Proceedings

1998 IEEE INTERNATIONAL CONFERENCE ON
ROBOTICS AND AUTOMATION

May 16 - 20, 1998
Katholieke Universiteit Leuven
Leuven, Belgium

Sponsored by

IEEE Robotics and Automation Society

Volume 3

Pages 1859 - 2770

ICRA-98 PROCEEDINGS

Additional copies may be ordered from:

IEEE Service Center
445 Hoes Lane
P. O. Box 1331
Piscataway, NJ 08855-1331 U.S.A.

IEEE Catalog Number 98CH36146
ISBN 0-7803-4300-X (Softbound)
ISBN 0-7803-4301-8 (Casebound)
ISBN 0-7803-4302-6 (Microfiche)
ISBN 0-7803-4758-7 CD-ROM
Library of Congress Catalog Number 90-640158
ISSN 1050-4729

Copyright and Reprint Permission:

Abstracting is permitted with credit to the source. Libraries are permitted to photocopy beyond the limit of U. S. copyright law for private use of patrons those articles in this volume that carry a code at the bottom of the first page, provided the per-copy fee indicated in the code is paid through Copyright Clearance Center, 27 Congress Street, Salem, MA 01970. Instructors are permitted to photocopy isolated articles for non-commercial classroom use without fee. For other copying, reprint, or republication permission, write to IEEE Copyrights Manager, IEEE Service Center, 445 Hoes Lane, P. O. Box 1331, Piscataway, NJ 08855-1331. All rights reserved. Copyright © 1998 by the Institute of Electrical and Electronics Engineers, Inc.

Other copying, reprint or republication requests should be addressed to: IEEE Copyrights Manager, IEEE Service Center, 445 Hoes Lane, P. O. Box 1331, Piscataway, NJ 08855-1331.

The papers in this book comprise the proceedings of the meeting mentioned on the cover and title page. They reflect the authors' opinions and, in the interests of timely dissemination, are published as presented and without change. Their inclusion in this publication does not necessarily constitute endorsement by the editors or the Institute of Electrical and Electronics Engineers, Inc.

Printed in the United States of America by OMNIPRESS.......The Proceedings Printer.

The Institute of Electrical and Electronics Engineers, Inc.

Foreword

The 1998 IEEE International Conference on Robotics and Automation, the 15th sponsored by the Robotics and Automation Society, is held in Leuven, an old university town situated 25 km from Brussels in the heart of Western Europe. It offers easy accessibility to attendees from all countries and provides the proper site for a major international Conference with a truly Pan-European flavor.

Leuven, former capital of the duchy of Brabant, is widely known as university town. The University, founded in 1425, within which the Conference takes place, is the oldest in the country and it has 27,000 students.

The Conference theme is **Intelligent Machines: from the Shop-Floor to Personal Robots**. Intelligent Machines are a new frontier for Advanced Robotics and Automation. They open a clear and exciting perspective with two challenges. First, the challenge and the opportunity to offer a solid ground for Robotics to emerge as an application-independent field, to construct it as a true scientific discipline based on a coherent set of concepts and formal tools. Second, the challenge and the opportunity to develop application-oriented robots and systems in a broad host of domains with important economic impacts and high social relevance. These domains range from new machinery on the shop-floor to novel real-world applications such as Tele-Operation in Remote and/or Hostile Sites, Field-Based Applications (mining, forestry, agriculture, ...), Service Robots including Public-Oriented needs and the highly demanding aspects related to Personal Robots to assist the aging and the impaired. Together with the newest results from the research community, the Conference emphasizes the importance of front-line Real-World Applications.

596 papers in 149 sessions were selected from the record number of 963 papers submitted by authors from 42 countries, corresponding to an overall breakdown of about one third from each of the general regions. Among these, five sessions are entirely devoted to papers motivated by problems of industrial relevance written, in most cases, by researchers from industry and nine "Special Sessions" are focused on specific topics with extra time allowed for in-depth discussion.

To complement this part of the technical program, several other events have been organized to favor open, broad discussion and exchanges within a unique audience of highly qualified researchers, industrial users and political decision makers. This includes three video sessions for presentation and discussion of the 27 segments selected for the Video Proceedings, every day key-note invited presentations, panels and round tables that address problems and issues that are important not only technically, but socially, economically, politically and even ethically. Special presentations on "Sojourner: the Robot of the Year 1997" and on new opportunities for research in Robotics are respectively set to entertain the Reception and the Banquet.

We are very grateful for all the hard work of the many people that endeavored to make ICRA'98 a success. We would like to emphasize the importance of the cooperative executive team-work distributed over four European countries - Belgium, France, Germany, Italy - and the United States and also to stress the fantastic amount of work done by the people of the Conference technical staffs. To all of them, our deepest thanks and gratitude.

Finally, a special thanks is owed to all the authors for contributing their research work, to all the members of the technical committees and the reviewers, the participants, the exhibitors, and very specially, for the support and the contribution of the European institutions and industrial companies that have well understood the role of Europe as a host. ICRA'98 owes its success to all.

Georges Giralt
General Chair

Paolo Dario
Program Chair

ORGANIZATION OF ICRA'98

SPONSORING ORGANIZATION
IEEE Robotics and Automation Society

CONFERENCE ORGANIZATION

General Chair

Georges Giralt, LAAS-CNRS, Toulouse, France

Honorary and Organizing Committee

Chair: Hendrik Van Brussel, Katholieke Universiteit Leuven, Belgium

International Program Committee

Program Chair: Paolo Dario, Scuola Superiore Sant'Anna, Pisa, Italy

Program Vice-Chairs
Vincent Hayward, McGill University, Canada
Bruno Siciliano, Università di Napoli Federico II, Italy
Kazuo Tanie, MEL, Japan
Harry Stephanou, RPI, USA

Industry-Oriented Program Committee
Chair: Rolf D. Schraft, IPA, Stuttgart, Germany

Special Sessions
Chair: Christian Laugier, INRIA, Rhone Alpes, France

Workshops and Tutorials
Chair: Ken Goldberg, University of California at Berkeley, USA

Video Proceedings Committee
Chair: Peter B. Luh, University of Connecticut, USA

IEEE Advisory and Coordinating Committee
Chair: Norman Caplan, NSF, USA

Finances
Chair: Steve Hsia, University of California at Davis, USA
Dick Klafter, Temple University, USA

Publications
Chair: Tzyh-Jong Tarn, Washington University, USA
C.S. George Lee, Purdue University, USA

Local Arrangements Committee
Chair: Joris De Schutter, Katholieke Universiteit Leuven, Belgium
Exhibits Chair: Herman Bruyninckx, Katholieke Universiteit Leuven, Belgium

Conference Secretary
Ralph. Sobek, IRIT, Toulouse, France

Honorary and Organizing Committee Members

Belgium:
   José Traest, Secretary-general of Fund for Scientific Research-Flanders (F.W.O.) in Belgium
   André Oosterlinck, Rector of the Katholieke Universiteit Leuven
European Commission:
   A. Garcia Arroyo, Director of Industrial and Materials Technologies, DG XII.
France:
   Jean-Jacques Gagnepain, Director of the Engineering Department, CNRS
   Bernard Larrouturou, President of INRIA
Germany:
   Gerd Hirzinger, DLR Oberpfaffenhofen
   Hartmut Weule, Research Director of DAIMLER
Italy:
   Umberto Cugini, Universit‡ di Parma
   Lorenzo Sciavicco, Terza Universit‡ di Roma
Russia:
   Dmitry E. Okhotsimsky, Keldysh Institute of Applied Mathematics
Spain:
   Gabriel Ferrate, Rector (President) of Universitat de Catalunya, Barcelona
   Juan Soto, President of HEWLETT-PACKARD-Spain
Sweden:
   Jan-Olof Eklundh, KHT, Stockholm
United Kingdom:
   Mike Brady, Oxford University

Program Committee

Europe/Africa/Middle East
   Hassane Alla, ENSIEG, St. Martin d'Hères, France
   Massimo Bergamasco, Scuola Sup. Sant'Anna, Pisa, Italy
   Antonio Bicchi, Università di Pisa, Italy
   Alicia Casals, ESAII, Barcelona, Spain
   Raja Chatila, LAAS - CNRS, Toulouse, France
   Stefano Chiaverini, Università degli Studi di Napoli, Italy
   Alessandro De Luca, Università Tor Vergata Roma, Italy
   Rüdiger Dillmann, Karlsruhe Univ., Germany
   Olav Egeland, Norwegian Univ., Trondheim, Norway
   Bernard Espiau, INRIA, Rhone Alpes, France
   Gerd Hirzinger, DLR, Wessling, Germany
   Okyay Kaynak, Bogazici University, Ankara, Turkey
   Wisama Khalil, Ecole Centrale de Nantes, France
   Krzysztof Kozlowski, Pozan Univ., Poland
   Jean Paul Laumond, LAAS-CNRS, Toulouse, France
   Jean Pierre Merlet, INRIA, Sophia Antipolis, France
   Friedrich Pfeiffer, Technical Univ. of Munich, Germany
   Vincent Rigaud, IFREMER, La Seyne sur Mer, France
   Elon Rimon, TECHNION, Haifa, Israel
   Giulio Sandini, Università degli Studi di Genova, Italy
   Paul Sharkey, Reading Univ., UK
   Sergey Sokolov, KIAM, Moscow, Russia
   Paul Taylor, Newcastle Univ., UK

USA/Canada
- Peter Allen, CEPSR, Columbia Univ., USA
- Haruhiko Asada, MIT, Cambridge, USA
- Antal Bejczy, JPL, Pasadena, USA
- Wayne Book, Georgia Inst. of Tech., Atlanta, USA
- Martin Buehler, McGill Univ., Montreal, Canada
- Joel Burdick, CALTECH, Pasadena, USA
- John Canny, Berkeley Univ., USA
- Hoda Elmaraghy, Univ. of Windsor, Canada
- Ron Fearing, California Univ., Berkeley, USA
- Eddie Grant, North Carolina Univ., Raleigh, USA
- Roderic Grupen, Massachusetts Univ., Amherst, USA
- Blake Hannaford, Washington Univ., Seattle, USA
- John Hollerbach, Univ. of Utah, Salt Lake City, USA
- Rob Howe, Harvard Univ., Cambridge, USA
- Dan Koditschek, Michigan Univ., Ann Arbor, USA
- Eric Krotkov, DARPA, Arlington, USA
- Vijay Kumar, Univ. of Pennsylvania, Philadelphia, USA
- Andrew Kusiak, Iowa Univ., Iowa City, USA
- Kok-Meng Lee, Georgia Inst. of Tech., Atlanta, USA
- Anthony Maciejewski, Purdue Univ., W. Lafayette, USA
- Dinesh Pai, Univ. of British Columbia, Vancouver, Ca
- Michael Peshkin, Northwestern Univ., Evanston, USA
- Kimon Valavanis, USL, Lafayette, USA
- Yuan Zheng, Ohio State Univ., Columbus, USA,

Asia/Oceania
- Fumihito Arai, Nagoya Univ., Japan
- Hirohoko Arai, Tsukuba Univ., Japan
- Suguru Arimoto, Ritsumeikan Univ., Shiga, Japan
- Minoru Asada, Osaka Univ., Japan
- Hajime Asama, RIKEN, Saitama, Japan
- Hyung Suck Cho, KAIST, Taejeon, Korea
- Peter Corke, CSIRO, Kenmore, Australia
- Li-Chen Fu, National Taiwan Univ., Taipei, Taiwan
- Masakatsu Fujie, HITACHI, Ibaraki, Japan
- Toshio Fukuda, Nagoya University, Japan
- Hideki Hashimoto, Tokyo Univ., Japan
- Katsushi Ikeuchi, Tokyo Univ., Japan
- Makoto Kaneko, Hiroshima Univ., Japan
- Norihisa Komoda, Osaka Univ., Japan
- Kazuhiro Kosuge, Tohoku University, Sendai, Japan
- Fumitoshi Matsuno, Tokyo Institute, Yokohama, Japan
- Mamoru Mitsuishi, Tokyo University, Japan
- Yoshihiko Nakamura, Tokyo University, Japan
- Sang-Rok OH, KIST, Seoul, Korea
- Isao Shimoyama, Tokyo University, Japan
- Masaru Uchiyama, Tohoku University, Sendai, Japan
- N. Viswanadham, Indian Inst. of Science, Bangalore, India
- Tsuneo Yoshikawa, Kyoto University, Japan

Industry-Oriented Committee Composition
- Carlisle, B., Adept, USA
- Cook, C., University of Wollongong, Australia
- Fabrizi, D., Italy
- Ferre, R., ICT-UPC, Spain
- Hägele, M., IPA Stuttgart, Germany
- Holland, St., GM, USA
- Kusuda, Y., Yaskawa, Japan
- Lacombe, J. L., Matra, France
- Martin, T., KfK Karlsruhe, Germany
- Schraft, R. D., IPA Stuttgart, Germany

Workshops and Tutorials Review Committee
- Stefano Caselli, U. Parma
- Mike Erdmann, CMU
- Dan Halperin, Tel Aviv University
- Hiro Hirukawa, ETL
- Jana Kosecka, UC Berkeley
- Jeff Trinkle, UT Austin
- Ian Walker, Rice University
- Randy Wilson, Kodak Research
- Shin'ichi Yuta, U. Tsukuba

Video Proceedings Committee Members
- Prof. David J. Cannon, The Pennsylvania State Univ., USA
- Prof. I-Ming Chen, Nanyang Technological Univ., Singapore
- Prof. Rajiv Dubey, University of Tennessee, USA
- Prof. Clement M. Gosselin, Universite Laval, Quebec, Canada
- Prof. Masayuki Inaba, Dept. of Mechano-Informatics, Japan
- Prof. Oussama Khatib, Stanford University, USA
- Prof. T. Kesavadas, SUNY, University at Buffalo, USA
- Ing. Giuseppe Mosci, Elsag Bailey, Italy

Conference Promotion Committee
- Gordon Dodds, Chair, The Queen's University of Belfast, UK
- Aydan Erkmen, Middle East Technical University, Turkey
- Vasek Hlavac, Czech Technical University, Czech Republic
- Liu Hsu, COPPE/UFRJ, Brazil
- Georges Kovacs, Hungarian Academy of Sciences, Hungary
- Adam Morecki, Warsaw Technical University, Poland
- Prof. Ivan Plander, Slovak Academy of Sciences, Slovakia
- George N. Saridis, Athens, Greece
- Helen C. Shen, Hong Kong University, Hong Kong
- Miomir Vukobratovic, Institut Mihajlo Pupin Beograd, Yu

Supporting institutions and industrial companies

The support of many important institutions and industrial companies has been precious to permit Europe, as a host and in the full spirit of the conference, to provide assistance for a broad participation from Eastern Europe and South Mediterranean rim countries.

Institutions

Industrial and technical societies

BULL	MATRA
CNES	SIEMENS
DAIMLER	TECNOSPAZIO
KUKA	THOMSON-CSF

Table of Contents

Foreword iii
Organizing Committee v
Author Index liii

Volume 1 (pages 1-926)

MOTION PLANNING I

Executing Motion Plans for Robots with Many Degrees of Freedom in Dynamic Environments 1
O. Brock and O. Khatib

A Modularized Sensitive Skin for Motion Planning in Uncertain Environments 7
D. Um, B. Stankovich, K. Giles, T. Hammond and V. Lumelsky

Collision-Tolerant Control Algorithm for Mobile Manipulator with Viscoelastic Passive Trunk 13
H-O. Lim, K. Yokoi, Q. Huang, S-R. Oh, A. Takanishi and K. Tanie

Planning Paths for a Flexible Surface Patch 21
C. Holleman, L.E. Kavraki and J. Warren

CAR-LIKE ROBOTS PLANNING AND CONTROL

Path Planning with Uncertainty for Car-Like Robots 27
T. Fraichard and R. Mermond

Obstacle Distances and Visibility for Car-Like Robots Moving Forward 33
J.P. Laumond, C. Nissoux and M. Vendittelli

Path Planning System for Car-Like Robot 40
B. Kreczmer

A Collision Checker for Car-Like Robots Coordination 46
T. Siméon, S. Leroy and J.P. Laumond

VR SIMULATORS

A Virtual Excavator for Controller Development and Evaluation 52
S.P. DiMaio, S.E. Salcudean, C. Reboulet, S. Tafazoli and K. Hashtrudi-Zaad

Creating Realistic Force Sensations in a Virtual Environment: Experimental System,
Fundamental Issues and Results 59
D.O. Popa and S.K. Singh

Internet-Based Remote Teleoperation 65
K. Brady and T.-J. Tarn

A Driving Simulator as a Virtual Reality Tool 71
W.-S. Lee, J.-H. Kim and J.-H. Cho

SERVICE ROBOTS

Bugs: Multiple Robots System Control Approach in Uxo Clearance and Minefield Countermeasure .... *
T.N. Nguyen, C. O'Donnell, C. Debolt and C. Freed

Adaptive Motion Generation with Exploring Behavior for Service Robots 77
J. Ota, M. Van der Loos and L. Leifer

Factory Automation Adapted for the Decommissioning of Nuclear Reprocessing Facilities 83
A.K. Bicknell and G. Hardey

A Hybrid Technique to Supply Indoor Service Robots 89
C. Wattanasin, Y. Aiyama, D. Kurabayashi, J. Ota and T. Arai

REDUNDANT MANIPULATORS

The Dynamic Manipulability Ellipsoid for Redundant Manipulators 95
P. Chiacchio and M. Concilio

The Structure of Time-Optimal Controls for Kinematically Redundant
Manipulators with End-Effector Path Constraints 101
M. Galicki

The Enhanced Compact QP Method for Redundant Manipulators Using Practical Inequality Constraints 107
K.C. Park, P.H. Chang and S.H. Kim

Real-Time Control of Redundant Robots Subject to Multiple Criteria 115
L. Li, W.A. Gruver, Q. Zhang and W. Chen

OUTDOOR APPLICATIONS

Multi-Resolution Planning for Earthmoving 121
S. Singh and H. Cannon

On the Development of a Real-Time Simulator for an Electro-hydraulic Forestry Machine 127
Y. Gonthier and E. Papadopoulos

Bilateral Matched Impedance Teleoperation with Application to Excavator Control 133
S.E. Salcudean, K. Hashtrudi-Zaad, S. Tafazoli, S.P. DiMaio and C. Reboulet

Planet Rover as an Object of the Engineering Design Work 140
A.L. Kemurdjian

WALKING MACHINES

Design and Development of a Legged Robot Research Platform JROB-1 146
S. Kagami, M. Kabasawa, K. Okada, T. Matsuki, Y. Matsumoto, A. Konno, M. Inaba and H. Inoue

Locomotion Controller for a Crab-Like Robot 152
W.C. Flannigan, G.M. Nelson and R.D. Quinn

Posture Control of a Cockroach-Like Robot 157
G.M. Nelson and R.D. Quinn

Computation of Walking Robots Movement Energy Expenditure 163
V.V. Zhoga

FRICTION AND BACKLASH

Nonlinear Friction Compensation Methods for an In-Parallel Actuated 6-DOF Manipulator 169
J.I. Song, Y.H. Choi, J.H. Shim, D.S. Kwon and H.S. Cho

Modeling of an Omni-Directional High Precision Friction Drive Positioning Stage 175
W.S. Chang and K. Youcef-Toumi

A Fuzzy System Compensator for Backlash 181
K.T. Woo, L-X. Wang, F.L. Lewis and Z.X. Li

Experimental Results with Observer-Based Nonlinear Compensation of Friction in a Positioning System 187
H. Henrichfreise and C. Witte

OPERATION AND CONTROL

A Robot Programming Environment Based on Free-Form CAD Modeling 194
C.G. Johnson and D. Marsh

Space Robot Autonomy Based on Distance Sensors 200
P. Bizzantino, M. De Bartolomei, G. Magnani and G. Visentin

A Control Architecture to Achieve Manipulation Task Goals for a Humanoid Robot 206
Y.-J. Cho, J.-M. Park, J. Park, S.-R. Oh and C.W. Lee

Synthesis of Impedance Control Laws at Higher Control Levels: Algorithms and Experiments 213
D. Surdilovic

ASSEMBLY PLANNING

Modeling and Controlling Variation in Mechanical Assemblies using State Transition Models 219
R. Mantripragada and D.E. Whitney

A Dynamic Programming Approach to a Reel Assignment Problem of a Surface
Mounting Machine in Printed Circuit Board Assembly 227
S. H. Lee, T.H. Park, B.H. Lee, W.H. Kwon and W. Kwon

Assembly Stability as a Constraint for Assembly Sequence Planning 233
H. Mosemann, F. Röhrdanz and F. Wahl

Micro Planning for Mechanical Assembly Operations 239
S.K. Gupta, C.J.J. Paredis and P.F. Brown

DESIGN TOOLS AND ENVIRONMENTS

A Virtual Design Environment using Evolutionary Agents 247
R. Subbu, C. Hocaoglu and A.C. Sanderson

On-Line Detection of Defects in Layered Manufacturing 254
T. Fang, I. Bakhadyrov, M.A. Jafari and G. Alpan

Design of Manufacturing Plant Layouts with Queuing Effects 260
S. Benjaafar

Interactive Virtual Factory for Design of a Shopfloor Using Single Cluster Analysis 266
L. Lefort and T. Kesavadas

MARS ROVERS

Maximum Likelihood Rover Localization by Matching Range Maps 272
C.F. Olson and L.H. Matthies

Physics-Based Planning for Planetary Exploration 278
S. Farritor, H. Hacot and S. Dubowsky

Mars Pathfinder Mission Internet-Based Operations Using WITS 284
P.G. Backes, K.S. Tso and G.K. Tharp

An Autonomous Path Planner Implemented on the Rocky7 Prototype Microrover 292
S.L. Laubach, J.W. Burdick and L. Matthies

VIRTUAL MANUFACTURING AND RAPID PROTOTYPING

Dextrous Exploration of a Virtual World for Improved Prototyping 298
D.G. Caldwell, C. Favede and N. Tsagarakis

Systematic Creation and Application of Virtual Factory with Object Oriented Concept 304
M.-H. Lin and L.-C. Fu

Concurrent Intelligent Rapid Prototyping System Framework 310
S. Liu and Z. Wang

Research on Improving Rapid Prototyping 314
Z. Wang, X. Wang and S. Liu

MOTION PLANNING II

Evolutionary Path Planning Using Multi-Resolution Path Representation 318
C. Hocaoglu and A.C. Sanderson

Automatic Generation of Sphere Hierachies from CAD Data 324
J. Pitt-Francis and R. Featherstone

6 DOF Path Planning in Dynamic Environments–A Parallel On-Line Approach 330
D. Henrich, C. Wurll and H. Wörn

Reactive Planning of Robot Arms in Single and Cooperative Tasks 336
K. Hamilton and G. Dodds

MOBILE ROBOT PATH PLANNING WITH SENSORS I

Towards Exact Localization without Explicit Localization with the Generalized Voronoi Graph 342
K. Nagatani, H. Choset and S. Thrun

Generalized Local Voronoi Diagram of Visible Region 349
R. Mahkovic and T. Slivnik

A_c*-DFS: an Algorithm for Minimizing Search Effort in Sensor Based Mobile Robot Navigation 356
L. Shmoulian and E. Rimon

A Sensory Uncertainty Field Model for Unknown and Non-Stationary Mobile Robot Environments 363
N.A. Vlassis and P. Tsanakas

HAPTIC DEVICES I

A Practical Measure of Dynamic Response of Haptic Devices 369
M. Moreyra and B. Hannaford

Haptic Manipulation of Virtual Mechanisms from Mechanical CAD Designs 375
A. Nahvi, D.D. Nelson, J.M. Hollerbach and D.E. Johnson

Haptic Interface for Virtual Reality Based Minimally Invasive Surgery Simulation 381
R. Baumann and R. Clavel

A Virtual Environment with Haptic Feedback for the Treatment of Motor Dexterity Disabilities 3721
G.M. Prisco, C.A. Avizzano, M. Calcara, S. Ciancio, S. Pinna and M. Bergamasco

HEALTH CARE AND WELFARE TECHNOLOGY

A Twenty-Four Hour Tele-Nursing System Using a Ring Sensor 387
B.-H. Yang, S. Rhee and H.H. Asada

Navigation Systems for Increasing the Autonomy and Security of Mobile Bases for Disabled People 393
S. Fioretti, T. Leo and S. Longhi

Docking Control of Holonomic Omnidirectional Vehicles with
Applications to a Hybrid Wheelchair/Bed System 399
S. Mascaro and H.H. Asada

Fast Range Image Segmentation for Servicing Robots 406
E. Natonek

Modular Robots

Why Snake Robots Need Torsion-Free Joints and How to Design Them 412
M. Nilsson

Traveling Wave Locomotion Hyper-Redundant Mobile Robot 418
G. Poi, C. Scarabeo and B. Allotta

The Self-Reconfiguring Robotic Molecule 424
K. Kotay, D. Rus, M. Vona and C. McGray

A 3-D Self-Reconfigurable Structure 432
S. Murata, H. Kurokawa, E. Yoshida, K. Tomita and S. Kokaji

Underwater Robots

Using Sonar in Terrain-Aided Underwater Navigation 440
P. Newman and H. Durrant-Whyte

Guidance and Control of Fish Robot with Apparatus of Pectoral Fin Motion 446
N. Kato and T. Inaba

Robust Nonlinear Control of an Underwater Vehicle/Manipulator System with Composite Dynamics 452
C. Canudas de Wit, E. Olguín Díaz and M. Perrier

Basic Research on Underwater Docking of Flexible Structures 458
K. Watanabe, H. Suzuki, Q. Tao and K. Yoshida

Control of Multilegged Robots

Forward Dynamics of Multilegged Vehicles using the Composite Rigid Body Method 464
S. McMillan and D.E. Orin

Solving the Optimal Force Distribution Problem in Multilegged Vehicles 471
J-S. Chen, F-T. Cheng, K-T. Yang, F-C. Kung and Y-Y. Sun

Quadratic Optimization of Force Distribution in Walking Machines 477
D.W. Marhefka and D.E. Orin

Gait Controllability for Legged Robots 484
B. Goodwine and J.W. Burdick

Elastic - Joint Manipulators

Key Issues in the Dynamic Control of Lightweight Robots for Space and Terrestrial Applications 490
J.-X. Shi, A. Albu-Schäffer and G. Hirzinger

Global Output Feedback Tracking Control for Rigid-Link Flexible-Joint Robots 498
W.E. Dixon, E. Zergeroglu, M.S. de Queiroz and D.M. Dawson

A General Algorithm for Dynamic Feedback Linearization of Robots with Elastic Joints 504
A. De Luca and P. Lucibello

Stable, On-Line Learning using CMACs for Neuroadaptive Tracking Control of Flexible-Joint Manipulators 511
C.J.B. Macnab and G.M.T. D'Eleuterio

ROBOT PROGRAMMING

Robot Programming by Demonstration–Selecting Optimal Event Paths 518
J. Chen and B. McCarragher

NEXUS: A Flexible, Efficient Robust Framework for Integrating Software
Components of a Robotic System 524
J.A. Fernandez and J. Gonzalez

Programming Groups of Local Models from Human Demonstration to
Create a Model for Robotic Assembly 530
M. Tsuda, H. Ogata and Y. Nanjo

Interactive Generation of Flexible Robot Programs 538
H. Friedrich, J. Holle and R. Dillmann

PARTS ORIENTATION

Determining Polygon Orientation using Model Based Force Interpretation 544
S. Rusaw, K. Gupta and S. Payandeh

Automatic Orienting of Polyhedra through Step Devices 550
R. Zhang and K. Gupta

Parts Orienting with Partial Sensor Information 557
S. Akella and M.T. Mason

Parts Orienting with Shape Uncertainty 565
S. Akella and M.T. Mason

SHOP FLOOR FLEXIBLE AUTOMATION

Accommodating FMS Operational Contingencies through Routing Flexibility 573
S.A. Reveliotis

Scalable and Maximally-Permissive Deadlock Avoidance for FMS 580
P. Kumar, K. Kothandaraman and P. Ferreira

Simulation as a Decision-Making Tool for Real-time Control of Flexible Manufacturing Systems 586
J.S. Smith and B.A. Peters

Flexible Routing and Deadlock Avoidance in Automated Manufacturing Systems 591
M. Lawley

PLANETARY ROVERS

The Atacama Desert Trek: Outcomes 597
D. Bapna, E. Rollins, J. Murphy, M. Maimone, W. Whittaker and D. Wettergreen

A Task Planner for the Computer-Aided Design of a Space-Lander Robot 605
M. Ghallab and J. Gout

NOMAD: A Demonstration of the Transforming Chassis 611
E. Rollins, J. Luntz, A. Foessel, B. Shamah and W. Whittaker

Robotic Deployment of Electro-Magnetic Sensors for Meteorite Search 618
L. Pedersen

VIDEO SESSION I – TELEOPERATION AND REHABILITATION ROBOTICS

Calibrated Synthetic Viewing .... **
W.S. Kim

Real-Time System for Virtually Touching Objects in the Real World Using a High Speed Active Vision System ...... **
T. Owaki, Y. Nakabo, A. Namiki, I. Ishii and M. Ishikawa

VR Training System for Crane Considering Adaptation to Operator's Skill .... **
M. Yoneda, F. Arai, T. Fukuda, K. Miyata and T. Naito

Facial 3-D Pose and Eye Gaze Direction Estimation Based on Robust Real-Time Face Tracking .... **
J. Heinzmann and A. Zelinsky

Piezohydraulic Parallel Micromanipulator .... **
P. Kallio, Q. Zhou, M. Lind and H.N. Koivo

Human-Machine Cooperative Telerobotics Using Uncertain Sensor and Model Data .... **
S. Everett and R. Dubey

Design and Virtual Prototyping of Rehabilitation Devices .... **
V. Krovi, I. Haulin, J. M. Vezien, I. Kakardiaris, R. Pito, R. Enciso, V. Kumar, G. K. Ananthasuresh and R. Bajcsy

A Robotic Orthosis Based on Mechanical Compatability with Humans .... **
S. Kawamura, Y. Hayakawa, M. Tamai, N. Kuribayashi and K. Suto

A Virtual Environment with Haptic Feedback for the Treatment of Motor Dexterity Disabilities .... **
G.M. Prisco, C.A. Avizzano, M. Calcara, S. Ciancio, S. Pinna and M. Bergamasco

PATH PLANNING I

Very Fast Collision Detection for Practical Motion Planning. Part I: The Spatial Representation 624
B. Martínez-Salvador, A.P. del Pobil and M. Pérez-Francisco

Choosing Good Distance Metrics and Local Planners for Probabilistic Roadmap Methods 630
N.M. Amato, O.B. Bayazit, L.K. Dale, C. Jones and D. Vallejo

Motion Planning with Uncertainty 638
H. Zhang, V. Kumar and J. Ostrowski

Very Fast Collision Detection for Practical Motion Planning. Part II: The Parallel Algorithm 644
M. Pérez-Francisco, A.P. del Pobil and B. Martínez-Salvador

MOBILE ROBOT PATH PLANNING WITH SENSORS II

Framed-Quadtree Path Planning for Mobile Robots Operating in Sparse Environments 650
A. Yahja, A. Stentz, S. Singh and B.L. Brumitt

Hierarchical Graph Search for Mobile Robot Path Planning 656
J. A. Fernandez and J. Gonzalez

Camera-based Observation of Obstacle Motions to Derive Statistical Data for Mobile Robot Motion Planning 662
E. Kruse and F.M. Wahl

Acquiring Mobile Robot Behaviors by Learning Trajectory Velocities with Multiple FAM Matrices 668
K. Ward and A. Zelinsky

TACTILE AND HAPTIC INTERFACES

Vibration Feedback Models for Virtual Environments 674
A.M. Okamura, J.T. Dennerlein and R.D. Howe

Tactile Feeling Display Based on Selective Stimulation to Skin Mechanoreceptors 680
H. Shinoda, N. Asamura and N. Tomori

Control of Dexterous Hand Master with Force Feedback 687
H.-P. Huang and Y.-F. Wei

A Tactile Display Using Human Characteristic of Sensory Fusion 693
J.-L. Wu, H. Sasaki and S. Kawamura

MEDICAL ROBOTICS

Stabilizer and Surgical Arm Design for Cardiac Surgery 699
T.J. Gilhuly, S.E. Salcudean, K. Ashe, S. Lichtenstein, P.D. Lawrence

Learning a Linear Association of Drilling Profiles in Stapedotomy Surgery 705
V.G. Kaburlasos, V. Petridis, P. Brett and D. Baker

Wavelet-Based Control of Penetration in a Mechatronic Drill for Orthopaedic Surgery 711
V. Colla and B. Allotta

Shared Control Framework Applied to a Robotic Aid for the Blind 717
P. Aigner and B. McCarragher

NONLINEAR MECHANICAL SYSTEMS

Observing Pose and Motion Through Contact 723
Y.-B. Jia and M. Erdmann

Dextrous Manipulation by Rolling and Finger Gaiting 730
L. Han and J.C. Trinkle

Optimal Planning of an Under-Actuated Planar Body Using Higher-Order Method 736
N. Faiz and S.K. Agrawal

Discretely Actuated Manipulator Workspace Generation using Numerical Convolution on the Euclidean Group 742
G.S. Chirikjian and I. Ebert-Uphoff

CONTROL OF UNDERWATER VEHICLES

A Fuzzy Model-Based Controller of an Underwater Robotic Vehicle Under the
Influence of Thruster Dynamics 750
W. Lee and G. Kang

Neural Network System for On-line Controller Adaptation and its Application to Underwater Robot 756
K. Ishii, T. Fujii and T. Ura

Successive Galerkin Approximations to the Nonlinear Optimal Control of an Underwater Robotic Vehicle 762
T.W. McLain and R.W. Beard

Task-Priority Redundancy Resolution for Underwater Vehicle-Manipulator Systems 768
G. Antonelli and S. Chiaverini

SPECIAL MOBILE MECHANISMS AND CONTROL

A Holonomic Omnidirectional Vehicle with a Reconfigurable Footprint
Mechanism and Its Application to Wheelchairs 774
M. Wada and H.H. Asada

The Bow Leg Hopping Robot 781
B. Brown and G. Zeglin

Experimental Implementation of a "Target Dynamics" Controller on a Two-link Brachiating Robot 787
J. Nakanishi, T. Fukuda and D. Koditschek

Control of a Bow Leg Hopping Robot 793
G. Zeglin and B. Brown

FLEXIBLE LINK MANIPULATORS I

Stable Inversion Control for Flexible Link Manipulators 799
A. De Luca, S. Panzieri and G. Ulivi

Inverse Dynamics Control of Flexible-Link Manipulators using Neural Networks 806
H. Talebi, R.V. Patel and K. Khorasani

Neural Network Model Based Control of a Flexible Link Manipulator 812
B. Song and A.J. Koivo

Adaptive Hybrid Force/Position Control of a Flexible Manipulator for Automated Deburring
with On-line Cutting Trajectory Modification 818
I-C. Lin and L.-C. Fu

PROGRAMMING AND CONTROL

A Target Approachable Force-Guided Control for Complex Assembly 826
S. Kang, M. Kim, C.W. Lee and K.-I. Lee

Hybrid Control as a Method for Robot Motion Programming 832
A.A. Rizzi

Physical Agent for Sensored Networked and Thinking Space 838
J-H. Lee, G. Appenzeller and H. Hashimoto

Message-Based Evaluation for High-Level Robot Control 844
C. Lee and Y. Xu

PART PUSHING/FEEDING

Dynamic Model for High-Speed Pushing as a Manipulator Operation 850
S. Su and I. Uzmay

A General Theory for Positioning and Orienting 2D Polygonal or
Curved Parts using Intelligent Motion Surfaces 856
M.G. Coutinho and P.M. Will

Flexible Part Feeder: Manipulating Parts on Conveyer Belt by Active Fence 863
A. Salvarinov and S. Payandeh

The Coulomb Pump: a Novel Parts Feeding Method Using a Horizontally-Vibrating Surface 869
D. Reznik and J. Canny

MODELLING, PLANNING AND INSPECTING IN INDUSTRIAL PLANTS

A Human Supervisory Approach to Modeling Industrial Scenes Using Geometric Primitives 875
J.P. Luck, C.Q. Little and R.S. Roberts

Planning Handling Operations in Changing Industrial Plants 881
M. Cherif and M. Vidal

Automatic Path Planning for Coordinate Measuring Machines 887
A. Limaiem and H.A. ElMaraghy

A New Algorithm for CAD-Directed CMM Dimensional Inspection 893
Y.-J. Lin and P. Murugappan

SPACE ROBOTS

Nonlinear Contact Control for Space Station Dexterous Arms 899
H. Seraji and R. Steele

Motion Estimation of an Unknown Rigid Body Rotating Freely in Zero Gravity
Based on Complex Spectrum of Position of a Point on the Body 907
H. Hirai, Y. Masutani and F. Miyazaki

Impact Analysis and Post-Impact Motion Control Issues of a Free-Floating Space
Robot Contacting a Tumbling Object 913
D. N. Nenchev and K. Yoshida

Momentum Control of a Tethered Space Robot through Tether Tension Control 920
M. Nohmi, D.N. Nenchev and M. Uchiyama

VIDEO SESSION II – HUMANOID ROBOTS AND MOBILE ROBOTS

Anthropomorphic Head-Eye Robot "WE-3R" .... **
A. Takanishi, S. Hirano, K. Sato, I. Kato and T. Otowa

A Human Symbiotic Humanoid: Hadaly-2 .... **
S. Hashimoto, S. Narita, K. Shirai, T. Takanishi, H. Kasahara, T. Kobayashi and S. Sugano

Intuitive Control of a Planar Bipedal Walking Robot .... **
J. Pratt and G. Pratt

Automatic Parallel Parking and Returning to Traffic Maneuvers .... **
I. Paromtchik and C. Laugier

Development of a Bipedal Humanoid Robot, WABIAN .... **
S. Hashimoto, S. Nrita, T. Kobayashi, A. Takanishi, J. Yamaguchi, P. Dario and H. Takonobu

Elements of Cooperative Behaviour in Autonomous Mobile Robots .... **
D. Jung, G. Cheng and A. Zelinsky

VIPER - Visual Position Estimation for Rovers .... **
F. Cozman, C.E. Guestrin and E. Krotkov

Posture Control of a Cockroach-Like Robot .... **
G.M. Nelson, R.J. Bachmann, R.D. Quinn, J.T. Watson and R.E. Ritzmann

A Hydrostatic Robot for Marine Applications .... **
R. Vaidyanathan, H.J. Chiel and R.D. Quinn

Volume 2 (pages 927-1858)

MOTION PLANNING III

Motion Planning for a 3-DOF Robot with a Passive Joint 927
K.M. Lynch, N. Shiroma, H. Arai and K. Tanie

Grasp Planning Algorithm for a Multifingered Hand-Arm Robot 933
N. Kawarazaki, T. Hasegawa and K. Nishihara

Cooperative Motion Planning for Grasp-Work Type Manipulators 940
G. Hirano, M. Yamamoto and A. Mohri

Minimum-time Open-Loop Smooth Control for Point-to-Point Motion in Vibratory Systems 946
A. Piazzi and A. Visioli

MOBILE ROBOT NAVIGATION I

Vision Based Navigation System by Variable Template Matching for Autonomous Mobile Robots 952
Y. Abe, M. Shikano, T. Fukuda, F. Arai and Y. Tanaka

Finding Landmarks for Mobile Robot Navigation 958
S. Thrun

Multiagent System with Event Driven Control for Autonomous Mobile Robot Navigation 964
R. C. Luo and T. M. Chen

A Multi-Loop Robust Navigation Architecture for Mobile Robots 970
J. Castro, V. Santos and M. I. Ribeiro

DYNAMIC SIMULATION AND HAPTIC SIMULATION

Analysis of Frictional Contact Models for Dynamic Simulation 976
P.R. Kraus, V. Kumar and P. Dupont

A Collision Model for Rigid and Deformable Bodies 982
A. Joukhadar, A. Deguet and C. Laugier

Real-Time Surgery Simulation with Haptic Feedback using Finite Elements 3739
S. Cotin and H. Delingette

Multirate Haptic Simulation Achieved by Coupling Finite Element Meshes through Norton Equivalents 989
O.R. Astley and V. Hayward

SURGICAL APPLICATIONS

A Robotics System for Stereotactic Neurosurgery and Its Clinical Application 995
M.D. Chen, T. Wang, Q.X. Zhang, Y. Zhang and Z.M. Tian

A PC-Based Workstation for Robotic Discectomy 1001
C. Casadei, P. Fiorini, S. Martelli, M. Montanari and A. Morri

Design of Haptic Interface Through Stiffness Modulation for Endosurgery: Theory and Experiments 1007
A. Faraz, S. Payandeh and A. Salvarinov

Remote Operation of a Micro-Surgical System 1013
M. Mitsuishi, Y. Iizuka, H. Watanabe, H. Hashizume and K. Fujiwara

MANIPULABILITY OF MULTIPLE ROBOT SYSTEMS

Kinematic Manipulability of General Constrained Rigid Multibody Systems 1020
J.T. Wen and L.S. Wilfinger

A Performance Index for Under-Actuated, Multi-Wire, Haptic Interfaces 1026
C. Melchiorri and G. Vassura

Manipulability and Singularity Analysis of Multiple Robot Systems: a Geometric Approach 1032
F.C. Park and J.W. Kim

Manipulability of Cooperating Robots with Passive Joints 1038
A. Bicchi and D. Prattichizzo

RECENT DEVELOPMENTS IN UNDERWATER ROBOTICS

Optimization of Configuration of Autonomous Underwater Vehicle for Inspection of Underwater Cables 1045
N. Kato, J. Kojima, Y. Kato, S. Matumoto and K. Asakawa

Experimental Study of Fault-Tolerant System Design for Underwater Robots 1051
K.C. Yang, J. Yuh and S.K. Choi

Force and Slip Sensing for a Dextrous Underwater Gripper 1057
D.J. O'Brien and D.M. Lane

Real-Time Estimation of Dominant Motion in Underwater Video Images for Dynamic Positioning 1063
F. Spindler and P. Bouthemy

CONTROL OF DYNAMICAL TASKS

Dynamic Modeling Approach to Gymnastic Coaching 1069
D. Nakawaki, S. Joo and F. Miyazaki

Experiments in Impulsive Manipulation 1077
W.H. Huang and M.T. Mason

A Volleyball Playing Robot 1083
H. Nakai, Y. Taniguchi, M. Uenohara, T. Yoshimi, H. Ogawa, F. Ozaki, J. Oaki, H. Sato,
Y. Asari, K. Maeda, H. Banba, T. Okada, K. Tatsuno, E. Tanaka, O. Yamaguchi and M. Tachimori

Stabilization of Systems with Changing Dynamics by Means of Switching 1090
M. Zefran and J.W. Burdick

FLEXIBLE LINK MANIPULATORS II

Consistent first and second Order Dynamic Model of Flexible Manipulators 1096
F. Boyer, N. Glandais and W. Khalil

An Efficient Motion Planning of Flexible Manipulator along Specified Path 1104
A. Mohri, P. Kumar Sarkar and M. Yamamoto

Hybrid Position/Force Control of Two Cooperative Flexible Manipulators Working in 3D Space 1110
M. Yamano, J.-S. Kim and M. Uchiyama

Parameters Identification of Flexible Robots 1116
Ph. Dépincé

BEHAVIOR - BASED SYSTEMS

Metrics for Evaluation of Behavior-Based Robotic Systems 1122
A.D. Mali and A. Mukerjee

Tradeoffs in Making the Behavior-Based Robotic Systems Goal-Directed 1128
A.D. Mali

Designing Stable Finite State Machine Behaviors Using Phase Plane Analysis and Variable Structire Control 1134
J.T. Feddema, R.D. Robinett and B.J. Driessen

Modelling and Realization of the Peg-in-Hole Task Based on Hidden Markov Model 1142
K. Itabashi, K. Hirana, T. Suzuki, S. Okuma and F. Fujiwara

ROBOTS FOR COST-EFFECTIVE MANUFACTURING

Robotic based Thermoplastic Fibre Placement Process 1148
M. Ahrens, V. Mallick and K. Parfrey

START: An Application Builder for Industrial Robotics 1154
E. Mazer, G. Boismain, J.M. Bonnet des Tuves, Y. Douillard,
S. Geoffroy, J.M. Dubourdieu, M. Tounsi and F. Verdot

Robotic Manipulation of Ophthalmic Lenses Assisted by a Dedicated Vision System 1160
X. Fernandez and J. Amat

Multimedia Communication Pendant for Sensor-Based Robotic Task Teaching by Sharing
Information - Modular Structure and Application to Sensing Systems 1166
Y. Nakamura, K. Kanayama and M. Mizukawa

CONTROL ISSUES IN MANUFACTURING

A State Variable Model for the Fluid Approximation of Flexible Manufacturing Systems 1172
F. Balduzzi and G. Menga

On the Synthesis of a Controllable Supervisor for Discrete Processes Modeled by Temporal Petri Nets 1179
S.I. Caramihai and H. Alla

A Three-Layer Workcell Control Architecture Design 1185
B. Bouzouia, F. Guerroumi and A. Boukhezar

Integrated Hybrid System Approach for Planning and Control of Concurrent Tasks in Manufacturing Systems 1192
M. Song, T-J. Tarn and N. Xi

MICROASSEMBLY AND MICRO-NANO MANIPULATION

Bio-Micromanipulation System for High Throughput Screening of Microbes in Microchannel 1198
K. Morishima, F. Arai, T. Fukuda, H. Matsuura and K. Yoshikawa

Parallel Microassembly with Electrostatic Force Fields 1204
K.-F. Böhringer, K. Goldberg, M. Cohn, R.D. Howe and A. Pisano

CAD-Driven Microassembly and Visual Servoing 1212
J.T. Feddema and R.W. Simon

Fusing Force and Vision Feedback for Micromanipulation 1220
Y. Zhou, B.J. Nelson and B. Vikramaditya

VIDEO SESSION III – ROBOTS IN GAMES, MANUFACTURING AUTOMATION, MANIPULATION AND CONTROL, AND ROBOTIC SUBSYSTEMS

Development of a Three Degree of Freedom Air Hockey Robot .... **
M Spong and B. Bishop

MiroSot Robot Soccer Systems .... **
J.H. Kim, H.S. Shim, M.J. Jung, H.S. Kim, I.H. Choy, K.C. Kim, D.H. Kim and Y.J. Kim

Solid Model Construction Using Meshes and Volumes .... **
P. Allen and M. Reed

CWRU Flexible Parts Feeding System .... **
G. Causey and R. Quinn

Technologies for Robust Agile Manufacturing .... **
R.D. Quinn, G.C. Causey, M.C. Birch, W.S. Newman, F.L. Merat,
M.S. Branicky, V.B. Velasco, Jr., N.A. Barendt, A. Podgurski, Y. Kim and J.Y. Jo

Two Industrial Robot Manipulators Rigidly Holding an Egg .... **
W.H. Zhu and J. De Schutter

Underactuated Robotic Hand .... **
C.M. Gosselin, T. Laliberte and E. Degoulange

A 3-D Deformable Shape Sensor .... **
D. Hristu, K. Morgansen, N. Ferrier and R.W. Brockett

Use of Sensors with Soft Contact Surfaces for Handling Flat Objects .... **
B. Borovac, L. Nagy, E. Begovic and M. Sabli

MOBILE ROBOTS I

Range and Pose Estimation for Visual Servoing of a Mobile Robot 1226
D. Jung, J. Heinzmann and A. Zelinsky

Reactive Navigation in Outdoor Environments Using Potential Fields 1232
H. Haddad, M. Khatib, S. Lacroix and R. Chatila

A Hybrid Collision Avoidance Method for Mobile Robots 1238
D. Fox, W. Burgard, S. Thrun and A.B. Cremers

Simultaneous Map Building and Localization for Mobile Robots: A Multisensor Fusion Approach 1244
J.A. Castellanos, J.M. Martínez, J. Neira and J.D. Tardós

MOBILE ROBOT NAVIGATION II

Route Presentation for Mobile Robot Navigation by Omnidirectional
Route Panorama Fourier Transformation 1250
Y. Yagi, S. Fujimura and M. Yachida

Algorithmic Navigation to Train Deictic Mobile Robot Operators 1256
M.E. Cleary and J.D. Crisman

A Graph-Based Exploration Strategy of Indoor Environments by an Autonomous Mobile Robot 1262
J.Y.-J. Hsu and L.-S. Hwang

A Path Following Controller for Wheeled Robots which Allows to Avoid Obstacles During Transition Phase 1269
P. Souères, T. Hamel and V. Cadenat

HUMAN TASK MODELS I

Virtual Lesson and its Application to Virtual Calligraphy System 1275
K. Henmi and T. Yoshikawa

Learning Force-Based Assembly Skills from Human Demonstration for
Execution in Unstructured Environments 1281
M. Skubic and R.A. Volz

Hand-in-Glove Human-Machine Interface and Interactive Control: Task Process
Modeling Using Dual Petri Nets 1289
S. Mascaro and H.H. Asada

Heterogeneous Function-Based Human/Robot Cooperations 1296
N. Xi and T.J. Tarn

HUMANOID ROBOTS

Development of a Remote-Brained Humanoid for Research on Whole Body Action 1302
F. Kanehiro, I. Mizuuchi, K. Koyasako, Y. Kakiuchi, M. Inaba and H. Inoue

Development of an Anthropomorphic Head-Eye System for a Humanoid Robot
-Realization of Human-Like Head-Eye Motion Using Eyelids Adjusting to Brightness 1308
A. Takanishi, S. Hirano and K. Sato

Design and Control of Mobile Manipulation System for Human Symbiotic Humanoid: Hadaly-2 1315
T. Morita, K. Shibuya and S. Sugano

The Development of Honda Humanoid Robot 1321
K. Hirai, M. Hirose, Y. Haikawa and T. Takenaka

MECHANISM DESIGN

Design Considerations of New Six Degrees-of-Freedom Parallel Robots 1327
N. Sima'an, D. Glozman and M. Shoham

The Minimum Form of Strength in Serial, Parallel and Bifurcated Manipulators 1334
R. O. Ambrose and M. A. Diftler

Design and Accuracy Evaluation of High Speed and High Precision Parallel Mechansm 1340
Y. Koseki, T. Arai, K. Sugimoto, T. Takatuji and M. Goto

Matrix Normalization for Optimal Robot Design 1346
L.J. Stocco, S.E. Salcudean and F. Sassani

VISUAL SERVOING

Positioning a Coarse-Calibrated Camera with Respect to an Unknown Object by 2D 1/2 Visual Servoing 1352
E. Malis, F. Chaumette and S. Boudet

Design of a Partitioned Visual Feedback Controller 1360
P.Y. Oh and P.K. Allen

What Can Be Done with an Uncalibrated Stereo System? 1366
J. Hespanha, Z. Dodds, G.D. Hager and A.S. Morse

Efficient Multi-strategic Hierarchical Motion Planning with Visual Servoing Constraints 1373
H. Sutanto and R. Sharma

BIPED LOCOMOTION

Attitude Control of a Biped Walking Robot Model with Circular Arced Soles Using a Gyroscope 1379
Y. Okuyama, A. Yabu and F. Takemori

Impactless Sagittal Gait of a Biped Robot During the Single Support Phase 1385
M. Rostami and G. Bessonnet

Dynamic Transition Simulation of a Walking Anthropomorphic Robot 1392
O. Bruneau, F. Ben Ouezdou and P. B. Wieber

Low Energy Cost Reference Trajectories for a Biped Robot 1398
C. Chevallereau, A. M. Formal'sky and B. Perrin

NEURAL NETWORKS I

Fusing a Hyper-Ellipsoid Clustering Kohonen Network with the Julier-Uhlmann-Kahlman Filter for
Autonomous Mobile Robot Map Building and Tracking 1405
J.A. Janét, M.W. White, M.G. Kay, J.C. Sutton, III and J.J. Brickley

A New Evolutionary Approach to Developing Neural Autonomous Agents 1411
J.-M. Yang, J.-T. Horng and C.-Y. Kao

Flexible Path Planning for Real-Time Applications using A*-Method and Neural RBF-Networks 1417
T. Frontzek, N. Goerke and R. Eckmiller

Integration of Knowledge-Based Systems and Neural Networks: Neuro-Expert Petri Net Models and Applications ... 1423
X.F. Zha, S.Y.E. Lim and S.C. Fok

IMPEDANCE AND FORCE CONTROL I

Impedance Control as Merging Mechanism for a Behavior-Based Architecture 1429
G. Beccari and S. Stramigioli

Impedance Matching for Evaluation of Dexterity in Execution of Robot Tasks 1435
S. Arimoto, S. Kawamura and H.-Y. Han

Motion/Force Decomposition of Redundant Manipulator and Its Application to Hybrid Impedance Control 1441
Y. Oh, W.-K. Chung, Y. Youm and I.-H. Suh

Robot Manipulator Hybrid Control for an Unknown Environment using Visco-Elastic Neural Networks 1447
K. Kiguchi and T. Fukuda

SMART ROBOT COMPONENTS

Gripper Design Guidelines for Modular Manufacturing 1453
G.C. Causey and R.D. Quinn

Industrial Exploitation of Computer Vision in Logistic Automation: Autonomous
Control of an Intelligent Forklift Truck 1459
G. Garibotto, S. Masciangelo, P. Bassino, C. Coelho, A. Pavan and M. Marson

Rover Continuous Path Planning using Merged Perceptions 3733
L. Rastel and M. Delpech

Active Laser Radar for High Performance Measurements 1465
J. Hancock, D. Langer, M. Hebert, R. Sullivan, D. Ingimarson, E. Hoffmann, M. Mettenleiter and C. Froehlich

ISSUES IN ASSEMBLY

A Flat Rigid Plate is a Universal Planar Manipulator 1471
D. Reznik and J. Canny

Complexity Reduction in Geometric Selective Disassembly using the Wave Propagation Abstraction 1478
H. Srinivasan and R. Gadh

An Analytic Approach to Assemblability Analysis 1484
S. Lee and C. Yi

Design for Tolerance of Electro-Mechanical Assemblies 1490
R. Sudarsan, Y. Narahari, K.W. Lyons, R.D. Sriram and M.R. Duffey

MICROACTUATOR CONTROL

Parallel Beam Micro Sensor/Actuator Unit Using PZT Thin Films and Its Application Examples 1498
T. Fukuda, H. Sato, F. Arai, H. Iwata and K. Itoigawa

Improvement of Control Method for Piezoelectric Actuator by Combining Induced
Charge Feedback with Inverse Transfer Function Compensation 1504
K. Furutani, M. Urushibata and N. Mohri

An Approach to Reduction of Hysteresis in Smart Materials 1510
J.M. Cruz-Hernandez and V. Hayward

Scale Effects and Thermal Considerations for Microactuators 1516
J. Peirs, D. Reynaerts and H. Van Brussel

COOPERATIVE ROBOTS I

Adaptive Control with Impedence of Cooperative Multi-Robot System 1522
A. Rodríguez-Angeles and V. Parra-Vega

Redundancy Optimization for Cooperating Manipulators Using Quadratic Inequality Constraints 1528
W. Kwon, B.H. Lee, W.H. Kwon, M.H. Choi and S.H. Lee

Experiments with Two Industrial Robot Manipulators Rigidly Holding an Egg 1534
W.-H. Zhu and J. De Schutter

A Control System for Cooperating Tentacle Robots 1540
M. Ivanescu and V. Stoian

MOBILE ROBOTS II

Probabilistic Mapping of an Environment by a Mobile Robot 1546
S. Thrun, D. Fox and W. Burgard

A Structured Dynamic Multi-Agent Architecture for Controlling Mobile Office-Conversant Robot 1552
H. Asoh, I. Hara and T. Matsui

Cooperative Behavior Acquisition in Multi Mobile Robots Environment by Reinforcement
Learning Based on State Vector Estimation 1558
E. Uchibe, M. Asada and K. Hosoda

GRAMMPS: A Generalized Mission Planner for Multiple Mobile Robots in Unstructured Environments 1564
B.L. Brumitt and A. Stentz

MOBILE ROBOT SHORT TERM NAVIGATION

VFH+: Reliable Obstacle Avoidance for Fast Mobile Robots 1572
I. Ulrich and J. Borenstein

"Where are you driving to?" Heading Direction for a Mobile Robot from Optical Flow 1578
A. Dev, B.J.A. Kröse and F.C.A. Groen

Feature Detection and Identification Using a Sonar-Array 1584
E.G. Araujo and R. Grupen

Wall Following Using Angle Information Measured by a Single Ultrasonic Transducer 1590
T. Yata, L. Kleeman and S. Yuta

ADVANCED TELEOPERATION

A 6 DOF Force-Reflecting Hand Controller Using the Fivebar Parallel Mechanism 1597
K.Y. Woo, B.D. Jin and D.S. Kwon

Learning Techniques in a Dataglove Based Telemanipulation System for the DLR Hand 1603
M. Fischer, P. van de Smagt and G. Hirzinger

Dimensional Analysis and Selective Distortion in Scaled Bilateral Telemanipulation 1609
M. Goldfarb

Human-Machine Cooperative Telerobotics Using Uncertain Sensor or Model Data 1615
S.E. Everett and R.V. Dubey

BIOMIMETIC DESIGN

A Graphical Method for Evaluating Static Characteristics of the Human Finger by Force Manipulability 1623
K. Hara, R. Yokogawa and A. Yokogawa

Design and Mechanics of an Antagonistic Biomimetic Actuator System 1629
R.M. Kolacinski and R.D. Quinn

Quantification of Masticatory Efficiency with a Mastication Robot 1635
H. Takanobu, T. Yajima, M. Nakazawa, A. Takanishi, K. Ohtsuki and M. Ohnishi

A Small-Sized Panoramic Scanning Visual Sensor Inspired by the Fly's Compound Eye 1641
K. Hoshino, F. Mura, H. Morii, K. Suematsu and I. Shimoyama

KINEMATICS OF HYPER REDUNDANT SYSTEMS

Inverse Kinematics for Modular Reconfigurable Robots 1647
I-M. Chen and G. Yang

Direct Kinematics of Manipulators with Hyper Degrees of Freedom and Frenet-Serret Formula 1653
H. Mochiyama, E. Shimemura and H. Kobayashi

Closed Form Solution of Forward Position Analysis for a 6 DOF 3-PPSP
Parallel Mechanism of General Geometry 1659
W.K. Kim, Y.K. Byun and H.S. Cho

Fast Estimation of the Kinematics of Parallel Modules of a Variable-Geometry-Truss
Manipulator Using Neural Networks 1665
K.E. Zanganeh and P.C. Hughes

VISION BASED TRACKING

Intelligent Robotic Manipulation with Hybrid Position/Force Control in a Uncalibrated Workspace 1671
D. Xiao, B.K. Ghosh, N. Xi and T.J. Tarn

Weighting Observations: The Use of Kinematic Models in Object Tracking 1677
K. Nickels and S. Hutchinson

Real Time Hand-Eye System: Interaction with Moving Objects 1683
D.E. Okhotsimsky, A.K. Platonov, I.R. Belousov, A.A. Boguslavsky,
S.N. Emelianov, V.V. Sazonov and S.M. Sokolov

Automatic Lane Following with a Single Camera 1689
S. Lee, K.S. Boo, D. Shin and D.H. Lee

LEGGED LOCOMOTION

A Strategy of Optimal Fault Tolerant Gait for the Hexapod Robot in Crab Walking 1695
J.-M. Yang and J.-H. Kim

Real-Time Dynamic Simulation of Quadruped Using Modified Velocity Transformation 1701
K-P. Lee, T-W. Koo and Y-S. Yoon

SCOUT: A Simple Quadruped that Walks, Climbs and Runs 1707
M. Buehler, R. Battaglia, A. Cocosco, G. Hawker, J. Sarkis and K. Yamazaki

Development of Quadruped Walking Robot with the Mission of Mine Detection and Removal 1713
S. Hirose and K. Kato

NEURAL NETWORKS II

Minimum Infinity-Norm Kinematic Solution for Redundant Robots Using Neural Networks 1719
H. Ding and S.K. Tso

A Neural Network Approach to Real-Time Trajectory Generation 1725
M. Meng and X. Yang

Analysis of Nonlinear Neural Network Impedance Force Control for Robot Manipulators 1731
S. Jung and T.C. Hsia

Visuo-Motor Coordination of a Robot Manipulator Based on Neural Networks 1737
L. Sun and C. Doeschner

IMPEDANCE AND FORCE CONTROL II

Impedance Based Combination of Visual and Force Control 1743
G. Morel, E. Malis and S. Boudet

On the Spatial Impedance Control of Gough-Stewart Platforms 1749
E.D. Fasse and C.M. Gosselin

Control of Moment and Orientation for a Robot Manipulator in Contact with a Compliant Environment 1755
C. Natale, B. Siciliano and L. Villani

Curvature in Force/Position Control 1761
J.M. Selig

MOBILE ROBOTS IN NON-MANUFACTURING FIELDS

Robot Driver for Guidance of Automatic Durability Road (ADR) Test Vehicles 1767
S. Shoval, J.P. Zyburt and D.W. Grimaudo

Experiences with the Development of a Robot for Smart Multisensoric Pipe Inspection 1773
H-B. Kuntze and H. Haffner

Field Test of Navigation System: Autonomous Cleaning in Supermarkets 1779
H. Endres, W. Feiten and G. Lawitzky

Undersea Robotics Activities in a Petroleum Company Research Center 1782
L.C.P. Messina and N.R.S. dos Reis

SCHEDULING

A Distributed Planning Network for Manufacturing Systems Management 1787
G.L. Kovács and I. Mezgár

An Efficient Search Algorithm for Deadlock-Free Scheduling in FMS Using Petri Nets 1793
I. Ben Abdallah, H.A. ElMaraghy and T. El Mekkawy

Lagrangian Relaxation Neural Networks for Job Shop Scheduling 1799
P.B. Luh, X. Zhao and Y. Wang

Dynamic Scheduling of Elevator System Over Hybrid Petri Net / Rule Modeling 1805
Y.-H. Huang and L.-C. Fu

MICROROBOTICS

Manipulating Biological and Mechanical Micro-Objects using LIGA-Microfabricated End-Effectors 1811
M.C. Carrozza, P. Dario, A. Menciassi and A. Fenu

Distributed Event-Based Control of Unifunctional Multiple Manipulator System 1817
K. Munawar and M. Uchiyama

A 3 DOF Piezohydraulic Parallel Micromanipulator 1823
P. Kallio, M. Lind, Q. Zhou and H.N. Koivo

Development of Underwater Microrobot using ICPF Actuator 1829
S. Guo, T. Fukuda, N. Kato and K. Oguro

COOPERATIVE ROBOTS II

Hybrid Position and Force Control of Two Industrial Robots Manipulating a
Flexible Sheet: Theory and Experiment 1835
D. Sun, J.K. Mills and Y. Liu

Human-Robots Collaboration System for Flexible Object Handling 1841
K. Kosuge, S. Hashimoto and H. Yoshida

Adaptive Learning Control of Robotic Systems with Model Uncertainties 1847
D. Sun and J.K. Mills

Rapid On-Line Learning of Compliant Motion for Two-Arm Coordination 1853
J. Zhang and M. Ferch

Volume 3 (pages 1859-2770)

MOBILE ROBOTS III

Visual Guidance of a Small Mobile Robot using Active, Biologically-Inspired, Eye Movements 1859
F. Mura and I. Shimoyama

Environmental Complexity Control for Vision-Based Learning Mobile Robot 1865
E. Uchibe, M. Asada and K. Hosoda

Unsupervised Learning to Recognize Environments from Behavior Sequences in a Mobile Robot 1871
S. Yamada and M. Murota

ROLLMOBS, a New Universal Wheel Concept 1877
L. Ferrière and B. Raucent

AUTOMATED CAR NAVIGATION

A New Satellite Selection Criterion for DGPS Using Two Low-Cost Receivers 1883
A. Pozo-Ruz, J.L. Martínez and A. García-Cerezo

Model-Based Car Tracking Integrated with a Road-Follower 1889
F. Dellaert, D. Pomerleau and C. Thorpe

Experiments in Autonomous Driving with Concurrent Goals and Multiple Vehicles 1895
B. L. Brumitt, M. Hebert and the CMU UGV Group

A Comparative Study of Vision-Based Lateral Control Strategies for Autonomous Highway Driving 1903
J. Kosecká, R. Blasi, C.J. Taylor and J. Malik

FORCE FEEDBACK CONTROL FOR TELEOPERATION

Towards Force-Reflecting Teleoperation Over the Internet 1909
G. Niemeyer and J.-J. E. Slotine

Gain-Scheduled Compensation for Time Delay of Bilateral Teleoperation Systems 1916
A. Sano, H. Fujimoto and M. Tanaka

Decoupling Control Based on Virtual Mechanisms for Telemanipulation 1924
A. Micaelli, C. Bidard and C. Andriot

Control of a Space Flexible Master-Slave Manipulator based on Parallel Compliance Models 1932
T. Komatsu and T. Akabane

CONTACT PROBLEMS

Modelling and Specification of Compliant Motions with Two and Three Contact Points 1938
H. Bruyninckx and J. De Schutter

The Instantaneous Kinematics of Manipulation 1944
L. Han and J. Trinkle

Contact Response Maps for Real Time Dynamic Simulation 1950
C. Ullrich and D.K. Pai

The Roles of Shape and Motion in Dynamic Manipulation: the Butterfly Example 1958
K.M. Lynch, N. Shiroma, H. Arai and K. Tanie

PARALLEL MECHANISMS

Working Modes and Aspects in Fully Parallel Manipulators 1964
D. Chablat and P. Wenger

The Isoconditioning Loci of A Class of Closed-Chain Manipulators 1970
D. Chablat, P. Wenger and J. Angeles

Efficient Computation of the Extremum of the Articular Velocities of a Parallel
Manipulator in a Translation Workspace 1976
J.-P. Merlet

Efficient Estimation of the Extremal Articular Forces of a Parallel Manipulator in a Translation Workspace 1982
J.-P. Merlet

DYNAMIC PROBLEMS IN VISUAL SERVOING

Dynamic Sensor Planning in Visual Servoing 1988
E. Marchand and G. Hager

Image-Based Visual Servoing by Integration of Dynamic Measurements 1994
A. Crétual and F. Chaumette

Tracking Adaptive Impedance Robot Control with Visual Feedback 2002
V. Mut, O. Nasisi, R. Carelli and B. Kuchen

6 DOF High Speed Dynamic Visual Servoing Using GPC Controllers 2008
J.A. Gangloff, M. de Mathelin and G. Abba

CONTROL ISSUES OF BIPEDAL WALKERS

Intuitive Control of a Planar Bipedal Walking Robot 2014
J. Pratt and G. Pratt

Realization of Dynamic Biped Walking Varying Joint Stiffness Using Antagonistic Driven Joints 2022
J. Yamaguchi, D. Nishino and A. Takanishi

Robust Biped Walking with Active Interaction Control between Foot and Ground 2030
Y. Fujimoto, S. Obata and A. Kawamura

Generation of Energy Optimal Complete Gait Cycles for Biped Robots 2036
L. Roussel, C. Canudas-de-Wit and A. Goswami

NEURAL AND SELF-ORGANIZING CONTROL

Neural Computation of the Equivalent Control in Sliding Mode For Robot Trajectory Control Applications 2042
M. Ertugrul and O. Kaynak

Neural Force Control (NFC) Applied to Industrial Manipulators in Interaction with Moving Rigid Objects 2048
M. Dapper, R. Maass, V. Zahn and R. Eckmiller

PSOM Network: Learning with Few Examples 2054
J.A. Walter

Organization and Reorganization of Autonomous Oceanographic Sample Networks 2060
R.M. Turner and E. H. Turner

MULTIFINGER HAND DESIGN

Optimal Design of a Five-Bar Finger with Redundant Actuation 2068
J.H. Lee, B.-J. Yi, S.-R. Oh and I.H. Suh

The Design and Development of the DIST-Hand Dextrous Gripper 2075
A. Caffaz and G. Cannata

DLR's Multisensory Articulated Hand - Part I: Hard- and Software Architecture 2081
J. Butterfass, G. Hirzinger, S. Knoch and H. Liu

DLR's Multisensory Articulated Hand - Part II: The Parallel Torque/Position Control System 2087
H. Liu, P. Meusel, J. Butterfass and G. Hirzinger

MANIPULATIVE ROBOTS IN NON-MANUFACTURING FIELDS

Redundancy Resolution of a Cartesian Space Operated Heavy Industrial Manipulator 2094
M. Honegger and A. Codourey

On-line Scheduling Algorithms for Improving Performance of Pick-and-Place
Operations on a Moving Conveyor Belt 2099
R. Mattone, L. Adduci and A. Wolf

Service Robots for Nuclear Safety: New Developments by Cybernetix 2106
J. Perret

Telerobotic System for Live Power Lines Maintenance: ROBTET 2110
L.F. Peñín, R. Aracil, M. Ferre, E. Pinto, M. Hernando and A. Barrientos

MULTI AGENTS IN MANUFACTURING

Development of a Distributed Object-Oriented System Framework for the
Computer-Integrated Manufacturing Execution System 2116
F-T. Cheng, E. Shen, J.-Y. Deng and K. Nguyen

Multi-agent Based Dynamic Scheduling for a Flexible Assembly System 2122
Y.-Y. Chen, L.-C. Fu and Y.-C. Chen

Multi-Agent System for Dynamic Scheduling and Control in Manufacturing Cells 2128
D. Ouelhadj, C. Hanach and B. Bouzouia

Multi-Agent Based Control Kernel for Flexible Automated Production System 2134
S.-H. Liu, L.-C. Fu and J.H. Yang

ACTUATORS AND SENSORS

Basic Study on a Magnetic Measurement for Balance Utilizing a Spherical Vessel 2140
T. Okada, K. Kimura and N. Mimura

Electrostrictive Polymer Artificial Muscle Actuators 2147
R. Kornbluh, R. Pelrine, J. Eckerle and J. Joseph

Development of a Distributed Actuation Device Consisting of Soft Gel Actuator Elements 2155
S. Tadokoro, S. Fuji, M. Fushimi, R. Kanno, T. Kimura, T. Takamori and K. Oguro

Preisach Model Identification of a Two-Wire SMA Actuator 2161
R.B. Gorbet, D.W.L. Wang and K.A. Morris

LEARNING

Learning by Biasing 2168
G. Hailu and G. Sommer

Eliminating Sensor Ambiguities via Recurrent Neural Networks in Sensor-Based Learning 2174
E. Cervera and A.P. del Pobil

Learning with Assistance based on Evolutionary Computation 2180
T. Omata

Multilayered Reinforcement Learning for Complicated Collision Avoidance Problems 2186
T. Fujii, Y. Arai, H. Asama and I. Endo

MOBILE ROBOT TRAJECTORY PLANNING

Motion Planning for a Mobile Manipulator Considering Stability and Task Constraints 2192
Q. Huang, S. Sugano and K. Tanie

Speed Planning and Generation Approach Based on the Path-Time Space for Mobile Robots 2199
V.F. Muñoz, A. Cruz and A. García-Cerezo

Accounting for Mobile Robot Dynamics in Sensor-Based Motion Planning: Experimental Results 2205
J.C. Alvarez, A. Shkel and V. Lumelsky

Motion Planning of a Wheeled Mobile Robot with Slip-Free Motion Capability on a Smooth Uneven Surface 3727
B.J. Choi and S.V. Sreenivasan

NAVIGATION

Controlling Sensory Perception for Indoor Navigation 2211
G.E. Hovland and B.J. McCarragher

The Detection of Faults in Navigation System. A Frequency Domain Approach 2217
S. Scheding, E.M. Nebot and H. Durrant-Whyte

Fault Detection and Identification in a Mobile Robot Using Multiple-Model Estimation 2223
S.I. Roumeliotis, G.S. Sukhatme and G.A. Bekey

AMADEUS: Mobile, Autonomous Decentralized Utility System for Indoor Transportation 2229
T. Kamada and K. Oikawa

HUMAN TASK MODELS II

On Discontinuous Human Control Strategies 2237
M. C. Nechyba and Y. Xu

Human Behavior Modeling in Master-Slave Teleoperation with Kinesthetic Feedback 2244
L. F. Peñín, A. Caballero, R. Aracil and A. Barrientos

Two Performances Measures for Evaluating Human Control Strategy 2250
J. Song, Y. Xu, M.C. Nechyba and Y. Yam

Towards Real-Time Robot Programming by Human Demonstration for 6D Force Controlled Actions 2256
Q. Wang and J. De Schutter

EXPERIMENTS ON TACTILE SENSING

Experiments in Synthetic Psychology for Tactile Perception in Robots: Step Towards
Implementing Humanoid Robots 2262
D. Taddeucci and P. Dario

Multifingered Robotic Hands: Contact Experiments using Tactile Sensors 2268
K.K. Choi, S.L. Jiang and Z.X. Li

Implementing Robotic Grasping Tasks Using a Biological Approach 2274
F. Leoni, M. Guerrini, C. Laschi, D. Taddeucci, P. Dario and A. Starita

Model and Processing of Whole-body Tactile Sensor Suit for Human-Robot Contact Interaction 2281
Y. Hoshino, M. Inaba and H. Inoue

PARALLEL MANIPULATORS

On the Design of Gravity-Compensated Six-Degree-Of-Freedom Parallel Mechanisms 2287
C.M. Gosselin and J. Wang

A Parallel x-y Manipulator with Actuation Redundancy for High-Speed and Active-Stiffness Applications 2295
S. Kock and W. Schumacher

Parallel Dynamics Computation and H-infinity Acceleration Control of Parallel
Manipulators for Acceleration Display 2301
K. Yamane, M. Okada, N. Komine and Y. Nakamura

Tracking Control of a Parallel Robot in the Task Space 2309
L. Beji, A. Abichou and M. Pascal

VISUAL SERVOING PERFORMANCE

Performance Evaluation of Vision-Based Control Tasks 2315
P. Krautgartner and M. Vincze

Performance and Sensitivity in Visual Servoing 2321
K. Hashimoto and T. Noritsugu

Experimental Evaluation of Fixed-Camera Direct Visual Controllers on a Direct-Drive Robot 2327
F. Reyes and R. Kelly

Visual Impedance Using 1ms Visual Feedback System 2333
Y. Nakabo and M. Ishikawa

MECHANICS OF SIMPLE BIPEDAL WALKERS

Active Leg Compliance for Passive Walking 2339
R.Q. van der Linde

The Motion of a Finite-Width Rimless Wheel in 3D 2345
A.C. Smith and M.D. Berkemeier

Speed, Efficiency and Stability of Small-Slope 2-D Passive Dynamic Bipedal Walking 2351
M. Garcia, A. Chatterjee and A. Ruina

A Design Method of Neural Oscillatory Networks for Generation of Humanoid Biped Walking Patterns 2357
M. Cao and A. Kawamura

INTELLIGENT CONTROL

A Model for the Organization Level of Intelligent Machines 2363
  M.N. Varvatsoulakis, G. Saridis and P.N. Paraskevopoulos

Stable Fuzzy Self-Tuning Computed-Torque Control of Robot Manipulators 2369
  M.A. Llama, V. Santibañez, R. Kelly and J. Flores

Sensor-Enhanced Robotic Cell Collaboration Using Shared Task Error Information 2375
  M. Motegi, T. Kakizaki and S.-Y. Muto

An Expert Opinion Approach to Tune Analytical Models of Nonlinear Systems 2383
  K.R. Chernyshov and F.F. Pashchenko

GRASPING FORMULATION

Virtual Truss Model for Characterization of Internal Forces for Multiple Finger Grasps 2389
  T. Yoshikawa

Generalized Stability of Compliant Grasps 2396
  H. Bruyninckx, S. Demey and V. Kumar

Grasping and Position Control for Multi-Fingered Robot Hands with Uncertain Jacobian Matrices 2403
  C.C. Cheah, H.Y. Han, S. Kawamura and S. Arimoto

Enveloping Grasp for Multiple Objects 2409
  K. Harada and M. Kaneko

CONSTRUCTION AND FIELD ROBOTICS

A Study of Autonomous Mobile System in Outdoor Environment (Part 2: Sign Guided
Autonomous Transportation System) 2416
  J. Takiguchi, K. Iwama, H. Sugie, M. Kato, T. Kiyonaga, T. Hashizume, F. Inoue, K. Yoshino and Y. Omote

Architecture of a GPS-Based Guiding System for Road Compaction 2422
  L-H. Pampagnin, F. Peyret and G. Garcia

Development of Automated Construction System for High-Rise Reinforced Concrete Buildings 2428
  K. Hamada, N. Furuya, Y. Inoue and T. Wakisaka

MONAI: An autonomous Navigation System for Mobile Robots 2434
  F. Carre, L. Gallo, B. Mazar, F. Megel and B. Serra

MACHINING

Fast Evaluation of Geometric Constraints for Bending Sequence Planning 2446
  M. Inui and H. Terakado

Automation of Chamfering by an Industrial Robot; for the Case of Machined
Hole on a Cylindrical Workpiece 2452
  N. Asakawa, K. Toda and Y. Takeuchi

An Improved Sculptured Part Surface Design with Jerk Continuity for a Smooth Machining 2458
T.S. Lee and Y.J. Lin

Accessibility Analysis in 5-Axis Machininh of Sculptured Surfaces 2464
A. Vafaeesefa and H.A. ElMaraghy

ACTUATOR CONTROL

Precise Position Control of Robot Arms using a Homogeneous ER Fluid 2470
N. Takesue, G. Zhang, J. Furusho and M. Sakaguchi

An Electrorheological Fluid Damper for Vibration Control 2476
J. Li and W.A. Gruver

Safety Oriented Mechanism and Control Using ER Fluid in the Joint 2482
F. Arai, A. Kawaji, T. Fukuda, H. Matsuura and H. Ota

3-DOF Closed-Loop Control for Planar Linear Motors 2488
A.E. Quaid and R.L. Hollis

FUZZY SYSTEMS

Cell Mapping Based Fuzzy Control of Car Parking 2494
M.C. Leu and T.Q. Kim

A Robust Model-Based Fuzzy-Logic Controller for Robot Manipulators 2500
M.R. Emami, A.A. Goldenberg and I.B. Türksen

Analysis of Linguistic Fuzzy Control for Curved-path-following Autonomous Vehicles 2506
Y.H. Fung and S.K. Tso

Fuzzy-Logic Dynamics Modeling of Robot Manipulators 2512
M.R. Emami, A.A. Goldenberg and I. Burhan Türksen

MOBILE ROBOT LOCALIZATION I

A Method for Tracking Pose of a Mobile Robot Equipped with a Scanning Laser Range Finder 2518
A. Dubrawski and B. Siemiatkowska

Mobile Robot Navigation Based on Vision and DGPS Information 2524
S. Kotani, K. Kaneko, T. Shinoda and H. Mori

Visual Place Recognition for Autonomous Robots 2530
H. D. Tagare, D. V. McDermott and H. Xiao

Homogeneous Neurolike Structures in Control Systems of Intelligent Mobile Robots 2536
I.A. Kaliaev

MOBILE ROBOT MAP BUILDING

Sonar Resolution-Based Environment Mapping 2541
L. Cahut, K.P. Valavanis and H. Deliç

Building Local Floor Map by Use of Ultrasonic and Omni-Directional Vision Sensor 2548
S.-C. Wei, Y. Yagi and M. Yachida

Map Building using Fuzzy ART, and Learning to Navigate a Mobile Robot on an Unknown World 2554
R. Araújo and A.T. de Almeida

Incremental Map Building for Mobile Robot Navigation in an Indoor Environment 2560
L. Delahoche, C. Pégard, M. Mouaddib and P. Vasseur

HAPTIC DEVICES II

Haptic Display for Object Grasping and Manipulating in Virtual Environment 2566
H. Maekawa and J.M. Hollerbach

Design of a Force Reflecting Master Arm and Master Hand Using Pneumatic Actuators 2574
S. Lee, S. Park, M. Kim and C.-W. Lee

Design of a Compact 6-DOF Haptic Interface 2580
Y. Tsumaki, H. Naruse, D.N. Nenchev and M. Uchiyama

Force Display System Using Particle-Type Electrorheological Fluids 2586
M. Sakaguchi and J. Furusho

SOCCER ROBOTS

Development of Self-Learning Vision-Based Mobile Robots for Acquiring Soccer Robots Behaviors 2592
T. Nakamura

Motion Control for Micro-Robots Playing Soccer Games 2599
S. Lee and J. Bautista

Sony Legged Robot for RoboCup Challenge 2605
H. Kitano, M. Fujita, S. Zrehen and K. Kageyama

Building Integrated Mobile Robots for Soccer Competition 2613
W.M. Shen, J. Adibi, R. Adobbati, B. Cho, A. Erdem, H. Moradi, B. Salemi and S. Tejada

UNDERACTUATED MANIPULATION

Time-Scaling Control of an Underactuated Manipulator 2619
H. Arai, K. Tanie and N. Shiroma

Adjustable Manipulability of Closed-Chain Mechanisms through Joint Freezing and Joint Unactuation 2627
S. Kim

Scaling Laws for Nonlinear Controllers of Dynamically Equivalent Rigid-Link Manipulators 2633
M. Ghanekar, D.W.L. Wang and G.R. Heppler

Robust Global Stabilization of the Underactuated 2-DOF Manipulator R2D1 2640
J. Mareczek, M. Buss and G. Schmidt

VISION-BASED CONTROL

Predictive Vision Based Control of High Speed Industrial Robot Paths 2646
F. Lange, P. Wunsch and G. Hirzinger

Tracking a Moving Target with Model Independent Visual Servoing: A Predictive Estimation Approach 2652
J.A. Piepmeier, G.V. McMurray and H. Lipkin

Toward Global Visual Servos and Estimators for Rigid Bodies 2658
N.J. Cowan and D.E. Koditschek

Toward 3D Uncalibrated Monocular Visual Servo 2664
B.E. Bishop and M.W. Spong

CONTROL OF NON-CONVENTIONAL LOCOMOTING MACHINES

Control of Autonomous Motion of Two-wheel Bycycle with Gyroscopic Stabilization 2670
A.V. Beznos, A.M. Formal'sky, E.V. Gurfinkel, D.N. Jicharev,
A.V. Lensky, K.V. Savitsky and L.S. Tchesalin

Toward the Control of a Multi-Jointed, Monoped Runner 2676
U. Saranli, W.J. Schwind and D.E. Koditschek

Dynamic Model of a Gyroscopic Wheel 2683
G.C. Nandy and Y. Xu

Omni-Directional Self-Propulsive Troweling Robot 2689
D.H. Shin, H. J. Kim, H.G. Lee and H.S. Kim

ADVANCED MANIPULATOR CONTROL

Frequency Modulation in Anthropomorphic Robots with Kinematic and Force Redundancies 2697
B.-J. Yi, S.-R. Oh, I. H. Suh and W.K. Kim

Fault-Tolerant Control and Optimal Operation of Redundant Robotic Manipulators 2703
A.J. Koivo and M. Ramos

Analytic Nonlinear H_∞ Optimal Control for Robotic Manipulators 2709
J. Park, W. Chung and Y. Youm

High Speed Tracking Control of Stewart Platform Manipulator via Enhanced Sliding Mode Control 2716
N.-I. Kim and C.-W. Lee

STATIC GRASP PLANNING

Caging Planar Objects with a Three-Finger One-Parameter Gripper 2722
C. Davidson and A. Blake

Geometric Formulation of Orientation Tolerances 2728
J.B. Gou, Y.X. Chu, H. Wu and Z.X. Li

Computing N-Finger Force-Closure Grasps on Polygonal Objects 2734
Y.-H. Liu

On Grasping and Manipulating Polygonal Objects with Disc-Shaped Robots in the Plane 2740
A. Sudsang and J. Ponce

MANUFACTURING

Schedule Execution Using Perturbation Analysis 2747
L. Bongaerts, H. Van Brussel and P. Valckenaers

A Dynamic Control Problem for a Two Part-Type Pull Manufacturing System 2753
F. Martinelli and P. Valigi

Multiple Control Policies for Two-Station Production Networks with Two Types of Parts using Fuzzy Logic 2759
R. Zhang and Y.A. Phillis

Analysis of Robot Motion Performance and Implications to Economy Principles 2765
S. Shoval, J. Rubinovitz and S. Nof

Volume 4 (pages 2771-3744)

3D SENSING

Flexible 3D Acquisition with a Monocular Camera 2771
M. Pollefeys, R. Koch, M. Vergauwen and L. Van Gool

A High Speed 3D Radar Scanner for Automation 2777
S.K. Boehmke, J. Bares, E. Mutschler and N.K. Lay

Zoom Tracking 2783
J.A. Fayman, O. Sudarsky and E. Rivlin

Rapid 3-D Digitizing and Tool Path Generation for Complex Shapes 2789
K.S. Kwok, C.S. Loucks and B.J. Driessen

SONAR SENSING

3D Object Localisation with a Binaural Sonarhead, Inspirations from Biology 2795
H. Peremans, A. Walker and J.C.T. Hallam

Perception of an Indoor Robot Workspace by Using CTFM Sonar Imaging 2801
Z. Politis and P. Probert

Mobile Robot Sonar Sensing with Pseudo-Random Codes 2807
K.-W. Jörg and M. Berg

Pipelined Sampling Techniques for Sonar Tracking Systems 2813
U.D. Hanebeck

MOBILE ROBOT LOCALIZATION II

Selection of Image Features for Robot Positioning using Mutual Information 2819
G. Wells and C. Torras

Toward Real-Time 2D Localization in Outdoor Environments 2827
A. Mallet and S. Lacroix

Continuous Localization Using Evidence Grids 2833
A.C. Schultz and W. Adams

Selecting Targets for Local Reference Frames 2840
S. Simhon and G. Dudek

DESIGN, PLANNING AND CONTROL IN MULTIPLE MOBILE ROBOT SYSTEMS

An Oscillation Analysis on Distributed Autonomous Robotic System 2846
T. Kaga and T. Fukuda

Multiple Mobile Robot Operation by Human 2852
A. Nakamura, S. Kakita, T. Arai, J. Beltrán-Escavy and J. Ota

Self-Organizing Collective Robots with Morphogenesis in a Vertical Plane 2858
K. Hosokawa, T. Tsujimori, T. Fujii, H. Kaetsu, H. Asama, Y. Kuroda and I. Endo

Controlling Formations of Multiple Mobile Robots 2864
J.P. Desai, J. Ostrowski and V. Kumar

TELEROBOT ARCHITECTURES

A Telerobotics System for Maintenance Tasks Integrating Planning Functions Based on Manipulation Skills 2870
T. Ogasawara, H. Hirukawa, K. Kitagaki, H. Onda, A. Nakamura and H. Tsukune

Detection of Discrepancies and Sensory-Based Recovery for Virtual Reality Based Telemanipulation Systems 2877
A. Kheddar, K. Tanie and P. Coiffet

Predictive Windows for Delay Compensation in Telepresence Applications 2884
J. Baldwin, A. Basu and H. Zhang

Augmentation of Safety in Teleoperation System for Intravascular Neurosurgery 2890
M. Tanimoto, F. Arai, T. Fukuda and M. Negoro

Multiple Robot Systems

Study on Cooperative Positioning System - Optimum Moving Strategies for CPS-III - 2896
R. Kurazume and S. Hirose

Manipulation of Multiple Objects by Two Manipulators 2904
Y. Aiyama, M. Minami and T. Arai

Optimization of Collision Free Trajectories in Multi-Robot System 2910
M. Mediavilla, J.C. Fraile, J.R. Perán and G.I. Dodds

Decentralized Control of Cooperating Mobile Manipulators 2916
T. Sugar and V. Kumar

Trajectories and Kinematics I

Two Methods for Interpolating Rigid Body Motions 2922
M. Zefran and V. Kumar

Isometric Visualization of Configuration Spaces of Two Degrees of Freedom Mechanisms 2928
G. Rodnay and E. Rimon

Removing the Singularities of Serial Manipulators by Transforming the Workspace 2935
J.E. Lloyd

Design and Kinematic Analysis of the Wire Parallel Mechanism for a Robot Pose Measurement 2941
J.W. Jeong, S.-H. Kim and Y.-K. Kwak

Hand-Eye Coordination

Improving Visually Servoed Disassembly Operations by Automatic Camera Placement 2947
F. Keçeci, M. Tonko, H.-H. Nagel and V. Gengenbach

Uncalibrated Hand-Eye Coordination with a Redundant Camera System 2953
C. Scheering and B. Kersting

The "Feature CMAC": a Neural-Network-Based Vision System for Robotic Control 2959
J. Carusone and G.M.T. D'Eleuterio

A Neuro-Fuzzy Solution for Fine-Motion Control Based on Vision and Force Sensors 2965
Y. von Collani, J. Zhang and A. Knoll

Nonholonomic Robots

A Global Approach for Motion Generation of Non-Holonomic Mobile Manipulators 2971
C. Perrier, P. Dauchez and F. Pierrot

Motion Control of the N.T.U.A. Robotic Snake on a Planar Surface 2977
K. Sarrigeorgidis and K.J. Kyriakopoulos

Sliding Mode Control of a Nonholonomic Wheeled Mobile Robot for Trajectory Tracking 2983
J.-M. Yang, I.-H. Choi and J.-H. Kim

Transportation of a Single Object by Two Decentralized-Controlled Nonholonomic Mobile Robots 2989
K. Kosuge, T. Oosumi, M. Satou, K. Chiba and K. Takeo

CONTROL

Compensation of Motor Torque Disturbances in Industrial Robots 2995
G. Ferretti, G. Magnani and P. Rocco

Adaptive Derivative Estimation for DSP-Based Acceleration Measurement 3001
O. Vainio

Analysis and Implementation of observers for Robotic Manipulators 3006
B. Bona and M. Indri

Disturbance Observer Based Force Control of Robot Manipulator without Force Sensor 3012
K.S. Eom, I.H. Suh, W.K. Chung and S.-R. Oh

DYNAMIC GRASPING

Vision-Guided Grasping of Unknown Objects for Service Robots 3018
P.J. Sanz, A.P. del Pobil, J.M. Inesta and G. Recatalá

Hybrid Closed-Loop Control of Robotic Hand Regrasping 3026
T. Schlegl and M. Buss

Biologically Inspired Robot Grasping Using Genetic Programming 3032
J.J. Fernandez and I.D. Walker

Transition Stability of Enveloped Objects 3040
M. Kaneko, M. Higashimori and T. Tsuji

MECHATRONIC SYSTEMS WITH NOVEL ACTUATOR APPLICATIONS

A Direct-Drive Pneumatic Stepping Motor for Robots: Designs for Pipe-Inspection
Microrobots and for Human-Care Robots 3047
K. Suzumori, K. Hori and T. Miyagawa

Pneumatic Muscle Actuator Technology a Light Weight Power System for a Humanoid Robot 3053
D.G. Caldwell, N. Tsagarakis, D. Badihi and G.A. Medrano-Cerda

A Rigid and Accurate Piezo-Stepper Based on Smooth Learning Hybrid Force-Position Controlled Clamping 3059
M. Versteyhe, D. Reynaerts and H. Van Brussel

Analysis of the Flight Performance of Small Magnetic Rotating Wings for Use in Microrobots 3065
N. Miki and I. Shimoyama

3D Modelling and Reconstruction

Range Data Merging for Probabilistic Octree Modeling of 3-D Workspaces 3071
P. Payeur, D. Laurendeau and C.M. Gosselin

An Efficient On-Line Algorithm for Direct Octree Construction from Range Images 3079
Y. Yu and K. Gupta

Autonomous Sensor Planning for 3D Reconstruction of Complex Objects from Range Images 3085
M.A. García, S. Velazquez, A.D. Sappa and L. Basañez

3D Scene Modelling and Curve-Based Localization in Natural Environments 3091
M. Devy and C. Parra

Sensor Systems

Determining the Value of Monitoring for Dynamic Monitor Selection 3097
T. Celinski and B. McCarragher

Instrumented Logical Sensor Systems-Practice 3103
M. Dekhil and T.C. Henderson

Integrated Precision 3-DOF Position Sensor for Planar Linear Motors 3109
Z.J. Butler, A.A. Rizzi and R.L. Hollis

Registering, Integrating and Building CAD Models from Range Data 3115
R. Yang and P.K. Allen

Localization for Mobile Robots

Position Estimation Using Principal Components of Range Data 3121
J.L. Crowley, F. Wallner and B. Schiele

Hybrid, High-Precision Localisation for the Mail Distributing Mobile Robot System MOPS 3129
K.O. Arras and S.J. Vestli

Mobile Robot Localization in Dynamic Environments using Places Recognition 3135
O. Aycard, P. Laroche and F. Charpillet

An Automatic Calibration Method for a Multisensor System: Application to a Mobile Robot Localization System ... 3141
H-J. von der Hardt, R. Husson and D. Wolf

Planning and Flexibility for Mobility

Hierarchical Path Planning on Probabilistically Labelled Polygons 3147
E. Piat and S. Lacroix

Local Path Re-Planning for Unforeseen Obstacle Avoidance by An Autonomous Sweeping Robot 3153
D. Kurabayashi, S. Koga, T. Arai, J. Ota, H. Asama and I. Endo

A Simple Space-Time-Symmetric Collision Avoidance Method for Autonomous Vehicles 3159
K. Matsumoto and M. Rude

Resource Modelling and Combination in Modular Robotics Systems 3167
J.A. Fryer and G.T. McKee

VISUAL VR INTERFACES

Designing Personal Tele-Embodiment 3173
E. Paulos and J. Canny

Guidance of Video Data Acquisition by Myoelectric Signals for Smart Human-Robot Interfaces 3179
O.A. Alsayegh and D.P. Brzakovic

Robotic Sightseeing - A Method for Automatically Creating Virtual Environments 3186
E. Bourque, G. Dudek and P. Ciaravola

Optimal 3D Viewing with Adaptive Stereo Displays 3192
S. Lee, S. Ro, J.-O. Park and C.-W. Lee

MULTIPLE AND MODULAR ROBOTS

Cooperative Control of Multiple Mobile Manipulators on Uneven Ground 3198
H. Osumi, M. Terasawa and H. Nojiri

A Cooperative Hunting Behavior by Mobile Robot Troops 3204
H. Yamaguchi

Evaluation on Flexibility of Swarm Intelligent System 3210
T. Fukuda, D. Funato, K. Sekiyama and F. Arai

Path Planning and Role Selection Mechanism for Soccer Robots 3216
J.-H. Kim, K.-C. Kim, D.-H. Kim, Y.-J. Kim and P. Vadakkepat

TRAJECTORIES AND KINEMATICS II

A Normal form Solution to the Singular Inverse Kinematic Problem for
Robotic Manipulators: The Quadratic Case 3222
K. Tchon and R. Muszynski

Generating Robust Trajectories in the Presence of Ordinary and Linear-Self-Motion Singularities 3228
J.E. Lloyd and V. Hayward

Alternative Computational Scheme of Manipulator Inverse Kinematics 3235
J. Lenarcic

Computation of Kinetostatic Performances of Robot Manipulators with Polytopes 3241
R. Finotello, T. Grasso, G. Rossi and A. Terribile

ACTIVE VISION

Spatial Attention and Saccadic Camera Motion 3247
J.J. Clark

Panoramic-Environmental Description as Robots' Visual Short-Term Memory 3253
K. Kayama, K. Nagashima, A. Konno, M. Inaba and H. Inoue

FOVEA: A Foveated Vergent Active Stereo System for Dynamic Three-Dimensional Scene Recovery 3259
W. Klarquist and A. Bovik

Person Tracking by Integrating Optical Flow and Uniform Brightness Regions 3267
T. Yamane, Y. Shirai and J. Miura

CONTROL OF NONHOLONOMIC ROBOTS I

Control of a Car-Like Robot Using a Dynamic Model 3273
M. Egerstedt, X. Hu and A. Stotsky

Robust Path-Following Control with Exponential Stability for Mobile Robots 3279
L.E. Aguilar, M.P. Souères, M. Courdesses and S. Fleury

Adaptive Motion Control of a Nonholonomic Vehicle 3285
S.V. Gusev, I.A. Makarov, I.E. Paromtchik, V.A. Yakubovich and C. Laugier

A Practical Approach to Feedback Control for a Mobile Robot with Trailer 3291
F. Lamiraux and J.P. Laumond

CONTROL AND IDENTIFICATION

Calibration of Coordinate System for Decentralized Coordinated Motion Control of Multiple Manipulators 3297
K. Kosuge, H. Seki and T. Oosumi

Experiment Design for Robot Dynamic Calibration 3303
G. Calafiore and M. Indri

INS-Based Identification of Quay-Crane Spreader Yaw 3310
M.A. Louda, D.C. Rye, M.W.M.G. Dissanayake and H.F. Durrant-Whyte

A Base Force/Torque Sensor Approach to Robot Manipulator Inertial Parameter Estimation 3316
G. Liu, K. Iagnemma, S. Dubowsky and G. Morel

GRASP OPTIMIZATION

Minimum-Deflection Grasps and Fixtures 3322
Q. Lin, J.W. Burdick and E. Rimon

Genetic Algorithm-Based Optimal Regrasping with the Anthrobot 5-Fingered Robot Hand 3329
A.M. Erkmen and M. Durna

Qualitative Test and Force Optimization of 3D Frictional Force-Closure Grasps Using Linear Programming 3335
Y.-H. Liu and M. Wang

Optimization of Robot Hand Power Grasps 3341
Y. Yu, K. Takeuchi and T. Yoshikawa

NEW TOOLS AND APPLICATIONS FOR AUTOMATION

Aligning Threaded Parts Using a Robot Hand 3348
M.A. Diftler and I.D. Walker

Automated Singulating System for Transfer of Live Broilers 3356
K.-M. Lee, R. Gogate and R. Carey

Microassembly Planning for Manufacturing by Flexible Microrobots 3362
S. Fatikow and R. Mounassypov

Nanorobotic Assembly of Two-Dimensional Structures 3368
A.A.G. Requicha, C. Baur, A. Bugacov, B.C. Gazen, B. Koel,
A. Madhukar, T.R. Ramachandran, R. Resch and P. Will

SHAPE RECOGNITION

Recognizing Surfaces Using Curve Invariants and Differential Properties of Curves and Surfaces 3375
D. Keren, E. Rivlin, I. Shimshoni and I. Weiss

Shape Recognition: A Fuzzy Approach 3382
S. Rolfes and M.J. Rendas

Object Skeletons from Sparse Shapes in Industrial Image Settings 3388
R. Singh, N.P. Papanikolopoulos and V. Cherkassky

3-D Object Recognition Using Projective Invariant Relationship by Single-View 3394
K.S. Roh, B.J. You and I.S. Kweon

SENSOR FUSION

On Design of Sequential Sensor Fusion System 3400
S. Emura and S. Tachi

A Hypothesis Testing Method for Multisensory Data Fusion 3407
X.-G. Wang, H. C. Shen and W.-H. Qian

A Decentralised Navigation Architecture 3413
M. Bozorg, E.M. Nebot and H.F. Durrant-Whyte

Triangulation based Fusion of Ultrasonic Sensor Data 3419
O. Wijk, P. Jensfelt and H.I. Christensen

LOCALIZATION USING GPS OR LANDMARKS

Landmark Perception Planning for Mobile Robot Localization 3425
J.M. Armingol, L. Moreno, A. de la Escalera and M.A. Salichs

Goal-Oriented Behaviour-Based Visual Navigation 3431
G. Cheng and A. Zelinsky

Achieving Integrity in an INS/GPS Navigation Loop for Autonomous Land Vehicle Applications 3437
S. Sukkarieh, E.M. Nebot and H.F. Durrant-Whyte

Positioning of Vehicle on Undulating Ground Using GPS and Dead Reckoning 3443
T. Aono, K. Fujii, S. Hatsumoto and T. Kamiya

ISSUES ON AUTONOMOUS VEHICLES

A Semi-Autonomous Robotic Airship for Environmental Monitoring Missions 3449
A. Elfes, S. Siqueira Bueno, M. Bergerman and J.G. Ramos, Jr.

Experimental Evaluation of a Fiber Optics Gyroscope for Improving
Dead-Reckoning Accuracy in Mobile Robots 3456
J. Borenstein

Airship Dynamic Modeling for Autonomous Operation 3462
S.B. Varella Gomes and J.G. Ramos, Jr.

Antenna Pointing for High Bandwidth Communications from Mobile Robots 3468
D. Bapna, E. Rollins, A. Foessel and R. Whittaker

HUMAN - ROBOT COEXISTENCE

Realization of Safety in a Coexistent Robotic System by Information Sharing 3474
Y. Wakita, S. Hirai, T. Hori, R. Takada and M. Kakikura

Human-Robot Coordination with Rotational Motion 3480
K.I. Kim and Y.F. Zheng

Development of an 8 DOF Robotic Orthosis for Assisting Human Upper Limb Motion 3486
K. Nagai, I. Nakanishi, H. Hanafusa, S. Kawamura, M. Makikawa and N. Tejima

EMG-based Human-Robot Interface for Rehabilitation Aid 3492
O. Fukuda, T. Tsuji, A. Ohtsuka and M. Kaneko

MULTIAGENT SYSTEMS

Experiments in Evolving Communicating Controllers for Teams of Robots 3498
I. Ashiru and C.A. Czarnecki

Principles of Minimal Control for Comprehensive Team Behavior 3504
B.B. Werger

Reactive Visual Control of Multiple Non-Holonomic Robotic Agents 3510
K. Han and M. Veloso

Development of a Hand-to-Hand Robot Based on Agent Network 3516
T. Suehiro, H. Takahashi and H. Yamakawa

ISSUES ON DYNAMICS

Swing Motion Control of Casting Manipulation (Experiment of Swing Motion Control) 3522
H. Arisumi, T. Kotoku and K. Komoriya

A Nonlinear Model for Harmonic Drive Friction and Compliance ..... *
H.D. Taghirad and P.R. Bélanger

Biped Robot Walking Using Gravity-Compensated Inverted Pendulum Mode and Computed Torque Control 3528
J.H. Park and K.D. Kim

Teleoperation System Via Computer Network for Dynamic Environment 3534
J. Kikuchi, K. Takeo and K. Kosuge

ISSUES IN VISION

Image-Based Manipulation Planning for Non-Rigid Objects 3540
P.W. Smith

Integrating Dependent Sensory Data 3546
A.C.S. Chung and H.C. Shen

Understanding Mechanism: From Images to Behaviors 3552
T. Dar, L. Joskowicz and E. Rivlin

Analysing Spatial Realizability of Line Drawings Through Edge-Concurrence Tests 3559
L. Ros and F. Thomas

CONTROL OF NON-HOLONOMIC ROBOTS II

Fourier Series based Method of Generating Continuous Controls for Driftless Nonholonomic Systems 3567
I. Duleba and J. Sówka

Asymptotic Stabilization of Multiple Nonholonomic Mobile Robots Forming Group Formations 3573
H. Yamaguchi and J.W. Burdick

Stabilization of the Acrobot via Iterative State Steering 3581
A. De Luca and G. Oriolo

Path Planner for Nonholonomic Mobile Robot with Fast Replaning Procedure 3588
L. Podsedkowski

Motion Control

Fundamental Control Concepts for Implementation of Transmission-Based Actuators in Robotics and Automation ... 3594
T.C. Widner and W.R. Hamel

A Class of Nonlinear PID Global Regulators for Robot Manipulators ... 3601
V. Santibañez and R. Kelly

Feedback Control for Robotic Manipulator with Uncertain Kinematics and Dynamics ... 3607
C.C. Cheah, S. Kawamura and S. Arimoto

Disturbance Attenuation and Load Decoupling with H∝ Positive Joint Torque Feedback ... 3613
F. Aghili, M. Buehler and J.M. Hollerbach

Strategic Grasping

Capturing Pyramidal-Like Objects ... 3619
M. Kaneko, M. Kessler, A. Weigl and H. Tolle

A Unified Distributed Cooperation Strategy for Multiple Object Handling Robots ... 3625
M.N. Ahmadabadi and N. Eiji

Coordinated Motion Generation and Real-Time Grasping Force Control for Multi-Fingered Manipulation ... 3631
Z.X. Li, Z. Qin, S. Jiang and L. Han

Controlling the Power Grasp with Incomplete Touch Sensor Information ... 3639
M.I. Vuskovic, G.R. Dunlop and K. Filali-Adib

Localization and Fixturing

A Geometric Approach of Form Tolerance Formulation and Evaluation ... 3646
J.B. Gou, Y.X. Chu and Z.X. Li

Localization Algorithms: Performance Evaluation and Reliability Analysis ... 3652
Y.X. Chu, J.B. Gou, H. Wu and Z.X. Li

Computer-Assisted Gripped and Fixture Customization Using Rapid-Prototyping Technology ... 3658
V.B. Velasco, Jr. and W.S. Newman

On the Hybrid Workpiece Localization/Envelopment Problems ... 3665
Y.X. Chu, J.B. Gou and Z.X. Li

Distance Computation

An Incremental Version of Growth Distance ... 3671
C-J. Ong and E. Huang

A Framework for Efficient Minimum Distance Computations ... 3678
D.E. Johnson and E. Cohen

Computing Distances between NURBS-defined Convex Objects 3685
C. Turnbull and S. Cameron

Design of a Collision Detection VLSI Processor Based on Minimization of Area-Time Products 3691
M. Hariyama and M. Kameyama

SENSOR-BASED NAVIGATION

Autonomous Land Vehicle Navigation Using Millimeter Wave Radar 3697
S. Clark and H. Durrant-Whyte

Autonomous Underground Navigation of an LHD Using a Combined ICP and EKF Approach 3703
R. Madhavan, M.W.M.G. Dissanayake and H.F. Durrant-Whyte

2 Dimensional Landmark-Based Position Estimation from a Single Image 3709
A.J. Muñoz and J. Gonzalez

Mobile Robot Exploration and Map-Building with Continuous Localization 3715
B. Yamauchi, A.C. Schultz and W. Adams

Author Index

Abba, G. ... 2008
Abdallah, I. Ben ... 1793
Abe, Y. ... 952
Abichou, A. ... 2309
Adams, W. ... 2833, 3715
Adduci, L. ... 2099
Adibi, J. ... 2613
Adobbati, R. ... 2613
Aghili, F. ... 3613
Agrawal, S.K. ... 736
Aguilar, L.E. ... 3279
Ahmadabadi, M.N. ... 3625
Ahrens, M. ... 1148
Aigner, P. ... 717
Aiyama, Y. ... 89, 2904
Akabane, T. ... 1932
Akella, S. ... 557, 565
Albu-Schäffer, A. ... 490
Alla, H. ... 1179
Allen, P.K. ... 1360, 3115
Allotta, B. ... 418, 711
Alpan G. ... 254
Alsayegh, O.A. ... 3179
Alvarez, J.C. ... 2205
Amat, J. ... 1160
Amato, N.M. ... 630
Ambrose, R. O. ... 1334
Andriot, C. ... 1924
Angeles, J. ... 1970
Antonelli, G. ... 768
Aono, T. ... 3443
Appenzeller, G. ... 838
Aracil, R. ... 2110, 2244
Arai, F. ... 952, 1198, 1498, 2482, 2890, 3210
Arai, H. ... 927, 1958, 2619
Arai, T. ... 89, 1340, 2852, 2904, 3153
Arai, Y. ... 2186
Araujo, E.G. ... 1584
Araújo, R. ... 2554
Arimoto, S. ... 1435, 2403, 3607
Arisumi, H. ... 3522
Armingol, J.M. ... 3425
Arras, K.O. ... 3129
Asada, H.H. ... 387, 399, 774, 1289
Asada, M. ... 1558, 1865
Asakawa, K. ... 1045
Asakawa, N. ... 2452
Asama, H. ... 2186, 2858, 3153
Asamura, N. ... 680
Asari, Y. ... 1083
Ashe, K. ... 699
Ashiru, I. ... 3498
Asoh, H. ... 1552
Astley, O.R. ... 989
Avizzano, C.A. ... 3721
Aycard, O. ... 3135
Backes, P.G. ... 284
Badihi, D. ... 3053
Baker, D. ... 705
Bakhadyrov, I. ... 254
Balduzzi, F. ... 1172
Baldwin, J. ... 2884
Banba, H. ... 1083
Bapna, D. ... 597, 3468
Bares, J. ... 2777
Barrientos, A. ... 2110, 2244
Basañez, L. ... 3085
Bassino, P. ... 1459
Basu, A. ... 2884
Battaglia, R. ... 1707
Baumann, R. ... 381
Baur, C. ... 3368
Bautista, J. ... 2599
Bayazit, O.B. ... 630
Beard, R.W. ... 762
Beccari, G. ... 1429
Beji, L. ... 2309
Bekey, G.A. ... 2223
Belousov, I.R. ... 1683
Beltrán-Escavy, J. ... 2852
Benjaafar, S. ... 260
Berg, M. ... 2807
Bergamasco, M. ... 3721
Bergerman, M. ... 3449
Berkemeier, M.D. ... 2345
Bessonnet, G. ... 1385
Beznos, A.V. ... 2670
Bicchi, A. ... 1038
Bicknell, A.K. ... 83
Bidard, C. ... 1924
Bishop, B.E. ... 2664
Bizzantino, P. ... 200
Blake, A. ... 2722
Blasi, R. ... 1903
Boehmke, S.K. ... 2777
Boguslavsky, A.A. ... 1683
Bohringer, K.-F. ... 1204
Boismain, G. ... 1154
Bona, B. ... 3006
Bongaerts, L. ... 2747
Bonnet des Tuves, J.M. ... 1154
Boo, K.S. ... 1689

Borenstein, J.	1572, 3456
Boudet, S.	1352, 1743
Boukhezar, A.	1185
Bourque, E.	3186
Bouthemy, P.	1063
Bouzouia, B.	1185, 2128
Bovik, A.	3259
Boyer, F.	1096
Bozorg, M.	3413
Brady, K.	65
Brett, P.	705
Brickley, J.J.	1405
Brock, O.	1
Brown, B.	781, 793
Brown, P.F.	239
Brumitt, B.L.	650, 1564, 1895
Bruneau, O.	1392
Bruyninckx, H.	1938, 2396
Brzakovic, D.P.	3179
Buehler, M.	1707, 3613
Bugacov, A.	3368
Burdick, J.W.	292, 484, 1090, 3322, 3573
Burgard, W.	1238, 1546
Buss, M.	2640, 3026
Butler, Z.J.	3109
Butterfass, J.	2081, 2087
Byun, Y.K.	1659
Caballero, A.	2244
Cadenat, V.	1269
Caffaz, A.	2075
Cahut, L.	2541
Calafiore, G.	3303
Calcara, M.	3721
Caldwell, D.G.	298, 3053
Cameron, S.	3685
Cannata, G.	2075
Cannon, H.	121
Canny, J.	869, 1471, 3173
Canudas-de-Wit, C.	452, 2036
Cao, M.	2357
Caramihai, S.I.	1179
Carelli, R.	2002
Carey, R.	3356
Carre, F.	2434
Carrozza, M.C.	1811
Carusone, J.	2959
Casadei, C.	1001
Castellanos, J.A.	1244
Castro, J.	970
Causey, G.C.	1453
Celinski, T.	3097
Cervera, E.	2174
Chablat, D.	1964, 1970
Chang, P.H.	107
Chang, W.S.	175
Charpillet, F.	3135
Chatila, R.	1232
Chatterjee, A.	2351
Chaumette, F.	1352, 1994
Cheah, C.C.	2403, 3607
Chen, I-M.	1647
Chen, J.	518
Chen, J.-S.	471
Chen, M.D.	995
Chen, T. M.	964
Chen, W.	115
Chen, Y.-C.	2122
Chen, Y.-Y.	2122
Cheng, F-T.	471, 2116
Cheng, G.	3431
Cherif, M.	881
Cherkassky, V.	3388
Chernyshov, K.R.	2383
Chevallereau, C.	1398
Chiacchio, P.	95
Chiaverini, S.	768
Chiba, K.	2989
Chirikjian, G.S.	742
Cho, B.	2613
Cho, H.S.	169, 1659
Cho, J.-H.	71
Cho, Y.-J.	206
Choi, B.J.	3727
Choi, I.-H.	2983
Choi, K.K.	2268
Choi, M.H.	1528
Choi, S.K.	1051
Choi, Y.H.	169
Choset, H.	342
Christensen, H.I.	3419
Chu, Y.X.	2728, 3646, 3652, 3665
Chung, A.C.S.	3546
Chung, W.	2709
Chung, W.-K.,	1441, 3012
Ciancio, S.	3721
Ciaravola, P.	3186
Clark, J.J.	3247
Clark, S.	3697
Clavel, R.	381
Cleary, M.E.	1256
Cocosco, A.	1707
Codourey, A.	2094
Coelho, C.	1459
Cohen, E.	3678
Cohn, M.	1204
Coiffet, P.	2877

Colla, V.	711	Driessen, B.J.	1134, 2789
Concilio, M.	95	Dubey, R.V.	1615
Cotin, S.	3739	Dubourdieu, J.M.	1154
Courdesses, M.	3279	Dubowsky, S.	278, 3316
Coutinho, M.G.	856	Dubrawski, A.	2518
Cowan, N.J.	2658	Dudek, G.	2840, 3186
Cremers, A.B.	1238	Duffey, M.R.	1490
Crétual, A.	1994	Duleba, I.	3567
Crisman, J.D.	1256	Dunlop, G.R.	3639
Crowley, J.L.	3121	Dupont, P.	976
Cruz, A.	2199	Durna, M.	3329
Cruz-Hernandez, J.M.	1510	Durrant-Whyte, H.F.	440, 2217, 3310, 3413, 3437, 3697, 3703
Czarnecki, C.A.	3498		
Dale, L.K.	630	D'Eleuterio, G.M.T.	511, 2959
Delpech, M.	3733	Ebert-Uphoff, I.	742
Dapper, M.	2048	Eckerle, J.	2147
Dar, T.	3552	Eckmiller, R.	1417, 2048
Dario, P.	1811, 2262, 2274	Egerstedt, M.	3273
Dauchez, P.	2971	Eiji, N.	3625
Davidson, C.	2722	Elfes, A.	3449
Dawson, D.M.	498	ElMaraghy, H.A.	887, 1793, 2464
de Almeida, A.T.	2554	El Mekkawy, T.	1793
De Bartolomei, M.	200	Emami, M.R.	2500, 2512
Deguet, A.	982	Emelianov, S.N.	1683
Dekhil, M.	3103	Emura, S.	3400
de la Escalera, A.	3425	Endo, I.	2186, 2858, 3153
Delahoche, L.	2560	Endres, H.	1779
Deliç, H.	2541	Eom, K.S.	3012
Delingette, H.	3739	Erdem, A.	2613
Dellaert, F.	1889	Erdmann, M.	723
del Pobil, A.P.	624, 644, 2174, 3018	Erkmen, A.M.	3329
De Luca, A.	504, 799, 3581	Ertugrul, M.	2042
de Mathelin, M.	2008	Everett, S.E.	1615
Demey, S.	2396	Faiz, N.	736
Deng, J.-Y.	2116	Fang, T.	254
Dennerlein, J.T.	674	Faraz, A.	1007
Dépincé, Ph.	1116	Farritor, S.	278
de Queiroz, M.S.	498	Fasse, E.D.	1749
Desai, J.P.	2864	Fatikow, S.	3362
De Schutter, J.	1534, 1938, 2256	Favede, C.	298
Dev, A.	1578	Fayman, J.A.	2783
Devy, M.	3091	Featherstone, R.	324
Diftler, M. A.	1334, 3348	Feddema, J.T.	1134, 1212
Dillmann, R.	538	Feiten, W.	1779
DiMaio, S.P.	52, 133	Fenu, A.	1811
Ding, H.	1719	Ferch, M.	1853
Dissanayake, M.W.M.G.	3310, 3703	Fernandez, J.A.	524, 656
Dixon, W.E.	498	Fernandez, J.J.	3032
Dodds, G.I.	336, 2910	Fernandez, X.	1160
Dodds, Z.	1366	Ferre, M.	2110
Doeschner, C.	1737	Ferreira, P.	580
dos Reis, N.R.S.	1782	Ferretti, G.	2995
Douillard, Y.	1154	Ferrière, L.	1877

Filali-Adib, K.	3639
Finotello, R.	3241
Fioretti, S.	393
Fiorini, P.	1001
Fischer, M.	1603
Flannigan, W.C.	152
Fleury, S.	3279
Flores, J.	2369
Foessel, A.	611, 3468
Fok, S.C.	1423
Formal'sky, A.M.	1398, 2670
Fox, D.	1238, 1546
Fraichard, T.	27
Fraile, J.C.	2910
Friedrich, H.	538
Froehlich, C.	1465
Frontzek, T.	1417
Fryer, J.A.	3167
Fu, L.-C.	304, 818, 1805, 2122, 2134
Fuji, S.	2155
Fujii, K.	3443
Fujii, T.	756, 2186, 2858
Fujimoto, H.	1916
Fujimoto, Y.	2030
Fujimura, S.	1250
Fujita, M.	2605
Fujiwara, F.	1142
Fujiwara, K.	1013
Fukuda, O.	3492
Fukuda, T.	787, 952, 1198, 1447, 1498, 1829, 2482, 2846, 2890, 3210
Funato, D.	3210
Fung, Y.H.	2506
Furusho, J.	2470, 2586
Furutani, K.	1504
Furuya, N.	2428
Fushimi, M.	2155
Gadh, R.	1478
Galicki, M.	101
Gallo, L.	2434
Gangloff, J.A.	2008
Garcia, G.	2422
Garcia, M.	2351, 3085
García-Cerezo, A.	1883, 2199
Garibotto, G.	1459
Gazen, B.C.	3368
Gengenbach, V.	2947
Geoffroy, S.	1154
Ghallab, M.	605
Ghanekar, M.	2633
Ghosh, B.K.	1671
Giles, K.	7
Gilhuly, T.J.	699
Glandais, N.	1096
Glozman, D.	1327
Goerke, N.	1417
Gogate, R.	3356
Goldberg, K.	1204
Goldenberg, A.A.	2500, 2512
Goldfarb, M.	1609
Gonthier, Y.	127
Gonzalez, J.	524, 656, 3709
Goodwine, B.	484
Gorbet, R.B.	2161
Gosselin, C.M.	1749, 2287, 3071
Goswami, A.	2036
Goto, M.	1340
Gou, J.B.	2728, 3646, 3652, 3665
Gout, J.	605
Grasso, T.	3241
Grimaudo, D.W.	1767
Groen, F.C.A.	1578
Grupen, R.	1584
Gruver, W.A.	115, 2476
Guerrini, M.	2274
Guerroumi, F.	1185
Guo, S.	1829
Gupta, K.	239, 544, 550, 3079
Gurfinkel, E.V.	2670
Gusev, S.V.	3285
Hacot, H.	278
Haddad, H.	1232
Haffner, H.	1773
Hager, G.D.	1366, 1988
Haikawa, Y.	1321
Hailu, G.	2168
Hallam, J.C.T.	2795
Hamada, K.	2428
Hamel, T.	1269
Hamel, W.R.	3594
Hamilton, K.	336
Hammond, T.	7
Han, H.-Y.	1435, 2403
Han, K.	3510
Han, L.	730, 1944, 3631
Hanach, C.	2128
Hanafusa, H.	3486
Hancock, J.	1465
Hanebeck, U.D.	2813
Hannaford, B.	369
Hara, I.	1552
Hara, K.	1623
Harada, K.	2409
Hardey, G.	83
Hariyama, M.	3691
Hasegawa, T.	933

Hashimoto, H.	838	Hughes, P.C.	1665
Hashimoto, K.	2321	Husson, R.	3141
Hashimoto, S.	1841	Hutchinson, S.	1677
Hashizume, H.	1013	Hwang, L.-S.	1262
Hashizume, T.	2416	Iagnemma, K.	3316
Hashtrudi-Zaad, K.	52, 133	Iizuka, Y.	1013
Hatsumoto, S.	3443	Inaba, M.	146, 1302, 2281, 3253
Hawker, G.	1707	Inaba, T.	446
Hayward, V.	989, 1510, 3228	Indri, M.	3006, 3303
Hebert, M.	1465, 1895	Ingimarson, D.	1465
Heinzmann, J.	1226	Inoue, F.	2416
Henderson, T.C.	3103	Inoue, H.	146, 1302, 2281, 3253
Henmi, K.	1275	Inoue, Y.	2428
Henrich, D.	330	Inui, M.	2446
Henrichfreise, H.	187	Iñesta, J.M.	3018
Heppler, G.R.	2633	Ishii, K.	756
Hernando, M.	2110	Ishikawa, M.	2333
Hespanha, J.	1366	Itabashi, K.	1142
Higashimori, M.	3040	Itoigawa, K.	1498
Hirai, H.	907	Ivanescu, M.	1540
Hirai, K.	1321	Iwama, K.	2416
Hirai, S.	3474	Iwata, H.	1498
Hirana, K.	1142	Jafari, M.A.	254
Hirano, G.	940	Janét, J.A.	1405
Hirano, S.	1308	Jensfelt, P.	3419
Hirose, M.	1321	Jeong, J.W.	2941
Hirose, S.	1713, 2896	Jia, Y.-B.	723
Hirukawa, H.	2870	Jiang, S.L.	2268, 3631
Hirzinger, G.	490, 1603, 2081, 2087, 2646	Jicharev, D.N.	2670
Hocaoglu, C.	247, 318	Jin, B.D.	1597
Hoffmann, E.	1465	Johnson, C.G.	194
Holle, J.	538	Johnson, D.E.	375, 3678
Holleman, C.	21	Jones, C.	630
Hollerbach, J.M.	375, 2566, 3613	Joo, S.	1069
Hollis, R.L.	2488, 3109	Jörg, K.-W.	2807
Honegger, M.	2094	Joseph, J.	2147
Hori, K.	3047	Joskowicz, L.	3552
Hori, T.	3474	Joukhadar, A.	982
Horng, J.-T.	1411	Jung, D.	1226
Hoshino, K.	1641	Jung, S.	1731
Hoshino, Y.	2281	Kabasawa, M.	146
Hosoda, K.	1558, 1865	Kaburlasos, V.G.	705
Hosokawa, K.	2858	Kaetsu, H.	2858
Hovland, G.E.	2211	Kaga, T.	2846
Howe, R.D.	674, 1204	Kagami, S.	146
Hsia, T.C.	1731	Kageyama, K.	2605
Hsu, J.Y.-J.	1262	Kakikura, M.	3474
Hu, X.	3273	Kakita, S.	2852
Huang, E.	3671	Kakiuchi, Y.	1302
Huang, H.-P.	687	Kakizaki, T.	2375
Huang, Q.	13, 2192	Kaliaev, I.A.	2536
Huang, W.H.	1077	Kallio, P.	1823
Huang, Y.-H.	1805	Kamada, T.	2229

Kameyama, M.	3691	Kimura, T.	2155
Kamiya, T.	3443	Kitagaki, K.	2870
Kanayama, K.	1166	Kitano, H.	2605
Kanehiro, F.	1302	Kiyonaga, T.	2416
Kaneko, K.	2524	Klarquist, W.	3259
Kaneko, M.	2409, 3040, 3492, 3619	Kleeman, L.	1590
Kang, G.	750	Knoch, S.	2081
Kang, S.	826	Knoll, A.	2965
Kanno, R.	2155	Kobayashi, H.	1653
Kao, C.-Y.	1411	Koch, R.	2771
Kato, K.	1713	Kock, S.	2295
Kato, M.	2416	Koditschek, D.E.	787, 2658, 2676
Kato, N.	446, 1045, 1829	Koel, B.	3368
Kato, Y.	1045	Koga, S.	3153
Kavraki, L.E.	21	Koivo, A.J.	812, 2703
Kawaji, A.	2482	Koivo, H.N.	1823
Kawamura, A.	2030, 2357	Kojima, J.	1045
Kawamura, S.	693, 1435, 2403, 3486, 3607	Kokaji, S.	432
Kawarazaki, N.	933	Kolacinski, R.M.	1629
Kay, M.G.	1405	Komatsu, T.	1932
Kayama, K.	3253	Komine, N.	2301
Kaynak, O.	2042	Komoriya, K.	3522
Keçeci, F.	2947	Konno, A.	146, 3253
Kelly, R.	2327, 2369, 3601	Koo, T-W.	1701
Kemurdjian, A.L.	140	Kornbluh, R.	2147
Keren, D.	3375	Kosecká, J.	1903
Kersting, B.	2953	Koseki, Y.	1340
Kesavadas, T.	266	Kosuge, K.	1841, 3297, 3534
Kessler, M.	3619	Kosuge, K.	2989
Khalil, W.	1096	Kotani, S.	2524
Khatib, M.	1232	Kotay, K.	424
Khatib, O.	1	Kothandaraman, K.	580
Kheddar, A.	2877	Kotoku, T.	3522
Khorasani, K.	806	Kovács, G.L.	1787
Kiguchi, K.	1447	Koyasako, K.	1302
Kikuchi, J.	3534	Kraus, P.R.	976
Kim, D.-H.	3216	Krautgartner, P.	2315
Kim, H.J.	2689	Kreczmer, B.	40
Kim, H.S.	2689	Kröse, B.J.A.	1578
Kim, J.-H.	71, 1695, 2983, 3216	Kruse, E.	662
Kim, J.-S.	1110	Kuchen, B.	2002
Kim, J.W.	1032	Kumar, P.	580
Kim, K.-C.	3216	Kumar, V.	638, 976, 2396, 2864, 2916, 2922
Kim, K.D.	3528	Kumar Sarkar, P.	1104
Kim, K.I.	3480	Kung, F-C.	471
Kim, M.	826, 2574	Kuntze, H-B.	1773
Kim, N.-I.	2716	Kurabayashi, D.	89, 3153
Kim, S.	2627	Kurazume, R.	2896
Kim, S.-H.	107, 2941	Kuroda, Y.	2858
Kim, T.Q.	2494	Kurokawa, H.	432
Kim, W.K.	1659, 2697	Kwak, Y.-K.	2941
Kim, Y.-J.	3216	Kweon, I.S.	3394
Kimura, K.	2140	Kwok, K.S.	2789

Kwon, D.S.	169, 1597	Lind, M.	1823
Kwon, W.	227, 1528	Lipkin, H.	2652
Kwon, W.H.	227, 1528	Little, C.Q.	875
Kyriakopoulos, K.J.	2977	Liu, G.	3316
Lacroix, S.	1232, 2827, 3147	Liu, H.	2081, 2087
Lamiraux, F.	3291	Liu, S.	310, 314
Lane, D.M.	1057	Liu, S.-H.	2134
Lange, F.	2646	Liu, Y.	1835
Langer, D.	1465	Liu, Y.-H.	2734, 3335
Laroche, P.	3135	Llama, M.A.	2369
Laschi, C.	2274	Lloyd, J.E.	2935, 3228
Laubach, S.L.	292	Longhi, S.	393
Laugier, C.	982, 3285	Loucks, C.S.	2789
Laumond, J.P.	33, 46, 3291	Louda, M.A.	3310
Laurendeau, D.	3071	Lucibello, P.	504
Lawitzky, G.	1779	Luck, J.P.	875
Lawley, M.	591	Luh, P.B.	1799
Lawrence, P.D.	699	Lumelsky, V.	7, 2205
Lay, N.K.	2777	Luntz, J.	611
Lee, B.H.	227, 1528	Luo, R. C.	964
Lee, C.	844	Lynch, K. M.	927, 1958
Lee, C.-W.	206, 826, 2574, 2716, 3192	Lyons, K.W.	1490
Lee, D.H.	1689	Maass, R.	2048
Lee, H.G.	2689	Macnab, C.J.B.	511
Lee, J.-H.	838, 2068	Madhavan, R.	3703
Lee, K.-I.	826	Madhukar, A.	3368
Lee, K.-M.	3356	Maeda, K.	1083
Lee, K.-P.	1701	Maekawa, H.	2566
Lee, S.	1484, 1689, 2574, 2599, 3192	Magnani, G.	200, 2995
Lee, S.H.	227, 1528	Mahkovic, R.	349
Lee, T.S.	2458	Maimone, M.	597
Lee, W.	750	Makarov, I.A.	3285
Lee, W.-S.	71	Makikawa, M.	3486
Lefort, L.	266	Mali, A.D.	1122, 1128
Leifer, L.	77	Malik, J.	1903
Lenarcic, J.	3235	Malis, E.	1352, 1743
Lensky, A.V.	2670	Mallet, A.	2827
Leo, T.	393	Mallick, V.	1148
Leoni, F.	2274	Mantripragada, R.	219
Leroy, S.	46	Marchand, E.	1988
Leu, M.C.	2494	Mareczek, J.	2640
Lewis, F.L.	181	Marhefka, D.W.	477
Li, J.	2476	Marsh, D.	194
Li, L.	115	Marson, M.	1459
Li, Z.X.	181, 2268, 2728, 3631, 3646, 3652, 3665	Martelli, S.	1001
Lichtenstein, S.	699	Martinelli, F.	2753
Lim, H.-O.	13	Martínez, J.L.	1883
Lim, S.Y.E.	1423	Martínez, J.M.	1244
Limaiem, A.	887	Martínez-Salvador, B.	624, 644
Lin, I.-C.	818	Mascaro, S.	399, 1289
Lin, M.-H.	304	Masciangelo, S.	1459
Lin, Q.	3322	Mason, M.T.	557, 565, 1077
Lin, Y.-J.	893, 2458	Masutani, Y.	907

Matsui, T.	1552	Morri, A.	1001
Matsuki, T.	146	Morris, K.A.	2161
Matsumoto, K.	3159	Morse, A.S.	1366
Matsumoto, Y.	146	Mosemann, H.	233
Matsuura, H.	1198, 2482	Motegi, M.	2375
Matthies, L.	272, 292	Mouaddib, M.	2560
Mattone, R.	2099	Mounassypov, R.	3362
Matumoto, S.	1045	Mukerjee, A.	1122
Mazar, B.	2434	Munawar, K.	1817
Mazer, E.	1154	Muñoz, A.J.	3709
McCarragher, B.	518, 717, 2211, 3097	Muñoz, V.F.	2199
McDermott, D.V.	2530	Mura, F.	1641, 1859
McGray, C.	424	Murata, S.	432
McKee, G.T.	3167	Murota, M.	1871
McLain, T.W.	762	Murphy, J.	597
McMillan, S.	464	Murugappan, P.	893
McMurray, G.V.	2652	Muszynski, R.	3222
Mediavilla, M.	2910	Mut, V.	2002
Medrano-Cerda, G.A.	3053	Muto, S.-Y.	2375
Megel, F.	2434	Mutschler, E.	2777
Melchiorri, C.	1026	Nagai, K.	3486
Menciassi, A.	1811	Nagashima, K.	3253
Meng, M.	1725	Nagatani, K.	342
Menga, G.	1172	Nagel, H.-H.	2947
Merlet, J.-P.	1976, 1982	Nahvi, A.	375
Mermond, R.	27	Nakabo, Y.	2333
Messina, L.C.P.	1782	Nakai, H.	1083
Mettenleiter, M.	1465	Nakamura, A.	2852, 2870
Meusel, P.	2087	Nakamura, T.	2592
Mezgár, I.	1787	Nakamura, Y.	1166, 2301
Micaelli, A.	1924	Nakanishi, I.	3486
Miki, N.	3065	Nakanishi, J.	787
Mills, J.K.	1835, 1847	Nakawaki, D.	1069
Mimura, N.	2140	Nakazawa, M.	1635
Minami, M.	2904	Nandy, G.C.	2683
Mitsuishi, M.	1013	Nanjo, Y.	530
Miura, J.	3267	Narahari, Y.	1490
Miyagawa, T.	3047	Naruse, H.	2580
Miyazaki, F.	907, 1069	Nasisi, O.	2002
Mizukawa, M.	1166	Natale, C.	1755
Mizuuchi, I.	1302	Natonek, E.	406
Mochiyama, H.	1653	Nebot, E.M.	2217, 3413, 3437
Mohri, A.	940, 1104	Nechyba, M.C.	2237, 2250
Mohri, N.	1504	Negoro, M.	2890
Montanari, M.	1001	Neira, J.	1244
Moradi, H.	2613	Nelson, B.J.	1220
Morel, G.	1743, 3316	Nelson, D.D.	375
Moreno, L.	3425	Nelson, G.M.	152, 157
Moreyra, M.	369	Nenchev, D.N.	913, 920, 2580
Mori, H.	2524	Newman, P.	440
Morii, H.	1641	Newman, W.S.	3658
Morishima, K.	1198	Nguyen, K.	2116
Morita, T.	1315	Nickels, K.	1677

Niemeyer, G.	1909
Nilsson, M.	412
Nishihara, K.	933
Nishino, D.	2022
Nissoux, C.	33
Nof, S.	2765
Nohmi, M.	920
Nojiri, H.	3198
Noritsugu, T.	2321
Oaki, J.	1083
Obata, S.	2030
Ogasawara, T.	2870
Ogata, H.	530
Ogawa, H.	1083
Oguro, K.	1829, 2155
Oh, P.Y.	1360
Oh, S.-R.	13, 206, 2068, 2697, 3012
Oh, Y.	1441
Ohnishi, M.	1635
Ohtsuka, A.	3492
Ohtsuki, K.	1635
Oikawa, K.	2229
Okada, K.	146
Okada, M.	2301
Okada, T.	1083, 2140
Okamura, A.M.	674
Okhotsimsky, D.E.	1683
Okuma, S.	1142
Okuyama, Y.	1379
Olguín Díaz, E.	452
Olson, C.F.	272
Omata, T.	2180
Omote, Y.	2416
Onda, H.	2870
Ong, C-J.	3671
Oosumi, T.	2989, 3297
Orin, D.E.	464, 477
Oriolo, G.	3581
Ostrowski, J.	638, 2864
Osumi, H.	3198
Ota, H.	2482
Ota, J.	77, 89, 2852, 3153
Ouelhadj, D.	2128
Ouezdou, F. Ben	1392
Ozaki, F.	1083
O'Brien, D.J.	1057
Pai, D.K.	1950
Pampagnin, L-H.	2422
Panzieri, S.	799
Papadopoulos, E.	127
Papanikolopoulos, N.P.	3388
Paraskevopoulos, P.N.	2363
Paredis, C.J.J.	239
Parfrey, K.	1148
Park, F.C.	1032
Park, J.	206, 2709
Park, J.-M.	206
Park, J.-O.	3192
Park, J.H.	3528
Park, K.C.	107
Park, S.	2574
Park, T.H.	227
Paromtchik, I.E.	3285
Parra, C.	3091
Parra-Vega, V.	1522
Pascal, M.	2309
Pashchenko, F.F.	2383
Patel, R.V.	806
Paulos, E.	3173
Pavan, A.	1459
Payandeh, S.	544, 863, 1007
Payeur, P.	3071
Pedersen, L.	618
Pégard, C.	2560
Peirs, J.	1516
Pelrine, R.	2147
Peñín, L.F.	2110, 2244
Perán, J.R.	2910
Peremans, H.	2795
Pérez-Francisco, M.	624, 644
Perret, J.	2106
Perrier, C.	2971
Perrier, M.	452
Perrin, B.	1398
Peters, B.A.	586
Petridis, V.	705
Peyret, F.	2422
Phillis, Y.A.	2759
Piat, E.	3147
Piazzi, A.	946
Piepmeier, J.A.	2652
Pierrot, F.	2971
Pinna, S.	3721
Pinto, E.	2110
Pisano, A.	1204
Pitt-Francis, J.	324
Platonov, A.K.	1683
Podsedkowski, L.	3588
Poi, G.	418
Politis, Z.	2801
Pollefeys, M.	2771
Pomerleau, D.	1889
Ponce, J.	2740
Popa, D.O.	59
Pozo-Ruz, A.	1883
Pratt, G.	2014

Pratt, J.	2014	Sanderson, A.C.	247, 318
Prattichizzo, D.	1038	Sano, A.	1916
Prisco, G.M.	3721	Santibañez, V.	2369, 3601
Probert, P.	2801	Santos, V.	970
Qian, W.-H.	3407	Sanz, P.J.	3018
Qin, Z.	3631	Sappa, A.D.	3085
Quaid, A.E.	2488	Saranli, U.	2676
Quinn, R. D.	152, 157, 1453, 1629	Saridis, G.	2363
Ramachandran, T.R.	3368	Sarkis, J.	1707
Ramos, M.	2703	Sarrigeorgidis, K.	2977
Ramos, Jr., J.G.	3449, 3462	Sasaki, H.	693
Rastel, L.	3733	Sassani, F.	1346
Raucent, B.	1877	Sato, H.	1083, 1498
Reboulet, C.	52, 133	Sato, K.	1308
Recatalá, G.	3018	Satou, M.	2989
Rendas, M.J.	3382	Savitsky, K.V.	2670
Requicha, A.A.G.	3368	Sazonov, V.V.	1683
Resch, R.	3368	Scarabeo, C.	418
Reveliotis, S.A.	573	Scheding, S.	2217
Reyes, F.	2327	Scheering, C.	2953
Reynaerts, D.	1516, 3059	Schiele, B.	3121
Reznik, D.	869, 1471	Schlegl, T.	3026
Rhee, S.	387	Schmidt, G.	2640
Ribeiro, M.I.	970	Schultz, A.C.	2833, 3715
Rimon, E.	356, 2928, 3322	Schumacher, W.	2295
Rivlin, E.	2783, 3375, 3552	Schwind, W.J.	2676
Rizzi, A.A.	832, 3109	Seki, H.	3297
Ro, S.	3192	Sekiyama, K.	3210
Roberts, R.S.	875	Selig, J.M.	1761
Robinett, R.D.	1134	Seraji, H.	899
Rocco, P.	2995	Serra, B.	2434
Rodnay, G.	2928	Shamah, B.	611
Rodríguez-Angeles, A.	1522	Sharma, R.	1373
Roh, K.S.	3394	Shen, E.	2116
Röhrdanz, F.	233	Shen, H.C.	3407, 3546
Rolfes, S.	3382	Shen, W.M.	2613
Rollins, E.	597, 611, 3468	Shi, J.-X.	490
Ros, L.	3559	Shibuya, K.	1315
Rossi, G.	3241	Shikano, M.	952
Rostami, M.	1385	Shim, J.H.	169
Roumeliotis, S.I.	2223	Shimemura, E.	1653
Roussel, L.	2036	Shimoyama, I.	1641, 1859, 3065
Rubinovitz, J.	2765	Shimshoni, I.	3375
Rude, M.	3159	Shin, D.	1689, 2689
Ruina, A.	2351	Shinoda, H.	680
Rus, D.	424	Shinoda, T.	2524
Rusaw, S.	544	Shirai, Y.	3267
Rye, D.C.	3310	Shiroma, N.	927, 1958, 2619
Sakaguchi, M.	2470, 2586	Shkel, A.	2205
Salcudean, S.E.	52, 133, 699, 1346	Shmoulian, L.	356
Salemi, B.	2613	Shoham, M.	1327
Salichs, M.A.	3425	Shoval, S.	1767, 2765
Salvarinov, A.	863, 1007	Siciliano, B.	1755

Siemiatkowska, B.	2518	Sun, L.	1737
Sima'an, N.	1327	Sun, Y-Y.	471
Siméon, T.	46	Surdilovic, D.	213
Simhon, S.	2840	Sutanto, H.	1373
Simon, R.W.	1212	Sutton, III, J.C.	1405
Singh, R.	3388	Suzuki, H.	458
Singh, S.	121, 650	Suzuki, T.	1142
Singh, S.K.	59	Suzumori, K.	3047
Siqueira Bueno, S.	3449	Tachi, S.	3400
Skubic, M.	1281	Tachimori, M.	1083
Slivnik, T.	349	Taddeucci, D.	2262, 2274
Slotine, J.-J.E.	1909	Tadokoro, S.	2155
Smith, A.C.	2345	Tafazoli, S.	52, 133
Smith, J.S.	586	Tagare, H.D.	2530
Smith, P.W.	3540	Takada, R.	3474
Sokolov, S.M.	1683	Takahashi, H.	3516
Sommer, G.	2168	Takamori, T.	2155
Song, B.	812	Takanishi, A.	13, 1308, 1635, 2022
Song, J.	2250	Takanobu, H.	1635
Song, J.I.	169	Takatuji, T.	1340
Song, M.	1192	Takemori, F.	1379
Souéres, M.P.	3279	Takenaka, T.	1321
Souères, P.	1269	Takeo, K.	2989, 3534
Sówka, J.	3567	Takesue, N.	2470
Spindler, F.	1063	Takeuchi, K.	3341
Spong, M.W.	2664	Takeuchi, Y.	2452
Sreenivasan, S.V.	3727	Takiguchi, J.	2416
Srinivasan, H.	1478	Talebi, H.	806
Sriram, R.D.	1490	Tanaka, E.	1083
Stankovich, B.	7	Tanaka, M.	1916
Starita, A.	2274	Tanaka, Y.	952
Steele, R.	899	Tanie, K.	13, 927, 1958, 2192, 2619, 2877
Stentz, A.	650, 1564	Taniguchi, Y.	1083
Stocco, L.J.	1346	Tanimoto, M.	2890
Stoian, V.	1540	Tao, Q.	458
Stotsky, A.	3273	Tardós, J.D.	1244
Stramigioli, S.	1429	Tarn, T.-J.	65, 1192, 1296, 1671
Su, S.	850	Tatsuno, K.	1083
Subbu, R.	247	Taylor, C.J.	1903
Sudarsan, R.	1490	Tchesalin, L.S.	2670
Sudarsky, O.	2783	Tchon, K.	3222
Sudsang, A.	2740	Tejada, S.	2613
Suehiro, T.	3516	Tejima, N.	3486
Suematsu, K.	1641	Terakado, H.	2446
Sugano, S.	1315, 2192	Terasawa, M.	3198
Sugar, T.	2916	Terribile, A.	3241
Sugie, H.	2416	Tharp, G.K.	284
Sugimoto, K.	1340	Thomas, F.	3559
Suh, I.-H.	1441, 2068, 2697, 3012	Thorpe, C.	1889
Sukhatme, G.S.	2223	Thrun, S.	342, 958, 1238, 1546
Sukkarieh, S.	3437	Tian, Z.M.	995
Sullivan, R.	1465	Toda, K.	2452
Sun, D.	1835, 1847	Tolle, H.	3619

Tomita, K.	432	Vidal, M.	881
Tomori, N.	680	Vikramaditya, B.	1220
Tonko, M.	2947	Villani, L.	1755
Torras, C.	2819	Vincze, M.	2315
Tounsi, M.	1154	Visentin, G.	200
Trinkle, J.	730, 1944	Visioli, A.	946
Tsagarakis, N.	298, 3053	Vlassis, N.A.	363
Tsanakas, P.	363	Volz, R.A.	1281
Tso, K.S.	284	Vona, M.	424
Tso, S.K.	1719, 2506	von Collani, Y.	2965
Tsuda, M.	530	von der Hardt, H-J.	3141
Tsuji, T.	3040, 3492	Vuskovic, M.I.	3639
Tsujimori, T.	2858	Wada, M.	774
Tsukune, H.	2870	Wahl, F.	233, 662
Tsumaki, Y.	2580	Wakisaka, T.	2428
Türksen, I.B.	2500, 2512	Wakita, Y.	3474
Turnbull, C.	3685	Walker, A.	2795
Turner, E. H.	2060	Walker, I.D.	3032, 3348
Turner, R.M.	2060	Wallner, F.	3121
Uchibe, E.	1558, 1865	Walter, J.A.	2054
Uchiyama, M.	920, 1110, 1817, 2580	Wang, D.W.L.	2161, 2633
Uenohara, M.	1083	Wang, J.	2287
Ulivi, G.	799	Wang, L-X.	181
Ullrich, C.	1950	Wang, M.	3335
Ulrich, I.	1572	Wang, Q.	2256
Um, D.	7	Wang, T.	995
Ura, T.	756	Wang, X.	314
Urushibata, M.	1504	Wang, X.-G.	3407
Uzmay, I.	850	Wang, Y.	1799
Vadakkepat, P.	3216	Wang, Z.	310, 314
Vafaeesefa, A.	2464	Ward, K.	668
Vainio, O.	3001	Warren, J.	21
Valavanis, K.P.	2541	Watanabe, H.	1013
Valckenaers, P.	2747	Watanabe, K.	458
Valigi, P.	2753	Wattanasin, C.	89
Vallejo, D.	630	Wei, S.-C.	2548
Van Brussel, H.	1516, 2747, 3059	Wei, Y.-F.	687
van der Linde, R.Q.	2339	Weigl, A.	3619
Van der Loos, M.	77	Weiss, I.	3375
van de Smagt, P.	1603	Wells, G.	2819
Van Gool, L.	2771	Wen, J.T.	1020
Varella Gomes, S.B.	3462	Wenger, P.	1964, 1970
Varvatsoulakis, M.N.	2363	Werger, B.B.	3504
Vasseur, P.	2560	Wettergreen, D.	597
Vassura, G.	1026	White, M.W.	1405
Velasco, Jr., V.B.	3658	Whitney, D.E.	219
Velazquez, S.	3085	Whittaker, R.	3468
Veloso, M.	3510	Whittaker, W.	597, 611
Vendittelli, M.	33	Widner, T.C.	3594
Verdot, F.	1154	Wieber, P.B.	1392
Vergauwen, M.	2771	Wijk, O.	3419
Versteyhe, M.	3059	Wilfinger, L.S.	1020
Vestli, S.J.	3129	Will, P.	856, 3368

Witte, C.	187	Yi, B.-J.	2068, 2697
Wöern, H.	330	Yi, C.	1484
Wolf, A.	2099	Yokogawa, A.	1623
Wolf, D.	3141	Yokogawa, R.	1623
Woo, K.T.	181	Yokoi, K.	13
Woo, K.Y.	1597	Yoon, Y.-S.	1701
Wu, H.	2728, 3652	Yoshida, E.	432
Wu, J.-L.	693	Yoshida, H.	1841
Wunsch, P.	2646	Yoshida, K.	458, 913
Wurll, C.	330	Yoshikawa, K.	1198
Xi, N.	1192, 1296, 1671	Yoshikawa, T.	1275, 2389, 3341
Xiao, D.	1671	Yoshimi, T.	1083
Xiao, H.	2530	Yoshino, K.	2416
Xu, Y.	844, 2237, 2250, 2683	You, B.J.	3394
Yabu, A.	1379	Youcef-Toumi, K.	175
Yachida, M.	1250, 2548	Youm, Y.	1441, 2709
Yagi, Y.	1250, 2548	Yu, Y.	3079, 3341
Yahja, A.	650	Yuh, J.	1051
Yajima, T.	1635	Yuta, S.	1590
Yakubovich, V.A.	3285	Zahn, V.	2048
Yam, Y.	2250	Zanganeh, K.E.	1665
Yamada, S.	1871	Zefran, M.	1090, 2922
Yamaguchi, H.	3204, 3573	Zeglin, G.	781, 793
Yamaguchi, J.	2022	Zelinsky, A.	668, 1226, 3431
Yamaguchi, O.	1083	Zergeroglu, E.	498
Yamakawa, H.	3516	Zha, X.F.	1423
Yamamoto, M.	940, 1104	Zhang, G.	2470
Yamane, K.	2301	Zhang, H.	638, 2884
Yamane, T.	3267	Zhang, J.	1853, 2965
Yamano, M.	1110	Zhang, Q.	115, 995
Yamauchi, B.	3715	Zhang, R.	550, 2759
Yamazaki, K.	1707	Zhang, Y.	995
Yang, B.-H.	387	Zhao, X.	1799
Yang, G.	1647	Zheng, Y.F.	3480
Yang, J.-M.	1411, 1695, 2983	Zhoga, V.V.	163
Yang, J.H.	2134	Zhou, Q.	1823
Yang, K.C.	1051	Zhou, Y.	1220
Yang, K-T.	471	Zhu, W.-H.	1534
Yang, R.	3115	Zrehen, S.	2605
Yang, X.	1725	Zyburt, J.P.	1767
Yata, T.	1590		

Visual Guidance of a Small Mobile Robot Using Active, Biologically-Inspired, Eye Movements

Fabrizio MURA & Isao SHIMOYAMA
The University of Tokyo, Dept. of Mechanical Engineering, Mechano-Informatics
7-3-1 Hongo, Bunkyo-ku, Tokyo 113, Japan. (fabrizio, isao)@leopard.t.u-tokyo.ac.jp

Abstract

In this paper we introduce and implement a new way of processing visual motion, which is directly inspired by the functional features of insects' visual systems. Rather than focusing on the computational performances of visual processing, alternative ways of detecting visual cues involving controlled eye movements are investigated. In this case, we show that the combination of retinal/eye movements significantly enhances the visual perception of motion for a sighted observer. The two kinds of eye movements introduced are: 1) **pulsed retinal scanning**, and 2) **visually-guided eye rotations**. Pulsed retinal scanning helps the eye to detect visual contrasts with enhanced reliability by increasing the S/N ratio of early visual signals. Rotations of the eye help a moving observer to maintain a continuous visual contact with obstacles while tracking their motion using pulses of retinal scanning. A direct demonstration of the performances has been implemented on a prototype eye, that succesfully guides the motion of HECTOR, a small mobile platform (diameter 0.12m; weight 0.75kg), autonomously and in real-time.

I. Introduction and motivation.

The age of digital computers has brought about consirerable progress to the field of robotic vision, raising new challenges in the fields of computational vision and machine vision. With a computer however, the overall performances of a robotic system are often restricted by the computational power of the digital system in use.

More recently, alternative approaches have been developed to alleviate the costs that an increase in the computational means entails. These approaches include; the concept of active vision; 'intelligent' silicon vision chips; biologically-inspired vision. Active vision is the process by which controlled movements of the visual sensor may enhance the quality and the amount of visual information, thereby releasing some of the burden in computational processing ([1]). Active vision systems have proven to be effective in overcoming the computational overload for robotic tasks including visually guided navigation and visual tracking ([2], [3], [15]). The design of intelligent vision chips focuses on the integration of visual processing together with visual sensing, typically supported by parallel circuit implementations. This line of approach is increasingly popular and has inspired the design of real-time, application-specific vision systems ([16], [13], [12]). There are increasing efforts towards the simulation and design of biologically-inspired visual systems, often in combination with the two aforementionned approaches ([6], [18], [19]). In particular the knowledge of the neural structure and function of insect visual systems have generated more cost-effective ways of processing visual information ([17], [19]). These recent developments have shown that real-time vision systems do not necessarily require large amounts of computation to operate efficiently. Furthermore, these approaches highlight the diversity of possible designs; there is no unique vision system and no single computer architecture which, if properly programmed, could perform a robotic visual task in the most reliable way.

The visual systems of walking/flying insects, as opposed to robotic visual systems operating with standard CCD devices, have developed remarkably versatile structures. The former use visual information in order to avoid obstacles, track moving tragets, stabilise flight, navigate towards a goal, land and take-off, escape from predators etc... Insect's visual sensors also share interesting properties: in the early stages of visual detection, analogue potentials propagate in continuous asynchronous flows to the higher order levels along parallel channels. The circuitry involved in the detection of visual motion is a network which closely integrates sensing and early-stage processing. This finely-tuned analogue pre-processing seems to be crucial for visual perception because it could ease off decision-making at higher processing level ([5], [9]). Eye movements also play an important role in insect visually-guided behaviour: rather than having to process a large amount of data, insects are able to control their eye movements in order to select the pertinent visual cues ([4], [10], [14]). The type of eye rotations for visual tracking proposed in this work does not rely on any memorisation of high-order image properties of the environment. Instead, visual tracking is the result of a feedback process by which the robot HECTOR maintains a continuous visual contact with the surrounding obstacles during self-motion. Each obstacle is described by the local variation in light intensity detected by the ACD circuit (Analog Contrast Detector).

Two particular eye movements of interest in our study are: pulsed retinal scanning movements for contrast detection and visually-guided eye movements for tracking. Scanning movements have been found to exist in the retina of many arthropods, with various types of scanning motion patterns ([8], [7]). These movements are generated by the contraction of muscles, that shift the orientation of the photoreceptors for small durations (typically several 100ms) and then bring them back to their initial position. The function of the small amplitude scanning movements is not yet clear, even though many observations tend to suggest that scanning could enhance the detection of moving objects such as obstacles, prey or predators. There is also some evidence that retinal scanning is a reflex-type mechanism which is involved in the early stages of visual processing. [17] have designed the first prototype of artificial scanning retina inspired by the fly's retinal scanning motion. Having only 24 pixels, the sensor is able to compute visual motion and extract the translatory optic flow in order to help a mobile platform steer its way. In [17], the S/N ratio of the visual inputs is advantagously enhanced by generating a smooth mechanical scanning. Scanning also allows the optic flow to be computed reliably, considering the low optical resolution of the eye (the inter-receptor separation across the eye is 3° ~3.5°). Additionally, smooth retinal scanning allows the visual system to perform straightforward self-calibration and auto-zero functions.

This paper is presented in the following order: section 2 addresses some key issues in motion perception for low resolution visual systems and presents the concept of fast retinal scanning; section 3 describes the hardware implementation of HECTOR's fast scanning retina and its overall performances; section 3.2 reports on the use of the scanning visual sensor in a simple obstacle avoidance strategy; section 4 summarizes the study and outines the advantages of pulsed retinal scanning and eye movements.

2. Perception of *visual motion*.

In the 2D environment such as presented in Figure 1a, light-contrasted features representing physical obstacles can be readily described by a set of points M1, M2, M3 on a plane surface. As a sighted observer 'O' moves in a straight horizontal line at the speed V, the contrast points {Mi} generate a field of *visual motion* while travelling across 'O''s retina. In the restricted case of a translatory motion, the angular velocity of the contrast points can be readily related to the observer's speed, the distance D and orientation φ of M seen from 'O''s line of move:

$$\Omega_t = V/D.\sin(\varphi) \qquad \text{equ (1)}$$

The angular velocity of M can be computed by solving the above equation, assuming that 'O' moves at a constant speed and M is a stationary point. The computation of Ω_t with regard to time on the y-axis (see Fig 1b) is carried out by varying the initial distance D_0 and keeping φ_0 fixed (in Fig 1c, by keeping D_0 fixed and varying the initial angle φ_0 respectively). The curve in Figure 1b, com-

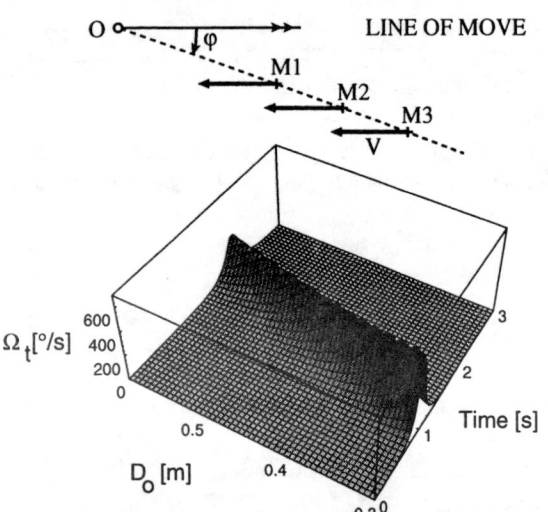
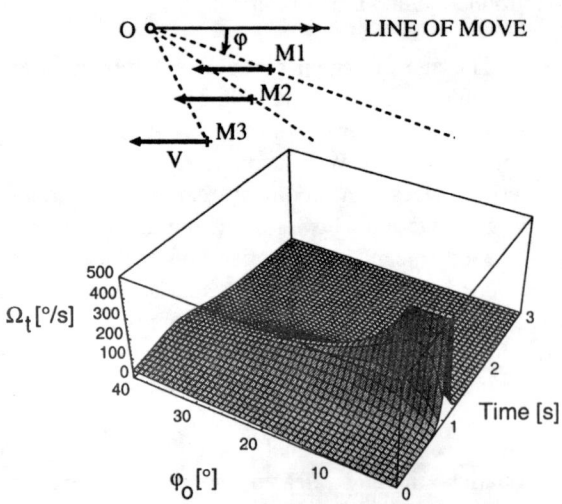

Figures 1a (top), 1b and 1c (bottom left and bottom right). 1a: (left and right) a sighted observer in 'O' moving at the speed V generates a field of *image motion* resulting from the induced movement (in the opposite direction) of the light-contrasted points {Mi} travelling accross 'O''s retina. 1b: as 'O' moves along, the angular velocity (z-axis) can be directly computed as a function of time (y-axis) and initial distance between 'O' and M (y-axis) by solving equation 1 (φ_0=10°; V=0.3m/s). 1c: the same computation is carried out by keeping D fixed and by varying the initial orientation of M (D_0=0.3m; V=0.3m/s)

puted with initial conditions ($\varphi_0=10°$; v=0.3m/s, D_0 varies), shows that the variations in distance for small initial orientations, have less influence on Ω_t than on the time-to-maximum of Ω_t (the time-to-maximum is the duration for which the contrast point M travels from φ_0 to $\varphi=90°$). This is particularly crucial for visual systems with a poor optical resolution; for example, with an interreceptor angular separation of 3° (as it is the case of the fly's eye) and 15°/s resolution in the measurement of Ω_t, a sighted observer would only be able to measure Ω_t correctly for D<0.2m. However, (see Fig 1c: D=0.3m; V=0.3m/s; φ_0 varies) the increase of φ_0 significantly affects Ω_t (see Fig 1c: D=0.3m; V=0.3m/s; φ_0 varies); the same low-resolution visual system would be able to measure Ω_t for obstacles oriented from $\varphi_0>15°$. For obstacles that are either too close to the line of move, or too far away from the observer, the drawback is that the low resolution visual system cannot always measure Ω_t on its own with sufficient accuracy. In the following study, we suggest that such visual systems can enhance their perception of visual motion by generating actively controlled eye movements.

2.1 Smooth retinal scanning.

In 1994, [17] built a retina that underwent a smooth, saw-tooth shaped scanning movement - adequately conditioned in amplitude and frequency - in order to induce an additive rotatory optic flow to the existing translatory component Ω_t. The total measured optic flow in the scanning retina was expressed as:

$$\Omega_T = \Omega_r + \Omega_t \qquad \text{equ (2)}$$

Ω_t was retrieved by subtracting the induced known rotatory component Ω_r from the measured total flow Ω_T. Ω_r was obtained by precisely calibrating the scanning actuator's motion:once calibrated the visual sensor was able to make correct measurements of the translatory optic flow given that $\Omega_t/\Omega_r \geq 0.15$ was satisfied (with a $\Delta\varphi \sim 3°$). Values of Ω_t that did not satisfy the preceding condition, were subject to fluctuations influenced by external factors (50Hz noise from artificial light sources and local variations in illumination), as well as by internal factors (individual efficiency of photodiodes and tolerances of the analog pre-processing circuits' discrete components).

2.2 Pulsed retinal scanning for contrast detection.

Instead of extracting small values of Ω_t by possibly increasing the performances of the hardware processing, we suggest using short, square wave pulses of retinal scanning to enhance the detection of slow moving obstacles. A square wave scanning signal will induce an Ω_r of much larger amplitude than the small Ω_t, as opposed to smooth retinal scanning (see § 2.1). In this case the visual signals can be used simply to detect and confirm the presence of a slow moving light contrast. Visual detection is increasingly reliable and is independent of the robot's movement. The detection of a slow-moving light contrast immediatly informs the robot about the orientation of the obstacle: by repeating the pulsed scanning periodically, the orientation of detection can also be very reliably updated. In this respect, the visual sensor is able to maintain a permanent visual contact with the oncoming obstacle, and eventually perceive its motion. Obstacles can also be detected whether the robot is stationary or whether in motion (with any additional parasitic motion, such as disturbances caused by an irregular terrain, wheel slippage or non-symetric performances of the wheels' actuators).

3. Hardware implementation of a fast scanning retina.

A fast retinal scanning eye was built and its performances evaluated. The experimental testbed comprises a single retina with 10 visual directions layed out on the horizontal plane, that occupy a total field of view of 30° (each pair of visual directions is separated by 3°, see Figure 2). The terminal end of each visual axis is a 250µm polished optic fibre connected to a phototransistor. With the 10 tips placed in the focal plane of a single biconvex lens (f/D=1), the optic fibres are bundled together, mounted and fixed on top of a sliding magnetic coil. The coil, actuated by the Lorentz force, oscillates around a permanent magnet with minimum mechanical friction. Under full load, a 500ms single-ended pulse of 80mA generates a motion pulse of 500µm of amplitude in the focal plane (average acceleration of 200mm/s^2).

The resulting visual signal is proportional to an average rotatory angular velocity of 150°/s, which significantly enhances the S/N ratio of signals at the earliest stages of processing. These visual signals are then fed in parallel through Analogue Contrast Detectors (see figure 3a): each ACD is composed of a pre-amplifier, a passive high-pass filter and a Schmidt trigger. Under a periodic fast scanning signal (see figure 3b and 3c: duration 140 ms; duty cycle of 0.5), a light contrast placed in the visual field of a single direction will periodically excite the high-pass filter and trigger a change of state in the ACD's output. However, if the contrast is removed the output of the ACD is a constant 20mV (see Fig. 3c).

visual information to the motor of the rotating eye so as to minimize the angular extent between the obstacle and the eye. Two important tasks that affect tracking in real-time are; 1) scanning and processing for extracting visual information; 2) computing the corrective feedback. The first task is mainly affected by the time delays of the Analogue Contrast Detectors (phototransduction time delay+high-pass filter time constant). The second task is affected by the delays in the corrective processing loop (control delay of the scanning actuator+sequential tracking algorithm+control delay of eye motor).

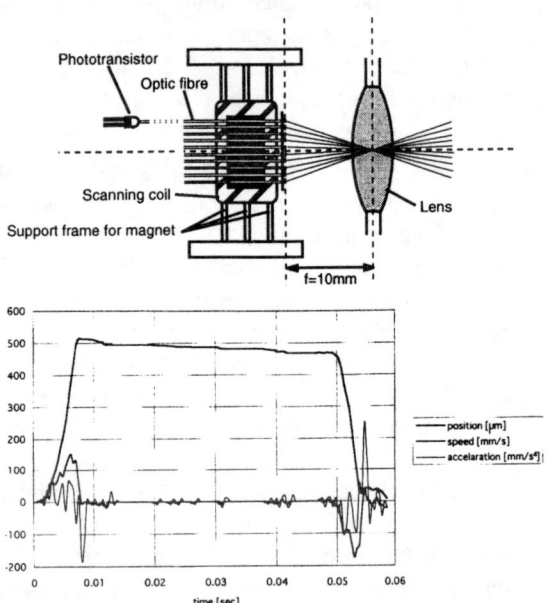

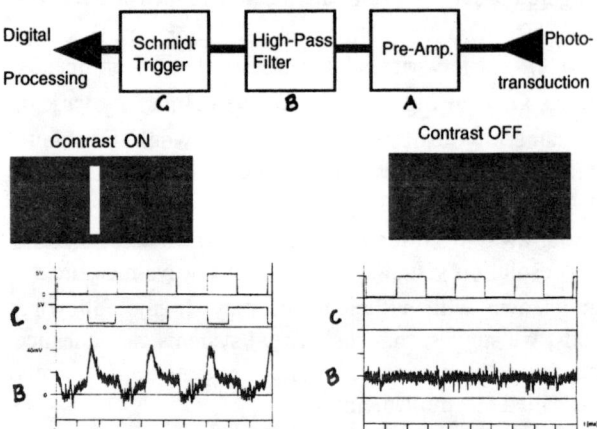

Figures 2a (top) and 2b (bottom). 2a: The fast scanning retina as seen from above. It is made of a bundle of 10 optic fibres attached to a magnetic coil that undergoes lateral motion in the focal plane (500um of amplitude). 2b: the scanning retina's position was recorded by observing motion under a high speed camera (frame rate of 750 images/s). The scanning control signal is a square pulse of 500ms at 80mA.

3.1 Tracking performances of the pulsed scanning retina.

A single pulsed scanning retina was tested onboard HECTOR's small mobile platform for visual tracking experiments (see Figure 5). The scanning retina is placed on top of a circular tower (diameter 5cm), rotated by a small DC motor at the maximum speed of 130°/s. The tower motor can either be controlled in continuous mode or in Pulse Width Modulation mode. This rotating eye, placed on top of a small mobile platform (diameter 120mm) has proprioceptive position sensors that detect 11 distinct orientations of the eye within 240°. At an instant t, the orientation of the eye can be evaluated by the combination of the proprioceptive intputs and the duration of motor activity. Furthermore, under PWM mode, the angular velocity of the rotating eye is obtained by the rate of control pulses that are sent to the eye motor.

The trigger outputs of the 10 ACDs are fed in parallel to a BASIC STAMP II microcontroller clocked at 20MHz. The Vision processor dispatches information to 2 other CPUs, the Rotating eye processor and the Platform control processor, accross a local network transmitting at 34Kbaud. Visual tracking of a light contrast essentially involves feeding back the pre-processed

Figure 3a (top), 3b(bottom, left) and 3c (bottom right). 3a: block diagram of the circuitry of the Analog Contrast Detector (ACD) responsible for the visual pre-processing. Each ACD has a pre-amplifier, a passive high-pass filter and a Schmidt trigger. 3b: when a single visual direction repeatedly scans in front of a black-white contrast, the ACD's output undergoes a change-of-state according to the periodic detection of the contrast. (scanning square wave of 140ms, duty cyle 0.5). 3c: when the light contrast is removed, the ACD's output remains unchanged.

In order to compare tracking of a scanning to that of a non-scanning retina, the platform is fixed at 35cm from a moving cart. The cart carries a single vertical cylinder (white colour; diameter 1.5cm) that moves in front of a black screen (800 lux illumination at the base of the cylinder). The cart is actuated horizontally at variable speeds along a straight rail (the cylinder's direction of movement is perpendicular to the eye's optical axis). To ensure stable visual tracking, the cycle time of the tracking loop must be small enough to compensate the disturbance caused by the continous movement of the cylinder. In PWM control mode, the visual tracking algorithm generates the following corrective eye rotation:

$$\tau(k+1) = P + V.\varphi_d(k) + \varphi_{pr}(k) \qquad \text{equ (3)}$$

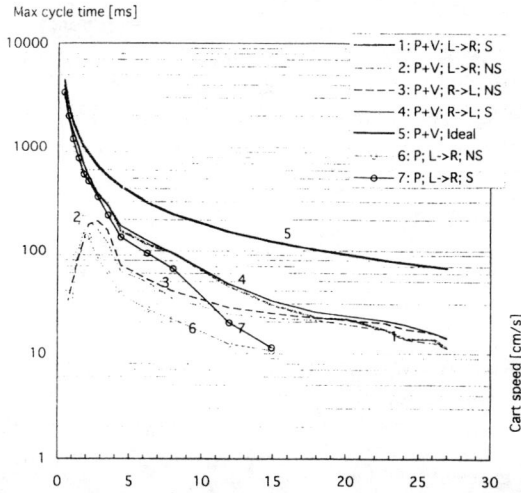

Figure 4. Comparative performances of the tracking system versus motion of a single obstacle using scanning (curves 2, 3, 6) and non-scanning (curves 1, 4, 7) (refer to section 3.1 for details).

Where τ is the duration of a control pulse, P is a corrective term in position, $V.\varphi_d(k)$ a corrective term in velocity ($\varphi_d(k)$ is the orientation of the contrast detected at time k), and $\varphi_{pr}(k)$ is an additional predictive correction.

For each value of the cylinder's speed, we measured and plotted the largest value of processing cycle time that ensures stable tracking (e.g., that allows the eye to maintain a visual contact with the moving cylinder while it travels for 80cm on the rail) (see Figure 4). A larger cycle time on the y-axis indicates a better performance: it means that the system is not constrained by a high feedback rate in order to ensure stable visual tracking. In the case of a mixed (position+velocity) correction, the maximum cycle time is measured for visual tracking of the cylinder moving in both directions (left to right (curves 1 and 2) and right to left (curves 3 and 4)). These curves are compared to the ideal tracking performances of a position+velocity correction (thick black curve). Firstly, whether the eye is scanning or not, tracking performances obtained with a mixed correction of (position+velocity), are superior to tracking with position correction only (Fig 4. For scanning, compare curves 1 and 4 (P+V correction) with curve 7 (P correction only). For non-scanning, compare curves 2 and 3 (P+V) with curve 6 (P only). Secondly, as the cart's speed decrases, tracking with scanning is relatively succesfull compared with non-scanning: the scanning eye is able to track the moving cylinder at very low speeds (less than 5 cm/s; see curves 1,4 and 7). The non-scanning eye must rely on higher feedback rates in order to ensure proper tracking at very low cart speeds (for speeds<3cm/s, see curves 2,3 and 6): also, for speeds lower than 1cm/s, the non-scanning eye loses track of the moving cylinder. The rela-

tively poorer performances are due to the fact that the non-scanning eye relies on the cylinder's motion as well as on its own rotations to detect visual motion. The ACD's enhanced temporal resolution result in the scanning eye's superior performance and the poor spatial resolution of the imaging system is overruled by the activity of the fast retinal scanning.

3.2 Obstacle avoidance with a single scanning retina.

To illustrate the importance of fast retinal scanning for robot navigation, the rotating scanning eye is assigned to guide HECTOR's movements by avoiding collisions in an unknown environment, made of stationary vertical cylinders. Obstacle avoidance is the result of a simple visuo-motor strategy embedded in the following sequence of reflex actions: 1) search for obstacles that are ahead; 2) confirm the detection by selecting and tracking the closest obstacle to the line of move; 3) start turning and monitor the eye movements of the approaching obstacle (angular speed and orientation); 4) if the angular velocity of the eye reaches a threshold or if the eye tracks the obstacle beyond a threshold orientation, then turn away a second time from the obstacle and reset the eye to its initial position.

Figure 5. The small mobile platform HECTOR. The scanning eye is mounted on a rotating tower.

Once detected, an oncoming obstacle is continuously tracked and the degree of confidence in tracking is monitored. As the confidence in tracking reaches a threshold, the robot turns away from the obstacle. Figure 6 displays a set of trajectories recorded in real-time, of HECTOR avoiding a single vertical obstacle placed in M (the robot's start point is at the far right). Typical trajectories display two distinct points of visually-guided turns. The first turning point is generated by the crossing of a threshold of the tracking confidence: the robot makes a single sharp turn which is followed by a large counter-rotation of the eye, so as to maintain a visual contact with the obstacle (the fast scanning and visual processing are inhibited during the large eye rotation). The robot then proceeds on a straight course while the eye tries to track the laterally travelling obstacle. If the tracking confidence is not maintained above the threshold, the eye stops tracking, returns to its resting frontal orientation and engages in a new search. however, if the speed of the eye exceeds the threshold then the robot makes a second safe turn away from the obstacle. At present, the robot's turning angle and turning direction are arbitrarily chosen and further work investigates a possible tuning of these parameters by environmental constraints as well as the robot's mission.

Figure 6. Recordings of the HECTOR's trajectory while avoiding a single obstacle placed in M (the robot's start point is at the far right. Black calibration bars: 10cm).

4. Summary and conclusion.

This work highlights the advantages of combined retinal and eye movements for the perception of visual motion. The fast retinal scanning and corrective eye rotations allow a single retina of 10 pixels with a resolution of 3° to perform complex visual tasks: search and detect obstacles, track them continously and guide the steps of a mobile robot in an unknown environment. These tasks are successfully implemented with cheap, low-energy consuming, off-the-shelf hardware: HECTOR's total power comsumption does not exceed 5W and the most expensive item, the DC eye motor, is worth 80 USD.

With equivalent imaging systems and digital hardware, no "passive" digital visual processing can match the performances of the combined "active" fast retinal scanning and visually-guided eye rotations. [11] have used a parallel tracking algorithm that runs in a 1ms visual loop (e. g., 30 times faster than HECTOR's tracking in figure 6) where visual tracking is computed on the basis of a spatio-temporal digital filtering process. The biologically-inspired function of retinal scanning applied to HECTOR's obstacle avoidance strategy is of significant importance in active vision where "acting in order to perceive" (e. g., using retinal scanning and eye rotations) is just as important as "perceiving in order to act" (e.g., maintaining a prolonged visual contact in order to avoid collisions).

Acknowlegment.

F Mura received a joint EU-JSPS post-doctoral fellowship for this research (grant award no PE95241).

References.

[1] Aloimonos Y (1993): Active Vision Revisited. In Active Perception, ed Aloimonos, Lawrence Erlbaum, 1-18.
[2] Appenzeller G, Weckesser P, Dillman R (1996): Active Parameter Control for the Low Level Vision System of a Mobile Robot. In proc. IROS'96, Osaka JP, 1256-1263.
[3] Batista J, Peixoto P, Araujo H (1997): Real-Time Visual Behaviors with a Binocular Active Vision System.In proc. IEEE R&A, Albuquerque NM, 3391-3396
[4] Collett TS (1978): Peering - a Locust Behaviour Pattern for Obtaining Motion Parallax Information. J Exp Biol, 76, 237-241.
[5] Franceschini N (1992): Sequence discriminating neural network in the eye of the fly. In Analysis and modelling of neural systems, ed Eeckman, Kluwer Academic, 189-197.
[6] Franceschini N, Pichon J-M, Blanes C (1992): From insect vision to robot vision. Phil Trans Roy Soc Lond B, 337, 283-294.
[7] Franceschini N, Chagneux R (1994): Retinal Movements in Freely Walking Flies. In Proc 22 Göttingen Neurobiology Conf, eds Elsner & Breer, 268.
[8] Gregory RL, Ross HE, Moray N (1964): the curious eye of Copilia. Nature,201, 1166-1168.
[9] Hausen K (1993): Decoding of Retinal Image Flow in Insects. In Visual Motion and its Role in the Stabilization of Gaze, eds Miles & Wallman, ch10, 203-235.
[10] Hengstenberg R (1984): Roll-Stabilization During Flight of the Blowfly's Head and Body by Mechanical and Visual Cues. In: Localization and Orientation in Biology and Engineering, eds Varju/Schnitzler, pp121-134.
[11] Ishii I, Nakabo Y, Ishikawa M (1996): Target Tracking algorithm for 1ms Visual Feedback System Using Massively Parallel Processing. In proc IEEE R&A, Minneapolis, Minnesota, 2309-2314.
[12] Koch C, Mathur B (1996) Neuromorphic vision chips. IEEE Spectrum May 1996, 38-46.
[13] Kyuma K. Lange E, Ohta J, Hermanns A, Banish B, Oita M (1994) Artificial retinas - fast, versatile image processors. Nature, vol 372, No. 6502, 197-198.
[14] Land M F(1993): How Animals Scan the Visual Environement. In Proc IEEE conf SMC, 144-149.
[15] LaValle S, Gonzalez-Banos H, Becker C, Latombe J-C (1997): Motion Strategies for Maintaining Visibility of a Moving Target. In proc. IEEE R&A, Albuquerque NM, 731-736.
[16] Mead C (1989): Analog VLSI and Neural Systems. Addison Wesley, Reading, Massachusetts.
[17] Mura F, Franceschini N (1996): Obstacle avoidance in a terrestrial mobile robot provided with a scanning retina. In proc IEEE Intelligent Vehicles'96, Tokyo, 47-52.
[18] Rind C (1997): Neural Circuits for Collision Avoidance., In proc IROS'97, Bio-Mechatronic Systems workshop.
[19] Torralba A, Herault J (1997): From Retinal Circuits To Motion Processing: a Neuromorphic Approach to Velocity Estimation. in ESANN'97, ed Verleysen D-Facto, Brussels.

Environmental Complexity Control for Vision-Based Learning Mobile Robot

Eiji Uchibe, Minoru Asada and Koh Hosoda
Dept. of Adaptive Machine Systems, Graduate School of Eng.,
Osaka University, Suita, Osaka 565-0871, Japan
uchibe@er.ams.eng.osaka-u.ac.jp

Abstract

This paper discusses how a robot can develop its state vector according to the complexity of the interactions with its environment. A method for controlling the complexity is proposed for a vision-based mobile robot of which task is to shoot a ball into a goal avoiding collisions with a goal keeper. First, we provide the most difficult situation (the maximum speed of the goal keeper with chasing-a-ball behavior), and the robot estimates the full set of state vectors with the order of the major vector components by a method of system identification. The environmental complexity is defined in terms of the speed of the goal keeper while the complexity of the state vector is the number of the dimensions of the state vector. According to the increase of the speed of the goal keeper, the dimension of the state vector is increased by taking a trade-off between the size of the state space (the dimension) and the learning time. Simulations are shown, and other issues for the complexity control are discussed.

1 Introduction

One of the ultimate goals of Robotics and AI is to realize autonomous agents that organize their own internal structure towards achieving their goals through interactions with dynamically changing environments. From a viewpoint of designing robots, there are two main issues to be considered:

- the design of the agent architecture by which a robot develop from the interaction with its environment to obtain the desired behaviors, and
- the policy how to provide the agent with tasks, situations, and environments so as to develop the robot.

The former has revealed the importance of "having bodies" and eventually also a view of the internal observer [7]. In [2], the first issue is focused and a discussion how the robot can develop from the interaction with its environment according to the increase of the complexity of its environment is given in the context of a vision based mobile robot of which task is to shoot a ball into a goal with/without a goal keeper. In this paper, we put more emphasis on the second issue, that is, how to control the environmental complexity so that the robot can efficiently improve its behaviors.

"Shaping by successive approximation" is a well-known technique in psychology of animal behavior [6]. A simple and straightforward analogy to this situation is to design a reward function to accelerate the reinforcement learning. However, this often requires *a priori* precise knowledge about the details of the relationship between the given task and the environment. Instead of providing such knowledge, an alternative called "Leaning from Easy Missions" (LEM) paradigm was proposed [3].

The basic idea of LEM can be extended to more complicated tasks, but more fundamental issues to be considered are how to define complexity of the task and the environment, and how to increase the complexity to develop robots. Since these issues are too difficult to deal with as general ones, a case study on a vision-based mobile robot is given in this paper where the environmental complexity is defined in the context of RoboCup Initiative [4] and a method to control the environmental complexity is proposed. First, we provide the most difficult situation, that is, the maximum speed of the goal keeper with chasing-a-ball behavior, and the robot estimates the full set of state vectors with the order of the vector components according to the contributions to reducing the estimation errors by a method of system identification. The environmental complexity is defined in terms of the speed of the goal keeper while the complexity of the state vector to cope with the environmental complexity is the number of the dimensions of the state vector. According to the increase of the speed of the goal keeper, the dimension of the state vector is increased by taking a trade-off between the size of the state space (the dimension) and the learning time.

The rest of the paper is organized as follows: first we give an overview of the whole learning system, and basics of the reinforcement learning, especially Q-learning, Next, a method for efficient learning and development coping with the increase of the task environment complexity is proposed. Then, an example task of shooting with avoiding a goal keeper is introduced. The proposed method is applied to scheduling the speed of the goal keeper for the efficient development of the learner that attempting at coping with new situations by adding a new axis in its state space. Finally, the preliminary experiments are shown, and other issues for the complexity control are discussed.

2 An Overview of The Whole System

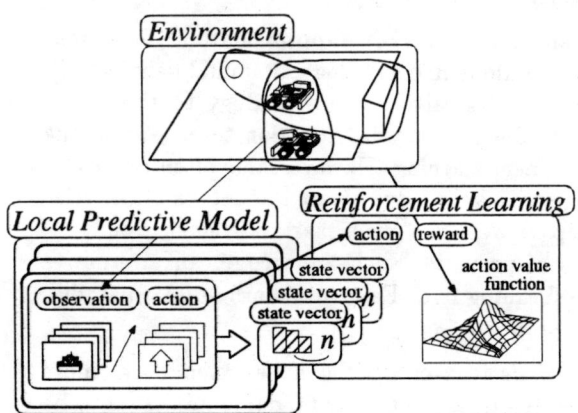

Figure 1: An overview of the whole system

Figure 1 shows an overview of the whole system consisting of a local predictive model and a learning architecture. The local predictive model outputs the state vector list in the order of the value of the estimated correlation coefficient with estimation errors. These state vectors are used to construct the state space for the reinforcement learning method to be applied in multi agent environment. About the details of the whole system, one can find other publications [9, 10]. Here, we focus on how to accelerate the Q-learning by appropriately increasing the environmental complexity. The rest of this section briefly explains the basics of state vector estimation and the reinforcement learning.

2.1 State Vector Estimation

In order to accelerate the learning according to the increase of the environmental complexity, it needs a mechanism to measure the complexity based on its experience. As such a mechanism, a local predictive model [9] is considered which estimates the relations between the learner's behaviors and the other agents through interactions (observation and action). In order to construct the local predictive model of other agents, Akaike's Information Criterion(AIC) [1] is applied to the result of Canonical Variate Analysis(CVA) [5]. We just briefly explained the method (for the details of the local predictive model, see [9, 10]).

CVA uses a discrete time, linear, state space model as follows:

$$\begin{aligned} \boldsymbol{x}(t+1) &= \boldsymbol{A}\boldsymbol{x}(t) + \boldsymbol{B}\boldsymbol{u}(t), \\ \boldsymbol{y}(t) &= \boldsymbol{C}\boldsymbol{x}(t) + \boldsymbol{D}\boldsymbol{u}(t), \end{aligned} \quad (1)$$

where $\boldsymbol{x}(t)$, $\boldsymbol{u}(t) \in \Re^m$ and $\boldsymbol{y}(t) \in \Re^q$ denote state vector, action code vector, and observation vector respectively. $\boldsymbol{A} \in \Re^{n \times n}$, $\boldsymbol{B} \in \Re^{n \times m}$, $\boldsymbol{C} \in \Re^{q \times n}$, and $\boldsymbol{D} \in \Re^{q \times m}$ represent matrices. CVA estimates a state vector $\boldsymbol{x}$ which is a linear combination of the previous observation and action sequences as follows:

$$\boldsymbol{x}(t) = [\boldsymbol{I}_n \; \boldsymbol{0}]\boldsymbol{U}\boldsymbol{p}(t), \quad (2)$$

where

$$\boldsymbol{p}(t) = [\boldsymbol{u}(t-1) \cdots \boldsymbol{u}(t-l) \, \boldsymbol{y}(t-1) \cdots \boldsymbol{y}(t-l)]^T,$$

and $\boldsymbol{U} \in \Re^{l(m+q) \times l(m+q)}$ is a matrix which is calculated by CVA.

2.2 Basics of Reinforcement Learning

After estimating the state space model given by Eq. (1), the agent begins to learn behaviors using a reinforcement learning method. Q learning [11] is a form of reinforcement learning based on stochastic dynamic programming. It provides robots with the capability of learning to act optimally in a Markovian environment.

In the previous section, appropriate dimension n of the state vector $\boldsymbol{x}(t)$ is determined, and the successive state is predicted. Therefore, we can regard an environment as Markovian. A simple version of Q learning algorithm is shown as follows:

1. Initialize $Q(x, u)$ to 0s for all combination of $\boldsymbol{X}$ and $\boldsymbol{U}$.

2. Perceive current state x.

3. Choose an action u according to the action value function.

4. Execute an action u in the environment. Let the next state be x' and immediate reward be r.

5. Update the action value function from x, u, x', and r,

$$Q_{t+1}(x,u) = (1-\alpha_t)Q_t(x,u) \\ + \alpha_t(r + \gamma \max_{u' \in U} Q_t(x',u')) \quad (3)$$

where α_t is a learning rate and γ is a fixed discounting factor between 0 and 1.

6. Return to 2.

3 The Method for Efficient Learning and Development

One can use all the state vectors to make the robot learn, but it would take enormously long time due to the large size of the state space. Instead of using the all vectors, one can start with a small size of the state vector set first and increase the dimension of the state space in the following stages. The action value function in the previous stage works as a priori knowledge so as to accelerate the learning. In order to transfer the knowledge smoothly, the state spaces in both the previous and current stages should be consistent with each other. Therefore, the robot should have a full list of the state vectors available in advance, and selects one among them at the periods when the robot no longer can cope with the changing environment with the current state vector set.

An algorithm to control the increase of the environmental complexity is given as follows:

1. Collect many sequences of data during action executions in the most complex task environment.

2. Construct the local predictive model to the data and output the state vector lists with estimation errors.

3. Set up the performance criterion.

4. Start with the minimum state vector set, say one or two dimensions for the lowest complexity of the task environment.

5. Keep the complexity until the robot learns the desired behavior (reach the performance criterion).

6. If the robot reaches the performance criterion, increase the complexity and return Step 5. Else, increase the dimension of the state space (add a new axis) and return Step 5.

As a learning method, we use modular reinforcement learning [8] based on Q-learning with the state space specified. The modular reinforcement learning can coordinate multiple behaviors (in the following, shooting behavior and avoiding one) taking account of a trade-off between the learning time and the performance.

4 Experimental Results

4.1 Task and Assumptions

We apply the proposed method to a simplified soccer game including two agents [8]. One is a learner to shoot a ball into a goal, and the other is a goal keeper of which speed is a control parameter in the environment complexity. Each agent has a single color TV camera and observes output vectors shown in Figure 2. The dimension of the observed vector about the ball, the goal, and the other robot are 4, 11, and 5 respectively.

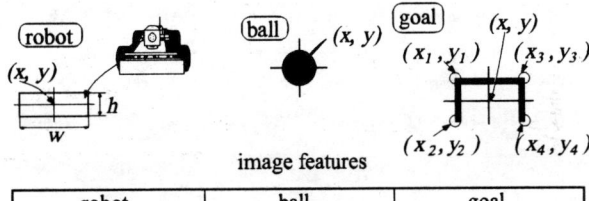

robot	ball	goal
area	area	area
center position	center position	center position
height	radius	4 corners
width		

Figure 2: Image features of the ball, goal, and agent

Two robots move around using a 4-wheel steering system. The effects of an action against the environment can be informed to the agent only through the visual information except the reward that is given by the environment (top down signal). Figure 3 shows a scene of two real robots and the environment. As motor commands, each agent has 7 actions such as go straight, turn right, turn left, stop, and go backward. Then, the input u is defined as the 2 dimensional vector as

$$\boldsymbol{u}^T = [v \ \phi], \quad v, \phi \in \{-1, 0, 1\},$$

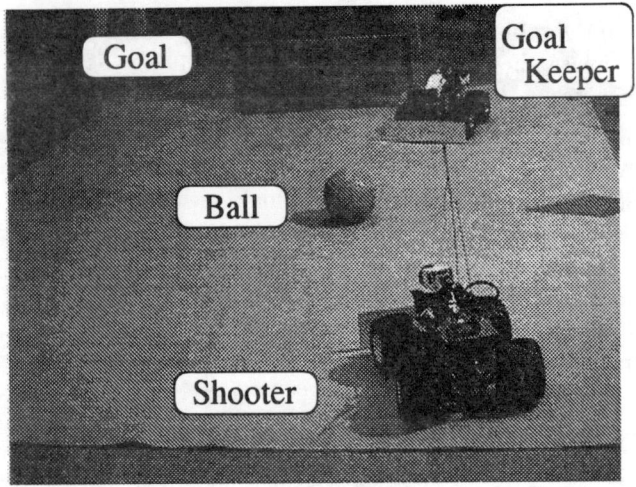

Figure 3: Two real robots and the environment

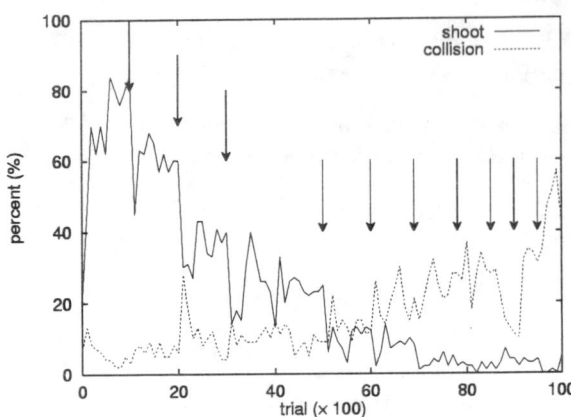

Figure 4: The dimension of the state vector n is one (the minimum dimension).

where v and ϕ are the velocity of motor and the angle of steering respectively and both of which are quantized.

We assume that the goal keeper has a basic behavior of moving to the ball, but its speed can be controlled as the complexity parameter.

4.2 Settings

We assign a reward value 1 when the ball was kicked into the goal or 0 otherwise. On the other hand, a reward value -0.3 is given to the robot when two robots make a collision between them. Discounting factor γ is 0.9.

To speed up the learning time, we select actions using the probability based on *semi uniform undirected exploration*. In this method, the learning agent executes random actions with a fixed probability. We set the probability of selecting a random action at 10 %.

4.3 Speed Control for the Goal Keeper with Fixed Dimension

At first, we demonstrate the experiments to control the complexity of the interactions in case of the fixed dimension of the estimated state vector about the goal keeper. The learning robot collects sequences of observation and action with the highest complexity, that is, the maximum speed $v_{\max}$ of the goal keeper, and applied the local predictive model to the obtained data. As a result, we obtained the list of the state vector for the goal keeper and others. The dimension of the estimated state vector of the goal keeper, the ball and the goal is 4, 4 and 2, respectively. The learning

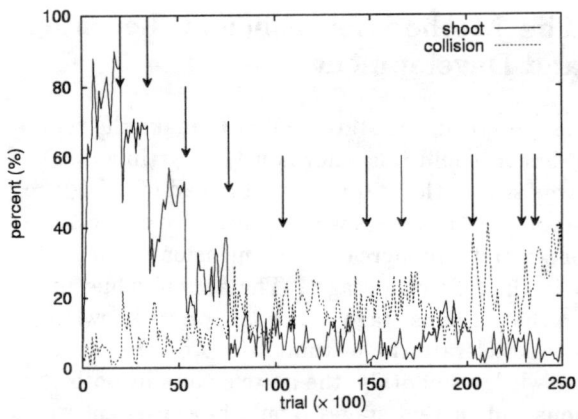

Figure 5: The dimension of the state vector n is 2.

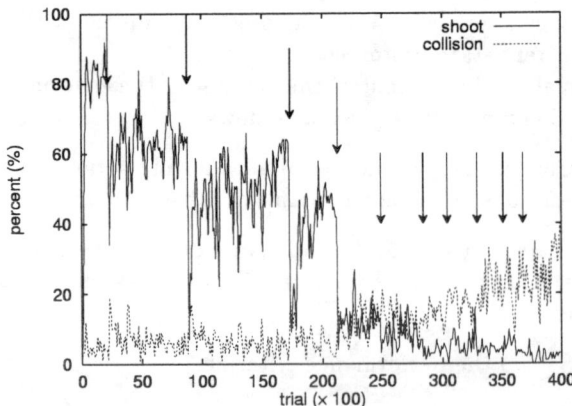

Figure 6: The dimension of the state vector n is 3.

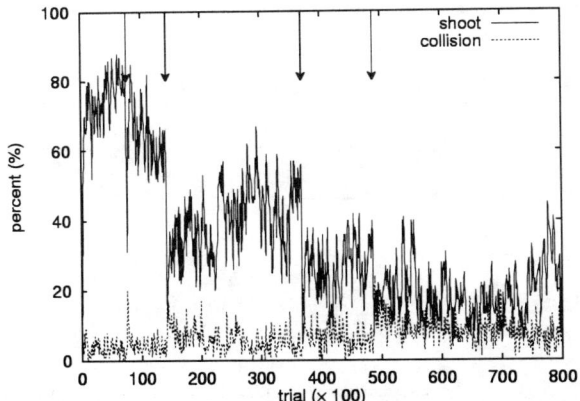

Figure 7: The dimension of the state vector n is 4 (the maximum dimension).

robot chooses the dimension of the only state vector about the goal keeper (other vectors are remained unchanged) which is estimated by the local predictive model to cope with the change of the complexity of the interaction.

Figures 4 ∼ 7 show graphs of the performance data (success rates of shooting and collision avoidance) in terms of the speed of the goal keeper with fixed dimension of the state vector for the goal keeper (from 1 to 4). The speed is increased when the robot achieves the pre-specified success rate (80%) or no improvement can be seen. The arrows show the time when the speed of the goal keeper is changed (10 % speed increase of the maximum motion speed v_{max} from 0 (stationary)). In spite of the number of dimensions, the best success rate of shooting is about 80 %. However it takes much time for learning agent to acquire the best performance when the dimension of the state space for the goal keeper increases. In Figure 7, the performance data until the speed of $0.4v_{max}$ is shown because of the space limit. As we can see from the Figures 4 ∼ 7,

- the success rate of shooting becomes worse when $v/v_{max} > 0.2$, and
- the collision rate is larger than success one of shooting when $v/v_{max} > 0.4$.

The learning agent has to take account of the trade off between shooting behavior and avoiding behavior while the goal keeper only pushes the ball. Therefore, the learning agent might not accomplish the shooting task if the goal keeper moves quickly.

4.4 Speed Control for the Goal Keeper with Variable Dimensions

Figure 8 shows the result of the speed control for the efficient learning. Short and long arrows indicate the times to increase the speed of the goal keeper and the dimension of the state vector, respectively. We set up 50% performance criterion by which the timing of the speed increase of the goal keeper is decided. Compared with Figures 4 ∼ 7, we may conclude that the fewer dimensions of the state space contribute to the reduction of the learning time but less performance and vice versa. For example, one dimensional state vector cannot cope with $0.2v_{max}$ while two dimensional state vector can not represent the situation with $0.3v_{max}$ for the learner to learn shooting behaviors. If we start with one dimension case and step up the dimension, we also give up $0.4v_{max}$ but with four dimensions the collision rate is much less than the success rate around 15,000 trials (See Figure 7.

Our proposed scheduling method can achieve the almost the same performance faster than the case of learning by the maximum dimension of the state vector from the beginning. We suppose that the reasons why our method can achieve the task faster are as follows. First, The time needed to acquire an optimal behaviors mainly depends on the size of the state space, which are determined by the dimension of the state vector estimated by the local predictive model. Our method assigns the appropriate dimension of the state vector according to the complexity while the full dimension of the state space (Figure 7) is redundant in the early state of learning. Second, since our proposed method utilizes the action value function which is previously acquired as the initial value, it can reduce the learning time. In other words, our method consider not only the size of the state space according to the complexity but also the initial values of the action value function which is usually initialized zeros. Finally, we show the example of an acquired behavior in Figure 9. The two lines emerged from the agent show its visual angle.

5 Discussion

We have shown the method of controlling the environmental complexity along with a simplified soccer task. There are two main issues to be considered. First, the number of control parameters is one in our experiments, but generally multiple, each of which is related to each other. Even in the example task, the

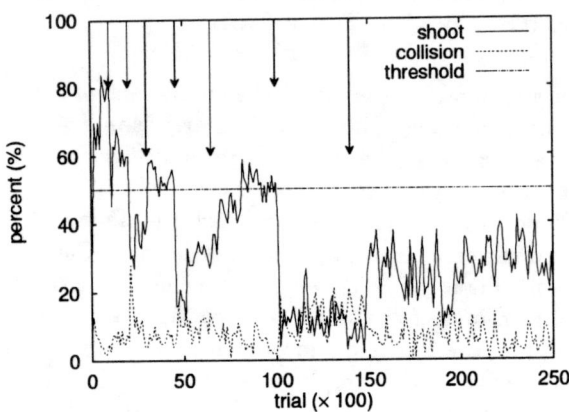

Figure 8: Result of the proposed method

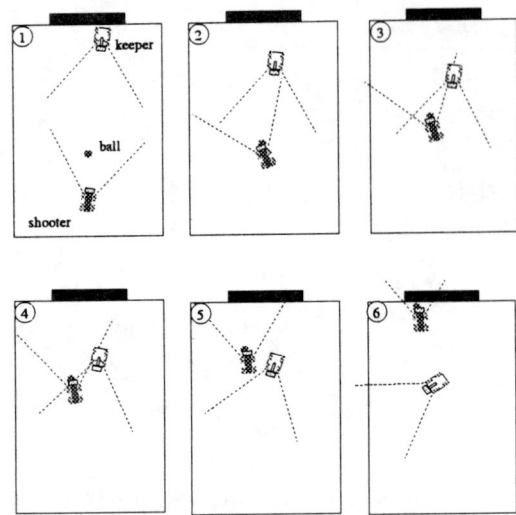

Figure 9: Visualization of the learning agent's policy at the end of a successful trial

speed of the learner, the dimensions of the state space, the resolution of the each dimension (fixed (3 partitions) in the experiments) and the initial configurations of the ball, the goal, the learner, and the goal keeper should be considered together with the speed of the goal keeper. In such a case, since designer cannot completely understand the relationships among them, it seems difficult to decide how to control the complexity completely.

Then, the second issue is revealed. To cope with unknown complexity, the robot should estimate the state vectors anytime when the task performance becomes worse. However, this causes inconsistency in state vector sets between the current and next learning stages. Therefore, the knowledge transfer is limited to the initial controller (action selection) and the robot needs much more memory and the learning time. Since this is against resource bounded condition, we should develop a new method which can take account of this trade-off.

References

[1] H. Akaike. A new look on the statistical model identification. *IEEE Trans. AC-19*, pp. 716–723, 1974.

[2] M. Asada. An agent and an environment: A view of "having bodies" – a case study on behavior learning for vision-based mobile robot –. In *Proc. of 1996 IROS Workshop on Towards Real Autonomy*, pp. 19–24, 1996.

[3] M. Asada, S. Noda, S. Tawaratumida, and K. Hosoda. Purposive behavior acquisition for a real robot by vision-based reinforcement learning. *Machine Learning*, 23:279–303, 1996.

[4] H. Kitano, M. Asada, Y. Kuniyoshi, I. Noda, E. Osawa, and H. Matsubara. Robocup a challenge problem for ai. *AI Magazine*, 18(1):73–85, 1997.

[5] W. E. Larimore. Canonical variate analysis in identification, filtering, and adaptive control. In *Proc. 29th IEEE Conference on Decision and Control*, pp. 596–604, Honolulu, Hawaii, December 1990.

[6] B. Schwartz. *Psychology of Learning and Behavior: Third Edition*. W. W. Norton, NY, London, 1989.

[7] J. Tani. Cognition of robots from dynamical systems perspective. In *Proc. of 1996 Workshop on Towards Real Autonomy*, pp. 51–59, 1996.

[8] E. Uchibe, M. Asada, and K. Hosoda. Behavior coordination for a mobile robot using modular reinforcement learning. In *Proc. of the 1996 IEEE/RSJ International Conference on Intelligent Robots and Systems*, pp. 1329–1336, 1996.

[9] E. Uchibe, M. Asada, and K. Hosoda. State space construction for behavior acquisition in multi agent environments with vision and action. In *Proc. of International Conference on Computer Vision*, pp. 870–875, 1998.

[10] E. Uchibe, M. Asada, and K. Hosoda. Cooperative behavior acquisition in multi mobile robots environment by reinforcement learning based on state vector estimation. In *Proc. of IEEE International Conference on Robotics and Automation*, 1998 (to appear).

[11] C. J. C. H. Watkins and P. Dayan. Technical note: Q-learning. *Machine Learning*, pp. 279–292, 1992.

Unsupervised Learning to Recognize Environments from Behavior Sequences in a Mobile Robot

Seiji Yamada   Morimichi Murota

CISS, IGSSE
Tokyo Institute of Technology
4259 Nagatsuta-cho, Midori-ku, Yokohama 226-0026, JAPAN
yamada@ymd.dis.titech.ac.jp
http://www.ymd.dis.titech.ac.jp/~yamada/

Abstract

In this paper, we describe development of a mobile robot which does unsupervised learning for recognizing environments from behavior sequences. Most studies on recognizing an environment have tried to build precise geometric maps with high sensitive and global sensors. However such precise and global information may not be obtained in real environments. Furthermore unsupervised-learning is necessary for recognition in unknown environments without help of a teacher. Thus we attempt to build a mobile robot which does unsupervised-learning to recognize environments with low sensitive and local sensors. The mobile robot is behavior-based and does wall-following in enclosures. Then the sequences of behaviors executed in each enclosure are transformed into input vectors for a self-organizing network. Learning without a teacher is done, and the robot becomes able to identify enclosures. Moreover we developed a method to identify environments independent of a start point using a partial sequence. We have fully implemented the system with a real mobile robot, and made experiments for evaluating the ability. As a result, we found out that the environment recognition was done well and our method was adaptive to noisy environments.

1 Introduction

In robotics research, most studies on recognizing an environment have tried to build a precise geometric map with high sensitive sensors [3]. However many natural agents like animals recognize their environments only with low sensitive sensors, and a geometric map may not be necessary. In view of engineering, it is important to develop a simple robot which is able to recognize environments only with inexpensive sensors because the cost is low. In many real environments like a dark room, the global information like vision may not be obtained.

Moreover, since a robot should be adaptive to unknown environments without help of a teacher, it needs to learn to recognize new environment by itself. Thus unsupervised learning is necessary.

Hence we attempt to build a mobile robot which does unsupervised-learning to recognize environments with low sensitive and local sensors. The robot is behavior-based and does wall-following in enclosures. Then the sequences of behaviors executed in each enclosure are obtained. The sequences are transformed into real-value vectors, and inputted to a Kohonen's self-organizing network. Learning without a teacher is done and a mobile robot becomes able to identify enclosures. Since we carefully define transformation from a behavior sequence into an input vector and the self-organizing network works well for generalizing data, the learned network is robust against noise like obstacles. Furthermore, for more robust recognition, we develop a method to identify environments independent of a start point using a partial sequence.

We have fully implemented the system using a real mobile robot with two infrared proximity sensors. For evaluating our approach, we made experiments including environments with obstacles as noise. As a result, we found out the environment recognition was done well and our method was adaptive to noisy environments.

Nehmzow and Smithers studied on recognizing corners in simple enclosures with a self-organizing network [6]. They used direction-duration pairs, which indicate the length of walls and shapes (convex or concave) of past corners, as an input vector to a self-organizing network. After learning, the network becomes able to identify corners. However the transformation from raw data to an input vectors is significantly sensitive to noise like small obstacles. We proposes another transformation which maintains better topology than their one. Furthermore we experimentally evaluate robustness of our method against obstacles.

Mataric represented an environment using automata consisting landmarks as nodes [5]. Though the representation is more robust than a geometric one, a mobile robot must segment raw data into landmarks and identify them. The segmentation and identification of landmarks is difficult for a robot only with low sensitive and local sensors. Thus, though her approach is similar to ours, it is not suitable for our purpose.

Tsuji and Li have done the excellent study on vision-based memorizing route scenes [7]. The mobile robot stores qualitative panoramic representation, and locates itself in the route by matching the memorized representation against that of the incoming scenes. If a robot has a vision system, their approach will be valid.

2 Overview

The overview of a whole system is shown in Fig.1. First a behavior-based mobile robot [2] goes round once in each of n enclosure by wall-following, and obtains n sequences of executed behaviors. Next the sequences (symbol lists) is transformed into the real-valued vectors, and they are given to a self-organizing network as input. Note that there is no teacher giving the correspondence from input vectors to enclosures. Learning is done on the network using the Kohonen's update rules. The n input vectors are repeatedly given to the network, and learning progresses. After learning, a winner node indicates one of the enclosures in which the robot did wall-following. The test data are given to the learned network, and the mobile robot identifies the test data with enclosures already trained. We mean this identification by *environment recognition*.

3 Wall-following by a behavior-based mobile robot

A mobile robot used in our research is shown in Fig.2(a). It has four infrared proximity sensors: two in front direction and other two in left and right direction on the both sides (top view in Fig.2(b)). Only two

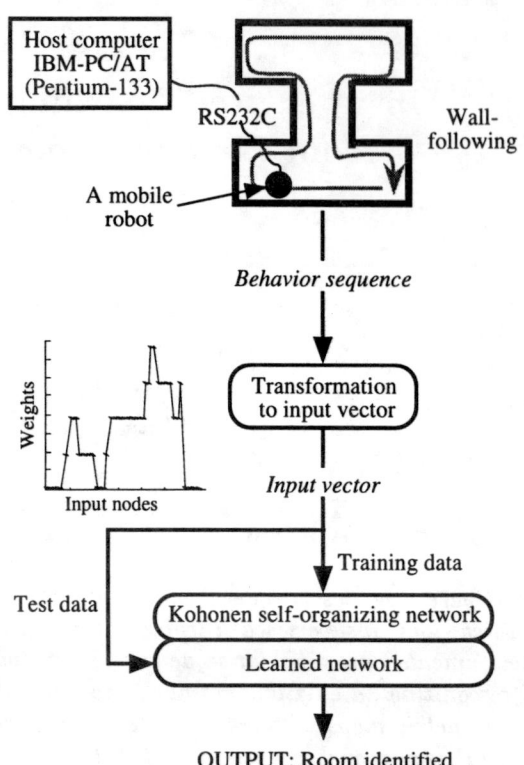

Figure 1 System overview

sensors on the left side are actually used since the wall-following is done clockwise. The mobile robot also has an orientation sensor for steering and an encoder for movement distance. Note that the infrared proximity sensors are low sensitive and obtain just local information within 20 cm. The actuators consists of two stepping motors for driving left and right wheels independently and a DC motor for a front steering wheel (Fig.2(b)).

Since the mobile robot needs to do wall-following even in an enclosure where some obstacles exist, we use the behavior-based approach [2] which is known robust against the change of an environment. The behavior-based approach also can controls a robot with low sensitive sensors, and have the advantage that the behavior sequence is invariant even when the geometric movement history of the mobile robot varies a little.

We use four behaviors (reactive rules) in the following for wall-following. The behaviors are periodically executed, and the mobile robot constantly goes forward.

Behavior-A (*turning in a concave corner*): *If* an obstacle within 10cm in the front and within 10cm on the left *then* turning 40° clockwise there.

Behavior-B (*turning in a convex corner*): *If* no

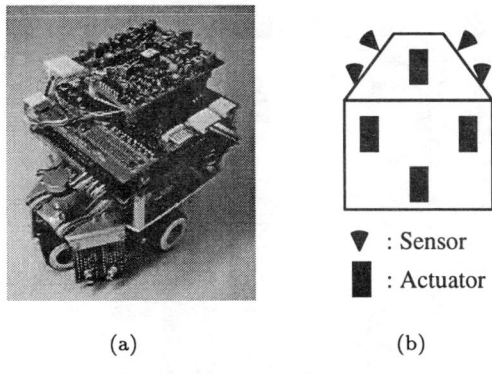

Figure 2 A mobile robot

obstacle within 5cm on the left and the right, and within 10cm in the front *then* turning 40° counterclockwise there.

Behavior-C (*following-1*): *If* an obstacle within 5cm on the left *then* steering 13.5° clockwise.

Behavior-D (*following-2*): *If* no obstacle within 5cm on the left *then* steering 13.5° counterclockwise.

We experimentally verified that a mobile robot turns well by executing the behavior A or B at corners. Although the above behaviors are very simple, a mobile robot is controlled well and smoothly follows walls.

4 Transformation from a behavior sequence into an input vector

By wall-following, a sequence of executed behaviors is obtained. It has information on the shape of the enclosure: the length of a continuous sequence of behavior C and D indicates the length of a wall, and behavior A and B indicate the existence of a concave corner and a convex corner respectively. Thus we consider a mobile robot can identify enclosures with the sequences.

Since a robot should be adaptive to unknown environments in which no other agents help it, it needs unsupervised-learning. Hence we introduce a Kohonen's self-organizing network for identifying enclosures. The behavior sequence is a list of symbols, thus we need to transform it into a real-valued vector as input to a self-organizing network. The transformation also needs to be robust against noise and easily computed.

We propose *BI-transformation*. Given a behavior sequence: $[r_1, r_2, \cdots, r_n]$ ($r_i \in \{A, B, C, D\}$) and an input vector: $\boldsymbol{I} = (v_1, v_2, \cdots, v_m)(n \leq m)$, the values of $\boldsymbol{I}$ are obtained using the following rules, where $v_1 = 0$.

1. If $r_i = $ A then $v_i = v_{i-1} + 1$.
2. If $r_i = $ B then $v_i = v_{i-1} - 1$.
3. If $r_i = $ C or D then $v_i = v_{i-1}$.
4. Otherwise $v_i = 0$ ($i > n$).

For example, Fig.5(a) shows an input vector transformed from a behavior sequence in a square enclosure. The x and y axis indicate dimensions and values of an input vector. For an input vector, the number of dimensions indicates the periphery length of an enclosure, a series of the same value indicates a wall, and the change of values stands for a corner. The increase and the decrease indicate a concave corner and a convex corner.

We explain the robustness of BI-transformation using an example shown in Fig.3. The movement history of a mobile robot and an input vector from the behavior sequences are indicated in Fig.3. Fig.3(a) stands for an enclosure without obstacle and Fig.3(b) stands for the same enclosure including a obstacle on a wall. Thus the mobile robot should identify them as the same one. The transformation needs to be defined so that the distance between the two input vectors may be small. Using the BI-transformation results in the input vectors as Fig.3. The size of shaded areas in Fig.3(b) indicates the difference, and it is relatively small. Note that the whole pattern of the input vector is not shifted. If a robot uses more rigid transformation in which an input vector is described with the length of walls, shapes (convex or concave) of past corners [6], the whole pattern will be shifted and the distance will be significantly large. Hence we consider BI-transformation is robust, and the effectiveness will be experimentally verified.

The idea of BI-transformation is similar to the *turning function* (or the chain coding)[1] in pattern recognition. In general, the turning function is extracted from the raw data like a bit image, while the input vector is directly obtained from a behavior sequence using BI-transformation.

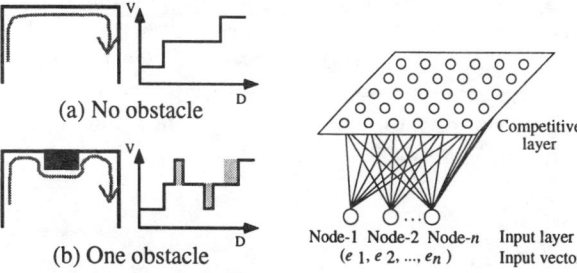

Figure 3 Robustness

Figure 4: Self-organizing network

5 Learning by a self-organizing network

In this section, we briefly explain a Kohonen's self-organizing network [4]. It clusters large dimensional input vectors by mapping them to small discrete vectors, and is used widely in pattern recognition and robotics. A self-organizing network is a two-layered network consists of an input layer and a competitive layer (Fig.4). Any input node is linked to all competitive nodes, and all links have weights. As an input vector is given, input nodes have values corresponding to the input vector, and competitive nodes has values which stand for the distance between their weights and the input vector. A winner node having the minimum distance is determined, and weights of the winner's neighbor nodes are updated.

Let an input vector and weights of links from all input nodes to a competitive node u_i be $\boldsymbol{E} = [e_1, e_2, \cdots, e_n]$ and $\boldsymbol{U}_i = [u_{i1}, u_{i2}, \cdots, u_{in}]$ respectively. First a self-organizing network computes the Euclidean distance between the input vector and competitive nodes.

After a *winner node* with the minimum distance is determined, the weights of winner's neighbor nodes are updated using the following formula.

$$u_{ij}^{\text{new}} = u_{ij}^{\text{old}} + \Delta u_{ij} \quad \Delta u_{ij} = \begin{cases} \alpha(e_j - u_{ij}) : i \text{ is neighbor} \\ 0 \quad\quad\quad\quad\; : \text{otherwise} \end{cases}$$

Update of the weights is done whenever a input vector is given. The learning rate and the size of the winner's neighborhood is usually decreased as the learning progresses. The learning is finished when no update is done.

Next the input vector is given as test data and the winner indicates the class including the input vector. The clustering is automatically done without a teacher.

In our research, the dimension of an input vector is set larger than the length of any behavior sequence. The competitive layer is one-dimensional because the computational complexity of learning with one-dimensional competitive layer is far less than that with a two-dimensional one. If we set sufficiently many nodes in an one-dimensional competitive layer, the resolution of classification will not worse than that of a two-dimensional one.

6 Experiments

Using the mobile robot mentioned earlier, we made experiments for evaluating the utility of our approach.

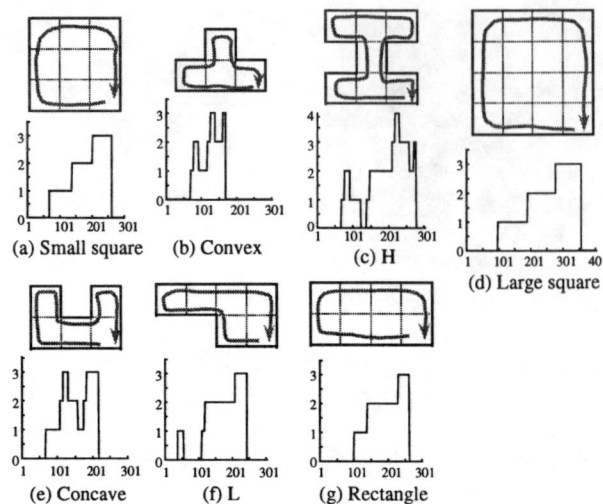

Figure 5 Seven enclosures

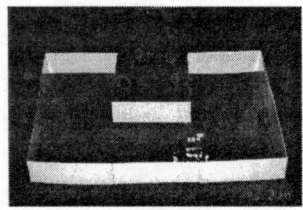

Figure 6 Experimental environments

As seeing from Fig.1, the system consists of a mobile robot and an IBM-PC/AT (Pentium 133MHz, 128M RAM) compatible personal computer as a host computer. All programming was done on the host computer using C^{++}. The program of wall-following was down-loaded into a mobile robot through RS232C interface, and a robot autonomously followed walls. After wall-following, the behavior sequences were sent to the host computer and the learning by a self-organizing network was done there.

6.1 Environment and a learning phase

Using white plastic boards, we built seven different enclosures in shape. Fig.5 shows the shapes of the enclosures and the input vectors obtained by wall-following in each enclosure. A mobile robot did wall-following ten times for each enclosure, and 70 behavior sequences were obtained in total. For each enclosure, a single behavior sequence was used as an input vector for learning, and other 6 sequences were used for testing. Fig.6 shows the scene that a mobile robot is doing wall-following.

The robot stops when it returns near a start point.

It can recognize the start location by dead reckoning using an orientation sensor and an encoder. A mobile robot, however, starts wall-following at the same position for the same enclosure. This restriction will be removed in §6.4.

The largest length of behavior sequences obtained from the enclosures was about 1400, and this was the minimum dimension of an input vector. Since the length was too large for a self-organizing network to learn tractably, all sequences were compressed into 1/4 and the length was 500 at most.

We constructed a self-organizing network consisting of 520 input nodes and 32 competitive nodes located in one dimension. The neighborhood in the competitive layer is defined with d nodes on both sides of the winner. The initial value d_0 and α_0 were set 5 and 0.2. We changed α and d as learning progresses. The initial weights of competitive nodes were set randomly within 1.5±0.15.

All of 70 behavior sequences were transformed into input vectors using BI-transformation. The seven training data (Fig.5) consisting of a single input vector for each of seven enclosures were randomly given to the self-organizing network until the total number becomes 4200, and learning was done.

When the learning began to converge, the particular nodes got to be winners frequently. We considered the winner nodes correspond to enclosures, and called them *r-nodes*. Hence the number of r-nodes is the number of enclosures recognized by a robot.

In the test phase, at every time a test input vector was given, a winner of r-nodes was determined. We considered the winner r-node corresponds to an enclosure in which the input vector was obtained.

6.2 Exp-1: Identifying the enclosures

After learning with training data, the test data (63 input vectors) were given to the learned self-organizing network. As a result, we found all the test data were correctly identified (the accuracy = 100%), and verified the utility of our approach. Note that the 10 input vectors for the identical enclosure are slightly different mutually because of noise like failure of executing behaviors.

Furthermore additional experiments were made using twelve enclosures in Fig.7. 120 action sequences (20 for each enclosure) were used for training and test examples. The network consisted of 1300 input nodes and 128 competitive nodes. As a result, the accuracy decreased to 76%.

Using the twelve enclosures, we made experiments for investigating the influence of self-organizing net-

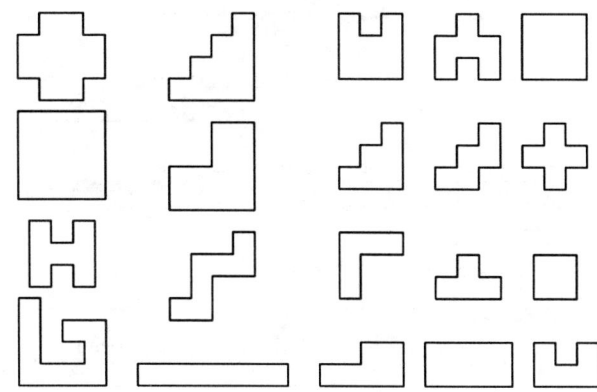

Figure 7 Twelve Enclosures

work's parameters on the accuracy. Setting the initial weight values: 0.5±10%, 1.5±10%, 4.5±10% and the structure of competitive layer: a line and a circle. As a result, the accuracy was 69% ∼ 74%, and the influence was considered slight.

6.3 Exp-2: Noisy environments

Though the test data used in Exp-1 included noise, it was not so much. In this experiment, we dealt with more noise like obstacles. We located obstacles in the seven enclosures, and the mobile robot did wall-following in the enclosures. The nine behavior sequences were obtained and transformed into input vectors. The input vectors (test data) were given to the self-organizing network which was trained in Exp-1.

As a result, five enclosures were correctly identified. Fig.8 shows success and failure examples. In Fig.8(a), the robot correctly recognized the enclosure including either a single obstacle or two small cubical obstacles on a wall. However, in Fig.8(b) the robot failed to recognize an enclosure including either two large obstacles or an obstacle located obliquely to the wall. The large or many obstacles changed the shape of enclosures from the original one, thus we consider the failure of recognition is natural.

6.4 Exp-3: Utilizing partial sequences

Although the experimental results were satisfactory, there are two important problems: a start point must be fixed in every trial, and a complete behavior sequence is necessary for recognition. The former makes recognition less robust, and the latter causes expensive recognition. Thus we developed a method to recognize environments with partial behavior se-

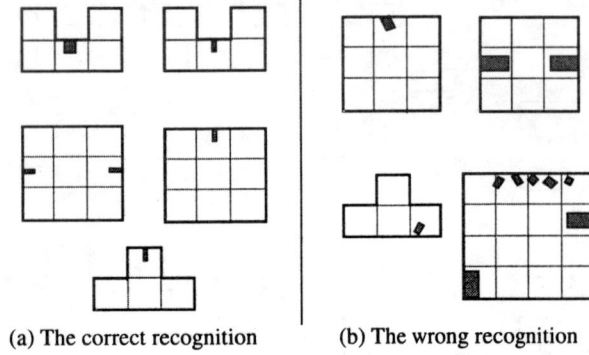

(a) The correct recognition (b) The wrong recognition

Figure 8 Enclosures with obstacles

quences independently of a start point. The following shows the procedure.

Training phase : Sift a behavior sequence so that the longest wall may be the head, and learn with it. After learning, r-nodes are obtained.

Testing phase : At every corner, the following is done. Make an input vector by BI-transformation, and obtain the partial input vector by shifting it so that the longest wall may be the head and setting value zero when a behavior was not executed yet. Compute the distance between the partial input vector and r-nodes. If the distance is sufficiently short, the environment is determined the r-node's environment.

Sifting an action sequence makes the recognition independent of a start point. Since the execution of a partial action sequence is more inexpensive than that of a complete one, the above procedure makes the recognition far more efficient.

We made experiments with the twelve enclosures in Exp-1. As a result, the accuracy and the partial ratio was 56.7% and 47.8%, where the partial ratio = $\frac{\text{The length of partial behavior sequences}}{\text{The length of complete behavior sequences}}$ (%).

7 Conclusion

We built a mobile robot which learns to recognize enclosures from behavior sequences without teaching. A self-organizing network was used for learning since it was able to generalize an input vector without a teacher and robust against noise. We also developed BI-transformation and a recognition method with a partial behavior sequence. The experiments using a real mobile robot were made for enclosures both with and without obstacles, and we verified the utility of our approach. Though this report may be preliminary, our approach is considered to contribute for designing a mobile robot which autonomously recognizes environments. While we obtained good experimental results, there are open problems like the followings.

- *Open environments*: Though we assume that an environment is a closed region, real environments may not be closed, e.g. a room with doors. The assumption is used only to terminate wall-following. Thus we need to develop a method to terminate robot's action in real environments.
- *Restriction on the shapes of enclosures*: The shapes of enclosures are restricted to rectangles with right corners. If an enclosure consists of curves like a circle, neither the behavior A nor B will be executed and a robot will not use information of corners. We need to improve the behaviors and BI-transformation.
- *Suitable behaviors for recognizing environments*: We used wall-following as behaviors for recognizing enclosures, however we do not consider it is best one. There may be a more robust and suitable behavior for characterizing environments. Currently we are developing evolutionary acquisition for suitable behaviors using genetic algorithm [8].

References

[1] E. M. Arkin, L. P. Chew, D. P. Huttenlocher, K. Kedem, and J. S. B. Mitchell. An efficiently computable metric for comparing polygonal shapes. *IEEE Transaction on PAMI*, 13(3):209–216, 1991.

[2] R. A. Brooks. A robust layered control system for a mobile robot. *IEEE Transaction on R&A*, 2(1):14–23, 1986.

[3] J. L. Crowly. Navigation of an intelligent mobile robot. *IEEE Transaction on R&A*, 1(1):31–41, 1985.

[4] T. Kohonen. *Self-Organization and Associative Memory*. Springer-Verlag, 1989.

[5] M. J. Mataric. Integration of representation into goal-driven behavior-based robot. *IEEE Transaction on R&A*, 8(3):14–23, 1992.

[6] U. Nehmzow and T. Smithers. Map-building using self-organizing networks in really useful robots. In *Proceedings of the First International Conference on Simulation of Adaptive Behavior*, pages 152–159, 1991.

[7] S. Tsuji and S. Li. Memorizing and representing route scenes. In *Proceedings of the Second International Conference on Simulation of Adaptive Behavior*, pages 225–232, 1992.

[8] S. Yamada. Learning behaviors for environment modeling by genetic algorithm. In *The First European Workshop on Evolutionary Robotics*, 1998. to appear.

ROLLMOBS, a new universal wheel concept

L. Ferrière and B. Raucent

Université catholique de Louvain
Department of Mechanical Engineering, 2 Place du Levant, 1348 Louvain-la-Neuve
e-mail : ferriere@prm.ucl.ac.be

Abstract

This paper presents a new family of omnidirectional and holonomic vehicles or, in short, omnimobile robots. The robot can perform simultaneously longitudinal, translational and rotational motions, providing full mobility. The concept is based on the original combination of a sphere driven by a classical universal wheel. The sphere in contact with the ground provides the robot high performances (load capacity, maximum surmountable bump height) and smooth motion (no vibrations). The universal wheel tread is not a tyre but consist on several freely rotating rollers that allow free motion of the sphere in the plane orthogonal to the roller axis.

1. Introduction

A large variety of wheeled mobile robots have been developed for industrial and personal purposes. For outdoor applications, four-wheeled car-like robots and robots equipped with caterpillars have traditionally been used, see for instance [1]. However, these systems are not well suited to indoor applications such as displacement in a congested room, corridor or workshop. Indeed, because of the kinematic constraints, these structures cannot move sideways. For such platforms, a parking operation requires several complex manoeuvres. This motion restriction is also a problem for part orientation in a stocking area and machine feeding as well as for any operation in a domestic environment such as cleaning, shopping, ... New platform principles have therefore been developed in research centres in order to improve motion capabilities.

When perfect mobility can be achieved, the robot is called an omnidirectional and holonomic robot or, in short, omnimobile robot. Those platforms can move instantaneously in any direction without reorientation of their wheels. It has been shown in [2] that, as regards the non slipping constraints relative to the wheels, omnimobility can be achieved using only off-centred or universal wheels.

Off-centred orientable wheels ("castor wheels") can rotate around a vertical axis which does not pass through the wheel centre (see Figure 1a). A robot equipped with three castor wheels is omnidirectional [3, 4]. But in order to avoid singularities, such as for a shopping trolley when the user wants to produce a transverse motion directly after a longitudinal one, the orientation of at least one wheel should be coordinated. This means that four motors are necessary. This complicates the robot control and, therefore, may lead to trajectory planning errors.

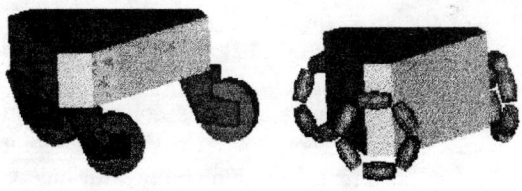

a. Off-centred wheels   b. Universal wheels
Figure 1. Omnimobile robots

Another solution is based on the "universal wheel" [5, 6], see Figure 1b. The wheel treads are not tyres but consist of several rollers whose axes are tangent to the wheel circumference, and free to rotate. As the shaft turns, the wheel is driven in a normal fashion in a direction perpendicular to the axis of the driven shaft. At the same time, the roller can rotate allowing a free motion perpendicular to the roller axis. An omnimobile platform can be designed by combining three or more such universal wheels.

The principal drawbacks of such a structure [7] are that the load capacity and the surmountable bump height (e.g. electrical cables, door sills, ...) are limited by the diameter of the roller and not, as for classical solutions, by the diameter of the wheel. Furthermore, this structure is sensitive to vibrations due to the successive shocks occurring each time the contact point passes from one roller to another.

To solve this problem and to improve the wheel performances, an orthogonal wheel composed of two truncated spheres placed side-by-side has been proposed in [8], see Figure 2. The wheel circumference appears to be continuous (see side-view in Figure 2), but the contact point moves from one roller to the other (each 90°). The rotations ω_i of the two spheres are mechanically coordinated. The wheels provide good performances and smooth motion but the bulkiness of the system is strongly increased.

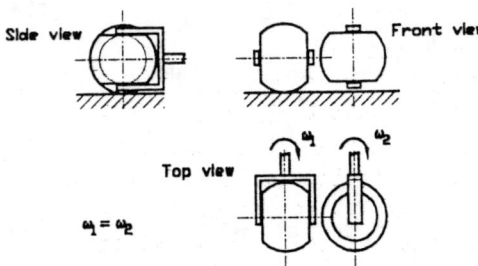

Figure 2. Orthogonal wheel

In the following section, we present a new spherical wheel assembly having the same kinematic properties as universal wheels. Furthermore, the new structure does not present the drawbacks of universal wheels (annoying vibrations, limited load and surmountable obstacle height) nor those of off-centred wheels (singular configurations). The wheel bulkiness is also reduced. This new design is protected by a patent [9]. The robot concept has been presented in [10] and [11]. We then show how a combination of three spherical wheels can be used to generate an omnimobile capability. In Section 3, we apply these concepts to the design of a prototype platform which provides omnimobility with independent longitudinal, translational and rotational degrees of freedom. Some experimental results illustrating these characteristics are presented in Section 4. Prospects are examined in Section 5 relative to improvements that will be required for a heavy-payload structure.

2. ROLLMOBS prototype

In a classical design, universal wheels are used to drive the robot and to carry it. As a consequence, the performances of the platform, such as the load capacity, are limited by the diameter of the roller in contact with the ground. The principle of the new design is to decouple the drive and load carrying functions. A spherical wheel is placed between the universal wheel and the ground. The performance of the structure now depends on the sphere diameter and no longer on the roller diameter.

In the MIT ball wheel mechanism [12], each ball wheel is held by a ring mechanism equipped with a set of rollers. The ring drives the sphere around an inclined axis so to develop a traction force between the ball and the floor. The rollers allow the sphere to rotate freely in the direction perpendicular to the active axis, thereby ensuring the robot's omnimobility.

Another solution is similar to the computer mouse but the rollers are replaced by universal wheels. In the case of a mobile robot, the roller should drive the sphere in order to move the structure. The driving torque is important. As a consequence, an important normal force needs to be applied between the roller and the sphere.

Use of a classical wheel is therefore not suitable because the friction forces prevent transverse motion, see Figure 3a. However, if a universal wheel is used, the free rotating roller in contact with the sphere allows free transverse motion, see Figure 3b. In the plane of the wheel, the sphere has a constrained motion which is controllable by the rotation of the wheel shaft, while the motion in the direction parallel to the shaft is unconstrained.

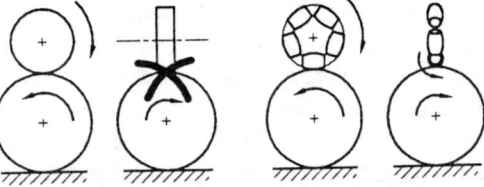

a. Classical wheel   b. Universal wheel
Figure 3. Drive devices

The second problem is to hold the spherical wheel in place and to allow a free rotation in any direction. Clearly, the sphere should be held at at least three points. The problem is thus to avoid friction at the contact points. Figure 4 presents two possible solutions.

A. Double universal wheels

B. Solid balls
Figure 4. Load carrying devices

Solution A uses non motorized double universal wheels. Solution B consists in using standard solid balls usually used in transfer devices. It should be noted that it is possible that dust will reduce the efficiency of the ball.

However the risk is lowered by using a proper cover for the sphere. Furthermore, since the contacting area between ball bearing and spherical wheel are significantly small, corresponding compressive forces tend to be extremely large. Solution B will be chosen for the first prototype but will be abandoned for the heavy structure because of its limited load capacity. Dedicated double wheels will be then used.

In case ROL1, presented in Figure 5a, the sphere is held in place by two solid balls and the universal wheel. The drive and load-carrying functions are thus coupled. Vertical vibrations are not completely eliminated but strongly reduced thanks to the use of small rollers which interpenetrate efficiently. This solution does not require the use of a suspension.

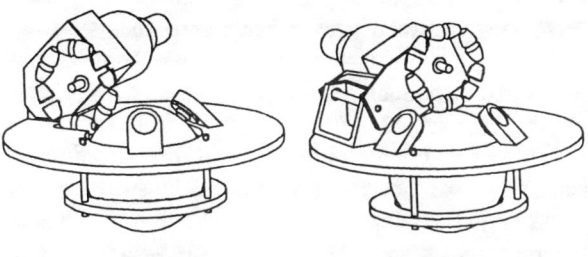

A. ROL1   B. ROL2
Figure 5. New wheel designs

For solution ROL2 presented in Figure 5b, the two functions are decoupled : three (or more) solid balls are used to carry the load and the universal wheel ensures only the drive function. However, this structure is hyperstatic and the force between the universal wheel and the sphere needs to be provided by means of a spring for example. In this case, the vertical vibration problem is totally eliminated.

Figure 6 shows the prototype that was built in order to validate the concept. The robot is equipped with 3 spheres made of polyurethane. The three assemblies are oriented at 120° from each other. Each sphere is held by 2 spherical ball bearings and 1 universal wheel. The driven wheel is made up of 12 half rollers. The chosen design is not optimal in the sense that spurious vibrations are not minimized but the wheel is easy to build. A list of the robot elements is summarized in Table 1. The greatest problem of sphere driving mechanism is friction force applied by the driven wheel. This force, depending on the position of the wheel on the sphere, increases when the wheel is close to the sphere equator. In case of ROLLMOBS prototype, the radial force on the wheel is about 33% of the robot total weight.

The power supplies and the computer that drive the robot are linked to the vehicle through a cable. For future structures both elements will be implemented on the robot. Typical DC motors drive the wheels. Tachometers are included on the main shafts. The controller sends its instructions to each motor at a rate of 5 ms. At present, there is no feedback to the controller, see Figure 7. The position and orientation of the robot are only calculated by dead reckoning.

- Platform dimension : ⌀ 500 mm ; weight : 25 kg
- Spheres : ⌀120 mm ; weight : 1 kg ; polyurethane
- Ball bearings : ⌀15 mm ; steel
- DC motors : 15 W ; 24 VDC ; tachogenerator
- Gearhead : 25:1 gear ratio ; 65% efficiency
  rated torque 1.45 Nm ; rated speed 60 rpm
- Universal wheels : ⌀90 mm ; radial force : 5 kg ;
  max. driven torque : 1.1 Nm ; friction coef. : 0.5
- Total weight : 15 kg

Table 1. System characteristics

Figure 6. ROLLMOBS prototype

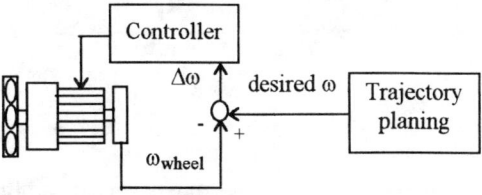

Figure 7. Current control block diagram

3. Kinematic relationships

The robot consists of a trolley whose frame, in the shape of an equilateral triangle, is equipped with 3 spheres placed at the vertices of the triangle.

Let us consider an inertial frame (I_1, I_2) in which the position of the frame centre of the robot is described by coordinates x and y. The angular position of a moving frame (X_1, X_2) attached to the robot is characterized by angle θ (see Figure 8). Let (ω_1, ω_2, ω_3) be the angular velocities of the three motors and (Ω_1, Ω_2, Ω_3) those of the spheres around an axis passing through the sphere centre and parallel to the wheel shaft.

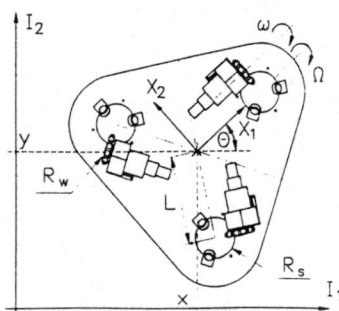

Figure 8. Robot coordinates

Let us define :

$$q = \begin{bmatrix} x & y & \theta \end{bmatrix}^T$$
$$\dot{\xi} = \begin{bmatrix} \omega_1 & \omega_2 & \omega_3 \end{bmatrix}^T \quad (1)$$
$$\dot{\psi} = \begin{bmatrix} \Omega_1 & \Omega_2 & \Omega_3 \end{bmatrix}^T$$

The kinematic constraints for each wheel correspond to a non slipping condition between the wheel and the sphere in the plane of the wheel, and to a non-slipping condition between the sphere and the ground. The kinematic model can then be written as :

$$J_1 R(\theta)\dot{q} + J_2 \dot{\psi} = 0 \quad (2)$$
$$J_3 \dot{\xi} = \dot{\psi} \quad (3)$$

with

$$J_1 = \begin{bmatrix} 0 & -1 & -L \\ \sqrt{3}/2 & 1/2 & -L \\ -\sqrt{3}/2 & 1/2 & -L \end{bmatrix}$$

$$J_2 = R_s, \quad J_3 = -\frac{R_w}{R_s} \quad (4)$$

$$\text{and } R(\theta) = \begin{bmatrix} \cos(\theta) & \sin(\theta) & 0 \\ -\sin(\theta) & \cos(\theta) & 0 \\ 0 & 0 & 1 \end{bmatrix}$$

L is the distance between the sphere centre and the frame centre. R_s is the sphere radius and R_w the wheel radius.

The kinematical constraints relate the angular velocities of the universal wheels ($\dot{\xi}$) to the angular position (θ) and velocity ($\dot{q}$) of the trolley. Considering equations (2), (3) and (4), we have :

$$\dot{\xi} = \frac{1}{R_w} J_1 R(\theta) \dot{q} \quad (5)$$

These constraints are not completely integrable : there is no analytical relation between ξ and q. This means that, at a given time t, q(t) and $\dot{q}$(t) do not depend only on ξ(t) and $\dot{\xi}$(t) but the complete evolution of $\xi(\tau)$ and $\dot{\xi}(\tau)$ for $\tau \leq t$. A control algorithm processing only ξ and $\dot{\xi}$ can provide only the regulation of the velocity ($\dot{q}$, θ) but not of the position x and y. Any error on ξ implies an irreversible position error on x and y.

Risk of slipping is important between the wheel and the sphere. This problem has been thoroughly studied in [11]. Using a good design, risk is considerably lowered. Unfortunately, that depends strongly on the friction coefficient between the wheel and the sphere. If conditions change (robot rolls on dust, water,...), the wheel may slip. Such a malfunction is not as yet detected on the current robot. On a future prototype, passive rollers coupled with encoders will be placed on the sphere in order to detect slipping. This will provide a feedback to the wheel controller and will allow to update the robot position and orientation. It will considerably increase the position dead reckoning.

The control algorithm is illustrated in Figure 9.

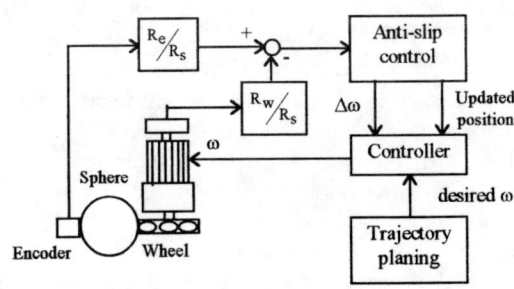

Figure 9. Future control algorithm

As yet, such a control algorithm that will minimize the error on the wheel velocities or, equivalently, on the robot speeds, has not been implemented on the prototype. More detailed results will be given during the oral presentation.

4. Experimental results

To demonstrate the operationality of the spherical wheel concept, a series of tests were performed. Different patterns were tested. Experimental results are shown in Figure 10, 11 and 12. A small light is mounted on the edge of the platform allowing to depict the actual path of the vehicle.

Figure 10 shows a 360° spin turn motion at a speed of 0.5 rad/s. Figure 11 illustrates the "crab-motion" capability of the robot. The robot follows a square (1 by 1 meter) at a speed of 0.2 m/s. Figure 12 shows the robot following a ⌀1 m circle with a linear speed of 0.3 m/s. Orientation is maintained during the rotation. Finally, Figure 13 shows a combination of a translational motion and a rotation.

In addition to this demonstration, the accuracy of the robot position is measured. The robot follows programmed trajectories depicted in Table 2 and returns back to its initial position. The final position of the robot centre is then compared with the initial one. In repeated experiments, the position error of the platform was found to be less than 2 % of the length of the trajectory.

n°	Description	Motion	Error
1	Straight line (2 x 2 m) backwards and forwards Speed 0.3 m/s		max : 1.4 %
2	Circle (Radius 1 m) constant orientation Speed 0.3 rad/s		max : 0.7 %
3	720° rotation Speed 0.9 rad/s		max : 1.7 %
4	Square (2 m square) constant orientation Speed 0.2 m/s		max : 1.25 %

Table 2. Programmed trajectories

5. Heavy platform

The small prototype has proved its performances in terms of accuracy and comfort (smooth motions, reduced vibrations). This encourages us to build another platform. The new robot will be autonomous (embarked batteries, PC on board) with a load capacity of 150 kg. Dead reckoning will be improved via passive rollers.

A new wheel design will be also tested. Indeed, in case of ROL1 (see Figure 5a), vibrations are not totally eliminated. In case of ROL2 (Figure 5b), the structure is hyperstatic and therefore requires a suspension. The idea is to eliminate spurious vibrations using the double wheel concept.

Unfortunately, classical double wheels cannot be used. Indeed, the contact point between the wheel and the sphere changes when the wheel shaft turns, producing undesirable vibrations (see Figure 14).

Figure 10. 360° Spin turn

Figure 11. Square

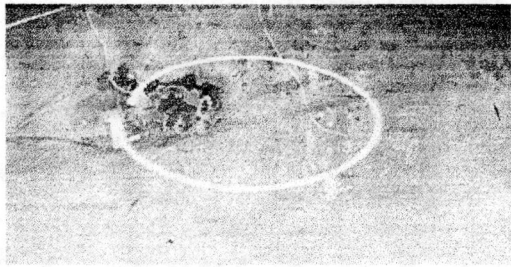

Figure 12. Circle

Figure 13. Spiral

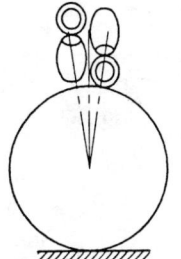

Figure 14. Double wheel drive

For this reason, we plan to try the "dedoubled wheels" assembly, see Figure 15. The two wheels equipped with few rollers (3/wheel) are placed on the sphere. Their axis are parallel and driven by the same actuator. There are no spurious torques and there is always a roller in contact with the sphere. Vibrations are then totally eliminated.

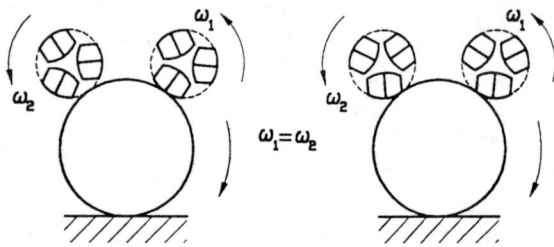

Figure 15. Dedoubled wheel concept

The two wheels rotations should be synchronized but numerous advantages ensue from the geometry of the wheel :

- increased load capacity
- continuous contact
- easy manufacturing of the rollers
- simplified component fitting (ball bearings)
- easy fixing of the rollers on the wheel

The heavy-payload structure will soon be manufactured.

Conclusions

An original spherical wheel has been presented. It is based on the mouse concept providing the robot with full mobility. Simultaneous longitudinal, translational and rotational motions are possible. The structure does not exhibit singular configurations such as for an omnimobile robot equipped with castor wheels. Comparing with robots equipped with classical universal wheels, performances are enhanced. Maximum surmountable bump height and load capacity are improved while spurious vibrations are totally eliminated or strongly reduced.

A prototype of such a platform has been constructed using three spherical wheels driven by a classical universal wheel. The mobile robot works well and proves its omnimobility. Trajectories are quite accurate despite lack of feedback control.

Further work will implement new features on the prototype such as optic encoders mounted on passive universal wheels. Additional comments will be made during the oral presentation.

The construction of a larger autonomous platform will be carried out. In this case, minor changes on the robot design should be taken into account in order to improve the robot load capacity and to eliminate all kinds of vibrations. The modelling of the robot performed in [13] will be helpful.

Acknowledgement

This work was sponsored by the Belgian Program on Interuniversity Attraction poles initiated by the Belgian State - Prime Minister's Office - Science Policy Programme (IUAP -24) and the Fonds National de la Recherche Scientifique (FNRS). The scientific responsibility rests with its authors.

references

[1] R. Jarvis, An Autonomous Heavy Duty Outdoor Robotic Tracked Vehicle, Proceedings of IROS'97, Grenoble (France), September 1997, pp. 352-359.

[2] G. Campion, G. Bastin and B. d'Andrea-Novel, Structural Properties and Classification of Kinematic and Dynamic Models of Wheeled Mobile Robot, IEEE Trans. on Robotics and Automation, Feb. 1996, vol. 12, pp. 47-62.

[3] L. Ferrière, B. Raucent and A. Fournier, Design of a Mobile Robot Equipped with Off-centred Orientable Wheels, book for the ERNET Workshop, Darmstadt (Germany), Sept. 1996, pp. 127-136.

[4] M. Wada and S. Mori, Holonomic and Omnidirectional Vehicle with Conventional Tires, Proceedings of ICRA'96, Minneapolis (USA), April 1996, pp. 3671-3676.

[5] R. E. Smith, Omnidirectional Vehicle Base, Int. Patent, WO 86/03132, June 5 1986.

[6] B. Raucent, P. Sente, H. Buyse and J.S. Samin, Robot Mobile à Trois Degrés de Liberté, Revue Générale de l'Electricité, n°6, 1990, pp. 35-43.

[7] L. Ferrière, B. Raucent and G. Campion, Design of Omnimobile Robot Wheels, Proceedings of the ICRA'96, Minneapolis (USA), April 1996, pp. 3664-3670.

[8] F. G. Pin and S. M. Killough, A New Family of Omnidirectional and Holonomic Wheeled Platforms for Mobile Robots, IEEE Trans. on Robotics and Automation, Vol. 10, n°4, August 1994, pp. 480-489.

[9] L. Ferrière, B. Raucent, Base Mobile Omnidirectionnelle, Belgian Patent.

[10] L. Ferrière, B. Raucent and J.-C. Samin, ROLLMOBS, A New Omnimobile Robot, ICRA'97 Workshop on Innovative Designs of Wheeled Mobile Robots, Albuquerque, April 1997.

[11] L. Ferrière, B. Raucent and J.-C. Samin, ROLLMOBS, A New Omnimobile Robot, Proceedings of IROS'97, Grenoble (France), September 1997, pp. 913-918.

[12] M. West and H. Asada, Design of Ball Wheel Vehicles with Full Mobility, Invariant Kinematics and Dynamics and Anti-slip Control, ASME Journal of Mechanical Design, 1994, pp. 377-384.

[13] L. Ferrière, P. Fisette, B. Raucent and B. Vaneghem, Contribution to the Modelling of a Mobile Robot Equipped with Universal Wheels, SYROCO'97 symposium, Nantes (France), Sept. 1997, pp. 715-722.

A NEW SATELLITE SELECTION CRITERION FOR DGPS USING TWO LOW-COST RECEIVERS.

A. Pozo-Ruz, J. L. Martínez and A. García-Cerezo.
Universidad de Málaga. Dep. Ingeniería de Sistemas y Automática.
E. T. S. I. Industriales. Plaza El Ejido, s/n. 29013- Málaga (Spain).
Phone: (+34) 5 2131418. Fax: (+34) 5 2131413. E-mail: anapozo@ctima.uma.es

Abstract: *This paper considers the employment of two low-cost GPS receivers for calculating positions based on a new satellite selection criterion. The parameters that decide the set of satellites used for positioning must take into account the coordinated work of both receivers. Thus, they are very different, or even contradictory to those conventionally used when the GPS receiver is working in single mode or in conjunction with a specialized fixed receiver. The method proposes the calculation of positions in the plane using the three highest satellites. The precision obtained with this technique is better than that provided when the receiver uses the conventional correction data.*

1. INTRODUCTION.

The Global Positioning System (GPS) is a navigation system consisting of a constellation of 24 satellites in 6 orbital planes inclined 55 degrees with respect to the plane of the Earth's equator. This system provides an accurate 3-dimensional position through trilateration techniques based on the time of flight of the radio signal transmited by the satellites (Sonnenberg, 1988).

However, civil applications of the GPS are not sufficiently precise, because of certain sources of error that affect the system. These errors can be divided into two groups: satellite geometry dependent and independent. Selective Availability (S/A), ephemeris error and receiver accuracy are independent of the satellite geometry. On the other hand, ionospheric delay, tropospheric delay and multipath reflections depend strongly on satellite geometry, especially the elevation angle (Herring, 1996).

S/A is the intentional changing of the satellites onboard clock introduced by the Defense Department of the U.S.A. in order to limit the civilian signal, which might be used by potentially hostile forces. S/A appears as a continuous variation in the user's position. S/A is the main source of error, and can produce errors of nearly one hundred metres. However, when averaged over a period of hours, the estimated position converges within metres of the exact position, because S/A has a correlation time of approximately 15 minutes (Bennet et al, 1995).

The GPS receiver relies on knowing exactly where the satellites are in space with respect to the center of the Earth. Errors in ephemeris limit GPS accuracy to approximately 5 meters at best. The receiver accuracy includes the precision of the onboard microprocessor, antenna/channel delays and the internal clock resolution.

Ionospheric effects are very much a function of satellite elevation angle, which determines the length of the signal path through the ionosphere, resulting in the shortest possible path when the satellite is directly overhead, reducing the signal delay (see fig. 1).

Tropospheric delays are caused by water vapor and other atmospheric effects. Due to the closer proximity of the troposphere relative to the receiver, the delay effects are even more dependent on satellite elevation than those associated with ionospheric delays. This dependence increases exponentially below an elevation angle of 15 degrees, because of the stronger concentration of water vapor at altitudes less than 12 km.

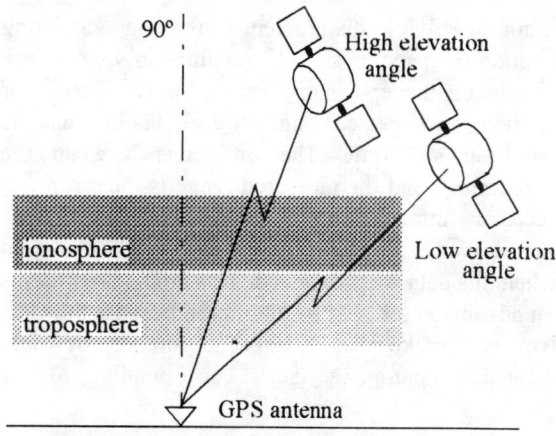

Figure 1. Satellite elevation angle.

If the signal from the satellite suffers multipath reflections, the time of flight of this signal is increased, so the GPS receiver calculates an erroneous position. It is very difficult to model multipath reflections properly because they heavily depend on the environment of the antenna such as buildings or water. But it is known that signals from a satellite with a lower elevation angle are much more affected by multipath reflections than those from one with a higher elevation angle.

GPS provides the position of an object in space (it does not give information about the orientation) when working in 3-D mode (Everett, 1995). However, errors in altitude are greater than in latitude and longitude, because the satellites are always over the receiver. Normally, errors in altitude are two or three times latitude and longitude errors.

All these problems can be partially solved by using differential techniques (DGPS). The differential GPS is a technique that employs two receivers and is based on the assumption that the satellites are so distant away in comparison to the separation between the receivers that they will be affected by the same errors. If one receiver is fixed at a precisely known location, its calculated position can be compared to the known position to generate a composite error vector using a fixed predefined code. This differential correction is then passed to the other receiver that accepts this data, reducing in this way the error in position.

The main drawback of generating the differential corrections using this technique is the ambiguity associated with the satellites which are used for the mobile receiver and the base receiver to generate the position. If both receivers select different combinations from the available satellite in sight, the differential corrections are not valid. Since most modern multichannel receivers employ dynamic algorithms for optimal satellite selection, this coordination between fixed and mobile receivers presents a real problem.

A method of solving this problem consists in calculating the location in space of all GPS satellites in view at any time by using the ephemeris. From this two pieces of information, an expected range to each satellite can be computed at any time. The difference between the computed range and the measured range is the error. So, the necessary information that must be transferred to the remote receiver is the correction for each satellite in sight and when this data was computed. This method represents a great advantage for DGPS techniques, because the GPS receiver can make use of this information by only choosing the appropiate data corresponding to the selected satellites.

The fixed DGPS reference station transmits these corrections several times a minute to any capable differential receiver employing a fixed code. Many commercial GPS receivers follow the RTCM SC-104 standard to transmit these correction messages (Radio Technical Commision for Maritime Services, 1990).

Since the price of a base station able to generate differential corrections is higher compared to the price of a receiver, some companies supply a service that consists in providing DGPS corrections, transmitting the error data from one or more reference base stations to several users economically and with good performance (Weber and Tiwari, 1995). These correction messages can be transmitted over a FM link or employing a special satellite (Huff, 1995). The accuracy strongly depends on the velocity of transmission, because of the rapid change in S/A errors, as well as the bandwidth usage, that dictates how much data can be sent in a given period of time.

Another technique, the carrier-phase-differential, achieves millimetre precision. This method is used, for example, in topographic applications, where high accuracy is required (Crane et al., 1995). However it is very expensive.

In this paper, a simple method for DGPS is proposed that employs two low-cost GPS receivers without using conventional differential messages (RTCM correction data), offering a good accuracy since a new satellite selection criterion is used. The efficiency of this technique has been demonstrated experimentally.

The paper is organized as follows. The new satellite selection criterion is explained in section 2. Section 3 describes the implementation of this technique. The experimental results are shown in section 4. Section 5 presents the conclusions and future works. The final sections are devoted to acknowledgments and references respectively.

2. THE PROPOSED METHOD.

The proposed method employs two low-cost GPS receivers, one of them being fixed in any location. The position of this fixed receiver is considered as the origin of the cartesian coordinate system. The position calculated by the fixed receiver is then transmitted to the mobile one, which calculates the difference between the two measurements, indicating the localization of the mobile receiver with respect to the origin.

But to be precise, it is necessary that the two receivers use the same set of satellites. However, this does not present a problem since both receivers are forced to employ the higher satellites only.

The next questions to solve are why to use the highest satellites and no other configurations such as the most equally spaced on the horizon or all common satellites in view for both receivers and how many satellites should be used for calculating positions.

Because the separation between both receivers is insignificant compared to the distance of the satellites in space, both GPS will use practically the same set of satellites in their calculations, differing perhaps in lower-angle satellites, not in the higher-angle ones.

Generally, Position Dilution of Precision in 3-D (PDOP) is used by GPS receivers as the automatic criterion for selecting satellites since it represents the geometrical contribution of observation errors on GPS positioning accuracy (Park et al., 1995).

GPS receivers allow the selection of 2-D or 3-D position if required. While a minimum of four satellites are needed to position in 3-D mode, only three are required to work in 2-D mode. Since in many applications the receiver moves on the ground, a two dimensional configuration has been adopted, so the PDOP criterion is substituted by the Horizontal Dilution of Precision (HDOP) criterion.

A low HDOP value means that the set of satellites employed for calculating positions has a good geometry with respect to the receiver.

A good geometry is understood as the satellites being more separate from each other on the horizon. For this reason, satellites with low elevation angles tend to be selected when the HDOP is used as the satellite selection criterion (see fig. 2).

However, satellites with low elevation angles produce larger ranging errors than others with higher elevation angles. Moreover, HDOP is defined under the assumption that the GPS receiver is working in single mode, without taking into account the coordination with other receivers for positioning.

f the GPS receiver is working in single mode, more than the minimum number of satellites required for positioning will be employed to increase the accuracy of the measurements. However, when the receiver works in coordination with another one, as in the proposed method, it is only necessary that the GPS receivers make the calculations with the minimum number of satellites to avoid ambiguities in the error obtained, that is, the contribution of each satellite to the error will be almost the same. Thus, the error will practically disappear when subtracting.

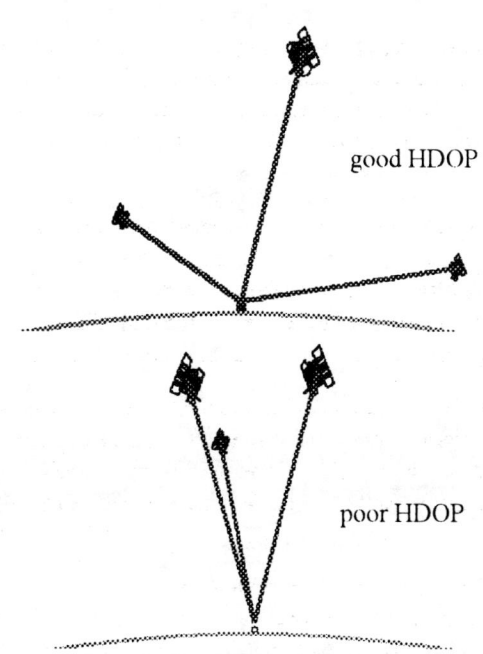

Figure 2. HDOP criterion.

Other configurations of three satellites that are not the three highest can also be employed, but they are rare and remain less time without change.

3. IMPLEMENTATION.

Two units of the GPS GN-74 model have been employed as low-cost GPS receivers (Furuno, 1995). This consists of a board of reduced dimensions conected to a PC through the serial port employing the RS-232 protocol (see fig. 3). It has 8 channels for tracking up to 12 satellites. This receiver uses the NMEA 0183 protocol for obtaining position, velocity, time and satellite information (National Marine Electronics Association, 1992).

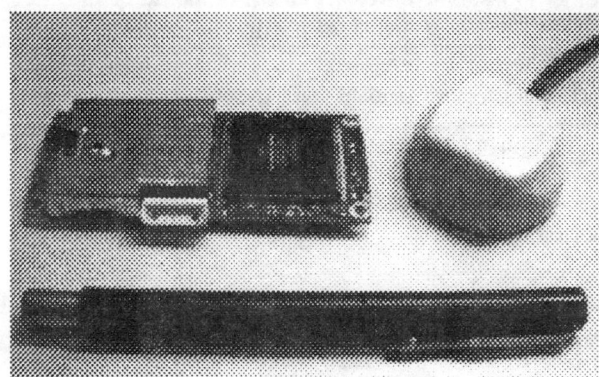

Figure 3. GPS GN-74 receiver.

The receiver must be considered as a black box that only permits communication through the NMEA protocol. Using this protocol it is possible to change some parameters of the GPS configuration, such as the satellite to be employed in the position calculations, but never modify the source program.

Due to this limitation, the algorithm proposes that a computer makes use of the NMEA messages to select the three highest satellites, masking the other ones in view (see fig. 4). In this way, the use of the same set of satellites during a determined period of time is guaranteed.

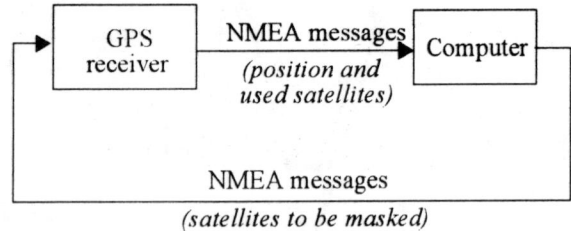

Figure 4. Communication between a GPS receiver a computer.

The main problem that has been found experimentally with the receivers is the maximun limit of HDOP that forces the receiver to stop in its calculations if it is overcome. Also, in order to converge the computations of both receivers with the same satellites, it is necessary to wait an initial time before data adquisition.

Moreover, the GN-74 does not mask a satellite that is not in view in that particular moment, thus after a period of time, a new satellite may enter. Since the satellite constellation can not be known in advance with this model, this problem must be solved by the computer.

4. EXPERIMENTAL RESULTS.

In order to taste the efficiency of the proposed method, a large collection of experiments has been made using different sets of satellites at different hours during a period of six months. Here some of them are presented.

The following experiment compares the accuracy obtained with the previously discussed satellite selection criterions: the three highest satellites and the conventional HDOP parameter.

Fig. 5 and 6 show the positions obtained during 30 minutes receiving data every two seconds, when one of the receivers was in a fixed location working in single mode. In these two cases, both receivers employ the same set of satellites. The data is plotted with respect to its mean value.

Fig. 5 was obtained when the HDOP criterion is employed in order to choose the satellites that provided the best accuracy, while fig. 6 shows the positions obtained when using the three highest satellites selection criterion. As can be observed, HDOP shows a better performance in single mode.

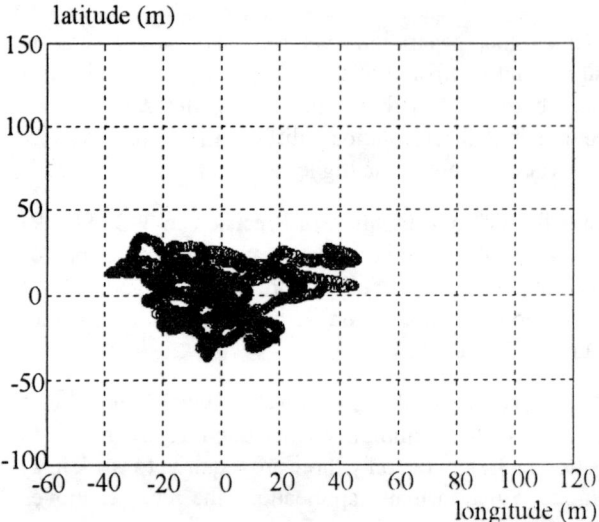

Figure 5. GPS data working in single mode with the HDOP criterion.

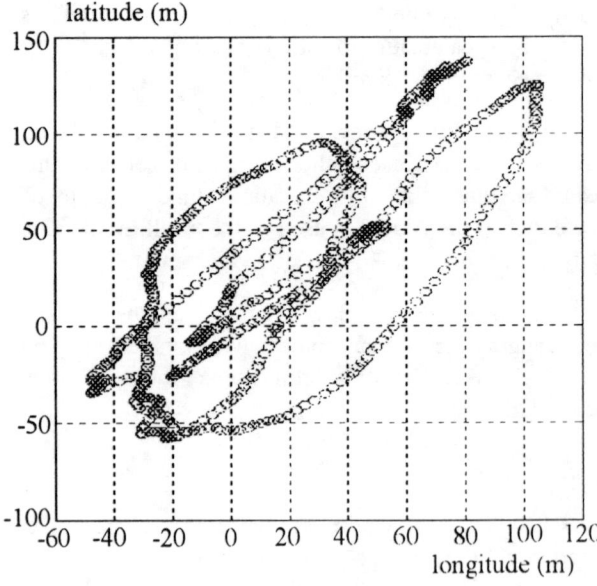

Figure 6. GPS data working in single mode positioning with the three highest satellites.

The subtraction of the data collected at the same time by the GPS receivers when they are in a fixed location and separated at a certain distance according to the HDOP criterion and the three highest satellites strategy can be observed in fig. 7 and 8. The data is also plotted with respect to its mean value.

As can be observed, the data obtained with the HDOP selection criterion has a higher deviation, even when both receivers use the same set of satellites.

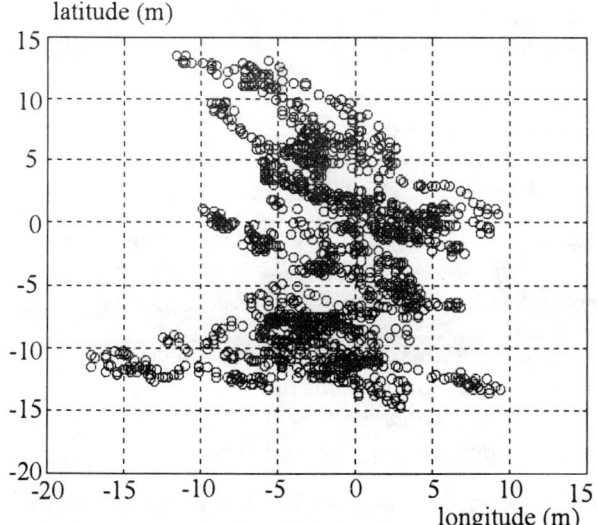

Figure 7. GPS data working in differential mode using the HDOP criterion.

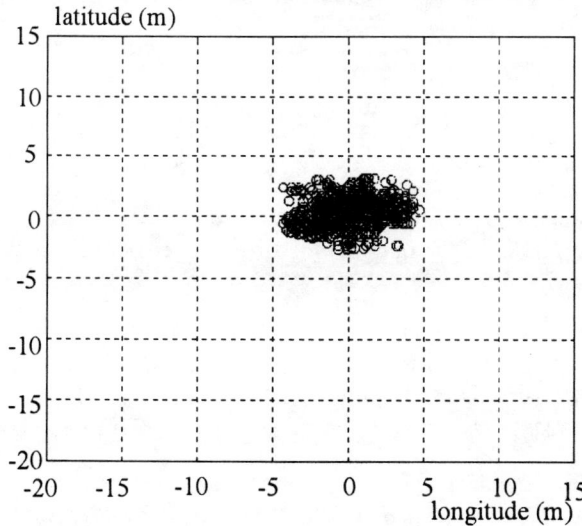

Figure 8. GPS data working in differential mode using the three highest satellites.

Moreover, the data obtained using the differential method proposed in this paper provides an accuracy that is better than provided by the GN-74 working in differential mode using RTCM data from a virtual base station (see fig. 9).

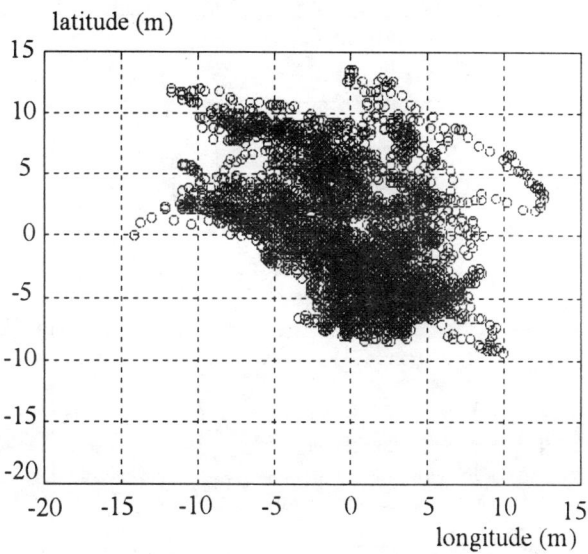

Figure 9. Collected data with the GPS GN-74 with RTCM data.

Table 1 shows for the above experiments the horizontal circular error probable (CEP) 50% and 95%, that represents, respectively, the radius of a circle within which the 50% and 95% of all measurements in longitude and latitude should be found.

GPS work mode	Criterion	Horizontal CEP (50%)	Horizontal CEP (95%)
Single	HDOP	21.2 m	40.5 m
Single	3-highest satellites	44.2 m	125.2 m
Differential	HDOP	7.6 m	14.2 m
Differential	3-highest satellites	1.88 m	3.5 m
Differential	RTCM	5.5 m	4.7 m

Table 1. Horizontal CEP.

Finally, the following experiment consists on testing the accuracy of the proposed technique when one of the receivers is moving. The recorded path when the receiver describes two times a rectangle of 13.5 x 26 metres is shown in fig. 10.

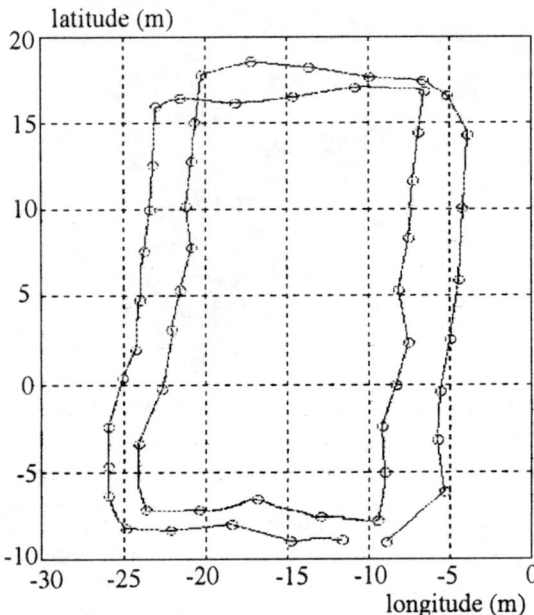

Figure 10. Movement test.

5. CONCLUSIONS.

The main contributions of the proposed method are the employment of cheap receivers, as well as the criterion for selecting satellites, which is very different, or even contradictory to those usually employed by GPS receivers. The reason is that the set of satellites chosen by this method takes into account that the two receivers are working together, rather than independently, so falling into the same errors with the exception of multipath reflections.

The position accuracy obtained is much better than the precision provide by the low cost receiver when working with RTCM data.

Essentially, this aproach does not make use of all available sources of information. If the user could manipulate the internal receiver's algorithm, then weighting functions or other heuristics could be considered.

The method is effective for corrections in local applications for short periods of time.

Future works include the integration of this method with other sensors in order to improve position data and its application to terrestrial navigation. Also, the study of the influence in the accuracy when changing one satellite to another will be considered.

The extension of the method to 3-D data with the selection of the four higher satellites will be also addressed, as well as the implementation of the method in other low cost GPS receivers.

6. ACKNOWLEDGMENTS.

This work has been partially supported by the C.I.C.Y.T. project TAP 96-1184-C04-02. The authors express their gratitude to Grafinta S.A., in particular to F. Mier and P. Mier for their useful hints.

The first author expresses her appreciation to the Fundacion Ramon Areces for her research grant.

7. REFERENCES.

Bennet S. M., D. E. Allen, W. Acker and R. Kidwell (1995). *"Blended GPS/DR Position Determination System"*. Andrew Technical Report.

Crane C. D., A. Rankin and D. G. Armstrong (1995). *"An Evaluation of INS and GPS for Autonomous Navigation"*. Proc. 2nd IFAC Conference on Intelligent Autonomous Vehicles 95, pp. 208-213. Espoo, Finland.

Everett H. R. (1995). *"Sensors for Mobile Robots: Theory and Applications"*. Ed. A. K. Peters.

Furuno Electric. Co. LTD. (1995). *"GPS Receiver Model GN-74. Technical Information"*.

Herring T. A. (1996). *"The Global Positioning System"*. Scientific American, pp. 32-38. February.

Huff M. K. (1995). *"Omnistar, A Versalite DGPS Positioning Tool"*.

National Marine Electronics Association (1992). *"NMEA-0183 Standard for Interfacing Marine Electronic Device Version 2.00"*.

Park C., I. Kim, J. G. Lee and G. Jee (1995). *"A Satellite Selection Incorporating the Effect of Elevation Angle in GPS Positioning"*. Control Engineering Practice, No. 12, pp. 1741-1746.

Radio Technical Commision for Marine Services (1990). *"RTCM Recommended Standards for Differential Navstar GPS Service Version 2.0 RTCM Special Committee No. 104"*. Washigton DC. U.S.A.

Sonnenberg G. J. (1988). *"The Global Positioning System"*. Radar and Electronic Navigation. Chapter 7. Ed. Butterworths.

Weber L. and A. Tiwari (1995). *"DGPS Architecture Based on Separating Error Components, Virtual Reference Station and FM Subcarrier Broadcast"*. Proc. 1995 ION Annual Meeting.

Model-Based Car Tracking Integrated with a Road-Follower

Frank Dellaert   Dean Pomerleau   Chuck Thorpe

Computer Science Department and The Robotics Institute
Carnegie Mellon University, Pittsburgh PA 15213

Abstract

This paper discusses how we integrated our 3D car tracking approach with the lane following module RALPH on the Navlab autonomous vehicles, obtaining a hybrid vision system that tracks both the road and cars better than those two systems in isolation. The tracking system brings precise and crisp measurements of the car in the image, and performs image stabilization. However, because it does not know about the yaw or lateral offset of the ego-vehicle, its curvature estimate can be misguided. RALPH takes a more global image processing approach and can provide this missing information, as well as a good estimate of curvature, so that the combined curvature estimate is superior to both taken in isolation. The additional information provided by RALPH also improves tracking performance, and allows us to estimate properties of the tracked car that were previously unobservable, in particular its in-lane displacement. Better car tracking, and a better idea of where the road is, gives us a substantial foundation on which to base other capabilities needed to realize fully autonomous vehicles.

1 Introduction

In this paper we show how we integrated our 3D car tracking approach with the lane following module on the Navlab autonomous vehicles, obtaining a hybrid vision system that tracks both the road and cars better than those two systems in isolation. The Navlabs are the experimental platforms for the Automated Highway System (AHS) research being conducted at CMU[1]. Several other groups have similar research programs, in particular Dickmanns' group at the Bundeswehr University in Munich [2]. Much of the research done on the Navlabs has been concentrated on road-following. The first convincingly successful system, ALVINN, consisted of a neural network that learned to predict steering direction from subsampled video images [3]. Since then, ALVINN has been superseded by RALPH, which uses a more domain specific method to estimate the curvature of the road and the lateral offset of the vehicle, which can then be used to calculate a steering command [4]. Although RALPH has been quite successful for road-following, it does not provide any information about the position or behavior of other cars on the road. This capability, *situational awareness*, enables autonomous vehicles to plan a course of action, ranging from matching the speed of a car ahead, to planning more complicated behaviors involving lane changes and overtaking other cars. To provide situational awareness, we recently proposed a model-based vision approach to car tracking, in which we estimate the 3D position and motion of a car by tracking a 2D bounding box in the video stream [5]. Since only line segments are tracked, the image processing involved is relatively simple, and the system can run at frame rate.

By combining the strengths of the tracking algorithm with the strengths of the RALPH road following module we can obtain an overall system which is superior to both approaches taken individually. The tracking algorithm provides very crisp measurements of the relative position and speed of the car that is being tracked, but could do even better if given an idea of exactly where the road is, as then it can form better expectations of how a tracked car will behave. Likewise, the road-follower could benefit from information provided by the tracker, since the position of the tracked vehicle provides an important clue as to where the road is. Having this additional estimate of curvature will be especially helpful in situations where RALPH traditionally has problems, e.g., when the flat earth assumption is violated. A principled way of combining the information provided by both techniques separately is by using an extended Kalman filter [6], and that is the way we will approach it here. Since the tracking algorithm relies heavily on a Kalman filter already, integrating the RALPH measurements in this framework comes rather naturally.

In the remainder of this paper we will give a brief overview of RALPH (Section 2) and the image processing used for the tracking (Section 3). Then, in Section 4, we describe the Kalman filter used for inte-

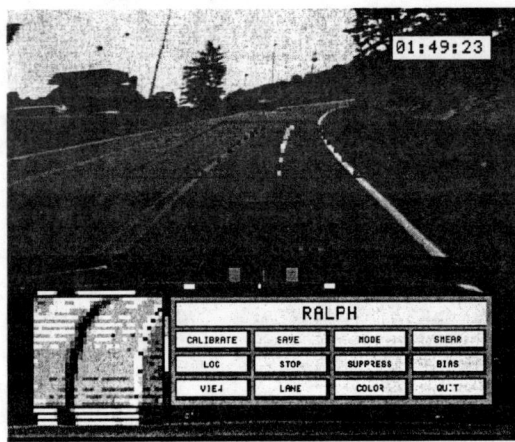

Figure 1: Screen shot of RALPH.

grating the two modules, i.e., we describe the model of how both cars relate to each other and the road, formulate the dynamics of this model and discuss the measurement equations that relate the two vision techniques to the model. Both qualitative and quantitative results are presented in Section 5 to establish the superiority of the resulting hybrid system to both modules in isolation. Finally, we conclude with a discussion and opportunities for future work in Section 6.

2 The RALPH Vision System

The RALPH vision system helps automobile drivers steer, by processing images of the road ahead to determine the road's curvature and the vehicle's position relative to the lane center. RALPH uses this information to either steer the vehicle autonomously, or warn the driver if he/she is steering inappropriately.

In order to locate the road ahead, RALPH first resamples a trapezoid shaped area in the video image to eliminate the effect of perspective (see Figure 1). RALPH then uses a template-based matching technique to find parallel image features in this perspective free image. These features can be as distinct as lane markings, or as subtle as the diffuse oil spots down the center of the lane left by previous vehicles. RALPH rapidly adapts to varying road appearance and changing environmental conditions by altering the features it utilizes to find the road. This rapid adaptation is accomplished in under one second.

3 Image Processing for Tracking

Although many features could be used to track a car in a sequence of video images, we track only the 2D bounding box around the car (Figure 2). When looking at the image of a car, you can see that in general it has strong edges on all sides where the image changes from the car to the background. This contour of strong edges is not always regularly shaped, but it can be approximated by a rectangle for a wide range of aspects. In the remainder of this section we will briefly discuss the image processing technique we use to track this 2D bounding box in a video stream. A more detailed treatment and a discussion of related work can be found in [5], where we also make the coupling with the Kalman filter more explicit. The overall approach is closest in spirit to the work of Schmid [7], although the underlying image processing is quite different.

We rely on an energy minimization technique to find the bounding box around the tracked car in each frame. An initial estimate is obtained by projecting the imaginary 3D bounding box around the car, as estimated by the Kalman filter (see below), into image space. As we are looking for a rectangular contour with strong edges on all sides, we maximize an objective function F that measures the strength of the edges around a particular contour. With (t, l) and (b, r) the top-left and bottom-right coordinates of the bounding box, respectively, and $I_v(I_u)$ the horizontal (vertical) gradient image, we define F as the contour integral of the average gradient perpendicular to the contour:

$$\begin{aligned} F &= \frac{1}{r-l}\int_l^r I_u(t,v)\,dv + \frac{1}{r-l}\int_l^r I_u(b,v)\,dv \\ &+ \frac{1}{b-t}\int_t^b I_v(u,l)\,du + \frac{1}{b-t}\int_t^b I_v(u,r)\,du \end{aligned}$$

We can relate this objective function to a Bayesian likelihood function by using the Gibbs/Boltzmann distribution [8, 9]. Thus, if we define $\alpha E_d = -F$ as the energy term we are trying to minimize, and $\mathbf{x} = [t\,b\,l\,r]^T$ as the 4-dimensional vector encoding the position of the bounding box, then the likelihood of bounding box $\mathbf{x}$ given the image $\mathbf{z}$ can be expressed as (where Z_d is a normalization factor):

$$P(\mathbf{z}|\mathbf{x}) = \frac{1}{Z_d}\exp[-\alpha E_d(\mathbf{x},\mathbf{z})] \qquad (1)$$

The advantage of working with this probabilistic interpretation is that we can conveniently introduce the notion of a Bayesian prior, and combine it by means of Bayes law with the likelihood term above. The result is a maximum a posteriori (MAP) estimate for the position of the bounding box. For example, when working in image space, the prior could be a Gaussian density centered around the previous position of the bounding box [5, 9].

Figure 2: Scaling a hill at $t = 765$.

Initialization of tracking is done automatically using the Candidate Selection and Search (CANSS) algorithm, which we discuss in detail elsewhere [5, 10]. It uses a Hough Transform on the image gradient to advance candidate image rows and columns that might contain edges of a car, after which a combinatorial search takes place for the bounding box most probably generated by a car. This search minimizes the same energy measure E_d that we use for tracking, but combines it with a prior probability distribution over likely bounding boxes.

4 Integration: the Kalman Filter

We use a Kalman-Bucy filter to accomplish three simultaneous goals: (a) extract useful information from the tracking process, (b) improve tracking performance by having better expectations of how the tracked car will behave, and (c) conveniently integrate the additional measurements provided by RALPH. A Kalman filter implements the iterative application of Bayes law under Gaussian white noise assumptions for both dynamic and measurement noise, and the use of linear dynamics and measurements [6]. It also propagates the conditional densities involved forward in time between measurements. We use an extended (continuous) Kalman-Bucy filter as both our system dynamics and measurements are non-linear.

4.1 System Dynamics

The filter has 14 state variables, chosen because they represent a quantity of interest, are needed to model the dynamics, or both. They can be partitioned in a natural way, which is how we will discuss them.

For the ego-vehicle B we model the forward velocity V_b, and furthermore the yaw ψ_b and lateral offset d_b, both with respect to the lane center. Thus, $\mathbf{x_b} = [V_b\ d_b\ \psi_b]^T$, where the state variables are related via the following vector differential equation:

$$\dot{\mathbf{x_b}} = \begin{bmatrix} \dot{V_b} \\ \dot{d_b} \\ \dot{\psi_b} \end{bmatrix} = \begin{bmatrix} 0 \\ -V_b sin(\psi_b) \\ -\psi_b/T_{\psi_b} \end{bmatrix} + \begin{bmatrix} w_{V_b} \\ 0 \\ w_{\psi_b} \end{bmatrix}$$

Here w_{V_b} and w_{ψ_b} are Gaussian white noise terms that constrain how much velocity and yaw can change over time. V_b is modeled as a random walk, and we use the dynamic noise term to constrain the acceleration within reasonable bounds. The yaw ψ_b is modeled as a zero-mean time-correlated noise, as we know that the mean yaw with respect to the road must be zero.

The lane is modeled as having constant curvature κ and width W_r. The time derivative of the curvature is modeled as a random walk, i.e., $\mathbf{x_r} = [\kappa\ \dot{\kappa}\ W_r]^T$ and

$$\dot{\mathbf{x_r}} = \begin{bmatrix} \dot{\kappa} \\ \dot{\kappa} \\ \dot{W_r} \end{bmatrix} = \begin{bmatrix} \dot{\kappa} \\ 0 \\ 0 \end{bmatrix} + \begin{bmatrix} 0 \\ w_{\dot{\kappa}} \\ w_{W_r} \end{bmatrix}$$

The position of the tracked vehicle C is given by its distance Y_c along the road, with lateral offset from the lane center d_c. In contrast to most work in car tracking, we do not make a flat earth assumption, but estimate the vertical coordinate Z_c of the tracked vehicle as well. We do require that the vehicle is in the same lane as the tracker, which we argue in detail elsewhere [5]. Lastly, we also model the width, length, and height of the vehicle as unknown constants, leading to the 7-variable state $\mathbf{x_c} = [Y_c\ V_c\ d_c\ Z_c\ W_c\ L_c\ H_c]^T$. V_c and Z_c are modeled as random walks, d_c as a time-correlated noise, and the distance Y_c changes in function of both V_b and V_c:

$$\dot{\mathbf{x_c}} = \begin{bmatrix} \dot{Y_c} \\ \dot{V_c} \\ \dot{d_c} \\ \dot{Z_c} \\ \dot{W_c} \\ \dot{L_c} \\ \dot{H_c} \end{bmatrix} = \begin{bmatrix} V_c - V_b \\ 0 \\ -d_c/T_{d_c} \\ 0 \\ 0 \\ 0 \\ 0 \end{bmatrix} + \begin{bmatrix} 0 \\ w_{V_c} \\ w_{d_c} \\ w_{Z_c} \\ 0 \\ 0 \\ 0 \end{bmatrix}$$

By including them as state variables, W_c, L_c, and H_c are iteratively estimated, based on the initial measurement output by the CANNS algorithm. The length L_c can only be observed when the tracked vehicle goes through a curve, but after it has been properly estimated it yields important information about the yaw of the tracked vehicle.

Finally, since the image measurement equation is very sensitive to changes in camera pitch, we estimate that on-line as well, in effect achieving image stabilization. We do that by means of the state variable α_n, a zero-mean time correlated noise which is added to the camera calibration baseline pitch. Thus:

$$\dot{\alpha_n} = -\alpha_n/T_{an} + w_{\alpha_n}$$

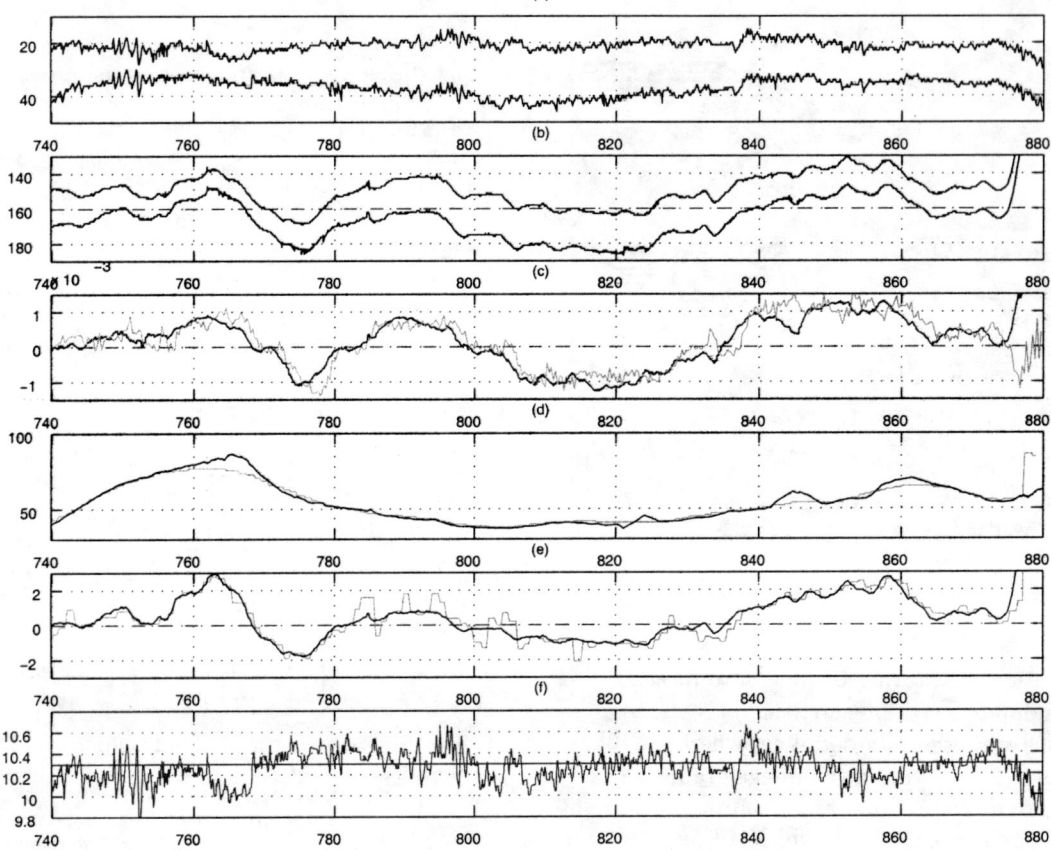

Figure 3: Combined system output. (a) pixel coordinates of top, bottom, and (b) left, right edges of tracked car; (c) curvature κ compared with the gyro (in gray); (d) distance Y_c to and (e) lateral position X_c (not d_c!) compared with radar (in gray); (f) pitch bounce estimate α_n

4.2 RALPH and Image Measurements

We integrate 3 quantities estimated by RALPH: road curvature κ, ego-vehicle lateral offset d_b, and lane width W_r. The measurement equation is trivial, as these are state variables of the filter. They are passed to the filter at a rate of 4 Hz, in addition to a measurement of the vehicle speed V_b obtained from GPS.

The image measurement equation simply consists of projecting the imaginary 3D bounding box around the tracked vehicle, as defined by the three state variables $[W_c\ L_c\ H_c]^T$, into the image. The 3D position and attitude of the car can be obtained directly from the filter, by combining the curvature estimate κ with the arclength Y_c, and the lateral offsets d_b and d_c. Taking into account the camera calibration and the current estimate of camera pitch, both the projection *and* its Jacobian are evaluated numerically using an OpenGL 3D graphics accelerator card. After the projection, it is a simple matter to determine the predicted 2D bounding box $[t\ d\ l\ r]^T$, which is used as an initial estimate for the image processing discussed above.

5 Results

Tracking results suggest excellent performance for the resulting integrated system, as will be illustrated here in both a qualitative and quantitative manner. In Figure 3 we present the output of the tracker on a long sequence of video recorded on the Navlab 8, an Oldsmobile Silhouette minivan, while it was being driven manually on I-79N near Pittsburgh. The images were taken early in the morning with the sun still low on the horizon, throwing long shadows of trees across the road. These shadows created strong distracting edges on the road surface. In addition, the terrain at that location was quite hilly, often violating the flat earth assumption for extended periods of time, in particular on hill crests.

The sequence shown in Figure 3 illustrates the strengths of our approach to car tracking, that have previously been discussed in [5]. Most notably, despite the challenging nature of the sequence, the system kept track at all times for the duration of the sequence. Since the processing was done at frame rate, and the sequence lasted 140 seconds, this represents

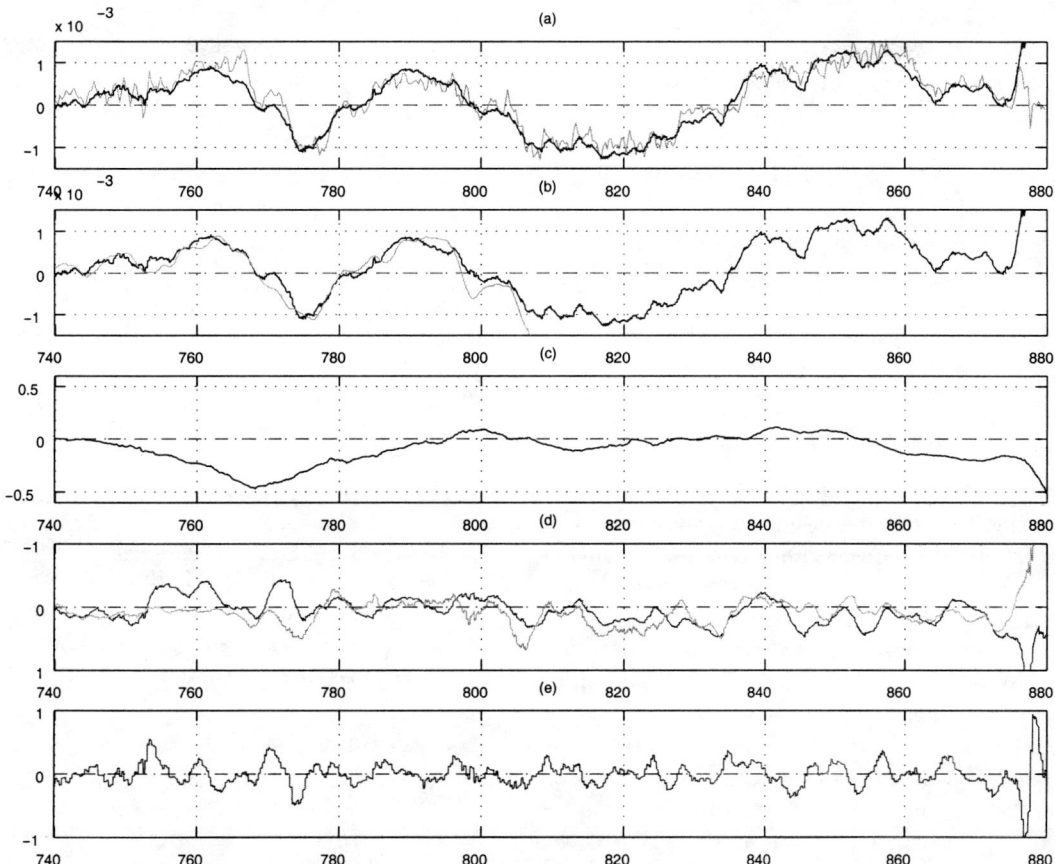

Figure 4: Benefits of integration. (a) curvature estimate κ compared with RALPH (in gray) (b) and with the tracker alone (in gray); (c) estimated target z coordinate Z_c; (d) lateral offset estimates d_b and d_c (in gray); (e) yaw estimate ψ_b

a continuous track over 4200 individual frames, while the distance covered was more than 4 km. In addition, the state estimates obtained by the tracker correspond closely to ground truth, as can be seen from Figure 3 (c) to (e), where the estimates of road curvature and relative target position are compared with recorded measurements from a yaw rate gyro and a Delco millimeter wave radar. The on-line pitch estimation, shown in panel (f), provides excellent image stabilization such that we do not lose track even when the Navlab drives over large bumps, e.g., the particularly challenging segment around $t = 750$. The extreme values at the end of the sequence occur because the Navlab exits the highway at that point.

The combination of RALPH and the tracker performs markedly better than RALPH in isolation, particularly where RALPH traditionally has problems, e.g., when the flat earth assumption is violated. This occurs for example at t=765, shown in Figure 2. Our tracking filter does not make a flat earth assumption: in Figure 4 you can see that the Z_c coordinate of the target is indeed negative at $t = 765$, while at the same time the grossly overestimated curvature output by RALPH is corrected by the tracker. Note that, as both the tracker and RALPH look ahead when estimating curvature, they will lead the gyro measurement when entering or leaving a curve. Finally, note that the filtered curvature estimate κ is a lot smoother than the raw RALPH measurement (panel a).

The integrated system also outperforms the tracker taken in isolation. To show this, we ran the tracker with the same settings but without the RALPH measurements, and compare its κ estimate with the integrated system in Figure 4 (b). Note that the comparison will not be not entirely fair, since in fact the tracker can be optimized to cope with the absence of RALPH's measurements. Nevertheless, the isolated tracker loses track about halfway through the sequence, which shows that RALPH helps the system keep track. More significant, however, is the large effect of ego-vehicle yaw ψ_b on the position of the tracked car in the image. Because RALPH has an accurate measurement of the lateral offset d_b, the integrated filter is now able to deduce ψ_b. For example, at $t = 770$ the estimated curvature is now zero, whereas the tracker in isolation mistakenly believed there was a curve to the right, because of the large yaw to the left (see panel e). Also important is that, for the isolated tracker to work at all, we needed to force both the ego-vehicle and target offsets d_b and d_c to zero, as they are now unobservable. This points out another advantage of bringing RALPH into the picture: cars significantly swerve in their lane, and this can corrupt the curvature estimate when assumed otherwise. But accurate estimates of in-lane displacements are also important to recognize lane changes and sudden maneuvers.

6 Conclusions and Future Work

We have shown how we integrated our tracking system with the RALPH road following module. The tracking system brings precise and crisp measurements of the car in the image, and performs image stabilization. However, because it does not know about the yaw or lateral offset of the ego-vehicle, its curvature estimate can be misguided. RALPH provides the missing information, as well as a good estimate of curvature, so that the combined curvature estimate is superior to both taken in isolation. RALPH also benefits from the capability of the tracker to cope with violations of the flat earth assumption, and is corrected appropriately when this is the case. The integration of these two systems substantially increased the overall system performance.

In the same way as we integrated RALPH, we could add in more measurements available from different sources. RALPH has already been integrated successfully with road map data in as yet unpublished work. We can very easily add this to our system. Other work in progress concentrates on tracking stationary points on and alongside the road, to obtain an accurate estimate of ego-motion. Lastly, although RALPH has been treated here as a black box, much of its internal processing could be integrated with the Kalman filter in a more direct fashion, such that we gain access to the probabilistic reasoning implicit in the filter.

There are also improvements to the tracker itself that we are considering. In particular, we would like to use machine learning to optimize which features we should be tracking, rather than hand tune the parameters or use ad-hoc techniques. For the bounding box feature this can be done readily by means of the likelihood energy term E_d, whose relationship to the image could be learned from training data. Multiple car tracking with occlusion reasoning, automatic error recovery and robust initialization are other opportunities for future work.

Better car tracking, and a better idea of where the road is, allows us to do more. Accurate estimates of ego-motion provide us with better information for controlling the vehicle. We should now be able to better recognize discrete events for what they are, e.g. lane changes or obstacle avoidance maneuvers of the car ahead. Using the accurate measurements of relative position and speed of other vehicles given by the system, we can predict ahead and construct a tactical plan to maneuver through traffic. In conclusion, the integration of two distinctly different vision systems into one hybrid system gives us a substantial foundation on which to base other capabilities needed to realize fully autonomous vehicles.

Acknowledgements

This work was supported in part by USDOT under Cooperative Agreement Number DTFH61-94-X-00001 as part of the National Automated Highway System Consortium, and by the National Highway Traffic Safety Administration (NHTSA) under contract DTNH22-93-C-07023.

References

[1] C. Thorpe, "Mixed traffic and automated highways," in *Proceedings of IEEE/RSJ International Conference on Intelligent Robots and Systems (IROS '97)*, vol. 2, (Grenoble, France), September 1997.

[2] E. Dickmanns, "Vehicles capable of dynamic vision," in *IJCAI-97, (Nagoya, Japan)*, 1997.

[3] D. Pomerleau, *Neural Network Perception for Mobile Robot Guidance*, Kluwer Academic Publishing, Boston, MA, 1994.

[4] D. Pomerleau and T. Jochem, "Rapidly adapting machine vision for automated vehicle steering," *IEEE Expert* 11, April 1996.

[5] F. Dellaert and C. Thorpe, "Robust car tracking using Kalman filtering and Bayesian templates," in *Proceedings of SPIE: Intelligent Transportation Systems*, vol. 3207, October 1997.

[6] P. Maybeck, *Stochastic Models, Estimation and Control*, vol. 1, Academic Press, New York, 1979.

[7] M. Schmid, "An approach to model-based 3-d recognition of vehicles in real time by machine vision," in *Proceedings of IEEE/RSJ International Conference on Intelligent Robots and Systems (IROS '94)*, vol. 3, (Munich, Germany), September 1994.

[8] D. Terzopoulos and R. Szeliski, "Tracking with Kalman snakes," in *Active Vision*, A.Blake and A.Yuille, eds., pp. 3–20, MIT Press, Cambridge, MA, 1992.

[9] A. L. Yuille, P. W. Hallinan, and D. S. Cohen, "Feature extraction from faces using deformable templates," *International Journal of Computer Vision* 8(2), pp. 99–111, 1992.

[10] F. Dellaert, "CANSS: A candidate selection and search algorithm to initialize car tracking," Tech. Rep. CMU-RI-TR-97-34, Robotics Institute, Carnegie Mellon University, 1997.

Proceedings of the 1998 IEEE
International Conference on Robotics & Automation
Leuven, Belgium • May 1998

Experiments in Autonomous Driving With Concurrent Goals And Multiple Vehicles

Barry Brumitt, Martial Hebert, and the CMU UGV Group

The Robotics Institute
Carnegie Mellon University
5000 Forbes Avenue
Pittsburgh PA 15213

Abstract[1]

In this paper, we report on experiments with a system for autonomously driving two vehicles based on complex mission specifications. We show that the system is able to plan local paths in obstacle fields based on sensor data, to plan and update global paths to goals based on frequent obstacle map updates, and to modify mission execution, e.g., the ordering of the goals, based on the updated paths to the goals.

Two recently developed sensors are used for obstacle detection: a high-speed laser range finder, and a video-rate stereo system. An updated version of a dynamic path planner, D*, is used for on-line computation of routes. A new mission planning and execution monitoring tool, GRAMMPS, is used for managing the allocation and ordering of goals between vehicles.

We report on experiments conducted in an outdoor test site with two HMMWVs. Implementation details and performance analysis, including failure modes, are described based on a series of twelve experiments, each over 1/2 km distance with up to nine goals.

This system is the first multi-vehicle and multi-goal system to be demonstrated in real, natural environments with this degree of generality.

The work reported here includes a number of results not previously published, including the use of a real-time stereo machine, a high-performance laser range finder, and the GRAMMPS planning system.

1 Introduction

Unmanned ground vehicles operate autonomously in natural, unstructured terrain. To accomplish this, they detect obstacles, plan paths, and build maps. In practical applications, multiple vehicles may operate simultaneously to carry out a common mission.

In an earlier paper [20], we described the first demonstration of an autonomous system with on-board, dynamic path planning combined with obstacle avoidance. At that time, the system was able to handle simple missions involving a single goal and a single vehicle. In realistic missions, however, several vehicles must coordinate their routes and complex mission plans may include multiple

---

1. The CMU UGV Group includes: Barry Brumitt, Peng Chang, Jim Frazier, John Hancock, Martial Hebert, Daniel Huber, Dirk Langer, Anthony Stentz, Chuck Thorpe, Todd Williamson, and Alex Yahja. This work was supported under DARPA/TACOM contract DAAE07-96-C-X075, entitled "Technology Enhancements for UGVs", monitored by U.S. Army TACOM.

goal locations. Once multiple vehicles and multiple goals are used, the system must be able to make complex decisions such as allocating goals between the vehicles and computing paths to many goals simultaneously in real time. Furthermore, the system must be able to accommodate a variety of different types of missions. For example, a mission might require all the vehicles to initially drive to a first goal location. e.g., a staging area, and then drive to a set of goals, e.g., observation points, in any order using any combination of vehicles while a different mission may require the vehicles to visit a set of goals in a specific sequence, e.g., to pick up supplies in a particular order.

In this paper, we report on experiments with a system that answers those needs by extending the capabilities described in [20]. The experiments were conducted with the two vehicles shown in Figure 1, operating in the environment shown in Figure 5. In the area of planning, the system described here demonstrates successful approaches to on-line route planning with multiple goals, on-line mission planning, e.g., allocation and ordering of goals, and to flexible mission specification.

Figure 1: The two HMMWVs used in the experiments.

Beyond the planning requirements, the system must be able to integrate information from the different sensors in order to provide a consistent map representation to the planners. In the experiments described in this paper, two different sensors were used: a new, high speed, laser range finder and a video-rate stereo machine. The experiments were also an opportunity to demonstrate those two state-of-the-art sensors and to demonstrate the ability of the system to accommodate different resolutions, fields of view, and ranges.

Although multi-vehicle planning systems have been reported by others [12][13][14], the experiments reported here are arguably the first demonstrations of vehicles with fully integrated systems operating in unstructured natural environments (see [5] for a survey of research in cooperative robotics.)

2 Architecture Overview

The system is divided into a mobility element, which is responsible for driving the vehicle around obstacles and for planning paths to goal points, and a mission element, which is responsible for specifying the goal points, monitoring the execution of the mission, and generating any necessary updates to the mission. Those elements are implemented as sets of decentralized, asynchronous modules communicating through dedicated communication links, using the IPT inter-process communication package. The complete system architecture is shown in Figure 2.

Mobility Element

The system architecture is based on the behavior-arbitration approach introduced in [17] and developed in the context of cross-country navigation systems in [18][11]. On each vehicle, a local obstacle avoidance module takes input from range sensors, constructs a local obstacle map, and outputs recommendations for steering and speed commands.

The local obstacle maps generated by the obstacle detection module are integrated into a global map maintained by a separate module called Intercom. In addition to maintaining the map for the vehicle on which it resides, Intercom is also responsible for exchanging map updates with the other vehicle. As a result, the two vehicles share the same map of the environment at all times.

The map representation is a large grid of cells labelled as obstacle or drivable. The map is used by a route planner, D*, to steer the vehicle toward goal locations. D* is a dynamic planner in that it is capable of continually updating the route to the goal based on update to the map from the obstacle detection module. Specifically, D* maintains an internal cost map in which the cost of driving from each cell to the goal is encoded; by modifying this cost map efficiently every time the map is updated, D* is able to compute the new optimal route to the goal. It has been shown in an earlier paper that D* is guaranteed to compute the optimal path to the given goal at all times [19].

Periodically, D* generates a set of steering recommendations based on the configuration of the cost map around the vehicle. More precisely, steering choices that guides the vehicle in directions with lower cost to the goal receive higher scores. A first algorithm for evaluating steering choices from cost maps was discussed in [20]. The algorithm used here is substantially different and leads to much smoother paths.

Recommendations from the local avoidance module and D* are lists of votes, one vote for each steering choice. Votes are normally continuous between 0 and 1, with 1 indicating a fully drivable arc. Either module can veto a steering choice by setting the vote to a veto value (-1). In particular, steering choices that would lead the vehicle outside of the field of view of the sensor are vetoed.

The votes are combined by an arbiter in order to generate a single driving command. This approach to command-based arbitration is based on the approach introduced in [18]. A typical snapshot of the steering arbiter operation is shown in Figure 3.

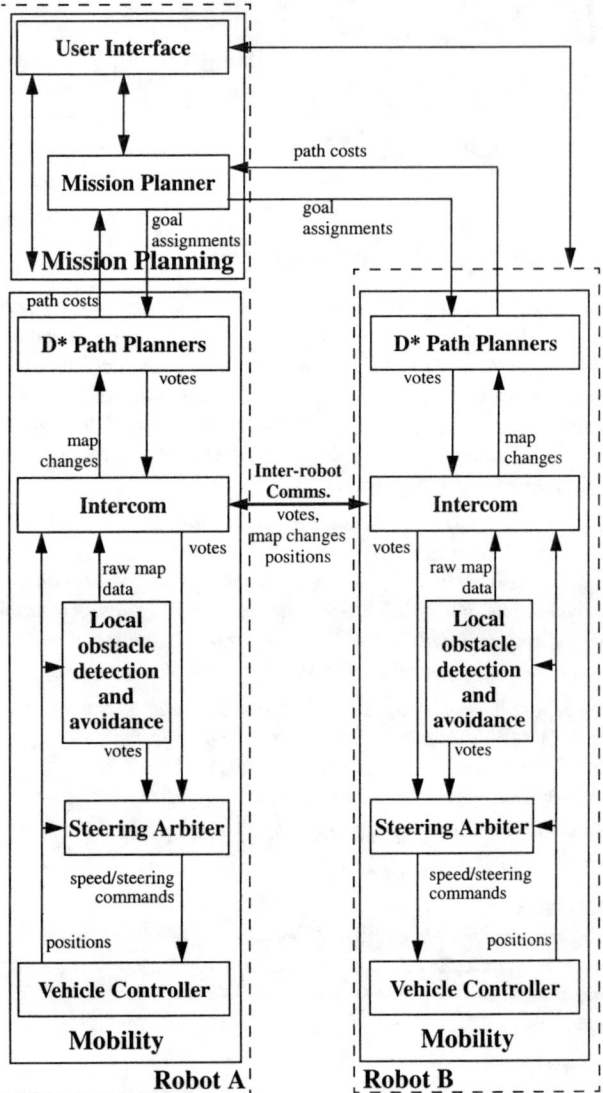

Figure 2: System architecture. The identical mobility systems reside on both vehicles. One of the robots supports the mission planning system. Mission planning could be moved to an off-vehicle base station if needed.

A long-standing issue with this approach is that a weighting scheme for combining votes between the two modules must be devised. Previous implementation of this approach used fixed weights, e.g., a large weight for local obstacle avoidance and a smaller weight for D*. The arbiter used in this paper incorporates a new approach for weight adaptation based on the local obstacle map. The weight adaptation algorithm is described in detail in Section 4.

Mission Planning Element

The main module in the mission planning element is the GRAMMPS (Generalized Robotic Autonomous Mobile Mission Planning System) mission planning and execution module which is responsible for assigning goals to vehicles based on a mission description and the current cost maps. GRAMMPS is also a dynamic planner in that it is capable of efficiently changing the order or allocation of goals between the vehicle as the global obstacle map is updated.

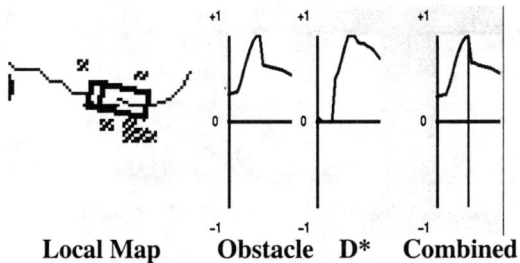

Figure 3: Arbitration results for the configuration shown at left. The votes from the obstacle avoidance module and D* are combined into a single distribution. The weight for D* is 8 times smaller than the weight for the obstacle avoidance module because the vehicle is in a cluttered area.

The basic approach to GRAMMPS was introduced in [3]. The basic concept is shown in Figure 4: A separate D* is associated with each goal location in each vehicle. All the D*'s use the same obstacle map in order to update their cost maps. Periodically, GRAMMPS uses the costs of all the robots to all the goals provided by D* in order to compute the optimal assignment of goals to robots. This description of GRAMMPS using separate D* modules for each robot and each goal is convenient but, in practice, a single route planning process is used on each vehicle to compute the costs to a subset of the goals. The details of the search algorithm currently used in GRAMMPS can be found in a companion paper [4]..

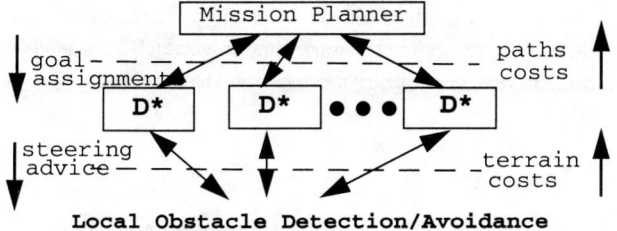

Figure 4: Conceptual view of GRAMMPS.

In this approach to mission planning and execution, all the modules run asynchronously. In particular, D* continuously updates its cost maps and generates command recommendations independently from GRAMMPS or the obstacle avoidance modules. Typically, updates from the mission planner, route planner, and local avoidance planner are issued at increasingly higher rates, e.g., 0.5 Hz, 1 Hz, and 2 Hz, respectively.

The second part of the mission planning element is a user interface in which mission descriptions can be built and sent to GRAMMPS. A general interpreted language is used for describing the missions. Robots may be instructed to visit specific goals in a given order, or to visit a set of goals in any arbitrary order. A goal may be assigned to a specific vehicle or may be visited by any of the two vehicles. The complete language specification is described in [4].

3 Experiments

The two-vehicle system described above was exercised in the field over a period of one month. Fragments of missions were first executed in order to evaluate the components. This section provides a detailed look at the execution of complex missions in real environments.

Twelve complete missions, i.e., missions in which the vehicles visit all the goals specified in the mission description, were recorded and analyzed in detail. Each mission involved driving each vehicle up to 600 meters and visiting up to nine goals. The average speed was 1 m/s in all the missions.

The mission descriptions used in the tests included a mix of three types of directives: drive a given vehicle to a specific goal or a set of goals in a specific order; drive each vehicle to a specific set of goals using any goal ordering; and drive to a set of goals using any goal ordering and any allocation of goals to vehicles.

For reasons of space, we can discuss only one mission in detail. Performance summaries over the whole set of experiments are provided in Section 5. This mission illustrates the basic capabilities of the mobility system such as goal re-ordering, route re-planning, and local obstacle avoidance. The example mission consists of eight goals. The mission description provided by the user is, in this case, a mix of mandatory orders, i.e., specific goals that must be visited by each vehicle, and optional orders, i.e., goals which can be visited in any order and by any vehicle as decided by the runtime system. In order to facilitate the description of the mission, each goal is designated by a name. In this example, the robots HMMWV2 and HMMWV1 are instructed to visit goals INTER and EDGE3, respectively; the remaining goals may be visited in any order.

During this experiment, HMMWV1 and HMMWV2 travelled 670m and 683m, respectively. A total of 592 obstacle cells were seen in the map, with an additional 1500 potential field cells assigned around the obstacle cells. The run took a total of 16 minutes.

The approximate locations of six of the goals are shown in Figure 5 overlaid on a mosaiced image of the test area. The final goal location EXEUNT is slightly outside of this image to the right. HMMWV2 starts on a narrow path as shown in Figure 5; HMMWV1 start position is close to the final goal EXEUNT.

In order to show the major stages of this mission, we include below displays obtained by replaying the data recorded during the run. In all those displays, HMMWV1's path and vehicle icon are drawn using shaded lines, while HMMWV2's are drawn using black lines. Two types of paths are drawn. The path behind the vehicle is the path actually driven by the vehicle under combined control of D* and local obstacle avoidance; the path ahead of the vehicle is the best path planned by D* given the current obstacle map. The obstacles are shown as grey squares and the goal locations are indicated by their names.

Figure 5: Actual position of the goals used in the mission described below; total distance travelled from starting point to final goal is approximately 500 meters

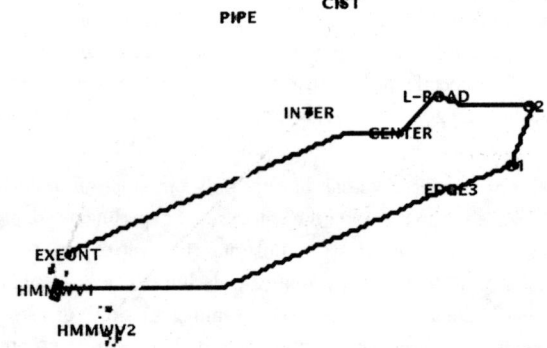

Initial configuration: Obstacles are known only in the immediate neighborhood of the vehicles at their starting positions. HMMWV2 plans its route from the starting location to the first goal, INTER, which is imposed by the mission plan, through two intermediate goals, ending at the final goal, EXEUNT. Similarly, HMMWV1's plan goes first to the mandatory goal EDGE3 and to four intermediate goals before going back to EXEUNT. GRAM-MPS generates the allocation of goals between the two vehicles.

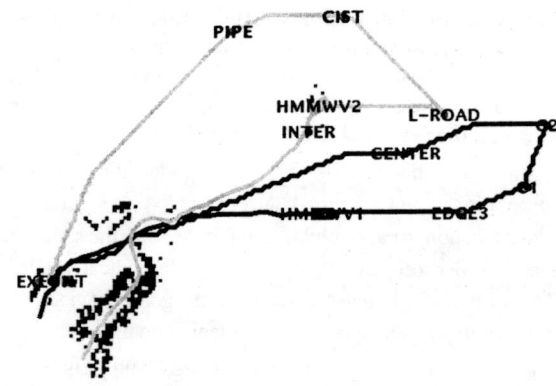

Change in goal allocation: Shortly after HMMWV2 reaches goal INTER, GRAMMPS decides that goal L-ROAD should be switched to HMMWV2's plan. This event occurs because newly detected obstacles just past INTER forced HMMWV2 to move closer to L-ROAD than had been initially planned. As the goal switching event occurs, the routes are immediately recomputed, leading to seamless transition between goal distributions.

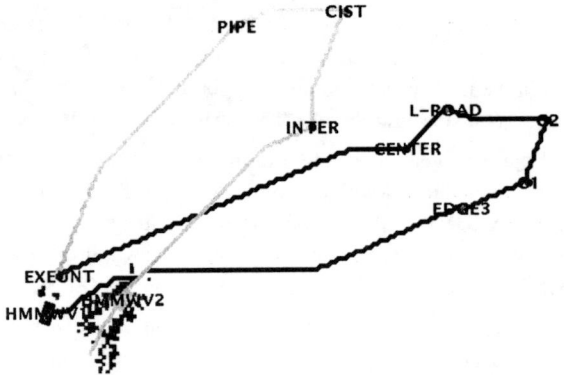

Mission start: HMMWV2 starts driving first, modifying its route based on the obstacle map accumulated since the start of the mission. At the same time, although it is not moving, HMMWV1 is updating its planned route based on the map built by HMMWV2. Specifically, the path to EDGE3 has now changed based on the new obstacles found by the laser system on HMMWV2.

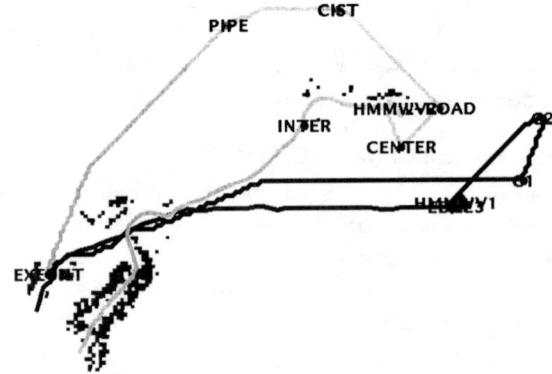

Change in goal ordering: As HMMWV2 is deflected from its path on its way to L-ROAD, the planner re-allocates goal CENTER to HMMWV2. As a result of this event, the ordering of goals for HMMWV1 is also changed; the plan now calls for C2 to be visited before C1.

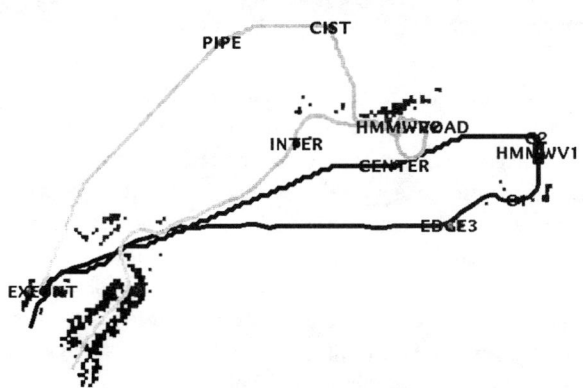

Correction: The planner has reverted its previous decision and restored the original goal allocation. At the same time, the ordering of goals for HMMWV1 has also been restored, as expected.

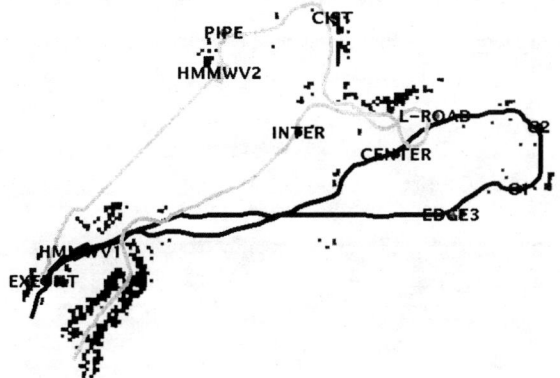

Driving back: Although it occupies a very cluttered area, HMMWV2 executes a path that leads it out of the area around L-ROAD (the edge of a forest.) This path is executed through the steering arbiter by combining the global optimal path to CIST and the local steering commands for obstacle avoidance. Because the speed arbiter reduced its speed in this area, HMMWV2 has just reached its last goal, PIPE, while HMMWV1 as is completing the mission. The route uses all the information accumulated since the beginning of the mission. For example, the path from PIPE to EXEUNT computed by D* uses the portion of the obstacle map constructed at the very beginning of the mission.

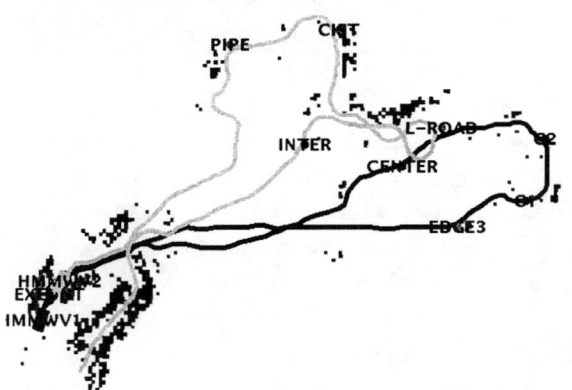

Mission completed: Both HMMWVs complete the mission. The paths shown here are optimal based on the information available at the time they were generated. Were these runs to be performed again, given the complete obstacle map shown above, the initial paths and goal ordering would be globally optimal and, therefore, would be very different from the ones shown above. For example, the map constructed between L-ROAD, INTER and CIST would prevent future mission from wasting time trying to cross directly from L-ROAD to CIST as HMMWV2's initial plan recommended. Instead, using this map, a planner would now generate the correct behavior, that is, the vehical would follow the tree line between L-ROAD and CIST.

4 Detailed Description

We describe below the details of the implementation of the system for the experiments described above. The purpose of this discussion is three-fold. First, we provide information on computation and communication requirements. Second, we describe as completely as possible the parameters used in the system, e.g., map resolution. Third, we emphasize algorithmic improvements over previous similar systems. For example, the D* vote generation and the steering arbitration algorithms described below are substantial improvements over previous implementations.

4.1 Communication

Communication bandwidth between the vehicles is always a serious concern with multi-agent systems because of limited bandwidths of radio systems. In these experiments, a FreeWave radio modem system was used. This system has a peak bandwidth of 110 Kbaud.

The communication rate between the vehicles was recorded and analyzed. The vehicles primarily exchange vote distributions and obstacles map updates. Vote distributions are updated at a maximum rate of 2Hz. The data rate for map updates can be measured in obstacle cells transmitted per second. The peak rate measured during the experiments was 80 cells per second. This rate was observed in the most cluttered areas of the test site. In those situations, the transmitted lists of obstacle cells are segmented into 50-cell packets.

The aggregate volume of transmission was always below 4800 baud. No compression algorithm was used. Straightforward approaches to data compression can be designed to reduce the data rate by a factor of 4. In particular, the encoding of cell coordinates and votes as floating-point numbers can be easily replaced by 1-byte encoding. This analysis shows that this system design is suitable for use with low-bandwidth communication hardware.

4.2 Sensing

A video rate stereo machine with a three-camera system is used on HMMWV1 for range sensing. The machine generates 100x256 range images at up to 15Hz. A new imaging laser range finder, used on HMMWV2, is capable of acquiring range data up to 50m in a $360°$ by $30°$ swath [7][8]. Because of mechanical limitations, the rotational speed of the scanner is set to a maximum of 2400Hz.

As a result, although a new laser design used in this sensor is capable of measuring points at 500kHz, the acquisition rate was limited to two images per second for 60x1000 images. More recent, faster, versions of the scanner are described in [7].

The two sensors have very different characteristics. In particular, the stereo system is faster but has a narrow field of view (approximately 50°; this was extended somewhat by using a pan control algorithm) while the laser range finder can sense all the way around the vehicle. In practice, the two sensors are complementary; stereo detects obstacles further ahead, while the laser range finder provides data in the periphery of the vehicle.

The use of very different sensors illustrates also the flexibility of the architecture. Both sensors used the obstacle detection and avoidance system described in [9][11], each with a different sensor model. Thanks to the use of the behavior-based approach, the change in sensor is limited to the obstacle detection module, and the rest of the system remains identical on both vehicles.

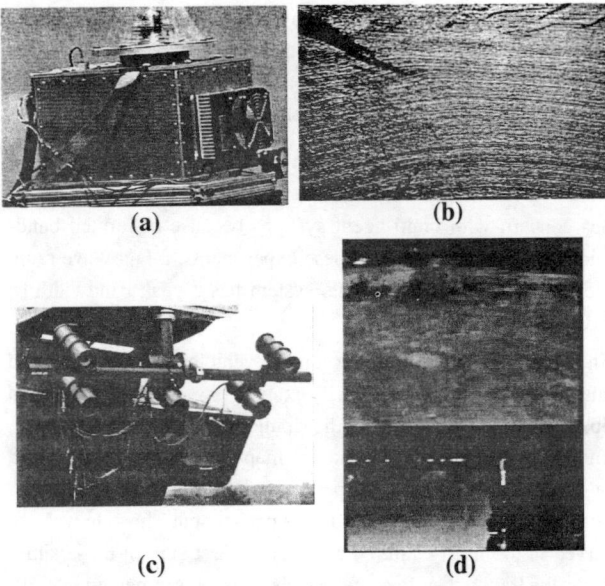

Figure 6: (a) Laser range finder; (b) 3-D display of data points from the laser range finder; (c) stereo cameras; (d) intensity image and corresponding disparity image from the stereo machine.

4.3 Planning and Map Management

As described above, the planning and map management responsibilities are distributed among three components, D*, Intercom, and GRAMMPS.

The D* route planner operated with a 1 meter resolution obstacle map in which obstacle cells are grown by a 2 meter potential field. Typical maps handled by D* are 500 meters on the side. D* is fast enough to update its internal cost map and to generate steering votes at 2Hz.

This vector of steering votes is computed by averaging the cost-to-goal of cells along each steering arc, weighted by the length of the segment of the arc contained in each cell. This approach is different from the one used in [20] and leads to substantially smoother vote distributions.

In addition to collecting map data and transmitting map data from one vehicle to the other, Intercom is also responsible for checking for interference between vehicles. Specifically, Intercom maintains a 10 meter buffer zone around each vehicle. One of the vehicles is stopped if the buffer zone is violated. The map management algorithm is temporarily suspended on the stopped vehicle so that the moving vehicle is not included in the map. More sophisticated algorithms can be found in the literature on collision avoidance in fleets of multiple robots [12][2]; we did not attempt to implement an optimal strategy for interference avoidance.

The GRAMMPS mission and execution planner updates the mission at 0.5 Hz or less. Updates consist of the ordering of goals to be visited by each vehicle. In the experiments described above, GRAMMPS resides on HMMWV1 and communicates the plan updates to the local D* module as well as to the system resident on HMMWV2. However, it is important to note that the mission planner, and its user interface, could reside at a base station outside of the vehicles.

4.4 Speed Control

Vehicle speed is controlled from three sources: mission directives may alter vehicle speed, for example by stopping at goals; maximum speed may be limited depending on the current steering radius in order to avoid tipover; and the speed should be adapted based on the complexity of the terrain, i.e., the more cluttered the terrain, the lower the speed.

In the current implementation, speed-based mission directives are only binary directives, that is, the vehicle is either stopped at a goal or waiting for the other vehicle, or it is driven at the speed commanded by the mobility system. Because of the low vehicle speeds used in the experiments, adaptation of speed as a function of turning radius was not necessary. The source of speed control was the obstacle avoidance system. The basic idea is that if the environment is cluttered, or, equivalently, if the steering votes have low values, then the speed should be reduced.

In the current speed control algorithm, the speed is the maximum desired speed multiplied by a speed factor s_O which is computed as: $s_O = v_{max} - \alpha(1 - r_{max})$, where v_{max} is the maximum steering vote in the current vote distribution, and r_{max} is the proportion of votes that are close to v_{max}. This algorithm controls speed with the desired behavior, as illustrated by the three main cases shown in Figure 7: If both v_{max} and r_{max} are large, the vehicle can maneuver over a wide range of arcs and, as a result, s_O is close to 1 and the commanded speed is close to the maximum speed. If v_{max} is high but r_{max} is small, the vehicle has a clear path but it has less room to maneuver and s_O is decreased. If both v_{max} and r_{max} are small, the vehicle is driving in a cluttered environment and s_O is decreased even further. Finally, this approach to speed forces the vehicle to stop if the environment is too cluttered, i.e., v_{max} and r_{max} are too small, even if no steering arcs are actually vetoed.

Experimentally, this approach to speed control allowed for fast driving across the obstacle-free areas of the map, e.g., in the segment between PIPE and EXEUNT in the earlier example, and to drive at reduced speed near obstacle regions such as the tree line near L-ROAD.

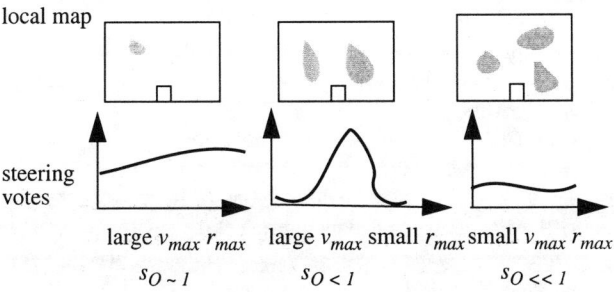

Figure 7: Speed adaptation in three cases: obstacle-free, narrow corridor, and cluttered environment; for each case, the local map is shown at the top and the corresponding steering vote distribution is shown at the bottom.

4.5 Steering Arbitration

The steering arbiter used 21 arcs regularly spaced between the minimum and maximum turning radii of -7m and 7m. New commands were issued at 2Hz based on the most recent sets of votes from the obstacle avoidance and D* modules.

This implementation of the steering arbiter is similar to the one originally described in [11]. However, as indicated above, a critical difference is the ability to adjust the relative weights of voting modules depending on the environment. Specifically, the votes are combined using a linear combination: $v_i = w_{D*} v_i^{D*} + w_O v_i^O$, where v_i^{D*} and v_i^O are the votes from D* and obstacle avoidance, respectively, and w_{D*} and w_O are the weights. If the vehicle is in an obstacle-free area, then D* should have control over the vehicle since the votes from the obstacle avoidance module do not carry any information except for the veto votes outside of the sensor's field of view. On the other hand, if the vehicle is driving in a cluttered area, then the contribution of D* should be decreased, or even eliminated, in order to give control of the vehicle to the obstacle avoidance module.

The idea of dynamic weight adaptation originates in the more general concept of adapting of the driving behavior based on the environment in a way similar to the speed adaptation algorithm described above. In fact, the weight adaptation algorithm used in those experiments is based on the speed factor, s_O generated by the obstacle avoidance module. More precisely, the weights are defined as $w_O = f(s_O)$, where f is a linearly decreasing function of s_O between a non-zero minimum value for w_O, to ensure some contribution of obstacle avoidance even if the speed is close to the maximum speed, and a maximum value less than 1, to ensure that D* introduces a small amount of bias toward the goal even in cluttered environments, as shown in Figure 8.

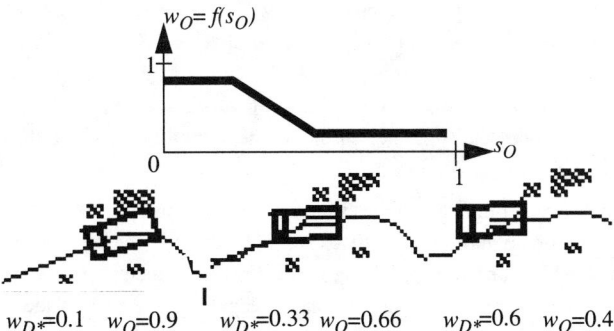

Figure 8: Adaptation of arbiter weight as function of the density of obstacles: (top) weight adaptation function; (bottom) example sequence in which w_{D*} increases as the vehicle drives away from the obstacles.

4.6 Position Estimation

Accurate position estimation is critical since, at all times, the vehicles need to know their positions with respect to each other and with respect to the map maintained by Intercom at all times. In exercising the system in real environments, it rapidly becomes clear that solutions that rely on single position sensors are not viable: GPS satellites may be occluded by terrain features; INS systems drift over time; and slippage makes encoder integration useless over any substantial amount of travel.

The solution chosen here is a tuned Kalman Filter that integrates data from a Novatel RT-20 differential GPS, the heading rate gyro from a Lear Astronautics MIAG INS system, and encoder readings. At low speeds, the filter integrates the readings from the heading rate gyro for heading estimation and GPS readings for position estimation; at higher speeds, the GPS readings are also used for estimating heading. Zero velocity updates are used whenever the vehicle is stopped, as measured by the encoders, in order to eliminate drift. This approach allows for seamless transition between different modes of estimation at different speeds.

5 Limitations and Failure Modes

As we indicated earlier, the two-vehicle navigation system was able to carry out complex missions in difficult, natural terrain. Although the system provides a solid base for the development of future autonomous systems, a number of limitations remain as can be seen in occasional failures of the system.

An important limitation has to do with the use of range sensors for obstacle detection to the exclusion of other types of sensors. The main problem here is that range sensors cannot discriminate between the changes in range due to vegetation and those due to actual obstacles. As a result, the vehicles are constantly avoiding small patches of vegetation which should not be reported as obstacles. In densely vegetated areas, the situation can degrade to the point that the system stops the vehicle and the mission fails.

This limitation is not an issue with the fundamental design of the system, but it is an issue of considerable importance for applying this work in any practical environment. The solution is to augment the sensor suite with a non-range sensor that can reliably discriminate vegetation. For example, we are currently working with a multispectral sensor which will permit real-time classification of vegetation. Details on the sensor and the algorithms can be found in [10].

A practical limitation of the system as presented here is that the vehicles do not have the ability to drive in reverse. More importantly, the arbiter and other parts of the system do not support reverse paths. As a result, a vehicle cannot escape if it is too far inside a cul-de-sac, even though it can recognize this occurance. Such a situation is unavoidable and leads to failure of the mission. Support for reverse driving must be added to the system in order to avoid this failure mode.

One drawback of an asynchronous, distributed system is the latency in communicating information between the components. If the maximum commanded vehicle speed is low, the latency and asynchronous nature of the system do not cause any significant problems. In fact, in the experiments described above, the system runs on a non-real time Unix OS. As vehicle speed increases, latency and unknown delays due to the OS become critical. Therefore, moving toward higher speeds will require porting the navigation system to a real-time operating system.

The last common failure mode is loss of position estimation, typically due to loss of GPS visibility. In fact, this was the most common failure mode in early experiments with the system. As we indicated earlier, the areas of the test site where GPS drop-outs were more frequent were known, thus allowing us to avoid this failure mode in subsequent missions. More generally, such systems will be truly widely applicable only when reliable positioning systems, both internal and external, are built.

6 Conclusion

We reported in this paper on experiments with a system for autonomously driving two vehicles based on complex mission specifications. We showed that the system is able to plan local paths in obstacle fields based on sensor data, to plan and update global paths to goals based on frequent obstacle map updates, and to modify mission execution, based on the updated paths to the goals. This system is the first multi-vehicle and multi-goal system demonstrated in real, natural environments with this degree of adaptation. Moreover, because it is built on the earlier design of distributed, behavior-based architectures, the system can be easily extended to accommodate new modules or additional vehicles.

In addition to the implementation and demonstration of the multi-vehicle/multi-goal capability, substantial enhancements were introduced in the various components of the system, including: adaptable weight in command arbitration; smooth estimation of vote vectors in D*; and use of two new sensors for obstacle detection.

References.

[1] Alami, R., et al. A General Framework for Multi-Robot Cooperation and its Implementation on a Set of Three Hilare Robots. *Experimental Robotics IV*. 1995. pp. 26-39.

[2] Arora, S. Polynomial-time Approximation Schemes for Euclidean TSP and other Geometric Problems. In *Proc. IEEE FOCS*. 1996. pp. 2-13.

[3] B. Brumitt. Dynamic Mission Planning for Multiple Mobile Robots. In *Proc. ICRA'96*. May 1996.

[4] B. Brumitt. GRAMMPS: A Mission and Execution Planner for Multiple Robots in Unstructured Environments. *To appear in ICRA'98*.

[5] Cao, U.Y., et al. *Cooperative Mobile Robotics: Antecedents and Directions*. Autonomous Robots 4. pp 7-27. 1997.

[6] Froelich, C., M. Mettenleiter, F. Haertl. Imaging laser radar (LIDAR) for high-speed monitoring of the environment. *SPIE Proceedings of the Intelligent Transportation Systems Conference*. October 1997.

[7] Froelich, C., J. Hancock, R. Sullivan, D. Langer. High-performance Imaging Laser Radar. *To Appear in ICRA'98*.

[8] Hancock, J., E. Hoffman, R. Sullivan, D. Ingimarson, D. Langer, M. Hebert. High-performance laser range scanner. *SPIE Proceedings of the Intelligent Transportation Systems Conference*. October 1997.

[9] Hebert, M. Pixel-Based Range Image Processing. In *Proc. ICRA'94*. May 1994.

[10] Huber, D., Denes, L., Hebert, M., Gottlieb, M., Kaminsky, B. A Spectro-polarimetric Imager for Intelligent Transportation Systems. In *Proc. SPIE Conference on Transportation Sensors. Pittsburgh*. October 1997.

[11] Langer, D., Rosenblatt, J., Hebert, M.. *A Behavior-Based Approach to the Autonomous Navigation Systems*. IEEE Transactions on Robotics and Automation, vol.10, no. 4. 1994.

[12] Le Pape, C. A Combination of Centralized and Distributed Methods for Multi-Agent Planning and Scheduling. In *Proc. ICRA*. pp. 488-493. 1990.

[13] Mackenzie, D.C., Arkin, R.A, Cameron, J.M. Multiagent Mission Specification and Execution. *Autonomous Robots 4*. pp. 29-52. 1997.

[14] Mataric, M. J. Reinforcement Learning in the Multi-Robot Domain. *Autonomous Robots 4*. pp.73-83. 1997.

[15] Morganthaler, M.K. UGV Mission Planning. *RSTA for the UGV*. O. Firschein and T. Strat Ed. Morgan Kaufman. 1997.

[16] Parker, L. *Heterogeneous Multi-Robot Cooperation*, Ph.D. Thesis, MIT, Feb. 1994.

[17] Rosenblatt, J.K. DAMN: A Distributed Architecture for Mobile Navigation. In *Proc. of the 1995 AAAI Spring Symposium*. H. Hexmoor & D. Kortenkamp (Eds.). AAAI Press, Menlo Park, CA. 1995.

[18] Rosenblatt, J.K., Thorpe, C.E. Combining Multiple Goals in a Behavior-Based Architecture. *Proceedings of IROS'95*. 1995.

[19] Stentz, A. Optimal and Efficient Path Planning for Unknown and Dynamic Environments. *International Journal of Robotics and Automation*. Vol. 10, No. 3. 1995.

[20] Stentz, A., Hebert, M.. *A complete navigation system for goal acquisition in unknown environments*. Autonomous Robots. Kluwer Academic Publishers. 1995.

A Comparative Study of Vision-Based Lateral Control Strategies for Autonomous Highway Driving

J. Košecká, R. Blasi, C. J. Taylor and J. Malik

Department of Electrical Engineering and Computer Sciences
University of California at Berkeley
Berkeley, CA 947, USA.
email: janak,blasirs,camillo,malik@cs.berkeley.edu

Abstract

This paper will present the results of a comparative study of a set of vision-based control strategies that have been applied to the problem of steering an autonomous vehicle along a highway. The aim of this work has been to further our understanding of the characteristics of various control laws that could be applied to this problem with a view to making informed design decisions. The control strategies that we explored include a lead lag control law, a full-state linear controller and input-output linearizing control law. Each of these control strategies was implemented and tested on our experimental vehicle, a Honda Accord LX, both with and without a curvature feedforward component.

1 Introduction

With the increasing speeds of modern microprocessors it has become ever more common for computer vision algorithms to find application in real-time control tasks. In particular, the problem of steering an autonomous vehicle along a highway using the output from one or more video cameras mounted inside the vehicle has been a popular target for researchers around the world and a number of groups have demonstrated impressive results on this control task. Dickmanns et. al. [2] developed a system that drove autonomously on the German Autobahn as early as 1985. The Navlab project at CMU has produced a number of successful visually guided autonomous vehicle systems. Other research groups include Ozguner et. al. at Ohio State [10], Broggi et al at the Universita' di Parma, Raviv and Herman at the National Institute of Standards [12] and Lockheed-Martin.

The goal of our research efforts in this field has been to understand the fundamental characteristics of this vision based control problem and to use this knowledge to design better control strategies. In [7] we presented an analysis of the problem of vision-based lateral control and investigated the effects of changing various important system parameters like the vehicle velocity, the lookahead range of the vision sensor and the processing delay associated with the perception and control system. We also described a static feedback strategy that enabled us to perform the lateral control task at highway speeds. We were able to verify the accuracy and efficacy of our modelling and control techniques on our experimental vehicle platform, a Honda Accord LX.

In this paper we present the results of a series of experiments that were designed to provide a systematic comparison of a number of control strategies. The aim of this work has been to further our understanding of the characteristics of various control laws that could be applied to this problem with a view to making informed design decisions. The control strategies that we explored include a lead lag control law, a full-state linear controller and input-output linearizing control law. Each of these control strategies was implemented and tested both with and without a curvature feedforward component.

Section 2 of this paper presents the basic equations that we have used to model the dynamics of our vehicle and our sensing system. Section 3 describes the design of the observer that we use to estimate the states of our system and the curvature of the roadway. Section 4 describes the various control strategies that we implemented on our experimental platform and section 5 presents the results of the experiments that we carried out with these controllers. Section 6 contains the conclusions that we have drawn from these experiments.

2 Modeling and Analysis

The dynamics of a passenger vehicle can be described by a detailed 6-DOF nonlinear model [11].

Since it is possible to decouple the longitudinal and lateral dynamics, a linearized model of the lateral vehicle dynamics is used for controller design. The linearized model of the vehicle retains only lateral and yaw dynamics, assumes small steering angles and a linear tire model, and is parameterized by the current longitudinal velocity. Coupling the two front wheels and two rear wheels together, the resulting bicycle model (Figure 1) is described by the following variables and parameters:

v linear velocity vector (v_x, v_y), v_x denotes speed

α_f, α_r side slip angles of the front and rear tires

$\dot{\psi}$ yaw rate

δ_f front wheel steering angle

δ commanded steering angle

m total mass of the vehicle

I_ψ total inertia vehicle around center of gravity (CG)

l_f, l_r distance of the front and rear axles from the CG

l distance between the front and the rear axle $l_f + l_r$

c_f, c_r cornering stiffness of the front and rear tires.

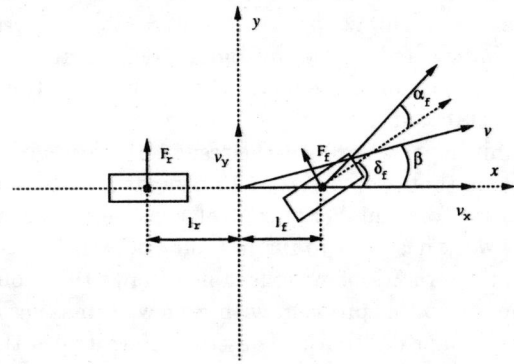

Figure 1: The motion of the vehicle is characterized by its velocity $v = (v_x, v_y)$ expressed in the vehicle's inertial frame of reference and its yaw rate $\dot{\psi}$. The forces acting on the front and rear wheels are F_f and F_r, respectively.

The lateral dynamics equations are obtained by computing the net lateral force and torque acting on the vehicle following Newton-Euler equations [1] and choosing $\dot{\psi}$ and v_y as state variables. The state equations have the following form:

$$\begin{bmatrix} \dot{v}_y \\ \ddot{\psi} \end{bmatrix} = \begin{bmatrix} -\frac{a_1}{mv_x} & \frac{-mv_x^2+a_2}{mv_x} \\ \frac{a_3}{I_\psi v_x} & -\frac{a_4}{I_\psi v_x} \end{bmatrix} \begin{bmatrix} v_y \\ \dot{\psi} \end{bmatrix} + \begin{bmatrix} b_1 \\ b_2 \end{bmatrix} \delta_f \quad (1)$$

where $a_1 = c_f + c_r$, $a_2 = c_r l_r - c_f l_f$, $a_3 = -l_f c_f + l_r c_r$, $a_4 = l_f^2 c_f + l_r^2 c_r$, $b_1 = \frac{c_f}{m}$ and $b_2 = \frac{l_f c_f}{I_\psi}$. The additional measurements provided by the vision system (see Figure 2) are:

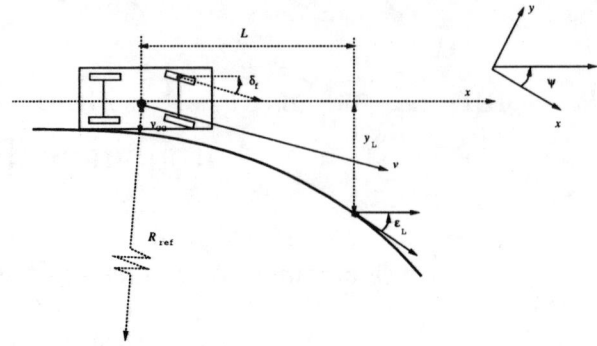

Figure 2: The vision system estimates the offset from the centerline y_L and the angle between the road tangent and heading of the vehicle ε_L at some lookahead distance L.

y_L the offset from the centerline at the lookahead,

ε_L the angle between the tangent to the road and the vehicle orientation

L denotes the lookahead distance of the vision system. The equations capturing the evolution of these measurements due to the motion of the car and changes in the road geometry are:

$$\dot{y}_L = v_x \varepsilon_L - v_y - \dot{\psi} L \quad (2)$$

$$\dot{\varepsilon}_L = v_x K_L - \dot{\psi} \quad (3)$$

We can combine the vehicle lateral dynamics and the vision dynamics into a single dynamical system of the form:

$$\dot{x} = A x + B u + E w$$
$$y = C x$$

with the state vector $x = [v_y, \dot{\psi}, y_L, \varepsilon_L]^T$, the output $y = [\dot{\psi}, y_L, \varepsilon_L]^T$ and control input $u = \delta_f$. The road curvature K_L enters the model as an exogenous disturbance signal $w = K_L$.

A block diagram of the overall system based on the state equations is shown in Figure 3. The transfer function $V_1(s)$ between the steering angle δ_f and offset at the lookahead y_L has the following form:

$$V_1(s) = \frac{1}{s^2} \frac{as^2 + bs + c}{ds^2 + es + f} \quad (4)$$

where the numerator is a function of both speed and lookahead distance and the denominator is parameterized by the speed of the car. $V_1(s)$ can be rewritten according to Figure 3 by singling out the vehicle dynamics in terms of $\ddot{y}_{CG}$ and $\dot{\psi}$ followed by the integrating action $1/s^2$:

$$V_1(s) = \frac{1}{s^2}(G(s) + L G_2(s)) \quad (5)$$

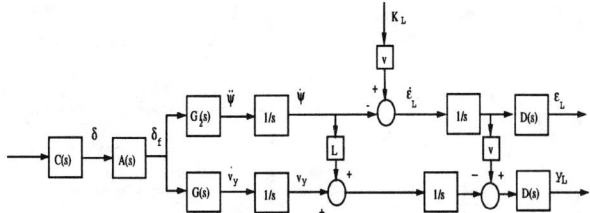

Figure 3: The block diagram of the overall system with the two outputs provided by the vision system.

where $G(s)$ and $G_2(s)$ are transfer functions between steering angle and lateral acceleration and yaw acceleration respectively. The actuator $A(s)$ is modeled as a low pass filter of the commanded steering angle δ and a pure time delay element $D(s) = e^{-T_d s}$ represents the latency T_d of the vision subsystem. In our system $T_d = 0.057$ s. The transfer function $C(s)$ corresponds to the controller to be designed. A more detailed analysis of how the behavior of this dynamic system changes as a function of important system parameters like, lookahead distance, processing delay and vehicle velocity can be found in [7].

3 Vision System

The vision-based lane tracking system used in our experiments is an improved version of the one presented in [13]. This system takes its input from a single forward-looking CCD video camera. It extracts potential lane markers from the input using a template-based scheme. It then finds the best linear fits to the left and right lane markers over a certain lookahead range through a variant of the Hough transform. From these measurements we can compute an estimate for the lateral position and orientation of the vehicle with respect to the roadway at a particular lookahead distance, L.

The vision system is implemented on an array of TMS320C40 digital signal processors which are hosted on the bus of an Intel-based industrial computer. The system processes images from the video camera at a rate of 30 frames per second.

4 Observer Design

In order to estimate the curvature of the roadway we have chosen to implement an observer based on a slightly simplified version of the systems state equations as shown in Equation (6). More specifically, in these equations we have chosen to neglect the vehicles lateral velocity, v_y.

$$\dot{x}' = A'(v_x)x' + B'\dot{\psi}$$

$$y' = C'x' \qquad (6)$$

where $x' = [y_L, \varepsilon_L, K_L]^T$, $y' = [y_L, \varepsilon_L]^T$. Note that the state vector x' includes the road curvature K_L. This differential equation can be converted to discrete time in the usual manner by assuming that the yaw rate, $\dot{\psi}$, is constant over the sampling interval T.

$$x'(k+1) = \Phi(v_x)x'(k) + \beta\dot{\psi} \qquad (7)$$

Equation (7) allows us to predict how the state of the system will evolve between sampling intervals.

Measurements are obtained from two sources: the vision system provides us with measurements of y_L and ε_L, while the on-board fiber optic gyro provides us with measurements of the yaw rate of the vehicle, $\dot{\psi}$. Our use of the yaw rate sensor measurements is analogous to the way in which information from the proprioceptive system is used in animate vision. The measurement vector y' is used to update an estimate for the state of the system $\hat{x}'$ as shown in the following equation:

$$\hat{x}'^+(k) = \hat{x}'^-(k) + L(y'(k) - C\hat{x}'^-(k)) \qquad (8)$$

where $\hat{x}'^-(k)$ and $\hat{x}'^+(k)$ denote the state estimate before and after the sensor update respectively.

The gain matrix L can be chosen in a number of ways [4], depending on the assumptions one makes about the availability of noise statistics and the criterion one chooses to optimize. In our case, the gain matrix was chosen to minimize the expected error of our estimate in the steady state using the function `dlqe` available in Matlab. The covariances of both the process and measurement noise were estimated by analyzing the data collected by our sensors during trial runs with the vehicle.

5 Controllers

The goal of all of the control schemes presented in the sequel is to regulate the offset at the lookahead, y_L, to zero. Passenger comfort is another important design criterion and this is typically expressed in terms of jerk, corresponding to the rate of change of acceleration. For a comfortable ride no frequency above 0.1-0.5 Hz should be amplified in the path to lateral acceleration [5]. Additional performance criteria may be specified in terms of the maximal allowable offset y_{Lmax} as a response to the step change in curvature and in terms of bandwidth requirements on the transfer function $F(s) = \frac{y_L(s)}{K_L(s)}$.

Lead-lag Control. Previous analysis [7] revealed that at speeds of up to 15 m/s with a lookahead

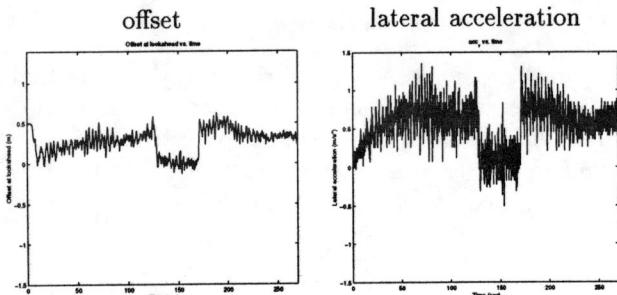
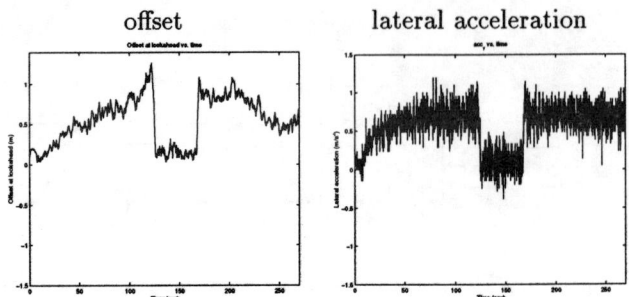

Figure 4: Lead-Lag controller. The offset was measured at the lookahead of 15 m. The performace of the lead-lag controler in terms of offset was superior compared to the other tested strategies.

Figure 5: Full State Feedback controller. The offset measured at the look-ahead distance of 6 m results in a less noisy measurements. The lateral acceleration profile degraded towards the end of the run due to the increase in linear velocity.

of around 10 meters one can guarantee satisfactory damping of the closed loop poles of $V_1(s)$ and compensate for the processing delay of the vision system using simple unity feedback control with proportional gain in the forward loop. As the velocity increases, the poles of the transfer function move toward the real axis and become more poorly damped which introduces additional phase lag in the frequency range 0.1-2 Hz. Since further increasing the lookahead does not improve the damping, gain compensation alone cannot achieve satisfactory performance. A natural choice for obtaining an additional phase lead in the frequency range 0.1-2 Hz would be to introduce some derivative action, however, in order to keep the bandwidth low an additional lag term is necessary. One satisfactory lead-lag controller has the following form:

$$C(s) = \frac{0.09s + 0.18}{0.025s^2 + 1.5s + 20} \quad (9)$$

where $C(s)$ is a lead network in series with a single pole. The above controller was designed for a velocity of 30 m/s (108 km/h, 65 mph), a lookahead of 15 m and 60 ms delay. The resulting closed loop system has a bandwidth of 0.45 Hz with a phase lead of 45° at the crossover frequency. A discretized version of the above controller taking into account the 33 ms sampling time of the vision system was used in our experiments. The performance of the controller is in Figure 4.

Since increasing the speed has a destabilizing effect on $V_1(s)$, designing the controller for the highest intended speed guarantees stability at lower speeds and achieves satisfactory ride quality. In order to tighten the tracking performance at lower speeds individual controllers can be designed for various speed ranges and gain scheduling techniques used to interpolate between them.

Full State Feedback. Given that the vehicle can be modeled as a linear dynamical system it seems natural to consider standard full state linear feedback laws of the form $u = Kx$. The controller was designed for velocity of 20 m/s and a lookahead of 6 meters. The gain matrix, K, was then chosen using pole placement techniques such that the two poles of the system that were originally at the origin were moved to a conjugate pair with a damping ratio $\xi = 0.707$ and a natural frequency $\omega_n = 0.989$ rad/s. The other two poles of the system were left unchanged. These pole locations were chosen so that the resulting system would satisfy our step response and bandwidth requirements.

In the resulting linear control law, the gain associated with the lateral velocity term v_y was small so we chose to neglect this component of the controller. Estimates for the remaining state variables, y_L, ε_L, and $\dot{\psi}$ are obtained from our observer and the yaw rate sensor. The offset and lateral acceleration profiles are in Figure 5.

Input-Output Linearization. Input-output linearization is typically used to linearize nonlinear systems by state feedback as described in [6]. The application of this technique to the bicycle model is not, strictly speaking, linearization by state feedback since the bicycle model is already linear. Nonetheless, this technique can be applied to render the model independent of the vehicles longitudinal velocity, v_x. In this case the feedback law has a zero cancelling effect instead of linearizing one and makes the vehicle dynamics poles unobservable. Given the bicycle model in the form $\dot{x} = f(x) + g(x)u$ consider the control law

$$u = \frac{1}{L_g L_f^1 h(x)}(-L_f^2 h(x) + u') \quad (10)$$

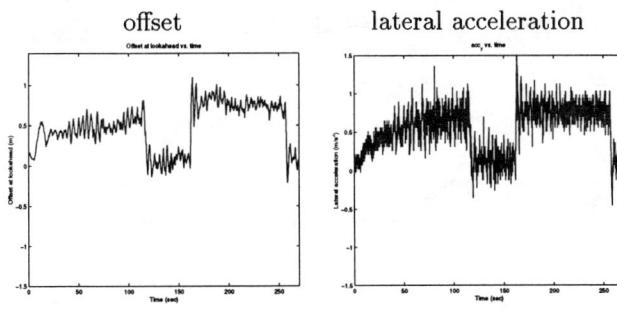

Figure 6: I/O linearized controller. The overshoots during the transitions are noticeable and performance slightly degrades with increasing speed towards the end of the test run.

where L_g^i denotes the i-th Lie derivative along g. For our particular example the control law becomes:

$$u = a \left(u' - \frac{(\frac{La_3}{I_\psi} - \frac{a_1}{m})v_y - (\frac{-La_4}{I_\psi} - \frac{a_2}{m})\dot{\psi}}{v_x} \right)$$

with
constants $a = 1/(-Lb_2 - b_1)$ and $a_1, a_2, a_3, a_4, b_1, b_2$ as defined in Equation 1. Employing this control law yields a second order equation of the form $\ddot{y} = u'$. Once the system has been reduced to this form we can employ the same lead-lag control law described previously to compute u' and stabilize the system as well as obtain the desired performance.

Feedforward Control The steady state steering input, δ_{ref}, that is required to track a reference K_{Lref} can be computed from the state equations by setting $[\dot{v}_y, \ddot{\psi}, \dot{y}_L, \dot{\varepsilon}_L]^T$ to 0.

$$\delta_{ref} = K_{ref} \left(l - \frac{(l_f c_f - l_r c_r)v_x^2 m}{c_r c_f l} \right). \quad (11)$$

This feedforward control component can be added to any of the control schemes that have been described. The feedforward control law allows the system to anticipate changes in curvature ahead of the car and improves the transient behavior of the vehicle when entering and exiting curves (see Figure 7). The effectiveness of the feedforward term will, of course, depend on the quality of the curvature estimates supplied by the observer.

6 Experimental Results

In order to evaluate different feedback and feedforward strategies we implemented them on our experimental vehicle and collected data from a number of trial runs. Our test track was a 7 mile oval and our experiments were run at speeds of approximately 75mph to simulate actual highway conditions. Each experimental trial lasted at least 5 minutes, long enough to explore how each controller fared on the straight sections, the curved sections and the transitions between them. Figures 4, 5, 6 describe performance of individual feedback control strategies and Figure 7 depicts the performance of the curvature estimator and some of the tested control strategies with the feedforward term.

7 Conclusions

The strategy behind the design of the feedback control laws was based on the observation that the behavior of our system was dominated by the two poles at the origin, since the other two poles are well behaved as long as the lookahead distance is large enough. This allowed us to design controllers for the highest intended operating velocity, which would operate satisfactorily in the whole range of lower velocities. However this approach sacrifices some performance criteria at lower velocities.

Our experiments indicate that all three of the feedback control strategies that we implemented provided acceptable performance on the lateral control task with the lead lag control law yielding the best tracking performance of the three. The data also shows that the curvature feedforward component definitely improves the tracking performance of all three control strategies. It allows the system to eliminate steady state tracking errors when following a curve and it minimizes the transient response of the system to changes in curvature. More detailed experimental evaluation of the control strategies in variety of weather and road conditions is necessary.

Acknowledgment. This research has been supported by Honda R&D North America Inc., Honda R&D Company Limited, Japan, PATH MOU257 and MURI program DAAH04-96-1-0341.

References

[1] R. S. Blasi. A study of lateral controllers for the stereo drive project. Master's thesis, Department of Computer Science, University of California at Berkeley, 1997.

[2] E. D. Dickmans and B. D. Mysliwetz. Recursive 3-D road and relative ego-state estimation. *IEEE*

Transactions on PAMI, 14(2):199–213, February 1992.

[3] B. Espiau, F. Chaumette, and P. Rives. A new approach to visual servoing in robotics. *IEEE Transactions on Robotics and Automation*, 8(3):313 – 326, June 1992.

[4] Arthur Gelb *et al*. *Applied optimal estimation*. MIT Press, 1994.

[5] J. Guldner, H.-S. Tan, and S. Patwarddhan. Analysis of automated steering control for highway vehicles with look-down lateral reference systems. *Vehicle System Dynamics (to appear)*, 1996.

[6] Alberto Isidori. *Nonlinear Control Systems*. Springer Verlag, 1989.

[7] J. Košecká, R. Blasi, C.J. Taylor, and J. Malik. Vision-based lateral control of vehicles. In *Proc. Intelligent Transportation Systems Conference, Boston*, 1997.

[8] M. F. Land and D. N. Lee. Where we look when we steer? *Nature*, 369(30), June 1994.

[9] Y. Ma, J. Košecká, and S. Sastry. Vision guided navigation for a nonholonomic mobile robot. In *Proceedings of CDC'97*, 1997.

[10] Ü. Özgüner, K. A. Ünyelioglu, and C. Hatipoğlu. Steering and Lane Change: A Working System. IEEE Conference on Intelligent Transportation Systems, Boston,1997.

[11] H. Peng. *Vehicle Lateral Control for Highway Automation*. PhD thesis, Department of Mechanical Engineering, University of California, Berkeley, 1992.

[12] M. Herman, M. Nashman, T. Hong, H. Schneiderman, D. Coombs, G.-S. Young, D. Raviv and A. J. Wavering Minimalist Vision for Navigation, *Visual Navigation: From Bilogical Systems to Unmanned Ground Vehicles*, Lawrence Erlbaum Associates, 1997, edt.: Y. Aloimonos, Mahwah, New Jersey.

[13] C. J. Taylor, J. Malik, and J. Weber. A realtime approach to stereopsis and lane-finding. In *Proceedings of the 1996 IEEE Intelligent Vehicles Symposium*, pages 207–213, Seikei University, Tokyo, Japan, September 19-20 1996.

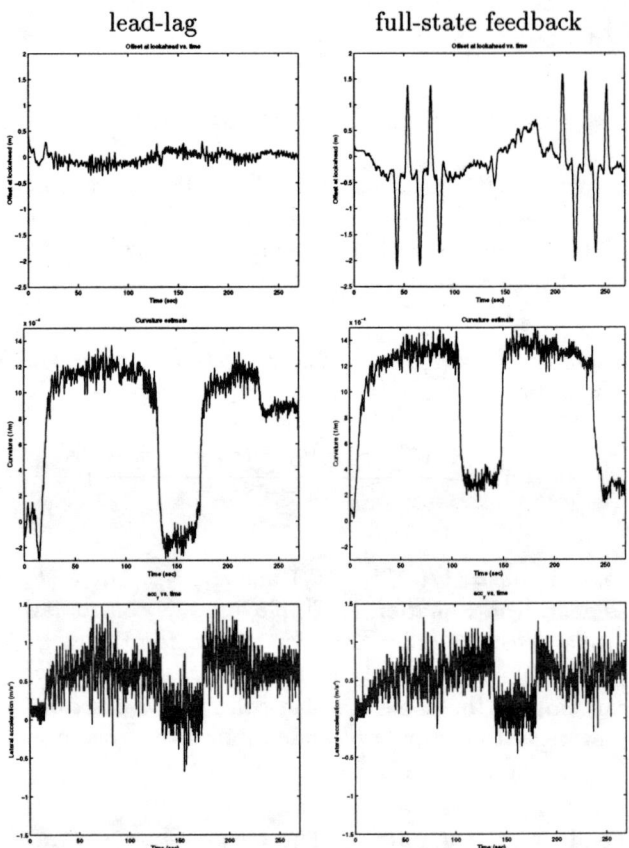

Figure 7: These plots demonstrate the effect of the feedforward control term on the overall tracking performance for two of the tested control strategies. The first row of plots indicates the tracking performance measured in terms of the offset at the lookahead, the second row depicts the curvature estimate used in the feedforward term, which was provided by the observer and the last row shows the lateral acceleration profiles. Notice that the steady state offset in the curved sections was essentially eliminated. The offset plots all exhibit a slight overshoot during transitions in curvature until the curvature estimates converge. In case of full state feedback controller the spikes in the offset measurements as well as in the lateral acceleration profile correspond to the lane change maneuvers which the vehicle performed at lower speeds.

Towards Force-Reflecting Teleoperation Over the Internet

Günter Niemeyer and Jean-Jacques E. Slotine

Nonlinear Systems Laboratory
Massachusetts Institute of Technology
Cambridge, MA 02139
gunter@ai.mit.edu, jjs@mit.edu

Abstract

This paper extends earlier results on stable force-reflecting teleoperation in the presence of significant time-delays to the case, frequent in practice, where the transmission delays are themselves varying with time in an unpredictable fashion. It shows that stability can be preserved through the systematic use of specially designed wave-variable filters. The resulting performance of the teleoperation system is illustrated in simulations, and is consistent with reasonable expectations on "ideal" behavior. The results may provide a practical tool for implementing force-reflecting teleoperation over the Internet.

1 Introduction

Teleoperation has enjoyed a rich history and has lead both to many practical applications and to a broad vision of interacting with environments far removed from the user (Sheridan, 1989). To provide a more complete interaction, force feedback is often included, since this information can considerably improve the user's ability to perform complex tasks (Sheridan, 1992).

However, by their very definition, teleoperation systems frequently experience significant time-delays in the communications between local and remote sites. Untreated, even small delays can lead to instability due to unwanted power generation in the communications. Through the introduction of the wave variable concept (Niemeyer and Slotine, 1991), based on a reformulation of the passivity formalism of (Anderson and Spong, 1989), systematic analysis tools can be developed to understand these problems. Moreover, *stable* force reflecting teleoperators can be designed and shaped to act as simple virtual tools when exposed to large delays. They are also transparent to the user when delays are below human reaction time (Niemeyer and Slotine, 1997a).

Existing results have concentrated on unknown but constant time delays. Here we extend these ideas to time-varying transmission delays. Two classes of practical problems come to mind. First, variable delays due to motion of the slave systems. For instance, space-based or underwater telerobotic applications involve moving vehicles and thus experience changing transmission times to and from the stationary operator. The resulting variations, however, are typically very slow and in practice can often be ignored.

Second, and significantly more interesting, are rapidly and possibly randomly varying transmission delays. This is the case, for instance, in satellite-based transmission through varying relay sites. Perhaps more intriguingly, this is also the case in the Internet, which has frequently been suggested as a means for creating teleoperation systems between a variety of remote sites, given the availability of ever less expensive force-reflecting interfaces. Information is transmitted in small packets and is routed in real-time through a possibly large number of intermediate stops. While average latencies may be low, the instantaneous delays may increase suddenly due to rerouting or other network traffic. In the extreme, the connection may be temporarily blocked. Such effects distort the signals (as seen with respect to time), can introduce high-frequency data, and can lead to instability if left untreated.

For example, examine in Figure 1 the observed trans-continental round trip delay times between MIT and California, a distance of roughly 4000 kilometers. The average latency of 0.1 seconds is comparable to the human reaction time, so that its effects can be made largely transparent to the user. However, the variation in the delay is strong, rapidly changing more than 50%, and contains many components near 10Hz. In a closed-loop system, such fluctuations may interact with the 0.1 second delay and cause stability problems.

The solution proposed in this paper exploits the

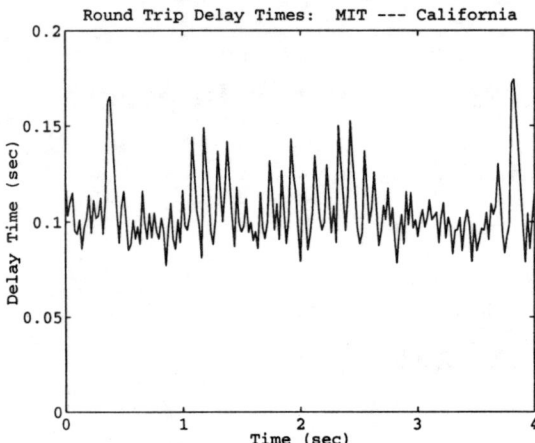

Figure 1: Observed round trip delay times between MIT and California, sending 50 data sets per second.

wave variable formalism to incorporate information as soon as it arrives, but modulating it if necessary so as to explicitly insure proper behavior. For a constant delay, the approach automatically simplifies to the standard wave variable method, which implicitly limits performance as needed to maintain stability at that delay (Niemeyer and Slotine, 1997b). For zero delay, it automatically simplifies to a classic force-reflecting configuration. As a result, the system feels soft when the instantaneous delay is large, then becomes crisp as the delay shrinks.

We first briefly review wave variables and their application to teleoperation in Section 2. We then detail the problems caused by variable time-delays and introduce this paper's suggested solution in Section 3. As the central part of the solution, the reconstruction filter is examined in more depth in Section 4. Finally, we demonstrate the complete system via simulation in Section 5, and offer some concluding remarks in Section 6.

2 Wave Variables

Wave variables are central to our developments and are briefly reviewed here. We refer the reader to (Niemeyer and Slotine, 1997a, 1997c) for a detailed discussion.

2.1 Definition

The key feature of wave variables is their encoding of velocity and force information. In particular, we define

$$\mathbf{u} = \frac{b\dot{\mathbf{x}} + \mathbf{F}}{\sqrt{2b}} \qquad \mathbf{v} = \frac{b\dot{\mathbf{x}} - \mathbf{F}}{\sqrt{2b}} \qquad (1)$$

Here $\mathbf{u}$ denotes the forward or right moving wave, and $\mathbf{v}$ denotes the backward of left moving wave. The characteristic wave impedance b is a positive constant or a symmetric positive definite matrix and assumes the role of a tuning parameter, which allows matching a controller to a particular environment or task.

To clarify this definition, consider the layout of a teleoperator system shown in Figure 2. At the local (master) side, the velocity $\dot{\mathbf{x}}_m$ and force $\mathbf{F}_m$ information is combined to provide a command wave signal $\mathbf{u}_m$, which reaches the remote (slave) side after a delay T. There it is decoded into a velocity $\dot{\mathbf{x}}_s$ or force $\mathbf{F}_s$ command. The same process returns information back to the master side.

Note that the inherent combination of velocity and force data makes the system well suited for interaction with unknown environments. Indeed it behaves like a force controller when in contact with a rigid object and like a motion controller when in free space. The parameter b also allows online trade-off between the two quantities, fine tuning the behavior.

A wave signal itself is then best described as a '*move or push*' command, where the sign determines the direction. The receiving side will then either move or apply forces depending on the current situation. The returning wave will have the opposite sign, i.e. 'push back', if no motion was possible. Or it will have the same sign, i.e. 'move with', if motion occurred.

2.2 Passivity

The original motivation for introducing wave variables, is their effect on passivity. Indeed the power input becomes

$$P_{in} = \dot{\mathbf{x}}^T \mathbf{F} = \tfrac{1}{2}\mathbf{u}^T\mathbf{u} - \tfrac{1}{2}\mathbf{v}^T\mathbf{v} \qquad (2)$$

where $\tfrac{1}{2}\mathbf{u}^T\mathbf{u}$ is the power flowing in the main forward direction and $\tfrac{1}{2}\mathbf{v}^T\mathbf{v}$ gives the power flowing back.

Now remember the condition for passivity, which requires more input energy than return energy

$$\int_0^t P_{in}\,d\tau = \int_0^t \dot{\mathbf{x}}^T \mathbf{F}\,d\tau \geq -E_{store}(0) \quad \forall t \geq 0 \qquad (3)$$

where $E_{store}(0)$ denotes the initial stored energy (Slotine and Li, 1991).

In the wave domain, this condition becomes

$$\int_0^t \tfrac{1}{2}\mathbf{v}^T\mathbf{v}\,d\tau \leq \int_0^t \tfrac{1}{2}\mathbf{u}^T\mathbf{u}\,d\tau + E_{store}(0) \quad \forall t \geq 0 \qquad (4)$$

Not surprisingly a system is passive if the energy in the returning wave $\mathbf{v}$ is limited to the energy provided by the ingoing wave $\mathbf{u}$ or stored initially.

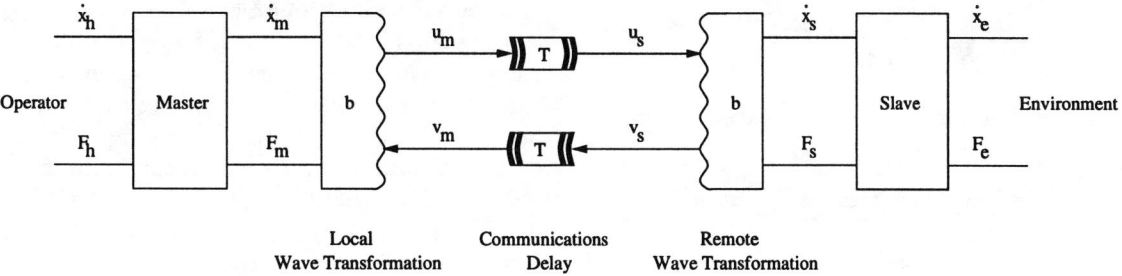

Figure 2: The wave based teleoperator transforms both local and remote information into wave variables before transmission to the other side.

Notice that each wave signal essentially contains its own power to execute the command, independent of its dual wave. In contrast, both power variables $(\dot{\mathbf{x}}, \mathbf{F})$ must be known in order to understand the power flow and passivity. It is this property and the new condition (4), which make wave variables robust to delays.

2.3 Time Delays

If each wave contains its own power and passivity only requires that the output or return wave be limited by the input or command wave regardless of phase, then delaying a wave signal does not alter passivity. Instead, it simply stores the energy in the wave for the delay time and releases it thereafter.

Consider again the basic teleoperator layout of Figure 2. Assume for now a constant delay in the communications between local and remote site. The total power input into the communications block at any point in time is given by

$$P_{in} = \dot{\mathbf{x}}_m^T \mathbf{F}_m - \dot{\mathbf{x}}_s^T \mathbf{F}_s \qquad (5)$$

where the minus sign appears because power is considered positive while flowing in the main direction from left to right.

Substituting the wave transformation equations, we can also compute this power input as

$$P_{in} = \tfrac{1}{2}\mathbf{u}_m^T\mathbf{u}_m - \tfrac{1}{2}\mathbf{v}_m^T\mathbf{v}_m - \tfrac{1}{2}\mathbf{u}_s^T\mathbf{u}_s + \tfrac{1}{2}\mathbf{v}_s^T\mathbf{v}_s \qquad (6)$$

where all variables are measured at the current time t.

But the communications transmits and delays the waves as

$$\mathbf{u}_s(t) = \mathbf{u}_m(t-T) \qquad (7a)$$
$$\mathbf{v}_m(t) = \mathbf{v}_s(t-T) \qquad (7b)$$

Substituting into (6) and integrating, we find that all input power is stored according to

$$\int_0^t P_{in}\,d\tau = E_{store}(t) = \int_{t-T}^t \tfrac{1}{2}\mathbf{u}_m^T\mathbf{u}_m + \tfrac{1}{2}\mathbf{v}_s^T\mathbf{v}_s\,d\tau \geq 0$$

assuming zero initial conditions. The wave energy in $\mathbf{u}_m$ and $\mathbf{v}_s$ is thus temporarily stored while the waves are in transit, making the communications not only passive but also lossless. This is independent of the delay time T, and does not require knowledge thereof.

2.4 Position Tracking

In the basic form, a wave based teleoperator transmits the wave signals, which encode both velocity and force, but do not contain any explicit position information. Position tracking is guaranteed only implicitly. Consider the deflection or position error between the master and slave sides.

$$\Delta\mathbf{x}(t) = \mathbf{x}_m(t) - \mathbf{x}_s(t) \qquad (8)$$
$$= \frac{1}{\sqrt{2b}}\int_0^t \mathbf{u}_m(\tau) + \mathbf{v}_m(\tau) - \mathbf{u}_s(\tau) - \mathbf{v}_s(\tau)\,d\tau$$

which is obtained by solving (1) for velocity and integrating. Substituting the delay equations (7), we have

$$\Delta\mathbf{x}(t) = \frac{1}{\sqrt{2b}}\int_{t-T}^t \mathbf{u}_m(\tau) - \mathbf{v}_s(\tau)\,d\tau \qquad (9)$$

which will reach zero when the system comes to rest and the wave commands are zero for the T seconds.

Besides requiring a constant delay, this argument is also questionable, as it assumes perfect numerical integration. Indeed an explicit position command for either master or slave can only be obtained by decoding the wave signals into a velocity command and integrating.

2.5 Wave Integrals

To avoid the numerical integration step and provide explicit position feedback, we can use and transmit the wave integrals in parallel with the wave signals themselves. Indeed, just as the wave signals encode

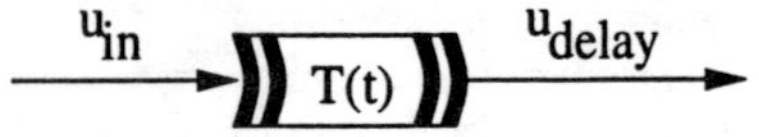

Figure 3: Variable time delay in the wave domain.

velocity and force, their integrals encode position and momentum information.

The integrated wave variables are defined as

$$\mathbf{U}(t) = \int_0^t \mathbf{u}(\tau)\,d\tau = \frac{b\mathbf{x} + \mathbf{p}}{\sqrt{2b}} \quad (10a)$$

$$\mathbf{V}(t) = \int_0^t \mathbf{v}(\tau)\,d\tau = \frac{b\mathbf{x} - \mathbf{p}}{\sqrt{2b}} \quad (10b)$$

where $\mathbf{x}$ denotes position and $\mathbf{p}$ denotes momentum, which is the integral of force

$$\mathbf{p} = \int_0^t \mathbf{F}\,d\tau \quad (11)$$

Note in many cases, we place little or no importance on the actual momentum value and it can often be eliminated from the system if so desired. Also the addition of wave integrals does not change the passivity arguments. It only provides an explicit feedback path for what is already theoretically guaranteed.

3 Variable Time Delays

We now focus our attention on variable delay as illustrated in Figure 3. Based on the previous discussion, we examine just an isolated single wave delay. The overall system remains passive if this element stays passive, i.e. if its output energy is limited by its input energy. And the position tracking is guaranteed its output wave integral tracks its input integral. As such the forward and return delays may be different and are handled separately by duplicating the following efforts for both transmission paths.

3.1 Untreated Variable Delays

Let us first understand the effect of a variable delay left untreated. Thus we use

$$u_{out}(t) = u_{in}(t - T(t)) = u_{in}(t_s(t)) \quad (12)$$

where t_s is the sample-time, for which the input value is currently presented at the output. The difference between the current time t and the corresponding input sample time $t_s(t)$ is the delay $T(t)$.

As the delay varies, the wave signal is distorted. Indeed, if the delay increases, the sample time changes only slowly and the input values are held longer. Hence the signal is stretched. In the extreme, if the sample time becomes constant and the delay grows as fast as time itself, the output also becomes constant. Note we assume that the order of the wave signal is preserved, i.e. that data arrives at the remote site in the same order it is transmitted. This implies that the sample time will never go backwards and the delay time can not increase faster than time itself,

$$\dot{T} \leq 1 \quad (13)$$

In contrast, if the delay shortens, the sample time and hence the output signal change more rapidly. In essence the signal is compressed. Here the extreme case can lead to shock-waves where multiple data samples arrive at the remote site simultaneously. This implies discontinuities and a jump in the output signal.

The delay variation and the corresponding changes in the wave signal may easily effect the system, if it is in any way correlated to the wave signal itself. For example, remember that the wave signal is interpreted as a 'push' command. If the signal is expanded during a positive push and compressed during a negative command, the output will be biased in the position direction.

More formally, both the wave integral, which determines position tracking, and the wave energy, which determines passivity, are no longer conserved.

$$U_{out}(t) = \int_0^t u_{out}(\tau)\,d\tau \neq U_{in}(t - T(t)) \quad (14a)$$

$$E_{out}(t) = \int_0^t u_{out}^2(\tau)\,d\tau \neq E_{in}(t - T(t)) \quad (14b)$$

Thus neither position tracking nor passivity are guaranteed. Also as part of the signal compression/expansion, the frequency content changes which may produce other unexpected effects.

3.2 Integral Transmissions

The above discussion illustrates that using the distorted wave signal based on (12) does not produce the desired results, as it does not preserve the wave integral or energy. As both of these quantities are central to the stability and performance of wave based systems, we propose the following solution, illustrated in Figure 4.

Instead of transmitting the wave signal itself through the delay and then integrating, transmit both the wave integral and wave energy explicitly:

$$U_{delay}(t) = U_{in}(t - T(t)) = \int_0^{t-T(t)} u_{in}(\tau)\,d\tau \quad (15a)$$

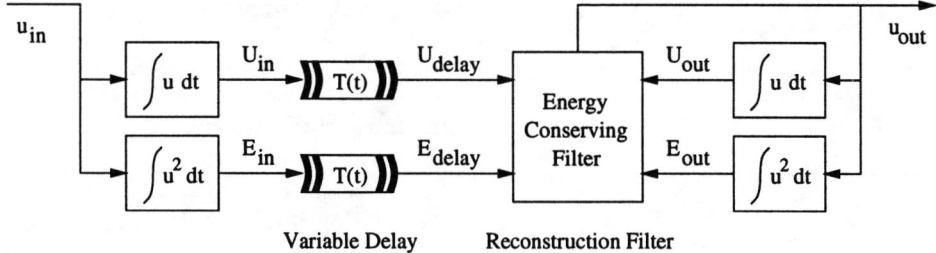

Figure 4: Transmitting the wave integral and energy, and reconstructing the output wave signal based on these quantities can overcome the damaging effects of variable delay distortions.

$$E_{delay}(t) = E_{in}(t-T(t)) = \int_0^{t-T(t)} u_{in}^2(\tau)\, d\tau \quad (15b)$$

The integration process thus remains consistent, though the resulting values are delayed. Also compute the equivalent quantities for the output wave signal $U_{out}(t)$ and $E_{out}(t)$ from (14). Then explicitly reconstruct the output wave signal such that its integral tracks the delayed input integral

$$U_{out}(t) \to U_{delay}(t) \quad (16a)$$

while using only the available energy

$$E_{out}(t) \leq E_{delay}(t) \quad (16b)$$

The following section details this process.

Such an explicit reconstruction has several advantages. First, passivity is guaranteed by the definition of the system and is independent of the actual delay and or fluctuations thereof. We can build a passive and stable teleoperator on top of such communications. Second, explicit use of the wave integrals provides explicit position feedback. To this end, the wave integrals should be computed directly from position measurements.

If the delay is constant, no distortion is present and the output of such a reconstruction filter should equal the delayed original wave input. But should the delay fluctuates, the input may change more rapidly than can be matched with the incoming energy. In such cases, the filter is forced to smooth the signal to conserve energy. Much like the wave based controllers, we see the system introduce an automatic performance limitation to remain passive.

4 Reconstruction Filter

Many alternatives are possible to the filter problem defined by (16). To better understand our solution, let us first examine the responses to impulse inputs. While the system is not linear, in the sense that the sum of two inputs does not necessarily produce the sum of the two individual outputs, such responses illustrate the basic behavior as a function of the system parameters. Also impulse inputs appear in real problems if transmission is temporarily blocked and the built-up data is released altogether.

4.1 Impulse Response

First define the wave 'distance to go' $U(t)$ and energy reserve $E(t)$ available to the filter as

$$U(t) = U_{delay}(t) - U_{out}(t) \quad (17a)$$
$$E(t) = E_{delay}(t) - E_{out}(t) \geq 0 \quad (17b)$$

Our solution takes the form

$$u_{out}(t) = \begin{cases} \alpha\, \dfrac{E(t)}{U(t)} & \text{if } U(t) \neq 0 \\ 0 & \text{if } U(t) = 0 \end{cases} \quad (18)$$

where α is a tunable parameter.

Such filters have several interesting properties. For zero input, they maintain the fixed ratio

$$\frac{E(t)}{U(t)^\alpha} = c \quad (19)$$

where c is a constant determined by initial conditions. Indeed,

$$\frac{d}{dt}\frac{E(t)}{U(t)^\alpha} = \frac{\dot{E}U - \alpha \dot{U} E}{U^{\alpha+1}} = \frac{-u_{out}^2 U + \alpha\, u_{out}\, E}{U^{\alpha+1}} = 0$$

Thus for an impulse response, $u_{out}(t)$ may be written as

$$u_{out}(t) = \begin{cases} \alpha\, c\, U(t)^{\alpha-1} & \text{if } U(t) \neq 0 \\ 0 & \text{if } U(t) = 0 \end{cases} \quad (20)$$

which is continuous at $U(t) = 0$ for $\alpha > 1$ and bounded for $\alpha = 1$. Furthermore we have

$$\frac{d}{dt} U(t) = -u_{out}(t) = -\alpha\, \frac{E(t)}{U(t)} = -\alpha\, c\, U(t)^{\alpha-1} \quad (21)$$

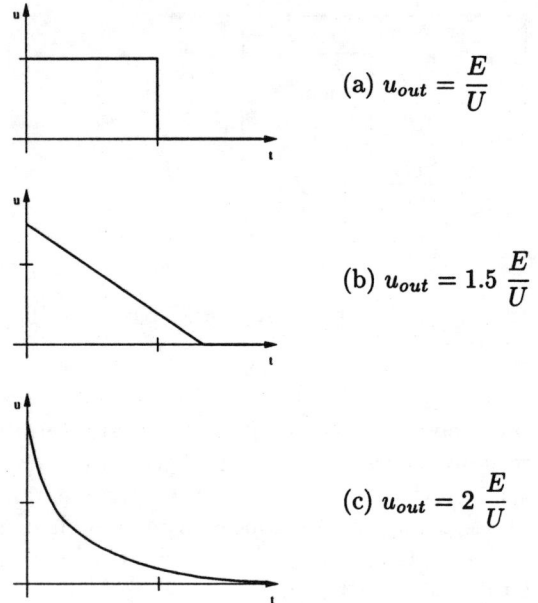

(a) $u_{out} = \dfrac{E}{U}$

(b) $u_{out} = 1.5\,\dfrac{E}{U}$

(c) $u_{out} = 2\,\dfrac{E}{U}$

Figure 5: Impulse responses for the reconstruction filter under a variety of parameter settings.

Consequently, both the distance to go $U(t)$ and the remaining energy $E(t)$ reach zero. Given the first order nature, $U(t)$ will not cross the origin during this convergence process and $E(t)$ remains positive.

Varying the parameter α selects how quick versus smooth the convergence process is. Figure 5 shows the impulse responses for the values of $\alpha = 1$, $\alpha = 1.5$, and $\alpha = 2$. In the first case, the response is constant until the goal is reached and all energy has been used. This provides the fastest time to reach the goal, but also contains discontinuities which may be disruptive in practice.

The response is linear for $\alpha = 1.5$ and exponential for $\alpha = 2$. We will constrain the following developments to the latter case. The response is given by

$$u(t) = \frac{2\,E_0}{U_0}\, e^{-\lambda t} \qquad (22)$$

where the bandwidth λ is determined as

$$\lambda = \frac{2\,E}{U^2} \qquad (23)$$

Notice that the speed of the response depends on the amount of energy E available for a given distance U. The more energy, the faster the response. This also suggests a practical advantage: To limit the frequency content coming out of the reconstruction filter, we can saturate the stored energy reserve via

$$E(t) \le \tfrac{1}{2}\, \lambda_{\max}\, U(t)^2 \qquad (24)$$

4.2 Discrete Implementation

In practice, with a changing input $U_{delay}(t)$, the filter will not reach zero but continue to track the input with the first order behavior of (21). Limiting the energy according to (24) further smoothes the signal.

Nevertheless, when $U(t)$ and $E(t)$ approach zero the division in (18) is hard to compute. In addition a finite sampling rate approximation may make $E(t)$ negative in violation of (16). We therefore suggest a discrete implementation which accounts for a finite sample rate.

Rather than approximating the continuous derivatives, we rederive the equations starting with the requirement

$$\frac{E_{n+1}}{U_{n+1}^2} = \frac{E_n}{U_n^2} = \text{constant}$$

for a zero input. This leads to the output

$$u_{out} = \begin{cases} 0 & \text{if } U = 0 \\[4pt] \dfrac{U}{\Delta t} & \text{if } U^2 \le E\,\Delta t \\[6pt] \dfrac{2EU}{U^2 + E\Delta t} & \text{if } U^2 > E\,\Delta t \end{cases} \qquad (25)$$

where Δt is the discrete sample time step.

The first case appears if there is no further energy or distance to go. The second case is new and accounts for the possibility that the convergence happens within one time step. Indeed in this case, explicitly reset $E = 0$ as the excess energy is unused. Finally the third case is adjusted to account for the zero order hold between samples.

5 Simulation Results

The resulting behavior is illustrated by the following simulation, which is chosen to be fairly extreme in its large and sudden delay variation. Consider a standard wave-based teleoperation setup. The remote (slave) side is contacting a rigid object. On the local (master) side, the operator is inputing a constant force together with some damping. The total delay is constant at 0.1 second, distributed between the forward and return path.

At time $t = 0.5$ the return transmission (from slave to master) is blocked until time $t = 1.0$. For a half second, no signals are received on the master side. The slave side is unaffected. Thereafter the communications is unblocked and the entire data is instantly

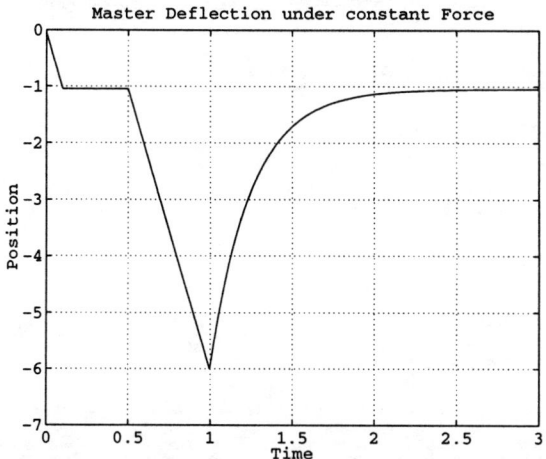

Figure 6: Master deflection under constant force and experiencing temporary transmission black out.

available. For the remainder of the time, the delay is again a constant 0.1 second.

Figure 6 shows the resulting master deflection. For the first 0.1 seconds, no signal is returning and the master is forced to deflect. After the regular delay time, the response from the remote site is available and the position holds constant, where the deflection is based on the equivalent system stiffness b/T. At $t = 0.5$, as the return signal disappears again, the position is once again forced to deflect to indicate the lack of knowledge. Finally, at $t = 1.0$ the situation returns to normal and the master can slowly reclaim its original position.

Note that the black-out time of 0.5 seconds directly translates to the time constant for the recovery. Shorter black-outs imply faster recoveries.

6 Concluding Remarks

The expansion and popularity of the Internet may allow force-reflecting teleoperation to achieve its full potential of letting users not only see, but physically interact with many distinct remote sites from the comfort of their local PC. The large fluctuations in time-delays for such network applications create unique and interesting problems, but we believe wave variables are well suited to these situations. With minimal computation and no advance knowledge of the transmission characteristics, wave variable filters can enable network based force-reflecting teleoperation, transparently during normal operation and degrading gracefully when the network is overloaded.

Acknowledgements

This report describes research done at the Nonlinear Systems Laboratory and the Artificial Intelligence Laboratory of the Massachusetts Institute of Technology. Support for the work has been provided in part by NASA/JPL.

References

Anderson, R. J. and Spong, M. W. (1989). Bilateral control of teleoperators with time delay. *IEEE Transactions on Automatic Control*, 34(5):494–501.

Niemeyer, G. (1996). *Using Wave Variables in Time Delayed Force Reflecting Teleoperation*. PhD thesis, MIT, Cambridge, MA.

Niemeyer, G. and Slotine, J.-J. E. (1991). Stable adaptive teleoperation. *IEEE Journal of Oceanographic Engineering*, 16(1):152–162.

Niemeyer, G. and Slotine, J.-J. E. (1997a). Designing force reflecting teleoperators with large time delays to appear as virtual tools. In *Proc. of 1997 IEEE Int. Conf. on Robotics and Automation*, pages 2212–2218.

Niemeyer, G. and Slotine, J.-J. E. (1997b). Telemanipulation with time delays. to be published in International Journal of Robotics Research.

Niemeyer, G. and Slotine, J.-J. E. (1997c). Using wave variables for system analysis and robot control. In *Proc. of 1997 IEEE Int. Conf. on Robotics and Automation*, pages 1619–1625.

Sheridan, T. B. (1989). Telerobotics. *Automatica*, 25(4):487–507.

Sheridan, T. B. (1992). *Telerobotics, Automation, and Human Supervisory Control*. The MIT Press, Cambridge, MA.

Slotine, J.-J. E. and Li, W. (1991). *Applied Nonlinear Control*. Prentice Hall, Englewood Cliffs, New Jersey.

Slotine, J.-J. E. and Niemeyer, G. (1993). Telerobotic system. U.S. Patent #5266875.

Gain-Scheduled Compensation for Time Delay of Bilateral Teleoperation Systems

Akihito Sano, Hideo Fujimoto and Masayuki Tanaka

Department of Mechanical Engineering
Nagoya Institute of Technology
Gokiso-cho, Showa-ku, Nagoya 466-8555, JAPAN

Abstract

In this study, the design of controllers for the master-slave system is discussed on the basis of the H_∞-optimal control theory. A few controllers that achieve given performance specifications for the free motion and the constrained motion are designed in order. Especially, by considering a variation of time delay, a time-varying controller incorporates to adjust to the current time delay is designed in the framework of the gain scheduling. This control strategy typically achieves higher performance in the face of large variations in operating conditions. The effectiveness of proposed method is confirmed by the experiments and the simulations.

1 Introduction

Recently, many new potential uses of advanced telerobotic systems have been explored, such as a micro-operation [1], a computer networked robotics and a teleoperation on a WWW browser. When the human being instructs, by operating the local master, the remote slave to execute the tasks, an instability phenomenon caused by a time delay exists in the communication channel between the master and the slave is one of the most serious problems [2].

In case of the operation with time delay on the order seconds, such as in undersea or Mars etc., Paul [3] proposed a teleprogramming theory in which the symbolic program commands are generated automatically based on the operator's interaction with a simulation of the remote site, and the slave attempts to execute each command immediately.

On the other hand, several papers have deal with the control issues [4]– [10]. Anderson and Spong [6] suggested a new communication architecture based on the scattering theory formalism to compensate for the time delay. In [7] a wave variable was utilized to characterize time delay systems and led to a new configuration for force reflecting teleoperation.

The H_∞ control theory and μ-analysis and synthesis are very effective for the design of bilateral controller which is required to the performance specifications in the frequency region [8], [9]. And, the feasibility of the given design specifications can be defined by checking whether or not a solution exists. Leung and Francis [10] modeled the time delay as a perturbation to the system and designed the system to be robust to such a perturbation by using μ-synthesis.

The business communication channels and not the exclusive channels undergo large variation of the time delay [11]. Therefore, it is impossible to achieve high performance over the entire operating range with a single robust LTI controller. Provided that the time delay is measured on-line, it is desirable to use controllers that incorporate such measurement. In this paper, the design of the time-varying H_∞ controller incorporates to adjust to the current time delay is discussed based on the framework of the gain scheduling. Futhermore, this study aims to develop the practical force reflecting teleoperators, and supposes the time delay on the order of several hundred milliseconds.

The paper is organized as follows. In Section 2, one-degree-of-freedom master and slave is introduced. And, the synthesis of the free motion controller is discussed. In Section 3, it is mentioned that the time delay is reflected as the perturbation to the system. Next, the design specifications such as the system should be stable for up to the prespecified amount of time delay are defined. In Section 4, the synthesis technique of the gain-scheduling is presented. The gain-scheduled H_∞ controller under the constrained motion is designed. In Section 5, to validate the design, the control experiments using the DD master-slave system and the computer simulations are confirmed.

2 Free Motion Control

2.1 DD Master-Slave System

In case of considering real applications, usage of a robot with multi-degree-of-freedom is very realistic. However, in a primary stage of the study, it is proper that a robot with single-degree-of-freedom is employed to avoid several difficult factors, such as an interference among each link. Figure 1 shows one-degree-of-freedom master and slave which has been developed to keep the analysis fairly simple.

The master and the slave are driven through the linear direct-drive motor, independently. This DD linear actuator has almost linear characteristic and is low friction (NSK Ltd., ML-YA series). A drive unit is used in the force control mode (analog input within ±10 V). A resolution of the position sensor (resolver) is 1 μm.

The weight of the master and the slave is 4.5 and 6.5 kg, respectively. The damping coefficient identified by experiments is 19.95 and 19.14 Ns/m. Two force sensors are able to measure an operation force and a reaction force from the slave environment. The stiffness of the environment can be changed by replacing several kinds of spring.

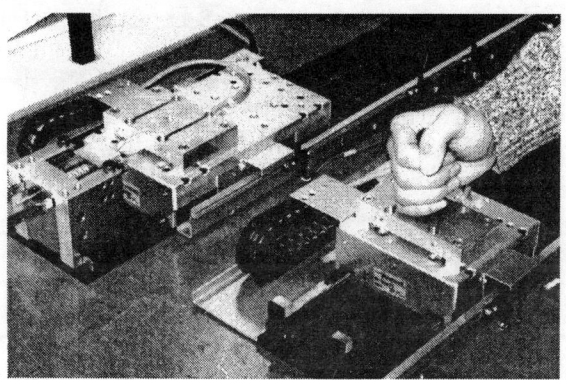

Fig.1: Master-slave system

2.2 Synthesis of Free Motion Controllers

Figure 2 illustrates a block diagram of the unilateral control system for free space motion. The transfer function of the master and the slave are represented by P_m and P_s. Thus

$$P_m = \frac{1}{M_m s + Q_m}, \quad P_s = \frac{1}{M_s s + Q_s} \quad (1)$$

In Fig.2, u_z and u_v are the applied forces on the master and the slave. v_m and v_s are the resultant velocities. In addition, f_m and f_d represent the force imposed on the master by the operator and the input disturbance, respectively.

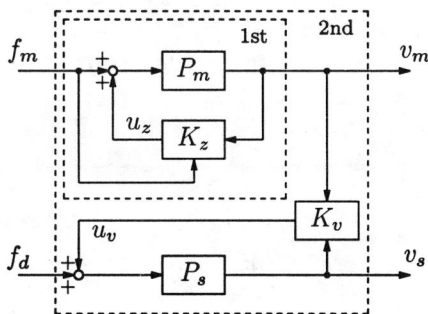

Fig.2: Unilateral control system

The controllers K_z and K_v are designed in this order to satisfy performance specifications for free motion which are summarized as follows [12]:

1. Minimizing the impedance error $v_m - Z_m^{-1} f_m$.

2. Tracking the slave velocity v_s to the master velocity v_m.

3. Prespecified saturation limits for the applied forces u_z and u_v.

By introducing the adjustable parameters α and β, the desired impedance is defined as follows:

$$Z_m = \alpha M_m s + \beta Q_m \quad (2)$$

In this paper, α and β are set to 1.0 and 2.0, respectively.

The H_∞ design procedure has been carried out using MATLAB. Figures 3 and 4 illustrate the gain characteristics of the controllers K_z of order 3 and K_v of order 4. The notations in the figure are given as follows:

$$u_z = \begin{bmatrix} K_{zfm} & K_{zvm} \end{bmatrix} \begin{bmatrix} f_m \\ v_m \end{bmatrix} \quad (3)$$

$$u_v = \begin{bmatrix} K_{vvm} & K_{vvs} \end{bmatrix} \begin{bmatrix} v_m \\ v_s \end{bmatrix} \quad (4)$$

Figure 5 shows the variation of position for the free motion when human operator exerts the force. In this

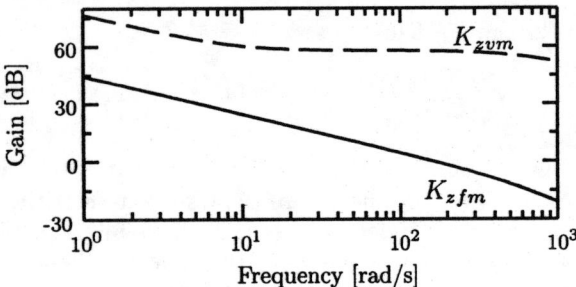

Fig.3: Gain characteristics of K_z

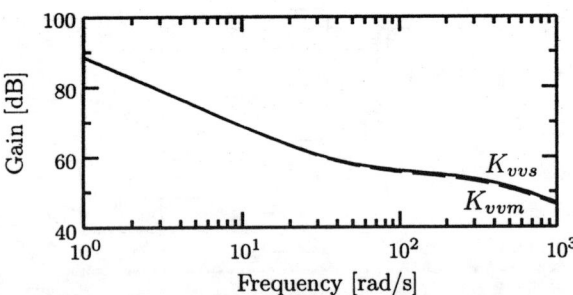

Fig.4: Gain characteristics of K_v

experiment, the time delay of 50 msec from the master to the slave was applied. As seen from this figure, the slave velocity $v_s(t)$ tracks the reference signal $v_{sd}(t)$ (i.e., $v_m(t-0.05)$). Such a time delay under the unilateral control does not affect the stability of the system.

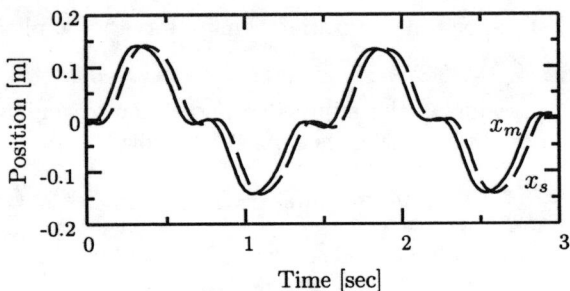

Fig.5: Variation of position (free motion)

3 Time Delay Compensation

3.1 Robust Stability for Time Delay

Figure 6 illustrates a block diagram of the bilateral control system with a communication channel. The blocks e^{-sT_1} and e^{-sT_2} represent the time delay from the slave to the master and vice-versa, respectively.

To simplify, the following blocks are introduced:

$$G_m = \frac{(K_{zfm}+I)P_m}{I-P_mK_{zvm}} \quad (5)$$

$$G_s = \frac{S_eP_sK_{vvm}}{(K_{vvs}-S_e)P_s-I} \quad (6)$$

where G_m represents the master that impedance is regulated by K_z. G_s represents the slave which is in contact with the (assumed known) external environment S_e (i.e., $S_e = \frac{K_e}{s}$, $K_e = 4.9 \times 10^3$) and whose velocity is controlled by K_v.

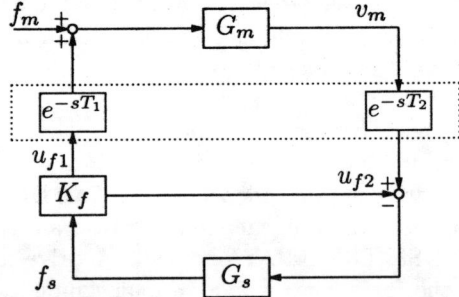

Fig.6: Bilateral control system with time delay

The constrained motion controller K_f will not affect the operation of the system designed for operation in free space [10]. Namely, for the case $f_s = 0$ (free motion), the output of K_f is also zero. This controller is designed to compensate the instability caused by the time delay exists in the communication channel between the master and the slave. Since it can be considered that the time delay reflects as the perturbation to the system, the compensation of time delay results in the robust stability problem of the system with the perturbation.

Leung et al. [10] have reconfigured the system so that the time delay is reflected as a perturbation Δ_T to the system. Let

$$\Delta_T(T) := e^{-sT} - 1 \quad (7)$$

In order to lump the delays into one block e^{-sT}, the left delay block e^{-sT_1} of Fig.6 is moved around the loop to the forward path of the loop. The block e^{-sT} represents the time delay of $T(=T_1+T_2)$ seconds.

Notice that the H_∞ norm $\|\Delta_T\|_\infty$ equals 2 for every $T > 0$, so it would be compensating for all perturbations of norm ≤ 2, not just the time delay. To reduce this conservatism, the perturbation Δ_T can be moved to surround the master system, G_m, Δ_m is defined as $\Delta_T G_m$. Since G_m is strictly proper by design in Section 2, there is a bandpass filter W_c such that

$$\|W_c^{-1}\Delta_m(T)\|_\infty < 1 \quad (8)$$

Finally, the system of Fig.6 can be redrawn as in Fig.7, since Δ_m can be expressed as $W_c W_c^{-1} \Delta_m$. Now, the perturbation is $W_c^{-1} \Delta_m$, and it has nom < 1. $W_c^{-1} \Delta_m$ has been replaced by Δ_s for the following discussion.

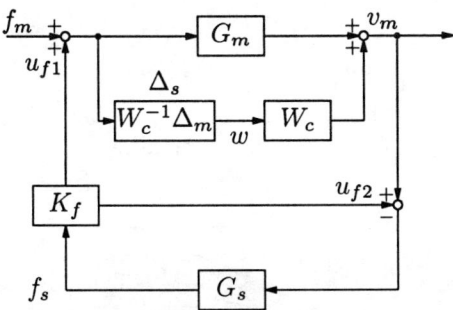

Fig.7: Perturbation model

Now, we check that the closed-loop system is stable for all time delays of $T \leq T_0$. T_0 is a prespecified time delay. Thus, the following condition is required

$$||\Delta_s(T_0)Q||_\infty < 1 \quad (9)$$

for stability of the system. Q is denoted as the transfer function from w to $f_m + u_{f1}$ as shown in Fig.7.

3.2 Design Specifications

The design specifications are taken to be as follows:

1. Prespecified limits for the control inputs u_{f1} and u_{f2}.

2. Tracking the slave velocity v_s to the master velocity v_m.

3. Minimizing the model-matching error $f_s - G_M f_m$.

4. Stabilizing for up to the prespecified amount of time delay.

where the model transfer function G_M mapping from f_m to f_s is chosen as follows:

$$G_M = \frac{f_s}{f_m} = -\frac{K_e}{\alpha M_m s^2 + \beta Q_m s + K_e} \quad (10)$$

Thus, for this design

$$z = \begin{bmatrix} W_m u_{f1} \\ W_s u_{f2} \\ W_v(v_s - v_m) \\ W_f(f_s - G_M f_m) \\ f_m + u_{f1} \end{bmatrix}, \quad w = \begin{bmatrix} f_m \\ W_c w \end{bmatrix} \quad (11)$$

By appropriate design of weighting functions (W_m, W_s,...), it is possible to design the controller that makes the norm from w to z less than or equal to 1.

When the time delay undergo large variation, it is impossible to achieve high performance over the entire operating range with a single robust LTI controller. Provided that the time delay is measured on-line, it is then desirable to use controllers that incorporate such measurement. For instance, the time delay T can be calculated on-line by the difference between the receiving time of the data and the time stamp on it. The time data which is stamped on the data u_{f1} at the slave site is copied into the data v_m at the master site, and is sent back immediately.

The design of the time-varying controller $K_f(T)$ incorporates to adjust to the current time delay $T(t)$ can be applicable to the framework of the gain scheduling which will be mentioned in next section.

4 Gain-Scheduled Compensation for Time Delay

4.1 Gain-Scheduling

Gain scheduling is a widely used technique for controlling certain classes of nonlinear or linear time-varying systems [13], [14]. The synthesis technique discussed below is applicable to time-varying and/or nonlinear systems whose linearized dynamics are reasonably well approximated by affine parameter-dependent plants (models) $G(.,p)$ with equations [15].

$$G(.,p) \begin{cases} \dot{x} = A(p)x + B_1(p)w + B_2 u \\ z = C_1(p)x + D_{11}(p)w + D_{12} u \\ y = C_2 x + D_{21} w + D_{22} u \end{cases} \quad (12)$$

where

$$p(t) = (p_1(t), \cdots, p_n(t)), \quad \underline{p_i} \leq p_i(t) \leq \bar{p_i}$$

is a time-varying vector of parameters and $A(\cdot)$, $B_1(\cdot)$, $C_1(\cdot)$, $D_{11}(\cdot)$ are affine functions of $p(t)$.

This control strategy typically achieves higher performance in the face of large variations in operating conditions. For this class of controllers, consider the following H_∞-like synthesis problem relative to the interconnection of Fig.8.

Design a gain-scheduled controller $K(.,p)$ satisfying the vertex property and such that

1. the closed-loop system is stable for all admissible parameter trajectories $p(t)$

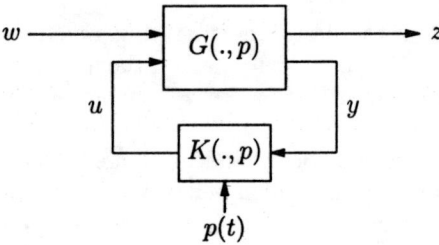

Fig.8: Gain–scheduled H_∞ problem

2. the worst-case closed-loop RMS gain from w to z does not exceed some level $\gamma > 0$

Furthermore, this synthesis problem can be reduced to the following LMI problem.

Find two symmetric matrices R and S such that

$$\left(\begin{array}{c|c}\mathcal{N}_R & 0 \\ \hline 0 & I\end{array}\right)^T \left(\begin{array}{cc|c}A_iR + RA_i^T & RC_{1i}^T & B_{1i} \\ C_{1i}R & -\gamma I & D_{11i} \\ \hline B_{1i}^T & D_{11i}^T & -\gamma I\end{array}\right) \left(\begin{array}{c|c}\mathcal{N}_R & 0 \\ \hline 0 & I\end{array}\right) < 0 \quad (13)$$

$$\left(\begin{array}{c|c}\mathcal{N}_S & 0 \\ \hline 0 & I\end{array}\right)^T \left(\begin{array}{cc|c}A_i^T S + SA_i & SB_{1i} & C_{1i}^T \\ B_{1i}^T S & -\gamma I & D_{11i}^T \\ \hline C_{1i} & D_{11i} & -\gamma I\end{array}\right) \left(\begin{array}{c|c}\mathcal{N}_S & 0 \\ \hline 0 & I\end{array}\right) < 0 \quad (14)$$

$$i = 1, \cdots, 2^n$$

$$\begin{pmatrix} R & I \\ I & S \end{pmatrix} \geq 0 \quad (15)$$

where

$$\begin{pmatrix} A_i & B_{1i} \\ C_{1i} & D_{11i} \end{pmatrix} := \begin{pmatrix} A(\Pi_i) & B_1(\Pi_i) \\ C_1(\Pi_i) & D_{11}(\Pi_i) \end{pmatrix} \quad (16)$$

and $\mathcal{N}_R$ and $\mathcal{N}_S$ are bases of the null spaces of (B_2^T, D_{12}^T) and (C_2, D_{21}), respectively.

If the parameter vector $p(t)$ takes values in a box of $\mathbf{R}^n$ (n is the number of parameters) with corners $\{\Pi_i\}_{i=1}^N$ ($N = 2^n$), given the convex decomposition $p(t) = \sum_{i=1}^N \lambda_i \Pi_i$ ($\lambda_i \geq 0$, $\sum_{i=1}^N \lambda_i = 1$) of the current parameter value $p(t)$, the values of $A_K(p), B_K(p), \cdots$ are derived from the values $A_K(\Pi_i), B_K(\Pi_i), \cdots$ at the corners of the parameter box by

$$\begin{pmatrix} A_K(p) & B_K(p) \\ C_K(p) & D_K(p) \end{pmatrix} = \sum_{i=1}^N \lambda_i \begin{pmatrix} A_K(\Pi_i) & B_K(\Pi_i) \\ C_K(\Pi_i) & D_K(\Pi_i) \end{pmatrix} \quad (17)$$

The controller state-space matrices at the operating point $p(t)$ are obtained by convex interpolation of the LTI vertex controllers.

4.2 Synthesis of GS-H_∞ Controller

The gain-scheduled H_∞ controller is designed as follows. For example, in this study, it is assumed that the time delay range in

$$0.1 \leq T(t) \leq 0.2 \quad (18)$$

The calculated perturbation $\Delta_m(T)$ is illustrated in Fig.9. Note that the apparent perturbation becomes small as the delay T becomes small.

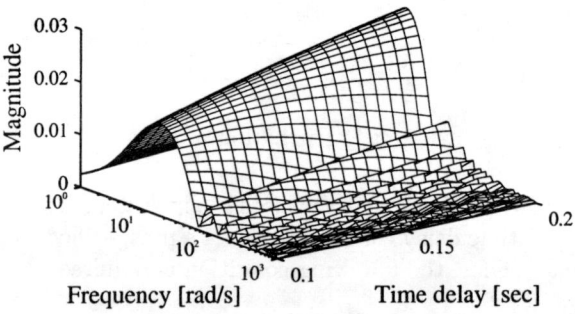

Fig.9: Perturbation $\Delta_m(T)$

Therefore, an adequate time-varying parameter p as a function of T can be included in the weight W_c on the perturbation Δ_m as follows:

$$W_c(p) = \frac{2.6s}{(4.5s + 39)(s + p(t))} \quad (19)$$

$$p(t) = \frac{17}{24}\frac{\pi}{T(t)} \quad (20)$$

where the range of p is given from Eq.(20) as follows:

$$11.1 \leq p(t) \leq 22.2 \quad (21)$$

Figure 10 shows $\Delta_m(0.1)$ and $W_c(22.2), W_c(11.1)$. In fact, it may be shown that this choice of W_c is clear the relation of Eq.(8).

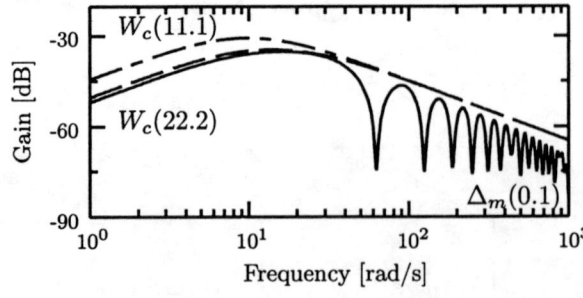

Fig.10: $\Delta_m(0.1)$, $W_c(22.2)$ and $W_c(11.1)$

Furthermore, the parameter p is also included in the weight W_s on the control input u_{f2} as follows:

$$W_s(p) = \frac{(4p(t) - 30) \times 10^3}{s + 10^3} \quad (22)$$

Note that the control input u_{f2} can be suppressed as the delay T becomes small. By appropriate design of W_s, we make the time delay stability margin be close upon 1 over the prespecified time delay range without being too conservative.

In above case, $A(p), C_1(p)$ of the generalized plant $G(.,p)$ are affine functions of time-varying parameter $p(t)$.

Other weight functions were chosen as follows:

$$W_m = \frac{10^4 s}{s + 10^6}, \quad W_v = \frac{\rho_v}{s + 10^{-4}}, \quad W_f = \frac{\rho_f}{s + 10^{-4}} \quad (23)$$

where ρ_v and ρ_f are free parameters. The weight W_v, W_f are set as a low-pass filter to accomplish the performane specifications. The parameters ρ_v, ρ_f are varied while checking whether or not a solution exists. And the W_m is set as a high-pass filter in order to suppress the gain of controller K_{f1}.

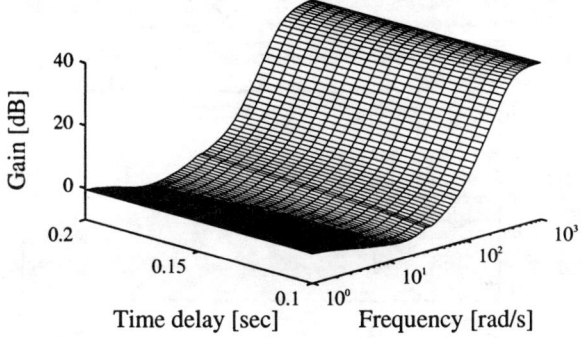

Fig.11: Gain characteristics of $K_{f1}(.,T)$

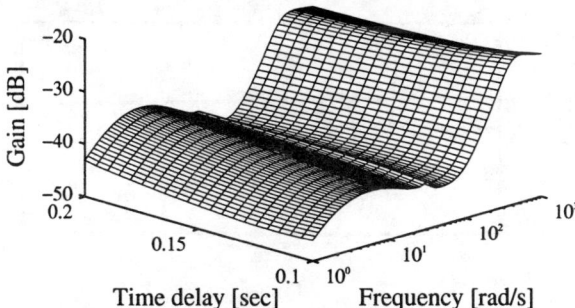

Fig.12: Gain characteristics of $K_{f2}(.,T)$

The H_∞ design technique may be performed utilizing the LMI Control Toolbox [15] in MATLAB. For given generalized plant $G(.,p)$, the controller $K_f(.,T)$ has been designed by using the gain-scheduling procedure.

Figures 11 and 12 illustrate the gain characteristics of controller $K_f(.,T)$ of order 17. The magnitude plot of $\Delta_s(T)Q$ is illustrated in Fig.13. Note, from this figure, that $\|\Delta_s(T)Q\|_\infty$ is within the specification and is close upon 1.

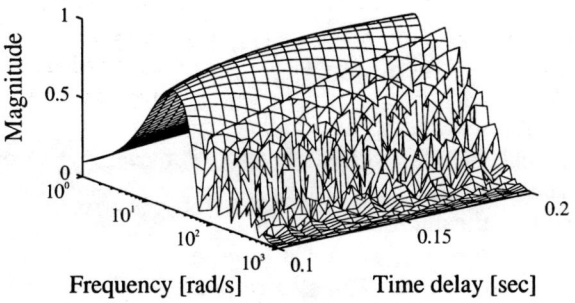

Fig.13: Time delay stability margin $|\Delta_s(T)Q|$

5 Experiments and Simulations

To validate the design, the control experiments using the DD master-slave system have been confirmed. This experiments were performed for the constrained motion where the slave is in contact with the virtual environment S_e. We apply a step input, $f_m(s) = 10/s$, to the system after 0.5 seconds from beginning of the experiment. The time delay is fixed at 100 msec.

The step response is illustrated in Figs.14 and 15. Figures 14 and 15 illustrate the variation of the velocity error and that of the force error, respectively. The experimental results are denoted as solid line, and for the reference, the simulation results are plotted as broken line.

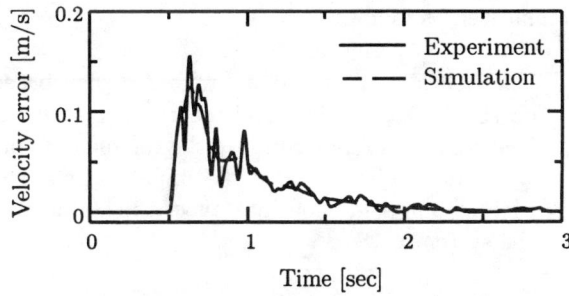

Fig.14: Step response ($v_m - v_s$, $T = 0.1$)

As seen from these figures, the closed-loop step response is good since the settling time is less than about

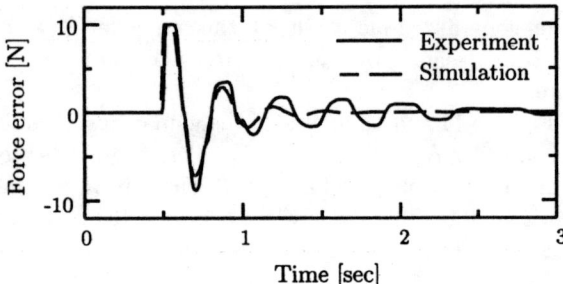

Fig.15: Step response ($f_m + f_s$, $T = 0.1$)

$2 \sim 2.5$ seconds. The stability was achieved for all time delays of less than 100 msec. Moreover, the error variation of the experiment coincides with that of the simulation comparatively.

Next, the time delay varies from 100 to 200 msec after 0.75 seconds. The step response is illustrated in Figs.16 and 17. The experimental result by usage of the gain-scheduled controller $K_f(T)$ which is designed in Section 4 are denoted in Fig.(a), and for the reference, the result by usage of the controller $K_f(0.2)$ which is fixed at the delay T of 200 msec are plotted in Fig.(b).

Note that although time delay undergo the quick variation, the system maintained stability. As seen from Figs.(a) and (b), it is possible to achieve superior performance properties over the prespecified range of time delay with gain-scheduled H_∞ controller.

6 Conclusions

In this paper, the gain-scheduled compensation for the time delay of the bilateral teleoperation systems has been proposed. The compensation of time delay results in the robust stability problem of the system with the perturbation. The results of this study are summarized as follows:

1. The free motion controllers were designed based on the H_∞-optimal control theory. The slave velocity has tracked accurately the reference signal with time delay. Under the unilateral control, such a time delay does not affect the stability of the system.

2. The adequate time-varying parameter p as a function of time delay T was included in the weight functions on the perturbation and on the control input. As the result, the design of the time-varying controller incorporates to adjust to the

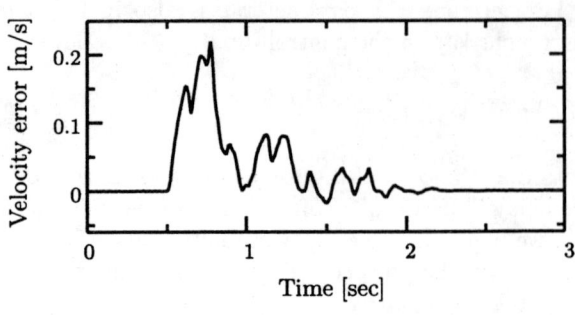

(a) $K_f(T)$

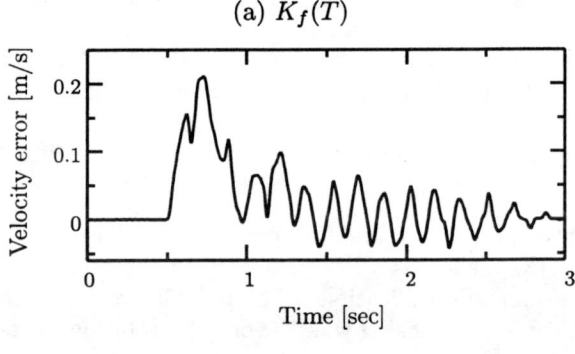

(b) $K_f(0.2)$

Fig.16: Step response ($v_m - v_s$, $T = 0.1 \to 0.2$)

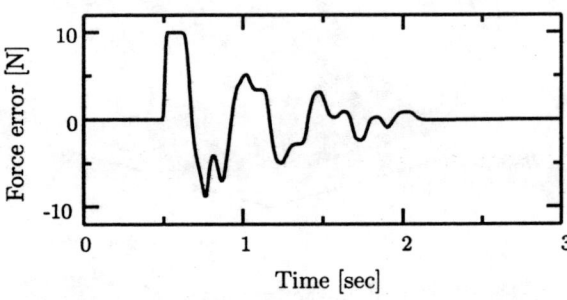

(a) $K_f(T)$

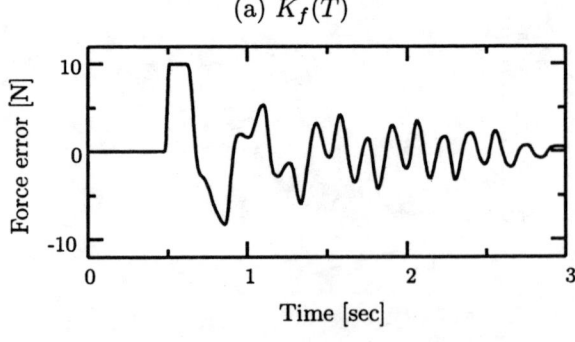

(b) $K_f(0.2)$

Fig.17: Step response ($f_m + f_s$, $T = 0.1 \to 0.2$)

current time delay T could be realized in the framework of the gain scheduling.

3. The closed-loop step response was good since the settling time is less than about 2 ~ 2.5 seconds. The stability was achieved for all time delays of less than 100 msec. Moreover, the error variation of the experiment was sufficiently close to that of the simulation.

4. Although time delay undergo the quick variation from 100 to 200 msec, the system maintained stability. Furthermore, it was possible to achieve superior performance properties over the prespecified range of time delay with gain-scheduled H_∞ controller.

Our future works are the discussion about the upper bound on T, and comparison between the proposed method and others. This work was supported in part by The Nitto Foundation.

References

[1] A. Sano, H. Fujimoto, and K. Kodani: "Development of Combined Master-Slave Tool for Medical Applications," *Proc. of the Japan/USA Symposium on Flexible Automation*, Vol.1, pp.245–250, 1996.

[2] S. Tachi and T. Sakaki: "Impedance Controlled Master Slave Manipulation System – Part I: Basic Concept and Application to the System with Time Delay (in Japanese)," *Journal of the Robotics Society of Japan*, Vol.8, No.3, pp.241–252, 1990.

[3] R.P. Paul, C.P. Sayers, and M.R. Stein: "The Theory of Teleprogramming," *Journal of The Robotics Society of Japan*, Vol.11, No.6, pp.782–787, 1993.

[4] Y. Yokokohji and T. Yoshikawa: "Bilateral Control of Master-Slave Manipulators for Ideal Kinesthetic Coupling – Formulation and Experiment," *IEEE Trans. on Robotics and Automation*, Vol.10, No.5, pp.605–619, 1994.

[5] T. Yoshikawa and J. Ueda: "Analysis and Control of Master-Slave Systems with Time Delay," *Proc. of the 1996 IEEE/RSJ Int. Conf. on Intelligent Robots and Systems*, No.3, pp.1366–1373, 1996.

[6] R.J. Anderson, and M.W. Spong: "Bilateral Control of Teleoperators with Time Delay," *IEEE Transactions on Automatic Control*, Vol.34, No.5, pp.494–501, 1989.

[7] G. Niemeyer, and J.J.E. Slotine: "Stable Adaptive Teleoperation," *IEEE Journal of Oceanic Engineering*, Vol.16, No.1, pp.152–162, 1991.

[8] H. Kazerooni, T.-I. Tsay, and K. Hollerbach: "A Controller Design Framework for Telerobotic Systems," *IEEE Transactions on Control Systems Technology*, Vo.1, No.1, pp.50–62, 1993.

[9] J. Yan and S.E. Salcudean: "Teleoperation Controller Design Using H_∞–Optimization with Application to Motion–Scaling," *IEEE Transactions on Control Systems Technology*, Vol.4, No.3, pp.244–258, 1996.

[10] G.M.H. Leung, B.A. Francis, and J. Apkarian: "Bilateral Controller for Teleoperators with Time Delay via μ–Synthesis," *IEEE Transactions on Robotics and Automation*, Vol.11, No.1, pp.105–116, 1995.

[11] K. Kosuge, H. Murayama, and K. Takeo: "Bilateral Feedback Control of Telemanipulators via Computer Network," *Proc. of the 1996 IEEE/RSJ Int. Conf. on Intelligent Robots and Systems*, No.3, pp.1380–1385, 1996.

[12] M. Sakaguchi, A. Sano, and J. Furusho: "Impedance Control of Robot Arms Based on H_∞ Control Theory (in Japanese)," *Trans. of the Japan Society of Mechanical Engineers*, Vol.62, No.596, C, pp.1503–1509, 1996.

[13] P. Apkarian, J.-M. Biannic, and P. Gahinet: "Self-Scheduled H_∞ Control of Missile via Linear Matrix Inequalities," *Journal of Guidance, Control, and Dynamics*, Vol.18, No.3, pp.532–538, 1995.

[14] H. Fujimoto, A. Sano, and T. Hanai: "Variable Impedance Control of Master Arm Based on Gain Scheduling (in Japanese)," *Trans. of the Japan Society of Mechanical Engineers*, Vol.63, No.609, C, pp.1632–1639, 1997.

[15] P. Gahinet, A. Nemirovski, A.J. Laub, and M. Chilali: LMI Control Toolbox, The MathWorks, Inc, 1995.

Decoupling Control Based on Virtual Mechanisms for Telemanipulation

Alain MICAELLI, Catherine BIDARD and Claude ANDRIOT

Sevice de Téléopération et de Robotique (STR), Commissariat à l'Energie Atomique (CEA)
BP 6, route de Panorama, 92265 Fontenay-aux-Roses Cedex, France

Abstract : Tuning a telemanipulation system is still a real challenge. The task model is often unknown, the system must remain stable for a wide range of tasks, configurations, for different modes which share in a different way all the system degrees of freedom (d.o.f). Thinking of numerous controllers which come up with all these different situations is quite unrealistic. In fact, a compromise between the number of control parameters and the system performances has to be found. Here, we propose a method for the design of a decoupling controller allowing different tunings along teleoperated d.o.f and locally controlled ones.

Keywords : telemanipulation, control design, dynamics decoupling, virtual mechanisms

1. INTRODUCTION

During the last ten years, CEA/STR has been involved in numerous telemanipulation projects devoted to nuclear applications (maintenance, intervention, dismantling,...). In order to help operators perform their tasks more efficiently, new assistance functions have been developed. How to share the control between an operator and the telemanipulation system has been one of the utmost problems to be solved.

This may be first seen as a task specification issue which has been formerly addressed by Mason [13] ; recent interesting works like [4] are based on his formulation. Lipkin [12], Doty [5], Selig [18], Featherstone [7], proposed some approaches in order to build frame and unit invariant specifications.

At this stage, a controller which is able to satisfy the tasks specifications must be designed. Following the initial formulation of Raibert and Craig [16], Khatib [11], for hybrid force/position control based on selection matrices, some improvements dealing with kinematic instabilities [8], non-invariance [7], or quite different control architectures [15], [17], have been proposed.

Initially devoted to robot control, these methods can be extended to teleoperation by considering that an operator may replace programmed commands along specified force or motion directions [3], [9].

Within this frame, CEA/STR recently developed and implemented a new specification and control approach based on *virtual mechanisms*. All the benefits of such an approach are well shown in [9], [2], [10]. As this method relies on physical considerations, it naturally leads to invariance, passivity (if desired), and a simplified controller design. Moreover, this approach can easily deal with very complex and non linear tasks.

In the frame of teleoperation, passivity relative to environment is of crucial importance, especially because environment is more often partially known and related tasks can only be approximately defined. This is the main reason why our approach is preferred to others mentioned before.

With respect to CEA previous work, this paper must be considered as an extension and presents a slight different organisation of the control in order to make the tuning easier and to improve the system behaviour. The main idea consists in decoupling the telemanipulation system in different subspaces where its dynamics and its behaviour are different, by using mechanical analogies.

The next section sketches the virtual mechanism approach. Then, our work objectives are stated and a first solution for a single manipulator is given. After some considerations about passivity and singularities, the case of a complete telemanipulation system is considered.

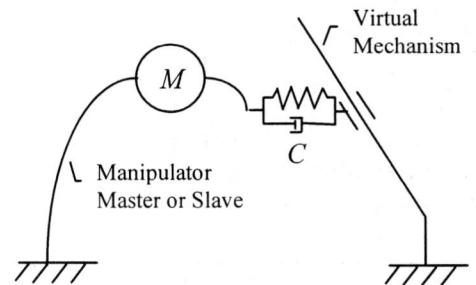

Figure 1 Virtual mechanism and robot controller C

2. VIRTUAL MECHANISM APPROACH

Basically, its main principle relies on a mechanical analogy which is presented on Figure 1.

The manipulator controller C must be considered as a passive six d.o.f controller in operational space which links the manipulator end effector to the virtual mechanism. Represented by a spring and a damper, this controller may be more involved, while still being passive.

Moreover, it will be supposed, for the following analysis, that Master and Slave sub-systems have been made locally backdriveable enough by an internal force-feedback loop.

The extension to a telemanipulation system can be explained, in a very simplified way, as the replication of Figure 1 for the other sub-system (Slave or Master). The overall system is then represented on Figure 2.

From Figure 2, it can be seen that controllers C_m and C_s, respectively designed for the Master and for the Slave sub-systems, have simultaneously to deal with virtually constrained and unconstrained directions along which the system dynamics are quite different. This may lead to a certain loss of performances which can be shown on the following simulated example.

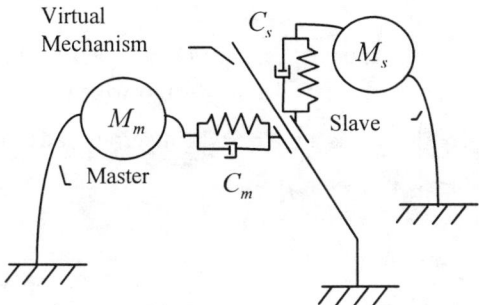

Figure 2 System and Virtual Mechanism

Example

Consider a simple teleoperation system consisting of :
- a 2 d.o.f. Cartesian Master arm, characterised by an isotropic inertia ;
- a 2 d.o.f. Cartesian Slave manipulator, also characterised by an isotropic inertia which is five times greater than the Master inertia ;
- a virtual mechanism realised by a free translation motion along one of the system Cartesian directions.

In a continuous ideal framework, controllers parameters would not be limited, but in real conditions, because of noise, filtering, sampling period, etc., a limited bandwidth must be considered. For a 30 rad/s bandwidth, if Master and Slave controllers are both position PD type (Proportional - Derivative) and both separately tuned for a 0.7 damping factor, then, along the unconstrained direction, the system behaviour can be characterised by the following plots on figure 3 and 4.

Figure 3 presents a bode plot which corresponds to the transfer function between Master and Slave velocities, when the Slave manipulator is kept free along the unconstrained direction. A substantial overshoot can be remarked at about 10 rad/s.

Figure 4 draws the Slave force response to a Master force step when the Slave manipulator is blocked along the virtual unconstrained direction. Again, a significant force overshoot can be observed.

With the control structure sketched on figure 2, the only way to solve the problem would be to make a compromise between virtually constrained and unconstrained directions. The following sections propose a method in order to avoid this compromise.

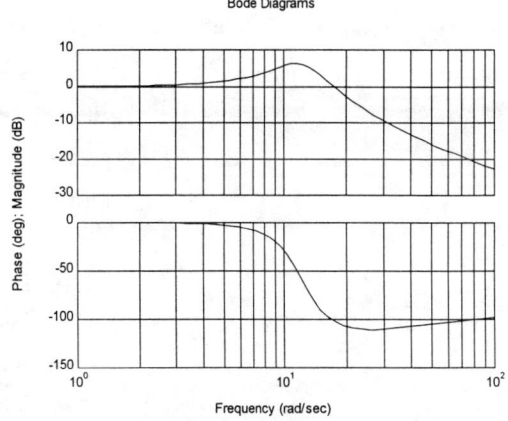

Figure 3 Slave velocity/Master velocity Transfer

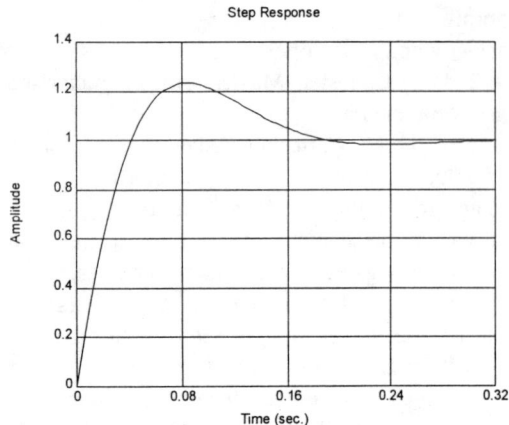

Figure 4 Slave force/Mater force step response

3. ONE MANIPULATOR CASE

3.1 PROBLEM STATEMENT

In order to improve the system behaviour along virtually constrained and unconstrained directions, we split each controller C_m and C_s into two new controllers as described by Figure 5 for the one manipulator case. In this way, optimising the system behaviour in different directions becomes feasible.

On this figure, a first virtual mechanism called $\mathcal{V}$ is related to the robot task (see[9]). Its end frame is defined by $X_v(q_v)$ relative to a fixed frame $X_v(0)$ (the base frame of virtual mechanism $\mathcal{V}$); q_v represents the general coordinates vector of this virtual mechanism. $X_v(q_v)$ can be expressed by three

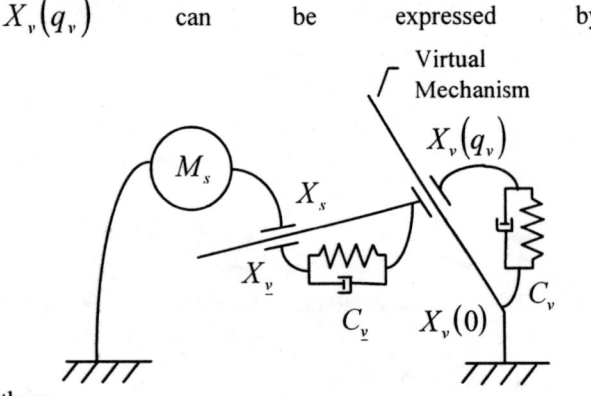

Figure 5 Building a decoupling controller

position coordinates in the Euclidian Space $E(3)$ and a rotation matrix in $SO(3)$. The robot behaviour along this virtual mechanism (consider that manipulator end frame satisfies $X_s = X_v$), is determined by controller C_v, which can be viewed as a six d.o.f. passive controller constrained by virtual mechanism $\mathcal{V}$. Application of C_v results in wrench $F_v(q_v)$ which only depends on q_v. $F_v(q_v)$ is expressed relatively to frame $X_v(0)$.

Now, rigidly attached to the end frame of $\mathcal{V}$, a second virtual mechanism, called $\underline{\mathcal{V}}$, completes $\mathcal{V}$, in order that $\mathcal{V}$ together with $\underline{\mathcal{V}}$ form a six d.o.f. virtual mechanism which is supposed to be non singular. When X_v is fixed, the manipulator behaviour is determined by controller $C_{\underline{v}}$ which creates a wrench between $X_{\underline{v}}(q_v, q_{\underline{v}})$ and $X_{\underline{v}}(q_v) = X_{\underline{v}}(q_v, 0)$. $C_{\underline{v}}$ is built so that it only depends on $q_{\underline{v}}$, generalised coordinates vector of $\underline{\mathcal{V}}$. In a way similar to C_v, $C_{\underline{v}}$ results in wrench $F_{\underline{v}}(q_{\underline{v}})$, which is expressed relatively to frame $X_{\underline{v}}(q_v, 0)$.

The velocity wrist V_s, associated to X_s and expressed relatively to the base frame of $\mathcal{V}$ is given by the following equation :

$$V_s = J_v(q_v)\dot{q}_v + Ad(q_v)J_{\underline{v}}(q_{\underline{v}})\dot{q}_{\underline{v}} \quad (1)$$

where Jacobian matrices J_v and $J_{\underline{v}}$ are related to mechanisms $\mathcal{V}$ and $\underline{\mathcal{V}}$ respectively. Ad, called adjoint transformation, is an inversible frame transformation for twists and is defined as follows (see [14], for instance) :

$$Ad(q_v) = \begin{bmatrix} R_v & \hat{p}_v R_v \\ 0 & R_v \end{bmatrix}$$

where R_v and p_v respectively specify the rotation matrix and the translation vector which are related to $X_v(q_v)$; $\hat{p}_v$ is a matrix which corresponds to the cross product with vector p_v :

$$\hat{p}_v(\cdot) = p_v \times (\cdot).$$

Here, Ad transforms twists from frame $X_v(q_v)$ to frame $X_v(0)$.

3.2 DECOUPLING

A simplified expression for the manipulator dynamics (relative to frame $X_v(0)$) is given by :

$$M_s \dot{V}_s = F_s \qquad (2)$$

where M_s is the manipulator inertia and may depend on its configuration.

With respect to $q = \begin{bmatrix} q_v \\ q_{\underline{v}} \end{bmatrix}$ and neglecting all quadratic terms in the velocity, equation (2) can be transformed into :

$$M_s \begin{bmatrix} J_v(q_v) & Ad(q_v)J_{\underline{v}}(q_{\underline{v}}) \end{bmatrix} \begin{bmatrix} \ddot{q}_v \\ \ddot{q}_{\underline{v}} \end{bmatrix} = F_s \qquad (3)$$

Now we consider the complete virtual mechanism $\mathcal{W}$ formed with $\mathcal{V}$ and $\underline{\mathcal{V}}$. If F_v and $F_{\underline{v}}$ were created by joint forces defined by :

$$\Gamma = \begin{bmatrix} \Gamma_v \\ \Gamma_{\underline{v}} \end{bmatrix}$$

in virtual mechanism $\mathcal{W}$, then, applying power conservation law, we would get :

$$\Gamma = \begin{bmatrix} J_v^T(q_v) \\ J_{\underline{v}}^T(q_{\underline{v}})Ad^T(q_v) \end{bmatrix} F_s \qquad (4)$$

Mixing equations (3) and (4), the following dynamic equation is obtained :

$$\begin{bmatrix} J_v^T M_s J_v & J_v^T M_s Ad J_{\underline{v}} \\ J_{\underline{v}}^T Ad^T M_s J_v & J_{\underline{v}}^T Ad^T M_s Ad J_{\underline{v}} \end{bmatrix} \begin{bmatrix} \ddot{q}_v \\ \ddot{q}_{\underline{v}} \end{bmatrix} = \begin{bmatrix} J_v^T \\ J_{\underline{v}}^T Ad^T \end{bmatrix} F_s \qquad (5)$$

A first decoupling condition is given by :

$$J_{\underline{v}}^T \left(Ad^T M_s J_v \right) = 0 \qquad (6)$$

Decoupling also needs the second member of equation (5) be written in this way :

$$\Gamma = \begin{bmatrix} \Gamma_v(q_v) \\ \Gamma_{\underline{v}}(q_{\underline{v}}) \end{bmatrix} \qquad (7)$$

(In equation (7), terms in parentheses may be completed with their derivatives, integrals,...).
In fact, this second decoupling condition is already satisfied because, as previously defined :
- F_v and C_v only depend on q_v, and,
- $F_{\underline{v}}$ and $C_{\underline{v}}$ only depend on $q_{\underline{v}}$.

Using the definition of F_v and $F_{\underline{v}}$, and applying again power conservation law, we get :

$$\Gamma = \begin{bmatrix} J_v^T F_v \\ J_{\underline{v}}^T F_{\underline{v}} \end{bmatrix} \qquad (8)$$

In equation (8), upper and lower parts of the second member only depend on q_v and $q_{\underline{v}}$ respectively.

3.3 CONTROL

Usually, equation (6) does not lead to a feasible virtual mechanism $\underline{\mathcal{V}}$. This fact relies upon integrability conditions (see [14]). However, if we suppose that the position errors are not to big, we may choose a local solution which will depend on q_v and the manipulator position.

The next step consists in determining $\mathcal{V}$ motion, or in other words, the dynamic equation for q_v estimation. This can be solved by using the manipulator position and velocity. Applying relation (6) to kinematic equation (1) leads to :

$$\dot{q}_v = \left(J_v^T M_s J_v \right)^{-1} J_v^T V_s \qquad (9)$$

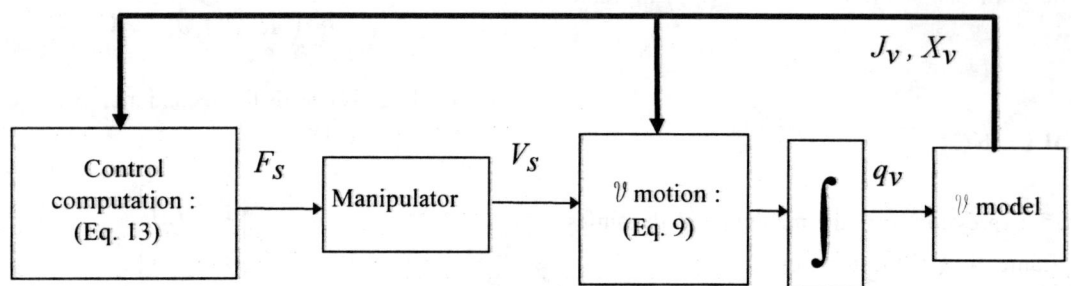

Figure 6 Implementation

The control itself, F_s, is obtained by the combining equations (4) and (8), i.e.:

$$F_s = \left[M_s J_v \left(J_v^T M_s J_v\right)^{-1} J_v^T\right] F_v +$$
$$\left[M_s Ad J_{\underline{v}} \left(J_{\underline{v}}^T Ad^T M_s Ad J_{\underline{v}}\right)^{-1} J_{\underline{v}}^T\right] F_{\underline{v}} \quad (10)$$
$$= \Pi F_v + \underline{\Pi} Ad^{-T} F_{\underline{v}}$$

The following relations can be verified:

$$\begin{cases} R(\Pi) = Ker(\underline{\Pi}) = Span(M_s J_v) \\ R(\underline{\Pi}) = Ker(\Pi) = Span(M_s Ad J_{\underline{v}}) \\ \Pi^2 = \Pi \\ \underline{\Pi}^2 = \underline{\Pi} \end{cases} \quad (11)$$

From these relations, it results that Π and $\underline{\Pi}$ are projectors and that:

$$\underline{\Pi} = Id - \Pi \quad (12)$$

Then, equation (10) can be simplified into:

$$F_s = \Pi F_v + (Id - \Pi) Ad^{-T} F_{\underline{v}} \quad (13)$$

Equation (13) means that there is no need at all for an explicit value of $J_{\underline{v}}$, and finally a rather simple expression is given for F_s.

For an implementation, a simplified block-diagram of the system and its controller is presented on Figure 6.

3.4 PASSIVITY

Proposition:
According to there definitions, if F_v and $F_{\underline{v}}$ result from passive controllers C_v and $C_{\underline{v}}$, then the overall controller is passive as well.

In order to prove this proposition, the following expression:

$$V_s^T F_s$$

is developed.

Applying equation (13), we get:

$$V_s^T F_s = V_s^T \Pi F_v + V_s^T (Id - \Pi) Ad^{-T} F_{\underline{v}} \quad (14)$$

From V_s definition (see equation (1)), and applying equation (9), then we have:

$$\Pi^T V_s = J_v \dot{q}_v = V_v \quad (15)$$

Combining equations (1) et (9) for the second term of (14) gives:

$$Ad^{-1}\left(Id - \Pi^T\right) V_s = V_{\underline{v}} \quad (16)$$

where $V_{\underline{v}}$ is the end frame velocity twist of $\mathcal{V}$ relative to frame $X_v(q_v)$.

Then, introducing (15) and (16) into (14), leads to the following relation:

$$V_s^T F_s = V_v^T F_v + \left(V_{\underline{v}}\right)^T F_{\underline{v}} \quad (17)$$

In equation (17), both of second member terms respectively express the power of two passive systems, if C_v and $C_{\underline{v}}$ are passive controllers. Then the overall controller is also passive.

Remark 1 :
If $\mathcal{V}$ could have been built in a global way, the passivity property would have been more obvious. As it is not the case here, a mathematical verification was necessary.

Remark 2 :
The passivity property does not depend on the M_s matrix used for Π computation. In other words, if the estimated manipulator inertia is different from the real one, the controlled system will not be decoupled anymore but will still remain passive.

Remark 3 :
Conventional inertial decoupling like computed torque does not lead to passivity whatever the estimated inertial matrix (see [1]).

3.5 ABOUT SINGULARITIES

Condition (6) also means that the local complete virtual mechanism $\mathcal{W}$ is non singular.

Indeed, consider the Jacobian matrix of the complete virtual mechanism :
$$J = \begin{bmatrix} J_v & AdJ_{\underline{v}} \end{bmatrix}$$
and take two vectors :
$$\begin{cases} x = J_v \lambda \\ y = AdJ_{\underline{v}} \mu \end{cases}$$

If we suppose that : $x = y$, then, according to (6), we have :
$$x^T M_s x = \mu^T J_{\underline{v}}^T \left(Ad^T M_s J_v \right) \lambda = 0$$

As M_s is an inertia matrix which is supposed to be positive definite, then : $x = y = 0$.

Finally, because $\mathcal{V}$ and $\underline{\mathcal{V}}$ were chosen non singular, the complete virtual mechanism is never singular as well.

4. TELEMANIPULATOR CASE

The extension of our approach to a master-slave system is depicted on Figure 7. $\mathcal{V}$ mechanism is now split into two identical virtual mechanisms $\mathcal{V}_m$ and $\mathcal{V}_s$. These two mechanisms are linked by a six d.o.f. controller C_v which creates a wrench F_v, function of $\begin{bmatrix} q_{vm} \\ q_{vs} \end{bmatrix}$ between. $X_{vs}(q_{vs})$ and $X_{vm}(q_{vm})$ relative to base frame $X_v(0)$. Master and Slave manipulators are respectively linked to $\mathcal{V}_m$ and $\mathcal{V}_s$ through $\underline{\mathcal{V}}_m$ and $\underline{\mathcal{V}}_s$ mechanisms, and six d.o.f. controllers $C_{\underline{v}m}$ and $C_{\underline{v}s}$ which only depend on $q_{\underline{v}m}$ ($\underline{\mathcal{V}}_m$ generalised coordinates), and $q_{\underline{v}s}$ ($\underline{\mathcal{V}}_s$ generalised coordinates) respectively.

The same approach as above will lead to similar formulas, i. e. :

- $F_s = \Pi_s F_{vs} + (Id - \Pi_s) Ad_s^{-T} F_{\underline{v}s}$
- $\Pi_s = M_s J_{vs} \left(J_{vs}^T M_s J_{vs} \right)^{-1} J_{vs}^T$
- $\dot{q}_{vs} = \left(J_{vs}^T M_s J_{vs} \right)^{-1} J_{vs}^T V_s$

for the Slave manipulator , and :

- $F_m = \Pi_m F_{vm} + (Id - \Pi_m) Ad_m^{-T} F_{\underline{v}m}$
- $\Pi_m = M_m J_{vm} \left(J_{vm}^T M_m J_{vm} \right)^{-1} J_{vm}^T$
- $\dot{q}_{vm} = \left(J_{vm}^T M_m J_{vm} \right)^{-1} J_{vm}^T V_m$

for the Master manipulator.

Referring to Figure 5, we have an additional relation :
$$F_{vs} = -F_{vm}$$

The controller architecture now consists of two subsystems similar to the one shown on Figure 6 and coupled with a common block which computes wrench
$$F_v = F_{vs} = -F_{vm}, \text{ knowing } \begin{bmatrix} q_{vm} \\ q_{vs} \end{bmatrix}.$$

In order to give an idea of the results which can be obtained, consider again the simple example presented above.

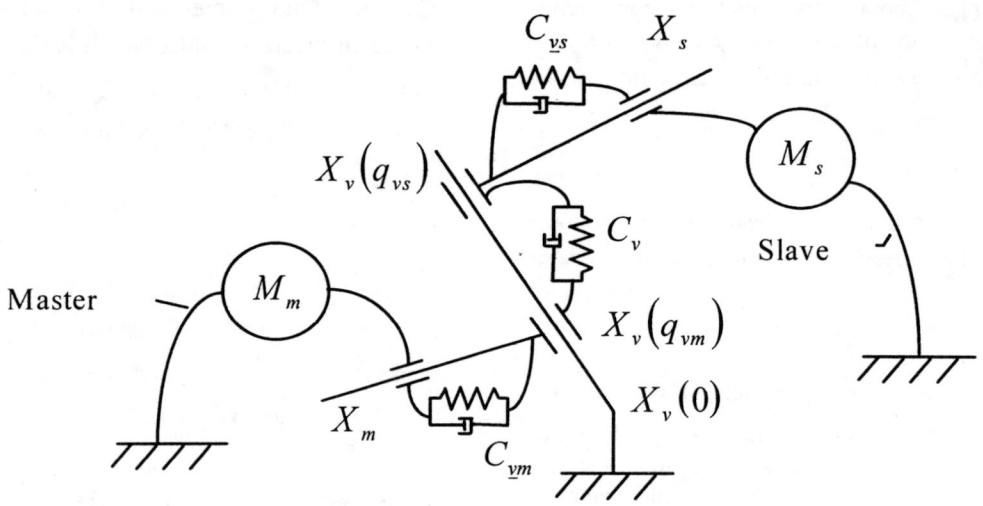

Figure 7 Building a decoupling Master-Slave Controller

Example

If the Master and Slave controllers are tuned in a same way as before, then, along the virtual constrained direction, the system behaviour will still be defined by independent Master and Slave behaviours characterised by a 0.7 damping ratio.

Proceeding as proposed in the previous paragraphs, it remains a new controller to tune along the virtual unconstrained direction. Choosing again a PD type controller and respecting the previous bandwidth limitation (30 rad/s), a significant improvement is obtained. On figures 8 and 9, the new system behaviour is presented and it can be compared to the previous obtained results on figures 3 and 4 respectively

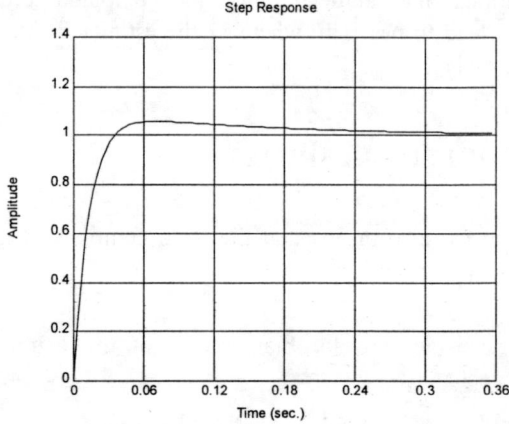

Figure 9 Slave force/Mater force step response

5. CONCLUSION

In this paper, we proposed a new control design approach in order to improve the behaviour of telemanipulation systems. The main idea of this approach is the design of three different controllers for three different systems which may work simultaneously :

- a Master manipulator alone (against virtual constraints) ;
- a Slave manipulator alone (against virtual constraints) ;
- a Master-Slave system (along the free virtual mechanism d.o.f.).

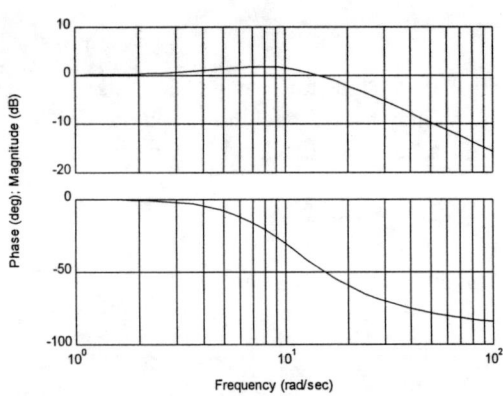

Figure 8 Slave velocity/Master velocity Transfer

The obtained theoretical result is rather simple and can easily be implemented.

However, because the analysis is based on perfect, rigid models which do not reflect the exact reality, and therefore, this work must be just considered as a guide to achieve better performances.

Nevertheless, the approach may be extended to more complicated dynamics than simple inertia.

The next step of our work will be to implement such a controller on a real Master-Slave System at CEA, and to qualify the performances improvements.

6. REFERENCES

[1] R. J. Anderson, Passive Computed Torque Algorithms for Robots, *Proceedings of the 28th Conference on Decision and Control*, Tampa, Florida, pp. 1638-1644, 1989.

[2] C. Andriot, Automatique des Systèmes Téléopérés avec Retour d'effort, Limitation des Performances, *Thèse de Doctorat, Spécialité Robotique*, Université de Paris 6, Octobre 1992.

[3] Y. Briere, *Téléopération en présence de retards : le concept de Téléopération Hybride Duale*, PhD., Ecole Nationale Supérieure de L'Aéronautique et de l'Espace, 1994.

[4] H. Bruyninckx, J. De Schutter, Specification of Force-Controlled Actions in the « Task Frame Formalism » - A Synthesis, *IEEE Transactions on Robotics and Automation*, vol. 12, no. 4, pp. 581-589, 1996.

[5] K. L. Doty, C. Melchiori, C. Bonivento, A Theory of Generalized Inverses Applied to Robotics, *The International Journal of Robotics Research*, vol. 12, no. 1, pp. 1-19, 1993.

[6] D. Escleine, F. Pierrot, P. Dauchez, P. Fraisse, Bilateral Teleoperation Based on External Force Control, *Preprints of the Fifth IFAC Symposium on Robot Control*, Nantes, vol. 3, pp. 625-630, 1997.

[7] R. Featherstone, S. Sonck, O. Khatib, A General Contact Model for Dynamically-Decoupled Force/Motion Control, *International Symposium on Experimental Robotics*, Barcelona, 1997.

[8] W. D. Fisher, M. S. Mujtaba, Hybrid Position/Force Control : A Correct Formulation, *The International Journal of Robotics Research*, vol. 11, no. 4, 1992.

[9] L. D. Joly, C. Andriot, V. Hayward, Mechanical Analogies in Hybrid Position/Force Control, *1997 IEEE International Conference on Robotics and Automation*.

[10] L. D. Joly, A. Micaelli, Hybrid Position/Force Control, Velocity Projection and Passivity, *Preprints of the 1997 IFAC Symposium on Robot Control*, pp. 345-351, Nantes, France.

[11] O. Khatib, A Unified Approach for Motion and Force Control of Robot Manipulators : The Operational Space Formulation, *IEEE Journal on Robotics and Automation*, vol. 3, no.1, pp. 43-53, 1987.

[12] H. Lipkin, J. Duffy, Hybrid Twist and Wrench Control for a Robotic Manipulator, *Journal of Mechanism, Transmissions, and Automation in Design*, vol.110, pp. 138-144, 1988.

[13] M. T. Mason, Compliance and Force Control for Computer Controlled Manipulators, *IEEE Transactions an Systems, Man, and Cybernetics*, Vol. SMC-11, no. 6, pp. 418-432, 1981.

[14] R. M. Murray, Z. Li, and S. S. Sastry, *A Mathematical Introduction to Robotic Manipulation*, CRC Press,1994.

[15] V. Perdereau, M. Drouin, A new Scheme for Hybrid Force-Position Control, *Robotica*, vol. 11, pp. 453-464, 1993.

[16] M. H. Raibert, J. J. Craig, Hybrid Position/Force Control of Manipulators, *Journal of Dynamic Systems, Measurement and Control*, vol. 102, pp. 126-133, 1981.

[17] L. Sciavicco, B. Siciliano, *Modeling and Control of Robot Manipulators*, McGraw Hill, 1996.

[18] J. M. Selig, P. R. McAree, A simple Approach to Invariant Hybrid Control, *Proceedings of the IEEE International Conference on Robotics and Automation*, Minneapolis, pp. 2238-2245, 1996.

Control of a Space Flexible Master-Slave Manipulator based on Parallel Compliance Models

Tadashi Komatsu

Department of. Mechanical Engineering
Kanto Gakuin University
Mutsuura 4834, Kanazawa
Yokohama 236-8501, Japan
tel&fax:+81-45-786-7749
e-mail:i02615@simail.ne.jp

Toshio Akabane

The Ashikaga Bank Ltd.

Abstract

This paper proposed a bilateral controller of a flexible master-slave manipulator(FMSM). A FMSM consists of a conventional compact rigid master arm and a flexible slave arm, and will be used in outer space in the future. This controller realizes stable compliance control and vibration control. The key idea is that this controller has dual compliance models.

1. Introduction

There are some conventional bilateral control methods for a master-slave manipulator (MSM)[1]-[5]. But in conventional MSM systems, all arms were controlled as rigid arms. On the other hand, when considering using a MSM in outer space, a slave arm will become a long and lightweight arm(see Fig.1). In this case, it is necessary to consider this arm as a flexible arm, because its elasticity decreases[7]. For this reason, a flexible master-slave manipulator (FMSM) system consisting of a conventional compact rigid master arm and a flexible slave arm has become necessary in this field.

There are two main problems for controlling a FMSM. The first is the vibration of a slave arm. In a FMSM, an operator controls the slave position by a master arm. In this case, when vibration occurs in a slave arm, it becomes difficult to detect the position of a slave arm. The second problem is the deformation of a flexible link. When an operator adds a large force to an object through a flexible slave arm, the link deforms or breaks easily. For these reasons, vibration control and precise control of the reflection force are necessary. Of course, many researchers have studied the vibration control of a flexible arm[7]-[8]. However, they dealt with only a flexible slave arm. On the other hand, a FMSM is the system in which an operator controls a flexible slave arm directly using a master arm. There are few studies of the control method and the control design method for this system.

For the reasons mentioned above, this paper presents an alternative control architecture of a FMSM based on the concept of dual compliance models following control with the vibration control. The key idea of this concept is to design each compliance model considering the elasticity of a master and slave arm. The proposed control architecture realizes the stable FMSM system with bilateral feedback. As the initial study, we consider one degree of freedom (d.o.f.) system.

2. Dynamics of the FMSM System

In this section, dynamics of a master and a flexible slave arm will be derived. Here, we consider a rigid master arm and a flexible slave arm with 1 d.o.f.(see Fig.2). Gravity is neglected because we are considering a space system. Also friction is neglected.

In general, dynamics of a rigid arm are written as follows;

$$I_m \ddot{\theta}_m = U_m - T_m, \qquad (1)$$

where, I_m is the inertia moment of the master arm, θ_m is the joint angle, U_m is the control torque of an actuator and T_m is the torque added by an operator.

On the other hand, we considered the first mode of the vibration in dynamics of a flexible arm. Here, the mass of a link was neglected and only the mass of a hand was considered. We used a quasi-static model and a virtual arm model[9], for example, which is represented by a dotted line in Fig.2.

Dynamics are written as follows;

$$I_s \ddot{\theta}_s = U_s - T_s. \quad (2)$$

Here, I_s is the inertia moment of a slave arm, θ_s is the joint angle of a virtual arm and written as follows;

$$\theta_s = \theta_{sr} + c(V/L), \quad (3)$$

where θ_{sr} is the joint angle of a slave arm, V is the deformation of the link tip and L is the link length. c denotes a scale coefficient. U_s is the control torque of an actuator and T_s is the torque added by the environment.

3. Control System using Parallel Compliance Models

There are two mechanical systems with quite different elasticities in a FMSM. Considering this situation, the newly proposed control system has two different virtual compliance models, one for a master arm and the other for a slave arm. Each arm is controlled independently along paths calculated by these models.

Figure 3 shows the configuration of the proposed control system. In this system, torque sensors are equipped near each actuator, i.e. this is the collocated system.

3.1 Control of a Master Manipulator

The torque of a master arm T_m and a slave arm T_s go into a controller of a master arm G_{mc} multiplying reflection coefficients R_{m1} and R_{m2}. In G_{mc}, the desired trajectory of a master arm is calculated by the compliance model as follows;

$$\theta_{md} = K_m(\Delta T_m + \int_0^t \Delta T_m dt), \quad (4)$$

where $\Delta T_m = -R_{m1}T_m - R_{m2}T_s$.

θ_{md} calculated in G_{mc} goes into a position controller of a master arm, and desired actuator torque is calculated. Here, the local PD control method was used

$$U_m = a_m(\theta_{md} - \theta_m) - b_m \dot{\theta}_m, \quad (5)$$

where a_m and b_m are the feedback gain.

3.2 Control of a Slave Arm

The torque of a master arm T_m and a slave arm T_s go into a controller of a slave arm G_{sc} multiplying reflection coefficients R_{s1} and R_{s2}. In G_{sc}, the desired trajectory of a slave arm is calculated by the compliance model as follows;

$$\theta_{sd} = K_s(\Delta T_s + \int_0^t \Delta T_s dt), \quad (6)$$

where $\Delta T_s = -R_{s1}T_m - R_{s2}T_s$.

θ_{sd} calculated in G_{sc} goes into a position controller of a slave arm;

$$U_s = a_s(\theta_{sd} - \theta_s) - b_s \dot{\theta}_s, \quad (7)$$

where a_s and b_s are the feedback gain.

We can obtain angle data θ_s by equation (3). The deformation V in equation (3) is calculated using data of a torque sensor, because we used a quasi-static model. θ_{sr} is obtained by an angle sensor. Therefore, θ_s includes the parameter V, and the vibration control is simultaneously realized by equation (7)[9].

3.3 Stability of the Proposed Control Method

In this section, the stability of the proposed control method is considered using the Lyapunov's direct method.

First, torques T_m and T_s are modeled as follows;

$$T_m = H_m(\theta_m - \theta_h), \quad (8)$$
$$T_s = H_s(\theta_s - \theta_r), \quad (9)$$

where H_m and H_s are the stiffness of the operator and the environment, and θ_h and θ_r are the position of the operator's hand and the object. Using equations from (1) to (9) without (3), state equations of the system are obtained as follows;

$$\begin{bmatrix} \dot{x}_1 \\ \dot{x}_2 \\ \dot{x}_3 \\ \dot{x}_4 \\ \dot{x}_5 \end{bmatrix} = \begin{bmatrix} 0 & 1 & 0 & 0 & 0 \\ -a_1-a_2-a_3 & -a_4 & -a_5 & 0 & a_2 \\ 0 & 0 & 0 & 1 & 0 \\ -a_6 & 0 & -a_7-a_8-a_9 & -a_{10} & \alpha a_8 \\ -a_{11} & 0 & -a_{12} & 0 & 0 \end{bmatrix} \begin{bmatrix} x_1 \\ x_2 \\ x_3 \\ x_4 \\ x_5 \end{bmatrix} \quad (10)$$

where
$x_1 = \theta_m, x_2 = \dot{\theta}_m, x_3 = \theta_s, x_4 = \dot{\theta}_s, x_5 = K_m \int_0^t \Delta T_m dt, \alpha K_m R_{m1} = K_s R_{s1}, \alpha K_m R_{m2} = K_s R_{s2}, \theta_h = \theta_r = 0, a_1 = a_m K_m R_{m1} H_m / I_m,$
$a_2 = a_m / I_m, a_3 = H_m / I_m, a_4 = b_m / I_m,$
$a_5 = a_m K_m R_{m2} H_s / I_m, a_6 = a_s K_s R_{s1} H_m / I_s,$
$a_7 = a_s R_{s2} K_s H_s / I_s, a_8 = a_s / I_s, a_9 = H_s / I_s,$
$a_{10} = b_s / I_s, a_{11} = K_m R_{m1} H_m,$
$a_{12} = K_m R_{m2} H_s.$

Here, we consider equation (11) as Lyapunov function;

$$v = \frac{n}{2}(x_1+x_2)^2 + \frac{p}{2}(x_3+x_4)^2 + \frac{q}{2}(x_5+\dot{x}_5)^2 + \frac{1}{2}\{n(a_1+a_2+a_3-1+a_4)-a_{11}/a_{12}\}x_1^2 + \frac{1}{2}\{p(a_7+a_8+a_9)+a_{10}/a_6-1/a_6-a_{12}/a_{11}\}x_3^2, \quad (11)$$

where
$n = 1/a_5, p = 1/a_6, q = a_2/(a_5 a_{11})$.

Then, using the Lyapunov's method, the following stability conditions are obtained from conditions of $v \rangle 0$ and $\dot{v} \langle 0$ concerning Lyapunov function and its differentiation along the solution trajectory of equation (10);

$R_{s1} R_{m2} = R_{m1} R_{s2},$
$b_m \rangle I_m,$
$b_s \rangle I_s. \quad (12)$

Therefore, by selecting parameters $R_{m1}, R_{m2}, R_{s1}, R_{s2}, b_m, b_s$ considering conditions of equation (12), the stability of the proposed control system is guaranteed.

4. Experiments

Next, we studied the effectiveness of the proposed method by experiments. Figures 4 and 5 show a master and a slave manipulator of an experimental system. A master arm is 1 m long and a slave arm is 1.5 m long. Each arm has 6 d.o.f., but only a shoulder joint and an upper arm were used for experiments. Joints were actuated by DC motors. Strain gages were equipped at the base of links for torque sensors. Potentiometers were used for angle sensors. A 32-bit computer system was used. The sampling time was 5 msec. A flexible link was made of stainless steels of diameter 6 mm.

First, the response of a slave arm following a master arm was analyzed. Figure 6 shows results using a conventional bilateral control of a force feedback method for a rigid arm. A time lag of 500 msec was appeared and vibration of 1 Hz occurred. On the other hand, Fig.7 shows results of the proposed method. Parameters in a controller were selected as follows;
R_{m1}=5, R_{m2}=10, K_m=2.0×10^{-2} rad/Nm,
a_m=5 Nm/rad, b_m=2 Nms/rad,
I_m=1.88 kgm^2,
R_{s1}=0.5, R_{s2}=1, K_s=0.2 rad/Nm,
a_s=12 Nm/rad, b_s=12 Nms/rad,
I_s=0.113 kgm^2, c=-4.

The slave arm followed the master arm with a little time lag and vibration was soon restrained. Because of the backlash, small vibration was remained. However, an overshoot of the slave angle appeared because of vibration control. It is the future theme to reduce this overshoot.

Next, force reflection tests were carried out. Figure 8 shows the case of R_{s1}=0.2 and R_{s2}=1. In this case, parameters in a controller were selected as follows;
R_{m1}=50, R_{m2}=250, K_m=9.0×10^{-3} rad/Nm,
a_m=200 Nm/rad, b_m=95 Nms/rad,
K_s=0.36 rad/Nm, a_s=20 Nm/rad,
b_s=12 Nms/rad, c=-8.

In these tests, we used a book with hard

covers as an object. Both joints moved a little within the limits of the backlash. The slave force became one-fifth of the master force stably. The desired responses of force reflection were obtained. Furthermore, Fig. 9 shows the results that the force of a slave was increased considering the elasticity of a flexible link. In this case, R_{s1} was set 0.5 and R_{s2} was set 1, and R_{m1} was set 50 and R_{m2} was set 100. Also, the desired result that the slave force became half of the master force was obtained. When we did same experiments using a conventional force feedback control method, the system became unstable.

These experimental results show that the effective bilateral control can be realized also for a FMSM using the proposed method.

6. Conclusions

This paper presents a newly developed bilateral controller of a FMSM using dual compliance models. The effectiveness of the proposed method was shown by experiments using an 1 d.o.f. system.

References

[1] T.Miyazaki and S.Hagihara, "Parallel control method for bilateral master-slave manipulator," J.Robotics Soc. Jpn., vol.7, no.5, pp.446-452, 1989(in Japanese).

[2] J.E.Colgate, "Power and impedance scaling in bilateral manipulator," Proc. of 1991 IEEE International conference on R & A, Sacramento, CA, pp.2292-2297, 1991.

[3] B.Hannaford, "A design framework for teleoperators with kinesthetic feedback," IEEE Trans. Rob. Auto, vol.5, no.4, pp.426-434, 1989.

[4] T.B.Sheridan, "MIT research in telerobotics," Proc. of the Workshop on Space Telerobotics, vol.2, pp.403-412, JPL Publication 87-13, 1987.

[5] N.Matsuhira, M.Asakura and H.Bamba, "Maneuverability of a master-slave manipulator with different configurations and its evaluation tests,"Advanced Robotics, vol.8, no.2, pp.185-202, 1994.

[6] V.Michtchenko, S.Kampen and F.Didot,"Extension of ERA capabilities by interaction with other automated facilities on the Russian segment of the ISS," Proc. i-SAIRAS'97, pp.243-248, 1997.

[7] R.H.Cannon, Jr. and E.Schmitz, "Initial experiments on the end-point control of a flexible one-link robot," Int. J. Robotics Research, vol.3, no.3, pp.62-75, 1984.

[8] W.J.Book, O.Maizza-Neto and D.E.Whitney, "Feedback control of two-beam, two joint systems with distributed flexibility," ASME J.DSMC, vol.97, no.4, pp.424-431, 1975.

[9] T.Komatsu,M.Uenohara,S.Iikura, H.Miura and I.Shimoyama, "Dynamic control for two-link flexible manipulator," Trans. JSME(C), vol.55, no.516, pp.2022-2028, 1988(in Japanese).

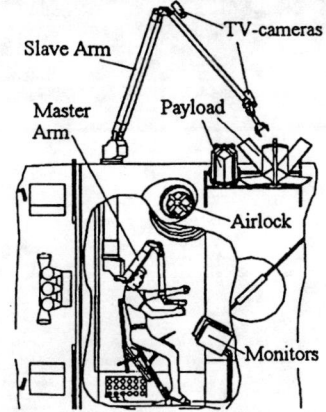

Fig.1 Configuration of space master-slave manipulator system[6].

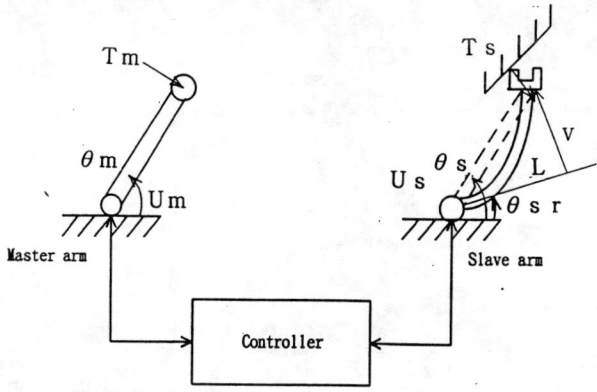

Fig.2 FMSM system with 1 d.o.f.

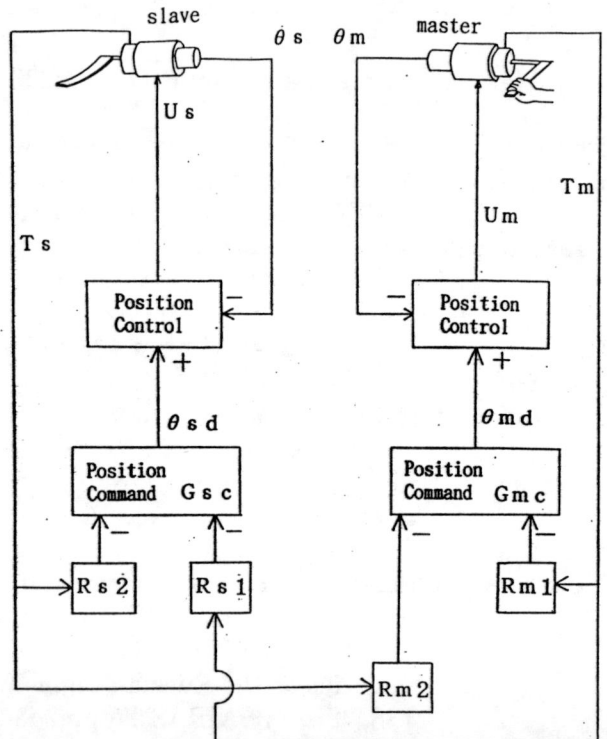

Fig.3 Dual compliance model.

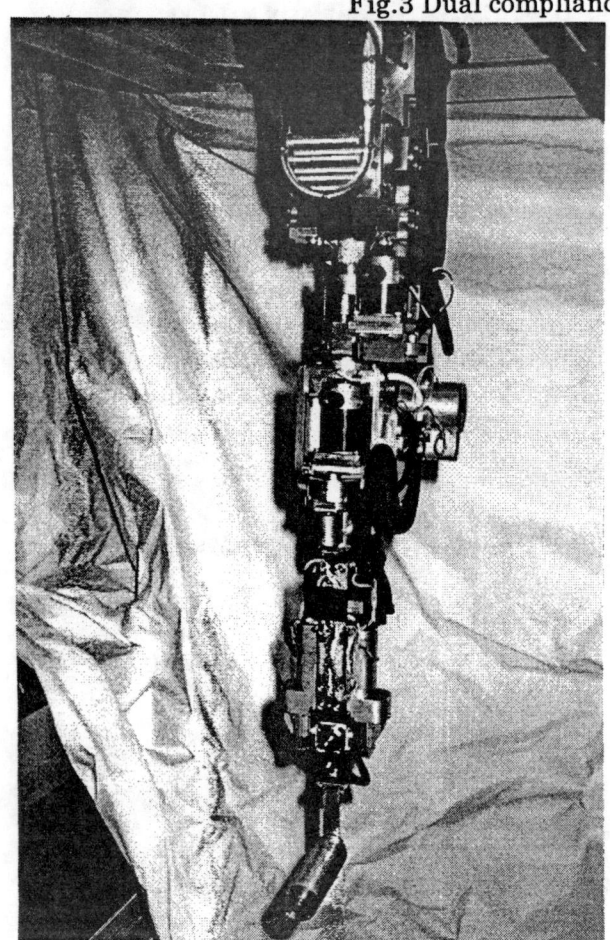

Fig.4 Photograph of the master arm.

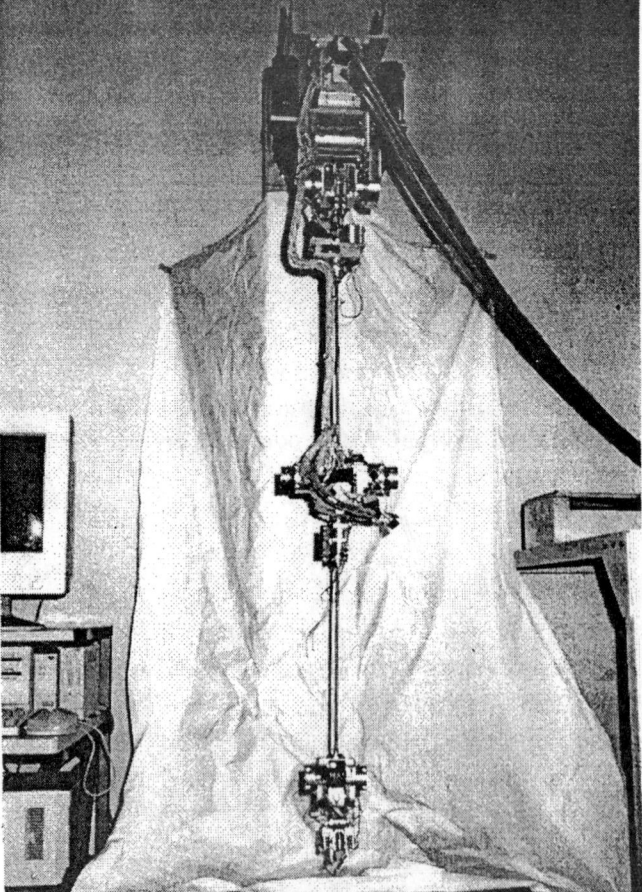

Fig.5 Photograph of the slave arm.

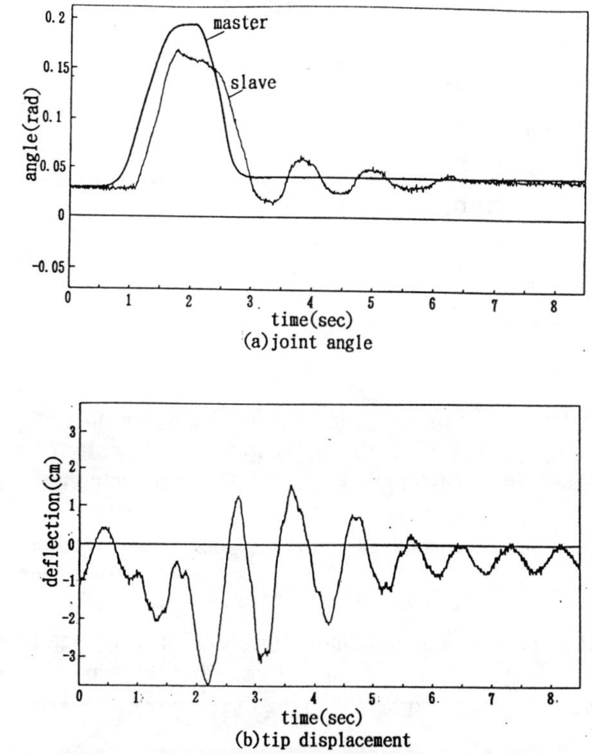

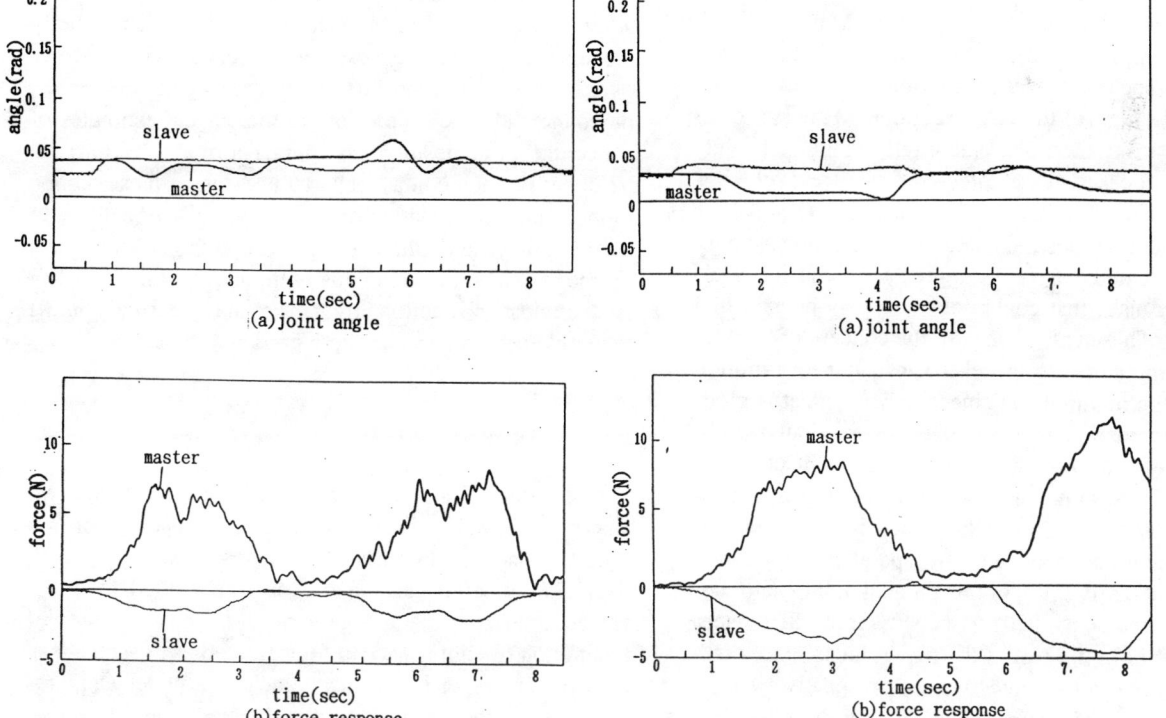

Fig.6 Experimental results of the conventional force feedback control method.

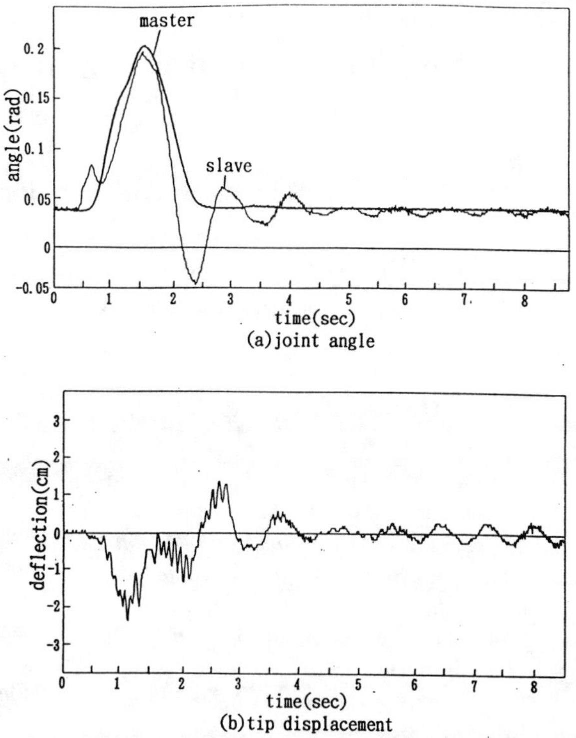

Fig.7 Experimental results of the proposed method.

Fig.8 Experimental results of the force reflection tests($R_{s1}=0.2, R_{s2}=1$).

Fig.9 Experimental results of the force reflection tests($R_{s1}=0.5, R_{s2}=1$).

Modelling and Specification of Compliant Motions with Two and Three Contact Points

Herman Bruyninckx, Joris De Schutter
Katholieke Universiteit Leuven
Department of Mechanical Engineering, Division PMA
Celestijnenlaan 300B, B-3001 Leuven, Belgium

Abstract

This paper describes the modelling and motion specification of compliant motion tasks with two or three contact points. These tasks cannot be done with Mason's classical "task frame" (TF) or "compliance frame" approach. Hence, a more flexible and versatile motion constraint model is introduced, that maintains most of the intuitiveness of the TF approach.

1 Introduction

The literature on force-controlled compliant motion most often uses (implicitly or explicitly) Mason's "Task Frame" (TF) approach, [5], to model the contact situation of the task, to specify the desired motion within this contact model, and to control the task execution. This approach is limited to tasks in which one single orthogonal reference frame suffices to model force-controlled and velocity-controlled directions of the motion constraint. (See [1] for more details.) Previous publications by the authors (e.g., [3]) have presented extensions to the TF approach that allow to tackle tasks that could not previously be executed successfully. These extensions use the concept of "virtual contact manipulators" to model the instantaneous motion freedom of the manipulated object: each contact is modelled by a kinematic chain that gives the manipulated object the same local motion freedom as the contact; if the total motion constraint consists of several contacts that act simultaneously on the manipulated object, this total motion constraint is hence modelled by a parallel manipulator. The kinetostatic properties of this parallel manipulator determine the force-controlled and velocity-controlled spaces at each instant. In the rest of the paper, these spaces will be called *wrench space* and *twist space*, respectively.

The advantage of this modelling approach is that it is completely general and independent of any coordinate representation. The resulting models, however, could lack the intuitiveness of the TF approach. Therefore, this paper gives ad hoc contact models for two frequently occurring contact situations (having two, respectively three, *vertex-surface* contacts between manipulated object and environment), in which the remaining degrees of freedom are defined in very intuitive and coordinate-independent ways. Coordinate expressions, however, are also given, such that implementation on a force-controlled robot system is straightforward. Section 2 describes the two-point contact situation, and Section 3 the three-point contact situation.

The contact models in this paper are basic to the motion specification and force control of every task that involves more than one single contact: *instantaneously* every contact is approximated by the position of the contact point and the direction of the contact normal, and this is exactly the situation where the presented models are valid. Moreover, *velocity-based* on-line identification of these contact parameters [1] (i.e., the errors in the current estimates of the contact point position and contact normal direction) can be done for each contact separately. Hence, these ad hoc models are very practical in two ways: (i) they allow to model and specify contact situations that the classical Task Frame formalism cannot cope with, and (ii) they simplify (without loss of functionality!) the general "virtual contact manipulator" approach in these particular cases.

2 Two-point contact

In this task, Fig. 1, the robot tool is in contact with two smoothly curved surfaces, which intersect each other in a "seam." Each contact is of the *vertex-surface* type, with five velocity-controlled and one force-controlled direction. Hence, the two contacts together reduce the dimension of the motion freedom space to four. Examples of such a task are: tracking pipes in chemical, nuclear or undersea plants (in this case, the "seam" exists in the model only); following a surface with a heavy tool that needs bracing on a support surface [6]; guidance of a welding torch or a glueing tool along a seam between two workpieces that have to be connected, etc.

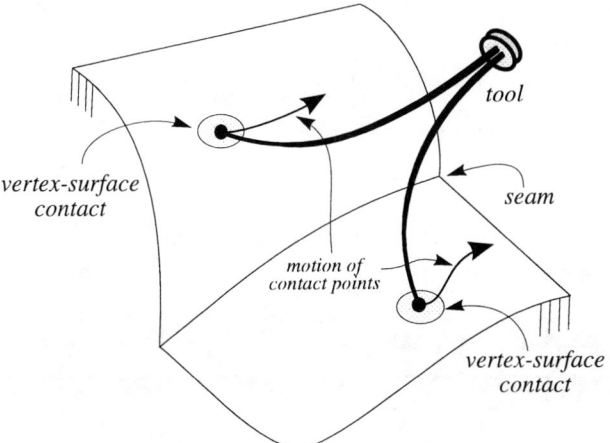

Figure 1: Seam following with two contact points.

2.1 Geometric parameters

The symbols $\{1\}$ and $\{2\}$ denote the two contact points, as well as their associated contact frames, Fig. 2. The *seam axis* is the intersection of the two tangent planes at $\{1\}$ and $\{2\}$. A *parallel plane* is each plane through a contact normal and parallel to the seam axis. A *perpendicular plane* is any plane perpendicular to the seam axis. The *seam angle* σ is the (free space) angle between the tangent planes. The contact points lie at distances d^1 and d^2, respectively, from the seam axis.

All these parameters can be calculated if the unit normal vectors e^1 and e^2 are known, as well as the vector p linking the two contact points. Expressed with respect to the reference frame $\{1\}$ of Fig. 3 this gives:

$$e^1 = \begin{bmatrix} 0 \\ 0 \\ 1 \end{bmatrix}, \; e^2 = \begin{bmatrix} e_x \\ e_y \\ e_z \end{bmatrix}, \; p = \begin{bmatrix} p_x \\ p_y \\ p_z \end{bmatrix}. \quad (1)$$

Then, the seam angle σ, the direction of the seam axis e^{sa}, the position vectors p_d^1 and p_d^2 of the points on the seam axis closest to $\{1\}$ and $\{2\}$, as well as the distances d^1 and d^2 from the contact points to the seam axis, are calculated as follows (the calculations are straightforward but rather tedious):

1. The seam angle σ:

$$\sigma = \arcsin\left(e^1 \cdot e^2\right) + \frac{\pi}{2} = \arcsin(e_z) + \frac{\pi}{2}.$$

2. The vector e^{sa} is the normalized cross product of e^1 and e^2:

$$e^{sa} = \frac{e^2 \times e^1}{|e^2 \times e^1|} = \frac{1}{\sqrt{e_x^2 + e_y^2}} \begin{bmatrix} e_y \\ -e_x \\ 0 \end{bmatrix}.$$

3. p_d^1 is the intersection of the tangent planes at $\{1\}$ and $\{2\}$, and the perpendicular plane through $\{1\}$:

$$p_d^1 = \frac{e_x p_x + e_y p_y + e_z p_z}{e_x^2 + e_y^2} \begin{bmatrix} e_x \\ e_y \\ 0 \end{bmatrix}.$$

Similarly, p_d^2 is the intersection of the tangent planes at $\{1\}$ and $\{2\}$, and the perpendicular plane through $\{2\}$:

$$p_d^2 = \begin{bmatrix} p_x \\ p_y \\ 0 \end{bmatrix} + \frac{e_z p_z}{e_x^2 + e_y^2} \begin{bmatrix} e_x \\ e_y \\ 0 \end{bmatrix}.$$

4. The distance d^1 is the length of p_d^1:

$$d^1 = \frac{|e_x p_x + e_y p_y + e_z p_z|}{\sqrt{e_x^2 + e_y^2}}.$$

The distance d^2 is the length of $p_d^2 - p$:

$$d^2 = |p_z| \sqrt{\frac{e_x^2 + e_y^2 + e_z^2}{e_x^2 + e_y^2}}.$$

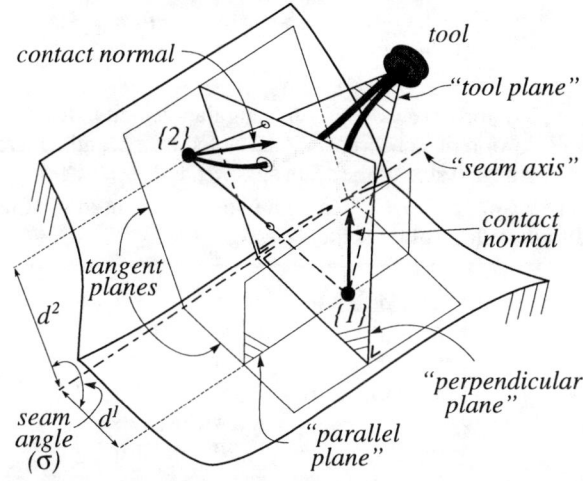

Figure 2: Two-point contact: geometric definitions.

The *tool plane* is defined as the plane through the two contact points and a third user-defined point p^t on the tool.

The tool plane has one coordinate-independent reference position when it lies in a perpendicular plane *and* $d^1 = d^2$. This situation is called the *symmetric* tool position. The tool looses one or two degrees of freedom if it is *parallel* to the seam, i.e., $d^1 = d^2 = 0$.

The contact frames at each of the two contact points have their Z axis along the contact normal. The X and Y axes are not uniquely determined geometrically; their direction can be freely chosen in the tangent plane. A contact frame is called *parallel* if its X axis is parallel to the seam axis, and the Y axis points towards the seam, Fig. 3.

2.2 Twist space basis

Each of the two contacts reduces the tool's motion freedom by one. With respect to the parallel reference frame in $\{1\}$, the bases for the five-dimensional twist space (i.e., the *Jacobian matrices* of the virtual manipulators at the two contacts) are:

$$\boldsymbol{J}^1 = \begin{bmatrix} 0 & 0 & 0 & 1 & 0 \\ 0 & 0 & 0 & 0 & 1 \\ 0 & 0 & 1 & 0 & 0 \\ 1 & 0 & 0 & 0 & 0 \\ 0 & 1 & 0 & 0 & 0 \\ 0 & 0 & 0 & 0 & 0 \end{bmatrix}, \quad (2)$$

$$\boldsymbol{J}^2 = \begin{bmatrix} 0 & 0 & 0 & 1 & 0 \\ 0 & 0 & -s_\sigma & 0 & -c_\sigma \\ 0 & 0 & -c_\sigma & 0 & s_\sigma \\ 1 & 0 & a & 0 & b \\ 0 & -c_\sigma & p_x c_\sigma & p_z & -p_x s_\sigma \\ 0 & s_\sigma & -p_x s_\sigma & -p_y & -p_x c_\sigma \end{bmatrix}. \quad (3)$$

The first three rows represent angular velocity, the last three rows represent translational velocity. c_σ and s_σ are the cosine and sine of the seam angle σ; $a = -p_y c_\sigma + p_z s_\sigma$, and $b = p_y s_\sigma + p_z c_\sigma$. Column two of $\boldsymbol{J}^2$ is used to simplify the other columns to:

$$\boldsymbol{J}^2 \cong \begin{bmatrix} 0 & 0 & 0 & 1 & 0 \\ 0 & 0 & -s_\sigma & 0 & -c_\sigma \\ 0 & 0 & -c_\sigma & 0 & s_\sigma \\ 1 & 0 & a & 0 & b \\ 0 & -c_\sigma & 0 & p_z - p_y \frac{c_\sigma}{s_\sigma} & -\frac{p_x}{s_\sigma} \\ 0 & s_\sigma & 0 & 0 & 0 \end{bmatrix}. \quad (4)$$

From this, it is clear that a basis for the twist space of the total constraint is found from either $\boldsymbol{J}^1$ or $\boldsymbol{J}^2$ with the second column removed. We define the total constraint's Jacobian matrix $\boldsymbol{J}$ as the matrix found from elementary column operations on $\boldsymbol{J}^2$:

$$\boldsymbol{J} \cong \begin{bmatrix} \boldsymbol{J}^2_4 & \boldsymbol{J}^2_3 & (s_\sigma \boldsymbol{J}^2_5 - c_\sigma \boldsymbol{J}^2_3 - p_y \boldsymbol{J}^2_1) & \boldsymbol{J}^2_1 \end{bmatrix}$$

$$= \begin{bmatrix} 1 & 0 & 0 & 0 \\ 0 & -s_\sigma & 0 & 0 \\ 0 & -c_\sigma & 1 & 0 \\ 0 & -p_y c_\sigma + p_z s_\sigma & 0 & 1 \\ p_z - p_y \frac{c_\sigma}{s_\sigma} & 0 & -p_x & 0 \\ 0 & 0 & 0 & 0 \end{bmatrix}. \quad (5)$$

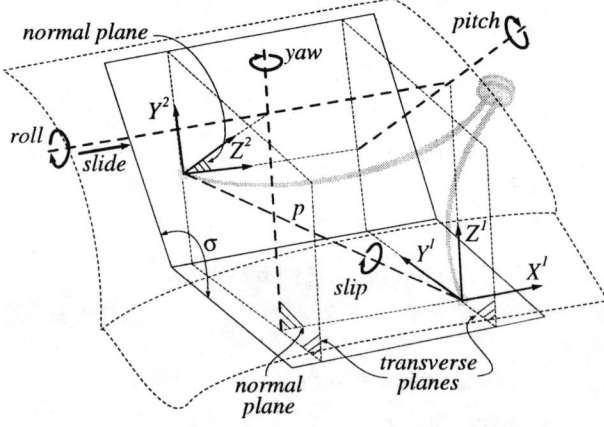

Figure 3: Two-point contact: *Roll, Pitch, Yaw, Slip,* and *Slide*.

These four remaining degrees-of-freedom represent rotations about, and translations along, geometrically defined lines, Fig. 3:

1. *Roll* is rotation about the intersection of the parallel planes through $\{1\}$ and $\{2\}$.

2. *Pitch* is rotation about the intersection of the perpendicular plane in $\{1\}$ and the parallel plane in $\{2\}$.

3. *Yaw* is the rotation about the intersection of the parallel plane in $\{1\}$ and the perpendicular plane in $\{2\}$.

4. *Slide* is translation in the direction of the seam axis. (It can also be considered as a rotation, i.e., about the intersection of the perpendicular planes in $\{1\}$ and $\{2\}$ which lies at infinity.)

Roll, Pitch and Yaw have (more or less) their original maritime interpretation if one looks at the tool as a "ship" travelling along the seam. Another coordinate-independent degree-of-freedom is the rotation about the line between the two contact points. This motion leaves the contact points unchanged on the environment; hence it is called *slip*. Slip is a linear combination of Roll, Pitch, Yaw and

Slide:
$$\begin{aligned} J^{slip} &= [p^T \; 0^T]^T \\ &= p_x J^{roll} - \frac{p_y}{s_\sigma} J^{pitch} + \left(p_z - p_y \frac{c_\sigma}{s_\sigma}\right) J^{yaw} \\ &\quad - (p_z s_\sigma - p_y c_\sigma) \, J^{slide}. \end{aligned}$$

2.3 Wrench space basis

The basis for the contact situation's wrench space (i.e., the force-controlled directions) is straightforward: it consists of unit forces along the contact normals in $\{1\}$ and $\{2\}$. The 6×2 matrix containing the coordinate expressions of these two forces is called the *wrench Jacobian matrix* and is denoted by G.

2.4 Comparison to TF

The differences between a classical Task Frame model and the Roll-Pitch-Yaw-Slide/Slip model are:

1. The lines on which the basis twists and wrenches are defined do not intersect in one point.

2. The bases are time-varying, i.e., the relative positions of the lines changes during the motion, due to the curvature of the contact surfaces.

The similarities are that:

1. An intuitive, geometric and hence coordinate-independent twist and wrench space model exists.

2. Only zero or infinite pitch screws are needed. At least in the *geometric* model, since, due to the non-intersecting axes of the geometric model, no coordinate representation exists in which the Jacobian matrices J and G also contain only zero or infinite pitch screws.

2.5 Motion specification

The previous paragraphs describe bases allowing to specify unambiguously the instantaneous twist of the tool and the desired ideal wrench on the tool. However, a human user might like more intuitive ways of specifying the instantaneous twist or the desired position. The following paragraphs describe two possible approaches, a *local* one and a *global* one.

The local specification approach follows the classical TF intuition: each individual contact gets its own TF, as if it were the only contact occurring on the manipulated tool. However, the user should not specify more than four independent motions in both TFs together. It is then the controller's job to translate this local specification in an instantaneous twist that does not violate the contact constraints. This translation can, e.g., be done with "projection matrices" on the instantaneous twist space basis, [2].

The global approach relies on a model of the remaining four motion degrees of freedom, for example *Roll-Pitch-Yaw-Slide* as described above. Then, the well-known Jacobian equation, $t = J\dot{q}$ applies, with J a basis of the twist space, and $\dot{q}$ the magnitudes of the Roll, Pitch, Yaw and Slide basis twists. The advantage is that, *by construction*, any specified twist will be compatible with the modelled constraint. However, the resulting motion of each individual contact point might be less intuitive than in the local approach. If the user prefers to specify the desired *position* of the tool, instead of the desired *instantaneous twist*, he could for example specify desired values for the following four geometrically determined distances: d_1, p_z^t, p_x^2, and the desired position along the seam. Again, the controller is responsible for transforming these four numbers into a resultant motion that is instantaneously compatible with the contacts.

3 Three-point contact

The general contact situation with three contact points, Fig. 4, has three degrees of motion freedom. The following paragraphs present an intuitive and coordinate-independent way to model the instantaneous degrees of freedom in this contact situation.

3.1 Geometric parameters

Instantaneously, the contact situation is determined by the tangent planes at the three contact points. As in the two-point contact case, the unit normal vector at contact point i is denoted by e^i. The geometric parameters defined in Sect. 2.1 exist in the three-point contact also, for each couple of contact points; the formulas to calculate these parameters remain unchanged. The notations, however, are slightly adapted, in order to discriminate the three possible combinations. For example, d_{13} denotes the distance between contact point 1 and the seam between the tangent planes in points 1 and 3; e^{13} is the unit vector parallel to this seam.

3.2 Twist space basis

A basis for the three-dimensional twist space can be chosen in many different ways. The following Jacobian

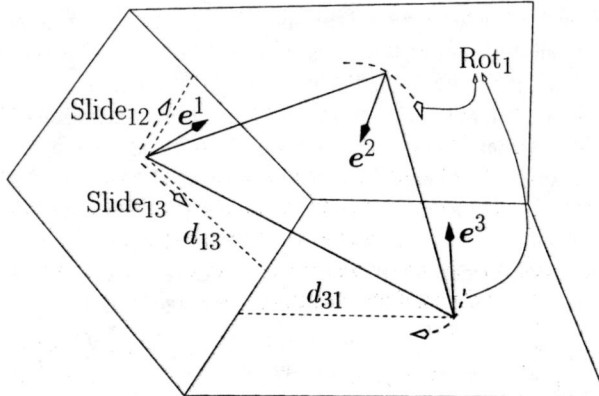

Figure 4: Three-point contact.

matrix has three basis twists that are an intuitively appealing extension to a classical Task Frame approach, Fig. 4:

$$J = [\text{Slide}_{ij}\ \text{Slide}_{ik}\ \text{Rot}_i]. \qquad (6)$$

Slide_{ij} is the translation of point i over its own tangent plane in the direction of the seam with point j; similarly for Slide_{ik}; Rot_i is the instantaneous pure rotation that leaves contact point i motionless, and moves the two other contact points in their local tangent planes. The basis in J can be used to specify the three available motion degrees of freedom by considering the motion of the contact point i only. Of course, all three basis motions must satisfy the instantaneous constraints. The following paragraphs explain how this is achieved:

1. **Slide_{ij}**. The seam between the contact points i and j is determined in exactly the same way as in the case of two-point contact. Hence, a corresponding "Roll" axis l_{ij} can be defined. Rotation about this axis makes the contact points i and j translate in their local tangent planes and perpendicular to the common seam axis. However, a pure rotation about this "Roll" axis is only possible if (i) the third tangent plane (i.e., the tangent plane at point k, $k \notin \{i,j\}$) is perpendicular to the two tangent planes that determine this "Roll" axis, or (ii) the third contact point k happens to lie on the "Roll" axis. Therefore, in general, a translational velocity v (Fig. 5) along the "Roll" axis should be added, in order to keep this third contact point k on its local tangent plane. Hence, the pure rotation "Roll" in the two-point contact case must be replaced by a *non-zero pitch screw* "Slide_{ij}" in the three-point contact case. The translational velocity component of this screw can be found as follows: $v_{ij,k}$ is the velocity of point k if it were to rotate about the "Roll" axis l_{ij}; v_k is the velocity in the tangent plane through k

that makes the point k follow the rotation about the "Roll" axis without leaving its instantaneous tangent plane; v_k is perpendicular to e^k (since it lies in the tangent plane) and to $d_{ij,k}$ (i.e., the direction vector through k and perpendicular to the "Roll" axis l_{ij}); the translational component v of the "Slide_{ij}" screw is parallel to l_{ij}, and proportional to the tangent of the angle between the vector $v_{ij,k}$ and the unit vector along v_k. All these vectors and angles can be calculated with simple vector calculus.

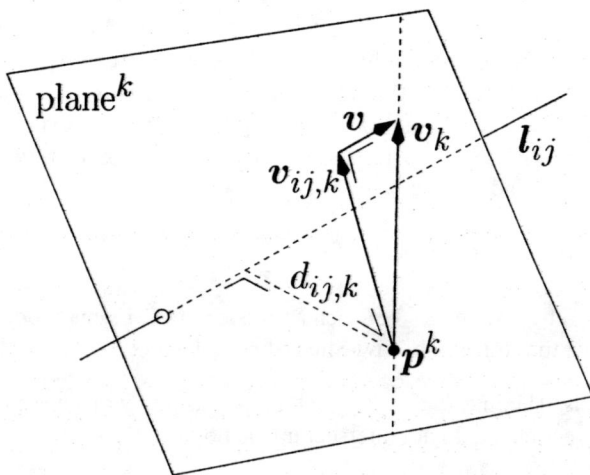

Figure 5: Three-point contact: velocity components due to slide.

2. **Rot_i**. Rotation about an axis through i moves the contact point j in the direction perpendicular to both $p^{i,j}$ (i.e., the vector from point i to point j, since point i remains motionless) and the normal direction e^j (since point j must move in its tangent plane). Hence, the axis of Rot_i goes through i and has direction vector $(p^{i,j} \times e^j) \times (p^{i,k} \times e^k)$.

3.3 Wrench space basis

The basis for the contact situation's wrench space is equally straightforward as in the two-point contact case: it consists of unit forces along the contact normals in $\{1\}$, $\{2\}$ and $\{3\}$. The wrench Jacobian matrix G is now a 6×3 matrix, containing the coordinate expressions of these three forces.

3.4 Comparison to TF

The differences between a classical Task Frame model and the above-described three-point contact model are:

1. The lines on which the basis twists and wrenches are defined do not intersect in one point.

2. The bases are time-varying, i.e., the relative positions of the lines change during the motion, due to the curvature of the contact surfaces.

3. The twist space cannot be spanned anymore by pure translations and/or pure rotations.

3.5 Motion specification

As in the two-point contact case, both *local* and *global* motion specifications are possible.

In the local approach, (part of) the motion of each contact point individually is specified as if no constraints were acting on the object. At the user level, the constraints are only taken into account by the requirement that one should not specify more than three independent velocity set-points. The others get "don't care" values that the robot controller must fill in in such a way that the resulting motion is compatible with the constraints.

In the global approach, the user constructs a compatible instantaneous motion by using, for example, the Jacobian matrix in Eq. (6). Alternatively, he can specify three desired distances of the contact points to some of the seams; for example, the distances of one of the three contact points to its two neighbouring seams, together with the distance of one of the other contact points to the tangent plane of the first contact point. The controller must again take care of the instantaneous motion interpolation required to reach the specified goal without violating the contact constraints. To this end, he can use the instantaneous twist space basis in Eq. (6).

4 Conclusions

This paper has described how classical Task Frame motion constraint modelling and motion specification procedures are extended to contact situations with two or three contact points. The presented approach is completely coordinate independent, and requires only the knowledge of the positions of the contact points as well as the contact normal directions in each of the points. The two presented contact models keep most of the intuitiveness of the Task Frame approach, but have nevertheless to compromise on two points: (i) some basis screws in the models are not pure translations or pure rotations, and (ii) the screw axes don't always intersect in one single point.

If the contact surfaces are curved, on-line "tracking" algorithms are required in order to be able to continuously update the contact normal directions during the motion of the contact points. This tracking can be done, for example, with the "velocity-based" tracking approach explained in [1, 4].

Acknowledgment

H. Bruyninckx is Postdoctoral Fellow of the Fund for Scientific Research–Flanders (F.W.O.) in Belgium. Financial support by the Belgian Programme on Inter-University Attraction Poles initiated by the Belgian State—Prime Minister's Office—Science Policy Programme (IUAP) is gratefully acknowledged.

References

[1] H. Bruyninckx and J. De Schutter. Specification of force-controlled actions in the "Task Frame Formalism": A survey. *IEEE Trans. Rob. Automation*, 12(5):581–589, 1996.

[2] H. Bruyninckx and J. De Schutter. Where does the Task Frame go? In *The Eighth International Symposium of Robotics Research (Preprints)*, pages 86–91, Hayama, Japan, 1997.

[3] H. Bruyninckx, S. Demey, S. Dutré, and J. De Schutter. Kinematic models for model based compliant motion in the presence of uncertainty. *Int. J. Robotics Research*, 14(5):465–482, 1995.

[4] J. De Schutter and H. Van Brussel. Compliant robot motion I. A formalism for specifying compliant motion tasks. *Int. J. Robotics Research*, 7(4):3–17, 1988.

[5] M. T. Mason. Compliance and force control for computer controlled manipulators. *IEEE Trans. on Systems, Man, and Cybernetics*, SMC-11(6):418–432, 1981.

[6] H. West and H. Asada. A method for the design of hybrid position/force controllers for manipulation constrained by contact with the environment. In *IEEE Int. Conf. Robotics and Automation*, pages 251–259, St. Louis, MS, 1985.

The Instantaneous Kinematics of Manipulation

L. Han and J.C. Trinkle *
Dept. of Computer Science
Texas A&M University
lihan, trink@cs.tamu.edu

Abstract

Dextrous manipulation planning is a problem of paramount importance in the study of multifingered robotic hands. In this paper, we show in general, that all system variables (the finger joint, object, and contact velocities) need to be included in the differential kinematic equation used for manipulation planning, even if the manipulation task is only specified in terms of the goal configuration of the object or the contacts only. The dextrous manipulation kinematics that relates the finger joint movements to object and contact movements is derived. With the results of inverse and forward instantaneous kinematics, we precisely formulate the problem of dextrous manipulation and cast it in a form suitable for integrating the relevant theory of contact kinematics, nonholonomic motion planning, and grasp stability to develop a general technique for dextrous manipulation planning with multifingered hands.

1 Introduction

Given an object to be manipulated by a robotic hand, the goal of dexterous manipulation planning algorithms is to generate finger joint trajectories that can drive the object to the desired configuration and/or achieve the desired grasp. There are 3 types of manipulation tasks for multifingered hand systems:

- Object Manipulation – achieve the desired object configuration without regard for contact locations;

- Grasp Adjustment – obtain desired contact locations without regard for object configuration;

- Dextrous Manipulation – achieve the goal configuration for the object and contact points simultaneously

Velocity kinematic relationships for the first two types of manipulation tasks have been derived previously in [11, 6], respectively. The relationship for the first type of task involves only the joint and object velocities, while that of the second type contains only the joint and contact velocities. These relationships could be used as the basis of manipulatoin planning algorithms, but they lead to certain difficulties that can be avoided by using the velocity relationship implied by the third type of task, which includs joint, object *and* contact velocities. For example, developing object manipulation without considering the contact variables, the contact locations may be undesirable, since the stability of the grasp may be lost, as shown in one example in section 3. On the other hand, planning without object velocity may give a solution that causes the manipulated object collide with finger links or other objects in a crowded environment. In general, all variables in the system (the finger joint vairables, the object configuration and contact locations) need to be considered when planning any three types of manipulation planing tasks defined above. Thus the corresponding velocities should appear in the kinematic equations used for planning.

In this paper, we derive in detail the *instantaneous manipulation kinematics* relating object *and* contact movement to finger joint movement, by incorporating closed kinematic chain constraints[8] and the physical constraints imposed by the contact models(e.g. sliding, rolling etc). The kinematic equation reveals the constraints on the feasible velocities of the finger joints, the object, and the contact points to maintain a grasp with a given contact mode. We discuss the existence and uniqueness of the kinematic solution for the dextrous manipulation problem. With the results of the forward and inverse instantaneous kinematics, we precisely formulate the problem of dextrous manipulation planning and cast it in a form suit-

---

*This research was supported by the National Science Foundation under grant IRI-9619850, the Amarillo National Resource Center for Plutoniumunder grant number UTA95-0278, and the Texas Higher Education Coordinating Board under grant number ATP-036327-017. Any findings, conclusions, or recommendations expressed herein are those of the authors and do not necessarily reflect the views of the funding agencies.

able for integrating the relevant theories of contact kinematics[7], nonholonomic motion planning [4, 9, 2] and grasp stability[10] to develop a general technique for dexterous manipulation with multifingered hands.

2 Kinematics of Manipulation

2.1 Mathematical Preliminaries

We denote by $p_{ab} \in \mathrm{R}^3$ and $R_{ab} \in SO(3)$ the position and orientation of a coordinate frame B relative to another coordinate frame A, and call $g_{ab} = (R_{ab}, p_{ab}) \in SE(3)$ the Euclidean transformation of B relative to A. The velocity of B relative to A is denoted by $V_{ab} = (v_{ab}, \omega_{ab}) \in \mathrm{R}^6$. The adjoint transformation $Ad_{g_{ab}} \in \mathrm{R}^{6\times 6}$ associated with g_{ab} is used to transform velocity between coordinate frames.

Consider two smooth rigid bodies, F and O, in contact. Let $\alpha_f = (u_f, v_f) \in \mathrm{R}^2$ and $\alpha_o = (u_o, v_o) \in \mathrm{R}^2$ be the local coordinates of the contact points on F and O, respectively, and the corresponding Gaussian frames be C_f and C_o. A contact configuration between two bodies is described by $\eta = (\alpha_o, \alpha_f, \psi) \in \mathrm{R}^5$, where ψ is the contact angle and is defined by the respective Gaussian frames C_f and C_o. $\dot{\eta}$ is called the *contact coordinate velocity*.

Denote the contact velocity of F relative to O in terms of the local Gaussian frames by

$$V_c \stackrel{\text{def}}{=} V_{of}^{c_f} = [v_x,\ v_y,\ v_z,\ \omega_x,\ \omega_y,\ \omega_z]^T$$

The contact is maintained if $v_z = 0$. The contact model will introduce more constraints, called *physical constraints*, on the contact velocity, e.g., rolling constraint requires v_x and v_y to be zero and pure rolling further requires ω_z to be zero.

The physical constraints will limit the admissible contact coordinate velocities. For example, from Montana's kinematics of pure rolling contacts[7], $(\dot{\alpha}_f, \dot{\psi})$ can be related to $\dot{\alpha}_o$ by

$$[\dot{\alpha}_f, \dot{\psi}]^T = J_{roll}\dot{\alpha}_o$$

where J_{roll} is a matrix in $R^{3\times 2}$, whose entries depend on the geometric parameters of the object and fingers. In general, the physical constraints on the contact coordinate velocities can be represented in matrix form:

$$\dot{\eta} = J_{g_c}\dot{\eta}_{g_c} \qquad (1)$$

where η_{g_c} represents the set of free parameters of change rate of contact coordinates and will be referred as the generalized contact coordinate velocity; J_{g_c} is the Jacobian mapping the generalized contact coordinate velocity to contact coordinate velocity. The equation (1) for the pure rolling contacts and general contacts,i.e. contacts that admit all possible contact velocities, are

$$\text{pure rolling: } \dot{\eta} = \begin{bmatrix} I^{2\times 2} & J_{roll} \end{bmatrix}^T \dot{\alpha}_o \qquad (2)$$
$$\text{general contacts:} \dot{\eta} = I^{5\times 5}\dot{\eta}$$

2.2 Kinematics of Multifingered Hands

For a multifingered robotic hand system, let P be the palm frame, O be the object frame, and F_i, be the frame of fingertip i. Denote the forward kinematic map and the Jacobian of finger i by

$$g_{pf_i}(\theta_i) \in SE(3), \qquad V_{pf_i} = J_{pf_i}(\theta)\dot{\theta}_i$$

where V_{pf_i} is the velocity of finger i with respect to the palm, $\theta_i = (\theta_{i1}, \cdots \theta_{in_i})$ is the joint variable vector of finger i and n_i is the number of joints of finger i. Let the contact configuration between the object and finger i be $\eta_i = (\alpha_{o_i}, \alpha_{f_i}, \psi_i)$. For an m-fingered hand, let $n = \sum_{i=1}^m n_i$ and

$$\theta = (\theta_1, \cdots \theta_m) \in \mathrm{R}^n, \qquad \eta = (\eta_1, \cdots \eta_m) \in \mathrm{R}^{5m}$$

Given $q_i = (\theta_i, \eta_i)$, the position and orientation of the object can be obtained by composing the forward kinematic map of finger i with a transformation defined by η_i,

$$g_{po} = g_{pf_i}(\theta_i) \cdot g_{f_io}(\eta_i) \qquad (3)$$

Differentiating equation (3) yields the velocity of the object with respect to the palm:

$$\begin{aligned} V_{po} &= \begin{bmatrix} Ad_{g_{f_io}^{-1}} J_{pf_i}(\theta_i) & J_{c_i}(\eta_i) \end{bmatrix} \begin{bmatrix} \dot{\theta}_i \\ \dot{\eta}_i \end{bmatrix} \\ &\stackrel{\text{def}}{=} J_i(q_i)\dot{q}_i. \end{aligned} \qquad (4)$$

$J_i(q_i) \in \mathrm{R}^{6\times(n_i+5)}$ relates the object velocity to the rate of change of the extended joint coordinates, $q_i = (\theta_i, \eta_i)$, and is referred to as the extended Jacobian of finger i.

By equating the right hand side of (3) and (4) for $i = 1, \cdots m$, we have the following closed-kinematic chain (or simply closure) constraints

$$\boxed{g_{po} = g_{pf_1}(\theta_1)g_{f_1o}(\eta_1) = \cdots = g_{pf_m}(\theta_m)g_{f_mo}(\eta_m)}$$
(5)

and

$$\boxed{V_{po} = J_1(q_1)\dot{q}_1 = \cdots = J_m(q_m)\dot{q}_m} \qquad (6)$$

Equations (5) and (6) are called the position and velocity closure constraints, respectively.

By substituting equation(1) to the equation (4), we can incorporate the physical constraints into the kinematic chain:

$$\begin{aligned} V_{po} &= \begin{bmatrix} Ad_{g_{f_i o}^{-1}} J_{pf_i} & J_{c_i} J_{g_{c_i}} \end{bmatrix} \begin{bmatrix} \dot{\theta}_i \\ \dot{\eta}_{g_{c_i}} \end{bmatrix} \\ &\stackrel{def}{=} \begin{bmatrix} Ad_{g_{f_i o}^{-1}} J_{pf_i} & \tilde{J}_{c_i} \end{bmatrix} \begin{bmatrix} \dot{\theta}_i \\ \dot{\eta}_{g_{c_i}} \end{bmatrix} \\ &\stackrel{def}{=} \tilde{J}_i(q_i) \begin{bmatrix} \dot{\theta}_i \\ \dot{\eta}_{g_{c_i}} \end{bmatrix} \end{aligned} \quad (7)$$

where $\tilde{J}_i$ is the extended Jacobian with respect to the generalized contact coordinate velocity.

The corresponding closed kinematic chain constraints are:

$$\boxed{V_{po} = \tilde{J}_1 \begin{bmatrix} \dot{\theta}_1 \\ \dot{\eta}_{g_{c_1}} \end{bmatrix} = \tilde{J}_2 \begin{bmatrix} \dot{\theta}_2 \\ \dot{\eta}_{g_{c_2}} \end{bmatrix} = \cdots = \tilde{J}_m \begin{bmatrix} \dot{\theta}_m \\ \dot{\eta}_{g_{c_m}} \end{bmatrix}} \quad (8)$$

Equation(8) incoporates the closed kinematic chain constraints and the physical constraints of the contacts. It will be satisified only if the contact mode of the grasp is maintained.

2.3 Instantaneous Manipulation Kinematics

By straight-forward algebraic manipulation of equation(7), we get

$$Ad_{g_{f_i o}} V_{po} - Ad_{g_{f_i o}} \tilde{J}_{c_i} \dot{\eta}_{g_{c_i}} = J_{pf_i} \dot{\theta}_i. \quad (9)$$

Stacking equation(9) for each finger, we can write the constraint for an m-fingered hand in matrix form that explicitly shows the dependence of the object and contact velocities on the finger joint velocities:

$$\boxed{J_{oc} \tilde{V}_{oc} = J_f \dot{\theta}} \quad (10)$$

where

$$J_{oc} = \begin{bmatrix} Ad_{g_{f_1 o}} & -Ad_{g_{f_1 o}} \tilde{J}_{c_1} & & 0 \\ \vdots & & \ddots & \\ Ad_{g_{f_m o}} & 0 & & -Ad_{g_{f_m o}} \tilde{J}_{c_m} \end{bmatrix}$$

$$\tilde{V}_{oc} = \begin{bmatrix} V_{po} \\ \dot{\eta}_{g_{c_1}} \\ \vdots \\ \dot{\eta}_{g_{c_m}} \end{bmatrix}, \quad J_f = \begin{bmatrix} J_{pf_1} & & 0 \\ & \ddots & \\ 0 & & J_{pf_m} \end{bmatrix}, \quad \dot{\theta} = \begin{bmatrix} \dot{\theta}_1 \\ \vdots \\ \dot{\theta}_m \end{bmatrix}$$

The sizes of $J_{oc}, \tilde{V}_{oc}, J_f$, and $\dot{\theta}$ are, respectively, $6m \times (6 + CDOF), (6 + CDOF) \times 1, 6m \times n$, and $n \times 1$.

$CDOF$ is the dimension of admissible contact velocity components and defined as follows:

$$CDOF = \sum_{i=1}^{m} dim(\dot{\eta}_{g_{c_i}}). \quad (11)$$

Pure rolling contacts have $dim(\dot{\eta}_{g_c}) = 2$; and general contacts have $dim(\dot{\eta}_{g_c}) = 5$.

From equation(10), we can extract the the dependency of the object velocity on the finger joint velocity and cast it to matrix form:

$$G^T V_{po} = J_h \dot{\theta} \quad (12)$$

Equation(12) is called fundamental grasp constraint in [9], in which G and J_h are called Grasp Map and Hand Jacobian, respectively. G is also known as the wrench matrix in [5].

3 Object Manipulation

The problem of object manipulation is to determine the velocity of the object relative to the palm V_{po}, without the concern for contact movement.

For an assumed set of contact locations and models, equation (12) constrains the choice of $\dot{\theta}$ for a given V_{po} or vice versa. If a solution doesn't exist, then the contact mode, positions, and/or velocity values determined by the planner must be changed. Once equation(12) is satisfied, the planner may continue to progress.

Unfortunately, the satisfaction of equation(12) is not sufficient for planning. Consider the current configuration under consideration by the planner. Suppose it is stable through *force closure* but the contacts are located such that some small movement could cause a loss of force closure. In this case, equation(12) is not desirable for planning, because velocities satisfying equation(12) could destabilize the grasp. For example: suppose a ball of radius 1 is grasped by two spherical fingertips of radius 0.2 and the contacts undergo pure rolling. The contact parameters, $(\alpha_f, \alpha_o, \psi)$, are $(0, 0, 0, 0, 0)$ and $(0, 0, 0, \frac{2\pi}{3}, 0)$. Suppose the coefficient of friciton is $\mu = \tan(30^\circ)$, then it can be shown that two contact points form a force closure grasp[10] but they are at the boundary: the grasp will not be force closure if two points move toward each other just a little bit. However, if the ball is rotated about the axis that is parallel to Z-axis and passes the point$(1,0,0)$, it can be shown that the fingertip velocities determined by the generalized inverse of J_h are $(0, 0, 0, 0, 0, 0)$ and $(-0.8660, -1.4423, 0, 0, 0, -0.2885)$. The contact velocities (w_x, w_y) are $(1, 0)$ and $(1.2885, 0)$, which will move two contact points closer to each other. Then the force closure grasp will be lost even if the equation (12) is satisified.

For the task of grasp adjustment, the kinematic equation that only includes the contact and object velocity can be derived from equation (10) and has similar problem as equation(12): it may give a solution of grasp adjustment that causes the object collide with obstacles. Therefore, in general, all system variables(the object, contact and finger joint variables) need to be considered for the manipulation planning even if the task only speicifies goal configurations for a subset of system variables, like in the cases of object manipulation and grasp adjustment. Next we will discuss in detail the dextrous manipulation kinematics which include all system states in the equation.

4 Dextrous Manipulation Kinematics

The kinematic problems to be solved are

- Forward Instantaneous Kinematics

  Based on the kinematic constraints(10), given joint velocity $\dot{\theta}$, are the object and contact velocity $\tilde{V}_{oc}$ uniquely determined?

  If so, the system is said to be **Kinematically-Determined**.

- Inverse Instantaneous Kinematics

  Given the desired object and contact velocity $\tilde{V}_{oc}$, is it possible to find appropriate finger joint velocity $\dot{\theta}$ to obtain such a trajectory?

  If the answer for the above question for any specified object and contact trajectory is yes, the system is said to be **Manipulable**

4.1 Inverse Instantaneous Kinematics

Given $\tilde{V}_{oc}$, a necessary condition for the existence of joint velocity $\dot{\theta}$ to satisfy equation(10) is

$$J_{oc}\tilde{V}_{oc} \in \Re(J_f). \quad (13)$$

where $\Re(J_f)$ is the range space of J_f.

If $\Re(J_{oc}) \subset \Re(J_f)$, then any value of $\tilde{V}_{oc}$ is feasible, such a system is called **Manipulable**. There are no constraints for a manipulable system on the instantaneous object and contact trajectory, since the finger joints can generate any contact and object velocity. A sufficient condition for a system to be manipulable is that all fingers have 6 joints and all finger configurations are nonsigular, since in this case $\Re(J_f) = \Re^{6m}$.

Denote by $V^\perp$ the orthogonal complement of a space V. Then since $(V^\perp)^\perp = V$, a condition equivalent to equation (13) is

$$J_{oc}\tilde{V}_{oc} \in ((\Re(J_f))^\perp)^\perp \quad (14)$$

Recall the fundamental theorem of linear algebra[12]: $N(A^T) = (\Re(A))^\perp$, where $N(A)$ and $\Re(A)$ denote null space and range space of matrix A.

Suppose the singular value decomposition of matrix J_f is

$$J_f = U \begin{bmatrix} \sigma_1 & & 0 & 0 \\ & \ddots & & \vdots \\ 0 & & \sigma_r & 0 \\ 0 & \cdots & 0 & 0 \end{bmatrix} V^T = U \begin{bmatrix} \Sigma & 0 \\ 0 & 0 \end{bmatrix} V^* \quad (15)$$

where U and V are orthogonal matrices of size $6m$ and n respectively, $\sigma_1 \ldots \sigma_r$ are the singular values of J_f, and r is the rank of J_f.

Suppose $U = [\ U_1 \ \ U_2\ ]$, $V = [\ V_1 \ \ V_2\ ]$ where $U_1 \in \Re^{6m \times r}, U_2 \in \Re^{6m \times (6m-r)}, V_1 \in \Re^{n \times r}, V_2 \in \Re^{n \times (n-r)}$. Then

$$R(J_f) = Span(U_1), \quad N(J_f^T) = Span(U_2)$$

Thus condition(14) can be rewritten as the following:

$$\boxed{U_2^T J_{oc} \tilde{V}_{oc} = 0} \quad (16)$$

i.e., $\tilde{V}_{oc} \in N(U_2^T J_{oc})$.

Put a basis of the null space of $U_2^T J_{oc}$ as columns to form matrix J_{ocg}, then the solution for equation(16) is

$$\tilde{V}_{oc} = J_{ocg} \tilde{V}_{ocg} \quad (17)$$

where $\tilde{V}_{ocg}$ represents the set of free parameters of object and contact coordinate velocities, and we will refer to $\tilde{V}_{ocg}$ as the generalized object and contact coordindate velocities.

Suppose the condition(13) is satisfied, then the necessary and sufficient condition to uniquely determine the finger joint velocity $\dot{\theta}$ is

$$rank(J_f) = dim(\dot{\theta}) = n. \quad (18)$$

When condition (18) is satisfied, the finger joint velocities can be determined using the generalized inverse of J_f in equation(10),

$$\dot{\theta} = (J_f)^+ J_{oc} \tilde{V}_{oc} = (J_f^T J_f)^{-1} J_f^T J_{oc} \tilde{V}_{oc}. \quad (19)$$

If condition (13) is satisfied, we can substitute equation(17) for $\tilde{V}_{oc}$ into above equation and get the explicit dependence of the finger joint velocity on the generalized object and contact coordinate velocity $\tilde{V}_{ocg}$:

$$\dot{\theta} = (J_f)^+ J_{oc} J_{ocg} \tilde{V}_{ocg} \quad (20)$$

4.2 Forward Instantaneous Kinematics

For a given joint velocity, $\dot{\theta}$, a necessary condition for the existence of $\tilde{V}_{oc}$ satisfying equation(10) is

$$J_f \dot{\theta} \in \Re(J_{oc}) \quad (21)$$

If the contacts are maintained and the contact models are correct, the above condition is automatically satisfied. A violation of the above condition indicates the joint velocities will cause a change in the the contact mode. When $\Re(J_f) \subset \Re(J_{oc})$, any value of the joint velocity is valid.

For the general case, we can follow the steps as we have done for the generalized object and contact velocities(13) and get similar expression as equation(17) for feasible $\dot{\theta}$:

$$\dot{\theta} = J_{fg_f} \dot{\theta}_{g_f} \quad (22)$$

where $\dot{\theta}_{g_f}$ is the real free parameters of feasible $\dot{\theta}$ and will be referred as generalized finger joint velocity.

Suppose condition(21) is satisfied. If J_{oc} has full column rank, then the object and generalized contact velocity can be uniquely determined by the generalized inverse of J_{oc}:

$$\tilde{V}_{oc} = (J_{oc})^+ J_f J_{fg_f} \dot{\theta}_{g_f} \stackrel{\text{def}}{=} J_{ocg_f} \dot{\theta}_{g_f}. \quad (23)$$

Otherwise, there is not a unique value for $\tilde{V}_{oc}$, but rather an infinite set of possible values.

For a kinematically-determined system, it is sufficient to use kinematic-based control to obtain a specified object/contact trajectory since actuating the finger joints to achieve the desired joint trajectories forces the object and contact velocities to be desired.

When the system is kinematically underdetermined, there are infinite solutions for equation (10). Then dynamic control needs be used to remove the ambiguity of the motion of the object and contact points. Therefore, dynamics can be thought of as additional constraints which could possibly fully determine the system motions.

Note that when the system is manipulable and kinematically determined, there will be no constraints of instantaneous manipulation planning in terms of the object and contact trajectory, and the kinematic-based control is sufficient to achieve the desired trajectory.

5 Dexterous Manipulation Planning

The objective of manipulation planning is to generate joint trajectories for the fingers so that the goal configuration of the object and/or contacts can be achieved, without dropping the object. The concerned state variables for dextrous manipulation are g_{po} and η. From the equations(1), we get

$$V_d \stackrel{\text{def}}{=} \begin{bmatrix} V_{po} \\ \dot{\eta}_1 \\ \vdots \\ \dot{\eta}_m \end{bmatrix} = \begin{bmatrix} I & 0 & \cdots & 0 \\ 0 & J_{gc_1} & \cdots & 0 \\ \vdots & \vdots & \ddots & \vdots \\ 0 & 0 & \cdots & J_{gc_m} \end{bmatrix} \begin{bmatrix} V_{po} \\ \dot{\eta}_{g_{c_1}} \\ \vdots \\ \dot{\eta}_{g_{c_m}} \end{bmatrix}$$

$$\stackrel{\text{def}}{=} J_d \tilde{V}_{oc} \quad (24)$$

If the system is manipulable, then there will be no constraints on the object and generalized contact coordinate velocities to be feasible. Thus we can use the above equation to do the manipulation planning directly with respect to $\tilde{V}_{oc}$. If all the contact points are general contacts, then $dim(V_d) = dim(\tilde{V}_{oc}) = 6 + 5m$, i.e. the DOF of velocity is equal to the dimension of concerned variables. While for pure rolling contact system, $dim(\tilde{V}_{oc}) = 6 + 2m < dim(V_d)$, the nonholonomic motion planning problem [9] arises.

For a general system without manipulability, we need to apply the constraints on $\tilde{V}_{oc}$ to equation(24) and further formulate the manipulation planning problem with respect to generalized object/contact velocity(17):

$$\boxed{V_d = J_d J_{ocg} \tilde{V}_{ocg}.} \quad (25)$$

The desired finger joint velocity can be obtained using inverse kinematic solution (20).

Also we can formulate the problem directly with respect to the generalized finger joint velocity (23):

$$\boxed{V_d = J_d J_{ocg_f} \dot{\theta}_{g_f}} \quad (26)$$

The corresponding finger joint velocity can be obtained using equation (22).

Treating $\tilde{V}_{ocg}$ and $\dot{\theta}_{g_f}$ as the control inputs for equations(25) and (26) respectively, systems (25) (26) are referred to as standard nonholonomic systems in [4, 9]. Thus we can use general *nonholonomic motion planning* techniques to generate a trajectory for V_d, and thus, achieve the object and grasp goal configurations simultaneously.

For a manipulation task which only specifies the goal configuration for the object, as we discussed in the previous section, we also need to consider the contact trajectory to maintain or optimize the grasp quality. Then we can further expand the original object manipulation task to (1) achieve the goal object configuration and (2) improve the quality of grasp. For the second objective, we need to define a measure of grasp quality and then use the gradient search to move the contact configuration to a locally optimal grasp. We have applied this methodology in the manipulation planning to two special but important manipulation cases: one flat finger rolling a ball on a plane and two flat fingertips manipulating a ball. The experimental results are reported in paper [3]. One particular point to notice is that a grasp is characterized by the contact points on the object $\{\alpha_{o_i}, i = 1...m\}$ and α'_{o_i} can be used as the generalized contact coordinate velocity for pure rolling contacts as indicated by equation(3). Therefore, if all contacts in a manipulation system are pure rolling and the system is manipulable, we can determine α'_{o_i} first by optimizing some grasp quality measure and then substitute it back to equation(24) to further determine V_d. As for the grasp adjustment, while only the goal contact points are specified, the object trajectory need to be collision free.

6 Conclusion

In this paper, we showed that in general all system variables(the velocities of the finger joints, the object and contact points) need be considered in manipulation planning even if the goal is only specified as a subset of the system states. We derived the *dextrous manipulation kinematics* which relates object *and* contact movement to finger joint movement. The existence and uniqueness of the solution for the kinematic equation of the dextrous manipulation were discussed. Using the results from forward and inverse manipulation kinematics, we precisely formulated the problem of dextrous manipulation planning and cast it in a form suitable for integrating the relevant theory of contact kinematics, nonholonomic motion planning, and grasp stability to develop a general technique for dexterous manipulation with multifingered robotic hands.

The current theory will be generalized to incorporate issues like workspace limits of hands,uncertainty and dynamic constraints. We are currently applying the analysis methods presented in paper[1] to study various properties of dextrous manipulation. While the instantaneous kinematics reveals the kinematic constraints clearly and is informative for local motion planning, we still need a representation of configuration space of the hand-object system and global motion planning techniques to enable us to implement automatic dextrous manipulation planning. This forms part of our ongoing research topics.

References

[1] A. Bicchi, C. Melchiorri, and D. Ablluchi. On the mobility and manipulability of general multiple limb robots. *IEEE Trans. on R. & A.*, 11(2):215–228, 1995.

[2] A. Cole, J. Hauser, and S. Sastry. Kinematics and control of a multifingered robot hand with rolling contact. *IEEE Transaction on Automatic Control*, 34(4), 1989.

[3] L. Han, Y.S. Guan, Z.X.Li, Q. Shi, and J.C. Trinkle. Dextrous manipulation with rolling contacts. In *Proc. of IEEE Intl. Conf. on Robotics and Automation*, 1997.

[4] Z. X. Li and J. Canny, editors. *Nonholonomic Motion Planning*. Kluwer Academic Publisher, 1993.

[5] M. Mason and K. Salisbury. *Robot hands and the mechanics of manipulation*. MIT Press, 1985.

[6] D. Montana. *Tactile sensing and kinematics of contact*. PhD thesis, Division of Applied Sciences, Harvard University, 1986.

[7] D. Montana. The kinematics of contact and grasp. *IJRR*, 7(3), 1988.

[8] D. Montana. The kinematics of multi-fingered manipulation. *IEEE Trans. on R. & A.*, 11(4):491–503, 1995.

[9] R. Murray, Z.X. Li, and S. Sastry. *A Mathematical Introduction to Robotic Manipulation*. CRC Press, 1994.

[10] V. Nguyen. The synthesis of force-closure grasps. Master's thesis, Department of Electrical Engineering and Computer Science, MIT, 1986.

[11] J. K. Salisbury. *Kinematics and Force Analysis of Articulated Hands*. PhD thesis, Dept. of Mechanical Engineering, Stanford University, 1982.

[12] G. Strang. *Linear Algebra and Its Applications*. Academic Press, Inc., Orlando, Florida 32887, 1980.

Contact Response Maps for Real Time Dynamic Simulation *

C. Ullrich[†]
Institute of Applied Mathematics
University of British Columbia
Vancouver, Canada
ullrich@cs.ubc.ca

D. K. Pai[‡]
Department of Computer Science
University of British Columbia
Vancouver, Canada
pai@cs.ubc.ca

Abstract

We describe the generation and use of "contact response maps" for real time dynamic simulation. Contact response maps are geometry and material dependent maps on physical objects which describe the surface tractions associated with local deformations during contact. We develop a technique for precomputing contact response maps for elastic bodies using the Boundary Element method to solve the corresponding plane strain problem. Such maps can then be used in a real-time simulation environment to accurately resolve collision dynamics.

1 Introduction

Much effort in the past few years has been focused on the simulation of contact and collision phenomena. However, accurate simulation at interactive rates remains challenging. Contact events are extremely fast, and delays in simulating the contact force and motion can lead to instabilities in a control algorithm or destroy the sense of realism in a virtual environment. Unfortunately, the contact response of real objects is very complex; even with somewhat idealized linear viscoelastic models of the materials, simulation requires the numerical solution of partial differential equations (PDEs) on domains with complex shapes. Therefore, much effort has been directed at developing empirical contact models that are sufficiently accurate, while being sufficiently fast for real time simulation.

We present a solution to this problem by introducing a data structure associated with each colliding body called the *contact response map*. The contact response map can produce a more accurate collision response than the rigid body models used in the literature, since it is computed from the governing partial differential equations of the body. On the other hand, the contact response map allows the collision response to be computed rapidly, at rates suitable for control and interactive simulation.

In this paper, we demonstrate this approach with a model of a two dimensional elastic solid, whose behavior is governed by Navier's equation (a system of hyperbolic PDEs). The model shows that the response of an elastic solid varies significantly with impact location. In addition, we describe an algorithm to make use of the precomputed step response functions in a real time dynamic simulator.

The remainder of the paper is organized as follows. The rest of section 1 discusses related work in the field of impact mechanics. Section 2 discusses the theory of elastic solids subject to impact. Section 3 quickly derives the boundary element method. Section 4 contains the numerical results and relevant notes. Section 5 presents a time stepping algorithm for dynamical simulation. In Section 6 we consider the storage issues for contact maps. Finally in section 7 we summarize the results to date and discuss some possible

---

*Supported in part by grants from NSERC and IRIS.
[†]M.Sc. student, Mathematics.
[‡]Associate Professor, Computer Science.

generalizations of this work.

1.1 Related Work

Most models intended for computationally demanding use, including the model in this paper, employ the rigid body assumption. Rigid bodies are idealized physical objects which deform only locally during impacts. Following [Cha97], rigid body impact models may be classified as impulse response rigid or force response rigid.

Impulse response rigid models consider impact events as occurring over an infinitely short period of time, resulting in a discontinuous change in the velocities of colliding bodies. Impulse models define a coefficient of restitution, e, relating post-impact to pre-impact states, quantifying the loss of kinetic energy due to a collision event. The coefficient was defined in terms of change in velocity by Newton, change in normal momentum [Kel86], [Bra89] or the change in normal work [Str90].

It is only recently that consistent algorithms for resolving impulse response rigid collisions were developed. In [WM87], the authors present a complete solution for two dimensional impulsive collision with friction. Three dimensional impulsive collision is complicated by the existence of multiple solutions, as discussed in Bhatt and Koechling [BK95]. A complete 3D impulsive collision algorithm is discussed in [Mir96]. Many other authors present analyses of the 3D impact problem [Bra89, Bar92, PG96]. Chatterjee [Cha97] examined a number of 3D models for energetic consistency and accuracy.

Force response rigid models attempt to model the forces produced during contact. Models in this category typically define a function $f(u, \dot{u})$ that gives force as a function of penetration depth u and velocity $\dot{u}$. Linear force functions are considered in [GPS94a], while more generalized models are discussed in [HC75]. The Hertz model, in which $f(u) \propto u^{\frac{3}{2}}$ can be derived from physical principles as in [Gol60].

Resolving force response rigid collisions is accomplished by enforcing the action-reaction principle (Newton's third law), friction constraints, and the force function. For most models this results in a system of non-linear algebraic equations to be solved at each time step. See [Gol60] or [GPS94a, GPS94b] for detailed algorithms.

Recent work, especially that of Stoianovici and Hurmuzlu [SH96], has shown the coefficient of restitution to be a geometry dependent phenomenon. In their study, a long thin steel bar (fig. 1) was dropped onto a massive block. The experimental setup allowed the angle of the bar to be varied so that impact occurred at various points around the rounded tip. Their results show the restitution factor varying over a range from .9 for normal impact to as low as .2 for a critical impact angle of about 66 degrees.

All collision models above all try to capture the impact behaviour of an object with a few simple numbers such as the coefficient of restitution, or spring and damping constants (for the linear force law). With the advent of computers that have large memories, we can consider a radically different approach in which we pre-compute a large number of collision responses for a given object using very accurate, but computationally expensive methods such as finite elements or boundary elements.

2 Theory

2.1 Elastic Approximation and Range of Validity

In this study, we consider only impacts that occur at relatively low velocities. In the case of contact between bodies moving slowly, there will be no permanent deformation at the point of contact, and we are justified in assuming that the solids may be modeled by the elastic constitutive equations.

A recent study by Lim and Stronge [LSar] used finite element calculations to bound the pre-impact velocity in terms of the yield strength of a material. They demonstrated that velocities in the 1-5 m/s range produce no plastic deformation for typical metals such as aluminum or steel.

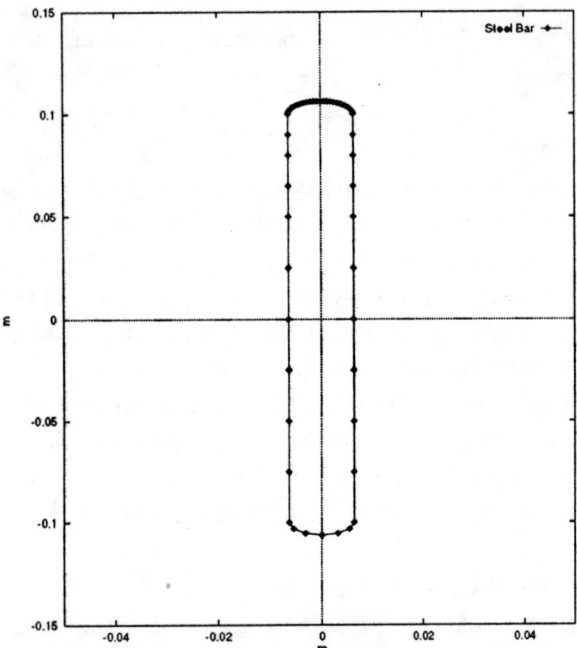

Figure 1: Steel Bar with Boundary Element Nodes shown. 56 nodes are shown, 26 along the bar length, and 28 on the tip. Nodes are more dense near the tip where we apply displacements.

2.2 Governing Equations

For simplicity we restrict our study to the 2D case. We need to model the response of an elastic solid to a contact event. In 2D, this is most easily done using the equations of plane-strain that model the response of an elastic solid which has negligible variation in the third axis (out of plane).

The field equations (Navier's equation) are written in component form as

$$\mu u_{i,jj} + (\lambda + \mu) u_{j,ji} + \rho b_i = \rho \ddot{u}_i \quad (1)$$

where $u \in \Omega$, u_i is the i'th component of the displacement, μ and λ are Lamé's constants, b_i is the i'th component of the body force, and ρ is the density. A comma denotes partial differentiation with respect to the space component, $\ddot{u}_i$ is the second time derivative of the i'th component of u. We apply the summation convention to repeated indices. Under the plane strain assumption, the indices i, j in Equation (1) run between 1 and 2. The equation is defined on $\Omega \subset \Re^2$.

2.3 Boundary Conditions

The field equations (1) give the generic behavior of an elastic body, but the boundary conditions specify a particular situation. An elastic body in free space not in contact with any other body has Neumann type boundary conditions. This is because the boundary is free to vibrate much like the free end of a string undergoing wave motion. Thus for all boundary points not in a contact situation (we define this region as Γ_1), we can say:

$$\frac{du}{dn} = 0,$$

where $u \in \Gamma_1$ and n is normal to the boundary Γ_1.

Now for a boundary segment which is in contact with an external body (Γ_2), we specify the displacement explicitly using a Dirichlet condition

$$u \cdot n = d_j,$$

where $u \in \Gamma_2$, and d_j is the j'th displacement depth (see below).

Given the field equations and boundary conditions, we can solve the well posed PDE using a variety of numerical techniques. A popular choice is the finite element method (FEM) in which the region Ω is discretized to solve the PDE. We consider instead the boundary element method which has a number of advantages over FEM, especially for elastodynamic contact problems.

3 Boundary Element Solution

There has been a great deal of work in the past two or three decades in the boundary element community on the solution of elastostatic and elastodynamic problems. Boundary element (BEM) solution techniques are ideally suited to contact problems because all of the unknowns are on the boundaries of the colliding bodies. In contrast, FEM (finite element) approaches deal with the entire body and solve for stresses and displacements at every point.

The basic idea of BEM is to formulate the partial differential equation as an integral equation. There are several ways of doing this for the Navier field equation. We briefly outline one approach and refer the interested reader to Dominguez [Dom93].

Denote the solution to Eq. 1 by u, with stress field σ and body forces b. Consider a second solution to the field equation (1) — with different initial conditions — for the same geometry, u^* with stress field σ^* and body forces b^*. The weighted residual of the two solutions is written as

$$\int_\Omega (\sigma_{kj,j} * u_k^*) d\Omega + \int_\Omega \rho(b_k * u_k^*) d\Omega - \int_\Omega \rho(\ddot{u}_k * u_k^*) d\Omega = 0, \qquad (2)$$

where * denotes a convolution product. Applying the divergence theorem, we can rewrite this as

$$\int_\Gamma (p_k * u_k^*) d\Gamma$$
$$+ \int_\Omega \rho(b_k * u_k^* + u_{0k}\dot{u}_k^* + \dot{u}_{0k}u_k^*) d\Omega$$
$$= \int_\Gamma (p_k^* * u_k) d\Gamma$$
$$+ \int_\Omega \rho(b_k^* * u_k + u_{0k}^*\dot{u}_k + \dot{u}_{0k}^*u_k) d\Omega, \qquad (3)$$

where $u_{0k} = u_k(x,0)$, $\dot{u}_{0k} = \dot{u}_k(x,0)$ are initial displacements and velocities respectively. Now, we generate the reciprocal solution u^* as a solution of the field equations when an impulsive load is applied at the point x^i in the direction l, i.e.,

$$\rho b_k^* = \delta(t)\delta(x - x^i)\delta_{lk}. \qquad (4)$$

Further, assume the initial conditions are identically zero and that the body forces of the primary solution may be neglected. Then we can write Eq. 3 as

$$u_l(x^i, t) = \int_\Gamma (u_{lk}^* * p_k) - (u_k * p_{lk}^*) d\Gamma, \qquad (5)$$

which gives the displacement at point x^i in direction l as a function of the boundary. Note that this same expression can be obtained from a variational principle (see [AP92]).

The integrals in Equation 5 are difficult or impossible to solve analytically. The numerical solution to these integrals constitutes the boundary element method.

Observe that we need only to discretize the boundary, which immediately reduces the dimension of the problem by one. In contrast, the FEM approach requires the domain Ω to be discretized. Typical BEM algorithms divide up the boundary into piecewise constant, linear or quadratic elements. Corners require special treatment. Also, it is not uncommon for the resulting integrals to be highly singular, in which case, the Cauchy principal value is used.

As a result of the discretization of the boundary and approximate evaluation of the integrals, we obtain a system of linear algebraic equations for the unknown boundary displacements or tractions. Although the system is much smaller than a similar FEM system, the matrix presents no special symmetry or sparseness properties and must be solved using dense matrix techniques.

When the integral (5) is discretized, the resulting linear equations can be rearranged so that for nodes on which displacements are given, the tractions are solved for and for nodes with given traction, the displacement is solved for. Hence for our boundary conditions, the BEM gives the surface traction (force response) for the point at which we specify the displacement. We write this as

$$f_j^A(t), \qquad (6)$$

which is the contact response at node j for body A. We store a discrete representation of $f_j^A(t)$ at a each boundary node j on a geometric model of body A. This is the contact response map.

4 Numerical Results

We considered a metal bar with the same dimensions as in [SH96], 20 cm in length with a diameter of 12.5mm and spherical ends (see fig. 1).

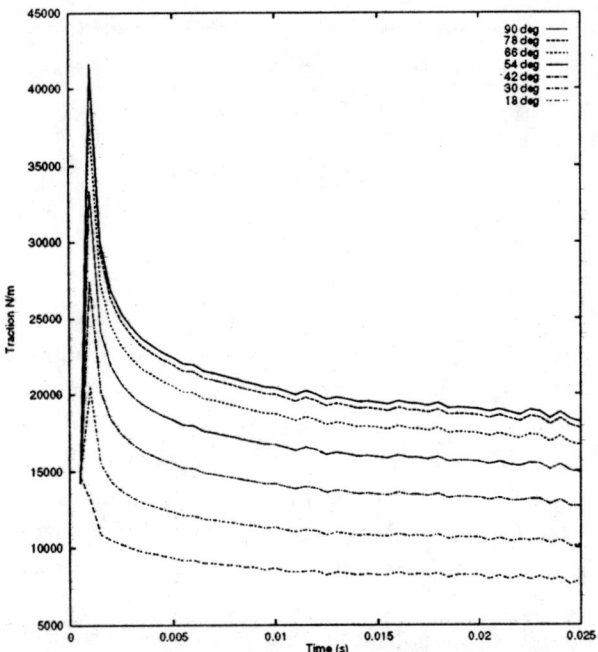

Figure 2: Step response curves for various impact angles at bar end. Angles are measured with respect to the horizontal axis of the bar in Figure 1.

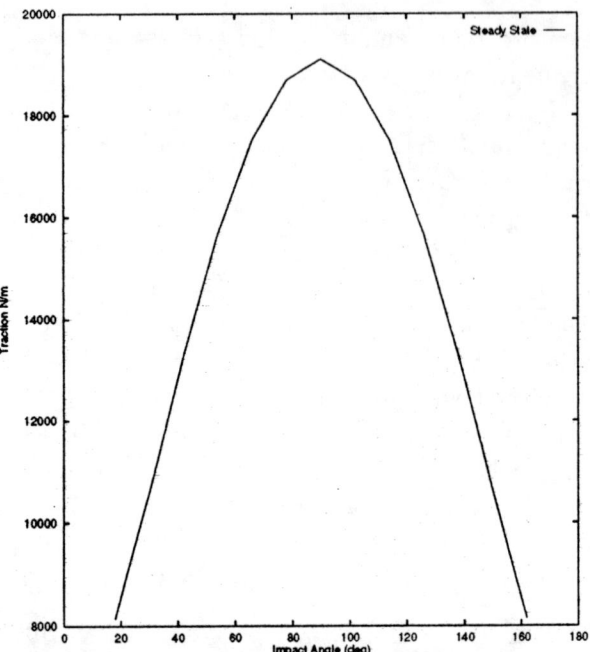

Figure 3: Steady state tractions giving the spring constant for each impact angle.

The boundary was discretized using 28 quadratic boundary elements, with three node points each. Using software developed by Dominguez [Dom93] for the simulation of elastodynamic systems, we computed the traction response at a number of node points to a step displacement. Calculations were performed on a Pentium II 266MHz computer, with each curve taking 3-4 minutes to generate. The results are shown in fig. 2 and clearly demonstrate the dependence on geometry of impact location.

The traction response curves shown in fig. 2 can be utilized for dynamic simulation in a variety of ways. Because we use a linear elastic model, the traction due to an arbitrary displacement trajectory $u(t)$ can be reconstructed from the contact response map by the superposition principle. We describe this in §5.

One could also use the contact response map to estimate parameters for simpler contact models. For instance, consider an approximate steady state response calculated from the tail end of the step response using an average (constant fit). In using only the steady state, we are effectively generating the proportionality ("spring") constant k in a force response rigid law $f(y) = -ky$. This law is only intended to demonstrate the simplest application of these results. Nevertheless, even this approximation demonstrates the geometry dependence of the contact response. This can be seen in Figure 3, which plots the effective spring constant k with respect to the contact position (angle) on the bar tip. A related map generated using only the steady state response is discussed in [TB93] where it is called the flexibility matrix.

More complex force models can be fit to BEM data by modifying the displacement function. The suitability of particular models is discussed in detail in [HC75].

5 Contact Simulation

We now describe an algorithm which uses the pre-computed boundary element surface tractions to generate collision forces during interactive simulations.

5.1 Action Reaction

The remaining requirement for a full solution to the contact problem is to enforce the action-reaction principle, or Newton's Third Law. As this is dependent on the exact contact configuration, it must be done during the simulation process. We state this requirement as: $f^A = -f^B$.

In our model, the response of the elastic solid is linear in the displacement depth, which allows us to write the force as a function of current depth and any residual effects from previous depths.

To simulate the contact, it is possible to define several numerical schemes for time stepping. We outline one scheme here; for simplicity we present the frictionless case — since the algorithm computes the normal contact force, one can extend this method to include Coulomb friction effects, as in [GPS94a].

We assume in this discussion that the elements in contact are already known, or are determined between each complete step of the following algorithm. It is not the purpose of this paper to discuss this problem, but an algorithm such as the enhanced gjk [Cam97] can be modified to get penetration depths for each element.

We first introduce some additional notation. Let the nominal, undeformed position of the contact point on body A be $\tilde{u}^A$, and let the actual position be u^A; then the penetration of A, δ^A, is defined as
$$\delta^A = \tilde{u}^A - u^A.$$

When the time needs to be specified, we will use the abbreviation $\delta_k^A = \delta^A(t_k)$ and the backward time difference operator Δ, defined as
$$\Delta \delta_k^A = \delta_k^A - \delta_{k-1}^A.$$

Finally, let $f_s(t)$ be the surface traction response to a unit step (of boundary node displacement) at $t = 0$.

At time t_k, we advance the simulation to time t_{k+1} as follows:

1. Compute the nominal forces on the body due to penetration upto time t_k.
$$\tilde{f}_{k+1}^A = \sum_{j=0}^{k} f_s^A(t_{k+1} - t_{k-j}) \Delta \delta_{k-j}^A.$$

   $\tilde{f}_{k+1}^B$ is computed similarly. We note that if the time steps are of equal size, this can be done efficiently as a digital filter applied to the penetration trajectory.

2. Integrate the equations of motion for each body (which may be part of a larger multi-body system), using an appropriate ODE or DAE integrator, and compute the nominal positions $\tilde{u}_{k+1}^A$ and $\tilde{u}_{k+1}^B$. Integration of multibody dynamics equations is treated extensively elsewhere in the literature (e.g., [APC97]). For adaptive stepsize integrators, additional bookkeeping needs to be done to ensure that the convolution in step 1 is correct.

3. Compute the total penetration along the contact surface normal n as
$$\delta_{k+1} = (\tilde{u}_{k+1}^A - \tilde{u}_{k+1}^B) \cdot n^B.$$

   Here n is pointed from B to A. We assume that the the total displacement of the bodies during contact is small relative to spacing of boundary element nodes.

4. Action-Reaction. Compute the penetration of each body so that it satisfies the kinematic constraint
$$\delta_{k+1}^A + \delta_{k+1}^B = \delta_{k+1} \quad (7)$$

   and Newton's third law
$$f_{k+1}^A = -f_{k+1}^B. \quad (8)$$

   Since
$$f_{k+1}^A = \tilde{f}_{k+1}^A + f_s^A(t_{k+1} - t_k) \Delta \delta_k^A,$$

   this is a linear equation in δ_{k+1}^A and δ_{k+1}^B. Therefore we can solve the Equations 7 and

8 simultaneously for the actual penetrations δ_{k+1}^A and δ_{k+1}^B.

If necessary, f_{k+1}^A and u_{k+1}^A can be computed from δ_{k+1}^A.

5. Advance time to t_{k+1}.

5.2 Sustained Contact, Multiple Contacts

Since the model we describe is force response rigid, sustained contact requires no special processing. This is a definite advantage over impulsive models which require additional calculations and assumptions to handle long term contact situations.

Force response models also have the advantage that multiple contacts are handled simply by superposition. In the impulse case, multiple contacts are known to cause (sometimes severe) uniqueness problems as discussed in [Bar92].

6 Storage considerations

Clearly this approach to dynamic simulation uses more memory per object than the coefficient of restitution approach [Cha97] or the spring-damper approach [GPS94a]. For each boundary element, we store N_f numbers to describe the force response ($N_f = 1$ for our steady state model). We can write the total storage per object as $N_b N_f$ floating point numbers. However, this is modest by the standards of today's computers. For example, in our study we used $N_b = 28$ boundary elements and the number of boundary points can often be reduced by considering object symmetries.

7 Conclusions

We presented a numerical technique to quantitatively generate the force response due to contact, given geometry and material parameters. Our approach is more accurate than impact laws currently used for real time simulation, since it is based on the PDEs of linear elastic solids. The PDEs are solved using well established boundary element method. Our approach captures experimentally observed impact phenomena such as the dependence of the contact response on the particular geometry of impacting objects. Given a computed force response, real time simulation becomes a possibility even for complex multibody systems. This will allow for the construction of detailed virtual environments that can be relied upon as faithful representations of the real world, useful in engineering, education and other fields.

In future work, we plan to extend this approach to other types of continuous media such as viscoelastic and anisotropic materials, and to demonstrate it in a 3D simulation environment being developed in our lab. We also plan to directly estimate the contact response map for more complex objects, from responses measured with our ACME facility.

References

[AP92] H. Antes and P. D. Panagiotopoulos. *The Boundary Integral Approach to Static and Dynamic Contact Problems*. Birkhauser, Basel - Boston - Berlin., 1992.

[APC97] U. Ascher, D. K. Pai, and B. Cloutier. Forward dynamics, elimination methods, and formulation stiffness in robot simulation. *International Journal of Robotics Research*, 16:6, December 1997. (to appear).

[Bar92] D. Baraff. *Dynamic Simulation of Non-Penetrating Rigid Bodies*. PhD thesis, Cornell University, 1992.

[BK95] V. Bhatt and J. Koechling. Three-dimensional frictional rigid-body impact. *ASME Journal of Applied Mechanics*, 62:893–898, 1995.

[Bra89] R. M. Brach. Rigid body collisions. *ASME Journal of Appied Mechanics*, 56:133–138, 1989.

[Cam97] S. Cameron. Enhancing gjk: Computing minimum and penetration distances between convex polyhedra. In *IEEE International Conference on Robotics and Automation*, pages 3112–3117, 1997.

[Cha97] A. Chatterjee. *Rigid Body Collisions: Some General Considerations, New Collision Laws, and Some Experimental Data*. PhD thesis, Cornell University, 1997.

[Dom93] J. Dominguez. *Boundary Elements in Dynamics*. Elsevier Science, Essex, 1993.

[Gol60] W. Goldsmith. *Impact: The theory and physical behavior of colliding solids.* Edward Arnold, London, 1960.

[GPS94a] S. Goyal, E. Pinson, and F. Sinden. Simulation of dynamics of interacting rigid bodies including friction i: General problem and contact model. *Engineering with Computers*, 10:162–174, 1994.

[GPS94b] S. Goyal, E. Pinson, and F. Sinden. Simulation of dynamics of interacting rigid bodies including friction ii: Software system design and implementation. *Engineering with Computers*, 10:175–195, 1994.

[HC75] K. H. Hunt and F.R.E. Crossley. Coefficient of restitution interpreted as damping in vibroimpact. *ASME Journal of Applied Mechanics*, 42:440–445, 1975.

[Kel86] J. B. Keller. Impact with friction. *ASME Journal of Applied Mechanics*, 53:1–4, 1986.

[LSar] C. T. Lim and W. J. Stronge. Normal elastic-plastic impact in plane strain. *Math and Computer Modeling*, to appear.

[Mir96] B. Mirtich. *Impulse-based Dynamic Simulation of Rigid Body Systems*. PhD thesis, UC Berkeley, 1996.

[PG96] F Pfeiffer and C Glocker. *Multibody Dynamics with Unilateral Contacts*. Wiley, New York, 1996.

[SH96] D. Stoianovici and Y. Hurmuzlu. A critical study of the applicability of rigid-body collision theory. *ASME Journal of Applied Mechanics*, 63:307–316, 1996.

[Str90] W. J. Stronge. Rigid body collisions with friction. *Proc. R. Soc. Lond. A*, 431:169–181, 1990.

[TB93] S. Takahashi and C. A. Brebbia. Elastic contact analysis with friction using boundary elements flexibility approach. In M. H. Aliabadi and Brebbia C. A., editors, *Computational Methods in Contact Mechanics*. Elsevier Science, Essex, 1993.

[WM87] Y. Wang and M. Mason. Modeling impact dynamics for robotic operations. In *IEEE International Conference on Robotics and Automation*, pages 678–685, 1987.

The Roles of Shape and Motion in Dynamic Manipulation:
The Butterfly Example

Kevin M. Lynch*   Naoji Shiroma†   Hirohiko Arai‡   Kazuo Tanie‡

*Mechanical Engineering Department
Northwestern University
Evanston, IL 60208 USA

†Institute of Engineering Mechanics
University of Tsukuba
1-1 Tennodai, Tsukuba, 305 Japan

‡Biorobotics Division
Mechanical Engineering Laboratory
Namiki 1-2, Tsukuba, 305 Japan

Abstract

We are studying a juggler's skill called the "butterfly." Starting with a ball resting on the palm of his/her open hand, a skilled juggler can accelerate and shape his/her hand so that the ball rolls up the fingers, over the top, and back down to the back of the hand. This paper describes a robotic implementation of the butterfly. The hand's shape and motion combine to effect the rolling motion of the ball, and we find that the shape and motion parameters enter the dynamic equations in a similar way. We define parameterized spaces of hand shapes and motions, and using a simulation based on the rolling equations, we identify shape and motion solutions that roll the ball from one side of the hand to the other. We describe an implementation of the butterfly on our planar dynamic manipulation testbed FLATLAND. *This example is our first step toward exploring the roles of shape and motion in dynamic manipulation.*

1 Introduction

The robot manipulation problem is to find a set of controls that map the current state of the world to a goal state, where this mapping $\mathcal{M}$ is governed by the laws of physics. Typically the robot controls are specified as motions or forces and the system is engineered to simplify the mapping $\mathcal{M}$, making it easier to construct robot plans. For instance, a grasped object tracks the motion of the hand, simplifying $\mathcal{M}$ to an identity relationship between hand and object motions.

For manipulation by pushing, throwing, tapping, rolling, and catching, however, the mapping $\mathcal{M}$ is a function of manipulator shape, compliance, friction, restitution, etc. If we treat these as design (control) variables, then it is possible to exploit the information embedded in $\mathcal{M}$ to simplify robot hardware. Instead of constructing powerful dexterous robots to force $\mathcal{M}$ to be a simple function, we can construct simple robots which work with, not against, the natural environmental dynamics.

In this paper we consider the relative roles of manipulator shape and motion in dynamic nonprehensile manipulation. By *nonprehensile* we mean that the manipulated object is not grasped. By *dynamic* we mean that the robot exploits dynamics to help control the motion of the part. An example of dynamic nonprehensile manipulation is shooting a basketball. As the hand moves, dynamic forces cause the ball to roll up the hand and off the fingertips into a free-flight trajectory toward the hoop. With dynamic nonprehensile ma-

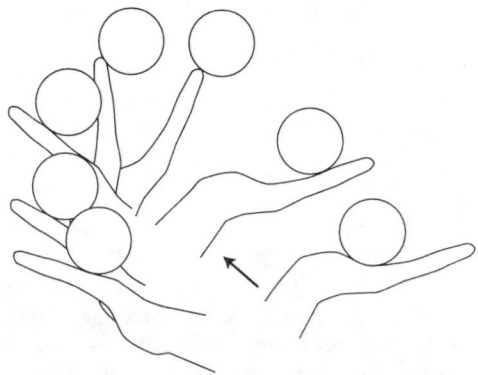

Figure 1: Sketch of a juggler doing a "butterfly."

nipulation, a robot can cause motion of the part relative to the end-effector, thereby controlling more part degrees-of-freedom (Lynch and Mason [9]).

This paper focuses on rolling manipulation, where an object rolls freely on the surface of the manipulator. We find that the derivative of the curvature of the manipulator surface and the acceleration of the manipulator enter into the dynamic equations of rolling in a similar way. This implies that, at least locally, we can trade freedom in the manipulator trajectory for freedom in the manipulator surface shape. To obtain a desired rolling motion of the object, we may be able to reduce the number of actuators required by properly designing the shape of the surface. This is a type of *dynamic cam*—we can use rolling dynamics and freedom in designing the shape of the manipulation surface to transform simple rotational or translational actuator motions to the desired motion of the object.

2 The Butterfly

We have begun our investigation into the roles of shape and motion in dynamic manipulation by examining a juggler's skill called the "butterfly" (Figure 1). Starting with a ball resting on the palm of his/her open hand, a skilled juggler can accelerate and shape his/her hand so the ball rolls up the fingers, over the top, and down to the back of the hand.

Our goal was to perform a planar version of this skill with a one joint robot, as shown in Figure 2. The design problem is to find a hand shape and 1 DOF revolute motion profile that rolls the disk from one side of the hand to the other while maintaining rolling contact at all times.

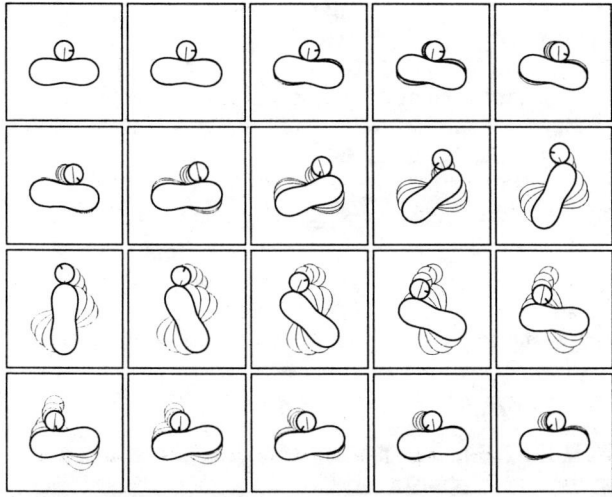

Figure 2: A simulated robotic butterfly. Gravity acts downward. The contact force is shown.

3 Dynamic Rolling Equations

In this section we present the dynamic equations of rolling between a moving surface (the "hand," which could be any manipulator surface) and an object rolling with point contact on the surface. We assume a hard contact model—no torque can be applied about the contact normal, and spin about the contact normal is not explicitly prohibited. Only slip is disallowed.

The kinematic equations of rolling have been derived previously in the context of rolling within a grasp. First-order analysis relating the relative velocity of two objects to the change in contact coordinates has been carried out by Kerr and Roth [7], Montana [10], and Cole et al. [3]. Sarkar et al. [11] built on Montana's work by deriving the second-order relationship between the relative motion of the contacting bodies and the acceleration of the contact coordinates. This information is used in the dynamic control of rolling motion in a grasp. Cai and Roth [2] derived the equations of motion in a manner allowing higher-order analysis.

Our derivation builds on this work, but differs in two ways. 1) The motion of the object relative to the hand is not specified, but rather is determined by the acceleration (and resultant contact forces) of the hand. The input to the equations of motion is the acceleration of the hand, and the outputs are the acceleration of the object and the acceleration of the contact coordinates. This is a second-order dynamic analysis. 2) Our development applies to both the planar and the spatial case. For this reason, we adopt notation similar to that of Cole et al. [3] (see Figure 3). In the spatial case, simple transformations provide the metric tensor, curvature tensor, and torsion form used by Montana [10].

The hand is a one-dimensional curve (planar case) or a two-dimensional surface (spatial case). The hand is locally parameterized by s_h, where $s_h \in \mathbf{R}$ in the planar case and $s_h \in \mathbf{R}^2$ in the spatial case. In a coordinate frame $\mathcal{F}_h$ attached to the hand, the contact position is given by $c_h(s_h)$, where $c_h \in \mathbf{R}^2$ in the planar case and $c_h \in \mathbf{R}^3$ in the spatial

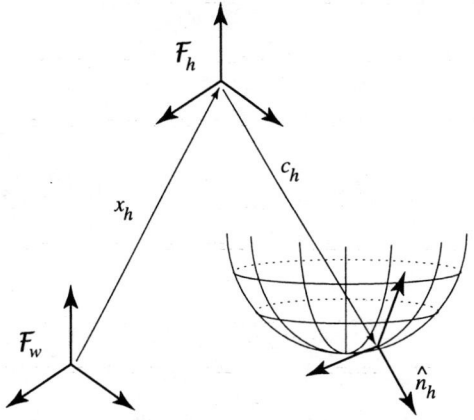

Figure 3: Coordinate conventions for the hand.

case. We define the 2x1 (or 3x2) Jacobian

$$J_h = \frac{\partial c_h}{\partial s_h}.$$

The outward-pointing unit contact normal at s_h is given by $\hat{n}_h(s_h)$. In the planar case, $\hat{n}_h$ is simply $J_h/\|J_h\|$ rotated by 90 degrees so that it is the outward-pointing normal. (Alternatively, considering $J_h/\|J_h\|$ as a 3-vector with zero third component, $\hat{n}_h$ is this vector crossed with a unit vector out of the plane.) In the spatial case, $\hat{n}_h$ is the normalized cross-product of the two columns of J_h. The 2x1 (or 3x2) Jacobian of $\hat{n}_h$ defines the curvature of the surface and is written

$$K_h = \frac{\partial \hat{n}_h}{\partial s_h}.$$

The contact point in a world frame $\mathcal{F}_w$ is given by

$$x_h + R_h c_h(s_h),$$

where x_h is the location of $\mathcal{F}_h$ in $\mathcal{F}_w$ and R_h is a rotation matrix ($x_h \in \mathbf{R}^n$, $R_h \in SO(n)$ where $n = 2$ in the planar case and $n = 3$ in the spatial case). The time derivatives of x_h are written $\dot{x}_h = v_h$ and $\ddot{x}_h = a_h$. The angular velocity and angular acceleration of $\mathcal{F}_h$ in $\mathcal{F}_w$ are written ω_h and α_h, respectively. In the planar case, we have

$$\dot{R}_h = (\omega_h \times) R_h = \begin{pmatrix} 0 & -\omega_h \\ \omega_h & 0 \end{pmatrix} R_h,$$

and in the spatial case, we have

$$\dot{R}_h = (\omega_h \times) R_h = \begin{pmatrix} 0 & -\omega_{h3} & \omega_{h2} \\ \omega_{h3} & 0 & -\omega_{h1} \\ -\omega_{h2} & \omega_{h1} & 0 \end{pmatrix} R_h.$$

The matrix $(\alpha_h \times)$ may be defined similarly to $(\omega_h \times)$.

We can make similar definitions for the object being manipulated, replacing the subscript "h" (for hand) with the subscript "o" (for object). The object frame $\mathcal{F}_o$ is fixed to the center of mass of the object. The mass of the object is m and its inertia matrix is I_o expressed in $\mathcal{F}_o$.

We are now ready to derive the rolling equations.

1) Contact position constraint. The contact points on the object and the hand must be coincident:

$$x_h + R_h c_h - (x_o + R_o c_o) = 0.$$

This constraint differentiates twice to yield the following two (three) linear equations in the planar (spatial) case.

$$a_h + (\alpha_h \times + \omega_h \times \omega_h \times) R_h c_h + 2\omega_h \times R_h J_h \dot{s}_h$$
$$+ R_h(J_h \ddot{s}_h + \dot{J}_h \dot{s}_h) - (a_o + (\alpha_o \times + \omega_o \times \omega_o \times) R_o c_o$$
$$+ 2\omega_o \times R_o J_o \dot{s}_o + R_o(J_o \ddot{s}_o + \dot{J}_o \dot{s}_o)) = 0. \quad (1)$$

2) Contact normal constraint. The unit contact normals must be opposite:

$$R_o \hat{n}_o + R_h \hat{n}_h = 0.$$

This constraint differentiates twice to yield the following two (three) linear equations in the planar (spatial) case. One of these equations is redundant because $\hat{n}_o$ and $\hat{n}_h$ are constrained to be unit.

$$(\alpha_o \times + \omega_o \times \omega_o \times) R_o \hat{n}_o + 2\omega_o \times R_o K_o \dot{s}_o$$
$$+ R_o(K_o \ddot{s}_o + \dot{K}_o \dot{s}_o) + (\alpha_h \times + \omega_h \times \omega_h \times) R_h \hat{n}_h$$
$$+ 2\omega_h \times R_h K_h \dot{s}_h + R_h(K_h \ddot{s}_h + \dot{K}_h \dot{s}_h) = 0. \quad (2)$$

3) Rolling constraint. To maintain rolling, the acceleration of the contact points on the object and the hand must be equal when projected to the contact tangent space. This yields one (two) linear equations in the planar (spatial) case.

$$R_h J_h^T (a_h + (\alpha_h \times + \omega_h \times \omega_h \times) R_h c_h$$
$$- a_o - (\alpha_o \times + \omega_o \times \omega_o \times) R_o c_o) = 0. \quad (3)$$

4) Force constraint. The contact force passes through the contact point. This yields one (three) linear equations in the planar (spatial) case:

$$\tau - r \times f = 0$$

or equivalently,

$$I_w \alpha_o + \omega_o \times I_w \omega_o - R_o c_o \times m a_o = 0. \quad (4)$$

$I_w = R_o I_o R_o^T$ is the object inertia matrix expressed in the world frame $\mathcal{F}_w$. The $\omega_o \times I_w \omega_o$ term vanishes in the planar case.

We would like to solve Equations (1)–(4) for a_o, α_o, $\ddot{s}_o$, and $\ddot{s}_h$ as a function of the state of the system and the inputs a_h and α_h. (We assume the input is the acceleration of the hand, not force/torque.) Rearranging the equations, we get the form

$$\mathbf{Ax} = \mathbf{b},$$

where $\mathbf{A}$ is 6x5 (11x10), $\mathbf{x}$ (the variables to be solved for) is a 5-vector (10-vector), and $\mathbf{b}$ is a 6-vector (11-vector) in the planar (spatial) case. These equations may be solved by premultiplying each side by $(\mathbf{A}^T \mathbf{A})^{-1} \mathbf{A}^T$, the pseudo-inverse of $\mathbf{A}$, allowing us to simulate dynamic nonprehensile rolling manipulation of smooth objects. In practice we also add a "snapping" routine to make sure the object stays on the surface of the hand. We have written a rolling simulator in C. The simulation enforces rolling contact; we check the contact force implied by a_o and α_o to see if contact and friction constraints are satisfied.

The curvature derivative shape information in the term $\dot{K}_h = \frac{\partial K_h}{\partial s_h} \dot{s}_h$ (Equation 2) gives us a way to design the shape of the hand to affect the rolling motion in a way similar to the acceleration of the hand. This curvature derivative information integrates to yield the shape of the hand.

4 Designing the Shape and Motion

Using the rolling simulation, we would like to design a hand shape and motion to solve the planar butterfly task. Below we describe a space of shapes and motions from which we will find a solution.

4.1 Shape Space

We would like a symmetric hand with a stable well (local minimum in a gravity field) when the disk is at the beginning or end of the roll, as in Figure 2. We chose simple polynomial functions satisfying these conditions:

$$x(s_h) = a + c s_h^2 + e s_h^4$$
$$y(s_h) = b s_h + d s_h^3 + f s_h^5,$$

where $s_h \in [-1, 1]$ describes the right half of the hand. (The other half is the mirror image.) Note that $x(s_h)$ and $y(s_h)$ are even and odd functions, respectively. The six coefficients must satisfy the two independent equations $x(-1) = x(1) = 0$ and $y'(-1) = y'(1) = 0$, where the constraint on $y'(\pm 1)$ forces the hand to be flat at these points. We chose $a = 18$, $b = -18$, leaving a two-dimensional shape space which can be parameterized by c and d (Figure 4). Essentially d controls the depth of the well and c controls whether or not the curvature is pushed out to the ends of the hand. For our experiments, we chose $a = 18.0$cm, $b = -18.0$cm, $c = -25.2$cm, $d = 21.6$cm, $e = 7.2$cm, and $f = -9.36$cm.

Two important points are worth noting. First, not all points in the shape space correspond to appropriate hands. Some choices of c and d yield self-intersecting curves, curves without wells, or wells with curvature too high to accommodate the disk (radius of 5.1cm in the experiments). Second, the curvature is discontinuous at $s_h = \pm 1$, because we have required only that $\hat{n}_h$ be continuous there. The rolling equations are ill-defined at $s_h = \pm 1$. This problem can be solved by increasing the order of the shape polynomials and forcing them to satisfy continuous curvature and curvature derivative constraints at $s_h = \pm 1$. Here we use the low order polynomials and simply take care in the simulation near these singularities.

4.2 Motion Space

We would like to find a symmetric hand motion that dips the hand to begin the disk rolling, rotates 180 degrees to roll

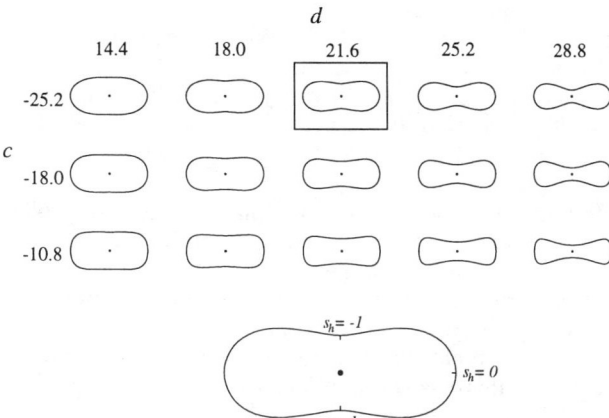

Figure 4: Some points in the two-dimensional hand shape space for $a = 18$, $b = -18$, and the shape we chose for our experiments.

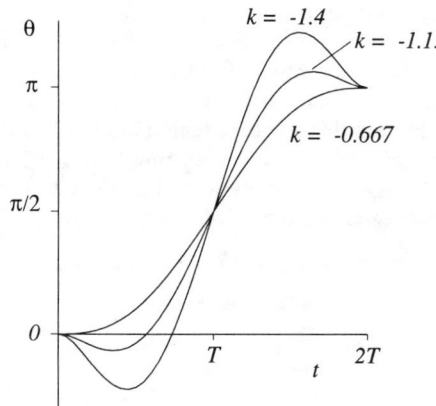

Figure 5: Example motion profiles for different values of k.

the disk to the other side of the hand, overshoots to stop the rolling, and finally settles in a horizontal position. We chose the following motion profile:

$$\theta(t) = \frac{\pi(j(t-T) + k(t-T)^3 + l(t-T)^5)}{2(jT + kT^3 + lT^5)} + \frac{\pi}{2}, t \in [0, 2T]$$

where $2T$ is the total time of the motion. This function rotates the hand from 0 to π. We require $\dot{\theta}(0) = \dot{\theta}(2T) = 0$ and we set $j = 1$. The remaining two-dimensional motion space can be parameterized by k and T, where k determines how far the hand initially dips to begin rolling of the disk. $k = -2/3$ yields zero dip, and the dip increases with increasingly negative values of k. Example motion profiles are shown in Figure 5.

4.3 Simulation

Using the simulation with the hand shape chosen in Section 4.1 and a disk of uniform mass and a radius of 5.1cm, we found a one-dimensional locus of solutions to the butterfly problem in the two-dimensional $k - 2T$ motion space. (For a given motion, we could similarly find a one-dimensional locus of solutions in the two-dimensional $c - d$ shape space.)

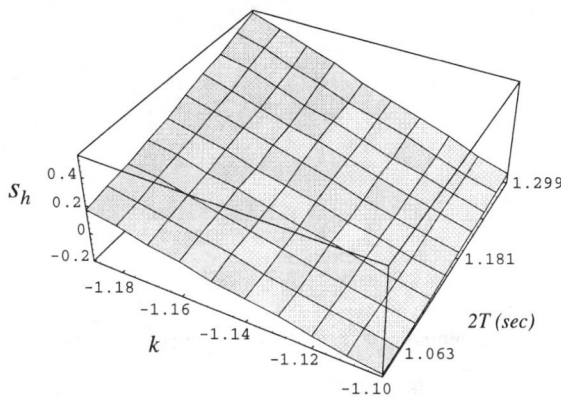

Figure 6: Surface plot of the contact parameter s_h at time T during the butterfly as a function of k and $2T$. The locus of $k - 2T$ solutions to the butterfly corresponds to the points where $s_h = 0$.

A "solution" is any motion that takes the disk to exactly zero velocity on the opposite side of the hand. Because the hand shape and the motion are symmetric, the motion of the disk on the hand is symmetric for any solution.

In a neighborhood of a solution in the $k - 2T$ space, increasing the initial dip (smaller k) or the motion time $2T$ causes the disk to roll further (past the goal position). Figure 6 shows the hand contact parameter s_h at time T during the butterfly as a function of k and $2T$. Because of the symmetry of solutions, a necessary condition for a solution motion is that $s_h = 0$ at time T. The smoothness and monotonicity of the s_h surface of Figure 6 allow us to quickly converge on the locus of points satisfying this condition.

In full gravity ($9.8 m/s^2$), the fastest solution which maintains contact at all times is $2T = 1.095s$, with $k = -1.1493$. At this speed, the contact force becomes zero when the disk reaches its apex. At any higher speed the hand will throw the disk. This is the time-optimal rolling solution in the $k - 2T$ motion space.

Figure 2 shows the solution for $2T = 1.181s$ and $k = -1.1393$. In this example, the contact force at the apex of the roll is 17.5% of the gravitational force. This motion requires a contact friction coefficient of 0.306 to maintain rolling contact. As we increase $2T$, the value of k that solves the butterfly problem increases, implying a smaller initial dip (Figure 7). The friction coefficient required to maintain rolling goes to zero as $2T$ goes to infinity. This is because there is a locus of (θ, s_h) equilibrium configurations connecting the start configuration $(0, -1)$ and the goal configuration $(\pi, 1)$ where the contact force is normal to the surface and through the disk center of mass. Some of these configurations (in the wells) are stable, others are unstable.

5 Experimental Testbed: FLATLAND

We implemented the butterfly on FLATLAND (Figure 8). We built FLATLAND to be a general testbed for experiments in planar dynamic manipulation. FLATLAND consists of a tiltable air table, allowing us to perform planar dynamic manipulation experiments with variable gravity; a set of ma-

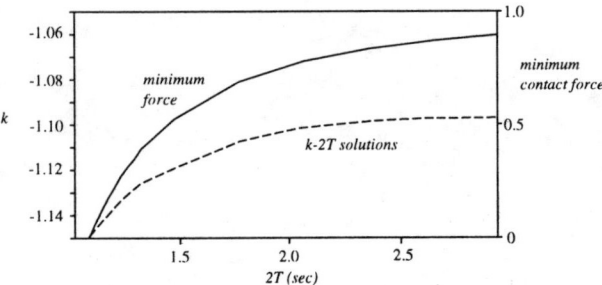

Figure 7: Plot of $k - 2T$ solutions to the butterfly problem and the corresponding minimum contact force as a ratio to the gravitational force.

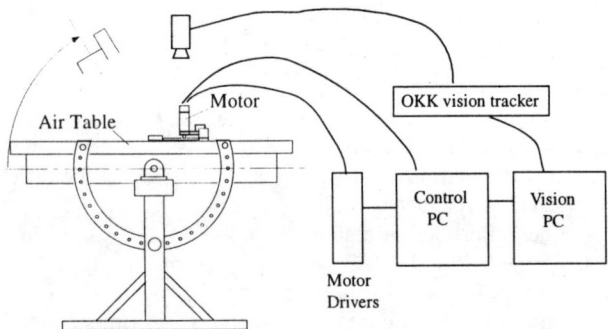

Figure 8: The FLATLAND experimental setup.

nipulator modules that can be configured as 1 or 2 DOF robots mounted on rails around or over the air table; and a 30 Hz OKK vision tracking system to track the motion of objects on the table. To the underside of each manipulator link is attached a *manipulation surface*. The manipulation surfaces are the parts of the robot actually making contact with the laminar objects floating on the table, and they can be changed easily to allow us to experiment with new manipulator shapes (Figure 9). The manipulators are controlled by a Pentium PC which receives vision data from an NEC PC98 which processes the data from the OKK vision tracker.

The hand was cut from aluminum by a CNC machine and attached to a manipulator module configured as a 1 DOF robot. The disk is made of plastic and is encircled by rubber O-rings to increase friction with the aluminum manipulation surface. We set the air table to a 5 degree angle, introducing a time-scaling $\kappa = \sqrt{1/\sin 5°} = 3.387$ from the full-gravity

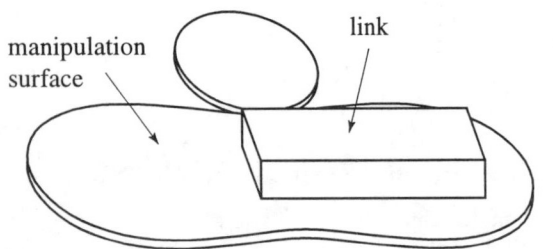

Figure 9: Manipulation surfaces attached to the robot links actually make contact with objects on the air table.

rolling solutions. (Increasing the time-scale permits more vision data during the roll.) The hand was made to follow the trajectory $2T = \kappa(1.181s) = 4.0s, k = -1.1393$, as shown in the simulation in Figure 2. Even without feedback, the butterfly often successfully rolled the disk to the goal well. By simply scaling $2T$, the butterfly was performed at different table angles.

The open-loop butterfly is not robust. Often the disk rolls too far, overshooting the final position, or does not roll far enough, never reaching the goal configuration. In both cases the hand drops the disk. To make execution robust, we implemented vision feedback control. A simple estimator was used to estimate the disk's position between vision data frames, and this data was used in a 1 kHz control loop. The control follows three stages:

1. Perform the initial dipping motion open-loop to get the disk rolling along the hand.

2. Once the disk has passed a certain point (typically $s_h > -0.9$), simply servo the hand toward the angle and angular velocity in the planned trajectory that corresponds to the disk's current position on the hand. In other words, the contact parameter s_h drives the hand's motion. During this stage s_h is monotonically increasing, so there is a one-to-one mapping between s_h and the planned manipulator angle and angular velocity. The control law is written

$$\ddot{\theta} = k_p(\theta_p(s_{h,a}) - \theta_a) + k_d(\dot{\theta}_p(s_{h,a}) - \dot{\theta}_a),$$

where $k_p, k_d > 0$, $s_{h,a}$ is the actual contact parameter (from vision feedback), θ_a is the actual hand angle, and $\theta_p(s_{h,a})$ is the planned hand angle when the disk is at $s_{h,a}$. (In our control system, the commanded acceleration $\ddot{\theta}$ is used to calculate a new reference position θ and velocity $\dot{\theta}$, which are then used in a PD controller to calculate joint torque. This approach suppresses nonlinearities in the actuator due to friction, etc.)

3. Once the disk has passed a certain point (typically $s_h > 0.9$), perform the final overshoot motion open-loop.

Despite the simplicity of the controller, it significantly stabilizes the rolling motion to small errors in initial conditions. Figure 10 shows experimental data for a butterfly under feedback control with $2T = 4.0s, k = -1.1393$. The θ and s_h trajectories are symmetric, but the execution time is extended to approximately $4.23s$. The controller slowed the motion of the hand in stage 2 to allow the disk more rolling time to compensate for errors.

If the rolling velocity after stage 1 is too large, the hand may throw the disk slightly when the disk reaches the top of the hand. (The butterfly often succeeds despite this, as the hand catches the disk and continues on the trajectory.) A slower nominal motion, which maintains a higher nominal contact force, could alleviate this problem. Another failure mode is when the disk does not have enough velocity from the initial dip to reach the top of the hand. We are currently investigating nonlinear feedback controllers based on the rolling dynamics to make the butterfly more robust.

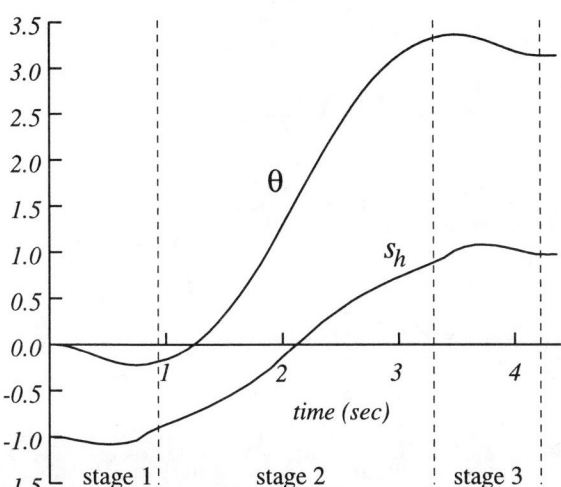

Figure 10: Experimental data for a butterfly with $2T = 4.0s$, $k = -1.1393$, and the table set at a $5°$ angle. The controller slows the execution of the hand trajectory during stage 2 to allow the disk more rolling time to compensate for errors. The total execution time is approximately $4.23s$.

6 Discussion

The butterfly system resembles a revolute two-joint robot with an unactuated second joint (Arai and Tachi [1]; Suzuki et al. [12]). Both systems are subject to a second-order nonholonomic constraint from a passive pivot joint. Three important differences are: 1) The pivot point is rolling along the surface of the manipulator with the butterfly. 2) The butterfly is performed in a gravity field. This allows equilibrium configurations to be stabilized, but trajectories cannot be time-scaled. The time scale is chosen by gravity. 3) There is a limited friction cone of contact forces that can be applied into the disk. Contact force constraints make the control problem particularly challenging.

We have broken the control of the butterfly into two stages, as suggested by the nonholonomic nature of the system: planning a nominal trajectory and feedback stabilization of that trajectory. Figure 6 suggests that gradient-descent approaches could be applicable to the trajectory planning problem; see, for example, (Divelbiss and Wen [4]; Fernandes et al. [5]; Lynch and Mason [8]). The optimization could also solve simultaneously for shape parameters according to some cost function weighting shape and motion. Since the motion of the disk is solved for by simulation, care must be taken to avoid numerical problems, especially if finite differences are used to approximate gradients.

7 Conclusion

We have derived the dynamic equations of nonprehensile rolling and used a simulation based on them to find shape and motion solutions to the planar butterfly problem. We have successfully implemented a solution on FLATLAND, our testbed for dynamic manipulation. Feedback control is used to stabilize the planned trajectories. Future work includes automatic trajectory planning for dynamic rolling manipulation; more robust nonlinear feedback control; designing manipulation surfaces to achieve a desired rolling motion with a low degree-of-freedom robot; and investigating the roles of shape and motion in other kinds of dynamics, such as impact in vibratory parts feeding.

Acknowledgments

This work was performed while the first author was an STA postdoctoral fellow at the Biorobotics Division of the Mechanical Engineering Laboratory. We thank the Science and Technology Agency of Japan and the Robotics Department of MEL for their support. We especially thank Garth Zeglin and Matt Mason for suggesting the butterfly problem. Their air table setup at Carnegie Mellon inspired many of the ideas used in FLATLAND.

References

[1] H. Arai and S. Tachi. Position control system of a two degree of freedom manipulator with a passive joint. *IEEE Transactions on Industrial Electronics*, 38(1):15–20, Feb. 1991.

[2] C. Cai and B. Roth. On the spatial motion of a rigid body with point contact. In *IEEE International Conference on Robotics and Automation*, pages 686–695, 1987.

[3] A. B. A. Cole, J. E. Hauser, and S. S. Sastry. Kinematics and control of multifingered hands with rolling contact. *IEEE Transactions on Automatic Control*, 34(4):398–404, Apr. 1989.

[4] A. W. Divelbiss and J. Wen. A global approach to nonholonomic motion planning. In *IEEE International Conference on Decision and Control*, pages 1597–1602, 1992.

[5] C. Fernandes, L. Gurvits, and Z. Li. Attitude control of a space platform/manipulator system using internal motion. *International Journal of Robotics Research*, 13(4):289–304, 1994.

[6] H. Hitakawa. Advanced parts orientation system has wide application. *Assembly Automation*, 8(3):147–150, 1988.

[7] J. Kerr and B. Roth. Analysis of multifingered hands. *International Journal of Robotics Research*, 4(4):3–17, 1986.

[8] K. M. Lynch and M. T. Mason. Dynamic underactuated nonprehensile manipulation. In *IEEE/RSJ International Conference on Intelligent Robots and Systems*, pages 889–896, 1996.

[9] K. M. Lynch and M. T. Mason. Dynamic manipulation with a one joint robot. In *IEEE International Conference on Robotics and Automation*, pages 359–366, 1997.

[10] D. J. Montana. The kinematics of contact and grasp. *International Journal of Robotics Research*, 7(3):17–32, June 1988.

[11] N. Sarkar, X. Yun, and V. Kumar. Control of contact interactions with acatastatic nonholonomic constraints. *International Journal of Robotics Research*, 16(3):357–374, June 1997.

[12] T. Suzuki, M. Koinuma, and Y. Nakamura. Chaos and nonlinear control of a nonholonomic free-joint manipulator. In *IEEE International Conference on Robotics and Automation*, pages 2668–2675, 1996.

Working modes and aspects in fully parallel manipulators

Damien Chablat   Philippe Wenger

Institut de Recherche en Cybernétique de Nantes
École Centrale de Nantes
1, rue de la Noë, 44321 Nantes, France
Damien.Chablat@lan.ec-nantes.fr Philippe.Wenger@lan.ec-nantes.fr

Abstract

The aim of this paper is to characterize the notion of aspect in the workspace and in the joint space for parallel manipulators. In opposite to the serial manipulators, the parallel manipulators can admit not only multiple inverse kinematic solutions, but also multiple direct kinematic solutions. The notion of aspect introduced for serial manipulators in [1], and redefined for parallel manipulators with only one inverse kinematic solution in [2], is redefined for general fully parallel manipulators. Two Jacobian matrices appear in the kinematic relations between the joint-rate and the Cartesian-velocity vectors, which are called the "inverse kinematics" and the "direct kinematics" matrices. The study of these matrices allow to respectively define the parallel and the serial singularities. The notion of working modes is introduced to separate inverse kinematic solutions. Thus, we can find out domains of the workspace and the joint space exempt of singularity. Application of this study is the moveability analysis in the workspace of the manipulator as well as path-planing and control. This study is illustrated in this paper with a RR-RRR planar parallel manipulator.

KEY WORDS : *Kinematics, Fully Parallel Manipulator, Aspects, Working modes, Singularity.*

1 Introduction

A well known feature of parallel manipulators is the existence of multiple solutions to the direct kinematic problem. That is, the mobile platform can admit several positions and orientations (or configurations) in the workspace for one given set of input joint values [3]. Moreover, parallel manipulators exist with multiple inverse kinematic solutions. This means that the mobile platform can admit several input joint values corresponding to one given configuration of the end-effector. To cope with the existence of multiple inverse kinematic solutions in *serial* manipulators, the notion of aspects was introduced in [1]. The aspects equal the maximal singularity-free domains in the joint space. For usual industrial serial manipulators, the aspects were found to be the maximal sets in the joint space where there is only one inverse kinematic solution.

A definition of the notion of aspect was given by [2] for parallel manipulators with only one inverse kinematic solution. These aspects were defined as the maximal singularity-free domains in the workspace. For instance, this definition can apply to the Stewart platform [4].

First of all, the working modes are introduced to allow the separation of the inverse kinematic solutions. Then, a general definition of the notion of aspect is given for all fully parallel manipulators. The new aspects are the maximal singularity-free domains of the Cartesian product of the workspace with the joint space.

A possible use of these aspects are the determination of the best working mode. It allows to achieve complex task in the workspace or to make path-planing without collision. As a matter of fact, currently, the parallel manipulators possessing multiple inverse kinematic solutions evolve only in one working mode. For a given working mode, the aspect associated is different. It is possible to choose one or several working modes to execute the tasks expected in the maximal workspace of the manipulator.

2 Preliminaries

In this paragraph, some definitions permitting to introduce the general notion of aspect are quoted.

2.1 The fully parallel manipulators

Definition 1 *A fully parallel manipulator is a mechanism that includes as many elementary kinematic* chains as the mobile platform does admit degrees of freedom. Moreover, every elementary kinematic chain possesses only one actuated joint (prismatic, pivot or kneecap). Besides, no segment of an elementary kinematic chain can be linked to more than two bodies [3].

In this study, kinematic chains are always independent. This condition is necessary to find the working modes. Also, the elementary kinematic chains can be called "legs of the manipulator" [5].

2.2 Kinematic relations

For a manipulator, the relation permitting the connection of input values ($\mathbf{q}$) with output values ($\mathbf{X}$) is the following

$$F(\mathbf{X}, \mathbf{q}) = 0 \qquad (1)$$

This definition can be applied to serial or parallel manipulators. Differentiating equation (1) with respect to time leads to the velocity model

$$\mathbf{At} + \mathbf{B\dot{q}} = 0 \qquad (2)$$

With

$$\mathbf{t} = \begin{bmatrix} w \\ \dot{\mathbf{c}} \end{bmatrix} \quad \text{For planar manipulators.}$$

$$\mathbf{t} = \begin{bmatrix} \mathbf{w} \end{bmatrix} \quad \text{For spherical manipulators.}$$

$$\mathbf{t} = \begin{bmatrix} \mathbf{w} \\ \dot{\mathbf{c}} \end{bmatrix} \quad \text{For spatial manipulators.}$$

Where w is the scalar angular-velocity and $\dot{\mathbf{c}}$ is the two-dimensional velocity vector of the operational point of the moving platform for the planar manipulator. For the spherical and the spatial manipulator, $\mathbf{w}$ is the three-dimensional angular velocity-vector of the moving platform. And $\dot{\mathbf{c}}$ is the three-dimensional velocity vector of the operational point of the moving platform for the spatial manipulator.

Moreover, $\mathbf{A}$ and $\mathbf{B}$ are respectively the direct-kinematics and the inverse-kinematics matrices of the manipulator. A singularity occurs whenever $\mathbf{A}$ or $\mathbf{B}$, (or both) that can no longer be inverted. Three types of singularities exist [6]:

$$\begin{aligned} det(\mathbf{A}) &= 0 \\ det(\mathbf{B}) &= 0 \\ det(\mathbf{A}) &= 0 \quad \text{and} \quad det(\mathbf{B}) = 0 \end{aligned}$$

2.3 Parallel singularities

Parallel singularities occur when the determinant of the direct kinematics matrix $\mathbf{A}$ vanishes. The corresponding singular configurations are located inside the workspace. They are particularly undesirable because the manipulator can not resist any force and control is lost.

2.4 Serial singularities

Serial singularities occur when the determinant of the inverse kinematics matrix $\mathbf{B}$ vanishes. When the manipulator is in such a singularity, there is a direction along which no Cartesian velocity can be produced.

2.5 Postures and assembling modes

The multiple inverse kinematic solutions induce multiple postures for each leg.

Definition 2 *A posture changing trajectory is equivalent to a trajectory between two inverse kinematic solutions.*

The multiple direct kinematic solutions induce multiple assembling modes for the mobile platform.

Definition 3 *An assembling mode changing trajectory is equivalent to a trajectory between two direct kinematic solutions.*

As an example, the 3-RRR planar parallel manipulator (the first joints are actuated joints), a posture changing trajectory exists between two inverse kinematic solutions (Fig. 1) and an assembling mode trajectory exits between two direct kinematic solutions (Fig. 2). In these trajectories, the mobile platform can meet a singular configuration.

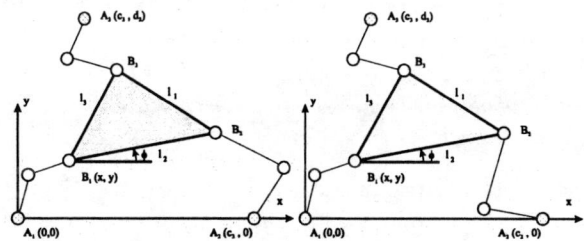

Figure 1: Two postures

2.6 Working Modes

The *working modes* are defined for fully parallel manipulators (Def. 1). From this definition, the inverse-kinematic matrix is always diagonal. For a manipulator with n degrees of freedom, the inverse kinematic matrix $\mathbf{B}$ is like in eq. (3). Each term $\mathbf{B}_{jj}$ is associated

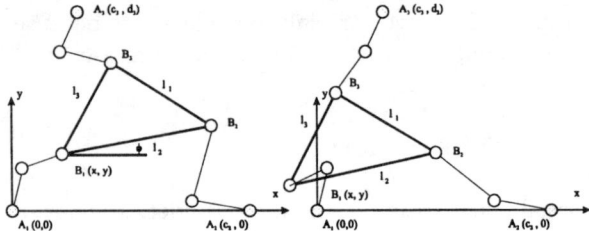

Figure 2: Two assembling modes

with one leg. Its vanishing induces the apparition of a serial singularity.

$$\mathbf{B} = \begin{bmatrix} \mathbf{B}_{11} & 0 & \cdots & \cdots & 0 \\ 0 & \ddots & & \ddots & \vdots \\ \vdots & & \mathbf{B}_{jj} & & \vdots \\ \vdots & \ddots & & \ddots & 0 \\ 0 & \cdots & \cdots & 0 & \mathbf{B}_{nn} \end{bmatrix} \quad (3)$$

Let W be the reachable workspace, that is, the set of all positions and orientations reachable by the moving platform ([7] and [8]). Let Q be the reachable joint space, that is, the set of all joint vectors reachable by actuated joints.

Definition 4 *A working mode, noted Mf_i, is the set of postures for which the sign of $\mathbf{B}_{jj}$ ($j = 1$ to n) does not change and $\mathbf{B}_{jj}$ does not vanish.*

$$Mf_i = \left\{ (\mathbf{X}, \mathbf{q}) \in W \cdot Q \setminus \begin{array}{c} sign(\mathbf{B}_{jj}) = constant \\ for (j = 1 \text{ to } n) \\ and \; det(\mathbf{B}) \neq 0 \end{array} \right\}$$
(4)

Therefore, the set of working modes ($Mf = \{Mf_i\}$, $i \in I$) is obtained while using all permutations of sign of each term $\mathbf{B}_{jj}$.

The Cartesian product of W by Q is noted $W \cdot Q$. According to the joint limit values, all working modes do not necessarily exist. Changing working mode is equivalent to changing the posture of one or several given legs. The working modes are defined in $W \cdot Q$ because the terms $\mathbf{B}_{jj}$ depend on both $\mathbf{X}$ and $\mathbf{q}$.

Theorem 1 *The working modes separate inverse kinematic solutions if and only if the legs are not cuspidal (see [9]).*

Proof 1 *If one leg is cuspidal then this leg can make a changing posture trajectory without meeting a serial singularity. In this case no $\mathbf{B}_{jj}$ vanishes during this trajectory. Reciprocally, if no leg is cuspidal, then the changing posture trajectory of one leg induces that some B_{jj} can vanish.*

In this study, the legs are not cuspidal so that the working modes allow the separation of the inverse kinematic solutions. The list of the most current non-cuspidal serial chains is given in [9].

Example 1 *For the robot Delta [10], a 3-dof manipulator (Fig. 3), there are 8 working modes (3 legs and 2 postures for each leg, with $2^3 = 8$ working modes). And for the Hexa robot [11], a 6-dof manipulator (Fig. 4), there are 64 working modes ($2^6 = 64$). For these manipulators, the serial singularities occur when one or more legs are outstretched.*

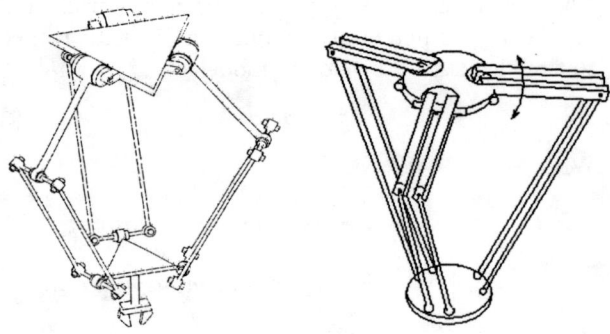

Figure 3: The Delta manipulator

Figure 4: The Hexa manipulator

2.7 Notion of aspect for fully parallel manipulators: General definition

The notion of aspect was introduced by [1] to cope with the existence of multiple inverse kinematic solutions in serial manipulators. Recently, the notion of aspect was defined for parallel manipulators with only one inverse kinematic solution [2] to cope with the existence of multiple direct kinematic solutions.

In this section, the notion of aspect is redefined formally for fully parallel manipulators with multiple inverse and direct kinematic solutions.

Definition 5 *The generalized aspects $\mathbf{A}_{ij}$ are defined as the maximal sets in $W \cdot Q$ so that*

- $\mathbf{A}_{ij} \subset W \cdot Q$;
- $\mathbf{A}_{ij}$ is connected;

- $\mathbf{A}_{ij} = \{(\mathbf{X}, \mathbf{q}) \in Mf_i \setminus det(\mathbf{A}) \neq 0\}$

In other words, the generalized aspects $\mathbf{A}_{ij}$ are the maximal singularity-free domains of the Cartesian product of the reachable workspace with the reachable joint space.

Definition 6 *The projection of the generalized aspects in the workspace yields the parallel aspects $\mathbf{WA}_{ij}$ so that*

- $\mathbf{WA}_{ij} \subset W$;
- $\mathbf{WA}_{ij}$ is connected.

The parallel aspects are the maximal singularity-free domains in the workspace for one given working mode.

Definition 7 *The projection of the generalized aspects in the joint space yields the serial aspects $\mathbf{QA}_{ij}$ so that*

- $\mathbf{QA}_{ij} \subset Q$;
- $\mathbf{QA}_{ij}$ is connected.

The serial aspects are the maximal singularity-free domains in the joint space for one given working mode.

3 A Two-DOF Closed-Chain Manipulator

For more legibility, a planar manipulator is used as illustrative example in this paper. This is a five-bar, revolute (R)-closed-loop linkage, as displayed in Fig. 5. The actuated joint variables are θ_1 and θ_2, while the Output values are the (x, y) coordinates of the revolute center P. The passive joints will always be assumed unlimited in this study. Lengths L_0, L_1, L_2, L_3, and L_4 define the geometry of this manipulator entirely. We assume here the dimensions $L_0 = 9$, $L_1 = 8$, $L_2 = 5$, $L_3 = 5$ and $L_4 = 8$, in certain units of length that we need not specify.

3.1 Kinematic Relations

The velocity $\dot{\mathbf{p}}$ of point P, of position vector $\mathbf{p}$, can be obtained in two different forms, depending on the direction in which the loop is traversed, namely,

$$\dot{\mathbf{p}} = \dot{\mathbf{c}} + \dot{\theta}_3 \mathbf{E}(\mathbf{p} - \mathbf{c}) \tag{5a}$$

$$\dot{\mathbf{p}} = \dot{\mathbf{d}} + \dot{\theta}_4 \mathbf{E}(\mathbf{p} - \mathbf{d}) \tag{5b}$$

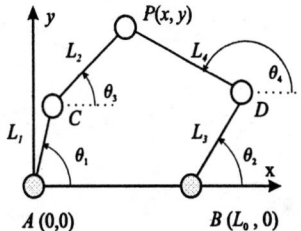

Figure 5: A two-dof closed-chain manipulator

with matrix $\mathbf{E}$ defined as

$$\mathbf{E} = \begin{bmatrix} 0 & -1 \\ 1 & 0 \end{bmatrix}$$

and $\mathbf{c}$ and $\mathbf{d}$ denoting the position vectors, in the frame indicated in Fig. 5, of points C and D, respectively. Furthermore, note that $\dot{\mathbf{c}}$ and $\dot{\mathbf{d}}$ are given by

$$\dot{\mathbf{c}} = \dot{\theta}_1 \mathbf{E}\mathbf{c}, \quad \dot{\mathbf{d}} = \dot{\theta}_2 \mathbf{E}(\mathbf{d} - \mathbf{b})$$

We would like to eliminate the two idle joint rates $\dot{\theta}_3$ and $\dot{\theta}_4$ from eqs.(5a) and (5b), which we do upon dot-multiplying the former by $\mathbf{p}-\mathbf{c}$ and the latter by $\mathbf{p}-\mathbf{d}$, thus obtaining

$$(\mathbf{p} - \mathbf{c})^T \dot{\mathbf{p}} = (\mathbf{p} - \mathbf{c})^T \dot{\mathbf{c}} \tag{6a}$$

$$(\mathbf{p} - \mathbf{d})^T \dot{\mathbf{p}} = (\mathbf{p} - \mathbf{d})^T \dot{\mathbf{d}} \tag{6b}$$

Equations (6a) and (6b) can now be cast in vector form, namely,

$$\mathbf{A}\dot{\mathbf{p}} = \mathbf{B}\dot{\boldsymbol{\theta}} \tag{7a}$$

with $\dot{\boldsymbol{\theta}}$ defined as the vector of actuated joint rates, of components $\dot{\theta}_1$ and $\dot{\theta}_2$. Moreover $\mathbf{A}$ and $\mathbf{B}$ are, respectively, the direct-kinematics and the inverse-kinematics matrices of the manipulator, defined as

$$\mathbf{A} = \begin{bmatrix} (\mathbf{p} - \mathbf{c})^T \\ (\mathbf{p} - \mathbf{d})^T \end{bmatrix} \tag{7b}$$

and

$$\mathbf{B} = \begin{bmatrix} L_1 L_2 \sin(\theta_3 - \theta_1) & 0 \\ 0 & L_3 L_4 \sin(\theta_4 - \theta_2) \end{bmatrix} \tag{7c}$$

3.2 Parallel singularities

For the manipulator studied, the parallel singularities occur whenever the points C, D, and P are aligned (6). Manipulator postures whereby $\theta_3 - \theta_4 = k\pi$ denote a singular matrix $\mathbf{A}$, and hence, define the boundary of the Joint space of the manipulator.

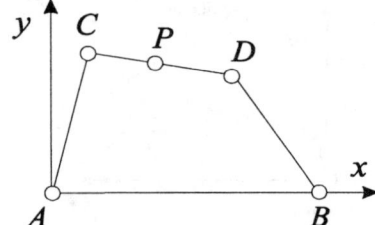

Figure 6: Example of parallel singularity

3.3 Serial singularities

For the manipulator at hand, the serial singularity occur whenever the points A, C, and P or the points B, D, and P are aligned (7). Manipulator postures whereby $\theta_3 - \theta_1 = k\pi$ or $\theta_4 - \theta_2 = k\pi$ denote a singular matrix **B**, and hence, define the boundary of the Cartesian workspace of the manipulator.

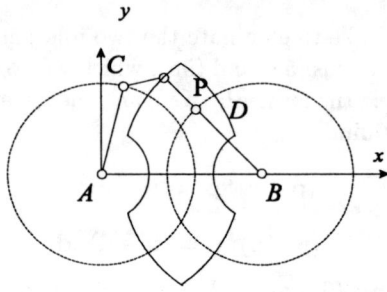

Figure 7: Example of serial singularity

3.4 The Working Mode

The manipulator under study has a diagonal inverse-kinematics matrix **B**, as shown in eq. (7c). There are four working modes, as depicted in Fig. 8. The different working modes in the Cartesian workspace and in the Joint space are displayed in figures 9, 10, 11 and 12.

3.5 The generalized aspects

For the manipulator at hand, the generalized aspects are defined with the definition 5. Figures 13-20 depict the different serial and parallel aspects obtained. Table 1 shows that there are 10 serial/parallel aspects for the manipulator (N and P stand for negative and positif, respectively).

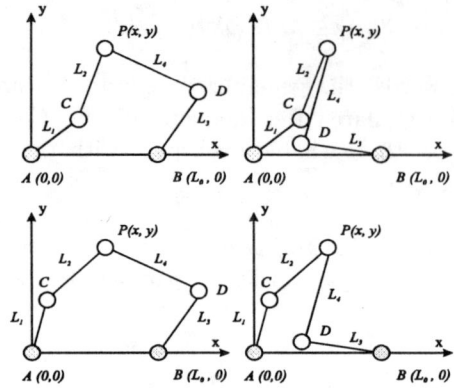

Figure 8: The four working modes

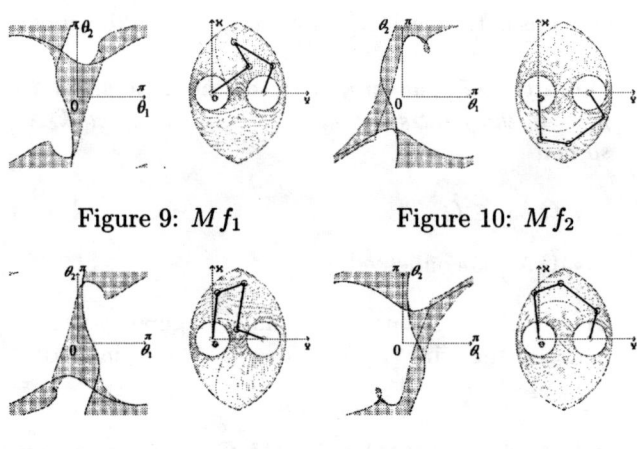

Figure 9: Mf_1 Figure 10: Mf_2

Figure 11: Mf_3 Figure 12: Mf_4

In opposite to the aspects defined by [1] or by [2], the generalized aspects are not disjoint. Only, the aspects belonging to the same working mode are disjoint. In the sample in which the manipulator has only one inverse kinematic solution, the notion of generalized aspect is equivalent to the notion of aspect given by [2]. Indeed, if the manipulator has only one working mode (like the 2-RPR planar manipulator), then the parallel aspects are disjoint and represent the maximal singularity-free domain in the Cartesian workspace.

4 Conclusions

In this paper, the notion of aspect was defined for parallel manipulators with multiple inverse and direct kinematic solutions. The working modes were introduced to define this notion. For one working mode, we can find out the maximal singularity-free domains of the Cartesian product of the workspace with the joint

Figures	13	14	15	16	17	18	19	20
$det(\mathbf{A})$	P	P	P	P	N	N	N	N
$\mathbf{B}_{11}$	P	P	N	N	P	P	N	N
$\mathbf{B}_{22}$	P	N	N	P	N	P	P	N
Nb of generalized aspects	1	1	1	2	1	2	1	1

Table 1: The generalized aspects

space. This work brings material to further investigations like trajectory planning and kinematic design which are the subject of current research work from the authors.

The generalized aspect are not the uniqueness domain in any case as it was shown in [2]. So, in a future study, the authors will define uniqueness domains for general fully parallel manipulators. In such domains, there are only one inverse and direct kinematic solution. These domains are of interest for the control of manipulator.

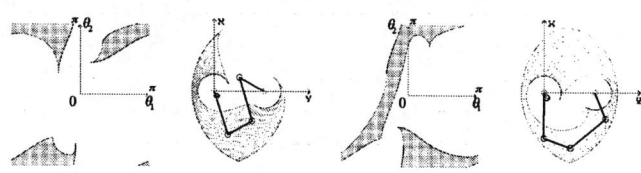

Figure 13: Aspect 1   Figure 14: Aspects 2

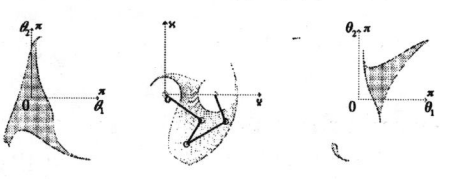

Figure 15: Aspect 3   Figure 16: Aspects 4 and 5

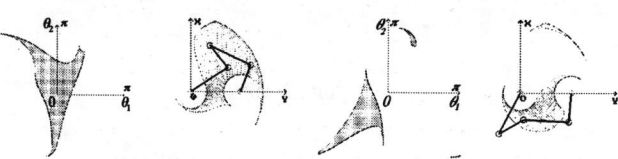

Figure 17: Aspect 6   Figure 18: Aspects 7 and 8

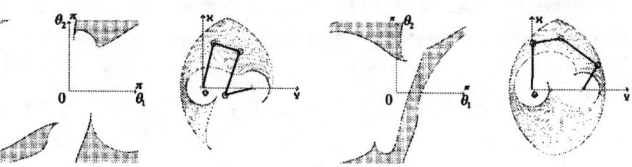

Figure 19: Aspect 9   Figure 20: Aspect 10

References

[1] Borrel, P. "A study of manipulator inverse kinematic solutions with application to trajectory planning and workspace determination" Proceeding IEEE International Conference on Robotic And Automation, pp 1180-1185, 1986.

[2] Wenger, Ph. and Chablat, D. "Uniqueness Domains in the Workspace of Parallel Manipulators" IFAC-SYROCO, Vol. 2, pp 431-436, 3-5 Sept., 1997, Nantes.

[3] Merlet J-P. "Les robots parallèles" HERMES, seconde édition, Paris, 1997.

[4] Stewart, D. "A platform with 6 degree of freedom" Proc. of the Institution of mechanical engineers, 180 (Part 1,15):371-386, 1965.

[5] Angeles, J. "Fundamentals of Robotic Mechanical Systems" SPRINGER 97.

[6] Gosselin, C. and Angeles, J. "Singularity analysis of closed-loop kinematic chains" IEEE Transactions On Robotics And Automation, Vol. 6, No. 3, June 1990.

[7] Kumar V. "Characterization of workspaces of parallel manipulators" ASME J. Mechanical Design, Vol. 114, pp 368-375, 1992.

[8] Pennock, G.R. and Kassner, D.J. "The workspace of a general geometry planar three-degree-of-freedom platform-type manipulator" ASME J. Mechanical Design, Vol. 115, pp 269-276, 1993.

[9] Wenger, Ph. "A classification of Manipulator Geometries Based on Singularity Avoidance Ability" ICAR'93, pp 649-654, Tokyo, Japan, Nov. 1993.

[10] Clavel, R. "A fast robot with parallel geometry" In 18th International Symposium on Industrial Robot, pages 91-100, Lausanne, 26-28 Avril 1988.

[11] Pierrot, F. "Robot Pleinement Parallèles Légers : Conception, Modélisation et Commande" Doctorat thesis, 1991, Montpellier.

The Isoconditioning Loci of A Class of Closed-Chain Manipulators

Damien Chablat   Philippe Wenger   Jorge Angeles[1]

Institut de Recherche en Cybernétique de Nantes
École Centrale de Nantes
1, rue de la Noë, 44321 Nantes, France
Damien.Chablat@lan.ec-nantes.fr   Philippe.Wenger@lan.ec-nantes.fr

[1]McGill Centre for Intelligent Machines and Department of Mechanical Engineering
McGill University, 817 Sherbrooke Street West
Montreal, Quebec, Canada H3A 2K6
Angeles@cim.mcgill.ca

Abstract

The subject of this paper is a special class of closed-chain manipulators. First, we analyze a family of two-degree-of-freedom (dof) five-bar planar linkages. Two Jacobian matrices appear in the kinematic relations between the joint-rate and the Cartesian-velocity vectors, which are called the "inverse kinematics" and the "direct kinematics" matrices. It is shown that the loci of points of the workspace where the condition number of the direct-kinematics matrix remains constant, i.e., the isoconditioning loci, are the coupler points of the four-bar linkage obtained upon locking the middle joint of the linkage. Furthermore, if the line of centers of the two actuated revolutes is used as the axis of a third actuated revolute, then a three-dof hybrid manipulator is obtained. The isoconditioning loci of this manipulator are surfaces of revolution generated by the isoconditioning curves of the two-dof manipulator, whose axis of symmetry is that of the third actuated revolute.

KEY WORDS : *Kinematics, Closed-Loop Manipulator, Hybrid manipulator, Isoconditioning surfaces, Singularity, Working Modes.*

1 Introduction

The aim of this paper is to study (a) a family of two-dof, five-bar planar linkages and (b) a derivative of this family, obtained when a third revolute is added in series to the above linkages, with the purpose of obtaining a three-dof manipulator. For the mechanical design of this class of manipulators, various features must be considered, e.g., the workspace volume, manipulability, and stiffness. The analysis of single-dof closed-loop chains is classical within the theory of machines and mechanisms [1]. The study of the workspace and the mobility of closed-loop manipulators, in turn, is given by Bajpai and Roth [2]. Gosselin [3], [4] conducted similar analyses for closed-loop manipulators with one single inverse kinematic solution on both a planar and a spatial mechanism. One important property of parallel manipulators is that they admit several solutions to both their inverse and their direct kinematics. This property leads to two types of singularities.

The singularities of these manipulators are correspondingly associated with two Jacobian matrices called here the "inverse kinematics" and the "direct kinematics" matrices. By means of the inverse kinematics matrix, we can define the "working mode" of the manipulator to separate the inverse kinematics solutions. It is useful to represent the manipulator in the workspace and to define its aspects in this workspace. The aspects of a manipulator are defined in [5]. Moreover, a novel three-dof hybrid manipulator is proposed, which is comparable to the one proposed by Bajpai and Roth [2]; ours is obtained as the series array of a one-revolute chain and the two-dof closed-chain manipulator described above. In this array, the axis of the former intersects the axes of the two actuated joints of the latter at right angles.

The proper operation of a manipulator depends first of foremost on its design; besides design, the operation depends on suitable trajectory-planning and control algorithms. In any event, a performance index needs be defined, whose minimization or maximization leads to an optimum operation. While various items come into play when assessing the operation of a manipulator, we focus here on issues pertaining to manipulability or dexterity. In this regard, we understand these terms in the sense of measures of distance to singularity, which brings us to the concept of condition number [6]. Here, we adopt the condition number of the underlying Jacobian matrices as a means to quantify distances to singularity. Furthermore, we derive the loci of points of the joint and Cartesian workspaces whereby the condition number of each of the Jacobian matrices remains constant. For the planar two-dof manipulators studied here, we term these loci the *isoconditioning curves*, while, for three-dof spatial manipulators, these curves become the *isoconditioning surfaces*.

2 A Two-DOF Closed-Chain Manipulator

The manipulator under study is a five-bar, revolute (R)-coupled linkage, as displayed in Fig. 1. The actuated joint variables are θ_1 and θ_2, while the Cartesian variables are the (x, y) coordinates of the revolute center P.

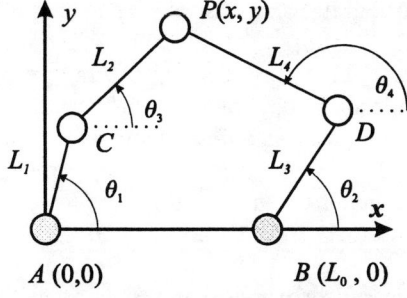

Figure 1: A two-dof closed-chain manipulator

Lenghts L_0, L_1, L_2, L_3, and L_4 define the geometry of this manipulator entirely. However, in this paper we focus on a symmetric manipulators, with $L_1 = L_3$ and $L_2 = L_4$. The symmetric architecture of the manipulator at hand is justified for general tasks. In manipulator design, then, one is interested in obtaining values of L_0, L_1, and L_2 that optimize a given objective function under some prescribed constraints.

2.1 Kinematic Relations

The velocity $\dot{\mathbf{p}}$ of point P, of position vector $\mathbf{p}$, can be obtained in two different forms, depending on the direction in which the loop is traversed, namely,

$$\dot{\mathbf{p}} = \dot{\mathbf{c}} + \dot{\theta}_3 \mathbf{E}(\mathbf{p} - \mathbf{c}) \tag{1a}$$

$$\dot{\mathbf{p}} = \dot{\mathbf{d}} + \dot{\theta}_4 \mathbf{E}(\mathbf{p} - \mathbf{d}) \tag{1b}$$

with matrix $\mathbf{E}$ defined as

$$\mathbf{E} = \begin{bmatrix} 0 & -1 \\ 1 & 0 \end{bmatrix}$$

and $\mathbf{c}$ and $\mathbf{d}$ denoting the position vectors, in the frame indicated in Fig. 1, of points C and D, respectively.

Furthermore, note that $\dot{\mathbf{c}}$ and $\dot{\mathbf{d}}$ are given by

$$\dot{\mathbf{c}} = \dot{\theta}_1 \mathbf{E} \mathbf{c}, \quad \dot{\mathbf{d}} = \dot{\theta}_2 \mathbf{E}(\mathbf{d} - \mathbf{b})$$

We would like to eliminate the two idle joint rates $\dot{\theta}_3$ and $\dot{\theta}_4$ from eqs.(1a) and (1b), which we do upon dot-multiplying the former by $\mathbf{p}-\mathbf{c}$ and the latter by $\mathbf{p}-\mathbf{d}$, thus obtaining

$$(\mathbf{p} - \mathbf{c})^T \dot{\mathbf{p}} = (\mathbf{p} - \mathbf{c})^T \dot{\mathbf{c}} \tag{2a}$$

$$(\mathbf{p} - \mathbf{d})^T \dot{\mathbf{p}} = (\mathbf{p} - \mathbf{d})^T \dot{\mathbf{d}} \tag{2b}$$

Equations (2a) and (2b) can now be cast in vector form, namely,

$$\mathbf{A}\dot{\mathbf{p}} = \mathbf{B}\dot{\boldsymbol{\theta}} \tag{3a}$$

with $\dot{\boldsymbol{\theta}}$ defined as the vector of actuated joint rates, of components $\dot{\theta}_1$ and $\dot{\theta}_2$. Moreover $\mathbf{A}$ and $\mathbf{B}$ are, respectively, the direct-kinematics and the inverse-kinematics matrices of the manipulator, defined as

$$\mathbf{A} = \begin{bmatrix} (\mathbf{p} - \mathbf{c})^T \\ (\mathbf{p} - \mathbf{d})^T \end{bmatrix} \tag{3b}$$

and

$$\mathbf{B} = L_1 L_2 \begin{bmatrix} \sin(\theta_3 - \theta_1) & 0 \\ 0 & \sin(\theta_4 - \theta_2) \end{bmatrix} \tag{3c}$$

3 The Isoconditioning Curves

We derive below the loci of equal condition number of the direct- and inverse-kinematics matrices. To do this, we first recall the definition of *condition number* of an $m \times n$ matrix $\mathbf{M}$, with $m \leq n$, $\kappa(\mathbf{M})$. This number can be defined in various ways; for our purposes, we define $\kappa(\mathbf{M})$ as the ratio of the largest, σ_l, to the smallest σ_s, singular values of $\mathbf{M}$, namely,

$$\kappa(\mathbf{M}) = \frac{\sigma_l}{\sigma_s} \tag{4}$$

The singular values $\{\sigma_k\}_1^m$ of matrix $\mathbf{M}$ are defined, in turn, as the square roots of the nonnegative eigenvalues of the positive-semidefinite $m \times m$ matrix $\mathbf{MM^T}$.

3.1 Direct-Kinematics Matrix

To calculate the condition number of matrix $\mathbf{A}$, we need the product $\mathbf{A}\mathbf{A}^T$, which we calculate below:

$$\mathbf{A}\mathbf{A}^T = L_2^2 \begin{bmatrix} 1 & \cos(\theta_3 - \theta_4) \\ \cos(\theta_3 - \theta_4) & 1 \end{bmatrix} \quad (5)$$

The eigenvalues α_1 and α_2 of the above product are given by:

$$\alpha_1 = 1 - \cos(\theta_3 - \theta_4), \quad \alpha_2 = 1 + \cos(\theta_3 - \theta_4) \quad (6)$$

and hence, the condition number of matrix $\mathbf{A}$ is

$$\kappa(\mathbf{A}) = \sqrt{\frac{\alpha_{max}}{\alpha_{min}}} \quad (7)$$

where

$$\alpha_{min} = 1 - |\cos(\theta_3 - \theta_4)|, \quad \alpha_{max} = 1 + |\cos(\theta_3 - \theta_4)| \quad (8)$$

Upon simplification,

$$\kappa(\mathbf{A}) = \frac{1}{|\tan((\theta_3 - \theta_4)/2)|} \quad (9)$$

In light of expression (9) for the condition number of the Jacobian matrix $\mathbf{A}$, it is apparent that $\kappa(\mathbf{A})$ attains its minimum of 1 when $|\theta_3 - \theta_4| = \pi/2$, the equality being understood *modulo* π. At the other end of the spectrum, $\kappa(\mathbf{A})$ tends to infinity when $\theta_3 - \theta_4 = k\pi$, for $k = 1, 2, \ldots$. When matrix $\mathbf{A}$ attains a condition number of unity, it is termed *isotropic*, its inversion being performed without any roundoff-error amplification. Manipulator postures for which condition $\theta_3 - \theta_4 = \pi/2$ holds are thus the most accurate for purposes of the direct kinematics of the manipulator. Correspondingly, the locus of points whereby matrix $\mathbf{A}$ is isotropic is called the *isotropy locus* in the Cartesian workspace.

On the other hand, manipulator postures whereby $\theta_3 - \theta_4 = k\pi$ denote a singular matrix $\mathbf{A}$. Such singularities occur at the boundary of the Joint space of the manipulator, and hence, the locus of P whereby these singularities occur, namely, the *singularity locus* in the Joint space, defines this boundary. Interestingly, isotropy can be obtained regardless of the dimensions of the manipulator, as long as i) it is symmetric and ii) $L_2 \neq 0$.

3.2 Inverse-Kinematics Matrix

By virtue of the diagonal form of matrix $\mathbf{B}$, its singular values, β_1 and β_2, are simply the absolute values of its diagonal entries, namely,

$$\beta_1 = |\sin(\theta_3 - \theta_1)|, \quad \beta_2 = |\sin(\theta_4 - \theta_2)| \quad (10)$$

The condition number κ of matrix $\mathbf{B}$ is thus

$$\kappa(\mathbf{B}) = \sqrt{\frac{\beta_{max}}{\beta_{min}}} \quad (11)$$

where, if $|\sin(\theta_3 - \theta_1)| < |\sin(\theta_4 - \theta_2)|$, then

$$\beta_{min} = |\sin(\theta_3 - \theta_1)|, \quad \beta_{max} = |\sin(\theta_4 - \theta_2)|; \quad (12)$$

else,

$$\beta_{min} = |\sin(\theta_4 - \theta_2)|, \quad \beta_{max} = |\sin(\theta_3 - \theta_1)|. \quad (13)$$

In light of expression (11) for the condition number of the Jacobian matrix $\mathbf{B}$, it is apparent that $\kappa(\mathbf{B})$ attains its minimum of 1 when $|\sin(\theta_3 - \theta_1)| = |\sin(\theta_4 - \theta_2)| \neq 0$. The locus of points where $\kappa(\mathbf{B}) = 1$, and hence, where $\mathbf{B}$ is isotropic, is called the *isotropy locus* of the manipulator in the joint space. At the other end of the spectrum, $\kappa(\mathbf{B})$ tends to infinity when $|\theta_3 - \theta_1| = k\pi$ or $|\theta_4 - \theta_2| = k\pi$, for $k = 1, 2, \ldots$, which denote singularities of $\mathbf{B}$. These singularities are associated with the inverse kinematics of the manipulator, and hence, lie within its Cartesian workspace, not at the boundary of this one. The singularity locus of $\mathbf{B}$ thus defines the Cartesian workspace of the manipulator. Therefore, the Cartesian workspace of the manipulator is bounded by the singularity locus of $\mathbf{B}$, i.e., the locus of points where $\kappa(\mathbf{B}) \to \infty$. Interestingly, $\mathbf{B}$ can be rendered isotropic regardless of the dimensions of the manipulator, as long as i) it is symmetric and ii) $L_1 \neq 0$ and $L_2 \neq 0$.

3.3 The Working Mode

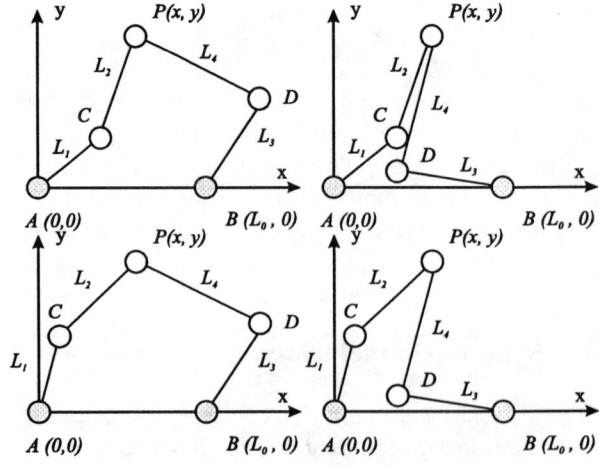

Figure 2: The four working modes

The manipulator under study has a diagonal inverse-kinematics matrix **B**, as shown in eq.(3c), the vanishing of one of its diagonal entries thus indicating the occurrence of a *serial singularity*. The set of manipulator postures free of this kind of singularity is termed a *working mode*. The different working modes are thus separated by a serial singularity, with a set of postures in different working modes corresponding to an inverse kinematics solution.

The formal definition of the working mode is detailed in [5]. For the manipulator at hand, there are four working modes, as depicted in Fig. 2.

3.4 Examples

We assume here the dimensions $L_0 = 6$, $L_1 = 8$, and $L_2 = 5$, in certain units of length that we need not specify.

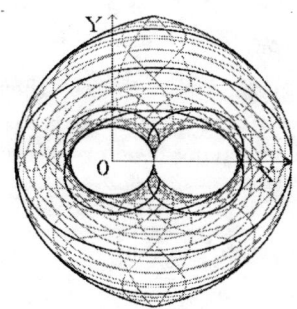 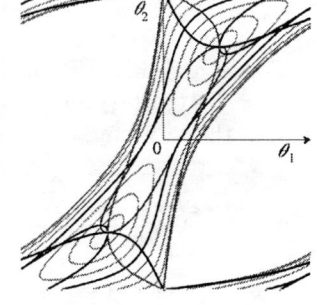

Figure 3: The isoconditioning curves in the Cartesian space

Figure 4: The isoconditioning curves in the joint space

The isoconditioning curves for the direct-kinematic matrix both in the Cartesian and in the joint spaces are displayed in Figs. 3 and 4, respectively. A better representation of isoconditioning curves can be obtained in the Cartesian space by displaying these curves for every working mode, which we do in Fig. 5.

In this figure, the isoconditioning curves are the coupler curves of the four-bar linkage derived upon locking the middle joint, of center $P(x, y)$, to yield a fixed value of $\theta_3 - \theta_4$. Each configuration where points C and D coincide leads to a singularity where the position of point P is not controllable.

4 A Three-DOF Hybrid Manipulator

Now we add one-dof to the manipulator of Fig. 1. We do this by allowing the overall two-dof manipula-

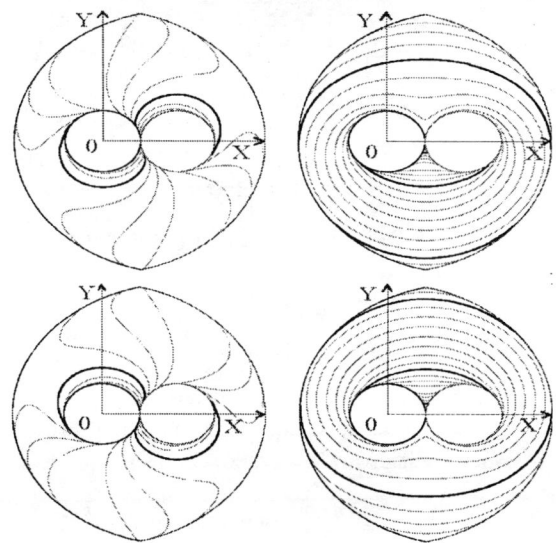

Figure 5: The four working modes and their isoconditioning curves in the Cartesian space

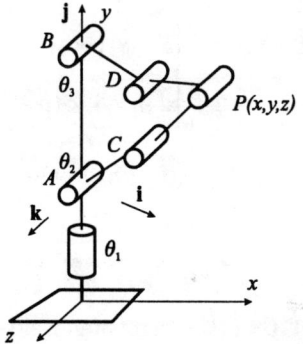

Figure 6: The three-dof hybrid manipulator

tor to rotate about line AB by means of a revolute coupling the fixed link of the above manipulator with the base of the new manipulator. We thus obtain the manipulator of Fig. 6.

4.1 Kinematic Relations

The velocity $\dot{\mathbf{p}}$ of point P can be obtained in two different forms, depending on the direction in which the loop is traversed, namely,

$$\dot{\mathbf{p}} = \dot{\mathbf{c}} + (\dot{\theta}_1 \mathbf{j} + \dot{\theta}_4 \mathbf{k}) \times (\mathbf{p} - \mathbf{c}) \quad (14a)$$

and

$$\dot{\mathbf{p}} = \dot{\mathbf{d}} + (\dot{\theta}_1 \mathbf{j} + \dot{\theta}_5 \mathbf{k}) \times (\mathbf{p} - \mathbf{d}) \quad (14b)$$

Upon dot-multiplying eq.(14a) by $(\mathbf{p} - \mathbf{c})$ and eq.(14b) by $(\mathbf{p} - \mathbf{d})$, we obtain two scalar equations free of θ_1 and the idle joint rates $\dot{\theta}_4$ and $\dot{\theta}_5$, i.e.,

$$(\mathbf{p} - \mathbf{c})^T \dot{\mathbf{p}} = (\mathbf{p} - \mathbf{c})^T \dot{\mathbf{c}} \quad (15)$$
$$(\mathbf{p} - \mathbf{d})^T \dot{\mathbf{p}} = (\mathbf{p} - \mathbf{d})^T \dot{\mathbf{d}} \quad (16)$$

Furthermore, we note that $\dot{\mathbf{c}}$ and $\dot{\mathbf{d}}$ are given by

$$\dot{\mathbf{c}} = (\dot{\theta}_1 \mathbf{j} + \dot{\theta}_2 \mathbf{k}) \times \mathbf{c} \quad (17)$$
$$\dot{\mathbf{d}} = (\dot{\theta}_1 \mathbf{j} + \dot{\theta}_3 \mathbf{k}) \times (\mathbf{d} - \mathbf{b}) \quad (18)$$

Substitution of the above two equations into eqs.(15 & 16), two kinematic relations between joint rates and Cartesian velocities are obtained, namely,

$$[(\mathbf{p} - \mathbf{c}) \times \mathbf{c}] \cdot \mathbf{k} \dot{\theta}_2 = (\mathbf{p} - \mathbf{c})^T \dot{\mathbf{p}} \quad (19)$$
$$[(\mathbf{p} - \mathbf{d}) \times (\mathbf{d} - \mathbf{b})] \cdot \mathbf{k} \dot{\theta}_3 = (\mathbf{p} - \mathbf{d})^T \dot{\mathbf{p}} \quad (20)$$

Moreover, upon dot-multiplying eqs.(14a & b) by $\mathbf{k}$, we obtain two expressions for the projection of $\dot{\mathbf{p}}$ onto the Z axis

$$\mathbf{k}^T \dot{\mathbf{p}} = \mathbf{k}^T \left[\dot{\mathbf{c}} + \dot{\theta}_1 \mathbf{j} \times (\mathbf{p} - \mathbf{c}) \right]$$
$$\mathbf{k}^T \dot{\mathbf{p}} = \mathbf{k}^T \left[\dot{\mathbf{d}} + \dot{\theta}_1 \mathbf{j} \times (\mathbf{p} - \mathbf{d}) \right]$$

which, in light of eqs.(17 & 18), readily reduce to

$$\mathbf{k}^T \dot{\mathbf{p}} = \mathbf{i^T p} \dot{\theta}_1$$
$$\mathbf{k}^T \dot{\mathbf{p}} = \mathbf{i^T p} \dot{\theta}_1$$

It is apparent that the right-hand sides of the two foregoing equations are identical, and hence, those two scalar equations lead to exactly the same relation, namely,

$$\mathbf{k}^T \dot{\mathbf{p}} = (\mathbf{i}^T \mathbf{p}) \dot{\theta}_1$$

It will prove useful to have the two sides of the above equation multiplied by L_2, and hence, that equation is equivalent to

$$L_2 \mathbf{k}^T \dot{\mathbf{p}} = L_2 (\mathbf{i}^T \mathbf{p}) \dot{\theta}_1 \quad (21)$$

In the next step, we assemble eqs.(19 & 20), which leads to an equation formally identical to eq.(3a), but with $\mathbf{A}$ and $\mathbf{B}$ defined now as 3×3 matrices, i.e.,

$$\mathbf{A} \equiv \begin{bmatrix} L_2 \mathbf{k}^T \\ (\mathbf{p} - \mathbf{c})^T \\ (\mathbf{p} - \mathbf{d})^T \end{bmatrix} \quad (23)$$

$$\mathbf{B} \equiv L_1 L_2 \begin{bmatrix} \sin\theta_2 + \lambda_1 \sin\theta_4 & 0 & 0 \\ 0 & \sin(\theta_2 - \theta_4) & 0 \\ 0 & 0 & \sin(\theta_3 - \theta_5) \end{bmatrix} \quad (24)$$

with λ_1 defined as $\lambda_1 \equiv L_2/L_1$, while vectors $\dot{\boldsymbol{\theta}}$ and $\dot{\mathbf{p}}$ are now given by

$$\dot{\boldsymbol{\theta}} \equiv \begin{bmatrix} \dot{\theta}_1 \\ \dot{\theta}_2 \\ \dot{\theta}_3 \end{bmatrix}, \quad \dot{\mathbf{p}} \equiv \begin{bmatrix} \dot{x} \\ \dot{y} \\ \dot{z} \end{bmatrix} \quad (25)$$

5 The Isoconditioning Surfaces

We conduct here the same analysis of Section 3.

5.1 The Direct-Kinematics Matrix

Apparently, matrix $\mathbf{A}$ in the 3-dof case has a structure similar to the corresponding matrix in the 2-dof case. Indeed, upon calculating $\mathbf{A}\mathbf{A}^T$ in the 3-dof case, we obtain

$$\mathbf{A}\mathbf{A}^T = L_2^2 \begin{bmatrix} 1 & 0 & 0 \\ 0 & 1 & \cos(\theta_4 - \theta_5) \\ 0 & \cos(\theta_4 - \theta_5) & 1 \end{bmatrix} \quad (26)$$

The eigenvalues of the foregoing matrix are, then, $\alpha_1 = 1 - |\cos(\theta_4 - \theta_5)|$, $\alpha_2 = 1$, and $\alpha_3 = 1 + |\cos(\theta_4 - \theta_5)|$, the foregoing eigenvalues having been ordered as

$$\alpha_1 \leq \alpha_2 \leq \alpha_3$$

The condition number of matrix $\mathbf{A}$ is thus

$$\kappa(\mathbf{A}) = \sqrt{\frac{1 + |\cos(\theta_4 - \theta_5)|}{1 - |\cos(\theta_4 - \theta_5)|}}$$

which can be further simplified to

$$\kappa(\mathbf{A}) = \frac{1}{|\tan((\theta_4 - \theta_5)/2)|} \quad (27)$$

Therefore, the condition number of the two direct-kinematics matrices, for the 2-dof and the 3-dof cases, coincide. However, the loci of isoconditioning points are now surfaces, because we have added one dof to the manipulator of Fig. 1. These loci are, in fact, surfaces of revolution generated by the isoconditioning curves of the 2-dof manipulator, when these are rotated about the axis of the first revolute. We represent the boundary of the workspace (Fig. 7).

5.2 The Inverse-Kinematics Matrix

Given the diagonal structure of matrix $\mathbf{B}$, its singular values are apparently, $\{ L_1 L_2 \beta_i \}_1^3$, with the definitions below:

$$\beta_1 = |\sin\theta_2 + \lambda_1 \sin\theta_4|,$$
$$\beta_2 = |\sin(\theta_2 - \theta_4)|,$$
$$\beta_3 = |\sin(\theta_3 - \theta_5)|$$

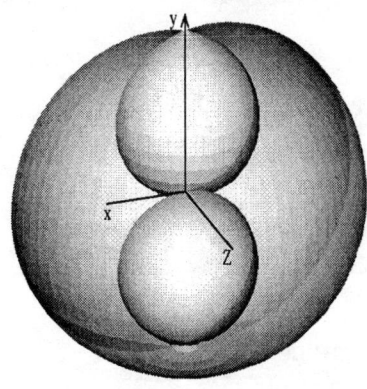

Figure 7: The boundary of the workspace

Therefore, the isoconditioning locus of $\mathbf{B}$ is determined by the relation

$$|\sin\theta_2 + \lambda_1 \sin\theta_4| = |\sin(\theta_2 - \theta_4)| = |\sin(\theta_3 - \theta_5)| \quad (28)$$

Notice that the distance d_1 of P to the Y axis is

$$d_1 = L_1 \sin\theta_2 + L_2 \sin\theta_4 = L_1 \beta_1 \quad (29)$$

Likewise, the distances d_2 and d_3 of P to the two axes of the other two actuated revolutes, i.e., those passing through A and B are, respectively,

$$d_2 = L_2 \beta_2 \quad (30)$$
$$d_3 = L_2 \beta_3 \quad (31)$$

It is now straightforward to realize that, for the case at hand, the locus of isotropic points of $\mathbf{B}$ are given by manipulator postures whereby P is equidistant from the three actuated revolute axes. Likewise, postures whereby point P lies on the Y axis are singular; at these postures, $\kappa(\mathbf{B})$ tends to infinity. Moreover, the inverse-kinematics singularities occur whenever any of the diagonal entries of $\mathbf{B}$ vanishes, i.e., when

$$d_1 = 0, \text{ or } \theta_2 = \theta_4 + k\pi, \text{ or } \theta_3 = \theta_5 + k\pi \quad (32)$$

for $k = 1, 2, \ldots$.

6 Conclusions

We have defined a new architecture of hybrid manipulators and derived the associated loci of isoconditioning points. Two Jacobian matrices were identified in the mapping of joint rates into Cartesian velocities, namely, the direct-kinematics and the inverse-kinematics matrices. Isoconditioning loci were defined for these matrices. Two special loci were discussed, namely, those pertaining to isotropy and to singularity, for each of these matrices.

The study has been conducted for three-dof-hybrid manipulators but applies to six-dof-hybrid manipulators with wrist as well.

The hybrid manipulators studied have interesting features like workspace and high dynamic performances, which are usually met separately in serial or parallel manipulators, respectively. Futher research work is being conducted by the authors on such hybrid manipulators with regard to their optimal design.

Acknowledgments

The third author acknowledges the support from the Natural Sciences and Engineering Research Council, of Canada, the Fonds pour la formation de chercheurs et l'aide à la recherche, of Quebec, and École Centrale de Nantes (ECN). The research reported here was conducted during a sojourn that this author spent at ECN's Institut de Recherche en Cybernétique de Nantes.

References

[1] Hunt, K. H. "Geometry of Mechanisms" Clarendon Press, Oxford, 1978.

[2] Bajpai, A. and Roth, B. "Workspace and mobility of a closed-loop manipulator" The International Journal of Robotics Research, Vol. 5, No. 2, 1986.

[3] Gosselin, C. "Stiffness mapping for parallel manipulators" IEEE Transactions On Robotics And Automation, Vol. 6, No. 3, June 1990.

[4] Gosselin, C. and Angeles, J. "Singularity analysis of closed-loop kinematic chains" IEEE Transactions On Robotics And Automation, Vol. 6, No. 3, June 1990.

[5] Chablat, D. and Wenger, Ph. "Working modes and aspects in fully parallel manipulators" to appear in Proc. IEEE International Conference of Robotic and Automation, Mai 1998.

[6] Golub, G. H. and Van Loan, C. F. "Matrix Computations" The Johns Hopkins University Press, Baltimore, 1989.

Efficient Computation of the extremum of the Articular Velocities of a Parallel Manipulator in a Translation Workspace.

Jean-Pierre MERLET
INRIA Sophia-Antipolis
BP 93, 06902 Sophia-Antipolis, France
E-mail: Jean-Pierre.Merlet@sophia.inria.fr

Abstract

This paper presents an efficient algorithm for computing, with a guaranteed error, the maximal and minimal articular velocities of a parallel manipulator so that whatever is the location of the end-effector in a given volume it may perform a motion at a given cartesian/angular velocity, under the assumption that its orientation is kept constant over the volume. This algorithm is much more faster and safe than the classical discretisation method.

1 Introduction

The design of a parallel manipulators involves various objectives which are either to be optimized or to be reached: positioning workspace, positioning accuracy, maximal articular forces over a given workspace etc.. One of these objectives may be the ability to perform a given cartesian/angular velocity of the end-effector for any of its position in a given workspace. It is then necessary to determine the maximal articular velocities of the robot for reaching this objective. Numerous papers have been devoted to the relations between articular velocities and cartesian/angular velocity of the end-effector [1],[2],[3], [4],[5] but none, to the best of the author knowledge, have addressed the problem of determining the extremum of the articular velocities when the end-effector is moving inside a given workspace.

In this paper we will consider the classical Gough type parallel manipulator [6] illustrated in figure 1. In this robot a base and a platform are connected through 6 extensible legs with ball-and-socket joints at each extremity.

We assume that the user is interested in the maximal and minimal velocities of the linear actuators

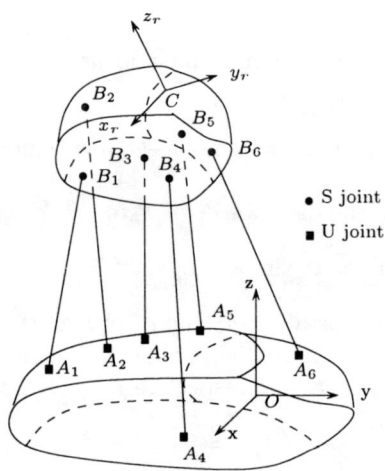

Figure 1: The classical Gough type parallel robot

needed to insure a given cartesian/angular velocity $\mathcal{V}$ of the end-effector, whatever the posture of the mobile platform in a given workspace is. The articular velocity vector $\dot{\rho}$ is related to the cartesian and angular velocity vector $\mathcal{V}$ of the end-effector by:

$$\dot{\rho} = J^{-1}(X)\mathcal{V} \qquad (1)$$

where X is the posture of the robot and J^{-1} is its inverse jacobian matrix. For a given cartesian/angular velocity $\mathcal{V}$ the articular velocities are therefore posture dependent. The purpose of this paper is to compute the maximal and minimal values of the components of the vector $\dot{\rho}$ for a given velocity vector $\mathcal{V}$ and for any posture X inside a given domain.

In the sequel C will denote the center of the moving platform, O the origin of the reference frame, A_i, B_i respectively the base and mobile joint centers of leg i. The vector $\mathcal{V}$ can be decomposed into a cartesian velocity vector V and an angular velocity vector Ω. We will assume that the orientation of the moving platform is constant (hence the vector CB_i is constant).

2 Extremal velocities for a segment

Let us assume that the point C is moving along a segment defined by two points M_1, M_2. Any position of C on the segment may be written as:

$$OC = OM_1 + \lambda M_1 M_2 \qquad (2)$$

where λ is a scalar in the range [0,1]. For a given velocity vector $\mathcal{V}$ the velocity of leg i is written as:

$$\dot{\rho}_i = J_i^{-1} \mathcal{V}$$

where J_i^{-1} is the ith row of the inverse jacobian matrix. It is well known that this row is:

$$\frac{A_i B_i}{\|A_i B_i\|}, \quad CB_i \times \frac{A_i B_i}{\|A_i B_i\|}$$

We notice immediately that for a given leg the articular velocity is not dependent upon the other leg velocities. Consequently we will drop the subscript in the sequel. We have:

$$\dot{\rho} = \frac{AB.V + (CB \times AB).\Omega}{\|AB\|} \qquad (3)$$

Note that AB may be written as:

$$AB = AO + OC + CB = AO + OM_1 + CB + \lambda M_1 M_2$$

In this equation the three first vectors are constant. Consequently we may write it in a simpler form as:

$$AB = U + \lambda M_1 M_2$$

The norm of this vector is therefore:

$$\|AB\| = \sqrt{\lambda^2 \|M_1 M_2\|^2 + 2\lambda U.M_1 M_2 + \|U\|^2}$$

Similarly we have:

$$CB \times AB = CB \times U + \lambda CB \times M_1 M_2$$

Using the previous equations (3) can be written in a simplified form:

$$\dot{\rho} = \frac{a_1 \lambda + a_2}{\sqrt{a_3 \lambda^2 + a_4 \lambda + a_5}} \qquad (4)$$

with

$$\begin{aligned}
a_1 &= M_1 M_2.V + (CB \times M_1 M_2).\Omega \\
a_2 &= U.V + (CB \times U).\Omega \quad a_3 = \|M_1 M_2\|^2 \\
a_4 &= 2\, U.M_1 M_2 \quad a_5 = \|U\|^2
\end{aligned}$$

The a_i are therefore constants which depend only upon the geometry of the robot, the imposed cartesian velocity and the segment extremities. Differentiating this expression with respect to λ leads to an expression whose numerator N is simplified to a linear function in λ. Consequently the minimal and maximal values of the articular velocity is obtained either for $\lambda = 0$ or $\lambda = 1$ or for the value which nullify N (if this value lie inside the range [0,1]).

3 Extremal velocities for a rectangle

Let us assume now that the workspace is defined by all the points inside a rectangle. Without loss of generality we may assume that any point in the rectangle is such that its coordinates verify

$$x_1 \leq x \leq x_2 \quad y_1 \leq y \leq y_2$$

Consequently we want to calculate the extremal values of $\dot{\rho}$ for any position of C verifying the previous constraints. This can be done using classical optimization techniques. We define two new variables α, β by:

$$\begin{aligned}
x &= x_1 + \frac{(1 + \sin \alpha)(x_2 - x_1)}{2} \\
y &= y_1 + \frac{(1 + \sin \beta)(y_2 - y_1)}{2}
\end{aligned} \qquad (5)$$

$\dot{\rho}$ is now a function of α, β whose extremal values satisfy necessarily the following equations:

$$\frac{\partial \dot{\rho}}{\partial \alpha} = 0 \quad \frac{\partial \dot{\rho}}{\partial \beta} = 0 \qquad (6)$$

It appears that equations (6) may be written as:

$$\cos(\alpha)\, F_1(\alpha, \beta) = 0 \quad \cos(\beta)\, F_2(\alpha, \beta) = 0 \qquad (7)$$

The solutions defined by $\alpha = \pm \frac{\pi}{2}$ and $\beta = \pm \frac{\pi}{2}$ correspond to an extremum on the edges of the rectangle, which can be computed using the previous section. The last remaining solutions are obtained when:

$$F_1(\alpha, \beta) = 0 \quad F_2(\alpha, \beta) = 0$$

which correspond to an extremum for a point inside the rectangle. In these equations the unknowns α, β appear via their sine only. F_1 is linear in $\sin \alpha$ and is solved for this unknown. The result is substituted into F_2 which become a third order polynomial in $\sin \beta$ only. By solving this equation the last remaining set of solutions of equations (6) are determined, which give the extremum inside the rectangle.. In summary the extremum on the edges are computed using the method described in the previous section and are compared to the extremum found for the inside of the rectangle to lead to the extremum for the whole rectangle.

4 Extremal velocities for a box

Let us assume now that the workspace is defined to be all the points inside a rectangular box. Without loss of generality we may assume that any point in the box is such that its coordinates verify

$$x_1 \leq x \leq x_2 \quad y_1 \leq y \leq y_2 \quad z_1 \leq z \leq z_2 \quad (8)$$

Consequently we want to calculate the extremal values of $\dot{\rho}$ for any position of C verifying (8).

4.1 The optimization approach

We define three new variables α, β, γ such that:

$$x = x_1 + \frac{1 + \sin\alpha(x_2 - x_1)}{2}$$
$$y = y_1 + \frac{(1 + \sin\beta(y_2 - y_1))}{2}$$
$$z = z_1 + \frac{1 + \sin\gamma(z_2 - z_1)}{2} \quad (9)$$

$\dot{\rho}$ is now a function of α, β, γ whose extremal values satisfy necessarily the following equations:

$$\frac{\partial \dot{\rho}}{\partial \alpha} = 0 \quad \frac{\partial \dot{\rho}}{\partial \beta} = 0 \quad \frac{\partial \dot{\rho}}{\partial \gamma} = 0$$

These three equations are transformed into algebraic equations using the classical half-angle tangent substitution. The resultants of two different pairs of equations lead to the two following equations:

$$\sum_{i,j=0,4} a_{ij} T_1^i T_2^j = 0 \quad \sum_{i,j=0,4} b_{ij} T_1^i T_2^j = 0$$

with $T_1 = tan(\alpha/2), T_2 = tan(\beta/2)$. Unfortunately we have not been able to solve this system.

At this point various options are possible: we may use a numerical method like continuation or intervals computing to solve numerically this set of equations. In our implementation we have chosen an alternative approach proposed in the next section.

4.2 An alternative approach

4.2.1 Principle

We intend to determine the maximal articular velocities in a box with a guaranteed error $\epsilon > 0$. The idea is to sweep the box by horizontal rectangles at various heights z, the difference of height between two successive rectangles being such that the difference of extremal articular velocities between the two rectangle does not exceed ϵ. Thus starting with the rectangle with the lowest z of the box, we will then determine the next z satisfying the previous constraint. The process will be repeated for the new rectangle until the height of the rectangle is greater or equal to the maximal z of the box. At this point the maximal velocity will have been determined with an error of at most ϵ.

4.3 Finding the increase of height

Let us assume that at the kth step of the algorithm the maximal velocity is $\dot{\rho}_k$ for the plane at height z_k. We want to determine the minimal increase of height z_Δ^2 such that

$$\dot{\rho}(z_k + z_\Delta^2) = \dot{\rho}_k + \epsilon \quad (10)$$

This may be seen a a classical optimization problem but the previous formulation is difficult to use in practice as $\dot{\rho}$ is a complex expression. But we have:

$$\dot{\rho} = \frac{F(z, \boldsymbol{\mathcal{V}})}{\rho(z)}$$

F and ρ^2 being algebraic functions of z. Thus equation (10) is transformed into:

$$F^2(z_k + z_\Delta^2) - (\dot{\rho}_k + \epsilon)^2 \rho^2(z_k + z_\Delta^2) = 0 \quad (11)$$

which has now an algebraic form. Now we have to solve the optimization problem of finding the minimal z_Δ^2 such that equation (11) is satisfied with the constraints on x, y defined in equations (9). We define the optimization function H as:

$$H = z_\Delta^2 + \mu(F^2(z_k + z_\Delta^2) - (\dot{\rho}_k + \epsilon)^2 \rho^2(z_k + z_\Delta^2))$$

in which the value of x, y have been substituted by equations (9). The minimum of z_Δ will be obtained by solving the system of equations:

$$\frac{\partial H}{\partial z_\Delta} = 0 \quad \frac{\partial H}{\partial \mu} = 0 \quad \frac{\partial H}{\partial \alpha} = 0 \quad \frac{\partial H}{\partial \beta} = 0$$

For each solution of this system it is necessary to check that the value of z_Δ does not lead to $\dot{\rho} = -\dot{\rho}_k - \epsilon$ which is also solution to equation (10). Due to the lack of space we will not present the details of the calculation but this problem is solved by manipulating a few sets of algebraic equations.

As for the determination of the minimal velocities a similar method is used with $\epsilon < 0$.

4.3.1 Computation time

Clearly the computation time of the previous algorithm is dependent upon the precision ϵ with which the extremal velocities are to be determined. Figure 2 presents the computation time as a function of the error ϵ, obtained with a SUN Ultra 1 workstation. It

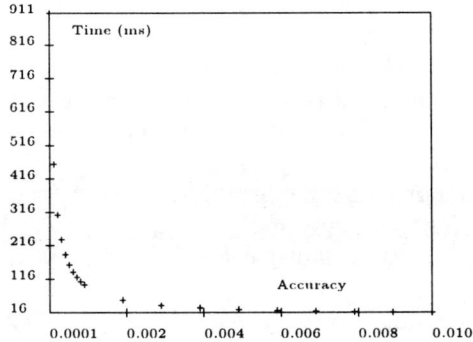

Figure 2: Computation time as a function of the precision on the value of the extremal velocities

may be seen that even with a high precision on the velocities the computation time is quite reasonable. Note that even with an accuracy of 0.01 the computation time of our algorithm is equal to the time needed to compute the articular velocities at less than 400 points in the box i.e. less than 8 points for each dimension: such limited number of points will never insure a correct estimation of the extremal velocities.

5 Maximal velocities in any volume

This method can be extended to more complex workspaces than a box. We will assume here that a workspace is described by a set of polygonal cross-sections. We will compute the extremal velocities for each volume defined between two successive cross-sections, then it will be easy to determine the extremal velocities for the whole workspace. A given volume will be decomposed into as many boxes as necessary until the velocities are determined with the desired accuracy. A list of box $\mathcal{B}$ is maintained during the algorithm: this list is initialized with the bounding box B_0 of the whole volume. Another list $\mathcal{L}$ will contain the current extremum of the articular velocities: this list is initialized by the extremum of the articular velocities computed over the vertices of the workspace boundary. At step k the algorithm perform the following operation:

1. if the box B_k is completely outside the volume we consider the next box in the list

2. if the box B_k is completely inside the volume we compute the extremal articular velocities for this box and update $\mathcal{L}$.

3. if the box B_k is partially inside the volume we compute the extremal articular velocities for this box.

   (a) if these extremum are within the range of $\mathcal{L}$ or the differences between the extremum are lower than ϵ we consider the next box in the list

   (b) otherwise the box is split into eight boxes by dividing each dimension of the box by 2. The resulting boxes are put at the end of the list and we consider the next box in the list.

The algorithm stop if there is no more box in the list. It enables to compute the extremal velocities for any type of workspace in a reasonable amount of time. For example we have considered the workspace defined by three square cross-sections: $z = 50, x \in [-10, 10], y \in [-10, 10]$, $z = 55, x \in [-5, 5], y \in [-5, 5]$, $z = 60, x \in [-10, 10], y \in [-10, 10]$ represented in figure 3.

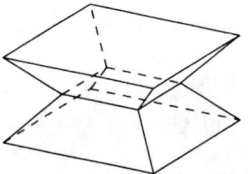

Figure 3: The test volume

The orientation is defined by the three Euler angles with value -60 degree, the desired cartesian/angular velocity is : -10, 0, -10, -10, 0 10. Table 1 indicates the computation time for various values of ϵ. We may

ϵ	1	0.5	0.1	0.05	0.01
Time (ms)	360	430	4260	12220	95500

Table 1: Computation time for a complex workspace as function of the desired accuracy ϵ on the articular velocities

note that the computation time for determining the velocities with an accuracy better than 1% is approximatively 4s.

6 Special case

In this special case we will assume that the angular velocity is equal to 0. Consequently the articular velocity is given by:

$$\dot{\rho} = V \cdot \frac{AB}{\|AB\|} \quad (12)$$

Let D_1 be the line going through A with vector V and D_2 be the line associated to the leg. If μ is the angle between these two lines we have:

$$\dot{\rho} = \|V\| \cos \mu \quad (13)$$

Hence if the line D_1 crosses the volume at a point M, then the maximal articular velocity will be $\|V\|$ if $AM.V$ is positive or the minimal articular velocity will be $-\|V\|$ if $AM.V$ is negative. In the sequel we will assume that $AM.V$ is positive (if this quantity is negative, then maximal has to be changed to minimal). If D_1 does not cross the volume the maximal velocity will be obtained for a location of C on one edge of the volume. Similarly the minimal articular velocity is also obtained for a location of C on one edge of the volume. Therefore for computing the maximal articular velocity we have to check if D_1 crosses the volume. We will consider here without loss of generality the volume between two cross-sections. D_1 may cross this volume either on a horizontal facet or on a side facet. An horizontal facet is defined by a polygon and an altitude z_i. For the two horizontal cross sections we compute the intersection of the line with the plane $z = z_i$ and check if the intersection point belongs to the polygon, in which case the maximal articular velocity is $\|V\|$.

A side facet is defined by two horizontal segments: the coordinates of the extremities of the first segment will be denoted $(x_1, y_1, z_1), (x_2, y_2, z_1)$ and $(x_3, y_3, z_3), (x_4, y_4, z_3)$ will denote the coordinates of the extremities of the second segment.

At a given altitude z the side facet is constituted of a segment whose start and end point X_k, X_{k+1} have the coordinates:

$$x_k = x_1 + (x_3 - x_1)(z - z_1)(z_3 - z_1) \quad (14)$$
$$y_k = y_1 + (y_3 - y_1)(z - z_1)(z_3 - z_1) \quad (15)$$
$$z_{k+1} = z_k = z \quad (16)$$
$$x_{k+1} = x_2 + (x_4 - x_2)(z - z_1)(z_3 - z_1) \quad (17)$$
$$y_{k+1} = y_2 + (y_4 - y_2)(z - z_1)(z_3 - z_1) \quad (18)$$

A point $M(x, y, z)$ which belongs to this segment has the coordinates:

$$x = x_k + \lambda(x_{k+1} - x_k) \quad (19)$$
$$y = y_k + \lambda(y_{k+1} - y_k) \quad (20)$$

with λ in the range [0,1]. If this point belongs to D_1 then:

$$AM \times V = 0 \quad (21)$$

This relation leads to 2 equations in the unknowns z, λ. If the vertical component of the velocity is equal to 0 then the z coordinate is the z coordinates of the point A and λ is obtained linearly from equation (21): if λ lie in the range [0,1], then we have intersection between D_1 and the volume.

If the vertical component of the velocity is not equal to 0 the resultant of two equations of (21) leads to a second order equation in z. This equation is solved and if it has a solution in the range $[z_1, z_2]$ the value of λ is computed and if λ lie in the range [0,1], then we have intersection between D_1 and the volume. With this method we are able to check if D_1 crosses the volume. If this is not the case and in order to compute the minimal articular velocities we have now to compute the extremal velocities for each edge of the volume. This is done by using the algorithm described in section 2.

Clearly this algorithm is very fast (less than 13 ms for the example presented above) and leads to the exact determination of the extremal articular velocities. Note that the algorithm can be trivially extended to workspace volumes described by spheres.

7 Articular workspace

Let us assume that the leg length have a minimal and a maximal values ρ_{min}, ρ_{max}. It may be of interest to compute the extremum of the articular velocities in the workspace defined by a constant orientation and any position of the platform which fulfill the constraints on the leg lengths: this workspace will be called the *articular workspace*. A simple adaptation of the previous algorithm enable to perform this task. Note first that a trivial algorithm enable to determine what are the extremum of each leg lengths while C moves in a given box: we will denote this algorithm $Max_\rho(B)$ where B is a box. Then notice that it is easy to determine a box which contain all the possible locations of the platform being given the extremal values of the leg lengths. We start the previous algorithm with this box. Then we have to change the inclusion test in the previous algorithm: a box will lie within the workspace if all the ranges given by $Max_\rho(B)$ lie within $[\rho_{min}, \rho_{max}]$ while a box will be completely outside the workspace if one of the ranges is outside the range $[\rho_{min}, \rho_{max}]$. In any other cases the box we assume that the box is partially within the workspace.

Strictly speaking this may be false: as $Max_\rho(B)$ gives the extremum of the leg lengths *independently* it may occur that there is no posture of the platform where *all* the leg lengths lie in the correct range at the same time, but as this type of box will be divided in smaller box during the process our assumption lie on the safe side.

Note also that another approach will be to use the algorithm described in [7] which enable to compute exact cross-sections of the workspace for a constant orientation and then use the algorithm described for the polyhedric workspace.

For an accuracy of 0.1 (0.01) the computation time is 29800 ms (67650 ms) and it is reduced to 1270 ms (1380 ms) if the angular velocities are equal to 0.

8 Another utility of the algorithm

Let us assume that the angular velocities are set to 0 and that the cartesian velocity is defined as a unit vector V. The algorithm will therefore compute the minimum and maximum of the quantity $AB.V/||AB||$ which is the cosine of the angle between the link direction and the vector V. Consequently we will get the minimum and maximum values of the angle of the passive joints with any fixed direction.

9 Extension to other types of parallel robots

The algorithm has been presented for the Gough-type parallel robot but may be extended for other types of parallel robots. Indeed it is well known that most of parallel robots have an inverse jacobian matrix of the same form as the one of the Gough-type robot. Therefore the principle of the algorithm will be similar. Consider for example the parallel robots with fixed leg lengths but whose A_i points moves on a line with unit vector u_i. The velocity $\dot{\gamma}_i$ of point A_i is related to the cartesian and angular velocities by [7]:

$$\dot{\gamma}_i = \frac{A_i B_i . V}{u_i . A_i B_i} + \frac{A_i B_i \times \Omega}{u_i . A_i B_i} \quad (22)$$

If C moves on a segment the derivative of the articular velocity with respect to λ is constant. Hence the minimal and maximal articular velocities will be obtained either for $\lambda = 0$ or $\lambda = 1$. If C moves into a horizontal rectangle we use equations (5) and the derivative of the articular velocity with respect to α, β have a similar form to (7).

10 Conclusion

An algorithm for computing the extremal articular velocities of a parallel robot whose end-effector must be able to perform a given translation/angular velocity over a whole workspace has been presented. It compute the extremal values of the articular velocity with a guaranteed error (without any error if the angular velocity is equal to zero) and is therefore safer than classical method relying on discretisation. Furthermore it is in general faster than the classical method. This algorithm can be used for the optimal design of parallel robots.

References

[1] Gosselin C. *Kinematic analysis optimization and programming of parallel robotic manipulators*. PhD thesis, McGill University, Montréal, June, 15, 1988.

[2] Ling S-H. and Huang M.Z. Kinestatic analysis of general parallel manipulators. In *ASME Mechanisms Design Conf.*, Minneapolis, September, 14-16, 1994.

[3] Martinez J.M.R. and Duffy J. A simple method for the velocity and acceleration analysis of in-parallel platforms. In *9th World Congress on the Theory of Machines and Mechanisms*, pages 842-846, Milan, August 30- September 2, 1995.

[4] Sorli M. and others . Mechanics of Turin parallel robot. In *9th World Congress on the Theory of Machines and Mechanisms*, pages 1880-1885, Milan, August 30- September 2, 1995.

[5] Zanganeh K.E. and Angeles J. Instantaneous kinematics and design of a novel redundant parallel manipulator. In *IEEE Int. Conf. on Robotics and Automation*, pages 3043-3048, San Diego, May, 8-13, 1994.

[6] Gough V.E. and Whitehall S.G. Universal tire test machine. In *Proceedings 9th Int. Technical Congress F.I.S.I.T.A.*, volume 117, pages 117-135, May 1962.

[7] Merlet J-P. *Les Robots parallèles*. Hermès, Paris, 1997.

Efficient estimation of the extremal articular forces of a parallel manipulator in a translation workspace

Jean-Pierre MERLET
INRIA Sophia-Antipolis
BP 93 06902 Sophia-Antipolis, France
E-mail: Jean-Pierre.Merlet@sophia.inria.fr

Abstract

In this paper we consider a classical Gough platform with extensible legs whose platform is submitted to a given load. This load induces forces in the linear actuators of the legs, these forces being dependent upon the posture of the platform, and we want to determine the extremal values of the articular forces when the platform is translating in a given 3D workspace (the orientation of the platform is assumed to be constant). We describe an efficient algorithm which enable to compute the extremal forces more efficiently than a discretisation method.

1 Introduction

In the design phase of a parallel manipulators it is extremely important to determine what will be the extremal articular forces of the platform induced by the presence of a given load on the moving platform. As an example we consider the classical Gough type parallel manipulator [1] illustrated in figure 1. In this robot a base and a platform are connected through 6 legs which have a ball-and-socket joint at each extremity. Linear actuators enable to change the leg lengths which in turn enable to control the position and orientation of the platform. If a given load is applied to the moving platform (for example the weight of some equipment) then each leg is submitted to a force acting along the leg axis. The values of these articular forces are position dependent and our purpose is to determine, for a given load, what will be the extremal values of the forces in the leg while the robot is moving in a given workspace. We will assume here that the workspace is only a translation workspace i.e. the orientation is kept constant and the workspace is defined by a 3D object which describe all the possible positions of the center C of the moving platform. The

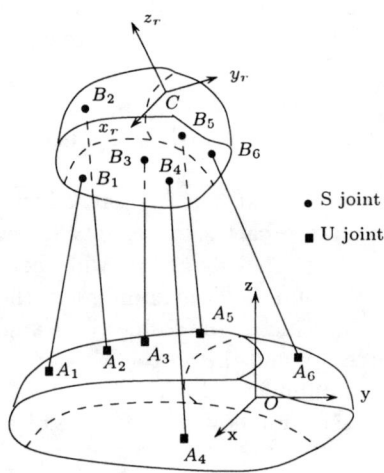

Figure 1: The classical Gough type parallel robot

coordinates of C in the reference frame (O, x, y, z) will be denoted x, y, z

Numerous papers have been devoted to the relations between articular forces and generalized forces acting on the platform [2, 3, 4, 5, 6, 7] but none, to the best of the author knowledge, have addressed our problem, which is however of practical importance. The computation time presented in this paper are established on a SUN Ultra 1 workstation.

2 Relation between the articular and generalized forces

Let $\mathcal{F}$ denote the generalized forces applied on the moving platform and τ the leg forces vector. It is well known that these quantities are related by:

$$\mathcal{F} = J^{-T}\tau \qquad (1)$$

where J^{-T} is the transpose of the inverse jacobian matrix of the robot, which is posture dependent. For the Gough platform a line J_i of the inverse jacobian matrix may be written as:

$$J_i = (\frac{A_iB_i}{||A_iB_i||} \quad CB_i \times \frac{A_iB_i}{||A_iB_i||}) \quad (2)$$

where $A_i(xa_i, ya_i, za_i), B_i(xb_i, yb_i, zb_i)$ are the extreme points of leg i. Note also that the leg length ρ_i is equal to $||A_iB_i||$. We define the matrix H_i as the matrix obtained by substituting the i-th column of the matrix J^{-T} by the vector $\mathcal{F}$. Equation (1) defining a linear system we get each component τ_i of τ by:

$$\tau_i = \frac{|H_i|}{|J^{-T}|} \quad (3)$$

Note that neither $|H_i|$ nor $|J^{-T}|$ are algebraic in terms of the coordinates of C as $||A_iB_i||$ appear in each matrix. A more convenient formulation will be presented now. Let the semi-inverse jacobian matrix J_s^{-1} be defined by the 6 lines J_s^i:

$$J_s^i = (A_iB_i \quad CB_i \times A_iB_i)$$

and let H_s^i be the matrix obtained by substituting the i-th column of J_s^{-T} by the vector $\mathcal{F}$. It is clear that:

$$|J^{-T}| = |J^{-1}| = \frac{|J_s^{-1}|}{\prod_{i=1}^{i=6} \rho_i} \quad (4)$$

By developing $|H_s^i|$ with respect to the i-th column we get:

$$|H_i| = \frac{|H_s^i|}{\prod_{j=1}^{j=6} \rho_j \quad j \neq i} \quad (5)$$

Note that both $|J_s^{-T}|, |H_s^i|$ are now algebraic in terms of the coordinates of C. Using these results and equation (3) we get:

$$\tau_i = \frac{\rho_i |H_s^i|}{|J_s^{-T}|} \quad (6)$$

Assume now that the coordinates of C are functions of a parameter r. The derivative D of τ_i with respect to r can be computed as:

$$D = \frac{\partial \tau_i}{\partial r} = \frac{(\frac{\partial \rho^2}{\partial r}|H_s^i| + 2\rho^2 \frac{\partial |H_s^i|}{\partial r})|J_s^{-T}| - 2\rho^2|H_s^i|\frac{\partial |J_s^{-T}|}{\partial r}}{2\rho|J_s^{-T}|^2} \quad (7)$$

Note that the numerator of this expression is algebraic in terms of r and that the denominator is strictly positive. We assume here that there is no singularity in the workspace of the robot, this being verified using a method which will be described in another paper. In the sequel we will present a method to compute the extremum of the articular force for one leg, the process being identical for each leg.

3 Segment workspace

We assume that C is moving on a given segment M_1M_2 and consequently we may write that:

$$OC = OM_1 + \lambda M_1M_2$$

where λ is a scalar in the range $[0,1]$. To determine the extremal value of the articular force as C moves on the segment it is sufficient to compute the roots λ_i of the polynomial equation defined by the numerator D_n of D and then to compute the value of the articular force for each location of C defined by the λ_i included in the range $[0,1]$ together with the force obtained for $\lambda = 0, 1$. Then the minimal and maximal values of these quantities are the extremal values of the articular force while C is moving on the segment. Let us study in more details the degree of D_n: $|J_s^{-T}|$ is usually a third order polynomial in λ, ρ^2 is a second order polynomial in λ while $|H_s^i|$ is a third order polynomial in λ. Consequently D_n will be a seventh-order polynomial in λ.

We consider now a special case where the base is planar ($za_i = 0$) and the segment is oriented along the x axis (or equivalently along the y axis as we may rotate freely the reference frame around the z axis) which will be useful in the sequel. In this case D_n is of degree 6 as $|J_s^{-T}|$ is of degree 2 in λ. Furthermore if the platform is parallel to the base ($zb_i = C^{te}$) then $|H_s^i|$ is of degree 1 in λ and it may be shown that the degree of D_n become 2 or 3 (this case will be denoted the *special case*).

The computation time of this procedure is about 55 ms for a general segment. If the base is planar and the segment is oriented along the x axis the computation time is 15 ms. Although this computation time may seem to be relatively high most of it is devoted to the computation of some constant coefficients which are to be computed whatever the workspace is: consequently for more complex workspace for which determination of the articular force extremum on segments is needed the computation time will be deeply reduced.

4 Horizontal rectangle workspace

Now we assume that the workspace for C is an horizontal rectangle defined by:

$$x_1 \leq x \leq x_2 \quad y_1 \leq y \leq y_2$$

Note that vertical rectangle workspace can be treated in the same manner with an appropriate change in the direction of the reference frame.

4.1 General case

A first approach to determine the extremal values of the articular force will be to define two auxiliary variables α, β such that $x = x_1 + (1+\sin\alpha)(x_2-x_1)/2$, $y = y_1 + (1+\sin\beta)(y_2-y_1)/2$. Equation (7) will then be used to obtain two constraint equations in the unknowns $\sin\alpha, \sin\beta$. Unfortunately the degree of these equations is high (7 or 8) and it is difficult to determine the solution of this system.

We use therefore another approach which is to compute the articular force with an accuracy at least better than a given constant ϵ. The idea is first to compute the articular force on the segment $x_1 \leq x \leq x_2, y = y_1$ using the result of the previous section. Let τ_{min}, τ_{max} be the force computed at this stage. We then investigate the minimal value of a variable y_Δ^2 such that on the segment $x_1 \leq x \leq x_2, y = y_1 + y_\Delta^2$ the corresponding articular force is equal to $\tau_{min} - \epsilon$ or $\tau_{max} + \epsilon$. In other words if $\tau_{min}^s(y), \tau_{max}^s(y)$ denote the minimal and maximal articular force on the segment $x_1 \leq x \leq x_2, y$ we have to find the minimal value of y_Δ^2 such that

$$\tau_{min}^s(y_1 + y_\Delta^2) = \tau_{min} - \epsilon \text{ or}$$
$$\tau_{max}^s(y_1 + y_\Delta^2) = \tau_{max} + \epsilon \quad (8)$$

If we define $x = x_1 + \lambda(x_2 - x_1)$ these equations are functions of y_Δ^2, λ. Here the value of y_Δ^2 is assumed to be small so that equations (8) may be developed at first order to become:

$$A_2(\lambda)y_\Delta^2 + A_0(\lambda) = 0$$

Thus the value of y_Δ^2 is obtained as a function of λ by:

$$y_\Delta^2 = -\frac{A_0(\lambda)}{A_2(\lambda)}$$

The derivative of y_Δ^2 with respect to λ is computed and lead to a 22nd order polynomial in λ. The value of y_Δ^2 is computed for each root of this polynomial in the range [0,1] together with the value at $\lambda = 0, 1$ and the minimal value of y_Δ^2 is retained. If this value is small the computation is supposed to be exact otherwise a fixed small value (0.3 in our current implementation) is assigned to y_Δ^2. The extremal articular force are computed on the segment $x_1 \leq x \leq x_2, y = y_1 + y_\Delta^2$ and the values of τ_{min}, τ_{max} are updated. This process is repeated until the value of y is equal to y_2.

In order to speed up this analysis we have computed the derivative of τ as function of y. Remember that the denominator of this derivative has a constant sign and that the numerator is algebraic in terms of x, y. As x, y are bounded a simple interval analysis enable to estimate what will be the minimum and maximum of this derivative. If they are found to be of constant sign then τ is monotonous with respect to y and the extremum of the articular force are obtained by computed the articular force on the two segments $x_1 \leq x \leq x_2, y = y_1$, $x_1 \leq x \leq x_2, y = y_2$.

4.2 Special case

As mentioned previously if both the base and the platform are planar we get a simplification in D_n which enable to use the optimization approach. In that case we may find the extremum of the articular forces by solving two pairs of second order equations.

For articular forces in the range of 200 and an accuracy of 1 the computation time is about 2500 ms for a general robot and 1600 ms if the base is planar. In the special case this time is reduced to 40 ms.

5 Box workspace

5.1 Principle

Now we assume that the workspace for C is a box defined by:

$$x_1 \leq x \leq x_2 \quad y_1 \leq y \leq y_2 \quad z_1 \leq z \leq z_2$$

In view of the previous section it is clear that the optimization approach cannot be used. Therefore we use a similar approach as for the rectangle workspace. We first compute the extremum articular force in the rectangle $x_1 \leq x \leq x_2, z = z_1$ using the result of the previous section. Let τ_{min}, τ_{max} be the current extremal articular forces. We then investigate the minimal value of a variable z_Δ^2 such that on the rectangle $x_1 \leq x \leq x_2, y_1 \leq y \leq y_2, z = z_1 + z_\Delta^2$ the corresponding articular force is equal to $\tau_{min} - \epsilon$ or $\tau_{max} + \epsilon$. In other words if $\tau_{min}^s(z), \tau_{max}^s(z)$ denote the minimal and maximal articular force on the rectangle $x_1 \leq x \leq x_2, y_1 \leq y \leq y_2, z$ we have to find the minimal value of z_Δ^2 such that:

$$\tau_{min}^s(z_1 + z_\Delta^2) = \tau_{min} - \epsilon \text{ or}$$
$$\tau_{max}^s(z_1 + z_\Delta^2) = \tau_{max} + \epsilon \quad (9)$$

If we define $x = x_1 + \lambda(x_2 - x_1)$, $y = y_1 + \mu(y_2 - y_1)$ these equations are functions of z_Δ^2, λ, μ. Here the value of z_Δ^2 is assumed to be small so that equations (9) may be developed at first order to become:

$$A_2(\lambda, \mu)z_\Delta^2 + A_0(\lambda, \mu) = 0$$

the value of z_Δ^2 is obtained as a function of λ, μ by:

$$z_\Delta^2 = -\frac{A_0(\lambda, \mu)}{A_2(\lambda, \mu)}$$

The minimal value of z_Δ^2 if obtained for λ, μ such that the derivatives of z_Δ^2 with respect to λ, μ vanish. As λ, μ belong to [0,1] these derivatives are approximated to the second order around the value 0, 0.5, 1. Hence for each case we get two second order equations in these variables and their resultant is a fourth order polynomial in λ. By solving this polynomial we get all the possible values of λ and for each λ we get corresponding values of μ. For each pair (λ, μ) we then compute the value of z_Δ^2 and we retain the smallest positive value. The value of z is updated to $z + z_\Delta^2$ if the computed value of z_Δ^2 is sufficiently small otherwise a small value is assigned to z_Δ^2. This process is repeated until the value of z is equal to z_2.

5.2 Speeding up the algorithm

In order to speed up this analysis we have computed the value of τ as function of x, y, z. Remember that τ_i may be written as $\rho_i |H_s^i|/|J_s^{-T}|$ where both $|H_s^i|, |J_s^{-T}|$ are algebraic in terms of x, y. As x, y are bounded a simple interval analysis enable to estimate what will be the minimum and maximum of $|H_s^i|/|J_s^{-T}|$. Another trivial algorithm enable to compute the maximal and minimal value of ρ_i in the rectangle. Consequently it is easy to find bounds on the value of τ_i in the rectangle at altitude z. If these bounds lie within the current range $[\tau_{min}, \tau_{max}]$, then we skip the computation of the current rectangle and compute a new value of z_Δ^2.

Similarly we compute the derivative of τ with respect to z: we get therefore an expression function of x, y, z. A simple analysis interval enable to compute the extremum values of this derivative and if the minimum and maximum are of same sign then τ is monotonous with respect to z and the extremum of the articular forces are obtained by computing the extremum for the rectangles at altitude z_1, z_2.

5.3 Computation time

For articular forces in the range of 200 and an accuracy of 1 the computation time vary from 6400 to 21000 ms for a general robot and from 3000 to 4000 ms if the base is planar (the difference between the two types of robot are mainly due to a more careful implementation of the case of the planar base). In the later case as the computation of the articular forces for one posture of the robot take about 0.5 ms a discrete method will have split each three main axis in 18 to 20 points (hence we may miss postures where important articular forces occur).

The sensitivity of the computation time to the accuracy with which the extremum of the articular forces are computed is presented in figure 2. It may be seen

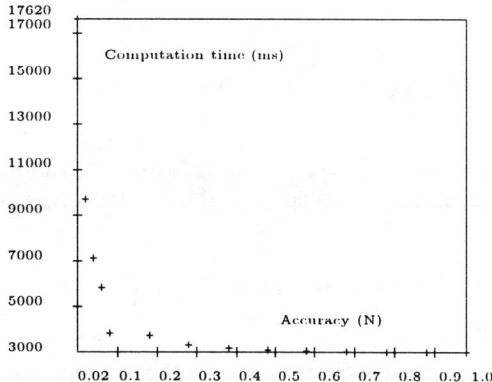

Figure 2: Computation time versus the desired accuracy for a box workspace and a robot with a planar base

that the computation time is prohibitive only for an accuracy which is far away from the usual one necessary for the design process.

5.4 Special case

In the special case the optimization approach can be used. The constraint equations can be reduced to solving a set of 4 univariate polynomials of degree 2,3,4,8. This enable to get the exact extremum and reduce the computation time to 120 to 200 ms.

6 Polyhedric workspace

The low computation time for the determination of the extremal articular forces when the workspace is a box suggests that this method can be extended to more complex workspaces. We will assume here that a workspace is described by a set of polygonal cross-sections. We will compute the extremal articular forces for each volume defined between two successive cross-sections, then it will be easy to determine the extremal articular forces for the whole workspace. A given volume will be decomposed into as many boxes as necessary until the articular forces are determined with the desired accuracy (note that it is trivial to determine if a box lie within the workspace). A list of box $\mathcal{B}$ is maintained during the algorithm: this list is initialized with the bounding box B_0 of the whole volume. A set of extremum articular forces is initialized by computing the articular forces at some vertex of the workspace. The range of these forces will be called the current articular forces range. At step k the algorithm perform the following operations:

1. if the box B_k is completely outside the volume we consider the next box in the list

2. if the box B_k lie completely within the workspace we compute the extremal articular forces for this box and update the current articular forces range

3. if the box B_k lie partially within the workspace we compute the extremal articular forces for this box

    (a) if these forces lie within the current articular forces range we consider the next box in the list

    (b) otherwise the box is split into eight boxes by dividing each dimension of the box by 2. The resulting boxes are put at the end of the list and we consider the next box in the list.

The algorithm stop if there is no more box in the list. Note that to speed up the process an heuristic is used: whenever the extremum for a box has to be computed we first estimate a bound on the value of the extremum by using an interval analysis similar to the analysis presented in the section devoted to the box workspace: if these bounds lie within the current articular forces range we skip the computation for this box and moves to the next box in the list.

The computation time is reasonable: for example we have considered the workspace defined by three square cross-sections: $z = 50, x \in [-10, 10], y \in [-10, 10]$, $z = 55, x \in [-5, 5], y \in [-5, 5]$, $z = 60, x \in [-10, 10], y \in [-10, 10]$ represented in figure 3. For a general robot the computation time vary from to 65s to 165s while when the base is planar the computation time vary from 7s to 9s (1.2s in the special case). The sensitivity of the computation time with respect to the accuracy is illustrated on figure 4. It may be seen that

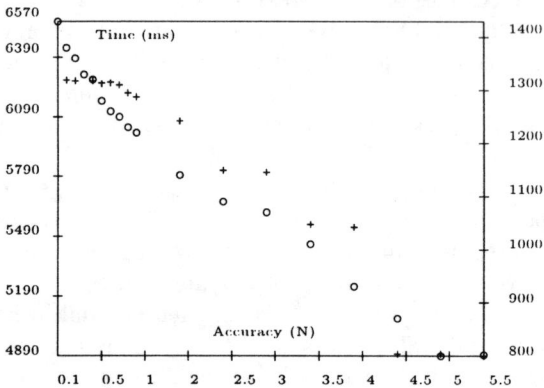

Figure 4: Computation time versus the desired accuracy for the test volume for a robot with a planar base (the crosses represent the general case while the circles represent the special case) .

even for an accuracy of 0.1 N the computation time is reasonably low at 6570 ms.

7 Articular workspace

Let us assume that the leg length have a minimal and a maximal values ρ_{min}, ρ_{max}. It may be of interest to compute the extremum of the articular forces in the workspace defined by a constant orientation and any position of the platform which fulfill the constraints on the leg lengths: this workspace will be called the *articular workspace*. A simple adaptation of the previous algorithm enable to perform this task. Note first that a trivial algorithm enable to determine what will the extremum of each leg lengths while C moves in a given box: we will denote this algorithm $Max_\rho(B)$ where B is a box. Then notice that it is easy to determine a box which contain all the possible locations of the platform being given the extremal values of the leg lengths. We start the previous algorithm with this box. Then we have to change the inclusion test in the previous algorithm: a box will lie within the workspace if all the ranges given by $Max_\rho(B)$ lie within $[\rho_{min}, \rho_{max}]$ while a box will be completely outside the workspace if one of the ranges is outside the range $[\rho_{min}, \rho_{max}]$. In any other cases we assume that the box is partially within

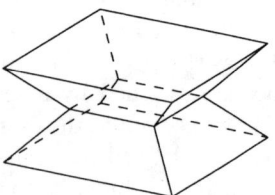

Figure 3: The test volume

the workspace. Strictly speaking this may be false: as $Max_\rho(B)$ gives the extremum of the leg lengths *independently* it may occur that there is no posture of the platform where *all* the leg lengths lie in the correct range at the same time, but as this type of box will be divided in smaller box during the process our assumption lie on the safe side.

In the implementation of this algorithm the following precautions have to be taken:

- if the leg lengths for a given box exceed by far the articular limits while still the box is partially inside the articular workspace it is better to split the box without computing the extremum of the articular forces for this box (to avoid computation for large boxes)

- as soon as all the dimensions of a box are quite small we consider that the articular forces in the box are given for the forces obtained at the center of the box. This avoid to create a large number of boxes in the case where the center of the box is on the boundary of the workspace

With these heuristics the computation time is reasonable: typically 4.5 mn for the full articular workspace for an accuracy of 1 (approximatively 2.5 mn in the special case) and 100 s for an accuracy of 5 (60s in the special case) but we still have the guarantee on the validity of the result. Note also that another approach will be to use the algorithm described in [8] which enable to compute exact cross-sections of the workspace for a constant orientation and then use the algorithm described for the polyhedric workspace.

8 Conclusion

The algorithm presented in this paper enable to compute efficiently one of the most important feature for the design of a parallel robot. Although the computation time may seem to be high it must be noted that for each type of workspace we have find numerous examples for which the computation time of a discretisation method necessary to determine the articular forces with the same level of accuracy exceed by far the computation time of our algorithm. It has also been noted that the discrepancy between the results of our algorithm and of a discretisation method with a similar computation time may reach up to 10 %.

Still the workspace we have been considering is only the translation workspace but a discretisation on the 3D orientation workspace will be by far less computer expensive than the discretisation on the full 6D workspace. This algorithm will be integrated in the near future in our design methodology DEMOCRAT for the design of parallel robot.

Note that this algorithm can also be extended to other mechanical architecture of fully-parallel 6 DOF robots as most of them have an inverse jacobian matrix similar to the matrix of the Gough platform.

References

[1] Gough V.E. and Whitehall S.G. Universal tire test machine. In *Proceedings 9th Int. Technical Congress F.I.S.I.T.A.*, volume 117, pages 117–135, May 1962.

[2] Agrawal S.K. and Roth B. Statics of in-parallel manipulator systems. *ASME J. of Mechanical Design*, 114:564–568, December 1992.

[3] Bryfogle M.D., Nguyen C.C., Zhou Z-l., and Antrazi S.S. A methodology for geometry design of closed kinematic chain mechanisms. In *IEEE Int. Conf. on Robotics and Automation*, pages 2974–2979, Albuquerque, April, 21-28, 1997.

[4] Duffy J. *Statics and Kinematics with Applications to Robotics*. Cambridge University Press, New-York, 1996.

[5] Kosuge K. and others . Input/output force analysis of parallel link manipulators. In *IEEE Int. Conf. on Robotics and Automation*, pages 714–719, Atlanta, May, 2-6, 1993.

[6] Orin D.E. and Oh S.Y. Control of force distribution in robotic mechanisms containing closed kinematic chains. *J. of Dyn. Syst. Meas. and Control*, 102:134–141, June 1981.

[7] Pang H. and Shahinpoor M. Analysis of static equilibrium of a parallel manipulator. *Robotica*, 11:433–443, 1993.

[8] Merlet J-P. Détermination de l'espace de travail d'un robot parallèle pour une orientation constante. *Mechanism and Machine Theory*, 29(8):1099–1113, November 1994.

Dynamic Sensor Planning in Visual Servoing

Éric Marchand[1] and Greg D. Hager
Dept. of Computer Science, Yale University
New Haven, CT 06520–8285
Email: marchand@irisa.fr, hager@cs.yale.edu

abstract

We present an approach to dynamic sensor planning problems in visual servoing. Specifically, one of the main problems in image-based visual servoing is to plan the camera trajectory in order to avoid undesired configurations (e.g., features out of view, collision with obstacles, ...). Our approach uses the robot redundancy and employs a control scheme based on the task function approach. It combines the regulation of the selected vision-based task with the minimization of a secondary cost function, which reflects given constraints on the manipulator trajectory. We describe how this methodology is applied to common problems in robotic vision: occlusion avoidance, field of view constraint and obstacle avoidance. We have demonstrated the validity of this approach with various experiments.

1 Overview

One of the key points of the perception action cycle is the automatic generation of the camera motion. Visual servoing [8][4][9] appears to be a very efficient approach to this problem. Although many of the theoretical control issues are now well known, the integration of visual servoing into complex robotics systems remains difficult for various reasons. Considering the case of an eye-in-hand architecture, planning camera trajectory remains an important issue. Indeed, if the control law computes a motion that leads the camera to undesired configurations (such as manipulator joint limits, occlusions or obstacles), visual servoing will fail. Control laws taking into account these "bad" configurations have thus to be considered.

We have chosen to build in avoidance of undesirable configurations using a control scheme based on the task function approach [14][4]. It combines the regulation of the vision-based task with the minimization of a cost function which reflects the constraints imposed on the trajectory. The visual task is considered as a primary and priority task. The cost function is then embedded in a secondary task which only the components which are compatible with the primary task are taken into account (*i.e.*, the minimization of the cost function is performed under the constraint that the visual task is realized). This cost function to be minimized is based on a measure of the risk of the occurrence of an undesired configuration. It must reach its maximal value when these configurations are likely to occur and its gradient must be equal to zero when the cost function reaches its minimal value [14]. In this paper, we applied the proposed methodology to various problems such as occlusions avoidance, constraints on the field of view (*i.e.*, keeping an object inside view), 3D contact in a cluttered environment (*i.e.*, obstacle avoidance). This method as been previously used for singularities and joint limits avoidance [11].

A similar approach has been proposed by Nelson and Khosla. It consists of minimizing an objective function which realizes a compromise between the visual task (a target tracking using a camera mounted on the end effector of a manipulator) and the avoidance of kinematic singularities, joint limits singularities but also with some other constraints on the field of view, the focus measure [12]. This function is used by exploiting the robot degrees of freedom which are redundant with respect to the visual task. However, the resulting camera motions can produce major perturbations in the visual servoing since they are generally not compatible with the regulation to zero of the selected image features.

The next section of this paper, taken from [11], recalls the application of the task function approach to visual servoing and the expression of the resulting control law. Section 3 describes the approach proposed to dynamic sensor planning. We finally present real time experimental results dealing with various robotic tasks. These results have been obtained using an eye-in-hand system composed of a camera mounted on the end-effector of a six d.o.f Zebra Zero robot.

2 Visual Servoing

The *image-based visual servoing* consists in specifying a task as the regulation in the image of a set of visual features[4][8]. Embedding visual servoing in the task function approach [14] allows us to take advantage of general results helpful for the analysis and the synthesis of efficient closed loop control schemes. A good review and introduction to visual servoing can be found in [9].

Let us denote $\underline{P}$ the current value of the set of selected visual features used in the visual servoing task and measured from the image at each iteration of the control law. To ensure the convergence of $\underline{P}$ to its desired value $\underline{P}_d$, we need to know the interaction matrix $L_{\underline{P}}^T$ defined by the classical

---

[1] Current address is Éric Marchand, IRISA - INRIA Rennes, Campus de Beaulieu, F-35042 Rennes Cedex, France.

equation [4]:
$$\dot{\underline{P}} = L_{\underline{P}}^T(\underline{P}, p) T_c \qquad (1)$$

where $\dot{\underline{P}}$ is the time variation of $\underline{P}$ due to the camera motion T_c. The parameters p involved in $L_{\underline{P}}^T$ represent the depth information between the considered objects and the camera frame.

A vision-based task $\underline{e}_1$ is defined by:
$$\underline{e}_1 = C(\underline{P} - \underline{P}_d) \qquad (2)$$

where C, called combination matrix, has to be chosen such that $C L_{\underline{P}}^T$ is full rank about the desired trajectory $q_r(t)$. It can be defined as $C = W L_{\underline{P}}^{T+}(\underline{P}_d, p_d)$. Assumptions on the shape and on the geometry of the considered objects in the scene have thus generally to be done in order to compute the desired values $\underline{P}_d$ and p_d. In that case, we set W as a full rank matrix such that $\text{Ker } W = \text{Ker } L_{\underline{P}}^T(\underline{P}_d, p_d)$.

If the vision-based task does not constrain all the n robot degrees of freedom, a secondary task $\underline{g}_s$ can also be performed and we obtain the following task function:
$$\underline{e} = W^+ \underline{e}_1 + (\mathbb{I}_n - W^+ W) \underline{g}_s^T \qquad (3)$$

where

- W^+ and $\mathbb{I}_n - W^+ W$ are two projection operators which guarantee that the camera motion due to the secondary task is compatible with the regulation of $\underline{P}$ to $\underline{P}_d$. Indeed, due to the choice of matrix W, $\mathbb{I}_n - W^+ W$ belongs to $\text{Ker } L_{\underline{P}}$, which means that the realization of the secondary task will have no effect on the vision-based task ($L_{\underline{P}}^T(\mathbb{I}_n - W^+ W) \underline{g}_s^T = 0$). On the other hand, if errors are introduced in $L_{\underline{P}}^T$, $\mathbb{I}_n - W^+ W$ no longer exactly belongs to $\text{Ker } L_{\underline{P}}$. This will induce perturbations on the visual task due to the secondary task. Let us finally note that, if the visual task constrains all the n degrees of freedom of the manipulator, we have $W = \mathbb{I}_n$, which leads to $\mathbb{I}_n - W^+ W = 0$. It is thus impossible in that case to consider any secondary task.

- $\underline{g}_s$ is the gradient of a cost function h_s to be minimized ($\underline{g}_s = \frac{\partial h_s}{\partial \bar{r}}$). This cost function is minimized under the constraint that $\underline{e}_1$ is realized.

In order to make $\underline{e}$ exponentially decrease and then behave like a first order decoupled system, we get:
$$T_c = -\lambda \underline{e} - W^+ \widehat{\frac{\partial \underline{e}_1}{\partial t}} - (\mathbb{I}_n - W^+ W) \frac{\partial \underline{g}_s^T}{\partial t} \qquad (4)$$

where:

- T_c is the camera velocity;
- λ is the proportional coefficient involved in the exponential convergence of $\underline{e}$;
- $\widehat{\frac{\partial \underline{e}_1}{\partial t}}$ represents an estimation of a possible autonomous target motion. If the scene is static, we can assume that $\frac{\partial \underline{e}_1}{\partial t} = \widehat{\frac{\partial \underline{e}_1}{\partial t}} = 0$.

3 Dynamic sensor planning

As already stated, when the vision-based task does not constrain all the six camera degrees of freedom, a secondary task can be combined with $\underline{e}_1$. Thus we can use the redundant degrees of freedom to propose a dynamic sensor planning strategy.

3.1 Avoiding occlusions

The main goal here is to avoid the occlusion of the target by static or moving (with unknown motion) objects. Therefore, the manipulator has to perform adequate motion in order to avoid the risk of occlusion while it ensures the desired constraints between the camera and the target (see Figure 1).

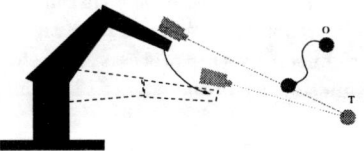

Figure 1: Reactive behavior for occlusion avoidance

Let us consider $\mathcal{O}$ the projection in the image of the set of objects in the scene which can possibly occlude the target T: $\mathcal{O} = \{O_1, \ldots O_n\}$. According to the presented methodology we have to define a function h_s which reaches its maximum when the target is occluded by another object of the scene. We thus define h_s as:
$$h_s = \frac{1}{2} \alpha \sum_{i=1}^{n} e^{-\beta (\|T - O_i\|^2)} \qquad (5)$$

where α and β are two scalar constants. α sets the amplitude of the control law due to the secondary task. The components of $\underline{g}_s$ and $\frac{\partial \underline{g}_s}{\partial t}$ involved in (4) are then:
$$\underline{g}_s = \frac{\partial h_s}{\partial \bar{r}} = \frac{\partial h_s}{\partial \underline{P}} \frac{\partial \underline{P}}{\partial \bar{r}}, \qquad \frac{\partial \underline{g}_s}{\partial t} = 0$$

Computing $\frac{\partial h_s}{\partial \underline{P}}$ is seldom difficult. $\frac{\partial \underline{P}}{\partial \bar{r}}$ is nothing but the interaction matrix $L_{\underline{P}}^T$ or image jacobian.

Let us consider the case of a single occluding object ; a generalization to multiple objects is straightforward. We want to see the target T at the center of the image. Thus we will consider the coordinates $\underline{P} = (X, Y)$ as its center of gravity. If we also consider the occluding object $\mathcal{O}$ by a point $\underline{P}_\mathcal{O} = (X_\mathcal{O}, Y_\mathcal{O})$, defined as the closest point of $\mathcal{O}$ to T, we have:
$$h_s = \frac{1}{2} \alpha e^{-\beta \|\underline{P} - \underline{P}_\mathcal{O}\|^2}$$

and

$$\underline{g}_s = \frac{\partial h_s}{\partial \underline{r}} = \frac{\partial h_s}{\partial X} \begin{pmatrix} -1/z_\mathcal{O} \\ 0 \\ X_\mathcal{O}/z_\mathcal{O} \\ X_\mathcal{O} Y_\mathcal{O} \\ -(1 + X_\mathcal{O}^2) \\ Y_\mathcal{O} \end{pmatrix} + \frac{\partial h_s}{\partial Y} \begin{pmatrix} 0 \\ -1/z_\mathcal{O} \\ Y_\mathcal{O}/z_\mathcal{O} \\ 1 + Y_\mathcal{O}^2 \\ -X_\mathcal{O} Y_\mathcal{O} \\ -X_\mathcal{O} \end{pmatrix}$$

with
$$\frac{\partial h_s}{\partial X} = -\alpha\beta(X - X_{\mathcal{O}})e^{-\beta\|\underline{P}-\underline{P}_{\mathcal{O}}\|^2}$$
and
$$\frac{\partial h_s}{\partial Y} = -\alpha\beta(Y - Y_{\mathcal{O}})e^{-\beta\|\underline{P}-\underline{P}_{\mathcal{O}}\|^2}$$

3.2 Field of view constraints

Let us consider that we want to keep a set $\mathcal{O}$ of objects in the field of view of the camera while ensuring a positioning with respect to another object T.

We have here to define a cost function h_s which is equal to 0 when $\mathcal{O}$ are located at the middle of the image and which is maximal near the border if the image. However, it is not always realistic with respect to the primary task to define h_s as a linear function of the distance to the center of the image. Furthermore, seeing $\mathcal{O}_i$ at the middle of the image is not required by this process. Therefore we do not use a linear function as proposed in [12] but the following cost function:

$$h_s = \sum_{i=1}^{n} h_s(O_i) \qquad (6)$$

with:
$$h_s(O_i) = \alpha e^{\beta(dc(O_i)^2)} \qquad (7)$$

where $dc(O_i)$ denotes the distance between O_i to the center of the image. Such a function increases quickly in the vicinity of the border of the image but reaches a nearly null value when O_i is located in a circle located around the center of the image and which radius can be easily tuned using β.

O_i can be either any object of the scene (not necessary related to the focused object T – see for example the results proposed in the next section) or any point of the edge of the focused object. It is possible to choose $\mathcal{O}$ as the set of points located on the edge of the focused object. It is thus possible to ensure that the whole object will be observed by the camera.

Note that the function h_s as proposed in (7) is not the only possible one. Another function (more similar to the one proposed by Nelson [12]) can be defined as:

$$h_s(O_i) = \begin{cases} \alpha\left(1 - \frac{db(O_i)}{dmin}\right) & \text{if } db(O_i) \leq dmin \\ 0 & \text{otherwise} \end{cases}$$

where $db(O_i)$ defines the distance between O_i to the nearest border of the image and $dmin$ is a predetermined threshold.

4 Trajectory planning

In the previous section, we have considered a single camera mounted on the end effector of a manipulator and the secondary task can be seen as a constraint introduced in the camera trajectory. In this section, we will consider a different problem: obstacle avoidance. To achieve this task, we cannot consider only one camera. A second motionless camera is added to the system and provides a global view of the scene by observing the gripper and the target (see Figure 2). Let us call $\mathcal{C}^l$, the camera mounted on the robot, the "local" camera and $\mathcal{C}^g$ the "global" camera. The main difference with the approach presented in the previous section is that the secondary task will no longer be used to *constrains* the camera trajectory but will *provide* to the system a collision free path.

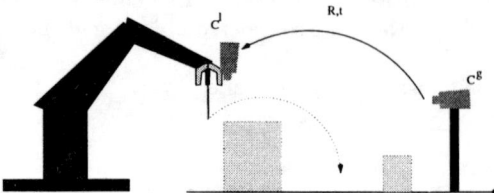

Figure 2: Two cameras system: a local and global view of the scene

4.1 Achieving contact...

The first task we want to perform is simply a point to point contact in 3D space. This kind of task can be achieved using the stereo visual servoing approach [6]. In our case, we will used two cameras as described above. To achieve the described task in an object free scene is straightforward. The primary task is to minimize the error between the current position of the target in the "local" image $\underline{P}^l$ and the position of the gripper $\underline{P}_d^l$ while the secondary task is nothing but the distance between gripper $\underline{P}^g$ and the target $\underline{P}_d^g$ in the "global" image. Note that $\underline{P}_d^l$ and $\underline{P}_d^g$ are fixed in the corresponding images.

$$h_s = \|\underline{P}^g - \underline{P}_d^g\|^2 \text{ and } \underline{g}_s = \frac{\partial h_s}{\partial \bar{r}} = \frac{\partial h_s}{\partial \underline{P}^g}\frac{\partial \underline{P}^g}{\partial \bar{r}}, \frac{\partial g_s}{\partial t} = 0$$

where $\frac{\partial \underline{P}^g}{\partial \bar{r}}$ defines the relation between the velocity of $\underline{P}^g$ in $\mathcal{C}^g$ and $\mathcal{C}^l$ velocity. It is given by:

$$\frac{\partial \underline{P}^g}{\partial \bar{r}} = L_{P^g}^T \begin{pmatrix} R & -R\left[sk(-R^T t)\right] \\ 0 & R \end{pmatrix}$$

where R and t are the rotation matrix and translation vector associated to the $\mathcal{C}^l$-to-$\mathcal{C}^g$ (or the gripper-to-$\mathcal{C}^g$) rigid transformation and $sk(a)$ is the skew-symmetric matrix associated with vector a. R and t are computed using the method proposed in [5].

4.2 ... in cluttered environment

The problem is quite different if we consider a cluttered environment. Let us actually consider not the whole robot but just the extremity of the gripper defined by a point. We will also consider that we have only 3 d.o.f in translation.

It is possible to modify the previous formulation (Section 4.1) in order to introduce in the cost function h_s a term which increases when the robot moves toward an object. However, if it is possible to ensure that the gripper will not encounter an obstacle, it is hardly possible to propose an object-free path toward the target (*i.e.*, between $\underline{P}^g$ and $\underline{P}_d^g$). To this purpose we propose to use a method derived from the potential fields methods: the navigation functions [10][13].

Let us define by $\mathcal{C}_{free}$ the set of object-free positions of the robot in the image: $\mathcal{C}_{free} = \mathcal{C} \setminus \bigcup_{i=1}^{n} \mathcal{B}_i$ where $\mathcal{B}_i$ is an obstacle and $\mathcal{C}$ is the configuration space. A navigation function is a potential function $U : \mathcal{C}_{free} \mapsto \mathbb{R}$ with a minimum located at the goal and whose domain of attraction includes

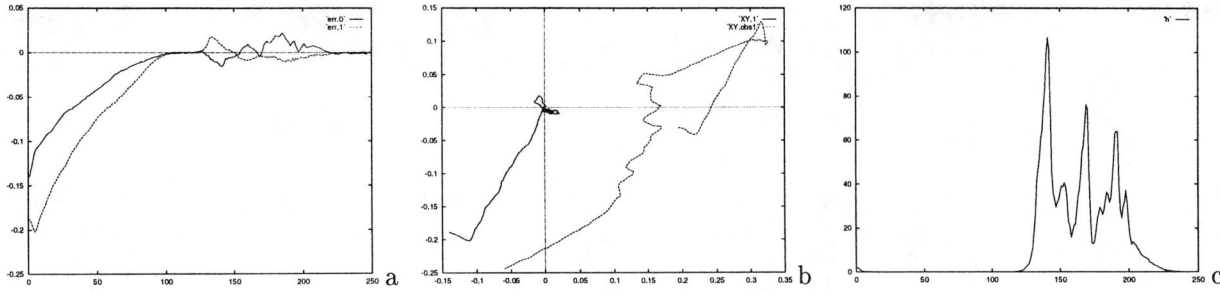

Figure 3: Positioning with respect to a point using an occlusion avoidance process (a) error between the current position of the point in the image and the desired position ($\underline{P} - \underline{P}_d$), (b) position of the two objects in the image, and (c) cost function h_s

the entire subset of $\mathcal{C}_{free}$ connected to the goal [10]. To compute this navigation function we have used the algorithm proposed in [1].

The resulting navigation function U is strictly decreasing and admits only one minimum located at the goal. Knowing U, the cost function can be defined as:

$$h_s = \beta U(X_d^g, Y_d^g)$$

where β is a scalar constant which sets the amplitude of the control law due to the secondary task.

The term $\partial h_s / \partial \underline{s}$ involved in the computation of g_s is merely the spatial gradients of h_s, i.e.:

$$\frac{\partial h_s}{\partial \underline{s}} = -\beta \begin{pmatrix} \nabla U_X \\ \nabla U_Y \end{pmatrix}$$

Using this formulation, the gripper will avoid any obstacle visible from the fixed camera. Furthermore if a path exists, then the specified task will be also achieved. If the obstacle and camera $\mathcal{C}_g$ are static then only the spatial gradients of U have to be recomputed. However, this approach can also deal with moving obstacles. In that case, the navigation function has to be recomputed at each iteration of the control law.

5 Experimental results

The method described above has been implemented on an experimental cell at Yale University. We have used a CCD camera mounted on the effector of a 6 d.o.f Zebra Zero robot arm. Image processing is performed at video rate using the XVision system [7].

5.1 Handling occlusions

In this experiment (as well as in the next one) we will consider a gaze control task. If $\underline{P} = (x, y)$ describes the position of the projection of the center of gravity of the "target", the goal (i.e. the primary task) is to observe this object at the center of the image: $\underline{P}_d = (0, 0)$. Only two degrees of freedom are necessary to perform the vision-based task, thus four motion components are redundant and can be used to avoid the undesirable configurations.

In this experiment the distance between the camera and the target is approximatively 400mm. An object is moving with an *a priori* unknown motion between the camera and the target in order to cause an occlusion.

Using the proposed occlusion avoidance process, when the distance in the image between the target and the occluding object decreases, the cost function h_s increases (see

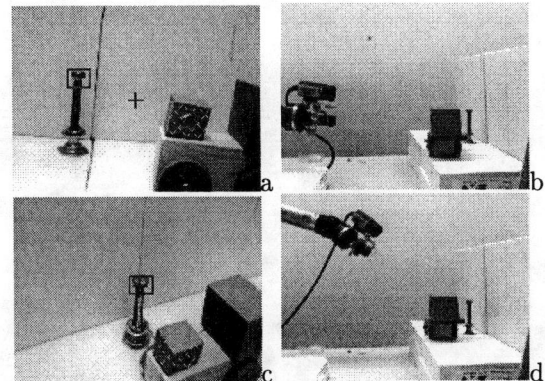

Figure 4: Positioning with respect to a point: (a) initial image acquired by the camera, (b) initial position of the camera with respect to the target (the screw) and the occluding object (the vehicle toy), (c) final image acquired by the camera, and (d) final position of the camera with respect to the target and the occluding object

Figure 3.c) and the other degrees of freedom are used to avoid the occlusion. When the occluding object moves away from the target, the cost function decreases (oscillations are due to a non-constant velocity of the occluding object).

We can observe that during the occlusion avoidance process, the vision based task is not perfectly achieved (see Figure 3.a). This is due to the fact that we do not have any information on the relative position between the camera and the target: the depth information z involved in the interaction matrix is unknown. As L_P^T cannot be updated at each iteration of the control law, assumptions have been made about the depth of the target.

5.2 Field of view

In this experiment we want to focus on a target (here the top hole on the cylinder) while keeping the trailer (see right of Figure 6) within the field of view of the camera. If no specific strategy is achieved, the vehicle moves out of the image during the camera motion due to the focusing task.

Figures 6.a and 6.b depict the initial image from the camera and an external view of the scene. Figures 6.c and 6.d show the final views. Figure 5.a shows the error between the current position of the target in the image and its desired position. Figure 5.b shows the position in the image of the focused target and the position of the tow which is the closest from the image border. Figure 5.c shows the value of the

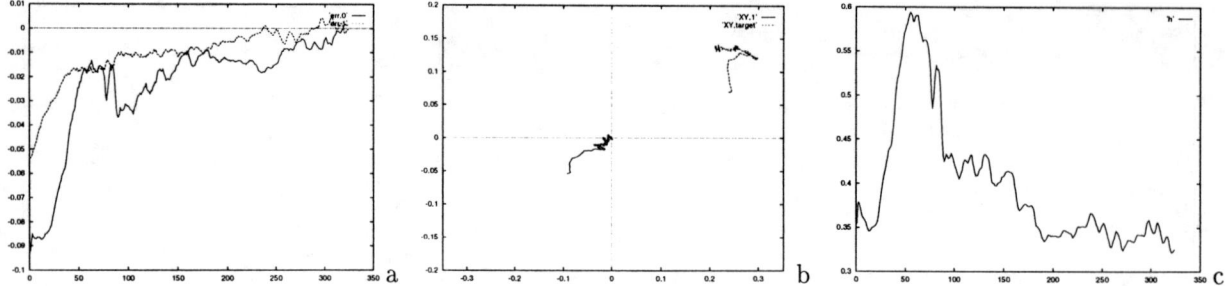

Figure 5: Field of view experiment (a) error $\underline{P} - \underline{P}_d$ (b) position of the focused target and of the vehicle tow in the image (c) cost function h_s

Figure 6: Field of view experiment

cost function h_s. Note that the system reacts rapidly to motion of the vehicle toward the border of the image. A motion along one of the previously unused translational degrees of freedom (x axis) is performed.

5.3 Avoiding obstacles

In this experiment we want to insert a 4mm wide screwdriver in a 5mm hole. It can be shown [6] that if the screwdriver and the hole are superimposed in the two images, then they are also superimposed in the 3D space. To achieve this task, we have used the three translational d.o.f of the robot. Introducing the rotational d.o.f into the process will require a more complex planning strategy to avoid the obstacles. Figure 9). As a result, in this experiment, the primary task will control 2 d.o.f (x and y in the mobile camera frame) while the secondary task controls the last one (z in this frame).

Figure 7 depicts the first and the last image acquired during the insertion process. The navigation function U is computed in the vicinity of group obstacle/target/robot using the method proposed in [1]. Figure 8 shows the resulting navigation function. Light areas correspond to high value of the navigation function while dark ones correspond to lower values. Obstacles are shown in black.

Figure 9.a depicts the error between the current and desired position in the image of the mobile camera. A small error (2 or 3 pixels which corresponds to a 1mm error in 3D space) can be observed due to a bad estimation of projection operators (W and $\mathbb{I} - W^+W$). Here again, the position of

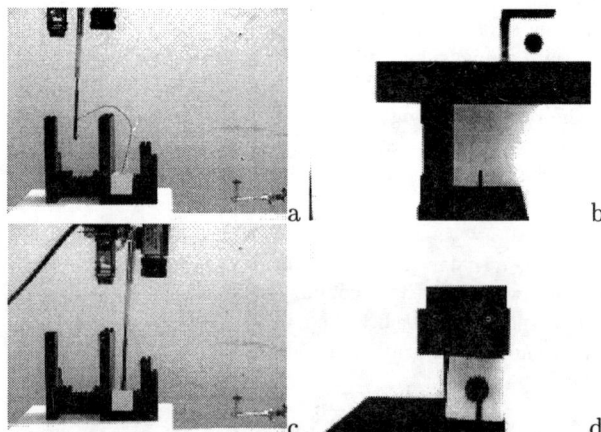

Figure 7: Insertion task in a cluttered environment: (a-b) initial image acquired by the cameras (a) global camera (b) local camera, and (c-d) final images. Note that the camera trajectory is depicted in image (a).

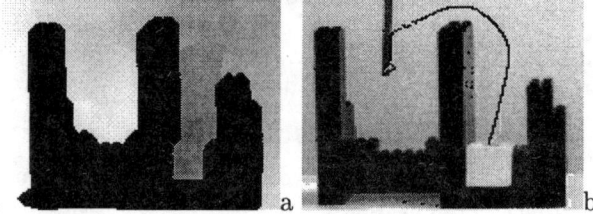

Figure 8: Obstacle avoidance (a) Navigation function (b) camera trajectory around the obstacle

the target w.r.t. the camera is unknown.

Figure 9.b depicts the error between the current and desired position of the screwdriver in the fix (global) image. In a first time (iterations 0–80) the error increases due to obstacle avoidance process (see robot trajectory on Figure 8.b). Figure 9.c depicts the cost function h_s which is the value of the navigation given the position of the robot.

6 Conclusion and future work

We have shown in various cases that it was possible to use redundancy to achieve dynamic sensor planning in visual servoing. Each time the constraints on the camera trajectory has been expressed as a function to be minimized with respect to the specified task. Results have been proposed for occlusion avoidance, field of view constraints and in a simple case for obstacle avoidance. Previous work has demonstrated

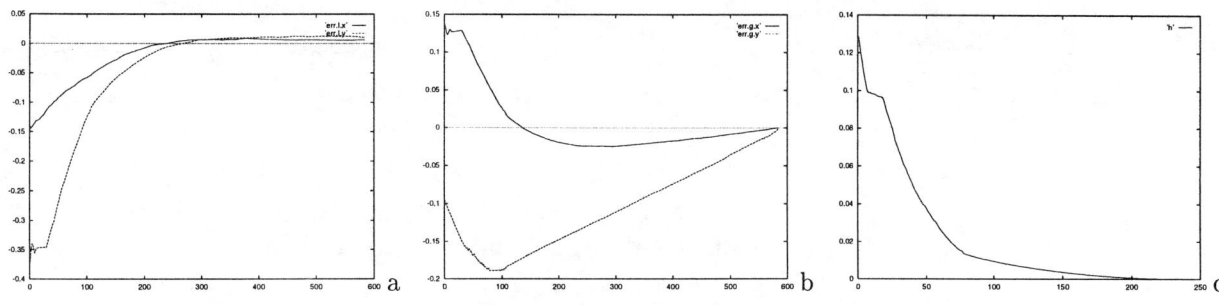

Figure 9: Obstacle avoidance experiment (a) error $\underline{P}^l - \underline{P}^l_d$ (b) error $\underline{P}^g - \underline{P}^g_d$ (c) cost function h_s

that the proposed method is able to deal with other problems such as singularities and joint limits avoidance [11] or for trajectory tracking[2]. Other tasks may be achieved such as introducing constraints on the measure of focus, resolvability (Nelson gives some cost functions for those purposes in [12]), visibility, resolution [3]. Most of these tasks have been solved in the case of static sensor planning (see for example [15]) and cost functions have been proposed. Using these cost functions within the framework proposed here is possible. However, here in each case, only image based control has been used. We have never used 3D information and our system has never been calibrated. To solve some others problems, more *a priori* informations (*e.g.*, a CAD of the scene) may be required.

Future work will be dedicated to incorporating an on-line estimation of depth of the objects using dynamic vision into the closed loop. Therefore we should be able to know if, in the occlusion avoidance problem, an object will actually occlude the target. Furthermore, this will give us a better estimation of the interaction matrix and thus of the projection operators. Other work has to be done dealing with the obstacle avoidance process. Obviously, the method presented here may fail (a path in the global image may not exist) and may not be optimal (a shorter path may exists using the other image). Future work will be devoted to determine a good path by computing new camera viewpoints. We will also consider the whole robot and be able to deal with motion along the rotation axes.

Acknowledgments

Éric Marchand was funded by INRIA (Institut National de la Recherche en Informatique et Automatique, France) under a postdoctoral fellowship. Work by Greg Hager was supported by NSF IRI-9420982. Authors whish to thank F. Chaumette for the discussion they had.

References

[1] J. Barraquand, B. Langlois, J.-C. Latombe. Numerical potential field techniques for robot path planning. *IEEE Transactions on Systems, Man, and Cybernetics*, 22(2):224–241, February 1992.

[2] F. Chaumette. Visual Servoing using image features defined upon geometrical primitives. *Conf on decision and Control*, Vol. 4, pp. 3782–3787, Orlando, December 1994.

[3] C.K. Cowan, P.D. Kovesi. Automatic sensor placement from vision task requirements. *IEEE Trans. on PAMI*, 10(3):407–416, May 1988.

[4] B. Espiau, F. Chaumette, P. Rives. A new approach to visual servoing in robotics. *IEEE Trans. on Robotics and Automation*, 8(3):313–326, June 1992.

[5] N. Fischler, R.C. Bolles. Random sample consensus: A paradigm for model fitting with application to image analysis and automated cartography. *Communication of the ACM*, 24(6):381–395, June 1981.

[6] G. Hager. A modular system for robust positioning using feedback from stereo vision. *IEEE Trans. on Robotics and Automation*, 13(4):582–595, August 1997.

[7] G. Hager, K. Toyama. The XVision system: A general-purpose substrate for portable real-time vision applications. *Computer Vision and Image Understanding*, 1998.

[8] K. Hashimoto, editor. *Visual Servoing : Real Time Control of Robot Manipulators Based on Visual Sensory Feedback*. World Scientific Series in Robotics and Automated Systems, Vol 7, World Scientific Press, Singapor, 1993.

[9] S. Hutchinson, G. Hager, P. Corke. A tutorial on visual servo control. *IEEE Trans. on Robotics and Automation*, 12(5):651–670, October 1996.

[10] J.C. Latombe. *Robot Motion Planning*. Kluwer Academic Publishers, 1991.

[11] E. Marchand, F. Chaumette, A. Rizzo. Using the task function approach to avoid robot joint limits and kinematic singularities in visual servoing. In *IEEE Int. Conf. on Intelligent Robots and Systems, IROS'96*, Vol. 3, pp. 1083–1090, Osaka, Japan, November 1996.

[12] B.J. Nelson, P.K. Khosla. Integrating sensor placement and visual tracking strategies. In *IEEE Int. Conf. Robotics and Automation*, Vol. 2, pp. 1351–1356, San Diego, May 1994.

[13] E. Rimon, D.E. Koditschek. Exact robot navigation using artificial potential functions. *IEEE Trans. on Robotics and Automation*, 8(5):501–518, October 1992.

[14] C. Samson, M. Le Borgne, B. Espiau. *Robot Control: the Task Function Approach*. Clarendon Press, Oxford, United Kingdom, 1991.

[15] K. Tarabanis, P.K. Allen, R. Tsai. A survey of sensor planning in computer vision. *IEEE Trans. on Robotics and Automation*, 11(1):86–104, February 1995.

Image-based visual servoing by integration of dynamic measurements

Armel Crétual François Chaumette
IRISA / INRIA Rennes
Campus de Beaulieu
35042 Rennes cedex, France
E-mail {acretual, chaumett}@irisa.fr

Abstract

Visual servoing based upon geometrical features such as image points coordinates is now well set on. Nevertheless, this approach has the drawback that it usually needs visual marks on the observed object to retrieve geometric features. The idea developed here is that these features can be retrieved by integrating dynamic ones, which can be estimated without any a priori knowledge of the scene. Thus, more realistic objects can be used to achieve vision-based control such as tracking and fixation tasks. We detail control laws concerning these two tasks, first using integration of speed in the image and then by direct regulation of these dynamic parameters. Results are finally presented and comparisons are made between the two types of control methods.

1 Introduction

The aim of visual servoing, as presented in [11, 13], is to control the robot displacements using visual features. One of the method used to complete such control laws is to apply the task function approach [19] to visual sensors and is based on the linear relation existing between image features variation and camera motion [9]. Geometric primitives have most often been used until now to complete robotic tasks such as positioning with respect to a given object. For example, visual marks are used in [9] to extract geometric features from the image, whereas [5, 7, 10] use informations from the object contour or particular features such as corners. Convergence is usually ensured and stable, at least when the initial position is in the 3D neighborhood of the desired position, and most of these applications run at video rate. The major problem encountered is that an a priori knowledge of the geometric features is needed.

Visual servoing based on dynamic features has recently been developed [8, 18, 21]. In this case, no knowledge of the observed pattern is necessary. The useful informations are extracted from 2D motion between two successive images. It can be seen that tasks such as tracking a moving object can be solved by this approach. To maintain the object at the same position in the image is equivalent to make its 2D projection keep a null speed.

Several papers deal with target tracking of mobile object. [3, 6, 12, 17] use visual marks to extract geometric information which are reinforced by an estimation of object speed in the image to compensate errors. Another method is used in [2, 15], but it is only able to track a small object. An affine model of 2D motion is computed between two successive images and the second image is compensated with the opposite motion. Threshold difference between this new image and the first one gives the position of the object, and the camera is controlled in pan and tilt so that this position stays at the image center. Finally, [1] uses a stereovision system to build a 3D model of the object motion in order to position the robot arm to grasp it.

This paper proposes a new approach to regulate speed in the image to zero. The problem of using directly dynamic visual features in the control loop is that, as the order of derivation is increased by one, there is generally no more a linear relation between features variations and camera motion. Furthermore, drift due to reacting time are not compensated. For these reasons, this paper develop the idea that position in the image can be retrieved by integration along time of speed in the image. Thus, visual servoing, as it is done with geometric features, can be used, but visual marks are no longer necessary. The principle of servoing by retrieving position from speed is quickly exposed in Section 2. Two applications are then presented and compared to methods using directly dynamic visual features in the control loop. We first describe the corresponding control laws, and then, display results obtained on our eye in hand 6 d.o.f. robotic system. The first application, detailed in Section 3, is the tracking of a mobile object using camera pan and tilt. The other one, in Section 4, corresponds to the positioning of the camera parallel to a plane coupled to a fixation task.

2 Image-based control from speed measurements

Our aim is to control the robot by classical image-based techniques but without having any a priori knowledge on the image content. The solution proposed is to retrieve geometric features by integrating dynamic measurements along time.

Let us call $s = (x, y)^T$, the 2D projection at time t of a 3D point M, and $\dot{s}$ its apparent speed in the image. s can obviously be retrieved knowing the projection s_0 at time 0 and the evolution of $\dot{s}$ along time, by:

$$s = s_0 + \int_O^t \dot{s}\, dt$$

This relation can be approximated under the following discrete form:

$$s = s_0 + \sum_{i=1}^{k} \dot{s}_i\, \delta t_i \qquad (1)$$

with $\dot{s}_i$ being the ith measurement of $\dot{s}$ and δt_i, the duration between (i-1)th and ith measurements.

The motion model used to approximate speed in the image is a simplified quadratic model with 8 parameters as below (see [8, 20]):

$$\begin{cases} \dot{x} &= a_1 + a_2 x + a_3 y + b_1 x^2 + b_2 xy \\ \dot{y} &= a_4 + a_5 x + a_6 y + b_3 y^2 + b_4 xy \end{cases} \qquad (2)$$

with

$$\begin{cases} a_1 = -v_x - \Omega_y & a_2 = \gamma_1 v_x + v_z & a_3 = \gamma_2 v_x + \Omega_z \\ a_4 = -v_y + \Omega_x & a_5 = \gamma_1 v_y - \Omega_z & a_6 = \gamma_2 v_y + v_z \\ b_1 = -\gamma_1 v_z - \Omega_y & b_2 = -\gamma_2 v_z + \Omega_x & b_3 = b_2\ \ b_4 = b_1 \end{cases}$$

where T and Ω respectively represent the translational and the rotational terms of the kinematic screw between the camera frame and the observed object frame, $(v_x, v_y, v_z) = \frac{1}{Z_p} T$, and $Z = Z_p + \gamma_1 X + \gamma_2 Y$ is the equation of the planar approximation of the object surface at the considered point expressed in the camera frame. The algorithm used to estimate parameters a_i and b_i is the RMRm (robust multi-resolution algorithm) developed in [16].

Of course, simpler models (constant, affine) can be used to estimate the position of the image center, which are simply deduced from the one presented above by identification with the corresponding parameters. In fact, there is a necessary compromise to find between precision of the estimation and duration of calculation.

The following block-diagram sums up the whole paragraph where s and s^*, respectively represent the current and desired position in the image, Ω the controlled rotation and a_i the dynamic visual features which give the 2D speed integrated to retrieve s.

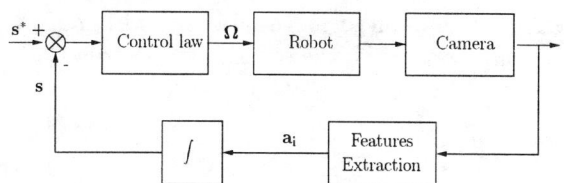

Figure 1: Block diagram: geometrical image-based control from dynamic visual features

3 Application to tracking

The principle of the tracking task is to control the camera pan and tilt such that a detected mobile object becomes projected at the center of the image or for want of anything better to keep it at the same position. We are not interested here in problems such as occlusions or multiple moving objects.

3.1 Detection of the mobile object

A step of detection of the mobile object has to be done first to obtain its initial projection mask on the image. As we do not use any a priori information on the target, this detection is performed using only the property that it is in motion. The camera remaining static until the mobile object is detected, the object projection location is determined by difference between two successive images. In practice, because of noise in the image, we use a local spatial (3 × 3 pixels) average of image intensities. Then, by considering a threshold difference between these two averaged images, we get a binary image separating moving zones from motionless ones. Under the hypothesis of a single moving object, it is easy to separate the mask of the mobile object from the background. The center of gravity of the mask gives the initial position which has to be regulated to zero, corresponding to s_0 in (2). Once the detection is done, s is obtained by integration of speed parameters given by the RMRm algorithm.

3.2 Associated control law

Having the estimation of the center of gravity (c.o.g.) of the target from (1), we thus can use a standard control law to complete the regulation of this estimated 2D position.

Let us consider the vector of error $s = (x, y)^T$. Using the fact that the camera motions are only the rotations around the x and y axes, we get from (2):

$$\dot{s} = L \begin{pmatrix} \Omega_{c,x} \\ \Omega_{c,y} \end{pmatrix} + \frac{\partial s}{\partial t} \text{ with } L = \begin{bmatrix} xy & (-1 - x^2) \\ (1 + y^2) & -xy \end{bmatrix}$$

where $\frac{\partial s}{\partial t}$ represents the 2D motion of the target and Ω_c the camera rotation.

Then, specifying an exponential decay with gain λ of the error s ($\dot{s} = -\lambda s$), the control law is given by:

$$\begin{pmatrix} \Omega_{c,x} \\ \Omega_{c,y} \end{pmatrix} = \frac{-\lambda}{1+x^2+y^2} \begin{pmatrix} y \\ -x \end{pmatrix} - L^{-1} \widehat{\frac{\partial s}{\partial t}}$$

The first term of this control law only allows to reach convergence when the observed object is motionless. To remove the tracking errors due to the object own motion, the second term has to be added and can be estimated by [4]:

$$\widehat{\frac{\partial s}{\partial t}} = \widehat{\dot{s}} - L\widehat{\Omega_c}$$

where the RMRm estimation algorithm gives $\widehat{\dot{s}}$ and $\widehat{\Omega_c}$ is the measured camera rotation. As done in [4], this estimation is filtered using a Kalman filter with a constant acceleration model with correlated noise.

3.3 Results

The tracking task has been tested on a 6 d.o.f. Cartesian robot cell, where the camera is mounted on the end-effector. 256 × 256 images were acquired by a SunVideo Board and treated on an UltraSpark station. As we do not have any exact measurement of the c.o.g. when using a real complex object, the control law has been first tested with a simple target from which we can extract geometric features. The position of the c.o.g. is unused in the control law but can be compared to its estimation from the motion parameters. Thus, we can observe the exact behavior of the control law. The object was a black surface where 4 white circles formed a square (see initial image on Fig 2(a)). An image processing running at video rate gives the position of the c.o.g. for each circle. Displacements of this four centers and measurement of time between two successive images give the speed for each. Applying a affine model of motion, in order to get a sufficient precision on constant parameters, leads to a linear system of 6 unknowns with 8 equations, which is solved by least square method. Finally, the c.o.g. of the square is given by the intersection of its two diagonals. Then, the experiment has been made with a 20 × 20 cm square from which no geometric features can be easily computed (see initial image on Fig 2(b)). In this case, to ensure a rate as closest as possible to the video rate, only the constant model of motion is used. With the object size such as it is obtained in this experiment, the rate reached is about 20 images per second.

The same conditions were taken for the two experiments, including initial positions of the target and of the camera. The target was translating along a rail alternatively to the right and to the left at the same speed, with a 4 seconds pause between the two motions. Accelerations and decelerations were 40 cm/s². The camera was about 1 m away from the object which was seen before its first motion, but did not necessarily appear at the center of the image. The images were 256 × 256 pixels. λ was chosen equal to 1.5. The successive motions are the following (where the number of iterations are approximative): it. 1 to 100, right move at 8 cm/s; it. 100 to 220, 4 seconds stop; it. 220 to 450, left move at 8 cm/s; it. 450 to 560, 4 seconds stop; it. 560 to 700, right move at 8 cm/s; it. 700 to 900, stop; it. 900 to 1000, right move at 30 cm/s; it. 1000 to 1100, 4 seconds stop, it. 1100 to 1220, left move at 30 cm/s; finally after iteration 1220, stop.

a b

Figure 2: Initial images. (a) Four points object (b) "Real" square

For the first experiment, we present in Fig. 3 the difference between the estimated position of the center of the object and the measured one. This error is always less than 0.5 pixel. Thus, as there is no drift due to the integration, we conclude we can trust the estimation of the c.o.g.. Furthermore, previous experiments have been made, in order to compare noise in estimations of the constant parameters of motion by the two methods presented above, with known motions of the observed object. It showed that noise is not greater with the RMRm algorithm than with the four point estimation. It is even sometimes lower, in particular when the object is motionless.

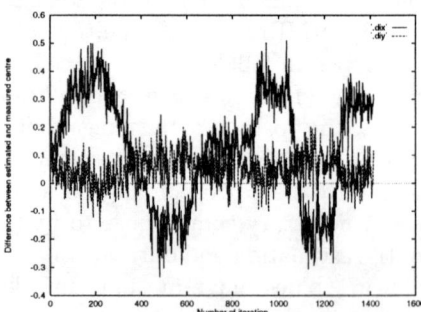

Figure 3: Four points experiment. Difference between the estimated displacement and the measured position of the center (in pixel)

The estimated displacement of the object center for the real square experiment, is displayed in Fig 4. This result is very similar to the one obtained with the four points experiment, and there is even less oscillations in the steps corresponding to the target stops. This experiment shows that convergence is well obtained for the initial error of about 40 pixels (it is brought to zero in less than 40 iterations even if the first motion of the object is on the opposite direction). At each abrupt change in the target motion (stop or start), there is an overrun due to the Kalman filter reacting time, but convergence is still obtained. This overrun of about 10 pixels is compensated in approximatively 20 iterations (0.8 seconds) when the speed is 8 cm/s (respectively 25 pixels and 50 iterations when the speed is 30 cm/s).

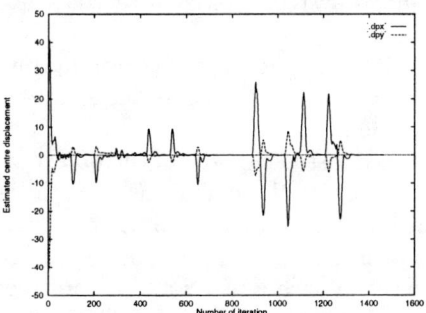

Figure 4: Square target experiment. Estimated displacement of the object center (in pixel)

The computed rotational velocities are displayed in (see Fig 5). A level of about 4 deg/s is necessary to track the 8 cm/s motion. In the case of the 30 cm/s motion, due to short time between start and stop of the object, we can note that Ω_x and Ω_y just reach their constant level of about 15 deg/s when the object stops, but no perturbation occurs in the regulation of the estimated c.o.g..

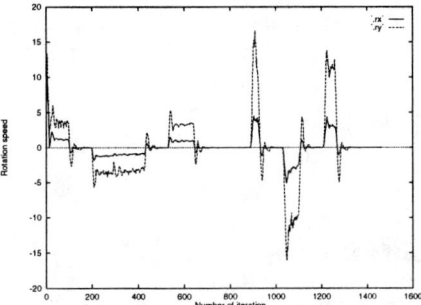

Figure 5: Square target experiment. Computed control law Ω_x and Ω_y (in deg/s)

3.4 Tracking with image motion based control

To see the interest of the previous method, a control directly based on image motion has been settled. The task consists now in trying to keep the object at the same position in the image by regulating to zero the constant terms of motion $s_1 = (a_1, a_4)^T$. Derivating s_1 along time leads to the interaction relation [20]:

$$\begin{pmatrix} \dot{a}_1 \\ \dot{a}_4 \end{pmatrix} = L \begin{pmatrix} \dot{\Omega}_{c,x} \\ \dot{\Omega}_{c,y} \end{pmatrix} + \frac{\partial s_1}{\partial t} \text{ with } L = \begin{bmatrix} 0 & -1 \\ 1 & 0 \end{bmatrix}$$

Applying the gradient type control, using a decreasing gain λ, the control law, expressed as angular acceleration, is then:

$$\begin{pmatrix} \dot{\Omega}_{c,x} \\ \dot{\Omega}_{c,y} \end{pmatrix} = -\lambda \begin{pmatrix} a_4 \\ -a_1 \end{pmatrix} - L^{-1} \widehat{\frac{\partial s_1}{\partial t}}$$

$\widehat{\frac{\partial s_1}{\partial t}}$ is estimated by $\frac{\widehat{s_1}_k - \widehat{s_1}_{k-1}}{\delta t}$ where indices k and $k-1$ stand for current and previous values and $\widehat{s_1} = (a_1 + \widehat{\Omega}_{c,y}, a_4 - \widehat{\Omega}_{c,x})^T$ which is the zero order of $\widehat{\frac{\partial s}{\partial t}}$ considered previously. Thus, it can be estimated with the same Kalman filter, only replacing $\widehat{s}$ by measure of $(\widehat{a}_1, \widehat{a}_4)^T$. Direct filtering of $\dot{s}_1$ is not accurate as the acceleration step is generally too short in front of the Kalman filter reacting time.

In the following summarizing block-diagram, s and s^* stand respectively for the current and desired values of the dynamic features $(a_1, a_4)^T$, which are given by the 2D-motion estimation algorithm and $\dot{\Omega}$ is the controlled rotational acceleration.

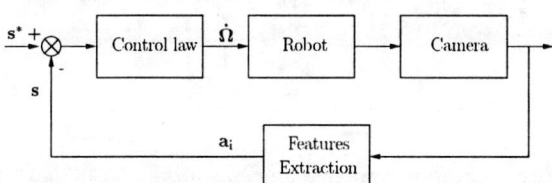

Figure 6: Block diagram: direct control of dynamic visual features

Let us note that in that case, the object can not be brought to the center of the image as we do not have anymore information on its position. Indeed, even if the detection step gives the first position of the c.o.g., trying to regulate it without any further information by an open loop would not be robust. Furthermore, if in the same time, the c.o.g. is still estimated and regulated to zero by rotational velocity, as regulation of s to zero is done by rotational acceleration, it would raise the problem, as in [14] of specifying the behavior of the control law by two different (and generally incompatible) means.

3.5 Results

The previous control law has been tested with the same initial condition and the same "real" target than in the previous results section. There, translational speeds were always 30 cm/s. First are presented the constant parameters of motion i.e. a_1 and a_4 (see Fig 7), the rotation acceleration computed by the control law (see Fig 8) and the estimated displacement of the object center (see Fig 9). The angular acceleration control law allows to fulfill the desired task which was to brought a_1 and a_4 to zero. It is accomplished in about 100 iterations after each abrupt change of motion, and it remains stable during the permanent running (during either a constant non-null speed motion or a motionless step), even if the computed parameters of control law are quite noisy. Finally, when the target becomes static again (between iterations 250 and 320, and between iterations 580 to 650), it appears a drift with respect to the initial position (1 or 2 pixels in the first stop, and about 10 to 15 in the second one). It is due to errors in the estimation of the object own acceleration. The accelerating step is very short (never more than 15 iterations), so the Kalman filter is not able to refine the estimated acceleration, and thus errors are not compensated.

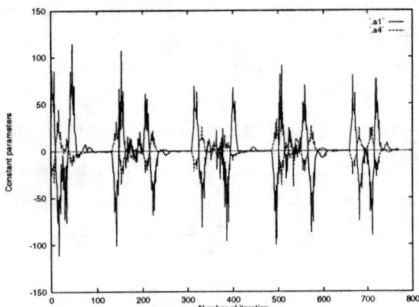

Figure 7: Square target experiment. Constant parameters of motion (in pixel/s)

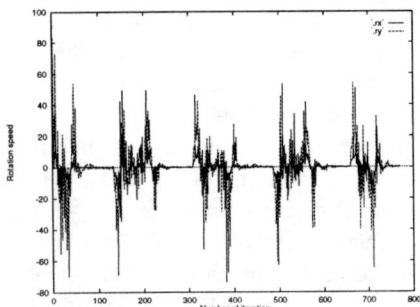

Figure 8: Square target experiment. Angular acceleration (in deg/s^2)

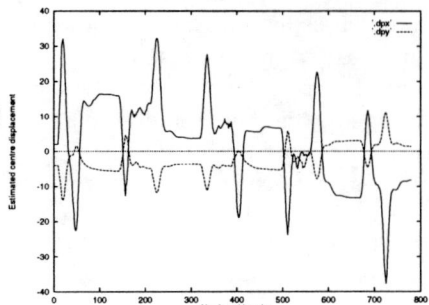

Figure 9: Square target experiment. Estimated displacement of the object center

3.6 Conclusion

The results presented here proves that tracking a real object, meaning without any visual marks, with an eye in hand system by visual servoing is possible whatever could its relative size be. It is solved by retrieving position of the object center by integration of its speed, and it runs close to video rate. On the contrary, dynamic control directly based on motion is unable to avoid a sensible drift of the object in the image. Furthermore, in the second case, the task aim was to keep the object in the same position. In most cases, a mobile object will appear at the border at the image, and keeping it in such a position does not seem to be very clever, while with the first control law it easily can be brought to the image center.

4 Application to fixation in an alignment task

The aim of our second task is to position the image plane parallel to an observed plane while ensuring a fixation task, such that $P_c = P_i$ (see Fig 10). Such a fixation is important in order that the object always appear in the image despite the rotational motion involved by the alignment.

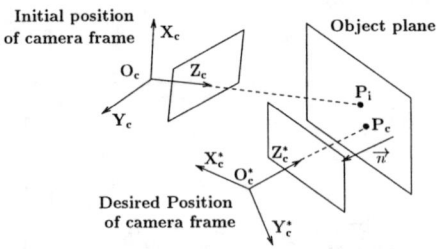

Figure 10: Task to be performed

4.1 Control laws

It can be denoted from (2) that the quadratic terms of the motion model show the terms γ_1 and γ_2 expressing the angular position of the observed plane relatively to the camera [8]. Thus, it appears that, with a non-null motion along the optical axis, the alignment task will be solved if and only if the following condition is respected:

$$\begin{pmatrix} b_1 + \Omega_y \\ b_2 - \Omega_x \end{pmatrix} = -\begin{pmatrix} \gamma_1 v_z \\ \gamma_2 v_z \end{pmatrix} = \begin{pmatrix} 0 \\ 0 \end{pmatrix}$$

Several approaches can be considered to solve the problem of adding a fixation task to the alignment one. In all the cases considered below, translation T_z along the optical axis is considered constant and rotation Ω_z around this axis is considered null, as it has no influence upon either the alignment nor the fixation. Furthermore, all control laws are established considering an exponential decay of the error.

The first approach is to consider that the image center will always correspond to the same point from the object if its apparent speed is zero, what means $(a_1, a_4) = (0,0)$ (see (2)). Then, a motion-based visual control law can be established considering the task constraining the following vector s to reach a zero value:
$$s = ((a_1, a_4, b_1 + \Omega_y, b_2 - \Omega_x)^T$$

Neglecting the second and upper order terms, the interaction relation linking the derivative $\dot{s}$ of s with the motion of the camera can be expressed under the following form [20]:

$$\dot{s} = \begin{bmatrix} -1/Z_p & 0 & 0 & -v_z \\ 0 & -1/Z_p & v_z & 0 \\ 0 & 0 & 0 & -v_z \\ 0 & 0 & v_z & 0 \end{bmatrix} \begin{pmatrix} \dot{T}_x \\ \dot{T}_y \\ \Omega_x \\ \Omega_y \end{pmatrix} + v_z \begin{pmatrix} -s_1 \\ -s_2 \\ s_3 \\ s_4 \end{pmatrix}$$
(3)

This leads to the the following rotational velocity and translational acceleration control law:

$$\begin{pmatrix} \dot{T}_x \\ \dot{T}_y \\ \Omega_x \\ \Omega_y \end{pmatrix} = -\begin{bmatrix} -T_z & 0 & T_z & 0 \\ 0 & -T_z & 0 & T_z \\ 0 & 0 & 0 & 1 \\ 0 & 0 & -1 & 0 \end{bmatrix} \begin{pmatrix} \frac{\lambda - v_z}{v_z} s_1 \\ s_2 \\ \frac{\lambda + v_z}{v_z} s_3 \\ s_4 \end{pmatrix}$$
(4)

A second approach, previously presented in [8], is to consider separately the alignment and the fixation tasks. Once the alignment control law is designed, the fixation can be obtained by direct compensation of the rotational motion using a translational one. It may be done by an open loop control in order to maintain a_1 and a_4 to zero using:

$$\begin{pmatrix} a_1 \\ a_4 \end{pmatrix} = 0 \Leftrightarrow \begin{cases} v_x = -\Omega_y \\ v_y = \Omega_x \end{cases}$$

Since Ω_x and Ω_y are controlled by the alignment task, T_x and T_y can be used for fixation and compensation of the rotational motion. However, this approach leads to an open control loop which does not take into account a possible error in the estimation of Z_p (needed to deduce T_x and T_y from v_x and v_y), nor than drift due to non zero affine and quadratic terms when the initial center does not appear at the image center anymore.

A third and new approach is to retrieve displacement $(x, y)^T$ due to rotational motion by integrating speed in the image, and to regulate it to zero by translational motion.

The vector of measures s for the alignment and fixation task can thus be chosen as:
$$s = (x, y, b_1 + \Omega_y, b_2 - \Omega_x)^T$$

In this case, the interaction relation between $\dot{s}$ and the camera motion is given by the following equation [8, 20] where the same approximation to the first order has been considered:

$$\dot{s} = \begin{bmatrix} -1/Z_p & 0 & 0 & -1 \\ 0 & -1/Z_p & 1 & 0 \\ 0 & 0 & 0 & -v_z \\ 0 & 0 & v_z & 0 \end{bmatrix} \begin{pmatrix} T_x \\ T_y \\ \Omega_x \\ \Omega_y \end{pmatrix} + v_z s$$

The corresponding control law is thus given by:

$$\begin{pmatrix} T_x \\ T_y \\ \Omega_x \\ \Omega_y \end{pmatrix} = \frac{\lambda + v_z}{v_z} \begin{bmatrix} -T_z & 0 & Z_p & 0 \\ 0 & -T_z & 0 & Z_p \\ 0 & 0 & 0 & 1 \\ 0 & 0 & -1 & 0 \end{bmatrix} s$$

This control law is similar to (4) (when replacing $\dot{T}$ by T and (a_1, a_4) with (x, y)). It can be easily explained noticing that the model of motion considered is $(\dot{x}, \dot{y})^T = (a_1, a_4)^T + v_z(x, y)^T$ and that $\dot{v}_z = -v_z^2$ [20]. Thus, relation (3) can also be deduced by derivating the previous one.

4.2 Results

These control laws have been tested on our experimental robotic cell. The initial image of the observed plane is presented in Fig 11. Experiments have been made with the same initial position for every case, where angular errors are about 30 degrees on each axes. As the estimation of the quadratic parameters of motion is quite costly, rate of control is about 1 Hz and the decreasing parameter λ was equal to 0.04. Results obtained for the alignment are displayed in [8]. We only present here results for the fixation.

Concerning the first control law, it appeared that trying to control by acceleration with such a rate is nearly impossible. Speed reaches a too great value before it can be updated by a new acceleration. Thus, we present results only for the second and third control laws. For the second one, the displacement of the image center along time has been estimated to be compared with the one obtained in the third case, but has not been taken into account in the control law. In each case, the number of iteration taken into account is the necessary one so that angular errors reach a continuous running at less that one degree.

Here, we present for each control law, the estimated drift of the initial center (see Fig. 12, 14) and a_1 and a_4 (see Fig. 13, 15).

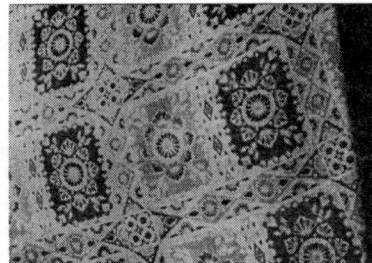

Figure 11: Alignment and fixation task: initial image

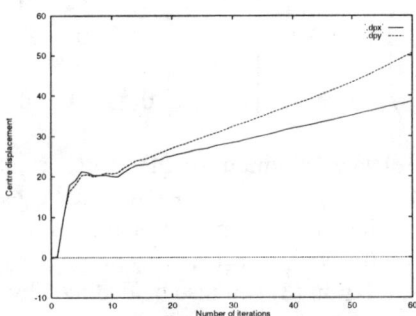

Figure 12: Open loop: Drift of initial centre (in pixel)

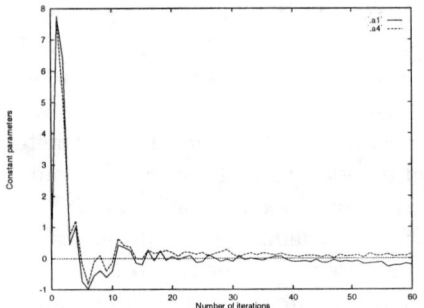

Figure 13: Open loop: Constant parameters

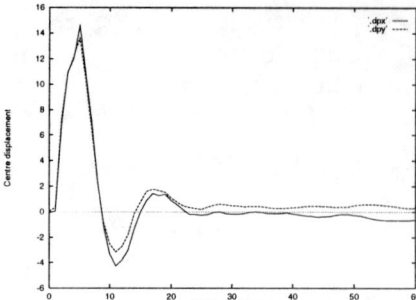

Figure 14: Closed loop: Drift of initial center (in pixel)

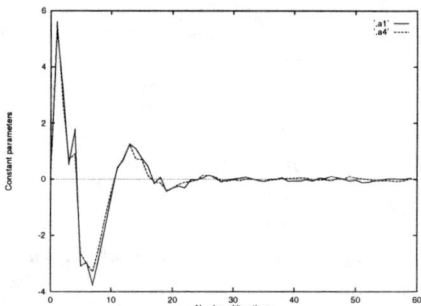

Figure 15: Closed loop: Constant parameters

The point projected at the center of the initial image has been "manually" retrieved on the final images for the two cases. In the open loop scheme, respective drifts on x and y axes are about 35 and 48 pixels whereas, in the closed loop one, they are only about 1 and 3 pixels. These results, similar to the estimated positions (which again validates the accuracy of the estimation scheme), proves that the second control law answers better to the desired behavior. This is due to the fact that the first control law only ensures the constant parameters to reach a zero value (which is done after the 15th iterations), but does not compensate drift of the center obtained during these first iterations. Then, after iteration 15, this center is subjected to the divergent motion brought by the translation along the optical axis. On the contrary, as the second control law is directly based on this estimated position, it is obviously brought to zero. There, we can notice that the constant parameters are also null at convergence, but drift appearing in the 5 first iterations is compensated in the following 5 ones.

5 Conclusion

The aim of this paper was to prove that image-based control can be done by integrating dynamic features. Usually, when measurements are integrated along time to estimate a value knowing its successive derivatives, a problem of bias in the estimation appears, due to the inevitable noises. Here, we showed that this problem does not appear. Tasks such as tracking a moving object or gazing at the same point of the scene when the camera is in motion, can be fulfilled using this technique. This is done by retrieving position from speed in the image and then applying control laws developed for image-based servoing.

References

[1] P.K. Allen, A. Timcenko, B. Yoshimi, and P. Michelman. Automated tracking and grasping of a moving object with a robotic hand-eye system. *IEEE Trans. on Robotics & Automation*, 9(2):152–165, Apr. 1993.

[2] M.G.P. Bartholomeus, B.J.A. Kröse, and A.J. Noest. A robust multi-resolution vision system for target tracking with a moving camera. In H. Wijshof, editor, *Computer Science in the Netherlands*, pages 52–63. CWI, Amsterdam, Nov. 1993.

[3] F. Bensalah and F. Chaumette. Compensation of abrupt motion changes in target tracking by visual servoing. In *IEEE Int. Conf. on Intelligent Robots and Systems*, volume 1, pages 181–187, Pittsburgh, Aug. 1995.

[4] F. Chaumette and A. Santos. Tracking a moving object by visual servoing. In *12th World Congress IFAC*, volume 9, pages 409–414, Sydney, Australia, July 1993.

[5] C. Colombo, E. Kruse, A.M. Sabatini, and P. Dario. Vision-based relative positionning through active fixation and contour tracking. In *Int. Symposium on Intelligent Robotic Systems*, Grenoble, July 1994.

[6] P.L. Corke and M.C. Good. Controller design for high performance visual servoing. In *12th World congress IFAC*, volume 9, pages 395–398, Sydney, Australia, July 1993.

[7] E. Coste-Manière, P. Couvignon, and P.K. Khosla. Visual servoing in the task-function framework: a contour following task. *Journal of Intelligent Robotic Systems*, 12(1):1–22, Jan. 1995.

[8] A. Crétual and F. Chaumette. Positionning a camera parallel to a plane using dynamic visual servoing. In *IEEE Int. Conf. on Intelligent Robots and Systems*, volume 1, pages 43–48, Grenoble, France, Sept. 1997.

[9] B. Espiau, F. Chaumette, and P. Rives. A new approach to visual servoing in robotics. *IEEE Trans. on Robotics & Automation*, 8(3):313–326, June 1992.

[10] G. Hager. The x-vision system: a general purpose substrate for real-time vision-based robotics. In *IEEE Workshop on Vision for Robots*, pages 56–63, Pittsburgh, USA, Aug. 1995.

[11] K. Hashimoto, editor. *Visual servoing. Real-time control of robot manipulators based on visual sensory feedback*. World scientific series in robotics and automated systems. World scientific, 1993.

[12] K. Hashimoto, T. Ebine, K. Sakamoto, and H. Kimura. Full 3D visual tracking with nonlinear model-based control. In *American Control Conference*, pages 3180–3185, San Francisco, California, June 1993.

[13] S. Hutchinson, G. Hager, and P.I. Corke. A tutorial on visual servo control. *IEEE Trans. on Robotics & Automation*, 12(5):651–670, Oct. 1996.

[14] P. Martinet, F. Berry, and J. Gallice. Use of first derivative of geometric features in visual servoing. In *IEEE Int. Conf. on Robotics & Automation*, volume 4, pages 3413–3419, Minneapolis, MN, Apr. 1996. IEEE.

[15] P. Nordlund and T. Uhlin. Closing the loop: detection and pursuit of a moving object by a moving observer. *Image and Vision Computing*, 14(4):265–275, May 1996.

[16] J.M. Odobez and P. Bouthemy. Robust multiresolution estimation of parametric motion models. *Journal of Visual Communication and Image Representation*, 6(4):348–365, Dec. 1995.

[17] N.P. Papanikolopoulos, B. Nelson, and P.K. Khosla. Six degree-of-freedom hand/eye visual tracking with uncertain parameters. *IEEE Trans. on Robotics & Automation*, 11(5):725–732, Oct. 1995.

[18] P. Questa, E. Grossmann, and G. Sandini. Camera self orientation and docking maneuver using normal flow. In *SPIE AeroSense'95*, Orlando, Florida, Apr. 1995.

[19] C. Samson, M. Le Borgne, and B. Espiau. *Robot control : the task function approach*. Oxford University Press, 1990.

[20] M. Subarrao and A. Waxman. Closed-form solutions to image equations for planar surface in motion. *Computer Vision, Graphics, and Image Processings*, 36(2):208–228, Nov. 1986.

[21] V. Sundareswaran, P. Bouthemy, and F. Chaumette. Exploiting image motion for active vision in a visual servoing framework. *International Journal of Robotics Research*, 15(6):629–645, Dec. 1996.

Tracking Adaptive Impedance Robot Control With Visual Feedback

Vicente Mut, Oscar Nasisi, Ricardo Carelli and Benjamín Kuchen
Instituto de Automática, Universidad Nacional de San Juan
San Juan, Argentina, 5400

Abstract

In this paper we propose a tracking adaptive impedance controller for robots with visual feedback. It is based on a generalized impedance concept where the sensed distance is introduced as a fictitious force to the control in order to avoid obstacles in restricted motion tasks. The controller is designed to compensate for full non linear robot dynamics. Robot parameters adjustment is introduced to reduce the sensibility of the controller design to dynamic uncertainties of the robot and the manipulated load. It is proved that the vision control errors are ultimately bounded in the image coordinate system. Simulations are carried out to evaluate the controller performance.

1 Introduction

The automation of tasks in the industry, those in which the robot interacts with the environment, needs the incorporation of sensors and the generation of control strategies which use this sensory information. Among controllers for constrained robot motion, the impedance control [11] is designed to regulate the manipulator's mechanical impedance, that is the dynamic relationship between the applied force and the motion error. Force sensors are required for sensing the interaction force.

The present work is related to the generalized impedance robot control using visual feedback in which a strategy of adaptive control is presented to deal with the problem of uncertainties in the manipulator's dynamics. In previous works [12] it is considered the concept of elasticity related to the distance sensing. In this work, the concept of impedance is generalized and the sensed distance is introduced like a fictitious force to the impedance control. The distance is assumed to be measured by means of the vision sensor. The tasks of the robotic manipulator with constrained motion can be classified into: with contact (force sensor) and without contact (distance sensor). Tasks without contact are emphasized, the first ones being reserved for the robot's eventual mechanical contacts with the environment. In tasks without contact, attractive surfaces (e.g. useful for profiles tracking) and refractory surfaces (e.g. to avoid obstacles) can be generated.

The use of visual information in the feedback loop represents an attractive solution to motion control of autonomous manipulators evolving in unstructured environments. In this context, motion robot control uses direct visual sensory information to achieve a desired relative position between the robot and -possible moving- object in the robot environment. Some solutions to this problem have been proposed [3] and [14] in which non-linear robot dynamics has not been considered for the controller design. These controllers can result in unsatisfactory control under high performance requirements, including high speed tasks and direct-drive robot actuators. In such cases, the robot dynamics has to be considered in the controller design, as partially done in [8] or fully included in [4]. These schemes assume an exact knowledge of the robot dynamics, and may result, in general, to be sensitive to robot model uncertainties as is presented in [1].

This paper presents an adaptive impedance robot controller in which signals are backfed directly from internal position and velocity sensors and visual information. The tracking impedance control errors are proved to be ultimately bounded [6]. The controller is based on inverse dynamics, the definition of a manifold in the error space [9], a desired impedance and a σ-modification type parameters update law. The control system proves useful to avoid obstacles when tracking an external object based on visual information.

2 Distance sensory feedback

To regulate the robot's mechanical impedance, it is necessary to measure the interaction force in the physical contact between the robot's end effector and the environment. In some applications this contact is not desirable. This is the case of obstacle avoidance or the follow contours. The robot should react with a compliant movement when an obstacle exists in the desired motion trajectory, without causing any physical contact. This can be achieved by using the impedance control, if it is considered a fictitious force F defined as a function of the sensed distance between the end effector and the surface. For example in [12] it is defined as,

$$F = \delta k(d - r) \quad (1)$$

with $\delta = 0$ if $\|d\| > \|r\|$ or $\delta = 1$ if $\|d\| < \|r\|$, where d is the sensed distance vector between the end effector and the obstacle. The distance d is calculated through an artificial vision algorithm. k is a constant with appropriate dimension and $r \in \mathbb{R}^n$ is a colinear vector with d representing the smaller acceptable approach from the robot to the surface. The definition given in equation (1) is adapted to develop a strategy to avoid collisions, because a refractory area is generated around the object. Other definitions of fictitious force are possible according to the particular application. For contour tracking, equation (1) should be used without the commutation function δ. This generates an appropriate fictitious force to define an attractive surface at $|r|$ distance.

3 Robot and Camera Models

In the absence of friction and other disturbances, the joint–space dynamics of an n–link manipulator can be written as,

$$H(q)\ddot{q} + C(q,\dot{q})\dot{q} + g(q) = \tau \quad (2)$$

where q is the $n \times 1$ vector of joint displacement, τ is the $n \times 1$ vector of applied joint torques, $H(q)$ is the $n \times n$ symetric positive definite manipulator inertia matrix, $C(q,\dot{q})\dot{q}$ is the $n \times 1$ vector of centripetal and Coriolis torques, $g(q)$ is the $n \times 1$ vector of gravitational torques.

The robot model (2) has some fundamental properties that can be exploited in the controller design [10].

Property 3.1 *Using a proper definition of matrix $C(q,\dot{q})$ (only the vector $C(q,\dot{q})\dot{q}$ is uniquely defined), matrices $H(q)$ and $C(q,\dot{q})$ in (2) satisfy*

$$z^T[\frac{d}{dt}H(q) - 2C(q,\dot{q})]z = 0 \quad \forall z \in \mathbb{R}^n \quad (3)$$

Property 3.2 *A part of the dynamic structure (2) is linear in terms of a suitably selected set of robot and load parameters, i.e.,*

$$H(q)\ddot{q} + C(q,\dot{q})\dot{q} + g(q) = \Omega(q,\dot{q},\ddot{q})\theta \quad (4)$$

where $\Omega(q,\dot{q},\ddot{q})$ is an $n \times p$ matrix and θ is an $p \times 1$ vector containing the selected set of robot and load parameters.

It is assumed that the robot is equipped with joint position and velocity sensors and a vision camera mounted in the robot hand. Following [4], let the camera and the object positions be, $s_c \in \mathbb{R}^n$ and $s_o \in \mathbb{R}^{m_o}$. For ideal perspective, the position of the object point in the image plane is,

$$\xi = \begin{bmatrix} x \\ y \end{bmatrix} = -\alpha \frac{f}{{}^C z_O} \begin{bmatrix} {}^C x_O \\ {}^C y_O \end{bmatrix} = i(s_c, s_o) \quad (5)$$

with f the focal length of the camera lens, α is the scaling factor in $pixels/m$ and $[{}^C x_O \; {}^C y_O \; {}^C z_O]^T$ the relative position vector of the object with respect to the camera coordinate system (${}^C z_O < 0$). This model can be extended to multiple feature points of the object to obtain an extended imaging model or model of the camera,

$$\xi = i^+(s_c, s_o) \quad (6)$$

with $\xi = [x_1 \, y_1 \cdots x_m \, y_m]^T$.

Noting that s_c depends on q, differentiating (6) it yields,

$$\dot{\xi} = J(q, s_o)\dot{q} + J_o(q, s_o)v_o \quad (7)$$

where $J \in \mathbb{R}^{2m \times n}$ is the Jacobian matrix obtained as,

$$J(q, s_o) = J_I(s_c, s_o) \begin{bmatrix} R_c(q) & 0 \\ 0 & R_c(q) \end{bmatrix} J_{geo}$$

where J_I is the image Jacobian matrix, J_{geo} is the geometric Jacobian matrix of the robot, [13]. Also, $J_o \in \mathbb{R}^{2m \times m_o}$ is the Jacobian matrix obtained as $J_o(q, s_o) = \partial i^+(s_c, s_o)/\partial s_o$ and v_o is the target velocity.

4 Adaptive Controller

The following assumptions are considered, similar to those in [4] and [7]:

Assumption 4.1 *There exists $q_d(t)$ such that the desired features vector ξ_d is achievable,*

$$\xi_d = i^+(s_c(q_d(t)), s_o(t)).$$

Assumption 4.2 *For the target path $s_o(t)$, there exists a neighborhood of q_d where J is bounded and invertible.*

Assumption 4.3 *The target velocity v_o is bounded.*

Now, we can formulate the tracking adaptive control problem for the robot with visual feedback.

4.1 Tracking control problem

Considering assumptions 4.1–4.3, desired features vector ξ_d, initial estimates of dynamic parameters θ in (3.2), initial estimates of target velocity $\hat{v}_o$ and its derivative $d\hat{v}_o/dt$, find a control law,

$$\tau = T(q, \dot{q}, \xi, \hat{\theta}, \hat{v}_o, d\hat{v}_o/dt, t), \qquad (8)$$

and a parameter update-law,

$$\frac{d}{dt}\hat{\theta} = \Theta(q, \dot{q}, \xi, \hat{\theta}, \hat{v}_o, d\hat{v}_o/dt, t) \qquad (9)$$

such that the control error in the image plane $\xi_d - \xi(t)$ is ultimately bounded by a sufficiently small ball B_r.

4.2 Generalized impedance control problem

The problem presented in the above subsection can be modified to include generalized impedance control objectives. An asymptotic impedance control objective can be defined as,

$$\xi_d - \xi(t) \to -(M_m p^2 + B_b p + K_k)^{-1} F(t) \qquad (10)$$

with $t \to \infty$, where $p = \frac{d(\cdot)}{dt}$ is the derivative operator, M_m, B_b and K_k are positive definite diagonal design matrices of order $2m \times 2m$. $F(t)$ is a fictitious force defined in the image plane.

The impedance error is defined now as,

$$\tilde{\xi} = (\xi_d + \xi_o(t)) - \xi(t) = \xi_r(t) - \xi(t) \qquad (11)$$

with $\xi_o(t) = -(M_m p^2 + B_b p + K_k)^{-1} F(t)$ and $\xi_r(t) = \xi_d + \xi_o(t)$. In this paper, instead of the ideal asymptotic convergence to zero (10), ultimately boundedness of $\tilde{\xi}(t)$ is considered as the control objective.

When there is no interaction between the robot an its environment, this problem reduces to the tracking one defined in subsection 4.1.

4.3 Control and Update Laws.

A manifold ν [9] is defined in the image error space as,

$$\nu = \frac{d\tilde{\xi}}{dt} + \Lambda\tilde{\xi}. \qquad (12)$$

Target velocity v_o and its derivative $\frac{dv_o(t)}{dt}$ can be estimated by means of a second order filter,

$$\begin{aligned}\hat{v}_o(t) &= \frac{b_o p}{p^2 + b_1 p + b_o} s_o(t) \\ \frac{d\hat{v}_o}{dt} &= \frac{b_o p^2}{p^2 + b_1 p + b_o} s_o(t).\end{aligned} \qquad (13)$$

Using $\hat{v}_o$ instead of v_o in equation (7), an estimate of the target velocity in the image plane $\dot{\xi}$ is given as, $\frac{d\hat{\xi}}{dt} = J\dot{q} + J_o \hat{v}_o$. Now, by substituting in (12), an estimate of ν is obtained,

$$\hat{\nu} = \frac{d\tilde{\hat{\xi}}}{dt} + \Lambda\tilde{\xi}. \qquad (14)$$

The following control law is proposed,

$$\tau = K\hat{\nu}' + \phi\hat{\theta} \qquad (15)$$

with,

$$\hat{\nu}' = J^{-1}\hat{\nu} = J^{-1}\dot{\xi}_o(t) - \dot{q} - J^{-1}J_o\hat{v}_o + J^{-1}\Lambda\tilde{\xi}, \qquad (16)$$

where, $\dot{\xi}_o(t) = -p(M_m p^2 + B_b p + K_k)^{-1}F(t)$ and $\phi(q, \dot{q}, \nu, \hat{v}_o, \frac{d\hat{v}_o}{dt})\hat{\theta}$,

$$\begin{aligned}\phi\hat{\theta} &= \hat{H}(q)\{\dot{J}^{-1}\hat{\nu} - J^{-1}\dot{J}\dot{q} - J^{-1}J_o\hat{v}_o + J^{-1}\ddot{\xi}_o \\ &- J^{-1}J_o\frac{d\hat{v}_o}{dt} - J^{-1}\Lambda J\dot{q} - J^{-1}\Lambda J_o\hat{v}_o \\ &+ J^{-1}\Lambda\dot{\xi}_o\} + \hat{C}(q,\dot{q})\{-J^{-1}J_o\hat{v}_o \\ &+ J^{-1}\Lambda\tilde{\xi} + J^{-1}\dot{\xi}_o\} + \hat{g}(q),\end{aligned} \qquad (17)$$

where: K and Λ are positive definite gain $(n \times n)$ matrices, $\hat{H}(q)$, $\hat{C}(q,\dot{q})$ and $\hat{g}(q)$ are the estimates of $H(q)$, $C(q,\dot{q})$ and $g(q)$ respectively. Parameterization of (15) is possible due to property 3.2.

To estimate θ, the following parameter update-law is considered, which is the σ-modification type [5],

$$\frac{d}{dt}\hat{\theta} = \Gamma\phi^T(q, \dot{q}, \hat{\nu}, \hat{v}_o, \frac{d\hat{v}_o}{dt})\hat{\nu}' - L\hat{\theta}, \qquad (18)$$

with Γ and L positive definite adaptation gain $p \times p$ matrices.

5 Stability analysis

Proposition 5.1 *Consider the control law (15) and the update law (18) in closed loop with the robot and camera models (2) and (6) with the assumptions 4.1–4.3. Then, there exists a neighborhood of q_d such that,*

a) $\tilde{\theta} = \theta - \hat{\theta} \in L_\infty^p$.

b) $\hat{\nu}' \in L_\infty^n \cap L_2^n$.

c) $\tilde{\xi}(t) = (\xi_r(t) - \xi(t))$ *is ultimately bounded.*

Prueba: The closed-loop system is obtained by combining (2) and (15),

$$K\hat{\nu}' + \phi\hat{\theta} = H\ddot{q} + C\dot{q} + g \qquad (19)$$

Using $\hat{\theta} = \theta - \tilde{\theta}$ and equations (12) and (18) it yields,
$$K\hat{\nu}' - \phi\tilde{\theta} + H\hat{D}\nu' + C\hat{\nu}' = 0 \quad (20)$$
where $\hat{D}\nu'$ is the estimate of the ν' time derivative. But, $\hat{D}\nu' = D\hat{\nu}' + \epsilon'$, with $\epsilon' = J^{-1}\Lambda J_o(v_o - \hat{v}_o)$, $v_o - \hat{v}_o = \epsilon_o$ the estimate error, and $D\hat{\nu}'$ the time derivative of $\hat{\nu}'$.

Then,
$$HD\hat{\nu}' = -(K+C)\hat{\nu}' + \phi\tilde{\theta} - \epsilon \quad (21)$$
where, $\epsilon = H\epsilon'$.

Considering the local non-negative function of time,
$$V = \frac{1}{2}\hat{\nu}'^T H\hat{\nu}' + \frac{1}{2}\tilde{\theta}^T \Gamma^{-1}\tilde{\theta} \quad (22)$$
whose time derivative along the trajectories of (21), considering the parameter update-law (18), is
$$\dot{V} = \hat{\nu}'^T[-(K+C)\hat{\nu}' + \phi\tilde{\theta} - \epsilon] + \frac{1}{2}\hat{\nu}'^T \dot{H}\hat{\nu}' + \tilde{\theta}\Gamma^{-1}[-\Gamma\phi^T\hat{\nu}' + L\hat{\theta}]. \quad (23)$$

Considering property 3.1, it results,
$$\dot{V} = -\hat{\nu}'^T K \hat{\nu}' - \tilde{\theta}^T \Gamma^{-1} L \tilde{\theta} - \hat{\nu}'\epsilon + \tilde{\theta}^T \Gamma^{-1} L \theta. \quad (24)$$

Using the expressions,
$$\|\tilde{\theta}\|\|\theta\| \leq \frac{1}{2}\frac{1}{\delta^2}\|\tilde{\theta}\|^2 + \frac{\delta^2}{2}\|\theta\|^2$$
$$\|\hat{\nu}'^T\|\|\epsilon\| \leq \frac{1}{2\eta^2}\|\hat{\nu}'\|^2 + \frac{\eta^2}{2}\|\epsilon\|^2 \quad (25)$$
with $\delta, \eta \in \mathbb{R}^+$, the equation (24) can be expressed as,
$$\dot{V} \leq -\mu_K\|\hat{\nu}'\|^2 - \mu_{\Gamma^{-1}L}\|\tilde{\theta}\|^2 + \|\hat{\nu}'\|\|\epsilon\| + \gamma_{\Gamma^{-1}L}\|\theta\|\|\tilde{\theta}\| \quad (26)$$
with $\mu_K = \sigma_{\min}(K)$, $\mu_{\Gamma^{-1}L} = \sigma_{\min}(\Gamma^{-1}L)$ and $\gamma_{\Gamma^{-1}L} = \sigma_{\max}(\Gamma^{-1}L)$, where σ denotes singular value.

Also,
$$\dot{V} = -\alpha_1\|\hat{\nu}'\|^2 - \alpha_2\|\tilde{\theta}\|^2 + \rho \quad (27)$$
where,
$$\alpha_1 = \mu_K - \frac{1}{2\eta^2} > 0$$
$$\alpha_2 = \mu_{\Gamma^{-1}L} - \frac{\gamma_{\Gamma^{-1}L}}{2\delta^2} > 0 \quad (28)$$
$$\rho = \gamma_{\Gamma^{-1}L}\frac{\delta^2}{2}\|\theta\|^2 + \frac{\eta^2}{2}\|\epsilon\|^2.$$

Equation (22 can be upper bounded by,
$$V = \beta_1\|\hat{\nu}'\|^2 + \beta_2\|\tilde{\theta}\|^2 \quad (29)$$
where, $\beta_1 = \frac{1}{2}\gamma_H$, with $\gamma_H = \sup_q[\sigma_{\max}(H)]$ and $\gamma_{\Gamma^{-1}} = \sigma_{\max}(\Gamma^{-1})$.

Then,
$$\dot{V} \leq -\gamma V + \rho \quad (30)$$
with, $\gamma = \min\{\frac{\alpha_1}{\beta_1}, \frac{\alpha_2}{\beta_2}\}$.

As ρ is bounded, equation (30) implies $\hat{\nu}' \in L_\infty^{2n}$, $\tilde{\theta} \in L_\infty^p$ which proves a) and b). Also, $x = (\hat{\nu}', \tilde{\theta})^T$ is ultimately bounded inside a ball $B(0, \mathbf{r})$. Now, from (12), $\hat{\nu} = J\hat{\nu}'$ and by assumption 4.2, $\hat{\nu} \in L_\infty^{2m}$. Note that, $\hat{\nu}$ can be expressed in terms of ν as,
$$\hat{\nu} = \frac{d\tilde{\xi}}{dt} + \Lambda\tilde{\xi} + J_o(v_o - \hat{v}_o) = \nu + J_o\epsilon_o. \quad (31)$$

As $J_o\epsilon_o$ is bounded, it means that ν is ultimately bounded. From the last equation, $\tilde{\xi} = O(\nu)$ where, $O(\cdot)$ is a linear operator with finite gain. Therefore, $\|\tilde{\xi}\| = \|O\|\|\nu\|$ then, as ν is ultimately bounded $\tilde{\xi}$ is also ultimately bounded, which proves c). Consequently $\xi_r(t) - \xi(t)$ is ultimately bounded. The last, implies that if the object is placed at a distance greater than $|\mathbf{r}|$ ($\xi_o \equiv 0$) it is verified that $\xi_d - \xi(t)$ is ultimately bounded, which implies that the tracking control objective is satisfied.

Remark 1: If more features than degrees of freedom of the robot are taken, a non-square Jacobian matrix is obtained. In this case a redefinition of ν as, $\nu = \frac{d(J^T\tilde{\xi})}{dt} + \Lambda(J^T\tilde{\xi})$, must be used. Reasoning in a similar way as in proposition above, it is possible to reach the same conclusions about the behaviour of the control system.

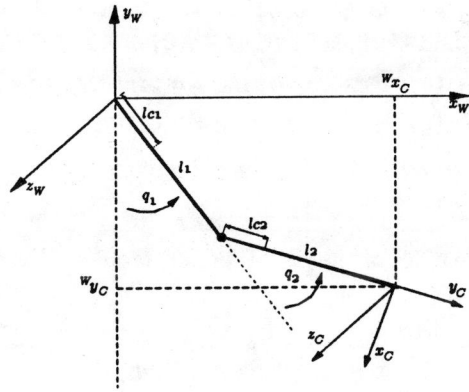

Figure 1: Direct drive 2 d.o.f. manipulator

Remark 2: It is possible to calculate a bound for the ultimately behaviour of the control errors. This bound of control errors depends on controller gains, the object velocity estimation error and the inertia matrix of the robot.

6 Simulations

Computer simulations have been carried out to show the stability and performance of the proposed tracking adaptive impedance controller. The manipulator used for the simulations is two–degree–of–freedom manipulator, as shown in Fig. 1. The meaning and numerical values of the symbols are listed in table 1.

Table 1: Manipulator parameters

	notation	value	units
Length link 1	l_1	0.45	m
Length link 2	l_2	0.55	m
C. of G., link 1	l_{c1}	0.091	m
C. of G., link 2 + camera	l_{c2}	0.105	m
Mass link 1	m_1	23.9	kg
Mass link 2 + camera	m_2	4.44	kg
Inertia link 1	I_1	1.27	Kg m^2
Inertia link 2 + camera	I_2	0.24	Kg m^2

The elements $H_{ij}(q)(i,j=1,2)$ of the inertia matrix $H(q)$ are,

$$H_{11}(q) = m_1 l_{c1}^2 + m_2 \left(l_1^2 + l_{c2}^2 + 2 l_1 l_{c2} \cos(q_2)\right) + I_1 + I_2$$
$$H_{12}(q) = m_2 \left(l_{c2}^2 + l_1 l_{c2} \cos(q_2)\right) + I_2$$
$$H_{21}(q) = m_2 \left(l_{c2}^2 + l_1 l_{c2} \cos(q_2)\right) + I_2$$
$$H_{22}(q) = m_2 l_{c2}^2 + I_2.$$

The elements $C_{ij}(q,\dot{q})(i,j=1,2)$ from the centrifugal and Coriolis matrix $C(q,\dot{q})$ are,

$$C_{11}(q,\dot{q}) = -m_2 l_1 l_{c2} \sin(q_2) \dot{q}_2$$
$$C_{12}(q,\dot{q}) = -m_2 l_1 l_{c2} \sin(q_2) (\dot{q}_1 + \dot{q}_2)$$
$$C_{21}(q,\dot{q}) = m_2 l_1 l_{c2} \sin(q_2) \dot{q}_1$$
$$C_{22}(q,\dot{q}) = 0.$$

The entries of the gravitational torque vector $g(q)$ are given by,

$$g_1(q) = (m_1 l_{c1} + m_2 l_1) g \sin(q_1) + m_2 l_{c2} g \sin(q_1 + q_2)$$
$$g_2(q) = m_2 l_{c2} g \sin(q_1 + q_2).$$

The numerical values of the camera model are *focal lenght* $f = 0.008\ m$ and the *scale factor* $\alpha = 72727\ pixels/m$.

The linear parameterization of equation (15) leads to a parameter vector

$$\theta = [m_1 l_{c1}^2 \quad m_1 l_{c1} \quad m_2 l_{c2}^2 \quad m_2 l_{c2} \quad m_2 \quad I_1 \quad I_2]^T.$$

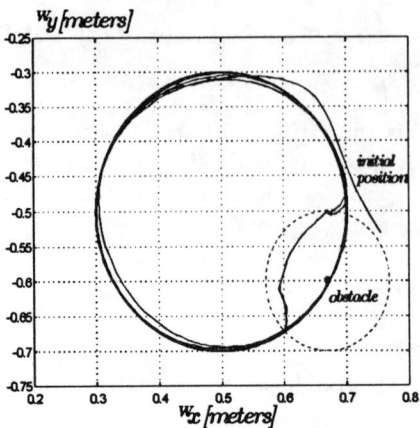

Figure 2: End–effector trajectory in the work space

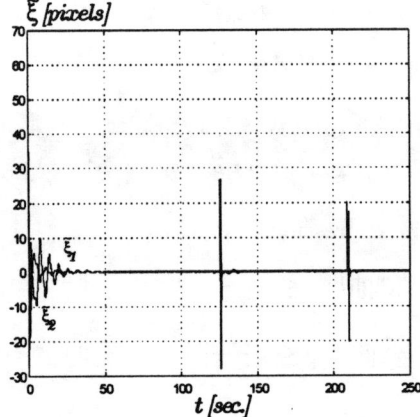

Figure 3: Evolution of the control error

For controller design it is assumed that the values are known with uncertainties of about 40%. Simulations are carried out using the following design parameters $\Lambda = diag(\lambda), \lambda_i = 15$, $K = diag(k), k_i = 100$ and $\Gamma = diag(\gamma), \gamma_i = 0.9$ and $L = diag(l_i)$, $l_i = .00025$.

The robot initial conditions are $q_1(0) = 30°$, $q_2 = 45°$, $\dot{q}_1 = 0$ and $\dot{q}_2 = 0$. The parameters of desired impedance are $M_m = diag(0.01)$, $B_b = diag(1)$ and $K_k = diag(30)$.

The trajectory of the object is a circle with radius $rd = 0.2m$, angular velocity $\omega = 1.05 rad/sec$ and the parameters of the second order filters are, $b_o = 4\ 10^4$, $b_1 = 400$.

The desired image feature point was set to $\xi_d = [0\ \ 0]^T$. The obstacle is considered circular and its center in $[0.67m \ \ -0.6m]^T$ within work space of the robot. The distance to which the fictitious force begin

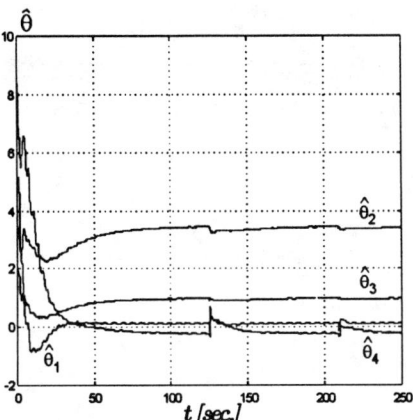

Figure 4: Evolution of the parameters

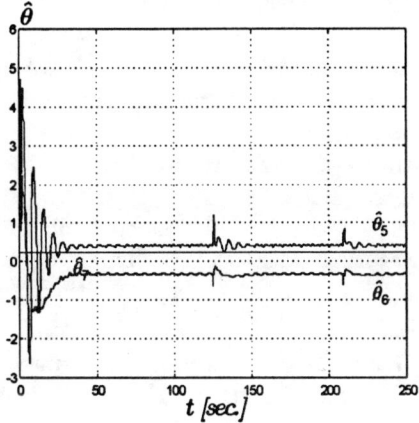

Figure 5: Evolution of the parameters

to act is $r = 0.1m$.

Simulation results are shown in figures 2 to 5. Figure 2 shows the end-effector trajectory in the work plane. Figure 3 shows evolutions of the control errors $\tilde{\xi}$. Finally, in the figures 4 and 5 are shown the paramaters of the robot.

7 Conclusions

In this paper, it has been presented a tracking adaptive impedance controller for robots with camera-in-hand configuration using visual feedback, in which a generalized impedance concept has been used in order to avoid obstacles. Full non-linear robot dynamics has been considered in the controller design. The control errors are proved to be ultimately bounded. Simulations illustrate the ability of the proposed controller to attain accurate control under dynamics uncertainties. Future work will solve the possible Jacobian singularities when object moves along a singular direction so it is not detectable by the camera.

References

[1] Carelli R., Nasisi O. and Kuchen B., "Adaptive robot control with visual feedback", *American Control Conference*, June 29 - July 1, 1994.

[2] Desoer C. and Vidyasagar M., "Feedback systems: Input-Output properties", *Academic Press*, 1975.

[3] Hashimoto K., Kimoto T., Ebine T. and Kimura H., "Manipulator control with image-based visual servo", *Proc. IEEE int. Conf. on Robotics and Automation*, Sacramento, California, 1991.

[4] Hashimoto K. and Kimura H., "Dynamic visual servoing with non-linear model based control", *IFAC World Congress*, Vol. 9, pp. 405-408, Sydney, 1993.

[5] Ioannou P. and Kokotovic P., "Adaptive systems with reduced models", *Springer Verlag*, New York, 1983.

[6] La Salle J. and Lefschetz S., "Stability by Lyapunov: Direct Method with Applications", *Academic Press*, Chapter 4, pp. 107-130, New York, 1961.

[7] Nasisi O., Carelli R. and Kuchen B., "Tracking adaptive control of robots with visual feedback", *Proceedings of the 13th World Congress (IFAC)*, San Francisco, USA, 1996.

[8] Papanikolopoulos N. P., Khosla P. K. and Kanade T., "Visual tracking of a moving target by a camera mounted on a robot: a combination of control and vision", *IEEE Trans. on Robotics and Automation*, Vol.9, n01, February, 1993.

[9] Slotine J. and Li N., "Adaptive manipulator control: a case of study", *Proc. IEEE Int. Conf. on Robotics and Automation*, Raleigh, N.C., April, 1987.

[10] Ortega R., Spong M., "Adaptive motion control of rigid robots: A tutorial", *Automatica*, Vol.25, n6, pp. 877-888, 1989.

[11] Hogan N., "Impedance control: an approach to manipulations, Parts I, II", *ASME Journal of Dynamics Systems, Meas. and Control*, Vol. 107, 1985.

[12] Sagues C., Montano L. and Neira J., "Guarded and compliant motions using force and proximity sensors", *International Workshop on Sensorial Integration for Industrial Robots*, pp. 274-280, Zaragosa, Spain, 1989.

[13] Sciavicco L. and Siciliano B., "Modeling and Control of Robot Manipulators", *Mc Graw-Hill Co.*, 1996.

[14] Weiss L. E., Sanderson A. C. and Newman C. P., "Dynamic sensor-based control of robots with visual feedback", *IEEE Journal of Robotics and Automation*, Vol. RA-3, no. 5, October, 1987.

6 DOF high speed dynamic visual servoing using GPC controllers

Jacques A. Gangloff, Michel de Mathelin*and Gabriel Abba

Strasbourg I University

Ecole Nationale Supérieure de Physique de Strasbourg (ENSPS)

Parc d'Innovation, Bd. Sébastien Brant, 67400 Illkirch, France

fax: +33 (0)3 88 65 54 89   e-mail: demath@hp1gra.u-strasbg.fr

Abstract

This paper presents a new way to model visual servoing in the case of a 6 DOF industrial manipulator. The manipulator with its actuators (DC motors), their current feedback loops and their velocity control loops, is modeled as a "virtual Cartesian device". The system with its visual feedback loop is decoupled to form 6 independent loops. Then, 6 Generalized Predictive Controllers (GPC) are implemented on line to take into account the dynamics of the manipulator. Simulations and experimental results show a drastic improvement in performance for a 6 DOF industrial manipulator in an eye-in-hand configuration compared to standard approaches neglecting these dynamics.

Introduction

Numerous visual servoing applications are described in a growing literature (see, e.g., [5]). In the case of 6 DOF manipulators, often the visual feedback computes set points for the joint-level controllers. In this case, the built-in joint position controllers of the manipulator are used as internal loops to stabilize the system. This avoids computing the Jacobian of the end-effector and dealing with kinematic singularities. However, in order to achieve a high performance visual feedback, velocity inputs should be directly applied to the joint-level velocity controllers, as it is done in this paper.

Indeed, our goal in this work is to track an elementary target (eight black dots on a white plane) as fast and as precisely as possible with a regular CCD camera and a 6 DOF industrial manipulator. The target complexity is deliberately reduced in order to minimize the time required by the image processing. However, more complex targets could be tracked by increasing the computing power of the hardware. For our target, the image processing rate is increased to 100 Hz. This is achieved by processing separately the upper and the lower half of the target image for each video raster. This sampling rate is fast enough to render significant the effect of the dynamics of the manipulator.

Some interesting control strategies have been previously tested in [8], [9], [4] and [3]. In [4], Hashimoto *et al.* describe a 6 DOF feature-based visual servoing with a LQ state feed-back controller where the states are the coordinates of the target points. They note that the system is uncontrollable because many states are linked together. Therefore, a controllable/uncontrollable mode decomposition is required in order to achieve their goal.

In [9], Papanikolopoulos *et al.* experiment various kinds of controllers for 2D tracking of an arbitrary 3D object: from the simplistic but very effective PI to the more complex LQG which gives good results in the case of noisy visual measurements. However, the dynamics of the manipulator are not taken into account.

In [3], Corke *et al.* introduce the manipulator's dynamics into the visual loop and use pole placement techniques to improve closed-loop performance for a pan/tilt camera system much simpler than a 6 DOF system.

In this paper, the manipulator's dynamics are taken into account to design the controllers of a 6 DOF visual servoing system. The six controlled outputs are the pose vector of the target in the camera frame which is measured by the camera through the image Jacobian. The robot with its built-in joint-level velocity controllers is modeled as a linear system whose dynamics are identified for different configurations. Then, Generalized Predictive Controllers (GPC) are designed for the all visual feedback loop.

The choice of GPC controllers was motivated by their good performance, their numerically stable algorithm, their robustness to noise and perturbations, and their fast real-time implementation.

---

*Author to whom all correspondence should be addressed

1 Open-loop modeling

1.1 The visual sensor

The camera on the end-effector is modeled by a perspective projection. Further, the distortion introduced by the lens is also added to the model. Thus, if $\mathbf{P_c} = [x\ y\ z]^T$ are the coordinates of a point with respect to the camera coordinate frame, then, the projection in the image plane of $\mathbf{P}_c$ has the following coordinates in pixels (see [10]):

$$\begin{bmatrix} u_d \\ v_d \end{bmatrix} = \frac{1}{zD} \begin{bmatrix} G_x x \\ G_y y \end{bmatrix} \quad (1)$$

where

$$D = 1 + K_d \left(u^2 + \left(\frac{G_y}{G_x}\right)^2 v^2 \right) \quad (2)$$

G_x and G_y are the magnification factors for the x and y directions, respectively, K_d is the distortion coefficient, and $[u\ v]^T$ are the coordinates of $\mathbf{P}_c$ if there is no distortion i.e.:

$$\begin{bmatrix} u \\ v \end{bmatrix} = \frac{1}{z} \begin{bmatrix} G_x x \\ G_y y \end{bmatrix} \quad (3)$$

In order to suppress distortion effects, the coordinates $[u_d\ v_d]^T$ of the projection of $\mathbf{P}_c$ in the image plane are corrected as follows

$$\begin{bmatrix} u_c \\ v_c \end{bmatrix} = \left(1 + K_d \left(u_d^2 + \left(\frac{G_y}{G_x}\right)^2 v_d^2 \right) \right) \begin{bmatrix} u_d \\ v_d \end{bmatrix} \quad (4)$$

where $[u_c\ v_c]^T$ are the corrected coordinates. Assuming that $u_d \approx u$ and $v_d \approx v$ (which is true for a small distortions), then $[u_c\ v_c]^T \approx [u\ v]^T$.

1.2 Computation of the target position

The target is made of eight black dots on a white background. The four upper dots and the four lower dots are forming two separate sub-targets, which are treated alternatively in the following manner. First, the outlines are detected with a Canny filter, then the center of each dot is found by computing its center of mass. The coordinates of the centers constitute the feature vector $\mathbf{f}$. Now, let $\dot{\mathbf{r}}$ be the velocity screw of the camera frame expressed in the coordinates of the camera, $\dot{\mathbf{r}} = [V_x\ V_y\ V_z\ \omega_x\ \omega_y\ \omega_z]^T$. Then, $\dot{\mathbf{r}}$ and $\dot{\mathbf{f}}$, the velocity of the features, are linked by the image Jacobian as follows (see Weiss et al. [11]):

$$\dot{\mathbf{f}} = \mathbf{J}_v(\mathbf{f})\,\dot{\mathbf{r}} \quad (5)$$

where $\mathbf{J}_v(\mathbf{f})$ is the image Jacobian at $\mathbf{f}$. If the feature is a point, i.e. $\mathbf{f} = [u\ v]^T$ with u and v the coordinates in pixels of the point in the image, then (see [11]), $\mathbf{J}_v(\mathbf{f}) =$

$$\begin{bmatrix} \frac{G_x}{z} & 0 & -\frac{u}{z} & -\frac{uv}{G_y} & \frac{G_x^2+u^2}{G_x} & -\frac{vG_x}{G_y} \\ 0 & \frac{G_y}{z} & -\frac{v}{z} & -\frac{G_y^2+v^2}{G_y} & \frac{uv}{G_x} & \frac{uG_y}{G_x} \end{bmatrix} \quad (6)$$

It should be noted from (6) that $\mathbf{J}_v$ not only depends on $\mathbf{f}$ but also on the depth coordinate, z. Therefore $\mathbf{J}_v$ cannot be computed without the knowledge of z. In absence of an accurate measurement of z, an estimation is used and the error made can be modeled as a gain uncertainty for the three translations. Now, let $\mathbf{f}$ consist of the coordinates of at least 4 points, i.e. the minimum number of points to have the translation and rotation of the object *uniquely* defined by the pseudo-inverse of $\mathbf{J}_v(\mathbf{f})$ (see [7]). Then, the Jacobian is built by stacking matrices like the one in (6) as many times as the number of points taken into account. Consequently, from (5), it follows that:

$$\dot{\mathbf{r}} = \mathbf{J}_v^+ \dot{\mathbf{f}} \quad (7)$$

where $\mathbf{J}_v^+ = \left(\mathbf{J}_v^T \mathbf{J}_v\right)^{-1} \mathbf{J}_v^T$ is the pseudo-inverse of $\mathbf{J}_v$.

1.3 The Jacobian of the camera

Let $\mathbf{q}_0$ be the current joints' position vector. And let $M_{bc}(\mathbf{q}_0)$ be the current homogeneous transformation between the base frame of the robot and the camera frame:

$$M_{bc}(\mathbf{q}_0) = \begin{pmatrix} R_{bc}(\mathbf{q}_0) & T_{bc}(\mathbf{q}_0) \\ O & 1 \end{pmatrix} \quad (8)$$

where $R_{bc}(\mathbf{q}_0)$ is a rotation matrix and $T_{bc}(\mathbf{q}_0)$ is a translation vector. Suppose that the joints' position has moved to $\mathbf{q}$. Then, the homogeneous transformation between the camera frame at $\mathbf{q}_0$ and the camera frame at $\mathbf{q}$ is given by:

$$M_{cc}(\mathbf{q}) = M_{bc}^{-1}(\mathbf{q}_0) \cdot M_{bc}(\mathbf{q}) \quad (9)$$

with

$$M_{cc}(\mathbf{q}) = \begin{pmatrix} R_{cc}(\mathbf{q}) & T_{cc}(\mathbf{q}) \\ O & 1 \end{pmatrix} \quad (10)$$

where R_{cc} is a rotation matrix and T_{cc} a translation vector. If,

$$T_{cc} = \begin{pmatrix} T_x \\ T_y \\ T_z \end{pmatrix} \quad R_{cc} = \begin{pmatrix} r_{11} & r_{12} & r_{13} \\ r_{21} & r_{22} & r_{23} \\ r_{31} & r_{32} & r_{33} \end{pmatrix} \quad (11)$$

using pitch, roll, yaw angles for orientation, the pose vector, $\mathbf{p}$, of the camera frame at $\mathbf{q}$ with respect to the camera frame at $\mathbf{q}_0$ is given by $\mathbf{p} = [T_x\ T_y\ T_z\ \Theta_p\ \Theta_r\ \Theta_y]^T$ with

$$\begin{aligned}
\Theta_r &= \arcsin(r_{13}) \\
\Theta_p &= \operatorname{arctan2}(-\frac{r_{23}}{\cos(\Theta_r)}, \frac{r_{33}}{\cos(\Theta_r)}) \\
\Theta_y &= \operatorname{arctan2}(-\frac{r_{12}}{\cos(\Theta_r)}, \frac{r_{11}}{\cos(\Theta_r)})
\end{aligned} \quad (12)$$

Note that $\cos(\Theta_r)$ is different from zero since Θ_r stays small. Then, the derivative of the pose vector, $\mathbf{p}$, is linked to the velocity screw of the camera by the following Jacobian, $\mathbf{J}_p$:

$$\dot{\mathbf{r}} = \mathbf{J}_p \dot{\mathbf{p}} \quad (13)$$

with (see [6]):

$$\mathbf{J}_p = \begin{bmatrix} I_3 & 0_3 \\ & 1 & 0 & \sin\Theta_r \\ 0_3 & 0 & \cos\Theta_p & -\cos\Theta_r \sin\Theta_p \\ & 0 & \sin\Theta_p & \cos\Theta_r \cos\Theta_p \end{bmatrix} \quad (14)$$

so that, $\mathbf{J}_p(\mathbf{q}_0) = I_6$ and $\dot{\mathbf{r}} = \dot{\mathbf{p}}$. Now, let's define $\mathbf{J}_c$ as the Jacobian of the camera linking $\dot{\mathbf{r}}$, the velocity screw of the camera, to $\dot{\mathbf{q}}$, the velocity of the joint coordinates:

$$\dot{\mathbf{r}} = \mathbf{J}_c \dot{\mathbf{q}} \quad (15)$$

with $\dot{\mathbf{q}} = [\dot{q}_1 ... \dot{q}_6]^T$. Since $\dot{\mathbf{r}}$ is expressed in the coordinates of the camera, the Jacobian of the camera must be derived with respect to the camera frame. Therefore, since $\dot{\mathbf{r}} = \dot{\mathbf{p}}$, the camera Jacobian, $\mathbf{J}_c$, can be derived from the following set of equations:

$$\mathbf{J}_c = \begin{pmatrix} \frac{\partial \mathbf{p}}{\partial q_1}(\mathbf{q}_0) & \cdots & \frac{\partial \mathbf{p}}{\partial q_6}(\mathbf{q}_0) \end{pmatrix} \quad (16)$$

In practice, $\mathbf{J}_c$ is computed by approximating $\frac{\partial \cdot}{\partial q_i}$ with $\frac{\Delta \cdot}{\Delta q_i}$ so that it can be done in real-time. Now, it becomes possible to control the translation and rotation speed of the camera along and around the axes of its coordinate frame. This can be done by computing $\mathbf{J}_c^{-1}$ and using

$$\dot{\mathbf{q}}^* = \mathbf{J}_c^{-1} \dot{\mathbf{r}}^* \quad (17)$$

as reference input to the joint velocity controllers of the robot where $\dot{\mathbf{r}}^*$ is the desired velocity screw computed by the controller from the measurements of $\Delta \mathbf{p}_t$, the displacement of the target from its desired position expressed as difference of pose vector with respect to the camera frame of coordinates at $\mathbf{q}_0$ (the desired pose vector $= 0$). Note that if $\mathbf{f}_0$ is the desired target position, then $\Delta \mathbf{f} = (\mathbf{f} - \mathbf{f}_0)$ is a feature vector error and, using a first order approximation, it holds from (7) and (13) that

$$\Delta \mathbf{p}_t = \mathbf{J}_p^{-1} \mathbf{J}_v^+ \Delta \mathbf{f} + \mathcal{O}(\Delta \mathbf{f}^2) \quad (18)$$

where $\mathbf{J}_v^+$ is computed at $\mathbf{f}_0$, $\mathbf{J}_p$ at $\mathbf{q}_0$ and $\mathcal{O}(\Delta \mathbf{f}^2)$ is a second order error term. Therefore, if the visual feedback loop is "fast enough" so that $\Delta \mathbf{f}$ remains small, then (18) gives a measurement of the small displacements of the target in the camera frame, i.e.:

$$\Delta \mathbf{p}_t \approx \mathbf{J}_v^+ (\mathbf{f} - \mathbf{f}_0) \quad (19)$$

1.4 Model of the robot dynamics

The robot is a 6 DOF industrial manipulator with 6 rotational joints powered by DC motors with individual current feedback loops and analog velocity feedback loops supplied by the manufacturer and a built-in mechanical gravity compensation device for the second and third axis. Therefore, around a given configuration, said $\mathbf{q}_0$, it can be shown experimentally that the dynamics of the robot with its actuators are almost decoupled and linear, from the joint velocity reference input signals, $\dot{\mathbf{q}}^*$, to the actual joint velocities, $\dot{\mathbf{q}}$. The transfer functions between $\dot{\mathbf{q}}^*$ and $\dot{\mathbf{q}}$ can be identified around different positions, $\mathbf{q}_0$, using classical identification techniques. From our experiments, except for the first joint velocity, $\dot{q}_1$, the other transfer functions are relatively independent of the configuration, $\mathbf{q}_0$. Let $F_i(z, \mathbf{q}_0)$, $i = 1, ..., 6$, be the six discrete time transfer functions around $\mathbf{q}_0$:

$$F_i(z) = \frac{\dot{q}_i(z)}{\dot{q}_i^*(z)} \quad i = 1, ..., 6 \quad (20)$$

where the F_i are normalized, by assuming that $\lim_{z \to 1} F_i(z) = 1$. Now suppose that

$$\mathbf{J}_c = \begin{pmatrix} J_{11} & \cdots & J_{16} \\ \vdots & \ddots & \vdots \\ J_{61} & \cdots & J_{66} \end{pmatrix} \quad (21)$$

and $$\mathbf{J}_c^{-1} = \begin{pmatrix} J_{11}^- & \cdots & J_{16}^- \\ \vdots & \ddots & \vdots \\ J_{61}^- & \cdots & J_{66}^- \end{pmatrix} \quad (22)$$

From (15), it comes that

$$\dot{r}_i = J_{i1}\dot{q}_1 + \cdots + J_{i6}\dot{q}_6 \quad (23)$$

Now, using (20),(23) can be rewritten as

$$\dot{r}_i(z) = J_{i1} F_1(z) \dot{q}_1^*(z) + \cdots \\ + J_{i6} F_6(z) \dot{q}_6^*(z) \quad (24)$$

Using (17) and (22) with (24), yields $\dot{r}_i(z) =$

$$\begin{array}{c}[J_{i1}F_1(z)J_{11}^- + \cdots + J_{i6}F_6(z)J_{61}^-]\dot{r}_1^*(z)\\+\\\vdots\\+\\{[J_{i1}F_1(z)J_{16}^- + \cdots + J_{i6}F_6(z)J_{66}^-]\dot{r}_6^*(z)}\end{array} \quad (25)$$

So that,

$$\dot{r}_i(z) = D_i(z)\dot{r}_i^*(z) + \sum_{k=1, k\neq i}^{6} C_{ik}(z)\dot{r}_k^*(z) \quad (26)$$

with
$$C_{ik}(z) = \sum_{l=1}^{6} J_{il}F_l(z)J_{lk}^- \quad (27)$$

$$D_i(z) = \sum_{l=1}^{6} J_{il}F_l(z)J_{li}^- \quad (28)$$

Since the $F_i(z)$ are normalized, it comes that

$$\lim_{z\to 1} D_i(z) = 1 \quad i = 1, ..., 6 \quad (29)$$

$$\lim_{z\to 1} C_{ik}(z) = 0 \quad k = 1, ..., 6 \quad k \neq i \quad (30)$$

Therefore, $\dot{\mathbf{r}}(z) = G(z)\dot{\mathbf{r}}^*(z)$ with

$$G(z) = \begin{bmatrix} D_1(z) & C_{12}(z) & \cdots & C_{16}(z) \\ C_{21}(z) & D_2(z) & \cdots & C_{26}(z) \\ \vdots & \vdots & \ddots & \vdots \\ C_{61}(z) & C_{62}(z) & \cdots & D_6(z) \end{bmatrix} \quad (31)$$

and $\lim_{z\to 1} G(z) = I_6$. The matrix, $G(z)$, is the transfer function of the so-called "Cartesian device" around $\mathbf{q}_0$ which is a linear MIMO system. Since the DC gain of $G(z)$ is the identity matrix, this MIMO Cartesian device can be considered in practice as six decoupled systems by neglecting the cross terms of $G(z)$.

2 Closing the loop

2.1 The whole visual feedback

A complete representation of the visual feedback is given in figure 1. The visual feedback tracks the reference features, $\mathbf{f}_0$, in the image plane. Further, $\Delta \mathbf{p}_t$ is the measurement in the camera frame of the current position of the target with respect to the reference position obtained by computing the image Jacobian (see (19)). Now $\Delta \mathbf{p}_t$ is fed back through $\mathbf{C}(z)$ which represents the controllers. Their output, $\dot{\mathbf{r}}^*$, contains Cartesian reference velocities which are converted into

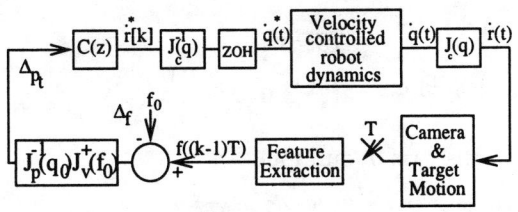

Figure 1: Model of the whole visual feedback.

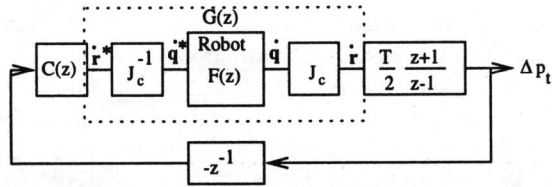

Figure 2: Linear dynamical model of the visual feedback.

joint velocities control inputs, $\dot{\mathbf{q}}^*$, with the camera Jacobian (see equation (17)). These control inputs are converted to analog voltages and passed as reference signals to the analog joint velocity controllers on the industrial robot. Then, a measurement of the target motion is made *via* the camera with a sampling time of $T = 10ms$. Indeed, by dividing the image and the target into two parts, the 50Hz sampling rate is multiplied by 2. Then, the feature extraction procedure gives the current position of the target in the image. The linear model of the whole visual feedback is given in figure 2. The block "Velocity controlled robot dynamics" is replaced by $F(z) = diag\{F_i(z)\}$ (see (20)). Then, the Cartesian device $G(z)$ is computed (see (31)). Further, the whole image processing is modeled by an integrator and a unit delay. This model is valid around the nominal position, $\mathbf{q}_0$, where the robot dynamics has been identified. If large motions are made, several models are identified and a gain scheduling strategy should be implemented.

2.2 The GPC controller

The Generalized Predictive Controller (GPC) was first introduced by Clarke *et al.* in [2]. One could define the GPC as a simpler version of the LQG controller. Both are based on a quadratic optimal control criterion which leads to a trade-off between performance and control energy. The GPC is sub-optimal in the sense that the criterion which must be minimized applies to a finite receding horizon. This horizon expands on future system outputs which must be predicted. Let the model of the system be given in the

ARIMAX form

$$A(q^{-1})y_t = B(q^{-1})u_{t-1} + \frac{C(q^{-1})}{\Delta(q^{-1})}\varepsilon_t \qquad (32)$$

with in the case of our application:

$$\frac{B(q^{-1})}{A(q^{-1})} = D_i(q^{-1})\frac{T}{2}\left(\frac{1+q^{-1}}{1-q^{-1}}\right) = P_i(q^{-1}) \qquad (33)$$

with D_i as in (28) and where ε_t represents the noise and the perturbations which affect the system. Since the spectral properties of ε_t are unknown, $C(q^{-1}) = 1$ is chosen. Selecting $\Delta(q^{-1}) = 1 - q^{-1}$ yields a controller that rejects step perturbations (see [1], for further details). First, let's solve recursively the following two Diophantine equations in E_j, F_j, G_j and H_j:

$$1 = E_j(q^{-1})A(q^{-1})\Delta(q^{-1}) + q^{-j}F_j(q^{-1}) \qquad (34)$$
$$E_j(q^{-1})B(q^{-1}) = G_j(q^{-1}) + q^{-j}H_j(q^{-1}) \qquad (35)$$

From (32) and (34) it can be deduce that:

$$y_{t+j} = \frac{F_j}{C}y_t + \frac{E_jB}{C}\Delta u_{t+j-1} + E_j\varepsilon_{t+j} \qquad (36)$$

For the sake of simplicity the arguments in q^{-1} have been omitted. Note that $\Delta u_t = u_t - u_{t-1}$ and that only the control input increments are considered in the design of this controller. Now, it is straightforward to show (see [1]) that $\hat{y}_{t+j}$, the minimum variance estimator of y_{t+j} is given by

$$\hat{y}_{t+j} = \frac{F_j}{C}y_t + \frac{E_jB}{C}\Delta u_{t+j-1} \qquad (37)$$

Then, using (35), the control inputs are split into past control inputs which are known at time t and future control inputs which have to be determined with the help of an optimal control criterion:

$$\hat{y}_{t+j} = G_j\Delta u_{t+j-1} + H_j\Delta u_{t-1} + F_jy_t \qquad (38)$$

The cost function to be minimized is $J(u,t) =$

$$E\left\{\sum_{j=1}^{N_p}[y_{t+j} - r_{t+j}]^2 + \lambda\sum_{j=1}^{N_u}[\Delta u_{t+j-1}]^2\right\} \qquad (39)$$

with $\lambda > 0$, $\Delta u_{t+j} = 0$ for $j \geq N_u$ and where $\mathbf{r} = [r_{t+1},...,r_{t+N_p}]^T$ are the future reference inputs. N_p is the prediction horizon and N_u is the control horizon. Let define,

$$\mathbf{F} = [H_1\Delta u_{t-1} + F_1y_t, ..., H_{N_p}\Delta u_{t-1} + F_{N_p}y_t]^T \qquad (40)$$

$$\tilde{\mathbf{u}} = [\Delta u_t, ..., \Delta u_{t+N_u-1}]^T \qquad (41)$$

$$\hat{\mathbf{y}} = [\hat{y}_{t+1}, ..., \hat{y}_{t+N_p}]^T \qquad (42)$$

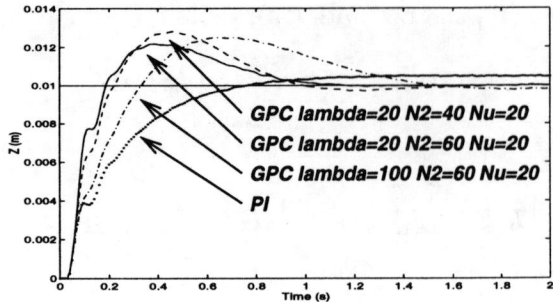

Figure 3: Step responses comparison.

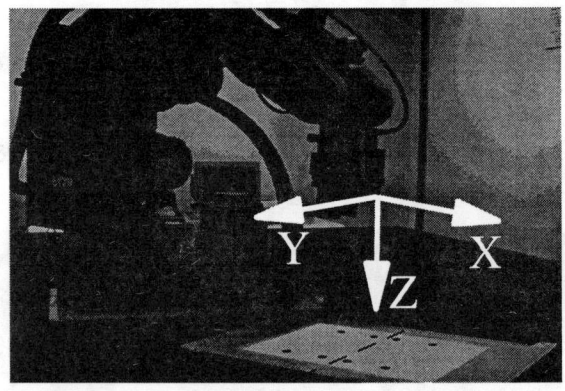

Figure 4: The robot with its camera.

Then, it follows that $\hat{\mathbf{y}} = \mathbf{\Gamma}\tilde{\mathbf{u}} + \mathbf{F}$ with

$$\mathbf{\Gamma} = \begin{pmatrix} g_0 & 0 & \cdots & 0 \\ g_1 & g_0 & \cdots & 0 \\ \vdots & \vdots & \ddots & \vdots \\ g_{N_u-1} & g_{N_u-2} & \cdots & g_0 \\ \vdots & \vdots & \cdots & \vdots \\ g_{N_p-1} & g_{N_p-2} & \cdots & g_{N_p-N_u} \end{pmatrix} \qquad (43)$$

and $G_j(q^{-1}) = \sum_{k=0}^{j-1} g_k q^{-1}$, so that (39) can be rewritten as $\mathbf{J} = (\hat{\mathbf{y}} - \mathbf{r})^T(\hat{\mathbf{y}} - \mathbf{r}) + \lambda\tilde{\mathbf{u}}^T\tilde{\mathbf{u}}$ whose minimum is obtained for $\tilde{\mathbf{u}} = (\mathbf{\Gamma}^T\mathbf{\Gamma} + \lambda\mathbf{I})^{-1}\mathbf{\Gamma}^T(\mathbf{r} - \mathbf{F})$.

3 Simulations and experiments

3.1 Simulation results

The robot model is obtained by identifying the robot joints dynamics with a high order. The camera model is derived from identification of the camera parameters and distortion effects. Figure 3 shows simulation results for a step added to the measurement of the third visual loop which controls the translation

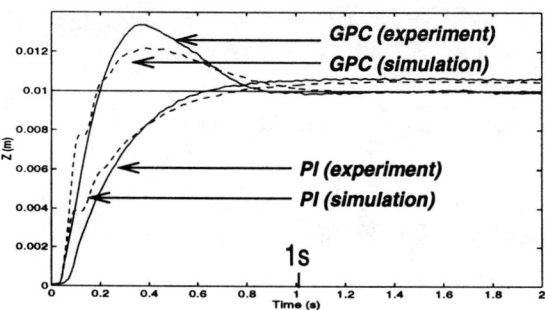

Figure 5: Comparison between PI and GPC.

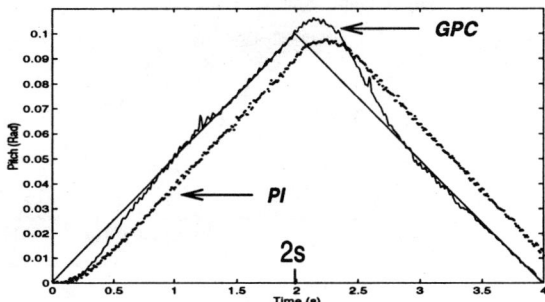

Figure 6: Tracking responses for the pitch angle.

along z. This "simulates" a step shift of the target along z. Four controllers: 3 GPC's and one PI, have been simulated. The PI has been tuned experimentally and its parameters have been adjusted in order to reach the best trade-off between speed and stability. The figure shows that: an increase of the design parameter λ induces a slowing down of the response and an increase of the overshoot. Furthermore, a decrease of the prediction horizon (N_p) induces larger overshoots but no slowing down of the response. Also, the PI is very fast at the beginning of the response but requires much more time than the GPC's to reach its final value. Indeed, for the sake of stability, the integral term has to be set to a small value. Consequently, by tuning λ, the performance of the system can be increased up or slowed down.

3.2 Experiments

A picture of the 6 DOF industrial manipulator is shown on figure 4. The target is made of 8 black dots. A dotted line makes a virtual separation between the two sub-targets. The three coordinates vectors are represented. The nominal robot position for which all tests are performed is represented in this picture. The values selected for the GPC controllers parameters are $\lambda = 20$, $N_p = 60$ and $N_u = 20$. A $1cm$ step function is added to the measurement. One can see on figure 5 the response in the Z direction. There is a good fit between experimental results and simulations. Furthermore, this figure shows a drastic improvement of the speed with the GPC controllers compare to the PI controllers. The rise time is about 0.15s for the GPC vs 0.4s for the PI. Figure 6 shows a tracking experiment for the pitch angle (around X). A $10cm$ height triangular function is added to the measurement to simulate a moving target at a constant velocity. This figure shows again the improvement achieved with the GPC.

References

[1] R. Bitmead, M. Gevers, and V. Wertz. *Adaptative optimal control, the thinking's man's GPC*. Prentice Hall, London, 1990.

[2] D.W. Clarke, C. Mohtadi, and P.S. Tuffs. Generalized predictive control - part 1. the basic algorithm. *Automatica*, 23:137–160, 1987.

[3] P.I. Corke and M.C. Good. Dynamic effects in visual closed-loop systems. *IEEE Transactions on Robotics and Automation*, 12(5):671–683, 1996.

[4] K. Hashimoto, T. Ebine, and H. Kimura. Visual servoing with hand-eye manipulator–optimal control approach. *IEEE Transactions on Robotics and Automation*, 12(5):766–774, 1996.

[5] S. Hutchinson, G.D. Hager, and P.I. Corke. A tutorial on visual servo control. *IEEE Transactions on Robotics and Automation*, 12(5):651–670, 1996.

[6] W. Khalil and E. Dombre. *Modélisation et commande des robots*. Hermès, Paris, 1988.

[7] H. Michel and P. Rives. Singularities in the determination of the situation of a robot effector from the perspective view of 3 points. *INRIA Research Reports*, 1850, 1993.

[8] N. P. Papanikolopoulos. *Controlled Active Vision*. PhD thesis, Carnegie Mellon University, 1992.

[9] N.P. Papanikolopoulos, P. K. Khosla, and T. Kanade. Visual tracking of a moving target by a camera mounted on a robot:combination of control and vision. *IEEE Transactions on Robotics and Automation*, 9(1):14–35, 1993.

[10] R.Y. Tsai. A versatile camera calibration technique for high-accuracy 3d machine vision metrology using off-the-shelf tv camera and lenses. *IEEE Journal of Robotics and Automation*, 3(4):323–344, 1987.

[11] L.E. Weiss and A.C. Sanderson. Dynamic sensor-based control of robots with visual feedback. *IEEE Journal of Robotics and Automation*, 3(5):404–417, 1987.

Intuitive Control of a Planar Bipedal Walking Robot

Jerry Pratt, Gill Pratt
MIT Leg Laboratory, Cambridge, MA 02139
http://www.leglab.ai.mit.edu/

Abstract

Bipedal robots are difficult to analyze mathematically. However, successful control strategies can be discovered using simple physical intuition and can be described in simple terms.

Five things have to happen for a planar bipedal robot to walk. Height has to be stabilized. Pitch has to be stabilized. Speed has to be stabilized. The swing leg has to move so that the feet are in locations which allow for the stability of height, pitch, and speed. Finally, transitions from support leg to support leg must occur at appropriate times. If these five objectives are achieved, the robot will walk.

A number of different intuitive control strategies can be used to achieve each of these five objectives. Further, each strategy can be implemented in a variety of ways. We present several strategies for each objective which we have implemented on a bipedal walking robot.

Using these simple intuitive strategies, we have compelled a seven link planar bipedal robot, called Spring Flamingo, to walk. The robot walks both slowly and quickly, walks over moderate obstacles, starts, and stops. Video, photographs, and more information on Spring Flamingo can be found at http://www.leglab.ai.mit.edu/

1 Introduction

Walking is a moderately easy task, and complex control techniques are not necessary to compel bipedal robots to walk. Instead, simple control strategies which can be explained in simple terms can be used.

Five things have to happen for a planar bipedal robot to walk. Height has to be stabilized. Pitch has to be stabilized. Speed has to be stabilized. The swing leg has to move so that the feet are in locations which allow for the stability of height, pitch, and speed. Finally, transitions from support leg to support leg must occur at appropriate instances. If these five objectives are achieved, then the robot will walk.

Any controller which results in stable walking must meet these five objectives. No matter what control technique is used, something must be happening which stabilizes height, pitch, and speed, swings the swing leg, and transitions from double to single and single to double support.

In this paper we describe intuitive control strategies which can be used to meet these five objectives and hence compel a robot to walk. We describe how these control strategies can be implemented with Virtual Model Control [12], a robot control language which itself appeals to intuition.

We introduce a simple technique, called the "virtual toe point" constraint for dealing with feet and actuated ankles. The virtual toe point is the point along the foot at which zero torques are commanded. This is similar to a static version of the zero moment point [16]. The virtual toe point constraint prevents the foot from over-rotating due to high ankle torques, while allowing the robot to go up on its toes in order to get an extended range of motion from the rear leg. The ability of the robot to go up on its toes allows it to walk with a straight leg which in turn increases energy efficiency.

We also present a simple scheme for dealing with force distribution when in the double support phase of walking. Instead of solving the force distribution problem exactly, we use an approximate force distribution parameter which can then be modified for speed control.

Using simple intuitive control strategies, the virtual toe point constraint, and the simple force distribution scheme, we have compelled a seven link planar bipedal robot to walk. The robot walks both slowly and quickly, walks over moderate obstacles, starts, and stops.

1.1 Intuitive Control

An intuitive controller is one which is based on human intuition of the system and an idea of what is going on. Descriptions of such controllers sound like something a coach would tell a player, like "Square up to the bucket, feet shoulder width apart, bend the knees, pretend you're shooting out of a telephone booth, and follow through." A robotic example might be to put a peg in a hole, "Push the peg and block together, slide the peg until part of it falls in the hole, line it up better, and wiggle it around until it is in."

Intuitive control is nothing new. A PD (Proportional-Derivative) controller is often described as a controller which pushes in the direction of the error and pushes back with increasing velocity to take some energy out and prevent the system from going too fast. Add the I (Integral) term and the controller keeps pushing harder and harder in the direction of the error until it finally goes away.

Raibert's hoppers use a simple 3-part intuitive controller [14]. The height, balance, and speed are stabilized using simple intuitive control laws which can be described in simple terms: to control height, energy is pumped into the leg spring when the leg is fully compressed; to control pitch, the body is servoed to be level to the ground when the stance leg is compressed; to control speed over a stride, the foot is placed further forward (to slow down) or further back (to speed up) from the neutral point in which speed is neither increased or decreased.

Intuitive controllers can be powerful, they are easy to apply, and by default they provide a high level of insight as to what is going on with the system and what is really important in the control of a robot. Unfortunately, intuitive controllers are often called "ad-hoc" or a "hack" because they are not mathematically based, and seldom rigorously proven to work. They are often considered to be "cheating" because parameters are usually hand tuned until desired performance is achieved.

However, using an intuitive controller does not preclude the use of mathematical analysis; if any controller for a system can be analyzed for stability, an intuitive one may be. Unfortunately, legged robots are complicated to analyze for several reasons which are independent of getting a legged robot to perform a task. Therefore, we believe that the limited capability of modern control theory and analysis should not be used as an excuse for the limited abilities of modern robots.

It is possible to use adaptive or learning techniques to automatically tune intuitive control parameters. When the adaptation or learning is complete it may then be possible to understand what was learned or why by looking at parameters whose effects can be understood. For example, suppose the single support to double support transition distance parameter is changed by a learning algorithm that is attempting to maximize efficiency. One may then draw insight into how that parameter affects efficiency.

In the following section we describe intuitive strategies that can be used to control bipedal walking robots. These strategies are easy to understand and easy to apply. Some of the strategies have been used to successfully compel Spring Flamingo, a planar bipedal robot, to walk.

2 Simple Intuitive Control Strategies for Bipedal Walking

We now describe simple intuitive control strategies which can be used to achieve the five objectives required for planar bipedal walking. These objectives are height stabilization, pitch stabilization, speed stabilization, swing leg placement, and support transitions.

There are a number of methods which can be used to implement the intuitive control strategies. These include inverse kinematics, high gain servos, feedforward control, impedance control, etc.

We use Virtual Model Control, a method which itself relies on intuition, to implement the control strategies. This control technique uses simulations of virtual mechanical components to generate real actuator torques (or forces). These joint torques create the same effect that the virtual components would have created, had they existed, thereby creating the illusion that the simulated components are connected to the real robot. Such components can include linear or non-linear springs, dampers, dashpots, masses, latches, bearings, potential and dissipative fields, or any other imaginable component.

In section 4 we discuss how the following control strategies are used to compel Spring Flamingo, a bipedal walking robot, to walk.

2.1 Height Stabilization

Stabilizing height is straightforward as long as we have a support leg firmly on the ground. This will be the case if the swing leg strategy and support leg strategies are working. There are many ways height can be stabilized. Two strategies are listed here.

1. Maintain a constant height above the ground.
2. Maintain a constant stance leg length.

Maintaining a constant height can be implemented with a virtual spring-damper mechanism attached between the ground and the robot's body. The spring set point will determine the height above ground that the robot maintains. The damper will cause oscillations about that height to decay. A virtual vertical force of the weight of the body can be used to allow for lower virtual spring constants and decrease the DC offset.

Maintaining a constant stance leg length can be done in a number of different ways. A simple method which does not require high gain feedback is to push upwardly on the body a little harder than gravity. This will cause the robot to increase its height until the knee hits its joint limit (knee cap).

The first strategy was used with Spring Turkey [12], our previous walking robot. We successfully applied both strategies on Spring Flamingo but embraced the second one because it is much more efficient and more similar to biological walking. By walking with fully extended legs, there is a low torque requirement on the knee. The robot will walk with somewhat of a compass gait, in which potential energy and kinetic energy are out of phase and hence total mechanical energy is nearly constant.

2.2 Pitch Stabilization

As in height stabilization, stabilizing pitch is straightforward as long as we have a support leg firmly on the ground. Two strategies for stabilizing pitch are

1. Maintain a level pitch.
2. Follow a pitch trajectory.

Both strategies require feedback to implement since the center of gravity is above the hip. If it were below, we could just rely on the natural dynamics which would already be stable. To control the pitch, we use a virtual torsional spring damper mechanism. If a level pitch trajectory is desired, then the virtual spring set point is held constant. If a pitch trajectory is desired (perhaps to help control speed, as described below), then the set point is changed to match the desired pitch position. We used the level pitch strategy in the control of Spring Flamingo.

2.3 Speed Stabilization

Most speed stabilization techniques for bipedal walking are discrete events and thus control the speed from stride to stride rather than throughout a given stride. In fact, for dynamic bipedal walking, it is impossible to arbitrarily control the forward speed during a stride since the center of mass projection lies outside the support foot polygon during much of the stride.

Five strategies for stabilizing speed are

1. Change stride length with speed.
2. Change transition events with speed.
3. Servo speed when the center of mass is over the support foot or when the robot is in double support.
4. Change the location of the body center of mass by pitching forward or backward.
5. Apply energy at strategic times.

The first two strategies (stride length and transitions) are discrete events which will stabilize the speed over a number of strides. By changing stride lengths and transition events one can change the percent of the stride in which the support leg is in front of the body and hence is slowing down the robot and the percent of the stride in which the support leg is behind the body and hence is speeding up the robot.

The third strategy (servo speed when possible) can be used when the center of mass is over the support foot polygon or when the robot is in double support. The placement of the virtual toe point can be used in single support by placing it forward, closer to the front of the foot, to slow the robot down and placing it backward, closer to the heel, to speed it up. Force distribution can be used during double support. More force can be applied by the rear leg to speed the robot up and more force can be applied by the front leg to slow the robot down.

The fourth strategy (pitching the body) can be used to speed the robot up by leaning forward or slow it down by leaning backward. This strategy is typically seen when a person leans into a hill when walking up it or leans back to brake when going down it.

The fifth strategy (applying energy at strategic times) is a bit trickier to apply. It may be possible to put energy into the system in one mode during part of the stride and later slosh it into forward speed. For example, it may be possible to increase the potential energy of the system by having the robot go up onto its toes earlier and then convert that energy into forward kinetic energy during a later portion of the stride.

The first three strategies were used in stabilizing Spring Flamingo's speed. The last two strategies are currently being investigated.

2.4 Swing Leg Placement

In order to walk successfully, the swing leg must swing quickly to its next support location. Fortunately, the exact placement is not important when walking on smooth ground.

Two possible strategies for swinging the swing leg are

1. Servo the swing leg, either as a function of time, or as a function of the other leg.
2. Let the swing leg swing passively, making sure it does not hit the ground.

Following a swing trajectory can be implemented with a virtual spring-damper mechanism attached between the body and the ankle of the robot. The spring set point can move along the trajectory, pulling the leg along to follow it.

The natural pendulum dynamics of the swing leg are exploited in the second strategy. The leg will naturally swing forward, as long as the foot clears the ground.

Both strategies were successfully applied to Spring Flamingo. However, at high speeds we had some trouble with the second strategy as one cannot rely completely on the natural dynamics of the swing leg but must get it started and stop it at the end. This proved challenging at high speeds and is currently being explored further.

2.5 Support Transitions

To continuously walk forward, the support legs must be alternated since a given leg can only support the body over a small range. For bipedal walking, double support to single support and single support to double support transitions must occur at appropriate times.

Three strategies for transitioning from double support to single support are

1. Transition to single support if the body is within a certain distance to the next support leg.
2. Transition to single support if the body is over a certain distance away from the previous support leg.
3. Transition to the single support leg after being in double support for a certain amount of time.

These strategies can be implemented by measuring joint angles and transitioning based on joint angle threshholds or by computing the kinematics of the robot and transitioning on center of mass to foot distance threshholds. A state machine can be used to keep track of what support state the robot is in.

The first strategy will ensure that the next support leg will have a long enough support time that the other leg will be able to swing through in time. The second strategy will ensure that the rear leg has enough range of motion that the robot does not have to drag its rear leg. The third strategy simply transitions to single support after a given time.

We used the first two strategies together in the control of Spring Flamingo. When either event happens, the robot transitions to single support.

Two strategies for transitioning from single support to double support are,

1. Transition to double support if the body is over a certain distance away from the support leg.
2. Transition to double support if the swing leg has swung beyond a certain position or has slowed below a certain speed.

The first strategy ensures that the robot will transition onto a new support leg before the body becomes too far from the current support leg. This will guarantee that the current support leg can safely support the body and hence stabilize height and pitch.

The second strategy will ensure that the next support leg has swung far enough that it is in a position to support the body when the time comes. This strategy also maximizes double support time as the robot will put its swing foot down as soon as it has swung through rather than wait for the transition distance event to occur.

We used the first strategy in the control of Spring Flamingo.

3 Virtual Actuator Implementation for a Planar Biped With Feet and Ankles

We use Virtual Actuators [13] and Virtual Model Control [12], two techniques based on intuition, to implement some of the strategies of the previous section. In this section we present the mathematics to implement virtual components on Spring Flamingo for the support leg in single support or both legs in double support, following the procedure described in [13].

3.1 Single Leg Implementation

Figure 1 shows a simple planar, five link, four joint, serial robot model that we use to represent a single leg of our walking robot. The toe joint and link do not exist on the real robot (Figure 2). They are used to represent the point on the foot in which no torque is applied. We refer to this as the "virtual toe point". It is similar to the center of pressure on the foot or the zero moment point [16] except that it is a commanded quantity, not a measured one, and is based on static, not dynamic, considerations.

The virtual toe point can be used for control in the following intuitive sense. If it is desired to accelerate the robot backward (or reduce the forward acceleration) one can move the virtual toe point forward on the foot. Similarly, if it is desired to accelerate the robot forward (or reduce the backward acceleration) one can move the virtual toe point backward toward the heel.

We wish to connect a virtual component between the virtual toe point frame, $\{A\}$, and the body frame, $\{B\}$. The angles θ_t, θ_a, θ_k, and θ_h are those of the virtual toe, ankle, knee, and hip. The upper link (femur) is of length L_2, the lower link (tibia) is of length L_1, and the height of the foot is L_h. We assume that the virtual toe is flat on the ground, so that $^O_A R = I$. The commanded distance to the virtual toe point from the ankle is L_{vtp}.

The forward kinematic map from frame $\{A\}$ to frame $\{B\}$ is as follows,

$$^A_B \vec{X} = \begin{bmatrix} x \\ z \\ \theta \end{bmatrix} = \begin{bmatrix} -L_{vtp} c_t - L_h s_t - L_1 s_{t+a} - L_2 s_{t+a+k} \\ -L_{vtp} s_t + L_h c_t + L_1 c_{t+a} + L_2 c_{t+a+k} \\ -\theta_h - \theta_k - \theta_a - \theta_t \end{bmatrix} \quad (1)$$

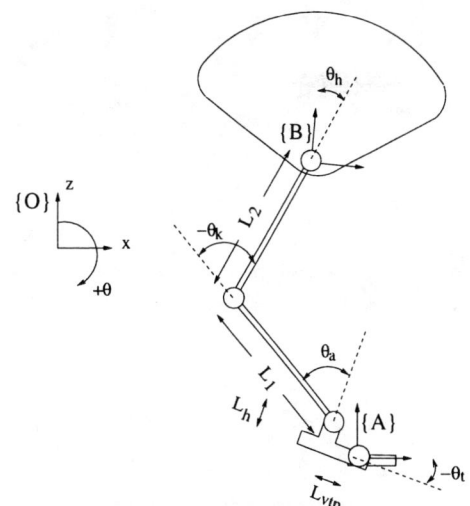

Figure 1: Single leg implementation. Reaction frame $\{A\}$ is assumed to be in the same orientation as reference frame $\{O\}$ so that $^O_A R = I$.

where $c_a = \cos(\theta_a)$, $s_a = \sin(\theta_a)$, etc. Partial differentiation produces the Jacobian,

$$^A_B J = \begin{bmatrix} J_{1,1} & J_{1,2} & J_{1,3} & 0 \\ J_{2,1} & J_{2,2} & J_{2,3} & 0 \\ -1 & -1 & -1 & -1 \end{bmatrix} \quad (2)$$

where,

$$J_{1,3} = -L_2 c_{t+a+k}$$
$$J_{1,2} = J_{1,3} - L_1 c_{t+a}$$
$$J_{1,1} = J_{1,2} + L_{vtp} s_t - L_h c_t$$
$$J_{2,3} = -L_2 s_{t+a+k}$$
$$J_{2,2} = J_{2,3} - L_1 s_{t+a}$$
$$J_{2,1} = J_{2,2} - L_{vtp} c_t - L_h s_t$$

The Jacobian relates the virtual velocity between frames A and B with the joint velocities,

$$^A_B \dot{\vec{X}} = {^A_B J} \, \dot{\vec{\Theta}} \quad (3)$$

and the virtual force to joint torque,

$$\vec{\tau} = (^A_B J)^T (^A_B \vec{F}) \quad (4)$$

where $\vec{\tau}$ is the joint torque vector and $\vec{F}$ is the virtual force vector.

Next we add the constraint of an unactuated toe, $\tau_t = 0$, since we desire zero actuated torque about the virtual toe point. This will constrain the direction in which virtual forces can be applied. With the virtual toe point constraint, Equation 4 is,

$$\begin{bmatrix} 0 \\ \tau_a \\ \tau_k \\ \tau_h \end{bmatrix} = \begin{bmatrix} J_{1,1} & J_{2,1} & -1 \\ J_{1,2} & J_{2,2} & -1 \\ J_{1,3} & J_{2,3} & -1 \\ 0 & 0 & -1 \end{bmatrix} \begin{bmatrix} f_x \\ f_z \\ f_\theta \end{bmatrix} \quad (5)$$

For our walking robot we are more concerned about applying forces in the vertical direction and torques about the body than we are concerned about applying horizontal forces. Therefore, we specify f_z and f_θ and solve for f_x

$$f_x = \begin{bmatrix} \frac{-J_{2,1}}{J_{1,1}} & \frac{1}{J_{1,1}} \end{bmatrix} \begin{bmatrix} f_z \\ f_\theta \end{bmatrix} \quad (6)$$

Plugging equation 6 back into equation 5, we get

$$\begin{bmatrix} \tau_a \\ \tau_k \\ \tau_h \end{bmatrix} = \begin{bmatrix} \frac{-J_{1,2}J_{2,1}}{J_{1,1}} + J_{2,2} & \frac{J_{1,2}}{J_{1,1}} - 1 \\ \frac{-J_{1,3}J_{2,1}}{J_{1,1}} + J_{2,3} & \frac{J_{1,3}}{J_{1,1}} - 1 \\ 0 & -1 \end{bmatrix} \begin{bmatrix} f_z \\ f_\theta \end{bmatrix} \quad (7)$$

Throughout the above derivation we have assumed that the toe is flat on the ground and that we can measure all angles. Because there is no toe (it is virtual) we cannot measure its angle with the ground. Instead we measure the body angle via a potentiometer on a boom or a gyroscope and compute what the toe angle would be if the toe was flat on the ground,

$$\theta_t = -\theta - \theta_h - \theta_k - \theta_a$$

We now have a simple set of equations for determining joint torques given virtual forces. These equations will be used in the next section in the control of a bipedal walking robot during the single support phase.

3.2 Dual Leg Implementation

In the previous subsection we discussed virtual actuator implementation for a single leg (when the robot is in single support). Here we examine the double support case.

As in [13] we could construct a constraint matrix and exactly solve for the force distribution between the two legs such that any arbitrary F_x, F_z, F_θ force vector could be commanded. However, we decide to use a simpler method because

- There is no solution when the feet are together as the constraint matrix is not invertible in such a configuration.
- It is unlikely that biological creatures exactly solve the force distribution problem.
- Solving the force distribution problem exactly is unnecessary.
- The method presented below appeals more to intuition.

Instead of solving the force distribution problem exactly we simply distribute the force between the two legs with force distribution parameter α such that,

$$\begin{bmatrix} f_z \\ f_\theta \end{bmatrix}_l = \alpha \begin{bmatrix} F_z \\ F_\theta \end{bmatrix}, \quad \begin{bmatrix} f_z \\ f_\theta \end{bmatrix}_r = (1-\alpha) \begin{bmatrix} F_z \\ F_\theta \end{bmatrix} \quad (8)$$

where $0 \leq \alpha \leq 1$

As in the single support case, we do not command forces in the x direction. Instead we command forces in the z and θ directions, decide the force distribution

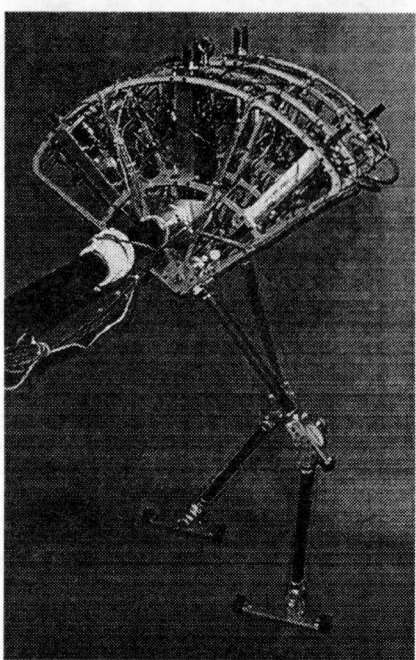

Figure 2: Spring Flamingo, a planar bipedal walking robot. There are six force controlled actuators attached to the body. Power is transmitted to the hips, knees, and ankles via cables. A boom prevents motion in the lateral, roll, and yaw directions.

between the two legs, and solve for the joint torques using Equation 7 for each leg.

We choose α such that the forces are divided between the legs in a natural way. If the robot's body is directly above the left leg, all forces are provided by the left leg ($\alpha = 1$). Similarly, if the robot's body is directly above the right leg, all forces will be provided by the right leg ($\alpha = 0$). If the robot's body is between the left and right legs, the forces will be divided with a linear relationship

$$\alpha = \frac{x_{right}}{x_{left} + x_{right}} \quad (9)$$

where $x_{left} \geq 0$ is the horizontal distance from the left leg to the body and $x_{right} \geq 0$ is the horizontal distance from the right leg to the body. If the legs are both close together, it is very much like the single support case and we simply set $\alpha = 0.5$, dividing the forces evenly between the two legs.

We can modify the force distribution parameter α for control in the following way. To accelerate forward, put more of the force distribution on the rear leg. To accelerate backward, put more of the force distribution on the front leg. For example we can use the simple control law $\alpha = \alpha_0 + b(\dot{x}_d - \dot{x})$ to help regulate velocity when the left leg is the rear leg.

In the next section we will use this control strategy to help regulate forward velocity during double support.

4 Intuitive Control Strategies Applied to a Bipedal Walking Robot

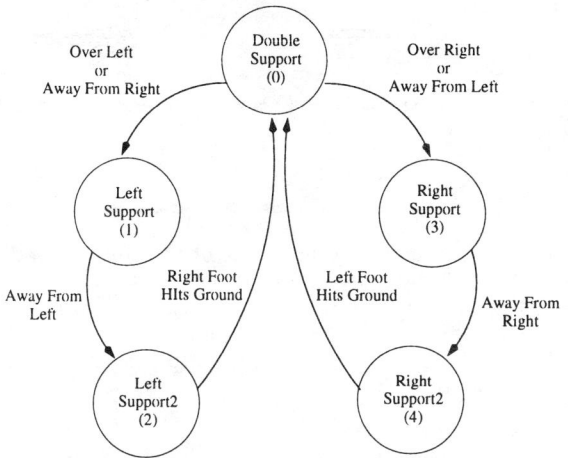

Figure 3: State machine used in Spring Flamingo's walking algorithm.

Table 1: Important Parameters for Spring Flamingo's Walking Algorithm.

Parameter	Value	Range
Height Control		
Virtual Z Anti-Gravity Force	110 N	90-120
Virtual Z Damper	$200 \frac{N}{m/s}$	100-250
Pitch Control		
Virtual Pitch Spring	$60 \frac{Nm}{rad}$	30-80
Virtual Pitch Damper	$10 \frac{Nm}{rad/s}$	4-15
Forward Speed Control		
Nominal Velocity	$0.4 \frac{m}{s}$	0.0-0.7
Virtual Toe Point Gain	$0.3 \frac{m}{m/s}$	0.0-0.5
Double Support Transfer Ratio Gain	$0.3 \frac{\%}{m/s}$	0.0-1.5
Double to Single Support Transition Distance Gain	$0.3 \frac{m}{m/s}$	0.0-0.5
Single to Double Support Transition Distance Gain	$0.3 \frac{m}{m/s}$	0.0-0.5
Swing Leg Control		
Virtual Swing Leg X Spring	$25 \frac{N}{m}$	10-40
Virtual Swing Leg X Damper	$3 \frac{N}{m/s}$	1-5
Virtual Swing Leg Z Spring	$150 \frac{N}{m}$	100-200
Virtual Swing Leg Z Damper	$8 \frac{N}{m/s}$	2-14
Support Transitions		
Double to Single Support Rear Leg Transition Distance	$0.26m$	0.20-0.30
Double to Single Support Front Leg Transition Distance	$0.05m$	0.01-0.10
Single to Double Support Transition Distance	$0.16m$	0.10-0.24
Nominal Stride Length	$0.36m$	0.24-0.42

Figure 2 is a photograph of Spring Flamingo, a planar bipedal walking robot. The robot has an actuated hip, knee, and ankle on each leg. An unactuated boom constrains Spring Flamingo's roll, yaw, and lateral motion, thereby reducing it to a planar robot. All of Spring Flamingo's motors are located in its upper body, with power being transmitted to the joints via cable drives. Series Elastic Actuation [11] is employed at each degree of freedom, allowing for accurate application of torques and a high degree of shock tolerance. The maximum torque that can be applied to the hips and ankles is approximately 18 Nm, while approximately 24 Nm can be applied to the knees. The force control bandwidth we achieve is approximately 20 Hz. Spring Flamingo weighs approximately 30 lbs (14 kg) and stands 3 ft (1 m) tall from floor to hip.

Potentiometers at the hips, knees, ankles, and boom measure joint angles and body pitch. Compression springs are used in the joint actuators to implement Series Elastic Actuation. Linear potentiometers measure the stretch in the springs.

4.1 Walking Algorithm

The intuitive control strategies used on Spring Flamingo are:

- Maintain a constant stance leg length by pushing up until hitting the knee cap.
- Maintain a constant level pitch using a virtual spring damper mechanism with constant set point.
- Transition from double support to single support if the body's x position becomes further than a certain distance from the rear foot or closer than a certain distance from the front foot.
- Transition from single support to double support if the body's x position becomes further than a certain distance from the support foot.
- Swing the non-stance leg so that the foot is roughly placed a nominal stride length away from the support foot when transitioning to double support.
- Increase the nominal stride length as the robot walks faster.

- Transition to double support later if the robot is walking too slowly or sooner if the robot is walking too quickly.
- Maintain the virtual toe point of the support foot approximately below the center of mass. Move it forward if walking too quickly or backward if walking too slowly.
- During double support put more of the load on the back leg if walking too slowly and more on the front leg if walking too quickly.

To implement Spring Flamingo's walking algorithm, we use a simple set of virtual components and a state machine shown in Figure 3.

The various virtual spring, damper, and force variables and walking parameters are chosen using physical insight and a manual search. Some of the parame-

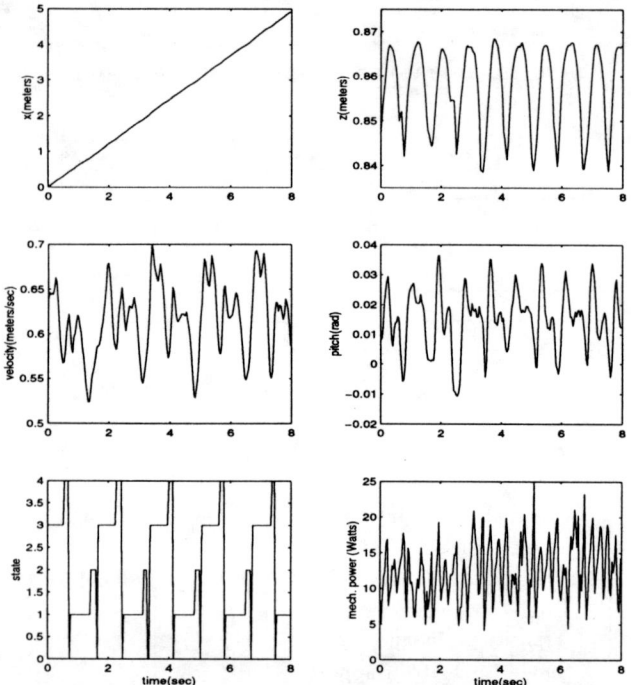

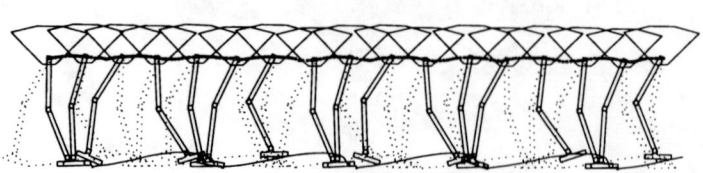

Figure 5: Elapsed time snapshot of the bipedal walking data in Figure 6. The drawings of the robot are spaced approximately 0.5 seconds apart. The left leg is dotted while the right leg is solid. Lines show the path of the tips of the feet and the hip trajectory. The robot walks from left to right.

Figure 4: Spring Flamingo walking data. Left graphs display, from top to bottom the horizontal position (x) and horizontal velocity of the body and the stat of the state machine. Right graphs display the vertical position (z) and pitch (θ) of the body and mechanical power being applied to the joints.

ters are listed in Table 1 along with their tuned values and reasonable ranges. This range represents the reasonable amount that the parameter can vary by while the robot remains well-behaved. For some parameters, the robot can continue to walk throughout this range. For others, the robot can not walk if the parameter is at the boundaries of the range. All the parameters can be individually varied by at least 10% from the tuned value without disrupting the stable walking.

The vertical force to control height is calculated to be a little larger than the weight of the robot. Many parameters are tuned by physically examining their effects (resistance to being pushed on, decay rate, etc.) until the desired effects are achieved and the robot walks successfully. These walking parameters consist of nominal stride length, transition distances, swing leg gains, and velocity gains on the transition distances, stride length, virtual toe points, and double support loading ratio. From the time the robot was built until the moment it could continuously walk, approximately 40 iterations were performed over a span of 3 weeks.

Spring Flamingo is initialized balancing with its feet together. It starts walking by lifting up one leg and transitioning into the single support phase. At no time is external intervention required. The robot stops by setting its desired velocity to zero after walking a given distance.

Figure 4 shows experimental data from Spring Flamingo while walking. The graphs on the left show (from top to bottom) the body's horizontal position (x), horizontal velocity, and state. The graphs on the right show the body's vertical position (z), pitch (θ), and the mechanical power being exerted at the joints. The mechanical power is computed as the sum of the absolute value of the torque times angular velocity at each joint.

The data in Figure 4 is plotted in graphical form in Figure 5. The snapshots in Figure 5 are approximately 0.5 seconds apart. Lines are drawn to show the path of the tips of the feet and the hip trajectory.

Spring Flamingo walked continuously at approximately 0.63 m/s. The data shows approximately 10 steps (left to right or right to left support transitions) in 8 sec, giving a step time of 0.8 seconds. The height fluctuated about 3 cm as the robot walked using a compass-like gait. The pitch was confined to ±0.04 radians (±2.1 deg). The mechanical power averaged about 15 watts. However, due to the inefficiencies of the motors, transmissions, and power electronics, the electrical power consumed is probably much higher.

4.2 Robustness of Walking Algorithm

The intuitive walking control algorithm discussed above is somewhat robust to external forces, rough terrain, and parameter variation. Spring Flamingo can be pushed fairly hard in either direction, temporarily changing its speed by about 25%, recovering to the original speed within a few steps. The robot can walk up and down slopes of approximately 5^0 without any change in the algorithm and without being informed of or detecting the presence of the slope. All of the control parameters can be individually changed by 10% or more while still maintaining stable walking.

The most common failure mode occurs when the robot is pushed too quickly and can not recover. The robot will typically take several short, choppy steps further increasing its speed and finally falling. A biological creature in this situation typically recovers by running a few steps and slowing down. Unfortunately, Spring Flamingo can not run and hence has no recourse when its speed increases above its natural walk to run transition speed.

4.3 Self-Stabilizing Speed

The above algorithm used several controllers to stabilize speed. In another algorithm, we successfully compelled Spring Flamingo to walk stably without any feedback on the forward speed. We used the speed control strategy "Take longer strides as the robot walks faster" but we never implicitly programmed it. Instead, we used low gains on the swing leg such that overshoot was significant and the natural dynamics of the system played a large roll in where the leg was placed. As the robot walked faster it naturally took longer strides without explicitly being told to. The speed-dependent stride lengths then self-stabilized the forward walking speed.

With this self-stabilizing speed algorithm, the robot walked continuously while being robust to external pushes. However, we could only get this algorithm to work for slow speed walking. Future work may focus on using a self-stabilizing speed algorithm at higher speeds.

5 Conclusions

Spring Flamingo walked continuously using a simple set of intuitive control strategies. These strategies are easy to develop, are easy to understand, and are easy to implement. In short, planar bipedal walking is easy to achieve despite being difficult to analyze mathematically.

By tuning the intuitive control parameters by hand, one gains insight into how the parameters relate to the resultant walking. This insight in turn helps speed up the tuning process. We are currently investigating the possibility of using learning or adaptive techniques to tune the parameters automatically.

Spring Flamingo gained several advantages over Spring Turkey [12] by the use of feet and actuated ankles. Since Spring Turkey had only point feet, it could not balance on one foot, had to walk with bent knees, and had large velocity fluctuations during each stride. Spring Flamingo exhibits smaller velocity fluctuations as it keeps its virtual toe point directly underneath when the center of mass passes over the support foot. It walks more efficiently as a compass gait with straight legs can be used without the worry of losing range of motion on the rear leg. Also, with feet and ankles, the robot is able to stand and balance on one leg.

Stable, robust, and efficient planar bipedal walking can be achieved using intuitive control strategies and intuitive control techniques. Spring Flamingo walked over moderate slopes with no change to its level ground algorithm. We are currently focusing on a few simple intuitive control strategies for dealing with more formidable slopes.

We are confident that we can develop similar strategies for three dimensional bipedal walking. We are currently developing such strategies and applying them to walking simulations.

References

[1] E. Dunn and R. Howe. Towards smooth bipedal walking. *IEEE Conference on Robotics and Automation*, pages 2489–2494, 1994.

[2] E. Dunn and R. Howe. Foot placement and velocity control in smooth bipedal walking. *IEEE Conference on Robotics and Automation*, pages 578–583, 1996.

[3] L. Jalics, H. Hemami, and B. Clymer. A control strategy for adaptive bipedal locomotion. *IEEE Conference on Robotics and Automation*, pages 563–569, 1996.

[4] S. Kajita and K.Tani. Experimental study of biped dynamic walking in the linear inverted pendulum mode. *IEEE Conference on Robotics and Automation*, pages 2885–2891, 1995.

[5] S. Kajita and K.Tani. Adaptive gait control of a biped robot based on realtime sensing of the ground profile. *IEEE Conference on Robotics and Automation*, pages 570–577, 1996.

[6] A. Kun and W. T. Miller. Adaptive dynamic balance of a biped robot using neural networks. *IEEE Conference on Robotics and Automation*, pages 240–245, 1996.

[7] Tad McGeer. Passive dynamic walking. *International Journal of Robotics Research*, 9(2):62–82, 1990.

[8] W. T. Miller. Real time neural network control of a biped walking robot. *IEEE Control Systems Magazine*, Feb:41–48, 1994.

[9] H. Miura and I. Shimoyama. Dynamic walk of a biped. *International Journal of Robotics Research*, 3(2):60–74, 1984.

[10] Simon Mochon and Thomas A. McMahon. Ballistic walking: An improved model. *Mathematical Biosciences*, 52:241–260, 1979.

[11] Gill A. Pratt and Matthew M. Williamson. Series elastic actuators. *IEEE International Conference on Intelligent Robots and Systems*, 1:399–406, 1995.

[12] J. Pratt, P. Dilworth, and G. Pratt. Virtual model control of a bipedal walking robot. *IEEE Conference on Robotics and Automation*, pages 193–198, 1997.

[13] J. Pratt, A. Torres, P. Dilworth, and G. Pratt. Virtual actuator control. *IEEE International Conference on Intelligent Robots and Systems*, pages 1219–1226, 1996.

[14] Marc H. Raibert. *Legged Robots That Balance*. MIT Press, Cambridge, MA, 1986.

[15] Ann L. Torres. Implementation of virtual model control on a walking hexapod. May 1996. Undergraduate Thesis, Massachusetts Institute of Technology.

[16] M. Vukobratovic, B. Borovac, D. Surla, and D. Stokic. *Biped Locomotion: Dynamics, Stability, Control, and Applications*. Springer-Verlag, Berlin, 1990.

[17] J. Yamaguchi, A. Takanishi, and I. Kato. Development of a biped walking robot adapting to a horizontally uneven surface. *IEEE International Conference on Intelligent Robots and Systems*, pages 1156–1163, 1994.

[18] K. Yi and Y. Zheng. Biped locomotion by reduced ankle power. *IEEE Conference on Robotics and Automation*, pages 584–589, 1996.

Realization of Dynamic Biped Walking Varying Joint Stiffness Using Antagonistic Driven Joints

*Jin'ichi Yamaguchi, **Daisuke Nishino *and* * ***Atsuo Takanishi

*Humanoid Research Laboratory, Advanced Research Institute for Science and Engineering, Waseda University
**Graduate School of Science and Engineering, Waseda University
***Department of Mechanical Engineering, School of Science and Engineering, Waseda University
3-4-1, Okubo, Shinjuku-ku, Tokyo, 169, Japan
Phone/Fax: +81-3-3208-8714   E-mail: yamajin@mn.waseda.ac.jp, PAH03322@niftyserve.or.jp
takanisi@mn.waseda.ac.jp

Abstract

In this paper, the authors introduce a life-size biped walking robot having antagonistic driven joints using a nonlinear spring mechanism and a dynamic biped walking control method using these joints.

In the current research concerning a biped walking robot, there is no developed example of a life-size biped walking robot with antagonistically driven joints by which the human musculo-skeletal system is imitated in lower limbs. Humans are considered to walk efficiently using the inertial energy and the potential energy of the lower limbs effectively, walk smoothly with less impact force when a foot lands and cope flexibly with the outside environment. The Human joint is driven by two or more muscle groups. Humans can vary the joint stiffness, using nonlinear spring characteristics possessed by the muscles themselves. These functions are indispensable for a humanoid. However, the biped walking robots developed previously have been unable to walk in this way.

Therefore, the authors developed a biped walking robot having antagonistic driven joints, and proposed a walking control method for dynamic biped walking that uses antagonistic driven joints to vary joint stiffness. The authors performed walking experiments using the biped walking robot and the control method. As a result, dynamic biped walking varying the joint stiffness using antagonistic driven joints was realized.

1. Introduction

In recent years, a lot of vigorous researches on biped walking robots have been carried out [1]-[7]. The authors and others are engaged in studies of biped walking robots, with "human form" as a key word, from the following two viewpoints. One is a viewpoint as a human science to elucidate the walking mechanism of humans from (robotics) engineering viewpoints. The other is a viewpoint towards the development of anthropomorphic robots called "humanoid" which will become human partners in the next century. And by now, the authors not only have realized a dynamic biped

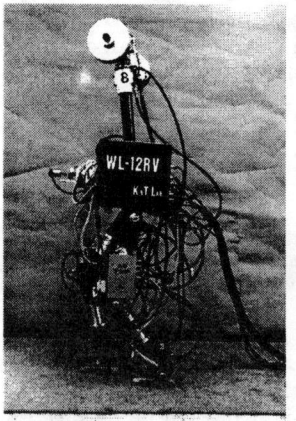
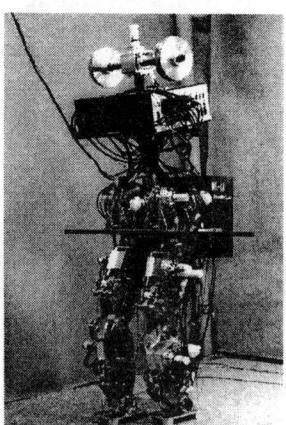

Fig. 1 WL-12RV   **Fig. 2** WL-13

walking on various known surfaces, but also have been improving its adaptability to the environment as a humans' living floor (unknown uneven surfaces) and so on [8]-[17].

However, in the current researches concerning a biped walking robot, there is no developed example of a life-size biped walking robot with the antagonistically driven joints by which the human musculo-skeletal system is imitated in lower limbs. As the biped walking robot WL-12(Waseda Leg-No.12 Refined V) shown in **Fig. 1** which the authors developed in 1992 or as ealier robots, one DOF(degree of freedom) is driven by one hydraulic rotary actuator. The drive system used a local position feedback system with a high gain ratio. Therefore, they had no capacity to vary joint stiffness in a stable way over a broad range. As a result, walking smoothly with less impact force and efficient walking using the inertial energy and the potential energy of the lower limbs could not be realized.

On the other hand, the human joint is driven by two or more antagonistic muscle groups having a nonlinear spring characteristic [18]. It is also reported that the muscular elasticity of cats when walking is adjusted at the swing and stance phase [19]. The tension and elasticity (stiffness) of the muscle are high in the stance phase, but low in the swing phase. As with cats, humans are considered to walk smoothly and efficiently using the hardware characteristics possessed by the muscles themselves. The authors consider that a stiffness control strategy using not only software but also mechanical hardware characteristics is very effective to realize a stable joint stiffness control varying quickly over a broad range.

The function of varying the joint stiffness is also an indispensable function for a humanoid which works cooperatively with humans. Therefore, in 1994, the authors began to make researches on a human-size biped walking robot with antagonistically driven joints by which the human musculo-skeletal system is imitated in lower limbs. In the same year, the authors contrived a straight type nonlinear spring mechanism which enables its spring coefficient to change from zero to infinity if we assume the spring coefficient of a wire is infinity. Then, the authors developed an anthropomorphic knee joint model WAK-1 (Waseda Anthropomorphic Knee-No.1) installed the mechanism and conducted joint driving tests using the model [20]. As a result, variable joint stiffness was realized. Based on the result of WAK-1, in 1995, the authors developed a bipedal walking robot WL-13 having antagonistic driven joints using a rotary type nonlinear spring mechanism which was a refined mechanism of the linear type [21]. Then, the authors realized a quasi-dynamic biped walking using antagonistic driven joints at the walking speed of 7.68 s/step with 0.1 m step length. In this walking performance, the stiffness of joints was not varied actively. The typical characteristics of the antagonistic driven joint using nonlinear spring mechanism is to vary the stiffness of joints over a broad range. Making the most of the characteristic, i.e. varying the stiffness of joints actively, enables dynamic walking with less energy consumption could be realized. But the authors did not achieve a dynamic biped walking varying the stiffness of joints.

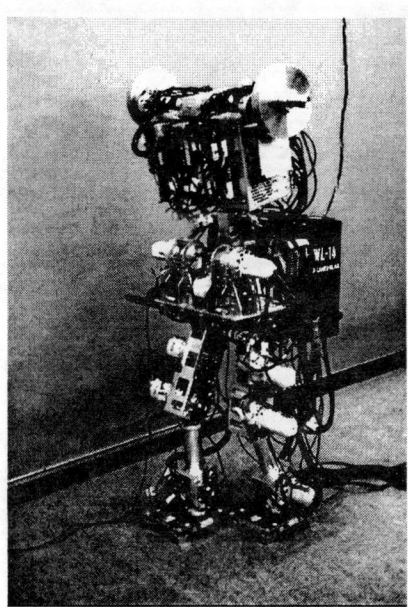

Fig. 3 Biped walking robot WL-14

By the way, in Japan and other developed countries, an increasingly aged society will require bipedal humanoid robots to assist humans. In order to conduct these practical activities in the human habitat, the robot must be wholly self-contained with complete control system and energy source and no need for external connections. Honda R & D Co. Ltd. developed the only example of that kind of humanoid robot having two legs and two arms [22]. But it is reported that the maximum battery operation time is only 15 minutes, not enough to conduct the practical activities at all. Therefore, developing a highly efficient joint driving mechanism and devising a control method for dynamic biped walking to use that mechanism would be an important step towards the realization of a practical bipedal humanoid robot.

There are two typical researches on efficient biped walking. One by McGeer, et al. studied passive walking using the effect of gravity, with a biped walking robot having no source of energy other than a downhill slope [23]. The role of the hip-mass is sometimes neglected in the work. In the other research, Kajita, et al. researched the potential energy conserving orbit of a bipedal walking robot [24]. They had their viewpoints upon a dynamics of an inverted pendulum model. But it presupposed the mass of lower limbs is small enough to ignore compared with the mass of torso. As mentioned above, no research had been done on developing a joint driving mechanism which varies its stiffness over a broad range imitating that of humans. Nor had work done to devise a control method for efficient biped walking by a human-size biped walking robot while considering the mass of the lower limbs.

From foregoing considerations, the purpose of this study was to develop a biped walking robot WL-14(Waseda Leg-No.14), which has antagonistic driven joints using a rotary type nonlinear spring mechanism, to devise a walking control method that controls the antagonistic driven joints and to realize an antagonistic driven type dynamic biped walking varying the stiffness of joints.

The authors performed walking experiments using the biped walking robot WL-14 and the control method. As a result, antagonistic driven type dynamic biped walking by varying the joint stiffness was realized. At the same time, it was confirmed that the energy consumption of the antagonistic driven joint at the swing phase was decreased to 75% of that in the case when the joint stiffness is not varied actively.

2. Biped walking robot WL-14

2.1 Machine model

The bipedal walking robot is shown in **Fig. 3**. The total weight of the machine model is 108.7 kg and the height in a static straight standing trunk position is 1.50 m. It is built mainly of extra-super-duralumin and GIGAS™ (YKK CORPORATION) for the pelvis. An assembly drawing of

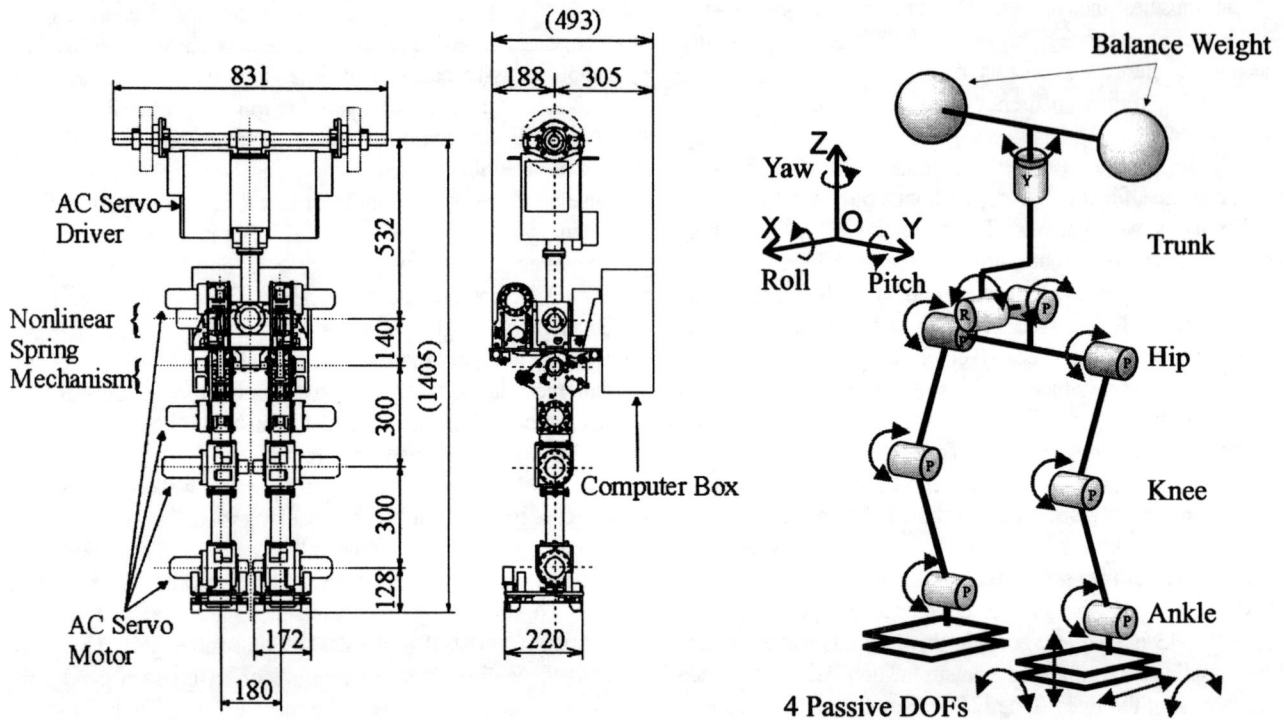

Fig. 4 Assembly drawing of WL-14

Fig. 5 Link structure of WL-14

WL-14 is illustrated in **Fig. 4** and the link structure and assignment of active DOF are illustrated in **Fig. 5**. The total active DOF of this machine is 9 DOF, consisting of 6 DOF for the pitch axis of the lower limbs and 3 DOF for the trunk (one each of the pitch axis, roll axis and yaw axis of the trunk). The trunk mechanism is able to compensate for the three-axis moment caused by the lower limbs motion. The one DOF of each hip joint is driven antagonistically by reeling two wires fixed to a joint in one end through a nonlinear spring mechanism using two AC servo motors. The other joint is driven directly through a reduction gear using one AC servo motor. Incidentally, the authors employed the antagonistic driven joints only for the hip joints for minimizing the inertial moment of the lower limbs and swinging the lower limb using inertial energy and potential energy.

For the biped walking control method to adapt to an unknown uneven surface and to increase the robustness under a program control, which is the same control method as WL-12RVII [15], two units of foot system WAF-3R (Waseda Anthropomorphic Foot-No.3 Refined) are installed on each foot. Then each foot has 4 passive DOF. The foot mechanism has two remarkable functions. One is the compensation mechanism of the model deviation (deviation between the simulated model of the walking system and the real world walking system). The other is the detection mechanism of model deviation.

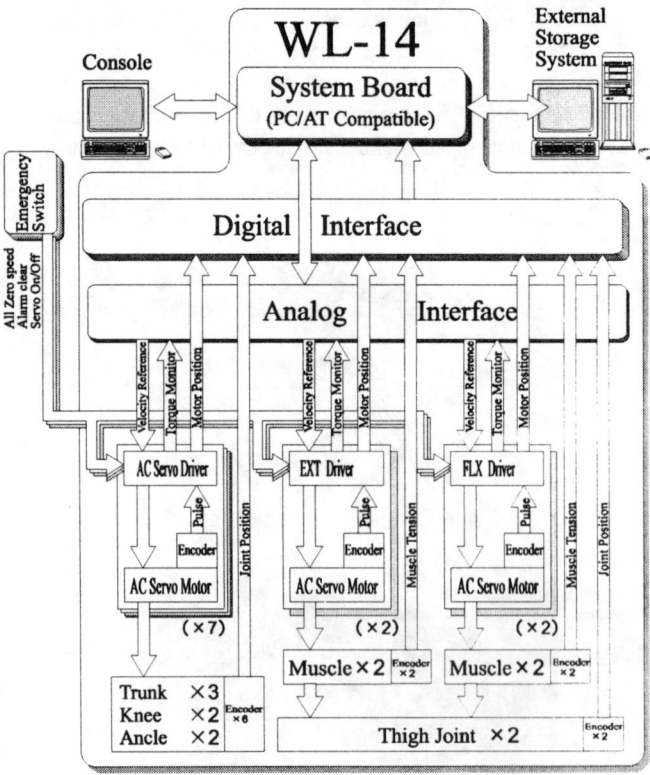

Fig. 6 Computer system structure of WL-14

The control system structure is illustrated in **Fig. 6**. The computer system which controls the machine model is mounted on the back of the waist and the servo driver modules are mounted on the upper part of the trunk. Thus the only external connection is to an outside electric power source. The computer system is controlled by a PC/AT compatible CPU board having an Intel Pentium processor 166 MHz. Four I/O boards (ISA bus) are installed in the computer system, one D/A conversion board (16ch, resolution 12bit at ±10V, settling time 3µs), one A/D conversion board (32ch, resolution 12bit at ±10V, settling time 6µs), one counter board (24ch, 24bit) and one Ethernet board. The 11 AC servo motors are centrally controlled by the computer system at the servo cycle of 1 ms.

2.2 Mechanism of antagonistic driven joints

The mechanism of antagonistic driven joints installed on the hip joints is composed of one free joint, two nonlinear spring mechanisms and two AC servo motors with reduction gears and driving pulleys to reel driving wires. Thus, the hip joints are driven antagonistically by reeling two wires fixed to a joint in one end through a nonlinear spring mechanism using two AC servo motors (**Fig. 7**). To use smaller pulleys for antagonistic driven joints and to allow more freedom of design, compared with a steel wire of the same strength, aramid fiber kevlar tape is used as driving wires for the mechanism.

The authors employed a rotary type nonlinear spring mechanism (**Fig. 8**) for the antagonistic driven system, which enables its spring coefficient to change from zero to infinity (assuming that the spring coefficient of a wire is infinity). The characteristic of the mechanism is that the positions of the free pulleys of the upper and lower mechanism are determined by the torque balance. Therefore, the authors can presume the driving wire tension by measuring the angle of the rotation of the rotary part (the spring angle). The movable spring angle range is 90 degree.

We developed the nonlinear spring mechanism under the following design conditions.
(1) In view of the control of the swinging leg landing, the stroke of the driving wire was set so that the time of the total spring coefficient going from zero to maximum is less than 150[ms] during active driving.
(2) At maximum tension, the total spring coefficient of the nonlinear spring mechanism becomes maximum, which is the value of the spring coefficient of the driving wire.
(3) When the flexion side muscle creates maximum tension antagonizing the extension side muscle, the total spring coefficient of the driving wire of the extension side is more than half of the maximum spring coefficient. (This is to prevent the joint stiffness from varying greatly in the direction of rotation at the time of antagonism.)

The following describes the problems of the antagonistic driven joint of WL-13 and the improvements on that in WL-14.

First, the initial linear spring tension of the nonlinear spring mechanism changed when the joints of WL-13 were driven at high tensions. To solve the problem, the authors installed a detection mechanism of the linear spring length and improved the fixation of the shaft adjusting the initial tension. Thus, the initial tension is constant for any walking experiments.

Second, because of the driving wire reeled one above

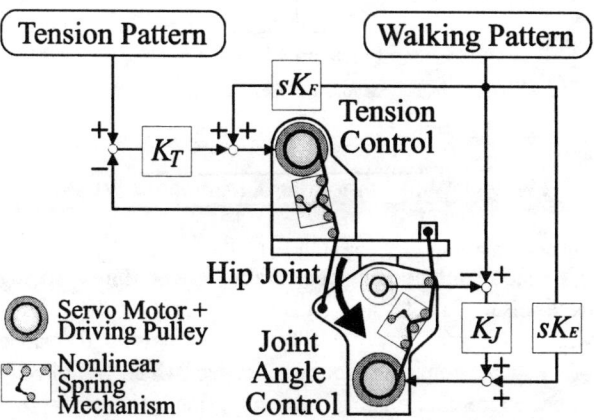

Fig. 7 Outline of the antagonistic driven joint system using non-linear spring mechanism

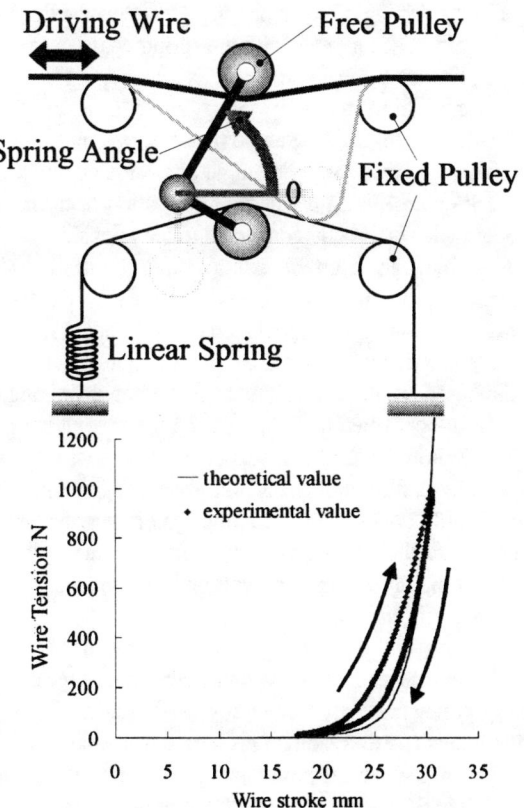

Fig. 8 Rotary type nonlinear spring mechanism

another by the driving pulley, a hysteresis occurred when the joints of WL-13 were driven at lower tensions. For this problem, it was difficult to make the positioning of the joint accurate enough to realize a dynamic biped walking using the control method of the antagonistic driven joint based on a feedback control. To solve the problem in WL-14, the driving wire was shortened to the minimum length necessary to drive the joint, so that the hysteresis was reduced and positioning became accurate enough for dynamic walking.

3. Walking control method

3.1 Outline of walking control method

The walking control method consists of four stabilization controls as follows, installing all algorithms of a biped walking control that the authors have already proposed in documents [9]-[17].
(1) Model based walking control (ZMP and yaw axis moment control)
(2) Robust walking using the compensation mechanism of the model deviation
(3) Model deviation compensative control
(4) Real-time control of ZMP and yaw axis moment (external force or torque compensative control)

The walking control method of WL-14 includes the two stabilization control of (1) (**Fig. 9**) and (2) above. The following describes the constitution.
<Before Walking>
I: Creating a preset walking pattern using the algorithm computing the trunk compensation trajectory
II: Creating a variable tension pattern for the antagonistic driven joints
<While Walking>
III: Program-controlling the biped walking robot to follow up the preset walking pattern during walking
IV: Program-controlling the antagonistic driven joint to follow up the variable tension pattern

To create the preset walking pattern of the biped walking robot, we employed the biped walking control method compensating for the three-axis moment by trunk motion as previously proposed by the authors [13].

3.2 Control method of antagonistic driven joint

The control method of the antagonistic driven joints for WL-14 is based on that of WL-13 which consists of two subroutines: a subroutine to control the joint angle and another subroutine to control the tension (stiffness) of the driving wire. The joints are thus cooperatively (antagonistically) driven. The antagonistic driven joints of WL-13 were driven only using the feedback control. But the control method for WL-14 uses a feedforward control to add the joint angular velocity references to the control input together with the feedback control above (**Fig. 7**). The

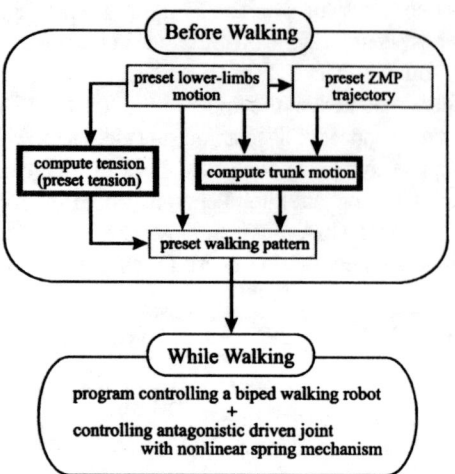

Fig. 9 Out-line of the control method for WL-14

following summarizes this control method.

1) Tension control using the rotary type nonlinear spring mechanism: The tension (stiffness) control of the driving wire is conducted by a feedback control using the angle information from the rotary encoder which is directly connected to the rotating section of the nonlinear spring mechanism and controlling the motor velocity.

Using the rotary type nonlinear spring mechanism, there is a nonlinear relation between the spring angle and the rotation angle of the driving motor. Therefore, by adjusting the tension control gain in accordance with the spring angle to cancel out the non-linearity, tension control is made possible over a broad range from the low spring angle (relaxed state of muscles) to the high spring angle (tense state of muscles).

2) Joint angle control: Joint angle control is conducted by a feedback control using the angle information from the rotary encoder which is directly connected to the joint shaft.

3) Feedforward control using joint angular velocity reference: When the joints are driven only by the feedback control of (1) and (2), the deviation of the joint angle becomes too large in dynamic walking, which makes it difficult to walk continuously. In this control method, the biped walking robot is program-controlled to follow up the preset walking pattern during walking, which makes the joint motion known. Therefore, controllability is improved by combining a feedforward control to add the joint angular velocity references to the angular velocity control inputs to the servo drivers of the driving motors with both of the tension control and joint angle control.

However, at the antagonistic driven joint in this machine model, the one end of the driving wire is connected to the driving pulley and the other end to the thigh (extension side) or the pelvis (flexion side). This structure causes a nonlinear variance in the virtual reduction ratio between the rotation angle of the driving motors and the joint angle. Therefore, the feedforward control inputs are calculated from the joint angular velocity references using the virtual reduction ratio corresponding to the joint angle.

4) Joint angle control for antagonistic joint drive: The joint angle is controlled by coordinating the two antagonistic driving motors by combining the control methods described above 1)-3). For this, a switch of the two feedback control will be conducted by varying the feedback control gain. The switch occurs when the deviation direction of the joint angle changes. Thus the motor on the drive direction side from the joint will conduct joint angle control following up the preset walking pattern. At the same time the other motor will conduct tension control following up the tension pattern.

However, frequent switches of the two control methods due to unknown external disturbances will destabilize the joint angle control. Therefore, unnecessary switches were prevented by providing hysteresis characteristics (as shown in **Fig. 10**) to the joint angle deviation when making decisions for the switch.

Also, the switch in the condition where is a large deviation between the spring angle of the joint angle control side immediately prior to changing to the tension control side and the spring angle reference will destabilize it. Because this kind of switch will cause excessive step-wise changes in the spring angle, that will result in step-wise disturbances in the antagonistic muscle on the opposite side. For this, the maximum spring angle deviation value is set for the switch and the switch does not take place when the spring deviation exceeds this value.

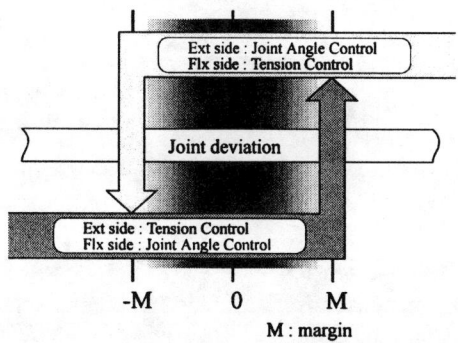

Fig. 10 Joint angle margin for switching control

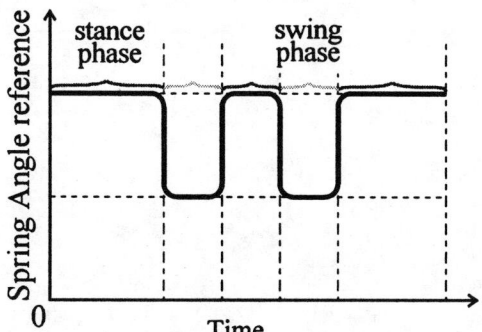

Fig. 11 Variable spring angle reference

5) Method for creating variable tension pattern: In this control method, as the first step for realizing a dynamic biped walking by varying the joint stiffness, the authors introduced a variable tension pattern in addition to a conventional preset walking pattern. As shown in **Fig. 11**, the method for creating a variable tension pattern is described below. First, certain values of the spring angle are set for stance and swing phases each other. The low spring angle reference is set in the swing phase where a high positioning accuracy in the joint angle and a high joint stiffness are not needed as in the stance phase. Second, high frequency elements of the initial reference that changes in steps are reduced by filtering. Also, stance-swing discriminating flags are added to make the judgment on stance-swing phase differences from the pattern.

4. Walking experiments

In order to actually prove the effectiveness of the biped walking system we developed and the walking control method we proposed, we performed walking experiments with the biped walking robot WL-14, using newly developed control software to antagonistically drive the joints.
(1) Walking experiment using constant tension pattern
(2) Walking experiment using variable tension pattern

The arrangement and the result of each walking experiment are described below.

4.1 Walking experiment using constant tension pattern

The arrangement of the walking experiment using a constant tension pattern is as follows.
(1) Creating a preset walking pattern.
(2) Program-controlling the biped walking robot using the preset walking pattern and the control software during walking. The tension (spring angle) references are constant through the whole range of walking.

As a result, we realized antagonistic driven type complete dynamic biped walking behaviors using the preset walking patterns at the parameters of 1.28-16.0 s/step, 0.10-0.20 m/step. In addition a complete dynamic biped walking backward was also realized.

4.2 Walking experiment using variable tension pattern

The arrangement of the walking experiment using a variable tension pattern is as follows.
(1) Creating a preset walking pattern and a variable tension pattern.
(2) Program-controlling the biped walking robot using the preset walking pattern, the variable tension pattern and the control software during walking.
(3) Acquiring the torque values which the driving motors generate using the torque monitoring function of AC servo

drivers to calculate the energy consumption of the antagonistic driven joints.

As a result, we realized antagonistic driven type dynamic biped walking varying the stiffness of the antagonistic driven joints using the preset walking pattern at the parameters of 1.28 s/step, 0.15 m/step, 6 step and a variable tension pattern.

The hip joint (antagonistic driven joint) angle responses during walking varying the joint stiffness are shown in **Fig. 12**. Even in the swing phase the follow-up characteristics are not much deteriorated, so the effectiveness of the combined feedforward control is confirmed.

The stiffness responses of driving wire are shown in **Fig. 13**. The stiffness values are divided by those of the kevlar tapes used as the driving wires. The stiffness value of the kevlar tape is the maximum stiffness value of the nonlinear spring mechanism.

The comparison of the consumption power of the antagonistic driven joints between using a constant tension pattern and a variable one is shown in **Fig. 14**. From this result it was confirmed that the energy consumption of the antagonistic driven joint at the swing phase when using the variable tension pattern was decreased to 75% of that when using the constant tension pattern (i.e. the joint stiffness is not varied actively). In this case, the energy consumption is calculated from the torque values and the angular velocity responses that the driving motors generated.

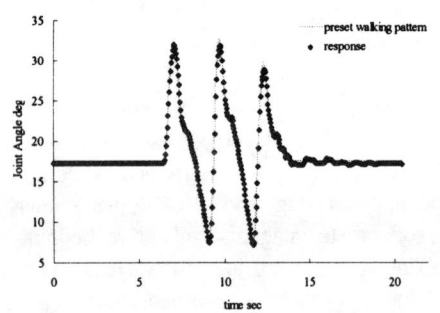
(a) Right hip joint

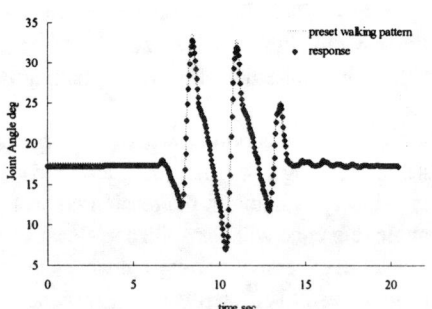
(b) Left hip joint

Fig. 12 Hip joint angle responses at walking experiment using variable tension pattern
Walking pattern: number of steps: 6, step length: 0.15m/step, step time: 1.28 sec/step

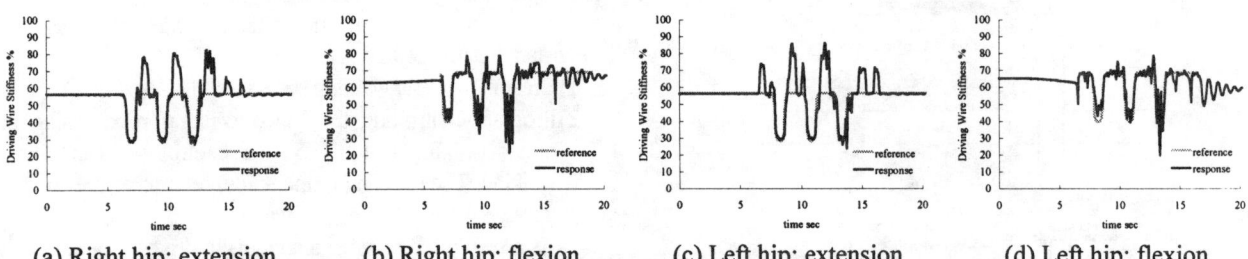

(a) Right hip: extension (b) Right hip: flexion (c) Left hip: extension (d) Left hip: flexion

Fig. 13 Driving wire stiffness responses at walking experiment using variable tension pattern
Walking pattern: number of steps: 6, step length: 0.15 m/step, step time: 1.28 sec/step
Maximum driving wire stiffness (100%): Extension side: 591 N/mm, Flexion side: 788 N/mm

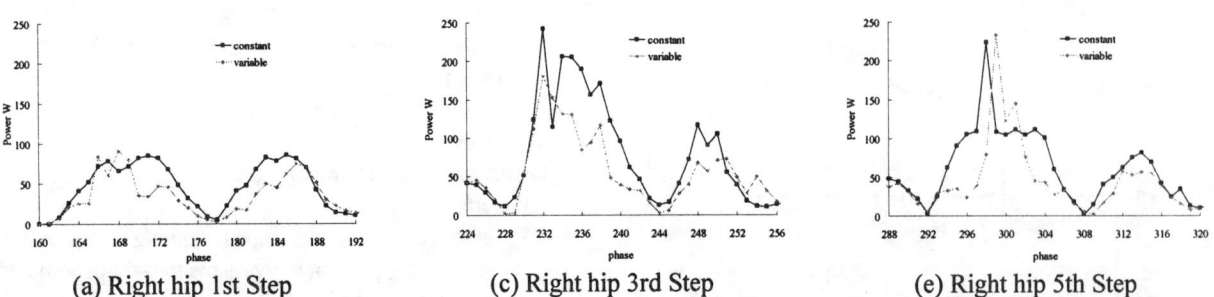

(a) Right hip 1st Step (c) Right hip 3rd Step (e) Right hip 5th Step

Fig. 14 Power of the hip joint motors in swing phase at walking experiment using variable tension pattern
Walking pattern: number of steps: 6, step length: 0.15 m/step, step time: 1.28 sec/step, phase time: 0.04 sec/phase

5. Conclusions

The purpose of this study was to develop a biped walking robot that has antagonistic driven joints using a rotary type nonlinear spring mechanism, to devise a walking control method that controls the antagonistic driven joints, and to realize antagonistic driven type dynamic biped walking varying the stiffness of joints.

The authors first developed a biped walking robot WL-14 having the antagonistic driven joints, and proposed a walking control method for antagonistic driven type dynamic biped walking varying the joint stiffness.

Next, we performed walking experiments using the biped walking robot and the control method. As a result, antagonistic driven type dynamic biped walking varying the stiffness of the joints was realized. At the same time, it was confirmed that the energy consumption of the antagonistic driven joint at the swing phase was decreased to 75% of that in the case that the joint stiffness is not varied actively.

In conclusion, the effectiveness of the walking control method proposed in this paper and the developed walking system have been experimentally supported.

For the next step of this study, we consider to develop a simulation software for the passive (zero muscle tension) drive of the antagonistic driven joints, and to install the control method of the realtime compensation for ZMP and yaw-axis moment in order to walk actually driving the antagonistic driven joints passively. In this way, we would like to realize antagonistic driven type dynamic biped walking by driving the joints in the lower range of tension and joint stiffness and efficiently using the inertial and potential energies of the lower limbs effectively.

Acknowledgments

This study has been conducted as a part of the project: Humanoid at HUREL (HUmanoid REsearch Laboratory), Advanced Research Institute for Science and Engineering, Waseda University. The authors would like to thank ATR, NAMCO Ltd., Nissan Motor Co., Ltd. and YASKAWA ELECTRIC Corp. for their cooperation in this study. A part of this study was done by the Japanese Grant-in-Aide for Science Research (No.07405012) and NEDO (New Energy and Industrial Technology Development Organization). The authors would also like to thank AMP Incorporated, Intel Corporation, Harmonic Drive Systems, Inc., Matsushita Electric Industrial Co., Ltd., NIPPON OIL COMPANY, LIMITED, OKINO Industries, Ltd., SANKYO FACTORY CORPORATION, TOKYO DEN-ON Co., Ltd., and YKK CORPORATION for supporting us in developing the hardware for a biped walking robot in the process of this study.

References

[1] L.Jalics, H.Hemami and Y.F.Zeng "Pattern Generation Using Coupled Oscillators for Robotic and biorobotic Adaptive Periodic Movement", *Proc. of the 1997 ICRA*, pp.179-184, 1997.

[2] J.Pratt, P.Dilworth and G.Pratt "Virtual Model Control of a Bipedal Walking Robot", *Proc. of the 1997 ICRA*, pp.193-198, 1997.

[3] K.Y.Yi "Locomotion of a biped robot with compliant ankle joints", *Proc. of the 1997 ICRA*, pp.199-204, 1997.

[4] B.Thuilot, A.Goswami and B.Espiau "Bifurcation and Chaos in a Simple Passive Bipedal Gait", *Proc. of the 1997 ICRA*, pp.792-798, 1997.

[5] D.G.Caldwell G.A.Medrano-Cerda and C.J.Bowler "Investigation of Bipedal Robot Locomotion using Pneumatic Muscle Actuators, *Proc. of the 1997 ICRA*, pp.799-804, 1997.

[6] A.Kun and W.T.Miller, III "Adaptive Dynamic Balance of a Biped Robot Using Neural Networks", *Proc. of the 1996 ICRA*, pp.240-245, 1996.

[7] S.Kajita and K.Tani "Adaptive Gait Control of a Biped Robot based on Realtime Sensing of the Ground Profile", Proc. of the 1996 ICRA, pp.570-577, 1996.

[8] A.Takanishi, M.Ishida, Y.Yamazaki and I.Kato "The Realization of Dynamic Walking by the Biped Walking Robot WL-10RD", *Proc. of the 1985 ICRA*, pp.459-466, 1985.

[9] A.Takanishi, Y.Egusa, M.Tochizawa, T.Takebayashi and I.Kato "Realization of Dynamic Walking Stabilized with Trunk Motion", *Proc. of ROMANCY7*, pp.68-79, 1988.

[10] A.Takanishi, L.Hun-ok, M.Tsuuda, and I.Kato "Realization of Dynamic Biped Walking Stabilized by Trunk Motion on a Sagitally Uneven Surface", *Proc. of IROS'90*, pp.323-329, 1990.

[11] A.Takanishi, M.Tochizawa, T.Takeya, H.Karaki and I.Kato "Realization of Dynamic Walking Stabilized by Trunk Motion Under Known External Force", (In Japanese), *Proc. of the 4th Symposium on Intelligent Mobile Robot*, pp.15-20, 1988.

[12] A.Takanishi, T.Takeya, H.Karaki M.Kumeta and I.Kato "A Control Method for Dynamic Walking under Unknown External Force", *Proc. of IROS'90*, pp.795-801, 1990.

[13] J.Yamaguchi, A.Takanishi and I.Kato "Development of a Biped Walking Robot Compensating for Three-Axis Moment by Trunk Motion", *Proc. of IROS'93*, pp.561-566, 1993.

[14] A.Takanishi, J.Yamaguchi, M.Iwata, S.Kasai, and T.Mizobuchi, "Study on Dynamic Turning of Biped Walking Robot", *Proceedings of The 72nd JSME Spring Annual Meeting*, pp.323-324, 1995.

[15] J.Yamaguchi, A.Takanishi and I.Kato, "Development of a Biped Walking Robot Adapting to a Horizontally Uneven Surface", *Proc. of the IROS'94*, pp.1156-1163, 1994.

[16] J.Yamaguchi, A.Takanishi, and I.Kato, "Experimental Development of a Foot Mechanism with Shock Absorbing Material for Acquisition of Landing Surface Position Information and Stabilization of Dynamic Biped Walking", *Proc. of the 1995 ICRA*, pp.2892-2899, 1995.

[17] J.Yamaguchi, A.Takanishi, and I.Kato, "Development of a Dynamic Walking System for Humanoid -Development of a Biped Walking Robot Adapting to the Humans' Living Floor-", *Proc. of the 1996 ICRA*, pp.232-239, 1996.

[18] H.Kusumoto, H.J.Park, M.Yoshida and K.Akazawa, "Simultaneous modulation of force generation and mechanical property of muscle in voluntary contraction", (in Japanese), *Biomechanisms 12*, pp.211-220, University of Tokyo Press, 1994.

[19] K.Akazawa, J.W.Aldridge, J.D.Steeves and R.B.Stein, "Modulation of stretch reflexes during Locomotion in the mesencephalic cat", *J.Physiol(1982)*, 329, pp.553-567, 1982.

[20] J.Yamaguchi, Y.Chujoh, D.Nishino, S.Inoue, G.Jin and A.Takanishi, "Development of an Anthropomorphic Biped Walking Robot -An Antagonistic Driven Joint Using Non-linear Spring Mechanism-", (in Japanese), *Proceedings of The 13th Annual Conference of Robotics Society of Japan*, pp.211-212, 1995.

[21] J.Yamaguchi and A.Takanishi "Development of a Biped Walking Robot Having Antagonistic Driven Joints Using Nonlinear Spring Mechanism", *Proc. of the 1997 ICRA*, pp.185-192, 1997.

[22] K.Hirai "Current and Future Perspective of Honda Humanoid Robot", *Proc. of the IROS'97*, pp.500-508, 1997.

[23] T.McGeer, "Passive Walking with Knees", *Proc of 1990 ICRA*, pp.1640-1645, 1990.

[24] S.Kajita and T.Yamaura and A.Kobayashi, "Dynamic Walk Control of a Biped Robot along a Potential Energy Conserving Orbit", *IEEE Trans. on R&A*, Vol.8, No.4, August, pp.431-438, 1992.

Robust Biped Walking with Active Interaction Control between Foot and Ground

Yasutaka Fujimoto
Dept. of System Design Eng.
Keio University
Yokohama 223 JAPAN
fujimoto@kawalab.dnj.ynu.ac.jp

Satoshi Obata   Atsuo Kawamura
Dept. of Elec. & Comp. Eng.
Yokohama National University
Yokohama 240 JAPAN
kawamura@kawalab.dnj.ynu.ac.jp

Abstract

This paper describes a biped walking control system based on the reactive force interaction control at the foothold. 1) robust control of reactive force/torque interaction at the foothold based on Cartesian space motion controller. 2) the posture control considering the physical constraints of the reactive force/torque at the foothold by quadratic programming. The proposed approach realizes the robust biped locomotion because the environmental interaction is directly controlled. The control is applied to the 20 axes simulation model, and the stable biped locomotion is realized even if unknown small slope exists. The stable attitude control is confirmed by 14-axis biped robot experiments.

1 Introduction

A number of biped walking systems have been proposed in the previous works[1]–[14]. The control objective of the biped walking is to carry the body of the robot in use of reactive force of foot with foot placement planning. The unknown disturbances are, however, exist in the terrain and also the reactive force is subject to nonlinear physical constraints such as *Zero Moment Point* conditions and friction conditions. Thus the conventional control systems calculate trajectories of joint angle or joint torque so as to approximately satisfy the stable contact condition[1]–[3]. The approximation, however, yields lack of walking robustness. The whole dynamic equation of the robot and the contact condition is considered in the generation of joints references in [3], but it is off-line type planning due to the complexity of dynamics of the biped robot. To improve the walking robustness, an adaptive method is introduced[7].

In this paper, robust approach is adopted. A hierarchical control system based on the robust reactive force control on the foothold and the force distribution system is proposed in order to improve the walking robustness. The physical constraints of the contact force on the foothold are precisely considered in the force distribution system. Then the robust force controller of the support foot locally suppresses unknown disturbances on the terrain.

First in this paper, the plant formulation are shown in the section 2. The control strategy with hierarchical system are described in the section 3. The proposed control system is applied to the 20 axes simulation model and the results of stable legged locomotion are shown in the section 4. The experimental results are shown in the section 5. The section 6 concludes this paper.

2 Model of Biped Robot

Consider the basic equation of the biped robot

$$\begin{bmatrix} H_{11} & H_{12} & H_{13} \\ H_{21} & H_{22} & H_{23} \\ H_{31} & H_{32} & H_{33} \end{bmatrix} \begin{bmatrix} \ddot{p}_0 \\ \dot{\omega}_0 \\ \ddot{q} \end{bmatrix} + \begin{bmatrix} b_1 \\ b_2 \\ b_3 \end{bmatrix} = \begin{bmatrix} 0 \\ 0 \\ \tau \end{bmatrix} + \begin{bmatrix} I_3 & o \\ [x_R \times] & I_3 \\ J_{R1}^T & J_{R2}^T \end{bmatrix} \begin{bmatrix} f_R \\ n_R \end{bmatrix} + \begin{bmatrix} I_3 & o \\ [x_L \times] & I_3 \\ J_{L1}^T & J_{L2}^T \end{bmatrix} \begin{bmatrix} f_L \\ n_L \end{bmatrix}$$ (1)

where

p_0 : 3×1 vector specifying position of the body
ω_0 : 3×1 vector specifying angular velocity of the body
q : $N \times 1$ vector specifying joint angle
τ : $N \times 1$ torque vector generated by actuator
f_L, f_R : 3×1 vector of reactive force at the center of left or right foot
n_L, n_R : 3×1 vector of reactive torque at the center of left or right foot

H_{ij} and b_i are inertia matrices and a non-linear term, respectively. J_{Ri}^T are transposal Jacobian matrices which transform reactive force and torque at the center of right foot into torque at joint coordinates. J_{Li}^T are those about left foot. x_R is a tip position of the right leg with respect to the origin of p_0 (see Fig. 1). x_L is that of the left leg. $[a \times]$ is a matrix representing a cross product, and I_n is a $n \times n$ identity matrix.

In addition, there are physical constraints on reactive force and torque. Let $f_R^T = [f_{Rx}\ f_{Ry}\ f_{Rz}]$, $n_R^T = [n_{Rx}\ n_{Ry}\ n_{Rz}]$, $f_L^T = [f_{Lx}\ f_{Ly}\ f_{Lz}]$, and $n_L^T = [n_{Lx}\ n_{Ly}\ n_{Lz}]$. The contact force and torque between foot and ground must satisfy the pressure condition, the friction condition, and the *Zero Moment Point* (ZMP) condition. First of all the normal component of the reactive force on the ground plane is not attractive but repulsive, which yield the following non-negative conditions.

$$f_{Rz} \geq 0, \quad f_{Lz} \geq 0 \quad (2)$$

The friction force, i. e., the tangent component of the reactive force on the ground plane always exists within the friction cone.

$$\sqrt{f_{Rx}^2 + f_{Ry}^2} \leq \mu f_{Rz} \quad \sqrt{f_{Lx}^2 + f_{Ly}^2} \leq \mu f_{Lz} \quad (3)$$
$$|n_{Rz}| \leq \mu' f_{Rz} \quad |n_{Lz}| \leq \mu' f_{Lz} \quad (4)$$

where μ and μ' are friction coefficients. It is possible to break out slips at the contact points when the equality in the (3)–(4) is realized.

The tangent component of the reactive torque at the center of the foot on the ground plane is also limited due to finiteness of the contact area.

$$|n_{Rx}| \leq d_y f_{Rz} \quad |n_{Lx}| \leq d_y f_{Lz} \quad (5)$$
$$|n_{Ry}| \leq d_x f_{Rz} \quad |n_{Ly}| \leq d_x f_{Lz} \quad (6)$$

where d_x and d_y are halves of length and width of the foot, respectively. (5)–(6) are equivalent to the *Zero Moment Point* conditions.

The first and second row of (1) represent the parallel and rotational motion of the body of the robot, respectively. The third row of (1) corresponds to the motion of the joint. Regarding the reactive force and torque f_R, f_L, n_R, n_L as an indirect control input, the position and attitude of the body p_0, ω_0 become controllable.

3 Hierarchical Control System

The control objective of the biped walking is to carry the body of the robot in use of reactive force of

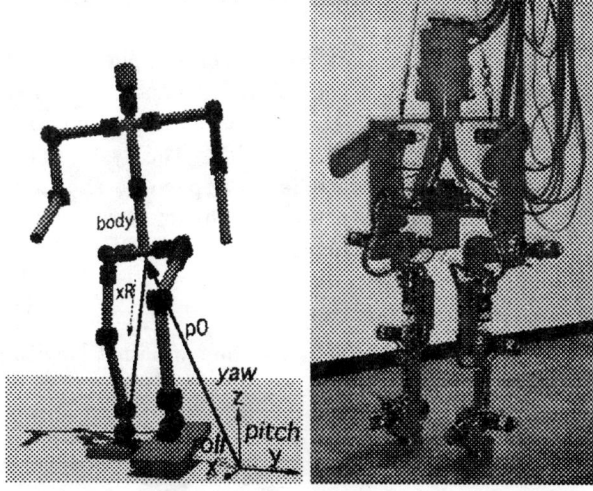

Figure 1: 20-axis simulation model and 14-axis real robot.

foot with foot placement planning. However, the relation between the input torque of the actuator τ and reactive force and torque f_{*i}, n_{*i} is hard to solve because the reactive force and torque depend on the non-linear static characteristics (2)–(6) and the unknown dynamic ones. Thus the conventional control systems calculate trajectory of joint angle or joint torque so as to approximately satisfy static characteristics (2)–(6). The approximation, however, yields lack of walking robustness.

In this paper, a control system based on the reactive force control and the force distribution system is proposed in order to improve the walking robustness, in which the physical constraints of the contact force are precisely considered. Fig. 2 shows the overview of the control system.

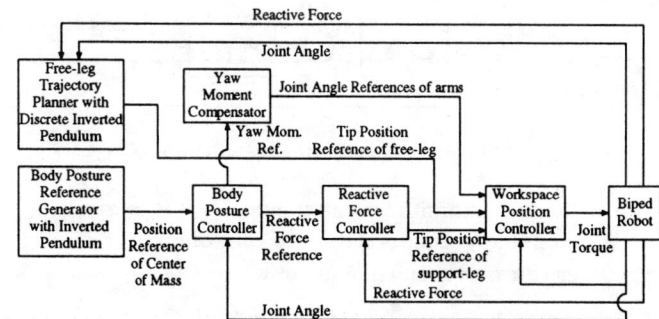

Figure 2: Biped walking control system.

3.1 Reactive Force Controller

The hybrid position/force control is applied to each leg, in which the force control mode is activated when the leg is in the support phase, otherwise the position control mode is activated. First in the system, the workspace position controller is applied as shown in Fig. 2 which consists of the H_∞ robust servo control with inertia compensation[16]

$$\boldsymbol{\tau} = \boldsymbol{H}_n(\boldsymbol{q})\boldsymbol{C}_q(s)(\boldsymbol{q}_r - \boldsymbol{q}) \quad (7)$$

and the inverse kinematics by the simplified Newton method

$$\boldsymbol{q}_r(t+T_s) = \boldsymbol{q}_r(t) + \begin{bmatrix} \boldsymbol{J}_R \\ \boldsymbol{J}_L \end{bmatrix}^{-1} \begin{bmatrix} \boldsymbol{g}_{Rx}(\boldsymbol{q}_r(t)) - \boldsymbol{x}_{Rr} \\ \boldsymbol{g}_{Lx}(\boldsymbol{q}_r(t)) - \boldsymbol{x}_{Lr} \end{bmatrix} (8)$$

where $\boldsymbol{q}_r$ is the reference of joint angles, $\boldsymbol{H}_n(\boldsymbol{q})$ is the nominal inertia matrix, $\boldsymbol{C}_q(s)$ is the H_∞ controller, T_s is the sampling period, $\boldsymbol{g}_{Rx}(\cdot)$ is the function of the right foot position given joint angles, and $\boldsymbol{x}_{Rr}$ is the position reference of the right foot.

Assuming the transfer characteristics of the Cartesian position control system is almost unity by the robust controller, the hybrid position/force controller is easily applied to the upper layer of the system. The following discussion in this section is in a case of right foot support. The force controller is simply given by

$$\boldsymbol{x}_{Rr} = \boldsymbol{C}_f(s)(\boldsymbol{f}_{Rr} - \boldsymbol{f}_R) \quad (9)$$

where $\boldsymbol{f}_R$ and $\boldsymbol{f}_{Rr}$ are the 6×1 force/torque vector and its reference on the right foot (the support foot), respectively. The configuration of the force control system is shown in Fig. 3. The plant system $\boldsymbol{P}_f(s)$ includes the dynamics of the environment and the Cartesian position control system, whose control input is $\boldsymbol{x}_{Rr}$, the Cartesian position reference of the support foot.

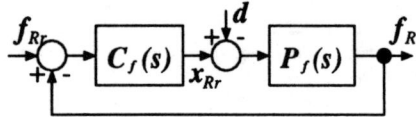

Figure 3: Force control system.

When the nominal plant model $\boldsymbol{P}_{fn}(s)$ is given, a very simple parameterization of the robust servo controller can be obtained as follows

$$\boldsymbol{C}_f(s) = \boldsymbol{P}_{fn}(s)^{-1}(\boldsymbol{I} - \boldsymbol{Q}_f(s))^{-1}\boldsymbol{Q}_f(s) \quad (10)$$

Here, $\boldsymbol{Q}_f(s)$ is the free parameter representing a complementary sensitivity function and is subject to $\boldsymbol{Q}_f(s)$, $\boldsymbol{P}_{fn}(s)^{-1}\boldsymbol{Q}_f(s)$, $\boldsymbol{P}_{fn}(s)(\boldsymbol{I}-\boldsymbol{Q}_f(s))\boldsymbol{d}(s)$, $(\boldsymbol{I}-\boldsymbol{Q}_f(s))\boldsymbol{r}(s) \in \boldsymbol{RH}_\infty$. Here, $\boldsymbol{RH}_\infty$ expresses a set of proper and stable transfer function matrices. These conditions are obtained from the internal stability and the output regulation.

In a case of the force control, the free parameter $\boldsymbol{Q}_f(s)$ and the nominal plant model $\boldsymbol{P}_{fn}(s)$ can be set $\boldsymbol{Q}_f(s) = \text{diag}\{Q_{f1}(s), \ldots, Q_{f6}(s)\}$ and $\boldsymbol{P}_{fn}(s) = \text{diag}\{P_{fn1}(s), \ldots, P_{fn6}(s)\}$ where

$$Q_{fi}(s) = \frac{3\tau_i^2 s^2 + 3\tau_i s + 1}{(\tau_i s + 1)^3}, \quad P_{fni}(s) = m_i s^2 + b_i s + k_i \quad (11)$$

$i = 1, 2, \ldots, 6$. Thus, the robust force controller is obtained from (10) as $\boldsymbol{C}_f(s) = \text{diag}\{C_{f1}(s), \ldots, C_{f6}(s)\}$ where

$$C_{fi}(s) = \frac{3\tau_i^2 s^2 + 3\tau_i s + 1}{\tau_i^3 s^3 (m_i s^2 + b_i s + k_i)} \quad (12)$$

3.2 Body Posture Controller

In this section, the method to control the parallel and rotational motion of the body is presented. The first row and second row of (1) can be transformed into the equation of parallel motion of the *Center Of Mass* (COM) and rotational one of the body.

$$\boldsymbol{M}\ddot{\boldsymbol{x}} + \boldsymbol{d} = \boldsymbol{u}_c = \boldsymbol{K}\boldsymbol{u} \quad (13)$$

where

$$\boldsymbol{x} = \begin{bmatrix} \boldsymbol{p}_c & \boldsymbol{\theta}_0 \end{bmatrix}^T \quad (14)$$

$$\boldsymbol{u} = \begin{bmatrix} \boldsymbol{u}_R & \boldsymbol{u}_L \end{bmatrix}^T \quad (15)$$

$$\boldsymbol{u}_R = \begin{bmatrix} \boldsymbol{f}_R & \boldsymbol{n}_R \end{bmatrix}^T \quad (16)$$

$$\boldsymbol{u}_L = \begin{bmatrix} \boldsymbol{f}_L & \boldsymbol{n}_L \end{bmatrix}^T \quad (17)$$

$$\boldsymbol{M} = \begin{bmatrix} m\boldsymbol{I}_3 & \boldsymbol{H}_{c12} \\ \boldsymbol{H}_{c21} & \boldsymbol{H}_{22} \end{bmatrix} \quad (18)$$

$$\boldsymbol{K} = \begin{bmatrix} \boldsymbol{I}_3 & \boldsymbol{0} & \boldsymbol{I}_3 & \boldsymbol{0} \\ [\boldsymbol{x}_{Rc} \times] & \boldsymbol{I}_3 & [\boldsymbol{x}_{Lc} \times] & \boldsymbol{I}_3 \end{bmatrix} \quad (19)$$

$$\boldsymbol{d} = \begin{bmatrix} mg & b_2 \end{bmatrix}^T \quad (20)$$

and $\boldsymbol{p}_c$ is COM of the robot. $\boldsymbol{x}_{Rc}$ is a tip position of the right leg with respect to the origin of COM $\boldsymbol{p}_c$. $\boldsymbol{x}_{Lc}$ is that of the left leg. m is total mass of the robot. $\boldsymbol{H}_{c12} = \boldsymbol{H}_{c21}^T$ is non-diagonal term of the inertia matrix.

The objective here is to make the COM of the robot and the attitude of the body converge its given reference trajectories with consideration of the physical constraints (2)–(6). The control input of this system is the reactive force reference which is realized by the force controller. While the degree-of-motion-freedom of (13) is 6, the degree of control input is time-variant, which becomes 12th in double support phase and 6th in single support phase.

The ideal force input at the COM of the robot and torque input around the body $\boldsymbol{u}_c^*$ is determined by the state feedback.

$$\boldsymbol{u}_c^* = \boldsymbol{M}[\boldsymbol{K}_p(\boldsymbol{x}^{ref}-\boldsymbol{x})+\boldsymbol{K}_d(\dot{\boldsymbol{x}}^{ref}-\dot{\boldsymbol{x}})+\ddot{\boldsymbol{x}}^{ref}]+\boldsymbol{d} \quad (21)$$

Due to the physical limitations (2)–(6), the ideal force $\boldsymbol{u}_c^*$ at the COM and torque at the body is not always realized by the reactive force and torque. Thus the following performance indices J_{main}, J_{sub} are introduced.

$$J_{main} = \frac{1}{2}(\boldsymbol{u}_c-\boldsymbol{u}_c^*)^T \boldsymbol{C}(\boldsymbol{u}_c-\boldsymbol{u}_c^*) \quad (22)$$

$$J_{sub} = \frac{1}{2}(\boldsymbol{u}_R-\boldsymbol{u}_L)^T(\boldsymbol{u}_R-\boldsymbol{u}_L) \quad (23)$$

The index J_{main} corresponds to the square error between the ideal force and torque and the realizable ones. The index J_{sub} corresponds to the square error between the force and torque of the left foot and those of the right one. The reactive force and torque input $\boldsymbol{u}$ is determined by quadratic programming, which minimizes the performance index under the linearized constraints of (2)–(6).

$$\min_{\boldsymbol{u}} \quad J_{main} + \epsilon J_{sub} \quad (24)$$

$$\text{subject to} \quad \boldsymbol{A}\boldsymbol{u} \leq \boldsymbol{b} \quad (25)$$

where ϵ is a small positive real number.

The main performance index J_{main} approaches the solution to the ideal force and torque $\boldsymbol{u}_c^*$ given by the state feedback. The sub performance index J_{sub} distributes the inner force and torque to the both foot in balance. Because ϵ is very small, the sub performance index does not almost have influence on the main performance index.

The optimization problem (24)–(25) is equivalent to the quadratic programming problem. The reactive force and torque reference can be obtained by solving a quadratic programming problem for each sampling period.

4 Simulations

The proposed control is applied to 20 axes human-type biped robot and is investigated by a precise simulator[15]. The parameters of the robot is shown in Table 1. The QP is solved by the algorithm in [18].

The snapshots of the simulation is shown in Fig. 4. The initial movement of COM is finished in $0 < t < 1$

Table 1: Parameters of biped robot.

parts	size [m]	weight [kg]
all	0.99	28.744
head	$0.14 \times 0.14 \times 0.14$ ($d \times w \times h$)	2.744
arm	0.3	3.5
body	0.4	8
thigh	0.2	2
shin	0.2	2
foot	0.2×0.1 ($d \times w$)	1.5

[s]. After that, the walking motion starts. Fig. 5 shows the trajectory of zero moment point (ZMP).

The walking motion becomes more robust when the yaw axis moment is compensated by the arm swing motion[17].

Walking on Unknown Slope Fig. 6 shows the biped walking simulation with slope environment whose information is not used in the controller. The controller used in the simulation is exactly as same as that of in the previous section. Thus the proposed control algorithm is robust against the environmental uncertainty.

The slope is set to 5 [deg] up, which is not used in the controller. Fig. 6 shows the trajectory of the ZMP. Due to the slope, the trajectory of the ZMP shifts to the heel in Fig. 6 (a) and (b), compared with the flat terrain case in Fig. 5 (a) and (b).

5 Experiments

Fig. 1 shows a photo of the real 14-axis biped walking robot developed in our lab. The specification of the robot is as follows.

- 6-degree-of-freedom for each leg.
- Dc servo motor with 50:1 harmonic gear.
- Rotary encoder sensor for each joints with resolution 4000 [pulse/rev] on motor shaft.
- 6DOF force/torque sensor on each ankle.
- 3DOF giro scope on body.
- Controller: DSP (TMS320C32-50MHz)

The robot has 6 joints for each leg so that the position and orientation of the foot can be chosen any posture in the 3-dimensional space. All calculation of the control is done by DSP board with TMS320C32-50MHz. The programs are written in the C language. The robot is about 1.2 [m] height and 20 [kg] weight. Fig. 7 shows the global system configuration. The host computer is used in cross compiling the DSP programs.

Figure 4: Snapshots of biped walking simulation.

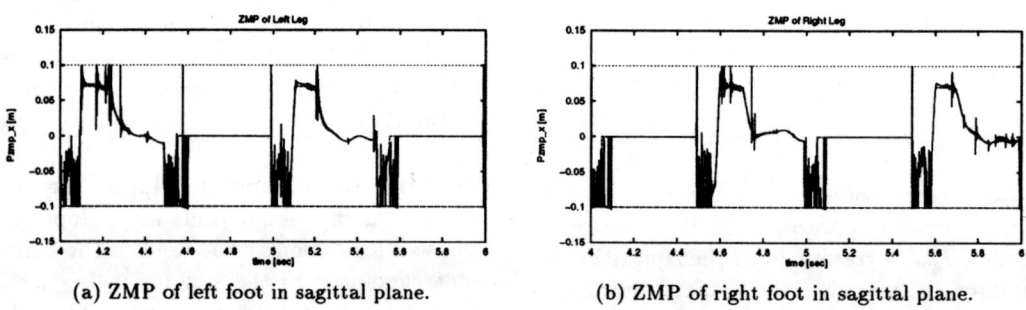

(a) ZMP of left foot in sagittal plane.

(b) ZMP of right foot in sagittal plane.

Figure 5: Trajectory of zero moment point in sagittal plane. (simulation)

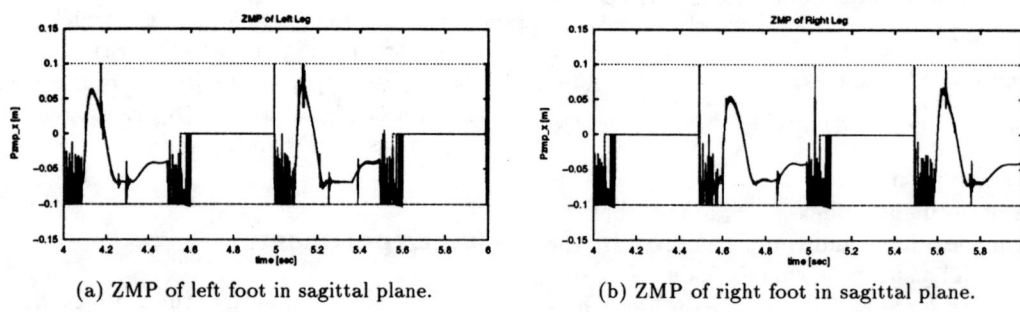

(a) ZMP of left foot in sagittal plane.

(b) ZMP of right foot in sagittal plane.

Figure 6: Trajectory of zero moment point in slope environment. (simulation)

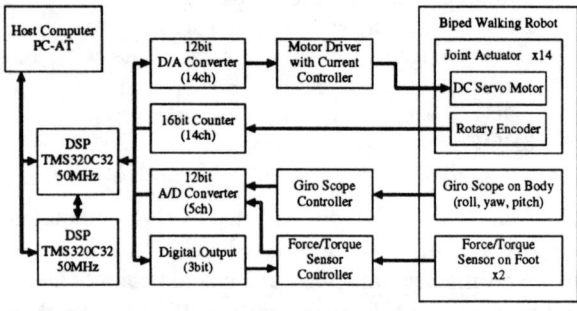

Figure 7: System configuration.

Implementation Aspects Although computing ability of recent micro processor progresses rapidly, it is not enough to implement all of the proposed control algorithm in real-time for the present. Especially the calculation of the quadratic programming costs very much. Here the simplified version of the force distribution instead of (24)–(25) is proposed as follows.

$$\boldsymbol{u} = \boldsymbol{W}^2 \boldsymbol{K}^T (\boldsymbol{K} \boldsymbol{W}^2 \boldsymbol{K}^T)^{-1} \boldsymbol{u}_c^* \qquad (26)$$

where $\boldsymbol{u}$ is the force reference of both feet and $\boldsymbol{u}_c^*$ is required force/torque at the body obtained by (21). This solution is the optimal in a sense that $\boldsymbol{f}_A$ has minimum square norm $\|\boldsymbol{W}^{-1}\boldsymbol{f}_A\|^2$. $\boldsymbol{W}$ is a weighting matrix $\boldsymbol{W} = \text{diag}\{w_1, w_2, \ldots, w_{12}\}$.

The stable posture control of the robot is shown as Fig. 8. The position of COM and the attitude of the body is well controlled within ± 0.03 [m] and ± 0.04

[rad] (= ± 2.3 [deg]) errors. Also ZMPs are used to control COM of the robot as shown in Fig. 9.

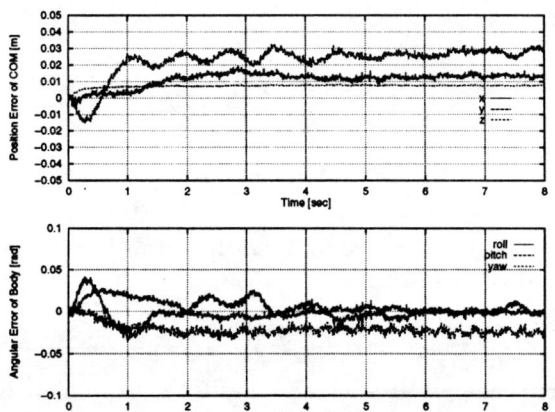

Figure 8: Error of COM position and body rotation.

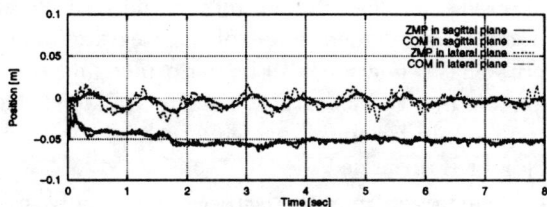

Figure 9: Measured ZMP in sagittal and lateral plane.

6 Conclusion

In this paper the following hierarchical control system is proposed. 1) robust control of reactive force/torque interaction at the foothold based on Cartesian space motion controller. 2) the posture control considering the physical constraints of the reactive force/torque on the foot by quadratic programming.

The proposed control system is applied to the 20-axis simulation model, and the stable biped locomotion is realized. The stable attitude control is confirmed by 14-axis biped robot experiments.

Finally, the authors would like to note that part of this research is carried with the subsidy of the Scientific Research Fund of the Ministry of Education.

References

[1] S. Kajita, T. Yamaura, and A. Kobayashi, "Dynamic Walking Control of a Biped Robot Along a Potential Energy Conserving Orbit," *IEEE Trans. RA*, vol. 8, no. 4, pp. 431–438, 1992.

[2] J. Furusho and A. Sano, "Sensor-Based Control of a Nine-Link Biped," *Int. J. Robotics Research*, vol. 9, no. 2, pp. 83–98, 1990.

[3] J. Yamaguchi, A. Takanishi and I. Kato, "Development of a Biped Walking Robot Compensating for Three-Axis Moment by Trunk Motion," *J. Robotics Society of Japan*, vol. 11, no. 4, pp.581–586, 1993. (in Japanese)

[4] M. H. Raibert, *Legged Robots That Balance*, Cambridge, MA, MIT Press, 1986.

[5] A. Kun and W. T. Miller, III, "Adaptive Dynamic Balance of a Biped Robot using Neural Networks," *Proc. IEEE Int. Conf. on RA*, pp. 240–245, 1996.

[6] J. K. Hodgins, "Three-Dimensional Human Running," *Proc. IEEE Int. Conf. on RA*, pp. 3271–3276, 1996.

[7] J. Yamaguchi, N. Kinoshita, A. Takanishi, and I. Kato, "Development of a Dynamic Biped Walking System for Humanoid, — Development of a Biped Walking Robot Adapting to the Human's Living Floor —," *Proc. IEEE Int. Conf. on RA*, pp. 232–239, 1996.

[8] S. Kajita and K. Tani, "Adaptive Gait Control of a Biped Robot based on Realtime Sensing of the Ground Profile," *Proc. IEEE Int. Conf. on RA*, pp. 570–577, 1996.

[9] H. Minakata and Y. Hori, "Development of Biped Bike Prototype 'Ostrich-I&II'," *Proc. Asian Control Conference*, vol. 3, pp. 319–322, 1997.

[10] A. W. Salatian, K. Y. Yi, and Y. F. Zheng, "Reinforcement Learning for a Biped Robot to Climb Sloping Surfaces," *J. Robotic Systems*, vol. 14, no. 4, pp. 283–296, 1997.

[11] S. Kawaji, N. Matsunaga and M. Arao, "Hierarchical Control of Biped Locomotion Robot", *Proc. IEEE Int. Workshop on Advanced Motion Control*, pp. 421–430, 1994.

[12] K. Sorao, T. Murakami, and K. Ohnishi, "A Unified Approach to ZMP and Gravity Center Control in Biped Dynamic Stable Walking," *Proc. IEEE/ASME Int. Conf. on Advanced Intelligent Mechatronics*, CD-ROM, 1997.

[13] T. Fukuda, Y. Komata, and T. Arakawa, "Stabilization Control of Biped Locomotion Robot based Learning with GAs having Self-adaptive Mutation and Recurrent Neural Networks," *Proc. IEEE Int. Conf. on RA*, pp. 217–222, 1997.

[14] J. Pratt, P. Dilworth, and G. Pratt, "Virtual Model Control of a Bipedal Walking Robot," *Proc. IEEE Int. Conf. on RA*, pp. 193–198, 1997.

[15] Y. Fujimoto and A. Kawamura, "Autonomous Control and 3D Dynamic Simulation of Biped Walking Robot Including Environmental Force Interaction," to appear in *IEEE RA Magazine*, June 1997.

[16] Y. Fujimoto and A. Kawamura, "An Inertia Fluctuation Insensitive Robust Control of Robot Manipulators Based on a Combination of Inertia Torque Computation Filter and H_∞ Control," *Trans. IEE of Japan*, vol. 117-D, no. 4, pp. 493–500, 1997. (in Japanese)

[17] Y. Fujimoto and A. Kawamura, "Robust Control of Biped Walking Robot with Yaw Moment Compensation by Arm Motion," *Proc. Asian Control Conference*, vol. 3, pp. 327–330, 1997.

[18] R. W. Cottle and G. B. Dantzig, "Complementary Pivot Theory of Mathematical Programming," *Linear Algebra and Its Applications*, vol. 1, pp. 103–125, 1968.

Generation of Energy Optimal Complete Gait Cycles for Biped Robots

L. Roussel, C. Canudas-de-Wit

Laboratoire d'Automatique de Grenoble
UMR-CNRS 5528, ENSIEG-INPG
B.P. 46, 38402, St Martin d'Hères
France

A. Goswami

INRIA Rhône-Alpes
655 ave. de l'Europe
38330 Montbonnot
France

Abstract

In this paper we address the problem of energy-optimal gait generation for biped robots. Using a simplified robot dynamics that ignores the effects of centripetal forces, we obtain unconstrained optimal trajectories generated by piecewise constant inputs. We study a complete gait cycle comprising single support, double support and the transition phases. The energy optimal gaits for different step lengths and velocities are compared with natural human gait.

1 Introduction

Recently, many studies have been devoted to locomotion, path planning and control of biped robots [3]-[6]. The main motivations for using walking robots rather than more conventional wheeled robots are their versatility in moving in unstructured and rough terrain and for their obstacle avoidance capabilities. In particular, bipeds robots, under suitable mechanical design, are potentially capable of producing gaits involving very little input energy (apart from the restitution energy needed to compensate for the losses due to friction and contact).

This last point has raised a lot of interest since the generation of low-energy trajectories for biped robots remains an open and non-trivial issue [3] [2] [8] [1]. From a practical point of view it seems reasonable to search for a trajectory that fulfills a certain objective in terms of the gait parameters (such as the walking velocity, the step length and the step frequency) while minimizing the input energy needed to produce such a gait.

The problem of determining whether or not these so called natural gaits exist is complex and only a few partial analytical results are available up to now, see [5]. This problem is equivalent to finding the joint velocity values before and after the swing and double support phases such that the desired gait cycle is reproduced. To this problem may be derived closed-form solutions provided that explicit integration of the support phase equations can be performed and that a reasonable model for contact losses is available. This is feasible only for simple systems such as the monopod [6], or if linearized model is considered.

The energy-optimal trajectories for highly non-linear equations of a complex robot may be found only numerically, and in general, they will be sub-optimal. Earlier investigations for numerical solutions approximated the joint trajectories to time-polynomials [3] [2], Fourier expansions [1], or a combination of both [8]. An exhaustive treatment of the application of the optimal programming to human locomotion is given in [4], where penalty functions are used to minimize the total mechanical work done. This technique is now superseded by new numerical optimization algorithms.

In this paper we propose an alternative method for energy optimal gait generation. The proposed approach searches for unconstrained trajectories (no particular time or frequency base function are chosen) generated by piecewise constant inputs.[1] A numerical study presented in [7] has shown that, for an equivalent amount of computational burden, this method provides motions with a lower input energy compared to polynomial or Fourier extensions. The solution is then found numerically after transforming the dynamical problem into a static one and solving it via a standard direct shooting optimization algorithm.

---

[1] Although we explicitly mention that our control input is piecewise constant, we need to remember that the numerical optimization always necessitates a discretization of the system equations resulting in piecewise constant control inputs.

2 Problem Formulation

A complete human gait cycle may be divided into two phases: the single support phase or the swing phase (one foot on the ground and the other foot swinging) and the double support phase (both feet on the ground). The transition from the single support to the double support phase, also called the *contact phase*, is associated with the heel of the front foot touching with the ground. The transition from the double support to the single support phase, also called the *take-off phase*, is caused when the toe of the rear foot leaves the ground. The dynamic equations of a robot consisting of all the described phases is composed of ordinary differential equations for the support phases and algebraic equations for the transition phases. Moreover, the robot's kinematic topology changes from the single support to the double support phase complicating further the differential equations.

It is not an easy task to choose a biped kinematics that captures the essence of the anthropomorphic gait while keeping the model reasonably simple to allow intuitive insights about its behavior. Admitting the fact that the simplifications may sacrifice some of the subtleties of human motion, we have converged upon a planar four degrees-of-freedom (DOF) biped mechanism as shown in Figure 1. For our study the following planar biped robot model will be used. The trunk mass m is located at the hip. a_1 and a_2 are the lengths of the shank and the thigh, respectively. Their masses are m_1 and m_2 respectively. See Figure 1 for a sketch of the model. We have assumed that in this model the trunk will be upright during the walk. This seems reasonable because the trunk's maximal excursion from the vertical axis is about 20mm at the pelvis point, see [8]. The foot of the swing leg is considered massless thereby obviating motor in the corresponding joint.

2.1 The Gait Phases

Mimicking the human gait phases, the dynamics of the biped robot can be decomposed into four different phases describes bellow.

• **Single Support Phase.** The dynamic equations, of this phase, derived by means of the familiar Euler-Lagrange formulation. Biped robot is modeled as a rotational joint open-chain manipulator in this phase.

$$\boldsymbol{H}(\boldsymbol{q})\ddot{\boldsymbol{q}} + \boldsymbol{C}(\boldsymbol{q},\dot{\boldsymbol{q}})\dot{\boldsymbol{q}} + \boldsymbol{g}(\boldsymbol{q}) = \boldsymbol{u}_{ss} \quad 0 < t < T_1 \quad (1)$$

where $\boldsymbol{q} \in R^4$ describes the generalized coordinates, $\boldsymbol{H}(\boldsymbol{q})$ is the inertia matrix, $\boldsymbol{C}(\boldsymbol{q},\dot{\boldsymbol{q}})$ is the matrix of centripetal acceleration and Coriolis terms, $\boldsymbol{g}(\boldsymbol{q})$ is the gravity vector, $\boldsymbol{u}_{ss}$ is the input torque vector during single support phase, and T_1 is the time duration of the swing phase.

• **Contact Phase.** During this phase, the robot configuration remains unchanged, $\boldsymbol{q} = const.$, while there is a discrete change in the joint velocities such that the swing foot stays on the ground after contact. The contact phase is assumed to be instantaneous and inelastic, and without sliping. Centripetal torques are assumed to be smaller than the impulsive forces and are neglected. Therefore, the dynamic equations can be integrated in order to establish the relationship between joint velocities just before and just after the impact:

$$\boldsymbol{H}(\boldsymbol{q})(\dot{\boldsymbol{q}}^+ - \dot{\boldsymbol{q}}^-) = \boldsymbol{I}_{cont} \quad t = T_1 \quad (2)$$

where $\dot{\boldsymbol{q}}^-$ and $\dot{\boldsymbol{q}}^+$ are respectively the joint velocity just before and just after the contact and $\boldsymbol{I}_{cont}$ is the impulse of the impact force which is active during the contact phase.

• **Double Support Phase.** As in the single support phase, the dynamic equations of the double support phase are also derived from the Euler-Lagrange formulation. The fact that the robot foot stays on the ground adds two supplementary constraints of the form $x_f = const$, $y_f = 0$, thereby reducing the admissible set of joint coordinates. The constraints are expressed as $\Phi(\boldsymbol{q}) = 0$, and the use of Lagrange multipliers, $\boldsymbol{\lambda}$, allow us to write the dynamic equations as,

$$\boldsymbol{H}(\boldsymbol{q})\ddot{\boldsymbol{q}} + \boldsymbol{C}(\boldsymbol{q},\dot{\boldsymbol{q}})\dot{\boldsymbol{q}} + \boldsymbol{g}(\boldsymbol{q}) = \boldsymbol{u}_{ds} + \boldsymbol{J}_\Phi^T \boldsymbol{\lambda} \quad T_1 < t < T \quad (3)$$

where $\boldsymbol{J}_\Phi = \frac{\partial \Phi}{\partial \boldsymbol{q}}$ is an $m \times n$ Jacobian matrix, $\boldsymbol{u}_{ds}$ is the input torque vector active during the double support phase, and T, the time of the complete gait cycle.

• **Take-Off Phase.** As in the contact phase, the robot configuration is constant while a change of the support foot reference appears. The joint velocity is discontinuous to bring the robot to the initial state of the following single support phase. This condition must be held to assure a cyclic walk.

$$\boldsymbol{H}(\boldsymbol{q}^+)\dot{\boldsymbol{q}}^+ - \boldsymbol{H}(\boldsymbol{q}^-)\dot{\boldsymbol{q}}^- = \boldsymbol{I}_{top} \quad t = T \quad (4)$$

$\boldsymbol{q}^-$ is the robot's joint coordinates just before the take-off and is computed with respect to a coordinate frame at the support foot for the previous swing phase whereas $\boldsymbol{q}^+$ is the robot's joint coordinates just after the take-off and is computed with respect to a coordinate frame at the new support foot.

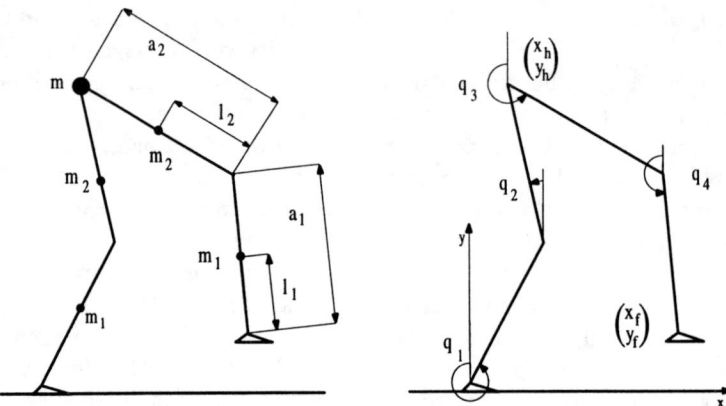

Figure 1: Simplified structure of the 4-DOF biped robot used for our study.

The complete dynamics of the robot is therefore given by the equations (1)-(4). Henceforth in the paper we will use a state-space description of the robot with the state vector $\boldsymbol{x} = [\boldsymbol{x}_1^T, \boldsymbol{x}_2^T]^T$ where $\boldsymbol{x}_1$ is the joint position vector and $\boldsymbol{x}_2$ is the joint velocity vector. In this description the complete dynamics of the robot (equations (1)-(4)) can be rewritten as

$$\begin{cases} \dot{\boldsymbol{x}} = f_{ss}(\boldsymbol{x}, \boldsymbol{u}_{ss}) & t \in [0, T_1[\\ \boldsymbol{x}^+ = \phi_{cont}(\boldsymbol{x}^-, \boldsymbol{u}_{cont}) & t = T_1 \\ \dot{\boldsymbol{x}} = f_{ds}(\boldsymbol{x}, \boldsymbol{u}_{ds}) & t \in]T_1, T[\\ \boldsymbol{x}^+ = \phi_{top}(\boldsymbol{x}^-, \boldsymbol{u}_{top}) & t = T \end{cases} \quad (5)$$

where $\boldsymbol{u}_{cont} = \boldsymbol{I}_{cont}/dt$ and dt is the impulsive time of the contact phase, and $\boldsymbol{u}_{top} = \boldsymbol{I}_{top}/dt$ and where the functions $f_{ss}, \phi_{cont}, f_{ds}, \phi_{top}$ can be obtained from the equations (1)-(4).

Due to the difficulties in the convergence of the optimization procedures with the complete dynamics, we introduce certain model simplifications. In the first place, our goal is to obtain optimal gaits to understand the walk phenomenon and for this, the simplifications are the follows. The inertia matrix is supposed to be constant and diagonal, and the centripetal accelerations are ignored. This assumption is reinforced by the fact that gear ratios of the D-C actuators are large enough so that coupling and position dependent terms of the inertia matrix can be ignored. The main nonlinearities considered by our model are the gravity terms. Then the dynamic model becomes:

$$\begin{cases} H\ddot{q} + g(q) = \boldsymbol{u}_{ss} & t \in [0, T_1[\\ H(\dot{q}^+ - \dot{q}^-) = \boldsymbol{u}_{cont} & t = T_1 \\ H\ddot{q} + g(q) = \boldsymbol{u}_{ds} + J_\Phi^T \lambda & t \in]T_1, T[\\ H(q^+)\dot{q}^+ - H(q^-)\dot{q}^- = \boldsymbol{u}_{top} & t = T \end{cases}$$

2.2 The Cost Function

The following cost function is used for our optimal control scheme:

$$\begin{aligned} J &= \int_0^{T_1} \boldsymbol{u}_{ss}^T \boldsymbol{u}_{ss}\, dt + \boldsymbol{I}_{cont}^T \boldsymbol{I}_{cont} \\ &+ \int_{T_1}^{T} \boldsymbol{u}_{ds}^T \boldsymbol{u}_{ds}\, dt + \boldsymbol{I}_{top}^T \boldsymbol{I}_{top} \end{aligned} \quad (6)$$

which quantifies the injected energy into the robot during a gait cycle. The injected energy to the robot is a reasonable criterion to minimize, especially for mobile robots that needs to carry their own power source.

The specified boundary conditions for the optimization are the initial and terminal joint angles of the single support phase $\boldsymbol{x}_1(0)$, $\boldsymbol{x}_1(T_1)$, and the final joint angle $\boldsymbol{x}_1(T)$ of the double support phase. We also specify T_1, and T, the time intervals of the swing phase and the total cycle time, respectively. This implicitly imposes an average progression speed for the robot. Please note that the three corresponding velocities $\boldsymbol{x}_2(0)$, $\boldsymbol{x}_2(T_1)$ and $\boldsymbol{x}_2(T)$ are free. This represents an additional degree of freedom giving the possibility for the optimization procedure to characterize minimum-energy trajectories. Let us define the variable $\boldsymbol{u}$ as:

$$\boldsymbol{u} = [\boldsymbol{u}_{ss}^T\ \boldsymbol{u}_{cont}^T\ \boldsymbol{u}_{ds}^T\ \boldsymbol{u}_{top}^T]^T \quad (7)$$

As a resume the optimal control problem can be states as:

Problem formulation 2.1 *Given the initial and final joint angles of the single support phase $\boldsymbol{x}_1(0) = \boldsymbol{x}_{1_0}$, $\boldsymbol{x}_1(T_1) = \boldsymbol{x}_{1_{T_1}}$, and the final joint angles $\boldsymbol{x}_1(T) = \boldsymbol{x}_{1_T}$ of the double support phase, and the time intervals T_1 and T, the problem is to find the optimal sequence $\boldsymbol{u}^\star(t)$, minimizing the cost function J (Eq 6), such that it steers the system (5) from $\boldsymbol{x}_{1_0}$ to $\boldsymbol{x}_{1_T}$.*

3 Optimization

The dynamic optimal control problem is next transformed into a static problem via a discretization as we describe subsequently.

3.1 The Piecewise Constant Method

We assume that the control input $u(t)$ is piecewise constant. Let N_1 be the number of time-intervals during the single support phase and N the total number of intervals during single and double support phases. The input control sequence is composed of the single support control phase sequence, the impact control, the double support phase control sequence and the take-off phase control. We now define $U \in R^{(4 \times N+2)}$ as the input control matrix:

$$U = [\underbrace{u_0, u_1, u_2, \ldots, u_{N_1-1}}_{U_{ss}}, u_{cont}, \underbrace{u_{N_1+1}, \ldots u_{N+1}}_{U_{ds}}, u_{top}]$$

Then, with this assumption, the cost function (6) can be reformulated as:

$$\begin{aligned} \mathcal{C} = & \sum_{k=0}^{N_1-1} u(k)^T u(k) \Delta t + u_{cont}^T u_{cont} \Delta t \\ & + \sum_{k=N_1+1}^{N+11} u(k)^T u(k) \Delta t + u_{top}^T u_{top} \Delta t \end{aligned} \quad (8)$$

The dynamic problem (2.1) will be transformed to a static one by approximating the derivative operator by the Euler's formula in the state equation during the two continuous time phases of the gait cycle, (1) (3). The two other phases are already discrete time representation and thus, this approximation has no influence on them.

• **Single support phase.** During this phase, the derivative operator $\dot{x}$ is approximated as follows:

$$x(\dot{k+1}) = \frac{x(k+1) - x(k)}{\Delta t} = f_{ss}(x(k), u(k))$$

It is then possible to write the state at the instant $k+1$ as a function of the previous state $x(k)$ and the input control $u(k)$. By induction, the final state $x(N_1)$ of the swing phase can be written as a function of the initial state $x(0)$ and the input control sequence of the single support phase U_{ss}:

$$\begin{aligned} x(N_1) &= F_{ss}(x(0), u(0), \ldots, u(N_1-1)) \\ &= F_{ss}(x(0), U_{ss}) \end{aligned} \quad (9)$$

where the operator F_{ss} is defined as:

$$F_{ss} = f_{ss} \circ f_{ss} \cdots \circ f_{ss}(x(0), u(0)) \quad (10)$$

• **Contact Phase.** By considering (2), which is already a discrete time equation, the state at the instant $N_1 + 1$ can be uniquely defined from the final state of the single support phase $x(N_1)$, which comes from (9), and the contact control $u_{cont} = u(N_1)$:

$$x(N_1 + 1) = \phi_{cont}(x(N_1), u(N_1)) \quad (11)$$

• **Double Support Phase.** In the same way as the swing phase, it is possible to express the final state $x(N)$ as a function of the initial state of this phase that is to say $x(N_1+1)$, which is obtained from (11), and the input control sequence of the double support phase U_{ds}:

$$\begin{aligned} x(N) &= F_{ds}(x(N_1+1), u(N_1+1), \cdots, u(N)) \\ &= F_{ds}(x(N_1+1), U_{ds}) \end{aligned} \quad (12)$$

where the operator F_{ds} is defined as:

$$F_{ds} = f_{ds} \circ f_{ds} \cdots \circ f_{ds}(x(0), u(0)) \quad (13)$$

• **Take-off Phase.** As in the impact phase, by considering (4), the state at the instant $N+1$ can be uniquely defined from the final state of the DS $x(N)$, which comes from (12), and the link control $u_{top} = u(N)$:

$$x(N+1) = \phi_{top}(x(N), u(N)) \quad (14)$$

The new problem formulation can be stated as the following static problem:

Problem 3.1 *Given the initial and final joint angles $x_1(0)$, $x_1(N_1)$ and $x_1(N)$ and the time intervals T_1 and T, the problem is to find the optimal value for $U^\star(t)$, minimizing the cost function $\mathcal{C}$ (8), such that it steers the system (5) from $x_1(0)$ to $x_1(N)$. Or equivalently:*

$$\begin{cases} \min_U \mathcal{C}(U) \\ under \begin{cases} x(N_1) &= F_{ss}(x(0), U_{ss}) \\ x(N_1+1) &= \phi_{cont}((x(N_1), u(N_1))) \\ x(N) &= F_{ds}(x(N_1+1), U_{ds}) \\ x(N+1) &= \phi_{top}((x(N), u(N))) \end{cases} \\ given \ (x_1(0), x_1(N_1), x_1(N), T_1, T, N_1, N) \end{cases}$$

In the following section, we describe the results obtained by simulation.

3.2 Optimal Energy Gaits

In Figure 2 we can see an example of optimal gait corresponding a step length $0.4\ m$ and $T = 1\ s$. This gait needs of $0.57\ Cal/m/kg$ with a walking velocity equal to $0.4\ m/s$. The gait of the biped is not similar during a complete gait cycle and during only a single

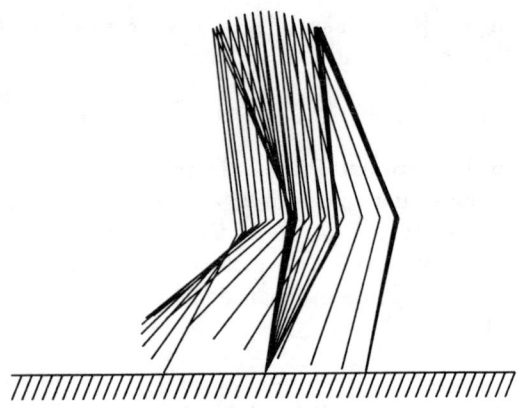

Figure 2: Optimal Gait obtained during a complete gait cycle with a step length $S = 0.4m$ and a time period $T = 1s$.

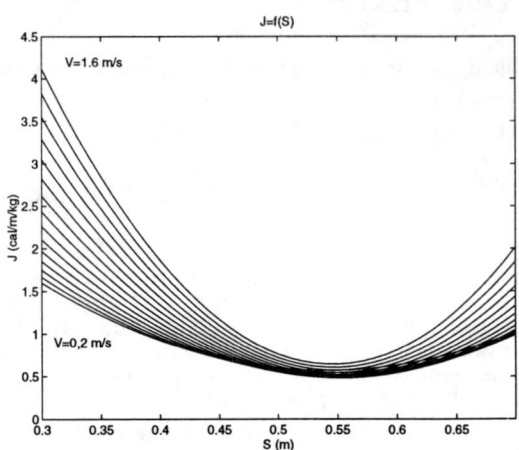

Figure 3: Optimal Criterion as a function of S for different values of V.

support phase [7] which produces ballistic gait. While during a complete gait cycle, the optimization takes all phases in consideration to find a minima, that is the reason why the foot of the swing is more "controlled" during the complete gait cycle in order to not penalize the cost function through the transition phases. The required energy to reach a step is more distribute during the complete cycle.

As the energy optimal gait generation is not sufficiently fast to be computed in real time, the aim of this study is to establish a database of pre-computed optimal gaits.

The walk parameters, the step length S, the time period T and as a consequence the walking velocity V, vary. The resulting cost criterion represents the energy required to perform one meter per kg. Figure 3 shows that each walking desired velocity corresponds to an optimal step length. It can be seen that, for all velocities, smaller steps correspond to higher energy consumption. It is essentially due to the greater number of steps required to cover the same distance and then the greater number of impact and take off phases involved. The main result which comes from this figure, it is that for all velocities, the optimal step length is the same, and it is approximately $0.54m$. That is to say, for each desired velocity, there exist a couple of optimal walking parameters S and T.

Figure 4, shows the cost function versus walking velocity for different step lengths, that for each step length, there is an optimal walking velocity. For human locomotion, this fact is characterized and this velocity is called "natural" walking velocity. If we are not constrained to walk at a certain velocity, this "natural" velocity is always adopted and then we choose to minimize our energy consumption. The profile of such human curve is shown in figure 5.

It is interesting to compare our results with those obtained from the study of normal human gaits. In the biomechanics literature, it is generally accepted that the normal human gait minimizes energy expenditure per unit body weight per distance although the function is rather flat at the minimum and a modification of the gait velocity does not significantly change the energy expenditure. According to [?] a grand average of most of the data available at that time gave a value of optimal walking speed of 80 m/min and at that velocity the human body spends 0.8 calorie of energy per meter of distance traveled per each kilogram of body weight. For the robot model considered in this paper we have approximatively one calorie of energy per meter.

A third thing in human gait is that people, in unforced walk, tends to use step lengths that are proportional to the cadence.

The figure 6 shows the percentage of consumed energy during each phase of a complete gait cycle. Firstly, the energy losses during the impact phase do not appear to be so important. It is due to the objective function which penalize the total amount of energy. Then, the best strategy (approach, procedure, scheme) is not necessarily find a ballistic swing phase if the impact phase losses a large quantity of energy due to the difference between the velocities just before and just after the impact. The optimization scheme tends to control all the swing phase in order to reduce the consumed energy during the impact: it is a compromise. After, it can be noted that the energy losses

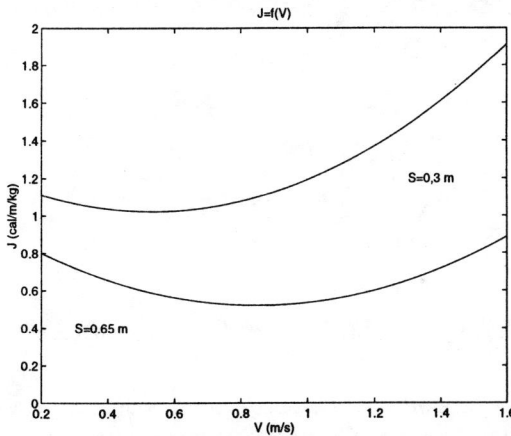

Figure 4: Optimal Criterion as a function of V for different values of S.

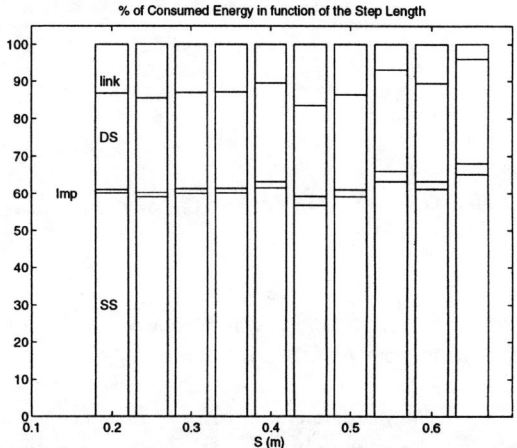

Figure 6: Repartition of Consumed Energy during the four phases.

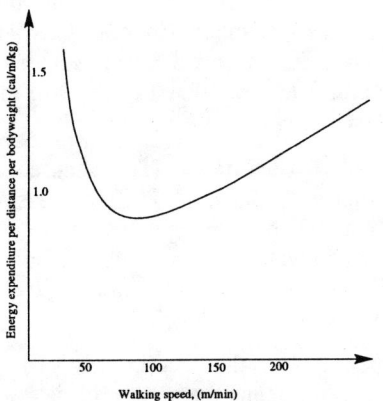

Figure 5: "Natural" speed in human case.

during the impact increase with the step length.

4 Conclusion

This article studies a complete gait cycle including the single support phase, the impact, the double support phases and the link phase. A generator of energy optimal gait is realized and some energizing results are compared to human results.

It is also interesting to study the potential benefits of introducing passive elements in the joint articulations [1], which appear as a natural way enhancing a passive walk.

References

[1] G. Cabodevilla, N. Chaillet, and G. Abba. Near optimal gait for a biped robot. In *Proc. of the AMS'95*, Karlsruhe, Germany, 1995.

[2] P.H. Channon, S.H. Hopkins, and D.T. Pham. Simulation and optimization of gait for a bipedal robot. *Math. Comput. Modelling*, 14:463–467, 1990.

[3] P.H. Channon, S.H. Hopkins, and D.T. Pham. Derivation of optimal walking motions for a bipedal walking robot. *Robotica*, 10:165–172, 1992.

[4] C.K. Chow and D.H. Jacobson. Studies of human locomotion via optimal programming. *Mathematical Biosciences*, 10:239–306, 1971.

[5] C. Francois and C. Samson. Running with constant energy. In *Proceedings of IEEE International Conference on Robotics and Automation*, pages 131–136, San Diego, USA, 1994.

[6] C. Francois and C. Samson. Energy efficient control of running legged robots. a case study: the planar one-legged hopper. Research Report 3027, INRIA, November 1996.

[7] L. Roussel, C. Canudas, and A. Goswami. Comparative study of method for energy-optimal gait generation for biped robots. In *International Conference on Informatics and Control*, pages 1205–1212, St Petersburg, June 1997.

[8] V. Yen and M.L. Nagurka. Suboptimal trajectory planning of a five-link human locomotion model. In *Biomechanics of Normal and Prosthetic Gait*, pages 17–22, Boston, MA, 1987. ASME Winter Annual Meeting.

Neural Computation of the Equivalent Control in Sliding Mode For Robot Trajectory Control Applications

Meliksah Ertugrul

TUBITAK Marmara Research Center, Robotics
and Automation Group, 41470, Gebze, TURKEY

Okyay Kaynak

M.Sc. Program in Mechatronics
National University of Singapore, SINGAPORE

Abstract: In the application of Sliding Mode Controllers, the main problem which is encountered is that a whole knowledge of the system dynamics (or inverse dynamics) and the system parameters is required to be able to compute the equivalent control. This is actually very rare in practice. In this paper, a feed-forward neural network is proposed to compute the equivalent control. The weights of the net are updated such that the additional control term of the sliding mode goes to zero. Experimental studies carried out on a direct drive arm indicate that the proposed approach is a good candidate for trajectory control applications.

1. Introduction

Variable Structure Controllers (VSC) with Sliding Mode Control (SMC) was firstly proposed in early 1950's. After the seventies, SMC became more popular and nowadays it enjoys a wide variety of application areas [1]. The main reason of this popularity is the attractive properties of SMC, such as good control performance for nonlinear systems, applicability to MIMO systems, and well established design criteria for discrete time systems[2]. The best property of the SMC is robustness. Loosely speaking, a system with a SMC is insensitive to parameter changes or external disturbances.

The SMC with discontinuous control suffers mainly from two disadvantages. The first one is the chattering, which is the high frequency oscillations of the controller output. The second one is the difficulty in the calculation of the equivalent control. A thorough knowledge of the plant dynamics is required for this purpose. In the literature, there are some suggestions to overcome these problems. For example, the use of a saturation function instead of the sign function is commonly suggested as a technique to eliminate chattering [3]. The use of an averaging filter is suggested in [4], for avoiding the difficulty in the calculation of the equivalent control.

One of the first robotics applications of Neural Networks (NN) is the *CMAC* (Cerebellar Model Articulation Controller) by Albus [13]. The main idea of the CMAC is the use of a look up table instead of solving analytic equations. He used three layer structure NN and the first connections are random while the second employs adjustable weights.

A popular topic on NN robot control is the use of NN to compute the *inverse dynamics* or *inverse kinematics* with learning. A few example can be given as Psaltis [14], Guez and Ahmad [15], Elsley [16], Grossberg and Kuperstein [17], Fukuda [18], G.A. Bekey [19].

Kawato and his coworkers developed *Feedback error learning* [6]. In this configuration, the output of a classical feedback controller (Constant Gain Proportional Controller) and a feed-forward NN are summed as control signal. The aim is to minimize the output of the feedback controller by updating the weights of NN.

Narendra and Parthasarathy [20] focused on *Model Reference Adaptive Control* (MRAC). In this application, the weights of the NN are updated such that the plant (robot) output converges to the model output. Tokita and his coworkers focused on a *self-tuning regulator* for force control of a robot manipulator [21].

In this paper, firstly, the relation between the inverse dynamics and the equivalent control is presented. In sliding mode, equivalent control has an effect identical to the inverse dynamics (computed torque). The motivation of showing the equivalence is to use a neural network to compute the equivalent control. Because, there are lots of NN applications in robotics to compute the inverse dynamics. Secondly, a two layer feed-forward NN is proposed to compute the equivalent control. Thirdly, an adaptation scheme is proposed to update the weights of NN.

The paper concludes with the presentation of some implementation results obtained for the control of a direct drive scara type robot.

2. Variable Structure Systems

In the application of Variable Structure System theory to the control of nonlinear processes it is argued that one only needs to drive the error to a "switching" or "sliding" surface, after which the system is in "sliding mode" and will not be affected by any modeling uncertainties and/or disturbances [1-2].

2.1 The System (Plant)

Consider a nonlinear, non-autonomous, multi-input multi-output system of the form,

$$x_i^{(k_i)} = f_i(X) + \sum_{j=1}^{m} b_{ij} u_j \quad (1)$$

where $x_i^{(k_i)}$ means the k_i^{th} derivative of x_i. Also, the vector U of components u_j is the control input vector and the state X is composed of the x_i's and their first (k_i-1) derivatives. Such systems are called square systems since they have as many control inputs as outputs x_i to be controlled [3]. The system can be written in a more compact form as letting

$$X = [x_1 \; \dot{x}_1 \; .... \; x_1^{k_1} \; .... \; x_m \; \dot{x}_m \; .... \; x_m^{k_m}] \quad (2)$$

$$U = [u_1 \; .... \; u_m]^T \quad (3)$$

and assume, X is (nx1). The system equation becomes,

$$\dot{X}(t) = F(X) + BU(t) \quad (4)$$

where B is (nxm) input gain matrix.

2.2 Sliding Surface

For the systems given in (4), generally, sliding surface, S, (mx1) is selected [4] as given below,

$$S(X,t) = G(X^d(t) - X(t)) = \phi(t) - S_a(X) \quad (5)$$

where, $\quad \phi(t) = G X^d(t) \; , \quad S_a(X) = G X(t) \quad (6)$

the time and the state dependent parts. Also X^d represents the desired (reference) state vector and G is (mxn) slope matrix of the sliding surface. Generally, the G matrix is selected such that the sliding surface function becomes,

$$S_i = \left(\frac{d}{dt} + \lambda_i\right)^{k_i - 1} e_i \quad (7)$$

where e_i is the error for x_i ($e_i = x_i^d - x_i$). Also λ_i's are selected as positive constants. Therefore e_i goes to zero when S_i equals to zero.

The aim in SMC is to force the system states to the sliding surface. Once the states are on the sliding surface, the system errors converge to zero with an error dynamics dictated by the matrix G.

2.3 Sliding Mode Controller Design

The method described in this section is based on the selection of a Lyapunov function [4]. The control should be chosen such that the candidate Lyapunov function satisfies Lyapunov stability criteria.

The Lyapunov function is selected as given below,

$$V(S) = \frac{S^T S}{2} \quad (8)$$

It can be noted that this function is positive definite. ($V(S=0) = 0$ and $V(S) > 0 \; \forall S \neq 0$)

It is aimed that the derivative of the Lyapunov function is negative definite. This can be assured if one can assure that

$$\frac{dV(S)}{dt} = -S^T D \, sign(S) \quad (9)$$

D is (mxm) positive definite diagonal gain matrix. $sign(S)$ means signum function is applied to each element of S,

$$sign(S) = [sign(S_1) \; ... \; sign(S_m)]^T \quad (10)$$

and $sign(S_i)$ is defined as,

$$sign(S_i) = \begin{cases} +1 & S_i > 0 \\ 0 & S_i = 0 \\ -1 & S_i < 0 \end{cases} \quad (11)$$

Taking the derivative of (8), and equating this to (9), one will obtain the following equation,

$$S^T \frac{dS}{dt} = -S^T D \, sign(S) \quad (12)$$

By taking the time derivative of (5) and using the plant equation,

$$\frac{dS}{dt} = \frac{d\phi}{dt} - \frac{\partial S_a}{\partial X} \frac{dX}{dt} = \frac{d\phi}{dt} - G(F(X) + BU) \quad (13)$$

is obtained. By putting (13) into (12), the control input signal can be obtained as,

$$U(t) = U_{eq}(t) + \Delta U(t) \quad (14)$$

where $U_{eq}(t)$ is the equivalent control and it is written as,

$$U_{eq}(t) = -(GB)^{-1}\left(GF(X) - \frac{d\phi(t)}{dt}\right) \quad (15)$$

and $\Delta U(t)$ is the additional control term and written as,

$$\Delta U(t) = (GB)^{-1} D \, sign(S) = K \, sign(S) \quad (16)$$

2.4 Chattering Elimination

The controller in (14) results with high frequency oscillations in its output, causes to a problem known as chattering. Chattering is undesirable because it can excite the high frequency dynamics of the system. To eliminate the chattering, it is suggested to use a saturation [3] or a shifted sigmoid function [4] instead of sign function. In this case, the additional control term is computed as,

$$\Delta U(t) = K h(S) \quad (17)$$

Where $h(.)$ is a shifted sigmoid function and defined as,

$$h(S_j) = \frac{2}{1 + e^{-S_j}} - 1 \quad (18)$$

3. Neural Computation of The Equivalent Control

In this part, firstly, the equivalence of the inverse dynamics and equivalent control will be presented.

As it is stated earlier, the sliding mode control converts the nth order system equations to the 1st order equations. The new dynamics equation while in sliding mode can be written as,

$$\frac{dS}{dt} = 0 \quad (19)$$

By solving (19) for desired control signal, one will obtain the following,

$$\frac{dS(X^d, U^d, t)}{dt} = 0 \quad (20)$$

With using (13), the equation (20) can be computed as,

$$\frac{dS}{dt} = \frac{d\phi}{dt} - G(F(X^d) + BU^d) = 0 \quad (21)$$

Solving (21) for desired control signal,

$$U^d = -(GB)^{-1}\left(GF(X^d) - \frac{d\phi(t)}{dt}\right) \quad (22)$$

is obtained. If the system is in the sliding mode, this means the system perfectly follows the desired trajectory, $F(X^d)$ term can be replaced with $F(X)$ which is the actual value. Then, the desired control signal becomes the equivalent control.

$$U^d \equiv U_{eq} = -(GB)^{-1}\left(GF(X) - \frac{d\phi(t)}{dt}\right) \quad (23)$$

As a conclusion, in the sliding mode, the equivalent control is an equivalence to the inverse dynamics.

3.2 NNs in inverse dynamics computation

There are mainly two approaches to learn the inverse dynamics of a robot with a NN [5]. The first one is called as "general learning" and it's an off-line learning and the second one is called as "specialized learning" and it's an on-line learning.

A very good example of inverse dynamics computation in robot control is "Feedback error learning" which is proposed by Kawato et al [6]. It is based on the NN realization of the computed torque plus a secondary PID controller which is one of the most used technique in robot control applications. In Kawato's structure, the output of the secondary controller is used as the error signal to update the weights of NN.

3.3 The motivation of U_{eq} computation by NN

If the knowledge of $F(X)$ and B matrices is very poor, then the equivalent control calculated will be too far off from the actual equivalent control. In other words, a whole knowledge of the system dynamics (or inverse dynamics) and the system parameters is required to be able to compute the equivalent control. This is actually very rare in practice. To be able to solve this problem, one can use a NN to compute the equivalent control. Hence, it is showed that the equivalent control is identical to the inverse dynamics in sliding mode, previously mentioned successful applications to compute the inverse dynamics motivate to use a NN in the computation of the equivalent control.

3.4 The structure of the proposed controller

In the proposed structure, the equivalent control term in sliding mode control is computed by a NN. The additional term in sliding mode is also computed and summed with the output of the NN to form the control signal. The output of the secondary controller is accepted as a measure of error to update the weights of the NN. The aim of the learning process of the NN is to minimize the output of the secondary controller. This is because, in sliding mode, equivalent control is enough to keep the system on the sliding surface and additional term is necessary to compensate the deviations from the surface. The overall system with the proposed controller is given in Fig. 1.

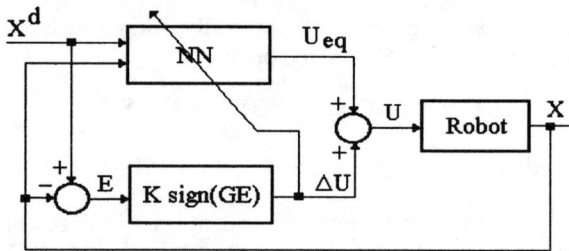

Figure 1. The overall system with proposed controller

3.5 NN structure to compute the equivalent control

The structure of NN is selected as two layer feed-forward network. The inputs and outputs of the network is selected from the equivalent control equation.

In the computation of the equivalent control, all the desired and actual states are used. This is obvious from (15). Therefore, the inputs to NN which will compute the equivalent control are the desired and the actual states. The selected NN has one hidden and one output layer. The number of neurons in the output layer are determined by the number of the actuators of the robot. In other words, it equals to the number of inputs of the robot. The number of neurons in the hidden layer should be selected such that the NN is capable to compute whole span of inverse dynamics. In practice, this can be selected as two times the number of neurons in the input layer. If any error occurs due to poor modeling, the secondary controller can compensate the error.

To be able to clarify the NN structure further, it will be better to explain it on a two degrees of freedom (DOF) robot manipulator. The states of the robot dynamics can be selected as angular positions and velocities. Therefore, the number of states will become four. In this case, the NN has eight inputs (four for actual states and four for desired states). Hence, the robot has two inputs, the number of neurons in the output layer of NN is two. The number of neurons in the hidden layer can be arbitrarily selected as sixteen which is two times the number of inputs. The structure of NN for a two DOF robot manipulator is presented in Fig. 2.

The inputs (designated as Z) to the net consist of desired and actual states ($Z = [(X^d)^T \; X^T]^T$). The net sum and the output of the hidden layer are designated as *Ynet* and *Yout*, respectively. The output layer has two neurons. The net sum and the output of the output layer are designated as *Unet* and *Ueq*, respectively. The values can be computed as,

$$Ynet_j = \sum_{i=1}^{8} Wz_{j,i} * Z_i \quad j=1..M \quad Yout_j = h(Ynet_j) \quad (24)$$

$$Unet_j = \sum_{i=1}^{M=16} Wy_{j,i} * Yout_i \quad j=1..2 \quad Ueq_j = h(Unet_j) \quad (25)$$

where $h(.)$ is an activation function. It is selected as a shifted sigmoid function as defined in (18).

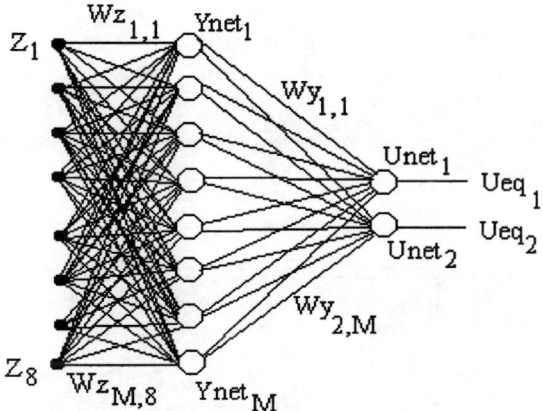

Figure 2. NN structure of U_{eq} for a two DOF robot

3.6 Weight Adaptation

The weight adaptation is based on a minimization of a cost function. The cost function is selected as,

$$E = \frac{1}{2} \sum_{j=1}^{2} \left(Ueq_j^d - Ueq_j\right)^2 \approx \frac{1}{2} \sum_{j=1}^{2} (\Delta U_j)^2 \quad (26)$$

Gradient Descent (or back propagation) for the output layer is used as,

$$\frac{dWy_{j,i}}{dt} = -\mu \frac{\partial E}{\partial Wy_{j,i}} = -\mu \frac{\partial E}{\partial Unet_j} \frac{\partial Unet_j}{\partial Wy_{j,i}} \quad (27)$$

$$\frac{dWy_{j,i}}{dt} = \mu \, \delta y_j \, Yout_i \quad (28)$$

where,

$$\delta y_j = -\frac{\partial E}{\partial Unet_j} = -\frac{\partial E}{\partial Ueq_j} \frac{\partial Ueq_j}{\partial Unet_j} = \left(Ueq_j^d - Ueq_j\right) h'(Unet_j)$$

$$\delta y_j = (\Delta U_j) h'(Unet_j) \quad (29)$$

The derivative of the shifted sigmoid f'n is computed as,

$$h'(Unet) = \left.\frac{dh(x)}{dx}\right|_{x=Unet} = \frac{1}{2}(1 - h^2(Unet)) = \frac{1}{2}(1 - Ueq^2) \quad (30)$$

Gradient descent for the hidden layer is computed as,

$$\frac{dWz_{j,i}}{dt} = -\mu \frac{\partial E}{\partial Wz_{j,i}} = \mu \, \delta z_j \, Z_i \quad (31)$$

where,

$$\delta z_j = \left(\sum_{k=1}^{2} \delta y_k \, Wy_{k,j}\right) h'(Ynet_j) \quad (32)$$

The most important point in this derivation is that the error between desired and actual equivalent control is replaced with the additional control term of the sliding mode control, as it is seen from (29)

4. Robotics Application

In order to study the performance of the proposed controller, extensive implementation studies are carried out on a two degrees of freedom, direct drive, scara type experimental manipulator, manufactured by Integrated Motion Corporation (see figure 3).

4.1 Robot Dynamics

The robot model is written as,

$$M(q)\ddot{q} + C(q,\dot{q})\dot{q} + f_c = \tau \quad (33)$$

The details of the dynamics can be found in [22]. The model in (33) can be written in the state-space form representation as,

$$\begin{bmatrix} \dot{x}_1 \\ \dot{x}_2 \end{bmatrix} = \begin{bmatrix} x_2 \\ -M^{-1}(Cx_2 + f_c) \end{bmatrix} + \begin{bmatrix} 0 \\ M^{-1} \end{bmatrix} u \quad (34)$$

where, $[x_1 \; x_2] = [q \; \dot{q}] = [\theta_1 \; \theta_2 \; \dot{\theta}_1 \; \dot{\theta}_2]$ and $u=\tau$

The equation (34) is in the form of (4), and the proposed method can be applied.

4.2 Experimental Results

In the experimental studies carried out, a desired end effector trajectory depicted in Fig. 4 is used. This corresponds to the state references shown in Fig. 5.

Instead of calculating the equivalent control, NN is used for the implementation studies. It was seen that with proper selection of gain matrices (G and D), the trajectory following performance is very good. Generally, G and D were selected as,

$$G = \begin{bmatrix} 5 & 0 & 1 & 0 \\ 0 & 5 & 0 & 1 \end{bmatrix} \text{ and } D = \begin{bmatrix} 2 & 0 \\ 0 & 2 \end{bmatrix}$$

In the weight adaptation process, the error which was used to adapt the weights (i.e. the additional control term(ΔU)) was accepted as the additional control term divided by the maximum control value. This is a protection against blowing the weights up. Also, the net output (U_{eq}) is multiplied by the maximum control value before applying the robot.

To eliminate chattering, shifted sigmoid function is used as described in (18). The adaptation rate (μ) is selected as 0.01.

The experimental results are presented from figs 6 to 8. The robot perfectly follows the desired trajectory.

Figure 3. The Experimental Direct Drive Scara Robot

5. Conclusions

In this paper, firstly, the SMC is designed by selecting a Lyapunov function. The design yields an equivalent control term plus an additional control term. Secondly, the relation between inverse dynamics of a robot and the equivalent control term is presented. As a result, the structures of the inverse dynamics and the equivalent control are equal. Thirdly, it is surveyed how the NNs are used to compute the inverse dynamics. Fourthly, a NN is proposed to compute the equivalent control.

The main contribution of the paper concerns the computation of the equivalent control term in the control equation with using neural networks. Its direct calculation

---

Box 1. The Application Algorithm

Step 1. Initialize
Set all weights of NN to small random values. Such as 0.05. Select a value for the adaptation rate as; $0<\mu<1$.

Step 2. Compute the hidden layer and net outputs
Using the equations (24) and (25) compute the net outputs to form the equivalent control

Step 3. Compute ΔU
Compute the additional control term (ΔU) as in (17).

Step 4. Apply control to robot
Sum the equivalent control and additional control term to form the control signal to be applied to the robot.

Step 5. Measure the outputs of the robot
Measure the angular positions and velocities of the robot.

Step 6. Update the weights of NN
Update the weights of NN as given below,

$Wy_{j,i}(t+1) = Wy_{j,i}(t) + \mu\, \delta y_j\, Yout_i$ j=1..2, i=1..M
$Wz_{j,i}(t+1) = Wz_{j,i}(t) + \mu\, \delta z_j\, Z_i$ j=1..M, i=1..8

where,

$\delta y_j = (\Delta U_j)(1 - Ueq_j^2)$ and $\delta z_j = \left(\sum_{k=1}^{2} \delta y_k Wy_{k,j}\right)\left(1 - Yout_j^2\right)$

Step 7. Repeat by going to step 2

---

requires a complete knowledge of the plant dynamics and the parameters, which is rarely the case in practice. There are some classical estimation techniques, but they are generally not easy to use. By showing the equivalence of the inverse dynamics and the equivalent control, a NN is proposed to compute the equivalent control.

The proposed method has the following advantages;

1. There is no need to know the inverse dynamics and the parameters to compute the equivalent control.

2. There is no need to compute the inertia (or inverse) matrix to compute the equivalent control.

3. No need to use the desired or actual acceleration term in NN. The most of the NN structures need it to compute the inverse dynamics.

4. It is a robust Neuro-Controller.

The experimental results presented in this paper indicate that the suggested approach has considerable advantages compared to the classical one and is capable of achieving a good chatter-free trajectory following performance without an exact knowledge of plant parameters. These characteristics make it a promising approach for motion control applications.

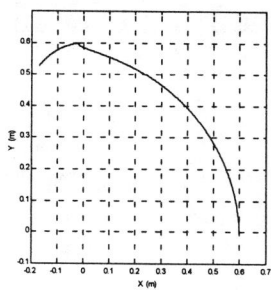 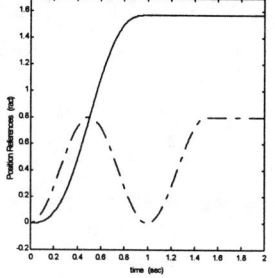

Figure 4. End effector Ref.   Figure 5. Angular Pos. Ref.

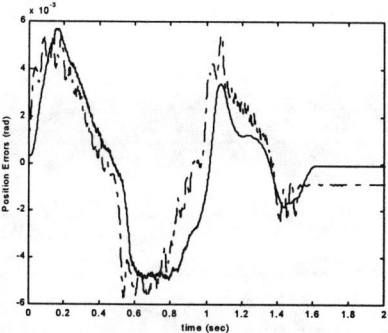

Figure 6. Angular Errors

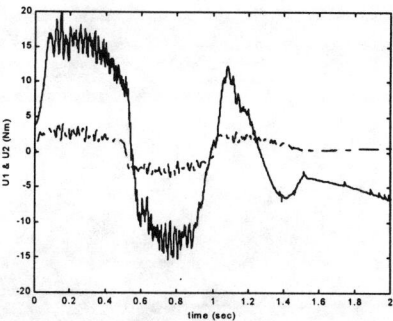

Figure 7. Controller Output 1&2

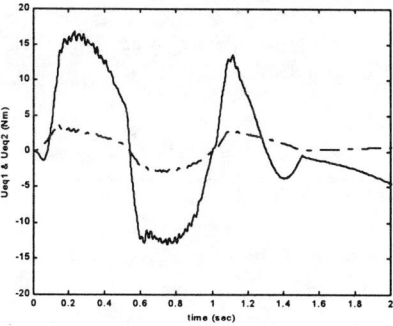

Figure 8. Equivalent Control 1&2

References

[1] Hung J. Y., *"Variable Structure Control: A survey"*, IEEE Transaction on industrial Electronics, Vol.40, no.1, February, 1993.

[2] Utkin, V.I., *"Sliding Modes in Control Optimization"*, Springer-Verlag, 1981.

[3] J. J. Slotine and W. Li, *"Applied Nonlinear Control"*, Prentice Hall, 1991.

[4] Ertugrul M., Kaynak O., Sabanovic A., Ohnishi K. *"A Generalized Approach For Lyapunov Design Of Sliding Mode Controllers For Motion Control Applications"*, Advanced Motion Control Conference, AMC'96, Mie, Japan, 1996.

[5] Tzafestas S.G., *"Neural Networks in Robotics: State of the Art"*, IEEE Int. Conf. On Industrial Electronics, 1995.

[6] Kawato M., Uno Y., Isobe M., Suzuki R. A., *"Hierarchical model for voluntary movement and with application to robotics"*, IEEE Control System Magazine, pp.8-16, April 1988.

[7] Fukuda T., and Shibata T., *"Theory and Applications of Neural networks for Industrial Control Systems"*, IEEE Trans. On Industrial Electronics, Vol.39, No.6, pp.472-489, 1992.

[8] Barto A.G., *"Neural Networks for Control"*, Chapter 1, pp.5-58. MIT Press, Cambridge, MA.

[9] Hunt K. J., Sbarbaro D., Zbikowski R. and Gawthrop P. J., *"Neural Networks for Control Systems-A Survey"*, Automatica Vol.28, No.6, pp1083-1112, 1992

[10] Ozaki T., Suzuki T., Furuhashi T., Okuma S., Uchikawa Y., *"Trajectory Control of Robotic Manipulators Using Neural Networks"*, IEEE transactions on Industrial Electronics, Vol. 38, No.3, June 1991.

[11] Lippmann R.P., *"An Introduction to computing with neural nets"*, IEEE ASSP magazine, April 1987.

[12] Bose N.K., Liang P., *"Neural network fundamentals with graphs, algorithms, and Applications"*, McGraw-Hill Int., 1996.

[13] J. Albus, *"A new approach to manipulator control: The cerebellar model articulation controller (CMAC)"*, J. Dyn. Syst. Meas. and Control, pp220-227, 1975.

[14] D. Psaltis, A. Sideris, A. Yamamura, *"Neural Controllers"*, Proc. IEEE Int. Conference on Neural Networks, 1987.

[15] A. Guez, Z. Ahmad, *"Solution to the inverse problem in robotics by neural networks"*, Proc. Int. Conference on Neural Networks, 1988.

[16] R. Elsley, *"A Learning architecture for control based on back-propagation neural-networks"*, IEEE Conf. on Neural Networks, Vol.2, pp.584-587, 1988.

[17] S. Grossberg and M. Kuperstein, *"Neural Dynamics of Adaptive Sensory-motor Control"*, Elmsford, NY:Pergamon Press, 1989.

[18] T. Fukuda et al, *"Neural Servo controller; Adaptation and Learning"*, Proc. IEEE Intl. Workshop Advanced Motion Contr., pp.107-115, 1990.

[19] G. A. Bekey, *"Robotics and Neural Networks"*, in Neural Networks for Signal Processing, B. Kosko, Prentice Hall, 1992.

[20] K. Narendra and K. Parthasarathy, *"Identification and control of dynamical systems using neural networks"*, IEEE Trans. Neural Networks, Vol.1, No.1, pp.4-27,1990.

[21] M. Tokita, T. Mitsuoka, T. Fukuda, and T. kurihara, *"Force control of robots by neural models: Control of one-dimensional manipulators"*, J. Japan Society of Robotics Engineers, vol..8-3, pp.52-59, 1989.

[22] *"Direct Drive Manipulator R&D package User Guide"*, Integrated Motion Incorporated, Berkeley CA.

Neural Force Control (NFC) Applied to Industrial Manipulators in Interaction with Moving Rigid Objects

M. Dapper, R. Maaß, V. Zahn, R. Eckmiller

University of Bonn, Dept. of Comp. Science VI
Römerstr. 164, D-53117 Bonn, F.R. Germany
Tel.: +49-228-73-4168, FAX: +49-228-73-4425
e-mail: dapper,maass,zahn,eckmiller@nero.uni-bonn.de

Abstract - *We developed a novel concept of hybrid (mixed) force/position control based on neural networks NFC (Neural Force Control) to significantly expand the range of manipulator applications. NFC includes neural approaches for complex robotic mappings such as inverse dynamics and kinematics: A Neural Dynamics Network (NDN), as the essential component of a computed torque controller, performs a fast and adaptive computation of the inverse manipulator model. The kinematic mappings are represented by a Neural Kinematics Network (NKN). The features of NKN provide singularity robustness and the handling of constraints in jointspace and Cartesian space (self collisions). To guarantee a tender impact while establishing contact between manipulator and surface, a cascaded velocity controller (CVC) is added to the NFC approach. Simulations for a 6DOF industrial manipulator have proved that the NFC concept is capable to manage various demanding tasks such as screw removal and surface tracking with high accuracy.*

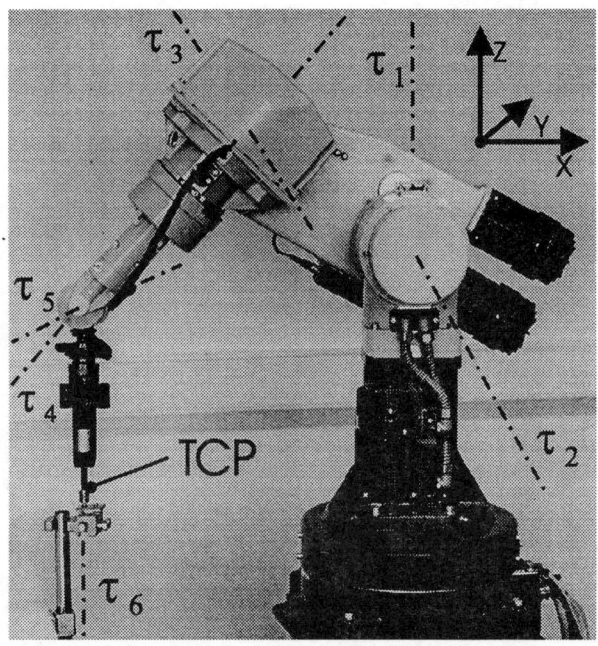

Fig. 1. 6 DOF manipulator (Siemens Manutec r2) with wrist force/torque sensor (ATI) and screwdriver tool as a testbed, where the simulated control laws will be implemented in the near future.

1 Introduction

Many robotic applications involve intentional interaction between the manipulator and the environment where contact stability to moving objects and surface tracking with defined contact force typify demanding tasks [6, 11, 12]. Establishing stable contact to rigid objects as well as tracking surfaces with unknown shape and elasticity are the prerequisites of a manipulator handling its environment naturally.

In several approaches of force/position control, the contact between the endeffector and the environment is assumed to be soft, realized by a flexible tool [2, 4, 7]. However, an artless tool and the environmental surface are often stiff, e.g. a screw driver and a metal surface like a screw head. At the contact point, the endeffector and the environmental surface will maintain their shape regardless of the forces exerted. Current technical approaches for hard contact are limited to special cases with a reduced number of degrees of freedom [5]. To make the endeffector of a manipulator follow in a stable way the edge or the surface of a workpiece while applying prescribed forces and torques we developed a novel control concept: Neural Force Control (NFC)[1].

---

[1] This work was supported in part by Federal Ministry for Education, Science, Research, and Technology (BMBF) under grant "DEMON"

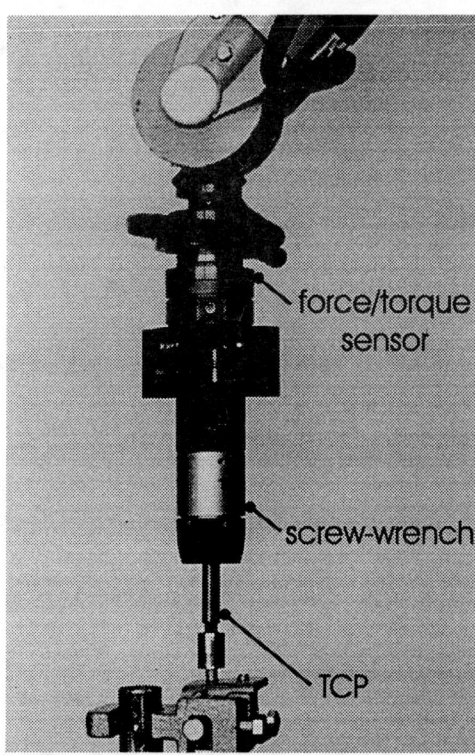

Fig. 2. Zoom of the manipulator's endeffector with 6 DOF force/torque sensor (ATI, FT Mini 100/5), to be controlled with the simulated control laws in the near future

2 Results

2.1 Control concept

NFC is a neural hybrid force/position controller [3, 9] based on the 'hard contact' approach. Containing the two modules NKN (fig. 3) and NDN (fig. 4), NFC (fig. 4) computes the control signals for the manipulator from the joint angle errors and the force errors at the endeffector. NKN delivers a fast computation of the singularity robust, differential kinematics (Jacobian matrix and its inverse), necessary for the NFC calculations. The inverse dynamics of the manipulator are performed by the NDN module. To establish stable contact to unknown surfaces or objects, CVC (fig. 6) is applied to the control concept to reach a tender impact. Fig. 1 shows the 6 DOF robot equipped with a 6 D-wrist force/torque sensor (Assurance Technologies Incorporation, FT Mini 100/5).

NKN: The NKN module (fig. 3) realizes the differential inverse kinematics (DIK) by calculating a modified Jacobian matrix $J_{Net}(\theta)$ and its inverse $J_{Net}^{-1}(\theta)$. NKN

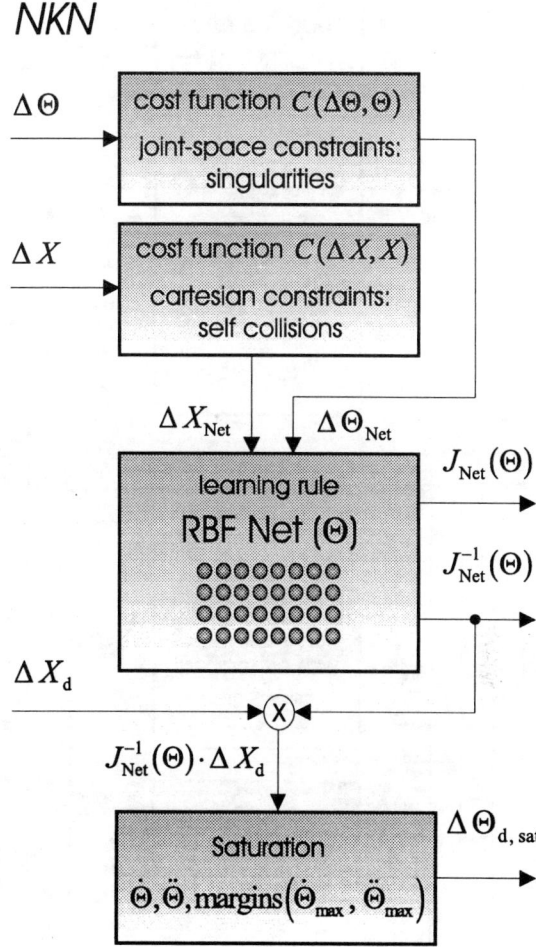

Fig. 3. Neural Kinematics Net (NKN) performs a fast computation of the Jacobian and the singularity robust (SR) inverse Jacobian of the manipulator. Cartesian constraints are taken into account.

meets two major requirements in robust robot control:

1. Singularities, appearing in the entire workspace, must not lead to undefined high joint angle velocities or tremendously increasing force errors in case of contact with surfaces. A strategy to maintain singularity robustness by using a modified inverse of the Jacobian matrix (SR-Inverse) has already been described [10]. This kind of inverse kinematics delivers continuous and feasible solutions at or in the neighborhood of singular positions. The SR-Inverse can be computed using singular value decomposition (SVD). To avoid the cumbersome SVD in the control loop, the SR-Inverse has been trained offline into a radial basis function type neu-

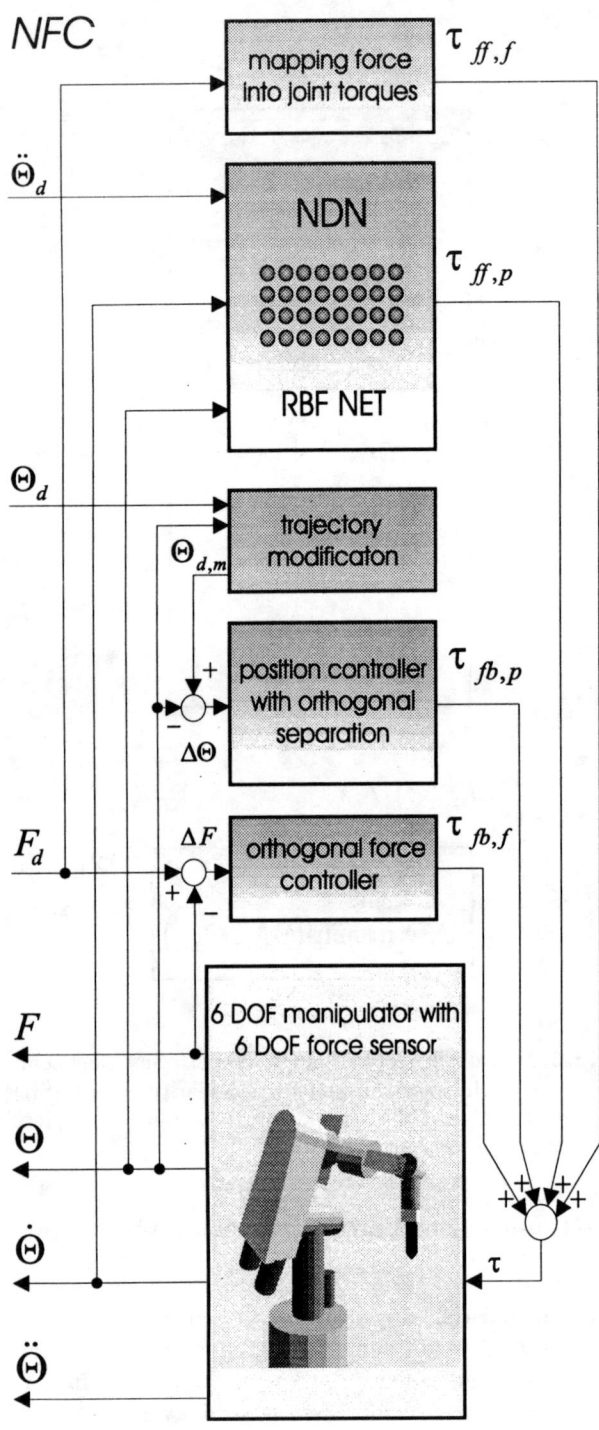

Fig. 4. Neural force/ position control system (NFC) inclusive 6DOF industrial manipulator with 2 ms sampling time in the control loop

static obstacles in workspace and joint angle margins. The implementation of the DIK as a NN includes the capability of modifying the inverse Jacobian such that critical positions cannot be reached.

The RBF net used in NKN consists of about 20,000 neurons. Its inner structure optimizes the total number of neurons by increasing the neuron density in critical regions (e.g. node allocation). The NN can be trained offline with the SR-Inverse of the Jacobian matrix as teacher input or online by using the variation of joint angle θ and cartesian vector x while tracking test movements. In the second approach the NN learns the DIK mappings only by using sensory input without any previous knowledge concerning the manipulator.

NDN: The NDN performs a nonlinear decoupling and feedback linearization of the manipulator's dynamics using the plants inverse dynamics model, trained into a neural network. NDN takes into account inertias, coriolis- and zentripetal effects as well as gravitation and various kinds of joint friction (coulomb-, stribeck- and viscose friction).

NDN is embedded into a neural hybrid force/position controlsystem NFC, achieving an orthogonal separation of force- and position constraints.

$$\tau_{ff,f} = \mathbf{J}^T(\boldsymbol{\theta}) \cdot \mathbf{S} \cdot \mathbf{F_d} \quad (1)$$

$$\tau_{ff,p} = \mathbf{M}(\boldsymbol{\theta})\ddot{\boldsymbol{\theta}} + \mathbf{C}(\boldsymbol{\theta},\dot{\boldsymbol{\theta}}) + \mathbf{G}(\boldsymbol{\theta}) + \mathbf{R}(\dot{\boldsymbol{\theta}}) \quad (2)$$

$$\tau_{fb,p} = \mathbf{J}^{-1}(\boldsymbol{\theta}) \cdot (\mathbf{I} - \mathbf{S}) \cdot \mathbf{J}(\boldsymbol{\theta}) \cdot \Delta\boldsymbol{\theta} \quad (3)$$

$$\tau_{fb,f} = \mathbf{J}^T(\boldsymbol{\theta}) \cdot \mathbf{S} \cdot \Delta\mathbf{F} \quad (4)$$

$$\tau = \tau_{ff,f} + \tau_{ff,p} + \tau_{fb,p} + \tau_{fb,f} \quad (5)$$

The most significant difference to common approaches is the realization of the inverse dynamics by neural networks [1, 8, 14]. The matrix elements of the inverse dynamics equation (eq. 2) like mass matrix $\mathbf{M}(\boldsymbol{\theta})$, coriolis coupling $\mathbf{C}(\boldsymbol{\theta},\dot{\boldsymbol{\theta}})$, gravitational influence $\mathbf{G}(\boldsymbol{\theta})$ and different types of angle velocity dependent frictions $\mathbf{R}(\dot{\boldsymbol{\theta}})$ (coulomb friction, stribeck friction and viscose friction) are represented by RBF nets respectively. The diagonal entries of the matrix $\mathbf{S}$ (eq. 1, 3, 4) associated with the components of the force to be controlled are chosen equal to one, and the remaining elements are zero. In the adaptation phase, the matrix $\mathbf{S}$ is completely set to zero, to turn off the force controller which allows to learn the invers model of the plant from the error $\Delta\tau$

ral network (NN).

2. Another feature of NKN is the handling of robot specific constraints as avoidance of self-collisions,

and the measured actual joint angles inclusive their first order- and second order derivations. During the adaptation phase, the robot executes a set of 500 accidentaly chosen fast point to point test movements, which excite nearly all mechanical frequencies of the plant, to get an optimal representation of the robots inverse dynamics.

The position controller in NFC is based on a differential inverse kinematics approach (DIK) combined with an orthogonal separation of force- and position constraints. As a linear approximation the DIK is only sufficiently correct when small inputs are used so NFC has to contain the function block **'trajectory modification'** (fig. 4), which performs the following incremental modification of the desired trajectory.

$$\theta_{dm,i} = \theta_{d,switch} + \theta_{T,i} + \theta_{F,i} \quad (6)$$

$$\theta_{T,i} = \theta_{T,i-1} + \mathbf{J_i^{-1}} \cdot [\mathbf{I} - \mathbf{S}] \cdot \mathbf{J_{d,i}} \cdot [\theta_{d,i} - \theta_{d,i-1}] \quad (7)$$

$$\theta_{F,i} = \theta_{F,i-1} + \mathbf{J_i^{-1}} \cdot [\mathbf{S}] \cdot \mathbf{J_i} \cdot [\theta_i - \theta_{i-1}] \quad (8)$$

The **trajectory modification** (eq. 6, 7, 8) becomes active when NFC switches from pure position control to hybrid force/position control by changing diagonal elements of the separation matrix **S** from zero to one. After that switching point $\theta_{d,switch}$ the trajectory is modified for each control step i by the equations (6) - (8) to reduce the control error, fed into the orthogonal position controller (eq. 3), to keep the differential inverse kinematics valid. At the switching control step $\theta_{T,i-1}$ and $\theta_{F,i-1}$ are initialized to zero. Fig. 5 shows an example why trajectory modification is necessary. The desired trajectory $\theta_\mathbf{d}$, representing a circle in the XY-plane has to be drawn on the unknown surface maintaining a defined TCP force in Z-direction. Thus, specified by the separation matrix **S**, the Z-coordinate of the TCP is not position controlled. The Z-component of the desired trajectory has to be adjusted (eq. 8) to the real trajectory produced by the force controller while the XY-components of $\theta_\mathbf{d}$ remain in their original shape (eq. 7).

CVC: A challenge in force/position control is the approaching phase while the TCP has not established contact to the unknown object yet and pure position control is switched to hybrid force/position control by modifying the separation matrix **S**. In this phase, the TCP velocity $\dot{\mathbf{X}}_\mathbf{sep}$ has to be bounded to a sufficient

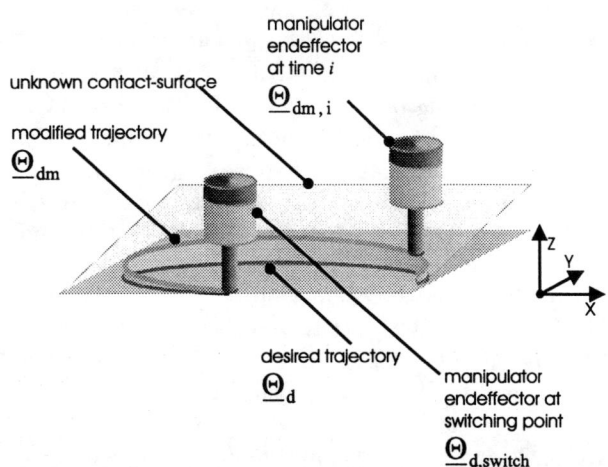

Fig. 5. Modification of the desired trajectory depending on the unknown surface to track

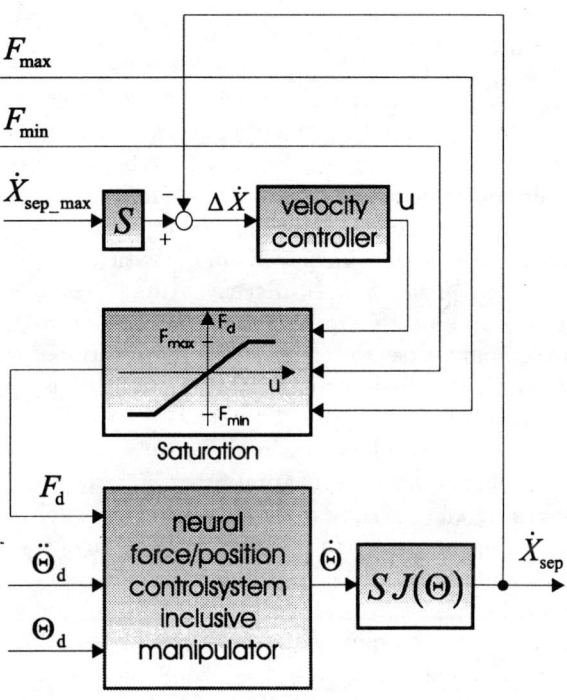

Fig. 6. Cascaded TCP-velocity controller (CVC) being active during approaching-phase for establishing contact to unknown object

small maximum value $\dot{\mathbf{X}}_\mathbf{sep,max}$ of e.g. 1.0 mm/sec to avoid damaging the manipulator or the force sensor during the impact into the unknown surface [13].

Having contacted the unknown object, the control system has got to change immediately from velocity control to force control, automatically performed by CVC (fig. 6). CVC consists of a velocity controller, saturation and mapping of joint angle velocities $\dot{\theta}$ to TCP velocities $\dot{\mathbf{X}}_{\mathbf{sep}}$. Having no contact to an object, the velocity controller changes the force $\mathbf{F_d}$, fed into the force/position control system, to obtain the desired TCP velocity $\dot{\mathbf{X}}_{\mathbf{sep,max}}$. After TCP acceleration and obtaining the desired velocity, $\mathbf{F_d}$ is reduced nearly to zero. Reaching the contact object, the actual TCP velocity is immediatly set to zero or a negative value (for a recoiling TCP) causing the velocity controller to run into the saturation raising $\mathbf{F_d}$ to the desired maximum value $\mathbf{F_{max}}$ of e.g. 100.0 N. Contact is established by a minimum impact effect independent of the distance between the TCP and the contact object when switching from position control to hybrid force/position control. The adequate value of $\dot{\mathbf{X}}_{\mathbf{sep,max}}$ for the actual manipulator configuration has to be estimated via an observer using hard contact simulations and adequate contact object constraints delivering the maximum impact force [15].

2.2 Simulation

Fig.7A shows the position of the metal screw to be contacted. After 1 second the manipulator control loop is switched from pure position control to hybrid force/position control whereas a force of 100 N in Z-direction is desired. 0.2 seconds later, the manipulator establishes contact to the unknown object with a maximum force overshoot of 20 percent. After 3 seconds the unknown object starts to move in Z-direction with 1 cm/sec and pushes the endeffector in Z-direction upwards while the desired contact force of 100 N has to be maintained (fig.7B).

Fig.7C shows examplary the simulated torques τ_2 and τ_3 of the manipulator (configuration as in fig.1) while establishing and exerting force in Z-direction. τ_2 has got a mostly negative value excited by the strong gravitational influence to be compensated while joint 3 exerts a mostly positive torque because link 3 rotates nearly its balance point and senses therefore only little gravitational influence and is completely used to generate the force at the endeffector.

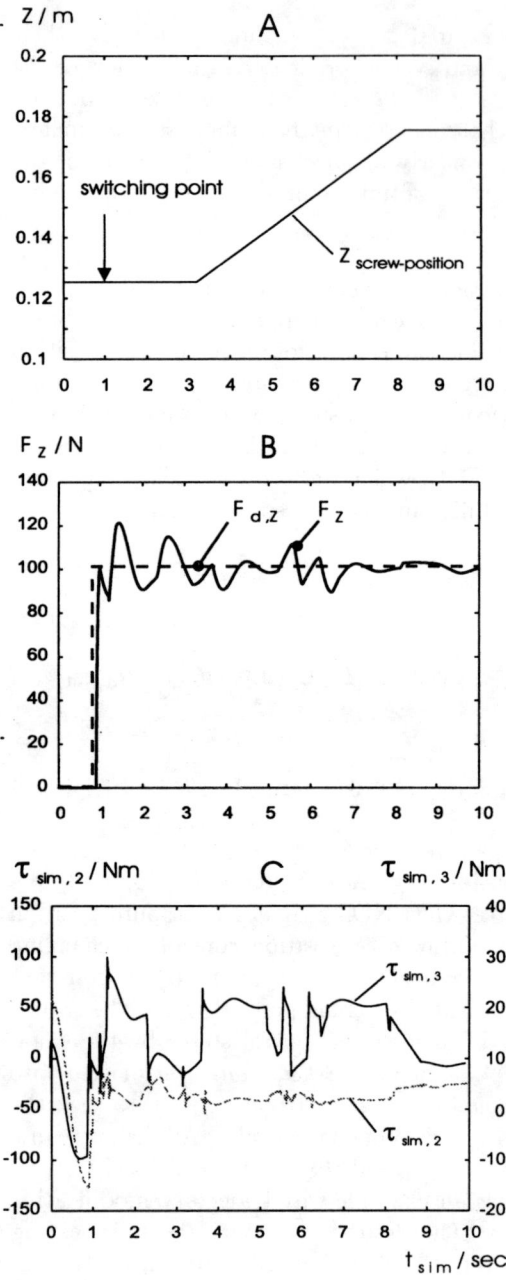

Fig. 7. A: Simulated screw-trajectory in Z-direction (slope 1cm per second). Components in X- and Y-direction remain constant. B: Simulated force exerted at endeffector while contacting unknown metal screw and while movement of the screw. C: Simulated torque of joint 2 and joint 3 while approaching and contacting object (metal screw) and switching from neural position control to neural hybrid force/position control.(2 ms sampling time in the simulated control loop)

3 Discussion

Simulations and first experiments have shown, that neural hybrid force/position control NFC allows contacting rigid objects without any need of flexiblity in the endeffector or in the manipulator construction. It has been generally recognized, that force control may cause unstable behaviour during contact with the environment.

The robustness of the closed loop controllers against disturbances is also very important for tasks like surface tracking, where the contact force has strong influence on the contact friction, disturbing the position controller [5].

A neural implementation of the differential inverse kinematics as it is done in the NKN concept has shown to be feasible. The advantage of this approach is the capability of an individual modification of the DIK to constraints in combination with a fast calculation necessary for a position separation in the realtime control loop.

4 Conclusions

NFC as a hybrid force control system with neural components combines the advantages of learning structures and advanced robot control theory. The high accuracy in force/position control using NN's offers a wide range of intentional interaction tasks of multi joint manipulators. In simulations we have demonstrated, that NFC allows the solution of various hard contact control problems including polishing, deburring, or disassembly, which require a continuous evaluation of the contact forces.

References

1. J.R. Beerhold. Stable adaptive closed-loop control of multi-joint robots using rbf-nets (in german). *at-Automatisierungstechnik*, 44(12):577–583, 1996.
2. G. Bonitz and T. Hsia. Robust internal force based impedance control for coordinating manipulators. In *IEEE Proc. International Conference on Robotics and Automation, Mineapolis, Minesota*, pages 622–528, 1996.
3. M. Dapper, R. Maaß, V. Zahn, and R. Eckmiller. Neural force control (nfc) for complex manipulator tasks. In *Springer Proc. Int. Conf. Artificial Neural Networks, ICANN97, Lausanne, October*, pages 787–792, 1997.
4. R.V. Dubey, T. F. Chan, and S. E. Everett. Variable damping impedance control of a bilateral telerobotic system. *IEEE Control Systems Magazine*, 17(1):37–45, 1997.
5. M.W. Dunnigan, D.M. Lane, A.C. Clegg, and I. Edwards. Hybrid position/force control of a hydraulic underwater manipulator. *Systems Engineering for Automation*, 143(2):145–151, 1996.
6. G. Ferretti, G. Magnani, and P. Rocco. On the stability of integral force control in case of contact with stiff surfaces. *Journal of Dynamic Systems, Measurement, and Control*, 117(4):547–553, 1995.
7. B. Heinrichs, N. Sepehri, and A. B. Thornton-Trump. Position-based impedance control of an industrial hydraulic manipulator. *IEEE Control Systems Magazine*, 17(1):46–52, 1997.
8. K. Kiguchi and T. Fukuda. Fuzzy neural friction compensation method of robot manipulation during position force control. In *IEEE Proc. International Conference on Robotics and Automation, Mineapolis, Minesota*, pages 372–377, 1996.
9. R. Maaß, V. Zahn, and R. Eckmiller. Neural force/position control in cartesian space for a 6 dof industrial robot: Concept and first results. In *IEEE Proc. Int. Conf. Neural Networks, ICNN97, Houston, June*, pages 1744–1748, 1997.
10. Y. Nakamura. *Advanced Robotics*. Addison-Wesley, 1991.
11. J. Steck, K. Rokhsaz, and S. P. Shue. Linear and neural network feedback for flight control decoupling. *IEEE Control Systems Magazine*, 16(4):22–30, 1996.
12. D. Surdilovic. Contact stability issues in position based impedance control: Theory and experiments. In *IEEE Proc. International Conference on Robotics and Automation, Mineapolis, Minesota*, pages 1675–1680, 1996.
13. T.J. Tarn, Y. Wu, N. Xi, and A. Isidori. Force regulation and contact transition control. *IEEE Control Systems Magazine*, 16(1):32–40, 1996.
14. D. A. White and D. A. Sofge. *Handbook of Intelligent Control, Neural, Fuzzy and Adaptive Approaches*. Van Nostrand Rheinhold, New York, 1992.
15. C.C. de Wit, B. Siciliano, and G. Bastin. *Theory of Robot Control*. Springer, 1997.

PSOM Network: Learning with Few Examples

Jörg A. Walter

Department of Computer Science · University of Bielefeld · D-33615 Bielefeld
Email: walter@techfak.uni-bielefeld.de

Abstract:

Precise sensorimotor mappings between various motor, joint, sensor, and abstract physical spaces are the basis for many robotics tasks. Their cheap construction is a challenge for adaptive and learning methods. However, the practical application of many neural networks suffer from the need of large amounts of training data, which makes the learning phase a costly operation – sometimes beyond reasonable bounds of cost and effort.

In this paper we discuss the "Parameterized Self-organizing Maps" (PSOM) as a learning method for rapidly creating high-dimensional, continuous mappings. By making use of available topological information the PSOM shows excellent generalization capabilities from a small set of training data. Unlike most other existing approaches that are limited to the representation of a input-output mappings, the PSOM provides as an important generalization a flexibly usable, *continuous associate memory*. This allows to represent several related mappings – coexisting in a single and coherent framework.

Task specifications for redundant manipulators often leave the problem of picking one action from a subspace of possible alternatives. The PSOM approach offers a flexible and compact form to select from various constraint and target functions previously associated.

We present application results for learning several kinematic relations of a hydraulic robot finger in a single PSOM module. Based on only 27 data points, the PSOM learns the inverse kinematic with a mean positioning accuracy of 1 % of the entire workspace. Another PSOM learns various ways to resolve the redundancy problem for positioning a 4 DOF manipulator.

1 Introduction

Many tasks in robotics require the availability of precise sensorimotor mappings – able to transform between various motor, joint, sensor, and abstract physical spaces. The construction of required relationship from empirical training data is a challenge for adaptive and learning methods. Unfortunately, many neural network approaches indeed require hundreds or thousands of examples and training steps. Since the acquisition of this data is related to cost and effort, this is a major obstacle for the practical application of those methods.

To make a learning system useful and efficient in robotics means that the learner provides *good generalization* capabilities – based on a *small* training data set, and uses a quick learning procedure without fragile learning parameters and without taking too much iteration time (with growing availability of computing power this need is increasingly relaxed, allowing also more elaborated algorithms to compete).

In this contribution we present the "Parameterized Self-Organizing Map" (PSOM) approach, which is particularly useful in situation where a high-dimensional, continuous mapping is desired. If information about the topological order of the training data is provided, or can be inferred, only a very small data set is required. In section 2 the PSOM algorithm is derived from Kohonen's Self-Organizing Map and the PSOM's *auto-associative* capabilities are presented.

In section 3 we report on a PSOM application for solving the forward and backward kinematics for a robot finger. If numerous degrees of freedom are available one has to pick a configuration of a continuous space of alternatives. Most solutions of this redundancy problem are based on some pseudo-inverse control (for a review see e.g. [2]). However a more flexible solution should offer a set of suitable action strategies and should offer to respond to different types of constraints. Sec. 4 suggests an *associative memory* that completes partial specification of a incomplete task specification as a natural solution. But in contrast to spin-glass type attractor networks, in robotics, we need a continuous attractor manifold instead of just isolated points. Here we show that a PSOM can provide the functionality of representing continuous relations in conjunction with a favorable flexibility to specify additional goals.

2 From SOMs to PSOMs

Kohonen [3] formulated the *Self-Organizing Map* (SOM) algorithm as a mathematical model of the self-organization of topographic maps, which are found in brains of higher animals. Fig. 1 illustrates a two-dimensional array A of processing units or formal "neurons". Each neuron has a reference vector $\mathbf{w_a}$ attached, which points in the embedding input space X. A presented input $\mathbf{x}$ will select that neuron $\mathbf{a}^*$ with $\mathbf{w_a}$ closest to the given input: $\mathbf{a}^* = \mathrm{argmin}_{\forall \mathbf{a} \in \mathbf{A}} \|\mathbf{w_{a'}} - \mathbf{x}\|$. This competitive mechanism tessellates the input space in *discrete* patches – the so-called *Voronoi cells* (see light gray border lines).

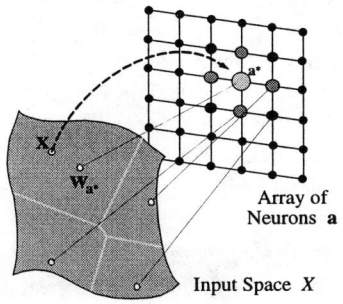

Figure 1: The "Self-Organizing Map" ("SOM") algorithm builds a topographic map and tessellates the input space to discrete Voronoi-cells.

The Kohonen learning rule (see e.g. [3, 6]) generates a dimension reducing, topographic mapping from a high-dimensional input space to a m-dimensional index space of neurons in the array S. Topographic order means that neighboring neurons are responsible for similar input situations $\in X$. The main features are: *(i)* the generation of the *topographic order* and, *(ii)* as a result, each learning step can profitably be shared among neighboring neurons, allowing to improve the convergence properties of the algorithm.

How can the SOM-network learn a smooth continuous input–output mapping? A simple strategy is the supervised teaching of a constant output value $y_\mathbf{a}$ (or vector $\mathbf{y_a}$) per neuron $\mathbf{a}$. The network output is then $F(\mathbf{x}) = y_{\mathbf{a}^*}$ of the winner neuron $\mathbf{a}^*$. The first improvement to increase the output precision is the introduction of a locally linear regression scheme for each neuron $\mathbf{a}$ and returning the "winner" neuron's output: $F(\mathbf{x}) = y_{\mathbf{a}^*} + \mathbf{B}_{\mathbf{a}^*}(\mathbf{x} - \mathbf{w}_{\mathbf{a}^*})$; i.e. a set of (hyper-) planes approximate the desired function. Unfortunately, in general the planes do not match at the borders of the Voronoi-cells, which may leave discontinuities in the overall mapping.

The PSOM concept [5] can be seen as the generalization of the SOM with the following three main extensions:

- the index space S in the Kohonen map is generalized to a *continuous mapping manifold* $S \in \mathbb{R}^m$.

- The embedding space $X = X^{in} \times X^{out} \subset \mathbb{R}^d$ is formed by the Cartesian product of the input space and output space.

- We define a *continuous mapping* $\mathbf{w}(\cdot) : \mathbf{s} \mapsto \mathbf{w}(\mathbf{s}) \in M \subset X$, where $\mathbf{s}$ varies continuously over $S \subseteq \mathbb{R}^m$.

We require that the *embedded manifold* M passes through all supporting reference vectors $\mathbf{w_a}$ and write $\mathbf{w}(\cdot) : S \to M \subset X$ as weighted sum:

$$\mathbf{w}(\mathbf{s}) = \sum_{\mathbf{a} \in \mathbf{A}} H_\mathbf{a}(\mathbf{s})\, \mathbf{w_a} \qquad (1)$$

This means that, we need a *"basis function"* $H_\mathbf{a}(\mathbf{s})$ for each formal neuron or "node", weighting the contribution of its reference vector (= initial "training point") $\mathbf{w_a}$. The $H_\mathbf{a}(\mathbf{s})$ depend on the location $\mathbf{s}$ relative to the node position $\mathbf{a}$, and also *all* other nodes $\mathbf{A}$ (however, we drop in our notation the dependency $H_\mathbf{a}(\mathbf{s}) = H_{\mathbf{a};\mathbf{A}}(\mathbf{s})$ on $\mathbf{A}$).

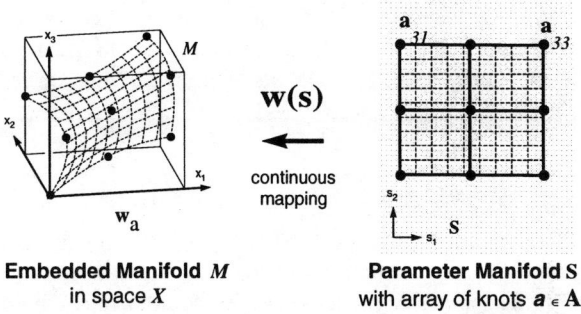

Embedded Manifold M in space X **Parameter Manifold** S with array of knots $\mathbf{a} \in \mathbf{A}$

Figure 2: The mapping $\mathbf{w}(\cdot) : S \to M \subset X$ builds a continuous image of the *right* side S in the embedding space X, as illustrated by the dotted test grid.

A suitable set of basis functions can be constructed in several ways but must meet two conditions: *(i)* the hyper-surface M shall pass through all desired support points (*orthonormality*), i.e. at those points, only the local node contributes $H_{\mathbf{a}_i}(\mathbf{a}_j) = \delta_{ij}$; $\forall\, \mathbf{a}_i, \mathbf{a}_j \in \mathbf{A}$; *(ii)* the sum of all contribution weights must be one: $\sum_{\mathbf{a} \in \mathbf{A}} H_\mathbf{a}(\mathbf{s}) = 1$, $\forall \mathbf{s}$ (*partition-of-unity*).

A simple construction of basis functions $H_\mathbf{a}(\mathbf{s})$ becomes possible when the topology of the given points is sufficiently regular. A particularly convenient situation arises for the case of a multidimensional rectangular grid. In this case, the set of functions $H_\mathbf{a}(\mathbf{s})$ can be constructed from products of one-dimensional Lagrange interpolation polynomials. See [7] for details.

Specifying for each training vector $\mathbf{w_a}$ a node location $\mathbf{a} \in \mathbf{A}$ introduces a *topological order* between the training points: training vectors assigned to nodes $\mathbf{a}$ and $\mathbf{a}'$, that are adjacent in the lattice $\mathbf{A}$, are perceived to have this specific neighborhood relation. The effect is important to note: it allows the PSOM to *draw extra curvature information* from the training set. Such information is not available within other techniques, such as the RBF approach and is the essential reason for the generalization capabilities of the PSOM (see below Fig. 5–7).

When M has been specified, the PSOM is used similar to the SOM: given an input vector $\mathbf{x}$, *(i)* find the best-match position $\mathbf{s}^*$ on the mapping manifold S by minimizing the distance function $dist(\cdot)$

$$\mathbf{s}^* = \mathbf{s}(\mathbf{x}) = \underset{\forall \mathbf{s} \in S}{\operatorname{argmin}} \; dist(\mathbf{w}(\mathbf{s}), \mathbf{x}). \quad (2)$$

(ii) The surface point $\mathbf{w}(\mathbf{s}^*)$ serves as the output of the PSOM in response to the input $\mathbf{x}$. The output $\mathbf{w}(\mathbf{s}^*)$ can be viewed as an *associative completion* of the input space component of $\mathbf{x}$ if the distance function $dist(\cdot)$ (in Eq. 2) is chosen as the Euclidean norm applied only to the input components of $\mathbf{x}$ (belonging to X^{in}). Thus, the function $dist(\cdot)$ actually selects the input subspace X^{in}, since for the determination of $\mathbf{s}^*$ (Eq. 2) and, as a consequence, of $\mathbf{w}(\mathbf{s}^*)$, only those components of $\mathbf{x}$ matter, that are regarded in the distance metric $dist(\cdot)$. A suitable definition is

$$dist(\mathbf{x}, \mathbf{x}') = \sum_{k=1}^{d} p_k (x_k - x'_k)^2. \quad (3)$$

which selects all components k with $p_k > 0$ as belonging to the input subspace; output are components k with $p_k = 0$. By changing the coefficients p_μ the PSOM the mapping direction can be (e.g.) reversed, as illustrated in Fig. 3.

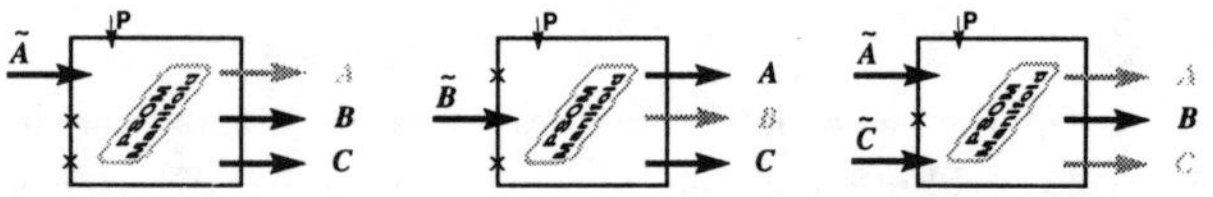

Figure 3: "Continuous associative memory" supports multiple mapping directions. The specified $\mathbf{P}$ vector selects different subspaces (here symbolized by $\tilde{A}$, $\tilde{B}$ and $\tilde{C}$) of the embedding space as inputs. Values of variables in the selected input subspaces are considered as "clamped" (indicated by a tilde) and determine the values found by the iterative least square minimization (Eq. 2). for the "best-match" vector $\mathbf{w}(\mathbf{s}^*)$. This provides an associative memory for the flexible representation of continuous relations.

The discrete best-match search in the standard SOM is now replaced by solving the continuous minimization problem for the determination of $\mathbf{s}^*$ in Eq. 2. A simple approach is to first perform the (SOM-like) discrete best-match search to find $\mathbf{s}_{start} = \mathbf{a}^*$ in the knot set $\mathbf{A}$, followed by an iterative procedure like the gradient descent. We found the Levenberg-Marquardt algorithm [4] best suited to find $\mathbf{s}^*$ in a couple of iterations.

In this scheme M can be viewed as a continuous attractor manifold with a recurrent dynamic. Since M contains the data set $\{\mathbf{w_a}\}$, any at least m-dimensional "fragment" of the data set will be attracted to the completion $\mathbf{w}$. Any other input will be attracted to an interpolation manifold point.

3 Application: The Robot Finger Kinematics

This section presents the results of applying the PSOM algorithm to the task of learning the kinematics of a 3 degree-of-freedom robot finger of a three-fingered modular hydraulic robot hand, developed by the Technical University of Munich. The finger is actuated by spring-loaded oil cylinders driven by a remote "base station" that provides the hydraulic pressure. Its mechanical design allows roughly the mobility of the human index finger, scaled up to 110%. A cardanic base joint (2 DOF) offers sidewards gyring of $\pm 15°$ and full adduction with two additional coupled joints (1 DOF). See Fig. 4.

In the case of the finger, there are several coordinate systems of interest, e.g. the joint angles $\vec{\theta}$, the cylinder piston positions $\vec{c}$, one or more finger tip coordinates $\vec{r}$, as well as further configuration dependent quantities, such as the Jacobian matrices J for force/moment transformations. All of these quantities can be simultaneously treated in one single PSOM allowing to map in multiple ways, as indicated in Fig. 3. Here we present results of the inverse kinematics, the classical hard part. When moving the three joints on a cubical $10 \times 10 \times 10$ grid within their maximal configuration space, the fingertip will trace out the "banana" grid displayed in Fig. 4 (confirm this workspace with your finger).

We exercised several PSOMs with $n \times n \times n$ nine dimensional data tuples $(\vec{\theta}, \vec{c}, \vec{r})$, all equidistantly sampled in $\vec{\theta}$. Fig. 5a–b depicts a $\vec{\theta}$ and an $\vec{r}$ projection of the smallest training set, $n = 3$.

To visualize the inverse kinematics ability, we ask the PSOM to back-transform a set of workspace points of known arrangement. In particular, the workspace filling "banana" set of Fig. 4 should yield a rectangular grid of $\vec{\theta}$. Fig. 5c–e displays the actual result. Distortions can be vi-

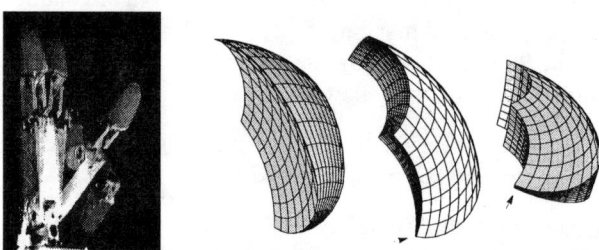

Figure 4: a–d: *(a)* stroboscopic image of one finger in a sequence of extreme joint positions. *(b–d)* Several perspectives of the workspace envelope $\vec{r}$, tracing out a cubical $10 \times 10 \times 10$ grid in the joint space $\vec{\theta}$. The arrow marks the fully adducted position, where one edge contracts to a tiny line.

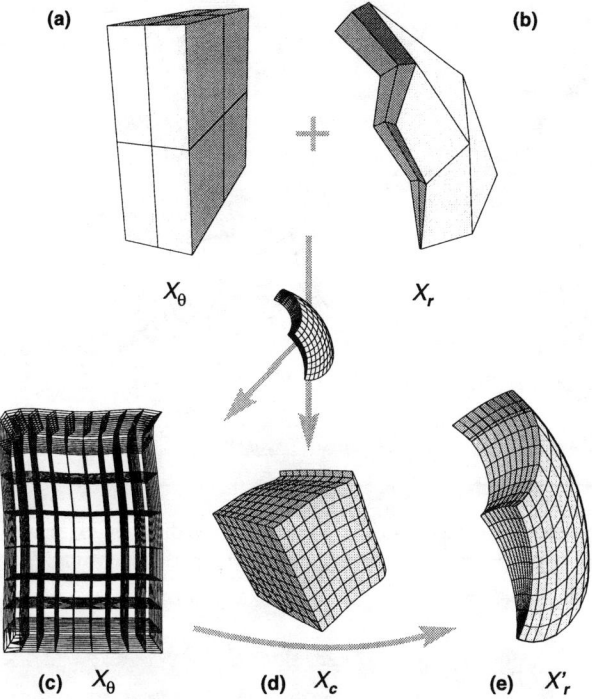

Figure 5: a–b and c–e; Training data set of 27 nine-dimensional points in X for the 3×3×3 PSOM, shown as perspective surface projections of the (a) joint angle $\vec{\theta}$ and (b) the corresponding Cartesian sub space. Following the lines connecting the training samples allows one to verify that the "banana" really possesses a cubical topology. (c–e) Inverse kinematic result using the grid test set displayed in Fig. 4. (c) projection of the joint angle space $\vec{\theta}$ (transparent); (d) the stroke position space $\vec{c}$; (e) the Cartesian space $\vec{r}'$, after back-transformation.

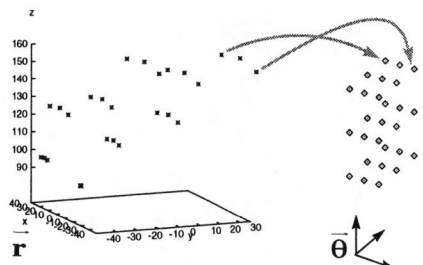

Figure 6: The 27 training data vectors for the Back-propagation networks: (left) in the input space $\vec{r}$ and (right) the corresponding target output values $\vec{\theta}$.

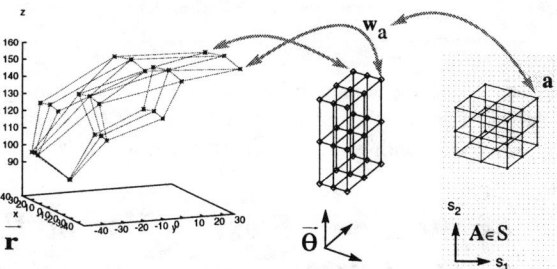

Figure 7: The same 27 training data vectors (cmp. Fig. 6) for the bi-directional PSOM mapping: (left) in the Cartesian space $\vec{r}$, (middle) the corresponding joint angle space $\vec{\theta}$. (Right:) The corresponding node locations $\mathbf{a} \in \mathbf{A}$ in the parameter manifold S. Neighboring nodes are connected by lines, which reveals now the "banana" structure on the left.

sually detected in the joint angle space (c), and the piston stoke space (d), but disappear after back-transforming the PSOM output to world coordinates (b). The reason is the peculiar structure; e.g. in areas close to the tip a certain angle error corresponds to a smaller Cartesian deviation than in other areas.

When measuring the mean Cartesian deviation we get an already satisfying result of 1.6 mm or **1.0 %** of the maximum workspace length of 160 mm (using a test set of 500 randomly chosen positions). In view of the extremely small training set displayed in Fig. 5a–b this appears to be a quite remarkable result.

Nevertheless the result can be further improved by supplying more training points. For a growing number of network nodes the "Local-PSOM" approach offers to keep the computational effort constant by applying the PSOM algorithm on a sub-grid, see [8, 7].

PSOM versus MLP: For comparison reasons, we employed the standard Multi-Layer-Perceptron with one and two hidden layers and linear units in the output layer. We found that this problem is not suitable for the MLP network. Even for larger training set sizes, we did not succeed in training them to a performance comparable to the PSOM network.

Why does the PSOM perform more than an order of magnitude better than the back-propagation algorithm? Fig. 6 shows the 27 training data pairs; on the left side, the Cartesian input space $\vec{r}$, one can recognize some zig-zag structure, but not much more. Fig. 7 depicts the PSOM situation: the PSOM gets the same data-pairs as training vectors — but additionally, it obtains the assignment to the node location $\mathbf{a}$ in the 3×3×3 node grid illustrated in Fig. 7. If neighboring nodes are connected by lines, it is easy to recognize the coarse "banana" shaped structure. Using the curvature information contained in the ordered data the PSOM could generalized to the reported positioning precision of 1%. This topological information is not available to other techniques, like the MLP or the radial basis approach.

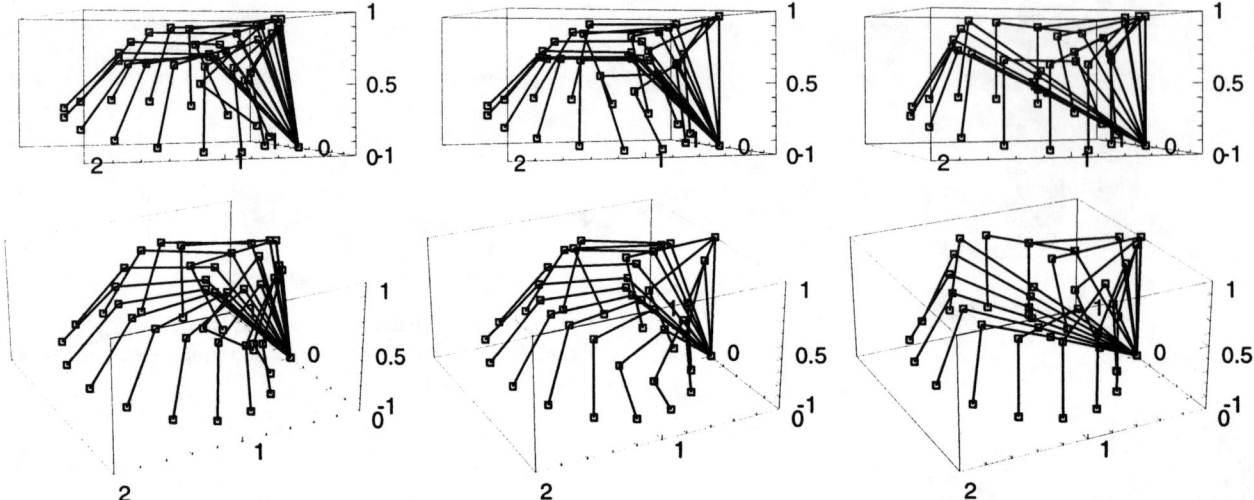

Figure 8: Tracking a point in a slanted elliptical path with a 4 DOF manipulator, using a 5×5×5×5 *PSOM* and three different extra target functions to resolve the redundancy problem: *(left)* with maximal similarity of the last two joints; *(middle)* with horizontal middle arm segment; *(rigth)* with vertical distal arm segment – as far as possible.

4 Application: Flexible Use of Redundant DOF

As mentioned earlier, in the presence of excess degrees of freedom one may specify extra constraints to determine a robot configuration. The question is how this can be done in a versatile manner? Here, the PSOM contributes an elegant way to offer several goal and contraint functions for flexible usage.

We illustrate the flexibility of the PSOM approach in the task to position a 4 DOF robot in 3D as depicted in Fig. 8. The simulated robot consists of a vertical revolute joint θ_1 and three horizontal ones $\theta_{2..4}$ with link lengths 0,1,0.7,0.6 l. For the construction of a PSOM, i.e. the configuration manifold of the arm, we choose an 5×5×5×5 grid covering the joint range $\vec{\theta} \in [0°, 180°] \times [0°, 120°] \times [-120°, 0°] \times [-120°, 0°]$. The embedding space X is spanned by $\mathbf{x} = (\theta_1, \theta_2, \theta_3, \theta_4, r_x, r_y, r_z, c_8, c_9, c_{10})^T$ and contains, similar to the example before, the angles $\theta_{1..4}$, the Cartesian position $\vec{r}$, and, here, three further parameters: c_8 is the difference $(\theta_4 - \theta_3)^2$, c_9 is the elevation angle of link-3 (relative to the horizontal), and c_{10} is the angle between distal link-4 and the vertical.

This allows to resolve the redundancy in various ways. For example, the goal can be to ...

(i) take the minimal joint motion from the current position to the specified position $\vec{r}$: all we need to do is to start the best-match search ($p_{5..7} = 1$) at the best-match position $\mathbf{s}^*_{curr}$ belonging to the current position, and the steepest gradient descent procedure will solve the problem;

(ii) keep joint $j \in \{2, 3, 4\}$ fixed: additionally specify θ_j and $p_j > 0$;

(iii) use similar adduction in the two distal joints (like the finger kinematics): by activating $p_8 > 0$ (to a small value e.g., 0.01) and setting $x_9 = 0$. Measuring the deviation for the inverse kinematics we find a mean value of 0.008 in the workspace; Fig. 8*(left)* depicts the solution for tracing a elliptical path for the end effector;

(iv) keep the middle segment horizontal: by specifying the target $x_9 = 0$ and $p_9 = 0.01$. Fig. 8*(middle)* reveals that this constraint can not be met in all cases. By setting p_9 to only a small value, as a "soft goal", the accuracy of the trajectory is not (significantly) compromised (see also below);

(v) approach vertically: after specifying $x_{10} = 0$, $p_{10} = 0.01$, Fig. 8*(right)* shows the stroboscopic tracking result.

For these different cases we do not need different networks, instead one single PSOM can be utilized. If one anticipates useful target functions, the embedding space can be augmented in advance, enabling to construct reconfigurable optimization modules. They are later activated on demand and show the desired performance. In conflicting situation, e.g. the distal reaching positions in the last example, a meaningful compromise is found. As shown in [7], the input selection coefficients can be made dynamical $p_\mu = p_\mu(t)$ during the iteration process. I.e., secondary goal functions are weighted by $p_\mu(t)$, starting at a small value, which decrease to zero. Primary positioning goals are not compromised and secondary goals satisfied as much as possible. This procedure allows to define priorities of goal functions, which are solved according to there rank.

5 Discussion and Conclusion

We presented the PSOM as a versatile module for learning continuous, high-dimensional mappings. As highlighted by the robot finger example, the PSOM draws its good generalization capabilities from curvature information available through the topological order of only a few reference vectors $\{\mathbf{w_a}\}$. This topological assignment can be learned by *(i)* Kohonen's SOM learning rule, or *(ii)* by construction – if the topological relation of the data is known. The first is an iterative learning method which requires more training data than the second. We find that in many robotic applications, the latter case can be realized by active, structured sampling of the training data – often without any extra cost. This can be viewed as a special mechanism for incorporating prior knowledge into the learning system. In the robot finger example a set of only 27 data points turns out sufficient to approximate the highly non-linear 3 D kinematics relation with remarkable precision. The inverse kinematics showed a mean positioning deviation of only 1 % of the entire workspace range.

Due to the compactness of the training set, the PSOM has some overlap with fuzzy networks (e.g. [10]): An expert defines a fuzzy class, assigns linguistic names (e.g. "left", "middle", "right stroke position"), and (initially) provides suitable output values. In the PSOM learning process, the grid node values **a** can be assigned to input-output pairs. Likewise, names can be alloted to support the interpretation of the learned knowledge.

The PSOM's associative mapping concept has various attractive properties. Several coordinate spaces can be maintained and learned simultaneously, as shown in the robot finger example. E.g., the *multi-way mapping* capability can solve the forward and inverse kinematics within the very same network. This simplifies learning and avoids worries about inconsistencies of separate learning modules. As pointed out by Kawato [1], the learning of bi-directional mappings is not only useful for the planning phase (action simulation), but also for bi-directional sensor–motor integrated control.

Another potential PSOM application is the representation of system states together with a set of values from different sensors. Here, the PSOM can serve as an integrated sensor data fusion mechanism. It allows not only to incrementally fuse available data, but also to deliver intermediate data predictions, usable for sensor guidance. For details we must refer to [7] where also examples of the full 6 DOF robot kinematics can be found together with responses to sudden changes in underlying mapping task. See [9] for several applications in visuo-motor coordination.

The input selection mechanism enables to easily add further, parameterized target functions. Those can be utilized to resolve redundancy problems when tasks are underspecified. Here the PSOM offers to build a battery of optimizer modules which can be learned within the same continuous associative memory. On demand, one can select or combine auxiliary goals by activating the components in the distance metric (p_k in Eq. 3). The continuous associative completion serves here as an elegant and compact mechanism to provide a variety of options and solutions.

References

[1] Mitsuo Kawato. Bi-directional neural network architecture in brain functions. In *Proc. Int. Conf. on Artificial Neural Networks (ICANN-95), Paris*, volume 1, pages 23–30, 1995.

[2] C.A. Klein and C.-H. Huang. Review of pseudoinverse control for use with kinematically redundant manipulators. *IEEE Trans. Sys. Man and Cybern.*, 13:245–250, 1983.

[3] Teuvo Kohonen. *Self-Organization and Associative Memory*. Springer Series in Information Sciences 8. Springer, Heidelberg, 1984.

[4] W. Press, B. Flannery, S. Teukolsky, and W. Vetterling. *Numerical Recipes in C – the Art of Scientific Computing*. Cambridge Univ. Press, 1988.

[5] Helge Ritter. Parametrized self-organizing maps. In S. Gielen and B. Kappen, editors, *Proc. Int. Conf. on Artificial Neural Networks (ICANN-93), Amsterdam*, pages 568–575. Springer Verlag, Berlin, 1993.

[6] Helge Ritter, Thomas Martinetz, and Klaus Schulten. *Neural Computation and Self-organizing Maps*. Addison Wesley, 1992.

[7] Jörg Walter. *Rapid Learning in Robotics*. Cuvillier Verlag Göttingen, 1996. also postscript http://www.techfak.uni-bielefeld.de/~walter/pub/.

[8] Jörg Walter and Helge Ritter. Local PSOMs and Chebyshev PSOMs – improving the parametrised self-organizing maps. In *Proc. Int. Conf. on Artificial Neural Networks (ICANN-95), Paris*, volume 1, pages 95–102, 1995.

[9] Jörg Walter and Helge Ritter. Rapid learning with parametrized self-organizing maps. *Neurocomputing*, 12:131–153, 1996.

[10] J. Zhang, Y. von Collani, and A. Knoll. On-line learning of b-spline fuzzy controller to acquire sensor-based assembly skills. In *Proc. Int. Conf. on Robotics and Automation (ICRA-97)*, pages 1418–1423, 1997.

Organization and Reorganization of Autonomous Oceanographic Sampling Networks*

Roy M. Turner & Elise H. Turner
Department of Computer Science
University of Maine
Orono, ME 04469
{rmt,eht}@umcs.maine.edu

Abstract

As robots and other autonomous agents become more widespread, there will be a greater need for them to work cooperatively to accomplish complex, long-duration tasks. Such groups of agents will often be highly heterogeneous and *open*: that is, agents will enter or leave the system while the task is being performed. This presents a formidable problem for organizing and controlling the agents. In this paper, we present an approach to this problem for autonomous oceanographic sampling networks (AOSNs), which are groups of robots and sensor platforms that cooperate to sample an area over a long period of time. The approach uses two organizations to control the agents. A *task-level organization* (TLO) controls the system during the actual mission, and a *meta-level organization* (MLO) is responsible for self-organization of the system, design of the TLO to fit the situation, and reorganization as necessary. This approach allows the TLO to be highly efficient, while allowing the MLO to give the overall system a great deal of flexibility in adapting to change.

I Introduction

Autonomous oceanographic sampling networks (AOSNs) [Curtin *et al.*, 1993] are multi-robot systems being developed to collect data from the ocean over long periods of time. During deployment, the AOSN may alter its task in response to previously-collected data or new instructions from scientists. An AOSN will be composed of a wide variety of components, including autonomous underwater vehicles (AUVs), remotely-operated vehicles (ROVs), and non-mobile instrument platforms. Because underwater robots are so expensive to develop, and because so few exist, components of the AOSN will include AUVs and ROVs that have been loaned to the AOSN by undersea robotics laboratories.

Like all multi-agent systems, the AOSN requires a mechanism for organizing its components. However, AOSNs have several characteristics which make existing approaches for organization inappropriate. We believe these characteristics will be shared by an increasing number of systems as more robots become available and are used to perform more complex missions over longer periods of time. These characteristics are:

Organization and reorganization must be autonomous. Finding the appropriate organization for a system requires knowledge of organizational theory as well as knowledge of the components of the system. Because the system will interact with "users" instead of "operators," we do not want to burden the user with having to define the organization. In addition, we would like the system to be able to operate without any user intervention. Since the situation may change and reorganization may be required at any time, the system must be able to reorganize autonomously.

The system will be composed of a wide variety of agents. The system will be composed of robots with a wide range of capabilities depending on the available robots and the tasks which the system is to perform. Consequently, there will be many more differences between robots than in most previous heterogeneous systems. In particular, robots will have different levels of intelligence: some will be able to reason about the organization while others will have only the intelligence required to perform some specific task.

The composition of the system will change. Because the system will be deployed for long periods, we expect robots to come and go. Robots will leave the system due to failure or because they are needed

---

*The authors would like to thank the Office of Naval Research for its support under contract N00014-89-J-3074. The authors would like to thank the University of Maine Cooperative Distributed Problem Solving research group and our colleagues at the Autonomous Undersea Systems Institute for helpful discussions on this work. We would also like to thank Steve Chappell and Charles Grunden for their work on the simulator.

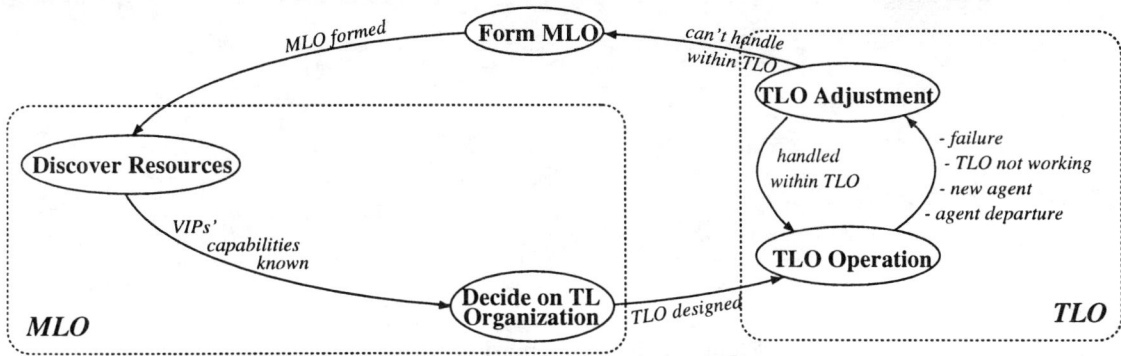

Figure 1: A two-level approach to AOSN control.

outside of the AOSN. Robots may also join the system when they are no longer needed elsewhere. In addition, new robots may be developed which are added to the system.

The tasks of the system may change. Whenever a system interacts with the real world, its task may change in response to unexpected features of the environment or in response to the system's own activities. A task may be successfully completed or the results of one task may cause another to be created. For example, exploratory robots may be able to determine when they have found something interesting and, based on that information, generate a task to explore that area in more detail. The user may also change the assigned task of the system. This is particularly likely for systems that are deployed for long periods, because the situation, the users' goals, or the users may change.

Communication is severely limited. In the ocean, acoustic communication is essentially the only option for communication that does not require breaking the surface. Current acoustic modems have baud rates significantly below that of off-the-shelf telephone modems. Many other domains and tasks also have communication constraints. For example, in covert operations, it is essential that communication be kept to a minimum. Even domains that have seemingly unlimited communication have some constraints on communication, including processing constraints which limit the amount of information that the agent can understand, as well as its ability to perform other tasks while communicating.

Excellent performance is expected. As robotics research advances, users will have elevated expectations about the performance of the system. Systems, at a minimum, will have to perform their task effectively. In addition, the systems should be efficient and reliable. For example, an AOSN will need to guarantee its users high levels of data quality and coverage. To these ends, there must be some global control to ensure efficiency and monitor expected results.

Existing CDPS approaches fall short on addressing one or more of these characteristics. For example, Partial Global Planning (PGP) [Durfee & Lesser, 1987] requires all agents to have sophisticated problem-solving abilities, and it does not take the sort of global perspective necessary to ensure data quality. The Contract Net Protocol (CNP) [Smith, 1980] also fails to take a global approach to designing the organization it creates, and it is not clear how well it handles the failure or exit of mid-level or top-level managers. Multiagent planning approaches [e.g., Cammarata et al., 1983; Georgeff, 1983], though taking a global perspective, do not cope well with the possibility of the planning agent(s) failing or leaving the system. Limited bandwidth is a problem for almost all approaches. In addition, most existing approaches suffer from a trade-off between flexibility and efficiency. Efficient approaches are not particularly adaptable when the situation changes, and flexible approaches are not as efficient as might be desired. A middle-ground approach is worse than either: the system will be neither adaptive enough nor efficient enough.

We have developed a two-level approach to controlling AOSNs to address these problems. Two organizations are used to control the agents. A *task-level organization* (TLO) controls the actual mission-related tasks, and a *meta-level organization* (MLO) is used to design as efficient a TLO as possible. Only the robots in the MLO are required to reason about the organization. The MLO self-organizes and can take a global perspective to design a TLO. The TLO then conducts the mission until there is a change beyond its capability to adapt. At that point, the MLO again

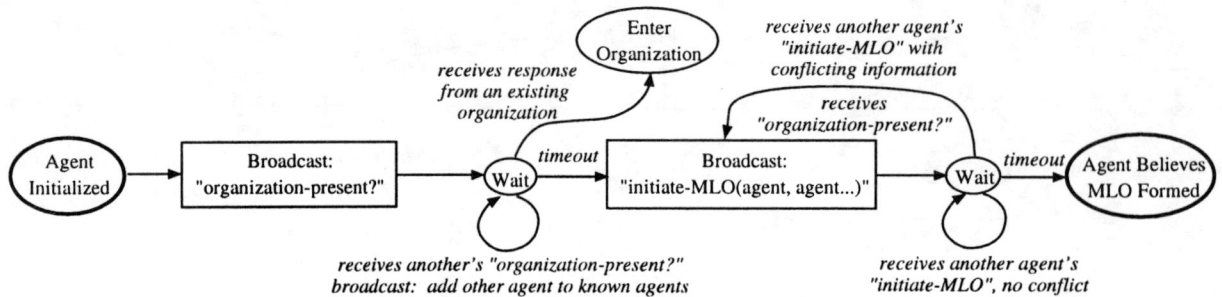

Figure 2: MLO formation protocol.

regains control to reorganize the system by repairing or designing a new TLO. This provides both efficiency and flexibility.

In the remainder of this paper, we first discuss the approach in general, then look more closely at task assignment and at how reorganization takes place in the system. We then describe the simulation testbed in which this work is being developed.

II A Two-Level Approach to Organizing CDPS Systems

Our overall approach is shown in Figure 1. The operation of the system is partitioned into two general phases, one controlled by the meta-level organization and one controlled by the task-level organization. Within these are smaller phases. All activities of the system and the agents are controlled by *protocols* that dictate acceptable actions given the situation and the phase the system is in. All vehicles and instrument platforms (*VIPs*) that participate in the system must abide by the protocols or a subset of the protocols appropriate for their level of participation. Thus those that can participate in the MLO must be able to follow most of the protocols, while those that can participate only in the TLO must follow the TLO protocols relevant to them.

A Meta-Level Organization

The meta-level organization is composed of those agents present that have the capability to reason about which organization is appropriate for the mission. This means that these agents, called *MLO agents*, must be fairly sophisticated. The MLO, then, will contain a subset of the VIPs present.

When first deployed, the task facing the AOSN's agents is to self-organize into a meta-level organization capable of designing the task-level organization. We assume that the system may not initially have much knowledge of itself. This could happen, for example, if the VIPs were deployed by air drop, in which case not all may survive, or if they were supposed to rendezvous at a location, in which case not all may have arrived. Thus, the MLO cannot be created *a priori*, nor should it rely on the presence of particular VIPs.

Consequently, our MLO formation protocol makes few assumptions about which agents will be present. The agents capable of participating in the MLO (called *MLO agents*) each carry out a simple sequence of actions designed to identify their peers. Figure 2 shows the protocol used for forming the MLO.[1] The result is a "flat" meta-level organization in which each member can communicate with all the others and no single agent is in charge. As long as there is a single MLO agent present, the MLO can form and the mission can continue.

The exchange of information specified by the MLO-formation protocol establishes common knowledge of the MLO's membership and location of the agents. The next task is to determine the total capabilities available to the AOSN to conduct its task. This is done by the MLO discovery protocol (see Figure 3), which guides the MLO agents in querying non-MLO VIPs about their abilities.

We treat each agent as a black box with advertised capabilities [see, e.g., Turner *et al.*, 1993]. This simplifies designing the TLO, as it abstracts away unnecessary detail about the specifics of the agent's implementation. It also facilitates interoperability of disparate agents, since a "vocabulary" of capabilities provides a common interface to the agents and a standard way of communicating about them. In many ways, the benefits are similar to those from object-oriented programming. This representation of agents means that, practically speaking, any VIP can participate in the system as long as it is capable of respond-

---

[1] In this and the other protocol diagrams, only the major pathways are shown for clarity.

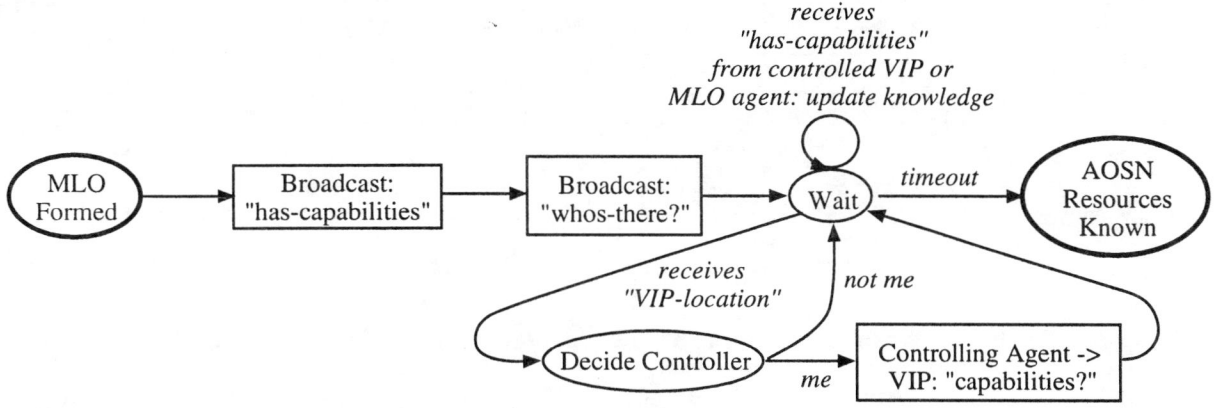

Figure 3: MLO discovery protocol.

ing appropriately to queries about its capabilities.

At the end of the protocol shown in Figure 3, the MLO knows all the capabilities of all agents in the system. To reduce bandwidth requirements, our current MLO organization distributes responsibility for this knowledge among the MLO agents. Each MLO agent is responsible for knowing about the capabilities of those VIPs closest to it.

At this point, the MLO can proceed to design the TLO. The current protocol, shown in Figure 4, calls for a single planner to be selected by convention from among the MLO agents. The planner gathers information from its peers about capabilities related to accomplishing the mission at hand, then it designs a task-level organization. In the future, we will examine how and when to distribute the planning task among multiple MLO agents.

We currently restrict the kinds of organizations considered for the TLO to be hierarchies. Hierarchical organizational structures have several benefits, including efficiency of communication [Malone, 1987], effectiveness in the face of uncertainty [Fox, 1981], and the possibility of global coherence of actions, since there is a top-level manager. Even with this restriction, however, the problem of designing a TLO is rich enough to be interesting, since hierarchies can vary along a number of axes, including the kind and timing of communication allowed, number of levels, and whether or not peers are allowed to cooperate outside the management structure. In the future, we plan to broaden our consideration to include other kinds of task-level organizational structures.

The planner designs the TLO by matching VIP capabilities to mission tasks, creating the hierarchy's management structure, and assigning VIPs to management roles, based again on their advertised capabilities. Our first implementation of task assignment used a very simple first-fit mechanism, and the assignment of management roles is done using simple heuristics (e.g., pick a manager from among the agents working on a task, if possible). The second version is based on constrained heuristic search [Fox et al., 1989], in which task assignment is treated as a constraint satisfaction problem. This is described in more detail in Section III. In the future, we intend to examine the feasibility of using an enhanced version of the distributed constrained heuristic search algorithm [Sycara et al., 1991] for task assignment to allow the entire MLO to participate. We will also extend the approach to organizational design in two ways. First, we are developing a representation for organizations that will allow an agent to analyze the situation, then quickly choose an organization type and specific parameters based on the situation's features. Second, we will ultimately examine the feasibility of incorporating management roles into the "tasks" assigned by CHS.

Once the TLO is designed, the planner informs the new managers of their roles and whom they control and tells the top-level manager to begin work. The planner then sends a message to its peers in the MLO informing them that the MLO is now dissolved.

B Task-Level Organization

Once formed, the task-level organization controls the AOSN until the mission is complete or there is a change severe enough to be beyond the TLO's ability to accommodate.

Figure 5 shows the major protocol used in hierarchical TLOs. When a TLO manager receives notification of its assignment, it analyzes itself and its subordinates to identify any slack resources (e.g., unused

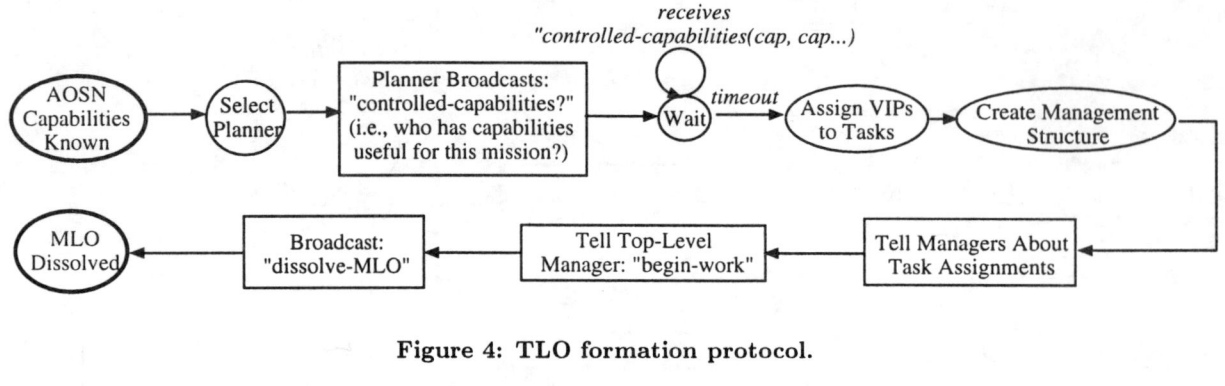

Figure 4: TLO formation protocol.

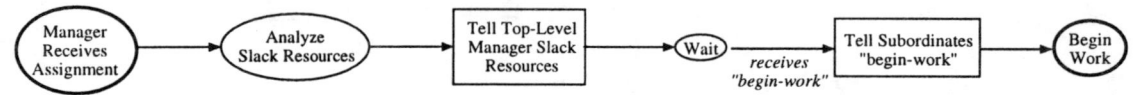

Figure 5: A TLO protocol.

capabilities). This information is passed to the TLO's top-level manager, which keeps a record of the slack capabilities and VIPs present in the system. This gives the TLO a limited ability to respond to changes in the task or its own composition, as discussed in Section IV. When the manager receives a "begin work" message, it passes it along to its subordinates (which pass it along to their subordinates, etc.) and begins work on its own task(s).

We have not yet defined protocols governing mission-related aspects of the AOSN control such as communication between manager and subordinate during work on tasks. To a large extent, this will depend on the specifics of the task, manager, and subordinate. Protocols will be needed, however, to allow the managers of the TLO to gather and maintain current information about the state of the TLO and the mission. This will be addressed in future work.

III Task Assignment

Task assignment identifies the VIP which will perform each required task. The task assignment algorithm receives a task decomposition tree as input. A task decomposition tree is a standard representation used by problem solvers that can be produced for a given mission by the planner selected for TLO formation. The tree represents all of the alternative methods for carrying out the assigned mission and the capabilities required by each alternative. Only a VIP that has some capability can be assigned to deliver that capability, and no VIP can be assigned tasks in excess of its resources.

Task assignment is currently implemented using a variant of *constrained heuristic search (CHS)* [Fox et al., 1989] that has been extended for task decomposition trees [Turner & Turner, in press]. CHS was chosen as the basis for our task assignment algorithm because it efficiently finds a solution if one exists. CHS ties constraint satisfaction to heuristic search by supplying operators which add constraints and variables to the constraint graph. The alternative methods for carrying out the task are selected using heuristics which indicate how likely it is that an assignment can be found within the resource constraints of the VIP. The selected capabilities are placed in a constraint graph as variables whose values are constrained by the resource limitations of the VIPs. Then constraint satisfaction techniques are used to find a solution.

IV Handling Change

Changes will occur during the operation of an AOSN that may require reorganizing the system. We are currently focusing on changes in composition of the AOSN. Agents will fail during long duration missions. In systems such as AOSNs, in which robots may be on loan to the system, agents will also need to exit the system as they are needed elsewhere. In addition, as robots become available, it is desirable to allow them to enter the system to augment its resources.

If an agent has time and sufficient knowledge, it may be able to gracefully leave the system by notifying the MLO or TLO, whichever currently exists. A protocol for doing this in the context of the TLO is shown in Figure 6. This protocol is specific for hierarchical TLOs and requires the top-level manager to maintain a list of slack resources present in the sys-

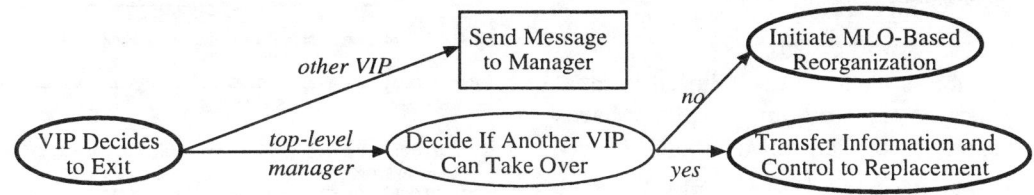

Figure 6: Exit protocol.

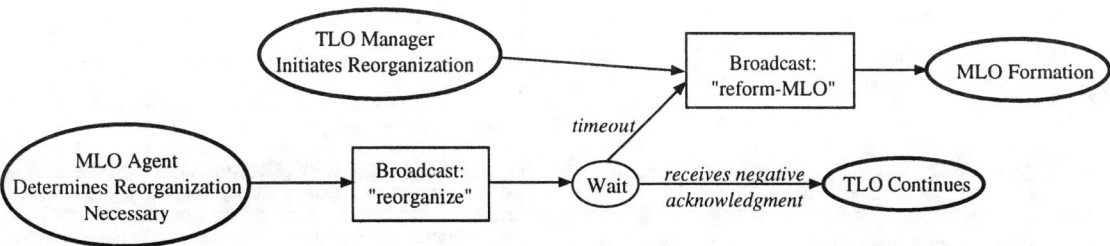

Figure 7: A reorganization protocol.

tem. If the top-level manager itself needs to leave, it can consult this list to determine if there is another agent present that can take over for it. If not, then the manager has exhausted the TLO's ability to adapt and initiates reorganization via re-forming an MLO, as discussed below. If the exiting agent is not the top-level manager, it signals its manager and, if that manager has knowledge about resources under its control that can be assigned to replace the agent, it replaces the agent and passes information up the hierarchy about the change. Otherwise, it notifies its own manager about the agent's intention to exit and lets that manager attempt to handle the change in the context of its more global viewpoint. Ultimately, if no other manager can handle the change, then the top-level manager will attempt to replace the exiting agent. If it cannot, it will initiate reorganization by the MLO.

It would be overly optimistic to assume that in all (or even most) cases, agents can exit gracefully. When an agent fails, it often will not have the time or the ability to signal its manager. In such cases, the agent's manager and peers will have to determine that the agent has failed. This is, in general, a very difficult problem that we will address in future work. Once failure is detected, however, it can be treated similarly to a more graceful exit.

The entry of new agents is also important. A new agent's arrival will not force the TLO or MLO to reorganize, but the availability of its capabilities will often provide an opportunity to achieve a task more effectively or to work on a task impossible to achieve previously. For example, at the time the TLO was created, perhaps the only vehicle present that could take data samples of a particular kind was one whose navigational precision was poor enough that some data points would be missed. Later, if a vehicle with the required sensor and a much better navigational system enters the AOSN, it would make sense to assign it to take over the old agent's task.

Our agent entry protocols are sufficiently complex, due to the need to handle agent entry in any phase of AOSN operation, that space does not permit a discussion here. They are, however, discussed elsewhere [Chappell *et al.*, 1997].

The task-level organization will not always be able to accommodate to agents entering or leaving, since it is meant to be efficient, not flexible. In these cases, the MLO regains control to reorganize the system. Other sorts of changes can also lead to reorganization. For example, the environment may change significantly, the mission can be changed by a user, or a task can fail. In these cases, some agent or agents in the TLO must detect the change. The agent can then either send a message up the hierarchy or attempt to initiate a reorganization without the participation of the top-level manager. The latter may be the case if the agent detecting the change has more information or is more intelligent than the top-level manager.

Figure 7 shows a reorganization protocol. When the top-level manager requests a reorganization, the MLO is immediately re-formed. Other agents follow a different protocol. First, they broadcast a message to ensure that all agents agree to the reorganization. If so, the MLO is re-formed. This protocol is tentative, and will be changed as work progresses.

MLO re-formation is identical to its initial formation. This is because the situation will likely have changed significantly since the last time the MLO existed. Consequently, we cannot rely on the presence of any of the MLO agents that participated previously. In the future, we will evaluate the usefulness of allowing the MLO to remain in existence while the TLO is controlling the system. This will likely decrease the time necessary to reorganize at the expense of increased bandwidth utilization during the mission.

V The Simulator

A simulation testbed was created in which to develop and evaluate the approach described above. Instead of building the usual sort of AI simulator, in which the decision-making processes of individual agents is simulated in detail, we chose to create a rule-based simulator that focuses on the aggregate properties of a group of agents following the protocols. This allows us to focus on the protocols rather than on how the agents are implemented.

This approach has several advantages. Development time is saved by not concerning ourselves with the internal decision-making of agents. Time that would normally be spent designing and programming the agents is instead available for developing and evaluating the protocols. This is also in keeping with the approach being simulated, since this does not overcommit to the kinds of agents that will be in the simulated system: agents are treated as black boxes with well-defined behavior (i.e., as governed by the prototocols). The rule-based approach makes it very easy to implement new protocols or change existing ones, since protocols are readily implemented as sets of rules. This allows rapid testing of new ideas in the simulator. A very important benefit of this kind of simulator is that we can concentrate on evaluating the protocols without worrying about how the particular decision-making processes of the agents impacted the results. This type of simulator also allows higher-fidelity rules and algorithms to replace lower-fidelity versions as they become available.

The simulator is written in the CLIPS [Giarratano, 1993] forward-chaining rule-based system. There are approximately 270 rules currently in the simulator that implement the protocol simulation itself, the environment, vehicle motion, and discrete-event simulation capabilities. In addition, the simulator includes C and Lisp code. The CLIPS and C portions of the simulator are available on the Web.[2] Figure 8 shows example output from the simulator.

```
;; MLO formation phase:
00:00:00 (SIM) new agent EAVE-Ariel is broadcasting organization-present?
    message.
00:00:00 (SIM) new agent EAVE-Arista is broadcasting organization-present?
    message.
;; [...]
00:00:05 (MLO) EAVE-Ariel: received organization-present? message from
    EAVE-Arista
00:00:30 (MLO) EAVE-Ariel believes it has waited long enough for replies.
00:00:30 (MLO) EAVE-Ariel is initiating MLO formation with agents =
    (EAVE-Arista EAVE-Ariel)
;; MLO discovery phase:
00:00:35 (MLO) Agents are attempting to discover other VIPs.
00:00:35 (MLO) EAVE-Arista --> EAVE-Ariel: I have capability(ies) (CDPS
    survey-side-scan-sonar transit search acoustic-link manage manage
    manage manage manage), and I'm at location (10 10 10).
;; [...]
00:00:40 (MLO) mooring-Able broadcasting: I am at (0 0 0).
00:00:40 (MLO) mooring-Baker broadcasting: I am at (200 0 0).
;; [...]
00:01:35 (MLO) Closest MLO agent EAVE-Ariel now controls mooring-Able
00:01:35 (MLO) EAVE-Ariel --> mooring-Able: tell me your capabilities.
;; [...]
00:01:40 (MLO) mooring-Able -> EAVE-Ariel: I have capabilities (radio
    acoustic-link LBL).
;; [...]
00:01:45 (MLO) MLO formation complete.
;; TLO design phase:
00:01:45 (MLO) Selecting EAVE-Ariel as planner (convention: first in MLO).
00:01:45 (MLO) Planner EAVE-Ariel querying others about capabilities
    they may contribute for tasks: (background-survey LBL3 LBL2 LBL1
    communication-relay convex).
;; [...]
00:02:00 (MLO) EAVE-Arista --> EAVE-Ariel: these agents may work: (AUV
    mooring-Charlie mooring-Baker mooring-Delta CONVEX-mooring
    EAVE-Arista).
;; [...]
00:02:05 (MLO) EAVE-Arista manages mooring-Able.
00:02:05 (MLO) EAVE-Arista manages EAVE-Ariel.
00:02:05 (MLO) EAVE-Arista manages mooring-Delta.
00:02:05 (MLO) EAVE-Arista manages mooring-Charlie.
00:02:05 (MLO) EAVE-Ariel manages CONVEX-mooring.
00:02:05 (MLO) EAVE-Ariel manages AUV.
00:02:05 (MLO) EAVE-Ariel manages mooring-Baker.
00:02:05 (MLO) mooring-Able is working on task LBL1.
00:02:05 (MLO) mooring-Baker is working on task LBL3.
00:02:05 (MLO) mooring-Charlie is working on task LBL2.
00:02:05 (MLO) CONVEX-mooring is working on task convex.
00:02:05 (MLO) mooring-Delta is working on task communication-relay.
00:02:05 (MLO) AUV is working on task background-survey.
00:02:05 (MLO) EAVE-Ariel -> all: dissolve meta-level organization.
```

Figure 8: Example simulator output.

VI Conclusion and Future Work

As robots and other autonomous agents become more widely used, there will be a greater need for them to work cooperatively to accomplish complex, long-duration tasks. Often these systems of agents will have characteristics and requirements similar to those of autonomous oceanographic sampling networks: autonomous organization/reorganization will be needed,

---
[2]http://cdps.umcs.maine.edu/MAUV

the system will be highly heterogeneous, its composition and tasks will change over time, communication bandwidth will be limited, and rigorous performance measures will have to be satisfied.

In this paper, we have discussed our approach to controlling such systems. We have developed a two-level organizational approach that addresses concerns of flexibility and efficiency without trading them off against one another. The task-level organization can be designed to be highly efficient, though inflexible, with the meta-level organization stepping in to reorganize when the situation changes.

We are currently developing and evaluating protocols that implement this approach. Future work will refine and extend the protocols based on the results of empirical simulation studies that are currently in progress. We will explore distributing the task assignment mechanisms across multiple MLO agents. Characteristics of organizational structures will be delineated, along with features of situations in which they are appropriate. This will allow the development of a mechanism for quickly selecting an appropriate organizational structure based on the current situation. In the longer term, we anticipate testing this approach in a fielded AOSN.

REFERENCES

Bond, A. & Gasser, L. (1988). *Readings in Distributed Artificial Intelligence.* Morgan Kaufmann, Los Altos, California.

Cammarata, S., McArthur, D., & Steeb, R. (1983). Strategies of cooperation in distributed problem solving. In *Proceedings of the 1983 International Joint Conference on Artificial Intelligence*, pages 767–770.

Chappell, S. G., Turner, R. M., Turner, E. H., & Grunden, C. M. (1997). Cooperative behavior in an autonomous oceanographic sampling network: MAUV Project update. In *Proceedings of the 10th International Symposium on Unmanned Untethered Submersible Technology (UUST)*.

Curtin, T., Bellingham, J., Catipovic, J., & Webb, D. (1993). Autonomous oceanographic sampling networks. *Oceanography*, 6(3).

Durfee, E. H. & Lesser, V. R. (1987). Using partial global plans to coordinate distributed problem solvers. In *Proceedings of the 1987 International Joint Conference on Artificial Intelligence*, pages 875–883.

Fox, M. S. (1981). An organizational view of distributed systems. *IEEE Transactions on Systems, Man and Cybernetics*, 11:70–80.

Fox, M. S., Sadeh, N., & Baykan, C. (1989). Constrained heuristic search. In *Proceedings of the Eleventh International Joint Conference on Artificial Intelligence (IJCAI-89)*.

Georgeff, M. P. (1983). Communication and interaction in multiagent planning. In *Proceedings of the 1983 Conference of the American Association for Artificial Intelligence*, pages 125–129. (Reprinted in [Bond & Gasser, 1988].)

Giarratano, J. C. (1993). *CLIPS User's Guide.* NASA, Information Systems Directorate, Software Technology Branch, Lyndon B. Johnson Space Center, Houston, TX.

Malone, T. W. (1987). Modeling coordination in organizations and markets. *Management Science*, 33(10):1317–1332.

Smith, R. (1980). The contract net protocol: High-level communication and control in a distributed problem solver. *IEEE Transactions on Computers*, C–29(12):1104–1113.

Sycara, K., Roth, S., Sadeh, N., & Fox, M. (1991). Distributed constrained heuristic search. *IEEE Transactions on Systems, Man, and Cybernetics*, 21(6):1446–1461.

Turner, E. H. & Turner, R. M. (in press). A constraint-based approach to assigning system components to tasks. To appear in the *Proceedings of the 11th International Conference on Industrial and Engineering Applications of Artificial Intelligence and Expert Systems*.

Turner, R. M., Blidberg, D. R., Chappell, S. G., & Jalbert, J. C. (1993). Generic behaviors: An approach to modularity in intelligent systems control. In *Proceedings of the 8th International Symposium on Unmanned Untethered Submersible Technology (AUV'93)*, Durham, New Hampshire.

Optimal Design of a Five-bar Finger with Redundant Actuation

Jae Hoon Lee[1], Byung-Ju Yi[1], Sang-Rok Oh[2], Il Hong Suh[3]

[1]Dept. of Control & Instrumentation Eng., Hanyang Univ. Korea
[2]Dept. of Electronics and Information Tech. KIST, Korea
[3]Dept. of Electronics Eng., Hanyang Univ. Korea

Abstract

In order to develop a human hand mechanism, a 5-bar finger with redundant actuation is suggested. Optimal sets of actuator locations and link lengths for the cases of minimum actuator, one, two, and three redundant actuators are obtained by employing a composite design index which simultaneously consider several performance indices such as workspace, isotropic index, and force transmission ratio. Eventually, several finger-configurations optimized for special performance indices are illustrated.

I. Introduction

Robot hands have been employed for fine motion control and assembling parts. Most of existing robot hands employ tendon-driven power transmission. However, frictions existing in the transmission line require more effort on control. In light of this fact, we propose a five-bar finger mechanism which is directly driven by ultra-sonic motor at joints of the mechanism. Since the five-bar finger mechanism has many potential joint locations for attaching actuators, redundant actuation mode can be achieved[1-2]. Redundant actuation prevails in general biomechanical systems, such as the human body, the bodies of mammals and insects. Redundant actuation can be also found in many robotic applications. They includes multiple arms, dual arms, multi-fingered hands, walking machines, and so on[4-6].

Redundant actuation can be easily explained in terms of mobility. When mobility of a system is greater than the degree-of-freedom, the system is called "a kinematically redundant system". On the other hand, when the number of actuators is greater than the mobility (this situation usually happens in a closed-chain system), the system is called "*redundantly actuated system*". For example, the mobility of the human upper-extremity (arm) can be considered as 7, while it has 29 human actuators (i.e, muscles)[9]. Accordingly, it has 22 redundant actuator.

The purpose of this paper is the optimum design and development of a five-bar finger employing redundant motors. Section 2 introduces the kinematic modeling for a five-bar finger. Optimal design for the five-bar finger is treated in section 3. According to the optimization result, a five-bar finger with two redundant actuators has been developed and explained in section 4. Finally, we draw conclusion.

2. Kinematic Modeling

2.1 Open-chain kinematics

Consider a 5-bar finger mechanism shown in Fig. 1. This system has one closed-kinematic chain. The closed-kinematic chain is formed by connecting the two open-chains at the given location of the second link of the left open-chain, as shown in Fig. 1. In order to enlarge the area encompassed by the finger, the folded-in configuration of the right open-chain is chosen. Since two chains of the 5-bar mechanism have a common kinematic relation at the end-point of the system, the components of the end-point vector u are described by

$$x = l_1 c_1 + l_2 c_{12} = l_3 c_3 + l_4 c_{34} + l_5 c_{345}, \quad (1)$$

$$y = l_1 s_1 + l_2 s_{12} = l_3 s_3 + l_4 s_{34} + l_5 s_{345}, \quad (2)$$

and

$$\Phi = \theta_1 + \theta_2 = \theta_3 + \theta_4 + \theta_5. \quad (3)$$

Adopting the standard Jacobian representation for the velocity of a vector of N dependent (output) parameters u in terms of a set of P independent input coordinates $_r\dot{\phi}$ of rth open-chain, one has

$$\dot{u} = [\,_r G^u_\phi\,]\,_r\dot{\phi}_a. \quad (4)$$

Here,

$$[\,_r G^u_\phi\,] = [\frac{\partial u}{\partial\,_r\phi_1}, \frac{\partial u}{\partial\,_r\phi_2}, \cdots, \frac{\partial u}{\partial\,_r\phi_P}] \quad (5)$$

is the Jacobian relating the coordinates u to $_r\dot{\phi}$, and is of dimension of $N\times P$, with the mth column being of dimension of $N\times 1$. Jacobians of the first and second open-chain, respectively, are given by

$$[_1G_\theta^u] = \begin{bmatrix} -(l_1s_1 + l_2s_{12}) & -(l_2s_{12}) \\ (l_1c_1 + l_2c_{12}) & (l_2c_{12}) \\ 1 & 1 \end{bmatrix}, \quad (6)$$

and

$$[_2G_\theta^u] = \begin{bmatrix} -l_3s_3 - l_4s_{34} - l_5s_{345} & -l_4s_{34} - l_5s_{345} & -l_5s_{345} \\ l_3c_3 + l_4c_{34} + l_5c_{345} & l_4c_{34} + l_5c_{345} & l_5c_{345} \\ 1 & 1 & 1 \end{bmatrix} \quad (7)$$

2.2 Internal kinematics for 5-bar Finger mechanism

Since the mobility of this mechanism is two, at least two actuators are required to control the mechanism. There exist several choices in the selection of independent joints (i.e., actuator locations). In general, the base joints have been chosen as the actuator locations in previously developed 5-bar systems, primarily to minimize the dynamic effect due to floating actuators. However, from a kinematic point of view, inclusion of one or two floating actuators may be promising. For example, a better manipulability, isotropy, or load handling capacity can be achieved by using a certain floating actuator[1]. An internal kinematic relationship between dependent joints and independent joints is required to deal with the problem addressed in the above.

The equivalent velocity relation is given by

$$\dot{u} = [_1G_\theta^u]_1\dot{\theta} = [_2G_\theta^u]_2\dot{\theta}. \quad (8)$$

Choosing the joints θ_1 and θ_3 as the independent joints (θ_a) and the joints θ_2, θ_4, and θ_5 as the dependent joints (θ_p), Eq. (8) can be rearranged according to the following form

$$[A]\dot{\theta}_p = [B]\dot{\theta}_a \quad (9)$$

where

$$[A] = [-[_1G_\theta^u]_{;2} \;\; [_2G_\theta^u]_{;2,3}], \quad (10)$$

$$[B] = [[_1G_\theta^u]_{;1} \;\; -[_2G_\theta^u]_{;1}], \quad (11)$$

$$\dot{\theta}_p = (\dot{\theta}_3 \;\; \dot{\theta}_4 \;\; \dot{\theta}_5)^T, \quad (12)$$

and

$$\dot{\theta}_a = (\dot{\theta}_1 \;\; \dot{\theta}_2)^T. \quad (13)$$

Now, premultiplying the inverse of the matrix $[A]$ to both sides of Eq. (8) yields

$$\dot{\theta}_p = [G_a^p]\dot{\theta}_a, \quad (14)$$

where $[G_a^p]$ denotes the first-order *KIC* matrix relating θ_p to θ_a.

According to the duality existing between the velocity vector and force vector, the force relation between the independent joints and the dependent joints is described by

$$T_a = [G_a^p]^T T_p. \quad (15)$$

Then, the effective load referenced to the independent joints is given by

$$T_a^* = T_a + [G_a^p]^T T_p = [G_a^\phi]^T T_\phi \quad (16)$$

where

$$[G_a^\phi] = \begin{bmatrix} I \\ [G_a^p] \end{bmatrix}, \quad (17)$$

$$T_a = (T_1 \;\; T_2)^T. \quad (18)$$

In Eq. (15), T_ϕ denotes a force vector consisting of T_a and the whole set or subset of the joint torque at the dependent joints.

2.3 Forward Kinematics for 5-bar mechanism

Since the joints ($_r\phi$) of the *rth* chain is composed of some of the independent and dependent joints, $_r\dot{\phi}$ can be expressed in terms of the independent joints by

$$_r\dot{\phi} = [^rG_a^\phi]\dot{\phi}_a \quad (19)$$

where the matrix $[^rG_a^\phi]$ is formed using elements of $[G_a^p]$ augmented with a unity in the *ith* row and *jth* column and with zeros in all other elements of the *ith* row if $_r\phi_i = \phi_{a_j}$. Thus, the forward kinematics for the common object is obtained by embedding the first-order internal KIC into one of the *rth* pseudo open-chain kinematic expressions as follows:

$$\dot{u} = [_rG_\phi^u]_r\dot{\phi} = [G_a^u]\dot{\phi}_a, \quad (20)$$

where the forward Jacobian is determined by

$$[G_a^u] = [_rG_\phi^u][^rG_a^\phi]. \quad (21)$$

3. Kinematic Optimal Design for Five-Bar Finger with Redundant Actuator

3.1 Optimization Methodology

To deal with a nonlinear optimization with constrains, three numerical methods are used. The exterior penalty function method is employed to transform the constrained optimization problem into an unconstrained optimal problem. Powell's method is applied to obtain an optimal solution for the unconstrained problem, and quadratic interpolation method is utilized for uni-directional minimization[3].

3.2 Kinematic Design Indices

Based on the effective force relationship between the operational force vector and the input force vector, the ratio of the 2-norm of the output load to that of the input load can be expressed as

$$\frac{\|T_u\|}{\|T_\phi\|} = \left\{ \frac{T_\phi^T [G_u^\phi][G_u^\phi]^T T_\phi}{T_\phi^T T_\phi} \right\}^{\frac{1}{2}}, \quad (22)$$

where $\|T_\phi\|$ and $\|T_u\|$ are defined as

$$\|T_\phi\|^2 = T_\phi^T T_\phi, \quad (23)$$
$$\|T_u\|^2 = T_u^T T_u. \quad (24)$$

Based on the Rayleigh quotient, the output bounds with respect to the input loads are given as

$$\sigma_{min} \|T_\phi\| \leq \|T_u\| \leq \sigma_{max} \|T_\phi\|, \quad (25)$$

where σ_{min} and σ_{max} are the square root of minimum and maximum singular values of $[G_u^\phi][G_u^\phi]^T$, respectively. Since the nonzero eigenvalues of $[G_u^\phi]^T[G_u^\phi]$ is the same as those of $[G_u^\phi][G_u^\phi]^T$, the nonzero eigenvalues are obtained in terms of $[G_u^\phi]^T[G_u^\phi]$, and these singular values are used in determining the bounds of the force transmission ratio. An alternative expression of Eq. (25) is

$$\frac{1}{\sigma_{max}} \leq \frac{\|T_\phi\|}{\|T_u\|} \leq \frac{1}{\sigma_{min}}, \quad (26)$$

where $\sigma_F (= \frac{1}{\sigma_{min}})$ is defined as the maximum force transmission ratios (actuator capacities for an unit operational load of $\|T_u\|$.

3.2.1 Single design index

The operating region or workspace of the five-bar finger will be characterized by a reachable workspace. Also, a manipulator should be designed so that it has well-conditioned workspace which allows its end-effector to move from one regular value to another without passing through a critical value (singularity). An isotropic index is a criterion to measure such phenomenon. The isotropic index, σ_I, is defined as

$$\sigma_I = \frac{\sigma_{min}}{\sigma_{max}}, \quad (27)$$

The global isotropic index is defined with respect to the entire workspace of the manipulator as

$$\Sigma_I = \frac{\int_W \sigma_I dW}{W}, \quad (28)$$

where the workspace of manipulators is denoted as

$$W = \int_W dW. \quad (29)$$

Maximum force transmission ratio is defined as the required actuator capacity for an unit operational load of $\|T_u\|$. The global maximum force transmission ratio is defined with respect to the entire workspace of the manipulator as

$$\Sigma_F = \frac{\int_W \sigma_F dW}{\int_W dW}. \quad (30)$$

The design of a manipulator system can be based on any particular criterion. However, the single criterion-based design does not provide sufficient control on the range of the design parameters involved. Therefore, multi-criteria based design has been proposed[8]. However, the previous multi-criteria methods such as weighed sum did not provide any systematic design procedure and flexibility in design. To consider these facts, we employ a composite design index[8].

3.2.2 Composite design index

As an initial step to this process, preferential information should be given to each design parameter and each design index. Then, each design index is transferred to common preference design domain which ranges from zero to one. Here, the preference given to each design criterion is very subjective to the designer. Preference can be given to each criterion by weighting. This provides flexibility in design. For σ_I, the best preference is given the minimum value, and the least preference is given the maximum value of the criterion. Then, the design index is transferred into common preference design domain as below

$$\widetilde{\Sigma}_I = \frac{\Sigma_I - \Sigma_{I min}}{\Sigma_{I max} - \Sigma_{I min}}, \quad (31)$$

where '~' implies that the index is transferred into the common preference design domain. Since workspace is also in favor of maximum value, the design index transferred into common preference design domain is given as

$$\widetilde{W} = \frac{W - W_{min}}{W_{max} - W_{min}}. \quad (32)$$

On the other hand, force transmission ratio is in favor of minimum value, the design index transferred into common preference design domain is given as

$$\widetilde{\Sigma}_F = \frac{\Sigma_{F max} - \Sigma_F}{\Sigma_{F max} - \Sigma_{F min}}. \quad (33)$$

Note that each composite design index is constructed such that a large value represents a better design. A set of optimal design parameters is obtained based on max-min principle[7]. Initially the minimum values among the design indices for all set of design parameters are obtained, and then a set of design parameters, which has the maximum of the minimum values, is chosen as the optimal set of design parameters. Based on this principle, the composite global design index(CGDI) is defined as the minimum value of the above mentioned design indices at a set of design parameters, and given as

$$CGDI = \min\{\hat{W}^\alpha, \hat{\Sigma}_I^\beta, \hat{\Sigma}_F^\gamma\}. \quad (34)$$

The upper Greek letters (α, β, etc) represent the degree of weighting, and usually large value implies large weighting, and usually large value implies large weighting. Now, a set of optimal design parameters is chosen as the set that has the maximum $CGDI$ among all $CGDI$'s calculated for all set of design parameters.

3.3 Kinematic Optimization

The link lengths and the base width of the five-bar mechanism can be cited as kinematic design parameters. Initially, we assume that the workspace of the five-bar mechanism is the first quadrant of the x-y plane. That is,

$$0.01m \leq x, y \leq 0.3m. \quad (35)$$

Now, kinematic constraints associated with these parameters are given as

$$l_1 + l_2 = 0.3m, \quad (36)$$

$$l_3, l_4 \geq 0.07m, l_5 \geq 0.02m, \quad (37)$$

where the sum of l_1 and l_2 are decided based on the range of the workspace. Also, l_3 and l_4 should be greater than the minimum link length which is decided based on the size of the transmission system embedded inside the link. l_5 requires a minimum length to attach a finger-tip at the end of the link.

Kinematic optimization for the five-bar mechanism has been performed for the case of $\alpha=1$, $\beta=4$, $\gamma=1$ in which a large weighting is given the isotropic index, and for the case of $\alpha=1$, $\beta=1$, $\gamma=4$ in which a large weighting is given the maximum force transmission ratio. In Table 1, simulation result for the case of minimum actuation is shown. Characteristics of the kinematic design indices resulting from the optimization procedure have been improved in comparison to those of non-optimized case in which all link lengths are chosen as unit length. Similar to Table 1, Table 2, Table 3, and Table 4 illustrate the simulation results for the case of one, two, and three redundant actuation, respectively. We can conclude that for minimum actuation case, actuation of the first and fourth joints(here, we denote it as 14) has best performance in both isotropic and maximum force transmission characteristics, and that for one redundant actuation case (i.e., three actuators), actuation of the first, fourth, and fifth joints(here, we denote it as 145) has the best performance in both characteristics, and that for two redundant actuation case (i.e., four actuators), actuation of the first, third, fourth, and fifth joints(here, we denote it as 1345) has the best performance in both characteristics. As the number of actuators increase, characteristics of kinematic isotropy and maximum force transmission ratio are enhanced except the case of full actuation(i.e., three redundant actuation) in which only the force transmission ratio is improved a little bit, while the isotropic characteristic deteriorates. Figure 2 illustrates optimal five-bar configurations for 14, 145, and 1345. The black dots denote the positions of actuators. Figure 3 and 4 represent the kinematic isotropic index for optimized and nonoptimized cases, respectively, and then Figure 5 and 6 represent the maximum force transmission ratio for optimized and nonoptimized cases, respectively. As expected, optimization results in reduction of maximum force transmission ratio and improvement of the kinematic isotropy throughout the workspace.

Though both kinematic isotropy and maximum force transmission ratio are considered in the above, maximum force transmission ratio is believed to be much important factor than kinematic isotropy and workspace because fingers in multi-fingerd hands usually require large payload, and is operated in a small workspace. Specifically, the value of force transmission ratio for 145 joints has been reduced as much as 15.3 percents of that of 14 joints, and the value of force transmission ratio for 1345 joints has been reduced as much as 37.3 percents of that of 134 joints, and the value of force transmission ratio for full actuation(i.e., 12345 joints) has been reduced much smaller that the two previous cases. Conclusively, two redundant actuation is suggested to enhance the force transmission ratio of the five-bar mechanism.

4. Development of Five-Bar Finger with Redundant Actuator

4.1 Structure of Five-Bar Finger

Figure 7 shows the prototype of the five-bar mechanism. According to the optimization result, four actuators are placed to 1345 joints. Each joint of the finger is driven by a compact actuator mechanism having ultrasonic motor and a gear set with potentiometer, and the system is controlled by VME Bus-based Control system. The ultra-sonic motors have high torque/size ratio as compared to DC motor with a similar size. The quantitative specifications of the ultrasonic motor are shown in Table 5. A gear transmission having about 15:1 speed reduction ratio is employed. Particularly, the gear transmission consisting of series of spur gears and the potentiometer, as shown in Fig. 8, are embedded inside the link, which yields compact and modular design of the finger mechanism.

5. Conclusions

In this paper, we proposed employment of redundant actuation in finger design on the purpose of enhancing the kinematic isotropic characteristic and maximum force transmission ratio of the finger mechanism. Using the concept of composite design index which allows multi-purpose and multi-variable optimization, optimal sets of actuator locations and link lengths for the cases of using minimum number of actuators, one-, two-, and three-redundant actuators are obtained. Three design indices such as workspace, isotropic index, and force transmission ratio were simultaneously optimized with consideration of their relative weighting factors. Eventually, several finger-configurations optimized for special performance index are suggested. Future work involves experimental work associated with internal force control[1,2] and development of a three-fingered hand made of five-bar finger mechanism.

Acknowledgment

This work has been supported by KIST 2000 Human robot program.

References

[1] Yi, B-J., Suh, I.H., and Oh, S-R., Analysis of A Five-Bar Finger Mechanism Having Redundant Actuators with Applications to Stiffness and Frequency Modulation. *IEEE Proceeding on Robotics and Automation Conference*, pp. 759-765. (1997).

[2] Yi, B-J., Oh, S-R., Suh, I.H., and You, B.J., Synthesis of Actively Adjustable Frequency Modulators : The Case for A Five-Bar Finger Mechanisms. IEEE/RSJ *Proceeding on IROS*. (1997).

[3] Thomas, M., Yuan-Chou, H.C., and Tesar, D., Optimal Actuator Sizing for Robotic Manipulators Based on Local Dynamic Criteria, Trans. on ASME Journal of Mechanisms, Transmissions, and Automation in Design, Vol. 107, pp. 163-169. (1985).

[4] Nakamura, Y. and Ghodoussi, M., Dynamic Computation of Closed-link Robot Mechanisms with Nonredundant and Redundant Actuators. *IEEE Journal of Robotics and Automation* **5**, 294-302 (1989).

[5] Kumar, V.J. and Gardner, J., Kinematics of Redundantly Actuated Closed-chain. *IEEE Journal of Robotics and Automation* **6**, 269-273 (1990).

[6] Kurz, R. and Hayward, W., Multipl-goal Kinematic Optimization of a Parallel Spherical Mechanism with Actuator Redundancy. *IEEE Journal of Robotics and Automation* **8**, 644-651 (1992).

[7] Terano, T., Asai, K., and Sugeno, M., Fuzzy Systems and Its Applications, 1st ed., Hardourt Brace Jovanovitch Publishers, San Diego.

[8] Lee, S.H., Yi, B-J., Kwak, Y.K., Optimal Kinematc Design Of An Anthropomorphic Robot Module With Redundant Actuators, *Mechatronics*, Vol. 7, No. 5, 443-464 (1997)

[9] Spence, P.A., *Basic human anatomy*. The Benjamin/Cummings Publishing Co. Inc. (1986).

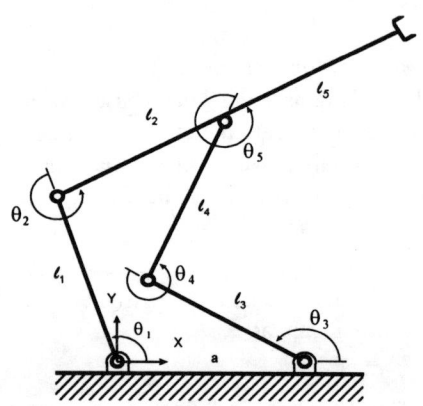

Figure 1. Five-bar Finger Mechanism

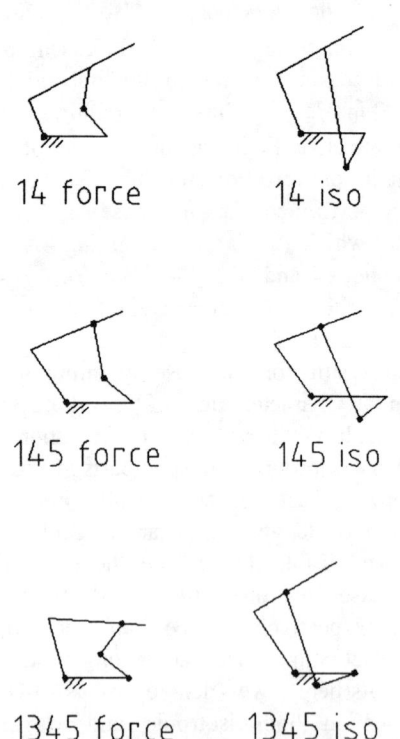

Figure 2. Optimal Finger Configurations

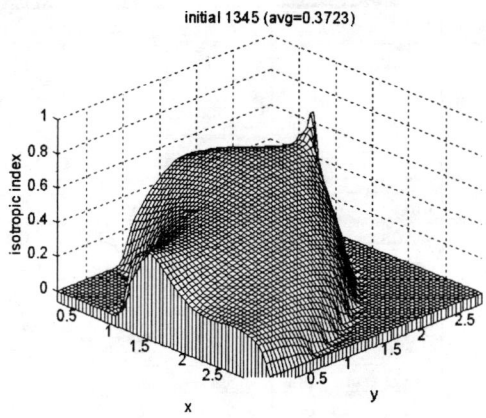

Figure 3. Kinematic Isotropic Index
(All links lengths are given unity)

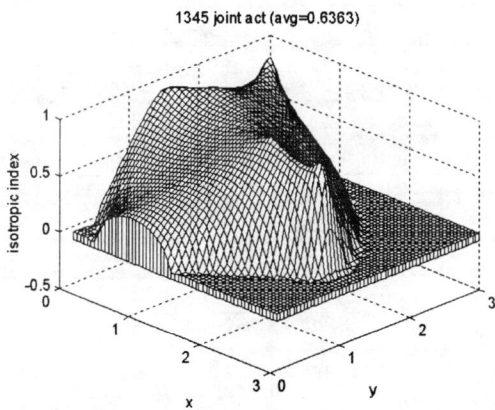

Figure 4. Kinematic Isotropic Index After Optimization

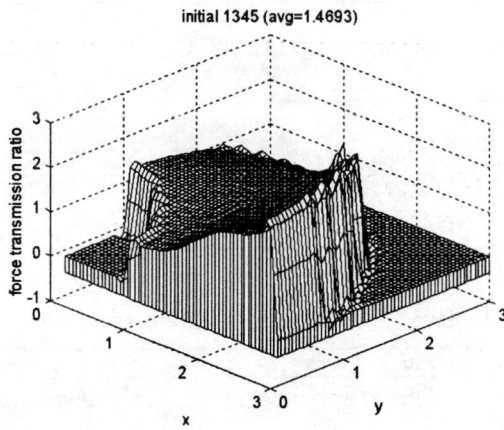

Figure 5. Max. Force Transmission Ratio
(All links lengths are given unity)

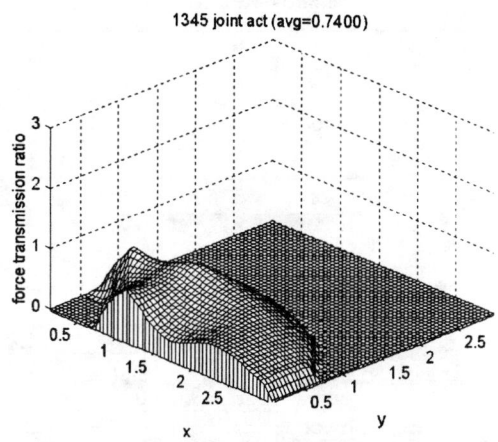

Figure 6. Max. Force Transmission Ratio After Optimization

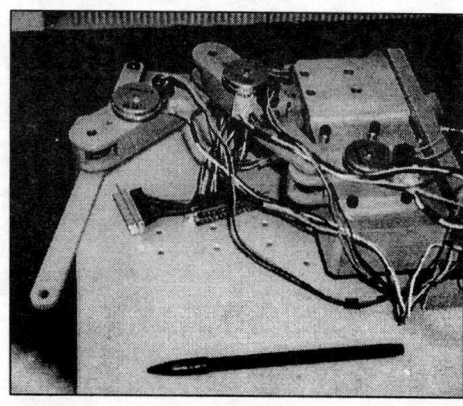

Figure 7. Proto-type of Five-Bar Finger Mechanism

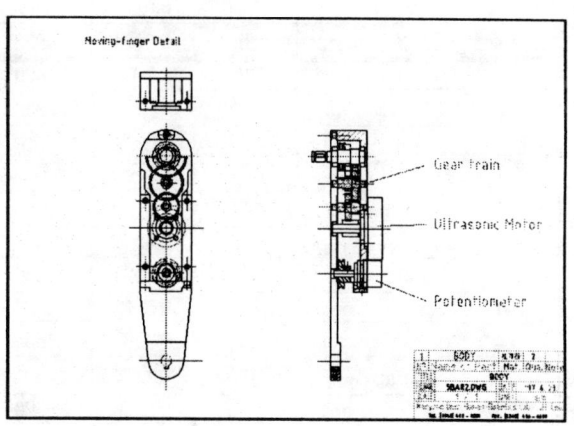

Figure 8. Gear Transmission System

Table 1. Optimization Result for Minimum Actuation

actuation	case	W_{area}	Σ_I	Σ_F
12	initial case	5.76	0.1970	2.8904
12	isotropic opt.	4.39	0.5454	1.9072
12	force trans. opt.	3.18	0.5054	1.8774
13	initial case	5.76	0.1757	4.5294
13	isotropic opt.	5.19	0.4990	2.2001
13	force trans. opt.	3.10	0.4328	1.9885
14	initial case	5.76	0.4468	1.9720
14	isotropic opt.	2.76	0.6676	1.2828
14	force trans. opt.	3.17	0.4759	1.1810
15	initial case	5.76	0.1325	6.5652
15	isotropic opt.	4.53	0.6258	2.5587
15	force trans. opt.	4.67	0.4989	1.9974
23	initial case	5.76	0.0620	7.9667
23	isotropic opt.	2.59	0.5179	1.8201
23	force trans. opt.	3.53	0.4959	1.7580
24	initial case	5.76	0.3508	2.3534
24	isotropic opt.	6.09	0.5368	1.8513
24	force trans. opt.	5.19	0.5230	1.6325
25	initial case	5.76	0.1424	5.2939
25	isotropic opt.	3.27	0.5544	1.5538
25	force trans. opt.	3.57	0.5029	1.5507
34	initial case	5.76	0.4129	2.2236
34	isotropic opt.	5.12	0.6967	1.9118
34	force trans. opt.	5.50	0.5583	1.5601
35	initial case	5.76	0.1109	7.5883
35	isotropic opt.	5.98	0.6155	2.1639
35	force trans. opt.	5.58	0.4624	2.0856
45	initial case	5.76	0.4373	2.5583
45	isotropic opt.	6.12	0.6058	1.5071
45	force trans. opt.	5.33	0.5869	1.3651

Table 2. Optimization Result for One-Redundant Actuation

actuation	case	W_{area}	Σ_I	Σ_F
123	initial case	5.76	0.2002	2.4400
123	isotropic opt.	4.92	0.5775	1.5406
123	force trans. opt.	2.49	0.5014	1.4832
124	initial case	5.76	0.3666	1.5429
124	isotropic opt.	6.43	0.5752	1.3838
124	force trans. opt.	3.70	0.5432	1.2076
125	initial case	5.76	0.2171	2.3956
125	isotropic opt.	3.24	0.5762	1.5985
125	force trans. opt.	4.14	0.4961	1.5166
134	initial case	5.76	0.3920	1.5632
134	isotropic opt.	2.78	0.5713	1.2562
134	force trans. opt.	4.30	0.4598	1.1634
135	initial case	5.76	0.1874	4.2490
135	isotropic opt.	1.99	0.5838	1.4320
135	force trans. opt.	5.42	0.5794	1.2906
145	initial case	5.76	0.4243	1.5554
145	isotropic opt.	4.75	0.6471	1.0702
145	force trans. opt.	4.40	0.5812	1.0008
234	initial case	5.76	0.3052	1.9618
234	isotropic opt.	6.05	0.5172	1.5416
234	force trans. opt.	4.07	0.4888	1.2885
235	initial case	5.76	0.1243	5.0690
235	isotropic opt.	1.79	0.4403	1.4665
235	force trans. opt.	3.50	0.4011	1.2577
245	initial case	5.76	0.3475	1.7242
245	isotropic opt.	6.11	0.5682	1.1147
245	force trans. opt.	3.55	0.5415	1.0291
345	initial case	5.76	0.3740	1.8138
345	isotropic opt.	6.17	0.6560	1.3368
345	force trans. opt.	4.37	0.4975	1.1462

Table 3. Optimization Result for Two-Redundant Actuation

actuation	case	W_{area}	Σ_I	Σ_F
2345	initial case	5.76	0.3034	1.6341
2345	isotropic opt.	4.03	0.6199	0.9745
2345	force trans. opt.	3.09	0.5088	0.7488
1345	initial case	5.76	0.3723	1.4693
1345	isotropic opt.	5.66	0.6363	1.0538
1345	force trans. opt.	3.13	0.5318	0.7400
1245	initial case	5.76	0.3597	1.3759
1245	isotropic opt.	4.57	0.5715	0.8527
1245	force trans. opt.	3.07	0.5416	0.7562
1235	initial case	5.76	0.2091	2.1979
1235	isotropic opt.	1.78	0.4913	1.2516
1235	force trans. opt.	4.56	0.4495	1.2347
1234	initial case	5.76	0.3336	1.4096
1234	isotropic opt.	4.44	0.6219	1.0100
1234	force trans. opt.	2.74	0.4566	0.8467

Table 4. Optimization Result for Three-Redundant Actuation

actuation	case	W_{area}	Σ_I	Σ_F
12345	initial case	5.76	0.3277	1.3235
12345	isotropic opt.	5.26	0.6397	0.8204
12345	force trans. opt.	3.29	0.5289	0.6996

Table 5. Specifications of Ultra-sonic Motor

Driving Frequency	50KHz
Drive Voltage	110Vrms
Maximum Torque	0.1Nm(1Kgf·cm)
Rated Speed	250rpm
Weight	20g

The Design and Development of the DIST-Hand Dextrous Gripper

Andrea Caffaz, Giorgio Cannata
DIST, University of Genova
Via Opera Pia 13, 16145 Genova, Italy
E-Mail: {caffaz, cannata}@dist.unige.it

Abstract

This paper presents the first proptoype of the DIST-Hand dextrous gripper. DIST-Hand is a *4-fingered* tendon driven device with *16* degrees of freedom, designed for experiments in the area of grasping control, and tele-manipulation. The current version of the gripper is lightweight and can be easely installed on the various existing robots. The paper outlines the kinematic and structural characteristics of the hand. Furthermore, some general methodological issues addressed during the design phase are discussed.

1 Introduction

The development of robot dextrous grippers is a very challenging endeavour which has been pursued by many researcher [3],[2], [5], [4], [6].

This paper briefly presents the development of the first prototype of the DIST-Hand dextrous gripper. The DIST-Hand is a 4 fingered mechanism with 16 degrees of freedom with a high degree of dexterity. The main goal pursued during the development of the DIST-Hand has been that of designing a small and lightweight dextrous gripper with anthropomorphic kinematics, which could be easily ported and installed even on small robot manipulators.

This paper outlines the characteristics of the hand and discuss some general design issues faced during the early stage of the development. In section 2, the description of the finger as the elementary building block of the hand is provided, and some analytical issues related with the dimensioning of its actuation system are discussed. Finally the integrated DIST-Hand is presented in section 3 and conclusions are drawn.

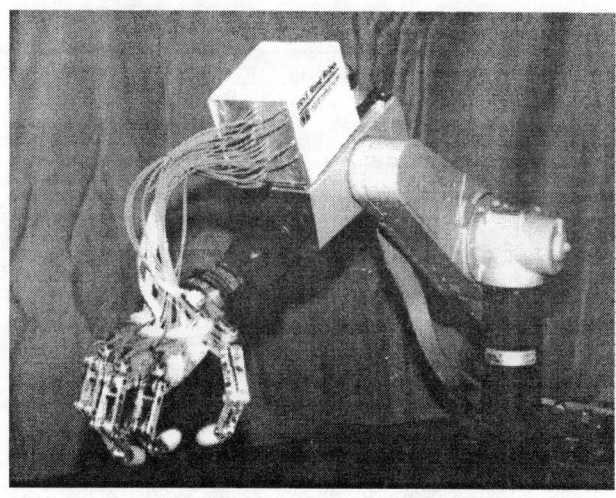

Figure 1: The DIST-Hand dextrous gripper mounted on a PUMA 260 robot arm.

2 Finger Design

The basic element designed for the development of the DIST-Hand is a 4 degrees of freedom finger. As the finger is the basic component of the hand in this paper we will mostly outline its characteristics and the methods followed during its design.

As it will appear in the forthcoming discussion, during the design of the finger (and of the DIST-Hand as well) a particular attention has been given to the possibility of using whenever possible cheap off-the shelf components. This allowed to keep the costs and the time of development of the hand at very low levels. The following sub-sections illustrate the characteristics of the finger.

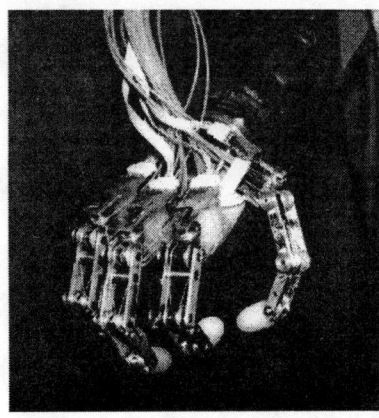

Figure 2: Detailed picture of the DIST-Hand.

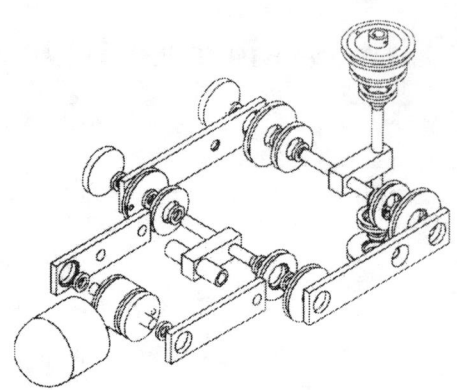

Figure 3: The mechanical drawing of the finger.

2.1 Finger Kinematics

The dimensions of the finger are close to that of a human one. The lengths of the various links, are listed in the following table

Link 1	Link 2	Link 3	Link 4
12 [mm]	42 [mm]	27.5 [mm]	28.5 [mm]

Each joint has a range of rotation which is larger than 90 deg and equivalent to that of a human hand.

The most important aspect considered during the kinematic and mechanical design of the finger has been related with the need of ensuring adequate mobility to the mechanism. The first two joints of the finger have orthogonal axes; in particular the first axis allows to rotate the *distal plane* (i.e. the plane where the finger is bended during actual operations). The distal plane is orthogonal the axes of the joints *2,3,4* and is a plane of symmetry of the finger. The distance between the first two axes as been kept as small as possible, to better emulate the finger kinematics.

The kinematic structure of the finger is sketched in figure 3. All the idle pulleys are mounted on miniature ball bearings as well as the joint supports in order to reduce the friction effects.

2.2 Finger Actuation

Each finger is actuated through 6 tendons ($\phi = 0.4$ [mm]) made of polyester, routed through pulleys and driven by 5 DC motors. In particular, tendons 5 and 6 driving the first joint are actuated by a single motor and are passively pre-tensioned.. The other 4 tendons are instead independently controlled by the remaining motors to drive the last 3 joints.

The actuation system has been carefully designed since it primarily affects the control performance of the robotic system. Three main problems have been addressed. Two of these have been mainly technological and involved basically the investigation of the most appropriate off-the-shelf components for the tendons, the tendons sheaths, and the motors.

The third problem has been instead mostly methodological and heavily influenced the final finger and hand design. This was the problem of designing the most appropriate routing of the tendons and dimensioning of the various pulleys.

As outlined by many authors (e.g. [2], [11] and particularly [1]) the problem of routing the tendons in mechanical hands is critical at least for two main reasons. First of all not all the tendon routings are admissible in order to generate arbitrary joint torques since tendons can only exert unidirectional forces. Caratheodory theorem establishes the minimum number of tendons needed, which is equal to $n+1$ where n is the number of joints [11], while Lee and Tsai [1] defined a procedure for the synthesis of admissible tendon routings.

Secondly, the mapping between tendon tensions (assumed to be massless and inelastic in the present discussion) and resultant joint torques is typically highly coupled thus making critical the problem of controlling the finger movements.

These two aspects have been carefully taken into account during the DIST-Hand design.

Consider a generic robotic hand where, with a little loss of generality, all the tendons are routed through pulleys. The mapping between joint torques τ and forces **f** applied by the tendons, is expressed by the following linear equation

$$\boldsymbol{\tau} = A^T \mathbf{f} \quad (1)$$

where $\mathbf{f} = [f_1, f_2, ..., f_k]^T$, k is the number of tendons, $\boldsymbol{\tau} = [\tau_1, \tau_2, ..., \tau_n]^T$, n is the number of joints. The matrix A^T is called *structure matrix*, it is constant and its entries (column-wise) are the radii of the pulleys through which each tendon is routed to generate any desired joint torque. The details on the structural characteristics of the *structure* matrix A^T can be found in [1]. To our purposes it is only relevant to remark that if one entry r_{ij} of the matrix is equal to 0 it means that the $j-th$ tendon is not routed through the $i-th$ joint and therefore all the entries r_{lj} with $l > j$ must be 0 as well. This property is the basic reason of the coupling between joints and tendons motions since diagonal or quasi-diagonal structure matrices cannot exist. On the other hand, Caratheodory theorem requires that the matrix A^T is rectangular, full-row rank and its null space must contain at least one vector with all positive (negative) components.

We have studied the properties of the matrix A^T not only in term of its *topological* properties as done in [1], but also as a function of its entries. In particular, we focused on the sensitivity of equation 1, and on the structure of the null space of A^T. Both these issues are, in our opinion, important in the area of control design although they are primarily related to mechanical design aspects.

Assume $\boldsymbol{\tau} = \boldsymbol{\tau}(q)$ to be a generic (joint level) control torque signal then it can be transformed into a *commanded* force signal $\mathbf{f}$ as follows

$$\mathbf{f} = A(A^T A)^{-1} \boldsymbol{\tau}(q) + \eta \mathbf{v} \quad (2)$$

where $\mathbf{v} \in \ker(A^T)$, $\mathbf{v} > 0$, and η is a positive coefficient (or function) ensuring that $\mathbf{f} > 0$ at each time instant. Then, the above control signal $\mathbf{f}$ can be fed into suitable servo level control loops.

Equation 2 is of interest here. The sensitivity of equation 1 is a measure of *regularity* of equation 2, therefore we expect to have low sensitivity in order to avoid sudden changes of $\mathbf{f}$ in response to small changes of $\boldsymbol{\tau}(q)$, and unavoidable uncertainties in the knowledge of the structural parameters (e.g., structure matrix coefficients, stiffnesses etc.). Furthermore, the vector $\eta \mathbf{v}$, which plays the role of a pretension term, should balanced, i.e. the magnitude of its components should be as much as possible the same, for reasons which will be made clearer below.

The sensitivity of equation (1) is measured by the condition number of the matrix A^T which can be defined as

$$\chi(A^T) = \frac{\sigma_{\max}(A^T)}{\sigma_{\min}(A^T)}$$

where $\sigma_{\min}$ and $\sigma_{\max}$ are the minimum and maximum singular values of A^T. It is important to remark that $\chi(A^T)$ represents the ratio of the maximum and minimum axes of the ellipsoid associated with the positive definite quadratic form $A^T A$. Minimizing the sensitivity is therefore equivalent to design a set of pulleys whose radii (as coefficients of the matrix A) allow to make the ellipsoid associated to $A^T A$ as close as possible to a sphere.

The most direct effect of reducing $\chi(A^T)$ and balancing $\mathbf{v}$ is that for norm bounded joint torques control signals such that $\|\boldsymbol{\tau}(q)\| < \tau_{\max}$ it is possible to reduce the pretensioning of the tendons, while still keeping their positivity.

It is now interesting to see if it is possible to design *optimal* structure matrices in the sense previously discussed. To this aim we can consider the following optimization problem

Problem 1: Minimize

$$J(A) = \chi(A^T) + \alpha \|\mathbf{v} - \mathbf{u}\|^2$$

subject to

$$A^T \mathbf{v} = \mathbf{0}$$
$$\|\mathbf{v}\| = 1$$

where $\mathbf{u}$ is a unit k-dimensional vector whose components are equal to $1/\sqrt{k}$, and $\alpha > 0$ a suitable weighting coefficient.

The above problem is not in general a trivial one, however, it is possible to construct, in simple but significant cases, optimal solutions. To this aim we must introduce some notations and some algebra results.

The routing of the tendons in a generic finger can be described by the $n-tuple$: $\rho = (\zeta_1, ..., \zeta_n)$ where ζ_i is the number of the tendons routed through the $i-th$ joint, or equivalently the number of non-zero entries in the $i-th$ row of the matrix A^T. More details about this definition can be found in [1].

Lemma 1 $\chi(A^T) = 1$ *if and only if* $A^T A = \gamma I$ *with* $\gamma > 0$

Then we can state the following proposition, keeping in mind that $J(A) \geq 1 \; \forall A$,

Proposition 2 *If the rows of A^T are orthonormal (or orthogonal and with the same norm) and $A^T \mathbf{u} = \mathbf{0}$ then $\min_A J(A) = 1$*

Proof. If the rows of A^T are orthonormal then $A^T A = \gamma I$ and by the above *Lemma* $\chi(A^T) = 1$. Furthermore, it is obvious that a well balanced vector $\mathbf{v} \in \ker(A^T)$ exists if $A^T \mathbf{u} = \mathbf{0}$, which is equivalent to say that the sum of the columns of A^T is equal to zero.

Consider now, as a simple but significant example a $n-dof$ finger actuated with $n+1$ tendons with routing $\rho = (n+1, n, n-1, ..., 2)$. In this case it can be easily verified that an optimal structure matrix (though not unique), in the sense of the *Problem 1* has the form

$$A^T = r \, diag\left[\frac{1}{\sqrt{n+n^2}}, ..., \frac{1}{\sqrt{6}}, \frac{1}{\sqrt{2}}\right] \cdot$$

$$\begin{bmatrix} 1 & 1 & 1 & 1 & 1 & -n \\ \cdots & \cdots & \cdots & \cdots & \cdots & \cdots \\ 1 & 1 & -2 & 0 & 0 & 0 \\ 1 & -1 & 0 & 0 & 0 & 0 \end{bmatrix}$$

with $r > 0$. It can be verified by inspection that the conditions stated in the above *Proposition* are fully met. The structure matrix has however quite an odd characteristic. In fact the radii of the pulleys are typically much larger at the distal joints than at the proximal ones. To clarify the situation consider a $4-dof$ finger, assuming $r = sqrt(2)$ [cm] the smaller pulleys have a radius which is about one third of the radius of the pulleys at the distal joint.

During the design of the DIST-Hand we faced the problems discussed above. As a matter of fact we modified the formulation of *Problem 1*, in order to keep into account kinematic, design and technical constraints.

Problem 2: Minimize

$$J(A) = \chi(A^T) + \|\mathbf{v} - \mathbf{u}\|^2$$

subject to

$$\begin{aligned} A^T \mathbf{v} &= \mathbf{0} \\ \|\mathbf{v}\| &= 1 \\ r_{ij} &> r_{\min} \quad \text{technological constraints} \\ r_{ij} &< r_{\max} \quad \text{design constraints} \\ f(r_{1j}, r_{2k}) &= 0 \quad \text{kinematic constraints} \\ \rho &= (6, 4, 4, 2) \end{aligned}$$

where $\mathbf{u}$ is a unit 6-dimensional vector whose components are equal to $1/\sqrt{6}$.

Remark 1 *The technological constraints were related to the components available off-the-shelf (e.g. micro ball-bearings), and to the workshop-floor requirements for the machining of the parts. The design constraints have been established in order to keep the size of the hand within the boundaries described in the previous sections. The kinematic constraints were related to specific components or parts, e.g. to leave sufficient clearance between the various parts as in the case of the cross-shaped joints 1 and 2 (see figure 3), or to keep the distance of the tendons over some given threshold and so on. The routing $\rho = (6, 4, 4, 2)$ has been adopted in part as a consequence of kinematic constraints, but also because it allows a symmetric routing of the tendons, and finally a better sensitivity as discussed in the following.*

Remark 2 *It must be remarked that for sake of simplicity the $5-th$ and $6-th$ tendons are driven by a single motor. This means that their pretensioning cannot be actively controlled, but must be mechanically preset. The adoption of a single motor for these tendons allowed to save in the final design of the hand a significant amount of space and weight.*

Remark 3 *In the present design $\chi(A^T) \cong 3.51$. However the adoption of smaller ball bearings (not available off the shelf at the time of the development of the present prototype) could lead to $\chi(A^T) \cong 2.34$, with a slight increasing of the radii of the distal joints pulleys. Finally it is interesting to notice that the better solution (subject to the same constraints) for a routing $\rho = (6, 4, 3, 2)$ lead to a value of sensitivity larger than 4.0, i.e. about 12% more the one reached in the present design.*

The routing of the tendons is sketched in figure 4.

2.3 Finger Joint Sensors

The tendons and relative sheaths produce elastic perturbations in the position of the finger which make critical the control of the fingers' motions using position and velocity feedback directly form the motor axes. Therefore, we have developed *ad hoc* rotation sensors mounted on each joint. Using these sensors it is possible to implement servo loops around the perturbations due to the elasticity and in part to friction.

The sensor is based on the use of a solid state Hall effect transducer as proposed also in [5] and [3]. The basic sensor design is sketched in figure 5. The sensor is contactless therefore it does not affect the motion of the joints. Furthermore, it has a significant immunity to noise with respect to other transducers of compa-

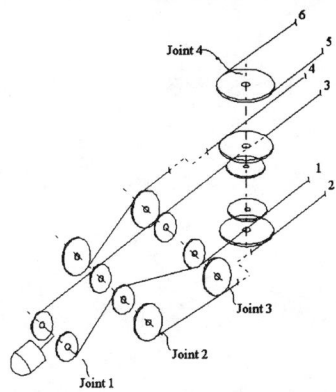

Figure 4: The tendons routing.

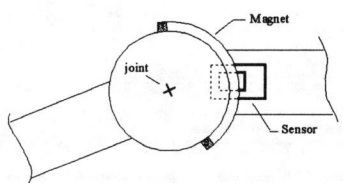

Figure 5: Sketch of the joint Hall effect position sensor.

rable size (e.g. micro/mini potentiometers, encoders).

The sensors output is a voltage $V = V(q)$, where q is the corresponding joint variable, with a S/N ratio of about 50dB, thus ensuring more than 8 bit resolution over a range of $110 deg$. The response curve of the sensor is mildly non linear, and strictly monotonically increasing, see figure 6.

3 Design of the Hand

The main goal of the DIST-Hand design has been that of developing an anthropomorphic gripper. The finger modules described in the previous sections and shown in figure 7 represent the basic building blocks.

A fundamental problem for developing a gripper with human like mobility is related with the fingers placement and in particular with the positioning of the thumb with respect to the palm. The position of the fingers on the supporting palm (except for the thumb) has been defined accordingly with tabulated anthropomorphic data [7]. The position of the thumb has been instead studied using a custom kinematic simulation tool. The simulator allowed to study the posture of the hand in response to various motion tasks involving the various fingers, using the techniques proposed in, [8]; this analysis allowed to study the co-ordinated motion of two or more fingers, with particular emphasis on the problem of determining the posture of the hand when the thumb tip is in contact with the other finger-tips.

Finally the hand has been assembled and tendons have been routed to the motor package formed by 20 motors. The motors are low-cost DC motors with reduction gears able to produce an output torque of about 2 $[Kg \cdot cm]$ and limited volume. The motor pack for a single finger has a volume lower than 100 $[cm^3]$. The hand and the 20 motors weigh less than $10 N$ and can be easily fitted on various robot arm. Figure 1, shows the DIST-Hand installed on a PUMA 260 arm. The mechanical interfaces are very simple and not *invasive* for the supporting arm. Although the tendons sheaths somehow constrain the admissible motions of the PUMA wrist the mobility of the whole hand is adequately large.

The actual size of the DIST-Hand is comparable with that of an *average* human hand in figure 8 Finally the functionality of the hand has been proven in simple position controlled task which have been presented in [9]

4 Conclusions

The design of the first prototype of the DIST-Hand dextrous gripper has been presented. The hand has a high degree of dexterity equivalent to that of other dextrous robot gripper already presented in the literature. However, some design issues followed during its development, to the best of the authors knowledge,

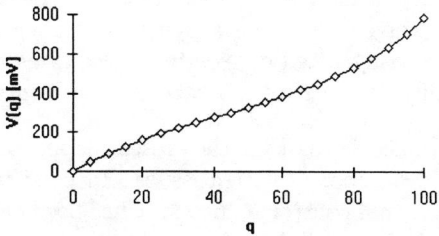

Figure 6: The response of the Hall effect sensor.

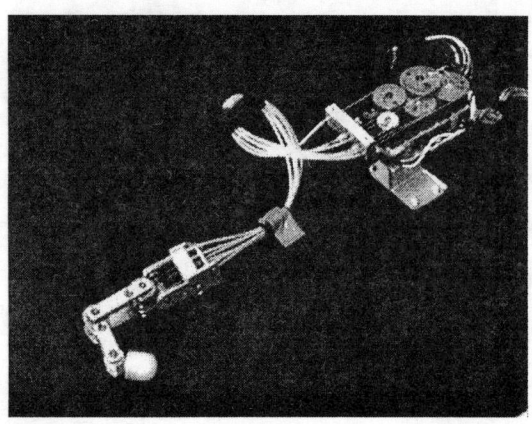

Figure 7: The modular finger. The mechanical structure of the finger and its related package of actuators are linked through flexible sheats.

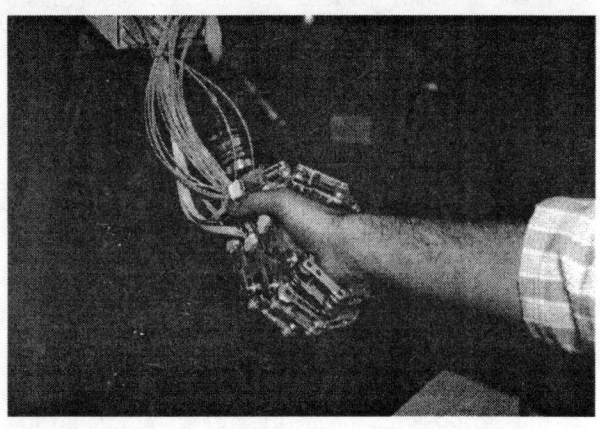

Figure 8: Comparison with a human hand

have not been fully addressed in the existing literature. Furthermore the design of the DIST-Hand has been conceived to obtain a lightweight device which could be easily fitted on existing robot arms and operationally used for tele-operation experiments.

The topics discussed in the present paper cover only a limited number of design issues. In particular the computational control architecture has been described in [10], while the design of a conductive rubber based distributed tactile sensor package to provide suitable force and contact feedback is currently under development.

5 Acknowledgments

This paper has been partially supported by the project *BRITE-Robotic NDT* funded by the European Commission.

References

[1] J. Lee, L. Tsai, "The Structural Synthesis of Tendon-Driven Manipulators having Pseudotriangular Structure Matrix", Int. Journ. of Robotic Research, vol. 10, n. 3, 1991, pp. 255-262

[2] K. Salisbury, M. mason, "Robot Hands and the Mechanics of Manipulation". MIT Press: Cambridge, MA. 1985.

[3] S.C. Jacobsen, J.E. Wood, D.F. Knutti, K.B. Biggers, "The Utah-MIT Dextrous Hand: Work in Progress", Int. Journ. of Robotic Research, vol. 3, n. 4, 1983, pp. 21-50.

[4] M.S. Ali, K.J. Kyriakopulos, H.E. Stephanou, "The Kinematics of the ANTHROBOT-2 Dextrous Hand", Proc. 1993 IEEE Int. Conf. on Robotics and Automation, Atlanta, May 1993, pp. 705-710.

[5] C. Bonivento, C. Melchiorri, "Towards Dextrous Manipulation with the UB-Hand II", Proc. 12th IFAC World Congress, Sydney (Australia), July 1993.

[6] J.D. Crisman, C. Kanojia, I. Zeid, "GRASPAR: A Flexible, Easely Controllable Robotic Hand", IEEE Robotics and Automation Magazine, vol. 3, n. 2, 1996, pp. 32-38.

[7] A. Farina, "Atlante di Anatomia Umana Descrittiva", Recordati. 1957 (in Italian)

[8] M. Aicardi, G. Cannata, G. Casalino, "Task Space Robot Control: Convergence Analysis and Gravity Compensation Via Integral Feedback", Proc. IEEE Conf. on Dec. and Control, Kobe (Japan), Dec. 1996.

[9] S. Bernieri, A. Caffaz, G. Cannata, G. Casalino, "The DIST-Hand Robot", IROS '97 Conf. Video Proceedings, Grenoble (France), September 1997.

[10] G. Cannata, G. Casalino, "Design of Task Level Robot Control Systems", ANIPLA Italian Conf., Torino (Italy), Nov. 1997, (in English).

[11] R.N. Murray, Z. Li, S.S. Sastry, "A Mathematical Introduction to Robotic Manipulation". CRC Press. 1994.

DLR's Multisensory Articulated Hand
Part I: Hard- and Software Architecture

J. Butterfass, G. Hirzinger, S. Knoch, H. Liu

DLR
German Aerospace Center
Institute of Robotics and System Dynamics
82234 Wessling, Germany
Joerg.Butterfass@dlr.de

Abstract

The main features of DLR's dextrous robot hand as a modular component of a complete robotics system are outlined in this paper. The application of robotics systems in unstructured servicing environments requires dextrous manipulation abilities and facilities to perform complex remote operations in a very flexible way. Therefore we have developed a multisensory articulated four finger hand, where all actuators are integrated in the hand's palm or the fingers directly. It is an integrated part of a complex light-weight manipulation system aiming at the development of robonauts for space.

After a brief description of the hand and it's sensorial equipment the hard- and software architecture is outlined with particular emphasis on flexibility and performance issues. The hand is typically controlled through a data glove for telemanipulation and skill-transfer purposes. Autonomous grasping and manipulation capabilities are currently under development.

Motivation

For many space operations, e.g. handling drawers, doors and bayonet closures in an internal lab environment, two finger grippers seem adequate and sufficient; the appropriate mechanical counterparts in the lab equipment are easily designed and realised even in a very late design stage. For more complex tasks however, future space robots need articulated multifingered hands.

In the past, impressive dextrous robot hands have been built [2][3][4][5][6][7][8][9]. However, all of them suffer from one main drawback: if the number of active degrees of freedom exceeds a fairly small number, there is no chance to integrate the actuators in the hand's wrist or palm if one wants to limit the size of the artificial hand to approximately 1.5 times the size of a human hand.

Thus, it was our declared goal to build a multisensory four finger hand with in total twelve degrees of freedom, where all actuators are integrated in the hand's palm or in the fingers directly.

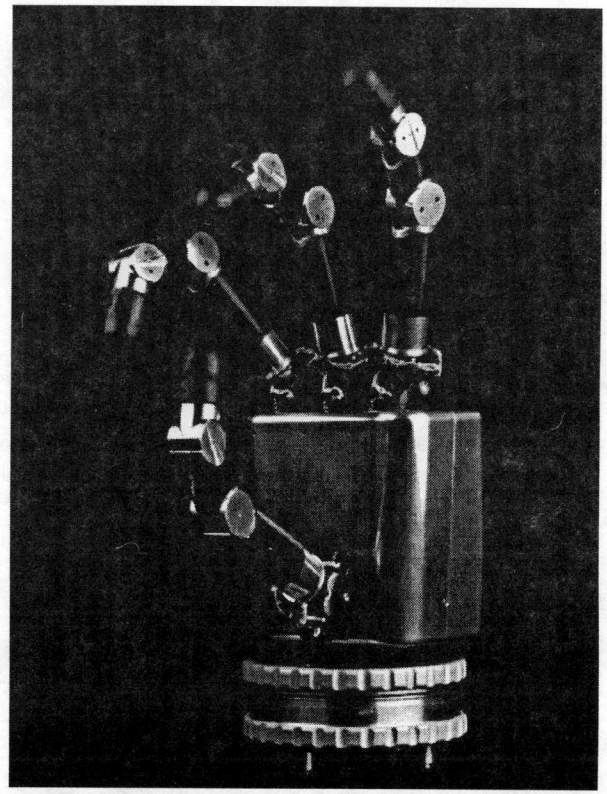

Fig. 1: DLR's Multisensory Articulated Hand mounted on a force torque sensor.

This became only feasible by using our specially designed miniaturised linear actuator.

Overall System Description

The DLR Hand, as shown in *Fig.1*, is a four fingered dextrous robot hand with a semi-anthropromorphic design. To achieve a high degree of modularity the hand consists of four identical fingers. The current arrangement shows three fingers and an opposing thumb (*Fig. 1*). Each finger shows up a two degrees of freedom base joint with intersecting axes for curling motion and for abduction/adduction driven by one actuator each. A third actuator is located in the proximal link actuating the medial link actively and, by coupling, the distal link passively.

The finger joints are actuated by specially designed linear actuators. Each linear actuator consists of a combination of a brushless DC motor with hollow shaft and DLR's miniaturised planetary roller spindle drive (*Fig. 2*) [10].

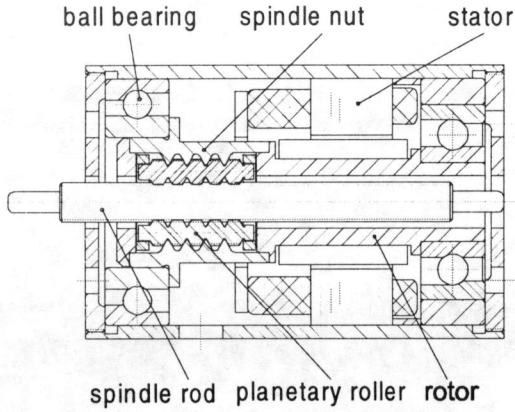

Fig. 2: Principal view of the DLR Linear Actuator.

With a cylindrical size of 21 mm in diameter and 33 mm in length this actuator is capable of applying a force of 150 N (*Table 1*). The 12 power converters needed to drive the 12 actuators are integrated in the hand's palm as well. Force transmission to the joints in the fingers is currently realised by SPECTRA® tendons.

DLR Linear Actuator	
diameter	21 mm
length	33 mm
mass	40 g
max. speed	50 mm/s
max. force	150 N

Table 1: Technical data of the DLR Linear Actuator.

Following our mechatronics design principles, the finger is equipped with various sensors, and literally any space in the finger is occupied by sensor signal processing electronics. Whenever possible the sensors are integrated into the mechanical structure to prevent them from being damaged (*Fig. 4 and 6*). Every finger unit with its 3 active degrees of freedom integrates 28 sensors. A small separate controller box houses the finger controllers coupled by a fiber optic link ring to any external workstation running the hand controller. The system is completed by a data glove with or without (*Fig. 3*) force feed-back and a SPACE MOUSE® as input device.

Fig. 3: Manipulation system consisting of DLR Hand and light weight robot, controlled by data glove and tracker.

Mechanical Structure

Each of the four identical fingers consists of two independent units, the base joint unit (*Fig. 4*) with two degrees of freedom realised in a cardanic manner and the finger unit (*Fig. 6*) with one actuator for two joints.

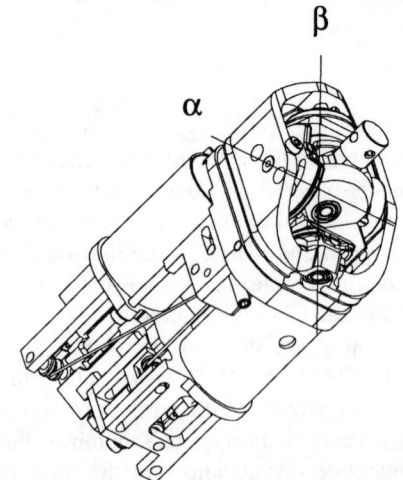

Fig. 4: Base joint unit with two actuators for two degrees of freedom.

Two actuators fixed in the base joint unit result in a slight kinematic coupling of the two axes. Due to mechanical constraints we are not able to measure the actual position of joint 2 (abduction/adduction) but of actuator 2. Thus we have to calculate the actual joint position from the measured position of joint 1 and actuator 2. Devoting the position of axis 1 α (*Fig. 4*) the position of actuator 2 β, the position γ of joint 2 is given as:

$$\gamma = sign(\beta)\arccos(\frac{1}{\sqrt{1+\cos^2(\alpha)\tan^2(\beta)}}).$$

On the other hand this means to achieve a desired position γ of axis 2, β has to be measured as:

$$\beta = \arctan(\frac{1}{\cos(\alpha)}\tan(\gamma)).$$

In case of axis 1 in its initial position ($\alpha=0$), there is no coupling. We observe the highest degree of coupling in the extreme positions of joint 1 and joint 2. In this case β deviates 9.23° from γ. Since the equation is known, the deviation can be compensated.

The range of motion for the joints is given as ±45° for joint 1 (curling motion of base joint), ±30° for joint 2 (abduction/adduction of base joint) and 105° for joint 3 and 110° for joint 4 respectively (both joints: curling motion of finger).

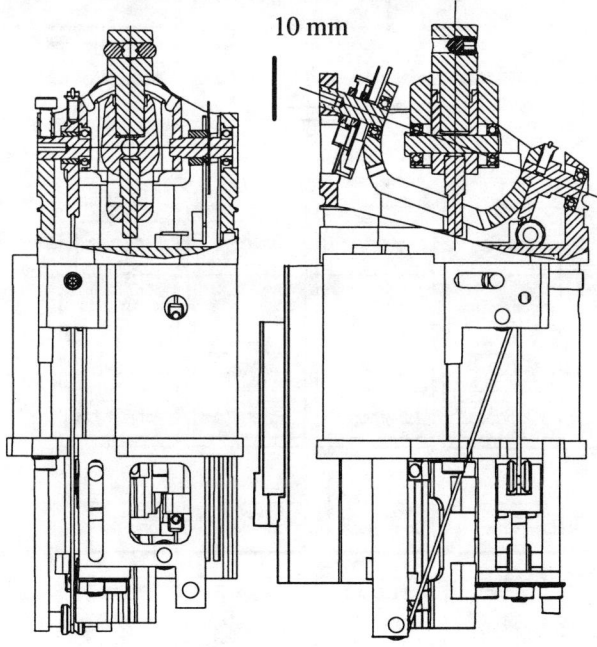

Fig. 5: Sectional view of base joint unit with two actuators and two joints from front (left) and side (right).

Sensorial Structure

A dextrous robot hand needs as a minimum a set of force and position sensors to enable control schemes like position control, force control and stiffness control. Special types of sensors add to this basic sensor equipment (*Table 2*).

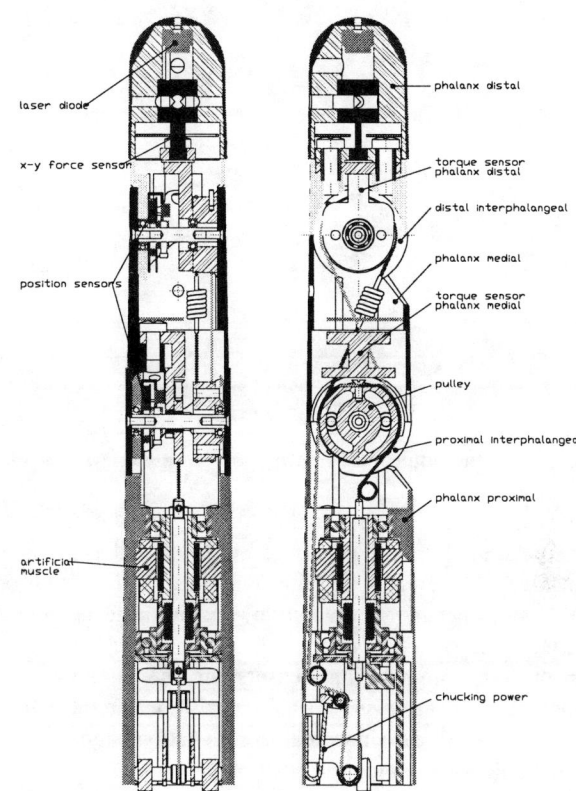

Fig. 6: Design of the finger unit with one integrated actuator and two coupled joints.

Beside conventional strain gauge based joint torque sensors in each joint a newly developed optical position sensor is integrated in every joint of the DLR Hand in order to meet the requirements. A separate joint angle sensor is necessary due to the presence of slippage in the planetary roller spindle drive and hysteresis of tendon transmission.

This optical joint position sensor (*Fig. 7*) fits in an almost human sized finger (*Fig. 6*).

The sensor is based on a one-dimensional PSD (Position Sensitive Device). This PSD is illuminated by an infrared LED via an etched measurement slot. Using an optimized PCB design exclusively equipped with tiny SMD items and the development of a circuit with a minimized number of items used it was possible to create an optical position sensor with excellent performance with respect to its size.

The sensor measures only 4.8 mm in thickness and 17 mm in diameter. Nevertheless a voltage regulator and the complete analogue conditioning circuit is included in the sensor itself. 10 Bit angular solution are achieved with a linearity error of less than 1 % using a supply voltage of 14 V.

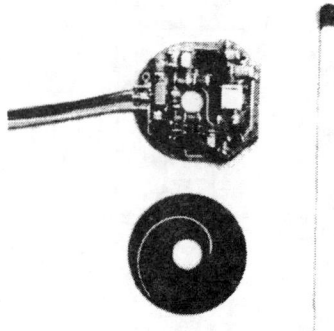

Fig. 7: Joint position sensor

By replacing the slice with the measurement slot the measurement range can easily be adapted to various requirements.

To increase the controllability of the actuators, speed sensors are important. Therefore we developed a so-called Tracking Converter providing us with a high resolution (3072 steps per motor revolution) position information of the motor's rotor. The sensor is based on linear Hall effect sensors as commutation sensors in the motors. The Tracking Converter is a small hardware circuit which converts the three sinusoidal commutation signals to a high resolution position information.

Sensor Type	count	Range/resolution
joint position	4	110° (8 bit), 90° (10 bit)
joint torque	5	1.8 Nm (9 bit)
Tactile sensors	4	0.5-10 N, 35 mN res.
rotor position	3	3072 steps per revolution
Temperature	5	0-100 °C, 0.5 °C res.
light barriers	6	
proj. laser diode	1	0.5 m

Table 2: Sensors of one finger

These three sensor types are essential for low level joint control. The hand is equipped with various other sensors as well.

- Tactile sensors cover each finger link. They consist of tactile foils detecting center and size of external forces applied to the fingers. The sensors are based on FSR (force sensing resistor) technology and arranged as XYZ pads [11]. With a cycle time of 2 ms they are also suitable for low level joint control.
- The finger tips provide a light projection laser diode to simplify image processing for a tiny stereo camera system integrated in the hand's palm. Due to the already mentioned modularity the finger tips might be easily exchanged with a version containing e.g. fiber optics. The two-axis torque sensor hereby serves as fast exchange adapter.
- Several temperature sensors and light barriers are present for security purposes.
- Additionally the hand is equipped with a six dimensional force torque sensor in the wrist.

Signal conditioning for the sensors in the finger is completely integrated in the finger unit.

Hand Controller Hardware Design

DLR's Dextrous Hand is controlled by a multiprocessor system.

The controller hardware design shows up a fully modular concept (*Fig. 8*). The control architecture is split into two levels, the global hand control level and the local finger control level. The global hand controller is externally located in a PC running a real time operating system.

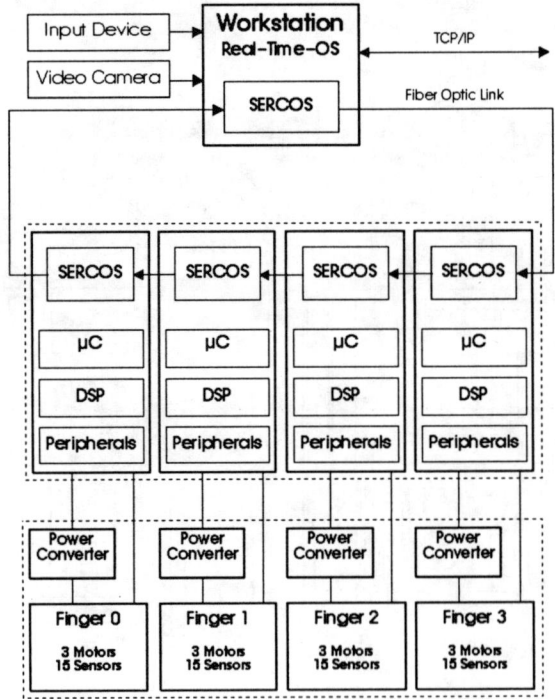

Fig. 8: Hand control system

Thus maximum flexibility is provided for implementing various hand control strategies.

In contrast to the widely used controller hardware designs the modular local finger controllers are located near the hand (*Fig. 9*) attached to the manipulator carrying the hand. This design became feasible due to the high degree of integration and miniaturisation of the finger controller hardware.

Global hand controllers and local finger controllers communicate via SERCOS (SErial Real Time COmmunication System) by fiber optic link. SERCOS is powerful enough to exchange all sensor information and control signal within 1 ms for all four fingers.

The main goals taken into consideration for this design are high flexibility, easy expandability and maximum computational performance.

The current finger controller solution is about 1.8 kg in weight and $12 \times 11 \times 22$ cm^3 in size for the whole hand with four fingers (*Fig. 9*). Thus it can easily be carried by common robots used with our dextrous hand. For the future additional miniaturisation is envisaged.

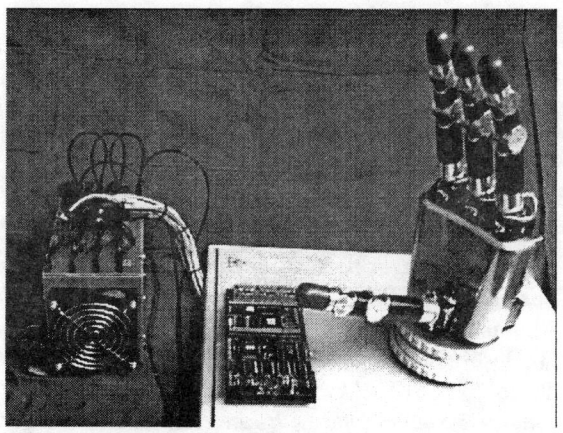

Fig. 9: Finger controller (left) and DLR Hand. In the middle one finger controller module is shown.

Due to the integration of the local finger controllers as well as the complete drive system including the power converters into the hand-arm system the cabling carried by the manipulator arm reduces to a fiber optic link ring and a four line power supply interface.

Finger Controller Hardware Architecture

For the control of each finger one finger controller module is necessary. The controller module (*Fig. 10*) is an independent subsystem of the hand system and receives commands from the global control level by SERCOS via fiber optic link.

One µ-controller per finger is responsible for the information management inside the controller module. An additional 60 MFLOPS floating-point DSP provides a control hardware with sufficient computational power to realise future control algorithms for the three motors of each finger. These finger controllers integrate all the peripherals shown in *Fig. 10* necessary to drive a complete finger.

Identical finger control modules are put together by a common power supply to form the complete control system for the local finger level. The modular design allows the use of up to four finger control modules with the current power supply design. In order to obtain a certain number of fingers for a fixed design a redesign of the power supply is sufficient to get the optimal solution in respect of size and weight.

Joint control strategies are presented in a complementary paper at this conference.

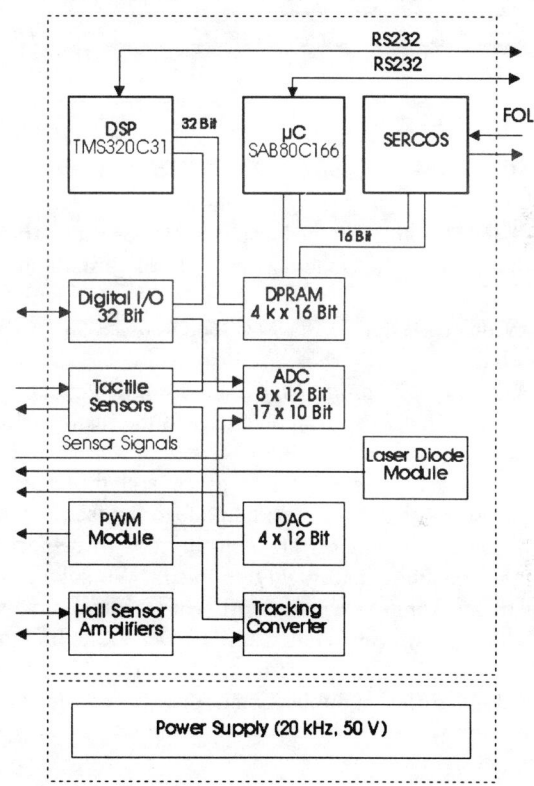

Fig. 10: Finger controller hardware

Software Structure

In terms of flexibility a modular software structure is essential for a research system like the hand system presented here.

Software is separated in the task level programming environment and the real time control environment (*Fig. 11*). The real time control environment contains the joint controllers running on the finger DSPs and the hand controller running a real time operating system.

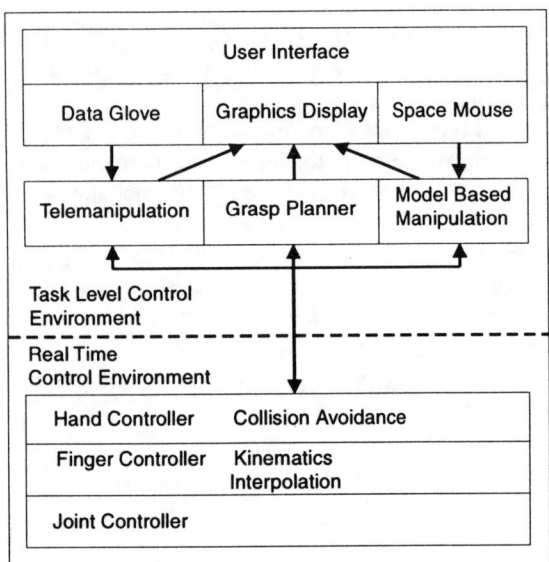

Fig. 11: Software structure

This hand controller is located between the finger controllers and the external task level programming environment and is responsible for kinematics calculation, cartesian stiffness calculation and trajectory interpolation.

Currently the task programming environment comprises collision avoidance, a grasp planner and the man machine interface.

The interface between the data glove and the artificial hand is a separate software module due to the fact, that the data glove measurements are non-linear with respect to the finger positions and highly coupled. We solved this problem by using a feed-forward neural network which learns the non-linear relationship between the fingertip positions and the data glove measurements. This will be outlined in a different publication at this conference.

Conclusion

The main features of a modular manipulation system with integrated actuators for dextrous manipulation in unstructured servicing environment has been presented. The presentation focused on the efficient control architecture.

It shows flexibility and computational power to meet even future requirements for autonomous manipulation.

Future activities will focus on the implementation of powerful grasping strategies, further miniaturisation and further improvement of mechanical and electrical reliability. The hand is supposed to become a central component for the development of future robonauts in space, but also aims at being used as a prosthesis at least long-term.

References

[1] G. Hirzinger, J. Butterfass, S. Knoch, H. Liu, *DLR'S Multisensory Articulated Hand,* Preprints Fifth Int. Symposium on Experimental Robotics, Barcelona, Spain, June 1997, pp. 28-34.

[2] S.C. Jacobsen, I.K. Iversen, D. Knutti, R.T. Johnson, K.B. Biggers, *Design of the Utah/MIT Dextrous Hand,* Proceedings IEEE Int. Conf. on Robotics and Automation, USA, 1986, pp. 1520-1532.

[3] C. Melchiorri, G. Vassura, *Mechanical and Control Features of the University of Bologna Hand Version 2,* Proc. IEEE Int. Conf. on Intelligent Robots and Systems, Raleigh, 1992, pp. 187-193.

[4] J.K. Salisbury, *Design and Control of an Articulated Hand,* Int. Symposium on Design and Synthesis, Tokio, 1984.

[5] B.M. Jau, *Man-Equivalent Telepresence Through Four Fingered Human-Like Hand System,* Proc. IEEE Conf. on Robotics and Automation, France, 1992, pp. 843-848.

[6] Y. Maeda, S. Tachi, A. Fujikawa, *Development of an Anthropomorphic Hand (Mark-1),* Proc. Int. Symp. on Industrial Robots, Tokyo, 1989, pp. 537-544.

[7] S. Sugano, I. Kato, *Wabot-2: Autonomous Robot with Dextrous Finger-Arm Coordination in Keyboard Performance,* Proc. IEEE Conf. on Robotics and Automation, Raleigh, 1987.

[8] H. Hashimoto, H. Ogawa, M. Obamam, T.Umeda, K. Tatuno, T. Kurukawa, *Development of a Multi-Fingered Robot Hand with Fingertip Tactile Sensors,* Proc. IEEE Conf. on Intelligent Robots and Systems, Yokohama, 1993, pp. 875-882.

[9] M. S. Ali, K.J. Kyriakopoulos, H.E. Stephanou, *The Kinematics of the Anthrobot-2 Dextrous Hand,* Proc. IEEE Conf. on Robotics and Automation, Atlanta, 1993, pp. 705-710.

[10] J. Dietrich, G. Hirzinger, B. Gombert, J. Schott, *On a Unified Concept for a New Generation of Light-Weight-Robots,* Proc. Int. Symp. on Experimental Robotics, Montreal, Canada, 1989.

[11] H. Liu, P. Meusel, G. Hirzinger, *Tactile Sensing System for the DLR Three-Finger Robot Hand,* IMEKO Technical Committee on Robotics, 1995, pp. 91-96.

[12] M.T. Mason, J.K. Salisbury, *Robot Hands and the Mechanics of Manipulation*, MIT Press, Cambridge, USA, 1985.

[13] S.T. Venkataramanan, T. Iberall, *Dextrous Robot Hands*, 1st ed., NY, Springer Verlag, 1990.

[14] T. Matsui, T. Omata, Y. Kuniyoshi, *Multi-Agent Architecture for Controlling a Multi-Fingered Robot,* Proc. IEEE Conf. on Intelligent Robots and Systems, Raleigh, 1992, pp. 182-186.

DLR's multisensory articulated Hand
Part II: The Parallel Torque/Position Control System

H. Liu P. Meusel J. Butterfass G. Hirzinger

DLR

German Aerospace Center
Institute of Robotics and System Dynamics, 82230 Wessling
e-mail: Hong.Liu@dlr.de

Abstract

This paper gives a brief description of feedback control systems engaged in DLR's recently developed multi-sensory 4 finger robot hand. The work is concentrated on constructing the dynamic model and the control strategy for one joint of the fingers. One goal is to make the hand follow a dataglove for fine manipulation tasks. Our proposed strategy for this task is parallel torque/position control; sliding mode control is realized for the robust trajectory tracking in free space; while impedance control is provided for compliance control in the constrained environment; and an easily-designed parallel observer is used for the switch between these two control modes during the transition from or to contact motion. Some experimental results show the effectiveness of proposed strategy for the pure position control, torque control, and the transition control.

1. Introduction

The DLR's multisensory articulated hand[1], as shown in Fig. 1, is a four fingered hand with in total twelve degrees of freedom. It has three fingers and an opposing thumb. The actuation system is uniformly based on Artificial Muscles®[1], a tiny linear electromechanical actuator integrating DLR's planetary roller screw-drive[2] with small brushless DC motor(BLDC), which are integrated in the hand's palm or in the fingers directly. Force transmission in the fingers is realized by special tendons made of highly molecular polyethylene. To achieve high degree of modularity, all four fingers are identical, and each has three active DOFs and integrates 28 sensors. The motions of middle phalanx and distal phalanx are not individually controllable; they are connected by means of tendons in such a way as to display motions similar to those of human fingers during grasping and are actuated only by one artificial muscle. The proximal joint has 2 degrees of freedom; one is for curling motion and another is for abduction/adduction motion.

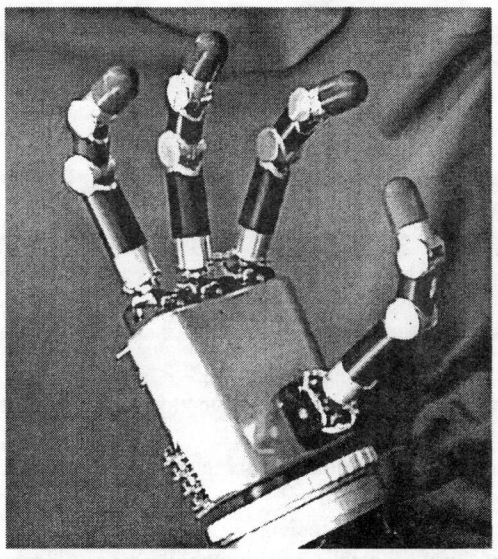

Fig. 1. The DLR's Multisensory Hand

There are two different position sensors for each active degree of freedom; one is a tracking converter for measuring the motor position on the basis of Hall sensors and another is the actual joint position sensor based on a one-dimensional PSD (Position Sensing Device), which is illuminated by an infrared LED via an etched spiral-type measurement slot. The effective combination between these two kind of sensors plays a key role in the joint position control in dealing with tendon hysteresis. Also, at each joint there is a torque sensor based on strain-gauges for accurate torque control. A more detailed description is given in Part I.

Our first approach was to make the hand to follow the desired states (positions, speeds, accelerations, and torques) commanded from higher levels, e.g., dataglove. This requires accurate tracking in free space, compliance in the constrained environment, and smooth transition between these two operational modes. The tracking control problem is to design a control scheme which generates the appropriate control signal so as to ensure that the joint angle follows any specified reference trajectory as closely

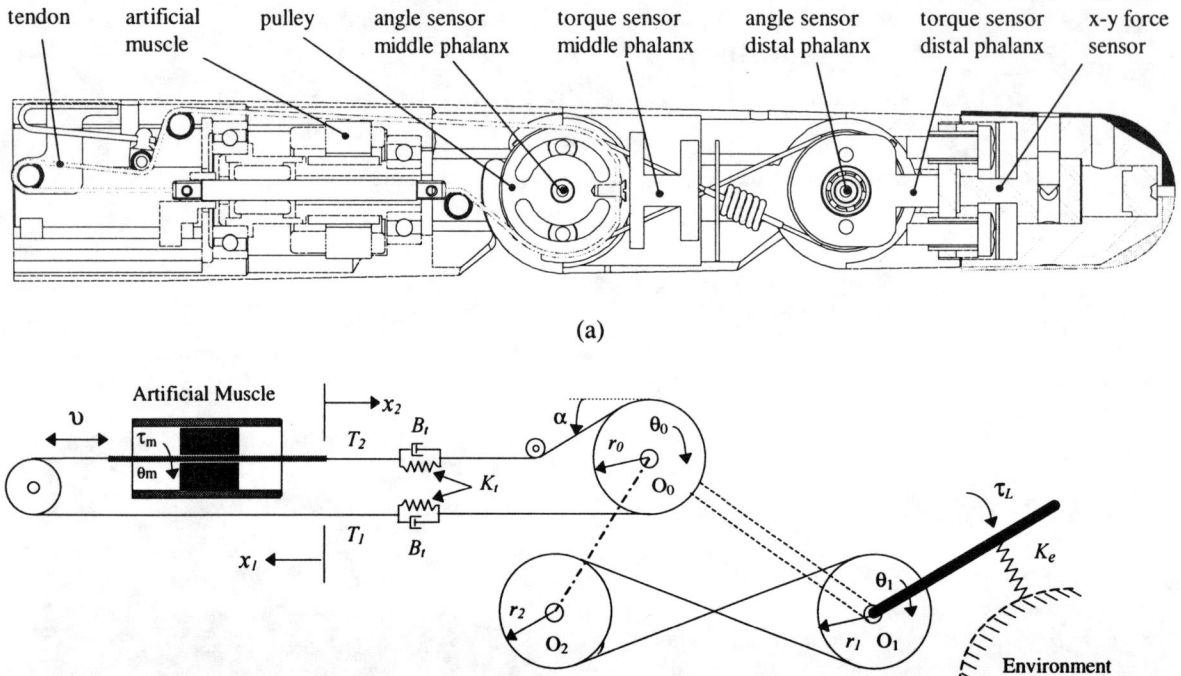

Fig.2 The Finger Mechanical Construction(a) and Its Schematic Diagram (b)

as possible. Compliance control is needed for fine manipulation and also for the protection of the hand itself. Transition control, called also impact control, is essentially the problem of making the energy conversion more effective and smooth. The well-known approaches to transition control include impedance control[3] and explicit force control[4]. Although the control schemes show good performance, this performance is achieved under fairly restrictive conditions. Common to the approaches is the requirement that the operating environment as well as the environmental interaction have to be very accurately modeled. This fact restricts their use in realistic operating environments. In this paper we propose a parallel torque/position control strategy for the pre-described task. The robust trajectory tracking in free space is implemented by sliding mode control; the compliant motion in the constrained environment is realized by using impedance control; and a parallel observer based on contact torques and system states has been built to determine the switch between these two kind of controllers for the transition control from or to contact motion.

2. Dynamic Model of a Joint

The actuation and transmission system of the third link of the DLR four-fingered hand is shown in Fig. 2(a). The main components of the single-joint model are : a) the BLDC-based artificial muscle; b) the tendon pulley power transmission system; and c) the joint itself. The artificial muscle is a linear actuator which translates the motor's rotation into axial movements based on DLR's planetary roller screw drive[2]. The tendon as a structure is represented by parallel spring and damping elements(K_{st}, B_{st}), which permit the structural oscillations between the actuator and load. A schematic diagram of the system is shown in Fig. 2(b). T_1 and T_2 represent the tension in the tendon at the appropriate points while T_{10} and T_{20} represent the corresponding pretensions. τ_L represents the external torque which results from the contact between the finger and its environment, and K_e is the environment stiffness. The system also exhibits a hysteresis effect due to the combination of Coulomb friction and tendon compliance. Fig.7(a) shows the experimental hysteresis measurements. Hystereses windup and winddown only occur during the initial startup, or result from direction changes. One may then write down the following equations describing the dynamics of the finger unit:

$$r_0 = r_2 = K_r r_1 \tag{1}$$

$$\theta_1(t) = n\theta_m(t), \theta_1(t) = K_r \theta_o(t) \tag{2}$$

$$T_1 = K_t(x_1(t) - r_0\theta_0(t)) + B_t(\dot{x}_1(t) - r_0\dot{\theta}_0(t)) + T_{10} \tag{3}$$

$$T_2 = K_t(x_2(t) - r_0\theta_0(t)) + B_t(\dot{x}_2(t) - r_0\dot{\theta}_0(t)) + T_{20} \tag{4}$$

$$\tau_m = K_I I_m \tag{5}$$

$$J_m\ddot{\theta}_m(t) + B_m\dot{\theta}_m(t) + T_f + T_g + n\tau_L = \tau_m \tag{6}$$

$$\tau_L + K_s\left(\theta_1(t) - x_1(t)/r_1\right) = 0 \quad (7)$$

where the remaining parameters are defined in Tabel 1.

3. Sliding Mode Position Control of a Joint

The theory and properties of sliding mode control[5] are well known in the automatic control field. The important features of the sliding mode controller which make it attractive for application in the power electronics area are a) high accuracy, b) fast dynamic response, c) good stability, d) simplicity of design and implementation, and, above all, e) robustness. Robustness, or low sensitivity to deviations in system parameters and external disturbances, is a very important index of the controller in industrial applications. Melchiorri[6] used the sliding mode technique for the position control of the University of Bologna (UB) hand, however he did not deal with the trajectory tracking problems. The goal of the research activity presented in this paper is to experimentally verify the feasibility of a sliding mode controller for the position tracking control of the third joint.

To design the position controller with sliding mode properties, the dynamic equation (6) is rewritten in terms of state space:

$$x_1 = \theta_m, \; x_2 = \dot{x}_1, \; x_3 = \dot{x}_2$$

$$\begin{bmatrix} \dot{x}_1 \\ \dot{x}_2 \\ \dot{x}_3 \end{bmatrix} = \begin{bmatrix} 0 & 1 & 0 \\ 0 & 0 & 1 \\ 0 & -\frac{B_m R_m}{J_m L_m} & -(\frac{B_m}{J_m} + \frac{R_m}{L_m}) \end{bmatrix} \begin{bmatrix} x_1 \\ x_2 \\ x_3 \end{bmatrix} + \begin{bmatrix} 0 \\ 0 \\ \frac{K_I R_m}{J_m L_m} \end{bmatrix} I_m$$

$$+ \begin{bmatrix} 0 \\ 0 \\ -\frac{R_m}{J_m L_m} \end{bmatrix} (T_f + T_g + n\tau_L) \quad (8)$$

Equation (8) is in the form of

$$\dot{X} = AX + bu + Dn \quad (9)$$

where the control u takes the form of the link current I_m, and the disturbance input n is the load torque $n\tau_L$, friction T_f, and gravitation T_g.

Let $R(t) = [\theta_d \; \dot{\theta}_d \; \ddot{\theta}_d]^T$ be a reference track vector, i.e. a vector containing the desired tracks of motor position, speed and acceleration, $X(t)$ is the system state vector. Let the error $E(t)$ be defined as follows:

$$E(t) = R(t) - X(t) \quad (10)$$

The switching surface s here is chosen as

$$s(t) = C \cdot E(t), \; C = [c_1 \; c_2 \; c_3], \; E(t) = [e_1 \; e_2 \; e_3]^T,$$

where C is a vector of weighting factors and $c_3 = 1$. The switching surface is defined as $s(t) = 0$.

The control law of the sliding mode control is as follows

$$u = \psi_1 e_1 + \psi_2 e_2 + d \, \text{sgn}(s) \quad (11)$$

where ψ_1 and ψ_2 are feedback gains of each state variable and d is the input gain for compensating disturbances. The term $d \, \text{sgn}(s)$ is a steady-state dither component that is used to remove the steady-state error. Note that parameters ψ_1, ψ_2 and d in equation (11) allow changes in the individual terms, whereas d allows changes common to all the terms. Parameters ψ_1 and ψ_2 are not fixed, but change discretely to maintain the system on the switching surface.

3.1 Conditions of existence

The existence conditions of sliding mode require that the state trajectories be always directed towards the sliding surface s. This is given in mathematical form as

$$\lim_{s \to o} s(t) \cdot \frac{ds(t)}{dt} \leq 0 \quad (12)$$

From equations (8), (10), and (11),

$$s\dot{s} = (c_1 \dot{e}_1 + c_2 \dot{e}_2 + \dot{e}_3)s$$

$$\approx \left(-c_1 c_2 + c_1(\frac{B_m}{J_m} + \frac{K_I}{J_m}) - \frac{K_I R_m}{J_m L_m}\psi_1\right)e_1 s$$

$$+ \left(-c_2\left(c_2 - \frac{B_m}{J_m} - \frac{K_I}{J_m}\right) + c_1 - \frac{B_m R_m}{J_m L_m} - \frac{K_I R_m}{J_m L_m}\psi_2\right)e_2 s$$

$$+ \left(\frac{R_m}{J_m L_m}(T_f + T_g + n\tau_L) + \dot{r}_3 + \left(\frac{B_m}{J_m} + \frac{R_m}{L_m}\right)r_3 + \frac{B_m R_m}{J_m L_m}r_2\right)s$$

$$- \frac{K_I R_m}{J_m L_m} d \, \text{sgn}(s) s$$

The existence condition of sliding mode given in (12) can be satisfied if

$$d > \frac{1}{K_I}\left(T_f + T_g + \tau_L\right)$$
$$+ \frac{J_m L_m}{K_I R_m}\left(\frac{B_m R_m}{J_m L_m}r_2 + \dot{r}_3 + \left(\frac{B_m}{J_m} + \frac{R_m}{L_m}\right)r_3\right) \quad (13)$$

$$\psi_1 = \begin{cases} \alpha_1 & \text{if } e_1 s > 0 \\ \beta_1 & \text{if } e_1 s > 0 \end{cases} \qquad (14)$$

$$\psi_2 = \begin{cases} \alpha_2 & \text{if } e_2 s > 0 \\ \beta_2 & \text{if } e_2 s > 0 \end{cases} \qquad (15)$$

where,

$$\alpha_1 = \max_t \left\{ \frac{J_m L_m}{K_I R_m} \left(-c_1 c_2 + c_1 \left(\frac{B_m}{J_m} + \frac{K_I}{J_m} \right) \right) \right\}$$

$$\beta_1 = \min_t \left\{ \frac{J_m L_m}{K_I R_m} \left(-c_1 c_2 + c_1 \left(\frac{B_m}{J_m} + \frac{K_I}{J_m} \right) \right) \right\}$$

$$\alpha_2 = \max_t \left\{ \frac{J_m L_m}{K_I R_m} \left(c_1 - c_2 \left(c_2 - \frac{B_m}{J_m} - \frac{K_I}{J_m} \right) - \frac{B_m R_m}{J_m L_m} \right) \right\}$$

$$\beta_2 = \min_t \left\{ \frac{J_m L_m}{K_I R_m} \left(c_1 - c_2 \left(c_2 - \frac{B_m}{J_m} - \frac{K_I}{J_m} \right) - \frac{B_m R_m}{J_m L_m} \right) \right\}$$

3.2 Conditions of stability

In order to make the motion in the sliding regime stable, the coefficient c_i needed for the design of a desired sliding mode cannot be chosen freely. Considering the motion on the sliding surface $s=0$, the characteristic equation is

$$\ddot{e}_1 + c_2 \dot{e}_1 + c_1 = 0 \qquad (16)$$

Since we do not want any overshoot, our choice of c_1 and c_2 must satisfy the condition:

$$c_2 \geq 2\sqrt{c_1} \; ; \; c_1, c_2 \geq 0 \qquad (17)$$

3.3 Conditions of hitting

In order to guarantee that the system hits the sliding surface s from any initial states, the following condition should be satisfied [5]

$$c_2 < \frac{B_m}{J_m} + \frac{R_m}{L_m} \qquad (18)$$

Equations (13), (14), (15), (17), and (18) together specify the bounds on the controller parameters for operation of the position tracking control system in sliding mode. From these equations it is clear that exact knowledge of system parameters is not necessarily needed for designing the sliding mode controller. It is sufficient to know the bounds on the system parameters for the design of the above controller.

3.4 Practical implementation of the SLM

The sliding mode controller is constructed using the position error, speed error, and acceleration error. The practical implementation is shown in Fig.3. The reference trajectory is produced through a fifth-order polynomial interpolation when the desired position and speed are given[6].

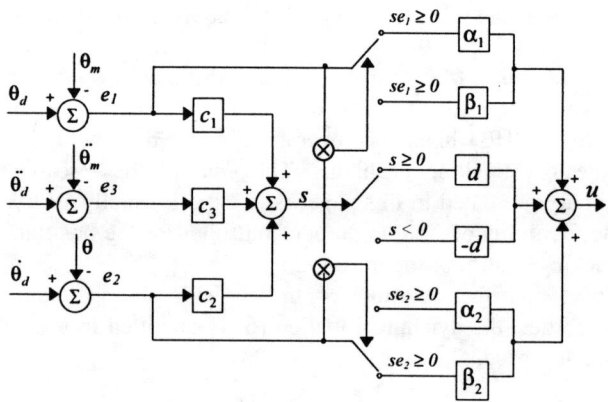

Fig.3 Block diagram of SLM position control of a joint

4. Impedance Joint Torque Control

Let J_d, B_d, K_d be the desired target impedance parameters of the robot finger; impedance control specifies this desired impedance relationship as a generalization of the second order dynamics of a damped spring:

$$J_d \ddot{\theta}_e + B_d \dot{\theta}_e + K_d \theta_e = \tau_{ext} \qquad (19)$$

where $\theta_e = \theta_d - \theta_m$ is the position error, while θ_d, θ_m, and τ_{ext} are the desired position, actual joint position, and the actual reaction force which the environment exerts on the robot finger, respectively. In order to keep the target impedance, one can deduce the following motor output torque by introducing (19) into (6):

$$\tau_m = J_m \left\{ \ddot{\theta}_d + J_d^{-1} \left[B_d (\dot{\theta}_d - \dot{\theta}_m) + K_d (\theta_d - \theta_m) - \tau_{ext} \right] \right\} \\ + B_m \dot{\theta}_m + T_f + T_g + n\tau_L \qquad (20)$$

This means that, with precise knowledge of finger dynamics and accurate sensors, one can achieve a perfect feedback linearization for driving torque calculation, and

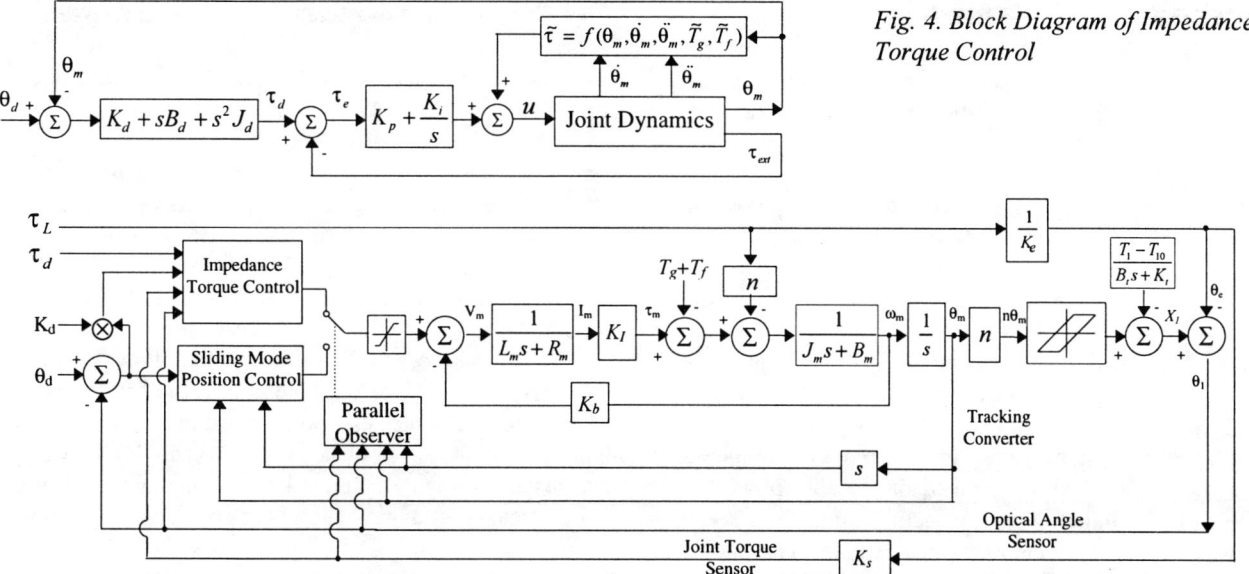

Fig. 4. Block Diagram of Impedance Torque Control

Fig. 5. Block diagram of the parallel torque/position control system

TABLE 1 : PARAMETERS OF ARTIFICIAL MUSCLE(AM20) AND TRANSMISSION SYSTEM

$K_I = 9.0 \times 10^{-3}$ Nm/A	torque constant		$t_E = 0.14$ ms	electrical time constant
$J_m = 1.62 \times 10^{-7}$ Kg.m^2	total inertia		$t_M = 4.17$ ms	mechanical time constant
$B_m = 1.11 \times 10^{-4}$ Nm/rad/s	damping coefficient		$K_s = 2.73 \times 10^1$ Nm/rad	stiffness of sensing joint
$K_b = 2.0 \times 10^{-2}$ V/rad/s	back EMF constant		T_1, T_2	tendon tension
$L_m = 1000.0$ μH	armature inductance		$K_t = 2.86 \times 10^4$ N/m	tendon stiffness constant
$R_m = 7.0$ Ω	ave. Terminal resistance		$B_t = 5.6 \times 10^{-4}$ Nm/rad/s	tendon damping constant
$n = 1/323$	reducer ratio		$K_r = 1.05$	pulley radii reducer ratio
I_m	motor link current		T_g	gravitational force
τ_m	motor output torque		T_f	frictional force

the finger will show up the desired impedance parameters -- J_d, B_d, K_d to the environment. However, in reality, the finger dynamics are not known precisely, tendon transmission brings some hysteresis, and the accuracy of the position and torque sensors are always affected by some noises. This means, practically, that it would be very difficult to realize a perfect linearization, and hence the desired impedance parameters can not be achieved. Alternatively, in order to keep the equality (19) as equal as possible, we can also introduce an explicit force control scheme, i.e., let

$$\tau_d = J_d(\ddot{\theta}_d - \ddot{\theta}_m) + B_d(\dot{\theta}_d - \dot{\theta}_m) + K_d(\theta_d - \theta_m) \quad (21)$$

where τ_d is the desired torque. let τ_e be the error function:

$$\tau_e = \tau_d - \tau_{ext} \quad (22)$$

Now we can introduce a simple PI control scheme with τ_e as input. If the τ_e converges to zero, the actual impedance parameters will converge to the desired values automatically. With the addition of the control signal from

PI and estimated finger dynamics, we can build an impedance controller shown in Fig.4. However the desired trajectory is not specified as a fixed function of time. Instead, it is a varying information from a high level, e.g. from a dataglove. In steady state, all measured and desired velocity and acceleration values are zero. This induces that the value of the steady state torque is the stiffness multiplied by the steady state deformation $(\theta_d - \theta_m)$, and the joint behaves like a programmable spring.

5. Parallel Torque/Position Control Strategy

Motions in free space and constrained environment belong to different operational modes with a qualitative change in the system model. In the previous section the sliding mode position control and impedance torque control have been introduced for these two motion modes. Some experimental results will be discussed in the next section. Pure position control systems will follow the commanded position trajectory while rejecting external forces which are considered as disturbances. Force control is introduced for the motion control in the constrained environment by tracking a dynamic relation between the active force and impedance. During the transition phase a large amount of kinematical energy need to be dissipated within a very short time. There exists a key problem of how to deal with the transition from free space to constrained environment or vice versa.

In this paper we propose a parallel torque/position control strategy for the transition phase, which is shown in Fig. 5. The strategy attempts to combine simplicity and robustness of the impedance control and sliding mode control with the ability of controlling both torque and position. The kernel of the strategy is the design of a parallel observer which determines which control mode should be active. The inputs to the observer are active torques, actual system states, and desired system states. Fig. 6 shows the block diagram of the parallel observer, where τ_{th} is the initial contact detecting torque and i, $i-1$ represent the current and previous events. The mode switching to contact motion is only determined by contact torque, while the recovering from contact motion to free motion is not only determined by τ_{th} but also dependent on whether the reference position changes. There is always one control mode active. The control signal exhibits a discontinuity only when the control mode has been changed. In order to make the transition phase as smooth as possible, a first-order low-pass filter is used for the control signal variations.

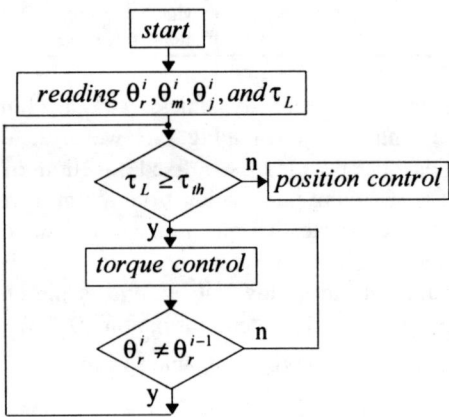

Fig. 6 Block diagram of parallel observer

6. Experimental Results

To verify the proposed control approach, a DSP(Digital Signal Processor)-based real-time control system has been built. The kernel of the system is a commercially available processor board with a TMS320C40. Some characteristics of this board are: 50 MHz clock(40 ns cycle time), floating point arithmetic unit, 768 Kbyte of fast static memory; and there are several I/O boards such as ADC, DAC, PIO which are installed in a separate box . All software development is done on the separate box which connects to a SUN workstation via ether net. The box takes care of the user interface, while most real-time computations and data I/O needed for the hand control are performed by several DSP boards.

Fig. 7(b) shows the sliding mode position control in free space, where there are two position sensors denoted by '1' and '2' which represent the motor position and actual joint position, respectively. Due to the hysteresis of tendon transmission and the slippage of the artificial muscle, there is always about 2° backlash between '1' and '2'. Our goal is to enable the motor to follow a desired trajectory as closely as possible and at the same time to achieve an accurate end-point position. Therefore, a dynamic variable reference trajectory is built, which changes according to the error between the motor position and the real joint angle. The result shows that the motor position is above the desired position of 20°, denoted as '3', but the optical joint angle sensor approaches '3' exactly.

Fig. 7(c) displays a torque step response based on impedance control. The experimental result shows the torque step response for seven different steps -10Nmm → -20Nmm → -30Nmm → -40Nmm → -30Nmm → -20Nmm → -10Nmm with 1 second interval. In this experiment, the stiffness $K_d=570$ *Nmm/rad*, damping $B_d=6000Nmm/rad/s$, and $J_d=0.0$. The overshoots for the loose and tension periods are nearly the same.

Fig. 7(d) shows a typical transition control between free space motion and constrained motion. The desired position denoted as '1' is 21°. The joint moves to this reference position at a speed of 50°/sec until it detects a contact. The control mode is switched automatically from position control to torque control when the contact torque is greater than the threshold value of $\tau_{th}=2.5Nmm$. The joint position denoted as '2' does not reach '1' but the torque approaches a desired value with a limited overshooting. It demonstrates the one side of the transition phase: from free space to constrained environment. By following the outputs of dataglove the joint can also move from contact phase to free motion with stability.

7. Conclusions

In this paper a dynamic model of the third joint of DLR's multisensory hand has been established. Sliding mode control has been successfully implemented in the tracking control of the joint in free space. The position error between the motor and active joint during slippage of artificial muscle and tendon transmission is always compensated in real-time to achieve a desired end-point position. A simplified impedance torque control with desired stiffness, damping, and mass has been also implemented for the contact phase. A parallel observer is used to switch between the two control modes for the transition control from or to contact motion, and the finger behaves like a programmable spring in steady state. The experimental results show the effectiveness of the proposed controller with the essential property of closed-loop system stability. It might serve as a general solution to the problem of impact control in realistic operating environments.

References:

[1] Hirzinger G., J. Butterfaß, S. Knoch, H. Liu, "DLR's Multisensory Articulated Hand", *Int. Symp. On Experimental Robotics, preprints,* pp.28-39, Barcelona, 1997

[2] J. Dietrich, G. Hirzinger, B. Gombert, J. Schott, "On a Unified Concept for a New Generation of Light-Weight-Robots", *Proc. Of the Conf. ISER, Int. Symp. On Experimental Robotics*, June 1989

[3] Hogan N., "Impedance Control: An Approach to Manipulator: Part I-III ", *Trans. ASME J. Dyn. Syst., Meas., Contr.*, Vol. 107, 1985, pp. 1-24.

[4] Volpe R., and P. Khosla, " A Theoretical and Experimental Investigation of Impact Control for Manipulators " *Int. J. Robot. Res.* 12(4), 1993, pp. 351-365.

[5] Utkin V.I., " Variable Structure Systems with Sliding Modes", *IEEE Trans. Automat. Contr.*, vol. AC-22, no. 2, pp. 212-222, 1977.

[6] C. Melchiorri and A. Tonielli, "Sliding Mode Control for a Robotic Hand", *Robotersysteme* 8, pp. 13-20, 1992.

[7] Paul. R, *Robot Manipulators: Mathematics, Programming and Control*. Cambridge, MA: M.I.T Press, 1981.

[8] Mills J.K., Lokhorst D.M., " Control of Robotic Manipulators During Genaral Task Execution: A Discontinuous Control Approach ", *Int. J. Robot. Res.* Vol. 12(2), 1993, pp. 146-163.

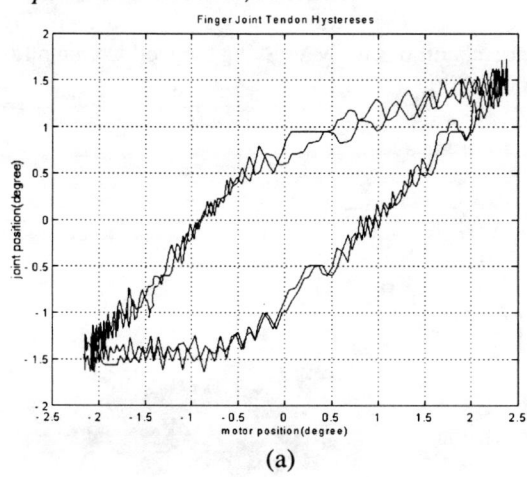

(a)

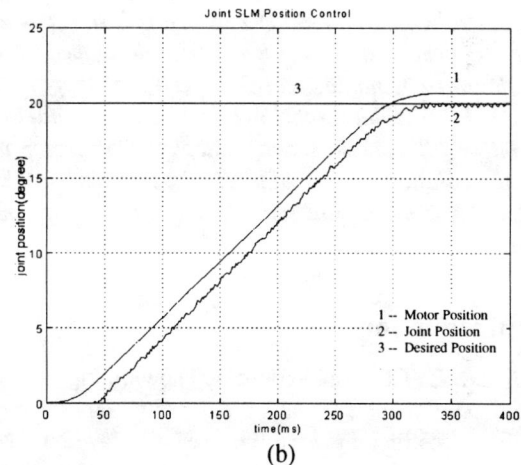

(b)

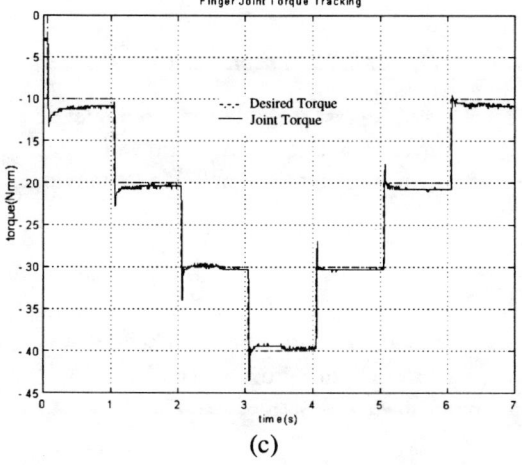

(c)

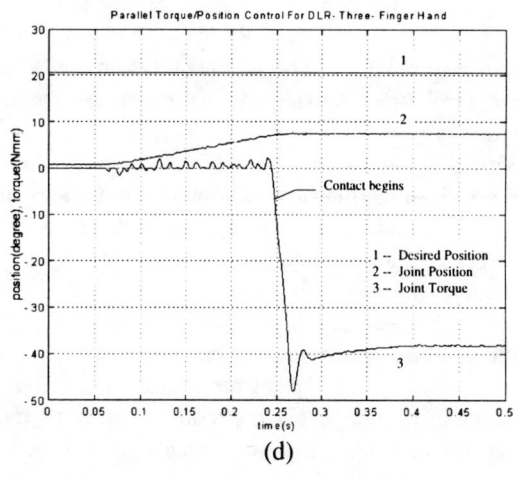

(d)

Fig.7. Some experiments on a finger joint (a) tendon hystereses , (b) joint position control, (c) joint torque tracking, (d) parallel torque/position control

Redundancy Resolution of a Cartesian Space Operated Heavy Industrial Manipulator

M. Honegger, A. Codourey
Institute of Robotics, ETH-Zürich, CLA, 8092 Zürich, Switzerland
Tel. +41 1 632 55 48, Fax +41 1 632 10 87
e-mail: honegger@ifr.mavt.ethz.ch

Abstract

In tunneling and mining construction work, heavy and large manipulators are used to spray liquid concrete on the walls. These manipulators are usually operated manually with simple units allowing to control all actuators independently. In cooperation between industry and university, a novel control system has been developed for the redundant heavy manipulator Robojet® that supports the operator by calculating the inverse kinematic model and controlling all actuators automatically. This paper presents the solution of the redundant inverse kinematics of the Robojet using the null-space method and its implementation in a control system.

1. Introduction

The MEYCO Robojet® sc-30 shown in figure 1 is a hydraulically actuated manipulator with application in tunneling construction work. It is used to spray liquid concrete on the walls of new tunnels. The design of this heavy and large manipulator with 8 degrees of freedom is ten years old, and the manipulator is being used worldwide. So far the manipulator has been operated manually with a simple control unit allowing to control the 8 actuators independently.

With this controller it is difficult to guide the jet along the wall of the tunnel while optimizing the spraying process and minimizing the losses of concrete. The operator must practice a long time in order to master the task in a satisfactory way.

We have developed a new control system that supports the operator in different ways. In one of the modes the operator can guide the jet directly in world-coordinates, using a space mouse, i.e. a 6 dof joystick [1]. In an automatic mode it is possible to scan the profile of the tunnel in a selected area using a laser scanner and to subsequently automatically control the distance and orientation between the jet and the wall. The operator needs only to guide the tool center point along the directions of the tunnel wall with the space mouse.

In both operation modes the movement given in cartesian coordinates has to be transformed in joint coordinates. The calculation of the redundant inverse kinematics and the control of the 8 hydraulic actuators is performed by the controller. The redundancies are solved either by using static conditions or by applying the well known null-space method [2].

Fig.1: MEYCO Robojet sc-30

Chapter 2 and 3 present the kinematic model of the robot and its inverse solution. The following chapters describe its implementation in a simulation system and in the controller of the real robot.

2. Kinematics

The manipulator is mounted on a vehicle that is not moving during the spraying process. The location of the tool is therefore always described with reference to the vehicle.

Figure 2 shows a sketch of the kinematics of the robot with its redundant degrees of freedom.

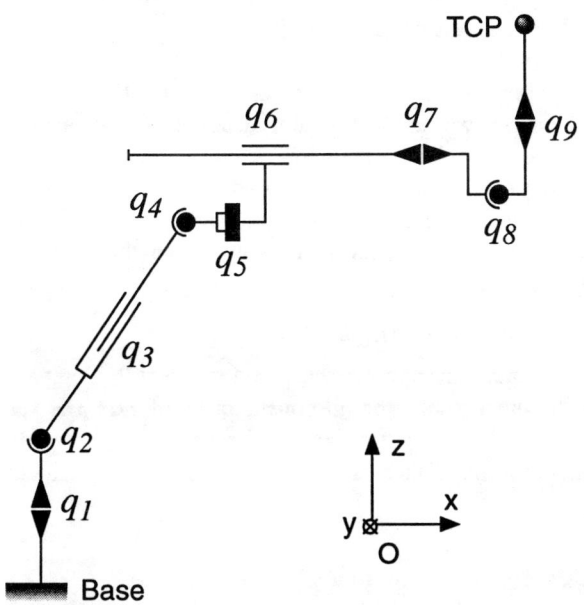

Fig.2: Kinematic model

The length of the manipulator when fully extended is bigger than 10m, thus allowing to spray in tunnels with a diameter of up to 25m without moving the vehicle.

All 9 joints are hydraulically actuated. Joint 9 is used for a small circling motion of the jet for a better distribution of the sprayed concrete. It has no effect on the calculation of the kinematic model. The joint angles involved in the calculation are thus

$$q = [q_1, q_2, q_3, q_4, q_5, q_6, q_7, q_8]^T. \quad (1)$$

The task requires the control of 5 dof of the jet, i.e. the position of the tool center point and 2 angles for the orientation. It is given by

$$x = [x, y, z, \alpha, \beta]^T. \quad (2)$$

The angle α describes the rotation about the x-axis of the reference frame O leading to a new rotated frame O'. β represents the rotation about the y-axis of the rotated frame O'. The calculation of the direct kinematic model $x=f(q)$ is straight forward using homogeneous transformation matrices. On the other hand calculating the inverse kinematics is not obvious. To solve the redundancies 3 constraints are required and no closed-form solution exists. A way to solve this problem is explained in the next section.

3. Redundancy Resolution

Many methods have been presented in the literature to solve this problem so far. In our case, two of them have been implemented and compared. The first one just uses some constraints for the position of the redundant axes. The second one is based on differential kinematics and uses the powerful null-space method.

3.1. Constraints on the position

As constraints for the redundant degrees of freedom, we used the following 3 static conditions for the variables:

$$q_5 + \frac{q_1}{3} = 0 \quad (3)$$

$$q_4 - \frac{1+q_2}{6} = 0 \quad (4)$$

$$q_3 = const \quad (5)$$

The selection of these equations can be understood as follows: In usual spraying tasks the position of the manipulator remains close to that of figure 2. In this configuration the axes 1 and 5, and 2 and 4 respectively, have a similar orientation in space. This redundancy can be used to increase the workspace of the manipulator using equations (3) and (4). The consumption of oil of the 3rd joint during large translational motions is very high and can be moved slowly only. To simplify the control it is kept constant during the automated tasks (equation (5)). Its static position is given manually before executing the task.

For five given pose coordinates in the operational space and the three additional conditions, the angles of all eight joints are fully determined and can be calculated. Due to the complicated kinematic structure of the robot there is however no closed-form solution for the inverse kinematic model. The joint angles are thus calculated numerically with the Newton-Raphson method.

If this method guarantees a large workspace, it doesn't allow high dynamics of operation because the velocity of the end-effector is limited by the slowest actuator. The null-space method is then used to solve this problem and explained in the following section.

3.2. Null-space method

Resolution of redundancy using the null-space method has already been presented many times in the literature. A good overview is given in [2] or [3]. It consists of resolving the redundancy by using the pseudoinverse of the Jacobian matrix of the manipulator for the movement of its TCP and the null-space of the Jacobian for internal movements. If $\dot{x}$ is the 5-dimensional velocity vector of the TCP and q, $\dot{q}$ the $n > 5$ dimensional vector of joint angles and velocity respectively, then

$$\dot{x} = J\dot{q}. \tag{6}$$

To solve this equation with respect to $\dot{q}$, it is necessary to use its pseudoinverse, which yields the following result:

$$\dot{q} = J^{\#}\dot{x} + (I - J^{\#}J)\phi \tag{7}$$

with

$$J^{\#} = J^{T}(JJ^{T})^{-1} \tag{8}$$

and where ϕ is an arbitrary joint velocity vector and $(I - J^{\#}J)\phi$ is its projection into the null-space of J. This corresponds actually to an internal motion of the manipulator which has no effect on the velocity of the end effector.

In order to optimize the motion of the manipulator with respect to some constraints, a weighted pseudoinverse can be used instead of the pseudoinverse. Equation (8) is then transformed to:

$$J^{\#}_{w} = W^{-1}J^{T}(JW^{-1}J^{T})^{-1} \tag{9}$$

which corresponds to the minimization of the cost function $\dot{q}^{T}W\dot{q}$ [5].

Both of these techniques have been combined and applied to the Robojet so that optimization of the oil consumption and resolution of the redundancy can be achieved simultaneously. This is obtained by using the following equation:

$$\dot{q} = J^{\#}_{w}\dot{x} + (I - J^{\#}J)\phi \tag{10}$$

3.2.1. Selection of the weighting matrix

In the case of the hydraulic manipulator Robojet, the objective is to minimize the oil consumption while moving. The oil consumption for each joint i is proportional to the velocity of the hydraulic cylinder, i.e.

$$Q_i = w_i \cdot |\dot{q}_i| \tag{11}$$

Minimizing the oil consumption of the manipulator can be obtained by minimizing the following cost function:

$$g(\dot{q}) = \sum \|w_i \cdot \dot{q}_i\|^2, \tag{12}$$

or equivalently

$$g(\dot{q}) = \dot{q}^{T} \cdot W \cdot \dot{q}, \tag{13}$$

where $W = diag(w_1^2, w_2^2, ..., w_8^2)$ is a diagonal matrix containing the squares of the consumption coefficients of each axis. The result of this optimization is actually given by equation (9).

However, in selecting the coefficients of the weighting matrix, we had to take into account that some joints can move quickly to their limits while others don't have enough oil to maintain the desired velocity. Thus, a fine parameter tuning and optimization was needed on the real manipulator. Our experiments have given the following matrix:

$$W = diag(37, 78, 100, 39, 25, 25, 12, 9) \tag{14}$$

A particular problem occured for joint 3. Compared to the other joints, the oil consumption of axis 3 is very high and thus the pump can not always deliver the needed amount of oil. This axis had to be weighted more than the others.

3.2.2. Null-space motion

As explained before, the use of the null-space method leads to internal motion of the manipulator without affecting the velocity of its end effector. The vector ϕ can thus be used to perform a secondary task such as keeping the axes of the manipulator as close as possible to the middle of their workspace [4], avoiding obstacles [5] or other kind of tasks. In our case, the secondary objective is to keep some relations between the axes constant so as to optimize dynamically the workspace of the manipulator. The same static constraint equations as before ((3) and (4)) have been chosen for axes 1 and 2. Axis 3 on the other hand is now constrained by the position of axis 6. The constraint equation is thus given by

$$\frac{q_6}{3} - \frac{q_3}{1.85} = 0. \tag{15}$$

Each of these constraints equations are used directly in the null-space by setting:

$$\phi = \begin{bmatrix} k_1\left(q_5 + \dfrac{q_1}{3}\right) \\ k_2\left(q_4 - \dfrac{1+q_2}{6}\right) \\ k_3\left(\dfrac{q_6}{3} - \dfrac{q_3}{1.85}\right) \\ 0 \\ 0 \\ 0 \\ 0 \\ 0 \end{bmatrix} \tag{16}$$

The internal velocity of each axis is thus adjusted until the constraints are satisfied. The coefficients $k_i, i = 1...3$ are used to tune the velocities of the internal motion.

4. Simulation

MOBILE, a simulation and 3D graphic animation software package [7] running on a SiliconGraphics workstation, is used to simulate the kinematics of the robot links and display its motions within a virtual tunnel (figure 3). The motions can be generated either by programming trajectories in a C++ program or online using a space mouse to guide the jet in world coordinates.

This simulation has been used to test the null-space algorithms before their implementation in the real application and also for the first tuning of the weighting matrix W and the parameters k_i for the internal motion.

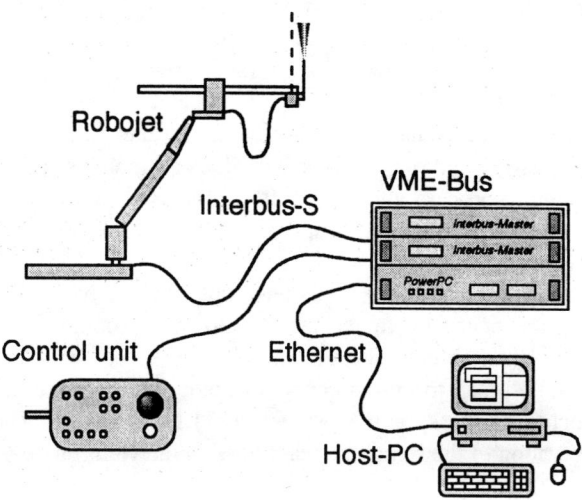

Fig.4: Hardware setup

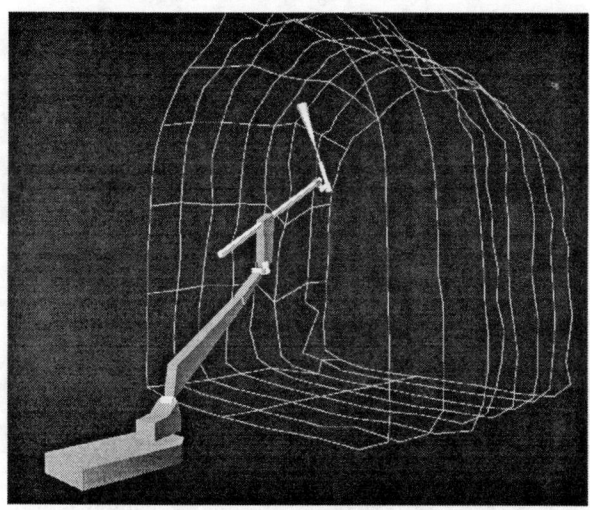

Fig.3: MOBILE simulation

As will be explained in the next section, the same space mouse is used for controlling the real robot. The 3D graphic animation can therefore also be used as a simulator for training purposes in a virtual reality environment.

5. Implementation and Results

In order to implement the new control system, the manipulator had to be equipped with encoders to measure the joint angles and with electrically controlled proportional valves for the hydraulic actuators. The sensors and actuators are connected to an Interbus-S peripheral bus system, that is controlled by a bus master board in a VME-bus chassis (see figure 4).

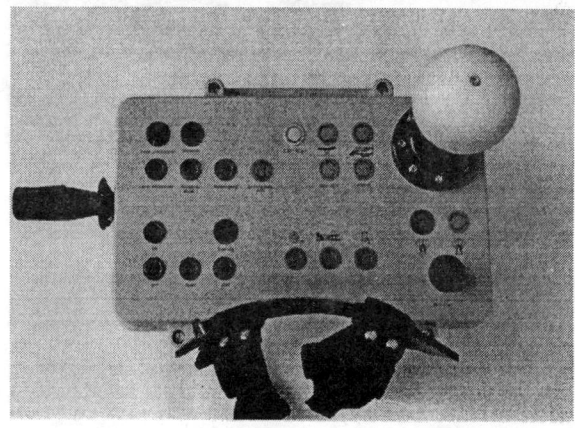

Fig.5: Control unit

The robot is operated through a control unit (figure 5) that allows to select different operating modes, to switch on the concrete pump, to start the measuring process and other functions. The chosen functions are confirmed by lighting the corresponding buttons. The digital signals from the control unit are transfered to and from the VME-bus system via a second Interbus-S connection. Both the space mouse that is integrated in the control unit and the laser scanner, that is mounted on the manipulator near the jet and is needed to scan the tunnel profile for the automatic operation, are connected to the processor board of the VME-bus system with a RS232 serial line.

The control software including the calculation of the inverse kinematics is running on a Motorola PowerPC 603 processor board and is programmed in the object-oriented real-time system Denia/XOberon [8], [9]. A Host-PC allows the programming and monitoring of the system. It can further be used to change some process parameters that can't be controlled from the operator's control unit, such as the distance between the jet and the wall during automatic spraying, the number of measurements in the scanning process and others. It is also used to visualize these parameters and to display a protocol of the complete operation cycle.

After first tests of the new controller we have seen that the operation of the manipulator has become much easier than before. The training phase was reduced from about 1 week to only a couple of hours. Furthermore the quality of the spraying process could be increased. The operation of the robot in a real tunnel is shown in figure 6.

Fig.6: Robojet in operation

Compared with the method described in section 3.1, the use of the null-space method has the advantage of simultaneously optimizing the robots workspace and its dynamic performance.

6. Conclusion

The Robojet is a redundant heavy manipulator that used to be operated with simple manual control units by skilled personel.

In this paper, we presented a new control system that supports the operator in his work in different ways. The operator can guide the tool in cartesian coordinates directly using a space mouse. The calculation of the redundant inverse kinematics and the control of the actuators is performed by the control system. The null-space method is used to achieve a big workspace and a high dynamic performance simultaneously.

A first version of this control system has been implemented on a manipulator and first tests in real tunnels have been realized. With this controller the operation of the manipulator was greatly simplified. It reduces the costs of the training and increases the quality of work.

7. Acknowledgments

We would like to thank A. Hess and M. Trigo for their valuable help in the implementation of the null-space method as well as R. Hueppi and R. Brega for their support in the development of the controller. Further thanks to O. Tschumi of MEYCO equipment for the valuable cooperation.

This project has been partially supported by the Swiss Commission of Technology and Innovation.

8. References

[1] Space Control GmbH, 82205 Gilching Germany, website: http://www.op.dlr.de/FF-DR-RS/SC/

[2] Nakamura, Y., Advanced Robotics, Redundancy and Optimization, Addison-Wesley, 1991.

[3] Hollerbach, J.M., Suh, K.C., Redundancy Resolution of Manipulators through Torque Optimization, A.I. Memo 882, Massachusetts Institute of Technology, January 1986.

[4] Liegeois, A., Automatic supervisory control of the configuration and behavior of multibody mechanisms, IEEE Trans. Systems, Man, Cybern., SMC-7, pp. 868-871, 1977.

[5] Maciejewski, A. A., and Klein, C. A., Obstacle avoidance for kinematically redundant manipulators in dynamically varying environments, Int. J. Robotics Research, 4 no. 3, pp. 109-117, 1985.

[6] Whitney, D.E., The mathematics of coordinated control of prosthetic arms and manipulators, ASME Journal of Dynamic Systems, Meaurement and Control 94(4), pp. 303-309.

[7] A. Kecskemethy: MOBILE - Users guide and reference manual, IMECH GmbH, Duisburg/Moers Germany, 1994.

[8] D. Diez, S. Vestli: D'nia, an object-oriented real-time system, Real-Time Magazine, August 1995.

[9] ETH Inst.of Robotics, 8092 Zurich Switzerland, website: http://www.ifr.mavt.ethz.ch

On-line scheduling algorithms for improving performance of pick-and-place operations on a moving conveyor belt

Raffaella Mattone [§,†]   Linda Adduci [†]   Andreas Wolf [§]

[§] Fraunhofer Institut für Produktionstechnik und Automatisierung (IPA)
Nobelstr. 12, 70569 Stuttgart, Germany

[†] Dipartimento di Informatica e Sistemistica
Università degli Studi di Roma "La Sapienza"
Via Eudossiana 18, 00184 Rome, Italy

{rfm,arw}@ipa.fhg.de

Abstract

In many industrial applications, robotic systems accomplish the task of sorting items on moving conveyor belts. The list of objects to be gripped can be viewed as a queue of clients waiting to be served. The main peculiarities of this queue are that the serving times of its elements vary in a dynamic way, and that any client has to be served before it exits the robot workspace. In most practical cases, a simple first-in-first-out (FIFO) rule can be used for scheduling the jobs in the queue, without dealing at all with the above issues. However, there are situations of industrial interest, as in the automatic sorting of wasted material, where the stochastic behavior of items flow gives rise to repeated overload situations, where the FIFO rule performs very inefficiently, requiring different scheduling strategies. In this paper, we propose two innovative on-line scheduling rules, based on suitable modifications of standard strategies for static queues, having the same complexity, but improved performance in the considered dynamic case. Simulation results confirm the validity of the proposed techniques.

1. Introduction

In many industrial applications, robotic systems accomplish the task of sorting items on moving conveyor belts. In most cases, the items flow can be controlled, and its parameters (distance between two consecutive items, d, and velocity of the belt, v_b) can be set in such a way that the maximum throughput is achieved, accordingly to system capabilities. In particular, the items are often equally spaced and v_b/d matches the minimum robot pick-and-place time. In this standard situation, a simple first-in-first-out (FIFO) rule is adopted as queue discipline for "serving" the items waiting to be gripped [1,2].

On the other hand, there are cases of interest where the items distribution cannot be fixed and only its stochastic description may be known. An example of this situation can be found in the sorting of recyclable waste, a procedure that is performed before the recycling process in order to separate different or undesired material fractions [3]. In this case, the distribution of items arrivals is usually characterized by a high variance of the distance between the incoming objects (see Fig. 1). This represents one of the main difficulties in obtaining an efficient, i.e., cost-effective, automatic system which is capable of assuring the minimum requested percentage of sorted material (72% according to German law). In particular, the commonly employed FIFO strategy may result in unacceptable values of the system *gripping rate* (number of picked items over the total) especially in overload cases, when the distance between consecutive objects is lower than the minimum d_0 allowing the robot to grip any item. Such overload situations might occur quite often if the average distance between the items is close to d_0, as in the case when the system *throughput* (gripped object per time unit) has also to be maximized. Thus, the need arises for different sequencing algorithms.

This planning process presents, however, some difficulties. The main issue to be addressed is related to the motion of the conveyor belt and to the fact that the robot picking time is, in general, not constant within its workspace. Thus, the *serving time* of the elements in the queue (i.e., the time needed to perform the pick-and-place "service" they require) varies with their moving position.

In particular, as will be shown, the evolution of the serving times is described by a discrete-time dynamic system. Furthermore, the objects in the queue are "impatient clients" [4], i.e., they can only be served within a definite time interval, after which they exit the workspace and are lost.

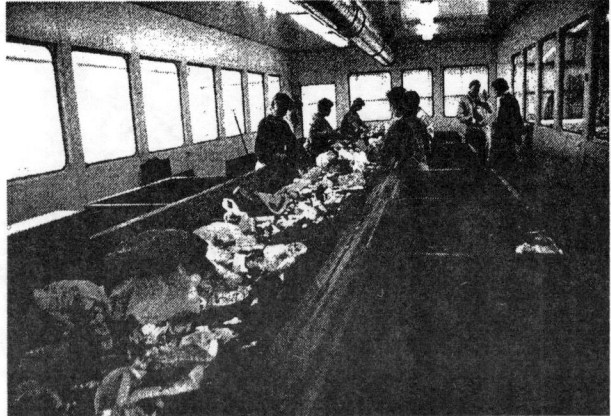

Fig. 1. A plant for the sorting of waste packaging

For this class of scheduling problems, specific solutions are not available, to our knowledge, in literature. In particular, most standard methodologies [5,6] and state-of-the-art techniques [7,8] do not deal with dynamic serving times, and/or with clients impatience.

In order to find a suitable on-line strategy, we start from standard scheduling rules for static queues, as FIFO and SPT (Shortest Processing Time) [9]. Then, we refine the above rules on the basis of local optima criteria and heuristic considerations on the peculiarity of the problem.

As a result, we propose here two innovative on-line algorithms, based on modifications of the FIFO and the SPT rules, having, respectively, the same complexity, but improved performance. Simulations have been carried out taking into account real data (velocity of the belt, workspace dimensions and gripping time function) from the robotic cell available at IPA (see Fig. 2) for experiments on the automatic sorting of recyclable waste [3].

The paper is organized as follows. In next section the problem is formulated, and some of its properties and issues are analyzed. In Sect. 3, the application of standard scheduling rules to dynamic queues is discussed, while Sect. 4 presents the modifications we propose to improve the performance of the above algorithms in the considered dynamic case. Simulation results are reported and discussed in Sect. 5. Some conclusions close the paper.

2. Problem formulation

Let us consider a queue of N elements representing objects on a moving conveyor belt, waiting to be gripped by a robot for being placed in a fixed position (*place point*, see Fig. 3). This queue is the result of the localisation process that, in the case of the waste sorting cell available at IPA, is performed by a human operator, by means of a touch screen system (see Fig. 2).

In order to focus on the main peculiarity of the problem, that is the dynamic evolution of the items serving times, we assume here that no new items enter the list after the beginning of the scheduling process, i.e., that at any step of the scheduling algorithm the items in the queue are those not yet processed in previous steps and still in the robot workspace.

As already introduced, the problem to be solved is finding a suitable scheduling rule for this queue, so that the smallest number of its elements are lost. In order to state it in a formal way, a description of the queue dynamics is needed.

2.1. State space description of the dynamic queue

As in the static case, the queue of items to be gripped can be described by the serving times t_i of its elements, where $i = 1, \ldots, N$ identifies the entering order of any element in the queue. However, these times are not constant in our case, because they depend on the moving positions of the items. Using a state space approach, the following equations can be used to describe their dynamic evolution

$$t_i(k+1) = time(x_i, y_i(k) + v_b \cdot u(k)),$$
$$y_i(k+1) = y_i(k) + v_b \cdot u(k),$$
$$t_i(1) = time(x_i, y_i(1)), \qquad y_i(1) = y_i^{start}, \quad (1)$$
$$u(k) \in \begin{cases} U(k), & U(k) \neq \varnothing, \\ \{0\}, & U(k) = \varnothing, \end{cases}$$

where k indicates the queue processing step, x_i (constant) and $y_i(k)$ are the X and Y coordinates (see Fig. 3) of i-th object at step k, $time(x, y)$ is the time needed by the robot to pick and place an item currently placed at point (x, y) on the moving conveyor belt, and input $u(k)$ is the pick-and-place time of the object selected at step k, or zero if no item can be gripped. While the queue is not empty, $u(k)$ belongs to the set $U(k)$ defined by

$$U(k) = \{t_i(k): t_i(k-1) \in U(k-1) / u(k-1),$$
$$\forall i: y_i(k) \leq Y_i^{max}\},$$
$$U(1) = \{t_i(1), \quad \forall i: y_i(1) \leq Y_i^{max}\},$$

where Y_i^{max} defines last admissible position for i-th item. The above equation states that the admissible objects are those not yet gripped, which are still in the workspace.

Then, the problem can be formulated as follows

$$\max k$$

$$\text{subject to } \begin{cases} (1), \\ u(k) \neq 0. \end{cases}$$

Fig. 2. The semiautomatic cell developed by IPA

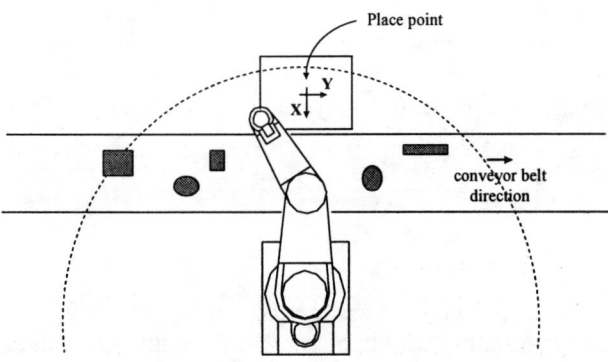

Fig. 3. Sketch of the robotic system for waste sorting

2.2. The gripping time function

A fundamental role in the analysis and solution of the described problem, is played by the function $time(x, y)$, describing the way the pick-and-place time varies within the robot workspace and, thus, governing the dynamics of the queue serving times. This function depends on the robot kinematics as well as on the conveyor belt velocity. In fact, it is bound not just to the time needed by the robot to reach the fixed point (x, y) (this value is given by the *static gripping-time function*), but to the time the robot needs to grip a moving object which is *currently* (i.e., at the moment the robot starts moving) at that point.

For any point (x, y), this time is the value of the *dynamic gripping-time function*, and can be computed as the minimum t_g satisfying

$$t_g \geq static\_time(x, y + v_b \cdot t_g), \quad (2)$$

where $static\_time(x, y)$ is the static gripping-time function. The above inequality states that the time needed by the robot to reach the gripping point has to be smaller or equal to the time it takes to the object to get there. Then, it is of course $time(x, y) = \min(2 \cdot t_g, t_g + T_{max})$.

The static gripping-time function depends on the robot kinematics. For a Scara robot (as that used in the cell available at IPA) and for a place point as in Fig. 3, symmetrically located w.r.t. the robot workspace, this function is approximately linear in the distance between the point (x, y) and the place point (see Fig. 4).

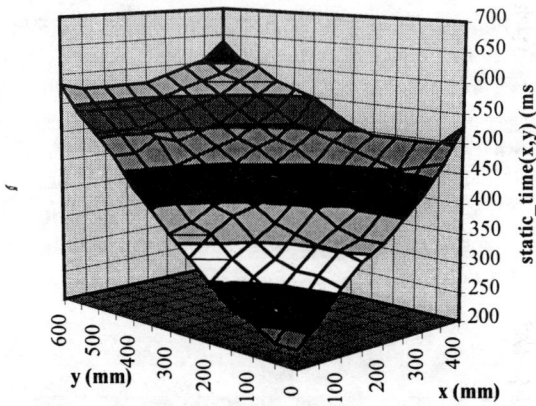

Fig. 4. Experimental behavior of the static gripping-time function for the Scara robot available at IPA

As, in the specific case we are considering, the x coordinate of each item remains constant (the conveyor moves in the Y direction), there is no loss of generality in assuming here $x_i = x = $ constant, $i = 1, ..., N$, so that the $static\_time$ function depends just on the y coordinate. Then, to our purposes, we can approximate the $static\_time$ function as being linear in $|y|$ within the robot workspace (defined by $|y| \leq Y_{static}^{max}$, see Fig. 5).

With these assumptions, the static gripping-time function can be characterized by the minimum gripping time T_{min} and by the slope coefficient $1/\alpha$, where

$$\alpha = \frac{Y_{static}^{max}}{T_{max} - T_{min}},$$

and T_{max} is the maximum gripping time within the robot workspace.

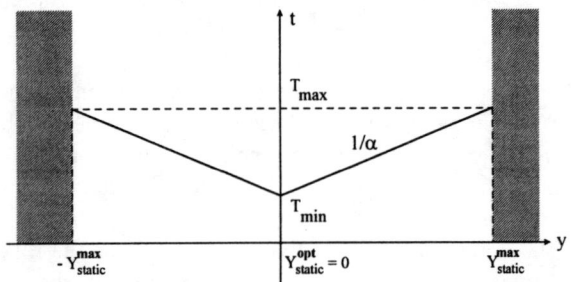

Fig. 5. The approximated static gripping-time function

The α parameter can be also given an interesting interpretation: it provides a measure of "how much it is possible to take advantage of the workspace extension, considering the longer time needed by the robot to reach distant points". We also call α *flexibility* of the kinematic device.

With the above structure for the static gripping-time function, and assuming $v_b < \alpha$, the dynamic gripping-time function can be easily obtained by solving inequality (2), and takes the form of Fig. 6 (if $v_b \geq \alpha$, then it is only defined for $y \leq -v_b \cdot T_{min}$).

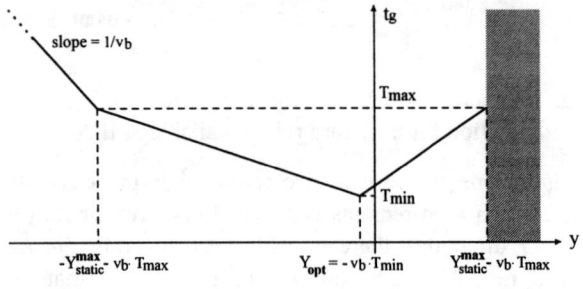

Fig. 6. Dynamic gripping time function corresponding to the static function of Fig. 5

In the real case, when the static gripping-time function cannot be approximated as in Fig. 5, inequality (2) may not have a closed form solution. In this case, a suitable approach seems to be the use of a look-up table of the gripping times, that can be built off-line and used on-line together with a suitable interpolation method. However, the monotonicity properties of the dynamic gripping-time function hold in general as shown in Fig. 6, so that any following consideration is also valid.

3. Standard scheduling rules

In this section, we discuss performance and limitations of two classical scheduling rules in the case of dynamic queues.

3.1. FIFO rule

This rule prescribes that the elements in the queue are served in the same order they enter it.

In terms of the dynamic system describing the queue, this rule implies that the input u in Eq. (1) is chosen accordingly to the following law

$$u(k) = t_j(k), \quad j = \min\{i: t_i(k) \in U(k)\}. \quad (3)$$

The main advantage of this scheduling algorithm is that its complexity is constant. In particular, it does not require the computation of the gripping times for the elements of the queue, resulting in a very low computational cost. Furthermore, in the case of underloaded systems, it can be proved to be optimum for many classes of scheduling problems (see, e.g., [10] where the problem of scheduling tasks in real-time systems is analyzed).

On the other hand, this algorithm proves to be inefficient in overload situations, as can be shown by studying the stability properties of system (1) subject to the input law (3). In particular, let us choose as system output at any step k the y coordinate of the scheduled item, namely

$$out(k) = y_j(k), \quad j: u(k) = t_j.$$

For any constant distance $d > d_0$ between the items, the above quantity has two possible equilibrium values out_{eq}, upstream and, respectively, downstream the minimum-gripping-time point Y_{opt} (i.e., $out_{eq}^{up} < Y_{opt}$, and $out_{eq}^{down} > Y_{opt}$), both characterized by $v_b \cdot time(out_{eq}) = d$. If the equilibrium is perturbed, e.g., by gripping one item a little bit over the point out_{eq} ($out(i) = y > out_{eq}$, for some i), different behaviors follow in the two cases, as one can easily realize by observing the shape of the dynamic gripping-time function. In the case of upstream equilibrium point out_{eq}^{up} the system asymptotically goes back to the previous equilibrium situation (which is therefore stable), while in the case of downstream equilibrium point out_{eq}^{down} the output becomes unstable, drifting to the limit of the workspace, characterized by the highest gripping times, and thus causing the loss of a large number of items.

Thus, the FIFO performs well just if the system works around a stable equilibrium point, and in underload situations. In fact, when the distance d approaches the minimum d_0 (overload situations), the work point of the system becomes closer to the limits of the stable area, so that even small variations in the distance between the objects can make the system unstable.

3.2. SPT rule

For static queues, it can be proved [9] that the minimum average waiting time is achieved by always serving first the client with the shortest processing time (SPT). The complexity of this algorithm is $O(N)$, being N the number of elements in the list, and the most expensive operation to be performed for any item is the computation of the gripping time.

In the dynamic case, even though the optimality properties of SPT do not hold anymore, it still provides a good heuristic rule. In fact, spending at each step the shortest time to grip an object, seems a good way to grip as many items as possible, and actually provides better results than FIFO (see simulation results in Sect. 5). On the other hand, this rule schedules the items on the mere basis of gripping time, not taking into any account the admissibility constraints on the objects positions. In particular, it does not allow spending "a little bit longer" time to grip an item that will otherwise exit the robot workspace. The example in Fig. 7 clarifies this situation. Object A is the fastest item to grip, but taking it will cause the loss of object B, while taking B first will still allow gripping A at next step. This suggests that the SPT rule behavior may also be improved in the dynamic case.

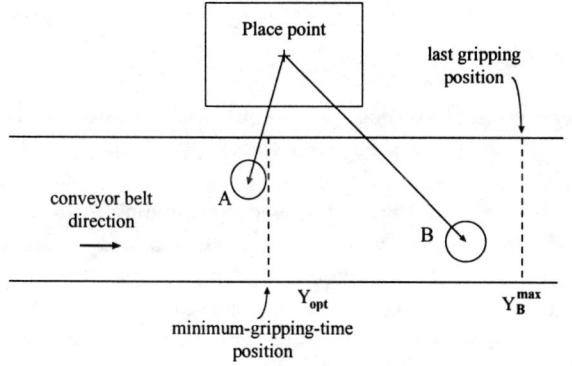

Fig. 7. A typical situation where SPT performance can be improved

4. Proposed rules

On the basis of above considerations on the application of standard rules, we propose here some modifications to improve the performance of the examined scheduling algorithms in the case of dynamic queues.

4.1. Improving FIFO performance

We have observed in Sect. 3.1 that the main limitation of FIFO rule is due to the fact that in overload situations it causes the system to become unstable and, consequently, to work around the workspace boundaries, where the gripping time is maximum. A simple way to overcome this problem, is to restrict the robot workspace to the set of its stable points, i.e., by considering not anymore admissible the objects gone beyond the minimum gripping time point. This results in the following scheduling rule:

Improved FIFO. *At each step k, take the first admissible item in the list that satisfies $y_i(k) \leq Y_{opt}$, being Y_{opt} the y coordinate of the minimum gripping time point.*

In terms of the input u to the dynamic system of Eq. (1), the above rule can be expressed as

$$u(k) = t_j(k), \quad j = \min\{i: \ t_i(k) \in U(k), \ y_i(k) \leq Y_{opt}\}.$$

For average objects distance close to the minimum admissible d_0, this modification makes the system work near to the minimum gripping time point, resulting in a gripping rate not much worse than that achieved by the SPT rule (see simulation results), though keeping the constant complexity of FIFO.

4.2 Improving SPT performance

In order to improve the performance of SPT rule, it is necessary to recognize list configurations as in Fig. 7, where taking other objects than the quickest one may increase the gripping rate. A reasonable way to get this behavior, seems to be choosing at each step the object that leaves the maximum number of admissible items in the workspace. In order to set the corresponding scheduling rule, we need the following definitions:

- Let O_1 be the shortest-gripping-time object in the list, n_1 the number of items being still admissible after taking O_1, and S_1 the set of items that will exit the workspace if O_1 is taken.

- If S_1 is not empty, let O_2 be the shortest-gripping-time object in S_1 and n_2 the number of items being still admissible after taking O_2.

- If $n_1 \geq n_2$, then set $O_{next} = O_1$, else $O_{next} = O_2$.

Then, it is straightforward to prove the following

Proposition (P1). *Choosing O_{next} as object to be gripped maximizes the number of items that will be still admissible at next step.*

The above proposition defines a scheduling rule, which is based on a local optimum criterion, and that we refer to as *Modified SPT*.

Unfortunately, the local improvement w.r.t. the SPT attained by this strategy, does not often correspond to better global performance (see simulation results, Fig. 9). This behavior can be easily understood by observing that having more admissible items does not imply that more items will be gripped. In particular, even though choosing O_2 leaves O_1 still admissible, this may be not anymore the quickest item at next step. Thus, if it is not lost, it will be gripped just before exiting the workspace, requiring a longer time than before. These considerations suggest to be more careful in evaluating the cases where taking other objects than the quickest may actually improve the global gripping rate.

A possible solution may be choosing O_2 as object to be gripped just if O_1 can be taken in a shorter time at next step. In order to translate this into a fast to evaluate rule, we proceed as follows. Let the *optimal gripping area* be defined as in Fig. 8. It is the conveyor region around the minimum-gripping-time point, whose boundaries are characterized by a constant gripping time $t_g = T^*$, and whose width is $L = v_b \cdot T_M$, where T_M is the longest pick-and-place time. Thus, any point in the optimal gripping area is characterized by a gripping time $t_g \leq T^*$. Then, the *upstream area* and the *downstream area* are, respectively, the admissible regions *before* and *after* the optimal gripping area (see Fig. 8).

It is clear that any item being in the upstream area at step k, may be gripped in a lower time at step $k+1$. Thus, we may decide to grip O_2 just if O_1 is in the upstream area. This modification to the previous rule restricts the number of cases when another object is chosen instead of the quickest one, so that the items which are not taken before entering the downstream area are very probably lost. This suggests to extend the research of O_2 to the whole set of items in this "bad" area, resulting in the following scheduling rule

Improved SPT. *Let O_1 be the shortest-gripping-time object in the list, and O_2 the shortest-gripping-time object among those in the downstream area. If O_1 is not in the upstream area, then take O_1, else take O_2.*

This rule has obviously the same complexity $O(N)$ of SPT, and has shown a global performance, in terms of gripping rate, better than SPT in all the performed simulations, even though the attained improvement is rarely larger than 2%.

Next section illustrates the gripping-rate results corresponding to all the examined strategies.

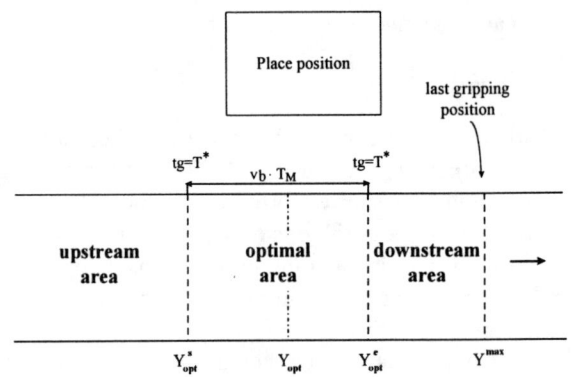

Fig. 8. Decomposition of robot workspace into areas of interest

5. Simulation results

In order to test the performance of the examined scheduling rules, simulations have been performed on lists having stochastic characteristics as close as possible to those of the wasted items in real sorting plants. In particular, the main peculiarity of the items flow, observed in real situations and reproduced here, is the cyclic presence of "large" and "small" groups of "near" objects (i.e., with average $d/d_0 < 1$) separated by "large" (i.e., greater than d_0) empty spaces. All the list parameters (number of items in large and small groups, distance between the items inside the groups, distance between the groups) have been assumed as being normally distributed stochastic variables. This basic structure of the list has been kept constant through all simulations, while the total average distance d between consecutive objects, and the value of the system flexibility α have been changed.

In Fig. 9 the gripping rates corresponding to lists with different average d/d_0 ratios and the same $\alpha = 300$ are visualized for the standard and proposed scheduling rules. The chosen values for d/d_0 are around the overload situation, which is the critical and thus most interesting one. The results show that any rule performs better for growing d/d_0 ratio (all the rules approach asymptotically the optimal 100% rate as the d/d_0 ratio increases), and confirm the better performance of the proposed rules w.r.t. the corresponding standard ones.

In Fig. 10 the performance is evaluated instead for varying values of α, and also the results of Modified SPT are reported. As can be noticed, the flexibility parameter also has, in general, an important influence on the gripping rate. The only rule for which this has shown to be not always true is the FIFO that, for the already discussed reasons, may perform better when the robot workspace (and thus α) is limited.

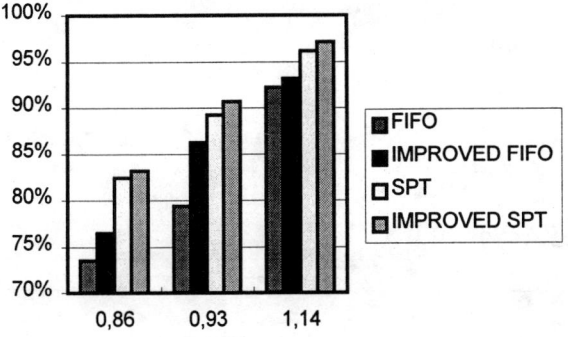

Fig. 9. Simulation results: Gripping rates of standard and proposed rules for representative lists with different average d/d_0 ratios, and $\alpha = 300$

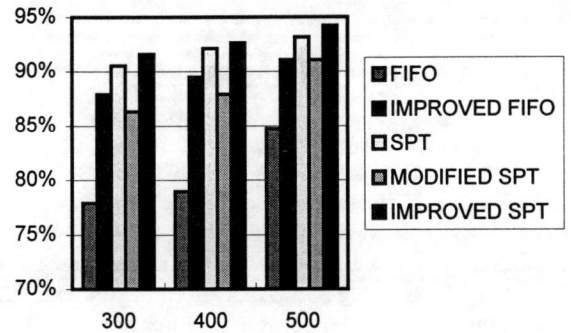

Fig. 10. Simulation results: Gripping rates of standard and proposed rules (including Modified SPT) for a representative list with $d/d_0 = 0.98$, for different values of α

6. Conclusions

In this paper, we have shown that maximizing the gripping rate of an automatic sorting system can be viewed as a scheduling problem for dynamic queues, i.e., queues where the serving times of the clients vary according to the evolution of a dynamic system. For this particular problem, we have also shown that standard scheduling rules as FIFO and SPT may behave poorly. On the base of local optimum criteria and heuristic considerations, we have then proposed some modifications to the above algorithms, so that two innovative on-line scheduling rules are obtained, having the same complexity of the considered standard ones, but improved performance in the case of dynamic queues. Simulations have confirmed the foreseen results.

References

[1] T. Li and J. C. Latombe, "On Line Manipulation Planning for Two Robot Arms in a Dynamic Environment," *IEEE Int. Conf. on Robotics and Automation*, Nagoya, J, pp. 1048-1055, 1995.

[2] G. Pardo-Castellote, S.A: Schneider, and R.H. Cannon Jr., "System design and interfaces for intelligent manufacturing workcell," *IEEE Int. Conf. on Robotics and Automation*, Nagoya, J, pp. 1105-1112, 1995.

[3] D. Schraft, A. Wolf, and S. Erhardt, "A Robot System for Automated Waste Sorting," *Int. Symp. on Industrial Robots*, Milano, I, pp. 205-209, 1996.

[4] D. Gross and C. M. Harris, "Fundamentals of Queueing theory," *Wiley series in Probability and Mathematical Statistics*, Wiley and sons, New York, 1974.

[5] D. Theune, "Robuste und effiziente Methoden zur Lösung von Wegprobleme," *Teubner-Texte zur Informatik*, Teubner, Stuttgart, 1995.

[6] K.L. Cooke and E. Halsey, "The shortest route through a network with time-dependent internodal transit times," *J. of Mathematical Analysis and Applications*, Vol. 14, pp. 493–498, 1966.

[7] G. Koren and D. Shasha, "Dover: An Optimal On-Line Scheduling Algorithm for Overloaded Uniprocessor Real-Time Systems," *SIAM J. on Computing*, Vol. 24, No. 2, pp. 318–339, 1995.

[8] G. Koole and M. Vrijenhoek, "Scheduling a repairman in a finite source system," *Mathematical Methods of Operations Research*, Vol. 44, pp. 333-344, 1996.

[9] L. E. Schrage, "A proof of the optimality of the shortest remaining time discipline," *J. of Operations Research*, Vol. 16, No. 3, pp. 687-690, 1968.

[10] M. L. Dertouzos, "Control Robotics: the procedural control of physical processes," *Proceedings of 1974 IFIF Congress*, pp. 807–813, 1974.

Service Robots for Nuclear Safety : New Developments by Cybernetix

Jérôme Perret

CYBERNETIX Établissement de Saclay
4, rue René Razel
91892 ORSAY Cedex
FRANCE
tel. : +33-1-69.35.68.15
fax. : +33-1-69.85.36.92
e-mail : rol.saclay@wanadoo.fr

Abstract

As nuclear powerplants age, maintenance tasks become more critical and more dangerous. In most Western countries today, and perhaps even more in the former East Block, the nuclear safety management becomes a major issue. In the next couple of years, strategic decisions have to be made regarding the nuclear waste management and the dismantling of old equipment.

Sadly, most of the dirty work is still done by humans and not by robots. The robotics community has achieved many breakthroughs in the last decade, thanks to the progress of computing power, but their impact has been small in the domain of nuclear intervention. Radiation tolerance implies heavy shielding, which in turn reduces the flexibility of the machine.

Looking ahead and striving for a better world, CYBERNETIX has been bringing together the latest advances in radiation-hard electronics, spread-spectrum transmissions and telecontrol technology, onto a single, modular, mobile platform. We present in this paper the state of the project and point out the many innovations which have been necessary. We conclude with a short survey of the market, and a tentative glimpse in the future.

1. Introduction

CYBERNETIX is a French company specialised in the design and manufacturing of service robots and advanced factory automation solutions. Its main domains of activity are the nuclear industry, civil security, off-shore, and floor cleaning. Its priviledged contact with the research community, through frequent co-operation for R&D projects and the funding of PhD students, makes it one of the industrial leaders in service robotics in the world.

In 1996, CYBERNETIX was chosen by KHG, a German company active in the field of nuclear safety, for the development of a new generation of intervention robots. The LMF system, to be delivered in 1998, will be the first fully 1Mrad radiation-tolerant robot able to operate inside a nuclear powerplant.

This paper describes the current state of the project.

2. Service robots for nuclear safety

The nuclear industry is a major user of robotics, in the form of telemanipulators. The whole nuclear fuel processing chain (from the first step - uranium ore - to the last - radioactive waste -) is dangerous for the workers' health, and has to be done in heavy-shielded cells, using remote handling techniques. Apart from the ore extraction in uranium mines, this process is well mastered.

But in most Western countries having a nuclear industry, as well as in the former East Block, nuclear infrastructures are now over 30 years old. Under irradiation, building materials age fast, and faster than architects had foreseen in the 60s. In order to insure nuclear safety, these buildings have to be inspected and maintained more often. When abnormal radiation levels are reported, then closer investigations have to be carried out in order to detect a potential leakage. Possibly, some repair work is executed, and eventually whole buildings will be dismantled.

This is a very hard work. Unlike in other industry sectors, it all has to be done from the inside, so that the minimum contamination can find its way out. It means carrying heavy tools (e.g. pneumatic drill) over long distances in small corridors, over very steep stairs (they were built at 45°, to save room), in a hot and moist atmosphere. But the worst is the ever-present threat of irradiation, meaning a short death or a long, painful disease.

It is unbearable that such a dirty work be done by a man, and yet it is very often the case. Robots are almost of no use at the job, because of their fragility and clumsiness. Very few systems are available on the market which are both radiation-tolerant and robust enough to do the task. But the need is enormous, and it can only increase in the next couple of years.

3. Main technical issues

The technical issues to be solved may not seem very exciting, but they are very challenging.

The first problem is the autonomy. In most cases, the mobile robot will have to carry its own source of energy, allowing it to work for several hours. Thermal motors give the best energy/mass ratio, but they use up the air in a closed room, and they need filters against contamination.

Stability is also a key issue : the system has to climb up and down stairs at 45°, which is not easy when carrying a heavy load. Legs would be very helpful, but they need a lot of power and are difficult to control.

Because the robot will be teleoperated, it needs a way to communicate with the outside. In such an environment, an umbilical is a real hindrance. But what other solutions do we have inside a building? Electromagnetic waves are reflected many times, delays add up so that radio transmission are almost useless.

Radiation tolerance is a real challenge. Most passive electronic components will hold, but every microchip has to be shielded with heavy, cumbersome sheets of metal.

The control software is not a critical issue, since the robot is teleoperated. Still, it is useful to relieve the operator as much as possible, in order to reduce his fatigue. Indeed, his job is very stressful, as he must work fast and without failures, in difficult conditions.

But perhaps the most critical issue is integration. Because of the complexity of the environment (narrow passages, frequent obstacles like naked pipes), the robot has to be small and manoeuvrable, while being able to carry and work with heavy tools.

4. The LMF project

In 1996, KHG chose CYBERNETIX for the development of its new nuclear intervention robot, the LMF. Able to climb up and down stairs at 45°, to turn about within a diameter of 1200 mm, to manipulate a mass of 100 kg and lift a glass without breaking it, it was to resist a constant radiation flow of 100 Gy/h during 100 hours.

Active in the field of nuclear safety, KHG owns a fleet of half a dozen robots, able to intervene anywhere in Germany and in Europe within a few days. KHG knew very well what a challenge it would be to build the LMF. CYBERNETIX was chosen because of its experience and tight contact with the research community, which promised an efficient, cost-effective service.

CYBERNETIX put the best of its technology into the project, using the latest advances in radiation-tolerant electronics to reduce the shielding mass, and bringing in the latest control software. Now the integration work is in progress, and we are confident that we will be successful in meeting both the challenging specifications and the 1998 deadline.

5. System architecture

5.1 Mobile platform

The LMF body is a traditional two-tracks platform, equipped with four tracked legs for increased stability while climbing stairs. It is able to cross a 400 mm wide trench or a 400 mm high step. The tracks are electrically-driven, but the legs are hydraulic for higher load capacity. The whole is driven by a battery pack at the bottom of the robot, giving it a two hours' autonomy. In case of a real emergency, the standard battery pack could be replaced with a not-rechargeable one, for six hours working capacity.

5.2 Telemanipulation

The LMF is equipped alternatively with a heavy-load hydraulic arm MAESTRO, or with a combination of a dextrous MA23M master/slave manipulator and a hydraulic left arm.

The MA23 is an electrical 6 d-o-f manipulator with a load capacity of 250 daN. All movements are driven over cables and tapes by motors situated on the arm base, which makes it light, fast and strong together. It is associated with a master arm with the same kinematics, suitable for force-feedback control.

The left arm is a SAMM by CYBERNETIX : a modular concept, based on standard actuators and links, configured here for 4 d-o-f.

The MAESTRO arm is a high-precision, heavy load hydraulic manipulator (see fig. 2). It is able to lift a 100 kg load at its full 2040 mm reach. We are developing a new control software including force-feedback control with a MA23 master arm, but it will not be available on the first version of the LMF.

Fig. 1 : The LMF robot.

5.3 Transmissions

The main transmissions are a development of LETI, a research laboratory in Grenoble, France. It is a spread-spectrum, direct sequence technology, radiation-hardened to 200 Gy, so that minimum shielding is necessary. It includes two JPEG-compressed video signals and the bi-directional control link at 256 kbits/second.

Alternatively, this communication medium can be replaced by a 100 m umbilical with a cable drum on the top of the LMF to allow for full turns.

The main transmissions are secured by a traditional radio link, called "safety link", also by CEA/LETI, which enables the operator to control the robot using a single video image, the only missing feature being the telemanipulation force feedback.

5.4 Control architecture

The low-level control is done on-board by a Siemens 88C166W microcontroller, which is also responsible for multiplexing/demultiplexing the data. It allows for reflex actions such as placing the LMF in a stand-by configuration in case of transmission loss.

The control station is driven by a real-time PC under VxWorks, running the control code for telemanipulation (see below). It is also responsible for the TELEVISE mode, which drives the stereo-head so that the gripper stays always in the centre of the video screen.

The supervision PC, under Windows NT, communicates with the former and controls the robot tracks and all other on-board services. It also drives the main MMI of the system.

Finally, a third PC called "virtual display" shows the current configuration of the robot and enables the operator to generate virtual side-views of the task. A specific algorithm is responsible for the detection of potential internal collisions between the platform, the legs and the manipulators.

5.5 Telecontrol

The telemanipulation programme includes the full features of the TAO 2000 project developed by the French Nuclear Agency CEA. It is a generic software for Cartesian-space and joint-space master/slave control with or without force feedback, with task-oriented d-o-f blocking and trajectory execution. The force feedback loop for MAESTRO is under validation by the CEA, but will not be available in time for the first version of the LMF. However, it is already fully operational for the MA23M master/slave manipulator.

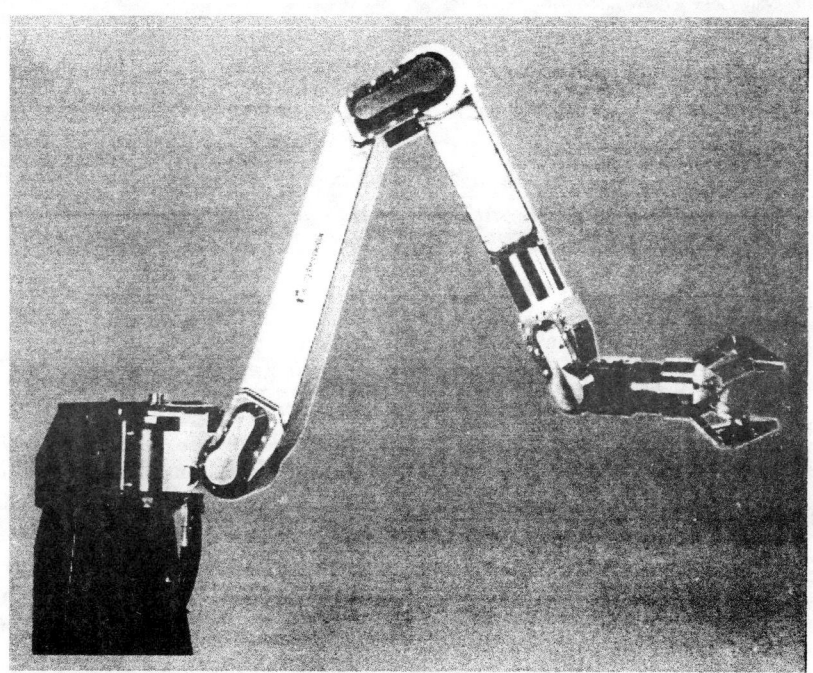
Fig. 2 : The MAESTRO manipulator.

6. State of the project

The project started in January 1997, for a delivery date end of 1998. By the time of the conference (June 1998), the integration phase should be almost completed, and the first results of the validation should be available.

7. Conclusion

Very few systems are available on the market today with sufficient radiation-tolerance and versatility for operational inspection, maintenance and intervention tasks in nuclear facilities. To add shielding on a standard vehicle is difficult, because of such problems as weight balance and heat exchange.

The users are still very few, partly because of the cost of new developments. It is foreseen that the demand will rapidly increase, as soon as reliable solutions become commercially available. After all, it is a matter of public safety, and we are confident that the nuclear industry will see the benefits of having a fleet of mobile robots ready to react to every emergency without risking human life.

The challenge of the LMF project is to integrate simultaneously so many new technologies into a single machine. We expect from KHG a tough validation campaing, and afterwards a long operational life for our robot. We will do our best to demonstrate that robotic technology today has become mature.

8. Acknowledgements

My special thanks go to Mr KRÜGER, NACHTIGAL, and GUSTMANN from the company KHG.

Some of the work described here uses algorithms and technology developed by the CEA/STR and CEA/LETI. Their contribution is acknowledged.

9. References

[1] R. Fournier, P. Gravez, P. Fontcuberta and C. Volle, *"MAESTRO Hydraulic Manipulator and its TAO 2000 Control System"*, in Proc. of RSTD'97, Alberta, 1997.

[2] J. Perret, *"Service Robots : a Manufacturer's Point of View"*, in Proc. of IARP'97, Genoa, Italy, Oct. 23-24, 1997.

TELEROBOTIC SYSTEM FOR LIVE POWER LINES MAINTENANCE: ROBTET

L.F. Peñín, R. Aracil, M. Ferre, E. Pinto, M. Hernando y A. Barrientos

Dpto. de Automática, Universidad Politécnica de Madrid (DISAM)
Jose Gutierrez Abascal, 2   Madrid 28006, SPAIN
Tel: +34-1-3363061   Fax: +34-1-5642961   E-mail penin@disam.upm.es

Abstract: Outage-free maintenance of electrical power lines has been conducted throughout the years in order to fulfill the increasing demand of uninterrupted power supply. In these techniques care regarding hazard of electric shock and falls must be exercised while the work is performed on or near energized elements. In order to increase the safety and comfort of the workers as well as the overall efficiency, a teleoperated system for live-line maintenance, called ROBTET, has been developed for the electrical utility IBERDROLA.

The ROBTET is described in detail in this paper. It has two hydraulic-driven master-slave teleoperated manipulators on top of an insulated boom over a truck. The operator commands the manipulators from a cabin on the truck via a pair of master arms, while he[1] receives visual feedback through a vision system. Multimedia display, voice commands, stereo vision and force-feedback are some of the new features implemented in order to achieve the required telepresence of the operator and hence increase the performance of the system.

1 Introduction

As a consequence of today's highly information-oriented society and increasing demand of electricity, uninterrupted power supply has become indispensable for most electric utilities. Outage-free maintenance techniques for overhead distribution power lines have been developed and used throughout the years by several companies in order to fulfill this requirement. For example, in Spain there is an experience of more than 25 years in performing this kind of work.

In the conventional maintenance techniques workers have to do their job on a live electrical power line, indirectly with various kinds of insulated hot-sticks or directly touching the line with rubber gloves from an insulated bucket. Therefore, work is performed in a hazardous environment with both the risk of electrical shock and the danger of falling from a high place. In addition, workers have to be very skilled and have to work cooperatively under very demanding tasks.

Since their first appearance in the 40's, many teleoperated systems have been developed and employed when dealing with non-structured environments and in applications where there is clear and unavoidable danger for the human operator [1].

---

[1] Throughout the paper, the pronoun 'he' is employed instead of 'she/he'.

It is true that live electrical lines maintenance encompass a series of operations (insulator string replacement, opening/closing bridges, etc.) made of highly standardized manipulation procedures, some of which are common to different operations. But the environment is very complex and variable and there is no standard set of equipment to work on. Therefore the use of an autonomous robotic device is not advisable [2]. Having again in mind the hazardous environment in which these operations take place, it is clear that some kind of telerobotic system can be introduced with great advantages.

Other utilities around the world have also been aware of the capabilities of the teleoperation concept for this application. Back in the mid-eighties, EPRI [3] in the United States and KEPCO [4] in Japan began the first developments in order to obtain a first prototype of a telerobotic system for live-line maintenance. Their approach was different, since EPRI tried to develop a system with the operator on the ground, while KEPCO prototype had the operator on a bucket near the line. In the following years several utilities began development of other systems [5]-[9], all of them following the KEPCO approach. Feasibility studies where also presented for the introduction of these kinds of systems in each region or company specific electrical network [10][11]. Recently, KEPCO began the

Figure 1 General view of the ROBTET

Table 1 Comparison between several modes of performing teleoperated tasks for live-line maintenance

Mode of operation	Human safety	Cost	Control complex	Protection	Productivity
Manual	↓↓	↑↑	↑↑	↓↓	↑
Direct Teleoperation	↑	↑	↑	↓	↓
Ground Teleoperation	↑↑	↑	↑	↑↑	↓
Semiautomatic Direct Tel.	↑	↓	↓	↓	↑
Semiautomatic Ground Tel.	↑↑	↓	↓	↑↑	↑
Automatic	↑↑	↓↓	↓↓	↑↑	↑

Note: The ↑ and ↓ symbols indicate a factor more or less favorable

development of a new prototype with the operator on the ground and with some semi-automatic features [12].
In 1990 Spanish utility IBERDROLA both with contracting company COBRA and DISAM (Polytechnic University of Madrid) began first studies in order to introduce robotic techniques into work on distribution hot-lines. These studies turned into a first laboratory prototype that showed the technical feasibility of the project and identified critical issues that needed special care. A new industrial prototype called ROBTET (ROBot para Trabajos en Tensión) began development in 1994. It has already been presented in several international conferences on live-line maintenance [13] [14] and is shown in Figure 1.
This paper presents a detailed description of the ROBTET, making special emphasis on the control architecture implemented and the solutions adopted regarding different designs issues. The ROBTET has several important innovations with respect to the preceding systems, especially on how to achieve the telepresence of the operator. These innovations will also be fully addressed in this paper.

2 Overview of telerobotic live-line maintenance

The introduction of telerobotic technology for live line maintenance gives the following advantages:

a) Ability to do and improve outage-free maintenance in countries with strict regulations regarding the interaction of humans with energized components.
b) Increase in the safety and comfort of the workers.
c) Decrease in the cost by eliminating the need for the operator to be working in a hazardous environment (wages, insurance, special equipment, etc.)
d) Able to work under moderate bad weather conditions
e) Decrease in labor requirements

Productivity depends on the type of telerobotic system selected, as pointed in [15]. All of them have in common the need of a truck with an insulated boom to reach the lines and the use of some kind of manipulator on its top to

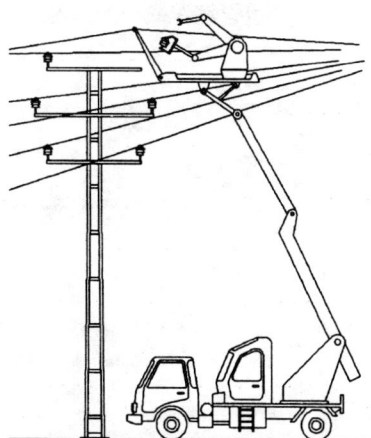

Figure 2 General scheme of the ROBTET

perform the operation. The main feature that characterizes each system is the operation mode. The type of control defines its practical and efficient use. Table 1 shows a comparison regarding various aspects of different modes of operation for this application.
Most existing prototypes work in manual direct teleoperation with the operator on top of the boom near the energized line. Instead, the ROBTET can be classified as almost semiautomatic with the operator on the ground. As seen in the table, this mode of operation gets de advantages (protection, human safety and workforce) of a ground system over a direct one, but limiting the disadvantages of control complexity and cost of the totally automatic one.
One of the basic themes while building the ROBTET has been to develop a system at maximum low-cost, using proven commercially available technology as a start off.

3 System description

The ROBTET has been designed to do maintenance and repair related tasks on the Spanish distribution network, which is rated up to 46 kV. It works in semiautomatic work with the operator giving commands from a cabin on the ground. It is capable of performing the following operations, which are by far the most usual ones in live-line work: Insulator string replacement, opening and closing of bridges, bypasses, and branch installation.
Figure 2 shows a general scheme of the main equipment present on the prototype. Its features are summarized on Table 2 and are fully explained below.

3.1 ROBTET equipment specification

Truck: is a 4x4 commercial truck of 5.5 ton. with an estimated weight allowance of 17 ton. It is 8 meters long and with a height of 3 m. The estimated power is 169 kW (230

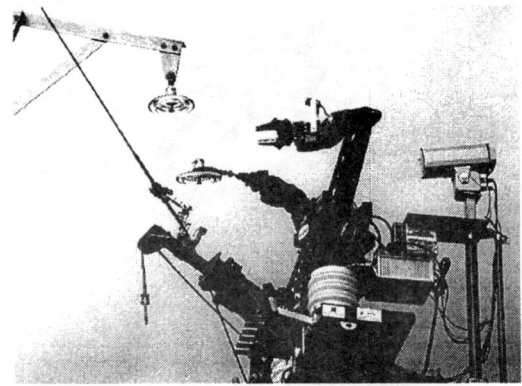

Figure 3 ROBTET remote platform

Figure 4 ROBTET control station

HP). It holds a hydraulic pump unit and an electric generator of 10 kW.

Boom: insulated telescopic boom with a maximum height of 15 m obtainable through the spanning of three sections. The last one has a dielectric strength of up to 69 kV. Its base is located on the truck chassis, next to the truck cabin. The truck hydraulically powers it and is switch controlled joint-by-joint through a specially designed control pendant situated on the operator cabin.

Rotating Platform: Located on the top of the boom. Has pan and tilt movements controlled from the control pendant. Holds two slave manipulators, an auxiliary jib, visual sensors, different tools, and an autonomous power supply with autonomy of 2 hours for the electric equipment on it. It is insulated from ground through the boom. Communication between the equipment present on the platform and the control cabin is done through several dedicated optic fiber cables going inside the boom. The platform also has two outlets for hydraulic tools. Figure 3 shows the equipment present on the platform.

Manipulators: there are two commercial master-slave systems with force reflection. The two slaves are on the rotating platform, are hydraulically powered and have 6 articulated degrees of freedom (DOF) plus grip action. Their maximum payload is 45 kg/arm and the net weight of each arm is of 60 kg. They are separated from the metallic platform by two specially designed insulators.

The two master arms located on the operator cabin are also articulated, have 6 DOF and resemble kinematically to the slaves, although they are a little smaller. They are powered with electric motors to implement a force-position bilateral control scheme [16].

Auxiliary Jib: It is placed next to the slave manipulators on the rotating platform. It is hydraulically powered and has three DOF: pan, tilt and extension of a telescopic link insulated up to 49 kV. It has a lifting capacity through a winch of 200 kg. The jib joints are switch controlled from the control pendant in the operator cabin. It is used both for sustaining the wire while the insulator replacement takes place, and to lift things from the ground.

Visual sensors: there are three different visual sensors located on the platform. First there is an overall-view camera with pan, tilt and zoom movements; all of them controlled through voice commands from the operator control station. It is located between both slaves a little behind them to have a good general view.

Second, there is stereo vision system with 3 DOF plus zoom. It is a commercial stereo head from HelpMate Robotics Inc. that continuously takes images of the environment. These images are processed in order to determine the position of each element of the working environment. The binocular system also makes use of an 810 nm laser scanner system of 500 mW to improve the segmentation of the cables during the image processing.

And finally, there is a small camera on the fifth link of the left slave manipulator. It has a fixed optic, and is used to be able, through the movement of the slave, to see areas of the environment not available to the overall camera.

Tools: the tools used by the ROBTET to do the maintenance tasks are similar to the tools used by the linemen. They are grabbed by the slaves and used in the same way as a human operator does. This is a very important fact, because there is a huge range of available tools already developed for every task and circumstance. The development of new special tools would have been very expen-

Table 2 Main equipment specification

Truck	4x4, 5.5 ton, 8 m long
Boom	15 m long, telescopic and with up to 69 kV of dielectric strength
Visual sensors	Stereo vision system (3-axes + zoom). Overall-view camera (2-axes + zoom)
Manipulators	Hydraulic, 7-function (6 axes + grip), articulated, masters with force-feedback, max. Payload 45 kg/arm, net weight 60 kg/arm
Jib	Hydraulic, 3 DOF, telescopic, lifting capacity: 200 kg; with winch
Rotating platform	Mounted on top of the boom. Holds the slave manipulators, the jib and the visual sensors.

sive. Only slight modifications on the tool grip and on the robot gripper itself had to be done.

Control station: the operator control station (Figure 4) is located in a closed cabin of 2.4 x 2.6 x 1.7 m^3 mounted on the truck chassis and behind the driver cabin. It has air conditioning, an anatomic chair and most of the commodities of an office environment. It also has lateral windows and a front-and-top window for the operator to see in every moment where the platform is.

The control station holds all the equipment required to conduct the operation of the system. But the operator only needs to allocate his attention in one multimedia interface, the two master arms and the control pendant for the movement of the boom, platform and jib. Since the operator's hands are on the master arms, communication with the interface is done through voice commands and speech synthesis.

4 How the ROBTET works

Seen from the outside, the ROBTET resembles very much to a conventional combination of truck plus insulated aerial lift used for manual live-line maintenance. The only apparent difference is the addition of the control cabin and the substitution of the bucket on the boom by a platform.

The operation begins by positioning the truck near the tower. The outriggers are then extracted to balance the truck in the rough terrain. Afterwards, the operator gets into the cabin and puts himself comfortable. After starting the system, the working procedure begins by indicating with friendly menus the type of operation to be conducted and the specific type of the different elements (insulator, wire, etc.) in the working environment. In the same manner, is necessary to validate the forbidden zones generated automatically by the system. These are zones around various elements of the environment in which the slaves should not enter in order to prevent damages or shortcuts.

The next step is to place the rotating platform using the control pendant on the best applicable position on the remote zone. Visual feedback to perform this task is both direct and through the overall-view camera.

Once the platform is placed and stabilized, the stereo visual sensors begin taking images of the environment, locating the positions of the various elements previously identified. This information updates a database with a computer model of the remote environment.

The operator can now take control of the system through the master arms. He begins operation, closing the control loop, through the use of various different views and visual aids on a multimedia interface. Figure 3 shows the ROBTET working on a line of 13 kV. Communication between computer and operator is done through voice

A collision detection algorithm that makes use of the computer model of the environment detects when the operator tries to move a slave to a forbidden zone. It then generates a virtual force on the respective master arm that pushes back the arm of the operator to a safe zone.

The operation is conducted following the same procedure as in manual maintenance with the use of hot-sticks, although in this case either manipulator grabs the conventional tools, both mechanic and hydraulic.

5 Control system

The main purpose of the ROBTET control system besides controlling the different equipment (robots, cameras, etc.) is to give the operator as much feeling of telepresence [1] as possible in order to perform the teleoperated task in a simple and comfortable way. Multimedia display, voice commands, stereo vision and force-feedback are some of the features implemented to achieve this goal. Additionally, other advanced characteristics, such as collision detection and virtual force generation, have been included to improve the safety and performance of the system.

Figure 5 shows a schematic of the hardware architecture implemented in the ROBTET. There are three different channels of communication between the distinct equipment: Ethernet 10baseT, optic fiber and analog.

A detailed description of the different elements of the control system will be given in the following paragraphs.

5.1 HIC (Human Interactive Computer)

Is a multimedia Indy workstation from SGI under IRIX. Implements the operators interface, and hold the sole display in which the operator attention is allocated. The display shows the views of the overall camera and the small camera on the left slave. It also shows a simulated scene of the remote environment that is continually (1 s) updated through the data obtained with the stereo vision head. Position of the slaves are also sampled and shown

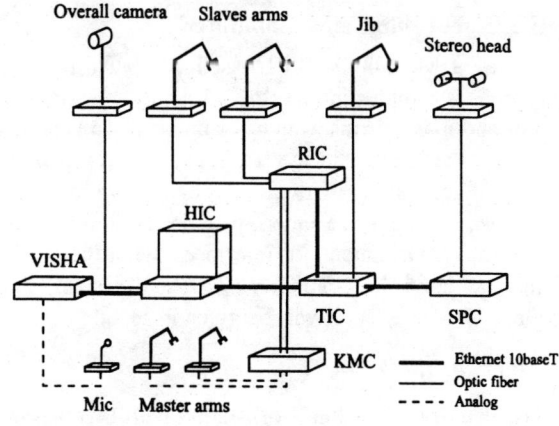

Figure 5 Hardware architecture

every 15 ms. Menus and information about the task are also displayed in a simple and compact way.

The simulator is primarily used to have always a complete view of the environment from any desired position. Work on the blending of real images with simulated ones is being conducted to check the model and to gain depth information. The simulator has been developed with OpenInventor, and is a general purpose one [17].

Communication with the interface is done through voice commands, especially useful to move the overall view camera to the required position and zoom as the task advances, since the operator has both hands occupied.

5.2 VISHA (Voice processing system)

Bus-PC based word-by-word voice recognition and processing system. It needs to be trained for each new operator and it has a 98% accuracy on a dictionary of 200 words. The VISHA gets the control commands of the operator from a microphone, translating them to a target language. This target language is the minimum set of actions that the teleoperation system can perform. They are passed to the HIC in order to implement them on the display (i.e. menus) or on the respective equipment.

5.3 SPC (Sensor Processing Computer)

Made of a Datacube mv200 image processing board based on a MC68000 and running under Lynx 2.0. It also has a floating point mp860 processor for complex mathematical computations. Its purpose is to update the database resident in the TIC with the model of the environment.

The SPC controls the binocular head and the laser scanner system. It gets the images from the stereo visual sensors, and processes them using matching algorithms that recover specific three dimensional data of the different elements in the environment (cables, fittings, insulators, etc.) [18]. This information is used to update the model database. The whole process is done every 1 second.

5.4 RIC (Robot Interactive Computer)

Consists of a MC68360 VME based computer running under OS-9. It samples in real time (5ms) the positions of the slave and master arms, which are updated via Ethernet in the database located in the TIC. The RIC is also responsible of generating a virtual force against the operator arm in case of the slave entering a declared forbidden zone of the environment. It interfaces with the KMC (commercial controller of the master-slave system) and with the TIC. More details will be given in section 6.

5.5 TIC (Task Interactive Computer)

The TIC is a Sparc 20 Sun Workstation with two processors running under Solaris 2.5. The TIC core is the environment model based database, which holds positional

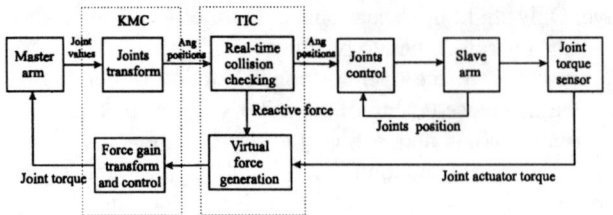

Figure 6 Slave arm control scheme

information of all the objects present in the environment. Data from the slaves comes from the RIC while the positions of the jib joints are directly sampled by the TIC. The location of all other relevant elements in the environment is updated by the SPC, as mentioned above. Objects in the database are modeled using rectangular polyhedra, generated by polygon linear swept, although boundary representation is used internally. The information present in the database is primarily used by the simulator in the HIC and by the collision detection module also present in the TIC. The collision detection module works in parallel with the virtual force generation module in order to improve the practical safety of the system. Due to its importance and for being a very advanced feature of the system, this part will be explained with more detail in next section.

6 Real-time virtual protection

The first procedure to do in any manual maintenance operation of live-lines is to put insulated covers on the tower and over the other two phases near the zone where the maintenance will take place. This time consuming operation is done to avoid the appearance of a shortcut due to a wrong maneuver of the operators, which could also be lethal to them.

In the ROBTET the placement of insulated covers has been substituted by a software protection. Whenever the operator moves one of the slaves to a previously declared forbidden zone, a collision detection algorithm will give the order to generate via the master a virtual force on the operator arm, not allowing to keep further with the movement. By this manner, shortcuts and any unwanted collision are immediately avoided. Figure 6 presents the master-slave bilateral control scheme employed with the inclusion of the corresponding modules for the implementation of software protection.

6.1 Collision detection algorithm

The collision detection algorithm employed is based on the successive approximation approach, in which the objects contour are successively refined as the algorithm progresses, as explained in [19], although we work with rectangular polyhedra. The detection algorithm is for static environments and in order to prevent the collision in the real environment the objects in the model have been extended a safety buffer. In a typical environment of 32 solids + two 6 DOF robots, the detection algorithm spends on the TIC machine an average of 1.5 ms. When an imminent collision is detected an artificial force is simulated on the master arms, substituting the real one, as expressed in Figure 6. This prevents the operator from reaching the obstacle, avoiding the collision. This module is totally transparent for the joint signals going to the slave.

6.2 Virtual force generation

Once a collision have been detected is necessary to compute the repelling virtual force to be applied to the master arm joints. First the repelling force is calculated for the slave's tip. The force is a non-linear function of the depth of the contact and its direction is calculated following the guidelines given in [20]. Through the use of the Jacobian, equivalent joint torque are obtained, which are later passed to the force gain transform and control module on the KMC, just as if these torques had come directly from the pressure sensors in the slave actuators.

The force in maintained as long as a collision is detected. Whenever the operator withdraws the slave from the forbidden zone this module becomes completely transparent.

7 Field trials

Up to date, several standard field tests have been carried out in collaboration with specialized linemen from COBRA. The tests performed to the prototype have been mainly in changing vertical insulator strings of the types 1507 and 1503 and putting bypasses. Time to change an insulator string during first trials has been around 15-20 min., very similar to the time spent by experience linemen using hot-sticks. During these trials, the multimedia display and the virtual protection mechanism have proven their usefulness and reliability.

8 Conclusions

This paper has presented a new teleoperated system for live-line maintenance called ROBTET, being designed to fulfill all the requirements for live-line maintenance of the Spanish power distribution (46 kV) network. With the operator on the ground, a great improvement in human safety has been introduced with respect to other similar systems in development. New advanced technologies, traditionally only applied in other technical fields, are being used and rationally integrated to greatly enhance the capabilities and performance of the basic teleoperation system. These technologies include 3D modeling, voice processing, collision avoidance, multimedia interface, telepresence, etc., having in mind what is going to be the future of everyday operation in hazardous environments.

9 Acknowledgments

The electric utility IBERDROLA, S.A. and the Spanish Ministry of Industry through OCIDE (PIE No. 132.198) supported the work presented in this paper. Special thanks are also due to A. Santamaría from IBERDROLA, M.A. Fernández and C. González from COBRA and F.M Sánchez and L.M. Jiménez from DISAM.

10 References

[1] T.B. Sheridan, *Telerobotics, Automation and Human Supervisory Control*, The MIT Press, Cambridge, 1992.
[2] J.E. McKenna, "Telerobotic Potential for Utility Applications", 2nd *Int. Conference on Line Maintenance, ICOLiM-94*, France, 1994.
[3] Elecric Power Research Institute (EPRI) "Live-Line Repair with Tomcat", *EPRI Journal*, July/August 1987.
[4] Y. Maruyama, Y. et al. (Kyushu Electric Power Co.), "A Hot-Line Manipulator Remotely Operated by the Operator on the Ground", *5th IEEE Int. Conf. on Transmission and Distribution Construction, Operation and Live-Line Maintenance ESMO-93*, 1993.
[5] T. Bennet (Pacific Gas & Electric Co.) "Creating Solutions. Utility company puts robotic arms to work", *Lift Equipment*, August-September issue, 1993.
[6] AICHI, "Aichi Manipulator system Design Feature", *Development Planning Department*, 1990.
[7] T. Ueno (Shikoku Electric Power Co.) "Development of Hot-Line Maintenance Robot", *Workshop on Robotized Hot-Line Maintenance*, Pisa University, Italy, 1988.
[8] M. Boyer (Hydro-Quebec Power Co.) "Telerobotics for Maintenance of Distribution Lines", *ICOLiM-94*, 1994.
[9] Yokoyama et al. (Tokyo Electric Power Co.) "Manipulator System for Constructing Overhead Distribution Lines", *5th ESMO-93*, 1993.
[10] J Lessard et al, " Study of remote control for Live Working on overhead distribution lines. Part I", *International Conference on Line Maintenance, ICOLiM-94*, 1994.
[11] Soler, R and Guillet, J, "Robotic Maintenance of the EDF transmission (63 to 400 kV) network: feasibility Study and effects on tower design". *5th IEEE Int. Conference on Transmission and Distribution Construction, Operation and Live-Line Maintenance ESMO-93*, 1993.
[12] Y. Maruyama et al, "MV overhead hot-line work robot", *3rd Int. Conference on Line Maintenance, ICOLiM-96*, Venice, Italy, 1996.
[13] R. Aracil et al, "ROBTET: A New Teleoperated System for Live-Line Maintenance", *6th IEEE ESMO-95*, 1995.
[14] R. Aracil, L.F. Peñín, M. Ferre and A. Barrientos, "ROBTET: Robot for live-line maintenance", *Int. Conf. on Live Maintenance (ICOLIM'96)*, Venice, October 1996.
[15] J Lessard et al, " Study of remote control for Live Working on overhead distribution lines. Part II", *International Conference on Line Maintenance, ICOLiM-94*, France, 1994.
[16] L.F. Peñín et al, "Fundamentals of a master-slave teleoperation system with a force-position bilateral control scheme", *IFAC's Int. Symposium on Robot Control, SYROCO'97*, Nantes, France, 1997.
[17] M. Ferre, R. Aracil, L.F. Peñín y A. Barrientos, "Multimedia Interfaces for Teleoperated Robots", *3rd SPIE Conference on Telemanipulator and Telepresence Technologies*, Boston, 1996.
[18] L.M Jiménez et al, "3D environment modeling for robot path-planning via stereo vision and active control of optic parameters", IV Spanish Artificial Vision Workshop, San Sebastian, Spain, 1995.
[19] S. Bonner and R.B. Kelley, "A Novel Representation for Planning 3-D Collision-Free Paths", *IEEE Trans. on Systems, Man and Cybernetics*, Vol. 20, No. 6, pp 1337-1351, 1990
[20] T. Kotoku et al, "Environment Modeling for the Interactive Display (EMID) used in Telebotics Systems", *IEEE/RJS IROS'91*, Osaka, Japan.

Proceedings of the 1998 IEEE
International Conference on Robotics & Automation
Leuven, Belgium • May 1998

Development of a Distributed Object-Oriented System Framework for the Computer-Integrated Manufacturing Execution System

Fan-Tien Cheng   Eric Shen   Jun-Yan Deng
Institute of Manufacturing Engineering
National Cheng Kung University
Tainan, Taiwan, R.O.C.

Kevin Nguyen
Mitta Technology Group Inc.
710 Lakeway Drive, Suite 100
Sunnyvale, CA 94086 U.S.A.

ABSTRACT

Today, most of the Manufacturing Execution Systems (MES) are large, monolithic, insufficiently configurable and difficult to modify. Using a distributed object-oriented technique, we will present a systematic approach to develop a computer-integrated MES Framework which is open, modularized, distributed, configurable, interoperable, collaborative, and maintainable. The CORBA infrastructure is adopted to develop this integratable MES. We also use OLE Automation and COM objects to construct sample applications. An example is shown to demonstrate the fruit of this systematic approach.

1 Introduction

A semiconductor manufacturing entity includes integration of the processing equipment with all of the supporting systems for product and process specification, production planning and scheduling, and material handling and tracking. IC fabrication is a very complex and capital-intensive manufacturing process [1].

In this paper, we will present a distributed object-oriented technique, to develop a computer-integrated MES Framework [2,3,5] which is open, modularized, distributed, configurable, interoperable, collaborative, and easy to maintain.

Our systematic approach is started with system analysis by collecting system requirements and analyzing domain knowledge. Our MES Framework is designed by the process of constructing abstract object model based on system requirements, partitioning application domain into components, identifying generic parts among components, defining framework inter-communication and messaging, and developing design patterns [11] for generic parts. Following the MES Framework design, various functional components can be designed by inheriting appropriate design patterns of the MES Framework's common components. An individual Application can be constructed by invoking corresponding methods of related components. At the final stage, the proposed MES can be integrated and tested.

The CORBA infrastructure [8,7] (see Fig. 1) is adopted to develop this integratable MES. Also, OLE Automation technique is applied to construct Applications (see Fig. 2). An example is shown to demonstrate the fruit of this systematic approach.

2 MES Development Procedure

As shown in Fig. 3, the MES development procedure consists of 5 steps: 1) system analysis; 2) MES Framework design; 3) component design; 4) Application construction; and 5) system integration and testing.

This work was supported by the National Science Council, Republic of China, under Contract NSC-87-2218-E-006-001.

The major purpose of the first step: system analysis is to collect all the related domain requirements and fully analyze these requirements such that the fullscale domain knowledge is obtained. According to this domain knowledge, system requirements, functions, constraints, and performance are studied and analyzed.

With the domain knowledge, the second step: MES Framework design can be commenced. The procedure of MES Framework design includes 1) construct abstract object model according to the domain knowledge; 2) partition application domain into components; 3) identify generic parts among components; 4) define framework messages; and finally 5) develop design patterns for generic parts.

After accomplishing the MES Framework design, the third step is the component design based on a design pattern; the fourth step is the Application construction which involves a variety of components; and the final fifth step is the system integration and testing.

In this paper, an integratable MES for the IC packaging factory is adopted as an example to demonstrate the MES Framework development procedure shown in Fig. 3. The detailed design will be explained in the following sections.

3 System Analysis

System analysis, the first step of the MES development procedure, is concerned with devising a precise, concise, understandable, and correct model of the real system. Before developing an MES, the developer must collect the domain requirements and the real-world environment in which it will operate. Then, he must examine and analyze these domain requirements such that all the important features and domain knowledge can be obtained.

Figure 4 shows a typical distributed architecture for an IC packaging manufacturing entity. As depicted in Fig. 4, system manager monitors and controls the status of the whole factory. Scheduler is in charge of scheduling and dispatching job assignments. Common database stores customer orders, job assignments, equipment status, recipes, bills of materials and other engineering data. Equipment manager controls and monitors equipment. Material manager handles the movement of AGV's, AS/RS, robots, and material.

In fact, a complete MES will also have the capabilities for work-in-process (WIP) tracking, statistical process control (SPC), etc. For lack of space and without loss of generality, these capabilities are not covered in this paper. After obtaining the result of system analysis as stated above, we are ready to commence the MES Framework design.

4 MES Framework Design

Our goal is to design an integratable MES by using the distributed object-oriented approach. Those basic foundations for distributed objects and framework [7] as well as the methodology of the Object Modeling Technique (OMT) [6] will be applied to achieve this goal.

The procedure for MES Framework design is composed of 5 steps (see Fig. 3):

1. Constructing abstract object model;
2. Partitioning application domain into components;
3. Identifying generic parts among components;
4. Defining framework messages; and
5. Developing design patterns for generic parts.

The details of these 5 steps are described below.

4.1 Constructing Abstract Object Model

As shown in Fig. 3, the abstract object model is constructed according to the domain knowledge obtained from system analysis. Using the system shown in Fig. 4 as an example, and considering the fact that an MES is composed of several functional modules that handle specifics like material, equipment, labor and planning [4], the abstract object model of an MES is constructed as in Fig. 5. The three key elements of a factory are Equipment, Material, and Labor. Each element is managed by its specific manager. All these three specific managers are controlled by System Manager. System Manager also dispatches orders to Scheduler. Scheduler dispatches jobs to Equipment, Material, and Labor Managers. As for the Common Database, it supports all of the objects to access data.

4.2 Partitioning Application Domain into Components

Our goal is to design an integratable MES which is highly distributed. Therefore, its application domain shall be partitioned systematically and methodologically.

The system shown in Fig. 5 may be partitioned into 6 components as depicted in Fig. 6. They are System Management, Scheduler, Common Database, Equipment Management, Material Management, and Labor Management Components.

System Management Component (which includes System Manager) is in charge of system-level management and services such as life-cycle services, collection services, and query services. Scheduler Component (which includes Scheduler) accepts orders from System Manager and conducts scheduling tasks, then dispatches jobs to Equipment Manager, Material Manager, and Labor Manager.

Equipment Management Component (which includes Equipment Manager and Equipment) manages process equipment; Material Management Component (which includes Material Manager and Material) controls the movements of materials, AGV's, AS/RS, and robots; Labor Management Component (which includes Labor Manager and Labor) handles labor. We define Equipment, Material, and Labor as system Resources. Therefore, these three Management Components are also considered as Resource Management Components.

4.3 Identifying Generic Parts among Components

Among the 6 components shown in Fig. 6, the Common Database Component represents a common facility of the CORBA infrastructure. As for other components indicated in Fig. 6, the structure of Equipment Management Component, Material Management Component, and Labor Management Component are quite similar. The structure of System Manager to control the other managers is similar to that of the Equipment Management Component. According to these observations, we identified generic parts among these four components and, further, we are able to propose a design pattern. As for the Scheduler Component, because its structure is different from those of the other components, a different design pattern is required.

4.4 Defining Framework Messages

The messages which enable interoperability and collaboration among all the components are termed framework messages. In order to maintain system uniformity, only framework messages are allowed to pass into and out of the MES Framework components.

According to the partition shown in Fig. 6, the framework messages are defined as in Fig. 7. System Manager invokes *Initiate*, *StartUp*, *ShutDown*, and *StandBy* methods of Resource Management Components, and Resource Management Components will reply *EventReport* and *AlarmErrorReport* messages.

Scheduler will accept *DispatchOrder* and *CancelOrder* from System Manager, and reply *OrderDoneReport* to it. Also, Scheduler will invoke *DispatchJob* and *CancelJob* methods of Resource Management Components and accept *JobDoneReport* from them. Among Resource Management Components, various demand-service messages are sent and their corresponding reply messages are received.

In addition to those messages described above, objects within each component also define several function-related-service messages in their IDL's to serve the other components. For example, System Manager provides a *CreateOrder* method in its IDL for the outside world to invoke such that an order can be created; and the Common Database Component provides *StoreData*, *RetrieveData*, *DeleteData*, and *ModifyData* methods to serve the other components.

Based upon the partition shown in Fig. 6 and the framework messages defined in Fig. 7, design patterns within the MES Framework can be developed. They are described in the next subsection.

4.5 Developing Design Patterns for Generic Parts

For lack of space, only the development of the Generic Component Design Pattern (GCDP) for Resource Management Components and System Management Component is explained here. Using the OMT methodology [6] this GCDP consists of object model, dynamic model, and functional model. Only the object model is described here.

In order to maximize the generality and reuse, we pay special attention to the common characteristics and behaviors among the components. By observing the object model shown in Fig. 6, we conclude that, the fundamental structure of a component is a System Manager controls several Component Managers and a Component Manager manages several Resources. Therefore, we propose the structure of the GCDP as in Fig. 8.

Base class is the superclass for Manager and Resource. It specifies the common attributes and operations for Manager and Resource to inherit. Among the attributes, *Owner* specifies its upper management class, *Name* and *Status* are self-descriptive, and *Capability* is optional and shows the functional abilities of this class. All of the operations in Base are designed for the purpose of system-management services.

Manager inherits all the attributes and operations of Base. Also, with the help of Resource Collection, Manager manages and serves several Resources. The operations in Manager are for the purposes of life-cycle services (*RegisterResource* and *RemoveResource*), resource-query service (*CheckResource*), and event-report services (*EventReport* and *AlarmErrorReport*). As for the operation of *GetResourceCollection*, it will get the object reference of Resource Collection for Manager. The purpose of Resource Collection is to help Manager accomplish the resource-collection services (*AddResource* and *DeleteResource*), resource-query services (*QueryResourceStatus*, *ListAllResource*, and *Verify*), and system-management service (*InitiateResource*).

Resource also inherits all the attributes and operations of Base. In addition, it adds an attribute: *Parameter* (to specify its own parameters) and 3 operations: *StartRunning*, *AbortRunning*, and *Reset*. These 3 operations are mainly for the purpose of handling job processes which will be described below.

Component Manager inherits all the attributes and operations of Manager. Besides, since Component Manager needs to handle job assignments and dispatches them to Resources for manufacturing processes, it adds three operations (*DispatchJob*, *CancelJob*, and *CompleteJob*) and includes the help of Job Handler to achieve those tasks. The relationship between Component Manager and Job Handler is similar to that between Manager and Resource Collection. Therefore, Component Manager needs the *GetJobHandler* operation to get the object reference of Job Handler. In order to help Component Manager handles job assignments, Job Handler is designed to have job-collection services (*AddJob* and *DeleteJob*), job-query service (*ListAllJob*), and parameter-setting service (*SetParameter*).

After all of the design patterns have been developed, the backbone of the MES Framework is established. The MES Framework architecture will be described next.

4.6 The MES Framework Architecture

The MES Framework architecture is shown in Fig. 9. The MES Framework is built on top of CORBA infrastructure which includes ORB, object services, and common facilities [8]. In this research, we treat the Common Database Component as one of the common facilities.

Since the fundamental framework messages and interfaces for the framework components have been considered in the various design patterns, each specific component shall select a proper design pattern to inherit and then include its own designated properties into the component. As such, the component can be integrated into the MES Framework easily and in a plug-and-play fashion.

The MES Framework provides suitable design patterns for the application component that is highly pluggable.

As shown on top of Fig. 9, an Application is constructed by involving several related components. The method for constructing an Application will be explained in Section 6. After finishing the design of the MES Framework, the next step is component design and implementation which will be described next.

5 Component Design and Implementation

The step for component design and implementation determines the full definitions of the classes and associations used in the implementation, as well as the interfaces and algorithms of the methods used to implement operations. As described in Section 4.6, a component shall inherit a proper design pattern and then include its own designated functions into the component.

By way of illustration, the procedures for designing and implementing System Management Component and Equipment Management Component will be selected and demonstrated in this section. To begin with, the GCDP developed in Section 4.5 will be implemented. Then, System Management Component and Equipment Management Component will be designed and implemented.

5.1 Implementation of Generic Component Design Pattern

After the processes of analysis and design as explained in Sections 3 and 4, we will have the object, dynamic, and functional models [6] of the GCDP, but the object model is the main framework around which the design and implementation is constructed. In fact, for this research the object model in Fig. 8 has already considered the necessities for converting the actions and activities of the dynamic model and the processes of the functional model into operations attached to classes in the object model itself.

With the operations in the object model being defined, their corresponding algorithms shall be designed for implementation. For lack of space, the details of the algorithms for all the operations are not illustrated here. As for the interfaces of the object model, they will be expressed in IDL.

Note that, only those attributes/operations of sever objects which are allowed to be accessed/invoked by the outside clients need to be defined in IDL. Therefore, the operations: *InitiateResource*, *AddResource*, and *DeleteResource* of Resource Collection and the operations: *AddJob*, *DeleteJob*, and *ListAllJob* of Job Handler are not shown in the IDL. As such, the IDL of GCDP is listed below:

```
interface Base {
    readonly attribute string Owner;
    readonly attribute string Name;
    readonly attribute short Status;
    readonly attribute long Capability;
    exception Reject { long ErrorNum; };
    void StartUp()        raises(Reject);
    void ShutDown()       raises(Reject);
    void StandBy()        raises(Reject);
    void Initiate (in string owner, in string name, in long capability)
                          raises(Reject);
    void StartOperation() raises(Reject);
};
interface Manager : Base {
    exception Reject { long ErrorNum; };
    exception Alarm { string Message; };
    void RegisterResource(in string name)    raises(Alarm);
    void RemoveResource(in string name)      raises(Alarm);
    boolean CheckResource(in short state);
    ResCollection GetResourceCollection();
    void EventReport(in string msg);
    void AlarmErrorReport(in string msg);
};
interface ResCollection {
    string ListAllResources();
    short QueryResStatus(in string name);
    boolean Verify(in string name);
};
interface ComMag : Manager {
    exception Reject { long ErrorNum; };
    void DispatchJob(in JobAssignment aJob)
         raises(Reject);
    void CancelJob(in string JobID)
         raises(Reject);
    void CompleteJob(in string JobID);
    JobHandler GetJobHandler();
};
interface JobHandler {
    string ListAllJobs();
};
interface Res : Base {
    readonly attribute long parameter;
    void StartRunning();
    void AbortRunning();
    void Reset();
};
```

For brevity, the concise model of the GCDP may be depicted as in Fig. 10.

5.2 Design and Implementation of System Management Component

The role of System Management is to provide system-management, life-cycle, component-manager-collection, component-manager-query, and event-report services to Component Managers. Besides, System Manager will create orders and dispatch them to Scheduler. Also, System Manager will accept order-done report from Scheduler and provide order status for other components to query. And, an order may be removed by System Manager.

Based on the functional requirements stated above, the System Management Component may be constructed as in Fig. 11 which uses GCDP as the foundation. Because the relationship between System Manager and Component Managers is similar with the relationship between Manager and Resources of GCDP, System Manager can be designed by inheriting the properties from Manager (of GCDP) and adding its own unique properties. As such, the IDL file for the System Management Component is shown below:

```
interface SysMag : Manager {
    string CreateOrder(in mfg_order_attribute aOrder);
    void OrderDoneReport(in string OrderNum, in short status);
    short QueryOrderStatus(in string OrderNum);
    void RemoveOrder(in string OrderNum);
};
```

5.3 Design and Implementation of Equipment Management Component

By the same token, as depicted in Fig. 12, the Equipment Management Component is designed by inheriting GCDP. The GCDP provides the basic behavior and fundamentals of a framework component. Therefore, by inheriting GCDP, the designer may not worry too much about the framework rule and can concentrate on the design of the specific properties belonging to the component itself. In the following section, Application construction with system integration and testing will be introduced.

6 Application Construction with System Integration and Testing

An Application is constructed by invoking corresponding methods of related components as shown at top of Fig. 9. In this research, the technique for CORBA integration with OLE/COM as shown in Fig. 2 is applied. In other words, Applications (client sides) are constructed by Microsoft's Visual Basic 4.0 [12] using the technique of OLE Automation and the components (server sides) are built by Microsoft's Visual C++ 4.0 [13] using Iona's Orbix 2.0.2 [14] which implements CORBA objects.

For the purpose of allowing Automation controllers [9,12] to view CORBA objects [14] as Automation objects [9,12] the correspondence (a gateway as shown in Fig. 2) between CORBA IDL and OLE Automation are defined [10]. For each IDL type, one or more corresponding OLE Automation type definitions as generated by the Orbix-OLE Wizard [10]. For example, an OLE Automation dual interface containing property and method definitions is generated for each IDL interface containing attribute and operation definitions. To invoke an operation on a remote CORBA object, an Automation controller simply invokes a method on the corresponding OLE Automation interface [10].

An example (called Application 1) for an IC wire bonding manufacturing process is adopted. For the purpose of demonstration and brevity, with this process, only a System Manager, a Scheduler, a Common Database, an Equipment Manger, and an Equipment of the IC packaging manufacturing entity shown in Fig. 4 are involved. For system bootstrap and resources monitoring purposes, an Application Console which serves as an OLE Automation controller is also included for the demonstration. Fig. 13 shows the message flow for the manufacturing process of Application 1. To begin with, all the components of the MES shall be bootstrapped, then an order will be created, and then the manufacturing process will be proceeded automatically following the message flow shown in Fig. 13.

The MES bootstrap procedure via Application Console is depicted in Fig. 14. For lack of space, the detailed explanation for this procedure is not shown here.

After the bootstrap procedure, all of the components in the MES will be in the *Operation* state and ready to receive an order.

Now, we may create an order on the Application Console, then the manufacturing process of Application 1 will be proceeded automatically according to the message flow shown in Fig. 13. Correspondingly, the framework messages for Application 1 are depicted in Fig. 15. The interaction of these framework messages have also been omitted here.

We may also use Application Console to construct generic Graphic User Interface (GUI) for querying and monitoring the status of all the Managers and Resources as well as all the jobs and orders. The approach is again to simply invoke the associated methods such as *QueryResourceStatus*, *ListAllResources*, *ListAllJobs*, etc.

Currently, this MES Framework Architecture and the demonstration programs of Application 1 have been successfully set up and running at the Factory Automation Laboratory of the Institute of Manufacturing, National Cheng Kung University, Tainan, Taiwan, Republic of China.

7 Summary and Conclusions

Applying the distributed object-oriented technique, a systematic approach for developing a computer-integrated MES Framework was proposed in this paper. Based on this MES Framework, an integratable MES which is open, distributed, interoperable and collaborative is achievable. Each component of the MES Framework was developed by inheriting a proper design pattern which is considered as the basic designs for architecture, framework messages, and interfaces of this component to interoperate and collaborate with the other components. The specific properties and implementation of the component can then be added into the component in a systematic approach. The component is integratable into the MES Framework in a plug-and-play fashion. Applications were built by invoking corresponding methods of related components with the techniques of OLE Automation and CORBA integration with OLE/COM. A design example was included in this paper. From this example, it is believed that the proposed systematic approach for developing an integratable MES is indeed a viable and efficient method.

References

[1] W. Maly, "Computer-Aided Design for VLSI Circuit Manufacturability," in *IEEE Proc. Design for Manufacturability*, 1993.

[2] D. Scott, "Comparative Advantage through Manufacturing Execution Systems," in *SEMICON Taiwan 96 IC Seminar*, pp. 227-236, Taipei, Taiwan, R.O.C., September 1996.

[3] K. Nguyen, "Flexible Computer Integrated Manufacturing Systems," in *SEMICON Taiwan 96 IC Seminar*, pp. 241-247, Taipei, Taiwan, R.O.C., September 1996.

[4] A. MacDonald, "MESs Help Drive Competitive Gains in Discrete Industries," *I&CS*, pp. 69-72, September 1993.

[5] J. McGehee, J. Hebley, and J. Mahaffey, "The MMST Computer-Integrated Manufacturing System Framework," *IEEE Transactions on Semiconductor Manufacturing*, Vol. 7, No. 2, pp. 107-115, May 1994.

[6] J. Rumbaugh, M. Blaha, W. Premerlani, F. Eddy, and W. Lorensen, *Object-Oriented Modeling and Design*, Englewood Cliffs, NJ: Prentice-Hall, Inc., 1991.

[7] R. Orfali, D. Harkey, and T. Edwards, *The Essential Distributed Objects Survival Guide*, New York: John Wiley and Sons, Inc., 1996.

[8] Object Management Group, *Common Object Request Broker: Architecture and Specification, Revision 2.0*, Framingham, MA: Object Management Group, 1995.

[9] K. Brockschmidt, *Inside OLE*, Redmond, Washington: Microsoft Press, 1995.

[10] *Orbix 2: Orbix Desktop for Windows User's Manual*, Cambridge, MA: IONA Technologies Ltd, October 1996.

[11] E. Gamma, R. Helm, R. Johnson, and J. Vlissides, *Design Patterns: Elements of Reusable Object-Oriented Software*, Reading, Massachusetts: Addison-Wesley, 1995.

[12] *Microsoft Visual Basic Tutorials Version 4.0*, Redmond, Washington: Microsoft Corporation, 1995.

[13] *Microsoft Visual C++ Tutorials Version 4.0*, Redmond, Washington: Microsoft Corporation, 1995.

[14] *Orbix 2: Programming Guide*, Cambridge, MA: IONA Technologies Ltd., October 1996.

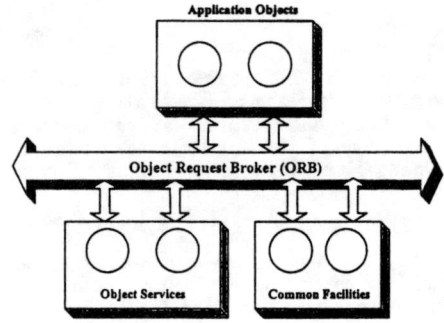

Figure 1: The Object Management Architecture

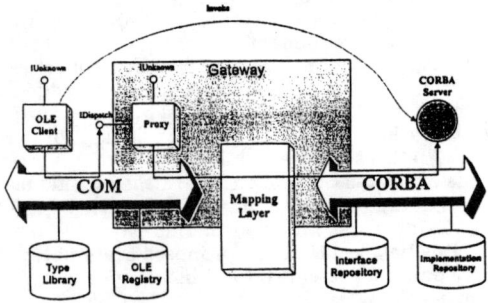

Figure 2: OLE-to-CORBA via a Dynamic Invocation Gateway

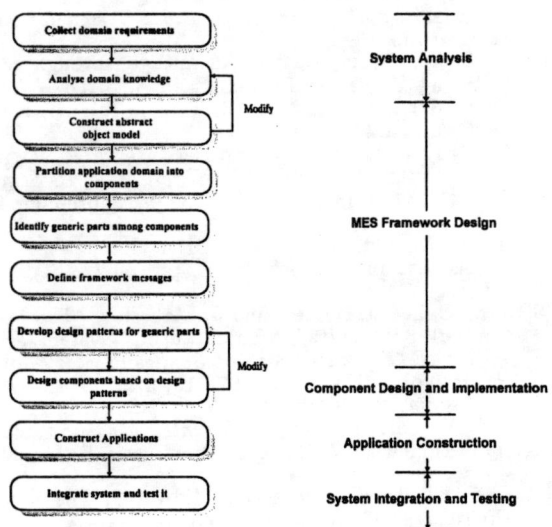

Figure 3: MES Development Procedure

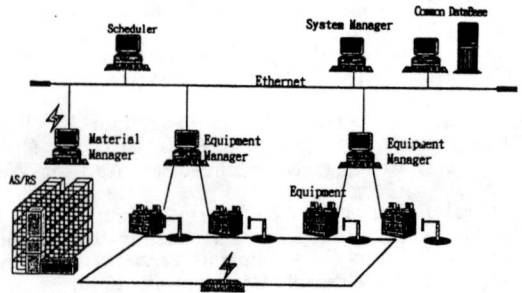

Figure 4: Typical Distributed Architecture for an IC Packaging Manufacturing Entity

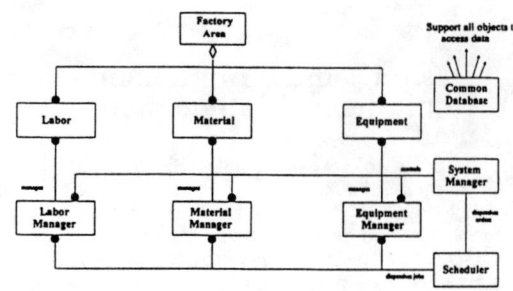

Figure 5: Abstract Object Model of an MES

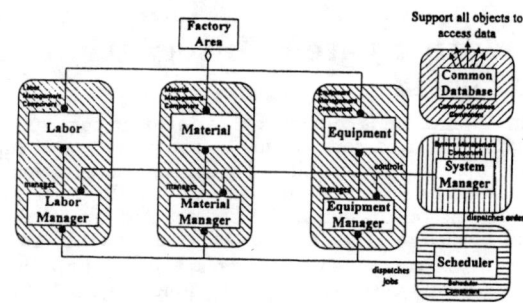

Figure 6: Partitioning Application Domain into Components and Identifying Generic Parts among Components

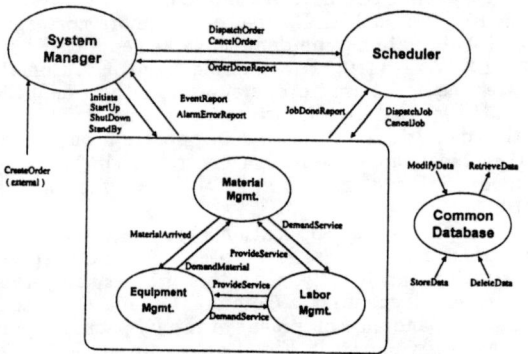

Figure 7: Defining Framework Messages

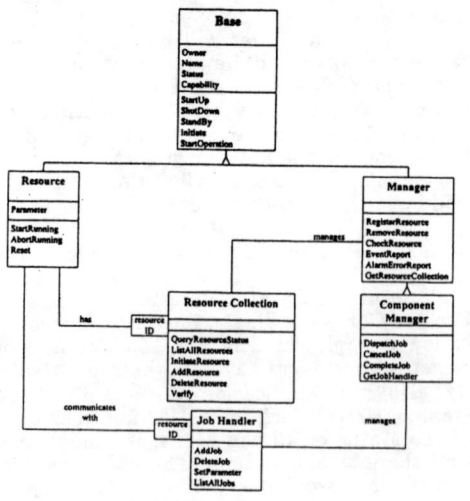

Figure 8: Object Model for the Generic Component Design Pattern

2120

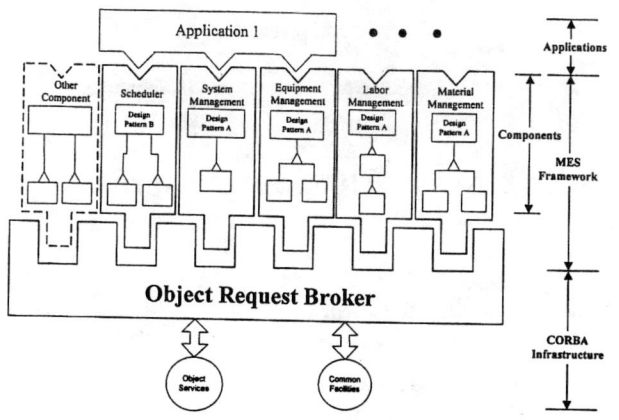

Figure 9: The MES Framework Architecture

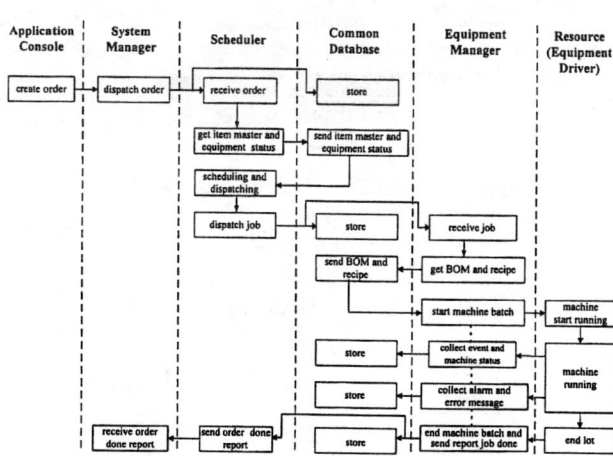

Figure 13: Application 1 Message Flow

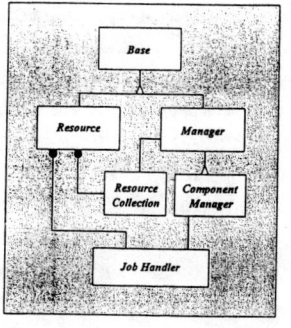

Figure 10: Concise Model of the Generic Component Design Pattern

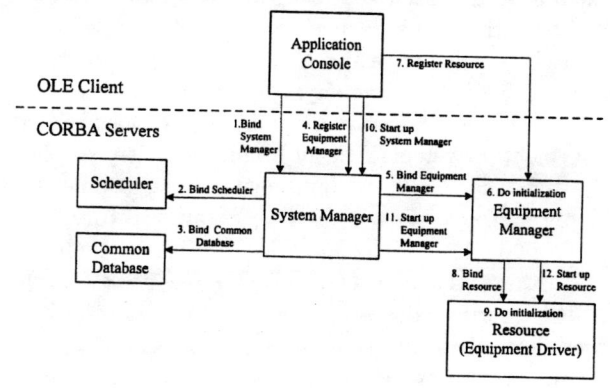

Figure 14: MES Bootstrap Procedure

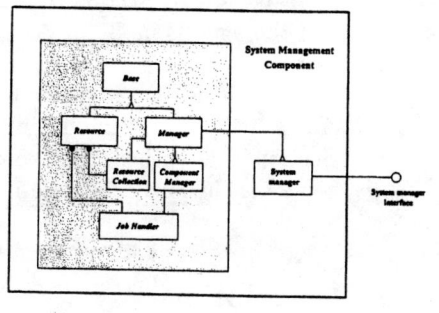

Figure 11: System Management Component

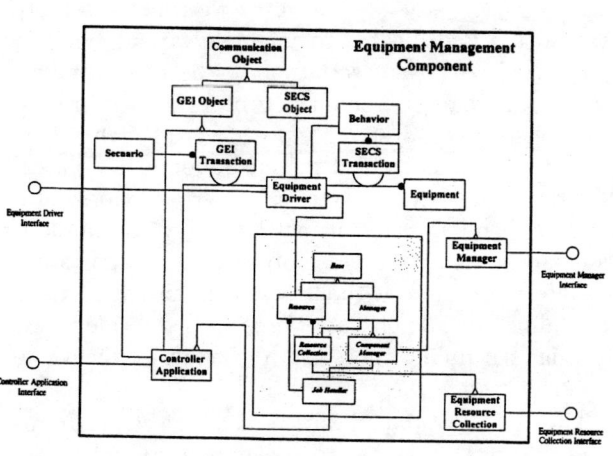

Figure 12: Equipment Management Component

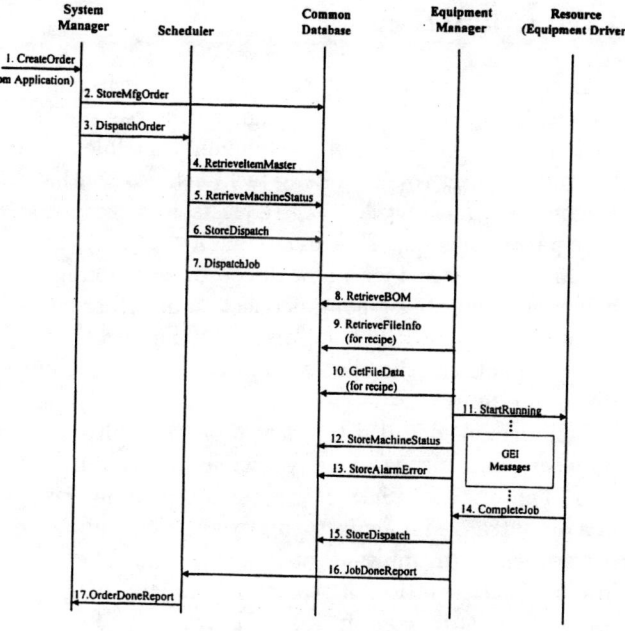

Figure 15: Application 1 Framework Messages Interaction Diagram

Multi-agent Based Dynamic Scheduling for a Flexible Assembly System

Yung-Yu Chen, Li-Chen Fu and Yu-Chien Chen
Dept. of Computer Science and Information Engineering
National Taiwan University, Taipei, Taiwan, R.O.C

Abstract

This paper proposes a multi-agent based dynamic scheduling approach for a flexible assembly system. We first introduce a flexible control system developed by IntelligentRobotics and Automation Laboratory in National Taiwan University. Based on that control system, the agents can communicate with each other conveniently. A generic agent architecture is proposed to model the pieces of equipment in the flexible assembly system. With a distributed architecture, the agents make their scheduling decisions using their local rule base. The agents acquire the resources following the distributed resource allocation protocol. The scheduling complexity is reduced to meet the real-time response requirement in the applications for flexible automated production. The present work is applied to the experimental robotized flexible assembly system in the above laboratory.

1 Introduction

A scheduler is called dynamic (on-line) if it makes its scheduling decisions at run time on the basis of the current requests for service. Dynamic schedulers are flexible to adapt to an evolving task scenario and have to consider only actual task requests and execution time parameters [2]. Research on multi-agent systems is mainly concerned with how to coordinate intelligent behavioral activities among a collection of autonomous agents [3]. The work in [3] discusses the negotiation among agents for the allocation of resource with time taken into account.

Jennings [4] detailed the design of the multi-agent structure and discusses the principle of handling errors. The work in [5] discusses the rescheduling issue in a decentralized manufacturing system. Manufacturing systems with multi-agent modelling can be found in [6, 7]. Resolution of real-time conflicts in multi-robot systems appears in [8]. The work in [9] proposes some criteria to set up a robotic assembly cell and analyzes the interaction among multiple robots.

Recently, agent-based approach is applied to robotic assembly environment [6, 7, 12, 13, 14]. The work [14] introduces an idea of cooperative action with group organization and a strategy for cooperative task processing using communication. There is an efficient negotiation algorithm using his proposed groupcast communication and a learning mechanism with reference to historical records on the past negotiation.

This paper consists of six sections. Section 2 describes a robotic assembly environment and introduces a general control system for that environment. The scheduling problem to be solved is also defined. Section 3 introduces an *agent architecture* and describes the functionality of the agent's module. Section 4 describes the strategy for resolving resource contention among agents. Section 5 describes the implementation of the dynamic scheduling for the flexible robotic assembly cell. Section 6 describes the experimental result. Finally, some conclusions are made in Section 7.

2 System Description

In our laboratory, we have a two-robot assembly system that is dedicated to assemble various types of mechanical parts sent serially into the conveyor belt by the part loader. There are two products currently assembled in this system, and each product has four parts that are assembled by the robot manipulator. The operations include vertical insertion, horizontal insertion, and rotation in assembling with the subassembly fixed at the assembly sites. The parts are fed into the system without a specific order, and the scheduling is made on-line. We proposed a rule based dynamic scheduling approach to schedule for the whole assembly system [17].

The cell is equipped with several pieces of hardware that work together to assemble parts. The brief description is given below:

- **Robot**: There are two robots in the system, namely **ADEPT** and **CRS**. Each of them is equipped with an automatic tool changer(**ATC**) in order to assemble different types of mechanical products. Each robot has a mounted CCD camera which serves the high-precision localization of the part. Also, each robot has its own force-torque sensor that can be used to correct the measurement error during assembly.

- **Part Loader**: It is composed of a Cartesian manipulator and a pallet that holds the parts waiting to be assembled. The Cartesian manipulator will pick up the parts from the pallete and put it on the coveyor belt.

- **CCD Camera**: We have one overhead CCD camera that can determine the type and orientation of the incoming part on the conveyor belt. Two Eye-In-Hand cameras are mounted on the robots, respectively.

- **Conveyor Belt**: It is responsible for carrying parts from outside into the cell.

- **Rotary Buffer**: This is used to temporarily store the incoming parts that are not suitable for immediate assembly. Both robots can access the buffer, but only one robot can be served at one time.

- **Proximity Sensors**: These are to detect the moving speed of each arriving part and act as anchors for the pick-up operation.

Based on the control kernal EMFAK [19], we can easily integrate the different pieces of euipment together. The scheduling problem is to provide a quick response and safety guarantee for the working pieces of equipment. We model each equipment as an agent. An agent has its own capability and thus the scheduling can be viewed as a group of schedulers working simutaneously. This is a distributed system architecture. The coordination among agents is crucial in making the entire system running smoothly.

3 Agent Architecture

There are different kinds of agents in the robotic assembly cell. They can communicate with each other through the communication center *EMFAK* [19]. Fig 1 shows the layout of the multi-agent based system.

Each agent is composed of two main processes which are manager process and controller process. A controller process is related to the domain level system,

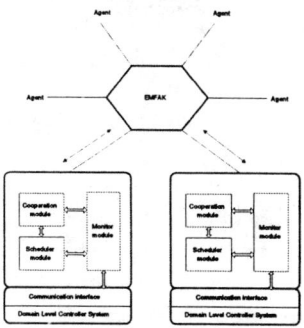

Figure 1: Muti-agent architecture

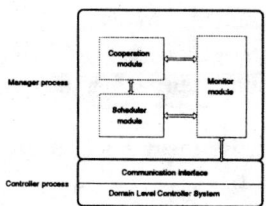

Figure 2: The agent architecture

which includes communication interface module and domain level controller module. Communication interface module is responsible for protocol conversion if controller moudule does not support the pre-selected communication protocol.

On the other hand, manager process is responsible for making the social contact with other agents and scheduling the local tasks. Its objective is to ensure that the agent's domain level activities are coordinated with the other agent and that its associated hardware runs efficiently.

The agent architecture is shown in Fig 2. The components of the generic agent architecture is described in the following sections.

3.1 Cooperation Module

This is an entity that handles the agent-to-agent coordination through message exchange. It is responsible for inter-agent communication. This module has two functional roles:

Message processing Since this module is directly connected to EMFAK, it is responsible for sending message on behalf of the agent and receiving message from the other one.

Coordination In the later section, the distributed resource allocation protocol will be described. Resource contention is the motivation for agent coordination. The agent's scheduler module is concentrated on making its local scheduling and depends on the cooperation module if it is in need of a resource managed by the other agent.

3.2 Scheduler Module

The function role of the scheduler module is to keep the controlled equipment working efficiently, and it will focus on the local affairs. If there is a need of the information or resource from the other agent, it will pass the request to the cooperation module to serve for it.

The agent has a degree of autonomy to make its scheduling decisions. We can view the scheduling problem as a single agent problem virtually. This is a divide-and-conquer approach, and scheduling becomes easier. The scheduler module searches for possible next moves with its local state taken into consideration, and it uses the cooperation module to cooperate with the other agent.

3.3 Monitor Module

This module is to monitor the status of tasks submitted from the scheduler module. It must detect whether the controller module finishes its work on time. Thus, a time-out mechanism is used to assure that the link between two components is active. There are two links in this architecture. They are the link between EMFAK and manager process and the link between manager process and controller process, respectively. For the link between EMFAK and manager process, the manager process periodically informs EMFAK of its liveness. EMFAK keeps a timer that is reset whenever EMFAK receives the liveness message. On the other hand, the other link is treated in another way. With the aid of the monitor module, the scheduler module can make decisions while the underlying domain level system is executing.

3.4 Communication Interface

It serves as the protocol conversion module that can translate different communication media into a preselected protocol. Currently we adopt TCP/IP as our standard protocol [20, 21]. In order to integrate different types of equipment, there should be a selected protocol for communication. If the underlying equipment does not follow that, the communication interface module will be a translator.

3.5 Domain Level Controller System

It is the actual task execution unit that waits for task given from the scheduler and then executes it. For example, they are programs that control robot movement or buffer rotation, etc. Domain level controller system directly controls the physical device. It is the entity that is dedicated to serve the request from the manager process. Basically, it possesses less intelligence and may be seen as a server that receives requests and offers the corresponding service.

4 Distributed Resource Allocation Protocol

In a decentralized environment, the resource allocation problem is complicated. We must consider the mutual exclusion problem when multiple agents want to access to the same resources. There should be a protocol to resolve the conflict on resource. Following the agent's architecture, the resource agent uses its manager process to allocate the resource. For clarity of discussion, the agent that needs the resources is mentioned as *consumer agent*. The consumer agent must get all the permission of all its required resources before utilizing them. Each consumer agent has its own priority that the resource agent uses to decide the precedence of the resource usage.

The consumer agent first sends requests to all relevent resource agents and the resource agent follows a decentralized protocol to allocate the resource. The resource agent has a request list that keeps the incoming request for the consumer agents. This protocol allows the consumer agent to try using mutiple resources at one request. The following sections introduce the distributed resource allocation protocol. The *consumer agent* executes the following protocol:

1. Send request messages to all needed resource agents.

2. Wait for each resource agents to respond with the status of reply.

    (a) If any reply is REJECT, **GOTO** 3.

    (b) If all replies are OK, **GOTO** 5.

3. Send all other resources an REJECTED message. The rejected agent's request will be removed from all resources' lists.

4. **GOTO** 1. (Try again)

5. Inform all resource agents that it is allowed to use the resources.
   (send the ACCEPTED messages to the resource agents)

6. Use the resources.

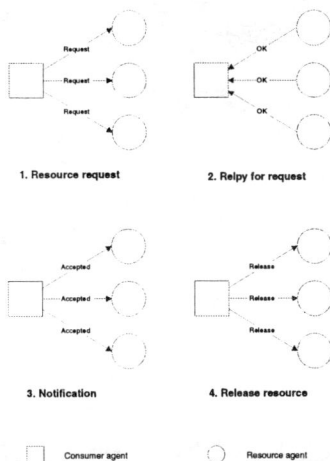

Figure 3: Distributed resource allocation protocol(1)

7. Release all resources.(Send RELEASE messages to resource agents)

The *resource agent* uses the following rules to make decision:

1. Reject all requests(send a REJECT message) when there is a request of higher priority already enqueued in the list or when the resource is *busy*. Otherwise, add the accepted request at the rear of the list.

2. Upon receipt of an REJECTED message, remove the sender's request from the list.

3. Send an OK message to the owner of the first request in the list or the owner of any request that is promoted to the front of the list when the requests in front of it is removed.

4. Upon receipt of an ACCEPTED message, send REJECT messages to the owner of all other requests in the list, empty the list, and mark the resource as *busy*.

5. Upon receipt of a RELEASE message, mark the resource as *free*.

Fig 3 illustrates the case that the consumer agent will get the resources it needs. In the first step, the consumer agent sends requests to all relevant resource agents. The resource agents all reply with OK messages to the consumer agent in the second step. The consumer agent notices the resource agents that it will use the resources in the third step. Finally, the consumer agent free its resources and send RELEASE message to the resource agents.

The protocol is flexible because it operates in a distributed manner. The new agents, whether consumer agents or resource agents, can be added into the system dynamically without interfering the running of the system. For one paticular agent, the arrival order for resource requesting messages does not affect the correctness of the protocol. Also, the message sendings of different consumer agents can be mixed. These above two characteristics can make the system have less assumption about the quality of the communication media and requirement of the communication protocol.

This protocol makes the resource allocation efficient. The resource is allocated to the consumer agent that owns the permissions of all its required resources. Once the resource is released, the contention for them starts again. With this way, resources are kept as busy as possible. The deadlock-free guarantee is proved. This property is important since it is concerned with safety issue and continuous running of the whole system.

5 Dynamic Scheduling in the Experimental FAS

Our experimental environment is a two-robot assembly cell which is dedicated to assembling various types of mechanical parts serially sent in through a conveyor belt. The cell is composed of several pieces of hardware.

Each product has four parts respectively. The first product is assembled with only vertical insertion operations. The second product includes more complex operations. To assemble the second part with the base part for the second product, the robot needs to do vertical insertion and then a rotation to fasten the part with the base part. Sixteen parts can be placed randomly in the pallete of the loader. The loader will load the part onto the conveyor belt one at a time on request.

We model our cell as multiple agents that work together. The modelling of the cell is depicted in Fig 4.

5.1 Error Recovery

Sometimes, there are machine failures during the assembly process. The proposed agent architecture can solve some of the failures. The manager process is assumed to be more reliable. The controller process is the main component that causes error for an agent.

Fig 5 shows the error recovery processs. There are two communication links in the manager process. The

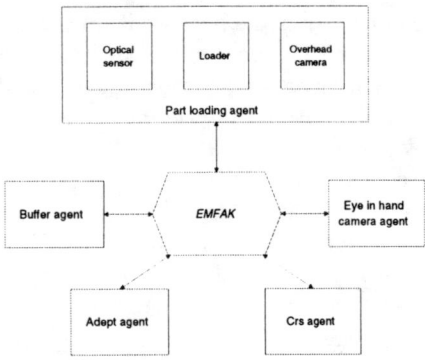

Figure 4: The agents in the FAS

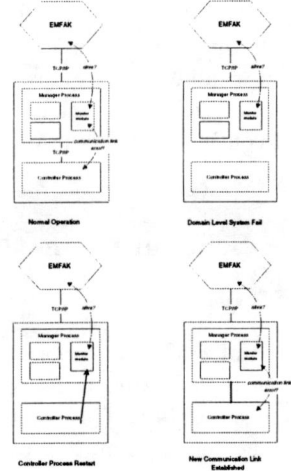

Figure 5: Error recovery for equipment failure

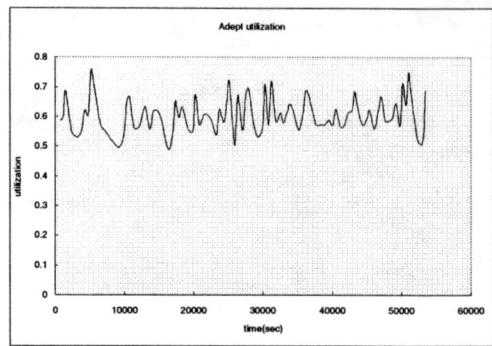

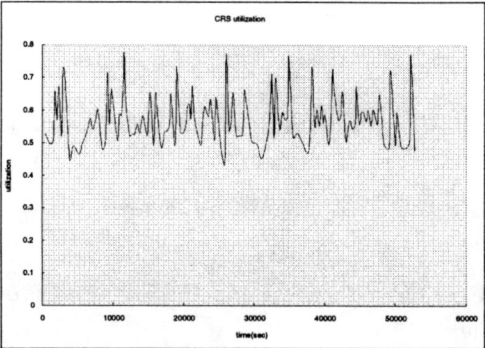

Figure 6: The utilization of the robots

monitor module detects the liveness of the link between the manager process and the controller process. When the physical equipment is faulty, we just shutdown the machine and kill its associated controller process. The manager process is still alive and will wait for the reconnection of the controller process once the physical equipment is recovered. The agent's internal state is kept and the agent restarts from the last kept status.

6 Experiment

The simulation is performed on the pseudo robotic assembly cell. We replace the physical equipment with software. In other words, the agents can be easily modified to work in the real environment. In the simulation, part loading machine loads part into the assembly cell randomly, and the robot will either assemble it or store it on buffer. There are total seven types of parts that may be fed into the cell. removed. The parameters that we will examine are the robot utilization, the histogram of the finished product, and buffer utilization. In this simulation, we give *Adept* a high priority than *CRS*.

6.1 Simulation result

We assume the robot operation time ranges from 2 to 12 seconds in this simulation, and the time depends on the complexity of that operation. The average operation time is less than 10 seconds.

In the two robot simulation environment, the utilization of the robot is between 0.5 and 0.8 or so as shown in Fig 6. There are resource contention between the two robots. For example, the robots compete for the common workspace, for the shared buffer, and for the shared *Eye-In-Hand* camera PC. Instead, if only one robot is in operation, the utilization of the robot will be above 0.8. The scheduling time is thus 1 to 2 seconds for each robot operation.

7 Conclusion

This paper is an extension of [17]. The relationship between the agents and communication backbone EMFAK is discussed in section 2. To meet the real-time

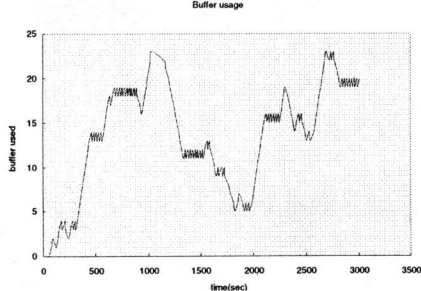

Figure 7: The buffer utilization

requirement in the robotic assembly environment, we use a decentralized approach to solve the scheduling problem. The basic idea is using the concept of agent which is an autonomous entity that can communicate with each other to achieve a coorinated behaviour. Section 3 introduces the agent architecture and the functionality of the agent's module. Section 3 also introduces the distributed resource allocation protocol that is used to allocate the resources among a group of distributed agents. The benefits of using multi-agent based modelling exhibits flexibility, robustness, and modularity. Section 4 describes the implementation of the agents for the flexible assembly cell. The experimental result is given in section 5.

References

[1] M. P. Groover, *Automation, Production Systems, and Computer Integrated Manufacturing*. Prentice-Hall International, 1987.

[2] H. Kopetz, *Scheduling*, ch. 18, pp. 491–509. Addison-Wesley publishing company, 1993.

[3] S. Karus, J. Wilkenfeld, and G. Zlotkin, "Multiagent negotiation under time constraints," *Artificial Intelligence*, vol. 75, pp. 297–345, 1995.

[4] N.R.Jennings, "Controlling cooperative problem solving in industrial multi-agent systems using joint intentions," *Artificial Intelligence*, vol. 75, pp. 195–240, 1995.

[5] T. K.Tsukada and K. G.Shin, "Priam: Polite rescheduler for intelligent automated manufacturing," *IEEE Transaction on Robotics and Automation*, vol. 12, no. 2, pp. 235–245, 1996.

[6] E. Oliveria, "Cooperative multi-agent system for an assembly robotics cell," *Robotics and Computer-Intergrated Manufacturing*, vol. 11, no. 4, pp. 311–317, 1994.

[7] R. J. Rabello and L.M.Camarinha-Matos, "Negotiation in multi-agent based dynamic scheduling," *Robotics and Computer-Intergrated Manufacturing*, vol. 11, no. 4, pp. 303–309, 1994.

[8] G. Cohen, "Concurrent system to resolve real-time conflicts in multi-robot systems," *Robotics and Computer-Intergrated Manufacturing*, vol. 8, no. 2, pp. 169–175, 1995.

[9] P. M. Pelagagge, G. Cardarelli, and M. Palumbo, "Some criteria to help the experimental setup of assembly cells with cooperating robots," *Robotics and Computer-Intergrated Manufacturing*, vol. 12, no. 2, 1996.

[10] Georgeff, "Communication and interaction in multi-agent planning," *Proceedings AAAI-83*, pp. 125–129, 1983.

[11] F.-Y. Wang and G. N. Saridis, "A coordination theory for intelligent machines," *Automatica*, pp. 833–844, 1990.

[12] J. S. Barsran, E. M. Petriu, and D. C. Petriu, "Flexible agent-based robotic assembly cell," *Proceedings IEEE International Conference on Robotics and Automation*, pp. 3461–3466, 1997.

[13] A. A. Rizzi, J. Gowdy, and R. L. Hollis, "Agile assembly architecture: An agent based approach to modular precision assembly systems," *Proceedings IEEE International Conference on Robotics and Automation*, pp. 1511–1516, 1997.

[14] H. ASAMA, K. OZAKI, Y. ISHIDA, K. YOKOTA, A. MATSUMOTO, H. KAETSU, and I. ENDO, "Collaborative team organization using communication in a decentralized robotic system," *Intelligent Robots and Systems*, pp. 816–823, 1994.

[15] R. G. SMITH, "The contract net protocol: High-level communication and control in a distributed problem solver," *IEEE Transactions on computers*, vol. C-29, no. 12, pp. 1104–1113, 1990.

[16] V. K. Garg and B. Waldecker, "Detection of weak unstable predicates in distributed programs," *IEEE Transcations On Parallel And Distributed Systems*, pp. 299–307, 1994.

[17] T.-S. Huang, L.-C. Fu, and Y.-Y. Chen, "Design and analysis of a dynamic scheduler for a flexbile assembly system," *Proceedings IEEE International Conference on Robotics and Automation*, pp. 3334–3339, 1997.

[18] C.-S. Jann and L.C.Fu, "Flexible control system for robot assembly automation," *Proceedings IEEE International Symposium on Assembly and Task Planning*, pp. 286–292, 1995.

[19] H.-S. Huang, L.-C. Fu, and J. Y. jen Hsu, "Rapid setup of system control in a flexible automated production systems," *Proceedings IEEE International Conference on Robotics and Automation*, pp. 1517–1522, 1997.

[20] D. E. Comer, *Internetworking With TCP/IP Vol 1: Principles, Protocols, and Architecture*. Prentice-Hall, 1991.

[21] W. R. Stevens, *UNIX network programming*. Prentice-Hall International, 1991.

[22] H. F. Wedde, B. Korel, S. Chen, D. C. Daniels, S. Nagaraj, and B. Santhanam, *Transparent Access to Large Files That Are Stored accross Sites*, ch. 9, pp. 490–510. IEEE Computer society press, 1994.

Multi-Agent System for Dynamic Scheduling and Control in Manufacturing Cells

D. Ouelhadj[1], C. Hanachi[2], B. Bouzouia[1]

1: Centre de Développement des Technologies Avancées. 128, Chemin Mohamed Gacem. El-Madania, Algiers, Algeria. Email: ouelhadj@hotmail.com
2: Université de Toulouse 1. Sciences sociales, Place Anatole France 31042 Toulouse cedex. E-mail : hanachi@univ-tlse1.fr

ABSTRACT

This paper describes an intelligent real time cell control architecture for flexible manufacturing systems. This architecture is based on the Multi-Agent concept of the Distributed Artificial Intelligence. We propose an actor architecture. This Multi-Agent prototype system involves the following functions :scheduling, dispatching, monitoring and error handling. A new contract net protocol for dynamic scheduling is presented. The purpose of this protocol is to dynamically assign operations to the resources of the Manufacturing system in order to accomplish the proposed tasks. This protocol is able to deal with exceptions.

1. INTRODUCTION

During the last years manufacturing systems have became increasingly complex. Competitive pressures are moving manufacturers toward shorter product cycles, lower inventories, higher equipment utilisation, and shorter lead times[21][22]. As a result, the problem of controlling a flexible manufacturing system grows in importance and has to be designed with an increased level of intelligence [23].

The function of a real time cell control system can vary depending on the size of a cell, its type and the degree of decision making capability given to a cell. The major functions of a cell control system include the need to schedule and monitor cell resources, and the ability to react to abnormal conditions or exceptions [8][14][16][26] [27].

Manufacturing, scheduling and control in real-time is an important and an increasingly popular research domain in a number of academic fields, including industrial engineering, artificial intelligence, and multi-agent systems.

Since manufacturing systems are often distributed in nature, both from the geographical and organisational point of view [6][8][13], decision making in manufacturing system control lends particularly well to distributed problem solving. Physically, the manufacturing system involves several resources (numeric control machines, robots, conveyors ...). From the logical point of view several tasks can be carried in parallel.

Due to these reasons, the framework of Distributed Artificial Intelligence, particularly multi-agent systems [10][11][12][15][17][18][28][13][20] seams more suitable for the dynamic control of manufacturing systems.

Multi-agent systems offer production systems that are decentralised rather than centralised, emergent rather than planned, and concurrent rather than sequential. Instead of centralising information and control, autonomous agent architectures recognise that data and control are distributed throughout the cell.

Yams [23][24] is the original implementation of multi-agent systems in manufacturing, which assigns an agent to each node in the control hierarchy (factory, workstation, machine). An agent at one level uses negotiation to identify agents under its control at the next lower level to whom to assign tasks. H.Lee [14] has proposed an architecture based on multi-blackboard/actor models that combines blackboard and actor models[1][2][9] to give a multi-blackboard/actor framework.

S.Devapriya [6] has developed a distributed intelligent system for FMS control based on the definition of a group of closely-coupled and cooperating cells where each cell is an intelligent problem solving agent. Carlos Ramos [25] has presented an architecture for the dynamic control and scheduling of manufacturing systems which use a new negotiation protocol suitable for the dynamic scheduling of tasks. This negotiation protocol is able to deal with exceptions. H.Norrie [20] has developed a multi-agent actor architecture for the integration of design, manufacturing and shop floor control activities.

This paper proposes a new architecture for an intelligent real time cell control system combining the advantages of the last three architectures cited above.

Section 2 presents the multi-agent control architecture. Section 3 details the functions of the multi-agent system. Finally, conclusions are presented in section 4.

2. THE MULTI-AGENT ARCHITECTURE FOR DYNAMIC CONTROL

The autonomous agent approach proposed replaces a centralised database and control computer with a network of agents, each endowed with a local view of its environment and the ability and authority to respond locally to that environment. The overall system performance is not globally planned, but emerges through the dynamic interaction of the agents in real time. Thus the system does not alternate between cycles of scheduling and execution. Rather the control emerges from the concurrent independent decisions of the local agents.

We propose an actor based framework which aims to the advantage of computational efficiency under steady conditions but flexibility in changing environments. The framework consists of several Agents which are actors[1], where each Agent is associated with a particular function. The control functions which are : scheduling, dispatching, monitoring and error handling are distributed over the Resource Agents.

This actor model aims to provide an architecture in which :

- different control functions can be performed in a distributed fashion;
- overall control can be achieved by passing appropriate messages between actors subsystems;
- negotiation, particularly contract net protocol is used to coordinate the agents.

The overall system architecture is shown in (fig.1) which contains the following agents:

Task Manager Agent : receives the new tasks to perform and creates their executing maps. Each task is a production routing which is an operation sequence.

We consider a single manufacturing facility consisting of M resssources that manufacture a set of part types. Each part is performed on a machine from the set M and has associated with it a deterministic processing time, which includes both the operation set up and runtimes.

Resource Agent (RA) : every resource in the cell (numerical control machines, industrial robots, tools, fixtures and control computers) is represented by an autonomous agent called Resource Agent. The resource agent represents the current situation of a resource : its status, and the activity to be carried out. The activity is a sequence of operations to be carried out which is represented into an agenda. The Resource Agent can be a Resource Manager that establishes the negotiation with the Resource Agents to select the appropriate resources that will carry out a specific task. It also renegotiates the operations of a resource in failure, not yet executed, with the Resource Agents that are able to execute them.

This agent is, also, responsible to perform the four control functions, which are :scheduling, dispatching, monitoring, and error handling.

Reactive Resource Agent : these reactive Agents are associated to sensors to inform the Resource Agents about the world status (parts, resources...).

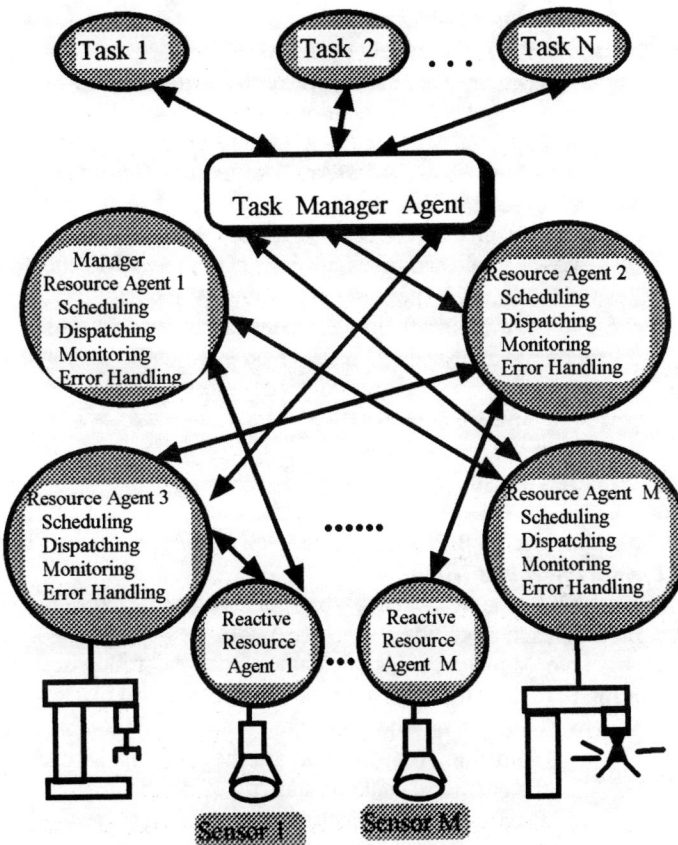

Fig 1: The multi-agent architecture for dynamic control

3. THE DYNAMIC CONTROL FUNCTIONS IN THE MULTI-AGENT SYSTEM

The multi-agent system includes the four principal functions mentionned above: scheduling, dispatching, monitoring, and error handling.

Scheduling
This control step assigns dynamically task operations to the cell resources that will execute them[5][7][13]. This assignation is accomplished within a contract net protocol [3][4][5][19][25]. The Resource Manager which is a Resource Agent is responsible of contracting the task operations. In order to guarantee the deadline, the negotiation between the Resource Manager and the Resource Agents is performed in backwards chaining.

The negotiation can be summarised in the following steps (fig.2).

The Task Manager Agent send a message to the Resource Manager as a request whose format is :

(Td, Lopr, Tres)

Where : **Td** is a task descriptor,
  Lopr is the list of operations,
  Tres is the list of resource types that can accomplish the operations.

This message sent by the Task Manager Agent to the Resource Manager and is referred to as the '*Operation Announcement Message*'*(OAM)*.

Once the set is received, the Resource Manager knows the sequence of operations to be negotiated with the Resource Agents. The last operation in the chain is chosen(backwards), then the constraints about this operation are analysed and a list of possible resources for the operation is constructed. The Resource Agents corresponding to this operation receive a message with the following format : **(Td, Opr)**,

Where : **Td** is the task descriptor,
  Opr is the operation itself.

This message will be referred as the '*Resource Request Announcement*'*(RRA)*.

All the Resource Agents, that had received request, propose execution bids and respond with messages to the Resource Manager. This messages have the following format : **(Td, Mintime, Maxtime, FreeTime, NbOp)**

Where : **Td** is the task descriptor,
  Mintime and **Maxtime** are the limits in which the operation could be started,
  FreeTime is the time free resource,
  NbOp is the number of operations to guarantee the deadline.

This message will be referred as to '*Resource Bid Message*'*(RBM)*.

After receiving all Resource Bid Messages, the Resource Manager decides for the resource to be contracted for a specific operation, by sending '*Acceptance Acknowledgement*'*(Acc.Ack)* messages to the chosen agents. If it is impossible to meet the deadline of an operation, its negotiation is finished without contract and a '*Non Acceptance*' message is sent to the Task Manager Agent. However, if all the operations of a task are contracted with success, an '*Acceptance*' message is sent by the Resource Manager to the Task Manager Agent that have proposed the task.

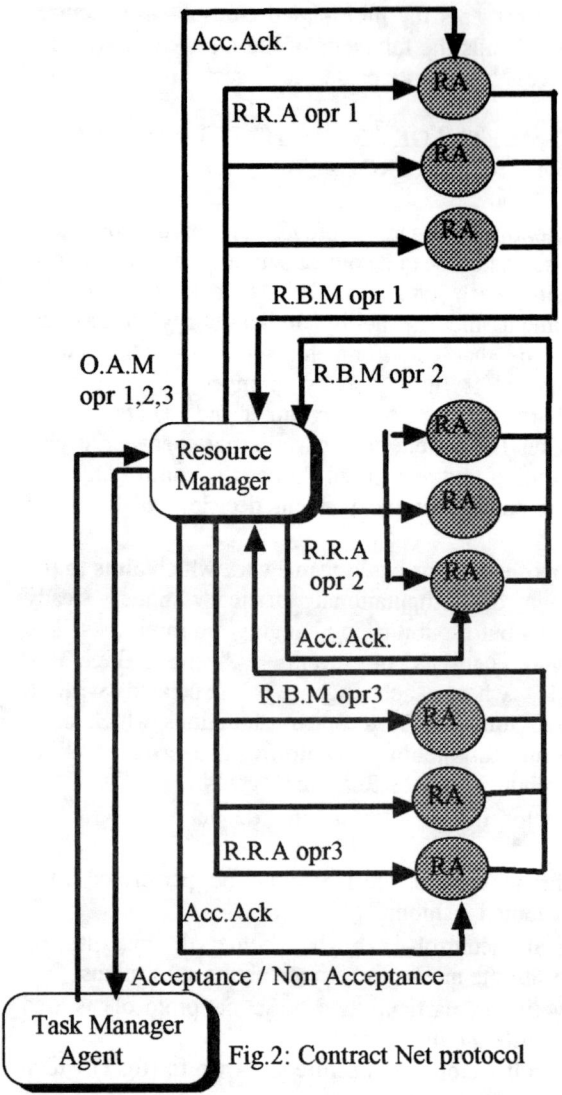

Fig.2: Contract Net protocol

Dispatching
The dispatching function is accomplished by the Resource Agents that send starting orders to their appropriate resources and wait for finish signals or abnormal events;

The execution of an operation can be summarised in the following points (fig.3) :
- extraction of an operation from the FIFO list;
- verification of its preconditions ;
- sending of starting orders ;
- handling of operation effects ;
- waiting of ending signal or abnormal.

The execution module of the Resource Agent extracts an operation from its FIFO list and sends its preconditions to the inference engine that verifies them, and responds with 'OK' if they are valid and by 'NOK' if they are not. In the favourable case (response with 'OK'), the operation is launched and a starting order, specifying the operation name and parameters, is sent to the resource interpreter.

When the operation is finished, the resource interpreter sends a signal to the execution module specifying the name of the terminated operation.

The Resource Agent takes in account the possible interruptions coming from the operator or from the error detection and recovery mechanism. The inference engine has to respect the due time of operation as well.

Monitoring and Error Handling
This function recognises and analyses the errors or exceptional conditions of the cell (machine in failure, operation in fault), and provides possible corrective actions to these problems.

When a Resource Agent executes an operation, the reactive agents associated to the sensory equipment(cameras) inform the Monitoring about the world status. This relevant information which concerns the status of operations, the parts and the resources is sent to the Error Handling to diagnose the errors by the use of production rules based on predicat logic. These rules have the following format:

If situation *then* diagnostic

Example :
For example the production rules related to the operation Take_Put (piece i, robot j, source, destination), which takes the 'piece i' by 'robot j' from 'source' to 'destination', are :

if not(free(robot j)) *and* (not(free(source)) *or* not(free(destination)) *then* camera in failure

if not(free(source)) *and* not(free(destination)) *then* camera in failure

if free(robot j) *and* not(free(source)) *and* free(destination) *then* robot j in failure

if not(free(robot j)) *and* free(source) *and* free(destination) *then* robot j in failure

if free(robot j) *and* free(source) *and* free(destination) *then* piece i falls from the robot j.

Once the errors are determined, they are sent to the Error Handling that will determine the corrective actions to take in order to recover these errors. The Error Handling uses production rules in reacting to these errors. These rules have the following format:

if diagnostic *then* actions

Example :
if numerical control machine (NC) in failure *then* transit to degraded mode, only the assembly line works, *and* call the operator.

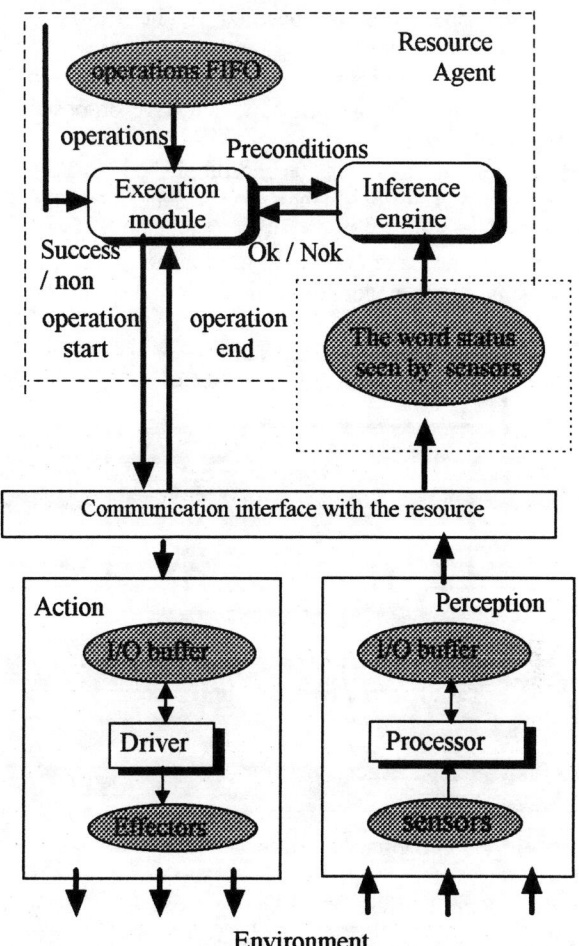

Fig.3: Operation execution mechanism

The corrective actions, which are not necessarily exclusive, are :

Urgent halt: it consists on stopping, momentarily, all the cell in order to protect its elements from possible deterioration. This happens when an abnormal presence of a non identified objects or a non authorised persons in the work area (area where robots work).

Transit to degraded mode : the characteristics of the cell are used, specially its flexibility, to continue the manufacturing process in spite of the errors or the failures. Some failures produce a partial halt of the cell.

The operator call: an operator is present in the cell and is solicited for possible repairing.

Renegotiation: the renegotiation protocol is illustrated in (fig.4). The Resource Agent related with the exception will send an '*Operation Failure*' message to the Resource Manager in which the amount of non accomplished operations is specified. The Resource Manager, as in a normal negotiation, proposes the operations to the Resource Agents that are able to execute them, and contracts those which have proposed the best bids of execution.

If the operations can not be accomplished by the resources, a *'task failure'* message is sent from the Resource Manager to the Task Manager Agent. Otherwise, a message is sent to inform that the operations have been negotiated with success.

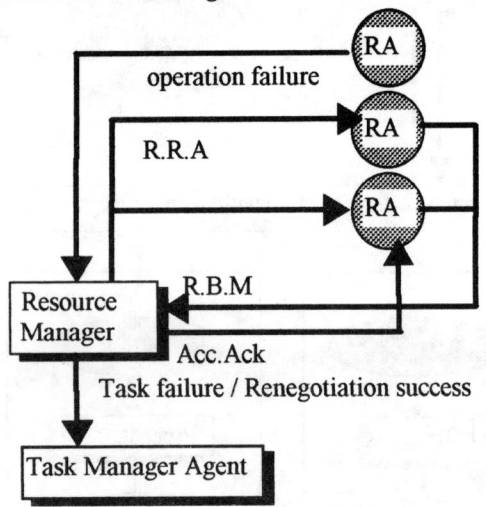

Fig.4: Renegotiation protocol

Operation leap (jump) : the operation in fault is ignored with all its drifted if does not perturb the manufacturing process.

4. CONCLUSION

This paper presents a multi-agent architecture for dynamic scheduling and control of manufacturing cells based on actor framawork. The main advantages of this architecture are: flexibility in more changing environment, increased fault-tolerance, inherent adaptability and reconfiguration, modular development and high parallelism degree. The four basic control functions, which are scheduling, dispatching, monitoring and error handling are distributed over all the agents in the system. There are also other functions, such as: cell initialisation and termination, communication and user interface. The use of contract net protocol in scheduling involves an architecture with a high performance.

The implementation of the system is in OOP(Object Oriented programming) with C++ under WINDOWS that allows multi-tasking.

5. REFERENCES

[1] Gul Agha. « ACTEURS: un modèle de calcul concurrent pour les systèmes distribués ». Traduction Meunier Nicolas. Coascas Patrick. 1993.

[2] Norman Carver & Victor Lesser. « The evolution of Blackboard control Architectures ». Cmpsci Technical Report 71-92. October 1992.

[3] S.Cammarata &D.MC Arthur & R.Steeb. » Strategies of cooperation in Distributed Problem Solving ». in Karlsruhe. Proceeding of 18 th International Joint Conference on Artificial Intelligence. Volume 2. Page 767_770. Germany. August 1993.

[4] V.Chevrier. « Coordination et structuration des échanges par négociation dans les systèmes multi-agents ». Journée Systèmes multi-agents. PRC-GDR. Intelligence Artificielle. Nancy. Decembre 1992.

[5] Thouraya Daouas & Khaled Ghedira & Jean Pierre Muller. « A distributed approach for the flow shop scheduling problem ». 3rd International Conference on Artificial Intelligence Applications. CAIRO Egypt. January. 1995.

[6] Dewasurendra S.Devapriya & Bernard Descotes & Pierre Lodet. « Distributed intelligence systems for FMS control using objects modelled with petri nets(SCOPE blackboard). IFAC symp on Distributed Intelligence Systems. 13-15. August 1991. Arlington Virginia.

[7] Touraya Douas & Khaled Ghedira & Jean Pierre Muller. « How to schedule a flow shop plant by agents ». Computer Science and Artificial Intelligence Institute. 1994.

[8] Neil A.Duffre & Kex S.Piper. « Non hierarchical control of a flexible manufacturing cell. Robotics and Computer Integrated Manufacturing. Vol 3. No2. Page 175_179.1987.

[9] Engelmorer.S & A.J.Morgan . « Blackboard systems : introduction in Blackboard systems ». Page 213_253. Also in : Blackboard Systems. R.Engelmore & T.Morgan. Addison Wesley. 1988.

[10] Jean Herceau & Jacques Ferber. « L'intelligence artificielle distribuée ». Recherche 233 Juin 1991 . Volume 22.

[11] J.Ferber & M.Ghallab. « Problématique des univers multi-agents ». Actes des Journées Nationales du PRC-IA. Mars 88.

[12] J.Ferber. « Les systèmes multi-agents : vers une Intelligence collective ». Inter- Edition. Paris 1995.

[13] Lamia Friha & Pauline Berry & Berthe Choueiry. « DISA : a distributed scheduler using abstractions ». Revue d'Intelligence Artificielle. Vol 11.N1.1997. Page 27_42.

[14] Peter O.Grady & Kwan H.Lee. « An intelligent cell control system for automated manufacturing ». Int Journal Production Res. 1988. Vol 26. N5. 845_861.

[15] Georgeff .M.P. « Communication and interaction in multi-agent planning ». Proceedings of the Third National Conference on Artificial Intelligence . AAAI83. Page 125_129. Also in : Readings in ditributed artificial intelligence. A.H.Bond & L.Gasser. Page 200_204. Morgan Kaufman Publishers. California. 1988.

[16] Yo Shio Kawauchi & Makoto Inaba & Toshio Fikuda. « CIRCA : a cooperative intelligent real time control Architecture». Robotics Man and Cybernitics. IEEE. 1993.

[17] Sofiane Labidi & Wided Lejouard.. « From ditributed intelligence to multi-agent systems ». INRIA Sophia Antipolis. 1993.

[18] Victor R.Lesser. « An overiew of DAI : Viewing ditributed AI as ditributed search ». Computer and Information Science. University of Massachusetts/Amherst. University Research Initiative grant number N00014_86_K_0764. 1992.

[19] Brigitte Laasri & Victor Lesser. « Negotiation and its role in cooperative ditributed problem solving ». Computer and Information Science Department University of Massachusetts. Coins Technical Report 90_39. May 1990.

[20] S.Balasubramanian & H.Norrie. « A multi-agent Design System Integrating Manufacturing and shop floor control ». Multi-agent Intelligent Design. February 97.

[21] H.Van Parunak. « Autonomous Agent Architectures : a non technical introduction ». Industrial Technology Institute .1994.

[22] H.Van Parunak. « MASCOT : a virtual factory for research and development in manufacturing scheduling and control ». Industrial Technology Institute. 1993.

[23] H.Van Parunak. « Applications of distributed Artificial Intelligence in industry ». Industrial Technology Institute. 1994.

[24] H.Van Parunak. « Implementing manufacturing agents ». Sponsored by the shop floor agents project of the National Center for Manufacturing Sciences in conjonction with PAAM'S 96. Westminster Central Hall. London UK. 25 April 1996.

[25] Carlos Ramos. « An architecture and a negotiation protocol for the dynamic scheduling of manufacturing systems ». IEEE Internatioanal Conference On Robotics and Automation. Vol 4. Page 8_13. May 1994.

[26] Peter.S.Vail. « Computer Integrated manufacturing ». New Hampshire Vocational Technical College. Manchester. New Hampshire. 1988.

[27] Peter.S.Vail. « System design Computer Integrated manufacturing. New Hampshire Vocational Technical Ccollege. Manchester. New Hampshire.1988.

[28] Micheal Wooldrige & Nicholas R.Jennings. « Intelligent agents : theory and practice ». 1995.

Multi-Agent Based Control Kernel for Flexible Automated Production System

Sung-Hahn Liu and Li-Chen Fu
Dept. of Computer Science and Information Engineering
National Taiwan University, Taipei, Taiwan, R.O.C

Jung-Hua Yang
Dept. of Electrical Engineering
Yung Ta Junior College of Technology & Commerce

February 26, 1998

Abstract

An Intelligent Automated Robotic Assembly System consists of several subsystems capable of providing dynamic interactions with the environment in order to accomplish a task properly. These subsystems perform various functions like data gathering, decision making, and task execution. Although a great deal of work has been done on individual subsystems, more attention must be given to the way how these subsystems are integrated so as to achieve the high efficiency of automated production. In this paper, we propose a cooperative multi-agent model of a shop floor control system architecture of robotic assembly atuomation and extend this model to all automated production system. Based on this model, we develop a control kernel named TOFAK(Task Oriented Flexible Automation Kernel) to support users to easily implement any shop floor control system. The by-product is to allow system designers to easily expand an existing system or to integrate several automation systems which are all controlled by TOFAK.

1 Introduction

Due to the rapid change in consumers world requirements, market flexibility has become one of the most important factors in manufacturing environment within the recent years. Large industial companies have realized that flexible production systems are capable of rapid adaptation to varying number and various kinds of products. In general, flexibility of a control system in automated systems is greatly emphasized nowadays because it can make the system more adaptable to various situations. One kind of flexibility is the capacity of on-line reconfiguration. In [10], a general control architecture for multiple vehicles is proposed and dynamic reconfiguration is allowed. Moreover, in [3] and [6], flexible control systems are implemented for flexible automated production systems. Recently, formal languages [7] are also adopted in the field of system control [5]. Hierarchical control methods [2, 3, 4, 5, 10] and objected-orient approaches [1, 6, 9] both contribute to this subject because they provide systematic ways to analyze and to build a control system, and then increase the reusablilty of the components in the programs. In [8], a multi-agent control system were introduced. A new distributed object model has been discussed recently, namely CORBA [11]. The new features of CORBA will be very helpful in integrating diverse production systems.

For the goal of integrating scheduling systems and operatioing systems on the shop floor, we here devise a flexible model of manufacturing system and a flexible control kernel for that model to solve those problems in the domain of shop floor control. The model is called Cooperative Multi-Agent Architecture (CMA) and the control kernel is called the Task Oriented Flexible Automation Kernel (TOFAK) .

2 CMA: Cooperative Multi-Agent Architecture

2.1 Agent Definition

In this model, the basic element is an agent, which is viewed as anything that can perceive its environment through sensors and can act on the environment through effectors.

In automated production systems, an agent can be regarded as a combination of software agent and robotic agent. That is, an agent is a program that has the ability of communicating with its environment and can also control real equipment to produce parts through physical link. In CMA, we define two kinds of agent : tack agent and communication agent. These two kinds of agent are described below.

- **Task Agent:** has both the ability to perform some tasks and the ability to communicate with domain server via TCP/IP.

- **Communication Agent:** links application software with different communication protocol (other than TCP/IP) to domain kernel.

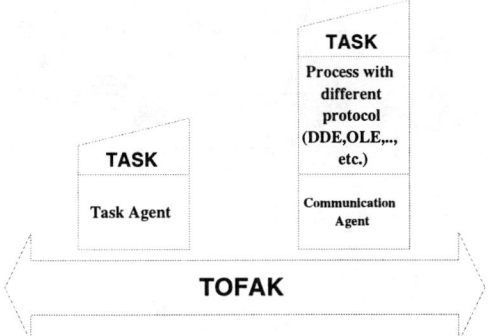

Figure 1: CMA Model

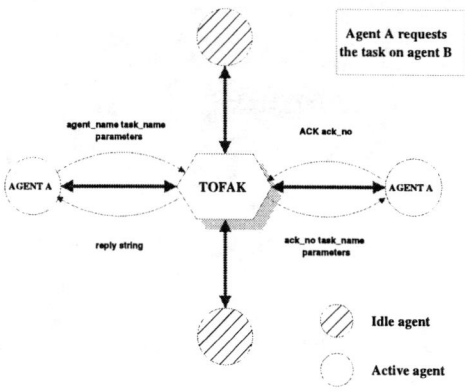

Figure 2: Task Execution Process

2.2 CMA Model

The main goal of this model is to provide a flexible manufacturing system. It is agreed that a successful automated production system should always try to allow use of different operating system (OS). For this reason, our model will allow applications running on different platforms. For example, application under DOS and application under UNIX can work together without any incompatibility. The whole architecture model is shown in Figure.1, which has two basic entities shown as follows:

- **TOFAK (Task Oriented Flexible Automation Kernel):** the task control kernel in the CMA model.

- **Agent:** a program which is able to perform some tasks and to connect the TOFAK.

By Figure.1, there are many agents working together, each agent has its own job function. An agent may need to handle the activity of physical robot arm, managing a vision system, or just a scheduler program. Thus, every agent performs some specific task and changes message with one another, of which all these efforts are to achieve the goal of a production system. As a result, communication network becomes indispensible due to the need. In this research work, the network protocol we choose is the TCP/IP, which is the most popular network protocol and is supported by most of vendors throughout the world.

Because of the above features of the CMA, we can easily integrate many small systems into a new one. In CMA, we can see that all agents are connected to the TOFAK and change messages transparently. The TOFAK is a broker based control kernel, which plays the most important part in this model. All tasks are sent to TOFAK first, and then TOFAK will decide how these tasks should be executed. Despite that, any agent does not need to know the addresses of other agents. Once an agent wants to communicate with another agent, it simply gives TOFAK the name of the target agent and all necessary parameters (Figure.2).

2.3 Useful Properties

In the definition of the CMA model, it can be found easy to integrate legacy systems into a new one, which complies with the CMA model. The legacy automation system can be a system without communication ability or a system with different communication protocol (COM,OLE,or RPC).

Besides, since TOFAK has communication ability, itself is also an agent. Becanse TOFAK can be treated like an agent, it can be connected to an upper level TOFAK, and hence can provide a hierarchical model of CMA, as shown in Figure.3.

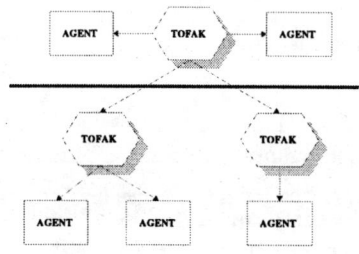

Figure 3: Hierarchical Model of CMA

As we have defined before, each agent should be associated with some kind of task. Therefore, TOFAK also has some task defined on it. An example of the task on TOFAK maybe is to report to the upper level system or to execute some orders from the upper level system.

3 TOFAK: Task Oriented Flexible Automation Kernel

Since the role of TOFAK in CMA model has been introduced in the previous section, a complete description of the architecture of TOFAK will be provided here. The basic architecture of TOFAK is illustrated in Figure 4.

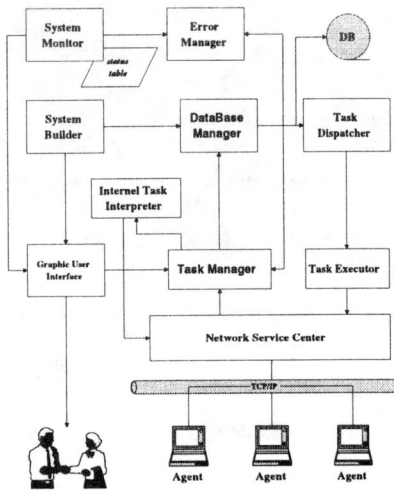

Figure 4: TOFAK Architecture

3.1 CMA Specification

The CMA specification contains all necessary data that TOFAK needs when building a brand new manufacturing environment. It is nothing but a text file such that anybody can tailor it for his own manufacturing cell. There are four sections in a CMA specification : Communication, Agent, Group and System_Task, which are respectively explained below.

Communication Section Because TOFAK is built on TCP/IP, there are certainly some information about the communication part that must be filled in inside the control kernel. For example, the socket port number and the name of this TOFAK.

Agent Section This section is focused on the issues like which agents will join the cell and what tasks can be provided by these agents. An agent which wants to enter this TOFAK will not be accepted if the name of this agent does not appear in this section. The task requested by some agent will be refused if the cell designer does not specify that task in the agent section. Therefore, this section allows one to layout his working environment flexibly in a transparent manner.

Group Section In TOFAK, the system designer can integrate several agents into one group and then can send message to these agents by sending message to this group. Agents in this TOFAK can send information to each group by using the internel task : SENDGROUP.

System_Task Section As has been pointed out earlier, a TOFAK can be an agent itself, and hence some task may be provided by it. This section will be used to implement those tasks if that will be the case. One can combine several internal tasks into a system task. To do so, one must first declare a unique task name and associate it with a list of internal tasks.

Status_report Section The cell designer can make a list of agent names on this section. Then, the system will automatically send agent status to every agent that is specified in this section whenever there is a status change in any of the agents.

3.2 Network Service Center

The Network Service Center (NSC) is the communication part in TOFAK. It takes responsibility of building connection among other components in a CMA environment. That is, it builds connections between agents and TOFAK and between the upper level TOFAK and the TOFAK itself. TOFAK sends and receives requests via NSC. There are two kinds of connections that need to be established. One is the registration request from an agent to TOFAK. When the agent undergoes the registration procedure, NSC will make a logical connection between the agent and TOFAK so that messages can be exchanged.

3.3 Internal Task Interpreter

There are two kinds of task in CMA environment. One is the agent task and the other is the internal task provided by kernel itself. When NSC receives a task request from the agent, it will be processed in two ways. If it is an internal task, then the NSC will send it into Internal Task Interpreter immediately. Otherwise, it will be sent into Task Manager instead.

3.4 Database Manager

Within our system, there are a large amount of data that need to be stored and retrieved such as task information, agent status, product working progress and error message. Since the data set will be very complex, huge and hard to handle, we prefer to design a special element for handling every data set we need in our system. This is what Database Manager needs to serve in TOFAK.

3.5 Task Manager

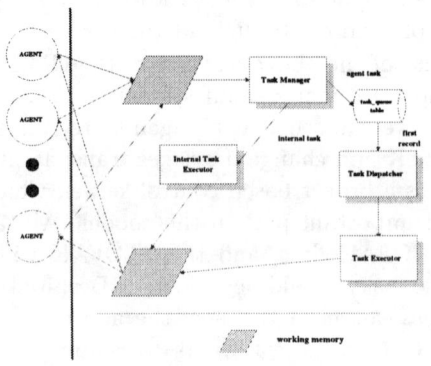

Figure 5: Task Manager

The major task of this component is that collecting all task request in this system and then check its validness and correctness, i.e., whether the task is clearly defined in CMA specification and whether the task carries the right parameters with it. If a task belongs to agent task type, the Task Manager will ask the Database Manager to add this task into the task table. All tasks in task table is indexed by its priority and would be retrived by Task Dispatcher one by one from top to end. The concept of Task Manager is shown in Figure.5.

3.6 Task Dispatcher

The Task Dispatcher takes a task from the task table maintained by the Database Manager if the task matchs these two condition: has the highest priority and the invoked agent is ready for serving this task. If the agent that provides this task is ready for accepting work, then it puts the task to the Task Executor for execution.

3.7 Task Executor

When there is an executable task coming, the Task Executor will find an agent which can perform this task and then invoke this task on remote agent. When the agent is executing this task, the Task Executor simply puts the task into running state and continues the next task invoking. Note that, because all the internal task is sent to the Internal Task Interpreter, the Task Executor only invokes the remote tasks. After an agent finished a task, it would send message to TOFAK to notify the Task Executor. It will move this task into finishing state and drop this task from the task table. The reply message will be sent to the request agent also.

3.8 System Monitor

When there is a task in this shop floor manufacturing system which must be executed, how do we know all resources are ready for it? We need a monitor mechanism to keep track of all system states. The System Monitor just plays this role in TOFAK. It can get the timely information from agents connected to this system and analyze all data captured. If there is a dangeous situation sensed by the System Monitor, it will ask the Error Manager to handle it.

3.9 Error Manager

In real world, there are many problems which may occur from time to time. For example, a collision between two robots occurs or some manufacturing machine is down. There should have some recovery methods provided by the shop floor control kernel. The Error Manager accepts these errors reported from outside agents and System Monitor. The concept of Error Manager is shown in Figure.6.

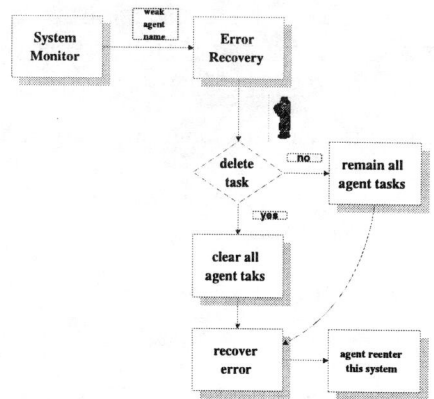

Figure 6: Error Manager Concept

3.10 Graphic User Interface

We provide a windows based interface in order to present how system works, how message is being exchanged, and what kind of problem exists. By this, we can communicate with the operator and get his feedback. The supervisory control with human operator can be implemented easily by using this interface. The difficulty as how to deal with unknown type of error can also be resolved by an operator through this interface.

3.11 Task Management

A task in TOFAK may be a remote agent task or an internal task. No matter what kind of task it will be, the task may stay in one of the five states, namely, new, ready, running, suspend and finishing, Figure 7 shows the state diagram of a task.

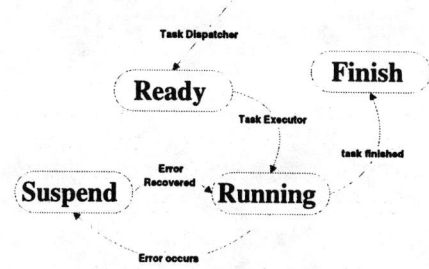

Figure 7: Task State Diagram.

- New : When a task is being created by the System Builder, it is in the new state. If there are some agents making a request on it, it enters the ready state.

- Ready : When a task is added on the task table and is waiting for invoking, it enters the ready state.

- Running : When a task is executed by the Task Executor, this task is in running state.

- Finish : When a task is ended normally, the task enters the finishing state. And, the task will be removed from the task table.

- Suspend : If there are some problems which occur during the stage of task execution, the task enters the suspend state and the Error Manager will handle the error.

3.12 Monitoring Mechanism

Within TOFAK, the function of System Monitor is to supply the necessary information to the agent controlling and task scheduling, so that they can carry out their respective tasks of planning and control. Thus, the role of the System Monitor is to make good use of real-time data collected from agents and internal state. The purpose of this monitor element is to make useful information for supporting system decision. There are three main activities of the System Monitor, namely, status capture, status analysis, and error finding.

3.13 Error Recovery Method

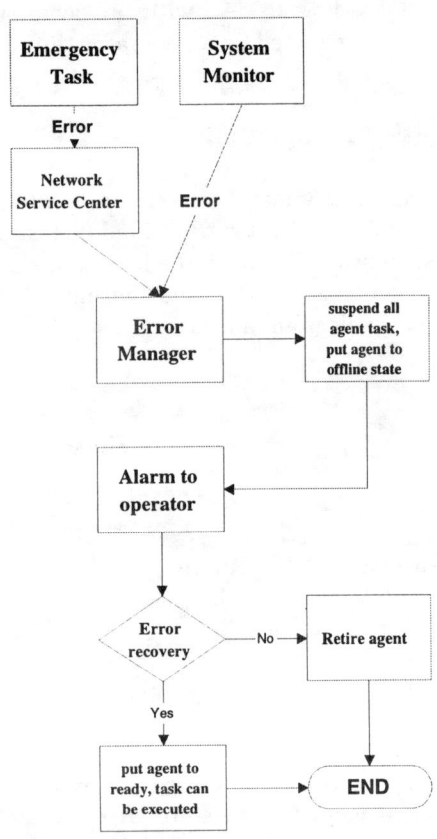

Figure 8: Error Recovery Method

In Figure 8, the main concept on error recovery method in this system is shown. When there are some errors which occur, the Error Manager (EM) will get the description of all these errors and then take the following steps to try to recover these errors.

1. Disable the agent associated with the error. The Error Manager will put this agent into dismissed state so that this agent will no longer be able to perform any task.

2. Send alarm to the operator and wait for operator to recover this error. By using Graphic User Interface, the Error Manager will send an alarm to human operator and inform him of the kind of error that happens to this system.

3. Do the right work according to the operator's reply. If the operator has completely recovered this error, the Error Manager should put this agent into ready state and let all tasks provided by this agent continue their running. But, if the operator reply that this error can not be recovered, then the Error Manager should disconnect this weak agent and delete all waiting tasks needed to be invoked on this agent.

4 Experiment

4.1 System Setup

Figure 9: Cell in Laboratory

In our laboratory, we have a two-robot assembly system that is dedicated to assemble various types of mechanical parts sent serially into the conveyor belt by the part loader as shown in Figure 9. There are two products currently assembled in this system, and each product has four parts that are assembled by the robot manipulator. The operations include vertical insertion, horizontal insertion, and rotation in assembling with the subassembly fixed at the assembly sites. The parts are fed into the system without a specific order, and the scheduling is made on-line. The cell is equipped with several pieces of hardware that work together to assemble parts, they included two robots, a part loader, several CCD cameras, a conveyor belt, a rotary buffer, and an assembly table with several kinds of fixture.

During operations of this assembly system, there are numerous interactions between different components using message passing. For example, when the optical sensor detects a part on the conveyor belt, it signals an interrupt and the associated interrupt service routine sends a message to the PC in charge of overhead camera to take a picture. The PC determines the part's type and orientation and sends a message to an assigned robot to pick up the part. This is a simple example that could appear in this assembly cell, and there are other similar activties concurrently taking place in the system.

4.2 CMA Model Specification

Since we have introduced the experiment environment in out labrotory, we want to make an example for demonstrating how to use the Cooperative Multi-Agent Architecture model and Task Oriented Flexible Automation Kernel. First, in this section, we will describ the CMA model for this case in detail.

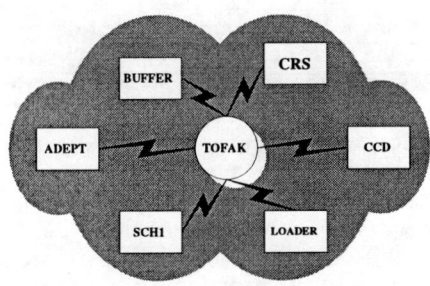

Figure 10: Example of Cooperative Multi-Agent Architecture

Figure.10 shows the abstract model view of whole assembly system. There are several agents designed for this cell. Each agent control one hardware in this cell and make a physical link with the hardware (RS-232 or one-bit signal port). Agents in this model have their own tasks to perform.

4.3 Results

After setting up all agents and the CMA specification file. This cell performs smooth assembly tasks without any problems. This control kernel help us to establish the full control of the cell. We can get runtime information via GUI of this kernel and perform the recovery procedure easily.

5 Conclusion

In this paper, we have proposed a multi-agent based model for an intelligent flexible automated production system. Under this model, every piece of equipment is given as an agent and communication among agents are through network using TCP/IP protocol. In order to realize such model, we further develop a task oriented flexible automation kenel (TOFAK) to establish the necessary message control platform. The present work is successfully demostrated in our intelligent robotic assembly cell in our laboratory. The results are considered extremely useful to expedite the process of creating the shop floor control for any automated production system.

References

[1] C. B. Basnet and J. H. Mize. An object-oriented framework for operating flexible manufacturing systems. In *Proceedings of International Conference on Object-oriented Manufacturing Systems*, pages 346–351, 1992.

[2] D. M. Dilts, N. P. Boyd, and H. H. Whorms. The evolution of control architectures for automated manufacturing systems. *Journal of Manufacturing Systems*, 10(1):79–93, 1991.

[3] Larry Jann and Li-Chen Fu. Flexible control system for robot assembly automation. Master's thesis, National Taiwan University, Department of Computer Science and Information Engineering, 1994.

[4] Yunho Jeon, Jungmin Park, Insub Song, Young-Jo Cho, and Sang-Rok Oh. An object-oriented implementation of behavior-based control architecture. In *IEEE Int. Conf. on Robotics and Automation*, pages 706–711, 1996.

[5] Sanjay B. Joshi, Erik G. Mettala, Jeffrey S. Smith, and Richard A. Wysk. Formal models for control of flexible manufacturing cells: Physical and system model. 11(4):558–570, August 1995.

[6] Li Lin, Masatoshi Wakabayashi, and Sadashiv Adiga. Object-oriented modeling and implementation of control software for a robotic flexible manufacturing cell. *Robotics and Computer-Integrated Manufacturing*, 11(1):1–12, 1994.

[7] John C. Martin. *Introduction to languages and the theory of computation*. McGRAW-HILL, 1991.

[8] Michel T. Martinez. Dynamic assembly sequence - a multi-agent control system. *IEEE Symposium on Emerging Technologies and Factory Automation*, 2:250–258, 1995.

[9] D. J. Miller and R. C. Lennox. An object-oriented environment for robot system architectures. *IEEE Control Systems*, 11(2):14–23, 1991.

[10] J. Borges Sousa and F. Lobo Pereira. A general control architecture for multiple vehicles. In *IEEE Int. Conf. on Robotics and Automation*, pages 692–697, 1996.

[11] Steve Vinoski. Corba:integrating diverse application within distributed heterogeneous environments. *IEEE Communication Magazine*, 1.14(2), February 1997.

Basic Study on a Magnetic Measurement for Balance Utilizing a Spherical Vessel

Tokuji Okada, Kuniyasu Kimura and Nobuharu Mimura

Faculty of Engineering, Niigata University
2-8050, Ikarashi, Niigata, 950-2102 JAPAN

Abstract

This paper proposes a magnetic measurement principle to develop a balance sensor. Basically, the sensor is composed of a spherical vessel sealing certain amount of liquid, a mushroom-shaped float and a permanent magnet. The circular plated magnet is fixed on the root of the float stem. According as the liquid moves in the vessel, the float changes the position of the magnet. And the Hall effect devices located around the vessel sense the change of balance with the aid of a signal processor. The density of the Hall effect devices and the algorithm for determining the direction of the resultant acceleration are considered. Also, the results of experiments for magnet's position and the angle of balance are shown. The proposed method is effective in detecting all directions of the resultant acceleration of motion and gravity with uniform resolution, since the sensor has no rotation axes.

1 Introduction

Such optical and magnetic sensors have been developed so far that can measure motional direction of a ball or magnet moving freely in the sealed spherical vessel. The sensor using a projected image[1] and the sensor using reed switches located around the spherical vessel sealing a magnetic float with magnetic fluid[2] are typical examples. These sensors are adequate to sense all directions, but rather complex in signal processing and not simple in structure.

A method for measuring magnitude or direction of a reactional force using springs are proposed by J. Bozicevic and N. E. Alexander[3~5]. From this motivation, a sphere of heavy weight is suspended by springs so that it can move proportionaly to the resultant of motional and gravitational forces, and the Hall effect devices are utilized to detect its displacement[6].

Magnetic and optical sensors utilizing a vessel are treated in [6~8]. These literatures have such characteristcs that the parts like a spherical ball or vessel arouse a person's interest in a lobster's equilibrium organ since it can sense omnidirectional information without installing specific axes as a sensing device.

In this paper, we describe the sensing structure for balance, the signal processing of electric signals produced by magnetic devices, the optimization of designing the sensor head, the measurement principle, and verification of the principle. In the first half, we introduce the basic scope of the sensor design and the simulation of the magnetic information processing. In the second half, we show the fabrication and assembly of parts for collecting data in basic experiments. Finally, we analyze the measurement error for future improvements.

2 Sensing Mechanism

Such a mechanism is considered that a magnet is sealed in the vessel with liquid so that it can move freely with keeping same distance from the interior wall of the vessel. Obviously, the magnet is fixed on the float. This mechanism is seen in the compass which is easily purchased from the market. The sensor based on this mechanism is available only in bright environment because we can not get directional information in the darkness. If we want to get the information, it is requested to produce electric signals. But we cannot find the compass which makes it possible to use in the darkness in the world at this moment. Even more such a senor is not found that gives electric wires for displaying the position or direction of the magnetic needle or rod of the compass in the brightness. In order to solve this problem, we discuss how the sensing mechanism might be devised and how the electric signals can be produced in the following.

Main parts of the sensing mechanism are divided into a sensor head and an information processing. The processing is enough to run a certain algorithm repetitively. This might be built in a smart module when its function is tested to be perfect. However, in our development process, a personal computer is used for easiness of refining. On the other hand, the sensor head is composed of three parts as shown in Fig.1. First one is a demagnetized spherical vessel of uniform thickness. Second one is a circular mushroom-shaped float. Third one is the Hall effect devices located at outside surface of the vessel with a certain density. From now, we express the Hall effect device and its output simply by Hall device and Hall output, respectively.

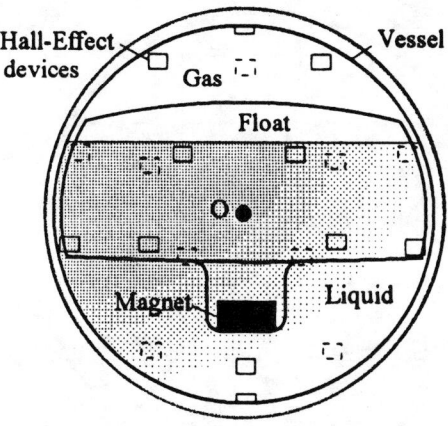

Fig.1 Interior sketch of the balance sensor.

The sensor head is compared to the equilibrium organ of a crayfish. That is, the spherical vessel and the magnet are supposed to be the statocystic and statolithic organs, respectively. The magnet can move freely along the inner wall of the vessel in the same way as the statolith behaves inside the statocyst. The information about the contact between the cilia and the statolith of the crayfish is collected by magnetic media since the magnet is floating without any contact in our design.

As a practical design, non-compressive and demagnetized alcohol is sealed in the vessel for floating up the magnet and for damping the swing motion of the magnet. The alcohol is poured in the vessel so that some amount of air rests and small gap remains between the float and the vessel interior. Therefore, the float can not only move along the inner wall but also rotate by itself in the vessel. Since the circular permanent magnet of a uniform shape is attached on the stem beneath the float, the magnet behaves as a heavy weight to find the direction of the resultant direction of the gravity and the external force operating on the vessel. The concentric spherical surface C in Fig.2 expresses the magnet's motion area. When the magnet's polar axis is normal to the float at its center, the magnet becomes quiet by directing the gravity direction in a steady state condition. In addition, the magnet becomes stable by indexing the pole of the earth as the compass when the polar axis is in the plane parallel to the float plane.

On the exterior of the vessel, Hall devices are fixed with a certain distance. Minimal number of the devices are four in general. Also, there is no rule to fix the devices on interior or exterior of the vessel. But exterior fixture makes the fabrication simple, in general. Fig.2 shows the case when the Hall devices H_i are located at vertices of the circumscribed tetrahedron to the vessel. Evidently, electric wires are connected to each of the devices to collect the Hall outputs which are proportional to the intensity of the magnetic field. The unified vessel with magnet, liquid, float, and Hall devices has no special rotation axis and can measure the intensity of the

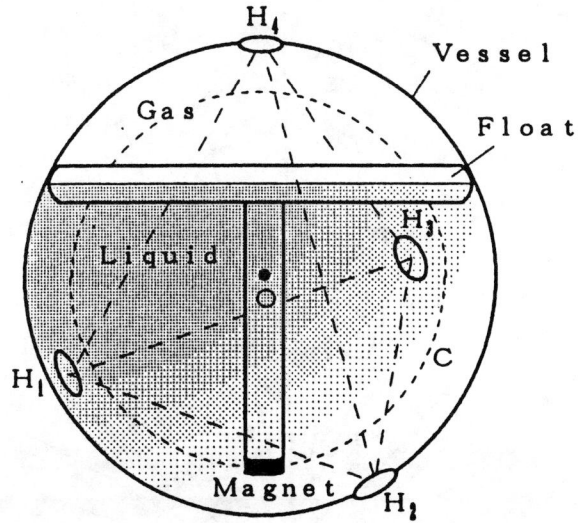

Fig.2 Hall devices located at vertices of the regular tetrahedron and the concentric surface C on which the magnet moves.

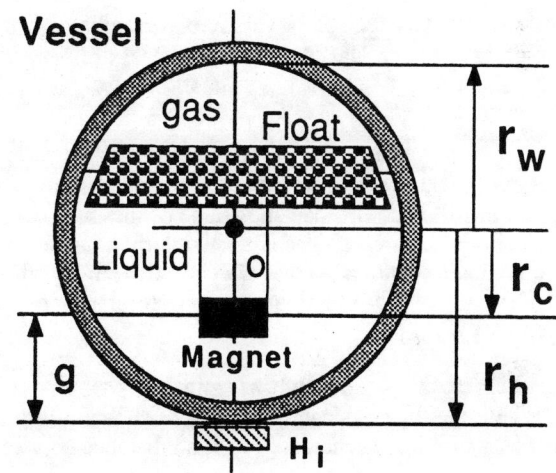

Fig.3 Physical parameters related to the sensor head.

magnetic field as if the Hall devices pursue the running magnet.

3 Measurement Principle

We describe how the Hall output is formulated and how the measurement is performed for sensing the balance.

3.1 Formulation of the Magnetic Potential

The permanent magnet beneath the float is supposed to be uniform and we take advantage of symmetry with respect to its axis. In the analysis of the magnetic field, it is important to extract the mathematical form expressing the magnetic potential in the coordinate system $O(x,y,z)$, where the polar axis of the magnet is in Z axis, and the magnetic surface is in X-Y plane. This

2141

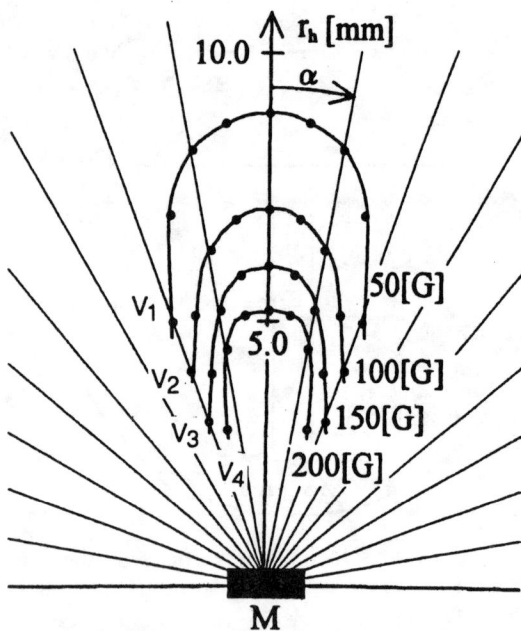

Fig.4 Characteristics of r_h versus α depending on the output V_i of the Hall device. Three cases of V_1, V_2, V_3 are treated, where rates of V_2/V_1, V_3/V_1 and V_4/V_1 are 2, 3, and 4, respectively.

analysis is basically common to that of the magnet shown in [6] since the size and shape of the magnet are quite similar. Principal idea is to divide the magnet into infinitesimal elements so that we can utilize integration formula. Detailed formulations are referred in the literature.

3.2 Calculation of the Magnet Position

We define the physical parameters as shown in Fig.3. That is, r_w; radius of the inner wall, r_c; distance between the magnet and the vessel center O, r_h; distance between the Hall device and the center O, g; gap between the magnet and the Hall device, r_m; radius of the magnet, h; thickness of the magnet. When the Hall output is given, same potential area is found from the analysis shown in section 3.1. The result is shown in Fig.4. Notice that the magnet moves always on the spherical surface C, then the magnet position is found on the common area between the surface C and the same magnetic potential surface which is determined uniquely based on the Hall output. We call the common area *potential circle* and express its radius by the symbol u_i. In the similar fashion, specific set of potential circle G_i and its radius u_i are determined to each of the Hall devices as shown in Fig.5.

If there is no symmetry around axes of the magnet and the Hall devices, the common area is not circular. But in general, most of the elements which are easily available in the market have the symmetry. Thus it is reasonable to take advantage of the symmetry.

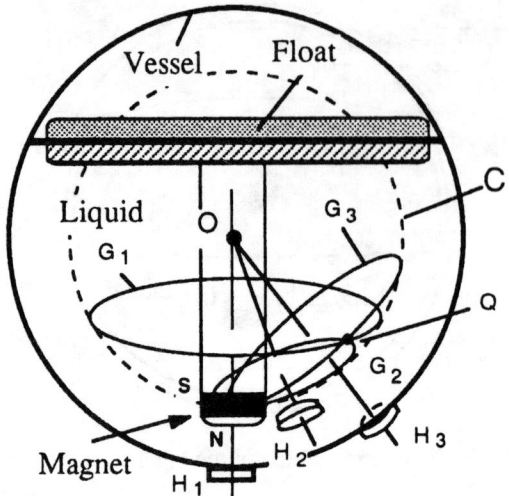

Fig.5 Circles expressing the magnetic potentials for each of the Hall devices.

Each of the potential circles which are obtained from the Hall devices indivisually means the magnet's location. And it is said that the magnet locates closer to the Hall devices when the magnitude of the radius of the potential circle becomes smaller. However, it is difficult to identify the exact position of the magnet. Therefore, we take note of three potential circles G_1, G_{i+1}, G_{i+2} which are collected by closely located Hall devices together. Their cross point is exactly the position of the magnet.

Since the consumption of the magnet intensity is negligible small and the electric characteristics of the Hall device is almost constant, the potential circle is calculated with high reproducibility while the magnet moves so randomly. Thus it becomes possible to determine the position of the magnet uniquely, say (X_m, Y_m, Z_m), by utilizing the three radii u_i (i=1,2,3) and geometrical information about the Hall devices on the spherical vessel. The magnet behaves as a heavey mass as mentioned before. Therefore, we can determine the direction of the external force operated to the vessel as the direction toward which the center of the vessel is observed from the center of the magnet. Also, we can recognize the direction of the gravity in a steady state, otherwise the direction is that of the resultant force of the gravitational and motional accerations.

3.3 Signal Processing

Since the magnet moves with keeping a small distance from the inner wall of the vessel and the Hall device is fixed on the vessel so that its sensitive area faces the vessel center, the radius of the potential circle u_i is a function of the parameter r_c in the form (see Fig.6).

$$u_i = r_c \sin \alpha \qquad (1)$$

The equation shows that the parameters u_i and α are transformable each other and we notice that α is enough

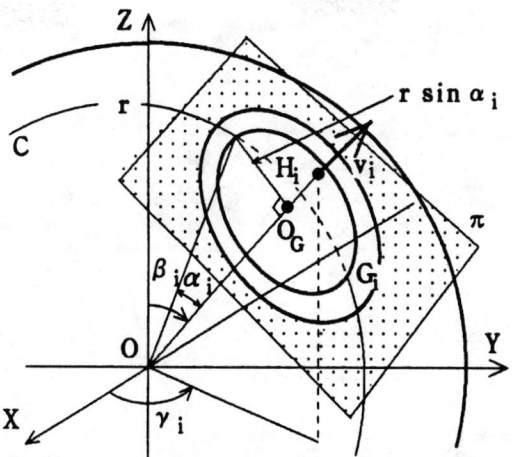

Fig.6 Relationship among α_i and the geometrical parameters β_i and γ_i regarding the Hall device H_i.

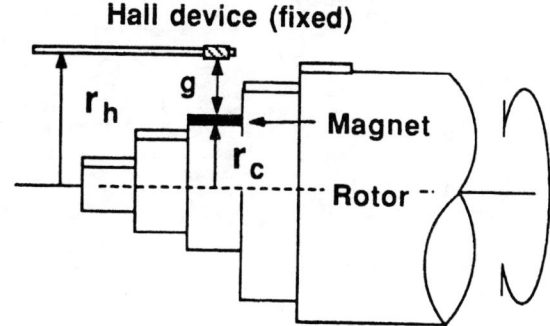

Fig.7 Magnetic measurement set without using the spherical vessel.

to evaluate u_i. In particular, it is difficult to express the same potential surface in a simple form when the surface becomes complicated (see Fig.4). This made us to change our thinking to calculate the position of the magnet not by the cross point of the three potential circles but by the cross point of the three planes, say π_i (i=1,2,3) corresponding to the potential circles G_i (i=1,2,3). This contributes to eliminate calculation of the radius u_i of the potential circle G_i. Suppose that the position of the Hall device H_i is expressed by the parameters β_i; deviation angle from Z axis, and γ_i; rotation angle around Z axis, then we obtain

$$\sin\beta_i\cos\gamma_i\, x + \sin\beta_i\sin\gamma_i\, y + \cos\beta_i\, z = r_c \cos\alpha_i \quad (2)$$

From (2) the form of the plane π_i with respect to the Hall device H_i is given and then the cramer equation makes it easy to find the magnet position (X_m, Y_m, Z_m). For instance, X_m is expressed as

$$X_m = \begin{bmatrix} r_c\cos\alpha_i & \sin\beta_i\sin\gamma_i & \cos\beta_i \\ r_c\cos\alpha_{i+1} & \sin\beta_{i+1}\sin\gamma_{i+1} & \cos\beta_{i+1} \\ r_c\cos\alpha_{i+2} & \sin\beta_{i+2}\sin\gamma_{i+2} & \cos\beta_{i+2} \end{bmatrix} / D \quad (3a)$$

where D is the determinant

$$D = \begin{bmatrix} \sin\beta_i\cos\gamma_i & \sin\beta_i\sin\gamma_i & \cos\beta_i \\ \sin\beta_{i+1}\cos\gamma_{i+1} & \sin\beta_{i+1}\sin\gamma_{i+1} & \cos\beta_{i+1} \\ \sin\beta_{i+2}\cos\gamma_{i+2} & \sin\beta_{i+2}\sin\gamma_{i+2} & \cos\beta_{i+2} \end{bmatrix} \quad (3b)$$

All of the Hall outputs are compared in their magnitudes to find three larger devices for practical use. That is, the smaller radius is selected. This is based on the fact that the larger signals make it possible to find the cross point among three potential circles reliably. The results of the calculation (X_m, Y_m, Z_m) is utilized to obtain

$$\theta = -\cos^{-1} Z_m / r_c \quad (4)$$
$$\phi = -\cos^{-1} X_m / (X_m^2 + Y_m^2)^{1/2} \quad (5)$$

where θ and ϕ are the rotation parameters around the axes of X and Y, related to the direction of the force to the vessel. These meanings are similar to those of the parameters β and γ as shown in Fig.6.

3.4 Simulation for α versus V Characteristics

Preparation of the program for calculating the α versus v characteristics is important to know how the error of the physical parameters makes worse the measurement accuracy and to optimize the number of the Hall devices. Therefore, we simulate the measurement by utilizing the results obtained by the calculation of the magnetic flux intensity.

We used an indexing equipment as shown in Fig.7 for setting the angle α in the experiment, in turn of the float. The Hall device is fixed at a certain distance from the axis of the indexer, while the magnet is rotated manually with keeping the parameter r_c constant. The data collected by using the experimental set is compared with the data obtained by the simulation program. The results are shown in Fig.8, where the parameter r_c is changed. Two figures (a) and (b) show simulated and experimental results, respectively. In Fig.8b, such symbols like ×, ●, ∨, ... are the experimental data. The curves are estimated results. From the two figures, it becomes clear that calculated and experimental results are quite close and also the simulation program is successful.

4 Basic Experiment

Implementation of the sensor head as a compact size without rotation axes, and the experiments by using a personal computer are described.

4.1 Fabrication of the Sensor Head

The sensor is designed based on the structure shown in Fig.1. We selected mechanical and electric parts and assembled them by hand into the sensor of outer diameter 30[mm]. Outlook of the sensor head in our assembling process is shown in Fig.9. Total weight is 10.2gram. Specifications of the sensor head is as follows:

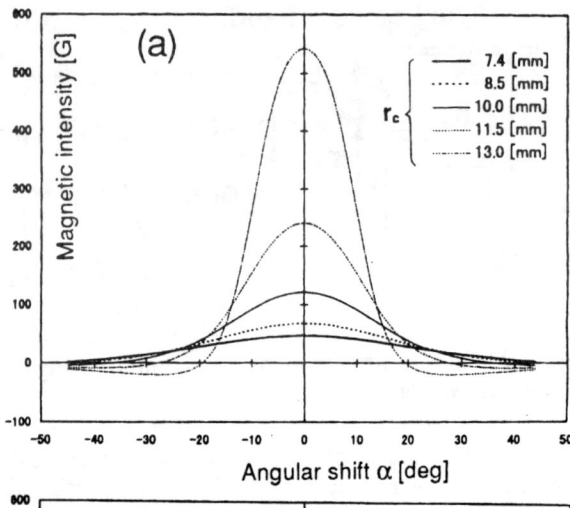

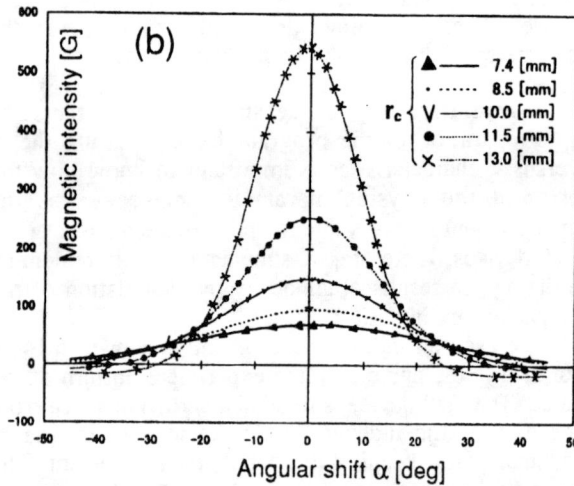

Fig.8 Relation of the parameters α versus r_c under the fixed values such that $r_w=12.7$, $r_h=13.5$, $r_m=2.5$, and $h=3$ in milimeters. (a) and (b) are simulated and experimental, respectively.

Permanent Nd-Fe-B magnet: $5^\phi \times 3^t$, 1.5gram

Plastic vessel (inner & outer diameters): 25.4^ϕ, 28^ϕ

Aluminum float
  (height, weight incl. magnet): 18.0mm, 3gram
  (cap diameter, thickness): 13.6-20^ϕ, 13.0mm
  (stem diameter, length): 7.5^ϕ, 5.0mm

Liquid (type, volume): volatile alcohol, 1.1cm^3

InSb Hall device (WxDxH): 2.1 x 2.1 x 0.6cm^3
(Asahi chemical Co. Ltd, HW-105C)

4.2 Symmetry of Magnet and Hall Device

It is needed to certify that the distribution of the magnet's intensity and the magnetic sensitivity of the Hall device are uniform around their axes. For the first purpose, the magnet is rotated around its axis by keeping axes of both the magnet and the Hall device parallel with a constant distance as shown in Fig.10a. For the second purpose, the axis of the magnet is rotated around the Hall axis as shown in Fig.10b.

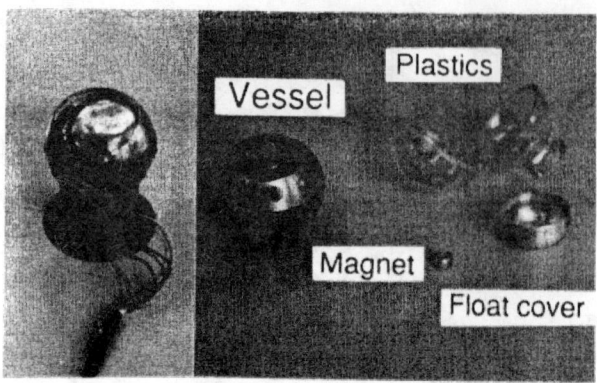

Fig.9 Outlook of the sensor head. Vessel before and after fixing the Hall devices are in (a) and (b), respectively.

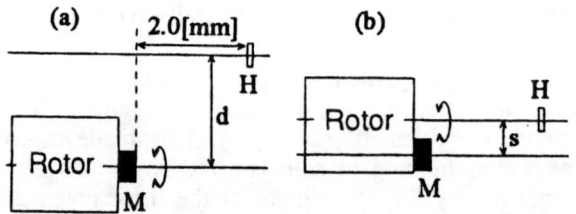

Fig.10 Conditional differences in the experiment regarding the arrangement of elements.

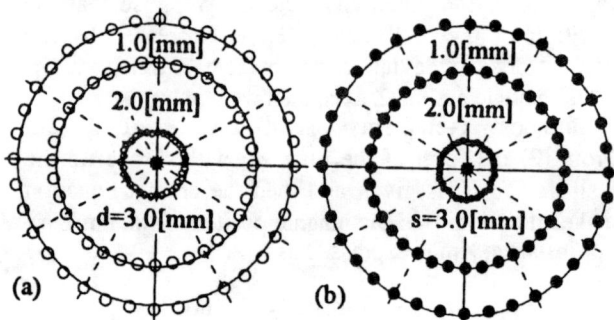

Fig.11 Axial symmetries for the magnet and Hall Effect devices are shown in (a) and (b), respectively.

Results of the experiment under such conditions that the axial distance 2.0[mm] between the magnet and the Hall device, radial distance d, and the radius s of the magnet rotation are shown in Fig.11, where d and s take the values of 1.0[mm], 2.0[mm] and 3.0[mm]. Figs.11a and 11b show the characteristics of the magnet and the Hall device, respectively. In both figures, Hall outputs are displayed in radial directions, i.e. 0.11[v], 0.38[v], and 0.55[v]. Evidently we can confirm that the magnet and the Hall device are almost symmetric around their axes.

4.3 Characteristics of α versus r_c

The change of the physical parameters like r_c, r_h, r_w and the radius r_m and thickness h have direct affects upon the characteristics α versus V. Then we estimate the

Table 1 Number of points for locating Hall devices with uniformity around the spherical vessel. Angular shift between two locations is shown in brackets.

Polyhedron	Surfaces	Vertexes	Edges
Tetrahedron	4 (109.5)	4 (109.5)	6 (36.8)
Hexahedron	6 (90.0)	8 (70.5)	12 (60.0)
Octahedron	8 (70.5)	6 (90.0)	12 (33.6)
Dodecahefron	12 (63.4)	20 (41.8)	30 (40.0)
Icosahedron	20 (41.8)	12 (63.4)	30 (26.9)

measurement resolution and the number of the Hall devices. Above all, the relation about α versus r_c is important since it is deeply concerned with optimization of the sensor design. Therefore, we pay attention to the parameter r_c for investigating the characteristics concerned with the angle α.

In the experiment we set the angle α as the angular displacement of the magnet attached on the body rotating around the line apart from the Hall device. And we get the Hall output V with the angular shift of the magnet from the direction toward the Hall device. When we change r_c under r_h=14.0[mm], the results shown in Fig.8b are obtained. From the figure, it is clear that the Hall output becomes large and the dynamic range of the angle α becomes narrow as the parameter r_c increases. Also it is clear that the angle α becomes wide as the parameter r_c decreases. Similarity in Figs.9a and 9b prove that the simulation and the experiment are successful.

In general, small Hall output V decreases the measurement sensitivity and small angles α makes the Hall device innactive. This makes us to use a lot of Hall devices. Notice that minimal number of the Hall devices is in proportion to r_c and inverse to $\tan(2\alpha)$, in general. The value of the parameter r_c will be determined depending on what we attach importance to, i.e. the size of the sensor head, the number of the Hall devices, or the measurement sensitivity.

Hall devices are recommended to be located at the exterior of the vessel with the same distance between two neighboring devices in order to make the measurement accuracy uniform in all directions. This makes us to locate the devices at points or directions which are seen in the center of planes, vertices, and the middle point of the edges of regular polyhedrons. These candidates are summarized in Table 1. Let suppose that $\alpha \leq \zeta$[rad], then a single device covers the small surface of the vessel which is roughly estimated as $\pi(r_h \sin \zeta)^2$. And the minimal number, say n, of the Hall devices is found from the following relation

$$n \geq 4 / \sin^2 \zeta \qquad (6)$$

Fig.12 shows the minimal number determined by using (6). For instance, 20 devices are needed at least and

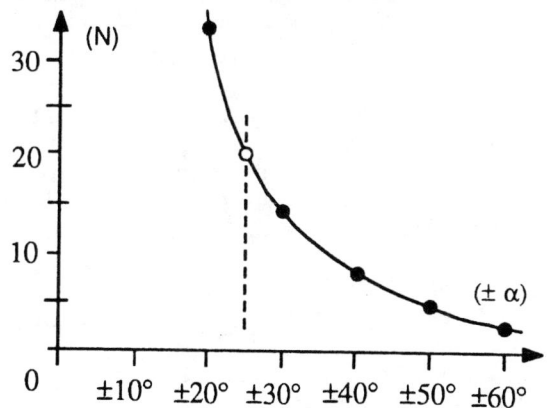

Fig.12 Minimal number of the Hall devices for covering all directions.

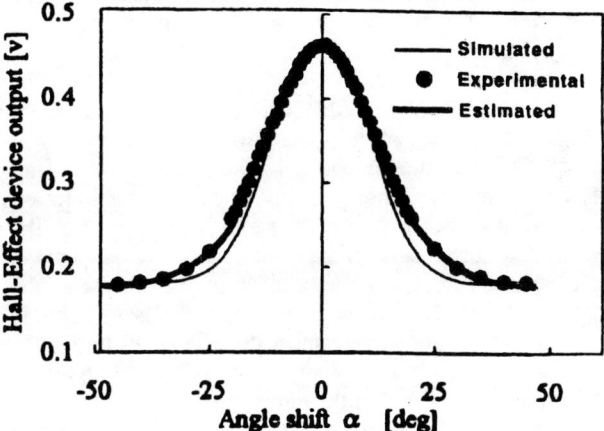

Fig.13 Experimental relations between α versus V under the specifications.

a regular dodecahedron and icosahedron are recommended for their arrangements. If the devices are coarsely arranged, the three potential circles happen to have no cross point, although three planes π_i (i=1,2,3) appear. Thus the number is recommended to be greater than the minimal number, also we must keep in mind that overestimated number makes the processing time heavy for selecting the three devices. In the specification shown in section 4.1, reasonable number is said to be 20 and the implemented sensor head has 20 devices in fact.

4.4 α - V Characteristics of the Sensor Head

When the sensor specifications are given, r_c and r_h become constant and the α - v characteristics is determined uniquely. And the experiment for collecting the data for the characteristics is performed. The result is shown in Fig.13. In the figure, full line, circle and bald line stand for the calculated, experimental and estimated data, respectively. The curve fitting is made by the LSM by assigning a 4th-degree equation.

It is clear from the figure that the curve has a single peak. Of course, in the large value of α which is not shown in Fig.13, the sign of V becomes inversed and the magnitude of V is quite small since the value of r_h is large. This is understandable by the fact that the magnet faces always the center of the vessel.

4.5 Results of the Practical Measurement

The sensor head is tested for the measurement of the overall direction of the external force operating on the vessel. Twenty signal channels are switched by the multiplexer board and the Hall outputs are converted through the A/D converter and processed by the microcomputer PC-98RS NEC. Good calibration showed us that the measurement error is less than 1.8 degrees in all directions.

5 Error Analysis

Errors are supposed to be caused by such factors like 1) rough accuracy in both machining and assembling the electrical and mechanical parts, 2) variations of the electrical characteristics of the Hall devices, and 3) environmental change of temperature, for instance. The first factor is detailed in dimensional error of the interior shape, unevenness of the vessel thickness, eccentricity of the float stem, dimensional error in fixing both the Hall devices and the magnet, unevenness of the junction between two hemispherical vessels, displacement of the magnet's mass center from the axis of the float, and so on. Second factor is mainly caused by the change of the temperature. That is, the Hall output tends to become smaller according as the temperature becomes high. The floating level will also change when liquid expands and contracts depending on the temperature, in general. The third is an external noise like the geomagnetic field. However the earth magnet is quite small as compared with our magnet giving flux density 3500 gauss at its surface. Thus the third error is negligible small.

To cope with the second error, the only thing we can do is to find and select the most appropriate devices of good characteristics. Non-linearity of the Hall device does not matter in our method since the actual relation of α versus V is collected in advance and referred in the experiment. In order to eliminate the first error, many improvements are considered. For instance, the position and direction errors of the Hall device fixation might be decreased by using special tools.

In the error analysis for the calculation of θ and ϕ, the errors $\pi/180$[rad] and $\pi/120$[rad] are estimated for β_i and γ_i, respectively. Notice that the small displacement of β and γ are expressed by $\Delta\beta$ and $\Delta\gamma$, respectively, then the real angular displacement $\Delta\lambda$ is

$$\Delta\lambda = \cos^{-1}(\cos\Delta\beta \cos\Delta\gamma) \qquad (7)$$

Above-mentioned errors are applied to (7) and we get $\Delta\lambda=\pi/100$[rad].

6 Concluding Remarks

The magnetic sensor has been proposed for the measurement of balance. This sensor uses a spherical vessel with the float fixing the permanent magnet, and Hall devices around the vessel. Hall outputs are transferred to the personal computer. The computer calculates the position of the magnet and gives the balance information since the magnet has a role of a heavy weight, that is, the overall direction-of-action operating on the vessel. After optimizing the sensor mechanism, it is designed and fabricated for demonstrating the measurement. In the experiment, basic characteristics of the magnet and the Hall devices are investigated. The experimental results are compared with the simulated data to show both results are quite similar. From the results of the experiment it is confirmed that the measurement error is less than 1.8 degrees in all directions. In the last, error analysis is performed to get some knowledge to improve the measurement acuracy. Actually, it is pointed out that the most effective treatment is in fixing the Hall devices on the vessel with preciseness with the aid of some special tools. The sensing mechanism might be revised to have the function as a compass too. These are our future works. The sensor is intended to be applied to the vehicle in pipe since the vehicle changes its attitude drastically against gravity. Also, some applications to evaluate angular motion of a human body are considered.

Acknowledgments

A part of this study was supported by the Nakatani Electronic Measuring Technology Association of Japan and also by the Japanese grant-in aid for scientific research.

References

[1] T. Okada. An optical sensor for measuring overall direction of action by using a projected image. *J. of Robotics Research*, vol.2, no.3, pp.32-45, 1983.

[2] Catalog of the TDK Co Ltd., Tilt Switch, 1989.

[3] J. Bozicevic. Theoretical foundation of a microprocessor based acceleration transducer. *Proc. of 9th IMEKO World Congress*, S-4.4, pp.1-10, 1982.

[4] N. E. Alexander, W. H. Thompson, and F. X. Mcnally. Spherical anemometer. United States Patent no. 2,959,052, Nov. 8, 1960.

[5] R. R. Segerdahl, and T. Erb. Device for measuring accelerations. United States Patent no. 3,713,343, Jan. 30, 1973.

[6] T. Okada and H. Tsutsui. Measurement mechanism of an overall direction-of-action sensor using a suspended weight. *Proc. of IFToMM-jc Int. Symp. on Theory of Machines and Mechanics*, pp.61-68, 1992

[7] T. Okada and H. Tsutsui. A fundamental study of measuring an overall direction-of-action using a suspended weight. *Proc. of IEEE/RSJ Int. Conf. on Intelligent Robots and Systems*, pp.1087-1094, 1992

[8] T. Okada, Z. Itou and S. Fujiwara. Measurement principles of an overall direction-of-action sensor using optical reflection on a mirror ball in a spherical vessel. *Proc. of the 4th Conf. on Intelligent Autonomous Systems*, pp.334-342, 1995

ELECTROSTRICTIVE POLYMER ARTIFICIAL MUSCLE ACTUATORS

Roy Kornbluh,[1] Ron Pelrine,[1] Joseph Eckerle,[1] Jose Joseph[2]

SRI International
333 Ravenswood Avenue, Menlo Park, California 94025

Abstract

Many new robotic and teleoperated applications require a high degree of mobility or dexterity that is difficult to achieve with current actuator technology. Natural muscle is an actuator that has many features, including high energy density, fast speed of response, and large stroke, that are desirable for such applications. The electrostriction of polymer dielectrics with compliant electrodes can be used in electrically controllable, muscle-like actuators. These electrostrictive polymer artificial muscle (EPAM) actuators can produce strains of up to 30% and pressures of up to 1.9 MPa. The measured specific energy achieved with polyurethane and silicone polymers exceeds that of electromagnetic, electrostatic, piezoelectric, and magnetostrictive actuators. A simple model using linear elastic theory can predict EPAM actuator performance from mechanical and electrical material properties and load conditions. A spherical joint for a highly articulated (snake-like) manipulator using EPAM actuator elements has been demonstrated. A rotary motor using EPAM actuator elements has been shown to produce a specific torque of 19 mNm/g and a specific power of 0.1 W/g. An improved EPAM motor could produce greater specific power and specific torque than could electric motors.

Introduction

Robots, manipulators and unmanned vehicles are increasingly proposed for use in field applications that require a high degree of mobility or dexterity. Such applications might require the ability to traverse difficult terrain or access and manipulate objects within heavily obstructed work spaces. These abilities place stringent requirements on actuator performance. Lightweight and compact actuators that offer sufficient force and stroke in a rapid and controllable manner are needed. For mobile applications where mission duration is an issue, it is also important that the actuators be energy efficient.

The biological world provides numerous examples of creatures that are physically capable of undertaking tasks similar to those proposed for robots and for teleoperated manipulators and vehicles. Indeed many robots and manipulators under development are based on designs inspired by nature. Insect-like legged robot platforms are used to traverse difficult terrain [Shastri 1997]. Highly-articulated snake-like manipulators can access confined areas [Hirose 1993]. Worm-like robots are used for pipe inspection and endoscopy [Aramaki et al. 1995]. The biological analogs to these examples all employ muscle as actuators. Muscle meets the stringent requirements of these difficult applications. Muscle is ubiquitous as an actuator throughout the higher orders of the animal kingdom. The performance of muscle is scale invariant, i.e., independent of size or mass. Thus, we find muscles with similar performance in applications as diverse as moving the legs of microscopic mites and lifting the trunk of an elephant. It follows that an actuator with muscle-like performance would be well suited to a wide variety of robots and unmanned vehicles.

Several researchers have noted the potential of "artificial muscle" actuators for robotic applications. DeRossi and Chiarelli [1994], Hunter and Lafontaine [1992], Pelrine, Eckerle, and Chiba [1992], and Kornbluh, Eckerle, and Andeen [1991] survey technologies used in artificial muscle actuators. These technologies include electromagnetics, mechano-chemical polymers, electrochemomechanical polymers (conducting polymers), piezoelectric and magnetostrictive materials, shape memory alloys and polymers, electrostatics, hydraulics and pneumatics, thermal expansion and thermal phase change, and fuel burning engines. All of these technologies are distinctly different from muscle in certain aspects of their performance and are therefore not well suited to certain applications that require muscle-like actuation.

Frequently, actuator requirements are defined in terms of stroke and force (or torque) requirements. However, the use of transmissions can trade off force, speed and stroke. For example, a small motor can be attached to a lead screw to produce a slow but high-force and high-stroke, linear actuator. Table 1 compares

---

[1] Advanced Automation Technology Center.
[2] Physical Electronics Laboratory.

actuation technologies, independent of any transmission systems. The metrics used in this comparison are energy density (energy output per unit volume) or specific energy (energy output per unit mass). These metrics describe how large or heavy an actuator would be have to be, to perform a given amount of work. When the speed of response of an actuator is an issue, its power density or specific power are also useful metrics for purposes of comparison. Power density and specific power can easily be calculated by multiplying the energy density or specific energy by the frequency of actuation (or by dividing these values by the speed of response).

Electrostrictive polymers and, in particular, electrostrictive polymers with low moduli of elasticity and highly compliant electrodes are used in a relatively new class of actuators that offer overall performance similar in some respects to that of biological muscle. This class of actuators is termed *electrostrictive polymer artificial muscle* (EPAM). Table 1 shows the performance of these actuators compared to that of several other electric actuation technologies and biological muscle. Note that EPAM technology does not provide the best performance according to any one metric. However, EPAM performance closely matches or exceeds that of biological muscle; it therefore follows that electrostrictive polymer actuators might be well suited to many robot, manipulator, and unmanned vehicle applications that require performance similar to that of biological creatures.

In the next section we briefly describe the operation of EPAM actuators, then experimental measurements of the electrostriction of several different polymer materials. Next we describe the potential use of electrostrictive polymer actuators in robots, manipulators, and unmanned vehicles, using two specific examples; a spherical joint actuator for a thin snake-like manipulator and a general purpose rotary motor. Each of these examples exploits the unique muscle-like features of EPAM actuators.

Principle of Operation

Figure 1 shows the principle of operation of an EPAM actuator. A film of an elastomeric polymer acts as an insulator or "dielectric" between two compliant electrodes. When a voltage is applied across the film, the unlike charges in each electrode attract each other, while the like charges in each electrode repel each other. The resulting forces compress the film in thickness and expand its area.

Table 1. Comparison of Actuator Technologies

Actuator Type (specific example)	Max Strain (%)	Max Pressure (MPa)	Max Energy Density (J/cm^3)	Max Efficiency (%)	Specific Density	Relative Speed (full cycle)
Electrostrictive Polymer Artificial Muscle[1]						
Silicone	32	0.21	0.034	90	1	Fast
Polyurethane	11	1.9	0.10	80	1	Fast
Electrostatic Devices (Integrated Force Array[2])	50	0.03	0.0015	>90	1	Fast
Electromagnetic (Voice Coil[3])	50	0.10	0.025	>90	8	Fast
Piezoelectric						
Ceramic PZT[4]	0.2	110	0.10	>90	7.7	Fast
Polymer (PVDF[5])	0.1	4.8	0.0024	90	1.8	Fast
Shape Memory Alloy (TiNi[6])	>5	>200	>5	<10	6.5	Slow
Shape Memory Polymer (Polyurethane[7])	100	4	2	<10	1	Slow
Thermal (Expansion[8])	1	78	0.4	<10	2.7	Slow
Electrochemo-mechanical Conducting Polymer (Polyaniline[9])	10	450	23	<1%	~1	Slow
Mechano-chemical Polymer/Gels (poly-electrolyte[10])	>40	0.3	0.06	30	~1	Slow
Magnetostrictive (Terfenol-D, Etrema Products[11])	0.2	70	0.025	60	9	Fast
Natural Muscle (Human Skeletal[12])	>40	0.35	0.07	>35	1	Med

1. Source: Pelrine, Kornbluh, and Joseph [1998].
2. Source: MCNC web site: http://www.mcnc.org/HTML/ETD/EMAD/ifa/ifa.html
3. These values are based on an array of 0.01 m thick voice coils, 50% conductor, 50% permanent magnet, 1 T magnetic field, resistivity of 2 ohm-cm, and 40,000 W/m^2 power dissipation.
4. PZT B, at maximum electric field of 4 V/μm, using data from Moulson and Herbet [1990], p. 293.
5. PVDF, at maximum electric field of 30 V/μm. Source: AMP literature, AMP Inc. Valley Forge, Pennsylvania, USA
6. Source: Hunter et al. [1991].
7. Source: Tobushi, Hayashi, and Kojima [1992].
8. Aluminum, using a temperature change of 500°C.
9. Source: Baughman et al. [1990].
10. Source: Hunter and Lafontaine [1992].
11. Source: Edge Technologies literature, Edge Technologies, Ames Iowa, USA
12. Source: Hunter and Lafontaine. [1992.

Most robotic actuators would be somewhat more complex than the basic element shown in Figure 1. Typically, several layers of polymer are stacked to produce sufficient force without the extremely high operating voltages that a single thick layer would require. Whether a single layer of polymer film or a multilayer stack is used, the basic element of Figure 1 can be incorporated into several different actuator configurations, several of which are shown in Figure 2. Note that the configurations are analogous to piezoelectric actuators. However, unlike piezoelectric devices, the stroke of EPAM actuators can be a significant fraction of their length (up to 30%). Thus, EPAM actuators can be used as linear actuators without the need for motion-amplifying transmissions. Also, unlike piezoelectric ceramics, the EPAM materials are extremely flexible and can be rolled into a cylindrical shape, as shown in Figure 2.

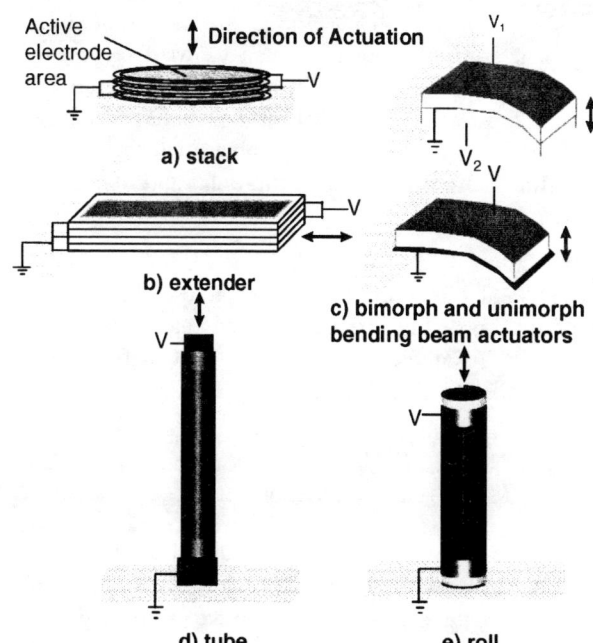

Figure 2. Possible Configurations of EPAM Actuators

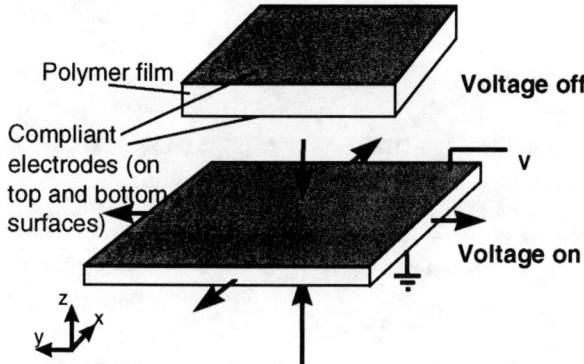

Figure 1. Principle of Operation of an EPAM Actuator

The performance and controllability of EPAM actuators can be predicted well by means of a relatively simple model. The effective compressive force per unit electrode area exerted by the electrodes on the polymer film can be calculated via the principal of virtual work. This approach assumes that the mechanical work done in deforming the dielectric to an infinitesimal degree is equal to the change in the electrical energy stored in the electric field of the device. The details of this derivation have been presented elsewhere [Pelrine, Kornbluh, and Joseph 1998]. If we assume that the electrodes are much more compliant than the polymer film itself, then this effective pressure, p, generated on the film can be expressed by

$$p = \varepsilon_r \varepsilon_o E^2 \qquad (1)$$

where ε_r is the relative dielectric constant of the polymer, ε_o is the permittivity of free space, and E is the electric field resulting from the voltage applied across the film.

Equation 1 is based on the assumption that the forces acting on the polymer film arise from the coulombic attraction of the free charges on the electrodes. Other researchers have proposed alternative mechanisms of electrostriction that generate forces in excess of those produced by coulombic attraction (e.g., Zhenyl et al. [1994]; Shkel and Klingenberg [1996]). However, these studies consider electrostriction at relatively low field levels. Our experiments have shown that at high electric fields (approaching the breakdown strength of the polymer film), coulombic forces dominate [Pelrine et al. 1997].

Equation 1 can be used to calculate the maximum force that material of a given cross-sectional area can produce with a given applied electric field. The amount of deformation of the EPAM material, which determines the stroke of the actuator, depends upon the loading on the actuator. Although the strains in the material can be quite large, we will approximate the materials as linearly elastic for purposes of illustration. In most of the actuator configurations shown in Figure 2, we can ignore mechanical constraints on the deformation of the film. In such cases, the strain in thickness of a single layer of film is

$$s_z = (-p - p_{z,load})/Y - 0.5 p_{x,load}/Y - 0.5 p_{y,load}/Y \qquad (2)$$

where p_{load} is the pressure on the EPAM material due to the load on the actuator, and Y is the Young's modulus of the material. We have assumed that the

2149

polymer is incompressible and thus has a Poisson's ratio of 0.5. Similar equations can be written for the strain in the plane of the film by using a generalized Hooke's law. A different set of equations could be developed for the partially constrained deformation of the film that occurs in unimorph and bimorph configurations.

By combining Equations 1 and 2 we can calculate the electric field that must be applied across the EPAM material in order to produce a given load at a given stroke. For example, the equation that defines the performance of a rolled actuator is

$$\Delta l = l\,(0.5 p - f_{load}/wt)/Y$$
$$= l\,(0.5\varepsilon_r \varepsilon_o E^2 - f_{load}/wt)/Y \qquad (3)$$

where l and w are the length and width of the film, t is the film thickness, f_{load} is the axial force pushing against the actuator, and Δl is the stroke of the actuator. This force vs. stroke performance of a rolled EPAM actuator material is shown graphically in Figure 3. To avoid buckling, a rolled actuator can be used with a return spring (or be antagonistically paired with another rolled actuator). Also shown in Figure 3 are the load line for a spring or opposing EPAM actuator, and a constant load, which would be experienced, for example, when lifting a weight.

The simplest type of loading involves an unloaded and unconstrained actuator. In this case (ignoring viscoelastic losses), all of the electrical energy that is converted to mechanical work causes deformation of the material itself. It may seem at first that this loading condition is not realistic for most robotic applications because the actuator cannot produce any external force. However, the energy of deformation is elastic, so that this energy could be recovered and later used for external work when the actuation voltage is removed. This loading condition is a good benchmark for comparisons of the actuation performance of different polymer materials and comparisons of EPAM with other actuation technologies. In the simple loading case we are discussing here, the corresponding volumetric strain energy density, e, of the deformed polymer material is

$$e = ps_t/2 = [\varepsilon_r \varepsilon_o E^2]^2 / 2Y = p^2/2Y \qquad (4)$$

Equations 3 and 4 can provide us with estimates of the force, stroke, and total energy output for an EPAM actuator of any given cross-sectional area and length. Dividing the volumetric strain energy density by the density of the polymer material gives the specific energy (energy per unit mass) of the material. The specific power is simply the specific energy multiplied by the rate of actuation.

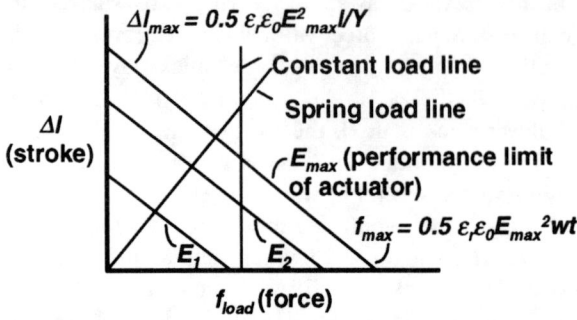

Figure 3. Rolled Actuator Force vs. Stroke

Performance and Fabrication of Electrostrictive Polymer Materials

We have identified a number of polymer materials that are capable of producing useful pressures, strains, and total energy densities. Table 2 is based on data from Pelrine, Kornbluh and Joseph [1998] and summarizes the experimentally measured maximum performance of some of these materials. As noted, the strains produced can be quite large. We have produced strains of > 30% in silicone rubber (polydimethylsiloxane) and strains of > 10% in a number of different polymers. The greatest pressure and energy density was achieved with polyurethane; however, such maximum values are not easily reproducible. Silicone rubber consistently gives a high energy density. The strains and pressures given in Table 2 should be considered the maximum values achievable under ideal conditions at the maximum sustainable electric field for each material. In practice, many factors will diminish actuator performance. These factors include viscoelastic losses in the polymer and electrodes, stiffness of the electrodes, and limitations on the applied field due to variations in the thickness or quality of the film.

The speed of response of an electrostrictive polymer actuator is limited at the most basic level only by the speed of sound across the polymer and the electrical impedance of the actuator and driving electronics [Pelrine, Kornbluh, and Joseph 1998]. We have measured pressure rise times of < 4 ms in linear actuators. We have also observed that EPAM actuators can produce sound at frequencies of at least 17 kHz; thus, the maximum speed of response may be < 1 ms. In many cases, the speed of response will be limited by the resonant modes of the actuator and driven mass. Note that in some applications, such as a rotary motor, it may be desirable to drive the actuator at resonance.

Resonant frequency can be simply calculated by using the spring rate found in Equation 3.

Table 2. Measured Electrostrictive Performance of Various Polymer Materials

Polymer (Specific type)	Energy Density (J/cm^3)	Pressure (MPa)	Strain (%)	Young's Modulus (MPa)	Electric Field (V/μm)
Polyurethane Deerfield PT6100S	0.10	1.9	11	17	160
Silicone Dow Corning Sylgard 186	0.034	0.21	32	0.7	144
Fluorosilicone Dow Corning 730	0.019	0.070	28	0.5	80
Fluoro-elastomer LaurenL143HC	0.0080	0.20	8	2.5	32
Polybutadiene Aldrich PBD	0.011	0.19	12	1.7	76
Isoprene Natural Rubber Latex	0.0052	0.094	11	0.85	67

EPAM actuator efficiency is quite high. The polymers generally have high volume resistivity, so losses during actuation are primarily due to viscoelastic damping in the polymer. These losses are generally around 5% for silicone rubber at low actuation frequencies. At a frequency of 200 Hz, these losses are approximately 20%. For many applications, efficiencies of > 80% should be feasible.

The most critical step in the fabrication of EPAM actuators is the fabrication of the polymer film. The film thickness must be uniform, in order to keep the electric field constant throughout the film and avoid areas where electrical breakdown would occur. The polymer films are typically fabricated by spin coating. Dip coating is used to produce tubular actuators. Films as thin as 2 μm have been produced although most actuators use films in the 10—100 μm range, requiring operating voltages of up to several thousand volts for maximum performance. The reduction of operating voltage by using thinner films is an area of ongoing research. Note, however, that the average electric current is extremely low (e.g., at 1 W of average power, the average current is just 0.5 mA for an operating voltage of 2,000 V). Such low current operation is inherently efficient because it results in lower losses, due to otimic heating (i^2r losses) in the wiring and electrodes.

A number of materials can be used to form the compliant electrodes. Powdered graphite that is brushed onto the film through a stencil offers sufficient conductivity but can slowly flake off the film. (Graphite was used for the measurements in Table 2.) Polymers filled with very fine conductive particles, such as carbon black, are more durable but constrain the motion with some added stiffness. These conductive polymer materials are applied by spraying or dipping.

Actuator Design and Performance

EPAM actuators can be used in robots in a variety of ways. The large stroke capability of EPAM actuators allow them to be used directly as linear actuators in much the same way as muscle is used in biological creatures. Figure 4 shows an example of this approach. Two rolled actuators are attached across the elbow joint of a 1:7 scale model of a human skeleton. While this example is intended mainly to demonstrate the similarities to biological muscle, the configuration is similar to that which might be used for applications such as moving the legs of insect-like robots or the fingers of a highly articulated hand. A typical rolled actuator for such applications has been fabricated, using silicone rubber as the electrostrictive polymer material. The actuator weighs 0.25 g (including connectors), and has an active length of 15 mm and a diameter of 2 mm. Despite its small mass and size, the actuator can produce more than 15 g of force and has a 1.5 mm stroke.

Spherical Joint Actuator

As noted above, a snake-like or serpentine manipulator poses difficult requirements for conventional actuator technologies. By analogy to snakes, worms, elephant trunks, or tentacles, muscle-like actuators are clearly appropriate for such a manipulator. A serpentine manipulator might be used to inspect and perform tasks in cluttered environments or reach inside an object for inspection. A typical serpentine manipulator consists of many links connected in series by spherical joints. The abilities of such a manipulator to reach around obstacles can be improved by maximizing the number of links, maximizing the range of motion of each link, and minimizing the length of each link, in order to increase the curvature that the manipulator can achieve. With a large number of actuators distributed along the length of the manipulator, it is important to minimize the mass of each actuator. Since it is also desirable to minimize the diameter of the manipulator, the size of the actuators is important as well. Due to their high specific energy and energy density, EPAM actuators are well suited to serpentine manipulators.

such a stroke. The manipulator shown in Figure 6 is tapered distally because the more distal joints do not need to support the weight of the links closer to the base.

If we assume that the cross-sectional area of each actuator element is r^2, then the maximum torque that can be produced at each joint is

$$T = 2 f_{load} r = 2pr^3 \qquad (6)$$

We can calculate the quasi-static lifting capabilities of a manipulator using EPAM technology, by applying Equation 6 and the data of Table 2 to a manipulator modeled as a straight beam supported at the base. For example, assuming that the mass of the manipulator is dominated by the mass of the actuators, a tapered manipulator 1 m long with a cross-sectional area of 25 cm^2 at the base could lift a payload of 100 g at full extension, using silicone rubber EPAM actuators. This calculation suggests that the EPAM technology is appropriate for serpentine manipulator tasks involving small reaction forces at the end effector, such as inspection.

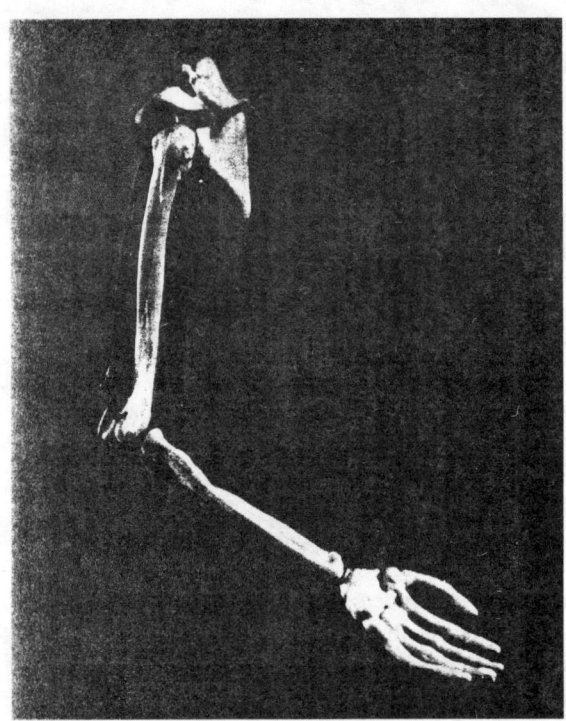

Figure 4. EPAM Roll Actuators Used as Linear Actuators on a Scale Model of a Human Arm

Figure 5 shows a spherical joint based on EPAM roll actuators. Three actuators are arranged in a triad about a central spine. The actuators are preloaded in tension. A pivot joint with a flexible shaft coupling (or a universal joint) allows the spine to bend in any direction and to resist torsion. Proportional control of the voltage to each actuator enables the pivot joint to bend in any direction. A minimum of three actuators are used to provide this motion capability, without the use of opposing springs. Thus, almost all of the mass in each link contributes to actuation. The use of three actuators also enables each link to be extended in length if the spine can telescope.

A manipulator configured as shown in Figure 5 can achieve a local radius of curvature, given by

$$R = \tfrac{1}{2}\, rl/\Delta l \qquad (5)$$

where r is the distance from the center of the spine to the attachment point of the actuator (the radius of the manipulator must be greater than r). The joint in Figure 5 is about 6 cm long and 4 cm in diameter. Figure 6 shows a kinematic model of a portion of a manipulator constructed with EPAM spherical joints. This figure illustrates the curvature that can be achieved if the stroke of each actuator is just 10% of its length. We have used a proof-of-concept joint with silicone rubber roll actuator elements to demonstrate

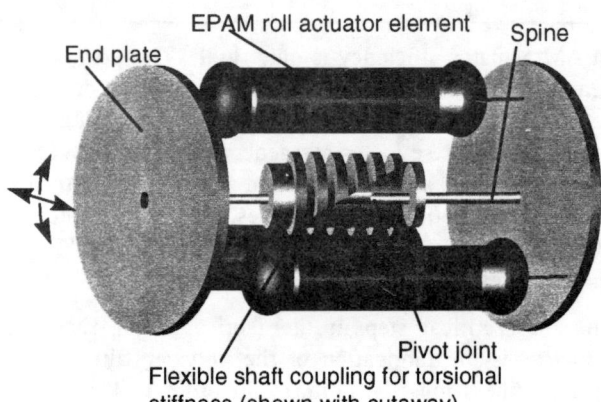

Figure 5. Spherical Joint Actuator for a Serpentine Manipulator

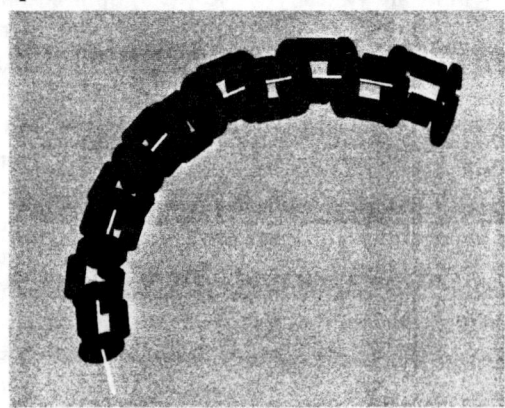

Figure 6. Kinematic Model of Serpentine Manipulator with Linked Spherical Joints

Rotary Motor

In certain applications, the deformation of the material in an EPAM actuator is not sufficient to produce the desired stroke. Other applications may require continuous rotary motion. A motor that uses repeated deformations of an EPAM actuator element can meet these requirements. The high specific energy of an EPAM actuator, combined with its rapid speed of response, suggests that the EPAM technology can produce motors with high specific power.

Figure 7 shows a simple rotary motor that converts the linear motion of an EPAM rolled actuator to rotary motion. The rolled actuator causes a rocker arm to reciprocate. The rocker arm is coupled to the output shaft through a one-way clutch that engages with the shaft in only one direction of rotation. Our proof-of-concept device uses a commercially available one-way roller clutch with a locking sprag mechanism. To enable the direction of rotation of the motor to be reversed, the clutch would have to be an active device such as an electromagnetic clutch or magnetic particle clutch; alternatively, a separate EPAM actuator element could be used to actuate a specially designed clutch. The motor requires a load on the output in excess of the frictional drag of the clutch, in order to produce motion in one direction. The output shaft can also be loaded by an inertial mass that produces motion when the motor is driven at higher frequencies. The output power is maximized by driving the motor at the frequency at which the EPAM element (coupled to the inertial mass) experiences its first longitudinal resonance.

Our proof-of-concept rotary motor was driven by a 0.25 g EPAM rolled actuator with an active length of 15 mm. The motor produced a maximum output speed of 110 rpm and a maximum torque of 1.5 mNm. The maximum output power was about 9 mW. This power was achieved by driving the EPAM element at its resonant frequency of 90 Hz. The mass of the active portion of the EPAM actuator element driving this motor was approximately 0.08 g. Therefore, the specific power of the EPAM actuator element is roughly 0.1 W/g. This value compares favorably with the power achieved by most electric motors. The best rare-earth magnet electric motors can produce a specific power of about 0.5 W/g. While the best electric motors exceed the performance of our proof-of-concept device, it should be noted that our motor is a simple design that is far from optimal.

Our measurements of the EP material capabilities indicate that we can expect more than an order of magnitude improvement as more is learned about the important design parameters and configurations. To estimate the maximum achievable performance, we note that the EPAM actuator element was operated at 90 Hz. Using the full 0.034 J/g maximum specific energy measured for the silicone material in the actuator element (and ignoring the mass of the electrodes and inactive portions of the device), the specific power at 90 Hz would be over 3 W/g. A thicker roll would have a higher resonant frequency, so that the maximum theoretical specific power could be even greater. Allowing for inefficiencies in the design, the mass of the inactive portions of the device, and a safety margin below the maximum output of the material, a specific power of 1.0 W/g should be achievable.

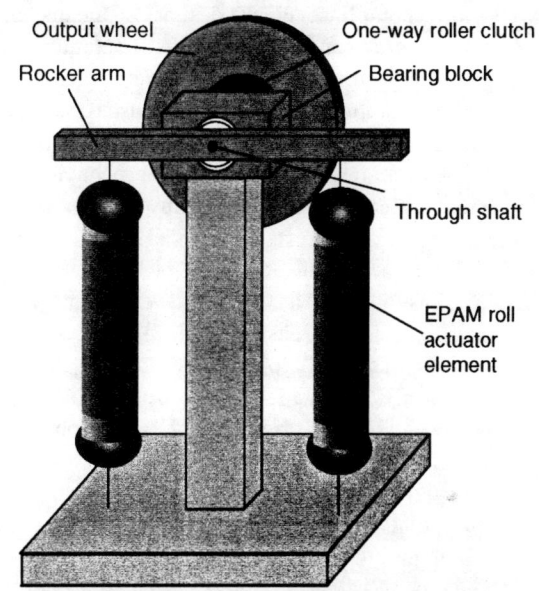

Note: Two EPAM elements are shown. One could be replaced by a passive spring element

Figure 7. Simple Rotary Motor Based on EPAM Actuator Elements

The specific torque of the motor is 19 mNm/g, considerably greater than that of most direct-drive electromagnetic motors and rotary actuators.

Rotary motors based on piezoelectrics are increasingly used in a variety of small mechanisms. Most piezoelectric motors drive the piezoelectric elements at ultrasonic frequencies. The specific power of piezoelectric motors can exceed that of electromagnetic devices and is comparable to that projected for an EPAM motor. The specific torque of these motors is also similar to that of our EPAM motor. While the performance of piezoelectric and EPAM motors is comparable, EPAM motors have some advantages for robotic applications. Since the EPAM elements can undergo large deformations, EPAM motors do not require the precision components that must be used in piezoelectric motors. Therefore, EPAM motors can be manufactured at lower cost from a greater variety of materials. Initial tests with rolled EPAM actuator suggest good long-term reliability and performance. Piezoelectric elements are, however, subject to fatigue failure as well as performance degradation due to aging of the material. Problems associated with cost and fatigue cracking of the piezoelectric elements have prevented the construction of large piezoelectric motors.

Summary and Conclusions

We have discussed the electrostriction of elastomeric polymers with compliant electrodes as a means of actuation. The performance of such EPAM actuators is unlike that of any other electrically powered actuator and is similar to that of biological muscle. These actuators are promising for many biologically inspired robotic and teleoperated applications. The behavior of EPAM actuators can be understood by means of a relatively simple electrostatic model in which electrostriction arises from the coulombic attraction of the free charges on the electrodes. Several materials demonstrate high energy density capabilities. Proof-of-concept devices were built to demonstrate the unique capabilities of EPAM actuators. A spherical joint actuator showed how a lightweight and compact actuator with large stroke capabilities could be used for highly articulated mechanisms. A rotary motor demonstrated that EPAM technology can produce a specific power comparable to that of electric motors. More development is needed, to improve the motor performance and design in order to develop a lightweight and low-cost motor with performance exceeding that of electromagnetic and piezoelectric motors. Such a motor would be expected to find widespread usage in robotics, teleoperation, and unmanned vehicles. Possible disadvantages of EPAM actuators include the need to operate at relatively high voltages and possible dynamic control issues for high-speed operation, due to the inherent compliance of the EPAM materials. Further research is also needed to produce larger actuators for applications requiring larger forces and motions.

Acknowledgments

Much of this work was supported by the U.S. Naval Explosive Ordnance Disposal Technology Division and the Office of Naval Research. The authors would also like to thank the many individuals at SRI whose work contributed to the results presented in this paper.

References

Aramaki, S., S. Kaneko, K. Arai, Y. Takahashi, H. Adachi, and K. Yanagisawa. 1995. "Tube Type Micro Manipulator Using Shape Memory Alloy (SMA)," *Proc. IEEE Sixth International Symposium on Micro Machine and Human Science*, Nagoya, Japan, pp. 115—120.

Baughman, R., L. Shacklette, R. Elsenbaumer, E. Pichta, and C. Becht. 1990. "Conducting Polymer Electromechanical Actuators," in *Conjugated Polymeric Materials: Opportunities in Electronics, Optoelectronics and Molecular Electronics*, eds. J.L. Bredas and R.R. Chance, Kluwer Academic Publishers, The Netherlands, pp. 559—582.

De Rossi, D., and P. Chiarelli. 1994. "Biomimetic Macromolecular Actuators," *Macro-Ion Characterization*, American Chemical Society Symposium Series Vol. 548, Ch. 40, pp. 517—530.

Hirose, S. 1993. *Biologically Inspired Robots: Snake-like Locomotors and Manipulators*, Oxford University Press, New York.

Hunter, I.W., and S. Lafontaine. 1992. "A Comparison of Muscle with Artificial Actuators," *Technical Digest of the IEEE Solid-State Sensor and Actuator Workshop*, Hilton Head, South Carolina, pp. 178—185.

Hunter, I., S. Lafontaine, J. Hollerbach, and P. Hunter. 1991. "Fast Reversible NiTi Fibers for Use in MicroRobotics," *Proc. 1991 IEEE Micro Electro Mechanical Systems— MEMS '91*, Nara, Japan, pp. 166—170.

Kornbluh, R., J. Eckerle, and G. Andeen. 1991. "Artificial Muscle: The Next Generation of Robotic Actuators," SME Paper MS91-331, presented at the Fourth World Conference of Robotics Research.

Pelrine, R., J. Eckerle, and S. Chiba. 1992. "Review of Artificial Muscle Approaches" (by invitation), in *Proc. Third International Symposium on Micro Machine and Human Science*, Nagoya, Japan.

Pelrine, R., R. Kornbluh, and J. Joseph. 1998. "Electrostriction of Polymer Dielectrics with Compliant Electrodes as a Means of Actuation," *Sensor and Actuators A: Physical 64*, pp. 77—85.

Pelrine, R, R. Kornbluh, J. Joseph, and S. Chiba. 1997. "Electrostriction of Polymer Films for Microactuators," *Proc. IEEE Tenth Annual International Workshop on Micro Electro Mechanical Systems*, Nagoya, Japan, pp. 238—243.

Shastri, S.V. 1997. "A biologically consistent model of legged locomotion gaits," *Biological Cybernetics*, Vol. 76, pp. 429—440.

Shkel, Y., and D. Klingenberg. 1996. "Material Parameters for Electrostriction," *Journal of Applied Physics*, Vol. 80(8), pp. 4566—4572.

Tobushi, H., S. Hayashi, and S. Kojima. 1992. "Mechanical Properties of Shape Memory Polymer of Polyurethane Series," *JSME International Journal*, Series I, Vol. 35, No. 3.

Zhenyl, M., J.I. Scheinbeim, J.W. Lee, and B.A. Newman. 1994. "High Field Electrostrictive Response of Polymers," *Journal of Polymer Sciences, Part B—Polymer Physics*, Vol. 32, pp. 2721—2731.

Development of a Distributed Actuation Device Consisting of Soft Gel Actuator Elements

Satoshi TADOKORO   Satoshi FUJI   Mitsuaki FUSHIMI
Ryu KANNO   Tetsuya KIMURA   Toshi TAKAMORI
Department of Computer and Systems Engineering, Kobe University
1-1 Rokkodai, Nada, Kobe 657 Japan
tel.: +81-78-803-1195, fax.: +81-78-803-1217, e-mail: tadokoro@in.kobe-u.ac.jp
Keisuke OGURO
Osaka National Research Institute, AIST

Abstract

It is desirable that soft objects like organs are manipulated by soft actuation devices. This paper presents a newly developed device using a large number of actuator elements made of a soft gel material, which applies a distributed driving force on objects. The material used in the device is ICPF and the structure of the elements is EFD. Experimental results showed that an object placed on such a device could move at a speed of 0.62 mm/sec. This paper refers also to fundamental experiments and a concept for a more efficient drive.

1 Introduction

Softness of end-effectors is important in manipulation of soft objects like organs, many food materials, etc.

This softness can be actualized using two approaches: 1) drive by hard actuators with soft attachments, and 2) direct drive by soft actuators by themselves. The former looks to be a sure method because present technologies cover much of the development. However, in order to create micromachines or compact machines like miniature robot hands, the limitations of the former are so strict and it is difficult to find a breakthrough. The latter has the problem that a readily available soft actuator material does not exist. However, the material revolution currently underway will surely result in the discovery of an appropriate material in near future. For these reasons, it is meaningful to study methodologies for effective use of such materials for manipulation before their discovery with an eye to future applications.

A promising candidate for such a soft actuator material is gel. Many gel materials for actuators have been studied up to the present. An ICPF (Ionic Conducting Polymer gel Film) [9] is a new material that is closest to satisfying the requirements for our applications.

Because such materials are soft, it is impossible to apply large forces/moments at only a few points on an object contrary to the case with conventional robot manipulation. At the same time, however, it is an advantage that large pressures cannot be applied actively or passively. So as not to detract from this feature, a

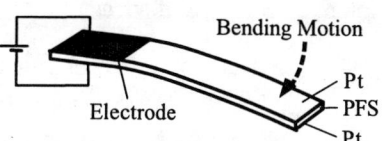

Fig. 1 ICPF actuator.

number of actuator elements should be distributedly used for applying the driving force.

The distributed drive is desirable also from the viewpoint of robust manipulation. Even if there are elements which cannot generate appropriate force, it is possible in principle for the other elements to compensate for them. This signifies insensitivity to environmental fluctuation. In human bodies, for examples, excretion of alien substances is performed by whipping motion of numerous cilia. Paramecia move by paddling their cilia. Centipedes crawl by the cooperative wavy motion of a number of legs. Any of these can accomplish their objectives robustly irrespective of environmental change.

There has been other research into distributed actuation. Böhringer fabricated multiple actuator elements on a silicon chip [1]. Brussel developed a 2D device driven by multiple piezoelectric ceramic elements [3]. Böhringer studied a plate with distributed vibration [2].

In this paper, a new distributed actuation device using a soft gel material is presented. The materials used in the studies up to the present date were not soft. It had been impossible to develop such a device because effective actuator materials had not been discovered, and the technology to utilize soft actuator materials had been insufficient. For these reasons, the development work described in this paper is completely original.

The material used in the device is ICPF and the actuator is composed of a structure containing EFD (Elliptical Friction Drive) actuator elements.

The prototype developed currently is not practical as a robotic device in the same sense as most micromachines. However, it is meaningful to demonstrate the

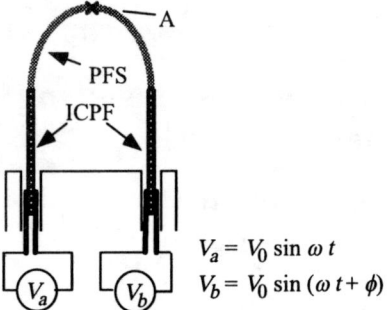

$V_a = V_0 \sin \omega t$
$V_b = V_0 \sin (\omega t + \phi)$

Fig. 2 Structure of an EFD actuator element.

possibility of manipulation using such materials and using this principle because the discovery of higher performance materials in the future would quickly lead to the development of practical devices.

2 ICPF Actuator and EFD Element

2.1 ICPF Actuator

The ICPF actuator is constructed of a novel high polymer composite discovered by Oguro in 1992.

This material is produced by chemically plating platinum onto perfluorosulfonic acid membrane (PFS). Applying a voltage to both platinum layers, using them as electrodes, causes the element to bend as shown in Fig. 1.

The characteristics of ICPF are as follows [7, 5].
- It moves in water or in a wet condition.
- The driving voltage is low (1.5 V).
- It responds to high frequency input (> 100 Hz).
- It is a soft material ($E = 2.2 \times 10^8$ Nm^2).
- Durability is high ($> 1 \times 10^5$ bending cycles) considering it is gel.

These characteristics are unique to the ICPF actuator. Application of this new material to micromachines or in living bodies is expected.

Its principle of motion has not been revealed until now. On characteristics of motion, a model consisting of 3 stages (electrical stage, stress generation stage, and mechanical stage) was proposed by Kanno [8, 6]. It approximates the characteristics well. Its nonlinear characteristics have not been revealed very well and future study is expected.

Some possibble applications of this material are as an active guide wire for a micro catheter system [4] and as an EFD actuator element [10] have been suggested.

2.2 EFD Actuator Element

The EFD is an actuator element which generates driving force by friction using ICPF actuators.

This element has the structure shown in Fig. 2. It has two ICPF parts with platinum plating for actuation and one PFS part without plating for connection. The whole structure is fixed so as to form the shape of an arch.

When sinusoidal voltages with a phase difference are applied to the two ICPFs, the excited sinusoidal bending motions of the ICPF also have phase difference. It

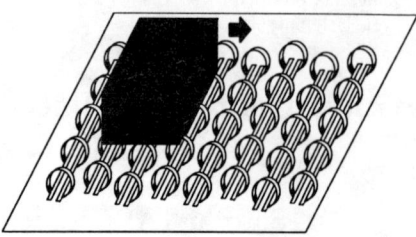

Fig. 3 Distributed actuation device consisting of multiple EFD elements.

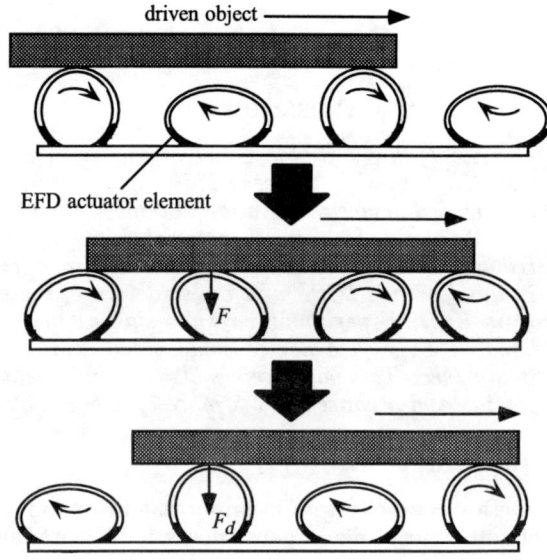

Fig. 4 Principle of distributed drive.

results in an elliptical motion in the top point A of the PFS.

3 Overview of the Distributed EFD Device

3.1 Principle of Drive

Figure 3 shows the developed distributed EFD device. A number of EFD elements cooperatively apply a driving force to an object.

The driving principle is shown in Fig. 4. Adjacent elements make elliptical motions with a phase difference of π (a two phase drive). On the planar contact face, a frictional force in the x direction is generated alternately by adjacent elements, and the object is driven.

3.2 Manufacturing Process

An ICPF actuator is produced by a processes consisting of surface roughening, adsorption of platinum, reduction and growth on a PFS membrane. A masking technique using a crepe paper tape with polyethylene coating can be used to form any arbitrary shape of ICPF on the PFS. This technique is called the pattern plating method. It is an essential technique for creating the various shapes in the gel material required for

Fig. 5 Photo of device developed.

Table 1 Mechanical design parameters of developed device.

Parameter	Value
Number of elements (N)	40
Thickness (t)	180 μm
Width (w)	1 mm
Length of ICPF part (l_I)	2 mm
Length of PFS part (l_P)	11 mm
Distance between supports (h)	7 mm
Angle of supports (θ)	180°
Element interval in the x direction (d_x)	14 mm
Element interval in the y direction (d_y)	4 mm

the actuator. It is important also for efficient supply of electricity.

The manufacturing process was as follows. First ICPF actuators were produced on the supports of a PFS membrane cut in a ladder shape. The ladder was formed into rings by rolling the membrane round a column in a hot water bath. The resultant multiple EFDs were fixed together on a plate. Then, they were wired electrically. Lastly, the shape of each element was adjusted.

Figure 5 shows a photo of the completed distributed actuator device. The mechanical design parameters are listed in Table 1.

The most serious problem in the manufacturing process was that the shape of each element varied widely. This gel material easily changes its size in response to external forces, moisture content, temperature, etc. It is however difficult to maintain constant conditions during the whole manufacturing process. Therefore, the hight of the elements was adjusted in the final stages of manufacture. The second problem was that the external electrodes must be pressed against the platinum surface using a constant pressure in order to keep stable voltage supply. The solution currently used is not effective enough, and more study is necessary. These problems are the most serious restriction in developing applications of this gel material.

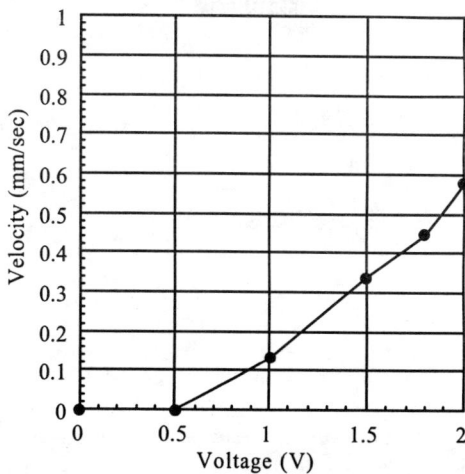

Fig. 6 Effect of input voltage amplitude on plate transfer velocity ($f = 2$ Hz, $\phi = 0.5\pi$ rad).

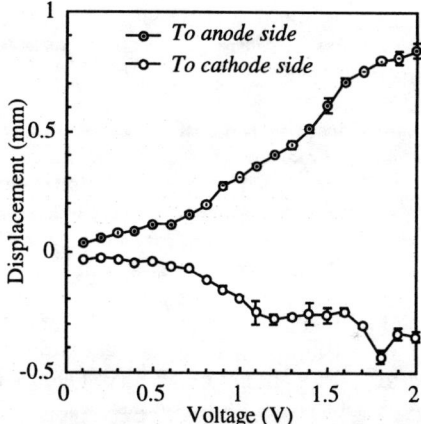

Fig. 7 Maximum displacement of ICPF actuator when applying a step voltage.

4 Experimental Results

For the experiments, a plate was placed on the device that had been manufactured was transferred on it. These were fundamental experiments for applications such as is shown in Fig. 13. In this case, the contact force was determined by the weight of the object.

The average contact force per element in the experimental results presented in the following results was 0.6 mN. All the experiments were performed in a water bath. The motion was measured by processing image data aquired with a TV camera.

4.1 Dependence of Motion upon Magnitude of Input Voltage

Figure 6 shows that the velocity of the plate varied with the amplitude of the sinusoidal voltage applied to each element. Under 0.5 V, motion was not observed. Over 1 V, the speed was proportional to the square of

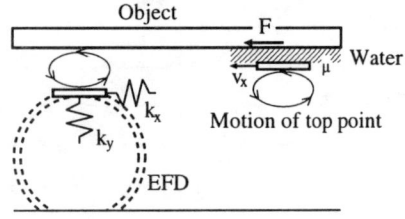

Fig. 8 Mechanism of drive.

Table 2 Compliance of one element (analytical result).

Direction	Compliance (Deflection by 0.6 mN)	
	Flat shape	Tall shape
y	0.5 m/N (0.3 mm)	0.6 m/N (0.4 mm)
x	0.8 m/N (0.5 mm)	2.7 m/N (1.6 mm)

the amplitude.

This curve is very similar to the relationship of the maximum displacement of the ICPF actuator to step input voltage shown in Fig. 7. This proves that the speed of this device is proportional to the displacement of the EFD elements.

The driving force is transmitted by viscous friction through a water layer. Therefore, the faster the elements in contact move in the x direction, the larger the driving force becomes.

It was assumed that the element was an elastic body with the shape of an arch, and the compliance in the x and the y directions was obtained. The results are shown in Table 2. The compliance varies according to the direction of the applied force. The compliance in the x direction is dependent on the EFD shape but the compliance in the y direction is constant.

For stability of repetitive contact and separation between the elements and the object, the active motion of the element must be larger than the deflection caused by contact force in the y direction. This is the reason why the velocity was 0 with a driving voltage lower than 0.5 V. In other words, the range of driving force where object transfer is possible is determined by the active displacement of the element in the y direction. When the y displacement is large, the drive becomes robust.

4.2 Dependence upon the Phase Difference between the Input Voltages

Figure 9 shows the experimental result obtained by varying the phase difference, ϕ, between the sinusoidal input voltages for each EFD element. The shape of the elliptical motion changes according to this phase difference. At the same time, the velocity of the plate changes.

Consideration of the analytical and experimental results showed that the elliptical motion in the x direction became large when the two phases were similar, and that the y displacement increased when the dif-

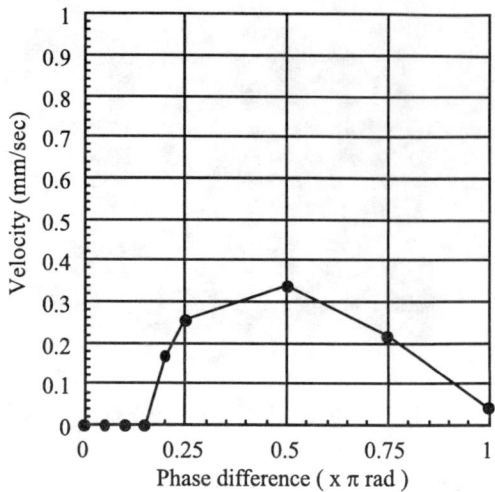

Fig. 9 Effect of phase difference between the input voltages on velocity of plate transfer ($f = 2$ Hz, $V_0 = 1.5$ V).

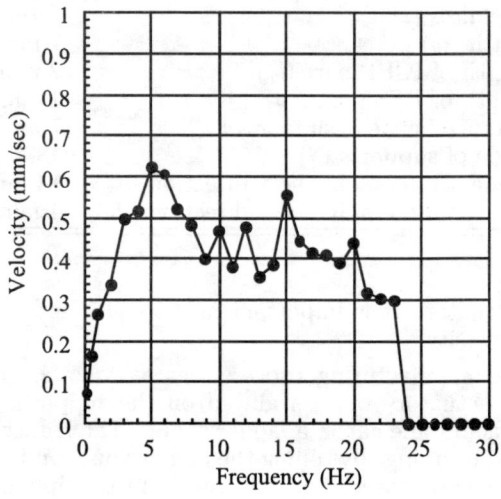

Fig. 10 Effect of input voltage frequency on velocity of plate transfer ($V_0 = 1.5$ V, $\phi = 0.5\pi$ rad).

ference between them was close to π. The speed of the object is determined by these two displacements. The optimal value in this experiment was found to be $\phi = \pi/2$.

4.3 Frequency Dependence

The results from varying the frequency of the sinusoidal input are shown in Fig. 10.

Analytical results, assuming elasticity of the material, indicated that the resonant frequencies of the element would be 5 Hz, 10 Hz and 16 Hz. The experimental curve showed a good correspondence because the speed was high at these frequencies. Viscous resistance of water was so small that complex flow near the device could be ignored. The major analysis error was caused by modeling error of the contact and the element shape.

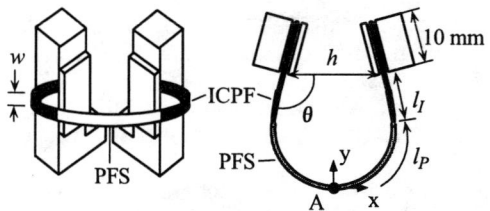

Fig. 11 Mechanical design parameters of an EFD element.

Under 3 Hz, the experimental result indicated that the velocity increased linearly with the frequency. In this range, the speed of the elements was proportional to the frequency because the effect of distributed flexibility was small.

The resonant frequencies vary by ±20% depending on the shape of the elements. These frequencies depend on the initial shape and the pressure exerted by the object being manipulated.

4.4 Manipulation of Very Soft Objects

These initial experimental results have demonstrated the capability to transfer objects with a rather hard surface. In its present form, the device cannot transfer very soft objects. This is because deformation of objects hinders the creation of the non-contact phase of the return motion of the elements. Therefore, if the objects are very soft, the device must be re-designed: 1) using broader elements, 2) arranging the elements more densely, 3) using better EFD design parameters for larger displacement in the y direction, and 4) using another actuator material with better performance characteristics.

5 On Optimal Design and Control

In this section optimal design and control of each element is discussed on the basis of the experimental results.

The parameters used for design and control of EFD are the following as shown in Fig. 11,

1. width between the supports of the two ICPF actuators (h mm),
2. angle between the two ICPFs (θ deg),
3. ratio of the lengths of the ICPF and PFS parts ($r = l_I/l_P$),
4. width of the element (w mm),
5. amplitude of the sinusoidal inputs (V_0 V),
6. phase difference between the sinusoidal inputs (ϕ rad),
7. frequency of the inputs (f Hz).

The following characteristics were determined from the experimental results.

1. The displacement in the y direction was smaller than that in the x direction. Therefore, design parameters should be selected considering the y displacement more.
2. On the displacement in the y direction, the best design parameters were with an the installation angle of $\theta \simeq 135°$ and an installation distance of $h \simeq 20$ mm, with $l_P \simeq 45$ mm and $l_I \simeq 5$ mm, as

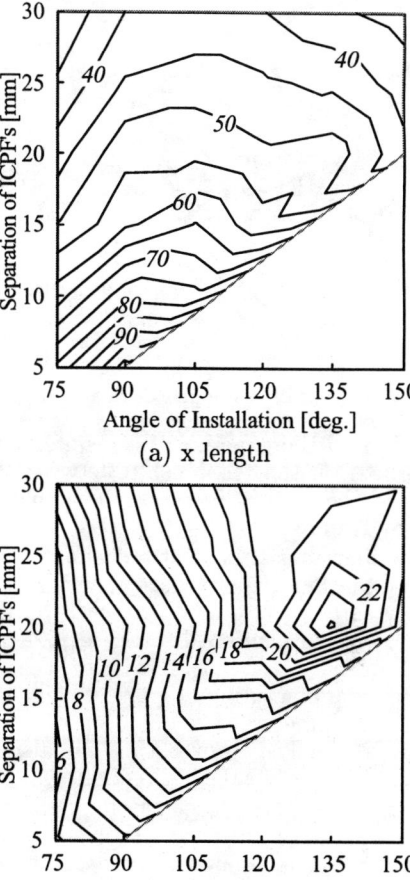

(a) x length

(b) y length

Fig. 12 Axis lengths of elliptical motion ($w = 4$ mm, $l_I = 5$ mm, $l_P = 45$ mm, $V_0 = 1.5$ V, $\phi = \pi/2$, $f = 1.0$ Hz).

shown in Fig. 12. When the ICPF length l_I was smaller, the optimal point shifted to a region of smaller installation angle.

3. On the displacement in the x direction, better results were obtained when the installation angle θ was small.
4. When the length of ICPF l_I was small, larger displacements were obtained. This is because PFS is more flexible than ICPF. At the same time, the motion approached a true circle.
5. Effect of the element width w was small.
6. Both displacements changed sinusoidally in response to changing the phase difference ϕ. The maximum in the y direction was obtained when $\phi = \pi$, and the optimum in the x direction was with $\phi = 0$.
7. The displacement was changed by the frequency f. At resonant frequencies, it became larger.
8. The amplitude of the input voltage V_0 also effected the displacement, the results corresponded to a curve of the second degree.

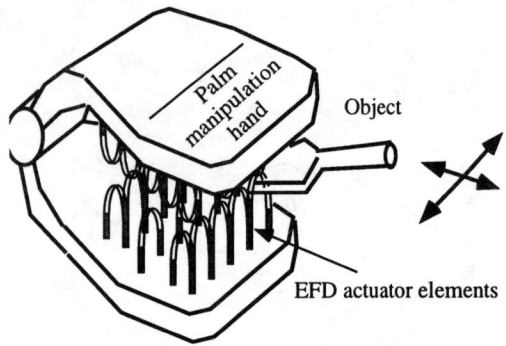

Fig. 13 Application of developed device to robotic hand.

In addition to the experimental results, simulation was performed on the relationship between these parameters and the motion, based on the actuator model proposed by Kanno. The results agreed qualitatively.

From the characteristics described above, it was concluded that the most appropriate structure for transferring very soft objects should have an ICPF length of 1.3 mm, an installation distance of 5.4 mm, and an installation angle of 135°. However, there remain problems in manufacturing at present, and these optimal design parameters have not been realized in the actual device. Analysis proved that the displacement becomes much larger in the y direction.

Increasing the input voltage can be used as a countermeasure. However there are problems over 1.5 V that gas bubbles are produced by electrolysis, which causes not only deterioration of the material but also danger of accumulated hydrogen and oxygen.

The optimal phase difference is near π for very soft objects and the best frequency is at the resonant frequency of 5 Hz.

In the transfer experiments, a method with reciprocal motion of two groups of elements was adopted. This two phase driving method is appropriate because the gap between the non-driving elements and the object becomes large. By adopting driving strategy with three or more phases, the gap becomes smaller so that the robustness is reduced, although the motion becomes smoother.

This element could be applied to a robot hand as shown in Fig. 13. Appropriate control of contact pressure is essential for stable manipulation of very soft objects.

6 Conclusions

This paper presented the following.
1. The development of a distributed driving device using a number of soft gel actuators.
2. Experimental results that demonstrated the ability to transfer objects.
3. Resultant characteristics that agreed with analysis and fundamental experiments.
4. In order to manipulate very soft objects, the optimal design and control of each EFD element has been discussed.

The following are the subject for future study.
1. Establishment of methodologies for precise manufacturing,
2. Modeling the transfer mechanism in detail, and
3. Applications to robotic hands.

Acknowledgements

This research was supported by the Scientific Research Fund of the Ministry of Education, Science and Culture (#07455113, #08555059 and #08750275). The authors appreciate Mr. Derek Ward for his contribution in preparing this manuscript.

References

[1] K.-F. Böhringer, B. R. Donald, R. Mihailovich, N. C. MacDonald, Sensorless manipulation using massively parallel microfabricated actuator arrays, Proc. 1994 IEEE Intl. Conf. on Robotics and Automation, pp. 826-833, 1994

[2] K.-F. Böhringer, V. Bhatt and K. Y. Goldberg, Sensorless manipulation using transverse vibrations of a plate, Proc. 1995 IEEE Intl. Conf. on Robotics and Automation, pp. 1989-1996, 1995

[3] H. V. Brussel, M. Versteyhe, D. Reynaerts, Development of a rigid and accurate positioning system based on a piezo-stepping principle, Proc. EUROMECH Colloquium 370, pp. 54-55, 1997

[4] S. Guo, T. Fukuda, K. Kosuge, F. Arai, K. Oguro, M. Negoro, Micro catheter system with active guide wire, Proc. 1995 IEEE Intl. Conf. on Robotics and Automation, pp. 79-84, 1995

[5] R. Kanno, A. Kurata, M. Hattori, S. Tadokoro, T. Takamori, Characteristics and modeling of ICPF actuator, Proc. Japan-USA Symp. on Flexible Automation '94, pp. 692-698, 1994

[6] R. Kanno, S. Tadokoro, T. Takamori, M. Hattori and K. Oguro, Linear approximate dynamic model of an ICPF (ionic conducting polymer gel film) actuator, Proc. 1996 IEEE Intl. Conf. on Robotics and Automation, pp. 219-225, 1996

[7] R. Kanno, S. Tadokoro, M. Hattori, T. Takamori and K. Oguro, Modeling of ICPF (ionic conducting polymer gel film) actuator, Part 1: fundamental characteristics and black-box modeling Trans. of the Japan Soc. of Mechanical Engineers (C), Vol. 62, No. 598, pp. 213-219, 1996

[8] R. Kanno, S. Tadokoro, M. Hattori, T. Takamori and K. Oguro, Modeling of ICPF (ionic conducting polymer gel film) actuator, Part 2: electrical characteristics and linear approximate model, Trans. of the Japan Soc. of Mechanical Engineers (C), Vol. 62, No. 601, pp. 3529-3535, 1996

[9] K. Oguro, Y. Kawami and H. Takenaka, Bending of an ion-conducting polymer film-electrode composite by an electric stimulus at low voltage, J. of Micromachine Society, Vol. 5, pp. 27-30, 1992

[10] S. Tadokoro, T. Murakami, S. Fuji, R. Kanno, M. Hattori and T. Takamori, An elliptic friction drive element using an ICPF (ionic conducting polymer gel film) actuator, IEEE Control Systems, 1997

Preisach Model Identification of a Two-Wire SMA Actuator

R.B. Gorbet
Systems Control Group
University of Toronto
Toronto, Canada M5S 3G4

D.W.L. Wang
Electrical & Computer Engineering
University of Waterloo
Waterloo, Canada N2L 3G1

K.A. Morris
Applied Mathematics
University of Waterloo
Waterloo, Canada N2L 3G1

Abstract

In recent years, the Preisach hysteresis model has emerged as the model of choice for the behaviour of many smart materials, such as shape memory alloys (SMA). This research treats the identification of Preisach models for a differential SMA actuator. The traditional identification technique is applied, and several models are derived for the actuator. It is seen that the classical Preisach model is able to model its behaviour. However, the results raise questions concerning the robustness of the traditional identification technique, as well as the conservative nature of current passivity-based control results for the Preisach model.

1 Introduction

The last decade has seen a growing interest in the application of so-called "smart materials" as actuators. The hysteretic nature of these materials has prompted renewed interest in general hysteresis models, and notably that originated by F. Preisach[13] in the 1930's. The generality of this model makes it conducive to the development of controller design and analysis techniques which benefit many different hysteretic systems, without concern for the underlying physical mechanism. The suitability of the model for the representation of SMA and piezoceramic hysteresis has been tested[5], and research has begun on control techniques based on passivity[3] and model inversion[4].

This work examines Preisach model identification of a two-wire differential SMA actuator, described in the remainder of the Introduction. In section 2, the Preisach model is presented and the identification technique of [10] is described. Section 3 details the identification process and the experimental verification of the resulting models, while section 4 contains a discussion of the results of this study.

1.1 Two-Wire SMA Actuator

The differential actuator is a natural configuration for achieving rotary motion. The mechanical details of the actuator are shown in Figure 1 (electrical connections are omitted). A 35cm length of 0.3mm SMA wire is anchored at its midpoint to an output shaft, creating two 15cm parallel lengths of wire. Angular position is read by a 2000 count/revolution shaft encoder. The wires have a common electrical connection at the shaft, and the two "free" ends have electrical lugs crimped over a knot tied in the wire. The actuator is operated by running current through either of the wires. Bipolar current is routed by two series diodes, so that positive current affects one wire while negative current commands affect the other. The convention adopted in this work is that negative temperatures refer to one wire, while positive values refer to the other.

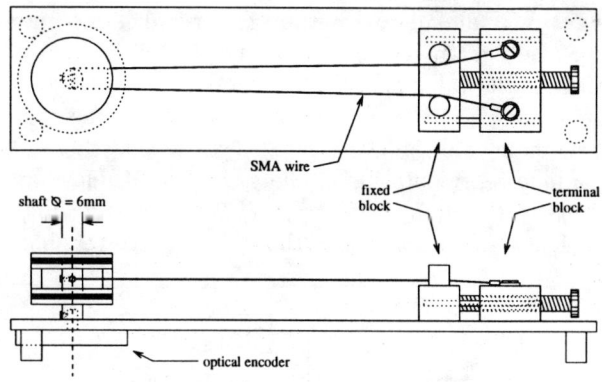

Figure 1: **Differential SMA Actuator**

The control current causes an increase in temperature of the active wire, leading to a phase change within the material which results in strain recovery. Since the structure is rigid, the total amount of strain in the wires is constant and the opposing wire will stretch, permitting rotation of the actuator. The amount of pre-strain which is recovered, and hence the rotation achieved, is dependent on the external load to which the actuator is subjected. Rotation in the opposite direction is achieved by heating the other wire. The measured hysteresis present in this actuator is shown in Figure 2. The method used to convert input current to wire temperature is described in Section 3.1.

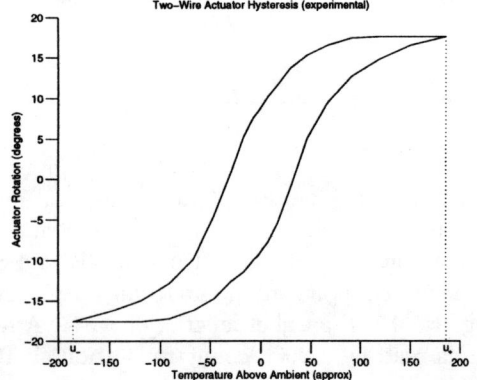

Figure 2: **Differential Actuator Hysteresis**

2 The Preisach Model

Historically, the Preisach model was developed to represent the hysteresis in magnetic materials, and assumes that the material is made up of individual dipoles. These dipoles are represented by hysteresis relays $\gamma_{\alpha\beta}$, and model output is a parallel summation of weighted relays. This is illustrated in Figure 3, where the value $\mu(\alpha, \beta)$ represents the weighting of the relay $\gamma_{\alpha\beta}$. Each relay is characterized by the pair of switching values (α,β), with $\alpha \geq \beta$, so that there is a unique representation of the collection of relays as points in the half-plane $\mathcal{P} = \{(\alpha,\beta)|\alpha \geq \beta\}$ (cf. Fig. 4). The vertical portions of the relays are irreversible:

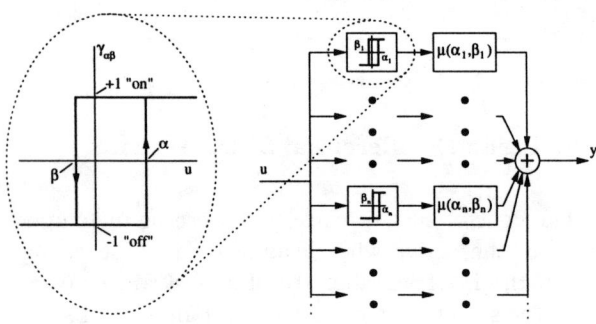

Figure 3: **Schematic of the Preisach Model**

they can only be traversed in one direction. The horizontal sections are reversible. Degenerate relays, those with $\alpha = \beta$, are fully reversible.

The behaviour of these relays, and hence the Preisach model, is only defined for continuous inputs u. As the input varies with time, each individual relay adjusts its output according to the current input value, and the weighted sum of all the relay outputs provides the overall system output (cf. Fig. 3)

$$y(t) = \iint_{\mathcal{P}} \mu(\alpha,\beta)[\gamma_{\alpha\beta}u](t)d\alpha d\beta. \qquad (1)$$

The collection of weights $\mu(\alpha,\beta)$ forms a weighting function $\mu : \mathcal{P} \mapsto \mathbf{R}$, which describes the relative contribution of each relay to the overall hysteresis.

2.1 The Preisach Plane

The region $\mathcal{P}$ is usually referred to as the Preisach plane. Every point in $\mathcal{P}$ represents a unique relay, and $\mathcal{P}$ is the support for the weighting function μ. The hysteresis in SMA has zero slope outside the hysteresis loop (cf. Fig. 2). There is no contribution to the output from inputs outside the domain of hysteretic behaviour (labeled $[u_-, u_+]$ in Fig. 2). Hence $\mu(\alpha,\beta) = 0$ outside the triangle corresponding to this domain, $\alpha = u_+$, $\beta = u_-$, $\alpha = \beta$. This can be thought of as restricting $\mathcal{P}$ to that triangle, as illustrated in Figure 4.

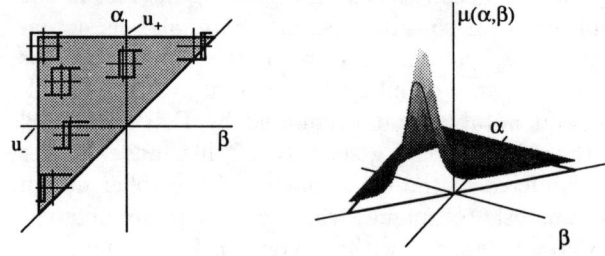

Figure 4: **The Preisach Plane**

The Preisach plane also provides a means for keeping track of the state of individual relays. First, the region $\mathcal{P}$ is divided into two time-varying regions, $\mathcal{P}_-$ and $\mathcal{P}_+$, defined as in [10]:

$$\mathcal{P}_-(t) \triangleq \{(\alpha,\beta) \in \mathcal{P} \mid \text{output of } \gamma_{\alpha\beta} \text{ at } t \text{ is } -1\}$$
$$\mathcal{P}_+(t) \triangleq \{(\alpha,\beta) \in \mathcal{P} \mid \text{output of } \gamma_{\alpha\beta} \text{ at } t \text{ is } +1\},$$

so that $\mathcal{P}_-(t) \cup \mathcal{P}_+(t) = \mathcal{P}$ at all times. It will become clear that each set is connected. The notation $\mathcal{P}_-$

and $\mathcal{P}_+$ will be used, with the time-dependence being implicit.

Now, consider an hysteretic system exposed to a monotonically increasing input, taking it from negative saturation to positive saturation along the major loop. In negative saturation, all relays are in the "-1" state and $\mathcal{P}_- = \mathcal{P}$, $\mathcal{P}_+ = \emptyset$ (Fig. 5a). As the input increases, it switches a relay $\gamma_{\alpha\beta}$ to "+1" as u increases past α; that relay now belongs to $\mathcal{P}_+$. The boundary between $\mathcal{P}_-$ and $\mathcal{P}_+$ can be represented as a horizontal line in the Preisach plane, moving up as the input increases (Fig. 5b), switching relays from $\mathcal{P}_-$ to $\mathcal{P}_+$ until the input stops increasing (Fig. 5c). If the input now reverses and begins to decrease monotonically, a vertical boundary is generated sweeping from right to left, switching a relay $\gamma_{\alpha\beta}$ from $\mathcal{P}_+$ to $\mathcal{P}_-$ as it passes $u = \beta$ (Fig. 5d). Further input reversals generate additional horizontal or vertical boundary segments.

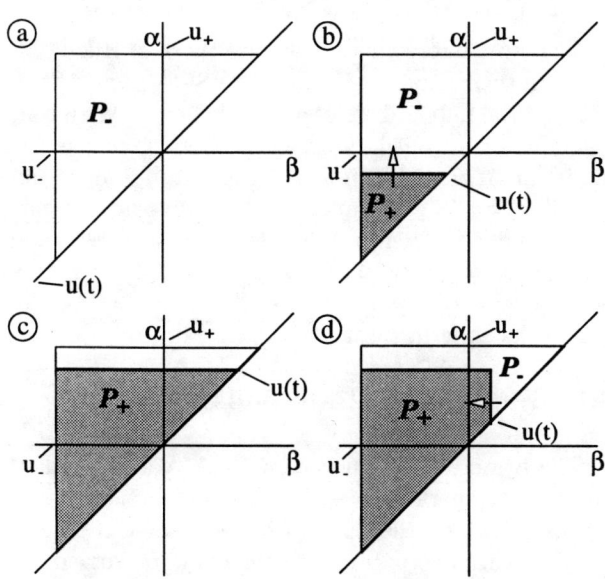

Figure 5: **Preisach Boundary Behaviour**

From the definitions of $\mathcal{P}_+$ and $\mathcal{P}_-$, the output equation (1) can be written as

$$y(t) = \iint_{\mathcal{P}_+} \mu(\alpha,\beta)d\alpha d\beta - \iint_{\mathcal{P}_-} \mu(\alpha,\beta)d\alpha d\beta$$

$$= 2\iint_{\mathcal{P}_+} \mu(\alpha,\beta)d\alpha d\beta - \iint_{\mathcal{P}} \mu(\alpha,\beta)d\alpha d\beta. \quad (2)$$

Since the boundary defines the region $\mathcal{P}_+$, knowledge of the boundary configuration at time t, along with the weighting function μ, is sufficient to determine $y(t)$.

Proposition 2.1 (Output Variation)
A monotonic change in input which causes the boundary to sweep out an area Ω from time t_1 to time t_2 results in an output variation

$$y(t_2) - y(t_1) = 2sgn[u(t_2) - u(t_1)] \iint_\Omega \mu(\alpha,\beta)d\alpha d\beta.$$

Proof
Consider the example of Figure 6. The input decreases from $u(t_1)$ to $u(t_2)$, removing the region Ω from $\mathcal{P}_+$ and adding it to $\mathcal{P}_-$.

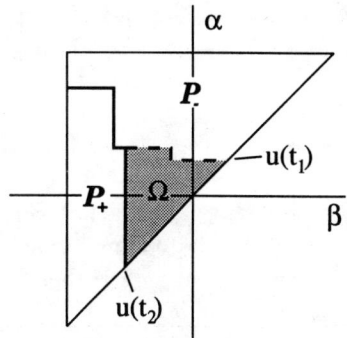

Figure 6: **Monotonic Input Decrease**

From equation (2), the difference in output is given by

$$y(t_2) - y(t_1) = 2\iint_{\mathcal{P}_+(t_2)} \mu(\alpha,\beta)d\alpha d\beta - 2\iint_{\mathcal{P}_+(t_1)} \mu(\alpha,\beta)d\alpha d\beta.$$

Since $P_+(t_1) = P_+(t_2) \cup \Omega$, this gives

$$y(t_2) - y(t_1) = -2\iint_\Omega \mu(\alpha,\beta)d\alpha d\beta.$$

Similarly, if u is increasing, it shifts the points of Ω from $\mathcal{P}_-$ to $\mathcal{P}_+$, and the sign of the output variation is reversed. ∎

2.2 Model Identification

In [10], Mayergoyz describes a technique to determine the Preisach weighting surface μ from experimental data. The identification involves the generation of several "first-order descending curves" (FOD), from which μ can be determined.

An FOD is generated by first bringing the input to negative saturation, followed by a monotonic increase to a value $u = \alpha_1$, then a decrease to $u = \beta_1$. The effect of this input on the Preisach plane is illustrated in Figure 7. The measured output values are labeled y_{α_1}, corresponding to $u = \alpha_1$, and $y_{\alpha_1\beta_1}$ when u has reached β_1.

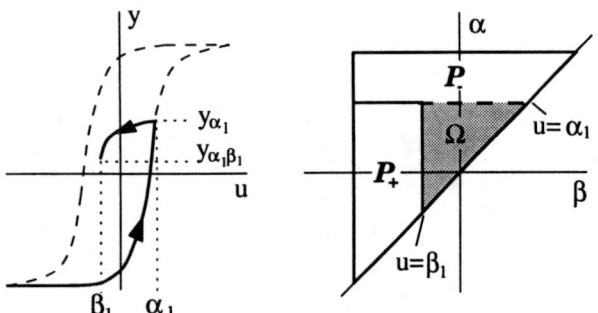

Figure 7: **Sample Identification Input**

After the input peaks at α_1, the decrease sweeps out Ω, generating the descending branch inside the major loop. From Proposition 2.1, the change in output along the descending branch is

$$y_{\alpha_1\beta_1} - y_{\alpha_1} = -2 \iint_\Omega \mu(\alpha,\beta) d\alpha d\beta.$$

Defining the function $F(\alpha_1, \beta_1) = y_{\alpha_1\beta_1} - y_{\alpha_1}$, rewriting the integral and then taking partial derivatives of both sides gives the following relationship for μ.

$$F(\alpha_1,\beta_1) = -2 \int_{\beta_1}^{\alpha_1}\int_\beta^{\alpha_1} \mu(\alpha,\beta) d\alpha d\beta$$
$$\frac{\partial^2}{\partial \alpha_1 \partial \beta_1} F(\alpha_1,\beta_1) = 2\mu(\alpha_1,\beta_1).$$

Then the value of the weighting function at any point $(\alpha_1, \beta_1) \in \mathcal{P}$ can be determined from

$$\mu(\alpha_1,\beta_1) = \frac{1}{2}\frac{\partial^2}{\partial \alpha_1 \partial \beta_1}F(\alpha_1,\beta_1) = \frac{1}{2}\frac{\partial^2 y_{\alpha_1\beta_1}}{\partial \alpha_1 \partial \beta_1}.$$

If the value $y_{\alpha\beta}$ could be identified for all points in $\mathcal{P}$ by collecting an infinite number of FOD measurements, it is clear on physical grounds that the surface $y(\alpha,\beta)$ formed of all points $y_{\alpha\beta}$ should be smooth. This surface could then be differentiated to obtain the weighting function,

$$\mu(\alpha,\beta) = \frac{1}{2}\frac{\partial^2 y(\alpha,\beta)}{\partial \alpha \partial \beta}. \quad (3)$$

More realistically, the hysteresis domain $[u_-, u_+]$ is divided in the ordered partition $\{u_i\}_{i=0,\ldots,n}$. FOD curves are then obtained for all pairs (u_i, u_j) with $j \leq i$, resulting in $\frac{n}{2}(n+3)$ FOD data points. A smooth approximation surface $\tilde{y}(\alpha,\beta)$ is then fit to these data points, and this surface is differentiated to obtain an approximate weighting surface $\tilde{\mu}$.

3 SMA Actuator Identification

Preisach models have been applied to SMA in the past. Ortín[12] identified a model for isothermal stress-strain hysteresis in a crystal of CuZnAl. A good quantitative match is observed between simulated output and experimental data. In [6], the author describes a complicated extension of the Preisach model. Four-parameter hysteresis kernels are used, rather than the two-parameter relays of the classical model. However, a technique for identifying this complex model is lacking, and only qualitative results are obtained.

The most interesting results are those of Hughes & Wen. In their experiments, a single SMA wire is fixed between the hub and tip of a flexible beam. Wire contraction is controlled by Joule-heating, current is the model input, and beam strain the output. In [4] an identification is performed using slowly-varying currents, and an attempt is made to fit a polynomial surface to the data. The polynomial fit fails mainly because low-order polynomials are unable to properly match the experimental FOD data.

3.1 Experimental Methods

The hysteresis which occurs in SMA is dependent on alloy temperature, and is not a function of the means by which the alloy is heated. In the case of the differential actuator, experiments were carried out using piece-wise constant currents and actuator position was measured in steady-state. The steady-state temperature of a wire carrying a constant current can be computed from the relationship

$$T_{ss} = \frac{R}{hA}i^2, \quad (4)$$

where R is resistance, A the surface area and h the coefficient of heat transfer to the environment[8, e.g.]. Temperatures in this paper should be considered approximate because of parameter variation, notably R (variation of 10%), during the SMA phase transformation. In this case, the average resistance of the austenite and martensite phases was used ($R = 2.0\Omega$). The value of h was taken from [9] ($h = 75\frac{W}{m^2C}$). The surface area is calculated at $A = 1.41 cm^2$.

3.2 FOD Data Collection

The range of wire currents is limited to $[-1, 1]$ Amps, corresponding to temperatures of approximately $[-185, 185]$ degrees above ambient. Recall that the sign of the temperature indicates which wire is being heated. This temperature range was partitioned into eleven sub-ranges $\{T_i\} = \{$-185,-145,-105,-75,-45, -15,15,45,75,105,145,185$\}$. The constant current required to achieve each T_i was computed from equation (4), and the piece-wise constant current input of Figure 8 was generated from these values. Each constant current was maintained until the position measurement had not changed for 25 seconds. At this point, it was considered that steady-state had been reached, the position was recorded and the next current step was applied. In this manner, eleven FOD curves were generated, and are shown in Figure 9. It is apparent from this data that the actuator was not fully exercised: higher temperatures should have been used. Because of this, the lower end of the FOD curves do not merge.

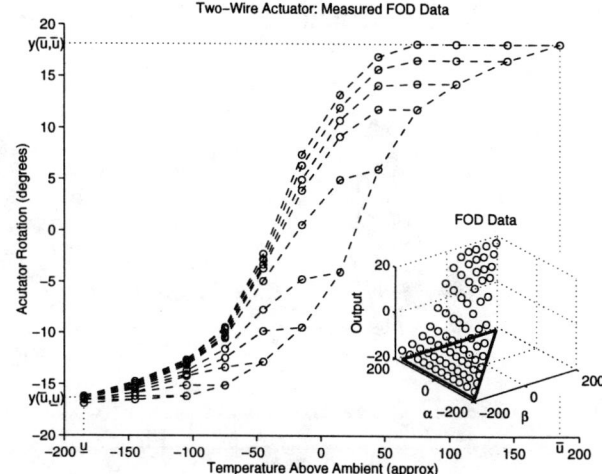

Figure 9: **Measured FOD Data**

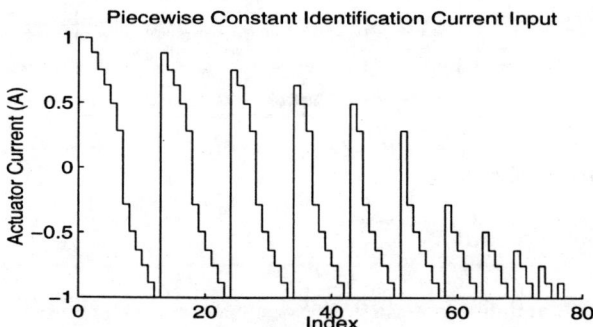

Figure 8: **Identification Current Input**

The inset of Figure 9 shows a three-dimensional plot of the FOD data. In order to determine the Preisach model weighting function, a surface $\tilde{y}(\alpha, \beta)$ is fit to this measured FOD data, then differentiated to obtain $\tilde{\mu}$.

3.3 Surface Fit

While higher order polynomial surfaces may effectively match the measured data, they may be highly oscillatory between points. This conflicts with the smooth behaviour observed experimentally, and has serious implications if the surface is to be differentiated to obtain $\tilde{\mu}$. In one of the initial works on modelling of SMA hysteresis[7], two curves of the form

$$y(u) = \frac{c_1}{1 + e^{-c_2(u+c_3)}} + c_4 \qquad (5)$$

are used to simulate the major loop of the hysteresis, suggesting that individual FOD curves might be well-approximated by this form. Equation (5) saturates asymptotically at both low and high values of u and is smooth in between, replicating the shape of the FOD curves. In this work, a three-dimensional surface based on (5) is used.

In order to determine an appropriate form for the candidate surface, a least squares fit of (5) was done for each individual FOD curve (fixed α). The variation of the curve parameters c_i with α was then used to extend the surface to three dimensions. The appropriate choice for $c_3(\alpha)$ being unclear, three different surfaces were tried, with constant, parabolic and exponential c_3 (see [2] for details).

The basic form of each candidate surface is

$$\tilde{y}(\alpha, \beta, c_3) = k \times \frac{[e^{-x_4(\underline{u}+c_3(\alpha))} - e^{-x_4(\beta+c_3(\alpha))}]}{[1 + e^{-x_2(\alpha+x_3)}][1 + e^{-x_4(\beta+c_3(\alpha))}]}$$
$$\times \frac{1}{[1 + e^{-x_4(\underline{u}+c_3(\alpha))}]} + y(\overline{u}, \underline{u}). \qquad (6)$$

The x_i are surface parameters. The constants $\overline{u}$ and $\underline{u}$ are the maximum and minimum input values used during the identification, $y(\overline{u}, \overline{u})$ and $y(\overline{u}, \underline{u})$ are the measured outputs at the top and bottom of the hysteresis curve, indicated in Figure 9. Since the full loop was not exercised, the lower branches don't merge, and the average measured value was used for $y(\overline{u}, \underline{u})$. The constant k is

$$k = [y(\overline{u}, \overline{u}) - y(\overline{u}, \underline{u})][1 + e^{-x_2(\overline{u}+x_3)}]$$
$$\times \frac{[1 + e^{-x_4(\overline{u}+c_3(\overline{u}))}][1 + e^{-x_4(\underline{u}+c_3(\overline{u}))}]}{[e^{-x_4(\underline{u}+c_3(\overline{u}))} - e^{-x_4(\overline{u}+c_3(\overline{u}))}]},$$

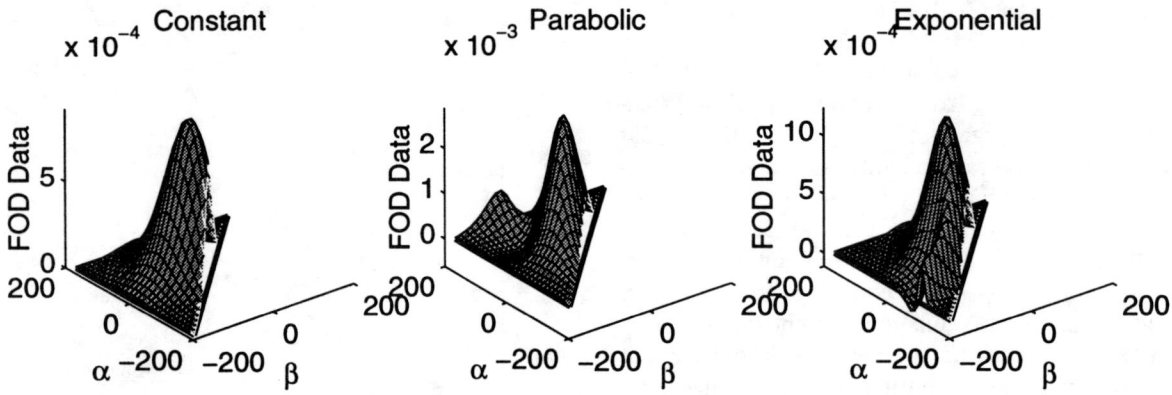

Figure 10: **Weighting Surfaces**

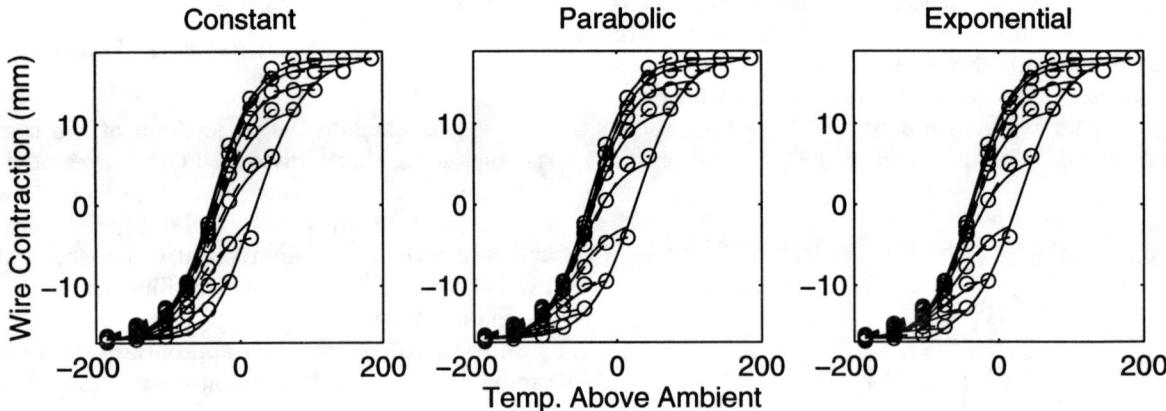

Figure 11: **Simulation Output**

chosen so that $\tilde{y}(\overline{u},\overline{u},c_3) = y(\overline{u},\overline{u})$ and $\tilde{y}(\overline{u},\underline{u},c_3) = y(\overline{u},\underline{u})$. In other words, the surface (6) is guaranteed to match the measured output saturation values of the hysteresis curve regardless of the choice of c_3.

The fitted surfaces $\tilde{y}$ vary in the choice of the function c_3. For the first, $\tilde{y}_c$, c_3 is chosen to be constant, $c_3(\alpha) = x_5$. A second candidate surface $\tilde{y}_p$ has a parabolic $c_3(\alpha) = x_5\alpha^2 + x_6\alpha + x_7$, while the third surface, $\tilde{y}_e$, has $c_3(\alpha) = x_5 e^{-x_6 \alpha} + x_7$.

The parameters x_i of the surfaces were optimized using a Nelder-Mead simplex algorithm[11] to minimize the error between the surface and the data in the inset of Figure 9. A modified least squares error function was used, weighting errors at surface points corresponding to the major loop more heavily than interior points, encouraging a match of major loop behaviour. The modified least squares error values were 9.59, 7.33 and 7.34 for $\tilde{y}_c$, $\tilde{y}_p$ and $\tilde{y}_e$ respectively. The results of the optimization are the parameter vectors

$$x_c = \begin{bmatrix} 0.026 & -16.773 & 0.033 & 38.473 \end{bmatrix}$$

$$x_p = \begin{bmatrix} 0.030 & -25.070 & 0.034 & 0.002 & -0.490 & 68.690 \end{bmatrix}$$
$$x_e = \begin{bmatrix} 0.029 & -24.277 & 0.034 & 27.277 & 0.021 & 36.027 \end{bmatrix}.$$

The approximate weighting functions $\tilde{\mu}_c$, $\tilde{\mu}_p$ and $\tilde{\mu}_e$ were obtained by differentiating equation (6) according to (3) and substituting these parameters. Plots of these weighting surfaces are shown in Figure 10.

3.4 Experimental Verification

In order to verify the models, the output to the identification input sequence was simulated for each of the surfaces. The results are shown in Figure 11. The measured FOD data is plotted in the figure and joined by dashed line segments to represent the experimental hysteresis. Since the Preisach model ensures that descending branches merge at $\underline{u}$, model output is smaller than experimental data in the lower ranges of input. This demonstrates the importance of exercising the full hysteretic range during the identification process.

The fit errors for $\tilde{y}_p$ and $\tilde{y}_e$ were very close, and smaller than that for $\tilde{y}_c$. This difference is manifested in the simulation results by a better match of lower minor loops.

4 Discussion & Conclusions

Differential Actuator Modelling
In the differential configuration, each SMA wire is subjected to a time-varying stress. As one wire contracts, the opposing wire is stretched, exerting more force on the wire being heated. In theory, this time-varying stress causes a bi-variate hysteresis behaviour to appear, with the shape of the $T - \varepsilon$ characteristic dependent on the stress. In this work, it was seen that the single-input Preisach model can represent the hysteresis in the two-wire actuator. In [2], a mechanical analysis of the actuator used for this study indicates that the variation in stress within the wire over the course of an actuator cycle is 7.8%. It is thought that this level of stress variation may not be enough to induce significant bi-variate behaviour in the actuator. This hypothesis remains to be verified in future research.

Implications for Passivity-Based Control
In [3], it is shown that if $\mu \geq 0$ then the associated Preisach model is passive from u to $\dot{y}$. This result is used to demonstrate stability of velocity control when the controller is strictly passive with finite gain. In the results of this work, only $\tilde{\mu}_c \geq 0$. However, the SMA actuator, which we expect to be passive based on physical intuition, is better modeled by $\tilde{\mu}_p$ or $\tilde{\mu}_e$. Each of these weighting functions has negative regions. This apparent sensitivity of actuator properties suggests the existence of a broader class of passive Preisach models, and a prime area of future research will be the extension of the passivity result.

Identification Technique
Finally, the fact that several models with different fundamental properties were derived emphasizes the need for more robust identification techniques. Deriving the weighting function by differentiating a fitted surface is inherently imprecise, and it would be preferable to fit μ directly. New identification techniques have recently been proposed[1], and their application to the differential actuator will provide an interesting comparative study.

References

[1] H.T. Banks, A.J. Kurdilla, and G. Webb. Identification of hysteretic control influence operators representing smart actuators: Convergent approximations. Technical Report CRSC-TR97-7, CRSC, NCSU, Raleigh, NC 27695-8205, April 1997.

[2] R.B. Gorbet. *Control of Hysteretic Systems with Preisach Representations*. PhD thesis, University of Waterloo, 1997.

[3] R.B. Gorbet, K.A. Morris, and D.W.L. Wang. Stability of control systems for the Preisach hysteresis model. *Journal of Engineering Design and Automation*, 1997. Special Issue, in press.

[4] D. Hughes and J.T. Wen. Preisach modeling and compensation for smart material hysteresis. In *Symposium on Active Materials and Smart Structures*, College Station, Texas, 1994.

[5] D. Hughes and J.T. Wen. Preisach modeling of piezoceramic and shape memory alloy hysteresis. In *IEEE Control Conference on Applications*, Albany, New York, 1995.

[6] Y. Huo. A mathematical model for the hysteresis in shape memory alloys. *Continuum Mechanics and Thermodynamics*, 1:283–303, 1989.

[7] K. Ikuta, M. Tsukamoto, and S. Hirose. Mathematical model and experimental verification of shape memory alloy for designing micro actuators. In *IEEE MicroElectroMechanical Systems Conference*, pages 103–108, 1991.

[8] F. Kreith. *Principles of Heat Transfer*. Intext Press, New York, 3 edition, 1973.

[9] Daniel R. Madill and David Wang. The modelling and L_2-stability of a shape memory alloy position control system. In *IEEE Conference on Robotics and Automation*, pages 293–299, 1994.

[10] I.D. Mayergoyz. *Mathematical Models of Hysteresis*. Springer-Verlag, New York, 1991.

[11] J.A. Nelder and R. Mead. A simplex method for function minimization. *Computer Journal*, 7:308–313, 1964.

[12] J. Ortín. Preisach modeling of hysteresis for a pseudoelastic Cu-Zn-Al single crystal. *Journal of Applied Physics*, 71(3):1454–1461, February 1992.

[13] F. Preisach. Uber die magnetische nachwirkung. *Zeitschrift fur Physik*, 94:277–302, 1935.

Learning by Biasing

G. Hailu, G. Sommer
Christian Albrechts University
Department of Cognitive Systems
Preusserstrasse 1-9, D-24105 Kiel, Germany

Abstract

In the quest for machines that are able to learn, reinforcement learning (RL) is found to be an appealing learning methodology. A known problem in this learning method, however, is that it takes too long before the robot learns to associate suitable situation - action pairs. Due to this problem, RL has remained applicable only to simple tasks and discrete environment. To accelerate the learning process to a level required by real robot tasks, the traditional learning architecture has to be modified. We propose a modified reinforcement based robot skill acquisition and adaptation architecture. The architecture has two components : a bias and a learning components. The bias component imparts to the learner coarse a priori knowledge about the task. Subsequently, the learner refines the acquired actions through reinforcement learning. We have validated the architecture and the learning algorithm on a simulated TRC mobile robot for a goal reaching task.

1 Introduction

Programming an autonomous robot to reliably carry out its task demands a complete knowledge of the task and the environment. Systems designed with complete knowledge are called *expert systems* and have no learning ability. Instead they are equipped with a large amount of data base that requires careful tuning. However, because of the complexity and uncertainty of the real world, it is prohibitive to create an expert system with large data base. Besides, it is argued that if the robot somehow possesses a *self-learning* ability, an enormous amount of human effort would be saved from tuning the data base.

In the past many machine learning techniques have been proposed. Most of the learning techniques assume the presence of teacher provided training instances in the form of stimuli and desired response. These types of learning techniques are known as *supervised learning* and successful applications have been booked in: function approximation, pattern recognition and, navigation of mobile robots [5, 12].

However, for many real world systems such as mobile robots working in dynamic environments, training instances in the form of stimuli and desired response are not easily available. Therefore, a robot has to learn for every stimulus the optimal response directly by interacting with its environment. This type of learning method falls into a class of learning methodology called *reinforcement learning* (RL). In RL the robot learns to associate the right responses to different stimuli of the world. It involves four components: the robot, its environment, a learner[a] (controller) to be trained and a trainer that provides *only* a scalar reinforcement signal.

Although RL method fits very nicely to robot learning, it is a slow learning process - it takes too long for the controller to converge toward the desired performance. There are many reasons [2, 8, 9] that contribute to the slow convergence of RL. The major one, however, is that the controller does not know beforehand where to search in action space for suitable reactions. This problem stems from the definition of RL: *reinforcement based learning robots learn by doing and do not require a teacher!* To overcome the problem, we have lifted up the above *unsupervised learning* restriction by providing the learner with a *bias component*. The bias component can be compared with a teacher in supervised learning. However, it does not supply the learner the desired response, hence we still demand the desired response to emerge from RL. Apart from accelerating the learning process, biasing enables the learner to avoid those actions that takes the robot to undesirable locations, thereby making the learning process safe [11].

---
[a] In this paper we use learner and controller interchangeably.

The paper is organized as follows. Section 2 presents briefly the architecture of the bias component, from which the controller gets a rough action. Section 3 describes the proposed learning architecture and adaptation algorithms. Section 4 presents the trajectory and the learning curves of the robot. At last, a conclusion is drawn from the experimental results.

2 Bias Component

The agent, for which the simulator is built, is a two wheeled $60cm$ square and $40cm$ high TRC mobile robot, Fig. 1. On the front periphery of the robot there are tactile and sonar sensors. The sonar sensors are programmed for a timeout distance of $2m$. Our simulator assumes sonar values are repeatable and motor actions are invertible. In addition, the simulated robot is given a capability to access global information - such as its position.

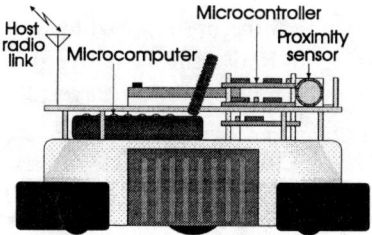

Figure 1. TRC robot

The architecture of the bias component is similar to [15] and is shown in Fig. 2. It consists of two purposive fuzzy behaviors: **obstacle avoidance** and **goal following**. As the purpose of the bias component is to deliver a rough estimate of the optimal action, it suffices to have few fuzzy rules and coarse input/output granulation levels.

To come up with few fuzzy rules, the sonars are first grouped into five regions: **right corner**, **right**, **front**, **left** and, **left corner** corresponding to their physical location on the TRC. Subsequently, from each region only the sonar that has the minimum reading is considered[b], i.e.,

$$D_j = \min_i(S_{i,j}) \quad i = 1, \ldots, N_j \; ; \; j = 1, \ldots, 5 \qquad (1)$$

where D_j is the depth value of region j, $S_{i,j}$ is the reading of sonar i located in region j and, N_j is the number of sonars in the region.

The obstacle avoidance behavior has a set of 32 fuzzy rules that are build by granulating the depth values

[b]On real robot this assumption does not hold true and another technique must be sought, see [6] for a possible method.

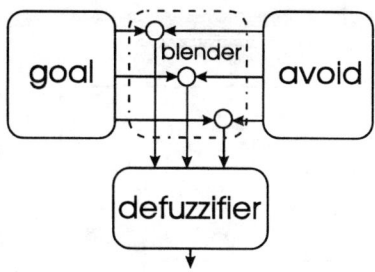

Figure 2. Bias component architecture

of each regions into two fuzzy sets: **n** (near) and **f** (far). Whereas the goal following behavior has a set of three fuzzy rules that are constructed by fuzzifying the acute angle θ between the robot heading and the vector connecting the current robot and goal locations into three fuzzy levels: **l** (left), **f** (front) and **r** (right). The output of the bias component is the turn rate[c] of the robot that is fuzzified into three fuzzy sets: **L** (Left), **F** (Front) and **R** (Right).

The obstacle avoidance and goal following behaviors output vectors $\boldsymbol{\alpha}_a$ and $\boldsymbol{\alpha}_g$ respectively, each with dimension three. The elements of the vectors indicate the activation levels of the output fuzzy sets. To combine the outputs of the behaviors a simple behavior blender with constant *desirability functions* $d_a = 0.9$ and $d_g = 0.1$ is employed. The blender fuses the outputs of each behaviors using Eqn. (2) and passes the fused vector $\boldsymbol{\alpha}_f$ to the defuzzifier, which decodes $\boldsymbol{\alpha}_f$ to a crisp value α using centroid technique.

$$\boldsymbol{\alpha}_f = d_g \boldsymbol{\alpha}_g + d_a \boldsymbol{\alpha}_a \qquad (2)$$

Note that when the robot is controlled by the bias component, most of the time it either collides or follows non-optimal trajectories. It is only for very simple tasks and carefully chosen desirability values that the bias component produces smooth and short trajectories. An example is shown in Fig. 3 where the robot has failed to follow optimal trajectory, though it reached the goal[d]. It is also very simple to find yet another example where the robot would fail to reach the goal point at all!

3 Learning Component

The task of the learner is to take the robot from a home location $\boldsymbol{p}_r(0)$ to some goal location $\boldsymbol{p}_g$. We assume that both the robot and goal locations are spec-

[c]The velocity of the robot is kept constant and only the turn rate is controlled within the range of $[-\pi/9, \pi/9]$ rad/sec.

[d]The bias component has been implemented on the real robot and the same result is obtained for the environment depicted.

2169

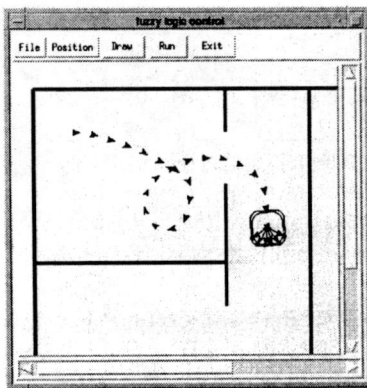

Figure 3. Non optimal trajectory

ified in a Cartesian coordinate system. Furthermore, at any time the robot determines its position by *dead reckoning* method.

The learner has eight inputs and one output. The first five inputs are depth information of each regions, Eqn. (1) and the remaining three inputs are the robot's current heading θ and position $p_r(t)$. However, before these inputs are applied to the learner they are converted to "primed" quantities by normalizing them appropriately. Hence the input, commonly called situation in connectionist, is characterized by a vector,

$$x = \left(\acute{D}_1, \acute{D}_2, \acute{D}_3, \acute{D}_4, \acute{D}_5, \acute{\theta}, \acute{x}_r, \acute{y}_r \right)^T \quad (3)$$

Similar to the bias component, the learning component maintains the vehicle velocity constant and controls the turn rate.

3.1 Trainer

Our trainer has two terms that penalize the learner immediately for every bad actions chosen. The first term f_1 penalizes the learner whenever the robot collides with or moves close to obstacles. If the robot collides, it is penalized by a fixed value, otherwise if the minimum depth reading is less than a certain threshold, the trainer penalizes the learner proportional to the inverse of the minimum reading with D_n as a proportional constant. Therefore, the term that teaches the robot to keep away from obstacles is :

$$f_1 = \begin{cases} -3 & : \text{if collision} \\ -D_n / \min_j(\acute{D}_j) & : \text{if close} \\ 0 & : \text{otherwise} \end{cases} \quad (4)$$

The other term f_2 teaches the robot how to *approach* a goal point. It computes first the acute angle θ between the robot heading and the vector connecting the current robot and goal location. Then as long as $|\theta| < \Theta$, for some positive Θ, the trainer penalizes the robot proportional to $|\theta|$. Beyond Θ, however, the robot is penalized as if a collision has occurred. This forces the robot to explore only the space which lies between $\pm\Theta$ from the goal direction, thereby bounding both the network size and the exploration space.

$$f_2 = \begin{cases} -|\theta|/\Theta & : \text{if} -\Theta \leq \theta \leq \Theta \\ -3 & : \text{otherwise} \end{cases} \quad (5)$$

The total immediate reinforcement r is the sum of the two terms, $r = f_1 + f_2$. Note that the trainer does not teach the robot directly how to *reach* the goal. It trains only how to approach (f_2) the goal without collision (f_1). Therefore, the above reinforce function presupposes that the environment satisfies the constraint that it has a free way (path) through which the robot can reach the goal without collision.

3.2 Learner

The learner architecture is a feed forward neural network consisting of RBF neurons in the hidden layer and a stochastic neuron in the output layer, Fig. 4. The architecture is an *actor-critic* type. The critic element is a one step ahead predictor of the expected future discounted sum of reinforcement values (utility). And the actor element is a multi-parameter stochastic unit that generates actions stochastically from a given distribution [4]. In the architecture, all neurons are tied up to the input and only a winning neuron is connected to the output. Each neuron represents a localized receptive field of width Σ that covers a hyper-sphere in the input space. In the present architecture the width of the receptive fields are all the same and kept fixed.

The learner is initially empty but grows gradually, similar to the work of [3], as it learns and explores the environment. When a new situation is presented to the learner, existing neurons (if any) compete to win the situation. If a winning neuron exists, it will be connected to the output layer to generate an action. Otherwise, a new neuron j is introduced and the following four learning parameters are attached to it: a) utility, u_j, b) prototypical action, p_j, c) output weight, w_j and, d) center position, c_j. Each of these parameters are initialized first and evolve later through RL.

The winning neuron generates an action by exploring only a restricted area around its prototypical action p_j. Restricted exploration accelerates the learning speed by focusing the search area to a fraction of the total action space. To enforce exploration, a *Gaussian stochastic unit* with parameters (μ, σ) is introduced at the output layer [14]. The parameters of the unit are directly determined from the learning parameters of

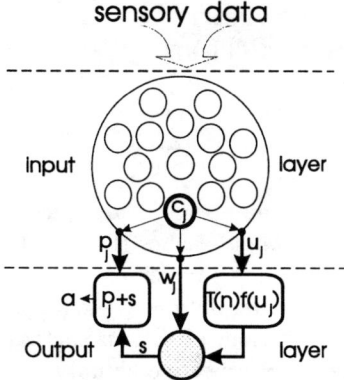

Figure 4. The architecture of the learner

the currently winning neuron using,

$$\mu = w_j \quad \text{and} \quad \sigma = T(n)f(u_j) \quad (6)$$

Where μ is the mean of the distribution, σ is the extent to which the stochastic unit searches for a better action, $T(n)$ is the search-range temperature, n is trial number, $f()$ is the logistic function that take values between [0, 1], Once the parameters are determined, the unit draws a random number $s = \mathcal{N}(\mu, \sigma)$ and generates the final action by modulating the prototypical action with the random number, i.e., $a = p_j + s$. The temperature T is cooled down[e], similar to [1], every time a trial[f] is started, so that the stochastic unit produces progressively deterministic actions.

Before learning starts the robot is located at origin $p_r(t = 0)$. At this location, the robot perceives a situation x. Since the learner is empty (has no neurons), it can not generalize the situation. Therefore, it invokes the bias component. Upon request, the bias component sends its action to the learner. The learner receives the action, adds a neuron, attaches the above learning parameters to the new neuron and initializes the parameters as described below.

To every new neuron j, the algorithm initializes the learning parameters as follows : the center position c_j is equated to the perceived situation x, the prototypical action p_j is set to the action received from the bias component, the utility u_j is estimated by computing the terms of the reinforce function for the current states of the robot and sonars readings and finally, the weight w_j is set to zero.

---

[e]Millán [10] has reported that his learner has determined suitable reactions without employing annealing techniques. However, this is only true if the learner starts near to the optimal actions and utility values - a case which is difficult to meet in general.

[f]A trial is a trajectory that starts at the home location and terminates when the robot collides or reaches the goal.

After initializing the parameters, the learner explores and generates action that moves the robot to a new location $p_r(t + 1)$. At this location, the trainer computes the immediate reinforcement $r(t + 1)$ for the action that brought the robot from $p_r(t)$ to $p_r(t + 1)$ and the robot perceives a new situation x. The new situation is presented to the learner that identifies first the winning neuron closest to the situation, i.e.,

$$winner = arg \min_i(d_i)$$
$$d_i = (c_i - x)^T(c_i - x) \quad (7)$$

If the distance of the winning neuron is larger than Σ, the situation is regarded as novel and the learner invokes the bias component and adds a neuron as discussed above. This way of adding neurons is called *distance driven*. Otherwise the situation is not new and can be generalized.

Next the learner adapts the learning parameters of the previous winning neuron using the immediate reinforcement received and the utility value of the current winning neuron. Thereafter, the learner explores and generates an action for the new situation. If the new action results in collision or takes the robot to the goal, the current trial is terminated, the robot is relocated to home location and, a new trial is started. Otherwise, adaptation and exploration continue until the robot collides or reaches the goal.

3.3 Adaptation

Before the learner generates an action for the present situation, it adapts the learning parameters of the previous winning neuron. Each of the learning parameters are adapted using different adaptation algorithms and error sources.

The utility value of the previous winning neuron $u_j(t)$ is updated by *temporal difference* (TD) method [13]. Assuming neuron i is the present winning neuron with an associated utility $u_i(t+1)$, $r(t+1)$ is the immediate reinforcement, and γ is a real value between [0, 1], the estimation error of u_j (commonly called TD error) between the estimates at $t + 1$ and t is,

$$\delta(t+1) = r(t+1) + \gamma u_i(t+1) - u_j(t) \quad (8)$$

During learning $\delta(t+1)$ is different from zero either because the utility values do not yet converge or the robot has chosen a non optimal action. If $\delta(t+1) < \Delta$, where Δ is some negative constant, then the situation at time t is incorrectly classified to neuron j. Because, even if the situation is close to neuron j as measured by Eqn. (7), it is found to have quite a different utility value from u_j. Therefore, the learner splits this situation from neuron j by creating and adding a new neuron at that situation. This is a second way of adding neuron and is called *error driven*, where the error is the TD error. Otherwise, if $\delta(t+1) > \Delta$, then u_j is adapted by:

$$u_j(t+1) = u_j(t) + \Delta u_j(t+1) \quad (9)$$

$$\Delta u_j(t+1) = \begin{cases} \eta_r\ \delta(t+1) & \delta(t+1) > 0 \\ \eta_p\ \delta(t+1) & \delta(t+1) < 0 \end{cases} \quad (10)$$

where η_r and η_p are two learning rates with $\eta_r > \eta_p$. The utility u_j is adapted less intensively when the TD error is negative than when it is positive. This is because a negative TD error is probably caused by bad action selection that results in a less utility estimate [11].

The output weight w_j directly controls the mean μ of the output stochastic unit and is updated in a direction that lies along the gradient of the expected utility. Williams' REINFORCE algorithm [14] is employed to update the weight w_j,

$$w_j(t+1) = w_j(t) + \Delta w_j(t+1) \quad (11)$$

$$\Delta w_j(t+1) = \begin{cases} \beta_r \delta(t+1)e_j & \delta(t+1) > 0 \\ \beta_p \delta(t+1)e_j & \delta(t+1) < 0 \end{cases} \quad (12)$$

where $\delta(t+1)$ is the TD error given by Eqn. (8) and e_j is the characteristic eligibility of w_j that measures how influential w_j was in determining the stochastic action [14]. Similar to utility update, the weight w_j is updated less intensively when the TD error is negative than when it is positive, i.e. $\beta_r > \beta_p$.

Finally, the center position c_j of the winning neuron is shifted toward x using,

$$\Delta c_j = \epsilon(x - c_j) \quad (13)$$

where ϵ is the learning rate. Our present architecture prevents neurons from collapsing in a region of high data density, since it activates only one neuron for every situation and the widths of all neurons are constant.

Every time the robot moves, the learner keeps track of the winning neuron $j(t)$, its associated utility value $u(j(t))$ and, the immediate reinforcement $r(t)$ along the trajectory. If a trajectory leads to the goal, the learning algorithm *back up* the utility values of all neurons that lie along this trajectory [7, 11]. While utility, output weight and, center position are adapted at each move, prototypical actions are *replaced* by the actual actions if the robot reaches the goal with the best total reinforcement. We define total reinforcement as the sum of immediate reinforcements the learner receives till the robot reaches the goal point.

$$R = \sum_{t=0}^{T} r(t) \quad (14)$$

Where T is total number of moves required to reach the goal. If the robot reaches the goal through a trajectory whose total reinforcement is greater than the maximum R_{max} so far obtained, the algorithm replaces the prototypical actions of all neurons that lie along the trajectory by their respective actual actions.

4 Results

We validate our architecture and learning algorithm on a simulator made for the TRC robot. The simulator has simplified dynamics and assumes noise free sensors. However, it takes into account the physical dimensions of the robot by reducing its size proportionally and places each sensors at the same locations as in the real robot. Besides, the fuzzy rules wired for the real robot are directly transfered without tampering to the simulator.

Figure 5 shows the final trajectory of the robot and Fig. 6 shows the number of neurons and total reinforcement value against the number of trials. It is observed that during the first eight trials the robot failed to reach the goal. This is not surprising, because the learner is empty and has to acquire enough situation-action pairs. Note that the total reinforcement Eqn.(14) is not defined if the robot fails to reach the goal, hence no data is available to plot.

At the ninth trial the robot reached the goal for the first time. It is during this trial that the learner received the lowest total reinforcement (Fig. 6b). Furthermore the path followed during this trial looks more of haphazard motion. After the robot has reached the goal at the ninth trail, it has chosen eight times non optimal actions that ultimately lead to real or virtual collisions (Fig. 6b). This is due to the exploratory nature of the actor element and is common in any reinforcement learning [11]. From the twelfth trial and afterwards the bias component has practically stopped intervening. Besides the size of network (Fig. 6a) has become saturated (between trials only few neurons are added). This indicates that the learner has already started operating in reinforcement mode.

After trial thirty the robot has visited the goal constantly, the total reinforcement remains stable within ± 3, except at trial 41, where the learner explore other actions from the currently known optimal values (Fig. 6b). In subsequent trials, however, it has

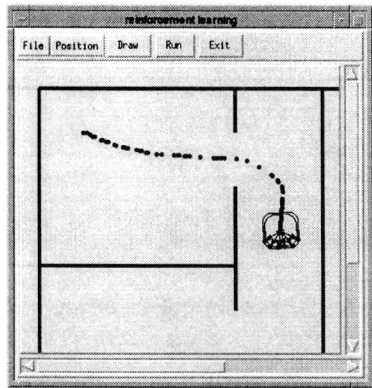

Figure 5. Robot trajectory

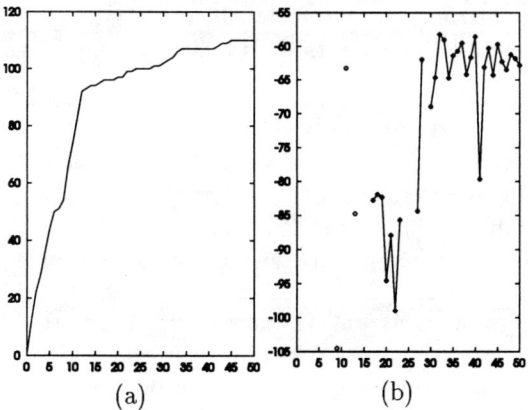

Figure 6. (a) Network size, (b) Reinforcement

quickly discovered its previous performance. The final result (Fig. 5) demonstrates that the robot has indeed adapted quickly the coarse and instinct skill acquired from the bias component to get smooth and planned like trajectory.

5 Conclusion

We have proposed a feasible robot learning architecture that learns quickly from reinforcement signal alone. The architecture has two components: a bias component and a learning component. Initially the bias component intervenes in the learning process frequently to resolve unknown situations. As learning proceeds, however, it stops intervening and the learner optimizes (refines) the acquired situation-action pairs using the reinforcement signal. The architecture has been tested on a simulated TRC mobile robot for a goal reaching task. The final planned like trajectory and the number of trials required validates our approach. Work is going on to transfer the obtained result on the real robot.

6 Acknowledgment

We would like to thank J. R. Millán for clarifying some points on reinforcement learning. The support given to the first author by DAAD under grant code 413/ETH-4-BOA is greatly acknowledged.

References

[1] Andrew G. Barto, Steven J. Bradtke, and Satinder P. Singh. *Learning to Act using Real Time Dynamic Programming.* University of Massachusetts, Amherst MA 01003, January 1993.

[2] Rodney A. Brooks and Maja J. Mataric. Real robots, real learning problem. *Robot Learning*, pages 193–214, 1993.

[3] Jörg Bruske and Gerald Sommer. Dynamic cell structure learns perfectly topology preserving map. *Neural Computation*, 7(4):834–846, 1995.

[4] Vijaykumar Gullapalli. A stochastic reinforcement learning algorithm for learning real valued function. *Neural Networks*, 3:671–692, 1990.

[5] Getachew Hailu. Distributed fuzzy and neural network based navigational behaviours. Technical Lab. Report H/696, CAU, Cognitive Systems Laboratory, 1996.

[6] Getachew Hailu, Jörg Bruske, and Gerald Sommer. Fuzzy logic control of a situated agent. In *Seventh International Fuzzy System Association - World Congress*, pages 494–500, Prague, June 1997.

[7] Sridhar Mahadevan and Jonathan Connell. Automatic programming of behavior-based robots using reinforcement learning. *Artificial Intelligence*, 55:311–365, 1992.

[8] Maja J. Mataric. *Interaction and Intelligent Behavior.* PhD thesis, Massachusetts Institute of Technology, Department of Electrical Engineering and Computer Science, May 1994.

[9] R. Andrew McCallum. Using transitional proximity for faster reinforcement learning. In *Machine Learning - Proceedings of the Ninth International Workshop (ML92)*, pages 316–321, Aberdeen, 1992.

[10] José R. Millán. Reinforcement learning of goal-directed obstacle-avoiding reaction strategies in an autonomous mobile robot. *Robotics and Autonomous Systems*, 15:275–299, 1995. Special Issue on Reinforcement Learning and Robotics.

[11] José R. Millán. Rapid, safe and incremental learning of navigation stratagies. *IEEE Transactions on Systems, Man, and Cybernetics*, 26(3):408–420, June 1996. Special Issue on Learning Autonomous Robots.

[12] Dean A. Pomerleau. *Neural Network Perception for Mobile Robot Guidance.* Kluwer Academic Publishers, 1993.

[13] Richard S. Sutton. Learning to predict by the methods of temporal differences. *Machine Learning*, 3(1):9–44, 1988.

[14] Ronald J. Williams. Simple statistical gradient-following algorithms for connectionist reinforcement learning. *Machine Learning*, 8:229–256, 1992.

[15] John Yen and Nathan Pfluger. A fuzzy logic based extension to payton and rosenblatt's command fusion method for mobile robot navigation. *IEEE Transactions on Systems, Man, and Cybernetics*, 25(6):971–977, June 1995.

Eliminating Sensor Ambiguities via Recurrent Neural Networks in Sensor-Based Learning

Enric Cervera and Angel P. del Pobil
Department of Computer Science
Jaume-I University
Castelló, Spain

Abstract

This paper presents a state identification approach which eliminates ambiguities caused by sensing and uncertainty. In manipulation tasks, the identification of contact states based on force and position sensing is affected by ambiguities. The approach uses recurrent neural networks to learn an internal representation of the finite state automata defined by the sensor patterns. The technique is demonstrated using a simulated learning task. State identification is combined with other techniques in a sensor-based learning architecture for robotic manipulation. Results are presented for simulated tasks. The system is able to manage ambiguous states which previously were impossible to learn.

1 Introduction

Despite the advances in robotics in recent years, the problem of uncertainty and adaptation to partially unknown environments is far to be solved in practice. The robot needs to be endowed with sensors for perceiving its environment, adapting to changes, and reducing the uncertainties.

State identification is a key step of the learning process, since any misunderstanding of the current situation of the system, the state, leads to the choice of wrong actions. In robotic manipulation, much work has been devoted to the identification of contact states [3, 4, 8]. Reinforcement learning in combination with recurrent neural networks have been used in [9]. Recurrent neural learning in mobile robot navigation has been explored in [10]. In real-world applications, ambiguities arise from partial observations of the physical state. In a previous work [2] we showed the inherent ambiguities of force sensing in a simple manipulation task. This paper presents a method for eliminating ambiguities which can be applied to a wide range of tasks, as long as sensor history provides enough information for an unambiguous identification of the states.

The learning architecture presented below combines a powerful state identification scheme with a reinforcement learning algorithm to allow the system to autonomously increase its skills by its own experience in the task.

2 Learning architecture

The proposed method is a framework which integrates robot sensing and different learning schemes in a modular architecture (Fig. 1).

This paper is mainly devoted to the state identification scheme. A new method for eliminating sensor ambiguities is shown. The classification of states is not directly observable a-priori in real-world tasks, so the performance of the state identification scheme is indirectly measured by the results of the learning process. The rest of the architecture and some experimental results on a manipulation task are described in [1].

Throughout this paper we will use the term *sensor measurement* to refer to any element in the finite set of outputs of the feature extraction module [1].

2.1 Recurrent neural networks

The problem faced by the robot is to infer the physical state of the world from the current and past observations. The underlying dynamics of the world are not known, and a model has to be inferred, based on the sequences of observations.

Elman [5] proposed a recurrent neural network architecture (Fig. 2) which has been proved to have as much computational power as a finite state machine. The input layer of an Elman network is divided into two parts: the true input units and the context units. The context units simply hold a copy of the activations of the hidden units from the previous time step. The modifiable connections are all feed-forward, and can be trained by conventional backpropagation methods. The recurrent connections from the hidden to the context layer are fixed. An Elman network can learn to mimic an existing Finite State Automaton (FSA), with different states of the hidden units representing the internal states of the automaton.

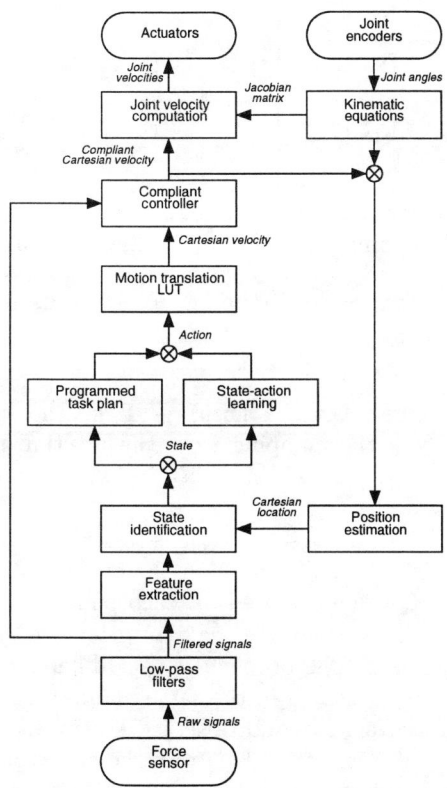

Figure 1: Sensor-based learning architecture

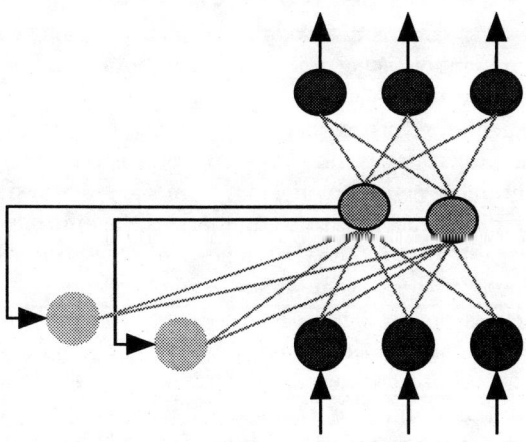

Figure 2: Schematic representation of an Elman network with 3 inputs, 2 hidden units and 3 outputs. Connections with the context units are fixed (black arrows). The rest of the connections (gray lines) are trainable

At each time step, the input of the network is the current sensor reading. The activation is recorded in the context units, so the output does not depend only on the current data but on the input in previous time steps. By means of recording sensor sequences and training a recurrent network for predicting the next sensor reading, the network learns an internal representation of the FSA defined by the sequences. This FSA can be extracted by different methods, determinized and, if necessary, minimized.

2.2 Reinforcement learning

Reinforcement learning (RL) is the learning of a mapping from situations to actions so as to maximize a scalar reward or reinforcement signal.

Q-learning [11] is a RL algorithm that can be used whenever there is no explicit model of the system and the cost structure. This algorithm learns the state-action pairs (called Q-values) which maximize a scalar reinforcement signal that will be received over time. In the simplest case, this measure is the sum of the future reinforcement values, and the objective is to learn an associative mapping that selects at each time step –as a function of the current state– an action that maximizes the expected sum of future reinforcement.

The Q-learning algorithm works by maintaining an estimate of the expected reinforcement for each state-action pair, and adjusting its values based on actions taken and reward received. This is done by using the difference between the immediate reward received plus the discounted value of the next state and the Q-value of the current state-action pair.

2.3 Exploration

During the learning process, two opposing objectives have to be combined. On the one hand, the environment must be sufficiently explored in order to find a (sub-)optimal controller. On the other hand, the environment must also be exploited during learning, i.e., experience gained during learning must also be considered for action selection, if one is interested in minimizing costs of learning. This trade-off between exploration and exploitation calls for efficient exploration capabilities, in order to maximize the effect of learning while minimizing the costs of exploration.

In *Boltzmann* exploration, the utility estimates (e.g. the $Q(i,u)$ values in Q-learning) are used for balancing exploitation and exploration. The probability of selecting an action u in state i is:

$$p(i,u) = \frac{\exp(\frac{Q(i,u)}{T})}{\sum_v \exp(\frac{Q(i,v)}{T})}$$

where T is a positive constant value, which controls the degree of randomness and is often referred to as

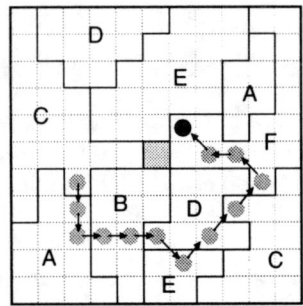

Figure 3: Example of motion in a grid world. The goal is the central gray cell. The sequence of sensor readings is: C A A B B D E D C F F F

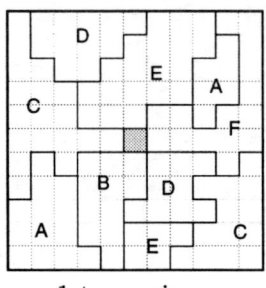 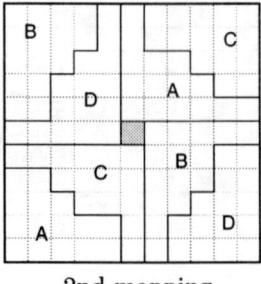

1st mapping　　　　　　　2nd mapping

Figure 4: Two sensor mappings for the sensor-based task

temperature. Its value is gradually decayed from an initial fixed value. When it is close to zero, exploration is turned off and the best action is always selected.

3 Simulation of a sensor-based task

The learning architecture is applied to a sensor-based task, using Q-learning and recurrent networks. A simple grid is chosen for illustrative purposes, but it could be any physical space, e.g. the contact space. Each cell in the grid produces a sensor reading, but the same measurement may be obtained at different locations. Due to this sensing ambiguity the learning task may turn out to be impossible.

The sensor measurement is the only information the robot is aware of. Each experiment begins at a different random position: prior to each step, the current sensor measurement is read and an action is chosen about the direction of motion. The robot can move from one cell to each of its 8-neighbors (an example trajectory is depicted in Fig. 3). A trial run ends when either the goal is attained, or when a predetermined number of steps have been taken.

After each step, the reinforcement algorithm updates its value function according to the reinforcement, which is constant and negative (penalty). The aim is to attain the goal with the minimum number of steps starting from a random location.

Two examples of sensor mappings used in the experiments are depicted in Fig. 4; in both cases there exist some sensing ambiguities. The second mapping is extremely ambiguous in the sense that, for each reading, there are two different regions which are located in opposite directions with regard to the goal.

Let (x,y) be the discrete location (which in fact is the *physical* state) of the robot with respect to a fixed frame of reference. Let (x_f, y_f) be the location of the goal.

Let $\Sigma = \{A, B, C, D, E, ...\}$ be a finite set of sensor measurements. Let $S : (x,y) \rightarrow \Sigma$ be the sensing function, i.e., the mapping from the position to the sensor measurement which is observed at such position (observed state).

Let A be the finite set of actions (discrete motions). If the decision is based only on the current observed state, one Q-value is stored for each pair (Σ, A), and the action whose value is maximum for the current state is chosen. The problem is that different physical states are observed in the same manner, i.e., their sensor measurements and observed states are equal. However, the best action to attain the goal from each state can be quite different.

If the current state is ambiguous, more information is needed in order to make a sensible choice. This information, in the absence of other sensors, can only be acquired by the analysis of past observations.

3.1 Finite state model of the task

During each run through the grid, a sequence of sensor symbols is generated, one symbol at each step. This sequence is a string $s \in \Sigma^*$, and the set of all possible generated strings is a language $L \subseteq \Sigma^*$. Since the number of states is finite, the language is regular, and the problem can be modeled by a finite state system. In such systems, the state summarizes the information concerning past inputs that is needed to determine the behavior of the system on subsequent inputs.

Let us define a finite state automaton (FSA) for that language by the 5-tuple $(Q, \Sigma, \delta, q_0, F)$, where Q is a finite set of states, Σ is the input alphabet, q_0 is the initial state, F is the set of final states, and δ is the transition function mapping $\delta : Q \times \Sigma \rightarrow Q$ [7].

Let $R = \{r_1, r_2, ..., r_n\}$ be the set of regions or clusters of grid cells such that all the cells share the same sensor measurement and it also holds that each region is connected. The goal defines a particular region r_f. Let us define a FSA over the set of regions:

- For each region $r \in R$, let $q[r] \in Q$ be a state of

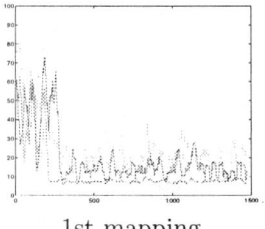
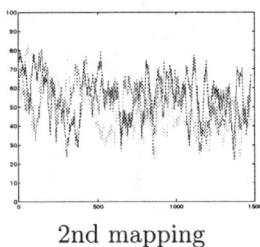

1st mapping   2nd mapping

Figure 5: Steps to goal vs. trials when using only the current sensor measurement

Results for the learning task which consider only the current sensor symbol as the state are depicted in Fig. 5. Three independent runs of 1500 trials each are shown. The plots represent the number of steps to attain the goal, with a limit of 100 steps if not found, versus the number of trials. A moving average window of 25 consecutive values is used to smooth the data.

In both mappings, Q-learning with Boltzmann exploration was used. Each trial starts at a random location and finishes either when the goal is attained or when 100 steps are carried out. In the first mapping, the algorithm converges to a good solution in one of the three trials; the other two show a performance increase affected by an oscillating behavior. The second mapping confirms the intuitive impossibility of learning any sensible action due to the completely ambiguous layout of the sensed states.

the automaton.

- Let q_0 be a distinct state of Q, not associated with any region.

- Let the alphabet Σ be the finite set of sensor measurements.

- Let $F = Q - \{q_0\}$, i.e. all the states except the initial one.

- The (non-deterministic) transition function δ is now defined as follows: $\forall r \neq r_f$
  - Define $\delta(q_0, \lambda) \to q[r]$
  - Define $\delta(q[r], S(r)) \to q[r]$
  - $\forall r' \in neighborhood(r)$, define $\delta(q[r], S(r)) \to q[r']$

The behavior of this FSA is defined by the topology of the grid world, i.e. the connections between regions of cells.

This methodology for building FSAs assumes a complete and perfect knowledge of the world, which is clearly an impossible assumption in any real task. More realistically, the system should build an internal representation of the FSA by means of an inference process, e.g. training a recurrent neural network with sample sequences, which can be obtained by random walks across the grid.

3.2 Sensing ambiguities

The presented approaches are illustrated with computer simulations of the sensor-based goal-finding task. Two possible sensor mappings are used in the simulations (Fig. 4). Each location of the grid is associated to a static sensor measurement, but two distinct isolated locations may produce the same sensor value.

Sensor measurements are ambiguous and the physical state of the system (in this particular case the position or whatever, e.g. the contact state) cannot be directly obtained from the current sensor signal.

3.3 FSA inference with recurrent nets

Previous results have demonstrated the need of taking the sequence of sensor readings into account to determine the state. The learning process is splitted into two phases:

1. Inference of FSA. Random walks are performed through the grid, recording the sensor sequences and training a recurrent network for predicting the next sensor reading. This network learns an internal representation of the FSA which is extracted with any suitable method, determinized and, if necessary, minimized.

2. Action learning. Once the FSA is extracted, the robot begins to learn the actions, with regard to the state determined by the FSA. The learning algorithm is the same that in the previous experiments; only the determination of the current state is different.

In the first mapping, the network has 6 input units (one for each symbol, using *1-of-N* codification scheme), 2 hidden units and 6 outputs. The training set consists of 100 random sequences of variable length (less than 30 symbols), and the network is trained to predict a symbol in its outputs based on the previous one in its inputs. A validation set of other 100 random sequences is used. The network is trained with the backpropagation algorithm with momentum for 150 epochs. The number of epochs was determined by the evolution of the squared error on the validation set. The learning rate is 0.04 and the momentum factor is 0.7. No exhaustive search for optimal parameters was made; we found that the trained network was not very sensitive to small changes in these parameters. The final Mean Square Error (MSE) was 0.54 for the

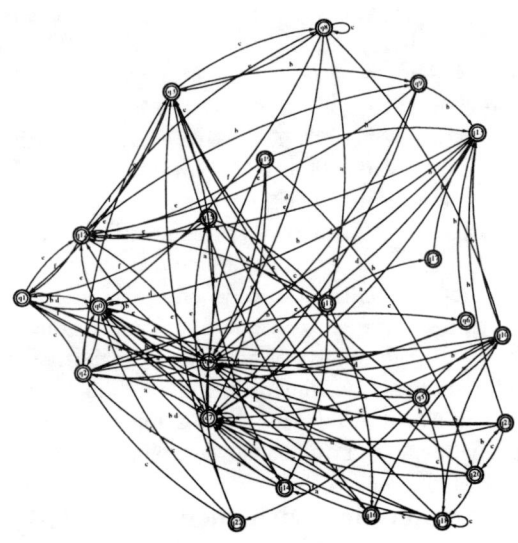

Figure 6: Minimized FSA extracted from an Elman network in the first mapping

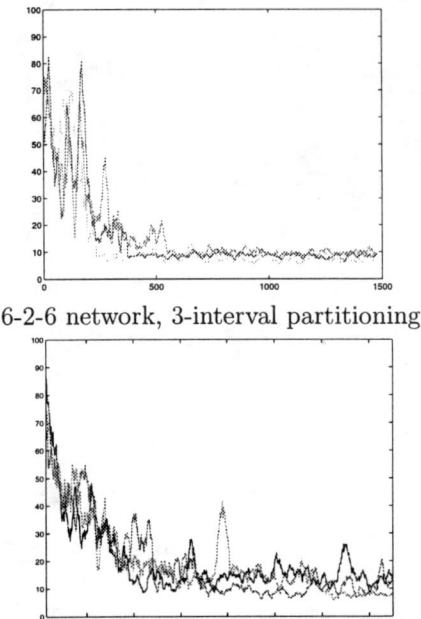

6-2-6 network, 3-interval partitioning

6-4-6 network, 3-interval partitioning

Figure 7: Steps to goal vs. trials using the FSA extracted from Elman networks for the first mapping

training set and 0.59 for the validation set. Additional hidden units did not improve the network performance very much, and were too complex for extracting the automata.

The FSA was extracted by *dynamic state partitioning* [6]. 100 random sequences of variable length (less than 100 symbols each) were presented to the network. The range of each hidden unit [0, 1] was divided into 6 equal intervals. The states of the FSA (up to 36) are the activated intervals and the transitions are defined by the symbols which cause the changes between intervals of activations. The extracted FSA was determinized and minimized to get a final FSA with 23 states and 113 transitions, which is depicted in Fig. 6.

Experimental results for the learning task with this FSA are shown in Fig. 7. The parameters of Q-learning and the exploration scheme are the same in all the experiments. The system achieves a good convergence in all the runs.

The choice of the number of hidden units and the number of intervals in state partitioning is relevant to the final performance. Unfortunately, there is no automatic determination of these values. Experimental results with other values are depicted in Fig. 7. The quality of the learning process depends on the extracted FSA, though a similar behavior is observable in these other cases.

In the second mapping, the Elman network had 4 inputs only (symbols A to D), 2 hidden units and 4 outputs. The training and validation set, and the training parameters were the same as before. The final MSE was 0.46 for the training set and 0.47 for the validation set. The FSA was extracted by partitioning each hidden unit in 5 intervals, and testing the networks with 100 sequences of length less than 30 symbols. The final minimized FSA with 13 states and 48 transitions is depicted in Fig. 8.

Experimental results for the learning task of the second mapping with this FSA are shown in Fig. 9. The parameters of Q-learning and the exploration scheme are the same than in the previous experiments. The system achieves a good convergence in all the runs, with a dramatic increase over the poor performance of the system based on current sensor information (Fig. 5).

4 Conclusion

A new method for state identification which eliminates sensing ambiguities has been presented. It has been experimentally demonstrated that the inference of FSA can help to improve dramatically the learning capabilities of a robot in a sensor-based task. The inferred FSA is able to learn an ambiguous task which is not solvable if only current measurements are considered.

Both the inference and the learning process are au-

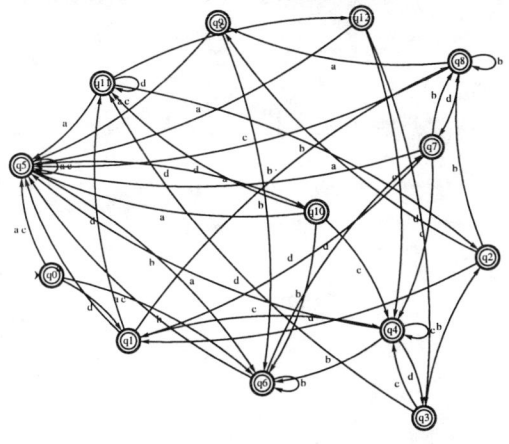

Figure 8: Minimized FSA extracted from an Elman network in the second mapping

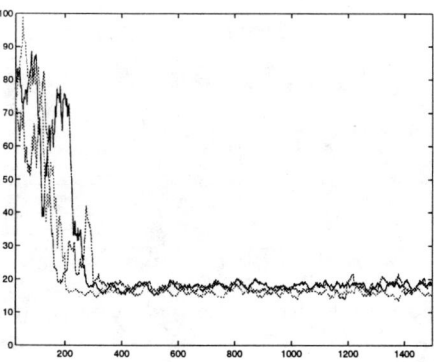

Figure 9: Steps to goal vs. trials using the FSA extracted a 4-2-4 Elman network, with 5-interval partitioning, in the second mapping

tomatically performed. However, a number of parameters have to be manually tuned by an external operator. An interesting extension would be to perform simultaneously both learning tasks: the FSA and the action learning. Future works includes the extension to the case where sensor reading itself is uncertain (i.e., it is probabilistically distributed), and an implementation of the scheme on real-world manipulation tasks which is under way.

Acknowledgments

This paper describes research done in the Robotic Intelligence Laboratory. Support for this laboratory is provided in part by the CICYT under project TAP95-0710, by the Generalitat Valenciana under project GV-2214/94, by the Fundació Caixa Castelló under PIA94-22, and by a scholarship of the FPI Program of the Spanish Department of Education and Science.

References

[1] E. Cervera and A. P. del Pobil. Programming and learning in real-world robotic tasks. In *Proceedings of the IEEE/RSJ International Conference on Intelligent Robots and Systems*, pages 471–476, 1997.

[2] E. Cervera, A. P. del Pobil, E. Marta, and M. A. Serna. Perception-based learning for motion in contact in task planning. *Journal of Intelligent and Robotic Systems*, 17:283–308, 1996.

[3] R. J. Desai and R. A. Volz. Identification and verification of termination conditions in fine motion in presence of sensor errors and geometric uncertainties. In *Proceedings of the IEEE International Conference on Robotics and Automation*, pages 800–807, 1989.

[4] S. Dutré, H. Bruyninckx, and J. de Schutter. Contact identification and monitoring based on energy. In *Proceedings of the IEEE International Conference on Robotics and Automation*, pages 1333–1338, 1996.

[5] J. L. Elman. Finding structure in time. *Cognitive Science*, 14:179–211, 1990.

[6] C. L. Giles, C. B. Miller, D. Chen, G. Z. Sun, H. H. Chen, and Y. C. Lee. Learning and extracting finite state automata with second-order recurrent neural networks. *Neural Computation*, 4:393–405, 1992.

[7] J. E. Hopcroft and J. D. Ullman. *Introduction to Automata Theory, Languages, and Computation*. Series in Computer Science. Addison-Wesley, 1979.

[8] G. E. Hovland and B. J. McCarragher. Frequency-domain force measurements for discrete event contact recognition. In *Proceedings of the IEEE International Conference on Robotics and Automation*, pages 1166–1171, 1996.

[9] L. J. Lin and T. Mitchell. Reinforcement learning with hidden states. In *Proc. of the Int. Conf. on Simulation of Adaptive Behavior: From Animals to Animats 2*, pages 271–280, 1992.

[10] J. Tani. Model-based learning for mobile robot navigation from the dynamical systems perspective. *IEEE Transactions on Systems, Man, and Cybernetics*, 26(3):421–436, 1996.

[11] C. J. C. H. Watkins and P. Dayan. Q-learning. *Machine Learning*, 8:279–292, 1992.

Learning with Assistance based on Evolutionary Computation

Toru Omata

Tokyo Institute of Technology
4259 Nagatsuta, Midoriku, Yokohama, Kanagawa, JAPAN

Abstract

This paper proposes a learning method which learns a motion by employing an assistance in order to simplify it. This learning is done from easy to difficult level. Using Genetic Algorithm, it searches for the parameters of a controller appropriate for controlling the motion to be learnt by gradually increasing difficulty, i.e., by gradually decreasing the degree of assistance. We show that this gradual search enables Genetic Algorithm to evolve a population of controllers efficiently by giving two examples: stable riding of a bicycle and stable controlling of a double inverted pendulum. A bicycle is much easier to control when it is running at a certain velocity. An initial velocity is given as assistance and it is decreased gradually. Similarly a double inverted pendulum is much easier to control when an upward force supports the distal end of the pendulum. The reduction rate of assistance is adjustable in accordance with the adaptability of a population to the reduction.

1 Introduction

Reinforcement learning is attracting research interests but its disadvantage is that it requires in general many trials and much time to learn. Learning from easy to difficult is a way to overcome this disadvantage. Bilchev *et al* propose Inductive Search which solves function optimization problems gradually from easy to difficult[1]. Yang *et al* propose Progressive Learning which learns a motion to be learnt from slow to fast and apply it to a peg insertion task[2]. Hikage *et al* propose Progressive Evolution in which evolution takes place stepwise to match environmental changes[3].

For mechanical systems, especially robots, assisting them mechanically with an external action is an efficient way to simplify a motion to be learnt. For example, a bicycle is difficult to control when it runs slowly. If one learns himself how to ride a bicycle which initially stands still, one must learn the difficult control at a slow speed first. But if someone else (typically parents teaching their children) pushes the bicycle to give an initial velocity as shown in Fig.1, the bicycle becomes much easier to control and the rider can learn the easier control at a certain speed first.

Figure 1: Practice of riding a bicycle initiated by a pusher

After the rider learns to ride a bicycle with a certain initial velocity, the assistant slightly decreases it. We expect that the rider eventually acquires a skill of riding a bicycle without assistance by gradually decreasing the initial velocity.

This paper uses evolutionary computation, especially Genetic Algorithm as a learning technique. GAs are widely used for function optimization but they are best suited for evolutionary systems[3] [6] [7]. Creatures have adapted themselves to environmental changes for a long time. They have acquired skills of motions and new species have emerged. Emergence of new species also causes environmental changes. Sims proposed a system for the evolution and co-evolution of virtual creatures that compete to each other[6]. Both the bodies and the brains of the virtual creatures are evolved simultaneously.

Changing the degree of assistance corresponds to actively changing the environment. This analogy motivates us to use Genetic Algorithm as a techniques to evolve controllers for the motion to be learnt.

This paper is organized as follows. Section 2 presents a GA based learning method which employs an assistance to simplify a motion to be learnt. Section 3 applies the method to stable riding of a bicycle and Section 4 to stable controlling of a double inverted pendulum. Section 5 presents a modified learning method which adjusts the reduction rate of assistance in accordance with the adaptability of controllers to the reduction.

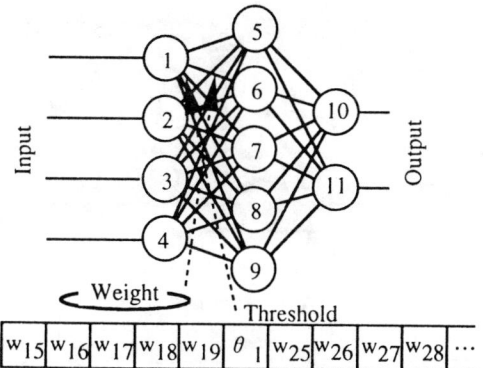

Figure 2: Neural network controller and coding

2 Learning with assistance

2.1 learning of a controller using GA

In this paper, we concern ourselves with parameter learning of a controller which controls a given motion. The inputs of the controller are sensor signals and the outputs are command signals to actuators. As a controller, a three layered neural network is employed simply because it can approximate any input-output mapping by adjusting its weight and threshold. Any other classes of functions can be employed if they can approximate any input-output mapping. We use GA to adjust the parameters of the NN controller. One dimensional array of the weights and thresholds of a NN controller forms the genotype of the GA as shown in Fig.2.

Much work has been done to improve the performance of NN and GA. For example, how to determine the best number of neurons has been a research topic in the studies of NN for a long time. Such results may help improve the performance of learning. This paper, however, avoids discussing general topics on NN and GA but concentrates in our own topic. The number of layers and the number of the neurons of the NN controller are fixed during learning. For the same reason, we use a standard GA as:
[Crossover]: one point crossover.
[Selection of crossover point]: random but does not cut the floating number of each weight or threshold.
[Selection of individuals]: roulette selection with elite preservation. An individual is selected as a parent with probability proportional to the fitness value FIT defined by

$$FIT = \begin{cases} (E_{max} - E)^2 & (E < E_{max}) \\ 0 & (E \geq E_{max}) \end{cases} \quad (1)$$

where E is a performance index. The smaller the E is, the better it is.
[Mutation]: replacement of randomly selected one of the weights or thresholds by a random value.

2.2 Assistance

If the motion to be learnt is difficult, a small set of NN controllers are suitable to control the motion. Simple GA is not adequate to search for the small solution set of controllers.

If an assistance simplifies the motion to be learnt and enlarges the solution set, it can be found more easily. The proposed learning method gradually searches for the solution set of controllers by gradually increasing difficulty, i.e., by gradually decreasing the degree of the assistance. Let $F \geq 0$ be the degree of an assistance. The learning procedure is described as follows.

1. Set the initial value of F and create an initial population of NN controllers randomly.

2. Evaluate the performance index E for all the NN controllers of the population. If E of the best n_{cri} controllers is smaller than E_{cri}, then the population is judged to learn the motion with the assistance of F and subtract ΔF from F.

3. If $F = 0$, then terminate. Otherwise create the next population by crossover and mutation and repeat Step 2.

Fig.3 schematizes this process. For this learning method to work, an assistance must enlarge the solution set of NN controllers and the difficulty of the motion must change continuously with the degree of assistance.

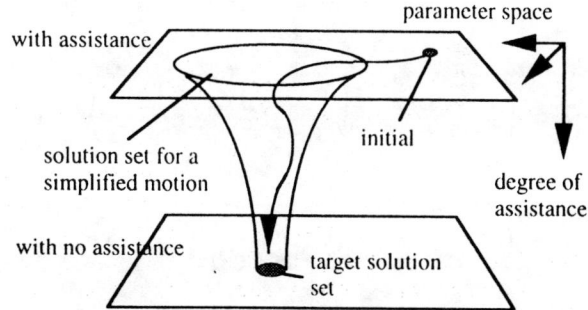

Figure 3: Schematized learning process

This paper is not concerned with *meta*-learning of searching for a good way of assistance for a motion but assumes that it is known *a prior*. This is first because it is not easy even for a man to find a good way of assistance and second because once it is found, it is not necessary to look for it any more. One might spend a long time to notice that giving an initial speed is a good way to assist riding of a bicycle if he/she did not know it.

3 Stable riding of a bicycle

This section applies the proposed learning with assistance to stable riding of a bicycle by giving an initial velocity as an assistance. Trajectory following is not considered and the workspace for the bicycle is assumed to be large enough. The goal is to be able to ride a bicycle even at a slow speed.

3.1 A model of the bicycle

For a dynamic model of a bicycle, we use the model described by nonlinear equations (2) studied by Getz[8](see Fig4). The model adopts the following simplifications and approximations.

1. All the mass of the bicycle is centralized at a point in the body.
2. The wheels have negligible inertial moments, mass, radii, and width.
3. The wheels roll without slippage on the ground.
4. Pedaling the bicycle to drive the rear wheel is considered to be equivalent to applying a force at the point mass.
5. The steering angle can track any desired input.

$$\tilde{M} \begin{pmatrix} \ddot{\alpha} \\ \dot{v}_r \end{pmatrix} = \tilde{F} + \tilde{B} \begin{pmatrix} w_\sigma \\ u_r \end{pmatrix} \quad (2)$$

where

$$\tilde{M} = \begin{pmatrix} p^2 & -cpc_\alpha\sigma \\ -cpc_\alpha\sigma & 1 + (c^2 + p^2 s_\alpha{}^2)\sigma^2 + 2p\sigma s_\alpha \end{pmatrix} \quad (3)$$

$$\tilde{F} = \begin{pmatrix} gps_\alpha + (1 + p\sigma s_\alpha)pc_\alpha\sigma v_r{}^2 \\ -(1 + p\sigma s_\alpha)2pc_\alpha\sigma v_r\dot{\alpha} - cps_\alpha\dot{\alpha}^2 \end{pmatrix} \quad (4)$$

$$\tilde{B} = \begin{pmatrix} cpc_\alpha v_r & 0 \\ -(c^2\sigma + ps_\alpha(1 + p\sigma s_\alpha))v_r & 1/m \end{pmatrix} \quad (5)$$

p	:	vertical position of the center of gravity (J=1.0[m])
c	:	horizontal position of the center of gravity from the rear wheel(=0.5[m])
b	:	distance between the front and rear wheels (J=1.0[m])
m	:	mass of the bicycle(=30.0[kg])
α	:	roll angle of the bicycle
v_r	:	velocity of the rear wheel
Φ	:	steering angle
σ	:	$= \tan(\Phi/b)$
w_σ	:	derivative of σ ($= \dot{\sigma}$)
u_r	:	force applied to the point mass
	:	by the rear wheel
s_α, c_α	:	$\sin\alpha, \cos\alpha$

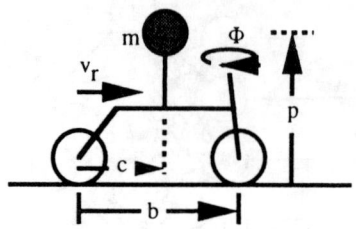

Figure 4: Bicycle model proposed by Getz

The first column of $\tilde{B}$ shows that the smaller the velocity v_r is, the less w_σ has an effect on the bicycle dynamics. When $v_r = 0$, w_σ has no effect on it. This model holds the characteristic of a bicycle that it is difficult to ride at a slow speed.

One often leans one's body to make a turn in the direction of the turn especially at a high speed. This motion is not important for stable riding at a slow speed. Thus we do not consider rider's swaying from side to side.

3.2 Learning of riding a bicycle with assistance

The inputs of the neural network for controlling the bicycle are the four states of the bicycle α, $\dot{\alpha}$, σ, and v_r. The outputs are the two control commands w_σ u_r which are amplified so that they range $-2\pi \leq w_\sigma \leq 2\pi$ [rad/s], and $-30.0 \leq u_r \leq 30.0[N]$, respectively. The negative value of u_r corresponds to braking the bicycle.

The NN controllers are evaluated by the performance index

$$E = \frac{1}{T}\int_0^T (Q_\alpha \alpha^2 + Q_v v_r{}^2)dt \quad (6)$$

where T is the duration of running till the bicycle falls. If it does not fall within T_{end}, it is equal to T_{end}.

This equation gives better evaluation when the bicycle remains as upright as possible and runs as slowly as possible. Without the second term, the bicycle might learn to run fast.

3.3 Simulation

The number of neurons at the hidden layer of the NN controller is five. As we mentioned, we do not search for a better number of the hidden neurons. From practical consideration, we limit the maximum velocity of the bicycle to 3.0[m/s]. Otherwise the bicycle would learn to run at a speed as fast as possible. For the same reason, the initial velocity given as an assistant is also limited to 3.0[m/s]. We disturb $\ddot{\alpha}, \dot{\sigma}$, and $\dot{v}_r$ within 3%. Table 1 shows the values of various parameters.

We did simulations of learning without assistance and learning with assistance ten times, respectively. Learning without assistance means the standard GA

Table 1: Parameters

ΔF	0.3[m/s]
n_{cri}	3
E_{cri}	0.1
Q_α : coefficient of α	3.0[rad^{-2}]
Q_v : coefficient of v_r	0.01[(m/s)$^{-2}$]
E_{max}	5.0
population size	30
preserved elite individuals	10
probability of mutation	0.03
range of initial value of α	$\pm 8.73 \times 10^{-2}$ [rad]
range of initial value of $\dot\alpha$	$\pm 8.73 \times 10^{-2}$ [rad/s]
range of initial value of δ	$\pm 8.73 \times 10^{-2}$ [rad]
T_{end}	5[s]
sampling period	0.01[s]

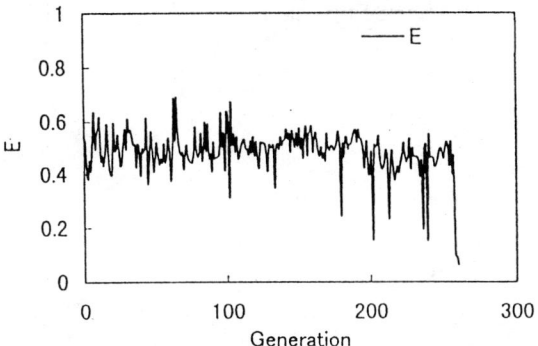

Figure 5: Learning without assistance

described in Section 2.1. Table 2 shows the numbers of generations required till the learning processes terminate ranked in increasing order.

Although the results range widely due to indeterminate factors such as mutation and disturbance, the average number of generations required by learning with assistance is one twelfth of that required by learning without assistance.

Table 2: The number of generations for obtaining stable riding of a bicycle

ranking	1	2	3	4	5
no assistance	60	101	114	136	147
with assistance	17	17	18	18	19

6	7	8	9	10	average
260	476	532	890	1000	372
23	26	31	35	117	31

Space does not allow us to show all the results in detail. So we select the sixth ranking simulations of learning with/without assistance as typical examples. Fig.5 shows the performance index E of the best controller versus generation for learning without assistance and Fig.6 for learning with assistance. For learning with assistance, the degree of assistance F is also shown. Note that the performance index E does not decrease monotonously in spite of the elite preservation. This is because the system is disturbed by random noises.

Fig.7 shows an example of bicycle riding acquired by learning with assistance. The bicycle makes turns

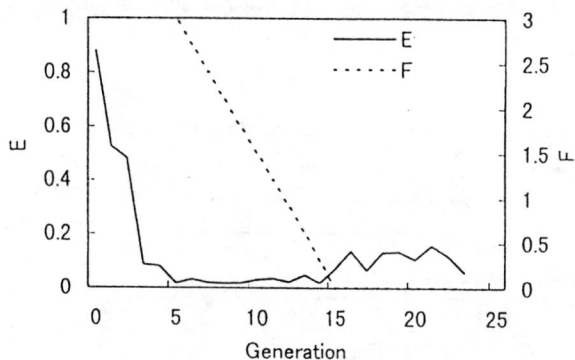

Figure 6: Learning with assistance $\Delta F = 0.3$

in order not to fall, which is often observed in human riding.

4 Stable controlling of a double inverted pendulum

This section applies the proposed learning with assistance to stable controlling of a double inverted pendulum as shown in Fig.8. An upward force exerted at the distal end of the upper link supports the pendulum as an assistance. The state variables of the double inverted pendulum are given by

$$\boldsymbol{X} = (x\ \theta_1\ \theta_2\ \dot x\ \dot\theta_1\ \dot\theta_2)^T \qquad (7)$$

where x is the position of the center of gravity of the lower pendulum, θ_1 and θ_2 the angles of the lower and upper pendulum.

NN controllers are evaluated by the performance index

$$E = \frac{1}{T}\int_0^T \boldsymbol{X}^T Q \boldsymbol{X} dt \qquad (8)$$

where Q is a weight matrix given by $Q = diag(1.5\ 3.0\ 3.0\ 0.3\ 0.6\ 0.6)$

The initial supporting force is set to 19[N], 96.8% of the gravitational force acting on the pendulum. The

Figure 7: Example of stable riding of a bicycle

Table 3: Parameters

ΔF	4.0[N]
n_{cri}	3
E_{cri}	0.03
E_{max}	5.0
population size	30
preserved elite individuals	10
probability of mutation	0.03
range of initial value of x	±0.3 [m]
range of initial value of θ_1	±0.05 [rad]
range of initial value of θ_2	±0.05 [rad]
range of initial value of $\dot{x}$	±0.06 [m/s]
range of initial value of $\dot{\theta}_1$	±0.1 [rad/s]
range of initial value of $\dot{\theta}_2$	±0.1 [rad/s]
T_{end}	5[s]
sampling period	0.01[s]

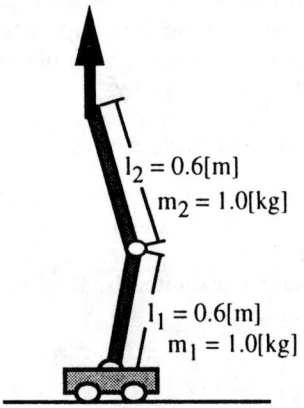

Figure 8: Double inverted pendulum with assistance of an upward force

Table 4: The number of generations for obtaining stable controlling of a double inverted pendulum

ranking	1	2	3	4	5
no assistance	81	95	184	195	196
$\Delta F = 4.0$	28	36	45	60	61
adaptive	40	47	54	65	70

6	7	8	9	10	average
233	248	271	311	345	216
63	64	73	78	80	59
70	79	81	84	93	68

cart is disturbed by a random force of maximum 0.1[N] in the lateral direction and the lower and upper links are disturbed by random torques of maximum 0.1[Nm]. The number of neurons at the hidden layer is five. Table 3 shows the values of various parameters The lengths and masses of the links are shown in Fig. 8.

We did simulations of learning without assistance and with assistance ten times, respectively. Table 4 shows the number of generations required till the learning processes terminates in increasing order. The last row is the results of learning with adaptively adjusted assistance which is discussed later in Section 5.

Like the previous example of bicycle riding, the results range widely due to indeterminate factors. The average number of generations required by learning with assistance is about one fourth of that required by learning without assistance.

We again select the sixth ranking simulations of learning with/without assistance as typical examples. Fig.9 shows the performance index E of the best controller versus generation for learning without assistance and Fig.10 for learning with assistance. For learning with assistance, the degree of assistance F is also shown.

5 Adaptive adjustment of learning speed

If learning with assistance reduces the degree of assistance too rapidly, the population of controllers can not adapt to the reduction. To cope with this inadaptation, we modify learning with assistance so that it preserves a population of NN controllers each time the generations alter, and it resumes the previous population when a new population can not adapt to the reduction.

The reduction rate ΔF is also reduced. But if ΔF is too small, learning with assistance progresses slowly. It is resumed after the inadaptation is settled.

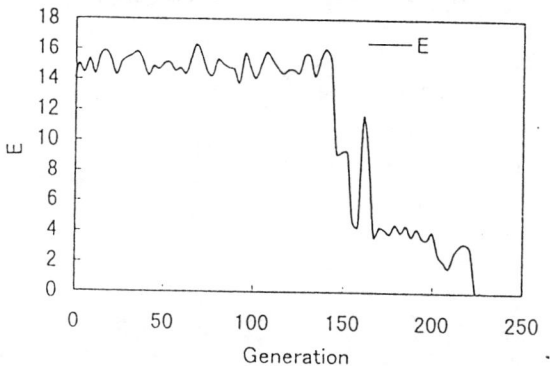

Figure 9: Learning without assistance

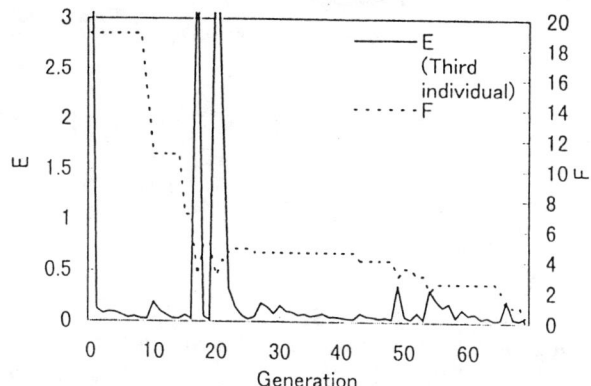

Figure 11: Learning with assistance adjusting ΔF adaptively

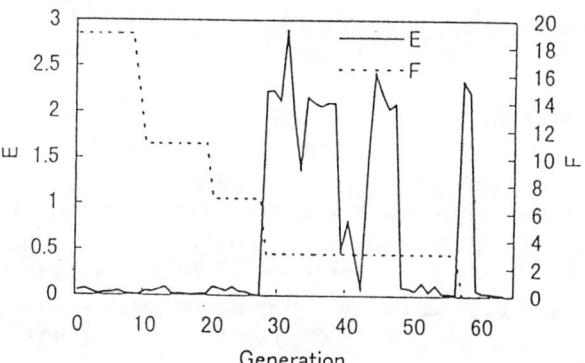

Figure 10: Learning with assistance $\Delta F = 4.0$

We did simulations of a double inverted pendulum ten times. A population of controllers is judged to adapt to the reduction of assistance if the performance index E of the best three controllers are smaller than 0.3. And the reduction rate ΔF is reduced by half when inadaptation occurs. It is doubled until it reaches its initial value after the inadaptation is settled.

The last row of Table 2 shows the numbers of generations required by the modified learning with assistance. They are a little more than those required by learning with assistance with ΔF being constant, 4.0. This is because the modified learning with assistance reduces the degree of assistance more carefully.

Fig.11 shows the performance index E of the third best controller for the sixth ranking simulation. Inadaptation is settled soon after it occurs.

6 Conclusion

This paper proposes learning with assistance which employs an assistance to simplify the motions to be learnt. The examples of bicycle riding and stable controlling of a double inverted pendulum show that the proposed learning method find an appropriate controller more efficiently than simple GA.

To apply this method to real robots is our future work. A robot helping another robot learn a motion is a new type of cooperation of two robots.

The proposed learning method, however, still requires many trials evaluating all the controllers in a population until it finds an appropriate controller. To reduce the number of trials in a real application, a controller should be evaluated on a model of a plant before on a real plant. We expect that an assistance also makes it easy to obtain a model of an unstable plant because it reduce the instability of the plant.

References

[1] G. Blichev adn I. Parmee, "Inductive Search", IEEE Int. Conf. on Evolutionary Computation, 832-836, 1996

[2] B. H. Yang and H. Asada, "Progressive Learning for Robotic Assembly: Learning Impedance with an Excitation Scheduling Method", IEEE Int. Conf. on Robotics and Automation, 2538-2544, 1995

[3] T. Hikage, H. Hemmi, and K. Shimohara, "Progressive Evolution Model Using a Hardware Evolution System", Proc. of 2nd Int. conf. of Artificial Life and Robots", 18-21, 1997

[4] D. E. Goldberg, "Genetic Algorithms in Search, Optimization and Machine Learning", Reading MA: Addison-Wesley, 1989

[5] T. Back, U. Hammel and H. P. Schwefel, "Evolutionary Computation: Comments on the History and Current State", IEEE Trans. on Evolutionary Computation, Vol. 1, No. 1, 3-17, 1997

[6] K. Sims, "Evolving 3D Morphology and Behavior by Competition", Artificial Life IV, The MIT press Cambridge, 28-39, 1994

[7] P. Bourgine and D. Snyers, "Lotka Voterra Coevolution at the Edge of Chaos", Artificial Evolution, Springer, 131-144, 1995

[8] N. H. Getz and J. E. Marsden, "Control for an Autonomous Bicycle", IEEE Int. Conf. on Robotics and Automation 1994.

Multilayered Reinforcement Learning for Complicated Collision Avoidance Problems

Teruo Fujii, Yoshikazu Arai, Hajime Asama, and Isao Endo

The Institute of Physical and Chemical Research (RIKEN)
2-1 Hirosawa Wako-shi Saitama, 351-01, JAPAN
{fujii, arai, asama, endo}@cel.riken.go.jp

Abstract

We have proposed the collision avoidance methods in a multirobot system based on the information exchanged by the "LOCISS: LOcally Communicable Infrared Sensory System", which is developed by the authors. One of the problems in the LOCISS based methods is that the number of situations which should be considered increases very much when the number of the robots and stationary obstacles in the working environment increases. In order to reduce the required computational power and memory capacity for such a large number of situations, we propose, in this paper, a multilayered reinforcement learning scheme to acquire appropriate collision avoidance behaviors. The feasibility and the performance of the proposed scheme is examined through the experiment using actual mobile robots.

1. Introduction

In order to achieve the attractive characteristics of multirobot systems [1, 2], a sophisticated mechanism for mutual collision avoidance should be implemented onto each robot as the primary autonomous functions. While robots' ability to recognize other robots and obstacles is totally dependent on their sensing capability. In order to make the recognition simple and efficient, we have developed a infrared sensory / communication system named "LOCISS: LOcally Communicable Infrared Sensory System [3]" by which relevant information for collision avoidance can be exchanged in a form of coded signals via infrared light emissions.

In the previous paper [4], we confirmed that simple collision avoidance can be realized by a rule based method using the LOCISS as far as the collision between two robots or one robot and one obstacle is considered. In this case, we can easily define the rules which determine what kind of behaviors should be taken according to the signals detected by the LOCISS. It is, however, difficult to extend this rule based approach to the situations in which three or more robots and obstacles appear, because it is almost impossible to define the large number of rules for such complicated situations. In order to solve such complicated situations usually appear in the working environments of multirobot systems, a reinforcement learning based method is introduced in [5] to install the mechanism for adaptive acquisition of the appropriate behavioral repertoire into each robot. While the basic potentials of the learning based method was verified, there still exist several problems when we consider the implementation of the method onto the real robots. One of the most serious problems among them is that the number of the possible situations becomes extremely large beyond the capacity of memories which can be mounted on an actual mobile robot, because of the combinatorial explosion of the sensory patterns exchanged by the LOCISS among multiple robots.

In order to reduce the number of combinations and to realize a feasible control mechanism which can be installed in a robot's onboard computer system, we propose, in this paper, a multilayered reinforcement learning scheme for acquisition of appropriate collision avoidance behaviors. A basic framework of the scheme along with a brief overview of the LOCISS system is described in the following chapters. And it is shown that a structured controller is constructed by the proposed learning scheme through numerical simulations. Finally the performance of the controller in a real world is examined through the experiment by implementing the controller onto the omni-directional mobile robots.

2. The LOCISS System

The LOCISS is basically a sensing system based on local communication with other robots using modulated infrared light [3]. Figure 1 shows the concept of the local communication using the LOCISS. By transmitting such data as self-ID, moving speed, etc. on the infrared signals,

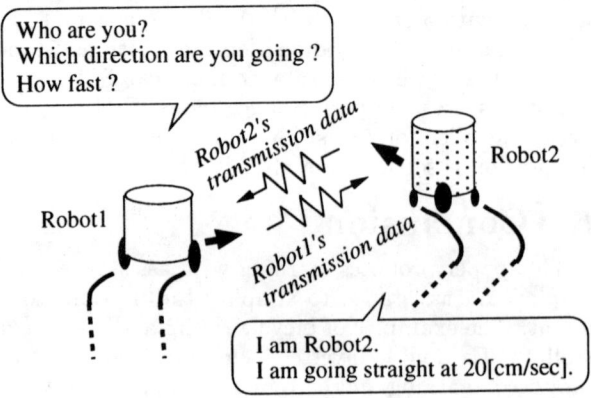

Figure 1 The Concept of the LOCISS

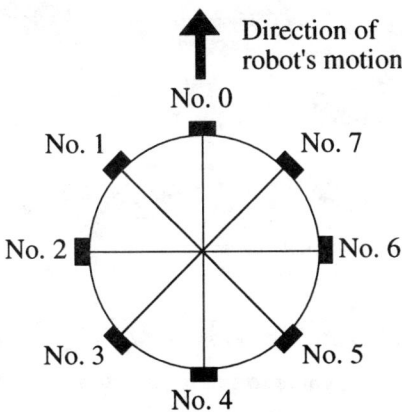

Figure 2 Configuration of the LOCISS

Figure 3 Mobile Robot with the LOCISS

those data can be exchanged among the robots exist in a short range. The source of the signals are easily distinguished by such kind of information. In the figure, the robot1 can immediately recognize robot2 by receiving the ID of the robot2. And when the robot1 detects the signal with the ID of its own, that means there is a stationary obstacle other than mobile robots.

Figure 2 shows a typical configuration of the LOCISS which consists of eight channels of transmitter/receiver devices mounted symmetrically along radial axes. Following the configuration the LOCISS system was fabricated and mounted on a top of an actual mobile robot (Fig.3). The communicable range is typically adjusted at around 1m. According to the direction of the robot's motion, each channel is numbered as shown in Fig.2. This number can be used as the direction index of the source of the received signal. As long as the number of robots was only two, collision avoidance between the robots could be easily accomplished by a rule matrix based method using the robot's ID and the direction indices [4].

3. Multilayered Reinforcement Learning

A learning based approach is one of the feasible methods to solve complicated problems for which a human designer is unable to define the appropriate rules or control laws in an explicit form. There are some classes of problems which are hardly solved by an analytical way. But most cases have at least one of the following three difficulties;
1) the number of possible situations to be considered is countable, but too large to prepare all the corresponding answers,
2) situation is unpredictable because random processes are essentially involved in the problem, and
3) the answer to the problem is not clearly known.

As the collision avoidance problem in multirobot systems, we assume that three or more robots and stationary obstacles exist in the working environment as shown in Fig.4. Each robot carrying the LOCISS system operates under its own mission, while avoiding collisions to each other by exchanging the information. The problem is so complicated that it may have all of above mentioned difficulties.

In order to address the difficulties 2) and 3), we introduced reinforcement learning approach in the previous paper [5]. In the real world applications, there must be unpredictable random noises and the answers to the complicated collision avoidance problems cannot be derived easily by human designers. It is, therefore, favorable that the answers to the problems can be automatically acquired through the learning process in the real or simulated world. In the acquired answers (behaviors), the characteristics of each robot's mechanisms and the random noises could be also automatically taken into account. But there is still the difficulty 1) which must be solved when we consider the implementation of the learning based method into the actual robot because only the CPUs and memories with limited capacity can be mounted on the robot. To avoid the combinatorial explosion of the number of possible situations, we propose here a new scheme named the "multilay-

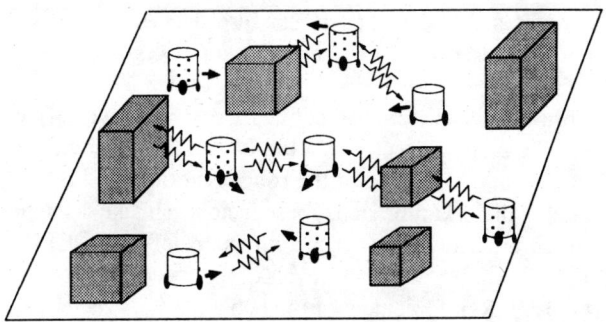

Figure 4 Complicated Collision Avoidance Problem

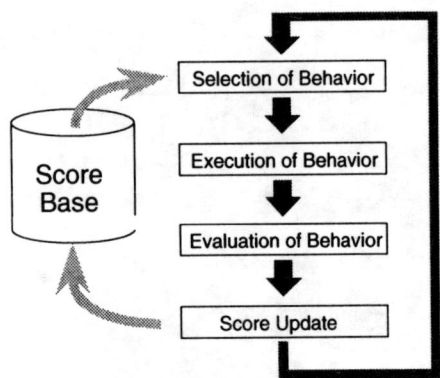

Figure 5 Procedure of Reinforcement Learning

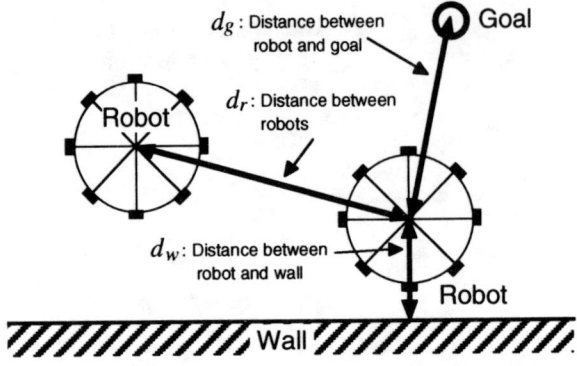

Figure 6 Evaluation of A Selected Behavior

ered reinforcement learning" as the LOCISS based collision avoidance strategy in multirobot environments.

3.1 Problem Definition

Figure 5 shows the procedure of the reinforcement learning which is a series of repetitive process executed once in every sampling cycle. A behavior is selected by the controller which involves a random selection process with weighted probability distribution. The probability distribution is tuned on the basis of the scores which are assigned to all the behaviors in the possible situations. For the working environment described in Fig.4, we evaluate the robot's behavior by weighted summation of three measures to calculate the score values, as follows (Fig.6);

$$E(t) = \alpha \Delta d_w(t) + \beta \Delta d_r(t) - \gamma \Delta d_g(t), \quad (1)$$

$$\begin{aligned}\Delta d_w(t) &= d_w(t) - d_w(t - \Delta t),\\ \Delta d_r(t) &= d_r(t) - d_r(t - \Delta t),\\ \Delta d_g(t) &= d_g(t) - d_g(t - \Delta t),\end{aligned} \quad (2)$$

where d_w denotes the distance between the robot and the stationary obstacle, d_r denotes the distance between two robots, and d_g denotes the distance from a robot to its goal. α, β, and γ are the weighing parameters. Note that when multiple robots and stationary obstacles are detected by the LOCISS simultaneously, these values should be the summation of the distances of all the pairs. According to this definition, when the robot move away from obstacles and other robots, the behavior gets a larger value. And when the robot get closer to its goal, the behavior also gets a larger value.

For the omni-directional mobile robot (Fig.3), we define here a behavior as a combination of two factors, i.e., the direction and the speed of the robot's motion, and is executed in the next time step. These factors can take discrete values which are defined by the directions of the LOCISS' channels $i=0,1\cdots 7$ and the levels of the speed $j=0, 10, 20, 30$ [cm/s]. Reinforcement of the behavior B_{ij} in the situation r is executed by updating the score S^r_{ij} as;

$$S^r_{ij(NEW)} = S^r_{ij(OLD)} + E(t) \quad (3)$$

The selection probability $p^r_{ij}(t)$ of a behavior B_{ij} in the situation r can be calculated as the normalized value of the score S^r_{ij};

$$p^r_{ij}(t) = \frac{S^r_{ij}}{\sum_{m=0}^{7}\sum_{n=0}^{30} S^r_{mn}} \quad (4)$$

3.2 The Controller's I/O

According to the above mentioned framework of the reinforcement learning, we can define the controller which outputs the direction $(0,1\cdots 7)$ and the speed $(0, 10, 20, 30$ [cm/s]$)$ of the robot to move when the inputs which express the surrounding situation are given (Fig.7). The situation is characterized by the robot's own status and the status of other robots and obstacles which can be exchanged by the LOCISS as shown in Fig.8. As the robot's own status, we should consider;
- direction of the goal $(0,1\cdots 7)$, and
- moving speed $(0, 10, 20, 30$ [cm/s]$)$.

By the local communication realized through the LOCISS, we can use;
- channel no. which detects the signal $(0,1\cdots 7)$,
- the received direction index $(0,1\cdots 7)$,
- the received robot ID (robot's own ID or others' ID), and
- the moving speed of another one $(0, 10, 20, 30$ [cm/s]$)$,

as the inputs to the controller.

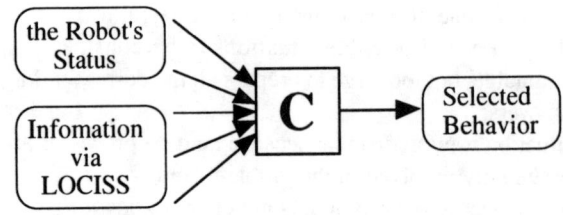

Figure 7 Definition of the Controller

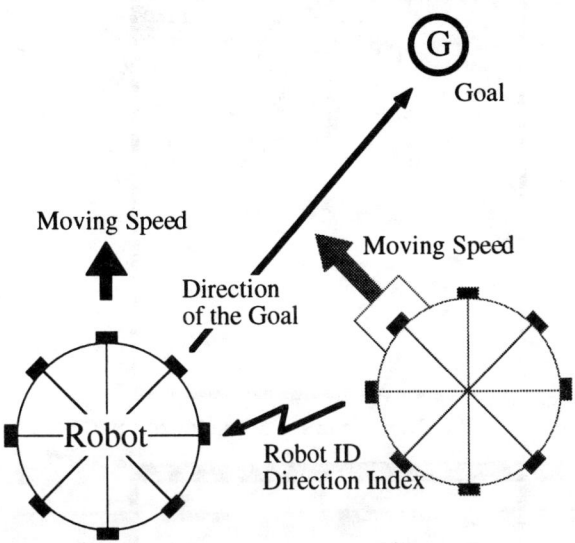

Figure 8 Information to Characterize the Situations

3.3 Multilayered Reinforcement Learning

Figure 9 shows the overall structure of the controller for the multilayered reinforcement learning which consists of four layers of modular controllers corresponding to the stages of reinforcement learning. The learning process is divided into four stages like a curriculum which starts with easier problems and proceed to more complicated ones;
1) learning the goal directed behaviors,
2) learning the avoiding behavior with a single object,
3) learning the avoiding behavior with multiple objects using subsets (three channels) of the LOCISS, and
4) learning with the full set of the LOCISS.

The stage 1) is to construct the controller C_g which outputs the goal directed behavior based on the moving speed and the direction of the goal. The stage 2) is to construct C_{o0} to C_{o7} (C_{oi}; i=0,1···7) which outputs the behavior to avoid the collision to a single robot or obstacle. As shown in Fig. 10, C_{oi} has a conditional switch in its structure according to the ID information received by the LOCISS. If the ID means the existence of other robots or obstacles, C^r_{oi} or C^o_{oi} works as the controller, respectively. If there is no signal for the channel i of the LOCISS, the outputs of the controller C_{oi} becomes the same as C_g (through). The inputs to the controllers C_{sj} (j=A, B, C) are the outputs of C_{oi} according to the grouping of the subset of the LOCISS as shown in Fig.11. The outputs of C_{sj} are passed to the controller C_f as its inputs. And the controller C_f outputs the resulted behavior which reflects all the modular controllers' decisions.

By structuring the process of reinforcement learning and the controller itself, the combinatorial number of situations and behaviors can be drastically reduced ($O(10^{15}) \rightarrow O(10^6)$ according to our calculation). Each layer of the controller is trained in each stage of the learning process. This learning scheme is, therefore, named "Multilayered Reinforcement Learning". The detailed procedure and the results of the learning process in each stage are described in the following chapters.

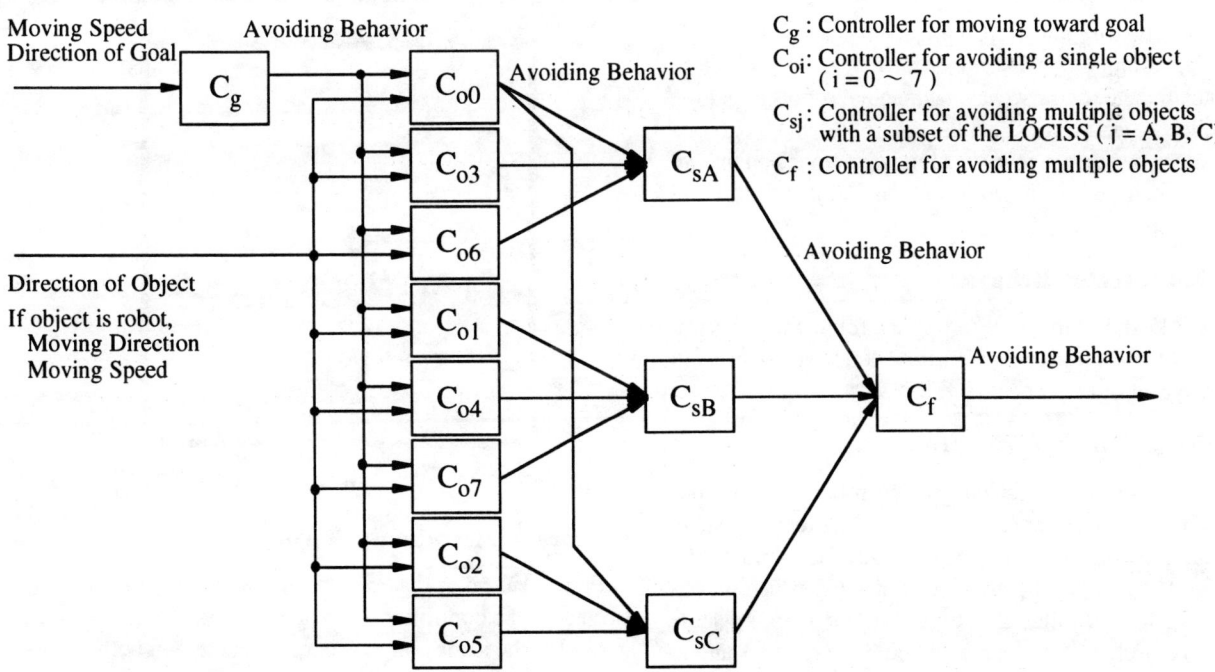

C_g : Controller for moving toward goal
C_{oi} : Controller for avoiding a single object (i = 0 ~ 7)
C_{sj} : Controller for avoiding multiple objects with a subset of the LOCISS (j = A, B, C)
C_f : Controller for avoiding multiple objects

Figure 9 Overall Structure of the Multilayered Controller

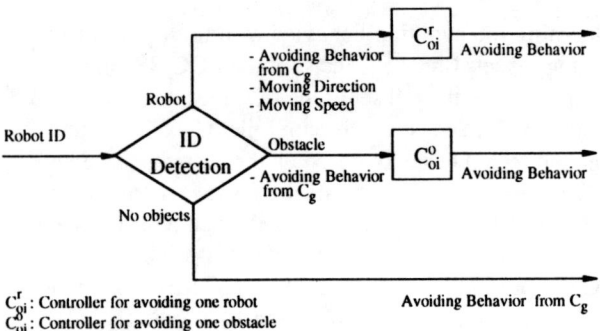

C^r_{oi}: Controller for avoiding one robot
C^o_{oi}: Controller for avoiding one obstacle

Figure 10 Substructure of the Controller C_{oi}

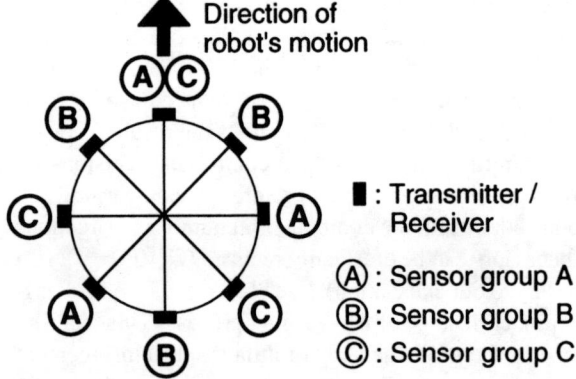

Figure 11 Definition of the Sensor Groups

4. Learning Process

We execute the multilayered learning process through numerical simulations at first, because it is impossible to execute all the necessary trials with actual robots to get the controller. After we construct the controller which exhibits satisfactory performance in the simulated environment, we can implement it into actual robots and adjust to the real world environment.

4.1 Goal Directed Behavior

Figure 12 shows the trajectory of the robot which is controlled by the controller Cg after the 1000 series of trials to learn the goal directed behavior.

4. 2 Single Object Avoidance

Figure 13 shows the trajectory of the robot which is controlled by the controller $C_g + C^o_{oi}$ during the learning stage 2). Wall avoiding behavior is successfully acquired by the controller C^o_{oi}. Figure 14 shows the trajectory which is controlled by the controller $C_g + C^r_{oi}$ during the learning stage 2). The controllers are mounted on both the robots 1 and 2. They become able to avoid mutually after 1000 series of trials.

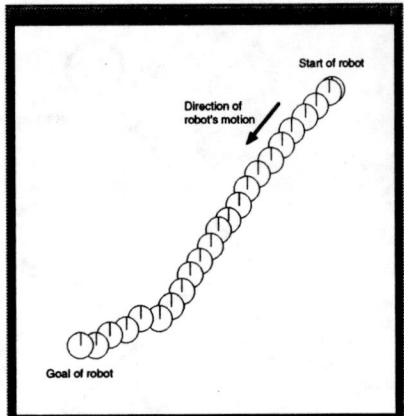

Figure 12 The Goal Directed Behavior of the Robot

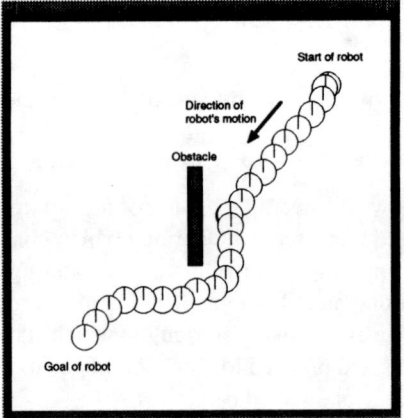

Figure 13 Avoidance of a Single Obstacle

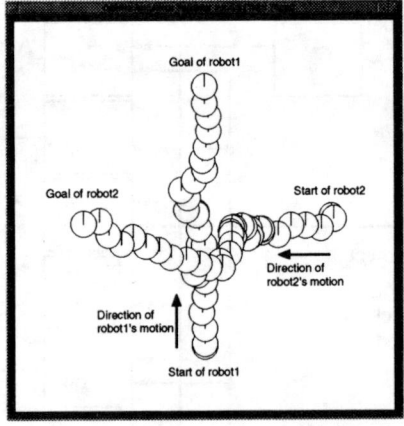

Figure 14 Avoidance of a Single Robot

4. 3 Learning by Sensor Groups

Figure 15 shows the trajectory of the robot which is controlled by the controller C_g through C_{sa} during the learning stage 3). The same controllers are mounted on both robots 1 and 2. While the robot 2 touches the obstacle a few times, it can be said that the behaviors to avoid the obstacle and other robots are partially realized.

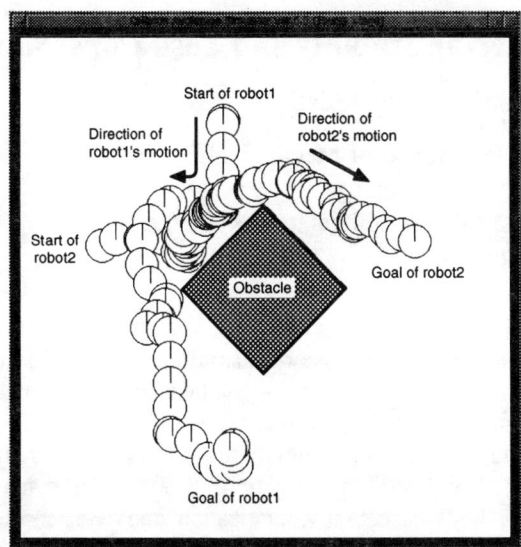

Figure 15 Avoidance of Multipe Robots and Obstacles

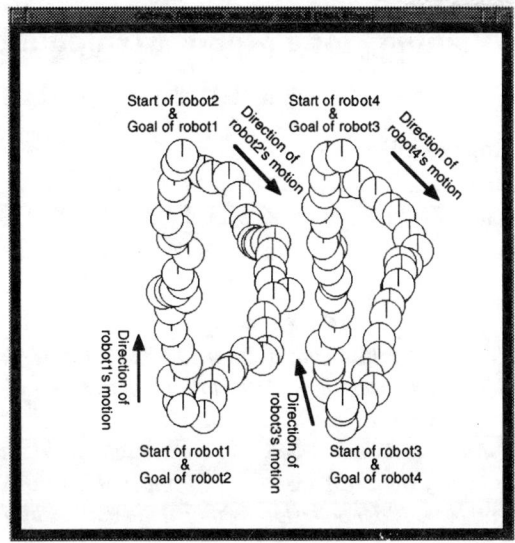

Figure 16 Avoidance of Multipe Robots and Obstacles

4. 4 Full Sensing

Figure 16 shows the trajectory of the robot which is controlled by the overall controller during the learning stage 4). Mutual avoidance among two or more robots are successfully executed. Usually it is difficult to design the built-in rules to realize this kind of behaviors. On the other hand, these behaviors have appeared automatically through the proposed learning process.

4.5 Experiment with actual robots

The constructed controller is implemented onto the omni-directional mobile robots and its performance is examined by experiments. Figure 17 shows the experimental result of collision avoidance. There are four robots and a wall in the working environment. The controller constructed through the learning process is implemented onto every robots. The robots exhibits excellent performance especially when the robot encounters complicated situations which are indicated by A and B in the Fig.17.

5. Conclusion

We proposed a new learning method called "Multilayered Reinforcement Learning" to realize versatile collision avoidance in a multirobot system using the LOCISS. The implemented controller showed excellent performance in experiments using actual robots. By introducing the new method, the required computational power and memory capacity can be drastically reduced. For future works, realization of on-line learning for various environments and types of robots (holonomic, non-holonomic,etc.) should be further discussed.

References

[1] H. Asama, et al., eds. "Distributed Autonomous Robotics Systems", Springer Tokyo (1994).
[2] H. Asama, et al., eds. "Distributed Autonomous Robotics Systems 2", Springer Tokyo (1996).
[3] S. Suzuki, et al., "An Infra-Red Sensory System with Local Communication for Cooperative Multiple Mobile Robots", Proc. IROS'95 (1995) pp.220-225
[4] Y. Arai, et al., "Collision Avoidance among Multiple Autonomous Mobile Robots using LOCISS", Proc. ICRA'96 (1996) pp.2091-2096
[5] Y. Arai, et al., "Adaptive Behavior Acquisition of Collision Avoidance among Multiple Autonomous Mobile Robots", Proc. IROS'97 (1997) pp.1762-1767

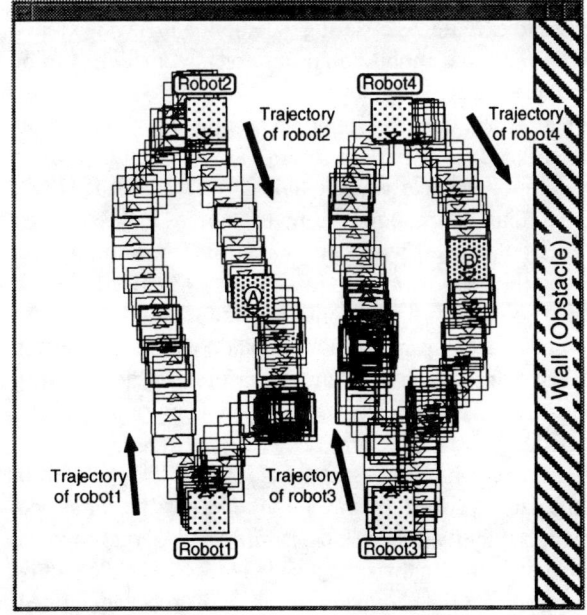

Figure 17 Experimental Result of Collision Avoidance

Motion Planning for a Mobile Manipulator Considering Stability and Task Constraints

Qiang HUANG*, Shigeki SUGANO**, and Kazuo TANIE*

* Department of Robotics
Mechanical Engineering Laboratory
1-2 Namiki, Tsukuba, Ibaraki 305, Japan
e-mail: huang@melcy.mel.go.jp, tanie@mel.go.jp

** Department of Mechanical Engineering
Waseda University
3-4-1 Okubo, Shinjuku, Tokyo 169, Japan
e-mail: sugano@cfi.waseda.ac.jp

Abstract

In order for a mobile manipulator to be used in areas such as offices and houses, the mobile platform must be small-sized. In the case of a small-sized platform, the mobile manipulator may fall down when moving at high speed, or executing tasks in the presence of disturbances. Therefore, it is necessary to consider both stabilization and manipulation simultaneously while coordinating vehicle motion and manipulator motion. In this paper, we propose a method for coordinating vehicle motion planning considering manipulator task constraints, and manipulator motion planning considering platform stability. Specifically, first, the optimal problem of vehicle motion is formulated, considering vehicle dynamics, manipulator workspace and system stability. Next, the manipulator motion is derived, considering stability compensation and manipulator configuration. Finally, the effectiveness of this method is demonstrated by simulation.

1 Introduction

Conventional industrial manipulators are fixed to factory floors and can only execute tasks within a limited workspace. In order to execute tasks during locomotion in a wide area, a vehicle-mounted mobile manipulator is considered to be effective.

Until now, one main topic on mobile manipulators has been to study optimal problems such as manipulator configuration, vehicle dynamics and obstacle avoidance using the redundancy of the mobile manipulator [1, 2]. Another topic is concerned with compensating for disturbances between the vehicle and the manipulator [3]. Recently, the method for integrating locomotion and manipulation was proposed to develop applications in outdoors or rough terrain [4]. Also, research on human safety for developing applications in human-robot collaboration environments was reported [5].

However, these investigations ignored the problem of system stability (overturn prevention). It is possible for a robot have high stability if the vehicle is sufficiently larger than the manipulator. But, in this case, a wide space for vehicle motion is necessary, applicable environments for the robot are limited, and energy consumption and mobility become problems. Considering these facts, especially for future applications in areas such as offices or houses, the vehicle must be miniaturized. In this case of a small-sized vehicle, it is necessary to consider the stability of the mobile manipulator.

Some researchers have studied the stability of a mobile manipulator, but most of them discuss only the static stability [6, 7]. Although some investigations [8, 9] concern dynamic stability, they consider only stabilization, and rarely consider manipulation. Since a mobile manipulator is required not only move stably but also execute tasks simultaneously, it is necessary to consider the compatibility of stabilization and manipulation.

In this paper, our objective is to derive coordinated motion so that the mobile manipulator can move stably and follow a given desired end-point trajectory (path, velocity) at an optimal configuration. First, in order to consider the compatibility of stabilization and manipulation, we propose a coordination algorithm in section 2. Next, in section 3, we formulate the vehicle motion, considering vehicle dynamics, manipulator workspace and system stability. Then, in section 4, we derive the redundant manipulator motion, considering stability compensation and manipulator configuration. Finally, we provide simulation results in section 5 and a conclusion in section 6.

2 Scheme of Coordinated Motion

When considering the compatibility of stabilization and manipulation, it is first necessary to maintain system stability. Then, based on the assurance of system stability, the mobile manipulator should execute tasks with an optimal configuration.

The vehicle motion mainly affects system stability: for example, the robot falls over easily if the velocity or the acceleration of the vehicle are large. Since the range of motion of the manipulator will not be large given an end-point trajectory, the manipulator has only limited ability to aid stability. Therefore, in order to ensure system stability, it is effective to plan vehicle motion first, then plan manipulator motion.

However, if considering only vehicle motion and system stability while planning vehicle motion, the manipulator configuration will not be suitable to execute end-point tasks; that is, manipulation will become difficult. Conversely, if considering only optimal configuration for executing end-

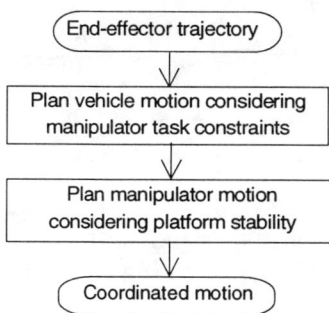

Fig. 1 Algorithm of coordination motion planning

point tasks while planning manipulator motion, the velocity and the acceleration of the vehicle may oscillate easily, and maintenance of system stability will become impossible.

Therefore, we propose the coordination algorithm as shown in Fig. 1. This algorithm consists of vehicle motion planning considering manipulator task constraints, and manipulator motion planning considering system stability. In the following, we discuss the vehicle motion and the manipulator motion, given the end-point trajectory (path, velocity).

3 Vehicle Motion Planning

Until now, although many investigations have studied vehicle motion or mobile robot motion, most of them concern only the vehicle's own factors, such as dynamic performance, obstacle avoidance and minimal distance. In the case of vehicle motion planning of a mobile manipulator, it is necessary to consider not only the vehicle's own factors but also factors such as manipulator configuration and system stability. In order to obtain such a vehicle motion, first plan the vehicle path mainly considering obstacle avoidance, shortest path and so on. Then, plan the vehicle motion (velocity, acceleration) along the planned path, considering vehicle dynamic performance, manipulator workspace and system stability.

We consider a vehicle with nonholonomic constraint. Since there are many investigations have studied such vehicle path planning [10, 11], we only discuss vehicle motion (velocity, acceleration) along the planned path. As mentioned by previous studies, we also use a vehicle path composed by arcs and line segments.

3.1 Minimum of Acceleration Sum

If the vehicle acceleration is small, the vehicle velocity will not oscillate largely, and the energy consumption of vehicle will be small. Therefore, the minimum of acceleration sum can be regarded as one of the characteristics of the vehicle dynamic performance. Let $P_v(t_i)=(x(t_i), y(t_i))$ denote one point on the path, and the minimum of cost function is denoted by the following equation [12]:

$$\text{Min.}\{G(x,y)\} = \text{Min.}\left\{\sum_{i=1}^{N}(\ddot{x}^2(t_i) + \ddot{y}^2(t_i))\right\} \quad (1)$$

where $\ddot{x}(t_i)$, $\ddot{y}(t_i)$ are the second derivatives of the position coordinates (x, y), t_i is the time of the vehicle at point $P_v(t_i)$, N is the number of sample points.

Dividing the whole length of the vehicle path into N equal parts, the relation between the time t_i and the time interval h_i required to move an equal length can be given as follows:

$$\begin{aligned} t_i - t_{i-1} &= h_i \\ t_i &= h_1 + h_2 + \cdots + h_i \end{aligned} \quad (2)$$

If the time interval h_i ($i = 1, 2, \cdots N$) is solved, the vehicle motion can be obtained. The problem of finding the time interval vector $H = (h_1, h_2 \cdots h_N)$ for minimizing the cost function can be result in the problem of nonlinear programming as follows:

$$\text{Min.}_{H \in B} G(H) = \text{Min.}\left\{\sum_{i=1}^{N}(\ddot{x}^2(h_1+h_2+\cdots h_i) + \ddot{y}^2(h_1+h_2+\cdots h_i))\right\} \quad (3)$$

$$B = (h_1, h_2, \cdots h_N)^T \bigg| h_i > 0, \sum_{i=1}^{N} h_i = T_0 \quad (4)$$

where T_0 denotes the fixed time to follow the given end-point trajectory.

3.2 Constraint in Manipulator Workspace

Within the manipulator's kinematic workspace, there are some regions in which the manipulator executes tasks easily, or with difficulty, or even impossibly. The region for the manipulator executing tasks easily is named the effective workspace. Therefore, it is necessary that the distance between the vehicle and the end-point is within the effective workspace while planning vehicle motion. This problem can result in a constraint on the time of vehicle moving the path.

Referring to Fig. 2, the reaching position of the vehicle is assumed as $P_v(t_i)$ at time t_i, and the reaching limit of the end-point is described by $P_m(T_a)$ and $P_m(T_b)$. If $t_i < T_a$, the end-point desired position $P_m(T_a - \Delta t_a)$, $(T_a - \Delta t_a = t_i$, $\Delta t_a > 0)$ is outside the effective workspace at time t_i. On the other hand, if $t_i > t_b$, the end-point desired position $P_m(T_b + \Delta t_b)$, $(T_b + \Delta t_b = t_i$, $\Delta t_b > 0)$ is outside the effective workspace at time t_i. Therefore, in order to follow the desired trajectory, the following constraint equation of the time t_i must be satisfied.

$$T_a \leq t_i \leq T_b \quad (5)$$

However, if we consider the constraint equation (5) of

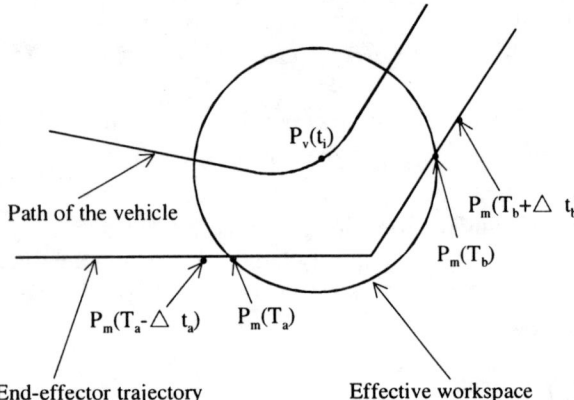

Fig. 2 Vehicle passing time and manipulator workspace

each point on the vehicle path when deriving vehicle motion based on equations (3) and (4), the number of constraints is infinite, and the nonlinear programming problem becomes difficult. On the other hand, we can consider constraint equations of some points. Vehicle velocity will not change violently in the case of minimizing acceleration sum based on equations (3) and (4). So, if the constraint equations of some points are satisfied, constraint equations of all points on the vehicle path will generally be satisfied.

3.3 Constraint in System Stability

As mentioned above, the manipulator has a limited ability to aid system stability, given the end-point trajectory. Therefore, in order to ensure system stability, the vehicle motion (velocity, acceleration) must be planned within this ability.

Since the minimum of acceleration sum is already considered in section 3.1, the acceleration will be continuous and small. Therefore, stability problems due to acceleration can be ignored. On the other hand, when the vehicle moves along arcs with a fixed turning radius, since the larger the velocity is the larger the centrifugal force becomes, the robot is easily destabilized. Therefore, the maximal velocity of arcs must be limited. Let V_c denote the limited velocity considering the manipulator ability to aid system stability [13], the time interval T_j of passing an arc is obtained by the following equation:

$$T_j \geq \frac{L_j}{V_c} \qquad (6)$$

where L_j is the length of arc j, $j = 1, 2, \cdots m$, and m denotes the number of arcs.

3.4 Solution of the Vehicle Motion

From constraints equations (2) (4) (5) (6), the whole path of the vehicle can be divided to k parts. The constraint equations of time interval vector H can be obtained as follows:

$$H \in B = \left\{ H = (h_1, h_2, \cdots h_N) \middle| h_i \geq 0, \sum_{i=1}^{n_1} h_i = T_1, \sum_{i=n_1+1}^{n_1+n_2} h_i = T_2, \cdots, \sum_{i=n_1+\cdots+n_{k-1}+1}^{n_1+\cdots+n_{k-1}+n_k} h_i = T_k, \sum_{j=1}^{k} T_j = T_0 \right\} \qquad (7)$$

where T_i ($i = 1, 2 \cdots k$) indicates the time interval required to move the length of the part i. n_i ($i = 1, 2 \cdots k$) indicates the integer of the length of part i divided by the equal length mentioned in section 3.1.

As mentioned above, the vehicle motion along the planned path considering the minimum of vehicle acceleration sum, manipulator's effective workspace and system stability, can result in the nonlinear programming problem of equation (3) based on constraint equation (7). The numerical solution of this nonlinear programming problem can be obtained by using a gradient projection method [14].

4 Manipulator Motion Planning

In this section, we discuss manipulator motion when the planned vehicle motion is executed. The manipulator consisting of an anthropomorphic arm and a turning trunk (Fig. 3) has two redundant degrees of freedom if considering only the wrist position (not including orientation).

With respect to manipulator redundancy, many methods have been proposed [15,16]. However, these

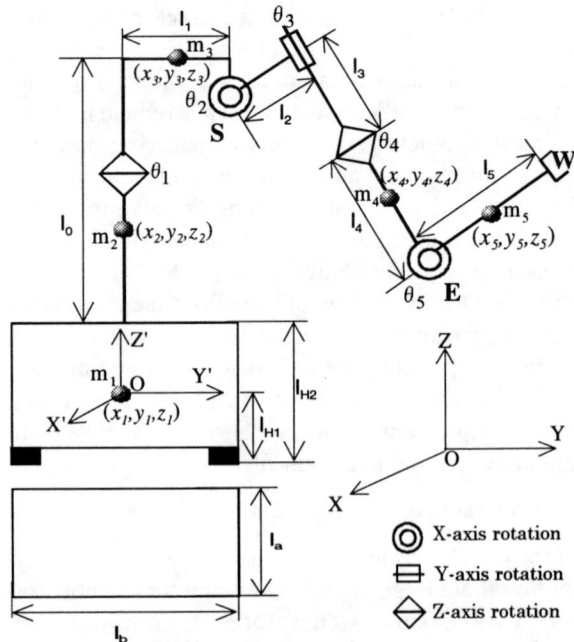

Fig. 3 Model of a mobile manipulator
S : shoulder position
E : elbow position
W : wrist position

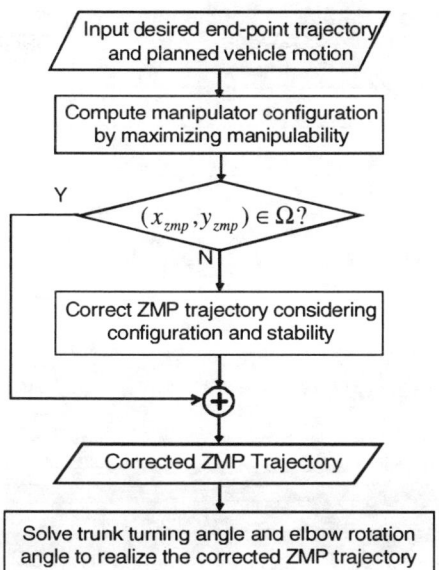

Fig. 4 Algorithm of manipulator motion
Ω : valid stable region

$$x_{zmp} = \frac{\sum_{i=1}^{5} m_i(\ddot{z}_i + g)x_i - \sum_{i=1}^{5} m_i \ddot{x}_i z_i}{\sum_{i=1}^{5} m_i(\ddot{z}_i + g)} \quad (8)$$

$$y_{zmp} = \frac{\sum_{i=1}^{5} m_i(\ddot{z}_i + g)y_i - \sum_{i=1}^{5} m_i \ddot{y}_i z_i}{\sum_{i=1}^{5} m_i(\ddot{z}_i + g)} \quad (9)$$

where m_i is the mass of link i, $(x_{zmp}, y_{zmp}, 0)$ is the coordinate of ZMP, g is the gravitational acceleration, and (x_i, y_i, z_i) is the coordinate of the mass center of link i on the absolute Cartesian coordinate system O-XYZ.

studies mainly concern manipulation or optimal configuration. Here, by using this redundancy, the mobile manipulator will follow the given trajectory with an optimal configuration when the robot is stable, and recover system stability when the robot is unstable.

In this study, in order to discuss dynamic stability considering environmental disturbances, we have already proposed the stability evaluation criterion [17] based on ZMP (Zero Moment Point) [18]. According to this stability evaluation, the ZMP must be inside the valid stable region to ensure system stability. In order to evaluate manipulator optimal configuration, the concept of manipulability [15] is used here. Therefore, the problem of the manipulator motion considering both system stability and manipulator configuration, will require that the ZMP must be inside the valid stable region and, simultaneously, the manipulability is maximal. To derive such a manipulator motion, the following algorithm is used.

First, decide the manipulator configuration by maximizing the manipulability when the planned vehicle motion is executed. Then, compute the ZMP trajectory in this case, and correct the ZMP trajectory considering the compatibility of stability and manipulability. Finally, solve the manipulator motion to realize the corrected ZMP trajectory (Fig. 4).

4.1 Consideration of Configuration and Stability

The manipulator configuration considering only manipulation can be obtained by maximizing the manipulability. In this case, the ZMP trajectory (in the following, this trajectory is called the un-corrected ZMP trajectory) can be computed as follows [17, 19]:

In order to move the un-corrected ZMP trajectory into the valid stable region, we can change only manipulator configuration. If the change of ZMP position is large, the manipulator configuration must change largely. On other hand, the un-corrected ZMP trajectory is the one with maximal manipulability. Therefore, in order to maintain system stability and simultaneously maximize manipulability, it is effective that the change of ZMP position is the smallest. Based on this fact, the point inside the valid stable region with the shortest distance to the ZMP outside the valid stable region, is regarded as the corrected ZMP position. For example, point A' is the corrected ZMP position corresponding to point A outside the valid stable region (Fig. 5).

4.2 Manipulator Posture Change

In general, there are two constraints between the corrected ZMP trajectory and the motion of the manipulator. However, it is difficult to directly get constraint equations for the two redundant joints of the manipulator. On the other hand, as shown in Fig. 3, the elbow is not constrained to a single position, given the specified positions of the shoulder and the wrist. So, we can introduce the elbow rotation angle which denotes the elbow rotating around the

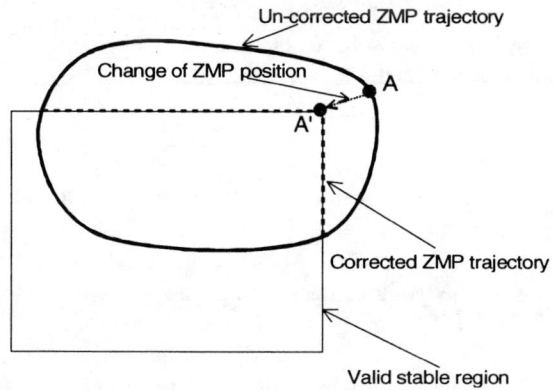

Fig. 5 ZMP path considering stability and configuration

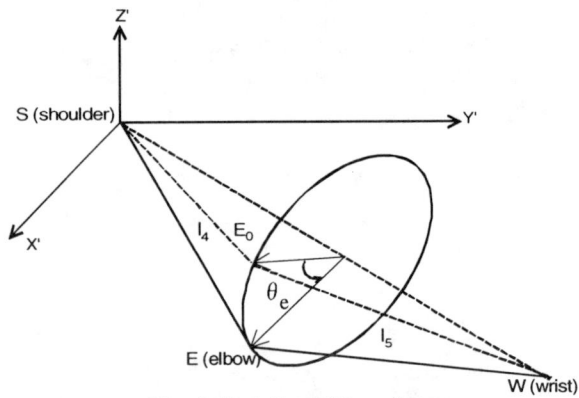

Fig. 6 Rotation of the elbow

axis of the straight line connecting the wrist and the shoulder (Fig. 6). By using the elbow rotation angle θ_e (Fig. 6) and the trunk turning angle θ_1 (fig. 3), we can formulate the two constraint equations.

The wrist position and the vehicle position are known according to the given trajectory and the planned vehicle motion. The shoulder coordinate can be denoted easily by the trunk turning angle. The elbow coordinate can be obtained with the following procedure.

Let $P_{E0} = [x_{e0}, y_{e0}, z_{e0}]$ be the position vector of the elbow at its lowest point, specified the positions of the shoulder and the wrist. By rotating the elbow (Fig. 6), the elbow coordinate can be given as:

$$P_E = E^{\theta_e} \cdot P_{E0} \qquad (10)$$

where E^{θ_e} denotes the transformation matrix.

From equations (8) (9) (10) and kinematic constraints, we can get the following two-rank nonlinear differential equations about the trunk turning angle and the elbow rotation angle.

$$\ddot{\theta}_1 = F_1\left(x_{zmp}, y_{zmp}, \theta_1, \theta_e, \dot{\theta}_1, \dot{\theta}_e\right) \qquad (11)$$

$$\ddot{\theta}_e = F_2\left(x_{zmp}, y_{zmp}, \theta_1, \theta_e, \dot{\theta}_1, \dot{\theta}_e\right) \qquad (12)$$

where F_1, F_2 denote functions on θ_1 and θ_e, (x_{zmp}, y_{zmp}) denote the ZMP on the corrected ZMP trajectory.

The numerical solutions to equations (11) and (12), can be obtained by the Runge-Kutta method.

5 Simulation

In simulation, parameters of the mobile manipulator (Fig. 3) are set according to Table 1. For convenience, the environmental disturbance is assumed a 200 [N] force adding on the trunk center, in this case the valid stable region is ± 0.135 [m] both in X-axis and in Y-axis.

Suppose that the given end-point trajectory is $O_m A_m B_m C_m$ with a uniform velocity of 1.0 [m/s], O_m(0.0, 0.0, 0.9), A_m(3.0, 0.0, 0.9), B_m(4.5, 2.6, 0.9), C_m(7.5, 2.6,

Table 1 Parameters of the mobile manipulator

	l_a	l_b	l_{H1}	l_{H2}	l_0
length (cm)	50	50	15	30	60
	l_1	l_2	l_3	l_4	l_5
	20	0	0	30	30
weight (kg)	m_1	m_2	m_3	m_4	m_5
	50	30	10	15	15

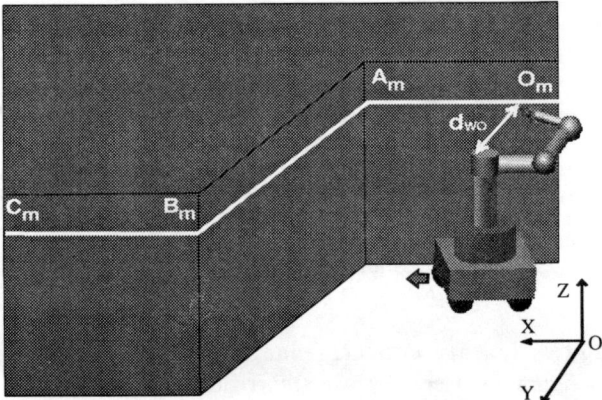

Fig. 7 Given end-point trajectory in simulation

0.9), on the world Cartesian coordinate system (Fig. 7).

The maximum distance of d_{wo} (Fig. 7) from the trunk to the end-point according to manipulator specification is 0.8 [m], and the distance d_{wo} considering manipulator effective workspace is assumed to be larger than 0.2 [m] and less than 0.6 [m]. The limited velocity of arcs is assumed to be 1.0 [m/s] when system stability can be maintained. Suppose that the initial point is S(0.0, 0.4, 0.0), the final point is G(7.5, 3.0, 0.0). The planned vehicle path composed of line segments and arcs is curve $SV_1V_2V_3V_4G$ shown in Fig. 8, and V_1, V_2, V_3, V_4 denote contact points between straight lines and circles.

In the following, we discuss the vehicle motion along the planned path, and manipulator posture change.

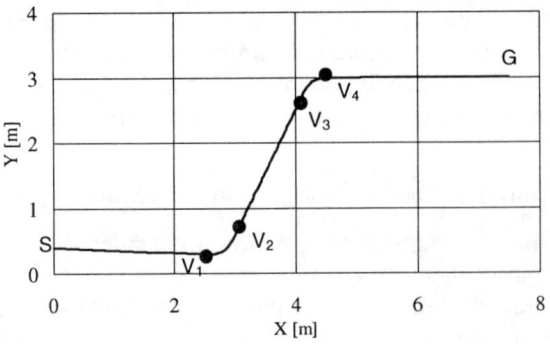

Fig. 8 Planned path composed by arcs and segments

5.1 Vehicle Motion

In order to illustrate the effectiveness of the proposed strategy for vehicle motion, we first consider the simulation example of vehicle motion considering only the minimum of acceleration sum.

5.1.1 Results of Considering Only Acceleration Sum

The dotted curves in Fig. 9 show the results of considering only the minimum of acceleration sum. Since the distance d_{wo} is larger than 0.6 [m] at some time (Fig. 9 (a)), the end-point cannot reach the given trajectory within the effective workspace, that is, the manipulator cannot follow the given trajectory at an optimal configuration. Since the vehicle velocity of arcs is larger than 1.0 [m/s] (Fig. 9 (b)), it is impossible to maintain system stability.

5.1.2 Results of Considering Acceleration and Workspace and Stability

The solid curves in Fig. 9 show the results of considering acceleration sum, effective workspace and system stability. Since the distance d_{wo} is always within the effective workspace (Fig. 9(a)), it is possible for the manipulator to follow the given trajectory at an optimal configuration. Since the vehicle velocity of arcs is less than 1.0 [m/s] (Fig. 9(b)), it is possible to maintain system stability.

5.2 Manipulator Motion

In order to illustrate the effectiveness of the proposed method for manipulator moiton, we first consider the simulation example of manipulator posture change considering only manipulator configuration.

5.2.1 Results of Considering Only Configuration

The dotted curves in Fig. 10 and Fig. 11 show the results of considering only manipulator configuration. In this case, the manipulator holds the configuration with the wrist directly in front of the shoulder and the elbow at almost its horizontal point (Fig. 10). This configuration is almost the same posture that human usually uses. However, since the Y_{zmp} between points V_1 and V_2 is outside the valid stable region (Fig. 11(b)), the mobile manipulator may tip over if affected by environmental disturbances.

5.2.2 Results of Considering Configuration and Stability

The solid curves in Fig. 10 and Fig. 11 show the results of considering both system stability and manipulator configuration. In this case, the elbow rotation angle is always the same as the case of considering only configuration (Fig. 10(a)), but the trunk turning angle is different (Fig. 10(b)). When the robot is stable during the time from points S to V_1, from points V_2 to G, the

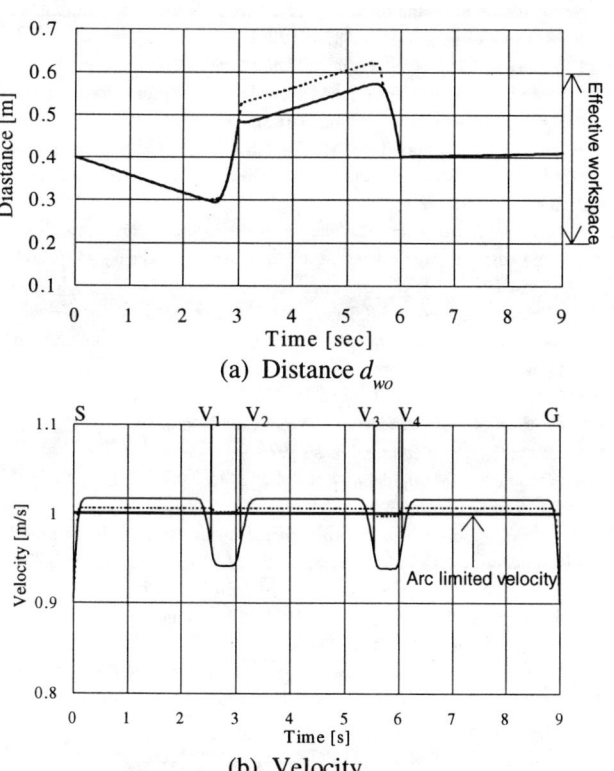

Fig. 9 Simulation results of the vehicle motion
· · · · · : the case of considering only acceleration
——— : the case of considering acceleration, workspace and stability

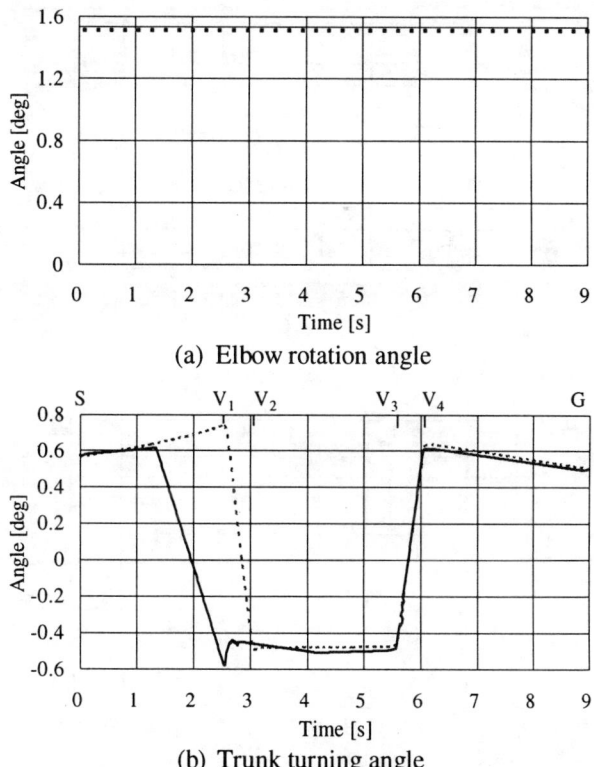

Fig. 10 Posture change of the manipulator
· · · · · : the case of considering only configuation
——— : the case of considering configuration and stability

manipulator holds almost the same configuration as the case of considering only manipulator configuration. But when the vehicle moves along the arc V_1V_2, the mobile manipulator might tip over to the outer side of the arc V_1V_2, if considering only manipulator configuration. Considering this, θ_1 of the result of the proposed method becomes smaller than the case of considering only configuration (Fig. 10(b)); that is, the trunk angle rotates to the inside of the arc V_1V_2. Therefore, the center of gravity of the whole system is moved to the inside of the arc V_1V_2, and system stability can be recovered. As shown as Fig. 11, the ZMP in this case is always inside the valid stable region.

6 Conclusion

In this paper, we proposed a method for coordinating vehicle motion planning considering manipulator task constraints, and manipulator motion planning considering platform stability. First, the optimal problem of vehicle motion is formulated, considering vehicle dynamics, manipulator workspace and system stability. Then, the arm posture change is derived, considering stability compensation and optimal configuration. Finally, the effectiveness of this method is suggested by simulation.

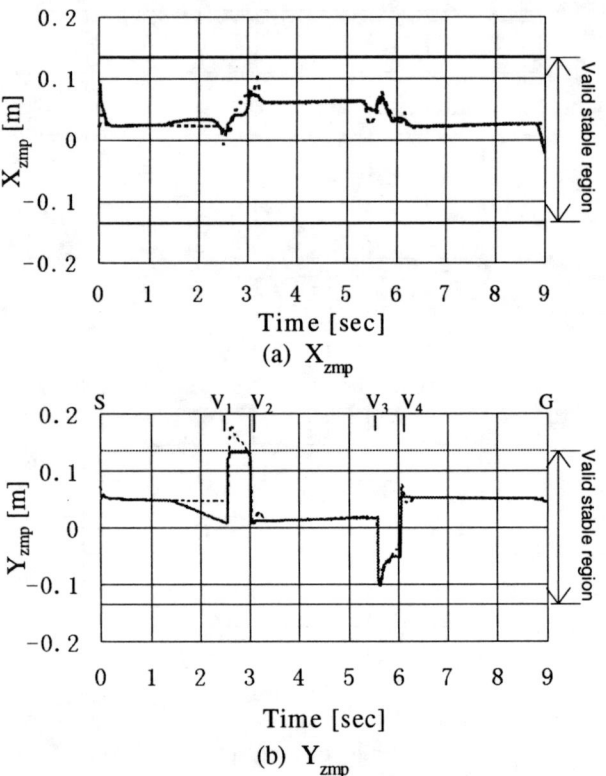

Fig. 11 ZMP trajectory
.... : the case of considering only configuration
___ : the case of considering configuration and stability

Reference

[1] Y. Yamamoto and X. Yun, "Coordinated Obstacle Avoidance of a Mobile Manipulator", Proc. IEEE Int. Conf. Robotics and Automation, pp. 2255-2260 (1995)

[2] W.F. Carriker, P.K. Khosla, and B.H. Krogh, "An Approach for Coordinating Mobility and Manipulation", Proc. IEEE Int. Conf. Robotics and Automation, pp. 59-63 (1989)

[3] N.A.M. Hootsmans and D. Dubowsky: "Large Motion Control of Mobile Manipulators Including Vehicle Suspension Characteristics", Proc. IEEE Int. Conf. Robotics and Automation, pp. 2236-2341 (1991)

[4] N. Koyachi, T. Arai, H. Adachi, K. Asami, and Y. Itoh, "Hexapod with Integrated Limb Mechanism of Leg and Arm", Porc. Int. Conf. Robotics and Automation, Japan, pp. 1952-1957 (1995)

[5] N. Chong, K. Yokoi, S. Oh, K. Tanie, "Position Control of a Collision-tolerant Passive Mobile Manipulator with Base Suspension Characteristics", Proc. IEEE Int. Conf. Robotics and Automation, pp. 594-599 (1997)

[6] D.A. Meessuri and E.E. Vance, "Automatic Body Regulation for Maintaining Stability of a Legged Vehicle During Rough-Terrain Locomotion", IEEE J. of Robotics and Automation, Vol. RA-1, No. 3, pp. 132-141 (1985)

[7] T. Fukuda and Y. Fujisawa, "Manipulator/Vehicle System for Man-Robot Cooperation," Proc. IEEE Int. Conf. Robotics and Automation, pp. 74-79 (1997)

[8] D.A. Rey and E.G. Papadopoulos, "On-line Automatic Tipover for a Mobile Manipulator", Proc. IEEE/RSJ Int. Conf. Intelligent Robotics and Automation, pp. 1273-1278 (1997)

[9] K. Yoneda and S. Hirose, "Tumble Stability Criterion of Integrated Locomotion and Manipulation", Proc. IEEE/RSJ Int. Conf. Intelligent Robots and Systems, pp. 870-876 (1996)

[10] L.E. Dubins, "On Curves of Minimal Length with a Constraint on Average Curvature, and with Prescribed Initial and Terminal Position and Tangents", American J. of Mathematics, Vol. 79, pp. 497-516 (1957)

[11] P. Jacobs, J.P. Laumond and M. Taix, "Efficient Motion Planners for Nonholonomic Mobile Robots", Proc. IEEE/RSJ Int. Conf. Intelligent Robots and Systems, pp. 1229-1235 (1991)

[12] S. P. Marin, "An Approach to Data Parametrization in Parametric Cubic Spline Interpolation Problems", J. of Approximation Theory 41, pp. 64-86 (1984)

[13] Q. Huang, S. Sugano, and K.Tanie, "Stability Compensation of a Mobile Manipulator by Manipulator Motion: Feasibility and Planning", Proc. IEEE/RSJ Int. Conf. Intelligent Robots and Systems, pp. 1285-1292 (1997)

[14] D.G. Luenberger, "Introduction to Linear and Nonlinear Programming", Addision-Wesley, Reading, Mass., 1973

[15] T. Yoshikawa, "Manipulability of Robotic Mechanisms", Proc. 2nd Int. Symp. Robotic Research, pp. 91-98 (1984)

[16] H. Seraji, "Task-Based Configuration Control of Redundant Manipulators", J. of Robotic Systems, Vol. 9, No. 3, pp. 411-451 (1992)

[17] S. Sugano, Q. Huang, and I. Kato, "Stability Criteria in Controlling Mobile Robotic Systems", Proc. IEEE/RSJ Int. Conf. Intelligent Robots and Systems, pp. 832-838 (1993)

[18] M. Vukobratovic and D. Juricic, "Contribution to the Synthesis of Biped Gait", IEEE Trans. on Bio-Medical Engineering, Vol. BME-16, No. 1, pp. 1-6, (1969)

[19] T. Takanishi, M. Ishida, Y. Yamazaki, and I. Kato, "Realization of Dynamic Walking Robot WL-10RD", Proc. IEEE Int. Conf. Robotics and Automation, pp. 459-466 (1985).

[20] O. Khatib, K. Yokoi, et.al, "Coordination and Decentralized Cooperation of Multiple Mobile Manipulators" J. of Robotic Systems, Vol. 13, No. 11, pp. 755-764 (1996)

SPEED PLANNING AND GENERATION APPROACH BASED ON THE PATH-TIME SPACE FOR MOBILE ROBOTS.

V. Muñoz, A. Cruz and A. García-Cerezo

Dpto. De Ingeniería de Sistemas y Automática. Universidad de Málaga. Plaza el Ejido s/n, 29013 Málaga (Spain).
Phone: (+34) 5 213-14-06; Fax: (+34) 5 213-14-13; E-mail: victor@ctima.uma.es

Abstract.

This paper presents a speed planning and generation algorithm for mobile robots that work under some kind of speed limitations. The proposed method takes the information about these speed constraints and the moving obstacles in the environment, and provides a safe speed profile which allows to build a trajectory that bypasses such obstacles. The method has been successfully tested in the RAM-2 mobile robot.

1. INTRODUCTION

The control motion system of a mobile robot drives its course by using the position and speed references (trajectory) computed by the planning system. Hence, this trajectory should be defined in such a way that provides a safe navigation with no collisions. Therefore, the obstacle avoidance is defined as the main function of the planning process.

The approach used for performing the above process, depends on the movement state of the obstacle. Some methods contemplate the obstacle's size and trajectory as entries to the local planner algorithm. Classic planning strategies can be adapted for this problem, such as visibility graphs [4][2], configuration space schemes [1][3] and potential fields [5].

The path-velocity decomposition is an efficient method for avoiding mobile obstacles by computing a safe speed function [4]. However, this methodology must be modified for providing the following features:

- First order continuity: The speed function and its first derivative should be continuous in order to reduce the speed tracking error.

- Admissibility from the point of view of the vehicle's kinematic and dynamic behaviour: The speed reference must be defined in such a way that allows the tracking system to follow the trajectory. Therefore, the kinematic and dynamic models of the robot must be taken into account in the speed function definition.

The method for avoiding moving obstacles, proposed in this paper, covers the above points by using piecewise cubic function, which is defined by studying other kind of speed restriction imposed by the vehicle's physical characteristics.

These limitations are commented in section 2. Section 3 is devoted to the previous work where the method presented in this paper leans on [7]. This work involves two stages: a speed planning process, which provides a speed plan that takes into account physical and operational speed limits, and a speed profile generation process, which builds the speed function needed to obtain the vehicle's trajectory. Section 4 details how the proposed method copes with the avoidance of moving obstacles. The speed planning made considering only kinematic and dynamic constraints is combined with the information about the mobile obstacles. In this way, a set of safety zones representing that speed planning, is built. Whenever a hazardous situation is detected, the original planning is modified in order to get a safe navigation. Implementation and experiments on the RAM-2 mobile robot are showed is section 5, and finally, section 6 presents the conclusions of this work.

2. SPEED PLANNING PROBLEM.

A robot path is defined as a set of evenly spaced postures $Q=\{q_1,...,q_m\}$, which are to be executed by the path tracking algorithm. A posture q_i is composed of five basic elements: x_i, y_i, θ_i, κ_i and s_i. The first two elements are position components, the third is the heading with respect to a global work frame, the fourth is the curvature component, and the last one is the distance along the path from the starting posture to the current one.

In order to convert a path Q into a trajectory $\tilde{Q}$ it is necessary to append a speed component to each posture of the path. In other words, the trajectory conversion process must turn each $q_i=(x_i,y_i,\theta_i,\kappa_i,s_i)$ into $\tilde{q}_i=(x_i,y_i,\theta_i,\kappa_i,s_i,v_i)$, where v_i is the posture speed component. This transformation is made by the definition of a parametric arc length speed function V(s). Such a curve is defined in the space-speed plane [10], where the upper speed limits for each posture q_i of the path Q are represented. These limits are obtained by taking into account the speed constraints introduced by the vehicle's features and operational speed limitations. Thus, V(s) is specified in such a way that it preserves all the posture speed limits, in order to obtain a speed profile with good tracking conditions. That means that V(s) must lie inside a safety area of the space-speed plane defined by the speed limits functions (see Fig. 1.).

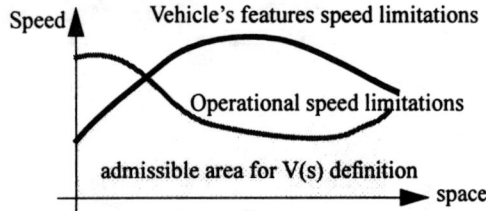

Fig. 1. Speed limitations in the space-speed plane.

The speed constraints considered at speed planning time are shown in table 1.

Table 1: Speed limits classification

Type	Constraint
Physical	Mechanical (ME)
	Kinematics (KI)
	Dynamic (DY)
Operational	Goal Point (DG)
	Known mobile obstacles (KM)

The physical speed restrictions group is due to the kinematic and dynamic behaviour of the mobile robot [9]. These restrictions impose a top speed and acceleration according to the peculiarities of the vehicle and the path to be followed.

The second group presents the external speed limitations, which arise because of performing the task in a real environment. In this way, some situations force to set a safe speed value to the vehicle in order to synchronize with other elements in the working environment, or even to stop before getting into a zone with collision hazard.

3. PREVIOUS WORK: SPEED FUNCTION V(s).

The speed function V(s) definition is made in two steps: speed planner and speed generation processes.

3.1. Speed planner process.

This stage chooses a set of control path postures $C=\{q^1, ..., q^p\}$ which divides the path into a set $S=\{S_1, ..., S_{p-1}\}$ of path segments, where S_i is composed of the path postures sequence between q^i and q^{i+1}. The choice of the elements which will be belong to C is made depending on the nature of the speed limitations:

- Kinematic and dynamic considerations are a function of the path curvature κ. Therefore, it is necessary to consider the variation law of this magnitude. In this way, the postures which have the local maximum or minimum curvature values are added to C set.

- Secondly, new path postures are chosen in order to satisfy the operational speed limitations. The closest posture to the goal point or known obstacle which satisfies the *distance to a goal point* speed limitation is selected.

A top speed v_i is assigned to each member q^i of C by using the minimum speed value provided by the speed limitations introduced by vehicle's features and operational speed constraints. This operation sets up a speed control set $V=\{0, v_1, ..., v_{p-1}, 0\}$ for the path Q, whose first component (always null) is the starting speed for S_1 and the remaining components v_i are speed boundary conditions between segments S_{i-1} and S_i. Sets S and V resulting from the speed planning process, are represented in the space-speed plane, and will be used by the speed profile generator process for building V(s).

3.2. Speed generation process.

The speed function V(s) is modelled as a set of space-time functions $\sigma_i(t)$, as it is shown in expression (1).

$$V(s) = \bigcup_{i=1}^{p} \frac{d}{dt}\sigma_i(\sigma_i^{-1}(s)) \qquad (1)$$

The ith component of this curve is assigned to the path segment S_i defined by the path postures sequence $\{q^i, ..., q^{i+1}\}$. Function $\sigma_i(t)$ is determined by the parameters set $\Pi_i=\{v_i, v_{i+1}, s_i, t_i\}$, where v_i and v_{i+1}, components of the set V, are the starting speed at q^i and the ending one at q^{i+1}; s_i is the segment length, and finally t_i is the navigation time assigned to the current segment which must be computed for $\sigma_i(t)$ definition.

The method evaluates $\sigma_i(t)$ by using two cubic polynomials $^1\sigma_i(t)$ and $^2\sigma_i(t)$, which are obtained from $\sigma_i(t)$:

$$\begin{aligned} ^1\sigma_i(t) &= {^1\alpha_{i0}}t^3 + {^1\alpha_{i1}}t^2 + {^1\alpha_{i2}}t + {^1\alpha_{i3}} \\ ^2\sigma_i(t) &= {^2\alpha_{i0}}t^3 + {^2\alpha_{i1}}t^2 + {^2\alpha_{i2}}t + {^2\alpha_{i3}} \end{aligned} \qquad (2)$$

Function $^1\sigma_i(t)$, with arc length ^1s_i, covers the first part of the segment, and $^2\sigma_i(t)$ the remaining length of S_i ($s_i - {^1s_i}$). In this way, by taking ^it_m as $t_i/2$ and ^iv_m as the average between v_i and v_{i+1}, the first function $^1\sigma_i(t)$ is defined by the set $\{v_i, {^iv_m}, {^1s_i}, {^it_m}\}$, and the second one $^2\sigma_i(t)$ by $\{^iv_m, v_{i+1}, {^2s_i}, {^it_m}\}$. Moreover, the shape detailed at expression (3) is imposed to their second derivatives.

$$^1\sigma''_i(0) = {}^2\sigma''_i({}^i t_m) = 0 \qquad {}^1\sigma''_i({}^i t_m) = {}^2\sigma''_i(0) \qquad (3)$$

These functions are detailed in matrix form as follows:

$$M \times \begin{bmatrix} {}^1\alpha_{i0} \\ {}^1\alpha_{i1} \\ {}^1\alpha_{i2} \\ {}^1\alpha_{i3} \end{bmatrix} = \begin{bmatrix} 0 \\ {}^1s_i \\ v_i \\ {}^i v_m \end{bmatrix} \qquad M \times \begin{bmatrix} {}^2\alpha_{i0} \\ {}^2\alpha_{i1} \\ {}^2\alpha_{i2} \\ {}^2\alpha_{i3} \end{bmatrix} = \begin{bmatrix} {}^1s_i \\ {}^2s_i \\ {}^i v_m \\ v_{i+1} \end{bmatrix} \qquad (4)$$

where M is defined by the expression:

$$M = \begin{bmatrix} 0 & 0 & 0 & 1 \\ ({}^i t_m)^3 & ({}^i t_m)^2 & {}^i t_m & 1 \\ 0 & 0 & 1 & 0 \\ 3({}^i t_m)^2 & 2({}^i t_m) & 1 & 0 \end{bmatrix} \qquad (5)$$

The values for the elements of these sets are:

$$^i t_m = \frac{s_i}{v_i + v_{i+1}} \qquad {}^i v_m = \frac{v_i + v_{i+1}}{2}$$
$$^1 s_i = \frac{s_i(5v_i + v_{i+1})}{6(v_i + v_{i+1})} \qquad {}^2 s_i = s_i - {}^1 s_i \qquad (6)$$

Thus, the relationship between the starting and ending speeds for keeping the a given top acceleration constraint a_{max} is defined by the following expression:

$$if(v_i < v_{i+1}) then$$
$$v_{i+1} \leq \sqrt{v_i^2 + a_{max} s_i}) \qquad else(v_{i+1} \geq \sqrt{v_i^2 - a_{max} s_i}) \qquad (7)$$

Finally, when the parameters sets Π_i have been determined via expressions (6) and (7), the speed profile V(s) is defined in the space-speed plane by using expression (1).

4. MOVING OBSTACLE AVOIDANCE METHOD.

The above speed planning V(s) is made in the speed-space phase plane, and it must be mapped into the path-time s × t space in order to detect the collisions with known mobile obstacles. The s × t space contains a set of rectangular forbidden regions, which represents the crossing points of the mobile obstacles trajectories through the robot path Q [4].

The speed planning is rendered in this space implicitly, by means of the space time functions $\sigma_i(t)$ (from i=1 to p). As it was stated in the previous section, every $\sigma_i(t)$ is defined by its parameters set $\Pi_i=\{v_i,v_{i+1},s_i,t_i\}$. Therefore, each $\sigma_i(t)$ is represented in the s × t by using its (s_i,t_i) elements. Furthermore, expressions at (6) show the relationship between these elements and the initial and final speed values of the segment (v_i and v_{i+1}).

In this way, the computed speed planning is safe when any $\sigma_i(t)$ does not touch any rectangular region. Let OB_j be a moving obstacle in the s × t space. The planned speed reference provides the maximum speed allowed by the vehicle's kinematic and dynamic features. Therefore, for avoiding the collision with OB_j, the robot must reduce its speed and let the moving obstacle cross the collision path point before the vehicle reaches it.

However, testing the collisions for every $\sigma_i(t)$ and all OB_j is a time-expensive process. Instead of this, the proposed method uses a set of hull-convex areas, called the safety zones Z_i, which encloses every $\sigma_i(t)$. Therefore, the proposed algorithm checks the collisions between every Z_i and OB_j, by using a geometric approach. Whenever this situation is detected, both $\sigma_i(t)$ and Z_i will be modified properly in the s × t space for avoiding the collision.

The building and modification of the safety zones will be developed in the following subsections.

4.1. Safety Zones

The building of the safety zone Z_i assigned to $\sigma_i(t)$ is based on the following lemma:

Lemma 1: The shape of the space-time function $\sigma_i(t)$ assigned to S_i is always either concave or convex. In other words, both $^1\sigma_i(t)$ and $^2\sigma_i(t)$ must be, simultaneously, concave or convex.

Proof:

Let $^1\sigma_i(t)$ and $^2\sigma_i(t)$ be the components of $\sigma_i(t)$. Their second derivative are defined in the following expression:

$$^1\sigma''_i(t) = 6\,{}^1\alpha_{i0}\,t$$
$$^2\sigma''_i(t) = 6\,{}^2\alpha_{i0}\,t + 2\,{}^2\alpha_{i1} \qquad (8)$$

where $^1\alpha_{i0}$ is obtained by solving the expression (4):

$$^1\alpha_{i0} = \frac{-(v_i - v_{i+1})(v_i + v_{i+1})^2}{s_i^2} \qquad (9)$$

In the same way,:

$$^2\alpha_{i0} = \frac{(v_i - v_{i+1})(v_i + v_{i+1})^2}{6 s_i^2}$$
$$^2\alpha_{i1} = \frac{-(v_i - v_{i+1})(v_i + v_{i+1})}{2 s_i} \qquad (10)$$

Concavity or convexity of these functions are studied by taking into account the sign of their second derivatives, which depends on the relationship between the starting and ending speed of S_i.

A increasing speed segment ($v_i<v_{i+1}$) implies that functions $^1\sigma_i(t)$ and $^2\sigma_i(t)$ are concave. On the other hand a decreasing speed segment ($v_i>v_{i+1}$) forces that derivatives are negative, and both subsegments are convex.

By taking into account the above lemma, the safety zone Z_i can be modelled as a triangle in the $s \times t$ space as it is shown in Fig. 2.

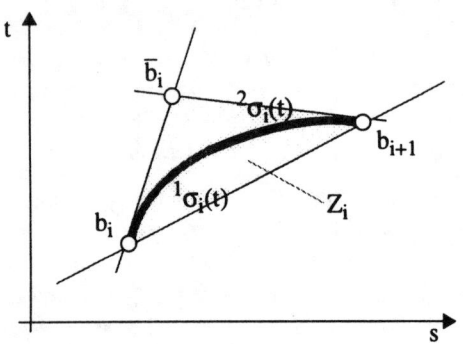

Fig. 2. The safety zone Z_i assigned to $\sigma_i(t)$.

This figure presents the safety zone as a shady triangle defined by the following lines:

- Line segment from the starting point b_i to the ending point b_{i+1}, which are defined as follows:

$$b_n = (gs_n, gt_n) = \left(\sum_{j=1}^{n} s_j, \sum_{j=1}^{n} t_j \right) \quad ; (s_j, t_j) \in \Pi_j \quad (11)$$

- Tangent line to $^1\sigma_i(t)$ at b_i (slope=v_i).
- Tangent line to $^2\sigma_i(t)$ at b_{i+1} (slope=v_{i+1}).

So, Z_i is defined by b_i, b_{i+1} and the intersection point $\bar{b}_i$ between the tangent lines.

4.2. Safety zones modification.

As it was stated before, the robot must decrease its speed when a collision between a safety zone Z_j and a forbidden region OB_i is detected. In this situation, the path segment S_j corresponding to Z_j is replaced with two new path segments SA_j and SB_j, in such a way that the obstacle OB_i is avoided by using the path-time point b_k (see Fig. 3.).

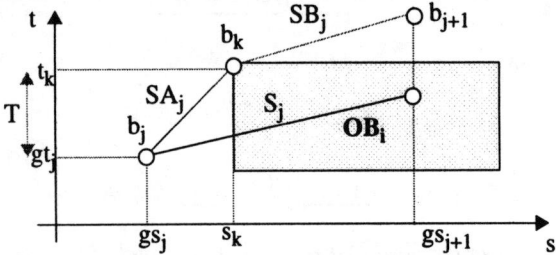

Fig. 3. Collision situation and path segment modification.

Let v_k be the speed reference at b_k. Therefore, the sets of path segments S and speed control set V are updated as shown in expression (12).

$$\begin{aligned} S &= \{S_1, ..., S_{j-1}, SA_j, SB_j, S_{j+1}, ..., S_p\} \\ V &= \{0, v_1, ..., v_j, v_k, v_{j+1}, ..., v_{p-1}, 0\} \end{aligned} \quad (12)$$

The parameters sets for SA_j and SB_j are defined by $\Pi_j^A = \{v_j, v_k, s_j^A, t_j^A\}$ and $\Pi_j^B = \{v_k, v_{j+1}, s_j^B, t_j^B\}$ where v_j and v_{j+1} are the planned speed references at b_j and b_{j+1}. In this way, the value for v_k is computed in order to travel path segment SA_j with a navigation time by using time expression at (6):

$$v_k = \frac{s_j^A}{T} - v_j \quad (13)$$

where $T=(t_k-gt_j)$. The direct consequence of inserting b_k with a speed reference v_k is the time displacement of point b_{j+1} in the $s \times t$ plane.

The speed components v_k and v_j for SA_j could not assure the acceleration constraint. Therefore, it is necessary to check this condition by using expression (7), and modify v_k when it is needed. If this new value for v_k does not guarantee a navigation time t_j^A for SA_j greater than avoidance time T, the current obstacle is not bypassed.

The correction of this fact is made via the iterative modification of the speed references for the previous segments. This operation is made by the back propagation of the time needed to avoid the obstacle which keeps the acceleration constraint.

This process is detailed in *AvoidMobileObstacle* algorithm which is composed of two different parts. The first one modifies set S by inserting the new path segments and computes the reference speed value v_k. Furthermore, this part calculates the time ΔT as the difference between the time T needed for avoiding the obstacle OB_i and the real time t_j^A for travelling SA_j imposed by the acceleration constraint. The second part (while loop) performs the time back propagation if it is needed.

Function AvoidMobileObstacle(S,V,OB$_i$, j)
   select S_j as the jth element of S.
   Divide S_j into SA_j and SB_j and update set S.
   Compute v_k by using expression (13) and insert in set V
   {V}=SpeedVerification(S,V)
   Compute navigation time t_j^A for SA_j with formula (6).
   $\Delta T=T- t_j^A$.
   while $\Delta T>0$ do
     Compute navigation time t_{j-1} for S_{j-1} with formula (6)
     required_time=$\Delta T+t_{j-1}$
     Compute v_j by using expression (13) with required_time.
     {V}=SpeedVerification(S,V)
     Compute navigation time t_{j-1} for S_{j-1} with formula (6).
     ΔT=required_time-t_{j-1}.
     j=j-1
   endWhile
   return S and V
endProcedure

4.3. Final algorithm.

The complete algorithm integrates both the speed planner and generation processes with the presented mobile obstacle avoidance method. The algorithm is divided into three stages: the speed planning process, the mobile obstacle avoidance and the speed profile generation.

The first step performs the speed planning, which involves the following operations. First, the path Q is divided into a set S of path segments and the speed control set V is computed. The components of this last set assure the acceleration constraint at expression (7).

The second stage is devoted to the moving obstacle avoidance. In this way, sets OB and Z are defined in the path-time space. After this, the algorithm tests the contacts between the elements of OB and Z. Whenever a collision is detected the sets S and V are modified for solving this situation. It is necessary to consider that avoiding a forbidden region OB_j might cause collisions between previous safety zones Z_k with k=0,...,i-1 and rectangles OB_l with l=0,...,j-1; so, the nested loops must be restarted if the function *AvoidMobileObstacle* modifies the previous speed planning.

Finally, the third stage builds the speed profile V(s) and merges this profile with path Q, in order to get the desired trajectory $\tilde{Q}$.

Procedure Path2Trajectory(Q,OT)
/* Speed planning process */
   $\tilde{Q}$ ={}
   {S}=DividePathIntoPathSegments(Q)
   {V}=ComputeSpeedControlSet(S)

/* Mobile obstacle avoidance */
   {OB}=ComputeObstaclesSet(S,OT)
   {Z}=ComputeSafetyZones(S,V)
   For each S_i belonging to S do
     For each OB_j belonging to OB do
       if Collision(Z_i,OB_j) then
         {V,S}=AvoidMobileObstacle(OB_j,S,V,i)
       end if
     end loop
   end loop

/* Speed profile generation and trajectoty building */
   For each S_i belonging to S do
     Compute $^1\sigma_i(t)$ and $^2\sigma_i(t)$
     $\sigma_i(t)$=Composition($^1\sigma_i(t),^2\sigma_i(t)$)
     $V_j(s)=\sigma_i'(\sigma_i^{-1}(s))$
     $\tilde{Q} = \tilde{Q} \cup \{S_i, V_i(s \in S_i)\}$
   end loop
End Procedure

---

5. IMPLEMENTATION AND EXPERIMENTS

The speed planning and profile generator algorithms have been integrated in the intelligent control architecture of RAM-2 [8].

Figures 4 to 6 show the results of the proposed method with the mobile robot RAM-2. The studied situation involves an environment with a moving obstacle. Figure 4 shows the path to be tracked by the robot in solid line and the trajectory followed by the obstacle in dotted line. Furthermore, the division of the path is marked with circular marks.

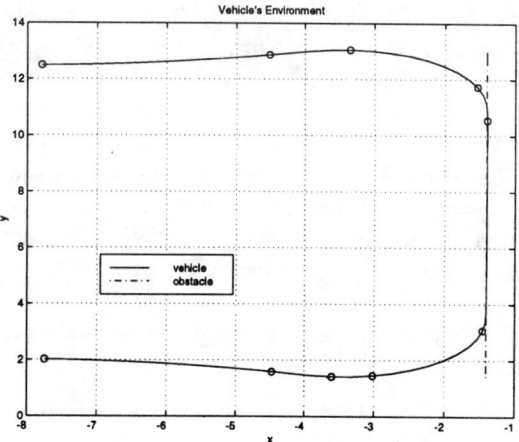

Fig. 4. Robot path and the moving obstacle.

The original speed planning provided by the speed planning process is represented, on space-speed plane, with a dotted line in figure 5. However, the moving obstacle crosses the robot path along its trajectory. This information is used for replanning the original speed profile, and obtaining a safe speed plannification for the mobile robot (solid line) which avoids the collision.

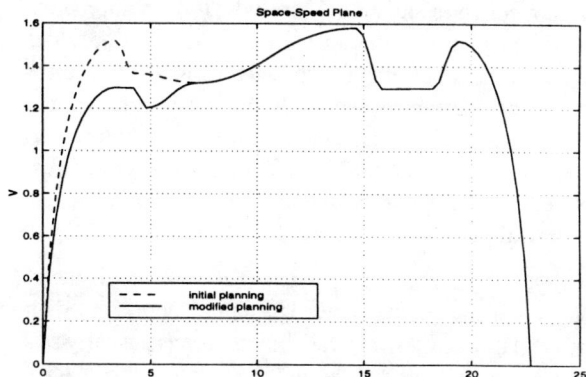

Fig. 5. Original speed planning and its modification for avoiding the mobile obstacle.

The above result is achieved via path-time representation (see Fig. 6.).

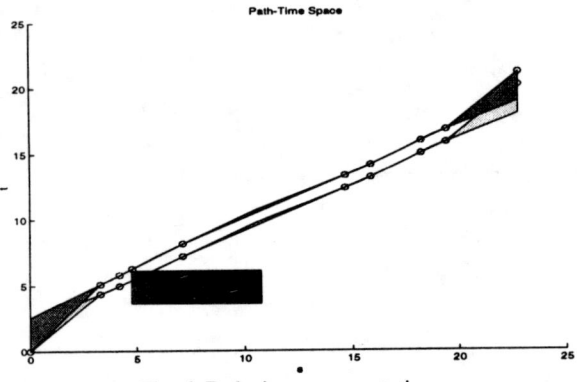

Fig. 6. Path-time representation.

This figure details the moving obstacle as a dark rectangle, and also contains the safety zones for every segment of the path as light shady triangles. The method detects the collision between the first safety zone and the obstacle, and eliminates this situation. The result of this action is shown as the set of dark shady triangles in Fig. 6., which includes the safety zone attached to the new segment needed to avoid the forbidden region. The new speed planning avoids the mobile obstacle and also considers the robot motion constraints due to its physical features.

6. CONCLUSIONS.

This paper presents a speed planning and generation algorithm for mobile robots that work under some kind of speed limitations. The proposed method takes the information about these speed constraints and the moving obstacles in its environment, and provides a safe speed profile which allows to build a trajectory that bypasses such obstacles. The algorithm applies three steps: i) speed planning, ii) mobile obstacles avoidance, and iii) speed profile generation.

The method modifies the speed planning when is required due to a collision situation. Moreover, it does not handle a visibility graph on the s × t space, and therefore does not use any kind of graph-search algorithm. The result is a speed profile which avoids the moving obstacles and takes into account other kind of speed constraints (i.e. kinematic and dynamic). This feature provides the real time execution for speed replanning which can be also made while the robot is tracking the path.

Finally, working in the presence of mobile obstacles demands a complete knowledge about them, i.e., their sizes and trajectories. Though this situation is usually real in multi-robot systems, the navigator must use a local speed planner which modifies the current speed reference using feedback control [6]. This action solves the uncertainties about the obstacle's trajectory, that the planner does not consider in planning time.

7. ACKNOWLEDGEMENTS

This work has been done within the framework of the projects TAP96-1184-C04-02 and TAP96-0763 of the C.I.C.Y.T. (Spain)

8. REFERENCES.

[1] Fujimura, K. and Samet, H. (1990) Motion planning in a Dynamic Domain. *Proc. IEEE Int. Conf. on Robotics & Automation*, pp. 324-330

[2] Gil de Lamadrid, J. (1994) Avoidance of Obstacles with Unknown Trajectories: Locally Optimal paths and Periodic Sensor Readings. *The International Journal of Robotics Research*, Vol 13,No 6, pp. 496-507.

[3] Griswold, N. C. and Eem, J. (1990) Control of Mobile Robots in the Presence of Moving Objects. IEEE Transactions on Robotics and Automation, Vol. 6, No. 2, pp. 263-268.

[4] Kant K., Zucker S. (1.986). Toward Efficient Trajectory Planning: The Path-Velocity Decomposition. The International Journal of Robotics Research, Vol 5, No. 3.

[5] Kyriakopoulos, K. J. and Saridis, G. N. (1993) An Integrated Collision Prediction and Avoidance Scheme for Mobile Robots in Non-Stationary Environments, Automatica, Vol. 29, No. 2

[6] Mandow A., Muñoz V., Fernandez R., García-Cerezo A. (1997). Dynamic Speed Planning for Safe Navigation. 1997 Proc. IEEE International Conference on Intelligent Robots and Systems.

[7] Muñoz V. (1995). Trajectory Planning for Mobile Robots. Ph. D. Thesis. University of Malaga (Spain).

[8] Ollero A., A. Simón, F. García, and V. E. Torres (1993). Integrated Mechanical Design of a New Mobile Robot. Proc. SICICA´92. Pergamon Press.

[9] Prado M., Simon A., Muñoz V., Ollero A. (1994). Autonomous Mobile Robot Dynamic Constraints due to Wheel-Ground Interaction. Proc. of 1994 European Robotics and Intelligent Systems Conference. Vol 1, pp 347-360. Malaga (Spain)

[10] Shiller Z., Gwo Y. (1991). Dynamic Planning of Autonomous Vehicles. IEEE Transactions on Robotics and Automation, Vol 7 No. 2.

Accounting for Mobile Robot Dynamics in Sensor-Based Motion Planning: Experimental Results*

J.C. Alvarez[†]   A. Shkel   V. Lumelsky

Robotics Laboratory
University of Wisconsin-Madison
Madison, Wisconsin 53706, USA

Abstract

The effect of robot dynamics on autonomous navigation is a key issue in sensor-based motion planning. In most of existing works, the solution is attempted by separating the planning and control into two sequential stages; as a minus, this may, for example, adversely effect the algorithm convergence. The strategy proposed in this paper solves the problem by combining motion planning and control within a single-stage procedure. The procedure exhibits good dynamic behavior, while providing safety (collision avoidance) and fast response. Results of testing the approach on a commercial Nomad-200 mobile robot are presented. Also discussed is the effect of model parameters on motion performance.

1 Introduction

The approaches to autonomous robot motion planning can be classified in terms of the amount of information available to the robot at each moment. In algorithms for planning with complete information, the workspace and the robot characteristics are assumed to be completely known *a priori*. This leads to the off-line calculation of the path, and reduces the problem to one in computational geometry [6]. The control is then a separate computational problem addressed within the classical control theory. This is in contrast with the paradigm of motion planning with incomplete information, where only a subset of the workspace sensed by the robot is known at each instant [8].

The motion of a physical robot is constrained by its kinematics, mechanical design, the properties of its drive system, and by its dynamics. As a result, a solely geometric solution to the motion planning problem is in general not feasible. For example, if a moving robot attempts to make a sharp turn, it can tip over. Among the related issues, the connection between robot dynamics and motion planning is perhaps the main one; it is also an active area of research. Considered approaches can be called *dynamic*, to distinguish them from the ones that deal solely with geometric and kinematic issues.

This paper addresses sensor-based motion planning with dynamics primarily in the context of physical reality and experimental verification. The algorithm presented below takes into account constraints imposed by the robot dynamics. An important property of the approach is that it combines the sensor-based planning and the control mechanism within a single-stage operation. This provides simplicity and preserves convergence. Parameters of the model are analyzed and identified from experimental data. Dealing with the system dynamics explicitly turns out to be crucial for assuring safety in real-time motion planning. Experiments, carried out on a commercial Nomad-200 mobile robot, demonstrate the improved system behavior under the combined single-stage planning/control procedure.

2 The approach

For the moment, consider the problem of sensor-based navigation as one consisting of two separate problems: a geometric task of path generation (call it Path Planner), for the robot to move in the workspace filled with obstacles, and a control task (call it Controller), which generates control commands to make

---

*The work was supported in part by the Office of Naval Research, grant 149710871, and by Sea Grant No. NA46RG048.

[†]On leave from the Systems Engineering & Automation Area, Electrical and Computer Engineering Dpt., University of Oviedo, Gijón 33208, Asturias, Spain

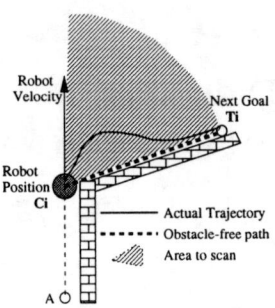

Figure 1: To handle system dynamics, planning and control must be tied together within a single-stage cycle.

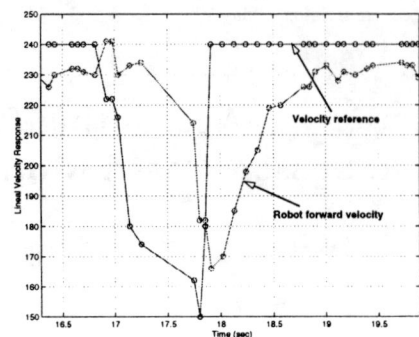

Figure 2: Forward velocity: experimental data for the (u_1, v) relation.

the robot follow that path. The input information of the Path Planner is robot current coordinates, C_i, and the description of the surrounding obstacles. Its output is an intermediate target point, T_i, and a straight-line path segment that leads to it. The Controller's input is the current state C_i, current velocity vector v_i, point T_i, and the path segment from the Path Planner which the Controller is expected to execute.

Notice a possible conflict, Figure 1. As the robot arrives at point C_i along the path AC_i, it decides on a new intermediate target, T_i, and a straight-line path segment to it. Since it arrives at C_i with a non-zero velocity, because of the system dynamics it cannot make a sharp turn suggested by the Path Planner. To preserve continuity in velocity, the real path must have a bulge around C_i (shown in Figure 1 in solid line). This curved path could in principle cut through an obstacle, which the Path Planner would not consider because it is off the intended straight line path. It would be much better if the operation of the Path Planner would directly account for the system dynamics. Besides planning the path proper, it would then also plan for the velocity and acceleration as well. In the example in Figure 1, if the robot expected an obstacle beyond point C_i, it would slow down just enough to decrease the bulge in the path and to keep closer to the line C_iT_i, while keeping the velocity at its optimum. This is the idea that inspired the strategy discussed in [10].

Under the proposed control, the robot always moves with the maximum velocity that is feasible under the circumstances. That is, if the intended path segment is a straight line, the robot will move with the maximum absolute speed – unless it senses an obstacle in this direction, in which case it slows down or/and plans a detour. If the intended path segment is curved, sensing is done within a prescribed sector; again, the velocity is adjusted so as to guarantee safety. The robot model is implicitly included in the selection of the sensor scanning sector. At all times, the robot must have a guarantee of emergency stopping path, in case an obstacle cannot be avoided in a smooth fashion.

3 Identification of Robot Dynamics

The Nomad–200 mobile robot used in this work has a synchro–drive mechanism with decoupled controls of forward velocity and orientation. We are interested in the transfer function relation between the inputs (u_1, u_2) and the robot actual motion (v, ω), corresponding to the forward velocity v, and angular (turning) velocity ω_1, respectively. We assume that the relations (u_1, v) and (u_2, ω) are mutually independent. Figure 2 shows the corresponding experimental data for the first relation. As shown, the command cycle period Δt is about 100 $msec$, which corresponds to about ten commands per second. In Figure 2 the value of the control u_1, responsible for the forward velocity motion v, changes between 15 and 24 $\frac{inch}{s}$. This produces a change in the robot actual velocity, with some time delay, and a certain acceleration rate. Similar behavior appears with the second pair (see [1]). For each pair there is an extra configurable parameter, the rate of change of velocity, set to $a = 10\frac{inch}{s^2}$ and $\alpha = 24\frac{deg}{s^2}$, respectively.

For many applications the system response can be explained with a model with constrained inputs (see, e.g. [9, 2]). This model has some range(s) within which the robot dynamics are not negligible. The robot motors, controlled by high frequency digital controllers, to set the velocities (v, ω) to the required values (u_1, u_2), work at high frequencies (1 kHz in the Nomad–200 case) under an independent microprocessor control [3]. Velocities generated by motion planning algorithms are usually commanded at lower rates

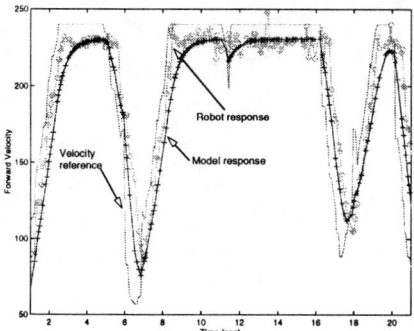

Figure 3: Comparison of (u_1, v) transfer function between the Nomad experimental data and model (4)

(for example, four commands per second [2]). This allows one to ignore the control loop dynamics with good precision, whenever the conditions above are fulfilled. When the planning frequency approaches the control frequency, the effect of the actual (rather than modelled) system dynamics is more pronounced, and larger deviations from the commanded values can be expected.

An alternative approach is to identify the input-output model from the actual robot response data. Along this line, one can estimate the system parameters under specific, experimental conditions, or carry out on-line recursive identification. To identify transfer functions between (u_1, v) and (u_2, ω), we need first to select a candidate model. The best fit, obtained with the least squares approximation of experimental data, corresponds to a first order model with unity delay (it will be used in the next section in the motion planning algorithm design):

$$G(z^{-1}) = \frac{b_1 z^{-1}}{1 + a_1 z^{-1}} \quad (1)$$

Here a_1, b_1 are the parameters to be identified. The comparison results between the experimental data and the model are shown in Figure 3. The real robot and the identified model are fed with the same references, and the responses are compared. From the identified model, the robot motion response for each control can be parameterized with (K, t_s, t_r), where K is the gain ratio, t_s the time to reach the reference (which relates to the maximum acceleration rate), and t_r is the time delay in the reference changes.

A constrained inputs model can be viewed as the discrete version of the differential equations

$$M\dot{v} = \tau_1, \qquad J\dot{\omega} = \tau_2 \quad (2)$$

where M is the robot mass and J represents the robot inertia distribution. Assuming that within the interval $\Delta t = t_1 - t_0$ the inputs $\frac{\tau_1}{M}, \frac{\tau_2}{J}$ are constant, simple integration of the model (2) produces,

$$v(t_1) = v(t_0) + \frac{\tau_1}{M} \cdot \Delta t \qquad \omega(t_1) = \omega(t_0) + \frac{\tau_2}{J} \cdot \Delta t \quad (3)$$

As mentioned above, from the control+planning point of view the main limitation of this model is that typical built-in digital motor controllers do not give one an ability of direct torque control. We can model the motor torque with the equations $\tau_1 = k_1(u_1 - v_k)$ and $\tau_2 = k_2(u_2 - \omega_k)$, with $k_1[\frac{Kg}{s}]$ and $k_2[Kg\frac{m}{s}]$ being dimensional constants. Our controls are now (u_1, u_2), and the corresponding motion equations are:

$$v(t_1) = v(t_0) + \frac{k_1}{M}(u_1 - v(t_0)) \cdot \Delta t \quad (4)$$

$$\omega(t_1) = \omega(t_0) + \frac{k_2}{J}(u_2 - w(t_0)) \cdot \Delta t \quad (5)$$

These relate directly to the difference equation identified in (1) from the experimental data, which can be considered an approximation of the differential equation

$$M\dot{v} + f_v = \bar{\tau}_1 \qquad J\dot{\omega} + f_\omega = \bar{\tau}_2 \quad (6)$$

with $f_v = k_1 v$ and $f_\omega = k_2 \omega$ being linear functions, and $\bar{\tau}_1 = k_1 u_1$, $\bar{\tau}_2 = k_2 u_2$. The relation between the identified parameters and this continuous model is given by:

$$\frac{k_1}{M} = \frac{1 + a_1}{\Delta t} \qquad k_1 = \frac{b_1}{\Delta t} \quad (7)$$

The second control is treated similarly. Model (6) had been used in robot motion control before [4]. Note that when $f_v = 0$ and $f_\omega = 0$, model (6) is reduced to (2).

4 The Algorithm

As explained in Section II, within one command cycle the navigation problem in our system is divided into two stages. First, the planner algorithm produces the next intermediate goal. Then the velocity controller, based on the sensor scan of the environment, decides what controls are to be applied in the next cycle so as to reach the intermediate goal in minimum time. The higher the command frequency, the better the system performance. (The maximum velocity achievable by Nomad-200, $v_{max} = 24[\frac{inch}{s}]$, requires about 10 control commands per second).

The planner can be implemented with any convergent sensor-based motion planning algorithm; in our experiments, the VisBug algorithm [7] has been used. The controller makes use of the scheme proposed in [10]; in brief, it operates as follows: 1) compute controls to reach the intermediate goal in minimum time, 2) scan the workspace to check if the computed controls guarantee an (emergency) stopping path in the next robot position, 3) if so, proceed at the maximum speed; otherwise, find suboptimal controls that guarantee the stopping path. The Safe-Scan Window (SSW) is the minimum sector that the robot sensors need to scan at each planning cycle in order to assure a safe stopping path if needed, no matter what the robot velocity and direction of motion are. When designing the SSW sector, the requirements to the motion stated above – high speed, safety and convergence – must be respected. Its design will depend on the robot motion equations.

The integration of these non–holonomic motion equations is not easy because of the coupling between velocity and orientation. Alternatively we can use a worst case analysis. To simplify the implementation, the same size area, based on the worst case, will be always scanned. This sector is found based on two canonical operations that cover the worst case of an emergency stop. These are the "Panic Stop" and the "Turn Panic Stop", adopting the nomenclature in [5]. The robot will scan an area big enough to assure that these two operations are always available.

The *Panic Stop operation* occurs when the robot initial conditions are $(v, \omega)_{PS} = (v_{max}, 0)$ and an emergency stop is required. This happens when the robot senses an obstacle while moving at the maximum speed along a straight line. Intuitively, the best controls to apply in order to stop here are $(u_1, u_2)_{PS} = (0, 0)$. The identified model (1) predicts a straight line stop path, with an exponential deceleration given by $v_k + a_1 v_{k-1} = 0$, or, using the model in (4):

$$v_k = \left(1 - \frac{k_1 \Delta t}{M}\right) v_{k-1} \qquad (8)$$

The distance r_d to stop must include the distance traversed due to the time delay; this leads to

$$r_d = t_r \cdot v_{max} + \sum_i^{t_s} v_i \Delta t = v_{max} \left(t_r + \frac{t_s}{1 - a_1}\right) \qquad (9)$$

Then, the linear velocity will decrease exponentially following the equation (6), giving $v = v_0 e^{-\frac{k_1}{M} t}$.

The *Turn Panic Stop operation* corresponds to an emergency stop when the robot is turning at its maximum speed. The corresponding conditions will then be $(v, \omega)_{TPS} = (v_{max}, \omega_{max})$. The necessary controls

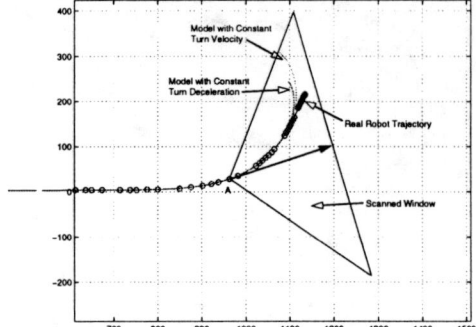

Figure 4: Turn Panic Stop operation.

will be again $(u_1, u_2)_{TPS} = (0, 0)$. From the robot motion model, the linear velocity will decrease exponentially, following the same equations as those for Panic Stop. The angular velocity has the same first-order dynamics $\omega = \omega_0 e^{-\frac{k_2}{J} t}$. An approximate solution is possible if we assume that, instead of $\omega = \omega_0 e^{-\frac{k_2}{J} t}$, ω is constant during the whole maneuver. Then the orientation will increase linearly, following $\theta = \omega_0 t$; call it the Constant Turn Velocity simplification. The stop trajectory describes an spiral curve:

$$\Delta x = v_0 (e^{-p_M \Delta t} \frac{-p_M \cos(\omega_0 \Delta t) + \omega_0 \sin(\omega_0 \Delta t)}{p_M^2 + \omega_0^2}$$
$$+ \frac{p_M}{p_M^2 + \omega_0^2}) \qquad (10)$$

$$\Delta y = v_0 (e^{-p_M \Delta t} \frac{-p_M \sin(\omega_0 \Delta t) - \omega_0 \cos(\omega_0 \Delta t)}{p_M^2 + \omega_0^2}$$
$$+ \frac{\omega_0}{p_M^2 + \omega_0^2}) \qquad (11)$$

where $p_M = \frac{k_1}{M}$. The portion of this spiral traversed by the robot depends on the time necessary to bring its forward velocity to zero. Note that with this model, only when $p_M = \frac{k_1}{M}$ is zero, (that is, the term f_v in (6) vanishes), we can say that the forward velocity remains constant, $v = v_0$. If the robot is able to turn fast enough, it can move inside the cone given by the worst case. This has been the case in the experiments with the Nomad: the control system demanded the opposite turn; the robot was able to produce it before stopping, moving towards the inside of the safe area.

From that, and given the Nomad's sonar sensing, a triangular scan window was selected as shown in Figure 4. Two parameters define the SSW area: height (d_f) and aperture (ρ). The SSW height (d_f) is defined by the Panic Stop operation. Let us call v_k the actual robot velocity, r_k the maximum distance traveled within a control cycle, and r_d the distance needed to stop with maximum brake effort in the Panic Stop operation. Using equation (9), the height condition is

$$r_k = v_k(\Delta t + t_r) \qquad d_f \geq r_k + r_d \qquad (12)$$

This is because the distance traversed within a control cycle depends only on the robot current status v_k. Using in these expressions the worst case $v_k = v_{max}$ fixes d_f: it now depends only on the robot model parameters and can be computed off-line.

The aperture ρ of the "Turn Panic Stop" spiral is obtained from equations (10,11). It should be sufficient to guarantee a stop maneuver inside the scanned area in the worst case of turning, when $(v_{max}, \omega_{max})_k$. Two parameters define the maximum curvature of turn with these velocities: the deceleration time until a complete stop t_{dec} (given by (10)), and the minimum turning radius r_{max} (assuming no deceleration, it is $r_{max} = \frac{v_k}{\omega_k}$; it can be calculated better from the model equations). Finally, the cone angle of the scanning cone must be enough to cover the robot dimension (a circle of radius r_r), and to satisfy another restriction: $d_f \tan \rho \geq r_r$. In our case the robot characteristics are

$$v_{max} = 24[\frac{in}{s}], \; a = 10[\frac{in}{s^2}], \; \omega_{max} = 50[\frac{deg}{s}], \; \alpha = 45[\frac{deg}{s^2}].$$

The scan window will be of dimensions $r_k = 100$, $r_d = 290$ (in tenths of an inch), and $\delta \approx 30$ degrees.

The velocity control strategy has to satisfy these considerations: 1) if there are no obstacles inside the scanned area, the robot should move at the maximum velocity, 2) if an obstacle appears inside the area, the robot should have enough time for a complete stop, no matter what its current velocity, 3) when an obstacle appears on the robot's way, it should reduce its velocity or/and plan a detour. Once a detour is safely complete, the robot will accelerate again to its maximum speed. If it needs to stop due to the obstacle, a special recovery maneuver is invoked to modify the path [10]. Intuitively, this strategy realizes an idea of maximum turning, by using the extreme values of the controls:

$$u_1 = K_v, \qquad u_2 = K_w(\delta - \theta). \qquad (13)$$

where δ is the angle between the current position and the intermediate goal T_i (in the robot local framework); K_v, K_w values depend on the sensory data about the environment. With no obstacles inside the scanned cone, $K_v = u_{1\,max}$ and K_w is chosen so as to produce maximum turning, $u_2 = u_{2\,max}$. If an object appears in the scanned area, the velocity is reduced linerary, with $K_v = u_{1\,max} \cdot \frac{d}{d_f}$, where d is the distance to the nearest obstacle inside the area, and d_f the length of the scanned window. The robot deceleration ability is accounted for in the value of d_f. If

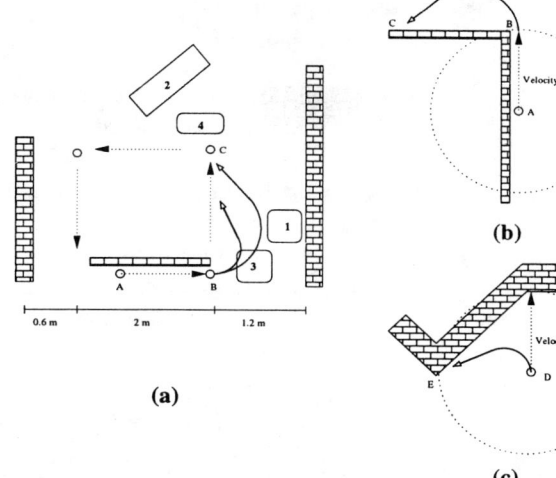

Figure 5: Experimental setup. a) The top view of the lab; objects 1,2,3,4 are additional obstacles. b,c) examples of local motion control.

it's clear that the deceleration is not sufficient to avoid the obstacle, the robot will come to a stop, and then initiate a detour. Such recovery is always possible for a synchro-drive robot.

5 Experimental Results

The experiments were carried out with a Nomad-200 robot, with the control algorithms implemented in the on-board 486–33MHz processor. The Safe–Scan window corresponds to a frontal sonar sensor. Based on the robot specifications, analysis above, and the energy profile of the Polaroid sonar sensor, the scan window aperture ρ was chosen at about 20 degrees. As the focus of the work is on the control strategy, some idiosyncrasies of a sonar sensor – such as false reflections from wall corners or from angular-shape obstacles – were avoided by a proper design of the test workspace.

A sketch of the lab top view is shown in Figure 5a. Obstacles are walls and other objects (chairs, table - objects 1,2,3,4). A sequence of global goals is given, which presents the corners of a 2 by 2 meter square (shown in Figure 5a by straight line arrows). Consider the moment when the robot (shown as a small circle) is moving at its maximum speed through the point A. Between points A and B only forward velocity is controlled. Given the robot model and sensory data, at each control cycle (with 20 cycles per second in our implementation) the motion planner defines a new intermediate goal T_i, which happens to be unchanged until the robot arrives in the vicinity of point B. At B, new intermediate goals appear (eventually this becomes point C), and new velocities are selected.

Note that in the corners of the square the robot moves forward while simultaneously turning towards its new intermediate target.

Tests were carried out with different combinations of obstacles, to study the system performance in various cluttered situations. Examples in Figures 5b,c show typical turning paths. The following table summarizes one set of experiments, aimed at analyzing the performance of the proposed strategy.

	Path Le.	Time	(v_m, a_m)	(w_m, α_m)
Exp 1	3200	25.78	(240,300)	(450,500)
Exp 2	3200	22.53	(240,a_M)	(450,α_M)
Exp 3	3505	24.77	(140,100)	(450,500)
Exp 4a	3422	22.92	(240,100)	(450,500)
Exp 4b	3802	22.57	(240,100)	(450,500)
Exp 4c	3990	22.88	(240,100)	(450,500)
Exp 4d	4450	22.68	(240,100)	(450,500)

Relation between the path length (in tenths of an inch) and motion time (in seconds) was studied under different velocity and acceleration patterns. In the first four experiments in the table above the environment was as shown in Figures 5a. In Experiment 1, a "stop & turn" strategy was used in order to reproduce the exact square path, Figure 5a. The robot was accelerated and decelerated along both control axes with accelerations a_m and α_m (in $\frac{0.1in}{s^2}$ and $\frac{0.1°}{s^2}$, respectively), and velocities v_m and ω_m. The task was completed in about 26 sec; as expected, the path length was that of the $2mx2m$ square, about 3200 of our $.1in$ units. The same strategy is used in Experiment 2, except the robot was forced to its acceleration limits ($a_M = 900$ and $\alpha_M = 2000$). This experiment represents the robot time performance limit, achieved at the expense of slippage, unstable motion and large localization errors. In Experiment 3 the velocity was controlled without any dynamic considerations, using the control law $v = K_v$ and $\omega = K_w(\delta - \theta)$, where values K_v and K_w are design parameters chosen empirically to maximize the robot speed.

In Experiments 4a to 4d the control strategy presented in this paper has been used. Compared to Experiments 1, 2, 3, the same task is finished in shorter time, in spite of smaller accelerations and smoother motion. Figure 6 shows the top view of the resulting path. Motion starts at the lower left corner of the path square. The corresponding velocity profiles are shown in Figure 7. Experiments 4b,c,d relate to the same task and the same control strategy as in Experiment 4a, carried out in increasingly simpler environments: Experiment 4b has no obstacle 4 in its environment, Experiment 4c – no obstacles 4 and 3,

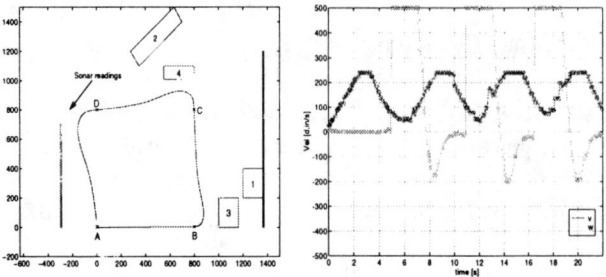

Figure 6: Exp.4a, cluttered environment; top view.

Figure 7: Experiment 4a: velocities (forward & ang.).

Experiment 4d – no obstacles 4, 3, 2. Overall, among the experiments shown in Table I, those based on the proposed strategy showed the best performance.

The experimental results suggest that the proposed approach works well within the range of speeds tested. It would be interesting to test it at higher speeds and accelerations; it is conceivable that further improvement in performance might be necessary. For such experiments, a robot with the range of velocities and accelerations wider than Nomad-200 would be needed.

References

[1] J. C. Alvarez, A. Shkel, and V. Lumelsky. Accounting for mobile robot dynamics in sensor–based motion planning. Tech.Rep. RL-97007, Rob.Lab., U.W.-Madison, jun 1997.

[2] D. Fox, W. Burgard, and S. Thrun. The dynamic window approach to collision avoidance. *IEEE Robotics and Automation Magazine*, april 1997.

[3] Galil Motion Control, Inc., Sunnyvale, CA 94086. *DMC-600 Series User Manual*, 1991.

[4] J. Guldner and V. I. Utkin. Tracking the gradient of artificial potential fields: sliding mode control for mobile robots. *Int. J. Control*, 63(3):417–432, 1996.

[5] A. Kelly and A. Stentz. Analysis of requirements for high speed rough terrain autonomous mobility. In *IEEE Int. Conf. on Rob. & Aut.*, pages 3318–25, 1997.

[6] Jean-Claude Latombe. *Robot Motion Planning.* Kluwer Academic Publishers, 1991.

[7] V. Lumelsky and T. Skewis. Incorporating range sensing in the robot navigation function. *IEEE Trans. Robotics and Automation*, 20(5):1059–1069, sep 1990.

[8] V. Lumelsky and A. Stepanov. Dynamic path planning for a mobile automaton with limited information on the environment. *IEEE Trans. Autom. Control*, 31(11), nov 1986.

[9] Nomadic Techs. *Nomad 200 Users's Guide*, 1996.

[10] A. Shkel and V. Lumelsky. The Jogger's problem: Control of dynamics in real-time motion planning. *Automatica*, 33(7):1219–1233, jul 1997.

Controlling Sensory Perception for Indoor Navigation

G.E. Hovland
Info. Tech. and Control Systems Division
ABB Corporate Research
Oslo, Norway

B.J. McCarragher
Department of Engineering
Faculty of Engineering and Info. Tech
The Australian National University
Canberra, Australia

Abstract

The problem of controlling sensory input and perception for use in mobile navigation is addressed in this paper. The proposed solution offers several advantages compared to existing methods in the literature. First, the proposed solution is based on a stochastic dynamic programming algorithm which guarantees cost-efficiency of the real-time sensory perception controller (SPC) by solving a constrained optimisation problem. Second, a new and unique discrete event model of mobile navigation is presented. The model is task-independent and can be used in a wide range of navigation problems. Third, the discrete events become a natural common representational format for the sensors, which further extends the applicability of the proposed solution. Fourth, the sensing aspect of discrete event systems has often been neglected in the literature. In this paper we present a unique approach to on-line discrete event identification.

1 Introduction

Sensory perception is one of the main bottlenecks in mobile navigation. Inaccurate sensors and world unpredictability frequently cause the failure of real-time mobile navigation systems. Single sensor systems provide high bandwidth solutions, but often lack the required robustness in noisy environments. Multiple sensor systems often have increased robustness, but may fail to meet response time requirements. Control of sensory perception aims at keeping the sensing costs low while increasing robustness by actively selecting different sensing techniques in real-time. In normal operation only a few sensing techniques are needed. Then, as the quality of the sensory perception decreases, additional sensors are utilised.

The most common approach to sensory perception in mobile navigation is to continuously use all available sensors. Different sources of information are combined into one representational format. This operation is often referred to as multi-sensor fusion, see [7, 9] for extensive surveys. As a robot's tasks become more complex and more numerous, the number of relevant features of the environment quickly exceeds the sensing and processing resources that are feasible to supply to a robot, [13]. Hence, the control of sensory perception is an important step towards designing autonomous robots in complex and uncertain environments.

In this paper we present a new and unique approach to dynamic sensing strategies. A sensory perception controller (SPC) using stochastic dynamic programming has been developed. To be of any practical value, the extra incurred overhead cost by the SPC can not outweigh the actual sensing costs. Otherwise, a more cost-efficient solution would be to use all available sensors and multi-sensor fusion techniques. Hence, computationally efficient methods such as dynamic programming are required. In [2, 12] brute-force and tree search methods using Bayesian decision theory are described. The improved performance of dynamic programming compared to these methods comes from solving multi-stage problems by analysing a sequence of simpler inductively defined single-stage problems.

The general problem of mobile robot navigation was summarised by [8] by three questions: "Where am I?", "Where am I going?" and "How should I get there?". The discrete event model provides an elegant framework for addressing these questions. The first question, which is the focus of this paper, is addressed by the real-time identification of discrete events. The other two questions are addressed by discrete event trajectory planning, see for example [10]. The discrete events become a natural common representational format for the sensors. A host of sensors such as tactile sensors, three-dimensional range sensors, force-torque sensors and acoustic sensors are currently being used in, for example, robotics and manufacturing applications and are well suited to discrete event identification. The sensory perception controller presented in this paper is not limited to mobile robot navigation problems. In particular, manufacturing, robotic assembly, communication networks, transportation systems and logistic systems all fall within the class of discrete event systems and have to deal with real-time discrete event identification.

A new and unique discrete event model of mobile navigation in structured environments is presented.

When the mobile unit is in contact with a surface, the free space motion becomes constrained. The discrete events are defined as changes in the mobile unit constraint equations. The advantage of such a model is a complete independency of given tasks. The successful completion of any given task becomes issues for the discrete event controller and the event trajectory planner. Hence, the proposed modelling of discrete events can be used in a wide range of applications.

Moreover, the sensing aspects are often neglected in discrete event systems research. The detection and identification of discrete events are assumed to be perfect. In this paper we present a new and unique approach to on-line discrete event identification in mobile navigation by using an active sensing architecture.

2 Discrete Event Formalism

The problem of mobile navigation is immense. Different strategies are required for static, dynamic, partially known and unknown environments. We focus our attention to the localisation problem in a planar static environment as shown in Figure 1. The proposed method handles noisy and imperfect sensing, but requires all surfaces in the world to be modelled. For simplicity, all surfaces in the model in Figure 1 are straight lines. More complex surfaces, for example circles and semi-circles, can easily be incorporated into the model. However, such surfaces complicate the event identification algorithms.

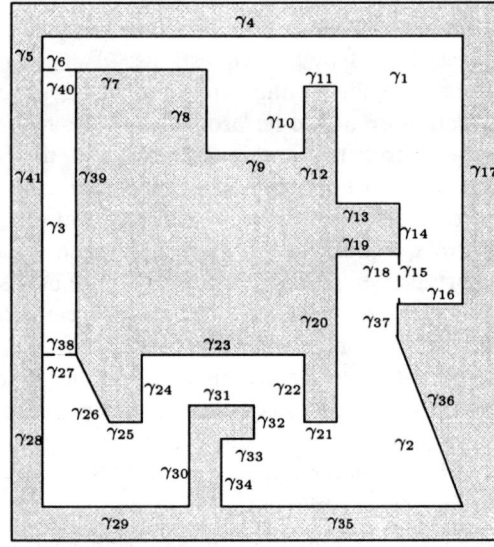

Figure 1: *Contact states for the navigation problem. The dotted lines describe doors. Both sides of the doors are modelled.*

We have chosen an environment consisting of three rooms, several walls and doors. The doors can be open or closed. There is no direct method for sensing whether a door is open or closed. This scenario allows us to demonstrate the ability of the SPC to recover from undetected events.

Each wall and door is modelled as a straight line. The mobile unit is modelled as a point (p_x, p_y) with a direction θ. The point (p_x, p_y) of the mobile unit can be in any of the free spaces or contact any of the walls and doors. The equation of motion of the mobile unit in free-space is generally written as

$$\dot{\mathbf{x}}(t) = f(\mathbf{x}(t), \mathbf{u}(t)) \quad (1)$$

where $\mathbf{x}(t)$ is the continuous time space vector, and $\mathbf{u}(t)$ is the input vector. As the mobile unit interacts with an inertially fixed environment, the free-space dynamics of equation (1) become constrained. For each point-surface contact, this constraint may be written as

$$g_j(\mathbf{x}(t)) = 0 \quad (2)$$

where g_j is the constraint equation for the jth point-surface contact. The distance function from the point to the surface is an ideal candidate for g_j. The distance functions can be constructed such that all $g_j(\mathbf{x})$ are positive inside the room and negative outside. Then, free-space motion in a room is constrained by the walls, ie.

$$\begin{aligned} i &= \mathrm{argmin}_j \quad |g_j(\mathbf{x}(t))| \quad \text{for all } j \in \mathcal{J} \\ g_i(\mathbf{x}(t)) &\geq 0 \end{aligned} \quad (3)$$

where $\mathcal{J}$ is the set of all constraints corresponding to a surface in the room. Equation (3) simply says that the mobile unit has to be on the room side of the closest surface. Since the mobile unit is modelled as a single point, only one constraint is active except for the corner points. Point-point contacts are not modelled. For many applications, point-point contacts occurring at corner points are short-lived and can be extremely difficult to detect and identify. A *discrete state* is defined as follows.

Definition 1 (Discrete State) *A discrete contact state γ_i is given by a unique constraint equation. For a point-surface contact (including walls and doors) the constraint is given by equation (2). For a no contact state (rooms) the free-space motion is given by equation (3).* □

The contact state network is shown in Figure 2. Note that the individual rooms are only connected through the door states. Hence, the addition of more rooms to the model is trivial. Also note that both sides of the

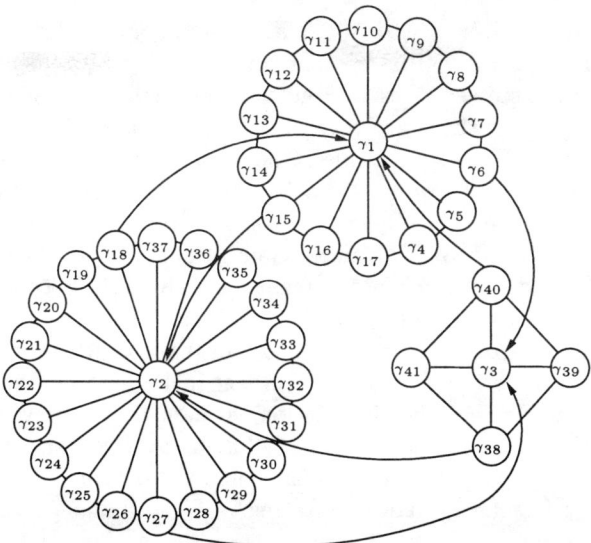

Figure 2: *Contact state network. The non-directional lines indicate bi-directional transitions between states. The directional lines indicate a transition from a door state to free space.*

doors are modelled as a discrete state. The constraints for the two sides of a door are different caused by the requirement of $g_j(\mathbf{x}) \geq 0$ on the room side of a surface.

A *discrete event* is now defined as follows.

Definition 2 (Discrete Event) *The mobile unit can be in any free space state or contact doors and walls. A discrete event is defined as a physical change of state $\gamma_i \to \gamma_j$ between the mobile unit and the planar environment.* □

3 Sensory Perception Controller

The sensory perception control problem is formulated as follows. Several monitors which outputs depend on the discrete events are available. Given an occurrence of a discrete event, select a sequential order in which monitors are consulted and a stopping rule such as to minimise the cost related to the error in classifying the event plus the costs of obtaining the monitor outputs. Each monitor is consulted at most once for each occurrence of discrete events.

The motivation for consulting each monitor at most once is the fact that discrete events are short-lived and the dynamics of the sensing signals are strongest immediately after the occurrence of the events. The most efficient event monitoring is achieved by analysing these early dynamic signals. The extra event information gained by analysing the measurements a long time after the event occurred is insignificant and usually no extra information is gained by consulting a monitor twice on the early dynamic sensor signals.

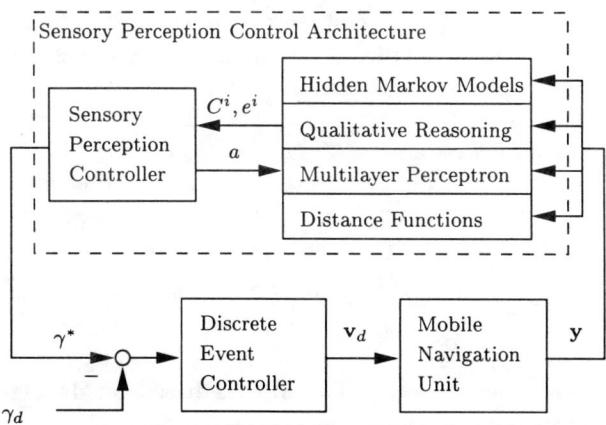

Figure 3: *Sensory perception control architecture in the discrete event framework. Monitor i recognises events e^i as they occur with confidence C^i. The action $\mathbf{a}$ from the SPC sequentially consults monitors to produce the final recognised state γ^*. γ_d is the discrete event controller reference command, $\mathbf{v}_d$ is the controller output and $\mathbf{y}$ is the continuous measurement vector.*

Figure 3 shows the block diagram of the discrete event control structure. The SPC is based on stochastic dynamic programming and a detailed description of the model is presented in [4]. Application of the SPC was demonstrated for a planar robotic assembly task by [5]. The perceptual capabilities of the discrete event system consist of several process monitors. Due to noisy measurements, model uncertainties or world unpredictability, the recognised event e^i may not correspond to the actual physical event e that occurred. A very important feature of a process monitor is its ability to indicate the confidence level C^i of the recognised event. A good process monitor produces low confidence levels for events recognised incorrectly and large confidence levels for events recognised correctly.

When a discrete event occurs, the sensory perception controller (SPC) has the option of sequentially consulting any of the process monitors. The SPC has two main objectives. First, the SPC must use the recognised events e^i and the corresponding confidence levels C^i efficiently to correctly recognise the Discrete Event System (DES) states. Even when some of the e^i are incorrect, a robust SPC is able to recognise the correct state. Second, the SPC must keep the perceptual costs low. In real-world applications computational resources are limited and have to be shared between the perceptual system and the other components.

The SPC maps the final recognised event e^* to a final recognised DES state γ^* which is sent to the discrete event controller. The discrete event controller changes its command vector $\mathbf{v}_d$ depending on γ^* and the desired state γ_d. The command vector depends on the process plant, but typically contains reference commands and parameters for low-level process plant controllers. The command vector $\mathbf{v}_d$ will drive the process plant until the next discrete event occurs. The sensory perception control system is then activated again.

A strict requirement of any SPC is low real-time overhead cost. The control of sensory perception becomes meaningless if the overhead cost outweighs the actual sensing costs. The SPC is based on dynamic programming which provides cost-efficient solutions to sequential multi-stage decision problems by analysing a sequence of simpler inductively defined single-stage problems. Moreover, the control of sensory perception fits particularly well in the discrete event control framework. Because of the discrete nature of events, the discrete event controller does not require feedback information continuously. Hence, valuable processing time is available for the analysis of sensory information between the occurrence of events.

4 Process Monitors

Four process monitors are used as shown in Figure 3. All the monitors recognise discrete events as they occur and produce real-time confidence level measurements. The *Distance Functions* monitor uses the definition of distance functions by [1]. The *Multilayer Perceptron* monitor is described in [6]. The *Qualitative Reasoning* monitor is described in [11] and the details of the *Hidden Markov Models* monitor are found in [3].

The required CPU times and the average recognition rates of the monitors in real-time operation were found from a sample set of 300 discrete events and are given in Table 1. Note the relatively low recognition rates

Process Monitor	CPU Sec.	Rec. %
1 – Distance Functions	0.11	72.8
2 – Multilayer Perceptron	0.61	76.0
3 – Qualitative Reasoning	0.09	66.0
4 – Hidden Markov Models	0.91	68.4

Table 1: *Required CPU times (CPU seconds) and average recognition rates of the process monitors in real-time operation.*

by monitor 3 and 4. These two monitors use the force/torque measurements only. In the experiments the force sensor is located in the centre of the maze and that there are several parallel surfaces, see Figure 4. This setup causes the measured planar forces and torque to be extremely similar for several of the discrete events. Hence, process monitoring using force only is difficult. We added Gaussian noise $\mathcal{N}(0,1)\,[cm]$ to the position measurements to reduce the recognition rates of the position based monitors. Further, the maze was moved $2cm$ (8% of maze width) in the P_x direction to make the model in Figure 1 inaccurate. In the following section we show that even with these limitations in the modelling and the sensors, the sensory perception controller is able to achieve high performance event monitoring.

5 Experiments

The experimental setup is shown in Figure 4. A Polhemus position sensor and stylus were used to obtain the planar position measurements, while a JR3 force/torque sensor was used to measure the planar forces and the torque. The Polhemus and the JR3 sensors were connected to a Motorola 68040 based VxWorks board. All the process monitors and the sensory perception controller were implemented on the VxWorks board for real-time control.

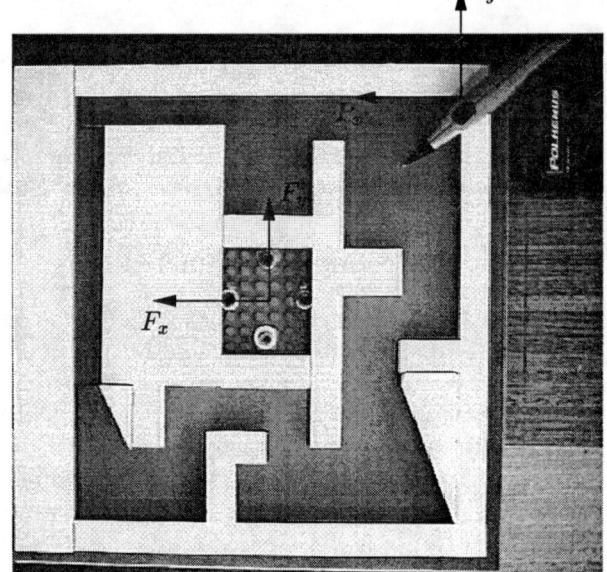

Figure 4: *Experimental setup consisting of maze, Polhemus position sensor and JR3 force/torque sensor (located under the maze). The outer dimensions of the rectangular maze is 25.6cm by 25.6cm.*

The sensory perception controller was tested on a sam-

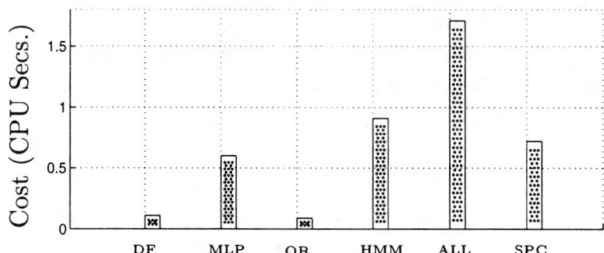

Figure 5: *Average costs of the process monitors and the sensory perception controller. DF is the distance functions process monitor, MLP is the multilayer perceptron, QR is the qualitative template matching method and HMM is the Hidden Markov Model method. ALL is a consensus-based method on all the monitors and SPC is the sensory perception controller.*

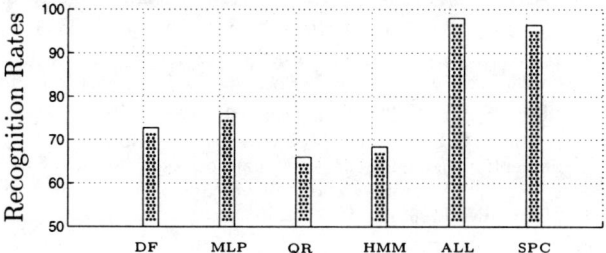

Figure 6: *Event recognition rates of the process monitors and the sensory perception controller.*

ple set of 300 discrete events; 100 discrete events occurring in each of the rooms defined by the free spaces γ_1, γ_2 and γ_3. Figures 5 and 6 show the average recognition rates and costs of 1) each individual process monitor, 2) all the monitors used all the time with consensus-based decision making and 3) the performance of the sensory perception controller. When all the monitors are used all the time, the consensus-based decision making is performed as follows

$$e^* = \mathrm{argmax}_{e_i} \sum_{j=1}^{4} S_j(e_i)$$
$$\gamma^* = \tau(e^*) \qquad (4)$$

where $S_j(e_i)$ is the score for event e_i from process monitor j and τ maps the recognised event e^* to a final recognised state γ^*. As expected, the highest recognition rate was achieved when using all the monitors all the time. However, the total cost of this method is considerably higher than the SPC method. Note also that the SPC method has a higher recognition rate than any individual process monitor.

The process monitors described in [3, 6, 11] use force measurements to detect event occurrences. When the mobile unit passes through an open door, for example $\gamma_1 \to \gamma_{15} \to \gamma_2$, the process monitors are unable to detect these events. Next we show how to recover from undetected events and the influence of undetected events on the performance of the sensory perception controller.

Process monitoring in the discrete event framework only evaluates events in the set of admissible discrete events. To be able to recover from undetected events, the set of admissible events must take into consideration possible future contact states that are not directly connected to the current contact state in the contact state network, Figure 2. When passing through an open door, two events are undetected, for example $\gamma_1 \to \gamma_{15}$ followed by $\gamma_{15} \to \gamma_2$. Hence, a look-ahead of three events is required to correctly classify new events after passing through the door. Let $\Gamma_{ij} = 1$ if $\gamma_i \to \gamma_j$ is an event defined the contact state network in Figure 2. Then, with a look-ahead of three events, the set of admissible discrete events in any contact state γ_i is given as follows.

$$\mathcal{E}_{\gamma_i} = \{\gamma_i \to \gamma_j, \gamma_j \to \gamma_k, \gamma_k \to \gamma_l\}$$
for all j, k, l such that $\Gamma_{ij} = \Gamma_{jk} = \Gamma_{kl} = 1$ \qquad (5)

Process Monitor	CPU Sec.	Rec. %
1 – Distance Functions	0.14	67.6
2 – Multilayer Perceptron	0.81	65.7
3 – Qualitative Reasoning	0.12	56.2
4 – Hidden Markov Models	1.16	60.1

Table 2: *Required CPU times (CPU seconds) and average recognition rates of the process monitors with a look-ahead of three events.*

The increased number of events in the set $\mathcal{E}_{\gamma_i}$ increases the costs and decreases the average recognition rates of the monitors as described in Table 2. One reason for the decrease in recognition rates of the force based monitors is the increased number of parallel surfaces to be considered. As mentioned earlier, the force/torque measurements are very similar for parallel surfaces. The reason for the lower recognition rate of the position based monitor is the fact that a larger number of events are possible for similar measurements. For example, the events $\gamma_1 \to \gamma_4$ and $\gamma_4 \to \gamma_1$ might have identical position measurements.

The sensory perception controller was tested on a sample set of 200 discrete events; 100 events occurring after passing through an open door and 100 events occurring with no doors involved. A look-ahead of three events was used by all four process monitors for all the events in the sample set. Again, the highest recognition rate (96.2%) was achieved when using all the monitors all the time. However, the cost of using all the monitors all the time is more than double the cost of the sensory perception controller. The recognition rate of the SPC was 91.4%. Again, the recognition rate of the SPC is higher than any individual monitor. To summarise, the SPC is able to recover from undetected events and still achieve relatively high recognition rates by increasing the set of admissible events.

For the particular contact state network in Figure 2 any contact state can be reached in maximum four events from any starting point. Even if the additional look-ahead required to cover the entire search space is only one, the increase in the size of the set of admissible events is large. For example, in γ_1 the set of admissible discrete events is almost doubled when the look-ahead is increased from three to the entire search space. In general, a trade-off has to be made between sensing costs and the additional robustness gained by a large look-ahead horizon.

6 Discussion and Conclusions

We have presented a new approach for controlling sensory perception in mobile navigation. The navigation problem is modelled as a discrete event system. The cost-efficient use of sensory perception reduces the need for mobile robots to carry powerful computational power. The cost of sensing is reduced compared to multi-sensor systems where all the sensors are used all the time. The experiments also demonstrate that the sensory perception controller achieves higher event recognition rates than any individual sensor.

Solving the task of of mobile robot navigation in an indoor environment range finders (such as lasers or optical range finders, ultrasonic sonars, etc.) are usually used. The incorporation of such sensors into the discrete event sensory perception framework requires the development of process monitors mapping the raw data into discrete events. In general, any new sensor can be made available to the SPC once a process monitor has been developed. The Hidden Markov Model and the Multilayer Perceptron are good candidates as they provide general methods for event recognition when faced with noisy and uncertain sensors.

The proposed method addresses sensing in the discrete event formalism. This area has often been neglected in discrete event research. We believe the incorporation of high performance discrete event recognition will improve the applicability of the discrete event theory in many areas, including mobile navigation.

Compared to traditional off-line sensor planning algorithms, the sensory perception controller takes noisy and imperfect sensing into account. Real-time measures of discrete event confidence levels are incorporated into the decision making. Sensory perception is often the main bottleneck in mobile robotic agents. The perceptual capabilities often limit the overall system's performance. The control of sensory perception is an important step towards robust agents operating in unpredictable real-world environments.

References

[1] Astuti, P. and B.J. McCarragher, "Sufficient Conditions for the Success of Robotic Assembly", *Proc. of the 1994 IEEE Intl. Conf. on Robotics and Automation*, San Diego, May, pp. 1693-1699.

[2] Hager, G.D., *Task-Directed Sensor Fusion and Planning: A Computational Approach*, Kluwer Academic Publ., 1990.

[3] Hovland, G.E. and B.J. McCarragher, "Frequency-Domain Force Measurements for Discrete Event Contact Recognition", *Proc. of the 1996 IEEE Intl. Conf. on Robotics and Automation*, Minneapolis, 22-28 April, pp. 1166-1171.

[4] Hovland, G.E. and B.J. McCarragher, "Control of Sensory Perception Using Stochastic Dynamic Programming", *Proc. of the 1st Australian Data Fusion Symposium*, Adelaide, 21-23 Nov. 1996, pp. 196-201.

[5] Hovland, G.E. and B.J. McCarragher, "Dynamic Sensor Selection for Robotic Systems", *Proc. of the 1997 IEEE Intl. Conf. on Robotics and Automation*, Albuquerque, 20-25 April, pp. 272-277.

[6] Hovland, G.E. and B.J. McCarragher, "Combining Force and Position Measurements for the Monitoring of Robotic Assembly", *Proc. of the IEEE/RSJ Intl. Conf. on Intelligent Robots and Systems*, Grenoble, 8-13 Sept. 1997.

[7] Kam, M., X. Zhu and P. Kalata, "Sensor Fusion for Mobile Robot Navigation", *The Proceedings of the IEEE*, Vol. 85, No. 1, 1997, pp. 108-119.

[8] Leonard, J. and H.F. Durrant-Whyte, "Application of Multi-Target Tracking to Sonar-Based Mobile Robot Navigation", *IEEE Trans. on Robotics and Automation*, Vol. 7, No. 3, 1991, pp. 376-382.

[9] Luo, R.C. and M.G. Kay, "Multisensor Integration and Fusion in Intelligent Systems", *IEEE Trans. on Systems, Man and Cybernetics*, Vol. 19, No. 5, 1989, pp. 901-931.

[10] McCarragher, B.J. and H. Asada, "The Discrete Event Modelling and Trajectory Planning of Robotic Assembly Tasks", *ASME Journal of Dynamic Systems, Measurements and Control*, Vol. 117, No. 3, 1995, pp.394-400.

[11] McCarragher, B.J. and H. Asada, "Qualitative Template Matching Using Dynamic Process Models for State Transition Recognition of Robotic Assembly," *ASME Journal of Dynamic Systems, Measurements and Control*, vol. 115, no. 2A, 1993, pp. 261-269.

[12] Sakaguchi, Y., "Haptic Sensing System with Active Perceptiob", *Advanced Robotics: The international journal of the Robotics Society of Japan*, Vol. 8, No. 3, 1994, pp.263-283.

[13] Simmons, R., "Robust Behaviour with Limited Resources", *AAAI Stanford Spring Symposium*, 1990.

The Detection of Faults in Navigation Systems: A Frequency Domain Approach

S. Scheding, E. Nebot and H. Durrant-Whyte
Department of Mechanical Engineering
The University of Sydney, NSW. 2006, Australia
e-mail: scheding/nebot/hugh @mech.eng.usyd.edu.au

Abstract—This paper provides an analysis of Kalman filter based systems with respect to fault detection. By using frequency domain techniques, a metric is developed that describes the detectability of a fault by showing how a fault is transmitted to the filter innovations (if at all). Through experiment, it is shown that redundancy must be employed for guaranteed detection of faults, and that unlike sensors should be used. Further, it is shown that modelling errors can be treated within the same framework as 'hard' actuator or sensor faults.

I. INTRODUCTION

This paper is concerned with the design and implementation of navigation systems for autonomous field robotics. As the demand for outdoor automation[1,2] increases, consideration needs to be paid to the reliability of such systems. For example, if a mining vehicle is automated, one can assume that at least in the initial stages of development, that it will interact in some way with conventional vehicles driven by human drivers.

To prevent a fault in the navigation system from causing major damage and harm (to itself or to others), there needs to be some mechanism by which faults are detected, and appropriate action taken.

A necessary part of the design of autonomous systems, therefore, is the inclusion of fault detection, and if possible identification algorithms which ensure the vehicle operates in a safe and reliable manner.

Existing Fault Detection and Identification (FDI) schemes usually fall into one of two categories [3];
- Model based [4–6], or
- Innovations (or residual) based [7]

Model based fault detection systems suffer from several drawbacks. The first is that a modelled fault may be similar or identical to a perfectly valid change in system parameters. The second drawback of model based FDI is that often (in the case of estimated or observed stochastic systems) the optimality of the estimator is compromised by the inclusion of a fault model during the no-fault condition which is clearly unwanted. Lastly, it is not unreasonable to expect that a fault may occur which does not closely match any of the modelled faults thus causing the FDI to fail.

The innovations of a filter (or observer) are the only *internal* performance metric of a filter, having known statistics when no fault has occurred. The difficulty in determining whether a fault has occured in a sensor from the innovation sequence arises when the filter *tracks* the fault rather than *rejecting* it, resulting in a state estimate that diverges from the true state, without changing the statistics of the innovations.

This paper develops a metric to determine the types of faults transmitted to the innovations. The reason that an innovations based approach is chosen is because no a priori assumptions are made about the fault.

Although the intended application of the theory presented in this paper is for the automation of large outdoor vehicles, the analysis provided is applicable to any linear system whose state is estimated by a Kalman filter.

Section II presents a brief introduction to the continuous time Kalman filter in both time and frequency domains. Sections III and IV examine the detectability of sensor and process faults respectively, using frequency domain classical control techniques. Experimental results are given in Sections V, and concluding remarks are made in Section VI.

II. BACKGROUND

Consider a linear system represented in state space form by the equations,

$$\begin{aligned}\dot{\mathbf{x}}(t) &= \mathbf{F}\mathbf{x}(t) + \mathbf{w}(t) \\ \mathbf{z}(t) &= \mathbf{H}\mathbf{x}(t) + \mathbf{v}(t)\end{aligned} \quad (1)$$

where $\mathbf{x}(t)$ is the state of the system at time t, $\mathbf{F}$ is the continuous time process model and $\mathbf{H}$ is the matrix mapping the observations to state space. The variables $\mathbf{w}(t)$ and $\mathbf{v}(t)$ are the process noise and sensor noise vectors respectively.

For a system in this form, the continuous time Kalman filter update equations are given by;

$$\dot{\hat{\mathbf{x}}}(t) = \mathbf{F}\hat{\mathbf{x}}(t) + \mathbf{K}[\mathbf{z}(t) - \mathbf{H}\hat{\mathbf{x}}(t)] \quad (2)$$

where $\mathbf{K}$ is the Kalman gain and the term $[\mathbf{z}(t) - \mathbf{H}\hat{\mathbf{x}}(t)]$ is defined as the filter innovations. This may be converted to Laplace space as,

$$s\hat{\mathbf{x}}(s) = \mathbf{F}\hat{\mathbf{x}}(s) + \mathbf{K}[\mathbf{z}(s) - \mathbf{H}\hat{\mathbf{x}}(s)] \quad (3)$$

giving the transfer between the estimate and the observations as,

$$\begin{aligned}\frac{\hat{\mathbf{x}}(s)}{\mathbf{z}(s)} &= [s\mathbf{I} - \mathbf{F} + \mathbf{KH}]^{-1}\mathbf{K} \\ &= G(s)\end{aligned} \quad (4)$$

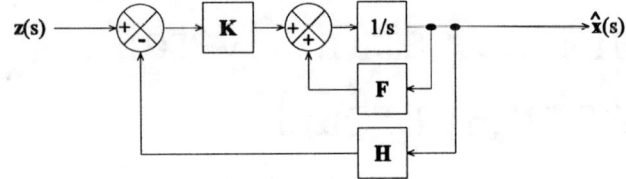

Fig. 1. Kalman Filter Block Diagram

which is constant once the filter has reached steady state.

Kalman filter operation may be seen graphically by the block diagram shown in Figure 1.

III. Sensor Fault Detectability

Under no-fault conditions,

$$\hat{\mathbf{x}}_{nf}(s) = G(s)\mathbf{z}_{nf}(s) \tag{5}$$

where the subscript nf denotes the no-fault condition, and the subscript f denotes a fault condition.

Now, consider an observation corrupted by a fault vector $\mathbf{f}_o(s)$,

$$\mathbf{z}_f(s) = \mathbf{z}_{nf}(s) + \mathbf{f}_o(s) \tag{6}$$

From equation 5, the state estimate becomes

$$\hat{\mathbf{x}}_f(s) = G(s)[\mathbf{z}_{nf}(s) + \mathbf{f}_o(s)] \tag{7}$$

The no-fault filter innovations are defined as

$$\nu_{nf}(s) = \mathbf{z}_{nf}(s) - \mathbf{H}\hat{\mathbf{x}}_{nf}(s) \tag{8}$$

so, under fault conditions,

$$\begin{aligned} \nu_f(s) &= \mathbf{z}_f(s) - \mathbf{H}\hat{\mathbf{x}}_f(s) \\ &= [\mathbf{z}_{nf}(s) + \mathbf{f}_o(s)] - \mathbf{H}[G(s)[\mathbf{z}_{nf}(s) + \mathbf{f}_o(s)]] \\ &= [\mathbf{z}_{nf}(s) - \mathbf{H}\hat{\mathbf{x}}_{nf}(s)] + [\mathbf{f}_o(s) - \mathbf{H}G(s)\mathbf{f}_o(s)] \\ &= \nu_{nf}(s) + [\mathbf{I} - \mathbf{H}G(s)]\mathbf{f}_o(s) \end{aligned} \tag{9}$$

Therefore, a sensor fault is detectable in the innovation sequence when the term $[\mathbf{I} - \mathbf{H}G(s)]\mathbf{f}_o(s)$ is non-zero. It is interesting to note that this has a direct analogy with the no-fault innovation which may be written as $[\mathbf{I} - \mathbf{H}G(s)]\mathbf{z}(s)$. It may therefore be thought of as representing the effects of the sensors' frequency content on the innovations - particularly when the frequency content is dissimilar as in the case when a sensor or sensor model is in fault.

IV. Process Fault Detectability

Figure 2 shows the continuous time Kalman filter with a fault $\mathbf{f_p}(s)$ injected at the process model. This model of process faults is proved valid by considering the summation block located directly after the gain block. The contribution of the process model loop is

$$\mathbf{F}\hat{\mathbf{x}}(s) + \mathbf{f}_p(s) \tag{10}$$

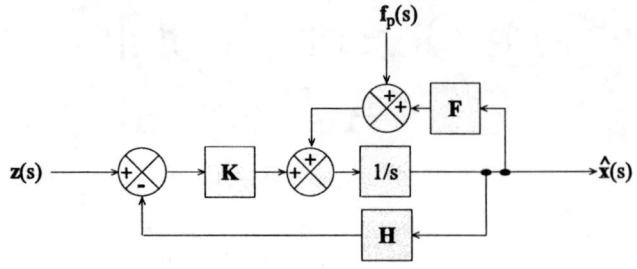

Fig. 2. Kalman Filter Block Diagram with Injected Process Fault

which may be rewritten as

$$\left[\mathbf{F} + \mathbf{f}_p(s)\hat{\mathbf{x}}(s)^{-1}\right]\hat{\mathbf{x}}(s) = [\mathbf{F} + \Delta\mathbf{F}(s)]\hat{\mathbf{x}}(s) \tag{11}$$

Showing that the injected fault $\mathbf{f}_p(s)$ is indeed representative of a fault $(\Delta\mathbf{F}(s))$ in the process model $\mathbf{F}$.

To analyse this system, the injected fault is considered a disturbance, and the analysis is performed using classical control techniques.

The transfer function from the disturbance input to the output may be derived as

$$\begin{aligned} \frac{\hat{\mathbf{x}}(s)}{\mathbf{f}_p(s)} &= [s\mathbf{I} - \mathbf{F} + \mathbf{KH}]^{-1} \\ &= G_{pf}(s) \end{aligned} \tag{12}$$

where the subscript pf denotes the transfer function with a process fault.

Therefore, under process fault conditions

$$\hat{\mathbf{x}}(s) = G_{pf}(s)\mathbf{f}_p(s) \tag{13}$$

The innovations with a process fault (and zero input) may now be defined as

$$\begin{aligned} \nu_{pf} &= -\mathbf{H}\hat{\mathbf{x}}(s) \\ &= -\mathbf{H}G_{pf}(s)\mathbf{f}_p(s) \end{aligned} \tag{14}$$

Now, in normal operation the input $\mathbf{z}(s)$ will be non zero. According to classical control techniques, the response of the system (or in this case the response of the innovations) will simply sum, as the system is linear. So, to find the response of the innovations subject to normal sensor operation together with a process fault (disturbance), the separate responses are simply added together. The innovations under process fault conditions become the no-fault innovations summed with the innovations due to a process fault as

$$\nu_{pf} = \nu_{nf} - \mathbf{H}G_{pf}(s)\mathbf{f}_p(s) \tag{15}$$

where the subscript nf denotes the no-fault condition.

Therefore, a process fault is detectable in the innovation sequence when the term $\mathbf{H}G_{pf}(s)\mathbf{f}_p(s)$ is non-zero. This means that only faults that are transmitted to observed states are detectable.

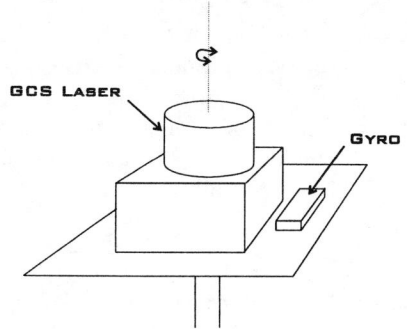

Fig. 3. The Experimental Setup

V. A Gyro-Laser Experiment

This section describes an experiment used to verify the theory presented previously in this paper. Equation 9 is used to show the detectability of sensor faults in the innovations sequence, and an example is given of an undetectable sensor fault. Process faults, described by Equation 15, are also shown to be detectable in one of the observed states.

A. Experimental Setup

The experiment used two reasonably high quality sensors, these were;
- A GCS[1] laser
- An Andrews fibre optic gyroscope

The GCS laser sensor operates by detecting the angle to a number of fixed beacons $\mathbf{B}_i = [X_i, Y_i]^T, i = 1, \ldots, N$. For this experiment, three beacons (actually retroreflective tape) were surveyed with a theodolite so their position was accurately known. The strips were placed roughly in an equilateral triangle with the experimental rig situated within the area described by the triangle.

The gyro used measures angular rate $\dot{\phi}$, but is subject to drift, particularly due to thermal influences. This sensor however has extremely good high frequency performance, being able to detect rotation rates of up to 100 degrees per second.

The sensors were mounted on a plate such that the axis of rotation of the GCS laser was approximately located over the center of the plate. The gyro was mounted at the periphery of the plate such that its sensitive axis was perpendicular to the surface of the plate. The plate itself was mounted on a stand such that it was free to rotate about its center. A schematic of the experimental setup may be seen in Figure 3.

The experiment itself was extremely simple. The platform was left for a period of time, usually 10 to 15 minutes, then the platform was rotated quickly through approximately 90 degrees. This process was repeated several times over the course of a single run. The entire experimented was run several times to ensure repeatability.

By combining the high frequency characteristics of the gyro and low frequency behaviour of the laser using a Kalman filter, the whole spectrum of platform rotation manouevres is able to be tracked.

The first set of results (Section V-C) for this experiment are for the nominal system, with no faults added. Section V-D shows results for the system in the presence of a laser bias.

B. Filter Design and Analysis

The orientation of the platform was modelled by a simple constant velocity model, allowing ϕ and $\dot{\phi}$ to be estimated. The gyro is known to drift, so a shaping filter must also be added. Also, to allow for small variations in position, x and y were also estimated.

The state vector for this system is therefore defined as

$$\mathbf{x}(t) = [x(t), y(t), \phi(t), \dot{\phi}(t), x_{sf}]^T \quad (16)$$

The continuous time linear process model may now be written as

$$\begin{bmatrix} \dot{x}(t) \\ \dot{y}(t) \\ \dot{\phi}(t) \\ \ddot{\phi}(t) \\ \dot{x}_{sf}(t) \end{bmatrix} = \begin{bmatrix} 0 & 0 & 0 & 0 & 0 \\ 0 & 0 & 0 & 0 & 0 \\ 0 & 0 & 0 & 1 & 0 \\ 0 & 0 & 0 & 0 & 0 \\ 0 & 0 & 0 & 0 & 0 \end{bmatrix} \begin{bmatrix} x(t) \\ y(t) \\ \phi(t) \\ \dot{\phi}(t) \\ x_{sf}(t) \end{bmatrix} + \begin{bmatrix} w_x(t) \\ w_y(t) \\ 0 \\ w_{\dot{\phi}}(t) \\ w_{sf}(t) \end{bmatrix} \quad (17)$$

Note that the shaping state x_{sf}, and the two position states x and y are all modelled as Brownian motion processes. The shaping state, however, is designed to reflect the coloured noise component of the gyro measurement, while the position state models are intended to reflect the uncertainty in the true state, and the rate (i.e. randomly) at which the true state is considered to vary.

The non-zero elements of the vector $[w_x, w_y, 0, w_{\dot{\phi}}, w_{sf}]^T$ are all assumed zero mean, uncorrelated gaussian sequences with strengths $\sigma_x^2, \sigma_y^2, \sigma_{\dot{\phi}}^2$ and σ_{sf}^2 respectively.

The gyro measurement equation is a linear combination of the angular rate $\dot{\phi}$ and the shaping state x_{sf}.

$$z_{gyro}(t) = \begin{bmatrix} 0 & 0 & 0 & 1 & 1 \end{bmatrix} \mathbf{x}(t) + [v_{gyro}(t)] \quad (18)$$

where the gyro white noise component v_{gyro} is assumed a zero-mean, uncorrelated gaussian sequence with strength σ_{gyro}^2.

The laser provides a nonlinear observation which may be considered to be of the form

$$\mathbf{z}(t) = \mathbf{h}(\mathbf{x}(t)) + \mathbf{v}(t) \quad (19)$$

The bearing to a beacon is given by $\arctan(\frac{Y_i - y(t)}{X_i - x(t)})$, however the platform is oriented in the direction ϕ, so for this system, the measurement equation for each beacon detected by the laser is given by the model

$$\mathbf{z}_\theta^i(t) = \left[\arctan(\frac{Y_i - y(t)}{X_i - x(t)}) - \phi(t) \right] + [v_\theta^i(t)] \quad (20)$$

where the laser observation noise v_θ^i is assumed to be identical for each beacon observed, uncorrelated, zero mean and gaussian with strength σ_θ^2.

[1] Guidance Control Systems

σ_x^2	$1^{-10}\ m^2$
σ_y^2	$1^{-10}\ m^2$
σ_ϕ^2	$4.9^{-7}\ rad^2 s^{-2}$
σ_{sf}^2	$1^{-10}\ rad^2 s^{-2}$
σ_{gyro}^2	$8.1^{-7}\ rad^2 s^{-2}$
σ_θ^2	$1^{-8}\ rad^2$

TABLE I
NOISE PARAMETERS

Throughout this experiment, the estimated noise strengths listed in the Table I were used.

Figures 4 and 5 show Bode Diagrams of the transfer functions ($G(s)$) from the laser and gyro respectively. The transfer functions are only shown for the states of most interest, ϕ, $\dot\phi$ and $x_s f$. Note that the transfer functions are not directly comparable, as the laser provides an observation of ϕ, while the gyro observes $\dot\phi$.

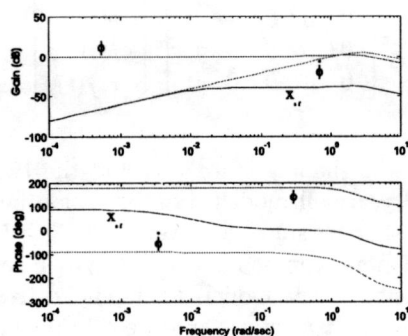

Fig. 4. Bode Diagram of the Transfer Functions from the Laser to the states ϕ, $\dot\phi$ and $x_s f$

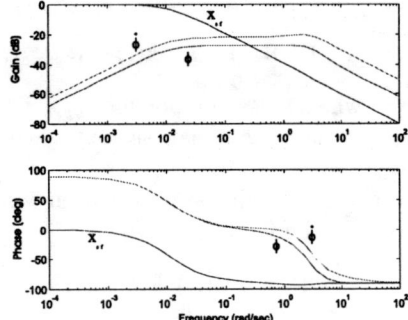

Fig. 5. Bode Diagram of the Transfer Functions from the Gyro to the states ϕ, $\dot\phi$ and $x_s f$

The fault plots[8] (a graphical representation of the term $[\mathbf{I} - \mathbf{H}G(s)]\mathbf{f}_o(s)$) for this system can be seen in Figures 6, 7, 8 and 9. These figures show that only very low frequency faults in either the gyro or the laser will go undetected in the innovations. However, if the Bode plot for the gyro is examined (Figure 5) at low frequencies, gyro information is transmitted to the shaping state. Therefore a fault in the gyro is either detectable, or transmitted to the shaping state. A fault in the gyro will not affect the states of interest ϕ or $\dot\phi$. A low frequency laser fault on the other hand *is* transmitted to the states of interest *and* is undetectable in the innovations. This represents a potentially catastrophic failure - one which is internally undetectable.

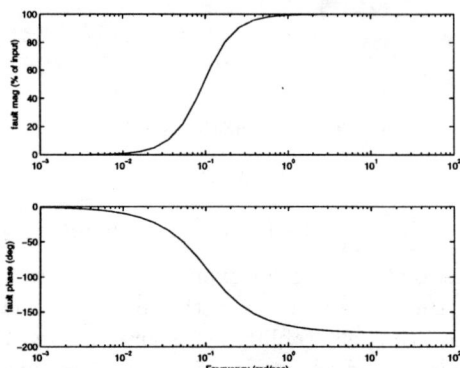

Fig. 6. Fault Plot - Percentage of Fault in Laser (and Corresponding Phase) Transmitted to the Laser Innovations

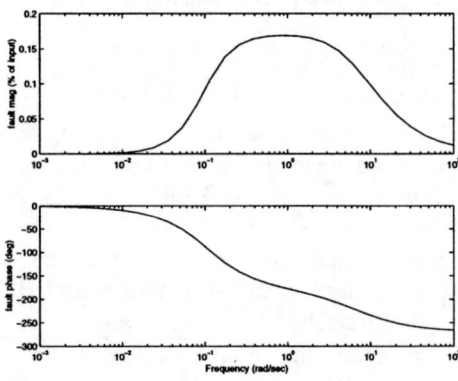

Fig. 7. Fault Plot - Percentage of Fault in Laser (and Corresponding Phase) Transmitted to the Gyro Innovations

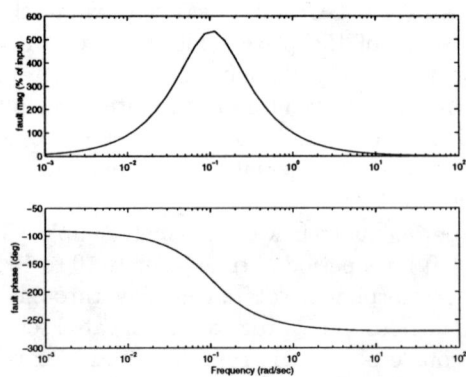

Fig. 8. Fault Plots - Percentage of Fault in Gyro (and Corresponding Phase) Transmitted to the Laser Innovations

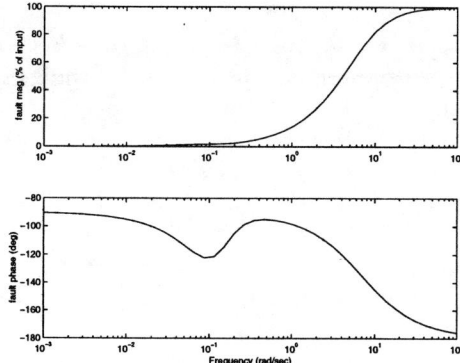

Fig. 9. Fault Plots - Percentage of Fault in Gyro (and Corresponding Phase) Transmitted to the Gyro Innovations

C. Results: The Nominal System

Figure 10(a) shows the orientation of the platform as estimated by the filter. This result agrees well with the experiment performed, the platform was stationary for relatively long periods of time, then rotated quickly by approximately 90 degrees. This result can be directly compared to that shown in Figure 10(b). This figure shows the effect of simply integrating the gyro measurements to obtain the platform's orientation. It can be seen that the integrated gyro drifts by approximately 90 degrees over the course of the experiment.

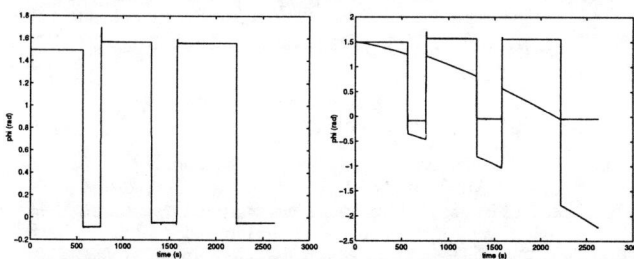

Fig. 10. (a) Estimated Orientation ϕ and (b) Estimated Orientation Overlayed with the Integrated Gyro Output

In principle, the platform rotates about the geometric center of the laser, and therefore the laser does not move in the xy plane. However, Figure 11 shows that indeed the laser moves in the order of a few millimetres during rotation. This small error can be attributed to poor manufacturing of the platform. This highlights the need for accurate models, if the xy position were not estimated, this small error would feed into the other states, perhaps causing filter divergence.

The spikes in these estimates, which correspond to the times at which the platform was rotated are an artifact of bearing only tracking. The laser is the only sensor which supplies information to the position states, allowing them to be estimated. The fact that the laser supplies bearing only, causes the filter to have a greater weight for the orientation state than for the position states. This causes the position states to take longer to converge to steady state than the orientation. In a system that is much more tightly coupled, for example by using a sensor that supplies range and bearing, one could reasonably expect the estimates of position to converge at a much greater rate.

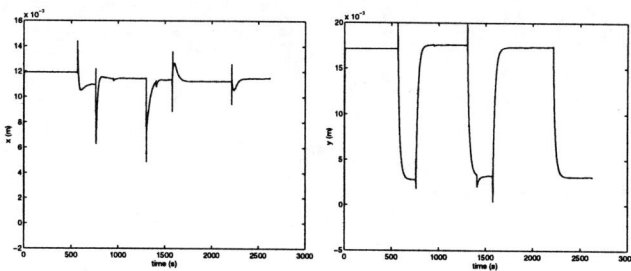

Fig. 11. Estimated Position (a) x and (b) y

Figure 12(a) shows the estimated shaping state x_{sf}. It can clearly be seen that the shaping state 'absorbs' the gyro drift. Again, there are spikes corresponding to the changes in orientation of the platform. In this case, *this is due to the process fault* (as can be seen and easily detected in the gyro innovation sequence shown in Figure 12(b). The process model is an extremely poor model of the system. There is no way for the process model to predict the onset of movement in the platform. The shaping state tries to compensate for some of this error (thus the spikes), but at steady state, only the gyro drift is estimated.

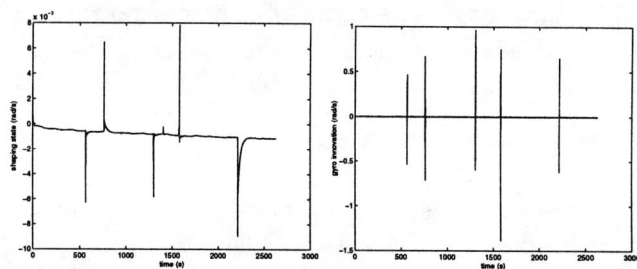

Fig. 12. (a) Estimated Shaping State x_{sf} and (b) Gyro Innovation Sequence

From the theory presented in Section IV, as both the orientation and the angular rate are being observed, any process faults occuring in either of those states is guaranteed be detectable in the innovations. This can clearly be seen in Figure 12(b) which shows the gyro innovation. The innovation is zero mean and white when the platform is stationary, indicating correct filter performance. However at the points where the platform is being rotated, the innovation sequence jumps, indicating a fault has occurred. The laser innovation sequence exhibits very similar behaviour.

Figures 13(a) and 13(b) show the innovations of the gyro and laser respectively during periods when the platform is stationary. Both innovation sequences appear to be unbiased and zero mean, indicating that when the platform is stationary, the filter performs correctly.

In summary, these results show that a constant velocity process model is not sufficient to accurately model the experimental rig. In steady state, the system performs extremely well, however when the platform was rotated sig-

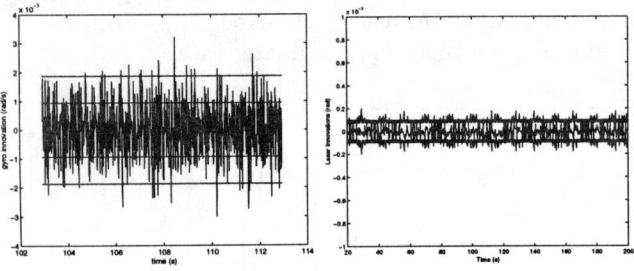

Fig. 13. (a) Gyro Innovation Sequence with 1σ and 2σ Bounds and (b) Laser Innovation Sequence with 2σ Bound

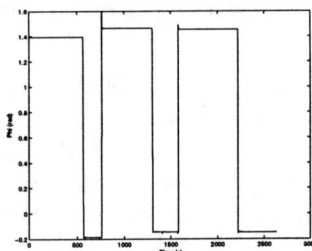

Fig. 14. Estimated Orientation ϕ

nificant error was induced, as evidenced by the innovations. Interestingly, although the model was not good enough to ensure filter consistency for the entire duration of the experiment, it *was* good enough to prevent the filter from diverging.

D. Results: in the Presence of Laser Bias

For this experiment a bias (an extremely low frequency fault) was added to the GCS laser of 0.1 radians (approximately 5.7 degrees). This bias simulates a misalignment of the sensor with respect to the platform. This is not an uncommon problem in the field, as it is extremely difficult to align a sensor such as the GCS laser to a tolerance that is smaller than the accuracy of the sensor.

Figure 14 shows the estimated orientation of the platform. As the bias is passed straight through to the estimate (from the Bode diagram in Figure 4), it is unsurprising that the estimated orientation appears to be consistently 0.1 radians offset from the unbiased case.

Figure 15 shows the innovation sequence for the gyro and laser respectively. Again, they appear unbiased and zero mean.

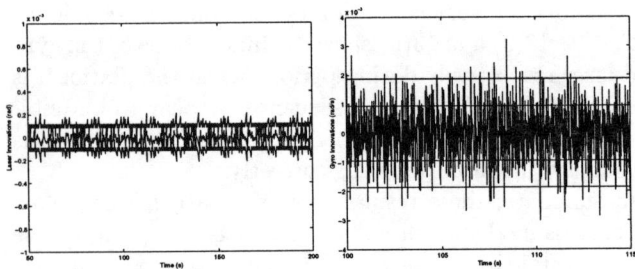

Fig. 15. (a) Laser Innovations and (b) Gyro Innovations

But, the system is at fault. As discussed in Section V-B very low frequency faults in the laser sensor will go undetected in the innovations, as evidenced by this example. The filter continues to produce consistent estimates with innovations that are white with zero mean. The estimates, however, *do not reflect the true state*. The only way to estimate this bias in a system such as this is to provide redundancy[8]. The term redundancy is used here to indicate multiple sensors that make comparable measurements - *not* identical sensors.

It should be noted that in this case the sensor did not fail in hardware. The simulated fault (misalignment) is one that is generally caused by human error. With redundancy, even faults such as these will be detectable.

VI. Conclusions

In this paper, a metric for determining the detectability of faults was developed using frequency domain techniques. This metric describes the percentage of the fault transmitted to the filter innovations versus the frequency content of the fault. Using this metric, it was shown that for a system fusing gyro and laser data, low frequency faults in the laser will be undetectable. This was verified by experiment. Further, process faults were also shown to be detectable as evidenced by the experiment.

These two examples demonstrate the utility of this approach. Faults are distinguished by their frequency content only. Therefore the source of the fault, be it in hardware or a fault in modelling, is irrelevant to this fault detection technique.

References

[1] S. Scheding, G. Dissanayake, E. M. Nebot, and H. F. Durrant-Whyte, "Slip Modelling and Aided Inertial Navigation of an LHD," in *IEEE Conference on Robotics and Automation*, 1997.

[2] H. F. Durrant-Whyte, "An Autonomous Guided Vehicle for Cargo Handling Applications," *International Journal of Robotics Research*, vol. 15, 1996.

[3] A. Willsky, "A Survey of Design Methods for Failure Detection in Dynamic Systems," *Automatica*, vol. 12, pp. 601–611, 1976.

[4] J. Gertler, "Survey of Model-Based Failure Detection and Isolation in Complex Plants," *IEEE Control Systems Magazine*, vol. 8, pp. 3–11, 1988.

[5] R. Isermann and P. Balle, "Trends in the Application of Model-Based Fault Detection and Diagnosis of Technical Processes," *Control Engineering Practise*, vol. 5, pp. 709–719, 1997.

[6] D. Lane and P. Maybeck, "Multiple Model Adaptive Estimation Applied to the Lambda URV for Failure Detection and Identification," in *Proc. 33rd Conf. Decision and Control*, 1994.

[7] E. Nebot, M. Karim, and J. Romagnoli, "Implementation of a Failure Detection-Identification Algorithm for Dynamical Systems," *J. Int. Soc. Mini and Microcomputers*, vol. 5, pp. 59–65, 1986.

[8] S. Scheding, E. Nebot, and H. Durrant-Whyte, "Fault Detection in the Frequency Domain: Designing Reliable Navigation Systems," in *Proceedings of Conference on Field and Service Robotics*, 1997, pp. 218–221.

[9] J.J. Leonard, *Directed Sonar Sensing for Mobile Robot Navigation*, Ph.D. thesis, University of Oxford, 1991.

Fault Detection and Identification in a Mobile Robot using Multiple-Model Estimation*

Stergios I. Roumeliotis, Gaurav S. Sukhatme[†] and George A. Bekey

stergios|gaurav|bekey@robotics.usc.edu

Department of Computer Science
Institute for Robotics and Intelligent Systems
University of Southern California
Los Angeles, CA 90089-0781

Abstract

This paper introduces a method to detect and identify faults in wheeled mobile robots. The idea behind the method is to use adaptive estimation to predict (in parallel) the outcome of several faults. Models of the system behavior under each type of fault are embedded in the various parallel estimators (each of which is a Kalman Filter). Each filter is thus tuned to a particular fault. Using its embedded model each filter predicts values for the sensor readings. The residual (the difference between the predicted and actual sensor reading) is an indicator of how well the filter is performing. A fault detection and identification module is responsible for processing the residual to decide which fault has occurred. As an example the method is implemented successfully on a Pioneer I robot. The paper concludes with a discussion of future work.

1 Introduction

Fault[1] tolerant behavior in mobile robots is desirable for a variety of reasons including safety and economics. Fault tolerant behavior refers to the autonomous detection and identification of faults as well as the ability to continue functioning after a fault has occurred. This paper deals with the first two components of the problem namely detection and identification. The ultimate goal of this work is to develop a methodology that allows mobile robots to autonomously detect, identify and rectify faults due to sensor failure, actuator failure as well as mechanical failure.

---

*This work is supported in part by JPL, Caltech under contract #959816 and DARPA under contract #F04701-97-C-0021

[†]contact author for correspondence

[1]In this paper fault and failure are used synonymously

Earlier work in the field (though not applied to mobile robots) is due to [6] and [15]. In [4] a network of adaptive virtual sensors is used to maintain reliable performance of a walking robot with many sensors, actuators and computers. The basic idea is that the virtual sensors reconfigure the way they use sensor information when a failure is detected. In [13] the authors investigate fault-tolerant techniques using redundant sets of control strategies. Recent work in generating fault residuals in robotics includes [14], a grid cell consistency measure based approach and [3], a observer-based approach to fault isolation in robot manipulators.

Kalman filtering [7], [9], [5] is a well known technique for state and parameter estimation. Kalman filtering is a recursive estimation procedure using sequential measurement data sets. Prior knowledge of the state (expressed by the covariance matrix) is improved at each step by taking the prior state estimates and new data for the subsequent state estimation.

Using a bank of Kalman filters was pioneered by Magill [8] who used a parallel structure of estimators in order to estimate a sampled stochastic process. Subsequently Athans et al. [1] used a bank of Kalman filters that provided state estimates to an equal number of LQG compensators to provide control over different operating regimes of an aircraft. Each estimator relied on a set of system equations linearized about a different operating point. Later Maybeck et al. [11] used the same technique (with an adaptive control strategy) to control F-15 aircraft. Further in [12], [10] the multiple model adaptive estimation (MMAE) technique was used to reliably detect and identify sensor and actuator failure for aircraft.

In recent years Kalman filter based localization has become common practice [2] in the robotics literature. Since the MMAE technique relies upon a bank of Kalman filters it seems natural to apply it to fault detection and identification in mobile robot systems. The basic philosophy of the method is to use **analytical redundancy** in the form of several system models (as opposed to say **hardware redundancy** which replicates hardware to identify a failure). A Kalman filter based framework provides a measure of the disparity (typically called a **residual**) between the measured sensor values and the values predicted by the model embedded within the filter. The residual is used in the filter to update the estimate and is an excellent indicator of failure. We demonstrate this fact using two failures as examples. A future challenge is to develop algorithms that can discriminate between the time profiles of multiple faults as they appear simultaneously in the residual.

The work reported in this paper is largely fault detection and a basic example of fault identification. The overall architecture is a two-component system comprised of 1. fault detection and identification and 2. fault accommodation. The architecture is ultimately intended for larger scale application. In this work four filters were designed and implemented. The filter bank used for fault detection and identification consisted of three of them (the first one was not used). The preliminary filter developed is an adaptive estimator used to estimate the values of two parameters. The basic observation is as follows: in the absence of any fault when the left and right wheels of the robot are commanded to rotate at fixed, equal speeds, the resultant trajectory is not a straight line. This is primarily due to two reasons 1. actuator mismatch and 2. unequal wheel radii. Measurements confirmed that the second factor was not significant. However, modeling actuator mismatch is quite complex. An alternative method is to build a filter with a kinematic model of the robot but to allow the radii of the two wheels to be parameters which can be estimated online. Even though the actual values of the two wheel radii are not (significantly) different the actuator mismatch is effectively modeled using them as parameters. Once the two radii parameters are learned they are used as base values in the other filters. The first filter in the bank uses two measurements from the wheel encoders, a chassis yaw rate and a kinematic model of the robot. This is called the base or nominal filter and it uses the learned radii values as constants in its design making it a good predictor of the nominal behavior of the robot. The second and third filters were each built with an embedded fault model. Filter two in the bank uses a reduced value for the left wheel radius as a model for a flat tire fault on the left wheel. Filter three in the bank uses a reduced value for the right wheel radius as a model for a flat tire fault on the right wheel.

Figure 1: The Pioneer I

In the experiments performed two faults are examined. The first (referred to henceforth as `right-tire-flat`) is a decrease in the radius of the right tire. The second fault (referred to henceforth as `periodic-bump`) was created by attaching a small object to the left wheel of the robot. A data stream from the robot sensors with these faults absent at first and present subsequently is used to to illustrate the response of the three filters. The rest of this paper is organized as follows: first the robot model and the filter residuals are presented. A block level view of the fault tolerant architecture is then discussed. This is followed by a summary of the experimental results, conclusions and a discussion of future work.

2 Robot Model

The Pioneer I used for experiments is a three wheeled robot shown in Figure 1. The front two wheels are actuated independently thereby enabling differential steering. The rear wheel is a passive castor. The kinematics of the Pioneer I are given in Equations 1-2.

$$v_L = r_L \dot{\theta}_L \qquad v_R = r_R \dot{\theta}_R \qquad (1)$$

$$\dot{\phi} = \frac{v_R - v_L}{a} \qquad v_{tot} = \frac{v_R + v_L}{2} \qquad (2)$$

where a is the axle length, r_L and r_R are the the radii of the left and right wheels respectively. The yaw rate of the robot in the x-y plane is denoted by $\dot{\phi}$ and the rotational speeds of the left and right wheels are denoted by $\omega_L = \dot{\theta}_L$ and $\omega_R = \dot{\theta}_R$. The linear speeds of the left and right wheels are denoted by v_L and v_R. The kinematic quantities are shown in Figure 2. The basic idea behind the approach used in this work is to do fault detection by processing the residual signature of the Kalman filter and fault identification by having a particular filter respond to its matching failure

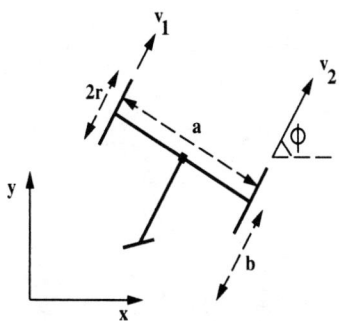

Figure 2: The Robot Kinematics

with minimal measurement residual. In the experiments reported here the measurement vector is composed of the two rotational speeds of the left and right wheels and the yaw rate of the chassis. The estimated measurement vector is denoted by $\hat{\mathbf{z}}$, the actual measurements from the sensors are denoted by $\mathbf{z}$ and the residual vector is denoted by $\mathbf{r}$. We have

$$\mathbf{z} = [\omega_L \; \omega_R \; \dot{\phi}]^T \quad \hat{\mathbf{z}} = [\hat{\omega}_L \; \hat{\omega}_R \; \hat{\dot{\phi}}]^T \quad \mathbf{r} = \mathbf{z} - \hat{\mathbf{z}} \quad (3)$$

The various filters developed in this work use the kinematics in Equations 1-2 and the measurements shown in Equation 3.

3 Fault Tolerant Architecture

The proposed architecture is depicted pictorially in Figure 3. The thrust of the current work is in the fault detection and identification modules as seen in the bold part of the figure. The control module is part of the future work. The first stage of the proposed approach is the detection of a fault. In the results presented here thresholding the residual is sufficient to detect a fault. The second stage of the proposed work is fault diagnosis. In the current work we are able to diagnose one fault apart from the other rather trivially since one of the filters designed minimizes the residual for a particular fault.

The bank of Kalman filters shown in Figure 3 is labeled using pairs of numbers (i, j). If the first number in the ordered pair is the same for two filters then the same set of kinematic equations (with different parameters) is used in both filters. All three filters used in the experiments reported here fall into that category. If the first entry is unequal the two filters have different kinematic equations of the system. Each filter produces a residual $\mathbf{r_{ij}}$. The nominal model results

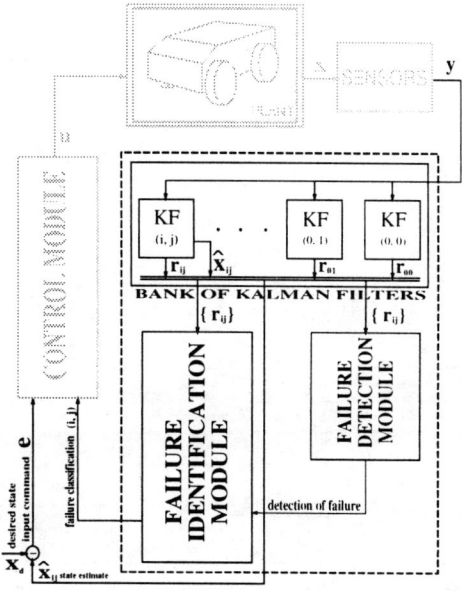

Figure 3: Failure Detection and Identification

in a residual of $\mathbf{r_{00}}$. All the residual vectors are fed into the FD (failure detection) and FI (failure identification) modules. The output of the FD module is a signal notifying a failure. When this is true the FI module becomes active. The FD module stays active even after a failure has been detected in case the failure spontaneously disappears or a new failure appears. A small portion of this scheme (a bank of three filters) is implemented here.

In the scheme proposed here the key insight is to use a bank of estimators. The fault detection/identification depends on the correct selection of the state estimator with the minimal residual. This estimator is the one which assumes currently sound knowledge of the system description, i.e. it has incorporated the failure effect in its structure. The FDI module thus provides a high quality estimate to the control module and allows for graceful degradation of the system. An increase in robustness is expected since the proposed scheme uses additional knowledge about expected failures. Thus it is capable of dealing with scenarios that in other schemes might be considered catastrophic.

The third stage in the process shown in Figure 3 is labeled control. This stage deals with modifying the control structure of the robot (fault accommodation) so that it can continue functioning after the fault has occurred.

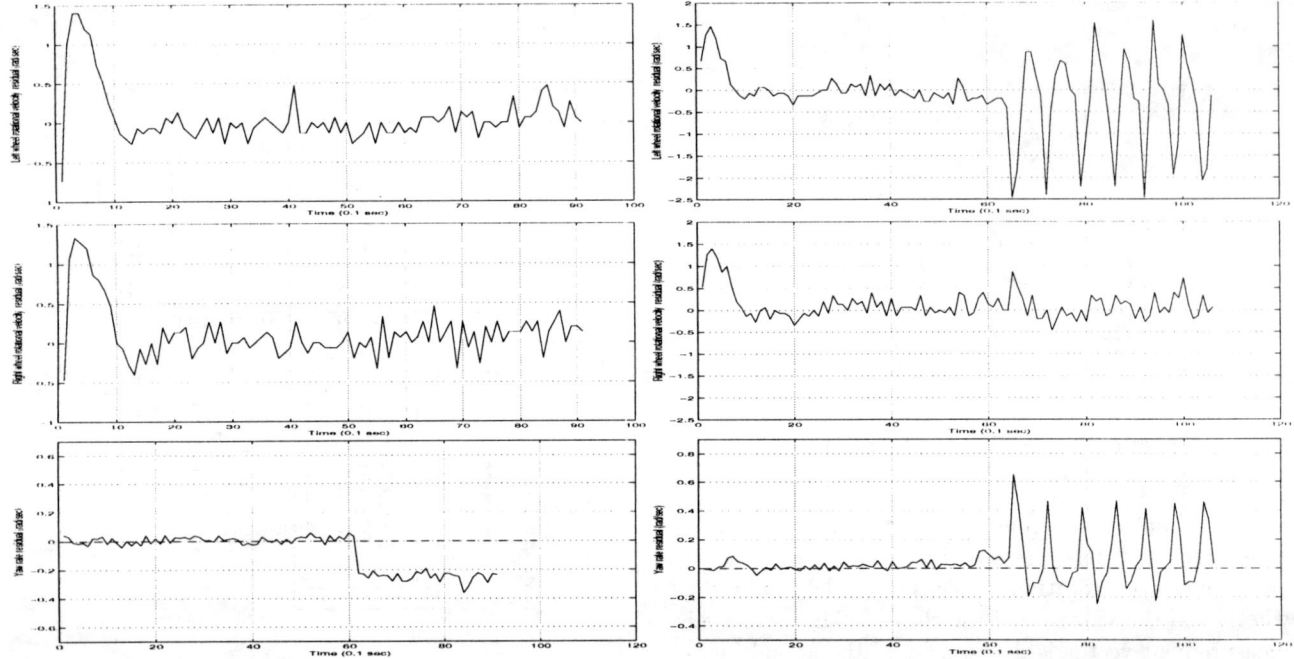

Figure 4: Each column of figures above shows the three **residuals for the nominal filter** (which has no failure model). The left column shows the residuals for the `right-tire-flat` fault. The right column shows the residuals for the `periodic-bump` fault.

4 Experimental Results

In the experiments reported here, two faults were considered and three filters were used in the filter bank to process the data stream from the robot's sensors as it transitioned from its nominal behavior to the faulty behavior. Each of the three filters used the three component measurement vector given in Equation 3 thus producing a three component residual vector.

Consider Figure 4. These are the residuals from the nominal filter (which has no failure model). The left column shows the residuals for the `right-tire-flat` fault. The right column shows the residuals for the `periodic-bump` fault. In the case of the `right-tire-flat` fault there is no observable change in the first two components of the residual since the (angular) rotational speeds of the wheels do not depend on their radii. However the vehicle does begin to yaw as seen by the yaw rate residual which jumps to a mean value well below zero at $t = 6$ s when the fault occurs. In the right column of Figure 4 one sees the residual response to the `periodic-bump` fault on the left wheel. The rotational velocity of the left wheel as well as the yaw rate residual respond with oscillations of large amplitude.

Consider the graphs shown in Figure 5. As before, the left column shows the residuals for the `right-tire-flat` fault and the right column shows the residuals for the `periodic-bump` fault. The filter is however tuned to the `right-tire-flat` fault and this is immediately seen in the yaw rate residual of the left column - the residual is high before the fault occurs ($t < 6$ s) and drops sharply to zero as soon as the fault occurs signaling the identification of this particular failure. The filter is not tuned to the `periodic-bump` fault but the residual signature clearly shows that the fault occurs at $t = 6.5$ s. The yaw rate residual switches from a steady bias of approximately 0.25 rad/sec to a high amplitude oscillation.

Lastly, consider Figure 6. The residuals in this figure are from the filter tuned to the fault where the left tire goes flat. This fault was **not** considered in the experiments. The filter residuals are again sensitive to both faults as seen in both the left and right columns. However in neither case is the filter able to bring the residual back to zero. This filter would be be able to zero out the residual (and thereby identify) for the case where the left tire goes flat.

In the examples given here the failure identification is done very simply using thresholding. Also the residual signature for the filter tuned to the `right-tire-flat` fault is distinctive. So the problem of detecting which fault has occurred is easily automated. Ideally one would like to design filters for every fault so that after a fault has occurred there is at least one filter whose state estimate is reliable and can be used for control. In this work we relax that condition.

5 Conclusion and Future Work

In this paper a multiple model based technique to detect and identify faults in mobile robotic systems was presented. The technique is based on using a bank of Kalman filters in parallel. Detection and identification of faults is done by analyzing the signature of the residual produced by each filter. In this paper we showed the application of this methodology to the case of a Pioneer I robot. Two faults were considered. The first was a flat tire and the second was an object stuck on the tire. The bank of filters was able to detect each of the two faults as evidenced in the residuals. Further, the flat tire fault is distinguishable by the low residual in the filter tuned to this particular fault. Without doing sophisticated residual analysis we are able to do detection and identification of two faults. It should be noted that the architecture presented here is generalizable to more faults and increasingly sophisticated filters and residual postprocessing. The method can easily be applied to other 3-wheeled mobile systems because it relies on a simple kinematic description.

In the future we plan to concentrate on these issues as well as sensor failure, actuator failure and other mechanical failures. Recovery from failure by altering the control strategy is also the subject of future work. The easiest (and least autonomous) solution is to stop, flag a fault and await human help. Other strategies include making guarded motions, changing the covariance matrices that characterize the sensors or switching to an entirely new control methodology.

Acknowledgments

The authors would like to thank Barry Werger and Dani Goldberg for help with data collection. The authors also thank S. Hayati, G. Rodriguez, R. Volpe, C. Weisbin and B. Wilcox for several useful discussions.

References

[1] M. Athans, D. Castanon, K-P Dunn, C. S. Greene, W. H. Lee, N. R. Sandell Jr., and A. S. Willsky. The stochastic control of the f-8c aircraft using a multiple model adaptive control (mmac) method- part i: Equilibrium flight. *IEEE Transactions on Automatic Control*, AC-22(5):768–780, October 1977.

[2] B. Barshan and H. F. Durrant-Whyte. Inertial navigation systems for mobile robots. *IEEE Transactions on Robotics and Automation*, 11:328–342, January 1995.

[3] F. Caccavale and I. Walker. Observer-based fault detection for robot manipulators. In *Proc. 1997 IEEE International Conference on Robotics and Automation*, pages 2881–2887, April 1997.

[4] C. Ferrell. Failure recognition and fault tolerance of an autonomous robot. *Adaptive Behavior*, 2(4):375–398, 1994.

[5] M. S. Grewal and A. P. Andrews. *Kalman Filtering, Theory and Practice*. Prentice Hall, 1993.

[6] D. T. Horak. Failure detection in dynamic systems with modeling errors. *AIAA Journal of Guidance, Control and Dynamics*, 11(6):508–516, Nov-Dec 1988.

[7] R. E. Kalman. A new approach to linear filtering and prediction problems. *ASME Journal of Basic Engineering*, 86:35–45, 1960.

[8] D. T. Magill. Optimal adaptive estimation of sampled stochastic processes. *IEEE Transactions on Automatic Control*, AC-10(4):434–439, 1965.

[9] P. S. Maybeck. *Stochastic Models, Estimation and Control*, volume 1. New York: Academic Press, 1979.

[10] P. S. Maybeck and P. D. Hanlon. Performance enhancement of a multiple model adaptive estimator. *IEEE Transactions on Aerospace and Electronic Systems*, 31(4):1240–1253, October 1995.

[11] P. S. Maybeck and D. L. Pogoda. Multiple model adaptive controller for the stol f-15 with sensor/actuator failures. In *Proceedings of the 20th Conference on Decision and Control*, pages 1566–1572, December 1989.

[12] T. E. Menke and P. S. Maybeck. Sensor/actuator failure detection in the vista f-16 by multiple model adaptive estimation. *IEEE Transactions on Aerospace and Electronic Systems*, 31(4):1218–1229, October 1995.

[13] D. Payton, D. Keirsey, D. Kimple, J. Krozel, and K. Rosenblatt. Do whatever works: A robust approach to fault-tolerant autonomous control. *Journal of Applied Intelligence*, 2:225–250, 1992.

[14] M. Soika. Grid based fault detection and calibration of sensors on mobile robots. In *Proc. 1997 IEEE International Conference on Robotics and Automation*, pages 2589–2594, April 1997.

[15] M. L. Visinsky, I. D. Walker, and J. R. Cavallaro. New dynamic model-based fault detection thresholds for robot manipulators. In *Proceedings of the 1994 IEEE International Conference on Robotics and Automation*, pages 1388–1395, 1994.

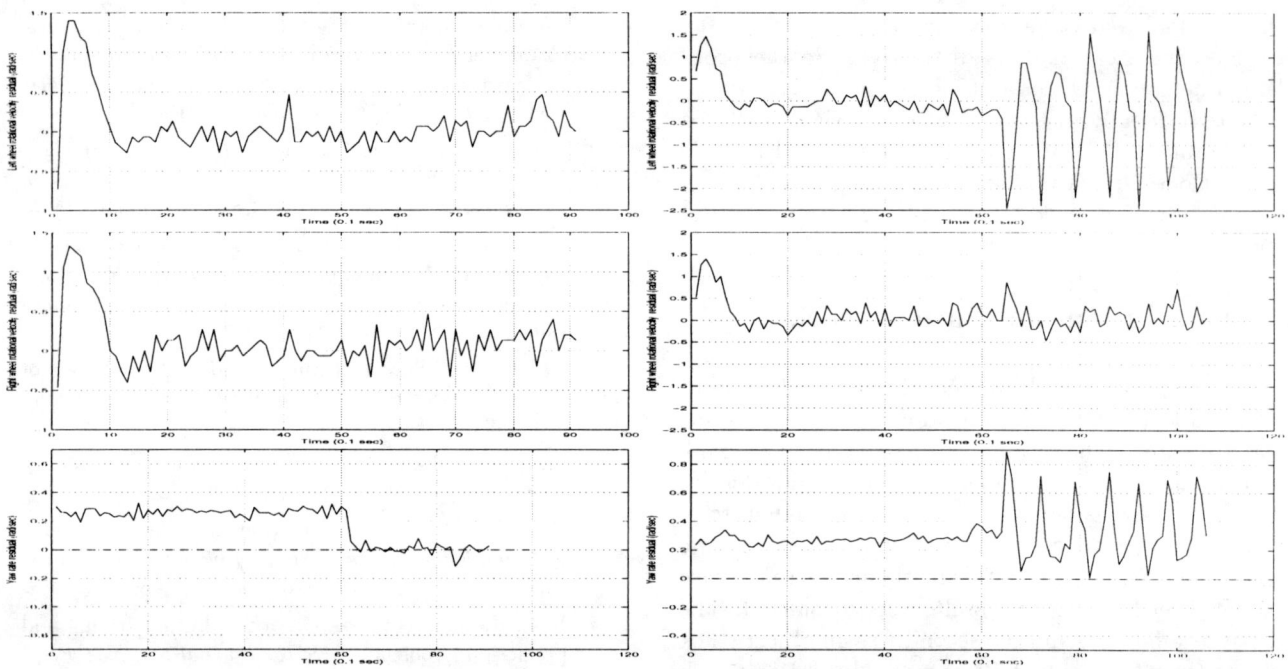

Figure 5: Each column of figures above shows the three **residuals for the filter which is tuned for the right tire flat model**. The left column shows the residuals for the `right-tire-flat` fault. The right column shows the residuals for the `periodic-bump` fault.

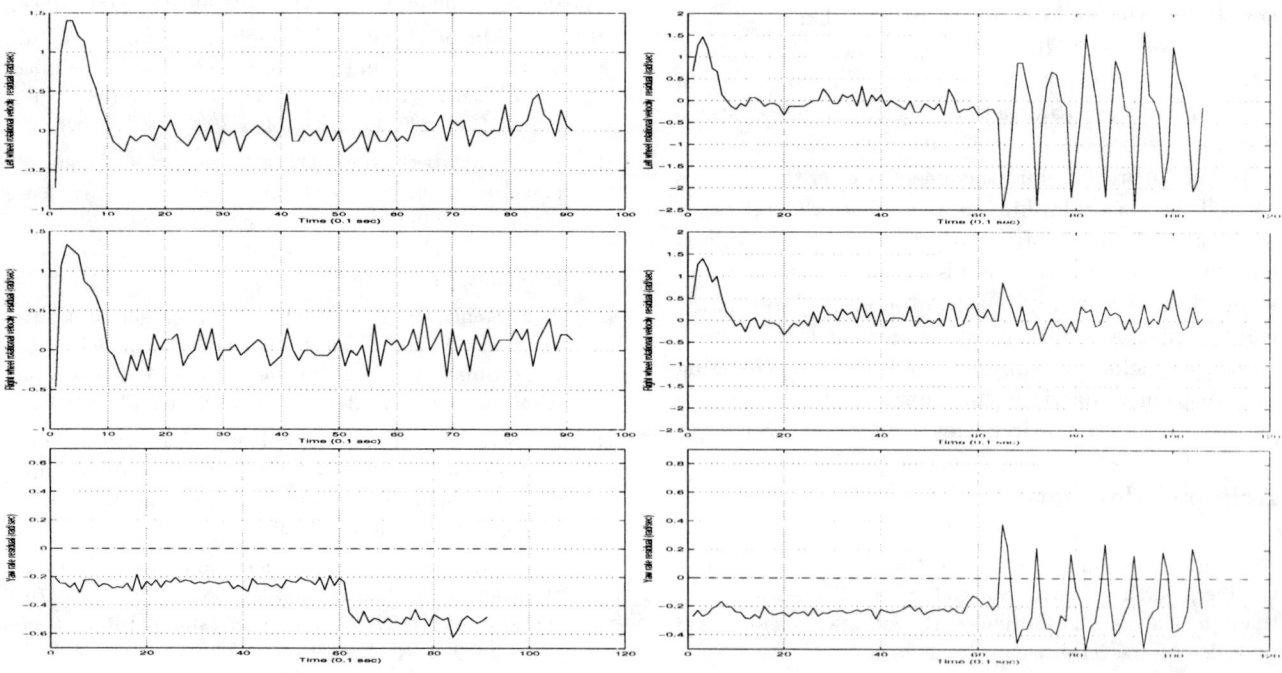

Figure 6: Each column of figures above shows the three **residuals for the filter which is tuned for the left tire flat model**. The left column shows the residuals for the `right-tire-flat` fault. The right column shows the residuals for the `periodic-bump` fault.

AMADEUS: A Mobile, Autonomous Decentralized Utility System for Indoor Transportation

Toru Kamada and Koichi Oikawa
Fujitsu Laboratories Ltd.
10-1 Morinosato-Wakamiya, Atsugi 243-01, Japan
E-mail: tkamada@flab.fujitsu.co.jp

Abstract

The authors have developed a mobile, autonomous decentralized utility system for indoor transportation, called AMADEUS. This paper explains how we put it to practical use. AMADEUS is a system designed around transportation agents. We eliminated centrally managed vehicle allocation and routing plans by implementing autonomous vehicle allocation negotiations and collision avoidance between agents. The transportation agent's collision avoidance function enables the bi-directional movement of vehicles on a single-track, thereby making efficient, space-saving transportation possible. This paper introduces the concept of AMADEUS and describes the transportation agent architecture by focusing on installing the collision avoidance function.

1. Introduction

In factories producing personal information units such as personal computers and portable telephones, production orders are currently increasing because of shorter product life cycles and reduced inventories. The increase in production orders has led to frequent changes in the types and amount of products manufactured by these factories.

To cope with this situation, factories are changing their production systems from conventional conveyer lines to production cells. In the cell production method, a work process is usually completed within a small U-shaped cell. Unlike conventional conveyer line methods, this method enables high productivity for multiple-product manufacturing and easy changes in the number and layout of cells that may occur due to changes in the scale of production.

To support the cell production method, a transportation system for supplying parts to and removing assemblies or products from the cells is required to:
- Link arbitrary cells with optimum routes at the appropriate times.
- Flexibly cope with changes in transportation environments such as changes in the number and layout of cells.

On a shop floor using the cell production method, cells are densely arranged and passages are narrow to minimize travel distances and the required floor space. A conventional automatic guided vehicle (AGV) transportation system used with the cell production method could only have a single-track and a fixed transportation direction. With this transportation system, vehicles can only follow a predetermined route, thus leading to low transportation efficiency. Moreover, conventional transportation systems must cope with potential changes in the number of AGVs and the transportation route, which require complex calculations [1,2] to allocate vehicles to the cells and determine vehicle routes. Modifying the complex management programs used for the transportation system requires enormous amounts of time and money.

Therefore, we focused our attention on an autonomous decentralized transportation system, because it features:
- AGVs having a collision avoidance (when passing each other) function on a single-track.
- AGVs that negotiate with cells regarding vehicle allocation.

This system supports bi-directional transportation, thus assuring a high transportation efficiency. Each AGV can autonomously move to a pre-negotiated target cell without colliding with other AGVs. This could eliminate conventional vehicle allocation and routing plans, allowing the centralized management system to be eliminated.

Results on the experimental performance of autonomous decentralized robot systems have been reported, but practical systems have been difficult to configure [3-5]. This is because before developing practical systems, it is necessary to make agents autonomous enough to cope with complicated, ever-changing real work environments, and to produce agents at a practical cost.

In other words, it is necessary to develop highly adaptive AGVs that can evade moving obstacles and are comparable in cost to conventional AGVs. The authors tried to solve this problem by introducing a behavior-based approach [6] to the hardware on the same functional level as that for ordinary AGVs, because this approach could be used to realize sophisticated behavior by means of simple low-priced sensors.

However, The attempt to install behavior in collision avoidance systems involves difficulties due to the coexistence of various types of target behavior in transportation work aspects. For this reason, the need arose for a new architecture that would prepare different behavioral styles for each aspect and provide a transition between behavioral styles.

Now that the concept of AMADEUS (A Mobile, Autonomous DEcentralized Utility System) has been covered, this paper will now describe the architecture used to make AGVs autonomous.

2. Concept of AMADEUS

Figure 1 shows the overview of AMADEUS. In this system, two different types of distributed agents, autonomous AGVs and cells, cooperate in transporting objects. There is no central function to manage and control these agents.

The major configuration and functions of AMADEUS are as follows:
1) AMADEUS consists of dynamic agents, called mobile agents, for transporting objects from cell to cell, and another type of agent, called cell agents, for supplying and removing objects to each cell.
2) Cell Agents and mobile agents communicate with one another via a contract net protocol [7] to allocate mobile agents to transportation tasks occurring in cells.
3) When a mobile agent is assigned a transportation task, it runs along a single-track in either direction to the target address. If it encounters an obstacle (such as a person, another mobile agent, or an object), it temporarily moves away from the track's guideline to pass the object. In areas where it is difficult to avoid collisions, such as in the vicinity of a cell station or a narrow passage, mobile agents negotiate with one another to agree on which will enter the area.

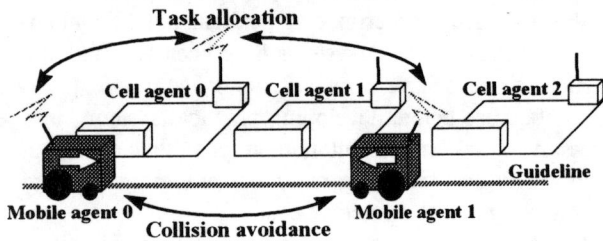

Figure 1. Concept of AMADEUS.

Implementing the configuration and functions stated above attains efficient, bi-directional transportation and eliminates the need to execute centrally managed vehicle allocation and routing plans.

2.1 Agents

The two types of agents that comprise AMADEUS have the following functions and roles:

1) Mobile agents

Mobile agents use a contract net protocol to negotiate with cell agents about vehicle allocation to acquire transportation tasks or target cell addresses. Once a mobile agent acquires a target address, it determines its direction of travel according to its current position indicated by a signpost on the floor and its own map and then starts moving toward the target cell.

If a mobile agent encounters an obstacle, it temporarily leaves the track's guideline to pass the obstacle. After it passes the obstacle, it returns to the guideline and proceeds. Moreover, a mobile agent can avoid collisions with other mobile agents by negotiated cooperation.

2) Cell agents

When a transportation request occurs in a cell, a cell agent uses contract net protocol to negotiate with mobile agents to determine vehicle allocation and assigns the transportation task to the unassigned mobile agent that will most quickly be able to receive the object to be transported.

2.2 Task allocation

The contract net protocol shown in Figure 2 is used to assign transportation tasks to mobile agents. It works as follows:

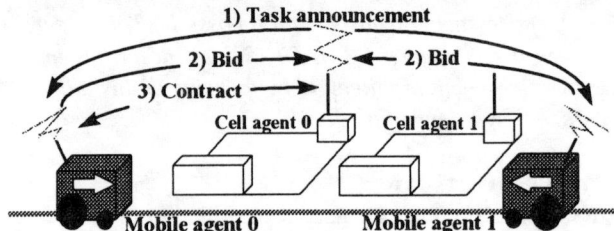

Figure 2. Task allocation with contract net protocol.

1) Task announcements by a cell agent

When a transportation request occurs in cell zero, as shown in Figure 2, the cell agent in charge of cell zero presents the transportation task to all mobile agents by radio.

2) Bids by mobile agents

For the presented transportation task, each mobile agent bids on the time required for it to get to cell zero by radio. If a mobile agent is not executing a transportation task, its bid is simply the time required for it to move from its current position to cell zero. If a mobile agent is currently executing a task, its bids is the total time required for it to complete the present task and that required for it to move from its completion position to cell zero.

3) Contract

Cell agent zero compares all bids to select the mobile agent that can satisfy the transportation request in the shortest time, now known as mobile agent zero, and concludes a transportation contract with mobile agent zero by sending address zero of cell zero to the mobile agent. If mobile agent zero is not executing a transportation task, it starts for address zero. If it is already executing a task, it finishes it before starting for address zero.

2.3 Collision avoidance

AMADEUS does not use centrally managed routing plans to avoid collisions between mobile agents. Instead, mobile agents autonomously avoid collisions using either of two methods. The first method is used in passageways that are wide enough for at least two mobile agents to pass. The second method is used in areas where it is difficult for two mobile agents to pass each other, such as in the vicinity of cell stations and in narrow passageways.

1) Passing each other

Each mobile agent is provided with a collision avoidance function to anticipate situations in which it may run across an oncoming mobile agent on a single-track's guideline. With the collision avoidance function, the mobile agent leaves the guideline temporarily to avoid collision with the oncoming mobile agent and, after the oncoming mobile agent has passed, returns to the guideline, as shown in Figure 3. The collision avoidance function is also effective when the mobile agent encounters a person or other obstacles. Mobile agents keep to the right or left (whichever is specified) to prevent deadlocks.

Research results on how to best enable mobile robots to pass each other have been reported [8,9]. However, there is no practical system that enables mobile objects to pass each other on a single-track guideline.

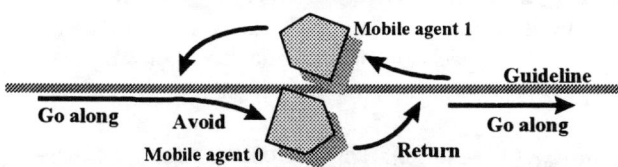

Figure 3. Passing each other on a single track guideline.

2) Negotiation

When mobile agents are in an area where it is difficult for them to pass each other, such as in the vicinity of a cell station or in a narrow passageway, they negotiate with each other by radio.

In the example shown in Figure 4, when mobile agent zero attempts to return from a cell station, it sends its address to other mobile agents. If mobile agent one receives the address of mobile agent zero, it reciprocates by sending its own address and direction of travel. On receiving this information, mobile agent zero recognizes that there is an oncoming mobile agent and waits at the cell station. Mobile agent one makes a similar decision. It memorizes that it is keeping mobile agent zero waiting. After passing an area of contention, mobile agent one allows mobile agent zero to enter the area by sending information by radio. Thus, collisions can be assuredly and efficiently avoided in areas of contention.

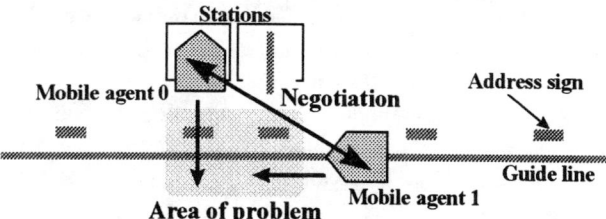

Figure 4. Negotiating in the vicinity of cell stations.

3. Mobile Agent Architecture

To realize collision avoidance between mobile agents as described in Section 2, it is necessary to respond to changes in situations flexibly and quickly. However, to produce mobile agents at a practical cost, we must select sensors with local sensing functions only. For mobile agents that cannot grasp their environment in a perspective to respond to unexpected changes in environments flexibly and quickly, it is necessary to act positively, precisely, and quickly by responding to the ever-changing situations around them. To meet this requirement, mobile agents use a behavior-based approach.

Behavior-based architecture concurrently processes behavior modules represented by finite state machines by setting up a suppressing relationship. So it is necessary to prepare behavior modules that suit each situation and set up a suppressing relationship specific to each behavior module. For a situation where collision avoidance is needed, for example, it is necessary to prepare behavior modules for avoiding collisions, for running along the guideline, and for returning to the guideline, and set up suppressing relationships according to the priority among these behavior modules.

In addition to the behavior styles described so far, mobile agents must have various other behavior styles that can support diverse aspects, such as a behavior style (a) for detecting a target address and entering the corresponding station, a behavior style (b) for receiving a new target address and leaving the station, and a behavior style (c) for detecting an intersection sign and moving in the target direction (see Figure 5).

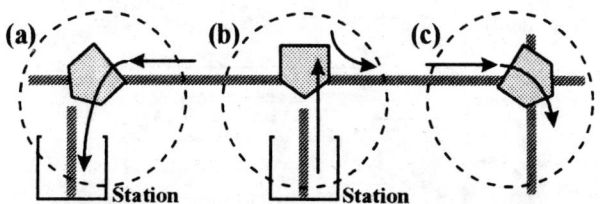

Figure 5. Behavior styles of mobile agent.

The types and number of behavior modules to be activated in each aspect vary from one behavior style to another. When a mobile agent is entering a station, for example, a behavior style for collision avoidance would detect the target station as an obstacle and try to divert the mobile agent from the station rather than causing it to approach the station. When entering a station, a mobile agent must use a behavior module for stopping instead of the collision avoidance behavior module.

To solve these problems, research on managing the actions of multiple mobile robots [10,11] has resulted in a hierarchical architecture in which high-order layers cooperate to solve problems. This architecture has a problem regarding real-time operation, however, because many steps are needed to exchange information among layers. Some other methods [12,13] group behavior modules for a particular aspect into a behavior set on a higher level and share behavior modules on a lower level by imposing suppressing relationships as required. If there are many behavior modules, however, these methods degrade in real-time response due to an increase in the number of execution steps, because finite state machines must be executed to suppress the operation of behavior modules on lower levels. In addition, low-level behavior modules that are seemingly shareable among different aspects are actually difficult to be shared in real environments, because it is necessary to precisely adjust the handling of sensor signals and the operation of actuators for each aspect.

The architecture we developed is based on the concept that the behavior module to be activated belongs to a specific aspect, and so sharing even low-level behavior modules would be difficult. Therefore, this architecture prepares behavior modules specific to a unit of behavior for each aspect and sets up a suppressing relationship specific among them. This unit of behavior is hereafter called a behavior style.

Figure 6 shows the mobile agent architecture. This architecture activates a behavior style that corresponds to a specific aspect according to communication between agents and information from sensors, by using a finite state machine called an activator. To be more specific, a transition among the states of the activator activates a behavior style. Once the behavior style is activated and achieves its goal, it becomes inactive by resetting its indicator, causing another behavior style to be activated.

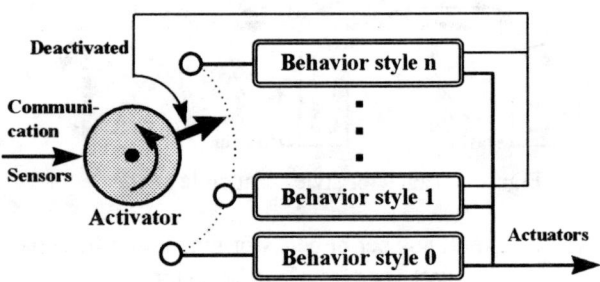

Figure 6. Architecture of mobile agent.

This way, each behavior style can be configured with a minimum number of behavior modules, thus improving its real-time characteristics. This method can also make it easier to create programs in a modular structure, thus enabling suppressing relationships to be set up among behavior modules for each behavior style without affecting other behavior styles. In addition, it is possible to tailor behavior modules for individual aspects, thereby avoiding functional trade-offs that would be result from sharing behavior modules among behavior styles.

3.1 Behavior styles

The mobile agent architecture features the use of multiple behavior styles, each of which is tailored to a specific aspect. No behavior module is shared among behavior styles. Instead, each behavior module has a completely independent structure. This configuration eliminates some control steps, thereby improving the real-time characteristics of the behavior style. Moreover, providing behavior modules specific to each behavior style enables the tailoring of each behavior module to meet the requirements of the corresponding behavior style. Figure 7 shows the major behavior styles necessary to AMADEUS.

1) Normal-running: In the normal-running behavior style, a mobile agent runs along the guideline and leaves it when an obstacle is encountered. After avoiding the obstacle, the mobile agent steers in the direction opposite to that taken when leaving the guideline according to its records, thereby moving back toward the guideline. Once the mobile agent returns to the guideline, it resumes its original course along the guideline (Figure 7a).

2) Approaching-station: The approaching-station behavior style is used when a mobile agent enters a station (Figure 5a). This behavior style is activated when a mobile agent detects a target address according to a signpost and diverts itself to the station. On detecting the guideline leading to the station, the mobile agent moves along the guideline and stops when it detects another signpost. When it completes its positioning, it resets the indicator of the activator to become inactive (Figure 7b).

3) Leaving-station: The leaving-station behavior style applies to when a mobile agent leaves a station (Figure 5b). On receiving a new target address, the mobile agent is activated and returns to the guideline. The mobile agent searches for the main guideline to return to it. On detecting the guideline, the mobile agent resets the indicator of the activator to become inactive (Figure 7c).

4) Changing-course: The changing-course behavior style is for when a mobile agent changes its course at an intersection (Figure 5c). Comparing information from an intersection sign with the target address to alter its course activates the mobile agent. After being diverted to a preset course, it proceeds. On detecting a guideline, the mobile agent resets the indicator of the activator to become inactive. If it encounters an obstacle before it finishes changing its course, the mobile agent goes back to avoid collision (Figure 7d).

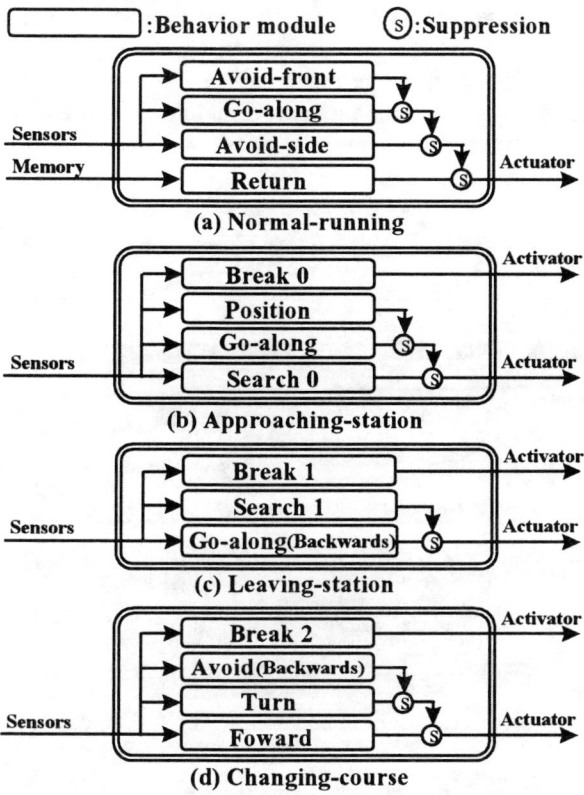

Figure 7. Behavior styles of mobile agent.

This way, the mobile agent architecture minimizes the number of behavior modules required for each aspect. In addition, when behavior modules respond to a real environment, their functions while basically similar, differ slightly for each behavior style, as shown with "Avoid" in Figures 7a and d. Therefore, it is difficult for behavior styles to share behavior modules. Hence, making each behavior style completely independent of each other is practical.

3.2 Behavior modules

A behavior module is the smallest functional unit in the mobile agent architecture. It is a behavior style component that determines the function of the behavior style.

How well a mobile agent can avoid collision depends on not only the performance and layout of its sensors and actuators but also on how well the behavior modules in it suit the goals of a particular behavior style and how easily they can be configured. The easier the configuration, the easier it becomes to maintain high real-time characteristics and to follow changes in situations. The mobile agent architecture can focus on the behavior type intended by each behavior style, therefore enabling the configuring of highly adaptable, efficient behavior modules.

Collision avoidance on a single-track guideline can be disassembled into four behavior types: go along, avoid front, avoid side, and return. These behavior modules can be represented with simple finite state machines. Each behavior module functions as described below:

1) Avoid-front: This behavior module is used when the mobile agent avoids an obstacle ahead. If the mobile agent detects an obstacle ahead, it moves over quickly. If it detects an obstacle in the left front, it moves over to the right. (This mode is used when the keep-to-the-right rule applies. It is possible to move to the left if so designed.) If the mobile agent detects an obstacle to the right or left toward the front, it moves forward slowly.

2) Go-along: This behavior module is used when the mobile agent is running along the guideline. If the mobile agent detects that it is deviating to the left from the guideline, it moves to the right. Similarly, if it detects that it is deviating to the right, it moves to the left. If it is not deviating to either side, it continues moving straight ahead.

3) Avoid-side: This behavior module is used when the mobile agent avoids an obstacle on its flank. If the mobile agent detects an obstacle on the left side, it moves over to the right.

4) Return: This behavior module is used when the mobile agent returns to the guideline. If the mobile agent remembers that it has just avoided a collision, it moves to the left. (This mode is used when the keep-to-the-right rule applies. It is possible to move to the right if so designed.)

The independence of the behavior styles is useful for setting up suppressing relationships among behavior modules. In the normal-running behavior style, it is only necessary to set up specific suppressing relationships among the behavior modules stated above, as shown in Figure 7a. When this method is applied, the mobile agent should follow the guideline via the go-along behavior module, avoid collisions via the avoid-front behavior module, and pass obstacles on the right side and return to the guideline via the avoid-side and return behavior modules before returning to follow the guideline via the go-along behavior module.

It is possible to configure behavior modules that belong to behavior styles other than the normal-running ones in a structure that meets the target behavior types of a specific behavior style and set up suppressing relationships among them.

A method has been proposed which can be used to set up suppressing relationships among behavior modules [14]. With the mobile agent architecture, however, designers can set up suppressing relationships among behavior modules easily, because the role of each behavior module is defined clearly and their suppressing relationships are self-explanatory.

4. Execution Example

This section gives an example of using AMADEUS for transportation among cells on a shop floor. The example illustrates how behavior style transition occurs according to the mobile agent architecture to implement collision avoidance (passing each other).

4.1 Equipment

AMADEUS is basically configured with mobile agents, cell agents, guidelines, signposts installed on the floor, and stations among which cargo is transported. The major structures of signposts installed on the floor and mobile agents are explained below.

1) Signposts on the floor

The signposts shown in Figure 8 are installed on the floor. The signposts are categorized into address signs (a), stop position signs (b), and intersection signs (c). The address signs (a) indicate street and block names. The intersection signs (c) indicate right- and left-side street names.

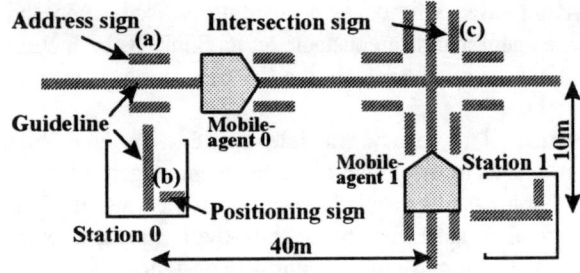

Figure 8. Signposts on the floor.

2) Mobile agent structure

Figure 9 shows the structure of a mobile agent. It measures 700 (W) x 850 (L) x 900 mm (H), and weighs 1000 N. It has right and left independent drive wheels and is guided using magnetic induction loops.

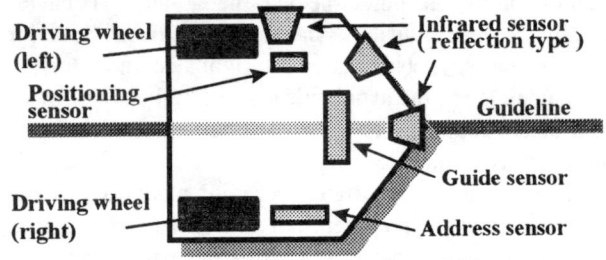

Figure 9. The outline structure of a mobile agent.

The guide sensor detects when the mobile agent deviates to the right or left from the guideline. The reflection-type infrared sensors are used to detect obstacles. The positions of stations and intersections are detected using address signs. Positioning the mobile agent within a station is performed using the positioning sensor to detect the deviation of the mobile agent from the stop position indicated on the floor. Control is carried out by representing behavior modules with finite state machines and executing them as one process in a multitasking mode.

4.2 Example of behavior style transition

Figure 10 shows an example of behavior style and module transition in the mobile agent architecture. This example illustrates how a mobile agent moves from station one to station zero. Steps 1) to 5) below sequentially describe how the behavior style transition occurs. Figure 11 shows how the mobile agent behaves in each behavior style.

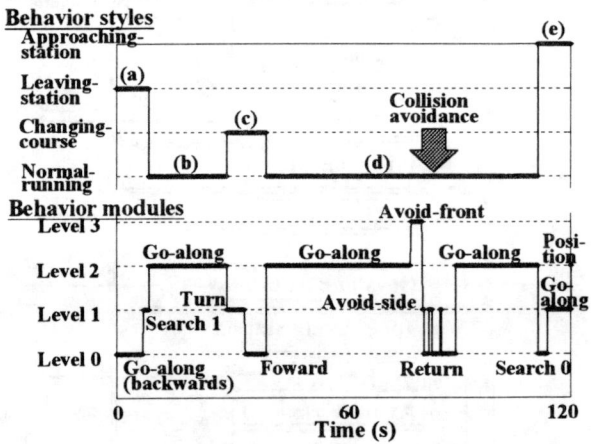

Figure 10. Transition of behavior styles and modules.

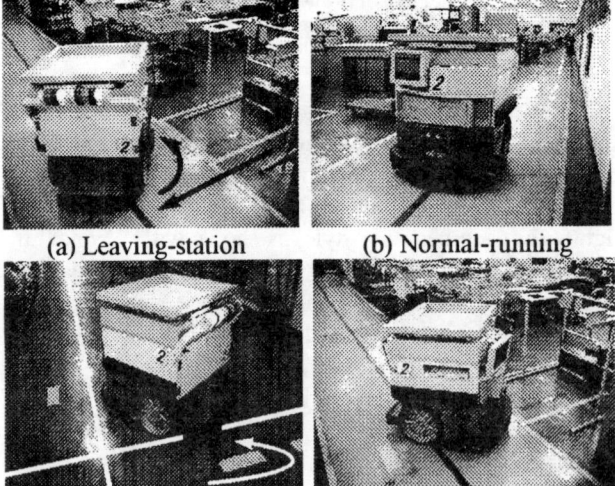

Figure 11. Behavior of mobile agent in each behavior style.

1) Leaving-station behavior style (Figure 10a): First, the mobile agent one (Figure 8) obtains a target address (Figure 8a) by negotiating vehicle allocation with the cell agent at station one (Figure 8). Then, the mobile agent leaves station one according to the leaving-station behavior style (Figure 11a).

2) **Normal-running behavior style** (Figure 10b): After leaving station one, the mobile agent searches for the guideline. When it detects the guideline, a behavior style transition of leaving-station to normal-running occurs. Following the normal-running behavior style, the mobile agent runs along the guideline toward an intersection (Figure 11b).

3) **Changing-course behavior style** (Figure 10c): When the mobile agent detects an intersection sign (Figure 8c), the behavior style changes from normal-running to changing-course. According to the changing-course behavior style, the mobile agent turns left at the intersection (Figure 11c). If the mobile agent encounters an oncoming mobile agent zero (Figure 8) when turning, it gives way to avoid a collision using the avoid-backward behavior module specific to the changing-course behavior style.

4) **Normal-running behavior style** (Figure 10d): After turning at the intersection, the mobile agent searches for the guideline. When it detects the guideline, a changing-course to normal-running behavior style transition occurs, according to which the mobile agent runs along the guideline toward station zero (Figure 8). If the mobile agent encounters an on-coming mobile agent zero when running along the guideline, both mobile agents carry out collision avoidance. In this case, each mobile agent only has to execute the avoid-front, go-along, avoid-side, and return behavior modules that form the normal-running behavior style (see Section 4.3 for details).

5) **Approaching-station behavior style** (Figure 10e): When a mobile agent detects a target address (Figure 8a), a normal-running to approaching-station transition occurs, according to which the mobile agent enters station zero. On detecting the stop position sign (Figure 8b), the mobile agent comes to stop (Figure 11d).

Use of the mobile agent architecture enables timely behavior style transitions so that the behavior style that is best suited for a specific aspect of transportation can be selected. This method executes ten or more conventional behavior modules in four or less (Figure10 Level 0 to 3) modules for each aspect of the transport process, thereby producing a high real-time response. Moreover, this method stipulates that only those behavior modules necessary for a specific behavior style be executed, making it possible to tailor behaviors for each aspect.

4.3 Example of collision avoidance

Figure 12 shows an example of how behavior module status transition occurs for causing mobile agents to avoid collision (passing each other) and how each mobile agent behaves. Figure 13 is a picture showing how collision is avoided. In this example, mobile agents run along a magnetic guideline at 0.5 m/s in opposite directions. The mobile agent architecture has selected the normal-running behavior style and caused only the related behavior modules to operate.

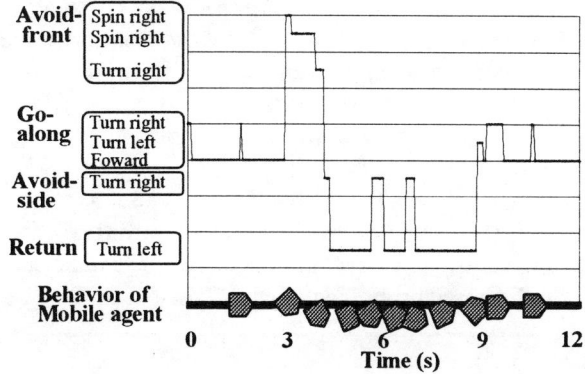

Figure 12. Transition of behavior modules and the behavior of mobile agent.

Figure 13. View of passing each other in each behavior module.

Both mobile agents first run along the guideline as directed by the go-along behavior module. When each mobile agent detects the other mobile agent, that is an oncoming mobile agent for each other, the avoid-front behavior module starts to operate and causes each mobile agent to move over to the right from the guideline (Figure 13a). Next, the avoid-side behavior module and the return behavior module, which is based on the memory that each mobile agent has moved over from the guideline, operate alternately. So, each mobile agent passes the other mobile agent on the right side by moving to the right then to the left (Figure 13b). After passing the oncoming mobile agent, each mobile agent moves toward the guideline as directed by the return behavior module (Figure 13c). After returning to the guideline, its memory that it has moved away from the guideline is erased. Next the go-along behavior module starts operating to cause each mobile agent to run along the guideline (Figure 13d).

In this example, the mobile agents can pass each other within 6 seconds, because a high real-time characteristic can be maintained for control using only the behavior modules of the normal-running behavior style. In addition, the mobile agents can easily avoid collisions only by temporarily moving away from the guideline no matter what the shape or behavior of the obstacle is, because the system shown in the example has improved its performance in avoiding collisions by making each behavior module specific to the normal-running behavior style.

5. Conclusions

We have developed an autonomous decentralized transportation system called AMADEUS, and applied it to transportation among cells on a shop floor. AMADEUS consists of autonomous transportation agents, or mobile agents, and agents in cells, or cell agents.

AMADEUS is characterized by vehicle allocation negotiated between agents and collision avoidance performed by mobile agents, which makes it possible to abolish centrally managed allocation and routing plans. To be more specific, a mobile agent moves to a target address obtained in negotiation for autonomous vehicle allocation with a cell agent without colliding with other mobile agents. This method has superseded the conventional central management system, which required high cost and long times to cope with changes in transportation environments. Moreover, the collision avoidance function of mobile agents enables them to run along a single-track guideline in opposite directions, resulting space-saving, efficient transportation.

One of the major challenging issues in realizing AMADEUS is how to install the collision avoidance function. A behavior-based approach has been used to control the mobile agents. Providing each mobile agent with a collision avoidance function would lead to an increased number of control steps and interference among behavior types because of their coexistence in various transportation aspects. To solve these problems, we have devised an architecture in which each behavior style is made of only the required basic units of functions it needs, and transitions occur among the behavior styles as required. With this architecture, it is only necessary to run an optimum basic behavior unit designed to realize the target behavior in each aspect. The best-suited behavior for each goal can be implemented without deteriorating the real-time characteristic for control in any aspect.

AMADEUS is now operating at Fujitsu's Kanuma Plant (6-1 Satsuki-cho, Kanuma 322, Japan).

References

[1] P.Pu, J.Hughes: "Integrating AGV Schedules in a Scheduling System for a Flexible Manufacturing Environment," in Proceeding of IEEE International Conference on Robotics and Automation, pp.3149-3154, 1994.

[2] O.M.Ulgen, P.Kedia: "Using Simulation of a Cellular Assembly Plant with Automatic Guided Vehicles," in Proceeding of Winter Simulation Conference, 1990.

[3] T.Fukuda, Y.Kawauchi, H.Asama: "Analysis and Evaluation of Cellular Robotics(CEBOT) as a Distributed Intelligent System by Communication Information Amount," in Proceeding of IROS'90, 1990.

[4] H.Asama, A.Matsumoto, Y.Ishida: "Design of Autonomous and Distributed Robot System : ACTRESS, " IEEE/RSJ Int. Conference on Intelligent Robots and Systems, pp.283-290, 1989.

[5] G.Lucarini M.Varoli, R.Cerutti, G.Sandini: "Simulation and HW Implementation," in Proceeding of IEEE International Conference on Robotics and Automation, pp.846-852, 1993.

[6] R.A.Brooks: "A Robust Layered Control System for a Mobile Robot," IEEE Journal of Robotics and Automation, vol.RA-2-1, pp.14-23, 1986.

[7] R.G.Smith: "The Contract Net Protocol: High-Level Communication and Control in a Distributed Problem Solver," IEEE Trans. on Computers, vol.29, no.12, pp. 1104-1113, 1980.

[8] T.Gomi, P.Volpe: "Collision Avoidance Using Behavioral-Based AI Techniques," in Proceeding of Intelligent Vehicles'93, 1993.

[9] T.Kamada: "A Behavior-based Automatic Guided Vehicle," in Proceeding of the 14th Conference of RSJ(Robotics Society of Japan), pp.133-134, 1996(in Japanese).

[10] F.R.Noreils: "Toward a Robot Architecture Integrating Cooperation between Mobile Robots: Application to Indoor Environment," The International Journal of Robotics Research, vol.12, No.1, pp.79-98, 1993

[11] P.Caloud, W.Choi, J.C.Latombe, C.L.Pade, M.Yim: "Indoor Automation with Many Mobile Robots," in Proceeding of the IEEE International Workshop on Intelligent Robots and Systems(IROS'90), pp.67-72, 1990

[12] L.E.Parker: "ALLIANCE: An Architecture for Fault Tolerant, Cooperative Control of Heterogeneous Mobile Robots," in Proceeding of Int. Conference on Intelligent Robots and Systems(IROS94), pp.776-783, 1994.

[13] L.E.Parker: "On the design of behavior-based multi-robot teams," Advanced Robotics, vol.10, no.6, pp.547-578, 1996.

[14] P.Maes: "The Dynamics of Action Selection," in Proceeding of Int. Joint Conference of Artificial Intelligence(IJCAI-89), pp.991-997, 1989.

On Discontinuous Human Control Strategies

Michael C. Nechyba[1] and Yangsheng Xu[1,2]

[1]The Robotics Institute, Carnegie Mellon University, Pittsburgh, PA 15213, USA
[2]Department of Mechanical and Automation Engineering, The Chinese University of Hong Kong, Hong Kong

Abstract

*Models of human control strategy (HCS), which accurately emulate dynamic human behavior, have far reaching potential in areas ranging from robotics to virtual reality to the intelligent vehicle highway project. A number of learning algorithms, including fuzzy logic, neural networks, and locally weighted regression exist for modeling **continuous** human control strategies. These algorithms, however, may not be well suited for modeling **discontinuous** human control strategies. Therefore, we propose a new stochastic, discontinuous modeling framework, for abstracting human control strategies, based on Hidden Markov Models. In this paper, we first describe the real-time driving simulator which we have developed for investigating human control strategies. Next, we demonstrate the shortcomings of a typical continuous modeling approach in modeling a discontinuous human control strategy. We then propose an HMM-based method of modeling discontinuous human control strategies, and show that the proposed controller overcomes these shortcomings and demonstrates greater fidelity to the human training data. We conclude the paper with further comparisons between the two competing modeling approaches.*

1. Introduction

In recent years, a number of different researchers have endeavored to abstract models of human skill directly from observed human input-output data (see [1] for an overview of the literature). Much of the work to date attempts to model human skill by *learning* the mapping from sensory inputs to control action outputs. Although the choice of learning algorithm varies, the most frequently used — including fuzzy logic, neural networks and locally weighted regression — are all examples of *continuous* function approximators (FAs). For each of these algorithms, control outputs are continuous and deterministic functions of model inputs.

Powerful as these may be, however, continuous learning algorithms may not be able to faithfully reproduce control strategies where discrete events or decisions introduce discontinuities in the input-output mapping. An example of this type of discontinuous control occurs in human driving, which requires control of (1) steering and (2) acceleration. While steering will tend to vary continuously with model inputs, acceleration control of the vehicle is decidedly discontinuous, since it involves explicit switching between the gas and brake pedals.

To adequately model such control behavior, we therefore propose a new stochastic, discontinuous learning algorithm, based on Hidden Markov Models. The proposed algorithm models possible control actions as individual HMMs. During run-time execution of the algorithm, a control action is then selected stochastically, as a function of both prior probabilities and posterior HMM-evaluation probabilities.

In this paper, we first describe the real-time graphic driving simulator for which we have recorded human control data, and for which we wish to abstract the corresponding driving control strategies. We then illustrate the difficulty of modeling a discontinuous control strategy using a continuous learning framework. Next, we propose a new HMM-based, discontinuous learning architecture for abstracting discontinuous human control strategies. We show that the resulting discontinuous modeling framework demonstrates better fidelity to the human training data than the continuous modeling approach. Finally, we offer some comparisons between the two learning architectures and suggest potential applications for each modeling approach.

2. Real-time driving simulator

Figure 1 shows the real-time graphic driving simulator which we have developed as an experimental platform for modeling human control strategies (HCS). In the simulator, the human operator has independent control over the steering of the car, the brake and the accelerator, although the simulator does not allow both the gas and brake pedals to be pushed at the same time. The state of the car is described by [2,3], $\{v_\xi, v_\eta, \omega\}$, where v_ξ is the lateral velocity of the car, v_η is the longitudinal velocity of the car and ω is the angular velocity of the car; the controls are given by,

$$-8000\text{N} \leq \alpha \leq 4000\text{N}, \qquad (1)$$

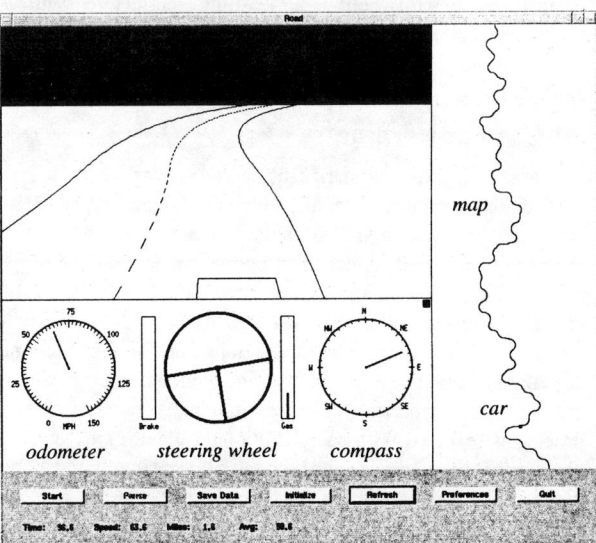

Fig. 1: The driving simulator gives the user a perspective preview of the road ahead. The user has independent controls of the steering, brake, and accelerator (gas).

$$-0.2\text{rad} \leq \delta \leq 0.2\text{rad}, \qquad (2)$$

where α is the user-applied longitudinal force on the front tires and δ is the user-applied steering angle.

Because of input device constraints, the force (or acceleration) control α is limited during each 1/50 second time step, based on its present value. If the gas pedal is currently being applied ($\alpha > 0$), then the operator can either increase or decrease the amount of applied force by a constant $\Delta\alpha_g = 200\text{N}$ or switch to braking. Similarly, if the brake pedal is currently being applied ($\alpha < 0$) the operator can either increase or decrease the applied force by a second constant $\Delta\alpha_b = 200\text{N}$ or switch to applying positive force. Thus, the $\Delta\alpha_g$ and $\Delta\alpha_b$ constants define the responsiveness of each pedal. In concise notation, denote $\alpha(k)$ as the current applied force and $\alpha(k+1)$ as the applied force for the next time step. Then, for $\alpha(k) \geq 0$,

$$\alpha(k+1) \in \{\alpha(k), \min(\alpha(k) + \Delta\alpha_g, 4000), \\ \max(\alpha(k) - \Delta\alpha_g, 0), -\Delta\alpha_b\} \qquad (3)$$

and for $\alpha(k) < 0$,

$$\alpha(k+1) \in \{\alpha(k), \max(\alpha(k) - \Delta\alpha_b, -8000), \\ \min(\alpha(k) + \Delta\alpha_b, 0), \Delta\alpha_g\} \qquad (4)$$

For the experiments in this paper, we collect human driving data across randomly generated roads like the 20km one shown in the map of Figure 1. The roads are described by a sequence of (1) straight-line segments and (2) circular arcs. The length of each straight-line segment, as well as the radius of curvature of each arc, lies between 100 and 200 meters. Finally, the visible horizon is set at 100m.

3. Continuous control

Below, we motivate the development of the discontinuous HMM-based learning architecture by first illustrating the learning problems that occur when attempting to model a discontinuous control strategy with a continuous learning architecture. While we choose *cascade neural network learning* for this purpose, we will show that the same problems would, in fact, be encountered by *any* continuous function approximator.

3.1 Cascade learning

Here, we briefly summarize the cascade neural network learning architecture. Further details, which are omitted for space reasons, may be found in [1,2,4]. Initially, there are no hidden units in the network, only direct input-output connections which are trained first. When no appreciable error reduction occurs, a first hidden unit is added to the network from a pool of *candidate* units, which are trained independently and in parallel with different random initial weights. Once installed, the hidden unit input weights are frozen, while the weights to the output units are retrained. This process is repeated with each additional hidden unit, which receives input connections from both the network inputs and all previous hidden units, resulting in a cascading structure.

In the experiments reported in this paper, we enhance the basic cascade learning framework in two ways: (1) we allow new hidden units to have variable activation functions [2], increasing the functional flexibility of the learning architecture; and (2) we train the neural network weights through node-decoupled extended Kalman filtering (NDEKF) [5], as opposed to gradient-descent techniques, such as quickprop or backpropagation. Both of these modifications have been shown to significantly improve learning speed and error convergence of the cascade learning architecture.

A necessary condition for successful learning is, of course, that the model be presented with those state and environmental variables upon which the human operator relies. Thus, the inputs to the cascade neural network should include, (1) current and previous state information $\{v_\xi, v_\eta, \omega\}$, (2) previous output (control) information $\{\delta, \alpha\}$, and (3) a description of the road visible from the current car position. More precisely, the network input vector $\zeta(k)$ at time step k is given by,

$$\{v_\xi(k-n_s), ..., v_\xi(k-1), v_\xi(k) \\ v_\eta(k-n_s), ..., v_\eta(k-1), v_\eta(k), \\ \omega(k-n_s), ..., \omega(k-1), \omega(k)\} \qquad (5)$$

$$\{\delta(k-n_c), ..., \delta(k-1), \delta(k) \\ \alpha(k-n_c), ..., \alpha(k-1), \alpha(k)\}, \qquad (6)$$

$$\{x_1(k), x_2(k), ..., x_{n_r}(k), y_1(k), y_2(k), ..., y_{n_r}(k)\}. \qquad (7)$$

where n_s is the length of the state histories and n_c is the length of the previous command histories presented to the network as input. For the road description, we partition the visible view of the road ahead into n_r equivalently spaced, body-relative (x, y) coordinates of the road median, and provide that sequence of coordinates as input to the network. Thus, the total number of inputs to the network are $3n_s + 2n_c + 2n_r$. The outputs of the cascade network are $\{\delta(k+1), \alpha(k+1)\}$, the steering and acceleration commands at the next time step, respectively.

3.2 Experiment

We ask Larry to drive over two different randomly generated 20km roads ρ_1 and ρ_2. A part of Larry's second run, each of which lasts about 10 minutes, is shown in Figure 2 below. Larry's driving behavior is representative of other runs recorded by him, as well as other individuals, in that (1) the steering control is reasonably continuous; (2) the acceleration control has significant discontinuities due to rapid switching between the brake and gas pedals; and (3) Larry manages to stay on the road (±5m deviation from the road median) for most of the run, with only a few brief off-road episodes in especially tight turns.

Now, we use Larry's first run (ρ_1) to train a cascade neural network, and reserve the second road ρ_2 for testing the network model. By searching the space of possible inputs parameterized by $\{n_s, n_c, n_r\}$, we arrive at the following suitable input space representation for Larry's cascade network model:

$$n_s = n_c = 6, n_r = 10 \qquad (8)$$

Thus, there are 50 inputs to the model. This input representation ensures that the cascade neural network forms a convergent control model. Figure 3 plots part of the neural network's driving control strategy over road ρ_2 for a linear cascade network (i.e. no hidden units), while Table 1 compares some aggregate statistics for Larry's second run and the linear model's run.

3.3 Discussion

From Figure 3 and Table 1, we make several observations. Most importantly, the linear model, despite the discontinuous acceleration command, is able to learn *something*; that is, the model keeps the vehicle on the road (except for one high-curvature turn that Larry himself was not able to handle properly). Not only that, but it does so at approximately the same average speed and lateral distance from the road median using a similar steering control strategy as Larry. In some respects, the model's control can even be considered superior to Larry's control. The model only rarely engages the brake, and maintains tighter lateral road position.

If we judge the model on how faithfully it reproduces Larry's acceleration control strategy, however, it rates significantly worse; that is, the model's acceleration control looks nothing like Larry's. Adding hidden units to impart nonlinearity to the model introduces additional high-frequency components to the control, but does not bring the model much closer to Larry's control strategy. Figure 4, for example, illustrates the model's acceleration control when two hidden units are introduced to the model.

To better appreciate what is happening, we would like to visualize how different input vectors in the training data map to different acceleration outputs. Since it is impossible to visualize a 50-dimensional input space, we decompose each of the input vectors $\zeta(k)$ in the training set into the principal components (PCs) for Larry's entire first run such that,

$$\zeta(k) = c_1^k \gamma_1 + c_2^k \gamma_2 + \ldots c_{50}^k \gamma_{50}, \qquad (9)$$

where γ_i is the PC corresponding to the ith largest eigenvalue σ_i. For Larry's control data, we have that,

$$|\sigma_2/\sigma_1| = 0.44, \ |\sigma_i/\sigma_1| \leq 0.05, \ i \in \{3, 4, \ldots, 50\} \qquad (10)$$

so that we can approximate the input vectors $\zeta(k)$ by,

Table 1: Statistical comparison

Road ρ_2	Larry	Linear model
v (mph)	71.9 ± 9.0	73.6 ± 2.5
d (m)	-0.72 ± 1.46	-1.00 ± 0.60
δ (rad)	± 0.094	± 0.068
α (N)	2240 ± 2620	1780 ± 760

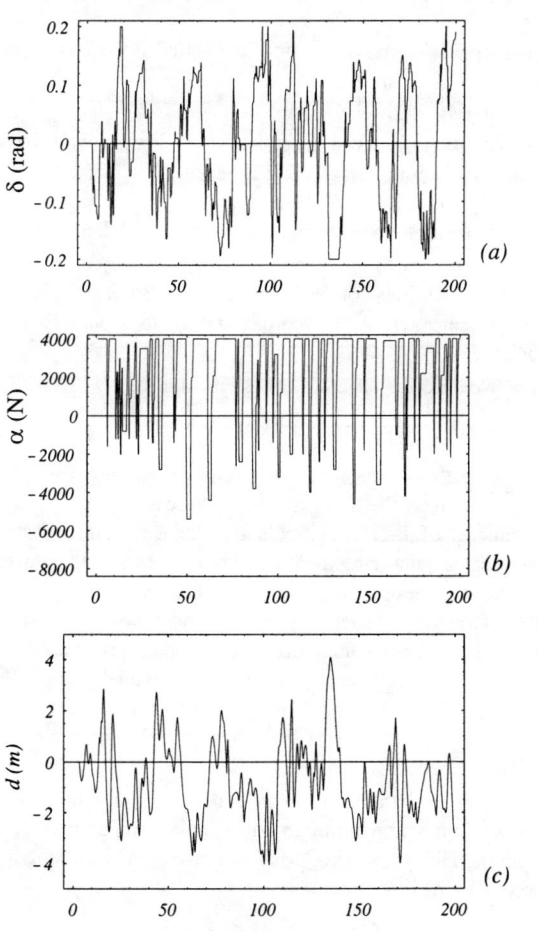

Fig. 2: Part of Larry's (a) steering command, (b) acceleration command and (c) resulting lateral distance from the road median. for the second road.

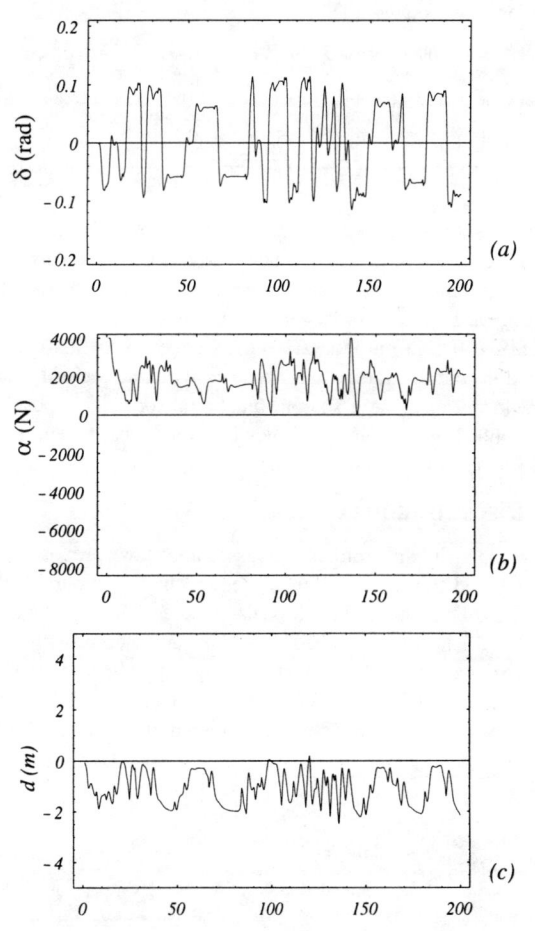

Fig. 3: Part of the linear model's (a) steering command, (b) acceleration command and (c) resulting lateral distance from the road median for the second road.

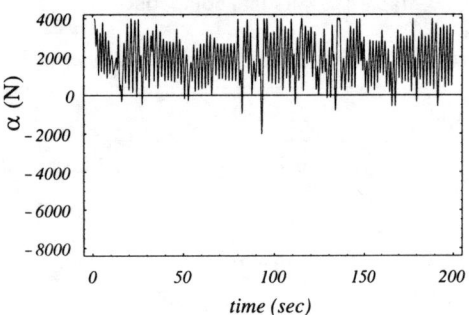

Fig. 4: Adding hidden units to the linear model does not bring the model significantly closer to Larry's acceleration control strategy.

$$\zeta(k) \approx c_1^k \gamma_1 + c_2^k \gamma_2. \tag{11}$$

Now, we can visualize the relative location of each $\zeta(k)$ by plotting its PC coefficients (c_1^k, c_2^k) in 2D space. Figure 5(a) and (b) show the results for $\alpha(k) < 0$ (brake), and $\alpha(k) > 0$ (gas), respectively. In each plot, we distinguish points by whether or not $\alpha(k+1)$ indicates a discontinuity (i.e. a switch between braking and accelerating) such that,

$$\alpha(k) < 0 \text{ and } \alpha(k+1) > 0 \quad [\text{Figure 5(a)}] \tag{12}$$

$$\alpha(k) > 0 \text{ and } \alpha(k+1) < 0 \quad [\text{Figure 5(b)}] \tag{13}$$

Those points that involve a switch are plotted in black, while a representative sample (20%) of the remaining points are plotted in grey.

We immediately observe from Figure 5 that — at least in the low-dimensional projection of the input vectors — the few training vectors that involve a switch overlap the many other vectors that do not. In other words, very similar input spaces can lead to radically different outputs $\alpha(k+1)$. Consequently, Larry's acceleration control strategy may not be easily expressible in a functional form, let alone a smooth functional form. This poses an impossible learning challenge not just for cascade neural networks, but *any* continuous function approximator.

4. Discontinuous control

To cope with the problems discussed above, we propose a new HMM-based framework for modeling discontinuous control strategies. Hidden Markov Models [6] are trainable statistical models which can be applied to model human control strategy, not as a deterministic functional mapping, but rather as a probabilistic relationship between sensory inputs and control actions outputs. They have previously been applied in a number areas,

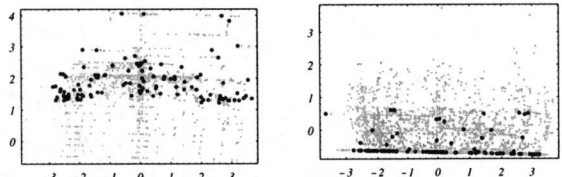

Fig. 5: Switching actions (black) significantly overlap other actions (grey) when the current applied force is (a) negative (brake), and (b) positive (gas).

including speech recognition [6, 7], modeling open-loop human actions [8], and analyzing similarity between human control strategies [9].

Although continuous and semi-continuous HMMs have been developed, discrete-output HMMs are often preferred in practice because of their relative computational simplicity and reduced sensitivity to initial parameter settings during training [6]. A discrete Hidden Markov Model consists of a set of n states, interconnected through probabilistic transitions, and is completely defined by the triplet, $\lambda = \{A, B, \pi\}$, where A is the probabilistic $n \times n$ state transition matrix, B is the $L \times n$ output probability matrix with L discrete output symbols, and π is the n-length initial state probability distribution vector. For an observation sequence O of discrete symbols, we can locally maximize $P(\lambda|O)$ (i.e. probability of model λ given observation sequence O) using the Baum-Welch Expectation-Maximization (EM) algorithm. We can also evaluate $P(O|\lambda)$ through the efficient Forward-Backward algorithm.

Figure 6 provides an overview of the resulting hybrid controller, where continuous outputs are modeled as before, and discontinuous outputs are modeled using the new framework. As shown in Figure 6, the discontinuous controller consists of three separate phases:

1. Input-space signals are first converted to an observation sequence of discrete symbols O^*, in preparation for Hidden Markov Model (HMM) evaluation.
2. The resulting observation sequence O^* is then evaluated on a bank of discrete-output HMMs, each of which represents a possible control action A_i and each of which has previously been trained on corresponding human control data.
3. Finally, the HMM evaluation probabilities are combined with prior probabilities for each action A_i to stochastically select and execute action A^* corresponding to input observation sequence O^*.

Below, we describe each of these steps in turn.

4.1 Signal-to-symbol conversion

In order to use discrete-output Hidden Markov Models, we must first convert the multi-dimensional real-valued input space, to a sequence of discrete symbols. At a minimum, this process involves vector quantizing the input-space vectors $\zeta(k)$ to discrete symbols. We choose the well-known LBG VQ algorithm [10], which iteratively generates vector codebooks of size 2^m, $m \in \{0, 1, ...\}$, and can be stopped at an appropriate level of discretization, as determined by the amount of available data. By optimizing the vector codebook on the human training data, we seek to minimize the amount of distortion introduced by the vector quantization process.

Once we have trained a vector codebook on all the input vectors $\zeta(k)$ in the human training data, and assuming that we want to train the Hidden Markov Models on sequences of length n_O, the sequence of input vectors,

$$\{\zeta(k - n_O + 1), \zeta(k - n_O + 2), ..., \zeta(k)\} \tag{14}$$

can be converted to,

$$O^k = \{o_1, o_2, ..., o_{n_O}\} \tag{15}$$

where o_i is the index of the codebook vector which minimizes the SSE distortion for $\zeta(k - n_o + i)$.

4.2 Stochastic, discontinuous controller

For the moment, assume that we wish to model a control task where at each time step k, we can choose one of N different discrete control actions A_i, $i \in \{1, ..., N\}$. Furthermore, assume that we have N groups of training observation sequences $\{O_i^1, O_i^2, ..., O_i^{n_i}\}$, $i \in \{1, ..., N\}$, where observation sequence O_i^j leads to control action A_i at the next time step. Then, using the Baum-Welch algorithm, we can train N different left-to-right HMMs λ_i in order to maximize,

$$\prod_{j=1}^{n_i} P(\lambda_i | O_i^j) \qquad (16)$$

for model λ_i. Given this bank of HMM models, and a new observation sequence O^*, we would now like to choose an appropriate action A^*. For each model, we can evaluate $P(O^* | \lambda_i)$ using the Forward-Backward algorithm. Since model λ_i corresponds to action A_i,

$$P(O^* | A_i) \equiv P(O^* | \lambda_i) \qquad (17)$$

By Bayes Rule,

$$P(A_i | O^*) = \frac{P(O^* | A_i) P(A_i)}{P(O^*)} \qquad (18)$$

where,

$$P(O^*) \equiv \sum_{i=1}^{N} P(O^* | A_i) P(A_i) \qquad (19)$$

and $P(A_i)$ represents the prior probability of selecting action A_i. We now propose the following stochastic control strategy:

$$A^* = A_i \text{ with probability } P(A_i | O^*) \qquad (20)$$

Thus, at each time step k, a control action is generated stochastically as a function of the current model inputs (O^*) and the prior likelihood of each action.

4.3 Acceleration control

For the acceleration control in our application, we have a total of $N = 8$ possible actions, as given in equations (3) and (4). When $\alpha(k) \geq 0$,

$$A_1 : \alpha(k+1) = \alpha(k) \qquad (21)$$

$$A_2 : \alpha(k+1) = \min(\alpha(k) + \Delta\alpha_g, 4000), \qquad (22)$$

$$A_3 : \alpha(k+1) = \max(\alpha(k) - \Delta\alpha_g, 0), \qquad (23)$$

$$A_4 : \alpha(k+1) = -\Delta\alpha_b, \qquad (24)$$

and when $\alpha(k) < 0$,

$$A_5 : \alpha(k+1) = \alpha(k) \qquad (25)$$

$$A_6 : \alpha(k+1) = \max(\alpha(k) - \Delta\alpha_b, -8000), \qquad (26)$$

$$A_7 : \alpha(k+1) = \min(\alpha(k) + \Delta\alpha_b, 0), \qquad (27)$$

$$A_8 : \alpha(k+1) = \Delta\alpha_g, \qquad (28)$$

Actions A_1 and A_5 correspond to no action for the next time step; actions A_2 and A_6 correspond to pressing harder on the currently active pedal; actions A_3 and A_7 correspond to easing off the currently active pedal; and actions A_4 and A_8 correspond to switching between the gas and brake pedals. We estimate the priors $P(A_i)$ by the frequency of occurrence of action A_i in the human control training data:

$$P(A_i) = n_i / \sum_{k=1}^{N} n_k \qquad (29)$$

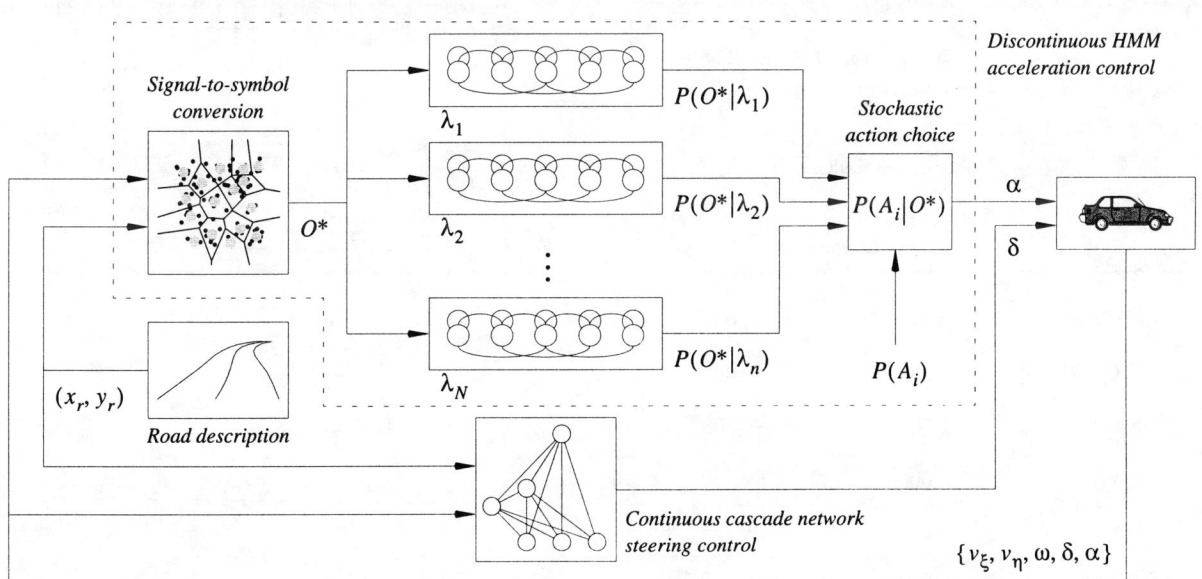

Fig. 6: Overall control structure. Steering is controlled by the cascade network, while the discontinuous acceleration command is controlled by the HMM-based controller (shaded box)

5. Experiment
5.1 Simulation

We once again try to model Larry's driving control strategy, only this time, we use the HMM-based approach to model the acceleration control α. The steering control δ is modeled by a cascade neural network as before (see Figure 6). Once again, we will use Larry's first run on road ρ_1 for training, and will reserve road ρ_2 for testing.

In processing Larry's control data for HMM training, we choose the following input representation and observation sequence length:

$$n_s = n_c = 1, n_r = 10, n_O = 6, \quad (30)$$

These choices ensure that approximately the same level of information is provided to the HMM controller as was provided to the cascade neural network in Section 3.2. The resulting input vectors $\zeta(k)$ are vector quantized to 512 codes, and three-state HMMs λ_i are trained for each possible action A_i.

Using equation (29), we calculate the priors $P(A_i)$ for Larry's first run as given in Table 2 below. We note from Table 2 that Larry never eases off either the gas or brake pedal, but rather opts to switch between braking and accelerating to achieve his desired control. Even so, the prior likelihood of a pedal switch is relatively low, which is part of the reason the neural network has difficulty modeling Larry's acceleration control strategy.

Figure 7, plots one of the stochastic HMM-based driving control strategies for road ρ_2, while Table 3 compares aggregate statistics for Larry's second run and the HMM controller's run.

5.2 Discussion

We see from Figure 7 that the stochastic HMM controller also appears to have learned a convergent control strategy. The big question is: Which controller, the continuous neural-network controller, or the discontinuous HMM controller, performs better? The answer to that question depends on what precisely is meant by "better."

If we evaluate the two controllers based on absolute performance criteria, the neural network controller probably performs better. It minimizes variations in the lateral position of the vehi-

Table 2: Prior probabilities

Road ρ_1	$\alpha(k) \geq 0$	$\alpha(k) < 0$
$P(A_1)$	0.819	0.000
$P(A_2)$	0.176	0.000
$P(A_3)$	0.000	0.000
$P(A_4)$	0.005	0.000
$P(A_5)$	0.000	0.730
$P(A_6)$	0.000	0.249
$P(A_7)$	0.000	0.000
$P(A_8)$	0.000	0.021

Table 3: Statistical comparison

Road ρ_2	Larry	HMM controller
v (mph)	71.9 ± 9.0	70.7 ± 8.1
d (m)	−0.72 ± 1.46	−1.38 ± 1.70
δ (rad)	±0.094	±0.081
α (N)	2240 ± 2620	1970 ± 2340

cle, conserves fuel by rarely "switching" to use the brake and averages a higher overall speed. By comparison, the HMM controller runs off the road more often and resorts to braking much more frequently than the neural network controller. Simply put, the neural network controller appears to be more stable than its HMM counterpart.

If, on the other hand, we evaluate the two controllers on how closely they approximate the operator's (Larry's) control strategy, the verdict changes drastically. As we have already noted previously, the neural network control does not look anything like

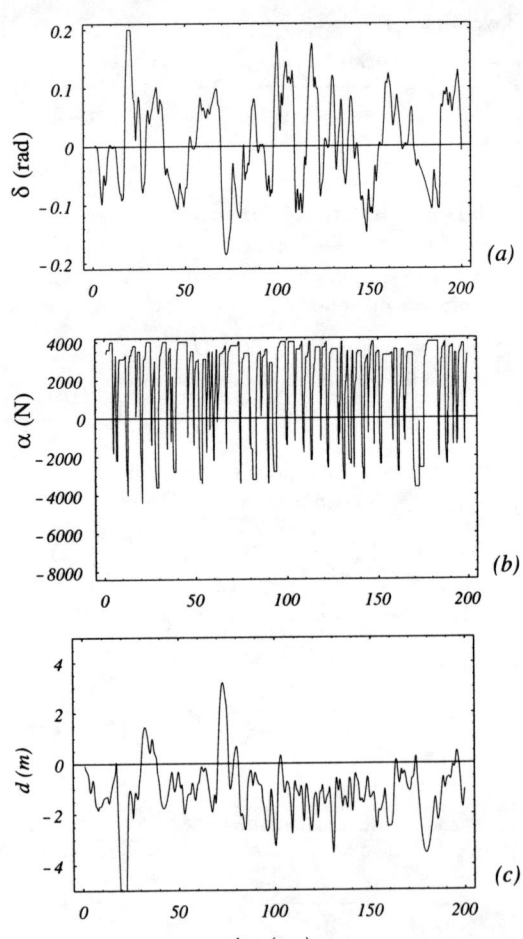

Fig. 7: Part of the HMM controller's (a) steering command, (b) acceleration command and (c) resulting lateral distance from the road median for the second road.

Larry's control; the HMM control trajectory appears to be a much better approximation of Larry's actual acceleration control strategy, including the consequent increases in the number of off-road incidents. We can quantify the degree of similarity to Larry's control strategy for each controller using a stochastic similarity measure σ which we have developed previously [9] for comparing different human control strategies. The similarity measure is capable of comparing stochastic, multi-dimensional trajectories and yields a value between 0 and 1, with larger values indicating greater similarity. For the similarity comparison here, we include all relevant state and control variables $\{v_\xi, v_\eta, \omega, \delta, \alpha\}$, and arrive at the following similarity values:

$$\sigma(Larry, HMM) = 0.628 \quad (31)$$

$$\sigma(Larry, NN) = 0.101 \quad (32)$$

Hence, the HMM controller shows greater fidelity to the source training data than does the neural network controller.

The most important reason behind the HMM controller's success is that it is able to successfully model the switching behavior between the gas and brake pedals as a probabilistic event. Because we train a *separate* HMM for each action A_i, this modeling approach does not encounter the same one-to-many mapping problem, illustrated in Figure 5, that the neural network encounters. The relatively few occurrences of switching are sufficient training data, since the switching HMMs λ_4 and λ_8 see *only* that data during HMM training. Including the priors $P(A_i)$ in the action selection criterion then ensures that the model is not overly biased towards switching.

Now, suppose the acceleration control α were not constrained by equations (3) and (4) and thus were not as readily expressible through discrete actions. For example, suppose that the separate gas and brake commands could change by an arbitrary amount for each time step, not just by $\Delta\alpha_g$ and $\Delta\alpha_b$. How would this change the proposed control framework? In such a case, two separate continuous controllers can be trained, one for controlling the gas pedal, and the other for controlling the brake pedal. Furthermore, the HMM controller's role would be reduced to simply regulating the switching behavior between the two pedals. Despite its reduced function, however, the general HMM framework would still retain a critical part in modeling the overall acceleration control accurately.

Of course, the HMM control framework does have some limitations in comparison to functional modeling approaches. Because we vector quantize the input space, the stable region of operation for the HMM controller is strictly limited by the range of the vector codebook. When, in rare circumstances, an unfortunate sequence of stochastic action selections causes the controller to stray too far from that limited codebook range, the VQ distortion dramatically increases, and the behavior of the controller becomes less predictable. In such cases, by monitoring the VQ distortion, it might be possible to temporarily suspend the HMM controller in favor of a neural network controller until the VQ distortion once again returns to an acceptable level. Alternatively, we are investigating whether semicontinuous HMMs [7], which model the VQ codebook as a family of Gaussian pdfs, would (1) improve performance, and (2) be computationally tractable in real-time execution.

A second limitation of the HMM approach is the inclusion of the prior probabilities $P(A_i)$ in the stochastic selection criterion. By including the priors, we are assuming an environment similar to the training environment. Radically different environmental conditions would presumably change the values of the priors, and therefore make the action selection criterion less valid.

6. Conclusion

In this paper we have developed a discontinuous modeling framework for abstracting discontinuous human control strategies, and have compared the proposed approach to a competing continuous learning architecture. Which control approach is preferred ultimately depends on the specific application for the HCS model. If the model is being developed towards the eventual control of a real robot or vehicle, then the continuous modeling approach might be preferred as a good starting point. Continuous models can operate for a larger range of inputs, show greater inherent stability, and lend themselves more readily to theoretical performance analysis. If, on the other hand, the model is being developed in order to simulate different human behaviors in a virtual reality simulation or game, then the discontinuous control approach might be preferred, since fidelity to the human training data and random variations in behavior would be the desired qualities of the HCS model. Thus, depending on the application, we believe a need exists for both types of modeling approaches.

References

[1] M. C. Nechyba and Y. Xu, "Human Control Strategy: Abstraction, Verification and Replication," *IEEE Control Systems Magazine*, vol. 17., no. 5, pp. 48-61, 1997.

[2] M. C. Nechyba and Y. Xu, "Learning and Transfer of Real-Time Human Control Strategies," to appear in *Journal of Advanced Computational Intelligence*, vol. 1, no. 2, 1997.

[3] H. Hatwal and E. C. Mikulcik, "Some Inverse Solutions to an Automobile Path-Tracking Problem with Input Control of Steering and Brakes," *Vehicle System Dynamics*, vol. 15, pp. 61-71, 1986.

[4] S. E. Fahlman, L. D. Baker and J. A. Boyan, "The Cascade 2 Learning Architecture," Technical Report, CMU-CS-TR-96-184, Carnegie Mellon University, 1996.

[5] M. C. Nechyba and Y. Xu, "Cascade Neural Networks with Node-Decoupled Extended Kalman Filtering," *Proc. IEEE Int. Symp. on Computational Intelligence in Robotics and Automation*, vol. 1, pp. 214-9, 1997.

[6] L. R. Rabiner, "A Tutorial on Hidden Markov Models and Selected Applications in Speech Recognition," *Proc. IEEE*, vol. 77, no. 2, pp. 257-86, 1989.

[7] X. D. Huang, Y. Ariki and M. A. Jack, *Hidden Markov Models for Speech Recognition*, Edinburgh Univ. Press, 1990.

[8] J. Yang, Y. Xu and C. S. Chen, "Human Action Learning Via Hidden Markov Model," *IEEE Trans. Systems, Man and Cybernetics, Part A*, vol. 27, no. 1, pp. 34-44, 1997.

[9] M. C. Nechyba and Y. Xu, "Stochastic Similarity for Validating Human Control Strategy Models," *Proc. IEEE Conf. on Robotics and Automation*, vol. 1, pp. 278-83, 1997.

[10] Y. Linde, A. Buzo and R. M. Gray, "An Algorithm for Vector Quantizer Design," *IEEE Trans. Communication*, vol. COM-28, no. 1, pp. 84-95, 1980.

HUMAN BEHAVIOR MODELING IN MASTER-SLAVE TELEOPERATION WITH KINESTHETIC FEEBACK

L.F. Peñín, A. Caballero, R. Aracil y A. Barrientos

Dpto. de Automática, Universidad Politécnica de Madrid (DISAM)
Jose Gutierrez Abascal, 2 Madrid 28006, SPAIN
Tel: +34-1-3363061   Fax: +34-1-5642961   E-mail penin@disam.upm.es

Abstract: In master-slave teleoperation systems the control loop is closed through the human operator. He[1] acts as a controller generating actuating signals based in visual and kinesthetic cues. It is then indispensable to have a model of his behavior in order to design and study this type of systems coherently. Up to now only simplified arm dynamic models have been included in the study of bilateral control schemes. This paper makes use of already existing human-in-the loop models, developed for aircraft piloting, in order to construct a complete and simple model of human behavior for master-slave teleoperation systems with visual and kinesthetic feedback. The model was validated through several experiments with two different master arms and under various circumstances of operation. It includes complete human dynamic behavior: both the generation of control actions and limb dynamics.

1 Introduction

One of the major differences between automatic and teleoperated systems is the inclusion of a human operator in the control process. This intervention is primarily due to the use of teleoperation systems in complex and non-structured environments, where human decision and response capacity upon unexpected events is essential.

The degree and mode of human intervention depends on the task and the operating circumstances. It can go from continuous manual control to supervisory control, passing through shared and traded control [1]. Manual control is by far the most common and is the one being considered in this study.

The significance of an operator in a master-slave teleoperation system lies in the fact that the control loop is closed through him. He acts as a controller generating actuating signals upon the master based on visual and kinesthetic feedback. Some definitions about his behavior can be found in the literature: "is the archetype hierarchical, adaptive, optimizing, decision making controller" [2]; "is an adaptive controller who learns from experience" [3]; "acts as an adaptive and robust controller" [4].

It is clear that inadequate control signals coming from the operator would make system response not acceptable, which in case there is kinesthetic feedback could also be harmful to the system or the operator himself. This fact is even more relevant in master-slave systems with time delay between the local and the remote zone [5].

Despite human's excellent ability for manual control, we are limited by a response dynamic, both in making the decision and in the manipulative movement. This dynamics greatly affects the whole performance of the teleoperation system. Lee [6] says that no model of a bilateral teleoperation system with some practical use can neglect the inclusion of a model of the operator.

Researchers in bilateral control have always follow this guideline [7][8], but up to date their model of the operator has been limited to the arm dynamics (mass, damping and stiffness). Only [6] tried to proposed a richer model including some decision behavior, but was oversimplified when practically used, becoming another mechanic model without information of how the operator generates the control actions from feedback information. In [9] a more detailed model is presented, but is too complex to be used with common bilateral systems models and does not takes into account kinesthetic feedback.

This paper makes use of already existing human-in-the loop models developed for modeling aircraft piloting, in order to develop a relatively simple and straightforward model that can be easily implemented in existing models of bilateral systems. The model includes both the generation of control actions and the dynamics of response of the arm. This model can be used to further study bilateral control schemes, human behavior in delayed systems or to help in the design of hand controllers.

Several experiments with two different master arms were conducted. With the data obtained the parameters of two adapted existing human-in-the loop models where identified and declared suitable for the modeling of human performance in teleoperation. One of the models was subsequently extended for the use with master-slave systems with kinesthetic feedback. This paper explains all this process in detail, but first it is necessary to give some introductory notes about human modeling.

2 Human modeling

First investigations in modeling human behavior began in the 50's, when Hick and Fitts [1] proposed respectively models to explain the time necessary to take decision and to actually implement it. But they where only valid for very simple tasks.

Interest in modeling the behavior of a human as an active feedback control device began during WWII, when studies where made to improve the performance of pilots. The *control theory paradigm* was then created. It says that a human controller can be represented as a set of linear, constant-coefficient differential equations [10].

Based in this paradigm different approaches were born. The first one generates a structural model in which different blocks try to identify each subsystem of the human operator senso-motor behavior [2][11]. It makes use of classic control theory and cause-effects laws. On the other hand there is an algorithmic approach [2][10][12] that uses optimal control theory. There are

---
[1] Throughout the paper the pronoun 'he' will be employed instead of she/he.

Table 1 Configuration of the experiments

Exp.	E1	E2	E3	E4	E5	E6	E7	E8	E9	E10	E11
HC	K	K	K	K	K	K	K	K	D	D	D
Joint	1	1	4	4	3	1	1	1	3	3	2
Disp.	D1	D1	D1	D1	D2	D1	D1	D1	D1	D1	D2
Target	S	C	S	C	S	S	S	S	S	C	S
# Op.	7	6	7	6	7	4	4	4	8	7	6
FFB	-	-	-	-	-	FE	FL	FR	-	-	-

also other approaches less extended, such as modeling through time series [3] or fuzzy logic [11][13].

Whatever the modeling mode selected is necessary to consider some general facts about human performance. First, his actuating bandwidth is around 5-10 Hz for self decided movements, although it gets reduced to 2 Hz when following random position signals [14]]. On the other hand kinesthetic feedback processing is over 20-30 Hz up to a maximum of 100 Hz [14].

The human operator is capable of varying at will his arm dynamic features, such as stiffness, inertia and viscous damping, although the variations in inertia are very low [6][2]. Modifying independently the values of torque generated and damping due to the co-activation of antagonistic muscles [15] (although stiffness and damping are correlated) allows to stabilize a system with a critical behavior [9][6][16]. Stiffness values from 2 to 400 Nm/rad can be selected for the elbow [17]. Humans can also sense vibrations up to 1 kHz of 1 μm [18] through several cutaneous mechano-receptors in the skin of the hand [14]. Some studies [9] [18] affirm that this tactile information is essential for the sequential coordination of activities during the manipulation of objects or tools, and therefore also for telemanipulation. But still is not clear the degree of influence of this low energy tactile information on the operator. In common teleoperation tasks, where not a high sensitive force feedback is expected, it is not considered significant.

3 Design of experiments

Several experiments have been conducted in order to find a model for human behavior in manual teleoperation. Generally, they consist in performing a 1 DOF given task on a display with a teleoperation master arm or hand-controller (HC). The task is mainly following a target circumference with another circumference whose position in the screen is directly related to the sampled position of the HC joint. The experiments differ in the type of HC employed, in the DOF of the HC used and in the movement of the target. The experiments can also be classified as with or without force feedback (FFB) on the master arm.

3.1 Hand Controllers (HC)

Two different HC have been employed. First there is the DISAMaster, developed in our lab and with 3 revolute DOF. The first and third one move on a horizontal plane, while the second one moves in a vertical plane. It has no FFB and the sampling period in the experiments for each joint angle is 100 ms. It has very low inertia and friction, and almost no viscous damping.

The second HC is from Kraft Telerobotics Inc. It has 6 revolute DOF joints, all of them with vertical movement, except the first and fourth one, which move in a horizontal plane. The first five

Figure 1 Operator performing an experiment with the Kraft HC

joints are actuated with AC motors in order to implement a force-position bilateral control scheme [19] in case it is used with the Kraft slave. The sampling period in the experiments for each joint angle is 60 ms. It has a relatively high inertia on the first joints and some stiction and viscous damping.

3.2 Displays

The displays are implemented on a SUN workstation screen. They consist of a target and follower circumferences moving over a predefined 1 DOF path. The follower position is given by the HC joint, while the target position is pre-programmed. The objective is to maintain the follower (of smaller diameter) inside the target as it moves. As the operator response only depends on the error between target and follower, the displays can be catalogued as compensatory [11].

There are two different classes of display depending on the type of the path. Display 1 (D1) path is a semi-circumference, and the follower position directly depends of the angle of the HC joint being sampled. This display is used for horizontal movements. Display 2 (D2) path is a vertical line, and the follower position is the projection of the HC handle, which describes a round path, over a vertical axis. This display is suitable for vertical movements.

3.3 Target movement

There are two types of target movements for the compensatory displays. First there is what we have called step target (S): the target appears on a given position, stays approximately 5 s before disappearing, and then appears in another given position, stays another 5 s and so on. This type of movement tries to simulate the smooth movement of a slave arm into a desired position in the environment, since 5 s is enough time to complete the movement with no need to hurry. The sequence of steps is the same for all the trials in one given experiment, and have the particularity of having the first and second half of the sequence identical.

Second, there is the continuos target (C) that moves continuously along the path following a predefined movement, apparently random for the operator. The movement was constructed by the summation of 13 sinusoids of 0.01, 0.015, 0.025, 0.040, 0.063, 0.1, 0.158, 0.251, 0.4, 0.63, 1, 1.58 and 2.51 Hz respectively. The magnitude of each sinusoid was chosen so that the magnitude spectrum of the target position appeared to be of a first order low-pass filter with cutoff frequency of 0.04 Hz. This was done to ease frequency analysis and corresponds to a typical validation set-up [20]

Table 2 RMS error between average curve and real data

Exp.	E1	E2	E3	E4	E6	E7	E9	E10
RMS	0.068	0.031	0.098	0.041	0.111	0.132	0.073	0.042

3.4 Force feedback (FFB)

In some of the experiments a force was applied to the operator arm through the Kraft HC. The force generated depended on the target and follower positions through three different functions. In the first one the torque applied on the joint was proportional (3.2 Nm/rad) to the error between target and follower (FE). It was a repulsive torque with a maximum of 10 Nm when the target is over the follower.

The second one generates a torque proportional (6.4 Nm/rad) to the distance between the follower and the left end of the path (FL). It is a repulsive torque with a maximum of 20 Nm when the target is on the left end. The third one works in the same way, having the right end as the maximum repulsive point (FR).

3.5 Operators

Nine different operators were randomly selected among the technical staff of the lab. They were male and young. Before doing the experiment they were allowed to train as much as they wanted with the system.

The experiments were conducted during several days at different times. The operators were told to put themselves as comfortable as possible and to perform the task relaxed. They were also told to move the follower smoothly in the step movements trying not to overshoot.

In the experiments with FFB they were told to move the HC as far as possible against the repulsive force but without forcing themselves or the system.

Depending on the set-up employed, 11 different experiments were conducted. The time of all the tasks was approximately 60 seconds. The most relevant are summarized in Table 1.

4 Data analysis and processing

Before trying to identify a model for the data obtained through the experiments, it is necessary first to process such amount of information in order to have simple coherent data to work with. In this sense spurious points due to error in the sampling or the recording were removed.

Each experiment generated several data curves, one for each operator. Since the target movement in a given experiment was always the same, it is possible to obtain an average response curve for each experiment. Visual inspection certifies that the average curve is a very accurate representation.

Table 2 shows the RMS error between the average curve and the experimental ones for each experiment over a 0 to π rad range of values. It is important to note that the average value of the follower for the different experiments with FFB (E6, E7) is not as good as it should. This is due to the different way of perceiving force of the operators and that the transmission of force of the Kraft HC is not as sensitive as would be desirable.

From now on further analysis will only make use of the average curves obtained. The proposed models will be validated comparing their output with the experimental average curves. The comparison will be done in the time-domain through visual inspection and RMS calculation, and in the frequency domain through spectral analysis [21].

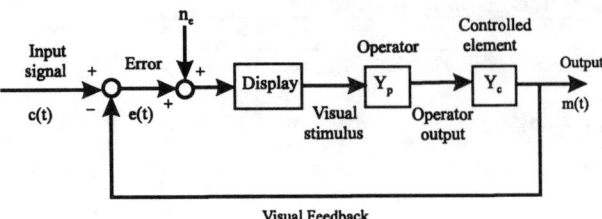

Figure 2 Manual control system in compensatory behavior

5 Cross-over model in teleoperation

McRuer in 1965 [2] modeled operator behavior in manual control loops such as the one shown in Figure 2. The operator works with a compensatory display in which he only sees the error between the target and the follower. His actions in order to minimize the error are to move an aircraft joystick (Y_c) with a certain dynamics, which in turn moves the follower. Remnant signal n_e will be neglected in our models.

His investigations resulted in finding that for a great variety of controlled elements dynamics and with continuos moving targets on a compensatory display, the operator behaves such as to always have:

$$G = Y_p Y_c = \frac{\omega_c e^{-\tau_e s}}{s} \qquad [1]$$

This is what is known as the crossover model and implies that the human plus the controlled element is an invariant. The crossover frequency (ω_c) and the time delay (τ_e) vary with the type of task and through training.

We have used this simple and well-accepted model to check the coherency of our experiments. Table 3 shows the values of the parameters of the crossover model identified for the respective experiments with both the continuous and the step-moving target. In the latter case the crossover hypothesis did not apply. To adapt the model the input is delayed τ_1 s, which is the time delay due to decision making every time the target appears in one position in the discrete movement. With this modification the crossover model has been extended to be used with movements not continuos.

Figure 3 shows the experimental follower position (gray) and the crossover model output (black) for E2. Figure 3 also shows the plot of the open-loop $H(j\omega)$ for the experimental data and the data predicted by the crossover model. It is seen that at low frequencies the phases differ slightly, which can be explained by some pursuit behavior developed during the experiment [11].

Table 3 Crossover parameters for experiments with a continuous and discrete moving target

	Continuos target			Step target		
	E2	E4	E10	E1	E3	E9
ω_c (rad/s)	1.92	3.02	2.74	0.82	1.05	1.06
τ_e (s)	0.3	0.24	0.23	0.28	0.19	0.24
τ_1 (s)	-	-	-	0.33	0.20	0.36

Despite the great utility of the crossover model is indeed very simple and it not behaves so well under changes in the set-up. Also, the two parameters are not enough to be able to explain completely human control behavior and adaptation. Also, it is not possible to construct a crossover model with FFB. A more complex model that makes use of the identification of the different subsystems in human senso-motor performance is needed. This is the structural model that will now be introduced.

6 The structural model without FFB

A structural model of the human operator in a compensatory manual control loop is offered in [10] [11] and is presented in Figure 4. It follows the crossover model but gives a more detailed representation of human operator dynamics. A brief description of each block will be provided here.

The structural model can be divided in two different parts: the Central Nervous System (CNS) and the Neuromuscular System (NS), which together correspond to the operator block Y_p of Figure 2. System error is presented to the operator through a display with no dynamics. A central time delay of τ_0 seconds accounts for the effects of latencies in the visual process sensing, motor nerve conduction times, etc. The blocks of the NS comprise the neuromuscular dynamics (Y_{pn}) of the particular limb driving the manipulator and a propioceptive feedback due to muscle spindles and tendons (Y_f). There is also a dynamics of the propioception associated with higher level signal processing that is represented by block Y_m in the CNS. The NS finally generates a force (δ_{nm}) that is applied to the controlled element (Y_c), the master arm in our case.

We want this model to predict human control movements in teleoperation related tasks. The most common type of movement is generated to move from one point to another more or less far away. The experiments with a step-moving target have been specially design to read this behavior and therefore are the ones

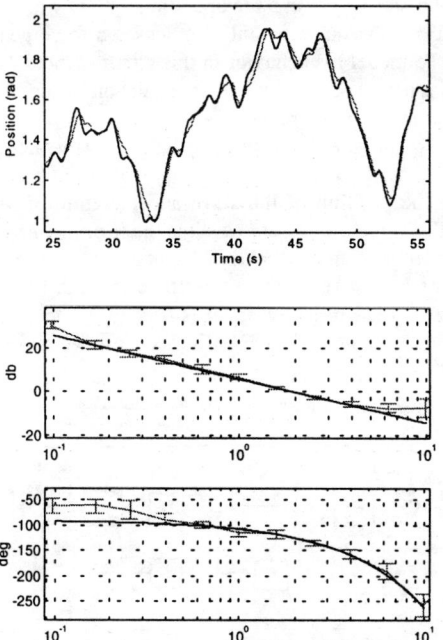

Figure 3 Experimental (gray) and crossover model (black) plots for E2

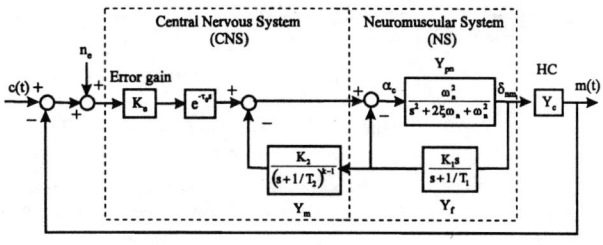

Figure 4 Structural model for compensatory behavior with no FFB

to be predicted by the model.

6.1 Parameters adjustment for experiment E1

The structural model comprises a great variety of parameters. First there is the transfer function representing the dynamics of the master arms Y_c. In a real teleoperation system it would also had in series the dynamic of the slave arm. In our case, the following transfer functions for the first joint of the DISAMaster and the Kraft master arm were identified experimentally:

$$Y_{c\,DISAM} = \frac{1}{0.15s^2 + s} \qquad Y_{c\,KRAFT} = \frac{1}{s^2 + 10s} \qquad [2]$$

Several trials using parameter values present in the literature [10][11] were conducted to predict the output of the experiments, but no reasonable results were obtained. It is necessary to account that the model was in its origin created for aircraft pilot modeling. Although it can be employed for other types of manipulation it has always been validated with continuous input signals, which is not our present case. So a specific parameter adjustment for this type of task is needed.

We follow the methodology proposed by Hess [11], taking as invariant for this type of experiments and HCs the following parameters:

$$T_1 = T_2 \quad K_1 = 1 \quad k = 2 \quad \omega_n = 10 \quad \xi = 0.707 \qquad [3]$$

As the movement of the target is discrete it is necessary to extend the model including a time delay block in the input representing the Hicks decision time τ_1, whose value was shown in Table 3 for some experiments. Table 4 shows the values identified for the rest of parameters and for each step of the target. Average and variance for all the experiment trials are also shown.

A linear correlation is detected between model gain K_e and the width of the step Δ_{step}. The relation is given by:

Table 4 Structural model parameters for E1

#	Step (rad)	K_e	τ_0	τ_1	$1/T_1$	K_2
2	1	8.900	0.185	0.250	17.80	8.00
3	-1.5	8.725	0.194	0.191	16.53	3.68
4	1.9	6.433	0.284	0.181	17.04	17.33
5	-1.4	8.795	0.020	0.369	14.45	16.44
6	2.1	6.273	0.087	0.390	10.34	16.26
7	-1	9.684	0.091	0.37	17.66	15.69
8	-1.2	8.394	0	0.36	13.36	16.37
Average	-	8.208	0.0836	0.335	14.88	15.08
variance	-	1.322	0.0874	0.081	2.85	4.30

$$K_e = 11.4159 - 2.3316 \times \Delta_{step} \quad \text{for } \Delta_{step} > 0 \quad [4]$$
$$K_e = 14.0639 + 4.1090 \times \Delta_{step} \quad \text{for } \Delta_{step} < 0$$

Note the asymmetry depending on the sign of the step. This corresponds with the human motion asymmetry of moving the arm (or wrist) to the right or to the left. For the rest of parameters the average is a good value. Figure 5 shows the experimental response and the model prediction for two sequential steps in the target movement during the performance of E1. The corresponding K_e for each step was used. The similarity between both curves is clear. The average value of K_e could have been used, and the similarity would have remained, as seen in the spectral plot of Figure 5 obtained with a fixed average value of K_e. It is clear that the accuracy of the model remains in the frequency domain.

The crossover frequency of the experiment is of about ω_c=0.797 rad/s, which is very similar to the one obtained through the crossover model: 0.82 rad/s (see Table 3). This is one of the conditions proposed by Hess [10] for building the structural model over the crossover one.

If we compare the open-loop transfer function Y_pY_c of the crossover model and the structural model in permanent state upon an impulse in the input, we have the following relation: $K_e=b_m\omega_c$, where b_m is the viscous damping of the master. This relation is congruent with the values already obtained and it allows to know the value of K_e from the simpler cross over model. For E1, we have $K_e=10\omega_c$.

Slight deviations from this rule in other experiments are because the same HC transfer function has been used for the different joints, which is just an approximation.

Table 5 Parameters of the structural model for different experiments

	E2	E4	E10	E1	E3	E5	E9
ω_c	1.92	3.02	2.74	0.82	1.05	0.94	1.06
K_e	20.98	31.07	2.22	8.20	10.60	10.21	1.02
$\tau_1(s)$	-	-	-	0.33	0.33	0.20	0.27
$\tau_0(s)$	0.08	0.08	0.08	0.08	0.08	0.073	0.08
$1/T_1$	79.00	77.10	365	14.88	56.00	63.74	70.00
K_2	15.08	15.08	15.08	15.08	15.08	15.08	15.08
ω_n	10	14.3	10	10.	14.3	10	10
ξ	0.707	0.707	0.707	0.707	0.707	0.707	0.707

6.2 Parameter adjustment for the rest of experiments

The same procedure done for E1 can be applied to the rest of experiments. The values obtained are shown in Table 5, where the value of K_e is the average. The relation between K_e and ω_c remains valid for all the experiments.

Parameters $1/T1$ and $1/T2$ are in straight relation with the adaptability of the human operator. It can be proven that as the task requires a more controlled approach less will be their value. Parameters τ_0, K_2 and ξ can be considered fixed for the kind of tasks we have experimented with.

Comparing the two HC it is seen that with the DISAMaster (E9, E10), (more simple to use and with almost no dynamics) the crossover frequency (and therefore K_e) is higher than with the Kraft HC ones (E1,E2), while the decision time delay is a little smaller. The same relation can be found when comparing experiments done with the arm (E1, E2) with the ones done with the wrist (E3, E4). Wrist movements, done in a more simple and controlled way, have higher crossover frequency and K_e but less decision-time. When moving with the wrist a slight change in ω_n was needed to have a good model.

7 The structural model with FFB

When the operator works in a teleoperation system with FBB he not only receives visual cues but also senses a force generated by the HC. To model his behavior in this circumstances we have constructed a new model, which is an extension of the previous one. It is presented in Figure 6.

There is a new variable in the system: the force generated by the actuators of the HC. It is not a real input because it generally depends on the position of the slave and therefore of the HC. Physically this force is generated against the force applied by the operator in order to move the HC dynamics (F_m). There is an inner loop of human reaction to this force through the propioceptors Y_f and Y_m. But a new input is needed in the perceptual system, which is expressed by F_{mp}. In this sense we make use of

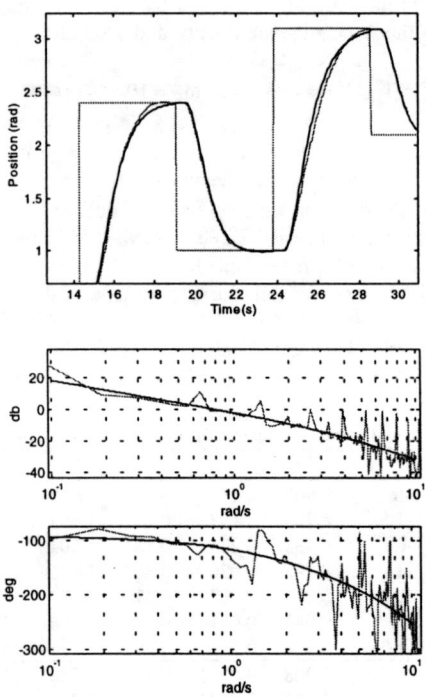

Figure 5 Experimental and structural model plots for E1

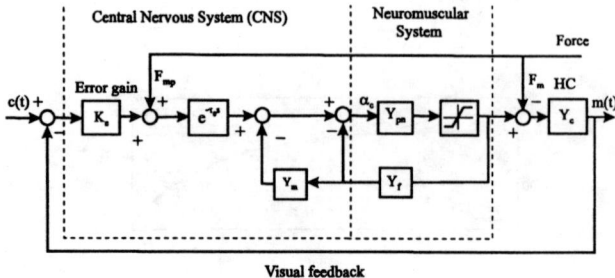

Figure 6 Structural model for compensatory behavior with FFB

the hypothesis that the human can integrate visual and force input in a coherent way without giving more importance to one or the other [6].

A saturation block is needed after the neuromuscular dynamics Y_{pn}, as there is maximum force that can be generated by the limb. The values of this block depend of the kind of movement, but for horizontal movements done with the arm semi-extended (hand at 50 cm of the shoulder) is around 15 Nm. Note that there is an asymmetry between movement to the left or to the right (15.5 vs 15 Nm). The values of the rest of the structural model without FFB remain.

Figure 7 shows the evolution of the position of the target, the follower and the model prediction for experiment E7 with the values of E1 shown in Table 5. Figure 7 also shows the corresponding frequency response. Similar results were obtained with the other type of FFB used in experiments E6 and E8.

The conclusions obtained through the analysis of the structural model with no FFB still apply, although it has to be considered that when force is introduced the variance of behavior of the operators increases significantly. Having in mind this fact the prediction of the model is very accurate and can be used extensively for the analysis of teleoperation systems with human manual control with kinesthetic feedback.

8 Conclusions

This paper has presented a new model for manual control of master-slave systems with kinesthetic feedback. The 1 DOF model was constructed from the basic crossover model and the structural model known in aircraft pilot modeling, being both extended for teleoperation. Parameters were identified for different circumstances through the use of the data supplied by several different experiments with two HC.

The model has proven very accurate for common tasks of moving the slave smoothly from one position to another, both with no FFB and FFB. In the latter case prediction is slightly worse due to the low repeatability of operator behavior when kinesthetic cues of high value are present.

The model is relatively simple and straightforward and can be easily implemented in existing models of bilateral control systems. Work has to be done to prove its validity with more complex movements and when more DOF are present. Also, studies have began to see the ability of the proposed human model to represent the effect of the human operator on stability.

9 References

[1] T.B. Sheridan, *Telerobotics, Automation and Human Supervisory Control*, The MIT Press, Cambridge, 1992.
[2] D. McRuer, "Human Dynamics in Man-Machine Systems", *Automática*, Vol. 16, pp. 237-253, 1980.
[3] F. Osafo-Charles et al, "Application of Time-Series Modeling to Human Operator Dynamics", *IEEE Trans. on Systems, Man and Cybernetics*, Vol. 10, No. 12, December, 1980.
[4] K.P. Chin and T.B. Sheridan, "The Effect of Force Feedback on Teleoperation", *Work with Computers: Organizational, Management, Stress and Health Aspects*, Elsevier, Amsterdam, 1989.
[5] T.B. Sheridan, "Space Teleoperation Through Time Delay: Review and Prognosis", *IEEE Trans. on R&A*, Vol. 9, No. 5, October, 1993.
[6] S. Lee & H.S. Lee, "Modeling, Design, and Evaluation of Advanced Teleoperator Control Systems with Short Time Delay", *IEEE Tran. on Robotics and Automation*, Vol. 9, No. 5, October 1993.
[7] Y. Yokokohji and T. Yoshikawa, "Bilateral Control of Master-Slave Manipulators for Ideal Kinesthetic Coupling - Formulation and Experiment", *IEEE Tran. on R&A*, Vol. 10, No. 5, October, 1994.
[8] D.A. Lawrence, "Designing Teleoperator Architectures for Transparency", *Proceedings of the 1992 IEEE ICRA'92*, France, May 1992.
[9] Van de Vegte et al, "Teleoperator Control Models: Effects of Time Delay and Imperfect System Knowledge", *IEEE Trans. on Systems, Man, and Cybernetics*, Vol. 20, No. 6, December 1990.
[10] R. Hess, "Human in the Loop Control", *The Control Handbook*, Editor W.S Levine, IEEE- CRC Press, 1996.
[11] R.A. Hess, "Pursuit Tracking and Higher Levels of Skill Development in the Human Pilot", *IEEE Transactions on Systems, Man and Cybernetics*, Vol. 11, No. 4, April 1981.
[12] D.L. Kleinman et al, "An Optimal Control Model of Human Response Part I: Theory and Validation", *Automatica*, Vol. 6, 1970.
[13] F. Wawak, A.M. Desodt y D. Jolly, "Fuzzy decision algorithm for man machine systems", *CIMNE*, Barcelona 1993.
[14] G. Burdea, *Force and Touch Feedback for Virtual Reality*, John Wiley & Sons, New York, 1996.
[15] N. Hogan, "Impedance Control: An approach to manipulation: Part I-Theory", *ASME Journal of Dynamic Systems, Measurement and Control*, Vol. 107, March 1985.
[16] W.S. Kim, B. Hannaford, A.K. Bejczy, "Force-Reflection and Shared Compliant Control in Operating Telemanipulators with Time Delay", *IEEE Tran.. on Robotics& Automation*, Vol. 8, No. 2, 1992.
[17] N. Hogan, "Controlling Impedance at the Man/Machine Interface", *IEEE Int. Conf. on Robotics and Automation*, Arizona, 1989.
[18] R.D. Howe y D.A. Kontarinis, "High-Frequency Force Information in Teleoperated Manipulation", *Proc. of the Third International Symposium on Experimental Robotics*, Springer-Verlag, 1993.
[19] L.F. Peñín et al, "Fundamentals of a master-slave teleoperation system with a force-position bilateral control scheme", *IFAC's Int. Symposium on Robot Control, SYROCO'97*, Nantes, France, 1997.
[20] W.S. Kim, et al, "A Comparison of Position and Rate Control for Telemanipulations with Consideration of Manipulator System Dynamics", *IEEE J. of Robotics & Automation*, Vol. 3, No. 5, 1987.
[21] P. Fiorini y A. Giancaspro, "A Procedure for the Frequency Analysis of Telerobotic Tasks Data", *IEEE/RJS IROS'92*, Raleigh, NC, 1992.

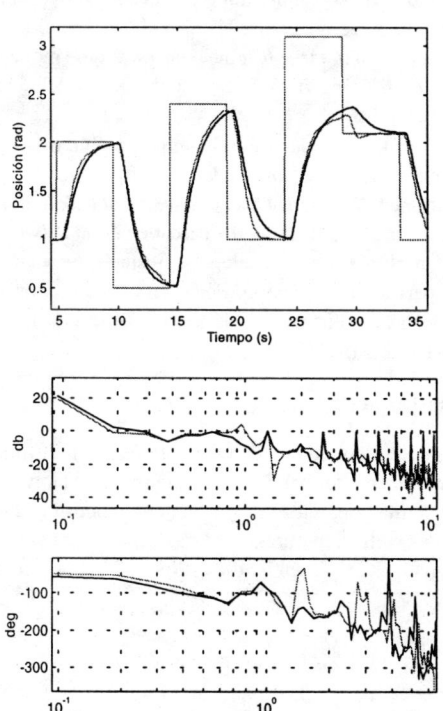

Figure 7 Experimental and structural model plots for E7

Two Performance Measures for Evaluating Human Control Strategy

Jingyan Song[1], Yangsheng Xu[2,3], Michael C. Nechyba[3] and Yeung Yam[2]

[1]Department of Systems Engineering and Engineering Management, The Chinese University of Hong Kong, Hong Kong
[2]Department of Mechanical and Automation Engineering, The Chinese University of Hong Kong, Hong Kong
[3]The Robotics Institute, Carnegie Mellon University, Pittsburgh, PA 15213, USA

Abstract

In the last few years, modeling dynamic human control strategy (HCS) is becoming an increasingly popular paradigm in a number of different research areas, such as the intelligent vehicle highway system, virtual reality and robotics. Usually, these models are derived empirically, rather than analytically, from real human input-output control data. As such, there is a great need to develop adequate performance criteria for these models, as few guarantees exist about their theoretical performance. It is our goal in this paper to develop several such criteria. In this paper, we first collect driving data from different individuals through a real-time graphic driving simulator. We then model each individual's control strategy through the flexible cascade neural network learning architecture. Next, we develop two performance measures for evaluating the resulting HCS models, one dealing with obstacle avoidance, the other with tight-turning behavior. Finally, we evaluate the relative skill of different HCS models through the proposed performance criteria.

1. Introduction

HCS models, which accurately emulate dynamic human behavior, find application in a number of research areas ranging from robotics to the intelligent vehicle highway system. Because human control strategy (HCS) is a dynamic, nonlinear stochastic process, developing good analytic models of human control strategies, tends to be difficult. Therefore, recent work in modeling HCS has focussed on learning empirical models, through, for example, fuzzy logic [1,2], and neural network techniques [3]. Since these HCS models are empirical, few if any guarantees exist about their theoretical performance. Thus, performance evaluation is an integral aspect of HCS modeling research, without which it is impossible to rank or prefer one HCS controller over another.

Skill or performance can be defined through a number of task-dependent as well as task-independent criteria. Some of these criteria may conflict with one another, and which is most appropriate for a given task depends in part on the specific goals of the overall task. Therefore, rather than examine performance evaluation in the abstract, we focus on one specific HCS namely, the task of human driving.

In this paper, we first record driving data from different individuals through a dynamic driving simulator. For each driver, we then train a HCS model using the flexible cascade neural network learning architecture. Because each of the different drivers exhibits a different style or control strategy, their respective models will likewise differ. It is the goal of this work to define performance criteria by which the driving models' performance can be evaluated and ranked.

In previous work, a stochastic similarity measure, which compares model-generated control trajectories to the original human training data, has been proposed for validating HCS models [4]. While this similarity measure can ensure that the neural network model adequately captures the driving characteristics of the human operator, it does not measure a particular model's skill or performance. In other words, it does not (nor can it) tell us which model is better or worse. In this paper, therefore, we propose two measures for evaluating the performance of HCS models.

For the task of driving, many candidate performance criteria, such as average speed, driving stability, driving safety and fuel efficiency, exist. Rather than select such specific criteria, however, we prefer to decompose the driving task into subtasks, and then define criteria that measure performance for the most meaningful subtasks. For example, we can view the driving task as a combination of the following subtasks: (1) driving along a straight road, (2) turning through a curve in the road and (3) avoiding obstacles. Of these subtasks, avoiding obstacles is perhaps the most useful one, as it can measure important characteristics of the HCS model, including its ability to change speeds quickly while maintaining vehicle safety and stability. Thus, the first performance criterion we develop is based on avoiding obstacles. Our second performance criterion investigates the HCS models' behavior in executing tight turns. Both of these test the behavior of the models outside the range of training data from which the models are learned.

In this paper, we first introduce the dynamic graphic driving simulator that we use to collect human data and from which the HCS models are trained. We then show how we model each individual's human control strategy using the cascade neural network learning architecture. Finally, we develop the obstacle avoidance and tight turning performance criteria. For each of the proposed criteria, we demonstrate their use on HCS models trained from different individuals' data.

2. Experimental setup

For this work, we collect human driving data from a real-time graphic simulator, whose interface is shown in Figure 1 below. In the simulator, the human operator has independent control of the vehicle's steering as well as the brake and gas pedals. The simulated vehicle's dynamics are given by the following second-order nonlinear model:

$$\ddot{\theta} = (l_f P_f \delta + l_f F_{\xi f} - l_r F_{\xi r})/I \qquad (1)$$

$$\dot{v}_\xi = (P_f \delta + F_{\xi f} + F_{\xi r})/m - v_\eta \dot{\theta} - (\text{sgn} v_\xi) c_D v_\xi^2 \qquad (2)$$

$$\dot{v}_\eta = (P_f + P_r - F_{\xi f} \delta)/m + v_\xi \dot{\theta} - (\text{sgn} v_\eta) c_D v_\eta^2 \qquad (3)$$

$$\begin{bmatrix} \dot{x} \\ \dot{y} \end{bmatrix} = \begin{bmatrix} \cos\theta & \sin\theta \\ -\sin\theta & \cos\theta \end{bmatrix} \begin{bmatrix} v_\xi \\ v_\eta \end{bmatrix}, \text{where,} \quad (4)$$

$\dot{\theta}$ = angular velocity of the car (5)

v_ξ = lateral velocity of the car, (6)

v_η = longitudinal velocity of the car, (7)

$$F_{\xi k} = \frac{\mu F_{zk}(\tilde{\alpha}_k - (\text{sgn}\delta)\tilde{\alpha}_k^2/3 + \tilde{\alpha}_k^3/27) \times}{\sqrt{1 - P_k^2/(\mu F_{zk})^2 + P_k^2/c_k^2}}, k \in \{f, r\} \quad (8)$$

$\tilde{\alpha}_k = c_k \alpha_k / (\mu F_{zk})$, $k \in \{f, r\}$, (9)

α_f = front tire slip angle = $\delta - (l_f \dot{\theta} + v_\xi)/v_\eta$, (10)

α_r = rear tire slip angle = $(l_r \dot{\theta} - v_\xi)/v_\eta$, (11)

$F_{zf} = (mgl_r - (P_f + P_r)h)/(l_f + l_r)$,

$F_{zf} = (mgl_f + (P_f + P_r)h)/(l_f + l_r)$, (12)

ξ = body-relative lateral axis,

η = body-relative longitudinal axis, (13)

c_f, c_r = 50000N/rad, 64000 N/rad, (14)

$c_D = 0.0005\text{m}^{-1}$, (15)

μ = coefficient of friction = 1, F_{jk} = frictional forces,

$j \in \{\xi, z\}$, $k \in \{f, r\}$ (16)

$$P_r = \begin{cases} 0, & P_f \geq 0 \\ k_b P_f, P_f < 0, k_b = 0.34 \end{cases}, m = 1500\text{kg},$$

$I = 2500\text{kg-m}^2$, $l_f = 1.25\text{m}$, $l_r = 1.5\text{m}$, $h = 0.5\text{m}$, (17)

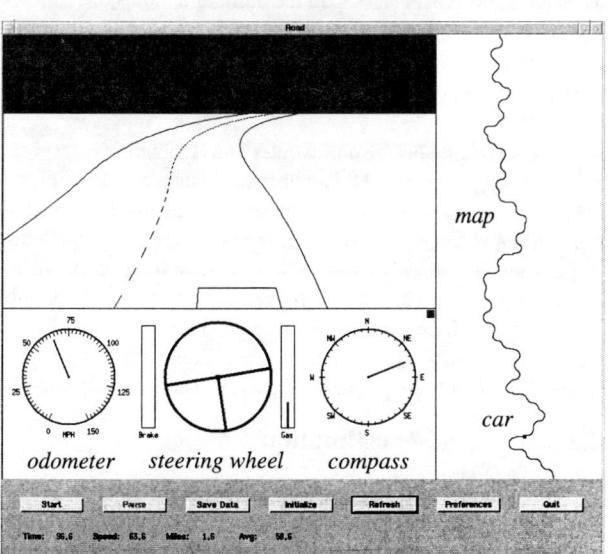

Fig. 1: The driving simulator gives the user a perspective preview of the road ahead. The user has independent controls of the steering, brake, and accelerator (gas).

and the controls are given by,

$-8000\text{N} \leq P_f \leq 4000\text{N}$ (18)

$-0.2\text{rad} \leq \delta \leq 0.2\text{rad}$ (19)

where P_f is the longitudinal force on the front tires, and δ is the steering angle.

Note that the separate brake and gas commands for the human are in fact the single P_f variable, where the sign indicates whether the brake or the gas is currently active. Each individual is asked to navigate across several randomly generated roads, which consist of a sequence of (1) straight-line segments, (2) left turns, and (3) right turns. The map in Figure 1, for example, illustrates one randomly generated 20km road for which human driving data was recorded. Each straight-line segment as well as the radius of curvature for each turn range in length between 100m and 200m. Nominally, the road is divided into two lanes, each of which has width $w_l = 10\text{m}$. The human operator's view of the road ahead is limited to 100m. Finally, the entire simulator is run at 50Hz.

3. HCS modeling

In this paper, we choose the flexible cascade neural network architecture with node-decoupled extended Kalman filtering (NDEKF) [5, 6] for modeling the human driving data. We prefer this learning architecture over others for a number of reasons. First, no *a priori* model structure is assumed; the neural network automatically adds hidden units to an initially minimal network as the training requires. Second, hidden unit activation functions are not constrained to be a particular type. Rather, for each new hidden unit, the incremental learning algorithm can select that functional form which maximally reduces the residual error over the training data. Typical alternatives to the standard sigmoidal function are sine, cosine, and the Gaussian function. Finally, it has been shown that node-decoupled extended Kalman filtering, a quadratically convergent alternative to slower gradient descent training algorithms (such as backpropagation or quickprop) fits well within the cascade learning framework and converges to good local minima with less computation [5].

The flexible functional form which cascade learning allows is ideal for abstracting human control strategies, since we know very little about the underlying structure of each individual's internal controller. By making as few *a priori* assumptions as possible in modeling the human driving data, we improve the likelihood that the learning algorithm will converge to a good model of the human control data.

In order for the learning algorithm to properly model each individual's human control strategy, the model must be presented with those state and environmental variables upon which the human operator relies. Thus, the inputs to the cascade neural network should include, (1) current and previous state information $\{v_\xi, v_\eta, \dot{\theta}\}$, (2) previous output (command) information $\{\delta, P_f\}$, and (3) a description of the road visible from the current car position. More precisely, the network inputs are,

$$\{v_\xi(k-n_s), ..., v_\xi(k-1), v_\xi(k)$$
$$v_\eta(k-n_s), ..., v_\eta(k-1), v_\eta(k), \quad (20)$$
$$\dot{\theta}(k-n_s), ..., \dot{\theta}(k-1), \dot{\theta}(k)\}$$

$$\{\delta(k-n_c), ..., \delta(k-1), \delta(k) \\ P_f(k-n_c), ..., P_f(k-1), P_f(k)\}, \quad (21)$$

$$\{x(1), x(2), ..., x(n_r), y(1), y(2), y(n_r)\}. \quad (22)$$

where n_s is the length of the state histories and n_c is the length of the previous command histories presented to the network as input. For the road description, we partition the visible view of the road ahead into n_r equivalently spaced, body-relative (x, y) coordinates of the road median, and provide that sequence of coordinates as input to the network. Thus, the total number of inputs to the network n_i are,

$$n_i = 3n_s + 2n_c + 2n_r \quad (23)$$

The two outputs of the cascade network are $\{\delta(k+1), P_f(k+1)\}$. For the system as a whole, the cascade neural network can be viewed as a feedback controller, whose two outputs control the driving of the vehicle. Figure 2 illustrates the overall structure of the model-vehicle system.

4. Obstacle avoidance performance criterion

In real driving, obstacles such as rocks and debris can unexpectedly obstruct a vehicle's path and force a driver to react rapidly. Thus, obstacle avoidance is one important performance criterion by which we can gauge a model's performance. In this section we develop and evaluate such criteria for different HCS models.

4.1 Virtual path equivalence

Since our HCS models receive only a description of the road ahead as input from the environment, we reformulate the task of obstacle avoidance as *virtual path following*. Assume that an obstacle appears τ meters ahead of the driver's current position. Furthermore, assume that this obstacle completely obstructs the entire width of the road ($2w$) and extends for d meters along the road. Then, rather than follow the path of the actual road, we wish the HCS model to follow a virtual path as illustrated in Figure 3. This virtual path consists of (1) two arcs with radius of curvature γ, which offset the road median laterally by $2w$, followed by (2)

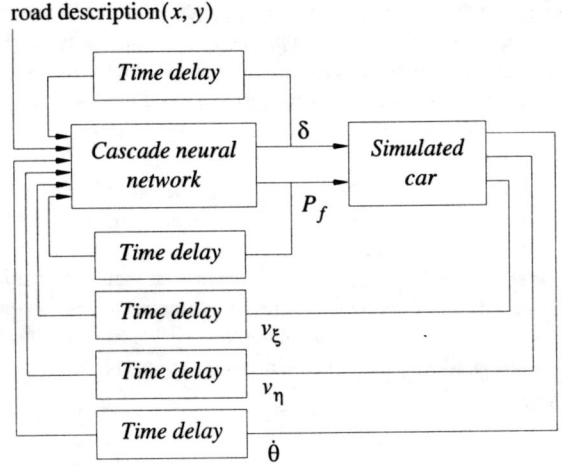

Fig. 2: The overall structure of the simulated system.

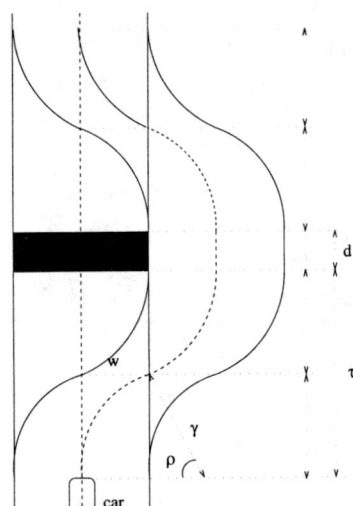

Fig. 3: Virtual path for obstacle avoidance.

a straight-line segment of length d, and (3) another two arcs with radius of curvature γ which return the road median to the original path. By analyzing the geometry of the virtual path, we can calculate the required radius of curvature γ of the virtual path segments in terms of the obstacle width $2w$ and the obstacle distance τ:

$$\gamma^2 = (\tau/2)^2 + (\gamma - w)^2 = \tau^2/2 + \gamma^2 - 2\gamma w + w^2 \quad (24)$$

$$\gamma = [\tau^2/(8w)] + w/2 \quad (25)$$

The corresponding sweep angle ρ for the curves is given by,

$$\rho = \sin^{-1}\left(\frac{\tau/2}{\gamma}\right) = \sin^{-1}\left[\tau/\left(\frac{\tau^2}{4w} + w\right)\right] \quad (26)$$

Consider an obstacle located $\tau = 60$m ahead of the driver's current position. For this obstacle distance and $w = 5$m, γ evaluates to 92.5m. This is less than the minimum radius of curvature (100m) that we allow for the roads over which we collect our human control data. Therefore, the obstacle avoidance task in part tests each HCS model's ability to operate safely outside the range of its training data.

As an example, Figure 4 illustrates one HCS model's response to the virtual path created by an obstacle distance of 60m. Figure 4(a) plots the vehicle's lateral distance from the road median through the virtual path. We observe that on the virtual path, the vehicle deviates sharply from the road median by over 4m. In addition, Figure 4(b) shows that the velocity of the car drops substantially from approximately 35m/sec to a low of about 23m/sec on the virtual path. The model's corresponding steering (δ) and force (P_f) outputs are plotted in Figure 4(c) and (d), respectively.

4.2 Lateral offset estimation

As we observed in Figure 4(a), a driving model may deviate significantly from the center of the road during the obstacle avoidance maneuver. Below, we derive the important relationship between the obstacle detection distance τ and a model's corresponding maximum lateral deviation ψ. First, we take N measurements of ψ for different values of τ, where we denote the ith measurement as (τ_i, ψ_i). Next, we assume a polynomial relationship of the form,

$$\begin{aligned}\psi_i &= \alpha_p \tau_i^p + \alpha_{p-1} \tau_i^{p-1} + \ldots + \alpha_1 \tau_i + \alpha_0 + e_i \\ &= \Gamma_i^T \alpha + e_i\end{aligned} \quad (27)$$

where the e_i are additive measurement error. Then, we can write,

$$\begin{aligned}\psi_1 &= \Gamma_1^T \alpha + e_1 \\ \psi_2 &= \Gamma_2^T \alpha + e_2 \\ &\ldots \\ \psi_N &= \Gamma_N^T \alpha + e_N\end{aligned} \quad (28)$$

or, in matrix notation,

$$\Psi = \Gamma \alpha + e, \text{ where,} \quad (29)$$

$$\Psi = [\psi_1, \psi_2, \ldots, \psi_N]^T, \quad (30)$$

is the observation vector,

$$\Gamma = [\Gamma_1, \Gamma_2, \ldots, \Gamma_N]^T, \quad (31)$$

is the regression matrix, and $e = [e_1, e_2, \ldots, e_N]$ is the error vector.

Assuming white noise properties for e ($E\{e\} = 0$ and $E\{e_i e_j\} = \sigma_e^2 \delta_{ij}$ for all i, j), we can minimize the least-squares error criterion,

$$V(\hat{\alpha}) = \frac{1}{2} \varepsilon^T \varepsilon = \frac{1}{2} \sum_{k=1}^{N} \varepsilon_k^2 = \frac{1}{2} (\Psi - \Gamma \hat{\alpha})^T (\Psi - \Gamma \hat{\alpha}), \quad (32)$$

with the optimal, unbiased estimate $\bar{\alpha}$,

$$\bar{\alpha} = (\Gamma^T \Gamma)^{-1} \Gamma^T \Psi, \quad (33)$$

assuming that $(\Gamma^T \Gamma)$ is invertible.

For example, consider the HCS model from Figure 4. We plot its measured ψ values for τ ranging from 20 to 100 meters in Figure 5. Superimposed on top of the measured data is the estimated fifth-order relationship ($p = 5$) between ψ and τ. We observe that the polynomial model fits the data closely and appears sufficient to express the relationship between ψ and τ.

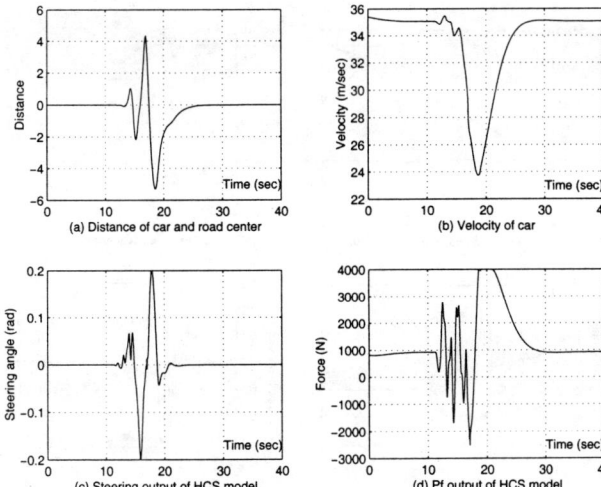

Fig. 4: Example of obstacle avoidance by a HCS model.

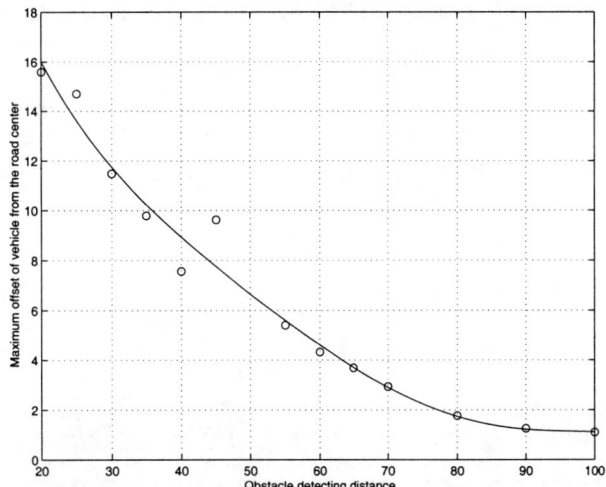

Fig. 5: Maximum lateral offset as a function of obstacle detection distance.

4.3 Obstacle avoidance threshold

We note from Figure 5 that as the obstacle detection distance decreases, the maximum lateral offset increases. Thus, for a given model and initial velocity $v_{initial}$, there exists a value τ_{min} below which the maximum offset error will exceed the lane width w_l. We define the driving control for obstacle distances above τ_{min} to be stable; likewise, we define the driving control to be unstable for obstacle distances below τ_{min}.

Now, define an obstacle avoidance performance criterion β,

$$\beta = \frac{\tau_{min}}{v_{initial}}, \quad (34)$$

where $v_{initial}$ is the velocity of the vehicle when the obstacle is first detected. The β criterion measures to what extent a given HCS model can avoid an obstacle while still controlling the vehicle in a stable manner. The normalization by $v_{initial}$ is required, because slower speeds increase the amount of time a driver has to react and therefore avoiding obstacles becomes that much easier.

Below, we calculate the β performance criterion for three HCS models, trained on real driving data from Tom, Dick, and Harry, respectively. Figure 6 plots ψ as a fifth-order function of τ for the three different models. From Figure 6, it is easy to approximate τ_{min} for each HCS model; thus, the corresponding β performance criterion for each model is,

$$\beta_{Tom} = 45/35 = 1.3 \quad (35)$$

$$\beta_{Dick} = 35/35 = 1.0 \quad (36)$$

$$\beta_{Harry} = 18/35 = 0.51 \quad (37)$$

Thus, as an obstacle avoider, Harry's model clearly outperforms Tom's and Dick's models, since β_{Harry} is the lowest performance measure for the three models.

4.4 Obstacle avoidance velocity loss

The performance criterion β measures the stability of a particular HCS model in avoiding an obstacle. It does not, however, directly measure how skillfully the model avoids the obstacle. Consider, for example, Figure 4(b). During the obstacle avoidance maneuver, the velocity of the vehicle drops sharply so that the model can adequately deal with the tight maneuvers required. Below, we define a performance criterion J which measures the distance lost as a result of this velocity drop:

$$J = \int_{t_0}^{t_f} |v_{initial} - v_{virtual}| dt, \quad (38)$$

where $v_{initial}$ is the velocity before obstacle detection, and $v_{virtual}$ is the time-dependent velocity during the obstacle avoidance maneuver.

Consider once again the three HCS models for Tom, Dick and Harry. Each model can successfully avoid the obstacle when τ ranges from 50 to 100 meters. Figure 7 plots J for this range of τ. Once again we observe that Harry's model performs best when evaluated with the J performance criterion, since its distance loss is smaller for each τ then either Tom's or Dick's model.

5. Tight turning

Here we analyze performance as a function of how well a particular HCS model is able to navigate tight turns. First, we define a special road connection consisting of two straight-line segments connected directly (without a transition arc segment) at an angle ζ. For small values of ζ, each HCS model will be able to successfully drive through the tight turn; for larger values of ζ, however, some models will fail to execute the turn properly by temporarily running off the road or losing complete sight of the road.

Figure 8 illustrates for example, how Harry's model transitions through a tight turn for $\zeta = 5\pi/36 \text{rad}$. Figure 8(a) plots the two straight-line segments connected at an angle ζ. The solid line describes the road median, while the dashed line describes the actual trajectory executed by Harry's HCS model. The length of the initial straight-line segment is chosen to be long enough (150m) to eliminate transients by allowing the model to settle into a stable state. This is equivalent to allowing the vehicle to drive on a straight road for a long period of time before the tight turn

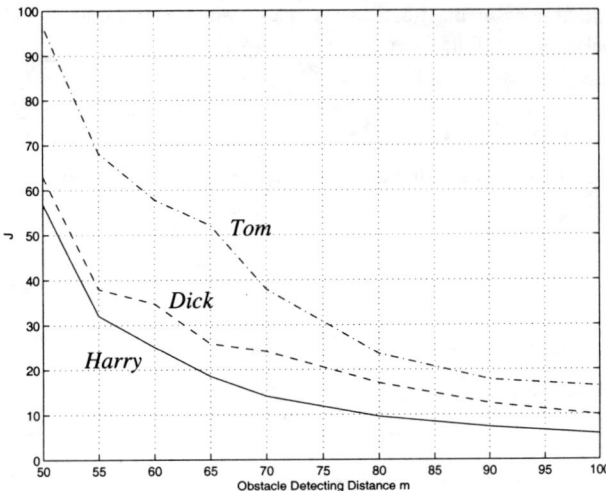

Fig. 7: Distance loss as a function of obstacle detection distance.

appears in the road. Figure 8(b) plots the lateral offset from the road median during the tight-turn maneuver. Here, Harry's model maximally deviates about 8m from the road center. Both before and after the turn, the lateral offset converges to zero. Figure 8(c) plots the commanded steering angle for Harry's HCS model, and Figure 8(d) plots the corresponding change in velocity.

Now, define the maximum lateral offset error corresponding to a tight turn with angle ζ to be ρ. We want to determine a functional relationship between ρ and ζ for a given HCS model. First, we take N measurements of ρ for different values of ζ, where we denote the ith measurement as (ζ_i, ρ_i). Then, similar to Section 4.2, we assume a polynomial relationship between ρ and ζ such that,

$$\rho_i = \alpha_p \zeta_i^p + \alpha_{p-1} \zeta_i^{p-1} + \ldots + \alpha_1 \zeta_i + \alpha_0 + e_i \quad (39)$$

The least-squares estimate of the model $\hat{\alpha}$ is given by,

$$\hat{\alpha} = (\hat{\zeta}^T \hat{\zeta})^{-1} \hat{\zeta} \hat{\rho}, \text{ where,} \quad (40)$$

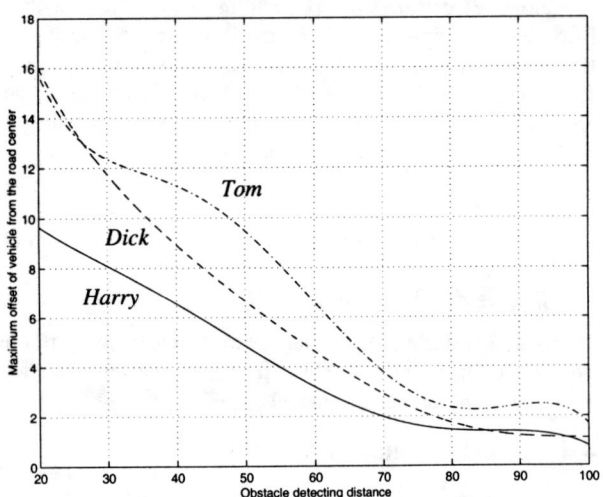

Fig. 6: Max. lateral offset for Dick's, Tom's and Harry's model.

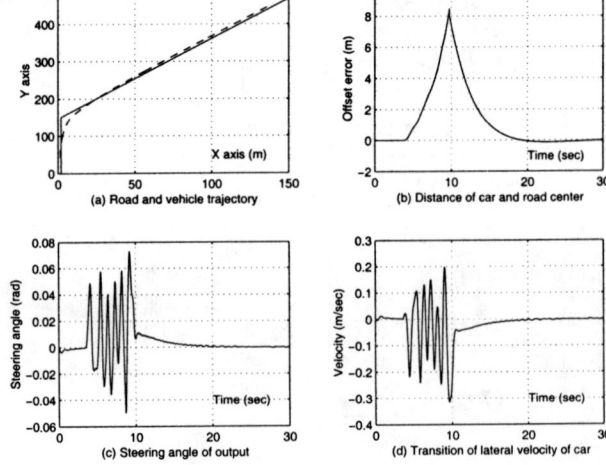

Fig. 8: Example model driving behavior through a tight turn.

$$\hat{\rho} = [\rho_1, \rho_2, ..., \rho_N]^T \quad (41)$$

$$\hat{\zeta} = \begin{bmatrix} \zeta_1^p & \zeta_1^{p-1} & \cdots & \zeta_1 & 1 \\ \zeta_2^p & \zeta_2^{p-1} & \cdots & \zeta_2 & 1 \\ \vdots & \vdots & \vdots & \vdots & \vdots \\ \zeta_N^p & \zeta_N^{p-1} & \cdots & \zeta_N & 1 \end{bmatrix} \quad (42)$$

$$\hat{\alpha} = [\alpha_p, \alpha_{p-1}, ..., \alpha_0]^T \quad (43)$$

For ζ ranging from $-4\pi/9\,\text{rad}$ to $4\pi/9\,\text{rad}$ and assuming a fifth-order model ($p = 5$), we arrive at the following estimate for Harry's model,

$$\rho = 2.78\zeta^5 - 0.584\zeta^4 - 0.599\zeta^3 - 4.286\zeta^2 + 11.68\zeta - 0.330 \quad (44)$$

and the following estimate for Dick's model,

$$\rho = -1.734\zeta^5 + 1.076\zeta^4 + 2.258\zeta^3 - 0.243\zeta^2 + 21.29\zeta - 0.679 \quad (45)$$

Equations (44) and (45) are plotted in Figure 9. For a given road width, we can determine the values of ζ for which each model stays on the road. For example, assume a road width of 20m. Then, the maximum allowable lateral offset is $\pm 10\text{m}$. From Figure 10 below, where the boundaries are explicitly drawn, we observe that Harry's model can execute tight turns from -0.65rad to 1.05rad, while Dick's model can only execute tight turns from -0.45rad to 0.48rad. Thus, Harry's model generates stable driving for wider range of conditions than does Dick's model.

We note that as a first-order approximation, we can define the tight-turning performance criterion J to be,

$$J = \alpha_1 \quad (46)$$

where α_1 is the linear coefficient in the fifth order model of equations (44) and (45), and smaller values of J indicate better performance. In that case,

$$J_{Harry} = 11.68 \text{ and } J_{Dick} = 21.29 \quad (47)$$

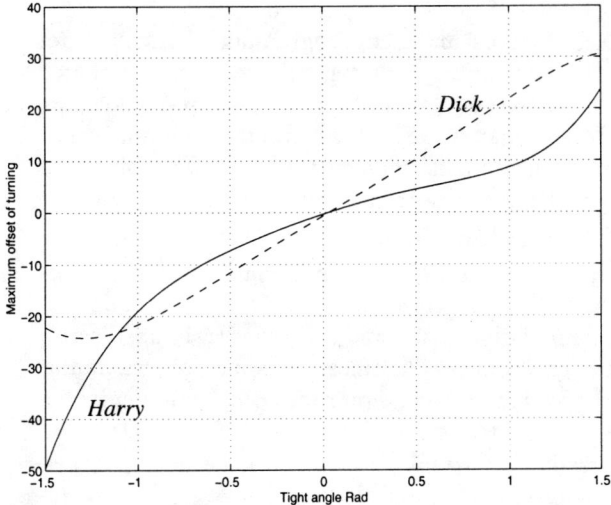

Fig. 9: Maximum lateral offset in tight turns for Dick's and Harry's model.

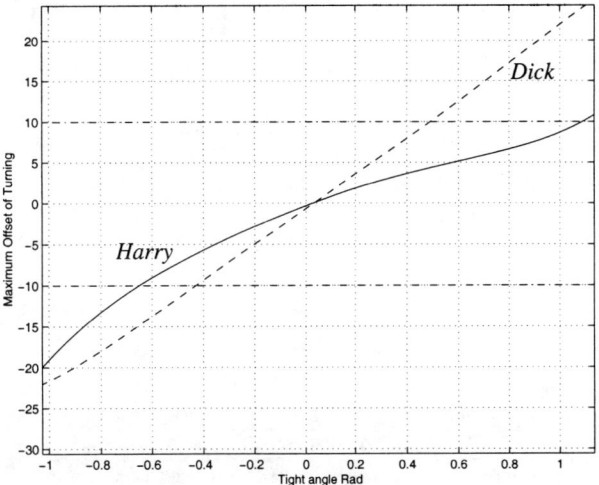

Fig. 10: Harry's model stays on the road for a greater range of tight turns.

6. Conclusion

Modeling human control strategy analytically is difficult at best. Therefore, an increasing number of researchers have resorted to empirical modeling of human control strategy as a viable alternative. This in turn requires that performance criteria be developed, since few if any theoretical guarantees exist for these models. In this paper, we develop several such criteria for the task of human driving, including obstacle avoidance and tight-turning performance criteria. We model human driving using the cascade neural network architecture, and evaluate the performance of driving models derived from different individuals using the developed performance criteria.

Acknowledgments

This work is supported in part by RGC Grant No. CUHK519/95E.

References

[1] M. Sugeno and T. Yasukawa, "A Fuzzy-Logic-Based Approach to Qualitative Modeling," *IEEE Transactions on Fuzzy Systems*, vol. 1, no. 1, 1993.

[2] U. Kramer, "On the Application of Fuzzy Sets to the Analysis of the System-Driver-Vehicle-Environment," *Automatica*, vol. 21, no. 1, pp. 101-7, 1985.

[3] M. C. Nechyba and Y. Xu, "Human Control Strategy: Abstraction, Verification and Replication," to appear in *IEEE Control Systems Magazine*, October 1997.

[4] M. C. Nechyba and Y. Xu, "Stochastic Similarity for Validating Human Control Strategy Models," *Proc. IEEE Conf. on Robotics and Automation*, vol. 1, pp. 278-83, 1997.

[5] M. C. Nechyba and Y. Xu, "Cascade Neural Networks with Node-Decoupled Extended Kalman Filtering," *Proc. IEEE Int. Symp. on Computational Intelligence in Robotics and Automation*, vol. 1, pp. 214-9, 1997.

[6] S. E. Fahlman, L. D. Baker and J. A. Boyan, "The Cascade 2 Learning Architecture," Technical Report, CMU-CS-TR-96-184, Carnegie Mellon University, 1996.

Towards Real-time Robot Programming by Human Demonstration for 6D Force Controlled Actions

Qi WANG and Joris DE SCHUTTER

Department of Mechanical Engineering, Katholieke Universiteit Leuven, Belgium
E-mail: Qi.Wang@mech.kuleuven.ac.be

Abstract

An approach for real-time robot programming by human demonstration for 6D force controlled actions is presented. A human operator utilises a joystick to guide a robot with a force sensor to execute a task including continuous contact between a manipulated object and an un-modelled environment. During the demonstration, the position, velocity and force of the manipulated object as well as the human commands via the joystick are recorded. In real-time, the recorded information is translated into a textual robot program providing more robust execution in the presence of uncertainties. This approach has three main features 1) on-line control type adjustment; 2) automatic subtask termination; 3) real-time program generation. Experiments show the potential industrial applicability.

1 Introduction

This work is about *robot programming by human demonstration*. A human operator uses a joystick to guide a robot with a force sensor to execute a task including continuous contact between a manipulated object and an un-modelled environment. This process is called human demonstration during which the position, velocity, and force of the manipulated object as well as the human commands via the joystick are continuously recorded. As soon as a subtask demonstration is finished, the recorded information is, in real-time, processed, analysed, and translated into a textual robot program, which provides more robust execution in the presence of uncertainties. Fig. 1 shows the overview of this programming approach.

1.1 Background

Force controlled action [2], also called *compliant/fine/constrained motion* [7, 3], refers to manipulation tasks involving continuous contact between manipulator and environment, and during the execution of which the end-effector trajectory is modified by the occurring contact forces. A well-known example is inserting a peg into a hole (so-called *peg-into-hole* assembly). A robot with force measurement and feedback is called a *force controlled robot*.

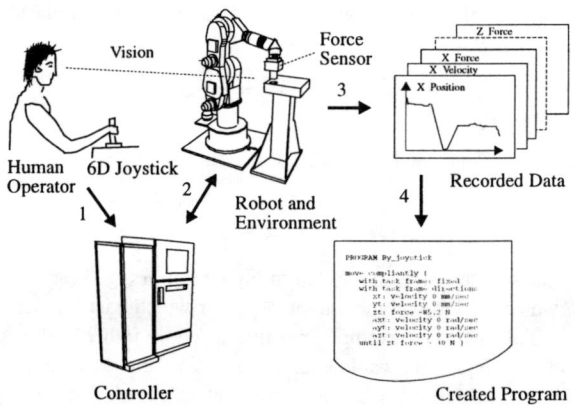

Figure 1: Real-time robot programming

Mason [7] proposes a *hybrid control* functional specification in which one can designate an appropriate control type, either position or force, to each of the coordinate axes of a so-called *task/compliance frame*. De Schutter and Van Brussel present an improved and operationalised *task frame formalism* [3] and a control approach based on *external force control loops* [4] closed around the robot positioning system. Recently, Bruyninckx and De Schutter [2] further synthesize the task frame formalism.

The aim of this research is to design intuitive, easy-to-use, flexible and fast programming strategies for force controlled actions which are usually executed under unexpected variations in workpiece location and/or geometry. For more robust program execution, in such a task, most of the subtask terminations are caused by force rather than position transients.

1.2 Previous Work

Human demonstration, mentioned in a lot of papers, has been utilised in visual understanding and grasping, skill transfer and controller training, discrete event and contact identification, robot programming, etc. This paper focuses on robot programming by human demonstration [5] for force controlled action.

Asada and Izumi [1] develop a method to generate a program for hybrid position/force control using a back-drivable robot guided by the human operator's hand directly. Delson and West [5] present a method in which an operator holds a particular device, that measures both the position and contact force, to perform the task by multiple demonstrations. Referring to their work, the authors have proposed an approach [10, 6] to derive programs for 3D translational motion.

The approach presented in this paper is different from the above work because of the real-time program generation and the 6D motion execution.

1.3 Problem Statement

Usually a human can not execute a task very consistently with a constant velocity or force, especially for unexperienced operators. For example, for the task of sliding the manipulated object on a table, it is not always easy to keep a constant contact force and a constant velocity. Of course, the more constant the force/velocity, the better the demonstration. In order to make the system easy-to-use for beginners, *Joystick Holding* is introduced to hold the joystick's output by pressing a particular button on the joystick as soon as the desired manipulation state is reached. After the current subtask demonstration is finished, re-pressing the button will turn off the holding state.

Joystick Holding highly improves the quality of human demonstration and liberates the operator when it is turned on. However a few hard problems appear when we extend the 3D translational motion to 6D motions. 1) The demonstration difficulty of some tasks; For instance, in the example of a vertex tool tracking a curved surface, it is very difficult, if not impossible, to demonstrate how the task frame follows the environment. If the operator does not use Joystick Holding, it is also hard to give a good demonstration. 2) The translation complexity; Because the signals used are position, velocity and force of the manipulated object, the techniques in [10] will be very complex for 6D motion.

In order to overcome these problems, a demonstration loop, called *Autopilot*, is closed around the original control system and the robot (Fig. 2) to support the following new features:

* **Control Type Adjustment**

As soon as Autopilot is turned on, it will hold the joystick's outputs while memorising and analysing the current manipulation state and the human joystick commands. If the current control types somewhat conflict with the environment, Autopilot will adjust the control types on-line to follow the environment.

* **Automatic Subtask Termination**

When Autopilot is on, the robot will automatically stop if the manipulated object gains a new contact or loses an existing contact.

* **Real-time Translation**

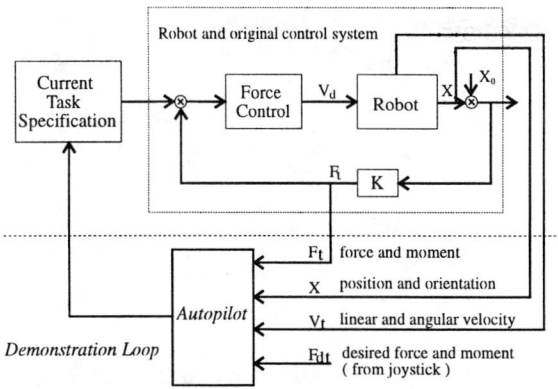

Figure 2: Overview of Autopilot

When the present subtask demonstration is finished, Autopilot will translate the recorded information into an explicit textual robot program in real-time.

On-line control type adjustment, automatic subtask termination, and real-time program translation are presented in Sections 2, 3, 4, respectively. The experimental results are given in Section 5.

2 Control Type Adjustment

The following simplifications are made: The world is ideal (the environment, robot and manipulated object are rigid and smooth; the controller, force sensor and joystick have ideal performances and no dynamics), and hence only *kinetostatic* concepts can be used (*twists* — instantaneous angular and linear velocities of the rigid manipulated object; *wrenches* — static contact forces, i.e., moment and linear forces; and the reciprocal relationship between both to describe any interaction [2]).

Our approach is based on motion-based damping force controllers — that transform force errors between the desired and measured forces into corrective velocities, e.g. the external force loop controller [4].

For demonstrating various tasks, the twist vector space consists of 0 task frame directions and the wrench vector space consists of 6 task frame directions. In each direction, the control type is *force*. The joystick output is transformed into the desired force in the task frame.

2.1 The Input of Autopilot

The input (Fig. 2), with respect to the task frame, includes: 1) measured forces and moments. 2) measured positions and orientations (of the task frame) with respect to the base frame. 3) measured linear and angular velocities. 4) desired forces and moments (from joystick).

2.2 The Relation of the Measurements

Applying a motion-based damping controller, we have: $V_t = k_a \times (F_{dt} - F_t)$ and $\omega_t = k_p \times (M_{dt} - M_t)$

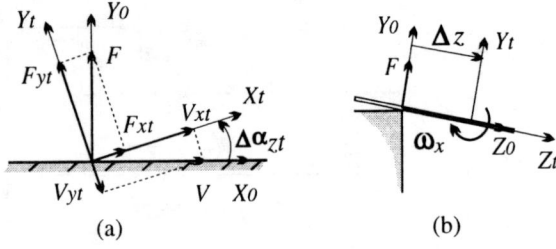

Figure 3: The polar and axial tracking

1) F_t/M_t: measured force/moment in the task frame. 2) V_t/ω_t: measured linear/angular velocity in the task frame. 3) F_{dt}/M_{dt}: desired force/moment in the task frame. 4) k_a/k_p: negative coefficient of the controller.

After Autopilot is turned on, F_{dt} and M_{dt} will be constants. According to the magnitudes of the measured forces/moments, the operator's control intent is interpreted as: if $F_t > F_{dt}/2$ or $M_t > M_{dt}/2$, *force* control; else *velocity* control.

2.3 Constraints

Constraints are divided into *unilateral* and *bilateral* constraints. A unilateral constraint can occur only in one direction of a task frame axis; A bilateral constraint occurs in both directions and a very small displacement may cause a sign change of the contact force/moment. Consequently, the stable velocity can occur only with a unilateral constraint.

2.4 Tracking

The purpose of tracking control is to let the task frame follow the environment continuously. Thus, tracking applies to both rotation and translation as shown below.

Because the normal direction(Y_0) of the environment does not align with the task frame Y_t (Fig. 3a), the contact force results in a force component F_{xt}. Thus, a polar tracking control around the Z_t axis is needed to align the task frame Y_t with the normal direction and make $F_{xt} = 0$. The tracking error is given by $\Delta\alpha_{zt} = \arctan(F_{xt}/F_{yt})$ (or $-\arctan(V_{yt}/V_{xt})$) [3].

In Fig. 3b, because the contact point deviates a certain distance from the task frame origin, the contact force in the Y_0 direction results in a coupled moment around the task frame X_t. Therefore, an axial tracking control along the Z_t axis is needed to move the origin of the task frame to the contact point and make $M_{xt} = 0$. The tracking error is given by $\Delta z = (M_{xt}/F_{yt})$ (or $-(V_{yt}/\omega_{xt})$) [9].

2.5 Control Type Identification

Considering all possible qualitative combinations of the measured velocity, force and desired force memorised as soon as Autopilot is turned on, 15 cases A1–A15 are listed for the control type in axial directions

(and 15 very similar cases P1–P15 exist for polar directions [9]).

	V_t	F_t	F_{dt}	Control type	Control type ID condition
A1)	0	0	$\pm f_d$	incorrect	
A2)	0	$\pm f$	$\pm f_d$	force $\pm f$	
A3)	0	$\mp f$	$\pm f_d$	incorrect	
A4)	$\mp v$	0	$\pm f_d$	velocity $\mp v$	
A5)	$\pm v$	0	$\pm f_d$	incorrect	
A6a)	$\mp v$	$\pm f$	$\pm f_d$	force $\pm f$	if $f > f_d/2$; tracking clue (TC)
A6b)				velocity $\mp v$	else
A7)	$\pm v$	$\mp f$	$\pm f_d$	incorrect	
A8a)	$\pm v$	$\pm f$	$\mp f_d$	force $\pm f$	if determined by tracking ID; TC
A8b)				velocity $\pm v$	default; TC
A9a)	$\pm v$	$\pm f$	$\pm f_d$	velocity $\pm v$	if determined by tracking ID; TC
A9b)				force $\pm f$	default; TC
A10a)	0	0	0	track	if this axial track is identified
A10b)				velocity 0	else (default)
A11a)	0	$\pm f$	0	incorrect	
A11b)				force 0	default; for bilateral constraint
A12	$\pm v$	0	0	incorrect	
A13)	$\mp v$	$\pm f$	0	incorrect	
A14a)	$\pm v$	$\pm f$	0	velocity $\pm v$	if determined by tracking ID; TC
A14b)				force $\pm f$	else-if determined by tracking ID; TC
A14c)				track	else-if this axial track is identified; TC
A14d)				force 0	else (default); TC
A15)	≈ 0	$\pm f$	0	force 0	for bilateral constraint

Under the simplifications in Section 2, cases A1,A3,A5,A7,A11a,A12,A13 can not happen. The control type of cases A2,A4,A10b,A15 is straightforward. In cases A8,A9,A14, a measured velocity and force appear at the same time along one axis with the same signs. This suggests that tracking is needed to solve the conflict between the task frame and the environment. We call this "tracking clue (TC)". Applying the result in Section 2.2, case 6 can be divided into A6a (a TC because a velocity exists along a force controlled direction) and A6b (this can be caused by sliding under friction, hence case A6b remains undecided about TC). The final control type will be determined during the tracking identification described in the next section. At this moment, the default will be used if multiple possibilities exist.

2.6 Tracking Identification

In the task frame formalism, a tracking control always involves 3 axes: the two axes containing the coupling and the tracking control axis. For polar tracking (for axial tracking, see [9]), there are 5 tracking identification templates:

Axis	V_t/ω_t	F_t/M_t	F_{dt}/M_{dt}	Control Type	Case
i	any	$\pm f_i$	$\pm f_{di}$	force $\pm f_i$	A6a, A2
j	$\pm v_j$	$\pm f_j$	f_{dj}	velocity $\pm v_j$	A8b, A9a
k	0	0	0	track (on velocity)	P10a
i	$\mp v_i$	any	$\pm f_{di}$	velocity $\mp v_i$	A6b, A4
j	$\pm v_j$	$\pm f_j$	f_{dj}	force $\pm f_j$	A8a, A9b
k	0	0	0	track (on velocity)	P10a
i	any	$\pm f_i$	$\pm f_{di}$	force $\pm f_i$	A6a, A2
j	$\pm v_j$	$\pm f_j$	0	velocity $\pm v_j$	A14a
k	0	0	0	track (on velocity)	P10a
i	$\mp v_i$	any	$\pm f_{di}$	velocity $\mp v_i$	A6b, A4
j	$\pm v_j$	$\pm f_j$	0	force $\pm f_j$	A14b

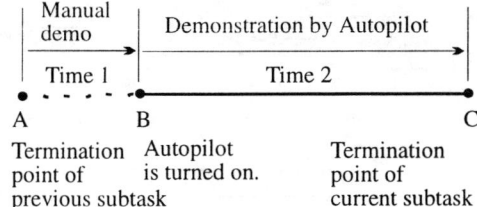

Figure 4: A subtask demonstration process

k	0	0	0	track (on velocity)	P10a
i	any	$\pm f_i$	$\pm f_{di}$	force $\pm f_i$	A6a
j	$\mp v_j$	0	$\pm f_{dj}$	velocity $\mp v_j$	A4
k	0	0	0	track (on velocity)	P10a

i and j are axial and k is polar. The italic control type results from the tracking identification.

If the operator wants a pen to slide on a table without aligning the table normal direction, he may turn on a *Orientation Holding* button which will disable the polar tracking identification and adjustment. Similarly for the axial tracking.

2.7 Type Adjustment Procedure

1) When Autopilot is turned on, the measured velocity, force and desired force are memorised. 2) Applying A1–A15 and P1–P15, determine the control type in each task frame axis. The default is used in case of multiple possibilities. 3) If there are tracking clues, the tracking identification templates are used to detect and determine the control type in each of the directions. 4) Finally the identified control types and the memorised parameters are applied to the current subtask in real-time.

3 Automatic Subtask Termination

Gaining/losing a contact corresponds to a sudden increase/decrease in the reaction force in a certain direction, as well as, dually, to a sudden decrease/increase of the velocity in this direction [2].

Based on the control types adjusted, and by comparing the measured and desired velocities and forces, Autopilot continuously monitors the manipulation state to detect any sudden changes. Contact-gain is detected only in velocity controlled directions, and contact-loss only in force controlled directions. When a contact-gain/contact-loss happens, Autopilot stops the current subtask demonstration and memorises the termination condition. Then the human operator may proceed to demonstrate the next subtask. For a termination condition concerning distance or time, the operator just turns off Autopilot manually when the desired distance or time has passed.

4 Real-time Translation

The following command modes are supported: 1) Point-to-point free space motion. 2) Holding the positions and the orientations in free space until a certain time has passed. 3) Compliant motion command with velocity control. 4) Compliant motion command in which all velocity controls are converted into force controls. 5) Keeping a contact until a certain time has passed.

Fig. 4 shows how to demonstrate a subtask using Autopilot. Table 1 indicates how to choose the command modes based on various combinations of the distance AB and BC in Fig. 4. After control type adjustment and automatic subtask termination, Autopilot will use the recorded information to generate a textual command corresponding to the demonstrated subtask in real-time.

In this approach, a new subtask is marked by turning on Autopilot. Similarly, a subtask end is marked by an automatic termination or manual turning off Autopilot.

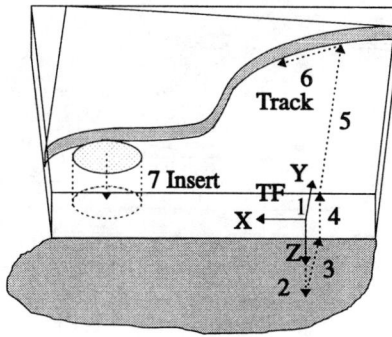

Figure 5: Peg-into-hole assembly

5 Experiments

The experimental system consists of a KUKA361 robot, a SCHUNK force sensor, a force control software named COMRADE [8], a digital joystick and a UNIX host. The operator can switch between several joystick work modes on-line: rotational-operation on/off; translational-operation on/off; dominant mode on/off — only the maximum of 6 joystick readings is valid.

Autopilot, written in parallel C, works at about 17–20 Hz (except the automatic subtask termination part at 80–100 Hz). It generates textual programs in the task specification language, supported by COMRADE, which is controller-independent and is based on the task frame formalism.

Some typical force controlled tasks as in [2] have been tested and executed. Here the first example is a peg-into-hole assembly (Fig. 5). The peg, with a 3mm/45deg chamfer, has a 50mm diameter and a 0.15mm clearance with the hole. The task includes the following subtasks: 1) Go to the initial position. 2) Go down until a contact is made. 3) Slide to the

AB	BC	Free space?	Command type during AB	Command type during BC
$\neq 0$	$\neq 0$	Yes	Point-to-point free space motion	Compliant motion using velocity control
		No	Compliant motion using velocity control	
$\neq 0$	0	Yes	Point-to-point free space motion	N/A
		No	Compliant motion using force control for velocity direction	N/A
0	$\neq 0$	Yes	N/A	N/A
		No	N/A	N/A
0	0	Yes	Holding position and orientation until time 2 passed	
		No	Keeping contact until time 2 passed	

Table 1: The command mode choice based on various combinations of the distance AB and BC in Fig. 4

front until the workpiece is hit. 4) Move upwards until the top edge is crossed. 5) Slide on the workpiece to the front until a contact is made. 6) Follow the curved wall to the left until the hole is reached. 7) Insert the peg. 8) Extract the peg. 9) Move a certain distance upwards.

First the operator guides the robot to the initial position, and then turns on the *programming* mode by pressing a button on the joystick. Then the operator presses the joystick downwards to cause the robot to move. When the instantaneous manipulation state is desired, the operator turns on Autopilot (and hence becomes free). After the control type identification, Autopilot adjusts the control type on-line while displaying the control types on the screen. After the peg contacts the table, Autopilot automatically stops meanwhile a message appears on the screen: force-contact-gain -Z. Then Autopilot generates this subtask program in much less than 1 ms.

Similarly, the operator continues until the last subtask is finished. During the programming, if a mistake/error is made, the operator may turn off the programming mode causing the current subtask to be cancelled. After the operator "manually" guides the robot back to the starting point of the subtask, he may turn on the programming mode again to re-demonstrate this subtask. Autopilot is only available when the programming mode is on. Programming the first example takes about 1 minute. After translation, for example, subtask 2, 6, 7 are listed as follows:

```
/* subtask 2 */
move compliantly {
xt :    velocity 0 mm/sec
yt :    velocity 0 mm/sec
zt :    velocity 36.8 mm/sec
axt :   velocity 0 rad/sec
ayt :   velocity 0 rad/sec
azt :   velocity 0 rad/sec
until zt force > -30 N }
/* subtask 6 */
move compliantly {
xt :    velocity 20.2 mm/sec
yt :    force -20.6 N
zt :    force -38.4 N
axt :   velocity 0 rad/sec
ayt :   velocity 0 rad/sec
azt :   track (on velocity)
until zt force > -3 N }
/* subtask 7 */
```

```
move compliantly {
xt :    force 0 N
yt :    force 0 N
zt :    force -95.3 N
axt :   force 0 Nm
ayt :   force 0 Nm
azt :   velocity 0 rad/sec
until zt force < -85.8 N }
```

In subtask 7, the operator inserts the peg until the bottom is reached, then turns on Autopilot and turns it off while keeping a big force. In this case, the command mode is (see table 1, with $AB \neq 0, BC = 0$, and not in free space): compliant motion using force control for velocity direction (a 90% contact force will be used as the termination threshold force). This assembly program has been tested on the robot and the peg is successfully inserted, even under a changed environment (displaced workpiece) because most terminations are caused by force transients. In fact, each subtask has been tested partly after Autopilot is on.

However, on the other hand, if the clearance between the peg and the hole is not small, and if the orientations of the peg and the hole match very well during demonstration, the identified control type in axt, for example, is often *velocity 0* rather than *force 0*. This is a problem of this approach, which is caused by the absence of an explicit model. When the variation of the hole orientation is not small, the *velocity 0* will cause a new execution to fail. There are several possible solutions. One is that continuous control type monitoring is used during subsequent executions. In fact, this involves qualitative on-line geometric identification. Another is that the operator might tell Autopilot an abstract word—"insert" by mouse, for instance, when he is free after turning on Autopilot. This will result in a very extensive user interface. Maybe the fastest solution is let the operator learn how to "explicitly" demonstrate the insertion clue for Autopilot by understanding how this un-modelled approach works. However, this is not suitable for an unexperienced user.

For the above example, another programming idea is tested: after guiding the peg very near to the hole "manually" from the initial position, the operator re-

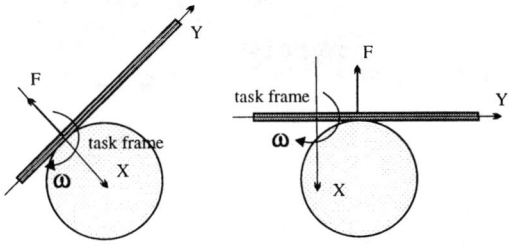

Figure 6: Rolling

leases the joystick, and turns Autopilot on and off, then demonstrates the insertion. After translation (see table 1, with $AB \neq 0, BC = 0$, and in free space), we get a point-to-point free space motion command from the initial position to the hole and a compliant inserting command. During the real execution, the peg is inserted successfully which means precise point-to-point free space motion is also supported in our approach. However, the variation of the hole location should be limited to the chamfer width.

The second example consists of rolling a narrow plate on a curved surface shown in Fig. 6. Autopilot generates this subtask as follows after the demonstration (see table 1, with $AB \neq 0, BC \neq 0$, and not in free space):

```
move compliantly {
  xt  :  force -30.6 N
  yt  :  track (on velocity)
  zt  :  velocity 0 mm/sec
  axt :  velocity 0 rad/sec
  ayt :  velocity 0 rad/sec
  azt :  velocity -0.26 rad/sec
  until relative time > 3.22 sec }
```

The translation is correct according to the task specification language. So far the re-execution of the generated task program is possible except for axial tracking control which is currently being implemented in COMRADE.

In addition, no force reflection is available on the joystick. If the operator wants a precise force, he may check the occurring forces on the screen before turning on Autopilot.

6 Summary

Real-time robot programming by human demonstration for 6D force controlled actions is presented. A human operator utilises a joystick to demonstrate a force controlled task in an un-modelled environment. During the demonstration, the position, velocity and force of the manipulated object and the human joystick commands are monitored by Autopilot which not only adjusts the control type on-line but generates the textual program in real-time. Two examples with several subtask programs are given. Experiments show the potential industrial applicability.

The main features are 1) semi-automatic demonstration; 2) real-time textual program generation; 3) full 6D compliant motion programming addressing the axial tracking; 4) five supported command modes (compliant motion using velocity control, precise point-to-point free space motion, ...) providing more robust execution in the presence of uncertainties.

Future work includes 1) testing a variety of force controlled actions as in [2] to refine this system; 2) extending COMRADE to support the axial tracking control completely; 3) designing a good user interface. Finally, this system should be integrated with our *programming by speech* system and visual system to give the user more options.

Acknowledgements

This work has been sponsored by the Belgian State Programme IUAP-50, IUAP 4/24. The first author has been supported by a K.U.Leuven Doctoral Fellowship.

Constructive comments and help by Prof. H. Van Brussel, H. Bruyninckx, W. Witvrouw, S. Graves, S. Dutré and other colleagues are gratefully acknowledged.

References

[1] H. Asada and H. Izumi, "Automatic program generation from teaching data for the hybrid control of robots," *IEEE Trans. on Robotics and Automation*, vol. 5, no. 2, pp. 163–173, April 1989.

[2] H. Bruyninckx and J. De Schutter, "Specification of force-controlled actions in the task frame formalism – a synthesis," *IEEE Trans. on Robotics and Automation*, vol. 12, no. 4, pp. 581–589, 1996.

[3] J. De Schutter and H. Van Brussel, "Compliant robot motion I: A formalism for specifying compliant motion tasks," *Int. J. of Robotics Research*, vol. 7, no. 4, pp. 3–17, 1988.

[4] J. De Schutter and H. Van Brussel, "Compliant robot motion II: A control approach based on external control loops," *Int. J. of Robotics Research*, vol. 7, no. 4, pp. 18–33, 1988.

[5] N. Delson and H. West, "Robot programming by human demonstration," In *Proc. of Japan/USA Symp. on Flexible Automation*, pp. 1387–1397, 1992.

[6] S. Graves, Q. Wang, W. Witvrouw, and J. De Schutter, "An environment for compliant motion programming by human demonstration," In *Proc. of IEEE/RSJ Int. Conf. on Intelligent Robots and Systems*, pp. 1579–1585, 1996.

[7] M. Mason, "Compliance and force control for computer controlled manipulators," *IEEE Trans. on Systems, Man, and Cybernetics*, vol. 11, no. 6, pp. 418–432, 1981.

[8] P. Van de Poel, W. Witvrouw, H. Bruyninckx, and J. De Schutter, "An environment for developing and optimising compliant robot motion tasks," In *Proc. of Int. Conf. on Advanced Robotics*, pp. 713–718, 1993.

[9] Q. Wang and J. De Schutter, "6d compliant motion programming by human demonstration in realtime," Technical report, Department of Mechanical Engineering, Katholieke Universiteit Leuven, Belgium, February 1997.

[10] Q. Wang, J. De Schutter, W. Witvrouw, and S. Graves, "Derivation of compliant motion programs based on human demonstration," In *Proc. of IEEE Int. Conf. on Robotics and Automation*, pp. 2616–2621, 1996.

Experiments in Synthetic Psychology for Tactile Perception in Robots: Steps Towards Implementing Humanoid Robots

Davide Taddeucci and Paolo Dario

ARTS Lab - Scuola Superiore Sant'Anna
via Carducci 40 - 56127 Pisa, Italy
Phone: +39-50-883400, Fax: +39-50-883402
E-mail: {davide, dario}@arts.sssup.it

Abstract

A robotic system which imitates the development process of stable grasping in infants is presented. The sensing devices of the system are tactile and visual sensors based on anthropomorphic design. The sensory data are processed and fused by the internal learning system based on artificial neural networks implementing the psychological aspects of the reinforcement learning paradigm. The design of the robotic system is modelled on the principles of synthetic psychology in order to obtain a human-like behavior, an important step towards "humanoids". Experiments show that the robot is able to find the best procedure to hold an object using the learned force with an average of 9 successful grasps over 10 trials.

1.1 Introduction

Recent studies in robotics and artificial intelligence have pointed out the limits of the traditional analytical approach: tasks like motor-sensory co-ordination, recognition and manipulation, which are natural and simple for humans, seem to be virtually unsolvable using current control methods. On the other hand, these human-like capabilities are crucial for those robots, like the service robot or the personal robot, whose main mission is to interact with humans in not-structured environments [1].

According to these considerations, the anthropomorphic approach to robot design and control seems particularly attractive: in fact, the ultimate goal of this approach should be to have anthropomorphic hardware controlled via anthropomorphic software. The design of anthropomorphic hardware is based on biological actuators and sensors (e.g. tactile, visual and olfactive) while the anthropomorphic software takes its inspiration from the replication of a simplified model of the human central nervous system (neural networks or NNs). The task to be performed by this software is essentially to integrate different sensory modalities in order to obtain new motor and cognitive behaviors using a dynamic, plastic and adaptive learning process.

The interest towards anthropomorphic solutions in robotics is demonstrated by increasing research activity on the development of humanoid robots. For example, a Symposium on Humanoid Robots [2] addressed specifically many different aspects of the design and development of anthropomorphic robots [3][4][5]. In our laboratory, research on anthropomorphic robots began first with the development of anthropomorphic hardware, such as skin-like tactile sensors and retina-like visual sensors: we performed experiments on tactile perception processed by NNs[6][7]; later, we investigate difficult tasks, such as cognitive [8] and the motor tasks [9] by fusing tactile and visual sensing at low level. While in [9] we presented a general model of artificial perception and sensory-motor co-ordination which included low level processing (for "unconscious" motor action) and high level processing (for tasks requiring higher levels of conscious cognitive effort), in the present paper we extend the low-level anthropomorphic approach to the case of more complex grasping procedures. The concept is to try to imitate the development process of stable grasping in infants. To this aim, at least two possible approaches can be followed according to an anthropomorphic perspective: the bottom-up approach, which tries to replicate the complex architecture of the cortical sensory-motor areas involved in the generation of the motor behavior, and the top-down approach, which makes use of psychology address. In our laboratory, we are pursuing both approaches. A separate paper [10] describes the first approach. In this paper we face the problem of sensory-motor co-ordination in manipulation by means of the principles of synthetic psychology. The assumption is that it is much more difficult to try to guess internal structure from the observation of the behavior than it is to create the structure that realizes the behavior itself.

2. Synthetic psychology & anthropomorphism

The term "synthetic psychology" was introduced by V. Braitenberg in 1984 [11] to describe the results of a series of experiments made on virtual and simple machines (the "vehicles") whose structure was in part inspired by the nervous system. From the point of view of an external observer, the movement of the vehicles exhibited a sort of human teleology: in fact, the vehicles behaved like they had complex goals and emotions (that the author called "love", "aggression", "foresight" and "optimism"). In this way, the author had induced a sort of "intention" (a "psychology") in the behavior of the small machines. In fact, in our laboratory, we implemented some of Braitenberg's concepts in a swarm of miniature robots [12].

The achievement of human-like psychology may be an important aspect to obtain true anthropomorphic robots that have to interact with humans. Human-like behavior (in addition to human-like shape) would help reducing the impact of the new synthetic creatures on human users.

According to the exact definition of anthropomorphism, the "internal" design of such a robot would have to be as adherent as possible to the human central nervous system, perhaps the most economic and efficient existing system dealing with knowledge, perception and intelligence. Encouraging results in this paradigm have been achieved by Fagg and Arbib [13], who modelled the role of the primate pre-motor cortex in triggering movements on the basis of visual stimuli and the performance curves obtained were qualitatively similar to those observed in animals.

Rucci, Tononi & Edelman [14] successfully synthesised the barn-owl audio-visual behavior reproducing the optic-tectum, a part of sensory-cortex of the rapacious.

The work described in this paper follows a simplified approach to the objective of replicating the behavior of complex human sensory-motor cortical areas involved in tactile perception. In fact, rather than simulating the low-level neuro-physiological aspects of such behavior (as in the previously cited papers), we only pursue a "mild" anthropomorphic approach to tactile perception by simulating the high level behavior of humans using NNs. In order to investigate the feasibility of this approach, we have performed experiments aimed at developing a particular robotic motor-behavior similar to the procedure of learning stable grasp in the child.

According to the observations of Piaget [15], during the development of psycho-sensory-motor co-ordination, the infant does not know an a-priori motor strategy suitable to lift up successfully an object; however he/she can learn it while interacting with the object itself. For example, if the force applied to the object by the hand is too light, the object will drop down while, if too strong, the object will be damaged. The child will verify the results of his/her efforts using essentially vision and touch: when the object fails in being lifted up, the infant is conscious of the failing since he/she sees the object onto the ground and feels it slipping out of the hand. According to the result of the applied motor procedure, the neuro-physiological hypothesis (confirmed in animal experiments) is that the synaptic weights of the neural structures involved in the generation of successful actions increase whereas they decrease in case of faults. In the experiment described below, we want to imitate almost exactly this learning scheme: visual and tactile sensors constitute the sensory system whereas NNs are the learning substrate.

2.1 Implementation of the synthetic behavior

The implementation of low level processing and learning is based on a *neural* approach integrated with the *reinforcement learning* paradigm.

The neural approach to low level processing is well suited for an anthropomorphic solution, for a number of reasons [16]:

1. the processing performed by NNs is achieved thanks to the connections in the nodes of the network, just as low level processing is performed in humans by physical connections between neurones;
2. information is processed by NNs in parallel, as sensory data are perceived simultaneously and fused in human multi-sensory perception;
3. one of the main features of NNs are the plasticity and adaptability of the structure, where connections among nodes can be modified dynamically, as they are reinforced or cancelled in the human brain;
4. NNs are generally redundant and guarantee the production of output values even if some units or connections are missing, by applying the same solution applied by Nature in animals, based on redundancy;
5. finally, the concept of learning is intrinsically related to artificial NNs, just as learning is intrinsic in the human mental processing.

Human beings can learn through a number of different paradigms; at present, though, a unified understanding of the problem has not been reached yet. Computational models of learning have been proposed which try to replicate some aspects of the animal learning process. Essentially, a rough taxonomy of computational learning comprises of *Supervised Learning*, *Unsupervised Learning* and *Reinforcement Learning*. *Supervised learning* is achieved when a teacher is available who specifies the input and the expected output the system must learn. During the *Unsupervised Learning*, instead, a system discovers regularities in the training phase and produce clusters of similar data in a self organised fashion. Finally, in the *Reinforcement Learning* the system has to emulate an expected behavior without a direct specification of the output but having as feedback from the environment only a scalar value representing how good was the action performed in response to a particular input.

We selected reinforcement learning as the paradigm for the implementation of the proposed model of learning, because of its strong analogy with the anthropomorphic psychological mechanism of correction or enhancement of behaviors.

Reinforcement Learning is "learning about, from, and while interacting with an environment in order to achieve a goal" [17]; in other words, it is a relatively direct model of the learning of humans and animals plausible both from a neuro-physiological (Hebbian rule [16]) and from a psychological point of view ("The law of effect" of Thorndike [18]).

In robotics, problems are sometimes defined in terms of abstract goals rather than sequential pair <perception,

action>. Reinforcement procedures usually fit this type of learning since they require only scalar values measuring the relative desirability of the state goodness [19]. The dynamic achieving of a robotic goal could be seldom attainable via an immediate reward (positive value), when the system has a good behavior, and a punishment (negative value), in case of wrong actions, since intermediate results are often difficult to quantify and only the initial and final phases are well defined.

3. The experiment

3.1 Experimental Scenario

The setup we used for our experiments is depicted in fig. 1.

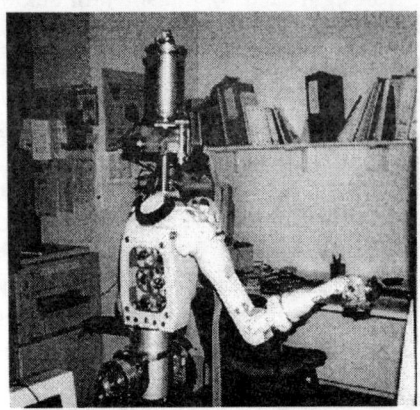

Fig. 2: The MOVAID robot

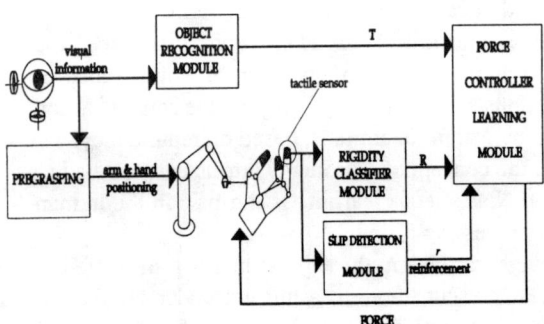

Fig. 1: Block diagram of the experimental setup

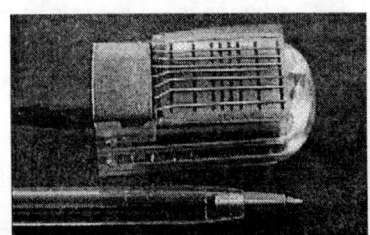

Fig. 3: The integrated miniature fingertip

The robotic platform used in the experiment is the MOVAID robot [20] (fig. 2), which includes anthropomorphic sensors (visual and tactile). The visual sensor is a CCD retina-like camera [21] mounted in the head of MOVAID with a pan/tilt control. The tactile sensor is the KIST sensor [22], which includes a 8x8 tactile array (for sensing the distribution of normal contact force), a piezo-electric dynamic sensor and a thermal sensor [fig. 3]. A characteristic of the tactile sensor array is its fovea-like geometry, leading to variable space resolution (1 mm at the center, 5 mm in the periphery).

The MOVAID manipulator comprises of an anthropomorphic 8 d.o.f. arm (DEXTER) and a 2 ½ d.o.f. prosthetic hand (MARCUS).

Typical "shopping bag" objects, including vegetables, fruits, cans and bottles have been used in the experiment.

3.2 Processing Algorithm

The following high-level algorithm describes the steps of the processing:

1) the image of the object to be grasped is captured with the camera;
2) the visual information is used by a PRE-GRASPING module [23] which, after extracting some features from the visual scene using a neuro-fuzzy module, calculates the position of the arm and of the hand suitable to pre-shape in a "good" way (according to a fuzzy definition);
3) in parallel, a visual module (OBJECT RECOGNITION MODULE) extracts four different shapes from the image (in the specific experiment we performed, the database we selected includes the following objects: bottle, large box, small box, circular fruits);
4) after an explorative curling movement of the fingers around the object, the RIGIDITY CLASSIFIER MODULE estimates the hardness of the object, by assigning it a number from 0 (soft) to 3 (hard);
5) the FORCE CONTROLLER LEARNING MODULE
   (a) receives in input the object type (T) and the rigidity score (R), while in output produces a scalar value representing the force of closure of the hand. Then,
   (b) the force computed by the neural network of the previous module is applied to the hand;
   (c) the SLIP DETECTION MODULE detects if the object is still in contact with the hand (reinforcement $r=1$) or if it has slipped down ($r=-1$) and then the reinforcement value r is backpropagated to the learning module;
   (d) the weights of the neural network of the learning module are updated according to the reinforcement learning rule;
6) The learning procedure starts again from the point 1) until a good grasping procedure is accomplished.

In the following sub-paragraphs, we illustrate in details the main modules of the algorithm.

3.3 Tactile Processing

The tactile data are used in the "rigidity recognition" neural network and in the "slip detection" module.

3.3.1 Rigidity classifier neural network

The task of this module is to classify the hardness of the object under tactile exploration. The subsystem learns how to determine the particular internal weights configuration by a supervised learning algorithm.

This module receives in input 4 subsequent tactile images acquired every 0.25 s during the explorative curling movements of the fingers around the object (duration: 1 s). From a qualitative analysis of a typical tactile sequence (fig. 4) the different behaviors of the tactile sensor for different objects is visible:
a) in case of soft object, the contact surface increases almost linearly, just like the pressure at the centre of the array;
b) in case of hard object, after the second tactile acquisition, images remain almost unchanged.

The neural network architecture is composed of 256 input nodes (64 tactile elements x 4 images), two levels of hidden layers (the first with 128 nodes and the following with 64 nodes), and an output level with 4 nodes. The training is performed with the classical backpropagation algorithm [24]. After training, each output node activates itself to classify the hardness of the explored object.

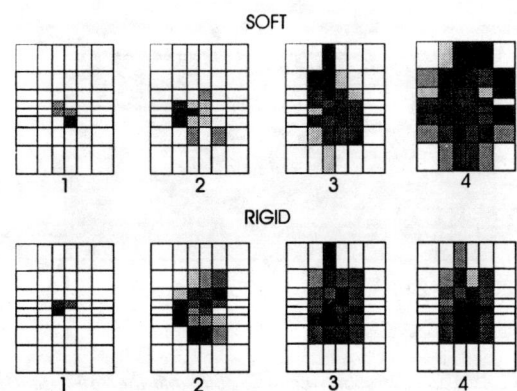

Fig. 4: Sequential acquisition of tactile data in soft and rigid objects

3.3.2 Slip detection module

The dynamic sensor is used in this module to detect the slipping of the object thus determining the reinforcement value (r) to be used in the force learning module. When the object in the hand is sliding off, this movement produces some micro-vibrations which however are also present for little adjustments of the object in the hand. A qualitative discrimination of the two different situations is thus needed. The criterion used to discriminate the slippage and to compute r is the following:

IF the micro-vibrations maintain themselves over a certain threshold continuously for a specified temporal range (both experimentally determined), THEN r=+1 ELSE r=-1.

In our experiment, the criterion adopted is useful only to detect a "dangerous" situation (object slipping), and is not intended to be a solution for the problem of "incipient slipping". Our approach is aimed at detecting the final situation of *object in the hand* or *object on the ground*, so the system cannot avoid the sudden falling of the object.

OBJECT TYPE	THRESHOLD	Temporal Range
Circle	128	0.7 s
Bottle	128	1.8 s
Large box	128	1.3 s
Thin box	128	0.7 s

Table 1: Threshold and temporal range values used to detect different object slippage

In Table 1, the values of threshold and temporal range determined experimentally are shown according to the different object type; the threshold value is coded with one byte and its average value (128) is sufficient to discriminate the 4 different shapes.

3.4 Visual Processing

Visual information is processed at an early stage in order to detect features relevant for further processing and integration with tactile data for learning. The module is able to classify the object during visual exploration according to four different pre-defined shapes (bottle, large box, thin box, circle).

Early vision processing can be divided into four phases:
1. space-variant image acquisition;
2. extraction of the foveal part of the image;
3. filtering and edge detection;
4. identification of the line approximating the edge segment focused onto the fovea.

The acquired grey scale image is filtered in order to reduce possible random noise with a smoothing algorithm; then edge detection is performed by means of a gradient-based algorithm.

The line approximating the edge is detected by applying the Hough method [9] on the edge image. The Hough method detects the parametric curves that cross the image points, and creates an accumulation array in which such curves are stored, together with a scoring value allowing the detection of the one best fitting the edge.

4. Results

Before performing the real learning procedure, the hand is set in a suitable position for grasping the object under examination. According to the definition of Iberall et al. [25], the hand has three different degrees of opening (tip, pad, palm). The output of the learning module is the force to be applied during the closure of the hand around the object. This force has to be "optimal" with respect to the aim of avoiding the object slippage onto the ground of the object itself.

The reinforcement learning paradigm has been implemented with NNs. The reinforcement value is a scalar value determined by the SLIP DETECTION MODULE: the value is +1 if the object is steadily grasped (as detected by the dynamic sensor in the fingertip); the value is –1 if it slips off. The internal configuration of the neural network will converge towards the one expressing the expected robotic behavior updating the synaptic weights of the learning module according to this value and repeating the entire learning scheme

The topology of the network is determined by 8 input nodes (4 for the TYPE OF THE OBJECT and 4 for the RIGIDITY), 6 hidden nodes and 4 output nodes. Each output node i represents a different degree of grasping force as measured by the tactile sensing array.

The learning rule chosen for the neural network is the classic backpropagation [2] modified according to the following reinforcement learning algorithm :

1) apply the input vector (T,R) to the neural network and compute the output value o_i of every output node i;
2) $a_i = o_i + rnd_i$, i = 1…p; p is the number of possible actions; o_i is the real output value of the network on unit i,; rnd_i is a small random value; a_i is the new value of the output node i;
3) choose the force$_{OUT}$ to apply to the hand according to the max value of a_i, i = 1 … p:
$$OUT = \max_i \{a_i\}$$
4) after applying force$_{OUT}$ to the hand, determine the reinforcement r by using the SLIP DETECTION MODULE;
5) update the weights according to the backpropagation rule considering as error vector the following one:

$$if \quad r = \begin{cases} +1 & e_i = \lambda_1[rnd_i *[(a_{OUT} - a_i)]^2 \\ -1 & e_i = \lambda_2[rnd_i *[1 - (a_{OUT} - a_i)]^2 \end{cases}$$

where λ_1 and λ_2 are real-valued constant learning parameters, with $0 < \lambda_2 < \lambda_1 < 1$;
6) repeat the entire learning procedure until a satisfactory robotic behavior is observed in grasping procedure.

The condition on the values of the constants λ_1 and λ_2 maximizes the effects of a right output specific node while introducing small negative changes in case of wrong behavior. In fact, in case of positive answer, we are sure about the right activation node, while in the opposite case we do not know which ones should have been active; so a mild punishment is introduced on all nodes.

The randomness introduced with the rnd_i factor allows the network to explore the search space, shaking the neural system from possible local minima (incorrect behavior).

It is important to underline that the correctness of the previous implementation of reinforcement learning is not demonstrated by a formal theory: the validity of this rule has been assessed only by experiments.

In the following tables, the average error of the learning network is shown during the first 500 grasping trials. The global (almost) monotonically decreasing trend of the learning curve, is visible in table 2.a.

Zooming in the range [200 … 250], as shown in in Table 2.b, the behavior of the curve appears to be more "randomly decreasing" than "monotonically decreasing"; this is due to the random multiplicative factor introduced in the learning rule.

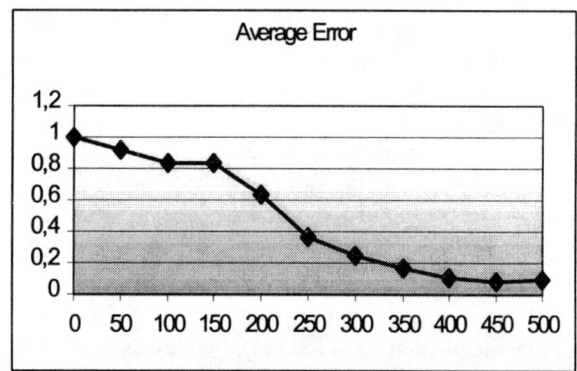

Table 2.a : Average error during the first 500 trials

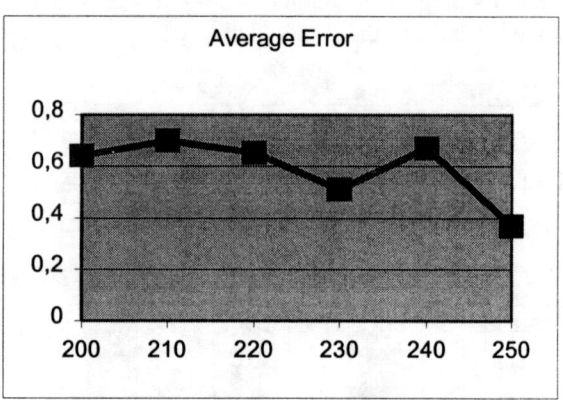

Table 2.b: Average error between the 200th and the 250th iteration

But, specifically, the robot system is able to find autonomously the appropriate procedure to hold an object using the learned force with an average of 9 successful grasps over 10 trials.

5. Conclusions and future work

In this paper, we described a robotic system that imitates the natural human-like behavior of motor strategy learning. The robotic system we have implemented takes a mild inspiration from biological structures, while the real anthropomorphism is only visible from a psychological perspective: an external observer looks at a robot that, after an initial randomness, learns from its faults to adopt the better strategy to firmly hold some objects (with a mean percentage of success of 87 grasps over 100 trials). The anthropomorphic behaviour is implemented following the reinforcement learning paradigm that, even if its main principle has generated controversies over the years, remains a considerable theory supported by many experimental trials. The neuro-biological mathematics of the reinforcement learning rule is very simple between small groups of our brain neurones; unfortunately at the moment, we do not have an exact representation of our cortex and of the complex links between the different parts of it. The result of this incomplete knowledge was that our artificial model have to be necessarily simplified, using a small number of neurones, and reducing significantly the complexity of the behaviour of the system. The reduction was made on the theory of emerging behaviors, according to which a set of units interacting among them exhibit a more complex global behaviour respect to the single one. If research efforts in robotics, computer science and neurobiology will be continue in a concerted and integrating manner towards the long term goal of interpreting and reproducing the human central nervous system, dramatic reduction of the complexity of intelligent dynamic systems could be achieved, leading to significant steps towards the development of the field of humanoid robots.

6. Bibliography

1. Bekey G., Crisman J. 1996, The 'grand challenge' for robotics and automation, Special Panel discussion, *IEEE Conference on Robotics and Automation*, Minneapolis, Minnesota, USA, April 22-28
2. *Proc. of the First International Symposium on HUmanoid RObots*, 1996, sponsored by Waseda University, Tokyo, Japan, October 30-31
3. Brooks R. 1996, Prospects for human level intelligence for humanoid robots, *Proc. of the First International Symposium on HUmanoid RObots*, Waseda University, Tokyo, Japan, October 30-31, 17-24
4. Lee C.W 1996 Project "CENTAUR" in mid-entry strategy – KIST 2000 human robot system project, *Proc. of the First International Symposium on HUmanoid RObots*, Waseda University, Tokyo, Japan, October 30-31, 73-82
5. Yamaguchi J. 1996, Development of a humanoid robot – Design of a biped walking robot having antagonist driven joints using nonlinear spring mechanism, *Proc. of the First International Symposium on HUmanoid RObots*, Waseda University, Tokyo, Japan, October 30-31, 102-110
6. Dario P., Rucci M. 1993, A neural network- based robotic system implementing recent biological theories on tactile perception, in *Proc. of 3rd International Symposium on Experimental Robotics*, Kyoto, Japan, 162-167
7. Rucci M., Dario P. 1994, Development of cutaneo-motor co-ordination in an autonomous robotic system, *Autonomous Robots*, 1:93-106
8. Taddeucci D, Laschi C, Lazzerini R., Magni R., Dario P., Starita A. 1997, An approach to integrated tactile perception, *Proc. of the 1997 International Conference on Robotics and Automation,* April 20-25,Albuquerque, Nex Mexico, USA
9. Laschi C., Taddeucci D., Dario P. 1997, An anthropomorphic model of sensory-motor co-ordination for robots,*Proc. of the Fifth Intenrational Symposium on Experimental Robotics,* June 15-18, Barcelona, Catalonia ES
10. Leoni F., Guerrini M., Laschi C., Taddeucci D., Dario P, 1997, Implementing robotic grasping tasks using a biological approach, submitted to the International Conference of Robotics and Automation ICRA '98, May 16-21, Leuven, Belgium
11. Braitenberg V. 1984, *Vehicles: experiments in synthetic psychology*, MIT Press, Cambridge, USA
12. P. Dario, F. Ribechini, V. Genovese, Sandini G. 1991, "Instinctive behaviors and personalities in societies of cellular robots", International Conference on Robotics and Automation ICRA '91, April 9-11, Sacramento, CA, USA
13. Fagg, A. H. and Arbib, M. A. 1992, A model of primate visual-motor conditional learning, *Adaptive Behaviours*, 1:3-37
14. Rucci M., Tononi G., Edelman G. M. 1997, Registration of neural maps through value-dependent learning: modeling the alignment of auditory and visual maps in the barn owl's optic tectum, *The Journal of Neuroscience*, 17(1):334-352
15. Piaget J. 1976, *The grasp of consciousness: action and concept in the young child*, Harvard University Press, Cambridge, MA, USA
16. Haykin S. 1994, *Neural Networks: A Comprehensive Foundation*, IEEE Comp. Spc. Press, McMillan
17. Sutton R.S. 1996, Reinforcement Learning, *NIPS Tutorial*, Dec. 2
18. Thorndike E. L. 1911, *Animal Intelligence*, Darien, CT
19. Meeden L.A. 1994, An incremental Approach to developing intelligent neural network controllers for robots
20. Dario P., Guglielmelli E., Laschi C., Guadagnini C., Pasquarelli G., Morana G. 1995, MOVAID: a new European joint project in the field of rehabilitation robotics, in *Proc. of the 7th International Conference on Advanced Robotics (ICAR '95)*, Sant Feliu de Guíxols, Spain, September 20-22, 51-59
21. Sandini G., Dario P., De Micheli M., Tistarelli M. 1993, Retina-like CCD sensor for active vision, in *Robots and Biological Systems*, Dario P., Sandini G., Aebischer P. Ed.s, Berlin-Heidelberg: Springer-Verlag, 553-570
22. Dario P., Lazzarini R., Magni R., Oh S.R. 1996, An integrated miniature fingertip sensor, in *Proc. of the 7th International Symposium on Micro Machine and Human Science (MHS '96)*, Nagoya, Japan, October 2-4
23. Cerbioni K., Colosimo C. 1997, Un sistema neuro-fuzzy per l'apprendimento e la pianificazione della presa in un manipolatore robotico, Master Thesis in Computer Science, University of Pise, to appear
24. Rumelhart E. E., Hinton G. E., Williams R.J. 1986, Learning internal representations by error propagation, in *Parallel and Distributed Processing: Explorations in the Microstructure of Cognition*, vol. 1, Cambridge, MA, MIT Press
25. Iberall T., Bingham G., Arbib M. A. 1986, Opposition space as a structuring concept for the analysis of skilled hands movements, in Heuer H., Fromm C. (eds), *Generations and modulations of action patterns*, Springer, Berlin, 158-174

Multifingered Robotic Hands: Contact Experiments using Tactile Sensors

K.K. Choi   S.L. Jiang   Z. Li

Department of Electrical and Electronic Engineering
Hong Kong University of Science and Technology
Clear Water Bay, Hong Kong, EMail: eezxli@ee.ust.hk

Abstract

Capacitive tactile sensors are constructed and installed to the fingers of the HKUST hands for measurement of position, force and direction of principle curvature of contact point. The hardware and software for signal processing are designed such that the contact information is sent to the motion control computer in real time. Experiments in rolling and sliding contact motions are then performed for testing the functionality of the tactile sensing system in motion control. The measurement of contact velocities obtained from the sensor is also compared with that calculated from the theoretical contact equations. This paper describes the tactile sensing system and the experimental result in contact motion control.

1 Introduction

Tactile sensor is a device for measuring contact position and contact force. Among the different methods of contact sensing [8], capacitive type is relatively robust, easy to construct and inexpensive. A typical block diagram is shown in figure 1. (See Fearing [1]). The actual constructed device consists of 16×16 capacitors, with an effective area of 14.3×14.3 cm^2.

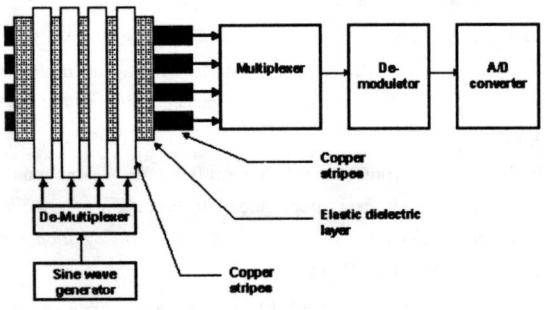

Figure 1: Schematic of Capacitive Tactile Sensor

2 Calculating Contact Information

The co-ordinate of contact point p_c is calculated by:

$$F = \sum_{p \in B_s}(I_p - I_{op})$$

$$p_c = \frac{\sum_{p \in B_s}(I_p - I_{op}).p}{F}$$

The symbols are defined as follow. S is the set $\{(x,y) \mid x,y = 0,\ldots,15\}$ of co-ordinates of the capacitors. p is the co-ordinate of a particular capacitor ($p \in S$). I_{op} is the amplitude of current flowing through capacitor C_p at no load. I_p is the measured amplitude of current flowing through capacitor C_p during operation. s is the point $\{ p \mid p \in S$ and $(I_p - I_{op})$ is maximum $\}$ which is the point with maximum increase in current amplitude. B_s is the 7×7 neighborhood of s. F is the contact force.

The summation is performed within B_s only in order to save computation time. In the above calculation, we have assumed a co-ordinate frame with the origin placed on the capacitor with index $(0,0)$. The x axis is parallel to the direction of copper stripes in the top layer and the y axis is perpendicular to the x axis. The distance between two adjacent capacitors which are located along a line parallel to either axis is one unit. Initial calibration shows that the relationship between F and the applied load is linear.

If the region of the object surface in contact with the sensor is round shaped, the set of capacitors providing active (above idle) signals will be located within an elliptic region. In the extreme case, if a sharp edge is in contact with the sensor, the set of capacitors providing active signals will be located along a line. The following procedure is used to calculate the parameters of the ellipse of the contact region. Let

$$
\begin{aligned}
N &= (cos\theta, sin\theta)^t = \text{Unit tangent vector} \\
&\quad \text{emerging from point } p_c \text{ of the sensor.} \\
\theta &= \text{Angle between } N \text{ and the } x \text{ axis.} \\
(x_c, y_c)^t &= \text{Co-ordinate of } p_c \\
(x, y)^t &= \text{Co-ordinate of } p
\end{aligned}
$$

Define

$$
\begin{aligned}
H &= \sum_{p \in B_s} (I_p - I_{op}).((p - p_c)^t.N)^2 \\
&= \sum_{p \in B_s} (I_p - I_{op}).((x - x_c).cos\theta + (y - y_c).sin\theta)^2 \\
&= A.cos^2\theta + 2B.sin\theta cos\theta + C.sin^2\theta
\end{aligned}
$$

where

$$
\begin{aligned}
A &= \sum_{p \in B_s} (I_p - I_{op}).(x - x_c)^2 \\
B &= \sum_{p \in B_s} (I_p - I_{op}).(x - x_c).(y - y_c) \\
C &= \sum_{p \in B_s} (I_p - I_{op}).(y - y_c)^2
\end{aligned}
$$

H is the weighted sum of square of projection onto N of vectors from p_c to mid-points of active capacitors within the contact region. If vector N points in the direction of the major or minor axis of the ellipse, then H should be either maximum or minimum. Setting $\frac{dH}{d\theta} = 0$ and solving for θ. we get

$$
\theta = arctan(\frac{2B}{A - C})
$$

The angles θ and $\theta + \pi/2$ are the direction angles of the major and minor axis of the elliptic contact region on the tactile sensor. Let H_{max} be the value of H of the major axis and H_{min} be that of the minor axis. Under the same applied load, a higher value of H_{max} and H_{min} means that the contact surface has smaller principle curvatures. In a typical experiment, spherical objects of different radius R are placed onto the sensor and pressed until the reading of F is 10,000. The value of H_{max} is then recorded. The relationship between H_{max} and R is plotted in figure 2. The spherical objects are not equally rigid and deform differently at the contact points. Hence, the effective radius are different from the measurement. This causes the fluctuation in the result.

The ratio $r = \frac{H_{max}}{H_{min}}$ can provide a rough idea of the shape of the object surface in contact. If $r = 1$, the contacting surface is similar to that of a sphere. A huge value of r means that the sensor is in contact with an edge. The angle θ of H_{max} is the direction angle of the major axis which can also be thought of as the direction of the minor principle curvature of the object

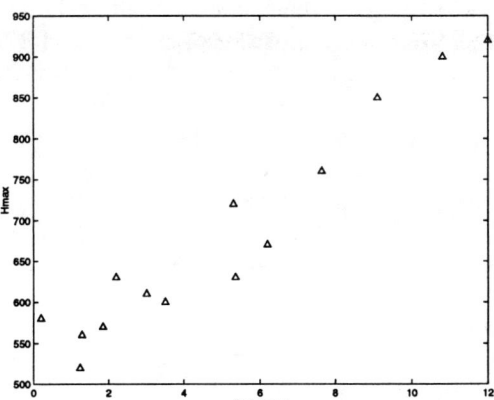

Figure 2: H_{max} and radius of spherical objects.

surface around the contact point. An experiment is performed in which an American football is placed onto a planar sensor and rotated horizontally. The change in calculated θ and the actual change in angle of rotation of the football is compared. The result is shown in figure 3. The accuracy of the calculated θ is good only if the contact area is large with significant number of active capacitors. In the experiment, extra weight (900 g) is added to the American football for increasing the contact force and the contact area.

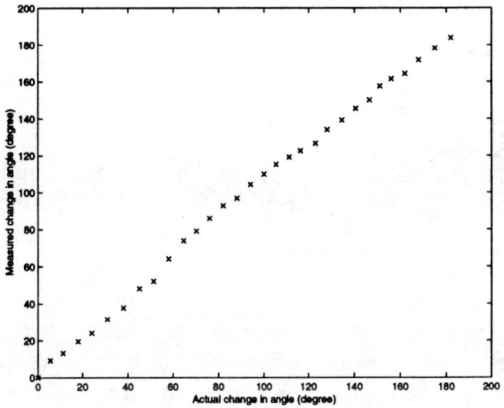

Figure 3: Measurement of American football rotation

3 Contact Experiment

3.1 Rolling Motion

An experiment is performed to test the co-ordination of robot motion in contact with curved surface using tactile sensing. An American football is placed on a table. One of the robot arm of the HKUST hands [9] with a flat tactile sensor installed in the tool frame is then programmed to touch the football. The tactile sensor then rolls in the direction of minor principle curvature of the American football. The co-ordinate of contact point and the direction of minor principle curvature measured from tactile sensor are used in real time for controlling the motion. The following is the algorithm of the rolling motion.

Initialize G_t with initial tool frame.

```
Repeat
{
  if ( F ∈ D )
  {
```
$$v = (\cos\theta \quad \sin\theta \quad 0)^t$$
$$n = (0 \quad 0 \quad 1)^t$$
$$h = v \times n = (-\sin\theta \quad \cos\theta \quad 0)^t$$
$$h' = (-\sin\theta \quad \cos\theta \quad 0 \quad 1)^t$$
$$\omega = G_t.G_s.h' = (\omega_x \quad \omega_y \quad \omega_z \quad 1)^t$$
$$\widehat{\xi} = \begin{pmatrix} 0 & -\omega_z & \omega_y & 0 \\ \omega_z & 0 & -\omega_x & 0 \\ -\omega_y & \omega_x & 0 & 0 \\ 0 & 0 & 0 & 0 \end{pmatrix}$$
$$G_t = G_t.G_s.e^{\widehat{\xi}\phi}.G_s^{-1}$$
```
  }
  else
  {
```
$$G_t = G_t + k_F.(F_d - F).G_t.G_s.(0 \quad 0 \quad 1 \quad 1)^t$$
```
  }
  Move to G_t;
  Delay 0.1 second;
}
```

The symbols are defined as follow. k_F is a constant. $\mathcal{D}$ is the range of desired contact force. F_d is the desired value of contact force. F is the measured value of contact force. θ is the measured angle between direction of minor principle curvature of object contact point and the x axis of contact frame on tactile sensor. v is the direction vector of object minor principle curvature in the sensor contact frame. n is the outward normal vector of the contact point on the sensor surface in the sensor contact frame. h is the axis of rotation around the contact point in the sensor frame. ω is the axis of rotation around the contact point in the world frame. ϕ is the angle of rotation in each step, which is set to $\pi/720$. G_s is the transformation matrix from the tool frame to the sensor contact frame. G_t is the transformation matrix from the world frame to the tool frame. It is also the target configuration of the tool frame ($G_s, G_t \in SE3$).

The robot repeatedly moves the tool frame G_t such that the sensor moves in the direction of the normal axis of contact until the contact force is within the desired range. Afterwards, the sensor will rotate around the contact point for a small angle. The axis of rotation h is the vector of major curvature on the object at the contact point. The mixed sequence of contact force adjustments and small rotations build up the rolling motion.

3.2 Contact Equations

In order to show how well the tactile sensor measures the dynamic contact point, the motion data is tested with the contact equations derived by Montana [7]. The contact equations are:

$$\dot{p}_s = M^{-1}(K_o + K_s)^{-1}\left(\begin{bmatrix} -\omega_y \\ \omega_x \end{bmatrix} - K_o \begin{bmatrix} v_x \\ v_y \end{bmatrix}\right)$$

$$\dot{p}_o = (K_o + K_s)^{-1}\left(\begin{bmatrix} -\omega_y \\ \omega_x \end{bmatrix} + K_s \begin{bmatrix} v_x \\ v_y \end{bmatrix}\right)$$

$$\dot{\psi} = T(K_o + K_s)^{-1}\left(\begin{bmatrix} -\omega_y \\ \omega_x \end{bmatrix} - K_o \begin{bmatrix} v_x \\ v_y \end{bmatrix}\right) + \omega_z$$

$$0 = v_z$$

The velocity $V_s = (\omega_x \quad \omega_y \quad \omega_z \quad v_x \quad v_y \quad v_z)^t$ is the body velocity of the sensor relative to the object in the contact frame. $\omega_s = (\omega_x \quad \omega_y \quad \omega_z)^t$ is the angular velocity of the sensor and $v_s = (v_x \quad v_y \quad v_z)^t$ is the linear velocity. K_o is the curvature form at the contact point of the object and K_s is the curvature form at the contact point of the sensor. Both of the curvature forms are relative to the contact frame. p_s is the contact co-ordinate of the sensor in the sensor frame and p_o is the contact co-ordinate of the object.

$$M = \begin{pmatrix} \|f_u\| & 0 \\ 0 & \|f_v\| \end{pmatrix}$$

$$T = \begin{pmatrix} \dfrac{f_v.f_{uu}}{\|f_u\|^2.\|f_v\|} & \dfrac{f_v.f_{uv}}{\|f_u\|.\|f_v\|^2} \end{pmatrix}$$

f_u and f_v are the vectors of co-ordinate axes of sensor frame at the contact point relative to the tool frame.

The tool frame is located at the centre of the sensor surface with its x and y axes parallel to the surface. We use a rectangular co-ordinate chart on the flat sensor surface for each contact point such that the co-ordinate of sensor contact point relative to the tool frame is $(u \quad v \quad 0)$. The orientation of the contact frame is always identical to that of the tool frame. f_u and f_v are then constant orthonormal unit vectors. Hence, M is the identity matrix. K_s and T are zero matrices and $\dot{\psi} = w_z$.

Imagine a plane passing through the two vertices of the American football. The plane intersect the football at a curve which is a portion of a circumference of a big circle. The tangent vector of the curve is always in the direction of minor principle curvature. The radius of the circle is estimated to be 180 mm and hence the minor principle curvature $k_{x'}$ of the American football is 1/180. Imagine again a line joining the two vertices of the football. A plane normal to the line will intersect the football in a smaller circle C. The major curvature is then estimated to be $1/r$ where r is the radius of the small circle C. We call this second curvature $k_{y'}$. The value of r is computed from the position of contact in the world frame.

In the contact equations, K_o is measured in the frame with the x and y axes in alignment with that of the sensor contact frame. Using the measured angle θ between the axis of the minor curvature and the x axis of the sensor contact frame, the object curvature form at each contact point is computed as follow.

$$B = \begin{pmatrix} \cos\theta & \sin\theta \\ -\sin\theta & \cos\theta \end{pmatrix}$$
$$K' = \begin{pmatrix} k_{x'} & 0 \\ 0 & k_{y'} \end{pmatrix}$$
$$K_o = B^t K' B$$

The data of sensor motion is sampled such that $G_t(t_n)$ is the tool frame relative to the world frame at time t_n where n is the sampling index. Let $p_s(t_n) = (p_{sx}(t_n), p_{sy}(t_n))$ be the measured sensor contact coordinate at time t_n. From $p_s(t_n)$, the sensor contact frame $G_s(t_n)$ relative to the tool frame at time t_n is given by:

$$G_s(t_n) = \begin{pmatrix} 1 & 0 & 0 & p_{sx}(t_n) \\ 0 & 1 & 0 & p_{sy}(t_n) \\ 0 & 0 & 1 & 0 \\ 0 & 0 & 0 & 1 \end{pmatrix}$$

The configuration $G(t_n)$ of the contact point relative to the world frame at time t_n is then given by $G(t_n) = G_t(t_n).G_s(t_n)$. At time t_{n+1}, the configuration of the original contact point on the sensor at time t_n will have moved to $G_p(t_{n+1})$ where $G_p(t_{n+1}) = G_t(t_{n+1}).G_s(t_n)$. The sensor body velocity $V_s(t_n)$ relative to the object in the contact frame at time t_n is then given by:

$$\widehat{V_s}(t_n) = G(t_n)^{-1} \frac{(G_p(t_{n+1}) - G(t_n))}{\delta t_n}$$
$$= \begin{pmatrix} 0 & -\omega_z & \omega_y & v_x \\ \omega_z & 0 & -\omega_x & v_y \\ -\omega_y & \omega_x & 0 & v_z \\ 0 & 0 & 0 & 0 \end{pmatrix}$$

where $\delta t_n = t_{n+1} - t_n$. Hence, the sensor velocity $V_s(t_n) = (\omega_x\ \omega_y\ \omega_z\ v_x\ v_y\ v_z)^t$ is computed. The velocity of object contact point co-ordinate is estimated by the actual distance traveled in the sensor frame, assuming that the object is static relative to the world frame. Let $\dot{p}_o(t_n) = (\dot{p}_{ox}(t_n)\ \dot{p}_{oy}(t_n))$ be the object contact velocity at time t_n. Let the x, y axes of $G(t_n)$ be $f_x(t_n), f_y(t_n)$ and the position vector of the origin of $G(t_n)$ be $q(t_n)$. Then,

$$\dot{p}_{ox}(t_n) = \frac{1}{\delta t_n}(q(t_{n+1}) - q(t_n))^t . f_x(t_n)$$
$$\dot{p}_{oy}(t_n) = \frac{1}{\delta t_n}(q(t_{n+1}) - q(t_n))^t . f_y(t_n)$$

For each sample, the velocity and curvature values are calculated assuming $\delta t_n = 1\ \forall n$. The values are then substituted into the contact equations for comparing with the expected $\dot{p}_o(t_n)$ and $\dot{p}_s(t_n)$. The result is shown in the following figures.

3.3 Rolling Contact

Figure 4 and figure 5 show the contact point velocities of the sensor ($\dot{p}_s$) and object ($\dot{p}_o$) in rolling motion. The graphs show close match of velocities measured from tactile sensor and that calculated from the contact equations.

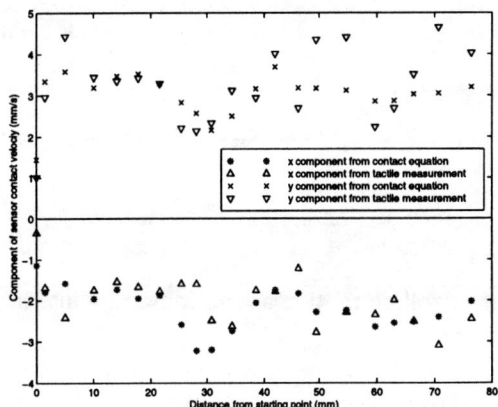

Figure 4: Sensor Contact velocity of rolling motion.

Notice that the object contact velocity is very close to the sensor contact velocity. This is a special case of rolling motion when the same chart are used in the sensor frame and the object frame at the contact. The measured v_z is very small when compared to the contact velocity. This also agrees with the contact equations.

Figure 6 shows the comparison of calculated $\dot{\psi}$ and measured $\dot{\psi}$ in rolling motion. In ideal rolling motion, $\dot{\psi}$ should be zero. The figure shows that the $\dot{\psi}$ measured from tactile sensor varies with a mean value close to the $\dot{\psi}$ calculated from the contact equations.

Figure 7 shows the loci of sensor contact point and object contact point in rolling motion. Both loci agree

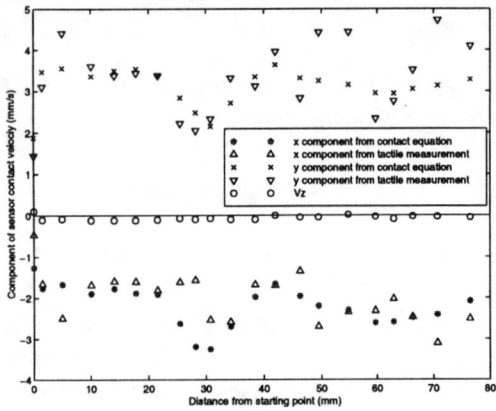

Figure 5: Object Contact velocity of rolling motion.

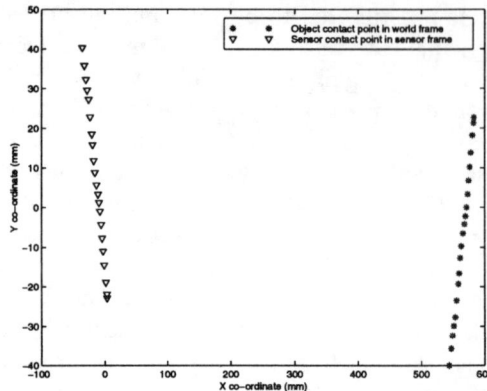

Figure 7: Loci of contact points in rolling motion.

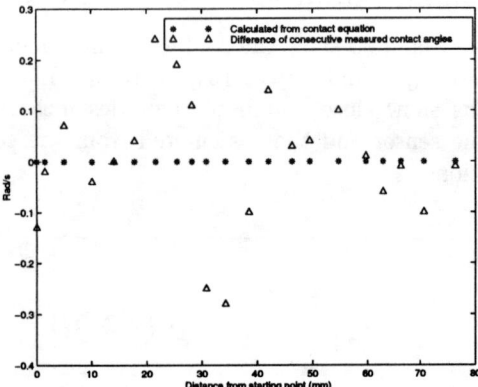

Figure 6: Contact angular velocity.

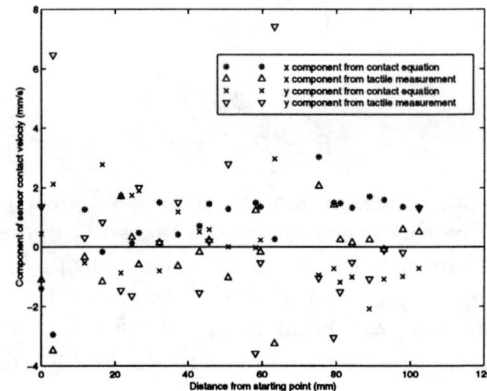

Figure 8: Sensor contact velocity of sliding motion.

with that of rolling motion as the contact points move in both cases.

3.4 Sliding Contact

Another similar experiment is performed in which the sensor is programmed to slide on the American football in the initial direction of minor principle curvature. The algorithm of motion control is similar to that of rolling with the addition of sensor contact point position adjustment in each loop. The ultimate sequence of tool frame configurations are sampled and recorded. The comparison of the motion with that of the contact equations is shown in the following figures.

Figure 8 and figure 9 show the comparison of calculated and measured contact point velocities in sliding motion. In figure 8, both the measured and calculated $\dot{p}_s$ vary around zero since $\dot{p}_s$ should be zero for ideal sliding motion.

Figure 9 shows the comparison of calculated $\dot{p}_o$ and measured $\dot{p}_o$ in sliding motion. The figure shows better match between the velocities computed from tactile measurement and that computed from contact equations. The actual motion consists of small sequential steps of rolling and sensor contact point adjustments. This causes extra interference to the motion of ideal sensor contact frames. Again, notice the small values of v_z which are in close match to the contact equations.

Figure 10 shows the comparison of calculated $\dot{\psi}$ and measured $\dot{\psi}$ in sliding motion. Similar to that of the rolling motion, the $\dot{\psi}$ from contact equation is near zero while that from tactile measurement varies with zero mean.

Figure 11 shows the loci of sensor contact point and object contact point in sliding motion. In the sliding motion, the sensor contact point oscillates around the initial contact location while the contact point moves along on the object surface.

The figures show that the measurements from the tactile sensor agree with that calculated from the Montana's contact equations.

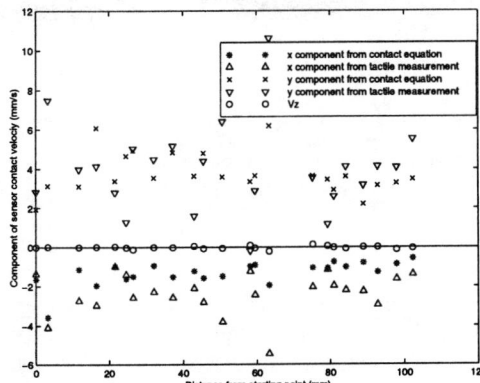

Figure 9: Object contact velocity of sliding motion.

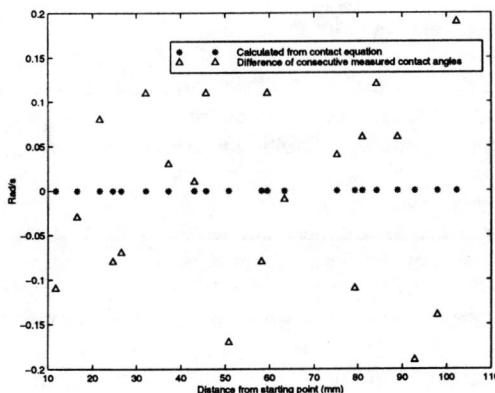

Figure 10: Contact angular velocity of sliding motion.

4 Conclusion

This paper presents the capacitive tactile sensor system of the HKUST hands. The signal processing operates in high speed which enables it to provide tactile information for real time motion control. Results of robot motion experiments in contact rolling and sliding utilizing the tactile signals are presented. The calculations of contact velocities based on the measurements from the tactile sensors agree with that derived from Montana's contact equations.

5 Acknowledgment

The authors would like to express their thanks to Professor R. Fearing and Mr. K. Kim of U.C. Berkeley who provided invaluable instructions and help in the construction of the tactile sensors.

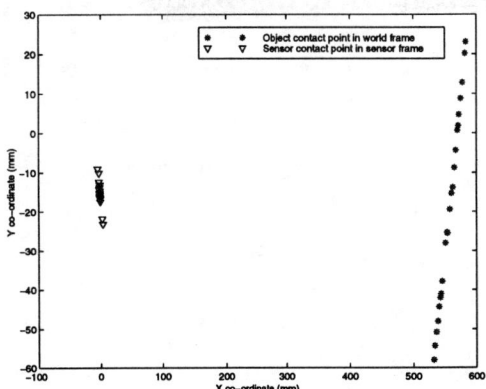

Figure 11: Loci of contact points in sliding motion.

References

[1] R.S. Fearing, "Tactile Sensing, Perception, and Shape Interpretation", PhD. dissertation, Department of Electrical Engineering, Stanford University, Dec 1987

[2] R.S. Fearing, "Some Experiments with Tactile Sensing During Grasping", in Proceeding IEEE International Conference, Robotics, Automation, April 1987, pp. 1637-1643.

[3] E. J. Nicolson, "Tactile Sensing and Control of a Planar Manipulator", PhD. dissertation, Department of Electrical Engineering and Computer Sciences, University of California, Berkeley, 1994.

[4] Richard M. Murray, Zexiang Li and S. Shankar Sastry, "A Mathematical Introduction to Robotic Manipulation", CRC Press, 1994.

[5] Zexiang Li, Ping Hsu and S. Shankar Sastry, "Grasping and Coordinated Manipulation by a Multifingered Robot Hand", The International Journal of Robotics Research, Vol 8, No. 4. Aug 1989.

[6] David J. Montana, "The Kinematics of Multi-fingered Manipulation", IEEE Transaction on Robotics and Automation, Vol 11, No. 4, Aug 1995.

[7] David J. Montana, "Tactile Sensing and the Kinematics of Contact", PhD. dissertation, Division of Applied Science, Harvard University, 1986.

[8] Howard R. Nicholls, "Advanced Tactile Sensing for Robotics", World Scientific, 1992.

[9] Y. S. Guan, Z. X. Li and Q. Shi, "Dextrous Manipulation with Rolling Contacts", ICRA' 1997, vol 2, pp. 992-997.

[10] Jae S. Son and Robert D. Howe, "Tactile Sensing and Stiffness Control with Multifingered Hands", ICRA' 1996, Vol 4, pp. 3228-3233.

[11] H. Zhang, H. Maekawa, K. Tanie, "Sensitivity Analysis and Experiments of Curvature Estimation Based on Rolling Contact", ICRA' 1996, Vol. 4, pp. 3514-3519.

Implementing Robotic Grasping Tasks Using a Biological Approach

Fabio Leoni, Massimo Guerrini, Cecilia Laschi, Davide Taddeucci, Paolo Dario, Antonina Starita°

ARTS Lab - Scuola Superiore Sant'Anna
via Carducci 40 - 56127 Pisa, Italy
Phone: +39-50-883400, Fax: +39-50-883402
E-mail: dario@arts.sssup.it

°Dipartimento di Informatica, Università degli Studi di Pisa
Corso Italia 40 - 56127 Pisa, Italy
Phone: +39-50-887215, Fax: +39-50-887226
E-mail: starita@di.unipi.it

Abstract

The capability of autonomously discovering relations between perceptual data and motor actions is crucial for the development of robust adaptive robotic systems intended to operate in a changing and unknown environment. In the case of robotic tactile perception, proper interaction between contact sensing and motor control is the basic step towards the execution of complex motor procedures such as grasping and manipulation. In this paper we propose an approach to the development of tactile-motor co-ordination in robotics, based on a neural model of the human tactile-motor system. The definition of such model is based on the features of biological systems as investigated by neuroscience. The autonomous development of tactile-motor co-ordination achieved through the implementation of the neural model is evaluated by experimental trials using a sensorised prosthetic hand and a robotic manipulator. The proposed neural network architecture linking changes in the sensed tactile pattern with the motor actions performed is described and experimental results are analysed and discussed.

1 Introduction

One of the main objectives of robotics research is to develop autonomous robots able to actively interact with an unknown environment. For this purpose robots should be able to perceive their surrounding environment, and to adapt their motor behavior accordingly. Though a variety of sensing devices exists, which are able to supply the required sensor capabilities to robots (force, proximity, tactile, visual and other sensor modalities), the real challenge is in the interpretation of the information gathered from the sensors and in its use for the execution of complex tasks.

As to tactile and haptic perception, which is the focus of the proposed work, first attempts to artificial perception and motor co-ordination in grasping and manipulation were mainly limited by the lack of suitable sensory tools. Recent advancements in sensing technology [1][2] allow a different approach to the problem, and allow to focus research on the issues related to multi-sensory integration, interpretation of sensory data, recognition and sensor-based control [3][4][5][6][7].

In this paper, we focus our interest on haptic perception and motor co-ordination by taking a biological approach. In fact, we believe that a proper comprehension of human biological structures and cognitive behaviour involved in the sense of touch and how these information are used to achieve proper and skilful manipulative capabilities is fundamental to design and develop a robotic system to be able to perform fine robotic manipulative tasks. Recently, the anthropomorphic or, more generally, the biological approach, well reproduced through neural-network-based algorithms, has shown to be successful, especially in applications in which robots are introduced in common human environments, such service and personal robotics [8][9].

In our lab, the anthropomorphic approach to sensory-motor co-ordination is being developed through two main lines [10]: on one side, by taking a top-down approach, we are focussing on the simulation of the psychological level, with no replication of the involved neural areas [11]; on the other hand, with a bottom-up approach, we try to obtain an anthropomorphic behavior by modelling and replicating the hierarchical architecture of the cortical sensory-motor areas.

This paper focuses on this second line of activity. Starting from the human model for tactile-motor co-ordination, we developed an artificial model which was implemented through a neural-network-based computing architecture. The implemented system was tested and evaluated in an experimental scenario including a robotic arm equipped with a three-fingered hand sensorised by a tactile array sensor on the thumb and force sensors on the fingertips. In particular, experimental trials aimed at proving the learning capability of the system and gave examples of application of the developed sensory-motor co-ordination in the generation of sequences of motor actions to find a target tactile information, with special focus on the procedure named tactile foveation in the text.

Experimental results show good learning capability of the system, which allow a high percentage (90% to 100%) of success in the system performance.

2 Structure and functionality of the model

2.1 The human tactile-motor co-ordination model

In humans, sensory-motor behaviour are generated by neural interactions occurring in different parts of the brain and connected to different sensory-motor structures, not fully known. Though many of the aspects of how the different perceptual data are associated into one environment perception, and how the different sensory structures are co-ordinated with the motor ones are still to be fully understood, neuro-physiological and psychological studies, together with the recent advancements in neuroscience, provide a consistent knowledge on the human tactile-motor co-ordination.

The different stimuli perceived by the human peripheral system through the different tactile receptors [12][13] are converted into neural signals and transmitted to the central nervous system (CNS) towards parallel paths (*somatic afferent paths*) so as to preserve, at this level, a complete representation and topographical organisation of the body area [14]. The cortical computation of tactile information occurs at specific levels of multiple areas of the CNS [15][16], which are hierarchically organised: each area receives information from the previous one towards neurones of cortical projections [17][18]. Such information is then integrated with the visual one and retransmitted to the frontal areas of the motor system to be used for driving intentional movements for reaching and grasping objects. On this basis, a model of the structures involved in the computation of somatic-sensory information has been developed, named *Simian Elaboration Model (SEM)*, as shown in Fig.1.

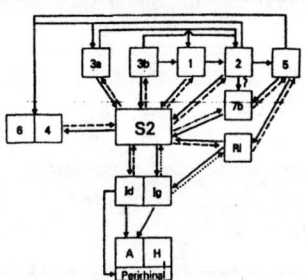

Fig.1. Corticortical connections of somatic-sensory area diagram (Friedman [17]).

Actually, motor control is directly involved in tactile perception, since touch is an intrinsically active sensorial modality, for the need of bringing the receptors in touch with explored surfaces, with proper positions and contact forces [19][20]. These explorative motor actions can be performed only if a strict relationship between tactile and motor modalities has been previously established. If at a peripheral level neuro-physiological aspects are prevalent, in this learning process also psychological mechanisms are deeply involved. One of the traditional psychological approaches to learning processes is the Piaget's [21], proposed for visual-motor co-ordination. Piaget studied children from a very early age and observed how, by making endogenous movements and gazing the resulting arm and hand spatial positions, they allow an auto-association to be created between visual and proprioceptive sensing.

2.2 The proposed artificial haptic-motor co-ordination model

Based on the described model for human tactile-motor co-ordination, the proposed artificial model includes the replication of the SEM and of the Piaget's learning paradigm. In fact, even though the tactile-motor co-ordination differs greatly from the visual-motor one, taking into consideration Rucci and Dario's suggestions [8], we propose to extend the circular reaction concept and apply it to the tactile context.

In our model, we identify the *sensory* and *motor* subsystems and, in analogy with cytoarchitectonic maps of the parietal cortex and to the hierarchical SEM, we further divide the sensory area into several sensory maps. The motor area should also be divided into the area 4 and the area 6 of the SEM, but according to the results of recent neuro-physiological researches, we focus on motor cortex functionality as a whole. The overall conceptual scheme is shown in Fig.2.

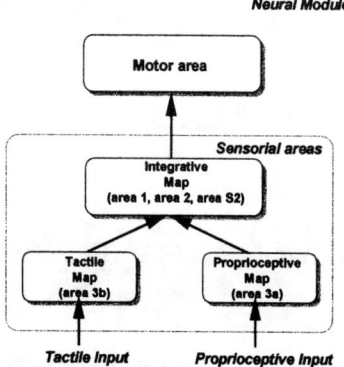

Fig.2. The conceptual scheme for the proposed artificial model of haptic-motor co-ordination

The *Sensor Area* reproduces a somaesthesic sense, several tactile inputs, and a kinaesthetic sense that can be learnt in an auto-organising way and can adjust to the input signal nature. This functionality allows to partly overcome the limits shown by other models for sensor and postural information mapping [9][22][23], in which the maps show a topographic order imposed at the beginning, on the basis of a priori knowledge of the whole possible input values variability [24]. According to the above mentioned needs and in analogy with the same somatic-sensorial cortex abilities for this area neural modelling, the sensor area was implemented through Kohonen's maps (SOFM) [25]. The integrative map is aimed at integrating and properly combining information coming from the two lower areas and then at transmitting them to the motor system for appropriate motor control computation. Due to the particular connection scheme of the SOFM structure of our model, we called it MSOFM (*Multilayered SOFM*).

The *Motor Area* is in charge of computation and realisation of motor control, by relating external inputs, internal state and pre-defined target. Motor controls are executed in endogenous way, so that they continuously require a dynamic control that is able to adjust the trajectory. That is why we decided to implement the motor area with Recurrent Neural Network (RNN) [26], showing good skills for dealing with dynamic process producing output value sequences for their internal state and input values.

The scheme of the proposed neural model is shown in Fig.3.

In the adaptation of the Piaget's scheme to haptic-motor co-ordination, we can take out relevant parameters of the circular reaction: visual target, spontaneous or endogenous movements and trajectory control. In order to apply this scheme to the tactile context, it is necessary to properly reproduce the above mentioned parameters.

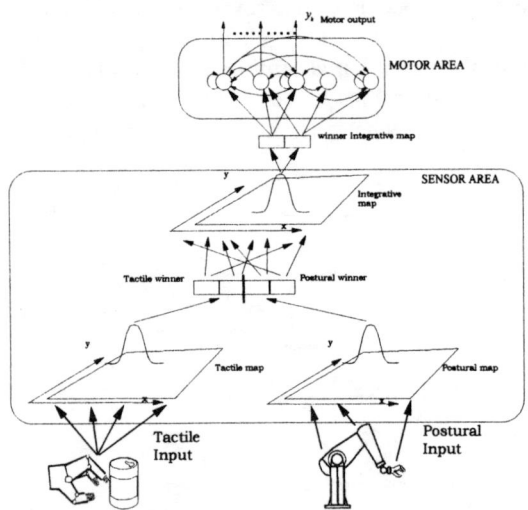

Fig.3. The proposed artificial neural model for haptic-motor co-ordination. Input vectors activate one unit for each map and the co-ordinates of these winner units represent the input vector for the integrative map. The co-ordinates of winner units set up by the integrative map are the input to the RNN of the motor area, while the motor control, in terms of parameters for the motor action, is its output

I. Tactile target. As for the visual pattern, the target perceived in the world reference system is mapped by means of suitable transformations in the retinal reference system. Tactile receptors cannot "see" the target but the fingers can touch the object thus starting up the tactile pattern in the tactile reference system.

II. Endogenous movements. Cerebral structures involved in movements are the same both in tactile and visual patterns. The difference is in the movement extents: in visual pattern the movements are wider and coarse, in tactile pattern they are shorter and finer.

III. Tactile trajectory. As for the visual pattern, Bullock and Grossberg [22] showed how an accurate motor synergism, through control mechanisms, can, rectify endogenous trajectory dynamically and in real-time. The control mechanism is named *planned and automatic control*. The planned control oversees the "quality" of the movements while the automatic control settles the variables of the mechanism necessary for the actual movement implementation, according to the system current condition. Both controls are implemented in the model and have the same control purpose over the trajectory, seen as a temporal sequence of arm and hand postures.

The learning paradigm proposed for the implementation of Piaget's scheme is the *Reinforcement Learning* [27].

The overall aim of the implemented system is to generate a sequence of endogenous movements leading to a given tactile target, on the basis of the initial tactile pattern and proprioceptive data. At a first level of detail the system scheme is given in Fig.4.

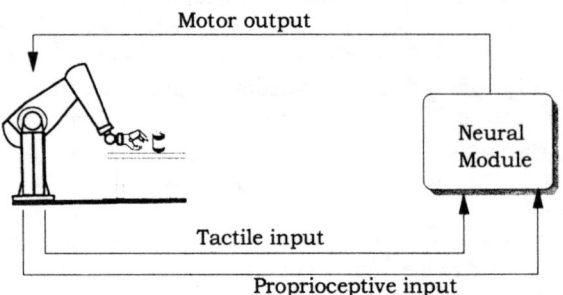

Fig.4. Interaction loop between the neural module and the robotic system. Postural and tactile information from arm and hand are transmitted to the neural module during grasping. The neural module elaborates the information and then transmits endogenous motor commands to the robotic controllers

If at time $t=0$ the initial tactile pattern is perceived, at each time $t>0$ the current tactile pattern and postural configuration start up one unit on the integrative map, whose co-ordinates become the input for the RNN of the motor area. According to the equations that rule its dynamics, the motor area computes the values of its output units as follows:

$$y_k(t+1) = f_k(\sum_{l \in U \cup I} w_{kl} z_l(t))$$

where,

$$z_k(t) = \begin{cases} x_k(t) & \text{if } k \in I \\ y_k(t) & \text{if } k \in U \end{cases}$$

with $f_k(\cdot)$ as output function, $x_k(t)$ as the input to the unit k-th, $y_k(t)$ as the output of the other recurrent units and w_{kl} as the usual component of weights matrix. Since at the time $t=0$ the weights are randomly initialised, the actual output generates an endogenous motor command, in terms of a vector coding the parameters of the relative movement that has to be imposed to the arm and hand joints; the components are calculated according to:

$$F(y_k) = \begin{cases} -1 & y_k < -\tau \\ 0 & |y_k| \leq \tau \\ +1 & y_k > \tau \end{cases} \quad \text{with } k = 0,...,n$$

where $F(y_k)$ is a threshold function and τ an appropriate threshold value. Therefore, the system takes an another postural configuration so that tactile sensors detect a new tactile image. In order to adjust the movement, it is necessary to compute the centroid distance between initial tactile pattern and target pattern and then, after the execution of the movement, to compute the distance of the new perceived pattern to the target. If the new distance results smaller than the previous one, the direction of the movement is considered "good" and it is *positively reinforced*; otherwise, it is *negatively reinforced*. An example of the implemented mechanism of reinforcement learning is shown in Fig.5.

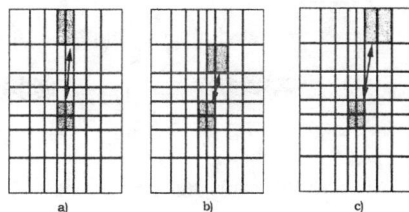

Fig.5. a) Distance between the initial tactile pattern and the target (the target is centred on the tactile image). b) Approaching of perceived tactile pattern to the target, corresponding to a positive reinforcement. c) Moving back of the tactile pattern from the target, corresponding to a negative reinforcement.

The equation for the calculation of the error is the following:

$$e_k(t) = \begin{cases} d_k(t) - y_k(t) & \text{if } k \in T(t) \\ 0 & \text{otherwise} \end{cases}$$

where $T(t)$ is the set of the output unit indexes and d is the output value expected for the k-th unit. The above described method avoid to calculate what the expected output d should be. If the movement, or rather the rotation direction of the corresponding joint, is in the right direction, the current output of the k-th unit has to be strengthened. Otherwise, if the movement goes to the wrong direction, the output must to be weakened. The equations are the following,

$$d_k(t) = \begin{cases} \text{sgn}(y_k(t)) & \text{if } \Delta(t+1) < \Delta(t) \\ -\text{sgn}(y_k(t)) & \text{if } \Delta(t+1) \geq \Delta(t) \end{cases}$$

$$e_k(t) = \begin{cases} \text{sgn}(y_k(t)) - y_k(t) & \text{se } \Delta(t+1) < \Delta(t) \\ -\text{sgn}(y_k(t)) - y_k(t) & \text{se } \Delta(t+1) \geq \Delta(t) \end{cases} \text{ with } k \in T(t)$$

where $e_k(t)$ is the error and $\Delta(t)$ the distance between tactile pattern perceived at time t and the target pattern.

3 Experimental Results

The implemented system was tested and evaluated through experimental trials aiming at verifying the learning capabilities and motor co-ordination functionality.

The experimental scenario comprises of:
- a PC (Pentium 90Mhz) to simulate the neural model, for elaborating tactile and postural information and driving the controllers for the arm-hand system;
- a PUMA 560 manipulator (6 degrees of freedom);
- a MARCUS hand (2 degrees of freedom), equipped with position and force sensors on the fingertips;
- a tactile array sensor [28], mounted on the thumb, presenting a space-variant fovea-like distribution of active site.

The fovea-like structure of the tactile sensor emulates primate's density of tactile units, in particular Slowly Adapting I (SAI) and Rapidly Adapting (RA)[12]. Fig.6 shows the MARCUS hand and the tactile array sensor.

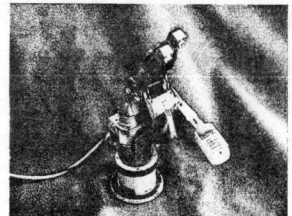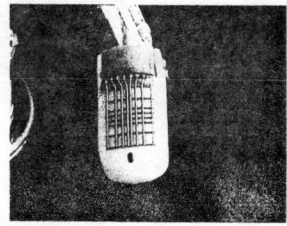

Fig.6. The MARCUS hand and the integrated fingertip including the tactile array sensor

The experimental task was to bring a arm-hand system in contact, with an object, by means of a touch and release motion, so that the distribution of pressure perceived by sensors on the fingers was centred on the tactile fovea. In fact, in this area humans have the higher density of tactile receptors thus offering a great number of tactile information, while, during the manipulative task, they bring the finger fovea in contact over the object surface. This task is referred to as *foveation task* and essentially consists in executing a sequence of fine adjusting movements. The foveation task has been successfully tested on two rigid and non-deformable objects with different geometric characteristics, with no assumptions on their shapes. The used objects are a cylindric metal can and a plastic orange, whose shape is approximately a sphere, as shown in Fig.7.

Fig.7. The objects to be grasped by the robotic system in the experimental trials.

3.1 Training of the System

Due to the different nature of the NN we have used for the sensory and motor areas, relating to the different supervised (SOFM) and unsupervised (RNN) paradigms, we had to train the NNs separately.

Training of Sensory Area

The training of the MSOM is essentially divided in two phases: the first one is for the creation of the training-set and the second is the real training of the MSOM. In building the training-set, we executed a specific procedure of pattern acquisition for the can and for the orange, aiming at getting a significant set of geometrical characteristics pattern for the objects. The procedure consists in bringing the MARCUS hand in touch with the object at different spatial locations, then making the hand close around the object and reading and storing the information provided by the position sensors of arm and hand and by the force sensors and the tactile array sensor. In this way, the training-set, TS, is a set of N couples of patterns so defined: $TS = \{(T_0, P_0), (T_1, P_1), ..., (T_{N-1}, P_{N-1})\}$

where P_i, with, $i = 0,..., N-1$ is the vector posture of the arm and the hand and T_i is the tactile information perceived by the sensors of the hand, pressure and tactile sensors, when the system is at position Pi. The number of the examples (couples) for the training-set is 224 for the can and 64 for the orange.

The training of the MSOM was executed in two different ways: the first way consisted in training each SOFM of the sensory area separately, that is, first the Tactile and Postural maps together and then the Integrative map with the co-ordinates of the winner units of each map for each example. The second way consisted in training the SOFMs contemporarily; both methods gave consistent results. There are no substantial differences in the two methods, as the location of the winner units for the same examples on the same maps for the two methods depends on the random initial value of the weights of the maps only. Only, the second method is faster, as shown by the results reported in the following table.

OBJECT	N° PATTERNS FOR TRAINING	N° ITERATIONS FOR SEPARATE TRAINING	N° ITERATIONS FOR PARALLEL TRAINING
ORANGE	64	25.000	13.000
CAN	224	33.000	18.000

The parameters of each SOFM are:
- the map dimension, as 25x25 units;
- the learning rate $\eta(n)$, as a linearly decreasing function of time and resembling the same standard value of a common SOFM [25];
- the neighbourhood function $\Lambda_{i(x)}$, as a square centred onto the winning unit with radius r initially set to the map dimension and set free to decrease linearly with the number of iterations down to zero (including the winner unit only).

After training, each map shows a topographical order that encodes its own processed information referred to the particular object. Fig.8 shows the distribution of the winner units after training for the Integrative map.

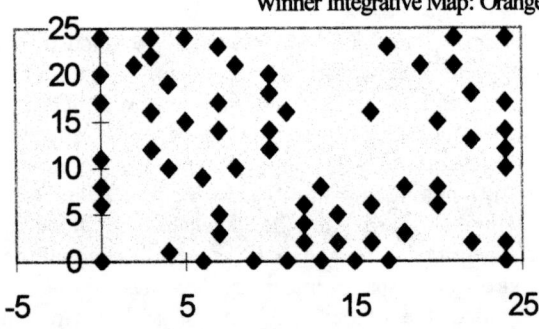

Fig.8. The resulting distribution of the winner units after training for the Integrative map, for the case of the orange. Some clusters evidently appear on the two-dimensional map, due to the homogeneity of the sensory and postural data in the case of a symmetric object.

Performance of Motor Area

The motor area does not need a separate phase of training because it has been implemented with a RNN that allows to change synaptic weights on-line. The performance of this area consists in applying the circular reaction scheme. After choosing a target tactile target pattern, the robotic arm is located so that the hand can close around the object at a random position. The weights of the motor area (a RNN with 25 recurrent units) are then randomly initialised in the range [-0.1, 0.1] to generate an endogenous movements. The output vector, with 8 components for the degrees of freedom of the system (6 for the arm and 2 for the hand), represents the relative movement for the robotic system. Consequently, this motor action brings the hand to another spatial location where to acquire new tactile and postural information. The new tactile information is elaborated to evaluate the new distance to the target and then used, according to previous discussion and equations shown in Section 2.2, to evaluate and reinforce the weights to compute and to refine the next movement. Fig.9. shows a sequence of tactile images acquired during an experimental session.

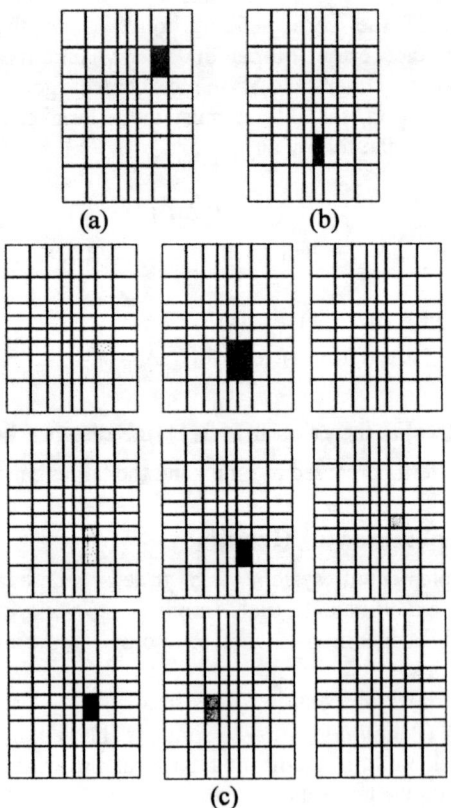

Fig.9. (a) The target tactile pattern; (b) the initial tactile image; (c) the sequence of tactile images during a typical operational sequence

The RNN is able to combine sensory information provided by the sensory area with the direction of movements and to develop a consistent relationship between these modalities. In fact, the correction of synaptic weights we have illustrated is able to bring the hand to a particular spatial position around the object so that the tactile sensor can perceive the specified target tactile pattern. This is also evident by comparing the graphical representations for the variation of the distance from the target and for the sequence of activation of the winner units on the integrative map, as shown in Fig.10.

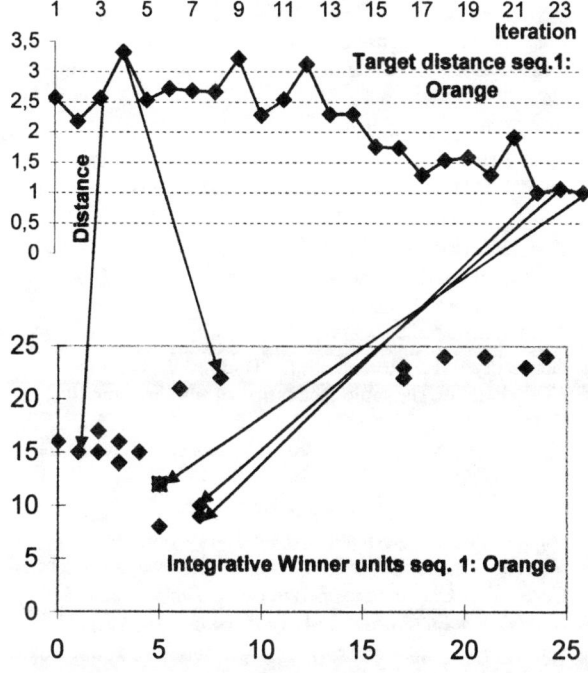

Fig.10. Variation of the distance of the tactile image from the target tactile pattern and sequence of activation of the winner units in the integrative map. The larger is the distance, the wider is the reinforcement and, consequently, the corresponding movement. The arrows point out the correspondence between relevant variation in the distance of the tactile image from the target and in the distance of the corresponding winner units in the integrative map.

When the system reaches the target tactile pattern, the experiment is repeated starting from another initial posture but keeping the target fixed, in order to cover the whole active area of the tactile sensor. For the orange, the total number of different initial postures is 20 and the experiment is repeated 15 times, whereas for the can we considered 30 different starting postures and the experiment is repeated 15 times, too.

Learning capability is shown by the smoothing of the graph of distance from the target pattern and the drastical reduction of the number of iterations (see Fig.11).

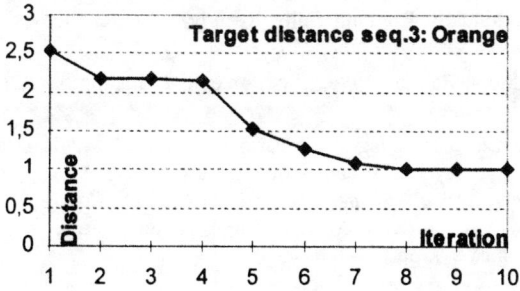

Fig.11. The graph of the variation of the distance of the tactile image from the target tactile pattern for the third sequence. In comparison with Fig.10 the line is much smoother, showing a reduction of error values and of consequent corrections, and the number of iterations for convergence decreased from 24 to 8.

The neural model presented has shown excellent learning capabilities: each requested sequence has been completed with success. The system has also shown a good rate of generalisation: afterwards, we have submitted the system new targets, different from those used in the first iterations, and the system reached the specified targets with a 100% percentage of success for the can and 90% for the orange. In Fig.12 some pictures of the robotic system during the attempts to reach targets for the orange and for the can are shown.

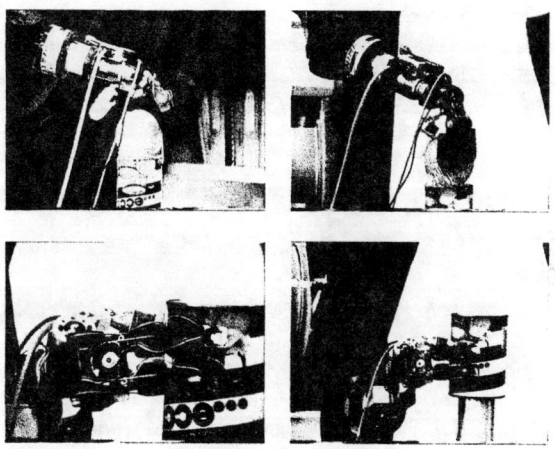

Fig.12 Phases of grasp adjustment.

4 Conclusions

The results obtained show the effectiveness of the proposed model, pointing out the importance of using neural network paradigms in developing sensory-motor co-ordination in robotic systems. In fact, neural networks, thanks to their learning capabilities, allow to solve perception problems without a priori knowledge of particular functionality.

The presented neural model has exhibited properties which are consistent with experimental results relating to biological control systems, thanks to the particular neural architecture, characterised by two systems linked in a hierarchical way and relying on different learning paradigms; the functionality and the nature of the sensory and motor areas of human tactile system, together with biological motivations, suggested us to use this type of coupling. The peculiar features of diagram connections of the SOFM sensory area, although being a significant simplification of reality, allow to capture the basic structures and some topic principles of biological systems. Another matter of some relevance, is that the executed movements are endogenous; the system is not forced to compute predetermined movements but only to focus on the target. The system generates motor commands on its own and corrects itself in the attempt of reaching the target. The correction algorithm of synaptic weights for motor control is obtained through the application of the circular reaction scheme introducing reinforcement learning; the system learns the "legal" movements to grasp the addressed object autonomously.

Finally, the system also allows to manage motor sequences dynamically and in real-time, thanks to the on-line weights correction technique peculiar for recurrent networks.

For future developments, we plan to test and improve the above mentioned method using the tactile array sensor on all the fingers of the MARCUS hand, to develop of further motor actions that involve pre-grasping and explorative procedures and to integrate haptic data with visual information.

5 References

[1] Dario P. 1989, Tactile sensing for robots: Present and future. In O. Khatib, J. Craig, and T. Lozano-Pérez, editors, *The Robotics Review 1*, chapter II, pp.133-146. MIT Press, Cambridge, Massachusetts.

[2] Howe R. D., Cutkosky M. R. 1992, Touch sensing for robotic manipulation and recognition. In O. Khatib, J. Craig, and T. Lozano-Pérez, editors, *The Robotics Review 2*, pp.55-112, MIT Press, Cambridge, Massachusetts, USA.

[3] Cutkosky M. R., Howe R. D. 1990, Human grasp choice and robotic grasp analysis. In S. T. Venkataraman and T. Liberall, editors, *Dextrous Robots Hands*, pp.5-31, Springer-Verlag.

[4] Allen P.K 1990, Mapping haptic exploratory procedures to multiple shape representations. In the *Proceedings of the 1990 IEEE International Conference on Robotics and Automation*, Cincinnati, Ohio, May 1990, pp.1679-1684.

[5] Fearing R. 1990, Tactile sensing for shape interpretation. In S. T. Venkataraman and T. Liberall, editors, *Dextrous Robots Hands*, chapter 10, pp.209-238, Springer-Verlag.

[6] Stansfield S. A. 1990, Haptic perception with an articulated, sensate robot hand", *Technical Report SAND90-0085*, Sandia National Laboratories, Albuquerque, New Mexico, March 1990.

[7] P. Dario, M. Rucci, C. Guadagnini, C. Laschi 1994, An investigation on a robot system for disassembly automation, in *IEEE/International Conference on Robotics and Automation*, San Diego, California, May 8-13.

[8] Rucci M., Dario P. 1994, Development of cutaneo-motor co-ordination in an autonomous robotic system, *Autonomous Robots*, 1:93-106.

[9] Kurpeinstein M. 1991, Infant neural controller for adaptive sensory-motor co-ordination, *Neural Networks*, 4:131-145.

[10] Laschi C., Taddeucci D., Dario P. 1997, An anthropomorphic model of sensory-motor co-ordination for robots, in *Proc. of the Fifth International Symposium on Experimental Robotics (ISER '97)*, June 15-18, Barcelona, Catalonia, Spain.

[11] Taddeucci D., Dario P. 1998, Experiments in Synthetic Psychology for tactile perceptions in robots: Steps towards implementing humanoid robots, submitted to *IEEE/International Conference on Robotics and Automation*, Leuven, Belgium, May 16-21.

[12] Jonhnson K. O., Hsiao S. S. 1992, Neural Mechanism of tactual form and texture perception, *Annual Review of Neurosciences*, 15:227-250.

[13] Albus J.S. 1981, Brains, Behavior and Robotics, *Byte Books*. Peterborough, N. H., USA.

[14] McClintic J.R 1988, Fisiologia del corpo umano, *Zanichelli Editore S.p.a.,* Bologna, Italy.

[15] Pons T. P., Garraghty P. E. and Mishkin M. 1992, Serial and Parallel Processing of Tactual Information in Somatosensory Cortex of Rhesus Monkeys, *Journal of Neurophysiology*, August, 68 (2):518-527.

[16] Kaas J.H. 1995, The reorganization of sensory and motor maps in adult mammals, *The Cognitive Neuroscience* pp.51-71, Cambridge: MIT Press

[17] Friedman D.P., Murray E.A. 1986, Thalamic Connectivity of the second somatosensory area and neighbouring somatosensory fields of the lateral sulcus of the monkey, *J. Comp. Neural* 252:348-373

[18] Garraghty P. E., Florence S. L., Tenhula W. N., Kaas J. H. 1991, Parallel Thalamic Activation of the First and Second Somato-sensory Areas in Prosimian Primates and Tree Shrews", *The Journal of Comparative Neurology*, 311:289-299.

[19] Burnod Y., Grandguillaume P., Otto I. et al. 1992, Visuomotor Transormations Underlying Arm Movements toward Visual Targets: A Neural Network Model of Cerebral Cortical Operations, *The Journal of Neurosciences*, April, 12 (4):1435-1453.

[20] Burnod Y., Guyot F., Otto I. et al. 1994, Computational Properties of the Cerebral Cortex to learn sensomotor programs, *Proceeding from Perception to Action Conference*, Lausanne, Switzerland, September 7-9, IEEE Computer Society Press, pp. 206-217.

[21] Piaget J. 1976, The Grasp of Consciousness: Action and Concept in the Young Child, *Harvad University Press*, Cambridge, MA, USA.

[22] Bullock D., Grossberg S. 1988, Neural dynamics of planned arm movements: Emergents invariants and speed-accuracy properties during trajectory formation, *Psychological Review*, 95:49-90.

[23] Thomas Miller W. III 1987, Real Time application of neural networks for Sensor Based Control of Robots with vision, *IEEE Transactions on Systems, Man and Cybernetics*, July-August, 19(4):825-831.

[24] Martinetz T. M., Ritter H. J., Shulten K. J 1990, Three-Dimensional Neural Net for Learning Visuomotor Co-ordination of a Robot Arm, *IEEE Transactions on Neural Networks*, March, 1(1):131-136.

[25] Kohonen T. 1984, Self-Organizing and Associative Memory, *Springer Series in Information Sciences*, vol.8.

[26] Williams R. J. and Zipser D. 1989, Learning Algorithm for Continually Running Fully Recurrent Neural Networks, *Neural Computation*, 1:270-280.

[27] Sutton R. S. 1996, Reinforcement Learning, *NIPS Tutorial*, Dec. 2.

[28] Lazzarini R., Magni R., Dario P. 1995. "A tactile array sensor layered in an artificial skin", in *Proc. of the International Conference on Intelligent Robots and Systems (IROS '95)*, Pittsburgh, Pennsylvania, USA, August 5-9, pp.114-119.

New Mexico, USA

Model and Processing of Whole-body Tactile Sensor Suit for Human-Robot Contact Interaction

Yukiko HOSHINO Masayuki INABA Hirochika INOUE
Dept. of Mechano-Informatics, Univ. of Tokyo,
7-3-1 Hongo Bunkyo-ku, Tokyo, 113 Japan

Abstract

The function that enables a robot to react when touched by a human is indispensable for a robot which enters our daily life. This paper describes the model and processing of full-body humanoid's tactile sensor suit for the human-robot interaction. The tactile sensor suit is first modeled to make a symbolized tactile data structure. Then conversion and compensation process is proposed to enable the robot to obtain accurate tactile data no matter what posture it takes.

Then we explain the robot system and two experiments using a 36 DOF full-body humanoid: the first is the acquiring maps to process the tactile data, and second is the conformation to a human's shape as a human-robot contact interaction with proposed sensor suit model and processing.

1 Introduction

We work on a robot which can work beside humans in our daily life. Our motivation is to build a robot which can have contact with a human, without being frightening, and can do some simple tasks.

To make such a robot, there are many research subjects, such as building sensors, developing a robot body, developing action decision process, realizing natural interaction between a human and a robot. There have been some works on humanoid robots which interact with a human [1–3]. However these researches deal mainly with audio and visual interaction, not contact interaction. Contact interaction between a human and a robot is indispensable, because a robot will surely have contact with a human when it enters our daily life. From this point of view, we set the robot which can react to human's contact as the goal of our research.

Figure 1 shows three examples of human-robot contact interaction which we are taking into consideration. The left one is the behavior of paying attention to a human's interruption. The robot which stays beside a human should react whenever a human touches. The middle one is intuitive motion teaching by holding a robot by its hands or feet. The right one is the behavior of the robot's hugging. When the robot detects that it is being held by a human, it moves its arms to conform to the human's body. To realize these kinds of interaction we mentioned above, the tactile sensor data from the surface of the whole body of the robot is important.

We developed a tactile sensor suit [4] which covers the whole body of a robot. It is made from electrically conductive fabric and strings and has enough flexibility. We implemented this tactile sensor suit for a child-size full-body humanoid [5] which has 160 sensing regions.

Data from distributed tactile sensor array have been processed in the similar manner as the image processing in several papers (e.g. [6, 7]). However the processing of the data from the tactile sensor suit all over the robot's body is more difficult since it has three dimensional structure and may have self-interference. Moreover, the tactile sensor suit is not fixed to the robot's body, and the correspondence of the tactile data to the robot's surface changes dynamically.

In this paper, we propose both the model of tactile sensor suit and the processing of the tactile data to realize the contact interaction between a robot and a human.

2 Tactile Sensor Suit

2.1 Features of Tactile Sensor Suit

Tactile sensor suit is a flexible sensor which covers the entire body of the robot. The tactile sensor suit has the following features.

Stable tactile data for a 3D curved surface are supplied by sensor units distributed all over the robot's body and sufficient number of binary data points. Flexibility ensures that the sensor suit does not disturb the robot's movement. The sensor suit is not attached to the robot's body, but is developed separately from it. This is termed "modularity" and enables re-formation of the tactile sensor suit for a de-

Figure 1: Human-robot contact interaction

sirable distribution. The wiring for data transmission runs through the suit, therefore the sensor suit can have a large amount of wiring without disturbing the robot's movement. However the correspondence of the tactile data to the robot's surface may change dynamically, and some processing is necessary to interpret the raw tactile data.

2.2 Issues of Tactile Data over the Entire Body

Issues concerning processing tactile data over the whole of the robot's body are:

1) How to represent the tactile data over the entire body of the robot

Raw tactile data over the entire body of the robot contains the large number of unstructured signals. Therefore mapping to a structured representation such as "somatotopic (skin-surface) map" [8] is indispensable. As the solution of this issue, the tactile data should be symbolized to the names of robot's body based on the modeling of the tactile sensor suit.

2) How to absorb the dynamic change of the correspondence of the tactile data to the robot's body

When the robot twists its joint, the correspondence of the sensor data to the robot's body changes dynamically. Therefore conversion and compensation is indispensable to absorb this changes. To solve this issue, a map which stores the rules of the change of correspondence is required.

3) How to detect changes caused by external interaction

A robot which has tactile sensors all over its body will detect changes caused by its own movements, which is termed "self-interference data". A robot which interacts with a human has to distinguish between the data caused by being touched from outside and the self-interference data. A map which stores the conditions of the occurrence of self-interference can be used to compensate the tactile data.

4) How to acquire the correspondence of the sensor suit to the robot's surface

As the distribution of sensor unit can easily be re-formed, the robot should be able to map the tactile data to its body automatically. It also has to acquire the mapping which is used to absorb dynamic changes in the correspondence of tactile data to the robot's body and in order to identify data caused by human interaction.

3 Modeling of Tactile Sensor Suit
3.1 Sensor Suit Model

Figure 2 shows the modeling of the tactile sensor suit. There are two cylinders for each arm, and the link at the base of each cylinder has 2 DOF. There are

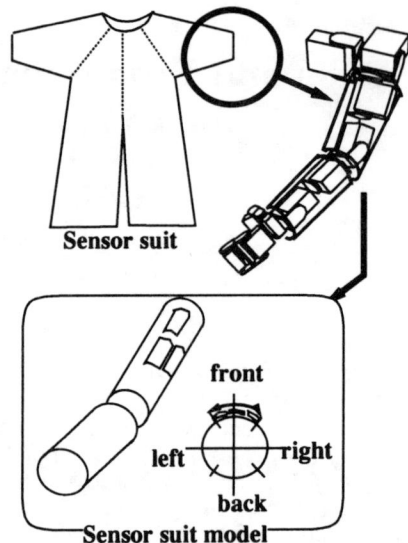

Figure 2: Modeling of sensor suit

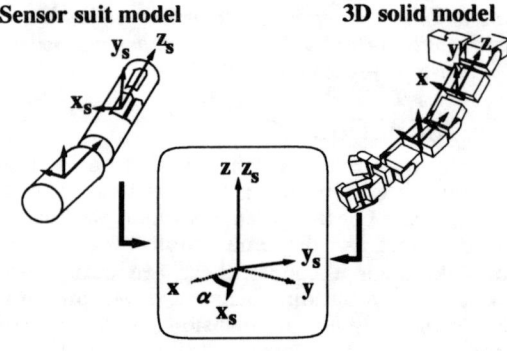

Figure 3: Two coordinate systems for the robot

patches for each sensor region on the cylinder, which represent the sensor units.

There are two coordinate systems for the robot as shown in Fig. 3: one is the c-ordinate system for the robot's body and the other is for Sensor Suit Model. The displacements of those coordinate systems (α in Fig. 3) change according to the robot's posture because the tactile sensor suit doesn't twist together with the robot's limbs.

3.2 Outline of Tactile Processing

Figure 4 shows the outline of tactile processing.

Conversion and compensation process are practiced with "Initial Maps". The processed sensor data is stored as "Symbolized Tactile Data", which is utilized to generate the robot's behavior.

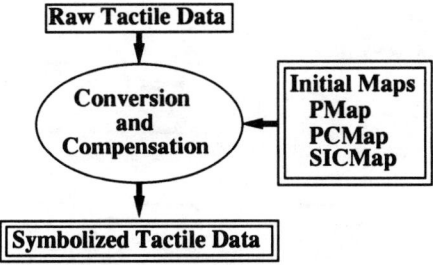

Figure 4: Outline of tactile processing

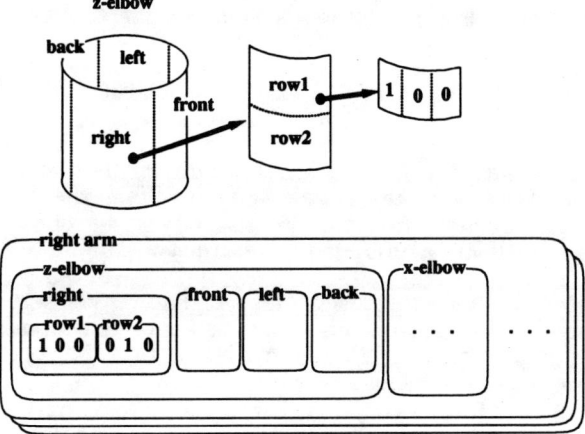

Figure 5: Data structure of Symbolized Tactile Data

3.3 Symbolized Tactile Data

Symbolized Tactile Data has data structure which is made by mapping the distributed tactile data onto the robot's surface. Figure 5 shows the structure. It has a nested structure: outermost is the limbs such as left arm, right arm. There are lists of links such as elbow, shoulder in it. Each list contains the direction lists, which are "right", "front", "left", and "back". Each direction list has lists which represent the rows from top to but tom. Each row data is the list of binary sensor data clockwise. The data structure of Symbolized Tactile Data reflects of robot's physical body structure.

4 Processing Using Sensor Suit Model
4.1 Initial Maps

Conversion and compensation process are practiced with "Initial Maps". The Initial Maps consist of three maps. The first map is "Sensor Position Map (PMap)" which represents the correspondence of the tactile data to the robot's body. The second one is "Position Compensation Map (PCMap)" which stores rules of the dynamic change of PMap. The third one is "Self-Interference Compensation Map (SICMap)".

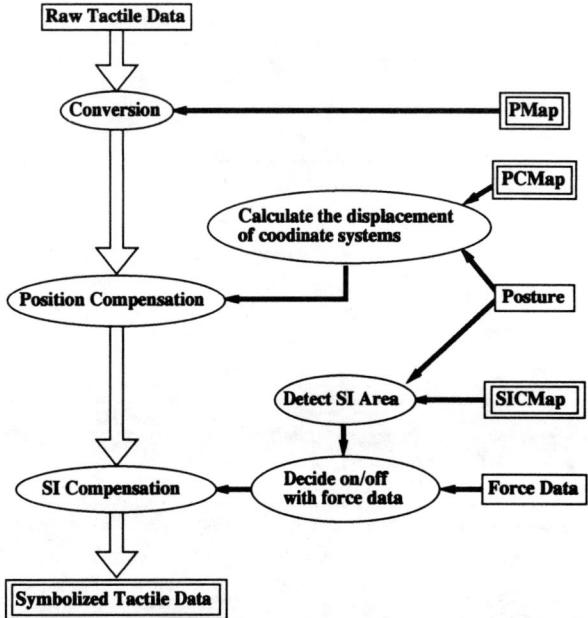

Figure 6: Conversion and compensation of tactile data

The self-interference data is related to the robot's posture, therefore this map has the relationship between the self-interference data and the angle of each robot's link and is utilized to compensate the self-interference.

4.2 Conversion and Compensation of Tactile Data

As the identifying process, conversion and compensation which is shown in Fig. 6 are practiced.

At the beginning of this process, raw tactile data are converted with PMap to the robot's surface. Next the robot compensates the displacement of the coordinate systems of the robot's body and the Sensor Suit Model. The displacement is calculated using PCMap and the robot's posture. At the end of the identifying process, the self-interference compensation is carried out. Using SICMap and the robot's posture, the robot detects self-interference areas in the neighborhood of the bending joints such as elbows, knees. Then the status of those areas are determined with another sensing channel such as force sensor.

As a result of this identifying process, the tactile data is converted to a structured form which can be accessed with a symbol such as "right arm, z-elbow, front", no matter what posture the robot takes.

4.3 Acquisition of the Initial Maps

The robot moves its own body to acquire the Initial Maps automatically. Figure 7 shows the process of acquiring the Initial Maps.

To acquire PMap, the robot calculates the posture to touch a certain point of its body using 3D solid model, and touches the spot to obtain the change of

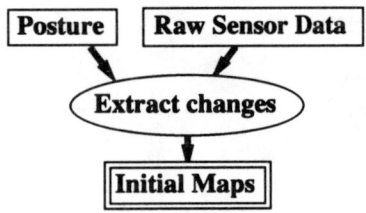

Figure 7: The flow of acquisition Initial Map

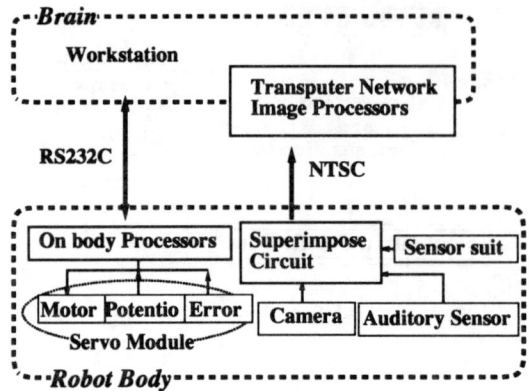

Figure 9: Robot system scheme

Actuators of this robot are commercial servo modules for radio-control. The robot has a stereo viewer to obtain binocular image with one CCD camera and binary auditory sensors to detect the direction of sound, and wears a tactile sensor suit to obtain the tactile information from all over its body. The robot is equipped with a on-body processor network [9] to control the servo modules. It is also utilized to obtain the value of potentiometers and error signals in each servo module. Accelerometer at the center of the robot's body gives the direction of gravity, so that the robot can detect the inclination of its own body.

Figure 9 shows an overview of the robot's system. The on-body processors obtain the reference angles for the servo modules from the workstation via RS232C, and send this data to each servo module. These processors also gather sensor data from each module, and send them to the workstation. As the image data is too large to be sent by the serial line, it is sent through another image line. The tactile data over the entire body of the robot are superimposed on the image, as shown in Fig.10, in what we call "Sensor Image" [10]. The rectangle area in the center of the Sensor Image represents raw tactile data area. A white dot in it indicates the presence of contact. The right map shows the correspondence of the tactile data to the robot's surface, which is stored in PMap.

5.2 Acquisition of PMap and SICMap by Self Movement

In this experiment, the robot acquires PMap by itself. The robot moves its body and touches every part of robot's body while checking for changes in raw tactile data. As a result, PMap is acquired. With regard to the areas where the robot cannot touch by itself, a human touches them to obtain PMap.

Figure 11 shows the snapshots in the experiment. The robot is checking of its left elbow. As the right figure shows, the sensor data changes after touching. The robot makes a correspondence of this changing area to the elbow in PMap.

The robot moves its right arm to acquire SICMap. Figure 12 shows the posture and the sensor data. The

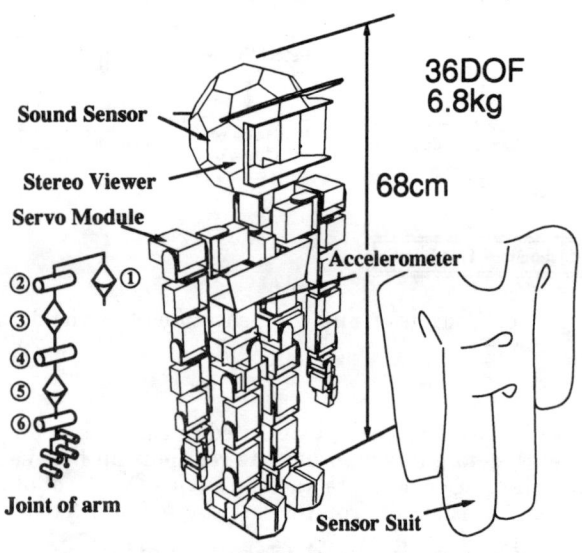

Figure 8: 36 DOF full-body humanoid

tactile data. Areas which the robot cannot touch by itself should be touched by a human.

To acquire SICMap, the robot detects the change of tactile data while moving a certain bending joint such as the elbows, and stores the relationship between the area of the change in sensor and the angle of that joint. The robot can also acquire PCMap with the same method.

5 Experiments with tactile sensor suit
5.1 Full-body Humanoid System

Fig. 8 shows the full-body humanoid robot, which is 68 cm tall with a weight of 6.6 kg and has 36 DOF.

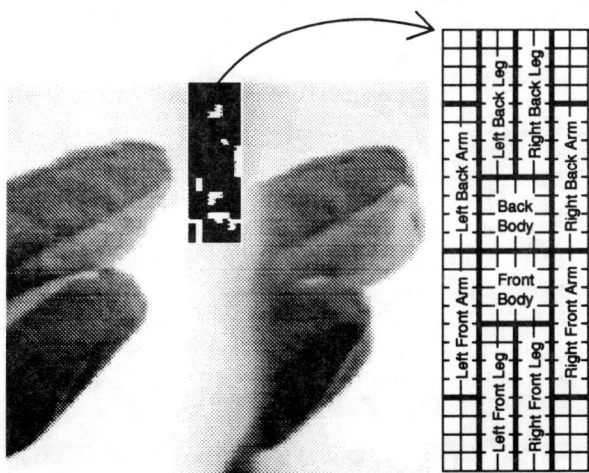

Figure 10: Sensor image

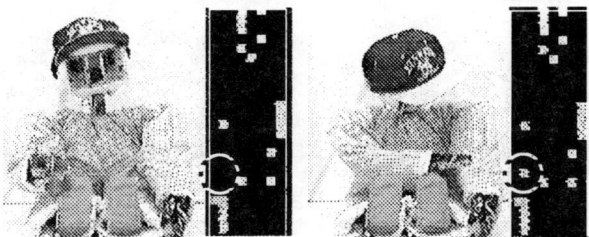

Figure 11: Acquisition of PMap by itself

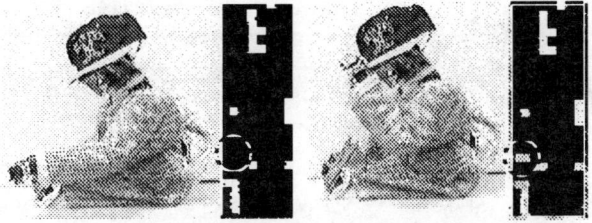

Figure 12: Acquisition of SICMap by itself

robot moves its elbow and self-interference areas appear at a certain angle of the joint.

5.3 Distinguish between External Interaction and Self-Interference

Using SICMap and the auxiliary force sensor, the robot can distinguish between external interaction and the self-interference. As shown in Fig.13, there are following three causes when tactile data at the elbow changes;

(a) external interaction

(b) self-interference

(c) both external interaction and self-interference

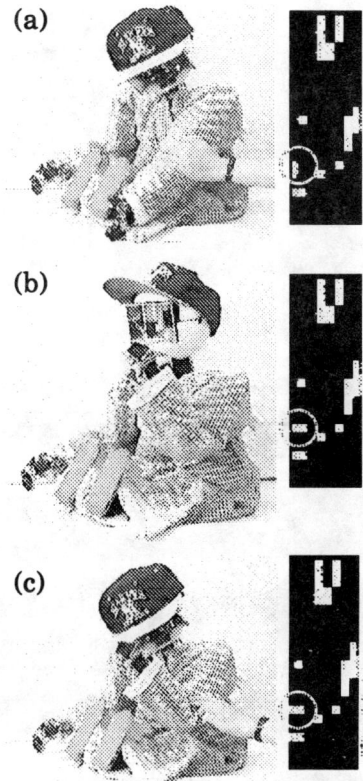

Figure 13: Recognition of external interaction

In case (a), the robot infers that no self-interference exists judging from its own posture because it is not bending its elbow. Therefore it determines that the change is caused by external interaction. In case (b), as the robot is bending its elbow, it infers that self-interference areas exist. Then it examines the force data, and determines that there is no external interaction since no external force data is detected. In case (c), the robot infers that self-interference areas exist, and recognizes the presence of external interaction, since external force is detected. In cases (a) and (c), the robot is turning to the elbow as a reaction to the detected external interaction.

5.4 Conforming to a Human's Shape

To conform to a human's shape, the robot has to twist some of the joints because the robot's arm has the arrangement of six joints shown in Fig.8. When the robot twists a joint, the correspondence of the sensor data to the robot's body changes. In this experiment, the compensation of the position using PCMap enables the robot to detect the accurate position of touched area.

Figure 14 shows the robot's movement. The robot closes the joint ① until it touches a human. Then it identifies the position of the touched point, and twists the joint ③ to enable the joint ④ to bend to the direction of touched area. Then it bends the joint ④ until

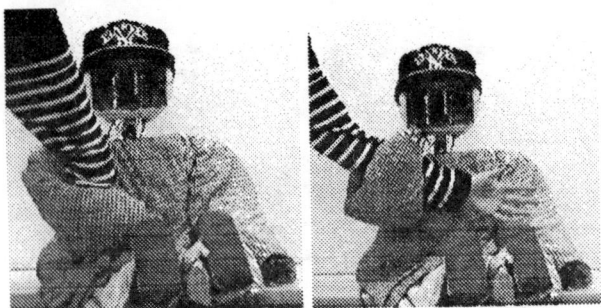

Figure 14: Conforming to a human's arm

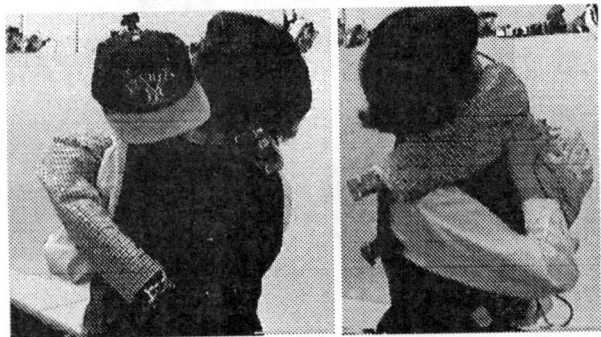

Figure 15: Conforming and holding on to a human's shape

it touches the human again. The robot detects the direction of the touched point and moves the joints ⑤ and ⑥ in the same manner, until it touches the human. In this experiment, a human touched two different areas of the robot's arm and those results of conforming behavior show that the change of the correspondence of the sensor data to the robot body was absorbed correctly. Figure 15 also shows the robot conforming to a human shape. The robot could hold on to a human with its arms.

6 Conclusions and Future Works

This paper described the modeling of the tactile sensor suit and conversion and compensation processing to symbolize the tactile data to obtain accurate tactile data. Then we presented two experiments to evaluate our approach. One is the acquisition of the Initial Maps and the other is conforming to a human.

The tactile data was symbolized to the names of robot's body based on the modeling of the tactile sensor suit to digest the large amount of raw signals. The PCMap enables the robot to absorb the displacement of the two coordinate systems, and SICMap also enables the robot and to distinguish the external interaction from the self-interference. In addition, methods for acquiring these maps are developed.

The goal of our research is to realize close interaction in daily life, as the conforming experiment shown an example. Therefore it is necessary for this system to be extended to the next stages of multi-sensor fusion, such as voice interaction where a human talks the robot to evaluate the robot reactions. The robot should also try to map the situation and behavior, and the robot has to recognize the situation and choose appropriate motion for that situation.

This research has been partly supported by grants of Grant-in-Aid for Scientific Research of the Ministry of Education, Science and Culture of Japan, and "Research for the Future" Program of the Japan Society for the Promotion of Science(JSPS-RFTF96P00801).

References

[1] Rodney A. Brooks, Cynthia Breazeal (Ferrell), Robert Irie, Charles C. Kemp, Matthew Marjanovic, Brian Scassellati, and Matthew Williamson. Alternate Essences of Intelligence. *Submitted to AAAI-98*, 1998.

[2] S.Hashimoto, S.Narita, K.Shirai, A.Takanishi, H.Kasahara, T.Kobarashi, and S.Sugano. A Human Symbiotic Humanoid Robot: Hadaly-2. In *Proc. 15th Annual Conference of Robotics Society of Japan (in Japanese)*, pp. 761–762, 1997.

[3] Atsushi Konno, Koichi Nagashima, Ryo Furukawa, Koichi Nishiwaki, Takuro Noda, Masayuki Inaba, and Hirochika Inoue. Development of a Humanoid Robot Saika. In *Proc. of Int. Conf on Intelligent Robots and Systems*, pp. 805–810, 1997.

[4] Masayuki INABA, Yukiko HOSHINO, Kenichiro NAGASAKA, Tatsuo NINOMIYA, Satoshi KAGAMI, and Hirochika INOUE. A Full-Body Tactile Sensor Suit Using Electrically Conductive Fabric and Strings. In *Proceedings of the 1996 IEEE/RSJ International Conference on Intellignet Robots and Systems*, pp. 450–457, 1996.

[5] Masayuki INABA, Takashi IGARASHI, Satoshi KAGAMI, and Hirochika INOUE. A 35 DOF Humanoid that can Coordinate Arms and Legs in Standing up, Reaching and Grasping an Object. In *Proceedings of the 1996 IEEE/RSJ International Conference on Intellignet Robots and Systems*, pp. 15–22, 1996.

[6] Howard R. Nicholls and Mark H. Lee. A Survey of Robot Tactile Sensing Technology. *The International Journal of Robotics Research*, Vol. 8, No. 3, pp. 3–30, 6 1989.

[7] Makoto Shimojo, Shigeru Sato, Yoshikazu Seki, and Akihiko Takahashi. A System for Simultaneous Measuring Grasping Posture and Pressure Distribution. In *IEEE International Conference on Robotics and Automation*, pp. 831–836, 1995.

[8] Leon D. Harmon. Automated Tactile Sensing. *The International Journal of Robotics Reserch*, Vol. 1, No. 2, pp. 3–32, 1982.

[9] Fumio KANEHIRO, Masayuki INABA, and Hirochika INOUE. Development of the Remote-Brained Robot which has a Nervous System by a On-Body LAN. In *Proc. 15th Annual Conference of Robotics Society of Japan (in Japanese)*, pp. 1023–1024, 1997.

[10] Masayuki INABA, Satoshi KAGAMI, Kazuhiko SAKAKI, Fumio KANEHIRO, and Hirochika INOUE. Vision-Based Multisensor Integration in Remote-Brained Robots. In *Proceedings of the 1994 IEEE International Conference on Multisensor Fusion and Integration for Intelligent Systems*, pp. 747–754, 1994.

On the design of gravity-compensated six-degree-of-freedom parallel mechanisms

Clément M. Gosselin and Jiegao Wang
Département de Génie Mécanique
Université Laval
Québec, Québec, Canada, G1K 7P4

Abstract

The design of gravity-compensated six-degree-of-freedom parallel mechanisms — or manipulators — with revolute actuators is studied in this paper. Two methods are studied for the static balancing of these mechanisms, namely, using counterweights and using springs. The first method leads to mechanisms with a stationary global center of mass while the second approach leads to mechanisms whose total potential energy (including the elastic potential energy stored in the springs as well as the gravitational potential energy) is constant. In both cases, the resulting mechanisms are fully compensated for gravity, i.e., the actuators do not contribute to supporting the weight of the moving links in any of the configurations of the mechanisms. The position vector of the global center of mass and the total potential energy of the manipulator are first expressed as functions of the position and orientation of the platform. Then, conditions for static balancing are derived from the resulting expressions. Finally, examples are given in order to illustrate the design methodologies.

1 Introduction

The balancing of mechanisms has been an important research topic for several decades (see for instance [1] for a literature review). A balanced mechanism leads to better dynamic characteristics and less vibrations caused by motion. Static and dynamic balancing of planar linkages has been studied extensively in the literature (see for instance [2, 3, 4, 5, 6]).

In the context of manipulators and motion simulation mechanisms, static balancing is defined as the set of conditions under which the weight of the links of the mechanism does not produce any torque (or force) at the actuators under static conditions, for any configuration of the manipulator or mechanism. This condition is also referred to as *gravity compensation*. Gravity-compensated serial manipulators have been designed in [7, 8, 9, 10, 11] using counterweights, springs and sometimes cams and/or pulleys. A hybrid direct-drive gravity-compensated manipulator has also been developed in [12]. Moreover, a general approach for the static balancing of planar linkages using springs has been presented in [13]. The balancing of spatial mechanisms has also been studied, for instance in [14] and [11].

However, to the knowledge of the authors, gravity-compensated spatial six-degree-of-freedom parallel manipulators or mechanisms cannot be found in the literature. Since spatial parallel mechanisms find more and more applications in robotics and flight simulation, their static balancing becomes an important issue. As mentioned above, a statically balanced parallel mechanism is one in which the actuators do not contribute to supporting the weight of the moving links, for any configuration. Hence, the actuators are used only to impart accelerations to the moving links, which leads to a reduction of the size and power of the actuators and results in the improvement of the accuracy of the control. In flight simulation, for instance, since the payload is very large (usually in the order of tons) and the motion of the platform of the mechanism is rather slow, the forces or torques exerted at the actuated joints are mainly due to the weight of the platform and links. Hence, if the mechanism is statically balanced, the actuating forces or torques will be greatly reduced, which will result in significant improvements of the control and energy efficiency. Finally, from a more general perspective, the design of a 'floating' gravity-compensated platform with six degrees of freedom may have several applications, including the simulation of space systems.

In this paper, the static balancing of spatial six-degree-of-freedom parallel mechanisms or manipulators with revolute actuators is addressed. Two approaches of static balancing are presented, namely, *i*) static balancing using counterweights and *ii*) using springs. When the mechanism is balanced using counterweights, a mechanism with a fixed global center of mass is obtained. In other words, the static balancing is achieved in any direction of the Cartesian space of the mechanism. This property is useful for applications in which the mechanism is needed to be statically balanced in all directions as for instance, when a system can be installed in different orientations with respect to the gravity vector. However, for some parallel mechanisms, static balancing with counterweights is difficult to realize. For example, in flight simulators, since the mass of the platform is very large, the counterweights required would be too large to be practical. Springs can be used in such instances. When springs are used, the total potential energy of the manipulator — gravitational and elastic — is set to be constant and the weight of the whole manipulator can be balanced with a much smaller total mass than when using counterweights, as pointed out in [13]. However, a mechanism which is statically balanced using springs will be statically balanced for only one direction and magnitude of the gravity vector, which may be unsuitable for some applications. Both methodologies are discussed in this paper.

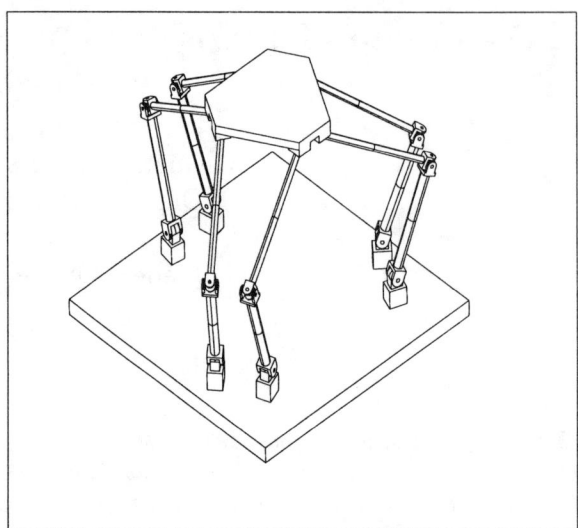

Figure 1: CAD model of a spatial six-degree-of-freedom parallel mechanism with revolute actuators.

2 Six-degree-of-freedom parallel mechanism with revolute actuators

A spatial six-degree-of-freedom parallel mechanism or manipulator with revolute actuators is illustrated in Figs. 1 and 2. It consists of six identical legs connecting the base to the platform. Each of these legs consists of an actuated revolute joint attached to the base, a first moving link, a passive Hooke joint, a second moving link and a passive spherical joint attached to the platform. A parallel manipulator of this type was described in [15]. The coordinate frame of the base, designated as the $O - x, y, z$ frame is fixed to the base with its Z-axis pointing vertically upward. Similarly, the moving coordinate frame $O' - x', y', z'$ is attached to the platform.

The Cartesian coordinates of the platform are given by the position of point O' with respect to the fixed frame, noted $\mathbf{p} = [x, y, z]^T$ and the orientation of the platform (orientation of frame $O' - x'y'z'$ with respect to the fixed frame), represented by matrix $\mathbf{Q}$, which

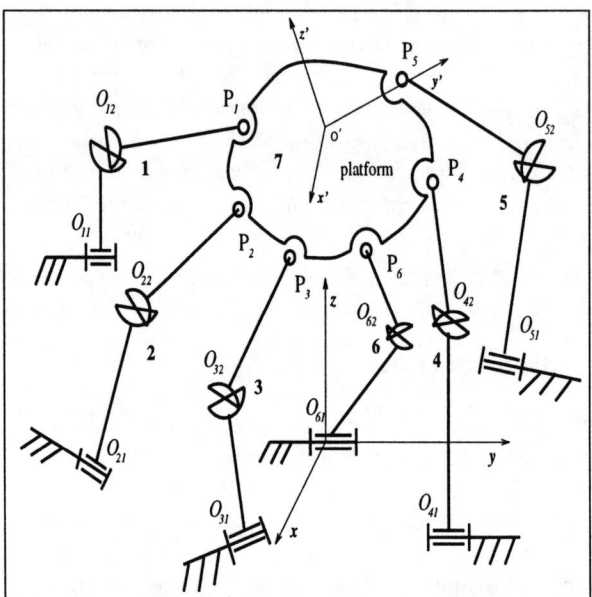

Figure 2: Schematic representation of a spatial six-degree-of-freedom parallel mechanism with revolute actuators.

can be written as

$$\mathbf{Q} = \begin{bmatrix} q_{11} & q_{12} & q_{13} \\ q_{21} & q_{22} & q_{23} \\ q_{31} & q_{32} & q_{33} \end{bmatrix} \quad (1)$$

where the entries can be expressed as functions of Euler angles, quadratic invariants, linear invariants or any other representation.

Finally, the coordinates of point P_i (Fig. 2) relative to the moving coordinate frame of the platform are noted (a_i, b_i, c_i) with $i = 1, \ldots, 6$.

3 Static balancing of spatial six-degree-of-freedom mechanisms

3.1 Derivation of the balancing conditions

The conditions for the static balancing of three types of spatial six-degree-of-freedom parallel mechanisms will now be derived using counterweights and then using springs.

Static balancing using counterweights consists in ensuring that the global center of mass of the mechanism remains fixed for any configuration of the mechanism. In other words, the resulting manipulator would be statically balanced for any direction of the gravity vector — and hence the weight of the mechanism does not have any effect on the actuators —, which is a desirable property for portable systems which may be mounted in different orientations.

On the other hand, static balancing using springs consists in ensuring that the total potential energy of the mechanism is kept constant, which also means that the weight of the mechanism does not have any effect on the actuators, but only for one direction of the gravity vector. Moreover, using this approach, the weight of the whole mechanism can be balanced with a much smaller total mass than when using counterweights.

The two links of the ith leg of the mechanism are represented in Fig. 3. A reference frame noted $O_{i1} - x_i, y_i, z_i$ is attached to the first link of the leg. Point O_{i1} is located at the center of the first revolute joint. The coordinates of point O_{i1} expressed in the base coordinate frame are (x_{io}, y_{io}, z_{io}), where $i = 1, \ldots, 6$. Moreover, the unit vectors defined in the direction of axes x_i, y_i and z_i are noted $\mathbf{x}_{i1}, \mathbf{y}_{i1}$ and $\mathbf{z}_{i1}$, respectively.

Vector $\mathbf{z}_{i1}$ is defined along the axis directed from point O_{i1} toward point O_{i2} while vector $\mathbf{x}_{i1}$ is defined

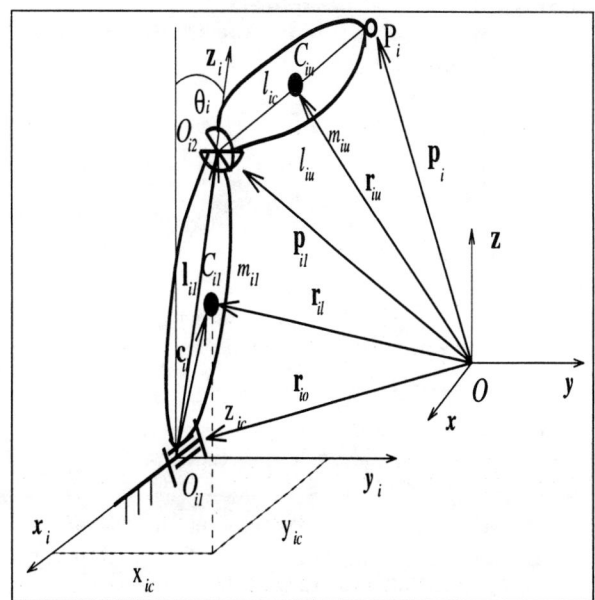

Figure 3: Geometry of the ith leg.

along the direction of the first revolute joint axis. Finally, vector $\mathbf{y}_{i1}$ is defined as

$$\mathbf{y}_{i1} = \frac{\mathbf{z}_{i1} \times \mathbf{x}_{i1}}{|\mathbf{z}_{i1} \times \mathbf{x}_{i1}|}, \quad i = 1, \ldots, 6 \quad (2)$$

Also, points C_{il} and C_{iu} denote respectively the center of mass of the lower and upper link of each leg.

Let θ_i be the joint variable associated with the first revolute joint of the ith leg and γ_i be the angle between the positive direction of the $\mathbf{x}$ axis of the base coordinate frame and the coordinate axis $\mathbf{x}_{i1}$, where it is assumed that vector $\mathbf{x}_{i1}$ is contained in the xy plane of the fixed reference frame (Fig. 3). One can write the rotation matrix giving the orientation of frame $O_{i1} - x_i, y_i, z_i$ with respect to the reference frame attached to the base as

$$\mathbf{Q}_{i1} = \begin{bmatrix} \cos\gamma_i & -\sin\gamma_i \cos\theta_i & \sin\gamma_i \sin\theta_i \\ \sin\gamma_i & \cos\gamma_i \cos\theta_i & -\cos\gamma_i \sin\theta_i \\ 0 & \sin\theta_i & \cos\theta_i \end{bmatrix}$$
$$i = 1, \ldots, 6 \quad (3)$$

Moreover, it is assumed that the center of mass of the second link of the ith leg lies on line $O_{i2}P_i$, as represented in Fig. 3. One can then write

$$\mathbf{p}_{i1} = \mathbf{r}_{io} + \mathbf{Q}_{i1}\mathbf{l}_{il}, \quad i = 1, \ldots, 6 \quad (4)$$

where $\mathbf{p}_{i1}$ and $\mathbf{r}_{io}$ are respectively the position vectors of points O_{i2} and O_{i1} expressed in the base coordinate frame, as represented in Fig. 3, while $\mathbf{l}_{il}$ is the vector pointing from O_{i1} to O_{i2} and expressed in the local coordinate frame, and

$$\mathbf{r}_{io} = \begin{bmatrix} x_{io} \\ y_{io} \\ z_{io} \end{bmatrix}, \quad \mathbf{p}_{i1} = \begin{bmatrix} x_{i1} \\ y_{i1} \\ z_{i1} \end{bmatrix}, \quad \mathbf{l}_{il} = \begin{bmatrix} 0 \\ 0 \\ l_{il} \end{bmatrix}$$
$$i = 1, \ldots, 6$$

where l_{il} is the distance from O_{i1} to O_{i2}.

Eq.(4) can be written in component form as

$$x_{i1} = x_{io} + l_{i1} \sin\gamma_i \sin\theta_i, \quad i = 1, \ldots, 6 \quad (5)$$
$$y_{i1} = y_{io} - l_{i1} \cos\gamma_i \sin\theta_i, \quad i = 1, \ldots, 6 \quad (6)$$
$$z_{i1} = z_{io} + l_{i1} \cos\theta_i, \quad i = 1, \ldots, 6 \quad (7)$$

Then, one can compute the position vector of the center of mass of the second link of the ith leg from the position vectors of points O_{i2} and P_i as

$$\mathbf{r}_{iu} = \mathbf{p}_i - \frac{l_{ic}}{l_{iu}}(\mathbf{p}_i - \mathbf{p}_{i1}), \quad i = 1, \ldots, 6 \quad (8)$$

where $\mathbf{r}_{iu}$ is the position vector of the center of mass of the upper link of the ith leg and where l_{iu} and l_{ic} are respectively the distance from O_{i2} to P_i and from O_{i2} to C_{iu}. Moreover, position vector $\mathbf{p}_i$ can be expressed as a function of the position and orientation of the platform, i.e.,

$$\mathbf{p}_i = \mathbf{p} + \mathbf{Q}\mathbf{p}'_i, \quad i = 1, \ldots, 6 \quad (9)$$

where

$$\mathbf{p} = \begin{bmatrix} x \\ y \\ z \end{bmatrix}, \quad \mathbf{p}'_i = \begin{bmatrix} a_i \\ b_i \\ c_i \end{bmatrix}, \quad i = 1, \ldots, 6 \quad (10)$$

The global center of mass of the mechanism, noted $\mathbf{r}$ can then be written as

$$M\mathbf{r} = m_p \mathbf{r}_p + \sum_{i=1}^{6}(m_{il}\mathbf{r}_{il} + m_{iu}\mathbf{r}_{iu}) \quad (11)$$

where M is the total mass of all moving links of the mechanism, m_p, m_{iu} and m_{il} are respectively the masses of the platform, the upper link and lower link of the ith leg, and

$$M = m_p + \sum_{i=1}^{6}(m_{il} + m_{iu}) \quad (12)$$

while $\mathbf{r}_p$ and $\mathbf{r}_{il}$ are respectively the position vectors of the center of mass of the platform of the mechanism and of the center of mass of the lower link of the ith leg, namely

$$\mathbf{r}_p = \mathbf{p} + \mathbf{Q}\mathbf{c}_p \quad (13)$$
$$\mathbf{r}_{il} = \mathbf{r}_{io} + \mathbf{Q}_{il}\mathbf{c}_{il}, \quad i = 1, \ldots, 6 \quad (14)$$

where $\mathbf{c}_p$ and $\mathbf{c}_{il}$ are the position vectors of the center of mass of the platform and of the lower links expressed in the local reference frame, and whose components are given as

$$\mathbf{c}_p = \begin{bmatrix} x_p \\ y_p \\ z_p \end{bmatrix}; \quad \mathbf{c}_{il} = \begin{bmatrix} x_{ic} \\ y_{ic} \\ z_{ic} \end{bmatrix}, \quad i = 1, \ldots, 6 \quad (15)$$

Substituting eqs.(8), (13) and (14) into eq.(11), one then obtains

$$M\mathbf{r} = \begin{bmatrix} r_x \\ r_y \\ r_z \end{bmatrix} \quad (16)$$

where

$$r_x = \sum_{i=1}^{6}(D_i \sin\gamma_i \sin\theta_i - D_{i+6} \sin\gamma_i \cos\theta_i)$$
$$+ D_{13}x + D_{14}q_{11} + D_{15}q_{12} + D_{16}q_{13} + D_{xo}$$
$$r_y = \sum_{i=1}^{6}(D_{i+6}\cos\gamma_i \cos\theta_i - D_i \cos\gamma_i \sin\theta_i)$$
$$+ D_{13}y + D_{14}q_{21} + D_{15}q_{22} + D_{16}q_{23} + D_{yo}$$
$$r_z = \sum_{i=1}^{6}(D_i \cos\theta_i + D_{i+6}\sin\theta_i)$$
$$+ D_{13}z + D_{14}q_{31} + D_{15}q_{32} + D_{16}q_{33} + D_{zo}$$

where D_{xo}, D_{yo} and D_{zo} are constant coefficients, and where

$$D_i = m_{il}z_{ic} + \frac{l_{il}}{l_{iu}}l_{ic}m_{iu}, \quad i = 1, \ldots, 6$$
$$D_{i+6} = m_{il}y_{ic}, \quad i = 1, \ldots, 6$$
$$D_{13} = m_p + \sum_{i=1}^{6}m_{iu}(1 - \frac{l_{ic}}{l_{iu}})$$
$$D_{14} = m_p x_p + \sum_{i=1}^{6}m_{iu}a_i(1 - \frac{l_{ic}}{l_{iu}})$$
$$D_{15} = m_p y_p + \sum_{i=1}^{6}m_{iu}b_i(1 - \frac{l_{ic}}{l_{iu}})$$
$$D_{16} = m_p z_p + \sum_{i=1}^{6}m_{iu}c_i(1 - \frac{l_{ic}}{l_{iu}})$$

In the above expressions for r_x, r_y and r_z, if the coefficients of the joint and Cartesian variables vanish, then the global center of mass of the mechanism will be fixed for any configuration of the mechanism. Hence, one obtains the conditions for static balancing as follows

$$D_i = 0, \quad i = 1, \ldots, 16 \qquad (17)$$

3.2 Example

An example is now given in order to illustrate the application of the balancing conditions to the type of mechanism described above. For this mechanism, let

$$m_p = 12, l_{iu} = l_{il} = 1 \, (i = 1, \ldots, 6)$$
$$a_1 = -0.5, \, b_1 = -0.5, \, c_1 = -0.3$$
$$a_2 = 0.5, \, b_2 = -0.5, \, c_2 = -0.3$$
$$a_3 = 0.5, \, b_3 = 0.5, \, c_3 = -0.3$$
$$a_4 = -0.5, \, b_4 = 0.5, \, c_4 = -0.3$$
$$a_5 = 0.5, \, b_5 = 0.0, \, c_5 = -0.3$$
$$a_6 = -0.5, \, b_6 = 0.0, \, c_6 = -0.3$$
$$x_{1o} = -1.5, \, y_{1o} = -1.5, \, z_{1o} = 0$$
$$x_{2o} = 1.5, \, y_{2o} = -1.5, \, z_{2o} = 0$$
$$x_{3o} = 1.5, \, y_{3o} = 1.5, \, z_{3o} = 0$$
$$x_{4o} = -1.5, \, y_{4o} = 1.5, \, z_{4o} = 0$$
$$x_{5o} = 1.5, \, y_{5o} = 0, \, z_{5o} = 0$$
$$x_{6o} = -1.5, \, y_{6o} = 0, \, z_{6o} = 0$$
$$\gamma_1 = \frac{\pi}{6}, \, \gamma_2 = -\frac{\pi}{6}, \, \gamma_3 = \frac{5\pi}{6}$$
$$\gamma_4 = \frac{3\pi}{6}, \, \gamma_5 = \frac{9\pi}{6}, \, \gamma_6 = \frac{7\pi}{6}$$

where the masses are given in kilograms and the lengths in meters.

¿From eqs.(17), one obtains

$$y_{ic} = 0, \, z_{ic} = -0.5 \, (\text{m})$$
$$m_{iu} = 4 \, (\text{kg}) \, m_{il} = 13 \, (\text{kg}) \, (i = 1, \ldots, 6)$$
$$x_p = 0 \, (\text{m}), \, y_p = 0, \, z_p = 0.3 \, (\text{m})$$

The balanced manipulator is represented schematically in Fig. 4. As can be realized from the numerical results and from the figure, large counterweights are necessary to balance the mechanism.

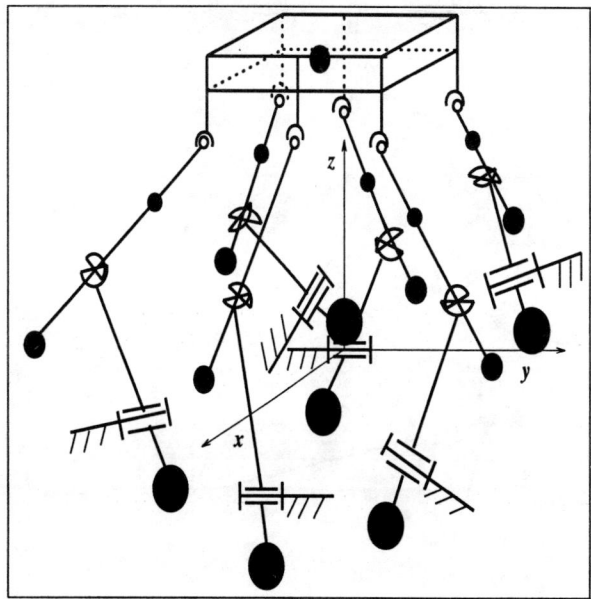

Figure 4: Complete balancing using counterweights.

4 Static balancing using springs

4.1 Kinematic architecture of the legs

In order to use springs to balance the manipulator, a special architecture (similar to what was used in [13]) is proposed for the legs. As represented in Fig. 5, a parallelogram four-bar linkage is used instead of the first link of the ith leg. This enables the attachment of a spring to the upper link of the leg and to a support which is maintained vertically. A spring is also attached to the parallelogram. The upper link of the leg is then mounted on a revolute joint with a horizontal axis which is in turn mounted on a revolute joint with a vertical axis. The latter two joint form a Hooke joint and therefore, the new architecture is kinematically equivalent to the previous one. However, the new architecture now allows the use of springs for the static balancing of the mechanism. Moreover, it is pointed out that the global center of mass of the parallelogram and the center of mass of the replaced first link of the ith leg can be handled similarly.

4.2 Derivation of the balancing conditions

The expression of the total potential energy of the mechanism including springs can be written as

$$V = V_w + V_s \qquad (18)$$

where V_w and V_s are respectively the gravitational potential energy and the elastic potential energy stored

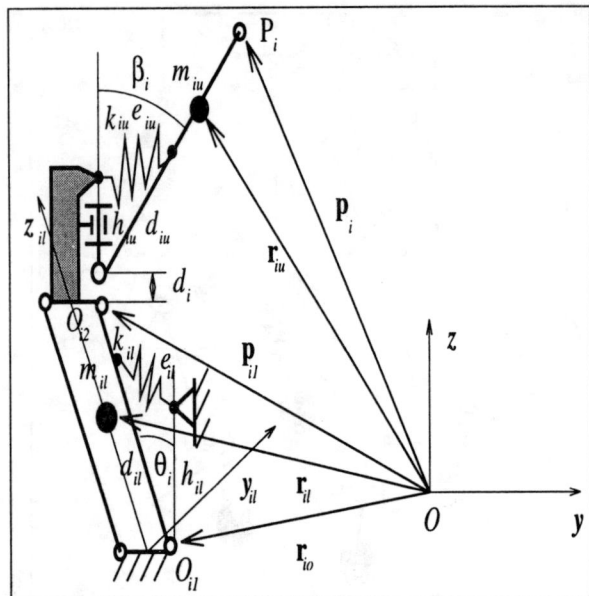

Figure 5: Geometry and kinematic architecture of the ith leg.

in the springs. Theses quantities can be written, for this mechanism, as

$$V_w = r_z g + D_c \quad (19)$$

$$V_s = \frac{1}{2}\sum_{i=1}^{6}(k_{il}e_{il}^2 + k_{iu}e_{iu}^2) \quad (20)$$

where r_z is defined in eq.(16), D_c is a constant which arises from the offset distance d_i (Fig. 5) and which can be written as $D_c = m_p g d_1 + \sum_{i=1}^{6}(m_i g d_i)$, g is the gravitational acceleration, k_{il} is the stiffness of the lower spring of the ith leg, e_{il} is the length of the lower spring of the ith leg, k_{iu} is the stiffness of the upper spring of the ith leg and e_{iu} is its length. It is assumed here that the undeformed length of the springs is equal to zero in order to obtain complete balancing [13]. As shown in [13], this condition can easily be met in a practical design. Using the law of cosines, the length of the springs can be written as

$$e_{il} = \sqrt{h_{il}^2 + d_{il}^2 - 2h_{il}d_{il}\cos\theta_i} \quad (21)$$

$$e_{iu} = \sqrt{h_{iu}^2 + d_{iu}^2 - 2h_{iu}d_{iu}\cos\beta_i} \quad (22)$$

$$i = 1,\ldots,6$$

where h_{il} and d_{il} are the distances from the revolute joint located at O_{i1} to the attachment points of the lower spring (Fig. 5) while h_{iu} and d_{iu} are the same distance for the upper spring. The cosine of angle β_i can be expressed as a function of angle θ_i as well as the position and orientation of the platform, i.e.,

$$\cos\beta_i = \frac{z_i - z_{i1} - d_i}{l_{iu}} \quad (23)$$

and where z_i and z_{i1} are respectively the third components of the position vectors $\mathbf{p}_i$ and $\mathbf{p}_{i1}$.

Substituting eqs.(19), (20) and (21)–(23) into eq.(18), one then obtains

$$\begin{aligned}V =& \sum_{i=1}^{6}[(D_i g - 2k_{il}h_{il}d_{il} - 2k_{iu}h_{iu}d_{iu}\frac{l_{il}}{l_{iu}})\cos\theta_i \\ &+ D_{i+6}g\sin\theta_i] + (D_{13}g - 2\sum_{i=1}^{6}\frac{k_{iu}h_{iu}d_{iu}}{l_{iu}})z \\ &+ (D_{14}g - 2\sum_{i=1}^{6}\frac{k_{iu}h_{iu}d_{iu}}{l_{iu}})q_{31} \\ &+ (D_{15}g - 2\sum_{i=1}^{6}\frac{k_{iu}h_{iu}d_{iu}}{l_{iu}})q_{32} \\ &+ (D_{16}g - 2\sum_{i=1}^{6}\frac{k_{iu}h_{iu}d_{iu}}{l_{iu}})q_{33} \\ &+ \frac{1}{2}\sum_{i=1}^{6}[k_{il}(h_{il}^2 + d_{il}^2) + k_{iu}(h_{iu}^2 + d_{iu}^2)] \\ &- \sum_{i=1}^{6}2\frac{k_{iu}h_{iu}d_{iu}}{l_{iu}} + D_{zo}g + D_c \end{aligned} \quad (24)$$

¿From eq.(24) one can finally obtain the conditions for the static balancing of the manipulator with springs as follows

$$D_i g - 2k_{il}h_{il}d_{il} - 2\frac{l_{il}}{l_{iu}}k_{iu}h_{iu}d_{iu} = 0 \quad (25)$$
$$i = 1,\ldots,6$$

$$D_{i+6} = 0 \quad (26)$$
$$i = 1,\ldots,6$$

$$D_{13}g - 2\sum_{i=1}^{6}\frac{k_{iu}h_{iu}d_{iu}}{l_{iu}} = 0 \quad (27)$$

$$D_{14}g - 2\sum_{i=1}^{6}\frac{k_{iu}h_{iu}d_{iu}}{l_{iu}} = 0 \quad (28)$$

$$D_{15}g - 2\sum_{i=1}^{6}\frac{k_{iu}h_{iu}d_{iu}}{l_{iu}} = 0 \quad (29)$$

$$D_{16}g - 2\sum_{i=1}^{6}\frac{k_{iu}h_{iu}d_{iu}}{l_{iu}} = 0 \quad (30)$$

4.3 Example

An example is now given in order to illustrate the application of the balancing conditions to this type of mechanism.

For the 6-dof manipulator with revolute actuators presented above, let

$m_p = 12, m_{iu} = m_{il} = 1.0, l_{il} = l_{iu} = 1.0 (i = 1, \ldots, 6)$

$$a_1 = -0.5, \; b_1 = -0.5, c_1 = -0.3$$
$$a_2 = 0.5, \; b_2 = -0.5, c_2 = -0.3$$
$$a_3 = 0.5, b_3 = 0.5, c_3 = -0.3$$
$$a_4 = -0.5, b_4 = 0.5, c_4 = -0.3$$
$$a_5 = 1.0, b_5 = 0, c_5 = -0.3$$
$$a_6 = -1.0, b_6 = 0, c_6 = -0.3$$
$$h_{il} = h_{iu} = 0.5, d_{il} = d_{lu} = 0.5$$
$$z_{ic} = l_{ic} = 0.5 \, (i = 1, \ldots, 6)$$
$$x_{1o} = -1.5, y_{1o} = -1.5, z_{1o} = 0$$
$$x_{2o} = 1.5, y_{2o} = -1.5, z_{2o} = 0$$
$$x_{3o} = 1.5, y_{3o} = 1.5, z_{3o} = 0$$
$$x_{4o} = -1.5, y_{4o} = 1.5, z_{4o} = 0$$
$$x_{5o} = 1.5, y_{5o} = 0, z_{5o} = 0$$
$$x_{6o} = -1.5, y_{6o} = 0, z_{6o} = 0$$
$$\gamma_1 = \frac{\pi}{6}, \gamma_2 = -\frac{\pi}{6}, \gamma_3 = \frac{5\pi}{6}$$
$$\gamma_4 = \frac{3\pi}{6}, \gamma_5 = \frac{9\pi}{6}, \gamma_6 = \frac{7\pi}{6}$$

where the masses are given in kilograms and the lengths in meters.

¿From eqs.(25)–(30) one obtains

$$z_{ic} = l_{ic} = 0.5 \, (\text{m}) \, (i = 1, \ldots, 6)$$
$$y_{ic} = 0, k_{iu} = 300 \, (\text{N/m}), k_{il} = 620 \, (\text{N/m}) \, (i = 1, \ldots, 6)$$
$$x_p = 0 \, (\text{m}), y_p = 0, z_p = 0.45 \, (\text{m})$$

The balanced mechanism is represented schematically in Fig. 6. Since each leg of the mechanism has an identical architecture, only one leg is represented in the figure.

As can be clearly seen from the numerical results and from the figure, the use of springs has allowed to eliminate the counterweights and the total moving mass is therefore much smaller (24 kg instead of 114 kg). If the mass of the platform is very large — as in flight simulators for instance — the use of springs would be much more practical than counterweights. However, the resulting mechanism will be statically balanced if and only if the gravity vector is aligned with the negative direction of the **z** axis of the fixed reference frame.

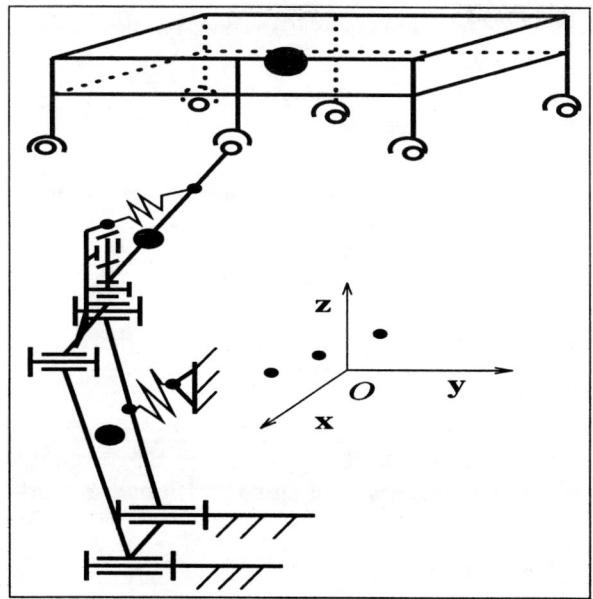

Figure 6: Balanced mechanism with springs.

5 Conclusion

The static balancing of spatial six-degree-of-freedom parallel mechanisms has been addressed in this paper. Two static balancing approaches, namely, with counterweights and with springs have been used. To this end, the expressions of the position vectors of the global center of mass and the potential energy of the mechanisms have been derived. The sets of equations of static balancing have finally been obtained from these expressions. Two examples have been given in order to illustrate the results. The examples are provided for illustrative purposes only. Indeed, it is clear, from the equations, that infinitely many statically balanced mechanisms exist, for each of the architectures studied here. Moreover, it is also found, by inspection of the equations, that balancing is always possible for any given value of the geometric parameters. This is an interesting result since it allows the kinematic design of a mechanism to be completed using any criterion and the balancing to be performed a *posteriori*.

It has been clearly shown that the types of spatial parallel six-degree-of-freedom mechanisms studied here can be statically balanced using either one of the two approaches presented in this paper. Each approach has its own advantages and is suitable to different applications. In all cases, the mechanisms obtained are perfectly balanced, i.e., no torque is required at the actuators to maintain the mechanism in static equilibrium for any configuration. Static balancing of spatial six-degree-of-freedom parallel mechanisms is of great interest and can be used in the design of mechanisms for robotics, flight simulators and several other applications involving large loads or the simulation of free-floating conditions.

Acknowledgements: The authors would like to acknowledge the financial support of the Natural Sciences and Engineering Research Council of Canada (NSERC) as well as the Fonds pour la Formation de Chercheurs et l'Aide à la Recherche du Québec (FCAR).

References

[1] G.G. Lowen, F.R. Tepper and R.S. Berkof, 'Balancing of linkages – an update', *Mechanism and Machine Theory*, Vol. 18, No. 3, pp. 213–220, 1983.

[2] E. N. Stevenson Jr., 'Balancing of machines', *ASME Journal of Engineering for Industry*, Vol. 95, No. 2, pp. 650–656, 1973.

[3] M. R. Smith, 'Optimal balancing of planar multibar linkages', *Proceedings of the 5th World Congress on the Theory of Machines and Mechanisms*, New-Castle-Upon-Tyne, pp. 142–149, 1975.

[4] C. Bagci, 'Shaking force balancing of planar linkages with force transmission irregularities using balancing idler loops', *Mechanism and Machine Theory*, Vol. 14, No. 4, pp. 267–284, 1979.

[5] F. Gao, 'Complete shaking force and shaking moment balancing of 17 types of eight-bar linkages only with revolute pairs', *Mechanism and Machine Theory*, Vol. 26, No. 2, pp. 179–206, 1991.

[6] Z. Ye and M.R. Smith, 'Complete balancing of planar linkages by an equivalence method', *Mechanism and Machine Theory*, Vol. 29, No. 5, pp. 701–712, 1994.

[7] R.H. Nathan, 'A constant force generation mechanism', *ASME Journal of Mechanisms, Transmissions, and Automation in Design*, Vol. 107, No. 4, pp. 508–512, 1985.

[8] J.M. Hervé, 'Device for counter-balancing the forces due to gravity in a robot arm', United States Patent 4,620,829, May 1986.

[9] D.A. Streit and B.J. Gilmore, 'Perfect spring equilibrators for rotatable bodies', *ASME Journal of Mechanisms, Transmissions, and Automation in Design*, Vol. 111, No. 4, pp. 451–458, 1989.

[10] N. Ulrich and V. Kumar, 'Passive mechanical gravity compensation for robot manipulators', Proceedings of the *IEEE International Conference on Robotics and Automation*, Sacramento, pp. 1536–1541, 1991.

[11] G.J. Walsh, D.A. Streit and B.J. Gilmore, 'Spatial spring equilibrator theory', *Mechanism and Machine Theory*, Vol. 26, No. 2, pp. 155–170, 1991.

[12] H. Kazerooni and S. Kim, 'A new architecture for direct drive robots', Proceedings of the *IEEE Int. Conference on Robotics and Automation*, Philadelphia, pp. 442–445, 1988.

[13] D.A. Streit and E. Shin, 'Equilibrators for planar linkages', Proceedings of the *ASME Mechanisms Conference*, Chicago, Vol. DE-25, pp. 21–28, 1990.

[14] C. Bagci,, 'Complete balancing of space mechanisms – shaking force balancing', *ASME Journal of Mechanisms, Transmissions, and Automation in Design*, Vol. 105, No. 12, pp. 609–616, 1983.

[15] R. Benea, 'Contribution à l'étude des robots pleinement parallèles de type 6R-RR-S', Ph.D. Thesis, Université de Savoie, France, 1996.

A Parallel x-y Manipulator with Actuation Redundancy for High-Speed and Active-Stiffness Applications

S. Kock, W. Schumacher
Institute of Control Engineering
Technical University Braunschweig, Germany

Abstract—A 2-d.o.f. parallel manipulator with actuation redundancy is examined for high-speed and stiffness-controlled operation. Advantages of actuation redundancy are outlined. The kinematics and singularity-free workspace of the manipulator are presented together with a force transmission analysis. Finally, a novel control scheme that guarantees a lower bound of the end-effector stiffness (LBSC) is presented. Simulation results are compared with a traditional control scheme for high-speed applications using the minimal 2-norm of actuator torques.

I. INTRODUCTION

Planar parallel manipulators have been studied extensively (e.g. [1], [2], [3]). They are generally constituted of three independent kinematic chains ('legs') with three independent 1 d.o.f. joints each, one end of the chains connected to a fixed base, the other to a common moving platform. Depending on the kind of joints in one chain (prismatic, P, or rotating, R), these manipulators can be classified as RRR, RPR, PPP etc. [1]. However, these manipulators show many singular configurations and a poor workspace, if physical joint limits come into play [4].

Redundancy can improve the abilities and performance of planar parallel robots in certain ways. At first, the term redundancy has to be specified.
Kinematic redundancy allows for minimizing joint speeds [5]. Apparently, to obtain an effect on all chains of a parallel robot, a high order redundancy is required.
Sensing redundancy can help to avoid uncertainties in the direct kinematics or to reduce the computational effort by adding information, thus reducing computation time in the control loop ([5], [6]).
Actuation redundancy results in an overconstrained system. This means that in general, assuming rigid joints and links, internal preloads can be generated. In case that one or more chains are in a singular position, extra actuation assures mobility of the manipulator. One way to use preloading is active stiffness control [7], another is force optimization [8].

In this paper, we examine an x-y RRR-type planar manipulator with a redundant branch and a redundant active joint and briefly describe the advantages of such a structure compared to a 2-d.o.f. manipulator without redundancy. This design is known as part of a small-scale haptic force-display application ([9], [10]). We show that a singularity-free workspace can be obtained, uncertainties in the direct kinematics can be removed, and high-speed as well as stiffness-controlled operation is possible in macro-scale. Although this manipulator is considered as a demonstrator for force and stiffness control strategies, it would be a versatile tool for pick-and-place as well as machining, clamping or bracing applications. Actuation redundancy promises improvements for 3-d.o.f. planar and higher d.o.f. spatial mechanisms as well and will experience growing interest in the future, calling for advanced control strategies.

II. MANIPULATOR

The manipulator is constituted of three actuated legs, as a regular planar parallel 3-d.o.f. manipulator. In this case, we chose an RRR-type mechanism with geared permanent magnet synchronous drives. Instead of a moving platform, the three legs meet at a center joint, where the tool would be mounted. We therefore lose a rotational degree of freedom, but we obtain an overconstrained linkage with one redundant actuator, gaining one degree of freedom to choose the combination of driving torques. Fig. 1 shows the manipulator in principle.

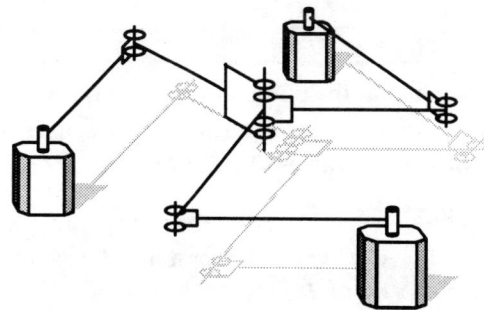

Fig. 1: Principle of redundant x-y manipulator

Extra redundancy can be added easily by spending more actuated legs. The center joint would of course become more sophisticated.

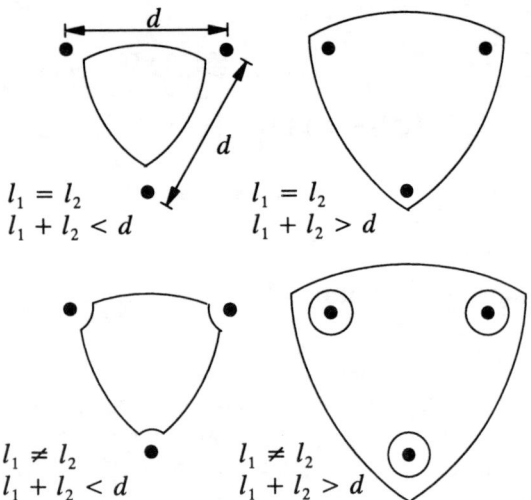

Fig. 4: Workspace for different link lengths

thors (e.g. [8], [9], [11], [12]). We therefore present the results for our manipulator rather then going into details.

If we define the 2x3 Jacobian matrix J as

$$J = \frac{\partial x_C}{\partial \Theta}, \quad (7)$$

where $\Theta = [\Theta_1 \; \Theta_2 \; \Theta_3]^T$ is the vector of the active joint angles, the relationship between the effective cartesian force $f = [f_x \; f_y]^T$ and the vector of driving torques $\tau = [\tau_1 \; \tau_2 \; \tau_3]^T$ becomes

$$f = J^{+T}\tau, \quad (8)$$

with $J^+ = \frac{\partial \Theta}{\partial x_C}$ being the pseudo-inverse of the Jacobian. Multiplying (8) from the right by $f^T = \tau^T J^+$ and taking the square root, the euclidian norm of the effective force at the end-effector is obtained:

$$|f| = \sqrt{f^T f} = \sqrt{\tau^T J^+ J^{+T} \tau}. \quad (9)$$

We can now define the force/torque transmission ratio

$$\frac{|f|}{|\tau|} = \frac{\sqrt{\tau^T J^+ J^{+T} \tau}}{\sqrt{\tau^T \tau}}. \quad (10)$$

It is known that the output bounds for $|f|$ with respect to the input loads $|\tau|$ are given by the square roots of the two singular values of the 2x2 matrix $J^{+T}J^+$:

$$\sigma_{min}|\tau| \leq |f| \leq \sigma_{max}|\tau|. \quad (11)$$

σ^2_{min} and σ^2_{max} can be obtained by a singular value decomposition (SVD) of $J^{+T}J^+$.

The ratio $D_l = \frac{\sigma_{min}}{\sigma_{max}}$ is a useful index for describing the local behavior of the manipulator, since it ranges from 1 (isotropy) to 0 (singularity). In literature, it is referred to as either local dexterity index [11] or force shape index [12]. Since the condition numbers of $J^{+T}J^+$ and JJ^T are identical [13], D_l can be computed from either J^+ or J, whatever seems more convenient. There are three ways for D_l to become zero, indicating a singularity. One is that $\sigma_{max} \to \infty$. This indicates a singularity of the 1st kind (undermobility), where J drops rank. Another is $\sigma_{min} = 0$ indicating a singularity of 2nd kind (overmobility) or a loss of rank of J^+. Both cases at the same time is known as a singularity of the 3rd kind.

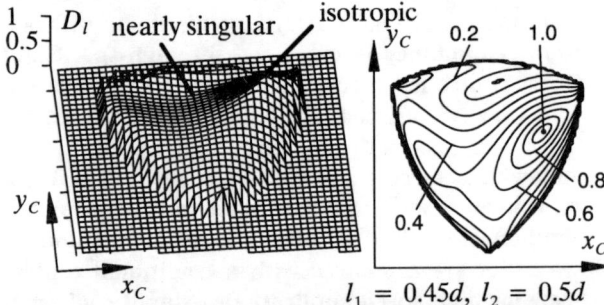

Fig. 5: Local dexterity plots of configuration f

We investigated the dexterity of our manipulator in configurations a and f of Fig. 3. Fig. 5 shows the local dexterity D_l over the workspace for configuration f, link lengths $l_1 = 0.45d$ and $l_2 = 0.5d$. There is an off-center isotropic position and a nearly singular position, which would become singular if $l_1 = l_2 = 0.5d$. This singularity is hypothetical because it would require joints B_2 and B_3 to have the same position. Nevertheless the local dexterity index approaches zero in this area.

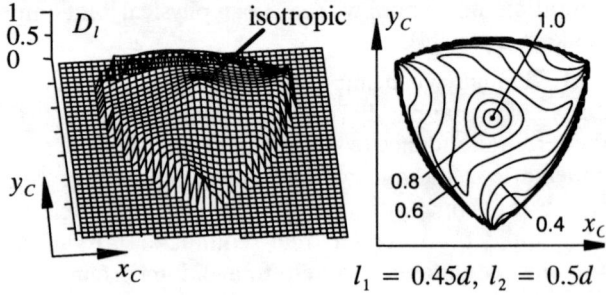

Fig. 6: Local dexterity plots of configuration a

Selecting configuration a of Fig. 3, the isotropy moves to the center of the workspace (see Fig. 6), and the dexterity index looks much more homogenous and somewhat symmetric throughout the area of interest. Singular or nearly singular positions are absent. We have hence proved that the inverse kinematic solutions a or e are the better choice from an isotropic force transmission standpoint, and that a singularity-free workspace can be obtained with actuation redundancy.

III. Kinematics

Fig. 2 shows the general coordinates. All chain links are assumed to have equal lengths

$$l_1 = \overline{A_1B_1} = \overline{A_2B_2} = \overline{A_3B_3} \text{ and}$$
$$l_2 = \overline{B_1C} = \overline{B_2C} = \overline{B_3C}$$

for ease of computation and manufacturing.

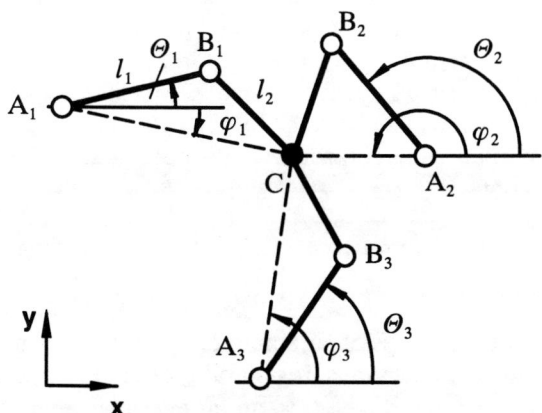

Fig. 2: Coordinates

A_i represents the ith actuated joint with coordinates $\boldsymbol{x}_{Ai} = [x_{Ai}\, y_{Ai}]^T$ with respect to the global cartesian reference frame. B_i represents the unactuated joint of leg i with coordinates $\boldsymbol{x}_{Bi} = [x_{Bi}\, y_{Bi}]^T$ and C the end-effector with $\boldsymbol{x}_C = [x_C\, y_C]^T$.

Direct kinematics

For computing the direct kinematics, we use

$$|\boldsymbol{x}_C - \boldsymbol{x}_{Bi}|^2 = l_2^2, \quad i = 1, 2, 3 \quad (1)$$

where
$$\boldsymbol{x}_{Bi} = \boldsymbol{x}_{Ai} + l_1 \begin{bmatrix} \cos\Theta_i \\ \sin\Theta_i \end{bmatrix}. \quad (2)$$

Θ_i is the angle of the ith actuated joint. Solving for x_C and y_C yields direct kinematics equations which use information about all three joint angles:

$$x_C = \quad (3)$$
$$\frac{|\boldsymbol{x}_{B1}|^2(y_{B2} - y_{B3}) + |\boldsymbol{x}_{B2}|^2(y_{B3} - y_{B1}) + |\boldsymbol{x}_{B3}|^2(y_{B1} - y_{B2})}{2[x_{B1}(y_{B2} - y_{B3}) + x_{B2}(y_{B3} - y_{B1}) + x_{B3}(y_{B1} - y_{B2})]}$$

$$y_C = \quad (4)$$
$$\frac{|\boldsymbol{x}_{B1}|^2(x_{B3} - x_{B2}) + |\boldsymbol{x}_{B2}|^2(x_{B1} - x_{B3}) + |\boldsymbol{x}_{B3}|^2(x_{B2} - x_{B1})}{2[x_{B1}(y_{B2} - y_{B3}) + x_{B2}(y_{B3} - y_{B1}) + x_{B3}(y_{B1} - y_{B2})]}$$

The reader should note that without the third leg, any 2-leg x-y manipulator would have two solutions for the direct kinematics, whereas this solution is unique.

Inverse kinematics

For the inverse kinematics, we obtain from any triangle A_iB_iC in Fig. 2:

$$l_2^2 = l_1^2 + \overline{A_iC}^2 - 2l_1\overline{A_iC}\cos(\Theta_i - \varphi_i), \quad (5)$$

where $\overline{A_iC}^2 = |\boldsymbol{x}_C - \boldsymbol{x}_{Ai}|^2$ and $\varphi_i = \arctan\frac{y_C - y_{Ai}}{x_C - x_{Ai}}$. Solving for Θ_i yields

$$\Theta_{i1,2}(\overline{A_iC}, \varphi_i) = \arccos\frac{l_1^2 - l_2^2 + \overline{A_iC}^2}{2l_1\overline{A_iC}} + \varphi_i. \quad (6)$$

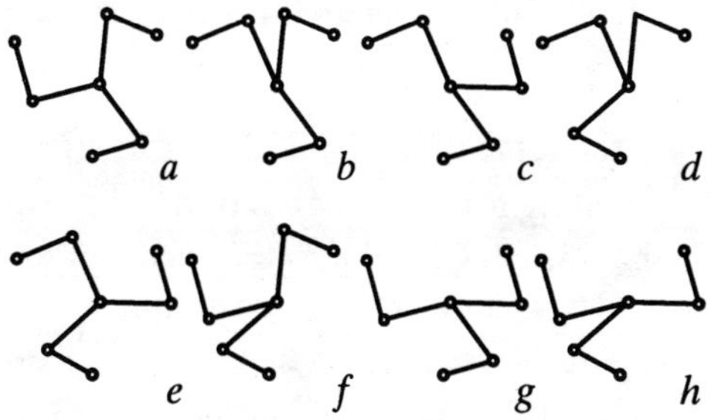

Fig. 3: All solutions of the inverse kinematics

Two solutions are obtained because of the uncertainty of the arccos function. Two solutions for each leg amount to $2^3 = 8$ solutions for the manipulator, as depicted in Fig. 3.

Configurations b - d and f - h are all kinematically equivalent. a and e are preferable because they have shown a more symmetric and isotropic force transmission throughout the workspace and no singularities, as shall be shown later.

IV. Workspace

To simplify our investigations, the active joints of the proposed manipulator are located on the corners of an equilateral triangle. Mechanical interferences between links and actuators are neglected in our examinations. The workspace is determined by three parameters: the distance d of the actuators and the link lengths l_1 and l_2. Fig. 4 shows workspace plots for $l_1 = l_2$ and the general case $l_1 \neq l_2$, given a fixed d.

The outer limit of the workspace is marked by the intersection of three circles of radius $l_1 + l_2$. If $l_1 \neq l_2$, three circular areas around the actuators with radius $|l_1 - l_2|$ remain unreachable.

Difficulties arise due to the fact that all three chains are in a plane, which means that they cannot move over one another. It has to be ensured by the design or the control that chains and actuators cannot collide, imposing constraints on the workspace.

V. Force Transmission

Force transmission in redundantly actuated parallel kinematic structures has been addressed by several au-

VI. MINIMAL NORM TORQUE VECTOR CONTROL (MNTC)

To reduce joint friction by internal forces at high speed operation, or to avoid jamming during point-to-point operation, a control is common which minimizes the internal forces of the manipulator. This is achieved by minimizing the euclidian norm (2-norm) of the vector $\boldsymbol{\tau}_c = [\tau_{c1}\ \tau_{c2}\ \tau_{c3}]^T$ of the commanded actuator torques, calling for an underlying torque controller.

The force equilibrium at joint C can be written as

$$f_I = f - f_{ext}, \qquad (12)$$

where f_I is the force of inertia of any mass attached to C, f the actuating force composed of the three driving forces in the links $\overline{B_iC}$, and f_{ext} the vector of external forces caused by a contact with the workpiece.

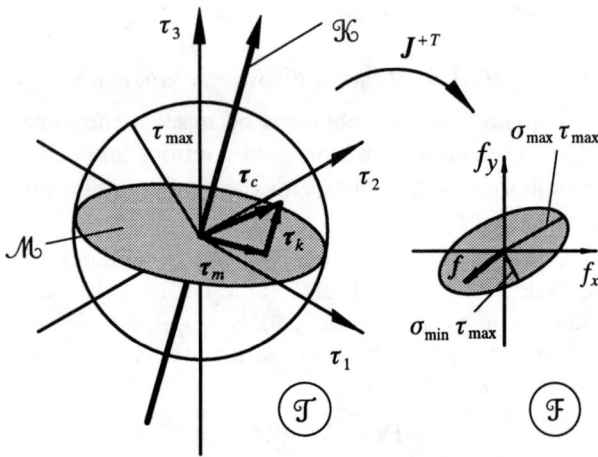

Fig. 7: Geometrical interpretation of mapping J^+

Let $\mathcal{T}$ be the 3-dimensional vector space of the actuator torques and $\mathcal{F}$ the 2-dimensional euclidean space of forces. Equation (8) can be written as

$$f = J^{+T}(\boldsymbol{\tau}_m + \boldsymbol{\tau}_k), \qquad (13)$$

where the vector of torques was split into a minimal-norm and a null space component. Since $\boldsymbol{\tau}_k$ is in the null space (kernel) of the linear mapping $J^{+T}: \mathcal{T} \to \mathcal{F}$, we know that

$$J^{+T}\boldsymbol{\tau}_k = 0 \text{ and } J^{+T}\boldsymbol{\tau}_m = f. \quad (14)$$

Thus, the vector space $\mathcal{T}$ can be divided into a 2-dimensional subspace $\mathcal{M}$ with minimum-norm vectors and a one-dimensional kernel $\mathcal{K}$ with torque vectors that preload the manipulator by generating internal forces, but do not apply any net force to the center joint C. Fig. 7 shows a geometrical interpretation.

The circular shaped subspace $\mathcal{M}$ with radius τ_{max} is mapped onto an ellipse in $\mathcal{F}$, whose radii are determined by the singular values of $J^{+T}J^+$ (see section 5). Any torque vector perpendicular to $\mathcal{M}$ belongs to the kernel of J^+ and is mapped onto zero output force.

Clearly, the inverse mapping $J^T: \mathcal{F} \to \mathcal{T}$ maps the ellipse onto $\mathcal{M}$, and therefore the equation $\boldsymbol{\tau}_m = J^T f_c$ automatically generates a minimal norm torque vector from any commanded cartesian force f_c. This can be utilized for a traditional cartesian minimal norm torque vector control, which we call MNTC, as depicted in Fig. 8. We will use this control for high-speed point-to-point operation.

VII. LOWER BOUND STIFFNESS CONTROL (LBSC)

High stiffness is of paramount importance for many machining applications. It generally describes the ratio of the restoring force Δf to an infinitesimal small displacement Δx_C of the end-effector ([7],[14]). The stiffness matrix K is then defined by

$$K = \lim_{\Delta x_C \to 0} \frac{\Delta f}{\Delta x_C} = \frac{\partial f}{\partial x_C}. \qquad (15)$$

Using (8), this can be rewritten as

$$K = \frac{\partial (J^{+T}\boldsymbol{\tau})}{\partial x_C} = H^T \boldsymbol{\tau} + J^{+T}\frac{\partial \boldsymbol{\tau}}{\partial x_C}. \qquad (16)$$

H^T is the 2x3x2 transpose of the Hessian matrix. With $\frac{\partial \boldsymbol{\tau}}{\partial x} = \frac{\partial \boldsymbol{\tau}}{\partial \boldsymbol{\Theta}} J^+$, one obtains

$$K = H^T \boldsymbol{\tau} + J^{+T}\frac{\partial \boldsymbol{\tau}}{\partial \boldsymbol{\Theta}} J^+. \qquad (17)$$

$\frac{\partial \boldsymbol{\tau}}{\partial \boldsymbol{\Theta}}$ is a diagonal matrix of the stiffness coefficients of the position-controlled actuators. Modeling all actuators as identical linear springs with stiffness k_a, equation (17) can be simplified:

$$K = H^T \boldsymbol{\tau} + k_a J^{+T}J^+. \qquad (18)$$

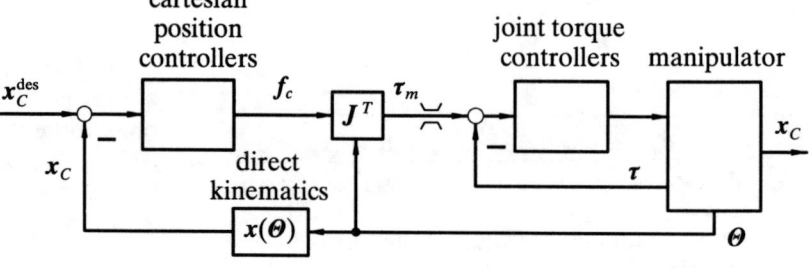

Fig. 8: Principle of point-to-point MNTC

The second term describes the "passive" stiffness of the manipulator, which is position-dependent and proportional to the actuators' stiffness. The first term can be used to affect the stiffness matrix by variation of the driving torques, and is therefore called "active" stiffness. Clearly, since we have one degree of actuation redundancy, we could arbitrarily vary one element of the 2x2 stiffness matrix

$$K = \begin{bmatrix} k_{xx} & k_{xy} \\ k_{yx} & k_{yy} \end{bmatrix} \quad (19)$$

by an appropriate choice of a null space torque vector – within the constraints imposed by the maximum actuator torques. Even better, we can guarantee a lower bound for the stiffness in all directions by controlling the smallest singular value of K, which determines the lower bound for the 2-norm of a force generated by a displacement Δx_C:

$$\sigma_{\min}(K)|\Delta x_C| \leq |\Delta f| \leq \sigma_{\max}(K)|\Delta x_C|. \quad (20)$$

Since K is positive semidefinite and symmetric, the singular values are identical to the eigenvalues, and the directions of maximum and minimum stiffness are therefore determined by the eigenvectors of K [14].

Let τ_{k0} span the kernel of J^{+T} and let τ_m be the minimal norm torque vector to satisfy (21). Then we have to scale τ_{k0} with a parameter β to adjust the smallest singular value of K to match the desired value $\sigma_{\min}^{des}$,

$$\sigma_{\min}(K) = \sigma_{\min}^{des}, \quad (22)$$

or $\sigma_{\min}(H^T(\tau_m + \beta\tau_{k0}) + k_a J^{+T}J^+) = \sigma_{\min}^{des}. \quad (23)$

At the moment, the solution of (23) is found numerically and used as a torque feedforward. Fig. 9 shows the proposed LBSC control scheme.

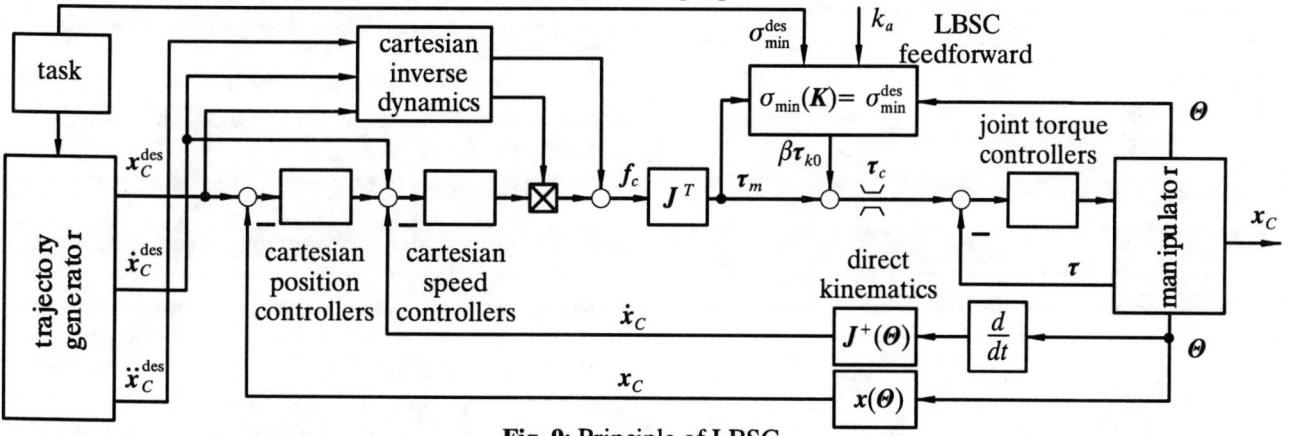

Fig. 9: Principle of LBSC

VIII. SIMULATION RESULTS

The effects of MNTC and LBSC commanded torque generation on stiffness parameters were verified by simulation. We simulated a constant-speed movement across the workspace along the line $\overline{DE}$ in Fig. 10. No external forces were applied at the end-effector. Fig. 11 shows the commanded torques along the trajectory using LBSC. Fig. 12 proves that a desired minimum stiffness $\sigma_{\min}^{des} = 5$ can be achieved. This means also, as can be seen in Fig. 13, that the main diagonal elements of the stiffness matrix stay beyond this lower bound at all times.

As a comparison, the same trajectory has been computed using minimal norm torques (MNTC), which of course means zero driving torques at constant speed if no external forces are applied. In Fig. 14, the minimum and maximum singular values of K along $\overline{DE}$ are plotted. They are comparably lower than in Fig. 12, where active stiffness is used. The same is true for the elements of the stiffness matrix in Fig. 15 as compared to Fig. 13.

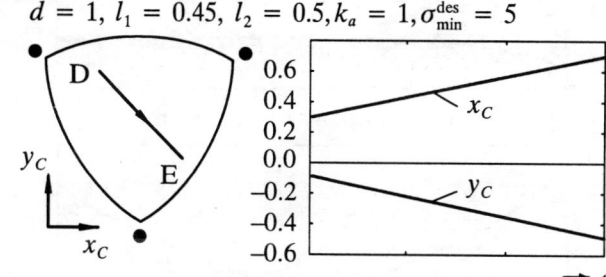

Fig. 10: Setup for simulation

IX. SUMMARY

In this paper, we have examined a 2-d.o.f. planar manipulator with actuation redundancy. The workspace as well as forward and inverse kinematics were investigated. Using a dexterity index, we have shown that a singularity-free and well-balanced workspace in terms of force transmission can be obtained, if a particular configuration of the inverse kinematics is chosen. A novel active-stiffness control scheme that guarantees a lower bound of the end-effector stiffness (LBSC) was proposed, verified, and compared with a traditional approach of minimizing the 2-norm of the torque vector (MNTC) in a computer simulation.

We have shown that actuation redundancy can be exploited for active stiffness and offers some good benefits for improving the behavior of parallel manipulators in terms of removing singularities and optimizing force transmission. We believe that spatial parallel manipulators will soon benefit from redundancy as well.

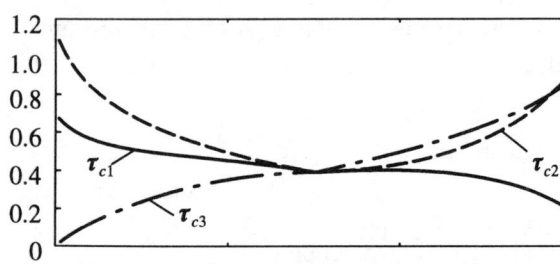

Fig. 11: LBSC – commanded torques

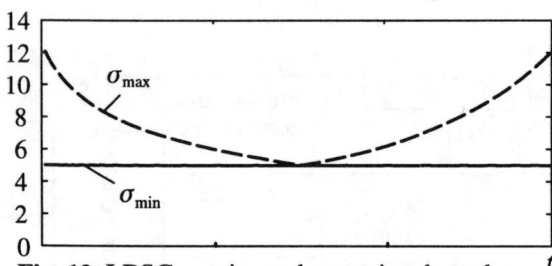

Fig. 12: LBSC – min. and max. singular value

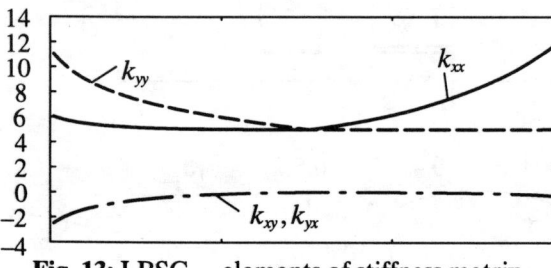

Fig. 13: LBSC – elements of stiffness matrix

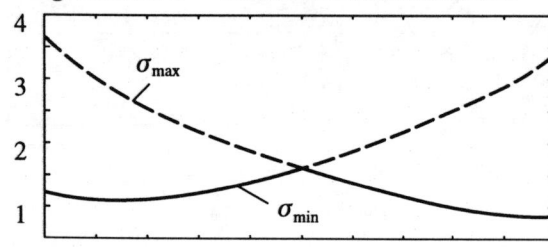

Fig. 14: MNTC – min. and max. singular value

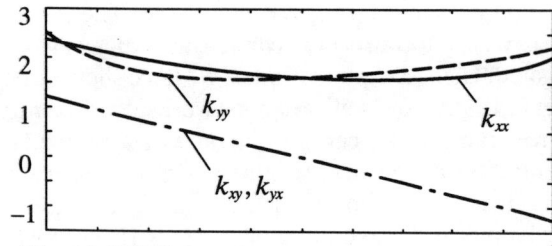

Fig. 15: MNTC – elements of stiffness matrix

References

[1] J.-P. Merlet. Direct Kinematics of Planar Parallel Manipulators. *Proc. 1996 IEEE Int. Conf. on Rob. and Autom.*, Minneapolis. pp. 3744-3749. Apr 1996

[2] C. M. Gosselin, J.-P. Merlet. The Direct Kinematics of Planar Parallel Manipulators: Special Architectures and Number of Solutions. *Mech. Mach. Theory*, Vol. 29, No. 8. pp. 1083-1097. 1994

[3] O. Ma, J. Angeles. Direct Kinematics and Dynamics of a Planar Three-Dof Parallel Manip. *ASME Des. and Autom. Conf.*, Chicago pp. 37-43. Sep 1990

[4] C. M. Gosselin, M. Jean. Determination of the Workspace of Planar Parallel Manipulators with Joint Limits. *Robotics and Autonomous Systems*, Vol. 17. pp. 129-138. 1996

[5] J.-P. Merlet. Redundant Parallel Manipulators. *Journal of Lab. Rob. & Autom.*, Vol 8 pp. 17-24. 1996

[6] J.-P. Merlet. Closed-form Resolution of the Direct Kinematics of Parallel Manipulators Using Extra Sensors Data. *Proc. 1993 IEEE Int. Conf. on Rob. and Automation*, Atlanta. pp. 200-204. May 1993

[7] B.-Y. Yi, R. A. Freeman, D. Tesar. Open-Loop Stiffness Control of Overconstrained Mechanisms/Robot Linkage Systems. *Proc. 1989 IEEE Int. Conf. on Robotics and Automation*, Scottsdale. pp. 1340-1345. May 1989

[8] M. A. Nahon, J. Angeles. Force Optimization in Redundantly-Actuated Closed Kinematic Chains. *Proc. 1989 IEEE Int. Conf. on Robotics and Automation*, Scottsdale. pp. 951-956. May 1989

[9] P. Buttolo, B. Hannaford. Advantages of Actuation Redundancy for the Design of Haptic Displays. *Proc. ASME 4th Annual Symp. on Haptic Interf. for Virtual Environm. and Teleoperation Systems*. DSC-Vol. 57-2, pp. 623-630. San Francisco, Nov 1995

[10] P. Buttolo, B. Hannaford. Pen-Based Force Display for Precision Manipulation in Virtual Environments. *Proc. IEEE Virtual Reality Annual Int'l Symp.* pp. 217-224. North Carolina, Mar 1995

[11] R. Kurtz, V. Hayward. Multiple-Goal Kinematic Optimization of a Parallel Spherical Mechanism with Actuator Redundancy. *IEEE Trans. on Rob. and Autom.*, Vol. 8, No. 5. pp. 644-651. Oct 1992

[12] B.-J. Yi, R. A. Freeman, D. Tesar. Force and Stiffness Transmission in Redundantly Actuated Mechanisms: The Case for a Spherical Shoulder Mechanism. *Robotics, Spatial Mechanisms and Mechanical Systems*, Vol. 45, pp. 163-172. Aug 1994

[13] R. S. Stroughton, T. Arai. A Modified Stewart Platform Manipulator with Improved Dexterity. *IEEE Trans. on Rob. and Autom.*, Vol. 9 No. 2. Apr 1993

[14] C. Gosselin. Stiffness Mapping for Parallel Manipulators. *IEEE Transactions on Robotics and Automation*, Vol. 6, No. 3. pp. 377-382. Jun 1990

Parallel Dynamics Computation and H_∞ Acceleration Control of Parallel Manipulators for Acceleration Display

K.Yamane, M.Okada, N.Komine* and Y. Nakamura
(E-mail: katz@ynl.t.u-tokyo.ac.jp)

Dept. of Mechano-Informatics, Univ. of Tokyo
7-3-1 Hongo Bunkyo-ku Tokyo,
113-0033 JAPAN

*Shimadzu Corporation
1 Nishi-no-kyo Kuwahara-chyo
Nakagyo-ku Kyoto, 604-8442 JAPAN

Abstract

In this paper, we propose a control scheme of parallel manipulators focusing on the accuracy of acceleration on endplate, which is an important factor when parallel manipulators are used as acceleration displays. We use two controllers — dynamic controller to achieve accuracy of position and to stabilize the system, and H_∞ controller to feedback the acceleration measured on the endplate. The main problem of dynamic control is computational complexity. In order to reduce computation time for inverse dynamics, parallel processing method called multi-thread pogramming is applied. H_∞ controller is added outside the closed loop of dynamic control to remove the vibration of the structure and the influence of modeling errors in dynamic controller.

1 Introduction

Acceleration display is one of many applications of parallel manipulators. Although the Stewart Platforms driven by hydraulic cylinders [1] have been widely used for simulators, it is difficult to control precisely such type of manipulators because of time delay and nonlinearity of the hydraulic actuators. Recently, parallel mechanisms using rotational and/or spherical joints (Figure 1), instead of sliding ones, are coming into use because servo motors are more suitable for precise control.

A problem of parallel mechanisms is that it is prohibitably costly to compute the dynamics and kinematics in real time. Therefore, in most cases, only simple position control is applied and parallel mechanisms are not used unfortunately to the best of their

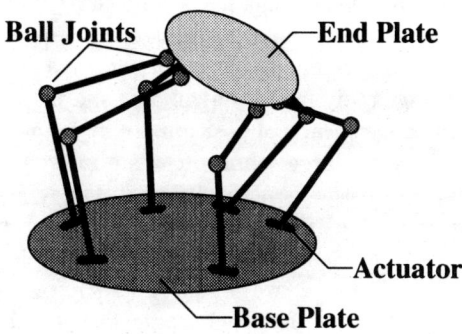

Figure 1: Parallel Mechanism

ability in high speed motion. There are two technical reasons for the difficulty: (1) computation of unmeasured passive joint angles, and (2) computation of dynamics due to closed kinematic chains.

One of the time-consuming steps of dynamic control is inverse dynamics. There have been basically two approaches to the problem in closed-link structures including parallel mechanisms. They resemble in that some joints of the closed-link structure are virtually cut and the structure is transformed into equivalent open-link tree structure. The earlier method, which uses Lagrange multiplier to compute the force acting at the virtually cut joints, is not computationally effective. The later approach [2], using the Jacobian of the free joints with respect to the actuated ones, is known to require much less computation. It is also known that some of the computation in inverse dynamics of parallel mechanisms can be computed in parallel, because of the parallelism of the structure [3]. Even now real-time dynamic control without approximated dynamical models are not yet ready for practical use.

Dynamic control is usually designed for the rigid body model of mechanism, wheras the real mechanism includes structural flexibility that causes unmodeled vibration on the endplate. The supression of such vibration is significant when it is used for an acceleration display. Designing a controller usually requires many trial-and-error steps. An useful method, however, has been developed, in which a controller is designed systematically based on the experimental results for identification [8, 9].

The main goals of this paper are: (1) to reduce the computation time for dynamic control by taking account of the mechanical parallelism, and (2) to design an acceleration feedback controller which realizes smooth acceleration on the endplate.

We first provide with the general computational algorithm of parallel mechanisms and show its parallelism. We then implement the algorithm using parallel processing technique. Experimental results show the advantage of dynamic control over simple position control in the accuracy of position. In high-speed motions, however, a large vibration was observed on the endplate due to the structural flexibility. Also modeling errors such as the frictions of the joints, are difficult to estimate. In order to make the acceleration smooth and accurate, we designed and implemented an H_∞ acceleration feedback controller for the dynamic control system. It was observed that the vibration and error of acceleration on the endplate was greatly reduced.

2 Dynamic Control Algorithm for Parallel Processing

The block diagram of dynamic control is shown in Figure 2, where θ_1 is the actuated-joint angles measured by the encoders, and $x_E, \dot{x}_E, \ddot{x}_E$ are the position, velocity, acceleration of the endplate, $\ddot{x}_{Ed}$ is the reference acceleration, $\hat{\ddot{x}}_{Ed}$ is the acceleration which the manipulator should generate, τ_1 is the torque required to generate $\hat{\ddot{x}}_{Ed}$, respectively. G is the whole acceleration-input-output system. For serial link manipulators, this is nothing but the resolved acceleration control[4].

Using the virtual cut algorithm for closed kinematic chain dynamics [3], the dynamic control for parallel manipulators requires the following subcomputations:

1. Angles of the actuated joints θ_1 are measured, and the velocity of those joints $\dot{\theta}_1$ can be computed by numerically differenciating θ_1.

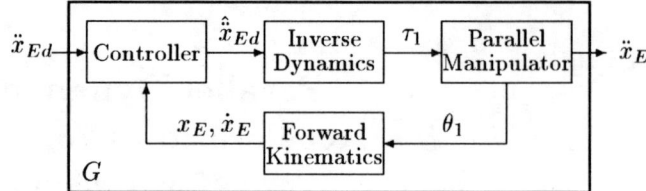

Figure 2: Block diagram of dynamic control

2. Compute the position of the endplate x_E (forward kinematics).

3. Compute the kinematics in parallel at each leg ($i = 1, 2, \ldots, 6$):

   (a) Compute the angles of the passive joints (inverse kinematics):
   $$\boldsymbol{\theta}_i = \boldsymbol{g}_i(\boldsymbol{x}_E) \qquad (1)$$
   where $\boldsymbol{g}_i(\boldsymbol{x}_i)$ is the function that gives all joint angles of each leg $\boldsymbol{\theta}_i$.

   (b) Compute the Jacobian matrix of each leg in parallel $\boldsymbol{J}_i = \partial \boldsymbol{x}_E/\partial \boldsymbol{\theta}_i$ and its inverse $\boldsymbol{G}_i = \boldsymbol{J}_i^{-1}$.

4. Form $\boldsymbol{S} = \partial \boldsymbol{\theta}_1/\partial \boldsymbol{x}_E$ by gathering all the elements of $\boldsymbol{G}_1, \boldsymbol{G}_2, \ldots, \boldsymbol{G}_6$ corresponding to the joints with actuators and compute its inverse $\boldsymbol{S}^{-1}$.

5. Compute velocity of the endplate:
   $$\dot{\boldsymbol{x}}_E = \boldsymbol{S}^{-1} \dot{\boldsymbol{\theta}}_1 \qquad (2)$$

6. Determine the acceleration of the endplate:
   $$\hat{\ddot{\boldsymbol{x}}}_E = \ddot{\boldsymbol{x}}_{Ed} + \boldsymbol{K}_D(\dot{\boldsymbol{x}}_{Ed} - \dot{\boldsymbol{x}}_E) + \boldsymbol{K}_P(\boldsymbol{x}_{Ed} - \boldsymbol{x}_E) \qquad (3)$$
   where $\ddot{\boldsymbol{x}}_{Ed}, \dot{\boldsymbol{x}}_{Ed}$ and $\boldsymbol{x}_{Ed}$ are the desired acceleration, velocity and position of the endplate, $\dot{\boldsymbol{x}}_E$ and $\boldsymbol{x}_E$ are the current accleration and position of the endplate, $\boldsymbol{K}_D$ and $\boldsymbol{K}_P$ are the feedback gains, respectively.

7. Here we virtually cut the joints between the endplate and the last link of all legs except for an arbitary leg, thus the closed-link structure is virtually transformed into an open-link structure, assuming all the 31 uncut joints are actuated. Let $\boldsymbol{\theta}_0$ be the angles of the uncut joints and $\boldsymbol{\tau}_0$ the torques.

8. Inverse dynamics is computed in parallel at each leg ($i = 1, 2, \ldots, 6$):

(a) Compute the velocities of all joints:
$$\dot{\boldsymbol{\theta}}_i = \boldsymbol{G}_i \dot{\boldsymbol{x}}_E \qquad (4)$$

and the accelerations:
$$\ddot{\boldsymbol{\theta}}_i = \boldsymbol{G}_i(\ddot{\hat{\boldsymbol{x}}}_E - \dot{\boldsymbol{J}}_i \dot{\boldsymbol{\theta}}_i) \qquad (5)$$

(b) Compute the joint torques of the virtual open-link structure $\boldsymbol{\tau}_i$ by applying Newton-Euler formulation.

9. Form $\boldsymbol{W} = \boldsymbol{\theta}_0/\boldsymbol{x}_E$ by gathering all the elements of $\boldsymbol{G}_i (i = 1\ldots 6)$ corresponding to the uncut joints.

10. Transform the torques to those of the actuated joints in the real closed-link structure:
$$\boldsymbol{\tau}_1 = \boldsymbol{S}^{-T}\boldsymbol{W}^T\boldsymbol{\tau}_0 \qquad (6)$$

where $\boldsymbol{\tau}_1$ is the actuator torques.

Among these steps, 3. and 8. can be computed in parallel for each leg. The computation time would be reduced by parallel computing. Practically, since the parallel-computation part is divided by serial computation of steps 4. to 7, the overhead due to switching between serial and parallel processing becomes so large as to make parallel processing meaningless. It is desirable that there exists only one segment of parallel computation in the whole sequence of the computation.

For this purpose, we see the six legs of the parallel mechanism as six independent serial manipulators and apply Unified Computation [5] to each leg. This can be realized by taking some simple points into account:

- Set the coordinate of the endlink of each serial leg coincident to that of the endplate of the parallel mechanism. Acceleration and velocity of the endlink coincide with those of the endplate. Position can be computed easily from that of the endplate by subtracting the offset of the leg base.

- Set the mass and inertia matrix of each endlink 0 and $\boldsymbol{O}$ respectively, except for an arbitrary leg, for which their real values are used.

The new algorithm is summarized as follows:

1. Measure $\boldsymbol{\theta}_1$ and compute $\dot{\boldsymbol{\theta}}_1$.

2. Compute the position of the endplate $\boldsymbol{x}_E$.

3. Compute $\ddot{\hat{\boldsymbol{x}}}_E$ by equation (3).

4. Compute the following steps in parallel for each leg ($i = 1, 2, \ldots, 6$):

   (a) Compute the desired position of the endlink $\boldsymbol{x}_{di}$:
   $$\boldsymbol{x}_{di} = \boldsymbol{B}_i^T \boldsymbol{x}_{Ed} - \boldsymbol{b}_i \qquad (7)$$
   where $\boldsymbol{B}_i$ and $\boldsymbol{b}_i$ are the attitude and position of the base coordinate of leg i, respectively.

   (b) Compute all joint angles $\boldsymbol{\theta}_i$ and their velocities $\dot{\boldsymbol{\theta}}_i$.

   (c) Compute the Jacobian matrix $\boldsymbol{J}_i$ and its inverse $\boldsymbol{G}_i$.

   (d) Compute the virtual torques $\boldsymbol{\tau}_i$ via Unified Computation [5], assuming all leg-joints are actuated. Note that inverse dynamics algorithm for serial-link manipulators can be applied without any modification.

5. Form $\boldsymbol{W}$ and $\boldsymbol{S}$ from $\boldsymbol{G}_1, \boldsymbol{G}_2, \ldots, \boldsymbol{G}_6$, and $\boldsymbol{\tau}_0$ from $\boldsymbol{\tau}_1, \boldsymbol{\tau}_2, \ldots, \boldsymbol{\tau}_6$.

6. Transform the computed virtual torques to those of actuated joints of real closed-link structure by equation (6).

In this algorithm, some values are redundantly computed at all legs. For example, six velocities of the endlinks are that of the endplate. This computation was done only once in the old algorithm. Although some duplicate computations are included, it is time-wise more efficient to unify the parallel-computed part and to eliminate overheads due to synchronization and switching from parallel to serial computations and vise versa. Figure 3 shows the concept of unifying the parallel-computation part.

3 Implementation of Dynamic Control Algorithm

3.1 Multi-thread Programming

There are many architechtures of multi-processor systems for robot control. For example one can connect some PC's, or select a special computer for I/O and interrupt handling to combine with a PC for other calculation. In general, PC's are not good at handling I/O and interrupts which are essencial to the real-time control. Using special computers that have higher real-time ability has a large advantage. Disadvantages of such systems are: (1) the programs will be

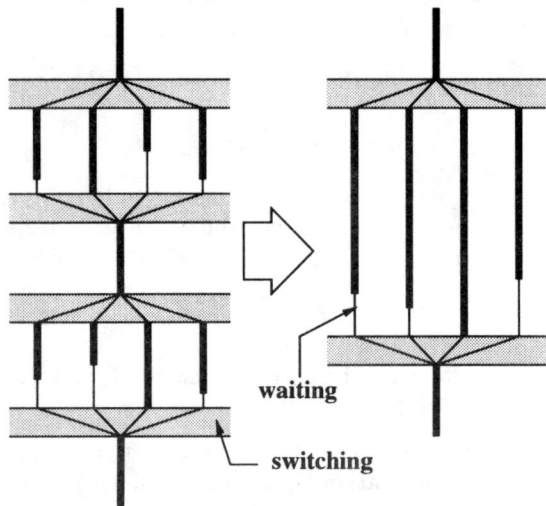

Figure 3: Unifying parallel-computation part

tailored to the system, and (2) programming parallel algorithm is usually difficult.

We chose a PC with four Pentium processors, WindowsNT for the operating system, and multi-thread programming[6] to implement parallel processing — all of these technologies are standard items for PC's, though not very suitable for real-time control of manipulators. The advantages of such a system are:

- Various procedures for parallel processing — synchronization, locking variables, and so on — are well supported by the operating system.

- The programs to be developed for the system will run on any other systems with the same operating system (even if there is only one processor).

A multi-thread program creates multiple *threads* in one process. Threads are small parts of the whole computation and there is no restriction on the number of threads. The operating system will distribute threads among the processors in such a way to equalize computation loads to some extent. Thus, parallel processing is realized.

3.2 The Parallel Manipulator System

Figure 4 shows the parallel manipulator developed and used in the experiments. The kinematics of this parallel manipulator was originally designed by Takeda et al. [7] We scaled up to make a platform for driving simulator. It has the ability of generating 2G with a load of 70kg.

Figure 4: Parallel Manipulator

We use a PC with four Pentium Pro 200MHz processors for parallel processing. The program is written in Microsoft Visual C++. WinRT is also used to drive I/O on WindowsNT. The servo motors are equipped with encoders to measure the input angles. Computed torques are input to the servo motors, which are run in torque control mode, through a D/A board. Acceleration on the endplate is also measured by the force sensor with a mass fixed at the center of the endplate.

3.3 Computation Time (Single Thread)

The computation time of each step in the algorithm when implemented in one thread is shown in table 1. The CPU is Pentium Pro 200MHz.

Although inputs from the counter and outputs to D/A can be handled in parallel, they are handled serially in a single thread. The whole computation time, if computed in a single thread, becomes 4.6msec, the sampling frequency 220Hz.

Table 1: Computation time in one thread

computation	time	remarks
forward kinematics	0.92msec	
read encoder counter	0.60msec	6 channels
inverse dynamics (1 leg)	0.43msec	2.58msec for 6 legs
torque transformation	0.19msec	
output through D/A	0.30msec	6 channels

3.4 Implementation by Multi-thread Programming

In a multi-thread program, a thread called *primary thread* is created at the beginning of the program. The other threads are created by this thread. Unfortunately one cannot predict when and which processor executes the thread. Therefore, we must synchronize the threads by using some methods — suspending / resuming threads, or setting / resetting events — to let threads run in the order we want. By setting event, for example, a thread can inform others that a sequence of computations has finished.

In our program, primary thread creates seven threads: one for forward kinematics and the other six for inverse dynamics of each leg. The task and procedure of each threads become as follows (see also Figure 5):

1. *Primary Thread* — Initialize parameters and I/O devices, and start the control loop. During motion, read the desired position, transform the torques, and output the torques to the D/A board.

    (a) Read desired position and wait for the forward kinematics computation to finish.

    (b) Resume the *Forward Kinematics Thread* and *Inverse Kinematics Threads* and wait for all inverse kinematics computations to finish.

    (c) Transform virtual torques into real torques, and output the result to the D/A board.

2. *Forward Kinematics Thread* — Read encoder counter board and compute current position of the endplate.

    (a) When the computation of forward kinematics finishes, set *Event1* and suspend itself.

    (b) Resumed by *Primary Thread*, return to (a).

3. *Inverse Dynamics Threads* — Compute the virtual torque of the open-link structure.

    (a) Resumed by *Primary Thread* when all the necessary information is ready.

    (b) Finishing the computation of inverse dynamics, suspend itself. The last one sets *Event2* before suspending.

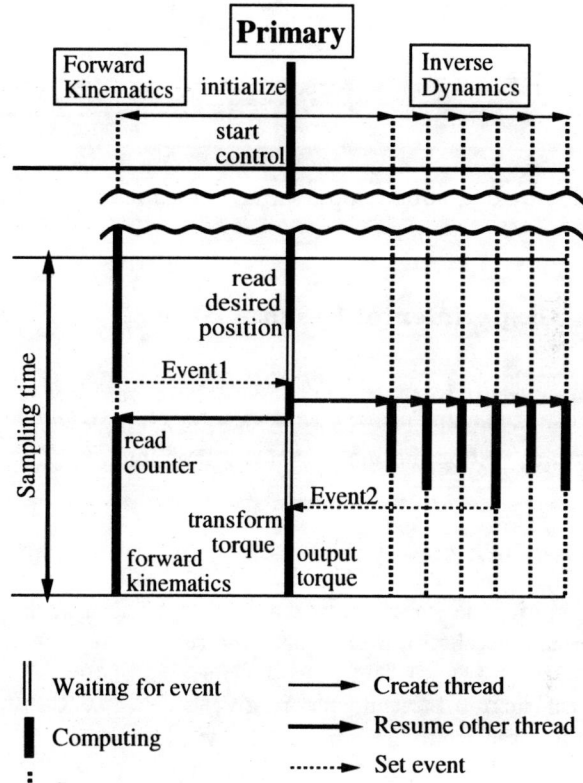

Figure 5: Scheduling of the threads

3.5 Computation Time (Multi Thread)

The sampling time was reduced by the parallel processing down to 2.0msec, which is less than half of the time when computed serially.

It is impossible to reduce the computation time to a quarter of the original even there are four processors because of two reasons: (1) not all four processors are always computing at one time, and (2) some idling time is unavoidable waiting for other thread to finish computation.

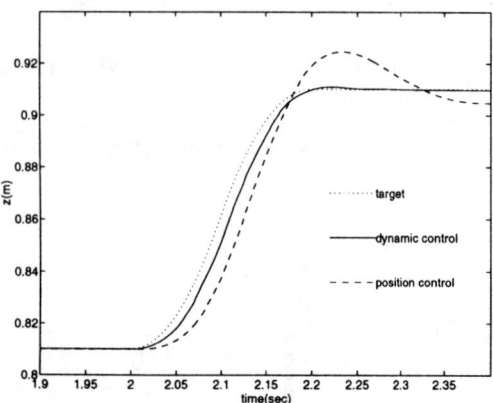

Figure 6: Endplate position in virtical motion

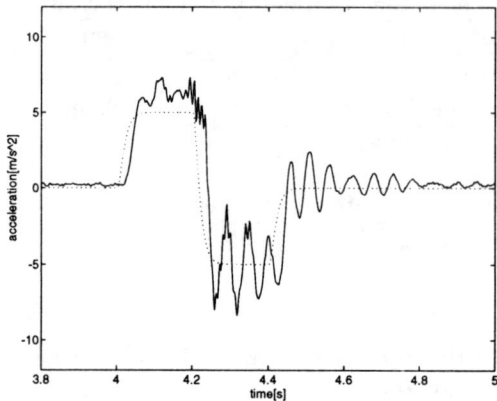

Figure 7: Acceleration measured on the endplate

3.6 Experimental Results

We executed experiments to see the advantage of dynamic control. The reference path is a simple constant acceleration motion in vertical direction.

3.6.1 Position

The motion shown in Figure 6 appeared when upward reference acceleration of $10m/s^2$ for the first 0.1 seconds, and downward reference acceleration of $10m/s^2$ for the next 0.1 seconds, were given. The position is computed from the input angles via forward kinematics computation. It is observed that timelag and overshoot are greatly reduced by dynamic control.

3.6.2 Acceleration on the Endplate

The real acceleration was measured on the endplate for constant acceleration motion as shown in Figure 7. Step reference acceleration of $5m/s^2$ was given for the first 0.2 seconds, $-5m/s^2$ for the next 0.2 seconds. Although dynamic control achives precise motion in terms of position, the acceleration measured on the endplate has large error and vibration due to mechanical flexibility of links of the parallel structure.

It is impossible to remove the mechanical vibration of the manipulator by this control method because it is not detected by the value of the encoder. The timelag, supposed to be caused by the friction of the joints, are also difficult to be estimated.

4 Acceleration Feedback Control

4.1 Joint Design of Identification and Controller

"Joint Design of Identification and Controller" [8, 9] is an approach to design a desirable controller which satisfies the control specification iteratively, based on H_∞ or H_2 control theory. Since the controller is designed systematically based on the experimental results using identification theory, this method can avoid trial-and-error steps.

In this paper, we identified the closed-loop system including dynamic controller and then designed an H_∞ controller because of the following reasons:

- By system identification, we can get a model of unmodeled dynamics such as the vibration due to the mechanical flexibility.

- As a system is stabilized with the dynamic controller, we can apply open-loop identification [1].

- Since H_∞ controller is designed in frequency domain, it is suitable for removing vibrations of known frequency.

---

[1] To identify a plant of an open-loop system, while closed-loop identification implies to identify a plant *inside* a closed loop. It is known that closed-loop one is much more difficult than open-loop one. Our problem is to identify the mechanism with dynamic controller, for which open-loop identification can be applied.

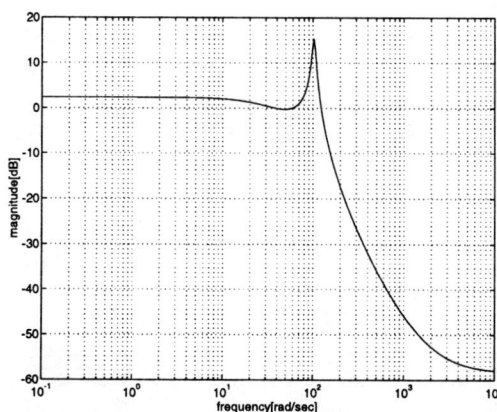

Figure 8: Bode plot of virtical direction system

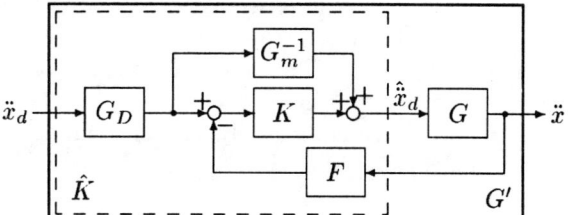

Figure 9: Acceleration feedback system

4.2 Designing Controller

It is desirable to identify the whole 6DOF system with 6 inputs and 6 outputs. In this case the designed controller will be high dimensional and require a large computational load. Regarding the cross terms as disturbance, we assume that six axes are independent of each others and identify the system of each axis separately. A rationale behind this assumption is the fact that the dynamic controller approximately decouples the system's behavior.

The controller is designed in the following steps:

1. System Identification
   First, we identify the system G as a linear time-invariant system G_m including the dynamic controller, whose input is the reference accleration and output is the measured acceleration of the endplate. Note that both input and output are acceleration. Here we identify the SISO system of six axes separately assuming all axes are independent of each others. For the identification an M-sequence signal [10] was used for reference. The model was assumed to be of the third order, which includes vibration of the second order and the first order delay. The Bode plot of the identified system of virtical direction is shown in Figure 8. There is a pole near the frequency of 100 rad/s as expected from Figure 7.

2. Controller Design

   Next, we design the controller K using the identified model G_m, which satisfies the following cost function J:

   $$J = \left\| \begin{array}{c} W_T K(I + G_m K)^{-1} G_m \\ W_S G_m (I + K G_m)^{-1} \end{array} \right\|_\infty < 1 \quad (8)$$

   where I is the unit matrix, W_T and W_S are the weighting functions. Model error Δ is assumed to be in the form of $G = G_m(1 + \Delta)$. Robustness against model error and low-sensitivity against noise are realized by W_T and W_S, respectively. W_T and W_S are obtained from the spectral analysis of the experimental results.

3. Adding the controller to the system
   The controller K is put into the closed-loop system shown in Figure 9, where G_D and F are the desired response of the system and the filter to remove noise of the sensor, and set as follows, respectively.

   $$G_D = \frac{100^2}{(s + 100)^2} \qquad F = \frac{1}{s + 300} \quad (9)$$

   Finally we obtain the controller $\hat{K}$, whose inputs are the original reference acceleration given by the user $\ddot{x}_d$ and the sensed acceleration $\ddot{x}$, output is the reference acceleration to the dynamic controller $\hat{\ddot{x}}_d$. If $\Delta = 0$, the transfer function from $\ddot{x}_d$ to $\ddot{x}$ becomes G_D.

4. Repeat 1.-3.
   When the system show unsatisfactory responses, we identify the closed-loop system as a new system G' and repeat the procedures of identification, designing and adding the new controller.

4.3 Experimental Results

We designed and implemented a controller in the way described in the previous section. The same reference acceleration as in Figure 7 was given to the new system with acceleration feedback. The acceleration measured on the endplate is shown in Figure 10.

Comparing to the result in dynamic control only (Figure 7), it is observed that the vibration and time-lag are greatly reduced. The influence of the cross terms with the other axes has turned out to be sufficiently small, which experimentally supports our assumption of independency of the six axes.

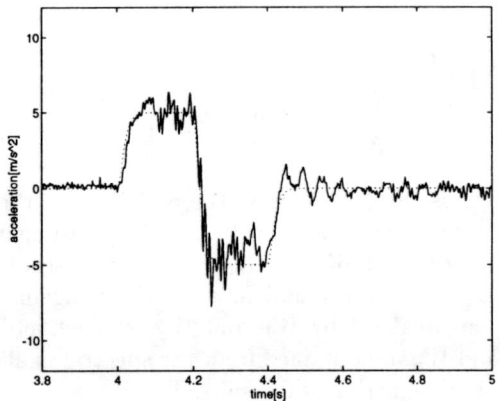

Figure 10: Measured acceleration with feedback

5 Conclusions

The following four conclusions concerning the control of parallel mechanisms for accleration displays were earned in this study:

1. A new dynamic control algorithm was established for parallel manipulators taking the mechanical parallelism into account. The algorithm is more efficient when computed in parallel.

2. The algorithm was implemented using multi-thread programming, and real-time dynamic control was realized without simplification.

3. In order to remove the vibration observed in high-speed motion, acceleration feedback controller was designed through identification and H_∞ control theory.

4. Experimental results illustrated that the whole system achives very smooth and accurate acceleration on the endplate.

Acknowledgments

This work was partially supported by the Ministry of Education, Culture, and Sports (the Grant in Aid of Scientific Research: Prototyping Research (B)(2) 06555065), the International Robotics and Factory Automation Center (Committe for High Speed Parallel Mechanisms, 1995-97), and the Hayao Nakayama Foundation for Science & Technology and Culture. The second author was supported by the "Reserach for the Future" Program from the Japan Society for the Promotion of Science (JSPS-RFTF 96P00801). The authors appreciate and acknowledge that all of these supports were essential for completing this work.

References

[1] Stewart, D.: "A Platform with Six Degrees of Freedom", Proceedings of the Institution of Mechanical Engineers, Vol.180 Pt.1 No.15, 1965-1966.

[2] Nakamura, Y. and Ghodoussi, M.: "Dynamics Computation of Closed-Link Robot Mechanisms with Nonredundant and Redundant Actuators", IEEE Transactions on Robotics and Automation Vol.5 No.3, 1989.

[3] Nakamura, Y.:"Dynamics Computation of Parallel Mechanisms", Journal of Robotics Society of Japan, Vol.10 No.6 pp.709-714, 1992 (in Japanese).

[4] Luh, J.Y.S., Walker, M.W. and Paul, R.P.C.: "Resolved Acceleration Control of Mechanical Manipulators", IEEE Trans. Automatic Control, pp.468-474, 1980.

[5] Nakamura, Y., Yokokohji, Y., Hanafusa, H. and Yoshikawa, T.: "Unified Computation of Kinematics and Dynamics for Robot Manipulators", Transactions of Society of Instrument and Control Engineers, Vol.23 No.5 pp.71-78, 1987 (in Japanese).

[6] Lewis, B. and Berg, D.J.:"Threads Primer", Sun Soft, Inc., 1996.

[7] Takeda, Y., Funahashi, H.:"A Development of a Spatial In-parallel Actuated Manipulator with Six Degree of Freedom with Consideration of Motion Transmissibility", Proceedings of 1993 Conference of Robotics Society of Japan, pp.853-856, 1993 (in Japanese).

[8] Sugie, T. and Okada, M.:"Iterative Controller Design Method on Closed-Loop Identification", Proceedings of the 3rd ECC, Vol.2, pp.1243-1248, 1995.

[9] Gevers, M.:"Towards a Design of Identification and Control?", Essays on Control, Trentelman, Willems Eds.(Birkhäuser), pp.111-151, 1993.

[10] Ljung, L.:"System Identification – Theory for the User", Prentice – Hall, 1987.

Tracking Control of a Parallel Robot in the Task Space

L. Beji *A. Abichou M. Pascal

Université d'Evry, CEMIF : Centre Etude en Mécanique d'Ile de France, 40 rue de Pelvoux 91020 EVRY Cedex.
*INAT, Institut National Agronomique de Tunis 43, Avenue Charles Nicolle 1082 Cité Mahrajène, Tunis.

Abstract

In this paper the tracking control problem of a parallel robot including the electrical actuator dynamics is addressed in the task space. For the electrically-actuated robots, we design a nonlinear control law in armatures' input voltages. The control technique consists of a cartesian tracking control and a force convergent control. The model obtained is in standard form to allow the application of singular perturbation methods. To validate the proposed corrective controller, passivity concept and singular perturbation techniques are combined successfully. Simulation results show a good behavior of the proposed task space tracking controller.

1. Introduction

The predicted performance of a robot manipulator will depend not only on the rigid dynamic part but also on the internal interactions due to the actuator dynamics. Furthermore, task space tracking control, which implies that all the control is done at the end effector level, will be maintained with respect to joint tracking control. Generally, resolved motion force control concept [8] is required to establish the adequate control law. The basic idea of this technique of control is to determine the applied torques to the joint actuators in order to perform the cartesian position control of the robot arm. It is essentially based on the relationship between the resolved force vector, and the joint torques at the joint actuators. Therefore the control technique consists of the cartesian position control and the force convergent control.

In the joint space, for a Rigid-Link Electrically-Driven (RLED) robot, Tarn et al [14] develop a feedback linearizing control. The RLED manipulator is transformed to a third-order dynamic model. Dawson et al [7] use the assumption of exact model knowledge and propose a corrective tracking controller for RLED robots. In the presence of unknown parameters of a RLED robot, Stepanenco et al [13] have designed an adaptive controller. The Robust tracking control of a RLED robot is addressed by Mahmoud [10] where a third-order differential model is considered. Recently the same author [11] proposes a robust tracking controller in the task space. For a hydraulically actuated serial robot in [1], the tracking control problem is addressed using singular perturbation methods, the hydraulic part was considered like a fast subsystem.

This work completes our recent results which are acheived in the joint space [3-4]. The organization of this paper is as follows. In the next section, the dynamic model of the mechanical and electrical parts are given in the task space. Some results about singular perturbation techniques are given in section 3. Section 4 deals with the designed task space tracking controller and the force convergent concept. To demonstrate the trajectory tracking performance and the robustness of the proposed controller, simulation tests are performed on our laboratory's six degrees of freedom and three limbs parallel robot. The application's results are presented in section 5.

2. Mechanical and Electrical Robot Models Design in the Task Space

In accordance with the modelisation theories of a closed kinematic chains mechanism [6], the robot mechanical part exhibits the following dynamic model :

$$M(q)\ddot{q} + B(q,\dot{q})\dot{q} + Q(q) = \tau \quad (1)$$

where $q \in R^n$ denotes the controlled variables (n denotes the end effector mobility), $\tau \in R^n$ denotes the vector of driving torques at joints, $M(q) \in R^{n \times n}$ denotes the inertia matrix, which is a symmetric positive definite matrix; $B(q,\dot{q})\dot{q} \in R^n$ represents the vector of centrifugal and Coriolis forces; $Q(q) \in R^n$ is the vector of gravitational forces.
The joints of the studied parallel robot are driven by DC-motors which are modelled by [3]:

$$L_{dc}\dot{\tau} + R_{dc}\tau + K_{dc}\dot{q} = u \quad (2)$$

where $R_{dc} = R(NK_t)^{-1}$; $K_{dc} = K_bN$; $L_{dc} = L(NK_t)^{-1}$. R and L are diagonal positives matrices of the armature resistance and inductance respectively. K_b is a constant diagonal matrix of the motors back emf. The input u is the armature voltages. $N \in R^{n \times n}$ is a diagonal matrix of the gear ratios ($N > 0$). $K_t \in R^{n \times n}$ ($K_t > 0$) is a diagonal matrix of the motor constants.

Now a task space rigid-robot dynamic model takes the following matrix form [12] :

$$A(X)\ddot{X} + C(X,\dot{X})\dot{X} + G(X) = F = \left(J^T\right)^{-1}\tau \quad (3)$$

where $\dot{X} = J\dot{q}$, $\ddot{X} = J\ddot{q} + \dot{J}\dot{q}$. $J \in R^{n \times n}$ is the manipulator task Jacobian matrix. X, $\dot{X}$, $\ddot{X}$ are n-vectors of the robot end effector Cartesian positions, velocities and accelerations. Analogous to the joint space quantities, $A(.)$

is the Cartesian mass matrix, $C(.,.)$ is a vector of velocity terms in Cartesian space, and $G(.)$ is a vector of gravity terms in Cartesian space. F is a force-torque vector developed at the end effector level. Moreover, using the transformations above, a relationship between the joint and the task space dynamic coefficients is given by:

$$A(X) = (J^T)^{-1} M(q) J^{-1}, \quad G(X) = (J^T)^{-1} Q(q),$$
$$C(X, \dot{X}) = (J^T)^{-1} B(q, \dot{q}) J^{-1} - (J^T)^{-1} M(q) J^{-1} \dot{J} J^{-1}$$

Now, the dynamic behavior of a RLED robot is governed by the following pair of equations:

$$\begin{cases} A(X)\ddot{X} + C(X,\dot{X})\dot{X} + G(X) = F \\ \varepsilon(J^T \dot{F} + \dot{J}^T F) = \alpha_1 J^{-1} \dot{X} + \alpha_2 J^T F + \alpha_3 u \end{cases} \quad (4a,b)$$

where $\varepsilon = L_M (N_M K_{tM})^{-1}$, $\alpha_1 = -N_M (L')^{-1} N' K_t' K_b N'$, $\alpha_2 = -(N_M K_{tM})^{-1} R(L')^{-1}$ are a constants diagonal matrices. $\alpha_3 = N' K_t' (L')^{-1}$ is a constant diagonal and reversible matrix. $L = L_M L'$ with L_M denotes the maximum value of L, etc. In (4) we have considered the fact that $\dot{\tau} = J^T \dot{F} + \dot{J}^T F$. ε is assumed to be a small scalar parameter.

The global system (4) will be referred to us as a singularly perturbed model. The control procedure which is based on force convergent method is organized in two steps. Firstly, the position control calculates the desired forces and moments to be applied to the end-effector in order to track a desired Cartesian trajectory. Secondly, the force convergent control determines the necessary joint torques to each actuator so that the end-effector can maintain the desired forces and moments obtained from the position control. The present analysis will be confirmed below.

3. Singular Perturbation Techniques

We consider the following nonlinear system:

$$\begin{cases} \dot{x} = f(x,y,\varepsilon,t) & x \in B_x \subset R^n, x(0) = x_0 \\ \varepsilon \dot{y} = g(x,y,\varepsilon,t) & y \in B_y \subset R^m, y(0) = y_0 \end{cases} \quad (5)$$

Further, for all $[t,x,y,\varepsilon] \in [0,\infty) \times B_x \times B_y \times [0,\varepsilon_0]$ we consider the following assumptions:

A1) The functions f and g are smooth enough with $f(0,0,\varepsilon,t) = 0$ and $g(0,0,\varepsilon,t) = 0$. Moreover, the equation $g(x,y,0,t) = 0$ has a unique real root $y = h(x,t)$ with $h(0,t) = 0$.
A2) f, g and their first partial derivatives with respect to x, y and ε are continuous and bounded.
A3) The function h and $\partial g(x,y,0,t)/\partial y$ have continuous first partial derivatives, $\partial f(x,h(x,t),0,t)/\partial x$ has bounded first partial derivatives with respect to x.
A4) The initial states x_0 and y_0 are regular with respect to ε.
A5) The origin of the reduced subsystem; $\dot{\bar{x}} = f(\bar{x}, h(\bar{x},t), 0, t)$ is exponentially stable.

A6) The origin of the following *boundary layer model* is exponentially stable uniformly with respect to (x,t); $d\hat{y}/d\varsigma = g(x, \hat{y}(\varsigma) + h(\bar{x},t), 0, t)$ where $\varsigma = t/\varepsilon$.

Theorem 1 [9]. *Suppose that Assumptions 1-6 hold, then the singular perturbation problem (5) has a unique solution $x(t,\varepsilon)$, $y(t,\varepsilon)$ defined for all $t \geq t_0 \geq 0$ with $x(t,\varepsilon) - \bar{x}(t) = O(\varepsilon)$ and $y(t,\varepsilon) - h(t,\bar{x}(t)) - \hat{y}(\varsigma) = O(\varepsilon)$ holds uniformly for $t \in [t_0, \infty)$, where $\bar{x}(t)$ and $\hat{y}(\varsigma)$ are the solutions of the reduced and boundary layer problems.*

4. Tracking Control of a RLED Robot in the Task Space

Using relation (3), the dynamic of the mechanical part can be transformed as:

$$A(X)\ddot{X} + C(X,\dot{X})\dot{X} + G(X) = (J^T)^{-1} \tau_d - (J^T)^{-1} \tilde{\tau} \quad (6)$$

where $\tilde{\tau} = \tau_d - \tau$. The $((J^T)^{-1} \tilde{\tau} = \tilde{F})$ term can be regarded as disturbance forces which can deteriorate the trajectory tracking of the robot terminal tool. $(J^T)^{-1} \tau_d = F_d$ is a reference Cartesian forces vector. Consequently, to achieve the tracking objectives, it is necessary to stabilize the mechanical part by the attenuation of the disturbance (i.e. $((J^T)^{-1}\tilde{\tau}) \to 0$) and to design an adequate auxiliary reference forces F_d.

We consider for the singularly perturbed system (4), the following state feedback control law in the input voltages:

$$u = \alpha_3^{-1} \dot{F}_d - \alpha_3^{-1}(I_d + \alpha_2 J^T) F - \alpha_3^{-1} \alpha_1 J^{-1} \dot{X} \quad (7)$$

where I_d denotes the identity matrix. We underline that the proposed controller depends on reference forces not yet specified and observed cartesian forces. The observed forces can be obtained using relation (4a). But to avoid the acceleration measurements and computational load, these forces can be estimated knowing the armature currents of the actuators. Now, referred to the passivity-based control concept [5], we propose the following robot end effector reference in forces and moments:

$$F_d = A(X)\ddot{X}_r + B(X,\dot{X})\dot{X}_r + G(X) - K_d s_x \quad (8)$$

with $s_x = \dot{e}_x + \Lambda e_x$, $e_x = X - X_d$, and $\ddot{X}_r = \ddot{X}_d - \Lambda \dot{e}_x$. K_d, Λ are constant diagonal matrices with positive-valued elements. X_d is a given task space trajectory that we wish the robot manipulator to track. It is assumed to be continuously-differentiable and uniformly-bounded function.

Theorem 2 *Consider for the global dynamic system (4) the feedback in armature voltages (7) and the passivity-based controller (8). Under the conditions $K_{d,m} > \frac{1}{2}$ and $\Lambda_m > \frac{1}{2}$, the tracking error e tends exponentially to zero. $K_{d,m}$ (resp. Λ_m) is the smallest eigenvalue of K_d (resp. Λ).*

Proof. Substituting (6) into (3), we get the following closed-loop global system:

$$\begin{cases} A(X)\ddot{X} + C(X,\dot{X})\dot{X} + G(X) = F_d - \tilde{F} \\ \varepsilon(J^T \dot{F} + \dot{J}^T F) = \tilde{F} \end{cases} \quad (9)$$

From (9) we can see that the error in forces $\tilde{F}$ is generated by the dynamic of the electrical part. With respect to the coordinate $\Theta = (e_x, s_x)^T$, system (10) can be reformulated into the form:

$$\begin{cases} \dot{\Theta} = f(\Theta, F, t) \\ \varepsilon(J^T \dot{F} + \dot{J}^T F) = g(\Theta, F, t) \end{cases} \quad (10)$$

where

$$f(\Theta, F, t) = \begin{pmatrix} s_x - \Lambda e_x \\ A^{-1}(e_x + X_d)\Big(F - C(e_x + X_d, \dot{e}_x + \dot{X}_d)(\dot{e}_x + \dot{X}_d) \\ -G(e_x + X_d)\Big) + \Lambda \dot{e}_x - \ddot{X}_d \end{pmatrix} \quad (11a)$$

and

$$g(\Theta, F, t) = F_d - F \quad (11b)$$

We assume that $(\Theta = 0, F = 0)$ is an equilibrium point of system (10). Main purposes is to prove that this equilibrium position is stable. Let us now verify the conditions of Theorem 1. The unique isolated root for the algebraic equation $g(\overline{\Theta}, \overline{F}, t) = 0$ is given by:

$$g(\overline{\Theta}, \overline{F}, t) = 0 \Leftrightarrow F_d - \overline{F} = 0 \Leftrightarrow \overline{F} = F_d = h(\Theta, t) \quad (12)$$

which imply that the robot model in closed loop (10) is in *standard form* [9] and the stability-based singular perturbation methods is available.
Functions f and g are assumed to be smooth enough to satisfy assumptions A1-4 of Theorem 1.
The slow subsystem is given by the following expression:

$$\dot{\overline{\Theta}} = f(\overline{\Theta}, h(\overline{\Theta}, t), t) \quad (13)$$

which is clearly the dynamic of the mechanical part having as input F_d. (for simplicity the slow variable $\overline{\Theta}$ will be replaced by Θ). Under (8) the dynamic behavior of the closed-loop slow subsystem is given by:

$$\begin{cases} \dot{e}_x = -\Lambda e_x + s_x \\ \dot{s}_x = A^{-1}(e_x + X_d)\big(-C(e_x + X_d, \dot{e}_x + \dot{X}_d)s_x - K_d s_x\big) \end{cases} \quad (14)$$

for which the Lyapunov function candidate is defined by:

$$V_x(\Theta) = \frac{1}{2}s_x^t A(X)s_x + \frac{1}{2}e_x^T e_x \quad (15)$$

It easy computation to show using the passivity property [5]: $\xi^T(\dot{A}(X) - 2C(X,\dot{X}))\xi = 0$ $(\forall \xi)$ and the conditions given by Theorem 2 that:

$$\dot{V}_x(\Theta) \leq -s_x^T(K_d - \frac{1}{2})s_x - e_x^T(\Lambda - \frac{1}{2})e_x \leq 0.$$

Then A5 is satisfied. The dynamic of the reduced fast part in closed loop is given by:

$$d\eta/d\varsigma = g(\Theta, \eta + h(\Theta,t), t) = F_d - (\eta + F_d) = -\eta \quad (16)$$

Then A6 is trivially satisfied. We can propose for (16) the Lyapunov function $W(\eta) = \frac{1}{2}\eta^T \eta$.

Remark 4. *By Theorem 1 we can assert that the tracking error in the task space remains in a small neighbourhood of zero which leads to a practical stability result. To achieve a complete stability result, it remains to prove that the global system is exponentially stable.*

We consider the composite Lyapunov function candidate

$$L(\Theta, \eta) = V(\Theta) + W(\eta) \quad (17)$$

The time-derivative of L along the trajectories of (9) yields:

$$\dot{L}(\Theta, \eta) = \frac{\partial V}{\partial \Theta} f(\Theta, h(\Theta,t)) + \frac{\partial V}{\partial \Theta}(f(\Theta, \eta + h(\Theta,t))$$
$$- f(\Theta, h(\Theta,t))) + \frac{1}{\varepsilon}\frac{\partial W}{\partial \eta} g(\Theta, \eta + h(\Theta,t))$$
$$- \frac{\partial W}{\partial \eta}\frac{\partial h}{\partial \Theta} f(\Theta, \eta + h(\Theta,t)) - \frac{\partial W}{\partial \eta}\frac{\partial h}{\partial t} \quad (18)$$

Remark 5. *An upper bound for all quantities in (15) can be easily achieved, except the term $\|(\partial W/\partial \eta)(\partial h/\partial t)\| \leq k\|\eta\|\|\partial h/\partial t\|$ (k is a constant). The biais is due to the presence of the unbounded term $\partial h(\Theta,t)/\partial t$.*

Lemma 1 *An upper bound for $h(\Theta,t)$ is given by:*
$$\|\partial h(\Theta,t)/\partial t\| \leq k_6 \|\Theta\|. \quad (30)$$

Proof. We recall the expression of $h(\Theta,t)$:
$$h(\Theta,t) = A(e_x + X_d)\ddot{X}_r + C(e_x + X_d, \dot{e}_x + \dot{X}_d)\dot{X}_r$$
$$+ G(e_x + X_d) - K_d s_x$$

where we have introduced $X = e_x + X_d$ and $\dot{X} = \dot{e}_x + \dot{X}_d$. The partial derivative of $h(\Theta,t)$ can be written as:

$$\frac{\partial h(\Theta,t)}{\partial t} = h_1(\Theta,t) + h_2(\Theta,t) \quad (17)$$

where

$$h_2(\Theta,t) = -\frac{\partial A}{\partial X}\dot{X}_d \Lambda \dot{e}_x - \left[\frac{\partial C}{\partial X}\dot{X}_d + \frac{\partial C}{\partial \dot{X}}\ddot{X}_d\right]\Lambda e_x \quad (18)$$

$$h_1(\Theta,t) = \frac{\partial A}{\partial X}\dot{X}_d \ddot{X}_d + A(X)\dddot{X}_d + \frac{\partial C}{\partial X}\ddot{X}_d^2$$
$$+ \frac{\partial C}{\partial \dot{X}}\ddot{X}_d \dot{X}_d + C(X,\dot{X})\ddot{X}_d + \frac{\partial G}{\partial X}\dot{X}_d \quad (19)$$

We assume that the dynamic coefficients A, C and G are C^∞ with respect to their arguments.

Given a desired task space trajectory as a continuously-differentiable and bounded function, and the fact that $\|e_x\| \le k_2 \|\Theta\|$ and $\|\dot{e}_x\| \le k_3 \|\Theta\|$, then the term $h_2(\Theta,t)$ satisfies the following inequality:

$$\|\partial h_2(\Theta,t)/\partial t\| \le k_4 \|\Theta\| \qquad (20)$$

Let us now examine the term $h_1(\Theta,t)$. By assumption A1 we have $z(t) \equiv A(X_d)\ddot{X}_d + C(X_d, \dot{X}_d)\dot{X}_d + G(X_d) = 0$. In fact this means that the trajectory tracking objectives in the task space are achieved, i.e. $\Theta = 0, F = 0$. Using this result and (19), after some arrangements $h_1(\Theta,t)$ can be rewritten as:

$$h_1(\Theta,t) - \frac{\partial z(t)}{\partial t} = y_1(\Theta,t)\dot{X}_d\ddot{X}_d + y_2(\Theta,t)\ddot{X}_d + y_3(\Theta,t)\dot{X}_d^2$$
$$+ y_4(\Theta,t)\ddot{X}_d\dot{X}_d + y_5(\Theta,t)\ddot{X}_d + y_6(\Theta,t)\dot{X}_d \qquad (37)$$

where

$$y_1(\Theta,t) = \frac{\partial A(e_x + X_d)}{\partial X} - \frac{\partial A(X_d)}{\partial X} \qquad (38)$$

$$y_2(\Theta,t) = A(e_x + X_d) - A(X_d) \qquad (39)$$

$$y_3(\Theta,t) = \frac{\partial C(e_x + X_d, \dot{e}_x + \dot{X}_d)}{\partial X} - \frac{\partial C(X_d, \dot{X}_d)}{\partial q} \qquad (40)$$

$$y_4(\Theta,t) = \frac{\partial C(e_x + X_d, \dot{e}_x + \dot{X}_d)}{\partial \dot{X}} - \frac{\partial C(X_d, \dot{X}_d)}{\partial \dot{X}} \qquad (41)$$

$$y_5(\Theta,t) = C(e_x + X_d, \dot{e}_x + \dot{X}_d) - C(X_d, \dot{X}_d) \qquad (42)$$

$$y_6(\Theta,t) = \frac{\partial G(e_x + X_d)}{\partial X} - \frac{\partial G(X_d)}{\partial X} \qquad (43)$$

Lemma 2 *Equation (37), consequently (34) is bounded in the following sens:*

$$\|h_1(\Theta,t) - \partial z(t)/\partial t\| = \|h_1(\Theta,t)\| \le k_9 \|\Theta\| \qquad (44)$$

Proof. For the proof see Appendix A. □

Consider the results of Lemma 2 (*i.e.* (33) and (44)), the proof of Lemma 1 is achieved. □

Therefore from Lemme 1, we can conclude that there exists ε^{**} leading to $\dot{L} \le 0$. Consequently, given a regular desired task space trajectory, the tracking error e_x of the robot terminal tool tends exponentially to zero. This complete the proof of theorem 2. □

5. Application and Simulation Results

To show the behavior of the proposed controller, the tests are performed on our laboratory electrically-driven 6 degrees of freedom parallel robot. The structure is composed by a top plate a fixed plat and three limbs. Each limb of the robot is controlled by a prismatic joint and an universal joint in which only one degree of freedom is motorized (figure 1). For further details, a complete geometric description of the robot is given in [3]. The desired trajectory was generated in the joint space under corrections in the task space as shown in figure 2. Simulation tests are obtained for initial conditions on task space tracking errors not equal to 0. The desired reference trajectory for each joint was presented in figure 3. The parameters of the DC-motors are shown in Table 1.

DC-motor parameters	RS110M (Prismatic joints)	RS240B (revolute-active joints)
K_t (Nm/A)	0.08368	0.068
R (Ω)	4.8	0.64
K_b (V/rd/s)	0.0242	0.0449
L (mH)	1.6	0.45
N	46.8	100

Table 1. Electrical parameters of the parallel robot

Simulations are performed with the numerical language ACSL. In order to evaluate the robustness of the task space tracking controller, we want to test a perturbed and non perturbed dynamic model but without compensation in control level. The perturbation is considered due only to frictions in joints.

For the case of a non perturbed model, the controller gains are given by: $\Lambda = diag\{30,30,30,25,25,25\}$ and $K_d = diag\{15,15,15,10,10,10\}$. Moreover fro a perturbed model the controller gains are changes as follows: $\Lambda = diag\{200,200,200,250,250,250\}$ and $K_d = diag\{25,25,25,30,30,30\}$.

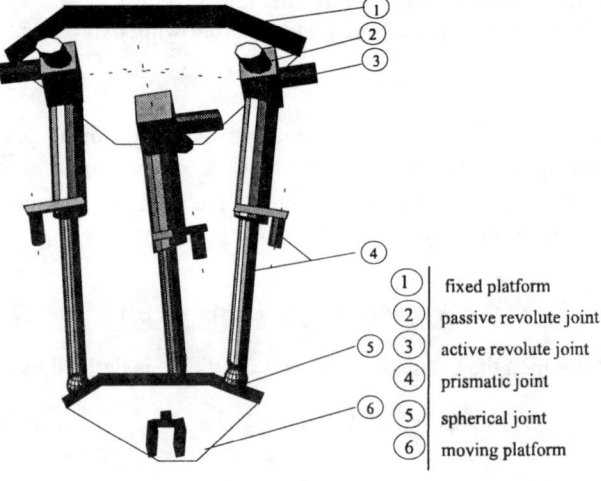

Figure 1. A spatial form of the parallel robot

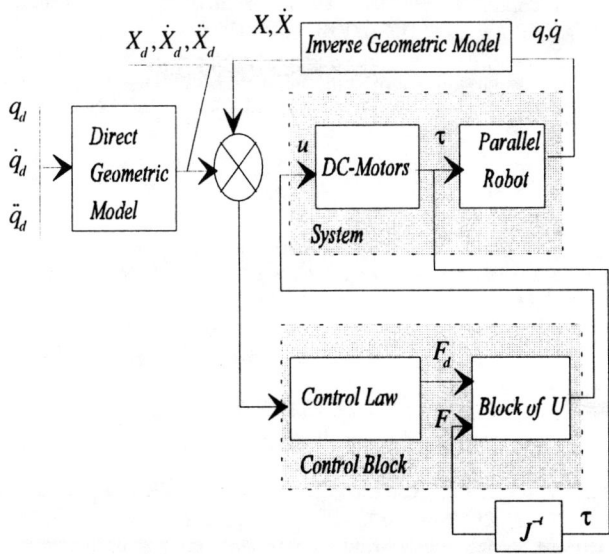

Figure 2. Block diagram of the system and the control block

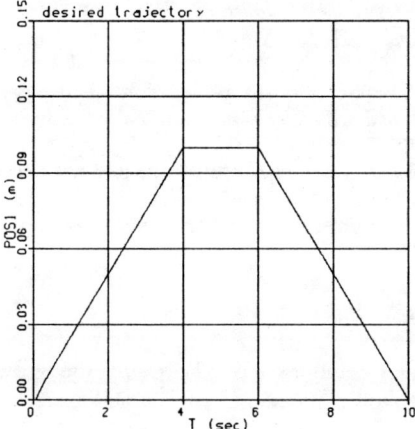

Figure 3. Desired joint trajectory

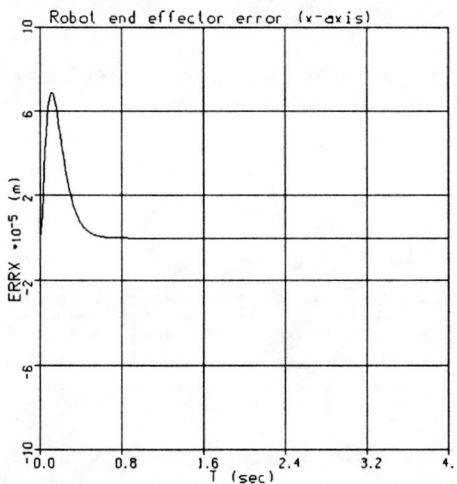

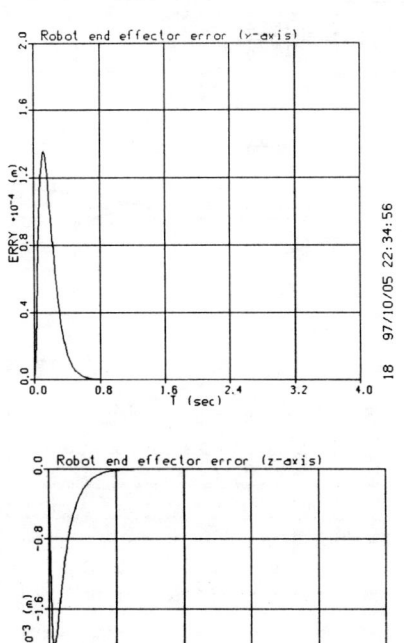

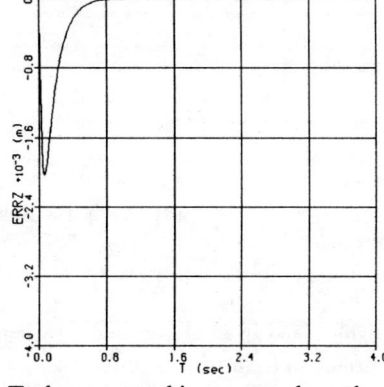

Figure 4. Task space tracking errors along the *x*, *y*, *z*-axis (without perturbed model)

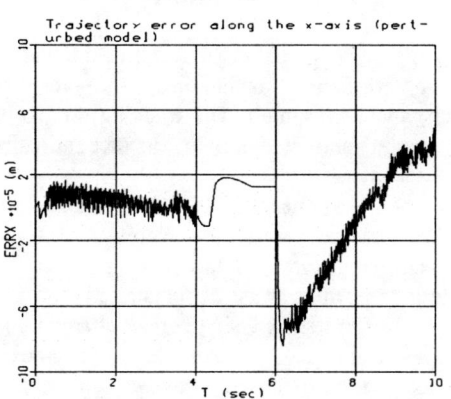

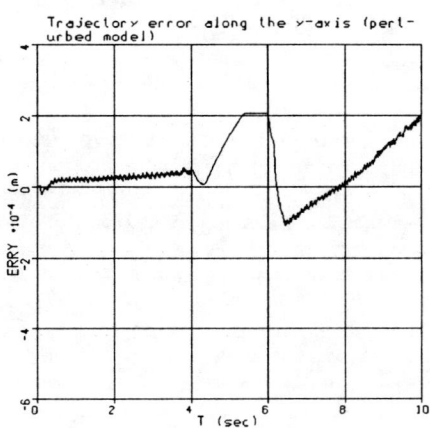

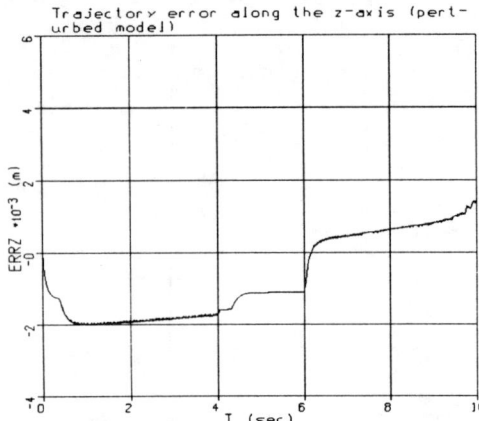

Figure 5. Task space tracking errors along the
x, y, z-axis (with perturbed model)

As shown in figure 4, in the case of a non perturbed model, the tracking errors along the different terminal tool axis are quite acceptable. In fact, this confirm the obtained theoritical stability results. Moreover we show that the observed forces and moments converge quickly to the proposed reference forces and moments which ensures the stability of the rigid robot part. As it is illustrated by figure 5, the tracking control of a perturbed robot model leads to a bounded end-effector tracking error. Necessarily the increase of the controller gains permits to decrease this error, but chattering was observed in control input. This result is illustrated in figure 5.

6. Conclusions

In this paper, a force convergent control method based on singular perturbation techniques and the passivity approach where presented. The tracking control law is a voltage input, and it takes in to account the robot manipulator dynamic as well as the internal actuators dynamic. An exponential stability results was also established under an exact model knowledge of the global system. The robustness of the task space controller in presence of frictions was demonstrated by simulation which was performed on a six degrees of freedom parallel robot. The results can be extended to various loading conditions and to electrically as well as hydraulically (where ε is related to the fluid compressibility) robot manipulators with and without kinematic loops.

7. References

[1] B. d'Andréa-Novel, M.A. Garnero; A. Abichou, Nonlinear control of a hydraulic robot using singular perturbations, *IEEE Trans, Systems Man, and Cybernetics*, San Diego, pp. 1932-1937, 1994.

[2] L. Beji, A. Abichou, P. Joli, M. Pascal, Nonlinear control of a parallel robot including motor dynamics, *Proc. of 11th. CISM-IFToMM Symposium, ROMANSY'96*, July 1-4, Udine, Italy, pp. 45-52, 1996.

[3] L. Beji, M. Pascal, P. Joli, Toward a minimal dynamic model of a 6-dof parallel robot, *Proc. of AMSE, 11th.*

Biennial Mechanism Conf., Sept.14-17, Sacramento, California, DETC'97/VIB-4224, 1997.

[4] L. Beji, A. Abichou, P. Joli, M. Pascal, Tracking control of a parallel robot with estimated state feedback, *5th. IFAC Symposium on Robot Control*, Nantes, Sept. 3-5, pp. 437-442, 1997.

[5] H. Berghuis, and H. Nijmeijer, A Passivity Approach to Controller-observer Design for Robots, *IEEE Trans. on Robotics and Automation*, Vol.9, pp. 740-754, 1993.

[6] C. Canudas de Wit, B. Siciliano, G. Bastin, Theory of Robot Control, *Springer Verlag*, London, 1996.

[7] D.M. Dawson, Z. Qu., J.J. Carroll, Tracking control of rigid-link electrically-driven robot manipulators, *International Journal of Control*, V.56, N°5, pp. 991-1006, 1992.

[8] K.S.FU, R.C.Gonzalez, C.S.G. Lee, Robotics: Control, Sensing, Vision, and Intelligence, *McGraw-Hill Book Company*, New York, 1987,

[9] H.K Khalil, Nonlinear Systems, *Macmillan Publishing Company*, NewYork, 1992.

[10] M.S Mahmoud, Robust control of robot arms including motor dynamics, *Int. Journal of Control*, Vol.58, N°4, pp.853-873, 1993.

[11] M.S Mahmoud, Robust Robot Tracking Controller in the Task Space, *IEEE Trans, Systems Man, and Cybernetics*, LILLE, pp. 637-642, 1996.

[12] J.J.-E. Slotine, W. Li, On the Adaptive Control of Robot Manipulators, *Int. Journal of Robotics Research*, Vol.6, pp. 49-59, 1987.

[13] Y. Stepanenco, C.-Y. Su, Adaptative Motion Control of Rigid-Link Electrically-Driven Robot Manipulators, *IEEE Conference on Decision and Control*, 1994, pp. 1050-4729.

[14] T.J Tarn, A.K. Bejczy, X. Yun, Z. Li, Effect of motor dynamics on nonlinear feedback robot arm control, *IEEE Trans. on Robotics and Automation*, Vol.7, pp. 114-122, 1991.

Appendix A. Proof of Lemma 2

We consider relations (37-43). Using the growth finite theorem an ultimate bound of $y_1(\Theta,t)$ and $y_2(\Theta,t)$ are given by :

$$\|y_1(\Theta,t)\| \leq \|\partial^2 A(X_d)/\partial X^2\| \|e_x\| \leq d_1 \|e_x\| \leq d_1 \|\Theta\| \quad (A1)$$

$$\|y_2(\Theta,t)\| \leq \|\partial A(X_d)/\partial X\| \|e_x\| \leq d_2 \|e_x\| \leq d_2 \|\Theta\| \quad (A2)$$

Moreover, we can reformulate relation (44) into the form:

$$y_5(\Theta,t) = C(e_x + X_d, \dot{X}_0) - C(e_x + X_d, \dot{X})$$
$$+ C(e_x + X_d, \dot{e}_x + \dot{X}_d) - C(e_x + X_d, \dot{X}_d)$$
$$+ C(e_x + X_d, \dot{X}_d) - C(X_d, \dot{X}_d) \quad (A3)$$

The expression of $y_3(\Theta,t)$ and $y_4(\Theta,t)$ can be trivially obtained by applying the operator $(\partial/\partial X)$ and $(\partial/\partial \dot{X})$ to (A3). Now, the growth finite theorem can be applied to bound the terms of $y_3(\Theta,t)$, $y_4(\Theta,t)$, $y_5(\Theta,t)$ and $y_6(\Theta,t)$.

The desired trajectory that we wish the robot terminal tool to track is considered bounded. Using (A1-3) and relation (37), $(h_1(\Theta,t) - \partial z(t)/\partial t)$ is ultimately bounded in the sens of (44). □

Performance Evaluation of Vision-Based Control Tasks

Peter Krautgartner   Markus Vincze

Institute of Flexible Automation, Vienna University of Technology
Gusshausstrasse 27-29, 1040 Vienna, Austria

pk@flexaut.tuwien.ac.at

Abstract

In this paper the tracking performance of vision-based control systems is evaluated and the optimal system configuration found. Configurations evaluated are serial or parallel image acquisition and processing, and pipeline processing. The basis of the optimization is the design of an optimal controller independent of the system configuration. Using this controller design a relation between system latency and maximum pixel error is derived. This relation is used to find maximum dynamic performance for the system configurations. The performance measure is the maximum velocity of the target in the image that can be tracked. The final comparison shows that processing in a pipeline obtains highest velocity due to high cycle rate of the system. The parameters for the point of maximum velocity are derived, i.e., the optimal number of steps in a pipeline.

1. Introduction

In recent years vision and robotics researchers built several active vision [Bajc88, OlCo91] and visual servoing systems [Hash93, HuHa96]. The principle is to use visual tracking to control motion. The basic difference to conventional control of motion is the additional latency in the feedback loop. Latency is introduced due to frame time, necessary to obtain and transfer an image or parts of an image, and by processing the image to obtain the control signal.

Processing the entire image is possible for simple tasks such as tracking a region of high contrast [Ande89, CoGo96]. Most tasks need sophisticated image processing to extract the target and research in robust vision will add to this additional request for computing power. To utilize these techniques within vision-based control, the general approach is to limit processing to small windows [GrMe96, HaTo96, RiKo96].

Latency has great impact on system dynamics. The dynamic performance of visual tracking is denoted by the velocity and acceleration the target can make without being lost. The standard methods to obtain high dynamic performance are filtering and prediction of target locations, e.g., after Kalman [WiWi96, WuHi97]. Another common procedure is to reduce processing time, and therefore latency, by careful system design. A common approach uses dedicated or parallel processors [LiBr95, WuHi97]. Another approach is processing in a pipeline, which nests several loops and obtains a small cycle time of the control system but large latencies [AlTi93, LiBr95].

The systematic investigation of dynamic performance of vision-based control was started in the excellent work of Corke [Cork96]. Corke investigated the demands on a controller for turning the wrist axes of a Puma robot in a fixation task. The dynamic performance of *visual tracking* (the vision processing in itself) has been first investigated in [ViWe97]. It is shown that there exists a fixed relationship between tracking velocity and window size, which is used to derive window size for maximum tracking performance. Starting from Corke's work we will extend this result on system design for vision-based control systems.

It is the goal of this paper to formalize the parameters of visual fixation control systems and to find the system design that gives best dynamic performance. The basic design cases are serial or parallel image acquisition and image processing and the effect of using a pipeline for processing. The systematic analysis of maximum performance relies on careful control loop design. The paper starts by outlining the basic control loop for vision-based systems (Section 2). A performance metric for visual fixation tasks is defined (Section 3), and the controller designed accordingly (Section 4). Using this controller the relation between maximum pixel error and latency is derived (Section 5). This relation is used to find the maximum dynamic performance for a given system design (Section 6) and finally we compare differ-ent designs to find the optimal architecture (Section 7).

2. Vision-Based Control System

From the view of control design the dynamics for steering a mechanism towards a pre-defined goal in the image (visual servoing) are the same as for centering a target in the image of a camera mounted on a mechanism (fixation). Fig. 1 shows the basic components of a vision-based control system.

The goal of fixation is to keep the target in the center of the image plane. As the target motion x_t is not directly measurable it has to be treated as a non-measurable disturbance input [Cork96]. The controlled variable is the image plane pixel error ΔX which should approach zero. As the vision sensor provides a position error which is transformed to a velocity demand by the controller, an integrator must generate a position setpoint for the control of the axis.

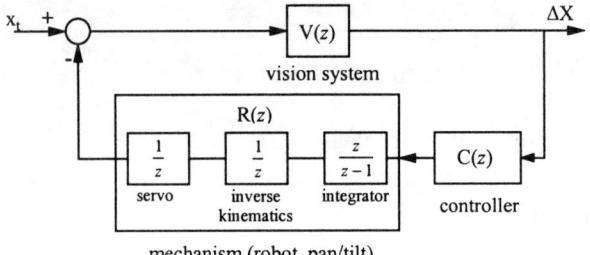

Figure 1: Block diagram of a vision-based control system.

The servo unit of the mechanism is ideally treated as a unit delay. This is justified by the assumption that the underlying position loop has a higher sampling rate than the vision system and therefore an interpolation of the position setpoint is possible. Corke showed [Cork96] that this simplification is appropriate and gives good results. One further unit delay represents the time needed for the integration, inverse kinematics, and eventual transfer times inherent in the feedback loop of the vision-based control system. When assigning these further delays to the mechanism, it results in a latency of the mechanism of two unit delays (see Fig. 1).

3. Performance Metric

Performance metrics often applied to dynamic control design are, for example settling time or overshoot of unit step responses. In case of fixation control, a more appropriate metric to evaluate the performance of the target motion is a ramp input, i.e., a constant velocity motion profile of the target. The rationale is that when trying to follow a moving target the critical value is the maximum deviation from the center of the image for not loosing the target within the window observed. Another requirement is to avoid heavy oscillations when following the target. Therefore the minimization of the mean pixel error within a given time interval is appropriate, that is

$$Q = \sqrt{\frac{\int_{t_1}^{t_2} (\Delta X)^2 dt}{t_2 - t_1}} \quad (1)$$

which constitutes the RMS pixel error, Q, over the time interval $[t_1, t_2]$.

Fig. 2 demonstrates the optimal controller design using this performance metric. One can clearly observe that the system is slightly oscillating, however shows moderate overshoot, that is small maximum pixel error in the image plane.

The small oscillations can be tolerated when dealing with fixation tasks but must be strictly avoided when focusing on assembly tasks. In assembly tasks velocity is however small and maximum tracking velocity is not the critical issue.

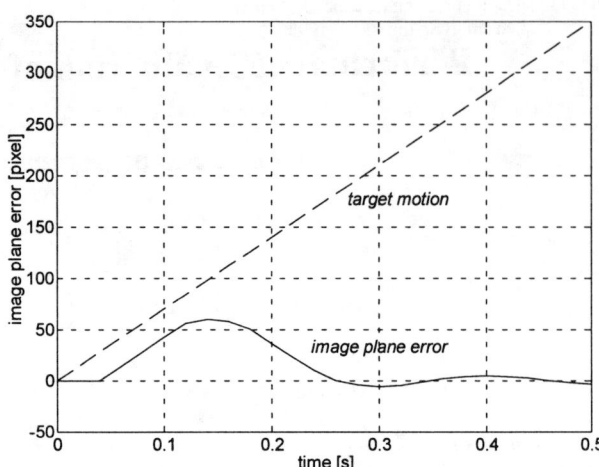

Figure 2: Dynamic response to a ramp input. Vision system: $V(z)=1/z^2$.

The assumption of constant target velocity is justified if the sampling interval is small enough. At each sampling step a changing target velocity stands for a step in acceleration which has to be tracked (see Fig. 2). This can be easily understood when the initial moment of target identification and fixation is taken into account: Assuming the target to move with constant velocity, at the first time instant the motion of the target is interpreted by the vision system as an instantaneous change in velocity, i.e., acceleration. Therefore the maximum pixel error is a measure for target acceleration and directly defines the lower limit for the window size necessary not to loose the target.

4. PID-Controller

4.1. Theoretical Considerations

In order to provide a ramp following behavior without any steady-state error a double integrator in the open loop transfer function is necessary, referred to as a Type 2 system in classical control theory. This can be accomplished by adding open-loop integrators to the consisting controlled system. Since the visually controlled system consists of just one integrator - as outlined in Section 2 - one further integrator has to be added for the steady-state tracking error to be zero. This can be achieved by a classical PID-controller with the discrete transfer function

$$C(z) = k_P + k_I \frac{zT}{z-1} + k_D \frac{z-1}{zT} \quad (2)$$

$$= \frac{z^2(k_P T + k_I T + k_D) - z(k_P T + 2k_D) + k_D}{z(z-1)T} \quad (3)$$

where k_P, k_I, and k_D are the proportional, integral and derivative gains respectively and T is the sampling time. The two compensator zeros can be chosen according to the dynamic behavior desired. The open-loop transfer function of the vision-based control task depicted in

Fig. 1 can now be written as

$$F_o(z) = V(z) \cdot C(z) \cdot R(z) \quad (4)$$

$$= \frac{k_C \cdot (z - z_1) \cdot (z - z_2)}{z^4 \cdot (z - 1)^2} \quad (5)$$

where $V(z)$, $C(z)$ and $R(z)$ are the transfer functions of the vision system, the compensator, and the controlled mechanism, respectively. k_C is the compensator gain and z_1 and z_2 constitute the compensator zeros which can be freely chosen. Here the vision system is assumed to result in two unit delays, one for acquiring the image and another one for processing the data.

4.2. Parameter Optimization

The optimal free parameters of a PID-controller, i.e., the compensator gain and the two zeros, are now investigated for different latencies in the control loop. The performance metric given in Section 3 is used to define the optimal controller.

The controller design is carried out with the aid of the root-locus method. The problem with a double integrator in the open-loop transfer function is the inherent tendency to become unstable. On the other hand this double integrator is necessary for ramp following behavior with a steady state error to become zero. The two zeros of the PID-compensator can now be used to bend the root locus inside the unit circle which corresponds to stability of the closed-loop system. A proper selection of the open-loop gain then yields the optimality of the closed-loop system with respect to performance metric of eq. (1).

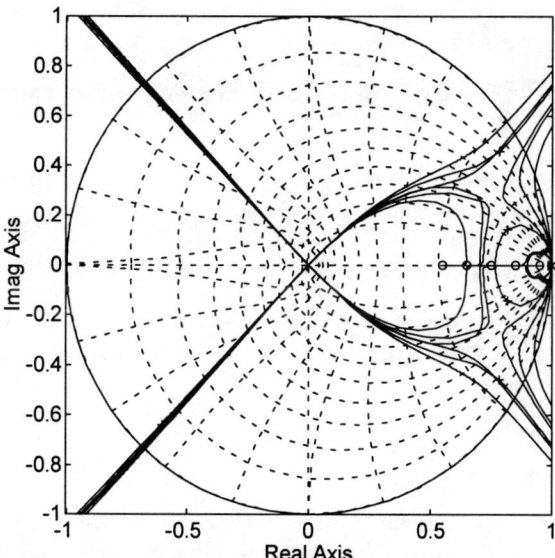

Figure 3: Root-locus plot of the fixation control:
vision system: $V(z) = 1/z^2$;
PID-controller: z_1=0.55, 0.65, z_2=0.75, 0.85, 0.95.

In contrast to the work in [Cork96] the controller is designed to obtain a closed-loop system without a dominant pole pair. This yields better maximum dynamic performance, that is a smaller maximum pixel error at the expense of small oscillations (see Fig. 2). Fig. 3 shows the root-locus arms for different combinations of the two controller zeros. The crosses mark the optimal open-loop gain for this setup. The disturbance behavior for the closed-loop system using this controller was shown in Fig. 2.

5. Image Plane Error vs. Latency

To derive an optimality criterion for the dynamic performance as a reaction to a given latency in the visual feedback loop, the relationship between latency and the resulting (optimal) image plane error has to be found.

Two different approximations have been investigated:

- The first approximation is based on the assumption that a P-controlled system leads to the same dynamic behavior, i.e., maximum image plane error ΔX as a reaction to a ramp input, as the PID-controlled system, provided that the damping factors D of both systems are the same. The difference between P- and PID-controlled system is that when using a P-controller there remains a steady-state error. The image plane error ΔX_I results from this steady-state error and the amount of overshoot of a P-controlled system and can be written as:

$$\Delta X_I = k \cdot T \cdot \frac{1}{k_R} \cdot \left(1 + e^{-\frac{\pi \cdot D}{\sqrt{1-D^2}}} \right) = k \cdot T \cdot C_I ; \quad (6)$$

where k_R is the gain of the P-controller, D is the damping factor, k is the rising factor of the ramp, and T is the sampling time.

- Another approximation results from the measured ramp response curves and is given by

$$\Delta X_{II} = k \cdot T \cdot (lat_{Mech} + C \cdot lat_{Vis}) = k \cdot T \cdot C_{II} ; \quad (7)$$

where lat_{Mech} and lat_{Vis} are the latencies of the mechanism and the vision system, respectively, denoted as multiples of the sampling time T. The constant factor C is ideally 1 for a controller which is able to instantaneously react to an error without any overshoot. The measurements show that a very good approximation is obtained for

$$C = 9/8 , \quad (8)$$

when using the optimal PID-controller selected in Section 4.

The two approximations differ in the boundary conditions:

If there exists one dominant closed-loop pole pair the first approximation is adequate (Fig. 4). When using this approximation for the estimation of the dynamic behavior, it has to be guaranteed that this assumption is met.

If there doesn't exist one dominant pole pair the second approximation with the constant C given in (8) is very precise, (Fig. 5). In case of one dominant pole pair this approximation fits well when the factor C of eq. (8) is changed to $^{12}/_{8}$. This means that maximum ΔX is larger, too (see eq. 7).

The difference between the approximations is depicted in Figs. 4 and 5, where the measured and approximated maximum ΔX are shown for varying latency in the control loop.

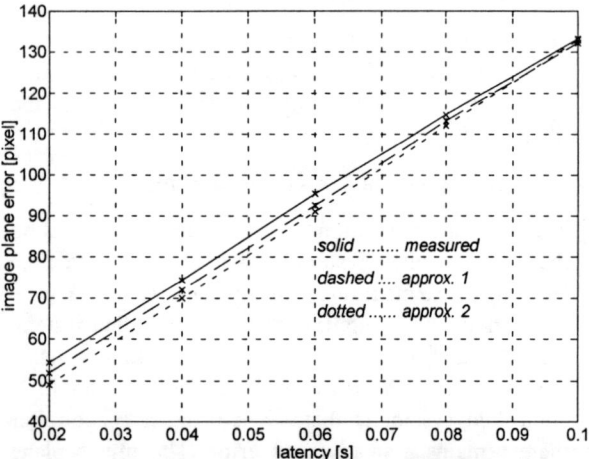

Figure 4: Measured and calculated image plane error for a system with one dominant pole pair; $C=^{12}/_{8}$.

The controller applied in Fig. 4 yields one dominant pole pair of the closed-loop system. Therefore approximation 1 yields better results than approximation 2. The results for the other case with no dominant pole pair are depicted in Fig. 5. Here approximation 2 is very exact, whereas the other one yields large deviations from the real values.

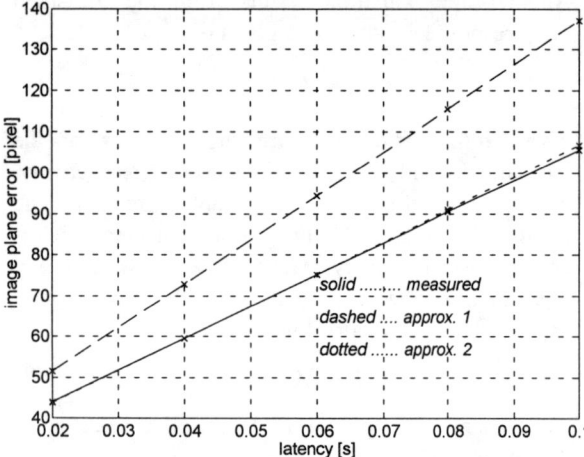

Figure 5: Measured and approximated image plane error for a system without dominant poles.

By comparing the measured curves (solid lines) in Figs. 4 and 5, we observe that the image plane error is much smaller for a controller which yields a closed-loop system without a dominant pole pair (Fig. 5). This supports the result of Section 4.2. We therefore select the optimal controller for the control system of eq. (5) which yields no dominant pole pair and the approximation 2 of eq. (7) to calculate the maximum pixel error.

6. Dynamic System Performance vs. Latency

In the last section the maximum pixel error was investigated as a function of latency. The pixel error was calculated for a given latency and the rising factor of the ramp, that is the velocity of the target. Now we relate dynamic performance to a given latency in the visual control loop. The question we want to answer is: *What is the maximum target velocity which can be tracked for a given window size?* Remember that window size determines processing time and therefore latency of the vision system.

In this section we consider the static case, that is visual tracking, which is defined as following an object within an image by permanently shifting a window within this image. For the static case there obviously exists a relationship between window size and processing time on the one hand and between window size and tracking velocity on the other hand.

An estimate valid for processing time t_p of common tracking techniques [ViWe97] states, that t_p is proportional to the number of pixels within the window. For a window size with the side length $2r$ processing time t_p is given by

$$t_p = 4 \cdot D \cdot r^2, \qquad (9)$$

where D describes the time necessary to evaluate one pixel. It is furthermore shown [ViWe96] that there exists a fixed relationship between tracking velocity v and window size (radius r), namely

$$v = \frac{r}{t_p} = \frac{r}{4 \cdot D \cdot r^2} = \frac{1}{4 \cdot D \cdot r} \qquad (10)$$

which is used to derive the window size for maximum tracking performance of visual tracking.

7. Comparison of Different Vision Strategies

Now the results of Section 6 for the static case will be applied to visual fixation control. The goal is to derive a relationship between tracking velocity and processing time like eq. (10) for the case of dynamic tracking control. The result we want to obtain is the tracking performance as a function of processing time.

To be able to compare different system configurations, the resulting differences regarding sampling time T must be investigated. The three different configurations which will be compared are parallel, serial, and pipeline processing. Fig. 6 shows the three cases pointing out image acquisition time t_{ac} and image processing time t_p. In each case arises a corresponding sampling time T of the discrete control system.

Timing:

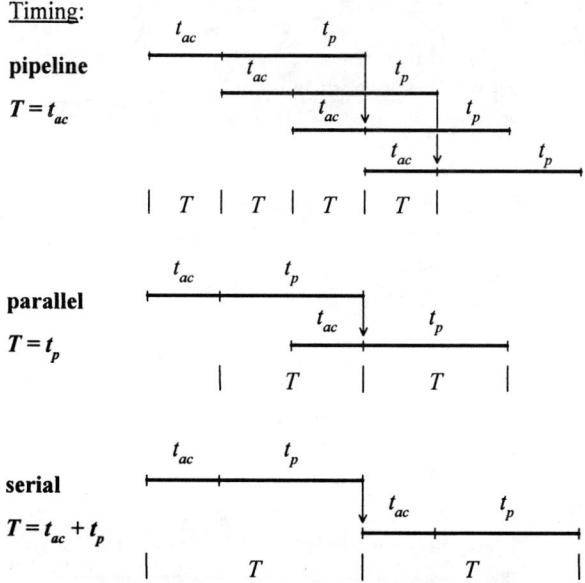

Figure 6: Parallel, serial and pipelining system.

We will now investigate the effects of latency on the three system designs. Eq. (7) is similar to eq. (10) when substituting v for k, that is replacing the ramp rising factor by the target velocity, and r (from eq.(10)) for ΔX. From eq. (7) we obtain as tracking velocity

$$v = \frac{r}{T \cdot C_{II}}. \quad (11)$$

Compared with eq. (10) for the static case, the velocity is decreased by the factor $1/C_{II}$ when including the dynamics of visual servoing. Using this equation together with the findings of Fig. 6, the basic expressions for the three system configurations can be derived.

The tracking velocity using a pipeline structure is

$$v_{pipe} = \frac{r}{t_{ac} \cdot C_{II}}. \quad (12)$$

With eq. (10), tracking velocity for the structure using parallel timing gives

$$v_{par} = \frac{r}{t_p \cdot C_{II}} = \frac{r}{4 \cdot D \cdot r^2 \cdot C_{II}} = \frac{1}{4 \cdot r \cdot D \cdot C_{II}}. \quad (13)$$

For the serial processing system follows

$$v_{ser} = \frac{r}{(t_{ac} + t_p) \cdot C_{II}} = \frac{r}{t_{ac} \cdot C_{II} + 4 \cdot r^2 \cdot D \cdot C_{II}}. \quad (14)$$

From eqs. (12) to (14) we observe that in all cases tracking velocity is a function of half the window side length r. Transforming identity (9) and substituting r into eqs. (12) to (14), further substituting the different values of C_{II} of eq. (7) for the corresponding system configuration we obtain tracking velocity as function of processing time t_p. Together with image the acquisition time t_{ac}, t_p determines the latency of the vision system.

Fig. 7 shows tracking velocity as a function of processing time in all three cases. Image acquisition time is 20ms and the entire latency of the mechanism is assumed to be two unit delays as derived in Section 2. Figs. 7 and 8 are examples of system configurations.

Many similar systems are possible depending on their mechanism and vision latencies.

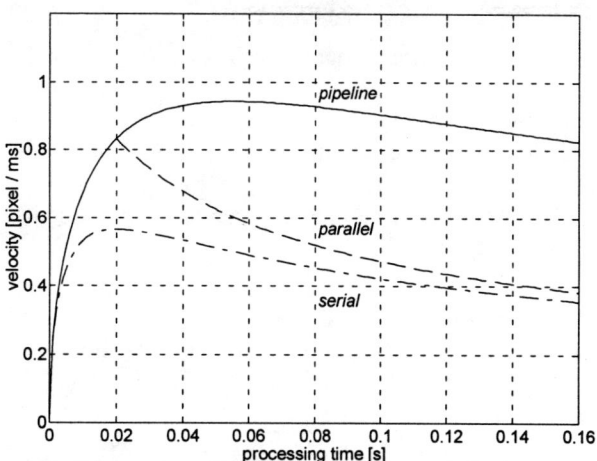

Figure 7: Tracking performance vs. processing time. t_{ac}=20ms, lat_{Mech}=2.

When transfer times and trajectory computation times can be neglected (= inverse kinematics in Fig. 1), the mechanism introduces only one unit delay. The effects on the dynamic performance are depicted in Fig. 8. Obviously velocity is better than in Fig. 7, since latency is smaller.

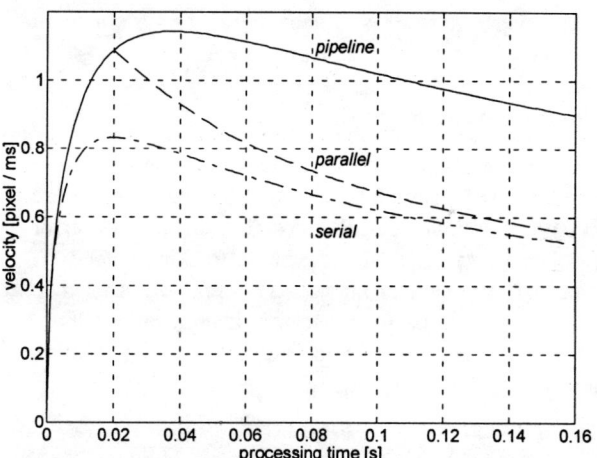

Figure 8: Tracking performance vs. processing time. t_{ac}=20ms, lat_{Mech}=1.

Concerning the various configurations we can always observe the following:

The sampling time T of the control circuit in a pipelining system equals the time required for taking one image t_{ac}, whereas in a parallel or serial system sampling time is a function of processing time t_p or of acquisition and processing time, respectively. This leads to a completely different dynamic behavior. Although vision latency in a pipeline system is very high, the vision data is available each sampling instant, thus providing an image plane pixel error signal to the PID-controller with high sampling rate. As a consequence the system is able to react much earlier to an image plane error than comparable parallel or even serial processing systems. In the latter two cases the performance is reduced by the lower sampling rate due to processing of the image. These effects are the reason for the different curves of tracking velocities shown in Figs. 7 and 8.

If processing time equals acquisition time, i.e., $t_p=t_{ac}=20$ms, and for t_p becoming even smaller the parallel case coincides with the pipeline case (see also Fig. 6). For very large processing times the performance of the parallel system comes close to the serial case. This is due to the fact that processing time is rather large in relation to acquisition time.

As Figs. 7 and 8 demonstrate, maximum tracking velocity is obtained for the pipeline structure. The point of maximum performance is obtained as follows. Substituting r from eq. (9) and C_{II} of eq. (7) into eq. (12) yields

$$v_{pipe} = \frac{\sqrt{t_p/(4 \cdot D)}}{t_{ac} \cdot (lat_{Mech} + C \cdot lat_{Vis})}. \quad (15)$$

After the differentiation with respect to t_p we obtain the maximum of tracking velocity for

$$t_p = t_{ac} \cdot \frac{lat_{Mech} + C}{C}, \quad (16)$$

which can be verified in Figs. 7 and 8. It shows that best dynamic performance expressed as maximum tracking velocity is obtained when using a pipeline structure with processing time as in eq. (16). This means that optimal tracking performance is obtained for a system with $(lat_{Mech}+C)/C$ steps of pipeline processing.

8. Conclusion and Future Work

The objective of this work is to find the optimal system layout for visual fixation control systems. First we outline a method of optimal controller design using an RMS error as performance metric. The maximum tracking velocity is proven to be related to latency. This relation is used to optimize dynamic performance for fixation tasks. Under all configurations investigated, the pipeline configuration yields highest tracking velocity.

The point of optimal performance specifies processing time for a given mechanism and corresponding latency introduced by this mechanism. When building the visual feedback system for this mechanism, the found processing time must be realized. The processing time fixes the window size where the target is searched for at each cycle. Higher computing power increases the window size (since processing one pixel needs less time) and therefore tracking performance. However, the point of optimal performance, that is, overall processing time remains unchanged. In this sense the system architecture is fixed. Adding more computer power for image processing increases performance without changing the optimal system architecture.

As future extensions of this work, two issues need to be tackled. First, all the investigations make the assumption that the dynamic behavior of the mechanism can be regarded as unit delay. This is an idealization and the effect of actual dynamics must be examined, for example the dynamics of a PT_2-system. Second, prediction is known to improve performance. The effect of prediction on coping with latency and the changes to optimal system architecture will be investigated.

Acknowledgments

This work is supported by the Austrian Science Foundation (FWF) in the Research Project PORTIME, contract number P11420-MAT.

References

[AlTi93] Allen, P.K., Timcenko, A., Yoshimi, B., Michelman, P.: Automated Tracking and Grasping of a Moving Object with a Robotic Hand-Eye System; IEEE Trans. RA Vol.9(2), pp.152-165, 1993.

[Ande89] Anderson, R.L.: Dynamic Sensing in a Ping-Pong Playing Robot; IEEE Trans. RA 5(6), pp.728-739, 1989.

[Bajc88] Bajcsy, R.: Active Perception; IEEE Proceedings 76(8), pp. 996-1006, 1988.

[Cork96] Corke, Peter I.: Visual Control of Robots: High Performance Visual Servoing, Research Studies Press (John Wiley), 1996.

[CoGo96] Corke, P.I., Good, M.C.: Dynamic Effects in Visual Closed-Loop Systems; IEEE Trans. on RA Vol.12(5), pp.671-683, 1996.

[GrMe96] Grosso, E., Metta, G., Oddera, A., Sandini, G.: Robust Visual Servoing in 3D Reaching Tasks; IEEE Trans./RA Vol.12(5), pp.671-683, 1996.

[Hash93] Hashimoto, K.: Visual Servoing; World Scientific, 1993.

[HaTo96] Hager, G.D., Toyama, K.: XVision: Combining Image Warping and Geometrical Constraints for Fast Visual Tracking, Proc. ECCV, pp. 507-517, 1996.

[HuHa96] Hutchinson, S., Hager, G.D., Corke, P.I.: A Tutorial on Visual Servo Control; IEEE Trans. on RA Vol.12(5), pp.651-670, 1996.

[LiBr95] Li, F., Brady, M., Hu, H.: Visual Guidance of an AGV; 7th Int. Symp. on Robotics Research, pp.403-415, 1995.

[OlCo91] Olson, T.J., Coombs, D.J.: Real-Time Vergence Control for Binocular Robots; Int. J. of Computer Vision Vol.7(1), pp.67-89, 1991.

[RiKo96] Rizzi, A.A., Koditschek, D.E.: An Active Visual Estimator for Dexterous Manipulation; IEEE Trans. RA Vol.12(5), pp.697-713, 1996.

[WiWi96] Wilson, W.J., Williams Hulls, C.C., Bell, G.S.: Relative End-Effector Control Using Cartesian Position Based Visual Servoing; IEEE Trans. RA Vol.12(5), pp.684-696, 1996.

[WuHi97] Wunsch, P., Hirzinger, G.: Real-Time Visual Tracking of 3-D Objects with Dynamic Handling of Occlusion; ICRA, pp.2868-2873, 1997.

[ViWe97] Vincze, M., Weiman, C.: *On Optimising Window Size for Visual Servoing;* ICRA, April 22-24, 1997.

Performance and Sensitivity in Visual Servoing

Koichi Hashimoto and Toshiro Noritsugu

Department of Mechanical Engineering
Okayama University
3-1-1 Tsushima-naka, Okayama 700 JAPAN

Abstract

This paper describes the relationship between the control performance and the number/configuration of the image features in feature-based visual servoing. The performance is evaluated by two ways: accuracy and speed. A quantitive definition of the sensitivity is given and the relationship among the sensitivity, the speed of convergence and the accuracy are discussed by using image Jacobian. It is proved that these performance indices are increased effectively by using a point with different height. Experiments on Puma 560 are given to show the validity of these performance measures.

1 Introduction

The performance of the feature-based visual servoing depends on the selection of image features. From an image recognition point of view, features must be robust and unique. On the other hand, the features must be sensitive against the object pose, i.e., the image features must change if the object position or orientation changes. Also the features must be controllable, i.e., the reference features must be selected so that they are attained by applying a sequence of control actions.

Feature selection problem has been discussed in e.g., [1, 2]. However they consider only minimum number of features and the effect of redundant features is not discussed. A lot of reference points are used to calibrate hand-eye systems [3, 4]. However, this is a result of least square estimation and the effect of feedback is not considered. In visual servoing, if the number of feature points are changed, one have to consider the change of dimension of feedback variable, the controllability and the closed loop stability. In consequence, the evaluation methods of calibration accuracy and visual servoing accuracy are different. For visual servo problem, two measures have been proposed so far. One is the resolvability introduced by Nelson and Khosla to find optimal sensor placement [5, 6]. And the other is the motion perceptibility proposed by Sharma and Hutchinson to solve the motion planning problem [7]. These indices are the same in a sense that they measures the norm of the feature change caused by unit motion of the object. The difference is that the resolvability is used to obtain directional properties for guiding the robot during the task execution but the perceptibility is a scalar quantity used to optimize the robot performance for an entire task [7].

This paper gives another quantitive measure, *sensitivity*, for controlling all the degree of freedom of a robot by visual servoing. The sensitivity is used to select the feature points so as to minimize the joint error and increases the response speed. Also discussed is the relations among the sensitivity and the control performance, i.e., accuracy and quickness. The Jacobian introduced by Weiss *et al.* [2], which is called *image Jacobian* in this paper, plays important roles. The sensitivity is defined by the smallest singular value of the image Jacobian.

It is shown, in this paper, that if the image Jacobian is not full rank, i.e., if the sensitivity is equal to zero, then the closed loop system becomes internally unstable. In other words, the input-output stability seems to be satisfied but the internal variable becomes unstable. On the other hand, if the sensitivity is increased, both the accuracy of the joint control and the quickness of the response are increased. Also proved is that adding proper features, i.e., using redundant features, increases the sensitivity and a feature point that has different height is effective to improve the sensitivity. Moreover, it is proved that the image Jacobian becomes full rank by using redundant features. To investigate the speed and accuracy of the redundant visual servo system, real time experiments on the PUMA 560 are carried out. Translational step response with three (which is minimum), four and five (which are redundant) features are examined. The results exhibit the quick and accurate performance of the visual servoing with redundant features.

2 Sensitivity

2.1 Definition

From the control theory point of view, robust and accurate performance can be expected by utilizing the measurements which contain rich information. For feature-based visual servo system we propose *sensitivity* for a measure of the richness. As discussed in [5] and [7] the image Jacobian can be used to evaluate the perceptibility of motion. To control m degree of freedom robot one need m features ($m/2$ feature points). If the features are under-observed, there always exists camera (equivalently, object) motions that are not perceptible by the observer. Thus we consider only for minimum and redundant cases ($m \leq 2n$).

Consider the singular value decomposition of image Jacobian

$$J = U\Sigma V^T \quad (1)$$

where

$$U = [u_1, \ldots, u_{2n}], \quad V = [v_1, \ldots, v_m],$$
$$\Sigma = \begin{bmatrix} \Sigma' \\ 0 \end{bmatrix}, \quad \Sigma' = \mathrm{diag}\{\sigma_1, \ldots \sigma_m\}. \quad (2)$$

The matrices U and V are orthogonal, u_i and v_j are, respectively, i-th left singular vector and j-th right singular vector, and σ_i is the i-th singular value. It is easy to verify that

$$Jv_i = \sigma_i u_i. \quad (3)$$

Thus the unit motion of joint angle $v_i = \Delta\theta, \|\Delta\theta\|_2 = 1$ is transformed to the feature motion in u_i direction and scaled by σ_i, i.e., $\sigma_i u_i = \Delta\xi$. Thus if the image Jacobian is not full rank, there exist joint motion that is not perceptible through the image features. Also, since the minimum singular value satisfies

$$\sigma_m = \sigma_{min} = \inf\{\|\Delta\xi\|_2 : \|\Delta\theta\|_2 = 1\}, \quad (4)$$

σ_m is the magnification for the most insensitive direction. Thus the minimum singular value is adequate for a measure of (worst case) *sensitivity*.

2.2 Internal Instability

Fig. 1 shows a block diagram of the visual servo system. The feature ξ is fed back and compared with the reference ξ_d. The feature error $e = \xi_d - \xi$ is multiplied by the controller gain K and generates the joint motion command $\Delta\theta$. The robot is driven by this

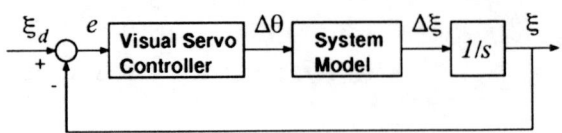

Figure 1: Closed Loop System

command and the camera moves. The camera motion generates the feature motion $\Delta\xi = J\Delta\theta$, which is integrated to yield the feature ξ. In this closed loop system, suppose that the minimum singular value σ_m is vanishingly small. If the generated joint command includes u_m direction component, this component is hardly reflected to image motion and the joints will keep moving in this direction. This means that joint variable θ (integration of joint displacement command $\Delta\theta$) inside the "System Model" box is internally unstable while the external signal ξ seems stable. We call this *internal instability*.

2.3 Improving Sensitivity

Feddema *et al.* selected feature points so as to minimize the condition number. Since the condition number is the ratio of maximum and minimum singular values ($\sigma_{max}/\sigma_{min}$), their approach is also appropriate for increasing the sensitivity. However we propose another approach. Suppose that there are n feature points and we add one more feature to the already existing features. Then we obtain the following theorem.

Theorem 1 *The sensitivity strictly increases by adding a feature point if and only if the minimum singular vector of the image Jacobian corresponding to the already existing points does not belong to the kernel of the image Jacobian corresponding to the newly added feature point.*

Proof: Let J_n and J_{n+1} be the image Jacobian with n and $n+1$ feature points. Since we have added a feature point to the already existing n feature points, we have

$$J_{n+1} = \begin{bmatrix} J_n \\ J^{(n+1)} \end{bmatrix} \quad (5)$$

where $J^{(n+1)}$ is the $2 \times m$ image Jacobian corresponding to the $(n+1)$-st feature point. Thus for any vector v, the following inequality holds

$$\|J_{n+1}v\|_2 \geq \|J_n v\|_2. \quad (6)$$

From the definition of minimum singular value (4), we have

$$\sigma_{min}(J_{n+1}) \geq \sigma_{min}(J_n). \quad (7)$$

The equality holds if and only if $J^{(n+1)}v_n = 0$, where v_n is the singular vector corresponding to the minimum singular value of J_n, i.e., the vector that attains the infimum of (4). ∎

Since the first three columns of $J_{image}^{(i)}$ are proportional to $1/Z_i$, changing the depth (Z_i) is effective to increasing the linear dependency of the image Jacobian.

2.4 Sensitivity and Response Speed

Let K be the controller gain matrix. Then the transfer function of the closed loop system (from ξ_d to θ) can be approximated by $(sI+JK)^{-1}K$. Most visual servoing schemes for redundant features [8, 9, 10] use the generalized inverse or transpose of the image Jacobian multiplied by a scaler k for the controller gain. Suppose that we fix the scalar k and add a feature point. If the generalized inverse is used, the closed loop poles do not depend on J. Thus the response speed will not be affected by the number of feature points. If the transpose is used, it is easy to show by doing mode decomposition that the closed loop poles are $-k\sigma_1^2, \ldots, -k\sigma_m^2$. Thus the slowest mode $(-k\sigma_m^2)$ will be fasten by increasing the sensitivity. For the case of the discrete time optimal controller [11, 12], one can not fix the controller K by adjusting Q and R because the system model depends on J and the dimension of the controller changes. However if Q, R are adjusted to yield the same matrix gain of the controllers $\|K\|_\infty$, it is confirmed by simulations that the slowest pole shifts to the origin in the complex plane by adding feature points (see Table 2).

3 Rank Condition

In this section we consider a six degree of freedom robot. Suppose that one want to control the position and orientation of the camera in 3D space. Then three features are necessary but not sufficient because there is a singular cylinder [13]. However, if one can use one more point on a plane that the three points lie on, then we have the following theorem.

Theorem 2 *Suppose that there are four points on a plane in 3D space and four feature points corresponding to these points are selected as the feature vector. Then the image Jacobian is full rank provided that the robot configuration is not singular and any three feature points out of four feature points are not collinear in the image plane.*

The proof is straightforward and omitted [14]. The assumption on feature point configuration is natural

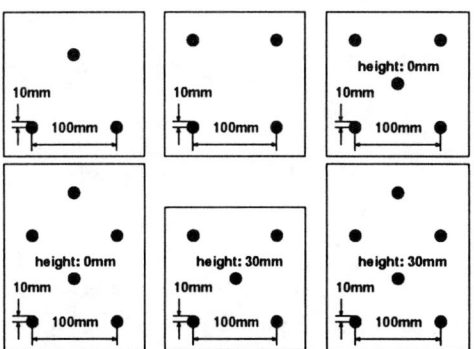

Figure 2: Configuration of Feature Points

Table 1: Minimum Singular Values

	a	b	c	d	e	f
p_1	0.00	0.36	0.38	0.43	2.52	2.56
p_2	0.29	0.58	0.59	0.60	2.63	2.63

because it is satisfied if the four feature points make a quadrangle (not necessarily square or rectangle but any four-sided plane) in the image plane.

4 Experiments

4.1 Sensitivity

Examples of smallest singular values for various feature sets are computed. As shown in Fig. 2, six cases with three to six feature points and two object position $p_1 = [-50, -50, -1000, 0, 0, 0]$ and $p_2 = [5, 10, -1000, -8, -5, -3]$ are tested (position is expressed in [mm] and orientation is expressed by Euler angles [degree]). The robot configuration is the same as the experiments (see the following section and Fig. 3). Table 1 shows the minimum singular values. For p_1, since the camera is on the singular cylinder, the sensitivity for three feature points is zero (a). The sensitivity is improved by increasing the number of feature points (b-f). Four feature points are necessary to guarantee the internal stability (b). Since the improvement is not very significant for five and six features on the same plane (c, d), one must consider the tradeoff between the performance and the image processing time. If a point with different height is available, the sensitivity is increased considerably (e, f). Even for the position p_2 which is outside of the singular cylinder, the sensitivity of three points is small. Four, five and six points give better sensitivity compared with p_1. By changing the height of the center

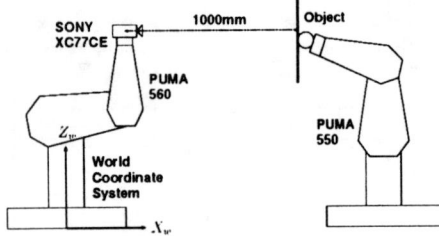

Figure 3: Robot Configuration and Object Position

point to 30 yields much better sensitivity but the difference between five and six is not significant.

4.2 Setup

Real time experiments were carried out on the visual feedback control system with a PUMA 560 to compare the performance within three feature sets (a, b, e). The objects are attached to a PUMA 550. The world coordinate system is at the base of the PUMA 560. A nominal camera position is in front of the plane on which the marks are and the distance is 1000mm (p_2). The nominal positions of the object and camera are shown in Fig. 3. The $X_w - Y_w - Z_w$ coordinate system is the world coordinate system. We carried out vertical step tests to verify that sensitivity is an appropriate measure to select the features.

4.3 Control Law

In this paper we adopt a discrete time optimal control law [12]. The discrete time state equation is $z_{k+1} = z_k + Bu_k$, where $z_k = J^T \xi_k, u_k = \Delta\theta_k$ and $B = J^T J$. Then, for positive definite matrices Q and R, the optimal control law is given by

$$u_k = -K_c J^T e_k, \quad K_c = (R + B^T PB)^{-1} B^T P \quad (8)$$

where $e_k = \xi_d - \xi_k$ and P is the positive solution of the Riccati equation $Q = PB(R + B^T PB)^{-1} B^T P$. The closed loop poles are $\det(zI - I + BK_c)$. The performance depends on the feature configuration as well as Q and R. To see the effect of feature configuration, we adjusted Q and R to obtain similar $\|K_c J^T\|_\infty$. The Q, R parameters and closed loop poles are shown in Table 2. Note that the slowest closed loop pole is shifted to the origin by increasing the number of feature points.

4.4 Vertical Step

The object is moved up 100mm in vertical axis Z_w. The camera moves to keep the features at the initial position. Thus the initial values and the reference values are the same. The object motion is considered as a disturbance for the plots of the features in the image

Table 2: Controller Design Parameters for Step Test

	3 points	4 points	5 points
Q	I	I	I
R	$2600I$	$5000I$	$100000I$
γ	10	10	10
$\|KJ_d\|$	0.0109	0.0117	0.0107
pole	0.998	0.995	0.967
	0.994	0.990	0.947
	0.900	0.900	0.900
	0.900	0.900	0.900
	0.900	0.900	0.900
	0.900	0.900	0.900

plane. On the other hand, the object motion becomes the step change of the reference position for the position of the camera in the world coordinate system. The reference orientation is the same as the initial orientation.

4.4.1 Three Points

Fig. 4 has six curves which show the x and y coordinates of the feature point in the image plane. The horizontal axis is the time. The curves disturbed largely are the y coordinates and the others are the x coordinates. They are almost stabilized in 6 seconds. Thus the response in the image plane is very good. However the plots in Fig. 5, which depicts the position errors of the camera in the world coordinate system $X_w - Y_w - Z_w$, is diverging. The features are kept in the neighborhood of the reference position due to the rotation of the robot wrist. Fig. 6 shows the orientation errors of the camera expressed in the Euler angles, say ψ, η, ϕ. The plot of ψ also becomes unstable. It shows that the camera keeps rotating. These plots exhibit the internal instability.

4.4.2 Four Points

Fig. 7 shows the response of the features in the image plane with four feature points. It takes 5 seconds to stabilize the disturbance. The response in the image plane is not improved very much compared with the response of three points. However, as shown in Fig. 8, the response of the camera position in the world coordinate system is stabilized, though it is sluggish. It takes over 20 seconds to stabilize the disturbance. The steady state errors are within 5mm. Thus the accuracy is fairly good. These plots show that the feature errors are reduced by the camera rotation as well as the camera translation in 5 seconds. After that, the

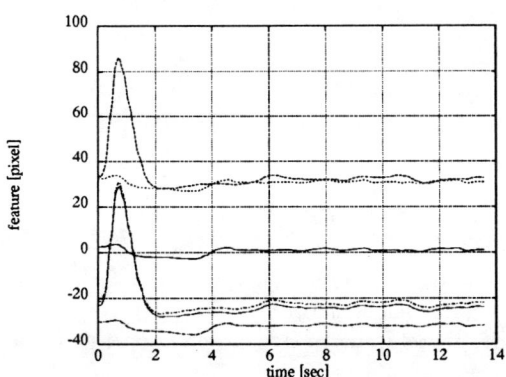

Figure 4: Response in Image Plane for 3 Points

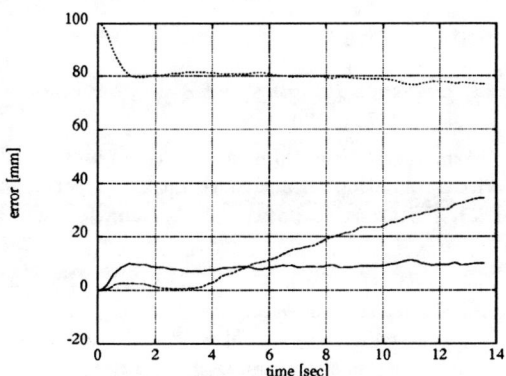

Figure 5: Error in 3D for 3 Points
— : X_w, - - - : Y_w, - - - : Z_w

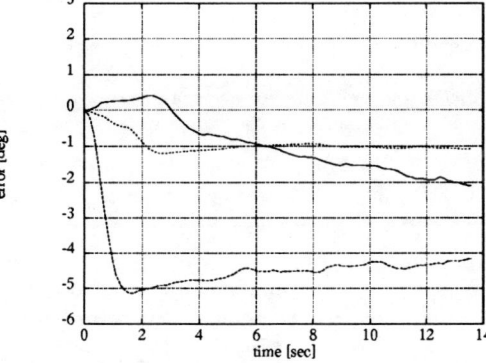

Figure 6: Orientation Error in 3D for 3 Points
— : ψ, - - - : η, - - - : ϕ

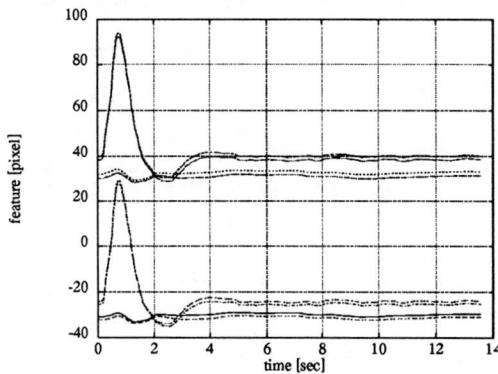

Figure 7: Response in Image Plane for 4 Points

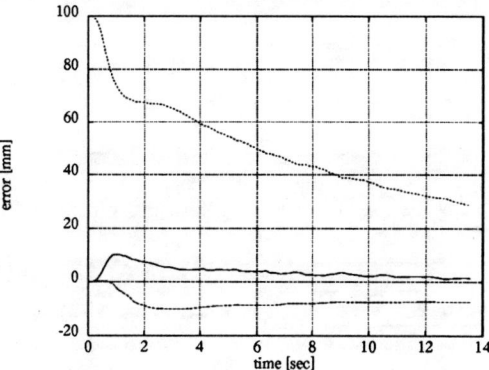

Figure 8: Error in 3D for 4 Points
— : X_w, - - - : Y_w, - - - : Z_w

orientation errors are gradually reduced, but the speed is slow because the sensitivity is small. The response in the image plane seems quick due to the fast poles (0.900) but the response is actually slow because of the very slow pole (0.995).

4.4.3 Five Points

Fig. 9 depicts the features in the image plane for the experiment with five points. The disturbance is stabilized in 5 seconds. The response in the image plane is similar to those with three and four points. Fig. 10 shows the response of the camera position in the world coordinate system. It is improved very much for both speed and accuracy because the sensitivity and the slowest pole are improved. The steady state errors are smaller than 5mm for all directions. These plots demonstrate the effectiveness of the redundant features for improvements of both speed and accuracy of the feature-based visual servoing.

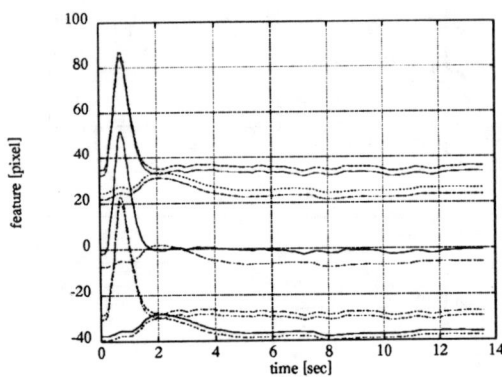

Figure 9: Response in Image Plane for 5 Points

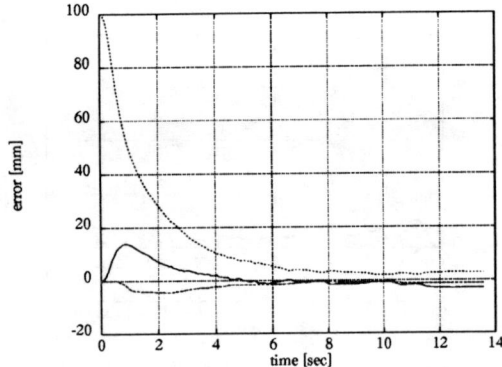

Figure 10: Error in 3D for 5 Points
—: X_w, - - -: Y_w, - - -: Z_w

5 Conclusions

Discussions on the performance improvement due to redundant features were presented. Real time experiments on PUMA 560 were carried out to evaluate the improvement of the accuracy and speed by utilizing the redundant features. The results have shown the quickly converging stable performance. The accuracy of the camera position control in the world coordinate system was increased by utilizing redundant features. Also the convergence speed was improved considerably by adding the extra feature point. Moreover, the experiments also verified that the minimum singular value of the extended image Jacobian plays an important role to evaluate the performance of the feature-based visual servoing.

References

[1] J. T. Feddema, C. S. G. Lee, and O. R. Michell, "Automatic selection of image features for visual servoing of a robot manipulator," in *IEEE Int. Conf. Robotics and Automation*, Scottsdale, Ariz., 1987, pp. 832–837.

[2] L. E. Weiss, A. C. Sanderson, and C. P. Newman, "Dynamic sensor-based control of robots with visual feedback," *IEEE J. Robotics and Automation*, vol. RA-3, no. 5, pp. 404–417, 1987.

[3] R. Y. Tsai, "A versatile camera calibration technique for high-accuracy 3D machine vision metrology using off-the-shelf TV cameras and lenses," *IEEE J. Robotics and Automation*, vol. RA-3, no. 4, pp. 323–354, 1987.

[4] R. Y. Tsai and R. K. Lenz, "A new technique for fully autonomous 3D robotic hand/eye calibration," *IEEE Trans. Robotics and Automation*, vol. 5, no. 3, pp. 345–358, 1989.

[5] B. J. Nelson and P. K. Khosla, "The resolvability ellipsoid for visual servoing," in *IEEE Conf. Computer Vision and Pattern Recognition*, 1994, pp. 829–832.

[6] B. J. Nelson and P. K. Khosla, "Force and vision resolvability for assimilating disparate sensory feedback," *Trans. on Robotics and Automation*, vol. 12, no. 5, pp. 714–731, 1996.

[7] R. Sharma and S. Hutchinson, "Motion perceptibility and its application to active vision-based servo control," *IEEE Trans. Robotics and Automation*, vol. 13, no. 4, pp. 607–617, 1997.

[8] F. Chaumette, P. Rives, and B. Espiau, "Positioning of a robot with respect to an object, tracking it and estimating its velocity by visual servoing," in *IEEE Int. Conf. Robotics and Automation*, Sacramento, Calif., 1991, pp. 2248–2253.

[9] B. Espiau, F. Chaumette, and P. Rives, "A new approach to visual servoing in robotics," *IEEE Trans. Robotics and Automation*, vol. 8, no. 3, pp. 313–326, 1992.

[10] K. Hashimoto et al., "Manipulator control with image-based visual servo," in *IEEE Int. Conf. Robotics and Automation*, Sacramento, Calif., 1991, pp. 2267–2272.

[11] K. Hashimoto et al., "Image-based dynamic visual servo for a hand-eye manipulator," in *MTNS-91*, Kobe, Japan, 1991, pp. 609–614.

[12] K. Hashimoto, T. Ebine, and K. Kimura, "Visual servoing with hand-eye manipulator —optimal control approach," *IEEE Trans. on Robotics and Automation*, vol. 12, no. 5, pp. 766–774, 1996.

[13] H. Michel and P. Rives, "Singularities in the determination of the situation of a robot effector from the perspective view of 3 points," Tech. Rep. n. 1850, INRIA, 1993.

[14] K. Hashimoto, A. Aoki, and T. Noritsugu, "Visual servoing with redundant features," *Journal of the Robotics Society of Japan*, vol. 16, no. 3, 1997.

Experimental Evaluation of Fixed–Camera Direct Visual Controllers on a Direct–Drive Robot [*]

Fernando Reyes
ECE–Universidad Autónoma de Puebla
Apdo. Postal 1651
Tel/FAX: +52 (22) 33 19 20
Puebla, Puebla, 72000, MEXICO.
e-mail: freyes@kim.ece.buap.mx

Rafael Kelly
CICESE Física Aplicada
Apdo. Postal 2615, Adm. 1
Ensenada, B. C., 22800, MEXICO.
FAX: +52 (61) 75 05 49
e-mail: rkelly@cicese.mx

Abstract

This paper addresses the visual servoing of robot manipulators in fixed-camera configuration. We present the experimental evaluation of three direct visual servo controllers on a planar two degrees of freedom direct-drive robot arm. These visual servo controllers belong to the family of transpose Jacobian-based control schemes yielding local asymptotic stability.

1 Introduction

Visual servoing deals with the posture control of an end–effector either relative to a world coordinate frame or relative to a target object by using real–time visual information. Visual information into feedback control loops represents an attractive solution to position and motion control of autonomous robot manipulators evolving in unstructured environments [1].

Although visual servoing has been an area of research over the last 30 years, attention to this subject has drastically grown in recent years (see [1, 2] for interesting reviews). Two types of visual servoing configurations can be recognized nowadays: fixed–camera (or static camera) and camera–in–hand (or eye–in–hand). In this paper we address the fixed–camera approach to visual servoing of planar robot manipulators with a single camera by assuming that the target posture is stationary (set–point regulation). The visual control problem considered is a subclass of the stereo visual servoing with moving targets (see [3, 4, 5, 6, 7, 8, 9, 10]).

This paper presents the experimental evaluation of three visual servo controllers belonging to the transpose Jacobian–based family [11, 12, 13]. The rationale behind this family of controllers is the transpose Jacobian philosophy introduced by Takegaki and Arimoto [14] to solve the regulation problem in Cartesian space. The common properties of the tested visual servo controllers are:

- Direct visual servo: The visual feedback is converted to joint torques instead of joint or Cartesian velocity inputs [15].

- "Endpoint–closed–loop" (ECL) systems: The vision system provides both the target and end–effector postures defined in terms of observable features rigidly attached to them [15].

- Image–based: Define servoing errors directly from the camera image.

Experimental results carried out on a vertical two degrees of freedom direct–drive arm allow to extract merits and drawbacks of each visual servo controller. The remaining of this paper is organized as follows. In next section we recall the robotic system model. Section 3 is devoted to describe three transpose Jacobian–based image–based direct visual controllers. Section 4 describes the experimental set-up. The experimental results are presented in Section 5. Finally, we give some conclusions in Section 6.

2 Robotic system model

The robotic system considered in this paper is composed by a planar robot manipulator and a vision system including a fixed camera as depicted in figure 1.

---

[*]Work partially supported by CONACyT–NSF grants No. 228050-5-C084A and IRI-9613737, project *Perception Systems for Robots-CYTED* and CONACyT I27218-A.

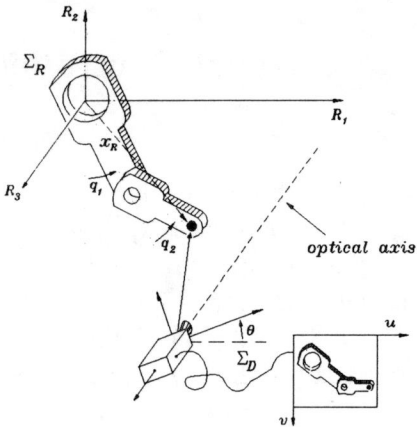

Figure 1: Robotic system.

2.1 Robot dynamics

In the absence of friction or other disturbances, the dynamics of a serial n-link rigid robot can be written as [16]:

$$M(q)\ddot{q} + C(q,\dot{q})\dot{q} + g(q) = \tau \qquad (1)$$

where q is the $n \times 1$ vector of joint displacements, $\dot{q}$ is the $n \times 1$ vector of joint velocities, τ is the $n \times 1$ vector of applied torques, $M(q)$ is the $n \times n$ symmetric positive definite manipulator inertia matrix, $C(q,\dot{q})$ is the $n \times n$ matrix of centripetal and Coriolis torques, and $g(q)$ is the $n \times 1$ vector of gravitational torques.

Let $\Sigma_R = \{R_1, R_2, R_3\}$ be a 3D right-hand Cartesian frame attached to the robot base. Consider the planar robot manipulator with two degrees of freedom moving in the plane R_1–R_2 as depicted in figure 1. The direct kinematics gives the position $x_R \in \mathbb{R}^2$ of the robot tip (end-effector) with respect to the robot coordinate frame ($x_{R3} = 0$) in terms of the joint positions $q \in \mathbb{R}^2$:

$$x_R = f(q) \qquad (2)$$

where $f : \mathbb{R}^2 \to \mathbb{R}^2$. The so-called analytical Jacobian matrix $J_A(q) \in \mathbb{R}^{2 \times 2}$ of the robot is defined from direct kinematics as

$$J_A(q) = \frac{\partial f}{\partial q}. \qquad (3)$$

2.2 Vision system model

A TV camera (CCD type) stationary with respect to Σ_R provides images of the whole robot workspace, including the robot end-effector and any other visible object as shown the Figure 1.

The position of the camera frame with respect to Σ_R is denoted by $o_C = [o_{C_1}, o_{C_2}, o_{C_3}]^T$. It is assumed that the camera frame possesses a rotation θ around axis R_3.

The image of the scene on the CCD is digitalized and transferred to the computer memory and displayed on the computer screen. We define the two dimensional computer image (screen) coordinate frame $\Sigma_D = \{u, v\}$. The origin of Σ_D is attached at the upper left corner of the computer screen while the axes u and v are selected parallel to the screen rows and columns respectively.

As defined previously, x_R represent the position of the robot tip with respect to the robot frame Σ_R. This description depends on the joint position q. The description of such a point in the computer image (screen) frame Σ_D denoted by $[u \;\; v]^T$ defines a mapping called the vision system model which includes a perspective projection and a rigid body transformation [4, 13]:

$$\begin{bmatrix} u \\ v \end{bmatrix} = \begin{bmatrix} -\alpha_u & 0 \\ 0 & \alpha_v \end{bmatrix} \begin{bmatrix} \frac{\lambda}{\lambda - o_{C_3}} R(\theta)^T \end{bmatrix}$$
$$\left[\begin{bmatrix} x_{R_1}(q) \\ x_{R_2}(q) \end{bmatrix} - \begin{bmatrix} o_{C_1} \\ o_{C_2} \end{bmatrix} \right] + o_I \right] + \begin{bmatrix} u_0 \\ v_0 \end{bmatrix}, \qquad (4)$$

$$R(\theta) = \begin{bmatrix} \cos(\theta) & -\sin(\theta) \\ \sin(\theta) & \cos(\theta) \end{bmatrix}, \qquad (5)$$

where $\alpha_u > 0, \alpha_v > 0$ are the scale factors in pixels/m, and u_0, v_0 denote the pixel position of the geometric center of plane CCD with respect to the system Σ_D, $R(\theta) \in SO(2)$ is the rotation matrix which represents the orientation of the camera with respect to the world frame Σ_R, $\lambda > 0$ is the focal length of the camera, and the position of the intersection of the optical axis with respect the geometric center of the plane CCD is denoted by o_I.

3 Image-based direct visual controllers

The robot task is specified in the image plane in terms of image features corresponding to observable points rigidly attached to the robot tip and target object (one point attached to each one). It is assumed that the target object resides in the plane R_1–R_2 but its position with respect to the robot and camera frames is unknown. Let $[u_d \;\; v_d]^T$ the description with respect to the computer image (screen) frame Σ_D of the target image feature corresponding to the attached point. Hereafter, $[u_d \;\; v_d]^T$ will be referred as the desired image feature vector which is constant because the target was assumed to be stationary.

The control problem it to design a controller to compute the applied torques τ such way that the image feature $[u \ v]^T$ corresponding to the point attached to the robot tip reaches the desired image feature $[u_d \ v_d]^T$ of the point attached to the target object. This formulation can be equivalently stated as driving the robot tip in such a way that the corresponding image feature $[u \ v]^T$ reaches a constant arbitrary point $[u_d \ v_d]^T$ into the computer image (screen) frame.

The image feature error is defined as

$$\begin{bmatrix} \tilde{u} \\ \tilde{v} \end{bmatrix} = \begin{bmatrix} u_d - u \\ v_d - v \end{bmatrix} \quad (6)$$

therefore, the control aim is to assure that $\lim_{t \to \infty} [\tilde{u}(t) \ \tilde{v}(t)]^T = \mathbf{0} \in \mathbb{R}^2$, at least for initial conditions $[\tilde{u}(0) \ \tilde{v}(0)]^T$ and $\dot{q}(0)$ sufficiently small.

In order to the control problem be solvable, we assume that there exists a joint configuration $q_d \in \mathbb{R}^2$ such that

$$\begin{bmatrix} u_d \\ v_d \end{bmatrix} = \begin{bmatrix} -\alpha_u & 0 \\ 0 & \alpha_v \end{bmatrix} \begin{bmatrix} \frac{\lambda}{\lambda - o_{C_3}} R(\theta)^T \\ \begin{bmatrix} \begin{bmatrix} x_{R_1}(q_d) \\ x_{R_2}(q_d) \end{bmatrix} - \begin{bmatrix} o_{C_1} \\ o_{C_2} \end{bmatrix} \end{bmatrix} + o_I \end{bmatrix} + \begin{bmatrix} u_0 \\ v_0 \end{bmatrix}.$$

Therefore, for analytical purposes, the image feature error (6) can be written as

$$\begin{bmatrix} \tilde{u}(q_d, q) \\ \tilde{v}(q_d, q) \end{bmatrix} = \frac{\lambda}{\lambda - o_{C_3}} \begin{bmatrix} -\alpha_u & 0 \\ 0 & \alpha_v \end{bmatrix} R(\theta)^T [f(q_d) - f(q)].$$

3.1 Visual servo controllers

In this paper we consider three direct visual servo controllers whose control laws are given by

$$\tau_1 = J_A(q)^T R(\theta) K_p \begin{bmatrix} \tilde{u} \\ -\tilde{v} \end{bmatrix} - K_v \dot{q} + g(q), \quad (7)$$

$$\tau_2 = J_A(q)^T K_p \tanh \left(\Lambda R(\theta) \begin{bmatrix} \tilde{u} \\ -\tilde{v} \end{bmatrix} \right) - K_v \dot{q} + g(q), \quad (8)$$

$$\tau_3 = J_A(q)^T R(\theta) K_p \tanh \left(\Lambda \begin{bmatrix} \tilde{u} \\ -\tilde{v} \end{bmatrix} \right) - K_v \dot{q} + g(q), \quad (9)$$

where $K_p, K_v, \Lambda \in \mathbb{R}^{2 \times 2}$ are diagonal positive definite matrices.

Visual servo controllers (7) and (8) have been already analyzed in [13] and [12], respectively. Their local asymptotic stability has been shown using the following Lyapunov functions

$$V_1(q, \dot{q}) = \frac{1}{2} \dot{q}^T M(q) \dot{q}$$
$$+ \frac{1}{2} \begin{bmatrix} \tilde{u}(q_d, q) \\ \tilde{v}(q_d, q) \end{bmatrix}^T K_p K^{-1} \begin{bmatrix} \tilde{u}(q_d, q) \\ \tilde{v}(q_d, q) \end{bmatrix},$$

$$V_2(q, \dot{q}) = \frac{1}{2} \dot{q}^T M(q) \dot{q} + \begin{bmatrix} \sqrt{\ln\{\cosh(\gamma_1 \tilde{x}_{R_1})\}} \\ \sqrt{\ln\{\cosh(\gamma_2 \tilde{x}_{R_2})\}} \end{bmatrix}^T$$
$$K_p \Gamma^{-1} \begin{bmatrix} \sqrt{\ln\{\cosh(\gamma_1 \tilde{x}_{R_1})\}} \\ \sqrt{\ln\{\cosh(\gamma_2 \tilde{x}_{R_2})\}} \end{bmatrix},$$

where $K = \text{diag}\left\{-\frac{\alpha_u \lambda}{\lambda - o_{C_3}}, -\frac{\alpha_v \lambda}{\lambda - o_{C_3}}\right\}$, and $\Gamma = \text{diag}\{\gamma_1, \gamma_2\} = K\Lambda$ are diagonal positive definite matrices, and $\tilde{x}_{R_i} = f_i(q_d) - f_i(q)$ with $i = 1, 2$.

Control law (9) is a slight modification to (8) proposed in this paper. The (local) asymptotic stability analysis can be performed via the following Lyapunov function candidate

$$V_3(q, \dot{q}) = \frac{1}{2} \dot{q}^T M(q) \dot{q} +$$
$$\begin{bmatrix} \sqrt{\ln\{\cosh(\gamma_1 [\cos(\theta) \tilde{x}_{R_1} + \sin(\theta) \tilde{x}_{R_2}])\}} \\ \sqrt{\ln\{\cosh(\gamma_2 [-\sin(\theta) \tilde{x}_{R_1} + \cos(\theta) \tilde{x}_{R_2}])\}} \end{bmatrix}^T$$
$$K_p \Gamma^{-1}$$
$$\begin{bmatrix} \sqrt{\ln\{\cosh(\gamma_1 [\cos(\theta) \tilde{x}_{R_1} + \sin(\theta) \tilde{x}_{R_2}])\}} \\ \sqrt{\ln\{\cosh(\gamma_2 [-\sin(\theta) \tilde{x}_{R_1} + \cos(\theta) \tilde{x}_{R_2}])\}} \end{bmatrix}.$$

4 Experimental set-up

We have designed and built at CICESE Research Center a planar direct drive robot arm with two degrees of freedom moving in the vertical plane (see Figure 2). It consists of two links actuated by brushless direct drive servo actuators to drive the joints without gear reduction. A motion control board based on a TMS320C31 is used to execute the control algorithm. This board is mounted in a PC 486 host computer which provides an environment for the experimental execution [17].

The vision system consists of a Panasonic GP-MF502 camera with a lens having a focal length $\lambda = 0.008$ [m] and a DT3851-4 frame processor board from Data Translation. A black disc was mounted on the robot end-effector, the centroid of disc was selected as the object feature point.

The CCD camera was placed in front of the robot arm but the reference system of the camera Σ_C is rotated an angle $\theta = 20$ [degrees] with respect to the system Σ_R. The camera frame Σ_C position with respect to the robot frame Σ_R was $o_C = [0.70, -0.48, 0.45]^T$ [m].

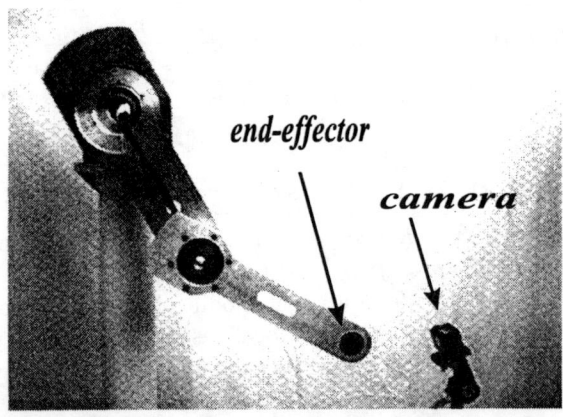

Figure 2: Robot arm

5 Experimental results

This section describes the experimental results obtained by testing the image–based controllers (7)–(9) on the direct drive robot manipulator.

We selected in all controllers the desired position in the image plane as $[u_d \ v_d]^T = [320 \ 180]^T$ [pixels] and the following initial configuration of robot $q_1(0) = 30$ [degrees] and $q_2(0) = 32$ [degrees], $[u(0) \ v(0)]^T = [359 \ 145]^T$ pixels and $\dot{q}(0) = \mathbf{0}$ [degrees/sec]. With this initial configuration the feature position error was $[\tilde{u}(0) \ \tilde{v}(0)]^T = [175, -179]^T$ [pixels].

Figures 3–4 shown the experimental results of the controller (7). The parameters of this controller were selected as $K_p = \text{diag}\{0.19, 0.0512\}$ [Nm/pixels^2] and $K_v = \text{diag}\{0.31, 0.045\}$ [Nm-sec/degrees]. Figure 3 depicts the time evolution of the feature error vector $[\tilde{u} \ \tilde{v}]^T$. The transient response is fast and it was around 1.5 sec. After transient, both components of the feature position error tend asymptotically to a small neighborhood of zero (-7 and -4 pixels, respectively). They remain stuck around zero because the presence of static friction at the arm joints.

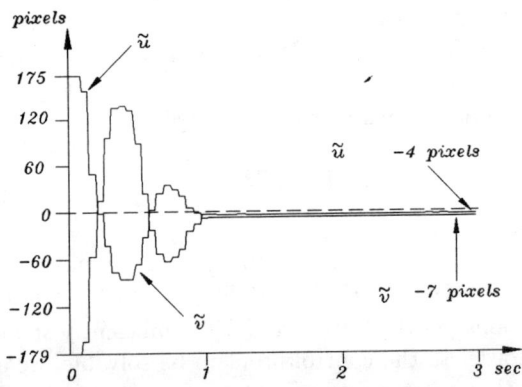

Figure 3: Feature position trajectory in image plane.

It is also interesting to describe the shape of image feature point looked on the image plane. Figure 4 depicts the trajectory of the robot manipulator end–effector saw in the image plane u–v. It can be observed that the end–effector convergence to a small neighborhood of the desired feature point. The oscillatory behavior may be due mainly to the friction at the joints and the delay in the image feature extraction, since the visual sampling period is larger than 33 msec., this is, between two image samples the arm moves in "open loop" with respect to image errors.

The experimental results for the controller (8) are shown in figures 5–6. The proportional and derivative gains were selected as $K_p = \text{diag}\{7, 6\}$ [Nm/pixels^2] and $K_v = \text{diag}\{0.35, 0.26\}$[Nm sec/degrees], respectively. While the matrix $\Lambda = \text{diag}\{0.8, 0.7\}$.

Figure 5 contains the evolution of the feature position error entries. The transient response was of 5 sec. The components of feature position error tend asymptotically toward a small region of -4 and -7 pixels, respectively. From experimental results of this controller, it can be observed that the oscillatory behavior is eliminated. The robot end–effector converges toward a small region of the desired point in the the image plan as shown in Figure 6.

The experimental results obtained with the proposed controller (9) are shown in Figures 7–8. The controller parameters were selected as $K_p =$

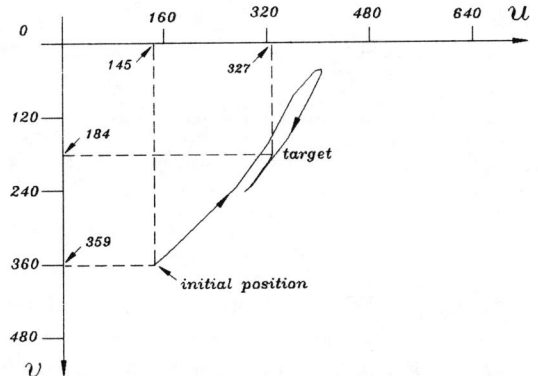

Figure 4: Path of the robot manipulator end–effector in the image plane.

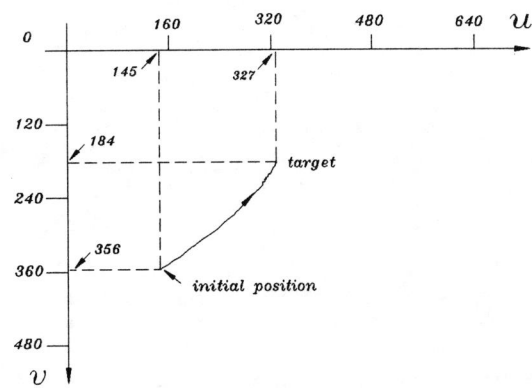

Figure 6: Path of the robot manipulator end–effector in the image plane for the controller (8).

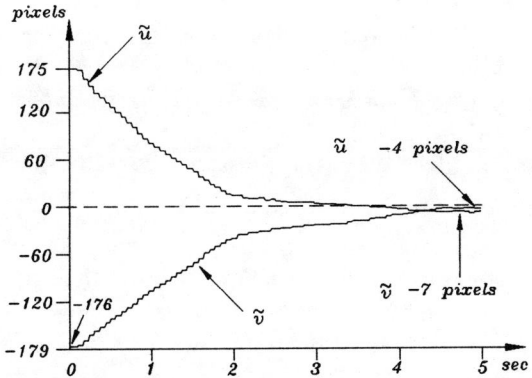

Figure 5: Feature position error in the plane image for controller (8).

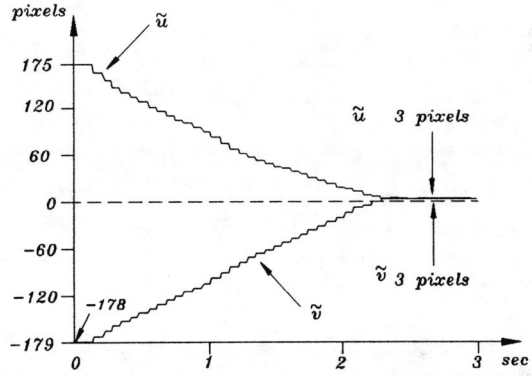

Figure 7: Feature position error in the image plane for the controller (9).

diag$\{7.2, 5\}$ [Nm/pixels^2], K_v = diag$\{0.31, 0.21\}$ [Nm sec/degrees] and Λ = diag$\{1, 1\}$.

Figure 7 contains the profiles of the components of the image feature error. The transient response was 2.5 sec without overshot. The two components of feature position error present a decreasing tendency and they remain in a neighborhood of 3 pixels around zero after 2 sec. The remaining value of the image feature error in steady state was due to the presence of static friction.

From experimental results of this controller, we can note that despite of the feature position error is usually large in the initial transient, the saturation function indirectly limits the applied torques, therefore the oscillatory behavior is eliminated completely in contrast with the response of controller (7).

The path that accomplishes the robot manipulator end–effector in the image plane is shown in figure 8 where the manipulator end-effector path converges to a small region of the desired target point.

6 Conclusions

In this paper we have presented an experimental evaluation of three direct visual controllers for fixed-camera configuration. These controllers yield locally asymptotically stable closed-loop systems. In practical implementation of the controllers, the image acquisition as well as the desired features are limited by the video rate. Experimental results showed that those visual servo controllers containing the smooth saturation function tanh($\cdot$) present the best behavior. However, the presence of static friction at the arm

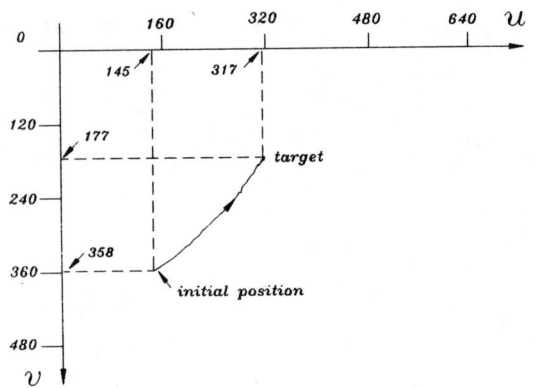

Figure 8: Path of the robot manipulator end-effector in the image plane for the controller (9).

joints produces steady state position errors.

References

[1] S. Hutchinson, G. D. Hager, and P. I. Corke, "A tutorial on visual servo control", *IEEE Transactions on Robotics and Automation*, Vol. 12, No. 5, pp. 651–670, October 1996.

[2] K. Hashimoto, *Visual servoing*, World Scientific Publishing, 1993.

[3] A. J. Koivo and N. Houshangi, "Real-time vision feedback for servoing robotic manipulators with self-tuning controller". *IEEE Tran. Syst. Man, Cybern.*, Vol. 21, No. 1, pp. 134–142, Jan./Feb. 1991.

[4] J. T. Feddema, C. S. G. Lee and O. R. Mitchell, "Weighted selection of image features for resolved rate visual feedback control", *IEEE Transactions on Robotics and Automation*, Vol. 7, No. 1, pp. 31–47, February 1991.

[5] B. Espiau, F. Chaumette, and P. Rives, "A new approach to visual servoing in robotics", *IEEE Trans. on Robotics and Automation*, Vol. 8, No. 3, pp. 313–326, June 1992.

[6] W. S. Wijesoma, D. F. H. Wolfe and R. J. Richards, "Eye to hand coordination for vision-guided robot control applications", *The International J. Robot Res.*, Vol. 12, No. 1, pp. 65–78, 1993.

[7] P. K. Allen, A. Timcenko, B. Yoshimi and P. Michelman, "Automated tracking and grasping of a moving object with a robotic hand-eye system", *IEEE Trans. on Robotics and Automation*, Vol. 9, No. 2, pp. 152–165, April 1993.

[8] M. Lei and B. K. Ghosh, "Visually guide robotic tracking and grasping of a moving object", *Proc. 32nd. Conf. on Decision and Control*, San Antonio, TX, pp. 1604–1609, Dec. 1993.

[9] G. D. Hager, W. C. Chang, and A. S. Morse, "Robot hand-eye coordination based on stereo vision". *IEEE Control System*, Vol. 15, No. 1, pp. 30–39, February 1995.

[10] B. J. Nelson, N. P. Papanikolopoulos and P. Khosla, "Robotic visual servoing and robotic assembly tasks", *IEEE Robotics and Automation Magazine*, Vol. 3, No. 2, pp. 23–31, June 1996.

[11] F. Miyazaki and Y. Masutani, "Robustness of sensory feedback control based on imperfect Jacobian", *Robotics Research: The Fifth International Symposium*, H. Miura and S. Arimoto, Eds. Cambridge, MA: MIT Press, pp. 201–208, 1990.

[12] R. Kelly, P. Shirkey and M. W. Spong, "Fixed camera visual servo control for planar robots". *Proceedings of the IEEE International Conference Robotics and Automation*, pp. Minneapolis, MN., Vol. 3, pp. 2463–2469, April 1996.

[13] R. Kelly, "Robust asymptotically stable visual servoing of planar robots", *IEEE Transactions on Robotics and Automation*, Vol. 12, No. 5, pp. 759–766, October 1996.

[14] M. Takegaki and S. Arimoto, "A new feedback method for dynamic control of manipulators", *ASME J. Dyn. Syst. Meas. Control*, Vol. 103, pp. 119–125, 1981.

[15] G. D. Hager, "A modular system for robust positioning using feedback from stereo vision", *IEEE Trans. on Robotics and Automation*, Vol. 13, No. 4, pp. 582–595, August 1997.

[16] M. Spong and M. Vidyasagar, *Robot Dynamics and Control*, John Wiley and Sons, New York, 1989.

[17] F. Reyes and R. Kelly, "A direct drive robot for control research", *Proc. IASTED International Conference, Applications of Control and Robotics*, Orlando, FL., pp. 181–184, January 1996.

Proceedings of the 1998 IEEE
International Conference on Robotics & Automation
Leuven, Belgium • May 1998

Visual Impedance Using 1ms Visual Feedback System

Yoshihiro Nakabo   and   Masatoshi Ishikawa

Department of Mathematical Engineering and Information Physics
University of Tokyo
7-3-1, Hongo, Bunkyo-ku, Tokyo 113-8656, Japan
nakabo@k2.t.u-tokyo.ac.jp   ishikawa@k2.t.u-tokyo.ac.jp

Abstract

We introduce visual impedance, a new scheme for vision based control which realizes task-level dynamical robot control using a 1ms visual feedback system. This method is simply described as applying image features to the impedance equation so that integration of a visual servo and a conventional servo system can be naturally accomplished. With visual impedance, an adaptive motion is obtained for real robot tasks in dynamically changing or unknown environments based on the framework of impedance control.

In such cases, very high rate visual feedback is necessary to control robot dynamics but most conventional vision systems using CCD cameras can never satisfy this condition because their sampling rate is limited by the video signal. To solve this problem, we developed a general-purpose vision chip SPE and the 1ms visual feedback system which can achieve an adequate servo rate to control dynamics.

In this paper, we first illustrate the concept of visual impedance. Then our 1ms visual feedback system for a robot control system is described. Last we show some experimental results with some real robot tasks.

1 Introduction

Task-level visual feedback is most effective when robots are to work adequately in dynamically changing or unknown environments. Much research has been widely made based on this. Recently realized direct vision-based control has been called visual servo [1, 2].

However, the robot systems developed in this research have not yet attained this aim. The problems are caused by a CCD camera used for capturing images in most systems. Because, using CCD cameras, images are scanned pixel by pixel and transmitted in the video signal so the video rate limits the image sampling rate up to the video field rate (60Hz in NTSC, 50Hz in PAL) even if fast image processing can be carried out. On the other hand it is generally accepted that a servo rate around 1kHz is needed to control robot dynamics. Compared to the robot dynamics the sampling rate of the conventional vision systems is too slow.

This limitation of the sampling rate leads most research attention to focus only on the problem of how the visual servo can be designed using, for example, prediction, known models and so on. It is rarely considered how the visual servo can be integrated into and fuse with the conventional robot system to apply it to real tasks. For this problem Castano and Hutchinson [3] proposed the concept of a visual compliance based on a framework of a hybrid vision/position control structure which lends itself to task-level specification of manipulation goals. However, in this method, directions of vision/position controls are restricted to strict orthogonal directions. And the visual servo is not used for dynamical control but only for position control in the image. Tuji et al. [4] realized a non-contact impedance control using visual information and they have controlled the robot dynamics. However, they use a position sensitive detector (PSD) as the vision sensor to obtain a high feedback rate. Thus the pattern information included in the image cannot be used. As we mentioned above, the limitation of the feedback rate in vision systems is the core problem which restricts the application of the visual servo to real robot tasks and the integration of the visual servo into conventional sensor feedback systems.

To solve this problem we developed a vision chip, the SPE (Sensory Processing Elements), in which all photo-detectors are directly connected to all processing elements. These pixels are integrated into one chip so that the bottleneck of the image transmission does not occurr. In addition we have developed the 1ms visual feedback system and demonstrated high speed visual tracking with a 1kHz feedback rate [5, 6]. Using this SPE chip, we can implement various kinds of image processing algorithms with far higher performance compared with conventional vision systems using CCD cameras.

In this paper, we propose the concept of visual impedance which realizes task-level visual feedback and dynamical robot control using the 1ms visual feedback system. With visual impedance, adaptive motion of the robot to the environment can be achieved through visually realized virtual contact based on the framework of a virtual impedance control.

2 Visual Impedance

2.1 Concept of visual impedance

Virtual impedance control is a method to regulate the mechanical impedance of a manipulator to a desired value according to a given task [7]. In this method, the control law applied to the manipulator is not only the desired trajectory but also adequate dynamics in which compliant motion with an environment can be achieved. For example, when a manipulator needs to contact with an object in an environment few errors in motion or models make serious disturbances. On the other hand, using impedance control, task-level feedback information from the force sensor is used to control dynamics of the manipulator so that these errors are absorbed by compliant motion.

To expand this method, we propose visual impedance in which the feature value extracted from the image is simply used in the equation of impedance control. Thus it can be applied to visually realized virtual contact instead of real contact. In this method, we assume an imaginary virtual surface on a real object calculated by real-time image processing. Inside this surface an adequate impedance is set in which, if the manipulator contacts with the virtual surface, an interaction is made that is similar to the real contact motion.

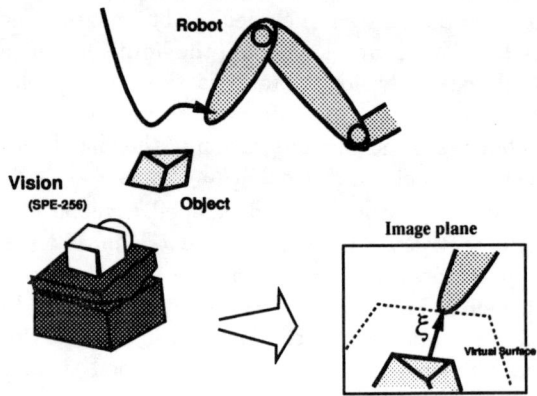

Figure 1 Visually realized virtual contact

2.2 Virtual contact

Now we define virtual contact between the robot and an environment expressed by the contact point P_r and contact vector ξ extracted by real-time image processing for resolved-rate motion control.

Let $C_o \subset \Re^2$ be the pattern of an edge on the object extracted from the image and $P = [X, Y]^T \in \Re^2$ be any point in the image plane except outer regions of the object, where (X, Y) are treated as coordinates of the point expressed in the image plane coordinate frame. The distance between P and the nearest point on the edge $x \in C_o$ can be expressed as follows:

$$\varphi = \min(|P - x|) \quad (\forall x \in C_o) \tag{1}$$

This $\varphi(P)$ describes the potential field and the value equals the distance from the object surface in the image plane. Now we call a set of points C_v a virtual surface where

$$C_v = \{P_v \mid \varphi(P_v) = c\} \quad (c : \text{const}) \tag{2}$$

which has constant distance from real surface of the object.

When the manipulator, in the image, comes inside the virtual surface we consider that a virtual contact has occurred. During the virtual contact, the vision system extracts the point P_r which describes the position of the manipulator and ξ is the contact vector defined as follows:

$$\xi = \begin{cases} -\nabla\varphi\,(c - \varphi) & \text{if} \quad c \geq \varphi \\ \mathbf{0} & \text{if} \quad c < \varphi \end{cases} \tag{3}$$

where

$$\varphi = \varphi(P_r). \tag{4}$$

Note that

$$|\nabla\varphi| = 1 \tag{5}$$

is used to define ξ in (3). As shown in Fig 2, the vector ξ is directed orthogonal and away from the real surface and its length is the distance from the virtual surface.

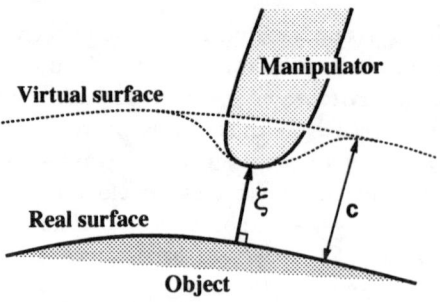

Figure 2 Contact vector ξ

2.3 Compliant motion by visual impedance

The following equation is applied to the extracted image feature ξ to obtain the desired dynamics.

$$M\ddot{\xi} + D\dot{\xi} - K\xi = \hat{F} \tag{6}$$

Where $M, D, K \in \Re^{2\times 2}$ are the desired inertia, viscosity and stiffness matrices of the virtual surface and $\hat{F} \in \Re^2$ is a vector of virtual external force defined only in the image plane.

Let us assume the manipulator to be rigid object. Therefor we have to determine a compliance center to calculate the effect of the manipulator from $\hat{F}$. It should be noted that this compliance center depends on the task so that we can design desired dynamics for compliant robot motion. To make it easy to explain, let the origin of the coordinate frame of the image plain be the compliance center. When the force $\hat{F}$ is exerted on the point P_r, following equation is obtained

$$\begin{aligned} F &= \frac{(\hat{F}\cdot P_r)}{|P_r|}P_r \\ M_z &= \frac{|\hat{F}-F|}{|P_r|} \end{aligned} \quad (7)$$

where the virtual force vector F is parallel to the image plane and virtual moment M_z is on the axis orthogonal to the image plane.

2.4 Visual impedance for image pattern

Next we expand this calculation of visual impedance to a binary image pattern. We add all the effects on the potential field φ by the manipulator to calculate the interaction between the object pattern and the manipulator pattern.

Let $C_r \subset \Re^2$ be an edge of the robot in the coordinate frame of the image plane and $P_r \in C_r$ a point on this edge. The expansion of (7) is given by

$$\begin{aligned} F &= \sum_{P\in C}\frac{(\hat{F}\cdot P)}{|P|}P \\ M_z &= \sum_{P\in C}\frac{|\hat{F}-F|}{|P|} \end{aligned} \quad (8)$$

where the P is all the points in C given by

$$\begin{aligned} C &= C_{v'} \bigcap C_r, \\ C_{v'} &= \{P_{v'} \mid \varphi(P_{v'}) \leq c\} \end{aligned} \quad (9)$$

Fig 3 shows a fitting motion using this method. The virtual external force exerted from several points realizes the desired dynamics dependent on the task.

2.5 Calculation of joint control torque

Last, we describe the control law based on the framework of image-based visual servo in which joint control torque is directly calculated from external torque and the moment obtained above. This torque realizes an additional motion of virtual contact as we noted at the begining of this section.

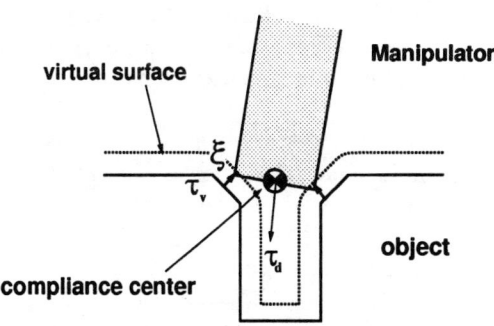

Figure 3 Visual impedance for fitting

First we redefine $F = [F_X, F_Y]^T$ in (8) by adding M_Z as follows

$$F = [F_X, F_Y, M_Z]^T \quad (10)$$

and similarly, the manipulator's position P by adding θ as follows

$$P = [X, Y, \theta]^T \quad (11)$$

where $\theta \in \Re$ describes the pose of the manipulator in the image plane. It is noted that now $F, P \in \Re^3$

Now we define the Jacobian matrix J including both image Jacobian J_{image} and robot Jacobian J_{robot} as follows

$$\begin{aligned} J &= \frac{\partial P}{\partial q} \\ (&= J_{image} J_{robot}) \end{aligned} \quad (12)$$

where $q \in \Re^l$ is the joint angle vector of an l-joint manipulator. To obtain J_{image} and J_{robot} proper models should be used for the detail (see [1, 2]).

At last, we obtain the control law of the manipulator as follows

$$\tau_v = J^T F \quad (13)$$

where $\tau_v \in \Re^l$ is a joint control torque vector of an l-joint manipulator.

As we proposed, the control law τ_v is obtained directly from image features ξ and P_r. Thus resolved-rate motion control using direct vision-based visual servo is realized in this scheme.

2.6 Integration of visual impedance and conventional control method

While the joint control torque τ_v is calculated as a control law, it is effected on the manipulator as if an external effect is caused by virtual contact. This leads us to obtain a simple control law for the total robot control law by adding the effect of virtual contact to the planned motion as follows

$$\tau = \tau_d + \tau_v \quad (14)$$

where $\tau_d \in \Re^l$ is the joint control torque calculated from another control law, like tracking objective trajectory and/or position based conventional virtual impedance.

The diagram of this total control law is shown in Fig (4). In this scheme, the robot is directly controlled

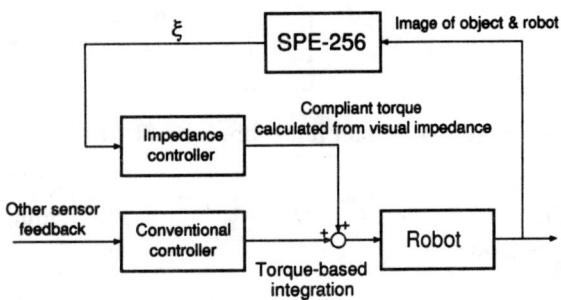

Figure 4 Blockdiagram of visual impedance control

both by vision and another sensor feedback. Therefore if the feedback rate of the visual servo is slower than others this difference in the rates produces a problem for controlling the robot dynamics.

However, in our system described in the next section, we realized visual feedback carried out every 1ms. Since this rate is fast enough for controlling the robot dynamics, we can simply integrate both visual servo and the other control law by summing each joint control torque as show in Fig (4).

3 1ms Visual Feedback System

In this section, we first discuss the necessity of high speed visual feedback. It is important to consider the performance of the vision system as the total system of controlling the robot. As the solution of this problem, we introduce our 1ms visual feedback system using the vision chip, SPE.

3.1 Servo rate of robot control

Considering the servo rate of the robot control, it is generally accepted that the lowest level servo rate T_s for the joint control of the manipulator should satisfy the following condition

$$T_s < T_m / 10 \quad (15)$$

where $1/T_m$ is the mechanical resonance frequency of the manipulator [8]. Therefore, it is necessary to control the manipulator with a feedback rate of $200 \sim 1\text{kHz}$, in other words, $1 \sim 5\text{ms}$ for the cycletime, since usual manipulators have mechanical resonance frequencies about $20 \sim 50\text{Hz}$. Needless to say, faster manipulations are usually desired at any cases.

From this point of view, conventional vision systems using CCD cameras are too slow for robot control, as we mentioned above. Therefore, in recent research of visual servoing, various kinds of methods are proposed to solve this problem using, for example, prediction, interpolation, assumption model of dynamics and so on. But all these methods leave out the most important aim of the visual information feedback, that is, to realize a robot system which can work in an unknown and dynamically changing environment. It is obvious that a high speed visual feedback system is needed to realize a robot which can work adequately in the real-world environment.

3.2 1ms visual feedback system

To solve the problem of the servo rate as discussed above we developed a 1ms visual feedback system using the vision chip SPE and realized high speed target tracking [5, 6]. We have applied this system to the robot control system and made several experiments which are described in section 4. The robot control system is shown in Fig 5 and a photograph is shown in Fig 6.

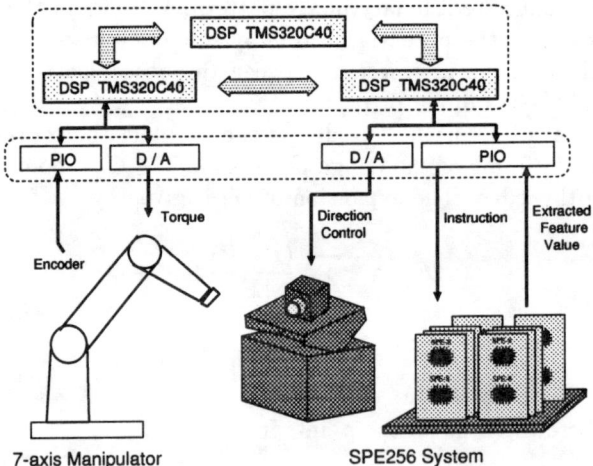

Figure 5 Robot control system using 1ms visual feedback system

3.3 Robot control system

The robot control system is constructed by a parallel architecture using several DSPs (TMS320C40) for high speed processing and communications and a sufficient numbers of input and output ports to avoid an I/O bottolneck, thus allowing the total system including vision system to achieve a 1ms cycle-time [9].

In this system the 7-axis manipulator is controlled and the two end joints of them are used in experiments. Also the vision system consists of a SPE-256

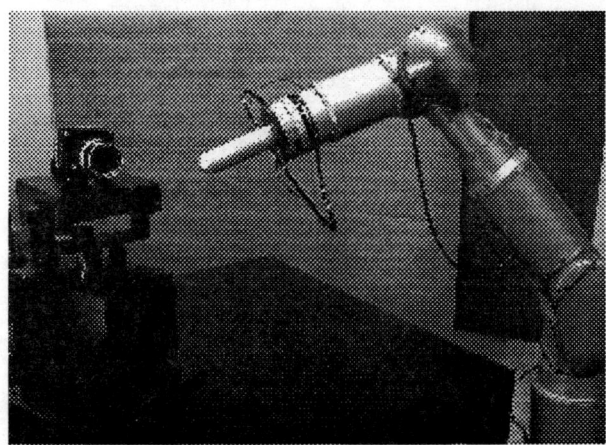

Figure 6 Overview of the system

system and an active vision system which has two DOF (pan and tilt) in its viewing direction.

3.4 SPE-256 system

As the vision system, we developed the SPE-256 system in which $16 \times 16 = 256$ processing elements(PEs) and a 16×16 array of PIN photo detectors(PDs) are directly connected. As shown in Fig 7, the PE in

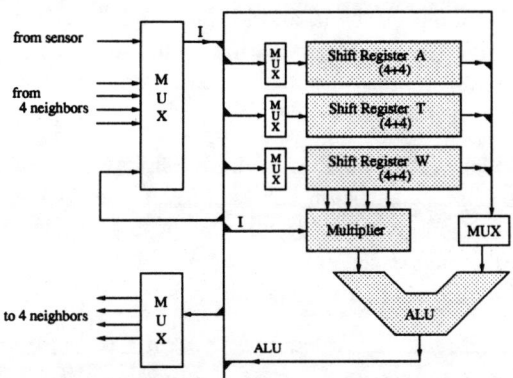

Figure 7 The architecture of SPE

the SPE-256 system is a 4-neighbor connected SIMD based parallel processor. Each PE has 32 bits of registers and a general-purpose processing unit allowing us to implement image processing algorithms to extract the image features.

This architecture is also compactly designed to integrate many processing elements on one-chip. Now we have developed a vision chip S^3PE which has successfully integrated both PEs and PDs in one-chip [10].

4 Experimental Results

In this section we show the experimental results applying visual impedance to collision avoidance and a fitting task.

4.1 Experimental setup

The experiments are carried out by the robot system as we described in section 3.

A white sphere is put on the end of the manipulator so the vision system can recognize the end-effector. Though the active vision system tracks this end-effector independently of the robot motion, the end-effector is always placed at the center of the image plane, since very fast target tracking has been achieved by the active vision system [5].

In both of the following experiments, the compliance center is placed on the center of the end-effector so that (7) can be simplified as $M_z = 0$, $\boldsymbol{F} = \hat{\boldsymbol{F}}$. And for J defined in (12) we set the directions of the motion of the manipulator parallel to the axes of the coordinate frame of the image plane so that we can obtain the simple Jacobian $J = \alpha I$, where α is a constant.

4.2 Collision avoidance

The experiments of collision avoidance are carried out to show the dynamical control of the robot motion using visual feedback. In these experiments the objective motion of the end-effector was given as constant torque in one direction, and contact with the virtual surface occurred.

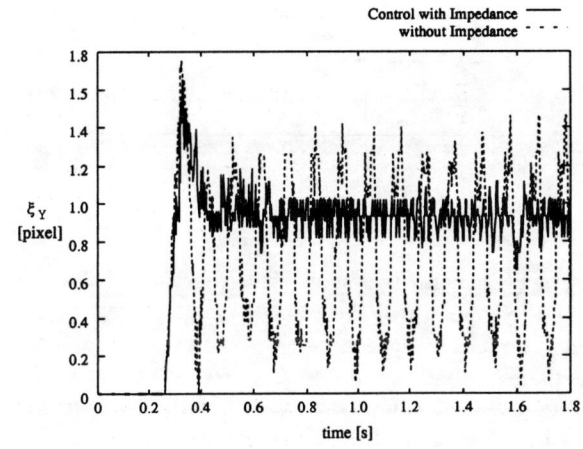

Figure 8 Collision avoidance with and without using visual impedance

In Fig 8 the lines show the trajectories of the end-effector whose direction is parallel to the Y axis. There is a balancing point at $\xi_Y = 1$. The solid line shows the trajectory using visual impedance in which tuned viscosity is applied as the impedance parameter. And the dotted line shows the trajectory without using visual impedance, using simple proportional control.

As shown in Fig 8, without using visual impedance,

even when the collision is avoided, the end-effector continues to vibrate. On the other hand, using visual impedance stable motion has been achieved due to controlling of the dynamics of the manipulator.

4.3 Fitting task

Now we show the experiments of a fitting motion as the robot task using visual impedance. For the fitting task it is necessary to realize compliant motion known as RCC (Remote Compliance Center) characteristics. With visual impedance, the pattern information extracted from an image is used to realize characteristics similar to the RCC.

In this experiment, the task is to push the sphere put on the end of the manipulator into a gap. Since the trajectory given first has an error, without any task-level feedback, it will not be accomplished. Using visual impedance the result as shown in Fig 9 is obtained.

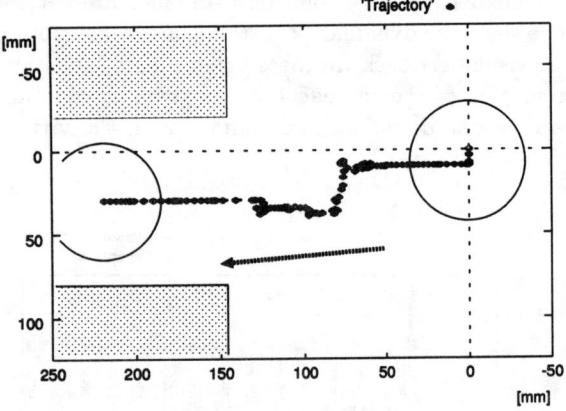

Figure 9 The fitting motion using visual impedance

The points in Fig 9 show the trajectory of the center of the sphere projected on the image plane sampled every 70ms. The pattern projected on the image plane has made virtual contact with the walls on both sides of the gap and compliant motion with the gap pattern has been achieved. This result shows that using visual impedance, task-level real-time feedback of visual information is accomplished to control the robot dynamics.

5 Conclusion

We proposed a concept of visual impedance in which task-level visual feedback is applied to the control of the dynamics of the manipulator. Since this concept is described as applying image features to the impedance equation, integration of visual servo and a conventional servo system can be naturally accomplished using the 1ms visual feedback system.

It is noted that virtual contact does not need real contact, but it can be with real contact. Our future research will apply the visual feedback system in combination with other sensors to the dynamical feedback control based on visual impedance.

References

[1] S. Hutchinsin, G. D. Hager, and P. I. Corke. "A tutorial on visual servo control." *IEEE Trans. Robotics and Automation*, Vol. 12, No. 5, pp. 651–670, 1996.

[2] K. Hashimoto and et al. *Visual Servoing*. World Scientific, 1993.

[3] A. Castano and S. Hutchinsin. "Visual compliance: Task-directed visual servo control." *IEEE Trans. Robotics and Automation*, Vol. 10, No. 3, pp. 334–342, 1994.

[4] T. Tsuji, H. Akamatsu, and M. Kaneo. "Non-contact impedance control for redundant manipulators using visual information." In *Proc. IEEE Int. Conf. on Robotics and Automation*, pp. 2571–2576, 1997.

[5] Y. Nakabo, I. Ishii, and M. Ishikawa. "High speed target tracking using 1ms visual feedback system." In *Video Proc. IEEE Int. Conf. on Robotics and Automation*, 1996.

[6] I. Ishii, Y. Nakabo, and M. Ishikawa. "Target tracking algorithm for 1ms visual feedback system using massively parallel processing vision." In *Proc. IEEE Int. Conf. on Robotics and Automation*, pp. 2309–2314, 1996.

[7] N. Hogan. "Impedance control: An approach to manipulation, part i, ii, iii,." *ASME journal of Dynamic Systems, Measurement, and Control*, Vol. 107, No. 1, pp. 1–24, 1985.

[8] R. P. Paul. *Robot Manipulators*. MIT Press, 1987.

[9] A. Namik and M. Ishikawa. "Optimal grasping using visual and tactile feedback." In *Proc. IEEE Int. Conf. on Multisensor Fusion and Integration for Intelligent Systems*, pp. 589–596, 1996.

[10] T. Komuro, I. Ishii, and M. Ishikawa. "Vision chip architecture using general-purpose processing elements for 1ms vision system." In *Proc. IEEE Int. Work. on Computer Architecture for Machine Perception*, pp. 276–279, 1997.

Active leg compliance for passive walking

Richard Quint van der Linde
Delft University of Technology
Mekelweg 2
2628CD DELFT, HOLLAND

R.Q.vanderLinde@wbmt.tudelft.nl

Abstract

Previous research has shown that passive (or ballistic) walking is an energetic efficient and mechanical cheap way of walking. Therefore ballistic walking would be suitable for applications in rehabilitation technology and autonomous robots.
Successful application would require a smooth hip trajectory in order to protect respectively the patient or electronics against large velocity changes due to ground collisions.
It is shown that an actively adjustable stance leg compliance in combination with a viscous damping can result in smaller hip velocity changes.

Keywords: Biped, Ballistic, Active compliance, Limit cycle

1. Introduction

In robotics movement is often achieved by actuators servod according to a defined trajectory. Many researches have shown that it is well possible to construct a bipedal walking motion by trajectory selection in combination with servo techniques [7, 8, 12], and even realize impactless walking [2, 4].
A cyclic motion like walking can also be realized by a mechanical system oscillating in its natural orbit. This is called passive (or ballistic) walking. Using the unforced oscillation implicates that the walking motion is not solely defined by the actuators, but largely by the passive inherent system properties. Actuation is mainly present to sustain the oscillation.
It has been shown that by choosing the right geometry of the robot a stable ballistic walking cycle can be constructed [5, 6, 10]. In these researches gravity was the main power supply.
The advantages of ballistic walkers are: a minimal energy consumption, mechanical simplicity, low weight, and 'natural' walking behaviour. These advantages make ballistic walking suitable for autonomous robots and rehabilitation technology. In both areas there is an inevitable energy restriction due to autonomous functioning. Also the second area is restricted to a mechanical compact design, and an as low-as-possible weight. Ballistic walking would be a suitable technology to overcome these restrictions.

However, in its turn ballistic walking also introduces several restrictions. Since the passive walking pattern is determined by the natural frequency of the mechanical system, changing the limit cycle can only be achieved by: 1. applying an additional force or torque, 2. actively changing the intrinsic system parameters. The first solution has been studied before [1, 6, 9, 11]. In this work the latter solution has been chosen.
The most plausible intrinsic parameter to vary is joint stiffness. Active adjustment of a joint stiffness offers the ability to 1. add or withdraw energy to or from the system, 2. modify the natural frequency 3. give the system a protection against high disturbance forces. If energy storing properties are to be used also, then the compliance must be passive. By phasically changing the joint stiffness a limit cycle can be realized [15].

Successful physical application of a walking motion requires a smooth hip trajectory. This in order to protect a patient (in the case of a prosthesis or orthosis) or electronics (in the case of a robot) against impulsive forces due to ground collisions. This effect can occur at heel contact, where the weight of a biped is transferred from one leg to another. Stiff legs imply an instantaneous velocity change due to an inelastic collision. If on the other hand the legs are compliant most energy can be stored during the landing and released during the push-off. This effect is often used in running Raibert-like hoppers [14]. The advantage of a short stance phase in combination with an aerial phase allows a simple bouncing effect. The efficiency of such a collision is nearly ideal, or nearly elastic.
If the bouncing effect is used for walking robots, the spring constant must be very low due to a long stance phase. This results in an undesired large leg shortening. It can be dealt with by applying locking mechanisms [3]. However these do not result in smooth hip trajectories, and are not biologically presumable.
In this paper it is shown that smooth hip trajectories for a phasically activated passive bipedal walker can be realized by regulating leg stiffness and leg damping. The values of these parameters define three different walking modes.

2. The model

2.1 Kinematics

The model is assumed to be according Fig. 1. It is a planar compass gait model, extended with an actively variable spring (C_{SL}) and a damper (β_{SL}) in the stance leg. Both legs have a mass (m) and inertia (I_L), and a foot radius (R). The mass (M_H) of the upper body is located between the hips and is a point mass. The unsprung leg length is L_0, and the distance between the hip and the leg center of gravity is p. The model has 3 degrees of freedom; stance leg angle φ_{SL}, trailing leg angle φ_{TL}, stance leg length L_{SL}.

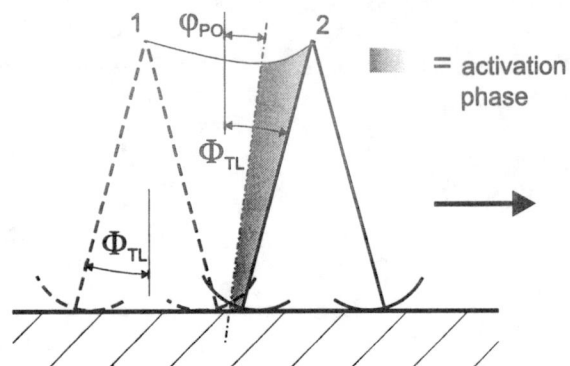

Fig. 2. *Complete walking cycle, with the gray area indicating the activation phase.*

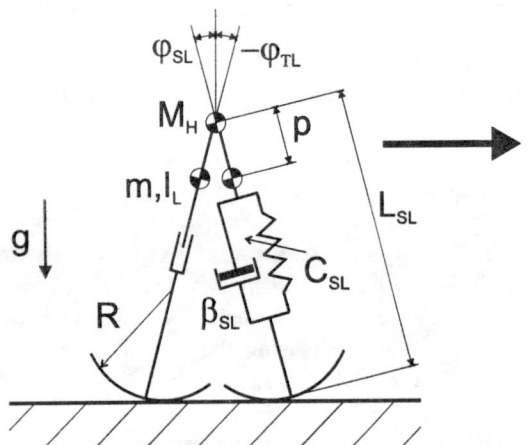

Fig. 1. *Model of the biped with actively variable leg stiffness and leg damping.*

The cycle begins when the foot of the trailing leg leaves the ground, and the system swings forward. The passive walker is activated by a stance leg stiffness change (dC), which is initiated by reaching a certain stance leg angle (φ_{PO}). This can be expressed by a dimensionless parameter :

$$\eta_{PO} = \frac{\varphi_{PO}}{\Phi_{TL}}, \qquad 0 \leq \eta_{PO} \leq 1 \qquad (1)$$

Where η_{PO} = dimensionless push-off parameter, φ_{PO} = push-off angle, and Φ_{TL} = trailing leg angle at push-off. The range of η_{PO} is limited, for the push-off can only occur after midstance and before foot contact. The activation phase ends when the stance phase ends.
When the foot of the trailing leg hits the floor the stance phase ends. This occurs when:

$$(L_0 - R)\cos\varphi_{TL} - (L_{SL} - R)\cos\varphi_{SL} = 0 \qquad (2)$$

Then the support leg becomes trailing leg and visa versa. After the transfer phase the next cycle begins. A complete cycle is illustrated in Fig. 2.

Furthermore the following assumptions are made:
1. The lower part of the legs is massless
2. The foot of the trailing leg retracts during mid stance, so it will not scuff the floor.
3. The support transfer is instantaneous, and the foot does not slip.
4. At collision angular and linear momentum are conserved.
5. The leg stiffness change is instantaneous

2.2 Equations

Based on the model (Fig. 1) and the preceding assumptions the governing equations can be derived. Mathematically there are two distinct phases: stance phase and transfer phase.

2.2.1 Dynamics stance phase

The support phase consists of two phases: ballistic and activation. During the ballistic phase the leg stiffness C_{SL} remains a nominal stiffness C_N. In the activation phase the leg stiffness is increased from C_N to C_N+dC. Both phases are governed by the same equations of motion, with different values for C_{SL}.
The equations for stance dynamics can be given by the following general equation of motion:

$$M(x)\ddot{x} + N(x,\dot{x})\dot{x} + G(x) + C(x) = \sum_{k=1}^{i} F_k \frac{\partial s_k}{\partial x} \qquad (3)$$

Where $x = [\varphi_{SL}\ \varphi_{TL}\ L_{SL}]^T$, $M(x)$ = 3 × 3 mass matrix, $N(x, \dot{x})$ = 3 × 3 coriolis matrix, $G(x)$ = 3 × 1 gravity matrix, $C(x)$ = 3 × 1 stiffness matrix, F_k = force k on state x. s_k = direction of F_k in point of application. The matrices M, N, G, C and the right hand expression are given as a function of system parameters in appendix A.

2.2.1 Transfer phase

The activation phase ends when the foot hits the floor. A non elastic collision occurs, during which impulse momentum is conserved. When the next cycle begins stance leg becomes trailing leg and visa versa. When the trailing leg becomes stance leg it will have unsprung length L_0. The state vector after collision can be written as:

$$x^+ = \begin{bmatrix} 0 & 1 & 0 \\ 1 & 0 & 0 \\ 0 & 0 & 1 \end{bmatrix} \cdot \begin{bmatrix} \varphi_{SL}^- \\ \varphi_{TL}^- \\ L_0 \end{bmatrix} \quad (4)$$

The impulse momentum equations can be written in the form:

$$Q^-(x)\dot{x}^- = Q^+(x)\dot{x}^+ \quad (5)$$

Where, the index "-" means before collision, the index "+" means after collision, and $Q(x) = (3\times3)$ matrix containing the impulse momentum equations. From (5) the speed and rates after collision ($\dot{x}^+$) can be derived, which will be the initial state vector for the next cycle. $Q^+(x)$, and $Q^-(x)$ are given as a function of system parameters in Appendix B.

3. Cycle analysis

Limit cycles can be found by applying return map analysis. A Poincaré map is defined by the a mapping function P which maps an initial state vector x onto its self after a complete cycle [13]. A periodic (or fixed) point can then be found finding the roots of the mapping equation:

$$P(x) - x = 0 \quad (6)$$

For the defined model the phase space is 6-dimensional. The end-of-step condition (2) causes 5 variables to be independent. So the mapping function is 5-dimensional. The following 5 parameters are chosen to be iterated $\dot{\varphi}_{SL}, \dot{\varphi}_{TL}, \dot{L}_{SL}, C_N$, and dC. Step width can then be given. If it is assumed that stance leg length at the end of the cycle equals L_0 then (2) is simplified by $\varphi_{SL} + \varphi_{TL} = 0$.

A Newton-Raphson method can now be used to iterate (6) until convergence. Therefore x is perturbed by Δx which results in:

$$x + \Delta x = P(x) + \nabla P_{X_0} \cdot \Delta x + \frac{\partial P}{\partial C_N} \Delta C_N + \frac{\partial P}{\partial dC} \Delta dC \quad (7)$$

Where ∇P_{X_0} = jacobian evaluated in X_0.

Rewriting this for the independent variables and iteration parameters results:

$$\begin{bmatrix} \Delta \dot{\varphi}_{SL} \\ \Delta \dot{\varphi}_{TL} \\ \Delta \dot{L}_{SL} \\ \Delta C_N \\ \Delta dC \end{bmatrix} = \left[\begin{pmatrix} \varphi_{SL} \\ L_{SL} \\ \dot{\varphi}_{SL} \\ \dot{\varphi}_{TL} \\ \dot{L}_{SL} \end{pmatrix} - P \begin{pmatrix} \varphi_{SL} \\ L_{SL} \\ \dot{\varphi}_{SL} \\ \dot{\varphi}_{TL} \\ \dot{L}_{SL} \end{pmatrix} \right] \cdot \left(\nabla S - \begin{bmatrix} 0 & 0 \\ I(3\times3) & 0 \end{bmatrix} \right)^{-1} \quad (8)$$

,with $\nabla S = \begin{bmatrix} \dfrac{\partial P_X}{\partial \dot{\varphi}_{SL}} & \dfrac{\partial P_X}{\partial \dot{\varphi}_{TL}} & \dfrac{\partial P_X}{\partial \dot{\varphi}_{SL}} & \dfrac{\partial P_X}{\partial C_N} & \dfrac{\partial P_X}{\partial dC} \end{bmatrix}$

4. Simulation results

Limit cycles were found according to the mentioned analysis. The simulations were performed for the following biped configuration parameters[1]: $L_0 = 0.655$ [m], $R = 0.2$ [m], $p = 0.282$ [m], $M_H = 1.53$ [kg], $m = 0.458$ [kg], $I_L = .0387$ [kgm^2]. These are parameters where stable passive walking can be expected [10].

The push-off parameter is chosen to be $\eta_{PO} = 0$. The dimensionless coefficient of damping is defined as

$$\beta_{SL} \equiv \kappa_{SL} / \sqrt{4 C_N \cdot M_T} \quad (9)$$

Where $M_T = 2m + M_H$ = total biped mass, κ_{SL} = stance leg damping constant [Ns/m]. Also a pre-tension U_0 is taken into account, caused by an agonist-antagonist construction [15]. In the plots U_0 is chosen to be a hypothetical 1 [m]. Later on it will be shown that it is energetically beneficial to decrease U_0, and increase dC.

Remarkable was the existence of three qualitatively different walking modes.

mode 1

A typical mode 1 cycle is characterized by the single up/down motion, illustrated in Fig. 3 for different settings of β_{SL}. Obviously the amplitude of the vertical body oscillation decreases with increase of β_{SL}. Since $\eta_{PO} = 0$ the maximum hip displacement is in midstance. The inter leg angle was chosen to be 2×0.23 [rad]. The gray line indicates the stiff legged compass gait hip trajectory.

---

[1] These parameters were largely based on the geometry of an existing prototype biped.

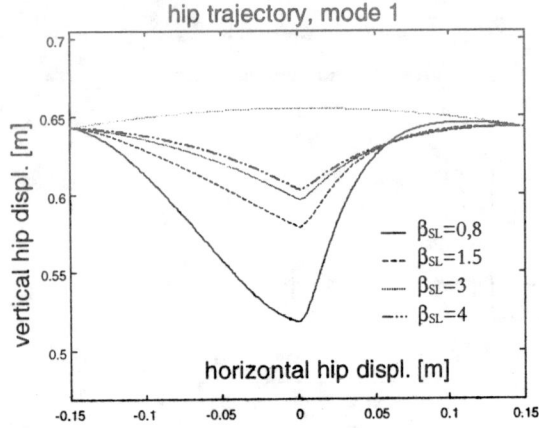

Fig. 3 *Mode 1 hip trajectory for different settings of stance leg damping:* $\beta_{SL} = .8, 1.5, 3, 4,$ *and* $4 < C_N \cdot L_0 / M_T g < 7$.

The leg stiffness range for mode 1 cycles in Fig. 3 is $4 < C_N \cdot L_0 / M_T g < 7$. Leg activation stiffness dC increases with an increase of leg damping β_{SL}, for more energy will be lost, so more energy must be added.

mode 2

A mode 2 cycle is characterized by the double up/down motion, illustrated in Fig. 4. The resulting walking motion looks rather funny.

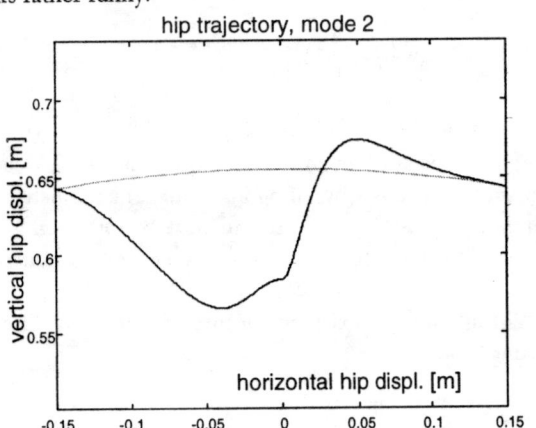

Fig. 4. *Mode 2 hip trajectory for* $\beta_{SL} = .2,$ *and* $C_N \cdot L_0 / M_T g = 10$.

The basin of attraction appeared to be very small for this mode. Cycles were found for $C_N \cdot L_0 / M_T g \approx 10$, and $\beta_{SL} \approx 0.2$. Also the cycles found were either slightly stable or slightly unstable. So this mode will most probably be very difficult to reproduce in practice.

mode 3

A mode 3 cycle is characterized by the presence of a damped vertical oscillation with small amplitude, illustrated in Fig. 5. The leg stiffness range for mode 3 cycles in Fig. 5 is $45 < C_N \cdot L_0 / M_T g < 76$. If damping is increased to overdamped the cycle changes into a mode 1 cycle with small steps.

The nominal leg compression in the ballistic phase is approximated by $M_T g / C_N$, which is the static spring compression.

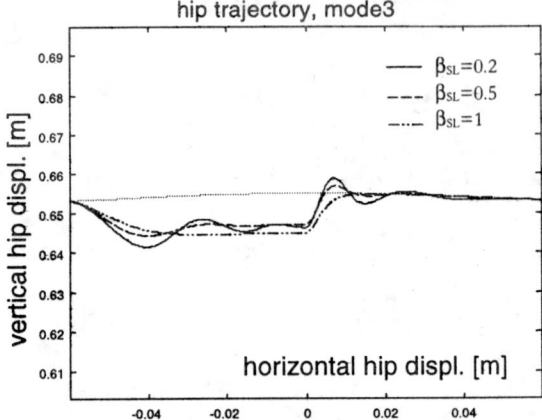

Fig. 5 *Mode 3 hip trajectory for different settings of stance leg damping:* $\beta_{SL} = .2, .5, 1,$ *and* $45 < C_N \cdot L_0 / M_T g < 76$.

A phase plane plot of the leg angle in one complete walking cycle (stance and swing) in the three different modes, and for a stiff legged biped model is given in Fig. 6. The stiff legged biped model is obtained by eliminating L_{SL} out of the state vector x, and rewriting the equations (3) and (5). Energy is added by instantaneous stance leg lengthening[2].

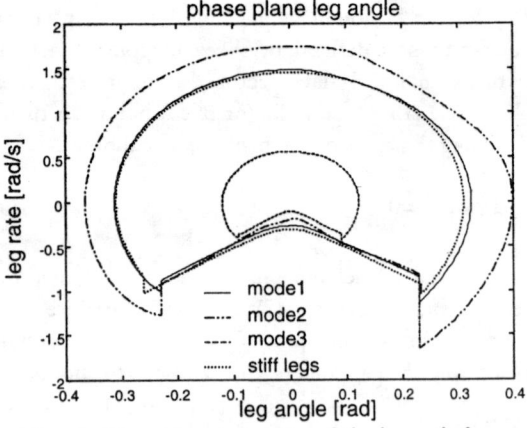

Fig. 6. *Phase plane trajectories of the leg angle for a complete walking cycle in three different modes and for a stiff legged biped model.*

The shapes of the phase plots in Fig. 6 do not differ a lot (except for the size). This can be expected, for leg angle behaviour will not be severely disrupted by a limited amount of leg lengthening.

---

[2] This is similar to McGeer's 'length cycling' [11]

A phase plot of the stance leg length for the three different modes is given in Fig. 7. For stiff legged walking the phase plane is not interesting, since stance leg length is assumed to change instantaneously. Fig. 7 clearly shows the difference between the three modes.

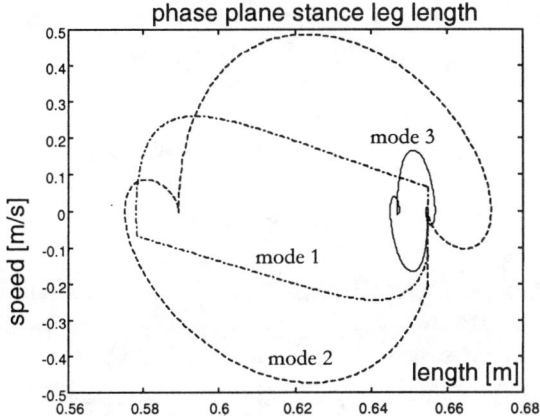

Fig. 7. *Phase plane trajectories of the leg length for a complete walking cycle in three different modes.*

Values for the vertical velocity jumps at heel contact on the hip (ΔV^z_{HIP}) are given in table 1. These values are for an equal stance leg angle $\Phi_{SL} = 0.23$ [rad].

Table 1. *Range of vertical velocity jumps of the hip at heel contact, for $\Phi_{SL} = 0.23$ [rad].*

	mode 1 [m/s]	mode 2 [m/s]	mode 3 [m/s]	stiff legs [m/s]
ΔV^z_{HIP}	0.0009 0.0084	0.013	0.005 0.007	0.22

The efficiency of transportation can be expressed by the specific resistance (SR). SR is defined as the work done divided by the product of weight and distance traveled. For the biped model the SR can be expressed as:

$$SR = \frac{energy\ input}{weight \times distance\ travelled} =$$

$$\frac{\frac{1}{2}(C_N + dC)\left(L_{SL}{}^A - L_0 - U_0\right)^2 - \frac{1}{2}C_N\left(L_{SL}{}^A - L_0\right)^2}{M_T g\{R(\varphi_{TL-} - \varphi_{SL-}) - (L_{SL-} - R)\sin\varphi_{SL-} + (L_0 - R)\sin\varphi_{TL-}\}} \quad (10)$$

Where, $L_{SL}{}^A$ = length of the stance leg on the moment of activation, and the index '-' means just before heel contact. In order to keep SR low, U_0 must be chosen as small as possible. In order to produce an appropriate push-off force, dC then needs to be increased. By varying the ratio of U_0 and dC the specific resistance could be varied between 500 and 0.05 ! So in order to keep efficiency high it is crucial to minimize the ratio U_0 / dC.

5. Conclusions and future work

Stable ballistic walking for a biped model with active leg stiffness and damping was achieved. Three different walking modes were found, determined by the values of leg stiffness and damping.

In spite of an additional dissipating effect of damping the specific resistance can be close to that of a stiff legged walker. This is because the energy loss at foot contact is smaller than is the case for stiff legged walkers, due to energy storage of the leg spring. As a result the vertical hip amplitudes can become larger, however velocity jumps at heel contact can be decreased by a factor 244.

Since the energy input is supplied by leg extension, the passive walker is able to walk on a horizontal floor. It is also expected that the implementation of leg stiffness makes simulation results versus practice more accurate with respect to collision behaviour. Since the leg can always chosen to be less stiff than the ground, properties of the surface can vary without having significant result on the walking cycle.

A disadvantage of leg compliance is the need for larger trailing leg shortening, since the vertical hip amplitude is much larger than is the case for stiff legged walking. It might therefore be interesting to study the effect of knees in the model.

In future research a force-velocity relationship will be implemented. If pneumatic McKibben muscles are used to generate the active stiffness, this relationship will be determined. This will relate damping (derivative of force to speed) and stiffness by muscle properties. If leg stiffness is varied mode switching between the three described modes can occur.

Currently a prototype biped is being built to study the practical effect of active leg stiffness.

References

[1] Ahmadi M., Buehler M., 'Stable control of a simulated one-legged running robot with hip and leg compliance'. *IEEE trans. on Rob. and Autom* (Feb. 1997), V.13(1), pp: 96-104.

[2] Blajer W., Schiehlen W., 'Walking without impacts as a motion/force control problem'. *J. Dyn. Syst. Meas. and Contr.* (Dec. 1992), 114: 660-665.

[3] Dhandapani S., Ogot M.M. 'Modeling of a leg system to illustrate the feasibility of energy recovery in walking machines'. *ASME, Advances in Design Automation* (1994) 69(2): 429-436.

[4] Dunn E.R., Howe R.D., 'Towards smooth bipedal walking'. *Proc. IEEE Int. Conf. on Rob. and Autom.* Vol. 3, p.p. 2489-2494. May 1994, San Diego.

[5] Garcia M., Chatterjee A., Ruina, A., Coleman, 'The simplest walking model: stability, complexity, and scaling'. accepted (4-16-'97) in *ASME Journ. Biom. Eng.*

[6] Goswami A., Espiau B., Keramane A., 'Limit cycles and their stability in a passive bipedal gait'. *Int. Conf. on Rob.&Autom*(1996), pp:246-251. Minneapolis, apr. '96.

[7] Grishin A.A., Formal'sky A.M., Lensky A.V., Zhitomirsky S.V., 'Dynamic walking of a vehicle with two telescopic legs controlled by two drives'. *Int. J. Robot. Res.* (1994), 13(2): 137-147.

[8] Kajita S., Yamaura T., Kobayashi A., 'Dynamic Walking Control of a Biped Robot Along a Potential Energy Conserving Orbit'. *IEEE Trans. on Robotics and Automation* (1992), 8 (4): 431-438.

[9] McGeer T., 'Passive bipedal running'. *Proc. R. Soc. Lond. B* (1990), 204:107-134.

[10] McGeer T., 'Passive dynamic walking'. *Int. J. Robot. Res* (1990), 9(2): 62-82.

[11] McGeer T., 'Dynamics and control of bipedal locomotion'. *J. theor. Biol.* (1993), 163: 277-314.

[12] Miura H., Simoyama I., 'Dynamic walk of a biped'. *Int. J. Robot. Res.* (1984), 3(2):60-74.

[13] Seydel, R., '*Practical bifurcation and stability analysis*'. New York 1994 (2nd ed.). Springer-Verlag. ISBN: 3-540-94316-1.

[14] Raibert M., *Legged robots that balance*. Cambridge, 1986. Mass.:MIT Press. ISBN: 0-262-18117-7.

[15] Van der Linde R.Q., 'Design, analysis and control of a low power joint for walking robots, by phasic activation of McKibben muscles'. subm. to *IEEE Rob. and Autom*.

Appendix A, body dynamics matrices

The general notation of the equations of motion for the model in the stance phase is given by (3). The equations of motion were derived using LaGrange's equation:

$$\frac{d}{dt}\left(\frac{\partial \wp}{\partial \dot{x}}\right) - \frac{\partial \wp}{\partial x} = \sum_{k=1}^{i} \vec{F}_k \frac{\partial \vec{s}_k}{\partial x}, \quad (A1)$$

Where $\wp = E_{KIN} - E_{POT}$ (the LaGrange operator)
For the kinetic energy (E_{KIN}) and the potential energy (E_{POT}) we can write,

$$E_{KIN} = \tfrac{1}{2}M_H\|V_H\|^2 + \tfrac{1}{2}m\|V_{SL}\|^2 + \tfrac{1}{2}m\|V_{TL}\|^2 + \tfrac{1}{2}I_L\|\dot{\varphi}_{SL}\|^2 + \tfrac{1}{2}I_L\|\dot{\varphi}_{TL}\|^2 \quad (A2)$$

$$E_{POT} = \mu g a \cdot \cos\varphi_{SL} - mgp \cdot \cos\varphi_{TL} + M_T R g + \tfrac{1}{2}C_N(L_{SL} - L_0)^2 + \tfrac{1}{2}dC(L_{SL} - L_0 - U_0)^2 \quad (A3)$$

$$\text{,with } \mu = M_H + \left(1 + \tfrac{b}{a}\right)m$$

Where V_N = speed vector of mass N, g = gravitation constant, U_0 = pre-tension.
Choosing a coordinate frame, expressing speeds (V) in x, substitute the result in A2 and A3, the matrices M, N, G, C, and D were derived by evaluating A1:

$M(x) = [3 \times 3]$, where
$M_{11} = 2\mu R a \cdot \cos\varphi_{SL} + \lambda(R^2 + a^2) + I_L$
$M_{21} = M_{21} = mp\{a \cdot \cos(\varphi_{SL} - \varphi_{TL}) + R \cdot \cos\varphi_{TL}\}$
$M_{31} = M_{13} = M_T R \cdot \sin\varphi_{SL}$, $M_{22} = mp^2 + I_L$
$M_{32} = M_{23} = -mp \cdot \sin(\varphi_{SL} - \varphi_{TL})$, $M_{33} = M_T$

$N(x, \dot{x}) = [3 \times 3]$, where
$N_{11} = 2L_{SL}\{a\mu + M_T R \cdot \cos\varphi_{SL}\} - \mu a R \dot{\varphi}_{SL} \cdot \sin\varphi_{SL}$
$N_{21} = mpa\dot{\varphi}_{SL} \cdot \sin(\varphi_{SL} - \varphi_{TL}) - mpL_{SL} \cdot \cos(\varphi_{SL} - \varphi_{TL})$
$N_{31} = -\mu a \dot{\varphi}_{SL}$
$N_{12} = mp\dot{\varphi}_{TL}\{R \cdot \sin\varphi_{SL} - a \cdot \sin(\varphi_{SL} - \varphi_{TL})\}$
$N_{32} = mp\dot{\varphi}_{TL} \cdot \cos(\varphi_{SL} - \varphi_{TL})$, $N_{22} = N_{13} = N_{33} = 0$
$N_{23} = -mp\dot{\varphi}_{SL} \cdot \cos(\varphi_{SL} - \varphi_{TL})$

$$G(x) = \begin{bmatrix} -\mu a g \cdot \sin\varphi_{SL} \\ mgp \cdot \sin\varphi_{TL} \\ M_T g \cdot \cos\varphi_{SL} \end{bmatrix}, \quad C(x) = \begin{bmatrix} C_H(\varphi_{SL} - \varphi_{TL}) \\ -C_H(\varphi_{SL} - \varphi_{TL}) \\ \overline{C_N(L_{SL} - L_0)} \\ +dC(L_{SL} - L_0 - U_0) \end{bmatrix}$$

with $b = L_{SL} - p - R$, $a = L_{SL} - R$, $M_T = M_H + 2m$,

$$\lambda = M_H + \left(1 + \tfrac{b^2}{a^2}\right)m$$

$$\sum_{k=1}^{i} \vec{F}_k \frac{\partial \vec{s}_k}{\partial x} = \begin{bmatrix} 0 & 0 & -\kappa_{SL} \cdot \dot{L} \end{bmatrix}^T$$

Appendix B, transition matrices

The transition matrices are given by:

$Q^-(x) = [3 \times 3]$, where
$Q^-_{11} = \{M_H a_-^2 + 2ma_-b_-\}\cos\alpha_- + R\{\mu_-a_- - mp\}\cos\varphi_{SL-} + R\mu_-a_- \cos\varphi_{TL-} + M_T R^2 - mpb_- + I_L$
$Q^-_{21} = -mp\{R \cdot \cos\varphi_{SL-} + b_-\}$
$Q^-_{31} = \mu_-a_- \cdot \sin\alpha_- + M_T R \cdot \sin\varphi_{TL-}$
$Q^-_{12} = -mp\{R \cdot \cos\varphi_{TL-} + b_-\}$, $Q^-_{22} = I_L$
$Q^-_{13} = M_T R \cdot \sin\varphi_{SL-} - \mu_-a_- \cdot \sin\alpha_-$
$Q^-_{23} = Q^-_{32} = 0$, $Q^-_{33} = M_T \cdot \cos\alpha_-$

$Q^+(x) = [3 \times 3]$, where
$Q^+_{11} = \lambda_+ a_+^2 + M_T R^2 + 2\mu_+a_+R \cdot \cos\varphi_+ - mp\{a_+ \cdot \cos\alpha_+ + R \cdot \cos\varphi_{T+}\} + I$
$Q^+_{21} = -mp\{a_+ \cdot \cos\alpha_+ + R \cdot \cos\varphi_{T+}\}$
$Q^+_{31} = M_T R \cdot \sin\varphi_{SL+}$
$Q^+_{12} = -mp\{a_+ \cdot \cos\alpha_+ + R \cdot \cos\varphi_{TL+} - p\}$
$Q^+_{22} = mp^2 + I_L$, $Q^+_{32} = Q^+_{23} = mp \cdot \sin\alpha_+$
$Q^+_{13} = M_T R \cdot \sin\varphi_{SL+} + mp \cdot \sin\alpha_+$, $Q^+_{33} = M_T$

Where, $\alpha_- = \varphi_{TL-} - \varphi_{SL-}$, $\alpha_+ = \varphi_{TL+} - \varphi_{SL+}$, the index '+' means post-transition, and the index '-' means pre-transition.

The motion of a finite-width rimless wheel in 3D

Adam C. Smith
acsmith@engc.bu.edu

Matthew D. Berkemeier
matthewb@gazelle.bu.edu

Department of Aerospace and Mechanical Engineering
Boston University, Boston, MA 02215

Abstract

We consider a new model for human (or biped robot) locomotion, consisting of a spoked rimless wheel of finite width rolling down a slope. In our model, consecutive spokes are on alternate sides of the wheel, and this models the finite leg separation in humans or robots. Full 3D motion is considered, in contrast to McGeer's 2D model. Numerical studies indicate that for a given slope, a single steady-state solution exists, and this corresponds to rolling straight down the slope at a particular average speed. Moreover, this equilibrium solution is asymptotically stable.

1 Introduction and Previous Work

Human walking is a seemingly complicated activity. Toe, ankle, knee, and hip joints, as well as numerous muscles, all participate in generating the motion. Nevertheless, several simple models have been shown to model important features of bipedal locomotion. Perhaps the simplest of these is a spoked wheel, such as an old-fashioned wagon wheel, with the rim removed [2]. This model captures many of the important features: As a rimless wheel rolls, its spokes impact the ground, just as feet impact during walking. For most of the motion, only one spoke is on the ground, just as one foot is on the ground during walking. When the next spoke touches the ground, there is a transition phase during which two spokes contact the ground simultaneously; this feature is also shared by human or biped robot walking.

McGeer [3] considered the motion of a rimless spoked wheel in two dimensions. In his work the wheel rolled down a slope of small angle γ with respect to the horizontal. This way, energy lost to spoke impacts was regained by traveling downhill. His analysis demonstrated that the rimless wheel had an equilibrium average speed which increased with slope. This speed was shown to be stable about the equilibrium. McGeer also extended his analysis to true bipeds with freely-swinging legs, which he showed are able to walk passively (i.e., without actuation) down shallow slopes in stable gaits. Analysis of limit cycles of McGeer-like bipeds was carried further by Thuilot, et.al. [5].

In previous work, we extended McGeer's analysis to a 2D model of quadrupedal walking [4]. Two rimless wheels modeled the front and rear pairs of legs, and a rigid connector modeled the body of a quadrupedal animal. Our analysis showed that the quadruped shared the speed stability that McGeer discovered for a single rimless wheel. More importantly, we found that the quadruped model walked significantly faster than a similarly configured biped model on a given slope, suggesting that quadrupedal walking may inherently be more efficient than bipedal walking.

Eventually, we wish to extend our quadruped model to include three-dimensional motions. As an intermediate step, we have begun to extend McGeer's analysis of a single rimless wheel to 3D motion. Coleman, Chatterjee, and Ruina [1] have recently performed just such an analysis with a flat or thin wheel. Our model is similar but has finite width. In our model, consecutive spokes are on alternate sides of the wheel, and this models the leg separation in humans or robots (Figure 1). Coleman, et. al. found that a thin rimless wheel rolling down a slope in 3D has a continuum of steady-state solutions, parameterized by heading direction. These solutions are asymptotically stable in the following sense: a perturbation from a steady-state solution will decay to a steady-state solution, but this solution may have a different heading than the original.

In contrast, our preliminary work indicates that a *finite-width* rimless wheel has only one steady-state motion: rolling straight down the hill. Furthermore, numerical studies indicate that this motion is asymptotically stable. We believe the existence of a single equilibrium in the return map to be an important advantage to our model. Moreover, the finite width of the wheel captures an important feature in human locomotion–the finite lateral separation of human legs.

The paper continues as follows: In Section 2, the model is described and the mathematics describing its

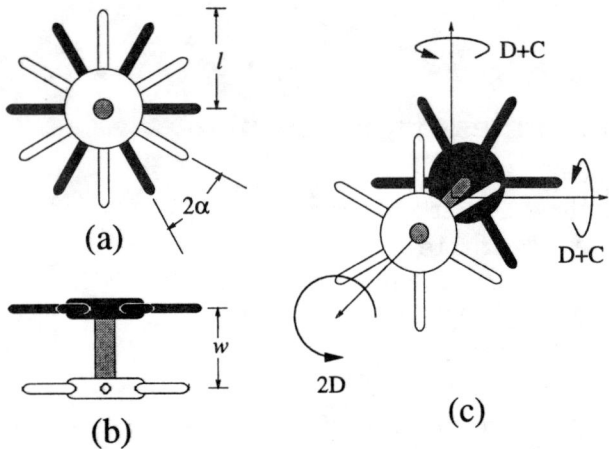

Figure 1: A rimless-wheel walker with spokes alternating between wheels. Diagrams (a) and (b) show dimensions and configuration. Diagram (c) shows inertial parameters.

motion are outlined. In Section 3, the methods used to analyze the system are explained, and some results of the analysis are given. Conclusions are presented in Section 4, and Section 5 outlines possible avenues of further research.

2 System Description

2.1 Parameters and Variables

Figure 1 shows the system investigated in this paper, referred to hereafter as the "walker". The walker can be thought of as consisting of two rimless wheels connected at their centers by a rigid shaft, making the walker a single rigid body. The rimless wheels are arranged such that their spokes are exactly out of alignment.

At any given point in time, the walker is standing on one of the spokes, known as the "stance" spoke, and is rotating about the single point of contact. This continues until the next spoke hits the ground, at which point support is instantaneously transferred to that spoke. Although it is possible for two consecutive stance spokes to be on the same wheel, this case is rare and is not considered here; in general, support spokes will alternate between the two wheels in a manner analogous to the alternating of support between two legs in a walking biped.

The physical parameters of the walker are:

- Leg length ℓ.
- Lateral wheel separation w.
- Angular spacing between legs 2α. Note that this is the spacing between two legs on opposite wheels, not between two legs on the same wheel.
- Total system mass m.
- Rolling moment of inertia about center of mass $2D$.
- Moment of inertia about axes perpendicular to the rolling axis $C + D$.

The environment provides two additional parameters:

- Ground slope γ.
- Gravitational acceleration g.

Note that by normalizing all equations with respect to m, ℓ, and g, we can reduce the system description to a set of five parameters:

$$\left\{\frac{w}{\ell}, \alpha, \frac{D}{m\ell^2}, \frac{C}{m\ell^2}, \gamma\right\}$$

Setting C and w to zero gives a wheel identical to that of Coleman, et. al. [1].

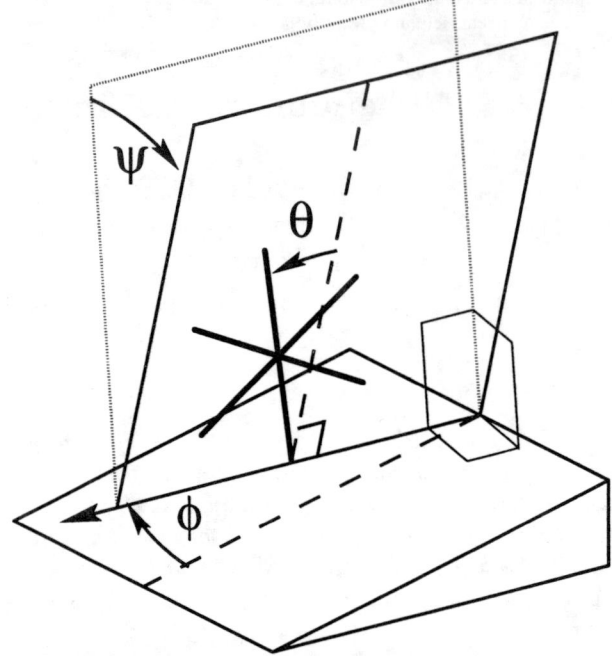

Figure 2: System state variables. For clarity, only the stance wheel is shown.

The state variables of the system, as shown in figure 2, are

- the "facing" angle ϕ, which is the heading of the walker, with $\phi = 0$ corresponding to facing straight downhill, and $\phi > 0$ indicating that the walker is facing "right" of downhill (from its point of view). This angle defines a line along the surface; the stance wheel is coplanar with this line.

- the "lean" angle ψ, which is the walker's lateral attitude with respect to the ground, with $\psi = 0$ corresponding to the case where the plane of the stance wheel is perpendicular to the ground, and $\psi > 0$ indicating that the walker is leaning to its left. (Note that ψ is measured with respect to the ground, which is tilted from the horizontal at an angle γ.)

- the "pitch" angle θ, which is the angle the stance leg makes with a line in the stance plane perpendicular to the line defined by ϕ. The case $\theta > 0$ indicates that the wheel is leaning forward, while $\theta < 0$ means that the walker is leaning backward. The angle θ increases throughout a step as the walker rolls forward. For a planar walker, such as those of McGeer or Coleman, et. al., θ will cycle from $-\alpha$ to $+\alpha$ over the course of a step; for a three-dimensional walker such as described here, start- and end-of-step values for θ will vary with ψ (subsection 2.3).

2.2 Equations of Motion

Since the system is passive, determination of the equations of motion between impacts is straightforward and can easily be accomplished by Lagrange's method. The result is a system of three second-order nonlinear differential equations in ϕ, ψ, and θ. Although easy to derive, the full equations are cumbersome, and are not presented here. They are of the form

$$\mathbf{M}(\Theta)\ddot{\Theta} + \mathbf{C}(\Theta, \dot{\Theta}) + \mathbf{N}(\Theta) = \mathbf{0}$$

where

$$\Theta = \begin{bmatrix} \psi \\ \phi \\ \theta \end{bmatrix}.$$

Thus, the system has a six-dimensional phase space with the coordinates $(\phi, \psi, \theta, \dot{\phi}, \dot{\psi}, \dot{\theta})$, and equations can be expressed as a system of six first-order differential equations, i.e.,

$$\dot{\mathbf{q}} = f(\mathbf{q}) \qquad (1)$$

where

$$\mathbf{q} = \begin{bmatrix} \phi & \psi & \theta & \dot{\phi} & \dot{\psi} & \dot{\theta} \end{bmatrix}^{\mathrm{T}}.$$

2.3 Initial and final conditions of a step

A step begins when the previous stance leg (the trailing leg) leaves the ground, and continues until the next stance leg makes contact with the ground. As previously stated, the initial and final values of θ for our walker will depend upon the lean angle ψ. Specifically,

$$\theta(0) = -\sin^{-1}\left(\frac{w \tan \psi(0)}{2l \sin \alpha}\right) - \alpha$$

$$\theta(\tau) = \sin^{-1}\left(\frac{w \tan \psi(\tau)}{2l \sin \alpha}\right) + \alpha \qquad (2)$$

where $t = 0$ at the beginning of the step, and $t = \tau$ at the end of the step.

2.4 Impact

The walker rolls under the influence of gravity, governed by the equations of motion, until the next leg hits the ground. The collision is modeled as instantaneous, and we assume the trailing leg leaves the ground in the same instant that the new stance leg makes contact. It is assumed that there is an instantaneous impulse applied at the point of impact, which causes a discontinuous change in velocity. It is also assumed that no such impulse occurs at the trailing leg. Angular momentum about the impact point will be conserved, since the impulse at the impact point is the only impulse applied during the instantaneous contact. Because the system's angular momentum about this point is a three-dimensional vector, this condition provides three equations that must be satisfied–enough to fully determine the three post-impact angular velocities. The collision can be modeled by

$$\mathbf{H}^{+} = \mathbf{H}^{-}$$

$$\mathbf{M}^{+}(\psi) \cdot \begin{bmatrix} \dot{\phi}^{+} \\ \dot{\psi}^{+} \\ \dot{\theta}^{+} \end{bmatrix} = \mathbf{M}^{-}(\psi) \cdot \begin{bmatrix} \dot{\phi}^{-} \\ \dot{\psi}^{-} \\ \dot{\theta}^{-} \end{bmatrix} \qquad (3)$$

where the superscripts "-" and "+" indicate pre- and post-impact values, respectively. Note that the pre- and post-impact mass matrices $\mathbf{M}^{-}$ and $\mathbf{M}^{+}$ are dependent on the orientation of the walker at impact. However, the facing angle ϕ is inconsequential during the impact, and the pitch angle θ at impact is a function of ψ, so the mass matrices effectively depend only on ψ.

3 Analysis

3.1 The step function

As stated above, the walker moves through a six-dimensional phase space. In order to investigate the properties of the system, we take a Poincaré section at the points where the wheel has just experienced a change of support from one spoke to the next. Since both the current and previous stance spokes must be in contact with the ground, θ is fixed as a function of ψ at these points. Thus, the state of the system at the

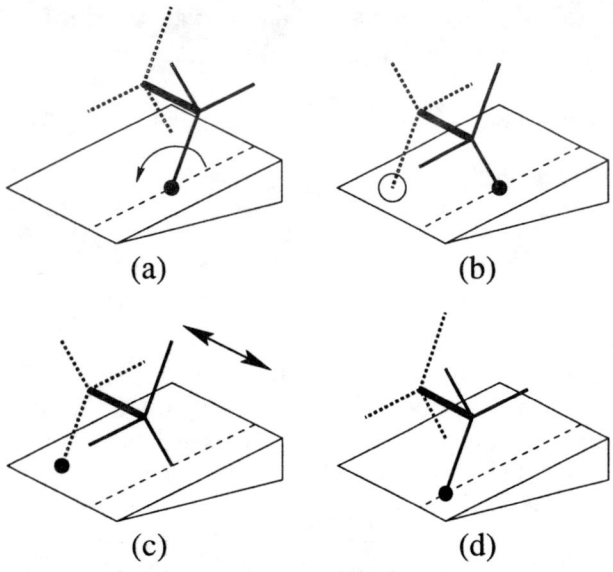

Figure 3: The evaluation of the step function for a system with six spokes ($\alpha = \pi/6$).

beginning of a step can be expressed in terms of the first five state variables. We define

$$\Phi_k = \begin{bmatrix} \dot{\phi}_0 & \dot{\psi}_0 & \dot{\theta}_0 & \phi_0 & \psi_0 \end{bmatrix}^\mathrm{T}_k$$

as the initial conditions of the system at the beginning of step k.

We define the step function S as the function that expresses the state of the system at the beginning of a step as a function of its state at the beginning of the previous step:

$$\Phi_{k+1} = S(\Phi_k) \qquad (4)$$

Figure 3 schematically illustrates the evaluation of the step function. The steps involved are as follows:

1. The equations of motion are integrated over time from the beginning of the step (figure 3a) until next stance foot contacts the ground (figure 3b).

2. The impact equations are applied to determine the angular velocities at the beginning of the next step (figure 3c).

3. A coordinate transformation is applied by which the system is reflected along the axis shown in figure 3c, allowing the same equations of motion to be used for the next step (figure 3d).

The step function is difficult to investigate qualitatively because steps k and $k+1$ will have stance legs on opposite sides of the body. In practice, it is easier to examine the function that takes the state of the system at the beginning of a step as its argument and returns the state of the system at the beginning of the *second* step following, at which point the stance leg will have returned to the same side of the body. This is simply S^2, the composition of the step function on itself.

$$\Phi_{k+2} = S^2(\Phi_k) = S(S(\Phi_k)) \qquad (5)$$

3.2 Finding Fixed points

In order to find a cyclic (equilibrium) gait of the system, we need to find a set of initial conditions for a step k that result in the walker having the same conditions at the beginning of step $k+2$. This is equivalent to finding a fixed point for S^2, an argument that returns itself:

$$\Phi^* = S^2(\Phi^*) \qquad (6)$$

Since the step function cannot be expressed in closed form, equation (6) must be solved numerically. The method used is Newton's method: If Φ_i is the ith guess at Φ^*, then

$$\Phi_{i+1} = \Phi_i - \left[\left. \frac{\mathbf{D}(S^2(\Phi) - \Phi)}{\mathbf{D}\Phi} \right|_{\Phi=\Phi_i} \right]^{-1} \cdot (S^2(\Phi_i) - \Phi_i) \qquad (7)$$

Solutions from [1] are used as starting guesses, and convergence to eight significant digits usually occurs within four or five iterations.

3.3 Equilibrium Gaits

Figure 4 shows an equilibrium gait for a typical walker found by the method described in subsection 3.2. The single-step period of this gait is 0.565. Note that ϕ and ψ oscillate around 0.

3.4 Stability

The stability of an equilibrium gait can be investigated by examining the linearization of S^2 about the fixed point; i.e., the matrix

$$\mathbf{S}^2_{\Phi^*} = \frac{\mathbf{D}S^2}{\mathbf{D}\Phi}(\Phi^*). \qquad (8)$$

If the eigenvalues of the $\mathbf{S}^2$ matrix at the fixed point all fall inside a unit circle centered at the origin, the fixed point will be asymptotically stable.

Table 1 shows the eigensystem of the $\mathbf{S}^2$ matrix for the gait of figure 4. The $\mathbf{S}^2$ matrix in this case has one real mode and two pairs of complex modes.

The real "speed" mode deals virtually exclusively with the rolling speed, $\dot{\theta}$, and represents convergence to the equilibrium speed for a given slope. This is the behavior exhibited by McGeer's 2D rolling wheel [3].

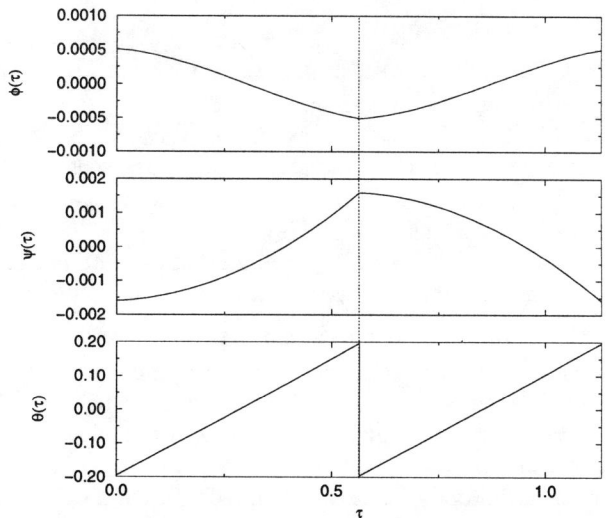

Figure 4: Equilibrium gait of a finite-width walker. The plot shows two steps; the discontinuity at the step boundary is due to the impulse applied at impact. Physical parameters used for this simulation were $w = 0.1\ell$, $\alpha = \pi/16$, $D = \frac{1}{4}m\ell^2$, $C = 0$, and $\gamma = 1/10$. Time is in units of $\sqrt{\ell/g}$.

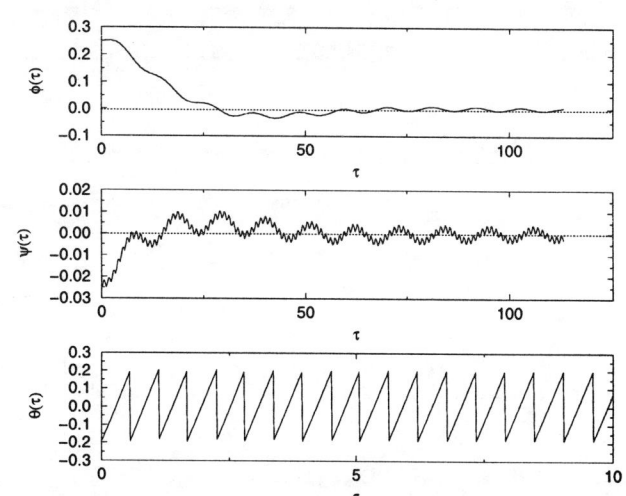

Figure 5: Results of a numerical simulation of an alternating-spoke walker with non-equilibrium initial conditions. Physical parameters used for this simulation were the same as those in figure 4. Time is in units of $\sqrt{\ell/g}$. Note that the time-axis of the lowest plot, $\theta(t)$ vs. t, differs from the other two.

Indeed, if this walker were projected into two dimensions and rolled down a 2D slope, as in McGeer's work, it would have an impact efficiency parameter of

$$\eta = \frac{\ell^2 \cos 2\alpha + 2D}{\ell^2 + 2D} = 0.9011$$

Deviations from equilibrium speed decay as η^{2n} over n steps. For two steps, $\eta^4 = 0.8120$, which agrees with the speed-mode eigenvalue of S^2 for our walker.

The two pairs of complex modes deal with the three-dimensional motion of the walker. The higher-frequency "stagger" mode has an eigenvalue of high magnitude, and thus disturbances in this mode will persist for an extended time, appearing as oscillations in ϕ and ψ with a period of $2\pi/0.650 = 9.67$ iterations of S^2, or about 19.3 steps. The lower-frequency "weave" mode is more heavily damped, and thus will only persist for a few oscillations before fading. Since all the eigenvalues have magnitude less than one, the system will be asymptotically stable at the fixed point.

Note the contrast with the planar walker of Coleman, et.al. [1], which on any given slope has a family of solutions parameterized by ϕ. Although a planar walker perturbed from a steady-state solution will asymptotically converge to a steady-state solution, the new solution may have a different heading than the old. In contrast, our studies indicate that a finite-width walker such as the one shown here has only one steady-state solution on a given slope, and, if perturbed, will return to that solution.

Figure 5 shows the trajectory of the walker over a period of 250 steps starting from a non-equilibrium state which seems to support the conclusions drawn from the analysis of the S^2 matrix. Note that the lightly-damped medium-frequency oscillations in ϕ and ψ have a period of about 11 time units, or about 19.5 steps using the equilibrium step period from the gait of figure 4. This figure agrees well with the estimate of the stagger-mode period derived above. Also, the lowest-frequency component of ϕ and ψ decays quickly and fades after only a single oscillation, agree-

Mode	Speed	Stagger		Weave	
		mag.	phase	mag.	phase
E-value	0.812	0.992	±0.650	0.940	±0.082
$\phi_0 - \phi_0^*$	-0.005	0.440	∓1.186	0.086	±0.452
$\dot\psi_0 - \dot\psi_0^*$	-0.033	0.220	±3.073	0.018	∓2.280
$\dot\theta_0 - \dot\theta_0^*$	0.999	0.006	∓1.142	0.003	±1.156
$\phi_0 - \phi_0^*$	-0.001	0.804	∓2.769	0.993	∓1.771
$\psi_0 - \psi_0^*$	0.000	0.334	±1.781	0.083	±2.410

Table 1: Step-to-step eigenvectors of the alternating-spoked rimless wheel for the gait shown in figure 4. Eigenvalues are given across the top line of the chart, with eigenvector components in columns beneath them.

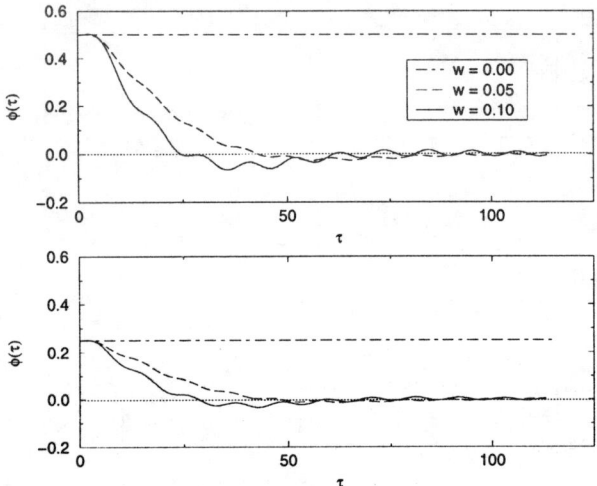

Figure 6: Comparison of convergence to equilibrium for walkers of different widths from initial heading angles of $\phi = 0.5$ (top) and $\phi = 0.25$ (bottom). Physical parameters for this simulation are the same as those of figure 4, with the exception of w. Time is in units of $\sqrt{\ell/g}$.

ing with the expected behavior of the weave mode. (The high-frequency activity visible in ψ occurs on the timescale of single steps, and is not captured in S.) The convergence of ϕ and ψ to zero supports the conclusion of asymptotic stability.

Figure 6 shows the change in walker behavior as the width w is increased. In both plots, all three walkers are given the same initial heading ϕ. The planar walker ($w = 0$) quickly finds a steady-state motion at a nonzero value of ϕ. The finite-width walkers, however, can only reach a dynamic equilibrium at $\phi = 0$, and they converge upon this value.

4 Conclusions

We have investigated a new model for bipedal walking, one that includes a finite lateral width, a significant feature not modeled in any previous rimless-wheel work. Numerical studies have shown that on any given slope, there is a steady-state solution corresponding to rolling straight down the slope. Furthermore, over a broad range of slopes these equilibria appear to be asymptotically stable. This stability may prove useful in designing biped robots, since asymptotically stable systems generally require only simple control.

5 Further Work

5.1 Analytical Approximations

Although the step function S cannot be solved in closed form, it may be useful to derive an analytical approximation. Coleman, et. al. [1] used a perturbation expansion to analyze their planar walker in three dimensions, and achieved good agreement between their analytical approximation and numerical results. A similar approach may prove fruitful here.

5.2 Extension to Quadrupedal Walking

In previous work, we have extended McGeer's two-dimensional rimless-wheel system to model quadrupedal walking [4]. This was accomplished by modeling a quadruped as a pair of bipeds connected by a rigid body, with one "biped" (a single rimless wheel) representing the front pair of legs and one representing the rear pair. A similar method could be used to model a quadruped in three dimensions.

References

[1] M. J. Coleman, A. Chatterjee, and A. Ruina. Motions of a rimless spoked wheel: a simple 3d system with impacts. *Dynamics and Stability of Systems*, (forthcoming).

[2] R. Margaria. *Biomechanics and Energetics of Muscular Exercise*. Clarendon Press, 1976.

[3] T. McGeer. Passive dynamic walking. *International Journal of Robotics Research*, 9(2):62–82, 1990.

[4] A. Smith and M. Berkemeier. Passive dynamic quadrupedal walking. In *Proceedings of the 1997 IEEE International Conference on Robotics and Automation*, 1997.

[5] B. Thuilot, A. Goswami, and B. Espiau. Bifurcation and chaos in a simple bipedal gait. In *Proceedings of the 1997 IEEE International Conference on Robotics and Automation*, 1997.

Speed, Efficiency, and Stability of Small-Slope 2-D Passive Dynamic Bipedal Walking

Mariano Garcia

Theoretical & Appl.Mech.
Cornell University
Ithaca, NY 14853

Anindya Chatterjee

Dept. Of Engr. Sci. & Mech.
Penn State University
University Park, PA 16802

Andy Ruina

Theoretical & Appl.Mech.
Cornell University
Ithaca, NY 14853

Abstract

This paper addresses some performance limits of the kneed and non-kneed passive-dynamic walking machines discovered by McGeer [10, 11]. Energetic *in*efficiency is measured by the slope γ needed to sustain gait, with $\gamma = 0$ being perfectly efficient. We show some necessary conditions on the walker mass distribution to achieve perfectly efficient walking. From our experience and study of a simpler model, only two gaits exist; the longer-step gait is stable at small enough slopes. Speed is regulated by energy dissipation. Dissipation can be dominated by a term proportional to speed^2 or a term proportional to speed^4 from normal foot collisions, depending on the gait, slope, and walker design. For special mass distributions of kneeless walkers, the long-step gaits are especially fast at small slopes. A period doubling route to chaos is numerically demonstrated for the kneed walker.

1 Introduction

This paper extends McGeer's work on passive-dynamic walking in the following ways that have not been described in previous publications: 1) near-zero slope walking is found for a class of kneeless and kneed 2-D walkers, 2) scaling laws are found for small slope walking for more than just the simplest walker, and 3) period doubling and chaos is found for kneed walking. The results may help those trying to build efficient robots, and to those trying to understand human walking. Energetic efficiency and speed maximization are obvious goals of both biological and artificial locomotion and transportation systems. Since animals and potentially-useful robot designs use legged walking motions it is interesting to consider the performance limits of such machines. In this paper, we address these questions in the context of two-dimensional

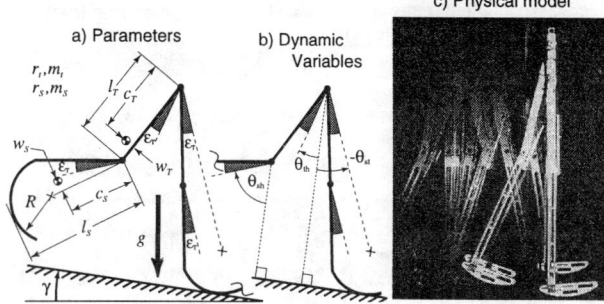

Figure 1: (a) Our description of McGeer's kneed walking model. Radii of gyration and masses of thigh and shank are denoted by r_t, m_t, r_s, and m_s, respectively. The circular-arc foot is centered at the +. ϵ_T is the angle between the stance thigh and the line connecting the hip to the foot center. A stop at each knee prevents hyperextension. (b) Dynamic variable values θ_{st}, θ_{th}, and θ_{sh} are measured from ground-normal to lines offset by ϵ_T from their respective segments. (c) A strobe photo of our physical model walking with dimensional parameters: $l_t = 0.35$m, $w_t = 0$m, $m_t = 2.345$kg, $r_t = 0.099$m, $c_t = 0.091$m, $l_s = 0.46$m, $w_s = 0.025$m, $m_s = 1.013$kg, $r_s = 0.197$m, $c_s = 0.17$m, $R = 0.2$m, $\gamma = 0.036$rad, $g = 9.81$m/s^2, $\epsilon_T = 0.097$rad.

passive-dynamic walking machines. We also describe some other newly found properties of these machines.

2 Passive Dynamic Walking Machines

Passive-dynamic walking machines that walk on shallow slopes were first designed, simulated and built by Tad McGeer [10, 11]. These machines consist of hinged rigid bodies that make collisional and rolling contact with a sloped, rigid ground surface. They are powered by gravity and have no control. The 2-D kneed walking machine we study here, essentially

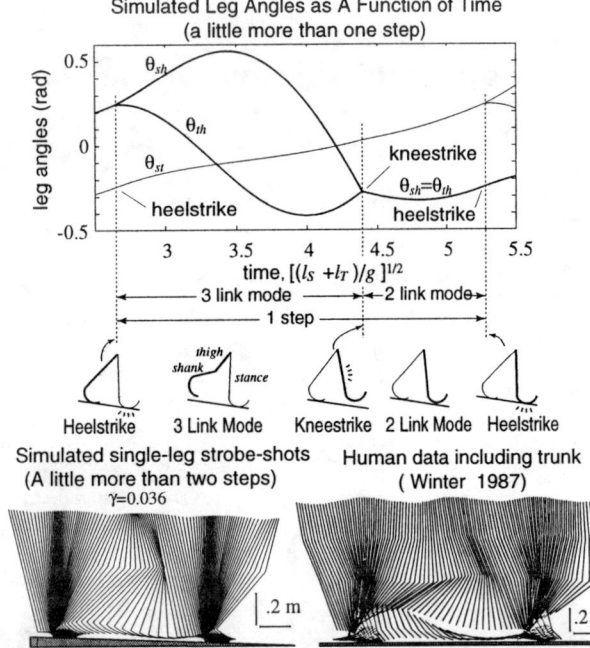

Figure 2: Simulated gait cycle (as per McGeer) of the walker in figure 1c. Angles of leg segments are shown from just before a heelstrike to just after the next heelstrike in a steady gait. The heavy line on the graph corresponds to the motion of the heavy-line leg on the small cartoon under the graph. At the start this is the stance leg, but it becomes the swing leg just after the first heelstrike, and again becomes the stance leg after the second heelstrike. The angular velocities of the joint segments have discontinuities at kneestrike and heelstrike, which appear as (barely visible) kinks in the curves. The strobe-like animation from the same simulation can be compared to measured human data (with a smaller scale and a longer stride).

a copy of McGeer's design, is shown schematically in figure 1. It consists of a *swing* leg (not in contact with the ground) and a *stance* leg (touching the ground), connected by a frictionless hinge at the hip. The non-kneed, or *kneeless* machines have the knee joint locked.

Figure 2 shows the simulated motion of such a machine.

The three features that make the McGeer-like models so intriguing for both robotics and the understanding of animal gait are these:

1. Existence of gait. A mechanism that resembles human legs in overall layout has an uncontrolled periodic motion that is rather anthropomorphic, as can be seen real models and simulations, or comparing simulated strobe data with human data (lower part of figure 2). Since passive walkers seem to be somehow close to human walkers, there is reasonable hope of learning something about human walking by studying these simpler passive models.

2. Efficiency. These machines can walk down shallow slopes. McGeer numerically found walking motions for slopes as low as about 0.005 radians and we will show here predictions of arbitrarily small slopes. Passive-dynamic based designs using other-than-gravity power schemes, e.g., toe-off, could have similarly high efficiencies. As argued clearly in, e.g.,[2], both evolutionary pressure and individual motivation push for high efficiency in animal locomotion.

3. Stability of gait. For certain parameter combinations, McGeer found *stable* limit cycle motion for both 2-D straight-legged and kneed walkers as [9] and [5] later repeated for some 2-D straight-legged walkers and we have repeated for kneed walkers (this work), and [3] found experimentally with a 3-D device. These stable motions indicate the possible role of passive-dynamics in *stabilizing* things which one might think need controlled stabilization.

It seems likely that the primary cost of locomotion is in the mechanical energy, and not the neural activity of control. It is thus natural to imagine that in evolution and learning of walking, a primary goal in perfecting walking motions would be energy-efficiency more than simplicity of control strategies. Such efficiency might even be achieved at the expense of passive stability, in contrast to item 3 above. Unstable limit cycle motions of mechanical systems can in principle be stabilized with 0+ energetic cost as has been addressed for a three-dimensional walking model by [4]. The tradeoffs, or lack thereof, between efficiency and stability for such systems are far from understood in these non-holonomic systems [12].

Method of analysis. The method used follows McGeer and is described in detail in, e.g., [5]. We find the Poincaré map for the change of the state of the walker in one step (starting just after heelstrike) by solving the Newton-Euler differential equations and collisional jump conditions numerically. Fixed points of this map are period 1 walking motions. We find both stable and unstable fixed points by numerical root finding. We evaluate the linearized stability of these gaits by numerically differentiating the Poincaré map. For general straight legged walkers the map of the fourth order system is three-dimensional (two for

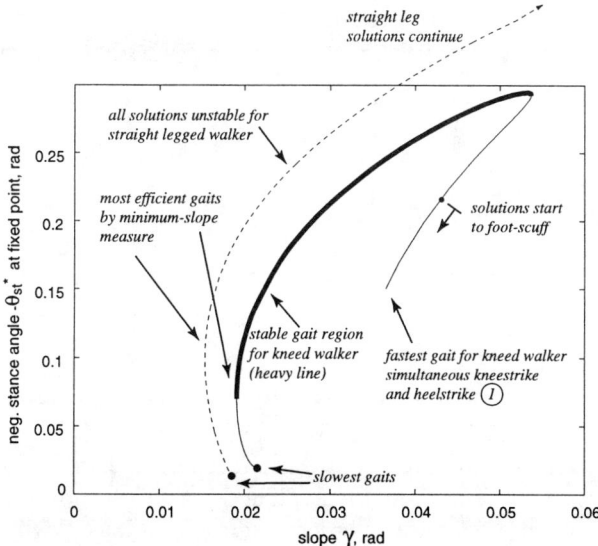

Figure 3: Calculated locus of solutions showing fixed points of stance angle as a function of slope for our physical kneed walking model (solid line) and for the same model but with the knees locked (dashed line). Each point on this graph is one periodic solution. The thick portion of the solid line denotes stable solutions for the kneed walker.

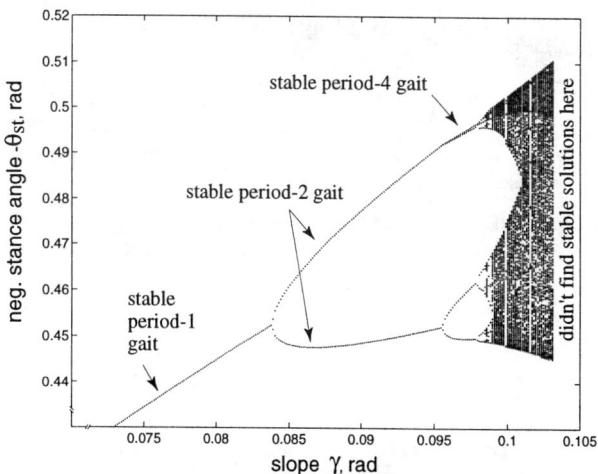

Figure 4: Period doubling route to chaos in *stable* kneed walking motions. The parameters are those of the zero-slope capable walker labeled (C) in figure 6.

some special cases). For the superficially 6th-order kneed walker the map is also only three-dimensional because the knee angle is not independent at heelstrike and because part of the swing-phase motion is only 4th order (Thus one of McGeer's numerically calculated eigenvalues near zero is actually zero). A plot of segment angles during a typical kneed gait cycle is shown in figure 2. For a straight-legged walker, the 3-link mode is absent, and $\theta_{sh} \equiv \theta_{th}$.

By assuming motions as described above, some of the periodic solutions we find might violate various physically-relevant inequality conditions [11]. We allow some of these violations since they could be circumvented by zero-energy-cost control action [6]. Allowed violations include scuffing of the swing foot (i.e., passing slightly underground when the two legs are parallel), unlocking torques at the stance knee, and hyperextension of the newly swinging leg.

3 Gaits of Generic Kneed and Straight-Leg Walkers

The number of gaits and their stability. Although the root finding involved in finding a gait cycle involves the solution of n equations in n unkowns (where n is the dimension of the return map) there is no *a priori* guarantee that any gait cycles will exist for a given passive-dynamic walking machine (i.e., a given set of masses, lengths, etc.), on a given slope γ. In practice, all searches with all designs have found either 0, 1, or 2 anthropomorphic solutions for given machine parameters and slope. Other non-anthropomorphic solutions may exist but are discounted. In these non-anthropomorphic solutions the leg swings forward and backward more than once, or the swing leg makes full revolutions.

Effect of varying slope. Figure 3 shows stance angle θ_{st} at the start of a gait cycle while slope is varied, for a given mass distribution, with and without knees. For both kneed and kneeless walkers there are slope γ regimes where there are either 0,1, or 2 solutions. Along the kneed curve, kneestrike occurs later and later in the step, until at one end of the locus of solutions (point 1), heelstrike and kneestrike occur simultaneously. At the other end of the locus of solutions the walker has just enough initial kinetic energy for the stance leg to make it past the vertical position. This is the slowest gait for these walkers. Neither of these walkers can walk at arbitrarily small slopes.

4 Chaos In A Kneed Walker

Like the simplest walker of [5], and the walkers of [13] and [9], kneed walkers can also exhibit period doubling and chaotic gait, as shown in figure 4.

5 Measures Of Performance

Since moving sideways in a gravitational field is workless (neglecting air friction), a rational dimensionless measure of work efficiency is somewhat problematic for locomotion on level ground. A natural measure of inefficiency, however, is the *specific cost of transport* η, (energy used)/(weight × distance travelled). It, as well as a few other reasonable measures of transport cost, reduces to the slope γ for small-slope passive-dynamic walking [6].

6 Walking At Near-Zero Slopes

Passive-dynamic walking at near-zero slopes has previously been demonstrated for the simplest walker [5]. Here we seek more general 2-D kneed and kneeless designs capable of zero-slope walking. Mathematical justifications for some of the arguments here can be found in [6].

Necessary Conditions on Mass-Distribution For Near-Zero-Slope Walking. Necessary conditions on the mass distribution for near-zero slope walkers are found as follows:

1. If walking motions do occur at very small slopes, these motions will be very slow [6]. The walker must be close to static equilibrium at all times. In the limit of zero slope, the walker configuration must approach a static equilibrium configuration. Thus the foot contact point must be where the foot-normal is directed towards the body center of mass.

2. At heelstrike both legs are straight and simultaneously touch the ground. As the slope (hence, step length) goes to zero, the spacing between the legs at this instant also goes to zero. In the limiting case, the foot contact point is seen to be that point on the foot which is farthest from the hip. Thus the normal to the foot contact point must pass through the hip.

3. From (1) and (2) the line from the hip through the body center of mass must intersect the foot curve normally at the nominal contact point at zero-slope walking. For circular feet this is equivalent to the collinearity of the center of mass of the whole body, the hip, and the foot center (see figure 5).

4. For the swing leg to be in static equilibrium in 3-link mode and to have zero knee-locking torque,

Conditions for Gait Solutions at Arbitrarily Small Slopes

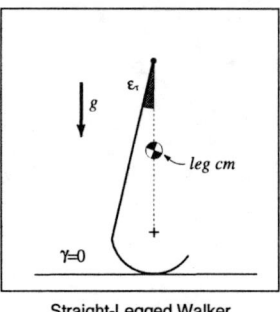
Straight-Legged Walker

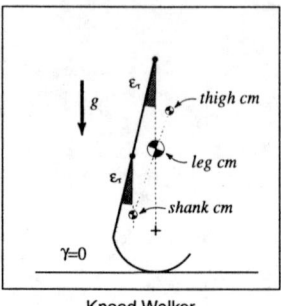
Kneed Walker

Figure 5: **Necessary conditions for near-zero slope walking.** The separations between the hips and the mass centers shown here are larger than those typical for our simulations. The necessary condition for elimination of first-order scuffing from knee flex is not shown.

the center of mass of the shank must lie directly under the knee, in the straight-leg configuration (see figure 5).

These necessary conditions on the mass distribution do not guarantee that near-zero slope walking solutions exist. In simulations we have found these conditions lead to zero-slope walking designs if the body center of mass is close to the hip. These conditions are reminiscent of the "straight" joints suggested by Alexander [1] as a means of achieving efficiency.

The Simplest Walking Model. Alexander's *minimal biped* [1] has a point-mass at the hip but no mass in the legs. Hence the minimal biped can be (and needs to be) supplemented with further stride length conditions. To make the minimal biped leg swing most simply deterministic, the simplest walker adds vanishing point-mass feet. The simplest walker is the minimal biped with miniscule feet. The simplest walker is also the limiting case of the 2-D McGeer straight leg walkers with no knees, a finite point-mass at the hip, and a vanishing point-mass at the point feet. The results presented here are generalizations of the results found for the simplest walker by Garcia *et al.* in [5].

The simplicity of the "simplest" walker is that: the non-dimensional form of the simplest walker has as its only free parameter the slope γ; the stance leg dynamics are decoupled from the swing leg motions; the Poincaré map is only two-dimensional and; the motion was only studied in a regime where most of the governing equations were linear.

By a mixture of analytic and numerical means the simplest walker was found to have two gaits at all small-enough slopes. Of these, the long-step gait is stable at small slopes ($\gamma < 0.015$), while the short-step gait is unstable at all slopes. The simplest walker was found to have near-zero slope walking with speed proportional to the cube root of the slope. It was found to have a period doubling route to chaos. This period doubling and chaos were discovered independently and reproduced (respectively) by [7, 9] in studies of a less extreme point foot model. For both gait cycles of the simplest walker the stance angle θ_{st} or step-length is proportional to $\gamma^{1/3}$ at small slopes, while the step periods tend to (different) nonzero constants. This implies a walking power consumption proportional to the fourth power of speed for small speeds, for *both* gaits. That is, power $\propto$ (speed)^4. This power scaling can be derived from Alexander's [2, 1] minimal biped results by assuming speed is proportional to step length (as it is for both small-slope gaits of the simplest walker).

An interesting feature of the long-period gait is that, in the limit $\gamma \to 0$, has a time-reversal symmetry. The configuration with both legs vertical is passed. Defining $t = 0$ for this configuration, $\theta(-t) = -\theta(t)$ for both legs. Equivalently, a movie of this walker shown backwards looks like a movie of the walker walking forwards in the opposite direction. A consequence of this symmetry is that the long period gait has no component of foot velocity tangent to the surface at heelstrike.

This symmetry is approximately observed even for the somewhat generic kneed walker of figure 2 (after averaging the shank and thigh motions, the shape of the curves is nearly preserved by 180 degree rotation of the graph).

Not quite the simplest walker. For the simplest walker, with negligible mass, the only kinetic energy lost is that of the hip. When the feet have finite mass, however, they also lose energy at heelstrike. If the striking foot hits the ground with no tangential velocity, its loss still scales as step length to the fourth power. If, however, the foot collision has a grazing component, then the energy lost scales as the step-length squared.

The point-foot walker still has two solutions at small slopes, even with non-negligible foot mass. To first order in the slope, the long-step solution has time-reversal symmetry, no tangential foot collisions, and step length proportional to the cube root of the slope. The short-step solution has some tangential component in foot collisions. Because this involves a finite mass colliding at a speed proportional to step length, the $\gamma \to 0$ motion has step length proportional to slope. Thus for the long-step gait power $\propto$ (speed)^4.

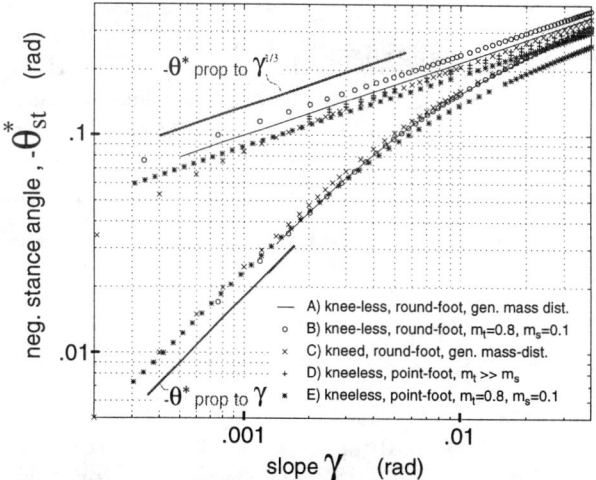

Figure 6: Low slope step-lengths of some zero-slope capable walkers vs slope γ. There are two gaits cycles at each slope, for each walker considered. Both step lengths for the simplest walker are proportional to $\gamma^{1/3}$ for small γ. The short-step gaits of the other walkers have step lengths proportional to γ for small γ. The long-step gaits for the other walkers have step lengths that are much longer than for the short-step gaits, approximately approaching $\gamma^{1/3}$ for larger γ. For a point-foot, kneeless walker with non-negligible foot mass, the step length of the long-step gait proportional to $\gamma^{1/3}$ for small γ.

For the short-step gait, power $\propto$ (speed)^2, at least at small speeds.

Scaling rule for more general straight-legged walkers. Near zero-slope the time period of steps for zero-slope capable walkers is asymptotically constant as $\gamma \to 0$. At steady walking the energy lost in collisions is balanced by the gravitational potential energy. The collision loss per step is proportional to the speed of the colliding foot contact point squared. The gravitational energy available is proportional to the product of step length and slope.

Taking account both the effects of the mass distribution on collisions and the kinematics of these walkers it is found that the dissipation per step is dominated by either $A\theta^{*2}$ tangential foot collision term or a $B\theta^{*4}$ normal foot collision term. This latter term is analogous to the point-mass collisional loss term for the point-mass-at-hub rimless wheel (or minimal biped) model of Alexander, but is derived for more general mass distributions of this 2 link mechanism [6]. Equating this loss with the gravitational energy available gives

$$A\theta^{*2} = \gamma\theta^* \quad \text{or} \quad B\theta^{*4} = \gamma\theta^*. \tag{1}$$

For zero-slope capable straight leg walkers, there is *no* tangential component of foot velocity at heelstrike for the long-period time-symmetric gait. Such walking motions follow the scaling rules found for the simplest walker [5]. Step length goes up with the cube root of slope. Power for walking increases with the fourth power of speed.

If the tangential foot velocity term is non-zero, the walkers have step length proportional to slope and the power for walking scales with the speed squared, at least as $\gamma \to 0$

Extension of scaling rule to kneed walkers. Kneed walkers dissipate kinetic energy in collisions at both heelstrike and at kneestrike. For heelstrike, the energy loss calculations described above still hold: the pre-collision velocities are determined from the straight-leg or 2 link configuration. The knee collision loss, for zero-slope walkers, scales with speed^2 and thus dominates at small enough slopes. However, the collisional loss of knee-strike is very small. In figure 6 it is only at the far left of the graph that the dominance of the knee-strike losses show as a switch from step length proportional to $\gamma^{1/3}$ to step length proportional to γ.

7 Conclusions

We have investigated the design of straight-legged and kneed passive-dynamic walkers that will walk at arbitrarily small slopes. At high speeds the power required for walking scales as v^4. At low speeds the scaling can be either with v^2 or v^4. On the one hand this shows how bad this kind of walking is when it is fast. On the other hand the v^4 scaling rule implies very small energy demand at low speeds. The essence of the scaling follows from the energy balance between collision losses and gravitational potential energy [5], rather like for the rimless wheel described in [10, 2, 1]. Perhaps one reason people in fact walk with higher frequencies at higher speeds, rather than to use a strictly pendulum-like fixed-period swinging motion, is to defeat the collisional losses which depend so strongly on step length.

References

[1] R. McN. Alexander. Energy-saving mechanisms in walking and running. *Journal of Experimental Biology*, 160:55–69, 1991.

[2] R. McN. Alexander. Simple models of human motion. *Applied Mechanics Review*, 48:461–469, 1995.

[3] Michael Coleman and Andy Ruina. An uncontrolled toy that can walk but cannot stand still. submitted to *Physical Review Letters*, 1997.

[4] J. Victoria Fowble and Arthur D. Kuo. Stability and control of passive locomotion in 3D. *Proceedings of the Conference on Biomechanics and Neural Control of Movement*, pages 28–29, 1996.

[5] Mariano Garcia, Anindya Chatterjee, Andy Ruina, and Michael Coleman. The simplest walking model: Stability, complexity, and scaling. Accepted for publication in the ASME Journal of Biomechanical Engineering, 1997.

[6] Mariano Garcia, Andy Ruina, and Anindya Chatterjee. Efficiency and stability of 2-d passive-dynamic kneed walking. An expanded version of this ICRA98 paper including rationalization of comments cited in this text. In preparation for ?, 1997.

[7] Ambarish Goswami, Bernard Espiau, and Ahmed Keramane. Limit cycles and their stability in a passive bipedal gait. *International Conference on Robotics and Automation*, 1996.

[8] Ambarish Goswami, Bernard Espiau, and Ahmed Keramane. Limit cycles in a passive compass gait biped and passivity-mimicking control laws. *Journal of Autonomous Robots*, 1997. In Press.

[9] Ambarish Goswami, Benoit Thuilot, and Bernard Espiau. Compass-like bipedal robot part I: Stability and bifurcation of passive gaits. *INRIA Research Report No. 2996*, 1996.

[10] Tad McGeer. Passive dynamic walking. *International Journal of Robotics Research*, 9:62–82, 1990.

[11] Tad McGeer. Passive walking with knees. *Proceedings of the IEEE Conference on Robotics and Automation*, 2:1640–1645, 1990.

[12] Andy Ruina. Non-holonomic stability sspects of piecewise holonomic systems. submitted to *Reports on Mathematical Physics*, 1997.

[13] Benoit Thuilot, Ambarish Goswami, and Bernard Espiau. Bifurcation and chaos in a simple passive bipedal gait. *IEEE International Conference on Robotics and Automation*, 1997.

# A Design Method of Neural Oscillatory Networks for Generation of Humanoid Biped Walking Patterns

Meifen Cao   Atsuo Kawamura

Department of Electrical and Computer Engineering, Yokohama National University, Japan
79-5 Tokiwadai, Hodogaya-ku, Yokohama 240, JAPAN

Abstract

The humanoid biped walking pattern attracts more and more attentions in both robot and rehabilitation fields. To express various human walking patterns in mathematical models is still under investigation. In this paper we orient towards analyzing the mechanisms of human walking and propose two design methods of neural oscillatory networks for generating various desired biped walking patterns of an 8 joints model in 3D plane. First, we discuss how the construction and the parameters of the network determine the inner state variable of each neuron in the network. Then we try to use two different methods to build up the suitable network which can generate the desired walking patterns. The one is to lower the system dimension to $3 \times n$ at first with considering some mechanisms in human walking, then to solve the connection weights of the network by the harmonic balance method. The other is to calculate all kinds of connections of the network by GA. With the results of simulations the methods proved to be effective.

1 Introduction

It is a dream for robot researchers to make biped robots walk like human beings. In a general walking control system, the first problem is the generation of the desired walking pattern which will be used as references of systems. On the other hand, a lot of different walking patterns have been analyzed by researchers of prosthetics. The generation of humanoid walking patterns indicates the potential applications both in robot and rehabilitation field.

It can be considered that walking patterns are mostly determined by joint angle trajectories. With the measured joint angle trajectories of various human walking, it was found that all of them are periodically oscillatory waves, and the frequency, amplitude and phase relationship between different joints make the different walking patterns.

On the other side, it is clarified by experiments of physiology that most of autonomic oscillatory activities of living organisms, such as, walking, swimming and breathing, are generated by rhythmic activities of the corresponding neural systems. To explain the mechanisms through which oscillations take place, some models have been proposed[1]– [4]. In these models, it is known that individual neural oscillators are coupled in such a way that they inhibit or stimulate each other, then the network makes an oscillation with a phase relation between neural oscillators. In such a case, the mutual coupled neural oscillators network is called neural oscillatory network.

Until now, various researches have been done in this field. For example, the symmetries of animal gaits were compared with the symmetry-breaking oscillation patterns[5]. The strategies in rhythm control were discussed. Using such mechanism, a model of the pattern change can be built in a quadruped locomotion [2]. Swimming patterns of aquatic animals and gait patterns of quadruped were generated by a cooperated oscillator[4]. It was also reported that a 4-neuron-oscillator-network was successfully used in the walking pattern generation of a 4-joint biped robot in the sagittal plane[6]. But it is not yet clarified how the neural oscillatory networks can be used to generate joint angle trajectories for biped walking with more joints. Also there is not yet a systematic design method of generating the desired biped walking patterns with neural oscillatory networks.

Two design methods were proposed in this paper. In the first one, we tried to investigate and analyze the mechanism of the human walking and constructed a network in which the way of stimulations between joints is considered most closely to that in human walking. The second one is a method of solving network's suitable connections for desired walking patterns by GA.

2 The Model of Neural Oscillatory Network

Although various mathematical models have been proposed to demonstrate neural rhythms, the essential common characteristics in every model are network properties and feedback(negative and positive)

properties. In this research, we adopt the following expression as a model of the network which has the mutual inhibitions between n neurons[2]. The dynamics of each neuron is expressed by equation(1) ~ (4). The a_{ij} represent the connections between neurons. When the network has proper connections(a_{ij}) and parameters(T_r, T_a, b, s_i), each x_i will oscillate with the same frequency and adequate phase shift. In the case of walking, it is assumed that a single x_i is dedicated to the trajectory for a specific joint and the mutual-coupled neural oscillators network is used for the whole joints (shown in Figure 1)

$$T_r \frac{dx_i}{dt} + x_i = -\sum_{j=1, j\neq i}^{n} a_{ij} y_j - b f_i + s_i \quad (1)$$

$$T_a \frac{df_i}{dt} + f_i = y_i \quad (2)$$

$$y_i = h(x_i) x_i \quad (3)$$

$$h(x_i) = \begin{cases} 1 & x_i \geq 0 \\ 0 & x_i < 0 \end{cases} \quad (4)$$

$(i = 1, 2, \cdots, n)$

x_i: the inner state variable of i-th neuron
y_i: the output variable of i-th neuron
f_i: the variable of fatigue strength of i-th neuron
b: the fatigue coefficient
s_i: the time-invariant input of i-th neuron
a_{ij}: the connection weight from j-th neuron to i-th neuron(inhibitory when $a_{ij} > 0$)
T_r, T_a: time constant
n: the number of neurons

The solution $x_i(t)$ is assumed as joint angle trajectories under the conditions below:

- only one leg supporting phase is assumed
- switching supporting leg when the step width is at largest in each step
- forward moving
- man-like walking

3 Network Design by Hamonic Balance Method

Consider the case of 8 joints biped walking model shown in Figure 2(a). The x_1, x_4 are hip joints, the x_5, x_6 are knee joints and the x_7, x_8 are arm joints in the sagittal plane. The x_2, x_3 are hip joints in lateral plane.

3.1 Structure of the Network

Because of 8 joints, a network with 8 neurons is needed, that is, there are at most 56 connection weights a_{ij} ($n \times n - n = 56$, $n = 8$). First, the parameters T_a, T_r, b and s_i are supposed to have given values(shown in Table.1(1)). Then, it is necessary but difficult to determine the suitable connections (a_{ij}) of the network. With the harmonic balance method[7] it is known that $3 \times n$ a_{ij} can be solved with given walking pattern parameters (including step length S,

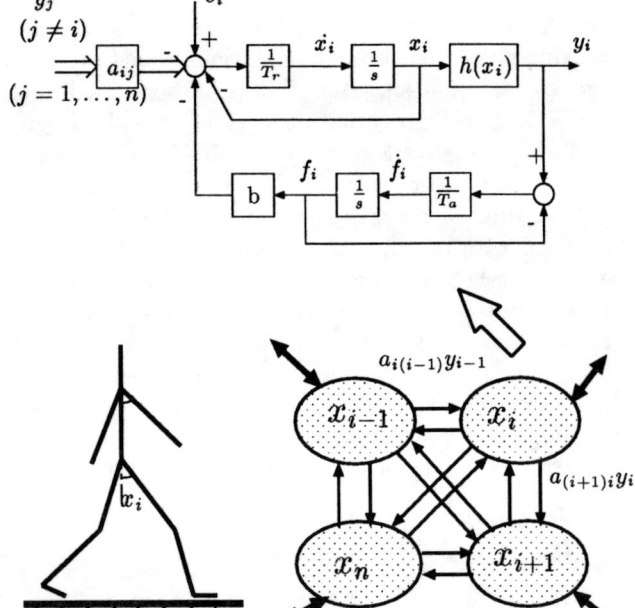

Figure 1: Neural oscillatory network used for biped walking

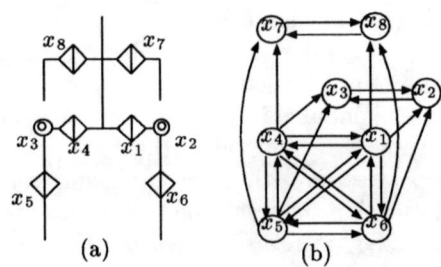

Figure 2: The structure of 8 joints and the network applied for it

step lift H, maximum slanting angles x_{2max}, x_{3max}, swing width of arms as_{2max}, as_{3max}[7]). Thus, networks with 24 connections in the structure are considered. With considering the mechanism in human walking the network shown in Figure 2(b) was built up, because it is known that in human walking the joint 8 is mostly stimulated by joint 1 and 6, while the joint 7 is mostly stimulated by joint 4 and 5. The symmetries between the right and the left side were considered at the same time. The Figure 3 is the oscillatory solutions using the above model with the given walking parameters shown in Table.1(2). The Figure 4 and Figure 5 are stick figures of the walking pattern in X-Y and 3D plane in use of the oscillatory solutions shown in Figure 3 as joint trajectories. Actually, the network shown in Figure 2(b) can generate various desired walking patterns of 8 joints.

3.2 Effects of T_a, T_r, b and s_i

From the above result, it is known that if the S, H, x_{2max}, x_{3max} and as_{2max}, as_{3max} are given, with the network in Figure 2(b) the connection weights a_{ij} can be calculated. Then the desired walking pattern can be generated with the oscillatory solutions obtained from solving equation(1) $\sim$ (4). So it can be concluded that the walking pattern is mostly decided by connection weights a_{ij}.

Here the effects of T_a, T_r, b and s_i will be discussed. In equation(1) f is the variable that represents the degree of fatigue or adaptation in the neuron, b is the fatigue weight. The b must be large enough to evoke the rhythm[2] because it is proportional to the rhythm frequency, but the large b will decline the amplitude too much. The effect on the walking pattern is that

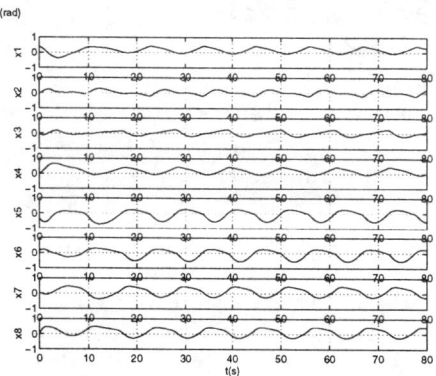

Figure 3: The oscillatory solution when the parameters as in Table.1

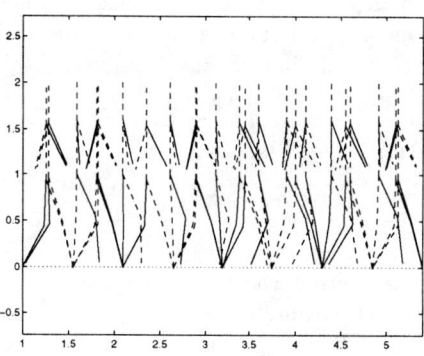

Figure 4: The stick graph of the walking pattern in X-Y plane

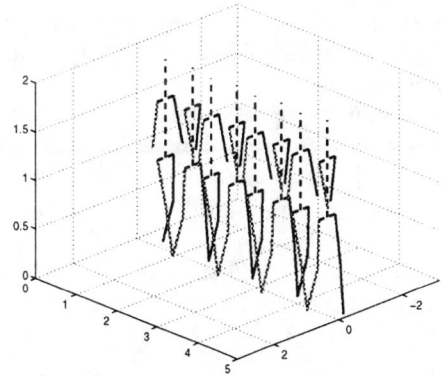

Figure 5: The walking posture in 3D plane

Table 1: Parameters in the simulations
(1). The given parameters in model

T_a	T_r	b	s_1, s_4	s_2, s_3	s_5, s_6	s_7, s_8
1	12	2.5	1	0.2	0.3	0.7

(2). The given parameters about walking pattern

$S(cm)$	$H(cm)$	$x_{2max}, x_{3max}(rad)$	as_{2max}, as_{3max}
60	5	$\frac{\pi}{12}$	$-(\frac{S}{2}-15), (\frac{S}{2}-5)$

(3). The calculated a_{ij}

i	a_{i1}	a_{i2}	a_{i3}	a_{i4}	a_{i5}	a_{i6}	a_{i7}	a_{i8}
1	0	0	0	1.42	0.13	0.76	0	0
2	0.89	0	0.04	0	0	-2.14	0	0
3	0	-0.43	0	-0.64	1.77	0	0	0
4	1.42	0	0	0	0.76	0.13	0	0
5	0.39	0	0	-0.45	0	2.34	0	0
6	-0.45	0	0	0.39	2.34	0	0	0
7	0	0	0	-0.74	0.35	0	0	1.19
8	-0.74	0	0	0	0	0.35	1.19	0

the step length will be shortened and step frequency will be enhanced.

The Figure 6 is the walking pattern in X-Y plane when the b is 3 times increased($b = 7.5$) and other parameters are the same as in Table.1. It was found that in the same time interval, the step number was increased but the step length was decreased too much.

A faster walking can be generated by decrease T_r or T_a. Because the time constants T_r or T_a can enhance the rhythm frequency but remains the pattern unchanged. The Figure 7 is the walking pattern in X-Y plane when the T_r is half of before($T_r = 0.5$) and other parameters are same as in Table.1. It was found that the walking pattern remained unchanged but the step frequency was enhanced.

The increase or decrease of input s_i induces no change in frequency but will increase or decrease the amplitude. The Figure 8 is the walking pattern in X-Y plane when all of s_i are two times enhanced and other parameters are the same as in Table.1. In this case, it was found that the step frequency remained unchanged but the step length was enhanced.

4 Network Design by GA

In section 3 we proposed a design method of constructing the network by lowering the system's dimension to $3 \times n$ with considering the mechanisms in human walking. From the above discussion, we know that the model parameters T_a, T_r, b and s_i decide the amplitude and frequency of the oscillations and the connection weights a_{ij} decide the pattern of the oscillations. In this section, we will propose a new design method with considering all possible connections by GA.

4.1 Description of the Problem

Let us consider the same case as in section 3. Here the parameters T_a, T_r, b and s_i are supposed to have given values as in section 3. There are totally 56 kinds of a_{ij} to be determined.

The most difficult problem here is to write the desired walking pattern into an objective function. we will focus on how to design the objective function or how to evaluate the properties of the network. In this paper we designed two fitness functions and some criterion rules according to the characteristics of the biped walking pattern which can be concluded as below:

- Periodical movement.
- Bilaterally symmetric.
- The phase differences between right and left joints are π. (but the phase difference between x_2 and

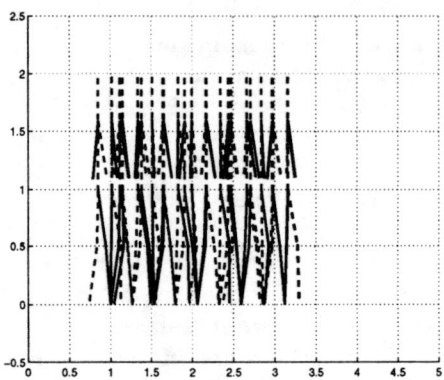

Figure 6: The walking pattern in X-Y plane when b=7.5 and other parameters are same as in Table.1

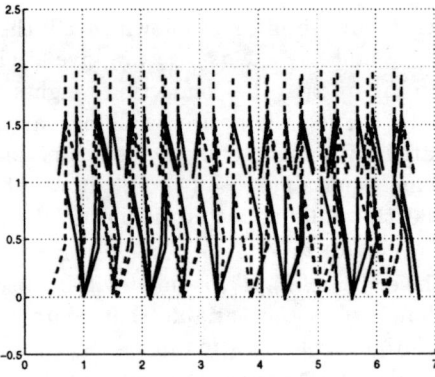

Figure 7: The walking pattern in X-Y plane when $T_r = 0.5$ and other parameters are same as in Table.1

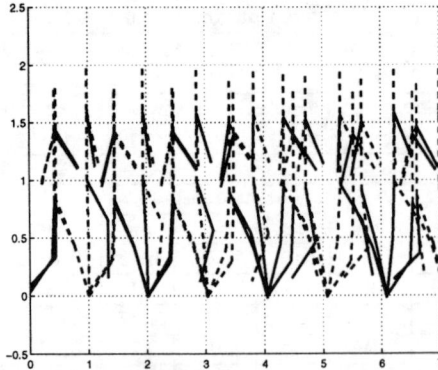

Figure 8: The walking pattern in X-Y plane when all of s_i were 2 times increased and other parameters are same as in Table.1

x_3 is supposed to near zero.) The phase differences between next joints determine the walking pattern.

- The knee joints bend forward.

- The walking pattern parameters, (such as the step length S, the step lift H, the maximum slanting angles x_{2max}, x_{3max} and the swing width of arms as_{2max}, as_{3max}[7]) decide the walking posture.

4.2 Design of the Objective Function

Let $y_i(k)$ is the FFT spectrum of $x_i(t)$. $yc_i(k)$ is the cosine component, $ys_i(k)$ is the sine component. ($k = 0, \cdots, m;\ i = 1, \cdots, n;\ m: data\ length;\ n: joint\ number$).

$$fitnessI = K1 \cdot \sum_{i=1}^{n} fit1_i$$
$$+ K2 \cdot \sum_{i=1}^{n}(C1_{imax} - \frac{fit2_i}{fit1_i})$$
$$+ K3 \cdot \sum_{i=1}^{n}(C2_{imax} - fit3_i) \quad (5)$$

$$fit1_i = \sqrt{yc_i^2(k_{imax}) + ys_i^2(k_{imax})} \quad (6)$$

$$fit2_i = \sqrt{yc_i^2(k_{imax}-1) + ys_i^2(k_{imax}-1)}$$
$$+ \sqrt{yc_i^2(k_{imax}+1) + ys_i^2(k_{imax}+1)} \quad (7)$$

$$fit3_i = k_{imax} - \overline{k_{max}} \quad (8)$$

In the above equations k_{imax} is the fundamental frequency of each $x_i(t)$, $\overline{k_{max}}$ is the average fundamental frequency of $x_i(t), (i = 1, \cdots, n)$. (6) means the amplitude of fundamental frequency component. (7) means that a kind of the amplitude of the side band of the main frequency in (6). The inner state variable solution $x_i(t)$ of each neuron in network will periodically oscillate with the same frequency when $fitnessI > thresholdI$.

$$fitnessII = |yc_1(k_{1max}) + yc_4(k_{4max})|$$
$$+ |ys_1(k_{1max}) + ys_4(k_{4max})|$$
$$+ |yc_6(k_{6max}) + yc_5(k_{5max})|$$
$$+ |ys_6(k_{6max}) + ys_5(k_{5max})|$$
$$+ |yc_7(k_{7max}) + yc_8(k_{8max})|$$
$$+ |ys_7(k_{7max}) + ys_8(k_{8max})|$$
$$+ |yc_2(k_{2max}) - yc_3(k_{3max})|$$
$$+ |ys_2(k_{2max}) - ys_3(k_{3max})|$$
$$(9)$$

When the $fitnessII$ is minimized the phase differences between joint 1 and 4, joint 6 and 5, joint 7 and 8 approach to π, and the phase difference between joint 2 and 3 approaches to $zero$.

The criterion for the knee joints must bend forward can be written as the following $ruleI$,

$$x_1(t) > x_6(t),\ x_4(t) > x_5(t)\quad for\ all\ t \quad (10)$$

The criterion about the walking pattern can be written as the following $ruleII, ruleIII$ and so on,

$$S = \max\{\sin x_6(k) + \sin x_1(k) - \sin x_4(k) - \sin x_5(k)\} \quad (11)$$

$$H = \max\{\cos x_6(k) + \cos x_1(k) - \cos x_4(k) - \cos x_5(k)\} \quad (12)$$

......

k is the sampling point.

4.3 The Simulation

- suppose the following finite binary string as one of chromosome:

$$< \underbrace{b_0 b_1 \cdots b_7}_{a_{12}} \underbrace{b_8 \cdots b_{15}}_{a_{13}} \cdots \cdots \underbrace{b_{440} \cdots b_{447}}_{a_{87}} >$$

- use two-points crossover with the probability of crossover $p_c = 0.99$.

- the probability of mutation $p_m = 0.02$.

- use roulette selection for reproduction.

The simulation flow chart is shown in Figure 9. The Figure 10 is the evolution process figure. When $population\ size = 30, K1 = 2, K2 = 1, K3 = 0.5, thresholdI = 35, S = 60cm, H = 5cm$. The $3th$ individual of the $74th$ generation has the adequate connection weights a_{ij}. The walking pattern generated by the network with such a_{ij} connections is shown in Figure 11 and Figure 12.

5 Conclusion

In this paper it has been discussed that when the neuron model is used for generation of the joint trajectories of desired walking patterns, the structure of the oscillatory network determines the patterns mostly and the model parameters T_a, T_r, b and s_i will change the frequency and amplitude of the oscillation. Thus if a desired walking pattern is wanted to be generated, first a proper network must be built up(connection weights a_{ij} must be calculated), then the oscillatory solutions can be obtained from equation(1) ∼ (4), and then the walking pattern can be generated by using the oscillatory solutions as joint angle trajectories.

When a faster walking pattern is requested, the time constants T_a or T_r must be decreased. When increases the input s_i, a larger step length walking pattern will be generated.

In this paper we proposed two methods for building up the suitable network. The one is to lower the system dimension to $3 \times n$ at first with considering some mechanisms in human walking and then to solve the suitable connection weights a_{ij} with the harmonic balance method. The other is to calculate all kinds of connections in the network by GA. With the results of simulations the methods proved to be effective.

References

[1] U.Bassler, "On the definition of central pattern generator and its sensory control," *Biol. Cybern.*, vol. 54, pp. 65–69, 1986.

[2] K. Matsuoka, "Mechanisms of frequency and pattern control in the neural rhythm generators," *Biol. Cybern.*, vol. 56, pp. 345–353, 1987.

[3] J. S. Bay and H. Hemami, "Modeling of a neural pattern generator with coupled nonlinear oscillators," *IEEE Trans. on Bio. Engineering*, vol. BME-34, no. 4, pp. 297–306, 1987.

[4] H. Yuasa and M. Ito, "Coordination of many oscillators and generation of locomotory patterns," *Biol. Cybern.*, vol. 63, pp. 177–184, 1990.

[5] J. J. Collins and I. N. Stewart, "Coupled nonlinear oscillators and the symmetries of animal gaits," *J. Nonlinear Sci.*, vol. 3, pp. 349–392, 1993.

[6] O. Katayama, Y. Kurematsu, and S. Kitamura, "Theoretical studies on neuro oscillator for application of biped locomotion," in *IEEE int. conf. on Robotics and Automation*, pp. 2871–2876, 1995.

[7] M. Cao and A. Kawamura, "Generation of humanoid biped walking pattern using neural oscillatory network," in *AIM'97*, p. 81 and CD ROM, 1997.

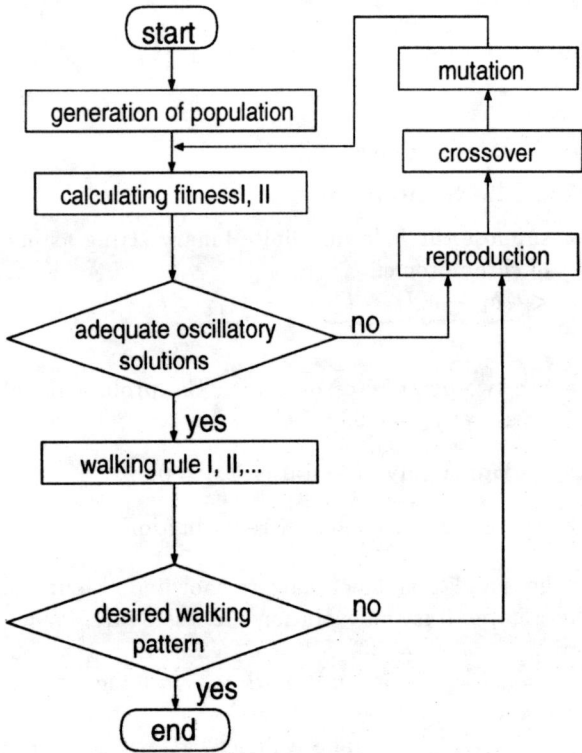

Figure 9: The flow of the simulation

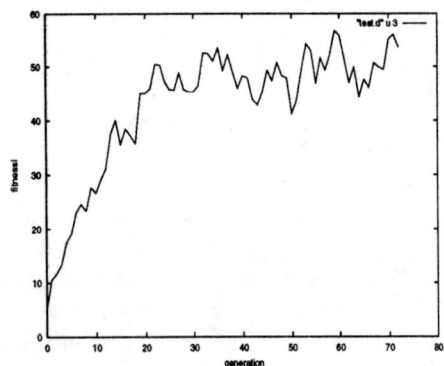

Figure 10: The result of evolution when population size = 30

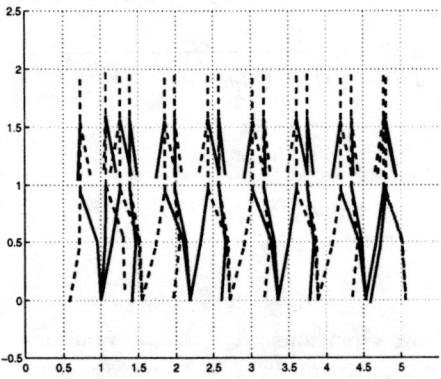

Figure 11: The stick graph of the walking pattern in X-Y plane when use the a_{ij} obtained from the evolutionary calculation

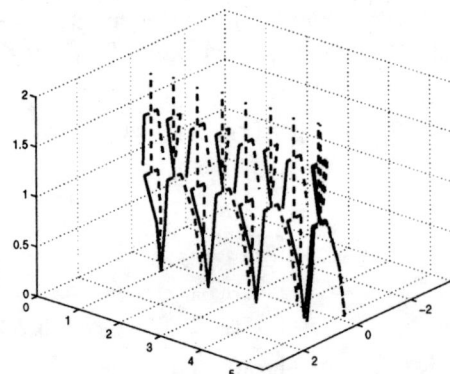

Figure 12: The walking posture in 3D plane when use the a_{ij} obtained from the evolutionary calculation

A Model for the Organization Level of Intelligent Machines

M.N.Varvatsoulakis
N.Technical University of Athens
Electrical Engineering Dept.
Zographou 15773-Athens-GREECE

G.Saridis
ECSE Department
Rensselaer Polytechnic Institute
Troy, NY 12180-3590

P.N.Paraskevopoulos
N.Technical University of Athens
Electrical Engineering Dept.
Zographou 15773-Athens-GREECE

Abstract - A model for the Organization level of a class of Intelligent Machines suitable for industrial applications is established. This model based on the theory of Hierarchically Intelligent Control Systems developed by Saridis combines the powerful high level decision making with advanced mathematical modeling and synthesis techniques of system theory and methods of dealing with imprecise or incomplete environment information.

1. Introduction

The evolution of control theory in the last few years points towards increasingly complex systems. Intelligent machines are based on the Principle of Increasing Precision with Increasing Intelligence to form an analytic methodology, using Entropy as a measure of performance. The original architecture represents a three level system, according to the principle, including Organization level, Coordination level and Execution level [Fig. 1].

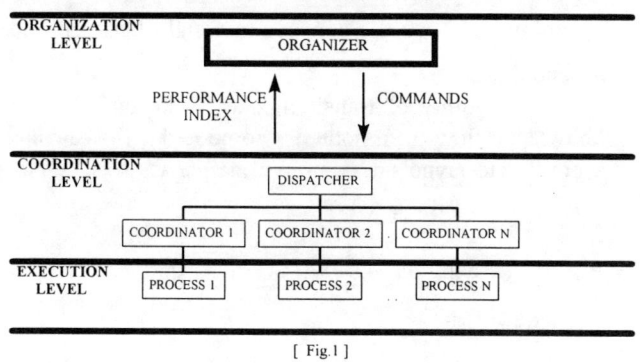

[Fig.1]

The Organization level is intended to perform such operations as planning and high level decision-making and may require large quantities of information processing but little or no precision.

In the present paper a new structure for this level is proposed [16], mainly oriented to manufacturing applications. The model requires the following capabilities:

• Task representation and processing ability which classifies the tasks in an hierarchical manner by defining initial, intermediate and final ones and in the sequel selects the appropriate control procedures to send as commands to the Coordination level.
• Learning ability which taking into account the performance indices from Coordination level improves the task sequence selection by reducing uncertainties in decision-making as more experience is obtained.

Since the late 60s, various strategies were proposed to address the control of complex learning systems. K.S.Fu [5],[6] was probably the first to write about Learning Control Systems and to coin as Intelligent Control Systems those systems of interdisciplinary nature, in the intersection of Artificial Intelligence and Automatic Control. An increasing number of other researchers have developed applications and theory in the new discipline by introducing new ideas such as neural nets, fuzzy logic, discrete event systems or hybrid systems.

The proposed model may be applied to manufacturing, robotic or other intelligent systems operating in unfamiliar environments.

The problem addressed in this paper is stated in Section 2. In Section 3 the model architecture is established and the algorithms are defined. In Section 4 a special case is examined in order to accelerate the convergence rate. Section 5 applies the theory to a manufacturing system. Finally, Section 6 summarizes the work and presents its major conclusions.

2. Problem Definition

A Boltzman-type Machine has been used for the structure that implements the Organization Level, developed by Moed and Saridis [3]. This machine would connect a finite number of nodes into sequential meaningful tasks, by minimizing at the first layer the total entropy of connections.

The functions of the Organizer, following the model of a knowledge-based system, comprise representation, abstract task planning (with minimal knowledge of the current environment), decision making and learning from experience. All those functions can be generated by a Boltzman-type Machine by considering a finite number of primitive elements at the nodes, constituting the basic actions and actors at the representing phase. Strings of these primitives are generated by the Boltzman-type Machine at the planning phase with the total entropy representing the cost of connections. The selection of the string with minimum entropy is the decision making process and the upgrading of the system parameters by rewarding the successful outcomes through feedback, is the learning procedure. The next to minimum entropy sequence may be selected as an alternate task in case of failure of the original, possibly due to unreliability of performance measures.

This hierarchical approach implies that the Organization level represents abstract activities and the performance cost after each execution is computed in the Coordinatation level. In this way the tasks generated are practically after some iterations independent of the current environment measures. However it is evident that the structure of the Coordination level dispatcher designed to interpret the Organizer strings and allocate commands among the coordinators is highly dependent on the natural language representing the sequence of the planned tasks.

3. Model Architecture

In what follows we propose a new analytic model [Fig.2] for the Organization level for which it is essential to derive the domain of the machine operation for a particular class of problems.

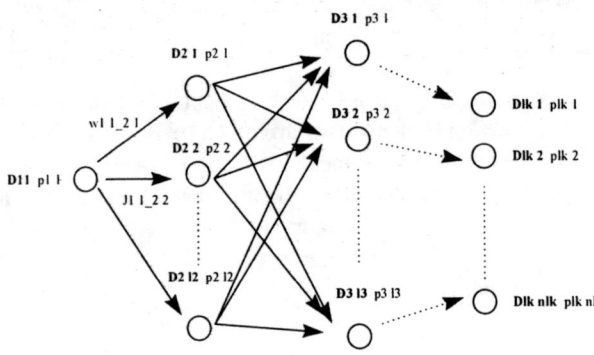

[Fig.2] Boltzman Machine for Organization Level

Assuming that there are uncertainties in the environment, one may define the following functionalities:

Task Representation is the association of a node to a number of tasks. A probability of activation is assigned to each node, a weight for the transfer between nodes and the entropy associated with it is calculated. In order to generate the required analytic model the following elements are defined:

-The ordered set of levels $L = (l_i)$, $i=1,..,k$ is the set of abstract primitive tasks of the machine and each one contains a number nl_i, $i=1,..,k$ of independent primitive nodes.

-The set of nodes $D = \{d_{11}, d_{12}, ..., d_{l_i nl_i}\}$, $i=1,..,k$ is the task domain of the machine and each node contains a number of independent primitive objects.

-The set $B \subseteq D$ contains the initial (starting) nodes.

-The set $S \subseteq D$ contains the final (ending) nodes.

-The set $Q = \{q_{11}, q_{12}, ..., q_{l_i nl_i}\}$, $i=1,..,k$ represent the state of events associated with each node D. The random variable q is binary (0,1) and indicates whether an event is inactive or active in a particular task.

-The set of probabilities $P = \{p_{11}, p_{12}, ..., p_{l_i nl_i}\}$, $i=1,..,k$ associated with the random variables q is defined as follows:

$P = \{p_{ij} = Pr(q_{ij} = 1), i = 1, .., l_k, j = 1, .., nl_i\}$

The probabilities p_{ij} are defined at the beginning of the representation stage according to previous experience.

Task Planning is the ordering of the activities. The ordering is obtained by properly concatenating for each level l_i the appropriate abstract primitive nodes d_{im}, $m=1,..,nl_i$ in order to form the right abstract activities.

The ordering is generated by a Boltzman Machine which measures the flow of knowledge $R_{d_{im}d_{i+1j}}$ transferred from node of level l_i to node of level l_{i+1} by

$$R_{d_{im}d_{i+1j}} = -\frac{1}{2} w_{d_{im}d_{i+1j}} q_{d_{im}} q_{d_{i+1j}}$$

where $w_{d_{im}d_{i+1j}} > 0$ the connection weight between nodes d_{im} and d_{i+1j}.

The probability of transition due to the uncertainty of knowledge flow from node d_{im} to node d_{i+1j}, is calculated according to Jayne's maximum principle as:

$$p_{d_{im}d_{i+1j}} = p(R_{d_{im}d_{i+1j}})$$

$$p_{d_{im}d_{i+1j}} = \exp(-a_{d_{im}} - \frac{1}{2} w_{d_{im}d_{i+1j}} q_{d_{im}} q_{d_{i+1j}})$$

Since

$$\sum_{j=1}^{nl_{i+1}} p_{d_{im}d_{i+1j}} = \sum_{j=1}^{nl_{i+1}} \exp(-a_{d_{im}} - \frac{1}{2} w_{d_{im}d_{i+1j}} q_{d_{im}} q_{d_{i+1j}}) = 1$$

we have $a_{d_{im}} = \ln \sum_{j=1}^{nl_{i+1}} \exp(-\frac{1}{2} w_{d_{im}d_{i+1j}} q_{d_{im}} q_{d_{i+1j}})$

The negative entropy in Shannon's sense of transfer d_{im} to d_{i+1j} is

$$H_{d_{im}d_{i+1j}} = -E[\ln p_{d_{im}d_{i+1j}}] = \bar{a}_{d_{im}} + \frac{1}{2} w_{d_{im}d_{i+1j}} p_{d_{im}} p_{d_{i+1j}}$$

with $\bar{a}_{d_{im}} = E[a_{d_{im}}] = \ln \sum_{j=1}^{nl_{i+1}} \exp(-\frac{1}{2} w_{d_{im}d_{i+1j}} p_{d_{im}} p_{d_{i+1j}})$

and defines the ordering of the concatenation of the nodes. It is evident that $H_{d_{im}d_{i+1j}}$ is an increasing function in terms of probabilities and weights of connected nodes and decreasing in terms of adjacent ones.

Decision Making is obtained by two alternatives:

-Starting by node d_{im} the connections to nodes d_{i+1j}, $j=1,..,l_{i+1}$ are searched until node j^* corresponding to the optimal transition is found. The search algorithm for optimal transition from node d_{im} to d_{i+1j} is

$$j^* = \arg\max_j H_{d_{im}d_{i+1j}} = \arg\max_j (\bar{a}_{d_{im}} + \frac{1}{2} w_{d_{im}d_{i+1j}} p_{d_{im}} p_{d_{i+1j}})$$

$$H_{d_{im}d_{i+1l}}^* = \max_j (\bar{a}_{d_{im}} + \frac{1}{2} w_{d_{im}d_{i+1j}} p_{d_{im}} p_{d_{i+1j}})$$

-By selecting the total maximum negative entropy at every transition which gives the optimum sequence of nodes to be selected. If we define as $S(f)$, $f = 1,..,k$ the array containing the selected nodes from each level the total maximum entropy of knowledge flow after $n \le k$ nodes is

$$H^*(n) = \max_{S(f), f=1,..,n} \sum_{i=1}^{n} (\bar{a}_{d_{iS(f)}} + \frac{1}{2} w_{d_{iS(f)}d_{i+1S(f+1)}} p_{d_{iS(f)}} p_{d_{i+1S(f+1)}})$$

Learning is obtained by feedback devices that upgrade the probabilities and the weights by evaluating the performance of the lower levels after each iteration.

The stochastic approximation reinforcement learning scheme used in this work is an extension of the algorithm proposed by Nicolic and Fu [5].

For every transition between nodes the performance index $J_{d_{iS(f)}d_{i+1S(f+1)}}(t)$ is estimated according to:

$$J_{d_{iS(f)}d_{i+1S(f+1)}}(t+1) = J_{d_{iS(f)}d_{i+1S(f+1)}}(t) +$$

$$\frac{1}{t+1}(J_{MRd_{iS(f)}d_{i+1S(f+1)}}(t+1) - J_{d_{iS(f)}d_{i+1S(f+1)}}(t)) \quad \text{where}$$

$J_{MRd_{iS(f)}d_{i+1S(f+1)}}(t)$ is the t measured value of $J_{d_{iS(f)}d_{i+1S(f+1)}}$ and $\lim_{t \to \infty} E[J_{MRd_{iS(f)i+1S(f+1)}}(t)] = J_{Sd_{iS(f)i+1S(f+1)}}$.

To update the probabilities and the weights Fu's stochastic approximation reinforcement learning scheme is also used:

$$p_{d_{iS(f)}}(t+1) = p_{d_{iS(f)}}(t) + \frac{1}{t+1}(p - p_{d_{iS(f)}}(t))$$

$$p_{d_{i+1S(f+1)}}(t+1) = p_{d_{i+1S(f+1)}}(t) + \frac{1}{t+1}(p - p_{d_{i+1S(f+1)}}(t))$$

$$w_{d_{iS(f)}d_{i+1S(f+1)}}(t+1) = w_{d_{iS(f)}d_{i+1S(f+1)}}(t) +$$

$$\frac{1}{t+1}(w - w_{d_{iS(f)}d_{i+1S(f+1)}}(t)) \quad \text{where}$$

$$p, w = \begin{cases} 1 & \text{if } \sum_{i=1,f=1}^{n} J_{d_{iS(f)}d_{i+1S(f+1)}}(t) = \min \sum J \\ 0 & \text{if } \sum_{i=1,f=1}^{n} J_{d_{iS(f)}d_{i+1S(f+1)}}(t) \ne \min \sum J \end{cases}$$

According to additional constraints imposed in system's operation it is possible to define two alternative learning schemes :

1. Learning scheme with full performance index observation: The system is forced to operate in all combinations of paths and performance measures are collected and compared at every iteration. In this way the performance estimates converge fast to their respective expected values and the algorithm converges fast to the optimum sequence. This learning scheme applies on systems with an initial learning phase when functioning is permitted in all combinations.

2. Learning scheme with operating performance index observation: After an initial phase where performance measures are collected for every combination of paths the system operates continuously and measures are collected only for the operating path. In this way the performance estimates converge slower to their respective expected values and so does the algorithm concerning the optimum sequence. This learning scheme applies on systems that evolve in parallel with algorithm execution.

Algorithm Evolution Index is defined in order to track convergence of the algorithm during execution.

The maximum entropy of the selected sequence is obtained when all probabilities and weights of its nodes are equal to 1 and all the others are equal to 0. In this way

$$H_{max}(n) = \frac{n-1}{2} + \ln[(e^{-1} + nl_2 - 1)(e^{-1} + nl_3 - 1)...(e^{-1} + nl_n - 1)]$$

The Algorithm Evolution Index (AEI) is defined as

$AEI = \dfrac{H(n)}{H_{max}(n)}$ and according to Learning $\lim_{t \to \infty} AEI = 1$

when the noise has been eliminated by the cost estimates and the optimum sequence is selected for a long number of continuous iterations.

4. System Configuration and Acceleration of Learning

The proposed general scheme treats equivalently all the possible paths. In this way the learning convergence rate depends on the number of levels and nodes and for some paths a long number of iterations may be needed until they are selected.

According to problem specifications and additional constraints imposed it is possible that dynamic system configuration may be needed by defining paths or groups of paths according to previous experience that will never meet specifications for optimum operation. In this way the learning procedure is modified as follows:

For every transition between nodes the performance index $J_{d_{iS(f)}d_{i+1S(f+1)}}(t)$ is estimated as previous according to :

$$J_{d_{iS(f)}d_{i+1S(f+1)}}(t+1) = J_{d_{iS(f)}d_{i+1S(f+1)}}(t) + \frac{1}{t+1}(J_{MRd_{iS(f)}d_{i+1S(f+1)}}(t+1) - J_{d_{iS(f)}d_{i+1S(f+1)}}(t))$$

After estimating $J_{d_{iS(f)}d_{i+1S(f+1)}}(t)$ the probabilities and the weights are updated as :

$$p_{diS(f)}(t+1) = p_{diS(f)}(t) + \frac{1}{t+1}(p - p_{diS(f)}(t))$$

$$p_{di+1S(f+1)}(t+1) = p_{di+1S(f+1)}(t) + \frac{1}{t+1}(p - p_{di+1S(f+1)}(t))$$

where

$$p = \begin{cases} \overline{p} & \text{if } \sum_{i=1,f=1}^{n} J_{d_{iS(f)}d_{i+1S(f+1)}}(t) = \min \sum J \\ 0 & \text{if } \sum_{i=1,f=1}^{n} J_{d_{iS(f)}d_{i+1S(f+1)}}(t) \neq \min \sum J \end{cases}$$

$$w_{d_{iS(f)}d_{i+1S(f+1)}}(t+1) = w_{d_{iS(f)}d_{i+1S(f+1)}}(t) + \frac{1}{t+1}(w - w_{d_{iS(f)}d_{i+1S(f+1)}}(t))$$

where

$$w = \begin{cases} \overline{W} & \text{if } \sum_{i=1,f=1}^{n} J_{d_{iS(f)}d_{i+1S(f+1)}}(t) = \min \sum J \\ 0 & \text{if } \sum_{i=1,f=1}^{n} J_{d_{iS(f)}d_{i+1S(f+1)}}(t) \neq \min \sum J \end{cases}$$

with $\overline{p} = \begin{cases} =0 & \text{if connected nodes belong to defined paths} \\ =1 & \text{otherwise} \end{cases}$

and $\overline{W} = \begin{cases} =0 & \text{if connected nodes belong to defined paths} \\ =1 & \text{otherwise} \end{cases}$

Paths including node with $\overline{p}=0$ or connected nodes with $\overline{W}=0$ are selected less frequently than the others or not at all during algorithm's evolution. In this way by properly defining $\overline{p}$, $\overline{W}$ system's configuration is possible contributing to algorithm's faster convergence.

5. Simulation Results

The operation and control of large scale manufacturing systems like production of rolling sheets in metal processing industries, is usually a difficult task because it involves several control alternatives in order to succeed an optimal policy. Such systems are very complex and the efficiency of the dynamic models which describe them is limited by the enormity of their dimensions. Process controllers based on expert systems have emerged in recent years and advantages in terms of efficiency and final goal success have been notified for many practical applications.

The production line, which is the basis of our case, consists of machines grouped in levels in a series configuration. Usual performance measures for line operation scheduling are processing times, throughput levels, gas consumption e.t.c. Parts flow from outside the system to a machine of the first level for processing then proceed to the second level and so forth until a machine of the last level, after which they exit the system. According to observed performance measures a cost is estimated for the processing in each machine. In the sequel the sum of the costs is calculated and compared to all the other costs in order to find the minimum, update the probabilities and converge to the optimum sequence of machines.

This special case study involves the simulation of a multi-level network representing the rolling plant of aluminum sheets in a typical metal processing industry [Fig.3].

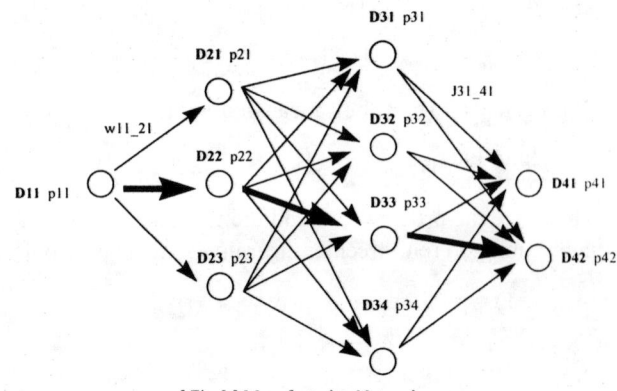

[Fig.3] Manufacturing Network

The rolling plant consumes the slabs produced in the casting plant and produces aluminum sheets of various dimensions. More precisely we have at each level:
1. Feeding Point of Hot Aluminum Slabs (dimensions around 500x1500x4000mm) [1].
2. Hot Rolling Mills fed with Slabs of thickness about 500mm and producing Sheet of thickness about 5mm [3].
3. Cold Rolling Mills fed with Sheet of thickness about 5mm and producing Sheet of thickness about 2mm [4].
4. Cold Rolling Mills fed with Sheet of thickness about 2mm and producing Sheet of thickness less than 1mm [2].

We have $D = (l_1, l_2, l_3, l_4)$ and $nl_1 = 1$, $nl_2 = 3$, $nl_3 = 4$, $nl_4 = 2$. In this way the total number of paths is 24 and for every case the Algorithm Evolution Index with the maximum number of iterations is presented in the following diagrams.

Case 1 : Equivalent Paths
Full Performance Index Observation

All weight and probability limits are equal to 1. The costs for all the paths are observed and estimated during every iteration. The optimum estimated cost in terms of the iterations compared with the next sub-optimum one is given in the diagram below. For this special case of network the maximum number of iterations to converge to the optimal sequence for a defined path is about 250.

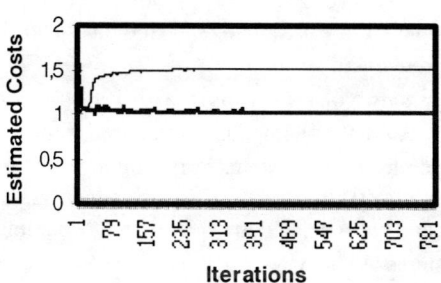

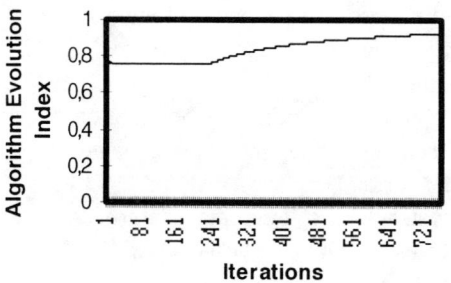

Case 2 : Equivalent Paths
Operating Performance Index Observation

All weight and probability limits are equal to 1. In this case only the cost for the selected path is observed and estimated during every iteration. For the other costs the estimations are based on last observed values. The optimum estimated cost in terms of the iterations compared with the next sub-optimum one is given in the diagram below. It is evident that convergence of cost to expected value needs more iterations than the previous case. For this special case of network the maximum number of iterations to converge to the optimal sequence for a defined path is about 400.

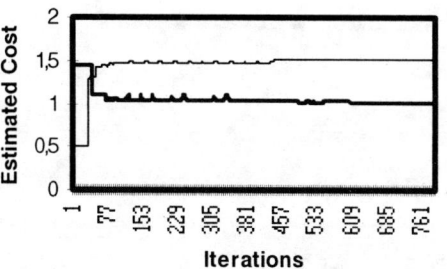

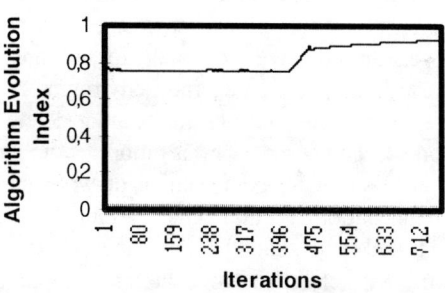

Case 3 : Configured Paths
Full Performance Index Observation

One weight limit is equal to 0. In this way the convergence rate of all the other paths is accelerated. The optimum estimated cost in terms of the iterations compared with the next sub-optimum one is given in the diagram below. For this special case of network the maximum number of iterations to converge to the optimal sequence for a defined path is about 40. Simulation results for networks with more levels and nodes show increment of the maximum number of iterations when Operating Performance Index Observation is obtained.

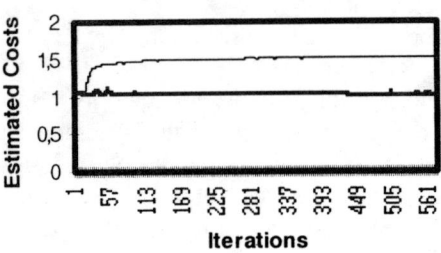

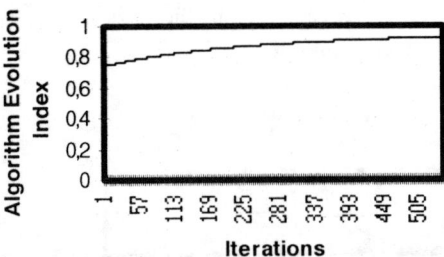

6. Conclusions

The architecture described in this paper is based on the architecture proposed by Saridis [1]. The details have been more clarified and more efficient internal structures have been used. The task representation provides the base for designing the task scheduling procedure and the learning algorithm gives an adaptive approach for finding the optimal task translation in the uncertain environment. The configurable structure adopted permits to add, delete and prioritize nodes. This approach is more suitable to industrial applications with fixed number of workstations. The main contribution is that this system has been successfully tested and that the resulting structure is extremely efficient, effective versatile as compared to other architectures.

On-going research [16] is trying to evaluate more complicated structures and establish a unified approach in dealing with discrete event control problems. Based upon modern automated processes, like flexible manufacturing systems, robotics and other advanced automation systems simulation studies should test the validity of the obtained results.

Acknowledgement

This work is financially supported by the Greek General Secretariat for Research and Technology (Ministry of Development) and the Greek State Scholarships Foundation.

References

[1] M.C.Moed and G.N.Saridis, "A Boltzman machine for the organization of intelligent machines", *IEEE Trans. SystemsManCybernetics,* vol.20, no.5, pp.1094-1102, 1990.

[2] G.N.Saridis and K.P.Valavanis, "Analytical Design of Intelligent Machines", *Automatica,* vol.24, no 2, pp.123-133, 1988.

[3] G.N.Saridis "Analytic Formulation of the Principle of Increasing Precision with Decreasing Intelligence for Intelligent Machines", *Automatica,* vol.25, no 3, pp.461-467, 1989.

[4] H.Lewis and C.H.Papadimitriou, "Elements of the theory of computation", Prentice-Hall Inc., 1981.

[5] Z.J.Nicolic and K.S.Fu, "An Algorithm for Learning without External Supervision and its application to Learning Control Systems", *IEEE Trans.Automatic Control,* vol.11, no.3, pp.414-422, 1966.

[6] K.S.Fu and Z.J.Nicolic, "On some reinforcement techniques and their relation to the stochastic approximation", *IEEE Trans.Automatic Control,* vol.11, no.2, pp.756-758, 1966.

[7] G.N.Saridis, "Intelligent Robotic Control", *IEEE Trans. Automatic Control,* vol.23, no.5, pp.547-557, 1983.

[8] Pedro U.Lima and G.N.Saridis, "Design of Intelligent Control Systems based on Hierarchical Stochastic Automata", World Scientific 1996.

[9] P.J.Ramadge and W.M.Wonham, "Supervisory control of a class of discrete event processes", *SIAM J.Contr.Optimization,* vol.25, no.1, pp.206-230, 1987.

[10] W.M.Wonham and P.J.Ramadge, "Modular supervisory control of discrete event systems", *Mathematics of Control, Signals and Systems,* vol.1, no.1, pp.13-30, 1988.

[11] F.Lin and W.M.Wonham, "Decentralized supervisory control of discrete event systems", *Information Sciences,* vol.44, pp.199-224, 1988.

[12] R.David and H.Alla, "Petri Nets for Modeling of Dynamic Systems-A Survey", *Automatica,* vol.30, no.2, pp.175-202, 1994.

[13] Jaynes E.T. "Information Theory and Statistical Mechanics", *Physical Review,* pp.106, 4, 1957.

[14] G.N.Saridis and J.H.Graham "Linguistic Decision Schemata for Intelligent Robots", *Automatica,* vol.20, no 1, pp.121-126, 1984.

[15] F.Y.Wang and G.N.Saridis "A Coordination Theory for Intelligent Machines", *Automatica,* vol.26, no 5, pp.833-844, 1990.

[16] M.N.Varvatsoulakis ''Learning Control of Discrete Event Processes'', Ph.D. Thesis, National Technical University of Athens, in preparation.

Stable Fuzzy Self–tuning Computed–torque Control of Robot Manipulators*

Miguel A. Llama†, Victor Santibañez†, Rafael Kelly‡, and Jesus Flores†

† Instituto Tecnológico de la Laguna, Apdo. Postal 49 Adm. 1
Torreón Coahuila, 27001, MEXICO
e–mail: *mllama@omega.itlaguna.edu.mx* and *vsantiba@omega.itlaguna.edu.mx*

‡ División de Física Aplicada, CICESE, Apdo. Postal 2615, Adm. 1
Ensenada, B. C., 22800, MEXICO
e–mail: *rkelly@cicese.mx*

Abstract

Computed–torque control is a well known motion control strategy for manipulators which ensures global asymptotic stability for fixed symmetric positive definite (Proportional and Derivative) gain matrices. In this paper we show that global asymptotic stability attribute also holds for a class of gain matrices depending on the manipulators state. This feature increases the potential of the computed–torque scheme to handle practical constraints in actual robots such as presence of friction in the joints and actuators with limited torque capabilities. We illustrate this potential by means of a fuzzy self–tuning algorithm to select the Proportional and Derivative gains according to the actual tracking position error. Experiments on a two degrees of freedom robot arm shown the usefulness of the proposed approach.

1 Introduction

Today a number of globally asymptotically stable motion control systems are available in the literature and discussed in robotics textbooks [1, 2, 3, 4]. Among them, the so–called *computed–torque* scheme is the simplest to understand. To say the truth, this is a control approach based on the feedback linearization technique which leads to a linear time–invariant closed–loop system. This is why this approach is also known as *inverse dynamics* control.

Our first contribution in this paper is to show that the computed–torque control scheme can also yield a globally asymptotically stable closed–loop system not only for constant positive definite gain matrices, but also for a class of manipulator state dependent gain matrices. This is a theoretical result with useful implications to handle real constraint of robot manipulators such as friction in the manipulator joints and torque capability limitations of their actuators.

The second contribution of this paper is the application of fuzzy logic to design a self–tuner for the computed–torque control taking into account specifications of allowable actuator torques limits and desired tracking accuracy in presence of friction. Although applications of fuzzy adaptation and self–tuning in robotics have been reported in the literature [5, 6, 7, 8], here we provide conditions for this fuzzy self–tuning algorithm to produce a globally asymptotically stable closed–loop system in the absence of friction in the manipulator joints. The performance of this control scheme is illustrated via experimental evaluation on a two degrees of freedom direct–drive vertical robot arm.

2 Robot dynamic model

In the absence of friction, the dynamics of a serial n–link rigid robot can be written as [2]:

$$M(q)\ddot{q} + C(q,\dot{q})\dot{q} + g(q) = \tau \qquad (1)$$

where q is the $n \times 1$ vector of joint displacements, $\dot{q}$ is the $n \times 1$ vector of joint velocities, τ is the $n \times 1$

---
*Work partially supported by COSNET and CONACyT (Mexico)

vector of applied torques by the actuators, $M(q)$ is the $n \times n$ symmetric positive definite manipulator inertia matrix, $C(q,\dot{q})\dot{q}$ is the $n \times 1$ vector of centripetal and Coriolis torques, and $g(q)$ is the $n \times 1$ vector of gravitational torques.

3 Computed-torque control with nonlinear gains

Consider the control law corresponding to a computed–torque scheme with nonlinear gain matrices:

$$\tau = M(q)[\ddot{q}_d + K_p(\tilde{q})\tilde{q} + K_v(\tilde{q},\dot{\tilde{q}})\dot{\tilde{q}}] + C(q,\dot{q})\dot{q} + g(q) \quad (2)$$

where $q_d, \dot{q}_d$ and $\ddot{q}_d$ are the $n \times 1$ vectors of desired position, velocity and acceleration respectively. The joint position error is denoted by the $n \times 1$ vector $\tilde{q} = q_d - q$ while $\dot{\tilde{q}} = \dot{q}_d - \dot{q}$ stands for the $n \times 1$ vector of velocity error. The standard computed–torque control has two parameters: K_p and K_v which are the $n \times n$ Proportional and Derivative gain matrices, but now $K_p(\tilde{q})$ and $K_v(\tilde{q},\dot{\tilde{q}})$ are suitable diagonal matrices functions whose entries are denoted by $k_{pi}(\tilde{q}_i)$ and $k_{vi}(\tilde{q}_i,\dot{\tilde{q}}_i)$ respectively. The practical usefulness of this control strategy will become clear later when a fuzzy self–tuning algorithm will be introduced.

The main stability result concerning the computed-torque scheme with nonlinear gains is stated in the following

Proposition 1. Consider the robot dynamic model (1) together with the computed–torque scheme with nonlinear gains (2). If there exists $\varepsilon > 0$ such that the nonlinear gains satisfy

- $k_{pi}(\tilde{q}_i) \geq \varepsilon$ for all $\tilde{q}_i \in \mathbb{R}$, and $i = 1, \cdots, n$
- $k_{vi}(\tilde{q}_i,\dot{\tilde{q}}_i) \geq \varepsilon$ for all $\tilde{q}_i, \dot{\tilde{q}}_i \in \mathbb{R}$, and $i = 1, \cdots, n$

then the closed-loop is globally asymptotically stable and the position tracking aim $\lim_{t \to \infty} q(t) = q_d(t)$ is achieved.

$\triangledown\triangledown\triangledown$

Proof. The closed-loop equation is obtained by combining the robot dynamic model (1) with the control law (2). This can be written as:

$$\frac{d}{dt}\begin{bmatrix}\tilde{q}\\ \dot{\tilde{q}}\end{bmatrix} = \begin{bmatrix}\dot{\tilde{q}}\\ -K_p(\tilde{q})\tilde{q} - K_v(\tilde{q},\dot{\tilde{q}})\dot{\tilde{q}}\end{bmatrix} \quad (3)$$

which is an autonomous nonlinear differential equation and the origin of the state space is the unique equilibrium.

To carry out the stability analysis we propose the following Lyapunov function candidate:

$$V(\tilde{q},\dot{\tilde{q}}) = \frac{1}{2}\dot{\tilde{q}}^T\dot{\tilde{q}} + \int_0^{\tilde{q}}\boldsymbol{\xi}^T K_p(\boldsymbol{\xi})\,d\boldsymbol{\xi} \quad (4)$$

where

$$\int_0^{\tilde{q}}\boldsymbol{\xi}^T K_p(\boldsymbol{\xi})\,d\boldsymbol{\xi} =$$

$$\int_0^{\tilde{q}_1}\xi_1^T k_{p1}(\xi_1)\,d\xi_1 + \cdots + \int_0^{\tilde{q}_n}\xi_n^T k_{pn}(\xi_n)\,d\xi_n.$$

Notice that since there exists a constant $\varepsilon > 0$ such that $k_{pi}(\tilde{q}_i) \geq \varepsilon$ for all $\tilde{q}_i \in \mathbb{R}$ and $i = 1, \cdots, n$, then

$$\int_0^{\tilde{q}_i}\xi_i k_{pi}(\xi_i)\,d\xi_i \geq \frac{1}{2}\varepsilon\,|\tilde{q}_i|^2, \quad (5)$$

and

$$\int_0^{\tilde{q}_i}\xi_i k_{pi}(\xi_i)\,d\xi_i \to \infty \text{ as } |\tilde{q}_i| \to \infty \quad (6)$$

which yields

$$\int_0^{\tilde{q}}\boldsymbol{\xi}^T K_p(\boldsymbol{\xi})d\boldsymbol{\xi} \geq \frac{1}{2}\varepsilon\,\|\tilde{q}\|^2 \quad (7)$$

and

$$\int_0^{\tilde{q}}\boldsymbol{\xi}^T K_p(\boldsymbol{\xi})d\boldsymbol{\xi} \to \infty \text{ as } \|\tilde{q}\| \to \infty. \quad (8)$$

This proves that $V(\tilde{q},\dot{\tilde{q}})$ introduced in (4) is a globally positive definite and radially unbounded function.

The time derivative of the Lyapunov function candidate is

$$\dot{V}(\tilde{q},\dot{\tilde{q}}) = \dot{\tilde{q}}^T\ddot{\tilde{q}} + \tilde{q}^T K_p(\tilde{q})\dot{\tilde{q}} \quad (9)$$

where we have used the Leibnitz' rule for differentiation of integrals. By substituting the closed-loop equation (3) in equation (9) we finally obtain:

$$\dot{V}(\tilde{q},\dot{\tilde{q}}) = -\dot{\tilde{q}}^T K_v(\tilde{q},\dot{\tilde{q}})\dot{\tilde{q}}.$$

Since by assumption $K_v(\tilde{q},\dot{\tilde{q}})$ is a diagonal positive definite matrix ($k_{vi}(\tilde{q}_i,\dot{\tilde{q}}_i) \geq \varepsilon > 0$), hence we have $\dot{V}(\tilde{q},\dot{\tilde{q}})$ is a globally negative semidefinite function. Thus by invoking the Lyapunov's direct method [10] we conclude stability of the closed–loop system.

In order to prove global asymptotic stability we exploit the autonomous nature of the closed–loop system (3) to apply the Krasovskii-LaSalle's theorem [10]. In the region

$$\Omega = \left\{\begin{bmatrix}\tilde{q}\\ \dot{\tilde{q}}\end{bmatrix} : \dot{V}(\tilde{q},\dot{\tilde{q}}) = 0\right\} = \left\{\begin{bmatrix}\tilde{q}\\ \dot{\tilde{q}}\end{bmatrix} = \begin{bmatrix}\tilde{q}\\ 0\end{bmatrix} \in \mathbb{R}^{2n}\right\}$$

the unique invariant is $[\tilde{\boldsymbol{q}}^T \ \dot{\tilde{\boldsymbol{q}}}^T]^T = \boldsymbol{0}$. Therefore invoking the Krasovskii–LaSalle's theorem we conclude that the origin of the state space is a globally asymptotically stable equilibrium of the closed–loop system (3).

◇

4 Fuzzy approach for self–tuning the computed–torque controller gains

Two important real constraint on robot manipulators are the friction in the manipulators joints and the technological limitation of torque (or force) capability in robot actuators. Static friction produces bias in positioning while torque capability reduces the class of desired position trajectories.

Fuzzy logic may be a suitable approach as a mechanism to determine the nonlinear Proportional and Derivative gains of the computed–torque control scheme because the input–output characteristics of Fuzzy Logic Systems could be easily suited in order to fulfill the stability requirements established in proposition 1, namely

- $k_{pi}(\tilde{q}_i) \geq \varepsilon$ for all $\tilde{q}_i \in \mathbb{R}$, and $i = 1, \cdots, n$
- $k_{vi}(\tilde{q}_i, \dot{\tilde{q}}_i) \geq \varepsilon$ for all $\tilde{q}_i, \dot{\tilde{q}}_i \in \mathbb{R}$, and $i = 1, \cdots, n$

for some $\varepsilon > 0$.

For this purpose, we define one conceptual Fuzzy Logic Tuner (FLT). Its goal is to tune the proportional gains $k_{pi}(\tilde{q}_i)$ and the derivative gains $k_{vi}(\tilde{q}_i)$ according to the input $|\tilde{q}_i|$. In summary, $2n$ elementary FLT will be involved in computation of n proportional gains and n derivative gains.

The design of the FLT follows steps similar to the fuzzy logic controllers [11, 12], starting from the creation of the knowledge base: Data base and rule base. In our application to robotics, the knowledge base is obtained from the experience of skilled control user of robot manipulators (the so-called "verbalization").

Having in mind the real-time implementation of the fuzzy self-tuning algorithm, a quite simple approach to design the FLT has been adopted. Let the conceptual FLT have an input $|x|$ and the corresponding output y. The universes of discourse of $|x|$, and y are partitioned into 3 fuzzy sets: B (Big), M (Medium), and S (Small) with each attribute being described by a membership function. We shall employ trapezoidal membership functions for input variables and singleton membership functions for output variables. In order to simplify notation, let us use the following convention. With reference to figure 1, the corresponding

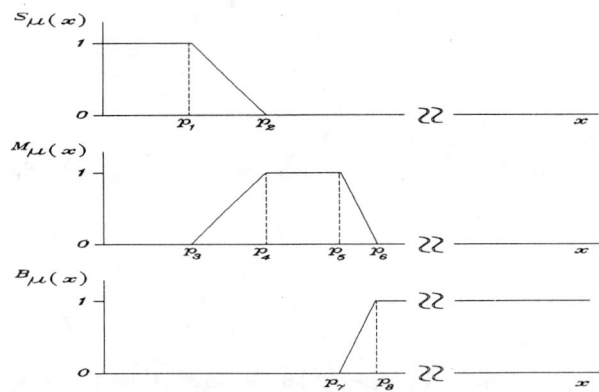

Figure 1: Input membership functions

Small, Medium and Big membership functions for the input variable x are denoted respectively by

$$^S\mu(|x|; p_1, p_2)$$
$$^M\mu(|x|; p_3, p_4, p_5, p_6)$$
$$^B\mu(|x|; p_7, p_8).$$

For convenience we define vector $\boldsymbol{\mu}(|x|)$ as

$$\boldsymbol{\mu}(|x|) = \begin{bmatrix} ^S\mu(|x|; p_1, p_2) \\ ^M\mu(|x|; p_3, p_4, p_5, p_6) \\ ^B\mu(|x|; p_7, p_8) \end{bmatrix}.$$

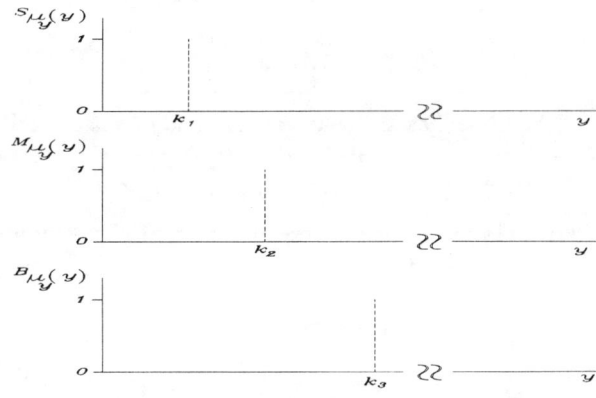

Figure 2: Output membership functions

With reference to the output variable y (see figure 2), the singleton membership functions corresponding to Small, Medium and Big are represented by

$$^S\mu_y(y;k_1)$$
$$^M\mu_y(y;k_2)$$
$$^B\mu_y(y;k_3).$$

It is also convenient for the sake of notation to define vector $\boldsymbol{\mu}_y(y)$ as

$$\boldsymbol{\mu}_y(y) = \begin{bmatrix} ^S\mu_y(y;k_1) \\ ^M\mu_y(y;k_2) \\ ^B\mu_y(y;k_3) \end{bmatrix}.$$

The rules in the rulebase are the following

IF $^S\mu(|x|;p_1,p_2)$ THEN $^B\mu_y(y;k_3)$

IF $^M\mu(|x|;p_3,p_4,p_5,p_6)$ THEN $^M\mu_y(y;k_2)$

IF $^B\mu(|x|;p_7,p_8)$ THEN $^S\mu_y(y;k_1)$.

Evaluation of the rules leads to

$$\boldsymbol{\mu}_y(y) = \begin{bmatrix} 0 & 0 & 1 \\ 0 & 1 & 0 \\ 1 & 0 & 0 \end{bmatrix} \boldsymbol{\mu}(|x|).$$

The defuzzification strategy chosen is a simplified version of the "center of gravity method":

$$y = [k_1 \ k_2 \ k_3] \frac{\boldsymbol{\mu}_y(|x|)}{\|\boldsymbol{\mu}_y(|x|)\|_1} \quad (10)$$

where $\|\cdot\|_1$ stands for the 1 norm. Variable y is the crisp value of the output of the FLS for some given crisp input $|x|$.

The described Fuzzy Logic Tuner (FLT) can be seen as a static mapping H defined by

$$H: \ \mathbb{R}_+ \to \mathbb{R}$$
$$|x| \mapsto y$$

This FLT has the feature that under weak conditions its output y is bounded away from a strictly positive constant. This is stated in the following

Property 1. Consider the described FLT. Assume that $\|\boldsymbol{\mu}(|x|)\|_1 > 0$ for all $x \in \mathbb{R}$, and $k_3 > k_2 > k_1 > 0$. Then

$$y(|x|) \geq k_1 > 0 \quad \forall \ x \in \mathbb{R}.$$

▽▽▽

Proof. Since by definition all entries of $\boldsymbol{\mu}_y(|x|)$ are nonnegative and by assumption $k_3 > k_2 > k_1 > 0$, then

$$[k_1 \ k_2 \ k_3]\boldsymbol{\mu}_y(|x|) \geq k_1 \|\boldsymbol{\mu}_y(|x|)\|_1.$$

Incorporating this expression in (10) we get the desired conclusion.

◇

Assumption $\|\boldsymbol{\mu}(|x|)\|_1 > 0$ means that the intersection of the input membership functions is nonempty for all input $|x|$. This is easy to check by testing the following inequalities

$$p_3 < p_2, \quad (11)$$
$$p_7 < p_6. \quad (12)$$

As previously described, the basic FLT is invoked to determine the proportional and derivative gains. Thus, a set of $2n$ FLTs are defined, that is

$$H_{k_{pi}}: \ \mathbb{R}_+ \to \mathbb{R}$$
$$|\tilde{q}_i| \mapsto k_{pi}$$

and

$$H_{k_{vi}}: \ \mathbb{R}_+ \to \mathbb{R}$$
$$|\tilde{q}_i| \mapsto k_{vi}$$

for $i = 1, \cdots, n$.

We are ready to present our main stability result.

Proposition 2. Consider the robot dynamic model (1) together with the computed–torque scheme (2). Let the proposed Fuzzy Logic Tuners be used to obtain the nonlinear gains k_{pi} and k_{vi}. If each FLT is designed assuming (11)–(12) and the fuzzy set supports for the output variables are strictly positive (i.e. the corresponding k_1, k_2 and k_3 are strictly positive), then the closed-loop system is globally asymptotically stable and the position tracking aim $\lim_{t\to\infty} \boldsymbol{q}(t) = \boldsymbol{q}_d(t)$ is achieved.

▽▽▽

Proof. In virtue of property 1, assumptions in proposition 2 mean that all gains are bounded away from a positive constant. This means that there exists $\varepsilon > 0$ such that

- $k_{pi}(\tilde{q}_i) \geq \varepsilon$ for all $\tilde{q}_i \in \mathbb{R}$, and $i = 1, \cdots, n$
- $k_{vi}(\tilde{q}_i, \dot{\tilde{q}}_i) \geq \varepsilon$ for all $\tilde{q}_i, \dot{\tilde{q}}_i \in \mathbb{R}$, and $i = 1, \cdots, n$.

Therefore, the desired conclusion follows straightforward from proposition 1.

◇

5 Experimental evaluation

Experiments on a two degrees of freedom direct-drive robot arm have been carried out in order to illustrate the performance of the proposed fuzzy self-tuning approach.

The dynamics (1) of this two degrees of freedom direct–drive robotic manipulator, is given by [13]

$$M(q) = \begin{bmatrix} 2.35 + 0.16\cos(q_2) & 0.10 + 0.08\cos(q_2) \\ 0.10 + 0.08\cos(q_2) & 0.10 \end{bmatrix}$$

$$C(q,\dot{q}) = \begin{bmatrix} -0.168\sin(q_2)\dot{q}_2 & -0.084\sin(q_2)\dot{q}_2 \\ 0.084\sin(q_2)\dot{q}_1 & 0 \end{bmatrix}$$

$$g(q) = 9.81 \begin{bmatrix} 3.921\sin(q_1) + 0.186\sin(q_1+q_2) \\ 0.186\sin(q_1+q_2) \end{bmatrix}$$

$$f(\dot{q}) = \begin{bmatrix} 2.288\dot{q}_1 \\ 0.175\dot{q}_2 \end{bmatrix}$$

where $f(\dot{q})$ has been incorporated into the model to include viscous friction. It is worth mentioning that Coulomb and static frictions are also present in the robot joints, however, we have decided to consider them as unmodeled dynamics.

The actuator saturations are

$$\tau_1^{max} = 150 \text{ [Nm]},$$
$$\tau_2^{max} = 15 \text{ [Nm]}.$$

The desired position trajectory q_d is given by

$$q_d = \begin{bmatrix} 0.78[1 - e^{-2.0\,t^3}] + 0.17[1 - e^{-2.0\,t^3}]\sin(\omega_1 t) \\ 1.04[1 - e^{-1.8\,t^3}] + 2.18[1 - e^{-1.8\,t^3}]\sin(\omega_2 t) \end{bmatrix}$$
$$+ \begin{bmatrix} 1.57 \\ 1.57 \end{bmatrix} \text{ [rad]} \qquad (13)$$

where $\omega_1 = 15$ rad/sec and $\omega_2 = 3.5$ rad/sec.

Through the experimental tests, the arm initial configuration was at rest in its vertical down position. This corresponds to $q(0) = 0$ and $\dot{q}(0) = 0$ and then $\tilde{q}(0) = \pi/2$ [rad] and $\dot{\tilde{q}}(0) = 0$.

5.1 Fuzzy self–tuning computed–torque control

The proportional and derivative gains have been on–line tuned by the FLTs described in the previous section.

Fuzzy partitions of the universes of discourse of the tracking errors $|\tilde{q}_1|$ and $|\tilde{q}_2|$ are characterized respectively by the sets

$$p_{\tilde{q}_1} = \{0.5, 1, 0.5, 1, 10, 15, 10, 15\} \text{ [deg]}$$
$$p_{\tilde{q}_2} = \{2, 4, 2, 4, 10, 15, 10, 15\} \text{ [deg]}$$

where $p_{\tilde{q}_i} = \{p_1, p_2, \cdots, p_8\}$ denotes the supports of the memership fuction of $\tilde{q}_i$ according to our convention.

The final partition of the universes of discourses for the proportional gains were

$$p_{k_{p1}} = \{9, 120, 500\} \text{ [sec}^{-2}\text{]},$$
$$p_{k_{p2}} = \{10, 1800, 5000\} \text{ [sec}^{-2}\text{]}.$$

The selection of the partition for the derivative gains k_{vi} were directly computed from the partition of the proportional gains in order to get ideally a critically damped second order linear system. This leads to the following partition of the universe of discourse

$$p_{k_{v1}} = \{6, 21.9, 44.7\} \text{ [sec}^{-1}\text{]},$$
$$p_{k_{v2}} = \{6.3, 84.8, 141.4\} \text{ [sec}^{-1}\text{]}.$$

Fuzzy partitions chosen ensure that the fuzzy self–tuners deliver proportional and derivative gains in agreement with conditions of Proposition 2; therefore, in case of absence of friction we conclude that the closed-loop system is asymptotically stable.

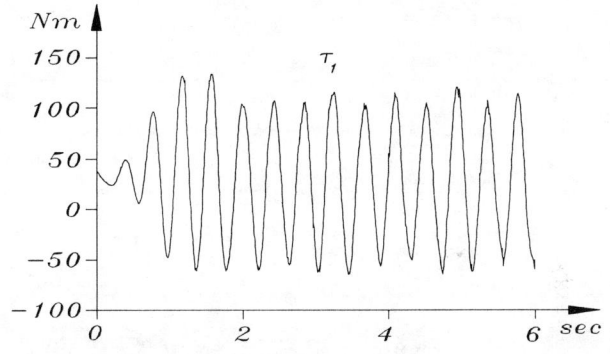

Figure 3: Applied torques

Figures 3–5 show the experimental results. Applied torque τ_1 and τ_2 are sketched in figures 3 and 4. As expected both signal remains within the prescribed allowable maximum torque for each actuator. The position tracking error $\tilde{q}$ whose components are shown in figure 5 are acceptable small.

6 Summary

The structure of the so-called computed–torque control scheme for robot motion control results useful to address real issues when its proportional and derivative gains can be allowed to depend in a nonlinear fashion on the robot state.

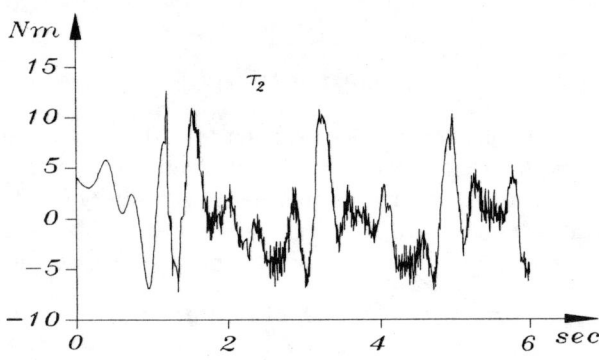

Figure 4: Applied torques

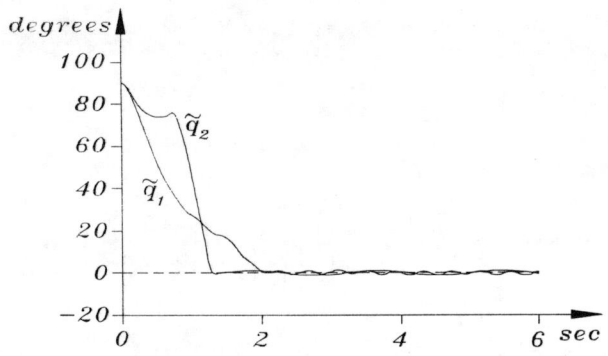

Figure 5: Position errors

We have characterized a class of nonlinear gains for which global asymptotic stability is preserved. Several techniques like gain scheduling, static neural networks or fuzzy logic, may be used to compute these gains according to practical specifications such as achievement of desired accuracy and avoiding to exploit the actuator beyond their torque capabilities.

We have shown the application of a gain tuning algorithm based on fuzzy logic. Experimental evaluation on a direct–drive robot arm illustrate the good performance of this approach.

References

[1] J.J. Craig, 1989, "Introduction to Robotics", 2nd ed., Addison–Wesley Pub.

[2] M.Spong and M. Vidyasagar, "Robot Dynamics and control", John Wiley and Sons, NY.,1989

[3] F. L .Lewis, C. T. Abdallah, D.M. Dawson, 1993, "Control of Robot Manipulators", MacMillan Publishing Company.

[4] L. Sciavicco, B. Siciliano, 1996, "Modeling and control of robot manipulators", McGraw–Hill Co.

[5] T. Fukuda and T. Shibata, 1992, "Hierarchical intelligent control for robotic motion by using Fuzzy, artificial intelligence, and Neural Networks", in Proc. Int. J. Conference on Neural Networks, Baltimore, MD, Vol. 1, pp. 269-274.

[6] J. Zhou and P. Coiffet, 1992, "Fuzzy control of robots", in Proc. lst. Int. Conf. on Fuzzy systems, San Diego, CA, pp. 1357-1364.

[7] J. M. Meslin, J. Zhou and P. Coiffet, 1993, "Fuzzy dynamic control of robot manipulators: A scheduling approach", in Proc. IEEE Int. Conf. on Syst. Man and Cybern., Le Tourquet, France, pp. 69-73.

[8] Sayyarrodsari, B. and A. Homaifar, "The role of Hierarchy in the design of fuzzy logic controllers", *IEEE Trans. on Systems, Man, and Cybernetics – Part B: Cybernetics*, Vol. 27, No. 1, pp. 108–118, 1997.

[9] R. Kelly, R. Haber, R. E. Haber, F. Reyes, "Lyapunov stable control of robot manipulators: A fuzzy self-tuning procedure", Intelligent Automation and Soft Computing, To appear, 1998.

[10] Vidyasagar M. (1993). "Nonlinear Systems Analysis", Prentice-Hall, Englewood Cliffs, NJ.

[11] Lee, C. C. "Fuzzy logic in control systems: Fuzzy logic controller – Part II", *IEEE Transactions on Systems, Man and Cybernetics*, Vol. 20, No. 2, pp. 419–435, 1990.

[12] Yager, R. R. and D. P. Filev, *Essentials of Fuzzy modeling and control*, John Wiley and Sons, NY, 1994.

[13] F. Reyes, R. Kelly, "Experimental evaluation of identification schemes on a direct–drive robot", Robotica, Vol. 15, Part 5, September–October 1997, pp. 563–571.

Sensor-Enhanced Robotic Cell Collaboration Using Shared Task Error Information

Manabu MOTEGI   Takao KAKIZAKI   Shin-yo MUTO

NTT Human Interface Laboratories
3-9-11 Midori-cho, Musashino-shi, Tokyo 180-8585, JAPAN

Abstract

A multiple-sensor-enhanced robot collaboration system has been developed for manufacturing applications. The system consists of basic robot modules, such as for locating and tracking. These modules have laser range finders and perform hybrid manufacturing tasks by working together as an advanced robotic cell. A hybrid task with uncertainties can be described by the task velocity and task error derived from the sensory and motion information on the robot modules. The modules send their task error and waiting time to each other via a collaboration network. They also modify their given motion sets in real time based on fuzzy rules in order to improve the accuracy and efficiency of the system. An experimental study of multi-station welding showed that this collaboration results in a satisfactory trade-off between accuracy and efficiency.

1 Introduction

Most industrial robots cannot communicate with other robots, and have difficulty performing a task including uncertainties due to workpiece position tolerances and shape differences. A number of sensor-based robotic systems have been investigated for performing tasks with uncertainties, such as arc-welding and automobile body sealing[1][2]. But most industrial tasks consist of not only a single task, but also related tasks, such as loading, positioning, and unloading. Performing such "hybrid tasks" by using one robot is unrealistic because these tasks include many work stages. Robotic cell systems and flexible manufacturing systems(FMS) have been developed, for performing hybrid tasks[3]. However, most of the developed systems use conventional teaching-playback robots, i.e., manipulators and related peripherals, so it is difficult to cope with uncertainties, such as workpiece position tolerances and shape differences between individual workpieces.

Visual sensor-enhanced robotic systems can track a target path with uncertainties and have been tested in factory applications. A path-tracking robot system can be considered to be a basic module having primitive autonomous characteristics. The next step is to integrate several basic modules into practical manufacturing systems. This will enable sensor-enhanced robots to collaborate in performing hybrid tasks that include uncertainties.

Many studies have been done on the design and control of multi-robot systems. Particular emphasis has been placed on multi-arm coordination control for manipulating an object by using an impedance-control scheme[4]. Multi-agent robot systems and distributed robotic systems have also been studied[5][6]. These studies, however, did not focus on practical industrial application. The planning, control, and interface of robotic-cell-type systems in dynamic environments have recently been studied[7][8]. However, the inherent sensing uncertainties encountered in practical robotic tasks were not considered.

We have developed a multiple-sensor-enhanced robot collaboration system for manufacturing applications. We integrated two sensor-enhanced robot modules into a robotic cell by using a local-area network. The robots collaborate to perform their tasks in a complementary manner. They do this by referring to the task error computed by each module from its sensory and motion information. Experimental evaluation of such a system constructed of two different-type modules performing a hybrid task showed that each robot efficiently and precisely worked together to perform the task.

2 Robot Cell Collaboration

2.1 Basic Concept

The basic concept of our proposed robotic-cell collaboration system for performing a hybrid task is shown in Fig. 1. A basic robot module (BRM) is a sensor-enhanced robot having the ability to communicate via a collaboration network interface. Each BRM performs a particular task that contributes to performing a hybrid

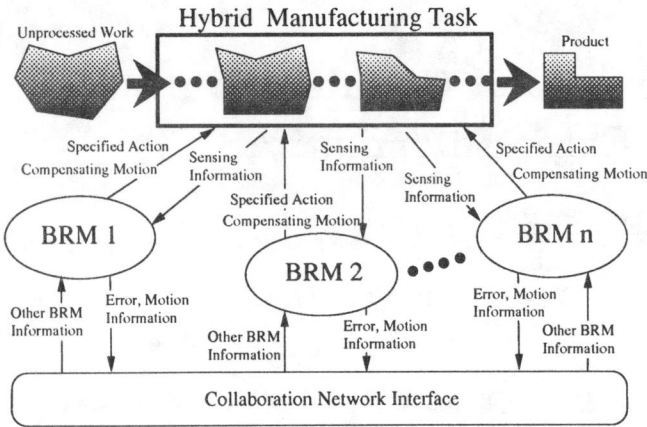

Figure 1: Concept of robotic-cell collaboration for manufacturing.

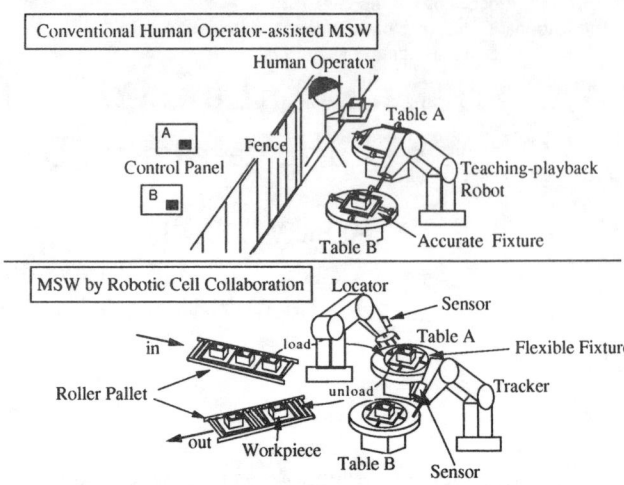

Figure 2: Multi-station welding (MSW) task used for experimental evaluation.

task. Each target workpiece has position and shape tolerances, and each BRM has a previously given motion-set and starting plan. Each BRM detects and estimates the workpiece tolerances, then modifies its previously given motion set so as to perform the task. To do this, the BRMs share their task error information via the collaboration network interface. They also share their motion information, i.e., velocity and status. Each BRM refers to the task error and motion information of the other BRMs and adjusts its own motion so as to improve task quality and efficiency. In this way, the desired trade-off between task error and task efficiency can be achieved.

The aim of robotic cell collaboration is to emulate the way people collaborate to perform complicated manufacturing tasks effectively.

2.2 Experimental Task

We used a multi-station welding (MSW) task to experimentally evaluate robotic cell collaboration (Fig. 2). In conventional MSW, a teaching-playback-robot welds in turn workpieces on positioning tables. While a workpiece is being welded, a human operator unloads the workpiece just completed and loads an unprocessed workpiece to be processed next. The operator must accurately position each workpiece on the table by using jig fixtures. In our experiment, a Locator module and a Tracker module were used instead of the human operator and the conventional welding robot. The Locator detects the workpiece position and orientation on the table and performs loading and unloading. "Locate" means estimating the workpiece position and orientation by using a sensor. The Tracker spatially tracks the path on the workpiece for arc welding.

No complicated or accurate fixtures are needed in this Locator and Tracker collaboration.

2.3 Basic Robot Module

We used a sensor-enhanced robot system consisting of a manipulator and a visual sensor for each BRM in the collaboration system. The sensor acquires the sectional data of the workpiece once each sampling period $\triangle \tau_s$ [9]. Then, the necessary feature point, such as a welding point, in the sensor coordinate system is computed by using a feature-recognition algorithm. The feature point in the robot coordinate system is then computed once each sampling period $\triangle \tau_r$ by taking account of the robot motion information.

The Locator and Tracker are constructed as BRMs in the collaboration system. As discussed below, both use visual sensory information to perform their task. The sensing performance may be affected by the robot motion speed because of the finite sampling periods ($\triangle \tau_s$ and $\triangle \tau_r$).

2.4 Motion Sets for Collaboration

Several motion sets must be prepared when an industrial robot system is used as the BRM. In the experiment discussed below, the following motion sets (Fig. 3) were used.

1. Locator motion sets
   Locate-All
   When no workpiece is on welding (or positioning) tables, the Locator loads an unprocessed workpiece from the pallet onto table A and locates it by using the sensor. It then loads another unprocessed workpiece onto table B and locates it.
   Locate-A
   After the Tracker finishes processing the workpiece on table A, the Locator unloads it. It then loads

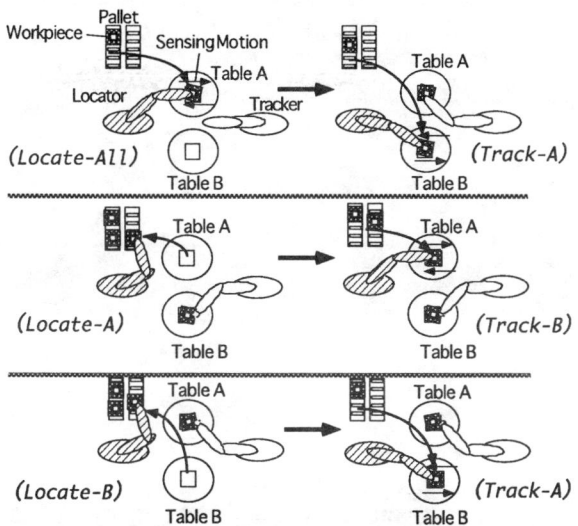

Figure 3: Motion-sets used for testing collaboration.

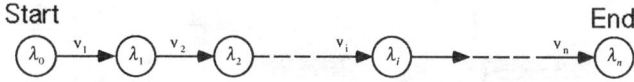

Figure 4: Robotic task flow using task velocity and task error.

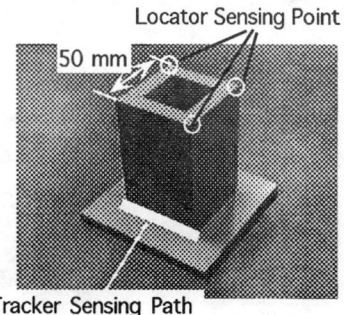

Figure 5: Tested workpiece.

an unprocessed workpiece onto table A and locates it.
Locate-B
The Locator performs the same task as *Locate-A* for the workpiece on table B.
2. Tracker motion-sets
Track-A
For the workpiece on table A, the Tracker performs the detection and welding tasks.
Track-B
The Tracker performs the same task as *Track-A* for the workpiece on table B.

Before a BRM begins executing its motion sets, the task parameters (motion velocity and starting position) are adjusted based on the task error information received from the other BRMs. By executing each BRM's task iteratively, robot cell collaboration for a hybrid task is performed.

2.5 Task Error and Task Velocity

Quick and highly accurate actions are most desirable in manufacturing; in other words, the objective is both efficiency and accuracy. We thus introduce the concepts of task velocity and task error. Task velocity represents efficiency, while task error represents a degradation in accuracy. There is a correlation between task velocity and task error. For example, if the task motion velocity or workpiece position tolerance increases, the task errors of the Tracker and Locator increase. The workpiece positioning tolerance depends on placement accuracy. In the following experimental case, the Locator releases the workpiece about 5 mm above the flexible fixture of the workpiece table. Therefore, the tolerance cannot be controlled by the BRM. The task velocity, however, can be controlled by the BRM because task velocity represents robot motion.

The MSW task is a hybrid task consisting of a chain of individual tasks: loading the workpiece, estimating its location, processing it, and unloading it. A schematic of the robotic task flow using task error and task velocity is shown in Fig. 4.

Task error λ_i due to the i-th task result can be expressed using a nonlinear function:

$$\lambda_i \cong f(v_i, \lambda_{i-1}), \qquad (1)$$

where v_i is the task velocity of the i-th task. While the value of each λ_i is not important, the value of λ_n affects the final result of the hybrid task, so it should be small.

A photograph of the workpiece we used is shown in Fig. 5. In the next two sections, we will discuss the relation-ship between task velocity and task error.

2.5.1 Locator

The Locator estimates the workpiece position and orientation by detecting its vertices sequentially. The vertices vector of the model workpiece, i.e., $\boldsymbol{P}^m = \{\boldsymbol{p}_1^m, \boldsymbol{p}_2^m \cdots \boldsymbol{p}_n^m\}$ is previously given. Initial robot teaching of the end-point position/orientation and velocity enables the BRM to acquire the vertices vector $\boldsymbol{P}^w = \{\boldsymbol{p}_1^w, \boldsymbol{p}_2^w \cdots \boldsymbol{p}_n^w\}$ of the unprocessed workpieces. After an unprocessed workpiece is loaded, the Locator starts its sensing motion based on the initial teaching data to determine the position/orientation tolerance of

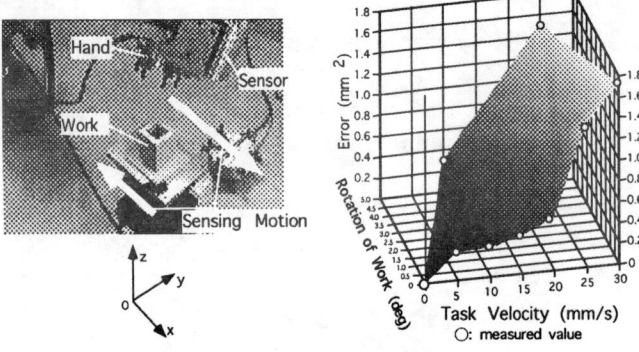

Figure 6: Locator task error due to task velocity and work positioning error.

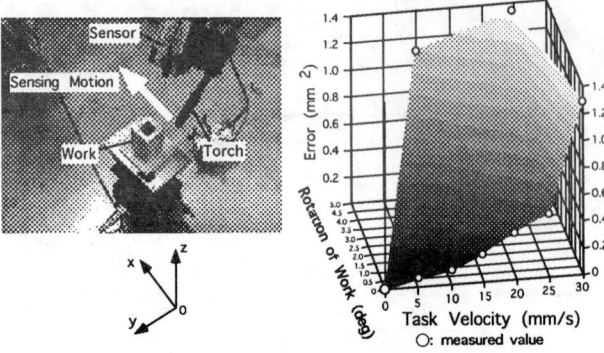

Figure 7: Tracker task error due to task velocity and work positioning error.

the workpiece. The correspondence between vertices in the model and vertices in the unprocessed workpiece are detecting order. The similarity transformation parameters ($H(R,t,c)$, R: rotation, t: translation, and c: scaling) from P^m to P^w are computed using least-squares estimation[10]. The Locator task error, λ_l, is defined as

$$\lambda_l(R, t, c) = \frac{1}{n} \sum \|P^w - (cRP^m + t)\|^2. \quad (2)$$

In our experiment, a minimum n of 3 was used for efficiency. The sensing motion velocity (average) was defined as the task velocity of the Locator. The effect of task velocity and workpiece position on the Locator task error is shown as the mean value of ten measurements in Fig. 6. The task error increases with the task velocity and with the workpiece position tolerance. The effect of task velocity on task error is due to the sampling period of the sensing, while the effect of the workpiece position tolerance on the task error is mainly due to the nature of the sensing algorithm used. The task error increases significantly when the workpiece has irregular deformations.

2.5.2 Tracker

The Tracker detects the seam-start point on a workpiece and continuously tracks the seam by means of visual sensing feedback. The welding task is completed when the seam-end point is detected. Similar to the Locator, a vector of the model path, $P_i^m = \{p_1^m, p_2^m \cdots p_i^m\}$, i.e., the seam start-point, the welding points along the seam, and the seam end-point of the model workpiece, is previously given. In the Locator/Tracker collaboration experiment discussed below, P_i^m is replaced by a new P_i^m based on the $H(R, t, c)$ similarity transformation parameters from the Locator.

For a simple straight seam path, the Tracker task error is given as follows. First, as in the Locator case, P_i^m is determined by sensing the seam of the model workpiece. The Tracker does this by slowly moving along the path. In a steady task flow, the Tracker moves at a speed higher than that needed for welding by detecting the actual path vector, P_j^w. Note that $i \neq j$ because the velocity is different. The Tracker then estimates the two lines ($Y_m(x_m)$ and $Y_w(x_w)$) corresponding to P_i^m and P_j^w by using the least-squares method. The corresponding relations between the seam-start point and the seam-end point on P_i^m and P_j^w are known. The unknown corresponding relations between the other points are assumed to be $x_m \simeq x_w$ for simplicity. The Tracker task error λ_t is defined as

$$\begin{aligned}\lambda_t &= \frac{1}{j} [\; (p_1^m - p_1^w)^2 \\ &+ \sum_{k=2}^{j-1} \{Y_m(x_{w,k}) - Y_w(x_{w,k})\}^2 \\ &+ \; (p_i^m - p_j^w)^2 \;]. \end{aligned} \quad (3)$$

The effect of task velocity and workpiece position on the Tracker task error is shown in Fig. 7. The Tracker task error increases with the task velocity and with the workpiece position tolerance. The relationship between the task error and the task velocity is similar to that for the Locator. This was expected because (a) both the Locator sensing and Tracker sensing are affected by the sensing motion speed and (b) both the Tracker and the Locator use the same recognition algorithm for their laser range sensor. From these results we can derive a simple but important principle: task quality is degraded if the task velocity is increased. This means there is a trade-off between task error and task velocity.

2.6 Collaboration Between Locator and Tracker

The required performance for a typical hybrid task can be expressed as follows:

$$\lambda_n \leq \lambda_{ref}, \quad (4)$$

$$\sum_{i=1}^{n} t_{vi} \leq T_{ref}, \quad (5)$$

where $t_{vi} = l_i/v_i$, v_i represents the task velocity and l_i represents the distance of the i-th individual task; λ_{ref} and T_{ref} are the reference values of the task error and total task execution time, respectively. These are previously given based on the user's task specifications.

To avoid colliding with BRM1, BRM2 must wait a moment before it performs the i-th task on a workpiece until BRM1 completes the $(i-1)$-th task on that workpiece. This waiting time wt_i is added to Eq. (5):

$$\sum_{i=1}^{n} (t_{vi} + wt_i) \leq T_{ref}. \quad (6)$$

Workpiece approach distance p is introduced as a variable parameter. For practical application, preventing collisions between the robot's end effector and the workpiece is important. Therefore, the BRM increases its p when the task errors of the other BRMs are relatively large. But in other words, l_i increases when a BRM increases p. Hence, task efficiency will be degraded, because the waiting time of the other BRMs will be lengthened. That is, there is a cause-and-effect relation between the task error and the task efficiency represented by the waiting time and the approach distance. A satisfactory trade-off between them is thus needed. Giving a general solution to this problem is difficult. To overcome this problem, we introduce fuzzy rules, which can handle vague information and easily represent the experience and knowledge of operators.

In our fuzzy-rule-based collaboration system(Fig. 8), the task velocity is adjusted based on fuzzy inference. The membership functions and fuzzy rules used for collaboration planning by the fuzzy inference engine are shown in Fig. 9. In the experiments, the same fuzzy rules were used for both the Tracker and the Locator. We did this because (a) if the chain of Locator tasks, i.e., loading, locating, and unloading is considered as one hybrid task, then $\lambda_n = \lambda_l$ in Eq. (4), and (b) if all the tasks of the robotic cell are considered as one hybrid task, then $\lambda_n = \lambda_t$ in Eq. (4). The fuzzy rules for each BRM are designed to reduce the task error of that BRM based on the waiting times and task errors of the other BRMs. In these fuzzy rules, the parameters on the left-hand side (input) are the deviations

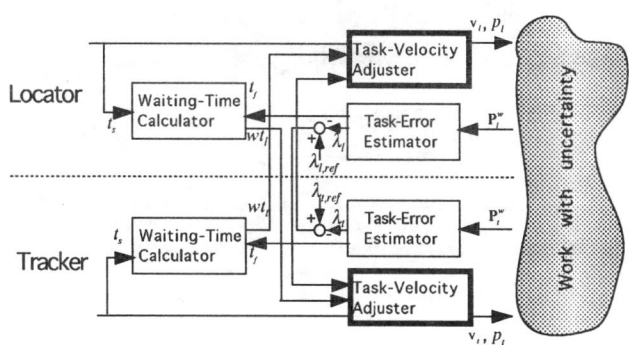

Figure 8: Concept of fuzzy-rule-based collaboration system.

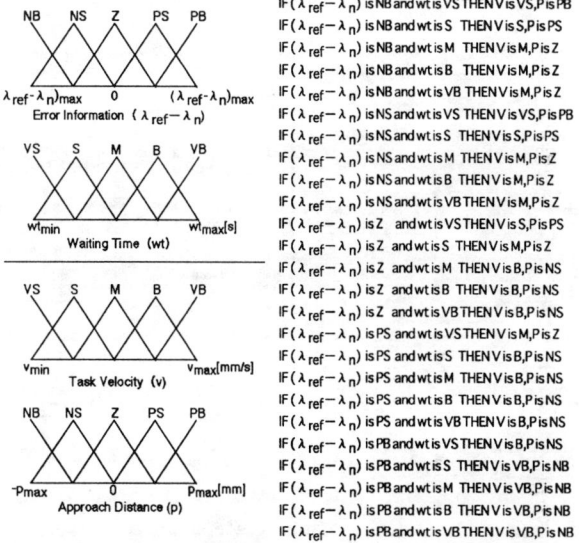

Figure 9: Membership functions and fuzzy rules for collaboration planning.

of the task errors of the other BRMs, $(\lambda_{ref} - \lambda_n)$, and the waiting times of the other BRM, wt. The parameters on the right-hand side (output) are task velocity v of the BRM itself and approach distance p, which represents the deviation from a standard position. A negative p means that the distance to the workpiece is short, and a positive p means that the distance to the workpiece is long. The membership functions on the left- and right-hand sides are triangular, and are the generally used functions. The min-max center of the gravity method is used to infer the velocity and work approach distance. For the experiment discussed below, the fuzzy rules and membership functions were determined based on previous experimental data.

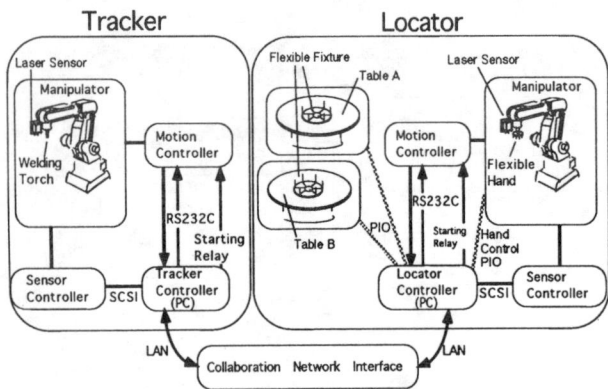

Figure 10: Collaboration system hardware used for experimental evaluation.

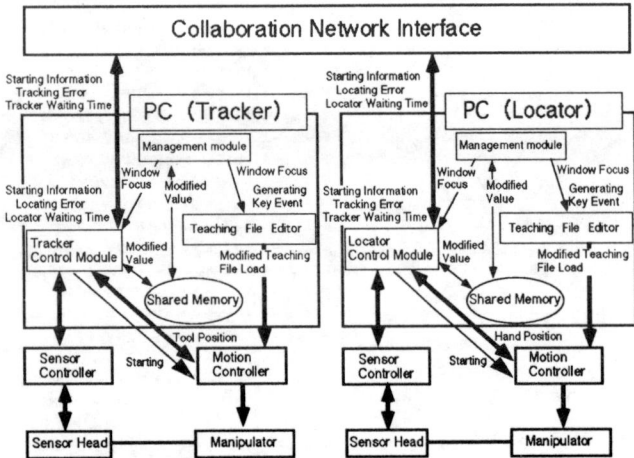

Figure 11: Collaboration system used for experimental evaluation.

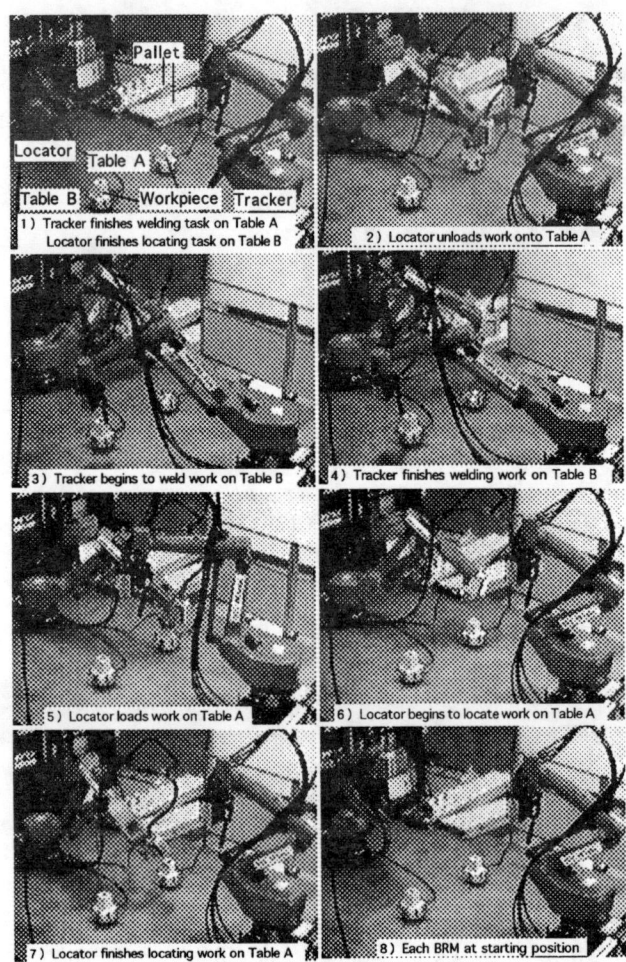

Figure 12: Multi-station welding tasks performed in experimental evaluation.

3 Experimental Evaluation

3.1 System Overview

As shown in Fig. 10, the Tracker consists of a PC (Gateway2000, P5-166), used to control the tracking, a 6-DOF industrial robot and its motion controller (Daihen, ALMEGA ROBOLAND), and a laser range sensor system (NTT Fanet, SH+). The Locator consists of the same devices, plus a pneumatic hand (Hamec, HM Hand B6072S) used as the robot's end effector and two large pneumatic hands (Hamec, HM Hand B6108S) used as a flexible fixture.

In the collaboration system (Fig. 11), the end-effector position and orientation information is sent from the motion controller to a PC, which is connected to the network interface. Through this network, each BRM shares its sensory and motion information. The teaching file is modified using an editing system (Daihen, ED-8700). Normally an operator would do this by keying in the changes. In our system, this editing work is done automatically by a management module that generates key-events. The modified teaching file is sent to the motion controller via a serial port.

As shown earlier, this system uses commercially available industrial robots and PCs, so it can be easily constructed.

3.2 Test Cases

Photographs depicting the steps in the MSW collaboration task are shown in Fig. 12. We tested three cases:

Case 1:
  –no collaboration
  –slow task velocity
  –invariable velocity($v_l = v_t = v_{min}$)

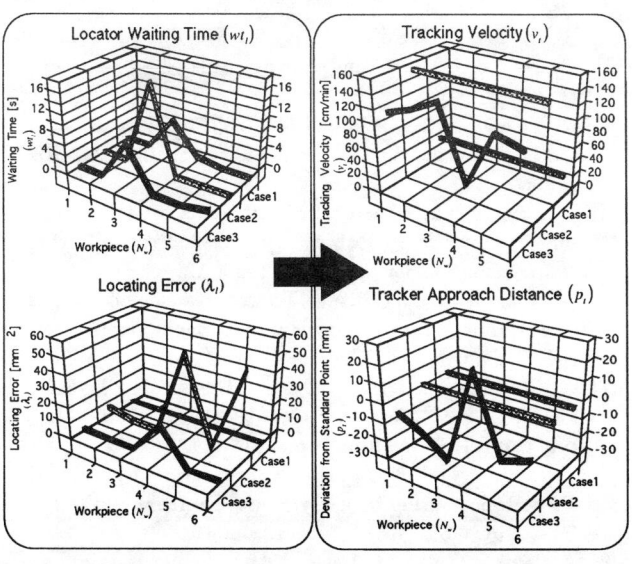

Figure 13: Effect of Locator performance on Tracker parameters.

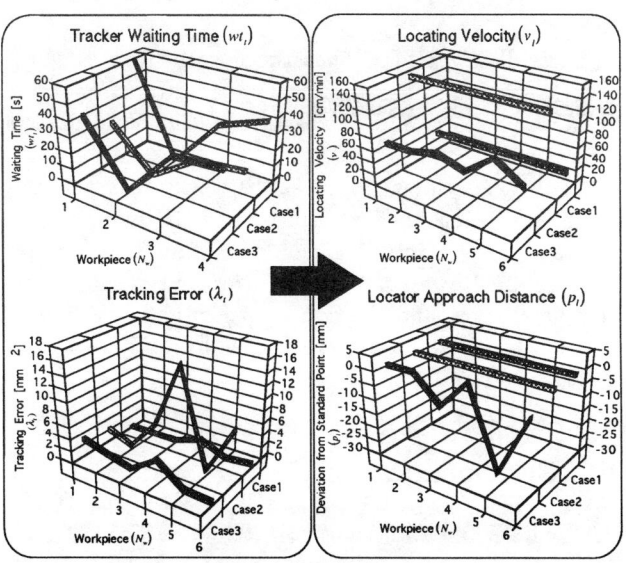

Figure 14: Effect of Tracker performance on Locator parameters.

Case 2:
- no collaboration
- fast task velocity
- invariable velocity ($v_l = v_t = v_{max}$)

Case 3:
- collaboration
- variable task velocity

For Case 3, as shown in Fig. 9, v_l, p_l and v_t, p_t vary in the range $v_{min}[\text{mm/s}] \leq v \leq v_{max}[\text{mm/s}]$, $-p_{max}[\text{mm}] \leq p \leq p_{max}[\text{mm}]$, and $wt_{max} = 10\text{s}$. Also, $v_{min} = 5\text{mm/s}$, $v_{max} = 25\text{mm/s}$, $p_{max} = 25\text{mm}$. For all cases, $v_l = 10\text{mm/s}$ for the first two workpieces,

The v_l is the locating velocity of the Locator, p_l is the approach distance of the Locator, v_t is the velocity of detecting and tracking of the Tracker, and p_t is the approach distance of the Tracker. The approach distance is the deviation from the standard position given by teaching.

In Cases 1 and 2, p_l and p_t were not variable. In all cases, motion velocity was set equal except for v_l, p_l and v_t, p_t.

3.3 Results and Discussion

The effect of Locator performance on the Tracker parameters is shown in Fig. 13. The left-hand side, i.e., the input to the Tracker's velocity adjuster, shows waiting time wt_l and locating task error λ_l, and the right-hand side, i.e., the output of the Tracker's velocity adjuster, shows task velocity v_t and approach distance p_t. The horizontal axes represent the workpieces.

The waiting time of the Locator is small in all cases, and is not remarkably different between cases. This was expected because the task execution time for each workpiece on the Locator is longer than that on the Tracker, because the tracking distance on the Tracker is shorter. Also, on workpieces $N_w \geq 3$, λ_l in Case 3 is smaller than that in Case 2 because λ_l depends on task velocity. In Case 3, when task error λ_l is small and waiting time wt_l is long, task velocity v_t is higher and approach distance p_t is shorter. That is, the Tracker made up for the effect of wt_l in the total task execution time. On the other hand, when λ_l is large and wt_l is small, v_t is lower and p_t is longer. That is, the Tracker made up for the effect of λ_l in task error λ_t. This compensation is most remarkable for workpiece 4. In this case, wt_l is relatively short and λ_l is relatively large. Hence, the Tracker's velocity adjuster determines that v_t is relatively short and p_t is relatively long.

The effect of Tracker performance on the Locator parameters is shown in Fig. 14. The left-hand side, i.e., the input to the Locator's velocity adjuster, shows waiting time wt_t and task error λ_t, and the right-hand side, i.e., the output of the Locator's velocity adjuster, shows task velocity v_l and approach distance p_l. The horizontal axes again represent the workpieces. Note that because the two fixtures are alternatively used in the MSW task and each fixture has different characteristics, v_l and p_l for the N_w-th workpiece is computed based on task error λ_t and waiting time wt_t for the $(N_w - 2)$-th workpiece. On workpieces $N_w \geq 3$ in Case 3, wt_t is smaller than that in Case 1, and λ_t is smaller than that in Case 2. In Case 3, on workpiece 3, –which is fixed by a flexible fixture on table A, wt_t is long and

λ_t is small. Therefore, when the Locator locates workpiece 5, which is the next workpiece after 3 on table A, it increases v_l and decreases p_l to make up for the effects of wt_t.

In this experiment, because the workpieces were different and each fixed workpiece had its own position tolerance, the results for each case cannot be strictly compared. However, the task error and waiting time for the Tracker, and the task error for the Locator were lower than those for the worst case by 31%, 79%, and 20%, respectively, on average.

4 Conclusion

We have experimentally investigated a cell-type collaboration system consisting of sensor-enhanced robots.

The basic concept of this robotic cell collaboration is that, each robot module refers to the task error and task velocity information of the other robot modules and modifies its own motion accordingly. A collaboration system constructed based on this concept achieved a satisfactory trade-off between efficiency and accuracy in performing a hybrid task.

Two basic robot modules were used in our test system: a Locator for estimating the position and orientation of the workpiece and a Tracker for tracking a welding seam path. We found that the task error tended to increase with the tolerance of the workpiece position and with the task velocity.

Examination of the relationship between the task error and the task velocity showed that better way to modify the task parameters is to use fuzzy rules.

We constructed an experimental system consisting of commercially available laser range finders, industrial robots, and PCs. Each robot module adjusts its motion to make up for the task errors and waiting times of the other modules. It therefore performed a hybrid task with a satisfactory trade-off between efficiency and accuracy. This proposed concept must be effective not only for simple robot cells but also for flexible manufacturing systems.

A remaining problem is self-tuning of the fuzzy rules[11]. Besides working on this, we will investigate flexible motion for irregular events, and extend the system to use more robots.

References

[1] T. Bamba, et al., "A Visual Seam Tracking System for Arc-welding Robots", Proc. of 13th ISIR, pp. 365-374, 1984.

[2] S. Sawano, et al., "A Sealing Robot System with Visual Seam Tracking", Proc. of '83 Int. Conf. on Adv. Rob., pp. 351-358, 1983.

[3] B. Irving, "Is Welding Ready for Flexible Manufacturing Systems?", Welding Journal, pp. 43-47, December 1966.

[4] K. Kosuge, et al., "Unified Control for Dynamic Cooperative Manipulation", Proc. of IROS, pp. 1042-1047, 1994.

[5] G. Beni, et al., "Theoretical Problems for the Realization of Distributed Robotic Systems", Proc. IEEE Int. Conf. Rob. Autom., pp. 1914-1920, 1991.

[6] H. Asama, et al.,"Mutual Transportation of Cooperative Mobile Robots Using Forklift Mechanisms", Proc. IEEE Int. Conf. Rob. Autom., pp. 1754-1759, 1996.

[7] G. P. Castellote, et al., "System Design and Interfaces for Intelligent Manufacturing Workcell", Proc. IEEE Int. Conf. Rob. Autom., pp. 1105-1112, 1995.

[8] T. Y. Li, et al., "On-Line Manipulation Planning for Two Robot Arms in a Dynamic Environment", Proc. IEEE Int. Conf. Rob. Autom., pp. 1048-1055, 1995.

[9] A. Ishii, et al., "A Laser Vision Sensor for an Arc-welding Robot", Proc. of 3rd France-Japan congress & 1st Europe-Asia congress on Mecatronics, Vol. 1, pp. 166-169, 1996.

[10] S. Umeyama, "Least-Squares Estimation of Transformation Parameters Between Two Point Patterns", IEEE Trans. Pattern Anal. Machine Intell., Vol. 13, No. 4, pp. 376-380, 1991.

[11] T. Fukuda, et al., "Fusion of Fuzzy, NN, GA to the Intelligent Robotics", Proc. IEEE Int. Conf. Syst. Man. Cybern., pp. 2892-2897, 1995.

An Expert Opinion Approach to Tune Analytical Models of Nonlinear Systems

K.R. Chernyshov
Institute of Control Sciences
Moscow, 117806

F.F. Pashchenko
Institute of Control Sciences
Moscow, 117806

Abstract

A general approach combining both analytical and expert knowledge-based techniques to identify a nonlinear system model is proposed. The approach leads to using a new type of stochastic dependence of random processes, involving expert opinion, and which is a generalization of the disperssional identification technique.

1 Introduction

A powerful extension of the conventional control theory techniques generated by involving knowledge-based approach leads to consideration of intelligent control systems. Generically, intelligent control system design implies solving problems corresponding to the following three hierarchical levels [1]. The upper level is a heuristic logical inference based on knowledge associated with the control process. The middle level corresponds to current information processing based on advanced schemes of experimental/sample data analysis. The lower level involves elaborating and transferring control signals.

Creating intelligent systems, or, generically, intelligent machines [2,3] which are able to function in a reliable and effective manner at various autonomous levels under known or unknown environment is an increasingly developed branch of modern control system theory. The branch has already involved a significant amount of research works [4]. Intelligent machines theory may be considered as a mathematical problem of obtaining a correct sequence of controls for a hierarchical system which meets to the principle of increasing precision with decreasing intelligence, with the sequence minimizing the total system entropy [2].

Advanced approaches to model the systems are presented in [1,4,5]. The approaches are represented by the following research branches [4]:

- design of planning and control systems, based on application of artificial intelligence and intelligent control in order to achieve autonomous performance [6-8];
- design of intelligent learning systems based on system theory [9,10];
- design of intelligent machines and systems based on Petri nets [11-13];
- task planning (at various abstract levels) using various approaches related to the artificial intelligence [14-20];
- expert control [21-23];
- modeling intelligent machines based on the principle of increasing precision with decreasing intelligence [2-4, 24-25].

Most of them assume deriving an analytical plant model and adaptation of the model. In another words, the identification problem takes an important place within the approaches, being a necessary preliminary step for solving a control problem when the plant model is unknown to some extent. In turn, body of knowledge about the model may vary from lack of information on the model structure at all to uncertainties in values of the model parameters.

Analytical models describing real-world processes have essential disadvantages which restrict branch practical applications [26]. It is caused by the circumstance that no model can properly handle all the factors affecting the system. To overcome the disadvantage some efforts have been taken. The efforts are based on heuristic fitting the system model in order to obtain the required properties of the model to be used [27-31]. The heuristic fitting provides matching of intuitive a priori assumptions on the system and determines the system model structure. Solving the problem requires using knowledge about the system, which enable one to handle the factors being omitted under conventional formal procedures.

2 Problem Statement

A model approximating the input/output behavior of a real-world dynamic system described, in general, in the form

$$F(y(t), x(s)) = 0, \quad (1)$$

will be searched as the following relationship

$$z(t) = L_{t,s} u(s). \quad (2)$$

In Eq. (1), $y(t)$ is an m-dimensional column-vector considered as the system output, $x(s)$ is an n-dimensional column-vector considered as the system input. In Eq. (2), $z(t)$ is an m-dimensional column-vector considered as the model output process, $u(s)$ is an n-dimensional column-vector considered as the model input process, $L_{t,s}$ is a linear dynamic operator subject to identification by observing processes $y(t)$ and $x(s)$. Processes $z(t)$ and $u(s)$ are considered as results of transforming processes $y(t)$ and $x(s)$ by some transformations B and C respectively, i.e.

$$z(t) = By(t), \quad u(s) = Cx(s). \quad (3)$$

In turn, within model (2) identification problem, transformations B and C are to be chosen to meet model (2) identification error $e(t)$,

$$e(t) = z(t) - L_{t,s} u(s), \quad (4)$$

to be acceptable in the sense of magnitude of the value $d(e(t))$, where $d(e(t))$ is a given measure calculated at $L_{t,s} = \hat{L}_{t,s}$. Again, $\hat{L}_{t,s}$ is determined from the condition

$$\hat{L}_{t,s} = \arg\inf_{L_{t,s}} d(e(t)). \quad (5)$$

Solving the problem formulated is constructed by combination of a formal analytical technique to determine $\hat{L}_{t,s}$ from Eq. (5) and heuristic methods to obtain transformations B and C in Eq. (3) based on involving knowledge about the investigated system.

A generalized scheme corresponding to the approach proposed is presented in the fig. 1 which is an analogy of that in [1]. The scheme is a hierarchical two-level structure. Its upper level combines data base and inference device, with lower level corresponding to using analytical identification algorithms. Basing on knowledge from the data base and magnitude of the value $d(e(t))$, the inference device forms transformations B and C in Eq. (3). At the lower level, under transformations B and C, operator $\hat{L}_{t,s}$ from Eq. (5) is obtained by formal analytical techniques, and magnitude $d(e(t))$ is calculated, corresponding to the $\hat{L}_{t,s}$, B and C.

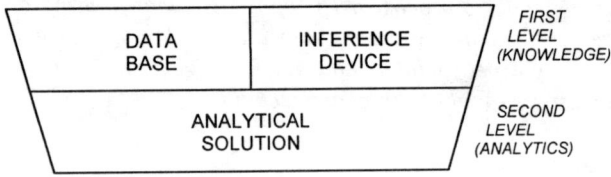

Fig. 1

Thus formulated, the problem statement reflects the following approach to system structure identification. Conventionally, unknown model structure identification is a direct search of an input/output relationship based on observing input and output system variables, with a corresponding scheme being presented in fig. 2.

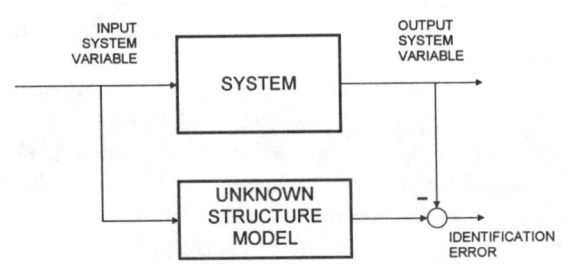

Fig. 2

Another way is to update the input and output variable by a manner when the system operator will take a priori given simple form, with identification of the operator analytical shape being implemented without further information on the system structure. Such an identification scheme is presented in fig. 3. In particular, within the Section, the model operator is searched as a linear transformation of the input model process into the output model process.

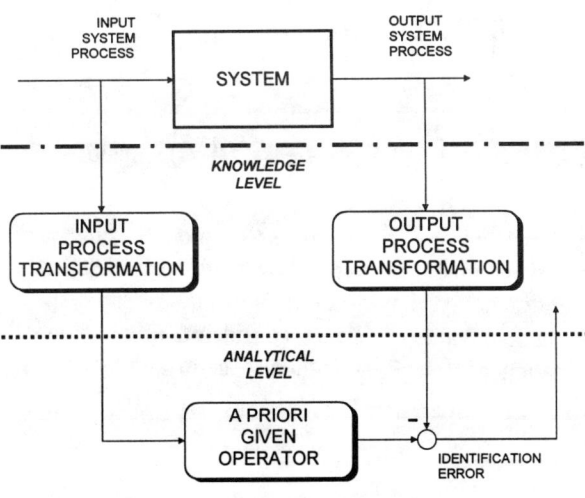

Fig. 3

Also, such an adopted approach reflects a principle of hierarchical system organization, which is known as increasing precision with decreasing intelligence [2,3].

3 Deriving SISO Model

Consider a single input / single output (SISO) model of form (2), in which the operator $L_{t,s}$ has the form

$$L_{t,s}u(s) = \int_{I_t} g(s)u(s)ds, \qquad (6)$$

with I_t being a time interval at which the system behavior is considered.

The model is natural to be considered as an approximation of a nonlinear dynamic system driven by an observable input scalar-valued process $x(s)$, and having corresponding output scalar-valued process $y(t)$. As a measure of $d(e(t))$ the mean-squared deviation

$$d(e(t)) = Me^2(t)$$

will be taken, where

$$e(t) = z(t) - \int_{I_t} g(s)u(s)ds,$$

and

$$\hat{g}(s) = \arg\inf_{g(s)} d(e(t)) \qquad (7)$$

is the desired estimate of the weight function of the linear integral operator in Eq. (6), subject to identification.

To argue the heuristic approach to determine processes $z(t)$ and $u(s)$ in the model given a reasoning based on the following analogies will be applied. As well known, from one hand side, the conditional mathematical expectation of a random value with respect to another one is the best approximation of a functional dependence between the variables. Again, within the intelligent system theory, values of the random process which is conditionally expected are natural to be considered as a kind of knowledge associated with the system, i.e. an expert opinion on values of $z(t)$ and $u(s)$ given $y(t)$ and $x(s)$ respectively. In turn, the marginal densities of $z(t)$ and $u(s)$ are natural to be assumed to meet Jaynes' maximum entropy principle [32].

From another hand side, there exists a terminological analogy within which the orthogonal projection of a random element onto a linear space formed by a set of random elements, being the best approximation of the element, is adopted to refer as conditional mathematical expectation in weak sense. In addition, for Gaussian random values, notions of the conditional mathematical expectation and the conditional mathematical expectation in weak sense coincide.

Thus, an intuitive solution is being appeared to use conditional mathematical expectation of $z(t)$ and $u(s)$ with respect to the corresponding processes to obtain the transformations B and C in Eq. (3), i.e.

$$By(t) = M\left(z(t)/y(t)\right),$$

$$Cx(s) = M\left(u(s)/x(s)\right),$$

with $M(\cdot/\cdot)$ standing for the conditional mathematical expectation.

Then, the system model takes the form

$$M\left(z(t)/y(t)\right) = \int_{I_t} g(s) M\left(u(s)/x(s)\right) ds. \quad (8)$$

Following the above reasoning, processes $z(t)$ and $u(s)$ in Eq. (8) have the maximum entropy marginal distribution densities $p(z)$ and $p(u)$ respectively, i.e. satisfying the conditions

$$z = -\mu_z - \ln p(z), \ u = -\mu_u - \ln p(u), \quad (9)$$

with the corresponding constants μ_z and μ_u.

Under Eq. (8), problem (7) is determined by the expression

$$\hat{g}(s) =$$
$$= \arg\inf_{g(s)} M\left(M\left(z(t)/y(t)\right) - \int_{I_t} g(s) M\left(u(s)/x(s)\right) ds \right)^2. \quad (10)$$

It follows from Eq. (10) that the desired weight function $\hat{g}(s)$ is to meet the following equation which extends the disperssional identification equation [33],

$$\kappa_{zuyx}(t,s) = \int_{I_t} g(\sigma) \kappa_{ux}(s,\sigma) d\sigma, \quad (11)$$

with

$$\kappa_{zuyx}(t,s) = M\left(\left(M\left(z(t)/y(t)\right) - M\{z(t)\} \right) \times \right.$$
$$\left. \times \left(M\left(u(s)/x(s)\right) - M\{u(s)\} \right) \right)$$

being the crossdisperssional function of the expert opinions $z(t)$ and $u(s)$ with respect to the random processes $y(t)$ and $x(s)$ correspondingly. The function is natural to be referred as *EO*(*expert opinion* based)-crossdisperssional function. Again,

$$\kappa_{ux}(s,v) = M\left(\left(M\left(u(s)/x(s)\right) - M\{u(s)\} \right) \times \right.$$
$$\left. \times \left(M\left(u(v)/x(v)\right) - M\{u(v)\} \right) \right),$$

is the *EO*-autodisperssional function of the expert opinion $u(s)$ with respect to the random process $x(s)$.

The technique to derive a multi input / multi output (MIMO) model for Eq. (2), Eq. (3) is equivalent to the scalar case.

By virtue of Eq. (9), Eq. (8) can be rewritten as

$$H\left(z(t)/y(t)\right) = \int_{I_t} g(s) H\left(u(s)/x(s)\right) ds,$$

with $\mu_z = \mu_u \int_{I_t} g(s) ds$. Here $H(\cdot/\cdot)$ stands for conditional entropy of the expert opinion $z(t)$ and $u(s)$ with respect to the corresponding random process, with the conditional entropy of a random value ξ with respect to a random value ζ being considered as conditional expectation of $(-\ln p(\xi))$ with respect to ζ. Again, the above *EO*-disperssional functions $\kappa_{zuyx}(t,s)$ and $\kappa_{ux}(s,v)$ take the form

$$\kappa_{zuyx}(t,s) = M\left(\left(H\left(z(t)/y(t)\right) - H\{z(t)\} \right) \times \right.$$
$$\left. \times \left(H\left(u(s)/x(s)\right) - H\{u(s)\} \right) \right),$$

$$\kappa_{ux}(s,v) = M\left(\left(H\left(u(s)/x(s)\right) - H\{u(s)\} \right) \times \right.$$
$$\left. \times \left(H\left(u(v)/x(v)\right) - H\{u(v)\} \right) \right).$$

Thus, the above equations establish an explicit link between disperssional and knowledge-based identification of an analytical model, with Eq. (11) being able to be rewritten in the form involving conditional entropies instead of conditional means.

4 Conclusions

A concept to improve nonlinear identification process by combining analytical and expert opinion based techniques has been proposed. Within the concept a nonlinear identification problem is considered as a two-step scheme meeting the principle of increasing precision with decreasing intelligence. From the above it follows that existence of a relevant knowledge is a necessary condition for the problem to be identified within the proposed scheme. In turn, relevance of the knowledge within the identification scheme is just a stochastic dependence of the expert opinions and the system input and output random processes. In turn, the *EO-*disperssional functions proposed enable one to construct a measure of relevance of the knowledge.

As well as continuous-time systems considered, the approach is applicable to discrete-time systems identification, with the linear integral operator being substituted by the corresponding discrete-time weight function.

Appendix: Disperssional functions

Within the disperssional identification techniques the following functions which describe nonlinear dependence between random processes are frequently used [33]:

- proper crossdisperssional function of random processes $y(t)$ and $x(s)$

$$\theta_{yx}(t,s) = M\left(M\left(y(t)/x(s)\right) - My(t)\right)^2;$$

- autodisperssional function

$$\theta_{xx}(t,s) = M\left(M\left(x(t)/x(s)\right) - Mx(t)\right)^2,$$

- generalized disperssional function of random processes $y(t)$, $z(v)$, and $x(s)$

$$\theta_{yzx}(t,v,s) = M\left(\left(M\left(y(t)/x(s)\right) - My(t)\right) \times \left(M\left(z(v)/x(s)\right) - Mz(v)\right)\right),$$

- disperssional *R*-function [34] of random processes $y(t)$, $z(v)$, $x(s)$, and $u(\tau)$

$$\theta_{yzxu}(t,v,s,u) = M\left(\left(M\left(y(t)/x(s)\right) - My(t)\right) \times \left(M\left(z(v)/u(\tau)\right) - Mz(v)\right)\right).$$

References

[1] R. Doraiswami, J. Jiang, "Performance monitoring in expert control systems", *Automatica*, Vol. 25, No 6, pp. 799-811, 1989.

[2] G.N. Saridis, "Analytical formulation of the principle of increasing precision with decreasing intelligence for intelligent machines", *Automatica*, Vol. 25, No 3, pp. 461-468, 1989.

[3] G.N. Saridis, K.P. Valavanis, "Analytical design of intelligent machines", *Automatica*, Vol. 24, No 2, pp. 123-133, 1988.

[4] K.P. Valavanis, H.M. Stellakis, "A general organizer model for robotic assemblies and intelligent robotic systems", *IEEE Trans. Syst. Man Cybern.*, Vol. SMC-21, No 2, pp. 303-317, 1991.

[5] M.G. Rodd, H.B. Verbruggen, A.J. Krijgsman, "Artificial intelligence in real-time control", *Engng. Applic. Artif. Intell.*, Vol. 5, No 5, pp. 385-399, 1992.

[6] K.M. Passino, P.J. Anstalkis, "A system and control theoretic respective on artificial intelligence planning systems", *Appl. Artificial Intell.*, Vol. 3, 1989.

[7] P.J. Antsalkis, K.M., Passino, S.J. Wang, "Towards intelligent autonomous control systems, architecture and fundamental issues", *J. Intelligent and Robotic Syst.*, Vol. 1, No 4. 1988.

[8] K.M. Passino, P.J. Anstalkis, "Fault detection and identification in an intelligent restructurable controller", *J. Intelligent and Robotic Syst.*, Vol. 1, 1988.

[9] A. Meystel, "Intelligent control in robotics", *J. Robotic Syst.*, Vol. 5, 1988.

[10] A. Meystel, Ed., *J. Intelligent and Robotic Syst., Special Issue on Intelligent Control.*, Vol. 2, No 2-3, 1989.

[11] H.J. Genrich, K. Lautenbach, "System modeling with high-level Petri nets", *Theoretical Comput. Sci.*, Vol. 13, 1981.

[12] T. Murata, "Petri nets: Properties, analysis and applications", *Proc. IEEE.*, Vol. 77, No 4, 1989.

[13] Wang Fei-Yue, G.N. Saridis, "Task translation and integration specification in intelligent machines", *IEEE Trans. Rob. Autom.*, Vol. 9, No 3, pp. 257-271, 1993.

[14] R.E. Fikes, N.J. Nilsson, "STRIPS: A new approach to the application of theorem proving to problem solving", *Artificial Intell.*, Vol. 2, 1971.

[15] E.D. Sacerdoti, "Planning in a hierarchy of abstraction spaces", *Artificial Intell.*, Vol. 5, 1974.

[16] M. Stefik, "Planning with constraints (MOLGEN: Part 1)", *Artificial Intell.*, Vol. 15, No 2, 1981.

[17] M. Stefik, "Planning and meta-planning: representation and plan generation (MOLGEN: Part 2)", *Artificial Intell.*, Vol. 15, No 2, 1981.

[18] D.E. Wilkins, "Recovering some execution errors in SIPE", *Computational Intell.*, Vol. 1, No 1.

[19] D.E. Wilkins, "Domain-independent planning: representation and plan generation", *Artificial Intell.*, Vol. 22, No 3, 1984.

[20] C. Tsatsoulis, R.L. Kashyap, "Case-based reasoning and learning in manufacturing with the TOLTEC planner", *IEEE Trans. Syst. Man Cybern.*, Vol. 23, No 4, pp. 1010-1023, 1993.

[21] K.J. Astrom, J.J. Anton, K.-E. Arzen, "Expert control", *Automatica*, Vol. 22, pp. 277-286, 1986.

[22] K.-E. Arzen, "An architecture for expert system based feedback control", *Automatica*, Vol. 25, No 6, pp. 813-827, 1989.

[23] K.J. Astrom, C.C. Hang, P. Persson, W.K. Ho, "Towards intelligent PID control", *Automatica*, Vol. 28, No 1, pp. 1-9, 1992.

[24] K.P. Valavanis, G.N. Saridis, "Information theoretic modeling of intelligent robotic systems", *IEEE Trans. Syst. Man Cybern.*, Vol. SMC-18, No 6, 1988.

[25] K.P. Valavanis, S.J. Carelo, "An efficient planning technique for robotic systems", *J. Intelligent and Robotic Syst.*, Vol. 3, No 4, 1990.

[26] R.S. Freedman, G.J. Tuzin, "A knowledge-based methodology for tuning analytical models", *IEEE Trans. Syst. Man Cybern.*, Vol. SMC-21, No 3, pp. 347-358, 1991.

[27] C. Apte, R. Dionne, "Building numerical sensitivity analysis using a knowledge-based approach", *IEEE Expert.*, Vol. 3, No 2, pp. 371-378, 1988.

[28] B. Fischoff, "Eliciting knowledge for analytical representation", *IEEE Trans. Syst. Man Cybern.*, Vol. SMC-19, No 3, pp. 448-461, 1989.

[29] W. Rouse, J. Hammer, C. Lewis, "On capturing human skills and knowledge: algorithmic approach to model identification", *IEEE Trans. Syst. Man Cybern.*, Vol. SMC-19, No 3, pp. 558-573. 1989.

[30] M. Lacy, "Artificial laboratories", *Artificial Intell. Mag.*, Vol. 10, No 2, pp. 43-48, 1989.

[31] H. Abelson, M. Eisenberg, M., Halfant, J. Katznelson, E. Sackes, G., Sussman, J. Wislom, K. Yip, "Intelligence in scientific computing", *Commun. ACM*, Vol. 32, No 5, pp. 546-561, 1989.

[32] E.T. Jaynes, "Information theory and statistical mechanics", *Physical Review*, Vol. 106, No 4, pp. 620-630, 1957.

[33] N.S. Rajbman, "Extensions to nonlinear and minimax approaches", *Trends and Progress in System Identification*, ed. P. Eykhoff, Pergamon Press, Oxford, 1981, pp. 185-237.

[34] I.S. Durgaryan, F.F. Pashchenko, "Identification of nonlinear plants with complex criteria", *Automation and Remote Control*, Vol.41, No 7, pp. 935-943, 1980.

Virtual Truss Model for Characterization of Internal Forces for Multiple Finger Grasps

Tsuneo YOSHIKAWA

Department of Mechanical Engineering
Kyoto University, Kyoto, 606 Japan
yoshi@mech.kyoto-u.ac.jp

Abstract — Internal force plays important roles in grasping and manipulation of objects by multiple fingers. It is well known that any internal force for three finger grasps with frictional point contacts can usually be decomposed into three pushing or pulling force components along the three lines joining the contact points. This paper studies the question of how far this intuitive decomposition can go for general multiple finger grasps. We propose a virtual truss model for the grasped object for characterizing the internal force and show that this characterization is valid for general multiple finger grasps under a non-degeneracy condition. We then discuss some degenerate cases and give a method of extending the virtual truss model to cope with these special cases. Based on this characterization, we give an explicit formulation of the categorization by Ponce et. al. of four finger grasps into concurrent, pencil, and regulus grasps. A method is also given for obtaining a non-redundant expression of all possible internal forces from this characterization.

1 Introduction

Grasping and manipulation of an object by multiple fingers have been studied by many researchers. One important issue in this research area is the characterization of internal forces, because it is a key step in dealing with various problems such as analysis of force closure grasp, determination of fingertip forces, optimization and planning of grasping and manipulation, analysis of power grasp, active fixtures, and so on [1]-[9].

Internal force has usually been characterized by the null space of a matrix that relates the vector of fingertip contact forces to the vector of resultant force. It is well known for three-fingered hands that three components of contact forces between three pairs of fingers are useful for synthesizing the fingertip contact forces for grasping and manipulation (see, for example, [4][5][6]). In [7] characterization of internal force in multiple whole-limb manipulation was studied.

In [8] and [9] it was shown using Grassmann geometry that four finger force-closure grasps can be classified into three categories: concurrent grasp, pencil grasp, and regulus grasp.

In [10] the virtual linkage model was proposed as a model of internal forces associated with the manipulation of an object grasped by multiple manipulators. This model consists of linearly actuated members and spherically actuated joints.

This paper proposes a model of internal forces for the case of multiple fingers grasping and manipulating a rigid object with frictional point contacts. Related to the well known fact that any internal force for three finger grasp can usually be decomposed into three pushing or pulling force components between three pairs of fingertips, an interesting question is how far this intuitive decomposition can go for general multiple finger grasps. As an answer to this question, we propose a model for the grasped object for characterizing the internal force, which we call the virtual truss model. It is shown that this model is valid for general multiple finger grasps under a non-degeneracy condition. We then discuss some degenerate cases and give a method of extending the virtual truss model to cope with these special cases. Based on this characterization, we give an explicit formulation of the categorization of four finger grasps by Ponce et. al. [8][9]. A method of obtaining non-redundant representation based on the model is also presented.

2 Multiple Finger Grasp
2.1 Contact Force and Resultant Force

We consider an n-fingered hand grasping and manipulating a rigid object in three-dimensional space as shown in Fig.1. Each finger is assumed to have a point contact with friction with the object and have enough degrees of freedom to exert force in any direction.

Let the contact point of finger i ($i = 1, 2, \cdots, n$) be denoted by C_i, its position vector expressed in a reference coordinate frame Σ_R be $r_i = [r_{ix}, r_{iy}, r_{iz}]^T$,

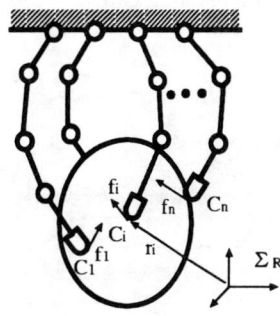

Figure 1: Object grasped by n-fingered hand

and the contact force vector be f_i.

Generally the relation between the resultant force and moment vector t (6-dimensional vector) and the contact force vector $f = [f_1^T, f_2^T, \cdots, f_n^T]^T$ (3n-dimensional vector) is given by

$$t = Gf \tag{1}$$

where G is an $6 \times 3n$ matrix defined by

$$G = \begin{bmatrix} I_3 & \cdots & I_3 \\ R_1 & \cdots & R_n \end{bmatrix} \tag{2}$$

Here I_3 is the 3-dimensional identity matrix and

$$R_i = \begin{bmatrix} 0 & -r_{iz} & r_{iy} \\ r_{iz} & 0 & -r_{ix} \\ -r_{iy} & r_{ix} & 0 \end{bmatrix} \tag{3}$$

Now a general solution of f to (1) will be given. We introduce a norm $\| t \|_{Mt} = (t^T M_t t)^{0.5}$ to the set of all t, and a norm $\| f \|_{Mf} = (f^T M_f f)^{0.5}$ to the set of all f. Weighting matrices M_t and M_f are arbitrarily selected positive symmetric matrices.* Using nonsingular matrices T_t and T_f satisfying $M_t = T_t^T T_t$ and $M_f = T_f^T T_f$, a general solution of (1) is given by

$$f = \hat{G}^+ t + [I - \hat{G}^+ G]\hat{k} \tag{4}$$

where $\hat{k}$ is a $3n$-dimensional arbitrary constant vector,

$$\hat{G}^+ = T_f^{-1}(T_t G T_f^{-1})^+ T_t \tag{5}$$

and "+" denotes the pseudo inverse matrix [13].

2.2 Internal Force

Equation (4) can be regarded as decomposing the fingertip contact forces into two components. The first term on the right hand side is a particular solution that represents the contact force component producing the resultant force and moment t that minimizes the norm $\| f \|_{Mf}$. The second term is a homogeneous solution representing the internal force that gives no effect on t.

Utilizing the fact $G\hat{G}^+ G = G$, we can easily show that the set of all internal forces (internal force subspace in the $3n$-dimensional contact force space) is given by

$$\{f \mid Gf = 0\} = \{f \mid f = [I - \hat{G}^+ G]\hat{k} \text{ for some } \hat{k}\} \tag{6}$$

Since $\text{rank}[I - \hat{G}^+ G] = 3n - \text{rank} G$, the expression $[I - \hat{G}^+ G]\hat{k}$ of the internal force is redundant for $n \geq 3$.

*When we need to change the reference coordinate frame and/or the units of the variables, we change the weighting matrices M_t and M_f accordingly, so that the solution is invariant [11]. Refer to [12] for discussion on the selection of the weighting matrices M_t and M_f.

In dealing with problems related to grasping and manipulation, it is desirable to have an expression of internal force that has a simple and easy physical meaning and to make its redundancy as small as possible. An approach that satisfies these requirements will be presented in the following sections.

It should be noted that for an internal force f_I to be realizable under the condition that the resultant force t should be applied on the object, the fingertip contact force $f = \hat{G}^+ t + f_I$ must satisfy the friction condition at each contact point, that is, each fingertip force must be pushing the object and the force should be within the friction cone [6]. Note also that, in the case of deciding whether a particular set of contact points can have an active (force) closure or not, we need to determine if there exists an internal force that satisfies the friction condition (see Proposition 6 of [14]).

3 Virtual Truss

In order to characterize the internal force, we propose the **virtual truss** as a model of the object grasped at n contact points. As shown in Fig.2, this truss model consists of skeletal line members connecting every pair of contact points, and each member is joined at its ends (that is, at the contact points) to other members by means of frictionless ball-and-socket joints. Hence the members can only resist against the axial force (tensile or compressive force).

In structural mechanics this kind of structure is known as a truss [15]. There are plane trusses in two-dimensional plane and space trusses in three-dimensional space. A truss is said to be rigid if its shape cannot be changed without changing the length of any of the members. The truss is said to be statically determinate if all the forces in the members can be determined by statics for any given forces at the joints that produce zero resultant force.

As will be discussed later, we may take away some members from the virtual truss model to make the truss statically determinate, or as long as the truss is rigid. We can also add some fictitious joints and corresponding line members to deal with some degenerate situations.

The virtual truss model is very similar to the virtual linkage model proposed in [10] in the sense that the objective is to characterize the internal force by a simple physical model. The differences of the truss model from the virtual linkage model are (i) the truss model is for multiple finger grasps whereas the virtual linkage model seems to be more suitable for multiple

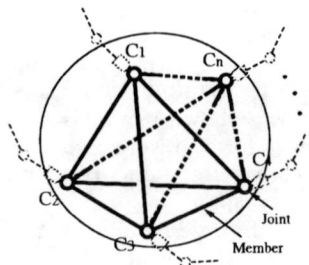

Figure 2: Virtual truss

hand grasps, (ii) the truss model is a passive structure that does not include any actuators, and (iii) there are a lot of research results available on the truss that may be helpful in studying the multiple finger grasps. Further comparison study of these two models should be done in the future.

4 Three-Fingered Hand

4.1 Non-Degenerate Case

We will first consider the case of a three-fingered hand in three-dimensional space. Under the assumption that the three contact points are not collinear, it is well known that the internal force can be expressed by a combination of pushing or pulling pairs of forces between the three pairs of fingers as shown in Fig.3(a). This corresponds to the following expression.

$$[I - G^+ G]\hat{k} = \begin{bmatrix} e_{12} & 0 & e_{13} \\ e_{21} & e_{23} & 0 \\ 0 & e_{32} & e_{31} \end{bmatrix} \begin{bmatrix} k_{12} \\ k_{23} \\ k_{31} \end{bmatrix} = Ek \quad (7)$$

where e_{ij} is the three-dimensional unit vector directing from C_i to C_j and k_{ij} is a real number representing the magnitude of the internal force component between C_i and C_j. Note that $e_{ji} = -e_{ij}$. The column vectors of E in (7) represent a set of basis vectors of the internal force subspace.

It is quite natural in this case to imagine the virtual truss model in Fig.3(b) and to consider the above three parameters k_{12}, k_{23}, and k_{31} as expressing the magnitude of compressive or tensile forces acting on the line members.

We can classify the internal forces into several modes based on whether the force acting on each member is a compressive force or a tensile force. Define $\alpha = [\alpha_{23}, \alpha_{31}, \alpha_{12}]$ where α_{ij} takes value $+1$ (-1) when k_{ij} is positive (negative, respectively), then any realizable internal force can be classified into four groups as shown in Fig.4 [6].

Note that almost the same argument holds also for three-fingered hands in two-dimensional plane.

4.2 Degenerate Case

We consider the case where the three contact points C_1, C_2, and C_3 are collinear, that is, on a straight line.

This degenerate case may not be useful in three-dimensional space because the object can freely rotate around the straight line L including the three contact

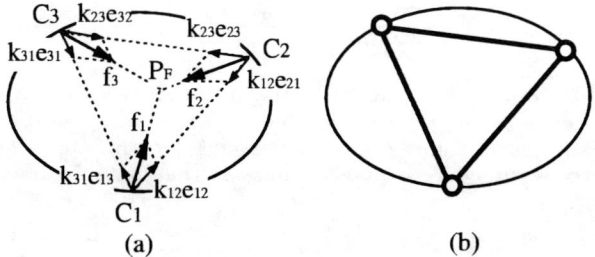

Figure 3: Representation of internal force and truss model for three-fingered hands

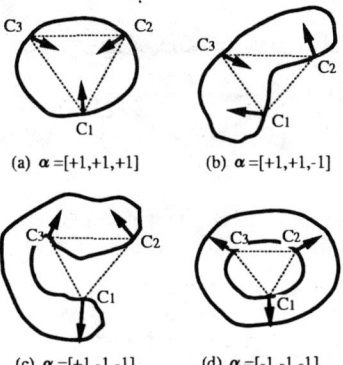

Figure 4: Mode of internal force for three-fingered hands

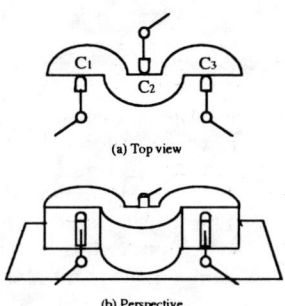

Figure 5: Collinear three contact points in two-dimensional space

points. It will be useful, however, for two-dimensional grasping in a plane. An example is shown in Fig.5.

A minimal dimensional representation of the internal force for this case is given by

$$\begin{bmatrix} e_{12} & 0 & d_{23}e_\eta \\ e_{21} & e_{23} & -(d_{12}+d_{23})e_\eta \\ 0 & e_{32} & d_{12}e_\eta \end{bmatrix} \begin{bmatrix} k_{12} \\ k_{23} \\ k_\eta \end{bmatrix} \quad (8)$$

where d_{12} is the distance between C_1 and C_2, d_{23} is that between C_2 and C_3, e_η is the unit vector on the horizontal plane and normal to the straight line L, and k_η is a real number. The column vectors of the matrix in (8) are a set of basis vectors of the internal force subspace (see Fig.6). Note that, although the first two basis vectors can be interpreted as the axial forces of the virtual truss, the last one cannot. Actually the last vector involves the three contact points and plays an important role in grasping.

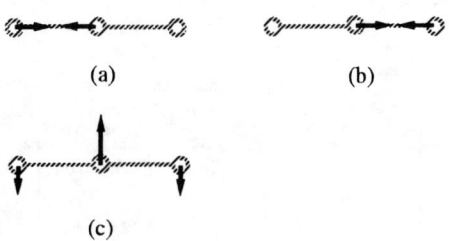

Figure 6: Basis vectors of internal force subspace

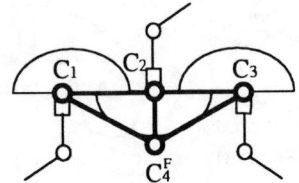

Figure 7: Augmented truss for collinear three contact points in two-dimensional space (top view)

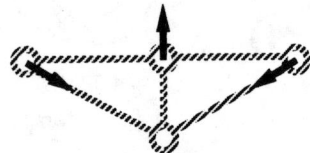

Figure 8: Basis vector corresponding to the third column vector

A method to express the last column vector by axial forces of members is to consider a fictitious contact point C_4^F that does not lie on line L, and add a joint and three line members to the original truss model as shown in Fig.7. This will be called the augmented virtual truss. Since the four points are not collinear, there exists a set of non-zero real numbers $\{q_1, q_2, q_3\}$ such that

$$e_{41}q_1 + e_{42}q_2 + e_{43}q_3 = 0 \tag{9}$$

On the other hand, the constraint at the fictitious contact point C_4^F is that the contact force at this point is zero:

$$e_{41}k_{41} + e_{42}k_{42} + e_{43}k_{43} = 0 \tag{10}$$

Using this augmented truss model, we can show that a representation of the internal force is given by

$$\begin{bmatrix} e_{12} & 0 & e_{14}q_1 \\ e_{21} & e_{23} & e_{24}q_2 \\ 0 & e_{32} & e_{34}q_3 \end{bmatrix} \begin{bmatrix} k_{12} \\ k_{23} \\ k_{14}/q_1 \end{bmatrix} \tag{11}$$

for any position of C_4^F as long as C_4^F is not on line L. The basis vector corresponding to the third column vector is shown in Fig.8. Note that Fig.6(c) corresponds to the case where C_4^F is taken at the infinity point on the straight line normal to line L.

5 Four-Fingered Hand

5.1 Non-Degenerate Case

We next consider the case of four-fingered hands. Under the condition that the four contact points are not coplanar, the truss model is unique and the internal force can be represented by (see Fig.9).

$$[I - G^+ G]\hat{k} = Ek \tag{12}$$

$$E = \begin{bmatrix} e_{12} & e_{13} & e_{14} & 0 & 0 & 0 \\ e_{21} & 0 & 0 & e_{23} & e_{24} & 0 \\ 0 & e_{31} & 0 & e_{32} & 0 & e_{34} \\ 0 & 0 & e_{41} & 0 & e_{42} & e_{43} \end{bmatrix} \tag{13}$$

$$k = [k_{12}\ k_{13}\ k_{14}\ k_{23}\ k_{24}\ k_{34}]^T \tag{14}$$

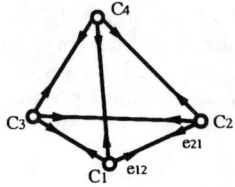

Figure 9: Truss model for four-fingered hands

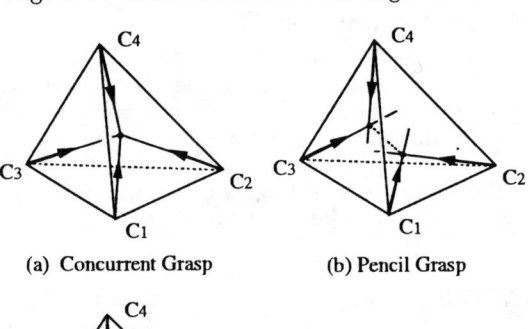

(a) Concurrent Grasp (b) Pencil Grasp

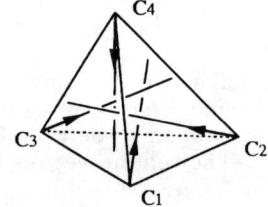

(c) Regulus Grasp

Figure 10: Categories of internal forces for 4-fingered hands

Ponce et al [8] have shown that four-finger force-closure grasps fall into three categories: concurrent, pencil, and regulus grasps as shown in Fig.10. Concurrent grasp is a grasp such that the four contact forces intersect at a point and their sum is zero. Pencil grasp is a grasp such that two contact forces intersect at a point and the remaining two contact forces also intersect at another point, and the sum of the former two forces and that of the latter two forces balance. Regulus grasp is a grasp such that the four contact forces do not intersect at all.

Using our characterization of the internal force (12), the above categorization can be formulated as follows. Let $\{i, j, h, k\} = \{1, 2, 3, 4\}$, $k_{ij} = k_{ji}$ and let the distance between C_i and C_j be $d_{ij}(= d_{ji})$. Then it can be shown that, for f_i and f_j to intersect, the parameters k_{ij} must satisfy

$$\frac{(k_{ih}/d_{ih})}{(k_{ik}/d_{ik})} = \frac{(k_{jh}/d_{jh})}{(k_{jk}/d_{jk})} \tag{15}$$

This condition means that the normalized ratio between k_{ih} and k_{ik} (normalized by the corresponding distances d_{ih} and d_{ik}, respectively) for the contact force f_i must be equal to that for the contact force f_j.

The proof of (15) is given as follows. Since the two forces f_i and f_j intersect, there exists a plane P including these forces(see Fig.11). Assume that this plane is not parallel to the line including C_h and C_k.

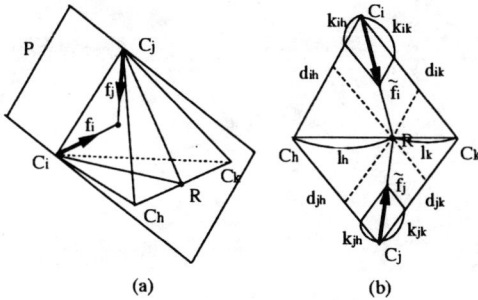

Figure 11: Triangles $C_iC_hC_k$ and $C_jC_hC_k$

Let the intersection point of the plane and the line be denoted by R.

Noting that $\boldsymbol{f}_i$ and $\boldsymbol{f}_j$ can be expressed as

$$\boldsymbol{f}_i = e_{ij}k_{ij} + e_{ih}k_{ih} + e_{ik}k_{ik} \tag{16}$$

$$\boldsymbol{f}_j = e_{ji}k_{ji} + e_{jh}k_{jh} + e_{jk}k_{jk} \tag{17}$$

we define new forces $\hat{\boldsymbol{f}}_i$ and $\hat{\boldsymbol{f}}_j$ by

$$\hat{\boldsymbol{f}}_i = e_{ih}k_{ih} + e_{ik}k_{ik} \tag{18}$$

$$\hat{\boldsymbol{f}}_j = e_{jh}k_{jh} + e_{jk}k_{jk} \tag{19}$$

Then we can show that these forces $\hat{\boldsymbol{f}}_i$ and $\hat{\boldsymbol{f}}_j$ also lie on this plane P.

Fig.11(b) shows the triangles $C_iC_hC_k$ and $C_jC_hC_k$ unfolded on a same plane. From the figure, it is straight forward to show that

$$k_{ih}/k_{ik} = (d_{ih}l_h)/(d_{ik}l_k) \tag{20}$$

holds for $\hat{\boldsymbol{f}}_i$. hSimilarly,

$$k_{jh}/k_{jk} = (d_{jh}l_h)/(d_{jk}l_k) \tag{21}$$

holds for $\hat{\boldsymbol{f}}_j$. From (20) and (21), we obtain (15). We can also show in a similar way that (15) holds in the case when the plane P is parallel to the line including C_h and C_k. This completes the proof.

Furthermore, we can show that, when (15) holds, the following equation must also hold.

$$\frac{(k_{hi}/d_{hi})}{(k_{hj}/d_{hj})} = \frac{(k_{ki}/d_{ki})}{(k_{kj}/d_{kj})} \tag{22}$$

This means that $\boldsymbol{f}_h$ and $\boldsymbol{f}_k$ must also intersect at a point[†].

Hence the case where (15) (and so (22)) holds for some $\{i,j,h,k\}$ corresponds to the pencil grasp. If we

---

[†] Equation (22) holds also for the case where f_h and f_k are parallel and in the same direction, and for the case where f_h and f_k are on the line going through C_h and C_k and in the opposite direction. We regard these cases as ones where the two forces intersect at an infinity point. The same argument holds for (15) too.

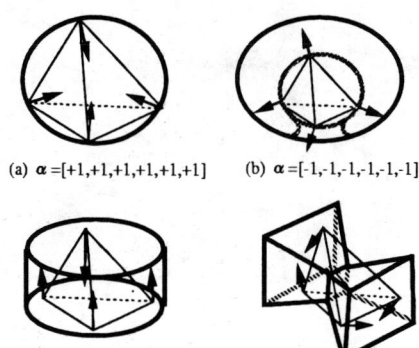

Figure 12: Mode of internal force for four-fingered hands

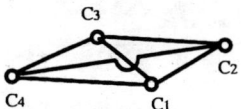

Figure 13: Truss model for coplanar four contact points

further assume that $\boldsymbol{f}_i$ and $\boldsymbol{f}_h$ intersect at a point, then, in a similar argument as above, $\boldsymbol{f}_j$ and $\boldsymbol{f}_k$ must also intersect. Since we assume that the four contact points are not coplanar, the four contact forces must intersect at a same point. Hence the case where (15) holds for any $\{i,j,h,k\}$ corresponds to the concurrent grasp. Finally the case where (15) does not hold for any $\{i,j,h,k\}$ corresponds to the regulus grasp

Note that the relation among these three grasps is now physically and mathematically very clear due to the formulation (12) of the internal force.

We can define the mode of internal force for four-fingered hands, in the same way as three-fingered hands, based on whether the force acting on each of the members of the truss is compressive force ($k_{ij} > 0$) or tensile force ($k_{ij} < 0$). The mode vector is defined by $\boldsymbol{\alpha} = [\alpha_{12}, \alpha_{13}, \alpha_{14}, \alpha_{23}, \alpha_{24}, \alpha_{34}]$. Fig.12 shows several kinds of the mode of internal force.

5.2 Degenerate Case

We consider the case where the four contact points are coplanar. For simplicity, we assume that no three points are collinear. The virtual truss is given in Fig.13. Matrix $\boldsymbol{G}$ is nonsingular (rank$\boldsymbol{G}=6$), but the truss is not rigid. To obtain a representation of the internal force, we place a fictitious joint (or contact point) C_5^F that is not coplanar with the original four contact points to obtain an augmented truss model as shown in Fig.14. Then there exists a set of four real numbers $\{q_1, q_2, q_3, q_4\}$ such that

$$e_{51}q_1 + e_{52}q_2 + e_{53}q_3 + e_{54}q_4 = 0 \tag{23}$$

and we can show that a representation of the internal

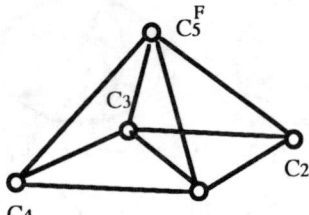

Figure 14: Augmented truss model for coplanar four contact points

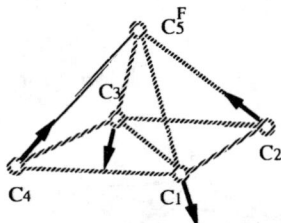

Figure 15: A basis vector involving four contact points

force is given by Ek where

$$E = \begin{bmatrix} e_{12} & e_{13} & e_{14} & 0 & 0 & e_{15}q_1 \\ e_{21} & 0 & 0 & e_{23} & 0 & e_{25}q_2 \\ 0 & e_{31} & 0 & e_{32} & e_{34} & e_{35}q_3 \\ 0 & 0 & e_{41} & 0 & e_{43} & e_{45}q_4 \end{bmatrix} \quad (24)$$

$$k = [k_{12}\ k_{13}\ k_{14}\ k_{23}\ k_{34}\ k_{15}/q_1]^T \quad (25)$$

Note hat the last column vector of E in (24) involves the four contact points as shown in Fig.15

6 Hands with Five or More Fingers

We now consider hands with five or more fingers. Note that when a human grasps an object with both hands, the number of fingers can be more than five.

We assume that the contact points are not coplanar as a whole. The same idea as in the four-fingered hands applies in this case too. One important difference from the case of four-fingered hands is that we have now some redundancy in the truss model. In general, for n-fingered hands ($n \geq 3$) grasping an object with non-degenerate contact points, the degree of freedom of the internal force is $(3n - 6)$ but the number of line members in the fully spanned truss model is $n(n-1)/2$. Since $(3n - 6) = n(n-1)/2$ holds for $n = 3, 4$, there is no redundancy in the case of three- and four-finger hands. Five-fingered hands have 1 degree of redundancy, and six-fingered hands have 3 degrees of redundancy. However, we can construct non-redundant truss models quite easily by taking out some line members from the fully spanned truss model to make it statically determinate. Hence we can obtain a non-redundant representation of the internal force. Fig.16 shows some examples of non-redundant truss model for five-fingered hands.

It should be noted that the fully spanned truss, although it is indeterminate, can be useful for the case where the contact points is time-varying.

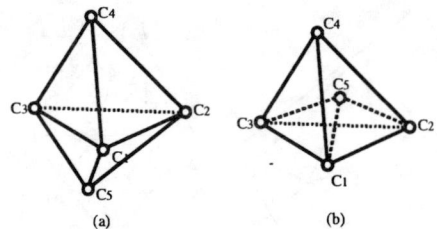

Figure 16: Truss models for five-fingered hands

Now we will prove that, for an n-finger hand ($n \geq 5$) grasping an object, any internal force can be represented as a sum of pushing or pulling internal forces acting on the members under the condition that the contact points are not coplanar as a whole.

Since the n contact points are not coplanar, we can number these contact points as $C_1, C_2, \cdots, C_n$ in such a way that

$$\text{rank}[e_{12}, e_{23}, e_{34}] = 3 \quad (26)$$

meaning that the first four contact points are not coplanar and form a non-degenerate tetrahedron. For any internal force $\{f_1, f_2, \cdots, f_n\}$ what we need to show is that there exists a set of real numbers $\{k_{ij}; i,j = 1, 2, \cdots, n; i \neq j; k_{ij} = k_{ji}\}$ such that

$$f_i = \sum_{j=1}^{n} e_{ij} k_{ij} \quad (27)$$

Since $\{f_1, f_2, \cdots, f_n\}$ is an internal force, the resultant force and moment are zero:

$$\sum_{i=1}^{n} f_i = 0, \quad \sum_{i=1}^{n} f_i \times r_i = 0 \quad (28)$$

From the assumption (26), we have

$$\text{rank}[e_{n1}, e_{n2}, \cdots, e_{n(n-1)}] = 3 \quad (29)$$

Hence there exist $\{k_{n1}, k_{n2}, \cdots, k_{n(n-1)}\}$ such that

$$f_n = \sum_{j=1}^{n-1} e_{nj} k_{nj} \quad (30)$$

Let

$$f_i^{(n-1)} = f_i + e_{ni} k_{ni} \quad (31)$$

then we have

$$\sum_{i=1}^{n-1} f_i^{(n-1)} = 0, \quad \sum_{i=1}^{n-1} f_i^{(n-1)} \times r_i = 0 \quad (32)$$

Hence $\{f_i^{(n-1)}, i = 1, 2, \cdots, n-1\}$ is an internal force acting on $(n-1)$ contact points $C_1, C_2, \cdots, C_{n-1}$.

Next we can show that there exist $\{k_{(n-1)1}, k_{(n-1)2}, \cdots, k_{(n-1)(n-2)}\}$ such that

$$f_{n-1}^{(n-1)} = \sum_{j=1}^{n-2} e_{(n-1)j} k_{(n-1)j} \quad (33)$$

Let
$$f_i^{(n-2)} = f_i^{(n-1)} + e_{(n-1)i}k_{(n-1)i} \qquad (34)$$

then we have

$$\sum_{i=1}^{n-2} f_i^{(n-2)} = 0, \quad \sum_{i=1}^{n-2} f_i^{(n-2)} \times r_i = 0 \qquad (35)$$

Repeating this procedure we obtain an internal force $\{f_i^{(3)}, i = 1, 2, 3\}$. It is also obvious that the three components of this internal force can be expressed as

$$f_1^{(3)} = e_{12}k_{12} + e_{13}k_{13} \qquad (36)$$

$$f_2^{(3)} = e_{23}k_{23} + e_{21}k_{21} \qquad (37)$$

$$f_3^{(3)} = e_{31}k_{31} + e_{32}k_{32} \qquad (38)$$

Using these coefficients k_{ij}, we finally obtain (27), completing the proof.

6.1 Non-Redundant Representation

A method of obtaining non-redundant representations based on the truss model in the previous section for the non-degenerate case will be given.

(i) Select a set of four contact points that is not coplanar or collinear and number them as C_1, C_2, C_3, and C_4. Number other contact points arbitrarily from C_5 through C_n.

(ii) Connect C_1, C_2, C_3, and C_4 by six line members as shown in Fig.9.

(iii) Let $i = 5$

(iv) Select a triangle of three line members among those obtained so far whose plane does not include C_i. Connect C_i and the three contact points of the selected triangle by three line members.

(v) If $i \leq n-1$, increase the value i by 1 and go to (iv).

(vi) Expression (27) with only k_{ij} corresponding to the members obtained in the previous steps is a non-redundant presentation of the internal force.

It is easy to show that this procedure gives those of Fig.16 for five-fingered hands.

7 Conclusions

The virtual truss model of the grasped object has been proposed in this paper for characterizing the internal force in multiple finger grasp. It has been shown that this truss model is valid under a non-degeneracy condition. Some degenerate cases have been discussed and it has been shown that there exist basis vectors of internal force subspace that involve three contact points in collinear case and four contact points in coplanar case. A method of extending the virtual truss model to cope with these special cases has been given. Based on this characterization, an explicit formulation has been derived of the categorization by Ponce et. al. of four finger grasps into concurrent, pencil, and regulus grasps. A method has also been given for obtaining a non-redundant expression of all possible internal forces.

References

[1] V. Nguyen: "Constructing Force-Closure Grasps," International Journal of Robotics Research, Vol. 7, No. 3, pp. 3–16, 1988.

[2] Y. Nakamura, K. Nagai, and T. Yoshikawa: "Dynamics and Stability in Coordination of Multiple Robotic Mechanisms," International Journal of Robotics Research, Vol. 8, No. 2, pp. 44–61, 1989.

[3] X. Markenscoff and C.H. Papadimitriou: "Optimum Grip of a Polygon," International Journal of Robotics Research, Vol. 8, No. 2, pp. 17–29, 1989.

[4] J. Kerr and B. Roth: "Analysis of Multifingered Hands," International Journal of Robotics Research, Vol. 4, No. 4, pp. 3–17, 1986.

[5] V. Kumar and K. Waldron: "Force Distribution in Closed Kinematic Chains," Proc. of IEEE Int. Conf. on Robotics and Automation, pp. 114-119, 1988.

[6] T. Yoshikawa and K. Nagai: "Manipulating and Grasping Forces in Manipulation by Multifingered Robot Hands," IEEE J. of Robotics and Automation, Vol. 7, No. 1, pp. 67–77, 1991.

[7] A. Bicchi, "Force Distribution in Multiple Whole-Limb Manipulation," Proc. of IEEE Int. Conf. on Robotics and Automation, pp. 196-201, 1993.

[8] J. Ponce, S. Sullivan, J-D. Boissonnat, and J-P. Merlet: "On Characterizing and Computing Three- and Four-Finger Force-Closure Grasps of Polyhedral Objects," Proc. of IEEE Int. Conf. on Robotics and Automation, pp. 821-827, 1993.

[9] A. Sudsang and J. Ponce: "New Techniques for Computing Four-Finger Force-Closure Grasps of Polyhedral Objects," Proc. of IEEE Int. Conf. on Robotics and Automation, pp. 1335–1360, 1995.

[10] D. Williams and O. Khatib: "The Virtual Linkage: A Model for Internal Forces in Multi-Grasp Manipulation," Proc. of IEEE Int. Conf. on Robotics and Automation, pp. 1025-1030, 1993.

[11] K.L. Doty, C. Melchiorri, and C. Bonivento: "A Theory of Generalized Inverses Applied to Robotics," International Journal of Robotics Research, Vol. 12, No. 1, pp. 1–19, 1993.

[12] T. Yoshikawa: "Hybrid Control Theory of Robot Manipulators," in Robotics Research, The Sixth International Symposium (eds: T.Kanade and R.Paul), The International Foundation for Robotics Research, Cambridge, USA, pp. 443-452, 1993.

[13] T. Yoshikawa. Foundations of Robotics, MIT Press, Cambridge, Mass., 1990.

[14] T. Yoshikawa, "Passive and Active Closures by Constraining Mechanisms," Proc. of IEEE Int. Conf. on Robotics and Automation, pp. 1477-1484, 1996.

[15] R.C. Coates, M.G. Coutie, and F.K. Kong. Structural Analysis, Thomas Nelson and Sons, Park Street, London, 1972.

Generalized Stability of Compliant Grasps

Herman Bruyninckx,* Sabine Demey†
Katholieke Universiteit Leuven
Department of Mechanical Engineering
Leuven, Belgium

Vijay Kumar
University of Pennsylvania
GRASP Lab
Philadelphia, U.S.A.

Abstract

We develop a geometric framework for the stability analysis of multifingered grasps and propose a measure of grasp stability for arbitrary perturbations and loading conditions. The measure requires a choice of metric on the group of rigid body displacements. We show that although the stability of a grasp itself does not depend on the choice of metric, comparison of the stability of different grasps depends on the metric. Finally, we provide some insight into the choice of metrics for stability analysis.

1 Introduction

The *stability* of a grasp or a fixture is one of the important aspects determining its "quality." Stability analysis determines where to place the fingers (fixture elements) on the object (workpiece), and how hard to squeeze, in order to keep this object rigidly grasped under the action of external disturbing forces. This paper addresses the development of a suitable measure of grasp stability using a differential geometric framework and identifies the important choices a user must make before using this measure.

Early work [1, 9] considered rigid point contacts and developed the concepts of *force closure* and *form closure*. Since all grasps with more than one frictional contact, and planar grasps with more than three (six for the spatial case) frictionless contacts are statically indeterminate, a rigid body analysis cannot be used for obtaining the force distribution. Further, "real" grasps have compliance, in the fingers, in the grasped object, and in the control system for the fingers. A knowledge of this compliance resolves this indeterminacy and allows us to determine the force distribution.

A model of the compliance of each contact allows us to characterize the stability of the system subject to small perturbations. The force-displacement characteristics at each contact can be described by a Cartesian stiffness matrix at that contact. The grasp can be considered a conservative system and the matrix of second partial derivatives of the associated potential energy gives us insight into the grasp stability. If this matrix, also called the grasp stiffness matrix [4], is positive definite, the grasp is stable in the sense of small displacements. A positive semi-definite matrix reveals that a higher-order analysis is necessary. An indefinite matrix shows that the grasp is unstable. Variants of this basic idea are used in definitions of first and second order stability [1, 4].

This paper addresses the assessment of the *quality* of a grasp. We assume that the Cartesian stiffness matrix for a grasp is given[1], and focus on the measure of stability. The measure of stability is based on the eigenvalue decomposition of the grasp stiffness matrix [6, 12]. This paper explains that such an eigenvalue decomposition is not uniquely defined, and some of the properties of this decomposition depend on arbitrary choices of metrics. Further, some of the properties may not be independent of the choice of reference frames.

The main results of this paper are easily obtained by formulating the problem in a differential geometric framework. The next section will discuss the geometric formulation of the Cartesian stiffness matrix and some basic results on grasp stability. Section 3 discusses metrics and eigenvalue decomposition of $(0, 2)$ tensors on $SE(3)$. The application of the ideas to develop a generalized measure of stability for grasps is presented in Section 6, with an example in Section 5.

---

*H. Bruyninckx is Postdoctoral Fellow of the Fund for Scientific Research–Flanders (F.W.O.) in Belgium. Financial support by the Belgian Programme on Inter-University Attraction Poles initiated by the Belgian State—Prime Minister's Office—Science Policy Programme (IUAP) is gratefully acknowledged.

†Currently at Materialise, N.V., Belgium.

[1]The computation of the grasp stiffness matrix from the contact model and from the control algorithms for the fingers is described, for example, in [1, 4].

2 The Cartesian stiffness matrix

In this section, we provide the basic differential geometric concepts needed to define the Cartesian stiffness matrix, and point out some properties that are useful for stability analysis.

It is well known that the stiffness matrix for a conservative system can be defined as the matrix of second partial derivatives of a twice-differentiable potential function. The Cartesian stiffness matrix is a $(0, 2)$ tensor defined on $SE(3)$, the group of all rigid body displacements. Since $SE(3)$ is *not* Euclidean, the definition of this matrix requires some basic ideas in differential geometry. These are summarized next. The unfamiliar reader is referred to [8, 13] for a more detailed account.

Preliminaries Any rigid body displacement can be represented by a 4×4 homogeneous transformation matrix that transforms points described in a body fixed reference frame {M} to an inertial reference frame {F}. The set of all displacements (homogeneous transformation matrices), form a matrix Lie group $SE(3)$, the special Euclidean group in three-dimensions [8]. Any element A of the group has the form:

$$A = \begin{bmatrix} R & d \\ 0 & 1 \end{bmatrix}, \quad (1)$$

where R is a 3×3 proper orthogonal matrix and $d \in \mathbb{R}^3$. On a Lie group, the tangent space at the group identity has the structure of a Lie algebra. The Lie algebra of $SE(3)$ is denoted by $se(3)$ and is given by:

$$se(3) = \left\{ \begin{bmatrix} \Omega & v \\ 0 & 0 \end{bmatrix}, \Omega \in \mathbb{R}^{3\times 3}, v \in \mathbb{R}^3, \Omega^T = -\Omega \right\}.$$

A 3×3 skew-symmetric matrix Ω can be uniquely identified with a vector $\omega \in \mathbb{R}^3$ so that for an arbitrary vector $x \in \mathbb{R}^3$, $\Omega x = \omega \times x$, where $\times$ is the cross product in $\mathbb{R}^3$. Each element $T \in se(3)$ can be thus identified with a *twist*, a 6×1 vector $t = \begin{bmatrix} v^T & \omega^T \end{bmatrix}^T$. We will use twists to represent *infinitesimal displacements* of a rigid body. If a and b are two twists, and T_a and T_b are their matrix representations, the Lie bracket of the two twists (sometimes called the motor product in the kinematics literature), $[T_a, T_b]$, is simply the matrix commutator $T_a T_b - T_b T_a$.

Every motion of a rigid body corresponds to a curve on $SE(3)$. The tangents to the curve define a vector field that represent the velocity at each point. When there are external forces and moments acting on a body, we can associate a generalized force at every point along the curve. The generalized force vector at a point is a co-vector that belongs to the cotangent space of the manifold at that point. The generalized force vectors define a *one form* on $SE(3)$.

In particular, we are interested in *left-invariant* or *right-invariant* vector fields. Left-invariant vector fields can be generated by left translating a twist vector, an element of the Lie algebra, so that at any point A, the vector X can be written as:

$$X_A = AT$$

where T is a twist. Similarly, right-invariant vector fields can be generated by right translation.

If a force f and a moment τ (about the origin of the body-fixed reference frame, {M}) act on a rigid body, we refer to the vector pair, $W = \{f, \tau\}$, as a *wrench*. If a wrench W acts on a rigid body that undergoes a twist, T, over a time interval Δt, the *work* done by the wrench is given by

$$\Delta E = (f^T v + \tau^T \omega)\Delta t,$$

which is a scalar. Wrenches therefore belong to the dual of the vector space of twists, $se^*(3)$. Just as twists can be used to generate left-invariant vector fields, any wrench can be used to generate a left-invariant one form. The quantity work is formally obtained by an *action*. The action $\phi(X)$ of the one-form ϕ on the vector field X is denoted by $\langle \phi, X \rangle$.

Geometry of $SE(3)$ If a smoothly varying, positive-definite, bilinear, symmetric form $\langle ., . \rangle$ is defined on the tangent space at each point on the manifold, we call it a Riemannian metric and we say the manifold is Riemannian. At every point, the bilinear form is an inner product on the tangent space at that point. If the form is nondegenerate but indefinite, the metric is called semi-Riemannian. It is well known that there is no *natural* Riemannian metric on $SE(3)$. In other words, there is no canonical way to define the distance between two sets of positions and orientations of a rigid body so that it is independent of the choice of inertial *and* body-fixed reference frames [7]. On the other hand, there is a natural, scale-dependent, left-invariant metric that can be obtained by embedding $SE(3)$ in $\mathbb{R}^{4\times 4}$. This metric depends on the choice of a length scale and the choice of the body fixed frame but is independent of the inertial reference frame [10].

No *natural identification* exists between $se(3)$ and its dual $se^*(3)$. Every such identification requires a *non-degenerate bilinear form* on $SE(3)$, [2, p. 59]. Since twists and wrenches are fundamentally different quantities that live in two different spaces, any attempt to identify them must rely on an arbitrary choice of the above-mentioned bilinear form.

Similarly, no *natural identification* exists between the tangent spaces or between cotangent spaces at two different points of $SE(3)$. Any such identification requires a rule for *parallel transport*, or an *affine connection*. An affine connection imposes additional structure on the manifold that allows us to compare a tangent or cotangent vector at a point $p \in SE(3)$ to a tangent or cotangent vector at another point $q \in SE(3)$. Thus, the derivative of a vector field or a one-form along a curve or in the direction of a vector field relies on the structure of an affine connection. The derivative of a vector field X in the direction of a tangent vector Y is called the *covariant derivative* $\nabla_Y X$, and depends on the connection.

Given any metric on SE(3) (or for that matter, any manifold) one can find a unique, symmetric connection that is compatible with the metric. This connection is called the *Levi-Civita*, or *Riemannian* connection of the metric.

Cartesian stiffness matrix Broadly speaking, a Cartesian stiffness matrix consists of components each of which describes how a component of a wrench acting on a rigid body changes as the body moves along a basis twist. This suggests that the wrench has to be differentiated along the left-invariant vector fields generated by the basis twists. However, a wrench belongs to the co-tangent space $se^*(3)$, the dual of the space of twists, and therefore, is only defined at a point (the identity element). It cannot be differentiated along a vector field (which is defined over the entire manifold). We must therefore formalize this notion of "changes in the components of a wrench" in terms of differentiation of the associated force one-form F. Further, the operation of differentiating a force one-form will necessitate the comparison of cotangent vectors at two nearby, but different points, before taking a limit. Thus a formal definition of a Cartesian stiffness matrix must entail a framework for differentiation on $SE(3)$, and a suitable recipe in terms of a derivative of a force one-form. As shown in [16], the elements of the matrix are given by:

$$\begin{aligned} K_{ij} &= \langle \nabla_{L_j} d\phi, L_i \rangle \\ &= L_j L_i(\phi) + L_k(\phi) \langle \nabla_{L_j} L_i, L_k \rangle \end{aligned} \quad (2)$$

where L_i and L_j are the ith and jth members of a basis of left-invariant vector fields and ϕ is the potential function. The matrix depends on the choice of basis vectors (including their ordering). Further, if at points away from equilibrium where ϕ is not stationary, the directional derivative $L_k(\phi)$ is nonzero and the components depend specifically on the affine connection [14, 16].

The study of the properties of Cartesian stiffness matrices [7, 12, 16] has produced a number of interesting results, that, on first sight, appear to contradict common engineering intuition. Some of these results are summarized below.

It is worth point out that in general, a Cartesian stiffness matrix cannot be diagonalized by a change of reference frame [7, 12]. It is therefore not always possible to find the *Remote Center of Compliance*, a concept used extensively in assembly and force control.

Another relevant property of the Cartesian stiffness matrix concerns its eigenvalues. The eigenvalues are not invariant under change of reference frame [12], however its signature matrix is. This is due to the fact that stiffness matrices transform according to *congruence* transformations and not *similarity* transformations.

3 Grasp stability

The previous section discussed the definition and the computation of the Cartesian stiffness matrix. For a given grasp, if the local geometry and the material properties of the contacting fingers and object are known, the stiffness matrix can be calculated either directly [4] or from the potential function for the grasp using Equation (2). In a condition of equilibrium, the potential energy is stationary and therefore, in Equation (2), the second term that consists of the first partial derivatives of the potential function, vanishes and the stiffness matrix is simply given by:

$$K_{ij} = L_j L_i(\phi).$$

Roughly speaking, a grasp is stable if a small disturbance twist on the system generates a restoring wrench that tends to bring the system back to its original configuration.

Definition 1 (Stability under given disturbance) *A grasp determined by a potential function is stable under a disturbance twist if this disturbance* **increases** *the potential energy in the grasp.*

Definition 2 (Stability) *A grasp determined by a potential function is stable if* **any** *disturbance twist increases the potential energy in the grasp.*

The stiffness K of the grasp transforms n infinitesimal displacement twist t into the *wrench* $w = \begin{bmatrix} f^T & \tau^T \end{bmatrix}^T = Kt$. Then, the grasp is stable if the work $w^T t$ is positive, i.e., the disturbance puts energy *into* the grasp instead of releasing energy from it. Since the work $w^T t$ can also be written as $t^T K t$, one most often uses the stability criterion from linear system theory: The system with system matrix K is stable if all of K's eigenvalues are positive. If one of more of the eigenvalues are zero and there are no negative eigenvalues, it is necessary to pursue a higher order analysis for determining the stability of the grasp. This case of marginal stability will not be considered here.

The general form of the eigenvalue decomposition for a stiffness matrix K (with eigenvalue λ and eigenvector t) is

$$Kt = \lambda t. \quad (3)$$

An important problem in grasp planning is to determine when a grasp is *more stable* than another grasp. We would like to define a measure of stability as follows:

Definition 3 (Measure of stability) *Grasp 1 is more stable than grasp 2 if the* **minimum** *restoring wrench over all unit twist disturbances is larger for grasp 1 than for grasp 2.*

It is clear that this definition requires a definition of the length of a twist vector and the length of a wrench. A more formal definition of these concepts is pursued later. But the motivation for the rest of the paper is very simple. If we assess stability in terms of the eigenvalues of the stiffness matrix, we would like to develop a stability measure that compares the smallest eigenvalues of competing grasps.

Some care is required when using the eigenvalues of a stiffness matrix K as basis for a quantitative stability analysis. Indeed, the following paragraphs show that these eigenvalues can change under a change of the reference frame with respect to which the twists and wrenches are expressed, as well as under a change of physical units. It is well-known that rigid body twists $t = [v^T \ \omega^T]^T$ change as follows under a change of world reference frame described by the homogeneous transformation matrix as in Eq. (1):

$$t' = \Gamma t = \begin{pmatrix} R & [d]R \\ 0 & R \end{pmatrix} t, \quad (4)$$

where $[d]$ denotes the 3×3 matrix representing the vector cross product with the three-vector d. Similarly, a wrench $w = [f^T \ \tau^T]^T$ transforms as

$$w' = \Gamma' w = \widetilde{\Delta}\Gamma\widetilde{\Delta} w, \quad \widetilde{\Delta} = \begin{pmatrix} 0_{3\times 3} & I_{3\times 3} \\ I_{3\times 3} & 0_{3\times 3} \end{pmatrix}. \quad (5)$$

The following properties follow straightforwardly:

$$\Gamma^{-1} = \widetilde{\Delta}\Gamma^T\widetilde{\Delta}, \quad \widetilde{\Delta}^T = \widetilde{\Delta}, \quad \widetilde{\Delta}\widetilde{\Delta} = I_{6\times 6}. \quad (6)$$

Hence, a change of reference frame transforms the stiffness matrix K in Eq. (3) into a matrix K' as follows:

$$\begin{aligned} w = Kt \Rightarrow w' &= \Gamma' K t \\ &= \Gamma' K \Gamma^{-1} t' \\ \Rightarrow K' &= \Gamma' K (\Gamma')^T. \end{aligned} \quad (7)$$

This is a so-called *congruence* transformation which preserves the *sign* of the eigenvalues λ, but not their values. On the other hand, if the alternative and equivalent twist representation $t = [\omega^T \ v^T]^T$ is used (i.e., the positions of linear and angular velocity are interchanged), the transformation of the eigenvalue problem

$$\widetilde{K}t = \lambda t, \quad \widetilde{K} = \widetilde{\Delta}K \quad (8)$$

yields

$$\widetilde{K}' = \Gamma'\widetilde{K}(\Gamma')^{-1}. \quad (9)$$

Hence, $\widetilde{K}$ is not symmetric, but transforms with a *similarity* transformation, which maintains the eigenvalues exactly. Moreover, a more careful inspection of Eqs (3) and (8) reveals that the eigenvalues of K have physical units (the first three having different units than the last three!), while the eigenvalues of $\widetilde{K}$ are dimensionless.

In summary, the eigendecomposition of a stiffness matrix is in general not invariant. However, the following invariant properties of parts of a stiffness matrix follow straightforwardly from the definitions above. Assume K is partitioned in 3×3 subblocks K_{11}, K_{12}, K_{21}, and K_{22}. Using the matrix inversion lemma and the fact that the inverse of a rotation matrix is its transpose, it is easy to prove that:

Proposition 1 *The eigenvalues of K_{11} are invariant.*

Proposition 2 *The eigenvalues of $K_{22} - K_{12}^T K_{11}^{-1} K_{12}$ are invariant.*

In summary, if all the eigenvalues of the Cartesian stiffness matrix of a grasp are positive, they will be positive regardless of changes in the body-fixed or inertial reference frames. However, a *quantitative* stability analysis of compliant grasps cannot be performed by simply looking at the eigenvalues of the grasp's stiffness matrix.

4 Generalized eigendecomposition

As announced in the previous section, grasp stability quantification can be based on an eigenvalue decomposition of the grasp's stiffness matrix. This Section explains that the corresponding (generalized) eigendecomposition (Sect. 4.1) is not uniquely defined, since it depends on the choice of a metric on $SE(3)$ (Sect. 4.2).

4.1 The eigenvalue problem

Since twists and wrenches live in different spaces, the correct way to define the eigenvalue problem for a stiffness mapping is as a *generalized* eigenvalue problem, [15]:

$$Kt = \lambda Mt. \quad (10)$$

The matrix M is the non-degenerate, bilinear form (i.e., the *metric*) required to *identify* the twist t with a wrench Mt, so that both sides of Eq. (10) lie in the same vector space, and hence an eigendecomposition can be stated unambiguously. Equation (10) is defined on $se(3)$, but it is straighforward to define a similar generalized eigendecomposition on the dual space $se^*(3)$:

$$Kt = \lambda Mt \Rightarrow Cw = \frac{1}{\lambda}M^{-1}w. \quad (11)$$

The 6×6 matrix C is the *compliance matrix* of the grasp, i.e., the inverse of the stiffness matrix K. The eigenvalue decompositions (10) and (11) do not suffer from any difficulties related to non-invariance with respect to rigid body transformations:

Proposition 3 *The eigenvalues of $M^{-1}K$ are invariant under changes of reference frame and changes in physical units.*

The proof for this is fairly straightforward. Intuitively, one can see that the changes in K (an $(0,2)$ tensor on $SE(3)$) due to a transformation are cancelled by the changes in M (another $(0,2)$ tensor on $SE(3)$).

4.2 Choice of metric

The choice of the metric M in Eqs (10) and (11) is *arbitrary*, and hence also any quantitative stability analysis. The following properties hold for any **symmetric, positive-definite** stiffness matrix K, and any **symmetric, non-degenerate** metric M:

Proposition 4 *All eigenvalues are real, [15, p. 37].*

Proposition 5 *The signature of $M^{-1}K$ is equal to the signature of M.*

Indeed, since K is positive-definite, its "square root" L (or Cholesky decomposition $K = LL^T$ exists. Multiplying $M^{-1}Kt = \lambda t$ by L^T gives $(L^T M^{-1} L)(L^T t) = \lambda(L^T t)$. The proposition follows from the fact that an eigendecomposition signature is invariant under congruence transformations and matrix inversion. Note that all eigenvalues are positive if M is positive-definite.

Proposition 6 $t^T M t$ *and λ have the same sign.*

Multiplying Eq. (10) from the left by t^T gives $t^T K t = \lambda t^T M t$. The left-hand side is a positive number, due to the positive-definiteness of K. Hence, the two factors on the right-hand side must have the same sign.

Proposition 7 *Eigenvectors t_1 and t_2 belonging to different non-zero eigenvalues λ_1 and λ_2 are M-orthogonal.*

Subtract $t_2^T M(K^{-1}Mt_1) = \frac{1}{\lambda_1} t_2^T M t_1$ from $(K^{-1}Mt_2)^T M t_1 = \frac{1}{\lambda_2} t_2^T M t_1$, and use the symmetry of K and M.

If M is not only symmetric and non-degenerate, but also **positive-definite** (e.g., a kinetic energy, or inertia matrix), the following additional property follows from Proposition 5:

Proposition 8 $M^{-1}K$ *is positive-definite if and only if K and M are positive-definite.*

4.2.1 Indefinite, non-degenerate metrics

Many papers dealing with the kinematics of grasping or force control (implicitly or explicitly) use a multiple of the reciprocity matrix $\widetilde{\Delta}$ (which is **indefinite**, but nondegenerate) to identify $se(3)$ and $se^*(3)$, e.g., [6, 12]:

$$M_{\widetilde{\Delta}} = \alpha \begin{pmatrix} 0_{3\times 3} & I_{3\times 3} \\ I_{3\times 3} & 0_{3\times 3} \end{pmatrix}. \quad (12)$$

This metric has some interesting properties:

1. $M_{\widetilde{\Delta}}$ is a *bi-invariant*, nondegerate, indefinite metric on $SE(3)$, [5, 7]. In other words, it is independent of transformations of both the inertial and the body-fixed reference frames.

2. $M_{\widetilde{\Delta}}$-orthogonality in Proposition 7 gives us the well-known condition for *reciprocity* of two screws [9].

3. The stiffness matrix eigenvalues defined by means of $M_{\widetilde{\Delta}}$ are dimensionless.

4. Special cases of stiffness matrix eigenvectors ("twist-compliant axes," "wrench-compliant axes") have been identified, [6, 12], depending on the parallelism of some of the eigenvectors' angular or linear three-vector parts.

$M_{\widetilde{\Delta}}$ has the less interesting property that three of its eigenvalues are negative, and three are positive, such that Proposition 8 does not apply. At first sight, this would mean that the stability of the grasp changes depending on the chosen identification metric, if one takes into account the sign criterion of the eigenvalues only. However, the physical stability criterion remains unchanged: the grasp is stable if any disturbance twist results in an increase of the potential energy of the system.

A second family of bi-invariant indefinite metrics are derived from a linear combination of the Killing and Klein forms [5], the so-called α-β metrics [16]:

$$M_{\alpha\beta} = \begin{pmatrix} 0_{3\times 3} & \beta I_{3\times 3} \\ \beta I_{3\times 3} & \alpha I_{3\times 3} \end{pmatrix}.$$

Their eigendecomposition properties are less straightforward, however, since they depend on the values of the scalars α and β.

4.3 Left-invariant metrics

There is a natural, scale-dependent, left-invariant metric that can be obtained by embedding $SE(3)$ in $\mathbb{R}^{4\times 4}$. This metric depends on the choice of a length scale and the choice of the body fixed frame but is independent of the inertial reference frame [10]. We will refer to this metric as M_k. It can be written in the form:

$$M_k = \begin{pmatrix} \gamma I_{3\times 3} & 0_{3\times 3} \\ 0_{3\times 3} & \alpha I_{3\times 3} \end{pmatrix}.$$

This is a special case of the so-called energy metric. The kinetic energy of a rigid body can be written in a reference frame aligned with its principal axes at the center of mass with the metric:

$$\begin{pmatrix} mI_{3\times 3} & 0_{3\times 3} \\ 0_{3\times 3} & H_{3\times 3} \end{pmatrix}.$$

where m is the mass of the rigid body and H is a diagonal matrix with the three moments of inertia along its diagonal. If the rigid body, for example, is a homogeneous sphere, we get the metric M_k.

5 Examples

First, we give the eigendecomposition for one single compliant element, using three different metrics. The compliance matrix $C = K^{-1}$ in this example comes from [11]:

$$C = \begin{pmatrix} 0.48 & 0 & 0 & 0 & 0 & 0 \\ 0 & 0.59 & 0 & 0 & 0 & 1.20 \\ 0 & 0 & 1.25 & 0 & -4.50 & 0 \\ 0 & 0 & 0 & 100 & 0 & 0 \\ 0 & 0 & -4.50 & 0 & 21.0 & 0 \\ 0 & 1.20 & 0 & 0 & 0 & 3.00 \end{pmatrix}.$$

The matrix is in MKS units and has been multiplied with 10^5. Wrench coordinates are $[f^T \ \tau^T]^T$, and twist coordinates are $[v^T \ \omega^T]^T$. The following three metrics are used as an example: $M_{\widetilde{\Delta}} = \widetilde{\Delta}$ (with appropriate units), $M_k = I$ (with appropriate units), and $M_{\alpha\beta}$ with $\alpha = 2Nm$ and $\beta = 1N$. The eigenvalues for the stiffness mapping $K = C^{-1}$ with these metrics are, respectively:

$$\lambda_{\widetilde{\Delta}} = \begin{pmatrix} .144 & -.144 & 1.579 & -1.579 & .450 & -.450 \end{pmatrix},$$

$$\lambda_k = \begin{pmatrix} .046 & 3.663 & 10.593 & .286 & 2.083 & .010 \end{pmatrix},$$

$$\lambda_{\alpha\beta} = \begin{pmatrix} -4.172 & .005 & -7.067 & -.183 & .164 & .0237 \end{pmatrix}.$$

As expected, $\lambda_{\widetilde{\Delta}}$ has three positive and three negative eigenvalues. Since all the eigenvalues in λ_k are positive the grasp is stable. However these eigenvalues depend on the choice of the metric. A more meaningful result might explicitly incorporate the inertia tensor of the grasped body.

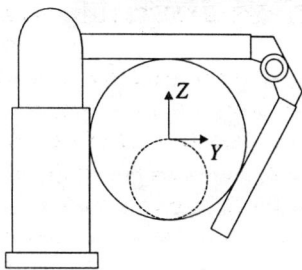

Figure 1: A full-envelope grasp by the WAM manipulator.

The second example (Fig. 1) is a whole arm grasp of a rigid cylinder. The Cartesian stiffness matrix K, expressed in the reference frame at the center of the cylinder, comes from the experimental results in [3]. The matrix models the elasticity of the "skin" of the manipulator at the contact points, as well as the elasticities in the actuation system of the WAM. In pounds and inches this matrix is

$$K = \begin{pmatrix} 378 & 0 & 0 & 0 & -866 & 1916 \\ 0 & 913 & -127 & 3488 & 0 & 0 \\ 0 & -127 & 413 & -2289 & 0 & 0 \\ 0 & 3488 & -2289 & 52455 & 0 & 0 \\ -866 & 0 & 0 & 0 & 6882 & -1506 \\ 1916 & 0 & 0 & 0 & -1506 & 14428 \end{pmatrix}.$$

We now choose two inertia metrics, expressed in the same frame as K. M_1 for a homogeneous mass distribution, with a total mass of 1.8 lbs, and M_2 for a mass distribution in which the same mass is contained in a cylinder of half the radius, with its center shifted over half the radius along the negative Z axis:

$$M_1 = diag(1.8, 1.8, 1.8, 44.1, 27.45, 27.45),$$

$$M_2 = \begin{pmatrix} 1.8 & 0 & 0 & 0 & 6.3 & 0 \\ 0 & 1.8 & 0 & -6.3 & 0 & 0 \\ 0 & 0 & 1.8 & 0 & 0 & 0 \\ 0 & -6.3 & 0 & 33.1 & 0 & 0 \\ 6.3 & 0 & 0 & 0 & 33.0 & 0 \\ 0 & 0 & 0 & 0 & 0 & 10.9 \end{pmatrix}.$$

The eigenvalues found from Eq. (10) are

$$\lambda_1 = \begin{pmatrix} 709 & \underline{23} & 254 & 1425 & 339 & 161 \end{pmatrix},$$
$$\lambda_2 = \begin{pmatrix} 2427 & \underline{15} & 701 & 8323 & 250 & 150 \end{pmatrix}.$$

Both metrics give positive eigenvalues (hence a stable grasp), and, as expected from physical intuition, the homogeneous mass distribution is more stable than the distribution with all mass below the center of the object (compare the underlined eigenvalues).

6 Generalized grasp stability

Taking into account the material presented in the previous sections, this section re-states grasp stability as introduced in Section 3. Two criteria are important for practical stability analysis of compliant grasps: (i) the analysis must be simple and efficient, and (ii) the results of the analysis must be invariant with respect to coordinate representation changes. Both criteria are satisfied by the *generalized eigenvalues* of stiffness matrices, using *positive-definite metrics* for the identification of the twist and wrench manifolds, $se(3)$ and $se^*(3)$, respectively. Indeed, the eigenvalues of $M^{-1}K$ are (i) dimensionless, (ii) positive if and only if the eigenvalues of K are positive (Proposition 8), and (iii) invariant under changes of the coordinate frame (Proposition 3). Hence, our stability definitions become:

Definition 4 (Generalized stability) *A compliant grasp is stable if all of its generalized eigenvalues are positive.*

Definition 5 (Generalized stability measure) *The smallest generalized eigenvalue is a quantitative measure of stability.*

Note that (i) the *value* of the smallest eigenvalue is not invariant under a change of identification metric, and (ii) it is in general not possible to come up with a *rank-ordering* of compliant grasps that is invariant under a change of identification metric. A physical interpretation of the stability definitions is straightforward. As shown in Section 4.2.2, the positive-definite metric M_k can be interpreted as an *kinetic energy metric*. Hence, the mass distribution of the grasped object influences the stability of the grasp. In the absence of information about the mass distribution, we can assume a mass distribution. In Example 1, $M_k = I$ was assumed. The results of the stability analysis depend, as expected, on this assumed mass distribution.

7 Conclusions

This paper has developed a geometric framework for the stability analysis of multifingered grasps. The major new concept introduced in this paper is the generalized eigendecomposition of the grasp's stiffness mapping. The generalization requires a choice of a metric on the group of rigid body displacements which enables the identification of twists with wrenches. The generalization eigendecomposition also leads to a new a measure of grasp stability for arbitrary perturbations and loading conditions. We show that although the stability of a grasp itself does not depend on the choice of metric, comparison of the stability of different grasps depends on the metric.

References

[1] M. R. Cutkosky. *Robotic grasping and fine manipulation*. Kluwer, Boston, MA, 1985.

[2] C. T. J. Dodson and T. Poston. *Tensor geometry: the geometric viewpoint and its uses*. Springer Verlag, 1991.

[3] W. S. Howard. *Stability of Grasped Objects Beyond Force Closure*. PhD Thesis, Univ. of Pennsylvania, Philadelphia, PA, 1995.

[4] W. S. Howard and V. Kumar. On the stability of grasped objects. *IEEE Trans. Rob. Automation*, 12(6):904–917, 1996.

[5] A. Karger and J. Novak. *Space kinematics and Lie groups*. Gordon and Breach, New York, NY, 1985.

[6] Q. Lin, J. Burdick, and E. Rimon. A quality measure for compliant grasps. In *IEEE Int. Conf. Robotics and Automation*, pages 86–92, Albuquerque, NM, 1997.

[7] J. Lončarić. *Geometrical Analysis of Compliant Mechanisms in Robotics*. PhD thesis, Harvard University, Cambridge, MA, 1985.

[8] R. M. Murray, Z. Li, and S. S. Sastry. *A Mathematical Introduction to Robotic Manipulation*. CRC Press, Boca Raton, FL, 1994.

[9] E. N. Ohwovoriole. Kinematics and friction in grasping by robotic hands. *Trans. ASME J. Mech. Transm. Automation Design*, 109:398–404, 1987.

[10] F. C. Park. Distance metrics on the rigid-body motions with applications to mechanism design. *Trans. ASME J. Mech. Design*, 117:48–54, 1995.

[11] T. Patterson and H. Lipkin. A classification of robot compliance. *Trans. ASME J. Mech. Design*, 115:581–584, 1993.

[12] T. Patterson and H. Lipkin. Structure of robot compliance. *Trans. ASME J. Mech. Design*, 115:576–580, 1993.

[13] B. F. Schutz. *Geometrical methods of mathematical physics*. Cambridge University Press, Cambridge, England, 1980.

[14] J. C. Simo. The (symmetric) Hessian for geometrically nonlinear models in solid mechanics: Intrinsic definition and geometric interpretation. *Comp. Methods Appl. Mech. Eng.*, 96:189–200, 1992.

[15] H. F. Weinberger. *Variational Methods for Eigenvalue Approximation*. SIAM, 1974.

[16] M. Žefran and V. Kumar. Affine connections for the Cartesian stiffness matrix. In *IEEE Int. Conf. Robotics and Automation*, pages 1376–1381, Albuquerque, NM, 1997.

Grasping and Position Control for Multi-fingered Robot Hands with Uncertain Jacobian Matrices

C. C. Cheah   H. Y. Han   S. Kawamura   S. Arimoto

Department of Robotics, Ritsumeikan University,
1916, Nojicho, Kusatsu, Shiga, 525 Japan

Abstract. Most research on multi-fingered robot control has assumed that the Jacobian matrices from joint space to task space is exactly known. This implies that the locations of contact points, geometry of the object, kinematics of the multi-fingered robot hands must be exactly known. In this paper, a task-space feedback control problem of multi-fingered robot hands with uncertain Jacobian matrices is formulated and solved. The stability and robustness of the proposed controllers to the uncertainties in Jacobian matrices are analyzed.

1 Introduction

Many research efforts have been devoted to the study of dextrous manipulation using multi-fingered robot hands [1]. Several control laws are proposed for dynamic control of multi-fingered robot hands [1-6]. In computed torque method [2-4], the exact model of the dynamics is used to compensate for the nonlinearity of the multi-fingered robot hands. To cope with the parametric uncertainties, model-based adaptive control using regressor matrix are proposed [5, 6]. However, such formulations require the exact knowledge of certain Jacobian matrices from the mapping of joint space to task space. In a sense, the locations of contact points, geometry of the object, kinematics of the multi-fingered robot hands are assumed to be exactly known and exact camera calibration is necessary when visual feedback is used. This limits the potential applications of multi-fingered robot hands because Jacobian matrices are usually uncertain in many applications. For example, in the case of fingers with surface and rolling contacts, the contact points are usually uncertain and changing during manipulation. Due to the depressions at the soft contact points, the kinematics of the fingers also become uncertain. As a result, previous research on control of multi-fingered robot hands has assumed point contacts with either fixed or rolling contact points.

In this paper, we propose task-space feedback control laws for setpoint control of multi-fingered robot hands with uncertain Jacobian matrices. The exact locations of the rolling motion of fingers on the object is not required in our approach since the main objective of object's manipulation is to move the object to a desired position rather than the finger tips. The uncertainties due to the surface and rolling contacts are treated as the uncertainties in Jacobian matrices. The fundamental problems of stability and robustness of the proposed controllers to uncertain Jacobian matrices are analyzed. Force and gravity regressors are proposed to compensate for the effect of uncertain Jacobian matrices and gravitational forces for a class uncertain Jacobian matrices. The updating of such regressors does not require the exact knowledge of the Jacobian matrices. We shall show that simple feedback control laws can effectively cope with the nonlinearity and uncertainty of the multi-fingered robot hand dynamics.

2 Dynamic Equations

Consider a set of k fingers holding an object in a n_o degree of freedom space as illustrated in Figures 1 and 2. Let $\sum$ denote the task or Cartesian reference frame and $\sum_o$ be the object coordinate frame fixed at the mass centre of the object and moving with the object. Let $\sum_{ci}$ be the contact point coordinate frame located at the contact point of the ith finger and let $\sum_{ei}$ be the a coordinate frame located on the ith finger as shown in Figure 2. The velocity vector v_o of the object in $\sum_o$ is related to the velocity vector v_{ci} at the contact point of the ith finger in $\sum_{ci}$ as $v_{ci} = J_{oi} v_o$, where $J_{oi} \in R^{n_o \times n_o}$ denotes a Jacobian matrix from $\sum_o$ to $\sum_{ci}$. The velocity vector v_{ci} in $\sum_{ci}$ is related to the velocity vector v_{ei} of the ith finger in $\sum_{ei}$ as $v_{ci} = J_{fi} v_{ei}$, where $J_{fi} \in R^{n_o \times n_o}$ denotes a Jacobian matrix from $\sum_{ei}$ to $\sum_{ci}$. Since v_{ei} in $\sum_{ei}$ is expressed as $v_{ei} = J_{ei} \dot{q}_i$, where $q_i \in R^{n_i}$ ($n_i \geq n_o$) is the joint coordinates of the ith finger and $J_{ei}(q_i) \in R^{n_o \times n_i}$ is the manipulator Jacobian matrix of $\sum_{ei}$ in q_i, we have $v_{ci} = J_{fi} J_{ei} \dot{q}_i$. The velocity of the joint variables $q \triangleq [q_1^T, \cdots, q_k^T]^T$ and the velocity of the object v_o are thereofre constrained by the following equation:

$$J_o v_o = J_e \dot{q}, \quad (1)$$

where $J_o = [J_{o1}^T, \cdots, J_{ok}^T]^T$ and $J_e = diag\{J_{f1} J_{e1}(q_1), \cdots, J_{fk} J_{ek}(q_k)\}$. In addition, let $x \in R^{n_o}$ denotes the position and orientation vector of $\sum_o$ in $\sum$, then the velocity vector $\dot{x}$ is related to the velocity vector as $v_o = J_v(x) \dot{x}$, where J_v is a non-singular Jacobian matrices from $\sum$ to $\sum_o$. Hence, the kinematic constraint between $\dot{x}$ and q is described by

$$J \dot{x} = J_e \dot{q}, \quad (2)$$

where $J = J_o J_v$.

The dynamic equation of the ith finger is described in the joint coordinates q_i as

$$M_i(q_i) \ddot{q}_i + (B_i + \tfrac{1}{2}\dot{M}_i(q_i) + S_i(q_i, \dot{q}_i)) \dot{q}_i + g_i(q_i) = \tau_i - J_{ei}^T J_{fi}^T f_{ei} \quad (3)$$

where $M_i(q_i) \in R^{n_i \times n_i}$ is the inertia matrix which is symmetric and positive definite for all q_i, $B_i \in R^{n_i \times n_i}$ denotes the viscous friction matrix, $g_i(q_i) \in R^{n_i}$ is the gravitational force, $\tau_i \in R^{n_i}$ denotes the control input, $f_{ei} \in R^{n_o}$ is the

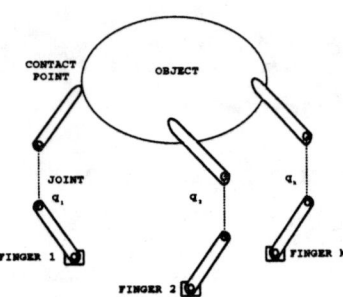

Figure 1: Multi-fingered Robot Holding an Object

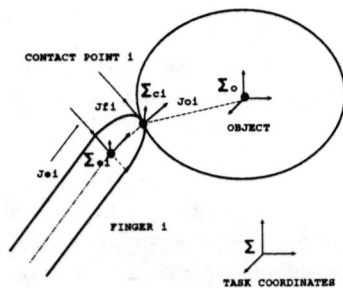

Figure 2: Definations of Coordinates

force/moment exerted on the object by the ith finger and $S_i(q_i, \dot{q}_i)$ is a skew-symmetric matrix expressed by

$$S_i(q_i,\dot{q}_i)\dot{q}_i = \frac{1}{2}\dot{M}_i(q_i)\dot{q}_i - \frac{1}{2}\{\frac{\partial}{\partial q_i}\dot{q}_i^T M_i(q_i)\dot{q}_i\}^T. \quad (4)$$

Clearly, the dynamic equation of the k fingers can then be described in q as

$$M(q)\ddot{q} + (B + \frac{1}{2}\dot{M}(q) + S(q,\dot{q}))\dot{q} + g(q) = \tau - J_e^T f_e, \quad (5)$$

where $M(q) = diag\{M_1(q_1),\cdots,M_k(q_k)\}$, $B = diag\{B_1,\cdots,B_k\}$, $g(q) = [g_1^T(q_1),\cdots,g_k^T(q_k)]^T$, $S(q,\dot{q}) = diag\{S_1(q_1,\dot{q}_1),\cdots,S_k(q_k,\dot{q}_k)\}$, $u = [u_1^T,\cdots,u_k^T]^T$, and $f_e = [f_{1e}^T,\cdots,f_{ke}^T]^T$. Similarly, $M(q)$ is symmetric and positive definite for all q and $S(q,\dot{q})$ is skew-symmetric. The equation of motion of the object can be written in $\sum$ as

$$M_o(x)\ddot{x} + (\frac{1}{2}\dot{M}_o(x) + S_o(x,\dot{x}))\dot{x} + g_o(x) = F, \quad (6)$$

where

$$F = J^T f_e, \quad (7)$$

is the total force exerted on the object by the fingers. From equatioin (7), f_e can be expressed as:

$$f_e = (J^T)^+ F + (I - (J^T)^+ J^T)\kappa, \quad (8)$$

where the pseudo-inverse of J^T is defined as $(J^T)^+ = J(J^T J)^{-1}$. Substituting equation (7) into equation (8), we have $(I - (J^T)^+ J^T)\kappa = (I - (J^T)^+ J^T)f_e$ and hence

$$f_e = (J^T)^+ F + Z f_e, \quad (9)$$

where $Z f_e$ is the internal force, $Z = I - (J^T)^+ J^T$ and $J^T Z = 0$. Since the pseudo-inverse of J^T is given as $J^+ = (J^T J)^{-1} J^T$, it follows that $(J^T)^+ = (J^+)^T$.

3 Robustness

As discussed in section 1, the Jacobian matrices $J = J_o J_v$, J_e and hence Z are uncertain in many practical applications such as using a soft-fingered robot hand. As a result, the internal force $Z f_e$ could not be calculated exactly too. It is therefore of theoretical and pratical importance to analyse the effects of such uncertainties on the object's stability. We assumed that the fingers are in contact with the object during manipulation. In order to maintain contact, the finger forces must satisfy a static frictional constraint. However, such constraint can be satisfied easily by using a soft-fingered robot hand with higher coefficient of static friction, as compared to fingers with rigid and point contacts.

Let us define a scalar potential function $S_i(\theta)$ and its derivative $s_i(\theta)$ as in [6] with the following properties:

(1) $S_i(\theta) > 0$ for $\theta \neq 0$ and $S_i(0) = 0$.

(2) $S_i(\theta)$ is twice continuously differentiable, and the derivative $s_i(\theta) = \frac{dS_i(\theta)}{d\theta}$ is strictly increasing in θ for $|\theta| < \gamma_i$ with some γ_i and saturated for $|\theta| \geq \gamma_i$, i.e. $s_i(\theta) = \pm s_i$ for $\theta \geq +\gamma_i$ and $\theta \leq -\gamma_i$ respectively where s_i is a positive constant.

(3) There are constants $\bar{c}_i > 0$, $d_i > 0$, $\bar{d}_i(> d_i) > 0$ such that,

$$\bar{d}_i s_i^2(\theta) \geq \theta s_i(\theta) \geq d_i s_i^2(\theta) > 0, \quad S_i(\theta) \geq \bar{c}_i s_i^2(\theta), \quad (10)$$

for $\theta \neq 0$. □

Let x_d denote a desired position and orientation of the object and f_d denote a desired grasping force, the control input is proposed as

$$\tau = \hat{J}_x^T(-K_p s(\triangle x) - K_v \dot{x} + \hat{g}_o(x)) + \hat{g}(q) + \hat{J}_m^T f_d, \quad (11)$$

where $\triangle x = x - x_d = (\triangle x_1,\cdots,\triangle x_{n_o})^T$, $s(\triangle x) = (s_1(\triangle x_1),\cdots,s_{n_o}(\triangle x_{n_o}))^T$, $\hat{J}_x = \hat{J}^+ \hat{J}_e$ and $\hat{J}_m = \hat{Z}^T \hat{J}_e$ are the estimated Jacobian matrices of $J_x = J^+ J_e$ and $J_m = Z^T J_e$ respectively, K_p, K_v are positive definite diagonal feedback gains for the position and velocity respectively, and $\hat{g}_o(x)$, $\hat{g}(q)$ are the estimates for $g_o(x)$, $g(q)$ respectively. We assumed that $\hat{J}_x$ and $\hat{J}_m$ are chosen so that

$$\|J_x^T - \hat{J}_x^T\| \leq p_1, \quad \|J_m^T - \hat{J}_m^T\| \leq p_2, \quad (12)$$

where p_1 and p_2 are constants to be defined.

Substituting equations (11) and (9) into equation (5) results in the following closed-loop equation

$$M(q)\ddot{q} + (B + \frac{1}{2}\dot{M}(q) + S(q,\dot{q}))\dot{q} + \hat{J}_x^T(K_p s(\triangle x) + K_v \dot{x} - \hat{g}_o(x)) + \delta g = -J_x^T(q)F + J_m^T \delta f - \delta_{Jm} f_d, \quad (13)$$

where $\delta g = g(q) - \hat{g}(q)$, $\delta f = f_d - f_e$ and $\delta_{Jm} = J_m^T - \hat{J}_m^T$. Let us define a vector given by:

$$y = \dot{q} + Q s(\triangle x) \quad (14)$$

where $Q^T = J^T(J_e^T)^+$ and $(J_e^T)^+ \triangleq (J_e J_e^T)^{-1} J_e$ such that $(J_e^T)^+ J_e^T = I$. Therefore, we note that

$$Q^T J_m^T = J^T(J_e^T)^+ J_e^T Z = J^T Z = 0,$$
$$Q^T J_x^T = J^T(J_e^T)^+ J_e^T (J^+)^T = (J^+ J)^T = I, \quad (15)$$

and

$$J_m \dot{q} = Z^T J_e \dot{q} = Z^T J \dot{x} = 0,$$
$$J_x \dot{q} = J^+ J_e \dot{q} = J^+ J \dot{x} = \dot{x}. \quad (16)$$

Taking inner product of y with equation (13) and using equations (6), (15) and (16) yields

$$\frac{d}{dt} V + W = 0, \quad (17)$$

where

$$V = \tfrac{1}{2} \dot{q}^T M(q) \dot{q} + \alpha s(\triangle x)^T Q^T M(q) \dot{q} + \tfrac{1}{2} \dot{x}^T M_o(x) \dot{x}$$
$$+ \alpha s(\triangle x)^T M_o(x) \dot{x} + \sum_{i=1}^{n_o} (k_{pi} + \alpha k_{vi}) S_i(\triangle x_i), \quad (18)$$

$$W = \dot{q}^T (J_x^T K_v J_x + B) \dot{q} + \alpha s(\triangle x)^T K_p s(\triangle x)$$
$$- \dot{q}^T (\delta_{Jx} K_v J_x) \dot{q} - \alpha s(\triangle x)^T (Q^T \delta_{Jx} K_p) s(\triangle x)$$
$$- \dot{q}^T (\delta_{Jx} K_p) s(\triangle x) - \alpha s(\triangle x)^T (Q^T \delta_{Jx} K_v J_x) \dot{q} + \alpha h + z^T \delta_1, (19)$$

$$h = s(\triangle x)^T Q^T (B - \tfrac{1}{2} \dot{M}(q) + S(q, \dot{q})) \dot{q}$$
$$- \dot{s}(\triangle x)^T Q^T M(q) \dot{q} - s(\triangle x)^T \dot{Q}^T M(q) \dot{q}$$
$$+ s(\triangle x)^T (-\tfrac{1}{2} \dot{M}_o(x) + S_o(x, \dot{x})) \dot{x} - \dot{s}(\triangle x)^T M_o(x) \dot{x}, (20)$$

and k_{pi}, k_{vi} denote the ith diagonal elements of K_p and K_v respectively, $\delta_{Jx} = J_x^T - \hat{J}_x^T$, $\delta_1 = \begin{bmatrix} I \\ \alpha Q^T \end{bmatrix} \delta$, $z = [\dot{q}^T, s(\triangle x)^T]^T$ and

$$\delta = \delta_{Jm} f_d + \delta_g + \delta_{Jx} g_o + \hat{J}_x^T (g_o - \hat{g}). \quad (21)$$

Since

$$\tfrac{1}{4} \dot{q}^T M(q) \dot{q} + \alpha s(\triangle x)^T Q^T M(q) \dot{q} + \tfrac{1}{4} \dot{x}^T M_o(x) \dot{x}$$
$$+ \alpha s(\triangle x)^T M_o(x) \dot{x} + \sum_{i=1}^{n_o} (k_{pi} + \alpha k_{vi}) S_i(\triangle x_i)$$
$$= \tfrac{1}{4} (\dot{q} + 2\alpha Q s(\triangle x))^T M(q) (\dot{q} + 2\alpha Q s(\triangle x))$$
$$+ \tfrac{1}{4} (\dot{x} + 2\alpha s(\triangle x))^T M_o(x) (\dot{x} + 2\alpha s(\triangle x))$$
$$- \alpha^2 s(\triangle x)^T (Q^T M(q) Q + M_o(x)) s(\triangle x)$$
$$+ \sum_{i=1}^{n_o} (k_{pi} + \alpha k_{vi}) S_i(\triangle x_i)$$
$$\geq \sum_{i=1}^{n_o} \{k_{pi} \bar{c}_i + \alpha(k_{vi} \bar{c}_i - \alpha \lambda_M)\} s_i^2(\triangle x_i), \quad (22)$$

where $\lambda_M \triangleq \lambda_{max}[Q^T M(q) Q + M_o(x)]$. Substituting into equation (18), we have

$$V \geq \tfrac{1}{4} \dot{q}^T M(q) \dot{q} + \tfrac{1}{4} \dot{x}^T M_o(x) \dot{x}$$
$$+ \sum_{i=1}^{n_o} \{k_{pi} \bar{c}_i + \alpha(k_{vi} \bar{c}_i - \alpha \lambda_M)\} s_i^2(\triangle x_i). \quad (23)$$

Hence, there exist an α and a K_v so that

$$k_{vi} \bar{c}_i - \alpha \lambda_M > 0, \quad (24)$$

and $V > 0$. Therefore, V represents a Lyapunov function candidate for the robustness problem of multi-fingered robot hands. From equation (19)

$$W \geq (\lambda_1 - p_1 b_{Jx} \lambda_{max}[K_v]) \|\dot{q}\|^2 + \alpha(\lambda_{min}[K_p]$$
$$- p_1 b_{QT} \lambda_{max}[K_p]) \|s(\triangle x)\|^2 - p_1 (\lambda_{max}[K_p] +$$
$$\alpha b_{QT} b_{Jx} \lambda_{max}[K_v]) \|\dot{q}\| \cdot \|s(\triangle x)\| - \alpha|h| - \|z\| \cdot \|\delta_1\|, (25)$$

where $\lambda_1 = \lambda_{min}[J_x^T K_v J_x + B]$, b_{Jx}, b_{QT} are the norm bounds for J_x, Q^T respectively. From equation (20), since $s(\triangle x)$ is bounded, there exist a constant c_0 such that [6]

$$\alpha|h| \leq \alpha c_0 \|\dot{q}\|^2. \quad (26)$$

Furthermore, note that

$$-\|s(\triangle x)\| \cdot \|\dot{q}\| \geq -\tfrac{1}{2}(\|s(\triangle x)\|^2 + \|\dot{q}\|^2). \quad (27)$$

Hence, substituting the above equations into equation (25) yields

$$W \geq \{\lambda_1 - p_1 k_1 - \alpha c_0\} \|\dot{q}\|^2$$
$$+ \{\alpha \lambda_{min}[K_p] - p_1 k_2\} \|s(\triangle x)\|^2 - \|z\| \cdot \|\delta_1\|$$
$$\geq l \|z\|^2 - \|z\| \|\delta_1\| \quad (28)$$

where $k_1 = b_{Jx} \lambda_{max}[K_v] + \tfrac{1}{2} \lambda_{max}[K_p] + \tfrac{1}{2} \alpha b_{QT} b_{Jx} \lambda_{max}[K_v]$ and $k_2 = \alpha b_{QT} \lambda_{max}[K_p] + \tfrac{1}{2} \lambda_{max}[K_p] + \tfrac{1}{2} \alpha b_{QT} b_{Jx} \lambda_{max}[K_v]$, $l = min\{\lambda_1 - p_1 k_1 - \alpha c_0, \alpha \lambda_{min}[K_p] - p_1 k_2\}$. Hence, K_v and K_p can be chosen so that

$$\lambda_1 - p_1 k_1 - \alpha c_0 > 0, \qquad \alpha \lambda_{min}[K_p] - p_1 k_2 > 0, \quad (29)$$

if p_1 is sufficiently small so that

$$p_1 < \min\{\frac{\lambda_1}{k_1}, \frac{\alpha \lambda_{min}[K_p]}{k_2}\}. \quad (30)$$

Therefore, W will be positive if

$$\|z\| > \frac{\|\delta_1\|}{l} \quad (31)$$

which causes $\frac{d}{dt} V$ to be negative (see equation (17)) and V to decrease. If V is decreased, then z must eventually decrease. However, if z decreases such that

$$\|z\| \leq \frac{\|\delta_1\|}{l}, \quad (32)$$

then $\frac{d}{dt} V$ may become positive and V will increase. The increase in V causes z to increase. However, when z is increased so that the inequality (31) is satisfied, z will start to decrease again. Therefore, z is bounded.

Differentiating equation (16) with respect to time, we have

$$J_m \ddot{q} + \dot{J}_m \dot{q} = 0, \quad (33)$$
$$J_x \ddot{q} + \dot{J}_x \dot{q} = \ddot{x}. \quad (34)$$

Substituting equations (6), (16) and (34) into (13), we have

$$[M(q) + J_x^T M_o(x) J_x] \ddot{q} + \{B + \tfrac{1}{2} \dot{M}(q) + S(q, \dot{q})$$
$$+ J_x^T M_o(x) \dot{J}_x + J_x^T (\tfrac{1}{2} \dot{M}_o(x) + S_o(x, \dot{x})) J_x\} \dot{q}$$
$$+ \hat{J}_x^T (K_p s(\triangle x) + K_v J_x \dot{q}) + \delta = J_m^T \delta f. \quad (35)$$

2405

Eliminating $\ddot{q}$ from equations (35) and (33) yields

$$J_m[M(q) + J_x^T M_o(x) J_x]^{-1} J_m^T \delta f = -\dot{J}_m \dot{q}$$
$$+ J_m[M(q) + J_x^T M_o(x) J_x]^{-1}\{(B + \tfrac{1}{2}\dot{M}(q) + S(q,\dot{q})$$
$$+ J_x^T(\tfrac{1}{2}\dot{M}_o(x) + S_o(x,\dot{x}))J_x)\dot{q}$$
$$+ \hat{J}_x^T(K_p s(\triangle x) + K_v J_x \dot{q}) + \delta\}. \qquad (36)$$

Hence δf is bounded since $\dot{q}$ and $s(\triangle x)$ are bounded.

The boundedness of the state z and the grasping force error δf is therefore specified by the following theorem:

Theorem 1. *The state error z and force error δf of the closed loop system described by equation (13) are bounded with uncertain Jacobian matrices $\hat{J}_x$ and $\hat{J}_m$ if the feedback gains K_p and K_v are chosen to satisfy conditions (24), (29) and (30) and $\hat{J}_x$ and $\hat{J}_m$ are chosen to satisfy condition (12).*

Remark 1. Note that z tends to zero as $\delta = \delta_{J_m} f_d + \delta g + \delta_{J_x} g_o + \hat{J}_x(g_o - \hat{g})$ tends to zero.

Remark 2. In a soft-fingered robot hand, the kinematics and Jacobian matrices are very complicated even when manipulating a very simple object [7]. This is because the contact points are changing during manipulation and hence the Jacobian matrices J_{fi} and J_{oi} are changing according to the geometry of the finger tips and object even when J_{ei} and J_v can be estimated with sufficient accuracy. Since the movements of the contact points are normally small variations of motion, by using the result of Theorem 1, it is possible to approximate the actual Jacobian matrices by using the Jacobian matrices of the initial contacts before manipulation if the errors due to other Jacobian matrices can be kept smaller.

4 Adaptive PD Control

In the previous section, the robustness of the multi-fingered robot system is analysed with a general class of Jacobian uncertainty. The result shows that if the uncertainty δ in equation (21) is small, then the position and force errors are small. As shown in equations (31) and (32), these errors can be reduced by increasing the feedback gains. However, a high feedback gain would be required if δ is large. In this section, we present an adaptive law to compensate for the effects of the uncertain Jacobian matrices and gravitional force for a class of Jacobian uncertainty such that:

$$\hat{J}_m \dot{q} = 0. \qquad (37)$$

where $\hat{J}_m = \hat{Z}^T \hat{J}_e$.

It is assumed that the the gravity terms can be completely characterised by a set of parameters $\theta = (\theta_1, \cdots, \theta_p)^T$ as

$$g(q) + J_x^T g_o(x) = Z_x(q,x)\theta, \qquad (38)$$

where $Z_x(q,x) \in R^{n \times p}$ is the gravity regressor matrix. Similarly, we assume the force can also be completely characterised by a set of parameters $\theta_f = (\theta_{f,1}, \cdots, \theta_{f,j})^T$ as

$$J_m^T f_e = J_e^T Z f_e = Z_m(q,x,f_e)\theta_f, \qquad (39)$$

where $Z_m(q,x,f_e) \in R^{n \times j}$ is the force regressor matrix. For example, if

$$J_m^T f_e = \begin{bmatrix} a_0 h_0(q) + a_1 h_1(q) & a_2 h_2(q) \\ a_3 h_3(q) & a_4 h_4(q) \end{bmatrix} \begin{bmatrix} f_{e1} \\ f_{e2} \end{bmatrix},$$

then $Z_m = \begin{bmatrix} h_0(q)f_{e1} & h_1(q)f_{e1} & h_3(q)f_{e2} & 0 \\ 0 & h_2(q)f_{e1} & 0 & h_4(q)f_{e2} \end{bmatrix}$

and $\theta_f = [a_0, a_1, a_2, a_3, a_4]^T$. Note that it is therefore implicitly assumed in equations (38) and (39) that the structure of the Jacobian matrices J_x and J_m are known so that the only uncertainties come from the parameters θ and θ_f.

In the presence of such uncertainties in the Jacobian matrices, we have

$$\hat{J}_m^T f_e = \hat{J}_e^T \hat{Z} f_e = Z_m(q,x,f_e)\hat{\theta}_f, \qquad (40)$$

and hence

$$(J_m^T - \hat{J}_m^T) f_e = Z_m(q,x,f_e)\bar{\theta}_f, \qquad (41)$$

where $\bar{\theta}_f = \hat{\theta}_f - \theta_f$ is an unknown parameter and its estimate is denoted by $\hat{\bar{\theta}}_f$.

The control input is proposed as

$$\tau = \hat{J}_x^T(-K_p s(\triangle x) - K_v \dot{x}) + \hat{J}_m^T f_d + \beta \hat{J}_m^T \int_0^t (f_d - f_e)d\tau$$
$$+ Z_m(q,x,f_e)\hat{\bar{\theta}}_f + Z_x(q,x)\hat{\theta}, \qquad (42)$$

$$\hat{\theta}(t) = \hat{\theta}(0) - L\int_0^t Z_x^T(q,x)(\dot{q} + \alpha \hat{Q} s(\triangle x))d\tau, \qquad (43)$$

$$\hat{\bar{\theta}}_f(t) = \hat{\bar{\theta}}_f(0) - \bar{L}\int_0^t Z_m^T(q,x,f_e)(\dot{q} + \alpha \hat{Q} s(\triangle x))d\tau, \qquad (44)$$

where

$$\hat{J}_x = \hat{J}^+ \hat{J}_e, \quad \hat{J}_m = \hat{Z}^T \hat{J}_e$$
$$\hat{J}^T \hat{Z} = 0, \qquad \hat{Q}^T = \hat{J}^T (\hat{J}_e^T)^+ \qquad (45)$$

and β is a constant, $Z_x(q,x)\hat{\theta} = \hat{g}(q) + \hat{J}_x \hat{g}(x)$, $\hat{\theta}(0)$, $\hat{\bar{\theta}}_f(0)$ are the initial estimations at $t=0$ and $L \in R^{p \times p}$, $\bar{L} \in R^{j \times j}$ are positive definite matrices. $(\hat{J}_e^T)^+ \triangleq (\hat{J}_e \hat{J}_e^T)^{-1} \hat{J}_e$, $\hat{J}^+ = (\hat{J}^T \hat{J})^{-1} \hat{J}^T$ such that $(\hat{J}_e^T)^+ \hat{J}_e^T = I$, $\hat{J}^+ \hat{J} = I$ respectively. Therefore, we note that

$$\hat{Q}^T \hat{J}_m^T = \hat{J}^T (\hat{J}_e^T)^+ \hat{J}_e^T \hat{Z}(x) = \hat{J}^T \hat{Z} = 0,$$
$$\hat{Q}^T \hat{J}_x^T = \hat{J}^T (\hat{J}_e^T)^+ \hat{J}_e^T (\hat{J}^+)^T = (\hat{J}^+ \hat{J})^T = I. \qquad (46)$$

It is assumed that $\hat{J}_x$ and $\hat{J}_m$ are chosen so that

$$\|J_x^T - \hat{J}_x^T\| \leq \bar{p}_1, \qquad \|J_m^T - \hat{J}_m^T\| \leq \bar{p}_2 \qquad (47)$$

and $\bar{p}_1$ and $\bar{p}_2$ are constants to be defined.

Substituting equations (42) and (9) into equation (5) results in the following closed-loop equation:

$$M(q)\ddot{q} + (B + \tfrac{1}{2}\dot{M}(q) + S(q,\dot{q}))\dot{q}$$
$$+ \hat{J}_x^T(K_p s(\triangle x) + K_v \dot{x}) + Z_x(q,x)\Delta\theta + Z_m(q,f_e)\Delta\theta_f$$
$$= -J_x^T(q)F + \hat{J}_m^T(f_d - f_e) + \beta \hat{J}_m^T \int_0^t (f_d - f_e)d\tau, \qquad (48)$$

where $\Delta\theta = \theta - \hat{\theta}$ and $\Delta\theta_f = \bar{\theta}_f - \hat{\bar{\theta}}_f$. Define $y_1 = \dot{q} + \hat{Q}s(\Delta x)$, taking inner product of y_1 with equation (48) and using equations (37), (46) and (6), we have

$$(\dot{q} + \alpha\hat{Q}s(\Delta x))^T\{M(q)\ddot{q} + (B + \tfrac{1}{2}\dot{M}(q) + S(q,\dot{q}))\dot{q}\}$$
$$+(\dot{q} + \alpha\hat{Q}s(\Delta x))^T\hat{J}_x^T(K_p s(\Delta x) + K_v \dot{x})$$
$$+(\dot{x} + \alpha J_x\hat{Q}s(\Delta x))^T\{M_o(x)\ddot{x} + (\tfrac{1}{2}\dot{M}_o(x) + S_o(x,\dot{x}))\dot{x}\}$$
$$+(\dot{q} + \alpha\hat{Q}s(\Delta x))^T(Z_x(q,x)\Delta\theta + Z_m(q,f_e)\Delta\theta_f) = 0. \quad (49)$$

Using equations (43) and (44), equation (49) can be written as:

$$\frac{d}{dt}V_1 + W_1 = 0, \quad (50)$$

where

$$V_1 = \tfrac{1}{2}\dot{q}^T M(q)\dot{q} + \alpha s(\Delta x)^T\hat{Q}^T M(q)\dot{q} + \tfrac{1}{2}\dot{x}^T M_o(x)\dot{x}$$
$$+\alpha s(\Delta x)^T\hat{Q}^T J_x M_o(x)\dot{x} + \sum_{i=1}^{n_o}(k_{pi} + \alpha k_{vi})S_i(\Delta x_i)$$
$$+\tfrac{1}{2}\Delta\theta L^{-1}\Delta\theta + \tfrac{1}{2}\Delta\theta_f \bar{L}^{-1}\Delta\theta_f \quad (51)$$

$$W_1 = \dot{q}^T(\hat{J}_x^T K_v J_x + B)\dot{q} + \alpha s(\Delta x)^T K_p s(\Delta x)$$
$$-\dot{q}^T(\hat{J}_x - J_x)K_p s(\Delta x) + \alpha h_1 \quad (52)$$

$$h_1 = s(\Delta x)^T\hat{Q}^T(B - \tfrac{1}{2}\dot{M}(q) + S(q,\dot{q}))\dot{q}$$
$$-\dot{s}(\Delta x)^T\hat{Q}^T M(q)\dot{q} - s(\Delta x)^T\dot{\hat{Q}}^T M(q)\dot{q}$$
$$+s(\Delta x)^T\hat{Q}^T J_x(-\tfrac{1}{2}\dot{M}_o(x) + S_o(x,\dot{x}))\dot{x}$$
$$-\dot{s}(\Delta x)^T\hat{Q}^T J_x M_o(x)\dot{x} - s(\Delta x)^T\dot{\hat{Q}}^T J_x M_o(x)\dot{x}$$
$$-s(\Delta x)^T\hat{Q}^T \dot{J}_x M_o(x)\dot{x}. \quad (53)$$

Similarly, since V_1 can be expressed as

$$V_1 = \tfrac{1}{4}\dot{q}^T M(q)\dot{q} + \tfrac{1}{4}\dot{x}^T M_o(x)\dot{x} + \sum_{i=1}^{n_o}(k_{pi} + \alpha k_{vi})S_i(\Delta x_i)$$
$$+\tfrac{1}{4}(\dot{q} + 2\alpha\hat{Q}s(\Delta x))^T M(q)(\dot{q} + 2\alpha\hat{Q}s(\Delta x))$$
$$+\tfrac{1}{4}(\dot{x} + 2\alpha s(\Delta x))^T M_o(x)(\dot{x} + 2\alpha s(\Delta x))$$
$$-\alpha^2 s(\Delta x)^T(\hat{Q}^T M(q)\hat{Q} + M_o(x))s(\Delta x)$$
$$+\tfrac{1}{2}\Delta\theta L^{-1}\Delta\theta + \tfrac{1}{2}\Delta\theta_f\bar{L}^{-1}\Delta\theta_f$$
$$\geq \tfrac{1}{4}\dot{q}^T M(q)\dot{q} + \tfrac{1}{4}\dot{x}^T M_o(x)\dot{x}$$
$$+\sum_{i=1}^{n_o}\{k_{pi}\bar{c}_i + \alpha(k_{vi}\bar{c}_i - \alpha\lambda_{\dot{M}})\}s_i^2(\Delta x_i)$$
$$+\tfrac{1}{2}\Delta\theta L^{-1}\Delta\theta + \tfrac{1}{2}\Delta\theta_f\bar{L}^{-1}\Delta\theta_f, \quad (54)$$

where $\lambda_{\dot{M}} \triangleq \lambda_{max}[\hat{Q}^T M(q)\hat{Q} + M_o(x)]$. Hence, there exist an α and K_v so that

$$k_{vi}\bar{c}_i - \alpha\lambda_M > 0, \quad (55)$$

and $V_1 > 0$. Therefore, the function V_1 represents a Lyapunov function candidate for the adaptive setpoint control problem of multi-fingered robot hands with uncertain Jacobian matrix. From equation (52)

$$W_1 = \dot{q}^T(\hat{J}_x^T K_v J_x + B)\dot{q} + \alpha s(\Delta x)^T K_p s(\Delta x)$$
$$+\dot{q}^T(\hat{J}_x - J_x)K_v J_x \dot{q} + \dot{q}^T(\hat{J}_x - J_x)K_p s(\Delta x) + \alpha h_1$$
$$\geq (\lambda_1 - \bar{p}_1 b_{J_x}\lambda_{max}[K_v])\|\dot{q}\|^2 + \alpha\lambda_{min}[K_p]\|s(\Delta x)\|^2$$
$$-\bar{p}_1\lambda_{max}[K_p]\|\dot{q}\|\cdot\|s(\Delta x)\| - \alpha|h_1| \quad (56)$$

where $\lambda_1 = \lambda_{min}[J_x^T K_v J_x + B]$. Note that, $-\|s(\Delta x)\|\cdot\|\dot{q}\| \geq -\tfrac{1}{2}(\|s(\Delta x)\|^2 + \|\dot{q}\|^2)$, and from equation (53)

$$\alpha|h_1| \leq \alpha\bar{c}_0\|\dot{q}\|^2, \quad (57)$$

since $s(\Delta x)$ is bounded. Hence, substituting into equation (56) yields

$$W_1 \geq \{\lambda_{max}[K_v]\bar{l}_1 - \alpha\bar{c}_0\}\|\dot{q}\|^2 + \lambda_{max}[K_p]\bar{l}_2\|s(\Delta x)\|^2. \quad (58)$$

where

$$\bar{l}_1 = \bar{\lambda}_1 - \tfrac{\bar{p}_1}{2}(2b_{J_x} + a_1), \quad \bar{l}_2 = \alpha\frac{\lambda_{min}[K_p]}{\lambda_{max}[K_p]} - \frac{\bar{p}_1}{2},$$
$$\bar{\lambda}_1 = \frac{\lambda_{min}[J_x^T K_v J_x + B]}{\lambda_{max}[K_v]}, \quad (59)$$

and $a_1 = \frac{\lambda_{max}[K_p]}{\lambda_{max}[K_v]}$. Hence if

$$\bar{p}_1 < \min\{\frac{2\bar{\lambda}_1}{(a_1 + 2b_{J_x})}, 2\alpha\frac{\lambda_{min}[K_p]}{\lambda_{max}[K_p]}\}, \quad (60)$$

then $\bar{l}_1 > 0$ and $\bar{l}_2 > 0$ and hence K_v can be chosen large enough so that

$$\bar{l}_1 - \frac{\alpha\bar{c}_0}{\lambda_{max}[K_v]} > 0, \quad (61)$$

and hence $W_1 \geq 0$.

An alternate condition for the bound $\bar{p}_1$ can be derived by expressing the first term of equation (52) in $\hat{J}_x$ so that

$$W_1 = \dot{q}^T(\hat{J}_x^T K_v \hat{J}_x + B)\dot{q} + \alpha s(\Delta x)^T K_p s(\Delta x)$$
$$+\dot{q}^T(J_x^T - \hat{J}_x^T)K_v\hat{J}_x\dot{q} - \dot{q}^T(\hat{J}_x - J_x)K_p s(\Delta x) + \alpha h_1 \quad (62)$$

Hence, condition (60) becomes

$$\bar{p}_1 < \min\{\frac{2\hat{\lambda}_1}{(a_1 + 2\hat{b}_{J_x})}, 2\alpha\frac{\lambda_{min}[K_p]}{\lambda_{max}[K_p]}\}, \quad (63)$$

where $\hat{\lambda}_1 = \frac{\lambda_{min}[\hat{J}_x^T K_v \hat{J}_x + B]}{\lambda_{max}[K_v]}$, and $\hat{b}_{J_x}$ is the norm bound for $\hat{J}_x$. This implies that the bound $\bar{p}_1$ can be represented by the actual or estimated Jacobian matrix whichever is larger. Note that in conditions (30) and (60) or (63), the viscous friction $B\dot{q}$ plays a role in ensuring the robustness of the uncertain Jacobian controller and increases the bound of the uncertainty. In pratice, such friction is always present in robots or can be added intentionally to the control input τ.

Figure 3 shows a graphical illustration of condition (60) or (63). If $\bar{p}_1$ is small, α can be chosen small and therefore a smaller controller gain is required as seen from equation (61). In addition, a wider range of a_1 can be chosen when $\bar{p}_1$ is small as illustrated in Figure 3. Conversely, if $\bar{p}_1$ is large, a larger controller gain is required and a narrower range of a_1 is allowed. Condition (60) therefore implies that if a_1 is increased, then the allowable bound $\bar{p}_1$ of the Jacobian uncertainty is reduced. Therefore, a_1 should be kept smaller so that the allowable bound of the Jacobian uncertainty is larger. This can be easily done by either reducing K_p or increasing K_v. Though the condition is a sufficient condition, it is reasonable because increasing K_p amplifies the estimated Jacobian $\hat{J}(q)$ and hence more accuracy on the estimation or

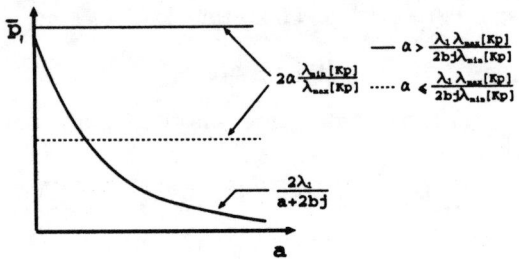

Figure 3: Variation of $\bar{p}_1$ with a

more damping is required. Note from equation (45) that $\bar{p}_2$ is denpendent on $\bar{p}_1$ since $\hat{J}_x = \hat{J}^+ \hat{J}_e$, $\hat{J}_m = \hat{Z}^T \hat{J}_e$ where $\hat{Z}$ is estimated according to the relationship $\hat{J}^T \hat{Z} = 0$.

The asymptotic stability of the equilibrium state $(x_d, 0)$ and the boundedness of the force is specified by the following Theorem:

Theorem 2. *The equilibrium state $(x_d, 0)$ of the closed loop system described by equations (48), (43) and (44) is asymptotically stable with uncertain Jacobian matrices $\hat{J}_m$ and $\hat{J}_x$ if the feedback gains K_p and K_v are chosen to satisfy conditions (55), (60) or (63), (61), and $\hat{J}_x$ and $\hat{J}_m$ are chosen to satisfy condition (47). In addition, the grasping force error δf is also bounded.*

Proof:
Since both V_1 and W_1 are positive definite, we have

$$\frac{d}{dt}V_1 = -W_1 \leq 0. \qquad (64)$$

Hence, V_1 is a Lyapunov function whose time derivative is negative in $s(\triangle x), \dot{q}$. This implies directly the asymptotic stability of the equilibrium state such that $x_d - x \to 0$, $\dot{q} \to 0$, when $t \to \infty$. Substituting equations (6), (16) and (34) into (48), we have

$$(M(q) + J_x^T M_o(x) J_x)\ddot{q} + (B + \tfrac{1}{2}\dot{M}(q) + S(q,\dot{q})$$
$$+ J_x^T M_o(x) \dot{J}_x + J_x^T(\tfrac{1}{2}\dot{M}_o(x) + S_o(x,\dot{x}))J_x)\dot{q} + \hat{J}_x^T(K_p s(\triangle x)$$
$$+ K_v J_x \dot{q}) + Z_x(q,x)\Delta\theta + Z_m(q, f_e)\Delta\theta_f$$
$$= \hat{J}_m^T(f_d - f) + \beta \hat{J}_m^T \int_0^t (f_d - f_e)d\tau. \qquad (65)$$

Hence the maximum invariant set of equation (65) satisfy

$$Z_x(q_d)\Delta\theta_\infty + Z_m(q_d, f_\infty)\Delta\theta_{f,\infty}$$
$$= \hat{J}_m^T(f_d - f_\infty) + \beta \hat{J}_m^T \int_0^t (f_d - f_\infty)d\tau, \qquad (66)$$

where $\Delta\theta_\infty = \theta - \hat{\theta}_\infty$. From equation (41), we have

$$Z_m(q_d, f_\infty)\Delta\theta_{f,\infty} = Z_m(q_d, f_\infty)\bar{\theta}_f - Z_m(q_d, f_\infty)\hat{\bar{\theta}}_{f,\infty}$$
$$= (J_1 - \hat{J}_{1,\infty})f_\infty, \qquad (67)$$

where $J_1 = J_m^T - \hat{J}_m^T$. Substituting into equation (66) yields

$$Z_x(q_d, x_d)\Delta\theta_\infty = \hat{J}_m^T f_d - (J_m^T - \hat{J}_{1,\infty}^T)f_\infty$$
$$+ \beta \hat{J}_m^T \int_0^t (f_d - f_\infty)d\tau, \qquad (68)$$

and hence $f_d - f_\infty$ is bounded because

$$Z_x(q_d, x_d)\Delta\theta_\infty + Z_m(q_d, f_d)\Delta\theta_{f,\infty}$$
$$= (J_m^T - \hat{J}_{1,\infty}^T)(f_d - f_\infty) + \beta \hat{J}_m^T \int_0^t (f_d - f_\infty)d\tau, \qquad (69)$$

where $Z_m(q_d, f_d)\Delta\theta_{f,\infty} = (J_1 - \hat{J}_{1,\infty})f_d$ △△△

Remark 3. It should be noted that in a vision-based system, the velocity $\dot{x}_I$ of a vector of image feature parameters x_I is related to the Cartesian coordiantes $\dot{x}$ as $\dot{x}_I = J_I \dot{x}$, where J_I is the image Jacobian matrix. Hence, if the only uncertainty comes from the image Jacobian matrix, the following control input can be used so that $z \to 0$ as $t \to \infty$

$$\tau = J_x^T \hat{J}_I^T(-K_p s(\triangle x)_I - K_v \dot{x}_I) + Z_x(q,x)\hat{\theta} + J_m^T f_d,$$
$$\hat{\theta} = \hat{\theta}(0) - L \int_0^t Z^T(\dot{q} + \alpha \hat{Q}_1 s(\triangle x))d\tau, \qquad (70)$$

where $s(\triangle x)_I = x_{I,d} - x_I$, $Z_x(q,x)\hat{\theta} = J_x^T \hat{g}_o(x) + \hat{g}(q)$ and $\hat{Q}_1^T = (\hat{J}_I^T)^+ J^T(J_e^T)^+$. The result follows directly from Theorems 1 and 2 by noting that δ is indepedent of J_I and

$$\hat{Q}_1^T \hat{J}_x^T \hat{J}_I^T = (\hat{J}_I^T)^+ J^T(J_e^T)^+ J_e^T(J^+)^T \hat{J}_I^T = I,$$
$$\hat{Q}_1^T J_m^T = (\hat{J}_I^T)^+ J^T(J_e^T)^+ J_e^T Z = (\hat{J}_I^T)^+ J^T Z = 0. \qquad (71)$$

This implies that the controller is robust to camera miscalibration.

Remark 4. The results presented in this paper can be similarly applied to the study of constrained robot system whose end effector is required to move on a holonomic constraint [6]. The stability problem of such constrained system has been studied by a number of researchers but with the assumption of exact constraint function [1]. Our result also provides an answer to the stability of constrained robot with uncertain constraint function.

5 Conclusion

We proposed simple task-space feedback control laws for setpoint control of multi-fingered robot hands. We have shown the robustness of the proposed controller to uncertainties in the Jacobian matrices. The positional error is bounded in the presence of uncertainties such as the exact locations of the contact points and the kinematics of the multi-fingered robot hands. Furthermore, this bound tends to zero as the uncertainties tend to zero. To compensate for the effects of the uncertainties, adaptive PD control using regressors are proposed for a class of Jacobian matrices.

References

[1] M. W Spong, F.L. Lewis and C.T. Abdallah, "Robot Control: Dynamics, Motion Planning and Analysis", IEEE Press, 1990.

[2] Z. Li, P. Hsu and S.S. Sastry, "Grasping and Coordinated Manipulation by a Multifingered Robot Hand", Int. J. of Robotics Res., vol 8, no. 4, pp. 33 – 50, 1989.

[3] A. Cole, J. Hauser and S.S. Sastry, " Kinematics and Control of Multifingered Robot Hands with Rolling Contact", IEEE Tans. on Automatic Control, vol 34, no. 4, pp. 398 – 404, 1989.

[4] K. Nagai and T. Yoshikawa, "Dynamic Manipulation/Grasping Control of Multifingered Robot Hands", IEEE Conf. on Robotics and Automation, Atlanta, USA, 1993, pp. 1027 – 1033.

[5] B. Yao and M. Tomizuka, "Adaptive Coordinated Control of Multiple Manipulators Handling a Constrained Object", IEEE Conf. on Robotics and Automation, Atlanta, USA, 1993, pp. 624 – 629.

[6] S. Arimoto, "Control Theory of Nonlinear Mechanical Systems - A Passivity-Based and Circuit-Theoretic Approach", Clarendon Press, Oxford, 1996.

[7] T. Nagashima, H. Seki and M. Takano, "Analysis and Simulation of Grasping/Manipulation by Multi-fingersurface", Mech. Mach. Theory, vol 32, no. 2, pp. 175 – 191, 1997

Enveloping Grasp for Multiple Objects

Kensuke Harada Makoto Kaneko

Industrial and Systems Engineering
Hiroshima University
Kagamiyama, Higashi-Hiroshima 739-8527, JAPAN

Abstract

This paper discusses the enveloping grasp of multiple objects under rolling contacts. We first provide a general mathematical formulation on the kinematic relationship for multiple objects enveloped by a multifingered robot hand, and then derive a condition for judging whether the rolling condition can be satisfied at each contact point. We also show a sufficient condition for rolling up general two objects grasped by a multifingered robot hand in contact with them. Finally, an experimental result is shown to confirm how easily two cylindrical objects can be enveloped by a simple grasping motion.

1 Introduction

Multifingered robot hands have a potential advantage to perform various skillful tasks like human hands. While much research has been done on multifingered robot hands, there are two basic grasp patterns. One is the finger tip grasp that emphasizes on dexterity and sensitivity [1, 2, 3], and the other is the enveloping grasp that provides highly stable grasp due to a large number of distributed contacts on the grasped object [4, 5, 6, 7, 8].

So far, most of works have implicitly assumed that a multifingered hand manipulates only one object. Under such a condition, they discussed several grasping issues, such as the stability of grasp, the analysis of contact force, the planning for manipulating an object and so forth. In this paper, we relax the assumption of single object, and discuss how to grasp multiple objects by a multifingered hand.

Suppose that a multifingered hand is grasping two objects by the finger tip grasp. Intuitively, we can imagine that it is difficult for such system to keep a stable grasp and the system will easily fail in grasping for a small disturbance. On the other hand, suppose that a multifingered hand is grasping two objects by enveloping grasp, as shown in Fig.1. It seems that the enveloping grasp can achieve this task even more easily than the finger tip grasp. We can find another advantage for enveloping two objects by a multifingered hand. Suppose that the friction between an object and the link surface is very significant. Under such a condition, since it is hard to lift up the object by slipping over the link surface, we have to provide an alternative scheme based on rolling contact. In such a case, one finger continuously pushes the object so that it may be rolled up over the surface of the other fingers. Generally, this motion planning is too complicated to be easily implemented to the actual system. However, for two objects satisfying the rolling contact each other, we can expect that a multifingered hand can easily achieve an enveloping grasp by simply pushing two links contacting with the objects, as shown in Fig.1. During the lifting phase, links and two objects behave as if they were just connected by mechanical gears. Due to this mechanical properties, achieving an enveloping grasp for two objects seems to be even easier than for a single object under a significant friction.

This work is motivated by these backgrounds. We first provide a general mathematical formulation on the kinematic relationship for multiple objects enveloped by a multifingered hand, and then discuss a condition whether each contact point can satisfy the rolling condition or not. We confirm the condition by a numerical example. We also show a sufficient condition for lifting up two objects by a simple pushing force exerted on the links contacting with them. Finally, an experimental result is shown to see how easily a multifingered robot hand can achieve an enveloping grasp for two cylindrical objects placed on a table.

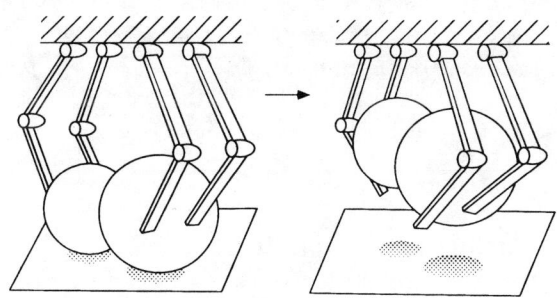

Fig. 1: Enveloping grasp of multiple objects

2 Related Works

Enveloping Grasp:
There have been a number of works concerning the enveloping grasp. Especially, Salisbury et al.[4] has

proposed the Whole-Arm Manipulation(WAM) capable of treating a big and heavy object by using one arm which allows multiple contacts with an object. Bicchi[5], Zhang et al.[6] and Omata et al.[7] analyzed the grasp force of the enveloping grasp. In our previous work, we have proposed the grasping strategies for achieving enveloping grasp work for cylindrical objects[8].

Grasp and Manipulation of Multiple Objects:
Dauchez et al.[10] and Kosuge et al.[11] used two manipulators holding two objects independently and tried to apply to an assembly task. However, they have not considered that two manipulators grasp and manipulate two common objects simultaneously. Recently, Aiyama et al.[9] studied a scheme for grasping multiple box type objects stably by using two manipulators. For an assembly task, Mattikalli et al.[12] proposed a stable alignments of multiple objects under the gravitational field. While these works treated multiple objects, they have not considered any manipulation of objects based on rolling contacts.

Grasp by Rolling Contacts:
Kerr et al.[1] and Montana[13] formulated the kinematics of manipulation of objects under rolling contacts with the fingertip. Li et al.[14] proposed a motion planning method with nonholonomic constraint. Howard et al.[15] and Maekawa et al.[16] studied the stiffness effect for the object motion with rolling. Cole et al.[2] and Paljug et al.[3] proposed a control scheme for the object motion.

Within our knowledge, this is the first challenge for enveloping multiple objects based on rolling contacts.

3 Modeling

Fig.2 shows the hand system enveloping m objects and n fingers, where finger j contacts with object i, and additionally object i has a common contact point with object l. Σ_R, Σ_{Bi} ($i = 1,\cdots,m$) and Σ_{Fjk} ($j = 1,\cdots,n$, $k = 1,\cdots,c_j$) denote the coordinate systems fixed at the base, at the center of gravity of the object i and at the finger link including the kth contact of finger j, respectively. Let p_{Bi} and R_{Bi} be the position vector and the rotation matrix of Σ_{Bi}, and p_{Fjk} and R_{Fjk} be those of Σ_{Fjk}, with respect to Σ_R, respectively. $^{Bi}p_{Cjk}$ and $^{Fjk}p_{Cjk}$ are the position vectors of the kth contact point of finger j with respect to Σ_{Bi} and Σ_{Fjk}, respectively. $^{Bi}p_{COt}(t = 1,\cdots,r)$ is the position vector of the common contact point between object i and object l with respect to Σ_{Bi}.

3.1 Constraint Condition

In this subsection, we derive the constraint condition. The contact point between object i and the kth contact of finger j can be expressed by Σ_{Bi} and Σ_{Fjk}. Similarly, the contact point between object i and object l can be expressed by Σ_{Bi} and Σ_{Bl}. As a result, we have the following relationships:

$$p_{Bi} + R_{Bi}{}^{Bi}p_{Cjk} = p_{Fjk} + R_{Fjk}{}^{Fjk}p_{Cjk}, \quad (1)$$

$$p_{Bi} + R_{Bi}{}^{Bi}p_{COt} = p_{Bl} + R_{Bl}{}^{Bl}p_{COt}. \quad (2)$$

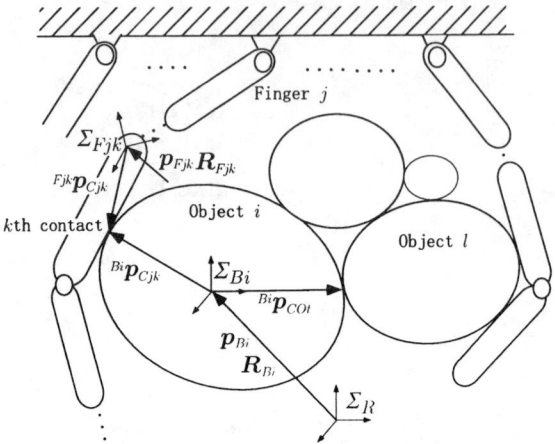

Fig. 2: Model of the system

Suppose that there is no slipping at each contact point. This means implicitly that we consider a sufficiently large coefficient of friction at each contact point. The small displacement of object i and the finger j should be same at the contact point. Also, the small displacement of two objects should be same at the common contact point between them. These discussions lead to the following equations[1]:

$$R_{Bi}{}^{Bi}\Delta p_{Cjk} = R_{Fjk}{}^{Fjk}\Delta p_{Cjk}, \quad (3)$$

$$R_{Bi}{}^{Bi}\Delta p_{COt} = R_{Bl}{}^{Bl}\Delta p_{COt}, \quad (4)$$

where $\Delta *$ denotes the finite differentiation of $*$. Substituting eq.(3) and eq.(4) into the differentiation of eq.(1) and eq.(2), respectively, we can derive

$$D_{Bjk}\begin{bmatrix}\Delta p_{Bi}\\ \Delta\phi_{Bi}\end{bmatrix} = D_{Fjk}\begin{bmatrix}\Delta p_{Fjk}\\ \Delta\phi_{Fjk}\end{bmatrix}, \quad (5)$$

$$D_{Oit}\begin{bmatrix}\Delta p_{Bi}\\ \Delta\phi_{Bi}\end{bmatrix} = D_{Olt}\begin{bmatrix}\Delta p_{Bl}\\ \Delta\phi_{Bl}\end{bmatrix}, \quad (6)$$

$$D_{Bjk} = [I \quad -(R_{Bi}{}^{Bi}p_{Cjk}\times)], \quad (7)$$

$$D_{Fjk} = [I \quad -(R_{Fjk}{}^{Fjk}p_{Cjk}\times)], \quad (8)$$

$$D_{Oit} = [I \quad -(R_{Bi}{}^{Bi}p_{COt}\times)], \quad (9)$$

where I, $(R_{Bi}{}^{Bi}p_{Cjk}\times)$, $(R_{Fjk}{}^{Fjk}p_{Cjk}\times)$ and $(R_{Bi}{}^{Bi}p_{COt}\times)$ denote the identity and the skew-symmetric matrices, respectively, and $\Delta\phi_{Bi}$ and $\Delta\phi_{Fjk}$ denote the small angular displacement vectors of Σ_{Bi} and Σ_{Fjk} with respect to Σ_R, respectively. Since the displacement of the finger link including the kth contact point of finger j can also be expressed by utilizing the joint displacement of finger j, we obtain the following relationships:

$$\begin{bmatrix}\Delta p_{Fjk}\\ \Delta\phi_{Fjk}\end{bmatrix} = J_{jk}\Delta\theta_j, \quad (10)$$

where J_{jk} is the jacobian matrix of the finger link with respect to the joint displacement. Substituting eq.(10) into eq.(5) and aggregating for $k = 1, \cdots, c_j$, the following equation is derived:

$$D_{Bj} \begin{bmatrix} \Delta p_{B1} \\ \Delta \phi_{B1} \\ \vdots \\ \Delta p_{Bm} \\ \Delta \phi_{Bm} \end{bmatrix} = D_{Fj} \Delta \theta_j. \quad (11)$$

where $D_{Fj} = \begin{bmatrix} D_{Fj1} J_{j1} \\ \vdots \\ D_{Fjc_j} J_{jc_j} \end{bmatrix}$, and D_{Bj} is the matrix whose components include D_{Bjk} ($k = 1, \cdots, c_j$). Aggregating eq.(6) for $t = 1, \cdots, r$, the following equation is derived:

$$D_O \begin{bmatrix} \Delta p_{B1} \\ \Delta \phi_{B1} \\ \vdots \\ \Delta p_{Bm} \\ \Delta \phi_{Bm} \end{bmatrix} = \mathbf{o}, \quad (12)$$

where D_O is the matrix whose components include D_{Oit} ($i = 1, \cdots, m$, $t = 1, \cdots, r$). Using eq.(11) and eq.(12), the following kinematic equation can be obtained:

$$\Omega(x, u) \Delta x = \mathbf{o}, \quad (13)$$

where

$$\Omega(x, u) = \begin{bmatrix} -D_{F1} & \mathbf{o} & \cdots & \cdots & \mathbf{o} & D_{B1} \\ \mathbf{o} & -D_{F2} & \mathbf{o} & \cdots & \mathbf{o} & D_{B2} \\ & & \cdots & \cdots & & \\ \mathbf{o} & \mathbf{o} & \cdots & \cdots & -D_{Fn} & D_{Bn} \\ \mathbf{o} & \cdots & \cdots & \cdots & \mathbf{o} & D_O \end{bmatrix},$$

$$\Delta x = [\Delta \theta_1^T \cdots \Delta \theta_n^T \Delta p_{B1}^T \Delta \phi_{B1}^T \cdots \Delta p_{Bm}^T \Delta \phi_{Bm}^T]^T,$$

$$u = [{}^{F11}p_{C11}^T {}^{Bi}p_{C11}^T \cdots {}^{Fn}p_{Cnc_n}^T {}^{Bi}p_{Cnc_n}^T]^T.$$

We note that eq.(13) shows the constraint condition for the system. In eq.(11), $\dim D_{Bj} = 3c_j \times 6m$ and $\dim D_{Fj} = 3c_j \times s_j$, where s_j denotes the number of joints of finger j. In eq.(12), $\dim D_O = 3r \times 6m$. Therefore, we have $\dim \Omega(x, u) = (3\sum_{j=1}^{n} c_j + 3r) \times (\sum_{j=1}^{n} s_j + 6m)$. Additionally, in eq.(13), u is the vector which represents the instantaneous position of contact during rolling motion.

3.2 Motion under Constraint

To evaluate the constraint condition (13), we define the variable α as follows[7]:

$$\alpha = (\sum_{j=1}^{n} s_j + 6m) - (3\sum_{j=1}^{n} c_j + 3r). \quad (14)$$

When $\alpha \leq 0$ and $\text{rank}\,\Omega(x, u) = \sum_{j=1}^{n} s_j + 6m$, the solution of eq.(13) becomes $\Delta x = \mathbf{o}$, which means that two objects are fully constrained. On the other hand, $\alpha > 0$ implies that there exists some possible degrees of freedom concerning velocity. Hereafter, we assume $\alpha > 0$. We further assume that

$$\text{rank}\,\Omega(x, u) = 3\sum_{j=1}^{n} c_j + 3r. \quad (15)$$

Let us now consider the motion under eqs.(13) and (15). We introduce a new vector $\Delta \eta = \Psi(x, u) \Delta x$ ($\eta \in R^\alpha$) such that $\text{rank} \begin{bmatrix} \Omega(x, u) \\ \Psi(x, u) \end{bmatrix} = \sum_{j=1}^{n} s_j + 6m$. By using the newly introduced $\Delta \eta$ and eq.(13), we obtain a set of equation as follows:

$$\begin{bmatrix} \mathbf{o} \\ \Delta \eta \end{bmatrix} = \begin{bmatrix} \Omega(x, u) \\ \Psi(x, u) \end{bmatrix} \Delta x. \quad (16)$$

We can regard that eq.(16) is a coordinate transformation between Δx and $\Delta \eta$. We note that the selection of $\Delta \eta$ is not unique. An example of selections of $\Delta \eta$ and $\Psi(x, u)$ will be shown in the numerical example. Due to the constrained conditions (13), only the α components of Δx can be utilized as independent parameters. Since $\begin{bmatrix} \Omega(x, u) \\ \Psi(x, u) \end{bmatrix}$ is nonsingular, we can obtain its inverse. Let $X(x, u)$ be the inverse matrix of $\begin{bmatrix} \Omega(x, u) \\ \Psi(x, u) \end{bmatrix}$. As a result, for a given $\Delta \eta$, Δx is uniquely determined by

$$\Delta x = X(x, u) \Delta \eta. \quad (17)$$

3.3 Displacement of Contact Point

In eq.(13), the vector of contact points is given by u. By extending the method proposed by Kerr et al.[1], we derive the velocity of the contact point as follows:

$$\Delta u = U(x, u) \Delta x. \quad (18)$$

Substituting eq.(17) into eq.(18), we have the following equation:

$$\Delta u = U(x, u) X(x, u) \Delta \eta. \quad (19)$$

Thus, for a given $\Delta \eta$, Δu is uniquely determined by using eq.(19). Along with eq.(17), all the motions of the system are determined.

Note that, in this section, while we deal with a 3D model, $\dim \Omega(x, u) = (2\sum_{j=1}^{n} c_j + 2r) \times (\sum_{j=1}^{n} s_j + 3m)$ for a 2D model, where α is given by

$$\alpha = (\sum_{j=1}^{n} s_j + 3m) - (2\sum_{j=1}^{n} c_j + 2r). \quad (20)$$

4 Judgment of Rolling Contacts

Using the results obtained in the previous section, we can derive a condition for judging whether the object can roll over the link surface or not. Using eq.(19), the velocity of the each contact point is rewritten as

$$\begin{bmatrix} ^{F11}\Delta p_{C11} \\ ^{Bi}\Delta p_{C11} \end{bmatrix} = B_{11}(x,u)\Delta\eta,$$

$$\vdots$$

$$\begin{bmatrix} ^{Fjk}\Delta p_{Cjk} \\ ^{Bi}\Delta p_{Cjk} \end{bmatrix} = B_{jk}(x,u)\Delta\eta,$$

$$\vdots$$

$$\begin{bmatrix} ^{Bi}\Delta p_{COt} \\ ^{Bl}\Delta p_{COt} \end{bmatrix} = B_{Ot}(x,u)\Delta\eta,$$

$$\vdots \tag{21}$$

where $UX = [B_{11}^T \cdots B_{nc_n}^T B_{O1}^T \cdots B_{Or}^T]^T$. In eq.(21), if the displacement of a contact point is 0 for any $\Delta\eta$, it is considered that the object cannot roll over the link surface at the point. Now we can find the following theorem:

[Theorem 1]
A necessary and sufficient condition for enabling the objects and contact links to satisfy the rolling condition at each contact point is

$$B_{jk}(x,u) \neq o \quad (j=1,\cdots,s_j \ \ k=1,\cdots,c_j), \tag{22}$$
$$B_{Ot}(x,u) \neq o \quad (t=1,\cdots,r). \tag{23}$$

[Proof]
Since sufficiency is obvious, we prove only the necessity. Suppose that the object 1 cannot roll at the 1st contact point of finger 1 even if $B_{11}(x,u) \neq o$. Since each velocity component of the left side in the first row of eq.(21) is 0, the first row of eq.(21) can be rewritten as

$$o = b_{111}\Delta\eta_1 + \cdots + b_{11\alpha}\Delta\eta_\alpha \tag{24}$$

where $B_{11}(x,u) = [b_{111} \cdots b_{11\alpha}]$. The elements of $\Delta\eta$ can be changed independently. Moreover, since $\Delta\eta$ is not included in the matrix $B_{11}(x,u)$. Thus, to satisfy eq.(24), it is needed that $b_{111} = \cdots = b_{11\alpha} = o$. The same condition is hold for another contact points. This holds the theorem. □

We note that the vector x is included in the matrices $B_{jk}(x,u)$ and $B_{Ot}(x,u)$. Moreover, although the matrices $B_{jk}(x,u)$ and $B_{Ot}(x,u)$ are composed of the matrix $X(x,u)$, it is generally hard to obtain $X(x,u)$ symbolically since the dimension of $X(x,u)$ is large.

5 Numerical Example

To confirm Theorem 1, we show a numerical example. Suppose that two cylindrical objects are enveloped by two planar fingers, as shown in Fig.3, where the finger 1 contacts with object 1 with two different points and finger 2 contacts with object 2 at one contact point. The constraint condition (13) is described as follows:

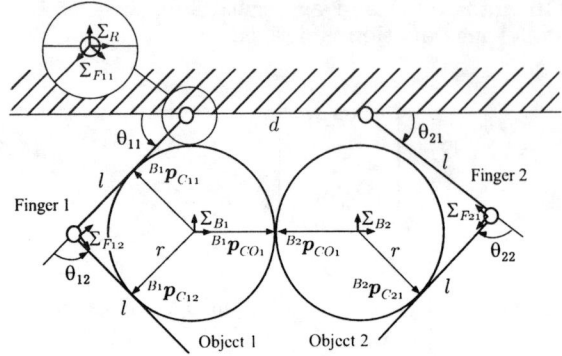

Fig. 3: 2 object system used in numerical example

$$\begin{bmatrix} -D_{F1} & o & D_{B1} \\ o & -D_{F2} & D_{B2} \\ o & o & D_O \end{bmatrix} \begin{bmatrix} \Delta\theta_1 \\ -- \\ \Delta\theta_2 \\ -- \\ \Delta p_{B1} \\ \Delta\phi_{B1} \\ \Delta p_{B2} \\ \Delta\phi_{B2} \end{bmatrix} = o, \tag{25}$$

where $\theta_1 = [\theta_{11}\ \theta_{12}]^T$, $\theta_2 = [\theta_{21}\ \theta_{22}]^T$, $p_{B1} = [x_{B1}\ y_{B2}]^T$ and $p_{B2} = [x_{B2}\ y_{B2}]^T$. The matrix of eq.(25) corresponds to $\Omega(x,u)$ in eq.(13). Since $\dim\Omega(x,u) = 8\times 10$ and $\text{rank}\Omega(x,u) = 8$, $\alpha = 2$ in eq.(20) and the condition of eq.(15) is satisfied. In $\dot{x}$, only the two elements can be independently chosen. In eq.(16), If we set η and Ψ as follows:

$$\eta = [\theta_{11}\ \theta_{21}]^T$$
$$\Psi = \begin{bmatrix} 1 & 0 & 0 & 0 & 0 & 0 & 0 & 0 & 0 & 0 \\ 0 & 0 & 1 & 0 & 0 & 0 & 0 & 0 & 0 & 0 \end{bmatrix},$$

$\begin{bmatrix} \Omega(x,u) \\ \Psi \end{bmatrix}$ becomes nonsingular. For numerical computation, we set $r = 1.0$, $l = 2.0$ and $d = 2.0$. The position and orientation of the object and the vector of the contact point are set as $p_{B1} = \begin{bmatrix} 0 \\ -\sqrt{2} \end{bmatrix}$, $R_{B1} = \begin{bmatrix} 1 & 0 \\ 0 & 1 \end{bmatrix}$, $p_{B2} = \begin{bmatrix} 2 \\ -\sqrt{2} \end{bmatrix}$, $R_{B2} = \begin{bmatrix} 1 & 0 \\ 0 & 1 \end{bmatrix}$, $^{B1}p_{C11} = \begin{bmatrix} -1/\sqrt{2} \\ 1/\sqrt{2} \end{bmatrix}$, $^{B1}p_{C12} = \begin{bmatrix} -1/\sqrt{2} \\ -1/\sqrt{2} \end{bmatrix}$ and $^{B2}p_{C21} = \begin{bmatrix} \cos(2\pi/9) \\ -\sin(2\pi/9) \end{bmatrix}$. As a result, the joint angles become $\theta_{11} = \pi/4$, $\theta_{12} = \pi/2$, $\theta_{21} = 0.3954$ and $\theta_{22} = 1.8735$. The values of the matrices $B_{11}(x,u)$, $B_{12}(x,u)$, $B_{21}(x,u)$ and $B_{O1}(x,u)$ in eq.(21) are computed by using MATLAB as follows:

$$B_{11}(x,u) = \begin{bmatrix} 0.0 & 0.1220\times 10^{-15} \\ 0.0 & 0.0 \\ 0.0 & -0.0862\times 10^{-15} \\ 0.0 & -0.0862\times 10^{-15} \end{bmatrix},$$

$$B_{12}(x,u) = \begin{bmatrix} 0.1148 \times 10^{-15} & -0.1220 \times 10^{-15} \\ 0.0 & 0.0 \\ 0.0812 \times 10^{-15} & -0.0862 \times 10^{-15} \\ -0.0812 \times 10^{-15} & 0.0862 \times 10^{-15} \end{bmatrix},$$

$$B_{21}(x,u) = \begin{bmatrix} 0.0 & 0.0 \\ -1.5783 & -1.1396 \\ 1.0116 & 0.7325 \\ 1.2056 & 0.8730 \end{bmatrix},$$

$$B_{O1}(x,u) = \begin{bmatrix} 0.0 & 0.0 \\ -0.9742 & -0.5405 \\ 0.0 & 0.0 \\ -0.9742 & -0.5405 \end{bmatrix}.$$

From this results, we can say that the matrices $B_{11}(x,u)$ and $B_{12}(x,u)$ are almost identical to the zero matrices since all components are either exactly zero or extremely small values. On the other hand, the matrices $B_{21}(x,u)$ and $B_{O1}(x,u)$ include enough large components. Thus, theorem 1 ensures that object 2 can keep rolling condition at both contact points while object 1 cannot.

6 Condition for Lifting up

In this section, we consider whether two objects can be lifted up by a simple pushing motion(Fig.1). As shown in Fig.4, the common tangential plane of two objects are defined as Π. The plane which is normal to Π and tangent to the gravity vector is defined as Γ. We consider the motion of the objects projected on Γ. The rolling condition is assumed to be satisfied at each

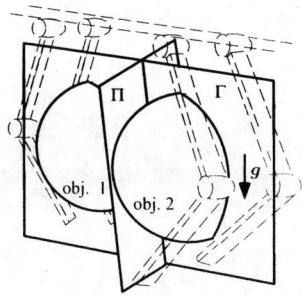

Fig. 4: Definition of the projection plane

contact point. The kinematic relationship between the objects projected on Γ is shown in Fig.5 where the suffix γ denotes a vector on the two dimensional plane Γ. For simplicity, we assume that an object contacts with one finger at one point or that contact points between an object and fingers are overlapped when they are projected on Γ. Now we provide the following definition.

[Definition]
When two objects rotate in opposite direction such that both center of gravity may be close to the palm, we call such a phase palm-reaching phase

For the objects being in palm-reaching phase, object 1 and object 2 have to rotate counter-clockwise and

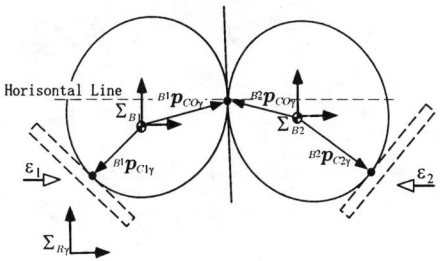

Fig. 5: The projected two object system

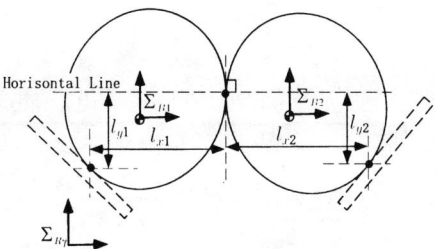

Fig. 6: Variables in eqs.(27),(28),(29) and (30)

clockwise, respectively. Here, we show a sufficient condition for achieving the palm-reaching phase.

[Theorem 2]
Consider that two objects are sandwiched by two finger links. A sufficient condition for achieving the palm-reaching phase by a horizontal pushing motion of finger links is that the contact points between an object and a finger link become lower than the horizontal line including the contact point between two objects.

[Proof]
By using eqs.(11) and (12), the following relation on Γ can be derived as

$$D_{F\gamma}\Delta\theta_\gamma = D_{B\gamma}\Delta p_{B\gamma}. \quad (26)$$

$D_{F\gamma}\Delta\theta_\gamma$ is defined as

$$D_{F\gamma}\Delta\theta_\gamma = [\epsilon_1 \; 0 \; -\epsilon_2 \; 0 \; 0 \; 0]^T,$$

where $\epsilon_1(>0)$ and $\epsilon_2(>0)$ are the displacements of finger links in the horizontal direction. $D_{B\gamma}$ and $\Delta p_{B\gamma}$ are also defined as

$$D_{B\gamma} = \begin{bmatrix} D_{B11\gamma} & \mathbf{0} \\ \mathbf{0} & D_{B21\gamma} \\ D_{O11\gamma} & -D_{O21\gamma} \end{bmatrix}$$

$$= \begin{bmatrix} 1 & 0 & -{}^{B1}y_{C11\gamma} & 0 & 0 & 0 \\ 0 & 1 & {}^{B1}x_{C11\gamma} & 0 & 0 & 0 \\ 0 & 0 & 0 & 1 & 0 & -{}^{B2}y_{C21\gamma} \\ 0 & 0 & 0 & 0 & 1 & {}^{B2}x_{C21\gamma} \\ 1 & 0 & -{}^{B1}y_{CO1\gamma} & -1 & 0 & {}^{B2}y_{CO1\gamma} \\ 0 & 1 & {}^{B1}x_{CO1\gamma} & 0 & -1 & -{}^{B2}x_{CO1\gamma} \end{bmatrix},$$

$$\Delta \boldsymbol{p}_{B\gamma} = [\Delta x_{B1\gamma}\ \Delta y_{B1\gamma}\ \Delta \phi_{B1\gamma}\ \Delta x_{B2\gamma}\ \Delta y_{B2\gamma}\ \Delta \phi_{B2\gamma}]^T,$$
$$^{B1}\boldsymbol{p}_{C11\gamma} = [^{B1}x_{C11\gamma}\ ^{B1}y_{C11\gamma}]^T,$$
$$^{B2}\boldsymbol{p}_{C21\gamma} = [^{B2}x_{C21\gamma}\ ^{B2}y_{C21\gamma}]^T,$$
$$^{Bi}\boldsymbol{p}_{CO1\gamma} = [^{Bi}x_{CO1\gamma}\ ^{Bi}y_{CO1\gamma}]^T\ (i=1,2).$$

As shown in Fig.6, when the contact points between the object and each finger link are lower than the horizontal line including the contact point between two objects, we can define

$$^{B1}y_{C11\gamma} - {}^{B1}y_{CO1\gamma} \overset{\triangle}{=} -l_{y1} < 0, \quad (27)$$
$$^{B2}y_{C21\gamma} - {}^{B2}y_{CO1\gamma} \overset{\triangle}{=} -l_{y2} < 0. \quad (28)$$

Moreover, from geometrical relationships, the following equations are obtained

$$^{B1}x_{C11\gamma} - {}^{B1}x_{CO1\gamma} \overset{\triangle}{=} -l_{x1} < 0, \quad (29)$$
$$^{B2}x_{C21\gamma} - {}^{B2}x_{CO1\gamma} \overset{\triangle}{=} l_{x2} > 0. \quad (30)$$

Using eqs.(27), (28), (28) and (29), eq.(26) can be solved for $\Delta \phi_{B1\gamma}$ and $\Delta \phi_{B2\gamma}$ as follows:

$$\Delta \phi_{B1\gamma} = (\epsilon_1 + \epsilon_2)l_{x2}/(l_{y1}l_{x2} + l_{x1}l_{y2}) > 0, \quad (31)$$
$$\Delta \phi_{B2\gamma} = -(\epsilon_1 + \epsilon_2)l_{x1}/(l_{y1}l_{x2} + l_{x1}l_{y2}) < 0. \quad (32)$$

From eqs.(31) and (32), we can show that object 1 rotates counter-clockwise and that object 2 rotates clockwise. This holds the theorem.

□

While the manipulation of multiple objects by rolling contact seems to be very difficult, the above theorem shows that the palm-reaching phase can be achieved by applying a simple pushing motion that can produce both ϵ_1 and ϵ_2.

7 Experiment

To verify how easily the sufficient condition can be satisfied, we execute an experiment by using Hiroshima-Hand. The Hiroshima-Hand is composed of three planar finger units, where each finger has three joints. Two cylindrical objects whose diameters are 33[mm] and 15[mm] are placed on the table. Fig.7 shows continuous photos during the experiment, where the two cylindrical objects on a table are gradually lifted up. After the fingers make contact with the objects, the fingers are driven by constant torque commands. The command torque for the left finger is 0.025[Nm], 0.025[Nm] and 0.0125[Nm] for joint 1, 2 and 3, respectively. Since there are two fingers in the right hand side, the command torque for the right fingers is chosen by the half of those of the left finger. As shown in the figure, although the fingers are driven by using the constant torque commands, the objects are lifted up easily and the enveloping grasp is achieved. It is generally difficult to achieve such a stable manipulation of two cylindrical objects by finger tip contacts.

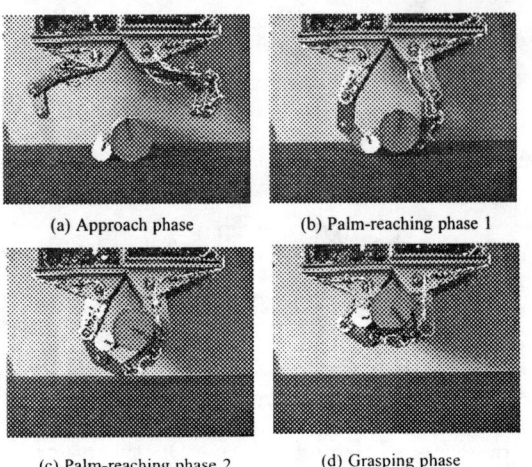

(a) Approach phase (b) Palm-reaching phase 1

(c) Palm-reaching phase 2 (d) Grasping phase

Fig. 7: Experimental results

This experimental result shows a potential advantage of enveloping grasp for manipulating multiple objects. Also, the theorem explains an essential principle behind the successful result for enveloping two cylindrical objects.

8 Conclusions

In this paper, we discussed the enveloping grasp for multiple objects. The n finger-m object system with rolling contact is modeled. We showed a necessary and sufficient condition for achieving a rolling motion at each contact point under multiple contacts (Theorem 1). A numerical example was shown to confirm the theorem 1. We also showed a sufficient condition for enabling the two objects to be lifted up(Theorem 2). An experiment was executed to show that an enveloping grasp of cylindrical objects can be achieved easily by utilizing a simple control scheme.

This work was supported by the Inter University Project(The Ministry of Education). We would like to express our sincere gratitude to Mr.Nophawit Thaiprasert who is a student of Hiroshima University for his help in the experiment.

References

[1] J.Kerr and B.Roth. Analysis of Multifingered Hands. *The Int. J. of Robotics Research, vol.4, no.4, pp.3-17,* 1986.

[2] A.B.A.Cole, J.E.Hauser, and S.S.Sastry. Kinematics and Control of Multifingered Hands with Rolling Contacts. *IEEE Trans. on Automatic Control, vol.34, no.4, pp.398-404,* 1989.

[3] E.Paljug, X.Yun, and V.Kumar. Control of Rolling Contacts in Multi-Arm Manipulation. *IEEE Trans. on Robotics and Automation, vol.10, no.4, pp.441-452,* 1994.

[4] K.Salisbury et al.. Preliminary Design of a Whole-Arm Manipulation Systems(WAMS). *Proc. of 1988*

[5] A.Bicchi. Force Distribution in Multiple Whole-Limb Manipulation. *Proc. of 1993 IEEE Int. Conf. on Robotics and Automation, pp.196-201*, 1993.

[6] X.-Y.Zhang, Y.Nakamura, K.Goda, and K.Yoshimoto. Robustness of Power Grasp. *Proc. of 1994 IEEE Int. Conf. on Robotics and Automation, pp.2828-2835*, 1994.

[7] T.Omata and K.Nagata. Rigid Body Analysis of the Indeterminate Grasp Force in Power Grasps. *Proc. of 1996 IEEE Int. Conf. on Robotics and Automation, pp.1787-1794*, 1996.

[8] M.Kaneko, Y.Hino, and T.Tsuji. On Three Phases for Achieving Enveloping Grasps. *Proc. of 1997 IEEE Int. Conf. on Robotics and Automation*, 1997.

[9] Y.Aiyama, M.Minami, and T.Arai. Operation of Multiple Objects by Dual Manipulator System. *Proc. of the JSME Conf. on Robotics-Mechatronics, pp.457-458*(in Japanese), 1997.

[10] P.Dauchez and X.Delebarre. Force-Controlled Assembly of two Objects with a Two-arm Robot. *Robotica, vol.9, pp.299-306*, 1991.

[11] K.Kosuge, M.Sakai, and K.Kanitani. Decentralized Coordinated Motion Control of Manipulators with Vision and Force Sensors. *Proc. of 1995 IEEE Int. Conf. on Robotics and Automation, pp.2456-2462*, 1995.

[12] R.Mattikalli, D.Baraff, P.Khosla, and B.Repetto. Gravitational Stability of Frictionless Assemblies. *IEEE Trans. on Robotics and Automation, vol.11, no.3, pp.374-388*, 1995.

[13] D.J.Montana. The Kinematics of Contact and Grasp. *The Int. J. of Robotics Research, vol.7, no.3, pp.17-32*, 1988.

[14] Z.Li and J.Canny. Motion of Two Rigid Bodies with Rolling Constraint. *IEEE Trans. on Robotics and Automation, vol.6, no.1, pp.62-72*, 1990.

[15] W.S.Howard and V.Kumar. Modeling and Analysis of Compliance and Stability of Enveloping Grasps. *Proc. of 1995 IEEE Int. Conf. on Robotics and Automation, pp.1367-1372*, 1995.

[16] H.Maekawa, K.Tanie, and K.Komoriya. Kinematics, Statics and Stiffness Effect of 3D Grasp by Multifingered Hand with Rolling Contact at the Fingertip. *Proc. of 1997 IEEE Int. Conf. on Robotics and Automation, pp.78-85*, 1997.

[17] M.Kaneko and T.Tsuji. Realization of Enveloping Grasp. *1997 IEEE Int. Conf. on Robotics and Automation(Video Proceeding)*, 1997.

Appendix

In the appendix, the vector of the point of contact u in eq.(13) is derived. First, the mathematical equations of the surfaces of the objects and the finger links are expressed as follows:

$$S_{Bi}(^{Bi}p_{Cjk}) = 0, \quad S_{Fjk}(^{Fjk}p_{Cjk}) = 0, \quad (33)$$

where $S_{Bi}(^{Bi}p_{Cjk}) < 0$ and $S_{Fjk}(^{Fjk}p_{Cjk}) < 0$ show the inside of the surfaces. Since the two surfaces touch at the point of contact, they are on opposite sides of a common tangent plane and thus must have equal and opposite outward unit normal vectors, i.e.,

$$R_{Bi}{}^{Bi}e_{Cjk} = -R_{Fjk}{}^{Fjk}e_{Cjk}, \quad (34)$$

where $^{Bi}e_{Cjk} = \frac{\partial S_{Bi}(^{Bi}p_{Cjk})/\partial ^{Bi}p_{Cjk}}{\|\partial S_{Bi}(^{Bi}p_{Cjk})/\partial ^{Bi}p_{Cjk}\|}$ and $^{Fjk}e_{Cjk} = \frac{\partial S_{Fjk}(^{Fjk}p_{Cjk})/\partial ^{Fjk}p_{Cjk}}{\|\partial S_{Fjk}(^{Fjk}p_{Cjk})/\partial ^{Fjk}p_{Cjk}\|}$. Differentiation of eq.(34) yields

$$\Delta\phi_{Bi} \times R_{Bi}{}^{Bi}e_{Cjk} + R_{Bi}A_{CBjk}\Delta^{Bi}p_{Cjk}$$
$$= -\Delta\phi_{Fjk} \times R_{Fjk}{}^{Fjk}e_{Cjk} - R_{Fjk}A_{CFjk}\Delta^{Fjk}p_{Cjk}, \quad (35)$$

where $A_{CBjk} = \partial^{Bi}e_{Cjk}/\partial^{Bi}p_{Cjk}^T$, $A_{CFjk} = \partial^{Fjk}e_{Cjk}/\partial^{Fjk}p_{Cjk}^T$. Since the contact coordinates $^{Bi}p_{Cjk}$ and $^{Fjk}p_{Cjk}$ are on the 2 dimensional surfaces, these become functions of 2 dimensional vectors[1] ξ_{CBjk} and ξ_{CFjk} [2] such as

$$^{Bi}p_{Cjk} = {}^{Bi}p_{Cjk}(\xi_{CBjk}), \quad (36)$$
$$^{Fjk}p_{Cjk} = {}^{Fjk}p_{Cjk}(\xi_{CFjk}), \quad (37)$$
$$\Delta^{Bi}p_{Cjk} = L_{CBjk}\Delta\xi_{CBjk}, \quad (38)$$
$$\Delta^{Fjk}p_{Cjk} = L_{CFjk}\Delta\xi_{CFjk}, \quad (39)$$

where $L_{CBjk} = \partial^{Bi}p_{Cjk}/\partial\xi_{CBjk}^T$, $L_{CFjk} = \partial^{Fjk}p_{Cjk}/\partial\xi_{CFjk}^T$. The following equations are derived from eqs.(35), (3), (38) and (39):

$$W_{Cjk}\begin{bmatrix}\Delta\xi_{CBjk}\\ \Delta\xi_{CFjk}\end{bmatrix} = Y_{jk}\begin{bmatrix}\Delta\phi_{Bi}\\ \Delta\phi_{Fjk}\end{bmatrix}, \quad (40)$$

where

$$W_{Cjk} = \begin{bmatrix} R_{Bi}L_{CBjk} & R_{Fjk}L_{CFjk}\\ R_{Bi}A_{CBjk}L_{CBjk} & R_{Fjk}A_{CFjk}L_{CFjk}\end{bmatrix},$$
$$Y_{jk} = \begin{bmatrix} 0 & 0\\ (R_{Bi}{}^{Bi}e_{Cjk}\times) & (R_{Fjk}{}^{Fjk}e_{Cjk}\times)\end{bmatrix}.$$

In eq.(40), the $\dim W_{Cjk} = 6 \times 4$. However, 4 lines of the matrix are independent because normal vectors are of unit length in eq.(34) and because the contact point must lie in the common tangent plane in eq.(3)[2]. Solving eq.(40) with respect to $\Delta\xi_{CBjk}$, $\Delta\xi_{CFjk}$, and substituting into eqs.(38) and (39), we can derive $\Delta^{Bi}p_{Cjk}$ and $\Delta^{Fjk}p_{Cjk}$. $\Delta^{Bi}p_{COt}$ can be derived similarly.

---
[1] 1 dimensional value for 1 dimensional line

A STUDY OF AUTONOMOUS MOBILE SYSTEM IN OUTDOOR ENVIRONMENT

(PART2 Sign Guided Autonomous Transportation System)

Jun-ichi Takiguchi[*1], Kiyoshi Iwama[*1], Hiroshi Sugie[*2]
Masanori Kato[*3], Takayuki Kiyonaga[*3], Takumi Hashizume[*3]
Fumihiro Inoue[*4], Kyoji Yoshino[*4], Yutaro Omote[*4]

*1 Mitsubishi Electric Corporation Kamakura Works, 325, Kamimachiya, Kamakura, Japan
*2 Mitsubishi Electric Corporation, Industrial Electronics & Sytem Laboratory,1-1,
    Tsukaguchi-honmachi 8-chome, Amagasaki-shi, Hyogo, Japan
*3 Waseda University, 17 Kikui-cho, Shinjuku-ku, Tokyo, Japan
*4 Obayashi Corporation, 640, Shimokiyoto 4-chome, Kiyose-shi, Tokyo, Japan

Abstract

This paper presents an autonomous mobile system in an outdoor environment especially applied to a construction site. The system consists of an autonomous vehicle and signs. The autonomous vehicle recognizes instructions given in the signs using a vision system, then it moves or handles materials according to the instruction. The system is adaptable to the changeable environment of the construction site because transportation paths can be easily changed by altering the contents or placement of the signs. Experimental results obtained in the outdoor environment show feasibility of the system.

1. Introduction

Recently, automated systems are becoming necessary in construction sites, inspection and surveillance operations and in workshops. But the current systems are not adaptable to these fields because of lack of adjustability to the changeable environment and lack of robustness towards disturbance. The authors aimed to develop a system which could be used in a complicated outdoor environment. As the first step, a high-resolution positioning system using dead-reckoning[1] and a high-resolution environmental map generating system using image processing[2] have been developed. As a second step, the authors have developed a more practical system which can be used in the actual field.

In the construction industry, production-efficiency needs to be improved to bring it close to the high level of manufacturing industry because of the increasing number of aged workers, needs for shortening the construction period, and the demand for cutting the cost. Especially in the transportation work on the finish-action-floor; the floor which is under interior processing, simple tools like hand trucks are still used predominantly. So a horizontal transportation system which can take the place of a hand truck is developed at the beginning. The proposed construction transportation system is shown in Fig.1.

In this paper, we consider the above issues and provide a mechanism and algorithm for the transportation system which can be used in the construction field, and evaluate the robustness, flexibility and utility of this system.

2. Outline of the system

An outline of the sign guided autonomous transportation system is shown in Fig.2. A conventional system like an AGV(Automatic Guided Vehicle) needs magnetic or reflecting tape on the floor for guidance. Therefore, in a construction site where the transportation course is often changed in short time, AGV systems have practical problems and so far

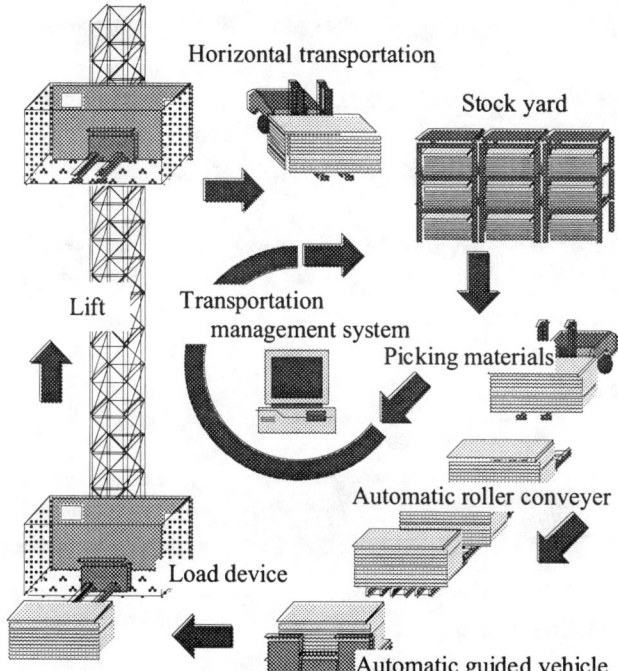

Fig.1 Automatic transportation system in a construction site

have been rarely used. In this system, the vehicle moves by tracking and recognizing instructions which are shown on the signs using a vision process. A route change and its behavior modification are made possible by changing the sign position and its content, therefore this system has sufficient utility for the construction site where changes in the environment occur frequently according to the construction progress.

A practical application in a construction site is as follows:. (A) The vehicle loaded with materials starts toward the first sign. (B) It turns left according to the left turn sign. (C)It turns at that point according to the right spin turn signal and goes back toward the unloading sign. (D) It unloads the materials. (E) It turns left. (F) It turns right. (G) It stops in front of the stop sign. After that, the next materials are loaded onto

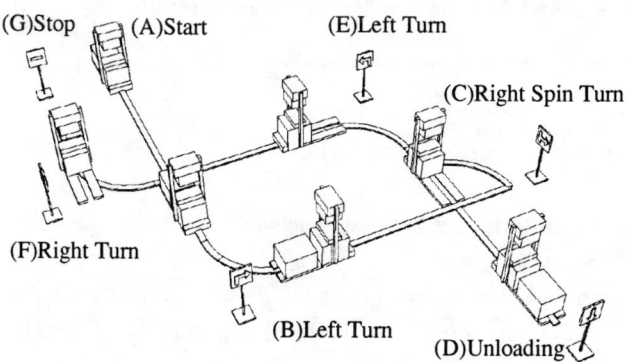

Fig.2 System outline

the vehicle by operators and the vehicle is taken to the start point. Then transportation is repeated.

3. Configuration of the system and control architecture

An appearance of the system is illustrated as Fig.3. This system; 2350mm length, 1043mm width, and 2050mm height contains the vehicle, the autonomous unit, and folks for lifting construction materials. The autonomous unit contains a SLR (Scan Laser Radar); measuring the distance from the vehicle to the sign, the deviation angle and the reflection ratio, a gyroscope; measuring the azimuth angle, a NIR(Near InfraRed) CCD camera, a vision processor, photoelectric sensors, and a pyroelectric NIR sensor; detecting a human being as an obstacle. The vehicle moves to the sign by measuring its position with the SLR, recognizes the instruction shown on the sign by using the vision system, and changes the path or unloads the materials at a fixed distance from the sign. This vehicle quickly slows down or stops when obstacles like a man are detected by the photoelectric sensor or bumper switch.

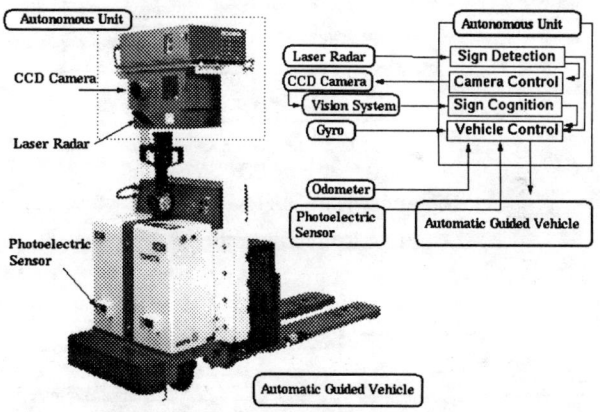

Fig.3 Appearance of the system and control architecture

4. Sign searching procedure
4.1 Sign searching algorithm

The signs are detected during two stages of a sign searching procedure using the SLR which radiates a pulse laser to an object and receives the reflection. The distance to the object is calculated from the time of flight. The intensity of the reflection is also measured. Several autonomous mobile systems using laser range finders for indoor[4] or for outdoor applications[5] have been developed.

At first, objects are distinguished according to their intensity of reflection. The signs include a material with high reflection ratio. Fig.4 shows the reflection ratios of several kinds of material which are typically found in construction sites. The ratio is normalized with reflection ratio of the sign, and the distinction threshold is set according to the distance to the object.

Secondly, the objects are distinguished according to their width. The measured data with high reflection ratios are plotted on a X-Y coordinate system as shown in Fig. 5. Data lying within 60 cm area perpendicular to the X axis is classified as a line, with the length of the line being the width of the sign plate. The tolerance of the sign plate width is decided according to distance to the object because the step scanning angle of the SLR is constant (0.15 degree resolution). The scanning angle range of the SLR is ± 6 degrees. As the SLR is mounted on a turret with ± 10 degree rotation angle, the total horizontal sign searching angle is ± 16 degrees.

4.2 Performance of sign searching

The distance between the SLR and the sign is given by equation (1). The width of the sign is given by equation (2). The direction for the center of the sign plate (deviation angle) is given by equation (3).

Fig. 6 shows width of sign plates as measured in the sign search experiment. When a sign is located within 30 m of the SLR, the length to the sign is measured with ±15 cm accuracy and the deviation angle is measured with ± 0.3 degree accuracy.

$$Distance = \sum_{n=1}^{r} Y(n) / (r + l - 1) \qquad (1)$$

$$\underset{r > l}{Width} = X(r) - X(l) \qquad (2)$$

$$\theta(m) = -6.0 + \frac{m}{N} \times 12.0 \qquad (3)$$

where r: Right Edge l: Left Edge
       m: Middle Point N: Data Number

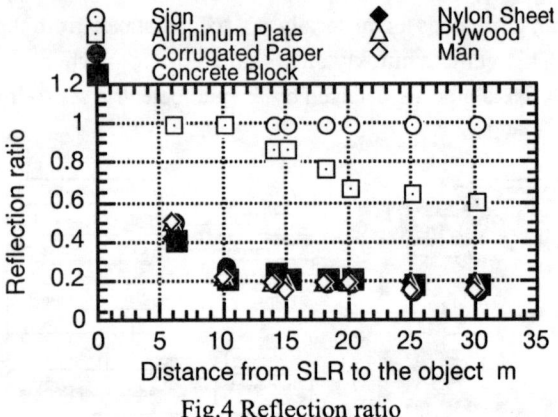

Fig.4 Reflection ratio

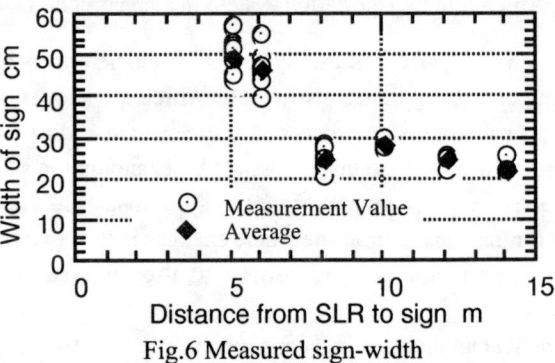

Fig.6 Measured sign-width

4.3 Sign searching experiment

The sign searching test reveals that this system can detect a sign which is located from 4 to 30 m away in real-time (processing time:165 ms) even if it is set up in a complicated background. Fig.7 shows the experimental results.

5. Sign recognition procedure
5.1 Sign recognition algorithm

The architecture of the sign recognition system is shown in Fig.8. The system is composed of photoelectric sensors, an electrically driven zoom lens, a NIR-CCD camera, a lens controller, and a NIR illuminator. First, using the distance measured by the SLR, the lens is focused on the sign, and is zoomed so that the sign is as large as the templates (64 pixels

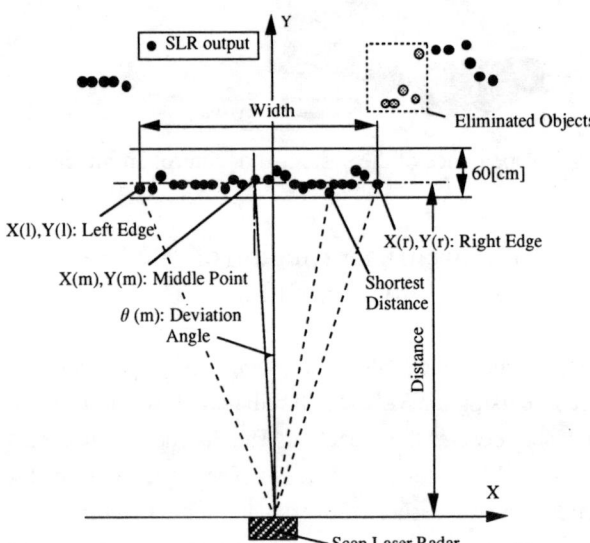

Fig.5 Sign-searching algorithm

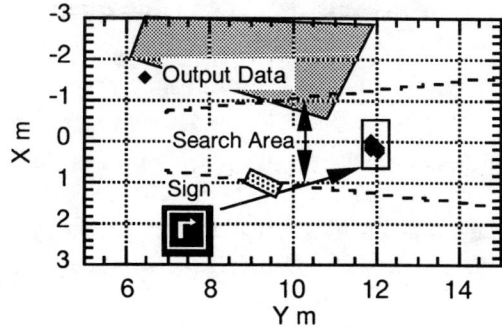

Fig.7 Sign searching result

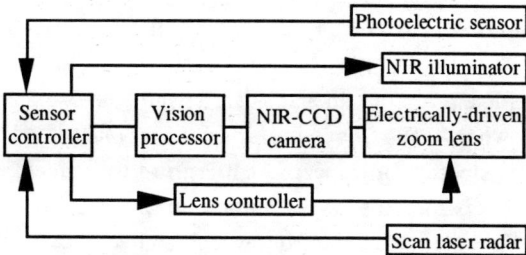

Fig.8 Configuration of sign recognition unit

by 64 pixels) in the image captured by the NIR-CCD camera. The iris is controlled so that the sign intensity can be always the same level. The flash of the NIR illuminator is synchronized with the shutter of the camera, since human eyes are not sensitive to the NIR, workers in the field are not dazzled by the illuminator. Then the resulting image is compared with the templates using a gray level matching method, and cross correlation between the image and each template are calculated[6]. The sign is identified with the template with which the image has the highest cross correlation. When the highest cross correlation is lower than threshold, the system does not identify the image to prevent misidentification.

5.2 Sign pattern and motion

The sign patterns are shown in Fig.9. They are a turn sign, a stop sign, a spin turn sign to turn the vehicle back toward an unloading place, and an unloading sign. Aimed for

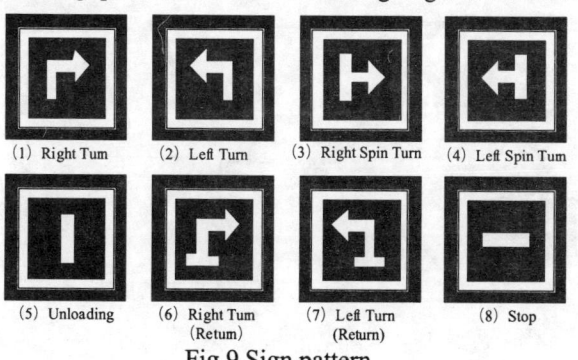

Fig.9 Sign pattern

misidentification prevention, there are two right signs: one that the vehicle obeys when driving to its destination and one for when it returns home position. Similarly, there are two left turn signs. The sign is composed of a black sheet with low reflection coefficient and white sheet with high reflection coefficient.

5.3 Performance of sign recognition

The result of right turn sign recognition is shown in Fig.10. Template numbers are as shown in Fig. 9. The sign recognition system calculates cross correlation between the captured image and 8 templates. This experiment resulted in that the image has the highest cross correlation with template No.1 (right turn) and the system recognized the sign correctly.

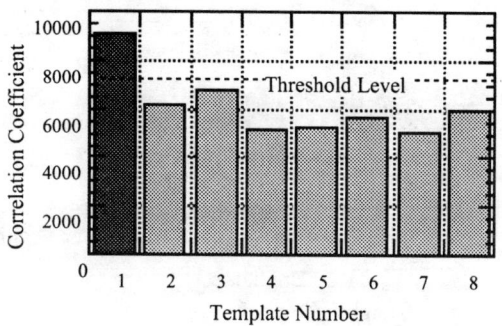

Fig.10 Sign recognition result

5.4 Experiment for evaluation of sign recognition in outdoor environment

An example of sign image in outdoor test field on fine day is shown in Fig.11. The white square in the image means corresponding area in the template. This experiment demonstrated in that the system recognized a sign correctly in spite of complicated background. The processing time was about 300 ms, under the condition that the search area was the entire visual field (512 pixels by 480 pixels) and that the number of template was 8. Fig.12 shows the comparison between the gray level histogram of the sign with a normal distribution curve to this histogram. The mean was 92.3 in 8 bit gray scale and standard deviation was 24.6. The mean of the gray level image captured at the same place in nighttime was 97.8. and standard deviation was 27.4. This result shows that there is scarcely any difference between daytime and nighttime.

Generally, performance of image processing declines according to the outside lighting conditions, such as reflected light in the morning or evening, partial shadow, fluctuation

Fig.11 NIR image in outdoor environment

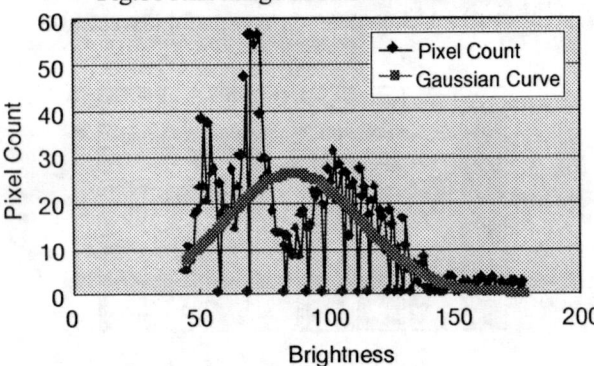

Fig.12 Sign brightness Gaussian curve

of light intensity, etc. As the transportation system is supposed to be used at finishing phase in building sites, the influence of direct or reflected sunshine from windows should be considered. Therefore this system uses a NIR-CCD camera, a NIR illuminator, and an iris control system. Due to the above countermeasures, the system can recognize the signs under an illumination disturbance outdoors, except for the situation in which the sun shines into the camera directly.

Fig. 13 shows maximum and minimum F-number with which the system could recognize the sign. F-number corresponds to brightness of the image and the image taken with small F-number is bright. The result shows that even if brightness of the image fluctuates the system can recognize the sign correctly. When the sign is at a distance less than 10 m, the system is influenced by light disturbance because of parallax of the lens and that of the illuminator. Robustness of the recognition procedure against light disturbance is confirmed through this experiment.

6. Motion control of autonomous vehicle

The vehicle has a driven and steered front wheel and two free rear wheels. The front wheel is driven at constant velocity and is steered on trapezoid pattern in turn maneuver. Fig.14 shows location of the vehicle in one second time steps, steering angle of the front wheel, and velocity of the front wheel in left turn maneuver. Fig. 15 shows those during a right spin turn maneuver. Accuracy of the distance between the sign and center line of the vehicle at the end of turn maneuver (Turn Length) was ±30 cm. Accuracy of the total rotation angle of the vehicle during a turn maneuver was ± 3.0 degrees.

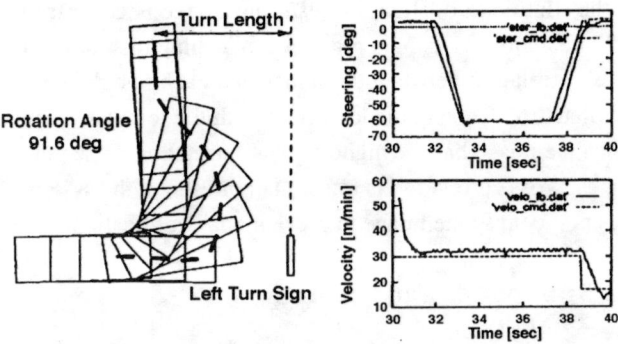

Fig.14 Vehicle movement, steering and velocity in left turn maneuver

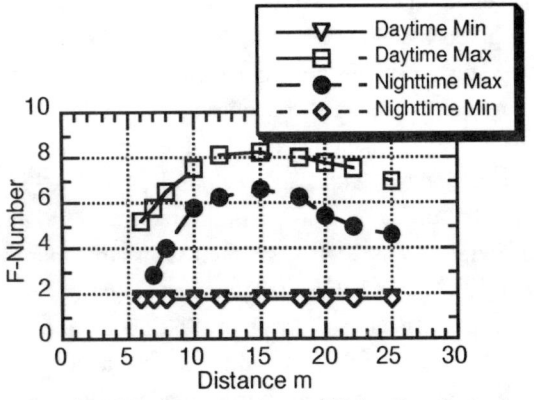

Fig.13 Robustness toward light disturbance

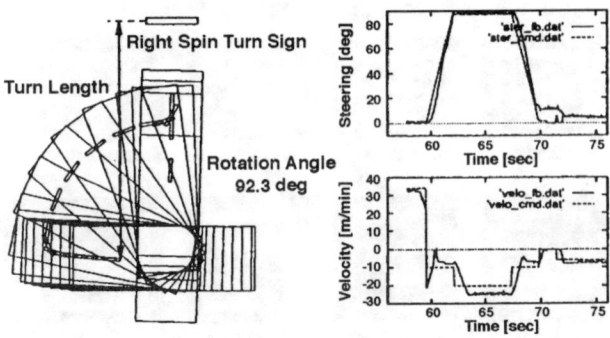

Fig.15 Vehicle movement, steering and velocity in right spin turn

7. Transportation experiment

An experimental result of transportation is shown in Fig.16. Location of the vehicle is calculated from the data of the rotary encoder attached to the front wheel and the data of the gyroscope by dead-reckoning, and calculated every two seconds. Although the experiment was performed in outdoor environment, the vehicle searched for signs and recognized instructions correctly and transported the material to the planned location. Thus transport ability of the system was confirmed. The trajectory of the vehicle was not straight because signs were not placed precisely. Even if workers place the signs coarsely on the floor, the vehicle can find the sign located within ±16 degrees of the heading of the vehicle.

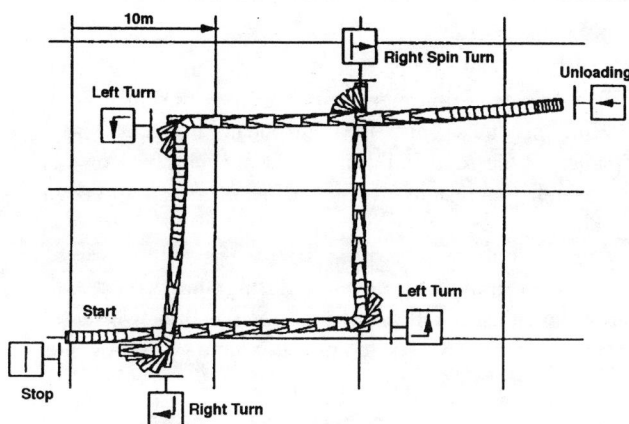

Fig.16 Vehicle movement in example transportation path

Fig.17 Practical experiment in Sapporo

open air confirmed transportation ability and accuracy of this system.

Now, the system is under a field test in a subway station construction site as shown in Fig.17. The system will be improved according to the results of the test.

8. Conclusion

In this paper the sign guided autonomous transportation system for construction site was outlined, and experiment results for evaluating efficiency of the system were investigated. The signs are detected through two stage of sign searching procedure using the SLR. The objects are distinguished according to their intensity of reflection, then they are distinguished according to their width. Reliability of the sign searching algorithm was confirmed through the experiments carried out with many objects. The signs are recognized through the sign searching procedure based on the gray level template matching method using a NIR-CCD camera, a NIR illuminator, and an iris control system. Robustness of the algorithm against light disturbance was proved through the experiments made under various lighting conditions in outdoor environment. Transportation experiments in the

Reference

[1] J.Takiguchi, T.Hashizume et.al, A study of autonomous mobile system in outdoor environment (Part1 A development of high-resolution-positioning system), 74th JSME, pp.391-392, 1996 (in Japanese)

[2] J.Takiguchi, T.Hashizume et.al, A study of autonomous mobile system in outdoor environment (Part2 A development of high-resolution environmental map generating system), 74th JSME, pp.393-394, 1996 (in Japanese)

[3] A.Ollero, A.Mandow, V.F.Munoz and J.Gomez de gabriel, Control architecture for mobile robot operation and navigation, Robotics & Computer-Integrated Manufacturing, Vol.11, No.4, pp.259-269, 1994

[4] J.Forsbeg, U.Larsson and A.Wernersson, Mobile Robot Navigation using the Range-Weighted Hough Transform, IEEE Robotics & Automation Magazine, pp.18-26, March 1995

[5] D.Langer, J..K.Rosenblatt and M.Heber, A Behavior-Based System for Navigation, IEEE Trans. RA, Vol.10, No.6, pp.776-783, 1994

[6] K.Sumi, M.Hashimoto, H.Okuda, Three-level Broad-Edge Matching based Real-time Robot Vision, Proc. '95 ICRA, pp.1416-1422, 1995

Proceedings of the 1998 IEEE
International Conference on Robotics & Automation
Leuven, Belgium • May 1998

Architecture of a GPS-based Guiding System for Road Compaction

Luc-Henri Pampagnin[1], François Peyret[2], Gaëtan Garcia[3]

[1] Cap Gemini France, ITMI division, BP 177, 61, chemin du vieux chêne, 38244 Meylan Cedex, France, phone : +33 (0) 4 76 41 40 00, fax : +33 (0) 4 76 41 28 05, e-mail : lpampagn@capgemini.fr

[2] Laboratoire Central des Ponts et Chaussées (LCPC), Site Robotics subdivision, BP 19 - 44340 - Bouguenais - France, phone : + 33 (0)2 40 84 59 40, fax : + 33 (0)2 40 84 59 92, E-mail : Francois.Peyret@lcpc.fr

[3] Institut de Recherche en Cybernétique de Nantes (IRCyN, UMR 6597), BP 92101, 44321 Nantes Cedex 3, France, phone: +33 (0)2 40 37 16 95, fax: +33 (0) 2 40 37 25 22, E-mail : Gaetan.Garcia@lan.ec-nantes.fr

1 Introduction

The CIRC (Computer Integrated Road Construction) project aims to develop precision systems for the real-time control of the positioning of road construction equipment [Peyret92].

CIRC products will provide integrated ready made solutions for operator support, for machine control and for quality assessment.

The CIRC project is supported by the European Commission (Brite-EuRam III N° BE-96-3039) and started at the beginning of 1997 for three years.

The project relies on two major technological approaches. The first one, for positioning, is the integration of GPS (Global Positioning System) technology. The other is the real-time use of CAD data on the machines as a reference geometrical data base for all the operations.

Due to the implementation of better resource management, CIRC systems will generate about a 5% gain on the budgets linked to compacting and paving. These savings permit a rapid return on investment.

Seven partners, among which are three industrial companies, are involved in the project, each providing complementary skills and abilities :

- CAP GEMINI France, ITMI division,
- Laboratoire Central des Ponts et Chaussées (LCPC), France,
- TEKLA OY, Finland,
- Karlsruhe University (IMB institute), Germany,
- University of East London (UEL), Great Britain,
- EUROVIA, France,
- National Land Survey (NLS), Sweden.

Alongside this consortium, an End-Users Club has been created. This group gathers several companies and organisations directly interested in the operation of the CIRC system. The role of this club is to participate in the specifications of requirements and facilitate the integration of CIRC products into the European market. Major companies of the road building world are members of this club, which represents over 50% of the European equipment manufacturing and over 30% of the European road construction.

The consortium will develop during the project two different versions of CIRC systems for the first two target machines : CIRCOM for compactors and CIRPAV for pavers.

2 Main objectives of CIRCOM

In road layers compaction, it is essential that the right level of energy should be transmitted to the material, with a uniform distribution. This energy, as far as the settings of the compactor do not vary, depends directly from the number of runs, or passes, of the machine (fig. 1). Until now, only the memory of the operator was used as a record to control the prescribed number of passes, inducing frequent defects in the achieved structure, given the extreme difficulty to perform this control in the site conditions.

So, the main objective of CIRCOM is to assist the driver in this task, so that he can perform the exact number of passes, at the right speed, everywhere on the surface to be compacted. From this improvement of quality in terms of level and uniformity of density, will result significant gains at the level of the life-time of the road, the operating time of the equipment and the saving of material.

To perform this task, it is essential to provide the operator and the system itself with an accurate and continuous location of the machine. "Accurate" in this case means 10 cm in both transversal and longitudinal directions. "Continuous" means that it should work everywhere on the site.

The second objective of CIRCOM is to record the actual work achieved by the compactor, in terms of trajectory and number of passes on every point, in order to feed the site data base and to perform a global quality control at the site level. So, the contractor should be able to get an immediate snapshot of the "as-built" in terms of compaction.

The main operational constraint of the CIRCOM system is at the level of the ergonomics of the man-machine interface that must absolutely remain very simple and user-friendly.

The feasibility of the CIRCOM concept had already been established, thanks to the demonstrator called MACC, which has been developed and patented by the LCPC. This demonstrator has been experimented using various positioning solutions, the last one, which has been chosen as the main positioning technology in the CIRC project, being real-time kinematic GPS, capable of centimetre accuracy [Froumentin95].

3 Overview of the CIRCOM system

The CIRCOM is designed as a hierarchical system. It is decomposed in three sub-systems split into modules. The 3 sub-systems are the following:
- ground sub-system (GSS),
- positioning sub-system (POS),
- on-board sub-system (OB).

The aim of the ground sub-system is to :
- provide the compactor with geometric data about the work-site, coming from CAD data, as well as guidelines for operation,
- compute compacting results and make some statistics about the work achieved.

The aim of the positioning sub-system is to locate precisely and in real-time the compactor by using Global Positioning System (GPS) technology as well as dead-reckoning sensors.

The role of the on-board sub-system embedded on the compactor is to :
- memorise and compute instruction data, position data, work done, and to
- manage a Man Machine Interface (MMI) which assists the driver in compacting.

From a hardware point of view, the CIRCOM comprises three units called "boxes" (fig. 2):
- the driver box includes the interfaces with the driver, i.e. the screen, the reduced keyboard and the PCMCIA disk drive. It is to be positioned in an ergonomic way and is removable.
- the computer box includes the acquisition and computation electronic boards as well as the power supply regulator. This unit is placed at a safe location in the compactor.
- the roof box includes the RTK GPS (Real Time Kinematic Global Positioning System) devices: processing unit, radio-modem and the antennas for GPS and radio-communication of position corrections.

Moreover, dead-reckoning sensors are connected to the computer box (see section 6).

4 Ground sub-system

The main challenge of the ground sub-system is to establish a link between the design and the construction of the road. Indeed, the data provided by the road CAD systems are directly processed in order to bring to the on-board sub-system the appropriate information about the work-site.

More precisely, the ground sub-system is in charge of the following functions.

1. to import information from the road CAD system
   CIRC systems are designed to import data from the main CAD software thanks a standard format called CIF (CIRC Input Format). Two major CAD software developers have already agreed to implement CIF output.
2. to plan the mission of the compactors
   This means that the GSS allows its operator to define the kind of compaction for each layer of the road, including the type of machine and the target number of passes.
3. to export the geographical database as well as the mission to the compactors
   This information is exported to the OB thanks to a large storage hard disk.
4. to import achieved work from the compactors
   The same medium allows to bring back to the GSS the data about the achieved work, mainly the number of passes per area of the road, for further exploitation.
5. to compute statistics about the achieved work
   The data about the performed work are processed in order to provide the contractor statistics about the work of each machine, the histogram of compaction of the road and so on. The GSS allows to display these results in a useful and friendly way.

The ground sub-system is a PC located in an office of the contractor. It is generally not on the work-site but several kilometres from it. The large storage disks are PCMCIA hard drives. Their credit card format allows them to be easily transported between the office and the work-site once or twice a week.

5 On-board sub-system

The OB is not the only part of the system located on-board the compactor. Actually, it gathers all the on-board functions except positioning. The main challenge of the OB is to implement a friendly user interface. Indeed, the drivers are not used to computers. The interface (fig. 3) is designed to offer an intuitive access to a restricted number of functions. Some simple rules are applied like "one function / one key" for inputs and "green means OK / red means problem" for outputs. From an external point of view, the most visible part of CIRCOM is this man-machine interface that helps the driver to precisely know the work already achieved and the work that remains thanks to a coloured map of the road.

More precisely, the functions implemented by the OB are the following:

1. to import geographical data and compacting instructions from the GSS
2. to compute the compacting status
   From the compacting point of view, the validity of passes of the compactor depends on several parameters including the vibration status and the speed. The vibration status is provided by a specific sensor.
3. to manage the compactor data-base
   This data-base stores the trajectory information provided by the positioning sub-system (POS) along with the compacting status.
4. to display the compaction map
   Thanks to the above information, it is possible to display a map of the road made of coloured tessels that indicate the current number of passes. This map is zoomable so that the driver can either check the joints between compaction stripes or locate his machine in whole work-site. This map also shows the current position of the compactor, by default in the centre of the scrolling map.
5. to display additional information including speed, vibration status and alarms.

From a hardware point of view, the OB is a PC104 computer running under Windows NT operating system with real-time extension.

6 Positioning sub-system

6.1 Problem statement

The localisation software is a key component of the CIRCOM system. In a first prototype version of such a system, localisation was performed by a RTK GPS unit alone. The main drawback of this solution was that, in some cases, no localisation data was available. Indeed, the GPS must be able to track at least four satellites to compute its location. In the presence of buildings or trees, the system may fail. Of course, when the machine passes under a bridge (a very common situation) or a tunnel, the system will not work.

In order to alleviate this problem, it was decided to use additional sensors to allow dead-reckoning localisation during these phases. What performances are desirable for this dead-reckoning ? The idea was to consider that the most common difficult situation would be that of passing under a bridge. In the worst case, this bridge supports a 2x2 lane road and crosses the work-site with a 45° angle. Taking into account the re-initialisation time of the GPS after a masking and the speed of the compactor, the distance travelled is approximately 100 m. Considering that the desired localisation precision is 0,2 m, the precision of the dead-reckoning localisation should be 0,2 %.

6.2 Difficulties of the situation

Odometry based on two fixed wheels is not feasible because there is no axis to which the speed is known to be perpendicular. Hence, the fixed wheels of the odometer would support lateral slipping, particularly during manoeuvres of the machine. In addition, due to the constraints of the work-site and to the high reliability demanded, the use of an external odometer [Ferrand92] is not acceptable. Moreover, the conditions of use are particularly difficult : the material on which the compactor moves is hot (more than 100° when laid on the road) and tends to stick to the roads. For these reasons, the sensor set-up includes a fibre-optic gyrometer to measure the rotation speed plus two sensors to measure the velocity : an encoder attached to the cylinder of the machine and a Doppler radar.

The reasons why two sensors are required are simple. At very low speeds, the Doppler radar will not work at all, and this happens often since the compactor repeatedly stops and inverts its speed (see a typical trajectory in the results). On the other hand, the encoder is not sufficient because, when the compactor is back to its normal speed, powerful vertical vibrations are applied to the cylinders for compacting, causing slippage ratios as high as 30%.

Another source of difficulty is that there is no simple kinematic model for the compactor. Since the "wheels" of the machine are cylinders (called drums), any non-straight trajectory results in slippage. If the cylinder is considered infinite, there is always one contact point of the cylinder with the ground for which the speed with respect to the ground is equal to zero, but it is not known and generally not constant.

Finally, it should be noted that what we are interested in is always absolute accuracy, i.e. accuracy of the position computed in a predefined absolute reference frame. This is in contrast to many studies about dead-reckoning precision. Indeed, in many cases, accuracy is measured by a test consisting in starting the mobile from a given position, performing a closed reference trajectory and measuring the distance from the return position to the initial position [Borenstein97]. In such a test, the initial heading error is implicitly equal to zero. In our case, the situation is different : we don't have any absolute heading measurement and, when a masking situation occurs, the current heading error will result in dead-reckoning localisation errors, even if the speed and rotation speed of the vehicle are measured perfectly later on.

6.3 Equations of the system

Since we do not have the possibility to measure the two components of the speed of the vehicle, we will suppose that the speed of the machine is always perpendicular to the axis of the front drum and we will set the Doppler sensor in such a way that it measures this component (supposing that its lateral sensitivity is null).

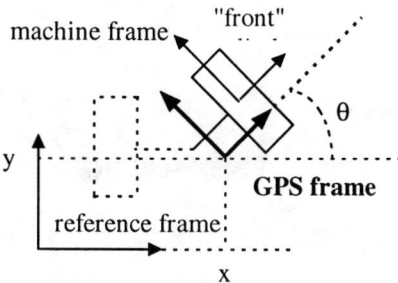

Fig. 4 : Notations.

Let T be the sampling period, and t_x, t_y the coordinates of the GPS frame in the machine frame. If v_k and ω_k are the speed and rotation speed of the 'front' cylinder of the machine, the coordinates of the GPS frame in the absolute reference frame at time k+1 are :

$$x_{k+1} = x_k + v_k T\cos\theta_k - \omega_k T(t_x \sin\theta_k + t_y \cos\theta_k) \quad (1)$$

$$y_{k+1} = y_k + v_k T\sin\theta_k + \omega_k T(t_x \cos\theta_k - t_y \sin\theta_k) \quad (2)$$

$$\theta_{k+1} = \theta_k + \omega_k T \quad (3)$$

These equations are the basis of the dead-reckoning localisation. They are used repeatedly (at 25 Hz) using the measurements of ω provided by the gyrometer and the speed provided by either the radar or the encoder. When a position measurement provided by the GPS is available (at 1 Hz), the current dead-reckoning estimated position is updated to prevent it from drifting. When the GPS is masked, the dead-reckoning algorithm is simply repeated. The updating of the estimated position is performed using an Extended Kalman Filter (EKF). We suppose that the GPS measures the position at the current time instant. In this case, since it is the GPS antenna that is localized, the measurement equations are particularly simple :

$$x\_gps_k = x_k \quad (4)$$

$$y\_gps_k = y_k \quad (5)$$

Let $U=(v,\omega)^t$, $Y=(x\_gps,y\_gps)^t$ and $X=(x,y,\theta)^t$. If we consider that the previous (ideal) equations are corrupted by additive noise, they can be written :

$$X_{k+1} = f(X_k, U_k) + \alpha_k \quad (6)$$

$$U_k^* = U_k + \gamma_k \quad (7)$$

$$Y_k = g(X_k) + \beta_k \quad (8)$$

We are in a context in which the EKF equations can be used. The standard equations must simply be adapted to the fact that the input U is corrupted by noise. Equations not reported here for the sake of brevity allow to calculate an estimated covariance matrix of the localisation error [Maybeck79].

In practice, the position calculated by the GPS is available with a constant time delay equal to N*T. If the inputs U of the last N periods are recorded, it is possible to make a GPS position correspond to the time instant at which it was valid. In this case, the equations are not modified but the position estimate is N*T late. This is the solution we chose since, in our case, the delay is only 70 ms. If, additionally, the states X and the corresponding covariance matrices P are also recorded it is possible, after each position update, to recalculate N prediction steps with eq. (6) to get an up-to-date position estimate.

6.4 Heading estimation

Since we do not measure θ, it is initialised by moving the vehicle in a roughly straight line and computing the heading using two sufficiently distant positions.

It can be shown that, although θ is not measured, it is (weakly) observable as soon as the speed is not equal to zero. The covariance matrix Q_γ of noise γ is of particular importance. The speed and angular speed measurement being independent, Q_γ is set to a diagonal matrix. The value of $Q_\gamma(2,2)$ controls how much the GPS position data is allowed to modify the dead-reckoning estimated heading. If this term is too large, the estimate of θ will tend to oscillate due to GPS position errors. In this case, the filter may enter mask periods with comparatively large heading errors (especially at low speeds). If too small, position data cannot sufficiently update the dead-reckoning estimate to prevent it from drifting.

We propose a methodology to tune this parameter, in spite of the fact that we do not have any external measurement of θ to which we could compare the estimate given by our filter. The idea is to use real data with GPS positions available and to do the following : for the same trajectory, simulate various time instants t, t+1, ..., t+n at which the masking phase begins. Start from a large value. If the position errors of the various dead-reckoning estimated trajectories are too different, lower the value ; if it takes too

long to compensate for an initial heading error, augment it
Continue until a correct compromise is found.

6.5 Results

Figures 5 to 8 show a typical trajectory of the compactor, the corresponding speed over the first 5 minutes, lateral error signals for masking situations starting at time instants 40 s to 50 s (11 curves) and, finally, the absolute error (solid line) and precision indicator (dotted line).

On figure 7, the 11 curves are nearly perfectly superimposed. The error changes sign when the speed is inverted because of our convention (lateral error is positive when the estimated position is on the left with respect to the real position when looking in the direction of the movement). Should the maskings occur in a curve, the various signals would have been distinguishable. Indeed, the variance of the noise on ω has been set to perform optimally in this type of situations since a straight line movement is the standard situation for the compactor.

Many tests have been performed on various trajectories. For all standard trajectories, the required precision of 0,2 m has been obtained for more than 100 m.

7 Development status

The CIRCOM is currently in the development phase. After the system design, each sub-system is designed in a detailed way, developed by the appropriate partners and individually accepted.

In the next steps, the CIRCOM will be tested following procedures :
1. the aim of the **experimental tests** is to verify the performances of some of the modules/sub-systems of the CIRCOM : robustness and noise immunity of the radio, accuracy of the positioning sub-system, ... They will be carried out on a real compactor on test tracks of the CER at Rouen, France.
2. the aim of the **validation tests** is to verify that the CIRCOM realises all the functions required by the user in a satisfactory way. They will be performed on a regular work-site from Eurovia. This last step is planned for June 1998.

8 Conclusions

At the moment we are writing this paper, the first one of the CIRC products is about to be presented at the biggest European construction machinery fair, the BAUMA, in Munich. This product will be the first totally integrated system, devoted to a civil engineering task, capable of managing the total process, from the design office up to the quality control of the work achieved.

From a scientific point of view, the development of the real-time positioning sub-system has undoubtedly been the one which has needed the greatest effort in terms of research, given the stringent specifications required. The result of this research is a novel device, using state-of-the-art GPS and optical fibre gyrometer as main sensors and advanced Kalman filtering techniques, for the first time on a civil engineering machine.

The research and developments inside the CIRC project will continue upon the second contemplated product which will be devoted to the profiling machines such as pavers or graders. For those machines, the most challenging functions will be again the positioning functions. This time, will be addressed the accurate 3D positioning of the tool, "accurate" meaning this time centimetre level for the height component.

9 References

[Borenstein97] J. Borenstein, H.R. Everett, L. Feng, D. Wehe, "Mobile robot positioning : Sensors and Techniques", Journal of Robotic Systems 14, n° 4, 1997, pp. 231-249.

[Ferrand92] A. Ferrand, "Localisation relative des robots mobiles. Conception et mise en oeuvre d'un odomètre indépendant de la structure de locomotion du robot", Revue d'Automatique et de Productique Appliquée, Vol. 5, n° 3, 1992, pp. 89-100.

[Maybeck79] P.S. Maybeck, "Stochastic models, estimation and control", Mathematics in Science and Engineering, vol. 141, Academic Press.

[Peyret92] F. Peyret & H. Philippe, "Towards Computer Integrated Road Construction", 9th ISARC, Tokyo, pp. 859-868, June 1992

[Froumentin95] M. Froumentin & F. Peyret, "An operator aiding system for compactors", 13th ISARC, Tokyo, June 1995.

Fig. 1: A compactor at work on a motorway work-site.

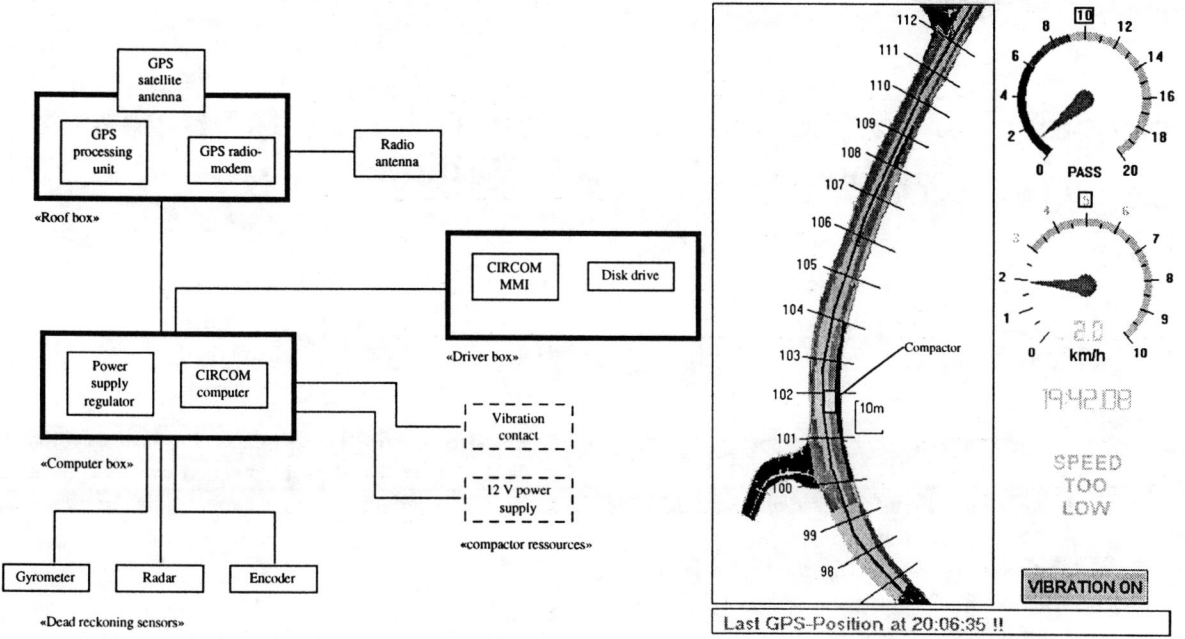

Fig. 2: Hardware architecture of CIRCOM.

Fig. 3: prototype of CIRCOM display.

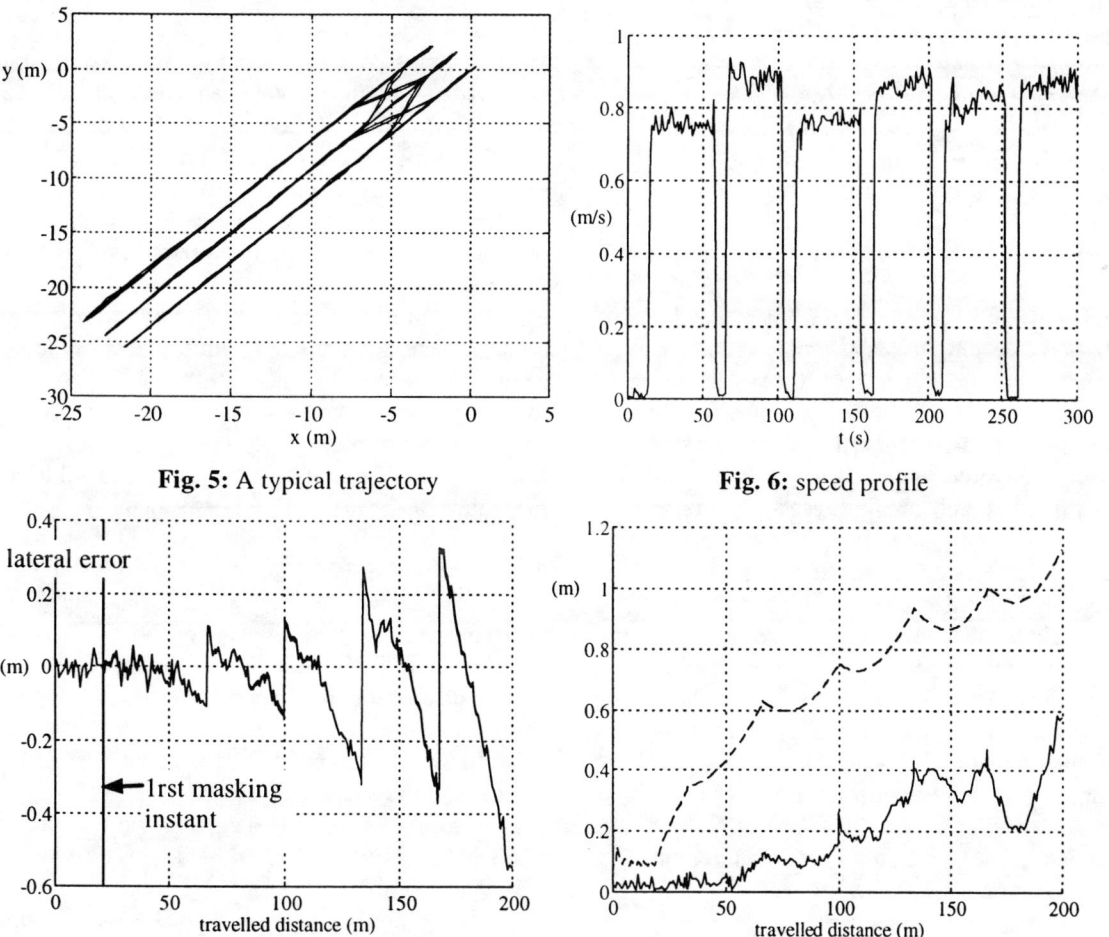

Fig. 5: A typical trajectory

Fig. 6: speed profile

Fig. 7: Lateral errors for 11 different masking instants

Fig. 8: absolute error and precision indicator

Development of Automated Construction System for High-rise Reinforced Concrete Buildings

Koji Hamada, Noriyuki Furuya, Yasuo Inoue
Technical Research Institute
Obayashi Corporation
4-640 Shimokiyoto, Kiyose-shi 204, Japan

Tatsuya Wakisaka
Building Construction Division
Obayashi Corporation
1-19-9 Tsutsumi-dori, Sumida-Ku 131, Japan

Abstract

An all-weather automated construction system has been developed to reduce the total cost of high-rise reinforced concrete building construction. It was applied for the first time in the world to the construction of a 26-story reinforced concrete condominium project in Chiba Prefecture in 1995.

This system incorporates four major elements: 1) a synchronously climbing all-weather temporary roof; 2) a parallel material delivery system; 3) prefabrication & unification of construction materials; and 4) a material management system.

It ensures good quality; improves the working and environmental conditions; reduces the construction period, manpower, and waste; and improves overall productivity.

1 Introduction

In Japan, the aging of skilled construction workers and the lack of younger ones is a steadily worsening problem. Further, the birth rate in Japan has declined in the past 20 years and the population of productive age (15-65) people is also steadily decreasing. Therefore, the lack of laborers will be a serious problem for the construction industry at the beginning of the 21st century. To solve this labor problem, it is necessary to raise labor productivity significantly and to improve the terms of employment and working conditions to attract young laborers.

With this background, in order to improve labor productivity, automated building construction systems that install an automation factory in the upper part of the building being constructed have been developed [1]. However, these automated building construction systems are applicable only to a building of steel structure. It is difficult to raise the level of automation for reinforced concrete structures compared with steel structures in terms of cost and effect. An automation plant can easily be set on a steel structure, but would be difficult to do on a reinforced concrete structure because the concrete must be left to harden.

An automated construction system for high-rise reinforced concrete buildings (hereinafter referred to as "BIG CANOPY") as described in this paper, has been developed to solve the above technical problems, and was used to build a 26-story reinforced concrete apartment house in the Tokyo Metropolitan area in 1995 for the first time in the world [2].

2 Features and construction progress

2.1 Main features of the system

In BIG CANOPY, as shown in **Fig. 1**, we set a parallel delivery system with three automated overhead cranes and one large construction lift under an all-weather synchronously climbing temporary roof frame. We also apply a material management system using a database linked with the CAD system, extensive prefabrication and unification of construction materials, and use versatile workers. The main features are as follows:

1. Improvement of productivity: The overhead crane is easier to operate compared with the tower crane, the parallel delivery system increases the efficiency of delivery and erection, and versatile workers can cooperate without wasting time.
2. Improvement of quality: Quality is improved by prefabrication, and by the all-weather temporary roof.
3. Short construction period: The period is shortened by the use of prefabrication and unification, stable processing by all-weather construction, and early commencement of the interior finishing work.
4. High degree of design freedom: As temporary posts are independent of the building, we can flexibly apply the system to various building shapes. According to statistics, this system can be applied to about 80% of high-rise reinforced concrete buildings.
5. Improvement of construction environment: Severe heat, wind and rain are moderated, and workers can work safely and comfortably under the temporary roof.
6. Safety of surroundings: The area of activity is compact, thus making the neighborhood safe.
7. Reduction of debris: Prefabrication and unification re-

duce debris.
8. Reduction of total cost: The above 1-7 reduce the overall cost.

2.2 Outline of construction progress

Figure 2 shows the construction progress when applying the system to a building site. This is described as follows:

1. Assembling the temporary roof frame: We assemble temporary roof frames and proceed with basement work under the roof.
2. Executing skeleton, finishing and equipment works: After raising the temporary roof two floors at a time, we perform skeleton work using an efficient delivery system under the roof, and start finishing and equipment work early to shorten the construction term.
3. Dismantling the temporary roof frame: After the skeleton work is completed, we lower the temporary roof onto the roof of the building and dismantle it. We separate the perimeter of the temporary roof frame and lower it in the reverse sequence to the ground and dismantle it safely there. We transport the temporary roof to the next construction project or leave it as, for instance, a heliport.

3 Key elements of the system

3.1 Synchronously climbing temporary roof

The synchronously climbing temporary roof consists of four tower crane posts erected independently outside of the building, climbing equipment, and temporary roof frame. We developed a central control unit that synchronously controls the four pairs of climbing equipment to keep the roof level while moving up and down. The oil hydraulic circuit and the electric control system are shown in **Fig. 3**. The features of the climbing device are as follows:

1. The difference between the cylinder strokes of the four climbing hydraulic jacks is controlled within 10 mm by a central control unit so that the roof can be kept horizontal while going up and down.

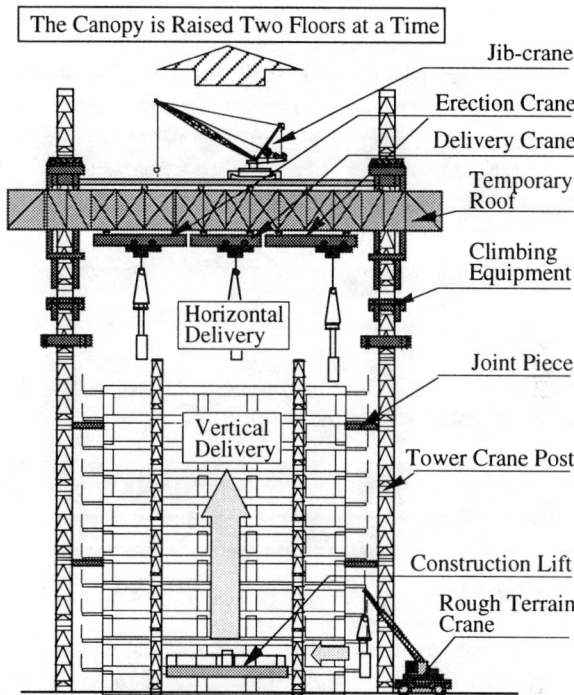

Figure 1: Parallel delivery system

Assembling the temporary roof frame   Executing skeleton, finishing and equipment works   Dismantling the temporary roof frame

Figure 2: From construction to dismantling of temporary roof

2. The control system consists of four modes combining manual/automatic control with a single interlock. When dismantling the roof, reverse climbing mode is used.
3. A worker directly operates a touch panel on the control unit's personal computer, and can watch the synchronously climbing situation on a monitor (**Fig. 4**).

The roof is raised two floors at a time. The up and down movement speed is 300 mm/min, and each 6-m (one post length) climb takes slightly less than one hour. The roof is raised during re-bar arrangement for slabs and concrete placing when the overhead cranes are not used.

The temporary roof is a shingle roof with folded thin steel plates on a steel frame truss structure and is about 50 m square. The entire weight, including the roof frame, climbing equipment, overhead cranes and jib-crane, is about 600 tons. The jib-crane set on the temporary roof is used for adding posts after climbing and for dismantling the temporary roof frame.

3.2 Parallel delivery system

The parallel delivery system consists of a construction lift and overhead cranes. The central overhead crane (delivery crane) removes a load from the lift and the right or left crane (erection crane) receives the hoist with a load from the delivery crane and erects it. These pieces of machinery reduce the waiting time of both workers and machines, and achieve efficient delivery and erection by simultaneous operation.

Before deciding the specifications of the parallel delivery system, we examined the cycle time of erecting each precast concrete (PC) member and the line balance of the lift and cranes by using a simulator on a computer (**Fig. 5**).

Hoist exchange between the delivery crane and erection crane is performed as shown in **Fig. 6**. This system is performed mostly manually with a wireless remote controller, but the traveling of the delivery crane and crane girder positioning between the delivery crane and erection crane are automated to reduce the work load on the operator. Members are delivered in the following order from the ground to erection location.

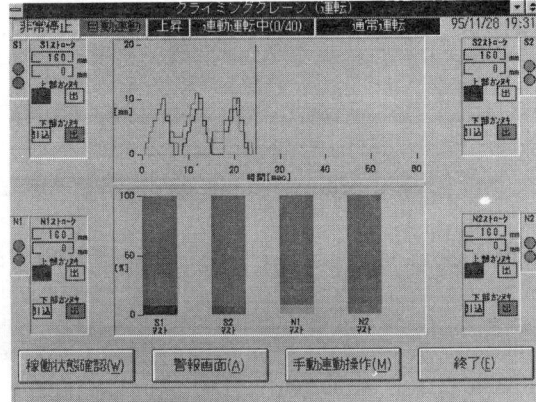

Figure 4: Control monitor for synchronously climbing equipment

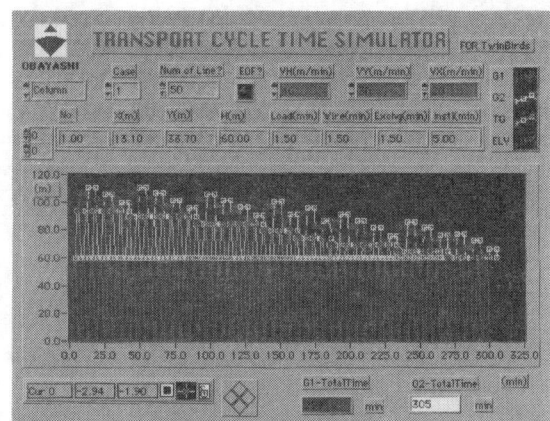

Figure 5: Transport cycle time simulator

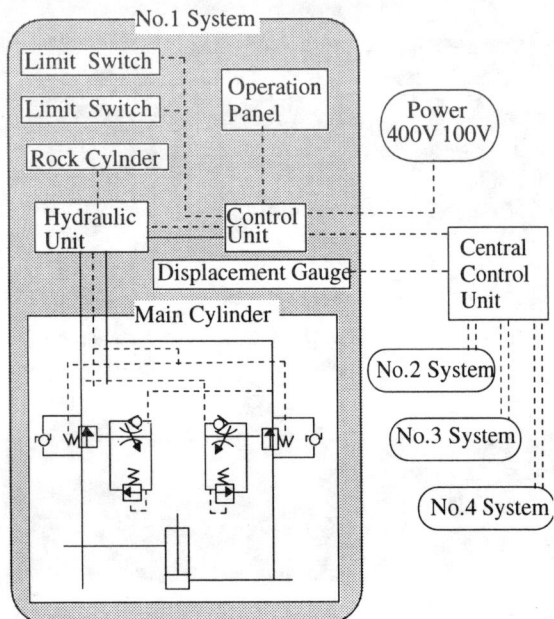

Figure 3: System of control and hydraulic circuit

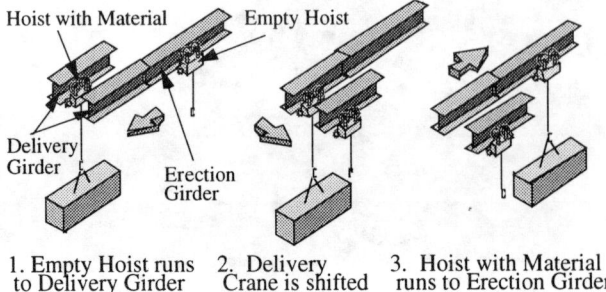

Figure 6: Hoist exchange system

1. The rough terrain crane discharges members carried to the site and loads them onto the lift.
2. The lift conveys them to the erection floor.
3. The delivery crane removes them from the lift.
4. The delivery crane travels to the waiting erection crane and stops at the correct position.
5. The exchange of hoists is done between the delivery crane and erection crane. Then the delivery crane travels back to the lift top and the loaded hoist traverses to the erection point.

We developed a suspender controlled by gyroscopic moments (GYAPTS) to control the rotation of building materials caused by wind and by inertia accompanied by crane movement [3]. There were two distinctive features in utilizing gyroscopic moment generated in the device: 1). by controlling the gyroscopic moments actively, the materials were turned with a quick response and correctly stopped at the target angle of the setting position, and 2). by using gyroscopic moment passively, the materials withstood external disturbances by generating an inverse gyroscopic moment. We introduced a suspender device equipped with GYAPTS for each hoist to prevent interference during hoist exchange and for positioning long PC members such as slab panels (**Fig. 7**) and balcony members. Since the operator of the overhead crane regulates the process by wireless remote control, danger to the worker is reduced and work efficiency is improved too.

The major specifications of the parallel delivery system are shown in **Table 1**.

3.3 Prefabrication and unification

We adopted the industrialized construction method using many PC members higher than the third floor, the cross section of the skeleton of which was standardized. Columns, walls and balconys are full PC members, and girders and slabs are half PC members as shown in **Fig. 8**. Concrete was placed in panel zones and the upper part of girders and slabs at the site. The prefabrication ratio was 71% of concrete volume, which is an extremely high ratio. We adopted grout connection through sleeves to join column bars, and joined screw bars with a coupler in panel zones in connecting girders.

3.4 Material management system

The objectives of the material management system were as follows:
- high quality planning of complicated material management;
- integration and use of actual material data;
- quick response to modification in delivery plans;
- labor saving.

The system consists of a material management database and five subsystems (**Fig. 9**). The hardware consists of a workstation, two computers, two bar code readers, a bar code printer and a page printer. The machines are linked by an Ethernet LAN in the construction site office.

The material management database linked with three-di-

Table 1: Specifications of lifting equipment

Name and Number of Device		Specification
Construction Lift	1	Loading Capacity :6t Winching up Speed :40m/min Control Type:Invertor
Hoist Exchanging Overhead Crane		Operation Type :Manual/Automatic Wireless Remote Control Control Type :Invertor
Delivery Crane	1	Maximum Traveling Speed :40m/min Suspended Capacity :7.5t
Erection Crane	2	Maximum Traveling Speed :30m/min Suspended Capacity :7.5t
Electric Hoist	3	Maximum Traversing Speed :33m/min Suspended Capacity :7.5t
Gyroscopic Suspender	3	Operation Type :Wireless Remote Control Weight:1100kg Rotating Drive:Gyroscopic Moment Inertia Moment of Load:25ton m^2

Figure 7: Installation of slab PC with GYAPTS

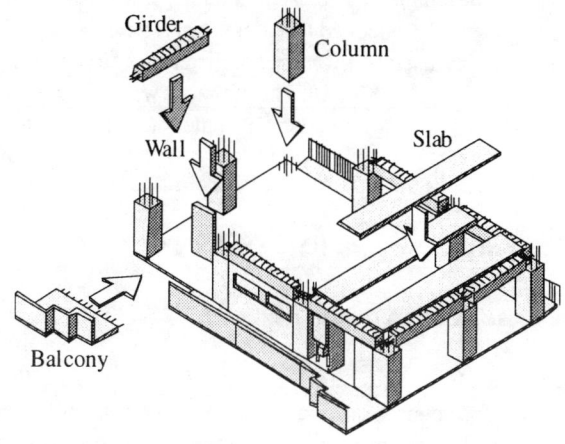

Figure 8: Main activities of PC construction

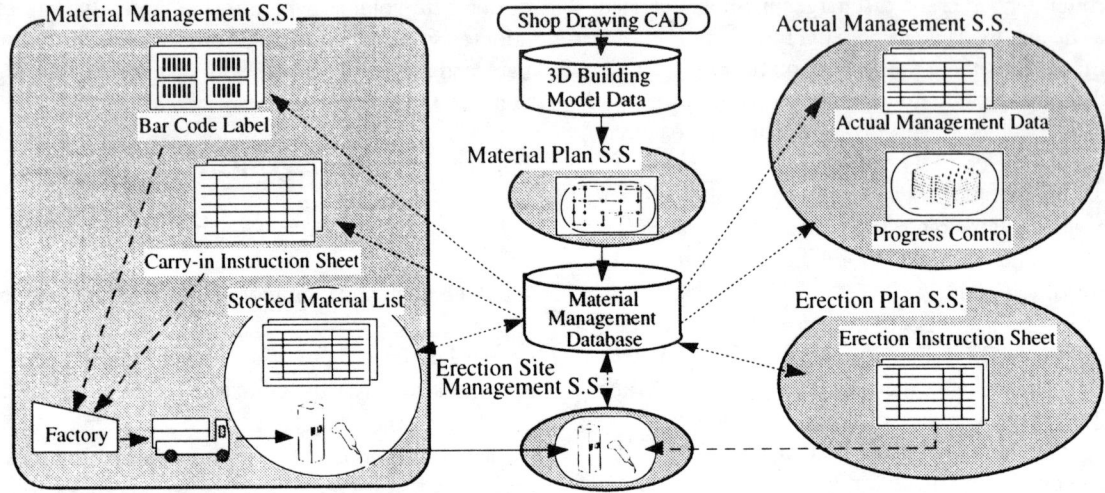

Figure 9: Material management system

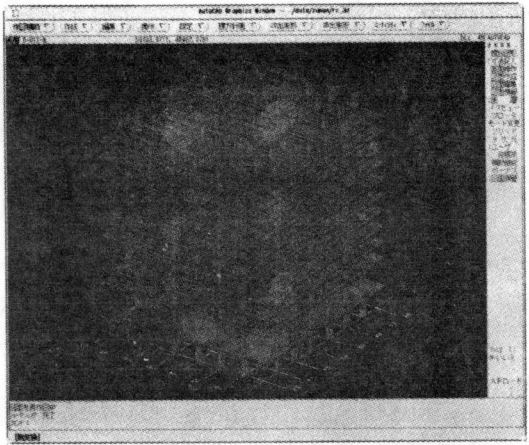

Figure 10: Progress control of PC member erection

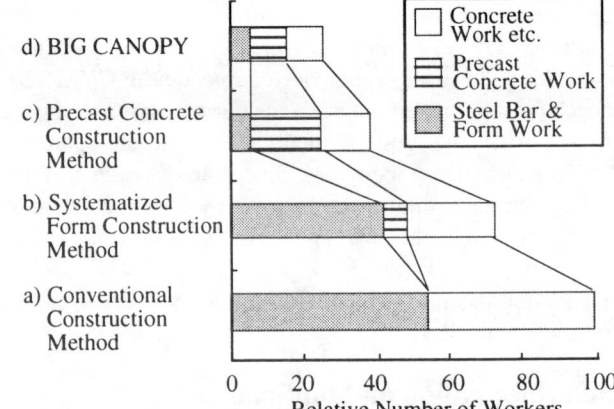

Figure 11: Productivity comparison among four construction methods

mensional CAD shop drawings was used to conduct all the planning from material delivery to erection and actual management. Attribute information, such as kind of member, shape, installation location and so forth, was input through CAD into the database. The material management was unified by using bar codes attached to the materials at the factory. The screen display used to monitor the progress of PC member erection is shown in **Fig. 10**. The manager was able to effectively accomplish various kinds of material management by utilizing CAD.

4 Benefits of using BIG CANOPY

4.1 Productivity improvement

Figure 11 compares the labor productivity of skeleton work for four construction methods for high-rise buildings.

The number of workers engaged in the skeleton work for BIG CANOPY was about 25% that of the conventional construction method, about 35% that of the systematic formwork construction method with metal form and about 65% that of the PC construction method. Labor saving is achieved by the simplification and standardization of work by prefabrication, the reduction in waiting time by an effective delivery system, and the use of versatile workers.

4.2 Reduction of construction period

We analyzed the relation between cycle time of the main PC members and the height of the erected floor from the actual data provided by the material management system. The results are shown in **Fig. 12**. The cycle time of the main PC members is the cumulative human working time (hanging and setting) which tends to decrease with learning and machine working time (lifting and delivering), which tends to increase with height. As the regression curve shows, we found that the cycle time tended to be the smallest at a height of about 60 m. Increasing the speed of delivery machines, especially of the lift, is expected to further reduce the cycle time.

Two tower cranes (200 tm) were initially planned to be used in this construction. Using the results of the above, we compared the erection time of PC members and found that the capability of the parallel delivery system was equal to the power of about 2.5 tower cranes as shown in **Fig. 13**. The reasons for this efficiency improvement are as follows:

· Overhead cranes are hardly affected by wind.
· The operator can precisely grasp the erection situation when operating the crane.
· By using a large construction lift, the wall and slab members can be lifted in large quantities.
· By using the delivery crane for erection work, three cranes can erect PC members simultaneously.

Consequently, skeleton work progressed faster than anticipated with the parallel delivery system, and work was not interrupted by the weather. As the temporary roof stabilized the skeleton work and allowed the finishing work to start early, the anticipated construction period by two tower cranes (28 months) was reduced by four months.

4.3 Improvement of work environment

The skeleton works under the temporary roof were analyzed from the ergonomics point of view during all seasons. From measurements of the surface temperature of workers' clothes and materials (rebars) under the temporary roof and outdoors in fine weather in the summer, there were differences of about 10°C maximum for workers' clothes and about 25°C maximum for rebars between under the temporary roof and outside. In outdoor work, workers' measured heart-rate was 134 beats/min at maximum with a mean value of 103 beats/min, and was about 80% and 40% higher compared with that at rest measured in the lunch break. On the other hand, when working under the temporary roof, the heart-rate was 108 beats/min at maximum with a mean value 89 beats/min, and was about 45% and 20% higher compared with that at rest.

Thus, the temporary roof of BIG CANOPY not only improved the working environment on rainy days, but also significantly reduced the physical load on workers on hot days.

5 Conclusion

BIG CANOPY is used mainly for the construction of high-rise reinforced concrete apartment buildings. However, there is the constraint that temporary construction costs must not increase greatly because of the low construction unit cost of reinforced concrete buildings. Accordingly, in our development we utilized the construction equipment that we already possessed and reduced the development cost by using commercial general-purpose equipment wherever pos-

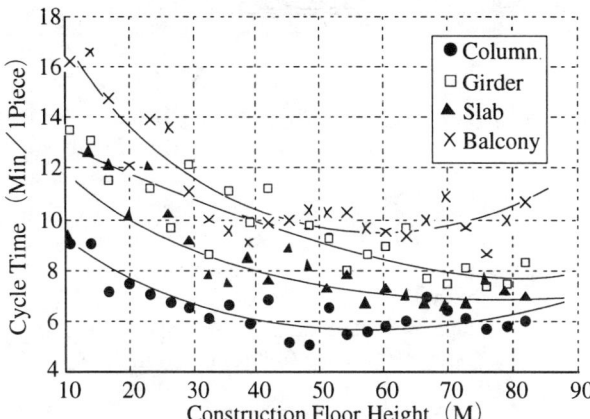

Figure 12: Cycle time of PC members

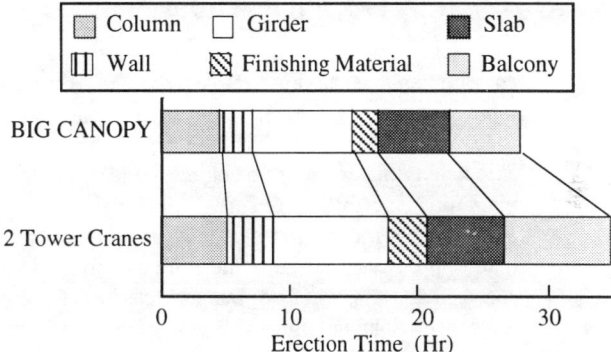

Figure 13: Efficiency comparison between two tower cranes and parallel delivery system

sible. We chose the optimal automation level of the system so that workers could use their skills, and examined the balance of cost and effect. We achieved very high productivity and reduced the construction period without increasing construction cost. This system increases the level of automation, adds a new function and is flexible according to the situation. BIG CANOPY is thus viable and has good potential.

BIG CANOPY has been used on another 20-story condominium project and two more new projects are under construction.

References

[1] R. Kudoh, "Implementation of an Automated Building Construction System", *Proc. of 13th International CIB World Building Congress*, 1995, pp17-28

[2] T. Wakisaka et al., "Automated Construction System for High-rise Reinforced Concrete Building", *Proc. of 14th ISARC*, 1997, pp111-118

[3] F. Inoue et al., "A Practical Development of the Suspender Device that Controls Load Rotation by Gyroscopic Moments", *Proc. of 14th ISARC*, 1997, pp486-494

MONAI : AN AUTONOMOUS NAVIGATION SYSTEM FOR MOBILE ROBOTS

Authors : F. CARRE L. GALLO B. MAZAR F. MEGEL B. SERRA

AEROSPATIALE Missiles, 1 rue Pablo Picasso 78114 Magny les Hameaux FRANCE

Email : laurent.gallo@missiles.aerospatiale.fr

Abstract : This paper presents the results of the simulation work concerning the development of a **MO**dule (for) **N**avigation (which is) **A**utonomous and **I**ntelligent (MONAI), made at AEROSPATIALE. Entirely based on vision channels, this system allows to achieve efficiently the navigation and guidance function of a rover, with a very favourable cost/effectiveness ratio. It is first dealt with the general principle of the navigation/guidance architecture, then with the detailed functions regarding landmark update, obstacle detection and motion assessment, with the first results from the simulations run.

1. INTRODUCTION

The navigation of autonomous robots in outside environments is an upstream study and research topic, considering the readiness of the techniques used, the complexity of the algorithms, the associated material architectures, and the low marketing efforts toward potential users. Autonomous mobile robotics may be of interest in many application ranges : the military, civilian or space fields. Owing to budget restrictions, the « Agence Spatiale Européenne » (the European Space Agency) does not give much credit to such a discipline. The Sojourner mission certainly brought about a fresh interest in that field and a revival of optimism, but concrete actions remain to be defined in Europe. As far as the civilian field is concerned, the markets remain to be opened. They concern the cleaning business, the logistics, airport surveillance operations, agriculture, or intelligent driving assistance. Such applications, however, require a mature and low cost technology, which is not very favourable in the present state of attainments and developments. There is the military market, however, which, taking into account the reduction of manpower and the fact that the armed forces are turning professional, constitutes probably an interesting source of development for mobile robotics. Finally, the duality of the military/space applications needs to be utilized in order to finalize the prospective studies inherent in the field of autonomous navigation of craft.

AEROSPATIALE has been working on that subject since 1996, and is developing an autonomous and intelligent navigation module (MONAI), transposable to any type of vehicle, in the above-mentioned fields of application. The work deal essentially with navigation and guidance systems based on vision channels. The challenge was to achieve both functions using only vision channels : visible or infrared cameras (passive and little expensive systems), and a compact image processing computer, fitted with several specific ASICs. The object of this paper is to explain the algorithms and the physical architecture implemented for the purpose of a feasibility demonstrator of the MONAI. After having covered rapidly the general functional architecture, we shall examine more particularly the original vision functions that have been, developed, namely the updating on landmarks, motion estimation and obstacle detection.

2. NAVIGATION AND GUIDANCE

The general philosopphy of our autonomous navigation system is based on a classical and determinist principle, contrary to the behavioural approaches (which can present, in other respects, advantages different from a conventional approach). The principle consists in maintaining an absolute position (or relative with respect to a map), and to guide the robot towards a point located within the same reference mark. The so called « global » guidance notion uses a « local » guidance function, allowing to avoid obstacles encountered along the advance course of the robot (which receives the orders from the global guidance), and allows also to take into account characteristic elements of the environment allowing to achieve a more « behavioural » guidance (road or edge following, detection and counting of intersections, etc ...).

The localisation function allows to estimate, at any time, the position of the robot, using various sensors providing enhancement of the measurement data or maintenance of an accuracy compatible with the requirement.

The « inertial unit » type sensors feature many advantages (compactness, accuracy), but feature regrettable disadvantages such as drifting with time, and very high cost. Therefore, the idea is to hybridize a very low cost sensor of that type (but which drifts very rapidly), with other sensors or

updating methods. Our approach consists in using a minimum class inertial unit, and updating it by :
- the motion estimation through image processing, and the detection of robot stopping (zero speed updating),
- aiming at and recognition of characteristic objects in the landscape during the progression of the robot, also called « landmarks ».

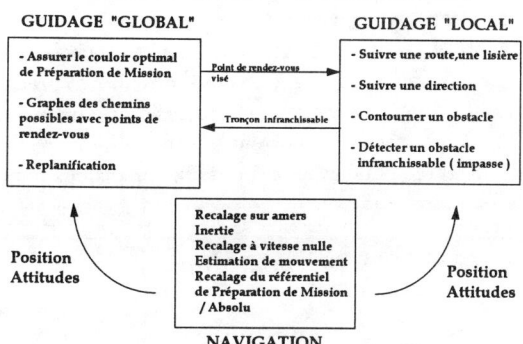

Studies are currently underway to demonstrate the feasibility of replacing an inertial unit by motion estimation through imagery.

2.1. Guidance principle

Going in an autonomous way from one point to another point while avoiding possible obstacles encountered during the progression creates in fact two problems of a very different nature:

Going from one point to another one is essentially a problem of a global nature :

A prior global knowledge of the situation is necessary. Such a knowledge is synthesized in the form of maps, accessible zones, prohibited areas... The algorithms of so-called 'global' guidance can cope with the problem.

The notion of unforeseen event is underlying. Obstacles may hinder the advance as initially planned because the data used for mission preparation may be of variable precision or may be sufficiently old so that the description of the situation is not correct any longer. So, the robot must face locally an obstacle 'which should not have been there'. The algorithms of so-called 'local'guidance can solve the problem.

Naturally, the two guidance principles function co-operatively, handing over to each other the control of the events, depending on the situation encountered. On the basis of an overall picture, the global guidance defines the main directions of advance and can therefore optimize an overall criterion. It is called path or route planning. Global guidance generates thus a set of way points, or even a precise trajectory. The local guidance is brought under control of those instructions, while recording information regarding the environment discerned by the robot via the sensors of the obstacle detection function. The local guidance allows therefore to reach the objectives determined by the global guidance while avoiding all the obstacles that were not known beforehand, but detected. This is known as « obstacle avoidance ».

The general principle that has been adopted consists in working on graphs. If the environment through which the robot moves is more or less known a priori (in the form of maps, aerial or satellite photographs, ground digital models), paths are then naturally apparent. It is therefore possible to extract a graph of the whole set of paths. It is implicitly understood that a path shown on the graph crosses an area that is passable for the robot, taking into account its kinematic capabilities and considering other criteria in connection the type of mission (such as discretion, survivability, amount of intelligence collected, radiofrequency or satellite visibility within the framework of a military mission), and free from obstacles.

Generating the graph of the possible paths is thus a high added value operation which must be achieved in mission preparation in a semi-automatic way, with the assistance of photo-analysis experts.

Each path appearing on the graph is designated in accordance with a number of criteria. Since the mission consists in going from the *Departure* point to the *Arrival* point, THE path that will meet the mission requirements at lesser cost should be extracted from the graph of the possible paths. Algorithm A* selected to perform this operation is a classical algorithm used for operational research.

Mission preparation transmits to the global guidance the graph and the list of the advance waypoints. During the progression, the global guidance manages the current waypoint to be aimed at, as a function of the current position of the robot, and delivers an order such as : direction to be followed, speed.

The local guidance will therefore try and achieve the course commanded by the global guidance algorithms. It will adapt locally the speed and advance direction of the robot as a function of obstacles unknown a priori, but detected by the sensors.

The principle on which the selected obstacle avoidance algorithm is based is the method called Vector Field Histogram (VFH) . Such a method consists in mapping the immediate environment discerned by the robot and in classifying as free / obstacle each cell of the map on the basis of the information supplied by the obstacle detection function. The only thing remaining to be done is to draw up the histogram of the density of obstacles, as a function of the lines of sight, and to steer in the direction where the density is below a preset threshold.

Two situations can occur :

- the density of obstacles detected in direction ψ_B 'recommended' by the global guidance is lower than the threshold; in that case the direction actually ordered to the control system is ψ_B. The local guidance must then confirm the set course, considering that there is no awkward, or even dangerous, obstacle nearby in that direction.

- the density of obstacles detected in direction ψ_B as 'recommended' by the global guidance is above the tolerance threshold; in that case the direction actually ordered to the control system is the direction closest to ψ_B in which the density of obstacles is below the threshold; *if there is such a direction!* In the opposite case (typically: detection of a cul-de-sac), initiative will be handed over to global guidance which must re-plan the mission : the point is to remove from the graph of the possible paths the one which frustrated the local guidance, and to look for another route to carry out the mission (provided that such a route exists; otherwise the mission will be unsuccessful).

Figure 1 details the structure of the simulation tool which has been implemented to evaluate the performance of the navigation/guidance channel.

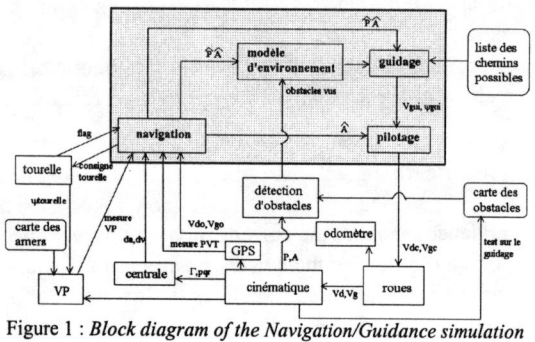

Figure 1 : *Block diagram of the Navigation/Guidance simulation tool*

3. The Vision channels

As mentioned before, the vision channels are more particularly concerned with :

- The « Update On Landmarks » (RA) function : the purpose of this function is to assess the feasibility, for a mobile robot, to update its position on the basis of the correspondence between an image and a 3D model, previously achieved off-line.

- The « Detection of Obstacles » (DO) function : this function is intended to assess the feasibility, for a mobile robot, to draw up a traversability map, where each element of the map bears attributes (height, slope, ...), and which is interpreted by the robot guidance algorithmics in terms of obstacles to be avoided.

- The « Estimation of Motions » (EM) function : on the basis of the 3D data made available at the output of the « Detection of Obstacles » function, the EM function is intended to only estimate through imagery the motion (translation, rotation) between two successive positions. The aim is twofold : estimate the motion speed to contribute to the navigation filter and make it possible to predict the position of the previous traversability map, and to merge it with the current traversability map; this is meant to obtain the best possible precisions in the calculated attributes.

3.1. The updating on Landmarks

3.1.1. General principles

On the basis of the following predefined requirements :
- using a system of passive stereoscopic sensors,
- using a 3D modeling of structured objects achieved during a preliminary phase of Mission Preparation (P.M.),

the main aims of the Updating on Landmarks function are as follows :
- recognition of a landmark described in the form of a 3D model projected into the images using a correlation technique,
- delivery, at the output of the image processing channels, of a landmark deviation measurement in connection with the line of sight, and the estimated conditions of photography.

The updating function is performed in three stages: « Generation of assumption - Verification - Estimation - 3D Position » :

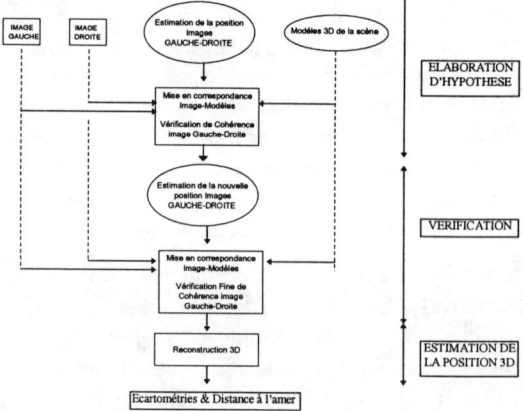

Figure 2 : *Block diagram of the Updating on Landmarks function*

3.1.2. Preparation of the 3D models

Among the main objects of the « Mission Preparation » function specific of the Updating on Landmarks function, the following can be noted:
- the definition of the Upadate points necessary for the resetting of the inertial navigation,
- for each updating point defined, the generation of the "3D environment model" necessary for the image-model correlation phase in the processing operations.

The description of the updating landmarks in the form of 3D environment models requires therefore source data of the site, which are : aerial or satellite images of the site, an adequate number of georeferenced points present in those images, completed plans of the site (optional).

Moreover, for performance evaluation of the image processing and in order to take into account the accuracy of the mission preparation data, each « reference 3D model » is downgraded in order to achieve so-called « aerial » 3D models (intermediate degradation) or « helios » 3D models (maximum degradation). These names assume that aerial image data (type IGN 1/25000) or satellite image data (hélios) are available during the mission preparation phase.

A single landmark is selected for all the trajectories. Only the angles and the distance at which the landmark is seen are different.

Test trajectories have been chosen arbitrarily in order to carry out a preliminary performance evaluation of the processing operations.

Three test trajectories are then defined along the 90°, 180° and 270° axes.

Three updating points at 55m, 75m, 95m from the landmark have been defined for each one of the trajectories.

Generation of the « 3D Edge models »

A refined geometrical description of each one of the updating landmarks is made based on its source data, so as to result in the achievement of the 3D models called « ground reference 3D models » (aerial or helios) the accuracy of which is consistent with the source data.

Then, as a function of the planned arrival axis, or more simply as a function of the predicted point of sight from which the the robot cameras will visualize the landmark during the mission, all the « ground reference 3D models » will be shaped in order to achieve models called "3D edge models ". Such a shaping consists in determining the model faces which will be visible (and consequently the the ones which will be hidden) in order to allocate visibility attributes to each constituent segment of the 3D model. Example of a perfect « 3D Edge model » for the 90° axis and a distance of 70-80m :

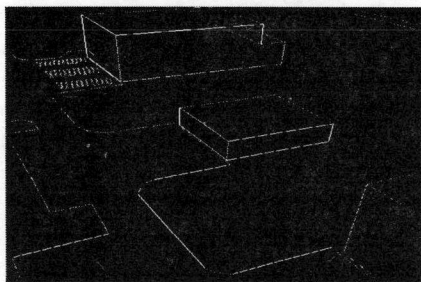

Figure 3 : 3D edge model

The attributes are used to project into the image the model segments which have an adequate percentage of visibility and make therefore a correlation phase (between the segments originating from the scenery image and the ones originating from the projected model) more reliable. Three types of "3D Edge models" are therefore available : perfect, aerial, helios.

3.1.3. Images - 3D models matching

The aim is to recognize the landmark within the image. The general principle of the processing operations is as follows :

On the basis of the position and the attitudes of the turret as estimated by the navigation filter, and of the predicted position of the landmark, a line of sight will be calculated allowing to point the turret in the direction of the landmark (fig. 4).

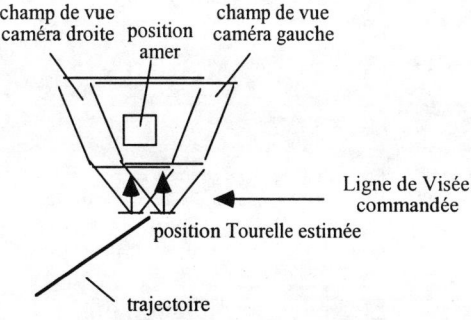

Figure 4 : Orientation of the line of sight

Then, a pair of LH and RH images of the landmark is generated (Figure 5).

LH Image   RH Image
Figure 5 : (Trajectory Axis 90° - Distance 70m)

The "3D edge models", generated for the estimated position, is then projected in both images and constitutes the predictive LH and RH images of the scenery (projected models segments (in green)).
At the same time, extraction of the contours is performed in both images (images segments).

Figure 6 : Example of a pair of stereoscopic images (Trajectory Axis 90° - Distance 70m) and of the extracted Contours (in red colour) :

In each one of the two images, a correlation operation is performed between the images segments and the projected models segments (figure 7). On completion of this operation, the position of the landmark is corrected in the LH and RH images.

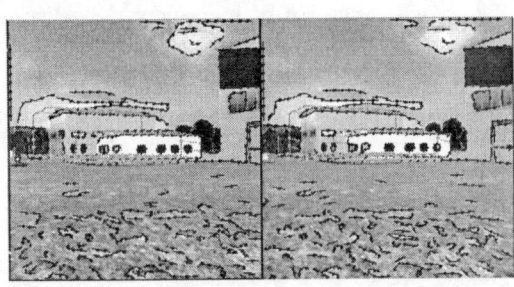

Figure 7 : Example of a pair of stereoscopic images (Trajectory Axis 90° - Distance 70m) and an image segments(red) - model segments (green) correlation; examples of matching arcs between the segments appear in light blue

3.1.4. Strategy : formulation of hypotheses - verification

The hypotheses formulation stage consists then in making sure that the results of the two correlations are « compatible »; the landmark point found in the RH image (LH resp.) is reprojected into the LH image (RH resp.). A measurement of the landmark localization coherence is made in both images. In the case of a coherence test, two new images are then required.
If the result of the coherence test is correct, then a « simplified » estimation is made of the photography conditions (C.P.D.V). of the RH images (LH resp.). The assumption verification stage consists in using the C.P.D.V. data obtained previously and in making sure that the results of the two correlations are still « compatible »; The Landmark point found in the RH image (LH resp.) is then reprojected into the LH image (RH resp.). A fine coherence measurement of the landmark localization is made in both images. In the case of a negative fine coherence test, two new images are then required . It is then possible to move on to the last stage of the processing operations : the stage of 3D Position Updating.

Figure 8 : Example of a correlation result for a pair of stereoscopic images (Trajectory, Axis 90° - Distance 70m); the « 3D Edge Model » is finally positioned in the images after the Assumptions Generation-verification steps; it appears in dark blue colour :

3.1.5. Performance Data

The performance of the Updating on Landmark function will be measured using two types of parameters :

- the first one consists in measuring the success percentage of the function,
- the second one consists in measuringthe accuracy of the error angles (Turret-Landmark) as obtained in the case when the function has not been declared faulty.

Performance evaluation has been achieved using the Bordeaux Issac synthetic image data base created by AEROSPATIALE. The test images have been generated for :
- 3 test trajectories (axis 90°, 180°, 270°);
- a distance ranging between 50m and 100m;
- 3 types of « 3D Edge models » : « perfect », « aerial », « helios »;
- 1 single landmark;
- a localization precision at 3 sigmas of 9m horizontally and 3 m vertically and 3° over all Euler's angles (heading, elevation and roll).
On the whole, this amounts to 4500 Updating Measurements performed using computer-generated images .

Percentage of success : results

The percentage of success of the Updating on Landmark function changes depending on the test trajectories : it depends on the arrival axis and on the type of « 3D Edge model » used. Figure 9 gives the percentage of success for the test trajectories.

	Axis 90°			Axis 180°			Axis 270°		
50 - 60 m	0.55			0.7	0.65	0.3	0.4		
70 - 80 m	0.7			0.6	0.5	0.4	0.6		
90 -100 m	0.66			0.6	0.6	0.38	0.3		

Figure 9 : % of success of the updating function

| Perfect 3D Edge models |
| Aerial 3D Edge models |
| Helios 3D Edge models |
| --- non considered cases |

Legend

The results obtained for axis 270° and a distance of 90-100m are poor because of the presence of trees in the data base, which obstructs the view for certain positions.

In the aggregate, the updating function is robust and the rate of success of the function is greater than 50%

Angular Update Accuracy : results

The accuracy of the Landmark-Turret deviation measurements changes according to the test trajectories : it depends on the axis of arrival and on the type of « 3D Edge models » used.

The measured values are the circular angle errors and elevations. These values characterize the precision with which the direction of the landmark is measured in the turret reference mark.

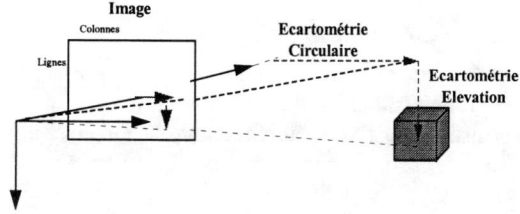

Figure 10 : definition of the circular and elevation angle errors

Figure 11 shows a graph representing the precision of the angle error measurements as obtained in azimuth and elevation as a function of the distance to the landmark for all statistical cases.

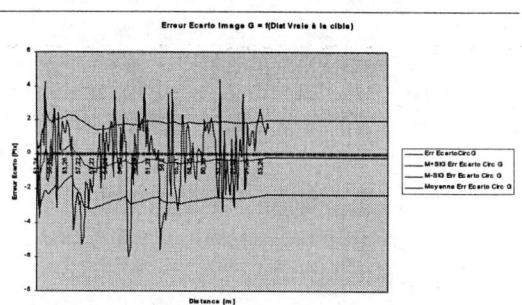

Figure 11 : angle error measurement in azimuth

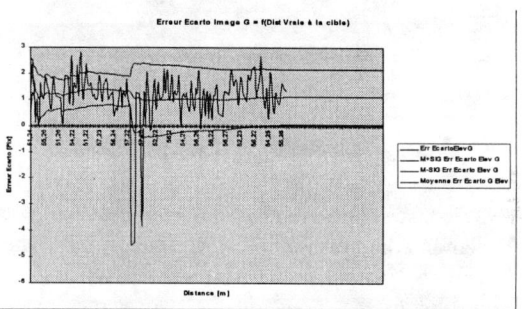

Figure 11(bis) : angle error measurement in elevation

They are characterized by a mean value and a standard deviation as a function of the distance and the axis of arrival. The errors in the deviation measurements are smaller in elevation. This is essentially due to the fact that, in the image, the model segments match together with less ambiguity along the vertical axis; the building roofs are discriminant segments which secure relative stability of the matchings along the vertica axis. Along the horizontal axis, the various vertical sides of the buildings maintain some ambiguity and are the main cause of the lack of precision in the measurements.

In short, all trajectories being taken together,
- the maximum deviation measurement error is :
  Circular : sigma = +/- 1.5°
  Elevation : sigma = +/- 1°
- the minimum percentage of success of the Updating function amounts to : 50%.

3.1.6. Estimation of the 3D position

Principle : The last stage of the processing operations consists in performing an Updating in 3D Position. This step consists in reconstructing 3D the point of the landmark observed using a stereoscopic vision technique. The pair of stereo cameras (Fig. 12) is supposed to be calibrated and the images are rectified to make up for the distortions due to the lenses.

The characteristics of the stereoscopic pair are selected in accordance with :
- the minimum size of the landmark,
- the navigation error (position and attitudes),
- the desired 3D reconstruction accuracy.

In order to achieve a 3D reconstruction accuracy close to one metre at 50m, the precision of the stereoscopic correlation will have to be at sub-pixel level.

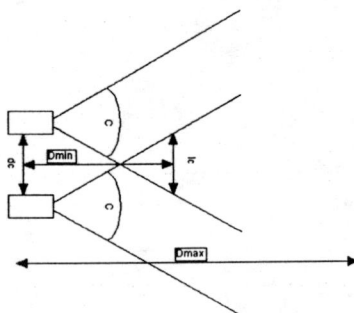

Fig 12: Geometry of the stereo system

The reconstruction principle consists in using , for the landmark point seen by the stereo pair, the difference in positions observed between the projection of that point on the two stereo image planes; at the end of the hypothesis Formulation-Verification stage, we have therefore 2 positions of the landmark aimpoint (1 in each LH and RH image). The variation between the 2 positions is called the disparity. In the case of a calibrated pair of stereo cameras and rectified images, there is a simple relation between distance Z of a point to one of the 2 cameras of the stereoscopic pair and disparity [D] : $Z = f \cdot Tx / D$. Calculations show that with a disparity error of 1 pixel, the reconstruction error amounts to nearly 9m at 50m ! The accuracy at the end of the hypothesis Formulation-Verification stage being on the order of a few pixels in each one of the LH-RH images , we must therefore select a technique allowing to achieve a sub-pixel accuracy in the disparity measurement in order to allow an accurate 3D reconstruction .

3.1.7. 3D Reconstruction

Using the position of the landmark given in the LH image, a small 7x7 image is extracted around that point. Then, using a correlation algorithm, the aim is to try and have the small image « coincide » about the same landmark point designated in the other image. The properties of the epipolar geometry (Calibrated cameras and rectified images) allow then, based on the disparity, to trace back to the 3D coordinates of each aim point. In order to make the calculation more robust and elimate matching errors, the same operation will be performed in a reciprocal manner, starting from the landmark point designated in the RH image.

A 3D filtering, consisting of a weighted average, is then performed on both reconstructed 3D points in order to obtain the combined 3D landmark point.

Performance data :

The performance data concerning the 3D reconstruction have been obtained only using sequences of computer-generated images, simulating perfectly calibrated caméras and perfectly rectified images .

The reconstruction accuracy of the position of the robot turret in the T.G.L. reference mark is variable in accordance with the test trajectories : it depends on the quality of the pre-positioning of the landmark as supplied by the Formulation-Verification phase.

Precision of Recons 3D	Axis 90°		Axis 180°		Axis 270°	
50-60m	-0.2m +/- 2.5m	0m +/- 1.7m	-0.1m +/- 1.9m	0.2m +/- 1.8m	0.1m +/- 2.0m	
70_80m	-0.9 +/- 3.3m	-1.5m +/- 2.6m	-1.4m +/- 2.5m	-1.6m +/- 1.8m	-1.0m +/- 2.6m	
90-100m	-0.2 +/- 2.6m	-0.25m +/- 2.8m	-0.4m +/- 2.3m	0;7m +/- 2.2m	-0.9m +/- 3.3m	

Figure 13 : Precisionsof 3D reconstruction

The 3D Reconstruction accuracies range between 1.7m and 3.3m depending on the axes of arrival and the distances (Fig. 13).

3.1.8. Real time installation

Real time installation of the Assumptions Formulation-Verification algoritms is currently underway. It is shown that it is possible to maintain a computing rate of about 1 Hertz on a an architecture consisting of an assembly of two PowerPC processors at 200 Mz.

The computing time planned for the 3D reconstruction phase negligible.

The algorithmics planned to perform an Update-on-Landmark function is therefore real-time compatible.

3.2. Detection of Obstacles

The aim of the DO function within the vision/perception module of a mobile robot is to construct in a fixed reference mark a traversability map of the environment perceived by the robot. Such a map will be made through a 3D reconstruction of the scenery as seen by a pair of stereoscopic sensors and through fusion of the various data acquired while the robot is moving.

The 3D reconstruction phase consists in reconstructing in three dimensions the scenery observed through a stereoscopic vision technique. The pair of sereo cameras is supposed to be calibrated and the images are rectified o make up for the distortions caused by the optical elements.

The reconstruction principle consists in calculating, for each point of space seen by the stereo pair, the displacement observed between their projection on the two stereo image planes. This value, called disparity, is calculated by a correlation algorithm which allows, for each point of an image of the stereo pair, to find the abscissa of the same point in the other image. The properties of the epipolar geometry allow then, based on the disparity, to trace back to the 3D coordinates of each aim point.

In order to make the calculation more robust and elimate matching errors, an additional check will be performed by means of a reciprocal correlation followed by a screening on the correlation score quality.

Finally, a sub-pixel accuracy of the disparity will be obtained for each point selected through parabolic approximation over 3 points about the correlation minimum values.

- The percentage of correlated points amounts to 70% approximately (the non-correlated points correspond to « mismatches »),
- The reconstruction errors in depth σz are less than 2.5 m between 0 and 30m: σz < 2.5m from 0 to 30m. This result shows that there is in fact a sub-pixel uncertainty σd concerning the disparities. Indeed, using relation $\sigma z = (z^2/dc*f)*\sigma d$ we obtain, at a distance of 30m, σd ≈ 0.4. We can therefore consider that under the same conditions, using a 512x512 pixels matrix, the reconstruction error between 0 and 30m will be then σz < 1.3m

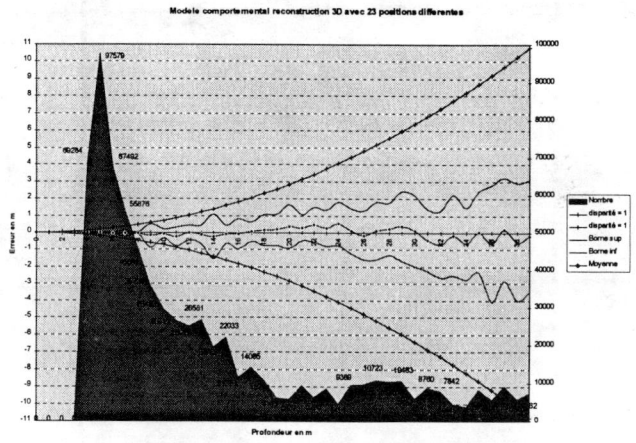

Figure 15 : Rreconstruction accuracy for a 256x256 pixel resolution image, field: 40°, and inter-cameras distance: 0,4m.

3.2.1. Local traversability map

The production of a local traversability map consists in meshing the terrain in front of the robot and in classifying each cell so constructed. The sizes of the cells may increase as a function of the distance to the robot in order to secure the presence of an adequate number of points in each cell.

The classification of the cells of the local map is done according to the following attributes : « obstacle », « non obstacle », « unknown ». This is achieved on the basis of a simplified modelling based on the number and the max., min., and mean heights of 3D points placed above each cell.

Such a modelling will produce good results on the simplifying assumption that the terrain is flat. A more sophisticated modelling must be envisaged when the terrain topology requires a finer classification of the cells (slope, step, ...). The meshing that has been adopted is made up of square cells, the smallest ones having a size of (50x50) cm. In this matter, the sizing factors are mainly the amount of data to be managed and the surface density of the reconstructed points (in connection with the configuration of the stereo pair).

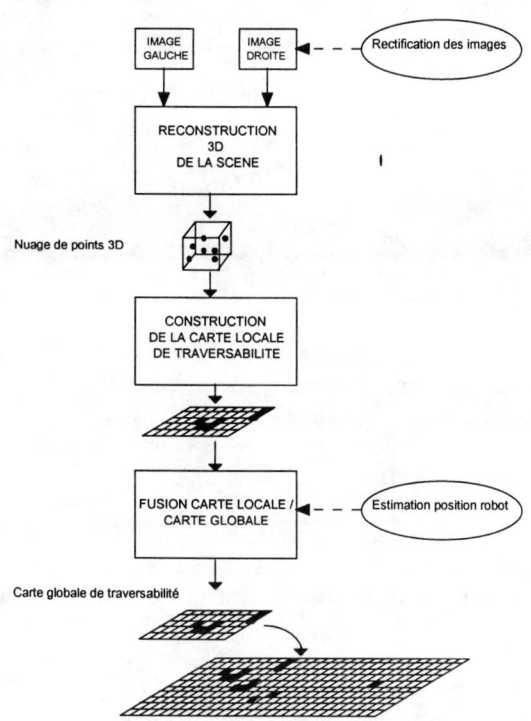

Fig. 14 : Block diagram of the DO processing operations

The performance data have been assessed on sequences of computer-generated images having a size of 256x256 pixels (Figure 15).

The detection and false alarm (FA) rate (i.e. of cells erroneously determined as obstacles) is evaluated quantitatively in the next paragraph below. From a qualitative point of view, figure 16 shows that such FAs (sky-blue colour) are localized aroud the obstacles, which is not in any case penalizing since the robot is programmed to avoid them. On the other hand, the « obstacle » cells that are not detected (red colour) are, for most of them, located in masked areas of the field of view of the two cameras.

3.2.2. Local and global maps fusion

The robot's navigation module takes its bearings within a fixed reference frame independent of the rover motions. The traversability maps as reconstructed locally by the robot must be therefore replaced into such a « global referential system ». Such a fusion allows also to reconstruct in a more reliable way the environment of the robot through a time screening of the various results obtained locally. The fusion can be done either at the level of the attributes of the cells, or direct at the level of the attributes of the 3D points. The method that has been adopted is based on the fusion of the local 3D points attributes with the attributes of the cells in the global map. In this context, the construction of a local map is interesting only tosuuply pinpoint data for the purpose of the robot's reflex actions (emergency stop, ...).

Figure 16 : Local traversability map

Figure 17 shows an estimation of the FAs space distribution with respect to the obstacles. The mean variation is of the same order of magnitude as the 3D points reconstruction accuracy ; at 30m from the robot, for instance, a mean variation below 2m can be observed.

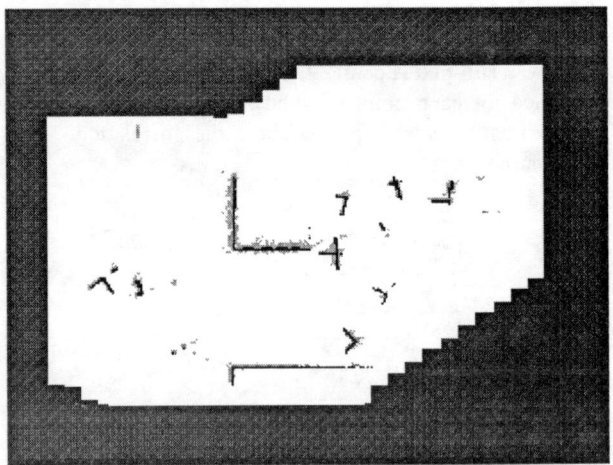

Fused global traversability map

Fusion of the local data requires that the robot position within the global (or local with respect to a reference position) reference frame should be known. The drift of the inertial sensors leads us to reset the robot position by means of vision techniques : updating on landmarks, motion estimation.

The performance data related to the making up of the global maps have been evaluated on sequences of computer-generated images, based on a precise knowledge of the robot travel (figure 18).

- The detection percentage, i.e. the number of obstacle-cells detected with respect to the total number of obstacle-cells, is estimated at 75 %.
- The FA rate , i.e. the ratio of the number of cells erroneously declared obstacles to the total number of cells declared obstacles, is estimated at 73 %. Those FAs are mainly due to reconstruction errors and are localized around the true positions of the obstacles.

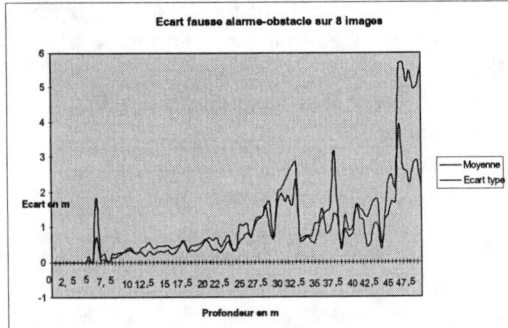

Figure 17 : Distance between false alarms and obstacles

- The percentage of error, i.e. the ratio of the number of wrongly classified cells to the total number of cells on the global map, is estimated at 3,7%.

3.3. Motion estimation

Determining the motion of the robot is a basic requirement for the navigation tasks and for the fusion of the environment perception maps. The two most simple techniques to acquire motion data are the use of odometers or of an inertial platform. Odometry is an intrisically little precise technique that is liable to significant errors when the grip of the wheels on the ground is not too good, which is often the case. The use of an inertial platform proves also unsatisfactory in the case of a mobile robot for which the point is to estimate motions of a few metres over periods of time of a few seconds: indeed, in such a case the time drift of an inertial platform is far from being negligible with respect to the robot natural motion.. Obtaining measurements with an adequate accuracy would require the use of types of platforms that are extremely expensive. These two types of sensors must be therefore supplemented by other processing operations so as to obtain an adequate motion measurement accuracy.

Using visual sensors, two types of position measurements can be made. The absolute position of the robot can be determined through recognition of the environment in which the robot is moving : this is the function performed by the Update-on-Landmark task presented above. This function offers the advantage of supplying an absolute position of the robot, but, on the other hand, it is necessary to have a model of the environment.

A second type of measurement can also be obtained through visual processing, the measurement of the relative motion between several positions of the robot (Figure 19). Such a processing offers the advantage of not requiring a prior knowledge of the environment. It can also be used both in a structured environment (buildings, ground installations) and in a non-structured environment (natural ground), which provides a very wide operating range.

The measurements supplied by the processing operations of motion estimation by vision have characteristics that are very different from those delivered by an inertial platform, more particularly in terms of drift : a motion measurement does not drift over time, as it is the case with an inertila platform, but only over the distance covered, since it is based on an estimation of the movement between two views of the same scene observed from two different positions.

The mobile robot is fitted with sensors based on a stereoscopic vision allowing it to have a 3D perception of its environment, more particularly for the measurement of the position of potential obstacles. The function of estimation of relative motions is based on the use of the same stereoscopic sensors. In addition to the fact that it does not require any additional sensor, such a function utilizes the results delivered by the 3D perception function operated for the detection of obstacles, which allows to limit the costs in terms of processing operations.

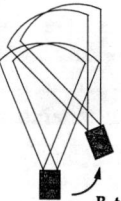

Figure 19 : Relative movement between two positions of the robot

The processing operations sequence used for motion computation is as follows (Figure 20) :

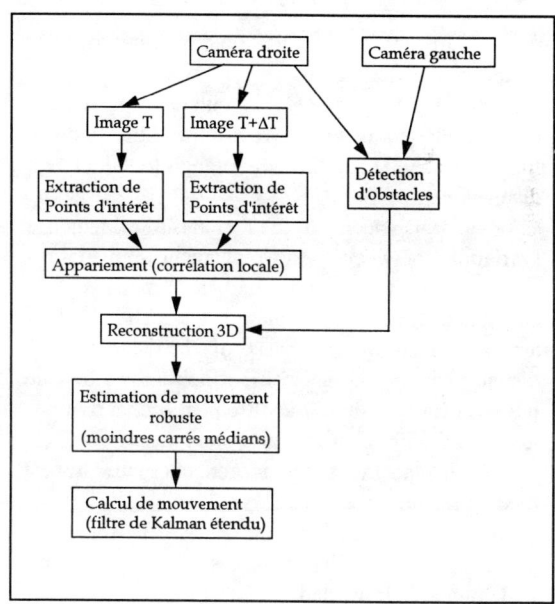

Figure 20 : Block diagram of the motion estimation function

1. An extraction of points of interest is performed independently on each one of the images taken by the RH camera in the successive positions. The aim of this step is to select a small number of well identified points, on which the following processing operations will be performed, which reduces considerably the complexity of the calculation.

2. A correlation is made between the points of

interest extracted at two successive positions. Such a correlation is performed through a simple correlation processing of the levels of grey in a window about the each point of interest. If a motion estimation is available (in the case when there is an inertial platform or odometers), it may be used to reduce the matching search envelope, and therefore the algorithmic complexity. At the end of this stage, there are consequently a series of *2D matches* available between two successive positions.

3. The results of the obstacle detection function (and more precisely of the 3D reconstruction processing operations) are used to obtain the the 3D coordinates of the matched points, as well as the associated co-variances. On completion of this stage, there are consequently a series of *2D matches* available between two successive positions, with a measurement of their uncertainties.

4. A robust motion estimation is performed using the so-called *least median squares* statistical technique and an extended kalman filter. The principle of such a method, which allows to obtain a robust calculation in the presence of a rate of erroneous data amounting to 50%,is as follows :

- A series of matching triplets is randomly selected. The number of selected triplets is determined in such a manner that the method guarantees a robust calculation it un calcul robuste in the presence of a rate of erroneous input data amounting to 50% (inaccurate matching).

- Each triplet is used to calculate a movement between the two positions (rotation and translation). Indeed, three 3D matches are enough to calculate a displacement.

- For each displacement, the residues of all matches (variations between the displacement applied to a point and the point with which it is matched) are calculated, then their median values. The residues are calculated on the basis of the Mahalanobis distance (statistical distance), which allows to take into account the mearurement uncertainties associated with the matching operations.

- The displacement which generates the lowest median residue is selected.

- All erroneous matching are detected by means of a threshold on the residues. The erroneous matching are eliminated.

- An extended Kalman filter is used to determine the displacement, based on the matches that have been kept by the robust estimation phase.

3.3.1. Performance data

The performance data of the motion estimation function have been evaluated over a complete trajectory of the robot, performed in computer-generated images . Each motion of the robot corresponds either to a 1 to 5 metre axial translation, or to a 10 to 20 degree rotation. The total displacement of the robot amounts to some 20 metres and some 60°.

An analysis of the results obtained (fig. 20 & 21) allows to bring out the following points :

The standard deviation of the axial translation measurement (direction of travel of the robot) is of the order of 5 cm per displacement, over the whole of the displacements. Such a standard deviation appears to be strongly connected with the amplitude of the displacement between two views. Such a behaviour is logical since a more important displacement necessitates the use of more distant points for the calculation of the movement. Since the uncertainty on the 3D points increases as the square of their distances, one can expect a quadratic degradation of the measurement of the motion over the distance travelled between two successive positions.

Between two successive measurement positions, a displacement below 2 m is desirable for an optimum measurement accuracy. In that case, the standard deviation of the axial translation measurement amounts to 1 cm, the mean error on a trajectory is 16 cm over 13 m.

The standard deviation of the lateral translation measurement is 4 mm per displacement (for displacements less than 5 m), the mean error on a 13m trajectory is negligible (3 mm).

The standard deviation of the rotation measurement is 0.02 degree per displacement, the mean error on a trajectory is 0.15 degrees on 64 degrees.

Axial transl. (m)	Rot (°)	Axial trans error (m)	Lateral translation error (m)	Rotation error (degrees)
2	0,0	0,027	-0,014	-0,053
2	0,0	0,024	-0,009	-0,041
0	11,3	0,000	0,004	0,000
0	14,7	-0,001	0,000	-0,034
0	19,0	0,001	0,003	-0,028
0	19,0	-0,003	0,002	-0,042
0	15,3	0,000	0,002	-0,026
0	10,7	0,003	0,002	-0,018
5	0,0	0,162	0,026	0,036
5	0,0	0,179	-0,009	0,005
0	-10,7	-0,001	0,000	0,011
0	-15,3	0,003	0,000	0,031
1	0,0	0,013	0,000	0,004
1	0,0	0,033	0,001	0,008
1	0,0	0,008	0,002	0,005
1	0,0	0,014	0,003	0,008
1	0,0	0,010	0,003	0,010
1	0,0	0,015	0,000	0,004
1	0,0	0,007	0,000	0,004
1	0,0	0,002	0,003	0,007
1	0,0	0,010	0,001	0,000

Standard deviation		0,050	0,007	0,024
Cumulat.	23	64,0	0,504	0,020	-0,108

Standard deviation (displa. <5m)		0,010	0,004	0,023
Cumulat (displa. <5m)	13	64,0	0,164	0,003	-0,149

Figure 20 : real translations and rotations measured along along a trajectory

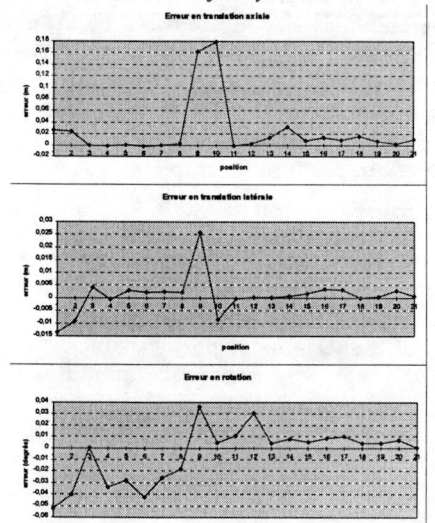

Figure 21 : Errors in the estimation of the translation and rotation along a trajectory

4. Conclusion

Navigation of autonomous robots by vision techniques is a promising approach in terms of cost/performance. The simulations of updating-on-ladmarks, detection of obstacles and motion estimation channels described in this paper are in the process of validation on real images, in open loop. The full validation of our approach will be performed on a small robot in order to demonstrate the feasibility of the autonomous navigation function on a computer architecture based on the PC/PCI standard. Tests on various scenery contexts (global ground, urban or country type environments) are envisaged to define the limits of performance and the optimum operating conditions for the MONAI.

5. BIBLIOGRAPHY

1. Vision stereoscopique et perception multisensorielle. Aplication ‡ la robotique mobile. **N.ayache. Inter-editions science informatique**

2. Rapport de recherche : real time correlation-based, stereo : algorithm, implementations and applications **o.faugeras , b.hotz , c. Proy . Inria**

3. Etude de technique de stereovision par correlation. Application au programme vehicule autonome planetaire (vap). **B.hotz. Cnes**

4. autocalibration d'un capteur stereoscopique. **Radu horaud, gravir-imag.**

5. A parallele stereo algorithme that produces dense depth maps and preserve image features. **P.fua. Inria**

6. Rapport de stage: stereoscopie **scherrer. Aerospatiale**

7. Quantitative and qualitative comparison of some area and feature-based stereo algorithme. **O.faugeras , b.hotz , p.fua , l.robert , m.thonnat , z.zhang. Inria**

8. Adaptive window algorithme for aerial image stereo. **Lotti, g. Giraudon. Inria**

9. Techniques de reconstruction 3d par stÈreovision. **P.bonnin. Credoc**

10. A reactive system for off-road navigation. **D.langer, j.k.rosenblatt, m.hebert. Robotic institut, pitsburg**

11. 3d autonomous navigation in a natural environnement. **F.nashashibi, p.fillatreau, t.simeon. Laas-cnrs**

12. The manhattan method: a fast cartesian elevation map reconstruction from range data. **P.ballard, f;vacherand. Leti / cea grenoble**

13. Autonomous navigation in outdoor ENVIRONNEMENT: ADAPTIVE APPROACH AND EXPERIMENT. **S.LACROIX, R.CHATILA, T.SIMEON. LAAS-CNRS**

Fast Evaluation of Geometric Constraints for Bending Sequence Planning

M. Inui and H. Terakado

Dept. of Systems Engineering
Ibaraki University
Naka-narusawa-cho 4-12-1
Hitachi-shi, Ibaraki 316-8511, JAPAN

Abstract

In the sheet metal part manufacturing, bending is the most commonly used to realize sufficient rigidity and to obtain a part of desired shape to perform a certain function. Feasible bending sequences of a sheet metal part can be determined by successively applying bending simulations to the geometric model of a blank. Unfortunately, this simulation based method is computationally expensive because enormous search-space must be explored if complex parts with many bending lines are given. Complex parts usually have rather small number of feasible bending sequences because other sequences cause self interferences or interferences with tools in the bending operations. Based on this characteristic of the sheet metal part, we develop a new method to accelerate the simulation based bending sequence planning. This method reduces the computation cost by efficiently detecting the interferences in the early stage of the search process. In the bending sequence planning, the same geometric computations for the interference detection are repeated many times. This redundancy is eliminated by saving and reusing the previous computation results. The proposed method is implemented and an experimental bending sequencing program is demonstrated.

1 Introduction

Blanks produced by cutting are used to manufacture sheet metal parts according to a series of forming operations, such as bending, flanging and hemming. In these operations, bending is the most commonly used to realize sufficient rigidity and to obtain a part of desired shape to perform a certain function. Bending operation usually involves putting a blank on the die of the bending machine, and then the punch goes down (or the die comes up) to bend the blank. Figure 1 illustrates a bending operation with a punch and a V-shaped die.

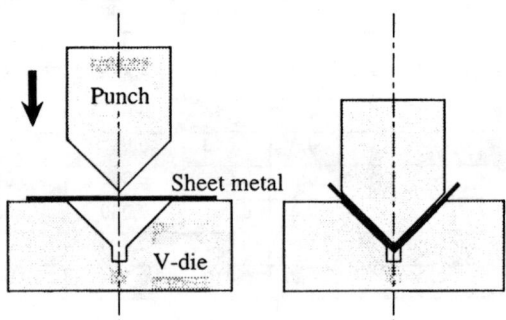

Figure 1: Bending operation with a punch and a V-shaped die.

Process planning of the bending operation, especially the determination of the bending sequence is important as a properly determined sequence is effective in reducing the manufacturing cost and time. Many CAD systems provide a function for modeling the sheet metal part. Several softwares have been developed for stamping and nesting the part to produce the blank [1][4]. On the other hand, most of the process planning tasks are still manually performed by manufacturing engineers. The number of possible bending sequences is very large even for a moderately complex part. A few days are often needed by a human expert to prepare a complete process plan.

The sequencing for bending sheet metal parts is highly geometry-dependent. A small modification of the part often causes completely different sequences. Therefore, using the variant approach to generate the bending sequence may not work. Feasible bending sequences of a sheet metal part can be determined by successively applying bending simulations to the geometric model of a blank. Unfortunately, this simulation based method is computationally expensive because enormous search-space must be explored if complex parts with many bending lines are given.

Only few studies have been reported for automating the bending sequence planning. Inui et al. originated the simulation based planning with the geometric model of a sheet metal part [3]. They proposed a tolerance analysis based method for the search-space reduction. Wang and Bourne developed another automatic process planning method [5]. They interpreted a sheet metal part as a set of features and used precedence heuristics and constraints associated with the features to help search for feasible bending sequences.

Complex parts usually have rather small number of feasible bending sequences because other sequences cause self interferences or interferences with tools in the bending operations. Based on this characteristic of the sheet metal part, we develop a new method to accelerate the simulation based bending sequence planning. This method reduces the computation cost by efficiently detecting the interferences in the early stage of the search process. In the bending sequence planning, the same geometric computations for the interference detection are repeated many times. Our method eliminates this redundancy by saving and reusing the previous computation results. Similar method is already adopted for the mechanical assembly planning. Successful results of the application are reported in [6].

The organization of the paper is as follows. After a brief explanation of the geometric model of the sheet metal part and its modification method in bending, outline of the simulation based bending sequence planning is explained in the third section. In the fourth section, some methods for accelerating the interference detection are described. Based on our developed algorithm, a bending sequencing program is implemented. Results of the computational experiments are given in the fifth section.

2 Sheet metal part model

2.1 Model representation

Bending has a distinct characteristic of stressing the sheet metal only at the cylindrical area of the bending radius [2]. During the bending, the localized area is stressed in tension on one surface and in compression on the other. The remaining metal is not stressed during the bending, and it retains its shape and contour unchanged. It is therefore possible to represent a sheet metal part as a set of stiff and thin solid models representing the unchanged regions, and bending axes representing imaginary centerlines of bending rotations of the thin solid models. In the following discussion, they are called plates and bends, respectively [3].

When a cross-section is made through the stressed area, the line of zero stress called the neutral axis is given. The neutral axis preserves the true representa-

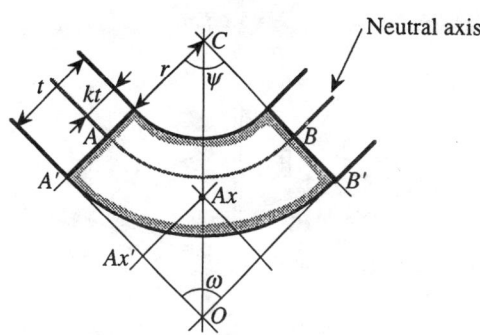

Figure 2: Definition of the centerline of a bending rotation.

tion of the original blank length. The imaginary centerline Ax of a bending rotation is determined based on the neutral axis. In the bent condition, distance between A' and Ax' in figure 2 becomes the half of the length of the neutral axis. By using the following geometric equations,

$$A'Ax' = (r + kt)\frac{\pi\psi}{360}$$
$$A'O = (r + t)\tan(\frac{\psi}{2})$$
$$CO = \frac{r+t}{\cos(\frac{\psi}{2})}$$

distance between C and Ax is derived as $CAx = CO\frac{A'Ax'}{A'O}$. A ratio k defining the neutral axis is usually around 0.3. The rotation angle $\psi = 0°$ corresponds to "flat" condition of two plates connecting to the bend, and $\psi = 180 - \omega°$ corresponds to "bent" condition where ω means the angle between the plates.

2.2 Model modification in bending

Topological relationships between bends $\mathcal{B} = \{B_i\}$ and plates $\mathcal{P} = \{P_i\}$ of a sheet metal part $\mathcal{S}$ compose a tree structure $\mathcal{S} = (\mathcal{P}, \mathcal{B})$ with the plates as nodes and the bends as links [5]. To each component plate P_i, we embed a local coordinate frame F_{P_i} representing the position and orientation of P_i with respect to the world coordinate frame F_W. Let $\mathbf{H}_{P_i}^W$ be a homogeneous transformation matrix which maps the axes of F_W to F_{P_i}.

Shape of a sheet metal part $\mathcal{S}$ in a certain bent condition can be represented by properly defining transformation matrices $\mathbf{H}_{P_i}^W$ for all component plates $P_i \in \mathcal{P}$. Consider a simple sheet metal part model with only two plates P_j and P_k linked by a bend B_i. A local coordinate frame F_{B_i} is fixed to each bend $B_i \in \mathcal{B}$ as its z axis becomes identical to the imaginary rotation axis. Let $\mathbf{H}_{B_i}^{P_j}$ and $\mathbf{H}_{B_i}^{P_k}$ be homogeneous transformation matrices that represent the position of F_{B_i} with respect to F_{P_j} and F_{P_k}. These two transformations can be computed using the CAD model of the blank.

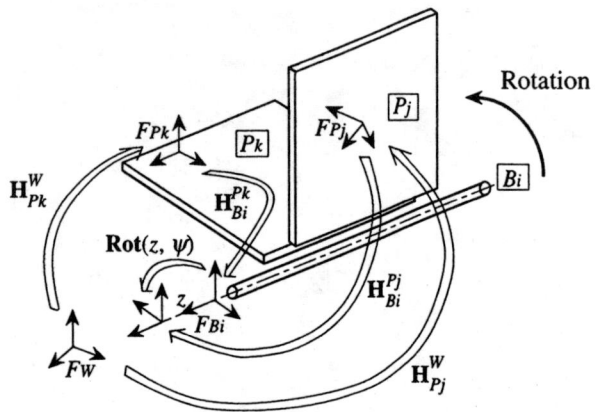

Figure 3: Spatial relationships between coordinate frames specified on a bend B_i and its connecting plates P_j and P_k.

When plates P_j and P_k are in the flat condition, the following equation is satisfied between the transformation matrices;

$$\mathbf{H}_{P_j}^{W}\ \mathbf{H}_{B_i}^{P_j} = \mathbf{H}_{P_k}^{W}\ \mathbf{H}_{B_i}^{P_k} \quad (1)$$

In our model, modification of the sheet metal part according to a bending operation on B_i is simulated by a rotation of one plate (for example P_j) around the z axis of frame F_{B_i} by $\psi = 180 - \omega°$. After the rotation, the following equation is satisfied (see figure 3);

$$\mathbf{H}_{P_j}^{W}\ \mathbf{H}_{B_i}^{P_j} = \mathbf{H}_{P_k}^{W}\ \mathbf{H}_{B_i}^{P_k}\ \mathbf{Rot}(z,\psi) \quad (2)$$

where $\mathbf{Rot}(z,\psi)$ means a rotational transformation around the z axis. Consider that one plate P_j (or P_k) is fixed somewhere in the world coordinate frame F_W and its corresponding transformation $\mathbf{H}_{P_j}^{W}$ (or $\mathbf{H}_{P_k}^{W}$) is given. Then the position of the other plate can be computed by using equation (1) if they are in the flat condition or equation (2) if they are in the bent condition.

Shape of a more complex sheet metal model $\mathcal{S} = (\mathcal{P}, \mathcal{B})$ can be computed in a similar manner. As the positional reference of $\mathcal{S}$, one component plate $P_i \in \mathcal{P}$ is selected and its coordinate frame F_{P_i} is fixed somewhere in F_W by assigning a proper matrix to $\mathbf{H}_{P_i}^{W}$. Then, all plates linked to P_i are checked and their positions are determined using the method described above. This positioning process is recursively applied to the plates until the positions of all component plates of $\mathcal{S}$ are determined.

3 Bending sequencing method

Feasible bending sequences of a part can be computed by successively applying bending simulations to the bends of its geometric model in the initial blank state. The same sequences can be derived by a backward search based method which investigates feasible "straightening" sequences of the part starting from the model in the final bent state. The reversal of the straightening sequence becomes a feasible bending sequence. Since plates of the model are more densely placed in the final state than in the initial blank state, more self interferences and interferences with tools tend to occur near the final state. The backward search based method is thus preferable because infeasible sequences can be detected and discarded in the early stage of the search process.

3.1 Backward search tree

The backward search process of feasible straightening sequences of a part with n bends $B_i, i \in [0, n-1]$ can be represented as a tree whose links correspond to straightening operations on bends. Figure 4 illustrates a part of a search tree for $n = 4$. Numbers appearing near links mean the subscripts of their corresponding bends B_i to be straightened. The root node represents the final shape of the sheet metal part. The other nodes represent intermediate shapes of the part after straightening such bends corresponding to the path from the root to the nodes. The terminal nodes thus represent the flat unbent shape.

The search starts from the root node and it follows the links connecting to the node downward. In the link following process, various geometric constraints concerning the bending operation corresponding to

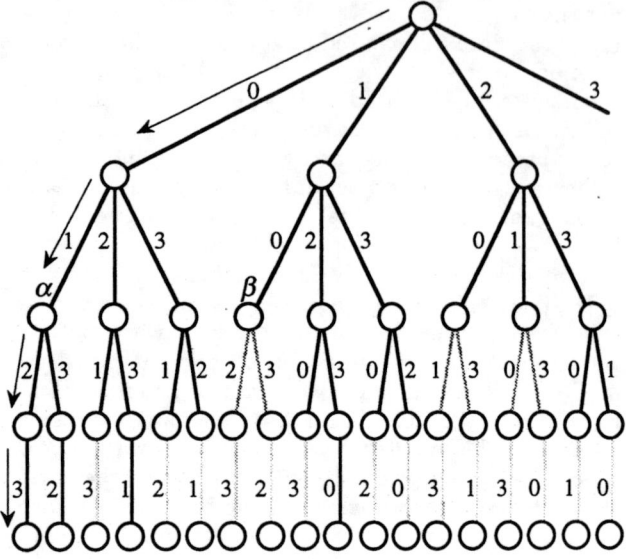

Figure 4: Part of a backward search tree of a sheet metal part with 4 bending lines.

the link are evaluated. If they are all satisfied, the sheet metal part model is straightened at the bend and the search proceeds, otherwise the following downward search is canceled. The search continues until it reaches a terminal node, and one feasible straightening sequence is obtained. In figure 4, the leftmost downward path represents a straightening sequence $B_0 \to B_1 \to B_2 \to B_3$. Its reversed sequence corresponds to a feasible bending sequence.

3.2 Geometric constraint evaluation

The following geometric constraints are evaluated in the search process to verify the applicability of a bending operation.

Tool accessibility: A pair of optimum punch and die to meet the bending requirements such as bending angle, bending radius and material is selected for each bend in advance. In the search process, accessibility of the selected tools are verified by placing their solid models at the pre-bending position and checking the interferences between the tool models and the sheet metal part model.

Self interferences: During the shape modification process in bending, some plates of the same part can have interferences. These self interferences are detected using sheet metal part models in the intermediate states corresponding to some discrete steps in the bending rotation.

Interferences with tools: For each discrete step in the bending rotation, interferences between the tool solid models and component plates of the intermediate state model are checked also.

4 Fast interference detection

The interference detection in the geometric constraint evaluation is the most time consuming task in the straightening sequence planning. We reduce the total computation time of the sequence planning by efficiently detecting the interferences. Some methods for accelerating the interference detection are discussed in this section.

4.1 Reusing previous detection results

In figure 4, node α in the search tree represents an intermediate state of a sheet metal part after straightening the bends in $B_0 \to B_1$ order. Node β corresponds to another state of the model after straightening the bends in $B_1 \to B_0$ order. Since the intermediate shape of a sheet metal part model depends only on the combination of the straightened bends and it does not depend on their order, these two nodes represent the same shape. Therefore, results of the geometric constraint evaluations corresponding to the downward links from these two nodes become the same.

We eliminate this redundancy by saving the constraint evaluation results at node α and reusing them when the geometrically same evaluation is needed again at node β. If we do not reuse the previous results, the constraint evaluation must be performed for all links of the tree visited in the search process. Some links in figure 4 are colored gray if they correspond to the bend straightening where reusing the previous constraint evaluation results is possible. Computations for such gray links become unnecessary if the previous results are reused.

4.1.1 Computation time analysis

The computation time of the simulation based sequence planning is proportional to the number of constraint evaluations performed in the search process. Without the saving and reusing mechanism, it is the same as the number of links of the search tree in the worst case. It becomes enormous for a complex part with many bending lines.

The computation time can be reduced to $O(n\,2^{n-1})$ (n means the number of bending lines) by reusing the results of the previous geometric constraint evaluations. Consider intermediate states of a sheet metal part whose i bends out of n are straightened in the prior search process. It has $\binom{n}{i}$ shape variations. The constraint evaluation is performed only once for each straightening operation of $n-i$ bends remained on the part. Results of the constraint evaluation are recorded in association with their corresponding intermediate shape. They are reused if the geometrically same straightening operations are repeated in the following search process. The total number of the constraint evaluations performed in the search process thus becomes

$$c = \sum_{i=0}^{n-1} \frac{n!}{(n-i)!\,i!}\,(n-i) = n\,2^{n-1}$$

$O(n\,2^{n-1})$ is still huge and the program with the saving and reusing mechanism is theoretically not efficient for large n. This improvement of the algorithm is, however, effective in the bending sequence planning purpose because most sheet metal parts have small number of bending lines. Many sample parts are investigated in our study and we find that the number of bending lines of them is usually less than 7. Complex parts with more than 10 bending lines are not found in our samples. Without the mechanism, 9 864 100 times of constraint evaluations are necessary for a part with 10 bending lines in the worst case. It can be reduced to $c = 5\,120$ by using the mechanism.

Manufacturing of parts with many bending lines is difficult and their feasible bending sequences are usually very limited. Therefore, the number of actual constraint evaluations is much smaller for most complex parts.

4.1.2 Implementation of the mechanism

Some implementation issues of the saving and reusing mechanism are briefly discussed. Results of the constraint evaluation are recorded in association with their corresponding intermediate shape of the sheet metal part. A unique bit sequence is generated for each intermediate shape as a key for recording and retrieving the results. For a part with n bending lines, a sequence of n bit length is prepared. A unique bit sequence can be determined based on the conditions of bends $B_i, i \in [0, n-1]$ of the intermediate state model. If B_i is straightened, 1 is set to the i-th bit of the sequence, otherwise 0 is set to the bit.

At each node in the search tree traversal, a bit sequence representing the intermediate state of the part corresponding to the node is generated. Before performing the geometric constraint evaluation, the database of the previous evaluation results is checked using the bit sequence as the key. If the previous results are retrieved, they are simply used without any real computations. Otherwise, interference detections are performed for evaluating the constraints. Evaluation results are then recorded with the key for the future retrieval.

4.2 Additional methods for the acceleration

The computation cost for the interference detection can be further reduced by using geometric characteristics of the sheet metal part and the bending operation.

- Bending a flat sheet metal does not cause any self interferences and interferences with tools. Therefore, geometric constraint evaluations are not necessary when only one bend is remained to straighten in the search process.

- A sheet metal part $S = (\mathcal{P}, \mathcal{B})$ is a tree with the bends as links. Consider two sub-trees $S_j = (\mathcal{P}_j, \mathcal{B}_j)$ and $S_k = (\mathcal{P}_k, \mathcal{B}_k)$ obtained by removing a link B_i from S. In bending B_i, plate groups $\mathcal{P}_j$ and $\mathcal{P}_k$ rotate around B_i while keeping the relative positions between plates in the same group. Therefore, self interferences of S occur only between a plate in $\mathcal{P}_j$ and another plate in $\mathcal{P}_k$.

- Figure 5 illustrates the shape modification of S according to a bending operation on B_i with a punch. A line pl in the figure represents the center plane of the punch. As shown in the figure, a

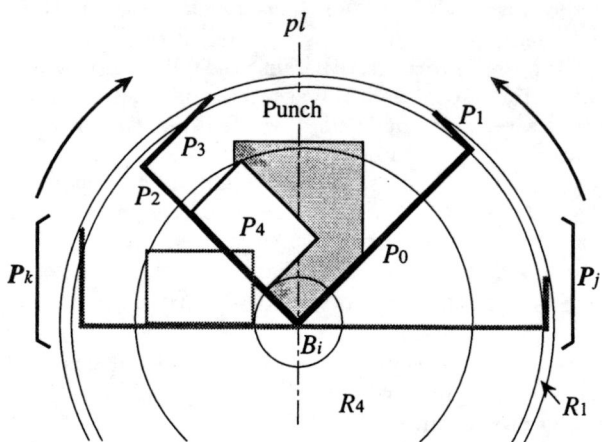

Figure 5: Motion of plates according to a bending operation on B_i.

plate group $\mathcal{P}_j$ and another group $\mathcal{P}_k$ approach pl from the opposite directions in rotating around B_i. Therefore, plates in $\mathcal{P}_j$ and plates in $\mathcal{P}_k$ that do not cross pl when B_i is completely bent do not have any mutual interferences in the bending process. In figure 5, plates $\{P_0, P_1\}$ in $\mathcal{P}_j$ and plates $\{P_2, P_3\}$ in $\mathcal{P}_k$ do not have interferences in bending B_i because they do not cross pl.

- According to the rotation of plates P_1 and P_4 around B_i, two ring-like regions R_1 and R_4 are swept by them as shown in figure 5. If these two regions do not overlap, their corresponding plates never collide in the bending rotation. Overlapping of the regions can be checked easily using their radii.

5 Computational experiments

The algorithm is implemented using C language. Feasible bending sequences of several sheet metal parts are computed with the program. All experiments are performed using SGI Indy workstation (CPU: R4600 100MHz, Memory: 64MB).

Figure 6 illustrates three "test piece" parts. Computation results are shown in table 1. The first column and the second column give the number of bending lines of the parts and the total number of computed feasible bending sequences. CPU seconds needed for the computation without the saving and reusing mechanism are given in the third column. They are reduced by using the mechanism as shown in the fourth column. The fifth column gives the reduction ratio.

As shown in the table, thousands of feasible bending sequences of complex sheet metal parts are computed in a practical speed for the interactive use.

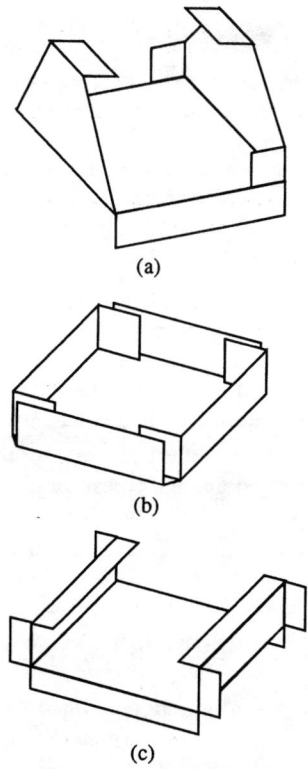

Figure 6: Example sheet metal parts.

ized. The proposed method is implemented and an experimental bending sequencing program is demonstrated. This program can compute all feasible bending sequences of complex parts in a practical speed for the interactive use.

The bending sequencing algorithm discussed in this paper will be used as a geometric engine of our future process planning system. The following enhancements are needed for the practical process planning.

- The bending sequencing program must be integrated with other modules for the process planning automation, for example the nesting program and the stamping planning program.

- Process planning experts elaborate the plan by considering various constraints concerning the manufacturing cost and the part precision. Introduction of some evaluation mechanisms of these factors is important.

Acknowledgments

This research work is financially supported by Amada Foundation for Metal Work Technology. We also thank Mr. Shin-ya Kawada of Tsuda Chemical Industries Co., Ltd. for his valuable comments on the sheet metal part manufacturing.

These results verify the effectiveness of the saving and reusing mechanism for the bending sequence planning.

Table 1: Computation results of feasible bending sequences of example parts.

	Number of bends	Number of seq.	Without reuse t_1	With reuse t_2	t_2/t_1
a	7	327	26.68	5.93	0.22
b	8	4704	386.65	23.75	0.061
c	10	1392	253.28	11.32	0.045

6 Summary and future research

A new method to accelerate the simulation based bending sequence planning is proposed. In the planning process, the same geometric computations for the interference detection are repeated many times. Our method eliminates this redundancy by saving and reusing the previous interference check results, and the substantial reduction of the computation time is real-

References

[1] K. H. Cho and K. Lee: Automatic Tool Selection for NC Turret Operation in Sheet Metal Stamping, Journal of Engineering for Industry, Vol. 116, 1994.

[2] D. F. Eary and E. A. Reed: Techniques of Pressworking: Sheet Metal, Prentice-Hall, Englewood Cliffs, 1958.

[3] M. Inui and F. Kimura: Design of Machining Processes with Dynamic Manipulation of Product Models, in Artificial Intelligence in Design, Springer-Verlag, London, 1991.

[4] R. Reich, J. B. Ochs and T. M. Ozsoy: Automated Flat Pattern Layout from Three-dimensional Wire-Frame Data, Journal of Engineering Design, Vol. 2, No. 3, 1991.

[5] C. H. Wang and D. A. Bourne: Using Features and Their Constraints to Aid Process Planning of Sheet Metal Parts, Proc. IEEE Int. Conf. Robotics and Automation, 1995.

[6] R. H. Wilson and J.-C. Latombe: Geometric Reasoning about Mechanical Assembly, in Algorithmic Foundations of Robotics, A K Peters, Wellesley, 1995.

Automation of Chamfering by an Industrial Robot; For the Case of Machined Hole on a Cylindrical Workpiece

Naoki Asakawa, Kenji Toda and Yoshimi Takeuchi

Department of Mechanical and Control Engineering
University of Electro-Communications, Chofu, Tokyo, 182-8585 Japan

Abstract

The study deals with the automatic chamfering for the case of a machined hole on a cylinder on the basis of CAD data, using an industrial robot. As a chamfering tool, a rotary-bar driven by an electric motor is mounted to the arm of the robot having six degrees of freedom in order to give an arbitrary position and attitude to the tool. The robot control command converted from the chamfering path is transmitted directly to the robot. From the experimental results, the system is found effective to remove a burr along the edge of a hole on a cylindrical metallic workpiece.

1 Introduction

The chamfering performed to deburr the edge of a workpiece after the drilling process is necessary to obtain workpieces with a smooth and clean edge. The requirement of automation of chamfering has strongly increased as the work is done in a contaminated environment. Therefore, industrial robots have been employed to automate the chamfering and / or deburring. However, the characteristics of burrs, whose shape are not necessarily constant, make it difficult to automate the operation.

A study copes with a variety of burr shapes by use of a laser-sensor to measure the shape in advance [1], and the others resolve the problem in terms of a feedback control and a compliance control to stabilize deburring condition [2, 3, 4, 5, 6]. These methods focused not on the path generation of itself but on the improvement of the tool path. While most of robots are controlled under a teaching-playback mode, some studies used CAD/CAM system to realize high-leveled automation [7, 8]. However, they did not consider the attitude of the robot arm, but considered the attitude of the tool. When an operation is performed with a multi-axis robot, it is important to keep the movement of the robot within the axis limit.

In the study, we treat a drilled hole on a cylindrical workpiece, frequently seen as cross-holes in a main spool of oil pressure parts, a hole on a cam shaft of a car and so on. Since the edge of the hole is not a simple circle but a 3-dimensional curved line, the chamfering has been conducted by skilled workers with a file, a whetstone or a rotary-bar.

To automate the chamfering, we introduce a robot with six-degrees-of-freedom on the basis of CAD/CAM system. As a result, the system is found effective to remove a burr along the edge of a hole on a cylindrical workpiece.

2 System Configuration

The system configuration is illustrated in Fig. 1. The robot having six degrees of freedom (Fanuc : Robot S-700), shown in Fig. 2, is used. The positioning accuracy of the robot is 0.2 mm, the load capacity is 300 N, and the arm is 2600 mm in length. As a chamfering tool, a motor with rotary-bar shown in Fig. 3 is attached to the robot arm. A touch sensor is mounted at the robot arm by a detachable holder as well as the chamfering tool to recognize the position and attitude of the workpiece. The workpiece shape is defined by use of 3D-CAD system (Ricoh : Design base). The chamfering path is generated by use of the our own CAM system on EWS (Sun : Sparc station), based on the CAD data. The robot control data generated on a personal computer (NEC : PC-98) with reference to the chamfering path are transferred to the robot controller through a RS-232-C. Then, the chamfering tool mounted to the robot arm starts chamfering.

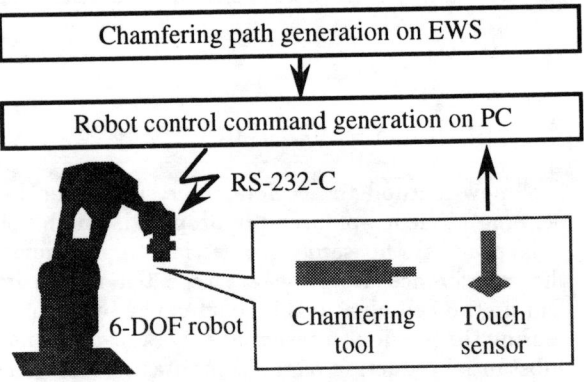

Fig. 1 Configuration of the system

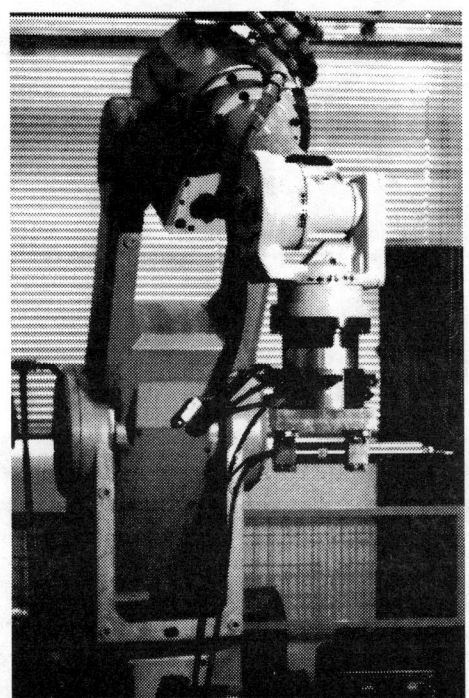

Fig. 2 Whole view of the robot used in the system

3 Basic Chamfering Model

3.1 Workpiece

Let us assume an edge with a right angle consisting of planes A and B, which is chamfered with 45 degree, as shown in Fig. 4. The symbols are as follows;

- P   Chamfering point
- N   Normal vector at each chamfering point
- F   Feed vector to the next chamfering point
- D   Tool axis vector

X_n means the n-th X in the figure. The parameters are called "workpiece surface data".

3.2 Chamfering Tool

The chamfering tool used in the system is modeled and

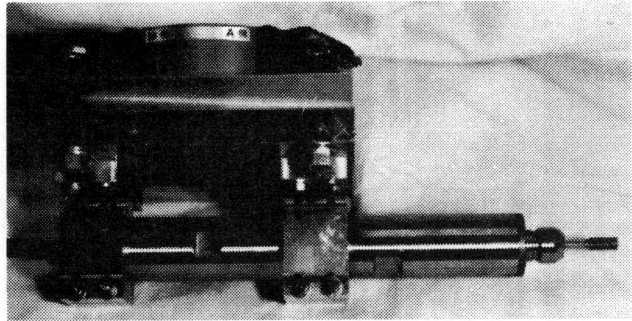

Fig. 3 Chamfering tool used in the system

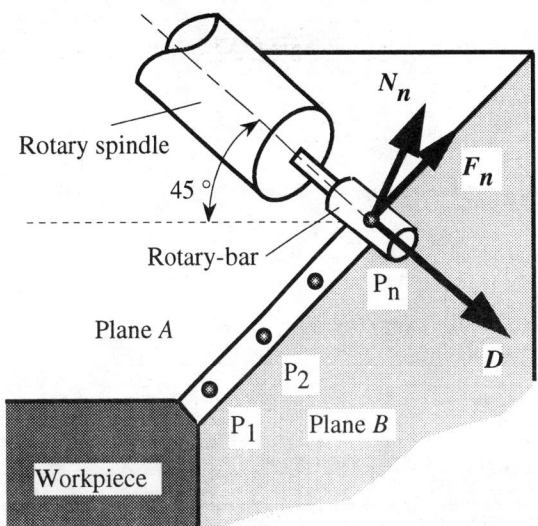

Fig. 4 Workpiece surface data on a basic chamfering model

attached to the robot as shown in Fig. 5, where

- D Tool axis vector
- T_f Functional tool vector
- T_g Geometrical tool vector

These parameters are called "tool attitude data". The tool vector D is parallel to the rotational axis of the chamfering tool, and is perpendicular to the vector N_{arm}, which is also parallel to the center axis of the robot arm.

Since the tool can machine a workpiece with any side part of the rotary-bar pushed against the workpiece, the functional tool vector T_f, representing the tool pushing direction, can be defined as an arbitrary vector within a plane M, whose normal is vector D. Considering the function of tool, the tool requires five degrees of freedoms.

On the other hand, the robot used in the system has six axes, which give the tool an arbitrary attitude. The robot attitude can be defined by appointing two vectors, the tool axis vector D and the geometrical tool vector T_g parallel to N_{arm}, considering the tool holder.

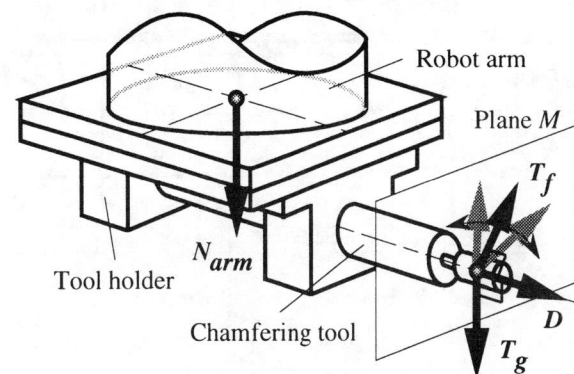

Fig. 5 Vectors for tool attitude

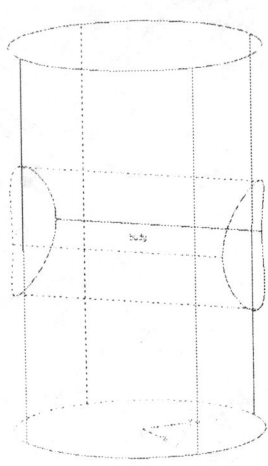

Fig. 6 CAD model of a workpiece

The tool path generation is to converte a series of workpiece surface data to a series of tool attitude ones.

4 Main-processor

4.1 Generation of Cylindrical Workpiece Surface Data

An example of the defined objective shape of cylindrical workpiece is shown in Fig. 6. The shape is a simplified model of a hole through a cylinder and is obtained by the subtractive operation between two cylinders on CAD system. Let us apply the above model to obtain the workpiece surface data for chamfering.

In Fig. 7, the cylinder A, the base cylinder, penetrates the cylinder B by drilling. The circle b is obtained as a cross section between the cylinder B and the plane L perpendicular

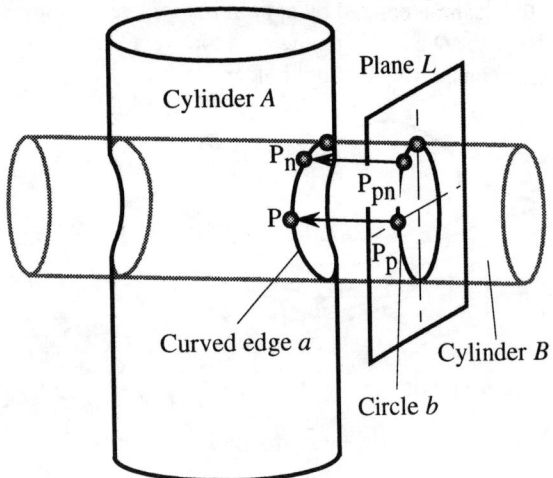

Fig. 7 A cylindrical workpiece model

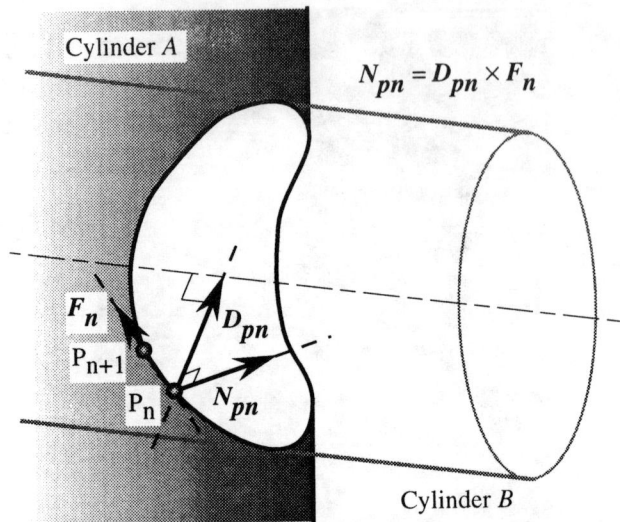

Fig. 8 Generation of chamfering points

to the cylinder B. The curved edge a is an edge of the cross section between the cylinder A and B. The circle b is equally divided by points P_p, depending on a chamfering condition. The chamfering point P is the projected one of P_p onto the curved edge a.

As shown in Fig. 8, D_{pn} and F_n are the vectors directing from P_n to the center axis of the cylinder B and from P_n to P_{n+1} respectively. An outer product of D_{pn} and F_n, corresponds to the normal vector of the plane A in Fig. 4. As shown in Fig. 9, the vectors D_n and N_n are obtained by revolving D_{pn} and N_{pn} around F_n by 45 degrees respectively. As a result, the vectors obtained correspond to the normal vector N, the feed vector F and the tool axis vector D in Fig. 4 at the chamfering point P. An example of the workpiece surface data displayed on CRT is shown in

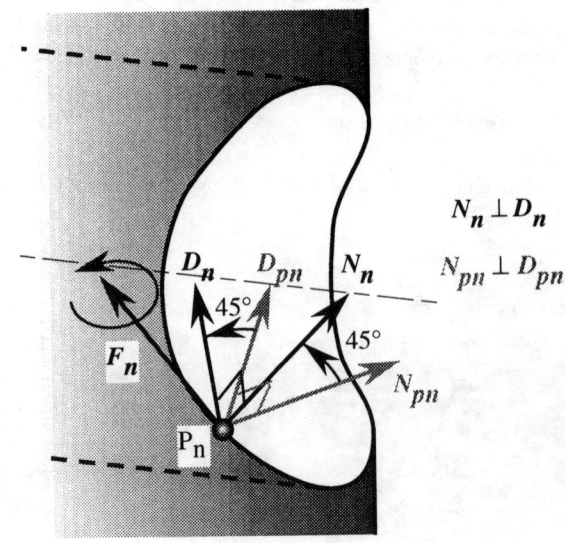

Fig. 9 Generation of tool vectors

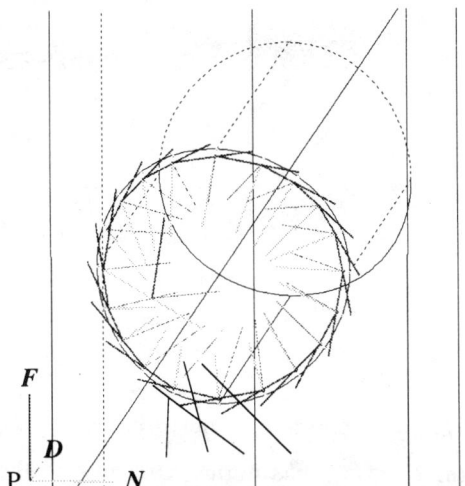

Fig. 10 Workpiece surface information

Fig. 10.

4.2 Conversion to the Tool Path

In this step, the workpiece surface data such as the chamfering point P, the normal vector N, the feed vector F and the tool axis vector D are converted to the tool attitude data such as the tool axis vector D, the geometrical tool vector T_g and the functional tool vector T_f. The vector D in the workpiece surface data is the same one in the tool attitude data. The vector T_f standing for the tool pushing direction is obtained as an inverse vector of N. The problem is how to relate T_f to T_g.

As a simple method, let us use T_f as T_g. The tool path generated by the method is shown in Fig. 11. With the method, however, the change in N directly influences T_g. When the robot is driven by T_g, as shown in Fig. 12, the change in the robot attitude becomes large, and makes the robot joints easily reach to the rotational angle limit. It is due to the direct generation of $T_g (=T_f)$ from an inverse

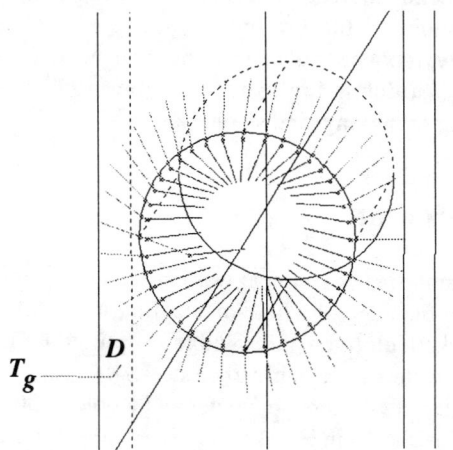

Fig. 11 Tool path based on the simple method

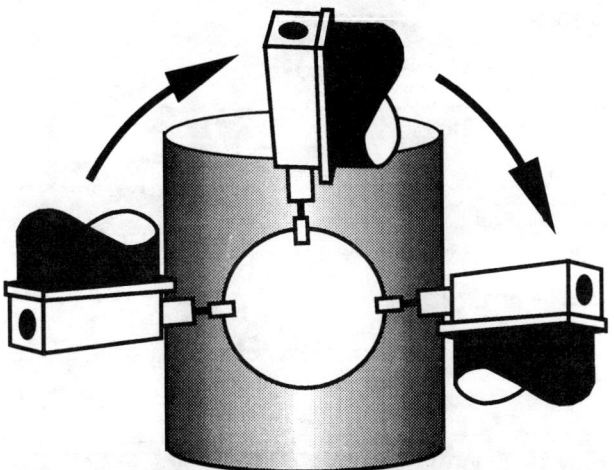

Fig. 12 Change in the robot attitude based on the simple method

vector of N.

As is mentioned above, T_f can be defined as an arbitrary vector within a plane having the vector T_f as its normal vector. When T_f is fixed, the tool can rotate around D. Therefore, T_g can arbitrarily be selected from T_f, which exists infinitely around D. This is because five degrees of freedom is enough for the tool though a robot has six degrees of freedom. In other words, a degree of freedom is redundant.

Based on the above characteristics, T_f can be converted to T_g, considering an attitude of the robot. The concept is realized by making selection of T_g so that T_f may make the changes in the robot attitude as small as possible.

Figure 13 illustrates the view of vector T_f from the center axis direction of the cylinder B. The angle between

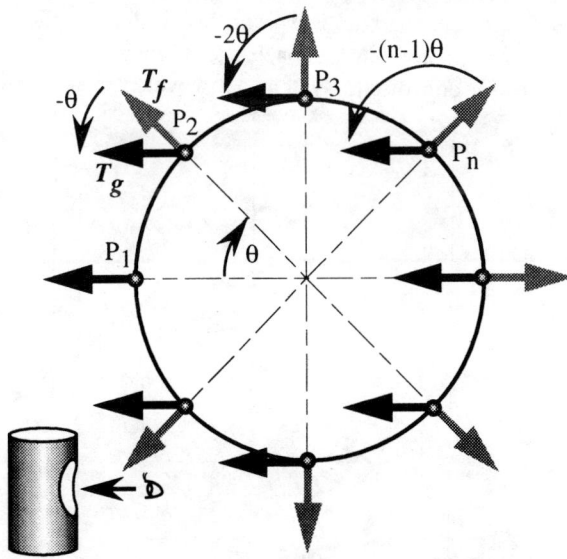

Fig. 13 Concept of a new method to generate the vector T_g from the vector T_f

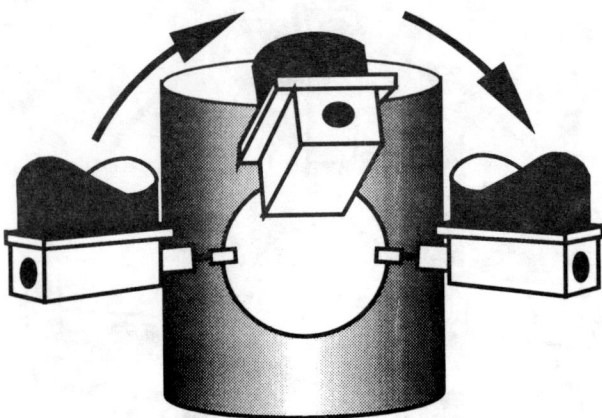

Fig. 14 Change in the robot attitude based on the new method

the horizontal line and T_f at each P_n, which is equally spaced round the circumference, increases from P_n to P_{n+1} by θ. Applying the characteristics, let us define T_g at P_2 by revolving T_f around D by $-\theta$. Then, it makes no influence to a chamfering condition. Besides, the tool can keep attitude near to the initial one by the rotation of $-(n-1)\theta$ at P_n.

As a result, the robot can be driven by the path with smaller changes in the attitude as shown in Fig. 14, and the tool attitude prevents the robot joints from reaching their limit of rotational angle. The tool path generated by the method is shown in Fig. 15.

5 Matching between the Coordinate Systems

The workpiece produced on the basis of CAD data can be placed arbitrarily on the table in front of the robot. A tool path generated in the workpiece coordinate system, as described in previous section, has to be converted to that in the robot coordinate system. However, the workpiece

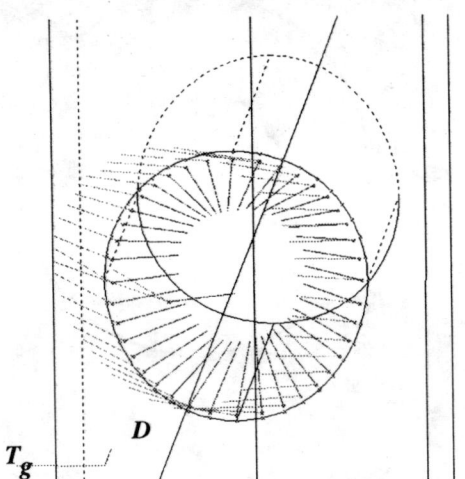

Fig. 15 Tool path based on the new method

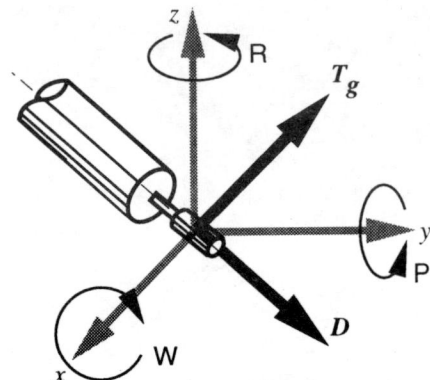

Fig. 16 Expression of tool attitude with angle

coordinate system has nothing to do with the robot coordinate one. The process, called "matching", relates these two coordinate systems each other. A touch sensor with a certain voltage, which is attached at the robot arm, makes contact with the workpiece, the voltage falls down to zero. This allows the robot to recognize the contact position. By use of the location information of the workpiece placed on the table, the tool position and the tool axis vector in the workpiece coordinate system is transformed to those in the robot coordinate system so that the system can generate the robot control commands necessary to control the robot movement.

6 Post-processor

The tool path is generated as a set of chamfering points and two vectors T_g and D, and is converted to the robot attitude expression since the robot used in the system has its own attitude expression. The attitude expression of robot, defined by W, P and R around x, y and z axes respectively, is shown in Fig. 16. Then, the extra tool path is added to the obtained tool path since the tool path generated in the previous section includes no path from the end of tool path to the beginning of the next one.

Finally, robot control commands are generated, taking account of a deburring speed and the mode such as chamfering or moving, and so on.

7 Experiment

A chamfering experiment was carried out to remove a burr along the edge of a hole on a cylindrical workpiece, which is difficult to chamfer with a simplified tool. The workpiece material is a plain carbon steel S45C, whose shape is a cylinder of 60 mm in diameter and a hole of 30mm in diameter, as shown in Fig. 17. The burr along the edge of the hole is about 1 mm in height. The chamfering conditions

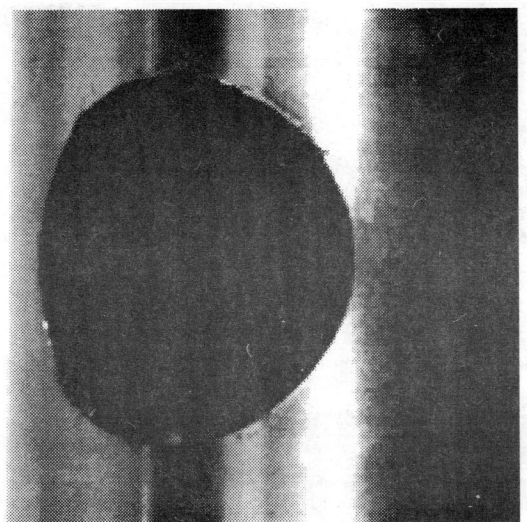

Fig. 17 Workpiece before chamfering

Fig. 19 Workpiece after chamfering

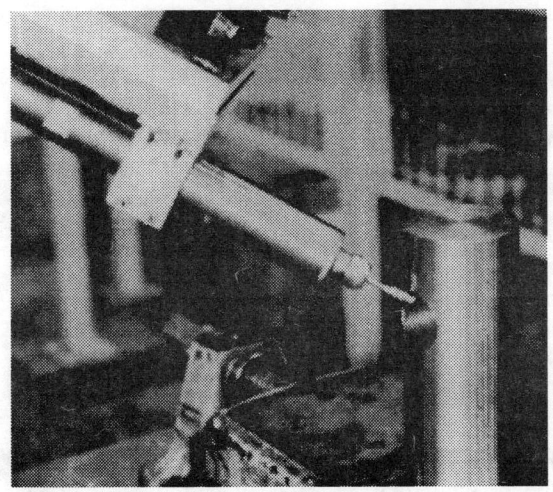

Fig. 18 Robot under chamfering

are follows; rotational speed of the tool : 35000 rpm, feed rate : 1.0mm/sec and depth of cut : 0.1 mm. The view of robot under chamfering is shown in Fig. 18. After several times repetition of the operation, the edge was chamfered by 1 mm in height as shown in Fig. 19, where the good chamfered edge can be seen.

8 Conclusion

The automatic chamfering system using an industrial robot is developed. The chamfering path is generated on the basis of CAD/CAM system. The system allows the chamfering tool to be set in a designated posture by 6-axis control. The system is experimentally found effective to remove a burr along the edge of a hole on a cylindrical workpiece considering robot arm attitude.

References

[1] G. Seliger, L. -H. Hsieh and G. Spur, Sensor-Aided Programming and Movement Adaptation for Robot-Guided Deburring of Castings, Annals of the CIRP, Vol.40/1 (1991) 487.

[2] H. Kazerooni, J. J. Bausch and B. M. Kramer, An Approach to Automated Deburring by Robot Manipulators, Trans. of the ASME, Vol.108, Dec. (1986) 354.

[3] T. M. Stepien, L. M. Sweet, M. C. Good and M. Tomizuka, Control of Tool/Workpiece Contact Force with Application to Robotic Deburring, IEEE Journal of Robotics and Automation, Vol.RA-3, No. 1 (1987) 7.

[4] M. A. Elbestawi, K. M. Yuen, A. K. Srivastava, and H. Dai, Adaptive Force Control for Robotic Disk Grinding, Annals of the CIRP, Vol.40/1 (1991) 391.

[5] M. G. Her and H. Kazerooni, Automated Robotic Deburring of Parts Using Compliance Control, Trans. of the ASME, Vol.113, Mar. (1991) 60.

[6] L. H. Chang and L. C. Fu, Nonlinear Adaptive Control of a Flexible Manipulatory for Automated Deburring, Proc. IEEE Int. Conf. on Robotics and Automation. (1997) 2844.

[7] G. M. Bone and M. A. Elbestawi, Sensing and Control for Automated Robotic Edge Deburring, Trans. of IEEE Industrial Electronics, Vol.41/2 (1994) 137.

[8] Y. Nonaka, S. Sakaue, Y. Yanagihara and K. Yokoshima, Development of an Impeller Grinding Robot System and a Gyro-moment Compensated Compliance Control, Proc. IEEE Int. Conf. on Robotics and Automation. (1995) 2084.

Proceedings of the 1998 IEEE
International Conference on Robotics & Automation
Leuven, Belgium • May 1998

An Improved Sculptured Part Surface Design with Jerk Continuity for a Smooth Machining

T.S. Lee and •Y.J. Lin
Department of Mechanical Engineering
University of Akron
Akron, OH 44325-3903
Phone: 330-972-5108 Fax: 330-972-6027
Email: *yl@uakron.edu*

Abstract: This paper presents an investigation of jerk continuity in milling operations for sculptured surface of parts. It has long been realized that chattering of machine tools during machining operations can cause detrimental effects on the quality of machined parts as well as on the cutting tool's life. One of the major reasons causing chattering is known to be rough transition of cutter acceleration changes during machining of different part surfaces. The problem becomes serious when machining sculptured surfaces of parts. In this work, an effective computer aided sculptured surface design technique is proposed. The ultimate goal is to achieve a smooth and near chattering-free machining for producing precision parts. The proposed surface design scheme models part's sculptured surfaces in such a way that it warrants a smooth "jerk" transition at the boundaries of common surface patches on the part. This results in a drastic reduction of large step changes of cutter accelerations during machining operations which will in turn eliminate a good portion of chattering effects. Three theorems concerning the necessary jerk continuity conditions for surface patches connections are developed and their proofs are presented. Examples of an airfoil and a concept car model are implemented using the proposed modeling approach to demonstrate its effectiveness.

1. Introduction

The ultimate goal of computer-aided free-form surface design is to use computer to creat mathematically defined smooth surfaces of products. Usually for the products with complex shapes, several patches of surfaces are required to be connected to generate the desired shape of the products. To ensure that these surface patches have smooth transitions at their common boundaries, static and dynamic continuity conditions at the boundaries of surface patches connection must be established and be satisfied.

In the past, creating a surface with G^1 (tangent planes) [1-9] and/or G^2 (curvature) [10-15] continuity condition at the common boundaries of the patches of the surface has been studied extensively. However, the higher-order continuity at the connecting boundaries of sculptured surface models which is deemed critical for smooth machining has usually been missing. When machining in a CAD/CAM enviroment, the lack of higher-order continuity in the part's sculptured surface patch model may cause large step changes of cutter accelerations as transiting from one cutting patch to another. This in turn may result in undesirable chattering effects during machining operations. It has been realized that chattering of cutting tools during machining can cause detrimental effects on the quality of machined parts as well as on cutting tool's lives. Therefore, chattering effects must be minimized, especially when the machined parts have sculptured surfaces.

In this paper, to achieve a smooth, chattering-free cutting an effective computer-aided sculptured part's surface design technique incorporating jerk continuity at the connecting boundaries of a composite sculptured surface patches is proposed. More specifically, the proposed technique models sculptured surfaces of prototypes in such a way that it warranys a smooth transition of the so-called "jerk" at the common surface patch boundary. As a result, more precise and smoother machining operations of parts having sculptured surfaces without being affected by chattering can be achieved. Three theorems along this line are derived and their proofs are presented in this paper. To implement the proposed sculptured surface design technique, Bezier's surface patch representation is employed with jerk continuity requirement. Finally, an airfoil and a concept car model are designed using the developed technique to demonstrate the effectiveness of the proposed sculptured surface design approach.

2. Dynamic Characteristics of a Milling Cutter on a Surface Patch

Figure 1 shows that $\vec{r}(u,v)$ is a parametric representation of a surface. In mathematical terms, it represents a function which maps a point in uv-space into a three-dimensional Euclidean space. At here, the parameters u,v are assumed to be valid in the interval of $0 \leq u, v \leq 1$. Mathematically, this interval can also be written as $u, v \in [0,1]$. This figure also shows that a curve, which is represented by $u(t), v(t)$, lying in the uv-plane. The image of the curve $u(t)$, $v(t)$ is described by $\vec{r}(t)$. Mathematically, $\vec{r}(t)$ can be interpreted as a curve lying on a surface. Physically, if parameter t is regarded as time variable, then the curve $\vec{r}(t)$ can be considered as a path traces out on the surface by a moving object (in this work it is a milling cutter) whose position at time $t = t_1$ is given by $\vec{r}(t_1)$.

From differential geometry, the unit tangent to the curve $\vec{T}$ is defined as derivative of $\vec{r}[s(t)]$ with respect to the arc length s of the curve. Thus, using the chain rule of calculus the velocity of the cutter becomes

$$\dot{\vec{r}} = \frac{d\vec{r}}{dt} = \frac{d\vec{r}}{ds}\frac{ds}{dt} = \dot{s}\vec{T} \quad (1)$$

where the magnitude of the velocity (it is also known as speed) is given by $\dot{s}$ and the velocity is pointed in the tangential direction of the curve.

Taking an additional time derivative of Equation (1) and applying the chain rule to the derivative term of the tangent vector $\vec{T}$ together with the first Serret-Frenet equation $\vec{T}' = \kappa \vec{N}$ [16], where the prime indicates derivative with respect to arc length s, gives

$$\ddot{\vec{r}} = \frac{d^2\vec{r}}{dt^2} = \ddot{s}\vec{T} + \dot{s}^2\kappa\vec{N} \quad (2)$$

in which κ is the curvature of the curve and $\vec{N}$ is the principal normal. The principal normal is orthogonal to tangential direction. In addition, it is lain in the same plane as that of unit tangent of the curve.

Differentiating $\ddot{\vec{r}}$ together with using second Serret-Frenet formula $\vec{N}' = \tau \vec{B} - \kappa \vec{T}$ [16] will yield the following jerk equation

$$\dddot{\vec{r}} = (\dddot{s} - \dot{s}^3\kappa^2)\vec{T} + (3\dot{s}\ddot{s}\kappa + \dot{s}^2\dot{\kappa})\vec{N} + \dot{s}^3\kappa\tau\vec{B} \quad (3)$$

where τ is the torsion of the curve, vector $\vec{B}$ is called binormal which is perpendicular to the osculating plane, the plane that contains both $\vec{T}$ and $\vec{N}$. Notice that the jerk equation not only depends on the magnitude of velocity and acceleration that the machine tool can generate, but also depends on the characteristics of the curve on the machined part's surface which in turn depends on the characteristics of the surface.

3. Condition for Jerk Continuity

Since mechanical parts usually have complex shapes, it is extremely difficult to represent them well using a single large surface. The logical way of solving this difficulty is to represent a complex shape with a composite of small simple patches. Unfortunately, these simple patches might have different surface characteristics. As a result the common boundary of any two adjacent patches might not be compatible, ending up with chattering of machine tool. In order to avoid chattering, it is clear that smooth transition of the jerk at the boundary is essential.

The parameter functions $\vec{a}(u,v)$ and $\vec{b}(x,y)$, as illustrated in Figure 2, denote two patches which are connected at their boundary. The parameters of the surfaces are varied from zero to one according to the directions of the arrows.

Before embarking to the task of deriving the necessary condition of jerk condition, a few terminologies of geometric modelling must be noted as follows:

(1) Two patches are said to have zeroth order geometric continuity G^0 at their common boundary when the position vectors of these patches along the boundary are identical. Mathematically, it can be expressed by

$$\vec{a}(1,v) = \vec{b}(0,y), \quad v = y \in [0,1] \quad (4)$$

(2) the first order geometric continuity condition G^1 at every point along the boundary of these patches is given by

$$\frac{\partial \vec{b}}{\partial x}(0,y) = p(v)\frac{\partial \vec{a}}{\partial u}(1,v) + q(v)\frac{\partial \vec{a}}{\partial v}(1,v) \quad (5)$$

Equation (5) implies that along the boundary the derivative of one of the patches is a linear combination of that of the other two lying on the same plane.

(3) two patches are said to be second order G^2 continuous when they have identical boundary curve (G^0 continuity) as well as G^1 continuity at their boundary. Moreover, they must have either identical osculating paraboloid, asymptotic direction, principal directions of curvature, or normal component of curvature (normal curvature) along the boundary. As a result, the second order geometric constraint G^2 is defined mathematically by

$$\frac{\partial^2 \vec{b}}{\partial x^2}(0,y) = p^2(v)\frac{\partial^2 \vec{a}}{\partial u^2}(1,v) + 2p(v)q(v)\frac{\partial^2 \vec{a}}{\partial u \partial v}(1,v) + q^2(v)\frac{\partial^2 \vec{a}}{\partial v^2}(1,v), \quad v = y \in [0,1] \quad (6)$$

In fact, from the physical point of view, the conditions of continuity of normals and normal curvatures at the common boundary of two surfaces implicitly imply the continuity of velocity and acceleration of the cutter, respectively.

Theorem 1: *The continuity of G^1 at the common boundary of two regular surfaces will guarantee the continuity of velocity of a cutter at the boundary.*

Proof: Let $\vec{a}(u(t),v(t))$ and $\vec{b}(x(t),y(t))$ be position vectors of a cutter on two regular contiguous surfaces. The velocities of the cutter on the surfaces are then obtained by taking time derivative of the position vectors. However, cross product the resultant velocity equations with $\partial \vec{a}/\partial v$ and $\partial \vec{b}/\partial y$ will lead to the equations of the normals of the surfaces

$$\frac{d\vec{a}}{dt} \times \frac{\partial \vec{a}}{\partial v} = \frac{\partial \vec{a}}{\partial u} \times \frac{\partial \vec{a}}{\partial v} \frac{du}{dt}$$
$$\frac{d\vec{b}}{dt} \times \frac{\partial \vec{b}}{\partial y} = \frac{\partial \vec{b}}{\partial x} \times \frac{\partial \vec{b}}{\partial y} \frac{dx}{dt} \quad (7)$$

From Equations (7) and (4) along with the fact that at the common boundary of the surfaces the normals of each surface is parallel to each other yields Equation (8). The constants C_1 and C_2 in this equation are defined as $C_1 = dx/dt|_{x=0}$, $C_2 = du/dt|_{u=1}$. If $\partial \vec{a}/\partial v$ is chosen to be zero, then the assumption that the surface is regular will be violated. Therefore, the other terms of the equation must be zero. This also implies that at the boundary of the surfaces the velocity at one surface is collinear to the other. In other words, the continuity of the normals at the boundary of the surfaces is implicitly meant the continuity of the velocities at the boundary.

$$\frac{\partial \vec{a}}{\partial u}(1,v) \times \frac{\partial \vec{a}}{\partial v}(1,v) = \frac{\partial \vec{b}}{\partial x}(0,y) \times \frac{\partial \vec{b}}{\partial y}(0,y)$$
$$\therefore \frac{1}{C_2}\frac{d\vec{a}}{dt}(1,v) - \frac{1}{C_1}\frac{d\vec{b}}{dt}(0,y) = 0 \text{ or } \frac{\partial \vec{a}}{\partial v}(1,v) = 0 \quad (8)$$

Theorem 2: *The continuity of G^2 at the common boundary of two regular surfaces will guarantee the continuity of the acceleration of a cutter at the boundary.*

Proof: Let $\vec{a}(u(t),v(t))$ and $\vec{b}(x(t),y(t))$ be position vectors of a cutter on two regular contiguous surfaces again. Applying Equation (2) to the position vectors $\vec{a}$ and $\vec{b}$ yields the accelerations of the cutter at the surfaces. Taking a dot product of the acceleration equation of the position vector $\vec{a}$ with the surface normal $\vec{n}_a$ and then notice that $\vec{T}_a$ and $\vec{n}_a$ are orthogonal as a result, the normal curvature equation of the surface patch becomes

$$\ddot{\vec{a}}.\vec{n}_a = \ddot{s}_a \vec{T}_a.\vec{n}_a + \dot{s}_a^2 \kappa^a \vec{N}_a.\vec{n}_a$$
$$= \dot{s}_a^2 \kappa_n^a \quad (9)$$

Notice that the definition of normal curvature $\kappa_n^a = \kappa^a \cos\theta$ from differential geometry has been applied to the equation in which θ is the angle between vectors $\vec{N}_a$ and $\vec{n}_a$. The normal curvature equation of surface $\vec{b}$ can be obtained similarly.

Applying the condition of continuity of normal curvatures of the two surface patches along their common boundary, one obtains

$$\kappa_n^a(1,v) = \kappa_n^b(0,y), \quad v = y \in [0,1]$$
$$\ddot{\vec{a}}(1,v).\vec{n}_a(1,v) = \ddot{\vec{b}}(0,y).\vec{n}_b(0,y), \quad v = y \in [0,1] \quad (10)$$

Now that G^1 condition will guarantee the coincidence of the surface normals of the patches at the boundary. Consequently, Equation (10) represents the condition of the continuity of acceleration of the cutter at the boundary of the connecting patches.

At here, two patches are called joining at jerk continuity if both patches meet G^0 and G^1 continuity at every point of their boundary. In addition, the normal curvature of each patch must be identical along the boundary. They also must have the same rate of change of normal curvatures.

The equation which describes characteristics of the rate of normal curvature $\dot{\kappa}_n$ of a surface can be obtained by taking dot product of the jerk equation, Equation (3), with the normal of the surface, which gives

$$\dddot{\vec{r}}.\vec{n} = (\dddot{s} - \dot{s}^3\kappa^2)\vec{T}.\vec{n} + (3\ddot{s}\dot{s}\kappa + \dot{s}^2\dot{\kappa})\vec{N}.\vec{n} + \dot{s}^3\kappa\tau\vec{B}.\vec{n}$$
$$= (3\ddot{s}\dot{s}\kappa + \dot{s}^2\dot{\kappa})\cos\theta + \dot{s}^3\kappa\tau\cos(\theta+90) \quad (11)$$
$$= 3\ddot{s}\dot{s}\kappa_n + \dot{s}^2\dot{\kappa}_n - \dot{s}^3\tau\kappa_g$$

This equation is obtained as a result of $\vec{T}$ being perpendicular to $\vec{n}$, and κ_n, κ_g and $\dot{\kappa}_n$ are defined as $\kappa_n = \kappa\cos\theta$, $\kappa_g = \kappa\sin\theta$ and $\dot{\kappa}_n = \dot{\kappa}\cos\theta$.

In the remaining of this section regarding derivation of necessary condition for smooth jerk transition, the indications of evaluation at the common boundary of the patches will be left out. Applying chain rule to $\vec{a}$ and $\vec{b}$ together with the acceleration continuity condition of Equation (10) as well as the constraint equations of G^0, G^1 and G^2 yields

$$\frac{du}{dt} = p\frac{dx}{dt}, \quad \frac{d^2v}{dt^2} = q\frac{d^2x}{dt^2} + 2\frac{dq}{dv}\frac{dx}{dt}\frac{dy}{dt} + \frac{d^2y}{dt^2}$$
$$\frac{dv}{dt} = q\frac{dx}{dt} + \frac{dy}{dt}, \quad \frac{d^2u}{dt^2} = p\frac{d^2x}{dt^2} + 2\frac{dp}{dv}\frac{dx}{dt}\frac{dy}{dt} \quad (12)$$

From the requirements that for smooth jerk transition from one patch to another the normal curvatures as well as the rate of normal curvatures of the patches must be the same, the jerk equation, Equation (11), can be simplified to

$$\dot{\kappa}_n^a = \dot{\kappa}_n^b, \quad \dddot{\vec{a}}.\vec{n}_a = \dddot{\vec{b}}.\vec{n}_b, \quad \therefore \dddot{\vec{a}} = \dddot{\vec{b}} \quad (13)$$

This is the equation that is going to be used to derive the necessary condition for jerk continuity. It is obtained by assuming the residue, the difference between constant multiplication of the torsion τ and the geodesic curvatues κ_g of the patches $\vec{b}$ and $\vec{a}$, is zero. In fact, as will be shown in Theorem 3, if one chooses jerk continuity condition as given in Equation (13) and let the constant terms C_1 and C_2 in Equation (8) be unity, then the residue will become zero.

Applying the chain rule to $\vec{a}$ and $\vec{b}$ together with Equations (12) and (13), the necessary condition for smooth jerk continuity along the common boundary of the patches is obtained as

$$\frac{\partial^3 \vec{b}}{\partial x^3}(0,y) = p^3(v)\frac{\partial^3 \vec{a}}{\partial u^3}(1,v) + 3p^2(v)q(v)\frac{\partial^3 \vec{a}}{\partial u^2 \partial v}(1,v) + 3p(v)q^2(v)\frac{\partial^3 \vec{a}}{\partial u \partial v^2}(1,v) + q^3(v)\frac{\partial^3 \vec{a}}{\partial v^3}(1,v) \quad (14)$$

Theorem 3: *If the constant coefficients in velocity continuity condition are chosen to be unity, then two regular surfaces which are osculating with the necessary jerk continuity condition have an identical surface curve torsion τ at their common boundary.*

Proof: Let $\vec{a}(u(t),v(t))$ and $\vec{b}(x(t),y(t))$ be defined as the position vectors of a cutter on two regular adjacent surfaces. From the differential geometry, the relationship between torsion and curvature of a curve on a surface (e.g. surface $\vec{a}$) is given by

$$\tau^a = (\kappa^a)^{-2}|\dot{\vec{a}}|^{-6}(\dot{\vec{a}}\cdot(\ddot{\vec{a}}\times\dddot{\vec{a}})) \quad (15)$$

But the normal curvature of the surface is related to the surface curve curvature by $\kappa_n^a = \kappa^a \cos\theta$ where $\cos\theta$ is defined as the dot product of the principal normal $\vec{N}_a$ of the surface curve with the normal $\vec{n}_a$ of the surface. It is noted that at the common boundary, the value of $\cos\theta$ of surface $\vec{a}$ is the same as that of surface $\vec{b}$. Therefore, using G^2 continuity condition, one obtains

$$\kappa_n^a(1,v) = \kappa_n^b(0,y), \quad \kappa^a(1,v)\cos\theta = \kappa^b(0,y)\cos\theta \quad (16)$$
$$\therefore \kappa^a(1,v) = \kappa^b(0,y), \quad v = y \in [0,1]$$

This equation is going to be used to show that when the constant coefficients of the velocity continuity condition are non-unity the surface curve torsion of surface $\vec{a}$ will be different in magnitude from that of surface $\vec{b}$ at their connected boundary.

Inserting surface curve torsion expressions as well as Theorems 1 and 2 in addition to Equation (13) into Equation (16) yields

$$(\tau^a)^{-1}|\dot{\vec{a}}|^{-6}(\dot{\vec{a}}\cdot(\ddot{\vec{a}}\times\dddot{\vec{a}})) = (\tau^b)^{-1}|\dot{\vec{b}}|^{-6}(\dot{\vec{b}}\cdot(\ddot{\vec{b}}\times\dddot{\vec{b}})) \quad (17)$$
$$\therefore \tau^a = C^5 \tau^b$$

The constant coefficient C is defined as the ratio of C_1 to C_2 which are the constant coefficients of the velocity continuity equation.

4. Patching Rectangular Bezier's Surfaces with Jerk Continuity

Let the patches $\vec{a}$ and $\vec{b}$ be represented by Bezier patch with degrees $k \times n$ and $n \times m$, respectively. Furthermore, assuming both of them are defined over a unit square. Therefore, in terms of Bezier's expression, the patches $\vec{a}$ and $\vec{b}$ can be expressed by

$$\vec{a}(u,v) = \sum_{i=0}^{k}\sum_{j=0}^{n} \vec{V}_{i,j}^{a} B_i^k(u) B_j^l(v), \quad u,v \in [0,1]$$
$$\vec{b}(x,y) = \sum_{i=0}^{m}\sum_{j=0}^{n} \vec{V}_{i,j}^{b} B_i^m(x) B_j^n(y), \quad x,y \in [0,1] \quad (18)$$

where $\vec{V}_{i,j}^{a}$ and $\vec{V}_{i,j}^{b}$ are the control vertices for the patches $\vec{a}$ and $\vec{b}$, respectively.

Substituting Equation (18) into the zero order geometric constraint G^0 equation yields

$$\sum_{j=0}^{n} \vec{V}_{k,j}^{a} B_j^n(v) = \sum_{j=0}^{n} \vec{V}_{0,j}^{b} B_j^n(y), \quad v = y \in [0,1] \quad (19)$$
$$\vec{V}_{k,j}^{a} = \vec{V}_{0,j}^{b}, \quad j = 0,1,2,\ldots,n$$

In terms, it just says that along the common boundary the last row of the vertices of patch $\vec{a}$ in the u-direction is coincided with the first row of the vertices of patch $\vec{b}$ in x-direction.

Using Equations (18) again, the G^1 constraint equation becomes

$$m\sum_{j=0}^{n} \Delta^{1,0}\vec{V}_{0,j}^{b} B_j^n(y) = kp(v)\sum_{j=0}^{n} \Delta^{1,0}\vec{V}_{k-1,j}^{a} B_j^n(v) + nq(v)\sum_{j=0}^{n-1} \Delta^{0,1}\vec{V}_{k,j}^{a} B_j^{n-1}(v), \quad v = y \in [0,1] \quad (20)$$

Notice that the left hand side of this equation is of degree n while the first and second terms on the right hand side are of degrees n and $n-1$, respectively. In order for the terms on both sides of the equation to have the same degree, $p(v)$ must assume a constant while $q(v)$ must be a linear combination of first degree of v and/or $(1-v)$. However, for generality the functions $p(v)$ and $q(v)$ are chosen to be

$$p(v) = \alpha, \quad q(v) = \eta + \beta(1-v) + \gamma v \quad (21)$$

where α, β, η and γ are constants. However, as a result of Theorem 3 the constant α must be equal to unity.

Using Equation (21) and definition of Bertein functions as well as elevation process on the Berstein functions with respect to parameter v, the Equation (20) becomes

$$m \Delta^{1,0} \vec{V}_{0,j}^{\,b} = \alpha k \Delta^{1,0} \vec{V}_{k-1,j}^{\,a} + (\beta + \eta)(n-j) \Delta^{0,1} \vec{V}_{k,j}^{\,a} + (\gamma + \eta) j \Delta^{0,1} \vec{V}_{k,j-1}^{\,a}, \; j=0,1,\cdots,n \quad (22)$$

Using the same procedures as that of obtaining G^1 constraint equation along common boundary of two rectangular Bezier surfaces, the G^2 continuity equation becomes

$$m(m-1) \Delta^{2,0} \vec{V}_{0,j}^{\,b} = \alpha^2 k(k-1) \Delta^{2,0} \vec{V}_{k-2,j}^{\,a} + 2\alpha k(\beta+\eta)(n-j) \Delta^{1,1} \vec{V}_{k-1,j}^{\,a} + 2\alpha k(\gamma+\eta) j \Delta^{1,1} \vec{V}_{k-1,j-1}^{\,a} + (\beta+\eta)^2 (n-j)(n-1-j) \Delta^{0,2} \vec{V}_{k,j}^{\,a} + 2(\beta+\eta)(\gamma+\eta)(n-j) j \Delta^{0,2} \vec{V}_{k,j-1}^{\,a} + (\gamma+\eta)^2 j(j-1) \Delta^{0,2} \vec{V}_{k,j-2}^{\,a}, \; j=0,1,\cdots,n \quad (23)$$

Similarly, the necessary condition of jerk continuity at the common boundary of two Bezier's surfaces are obtained as shown in Equation (24). The equation shows that in order to find the fourth row of control vertices of patch $\vec{b}$ one needs to have information regarding the last four rows of control vertices of patch $\vec{a}$. In addition, it also requires information concerning the first three rows of control vertices of patch $\vec{b}$. However, the first row of control vertices of patch $\vec{b}$ can be obtained from requirement of G^0 continuity at the common boundary. The other two rows of control vertices of patch $\vec{b}$ can be calculated from G^1 and G^2 continuity equations.

$$m(m-1)(m-2) \Delta^{3,0} \vec{V}_{0,j}^{\,b} = \alpha C_1 (k-2) \Delta^{3,0} \vec{V}_{k-3,j}^{\,a} + 3 C_1 (n-j)(\beta+\eta) \Delta^{2,1} \vec{V}_{k-2,j}^{\,a} + 3 C_1 j (\gamma+\eta) \Delta^{2,1} \vec{V}_{k-2,j-1}^{\,a} + 3 C_2 (\beta+\eta) (n-1-j) \Delta^{1,2} \vec{V}_{k-1,j}^{\,a} + 6 C_2 j (\gamma+\eta) \Delta^{1,2} \vec{V}_{k-1,j-1}^{\,a} + 3 C_4 \alpha k \Delta^{1,2} \vec{V}_{k-1,j-2}^{\,a} + (n-2-j) C_3 (\beta+\eta) \Delta^{0,3} \vec{V}_{k,j}^{\,a} + 3 C_3 j (\gamma+\eta) \Delta^{0,3} \vec{V}_{k,j-1}^{\,a} + 3 C_4 (n-j)(\beta+\eta) \Delta^{0,3} \vec{V}_{k,j-2}^{\,a} + (j-2) C_4 (\gamma+\eta) \Delta^{0,3} \vec{V}_{k,j-3}^{\,a}, \; j=0,1,\cdots,n \quad (24)$$

where the terms C_1, C_2, C_3 and C_4 in this equation are given as

$$C_1 = \alpha^2 k(k-1), \; C_2 = \alpha k(n-j)(\beta+\eta)$$
$$C_3 = (n-j)(n-1-j)(\beta+\eta)^2, \; C_4 = j(j-1)(\gamma+\eta)^2$$

5. Implementation of Proposed Methodology

The result of theoretical derivation of jerk continuity condition has been converted into C program and run on a windows-based computer. Using the proposed design methodology, two practical mechanical parts with jerk continuity surfaces have been designed. They are an airfoil and a concep car body.

Figure 3 shows the shape of an airfoil which is made up of two surface patches. The top conjunction of the surfaces is connected with the "jerk" continuity condition while the other conjunction of the airfoil is joint with G^0 continuity condition. As can be seen in Figure 3, the master patch of the airfoil is shaded and the other surface patch is generated as result of the requirement of "jerk" and G^0 continuity conditions, respectively.

Figure 4 illustrates a car body profile which is design and constructed by using a series of surface patches. Each boundary of the patches is connected with the "jerk" continuity condition. Again the master surface patch is shaded to distinguish it from the patches that are obtained by implementing the "jerk" continuity condition.

6. Conclusion

This paper raised a crucial issue of "jerk" continuity requirement and delved into its technological significance in mechanical part's sculptured surface design and the corresponding machining operations. The investigation advanced the conventional way of constructing composite sculptured surfaces of mechanical parts from approach based on a geometric point of view, to that based on a dynamics point of view.

The study began with derivations of dynamic characteristics of a typical milling cutter on a mechanical part's surface. The finding showed that the "jerk" constraint not only depends on the magnitudes of the velocity and acceleration of the cutter, but also on the part's surface characteristics. This finding justified that the "jerk" continuity is a necessary condition to be satisfied to ensure the machining continuity without encountering significant step acceleration changes of the cutter at the common boundaries of a machined part's composite sculptured surfaces transition. It also implied that the undesirable chattering effects during machining of sculptured parts would be greatly reduced and the precision of machining results would be increased if the jerk continuity requirement could be guaranteed.

With regard to the so-called "jerk" continuity constraint, three theorems required for employing the jerk constraint equation have been developed and proved thoroughly.

To implement the proposed sculptured surface design technique, Bezier's surface patch representation was employed and incorporated with the "jerk" continuity requirement. An airfoil model and a concept car body were designed using the developed technique which have demonstrate the effectiveness of the proposed approach.

Acknowledgement

This work was partially supported by SME foundation Grant #596-2205 and the Ohio Board of Regents' Research Challenge Fund #R3755. The funding was gratefully acknowledged.

References

1. Hosaka, M., Kimura, F., "Synthesis methods of curves and surfaces in interactive CAD", IEEE Int Conf on Interactive Techniques in CAD, 1978, pp. 151-156
2. Faux, I.D., Pratt, M.J., Computational geometry for design and manufacture, Ellis Horwood Limited, 1979
3. Farin, G., "A construction for visual C^1 continuity of polynomial surface patches", Computer Graphics and Image Processing, Vol. 20, 1982, pp. 272-282
4. Beeker, E., "Smoothing of shapes designed with free-form surfaces", CAD, Vol. 18, No. 4, May 1986, pp. 224-232
5. Liu, D., Hoschek, J., "GC^1 continuity conditions between adjacent rectangular and triangular Bezier surface patches", CAD, Vol. 21, No. 4, May 1989, pp. 194-200
6. Du, W.H., Schmitt, F.J.M., "On the G^1 continuity of piecewise Bezier surfaces: a review with new results", CAD, Vol. 22, No. 9, November 1990, pp. 556-573
7. Serraga, R.F., "G^1 interpolation of generally unrestricted cubic Bezier curves", Computer Aided Geometric Design, Vol. 4, No. 1, 1987, pp. 23-39
8. Choi, B.K., Surface modelling for CAD/CAM, Elsevier Science Publishing Company, 1991
9. Liu, Q., Sun, T.C., "G^1 interpolation of mesh curves", CAD, Vol. 26, No. 4, April 1994, pp. 259-267
10. Boehm, W., "Visual continuity", CAD, Vol. 20, No. 6, July/August 1988, pp. 307-311
11. Veron, M., Ris, G., Musse, J.P., "Continuity of biparametric surface patches", CAD, Vol. 8, No. 4, October 1976, pp. 267-273
12. Kahmann, J., "Continuity of curvature between adjacent Bezier patches", Pro Surf Comp Aided Geometric Design, 1982, pp. 65-75
13. Jones, A.K., "Nonrectangular surface patches with curvature continuity", CAD, Vol. 20, No. 6, July/August 1988, pp. 325-335
14. Du, W.H., Schmitt, F.J.M., "On the G^2 continuity of piecewise parametric surfaces", Pro Math Methods Comp Aided Geometric Design II, 1992, pp. 197-207
15. Li, H., Liu, S.Q., "Local interpolation of curvature-continuous surfaces", CAD, Vol. 24, No. 9, September 1992, pp. 491-503
16. Lipschutz, M., Schaum's outline: differential geometry, McGraw-Hill Inc., 1969

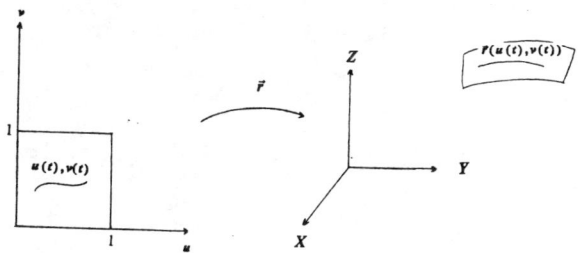

Fig. 1 Parametric surface with a surface curve

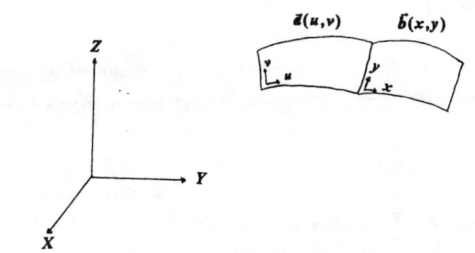

Fig. 2 Two patches connected at common boundary

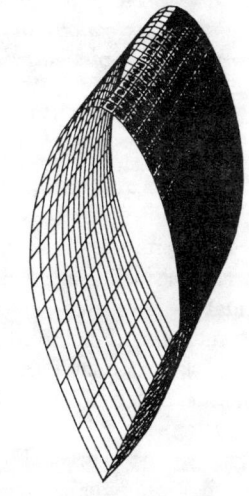

Fig. 3 Airfoil with "jerk" continuity at the top part

Fig. 4 Car body profile with "jerk" continuity

Accessibility Analysis in 5-axis Machining of Sculptured Surfaces

Abbas Vafaeesefa, Ph.D. Candidate
Hoda A. ElMaraghy, Ph.D., P.Eng.

Industrial and Manufacturing Systems Engineering,
University of Windsor, Windsor, Ontario, N9B 3P4

Abstract

In this paper, an approach for evaluating the global accessibility for 5-axis machining of sculptured surfaces is presented. The objective is to develop an algorithm which can determine all the feasible orientations along which the tool can reach a specified point without colliding with the workpiece. It is based on the projection of features, which define the unfeasible domains, on a unit sphere centered at the target point. The local curvature of the surface, the shape and the size of the tool and other obstacles (i.e. clamps and fixtures) are taken into consideration. The method also has potential applications in fields such as workpiece set-up, inspection planning and robotics manipulation.

1. Introduction

Five-axis machines have recently been used to produce the complex sculptured surfaces encountered in turbine blades, aerospace parts, dies and molds. Five-axis milling generally generates a smaller cusp height than 3-axis milling, when a toroidal cutter is used with an appropriate direction. Therefore, cutting and polishing times in this machining process are shorter than those in the 3-axis machining. The orientation of the milling cutter to produce minimum cusp height on the machined surfaces was discussed by Mullins [1] and Jensen [2]. The surface principal curvatures can locally define the shape of the surface. The position and orientation of the tool are determined by maximum matching of the tool curvature and surface curvature at the contact point. However, there is no guarantee that the derived tool orientation is free from collision with other portion of the surface.

Accessibility determines a bounded space in which a tool can maneuver without colliding within the workpiece. There are two ways to analyze the accessibility of a workpiece: 1) entire feature accessibility, where the accessibility of the whole workpiece is evaluated [3-7], and 2) point accessibility, where the directions along which the tool can reach a specified point are examined [7, 8]. Much research has been conducted to determine the optimized workpiece orientation(s) based on the first approach. Different classes of geometric algorithms are induced on the unit sphere based on observations made on the geometry of the cutting tools and degrees of freedom in 3, 4, and 5-axis numerical control machines [3,4]. Determining the Gaussain map of parametric surfaces, the visibility map which represents the entire feature accessibility is constructed. Tseng and Joshi [5] proposed a method that subdivides the 2-D Bezier curves and uses the tangential vectors to determine the entire feature accessibility. This method determines only the workpiece orientation in 2-D when the part is machined in 3-axis machining. Elber [6] used the hidden line and surface removal algorithms to evaluate the accessibility in 5-axis end-mill machining. The proposed algorithm computes the mapping that reduces the 5-axis accessibility to a 3-axis visibility problem. The algorithm is restricted to convex parametric surfaces obstructed by other surfaces, like a handle.

A technique for computing the feasible probe orientation at a specified point during CMM inspecting planning was discussed by Spyridi and Reoquicha [7]. They used algorithms based primarily on the computation of Gaussain images and Minkowski (sweeping) operation. The accessibility cone (AC) was defined as a place where the tool is safe from collision. AC is analyzed at two levels: 1) the local accessibility cone (LAC) that includes only the characteristic of the feature and the area in the immediate neighborhood of the point being measured, and 2) the global accessibility cone (GAC) that involves interference from other features of the part, and also possible interference from the fixtures, machine, etc.

Lim and Menq [8] extended Spyridi's approach to compute LAC and GAC on discrete points. Some simplifications, such as converting a 3-D accessibility cone into a 2-D accessibility map, were implemented to cut down the expensive computation time. To define the GAC at each measurement point, all possible tool directions are checked by tracing algorithms to determine their feasibility. Limaiem and ElMaraghy [9] proposed a method based on the intersection of concentric spherical shells centered at the measurement point. Geometric transformations (spherical scaling and solid intersection) are applied to the spherical shells and its thickness is reduced at each iteration. At the limit, the surface shell represents the accessibility domain.

The determination of accessibility for 5-axis sculptured surface machining is more complex than for inspection, because the cutter orientation need not be constant. The orientation changes to provide the required surface finish. Another difficulty is that milling tools have a cylindrical shape with a considerable radius which should be considered when determining the local accessibility. The local accessibility changes depending on the shape and the radius of the tool, surface curvature, and the machining direction. The objective of this paper is to develop a method that can efficiently determine all feasible tool orientations at a specified point for 5-axis sculpture surface machining, while taking the above factors into consideration.

We are aware of the recent work done by Morishige et al. [10], which has some similarities with our approach. However, we are presenting a general approach which can be used not only in 5-axis machining, but also in CMM inspection planning and robotics. Secondly, the tool orientation is not set in advance, rather it is defined when the domain accessibility is well defined. Therefore, intensive computation for correcting the tool orientation is not required. Finally, toroidal tool (as a general tool) is considered in this study and not only polyhedral objects, but also the surface, tool geometry, surface curvature, and the machining direction are considered when defining the feasible tool direction.

2. Accessibility Analysis

The accessibility analysis in 5-axis machining can be divided into four sections:

1. Feature accessibility that deals with surfaces, clamps, and fixtures.
2. Local accessibility that is defined initially by the tangent plane and then is changed depending on tool diameter, surface curvature and machining direction.
3. Machine accessibility which defines the maximum orientations that the machine can provide.
4. Global accessibility that is the intersection of all the above accessibilities, at the specified point.

In order to apply the proposed approach to sculptured surfaces, they should be subdivided into discrete sample points to create the triangular facets. A facet is considered to be the approximation of a small local surface region. Sculptured surfaces are usually described by implicit parametric representation $S(u,v)$. A simple method for sculptured surface approximation is to take discrete points with constant u/v intervals from the iso-parametric curves. This creates an approximate polyhedral model close enough to the nominal sculptured surface.

2.1. Feature Accessibility

The objective is to find directions that will permit access to the point of interest without interfering with neighboring obstacles. An optical analogy to this problem is to find the light regions on the unit sphere centered at the considered point, when a point light source (S) exists (see Fig. 1). The light rays are obstructed by the features surrounding the point S. By projecting these features on a unit sphere and subtracting them from all possible domains, the accessible domain at point S (or point accessibility cone) can be determined.

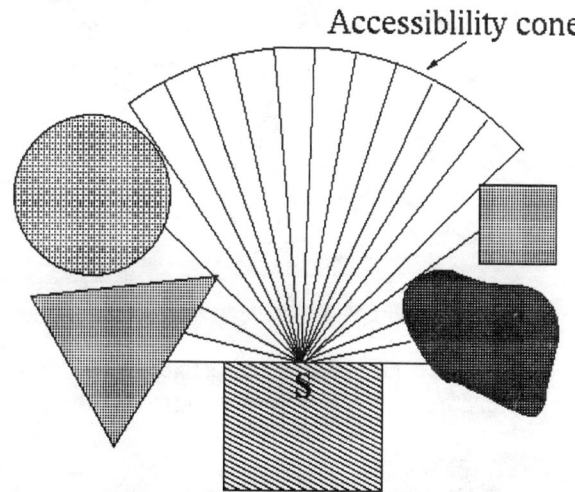

Fig. 1 Optical analogy

Point accessibility cone (PAC)

Since we are looking for the accessibility domain at a specified point on the feature, the local accessibility is the first constraint that limits the tool angle. The normal plane at the analyzed point divides the unit sphere into two hemispheres. The one that contains the normal vector considered as the initial local accessibility. Let N_c be the unit normal vector at point C, the unit hemisphere ϕ which defines the local accessibility can be expressed by:

$$\phi = \{l | <(l, N_c) \leq \pi/2\} \quad (1)$$

Equation (1) simply states that in order to make contact with point C, the tool must approach from an angle no greater than 90° from the normal vector at the contact point.

Note that the local accessibility ϕ is the initial visibility area, some of which may be interfered with by other features. To find the projections of obstacles on the unit hemisphere, the features are defined as a set of planar facets $\Delta_i(u)$, $i \in [1:n]$ (see Fig. 2). The PAC is represented by:

$$PAC = \phi - \left(\bigcup_i \{\Delta_i^p(u), i \in [1:n]\}\right) \quad (2)$$

Where (-) indicates the difference operation, and $\Delta_i^p(u)$ is the projection of the ith facet $\Delta_i(u)$ on ϕ. To reduce the computation time, the facets that are visible from the C point of view are considered.

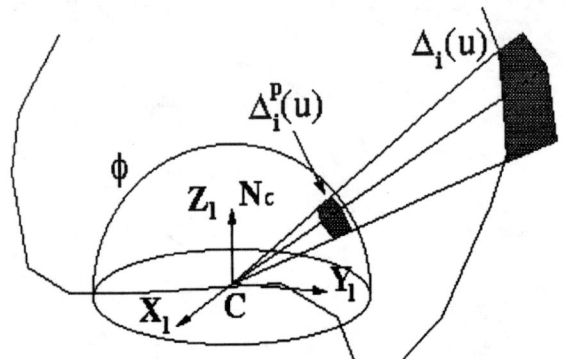

Fig. 2 Feature projection

Let $P_i(u)$ be the polygon that defines the *ith* facet $\Delta_i(u)$. Then, $\Delta_i^p(u)$ is given by the normalized points:

$$p_i^p(u) = \frac{C - p_i(u)}{|C - p_i(u)|} \quad (3)$$

The local coordinate on the normal plane is defined arbitrarily as long as the Z_l-axis in the local coordinates is along the normal direction (see Fig. 2). One can define the local coordinate system on the surface by:

$Z_l = N_c$
$X_l = S_u$
$Y_l = Z_l \times X_l$

where S_u is the tangent vector along the *u*-isoparametric curve of the surface. These vectors construct a homogenous transformation matrix M.

$$M = \begin{bmatrix} X_l^x & Y_l^x & Z_l^x & 0 \\ X_l^y & Y_l^y & Z_l^y & 0 \\ X_l^z & Y_l^z & Z_l^z & 0 \\ 0 & 0 & 0 & 1 \end{bmatrix}$$

The projected points $P_i^p(u)$ of facet $\Delta_i^p(u)$ are now transformed to the local coordinate system by:

$$P_i^{pt}(u) = M^T \cdot p_i^p(u) \quad (4)$$

where M^T is the transpose of the transformation matrix M. To determine the union of the projected facets $\Delta_j^p(u)$ (Eq. 2), they are transformed into two-dimensional coordinate systems on X_l-Y_l by the following expression (see Fig. 3):

$$q_i(u) = \frac{1 - p_{iz}^{pt}(u)}{\sqrt{1 - p_{iz}^{pt}(u)^2}} \begin{bmatrix} p_{ix}^{pt}(u) \\ p_{iy}^{pt}(u) \end{bmatrix} \quad (5)$$

The directions of vectors $q_i(u)$ on the X_l-Y_l plane are the same as $P_i^{pt}(u)$, but lengths of the vectors $q_i(u)$ are defined as $(1-P_{iz}{}^{pt}(u))$. Note that some of the facets' projections may lie partially inside and partially outside of ϕ. These projections are cut-off at the boundary before being mapped on 2-D.

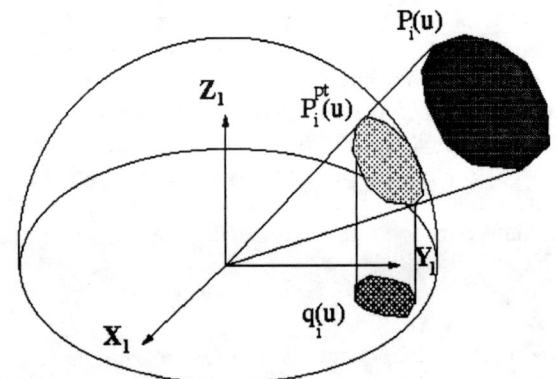

Fig. 3 Projection on 2-D

The initial local accessibility ϕ is also transformed to 2-D polygon on X_l-Y_l plane by the following expression:

$$LA(v) = \begin{bmatrix} \cos(2\pi v) \\ \sin(2\pi v) \end{bmatrix}, \quad v \in [0{:}1] \quad (6)$$

The above equation simply presents a circle. Both $q_i(u)$ and $LA(v)$ are presented as a polygon in two dimensions. By subtracting the union of polygons $q_i(u)$ from the local accessibility $LA(v)$, the resulting polygon $FP(s)$ defines the PAC in 2-D:

$$FP(s) = LA(v) - \left(\bigcup_i \{q_i(u), \ i \in [1{:}n]\} \right) \quad (7)$$

The above equation can be solved by the existing 2-D polygon Boolean operation technique. Without considering the tool geometry and surface curvature (i.e. in CMM inspection), a given tool direction at point C is feasible, if its projection on the local coordinate (equation 4 and 5) lies inside or on the boundary of the polygon $FP(s)$.

In order to define PAC, the polygon $FP(s)$ is transformed to the global three dimensions coordinate system by:

$$PAC(s) = M \cdot \begin{bmatrix} \dfrac{FP_x^p(s)\cdot\sqrt{1-R(s)^2}}{1-R(s)} \\ \dfrac{FP_y^p(s)\cdot\sqrt{1-R(s)^2}}{1-R(s)} \\ R(s) \end{bmatrix} \quad (8)$$

where

$$R(s) = 1 - \sqrt{FP_x^p(v)^2 + FP_y^p(v)^2} \quad (9)$$

2.2. Local Accessibility

The local accessibility defined in the last section does not consider the tool geometry and local surface properties which are important in avoiding gouging in 5-axis machining. In 5-axis machining, the tool orientation, in the local coordinate system, is changed by inclining the toroidal cutter by λ and tilting it by ω to get better surface finish (see Fig. 4). The tool orientation affects the effective tool radii in both Z_l-Y_l and Z_l-X_l planes. It is obvious that the surface curvature radii should not exceed the effective tool radii in these planes.

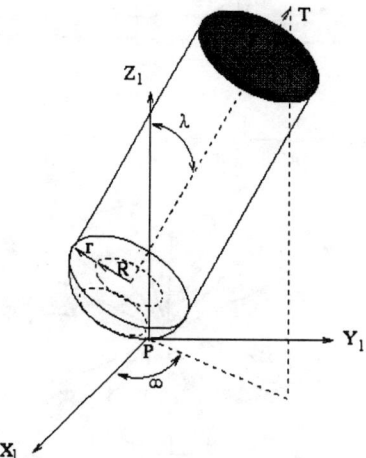

Fig. 4 Torus-shaped cutter in the local coordinate system

The principal curvatures of the toroidal cutter can be defined as:

$$k_1 = \frac{\sin\lambda}{R + r\sin\lambda}, \quad k_2 = \frac{1}{r} \quad 0 \le \lambda \le \frac{\pi}{2} \quad (10)$$

where R is the radius of the bottom portion and r is the fillet radius of the torus-shaped cutter. The effective tool radius in Z_l-Y_l plane is calculated by:

$$R_{eff}^x = \frac{1}{k_1\cos(\omega)^2 + k_2\sin(\omega)^2} \quad -\frac{\pi}{2} \le \omega \le \frac{\pi}{2} \quad (11)$$

and in Z_l-X_l plane by:

$$R_{eff}^y = \frac{1}{k_1\cos(\omega+\frac{\pi}{2})^2 + k_2\sin(\omega+\frac{\pi}{2})^2} \quad (12)$$

where ω is the rotation of the tool around the Z_l axis (assuming that X_l is the machining direction).

A set of angles (λ_i, ω_j) are created, and the effective tool radii of both planes are then calculated in these angles. Knowing the surface curvature radii of the surface in these planes, the inaccessible domain, where the effective tool radii are greater than that of the surface, can be defined. The boundary where the tool and surface radii are equivalent creates the local inaccessible boundary.

2.3. Machine Accessibility

The structure of the 5-axis control machining centers is composed of three translational movements along the **X**, **Y**, and **Z**-axes and two movements of rotation and tilt. The possible orientations of the tool are generally restricted depending on the type of milling machine. This restriction should be taken into account when determining the final global accessibility of a specified point.

2.4. Global Accessibility

The global accessibility is defined as a portion of the feature accessibility which is not part of the local inaccessibility domain and is within the machine accessibility domain. A given tool direction (λ_c, ω_c) at point C is feasible, if its projection lies inside or on the boundary of the global accessibility region. Note that the milling tool is not dimensionless but has a cylindrical shape of a considerable radius that should also be considered when determining the global tool accessibility.

3. Examples

The proposed algorithm was used to determine the feasible accessibility domains of two selected points on a sculptured surface. The complex sculptured surface shown in Fig. 5 was approximated by a set of triangles. The feature accessibility cones of these points are shown in Fig. 6 (elevated vertically for clarity). The global accessibility domains of the point used in Fig. 6a with the tool radii R=5 and r=3 is shown in Fig. 7. The global accessibility domain of the same point with different tool radii R=8 and r=5 is also shown in Fig. 8. The surface curvature radii in this example are 39.71 and 10.53 in X_l and Y_l direction respectively. To demonstrate the application of this method in inspection planning, the accessibility domains of six measurement points on the workpiece shown in Fig. 9 have been generated. The accessibility domains of points P5 and P6 are half spheres oriented along the surface normal direction. The point accessibility cones of points P1, P2, P3, P4, and the accessibility domain of P3 in 2-D are shown in Fig. 10.

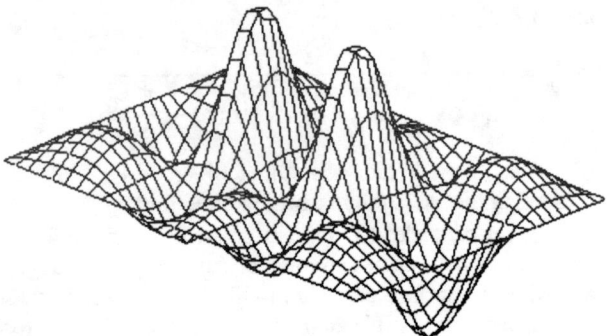

Fig. 5 Sculptured surface

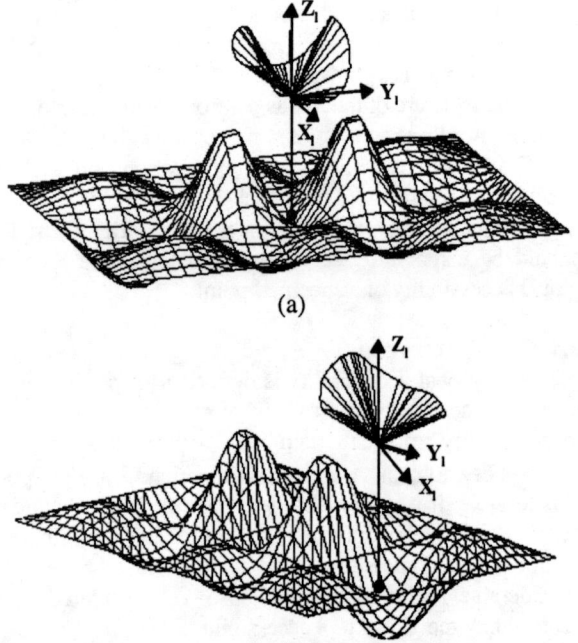

Fig. 6 Point accessibility cones on the sculptured surface

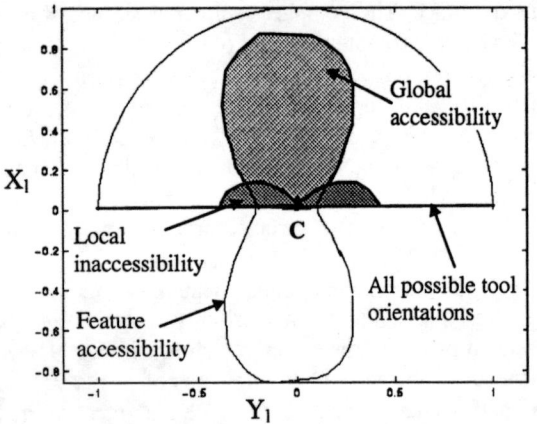

Fig. 7 Global accessibility domain with tool radii R=5 and r=3.

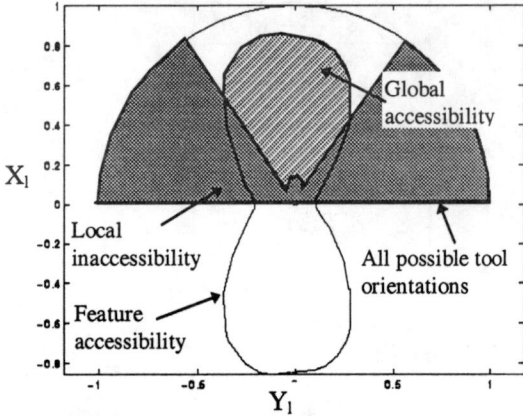

Fig 8 Global accessibility domain with tool radii R=8 and r=3.

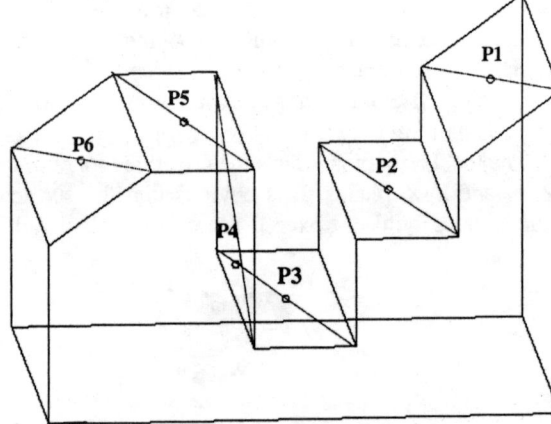

Fig. 9 Part example and the inspection points

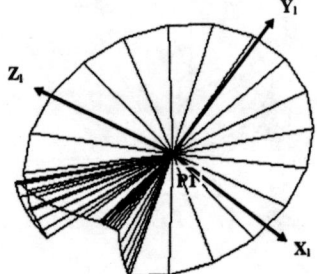

Fig. 10a Point accessibility cone of point **P1** in Fig. 9

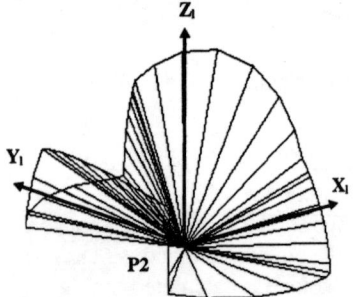

Fig. 10b Point accessibility cone of point **P2** in Fig. 9

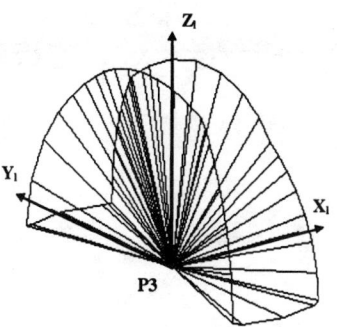

Fig. 10c Point accessibility cone of point **P3** in Fig. 9

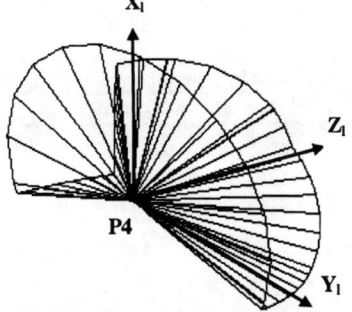

Fig. 10d Point accessibility cone of point **P4** in Fig. 9

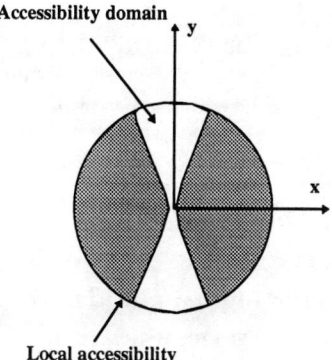

Fig. 10e Point accessibility cone in 2-D of point **P3** in Fig. 9

Conclusion

An efficient approach for analyzing the global feasibility of the tool orientation in 5-axis sculptured surface machining is presented. Unlike the other methods, the tool geometry, the local shape of the surface and other obstacles (i.e., clamps, and fixtures) can be taken into consideration. This approach is not limited to a particular surface or solid model representation and has potential applications in other fields such as workpiece set-up, CMM inspection planning, for mechanical parts and robotics manipulation.

References

1. Mullins, S. H., Jensen, C.G., and Anderson, D.C., 1993, "Scallop elimination based on precise 5-axis tool placement, orientation, and step over calculations", Advances in Design Automation, De-Vol 65-2, 534-544.

2. Jensen, C.G. and Anderson, D.C., 1993, "Accurate tool placement and orientation for finish surface machining", Journal of Design and Manufacturing, 3, 251-261.

3. Haghpassand, K., Oliver, J.H., 1995, "Computational Geometry for Optimal Workpiece Orientation", Transactions of the ASME Journal of Mechanical Engineering, Vol. 117, 329-335.

4. Tang, K., Woo, T., Gan, J., 1992, "Maximum Intersection of Spherical Polygons and Workpiece Orientation for 4- and 5-axis Machining", Transactions of the ASME Journal of Mechanical Engineering, Vol. 114, 477-485.

5. Tseng, Y. J., Joshi, S., 1991, "Determining Feasible Tool-Approach Directions for Machining Bezier Curves and Surfaces", Computer Aided Design, Vol. 23, No. 5, 367-379.

6. Elber, G., 1994, "Accessibility in 5-Axis Milling Environment", Computer-Aided Design, Vol. 26, No. 11, 796-802.

7. Spyridi, A. J., Reoquicha, A.A.G., 1990, "Accessibility Analysis for Automatic Inspection of Mechanical Parts by Coordinate Measuring Machines", IEEE International Conference on Robotic and Automation, 1284-189.

8. Lim, C.P., Menq, C. A., 1994, "CMM Feature Accessibility and Path Generation", Int., J., Prod., Res., Vol. 32, No. 3, 597-618.

9. Limaien, A. and ElMaraghy, H.A., 1997, "A General Method for Analyzing of Features Using Concentric Spherical Shells", Int. J. Adv. Manuf. Technol., Vol. 13, 101-108.

10. Marishige, K., Kase, K., Takeuchi, Y., 1997, "Collision-Free Tool Path Generation Using 2-Dimensional C-Space for 5-Axis Control Machining,", Int. J. Adv. Manuf. Technol., 13, 393-400.

Precise Position Control of Robot Arms Using a Homogeneous ER Fluid

Naoyuki TAKESUE, Guoguang ZHANG, Junji FURUSHO and Masamichi SAKAGUCHI

Department of Computer-Controlled Mechanical Systems,
Graduate School of Engineering, Osaka University,
2-1 Yamadaoka, Suita, Osaka 565-0871, JAPAN

Abstract

A semiclosed-loop control which utilizes the signal of an encoder mounted on a servomotor is adopted in the control of industrial robots. In this method, the position of the end-effector is not controlled very accurately. Therefore, we aim to apply a closed-loop control which uses the positional signal measured directly at the end-effector. Because of the flexibility of the driving system, however, it is very difficult to control the arm based on this method. In this study, a robot arm is developed which has a variable damper using an electrorheological fluid (ER fluid). It is shown that the ER damper is very effective for the precise position control based on a closed-loop control method.

1 Introduction

In industrial robots, each joint is driven through a reduction gear unit such as harmonic drive gear. Because of the elasticity of the driving systems including these reduction gear units, serious problems of vibratory behavior are being caused [1]-[3].

Most industrial robots operating at the present time use a semiclosed-loop control method for their control. This is a method whereby the feedback control of the driving motor for each joint is executed by using only signals from a sensor, such as an encoder, attached on that motor. The good point of this control method lies in the fact that the stability of the control system is assured even when its driving transmission system is elastic. However, a semiclosed-loop control method achieves only the control of the angle of a motor, so it is not clear whether the end-effector is precisely positioned or not. It is thought, in terms of improving the positioning accuracy, that the use of a closed-loop control method will be effective since this method applies the information on the relative position between a robot end-effector and an object directly. In this case, however, industrial robots, which have elasticity in their driving systems, may become unstable easily because the positional relationship between its sensor and actuator is non-collocation.

This study, which is aimed at realizing high-speed and high-precision closed-loop control, is intended to use an electrorheological fluid (to be abbreviated as ER fluid) that is currently attracting attention as an intelligent fluid [4],[5].

In Chapter 2, a brief introduction is given with respect to two types of ER fluids and the development of ER damper. Chapter 3 will describe the experimental apparatus consisting of an ER damper and a one-link robot arm and then discuss its mathematical model. In addtion, the effects of an ER damper to the resonance/antiresonance characteristics of 2-inertia systems are discussed. In Chapter 4, the design of the hierarchical control system is discussed. Firstly, the design of the inner loop controller for motor velocity control is presented. As an outer loop controller, an H_∞ controller is obtained by using a mixed sensitivity design method of the robust control theory. Then it is shown that the ER damper makes the design of a high gain controller possible while maintaining robust stability. In Chapter 5, the effectiveness of the proposed control method is ensured by experiments.

2 Variable Damper Using a Homogeneous ER Fluid

2.1 Electrorheological Fluids

An ER fluid is a substance which changes its apparent viscosity (rheological characteristics) according to the strength of an applied electric field [4],[5]. The ER fluids currently being developed may be classified into two types in terms of their characteristics: particle-type ER fluid and homogeneous ER fluid.

<u>Particle-type ER fluid</u>   ER fluids of this type are colloidal fluids which are solvents containing dispersed particles, and show their characteristics as shown in Fig.1 (a). The characteristics change according to the electric field E as shown in the diagram. It is clear from this diagram that the ER fluids exhibit the characteristics of a Newtonian fluid when no electric field is applied but that they reveal the characteristics of a Bingham fluid when an electric field is applied.

<u>Homogeneous ER fluid</u>   ER fluids of this type have been developed by using liquid crystalline polymers [6], and exhibit their characteristics as shown in Fig.1 (b). Shearing stress nearly proportional to shearing rate is

generated, and its slope, namely viscosity can be controlled by an electric field. As a result, it is possible to acquire a mechanical control force proportional to the speed under a constant electric field and to mechanically realize what is equivalent to the so-called differential control.

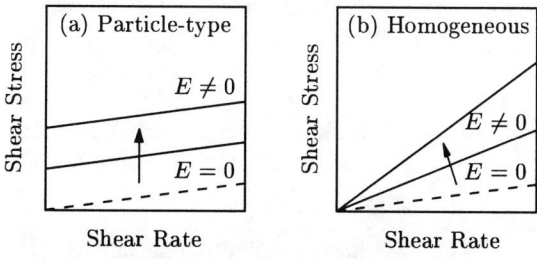

Fig.1: Two types of ER fluids

2.2 ER damper

J.Li, et.al. developed a damper using an ER fluid for robots to control the elastic vibration [7]. In their study, a particle-type ER fluid which has characteristics like Coulomb friction and exhibits nonlinear behavior, is used for the ER damper.

For developing an ER damper, we use a homogeneous ER fluid which nearly exhibits linear characteristics and is recently developed using liquid crystalline polymers [6]. Figure 2 shows a variable damper using a homogeneous ER fluid. We reported that this type of variable damper was effective for vibration control of robot arms [8].

The damper is structured in such a way that the ER fluid is filled into the ring-shaped grooves cut in the metallic ring(A) and this ring is coupled with another metallic ring(B) also having ring-shaped grooves. The gap between these metallic rings (electrodes) is 0.5mm. For high-voltage supply, a power source with a maximum output voltage of 1kV is used. Since an electric field is applied to the electrodes, they are each insulated with the use of an engineering plastic.

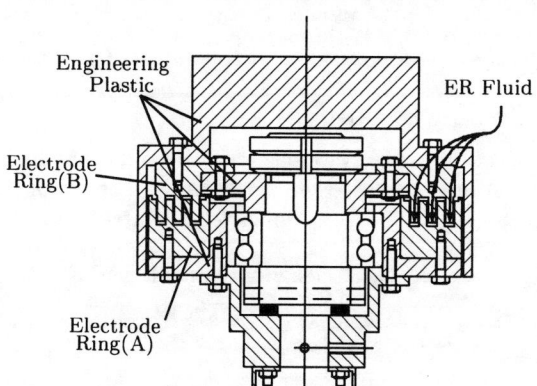

Fig.2: Variable damper using ER fluid

3 Experimental apparatus and its mathematical model

3.1 Experimental system

Figure 3 shows the one-link robot arm used in this study. The arm is driven by a DC servomotor via a harmonic drive gear. The developed variable damper has its lower end fixed onto the arm and its upper end onto the outer frame. In order to measure the relative position of the end-effector and an object, a position-measuring laser sensor is used.

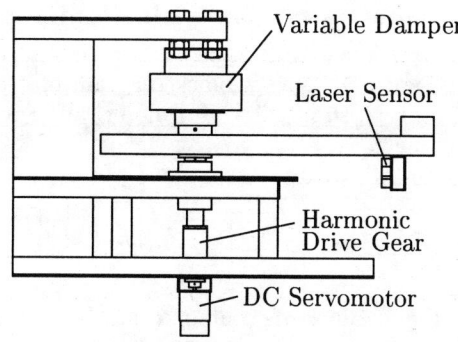

Fig.3: Experimental setup

3.2 Mathematical model

A reduction gear unit that drives articular joint has elasticity [1]-[3]. So, the one-link manipulator used for this study is modeled in the form of an arm and a motor coupled with a spring (Fig.4). Then the equation of motion of the robot arm is given as follows:

$$\begin{cases} J_m \ddot{\theta}_m = -K_j(\theta_m - \theta_a) + T_m \\ J_a \ddot{\theta}_a = K_j(\theta_m - \theta_a) - Q_a \dot{\theta}_a \end{cases} \quad (1)$$

where

θ_m : rotational angle of the motor (rad)
θ_a : rotational angle of the arm (rad)
J_m : moment of inertia of the motor (kg·m^2)
J_a : moment of inertia of the arm (kg·m^2)
K_j : coefficient of elasticity (N·m/rad)
Q_a : coefficient of viscous friction of the arm (N·m·s/rad)
T_m : motor torque (N·m)

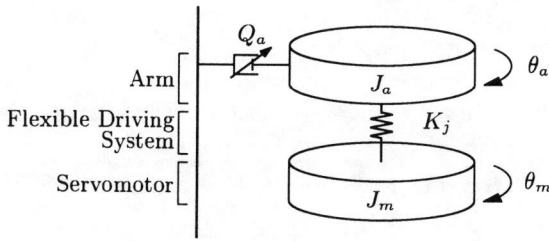

Fig.4: Model of flexible driving system

Q_a is given as follows:

$$Q_a = \bar{Q}_a + C_a \eta \quad (2)$$

where
- $\bar{Q}_a$: coefficient of viscous friction at the arm side of the harmonic drive (N·m·s/rad)
- C_a : coefficient determined by the shape and dimension of the damper (m³/rad)
- η : coefficient of viscosity of the ER fluid (Pa·s)

It follows then that Q_a can be varied by changing the applied electric field E.

3.3 2-inertia system with ER damper

The developed ER damper is attached to the arm joint. Figure 5 shows a block diagram of a one-link robot arm which consists of a moter, an arm and an ER damper.

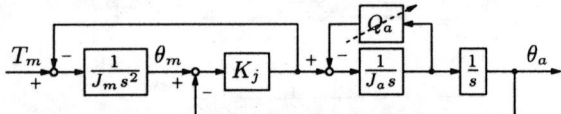

Fig.5: Block diagram of 2-inertia system

The 2-inertia system in Fig.5 can be expressed with an equivalent block diagram in Fig.6.

$$T_m \rightarrow \boxed{\frac{s^2+2\zeta_a\omega_z s+\omega_z^2}{f_o(s)}} \xrightarrow{\theta_m} \boxed{\frac{\omega_z^2}{s^2+2\zeta_a\omega_z s+\omega_z^2}} \rightarrow \theta_a$$

Fig.6: Transfer function description of Fig.5

where

$$f_o(s) = J_m s^2(s^2 + \omega_p^2) + 2\zeta_a\omega_z s(J_m s^2 + K_j) \quad (3)$$

$$\omega_z = \sqrt{\frac{K_j}{J_a}}, \ \zeta_a = \frac{Q_a}{2\omega_z}\frac{1}{J_a}, \ \omega_p = \omega_z\sqrt{1+\frac{J_a}{J_m}} \quad (4)$$

In the above equations, ω_z is an antiresonance frequency of the 2-inertia system, and ζ_a is a damping coefficient of the arm part. ζ_a can be changed by controlling the applied electric field E to the ER damper. The gain characteristics of the transfer function from T_m to θ_m in the left side of Fig.6 have notch at the antiresonance frequency ω_z when ζ_a is small enough. In the meanwhile, ω_p is a resonance frequency. As seen from Eq.(4), it holds that $\omega_z < \omega_p$.

On the other hand, it is seen from Fig.6 that there exists no zeros with the transfer function from T_m to θ_a since the zeros are canceled owing to the structure itself of the flexible joint in the arm side.

3.4 Experiments using ER damper

We conducted the open loop frequency response experiments by using a new homogeneous ER fluid which is recently developed. The results of the frequency response experiments from T_m to θ_m are shown in Fig.7.

In the diagram, the ● marks indicate the case where none of electric field is applied, whereas the ○ marks denote the case where an electric field is applied.

Figure 7 shows that when no electric field is applied, the damping coefficient ζ_a is low (approximately 0.1) and therefore the 2-inertia system has the antiresonance (corresponding to the notch) and resonance (corresponding to the peak)properties in the gain characteristics of the motor part distinctly. On the other hand if an electric field is applied, then the notch and the peak almost disappear. This might be derived from the reason that the damping coefficient ζ_a is made greater due to application of the electric field.

The results of the frequency response experiments from T_m to θ_a are shown in Fig.8. Figure 8 shows that the gain characteristics have a peak when no electric field is applied, but the peak disappears when an electric field is applied.

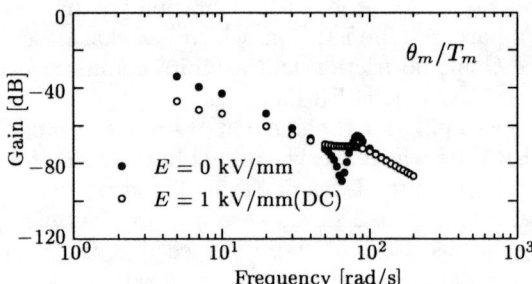

Fig.7: Frequency response of θ_m/T_m

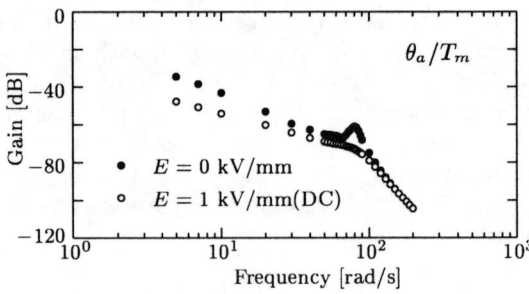

Fig.8: Frequency response of θ_a/T_m

4 Design of control system

The total control system is structured hierarchically as shown in Fig.9. At the lower level the speed of the motor is controlled as an inner loop control, while at

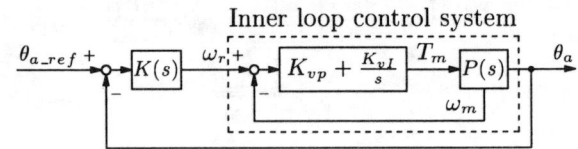

Fig.9: Total control system

the upper level the reference input to the velocity control loop ω_r is determined with the use of information from the laser distance sensor.

4.1 Inner loop control
4.1.1 Design of velocity control

Figure 10 shows the system which carried out the velocity feedback as the inner loop control.

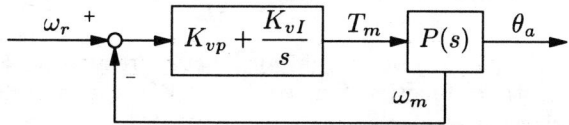

Fig.10: Inner loop control system

In this diagram, K_{vp} is a proportional feedback gain and K_{vI} is an integral feedback gain for the angular velocity control of the motor. As seen from the discussion in Section 3.4, the damping coefficient ζ_a becomes large when an electric field is applied. So, the velocity control system can be stabilized by applying an electric field even when the servo gain K_{vI} is high.

In this paper, the servo gain K_{vI} is determined as follows:

$$K_{vI} = \begin{cases} K_j/2 & (E = 0 \text{ kV/mm}) \\ K_j & (E = 1 \text{ kV/mm}) \end{cases} \quad (5)$$

Figure 11 shows the gain characteristics of the frequency response experiments ω_a/ω_r, where ω_a is the angular velocity of the arm. The ● marks indicate the case where none of electric field is applied, whereas the ○ marks indicate the case where an electric field is applied. As seen from this diagram, the gain characteristics of ω_a/ω_r has no peak when an electric field is applied, even though the higher feedback gain is used in this case as compared with the case where none electric field is applied.

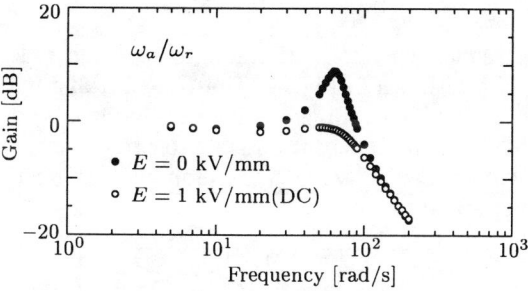

Fig.11: Frequency response of velocity control

4.1.2 Nominal model

From the results of the frequency response experiments, the transfer functions θ_a/ω_r are identified as in the following. The following equation is obtained from the data when no electric field is applied to the ER damper (Fig.12).

$$P_{0\_0} = \frac{1.389 \times 10^6 (s + 4)}{s(s + 3.92)(s + 306.9)(s^2 + 19.03s + 4618)} \quad (6)$$

Next, the data for the case where an electric field is applied yield the following equation (Fig.13).

$$P_{0\_1} = \frac{1.389 \times 10^6 (s + 8)}{s(s + 6.57)(s + 335.4)(s^2 + 74.65s + 5042)} \quad (7)$$

By using these equations as nominal models $P_{0\_i}(s)$ of the control objects, their respective outer loop controllers are designed.

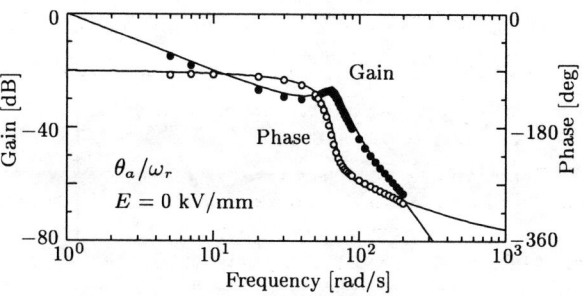

Fig.12: Frequency response of identified model P_{0_0}

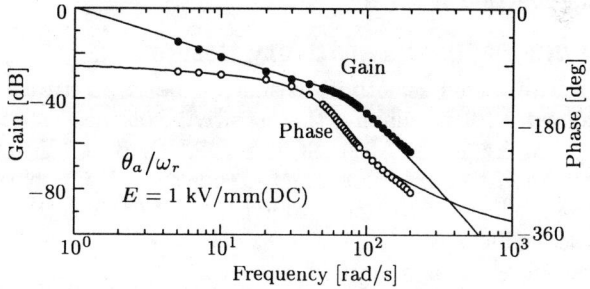

Fig.13: Frequency response of identified model P_{0_1}

4.2 Design of outer loop control

In this section, we design controllers for the outer loop by applying the H_∞ control theory.

4.2.1 Uncertainty of the model

The uncertainty of the model is to be expressed by using a multiplicative perturbation $\Delta(s)$ as follows:

$$P(s) = \{1 + \Delta(s)\} P_0(s) \quad (8)$$

Due to the factors, such as lost motion and solid friction in the harmonic drive, the control object possesses nonlinear characteristics, and the frequency response changes depending on the size of the input amplitude. Taking these characteristics into consideration, a number of frequency response experiments were carried out by putting as input amplitude 25, 50, ... ,150 percent of the value of the input amplitude used in obtaining the nominal model; and the maximum of the absolute value of the multiplicative perturbation $\Delta_{max}(j\omega)$ was plotted in Fig.14. Figure 14 (a) refers to the case in which none of electric field is applied, while Fig.14 (b) represents the case where an electric field is applied.

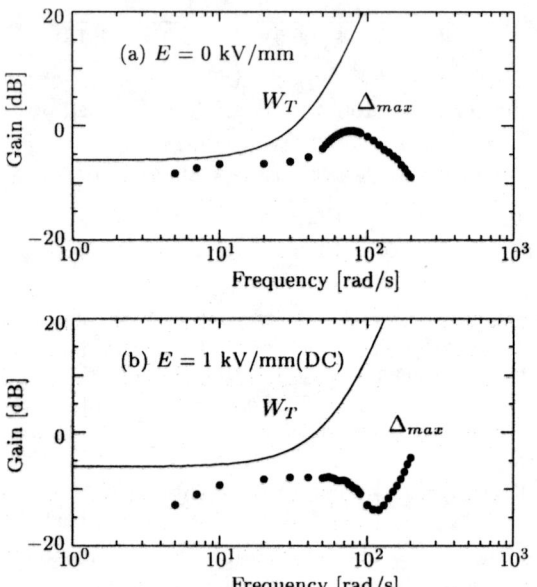

Fig.14: Maximum of perturbation Δ_{max} and weighting function W_T

4.2.2 Mixed sensitivety design

The mixed sensitivity design in the H_∞ control is aimed both at making the sensitivity of the system at low frequency as small as possible and at raising the robust stability at a high-frequency region where the modeling error tends to get large. In this section, the feedback controller $K(s)$ is designed by the mixed sensitivity design method.

First, consider the robust stability. It is assumed that the perturbation $\Delta(s)$ is suppressed by a weighting function W_T as follows:

$$|\Delta(j\omega)| < |W_T(j\omega)| \quad \forall \omega \quad (9)$$

When the closed-loop system becomes stable with respect to an arbitrary Δ which satisfies the above equation, it is said that the robust stability of the closed-loop system holds. The necessary and sufficient condition for robust stability is given as follows:

$$\|W_T T\|_\infty < I \quad (10)$$

where

$$T = P_0 K(I + P_0 K)^{-1} \quad (11)$$

As the weighting function $W_T(s)$ which suppresses the perturbation, the following equation were chosen for cases where no electric field is applied and for cases where an electric field is applied, respectively:

$$W_T(s) = \frac{(1/50s + 1)^4}{2} \quad (E = 0 \text{ kV/mm}) \quad (12)$$

$$W_T(s) = \frac{(1/70s + 1)^4}{2} \quad (E = 1 \text{ kV/mm}) \quad (13)$$

The way W_T suppresses the perturbation of the model is indicated in Fig.14. As is clear from Fig.14, the cases where no electric field is applied to the ER damper have a large perturbation peak and therefore make it more difficult to set up $W_T(s)$.

Next, consider the sensitivity minimization. The sensitivity function S may be defined as follows:

$$S = (I + P_0 K)^{-1} \quad (14)$$

By putting the weighting function relative to the sensitivity function S as $W_S(s) = \rho \bar{W}_S(s)$ and configuring the controller so that

$$\|W_S S\|_\infty < I \quad (15)$$

holds for as large ρ as possible, the minimization of the sensitivity is attempted.

The preceding robust stabilization and the sensitivity minimization are simultaneously satisfied, when the following relation holds (mixed sensitivity problem):

$$\left\| \begin{bmatrix} W_S S \\ W_T T \end{bmatrix} \right\|_\infty < I \quad (16)$$

Here, we choose $W_S(s) = \rho/s$ and adopt the maximum ρ within a range where the mixed sensitivity problem can be solved. By computing the problem, the H_∞ controllers can be obtained as follows:

The case where no electric field is applied:

$$K_0(s) = \frac{5702.2(s + 252.5)(s^2 + 19.7s + 4649)}{(s + 588.7)(s + 89.05)(s^2 + 107.2s + 7648)} \quad (17)$$

The case where an electric field is applied:

$$K_1(s) = \frac{13466(s + 4.615)(s^2 + 78.84s + 5805)}{(s + 233.4)(s + 5.497)(s^2 + 117.2s + 12310)} \quad (18)$$

The frequency response characteristics of controllers obtained are shown in Fig.15. The H_∞ controller with an electric field has a higher gain. In other words, the use of an ER fluid makes it possible to carry out high-gain feedback control while maintaining a high degree of robust stability.

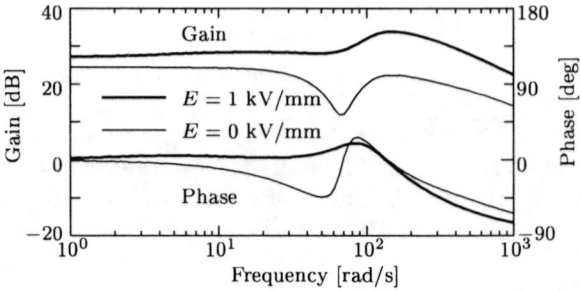

Fig.15: Frequency responses of H_∞ controllers

5 Experiments

Figure 16 (a) shows the result of an experiment for a case where no ER damper is used. The response characteristic is vibratory due to the flexibility of the driving system. Figure 16 (b) shows the result of an experiment for a case with ER damper where no electric field is applied. Figure 16 (c) indicates the response for a case with ER damper where an electric field is applied. As seen from the comparison of these diagrams, a control system which gives exceedingly good response shown in Fig.16 (c) can be constructed by ER damper.

Figure 17 shows the frequency response of a closed-loop system configured by using the H_∞ controller for a case where an electric field is applied. The input is the reference arm angle θ_{a_ref}, while the output is the arm angle θ_a. As is clear from the diagram, the bandwidth spans about 50 rad/s.

6 Conclusions

In order to achieve the high-speed and high-precision positioning for a robot arm, a control method using a homogeneous ER fluid has been proposed. And its effectiveness has been verified by experiments. The obtained results may be summarized as follows:

1. A robot system was developed which has a variable damper using a homogeneous ER fluid. For the developed system, it is shown that the influence of the resonance and antiresonance of a 2-inertia system is lowered by applying an electric field to the ER damper.

2. The design of inner velocity control was presented. It is possible to realize high-gain feedback by applying an electric field.

3. In order to maintain the robust stability, an H_∞ controller was designed by a mixed sensitivity design method of the robust control theory. As a result, it was shown that the applying an electric field makes high-gain position control possible. Finally, control experiments were carried out, and good responses were obtained.

Acknoeledgments

The first author is supported by the JSPS Research Fellowships for Young Scientists(No.8721). This study is supported by the Grant in Aid of Scientific Research from the Ministry of Culture, Sports and Education (No.09650282).

References

[1] M.C.Good, L.M.Sweet, "Dynamic Models for Control System Design of Integrated Robot and Drive Systems", ASME J.Dyn.Syst.Meas.Contr., Vol.107, No.1, pp.53-60 (1985)

[2] J.Furusho, H.Nagao and M.Naruse, "Multivariable Root Loci of Control Systems of Robot Manipulators with Flexible Driving Systems", JSME Int.J., Series III, Vol.35, No.1, pp.65-73 (1992)

[3] J.Y.S.Luh, "Conventional Controller Design for Industrial Robots – A Tutorial", IEEE Trans. on Sys. Man and Cyb.,Vol.SMC-13, No.3, pp.298-316 (1983)

[4] T.C.Jordan and M.T.Shaw, "Electrorheology (Review Article)", IEEE Trans. on Electrical Insulation, Vol.24, pp.849-878 (1989)

[5] J.Furusho, "Control of Mechatronic Systems Using Electrorheological Fluids (Review Article)", (In Japanese), J. of SICE, Vol.34, pp.687-691 (1995)

[6] A.Inoue and S.Maniwa, "Electrorheological Effect of Liquid Cryatalline Polymers", J. Appl. Polym. Scie., Vol.55, pp.113-118 (1995)

[7] J.Li, D.Jin and X.Zhang, "An Electrorheological Fluid Damper for Robots", Proc.Int.Conf.of Robotics and Autom., Vol.3, pp.2631-2636 (1995)

[8] J.Furusho, G.Zhang and M.Sakaguchi, "Vibration Suppression Control of Robot Arms Using a Homogeneous-Type Electrorheological Fluid", Proc.Int.Conf.of Robotics and Autom., Vol.4, pp.3441-3448 (1997)

[9] R.P.Paul, "Robot Manipulators", The MIT Press (1981)

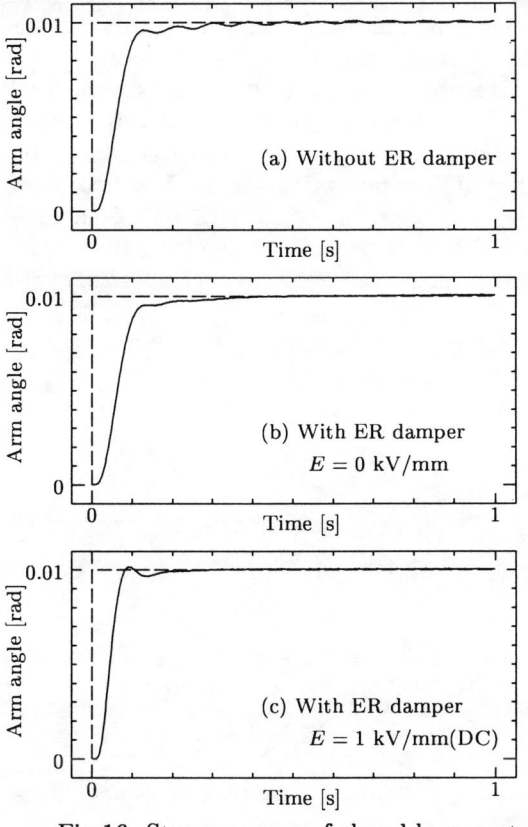

Fig.16: Step responses of closed-loop system

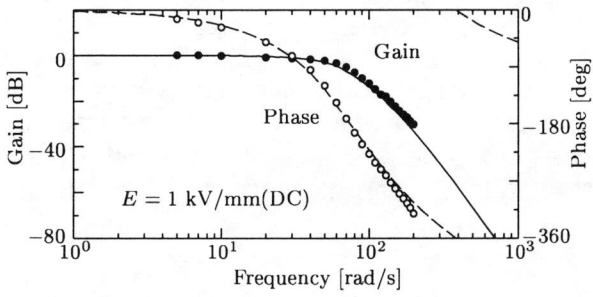

Fig.17: Frequency response of closed-loop system

An Electrorheological Fluid Damper for Vibration Control

Jianjun Li and William A. Gruver

Intelligent Robotics and Manufacturing System Laboratory
School of Engineering Science
Simon Fraser University
Burnaby, BC V5A 1S6 Canada

jianjun@cs.sfu.ca gruver@cs.sfu.ca

Abstract

A new electrorheological fluid joint damper based on orifice restriction has been developed for semi-active vibration control. The damper is capable of providing a continuously variable damping torque in response to an electric field. A mathematical model of the damper is developed and the influence of model parameters on the damping torque is analyzed. Performance tests show that the magnitude of the output torque increases 1.6 times within 10 milliseconds.

1 Introduction

Methods to control vibrations can be classified as passive, semi-active and active [1]. In passive devices, called dynamic absorbers, control is obtained by combining masses, springs and dampers. Since it has no sensory feedback, however, a dynamic absorber cannot compensate for undesirable vibrations of the controlled system. Active control is based on instrumenting a dynamic absorber with external sensors and actuators. Another approach is based on altering the characteristics of the spring or the damper of the passive dynamic absorber to achieve optimal operating conditions. This concept, called semi-active control, retains the benefits of an active system without requiring additional power requirements and system complexity. However, the low response characteristics of the springs or the dampers of the system often restrict the use of semi-active control.

In this research, we present an electrorheological fluid damper ideally suited for use in semi-active vibration control of robotic systems.

2 Electrorheological Fluids

Electrorheological (ER) fluids are suspensions of fine particles in liquids such as non-conducting oils. When subjected to an electric field, these fluids become a gel-like solid due to the suspended particles becoming polarized and aligned into chains along the direction of the field. Because these chains resist shear along a direction vertical to the field, the liquid reacts like a solid. When the field is removed, the material reverts back to a liquid state within milliseconds. Moreover, the degree of gelling is proportional to the field density, so, by varying the voltage, any rheological state from liquid to solid to liquid can be smoothly and instantly selected. This phenomenon is called the *electrorheological effect*.

As shown in Fig.1, in the absence of an electric field, an ER fluid exhibits Newtonian flow whereby the

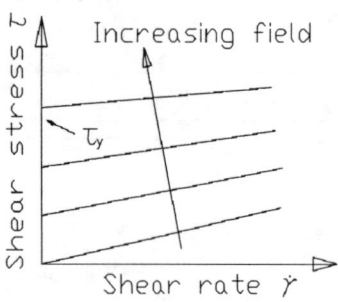

Fig. 1. Stress-shear rate behavior of an ER fluid

shear stress is directly proportional to the shear rate. When an electric field is applied, a yield stress phenomenon appears and no shearing takes place until the shear stress exceeds a minimum yield value that increases with the field density, i.e., the fluid appears to behave like a Bingham plastic. The shear stress is given by [2]

$$\tau = \tau_y + \mu \dot{\gamma} \qquad (1)$$

where τ_y is the yield stress, μ is the plastic viscosity, and $\dot{\gamma}$ is the shear rate.

The sensitivity of ER fluids to an electric field means that they can be directly controlled by a computer, eliminating the need for intermediate machinery. Hence, devices employing these fluids could meet the need for systems capable of responding to electrical signals faster and more precisely than existing technologies [2], [3], [8].

3 Electrorheological Damper

Previously, the authors described the design and analysis of an ER damper based on clearance restriction [4]. In this research an improved design is presented for an ER damper using orifice restriction. As shown in Fig.2, a cylindrical, outer housing is divided into two chambers by a separating web and a rotating vane fixed to the joint axis. The chambers are filled with an ER fluid. On the separating web is an orifice. Two electrodes are arranged to face each other across the orifice. When a high potential difference is applied across the two electrodes, an *ER effect* occurs in the fluid passing through the orifice.

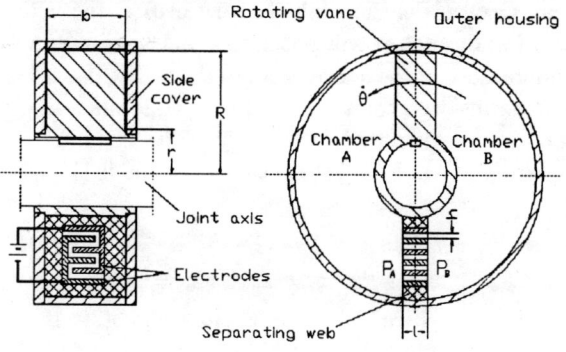

Fig. 2. Structure of the ER damper

The operational principle for the damper is as follows. As the vane rotates, the volume of fluid in both chambers changes. Due to the orifice restriction, the net pressure on the vane increases and the resulting pressure drop will cause a damping torque to oppose rotation. By varying the voltage applied across the electrodes, we can control the flow resistance of the fluid passing through the orifice. By this means, the output damping torque of ER damper can be controlled.

4 Model of the ER Damper

4.1 Mathematical model

To develop a mathematical model of the proposed damper, we make the following assumptions:

(1) The fluid is incompressible and the flow through the orifice is laminar;

(2) The pressure in each chamber is uniform;

(3) The theoretical model of ER fluids is Bingham plastic.

Under these assumptions, the output damping torque of ER damper can be expressed as [5]

$$M_R = M_P + M_f \qquad (2)$$

where M_p is the torque caused by the pressure drop acting on the vane, and M_f is the torque caused by Coulomb friction between the surfaces of rotating vane and two side covers.

As shown in Fig.3.

$$M_P = \int_{S(ABCD)} \Delta P \cdot r(x) \cdot ds \qquad (3)$$

where ΔP is the pressure drop across the chambers.

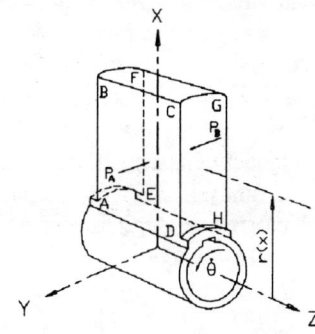

Fig. 3. Forces applied on the vane

Since M_f depends on the machining and assembly tolerances of the ER damper, it is hard to estimate. After the ER damper is fabricated and the ER fluid is selected, however, the variation of M_f is negligible. Here, it is treated as a system dependent constant that could be obtained by experiment. Therefore, the key issue in the model is to calculate the pressure drop across the chambers.

4.1.1 Pressure drop without an electric field

When no electric field is applied to the orifice, the ER damper behaves like an ordinary orifice restriction damper. Its pressure drop consists of two parts: the first part is the *restriction pressure drop*, caused by the orifice restriction; the second part, caused by the sudden changes in flow cross section and direction, and the effect of fluid hammer, is called the *impact pressure drop*.

The *restriction pressure drop* could be calculated from viscous flow theory. As shown in Fig.4, the rate of viscous steady flow passing through the plates is [6]:

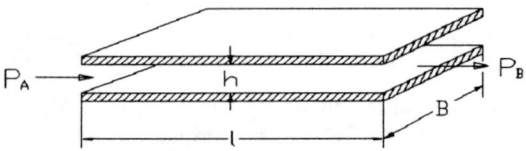

Fig. 4. Viscous flow through parallel plates

$$Q = \frac{(P_A - P_B)h^3 B}{12\mu l} \quad (4)$$

where P_A and P_B are the inlet and outlet pressures, respectively; h is the separation of the plates; B is the overlapping width of the plates; l is the overlapping length of the plates; and μ is the fluid viscosity.

Eq. (4) is a *laminar flow* model of viscous flow through parallel plates. Based on analysis [5], the *Reynolds number* of the ER fluid passing through the orifice is less than 20. Therefore, the flow is *laminar*. Since about a third of the particles are suspended in the fluid, the actual flow resistance of the ER fluid is much larger than that of ordinary viscous fluids. We introduce an ER fluid material related flow coefficient C_0 into Eq. (4)

$$Q = \frac{C_0(P_A - P_B)h^3 B}{12\mu l} \quad (5)$$

where C_0 (<1) decreases with an increase of the volume of the particles and their mass density in the ER fluid.

Thus, the pressure drop caused by orifice restriction is

$$\Delta P_{rstr0} = \frac{12\mu l Q}{C_0 h^3 B} \quad (6)$$

where B is the length of the middle axis line of the Z-shaped orifice, and

$$Q = \frac{1}{2}(R^2 - r^2)b\dot{\theta} \quad (7)$$

The *impact pressure drop*, which is proportional to the square of *flow speed*, could be estimated from tabular data [6]. Here, by means of an influence coefficient, the *impact pressure drop* is expressed as a function of the *restriction pressure drop* ΔP_{rstr0} and the *flow speed* $\dot{\theta}$. Therefore, we obtain a pressure drop in the chambers in the absence of an electric field

$$\Delta P_0 = (1 + \alpha\dot{\theta})\Delta P_{rstr0} \quad (8)$$

where α is the influence coefficient of the *impact pressure drop*. It depends on the structure and geometric size of the ER damper.

4.1.2 Pressure drop with an electric field

When an electric field is applied at the orifice, the ER fluid passing through it will generate a yield stress τ_y to resist liquid flow. If the shear stress in the orifice caused by the pressure drop cannot exceed the yield value, the vane and joint will remain stationary. The maximum pressure drop that the yield stress can withstand is [7]

$$\Delta P_E \cdot S = \tau_y \cdot A \quad (9)$$

where S is the orifice compression area and A is the shear area surrounding the orifice. Thus, we obtain

$$\Delta P_E = \frac{2l}{h}\tau_y \quad (10)$$

The influence of the *ER effect* corresponds to an On/Off valve. If the pressure drop across the chambers is $\Delta P = \Delta P_E$, the orifice will be closed.

When $\Delta P > \Delta P_E$, the fluid in the compression chamber could pass through the orifice into the other chamber. Due to the *ER effect*, the flow resistance of the orifice increases. Therefore, under the same pressure drop, the rate of flow at the orifice is less than that without the *ER effect*. The rate of flow, Eq. (5), becomes

$$Q = \frac{C_E C_0 (\Delta P_{rstr} - \Delta P_E) h^3 B}{12 \mu l} \quad (11)$$

where C_E is the *ER effect* flow coefficient which decreases with an increase of yield stress τ_y and can be approximately expressed as

$$C_E = (1 + \tau_y / \tau_{max})^{-1} \quad (12)$$

where τ_{max} is the maximum yield stress of the ER fluid.

From Eqs. (11) and (12) we obtain the *restriction pressure drop*

$$\Delta P_{rstr} = \frac{12 \mu l Q}{C_E C_0 h^3 B} + \Delta P_E \quad (13)$$

Similarly, considering the influence of the *impact pressure drop* we obtain the pressure drop across the chambers,

$$\Delta P = (1 + \alpha \dot{\theta}) \Delta P_{rstr} \quad (14)$$

4.1.3 Damping torque of the ER damper

By integrating Eq. (3) we obtain

$$M_P = \int_{s(ABCD)} \Delta P \, r(x) ds = \int_r^R \Delta P b x \, dx$$

$$= \frac{1}{2}(R^2 - r^2) b \Delta P = \frac{bl(R^2 - r^2)}{h}(1 + \alpha \dot{\theta}) \cdot$$

$$\left[\tau_y + \frac{3b(R^2 - r^2)}{C_0 h^2 B}\left(1 + \frac{\tau_y}{\tau_{max}}\right) \mu \dot{\theta} \right]$$

$$(15)$$

Substituting Eq. (15) into Eq. (2) we obtain the *damping torque* of the ER damper.

The relative parameters of the prototype ER damper are shown in Table 1.

Table 1. Parameters of the ER damper

R	r	b	B	l
27.5mm	10.5mm	13.9mm	16mm	11mm

h	C_0	α	τ_{max}	μ
0.5mm	0.5	3.1	6.6KPa	140mPa·s

4.2 Performance analysis

The damping torque of the ER damper can also be expressed as

$$M_R = M_0 + M_E + M_f \quad (16)$$

From Eq. (2), $M_0 + M_E = M_P$, where M_0 is the viscous damping of the damper in the absence of an electric field; M_E is the damping torque caused by the ER effect, called the *ER effect torque*; M_f is the Coulomb frictional torque caused by the friction between the surfaces of the rotating vane and the side covers.

In Eq. (16) the controllable part of the output torque of the damper is the *ER effect torque* M_E. Only when M_E is sufficiently large can the ER damper be useful. To evaluate the effect of M_E we compare the output torque of the ER damper with and without an applied electric field. The relative controllable range of the output damping torque of the ER damper is

$$K = \frac{M_0 + M_E + M_f}{M_0 + M_f} = 1 + \frac{M_E}{M_0 + M_f} \quad (17)$$

From Eq. (17) we see that the presence of the Coulomb frictional torque M_f decreases the relative controllable range of the ER damper. For reasons described in Section 4.1, we shall not explicitly consider the effect of M_f. Thus, we replace Eq. (17) by

$$K = \frac{M_P}{M_0} \quad (18)$$

$$M_E = M_P - M_0 \quad (19)$$

From Eqs. (18), (15), (14) and (8) we obtain

$$K = \frac{\Delta P}{\Delta P_0} = \frac{\Delta P_{rstr}}{\Delta P_{rstr0}}$$

$$= 1 + \left[\frac{1}{\tau_{max}} + \frac{C_0 h^2 B}{3\mu(R^2-r^2)b\dot{\theta}}\right]\tau_y$$

$$= 1 + (1.52 + 5.3/\dot{\theta}) \times 10^{-4}\tau_y \qquad (20)$$

From Eqs. (19), (15), (14) and (8) we obtain

$$M_E = \frac{bl(R^2-r^2)}{h}\left[1 + \frac{3\mu(R^2-r^2)b\dot{\theta}}{\tau_{max}C_0h^2B}\right](1+\alpha\dot{\theta})\tau_y$$

$$= (1.98 + 0.564\dot{\theta})(1 + 3.1\dot{\theta})10^{-4}\tau_y \qquad (21)$$

Eqs. (20) and (21) clearly show that an increase in the yield stress of the ER fluid is the key factor improve the performance of the damper. A decrease in the orifice separation is the most effective method to increase the output torque of the damper; and an increase of vane speed will cause the relative controllable range of the damper to decrease and increase its controllable range.

When $\tau_y = 4Kpa$ and $\dot{\theta} = 1/sec$, From Eqs. (20), (21) and (19) we obtain $K = 3.73$, $M_E = 4.17Nm$ and $M_R = 5.71Nm$.

5 Performance Test

To verify the mathematical model of the ER damper we tested its performance [5]. The ER fluid was obtained from ER Fluid Development Ltd., Sheffield, United Kingdom. The density of this fluid is $1.4g/mm^3$. The operation temperature was $25C^0 \sim 35C^0$. During the test we applied an electric field from *zero* to *5Kv/mm* and the speed of the rotating vane varied from *0.35/sec.* to *1.0/sec*. The yield stress of the ER fluid is shown in Table 2.

Table 2. Yield stress of the ER fluid

Electric field density (Kv/mm)	0	1	2	3	4	5
Yield stress (Kpa)	0	0.6	1.5	2.5	3.3	4.0

The results of the performance test shown in Table 3 and Fig. 5 confirmed the theoretical analysis. When the speed of the rotating vane is fixed, the output torque is proportional to the field density (yield stress); when the field density is fixed, the output torque is proportional to the speed of the rotating vane.

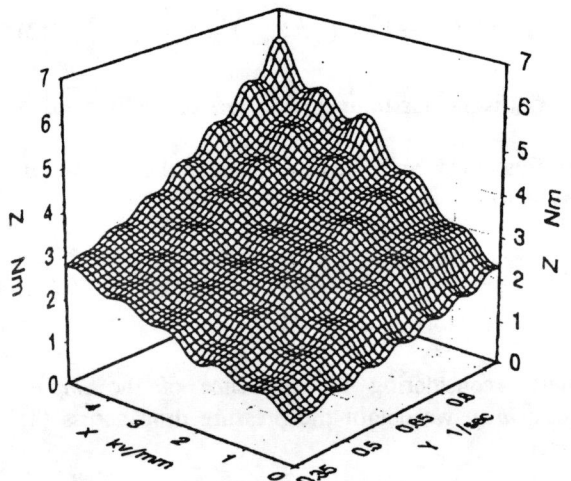

Fig. 5. Performance of the ER damper

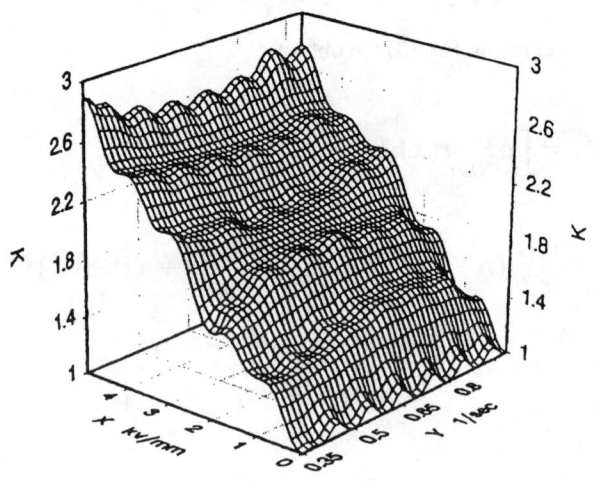

Fig. 6. Influence of electric field density on output torque

When the field density is *5Kv/mm*, K attains a minimum value of *2.6* (Fig. 6). The actual value of K is lower than its predicted value because the Coulomb frictional torque M_f was not included in the theoretical calculation (M_f is about $0.7Nm$). Imprecision in machining and assembly precision of the damper caused an increase in M_f. In spite of this difficulty, the influence of the ER effect is remarkable. By controlling the voltage applied at the orifice, the output torque increases 1.6 times within 10 milliseconds.

To verify the validity of the theoretical model, we substituted the test parameters in Table 3 and vane speeds and applied field strengths (converting them into yield stresses by means of Table 2) into Eq. (15). By this means we obtained the theoretical output torque which is shown in Table 4 (Coulomb frictional torque $M_f = 0.7Nm$ has been included in the results). Comparing the results in Table 3 and Table 4, the largest error is less than 21%, thereby confirming the theoretical model.

Table 3. Test output torque of the ER damper

Output Torque(Nm)		Applied field density (Kv/mm)					
		0	1	2	3	4	5
Vane Speed (rad /sec)	0.35	0.96	1.35	1.53	2.09	2.37	2.79
	0.5	1.19	1.67	2.25	2.50	2.87	3.24
	0.75	1.69	2.67	3.22	3.39	3.94	4.48
	1.0	2.29	3.00	3.94	5.02	5.38	6.34

Table 4. Theoretical output torque of the ER damper

Output Torque(Nm)		Applied field density (Kv/mm)					
		0	1	2	3	4	5
Vane Speed (rad /sec)	0.35	0.97	1.24	1.65	2.11	2.47	2.79
	0.5	1.17	1.52	2.04	2.62	3.08	3.13
	0.75	1.63	2.11	2.83	3.63	4.26	4.36
	1.0	2.23	2.85	3.79	4.83	5.67	5.83

Since the conductivity of the ER fluid is very low, the maximum current needed by the prototype damper does not exceed $1mA$. Therefore, the power consumption of the ER damper is very low.

6 Conclusions

We have described an electrorheological fluid rotary damper based on orifice restriction. From this research we conclude the following:

- The ability of ER fluids to respond rapidly to an electric field makes the ER damper ideally suited for use in semi-active vibration control systems.

- The torque generated by the damper consists of torques due to viscous damping, the *ER effect,* and Coulomb friction. When the geometry of the damper and the ER fluid are selected, the magnitude of viscous damping torque depends on the speed of the rotating vane; the *ER effect torque* depends on the applied field density; and the Coulomb frictional torque is a system dependent constant. Increasing the yield stress of the ER fluid is a key factor to improve the performance of the damper; and reducing the orifice separation is the most effective method to increase the output torque of the damper.

- The ER effect torque can be controlled by adjusting the voltage applied at the orifice. By this means the magnitude of the output torque increases 1.6 times within 10 milliseconds.

References

[1] Seto, K., "Trends on active vibrations control in Japan," *First International Conference on Motion and Vibration Control*, Yokohama, Sept. 1992, pp. 1-11.

[2] Burchill, P.J., "Electrorheological fluids and their application," *Materials Forum*, Vol. 15 No. 3, 1991, pp. 197-204.

[3] Goldstein, G., "Electrorheological fluids: applications begin to gel," *ASME Mechanical Engineering*, Vol. 112, Oct. 1990, pp. 48-52.

[4] Li, J., Jin, D., Zhang, X., Zhang, J., and Gruver, W. A., "An electrorheological fluid damper for robots," *1995 IEEE International Conference on Robotics and Automation*, May 1995, pp. 2631-2637.

[5] Li, J., "A study on the dynamic performance of a light-weight robot and a preliminary approach for a health care robot," *Ph.D. Thesis*, Department of Precision Instruments, Tsinghua University, Beijing, China, July 1996.

[6] Yeaple, F., *Fluid power design handbook*, Second Edition, Marcel Dekker., 1990.

[7] Bullough W. A., Peel, D. J., "Development of electrorheological fluids for application on mobile construction industry machinery," *Chinese Journal of Mechanical Engineering*, Vol. 23 No. 4, Dec. 1987, pp. 51-58.

[8] Kenneth, J. K., "Putting ER fluids to work", *Machine Design*, Vol. 63, No. 9, May 1991, pp. 52-53, 56-57, 60.

Safety Oriented Mechanism and Control Using ER Fluid in the Joint

Fumihito Arai*, Akiko Kawaji**, Toshio Fukuda***,
Hideo Matsuura* and Hiroshi Ota**

*Department of Micro System Engineering, Graduate School of Engineering
**Department of Mechanical Engineering, Graduate School of Engineering
***Center for Cooperative Research in Advanced Science & Technology
Nagoya University
Furo-cho 1, Chikusa-ku, Nagoya 464-01, JAPAN
E-mail: kawaji@robo.mein.nagoya-u.ac.jp

Abstract

Working robots with agents in the same place has been become known by the expansion of the use of a remote control robot. For that reason, we need the safety system which prevents a robot from a collision with an agent. An ER-Joint realizes high-speed response and force-limiting mechanism at the same time. We apply these properties to an autonomic safety mechanism to prevent from a collision accident. Then, we prove the stability of this system in the case of normal operations by Lyapunov's direct method, and actually apply the system to a micro-injector in order to show the effectiveness of the mechanism.

1 Introduction

The history of the research on a master-slave remote control system is long. But surprisingly, the fundamental general idea of this system has hardly changed from the beginning when the theory of the bilateral control system was presented in 1952 [1]. As for former researches of this system, reproducing more realistic replica, just like touching the environment directly of a slave robot on master side, was ideal to improve performance of operation. Therefore, it has become the most important subject of the research to realize this feeling. However, reproduction of the well-made replica by a lot of method, such as utilization of an acceleration signal bring about system complecations. Even if it is possible to make the system cleverly, operator's skill is needed in a difficult task after all. And it isn't necessarily connected with the increase in efficiency of the work. On the other hand, work with people or agents in the same place has been realized recently. In this environment, there is the possibility that the robot contacts with them. For that reason, the safety mechanism which prevents the robot from a drastic collision with the agents in the surrounding is indispensable. To satisfy these requirements, a human-machine cooperative work, adopting Shared Autonomy [2] into the master-slave remote control system, and combining simple bilateral control and a slave robot with a high autonomy function in other words, is the most effective.

In this paper, the safety in this work is investigated. Furthermore, as a part of the human-machine cooperative system realization, a mechanism is added to slave side like the autonomous machinery of the safety contact by the "ER-Joint" [3]. It utilizes the ability of quick response and the Bingham characteristic of the ER fluid [4]. This system prevents the accident when an operator makes an error or a machine goes out of control. And then, we demonstrate that the system can be controled stably in the usual work except for the time of danger avoidance by the Lyapunov's direct method. Finally, the danger avoidance system is actually applied to a micro-injector, the effectiveness of which system is shown.

2 Safety Condition in the Human-Machine Cooperative Work

Construction of the safety working system [5] has been realized by Sugimoto using two methods which are Interlock and Fail-Safe from about 1960's, as for the viewpoint which the safety means that a machine doesn't exist at the same time with a person in the group working space. What is more, safety conditions at the time when a robot and a human being exist simultaneously in a space and the force limiting equipment [6] which realizes the conditions were proposed in the recent research. What has been shown in the latest study, the safety condition S when a slave robot touches a person or an object is shown as the next formula with the interference force f formed to each other and the allowable marginal value f_H of the contact force.

$$S = \begin{matrix} 1 & : & f_H \geq f \\ 0 & : & f_H < f \end{matrix} \right\} \quad (1)$$

The confirmation of the situation that the interference force is less than f_H has to precede danger avoidance. However, the principle of the safety isn't concluded by the danger detection with the only force sensor that will go wrong possibly. Consequently, the mechanism which restricts the force output in the construction of the robot is necessary to make a safety system. To increase the efficiency of the work, the high rigidity and the repression of the vibration are demanded persistently within the range of f_H which satisfies the safety condition.

The structure which has a specific characteristic of threshold value is necessary for a force limiting equipment. The ER-Joint using the particle type of the ER fluid which has this characteristic is very effective as the equipment. But then, the general force limiting equipment isn't actually sufficient to work safely because it is weak against the shock acceleration. ER-joint, by contrast, functions as a stopping mechanism based on the information of the force sensor with high-speed response, too. We show how to build these double safety mechanisms at the same time using the ER-Joint in the following.

3 Safety Mechanism by ER fluid

The viscosity of ER fluid (Electro-rheological fluid) changes by an electric field. ER fluids are divided into two types roughly. One of them is the particle type which has Bingham characteristic, and the other is the liquid crystal type which has Newton behavior.

ER-Joint was made using the particle type fluid which has the quickness (less than several microseconds) of response concerning the voltage given between electrodes and the Bingham characteristic.

3.1 ER-Joint

Force transmission of the mechanical system arises between the links or the link and the environment. Then it is important to adjust the mechanical impedance [7] as desired to improve the performance of the mechanical system. How to adjust the mechanical impedance is classfied as [A] passive impedance method and [B] active impedance method. The active impedance method is superior in the point of adjustable range, but it requires actuator and sensor to control the force. In the field of robotic manipulator control, to realize active impedance method, there have been proposed (i) an approach based on the feedback control and (ii) that based on adjustment of the mechanical impedance. The first method has an disadvantage that the actuator size becomes big to reduce the settling time. The second method has an disadvantage that the mechanism becomes complicated.

On the other hand, the ER fluid has function to change damping property. It is impossible for the ER fluid itself to change the mechanical impedance arbitrarily. However, it has great advantage of quick response and easiness of miniaturization. To make use of these characteristics, we can improve the mechanical impedance control system. So, we proposed the ER-Joint. The ER fluid is filled in the joint. The damping property is changed by applying the voltage between the electrodes. Here, the role of the ER fluid is to change the damping property. It is impossible to control the restoring force by itself. The joint needs spring property to some extent. To change the spring property of the ER-Joint, we proposed (a) a method to use passive spring (passive type) and (b) that to change the spring property by the active control of the actuator (active type). Figure 1 shows the conceptual figure of the ER-Joint.

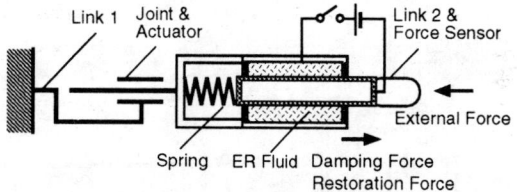

(a) Passive Type (Passive spring is used together.)

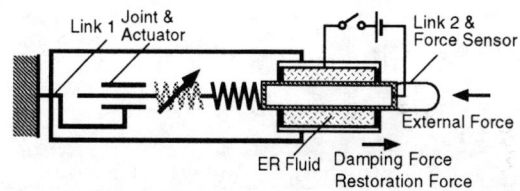

(b) Active Type (Actuator control is used together.)
Fig.1 Conceputual Figure of the ER-Joint

3.2 Shock Absorption Mechanism using ER-Joint

Figure 2 shows the ER-Joint with one degree of freedom. This machinery makes the double safety mechanism come into being at the same time.

Fig.2 ER-Joint

ER-Joint is fixed by an electric field exerted on the fluid in advance. When the force sensor furnished on the joint tip is normal, in case the force goes over allowable limited value f_S which is prescribed in the range that a safety collision is possible is detected, the power supply is cut off at once. As a consequence, the piston part of the ER-joint is shortened. It prevents the manipulator from a collision accident. Even if a trouble occurs in the force sensor, the principle of the safety is kept at the worst by restricting the volume of the initial voltage given to the ER fluid, which is sheared when the force gets over the threshold value f_H.

4 Modeling of the Control System

Think about the dynamics model, which one degree of freedom manipulator connected with the ER-Joint collides with the target that its movement characteristics are unknown, in figure 3.

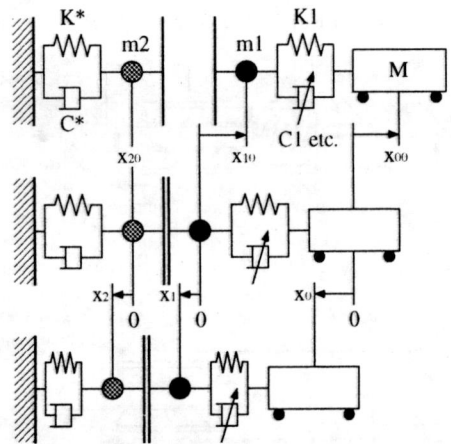

Fig.3 Mathematical Model for Control

4.1 Assumption

The following assumptions [8] are given.

Assumption 1. The manipulator and the force sensor of the tip are modeled on the mass-spring-damper system taken account of the resisting force which arises between the cylinder and the piston of the ER-Joint. The control input which is added to the system are independent of these parameters.

Assumption 2. The target is the linear system consists of mass, spring, and damper.

Assumption 3. The force of a collision is caused by inroads of the manipulator tip into the target. The force is modeled as Hertz shape multiplied by the parameter shown effect on attenuation.

Assumption 4. The power caused by a collision is only a positive (the condition which exerts the rebounding force each other).

4.2 Force of Resistance of the ER-Joint

We define the power of resistance f_{er} formed from the ER-Joint here.

Shear rate $\dot{\gamma}$ between the cylinder and the piston of the ER-Joint is given to the following.

$$\dot{\gamma} = \frac{\nu}{h} = \frac{\dot{x}_1 - \dot{x}_0}{h} \tag{2}$$

When there isn't an electric field, the shear stress of the ER fluid acts like a Newton fluid in proportion to $\dot{\gamma}$. On the other hand, the fixed stress σ_y is added to this under an electric field; accordingly, the fluid has a Bingham characteristic. Moreover, when it is compared with σ_y, increase in the term which is proportional to $\dot{\gamma}$ is minute. It is thus reasonable to suppose that a particle type of ER fluid has a characteristic of threshold value. σ_y is called the yield stress in this meaning. The shear stress σ can be defined as equation (3) because this yield stress is in proportion to square of the voltage E applied to ER-Joint in most of the particle fluids [9].

$$\begin{aligned} \sigma &= \sigma_y + \eta\dot{\gamma} \\ &= sgn(\dot{x}_1 - \dot{x}_0)\zeta E^2 + \eta\frac{\dot{x}_1 - \dot{x}_0}{h} \end{aligned} \tag{3}$$

$$sgn(\dot{x}_1 - \dot{x}_0) = \begin{cases} 1 & (\dot{x}_1 - \dot{x}_0 > 0) \\ 0 & (\dot{x}_1 - \dot{x}_0 = 0) \\ -1 & (\dot{x}_1 - \dot{x}_0 < 0) \end{cases} \tag{4}$$

Therefore, f_{er} is given as equation (5).

$$f_{er} = \left\{ sgn(\dot{x}_1 - \dot{x}_0)\zeta E^2 + \eta\frac{\dot{x}_1 - \dot{x}_0}{h} \right\} A \tag{5}$$

Variables and coefficients are defined as follows.

ν	:	Relative velocity between the cylinder and the piston of ER-Joint
h	:	Distance between the pole plates filled with ER fluid
x_0	:	Position of the manipulator
x_1	:	Position at the tip of the force sensor
E	:	Voltage applied to ER fluid
A	:	Area of a pole plate
ζ	:	Proportion coefficient toward the voltage
η	:	Proportion coefficient toward the shear rate

4.3 Equation of Motion

The equations of motion concerning three masses are made up with reference to the resisting force formed from the ER-Joint.

$$\begin{aligned} M\ddot{x}_0 = &-K_1(x_0 - x_1) - C_1(\dot{x}_0 - \dot{x}_1) \\ &- \left\{ sgn(\dot{x}_0 - \dot{x}_1)\zeta E^2 + \eta\frac{\dot{x}_0 - \dot{x}_1}{h} \right\} A + f \end{aligned} \tag{6}$$

$$\begin{aligned} m_1\ddot{x}_1 = &-K_1(x_1 - x_0) - C_1(\dot{x}_1 - \dot{x}_0) \\ &- \left\{ sgn(\dot{x}_1 - \dot{x}_0)\zeta E^2 + \eta\frac{\dot{x}_1 - \dot{x}_0}{h} \right\} A - f_w \end{aligned} \tag{7}$$

$$m_2\ddot{x}_2 = -K^* x_2 - C^* \dot{x}_2 + f_w \qquad (8)$$

The meaning of each character in this equation is the following.

- M : Mass of the manipulator with the Cylinder part of ER-Joint
- m_1 : Mass of the force sensor with the piston part of ER-Joint
- m_2 : Mass of the target
- x_2 : Position of the target
- f : Control input to the actuator of the manipulator
- f_w : Collision power
- K_1, C_1 : Spring fixed number and a viscous coefficient between M and m_1
- K^*, C^* : Spring fixed number and a viscous coefficient of m_2

The control input f is shown as equation (9).

$$\begin{aligned} f &= -K_p(x_0 - x_d) - K_v \dot{x}_0 \\ &= -K_p x_0 - K_v \dot{x}_0 \end{aligned} \qquad (9)$$

- $x_d (= 0)$: Target of the position
- K_p : Proportional gain concerning the position
- K_v : Gain of velocity

4.4 Collision Force

The collision force is represented by the next equation.

$$f_w = \begin{cases} (1 + p\dot{u})Hu^{3/2} & (u \geq 0 \text{ and } \dot{u} \geq -1/p) \\ 0 & (\text{the others}) \end{cases} \qquad (10)$$

where u and $\dot{u}$ are given by

$$u = x_1 - x_2, \quad \dot{u} = \dot{x}_1 - \dot{x}_2 \qquad (11)$$

In other words, u is the relative position between the target and the tip of the manipulator, and $\dot{u}$ means the relative velocity of the tip with the target. p is the parameter of an energy loss. A perfect elastic collision takes place at the time of $p = 0$. On the other hand, a perfect one occurs at the time of $p > 0$. And, H is a great positive number based on the theory of Hertz.

$$p \geq 0, \quad H > 0 \qquad (12)$$

In this study, we also have to model the force during the collision so that we may assume that the collision time can be measured finitely, when the system stability is investigated. Here we adopted equation (10) [10] thought to be the most suitable. We think equation (13) contacting condition, and the other cases non-contacting conditions by the assumption 4.

$$u \geq 0 \text{ and } \dot{u} \geq -1/p \qquad (13)$$

4.5 Voltage applied to ER-Joint

Until the detection value f_o of the force pressed against the target by the force sensor gets over the allowable marginal value f_S of the contacting force, the voltage E applied to the ER-Joint are fixed to do the ordinary work. Then, once f_o gets over f_S, E is turned off until the safety confirmation of the system is finished.

$$E = \begin{cases} E_0 & (\text{before becoming } f_o \geq f_S) \\ 0 & (\text{after } f_o \geq f_S) \end{cases} \qquad (14)$$

E_0 : initial voltage applied to ER-Joint

Here E_0 is expressed in equation (15) by threshold value f_H of the force limiting equipment.

$$E_0 = \sqrt{\frac{f_H}{\zeta A}} \qquad (15)$$

5 The Stability Analysis of the System

As for the danger avoidance in the state of emergency, safety of the system has to be secured first of all. ER-Joint can fulfill the role in the emergency. On the other hand, this system must be able to control stably in the usual work. Here, we show the stability of the system when the detection value f_o of the force caused by a collision is within the allowable limited value f_S by using the stability analysis of Lyapunov.

We define a non-linear scalar function as equation (16).

$$\begin{aligned} V = &\tfrac{1}{2}\{M\dot{x}_0^2 + m_1 \dot{x}_1^2 + m_2 \dot{x}_2^2 \\ &+ K_1(x_1 - x_0)^2 + K_p x_0^2 \\ &+ K^* x_2^2\} + \Delta k \cdot u^2 \end{aligned} \qquad (16)$$

where Δk is expressed as follows.

$$\Delta k = \begin{cases} \tfrac{2}{5}H u^{1/2} & (u \geq 0 \text{ and } \dot{u} \geq -1/p) \\ 0 & (\text{the others}) \end{cases} \qquad (17)$$

The scalar function V consists of the clauses which have the form of the perfect square, and it is a non-negative. Differentiate this.

$$\begin{aligned} \dot{V} = & M\dot{x}_0 \ddot{x}_0 + m_1 \dot{x}_1 \ddot{x}_1 + m_2 \dot{x}_2 \ddot{x}_2 \\ &+ K_1(x_1 - x_0)(\dot{x}_1 - \dot{x}_0) + K_p x_0 \dot{x}_0 \\ &+ K^* x_2 \dot{x}_2 + \tfrac{5}{2}\Delta k \cdot u \dot{u} \\ = & -C_1(\dot{x}_1 - \dot{x}_0)^2 \\ &- (\dot{x}_1 - \dot{x}_0) sgn(\dot{x}_1 - \dot{x}_0)\zeta E^2 A \\ &- \eta \tfrac{A}{h}(\dot{x}_1 - \dot{x}_0)^2 - K_v \dot{x}_0^2 - C^* \dot{x}_2^2 \\ &- (\dot{x}_1 - \dot{x}_2)f_w + \tfrac{5}{2}\Delta k \cdot u \dot{u} \end{aligned} \qquad (18)$$

Let us consider a non-contacting condition and a contacting condition separately.

5.1 Case of the Non-Contacting Condition

Equation (18) becomes (20) because f_w and Δk is shown by equation (19) in a case of $u < 0$ or $\dot{u} < -1/p$.

$$f_w = 0, \ \Delta k = 0 \tag{19}$$

$$\begin{aligned}\dot{V} &= -C_1(\dot{x}_1 - \dot{x}_0)^2 \\ &\quad -(\dot{x}_1 - \dot{x}_0)sgn(\dot{x}_1 - \dot{x}_0)\zeta E^2 A \\ &\quad -\eta\frac{A}{h}(\dot{x}_1 - \dot{x}_0)^2 - K_v\dot{x}_0^2 - C^*\dot{x}_2^2\end{aligned} \tag{20}$$

The 2nd clause of the right side of this equation is a non-plus because $\dot{x}_1 - \dot{x}_0$ has the sign which is always the same as of $sgn(\dot{x}_1 - \dot{x}_0)$ and this clause consists of these and a positive coefficient and a minus sign. Moreover, the other clause also contain a minus sign, a positive coefficient and forms of the perfect square, so it is a non-plus obviously. Therefore, equation (20) becomes (21).

$$\dot{V} \leq 0 \tag{21}$$

5.2 Case of the Contacting Condition

In the case of $u \geq 0$ and $\dot{u} \geq -1/p$, equation (18) becomes (23) because f_w and Δk is expressed as (22).

$$\begin{aligned}f_w &= (1 + p\dot{u})Hu^{3/2} \\ \Delta k &= \tfrac{2}{5}Hu^{1/2}\end{aligned} \tag{22}$$

$$\begin{aligned}\dot{V} &= -C_1(\dot{x}_1 - \dot{x}_0)^2 \\ &\quad -(\dot{x}_1 - \dot{x}_0)sgn(\dot{x}_1 - \dot{x}_0)\zeta E^2 A \\ &\quad -\eta\frac{A}{h}(\dot{x}_1 - \dot{x}_0)^2 - K_v\dot{x}_0^2 - C^*\dot{x}_2^2 \\ &\quad -(\dot{x}_1 - \dot{x}_2)(1 + p\dot{u})Hu^{3/2} \\ &\quad + \tfrac{5}{2}\cdot\tfrac{2}{5}Hu^{3/2}\dot{u} \\ &= -C_1(\dot{x}_1 - \dot{x}_0)^2 \\ &\quad -(\dot{x}_1 - \dot{x}_0)sgn(\dot{x}_1 - \dot{x}_0)\zeta E^2 A \\ &\quad -\eta\frac{A}{h}(\dot{x}_1 - \dot{x}_0)^2 - K_v\dot{x}_0^2 - C^*\dot{x}_2^2 \\ &\quad -pHu^{3/2}\dot{u}^2\end{aligned} \tag{23}$$

The 6th clause of equation (23) is a non-plus on the grounds that it is based on $u \geq 0$ and has the form of the perfect square of $\dot{u}$. Equation (24) comes into being in the same way as the case of the non-contacting condition.

$$\dot{V} \leq 0 \tag{24}$$

Hence V became Lyapunov function, and the stability of this safty system which contains a collision is shown from the above. Incidentally, it has already been shown that there is no problem concerning the continuity and the differentiability of this Lyapunov function by Shoji et al. [11].

6 Experiment system

The experiment of the safety mechanism using ER-Joint was done by applying to micro-injection. As common knowledge, the micro-injection is the work in which we inject pharmaceutical liquid into a cell, absorb the body fluid inside a cell, hold an ovule by suction. The tip of the micro-injecter is made of a very fragile glass tube generally. Hence it is sometimes damaged by the collision with the environment due to a little mis-operation. And then, the actuator used for micro-manipulation is a stepping motor and so on to satisfy the requirements of the precision and the movable range; however, the responce of this actuator is slow. For this reason, it is considered that the safety mechanism using the ER-Joint is effective to prevent the tip of micro-manipulator from being destroyed. In addition, we can introduce the safety mechanism into the micro-territory well, since there is hardly any influence of scaling on the ER-Joint [3].

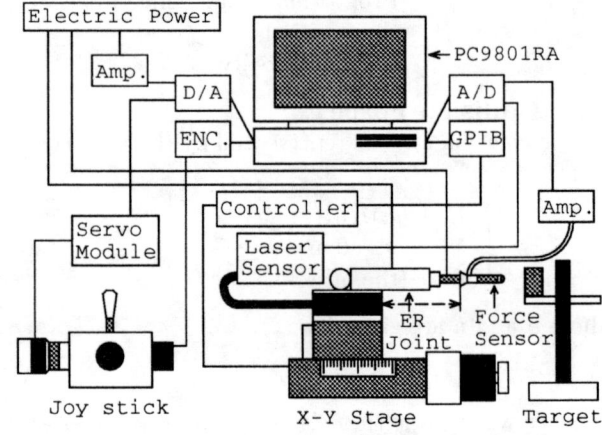

Fig.4 System Configuration

We show the configration of this experiment system in figure 4. The slave side which consists of the ER-Joint and the stepping motor in the X-Y stage has 1 degree of freedom, and is installed a force sensor [12] and a laser displacement sensor in the tip part of the ER-Joint. The master side used a joystick also has 1 degree of freedom, and a encoder and a motor are built in this equipment. While we can expect the stepping motor of the precision in micro-order, we are not able to remove the problem of slow response, since the motor is originally a position control type despite realization of tentative velocity control in this system. We send on-off information to the power supply of the voltage applied to ER-Joint by the value of the force sensor.

7 Experiment

We experiment with ER-Joint practically to confirm the quickness of the responce velocity. Figure 5 shows the output of the force sensor and the laser displacement meter when the ER-Joint is contacted with the target in 1000 pps (= 1mm/s) and $f_S = 5$ gf ($= 4.90 \times 10^{-2}$N). We see from (a) that in the case of 0V, the output of the force sensor remains within the range of f_S, and decreases after one value increasing since the threshold value is lower than the force of collision. In (b), noteworthy change isn't seen by the graph because the value of the force sensor doesn't get over f_S though the threshold value rises. The voltage becomes 0 when the force gets over f_S in (c). The responce time is less than 30msec even if the sampling time of front and back is included. While the stepping motor of the X-Y stage takes the responce time of about 300msec, ER-Joint can react at the time of 1/10 and under.

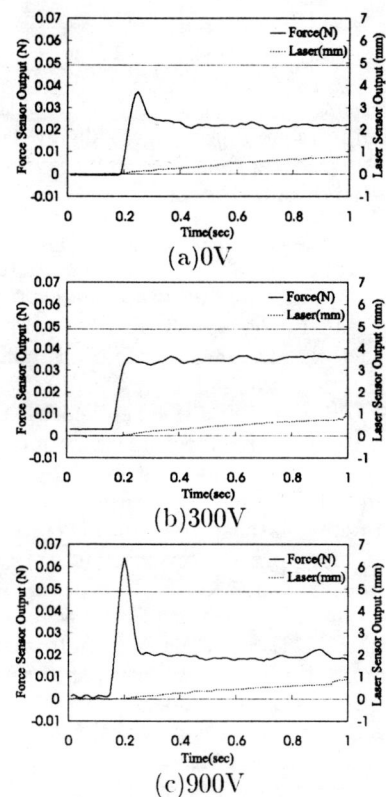

Fig.5 Response of ER-Joint

8 Summary

In this paper, we argued the need of the human-machine cooperative system first. Then we examined about the safety condition in the cooperative work in the 2nd chapter as a part of the realization. In the 3rd chapter we proposed to add the autonomy function using ER-Joint which realizes a double safety mechanism at the same time on slave side. Moreover, we showed the control model and the stability of the system in the 4th and 5th chapter. Finally we applied the system to micro-manipulation, and experimented to confirm its validity in the 6th and 7th chapters.

References

[1] Raymond C. Goertz and Frank Bevilacqua, "A Force-Reflecting Positional Servamechanism" Nucleonics, Vol.10, No11, pp.43-45, (1952).

[2] Shigeoki Hirai, "A Theoretical View of Shared Autonomy" Journal of the Robotics Society of Japan (in Japanese), 11-9, pp.822-827 (1993).

[3] Fumihito Arai, et al, "Bio-Micro-Manipulation (New Direction for Operation Improvement" Proceedings of the IEEE/RSJ International Conference on IROS, Volume 3, pp.1300-1305, (1997).

[4] W. M. Winslow, "Induced Fibration of Suspensions", Journal of Applied Physics, 20, pp.1137-1140, (1949).

[5] Masayoshi Sakai, et al., "Logic and Method of Safety in Controlling a Power Press" JSME International Journal, Series C, Vol36, No.4, (1993).

[6] Thuyoshi Saito and Noboru Sugimoto, "Basic Requirements and Construction for Safe Robots", JSME Annual Conference on Robotics and Mechatronics (in Japanese), 96-17, Vol.B, pp.1173-1176, (1996).

[7] Toshio Morita and Shigeoki Sugano, "Design and Development of a new Robot Joint using a Mechanical Impedance Adjuster", IEEE Internal Conference on Robotics and Automation, 95, Vol.3, pp.2469-2475, (1996).

[8] Takanori Shibata, et al, "Sencing and Control for Robotic Motions by Neural Network", Math. Modelling and Sci. Computing, Vol.1, No.3-4, pp.247-262, (1993).

[9] Howard See and Masao Doi, "Shear resistance of electrorheological fluids under time-varying electric fields", Journal of Rheology, 36(6), pp.1143-1163, (1992)

[10] Takafumi Fujita and Shinobu Hattori, "Periodic Vibration and Impact Characteristics of a Nonlinear System with Collision", Bulletin of the Japan Society of Mechanical Engineers, 23-177, pp.409-418, (1980)

[11] Yasumasa Shoji, Makoto Inaba, Toshio Fukuda, "Impact Control of Grasping", IEEE Transactions on Industrial Electronics, Vol.38, No3, pp.187-194, (1991).

[12] Mitsutaka Tanimoto, et al., "Micro Force Sensor for Intravascular Neurosurgery", IEEE Internal Conference on Robotics and Automation, 95, Vol.2, pp.1561-1566, (1997).

3-DOF Closed-loop Control for Planar Linear Motors

Arthur E. Quaid and Ralph L. Hollis

The Robotics Institute, Carnegie Mellon University
{aquaid, rhollis}@ri.cmu.edu

Abstract

Planar linear motors (Sawyer motors) have been used in industry as open-loop stepping motors, but their robustness and versatility has been limited. Using a sensor recently integrated into such a motor, a closed-loop 3-DOF controller has been implemented. The software-based control system consists of a commutator for computing amplifier currents from actuator forces, a force resolution function for solving the redundant actuation and saturation problems, and an observer for producing a velocity estimate, together with a PID controller. Experiments are performed using a 2-axis laser interferometer to show that the controller has sub-micron resolution, 2 μm peak-to-peak repeatability, and settling times after trajectories of about 20 ms. Limitations of the PID controller are discussed and ideas for improvements are presented.

1 Introduction

Commercial planar linear motors are available that have micron-level precision over meter-sized planar workspaces. They can move with velocities of several meters per second and accelerations of several g's. However, due to the lack of a suitable position sensor, they have been operated as open-loop stepping motors. It was recognized early that performance could be enhanced through sensing [1], but early attempts to develop sensors were not very successful.

Recently, prototype sensors have been designed and developed. In [2] a 1-DOF magnetic platen sensor was mounted on an outrigger off a commercial planar motor, and a PID controller was implemented, although few results are presented. In [3] a 1-DOF sensor of a different design was similarly mounted, and preliminary results for PD control suggested improved resolution, stiffness, and settling times.

Our group has recently completed development of a compact 3-DOF magnetic platen sensor. It has been integrated into the center of a commercial planar linear motor, in space that was otherwise wasted, and has a linear resolution (1σ) of 0.2 μm and an angular resolution of 0.0014°. With sensing (described in a companion paper [4]) and closed-loop control (the subject of this paper), planar linear motors have the potential for much wider application. Guarded moves, fault detection, automatic registration, compliant motions, and cooperative manipulation are some of the

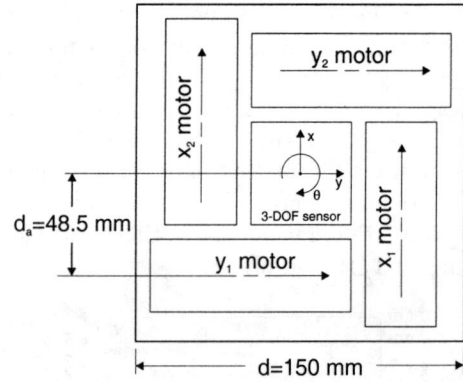

Figure 1: Schematic view of the underside of the forcer. Four linear motors combine to provide a 3-DOF force capability. An integral platen sensor provides 3-DOF sensing for closed-loop control.

abilities possible with closed-loop control. In [5], an adaptive controller for a 3-DOF planar motor is presented with simulation results, although sensor noise and actuator saturation appear to be neglected.

In more detail, planar linear motors consist of a moving *forcer* that translates in two directions on a passive steel *platen* stator surface etched with a waffle-iron type pattern. The forcers are supported by a 12-15 μm thick air bearing pre-loaded by permanent magnets, and require a tether to supply air and power. These forcers and platens are available commercially.

The particular forcer examined in this paper is shown schematically in Fig. 1. Two pairs of *motors*[1] mounted orthogonally generate balanced forces about the center of mass. Each of the four motors consists of a stack of laminations and two coils, shown schematically in Fig. 2. The motors operate on a flux-steering principle, with the coil currents acting to switch the permanent magnetic flux from one set of poles to the other. The poles with the most flux tend to align themselves with the platen teeth, so that by activating the poles in the proper order, a stepping motion is achieved. The coil currents can also be *microstepped* by applying a sine wave to one coil and a cosine wave to the other. A more detailed presentation of open-loop operation can be found in [6].

---

[1] Here, *motor* refers to one of the four actuators on the forcer, and *planar linear motor* refers to the entire device.

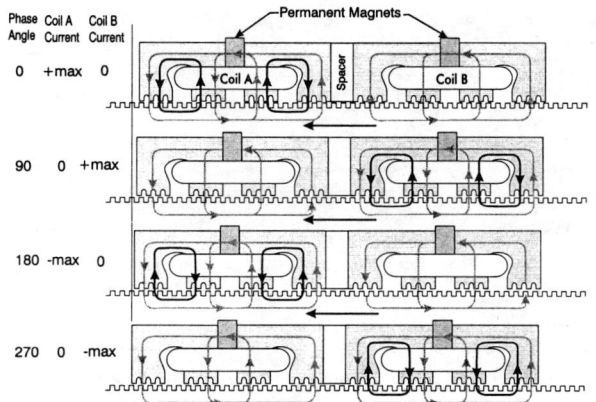

Figure 2: Basic linear motor operation: Currents in the motor coils generate magnetic flux (dark flux path) that sums with the permanent magnet flux (light flux path) to produce forces.

The next section describes the software components of the control system. Section 3 presents the hardware system and the experimental results. The paper concludes with a summary of the results and a discussion of the limitations of the implemented controller.

2 Controller Formulation

A block diagram of the system is shown in Fig. 3. There are four software blocks required. The *estimator* is used to compute a velocity and position estimate based on the sensor position output and the motor dynamics. The *controller* computes a wrench ($\mathbf{w} = [f_x\ f_y\ \tau]^T$) command to be generated by the forcer, based on the estimator outputs and desired trajectory. The *force resolution* block is needed to handle the redundancy of the force generation, as four actuators are used to generate a 3×1 wrench. Finally, the *commutator* computes currents to send to the motor coils based on the forcer position and the desired forces.

It will simplify things to specify wrenches and positions at the center of mass or center of actuation for different blocks. The wrench at the center of actuation, $\mathbf{w}_{ca}$, and the wrench at the center of mass, $\mathbf{w}_{cm}$, are related by:

$$\mathbf{w}_{ca} = \begin{bmatrix} 1 & 0 & 0 \\ 0 & 1 & 0 \\ -p_y & p_x & 1 \end{bmatrix} \mathbf{w}_{cm}, \quad (1)$$

where $\mathbf{p} = [p_x\ p_y]^T$ is the location of the center of mass of the forcer expressed in the coordinate system shown in Fig. 1. Similarly, the positions and velocities are related by:

$$\begin{bmatrix} x \\ y \\ \theta \end{bmatrix}_{ca} = \begin{bmatrix} 1 & 0 & p_y \\ 0 & 1 & -p_x \\ 0 & 0 & 1 \end{bmatrix} \begin{bmatrix} x \\ y \\ \theta \end{bmatrix}_{cm} - \begin{bmatrix} p_x \\ p_y \\ 0 \end{bmatrix}, \text{ and} \quad (2)$$

$$\begin{bmatrix} \dot{x} \\ \dot{y} \\ \dot{\theta} \end{bmatrix}_{ca} = \begin{bmatrix} 1 & 0 & p_y \\ 0 & 1 & -p_x \\ 0 & 0 & 1 \end{bmatrix} \begin{bmatrix} \dot{x} \\ \dot{y} \\ \dot{\theta} \end{bmatrix}_{cm}, \quad (3)$$

where the small rotation range of the motor justifies linearization of the equations about the zero angle.

In this remainder of this section, each software block is examined, with relevant motor modeling introduced as needed.

2.1 Commutation

Derivation of the commutation functions involves first finding a suitable model of the force generated by a single motor segment, given the amplifier inputs. An inversion of this model is then used for commutation.

As described more fully in [7], each linear motor segment generates forces according to the equation:

$$\begin{aligned} f_{x1} &\triangleq f(i_{x1}, x_1, \psi_{x1}) \\ &= k_a(i_{x1}, x_1, \psi_{x1}) \sin\left(\frac{2\pi}{p} x_1 - \psi_{x1}\right), \end{aligned} \quad (4)$$

where x_1 is the motor position in its direction of force generation, p is the pitch of the motor, and ψ_{x1} is the motor phase commanded by the amplifier. k_a is a proportionality factor which depends largely on amplifier current, but also on the motor position and skew angle. In this work, all experiments are performed with small ($\ll 1°$) rotations and are not overly sensitive to force ripple. Thus, it is sufficient to ignore the dependence on position and angle. Then, k_a is a linear function of only the amplifier current i_{x1}, and (4) becomes:

$$f_{x1} = k i_{x1} \sin\left(\frac{2\pi}{p} x_1 - \psi_{x1}\right). \quad (5)$$

The commutator needs to find an i_{x1} and ψ_{x1} that generate a commanded force f_{x1} by inverting (5). There are an infinite number of solutions, but as introduced in [7], two interesting possibilities are a *fixed amplitude* solution, where i_{x1} is set to a constant value and ψ_{x1} is varied according to f_{x1}, and a *fixed phase* solution, where ψ_{x1} is chosen constant relative to the motor position and i_{x1} is varied. In this work, the *fixed phase* approach is used, allowing the amplifier currents to be zero when f_{x1} is zero, which reduces thermal effects. With this approach, the commutator chooses amplifier inputs according to the equations:

$$i_{x1} = \frac{f_{x1}}{k}, \text{ and} \quad (6)$$

$$\psi_{x1} = \frac{2\pi}{p} x_1 - \frac{\pi}{2}. \quad (7)$$

The position x_1 is computed based on the sensed forcer position. At high speeds, it is also important to add *phase advance* to compensate for the latency between the sensor reading and the control output [7].

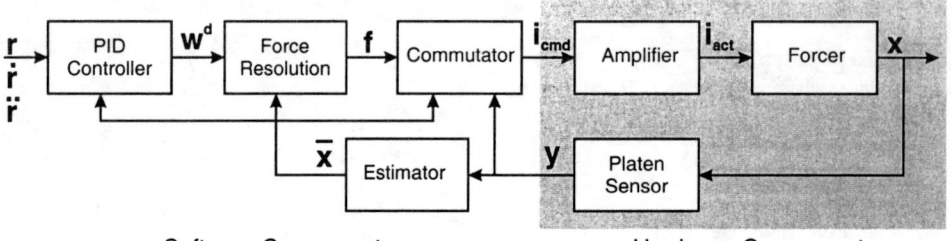

Figure 3: Controller block diagram

This term requires a velocity estimate. The position of each linear motor can then be computed:

$$\begin{bmatrix} x_1 \\ x_2 \\ y_1 \\ y_2 \end{bmatrix} = \begin{bmatrix} 1 & 0 & -d_a \\ 1 & 0 & d_a \\ 0 & 1 & -d_a \\ 0 & 1 & d_a \end{bmatrix} \left(\mathbf{y}_{ca} + t_a \begin{bmatrix} \bar{\dot{x}} \\ \bar{\dot{y}} \\ \bar{\dot{\theta}} \end{bmatrix}_{ca} \right), \quad (8)$$

where t_a is the phase advance time, $[\bar{\dot{x}}\ \bar{\dot{y}}\ \bar{\dot{\theta}}]^T$ is the *estimated* forcer velocity, and $\mathbf{y}$ is the *sensed* forcer position, both expressed at the center of actuation.

2.2 Force resolution

The force resolution function must consider the force kinematics and force saturation properties of the forcer. The force kinematics are given by:

$$\mathbf{w}_{ca}^d = \begin{bmatrix} f_x^d \\ f_y^d \\ \tau^d \end{bmatrix}_{ca} = \begin{bmatrix} 1 & 1 & 0 & 0 \\ 0 & 0 & 1 & 1 \\ -d_a & d_a & -d_a & d_a \end{bmatrix} \begin{bmatrix} f_{x1} \\ f_{x2} \\ f_{y1} \\ f_{y2} \end{bmatrix}, (9)$$

where $\mathbf{w}_{ca}^d$ is a wrench applied at the center of the forcer, and d_a is the distance from the center of the motor to the center of the forcer, as shown in Fig. 1.

The force saturation properties of the motor sections are determined by their maximum rated current: $f_{max} = k i_{max}$. If each of the four motor sections can generate up to a maximum force f_{max}, the wrench saturation constraints for the forcer are:

$$|f_x^d| \leq 2f_{max}, \quad (10)$$
$$|f_y^d| \leq 2f_{max}, \text{and} \quad (11)$$
$$|\tau^d| + d_a \left(|f_x^d| + |f_y^d| \right) \leq 4f_{max} d_a. \quad (12)$$

The *force resolution problem* is to find a suitable solution to (9) subject to these constraints.

It is helpful to consider this problem geometrically. By eliminating the absolute values, these equations can be expanded into 12 inequalities linear in the wrench parameters f_x, f_y, and τ. These constraints combine to form a *wrench envelope* $\mathcal{E}$ that can be represented in $I\!R^3$ as a rhombic dodecahedron, depicted

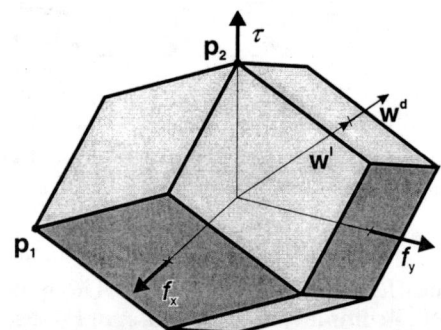

Figure 4: A *wrench envelope* represents the force/torque limits of the forcer. The darker faces result from constraint equations (10,11), and the lighter faces result from constraint equation (12).

in Fig. 4. For example, at point $\mathbf{p}_1$, the forcer is generating the maximum possible force in both the $+x$ and $-y$ directions and cannot generate any torque. Alternatively, at point $\mathbf{p}_2$, the forcer is using its full capabilities to generate torque, and cannot generate any forces.

The input to the force resolution function is the desired wrench $\mathbf{w}_{ca}^d$, which may be outside, inside or on $\mathcal{E}$ leading to over-, under-, and uniquely-constrained cases.

In the under-constrained case, an infinite number of solutions can be found that satisfy (9)-(12). Here, we choose the solution:

$$\begin{bmatrix} f_{x1} \\ f_{x2} \\ f_{y1} \\ f_{y2} \end{bmatrix} = \begin{bmatrix} \frac{1}{2} & 0 & -\frac{a}{a+b}\frac{1}{2d_a} \\ \frac{1}{2} & 0 & \frac{a}{a+b}\frac{1}{2d_a} \\ 0 & \frac{1}{2} & -\frac{b}{a+b}\frac{1}{2d_a} \\ 0 & \frac{1}{2} & \frac{b}{a+b}\frac{1}{2d_a} \end{bmatrix} \begin{bmatrix} f_x^d \\ f_y^d \\ \tau^d \end{bmatrix}_{ca}, \quad (13)$$

where $a = 2f_{max} - |f_x^d|$ and $b = 2f_{max} - |f_y^d|$. Note that a and b are a measure of the remaining force capability of the forcer in the x and y directions. Note that a or b are both non-negative, and because the wrench is inside the force envelope, at least one is non-zero. Therefore, (13) will always be defined, and (by examining the derivative) can also be shown to be continuous.

If the desired wrench lies on envelope $\mathcal{E}$, there is a single solution. This solution is identical to the under-constrained case except that it becomes undefined when $a = b = 0$. In this special case the desired wrench is at one of the corners in the f_x, f_y plane (e.g. $\mathbf{p}_1$) in Fig. 4, so that $\tau = 0$, and (13) is replaced by:

$$\begin{bmatrix} f_{x1} \\ f_{x2} \\ f_{y1} \\ f_{y2} \end{bmatrix} = \begin{bmatrix} 1/2 & 0 & 0 \\ 1/2 & 0 & 0 \\ 0 & 1/2 & 0 \\ 0 & 1/2 & 0 \end{bmatrix} \begin{bmatrix} f_x^d \\ f_y^d \\ \tau^d \end{bmatrix}_{ca}^d . \quad (14)$$

In the over-constrained case, $\mathbf{w}_{ca}^d$ lies outside the wrench envelope. In this case, the forcer is saturated, and there are no solutions to (9) that satisfy the constraints. However, if we redefine the problem for this case to be mapping $\mathbf{w}_{ca}^d$ back onto the wrench envelope, there are once again infinite solutions. One simple solution is to linearly scale the desired wrench vector back to the point where it pierces the wrench envelope. For example, $\mathbf{w}_{ca}^d$ in Fig. 4 is mapped to $\mathbf{w}_{ca}^l$. They are related by a scale factor s_l:

$$s_l = \max\left(\frac{|f_x^d|}{2f_{max}}, \frac{|f_y^d|}{2f_{max}}, \frac{|f_x^d| + |f_y^d| + |\tau^d|/d_a}{4f_{max}} \right), \quad (15)$$

such that $\mathbf{w}_{ca}^l = \mathbf{w}_{ca}^d / s_l$.

Although this solution has been implemented, the experiments are designed to avoid saturation cases. Saturation in general is a difficult non-linearity to deal with effectively. In this case it also acts to couple the axes, requiring a more complicated controller design.

2.3 Control

Given the commutator and force resolution functions, the controller can be designed around a simple linear model. The mechanical dynamics of the forcer are simply those of a mass moving in the plane:

$$\mathbf{w}_{cm} = \begin{bmatrix} m & 0 & 0 \\ 0 & m & 0 \\ 0 & 0 & I_z \end{bmatrix} \begin{bmatrix} \ddot{x} \\ \ddot{y} \\ \ddot{\theta} \end{bmatrix} + \begin{bmatrix} b_v(\dot{x}) & 0 & 0 \\ 0 & b_v(\dot{y}) & 0 \\ 0 & 0 & b_\omega(\dot{\theta}) \end{bmatrix}, \quad (16)$$

where m is the forcer mass and I_z is the rotational inertia. The b_v and b_ω functions model the eddy-current damping, but are neglected for simplicity and because the damping is negligible relative to the amount of added controller damping. To express the dynamics in this decoupled form, I_z and $\mathbf{w}_{cm}$ are defined relative to the center of mass of the forcer. The gains of a PID control law,

$$u = K\left(e + T_d \dot{e} + \int \frac{1}{T_i} e \right), \quad (17)$$

can then be chosen independently for each axis. Feed-forward acceleration terms from Eq. 16 are also added to the controller forces.

2.4 State estimation

The estimator exploits the linear dynamic model of the forcer to produce a filtered position and velocity signal without excessive lag. The discrete time estimator takes the form:

$$\bar{\mathbf{x}}_{cm}(k+1) = \begin{bmatrix} \boldsymbol{\Phi} & 0 & 0 \\ 0 & \boldsymbol{\Phi} & 0 \\ 0 & 0 & \boldsymbol{\Phi} \end{bmatrix} \bar{\mathbf{x}}_{cm}(k) +$$

$$\begin{bmatrix} \mathbf{L} & 0 & 0 \\ 0 & \mathbf{L} & 0 \\ 0 & 0 & \mathbf{L} \end{bmatrix} \tilde{\mathbf{y}}_{cm}(k) + \begin{bmatrix} \frac{T^2}{2m} & 0 & 0 \\ \frac{T}{m} & 0 & 0 \\ 0 & \frac{T^2}{2m} & 0 \\ 0 & \frac{T}{m} & 0 \\ 0 & 0 & \frac{T^2}{2I_z} \\ 0 & 0 & \frac{T}{I_z} \end{bmatrix} \mathbf{w}_{cm}^l (18)$$

with

$$\boldsymbol{\Phi} = \begin{bmatrix} 1 & T \\ 0 & 1 \end{bmatrix}, \quad \mathbf{L} = \begin{bmatrix} l_1 & 0 \\ 0 & l_2 \end{bmatrix}, \text{ and}$$

$$\tilde{\mathbf{y}}_{cm}(k) = \mathbf{y}_{cm}(k) - \begin{bmatrix} 1 & 0 & 0 & 0 & 0 & 0 \\ 0 & 0 & 1 & 0 & 0 & 0 \\ 0 & 0 & 0 & 0 & 1 & 0 \end{bmatrix} \bar{\mathbf{x}}_{cm}(k).$$

Here, $\mathbf{y}_{cm}(k) = [x\ y\ \theta]^T$ is the sensor position output at time k, $\bar{\mathbf{x}}_{cm}(k) = [\bar{x}\ \dot{\bar{x}}\ \bar{y}\ \dot{\bar{y}}\ \bar{\theta}\ \dot{\bar{\theta}}]^T$ is the state estimate at time k, and T is the sample time.

Note that the estimator is decoupled and the estimator gains l_1 and l_2 can be computed based on the desired estimator pole locations using well-known pole-placement techniques (i.e. MATLAB's[2] `PLACE` command).

3 Experimental results

The software blocks in Fig. 3 are implemented on a Motorola PowerPC 133 MHz computer running the LynxOS real-time operating system. I/O hardware consist of a number of Industry Pack (IP) credit-card sized modules on an ISA bus carrier. The computer and I/O hardware fit in a standard mini-tower PC case, and connect to the planar linear motor through a tether. The planar linear motor consists of a Normag platen and a modified Normag 2-phase forcer with a 1.016 mm pitch, 60 N nominal static force, 1.4 Kg mass, and 4 A peak operating current. The test setup includes a Zygo 2-axis laser interferometer, which can measure the differential skew angle and one translational axis of the forcer, providing a position measurement independent of the magnetic platen sensor.

The software is structured with a single high-priority thread running at 3500 Hz that includes the commutation, sensor I/O, force resolution, controller,

---
[2]Product of MathWorks, Inc.

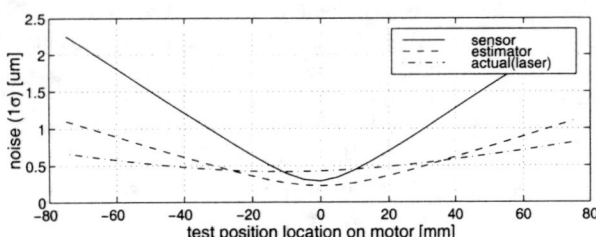

Figure 5: The controller regulates points near the sensor better than points on the periphery of the forcer.

observer, and trajectory generator functions. This set of functions takes approximately 150 μs to complete, leaving some time for lower-priority user interface and network communications threads to execute. The fast update rate is necessary to provide a reasonable number of updates per pitch. With a forcer pitch of 1.016 mm, and a peak speed above 1 m/s, even with 3500 Hz updates there may may be less than four updates per pitch. However, a more complicated controller can still be implemented by separating the commutator and controller into separate threads, with the controller running at a slower rate than the commutator.

3.1 Resolution tests:

To characterize the position resolution under closed-loop control, the PID controller was used to regulate the forcer to zero position and angle. The gains of the controller were $K = 220$ N/mm, $T_d = 0.0053$ s, and $T_i = 0.028$ s. The controller poles are underdamped with natural frequency of 40 Hz. Estimator poles were placed at 80 Hz, which was near the upper limit for the controller rate of 3500 Hz. Above this rate, an unacceptable level of noise from the sensor was passed into the controller, causing the forcer to be *audibly* noisy.

Readings for the x translation and θ skew angle from the platen sensor, estimator, and laser interferometer were recorded at 3500 Hz for 1000 samples. The amount of motion in the x direction of a particular point on the forcer was computed using the simple differential kinematic equation:

$$\delta x_t = \delta x - \delta \theta \, y_t, \qquad (19)$$

where δx_t is the x differential motion of a test point (x_t, y_t) on the forcer given a differential motion at the middle of the forcer of δx and $\delta \theta$. Figure 5 shows the standard deviation of δx_t as recorded by the sensor, estimator, and laser interferometer as y_t is varied. Note that points with $|y_t| = 75$ mm correspond to the edge of the forcer. Because the sensor measures angle by differencing two parallel position measurements that are close together (see [4]), there is a low-noise *sweet spot* in the middle of the forcer, where the sensor is located. However, Fig. 5 indicates that even at the edge of the forcer, the controller maintains micron-level resolution (1σ), which is sufficient for many applications.

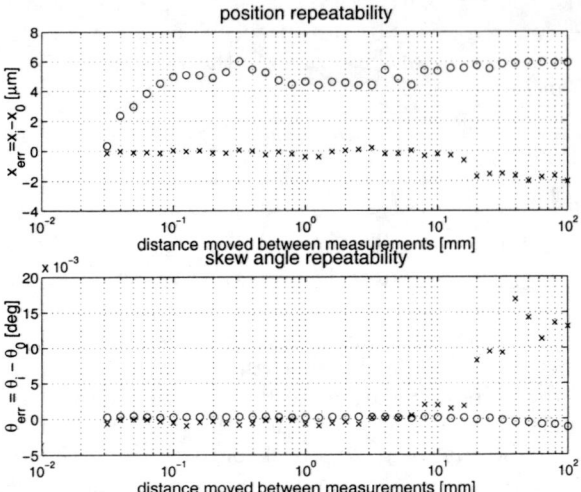

Figure 6: Bi-directional repeatability: errors in moving to a reference position from two approach directions are shown for open-loop (circles) and PID control (crosses) for a range of motion distances.

3.2 Repeatability

Repeatability is the ability of an actuator to return to the same position. The difference in the forcer positions after moving to a reference location from two different directions was measured with the laser interferometer. This process was repeated with varying move distances. The controller was started several minutes before testing began, and the test was designed to be completed in under a minute to minimize thermal effects. The crosses in Fig. 6 show this bi-directional position repeatability for the PID controller to be under 2.5 μm peak-to-peak, and the skew angle repeatability to be under 0.02° peak-to-peak over 36 motions. Note that the error increases at a travel distance of about $20mm$, which is when the linear motors start to overlap the reference position, so it appears likely that the motors are leaving a residual magnetic field in the platen that is slightly affecting the sensor operation. The exact mechanism for this interaction is under investigation. For comparison, the bi-directional repeatability tests are repeated under open-loop control. The error here is probably due to a combination of tether disturbance forces and the same magnetic hysteresis that affects the sensor.

3.3 Trajectory commands

The closed-loop controller was used to track a trajectory with a bang-bang acceleration of 10 m/s^2, maximum velocity of 0.8 m/s, and a position change of 0.1 m. Integral gains were disabled for this experiment to prevent integral windup during the motion.[3]

As shown in the dark traces in Fig. 7, the PD controller tracks the trajectory to within 50 μm, and settles to 1 μm within 20 ms. The tracking error comes

---
[3] Gain-scheduling could be used to re-enable the integral gain near the end of the trajectory to correct the steady-state errors.

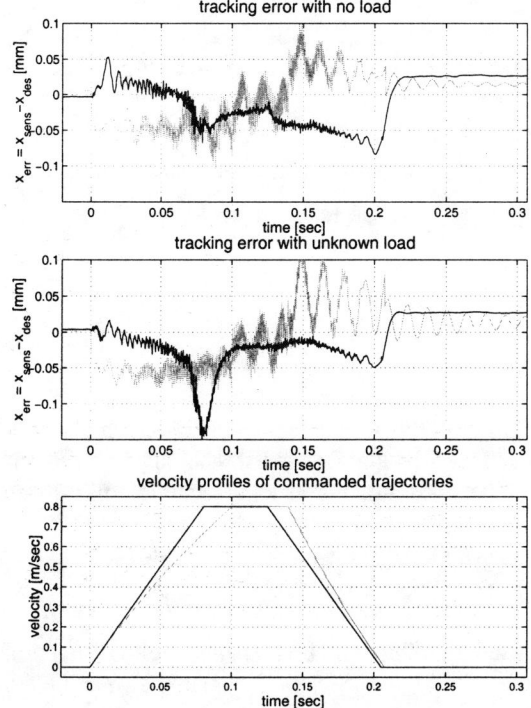

Figure 7: Trajectory tracking: tracking errors are shown for PD (dark trace) and open-loop (light trace) control given no load (upper plot) and an "unknown" load of 240 g (middle plot). The lower plot shows the velocity profile of the trajectories.

from motor modeling errors (i.e. eddy-current damping, errors in the mass or peak force model parameters, or unmodeled actuator nonlinearities.). The tracking error is also shown when a mass of 240 g is attached to the edge of the forcer, causing errors in in the mass, inertia, and center of mass of the model. Even in this case, the controller settles just as well, although the tracking error increases significantly during one part of the trajectory.

For comparison, the tracking experiment was repeated under open-loop control. However, it would be unfair to have the open-loop controller attempt to track the same bang-bang acceleration trajectories. Instead, the trajectory described in [8] was implemented. This trajectory is identical to the bang-bang acceleration trajectory, except for *burst* (step changes applied to the desired position) and *acceleration rolloff* as the velocity increases. These changes are designed so that the open-loop forcer dynamics will not be excited and, in the ideal case, there will be a constant tracking error during each phase (acceleration, slew, and deceleration) of the trajectory.

After a reasonable amount of tuning of the trajectory, tracking errors were as shown in the light traces of the top plot of Fig. 7. Although the tracking error levels are reasonably good, there is a significant oscillation, indicating that some dynamic parameters had errors or there were unmodeled nonlinearities. This oscillation grew worse when the extra load was added, as shown in the middle plot. Furthermore, in both cases, the open-loop controller takes much longer to settle at the end of the trajectory.

4 Conclusions

This work has demonstrated 3-DOF closed-loop control of a planar linear motor using an integrated platen position sensor. Experiments indicate submicron resolution and repeatability, and improved tracking and settling time relative to open-loop control. The *force resolution* problem was identified as a consequence of the actuator redundancy and a solution was presented.

Limitations were found with the PID controller. Trajectories with accelerations or velocities near the limits of the forcer could not be reliably performed due to the non-linear and coupling effects of actuator saturation. Also, at high velocities, disturbance torques (in practice, disturbances from the tether were most common) may momentarily saturate the torque capability of the motor. To reject this disturbance, a PD or PID controller would need an unrealistically high bandwidth. We are presently exploring the use of non-linear switching controllers to address these problems.

Acknowledgements

This work is supported in part by NSF grants DMI-9523156 and DMI-9527190, and the CMU Engineering Design Research Center. Quaid was supported by an AT&T Fellowship. We acknowledge Yangsheng Xu's early participation in this work, and Alfred Rizzi for work on the real-time computing hardware and operating system and for many helpful discussions. Finally, this work would not be possible without the efforts of Zack Butler in development of the platen sensor.

References

[1] B. A. Sawyer, "Linear magnetic drive system." U. S. Patent 3,735,231, May 22 1973.

[2] F. Y. Wong, H. Schulze-Lauen, and K. Youcef-Toumi, "Modelling and digital servo control of a two-axis linear motor," in *Proc. American Control Conference*, pp. 3659–3663, June 1995.

[3] J. Ish-Shalom, "Sub-micron large motion multi-robot planar motion system," in *Proc. 7th Int'l Symp. on Robotics Research*, pp. 582–595, October 1995.

[4] Z. Butler, A. Rizzi, and R. Hollis, "Integrated precision 3-DOF position sensor for planar linear motors," in *Proc. IEEE Int'l Conf. on Robotics and Automation*, May 1998.

[5] F. Khorrami, H. Melkote, and J. Ish-Shalom, "Advanced control system design for high speed ultra accurate manufacturing systems," in *Proc. of the American Control Conference*, pp. 164–165, June 1997.

[6] E. R. Pelta, "Precise positioning without geartrains," *Machine Design*, pp. 79–83, April 1987.

[7] A. E. Quaid, Y. Xu, and R. L. Hollis, "Force characterization and commutation of planar linear motors," in *Proc. IEEE Int'l Conf. on Robotics and Automation*, April 1997.

[8] J. I. Nordquist and P. M. Smit, "A motion-control system for (linear) stepper motors," in *Proceedings of the Fourteenth Annual Symposium on Incremental Motion Control Systems and Devices*, pp. 215–231, 1985.

Cell Mapping Based Fuzzy Control of Car Parking

Ming C. Leu[1] and Tea-Quin Kim[2]

1. National Science Foundation, Arlington, Virginia, U.S.A., on leave from New Jersey Institute of Technology, Newark, New Jersey, U.S.A.

2. Samsung Inc., Seoul, KOREA.

Abstract – This paper describes the development of a near-optimal fuzzy controller for maneuvering a car in a parking lot. To generate the rules of the fuzzy controller, a cell mapping method is utilized to systematically generate near-optimal trajectories for all possible initial states in the parking lot. Based on the input-output relations of these trajectories, which represent the states and controls of the corresponding cells, a set of fuzzy rules are generated automatically. In order to result in a small number of fuzzy rules from the large amount of numerical information generated by cell mapping, grouping of trajectories is proposed and each rule applies to the cells in one group. This reduces substantially the number of rules in the fuzzy controller compared with establishing the rules directly using the control data of individual cells.

I. INTRODUCTION

Fuzzy control is one of the most successful application areas of the fuzzy theory invented by Zadeh [1]. It is effective for solving control problems which are difficult or even impossible to solve by developing precise mathematical models. To construct a fuzzy controller, the generation of fuzzy rules is the main issue. Fuzzy rules are often generated by extracting knowledge from skillful human operators.

The application of fuzzy control for a car parking control problem was introduced by Sugeno et al. [2]. Several methods have since been proposed for the generation of fuzzy rules applicable to solving this problem. Kosko [3] developed a space clustering method to obtain fuzzy rules with the fuzzy associative memory. Wang and Mendel [4] proposed a fuzzy rule generation method by assigning a degree to each rule from input and output sample data. Lin and Lee [5] constructed the input-ouput mapping using a neural network. These research works were all based on the numerical input and output data obtained by a human expert with lots of experiments for the states that the car encounters.

Cell mapping can generate global optimal controls for nonlinear dynamic systems. This approach involves dividing the continuous state space into finite discrete cells. Hsu [6] developed the original cell-to-cell mapping method which first generates all possible paths and then searches for optimal paths with a systematic search algorithm. A modified method developed by Zhu [9] was used by Zhu and Leu [10] to solve the optimal trajectory planning problem for industrial manipulators. To implement, at each sampling time a system state is identified and the cell corresponding to this state is checked against the controller table to look for the control to use in the next time interval. A problem with the table-based control is that the table can be very large, especially if the system has many inputs and outputs. Table based control may also give a bumpy response as the controller jumps from one table value to another.

From the perspective of fuzzy control, the optimal control strategy obtained from cell mapping can be incorporated into the fuzzy controller. The near-optimality is realized by deriving fuzzy rules based on the numerical data of optimal cell mapping. A fuzzy controller constructed from the cell mapping data provides a promising solution for the global optimal control of nonlinear dynamic systems. In our work described herein, the fuzzy controller construction is based on cell mapping generated optimal control data of a dynamic system, instead of using data from human experiments.

To construct a fuzzy controller based on the numerical information of the optimal cell mapping data, there are two main issues: first, how to design the rules of the fuzzy controller; second, how to manage the great amount of numerical information generated by cell mapping. Kang and Vachtsevanos [7] generated fuzzy rules based on cell mapping with a tree search approach. In their method, each cell of the state space is used as a fuzzy region, i.e. a region for a fuzzy rule. However, the number of fuzzy rules can be enormous if the amount of optimal cell mapping data is very large. It is desirable to have a fuzzy controller with a small number of fuzzy rules for the entire state space.

In this paper, a cell mapping based systematic method is developed to construct a fuzzy controller for car parking control. Near-optimal car trajectories are created from the cell mapping data, and trajectories with similar features are collected to form groups. Fuzzy control rules and membership functions are then expressed with respect to the trajectory groups instead of individual cells. The developed fuzzy controller is shown in simulation to have a performance similar to that of the table-based controller generated from optimal cell mapping.

II. Cell Mapping

To generate the global optimal trajectory, the description of trajectories (or paths) in task space is converted to a description in discrete cell state space. In the discrete cell state space, a cell mapping algorithm [6, 9] plans the optimal trajectory with the dynamic equations of a system and its constraints. The cell mapping algorithm constructs tabular numerical information which represents trajectories and their control actions.

To generate global optimal trajectories of a dynamic system, the continuous state space is divided into finite discrete cells. Then, the behavior of the dynamic system is transformed into a description in the discrete cell space. The discrete cells are in the form of rectangular shape. To construct the discrete cell space, first each axis of the continuous space is divided into a number of intervals, with size h_i for axis x_i. An interval is denoted by an integer z_i, i.e.

$$\left(z_i - \tfrac{1}{2}\right) h_i \le x_i \le \left(z_i + \tfrac{1}{2}\right) h_i \tag{1}$$

The n-tuple $(z_1,...,z_n)$ is called a cell and is denoted by z. n is the dimension of the cell space; for example, n is 2 for (x, y) and is 3 for (x, y, θ).

Let the control output from actuators be a p-dimensional vector, u, given by:

$$u = [u_1, u_2, ..., u_p] \tag{2}$$

The differential equation of an n-dimensional dynamic system can be generally expressed as

$$\dot{x} = f(x, u(t), t) \tag{3}$$

where $x \in R^n$ and $t \in R$. By solving this set of differential equations, a point mapping can be established in the form of

$$x(k+1) = x(k) + \int_{t_k}^{t_{k+1}} f(x(k), u(k)) dt$$
$$k = 1,..,N \tag{4}$$

A cell mapping can be constructed from a point mapping and written in the form

$$z(k+1) = H(z(k), u(k), t(k)) \tag{5}$$

where $z \in I^n$ is an n-tuple of integers.

The dynamics of the system are expressed in terms of cell-to-cell mappings. Determining the image of a cell can proceed as follows: For a given cell $z(k)$, first find the coordinates of its center $x(k)$. Under control $u(k)$ and time duration $t(k)$, $x(k+1)$ is determined as the image of $x(k)$ by point mapping. If the corresponding cell of point $x(k+1)$ is $z(k+1)$, then $z(k+1)$ is the image cell of $z(k)$.

The functional to be minimized for finding an optimal path assumes the form

$$J = \sum_{k=1}^{m} (c_k \int_{t_k}^{t_{k+1}} f_0(x_k, u_k) dt + t_k) \tag{6}$$

where k is a number used to indicate the mapping step, m is the total number of steps from the initial cell to the target cell, and c_k is a coefficient used to weight the two terms in J. For time optimal control, all c_k's are zero. If it is important to save control energy, c_k's should be large.

After the discrete cell state space is constructed and the goal state (target) given, optimal cell-to-cell mappings are performed to search for optimal trajectories from various initial states to the target in the discrete cell state space satisfying all the conditions and dynamic equations.

III. Fuzzy Systems

A fuzzy controller consists of a set of fuzzy control rules processed with a system of logic inference. The fuzzy rule base consists of a collection of rules which represents all possible situations and their corresponding control actions. The logic inference [11] converts the linguistic labels of the rule base into a final crisp control action. Figure 1 illustrates the structure of a rule-based fuzzy controller.

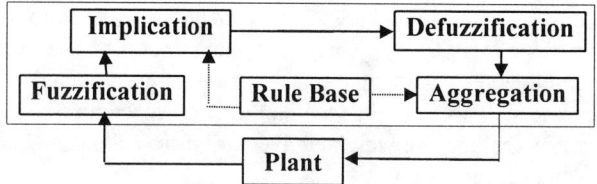

Fig. 1 Structure of rule-based fuzzy controller.

The rules in a fuzzy controller take the following form:

If <condition>, then <control action> (7)

The <condition> is the premise part and the <control action> is the consequence part. For example, a fuzzy control rule to steer a vehicle toward a parking target may be: If Parking Target x is 'ahead,' then steering angle y is 'straight.' The terms x and y are referred to as the state variable and control variable, respectively. The terms "ahead" and "straight" are linguistic labels. The adjectives such as (ahead, straight) in the rules do not correspond to precisely defined ranges.

A fuzzy variable is a linguistic term (or label) associated with a membership function. The fuzzy rule base represents qualitative knowledge in a fashion which can be interpreted by a fuzzy logic process. Among the common rule-based representations, one is the Mamdani type:

R_1: if x_1 is F_1^1 and x_2 is F_2^1 ... x_n is F_n^1, then y is A_1;

R_2: if x_1 is F_1^2 and x_2 is F_2^2 ... x_n is F_n^2, then y is A_2;

...

R_l: if x_1 is F_1^l and x_2 is F_2^l ... x_n is F_n^l, then y is A_l;

(8)

where F_1^i, F_2^i,..., and A_i are premise and consequence linguistic labels, respectively.

The process of fuzzification is assigning a "belief" value to a real-valued input variable with respect to each membership function. It maps each coordinate x_i of a crisp point x into a fuzzy set. Implication maps the belief values of the premise part to the output of the fuzzy logic system, which is the consequence part. Each fuzzy IF-THEN rule defines a fuzzy implication. Many fuzzy implication rules have been proposed in the fuzzy logic literature. Two commonly used fuzzy implication rules are the Min-operation implication and Product-operation implication. Defuzzification generates real valued controls. There are many defuzzification methods. Among them the center of gravity method and the mean of maximum method are used frequently. Since multiple rules of a fuzzy control system can be active simultaneously, all of the active rules are combined to create the final result. One simple way to combine the rules is taking the weighted average of the outputs. This is the aggregation step.

IV. CONSTRUCTION OF NEAR-OPTIMAL FUZZY CONTROLLER

We present in this section a method which uses the numerical data obtained from optimal cell-to-cell mapping to efficiently generate fuzzy rules for constructing a near-optimal fuzzy controller. We first discuss how to group trajectories for this purpose.

IV.1 Defining Trajectories

To define trajectories from the optimal control table and group them based on similarity features, we introduce the concept of initial cell, simple cell, and merged cell. A cell z_i is an initial cell (IC) if there exists no z_j such that $H(z_j) = z_i$ and $z_i \neq z_j$ for $j=1,...,N_{c_i}$. The cell z_i is a simple cell (SC) if there exists only one z_j such that $H(z_j) = z_i$ and $z_i \neq z_j$. If there exists more than one z_j, then z_i is a merged cell (MC). A trajectory is a set of connected cells, each evolves either from an initial cell to a merged cell or from a merged cell to another merged cell.

The following procedure is used to derive trajectories in a cell state space for the purpose of grouping:

Step 1: Each cell in the set $\{z_1, z_2, ..., z_r\}$ is assigned as an IC, SC, or MC.

Step 2: Construct the set $\Re$ consisting of IC's and MC's.

Step 3: From any cell $S_i \in \Re$, various cells are linked according to Eq. (5) until $S_j \in \Re$, $S_i \neq S_j$, is encountered. This forms one trajectory from S_i to S_j.

Step 4: Step 3 is performed repetitively for every element in $\Re$.

For illustration, the gray rectangles in Figure 2 are initial cells, the white rectangles are simple cells, and the black rectangles are merged cells. There are four trajectories in Figure 2: one from IC_i to MC_p, one from IC_j to MC_h, one from IC_k to MC_h, and one from MC_h to MC_p.

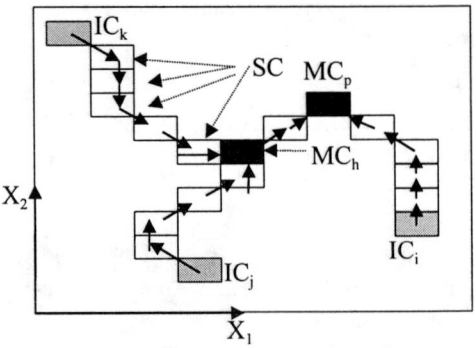

Fig. 2 Illustration of simple cells, initial cells, merged cells, and trajectories on cell state space.

IV.2 Features of Trajectories

To group similar trajectories, we consider features which distinguish trajectories by the overall locations and controls of their cells. If the difference between the overall locations and controls of the individual cells of two trajectories is small enough, these two trajectories are said to be similar. Since two trajectories generally have different numbers of cells, it is not possible to compare the locations and controls of their individual cells. We thus use the mean location and the mean control of the cells for each trajectory. Let T_a and T_b be the two trajectories under examination. The features of them are the mean locations, $\bar{x}_a$ and $\bar{x}_b$, and the mean controls, $\bar{u}_a$ and $\bar{u}_b$. Based on these features, we can compute the difference between the two trajectories as follows:

$$\eta_1 = |\bar{x}_a - \bar{x}_b| \tag{9}$$

$$\eta_2 = |\bar{u}_a - \bar{u}_b| \tag{10}$$

$$\beta = \gamma_1 \eta_1 + \gamma_2 \eta_2 \tag{11}$$

The values of γ_1 and γ_2 represent the weights of individual features and they can be chosen depending on the relative importance of the features.

IV.3 Grouping Similar Trajectories

The basis for grouping similar trajectories is Equation (11). The grouping of trajectories is achieved by beginning with empty group lists. The first trajectory to be classified forms the first group. Assume that second trajectory considered for grouping has a feature difference of β from the trajectory in the first group. If this difference is less than or equal to the predefined threshold β_T then the considered trajectory is placed in the first group. Otherwise it forms another group which is the second group. New groups are generated whenever the minimum feature difference between a newly considered trajectory and the trajectories of all established groups exceeds the threshold. The process continues until all the trajectories in the cell space have been categorized.

IV.4 Fuzzy Rule Base and Membership Functions

The main purpose of grouping trajectories is to obtain simplified state and control information for the generation of fuzzy rules and membership functions. Cell states are in the IF-part and control levels are in the THEN-part of a fuzzy rule. There could be different ways of using the cell states and control levels of a trajectory group. An intuitive one is to use a representative trajectory of a group, such as the first trajectory or the longest trajectory of a group. Another is to consider all cells of a group and extract their statistical properties, which may include the mean location of the cells, its standard deviation, mean control, etc. Based on these statistical properties, fuzzy membership functions corresponding to the various groups can be created.

In this study, we generate the fuzzy rules and membership functions based on the statistical properties of the trajectory groups. If there are k groups, G_j ($j = 1,.., k$), then the number of fuzzy rules is k. The membership functions we use have trapezoidal or triangular shape. Figure 3 illustrates the generation of fuzzy membership functions which are projected onto input axes. The projections of a group onto the input axes become the fuzzy membership functions of the IF-part. The mean location, $m_{xi,j}$, obtained by calculating the coordinates of all cells of group G_j is the middle point of the membership function of this group. The standard deviation, $\sigma_{xi,j}$, of the mean location of the group is the width of the fuzzy membership function. Based on these two statistical values, a set of fuzzy premises can be created for the fuzzy rule base. The membership functions for the fuzzy consequences are similarly created.

After designing the fuzzy rule base and membership functions, a global fuzzy controller is constructed. Constructing the global fuzzy controller consists of fuzzification, implication, defuzzification, and aggregation, as described previously. In the simulation described in Section V, the fuzzy control system is constructed using the following: 1) the rule base has the Mamdani type, 2) the implication uses the Min-operation, 3) the defuzzification is based on the center of gravity method, and 4) the aggregation uses the weighted average of the outputs.

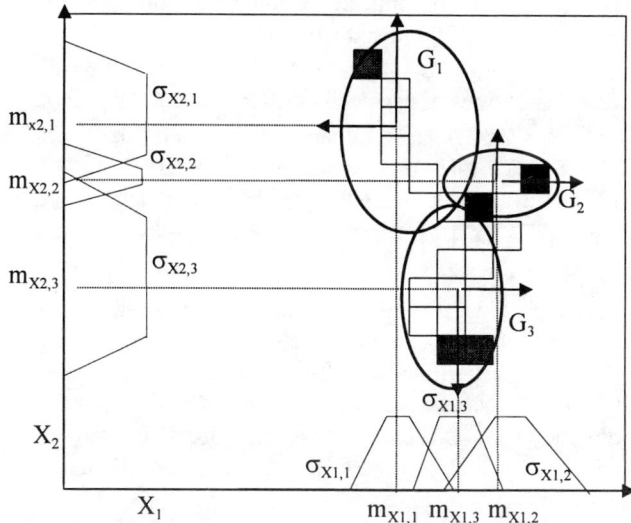

Fig. 3 Each group defines a fuzzy rule region and its membership functions on the input axes.

V. FUZZY CAR PARKING CONTROL

V.1 The Car Parking Control Problem

The car parking control problem is illustrated in Figure 4 which shows a car and its parking lot. The state variables representing the car are x, y, and ϕ, which are the Cartesian coordinates of the center of the rear wheels and the angle of the vehicle with respect to the horizontal axis. The objective of the control is to move the center of the rear wheels, (x, y), to the target, (x_d, y_d), with the vehicle perpendicular to the dock, i.e. $\phi_d = \pi/2$, at the target position.

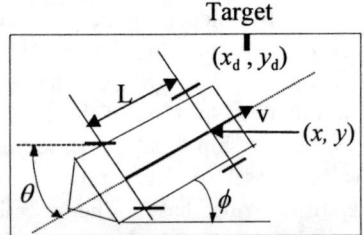

Fig. 4 The car and its parking lot.

Let θ denote the steering angle and v denote the speed of the rear center point of the vehicle. The equations of motion are:

$$\dot{x} = v\cos\phi, \quad \dot{y} = v\sin\phi, \quad \dot{\phi} = \frac{v}{L}\tan\theta \qquad (12)$$

where L is the distance between the center of the front wheel and the center of the rear wheels.

V.2 Generation of Optimal Trajectories by Cell Mapping

To generate optimal trajectories, the region of interest in the state space is taken to be $[x_{min}, x_{max}] \times [y_{min}, y_{max}] \times [\phi_{min}, \phi_{max}] = [0, 12] \times [0, 6] \times [-\pi/2, 3\pi/2]$. The dimensions of each cell are: $\Delta x = 12/27$, $\Delta y = 6/14$, and $\Delta\phi = 2\pi/27$. The total number of cells is 10,206 ($N_c =27*14*27$). The cell coordinates of the target are [13, 13, 13], which represent the physical coordinates of $(x, y, \phi) = [6, 3, \pi/2]$.

In the sorting of cell mappings for optimal trajectories, we use minimization of path length as the objective function. Only one switching of the car moving direction is allowed in this example study. And when it is necessary for the car to switch its moving direction, the switching is always from forward to backward direction because of the prespecified car orientation at the target.

In the cell mapping, first only backward movements of the car are allowed, until all the cells in the cell space are processed. Among the processed cells, the cells are (target) reachable cells if their trajectories are connected to the target. The other cells are unreachable cells. The trajectories of the unreachable cells are not connected to the target but to the sink cell (the entire region outside the domain region). If there are unreachable cells after the process of allowing only backward movements, then forward movements of the car are allowed to connect previously unreachable cells with previously reachable cells.

After the cell mapping procedure is completed, a total of 5,229 cells are found as reachable cells to the target when one switching of the vehicle moving direction is allowed. Among them, 2,371 cells are reachable cells to the target with only backward movements.

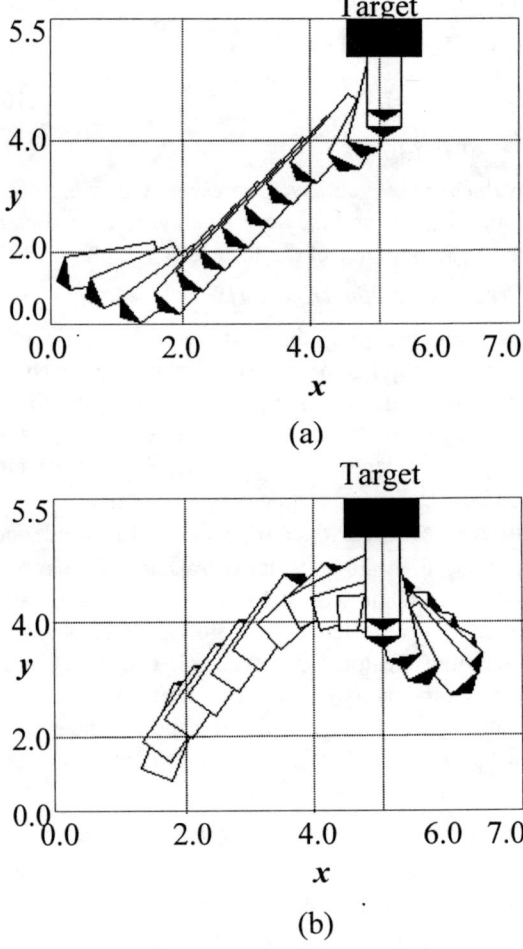

Fig. 5 Car paths for 2 different initial locations, with the same position but different orientations.

Figure 5 shows the car trajectories for 2 different initial locations. These locations are set such that the vehicle is initially close to the boundary of the loading dock. At the position of Figure 5 (a), where $x = 1.6$, $y = 1.6$, and $\phi = 5$ degrees, the car successfully arrives at the target with only backward movements. At the position of Figure 5 (b), where $x = 1.6$, $y = 1.6$, and $\phi = 260$ degrees, the car initially moves forward 14 steps to avoid collision with the left wall of the loading dock and then changes its moving direction from forward to backward.

V.3 Grouping of Trajectories and the Generated Fuzzy Controller

The total number of trajectory groups is 68, obtained using the trajectory determining procedure previously described, with the threshold value β_T set at 3.0. Among the 68 groups, the 6 groups of the longest trajectories are shown in Figure 6.

Figure 7 compares the simulated trajectories using the table-based controller with those using the fuzzy controller for the car parking problem. There are 14 different trajectories starting from different locations in the parking lot. It can be seen that the result of the fuzzy controller approximates that of the table-based optimal controller, despite that only 68 rules are used in the fuzzy controller while a data table for 2,371 cells are used in the table-based controller.

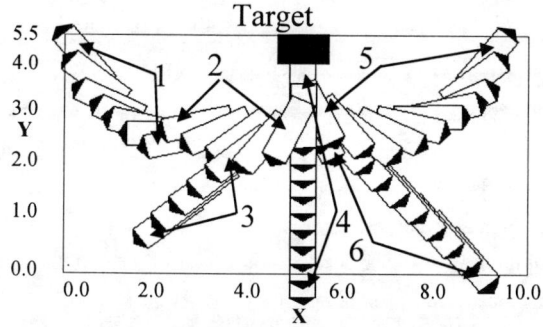

Fig. 6 Representative groups of the six longest trajectories.

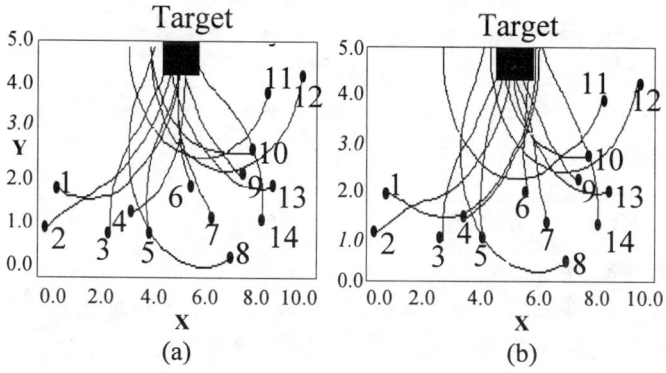

Fig. 7 Simulated trajectories using (a) the table-based controller and (b) the fuzzy controller.

VI. CONCLUSION

A cell mapping based method has been developed to systematically generate the rules of a near-optimal fuzzy controller for autonomous car parking. In this method, the optimal trajectories in the state space are grouped using the data obtained from cell mapping. The fuzzy rules and membership functions are then generated using the statistical properties of the individual trajectory groups. This method significantly reduces the number of rules in the fuzzy controller compared with generation of fuzzy rules directly using the control data of all individual cells. Simulation results show that the performance of the near-optimal fuzzy controller approximates that of the table-based optimal controller, despite a small number of rules used in the fuzzy controller compared with a large amount of cell data used in the table-based controller.

REFERENCES

[1] L.A. Zadeh, "Fuzzy sets," *Information and Control.*, vol. 8, pp. 338-353, 1965.

[2] M. Sugeno and M. Nishida, "Fuzzy control of model car," *Fuzzy Sets and Systems.*, vol. 16, pp. 103-113, 1985.

[3] B. Kosko, *Neural networks and fuzzy systems.* Prentice Hall, Englewood Cliffs, New Jersey, 1992.

[4] L.X. Wang and J. M. Mendel, "Generating fuzzy rules from numerical data from examples," *IEEE Trans. Syst., Man, Cybern.*, vol. 22, no. 6, pp. 1414-1427, 1992.

[5] C.T. Lin and C.S.G. Lee, "Neural-network based fuzzy logic control and decision system," *IEEE Trans. Comput.*, vol. 40, no. 12, pp. 1320-1336, 1991.

[6] C.S. Hsu, "A discrete method of optimal control based upon the cell state space concept," *J. of Appl. Mech.*, vol. 46, pp. 547-569, 1985.

[7] H. Kang and G.V. Vachtsevanos, "Nonlinear fuzzy control based on the vector fields of the phase portrait assignment algorithm," *Proc. of American Control Conf.*, 1990.

[8] T.Q. Kim, Global Fuzzy Control based on Cell Mapping, Ph. D. Dissertation, New Jersey Institute of Technology, Newark, New Jersey, 1996.

[9] W.H. Zhu, "An applied cell mapping method for optimal control systems," *J. of Opt. Theo. Appl.*, vol. 60, no. 3, 1989.

[10] W.H. Zhu and M. C. Leu, "Planning optimal robot trajectories by cell mapping," in *Proc. of IEEE Int. Conf. on Robotics and Automation*, 1990.

[11] Integrated Systems, Inc., SystemBuild/Workstation Real-Time Fuzzy Logic Block User's Guide, 1991.

A Robust Model-Based Fuzzy-Logic Controller for Robot Manipulators

Mohammad R. Emami
Robotics and Automation Laboratory
Dept. of Mechanical & Industrial Eng.
University of Toronto
emami@mie.utoronto.ca

Andrew A. Goldenberg
Robotics and Automation Laboratory
Dept. of Mechanical & Industrial Eng.
University of Toronto
golden@mie.utoronto.ca

I. Burhan Türksen
Intelligent Fuzzy Systems Laboratory
Dept. of Mechanical & Industrial Eng.
University of Toronto
turksen@mie.utoronto.ca

Abstract

This paper represents a fuzzy-logic control structure, containing the fuzzy-logic model of the dynamic system and fuzzy control rules that ensure the stability and robust performance. The robust fuzzy rules are designed based on a generalized formulation of sliding mode control for a class of nonlinear muti-input multi-output systems. The proposed fuzzy-logic control scheme was applied to the trajectory control of a four degree-of-freedom robot, and compared with the high-gain PID controllers. A superior tracking performance was achieved.

1 Introduction

New efforts have been made Recently to investigate the connection between Fuzzy Logic Control (FLC) and variable structure control [9, 7, 8 and 14]. It has been concluded that, due to partitioning the input-output space, the FLC is a qualitative extension of the sliding mode control. Some guidelines have been specified to derive the fuzzy IF-THEN control rules based on the theory of variable structure systems for the case of single-input single-output systems [10]. For Multi-Input Multi-Output (MIMO) nonlinear systems, due to the state interactions, more information about the system is required leading to a model-based fuzzy-logic control approach that is our focus in this paper. In [3, 4], a systematic methodology has been proposed for developing the fuzzy-logic model of complex systems. As a continuation, in this paper, we use the fuzzy-logic model of the nonlinear system for control tasks. A few researchers have also attempted to apply the fuzzy sliding mode control approach to robot manipulators [2, 13, 1]. Despite successful results, the lack of a systematic approach to design and analysis of the FLC, based on the sliding mode control theory, is observed.

2 The Proposed FLC Structure

Figure 1 illustrates the proposed structure of the FLC for a nonlinear MIMO second order system. The controller consists of two parts. In the first part, a set of fuzzy IF-THEN rules expresses the dynamic behavior of the system. The second part of the FLC consists of "*decoupled*" robust fuzzy IF-THEN rules for each state independently, in order to guarantee system stability, and to ensure achievement of the desired performance.

A systematic methodology of design and analysis requires the following steps:

i) Development of a fuzzy-logic model that encapsulates the basic knowledge of the system characteristics in the form of fuzzy IF-THEN rules. This task has been accomplished in [5,6].

ii) Design of the robust fuzzy control IF-THEN rules for each system state independently, and proof of the stability and completeness of the entire control structure. This step is the subject of the present paper.

3 A Generalized Formulation of the Sliding Mode Control

Without loss of generality, we consider second order dynamic systems with n system states and n input variables:

$$\ddot{\mathbf{q}} = \mathbf{f}(\mathbf{q},\dot{\mathbf{q}};t) + \mathbf{B}(\mathbf{q},\dot{\mathbf{q}};t)\mathbf{u}(t) = \mathbf{G}(\mathbf{q},\dot{\mathbf{q}},\mathbf{u};t) \quad (1)$$

where, $\mathbf{q} = [q_1, q_2, ..., q_n]^T$, $\dot{\mathbf{q}} = [\dot{q}_1, \dot{q}_2, ..., \dot{q}_n]^T$ and $\ddot{\mathbf{q}} = [\ddot{q}_1, \ddot{q}_2, ..., \ddot{q}_n]^T$ are vectors of system states, state velocities and accelerations, respectively. Function $\mathbf{f} \in R^n$ is a nonlinear vector function that represents system dynamics including system uncertainty and disturbances, and $\mathbf{B} \in R^{n \times n}$ is the nonlinear control gain matrix. Vector $\mathbf{u} \in R^n$ is the control input vector. The acceleration vector of the entire system can be considered as a nonlinear vector function $\mathbf{G} \in R^n$.

We assume that the system is controllable, and the control gain matrix B is positive definite, non-singular and bounded, i.e.,

$$\forall t \geq 0 \;;\; \forall \mathbf{q}, \dot{\mathbf{q}} \in R^n : \underline{b}\mathbf{I} \leq \mathbf{B}(\mathbf{q},\dot{\mathbf{q}};t) \leq \overline{b}(\mathbf{q},\dot{\mathbf{q}};t)\mathbf{I} \quad (2)$$

where $\underline{b}$ is a positive constant and $\overline{b}(\mathbf{q},\dot{\mathbf{q}};t)$ is a positive definite function and I is the n×n unity matrix. Based on the above assumptions, the inverse dynamics of the system can be presented as:

$$\mathbf{u} = \mathbf{M}(\mathbf{q},\dot{\mathbf{q}};t)\ddot{\mathbf{q}} + \mathbf{h}(\mathbf{q},\dot{\mathbf{q}};t) = \mathbf{F}(\mathbf{q},\dot{\mathbf{q}},\ddot{\mathbf{q}};t) \quad (3)$$

with $\mathbf{M} = \mathbf{B}^{-1}$, $\mathbf{h} = -\mathbf{B}^{-1}\mathbf{f}$, and

$$\forall t \geq 0 \;;\; \forall \mathbf{q}, \dot{\mathbf{q}} \in R^n : \underline{m}\mathbf{I} \leq \mathbf{M}(\mathbf{q},\dot{\mathbf{q}};t) \leq \overline{m}(\mathbf{q},\dot{\mathbf{q}};t)\mathbf{I} \quad (4)$$

where $\underline{m} = 1/\overline{b}$ and $\overline{m} = 1/\underline{b}$.

The control task is to follow a desired $\mathbf{q}_d$ and $\dot{\mathbf{q}}_d$ in the presence of system parameter variation and uncertainty. The tracking error $\mathbf{e} = \mathbf{q} - \mathbf{q}_d$ and the rate of error $\dot{\mathbf{e}} = \dot{\mathbf{q}} - \dot{\mathbf{q}}_d$ are to be observed. We define a generalized error vector as:

$$\mathbf{s} = \dot{\mathbf{e}} + 2\Lambda\mathbf{e} + \Lambda^2 \int_0^t \mathbf{e}\,dt \quad (5)$$

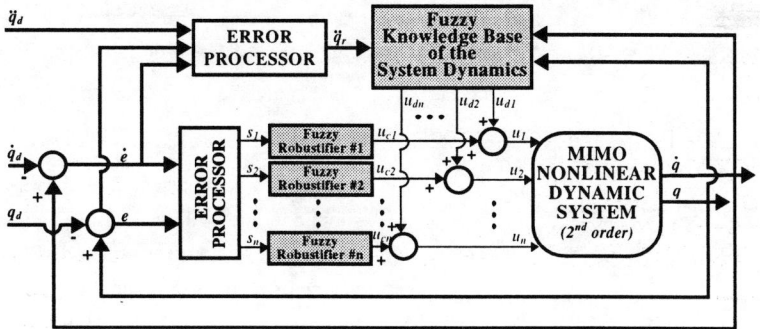

Figure 1: *The structure of the proposed fuzzy-logic control system*

where matrix Λ is an n×n positive definite diagonal matrix. The tracking control problem can be formulated as keeping the error vector **e** on the sliding surface defined as follows:

$$\dot{\mathbf{s}} = \ddot{\mathbf{e}} + 2\Lambda\dot{\mathbf{e}} + \Lambda^2 \mathbf{e} = 0 \quad (6)$$

Asymptotic convergence of system states to a small neighborhood of their corresponding switching surfaces is achieved if the control law **u** is designed such that for each state a Lyapunov-like condition for system stability holds [11]:

$$\frac{1}{2}\frac{d}{dt}\left(s_i^2\right) \leq -\eta_i\left(|s_i| - \Phi_i\right); \; \eta_i, \Phi_i > 0 \; ; \; i = 1, 2, \ldots, n \quad (7)$$

or in sum:

$$\frac{1}{2}\frac{d}{dt}\left(\mathbf{s}^T\mathbf{s}\right) = \mathbf{s}^T\dot{\mathbf{s}} \leq -\sum_{i=1}^{n}\eta_i\left(|s_i| - \Phi_i\right); \; \eta_i, \Phi_i > 0 \quad (8)$$

The parameter Φ_i is the thickness of the boundary layer and η_i is a design parameter that sets the time the system trajectory requires to reach the boundary layer from an outside initial condition.

From equations (1) and (6), the dynamics of the generalized error vector **s** is determined as:

$$\dot{\mathbf{s}} = \mathbf{M}^{-1}\left[\mathbf{u} - \left[\mathbf{M}\ddot{\mathbf{q}}_r + \mathbf{h}\right]\right] \quad (9)$$

where the "*reference*" acceleration $\ddot{\mathbf{q}}_r$ is defined as:

$$\ddot{\mathbf{q}}_r = \ddot{\mathbf{q}}_d - 2\Lambda\dot{\mathbf{e}} - \Lambda^2 \mathbf{e} \quad (10)$$

The control input **u** is assigned as:

$$\mathbf{u} = \hat{\mathbf{u}}_d + \mathbf{u}_c \quad (11)$$

where $\hat{\mathbf{u}}_d$ is the desired control calculated by the inverse dynamics model of the system:

$$\hat{\mathbf{u}}_d = \hat{\mathbf{M}}(\mathbf{q},\dot{\mathbf{q}};t)\ddot{\mathbf{q}}_r + \hat{\mathbf{h}}(\mathbf{q},\dot{\mathbf{q}};t) = \hat{\mathbf{F}}(\mathbf{q},\dot{\mathbf{q}},\ddot{\mathbf{q}}_r;t) \quad (12)$$

and $\mathbf{u}_c$ is the compensation part (robust control term) due to the model uncertainty. Equation (9) can then be rewritten as:

$$\dot{\mathbf{s}} = \mathbf{M}^{-1}\mathbf{u}_c + \mathbf{M}^{-1}\Delta\mathbf{F} \quad (13)$$

where $\Delta\mathbf{F}$ is the error vector of the inverse dynamics model and is bounded as:

$$\left\|\Delta\mathbf{F}(\mathbf{q},\dot{\mathbf{q}},\ddot{\mathbf{q}}_r;t)\right\| \leq \rho(\mathbf{q},\dot{\mathbf{q}},\ddot{\mathbf{q}}_r;t) < \infty \quad (14)$$

Consequently, the left hand side of the sliding condition (8) becomes:

$$\mathbf{s}^T\dot{\mathbf{s}} = \mathbf{s}^T\mathbf{M}^{-1}\mathbf{u}_c + \mathbf{s}^T\mathbf{M}^{-1}\Delta\mathbf{F} \quad (15)$$

By applying the following inequalities [6]:

$$\mathbf{s}^T\mathbf{M}^{-1}\Delta\mathbf{F} \leq \frac{1}{\underline{m}}\rho\|\mathbf{s}\| \; ; \quad \text{and} \quad \|\mathbf{s}\| \leq \sum_{i=1}^{n}|s_i| \quad (16)$$

we can deduce:

$$\mathbf{s}^T\dot{\mathbf{s}} \leq \frac{1}{\underline{m}}\rho\sum_{i=1}^{n}|s_i| + \mathbf{s}^T\mathbf{M}^{-1}\mathbf{u}_c \quad (17)$$

In order to satisfy the sliding condition (8), we should choose a continuous $\mathbf{u}_c$ such that:

$$\mathbf{s}^T\mathbf{M}^{-1}\mathbf{u}_c \leq -\sum_{i=1}^{n}\left[\left(\frac{\rho}{\underline{m}} + \eta_i\right)|s_i| - \eta_i\Phi_i\right]. \quad (18)$$

THEOREM [6]: The inequality (18) holds if the following conditions are satisfied for the robust control term u_{ci} of each state i (i=1,2,...,n) independently:

a) $u_{ci}(s_i)$ is continuous;

b) $u_{ci}(s_i)$ is monotonic and decreasing for $0 < |s_i| < \Phi_i$;

c) $u_{ci}(0) = 0$; (19)

d)
$$\begin{cases} \text{if } s_i > 0 \Rightarrow u_{ci} < -\left[\overline{m}\left(\frac{\rho}{\underline{m}} + \eta_i\right) - \overline{m}\frac{\eta_i\Phi_i}{|s_i|}\right] \\ \text{if } s_i < 0 \Rightarrow u_{ci} > \left[\overline{m}\left(\frac{\rho}{\underline{m}} + \eta_i\right) - \overline{m}\frac{\eta_i\Phi_i}{|s_i|}\right] \end{cases}$$

Figure 2 shows the domain in which each u_{ci} can compensate for system uncertainties. We call this domain the "*Robustness Region*". This region and properties specified in (19) help us assign the robust control term u_{ci} for each state, independently.

In our methodology, the control terms u_{ci} are produced by suitable fuzzy IF-THEN rules. The procedure is as follows: we consider the nonlinear MIMO system (1) with the following inverse dynamics: $\mathbf{u} = \mathbf{F}(\mathbf{q},\dot{\mathbf{q}},\ddot{\mathbf{q}};t)$. (20)

First, we generate the fuzzy-logic inverse dynamics model of the system from the experimental input-output data as:

$$\hat{\mathbf{u}}_d = \hat{\mathbf{F}}_{fuzz}(\mathbf{q},\dot{\mathbf{q}},\ddot{\mathbf{q}}_r;t), \quad (21)$$

having a known bounded error $\Delta\mathbf{F}$. Then the control input is of

the form:
$$\mathbf{u} = \hat{\mathbf{F}}_{fuzz}(\mathbf{q},\dot{\mathbf{q}},\ddot{\mathbf{q}}_r;t) + \mathbf{u}_c \quad (22)$$
where, $\ddot{\mathbf{q}}_r$ is defined by equation (10). The general conditions of $\mathbf{u}_c$ are expressed as properties (19). The robustness region depends on design parameters λ_i, η_i, and Φ_i, and system parameters $\underline{m}$, $\overline{m}$, and ρ that are defined as follows:

$$\begin{cases} \rho(\mathbf{q},\dot{\mathbf{q}},\ddot{\mathbf{q}}_r;t) \geq \|\Delta \mathbf{F}(\mathbf{q},\dot{\mathbf{q}},\ddot{\mathbf{q}}_r;t)\| & (23) \\ \underline{m} \leq \dfrac{1}{\|\ddot{\mathbf{q}}\|^2}\left[\ddot{\mathbf{q}}^T\left[\hat{F}_{fuzz}(\mathbf{q},\dot{\mathbf{q}},\ddot{\mathbf{q}};t) - \hat{F}_{fuzz}(\mathbf{q},\dot{\mathbf{q}},0;t)\right]\right] \leq \overline{m}(\mathbf{q},\dot{\mathbf{q}},\ddot{\mathbf{q}};t) & (24) \end{cases}$$

It is noted that in situations where $\ddot{\mathbf{q}}$ is zero, the inequalities (24) become ambiguous, illustrating the fact that these trajectory points do not give any information about the system inertia, and hence must be removed from the data set.

4 Design of the Robust Fuzzy Rules

From Figure 2, the characteristic relationship between u_{ci} and s_i can be qualitatively expressed as: "*u_{ci} is inversely as large as s_i within certain limits*". We interpret the above characteristic by the following seven IF-THEN rules :

$$\begin{cases} \text{IF } s \text{ is } Positive\ Big\ (PB), & \text{THEN } u_c \text{ is } Negative\ Big, \\ \text{IF } s \text{ is } Positive\ Medium\ (PM), & \text{THEN } u_c \text{ is } Negative\ Medium, \\ \text{IF } s \text{ is } Positive\ Small\ (PS), & \text{THEN } u_c \text{ is } Negative\ Small, \\ \text{IF } s \text{ is } Almost\ zero\ (AZ), & \text{THEN } u_c \text{ is } Almost\ Zero, \\ \text{IF } s \text{ is } Negative\ Small\ (NS), & \text{THEN } u_c \text{ is } Positive\ Small, \\ \text{IF } s \text{ is } Negative\ Medium\ (NM), & \text{THEN } u_c \text{ is } Positive\ Medium, \\ \text{IF } s \text{ is } Negative\ Large\ (NL), & \text{THEN } u_c \text{ is } Positive\ Big. \end{cases} \quad (25)$$

By using the modified Sugeno's reasoning formulation [12], given the input s_i, the crisp output u_{ci} is obtained as:

$$u_{ci} = \sum_{k=1}^{7} A_{ik}(s_i) b_{ik} \bigg/ \sum_{k=1}^{7} A_{ik}(s_i) \quad (26)$$

where, $A_{ik}(s_i)$ is the membership function of s_i in the antecedent fuzzy set of the kth rule, and b_{ik} is the centroid of the consequent fuzzy set of the kth rule.

The goal is to assign suitable membership functions for generating the robust control rules such that properties (19) are fulfilled. Considering the input membership functions shown in Figure 3, and seven output fuzzy set centroids b_i^1 for "*Almost Zero*" and b_i^2, b_i^3, b_i^4 for "*Positive Small, Medium, Big*", and $\underline{b}_i^2, \underline{b}_i^3, \underline{b}_i^4$ for "*Negative Small, Medium, Big*", the s_i-u_{ci} relation can be represented as shown in Figure 4. For the sake of simplicity and without loss of generality, the input membership functions are arranged such that they always overlap at the degree of membership equal to 0.5. Therefore, for each input s_i, two rules are fired at most. Furthermore, a symmetric behavior for $u_{ci}(s_i)$ is assumed. Hence,

$$\begin{aligned} \underline{a}_i^2 = -a_i^2 \ ;\ \underline{a}_i^3 = -a_i^3 \ ;\ \underline{a}_i^4 = -a_i^4 \ ; \\ \underline{b}_i^2 = -a_i^2 \ ;\ \underline{b}_i^3 = -b_i^3 \ ;\ \underline{b}_i^4 = -b_i^4 \ . \end{aligned} \quad (27)$$

Also, From Figure 4 and conditions (19): $a_i^1 = b_i^1 = 0 \quad (28)$

By using the inference formulation (26) and membership functions shown in Figure 3, a piece-wise linear characteristic is produced for the robust fuzzy control function of each system state i (i=1,2,...,n), that can be formulated as follows:
for

$$a_i^j \leq |s_i| < a_i^{j+1} \Rightarrow u_{ci} = -\frac{K_i^j}{\varphi_i^j} s_i + \overline{u}_{ci}^j\ \text{sgn}(s_i);\ j=1,2,3 \quad (29)$$

where, $K_i^j = b_i^{j+1} - b_i^j$; $\varphi_i^j = a_i^{j+1} - a_i^j$, and

$$\overline{u}_{ci}^j = \begin{cases} -\sum_{\sigma=1}^{j-1} K_i^\sigma + \dfrac{K_i^j}{\varphi_i^j} \sum_{\sigma=1}^{j-1} \varphi_i^\sigma & j=2,3 \\ 0 & j=1 \end{cases} \quad (30)$$

From equations (13) and (29), the dynamic behavior of the generalized error vector **s** can be rewritten as (i=1,...,n ; j=1,...,n):

$$\dot{s}_i + B_{ii}\frac{K_i^j}{\varphi_i^j}s_i = B_{ii}\overline{u}_{ci}^j\ \text{sgn}(s_i) + \sum_{\substack{k=1 \\ k \neq i}}^{n} B_{ik} u_{ck} + \sum_{k=1}^{n} B_{ik}[\Delta F]_k \quad (31)$$

Equation (31) represents the behavior of a state-dependent first order filter with corner frequency equal to $B_{ii} K_i^j / \varphi_i^j$.

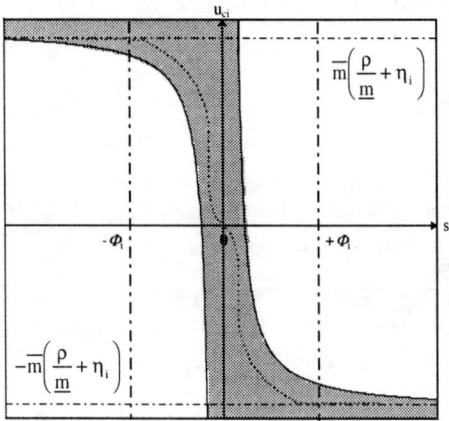

Figure 2 : *The specified domain for robust control term u_{ci}*

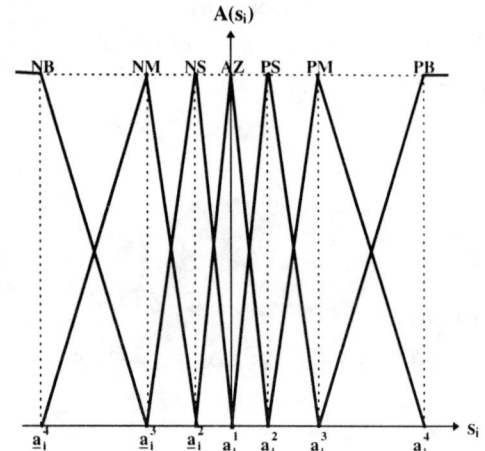

Figure 3 : *Membership functions of the generalized error s_i for robust control rules*

and system uncertainty and state interaction dynamics are inputs to the filter. A suitable selection of the corner frequency for such a filter is that:

$$B_{ii}\frac{K_i^j}{\varphi_i^j} \le \lambda_i \quad (32)$$

where λ_i is the lower band of the unmodeled frequencies. Since $\overline{b} = 1/\underline{m}$ is an upper value of the gain matrix B, a reliable break away frequency for the filter (31) can be assigned such that:

$$\frac{K_i^j}{\varphi_i^j} \le \underline{m}\lambda_i \quad (33)$$

However, within the boundary layer, the unmodeled frequencies can affect system performance only when s_i is close to zero, i.e., for the first segment where $|s_i| \le a_i^2$. Therefore, for each state i, parameters a_i^2 and b_i^2 should be selected such that:

$$\frac{b_i^2}{a_i^2} \le \underline{m}\lambda_i. \quad (34)$$

For a larger distance between the state and the switching line, since the unmodeled frequencies and state interactions can not change the sign of the control input we are able to assign higher break away frequencies which provide better control and consequently, faster response without any performance degradation. Therefore, parameters a_i^3 and b_i^3 are designed such that:

$$\frac{b_i^3 - b_i^2}{a_i^3 - a_i^2} \ge \frac{b_i^2}{a_i^2} \quad (35)$$

Similar to the sliding mode control, the tracking quality is guaranteed by choosing:

$$\frac{K_i^3}{\varphi_i^3} = \frac{b_i^4}{a_i^4} = \underline{m}\lambda_i \quad (36)$$

Outside the boundary layer, the maximum value of the robust control is assigned to be the lower bound of the "robustness region", therefore, from (19):

$$b_i^4 = \overline{m}\left(\frac{\rho}{\underline{m}} + \eta_i\right), \quad (37)$$

and from (36) and (37), the boundary layer thickness $\Phi_i = a_i^4$ is specified as:

$$\Phi_i = a_i^4 = \frac{\overline{m}}{\underline{m}}\left(\frac{\rho}{\underline{m}} + \eta_i\right)\frac{1}{\lambda_i}. \quad (38)$$

Equations (27), (28), (34), (35), (36), (37) and (38) assign the membership parameters of robust fuzzy IF-THEN rules (25). Obviously, these parameters depend on the design parameters λ_i and η_i. The natural frequency λ_i specifies the rate of convergence on the sliding surface, and it should be less than the minimum frequency associated with the largest unmodeled time delay τ_m and the frequency associated with the sampling rate τ_s. A suggested criterion for selecting λ_i is [11]:

$$\lambda \le \min\left(\frac{1}{3\tau_m}, \frac{1}{5\tau_s}\right) \quad (39)$$

5 Application to Robot Manipulators

The proposed fuzzy-logic control structure was applied to a 4 d.o.f. robot manipulator shown in Figure 5. In this application, the fuzzy-logic inverse dynamics model of the robot was first obtained from the input-output data through a systematic fuzzy-logic modeling methodology [6]. The model error vector was then calculated from the experimental data. Figure 6 shows the error norm of the test data set. The value of ρ was assigned as the upper bound of the error norm. The inertia parameters $\underline{m}$ and $\overline{m}$ were also obtained as the lower and upper bound of the inertia value m calculated from the fuzzy-logic model as follows:

$$m = \frac{\ddot{q}^T\left[F(q,\dot{q},\ddot{q}) - F(q,\dot{q},0)\right]}{\|\ddot{q}\|^2} \quad (40)$$

Figure 7 shows the inertia values for the experimental data and its lower and upper limits. For each joint i ($i=1,2,3,4$), parameters λ_i, η_i and Φ_i were designed as explained in Section 4.

The performance of the proposed fuzzy-logic control was compared with that of the high-gain PID control for different trajectories such as random, sinusoidal and step. The PID gains were designed and tuned for each trajectory separately in order to provide the high gain performance, while the design parameters of the proposed controller were fixed for all

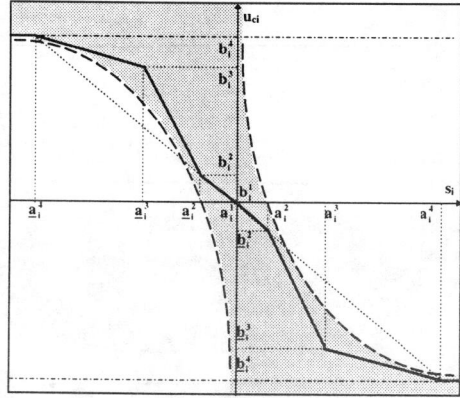

Figure 4 : *The robust fuzzy control characteristics*

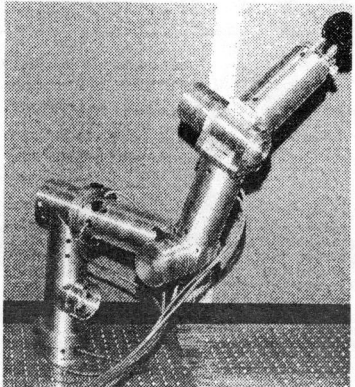

Figure 5 : *Desired configuration of the IRIS arm*

trajectories. The FLC controller outperformed the servo controller for all trajectories, both in terms of joint displacement and velocity tracking. Figure 8 illustrates the tracking performance of the first two joints for a typical sinusoidal trajectory.

6 Conclusion

In this paper, we introduced a fuzzy-logic control structure for a class of nonlinear MIMO complex systems. The core of this structure is the knowledge of the system dynamics in the form of fuzzy IF-THEN rules, and additional robust fuzzy rules are applied to further compensate system uncertainty and knowledge incompleteness. In order to prove the stability and robustness of the proposed control system, and provide systematic guidelines for designing the robust fuzzy rules, we developed a generalized formulation of sliding mode control for a class of nonlinear MIMO systems. This formulation has two distinct features. First, it is possible to apply this formulation while considering the system as a *"black box"* without any need to identify the internal parameters or to assume specific properties. Second, it is possible to design the robust control commands for each system state independently while the stability and robustness of the entire system is guaranteed. As a result, for each state, a *"robustness region"* was defined within which suitable fuzzy rules can be designed. The results of the developed formulation led us to guidelines for designing the robust fuzzy IF-THEN rules, that are simple and efficient. Application of the proposed FLC structure to a four degree-of-freedom robot manipulator with the superior tracking performance than that of the high-gain servo controllers for different trajectories is an illustration of its applicability and efficiency. The proposed structure behaves like complicated classical model-based algorithms while, due to the simplicity of its structure, it is as applicable as a servo controller.

References

[1] P. Begon, F. Pierrot, P. Dauchez, "Fuzzy sliding mode control of a fast parallel robot", *Proc. IEEE Int. Conf. Robotics and Automation*, pp. 1178-1183, 1995.

[2] J.S. Chen, C.S. Liu, Y.W. Wang, "Control of robot manipulator using a fuzzy model-based sliding mode control scheme", *Proc. the 33rd IEEE Conf. Decision and Control*, Lake Buena Vista, FL, USA, Vol. 4, pp. 3506-3511, December 1994.

[3] M.R. Emami, I.B. Turksen, A.A. Goldenberg, "An improved fuzzy-logic modeling: Part I: Inference mechanism", *Proc. North American Fuzzy Information Society NAFIPS*, University of Berkeley, California, USA, June 1996.

[4] M.R. Emami, I.B. Turksen, A.A. Goldenberg, "An improved fuzzy-logic modeling: Part II: System identification", *Proc. North American Fuzzy Information Society NAFIPS*, University of Berkeley, California, USA, June 1996.

[5] M.R. Emami, I.B. Turksen, A.A. Goldenberg, "Development of a systematic methodology for fuzzy-logic modeling", to be published in *IEEE Trans. Fuzzy Systems*, 1998.

[6] M.R. Emami, "Systematic Methodology of Fuzzy-Logic Modeling and Control and Application to Robotics", *Ph.D. Thesis*, Department of Mechanical and Industrial Engineering, University of Toronto, Toronto, 1997.

[7] D.P. Filev, R.R. Yager, "Three models of fuzzy logic controllers", *Cybernetics and Systems*, Vol. 24, pp. 91-114, 1993.

[8] M.B. Ghalia, A.T. Alouani, "Sliding mode control synthesis using fuzzy logic", *Proc. The American Control Conference*, Seattle; Washington, pp. 1528-1532, 1995.

[9] S. Kawaji, M. Matsunaga, "Fuzzy control of VSS type and its robustness", *Proc. the 3rd IFSA Congress*, Brussels, pp.81-88, 1991.

[10] R. Palm, "Sliding mode fuzzy control", *Proc. the First Int. Conf. Fuzzy Systems*, San Diego, pp. 519-526, 1992.

[11] J.J. Slotine, W. Li, *"Applied Nonlinear Control"*, Prentice Hall:Engle-Wood Cliffs, NJ, 1990.

[12] M. Sugeno, T. Yasukawa, "A fuzzy-logic-based approach to qualitative modeling", *IEEE Tran. Fuzzy Systems*, No. 1, pp. 7-31, 1993.

[13] T.I.J. Tsay, J.H. Huang, "Robust nonlinear control of robot manipulators", *Proc. IEEE Int. Conf. Robotics and Automation*, pp. 2083-2088, 1994.

[14] J.C. Wu, T.S. Liu, "A sliding-mode approach to fuzzy control design", *IEEE Tran. Control Systems and Technology*, Vol. 4, No. 2, 1996.

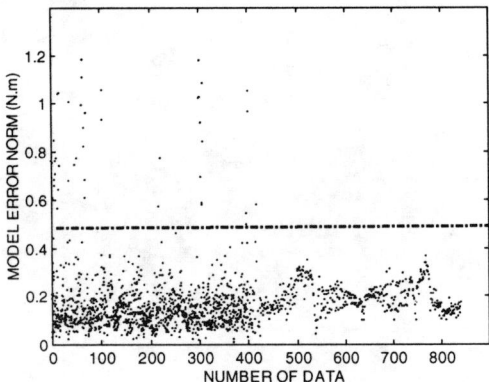

Figure 6 : *Norm of ΔF for the test data set*

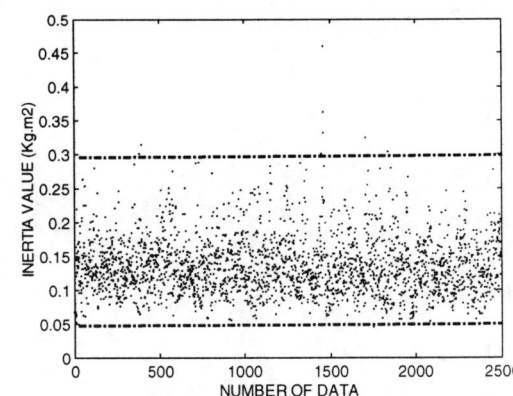

Figure 7 : *The value of m for some experimental data*

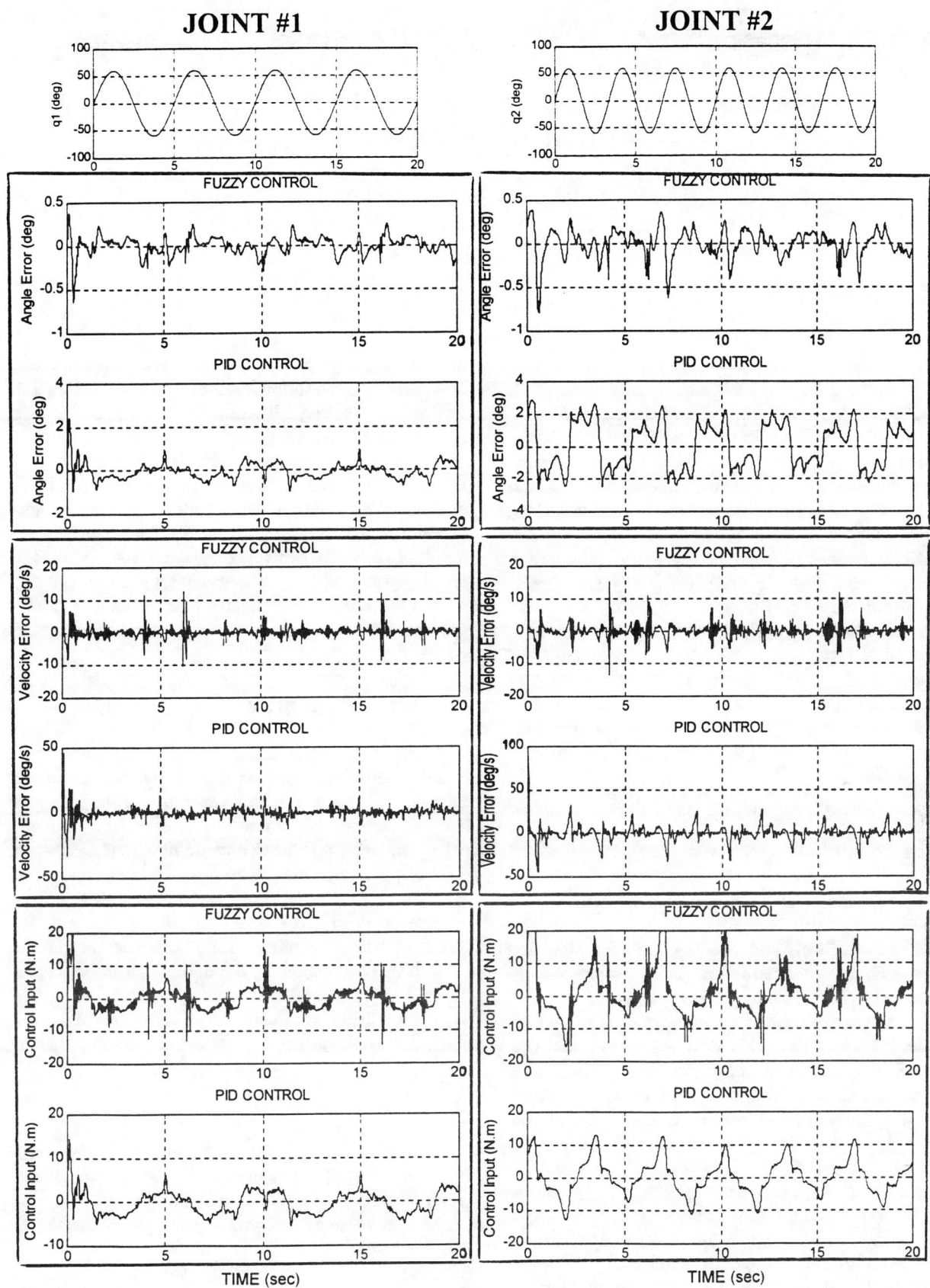

Figure 8: *Comparison of the proposed fuzzy control and PID control of the IRIS arm for **sinusoidal trajectory***

Analysis of Linguistic Fuzzy Control for Curved-path-following Autonomous Vehicles

Y. H. Fung and S. K. Tso

Centre for Intelligent Design, Automation and Manufacturing
City University of Hong Kong, HONG KONG

Abstract

Linear controllers perform well for straight-line path-following autonomous vehicles. For curved paths, significant large errors may result. These can be improved using two linguistic fuzzy-logic controllers (FLCs) even in the presence of inaccurate measurement of path curvature. One of the FLCs is used for accounting for the path curvature. Using a recently-developed linear-to-fuzzy (LIN2FUZ) algorithm, the synthesis of the linguistic FLCs is readily achieved. Also using another recently-developed equivalent transformation (ET) algorithm, the stability analysis of the linguistic FLCs can be carried out based on a Lyapunov function which is obtained using the linear-matrix-inequality (LMI) method. Experimental results are presented to illustrate the effectiveness of the proposed FLCs.

1 Introduction

In autonomous vehicles, after the information of the surrounding environments has been perceived and processed, desired paths are generated by path planners [4, 10]. The desired paths are usually expressed in terms of straight-line and curved segments. The vehicles are usually required to follow the desired paths as quickly as possible. This kind of task is called the path-following control problem [5].

Although this problem has been previously tackled in many autonomous guided vehicles (AGVs), the linear control laws developed are often based on the assumption of straight-line paths [6]. If the forward speed of the AGVs used in the manufacturing environment is slow and the radius of curvature of the paths is large, the linear control is quite acceptable. However, the performance degrades significantly for highly curved paths. In order to deal with this problem, a new set of control laws is found desirable [1, 2].

In this paper, the developed linear control laws is converted into FLC laws via a recently-developed LIN2FUZ (linear-to-fuzzy) conversion algorithm [8]. Performance of the FLC laws is then improved with suitable heuristics applied.

The developed linguistic FLCs can be converted into another form which is amenable to stability analysis using another recently-developed equivalent-transformation (ET) algorithm [9]. The error convergence of the fuzzy-controlled systems is guaranteed in the Lyapunov sense. Experimental results are presented to illustrate the effectiveness of the methodologies introduced.

2 Error dynamics

In Fig. 1, an autonomous vehicle centred at point F, is required to follow a desired path. The vehicle moves at a constant speed ν with a yaw rate $\dot{\phi}$. x-O-y denotes an inertia reference frame. At the arbitrary time t, when the vehicle is located at point F, $x(t), y(t)$, and $\phi(t)$ respectively denote the position coordinates and orientation of the vehicle.

The position error is defined as the shortest distance between the reference point of the vehicle and an arbitrary point along the desired path. In Fig. 1, point E is the shortest-distance point whose position and orientation are denoted by x_d, y_d, and ϕ_d. Let $\epsilon_d(t)$ denote the position error at time t, which is given by

$$\epsilon_d(t) = \overline{EF} \qquad (1)$$

The orientation error is defined as the acute angle between the line tangent to the desired point E and the line tangent to the longitudinal motion of the vehicle. The orientation error at time t is given by

$$\epsilon_\phi(t) = \phi_d(t) - \phi(t) \qquad (2)$$

From time t to time $t + \Delta t$, let the vehicle move from point F to point G'. The desired path actually

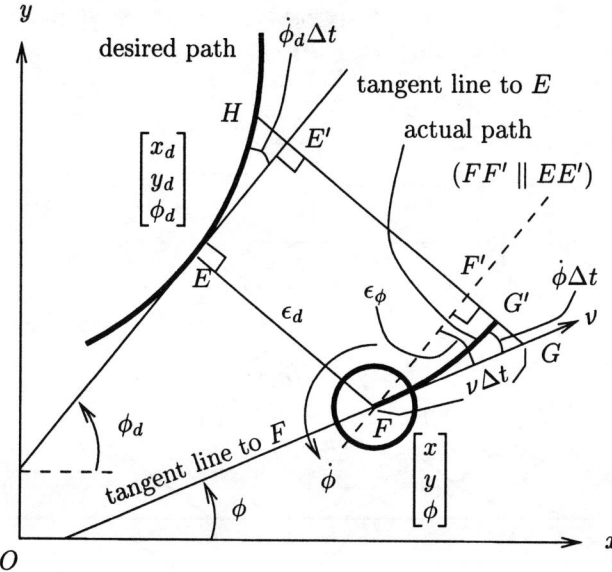

Figure 1: Definition of position and orientation errors

is described by EH. For a very short period of time Δt, point H is very close to point E and point G' is also close to point F. Hence, the following situations hold approximately: (i) the lines $\overline{HG} \parallel \overline{EF}$; (ii) since $\angle FEE'$ is a right-angle and $\overline{HG} \parallel \overline{EF}$, then $\angle EE'F'$ is a right-angle; and (iii) the paths EH and FG' are treated as straight lines.

The position error at time $t + \Delta t$ is given by

$$\epsilon_d(t+\Delta t) = \overline{HG'} = \overline{HE'} + \overline{E'F'} + \overline{F'G'} \quad (3)$$

The length of FG, the incremental forward distance moved by the vehicle, is given by $FG = \nu \Delta t$. In $\triangle FF'G$, $\overline{FF'}$ is given by

$$\overline{FF'} = \nu \Delta t \cos \epsilon_\phi \quad (4)$$

In $\triangle FF'G'$, $\overline{F'G'}$ is given by

$$\overline{F'G'} = \overline{FF'} \tan(\epsilon_\phi - \dot{\phi}\Delta t)$$
$$= \nu \Delta t \cos \epsilon_\phi \tan(\epsilon_\phi - \dot{\phi}\Delta t) \quad (5)$$

In $\triangle EE'H$, $\overline{HE'}$ is given by

$$\overline{HE'} = \overline{EE'} \tan(\dot{\phi}_d \Delta t) = \overline{FF'} \tan(\dot{\phi}_d \Delta t)$$
$$= \nu \Delta t \cos \epsilon_\phi \tan(\dot{\phi}_d \Delta t) \quad (6)$$

Since $EE'F'F$ is a rectangle, $E'F'$ is given by

$$\overline{E'F'} = \overline{EF} = \epsilon_d(t) \quad (7)$$

Substituting (6), (7), and (5) into (3) and taking the limit as $\Delta t \to 0$, we have

$$\dot{\epsilon}_d = \lim_{\Delta t \to 0} \frac{\epsilon_d(t+\Delta) - \epsilon_d(t)}{\Delta t}$$
$$= \lim_{\Delta t \to 0} \nu \cos \epsilon_\phi \tan(\dot{\phi}_d \Delta t) + \nu \cos \epsilon_\phi \tan(\epsilon_\phi - \dot{\phi}\Delta t)$$
$$= \nu \sin \epsilon_\phi \quad (8)$$

Differentiating (2) with respect to time gives

$$\dot{\epsilon}_\phi = \dot{\phi}_d - \dot{\phi} \quad (9)$$

3 Linear control

In order to drive the errors to zero, let

$$\dot{\phi} = \delta f_s(\epsilon_d, \epsilon_\phi) + \delta f_c(\epsilon_d, \epsilon_\phi) \quad (10)$$

where δf_s is used to drive ϵ_d and ϵ_ϕ towards zero in the case of straight-line paths; and δf_c is used to compensate for the path curvature.

Substituting (10) into (9) yields

$$\dot{\epsilon}_\phi = (\dot{\phi}_d - \delta f_c) - \delta f_s \quad (11)$$

Let the linear control laws for the curved paths be given by [1, 6]

$$\nu = k_3 \quad (12)$$
$$\delta f_s = k_1 \epsilon_d + k_2 \epsilon_\phi \quad (13)$$
$$\delta f_c = \dot{\phi}_d \quad (14)$$

where k_1, k_2, and k_3 are constants.

Substituting (12) into (8), and (13) and (14) into (11), we have

$$\dot{\epsilon}_d = k_3 \sin \epsilon_\phi \quad (15)$$
$$\dot{\epsilon}_\phi = -k_1 \epsilon_d - k_2 \epsilon_\phi \quad (16)$$

Linearising (15) in the neigbourhood of the zero-error equilibrium point, (15) and (16) become

$$\begin{bmatrix} \dot{\epsilon}_d \\ \dot{\epsilon}_\phi \end{bmatrix} = \begin{bmatrix} 0 & k_3 \\ -k_1 & -k_2 \end{bmatrix} \begin{bmatrix} \epsilon_d \\ \epsilon_\phi \end{bmatrix} \quad (17)$$

3.1 Design of linear controller

Making $k_1 = k_2$, and taking $k_3 = 0.125$ and $\zeta = 1$ for the second-order system (17), we have $k_1 = k_2 = 0.5$. (Other control laws may be designed as preferred.) The linear control law for δf_s in (13) is given by

$$\delta f_s = 0.5\epsilon_d + 0.5\epsilon_\phi \quad (18)$$

It is supposed that the autonomous vehicle is required to follow a circle with a radius ρ of 0.5 meter as shown in Fig. 7, and a linear control law for δf_c in (14) is thus given by

$$\delta f_c = \dot{\phi}_d = \frac{\nu}{\rho} = \frac{k_3}{\rho} = \frac{0.125}{0.5} = 0.25 \qquad (19)$$

4 Fuzzy control

With the application of the LIN2FUZ algorithm introduced in [8], the development time in the design of FLCs can be significantly reduced. In LIN2FUZ, the fuzzy reasoning method used is the product-sum-gravity type [7].

Two linguistic FLCs denoted by FLC_s and FLC_c are synthesised based on the linear control laws given by δf_s and δf_c, respectively. Let R_{ϵ_d}, R_{ϵ_ϕ}, and $R_{\delta f_s}$ denote respectively the universes of discourse of the antecedents ϵ_d and ϵ_ϕ, and the consequent δf_s.

4.1 Design of FLC_s

Taking $R_{\delta f_s} = \pm 1$ (as obtained from experiments such that this value is close to the maximum value of δf_s); and using the LIN2FUZ algorithm, the operating range R_{ϵ_d} of the antecedent ϵ_d of FLC_s is given by

$$R_{\epsilon_d} = \frac{1}{2} \cdot \frac{R_{\delta f}}{k_1} = \frac{1}{2} \cdot \frac{\pm 1}{0.5} = \pm 1 \qquad (20)$$

Similarly, $R_{\epsilon_\phi} = \pm 1$.

Figs. 2 and 3 show the distribution of membership functions for ϵ_d and ϵ_ϕ, respectively. Five linguistic labels $A_{-2}, \ldots, A_2$ and $B_{-2}, \ldots, B_2$ are employed in each distribution. Based on the LIN2FUZ algorithm [8], nine linguistic labels for the consequent δf_s are formed, namely $C_{-4}, \ldots, C_4$. The distribution of membership functions for δf_s is shown in Fig. 4. The fuzzy rules are given by [8]

$$\text{IF } \epsilon_d \in A_i \text{ AND } \epsilon_\phi \in B_j \text{ THEN } \delta f_s \in C_k \qquad (21)$$

where $k = i + j$ for $i, j = -2, \ldots, 2$. With suitable heuristics applied to modify the fuzzy rules for curved paths, Table 1 shows the revised fuzzy rules for δf_s. If ϵ_d is positive, then a larger control effort is applied. The revised fuzzy rules are 'boxed' for easy identification. Fig. 2 is modified to Fig. 5 by reducing R_{ϵ_d} from ± 1 to ± 0.5 to increase the gain for ϵ_d.

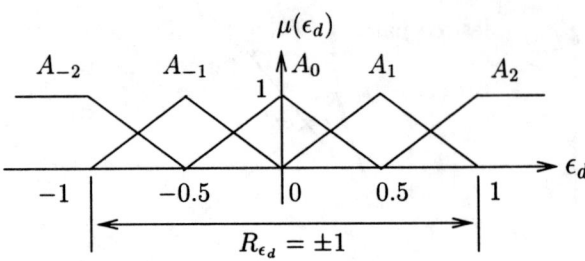

Figure 2: Distribution of membership functions for ϵ_d

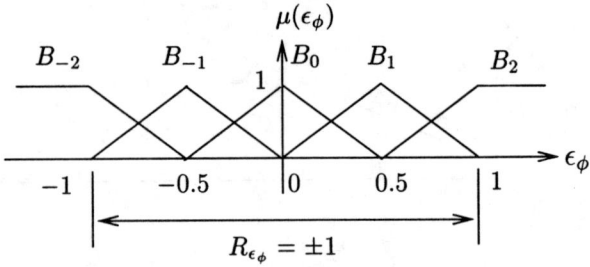

Figure 3: Distribution of membership functions for ϵ_ϕ

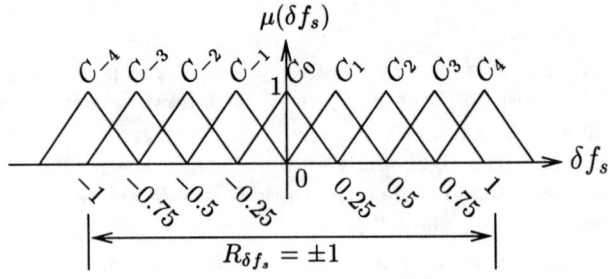

Figure 4: Distribution of membership functions for δf_s used in FLC_s

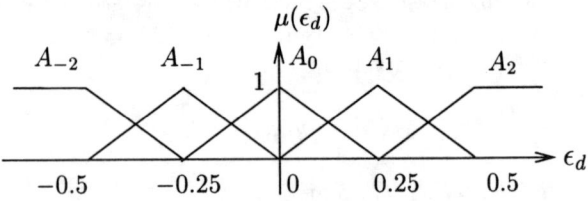

Figure 5: Distribution of membership functions for ϵ_d

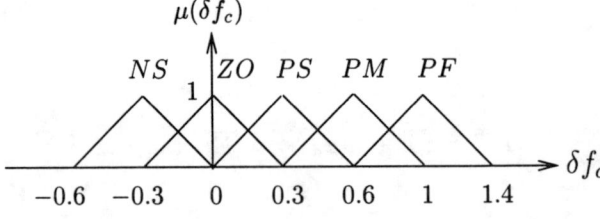

Figure 6: Distribution of membership functions for δf_c used in FLC_c

Table 1: Fuzzy rules for FLC_s

	B_{-2}	B_{-1}	B_0	B_1	B_2	(ϵ_ϕ)
A_{-2}	C_{-4}	C_{-3}	C_{-2}	C_{-1}	C_0	
A_{-1}	C_{-3}	C_{-2}	C_{-1}	C_0	C_1	
A_0	C_{-2}	C_{-1}	C_0	C_1	C_2	
A_1	C_0	C_1	C_2	C_4	C_4	
A_2	C_1	C_2	C_3	C_4	C_4	
(ϵ_d)						(δf_s)

Table 2: Fuzzy rules for FLC_c

	B_{-2}	B_{-1}	B_0	B_1	B_2	(ϵ_ϕ)
A_{-2}	NS	ZO	ZO	ZO	ZO	
A_{-1}	NS	ZO	ZO	ZO	ZO	
A_0	NS	NS	PF	PF	PF	
A_1	ZO	ZO	PF	PM	PS	
A_2	ZO	ZO	PF	PS	PS	
(ϵ_d)						(δf_c)

4.2 Design of FLC_c

The design of the FLC for δf_c is slightly different from that of the FLC_s because δf_c is given instead by

$$\delta f_c = \hat{\dot{\phi}}_d \times FLC_c(\epsilon_d, \epsilon_\phi) \quad (22)$$

where $\hat{\dot{\phi}}_d$ denotes the estimate of $\dot{\phi}_d$. The antecedents of FLC_c are the same as that used in FLC_s. ($R_{\epsilon_d} = R_{\epsilon_\phi} = \pm 1$.) The distribution of membership functions for the consequent δf_c is depicted in Fig. 6. Five linguistic labels are used to represent the speed of the vehicle, namely NS (negative slow), Z0 (zero), PS (positive slow), PM (positive medium), and PF (positive full). The heuristics used in building the fuzzy rules in Table 2 suggests that if ϵ_ϕ is zero and ϵ_d is positive, full compensation (PF) is required. Along the column B_{-2}, if ϵ_d is negative, NS is applied to drive the vehicle away from the centre of the circle. If ϵ_d and ϵ_ϕ are large, only positive medium or small compensation for path curvature is applied because the coarse control is given by the FLC_s.

5 Error convergence analysis

Let $\mathbf{x} = [x_1, x_2]^T = [\epsilon_d, \epsilon_\phi]^T$. Replacing the linear control laws δf_s and δf_c in (11) by the corresponding FLC_s and FLC_c, (8) and (11) become

$$\dot{x}_1 = \nu \sin x_2 \quad (23)$$

$$\dot{x}_2 = \dot{\phi}_d - \hat{\dot{\phi}}_d \times FLC_c - FLC_s \quad (24)$$

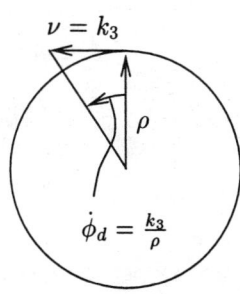

Figure 7: Curvature compensation

Figure 8: Laboratory platform

Let $FLC_s = k'_1 x_1 + k'_2 x_2$ and $\nu \sin x_2 = k'_3 x_2$. Using the ET algorithm [9], the lower and upper bounds of the nonlinear parameters k'_1, k'_2, and k'_3 are obtained and given by the sets $S_1 = \{0.537, 2.197\}$, $S_2 = \{0.230, 1.139\}$, and $S_3 = \{0.105, 0.125\}$, respectively. Let

$$g_1(\mathbf{x}) = \dot{\phi}_d - \hat{\dot{\phi}}_d \times FLC_c \quad (25)$$

(23) and (24) become

$$\dot{\mathbf{x}} = \mathbf{A}_i \mathbf{x} + g_1(\mathbf{x}) \quad (26)$$

where

$$\mathbf{A}_i = \begin{bmatrix} 0 & k'_3 \\ -k'_1 & -k'_2 \end{bmatrix} \quad (27)$$

with $k'_j \in S_j$, for $j = 1, 2, 3$. Hence there are eight combinations for $\mathbf{A}_i$ for $i = 1, 2, \ldots, 8$.

Let the Lyapunov function candidate be given by

$$V = \mathbf{x}^T \mathbf{P} \mathbf{x} \quad (28)$$

where $\mathbf{P}$ is a symmetric positive-definite matrix.

Differentiating V with respect to time yields

$$\dot{V} = \dot{\mathbf{x}}^T \mathbf{P} \mathbf{x} + \mathbf{x}^T \mathbf{P} \dot{\mathbf{x}} \quad (29)$$

Let there be constants γ and β satisfying

$$g_1(\mathbf{x}) \leq \gamma \|\mathbf{x}\| + \beta, \quad \forall x_1 \in R_{\epsilon_d}, x_2 \in R_{\epsilon_\phi} \quad (30)$$

where β may represent the offset when $\mathbf{x} = 0$ and/or the measurement error in $\hat{\dot{\phi}}_d$.

Substituting (26) into (29), we have

$$\dot{V}_i = \mathbf{x}^T (\mathbf{A}_i^T \mathbf{P} + \mathbf{P} \mathbf{A}_i) \mathbf{x} + 2 \mathbf{x}^T \mathbf{P} g_1(\mathbf{x}) \quad (31)$$

where $i = 1, \ldots, 8$.

Suppose that there exists a positive-definite matrix $\mathbf{Q}$ satisfying

$$\mathbf{x}^T(\mathbf{A}_i^T\mathbf{P} + \mathbf{P}\mathbf{A}_i)\mathbf{x} \leq -\mathbf{x}^T\mathbf{Q}\mathbf{x}, \quad i = 1, \ldots, 8 \quad (32)$$

For $\dot{V}_i < 0$, $i = 1, \ldots, 8$, substituting (32), (30) into (31), we have

$$-\lambda_{\min}(\mathbf{Q})\|\mathbf{x}\|^2 + 2\|\mathbf{x}\|\lambda_{\max}(\mathbf{P})(\gamma\|\mathbf{x}\| + \beta) \leq 0 \quad (33)$$

where $\lambda_{\min}(\mathbf{M})$ and $\lambda_{\max}(\mathbf{M})$ denote respectively the minimum and maximum eigenvalues of a symmetric matrix $\mathbf{M}$, and $\|\mathbf{x}\|$ denotes the Euclidean norm of $\mathbf{x}$.

For $\beta \neq 0$ and $\lambda_{\min}(\mathbf{Q}) > 2\gamma\lambda_{\max}(\mathbf{P})$, provided that

$$\|\mathbf{x}\| \geq \frac{2\lambda_{\max}(\mathbf{P})\beta}{\lambda_{\min}(\mathbf{Q}) - 2\gamma\lambda_{\max}(\mathbf{P})} \quad (34)$$

then $\mathbf{x}$ is bounded.

The inequality (34) may lead to conservative results because γ is not uncertain but may be made smaller if the state-variable norm is known to be smaller.

5.1 Linear matrix inequality (LMI)

In order to obtain a less conservative result, γ and β are assumed to be zero for the time being. (31) becomes

$$\dot{V}_i = \mathbf{x}^T(\mathbf{A}_i^T\mathbf{P} + \mathbf{P}\mathbf{A}_i)\mathbf{x}, \quad i = 1, \ldots, 8 \quad (35)$$

For all $x_1 \in R_{\epsilon_d}$, and $x_2 \in R_{\epsilon_\phi}$, if $\dot{V} < 0$ in (35), then the following LMIs must hold:

$$\mathbf{A}_i^T\mathbf{P} + \mathbf{P}\mathbf{A}_i < 0, \quad i = 1, \ldots, 8 \quad (36)$$

where '<' denotes that each LMI is negative definite.

Using toolboxes [3, 11] for solving the LMIs gives

$$\mathbf{P} = \begin{bmatrix} 2.0979 & 0.2354 \\ 0.2354 & 0.1827 \end{bmatrix} \quad (37)$$

which satisfies (36) except for $i = 1$ or $i = 5$. It so happens that (37) satisfies (29) which is the objective for seeking the matrix $\mathbf{P}$ using the LMI method.

For all $x_1 \in R_{\epsilon_d}$, and $x_2 \in R_{\epsilon_\phi}$, substituting (37) into (29) gives Fig. 9 where $\dot{\mathbf{x}}$ are given by (23) and (24), and $\hat{\dot{\phi}}_d = \dot{\phi}_d$. Since $V > 0$ and $\dot{V} < 0$ for all $x_1 \in R_{\epsilon_d}$, and $x_2 \in R_{\epsilon_\phi}$ but $V = \dot{V} = 0$ at $x_1, x_2 = 0$, hence $\lim_{t\to\infty} \mathbf{x}(t) = 0$

In the case of inaccurate path-curvature measurement, i.e., $\hat{\dot{\phi}}_d \neq \dot{\phi}_d$, then $\gamma, \beta \neq 0$ at $\mathbf{x} = 0$. The errors then converge to a limited circle which can be determined by plotting $\dot{V}$ using the inaccurate path-curvature measurement.

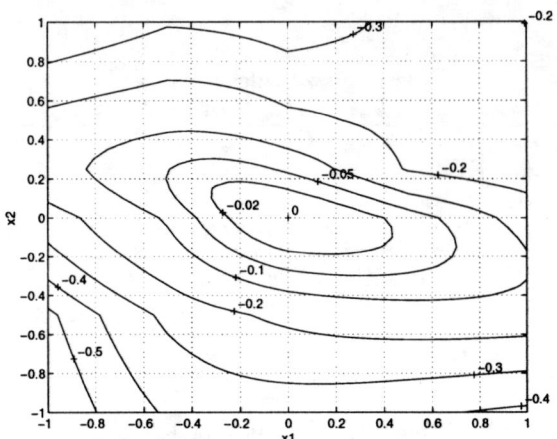

Figure 9: The contour plot of $\dot{V}$

6 Experimental results

A laboratory vehicle shown Fig. 8 is employed to implement both the linear and fuzzy controllers. The actual radius of curvature of the circle is $\rho = 0.5$m (constant). Three different cases representing the uncertainties of the curvature information are presented.

Case I, the estimated $\hat{\rho} = \infty$, as assumed for a straight line, i.e., no curvature compensation is applied. Figs. 10 and 11 show the experimental results of the linear controller and the fuzzy controller, respectively. Both are not very effective but the fuzzy controller is clearly better than the linear controller even in the absence of curvature compensation.

Case II, the estimated $\hat{\rho} = 1$m which is twice the true value. Figs. 12 and 13 show the experimental results using linear controller and fuzzy controllers, respectively. With some curvature compensation applied, the performance of the linear controller is improved but is not as good as that of the fuzzy controller.

Case III, the estimated $\hat{\rho} = \rho = 0.5$m, i.e., the path curvature is fully compensated. Figs. 14 and 15 show the experimental results of the linear controller and fuzzy controller, respectively. Both controllers perform well but the fuzzy controller can get the vehicle into the track of the circular path more quickly.

The experimental results confirm that the FLC outperforms the linear controller both in the presence and in the absence of the desired curvature information.

7 Conclusion

Based on the experimental results presented in this paper, the error dynamics of autonomous vehicles trying to follow curved paths is confirmed. The introduced curvature-compensation term is very effective provided that accurate path-curvature measurement is obtainable.

Based on the LIN2FUZ algorithm, the synthesis of the nonlinear linguistic FLCs can be easily achieved. Using the ET algorithm, the linguistic FLCs can be converted into another form which is amenable to stability analysis. The error convergence of the linguistic FLCs is guaranteed in the Lyapunov sense. The Lyapunov function is generated via the linear-matrix-inequality (LMI) method.

Given the Lyapunov function, the performance of the linguistic fuzzy-controlled systems for a larger path curvature can be further improved by the modification of the fuzzy rules, distributions of antecedent and/or consequent membership functions using neural networks or genetic algorithms, while still maintaining stability in the process. The issue about improving the robustness of the FLC in the face of measurement uncertainty will be separately treated.

References

[1] Y. H. Fung and S. K. Tso. AGV controller for arbitrary curvature negotiation. In *Proceedings of CAI Symposium '95 on Applications in Control and Automation*, pages 22–27, Hong Kong, 1995.

[2] Y. H. Fung and S. K. Tso. Performance improvement of differential-wheel-drive AGVs with curvature compensation. In *Proceedings of Second International Conference on Mechatronics and Machine Vision in Practice*, pages 169–174, Hong Kong, September 1995. City University of Hong Kong.

[3] P. Gahinet, A. Nemirovski, A. J. Laub, and M. Chilali. *LMI Control Toolbox: For use with MATLAB*. The MathWorks, Inc., May 1995.

[4] Y. B. Lee and K. D. Jang. The fuzzy system for velocity and curved-path control of an autonomous guided vehicle. In *Second Asian Conference on Computer Vision. Proceedings (1995)*, volume 3, pages 315–319, Singapore, December 1995.

[5] M. G. Mehrabi. *Path Tracking Control of Automated Vehicles: Theory and Experiment*. PhD thesis, Concordia University, Montreal, Quebec, Canada, September 1993.

[6] M. G. Mehrabi, A. Hemami, and R. M. H. Cheng. Analysis of steering control in vehicles with two independent left and right wheels. In *Proceedings of the Fifth International Conference on Advanced Robotics '91 ICAR*, volume 2, pages 1634–1637, Pisa, Italy, 1991.

[7] M. Mizumoto. Realization of PID controls by fuzzy control methods. In *Proceedings of IEEE International Conference on Fuzzy Systems*, pages 709–715, San Deigo, USA, 1992.

[8] S. K. Tso and Y. H. Fung. Methodological development of fuzzy-logic controllers from multivariable linear control. *IEEE Transactions on Systems, Man, and Cybernetics, Part B: Cybernetics*, 27(3):566–572, June 1997.

[9] S. K. Tso and Y. H. Fung. Synthesis and stability analysis of linguistic fuzzy controlled systems. In *Proceeding of 1998 IEEE International Conference on Fuzzy Systems*, Anchorage, Alaska, May 1998. (Accepted for presentation).

[10] P. G. Tzionas, A. Thanailakis, and P. G. Tsalides. Collision-free path planning for a diamond-shaped robot using two-dimensional cellular automata. *IEEE Transactions on Robotics and Automation*, 13(2):237–250, April 1997.

[11] L. Vandenberghe and S. Boyd. Software for semidefinite programming, user's guide. Available via anonymous ftp to isl.stanford.edu under /pub/boyd/semidef_prog, December 1994.

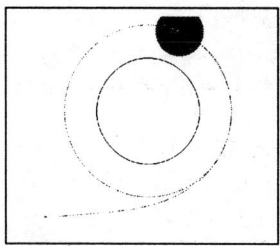
Figure 10: $\hat{\rho} = \infty$ m in linear control

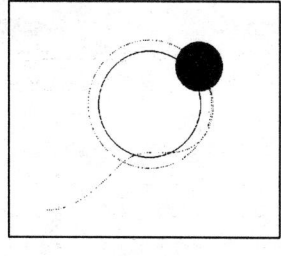
Figure 11: $\hat{\rho} = \infty$ m in fuzzy control

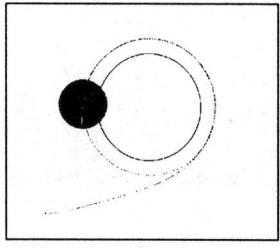
Figure 12: $\hat{\rho} = 1$ m in linear control

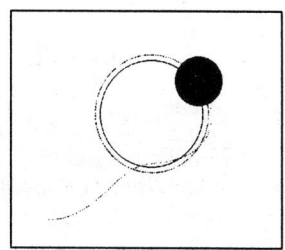
Figure 13: $\hat{\rho} = 1$ m in fuzzy control

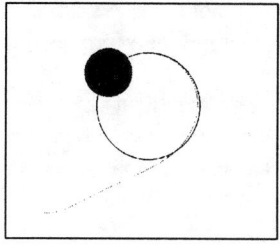
Figure 14: $\hat{\rho} = 0.5$ m in linear control

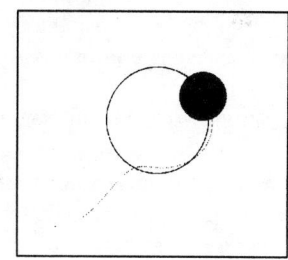
Figure 15: $\hat{\rho} = 0.5$ m in fuzzy control

Fuzzy-Logic Dynamics Modeling of Robot Manipulators

Mohammad R. Emami
Robotics and Automation Laboratory
Dept. of Mechanical & Industrial Eng.
University of Toronto
emami@mie.utoronto.ca

Andrew A. Goldenberg
Robotics and Automation Laboratory
Dept. of Mechanical & Industrial Eng.
University of Toronto
golden@mie.utoronto.ca

I. Burhan Türksen
Intelligent Fuzzy Systems Laboratory
Dept. of Mechanical & Industrial Eng.
University of Toronto
turksen@mie.utoronto.ca

Abstract

A systematic approach of fuzzy-logic modeling from the system input-output data is applied to modeling dynamics behavior of a four degree-of-freedom robot manipulator. The output of the fuzzy model is compared with the experimental data and with that of an analytical simulation. The simplicity of the fuzzy model and its improved accuracy is significant.

1 Introduction

Advanced control schemes for robot manipulators often involve computation of the joint actuator torque from its motion by using a dynamics model. The dynamics model is expected to properly predict the joint torques for different trajectories and load conditions, in real-time. Obtaining such a model through analytical methods involves two major burdens: **(i)** it is generally difficult to accurately model some complex phenomena such as backlash, flexibility, friction, etc; and **(ii)** even if we formulate these effects, many internal parameters should be accurately identified in advance, and this requires an exhaustive amount of specifically designed experiments. Further, the resulting model would be complicated and still hard to apply in real-time. In this research, through a systematic methodology, the fuzzy-logic inverse dynamics model of a four degree-of-freedom robot manipulator is developed directly from the input-output data. In the proposed approach, both the model structure and its parameters are identified from the input-output data. The simplicity in terms of system presentation and model computation effort, and the fuzzy model capability of capturing the complicated system behavior is significant.

The linguistic approach of system modeling that, similar to human thinking, introduces *"fuzziness"* into system theory, was initially proposed by Zadeh [1], and further developed into a new class of systems called *"fuzzy systems"* by Tong [2], Pedrycz [3], Sugeno et al. [4], and Yager et al. [5].

IF x_{s1} is B_{s11} AND ... AND x_{sr} is B_{s1r} THEN y_s is D_{s1}
ALSO
....
ALSO
IF x_{s1} is B_{sn1} AND ... AND x_{sr} is B_{snr} THEN y_s is D_{sn}

Figure 1: *The fuzzy-logic model of the MISO system s*

Generally, the inverse dynamics model of an S degree-of-freedom (d.o.f.) manipulator can be represented as:

$$y_s = F_s(q, \dot{q}, \ddot{q}, \xi) \qquad s = 1, 2, \cdots, S \qquad (1)$$

where $q(\dot{q}, \ddot{q})$ is the vector of joint displacements (velocities, accelerations), y_s is the torque applied on the sth joint and ξ is the vector of robot kinematic and dynamic parameters. The function F_s is highly nonlinear, and contains dynamic coupling, gravity, joint friction, flexibility and backlash, load variation and other effects. From the fuzzy-logic point of view, the encoded knowledge of dynamics behavior of the robot manipulator can be interpreted by S fuzzy models. Each model expresses the variation of one joint torque as a result of the motion of all joints. In other words, each model consists of IF-THEN rules with several antecedent variables, i.e., joint displacements, velocities and accelerations, and single consequent variable, i.e., joint torque. Figure 1 represents such a model for joint s $(s = 1, 2, \cdots, S)$ with n rules, where $x_{s1}, x_{s2}, \ldots, x_{sr}$ are input variables, and y_s is the output, B_{sij} and D_{si} $(i = 1, \cdots, n; j = 1, \cdots, r)$ are fuzzy sets representing the input and output membership functions, respectively. For each joint, the applied torque could be influenced by the dynamics of all joints, and hence $3 \times S$ input candidates can be considered. However, the number of practically *"significant"* inputs r may be less, depending on the joint and robot configuration. The process of fuzzy modeling is illustrated in Figure 2. A systematic approach of fuzzy modeling requires formal techniques for each of the blocks of the flowchart.

In this paper, we briefly review the systematic approach of fuzzy-logic modeling, and then apply the methodology to dynamics modeling of a 4 d.o.f. robot manipulator. The fuzzy model output and analytical simulation results are compared with the experimental data, and concluding remarks are made at the end.

2 A Review of the Systematic Methodology of Fuzzy-Logic System Modeling

The goal of the systematic approach is to improve the objectivity of fuzzy modeling by developing appropriate formulations and criteria to specify those features of the

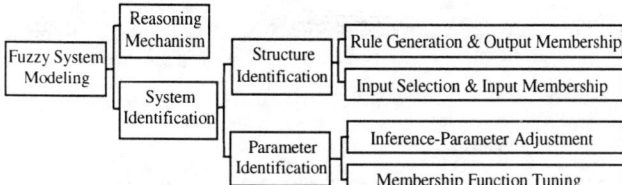

Figure 2: *Flow chart of fuzzy system modeling*

model that are usually assigned heuristically in the fuzzy modeling approaches. As a result, unlike *ad hoc* fuzzy modeling techniques that are mainly based on the expert knowledge, the proposed methodology merely exploits input-output data to extract some information about system characteristics. The reader is referred to [6,7] for more details.

2.1 Reasoning Mechanism

The proposed methodology considers the inference mechanism as an "*identifiable*" object of fuzzy systems. We developed a unified parameterized formulation for the reasoning process as follows:

For the crisp input $x^* = (x_1^*, x_2^*, \ldots, x_r^*)$, the fuzzy output of the system introduced in Figure 1 is obtained as (the index s is removed for convenience):

$$E(y) = \beta\left(1 - S_p\left[T_p[\tau_1(x^*)\overline{D}_1(y)], \ldots, T_p[\tau_n(x^*)\overline{D}_n(y)]\right]\right) + (1-\beta)S_p\left[T_p[\tau_1(x^*)D_1(y)], \ldots, T_p[\tau_n(x^*)D_n(y)]\right] \quad (2)$$

where τ_i called "*rule degree of firing*" is computed as:

$$\tau_i(x^*) = T_q\left(B_{i1}(x_1^*), B_{i2}(x_2^*), \ldots, B_{ir}(x_r^*)\right) \quad (3)$$

and
$$\overline{D}_i(y) = 1 - D_i(y) \quad (4)$$

In equation 2, S_p is the *n*-ary *t*-conorm operator computed as:

$$S_p(a_1, a_2, \ldots, a_n) = \quad p > 0 \quad (5)$$
$$\left[a_1^p + (1-a_1^p)\left[a_2^p + (1-a_2^p)\left[\cdots\left[a_{n-2}^p + (1-a_{n-2}^p)\left[a_{n-1}^p + (1-a_{n-1}^p)a_n^p\right]\right]\cdots\right]\right]\right]^{1/p}$$

The operator T_χ ($\chi = p, q$) is the *n*-ary *t*-norm operator that is calculated as:

$$T_\chi(a_1, a_2, \cdots, a_n) = 1 - S_\chi\left((1-a_1), (1-a_2), \cdots, (1-a_n)\right) \quad (6)$$

Equation 2 is a linear combination of two extreme reasoning approaches, Mamdani's and Logical, with adjustable parameters [7].

The crisp output is then obtained by using the *Basic Defuzzification Distribution* method as follows [5]:

$$y^* = \frac{\int_{y_0}^{y_1} y[E(y)]^\alpha dy}{\int_{y_0}^{y_1} [E(y)]^\alpha dy} \quad 0 \leq \alpha < \infty \quad (7)$$

In the above reasoning formulation, i.e., equations 2 and 7, four reasoning parameters p, q, α, and β are introduced whose variation will cause a continuous range of variation for the reasoning mechanism. Consequently, unlike traditional approach of selecting the inference mechanism *a priori*, the optimum reasoning mechanism would be identified for the system by adjusting the above parameters based on the input-output data.

2.2 Fuzzy Structure Identification

Fuzzy structure identification is to assign the optimum number of rules, significant input variables, input and output membership functions, and the amount of overlap between the membership functions required for the fuzzy model. These are briefly discussed in the following two sections.

2.2.1 Rule Generation, Output Membership Functions

An intuitive approach of objective rule generation is based upon clustering of input-output data. However, in the proposed methodology, we first cluster only the output space, and then derive the input space fuzzy partition by projecting the output space partition on each input space, separately. Simplicity and applicability, particularly for systems with a large number of input variables, are the main advantages of this approach.

The output fuzzy clusters are carried out by the Fuzzy C-Means (FCM) algorithm [8]. The idea of fuzzy clustering is to divide the output data into fuzzy partitions that overlap with each other. Therefore, the containment of each datum y_k to each cluster i with a center v_i is defined by a membership grade u_{ik} in [0,1]. The membership grades and cluster centers are obtained through an iterative procedure as follows:

$$u_{ik,t} = \left[\sum_{j=1}^{c}\left(\frac{\sqrt{y_k - v_{i,t-1}}}{\sqrt{y_k - v_{j,t-1}}}\right)^{\frac{2}{m-1}}\right]^{-1} \quad (8)$$

$$v_{i,t} = \frac{\sum_{k=1}^{N}(u_{ik,t})^m y_k}{\sum_{k=1}^{N}(u_{ik,t})^m} \quad (9)$$

where N is the number of data, and $u_{ik,t}$ and $v_{i,t}$ are the membership grade of output y_k in the cluster i, and the center of cluster i, respectively, at the tth iteration. Three crucial pieces of information are required *a priori* for constructing the suitable partition from the data, i.e., **(i)** adequate number of rules for expressing the system behavior which, in most cases, equals to the number of clusters c; **(ii)** the order of fuzziness of the system model that is the overlap of the fuzzy clusters, and is adjusted by parameter m called "*weighting exponent*"; and **(iii)** suitable initial location of the cluster centers which affects the model formation.

Specification of the Number of Rules: The following cluster validity index is developed for assigning the optimum number of output clusters, and hence number of rules, in the fuzzy model:

$$s_{cs} = \sum_{k=1}^{N}\sum_{i=1}^{c}(u_{ik})^m\left((y_k-v_i)^2-(v_i-\bar{v})^2\right) \quad (10)$$

where, $\bar{v}$, is a weighted mean of data considering their membership to each of the clusters defined as:

$$\bar{v} = \frac{1}{\sum_{i=1}^{c}\sum_{k=1}^{N}(u_{ik})^m}\sum_{k=1}^{N}\sum_{i=1}^{c}(u_{ik})^m y_k \quad (11)$$

Minimization of s_{cs} will increase the compactness of clusters and the separation between them, at the same time. Hence, the optimum number of clusters c corresponds to minimum s_{cs}. In most cases, c is equal to the number of rules n for the fuzzy model.

Specification of the Order of Fuzziness: The weighting exponent m controls the extent of membership sharing between the output fuzzy clusters in the data set. In the range of $(1,\infty)$, the larger m is, the *"fuzzier"* are the membership assignments to each data point. For selecting m, The following index is developed:

$$S_T = \sum_{k=1}^{N}\left(\sum_{i=1}^{c}(u_{ik})^m\right)(y_k-\bar{v})^2 \quad (12)$$

An appropriate value for m is what makes S_T equal to a constant parameter $K/2$, where K is defined as:

$$K = \sum_{k=1}^{N}\left[\left(y_k-\frac{1}{N}\sum_{j=1}^{N}y\right)^2\right] \quad (13)$$

The Initial Cluster Centers: The initial cluster centers for the FCM algorithm are assigned through the hard clustering techniques. This approach provides more efficient strategy compared to the previous approach of randomly selecting the initial values [7].

2.2.2 Input Selection and Input Membership Function Assignment

In order to identify the significant input variables among a finite number of candidates, we first project the output clusters onto the space of each of the input candidates. As a result, for each input candidate x_j, the membership functions $\hat{B}_{ij}\;(i=1,2,\cdots,n)$ are formed. Then, we calculate the following index:

$$\pi_j = \prod_{i=1}^{n}\frac{\Gamma_{ij}}{\Gamma_j} \qquad j=1,2,\cdots,\hat{r} \quad (14)$$

where Γ_{ij} is the range in which the membership function $\hat{B}_{ij}$ is equal to one, Γ_i is the entire range of x_j, n is the number of rules, and $\hat{r}$ is the number of input candidates. Less π_j illustrates more dominant variable x_j, and hence, significant variables are selected among those that produce less π.

The convex membership functions B_{ij} for significant inputs $x_j\;(j=1,2,\cdots,r)$ are then formed by using the range Γ_{ij} and performing *"fuzzy line clustering"* as described in details in [6].

2.3 Fuzzy Parameter Identification

Optimum values of the inference parameters (p, q, α, and β) are identified through a nonlinear constrained optimization problem, by minimizing:

$$PI(p,q,\alpha,\beta) = \sum_{k=1}^{N}(y_k-\hat{y}_k)^2 \Big/ N \quad (15)$$

subject to the following constraints:

$$0 < p,\,q < \infty \;\&\; 0 < \alpha < \infty \;\&\; 0 < \beta < 1 \quad (16)$$

where, y_k is the kth actual output and $\hat{y}_k$ is the kth model output.

The input-output membership functions that have already been identified in the structure identification phase are approximated by trapezoidal functions, and then an incremental tuning procedure is applied to adjust the membership function parameters based on the tuning data set and the performance index defined by equation 15 [6].

3 Application to a 4 d.o.f. Robot Manipulator

3.1 The Experimental Setup

The systematic fuzzy-logic modeling approach was applied to a 4 d.o.f. robot manipulator which is a part of the IRIS facility. This facility is a versatile, reconfigurable and expandable setup composed of several robot arms that can be easily disassembled and reassembled to provide a multitude of configurations. The basic element of the system is the joint module that has its own input and output link. Each module is equipped with a brushless DC motor coupled with harmonic drive gear and instrumented with an optical encoder to measure the rotor angular displacement and a tension-compression load cell torque sensor to measure the applied torque on the joint. The setup is controlled by a distributed computer system based on AMD 29050 RISC processor tightly coupled with the host computer based on 50 MHz Intel 80486 processor accelerated with a large cache memory. A fast parallel I/O system allows up to 5 KHz sampling rate. Modularity of the joints enables the user to arrange various configurations. Figure 3 shows the specified configuration of the IRIS arm for our modeling purposes.

3.2 Test Plan and Data Acquisition

A crucial phase of any system identification approach is planning appropriate experiments, as the model accuracy and robustness critically depends on the data obtained through the experiments. This situation is even more critical in *"black-box"* approaches such as ours. In this application, the manipulator system has 12 input

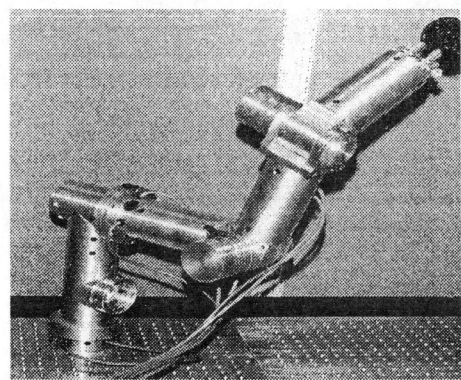

Figure 3 : *The desired configuration of IRIS arm*

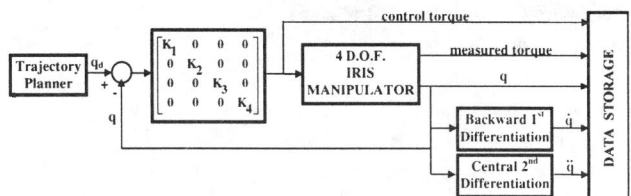

Figure 4 : *Block diagram for modeling experiments*

candidates, i.e., joint displacements, velocities, and accelerations, and 4 output variables as joint torques. The fuzzy-logic model of each joint is built up separately by considering the effect of other joints dynamics. Our plan is to drive the manipulator joints along different trajectories while the end-effector carries various amounts of load. A simple control feedback was designed to ensure a stable performance in all cases as shown in Figure 4. A user shell was also programmed on the host computer to calculate joint velocities and accelerations on line by using backward and central difference formulations, respectively.

Although designing the experiments is problem dependent, some general rules must be examined to guarantee the validity of the information obtained from the "*proper*" experiments, as follows:

CAUSALITY: The input and output variables must be assigned so that the former causes the latter. In a closed-loop experiment, there are obviously causal dependencies in both directions. However, if the control signal is generated by a computer, as in our case, then due to the physical system delay, the consequence of "*system state*" and "*control signal generation*" is closer in time than the consequence of the "*control signal*" and the new "*system state*". Hence, it is more appropriate to consider the system state (joint displacements, velocities and accelerations) as the input, and the control signal (joint control torque) as the output. This is already defined as the "*inverse dynamics model*" of the system. The causality condition also implies that the measured joint torque can not be considered as an identification parameter, since it is neither a cause nor an independent effect.

SUFFICIENT EXCITATION: Experiments should excite at least all modes of the system that may be excited when the model is used, in the same range of variation. For robot manipulators, random joint trajectories that cover the desired range of input/output parameters are suggested as "*proper*" input signals [9], provided that joint velocity and acceleration do not exceed the physical limitations. In our application, we generated these trajectories by an interactive program prepared in MATLAB environment. Moreover, in order to enrich the obtained information, we also tested the manipulator for several step joint trajectories and sinusoidal trajectories with different frequencies (due to the system bandwidth) and maximum desired amplitude. Experiments were performed under three different loading conditions, i.e., light, moderate and heavy loads. The load effect is considered as a disturbance to the system, and the fuzzy-logic model is expected to be robust enough due to various loading conditions.

REPEATABILITY: Another condition is required to ensure that the same system output (within an acceptable range of disturbance) is obtained from repeated experiments. In our application, reproducibility tests were performed, and as a result, joint displacements, velocities and control torque signals were found quite repeatable. However, as expected, the noise spectrum of the calculated joint accelerations was high, and hence some post-processing was required for filtering the signal.

SEPARABILITY: The input signal to the system should be independently generated, or at most, be influenced by the past and present system output only. A closed-loop experiment does satisfy this condition. It should be verified however that the channels transmitting input and output data do not have any cross-interference. This condition was also examined closely in our application.

3.3 Data Processing and Data Selection

One of the advantages of the proposed methodology is that, unlike classical approaches, there is no assumption of zero-mean signal; hence there is no need to remove the trend and outliers in the data. Furthermore, scaling is not required for our algorithm, either. However, the following data processing steps are useful for the modeling tasks:

Sampling Frequency: A lower band on the sampling frequency is obtained from Shannon's sampling theorem, which states that in order to recover a continuous-time signal exactly, the sampling frequency ω_s should be chosen such that the signal does not contain any useful frequencies above the Nyquist frequency $\omega_s/2$. By considering the power spectral density of the measured joint torques, the highest useful bandwidth can be obtained. As a result of this analysis, Nyquist frequency around $\omega_s/2 = [20\text{-}30]$ Hz determines the width of the meaningful bandwidth in our experiments.

Low-Pass Filtering: The second derivative of joint displacement signal contains a wide-band noise. Using input data with high bandwidth for system identification makes the model too sensitive with a higher error variance. Therefore, it is desirable to filter the acceleration signal within an appropriate range of frequency. The measured joint torque signal can be a useful hint to specify the maximum meaningful frequency of the acceleration of each joint. What the torque sensor measures is the torque load applied on each joint as a result of the dynamics of all joints. Hence, this signal contains the effect of joint accelerations, as well. As a result, spectral analysis of the measured torque signal provides a suitable cut-off frequency for filtering the calculated joint accelerations. We used a low-pass digital *Butterworth* filter to eliminate the effect of higher frequencies. This filter is characterized by a magnitude response that is maximally flat in the pass-band and monotonic overall. Monotonicity and smooth behavior are the advantages of this type of filter over its counterparts [10].

Data Selection and Categorization: The sampling frequency for system identification is not necessarily as high as the sampling frequency for data collection. In fact, as discussed before, in order to reduce the sensitivity of the model to noise, it is recommended to choose a sampling frequency less than ω_s for the identification. Also, it is desirable to reduce the amount of data used for the identification. For reduction of sampling frequency, first, we implemented a *Butterworth* low-pass filter with cut-off frequency around Nyquist frequency $\omega_s/2$ in order to prevent aliasing effects. Then we reduced the number of data carefully such that the main information contained in the signal was not damaged. This procedure depends on the signal and the applied trajectory. The resultaning data from all experiments were finally combined and categorized into three different sets:

(i) **Training Set:** for constructing the fuzzy model structure;
(ii) **Tuning Set:** for adjusting the inference and membership function parameters.
(iii) **Testing Set:** for examining the validity of the model.

4 Comparison with the Analytical Models and Conclusions

The fuzzy-logic model of the IRIS arm was constructed from the input-output data. Figure 4 illustrates these models for joints #1 and #2. In parallel, an analytical simulation of the manipulator was also prepared that contains the effects of nonlinear dynamics interaction, motor and harmonic drive dynamics, friction, and joint flexibility [11]. Figure 5 shows the results of the fuzzy and analytical models of the first two joints for a testing sinusoidal trajectory under a moderate loading condition, compared to the experimental data. A better performance of the fuzzy model is clearly observable for all joints. The same performance is also observed for different trajectories and loading conditions. While the fuzzy-logic model performs almost uniformly for all joints in different situations, the outcome of the analytical model critically depends on how we model various physical effects in the system. For instance, the behavior of joint #2 is more complicated due to the weight effect and undesired backlash and friction. These phenomena are hardly captured by analytical formulation, and therefore, simulation performance for these joints is degraded.

This application shows that the fuzzy-logic modeling paradigm can provide a strong potential for interpreting systems specially those that are too complicated to be modeled by analytical methods. Through this application, we demonstrated that although "*approximation*" is inherited in fuzzy modeling, based on a firm theoretical background, we can achieve more accuracy and better performance than what analytical approaches offer, without sacrificing the simplicity and applicability. We conclude that our expectations from analytical approaches should be limited by the degree of complexity of the system. Beyond a threshold, it is required to employ new paradigms such as fuzzy-logic approach, if we demand both simple and relevant interpretation, and fair accuracy and satisfactory performance, at the same time.

References

[1] L. Zadeh, "Outline of a new approach to the analysis of complex systems and decision processes", *IEEE Trans. Systems, Man, and Cybernetics*, SMC-3, pp. 28-44, 1973.

[2] R.M. Tong, "The construction and evaluation of fuzzy models": M.M Gupta., R.K. Regade, R.R. Yager (eds.), "*Advances in Fuzzy Set Theory and Applications*", North-Holland, Amsterdam, pp. 559-576, 1979.

[3] W.Pedrycz, "Identification in fuzzy systems", *IEEE Trans. on Systems, Man, and Cybernetics*, No. 14, pp. 361-366, 1984.

[4] M. Sugeno, T. Yasukawa, "A fuzzy-logic-based approach to qualitative modeling", *IEEE Trans. on Fuzzy Systems*, No. 1, pp. 7-31, 1993.

[5] R.R.Yager, D.P. Filev, "*Essentials of Fuzzy Modeling and Control*", John Wiley and Sons, 1994.

[6] M.R. Emami, Turksen I.B., Goldenberg A.A., "Development of a systematic methodology for fuzzy-logic modeling", to be published in *IEEE Trans. Fuzzy Systems*, 1998.

[7] M.R.Emami, "*Systematic Methodology of Fuzzy-Logic Modeling and Control and Application to Robotics*", Ph.D. Thesis, Department of Mechanical and Industrial Engineering, University of Toronto, Toronto, 1996.

[8] J.C. Bezdek, "*Pattern Recognition with Fuzzy Objective Function Algorithms*", Plenum Press, NY, 1981.

[9] M. Gautier, W. Khalil, "Exciting trajectories for the identification of base inertial parameters of robots", *Int. J. Robotics Research*, Vol. 11, No. 4, pp. 362-375, 1992.

[10] J.N. Little, L. Shure, "*Signal Processing Toolbox for Use with MATLAB*", The MathWorks Inc., 1992.

[11] M.R. Emami, "Simulation of the IRIS arm", *Internal Report*, Robotics and Automation Laboratory, U. of Toronto, 1995.

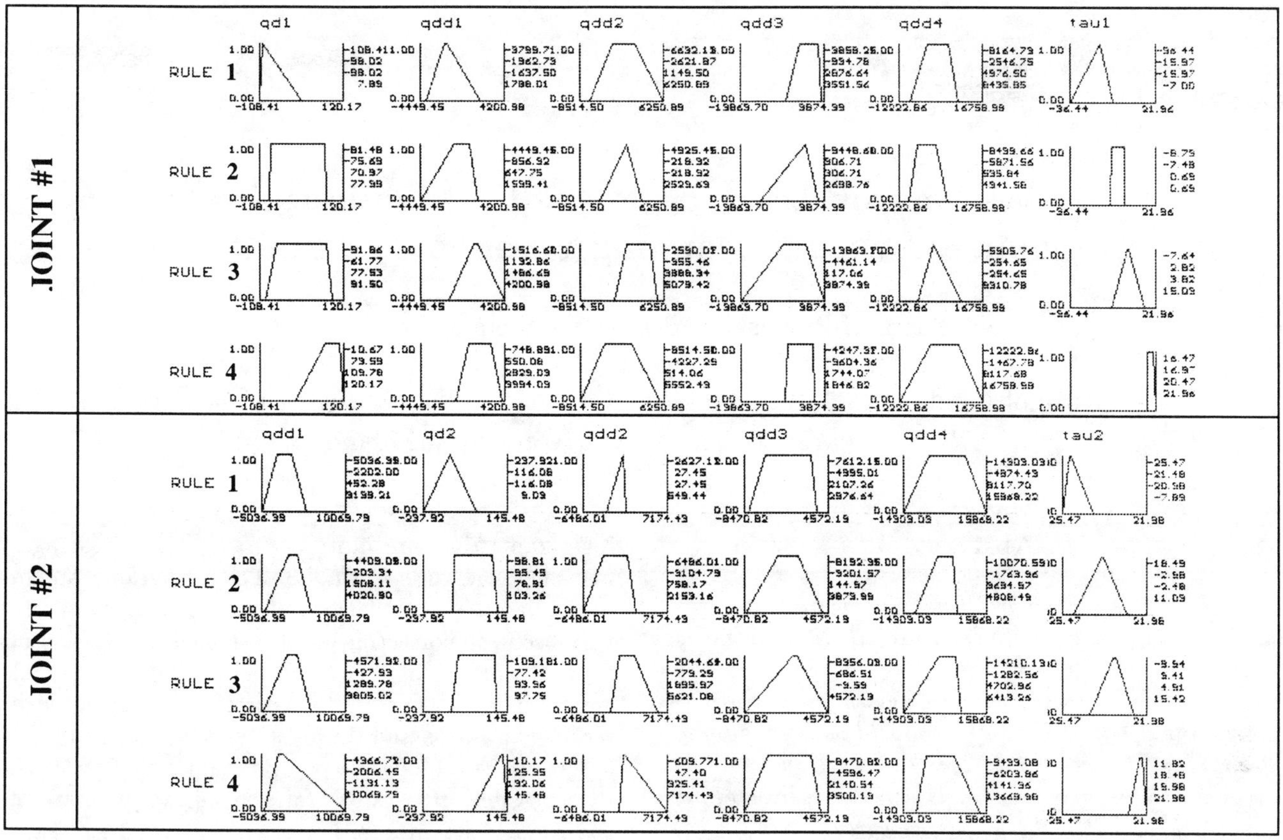

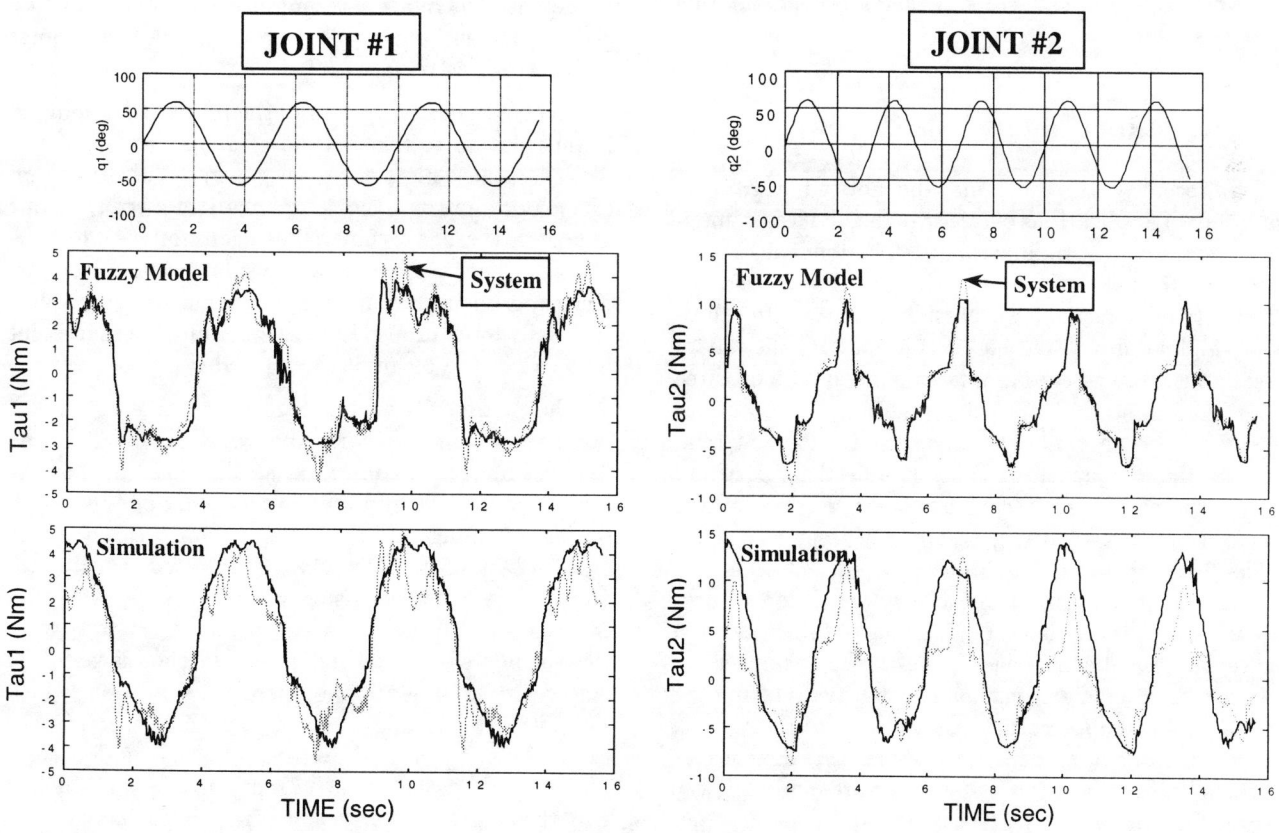

Figure 5: *Fuzzy-logic model of the first two joints of the IRIS arm*

Figure 6: *Comparison of the fuzzy model and simulation joint torque with the experimental data for the first two joints of the IRIS arm*

A Method for Tracking Pose of a Mobile Robot Equipped with a Scanning Laser Range Finder

Artur Dubrawski and Barbara Siemiątkowska

Institute of Fundamental Technological Research, Polish Academy of Science
21 Świętokrzyska Str., 00-049 Warszawa, Poland

Abstract

One of the essential problems in navigation of a mobile robot is to accurately determine its location using data obtained with range sensors. In this paper we present a novel approach to tracking changes of orientation and position of a robot equipped with a scanning laser range finder and designed to work in partially structured environments. It is shown experimentally on a real robot that the proposed approach is more robust against sensory noise than the method of angle histograms, which is often used to track orientation and position of indoor vehicles equipped with scanning range sensors.

1 Introduction

In recent years there may be noticed a growing number of applications of laser ranging devices in navigation systems of mobile robots. The tendency stems from useful characteristics of the laser-based distance measurement technology, particularly from its relatively high accuracy and high resolution. By the way of scanning a laser beam or a laser plane in two or three dimensions, state-of-the-art range finders are able to deliver huge amounts of fairly accurate data at relatively high frequencies. There is a need for fast and practical processing techniques, capable of handling massive data streams, generated by laser range finders, in real time and in a memory-efficient manner.

In this paper we introduce a new method of processing two-dimensional range scans obtained using a rotating laser beam device. The method searches for segments of colinear readouts by the way of finding groups of neighboring data that belong to lines which bear the same orientation. The extracted colinear segments are then used in a search for predominant wall directions in the scene. With this approach it is possible to very inexpensively generate a sort of angle histograms that are quite accurate and robust against range measurement noise. The histograms, and in particular cross-correlations of the pairs of them, are very useful in correcting dead reckoned orientation of a robot working in structured or partially structured environments. Knowing the correct orientation, dead reckoned position and the locations of the colinear segments oriented along the predominant wall directions, it is possible to obtain a fairly accurate estimate of the robot's actual position in respect to the objects of the scene. Correct orientation and position estimates are crucial for a successful performance of other range-based navigation techniques, such as occupancy grid-based mapping and planning, which were topics of our previous research [4].

The basic idea lying behind the presented technique is very similar to the method of *angle histograms* [5]. However, those histograms are generated directly from (x,y) coordinates of the components of a scan. In our approach the histograms reflect cummulative lengths of wall segments detected along subsequent directions. The experimental results show that our method is substantially more robust against inevitable measurement noise, than the technique described in [5].

The method presented here is to some extent related to Hough transform, known most of all in the field of image processing, but already used in the context of navigation of a mobile robot equipped with an optical scanning range sensor [2]. Our technique also searches for lines which share a selected orientation. But, unlike Hough transform which usually works on 2D rasterized images and is typically used to extract line segments in 2D (as in [2]), our method is very well suited to one-dimensional polar representation, which is natural for an image obtained with a laser scanner in 2D. Due to that and because of a very modest complexity, our method can be much faster when detecting the predominant wall directions in the laser scans, than its counterparts based on Hough transform.

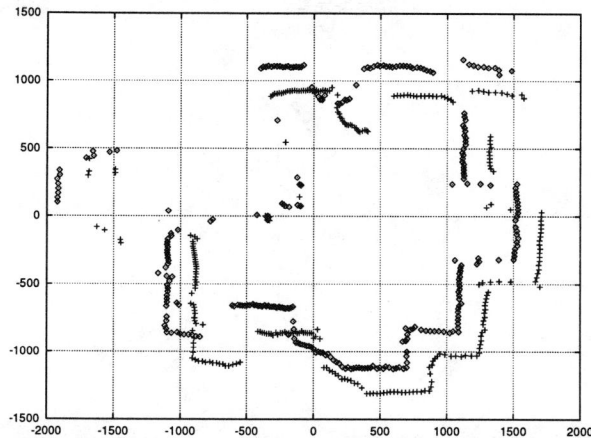

Figure 1: A pair of sample 2D laser rangefinder scans of an indoor environment plotted in the robot's coordinates (dimensions in millimeters).

The presented technique relies on a local and parallel data processing scheme and thus it may be implemented in a form of a cellular neural network and run in a massively parallel fashion to additionally accelerate the eventual realization. We propose to call the particular one-dimensional cellular neural architecture a *circular cyclic network* [1].

2 Fundamentals of the method

A scanning range sensor collects *scans*, ie. sets of m readouts $\{R_i, \varphi_i\}$, where R_i represents an observed distance to an object placed in the way of the laser beam in the direction determined by a scanning angle φ_i. The scanning angle takes m discrete values ranging from 0 to 360 degrees[2], so that $i = 1, \ldots, m$ and $\Delta\varphi = \varphi_{i+1} - \varphi_i = const$, and the indices i are additive modulo m. Scans like that may be visualized in the (x,y) coordinates of the robot's environment in a form shown in Fig. 1.

We may write a normal equation of a line in the plane (x,y) oriented along the direction α as follows:

$$x \cdot sin\alpha - y \cdot cos\alpha + c = 0 \qquad (1)$$

where $|c|$ determines the distance between the line and the origin ($x = 0, y = 0$). By taking into account

[1] This architecture foolows the definition of cellular neural networks given in [1] with one exception: it's layers are one-dimensional structures. Nb. most cellular networks (including the one described in this paper) do not learn in a sense of other neural structures. Here, node-to-node connection strenghts are fixed at a design stage, similarly to the human retina.

[2] Practically, the working range of the scanning angle may be, and often is, limited and does not cover the full horizon.

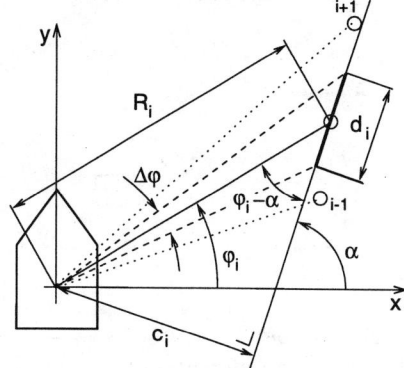

Figure 2: Readout $\{R_i, \varphi_i\}$ is a candidate member of a segment of three colinear readouts ($n=1$) oriented along a direction α. Unit length d_i contributed by this readout to the total length of the segment depends on the perpendicular distance c_i, and the angle at which this part of the segment is seen, $\varphi_i - \alpha$.

that $x_i = R_i \cdot \cos\varphi_i$ and $y_i = R_i \cdot \sin\varphi_i$ we may determine the distance c_i for the line given by (1) which crosses a given point (x_i, y_i):

$$c_i = R_i \cdot sin(\varphi_i - \alpha) \qquad (2)$$

Two points $\{R_i, \varphi_i\}$ and $\{R_j, \varphi_j\}$ belong to the same line oriented along α if $c_i = c_j$.

The above formulas compose the essential idea for testing colinearity of the neighboring range readouts in the presented method (Fig. 2). We may say that the readout $\{R_i, \varphi_i\}$ belongs to a line segment oriented along a specified direction α if it and its $2n$ neighbors fulfill the following simple condition:

$$\bigvee_{i-n \leq k \leq i+n} |c_k - 0.5 \cdot (c_{i-n} + c_{i+n})| \leq \epsilon \qquad (3)$$

So, for the colineraity check purpose, it is sufficient to compare the neighboring c-values computed for a praticular α using the set of current readouts $\{R_i, \varphi_i\}$.

Each readout, which belongs to a segment composed of consecutive colinear readouts, contributes a certain unit length d_i to the cummulative length of the segment (Fig. 2):

$$d_i = \left| \frac{c_i \cdot \sin \Delta\varphi}{\sin^2(\varphi_i - \alpha) - \sin^2(0.5\Delta\varphi)} \right| \qquad (4)$$

Taking into account that in our system $\Delta\varphi = 1°$ we may ignore $\sin^2(0.5\Delta\varphi)$, as a small of a higher order than the other components of the formula. Furthermore, $\sin \Delta\varphi$ is a multiplicative system constant, so we can ignore it when comparing different values of d

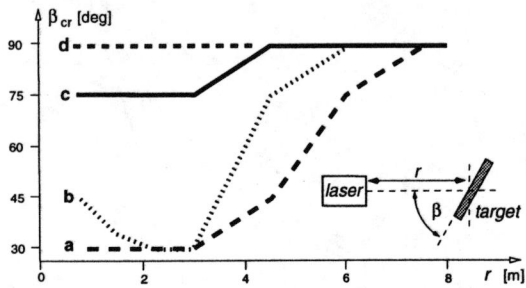

Figure 3: Critical detection angle ($\beta = \varphi - \alpha$) vs. measured distance for different target materials (LightRanger): (a) white paper, (b) cardboard, (c) polished furniture, (d) metal plate.

or linear combinations of d's, and consider d_i proportional to:

$$d_i \sim \frac{|c_i|}{\sin^2(\varphi_i - \alpha)} \quad (5)$$

The formula (5) becomes unstable at small values of $\varphi_i - \alpha$, that is at sharp angles of observation. However, typical optical range finding devices use safe low power lasers, which due to the energy dissipation tend to produce erroneous outputs at the sharp angles to the detected objects. It is practical to ignore the readouts which may belong to the colinear segments observed at the too sharp angles. The critical detection angle depends on the distance and on the reflexivity of the measured surfaces. In case of our experimental setup it is reasonable to skip all the readouts R_i observed at the angles $\varphi_i - \alpha$ sharper than 22.5° (see Fig. 3).

Now, with the formulas (2), (3) and (5) we can iterate through the values of $\alpha \in (0°, 180°)$, computing for each of them sums D of the unit lenghts d_i related to the readouts recognized as the members of the detected colinear segments. The histograms $D(\alpha)$ will have maxima at the orientations α corresponding to the predominant wall directions, if such directions exist. By using approximate formula (5) instead of the exact form (4) we only scale the histogram's amplitude, while the locations of the maxima remain intact. Formulas (2), (3) and (5) are very inexpensive to process. To do so one needs to access only a local information and the value of α.

3 Neural implementation

In the neural implementation of the described processing scheme (Fig. 4) the first ring-shaped input layer is composed of processing units which model the equation (2). Each of the m units receives a separate input signal R_i and a common input α. The φ_i's are hardware dependent and may be assigned constant values, one for each of the input layer's units. The second layer is composed of m laterally interconnected neurons which receive signals proportional to c_i's computed by the units of the input layer. The neurons take into account signals received from their immediate neighbors[3] and check the condition (3). A second layer neuron returns c_i if its colinearity condition is met and 0 otherwise. The third layer implements the right hand side of the formula (5). Similarly to the case of the input layer, every neuron in the third layer receives a common signal α, and has encoded an unique value of φ_i. The outputs, proportional to the unit segment lenghts, are then summed up to produce a component of the histogram for the given α.

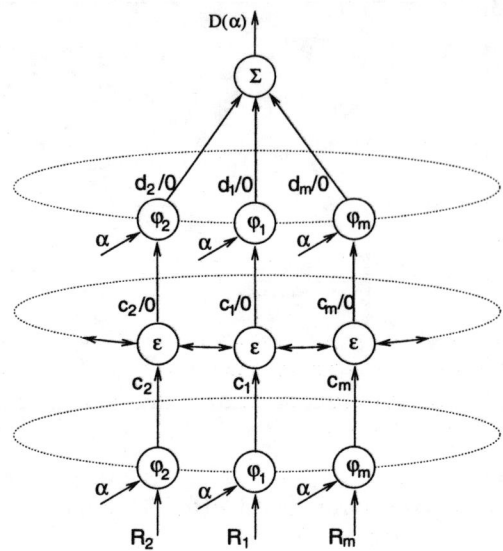

Figure 4: Cyclic cellular neural network in a setup suitable for processing laser range scans in order to obtain segments length vs. orientation histograms.

To perform a complete verification we conduct the described processing cycle for all the range of possible line segments orientations, ie. from 0 to +180 degrees in the robot's coordinates. For each α from that range we sum up d_i's present on the outputs of the third layer of neurons, and then we plot these sums, denoted $D(\alpha)$, against α's. In such an inexpensive way we obtain a sort of an angle histogram, which actually represents the cummulative dimensions of scene segments oriented in particular directions as seen by the sensor. The histograms, shown in the top graph in Fig. 5 for $\epsilon = 10$mm and several values of n, clearly

---

[3] Of course in general the range of lateral connections or equivalently the neighborhood size may be larger than $n = 1$.

position their extrema at the locations of the most and the least predominant directions of the scene segments. In a typical structured or partially structured indoor environment there are usually two maxima, which correspond to (most often) roughly perpendicular directions.

4 Tracking pose changes

If we compared two scans taken at nearby locations, we could expect them to reveal changes in position and orientation of the viewpoint in respect to the elements of the scene. The difference in α's which correspond to the maxima of two histograms computed for consecutive snapshots taken enroute is roughly equivalent to the robot's orientation change if the scans were being collected sufficiently often. Watching changes of α_{max} is an easy way of keeping track of the robot's orientation changes, but it does not provide a perfect accuracy since the locations of the maxima may change with the values of ϵ and n. Much more robust way is to compute cross-correlations of the histograms obtained for the pairs of consecutive scans. The change of the robot's orientation in respect to the environment configuration, $\Delta\alpha$, can be accurately approximated then by the location of the maximum of the cross-correlation curve (Fig. 6).

Classical approaches to position tracking [3] assume watching displacements of the robot in respect to the selected characteristic points of the scene, such as centers of the longest colinear segments. In our approach we focus on all the eligible readouts of two consecutive scans that correspond to the original predominant wall direction. If α is the predominant direction of the first scan (or its estimate provided by the odometry and/or the model of the environment), $\Delta\alpha$ is the orientation change estimated by cross-correlating the histograms obtained for the two scans, $C_1(\alpha)$ and $C_2(\alpha + \Delta\alpha)$ are respective vectors of c-values computed according to the formula (2), then we can estimate the robot's translation in the direction perpendicular to α by the way of comparing selected components of $C_1(\alpha)$ and $C_2(\alpha + \Delta\alpha)$. The selection accepts only these pairs of readouts for which both $c_{1i}(\alpha)$ and $c_{2j}(\alpha + \Delta\alpha)$ fulfill the colinearity criterion (3), $\varphi_j = \varphi_i + \Delta\alpha$, and the following eligibility requirement is satified:

$$| c_{1i}(\alpha) + \tilde{s}(\alpha) - c_{2j}(\alpha + \Delta\alpha) |\leq \delta \qquad (6)$$

where $\tilde{s}(\alpha)$ is a projection of a (for instance dead reckoned) translation onto the direction perpendicular to α (ie. the sum $c_{1i}(\alpha) + \tilde{s}(\alpha)$ is a prediction of $c_{2j}(\alpha + \Delta\alpha)$), and δ is a tolerance factor. Then if Ω is a set of pairs of indices $\{i, j\}$ for which $c_{1i}(\alpha)$ and $c_{2j}(\alpha + \Delta\alpha)$ meet criteria (3) and (6), and the power of the set $| \Omega |\neq 0$, the estimated displacement of the robot along the direction perpendicular to α may be computed as follows:

$$s(\alpha) = \frac{1}{| \Omega |} \cdot \sum_{\{i,j\}\in\Omega} (c_{1i}(\alpha) - c_{2j}(\alpha + \Delta\alpha)) \qquad (7)$$

The estimated displacement along the secondary predominant direction (if such a direction exists) may be computed analogically. The corrections $\Delta\alpha$ and $s(\alpha) - \tilde{s}(\alpha)$ are computed for a pair of consecutive scans under the assumption that the location of the robot and the environment configuration have not been changed too much in between the data collection cycles. Such an assumption is neccesary to secure a correspondence of the predominant wall directions α and $\alpha + \Delta\alpha$. Although we do not discuss it here, the immediate corrections obtained in the presented way can be directly used in a predictive filtering technique (such as Kalman filter for instance) to additionally keep track of the accuracy of the position and orientation estimates.

5 Experiments

The experiments described here were performed using HelpMate's LightRanger 3D scanning laser rangefinder set up to work in a 2D mode. It was mounted on RWI's Pioneer-1 robot. The environment was a realistic indoor scene, composed of the wall segments made of various materials, painted in various colors and having various surface textures and geometries.

Satisfactory performance of the presented method was observed for all of several hundred tested scans, which contained a certain amount of noise varying from segment to segment because of differences in material, color and type of surface. It may be noticed in the graph presenting obtained histograms (top plot in Fig. 5), that the shape of the plot depends on the selected neighborhood size n. However, the locations of global maxima do not change very much with n. Similar observation can be made with regard to the colinearity tolerance ϵ, which may take even as restrictive values as $\pm 3mm$, and still histograms of a sensible geometry could be obtained for typical indoor environments[4].

---

[4] We suspect, however, that there exist some environment dependent fenomena which may influence the relationship be-

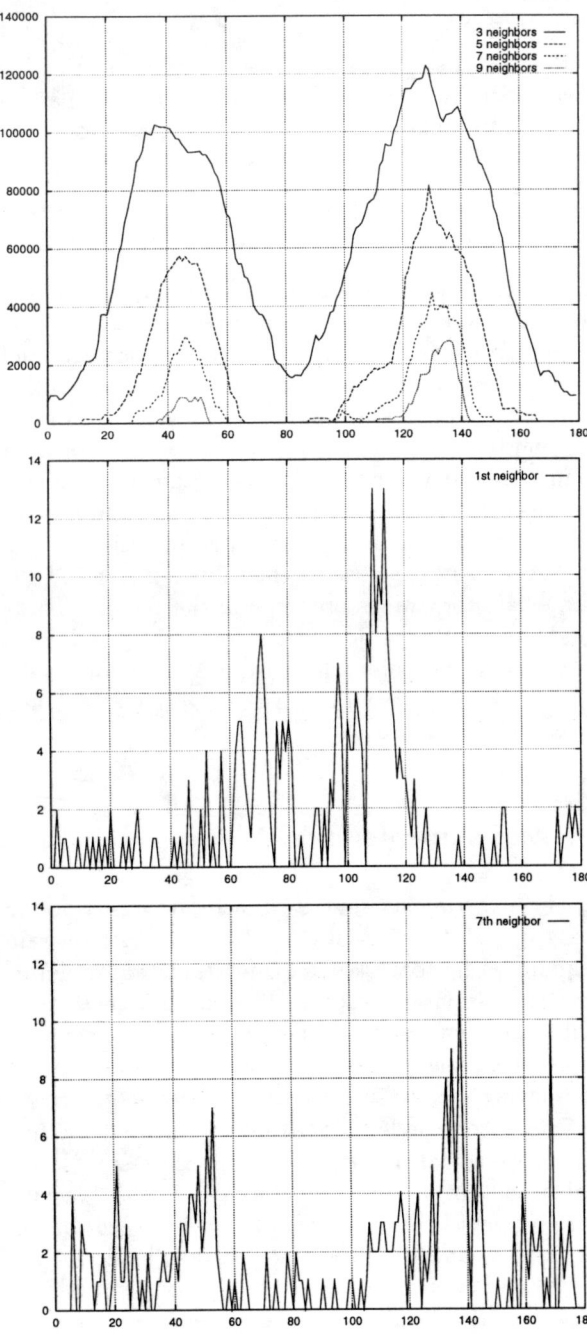

Figure 5: The $D(\alpha)$ histograms obtained for the first of the scans from Fig. 1 with n equal 1, 2, 3 and 4, ie. respectively 3, 5, 7, and 9 consecutive readouts were taken into account for colinearity test purposes (top) and the angle histograms computed with the method described in [5] for the immediate neighborhood (middle) and with the neigborhood of size 7 (bottom).

In the remaining graphs of Fig. 5 the angle histograms obtained with the method described in [5] are shown. The middle plot is a result of a direct application of the cited technique. It is not hard to notice that the displayed maximum of the histogram is shifted by about 30° away from the actual orientation of the most predominant direction of the segments. It is a result of taking into account simply pairs of consecutive readouts expressed in (x,y) coordinates when computing the histogram. A quite common noise in distance measurements can seriously disturb such a technique of the orientation estimation. By increasing the range of differentiantion (in the presented example from 1 to 7) one may obtain a much more correct histogram (bottom graph in Fig. 5), but still it is not as smooth and not as convincing as the ones obtained with the method introduced in this paper.

In the Fig. 1 the scans taken at two consecutive data collection points are shown. The robot traversed the scene heading in a direction of -45° (towards the lower right corner of the graph). According to the odometry, the robot's orientation was kept constant and location changed by +200mm along the x axis of the global coordinate system (the horizontal axis of the graph) and by -200mm along the y axis (the vertical axis). In Fig. 1 the locations of the readouts taken at the initial position are depicted as crosses, and the new scan is represented with squares.

The $D(\alpha)$ histograms were computed for the pair of scans with $n=2$ (ie. with 5 neighbors) and $\epsilon=10$mm. The starting configuration's histogram revealed its maxima, ie. the predominant wall directions, along $\alpha=129°$ and $\alpha=47°$ in the robot's coordinate system. The normalized cross-correlation of the two histograms is shown in Fig. 6 as a solid line. It has a maximum at $\Delta\alpha = 178°$, ie. the new scan was rotated by $-2°$ in respect to the starting configuration. It was observed that the cross-correlation maxima locations do not change by more than 1° when changing the values of ϵ and n within reasonable boundaries ($1 \leq n \leq 4$ and $5\text{mm} \leq \epsilon \leq 20\text{mm}$). Fig. 6 depicts also a cross-correlation of the histograms obtained with the original method implemented after [5] (with the neighborhood extended to 7 though). It is easy to notice how much smoother is the result obtained with the technique presented in this paper.

There is an additional very useful aspect in computing cross-correlations of the histograms. By me-

---

tween the colinearity tolerance ϵ, the neighborhood size n, and the histogram shape. The ongoing research focuses, among the others, on the automatic calibration of ϵ and n given the environmental conditions.

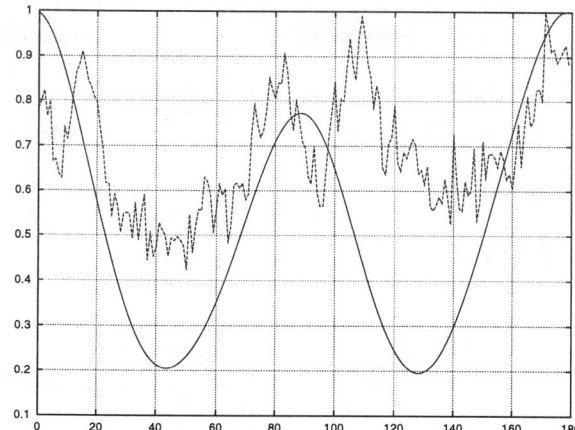

Figure 6: Normalized cross-correlations of the histograms obtained for the two consecutive scans (Fig. 1). Solid line is the result of our method ($n=2$, $\epsilon=10$mm). Dashed line was computed with the method of angle histograms [5] with the neighorhood size 7.

asuring the amplitude of the obtained normalized cross-correlation we may score the confidence in the estimated change of the robot's orientation. The larger the amplitude the more certain is the prediction. If for instance the robot was placed in an oval-shaped environment (with no particular prediominant wall directions present), the respective cross-correlations would look quite flat in the graph. Similar effects should be expected in case if a major change of a rectangular-like environment configuration occured in between the data collection cycles, or if the sensor lost a visual contact with an important part of the scene because of the limited range of the scanning angles. The consecutive histograms would be then hardly correlated, and the normalized cross-correlation amplitude would be small. This way the applicability of the presented method of tracking orientation may be automatically verified, and potentially erroneous predictions can be filtered out on the spot.

For the scans from Fig. 1, 65 pairs of readouts passed the eligibility test (6) for $\delta=50$mm, and were used for computing the robot's displacement estimates. The estimated displacement in the x direction was approx. 189mm (odometry reported 200mm, correction of -5.5%), and along the y axis -237mm (odometry reported -200mm, correction of -18.5%).

A series of extensive tests of the presented method in various natural indoor environments has been already carried out, and further experiments are being conducted. The obtained results have been very encouraging so far, especially in the orientation channel.

The errors have never exceeded $\pm 3°$ and most often remained within a range of $\pm 1.5°$. The position estimation inaccuracy was most of the time kept below 5% of the distance traveled, which is a decent result, but the authors intend to improve it in a close future.

6 Conclusion

In this paper we presented a novel approach to estimation of orientation and position of a mobile robot working in a partially structured environment. The method is especially suitable for use with optical scanning range devices. The technique relies on a local information processing scheme and it may be implemented in a form of a neural network. Experiments performed with a real robotic vehicle in a realistic indoor environment revealed the method's computational efficiency and robustenss against distance measurement noise. The results clearly outperformed those obtained with a more popular alternative approach, ie. the angle histograms. The presented method may serve as a useful add-on to the existing autonomous navigation systems, and it bears a big promise of increasing accuracy and efficiency of navigation, especially in realistic, known or unknown, partially structured environments.

Ongoing and future works include incorporation of a predictive filtering scheme to maintain a probabilistic model of the position and orientation estimates; and research towards experimental optimization and automatic calibration of the parameters ϵ, n and δ, in respect to the characteristics of the robot's workspace.

References

[1] Chua L.O., Roska T. The CNN Paradigm, *IEEE Trans. on Circuits and Systems-I*, 40, 147–156, 1993.

[2] Forsberg J., Larsson U., Ahman P., Wernersson A. Navigation in Cluttered Rooms Using a Range Measuring Laser and the Hough Transform, *Intelligent Autonomous Systems IAS-3*, 248–257, Pittsburgh, 1993.

[3] Leonard J.J., Durrant-Whyte H.F. *Direct Sonar Sensing for Mobile Robot Navigation*, Kluwer AP, 1992.

[4] Siemiątkowska B. Cellular Neural Network for Path Planning, *2nd Int. Symp. on Inteligent Robotic Systems SIRS'94*, 125–130, Grenoble, France, July 1994.

[5] Weiss G., Wetzler C., von Puttkamer E. Keeping Track of Position and Orientation of Moving Indoor Systems by Correlation of Range-Finder Scans, *IEEE/RSJ Int. Conf. on Intelligent Robots and Systems*, 595–601, Munich, Germany, Sept. 1994.

Mobile Robot Navigation Based on Vision and DGPS Information

S. Kotani, K. Kaneko, T. Shinoda & H. Mori
Faculty of Engineering, Yamanashi University
Takeda-4, Kofu, 400 Japan

Abstract

This paper describes a navigation system for an autonomous mobile robot in outdoor environments. The robot uses vision to detect landmarks and DGPS information to determine the robot's initial position and orientation. The vision system detects landmarks in the environment by referring to an environment model. As the robot moves, it estimates its position by conventional dead-reckoning, and matches up the landmarks with the environment model in order to reduce the error in the robot's position estimate. The robot's initial position and orientation are calculated from the coordinate values of the first and second locations which are acquired by DGPS. Subsequent orientations and positions are derived from map matching. We implemented the system on a mobile robot "Harunobu-6". Experimental results in real environments have showed the effectiveness of our proposed navigation methods.

1 Introduction

When a mobile robot navigates safely and effectively, the robot needs an environment model from a start to a goal. There are different types of map are presented in a hierarchy. The types are depend on its environment and various capabilities of the robot [1, 2]. When a robot does not know its initial position and orientation, a cost that a robot finds autonomously its position and orientation is very high.

GPS (Global Positioning System) is one of the solution of this problem. But GPS is inaccuracy for navigating autonomous mobile robots. If the robot knows initial position and orientation, the robot may reach a destination using dead_reckoning and environment maps. But we must consider an error distribution of the environment map and an error in estimated position by dead_reckoning of the robot increases with the traveling distance. Therefore the robot must compensate for these errors by map matching method using external sensors. There are many researches for landmark map matching using vision, sonar, laser range finders and so on [3, 4]. In outdoor environments, it is very difficult to match landmarks with environment map using vision. Asada and Shirai[5] proposed an effective approach using color image and knowledge. But they did not implement the method on robots.

We have developed outdoor autonomous mobile robots since 1982 [6, 7]. In this paper, target environments are outdoor environments with much noise and variability associated with dynamic worlds. As for an error correction of dead_reckoning, we use map matching method using γ–ω Hough transformation [8] which is robust for noise. Landmarks are points, lines and curves of objects in outdoor environments. As for an decision of its initial position and orientation, we use DGPS (Differential Global Positioning System). We describe configurations of our autonomous mobile robot and navigation method in detail. Experimental results for several environments have proved the validity of the methods.

2 System Components
2.1 Hardware

Fig. 1 shows "Harunobu-6", the autonomous mobile robot that is developed in 1996. The physical component is a motorized wheelchair (Suzuki Co. Ltd. MOTOR CHAIR MC-13). It has two drive wheels and two free casters. Table 1 shows the specifications of the components. Harunobu–6 can travel autonomously for about 3 hours with a computer system, a sensory system and batteries. Shaft encoders with a precision of 500 steps per resolution determine the distance traveled to a precision of 1.5 mm. Harunobu-6 can perform a dead reckoning using the shaft encoders and optical–fiber–gyro (HITACHI Densen Co. Ltd. OFG-3) information. A vision sensor (SONY Co. Ltd. Color CCD Camera XC–711, ASAHI Co. Ltd. Auto Iris Lens C814BEX-2) is mounted which can be seen in Fig. 1 on the top front. As the vision sensor is rotated by a servo motor, Harunobu–6 can observe the area around the circumstance of its own.

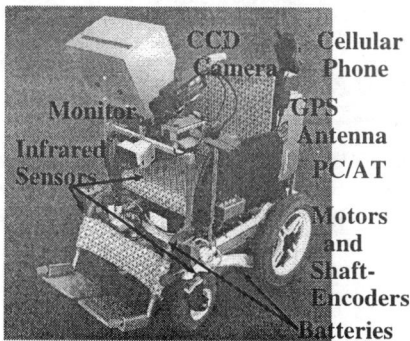

Figure 1: Mobile Robot "HARUNOBU-6"

Table 1: Component Specifications of the robot

Components	Specifications
Dimensions [L, W, H]	1170 x 600 x 860 [mm]
Weight (with batteries)	76 [kg]
Motor Power	24 [V] 140 [W] x 2
Maximum Speed	6.0 [Km/h]
Minimum Rotation Radius	560 [mm]
Maximum Travel Time	3 [hours]

Three infrared sensors in front of the robot are used to detect and avoid obstacles. As two of them are rotated by a servo motor, Harunobu-6 can detect precise obstacles position and can search free space. Harunobu-6 has one more sensor, in addition to conventional sensors. A GPS sensor is used to detect the robot's initial position and orientation.

The computer system is a PC/AT with a clock speed 120 MHz Pentium. The system has an UPP (Universal Pulse Processor) card which is developed our laboratory and an image processing card (HITACHI Co. Ltd. IP-2000).

2.2 Software

Fig. 2 is a schematic representation of "Harunobu-6" control architecture. The arrows in Fig. 2 show the main lines of communication.

An operating system is an μC/OS, which is a preemptive real time multi tasks kernel. The system executes **motion control**, **sensor processing**, **position estimation**, **path planning** and **image processing** tasks in real time. Especially an image processing task needs much processing time, so we use a hardware architecture of the IP-2000 image processing card as much as possible.

The details of these tasks are as follows:

Motion Control Task, Position Estimation Task These tasks executes movement control and position estimation using sensor information and map data.

Sensor Control Task This task gets various sensor data, DGPS data are used in order to calculate robot's initial position and orientation, shaft-encoders data and optical fiber gyro data are used in order to execute dead reckoning, and infrared sensor data are used in order to detect and avoid obstacles.

Image Processing Task This task detects and follows lines and matches up the landmarks with the environment model.

Path Planning Task This task plans the path using a map database and estimated robot's position and obstacle information.

2.3 Environment Model

The robot has a "full metric model" in an environment. Fig. 3 is a schematic representation of our environment map. Paths are expressed networks which are specified in a fixed coordinates. The networks consists of nodes and arcs. Locations of landmarks are specified in the coordinates. The landmarks have some features. The networks data is almost the same road information as that for digital car navigation maps. The landmarks data are added by human beings.

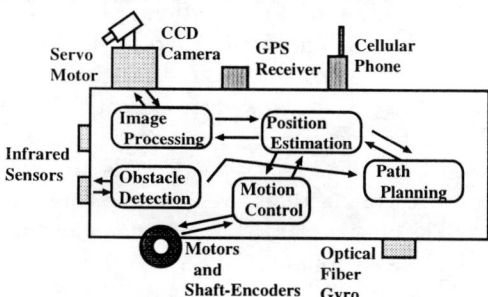

Figure 2: A Schematic Representation of the Robot

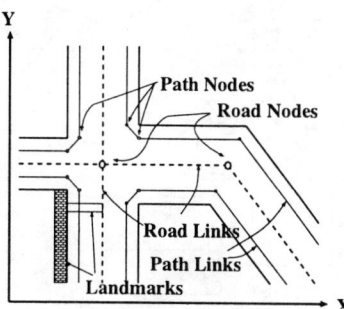

Figure 3: A Schematic Representation of our Environment Model

3 Decision Method of Robot's Initial Position and Orientation

3.1 Principle and Configuration of DGPS

GPS was developed by the U.S. Military to provide highly accurate position, velocity and time information to users anywhere on the earth. It is said that the horizontal position accuracy is 21 [m]. However the military introducing random timing errors, called Selective Availability, "SA", so that the accuracy is only 100 [m].

DGPS has been developed to compensate for the SA errors and timing errors. In this system, the correction data are transmitted from a central base station on the ground. As a result, the accuracy becomes just less than 5 [m]. Our DGPS system specifications are shown in Table 2. Our DGPS system configurations are shown in Fig. 4. All devices within dotted rectangle are mounted on the robot.

Table 2: DGPS Receiver Specifications and Services

Interface	RS-232C
Dimensions	84 x 145 x 44[mm]
Weight	525[g]
Base Location	Yokohama, Kanagawa
	Antenna1 (WGP-84)
Location	N. 35°32′09.070971
	E.139°41′03.095701
	H. 54.177[m]
Correction Data	RTCM-104 Message Type 1,2,3
Accuracy	5 [m] within 300[km] from Base
Available Time	24 [hours/day]

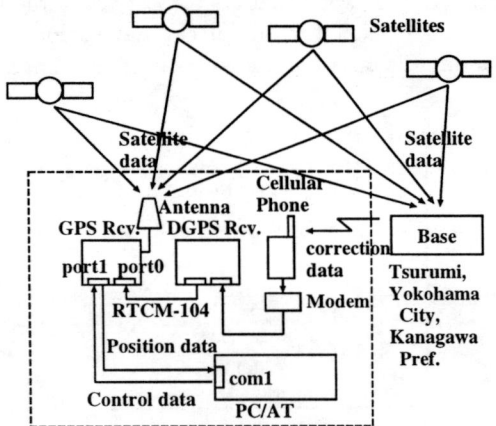

Figure 4: DGPS System Configuration

3.2 Decision Method

Decision outlines of the robot's initial position and orientation are as follows:

1. **Power On:**
   The robot does not know its position and orientation at all. The GPS starts, consequently the robot acquires correction data for DGPS compensation by a cellular phone.

2. **GPS mode ⇒ DGPS mode:**
   The robot gets a current position $L1(X1, Y1)$ by DGPS.

3. **Getting the second point information:**
   The robot searches area free from obstacles by Infra-red sensors. If the robot searches for the area, the robot moves to the free area and after that the robot gets a current position $L2(X2, Y2)$ by DGPS. Now the robot knows two points information, the robot can calculate its orientation. As the robot gets its orientation and position, navigation system of the robot plans the paths to a destination. The robot repeats landmark matching by vision in order to compensate for an error which is caused by dead_reckoning until the robot will reach the destination.

4 Localisation

The dead-reckoning is based only on odometry. Our robot estimates and corrects its position using Optical Fiber Gyro information and vision information. The technique is based on the Extended Kalman Filter using a robot model and a measurement model.

It is important technique to match up landmarks in an environment with the environment model. The robot detects line landmarks [3] in environments by vision. The detected line landmark is matched up with the environment model. As a result, The robot can compensate for an error which is caused by dead_reckoning. The method in order to correct the robot's position and orientation is based on Komoriya's method [3] The method in order to detect line landmarks is based on *HOUGH* Transformation [8] which is robust in outdoor environments. Moreover the robot detects curves, too. It is for reason that the area where the robot curves tends to cause an error dead_reckoning than the area the robot moves straight. The outlines for detecting and following lines and curves are as follows: The robot sets one small window. The area which is set is estimated area by environment model. While the robot successes detection of the line or the arc that are satisfied with some conditions, for example, the length and the direction, the robot sets small windows at the elongated area of the line or the curve. Fig. 5 shows an example of a curve detection and following. The average processing time is 250 [msec] using IP-2000 Image Processing Card and PC's CPU. The processing time depends on the individual images, but 95% of all images were processed within 350 [msec]

Figure 5: Example of a Line, a Curve Detection

4.1 Correction Method at Lines, Curves

We define the line model and the curve model as shown in Fig. 6. We get the distance **r** and α from equation (1).

$$\begin{pmatrix} r \\ \alpha \end{pmatrix} = \begin{pmatrix} \frac{ax_t + by_t + c}{\sqrt{a^2 + b^2}} \\ \varphi - \theta_t + \frac{\pi}{2} \end{pmatrix} \quad (1)$$

The landmarks are not only straight lines but also the curved lines. We get the distance **r** and α from equation (2).

$$\begin{pmatrix} r \\ \alpha \end{pmatrix} = \begin{pmatrix} \sqrt{(x_t - a)^2 + (y_t - b)^2} - R \\ \tan^{-1} \frac{y_t - b}{x_t - a} - \theta_t \end{pmatrix} \quad (2)$$

where

Line Model: $ax + by + c = 0$
Curve Model: $(x - a)^2 + (y - b)^2 = R^2$
(x_t, y_t): the position of the robot
θ_t: the direction of the robot
r: the distance from the robot to the tangent
α: the direction from the robot to the tangent

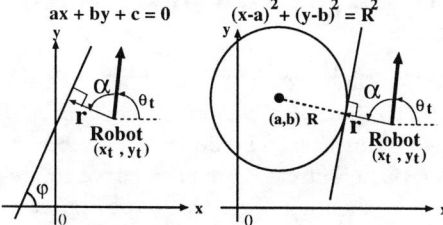

Figure 6: Relationship Between a Robot and a Curve

5 Experimental Results

5.1 DGPS evaluation

Fig. 7 and Fig. 8 show the results of our DGPS evaluation. The robot stopped at the point A (5 [min]) → B (30 [sec]) → C (30 [sec]) → D (30 [sec]) → E (30 [sec]) individually. After that robot moved.

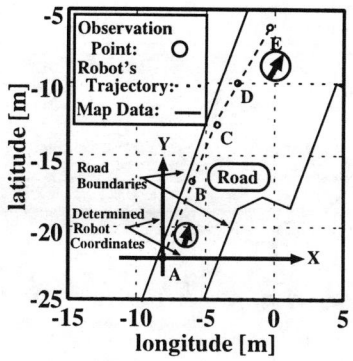

Figure 7: Decided Coordinates and Initial Position

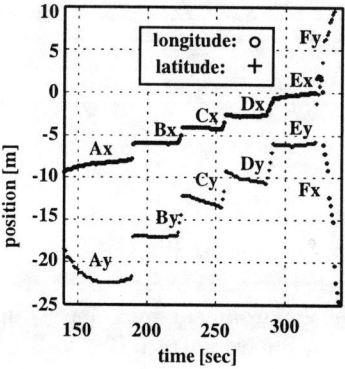

Figure 8: Position Trajectories of a robot

Fig. 8 shows each (x, y) value of the robot's trajectories. Fig. 8 and Fig. 7 show the global coordinates at our university map. At the point "A", the **SA**'s influence was still noticeable, but after that DGPS data were very stable. Based on this experimental results we may say that it is effective that the method for determining robot's initial position and orientation using DGPS.

Fig. 9 shows the robot's trajectories of dead_reckoning data based on the DGPS initial position and orientation. The thick line shows the estimated robot's position. The thin line shows our university map. The small squares show the DGPS data. The robot started at "A". DGPS data were not stable. At parts "P" and "Q", the **SA**'s influence was still noticeable, but after that DGPS data were very stable. The shaded portion shows the area of Fig. 7. The travel distance was

1,179 [m] and the travel time was 40 minutes. The trajectories of the robot were fairy good, but the data showed that the robot coursed out (See Fig. 9 "R"). As for a direction of the robot, the data is more or less correct. All data were shifted upper right direction, because of errors in the initial position estimate. The cause of the shift is the allocations of satellites. If we got the good allocation of the satellites, the trajectories became more correct.

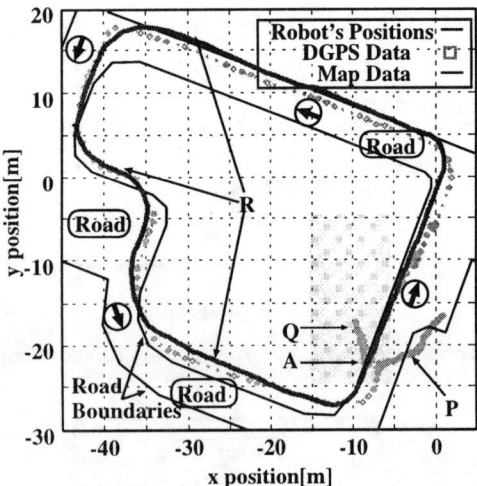

Figure 9: Robot Trajectories based on DGPS Initial Position

Fig. 10 shows an example of good and bad allocation of the satellites.
We can not expect the satellites which are always good allocation. Therefore the robot needs to detect landmarks in the environment and match up the landmarks with the environment model.

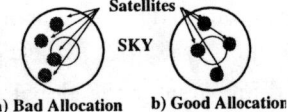

Figure 10: Examples of the Satellites Allocation

5.2 Error Correction with Line Landmark

The speed of the autonomous mobile robot, Harunobu-6, is 20 [cm/sec] when the robot moves along detected lines by vision and 15 [cm/sec] when the robot moves dead_reckoning. The robot stops in order to improve of an image processing accuracy and in order to avoid an influence of noise when the robot matches up landmarks with environment model. The robot takes images of eight times. Each images are taken from different direction. The position of the landmark is the mean of the eight images. Fig. 11 shows this experimental map. The robot started from "Start" and reached to "Goal". The travel distance was 60 [m]. The robot used a line (near "A") and a line (near "B") in order to correct the robot's position. The dotted lines show the trajectories of the robot. Fig. 12 shows the magnified area of Fig. 11's shaded portion.

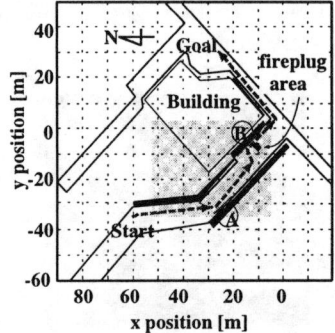

Figure 11: Example of a Route Map

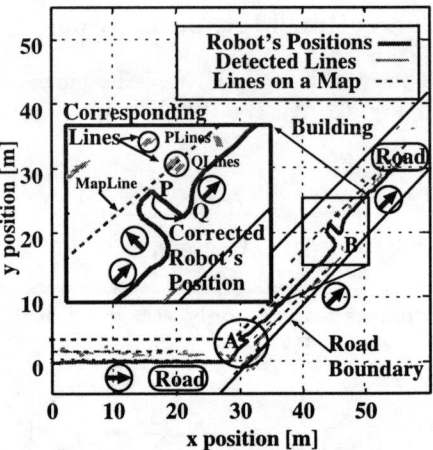

Figure 12: Details of a Corrected Position

The thick line shows the robot's position. The dotted lines shows the lines (landmarks) on environment map. The thin lines show the detected lines by vision. The robot matched up detected lines with environment model (See Fig. 12 "Corresponding Lines"), and corrected the robot's position from "P" to "Q" (See Fig. 12 "Corrected Robot's Position")

5.3 Error Correction with Curved Landmark

Fig. 13 shows this experimental map. The travel distance was 180 [m]. The thick solid lines show detected lines by vision. The thin solid lines show robot's position. The dotted lines show the map data. The dotted circles show the curve model.
Fig. 14 shows the magnified area of of Fig. 13's shaded portion. Ellipses show single-σ ellipses at each viewpoint represent the positional uncertainty. The ellipses are expanded into the size of 100 times for the

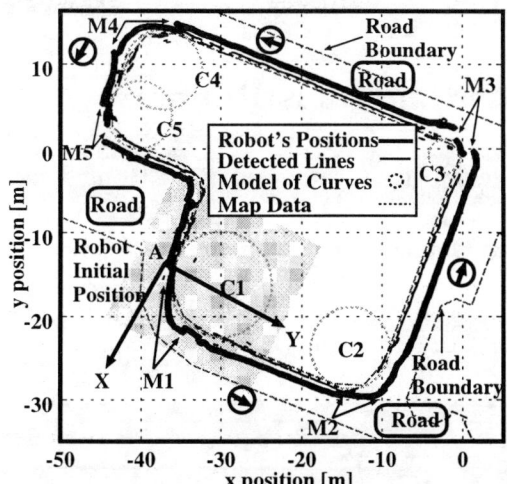

Figure 13: Trajectories at Yamanashi University confirmation.

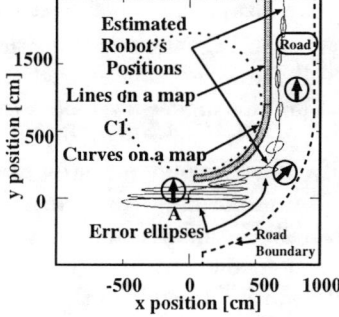

Figure 14: Changing of Error Ellipses at an around Start Point

It is understand that the ellipses becomes small to the center of the circular arc in accordance with a robot's being moved. Generally speaking, the amount of compensation in the straight line direction is small by the line landmark. However, the robot moved along the circular arc. As a result, the robot accomplished high precision compensation.

6　Conclusions and Future works

In this paper, we proposed the system configurations of our autonomous mobile robot and the method of the navigation in outdoor environments. We implemented the method on our autonomous mobile robot and we evaluated the method in long travel distance experiment. The experimental results in real outdoor environments showed the robustness and effectiveness of our proposed system and the method. The robot corrected its position error which was caused by dead_reckoning to match up landmarks which were detected by vision with environment model. We used Hough Transformation method in order to detect lines and curves for noisy environment. As for an disadvantage of processing speed of the method, we set a small size window dynamically based on environment model, consequently followed the elongated line or curve. If there are multi lines or curves in that environment, we used intensity information of the both side's edge of these lines and curves. As a result, the robot could select the correct line or curve from many lines or curves. Moreover we proposed the method in order to determine the robot's initial position and orientation. We implemented the method on the robot. We accomplished high accuracy positioning.

As for future works, we have two difficult problems. The first is an environment map-building problem. We need high cost in order to build our environment map. A precise "full metric map" needs accurate measurement. Moreover we must add information and attributes of landmarks by human beings.

The second is an DGPS problem. In city center, there are many high buildings, therefore the robot does not always acquire GPS information. We think that sensor fusion technique needs in order to solve this problem.

References

[1] B.J.Kuipers and Y.T.Byun: "A Robust, Qualitative Method for Spatial Learning", Proc. of AAAI-88, pp. 774–779, 1988.

[2] D. Lee : "The Map-Building and Exploration Strategies of a Simple Sonar-Equipped Mobile Robot", Cambridge University Press, 1996.

[3] K. Komoriya, H. Ohyama and K.Tani: "Planning of Landmark Measurement for the Navigation of a Mobile Robot", JRSJ, Vol.11, No.4, pp.533–540, 1993 (in Japanese).

[4] K. Nagatani and S. Yuta: "Path and Sensing Point Planning for Mobile Robot Navigation to Minmize the Risk of Collision", Proc. of IEEE/RSJ IROS'93, pp.2198–2203, 1993.

[5] A. Okamoto, Y. Shirai and M. Asada: "Integration of color and range data for three-dimensional scene description ", IEICE, Vol.E76-D, no.4, p.501-506, 1993.

[6] H. Mori: "A Mobile Robot Strategy Stereotyped Motion by Sign Pattern", in Robotic Research the Fifth Int. Symp., H. Miura and S.Arimto, Eds. Cambridge, MA, M.I.T. Press, pp.161–171, 1990.

[7] S. Kotani, H. Mori and N. Kiyohiro: "Development of the robotic travel aid "HITOMI"", Robotics and Autonomous Systems, Vol.17, pp.119–128, 1996.

[8] T. Wada and T. Matsuyama: "$\gamma - \omega$ Hough Transforma – estimation of Quantization Noise and Linearrization of Voting Curves in $\rho - \theta$ Parameter Space", proc. of 11th IAPR Int. Conf. Pattern Recognition, Vol.3, pp.272-275, 1992.

Visual Place Recognition for Autonomous Robots

Hemant D. Tagare

Department of Radiology
& Computer Science Department
Yale University
New Haven, CT 06520
USA

Drew McDermott

Computer Science Department

Yale University
New Haven, CT 06520
USA

Hong Xiao

Computer Science Department

Yale University
New Haven, CT 06520
USA

Abstract

The problem of place recognition is central to robot map learning. A robot needs to be able to recognize when it has returned to a previously visited place, or at least to be able to estimate the likelihood that it has been at a place before. Our approach is to compare images taken at two places, using a stochastic model of changes due to shift, zoom, and occlusion to predict the probability that one of them could be a perturbation of the other. We have performed experiments to gather the value of a χ^2 statistic applied to image matches from a variety of indoor locations. Image pairs gathered from nearby locations generate low χ^2 values, and images gathered from different locations generate high values. The rate of false positive and false negative matches is low.

1 Introduction

This paper presents a new visual place recognition algorithm. The algorithm accepts two images (one new + one old), and decides whether they are images of the same indoor environment taken from nearby positions. This information can be used (along with other information) to decide whether a robot is close to a place it visited before. We call the image comparison problem *Visual Place Recognition*. Although we discuss the more general problem below, visual place recognition is what is discussed in detail in this paper. Its success is judged by the statistical misclassification rates (false positive and false negative rates) against ground truth in real-world experiments. These values suggest that an autonomous mobile robot can use the algorithm to recognize whether it is close to a location that it had visited in the past.

As mentioned above, visual place recognition is only a part of the larger problem of map building. We intend to use this algorithm in the context of a map-learning algorithm that represents maps as networks of places with probabilistic transitions between them [10, 6, 14]. However, space constraints prevent us from discussing this higher-level algorithm in detail. Suffice it to say that we plan to use visual place recognition to verify that the robot is near a previously visited location, thus allowing it to correct accumulated odometric error since that previous visit.

We have focused on matching images directly, as opposed to attempting to recover the 3-d structure of the world and matching the structures recovered from two different images. Our algorithm will eventually attempt to reconstruct the shape of the place network, but not the shape of places the robot sees.

All image-based techniques can be classified according to how they address three fundamental issues: how they estimate the *correspondence* between the images, how they estimate the *domain* of the correspondence, and the relation they assume between image gray levels at the corresponding pixels.

The three issues are illustrated in figure 1 which shows the robot with a pin-hole camera obtaining an image I from place P and an image I' from a nearby place P'.[1] The *correspondence* between I and I' is a pairing of the pixels of the two images such that a pixel-pair images the same point of the external world. A particular pixel-pair in the correspondence is illustrated in figure 1. Since, in general, the camera axis at P and P' may not be aligned, a precise estimate of cor-

---
[1] The points are drawn as if the camera is oriented in completely different directions at the two points; this is simply because the diagram would be too cluttered if we drew the two points approximately coinciding in position, which is the case we are really interested in.

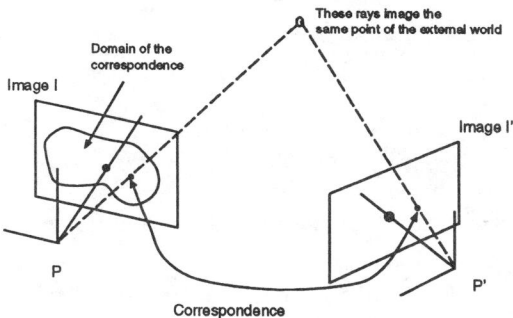

Figure 1: Correspondence and its domain.

respondence has to allow for projective transformation under arbitrary rotation of the camera axis. Further, a part of the external world which is visible from P may be occluded from P', every pixel need not have a corresponding pixel in I' (and vice versa). The subset of the pixels at P' that have corresponding pixels in P is called the *domain* of the correspondence. Any estimate of the correspondence requires an estimate of its domain.

Finally, if P and P' are close to each other, the corresponding pixels of the two images view the external world from similar directions. This implies that the amount of change in gray levels at corresponding pixels is small. The exact nature of the relation is the third issue in image comparison.

The simplest image-based algorithms use correlation to compare the two images [8], which makes sense only if the correspondence is the identity function. More sophisticated approaches attempt to estimate the correspondence during image comparison. These approaches tend to be expensive, and usually require further simplifying assumptions.

Although our approach is also image-based and ignores the three-dimensional geometry of the external world, it differs significantly from previous approaches. Instead of calculating the actual three-dimensional geometry at P or at P' we assume that the geometry belongs to a class of geometries (i.e. it is a stochastic process). We use knowledge only of the class of geometries and make no attempt to identify which member of the class occurs in a given situation. The class of geometries is specified as non-informatively as possible (analogous to models for white noise). Finally, we pose the recognition problem in the framework of statistical decision theory.

2 Related Work

The basic idea of treating maps as networks of places is from [10]. It has recently been refined by treating the networks as Markov decision processes with probabilistic transitions [14]. In [15], nodes are obstacle boundaries from which landmarks are visible.

Most workers in this area have used sonar as their main sensor for place recognition Others have used data from cameras, including [9, 8] and especially [13, 15, 4, 5, 6].

The major competing paradigm for map representation is the use of certainty grids, in which space is resolved into a grid of squares, and sonar data are used to assign a probability of occupancy to each square [3, 2]. Data from stereo vision can be used in a similar way [12].

Matching panoramic views (as we describe below) to achieve place recognition has been pursued before. Hong et al. [7] describe a system for taking a one-dimensional panorama for use in robot localization. It uses a spherical mirror to produce a 360-degree image of a location. One slice through this image remains at the same height as the robot moves around. This slice is used to produce a one-dimensional list of features used as landmarks. Our approach uses a full two-dimensional panorama. In [16] there is described a scheme for taking a panorama of what the robot sees to its *side*, and looking for features to use to retrace the same route. The panorama is obtained as a result of translation and does not model either perspective transformation or camera-axis rotation, as we do, albeit probabilistically.

3 Our Approach

Using the notation developed in the last section, we consider a location $\mathcal{L}$ randomly chosen in the neighborhood of P', and ask what the chances are that P is one such $\mathcal{L}$. To answer this question, we need to have a stochastic model of how images are perturbed by small camera motions.

Before we provide that answer, we point out that it may seem backward to measure the probability of generating P, the old image, as a perturbation of P', the new one. It should be clear, however, that the two derivations are symmetric. There are two technical reasons for doing it in this direction. One can be explained only after we present our algorithm. The other we explain now. The largest perturbation to contend with is camera rotation around the y (vertical) axis. Even a slight rotation between the pose at P and the pose at P' can cause the overlap between the two images to be small. To avoid this problem, we have the robot collect a *panorama* at a location it may want to match later. This is a "virtual wide-angle" image obtained by gluing several ordinary images together. Now when we collect a new image, I', at a place P', all we need to do is find the slice of the panorama taken at P that matches best. In what follows, we will treat image I as such a slice. This eliminates the registration problem. Of course, most of the correspondence problem remains, because different parts of the image will have shifted by different amounts, and some (the parts outside the domain Ω) are completely lost.

One consequence of the use of panoramas is that we must use spherical coordinates for our images, so that camera rotations are simple x-coordinate translations. Space does not allow us to go into details in this paper, but in the rest of the paper we will use longitude and latitude (e and l) instead of x and y.

We now present our stochastic model of image perturbation. We are letting $\mathcal{L}$ be a random variable that ranges over poses at places near to P', all of which we assume have a z (depth) axis parallel to that at P'. Every point of the external world that is visible from $\mathcal{L}$ is visible from P' except for unpredictable occlusions. We use C for the correspondence from R, the image domain at $\mathcal{L}$, to R', the domain at P'. Hence we derive the following expression for the *warped image process* at $\mathcal{L}$:

$$\tilde{\mathcal{I}}(e,l) = I'(C(e,l)) + n(e,l) \qquad (e,l) \in R$$

As explained in Section 1, occlusions limit the domain Ω of the correspondence; outside of Ω image pixels are provided by an unknown process E. Hence the *observed image process* at $\mathcal{L}$ can be written as

$$W_\Omega \tilde{\mathcal{I}} + (1 - W_\Omega) E,$$

where W_Ω is the window function

$$W_\Omega(u) = \begin{cases} 1 & \text{if } u \in \Omega \\ 0 & \text{otherwise,} \end{cases}$$

and, $E()$ is a stochastic process which describes the unknown image in $R - \Omega$.

Sampling the image process $\mathcal{I}$ on a discrete grid of pixels allows us to write down a joint probability distribution for it

$$p_{\mathcal{I},\Omega,E}(\mathcal{I},\Omega,E) = p_{\mathcal{I}|\Omega,E}(\mathcal{I}) p_\Omega(\Omega) p_E(E),$$

where, p_Ω and p_E are prior distributions on Ω and E (and where we have taken advantage of the fact that Ω and E are independently distributed).

Turning to the set of images obtained at P, we let the relative orientation between the coordinate system at P and $\mathcal{L}$ be ϕ. As explained above, the image I at P is actually chosen by selecting a slice of a panorama built from several images $I_1, \cdots, I_n$ obtained at P. Slices correspond to y-axis rotations of angle ϕ, so we denote them by I_ϕ. We use a factor m models the aperture and overall illumination change.

To proceed in a strictly Bayesian manner, we would evaluate the likelihood that some interpolated image at P comes from the observed image process at $\mathcal{L}$; and if the likelihood is high enough, we can declare that P is in fact $\mathcal{L}$. However, although this alternative is theoretically appealing, it is practically infeasible because the likelihood equations do not have a closed-form solution and are expensive to evaluate numerically. To obtain a more practical decision procedure, we make a key simplification: Rather than think about the correspondence function C as a whole, we think about the probability of individual pixel-to-pixel correspondences. In principle, we can eventually recover the distribution of entire correspondence by considering pairs of pixels, triples, and so forth. But we get good results by considering the simplest possible approximation, namely, the distribution for a single pixel.

We therefore ask the question, for a given pixel (e,l) of the image I, what range of pixels might it correspond to in the image I'? Assuming that the location of P' with respect to $\mathcal{L}$ is (Δ_x, Δ_z) and that the external point is at a distance r from $\mathcal{L}$, the correspondence function $C(e,l) = (e',l')$ can be written as shown in Equation 1. (In this equation f is the focal length of the camera.)

Notice that the correspondence function depends only on the *relative displacements* $\frac{\Delta x}{r}$ and $\frac{\Delta z}{r}$, and in the limit $\frac{\Delta x}{r}, \frac{\Delta z}{r} \to 0$, the correspondence function tends to the identity function. To get a sense of how the correspondence function departs from identity, it is useful to plot the ranges of (e',l') for a given range of $\frac{\Delta x}{r}$ and $\frac{\Delta z}{r}$ as a function of (e,l). Figure 2 shows this plot. The figure was constructed as follows. First the relative displacements were set to zero so that the extended correspondence function was identity. The images of (e,l) at $22.5^\circ f$ increments were plotted as cross hairs. Next the relative displacements $(\frac{\Delta x}{r}, \frac{\Delta z}{r})$ were varied in the unit square $[-0.2, 0.2] \times [-0.2, 0.2]$ and the extreme values of the extended correspondence were plotted as a curve surrounding the cross hairs. These are the bent quadrilaterals in figure 2. They show the region within which the extended correspondence varies as the distance r of the imaged point varies from $5.0 \max(\Delta x, \Delta z)$ to infinity. We will

$$e' = f \tan^{-1} \left(\frac{\sin \frac{e}{f} \sin \frac{l}{f} - \frac{\Delta x}{r}}{\cos \frac{e}{f} \sin \frac{l}{f} - \frac{\Delta z}{r}} \right), \quad (1)$$

$$l' = f \cos^{-1} \left(\frac{\cos \frac{l}{f}}{\sqrt{\cos^2 \frac{l}{f} + (\sin \frac{l}{f} \sin \frac{e}{f} - \frac{\Delta x}{r})^2 + (\sin \frac{l}{f} \cos \frac{e}{f} - \frac{\Delta z}{r})^2}} \right).$$

use the notation $P(e,l)$ to refer to the subset of points in I' that might be mapped to (e,l) in I.

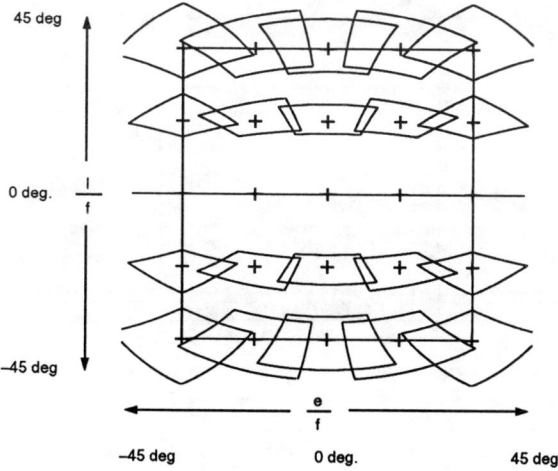

Figure 2: Limits of the correspondence process.

To proceed, we have to specify the distribution of the displacement $\Delta x, \Delta y$ and of the depth r and calculate the statistics of correspondence function via the approximation of equation (1). Since we are not willing to assume anything about the three-dimensional structure of the world, the most conservative approach is to (1) assume that the correspondence process maps (e,l) uniformly from $P(e,l)$, and (2) ignore all correlations, i.e., to assume that the mappings of (e_1, l_1) and (e_2, l_2) are statistically independent when $(e_1, l_1) \neq (e_2, l_2)$.

Given these independence assumptions, the probability that a pixel (e,l) actually corresponds to some pixel in $P(e,l)$ depends on the distribution of gray levels in $P(e,l)$. Although more complex models are possible, we assume that the gray levels have a Gaussian distribution, and hence that the probability is

$$P_{(e,l)}(I(e,l)) = \frac{1}{\sqrt{2\pi\sigma^2(e,l)}} \exp\left(-\frac{(I(e,l) - \mu(e,l))^2}{2\sigma^2(e,l)} \right)$$

where $\mu(e,l)$ is the average gray level over the parallelogram $P(e,l)$, and $\sigma^2(e,l)$ is the variance of the gray levels over $P(e,l)$. The probability of the entire unoccluded image is then simply

$$P_{\tilde{\mathcal{I}}}(\tilde{\mathcal{I}}|\Omega) = \prod_{(e,l) \in \Omega} P_{(e,l)}(I(e,l))$$

where Ω is a suitably discretized subset of the image. The MAP estimates of Ω, E, ϕ, m are obtained by maximizing the log likelihood.

There is no closed form solution to the maximization and we choose a numerical method for calculating the maximum. The numerical procedure uses a multi-resolution strategy. It is called with the limits of the search range of ϕ and an increment $\Delta \phi$. The procedure iterates through all feasible values of ϕ and finds the maximum log likelihood, say at ϕ^*. The procedure is recursively called with the the search limits set of $\phi^* \pm \Delta \phi$ and increment set to $(\Delta \phi)/2$. The recursion stops when the increment corresponds to one pixel – at which point the the maximum of the log likelihood found by the procedure is returned.

During the search for the maximum likelihood, the maximizing values of Ω and m are found by an iterative coordinate ascent procedure. This procedure alternately maximizes Ω and m. For a given interpolated image and m, the correspondence domain Ω which maximizes the log likelihood is easily found using the standard method [1], namely, removing all pixels the Gaussian probability of whose gray levels are low. Having chosen Ω, we pick m (the aperture factor) by estimating the mean illumination over the set Ω.

At the end of the MAP procedure outlined above, we have the MAP estimates ϕ^*, Ω^*, m^*. Once we have the MAP estimates, we can evaluate the similarity of the interpolated image $I = T(\phi^*, I_1, \cdots, I_n)$ and the warped image process by calculating the standard χ^2-statistic:

$$\chi^2 = \frac{1}{|\Omega^*|} \Sigma_{r,s \in \Omega^*} \frac{(I(r,s) - \mu(r,s))^2}{\sigma^2(r,s)},$$

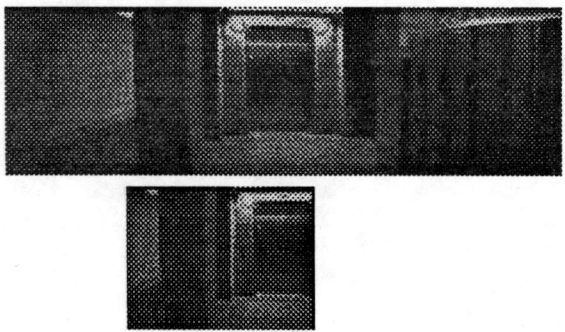

Figure 3: An example panorama and matching image

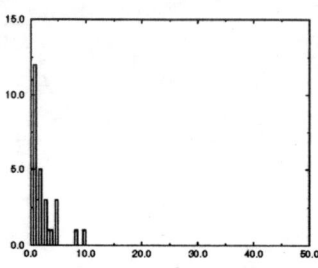

Figure 4: Histogram of χ^2 values for 41 matches

where, $|\Omega^*|$ is the area of the estimated correspondence domain in pixels.

We expect to get low values of χ when the images $I_1, \cdots, I_n$ and I' are obtained from nearby locations and "look similar." Conversely, we can expect to get high values of χ when images are dissimilar.

At this point we can explain the second technical reason for inquiring whether P is derived from P' rather than the other way around. Typically our map-learning algorithm will acquire a single new image, and then check it against a set of candidate stored panoramas. Each such check requires computing, for each pixel (e, l), the probability of its gray level assuming that it was drawn from a Gaussian distribution with the mean and variance $\mu(e, l)$ and $\sigma^2(e, l)$, as discussed above. That requires computing two derived arrays μ and σ^2, which we call the *mean image* and *variance image*. These are arrays can be computed for I' once and for all, and then used repeatedly as we compare I' to each panorama in turn.

4 Experiments

In our experiments, we created three panoramas from three different locations on the ground floor of the Watson building of the Yale Computer Science Department. Each panorama was constructed by taking 9 shots as the camera rotated, then gluing them together. In addition, we took 14 more images from points near the original camera position of one of the panoramas. Figure 3 shows one of the panoramas and one of the images, from the set of 14, that matched this panorama.

Each image was matched twice, once to a panorama taken near where it was taken, once to a different panorama. Running on a Sparc 2 the program, written

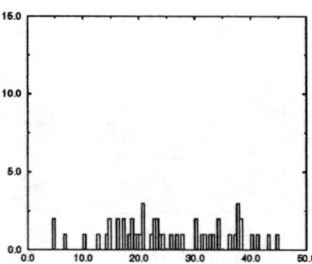

Figure 5: Histogram of χ^2 values for 54 mismatches

in C, took 3 seconds to produce the mean and variance images for the image, and 3 seconds to match it to each panorama. We computed the χ^2 statistic for each comparison. Figure 4 shows the distribution for the case where the two images were taken from nearby locations. Figure 5 shows the distribution for the other case. Obviously, the values are much closer to zero for the first case. If we classify all comparisons with a $\chi^2 \leq 10$ as "matches" and all those with $\chi^2 > 10$ as "mismatches," then the false positive rate is 5.6%, and the false negative rate is 0%.

Obviously, further work is required to flesh these statistics out. We intend to rerun the same comparisons again using a simple correlation test to see if our technique is more reliable, and if so by how much. We have also done some preliminary experiments to find the area over which the camera can be moved while preserving the criterion that $\chi^2 < 10$. The area depends on the depth of objects in the scene, but it appears that, when objects are two meters or more away, the camera can be moved over an area of about one square meter without causing a mismatch.

5 Conclusions

A promising approach to place recognition is to employ a stochastic model of image perturbations in order to decide if one image of a place could be produced by moving the camera within an area near where a previous image was taken. The model is based on the assumption that the only motions of the camera are rotation around the y axis, plus translation in the x-z plane. The resulting set of image perturbations includes large horizontal shifts, scale changes, and unpredictable occlusions.

We factor all of these into a model of where pixels in the perturbed image can come from in the unperturbed image. For each pixel, there is an area in the unperturbed image it might have been drawn from, and we can use a simple Gaussian distribution to estimate the probability that it actually did come from there. We take the unperturbed image to be the one just taken, and the perturbed image to be a slice of a panorama from a previously visited location. Taking panorama slices solves the problem of sensitivity to y-axis rotation. Using the new image as the unperturbed image allows us to compute the mean and variance of the unperturbed image just once before matching it to several candidates.

Preliminary experiments show that the method does a good job of distinguishing previously visited locations from new ones, at least in indoor environments. Further work includes doing detailed comparison with other methods, investigation of how far the camera can be moved before a match breaks down, and how the method can be adapted to outdoor conditions and other conditions where lighting is more variable.

References

[1] Andrew Blake and Andrew Zisserman. *Visual Reconstruction*. MIT Press, 1987.

[2] Joachim Buhmann, Wolfram Burgard, Armin B. Cremers, Dieter Fox, Thomas Hofmann, Frank E. Schneider, Jiannis Strikos, and Sebastian Thrun. The mobile robot Rhino. *AI Magazine*, 16(2):31–38, 1995.

[3] Alberto Elfes. Using occupancy grids for mobile robot perception and navigation. *IEEE Computer*, pages 46–58, 1989. special issue on Autonomous Intelligent Machines, June.

[4] Sean Engelson. Passive Map Learning and Visual Place Recognition. Technical Report 1032, Yale Computer Science Department, 1994.

[5] Sean P. Engelson and Drew V. McDermott. Image signatures for place recognition and map construction. In *Proceedings of SPIE Symposium on Intelligent Robotic Systems, Sensor Fusion IV*, 1991.

[6] Sean P. Engelson and Drew V. McDermott. Error correction in mobile robot map learning. In *Proc. IEEE Conf. on Robotics and Automation*, pages 2555–2560, 1992.

[7] Jiawei Hong, Xianan Tan, Brian Pinette, Richard Weiss, and Edward M. Riseman. Image-based homing. In *Proc. IEEE Conf. on Robotics and Automation*, pages 620–625, 1991.

[8] Ian Horswill. *Specialization of Perceptual Processes*. PhD thesis, 1993.

[9] David Kortenkamp, L. Douglas Baker, and Terry Weymouth. Using gateways to build a route map. In *Proc. IEEE/RSJ Int'l. Workshop on Intelligent Robots and Systems*, 1992.

[10] Benjamin Kuipers and Yung tai Byun. A robot exploration and mapping strategy based on a semantic hierarchy of spatial representations. *Robotics and Autonomous Systems*, 8:47–63, 1991.

[11] Don Murray and Cullen Jennings. Stereo vision based mapping and navigation for mobile robots. In *Proc. IEE Int'l Conf. on Robotics and Automation*, pages 1694–1699, 1997.

[12] Randall Nelson. *Visual Navigation*. PhD thesis, 1989.

[13] Reid Simmons and Sven Koenig. Probabilistic navigation in partially observable environments. In *Proc. Ijcai*, 1995.

[14] Camillo J. Taylor and David J. Kriegman. Exploration strategies for mobile robots. In *Proc. IEEE Int'l. Conf. on Robots and Automation*, 1993.

[15] Jian YuZheng, Matthew Barth, and Saburo Tsuji. Autonomous landmark selection for route recognition by a mobile robot. In *Proc. IEEE Conf. on Robotics and Automation*, pages 2004–2009, 1991.

HOMOGENEOUS NEUROLIKE STRUCTURES IN CONTROL SYSTEMS OF INTELLIGENT MOBILE ROBOTS

I. A. Kaliaev

Department of Robots Control Systems
Scientific Research Institute of Multiprocessor Computing Systems
Taganrog, 2 Chekhov ST., 347928, Russia

Abstract

The paper deal with the problem of collision-free mobile robot motion in complex real environment without human control. A noncomputing method, oriented on realization by means of homogeneous neurolike structures (HNS), for such type of problems solution in real time is suggested. The principles of HNS function and structure of mobile robot control system on its basis are designed. Tht practical realizations of intelligent mobile robot control system on the basis of HNS are described.

1 Introduction

Today a problem of creation of intelligent mobile robots, which can function independently in complex real environment without human control, has a great significance in different branches, for example, in military, in atomic engineering for work in red zone, in space researches and etc. In the first place the complexity of such robots creation connectes with the problem of on board control systems construction, which can solve in real time huge complex of intelligent tasks. The problem of robot collision-free motion planning and control for achievement of target in beforehand unknown environment takes one of the central places among this complex of tasks. As far as robot environment is apriori unknown it is impossible beforehand to determine and load into robot memory its movement control programs in all possible situations. Therefore, the robot control system must automatically form the current motion control for achivement the target state by optimum way in the rate of environment changes, that is in real time. Such type of problems belongs to class of so called position optimum control problems. The position strategy permits to form the object control a posteriori, correcting it on the basis of additional information, received during motion. The problem of position optimum control has the important practical significance especially in the case, when the object moves under beforehand unknown conditions and influences. The main complexity of position optimum control problems consist in the fact that their solution must be executed in real time of object movement. This requirement hampers the solution of such problems by standard computers with consecutive processing of information. It is explained by difference between sequential way of information processing in computer and parallel nature of position optimum control problems, which sense consists in choosing of one optimum solution among a set of possible alternatives. At the same time the human brain easily solves such type of problems. This fact suggests an idea of using neurolike structures for robot collision-free motion control in real time.

2 The principles of homogeneous neurolike structure organization

The method of position optimum control problem reduction to the problem of extermal path construction on the graph-model of the object state space is proposed in [1,2]. The essense of the approach is the following. The object states space $\{Y\}$ is covered by the regular graph $G(Q, X)$, whose vertices $q \in Q$ correspond to discrete points of the space $\{Y\}$ and arcs $x(q_j, q_{j+1}) \in X$ connect the vertices, corresponding to the neigbour discrete points Y_j and Y_{j+1} of the space $\{Y\}$. The weight $\gamma(q_j, q_{j+1})$, equaled to the increment of optimization functional on the transition between points Y_j and Y_{j+1}, is assigned to the arc $x(q_j, q_{j+1})$. Besides, the vertices q_0 and q_k, whose coordinates are defined by the current and target object states, are selected. Moreover, it is necessary to define a set of possible vertices Q_A and set of possible arcs X_A, sytisfying to the system of limitations, put on object states and controls.

As a result of such construction we receive the discrete model of the object states space, represented in the form of a homogeneous graph $G(Q, X)$. By using this model the problem of object position optimum control is changed by the problem of path construction on the graph $G(Q,X)$ between the vertices q_0 and q_k. This path must pass only through a set of the possible vertices Q_A and a set of the possible arcs X_A and must have the extremum sum weight of the arcs, belonged to it (Fig.1). It is shown [1,2] that initial arc of this path defines the current vector of optimum control for object collision-free movement to target. The fact that the problem of optimum position control may be reduced to the task of the extreme path construction on the graph-model allows to use homogeneous neurolike structures (HNS) with parallel information processing to solve it.

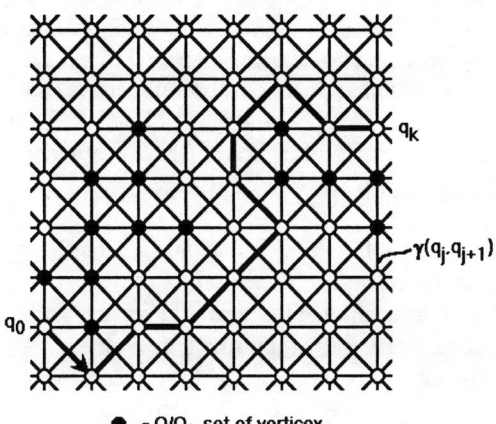

Figure 1: Graph-model of object state space

Let we have a neurolike structure topologically similar to the graph-model of the object states space. Each structure neuroprocessor must correspond to vertix of the graph-model, moreover, if two vertices of the graph are connected by the arc then the neuroprocessors, corresponding to them, are connected by the link. Besides, let the time delay of the signal passing between the neuroprocessor input and its output is proportional to the weight of corresponding arc of the graph. If the arc belongs to the set of impossible arcs X/X_A, then the signal passing through this link is blocked. Similarly, all the link, connecting the neuroprocessor with the neighbour neuroprocessors, are blocked too, if corresponding vertex belongs to the set of the impossible vertices Q/Q_A of graph-model $G(Q,X)$.

If in such structure we shall generate a signal in the neuroprocessor, corresponding to vertex q_k, then this signal will be propagated along the whole structure. In this case the time, during which this signal will reach the inputs of neuroprocessor, corresponding to the vertex q_0, will be proportional to the weight of the extremum path in the graph between the vertices q_0 and q_k. Moreover, the link, by which the first of the signals will come to the neuroprocessor q_0, defines the initial arc of this path.

Analysing the above procedure we may define the following simple functions which each neuroprocessors of such HNS must realize.

1. If the neuroprocessor corresponds to the vertex q_k, then the signal of wave propagation is formed in it.

2. If the signal has come to the neuroprocessor then it is transfered to the neighbour neuroprocessor with the time delay proportionally to weight of the corresponding arc in case when this arc belongs to the set of possible arcs X_A and corresponding vertex belongs to the set of possible vertices Q_A. Otherwise it is not transfered.

3. In the neuroprocessor q_0 the link is fixed along which the first signal has come to it. The scheme of HNS neuroprocessor, represented on fig.2, realizes all this functions. Here h_i and $h'_i (i = 1,2,...,l)$ - inputs and outputs of wave propagation signals, α_0 and α_k- inputs of correspondence to vertices q_0 or q_k, α_A - input of correspondence to set of vertices Q/Q_A, $h_i^* (i = 1,2,...,l)$ - outputs for fixation of link index, through which the first signal has come to neuroprocessor, if this neuroprocessor corresponds to vertex q_0.

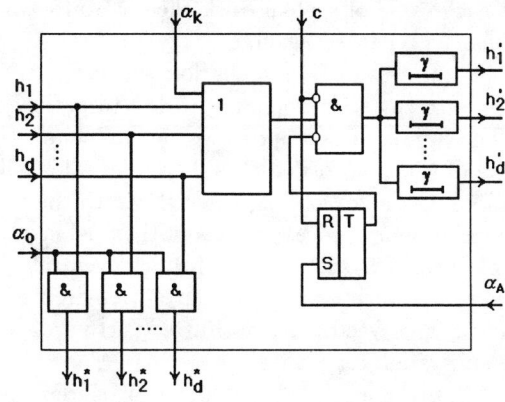

Figure 2: Scheme of HNS neuroprocessor

The proposed HNS has a high vitality, because it ensures the problem solutions when some neuroprocessors or the links between them break down. Really, the break down of some neuroprocessors or links is equivalently to narrowing of allowable conditions area, determinated by system of limitation. In this case the structure ensures the problem solution on remaining part of neuroprocessors and links.

3 Organizations of mobile robot motion control system on the bases of HNS

The structure of mobile robot collision-free motion control system, based of HNS, is shown on fig.3. The information about robot current state and current condition of environment is received with help of navigating subsystems and information subsystem. On the base of this information the graph-model $G(Q, X)$ of robot state space $\{Y\}$ is formed by means of model formation subsystem, i.e. the weight of arcs and sets of possible vertices and possible arcs of graph-model $G(Q, X)$ are determined. Besides with the help of this subsystem the verticex q_0 and q_k, corresponding to current and target robot states are chosen. Note, that the target robot state can be set by operator or systems of top level.

The graph-model $G(Q, X)$, formed by means of model formation subsystem, is reflected further in HNS, i.e. the links time delay and signal passing through neuroprocessors are programed in HNS. If this graph-model does not change during work of system and only the boundary conditions, determined by current and target robot states, are varied, then HNS can be programed once on the initial stage. If it is not so, then in each cycle of system work the HNS programming must be repeated again.

Further the wave of signals is formed in the HNS neuroprocessor q_k, corresponding to the target robot state. The inputs of neuroprocessor q_0 are connected to inputs of solution formation subsystem. This subsystem fixes the index of link, through which the first signal reaches neuroprocessor q_0, and then defines the vector of current motion control for achievement the target state by optimum way in current environment conditions. This vector is transmitted further on executive subsystem for realization.

After this the cycle of system work is repeated anew, with new information about current robot state and current condition of environment and so on, until the current robot state will not coincide with target state.

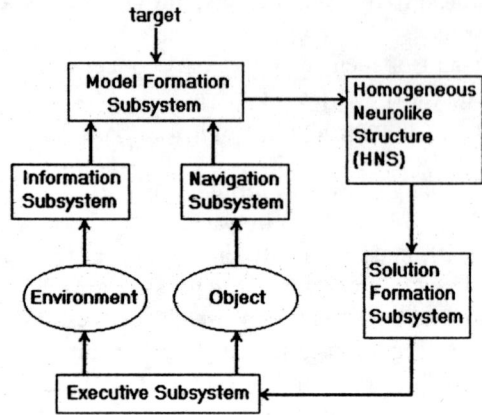

Figure 3: Structure of mobile robot control system

4 Experimental results

The theoretical researches have formed the basis for creation of series of experimental intelligent mobile robots (IMR), intended for exploration of Solar system planets surface, in particular Mars. This researches was carried out in frameworks of Russian space program.

The feature of such robots functioning consists in the large time of information exchange between robot and operator (more than 40 minutes if robot works on Mars). Therefore the robot must have the units for automatic motion control for achievement target without operator help. As far as the environment of robot motion is beforehand unknown, then it is impossible beforehand to plan and to keep in robot memory all variants of motion to target. Therefore the problem of position optimum control of robot motion to target in current situation is appeared. In spite of this, the rigid demands are made to the robot control system, such as:

- compactness, giving the possibility of system installation on robot chassis;
- survivability, ensuring preservation of system function in conditions of repair impossibility;
- superhigh performance, ensuring of robot motion control in real time of situation change (or new information reception about situation).

The control system, based on HNS, answers to all these demands.

With the purpose of compact realization of HNS the special element base, including inself the chip of HNS fragment and multichips module was developed and created. The high homogeneity and simplicity of HNS neuroprocessors permit to put on one chip the large number of such neuroprocessors. Created chip of HNS fragment contains 128 neuroprocessors. Moreover the opportunity of such chips connection one with another with the purpose of HNS size incrementation is provided. Multichips module contains 8 chips of HNS fragment in self (1024 neuroprocessors at whole) and the opportunity of such modules connection one with another is also provided.

The scheme of IMR planet-rover control system is shown in fig.4. It contains two processor modules (PM1 and PM2), serving for processing of sensor information from range-finder and TV-camera; processor module (PM3) for processing navigating information; processor module (PM4), serving for IMR movement planning and processor module (PM5), forming control effect on IMR chassis drives for realization of constructed movement with account of chassis dynamic properties. The coordination of all processor modul work is executed with help of central processor module. The functions of accumulation and correction of robot knowledge base about environments are entrusted on central processor too.

The system feature consists in the fact that the processor module PM4, used for task solving of IMR movement planning to target, is constructed on the base of homogeneous neurolike structure (HNS).

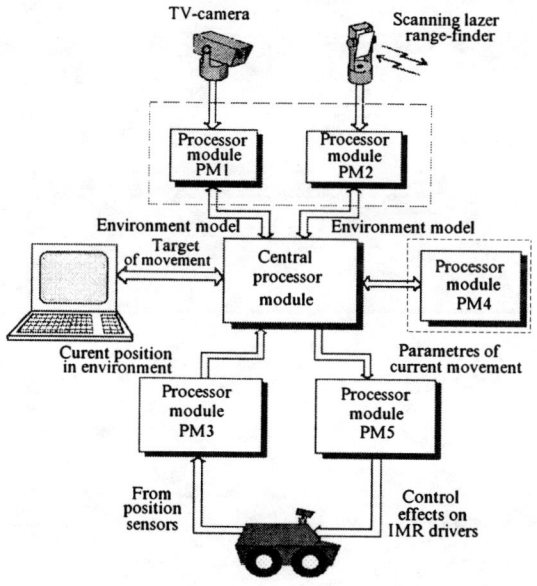

Figure 4: Structure of IMR control system

The HNS, which is used in processor module PM4, contains 4096 elementary neuroprocessors, incorporated in common solution field and realized by means of 32 chips of HNS fragment. With help of HNS the all possible variants of IMR movement trajectory to target are analysed in parallel and the optimum one is chosen on the base of current information about environment, accumulated to current time moment in robot knowledge base.

As a whole the system works as follows. On the basis of information, which is received in current moment of time by means of sensor (scanning laser rangefinder or TV-camera), corresponding processor module builds the model of visible environment zone. The integrated sign, determined difficulty its passing for robot, is put in correspondance for each section of this model. After this sensor processor module forms the request in central processor, which read the constructed model and with help of special algorithm put it over accumulated pivviously the hierarchical environment model, connected with current IMR position and made up its knowledge base about environment.

After this central processor passes to main processing procedure, which consists in following. The model of environment, accumulated in knowledge base, is reflected in HNS, where the all possible variants of IMR movement trajectory to target are analyzed and is chosen the optimum one, information about which is transmitted back in central processor module. On the basis of this information central processor calculates the parameters of current IMR movement alone the chosen trajectory with account of chassis dinamic charactiristics. These parameters are transmitted further to processor module PM5, which is used for control effects of IMR chassis drives for realization of current movement parametres. The procedure of planning optimal trajectory of movement and formation control effects on IMR chassis drivers is repeated periodically through each 0,2 sec., every time on the base of new information, accumulatad to this moment in knowledge base, that provides the collision free IMR movement through previously unknoun real environment with the speed up to 10km / hour.

The experiments have shown the efficiency of intelligent mobile robot control systems construction on the bases of HNS. Firstly, they are compact, that permits to put them on robot chassi. Secondly, they have large level of survivability as far as they may function when some neuroprocessors and links disable. At last, thirdly, they have of superhigh performance, achieved by use of nonnumerical method of solution, that provides the possibility of collision-free robot function in complex real envinronment without human control.

Acknowledgments

The suggested approach can be used for creation of mobile robotics systems, intended for function in beforehand unknown, real environment without human control, for example, in cosmic or underwater, in zone of radioactive and chemical pollutions, for inspection of dangerously explosive zones and objects, for realization of rescue works and so on, as well as industrial and domestic intelligent robots.

References

[1] I. A. Kaliaev. Homogeneous Neurolike Structures for Optimization Variational Problem Solving.- PARLE'93. Parallel Architectures and Languages Europe. 5-th Inter. Conf., Munich, Germany, June, pp. 438-451, 1993.

[2] I. A. Kaliaev. Homogeneous Neurolike Structures for Solving the Optimum Control Problems.- PARCELLA'94. Proc. of the VI Inter.Workshop by Cellural Automata and Arrays. Potsdam, Germany, Sept.21-23, pp.187-195, 1994.

Sonar Resolution-Based Environment Mapping

Leyla Cahut[*], Kimon P. Valavanis

Robotics and Automation Laboratory
Center for Advanced Computer Studies
University of Southwestern Louisiana
Lafayette, Louisiana 70504, USA

Hakan Deliç

Signal and Image Processing Laboratory
Department of Electrical Engineering
Boğaziçi University
Bebek 80815 Istanbul, Turkey

Abstract

Ultrasonic range sensors are used to obtain the information required for collision-free navigation of a mobile robot in a semi-structured or unstructured environment. A set of range readings from a ring of sonars are correlated to acquire a 2-D map of the robot's environment. The map is continuously enhanced via novel matching and update algorithms as new data are collected while the robot is in motion. The algorithm utilizes confidence measures that are directly obtained from the sonar's resolution or accuracy. There are no additional modeling assumptions and the approach is robust. The algorithms are experimentally tested on the Nomad 200 mobile robot and Nomad's Cognos Software Development Package.

1 Introduction

Feature-based environment mapping has been preferred for collision-free navigation and localization purposes due to its accurate representation of objects [1, 6]. Particularly, systems that rely solely on ultrasonic range sensors (sonars) have become popular since sonar offers a low-cost and relatively accurate alternative to other sensing types [7]. For instance, in [3], the environment map is a two-dimensional depiction where surfaces and objects are represented as connected sequences of line segments. The uncertainty in data is taken into account by maintaining a confidence measure with each line segment.

Unfortunately, the tolerances employed for matching line segments in [3] were set in an ad hoc manner. Variance and covariance expressions derived through Kalman filtering of the robot's position were used in [4]. However, as experimental observations indicate [5, 8], sensor noise and errors can be accurately modeled by a class of probability distributions, calling for robust statistical procedures, whereas Kalman filtering assumes strict Gaussianity for both measurement and system noise.

In this paper, we consider a mapping algorithm where the distance information returned by the sonar sensors are transformed into points in the Cartesian coordinate system. The points are grouped into linear clusters, and recursive line fitting is subsequently applied to each cluster. Once an initial map is thus obtained, an enhanced map of the environment is maintained through continuous, joint matching and update processes as new data are collected. Two consecutive sonar readings are said to "match" if they originate from the same location. Then a proper "update" procedure fuses the readings in order to enhance the available map.

We propose sonar accuracy-based tests for matching line segments. That is, the test tolerances become functions of sensor specifications and error performance rather than ad hoc settings. In particular, the Polaroid 6500 sensor, used extensively in the experiments, has ±1% accuracy over its entire range [9], which we take advantage of to define a ball of uncertainty around each sonar reading. The resulting map enhancement procedure is adaptive and makes no a priori modeling assumptions.

2 Sonar Uncertainty

Ignoring errors caused by scattering, the sonar beamwidth and resolution are the sources of uncertainty. The Polaroid 6500 sonar manual claims ±1% resolution (accuracy). However, in experiments and simulations, errors of magnitude up to 3% were observed for slightly tilted objects [2] owing to more pronounced scattering. Assuming that a resolution of $\pm\xi\%$ creates a linear uncertainty of $\pm\xi D_1/100$ along the signal direction, where D_1 is the range read-

---

[*]L. Cahut is now with Profilo Communications Technologies-PROTEK Inc., Mecidiyeköy 80384 Istanbul, Turkey.

ing, a second range reading D_2, obtained by perhaps some other sonar on the ring, should lie at most $\xi(D_1 + D_2)/100$ apart from the first point if the difference is to be attributed to sonar resolution. Thus, we define a resolution-based uncertainty region as a ball of radius $\xi D_1/100$ around the point-coordinate (x_1, y_1) corresponding to the range reading D_1:

$$B_{D_1} = \{(x, y): \sqrt{(x-x_1)^2 + (y-y_1)^2} \leq \xi D_1/100\}.$$

The ball B_{D_1} may be centered at any point on the arc of uncertainty due to the 25° beamwidth of the Polaroid 6500 sonar. The uncertainty region U_1 is the union of balls:

$$U_1 = \bigcup_{\{B_{D_i}(x_1, y_1) \text{ is on the arc}\}} B_{D_1}.$$

If the uncertainty region of two consecutive readings, D_1 and D_2 intersect, then any discrepancy in the point coordinates can be accounted for by the resolution or beamwidth of the Polaroid 6500 sonar, and hence we can assert that there is a match between the two points. Defining I_{D_1, D_2} to be the indicator function of the event of a match between the points corresponding to the range readings D_1 and D_2, the test can be described as follows:

$$I_{D_1, D_2} = \begin{cases} 1, & \text{if } U_1 \cap U_2 \neq \emptyset \\ 0, & \text{if } U_1 \cap U_2 = \emptyset. \end{cases}$$

The inherent challenge in resolution-based matching described above is the enormous amount of computation required to search for ball intersections corresponding to each center point located on the arcs of uncertainty. On the other hand, there is no satisfactory recipe for extracting the exact point of reflection on the 25°–arc from the range reading information. In fact, we strongly believe that issues such as arc uncertainty and backscattering must be tackled at the sonar signal processing level where raw data can be manipulated.

Given the observations in the preceding paragraph, we shall assume in this paper that any sonar reading comes from the middle of the 25°–arc. The map errors that might be caused by this simplistic approach will be temporary since more reliable data will be collected as the robot moves in the direction of the source of the erroneous reading.

Since midpoints of arcs are assumed as sources of reflection, the matching operation reduces to searching for the intersection of two balls, each centered at the points that correspond to two consecutive range readings. That is, the test becomes

$$I_{D_1, D_2} = \begin{cases} 1, & \text{if } B_{D_1} \cap B_{D_2} \neq \emptyset \\ 0, & \text{if } B_{D_1} \cap B_{D_2} = \emptyset. \end{cases}$$

This sonar accuracy-based approach, to test if two range readings emanate from the same point on the same object with $\xi = 3$, will form the basis for developing matching procedures between line segments in the next section.

3 Sonar Resolution-Based Map Generation

The process of generating and improving a sonar map consists of three stages: 1) clustering or grouping the sonar data; 2) processing the grouped information, and; 3) map enhancement.

3.1 Clustering the Sonar Data

The range readings received from the sonars are checked against the threshold d_{max}, where d_{max} is chosen to be 255 inches [2], [9]. Any depth reading that exceeds this threshold is discarded due to the fact that the signal–to–noise ratio (SNR) becomes unacceptable at such distances. The range readings which are below the threshold are transformed to Cartesian coordinates:

$$x_k = D\cos(\alpha_k \pm \beta + \gamma) + r\cos(\alpha_k + \gamma) + R_x,$$

$$y_k = D\sin(\alpha_k \pm \beta + \gamma) + r\sin(\alpha_k + \gamma) + R_y,$$

where D is the depth reading received from a sonar, β is an angle within the sonar's beam where the echo is received ($|\beta| \leq 12.5°$), r is the robot's radius, α_k is sonar k's orientation with respect to sonar 0 ($\alpha_k = 22.5k$), γ is sonar 0's angular position on the sonar ring with respect to the x-axis, R_x and R_y denote robot's position in the world coordinate frame. For the reasons stated in Section 2, β is taken to be 0°.

Once the range readings are converted to Cartesian coordinates, they are clustered such that each group represents a single face of an object. An outermost point of an object face is found when the distance between two neighboring points is greater than a preset threshold t_{gap}. If no gap is detected (t_{gap} is not exceeded), the distance between the next pair of neighboring points is checked. The clustering operation continues until all available points are exhausted.

3.2 Processing the Cluster Information

The clustered points are processed through recursive line fitting procedure. This part of the proposed methodology follows the same path as Crowley's [3].

3.3 Map Enhancement

The initial environmental map created needs to be improved as the robot continues navigating in the environment. The update of the map is done by fusing the new information with the existing knowledge and involves three steps. <u>Matching process</u>: Each new line segment is compared to the existing line segments to determine if there is a match. If no match is found, the new line segment is either dismissed, or it is kept as a new object face. <u>Update process</u>: If a match is found between two line segments, the higher confidence end-points are employed to form the line equation of the updated segment. The procedure is finalized by projecting the outermost of the four points (from the two lines segments) onto the updated line equation and extending the line segment to the projected point. <u>Correlation process</u>: If the matching process determines that a new line segment represents a new object face, this process checks to see if the line segments might be treated as extensions of one another. The correlation process smoothens the map but does not add any crucial information for navigation purposes. Therefore, we only furnish the procedure in the next section without providing experimental results. For results, see [2].

3.3.1 Matching Process

Let l_e and l_n denote line segments in the existing and new maps, respectively, that are candidates for a match. Two line segments are declared to have a match if all of the following three tests are satisfied in sequence.

- <u>Orientation test</u>: Difference between slopes of l_e and l_n should be small. This translates to requiring the difference in angular orientation with respect to some reference axis to be less than some threshold. Typically, this threshold is chosen to be 15° [3].

- <u>Colinearity test</u>: In addition to the orientation test, the distance, d, from the mid-point of l_n to the line equation of l_e should be smaller than the threshold defined below:

$$|d| \leq 0.03 D_1 + 0.015(D_2 + D_3).$$

The first and second terms on the right hand side of the equation account for the accuracy of the range reading D_1 and the mid-point between D_2 and D_3 respectively.

- Overlap test: This stage of the matching process consists of two steps: Mid-point overlap check (MPC), and end-point overlap check (EPC). MPC is satisfied if the projection of the mid-point of l_n intersects l_e. This check is repeated for all the existing lines until a match is found. If MPC fails with all the available lines, then we seek to find out if there is any overlap at all by initiating EPC, which is a check for partial overlap. EPC passes if one of the end-points of l_n intersects l_e when projected onto l_e.

The first two tests jointly ensure that the line segments are aligned along a common line equation. Failure of the first test implies mismatch in angular orientation. If the orientations match but the second test fails, then the two line segments are sufficiently "parallel" to one another (in the sense of the test) but too distant. Once the orientation and alignment match is confirmed, the third test checks for overlap. The possibilities are as follows:

1. MPC passes: The new line segment is about parallel to and near an existing line segment, and the two lines have at least 50% overlap. Thus a "good" match is achieved. If MPC fails for an existing line segment, we repeat the test for the neighboring line segments in the map next until a match is found. If MPC fails for all the existing line segments, the matching process is continued with the EPC test between the existing line segments possessing the outermost points of the object face and the new line segment.

2. MPC fails, EPC passes: There is "partial" match with one of the existing lines. Since this match accounts for less than 50% overlap, we declare that the new line segment is an extension of the existing one.

3. MPC and EPC both fail: There are two approaches that can be taken. Either the new line segment is considered as a newly detected object or object face and plotted on the map as an entirely separate line, or its proximity to the existing lines is checked and a decision is made whether the new line segment is to be connected to the nearest existing line segment.

If any one of the three tests fails, then the new line segment is assumed to represent a new object face. The only exception occurs when the orientation test fails but the colinearity and overlap tests pass for any of the existing line segments. This anomaly might stem from erroneous data and hence the new line segment is dismissed. It is now clear that the matching process first seeks for a match, then for dismissal. Only after it is neither matched nor dismissed, is a new line segment kept as a new object face.

The matching process will work only if the distance traveled between two consecutive snapshots is small enough that the signal-to-noise ratios of the corresponding readings in the two data sets are comparable. Otherwise, a false inconsistency will occur due to the significant improvement in the quality of the range readings as the robot moves much closer to the same object, or vice versa. In our experiments, consistently accurate matches were found for distances up to 10 inches traveled between two consecutive snapshots.

3.3.2 Update Process

Once all three tests pass, updating the line segment will be performed between two consecutive snapshots. There are two goals that we try to achieve simultaneously: given the four points, draw the longest possible line segment satisfying the line equation given by the two highest-confidence end-points. To that end, between the two sets of matching end-points, the ones with the higher confidence are kept to determine the equation of the updated line segment, l_u. In other words, the confidence in a point is inversely proportional to its corresponding range reading. If any of the lower-confidence points lies outside the segment bounded by the higher-confidence points, then the former point is projected to intersect l_u, and subsequently, l_u is extended to the point of intersection. This way a worst case scenario is presented to the navigation unit, to avoid collision with an obstacle that is not yet well-observed [2].

Let an existing line segment be defined between the left end-point (x_o^L, y_o^L) and the right end-point (x_o^R, y_o^R). Likewise, suppose the new line segment be between the left and right end-points (x_n^L, y_n^L) and (x_n^R, y_n^R), respectively. The update procedure takes the following course depending on the outcomes of MPC and/or EPC:

1. MPC passes: First the higher confidence end-points are determined. Let $D[(\cdot, \cdot)]$ be the range reading corresponding to a point. Then the end points (x_u^L, y_u^L) and (y_u^R, y_u^R) of the updated line equation are such that

$$D[(x_u^L, y_u^L)] = \min\{D[(x_o^L, y_o^L)], D[(x_n^L, y_n^L)]\},$$

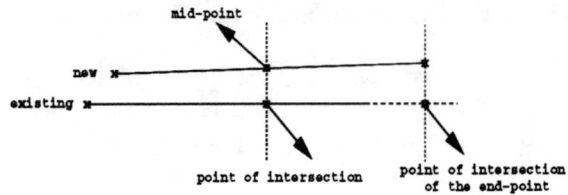

Figure 1: Projection of one of the endpoints lying outside the old line segment.

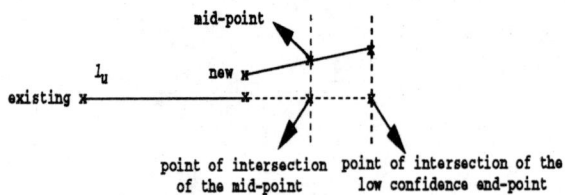

Figure 2: Projection of the right end-point of the new line segment on the updated line segment.

$$D[(x_u^R, y_u^R)] = \min\{D[(x_o^R, y_o^R)], D[(x_n^R, y_n^R)]\},$$

respectively. The points (x_u^L, y_u^L) and (x_u^R, y_u^R) determine the line equation l_u will lie on. Next we project the lower confidence end-points onto the line equation of l_u. If a projection does not intersect the updated line segment, then a new line is drawn between the updated end-points and the point of intersection (POI) of that projection and the line equation (see Figure 1). Otherwise, (x_u^L, y_u^L)–(x_u^R, y_u^R) defines the updated line segment. Thus the longest possible line segment satisfying the highest-confidence line equation is obtained given the four points (x_o^L, y_o^L), (x_o^R, y_o^R), (x_n^L, y_n^L) and (x_n^R, y_n^R).

2. MPC fails, EPC passes: Without loss of generality, suppose that EPC passed for the left end-point only. Then, we keep either the left end-point of the new line segment or the right end-point of the existing line segment as a middle-point (x_u^M, y_u^M) of l_u such that

$$D[(x_u^M, y_u^M)] = \min\{D[(x_o^R, y_o^R)], D[(x_n^L, y_n^L)]\}.$$

Of the two end-points (x_o^L, y_o^L) and (x_n^R, y_n^R), the one with the higher confidence is kept as the end-point of l_u, and the one with the lower confidence is projected on the line equation of l_u formed by (x_u^M, y_u^M) and the higher confidence end-point (see Figure 2). Thus,

$$(x_u^L, y_u^L) = \begin{cases} (x_o^L, y_o^L) \text{ if } D[(x_o^L, y_o^L)] \leq D[(x_n^R, y_n^R)] \\ \text{POI of } (x_o^L, y_o^L) \text{ and } l_u \text{ if else,} \end{cases}$$

$$(x_u^R, y_u^R) = \begin{cases} (x_n^R, y_n^R) \text{ if } D[(x_n^R, y_n^R)] \leq D[(x_o^L, y_o^L)] \\ \text{POI of } (x_n^R, y_n^R) \text{ and } l_u \text{ if else.} \end{cases}$$

As a result of update, the object face is represented by a set of existing and new points. Using these points, recursive line fitting is applied again to refine the sonar profile of the object face.

3.3.3 Correlation Process

One approach, if there is no match, is to assume that the new line segment could be representing another face of some object previously detected. The correlation algorithm described below is applied to append the new unmatched face to the existing object shape.

The correlation is performed when the minimum distance between the end-points of the new line segments and existing line segments is found to be less than or equal to some preset threshold t_{gap}. Define $L_{\text{existing}} = \{l_1, \ldots, l_p\}$ and $L_{\text{new}} = \{l_{p+1}, \ldots, l_{p+q}\}$ as the set of existing line segments belonging to all the detected object faces and the set of new lines forming some new object face, respectively. Let $d_i(L_{\text{existing}}, l_i)$, $l_i \in L_{\text{new}}$, denote the minimum distance from any end-point of l_i to any of the end-points of the lines in L_{existing}. Furthermore, define:

$$d_{min} = \min_{l_i \in L_{\text{new}}} d_i(L_{\text{existing}}, l_i).$$

If $d_{min} \leq t_{\text{gap}}$, then the set of line segments in L_{new} are appended to the line

$$l^* = \arg \min_{l_i \in L_{\text{new}}} d_i(L_{\text{existing}}, l_i).$$

If the minimum d_{min} exists, it is determined that the lines in L_{new} represent another face of some existing object; in particular the new line whose end-point is closest to l^* should form a corner with l^*. If a corner is not formed, the end-points closest to each other, and the lines that these points are associated with are interpolated. This process is repeated until all the unmatched new object faces in L_{new} above are exhausted. The new line obtained as a result of interpolation forms a corner between the closest two lines on different faces of same object (see Figure 3).

If the line segments in L_{new} neither match nor correlate with the existing line segments, then L_{new} is distinguished to be a distinct object face.

4 Simulation Results And Experiments

The developed algorithms have been tested on the Nomad 200 mobile robot which is endowed with a ring

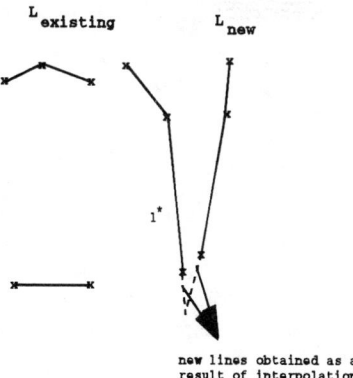

Figure 3: Correlating a new line segment with an existing line segment.

of sixteen sonar sensors. The robot was placed in a rectangular room of 305×154 inches, called the "Blue Room" because of the blue styrofoam walls forming the borders of the test area. The minimum gap size, t_{gap}, is taken to be twice the robot's diameter including the bumper, ensuring for passage through an opening under the worst case scenario where the neighboring sonars give identical readings and the reflections occur from the opposing ends of the neighboring sonar arcs [2]. For the Nomad 200, $t_{\text{gap}} = 2(2r + 3) = 42$ inches (3 inches account for the bumper circling Nomad 200). When both leftmost and rightmost vertices are found, the leftmost point, the intermediate points and the rightmost point are grouped together to constitute a face of the object.

In Figures 4 and 5, the robot is depicted within the mapped 2-D contours of the Blue Room. In the experiments, the robot was first positioned at various locations inside the Blue Room. Then, a snapshot of sixteen sonar readings were collected. The robot was moved 5 inches to the right, and a second set of data were recorded from the sonars. Each time a snapshot was obtained, a map was drawn using the recursive line fitting procedure. In the end, the two sets of points were fused through the matching and update algorithms of Section 3. As can be seen in Figure 4, as the robot moves to the right, it sees more of the left wall accounting for the greater coverage in the latter map. The line segments with points of higher confidence are kept (see Figure 5). Notice that the upper wall remains at about the same distance as the robot moves to the right, and both snapshots give identical line segments.

A case where the sonar returns unreliable data is shown in Figure 6 (upper right corner of the room). The mapping algorithm successfully draws the best

Figure 4: Environment maps derived from the first and second snapshots using recursive line fitting. The Nomad 200 mobile robot moved 5 inches to the right between the snapshots. (Red: first snapshot. Blue: second snapshot.)

Figure 5: Enhanced environment map obtained by fusing the first and second snapshots in Figure 4.

contours based on the available data as seen in Figure 7.

A common observation from these experiments is that in a large room, the sonar ring is not capable of detecting all the walls. Some sections of the Blue Room were either too far from the mobile robot, or the sonic beams were scattered due to the severe angle of incidence yielding the walls undetectable. In such instances, the robot has to navigate to different locations within the room for maximum coverage.

The algorithms developed in this paper work even when the robot is closely surrounded by objects (e.g. a tight room with a tiny opening). As seen in Figures 8 and 9, a complete map is successfully drawn and updated. More complex floor plans were considered using Nomad's Cognos Software Development Package in citeleyla.

Figure 6: Environment maps derived from the first and second snapshots using recursive line fitting. The Nomad 200 mobile robot moved 5 inches to the right between the snapshots. (Red: first snapshot. Blue: second snapshot.)

Figure 7: Enhanced environment map obtained by fusing the first and second snapshots.

Figure 8: Environment maps derived from the first and second snapshots using recursive line fitting, superimposed on the floor plan. The simulated robot, which is surrounded all around by objects, moved 5 inches to the right between the snapshots. Nomad's Cognos Software Development Package was used in this experiment. (Red: first snapshot. Blue: second snapshot.)

5 Discussion and Conclusions

In this paper, we introduced a novel matching and update processes which are based on the sonar accuracy or resolution. When consecutive snapshots are taken within 1–5 inches difference between robot's previous and current location, it is expected that the readings should fall within a ball of radius of $\pm 1 - 3\%$ times the previous range reading by virtue of the expected sonar accuracy specifications. Any discrepancy beyond this is attributed to erroneous data or the detection of new object faces. Two-dimensional representations obtained as a result of resolution-based mapping are reliable since the thresholds are based on the accuracy range. Another advantage of tying decision thresholds to sonar resolution is that confidence in the sensor readings becomes a function of the robot-to-object distance. This quality is intuitively pleasing since signal attenuation increases with distance.

The mapping and enhancement procedures developed in this paper have a number of advantages. The initial map generated by the first set of sonar data serves as the local model of the environment, and no a priori information is needed. This approach makes the method robust because in semi-structured and unstructured environments, where the local model based on a floor plan may not always be relevant. A typical example is an office room with people walking around.

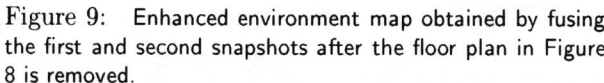

Figure 9: Enhanced environment map obtained by fusing the first and second snapshots after the floor plan in Figure 8 is removed.

way are adaptive, depending on the sonar-to-object distance.

The limitation of the matching and update processes embedded in the mapping algorithm is that it requires continuous data acquisition with a robot displacement of at most 10 inches between two consecutive snapshots. With the advent of high-speed microprocessors, the data acquisition and processing rates are already fast enough to satisfy this constraint.

The algorithms described in this paper have no provision to deal with erroneous data resulting from scattering or sensor uncertainties due to the 25° beamwidth of the Polaroid 6500 transducer. In order to overcome these problems and achieve a more accurate map of the environment, sonar signal processing of the raw data is necessary prior to data fusion. The actual TOF readings should be adjusted depending on the type of surfaces (i.e., diffracting or reflecting). Furthermore, multiple-sensor integration of visual, x-ray, tactile, etc. sensors will robustify the fusion process by compensating for the inherent weakness of each type of sensors with the information provided by the others. Our method is universal since it can be applied to range readings obtained from any kind of sensors.

Acknowledgment

This work has been partially supported by NSF Research Grants BES-9506771 and BES-9712565.

References

[1] J. Borenstein, H. R. Everett, and L. Feng, *Navigating Mobile Robots: Systems and Techniques*, Wellesley, Massachussetts: A K Peters, 1996.

[2] L. Cahut, "Feature-Based Environment Mapping Using Sonar Sensors", M.S. Thesis, University of Southwestern Louisiana, Lafayette, Louisiana, December 1997.

[3] J. L. Crowley, "Navigation for an Intelligent Mobile Robot", *IEEE Journal of Robotics and Automation*, vol. 1, 1985, pp. 31-41.

[4] J. L. Crowley, "World Modeling and Position Estimation for a Mobile Robot Using Ultrasonic Ranging", *Proceedings of the IEEE International Conference on Robotics and Automation*, Scottsdale, Arizona, May 1989 pp. 674-680.

[5] H. F. Durrant-Whyte, "Sensor Models and Multisensor Integration", *International Journal of Robotics Research*, vol. 7, 1988, pp. 73-89.

[6] D. Lee, *The Map-Building and Exploration Strategies of a Simple Sonar-Equipped Mobile Robot*, Cambridge, United Kingdom: Cambridge University Press, 1996.

[7] J. J. Leonard, H. F. Durrant-Whyte, *Directed Sonar Sensing for Mobile Robot Navigation*, Boston, Massachussetts: Kluwer Academic Publishers, 1992.

[8] R. Mandelbaum, G. Kamberova, and M. Mintz, "Statistical Decision Theory for Mobile Robotics: Theory and Application", *Proceedings of the Conference on Multisensor Fusion and Integration*, Washington, DC, December 1985, pp. 17-24.

[9] The Nomad 200 User's Manual, Nomadic Technologies Inc., Mountain View, California, 1996.

Building Local Floor Map by Use of Ultrasonic and Omni-directional Vision Sensor

Shih-Chieh Wei, Yasushi Yagi, Masahiko Yachida
Email: {seke,y-yagi,yachida}@sys.es.osaka-u.ac.jp
Department of Systems and Human Science
Graduate School of Engineering Science, Osaka University, Japan

Abstract

In this paper, we propose a new fusion approach which uses the ultrasonic sensor aided by an omni-directional vision sensor to give a grid based free space around the robot. By use of the ultrasonic sensor, the robot can obtain a conservative range information based on our nearby range filtering method. This filtering can give a more reliable result considering the sensor's problem of specular reflection. Also, by use of the special omni-directional vision sensor we developed, the color and edge information can be obtained in a single picture and mapped to the ground plane by the inverse perspective transformation. Thus the range, color and edge information can all be expressed on a metric grid-based representation which forms the basis of our radial fusion processing. Results in an indoor cluttered environment are given which show the usefulness of our proposed sensor fusion approach.

1. Introduction

For robot navigation, it is important to build a local map from sensor input to capture the environment around the robot. Based on this local map, obstacles can be avoided and free space can be followed.

However, due to the cost and limitation of various sensors, a single sensor is generally not sufficient to provide a satisfying result. Thus sensor fusion has been widely used to enhance the map precision [1]. Among the choices, the ultrasonic sensor is often used with the vision sensor [2,3]. This is due to the fact that the ultrasonic sensor is fast at acquisition of the environment model, though it's poor at angle resolution and error-prone as specular reflection occurs [4]. On the other hand, the vision sensor is good in resolution, though it's slow at image processing, and suffers from the varying lighting condition in the environment. In map building, by combining these two sensors, it allows the exploitation of the vision's area or edge information with the readily available ultrasonic range information. However, in applying the two sensors there were the following difficulties.

For the ultrasonic sensor, the obtained range information is known to have the problem of specular reflection and wide beam width. Approaches like discarding range readings above a certain threshold, or fusion of ultrasonic readings at various robot positions have been proposed to enhance the range reliability [5]. But depending on the thresholds, the discarding of longer range data tends to lose useful information. For a passive vision intensity sensor, the obtained area or edge information is subject to the lighting condition in the environment. Also, for vision sensors, a panoramic vision sensor is more desired in building a map around the robot.

In map building, fusion of multiple sensor input is often done on a feature or a grid basis [6]. A *feature based fusion* consists in identification of physical features supported by multiple sensor readings. The correspondence of sensor readings to a physical feature is often established by feature tracking [3] or a time-warping matching function [2] in robot motion. In contrast, a *grid based fusion* consists in storing all sensor readings on a common metric reference. The correspondence of readings to grids is established by this metric reference and thus affected only by the precision of sensor calibration at initialization. In general it can be observed that feature based correspondence is not easy to find at robot initialization while grid based correspondence can be established all the time. Traditionally the feature map built by the feature based fusion is often used in robot localization, and the grid map built by the grid based fusion used in obstacle avoidance. This is due to the fact that the location of features in a feature map can directly be used to compute the robot position by triangulation, while the grid occupancy information in a grid map can directly provide the bumping information along the grids in the robot's motion direction. However, in recent years, by use of moment of area matching [7] on the grid representation, it has been shown that localization is also possible on a grid map.

To build the floor map, we have built a system based on an omni-directional vision sensor and the ultrasonic sensor [8]. The omni-directional vision sensor we use can get a less distorted and faster scan of the surrounding floor [9] than traditional vision sensors. Also, the range filtering method we use on the ultrasonic range data can provide a more reliable initial estimation of the free space.

In this paper, we refine our fusion algorithm further by use of a radial fusion approach. Free index functions obtained from the ultrasonic and vision input are fused along the radial directions to give a combined safety index for each point in the grid representation. The finally obtained free space in the floor map can then be used for either obstacle avoidance or robot relocation by the methods mentioned above.

As related work, Bang, et al [2] and Yagi, et al [3] also integrate the sensor input from a conic omni-directional vision sensor and the ultrasonic sensor for building the local map. For Bang, the ultrasonic sensor gives the range, and the vision sensor the azimuth of the edge feature; for Yagi, the vision sensor gives the azimuth of the edge, and the ultrasonic sensor the confirmation of free space between two vertical edges. However, Bang assumed that the map is given and did not deal with map building. Also for both they did not deal with the ultrasonic specular reflection problem.

2. System overview

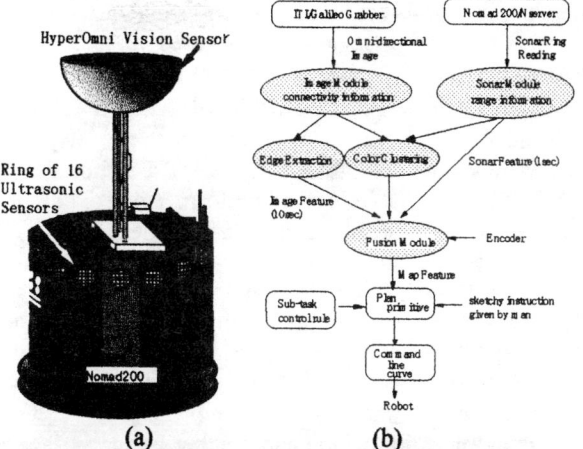

Figure 1. (a) Our vision sensor atop the Nomad 200. (b) Block diagram of the floor map building process

As shown in Figure 1(a), the system is built on a Nomad robot which has a ring of 16 ultrasonic sensors on the periphery of the upper body. Each ultrasonic sensor is 22.5 degrees apart from nearby sensors, and has a beam width of 22.5 degrees. Atop the robot body is a hyperboloidal omni-directional vision sensor [9]. The sensor captures the floor image and via the UHF antenna transmits it to a Sun 4 workstation where the image processing is done.

The block diagram of the system is shown in Figure 1(b). There are two major modules doing on the respective ultrasonic and vision sensor input before the final local map fusion. The *image module* gets input of an omni-directional image and corrects the distortion by inverse perspective transformation. The distortion corrected image is then fed into the *edge extraction* and *color clustering* modules to obtain the edge map, and the color map, respectively. These two maps reflect the connectivity information about the robot's surrounding environment. The *sonar module* gets input of 16 sonar ring readings, and produces a sonar map. The sonar map reflects the range information about the robot's environment. From the sonar map, an initial area of the surrounding free space is also estimated and used by the color clustering module to give an initial estimation of the surrounding floor color.

The *local map module* does the main fusion process for the 3 separate maps previously obtained. The fusion result represents the environment model of the robot. When the robot moves, the maps are shifted with encoder information to keep in validity. Other modules like *plan primitive* and *command modules* can be added for robot control.

In terms of fusion, there are three levels of sensor fusion involved in these modules. The sonar module does the fusion of 16 ultrasonic sensors to produce the echo map; the color clustering modules fuses with an initial free space information from the echo map to produce a uniform color map; and finally the local map module fuses the echo, color and edge map from different sensors to produce the final free space around the robot. We will introduce the fusion process in Section 3, 4 and 5 respectively.

3. Floor map from ultrasonic sensor

The ultrasonic sensor is known to have measurement error due to (1) wide beam width, and (2) specular reflection when the incident echo is over the critical angle of reflection of the wall. Wide beam width degrades the map's angular resolution while specular reflection defies the sensor's detection of a nearby obstacle. In this work, we try to enhance this poor angular resolution by incorporating the visual information. We also try to enhance the reliability of the range data by a nearby range filtering method which we will introduce below. With this filtering, a more reliable initial area of the surrounding space can be estimated and used in the color clustering module later.

3.1. The nearby range filtering

An ultrasonic sensor often expects echo return from its main lobe energy which travels along the same direction as the transducer axis. So the sensor can measure the distance of an object it is pointing to by the time of flight it takes the echo to travel to and fro [4]. However, there might be cases of specular reflection where the main lobe energy is not echoed back. Instead, it's the side lobe energy which gets reflected back whose traveling direction is different from the sensor's pointing direction. It is therefore difficult for a single sensor to make sure that specular reflection has occurred or not and thus be affirmative about its detected range value. However, by fusion of properly spaced sonar ring sensors, we find that it is possible to make a conservative evaluation of the returned range value by use of the minimum d_{min} of its nearby range values if they exist. That is,

$$d_{min} = \min(Range_{left}, Range_{self}, Range_{right})$$

A similar filtering approach can be found in Lee et al's work [6]. False reflection due to multiple reflections is not considered in this paper as a proper firing sequence of the sensors can avoid the interference from other sensors [10]

In the robot we use in this work, the beam width of each ultrasonic sensor is 22.5 degrees, which is equal to the angular separation between neighboring ultrasonic sensors on the robot periphery. This configuration ensures that for each reflected echo, there could be either an actual obstacle in the echo direction at the echo distance (due to main lobe reflection), or an obstacle in the nearby echo directions at nearer than the echo distance (due to side lobe reflection). This justifies the premise of the ultrasonic free index function which we will describe below.

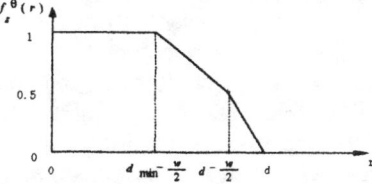

Figure 2. The ultrasonic free index function along a radial line in an ultrasonic beam cone

3.2. Floor map from the range information

To deal with the specular reflection problem, we propose an *ultrasonic free index function* $f_s^\theta(r)$, which defines the safety index for an area at each radial distance from the ultrasonic sensor as follows:

$$f_s^\theta(r) = \begin{cases} 0 & d < r \\ 0.5 - (r - (d - \frac{w}{2}))/(\frac{w}{2}) * 0.5 & d - \frac{w}{2} < r < d \\ 1 - (r - d_{min})/(d - \frac{w}{2} - d_{min}) * 0.5 & d_{min} - \frac{w}{2} < r < d - \frac{w}{2} \\ 1 & 0 < r < d_{min} - \frac{w}{2} \end{cases}$$

where r is the distance from the robot fringe where the sensor is installed; d the range value of the sensor; d_{min} the shortest range value among the neighboring 3 range values defined earlier; and w the robot diameter.

As shown in Figure 2, to each detected range value d, the safety index within the sensor's beam cone starts from 1 at the robot fringe, and begins to decrease at the minimum nearby range distance at d_{min} minus the robot radius. The free index function drops to 0.5 at the robot radius from the detected range d, and drops to 0 at d where a candidate obstacle lies.

The echo map around the robot is then obtained from fusion of the safety index function from the 16 ultrasonic sensors. In the echo map, the area of safety index 1 is later used for estimation of the initial floor color in Section 4.1. And in the final fusion in Section 5, the safety index is used to provide an average safety index in a radial region segmented by the visual information as described below.

4. Floor map from omni-directional vision sensor

The CCD camera has a limited angle of view. To capture a broader view in a single image, Yamazawa et al [9] designed an omni-directional vision sensor. The camera is fixed above the robot, and looks up to a fixed hyperboloidal mirror facing downward. Due to the geometrical property of the hyperbolic curve, light going toward the focal point of the upper hyperboloidal mirror will be reflected toward the focal center of the camera fixed at the other focal center of the hyperbolic curve below. The taken picture can thus be transformed to a floor map image as if seen by a camera looking down at the floor.

Note that the transformation is done based on the assumption that all obstacles lie on the ground plane. For those obstacles which stand above the ground plane, their upper body will appear elongated in the radial direction and lie farther than their lower body on the floor. This will not pose a problem to our map building as we are only interested in the free space region of the floor. However, to have this extracted free space useful, we do require that each obstacle extends its body to the ground plane which is almost always fulfilled in the indoor environment setting in concern. After the transformation, the obtained floor image is then used to extract the connectivity information below. Note that in the extraction process, the same ultrasonic fan region can be used to localize the processing from non-local environment factors.

4.1. Floor map from color information

From the floor map image, we extract two maps, the uniform color map, and the edge map. The uniform color map is composed of areas of uniform color. The areas could be the floor, the obstacle, or simply the shadow.

To extract the uniform color areas, we use the k-mean clustering to do the color segmentation. The clustering is expected to accommodate the minor color variance in the floor area. Since we are only interested in the floor region surrounding the robot, we fix to classify each fan region around the robot into 3 clusters. Each cluster represents a uniform color region which could either be the surface of the floor, the obstacle, or the shadow. In selection of the color space for clustering, we find that while YIQ space is good for discriminating the different surfaces in the image, it is not good as a clustering space. Among the color spaces we tried, we find that the (L,a,b) color space gives a generally good result in extraction of uniform color regions. In some well-lit environment, use of the saturation axis might also give good segmentation result. Therefore we choose the (L,a,b) space for clustering in general.

As we could obtain a good initial estimation of the floor region from the filtered ultrasonic range data, we compute the average color of this area as the initial central color for the floor cluster. For the initial central color of other clusters, we choose the (L,a,b) color in the fan region which has the maximum or minimum Y value (intensity) in the YIQ color space. In general, the color which has the minimum Y value can also be replaced by the dark central robot body area which is known in advance in calibration. The resultant color connectivity map contains the color region labels for different uniform color regions.

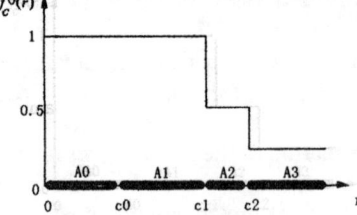

Figure 3. The color free index function along a radial line in the uniform color map

The obtained uniform color region can provide the free space information based on its color as well as distance from the robot. Like the ultrasonic free index function in Section 3 we define a *color free index function* $f_c^\theta(r)$ along the radial direction to give the safety index based on the adjacency of the uniform color region from the robot. The color free index function starts from 1 in the robot fringe and decreases by half upon entering a new color region. In Figure 3, a typical color free index function is shown. Region *A0-A3* are the four uniform color regions as a result of the clustering. Region $A0$ corresponds to the robot body. The safety index decreases by half at the uniform color region boundaries ($c1,c2$) excepting the initial robot boundary ($c0$). The safety index can be used to evaluate the segmented region aided by the ultrasonic and edge information later.

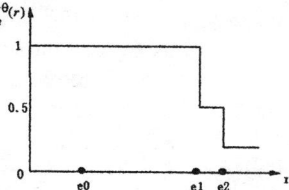

Figure 4. The edge free index function along a radial line in the edge map

4.2. Floor map from edge information

The edge map is defined by the Laplacian filter on the floor image whose response is over a given threshold. Isolated noise in the obtained binary edge is deleted by 8-neighborhood mask operation. The resultant edge map is then expanded and shrunk by an 8-neighborhood operation to result in better connectivity. The edge could be the boundary of a specular light on the floor or the obstacle.

As in the case of the color map, from this edge map, we can define the *edge free index function* $f_e^\theta(r)$ along a radial line which gives the safety index based on the edge information. The function starts from 1 in the outer rim of the robot and decreases by half upon finding a new edge in the radial direction. In Figure 4, a typical edge free index function is shown. $e0$-$e2$ are the edges along the radial direction. $e0$ often corresponds to the robot fringe and thus the index starts to drop from $e1$ and so on. The free index can be used to evaluate the segmented region aided by the ultrasonic and color information later.

5. Free space from fusing the sonar, color and edge maps

After obtaining the various maps from different sensors, we are in a position to fuse these 3 types of information to obtain the final free space as follows.

5.1 Choice of fusion representation

By the aid of a common metric reference in sensor calibration, grid based representation is good for fusion of heterogeneous sensors. However, the grid can be based on Cartesian or polar coordinates which give uniform and no-uniform resolution respectively. In this paper, a finer polar grid at origin is not important to us as no sensor data is available at the robot center. So in implementation, we use the Cartesian grid and compute the value of all the free index function on it for radial fusion. In the following sections, for each grid around the robot, a final safety index will be computed based on the respective free index functions $f_s^\theta(r)$ $f_e^\theta(r)$ and $f_c^\theta(r)$ defined earlier.

5.2 Fusion of the various information from the two different sensors

In extraction of free space, we find that the range information in the echo map seldom corresponds to true physical boundaries due to the wide beam width of the ultrasonic sensor. Instead, it's in the uniform color region boundaries or edges where physical boundaries often lie. So we take the approach to divide the extraction process into a radial region segmentation step and a segmented region evaluation step. In the segmentation step, only the color and edge information is used while in the evaluation step, a weighted sum of the 3 respective information $f_s^\theta(r)$, $f_e^\theta(r)$ and $f_c^\theta(r)$ is computed.

To illustrate these two steps, we assume that on a given radial direction we have obtained the 3 free index functions as shown in Figure 2, 3, and 4 respectively.

- **Radial region segmentation from the visual information:**

To determine the boundary of the new regions, we simply collect boundaries from the edge elements in the edge map and boundaries from the color regions in the uniform color map by

$$\{b_k\} = \{e_i\} \cup \{c_j\}$$

where b_k is the boundary point of the new regions along the given radial line; e_i the edge elements along the radial line in the edge map; and c_j the boundary points of the color regions along the radial line in the uniform color map.

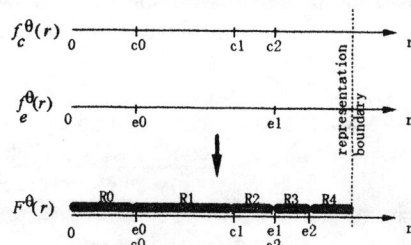

Figure 5. Radial 1-D segmentation by combining the uniform color and edge information

With the new boundaries b_k as delimiters, the new segmented regions can be defined. As shown in Figure 5, the result of the union operation is the 4 boundary points where $b0=e0=c0$, $b1=c1$, $b2=e1=c2$, and $b3=e2$. Among the resultant 5 regions $R0$-$R4$, $R0$ is often ignored as it corresponds to the robot body or the camera itself as seen in the vision input. There is no floor data derivable from this region. Radial segmentation in this way keeps all the boundary points of the edge and uniform color map intact, and thus introduces less artificial boundaries which arise when ultrasonic range boundaries are used.

- **Safety index evaluation for each segmented region:**

For each segmented region R in a radial line, we compute the average combined weighted safety index in it by

$$\frac{1}{N_R}\sum_{r\in R}F^\theta(r) = \frac{1}{N_R}\sum_{r\in R}[w_s\cdot f_s^\theta(r)+w_e\cdot f_e^\theta(r)+w_c\cdot f_c^\theta(r)]$$

where $f_s^\theta(r)$, $f_e^\theta(r)$, and $f_c^\theta(r)$ are the free index functions obtained earlier; w_s, w_e and w_c are the respective ultrasonic, color, and edge weights for the 3 free index functions; R the segmented region; and N_R the number of elements in the representation of the region R.

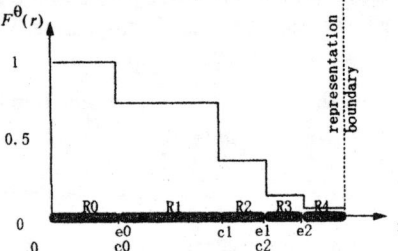

Figure 6. The resultant safety index in each segmented radial region along a radial line

In the above example, the result of the average safety index in each segmented region $R0..R4$ is shown in Figure 6. Note that the weights are empirically determined for now. But in general, the ultrasonic weight should be higher than the other two weights as it provides a different dimension of information than the visual ones.

5.3 Noise reduction

To reduce the noise, we produce a mask by the morphological *close* operation (dilation after erosion), and retain only those safety index values in the mask. We deemed those protruding parts below the size of the robot as noise. Even if there is a real narrow passage in those masked out regions, it will not cause trouble to us as in the map building we are only concerned with those areas reachable by the robot. We finally take a threshold on the masking result and produce the final free space. The threshold is empirically determined for now.

6. Experiment result

In the inverse perspective transformation, the omni-directional vision sensor is calibrated such that each pixel is about 1cm wide in the transformed image, and that the robot takes the shape of a circle with a radius of about 70 pixels in the transformed image. The transformed image map is 640 x 480 pixels in size.

The transformed image map is updated once per 10 sec, while the echo map is updated once per sec. The main bottleneck in the processing is the transformation speed.

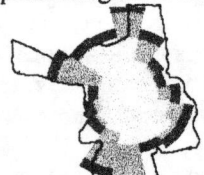

Figure 7. The sonar map after fusing the 16 sensors

From Figure 7 through Figure 11, we show the fusion process for the robot in an indoor cluttered environment.

The space in this experiment (Cf. Figure 8) is about 3.2m x 2.6m which is mostly surrounded by computer desks, chairs, and partitions, with passages in the upper, left upper, and the lower corners. The hand-drawn white curve denotes the free space we want to extract in the end.

Figure 7 shows the sonar map from the ultrasonic sensor. This is the result of combining the range data from the 16 sensors around the robot. In each ultrasonic beam cone, there are 3 regions marked by different colors. The outermost dark gray region shows the region whose safety index is below 0.5 in Figure 2. The outer rim of this region indicates the echo return distance. The innermost light gray region next to the central white circle (robot body) shows the region whose safety index is 1 in Figure 2. As a result of the range filtering, the outer rim of this region is computed as the minimum of the nearby three range readings minus the radius of the robot body. Between the previous two regions, the medium gray region shows the region whose safety index is between 0.5 and 1. The clipping in the upper and lower beam cones results from the limited size of the map representation. As before, the hand-drawn black curve denotes the actual free space in the environment. The difference in the echo map with the actual free space can be attributed to the wide beam width of the ultrasonic sensors.

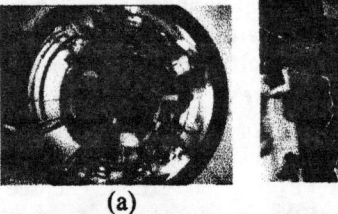 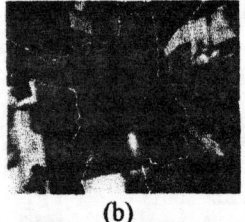

(a) (b)

Figure 8.(a) The raw image from the omni-directional vision sensor; (b) The floor map after inverse perspective transformation.

Figure 8(a) shows the raw image from the omni-directional vision sensor. The central black circle is the camera atop the robot. The circular robot body is occluded by the camera on top. Note that due to the hyperboloidal geometry of the reflecting mirror, the image distortion becomes more obvious near the periphery of the image. Figure 8(b) shows the floor map after inverse perspective transformation of the left image. Note that with the transformation the resolution downgrades seriously at larger distances from the camera center.

 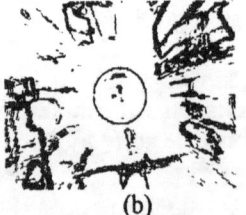

(a) (b)

Figure 9.(a) The uniform color areas after doing k-mean (k=3) color clustering; (b) The edge map after expansion and shrinking of the Laplacian filtered floor image

Figure 9(a) shows the uniform color regions after doing k-mean color clustering on the floor. Note that we fix the number of clusters to 3, intending to differentiate only the robot (medium gray), the surrounding floor (dark gray), and others (medium and bright gray). The trash can in the upper central region of the floor is mis-categorized as floor. This is caused by the few number of clusters we use in the clustering. However it can be remedied by later fusion processing as shown below.

Figure 9(b) shows the edge map after expansion and shrinking of the Laplacian filtered floor image. Dark pixels denote candidate edge cells. From this edge map, we extract the connectivity information by giving a high free index near the robot, and decrease it by half upon finding a new edge pixel radially.

Figure 10. (a) Initial fusion result after combining the echo, color, and edge information; (b) The mask for noise reduction after doing a morphological close operation.

In Figure 10 (a), the fusion result after combining the echo, color, and edge information is shown. The free index decreases according to increasing distance from the robot. Note that due to the edge information the result approximates the real floor boundary better at some fan regions, while over-extrapolating at those unclosed edge directions. Compared with Figure 9(a), we observe that some shadow area on the floor cannot be properly categorized as floor. This is due to the fact that we fix the cluster number to 3.

As postprocessing, to remove the protruding noise from the extracted free space, we produce a mask by using a morphological closing operation with a circular structure element about the size of the robot (56 pixels in diameter). The generated mask is shown in Figure 10 (b).

 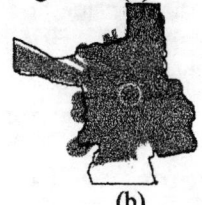

Figure 11. (a) The masking result of Figure 10(a); (b) The final thresholded free space.

With this masking, the noise reduced safety index map is shown in Figure 11(a). With a proper thresholding on the masking result, we obtain the final binary free space in Figure 11(b). Note that compared with Figure 9(a), in the final result, the original specular light area is not seen due to the aid of the ultrasonic information. Also while the trash can in the upper central floor region is mis-categorized as floor in Figure 9(a), it is remedied by the aid of the lower free index from the edge information. The over extrapolated regions in Figure 10 (a) is cleaned off by a morphology generated mask. For comparison, the original hand-drawn free space boundary is also shown in Figure 11. It can be seen that most of the actual free space is correctly extracted in the final thresholded map. The reason that the free space in the lower part of the map fails to be extracted can be attributed to its great color difference with the general floor color around the robot.

7. Conclusion

We built the local floor map for an indoor environment based on the range information from the ultrasonic sensor, as well as the uniform color and edge information from an omni-directional vision sensor. A nearby range filtering method is proposed to address the specular reflection issue. The clustering of uniform color areas is used to allow for less noise sensitive identification of the floor region. Finally a safety index on the grid representation is used as the basis for fusion of information from the two heterogeneous sensors. In the experiment result, we have shown the effectiveness of the fusion result in a cluttered indoor environment.

As future work, the various weights of sensor input and thresholds in the fusion process should be automated. Also we are considering incorporating the back projection of the 3D reflected ultrasonic cone onto the 2-D vision map for inspection of the validity of the range reading.

Reference

[1] Luo, R.C. and Kay, M.G., *Multisensor integration and fusion in intelligent systems*, IEEE Transactions on Systems, Man, and Cybernetics, 19:5, Sep/Oct 1989.

[2] Bang, S.W., Yu, W., and Chung, M.J., *Sensor-based local homing using omnidirectional range and intensity sensing system for indoor mobile robot navigation*, pp542-548, IROS 95.

[3] Yagi, Y., Nishizawa, Y., and Yachida, M., *Determination of free space by integrating omni-directional sensor COPIS and ultrasonic sensor*, pp877-882, International Conference on Advanced Mechatronics, 1993.

[4] Leonard, J.J., and Durrant-Whyte, H.F., *Dynamic map building for an autonomous mobile robot*, International Journal of Robotics Research, 11:4, August 1992.

[5] Elfes, A., *Sonar-based real-world mapping and navigation*, IEEE Journal of Robotics and Automation, RA-3:3, June 1987.

[6] Lee, D., and Recce, M., *Quantitative evaluation of the exploration strategies of a mobile robot*, International Journal of Robotic Research, 16:4:413-447, August 1997.

[7] Courtney, J.D., Jain, A.K., *Mobile robot localization via classification of multisensor maps*, pp1672-1678, ICRA 94.

[8] Wei, S.C., Yagi, Y., and Yachida, M., *On-line map building based on ultrasonic and image sensor*, pp1601-1605, International Conference on Systems, Man, and Cybernetics, 1996.

[9] Yamazawa, K., Yagi, Y., and Yachida, M., *Obstacle detection with omnidirectional image sensor Hyperomni Vision*, pp1062-1067, ICRA 95.

[10] *Nomad 200 User's Guide*, Nomadic Technologies, Inc. 1993.

Map Building Using Fuzzy ART, and Learning to Navigate a Mobile Robot on an Unknown World

Rui Araújo, and Aníbal T. de Almeida

Institute for Systems and Robotics (ISR); and
Electrical Engineering Department; University of Coimbra;
Pólo II, Pinhal de Marrocos; 3030 Coimbra - Portugal; Email: rui@isr.uc.pt

Abstract

This paper introduces a new approach, based on the application of the Fuzzy ART neural architecture, for on-line map building from actual sensor data collected with a mobile robot. This method is then integrated, as a complement, on the parti-game learning approach, allowing the system to make a more efficient use of collected sensor information. Also, a predictive on-line trajectory filtering method, is introduced on the learning approach. Instead of having a mechanical device (the robot) moving to search the world, the idea is to have the system analysing trajectories in a predictive mode, by taking advantage of the improved world model. The real robot will only move to try trajectories that have been predicted to be successful, allowing lower exploration costs. This results on an overall new and powerful method for simultaneous and cooperative construction of a world model, and learning to navigate from an initial position to a goal region on an unknown world. It is assumed that the robot knows its own current world location. It is additionally assumed that the mobile robot is able to perform sensor-based obstacle detection (not avoidance), and straight-line motions. Results of experiments with a real Nomad 200 mobile robot will be presented, demonstrating the effectiveness of the proposed methods.

1 Introduction

It is important for an autonomous mobile robot to be able to navigate on unknown environments, where the location, shape and size of obstacles is unknown, and where there is no map or model of the world initially available. In fact, it is difficult to provide the robot control system with a global map model of its world. This, may easily become a tedious and time consuming programming task. In addition, robot programming and control architectures must be equipped to face unstructured environments, which may be partially or totally unknown at programming time. A variety of approaches to motion planning, such as road-map, cell decomposition, and potential field methods (see [9] for an overview and further references) have been proposed. However, few methods have been able to fully cope with the above problem of model building and learning to navigate.

Both simulation [2], and real-robot experiments [1], [3], have demonstrated the parti-game multiresolution cell-based learning approach [10], as a powerful method for the simultaneous learning of a world model, and learning to navigate a mobile robot from a specified initial position to a known goal region on an unknown world. In [2], it is shown that the constructed world model is general-purpose, in the sense that its usefulness is not restricted to be used on self-learning a particular path, but is valuable for learning paths with different (Start,Goal) pairs. Some forgetting of accumulated world knowledge takes place, when cell-splitting occurs on the system [10], [3]. This leads to additional, time-consuming, exploration with the robot. In [1], modifications on the original parti-game cell-splitting strategy [10] were introduced to improve its operation by decreasing the exploration and modelling efforts. However, most of the received sensor information is lost, not being explicitly integrated into the constructed world model, and not used to plan robot trajectories.

In this paper we introduce, and demonstrate the effectiveness, of a new approach for sensor-based map building with geometric primitives, that is based on the application of the Fuzzy ART neural architecture [6]. This approach is then integrated, as a complement, in the original parti-game learning approach, resulting on a new method, allowing the system to make a more efficient use of collected sensory information for simultaneous and cooperative construction of a world model and learning to navigate from an initial location to a goal region on an unknown world. In this context, a predictive on-line trajectory filtering method is introduced on the learning approach, allowing a very significant reduction on the time-consuming exploration effort that is associated with searching the world with a real robot. Instead of having a mechanical device (the robot) searching the world, the idea is to have the system analysing trajectories in a predictive mode, by taking advantage of the improved world model. The real robot will only move to try a trajectory that has been predicted to be successful.

The organisation of the paper is as follows. Section 2 summarises the basic learning architecture. Section 3 presents the new method for sensor-based map building, that in section 4 is integrated on the parti-game algorithm, yielding an improved approach to navigate a mobile robot. Section 5 presents experimental navigation results with a real Nomad 200 mobile robot. Finally in section 6 we make some concluding remarks.

2 Learning Architecture

In this section we briefly discuss the parti-game learning approach [10] that constitutes the original core of the method we use for learning to navigate a mobile robot. With the method, the robot can simultaneously, learn a kind of map of its environment, and learn to navigate to the goal on an unknown world, having the predefined abilities of doing straight-line motion to a specified position in the world, and obstacle detection (not avoidance) using its own distance sensors. The system also requires the knowledge of the robot current position. However, in this paper we do not deeply address the problem of mobile robot localisation. We simply use accumulation of encoder information to perform robot localisation. Even though this simple approach induces errors, it was sufficient to validate the learning approach in our experiments. The two concurrent learning abilities may be seen as cooperating and enhancing each other in order to improve the overall system performance. For a more extensive discussion see [10], [3], [2].

The parti-game algorithm is based on a selective and iterative partitioning of the state-space. It is a multiresolution approach, beginning with a large partition, and then increasing resolution by subdividing the state-space (see Fig. 4) where the learner predicts that a higher resolution is needed. In order to reach the goal, the mobile robot path is planned to traverse a sequence of cells. The ability of straight-line motion is used as a greedy controller to move from one cell to the next cell on the path. This request to move to the next cell on the path (which is a neighbouring cell) may fail – usually due to an unexpected obstacle that is detected to be obstructing the robot path. A database of cell-outcomes, observed when the system aims at a new cell, is memorised and maintained in real-time. The information in this database may be seen as being organised on a graph data structure. In addition, cells are simultaneously organised in a kd-tree [7], for fast state-to-cell mapping. The database is in turn used to plan the sequence of cells to reach the goal cell, using a game-like minimax shortest path approach. In fact the next-cell outcomes observed as a result of a cell-aim (which is not guaranteed to succeed) may be viewed as "moves" available to an imaginary adversary that would be working against our objective of reaching the next cell, and ultimately reach the goal. The next cell on the path is chosen taking into account a worst case assumption, i.e. we imagine that for each cell we may aim, the adversary is able to place us on the worst position on the current cell such that the next cell that results from the aim is also the worst. In this way we always aim at the neighbouring cell with the best worst-outcome. For this purpose, a minimax problem is solved using Dynamic Programming methods [5]. Spatial resolution is robustly chosen using a game-theoretic cell-splitting criterion. Cells are split when the robot is caught on a losing cell - a cell for which the distance to the goal cell is ∞. Intuitively this means that for each sequence of cell-aims we may choose, our "adversary" may "respond" with cell outcomes that permanently prevent us from reaching the goal, i.e. for the current resolution, the game of arriving at the goal cell is lost. In those situations, as explained in [10], [3], [1], cells in the neighbourhood between losing and non-losing cells are split.

Algorithms 1, and 2 (Fig. 1) describe the overall parti-game learning method. In those figures, NEIGHS(i) represents the set of (cell-) neighbours of cell i, OUTCOMES(i, j) is the set of cells that were previously observed to be attained when the system was on cell i and aimed cell at j, P is the world partition, and D is a database of observed outcomes, represented as a set of triplets of the form: (starting-cell, aimed-cell, actually-attained-cell). **Algorithm 1** (see Fig. 1) keeps applying the local greedy controller, aiming at the next cell, on the "minimax shortest path" to the goal, and accumulating observed cell-aim outcome-experience, until

---

ALGORITHM 1
REPEAT FOREVER
1. **FOR** each cell i and each neighbour $j \in$ **NEIGHS**(i), compute the **OUTCOMES**(i, j) in the following way:
   1.1 **IF** there exists some k' for which $(i, j, k') \in D$
       THEN: **OUTCOMES**(i, j) = $\{ k \mid (i, j, k) \in D \}$
   1.2 **ELSE**, use the optimistic assumption in the absence of experience: **OUTCOMES**(i, j) = $\{j\}$
2. Compute $J_{WC}(i')$ for each cell using minimax.
3. Let $i :=$ the cell containing the current real-valued state s.
4. **IF** $i =$ GOAL **THEN** exit, signalling SUCCESS.
5. **IF** $J_{WC}(i) = \infty$ **THEN** exit, signalling FAILURE.
6. **ELSE**
   6.1 Let $j := \underset{j' \in \text{NEIGHS}(i)}{\text{argmin}} \underset{k \in \text{OUTCOMES}(i,j')}{\max} J_{WC}(k)$
   6.2 **WHILE** (not stuck and s is still in cell i)
       6.2.1 Actuate local greedy controller aiming at j.
       6.2.2 $s :=$ new real-valued state.
   6.3 Let $i_{new} :=$ the identifier of the cell containing s.
   6.4 $D := D \cup \{(i, j, i_{new})\}$
LOOP

ALGORITHM 2
WHILE (s is not in the goal cell)
1. Run Algorithm 1 on s and P. Algorithm 1 returns the updated database D, the new real-valued state s, and the success/failure signal.
2. **IF** FAILURE was signalled **THEN**
   2.1 Let $Q :=$ All losing cells in P ($J_{WC} = \infty$).
   2.2 Let $Q' :=$ The members of Q who have any non-losing neighbours.
   2.3 Let $Q'' := Q'$ and all non-losing neighbours of Q'.
   2.4 Split each cell of Q'' in half along its longest axis producing a new set R, of twice the cardinality.
   2.5 $P := P + R - Q''$
   2.6 Recompute all new neighbour relations, and delete from the database D, those triplets that contain a member of Q'' as a start point, an aim-for, or an actual outcome.
LOOP

Figure 1: Algorithm 1, and top-level Algorithm 2.

either the robot is caught on a losing cell ($J_{WC} = \infty$), or reaches the goal cell. The top-level **Algorithm 2** (see Fig. 1), is responsible for subdiving a selected set of cells whenever the robot is caught on a losing cell.

3 Map Building

This section introduces the Fuzzy ART architecture [6] as a new approach for map building, based on geometric primitives. A map building algorithm should ideally have a set of characteristics. **(1)** The Fuzzy ART model allows self-organisation: to be autonomous, the mobile robot must organise in a useful way, the sensor data it collects from the environment. **(2)** Multifunctionality: for representing the environment, we want a compact model that allows for efficient sensor-based map-building, motion planning, self-referencing, etc. The application of the Fuzzy ART model for map-building is discussed in sections 3.1 and 5, and an example of its usefulness and application in motion planning is shown in sections 4 and 5. Under the algorithmic point of view, if possible, we also want a similar model to be useful for other pattern recognition, control, and reasoning problems. **(3)** Updatability: the model should be easy to update according to new information arriving from sensors: the Fuzzy ART model can be updated by learning each isolated data point as it is received on-line, with the same result as if the update were made in conjunction with a set of other data points - model update is made on a point by point basis not requiring the simultaneous consideration of a, possibly large, set of data points. This is a significant convenience, allowing the robot to use new sensor data as soon as it arrives, thus enabling other system components, such as path planning and localisation, to take advantage of an updated model as soon as possible. See section 5 for a further discussion on the updatability of the Fuzzy ART model. The Fuzzy ART model enables (4) a compact geometric representation allowing small data requirements, and (5) low computational complexity. (6) Unlimited dimensions: the Fuzzy ART model is easy to extend to the modelling of data that is represented in higher dimensions (e.g. mapping of R^3 objects) without adversely impacting on the data size or complexity.

Grid-based certainty maps are widely used to store and maintain occupancy information because they are easy to build and maintain [12]. However, it is difficult to select a resolution for the grid, that is suitable for representing, and to serve as a basis for reasoning on, the entire world. A very localised feature of the world may impose a very high (constant-)resolution grid over the entire state-space. This implies higher data requirements, and induces excessive detail on world modelling, on reasoning (higher computational costs), and on the paths that result from reasoning under such a model. The difficulties on the direct application of grid-based models on localisation have also been pointed out [11]. Geometric representations (e.g. [8], [11]), on the other hand, have been difficult to build, but are significantly more compact, less complex, and fully applicable to high- and low-level motion planning (e.g. this paper) and localisation approaches (e.g. [11]). With higher dimensions the geometric model data requirements become exponentially smaller than the requirements of constant-resolution cellular models.

3.1 Map Building with Fuzzy ART

Other works, e.g. [8], [11], have used different methods to extract geometric primitives. In this subsection, we give a compact overview of the Fuzzy ART learning architecture [6], and discuss its application to map building. With the approach we are able to extract a set of (hyper-) rectangles, whose union represents occupied space, where sensor data points associated with objects have been perceived - a kind of unsupervised clustering.

A Fuzzy ART system includes a field F_0, of nodes representing a current input vector; a field, F_1, that receives both bottom-up input from F_0, and top-down input from a field, F_2, that represents the active code, or category (Fig. 2(a)). The F_0 activity vector is denoted by $I = (I_1, \cdots, I_M)$, with $I_i \in [0,1], i = 1, \cdots, M$. The F_1 and F_2 activity vectors are respectively denoted by $y_1 = (y_{11}, \cdots, y_{1M})$ and $y_2 = (y_{21}, \cdots, y_{2N})$. The number of nodes in each field is arbitrary. Associated with each F_2 category node j ($j = 1, \cdots, N$) is a vector $w_j = (w_{j1}, \cdots, w_{jM})$ of adaptive weights, or LTM traces. Initially weights are set to $w_{j1}(0) = \cdots = w_{jM}(0) = 1$, and all categories are said to be *uncommitted*. After a category is selected for coding it becomes *committed*. The Fuzzy ART operation is controlled by a choice pa-

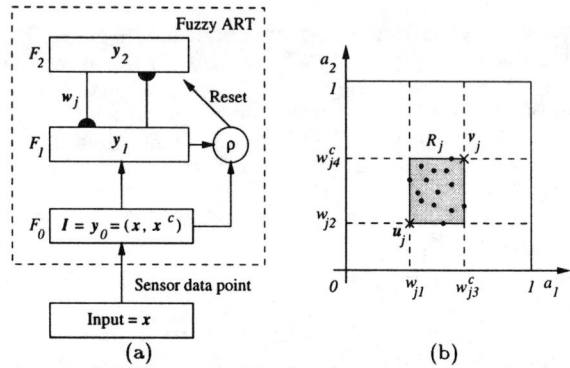

Figure 2: (a) Fuzzy ART neural architecture. (b) Rectangle associated to category j.

rameter $\alpha > 0$, a learning rate parameter $\beta \in [0,1]$, and a vigilance parameter $\rho \in [0,1]$. For each presentation of input I, and F_2 node j, a *choice function* is defined by $T_j(I) = |I \wedge w_j|/(\alpha + |w_j|)$, where, for any M-dimensional vectors p and q, '$\wedge$' denotes the 'min' version of the fuzzy AND operator defined by $(p \wedge q)_i = \min(p_i, q_i)$, and '$|\cdot|$' denotes the norm defined by $|p| = \sum_{i=1}^{M} |p_i|$. For notational simplicity, $T_j(I)$ is often written as T_j when input category I is fixed.

The system is said to make a *category choice* when at most one F_2 node can become active at a given time. The category choice is indexed by J, where $T_J = \max\{T_j : j = 1, \cdots, N\}$. If more than one T_j is maximal, the category j with the smallest index is chosen. In particular, nodes become committed in order $j = 1, 2, 3, \cdots$. When the Jth F_2 category is chosen, $y_{2J} = 1$, $y_{2j} = 0$ for $j \neq J$, and the F_1 activity vector is given by $y_1 = I \wedge w_J$. Resonance occurs if the *match function*, $|I \wedge w_J|/|I|$, of the chosen category meets the following vigilance criterion: $|y_1| = |I \wedge w_J| \geq \rho|I|$. If so, then learning takes place as defined below. *Mismatch reset* occurs if $|y_1| = |I \wedge w_J| < \rho|I|$. In this situation, the match function T_J is set to 0 for the duration of the current input presentation to avoid the persistent selection of the same category during search. A new index J maximising the choice function is chosen, and this search process continues until the chosen J leads to resonance. Once search ends, *learning* takes place by updating weight vector J according to the following equation: $w_J^{(new)} = \beta(I \wedge w_J^{(old)}) + (1-\beta)w_J^{(old)}$. By definition *fast learning* corresponds to setting $\beta = 1$. To avoid proliferation of F_2 categories, a *complement coding* input normalisation rule is used. With complement coding, if the input is an M-dimensional vector x (in our case, a sensor data point), then field F_0 receives the $2M$-dimensional vector $I = (x, x^c) = (x_1, \cdots, x_M, x_1^c, \cdots, x_M^c)$, where the complement of x is denoted by x^c, with $x_i^c = 1 - x_i$. The weight vector, w_j, can also be written in complement coding form: $w_j = (u_j, v_j^c)$, where u_j, and v_j are M-dimensional vectors. Let an (hyper-) rectangle R_j be defined by two of its corners (in diagonal) as illustrated in Figure 2(b). The size of R_j is defined as $|R_j| = |v_j - u_j|$, which in the 2D case, is equal to the sum of the height and width. A Fuzzy ART system with complement coding, fast learning, and constant vigilance forms hyperrectangular categories, R_j, that grow monotonically in all dimensions, and converge to limits in response to an arbitrary sequence of input vectors [6]. Rectangle, R_j, includes/represents the set of all data points which have activated Fuzzy ART category j without reset [6]. Additionally [6], the maximum size of the rectangles R_j can be controlled with the vigilance parameter: $|R_j| \leq (1-\rho)M$. Recall that input data, I, presented to Fuzzy ART must have components satisfying the condition $I_i \in [0,1]$. However, if we have sensor data that is assumed to initially belong to one (any) axis-aligned hyperrectangle, then a linear transformation enables the satisfaction of this condition.

3.2 Sensor Data Filtering

Due to the accumulating nature of the Fuzzy ART system, when applying it to modelling real sensor data, it is useful to perform some prior filtering for removing noisy exemplars. In our implementation, we have used two filtering operations on sensor data points. First experience with the infrared range sensors we have used, shows that above a certain limit, distance readings were not very reliable, and thus were rejected. A second filtering operation, probably more generally applicable to other types of sensors, was performed. Let S be a set of sensor points. A point x is rejected, if no other data point is found inside a circle of radius r_f and centre at x. Since this second operation requires the presence of a set of points S, we are not able to take full advantage of the isolated-point learning capability of Fuzzy ART. However, excellent results were obtained with small data sets, S, composed of points coming from a number of as low as two consecutive sensor-ring scans.

4 Predictive On-line Planning

In our previous work we demonstrated the application of the algorithm of section 2, to navigate a mobile robot [3], [1]. However, this work also enabled the identification and understanding of some aspects where this powerful method could be improved. Two comments emerge in this context. First, the parti-game model is based on its multiresolution partition, and on its aims-outcomes database (D) that summarises the information collected while exploring the world. The system gets rid of almost all sensor information received from the environment and, in spite of its clear usefulness, the information maintained on database D is somewhat indirect, scarce, and implicit, in its description of the world. The second comment is that D must be subject to some forgetting when cell-splitting takes place (cf. point 2.6 of Algorithm 2 - Fig. 1). In fact when a set of cells is split, the associated aim-outcome information is not directly inheritable from "parent-cells" to "son-cells". This constitutes one of the basic ideas of the parti-game operation: e.g. in spite of an aim-failure between two "parent-cells", it is quite possible to have aim-success (and the method is searching for it) between two corresponding "son-cells". But from an external point of view, this induces redundant exploration.

These two comments have motivated two developments on the navigation architecture. First, a new map building method, based on the Fuzzy ART model, and making better use of the received sensor information,

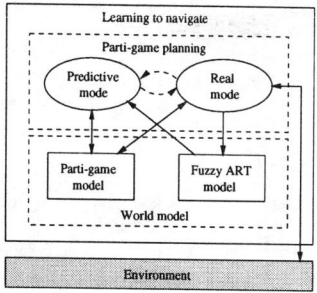

Figure 3: Predictive on-line trajectory filtering.

was developed (section 3), and integrated in the parti-game system of section 2, for improving its world model. Second, the parti-game learning approach was extended by the introduction of a method for Predictive On-line Trajectory Filtering (POTF), allowing a very significant reduction on the time-consuming exploration effort that is associated with searching the world with a real robot. Instead of having a mechanical robot exploring the world, the idea is to have the system analysing trajectories in a *predictive mode*, by taking advantage of the improved world model. The real robot will only move to explore a planned trajectory, when the system is on *real mode*, and the system will enter this mode, only after a *predictive success* has occurred. In real mode, obstacle detection is performed using the real distance sensors of the robot. In predictive mode, on the other hand, exploration trajectories have an on-line predictive/simulation nature not involving any real-robot motion, and obstacle detection is performed using the Fuzzy ART world model, possibly enlarging the rectangles by a percentage of the robot radius in a border gap. In both modes, path planning is performed using the parti-game approach, with the parti-game model (partition, and aim-outcomes database D) being incrementally updated, according to the results of both predictive and real exploration. However, only in real mode is the Fuzzy ART model incrementally updated, because only in this mode is real sensor data available for this purpose.

As already described, one of the main ideas of the method, is to reduce real-robot exploration by giving priority to predictive exploration. However, the "*extent*" of the predictive effort may be controlled by configuring the exigency level of the "*predictive success*" condition that is used to trigger the transition from predictive mode to real mode. Two options may used to establish this condition: a predictive success may be said to occur when, (1) N consecutive predictive cell-aim successes (or the predictive arrival at the goal cell whichever comes first), or (2) a predictive arrival at the goal cell, take(s) place after starting from the current robot location. Also, the "*frequency*" of predictive effort may be controlled, by configuring the condition that is used to trigger the transition from real mode to predictive mode. The system always starts in predictive mode, and the following four options (listed in increasing order of predictive frequency) may be used: enter predictive mode (1) after cell splitting that takes place when the robot is caught on a losing cell, or (2) at the end of every failed cell-aim, or (3) at the end of every cell-aim, or (4) at the end of every motion sampling interval. Figure 3 illustrates the ideas introduced in this section.

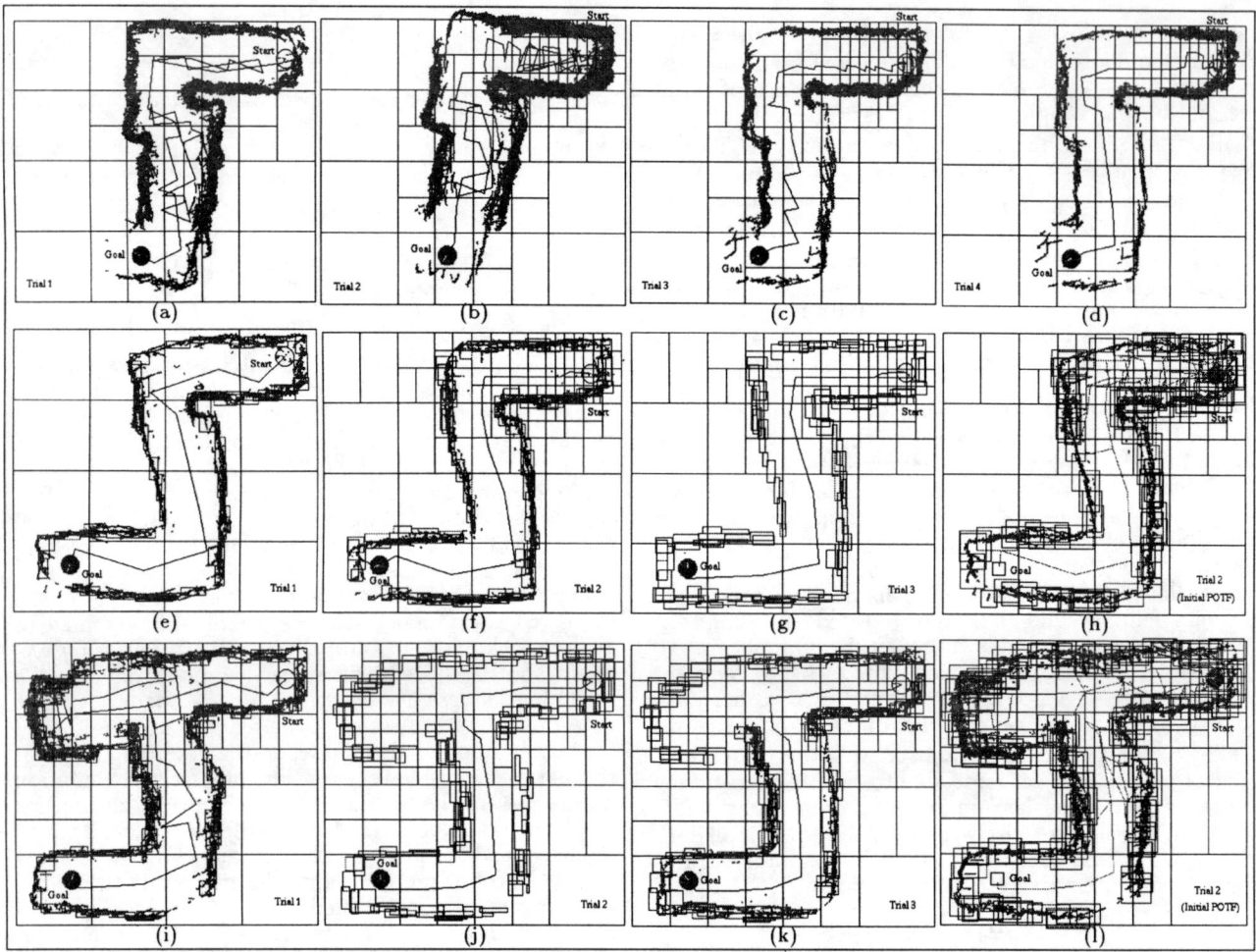

Figure 4: Experiment 1: (a) Trial 1, (b) Trial 2, (c) Trial 3, (d) Trial 4. Experiment 2: (e) Trial 1, (f) Trial 2, (g) Trial 3, (h) Trial 2 (Initial POTF). Experiment 3: (i) Trial 1, (j) Trial 2, (k) Trial 3, (l) Trial 2 (Initial POTF).

5 Experimental Results and Discussion

The methods presented in this paper have been implemented on a zero turning radius real Nomad 200 mobile robot which includes, 3 wheels, and 16 infrared range sensors (equally spaced around its body) that were used for obstacle detection.

In this section we present results of 3 experiments. Each experiment was organised as a sequence of trials to navigate, in a world with obstacles, from a start location to a goal region. The first trial starts with no model of the world. Subsequent trials start with, and build upon, the world model that was learned until the end of the previous trial. The first experiment (Figs. 4(a)-(d)) is included for comparative purposes, and was discussed in [3]. In this experiment, the method of section 2 was used without the integration of the Fuzzy ART model (section 3), or the use of POTF (section 4). We do not deeply address the problem of mobile robot localisation. We simply used accumulation of encoder information to perform robot localisation, with localisation accumulators being set to correct values at the beginning of each trial. Even though this simple approach induces localisation errors, it was sufficient to demonstrate the mobile robot map-building and navigation methods of the previous sections. However, it should be interesting to improve the methods, in order to make them more robust to uncertainty in localisation. Figure 4 includes: infrared information (not in (g), (j)), robot trajectories (not in (h), (l)), state-space partition at the end of trials, Fuzzy ART model (not in (a)-(d)), and the predictive trajectories at the beginning of trials (only in (h), (l)). In all the experiments of figure 4, the dimensions of the state-space were $7.42\,m \times 6.73\,m$.

Experiment 2 was similar to experiment 1, except that: (1) the path to the goal was somewhat greater with the goal region set at $1.7\,m$ to the left, of its location on experiment 1, and (2) the Fuzzy ART map building approach of section 3 was activated, and used to make predictive on-line trajectory filtering (POTF - section 4). In figures 4(e)-(g) the efficient trajectories of the first 3 trials of experiment 2 can be observed, with very direct navigation to the goal starting from the very first trial. These results demonstrate that the introduction of the new methods of sections 3, and 4, lead to a new, and very effective, approach for simultaneous model building, and learning to navigate a mobile robot on an unknown world. Additionally, from a relative point of view, when comparing with experiment 1 (Figs. 4(a)-(d)), the improvements are very significant. In figures 4(e)-(g), it can also be observed how the Fuzzy ART approach of section 3, was able to create a geometric-primitive-based map with the location of objects as they were perceived by the infrared range sensors. In both experiments 2 and 3, sensor data filtering (section 3.2) was performed with sensor readings above 13 inches being rejected, and $r_f = 67\,mm$. On trial 2, a few rectangles are present in places, that are slightly apart from where the infrared sensors have perceived objects on this trial. There are 2 reasons for this: (1) slight differences in localisation positions (e.g. from trial to trial) imply different locations for the perceived objects, and (2) the Fuzzy ART model has an accumulative nature, and currently is not able to detect and remove primitives from places where no objects are perceived anymore. In both experiments 2 and 3, the system was configured to enter predictive mode (section 4) at the end of every cell-aim, and to signal a *predictive success* after a predictive arrival at the goal cell. Also, a border gap of 80% of the robot radius was used on the Fuzzy ART rectangles. Figure 4(h) illustrates the predictive trajectories that were analysed at the beginning of trial 2, even before any real robot motion. Note that, since the robot enters predictive mode at the end of cell-aims, this is just a part of the predictive trajectories analysed during trial 2. As can be seen, a considerable amount of predictive exploration takes place. This enables a significant decrease of the exploration effort that would have to be done with the real mechanical robot (a more time consuming operation). Also, the lower amounts exploration motion enabled by the introduction of POTF, lead to a reduction on the severity of the localisation error problem without, however, solving it. In fact it may be seen that, when comparing with trials 1 and 2 of Experiment 1 (Figs. 4(a),(b)), the localisation errors have clearly decreased in Experiments 2 and 3.

Experiment 3 (Figs. 4(i)-(l)), used the same navigation controller that was used in experiment 2, and is an additional experimental evidence of the effectiveness of the overall navigation method that was introduced in this paper. Note how the system is able to backtrack from the dead end at the upper-left corner of the world, and subsequently this area does not need to be visited by the real robot anymore. An effective use of the available sensor information is made, which leads to efficient navigation trajectories from the very first trial (see Figs. 4(i)-(k)). The predictive trajectories at the beginning of trial 2 are again shown in figure 4(l). Further discussion of the results of experiment 3 would be very similar to the discussion of experiment 2. Finally, we remark that, in all experiments, the time expended in computational costs is only a small fraction of the total operation time that includes the sampling intervals when the mechanical robot was moving. A quantitative evaluation of the effectiveness of the introduction of POTF will be presented in [4].

Future work on Fuzzy ART map building includes relevant aspects of model updatability which, at present state, have not yet been considered: the division, and pruning (an obstacle may be no longer present), of geometric primitives in order to better model the world data. However, from a structural point of view, it will not be difficult to delete geometric primitives. Thus, we are optimistic on the feasibility of overcoming the above two aspects (especially pruning), provided that suitable tests are integrated to detect the two situations. This optimism is further supported by the fact that the Fuzzy ART world model may be used to make (sensor) measurement predictions, and those predictions may be compared with real sensor readings. This would be especially useful in non-static worlds. See section 3 for the discussion on another aspect of updatability, where Fuzzy ART is clearly strong. Exploring the application of the Fuzzy ART world model for place recognition and localisation is another line of future research.

6 Conclusion

A new approach, based on the Fuzzy ART neural architecture, has been introduced for on-line map building from actual sensor data. This method was then integrated, as a complement, on the parti-game learning approach, allowing the system to make a more efficient use of sensor information. Also, a predictive on-line trajectory filtering method, was introduced on the learning approach. This resulted on an overall new and powerful method for simultaneous/cooperative construction of a world model, and learning to navigate from an initial position to a goal region on an unknown world. Results of experiments demonstrated the application of the learning approach to a real mobile robot.

References

[1] R. Araújo and A. T. de Almeida. "Exploration-based path-learning by a mobile robot on an unknown world." *Prep. Fifth Int. Symp. on Experimental Robotics (ISER'97)*. Barcelona, Spain, pp. 551–562. June 1997.

[2] R. Araújo and A. T. de Almeida. "Mobile robot path-learning to separate goals on an unknown world." *Proc. IEEE Int. Conf. on Intell. Engineering Systems (INES'97)*. Budapest, Hungary, pp. 265–270. Sept. 1997.

[3] R. Araújo and A. T. de Almeida. "Sensor-based learning of environment model and path planning with a nomad 200 mobile robot." *Proc. IEEE/RSJ Int. Conf. on Intell. Robots and Systems (IROS'97)*. Grenoble, France, pp. 539–544. Sept. 1997.

[4] R. Araújo and A. T. de Almeida. "Integrating a geometric-primitive map into a multiresolution motion planning approach." *Submitted for presentation at the IEEE Int. Workshop on Advanced Motion Control (AMC'98)*. Coimbra, Portugal. June/July 1998.

[5] D. P. Bertsekas. *Dynamic Programming: Deterministic and Stochastic Models*. Englewood Cliffs, NJ: Prentice-Hall, Inc. 1987.

[6] G. A. Carpenter, S. Grossberg, and D. B. Rosen. "Fuzzy ART: Fast stable learning and categorization of analog patterns by an adaptive resonance system." *Neural Networks*, vol. 4, no. 6, pp. 759–771. 1991.

[7] J. H. Friedman, J. L. Bentley, and R. A. Finkel. "An algorithm for finding best matches in logarithmic expected time." *ACM Trans. on Mathematical Software*, vol. 3, no. 3, pp. 209–226. Sept. 1977.

[8] J. A. Janét, S. M. Scoggins, M. W. White, I. J. C. Sutton, E. Grant, and W. E. Snyder. "Self-organising geometric certainty maps: A compact and multifunctional approach to map building, place recognition and motion planning." *Proc. IEEE Int. Conf. on Robotics and Automation (ICRA'97)*. Albuquerque, New Mexico, pp. 3421–3426. April 1997.

[9] J.-C. Latombe. *Robot Motion Planning*. Boston, NJ, USA: Kluwer Academic Publishers. 1991.

[10] A. W. Moore and C. G. Atkeson. "The parti-game algorithm for variable resolution reinforcement learning in multidimensional state-spaces." *Machine Learning*, vol. 21, no. 3, pp. 199–233. Dec. 1995.

[11] J. Vandorpe, H. V. Brussel, J. D. Shutter, H. Xu, and R. Moreas. "Positioning of the mobile robot lias with line segments extracted from 2d range finder using total least squares." *Prep. Fifth Int. Symp. on Experimental Robotics (ISER'97)*. Barcelona, Spain, pp. 309–320. June 1997.

[12] B. Yamauchi. "Mobile robot localisation in dynamic environments using dead reckoning and evidence grids." *Proc. IEEE Int. Conf. on Robotics and Automation (ICRA'96)*. Minneapolis, Minnesota, pp. 1401–1406. April 1996.

Incremental Map Building for Mobile Robot Navigation in an Indoor Environment

Laurent Delahoche, Claude Pégard, El Mustapha Mouaddib, Pascal Vasseur

GRACSY
Groupe de Recherche sur l'Analyse et la Commande des Systèmes
Equipe perception en robotique
Université de Picardie Jules Verne
7, Rue du moulin neuf, 80000 Amiens, France

Abstract

In this article we present a navigation system allowing a mobile robot to be localized in an indoor environment which is only partially known. This system integrates an environment map updating module allowing the mobile robot to estimate the position of new vertical landmarks along its path. An Extended Kalman filter is used on the one hand to estimate the mobile robot position and on the other hand to extract observations which will be used to determine the positions of unlisted landmarks. The integration of new landmarks into the environment global map is managed from the covariance matrix associated with each unlisted landmark. We present the experimental results we have got with SARAH, our mobile robot.

1 Introduction

This article describes the problem of the robot navigation in an environment which is only partially known. In some cases it is difficult to establish a precise representation of the robot evolution world. It is then interesting to enlarge dynamically the robot environment map along its path. The safety aspect in the navigation will then directly depend on the accuracy and the coherence of this dynamic building.

For a dynamic map building it is necessary to localize the robot in relation with already known elements. Localization and environment map updating are thus narrowly linked. This is mainly what Crowley has pointed out by using the Kalman Filtering technique to build the environment global map and to localize the robot [1]. The fusion of dead reckoning and ultrasonic data is thus realized. Kalman Filtering is also used by Leonard and Durrant-Whyte [2] to realize the dynamic building of the environment map. The landmarks dealt with are corners, planes and cylinders and are detected thanks to a specific ultrasonic signature : the RCDs (Region of Constant Depth). Elfes uses ultrasonic data in another approach : he represents the environment under the shape of two dimensional grid [7]. This grid is composed of cells which can be found in three different states : empty, occupied or unknown. This type of representation is also used by Boreinstein in [5]. But the ultrasonic data remain relatively complex to exploit mainly because of the lack of directivity from the sensors. Using other systems of perception such as the laser or the vision is a more reliable way of building the environment map. A rotating laser rangefinder is used by Gonzalez who builds an environment local map where the geometric primitives managed are straight line segments [4]. In [6], a laser rangefinder is used to build a map represented as stochastic obstacle regions equipped with their own stochastic variables such as mean, variance and eigenvalues. As the laser is today a costly material, the vision seems to be an interesting alternative to deal with the problem of the world modeling.

The tridimensional vision linked to the Kalman Filter is used in [8] to update the map of an outdoor environment. Then the representation of the object of the map is tridimensional. As for Yagi, he uses the omnidirectional vision with his COPIS system to build a two dimensional map of free spaces and to estimate the robot position [3].

This article presents a navigation system allowing a mobile robot to update dynamically its environment map which is only partially known. This environment map is used to localize the robot all along its mission. We propose an approach based on the exploitation of the data given by an omnidirectional vision system and by an odometer. The robot position estimation and the map updating are based on the use of an Extended Kalman Filter.

2 The perception system

To localize our mobile robot we use an omnidirectional vision system composed of two parts : a CCD camera and a conic mirror which allows us to get a panoramic projection of the robot evolution world. These two elements are separated by a glass support and are in line. This type of perception system is also used by Yagi with the COPIS system [3].

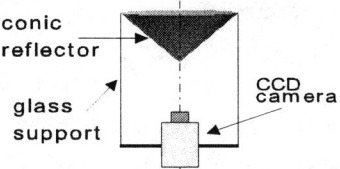

Figure 1: The omnidirectional perception system

Finally this perception system allows to get a two dimensional sensorial model of the robot evolution world on a 360 degrees vision field. In this sensorial model the vertical objects of the environment can be characterized by a set of radial straight lines converging to the center of the cone, i.e. to the center of the image (figure 2).

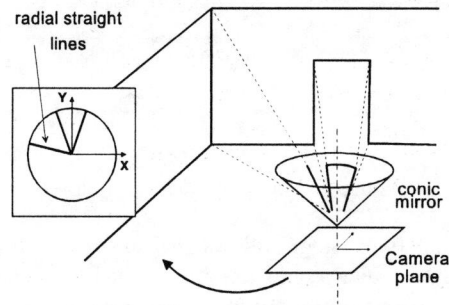

Figure 2: Properties of the omnidirectional vision sensor

The extraction of the radial straight lines characterizing the vertical landmarks of the sensorial model is based on the gradient treatment. The extracting algorithm is performed in two stages [9] :
- applying a 3x3 Sobel filter followed by a binarization,
- applying a simplified Hough transform to group the points forming the same radial straight line (figure 3).

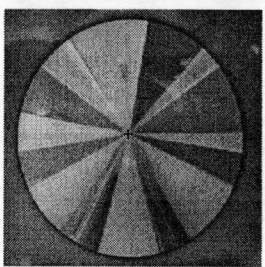 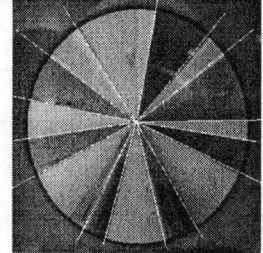

Figure 3: Initial image and primitives extraction result

The radial straight lines, allow to get the azimuth angles of the different vertical landmarks. The r azimuth angles ($\phi_k^1,..,\phi_k^r$) for an acquisition k are our observation set [9].

3 Dynamic localization algorithm

When starting its mission the mobile robot has only a partial knowledge of the environment map. Its localization will then be made according to the known landmarks $B_{1,...,Nb}$ determined in the environment reference $R_E=\{O_E,X_E,Y_E\}$. N_b represents the total number of known landmarks listed in R_E.

The estimation of the robot position has been made through a dynamic approach based on the fusion of the data given by an odometer and by our panoramic vision system [9]. To realize the data fusion we used an Extended Kalman Filter which allows us to treat two kinds of problems in our application :
- taking into account the safety aspect linked to the navigation : the Extended Kalman Filter allows to minimize the covariance estimation.
- differentiating the sensorial data used to localize the robot from those used to update the map (figure 4).

The robot localization mainly rests on associating the azimuth angles given by our omnidirectional sensor and the theoretical beacons listed in the environment map. The azimuth angle observed which cannot be matched to localize the robot will be used to determine the positions of unlisted landmarks (figure 4).

The state formalism of the system is made of two equations ((1) and (2)). Equation (1) is recurrent and describes the evolution of the robot configuration in relation to the command u_k and the disturbance v_k.

$$X_{k+1} = f_k(X_k, u_k) + v_k \quad \text{with} \quad v_k \sim N(0, Q_k) \quad (1)$$

where $f_k(X_k, u_k)$ is a nonlinear function.

We use notation $v_k \sim N(0, Q_k)$ to indicate that v_k is a zero mean blank Gaussian noise with variance Q_k.

The equation (2) called observation equation, allows us to give the observation measures in relation to the matched beacons, from the mobile robot configuration.

$$Z_{k+1} = h_{k+1}(X_{k+1}, B_{1,...,n}) + w_{k+1} \quad (2)$$

with $w_k \sim N(0, R_k)$, and where n is the number of matched beacons B at point X_{k+1} and at time $k+1$.

The nonlinear function $h_{k+1}(.,.)$ of the equation (2) expresses the link between the measures Z_{k+1} in relation to the beacons $B_{1,...,n}$ and the robot's configuration X_{k+1}. This relation is given by the equation (3) :

$$\phi_i = \arctan\left(\frac{yc_k - y_i}{xc_k - x_i}\right) - \theta_k \quad (3)$$

where $X_k = [xc_k, yc_k, \theta_k]^T$ is the robot's configuration in the environment reference, ϕ_i are the observed azimuth angle linked to the known landmark positions (x_i, y_i).

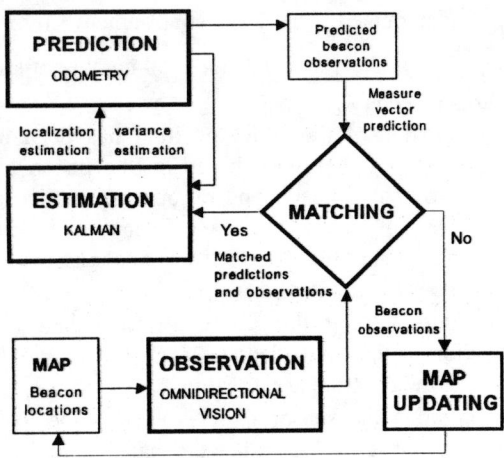

Figure 4: Navigation algorithm

The navigation algorithm (figure 4) we have developed [9] can be split out into several steps linked to the Extended Kalman Filtering [2]. The first one consists in *predicting* the current robot's configuration with the dead-reckoning data. The predicted configuration of X_{k+1} is thus given by the recurring equation (4):

$$\hat{X}_{k+1\setminus k} = f_k(\hat{X}_{k\setminus k}, u_k) \quad (4)$$

where $\hat{X}_{k+1\setminus k}$ will correspond to X predicted at time $k+1$, k being known. Moreover, the prediction stage allows us to get the covariance matrix $P_{k+1\setminus k}$ linked with the predicted position $\hat{X}_{k+1\setminus k}$. A predictive measure vector (figure 4) is determined before the matching stage in relation with the predicted configuration $\hat{X}_{k+1\setminus k}$ given by dead-reckoning. It is made up by the set of azimuth angles which can be seen in theory on the environment map. This predictive measure vector will be then matched with the observation vector during the matching stage.

In the second step when the data coming from the omnidirectional vision system are available (*observation*), they are matched (*matching*) with the environment theoretical model (figure 5).

The observation stage gives a measure vector Z^*_{k+1} coming from the real configuration X_{k+1} of the robot at the time $k+1$ which we will try to approach as best we can with $\hat{X}_{k+1\setminus k+1}$ during the estimation stage:

$$Z^*_{k+1} \cong h_{k+1}(\hat{X}_{k+1\setminus k}, B_{1,\ldots,n}) + H_{k+1}(X_{k+1} - \hat{X}_{k+1\setminus k}) + w_{k+1} \quad (5)$$

where w_{k+1} is a noise disturbance defined with the equation (2) and $H_{k+1} = \left(\dfrac{\partial h}{\partial X}\right)_{X=\hat{X}_{k+1\setminus k}}$.

The matching stage consists in associating each measure $z^*_{j,k+1}$, $j = 1,\ldots,n$ of the observation Z^*_{k+1} with a landmark $B_{1,\ldots,Nb}$ known in the absolute reference R_E. Finally, it consists in comparing the prediction $(\hat{Z}_{k+1}, B_{1,\ldots,p})$ and the observation Z^*_{k+1} (figure 5).

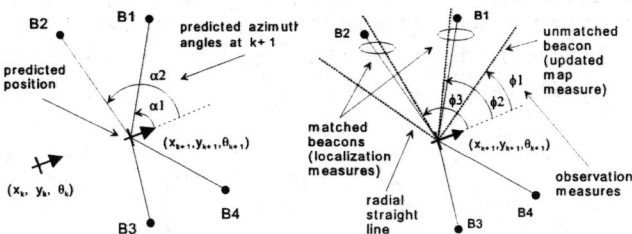

Figure 5: Matching stage principle

The matching stage allows us to obtain a set of observation which could not be matched at the time k. These measures will be used to determine the position of unlisted landmarks Bnr_i.

The last step of the localization algorithm consists in computing the Extended Kalman Filtering equations and estimating the best real configuration of the robot from the available data. This stage gives the robot configuration estimation $\hat{X}_{k+1\setminus k+1}$ and its associated covariance matrix $P_{k+1\setminus k+1}$. The uncertainty domain linked to the position estimation will be shown by an uncertainty ellipse defined from the matrix $P_{k+1\setminus k+1}$ [9].

4 Unlisted landmark treatment

From several paths of the robot we could note that the unmatched measures are generally generated by vertical landmarks which are not listed on the map. Our incremental map updating algorithm is based on the exploitation of these unmatched angular measures. Finally the data which allows to estimate the landmark positions to an acquisition k are:

- the robot configuration $[xc_k \ yc_k \ \theta c_k]^T$
- The azimuth angles $(\phi^1_k,\ldots,\phi^r_k)$ of the unlisted landmarks in the robot reference. r is the number of observed unmatched landmarks to an acquisition k.

The relation linking these parameters is identical to the one used to estimate the robot configuration (equation (3)). When the observations on the same unlisted landmark Bnr_i are available, its coordinates (xb^i, yb^i) in the map reference are directly got from the following equation (figure 6) :

$$\tan(\theta c_k + \phi_k^i) = \frac{yc_k - yb^i}{xc_k - xb^i} \quad (6)$$

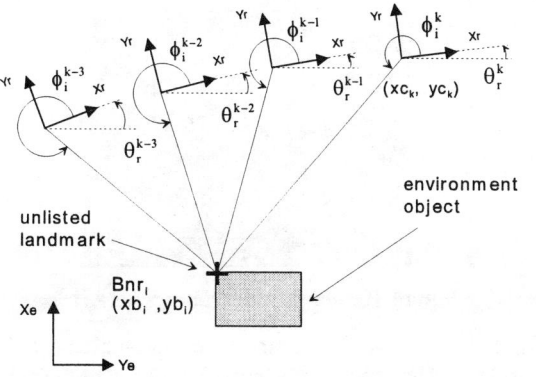

Figure 6: Position estimation of an unlisted landmark Bnr_i from several observations

This equation system is overdetermined. To estimate the parameters (xb^i, yb^i) incrementally we use the recursive least squares method :

$$\begin{aligned}
\hat{\underline{a}}_k &= \hat{\underline{a}}_{k-1} + K_k(y_k - \underline{x}_k^T \hat{\underline{a}}_{k-1}) \\
P_k &= P_{k-1} - K_k \underline{x}_k^T P_{k-1} \\
K_k &= P_{k-1}\underline{x}_k(1 + \underline{x}_k^T P_{k-1} \underline{x}_k)^{-1}
\end{aligned} \quad (7)$$

where a_k is the state vector, y_k the observation vector and x_k represents the known parameters. To apply the recurrence equations (7) it is necessary to express the equation (6) under the following form :

$$\underline{x}_i^T \underline{a} = y_i \quad (8)$$

which gives us with the equation (9) :

$$\begin{bmatrix} \tan(\alpha_k^i) & -1 \end{bmatrix} \begin{bmatrix} xb^i \\ yb^i \end{bmatrix} = xc_k \tan(\alpha_k^i) - yc_k \quad (9)$$

where $\alpha_k^i = \theta c_k + \phi_k^i$

Moreover, the recursive least squares method allows us to obtain an estimation of the uncertainty domain associated with a landmark position $\begin{bmatrix} xb^i & yb^i \end{bmatrix}^T$.

To estimate the position of a landmark, it is necessary to match the unmatched measures of an acquisition $k-1$ with those of an acquisition k. To solve this problem, we have developed a method based on the azimuth angle evolution along the robot path. On figure 7 we can notice that if the cartesian distances separating each omnidirectional acquisition are not too big (approximately 50 cm) the azimuth angles characterizing a same landmark do not vary much from an acquisition $k-1$ to an acquisition k.

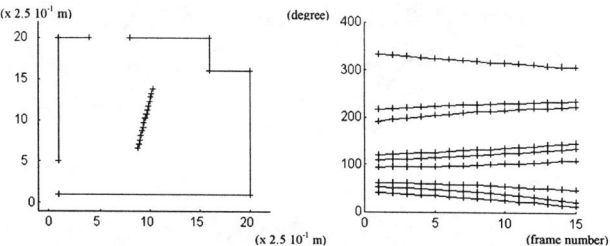

Figure 7 : azimuth angle diagram on 15 acquisitions (right) for a robot path in a structured environment (left)

The grouping of these unmatched angular measures from an acquisition $k-1$ to an acquisition k will be made directly in relation with a threshold level :

```
If |φ_{k-1}^i - φ_k^j| < ε  Then
    φ_k^i is matched with φ_{k+1}^j
End If
```

A grouping of unmatched observations of several acquisitions ($\phi_k^i, .., \phi_{k+n}^h$) will allow us to incrementally estimate the position of an unlisted landmark and its uncertainty domain.

5 Map updating algorithm

The map updating algorithm allows to manage the coherence of object insertions in the map. Managing insertions is mainly based on distinguishing two cases :
- it is necessary to initialize a new unlisted landmark,
- it is necessary to estimate the parameters of an already created unlisted landmark.

Following an occlusion, an unlisted landmark can be hidden on a part of the path and observed again thanks to omnidirectional vision. This relatively frequent case implies that an unmatched angular measure ϕ_k^j cannot be associated with a measure ϕ_{k-1}^i, but with previous measures on the other hand. In this case, we must not initialize a new beacon but estimate the parameters of an already created unlisted landmark Bnr_h. To associate the measure ϕ_k^j with a landmark Bnr_h whose position has already been estimated, we compare the value of the angular measure ϕ_k^j with the angles created by all newly inserted landmarks of the map $M_{bnr}(k)$ (figure 8) :
- the angles β_i created by the newly created unlisted landmarks are calculated in relation to the robot position at the acquisition k.

- These angles β_i are compared to the measure ϕ_k^j which could not be matched. If an association is possible then the parameters of the landmark Bnr_i are estimated with the measure ϕ_k^j. In the opposite case the measure ϕ_k^j is memorized to be used later (figure 9).

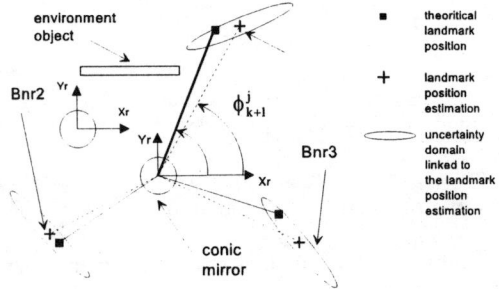

Figure 8: Treatment of the landmark fusion problem

The map updating algorithm is described with the following scheme :

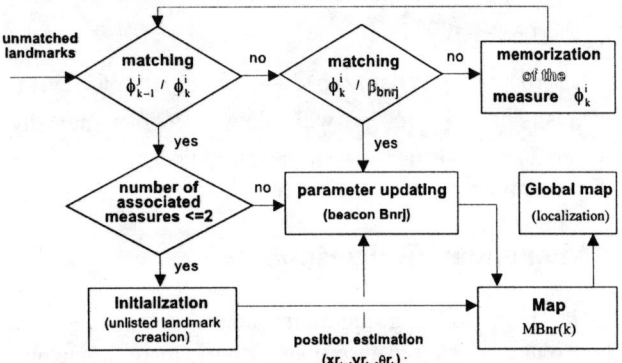

Figure 9: Map updating algorithm

At an acquisition k, all the unlisted beacons are managed in the same map M_{bnr} :

$$M_{Bnr}(k) = \{ \hat{\underline{a}}_k^i , P_k^i \mid 1 \leq i \leq r_k \}$$

where r_k is the number of unlisted landmarks, $\hat{\underline{a}}_k$ the landmark position estimations, P_k its covariance matrix.

The credibility linked to the new landmark position estimation is managed in relation to the surface of its uncertainty domain. It is an equiprobability ellipse whose parameters are defined from the covariance matrix P_k. The bigger the number of observations on a landmark Bnr_i is, the smaller its associated uncertainty domain and the more credibility it is possible to give to this new landmark.

An unlisted landmark is set in the environment global map and used to localize the robot only when the surface of its uncertainty domain is inferior to a reference surface S_{ref}.

6 Experimental results

We have set up our algorithm on SARAH our experimental platform. This mobile robot is equipped with an omnidirectional vision system and two odometers, one on each wheel (figure 10).

Figure 10: Our mobile robot SARAH

The robot evolved in a structured indoor environment (figure 11). The map of this environment is only partially known. To start with, it is constituted with the position of some vertical landmarks (figure 11).

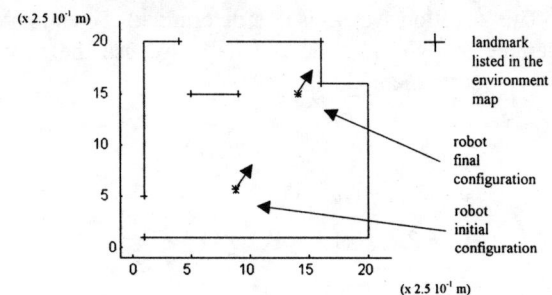

Figure 11: Experimental environment

In our experiments, the robot goes along a straight path (figure 11), during which we have made 20 exteroceptive acquisitions. On the following figures we show the evolution of the position estimation (characterized by a cross) and their associated uncertainty domain characterized by an ellipse (left inside figure). Moreover, we present the angular measures evolution curves. On these curves, the unmatched measures are shown with an asterisk and the ones used to localize the robot with a cross. The measures grouped by the updating algorithm are linked by a line.

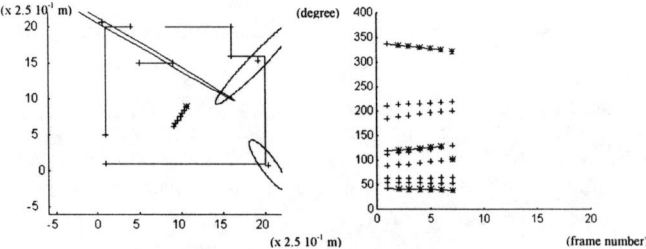

Landmark position estimations for acquisition k=7

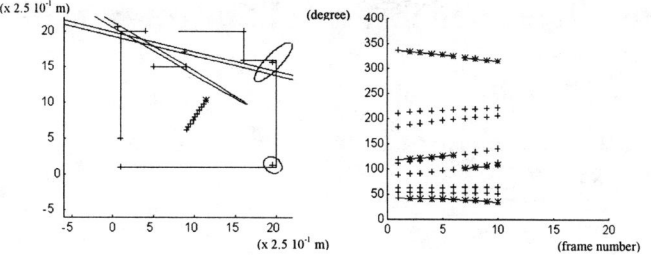
Landmark position estimations for acquisition k=10

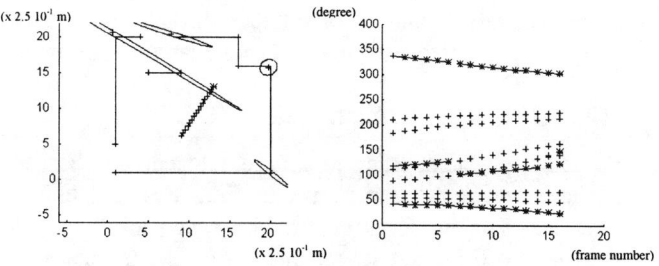
Landmark position estimations for acquisition k=16

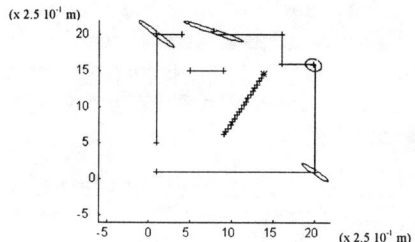
Final estimation of the 4 unlisted landmarks

Figure 12: Estimation of the landmark positions and their uncertainty domain during the mission

We can note that the uncertainty domain linked to the position estimation grows smaller along the observation. We can also note that the coherence of the unlisted landmarks map is correctly settled all along the robot path. The management of occultation (figure 12) is correctly done and the evolution of the new object creation number is relatively steady (figure 12).

We have observed on several paths a typical distance error between the real positions of the unlisted landmarks and their position estimations equal to 10 cm.

Finally along its path, the robot localizes itself in relation with an a priori known map but could also update this map with two of its four new unlisted landmarks.

7 Conclusion

Along this study we have developed a navigation system allowing a robot to carry out elementary missions in an only partially known environment. The robot position estimation is given by an Extended Kalman Filter. Using the Kalman Filtering allows us to differentiate the matched sensorial data used to localize the robot, from the unmatched, which will be used to determine the position of the new landmarks. This position is estimated incrementally thanks to the recursive least squares method. The advantage of our approach is to propose relatively a reliable updated method. Thanks to a coherence management of the uncertainty domain linked to each newly created landmark, we can manage in a reliable way the integration of these landmarks into the environment global map. Thus the robot can navigate safely in even partially defined environment zones.

Acknowledgments

This work was supported in part by the "Conseil Régional de Picardie" under the project "Diva Pole".

References

[1] J.L. Crowley, "World modelling and position estimation for a mobile robot using ultrasonic ranging", *Proc. of Int. Conf. on Robotics and Automation*, pp. 674-680, May 1989.

[2] J.J. Leonard, H.F. Durrant-Whyte, "Dynamic map building for an autonomous mobile robot", *The International Journal of Robotics Research*, Vol. 11 n°4, August 1992.

[3] Y. Yagi, Y. Nishizawa, M. Yachida, "Map-based navigation for a mobile robot with omnidirectional image sensor COPIS", *IEEE Trans. on Robotics. and Automation*, Vol. 11, pp. 634-648, October 1995.

[4] J. Gonzalez, A. Ollero, A. Reina, "Map building for a mobile robot equipped with a 2D laser rangefinder", *Proc. of Int. Conf. On Robotics and Automation*, pp. 1904-1909, May 1994.

[5] J. Boreinstein, Y. Koren, "Histogrammic in-motion mapping for mobile robot obstacle avoidance", *IEEE Trans. on robotics and automation*, Vol. 7, N°4, pp. 1688-1693, August 1991.

[6] Y. D. Kwon, J. S. Lee, "A stochastic environment modelling method for mobile robot by using 2D laser scanner", *Proc. Of Int. Conf. On Robotics and Automation*, pp. 1688-1693, April 1997.

[7] A. Elfes, "Sonar-based real world mapping and navigation", *IEEE Journal of robotics and automation*, Vol. RA-3, N°3, pp. 249-265, June 1987.

[8] S. Betgé-Brezetz, P. Hébert, R. Chatila, M. Devy, "Uncertain map making in natural environments", *Proc. of Int. Conf. On Robotics and Automation*, pp. 1049-1053, April 1996.

[9] L. Delahoche, C. Pégard, B. Marhic, P. Vasseur, "A navigation system based on an omnidirectional vision sensor", *Proc. Of Int. Conf. on Intelligent robots and Systems IROS'97*, pp. 718-724, September 1997.

Haptic Display for Object Grasping and Manipulating in Virtual Environment

Hitoshi Maekawa　　John M. Hollerbach

Department of Computer Science, University of Utah
Salt Lake City, Utah 84112 USA

Abstract

A haptic display for grasping and manipulating virtual objects in a CAD environment is investigated for the development of rapid prototyping technology. The operator receives the sensation of contacting and tracing of the surface of the virtual object, and of grasping and manipulating the object from the haptic display. After the control of the haptic display is formulated, it is implemented on the Sarcos Dexterous Arm Master. The proposed haptic display is experimentally confirmed to provide realistic sensation that enables the operator to grasp and manipulate the virtual object easily as intended.

1. Introduction

In this paper, a haptic display is presented for grasping and manipulating of virtual objects designed in an advanced CAD modeling system, Utah's *Alpha_1* [4]. The final goal of this research is a realization of realistic haptic sensations for interactive rapid prototyping in a virtual CAD environment. The hope is to reduce the duration of the cyclic iteration of design, evaluation and modification of prototypes. A fundamental task is considered of grasping of a virtual object by two fingers, manipulating it to a desired position and orientation and releasing the object at a new location.

In order to apply the sensation of grasping and manipulating the virtual object to the operator, the haptic display device should provide force-controllable degrees of freedom at the arm and hand as well as large workspace where the object is manipulated. Additionally, both of the following two forces, namely, the external force and the internal force need to be generated:

External force
The net force-moment applied to the operator that does not cancel in the operator's body. This force is caused by an object's gravity load, inertia, contact with the environment, and so on.

Internal force
The force applied at the fingers that cancels together in operator's body. This force is caused by squeezing the object by the fingers.

However, in reviewing conventional haptic display devices, they do not satisfy the requirements described above. Namely, a conventional haptic display device does not provide a large workspace [7] for manipulation or is capable of generating only one of the external force [1][3][5] or the internal force [2].

For this problem, the authors employed the Sarcos Dexterous Arm Master shown in **Fig. 1** as the haptic interface. The master arm provides seven joints at the arm, one at the index finger and two at the thumb. The operator grabs the hand rest at the wrist and inserts the index finger and thumb into the finger attachment while interacting with the master arm. Since the joints of the master arm are capable of force control, it is possible to apply specified forces at the wrist, index and thumb of the operator.

2. Fundamental formulae for virtual grasping and manipulating

In the proposed haptic display, the following assumptions are made for providing the sensation of grasping and manipulating to the operator by the master arm:

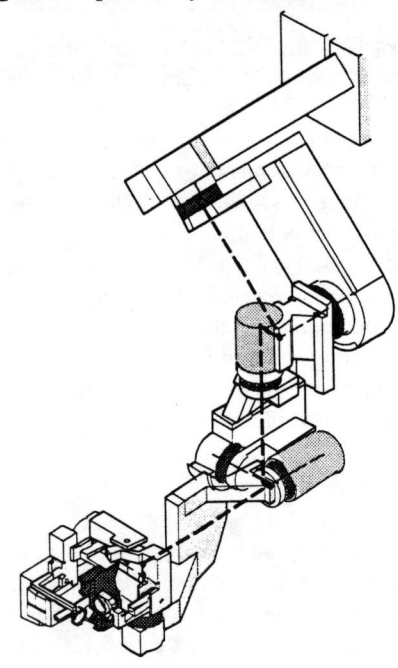

Fig. 1 Sarcos Master Dexterous Arm.

1. Only two points, at the tips of the index finger and the thumb of the master arm, interact with the virtual object.
2. The object is fixed in the absolute coordinate system when it is not grasped by the master arm.
3. When only one fingertip makes contact on the object, the contact is frictionless.
4. When an object is grasped by two fingers, each fingertip sticks on the object surface without slip.
5. Although the object is grasped only by two fingers, its rotation around the axis that connects two fingertips is constrained.
6. While manipulating the object, its dynamics due to inertia is omitted.

As a preparation for the formulation, two coordinate systems, namely the absolute coordinate system o^a-$x^a y^a z^a$ and the object coordinate system o^o-$x^o y^o z^o$ fixed on the object, are defined as shown in Fig. 2. The superscripts a and o represent the parameters described in the absolute and object coordinate systems, respectively. Also the subscripts i, t, w and o represent the parameters for index, thumb, wrist and object, respectively.

The positions of the tip of the index finger, the tip of the thumb and the wrist of the master arm are represented in the absolute coordinate system by the vectors x_i^a, x_t^a, $x_w^a \in \Re^3$, respectively. The object position is located by the vector $x_o^a \in \Re^3$ to the origin of the object coordinate system o^o. The object orientation is represented by vector $\xi_o^a \in \Re^3$, which consists of roll, pitch and yaw (RPY) angles of the object coordinate system relative to the absolute coordinate system. The orientation of the wrist of the master arm is represented by RPY angles $\xi_w^a \in \Re^3$ relative to the absolute coordinate system.

The rotation matrix $R(\xi_o^a) \in \Re^{3 \times 3}$ that converts the vector described in the absolute coordinate system to the corresponding vector described in the object coordinate system is defined as follows:

$$R(\xi_o^a) = \begin{bmatrix} c_y c_z & c_x s_z + s_x s_y c_z & s_x s_z - c_x s_y c_z \\ -c_y s_z & c_x c_z - s_x s_y s_z & s_x c_z + c_x s_y s_z \\ s_y & -s_x c_y & c_x c_y \end{bmatrix} \quad (1)$$

where,

$$\xi_o^a = \begin{bmatrix} \xi_{o\,x}^a & \xi_{o\,y}^a & \xi_{o\,z}^a \end{bmatrix}^T$$
$$c_x = \cos \xi_{o\,x}^a, \quad c_y = \cos \xi_{o\,y}^a, \quad c_z = \cos \xi_{o\,z}^a$$
$$s_x = \sin \xi_{o\,x}^a, \quad s_y = \sin \xi_{o\,y}^a, \quad s_z = \sin \xi_{o\,z}^a$$

Consequently, the positions of the tip of the index finger and of the thumb x_i^o, $x_t^o \in \Re^3$ in the object coordinate system are described as:

$$x_i^o = R(\xi_o^a)(x_i^a - x_o^a) \quad (2)$$
$$x_t^o = R(\xi_o^a)(x_t^a - x_o^a) \quad (3)$$

When a fingertip of the master arm collides with the virtual object, the penetration of the fingertip into the object is calculated. The penetration is the minimum distance between the fingertip and the object surface. Therefore, the penetration is directed toward the surface normal of the object.

For the index finger and thumb, the penetrations of the fingertips into the object in object coordinate system p_i^o, $p_t^o \in \Re^3$ are determined from the fingertip position x_i^o, x_t^o relative to the object coordinate system by the function P_o that is defined based on the geometry of the object surface as:

$$p_i^o = P_o(x_i^o) \quad (4)$$
$$p_t^o = P_o(x_t^o) \quad (5)$$

The penetration vector of the index finger and thumb relative to the object coordinate system is converted to that in the absolute coordinate system p_i^a, $p_t^a \in \Re^3$ by rotational transformation between coordinate systems as:

$$p_i^a = R^{-1}(\xi_o^a) p_i^o \quad (6)$$
$$p_t^a = R^{-1}(\xi_o^a) p_t^o \quad (7)$$

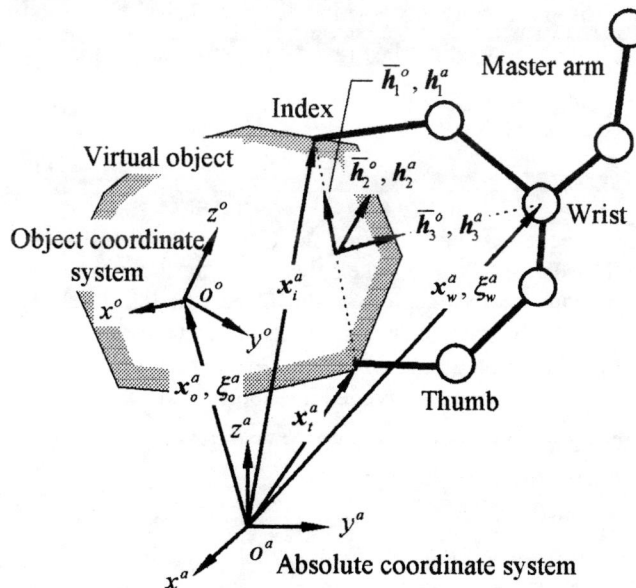

Fig. 2 Coordinate system describing the position, orientation of master arm and object.

3. Haptic display for object grasping and manipulating

The state transition diagram of the haptic display is shown in **Fig. 3**. The control consists of five states, namely FREE, INDEX CONTACT, THUMB CONTACT, GRASP and MANIPULATE. In order to create a haptic sensation of object grasping and manipulating, the posture of the operator's hand and arm is measured from the joint angles of the master arm. The joint torque is controlled according to the current state and the geometric relation between the master arm and virtual object as described next.

FREE state
In this state, no interaction exists between the operator and the object. When the tips of both the index finger and the thumb of the master arm are positioned outside the virtual object, the control is in the FREE state. No force or moment is applied to the operator while controlling the joint torques of the master arm $\tau^j \in \Re^N$ (N: Number of joints) to only counteract the gravity loading on the master arm:

$$\tau^j = \tau_g^j(\theta^j) \tag{8}$$

where $\tau_g^j \in \Re^N$ is the joint torque for gravity compensation of the master arm which is determined from the joint position $\theta^j \in \Re^N$.

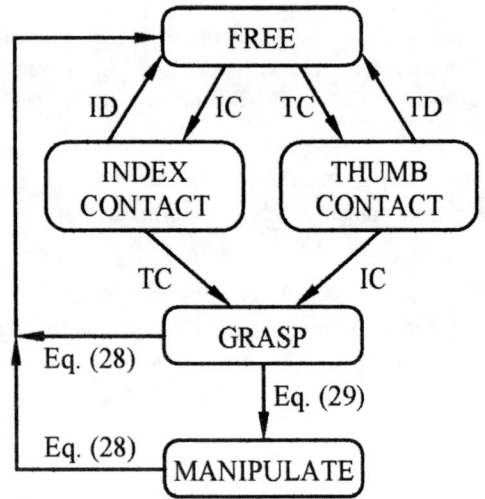

IC: Index contact on object
ID: Index detach from object
TC: Thumb contact on object
TD: Thumb detach from object
Eq. (28): Fingertip separates
Eq. (29): Reactive force counteracts the gravity

Fig. 3 State transition of the haptic display.

INDEX CONTACT, THUMB CONTACT state
In these states, the operator touches and traces the object surface with one fingertip while feeling frictionless contact. The object is still fixed in the absolute coordinate system. When the tip of the index finger moves into the internal region of the virtual object, the control transits from the FREE to the INDEX CONTACT state. Based on the contact model that contains nonlinear stiffness and damping [6], the contacting force at the index finger $f_i^a \in \Re^3$ that realizes the sensation of stable contact is exerted according to the penetration of the tip of the index finger p_i^a into the virtual object as follows:

$$f_i^a = -\left(K_c + B_c \frac{\mathrm{d}|p_i^a|}{\mathrm{d}t}\right)\sqrt{|p_i^a|}\frac{p_i^a}{|p_i^a|} \tag{9}$$

where K_c is a constant specifying the nonlinear stiffness of the contact that the force is proportional to the square-root of the penetration. On the other hand, B_c is a constant for the nonlinear damping that the force is proportional to the product of the differentiation and the square-root of the penetration. Since the contacting force is exerted in the same direction of the penetration of the fingertip that is parallel to the surface normal of the virtual object, the operator senses frictionless contact while tracing the surface of the object with the index finger.

The contact force at the index finger is converted to the corresponding joint torque of the master arm τ^j by the Jacobian matrix $J_i \in \Re^{3 \times N}$, which relates the deviation of the index fingertip position x_i^a and the joint position θ^j while appending the gravity compensation torque:

$$\tau^j = J_i^T f_i^a + \tau_g^j(\theta^j) \tag{10}$$

$$J_i = \frac{\partial x_i^a}{\partial \theta^j} \tag{11}$$

On the other hand, the state transits from FREE to THUMB CONTACT when the tip of the thumb moves into the internal region of the object. In the THUMB CONTACT state, as well in the INDEX CONTACT state, the contact force at the thumb is determined according to its penetration and converted into joint torque.

The control transits to the FREE state when the fingertip moves outside of the object.

GRASP state
In this state, both the index finger and the thumb stick on the virtual object without slip and grasp it. However, the object is constrained by virtual springs that suspend the object in the absolute coordinate system. When the other

fingertip that is separated from the object at INDEX CONTACT or THUMB CONTACT state comes into the internal region of the virtual object, the control transits to the GRASP state. At the moment of transition to the GRASP state, the position of the index finger, thumb, and wrist, and the position and orientation of the object for the absolute coordinate system is preserved as $\bar{x}_i^a$, $\bar{x}_t^a$, $\bar{x}_w^a$, $\bar{x}_o^a$, $\bar{\xi}_o^a \in \Re^3$, respectively. Also, the following vectors $\bar{h}_1^o$, $\bar{h}_2^o$, $\bar{h}_3^o \in \Re^3$ relative to the object coordinate system are calculated and preserved.

$$\bar{h}_1^o = R(\bar{\xi}_o^a) \frac{\bar{x}_i^a - \bar{x}_t^a}{|\bar{x}_i^a - \bar{x}_t^a|} \tag{12}$$

$$\bar{h}_2^o = \frac{\bar{h}_1^o \times R(\bar{\xi}_o^a)(\bar{x}_i^a - \bar{x}_w^a)}{|\bar{h}_1^o \times R(\bar{\xi}_o^a)(\bar{x}_i^a - \bar{x}_w^a)|} \tag{13}$$

$$\bar{h}_3^o = \bar{h}_1^o \times \bar{h}_2^o \qquad (|\bar{h}_1^o|=|\bar{h}_2^o|=|\bar{h}_3^o|=1) \tag{14}$$

As illustrated in Fig. 2, $\bar{h}_1^o$, $\bar{h}_2^o$ and $\bar{h}_3^o$ are orthogonal to each other. Vector $\bar{h}_1^o$ is aimed toward the tip of the index finger from that of the thumb. Vector $\bar{h}_3^o$ is in the plane that contains the tip of the index finger, thumb and wrist.

Additionally, the position of the midpoint between the fingertips relative to the object coordinate system is preserved as $\bar{x}_m^o \in \Re^3$:

$$\bar{x}_m^o = R(\bar{\xi}_o^a)\left(\frac{\bar{x}_i^a + \bar{x}_t^a}{2} - \bar{x}_o^a\right) \tag{15}$$

While in the GRASP or MANIPULATE state, the position and orientation of the virtual object are determined according to the motion of the master arm so that $\bar{h}_1^o$ to $\bar{h}_3^o$ and $\bar{x}_m^o$ is kept constant in the object coordinate system. As a result, the rotation of the object around $\bar{h}_1^o$ is artificially constrained although the object is grasped by only two fingers.

After preserving the above parameters at the transition, the position and orientation of the virtual object are determined according to the configuration of the master arm while it is moved by the operator. At first, vectors h_1^a, h_2^a, $h_3^a \in \Re^3$, which correspond to $\bar{h}_1^o$ to $\bar{h}_3^o$ but relative to the absolute coordinate system, are determined from the position of the tips of the index finger x_i^a and the thumb x_t^a and of the wrist x_w^a as:

$$h_1^a = \frac{x_i^a - x_t^a}{|x_i^a - x_t^a|} \tag{16}$$

$$h_2^a = \frac{h_1^a \times (x_i^a - x_w^a)}{|h_1^a \times (x_i^a - x_w^a)|} \tag{17}$$

$$h_3^a = h_1^a \times h_2^a \qquad (|h_1^a|=|h_2^a|=|h_3^a|=1) \tag{18}$$

The constraint that vectors h_1^a to h_3^a should coincide with $\bar{h}_1^o$ to $\bar{h}_3^o$ in the object coordinate system is expressed as:

$$R(\xi_o^a)\begin{bmatrix} h_1^a & h_2^a & h_3^a \end{bmatrix} = \begin{bmatrix} \bar{h}_1^o & \bar{h}_2^o & \bar{h}_3^o \end{bmatrix} \tag{19}$$

From this equation, the rotation matrix $R(\xi_o^a)$ between the absolute and object coordinate systems is given as:

$$R(\xi_o^a) = \begin{bmatrix} \bar{h}_1^o & \bar{h}_2^o & \bar{h}_3^o \end{bmatrix} \begin{bmatrix} h_1^a & h_2^a & h_3^a \end{bmatrix}^{-1} \tag{20}$$

Reviewing Eq. (1), the orientation of the object ξ_o^a is determined as follows, where R_{mn} represents the element of matrix $R(\xi_o^a)$ at the m-th column and n-th row.

$$\xi_o^a = \begin{bmatrix} \mathrm{atan2}(-R_{32}, R_{33}) \\ \mathrm{atan2}\left(R_{31}, \frac{R_{11}}{\cos(\mathrm{atan2}(-R_{21}, R_{11}))}\right) \\ \mathrm{atan2}(-R_{21}, R_{11}) \end{bmatrix} \tag{21}$$

The object position x_o^a is determined from the position of the fingertips x_i^a, x_t^a, the object orientation ξ_o^a, and the position of the midpoint of the fingertips $\bar{x}_m^o$ preserved at the transition to the GRASP state using Eq. (15):

$$x_o^a = \frac{x_i^a + x_t^a}{2} - R^{-1}(\xi_o^a)\bar{x}_m^o \tag{22}$$

After determining the position and orientation of the object through this process, the force and moment that apply the sensation of grasping to the operator are determined.

Since the object is constrained by virtual springs, the reactive force and moment are exerted at the wrist of the master arm. The force and moment at the wrist f_w^a, $m_w^a \in \Re^3$ relative to the absolute coordinate system are made proportional to the translation and rotation of the object from the initial condition at the transition to the GRASP state:

$$\begin{bmatrix} f_w^a \\ m_w^a \end{bmatrix} = -\begin{bmatrix} K_{ox}(x_o^a - \bar{x}_o^a) \\ K_{o\xi}(\xi_o^a - \bar{\xi}_o^a) \end{bmatrix} \tag{23}$$

In above equation, K_{ox} and $K_{o\xi}$ specify the translational and rotational stiffnesses that constrain the object motion.

Besides the force and moment exerted at the wrist, a grasping force is exerted between the tips of the index finger and the thumb in order to provide the sensation of squeezing the object. As well as the contact force exerted at the INDEX CONTACT and THUMB CONTACT states, the grasping force f_g that consists of nonlinear stiffness and damping specified by K_g and B_g, respectively, is exerted according to the distance between the fingertips $|x_i^a - x_t^a|$ as:

$$f_g = \left(K_g - B_g \frac{d|x_i^a - x_t^a|}{dt}\right)\sqrt{|\bar{x}_i^a - \bar{x}_t^a| + d_h - |x_i^a - x_t^a|} \quad (24)$$

The grasping force that varies for quasi-static change of the distance between the fingertips is illustrated in Fig. 4. When the control transits to the GRASP state at A where the fingertip distance is $|\bar{x}_i^a - \bar{x}_t^a|$, the grasping force is immediately exerted (A→B). While in the GRASP state, the grasping force increases nonlinearly as the virtual object is squeezed (B→C). In case the object is going to be released, the grasping force decreases to zero as the fingertips separate (C→B→D→E). The control transits to the FREE state at D and the object is considered to be released. Due to the hysteresis of the grasp d_h (A-D), the control will not transit from FREE to GRASP state again until the operator squeezes the fingers to A. Therefore, undesired frequent transition between FREE and GRASP state are prevented.

Both the reactive force and moment at the wrist and the grasping force at the fingers are converted to corresponding joint torques by the Jacobian matrix $J_w \in \Re^{6 \times N}$ for the motion of the wrist and $J_g \in \Re^{1 \times N}$ for the distance between fingertips as:

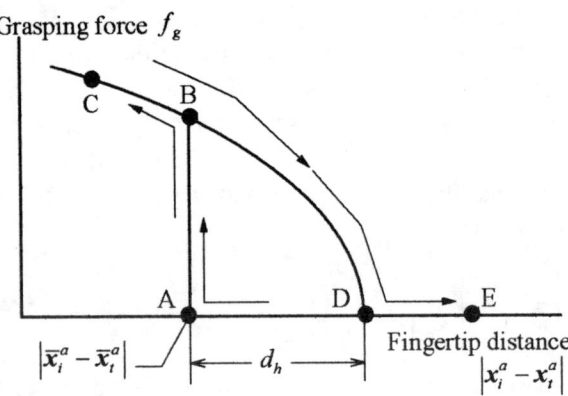

Fig. 4 Quasi-static change of grasping force.

$$\tau^j = J_w^T \begin{bmatrix} f_w^a \\ m_w^a \end{bmatrix} + J_g^T f_g + \tau_g^j(\theta^j) \quad (25)$$

$$J_w = \frac{\partial \begin{bmatrix} x_w^a \\ \xi_w^a \end{bmatrix}}{\partial \theta^j} \quad (26)$$

$$J_g = \frac{\partial |x_i^a - x_t^a|}{\partial \theta^j} \quad (27)$$

The GRASP state transits to the FREE state when the distance between the fingertips increases beyond the initial distance plus the hysterisis for grasping (D in Fig. 4) as:

$$|x_i^a - x_t^a| \geq |\bar{x}_i^a - \bar{x}_t^a| + d_h \quad (28)$$

In this case, the object returns to its initial position and orientation $\bar{x}_o^a$, $\bar{\xi}_o^a$ preserved at the transition to the GRASP state.

On the other hand, the GRASP state transits to the MANIPULATE state when the reactive force produced by the stiffness increases to counteract the gravity of the object as:

$$f_w^a \cdot g^a \leq -M_o |g^a|^2 \quad (29)$$

where $g^a \in \Re^3$, M_o represent the gravity acceleration relative to the absolute coordinate system and the mass of the virtual object, respectively. Since the gravity load of the object is applied to the operator in the MANIPULATE state as mentioned later, the vertical force applied to the operator varies continuously at the transition from GRASP to MANIPULATE states.

MANIPULATE state

In this state, the operator can freely manipulate the object without any constraint while sensing the gravity load of the virtual object. Instead of the reactive force-moment exerted by the stiffness in the GRASP state, the force-moment due to the gravity load of the object is exerted at the wrist as:

$$\begin{bmatrix} f_w^a \\ m_w^a \end{bmatrix} = \begin{bmatrix} M_o g^a \\ [x_o^a + R^{-1}(\xi_o^a)x_g^o - x_w^a] \times M_o g^a \end{bmatrix} \quad (30)$$

where $x_g^o \in \Re^3$ represents the position of the object's center of gravity relative to the object coordinate system. The inertia force-moment exerted by the object acceleration is currently omitted to simplify the control.

Similar to the GRASP state, the position and orientation of the object is determined according to the configuration of the master arm through the same process of Eqs. (16)

to (22). Also, the control transits to the FREE state and the object is located at a new position and orientation when the fingertips separate over the threshold in Eq. (28).

4. Implementation

The haptic display is implemented on the system shown in Fig. 5 consisting of the Sarcos Dexterous Arm Master, two single board computers (Motorola 68040 and PowerPC 604e) and an SGI graphics workstation. The single board computers hosted on a VME bus communicate together through shared memory.

The Sarcos Dexterous Arm Master has ten joints, each of which is equipped with a hydraulic actuator, a potentiometer for position sensor and load cell for torque sensor.

The Motorola 68040 manages the signal I/O and joint torque control of the master arm. The joint position and torque measured by sensors are acquired through 12 bit A/D converters. The acquired sensory data are written to the shared memory to be read by the PowerPC 604e. The desired joint torque is written to the shared memory by the PowerPC 604e, and the servo valves are controlled through 12 bit D/A converters so that the joint torque is set to the desired value.

The PowerPC 604e executes the main body of the haptic display. For both single board computers, the ControlShell (Real-Time Innovations, Inc.) object-oriented real-time software package that runs on VxWorks® (Wind River Systems, Inc.) real-time kernel and development environment is employed. The sampling rate of the control is 1920Hz.

The PowerPC 604e transmits the configuration of the master arm and the position and orientation of the virtual object to the SGI graphics workstation through the Myrinet local area network (Myricom, Inc.). On the workstation, the *Alpha_1* OPEN-GL viewer draws the solid model of the master arm and the object as the visual display to the operator. The transmission and refresh rate of the visual display is 32Hz, which is fast enough compared to the scanning rate of the CRT display.

On the other hand, previously recorded sounds that correspond to each transition are replayed when the control transits to a new state. This audio feedback assists the operator for recognizing the state transition while touching, grasping and manipulating the virtual object.

5. Experimental results

A model of a cylinder (diameter: 0.1m, length: 0.4m, mass: 2kg) is created as the object in the virtual environment to be grasped and manipulated. The parameters for stiffness and damping for contacting and grasping is set as:

$$K_c = 130\text{N/m}^{0.5}, \quad B_c = 100\text{Ns/m}^{1.5}$$
$$K_g = 1000\text{N/m}^{0.5}, \quad B_g = 800\text{Ns/m}^{1.5}$$

In the FREE state shown in Fig. 6 (a), the operator can freely move the hand and arm while the gravity loading on the master arm is compensated until the tip of the index finger or the thumb collides with the virtual cylinder.

In the INDEX CONTACT state as in Fig. 6 (b) and the THUMB CONTACT state as in Fig. 6 (c), the operator feels the contact on the cylinder by one fingertip. The contact is stable without any undesired vibration since sufficient damping is provided. The operator can easily recognize the shape of the cylinder by tracing its surface while receiving the contact force at the fingertip.

The control transits to the GRASP state as shown in Fig. 6 (d) when the operator contacts the cylinder with the tips of both the index finger and the thumb. The steep increase of the grasping force at the transition (A→B in Fig. 4) applies significant sensation of grasp to the operator. The operator feels as if the cylinder is suspended by translational and rotational springs since the reactive force-moment is applied at the wrist. Also the operator receives the sensation of squeezing the cylinder since the grasping force is exerted at the fingers.

As the operator lifts the cylinder, the reaction force directed downward increases as the cylinder that is constrained by the stiffness moves upward. When the reaction force increases to counteract the gravity load of the cylinder, the control transits to the MANIPULATE state. This transition applies a natural sensation to the

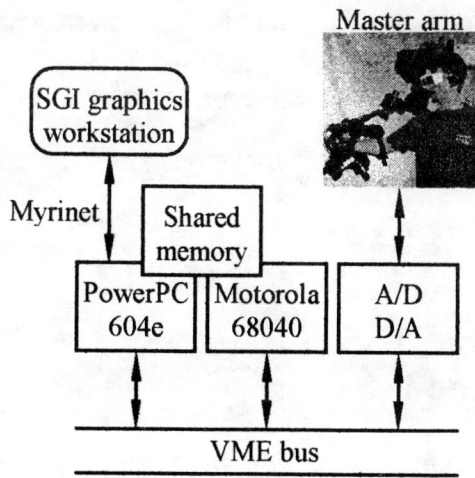

Fig. 5 Haptic-visual display system.

operator because the vertical force applied by the master arm varies continuously.

The operator can manipulate the cylinder freely in the MANIPULATE state as shown in **Fig. 6 (e)** while feeling both the gravity load of the cylinder at the wrist and the grasping force at the fingers. The operator can sense the position of the cylinder's center of gravity since the moment is determined according to it. As a result, when the operator grasps the eccentric part of the cylinder, it can be recognized and the eccentricity of the gravity load is sensed to vary while the operator rotates the grasped cylinder.

After the cylinder is manipulated to a desired position and orientation, the operator can place it there by releasing the hand as shown in **Fig. 6 (f)**. As a whole, the operator receives a natural sensation of grasping and manipulating the cylinder easily as intended.

Although the sounds played for the audio feedback are not precisely synthesized based on the mechanical model of contacting, grasping and manipulating, it assists the operator for recognizing the state transition as well as the force feedback.

6. Conclusion

In this paper, a haptic display for object grasping and manipulating is proposed as a fundamental technique for rapid prototyping in a virtual environment. The haptic display is implemented on the Sarcos Dexterous Arm Master and experimentally confirmed to provide a realistic sensation of grasping and manipulating that enables the operator to manipulate the virtual object easily as desired.

Since the final aim of this research is a realization of haptic display in a CAD environment, it is important how realistic is the haptic sensation that the operator receives. Although such a quantitative evaluation from a psychological viewpoint is not discussed in this paper, it will be investigated in the future.

In addition to the virtual grasping and manipulating achieved here, more complicated tasks such as assembling multiple parts, checking the interference, confirming the motion of the mechanism and so on would be required for advanced rapid prototyping. The realization of a haptic display capable of providing the sensation of such tasks will be a next target.

(a) FREE state at the beginning.

(b) INDEX CONTACT state.

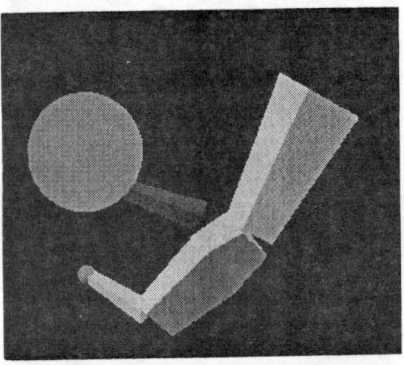
(c) THUMB CONTACT state.

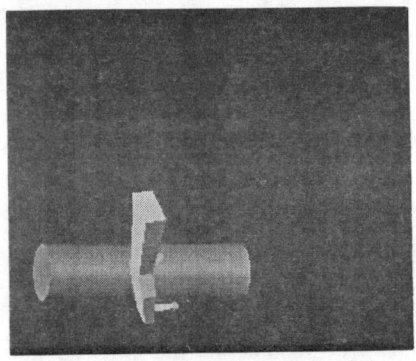
(d) GRASP state.

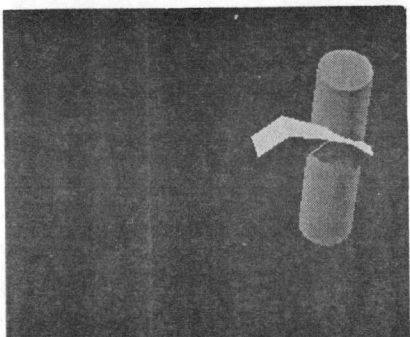
(e) MANIPULATE state.

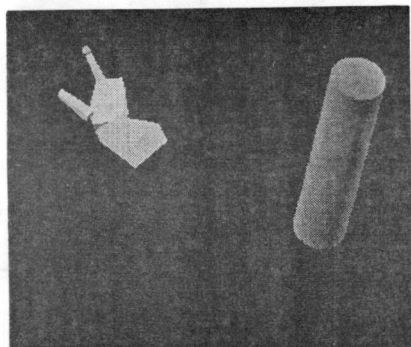
(f) FREE state at the end.

Fig. 6 Experimental grasping and manipulating of cylinder.

Acknowledgments

Support for this research was provided by NSF Grant MIP-9420352. The first author visited The University of Utah while the fellowship program provided by Science and Technology Agency, Japan. The authors sincerely thank Rodney Freier, Don Nelson and Thomas V. Thompson II, for their contribution in implementing the haptic-visual-audio display.

References

[1] Agronin, M. L., The Design of a Nine-String Six-Degree-of-Freedom Force-Feedback Joystick for Telemanipulation, Proc. of NASA Workshop on Space Telerobotics, pp. 341-348, 1987.

[2] Bergamasco, M., B. Allota, L. Bosio, L. Ferretti, G. Parrini, G. M. Prisco, F. Salsedo, and G. Sartini, An Arm Exoskeleton System for Teleoperation and Virtual Environments Applications, Proc. of IEEE Int. Conf. on Robotics and Automation, pp. 1449-1454, 1994.

[3] Brooks, F. P. Jr., M. Ouh-Young, J. J. Batter, and P. J. Kilpatrick, Project GROPE-Haptic Displays for Scientific Visualization, Computer Graphics, vol. 24, no. 4, pp. 177-185, 1990.

[4] Hollerbach, J. M., E. Cohen, W. Thompson, R. Freier, D. Johnson, A. Nahvi, D. Nelson, T. V. Thompson II, and S. C. Jacobsen, Haptic Interfacing for Virtual Prototyping of Mechanical CAD Designs, Proc. of ASME Design for Manufacturing Symp., 1997.

[5] Iwata, H., Artificial Reality with Force-Feedback: Development of Desktop Virtual Space with Compact Master Manipulator, Computer Graphics, vol. 24, no. 4, pp. 165-170, 1990.

[6] Marhefka, D. W. and D. E. Orin, Simulation of Contact Using a Nonlinear Damping Model, Proc. of IEEE Int. Conf. on Robotics and Automation, pp. 1662-1668, 1996.

[7] Yoshikawa, T. and H. Ueda, Haptic Virtual Reality: Display of Operating Feel of Dynamic Virtual Objects, Proc. of Int. Symp. on Robotics Research, pp. 214-221, 1995.

Design of a Force Reflecting Master Arm and Master Hand using Pneumatic Actuators

Sooyong Lee, Sangmin Park, Munsang Kim, Chong-Won Lee

Korea Institute of Science and Technology
Advanced Robotics Research Center
Seoul, KOREA
gemma@kistmail.kist.re.kr

Abstract

A lot of researches have been done in the teleoperation field. In order to control a robot from remote site, simple position teaching devices were developed followed by the advanced force feedback devices. Most of the force feedback master devices are as big as the slave robot and equipped with heavy actuators.

A new teleoperation device is presented in this paper. A simple exo-skeleton type master arm is designed based on the kinematic analysis of the human arm. For force reflection, pneumatic actuators are used. This device is integrated with KIST Humanoid robot as well as graphic simulator, so that the master arm can be used for teaching the robot or as the virtual reality device integrated with the graphic simulator. A force reflecting master hand was also developed with similar design to the master arm. These force reflecting master arm and hand are very light and compact so that a human can easily wear while he feels the same force as the robot does while giving position command to the slave robot.

1. Introduction

Robots don't have sufficient capabilities to perform complex tasks unless it is completely autonomous. Numerous studies have been performed over years on the robot teleoperation so that a supervisor can command the robot either from remote site monitoring the interaction of the robot with environments [Sheridan, 1992]. In addition to teaching capability, the operator can have more realistic interaction with the environment by providing feedback. There is an obvious need to use feedback from widely differing sensors in order to properly monitor various aspects of the task, particularly for complex tasks. For teleoperated object manipulation, force and vision feedback are the most important sensing modes. Vision is useful for aligning objects, while force ensures reasonable contact forces are maintained as parts mating occurs.

In force reflecting master-slave systems, forces are measured at the slave end which are transmitted to the master system. Since the pioneering work of [Goertz, 1964], who developed several teleoperation systems for nuclear application, a number of similar systems have been proposed. It has also been shown that force reflection can be applied to a rate-control joystick [Lynch, 1972]. A survey of early master hand controllers for teleoperation has been provided by [Vertut, 1977]. Using coordinate transformation, it has been shown that master and slave need not have the same kinematics if force reflection is to be used. [Corker, Bejczy, 1985] used the Salisbury/JPL master arm to show this. In recent years, the feedback of tactile and kinesthetic sensory input has been discovered for the application in virtual reality technology. The interest in force feedback systems for virtual reality applications has led to the development of many systems, ranging from force reflecting joysticks and whole force feedback arm-exoskeletons [Hill, 1995]. Most of the exoskeleton type master arms have a similar kinematic design to that of the slave arm, with actuators, usually electric motors, thus making the master arm bulky and heavy.

In this paper, we present a new design of master arm master hand. Our goal is to make it compact and light while providing sufficient position measurement and force reflection capabilities. In the following section, the design procedure is described. For the master arm, it is not necessary to have the same kinematics as the slave arm. Therefore, we reduced the degree of freedom of master arm, and the number of actuating joints. After several iterative modifications of the design, the master arm and hand design is finalized. The pneumatic actuators are used and analytic model is derived for precise force control in section 3. In section 4, the integrations of master arm/hand with KIST humanoid robot and virtual environment are described.

2. Kinematic Design of Master Arm and Hand

Our goals in designing master arm are as follows. First, the master arm should follow the motion of human arm without any resistance due to the kinematic constraints. Secondly, the number of actuating joints is to be as small as possible while compromising the possible force reflection space for compact and simple design. Finally, the actuator link parameters should be carefully chosen to satisfy the desired range of each joint. Figure 1 shows the design procedure considering these three goals. If the final design doesn't satisfy our goal, previous parameters selections, which are actuator link parameters, actuating joints, or kinematic parameters of the arm, should be modified.

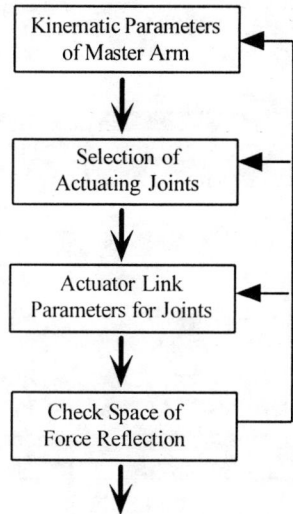

Figure 1. Design Procedure

The selection of parameters of master arm is mainly based on the analysis of human arm movement. The base of this device is to be attached to the back of operators. Our final design has 7 rotational joints and 2 prismatic joints excluding wrist as shown in Figure 2.

D-H notation of this kinematics is shown in Table 1. These are decided based on the analysis of human arm.

Link	θ	d	a	α
1	θ_1	0	0.12	0
2	θ_2	0	-0.009	$\frac{\pi}{2}$
3	θ_3	0	0	$-\frac{\pi}{2}$
4	0	d_4	0	$\frac{\pi}{2}$
5	θ_5	0	0	$-\frac{\pi}{2}$
6	θ_6	0	0	$\frac{\pi}{2}$
7	θ_7	-0.002	0	$-\frac{\pi}{2}$
8	θ_8	0	0	$\frac{\pi}{2}$
9	0	d_9	0	$-\frac{\pi}{2}$

Table 1. D-H Parameters of Master Arm

The desired range of each joint has been based on human motion. For instance, we neglected the movement of joint 7 because twisting of lower arm is not common, and human usually twists his wrist. Joint 6 is neglected because this motion is not usually possible. Two prismatic joints are used for joint 4 and joint 9. Human doesn't have this axis movement, but these prismatic joints are used to constrain motion taking advantage of the offsets from the arm. Both for rotational and prismatic joints, the same type of pneumatic actuator is used. For the prismatic joint, the actuator is attached in parallel, and for a rotational joint, the actuator's linear force is converted to the joint torque using the link mechanism as shown in Figure 3. The range of desired joint rotation angle is to be considered in designing these links.

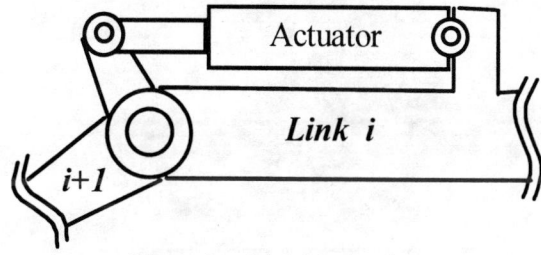

Figure 3. Link Mechanism

These selected parameters are verified using graphic simulator as shown in Figure 4.

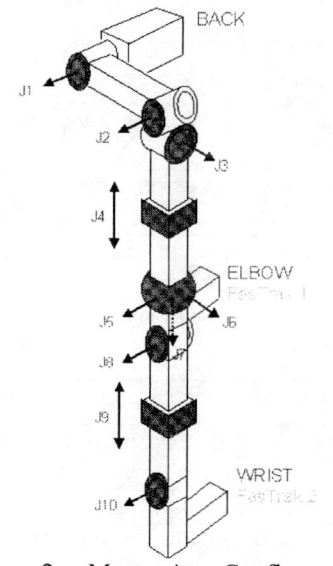

Figure 2. Master Arm Configuration

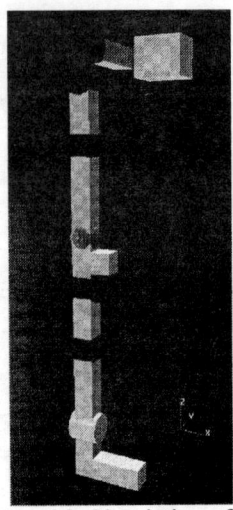

Figure 4. Graphic Simulation of Master Arm

With the master arm's kinematic parameters and link parameters selected, force analysis is done from eq. (1)

$$\tau = J^t(\theta)F \quad (1)$$

where τ is joint space torque, and F is force in Cartesian space. Jacobian matrix, J is calculated from D-H parameters in Table 1. Due to the fewer number of actuators than the number of joints, the force field includes a null space depending on the configuration θ and actuating joints. The desired design goal is to have a uniformly distributed force reflection space. Based on the force analysis, design parameters are modified and the final design is shown in Figure 5.

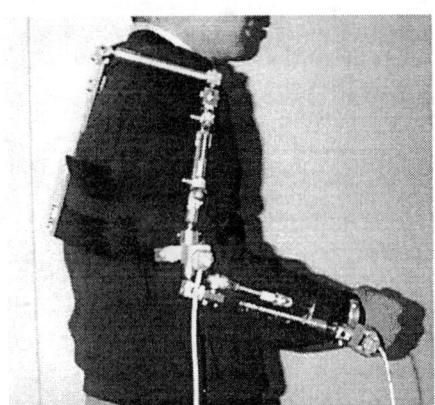

Figure 5. Master Arm with Position Sensor

Figure 5 shows two position sensors attached at the elbow and close to the wrist. From these three dimensional position/orientation sensors, the exact configuration of the master arm is provided.

The master hand has simpler design than that of the master arm. Currently, it is composed of three fingers, each of which has two degrees-of-freedom (DOF). Fingers have rotations in the same plane without base rotation. Human can feel virtual volume in his hand with force reflection from 3x2 DOF master hand. Figure 6. shows one 2 DOF finger attached to the human finger. The potentiometer is used for measurement of each joint's angle.

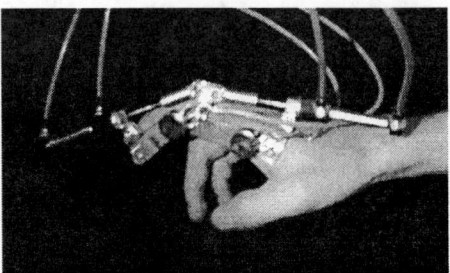

Figure 6. A Master Finger

The same number of actuators as DOF are used for the master hand with simple two link mechanism. However, the link parameters are chosen based on graphic simulation as the master arm.

3. Control of the Pneumatic Actuators

Unlike electric motor, pneumatic actuators have nonlinear characteristics. An analytic model is derived for the pressure control valve and cylinder, and then this model is linearized and simplified. Parameters for one of the valves and cylinders used for the master arm are shown in table 2.

M_v	mass of pressure control valve	$0.05\ kg$
B_v	damping constant of pressure control valve	$40.04\ Nm/sec$
K_v	spring constant of pressure control valve	$89090\ N/m$
R	air constant	0.287
k	air specific heat constant	1.4
T_1	temperature inside cylinder	$300K$
K_g	flow gain of pressure control valve	$-2.5\ kg/m$
K_c	flow pressure gain of pressure control valve	0.0
A_v	area of pilot at pressure control valve	$4.714 \times 10^{-5}\ m^2$
A_p	cross section area of cylinder	$7.85 \times 10^{-3}\ m^2$
V	volume of cylinder chamber	$3.9 \times 10^{-5}\ m^3$
M_p	mass of cylinder spool	$0.5\ kg$
B_p	damping constant inside cylinder	$5.8\ Nm/sec$

Table 2. Pneumatic System Parameters

The analytic model of the pressure control valve is

derived, followed by that of the pneumatic cylinder. From air dynamics, the mass flow rate of air through orifice, $\dot{m}$ is

$$\dot{m} = cA_o(x_v)\sqrt{\tfrac{2k}{k+1}P_s\rho_1\left[\left(\tfrac{P_1}{P_s}\right)^{\tfrac{2}{k}} - \left(\tfrac{P_1}{P_s}\right)^{\tfrac{k+1}{k}}\right]} \quad (2)$$

where c is discharge constant, P_1 is outlet pressure and P_s is supply pressure.

$$A_o(x_v) = \omega(x_{v.\max} - x_v) \quad (3)$$

where ω is the area gradient. Specific heat ratio k is defined as

$$k = \tfrac{C_p}{C_v} \quad (4)$$

and usually 1.4 for air. Dynamic model of valve spool is

$$M_v\ddot{x}_v + B_v\dot{x}_v + K_v x_v = P_1 A_v - K_v x_o \quad (5)$$

where x_o is spring preload. Assuming the inlet air flow rate $\dot{m}_1$ is equal to $\dot{m}$

$$\dot{m}_1 = \dot{m} = c\omega(x_{v.\max} - x_v)\sqrt{\tfrac{2k}{k+1}P_s\rho_1\left[\left(\tfrac{P_1}{P_s}\right)^{\tfrac{2}{k}} - \left(\tfrac{P_1}{P_s}\right)^{\tfrac{k+1}{k}}\right]} \quad (6)$$

Modeling of pneumatic cylinder is as follows. Applying continuity equation to the control volume where air flows in,

$$\dot{m}_1 = \frac{d}{dt}(\rho_1 V_1) \quad (7)$$

for ideal gas

$$\rho_1 = \frac{P_1}{RT} \quad (8)$$

from eq. (7) and eq. (8),

$$\dot{m}_1 = \frac{1}{R}\frac{d}{dt}\left(\frac{P_1 V_1}{T_1}\right) \quad (9)$$

Applying energy conservation law,

$$-C_p T_1 \dot{m}_1 - P_1 \tfrac{dV_1}{dt} + \tfrac{dQ_1}{dt} = \tfrac{d}{dt}(C_v \rho_1 V_1 P_1 T_1) \quad (10)$$

where C_p is constant pressure specific heat constant, and C_v is constant volume specific heat constant. From eq. (9) and eq. (10),

$$\tfrac{C_v}{R}\tfrac{d}{dt}(P_1 V_1) = C_p T_1 \dot{m}_1 - P_1 \tfrac{dV_1}{dt} + \tfrac{dQ_1}{dt} \quad (11)$$

Similarly, for the control volume where air flow out

$$\tfrac{dP_1}{dt} = \tfrac{k}{V_1}(T_1 R \dot{m}_1 - P_1 \tfrac{dV_1}{dt}) \quad (12)$$

$$\tfrac{dP_2}{dt} = \tfrac{k}{V_2}(-T_2 R \dot{m}_2 - P_2 \tfrac{dV_2}{dt}) \quad (13)$$

mass flow rate from the cylinder due to the pressure difference P_1 and P_2 is

$$\dot{m}_2 = cA_o\sqrt{\tfrac{2k}{k+1}P_2\rho_2\left[\left(\tfrac{P_1}{P_2}\right)^{\tfrac{2}{k}} - \left(\tfrac{P_1}{P_2}\right)^{\tfrac{k+1}{k}}\right]} \quad (14)$$

and force generated is

$$F = A_p(P_1 - P_2) - B\dot{x}_p = M\ddot{x}_p \quad (15)$$

using parameter values shown in table 2, and linearizing around operating range, the transfer function between displacement input to the pressure control valve to output force is,

$$G(s) = \frac{N(s)}{D(s)} \quad (16)$$

where,

$$N(s) = 1.15 \times 10^{-5} s(4.7 \times 10^{-4} s - 2.5)(0.5s^2 + 5.8s + 378.6)$$
$$D(s) = (5.7 \times 10^{-4} s^2 + 0.34s + 570) \times$$
$$(2.7 \times 10^{-9} s^5 + 3.5s^4 + 0.35s^3 + 58s^2 + 741s + 1.3 \times 10^5)$$

Pressure is controlled by changing displacement of the pressure control valve, and this displacement is controlled by the position controller. Including this position controller and PD pressure controller, and neglecting small higher order terms, the total transfer function can be represented as the second order system. The closed loop frequency response is shown in figure 7.

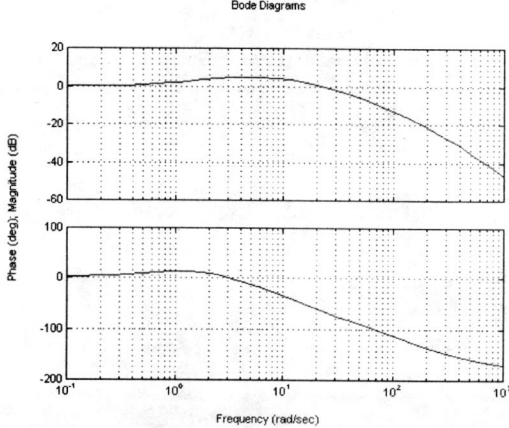

Figure 7. Frequency Response of Closed Loop System

From experiments, the force response to step command is shown in Figure 8.

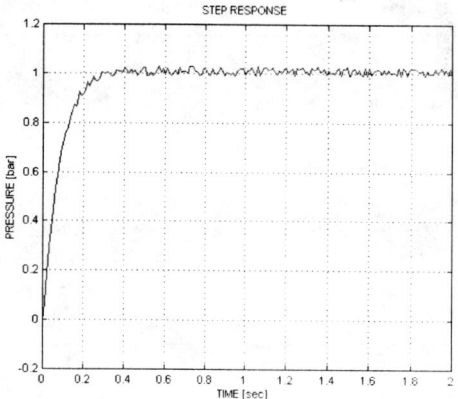

Figure 8. Measured Force Step Response

This force is converted to joint level torque through link mechanism.

4. Integration with KIST Humanoid Robot and Virtual Environment

These master arm and master hand are integrated with real robot and virtual environment. Each device is attached to the operator and trajectory commands are generated as the operator moves his arm and hand. At the same time, force information is fed back to the master arm/hand controller for force reflection. Therefore these devices are to be used for position teaching with force reflection.

KIST is developing Humanoid robot which has two arms, two hands, four legs, and vision as shown in Figure 9.

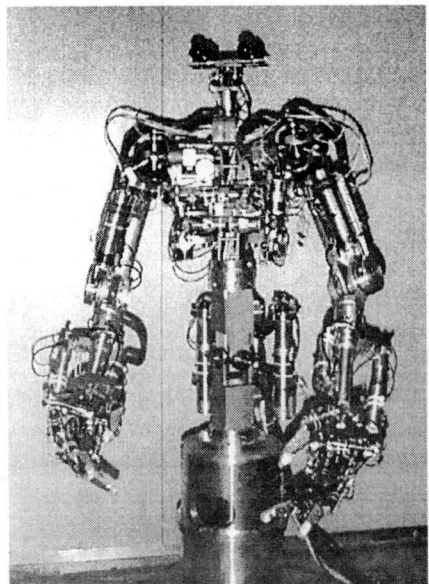

Figure 9. KIST Humanoid Robot

Each arm has 9 degrees of freedom (DOF), and each hand 10 DOF as well as 3 DOF for waist and 2 DOF for its neck. Each joint is driven by linear or rotary actuator with joint force/torque sensors. This design is one of the closest to human. The position controller is made for this robot, and can be easily integrated with master devices. Figure 10. shows information flow between the robot and master arm/hand. As human moves his arm, current position in Cartesian space is measured and inverse kinematics routine solves joint space command which are fed to the robot position controller. For master hand, the measured 6 joint angles are converted to Cartesian position and then fed to robot hand controller. Force imposed to the robot is measured using Force/torque sensor attached at its wrists or torque sensors at each joint. This force is converted to the force command to master arm/hand from Eq. (1).

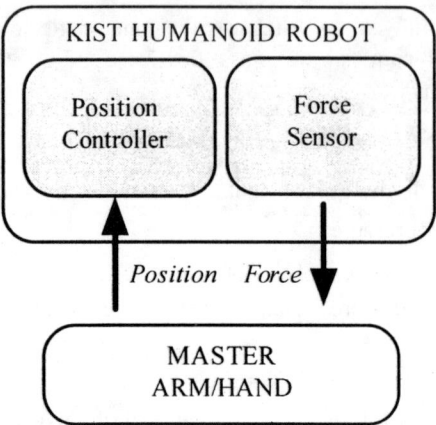

Figure 10. Integration with Robot

This master arm/hand are integrated with virtual environment using graphics, too. The master arm, and hand are modeled graphically as shown in Figure 11, and Figure 12 on Silicon Graphics Workstation. This graphics simulator is interfaced with master arm/hand controller. In this case, we can freely design the force as the virtual robot interacts with environment. For example, as the robot contacts with a wall, the force is calculated from the wall stiffness once it hits so that the human who wears the master arm feels such as he contacts a wall. Similarly, force reflection can make the human can feel as if he grasps an objects in his hand with a particular shape.

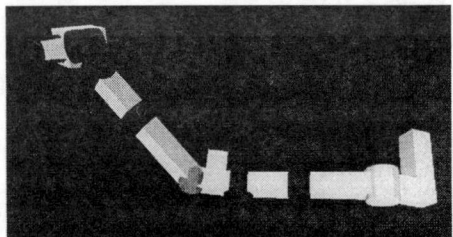

Figure 11. Graphic Display of Master Arm

Figure 12. Graphic Display of Master Hand

Figure 13. shows the similar integration with the robot. The position and force information is exchanged between the master arm/hand controller and virtual environment. Position controller is replaced with the graphic display and force sensor with force command generated based on desired environment.

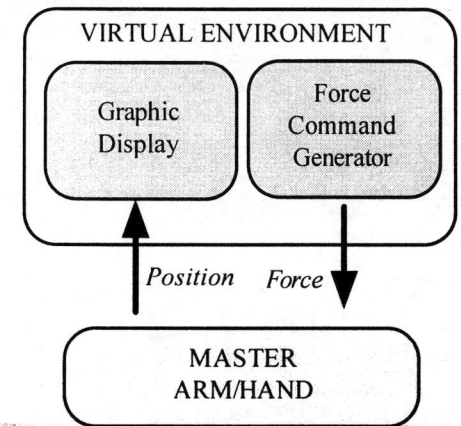

Figure 13. Integration with Virtual Environment

5. Conclusion

In this paper, a new design for a force reflecting master arm and master hand is presented. Somewhat different from conventional master devices, pneumatic actuators are used to make them very light and compact. Iterative optimal design procedures are done based on the kinematic analysis and graphic simulation so that the only minimum number of actuators are to be used. The closed loop controller is developed for the force control based on analytic model of the pneumatic system. This devices are interfaced with KIST humanoid robot as well as graphic virtual environment. This can be used as a teaching device for the precise robot teleoperation with force reflection and as a virtual environment device integrated with graphics.

References

[Corker, Bejczy, 1985] K. Corker, A.K. Bejczy, "Recent advances in telepresence technology development", Prceedings of 22nd Space Congress, Kennedy Space Center, 1985

[Goertz, 1964] R. Goertz, "Manipulator Systems Development at ANL", proc. 12th Conf. On Remote Systems Technology, American Nuclear Society, vol. 12, 1964

[Hill, 1995] J.W. Hill, J.F. Jensen, P.S. Green, A.S. Shah, "Two-Handled Telepresence Surgery Demonstration Systems", Proc. ANS Sixth Annual Topical Meeting on Robotics and Remote Systems, 1995

[Lynch, 1972] P.M. Lynch, "Rate Control of Remote Manipulators with Force Feedback", SM Thesys, MIT, 1972

[Salisbury, 1981] J.K. Salisbury, "Articulated hands : Force control and Kinematic Issues", Joint Automatic Control Conference, 1981

[Sheridan, 1992] T.B. Sheridan, "Telerobotics, Automation and Human Supervisory Control", The MIT Press, Cambridge, MA

[Vertut, 1977] J. Vertut, "Control of Master Slave Manipulators and Force Feedback", Proc. Conference on Joint Automatic Control, 1977

Design of a Compact 6-DOF Haptic Interface

Y. Tsumaki*, H. Naruse*, D. N. Nenchev** and M. Uchiyama*

*Dept. of Aeronautics and Space Eng.
Graduate School of Engineering
Tohoku University
Aramaki-aza-Aoba, Aoba-ku,
Sendai 980-8579, Japan

**Dept. of Mechanical and Production Eng.
Faculty of Engineering
Niigata University
8050 Ikarashi 2
Niigata 950-2181, Japan

Abstract

In this paper we propose a new compact 6-DOF haptic interface with large workspace. It contains a newly developed five bar spatial gimbal mechanism for orientation, placed on a modified Delta parallel-link mechanism. The motion range of each axis of the five bar mechanism is over ±70 degrees. Quick motions can be realized easily due to the parallelism inherent to both the modified Delta substructure and the five bar substructure.

1. Introduction

Recently, the application of virtual reality is expanding fast in various areas: medical, teleoperation, welfare, amusement, etc. Virtual reality is based mainly on three techniques: vision (computer graphics), force and haptic display, and audio. Computer graphics has made rapid progress owing to the remarkable achievements of computer technology during the last decade. On the other hand, progress in the field of force and haptic displays is much slower. It should be noted that high-quality virtual reality can be achieved only through integration of the different techniques, matured to a considerable level. This means that the development of high-performance haptic displays is highly desirable.

The history of haptic interfaces dates back to the 1950s, when a master-slave system was proposed by Goertz [1]. Since then, a number of master devices have been developed. The main desirable characteristics are quick response and a wide workspace. In addition, compactness is also desirable, since the haptic interface should be treated in a way similar to a conventional computer mouse.

An exoskeleton type master arm is one possible solution for displaying arm motion within a large workspace. However, practical realization is difficult [2], and besides, one cannot expect quick motion. Quick motion could be achieved with a parallel mechanism since the mass of the end-effector can be reduced [3]. PHANToM, developed by Massie and Salisbury, can be considered one of the most excellent haptic displays [4]. Unfortunately, the three DOF of the device are not enough to display the real phenomenon. In addition, hard contact feeling cannot be achieved with this device. The problem of hard contact dis-

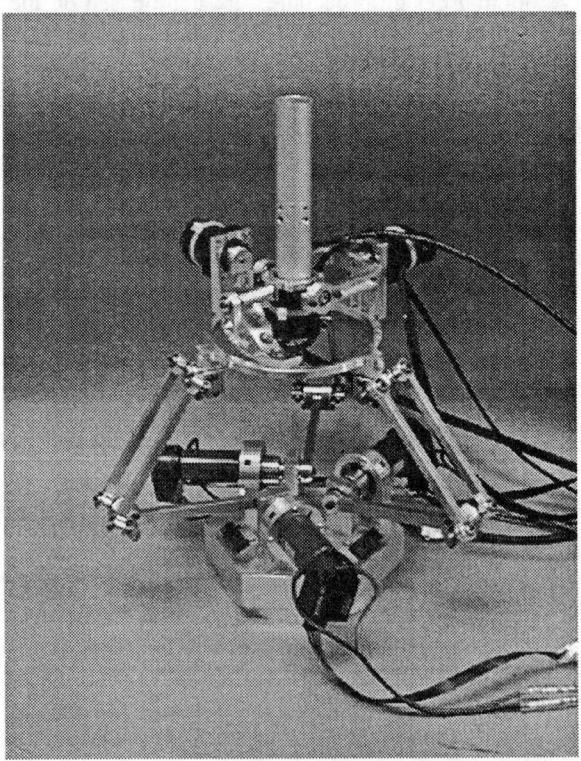

Figure 1. Overview of the haptic interface.

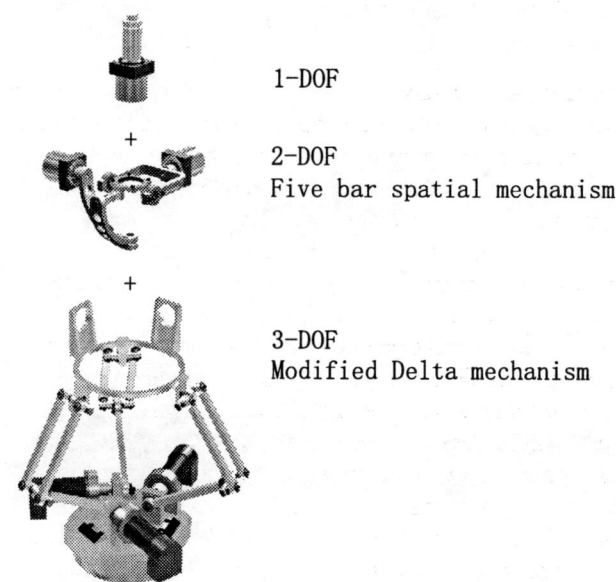

Figure 2. Distribution of degrees of freedom.

play can be alleviated by employing the nonholonomic haptic display principle proposed by Colgate et al. [5]. Devices with higher degrees of freedom based on this principle would be, however, difficult to realize.

In terms of quick motion, the magnetic levitation haptic interface of Berkelman et al. [6] is very appealing, but its workspace is small. Quick motion can be also achieved through a wired system, e.g. SPIDER [7]. Such devices occupy, however, a large space. It is possible to built a compact haptic device with the help of a 6-DOF parallel mechanism, as shown by Long [8]. But in general, fully 6-DOF parallel mechanisms (e.g. Stewart Platform [9], Pantograph Linkage [10], HEXA [11]) are characterized with restricted workspace for orientation. To increase the workspace, parallel mechanisms with kinematic redundancy can be employed [12], [13].

Concluding this brief analysis we note that, although a variety of haptic interfaces have been developed so far, a high-performance, compact 6-DOF device is still missing. In this paper, we propose such a device characterized by a relatively large workspace, quick motion and compactness. In addition, our device should be able also to realize hard contact feeling, using its mechanical characteristics.

2. Design of the New Haptic Interface

An ideal haptic interface should:

1. be able to display motion in 6-DOF,
2. be able to realize quick motion,
3. have a large enough workspace, and
4. be compact.

The first requirement is needed to display rigid body motion, not only the motion of a particle. The second one is needed to display mainly a free motion of the object. In addition, the ability of quick motion yields also a possibility of texture representation. We note that the texture of an object can be represented by vibration [14]. The third requirement is needed to avoid saturation in terms of position/orientation. The final requirement is needed to satisfy the desktop environment constraints. Furthermore, a compact haptic interface has the potential of a new input/output device which can determine a new and appealing way of interaction between computer and human.

To realize such a new haptic interface, we decided to use parallel mechanisms. A parallel mechanism has the ability of high-speed, high-power and high-precision motion. Especially, the HEXA type parallel robot can realize very fast motions [11], and is quite suitable for compact design. However, a 6-DOF parallel mechanism has a lot of singularities in its workspace. As a result, its workspace for orientation is too small for a haptic interface. Furthermore, we should point out that its kinematics are really complicated. The direct kinematics has not been solved analytically yet.

The above reasons lead us to adopt a hybrid parallel mechanism with decoupled substructures for position and orientation. We use a 3-DOF modified Delta mechanism as a positioning subsystem, and develop a five bar orientation subsystem. A significant advantage of the Delta mechanism is the lack of singularities in its workspace [15]. Furthermore, the mechanism can realize fast and high-power motions within a wide workspace. On the other hand, the orientation subsystem we propose consist of a 1-DOF yaw axis mechanism and a 2-DOF parallel gimbal mechanism for roll and pitch. The latter is a five bar spatial mechanism which does not have any singularities within the workspace. The details of the mechanism will be addressed in the next section.

To realize the compactness requirement, we use motors with high-reduction harmonic gears, without any backlash. A possible concern here is the lower stiffness due to the harmonic gear. This, however, can be compensated by the higher stiffness inherent to the parallel structure. In addition, a compact six axis force/torque sensor will be used to compensate the non-backdrivability feature of the high-reduction gears. The weight of this sensor, developed by BL auto tech Ltd. (the NANO sensor), is just 70 g. There is

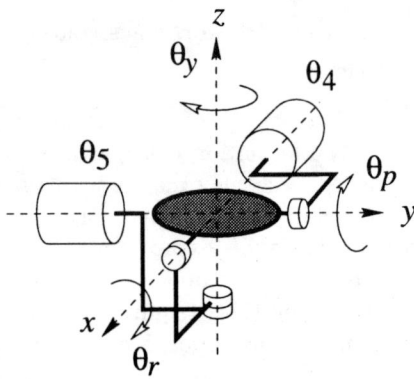

Figure 3. Five bar spatial gimbal mechanism.

an additional advantage in using the sensor. Without a force/torque sensor, a special isotropic design of the haptic interface would be needed to ensure uniform force controllability in each direction. Unfortunately, it is difficult to design a 6-DOF isotropic mechanism with good backdrivability. Thus, using a force/torque sensor in this case seems to pay off. Finally, to ensure the desired hard contact feeling ability, we will employ a recently proposed method [16] which makes effective use of hardware friction.

An overview of the new haptic interface and the distribution of its degrees of freedom are shown in Fig. 1 and 2, respectively. We consider the following target specification for the new haptic interface:

- 6 degrees of freedom,
- wide workspace (positioning: sphere with 75 mm radius, orientation: more than ±70 degree for each axis),
- presentability of high frequency vibration (amplitude: 0.1 mm, frequency: 100 Hz),
- maximum end-effector forces: more than 10 N,
- as compact as possible.

3. The Five Bar Spatial Gimbal Mechanism

The orientation subsystem should be based on a compact and light mechanism which can realize quick motion within a wide workspace. Especially, it should be as light as possible, because the weight of the subsystem influences significantly the performance of the positioning subsystem. The "Agile Eye" mechanism is a well known parallel mechanism for 3-DOF orientational motion [17]. This mechanism has enough workspace (±70 degrees) for the roll and pitch axes, but the range of the yaw axis is just about ±30 de-

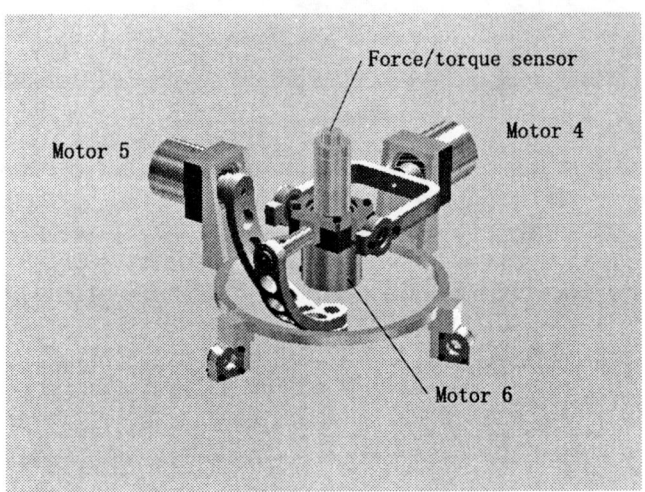

Figure 4. Some details of the orientation subsystem.

grees. On the other hand, a conventional gimbal mechanism has a wide workspace. Unfortunately its ability for quick motion is restricted due to the serial link arrangement. In addition, such a mechanism would become too heavy.

To satisfy the requirements, we adopt a hybrid orientation subsystem. A 1-DOF yaw axis and a five bar spatial gimbal mechanism are combined. Figure 3 shows the kinematic structure of the five bar spatial gimbal mechanism. The main features of this system are described as follows:

- light weight can be achieved;
- wide workspace can be established for each axis;
- quick motion, high power and high stiffness are expected;
- the kinematics are quite simple.

The direct kinematics of this mechanism are:

$$\begin{pmatrix} \theta_r \\ \theta_p \end{pmatrix} = \begin{pmatrix} \theta_4 \\ \mathrm{atan2}(s_5, c_4 c_5) \end{pmatrix}, \qquad (1)$$

where θ_r and θ_p denote the roll and the pitch angle, respectively, $s_i = \sin\theta_i, c_i = \cos\theta_i$, θ_i is the joint angle of motor i. The Jacobian matrix and its determinant can be obtained as:

$$\boldsymbol{J} = \begin{pmatrix} 1 & 0 \\ \dfrac{s_4 s_5 c_5}{s_5^2 + c_4^2 c_5^2} & \dfrac{c_4}{s_5^2 + c_4^2 c_5^2} \end{pmatrix}, \qquad (2)$$

and

$$\det \boldsymbol{J} = \frac{c_4}{s_5^2 + c_4^2 c_5^2}, \qquad (3)$$

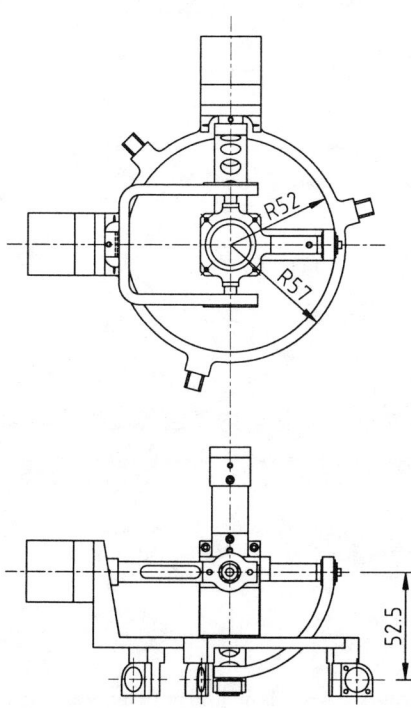

Figure 5. Assembly drafting of the orientation subsystem.

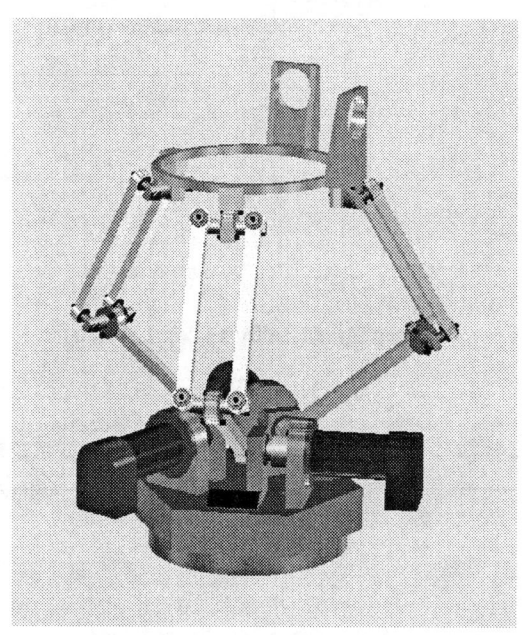

Figure 6. Overview of the modified Delta mechanism.

respectively. It is clear that this mechanism becomes singular when $c_4 = 0$. Namely, the singularity appears at $\theta_4 = \pm\frac{\pi}{2}$. However, this singularity is placed out of the feasible workspace. Obviously, the calculation cost of the kinematics is very low. We note also that the motion range of the yaw axis is unlimited since it is totally decoupled from the assembly of the five bar mechanism. If we consider the wiring of the force/torque sensor, then the motion range of each axis would be about ±70 degrees. To realize the requirement for low weight, we use very compact motors developed by Yasukawa Electric corporation, specially for use in robotic fingers. The weight of each motor is 70 [g], the harmonic gear (gear ratio: 1/80) inclusively. The nominal torque and the maximum velocity are 0.71 [Nm] and 22.5 [rpm], respectively. Details of the orientation subsystem are shown in Fig. 4.

We must note that the joint axes of the mechanism have to intersect at one point (at the center of the gimbal mechanism). To satisfy this requirement, the base of the mechanism was machined out of a single material block. Furthermore, adjustment mechanisms are also provided. The assembly drawing of this system is shown in Fig. 5.

4. The Modified Delta Mechanism

The modified Delta mechanism is used as a positioning subsystem (Fig. 6). The conventional Delta mechanism incorporates ball joints, thus the structure is simple and of light weight. For our purpose, however, these advantages of the ball-joint based design are not so significant. One reason is that the weight of a ball joint is not as small, when compared to the total weight of a compact haptic interface. Furthermore, the mechanical limits in the motion range of the ball joints restrict the workspace significantly. Therefore, we will employ conventional bearings in the passive joints. Usually, the angle of inclination of a ball joint is ±25 degree. On the other hand, a bearing system can incline about ±70 degree. This difference is illustrated in Fig. 7. Furthermore, we modified arrangement of freedoms to decrease the number of bearings (see Fig. 8). In this mechanism, each part of the rods must be a planar parallel link mechanism. As a result, a wide workspace can be obtained with a fewer number of bearings. The disadvantage of this mechanism is that each rod is subjected to torsion. Hence, the rods must be designed to withstand such torsion. Note that in the case of the conventional Delta the rods are not subjected to torsion. Nevertheless, we prefer to employ the modified Delta mechanism described above, due to its wide workspace. The workspace of our positioning subsystem and the assembly drawing are shown in Figs. 9 and 10, respectively.

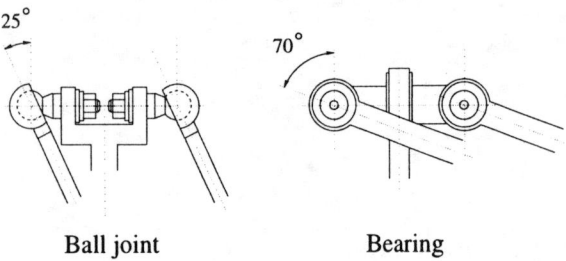

Figure 7. Illustrating the difference in the mechanical joint limits.

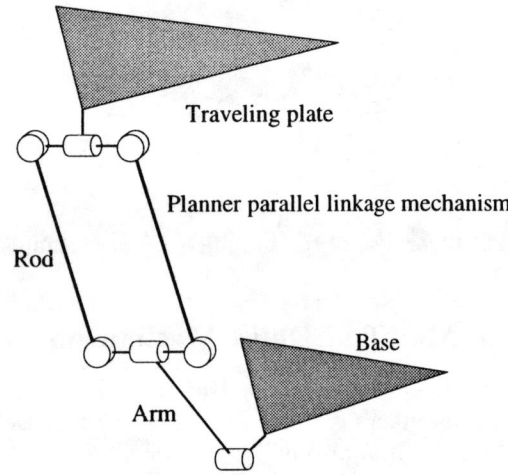

Figure 8. Arrangement of freedom for modified DELTA mechanisms.

5. Conclusion

A new compact haptic interface mechanism for high performance is designed. This mechanism ensures wide workspace, quick motion and compactness. Especially, the workspace for orientation is quite large. To realize such a system, we designed a five bar spatial gimbal mechanism and a modified Delta mechanism. Our next goal is to evaluate the performance of the device in a real teleoperation system.

Acknowledgment

The authors wish to express their gratitude to Mr. Abe, Mr. Yaegashi and Mr. Kawabata, technical staff of Tohoku University, for their great help in producing the haptic interface. This research was supported by the Grant-in-aid for Encouragement of Young Scientists from the Ministry of Education, Science, Sports and Culture (#09750283).

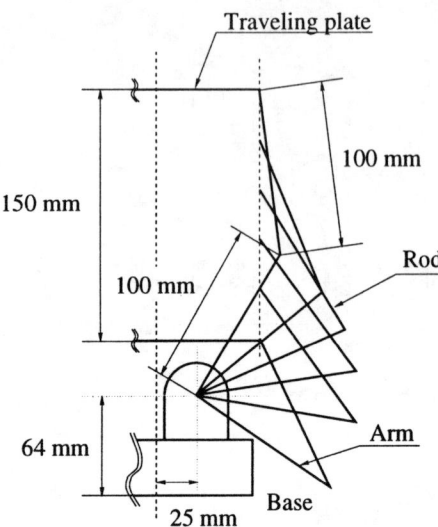

Figure 9. Workspace of the modified Delta mechanism.

References

[1] R. C. Goertz, "Fundamentals of General-Purpose Remote Manipulators," *Nucleonics*, Vol. 10, No. 11, pp. 36–42, 1952.

[2] M. Bergamasco, et al., "An Arm Exoskeleton System for Teleoperation and Virtual Environments Applications," *Proc. 1994 IEEE Int. Conf. on Robotics and Automation*, San Diego, California, pp. 1449–1454, 1994.

[3] R. Hui et al., "Mechanisms for Haptic Feedback," *Proc. 1995 IEEE Int. Conf. on Robotics and Automation*, Nagoya, Japan, pp. 2138–2143, 1995.

[4] T. H. Massie and J. K. Salisbury, "The PHANToM Haptic Interface: A Device for Probing Virtual Objects," *Proc. of the 1994 ASME Int. Mechanical Engineering Exposition and Congress*, Chicago, Illinois, pp. 295–302, 1994.

[5] J. E. Colgate, M. A. Peshkin and W. Wannasuphoprasit, "Nonholonomic Haptic Display," *Proc. 1996 IEEE Int. Conf. on Robotics and Automation*, Minneapolis, Minnesota, pp. 539–544, 1996.

[6] P. J. Berkelman, R. L. Hollis and S. E. Salcudean, "Interacting with Virtual Environments Using a Magnetic Levitation Haptic Interface," *Proc. of the IEEE/RSJ Int. Conf. on Intelligent Robots and Systems*, Pittsburgh, Pennsylvania, pp. 117–122, 1995.

[7] Y. Hirata and M. Sato, "3-Dimensional Interface Device for Virtual Workspace," *Proc. of the IEEE/RSJ Int. Conf. on Intelligent Robots and Systems*, Raleigh, North Carolina, pp. 889–896, 1992.

[8] G. L. Long and C. L. Collins, "A Pantograph Linkage Parallel Platform Master Hand Controller for Force-Reflection," *Proc. 1992 IEEE Int. Conf. on Robotics and Automation*, Nice, France, pp. 390–395, 1992.

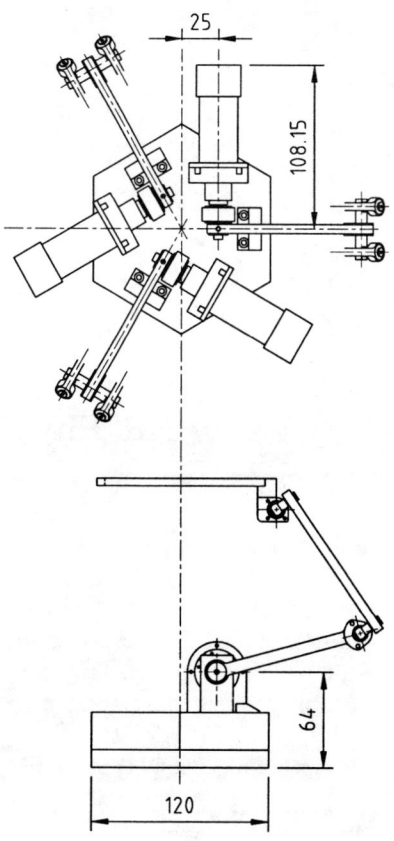

Figure 10. Assembly drafting of the positioning subsystem.

[9] D. Stewart, "A Platform with Six Degrees of Freedom," *Proc. of the Institution of Mechanical Engineers 1965–1966*, Vol. 180, Part 1, No. 15, pp. 371–386, 1965.

[10] H. Inoue, Y. Tsusaka and T. Fukuizumi, "Parallel Manipulator," *Robotics Research, The Third International Symposium*, O. Faugeras and G. Giralt (Eds), The MIT Press, pp. 321–327, 1986.

[11] F. Pierrot, M. Uchiyama, P. Dauchez and A. Fournier, "A New Design of a 6-DOF Parallel Robot," *J. of Robotics and Mechatronics*, Vol. 2, No. 4, pp. 308–315, 1991.

[12] T. Asano, H. Yano and H. Iwata, "Basic Technology of Simulation System for Laparoscopic Surgery in Virtual Environment with Force Display," *Medicine Meets Virtual Reality, Global Healthcare Grid, Studies in Health Technology and Informatics, IOS Press*, Vol. 39, pp. 207–215, 1997.

[13] V. Hayward, "Toward a Seven Axis Haptic Device, " *Proc. of the IEEE/RSJ Int. Conf. on Intelligent Robots and Systems*, Pittsburgh, Pennsylvania, pp. 133–139, 1995.

[14] Y. Ikei, K. Wakamatsu and S. Fukuda, "Texture Presentation by Vibratory Tactile Display," *Proc. 1997 IEEE Virtual Reality Annual Int. Symp.*, Albuquerque, New Mexico, pp. 199–205, 1997.

[15] R. Clavel, "Une Nouvelle Structure de Manipulateur pour la Robotique Légère," *APII*, Vol. 23, pp. 501–519, 1989 (in French).

[16] Y. Tsumaki and M. Uchiyama, "A Model-Based Space Teleoperation System with Robustness against Modeling Errors," *Proc. 1997 IEEE Int. Conf. on Robotics and Automation*, Albuquerque, New Mexico, pp. 1594–1599, 1997.

[17] C. M. Gosselin and J. F. Hamel, "Development and Experimentation of a Fast Three-Degree-of-Freedom Spherical Parallel Manipulator," *Proc. of the First World Automation Congress*, Maui, Hawaii, TSI Press, Vol. 2, pp. 229–234.

Force Display System Using Particle-Type Electrorheological Fluids

Masamichi SAKAGUCHI   Junji FURUSHO

Department of Computer-Controlled Mechanical Systems,
Graduate School of Engineering, Osaka University
2-1 Yamadaoka, Suita, Osaka 565-0871, JAPAN
E-mail :{saka, furusho}@mech.eng.osaka-u.ac.jp

Abstract

We developed ER actuators with low inertia. ER actuator is a torque-controllable clutch which uses an electrorheological fluid. It is shown that this actuator has good properties for force display device, physical therapy treatment, etc. Then, we developed new force display devices for virtual reality by using the developed ER actuators.

1 Introduction

In recent years, what is called an ER fluid (Electrorheological Fluid) has been attracting attention [1] - [6]. An ER fluid has its rheology varied by an electric field. The existence of a large variety of fluids having such characteristics has been known, and research has been carried out actively on their applications in various types of mechanical systems.

Force display systems are expected to have much effects and advantages as means of human interface [7]. Force display technology works by using mechanical actuators to apply forces to the user. These forces are computed in real time by simulating the physics of the user's virtual world, and then sent to the actuators so that the user feels them.

In this paper, the properties of a clutch using a particle-type ER fluids are presented. The development of a low-inertia actuator consisting of a couple of ER clutch and their driving system is discussed, and a control method of the developed ER actuator is introduced. Then, the merits of ER actuators in application to human interface such as force display devices in virtual reality are discussed. Closed-link-type and parallel-link-type force display devices are developed by using ER actuators. The force display experiments in virtual environments are also conducted by using these devices. Lastly, the proposal of a new device using ER actuators is also presented.

2 Clutch Using ER Fluids

2.1 Electrorheological Fluids

An ER fluid is a substance which changes its apparent viscosity (rheological characteristic) according to the strength of the electric field immersing it, and its main features include a fast response speed and extremely low power consumption, among others. The ER fluids currently being developed may be classified into two types in terms of their characteristics: particle-type and homogeneous-type.

(a) Particle-type ER fluid ER fluids of this type are colloidal fluids which are solvents containing dispersed particles (Fig.1). They show characteristics as shown in Fig.2 (a). The horizontal axis represents

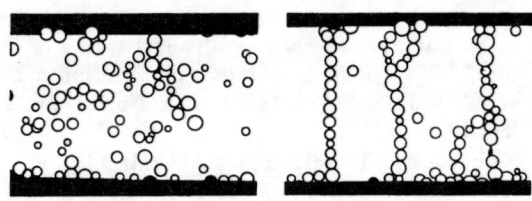

Fig.1: Behavior of Particle-Type ER Fluid

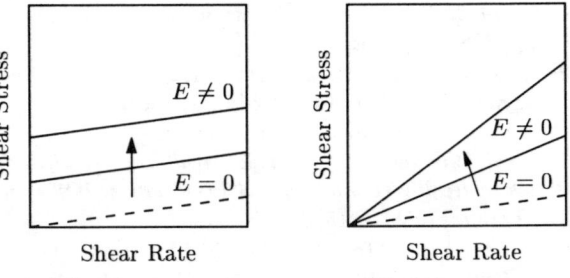

Fig.2: Characteristics of Particle-Type and Homogeneous ER Fluids

shearing speed, while the vertical axis indicates shearing stress; and the characteristics change according to the strength of the electric field E as shown in the diagrams. It is clear from those diagrams that the ER fluids exhibit the characteristics of a Newtonian fluid when no electric field E is present but that they reveal the characteristics of a Bingham fluid when an electric field is applied. In other words, when a voltage is applied, these ER fluids behave very much like Coulomb friction.

(b) Homogeneous ER fluid   Fluids of this type have been developed by using lowmolecule liquid crystal or macromolecular liquid crystal [8], and exhibit characteristics as shown in Fig.2 (b). Shearing stress nearly proportional to shearing speed is generated, and its slope, namely viscosity can be controlled by the electric field. As a result, it is possible to acquire a mechanical control force proportional to the speed under a constant electric field and to mechanically realize what is equivalent to the so-called differential control.

We have been studying the precise position control and the vibration control of robot arms using a homogeneous ER fluid [9]. In this study, an actuator using a particle-type ER fluid is developed by making use of characteristics such as the capability to control force directly by means of an electric field.

2.2 Principle of ER Clutch

Figure 3 shows the principle of ER clutch. The ER fluid is filled into the gap between the input rotational cylinder and output rotational cylinder. Both the input cylinder and the output cylinder serve as electrodes. The input rotational cylinder are driven in constant velocity.

If the voltage is applied to the electrodes, the shear stress is generated between the ER fluid and the surface of the rotational cylinders, then output torque is created. The output torque is controlled by the electric field between the electrodes. The system consisting of ER clutch and the driving system of the input rotational cylinder is called ER actuator.

2.3 Properties of ER Actuator

Figure 4 shows the step responses of output torque to the change of the applied voltage. As seen from the figure, the time constant of force response is $3 \sim 4 [m\,sec]$. The ER clutch has very quick response property, compared with powder clutches.

Figure 5 shows the response of the output torque in the case that the rotational speed of the input cylinder is suddenly changed. As seen from the figure, the output torque is hardly affected by the change of the rotational speed of the input cylinder. So, the rotational speed of the input cylinder does not need to be controlled exactly.

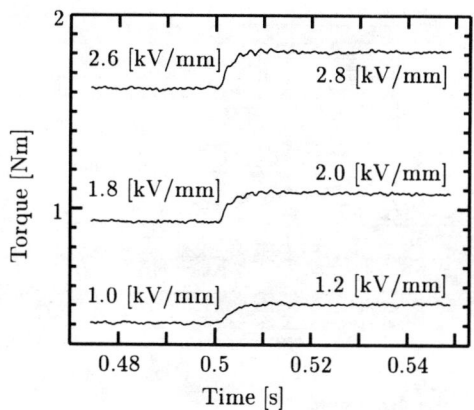

Fig.4: Step Response of ER Actuator

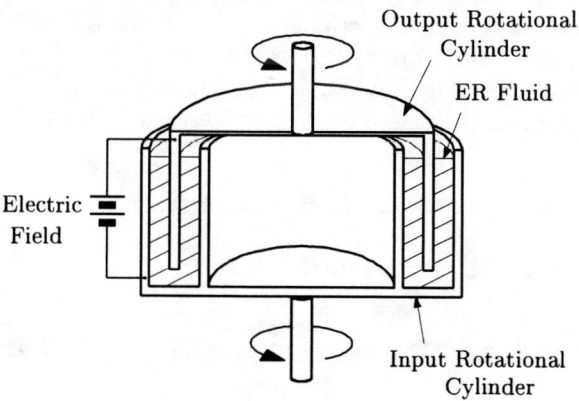

Fig.3: Principle of ER Clutch

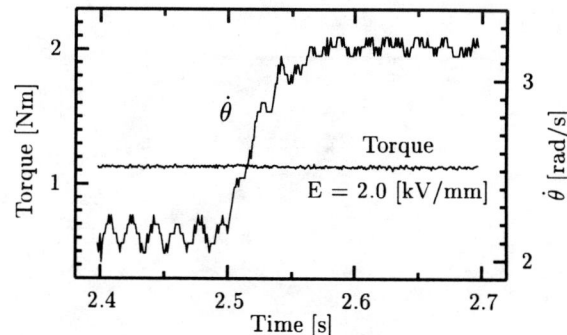

Fig.5: Effect of Velocity Variation of Input Cylinder

Table1: Comparison of Torque/Inertia Ratio

Actuator	Torque/Inertia Ratio	Response Time of ER Fluid and Powder
ER Actuator	$2 \times 10^3 \sim 5 \times 10^5$	$3 \sim 5$ [ms]
Powder Clutch	$3 \times 10^4 \sim 5 \times 10^5$	$30 \sim 300$ [ms]

3 Development of ER Actuators with Low-Inertia Property

The sectional view and the photograph of a developed ER actuator which is capable of producing output torque in two directions are given in Figs 6 and 7. The upper and lower input rotational parts which possess cylindrical structure are driven in different direction of rotation by an electric motor and gear system. The input torque is transferred to the rotating cylindrical section of the output axis via the particle-type ER fluid filled in the rotating cylinder. Both the input axis cylinders and the output axis cylinder serve as electrodes, and output torque is controlled by the electric field applied between the electrodes. The output cylinder is made of aluminum alloy in order to reduce the moment of inertia of the output axis.

The ER actuator has the following characteristics:

(1) As the ER fluid acts on the specific characteristics of the system consisting of motor and reducer as if it were a mask, the characteristics of the ER actuator are precisely those of the ER fluid. As a result, stable and quick torque control in a wide range is made possible by the electric field.

(2) The moment of inertia of the ER actuator is almost that of the output axis cylinder section which is structured simply, so that by using a lightweight and strong material, the moment of inertia can be made very small. As a result, it is easy to obtain a large torque/inertial ratio.

The torque/inertia ratio of this actuator is higher than that of the DD motor and its response characteristic is quite good. ER actuators make use of the surface friction effect, and the smaller their size the higher their torque/inertia ratio will be. When comparing with powder clutches in which the torque/inertia ratio is considered to be extremely high, the torque/inertia ratio of small-size ER actuators is not inferior. Considering the slow response property of powders in powder clutches, it can be stated that ER actuators are devices with extremely high response characteristics (Table 1).

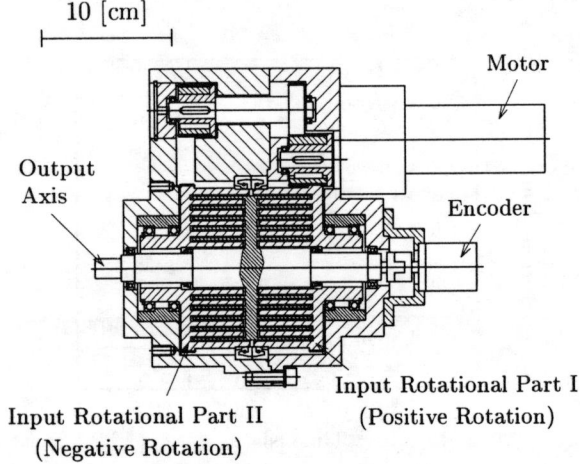

Fig.6: Sectional View of ER Actuator

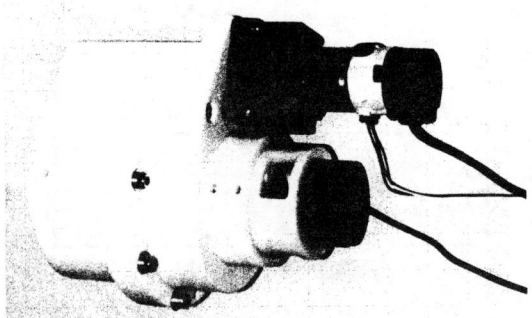

Fig.7: ER Actuator

4 Control of ER Actuator

4.1 Push-Pull Control and Linearization

In order to produce both the clockwise and counterclockwise torques in ER actuator, the push-pull-type control system must be adopted. Figure 8 shows the control system. Two high voltage amplifiers are used to control the electric field of ER fluid filled in Input

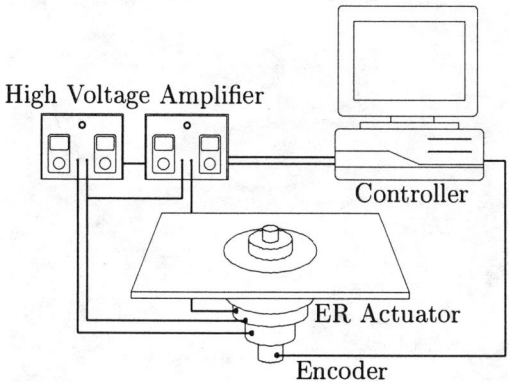

Fig.8: Experimental System of ER Actuator

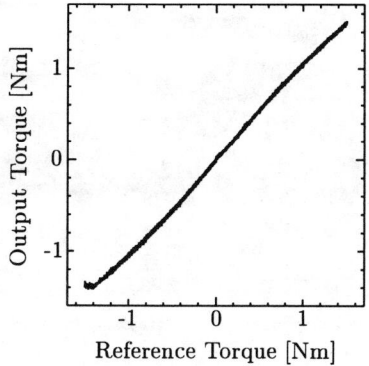

Fig.9: Input-Output Property of the Compensated ER Actuator

Rotational Part I and II. The linearization of the non-linear property of ER actuator is implemented in the computer. The linearized and push-pull-controlled ER actuator has the property shown in Fig.9.

4.2 Feedback Control

Figure 10 shows the results of frequency response experiment from the reference input torque to the output shaft angle. Figure 11 shows the result of a step

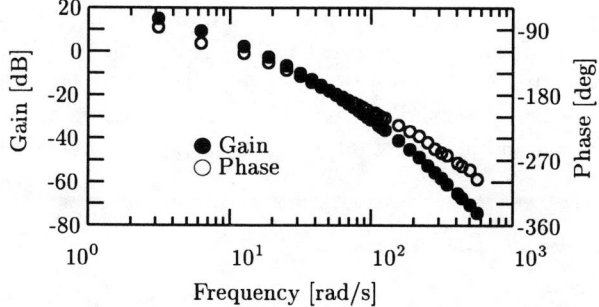

Fig.10: Frequency Response in Position Control

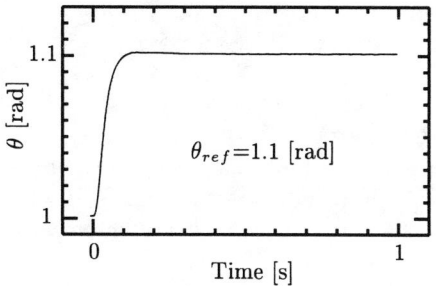

Fig.11: Step Response in Position Control

response experiment in which PID is used as its position feedback controller. As seen from the figure, the proposed control method gives a good response.

5 Merits of ER Actuators in Application to Human Interface System

We discuss the merits of ER actuators in application to human interface such as a force display device, a master arm of master-slave system and a device of physical therapy. From the viewpoints of safety as human interface device, ER actuators have the following merits.

(1) An ER actuator is a clutch in which the transmission torque is controlled by the electric field applied to ER fluid. Then, the maximum speed of output shaft is restricted by the rotational speed of input shaft. Therefore, when the rotational speed of input shaft of ER actuator is set slow, a human interface system using this actuator is safe for human operation.

(2) The moment of inertia of the output axis of ER actuator can be made very small. So, if the link system of the force display device using ER actuator is made by light materials, the impact force to human beings can be reduced in the case of unexpected accidents.

6 Force Display Device

6.1 High Fidelity Property in Force Presentation

Since the moment of inertia of the ER actuator can be made exceedingly small; so, in case that it is to be used as a force display device, it can be manipulated by any operator with ease. In addition, force display devices using ER actuators can give force sense with high fidelity as shown in the following.

(1) Quick force response property originated from the low inertia of ER actuator and the rapid response of ER fluids makes the force presentation with high fidelity possible. So, the impact force with hard object can be presented which contains high frequency signal.

(2) Simple system configuration of force display device is possible when ER actuators are used. The output torque of the ER actuator is sufficiently large, so direct drive system can be constructed. Reduction gear units with little backlash have large friction and bad backward-drivability. System configuration without such reduction gear units brings high fidelity property.

6.2 Development of Force Display Devices

Figure 12 shows a closed-link-type force display device developed in this study. This device uses the ER actuators shown in Figs 6 and 7. This device can reproduce a sensation of free movement inside a two-dimensional plane, a sensation in contact with an obstacle such as a wall and sensations of many kinds of physical systems.

Figure 13 shows a parallel-link-type force display system which was developed recently. This force dis-

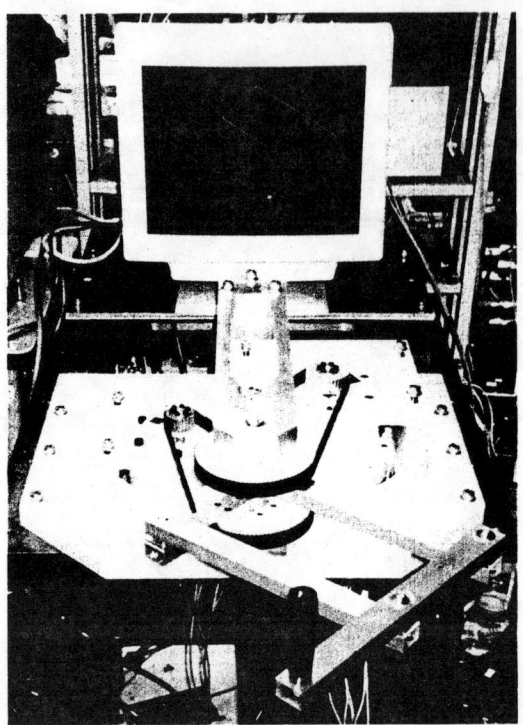

Fig.13: Parallel-Link-Type Force Display Device

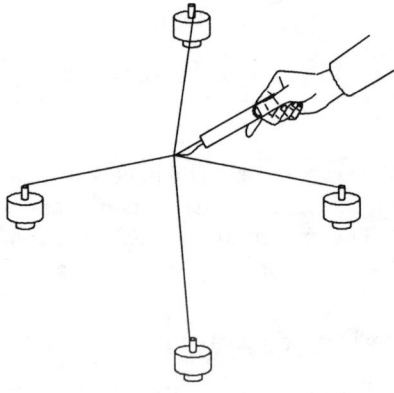

Fig.14: Force Display System for Virtual Surgery

play system can produce at least the force of 50 [N] in all directions within the rectangular area of 30×50[cm], when the shear stress in ER fluid is 2,000 [Pa].

Figure 14 shows a conceptual sketch of the force display system for virtual surgery. We are now developing small ER actuators with high performance for such system.

6.3 Force Display Experiments

By using the developed force display system, several force presentation experiments have been conducted.

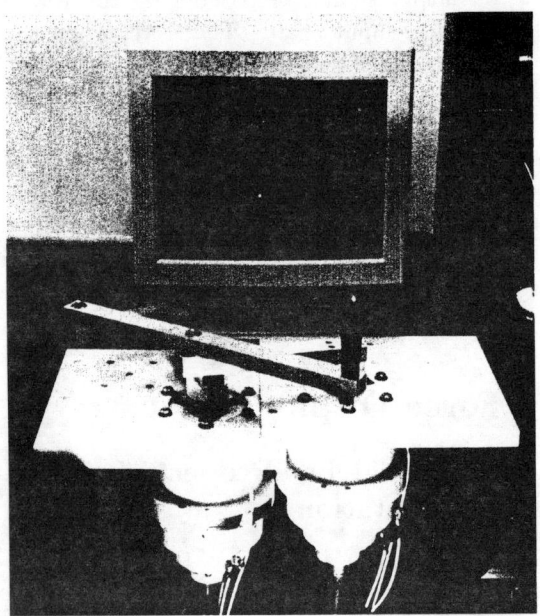

Fig.12: Closed-Link-Type Force Display Device

In the experiments of the virtual wall, the sense of smooth wall and sharp edge was clearly presented. A few games with force sense such as Hockey Game with impact force sensation were also developed.

The basis of force presentation is the presentation of the sense of spring, viscosity and inertia. Figure 15 shows result of force presentation experiment of a virtual two-dimensional linear spring. Figure 16 shows a variation of resistance force F_v in the experiment of a virtual viscous field.

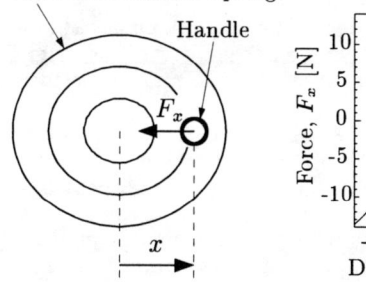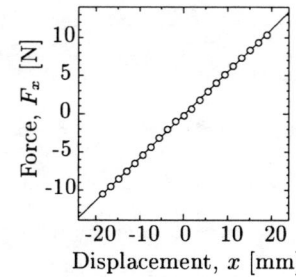

Fig.15: Force Presentation of Two-Dimensional Spring (Experiment)

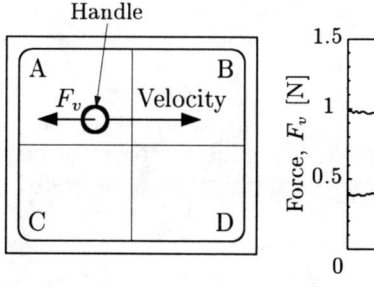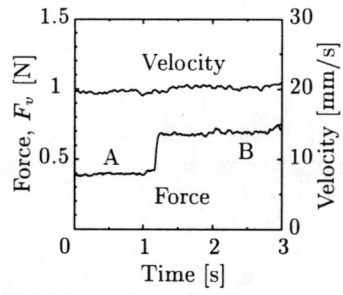

Fig.16: Force Presentation of Viscous Fields. Area A, B, C and D have different viscosity. When the handle is moved from area A into area B, the resistance force F_v suddenly changes on the boundary. (Experiment)

6.4 Use in Rehabilitation

From the view point of mechanical safety as human interface, ER actuators have some merits as discussed in Section 5. Then, we are examining the application of the developed force display system to rehabilitation by a virtual therapist. This system is also useful to estimate the control function of muscle for the purpose of rehabilitation.

7 Conclusions

(1) An ER actuator was developed having a lower moment of inertia and a higher torque/inertia ratio than conventional motors.

(2) Closed-link-type and parallel-link-type force presentation devices were developed by using ER actuators.

(3) The merits of ER actuators in application to human interface system was discussed from the viewpoint of mechanical safety. The high fidelity property of ER actuators in force presentation was also discussed.

(4) Several kinds of force display experiments were conducted, and the application to rehabilitation was examined.

References

[1] J. Furusho and M. Sakaguchi, "Active Control Using Electrorheological Fluids (Review Article)", *Japanese J. of Tribology (Published by Allerton Press)*, **Vol.41**, No.6 (1996), pp.645-654 (in English).

[2] M. Sakaguchi, J. Furusho and Z. Wei, "Application of Actuators Using Particle-Type ER Fluids for Force Display Devices, Master-Slave Systems, and Medical Treatments and Welfare", *Proc. of Japan/USA Symp. on Flexible Automation*, **Vol.1** (1996), pp.241-244.

[3] T. C. Jordan and M. T. Shaw, "Electrorheology (Review Paper)", *IEEE Trans. on Electrical Insulation*, **Vol.24** (1989), pp.849-878.

[4] *Proc. of 5th Int. Conf. on Electro-Rheological Fluids, Magneto-Rheological Suspensions and Associated Technology* (Ed. by W. A. Bullough), World Scientific (1996).

[5] J.Furusho, "Control of Mechatronics Systems Using Electrorheological Fluids (Review Paper)", *J. of SICE*, **Vol.34**, No.9 (1995), pp.687-691 (in Japanese).

[6] J. Furusho and M. Sakaguchi, "New Actuators Using ER Fluids (Review Paper)", *J. of the Robotics Society of Japan*, **Vol.15**, No.3 (1997), pp.323-325 (in Japanese).

[7] M. Minsky, M. Ouh-young, O. Steele, F. P. Brooks, Jr. , M. Behensky, "Feeling and Seeing: Issues in Force Display", *Computer Graphics*, **Vol.24**, No.2 (1990), pp.235-243 (ACM SIGGRAPH).

[8] A. Inoue, "Trends of Homogeneous ER Fluid Development (Review Article)", *J. of SICE*, **Vol. 34**, No.9 (1995), pp. 698-701 (in Japanese).

[9] J. Furusho, G. Zhang and M. Sakaguchi, "Vibration Suppression Control of Robot Arms Using a Homogeneous-Type Electrorheological Fluid", *Proc. of IEEE ICRA'97* (1997), pp.3441-3448.

Development of Self-Learning Vision-Based Mobile Robots for Acquiring Soccer Robots Behaviors.

Takayuki Nakamura

Nara Inst. of Science and Technology Dept. of Information Systems
8916-5, Takayama-cho, Ikoma, Nara 630-01, Japan
takayuki@is.aist-nara.ac.jp

Abstract

An input generalization problem is one of the most important ones in applying reinforcement learning to real robot tasks. To cope with this problem, we propose a self-partitioning state space algorithm which can make non-uniform quantization of the state space. To show that our algorithm has generalization capability, we apply our method to two tasks in which a soccer robot shoots a ball into a goal and prevent a ball from entering a goal. To show the validity of this method, the experimental results for computer simulation and a real robot are shown.

1 Introduction

Recently, many researchers in robotics [1] have paid much attention to reinforcement learning methods by which adaptive, reflexive and purposive behavior of robots can be acquired without modeling its environment and its kinematic parameters. A problem in applying reinforcement learning methods to real robot tasks which have continuous state space is that the value function [1] should be able to represent the value in terms of infinitely many state and action pairs. For this reason, function approximators are used to represent the value function when a closed-form solution of the optimal policy is not available.

One approach that have been used to represent the value function is to quantize the state and action spaces into a finite number of cells and collect rewards and punishments in terms of all states and actions. This is one of the simplest forms of generalization in which all the states and actions within a cell have the same value. In this way, the value function is approximated as a table in which each cell has a specific value (e.g., [1]). However, there is a compromise between the efficiency and accuracy of this table. In order to achieve accuracy, the cell size should be small to provide enough resolution to approximate the value function. But as the cell size gets smaller, the number of cells required to cover the entire state and action spaces grows exponentially, which causes the efficiency of the learning algorithm to become worse because more data is required to estimate the value for all cells. Chapman et. al [2] proposed an input generalization method which splits an input vector consisting of a bit sequence of the states based on the already structured actions such as "shoot a ghost" and "avoid an obstacle." However, the original states have been already abstracted, and therefore it seems difficult to be applied to the continuous raw sensor space of real world. Moore et. al [3] proposed a method to resolve the problem of learning to achieve given tasks in deterministic high-dimensional continuous spaces. It divides the continuous state space into cells such that in each cell the actions available may be aiming at the neighboring cells. This aiming is accomplished by a local controller, which must be provided as a prior knowledge of the given task in advance. The graph of cell transitions is solved for shortest paths in an online incremental manner, but a minimax criterion is used to detect when a group of cells is too coarse to prevent movement between obstacles or to avoid limit cycles. The offending cells are split to higher resolution. However, the restriction of this method to deterministic environments might limit its applicability since the real environment is often non-deterministic.

Another approach for representing the value function is to use other types of function approximators, such as neural networks (e.g., [4]), statistical models [5, 6, 7, 8] and so on. Boyan and Moore [9] used local memory-based methods in conjunction with value iteration; Lin [4] used backpropagation networks for Q-learning; Watkins [10] used CMAC for Q-learning; Tesauro [11] used backpropagation for learning the value function in backgammon. Asada et al. [7] used a concentration ellipsoid as a model of cluster (state) of input vectors, inside which a uniform distribution is assumed. They define a state as a cluster of input vectors from which the robot can reach the goal state or the state already obtained by a sequence of one kind action primitive regardless of its length. However, actual distributions are not always uniform.

This paper propose a new method for incrementally dividing a multidimensional continuous state space into some discrete states. This method recursively splits its continuous state space into some coarse spaces called tentative states. It begins by supposing that such tentative states are regarded as the states for Q learning. In parallel with the learning, it collects statistical evidence regarding immediate rewards r and Q values within this tentative state space. When a tentative state is considered to be *relevant* to the given

---
[1]*value function* is a prediction of the return available from each state and is important because the robot can use it to decide a next action. See [14] for more details.

task by the statistical test for r and Q values based on a minimum description length (hereafter, MDL) criterion [12], this coarse state is divided into finer states. These procedures can make non-uniform quantization of the state space. Our method can be applied to non-deterministic domain because the Q-learning is used to find out the optimal policy for accomplishing the given task.

The remainder of this article is structured as follows: In the next section, we describe our method to automatically quantize the sensor spaces. In section 3, we show the results of the experiments with computer simulation and real robot in which a vision-based mobile robot tries to shoot a ball into a goal and tries to prevent a ball from entering a goal. Finally, we give concluding remarks.

2 Self-Partitioning State Space Algorithm Based on MDL Criterion

2.1 Details of Our Algorithm

Here, we define sensor inputs, actions and rewards as follows:

- Sensor input d is described by a N dimensional vector $d = (d_1, d_2, \cdots, d_N)$, each component $d_i (i = 1 \sim N)$ of which represents the measurement provided by the sensor i. The continuous value d_i is provided by the sensor i. Its range $Range(d_i)$ is known in advance. Based on $Range(d_i)$, a measurement d_i is normalized in such a way that d_i can take values in the semi open interval $[0, 1)$.

- The agent has a set A of possible actions a_j, $j = 1 \sim M$. Such a set is called the action space.

- One of the discrete rewards $r = r_k$, $k = 1 \sim C$ is given to the agent depending on the evaluation of the action taken at a state.

Our method utilizes a hierarchical segment tree in order to represent the non-uniform partitioning of the state space which consists of N dimensional input vector. This representation has an advantage for approximating the non-uniform distribution of sample data. The inner node at i th depth in the j th level keeps the range $b_i(j) = [t_i^{low}, t_i^{high})$ of a measurement provided by each sensor i. (Actually, j corresponds to the number of iteration of this algorithm.) At each inner node in the j the level, the range of a measurement is partitioned into two equal intervals $b_i^0(j) = [t_i^{low}, (t_i^{low} + t_i^{high})/2)$ and $b_i^1(j) = [(t_i^{low} + t_i^{high})/2, t_i^{high})$. For example, initially $j = 0$, the range of each dimension i is divided into two equal intervals $b_i^0(0) = [0.0, 0.5)$ and $b_i^1(0) = [0.5, 1.0)$. When sensor input vector d has N dimensions, a segment tree whose depth is N is built (see Fig.1). The leaf node corresponds to the result of classification for observed sensor input vector d. As a result, 2^N leaf nodes are generated. These leaf nodes can represent the situations in the agent's environment. The state space represented by the leaf nodes is called

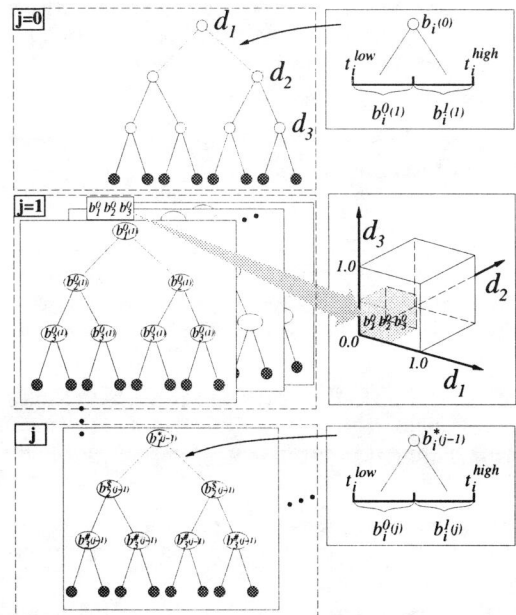

Figure 1: Representation of state space by a hierarchical segment tree

"tentative state space" TS. Let ts_k, $k = 1 \sim 2^N$ be the component of the tentative state space which is called "tentative state."

Our algorithm works as follows:

1. It starts by assuming that the entire environment is as if it were one state. Initially, the total number of the states $|S|$ and the tentative states $|TS|$ are one and 2^N, respectively.

2. Based on TS, our algorithm begins Q-learning. In parallel with this process, it gathers statistics in terms of $r(a_i|ts_k = on)$, $r(a_i|ts_k = off)$, $Q(a_i|ts_k = on)$ and $Q(a_i|ts_k = off)$, which indicate immediate rewards r and discounted future rewards Q in case that individual state is "on" or "off," respectively. In this work, it is supposed that if a N dimensional sensor vector d is classified into a leaf node ts_k, the condition of this node ts_k is regarded as "on," otherwise (this means the case that d is classified into the leaf node except ts_k), it is regarded as "off."

3. After Q-learning based on TS is converging, our algorithm asks the question whether there are some states in TS such that the r and Q for states "on" are significantly different from such values for states "off." When the distributions of statistics of ts_k in case of "on" and "off" are different, it is determined that ts_k is *relevant* to the given task. In order to discover the difference between two distributions, our algorithm performs the statistical test based on a MDL criterion. In the section 2.2.1 and 2.2.2, these procedures are explained.

2593

4. (a) If there is the state ts'_k adjoining the state ts_k which is shown to be relevant such that the statistical characteristic of Q and actions assigned at the adjoining state are same, merge these two states into one state.

   (b) Otherwise, skip this step.

5. Each leaf nodes ts_k is represented by a combination of intervals each of which corresponds to the range of a measurement provided by each sensor i. These intervals in ts_k which is shown to be relevant are bisected. As a result, in terms of one ts_k, 2^N leaf nodes are generated and correspond to tentative states. These tentative states are regarded as the states in Q-learning at the next iteration.

6. Until our algorithm can't find out any relevant leaf nodes, the procedures 2 ~ are repeated. Finally, a hierarchical segment tree is constructed to represent the partitioning of the state space for achievement of a given task.

After the learning, based on Q stored at leaf nodes, the agent takes actions for accomplishing the given task.

2.2 The Relevance Test Based on MDL Criterion

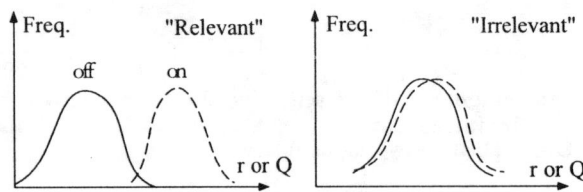

Figure 2: Criterion for determining the relevance of the state

Here, we explain how to determine whether a state is *relevant* to the task or not. **Fig. 2** shows the difference between the distributions of r or Q values regarding to the state ts_k in case ts_k is *relevant* or *irrelevant*. As shown in left side of this figure, when two peaks of the distributions of r or Q values, which correspond to pair of $r(a_i|ts_k = on)$ and $r(a_i|ts_k = off)$, or pair of $Q(a_i|ts_k = on)$ and $Q(a_i|ts_k = off)$), can be clearly discriminated, it is supposed that ts_k is relevant to the given task because the such state affects the value of state more heavily than the other states does, therefore, it affects how the robot should act at next time step. On the contrary, in case two peaks are ambiguous as shown in right side of this figure, it is considered that ts_k is *irrelevant* to the given task.

Actually, we perform the statistical test with respect to r and Q values based on a MDL criterion [12] in order to distinguish the distribution of such reinforcement values.

2.2.1 The Statistical Test for r Value

Since the immediate reward r_j value is given at each trial among one of C mutually exclusive rewards r_j $j = 1, 2, \cdots C$, the distribution of r_j in the state ts_k follows a multinominal distribution. Supposing **Tab. 1** shows the distribution of immediate rewards in case the

Table 1: The distribution of rewards in ts_k

Rewards	On	Off
R_1	$n(On, R_1)$	$n(Off, R_1)$
R_2	$n(On, R_2)$	$n(Off, R_2)$
$\vdots$	$\vdots$	$\vdots$
R_C	$n(On, R_C)$	$n(Off, R_C)$
	$n(On)$	$n(Off)$

$n(i)$: the frequency of sample data in the state i $(i = 1, \cdots, S)$

$n(i, r_j)$: the frequency of reward r_j $(j = 1, \cdots, C)$ given in the state i

$p(r_j|i)$: the probability that reward r_j is given in the state i

$$\sum_{j=1}^{C} n(i, r_j) = n(i), \quad \sum_{j=1}^{C} p(r_j|i) = 1, \quad (i = 1, \cdots S).$$

state ts_k is "*on*" or "*off*," our algorithm tries to find the difference between the distributions of rewards in two cases of ts_k "on" and "off" based on this table.

The probability $P(\{n(i, r_j)\}|\{p(r_j|i)\})$ that the distribution of the immediate rewards is acquired as shown in **Tab. 1** $\{n(i, r_j)\}$, $(i = 1 \cdots, S, j = 1, \cdots, C)$, can be described by

$$P(\{n(i, r_j)\}|\{p(r_j|i)\})$$
$$= \prod_{i=1}^{S} \left\{ \frac{n(i)!}{\prod_{j=1}^{C} n(i, r_j)!} \prod_{j=1}^{C} p(r_j|i)^{n(i, r_j)} \right\}.$$

The likelihood function L of this multinominal distribution can be written as follows:

$$L(\{p(r_j|i)\}) = \sum_{i=1}^{S} \sum_{j=1}^{C} n(i, r_j) \log p(r_j|i) + K,$$

where $K = \log \left\{ \frac{\prod_{i=1}^{S} n(i)!}{\prod_{i=1}^{S} \prod_{j=1}^{C} n(i, r_j)!} \right\}$.

When two multinominal distributions in case each tentative state ts_k is "on" and "off" can be considered to be same, the probability $p(r_j|i)$ that the immediate reward r_j is given in the state i can be modeled as follows:

M1: $p(r_j|i) = \theta(r_j)$ $(i = 1, \cdots, S, j = 1, \cdots, C)$.

Furthermore, its likelihood function L_1 can be written by

$$L_1(\{\theta(r_j)\}) = K + \sum_{j=1}^{C} \left[\left\{ \sum_{i=1}^{S} n(i, r_j) \right\} \log \theta(r_j) \right].$$

Therefore, the maximum likelihood ML_1 can be written by

$$ML_1 = L(\theta(\hat{r_j})) = L\left(\frac{\sum_{i=1}^{S} n(i, r_j)}{\sum_{i=1}^{S} n(i)} \right).$$

where $\theta(\hat{r_j})$ is the maximum likelihood estimation of $\theta(r_j)$. Therefore, the description length $l_{MDL}(M1)$ of the model $M1$ based on MDL principle is

$$l_{MDL}(M1) = -ML_1 + \frac{C-1}{2} \log \sum_{i=1}^{S} n(i).$$

On the contrary, when two multinominal distributions in case of "on" and "off" can be considered to be different, the probability $p(r_j|i)$ can be modeled as follows:

M2: $p(r_j|i) = \theta(r_j|i)$ $(i=1,\cdots,S, \quad j=1,\cdots,C)$.

Furthermore, its likelihood function L_2 can be written by

$$L_2(\{\theta(r_j|i)\}) = K + \sum_{i=1}^{S}\sum_{j=1}^{C} n(i,r_j)\log\theta(r_j|i).$$

In the same way, the maximum likelihood ML_2 can be written by

$$ML_2 = L(\theta(\hat{r}_j)) = L\left(\frac{n(i,r_j)}{n(i)}\right).$$

Therefore, the description length $l_{MDL}(M2)$ of the model $M2$ based on MDL principle is

$$l_{MDL}(M2) = -ML_2 + \frac{S(C-1)}{2}\log\sum_{i=1}^{S}n(i).$$

Based on MDL criterion, we can suppose that discovering the difference between two distributions is equivalent to determining which model is appropriate for representing the distribution of data. In this work, the difference between the distributions is found based on the following conditions: If $l_{MDL}(M1) > l_{MDL}(M2)$, two distributions are different. Otherwise, two distributions are same.

2.2.2 The Statistical Test for Q Value

In order to distinguish the distribution of sampled data of Q values, we perform the statistical test based on a MDL criterion. Let $\boldsymbol{x}^n$ and $\boldsymbol{y}^m$ be the sample data $(x_1, x_2\cdots,x_n)$ and $(y_1, y_2\cdots,y_m)$, respectively. $\boldsymbol{x}^n$ and $\boldsymbol{y}^m$ indicate a history of $Q(a_i|ts_k = on)$ and $Q(a_i|ts_k = off)$, respectively. We'd like to know whether these two sample data $\boldsymbol{x}^n$ and $\boldsymbol{y}^m$ come from the two different distributions or the same distribution. Here, we assume the following two model for the distribution of sampled data are $M1$ based on one normal distribution and $M2$ based on two normal distributions.

M1: $\quad N(\boldsymbol{\mu}, \boldsymbol{\sigma}^2)$
M2: $\quad N(\boldsymbol{\mu}_1, \boldsymbol{\sigma}'^2), N(\boldsymbol{\mu}_2, \boldsymbol{\sigma}'^2)$

The normal distribution with mean μ and variance σ is defined by $f(x:\mu,\sigma^2) = \frac{1}{\sqrt{2\pi\sigma^2}}\exp\left\{\frac{-(x-\mu)^2}{2\sigma^2}\right\}$. If both $\boldsymbol{x}^n$ and $\boldsymbol{y}^m$ follow the model $M1$, the probabilistic density function of these sampled data can be written by $\Pi_{i=1}^n f(x_i:\mu,\sigma^2) + \Pi_{i=1}^m f(y_i:\mu,\sigma^2)$. The likelihood function LL_1 of the model $M1$ can be written by

$$LL_1 = \sum_{i=1}^{n}\log f(x_i:\mu,\sigma^2) + \sum_{i=1}^{m}\log f(y_i:\mu,\sigma^2).$$

Therefore, the maximum likelihood MLL_1 can be derived as follows:

$$MLL_1 = -\frac{n+m}{2}(1+\log 2\pi\hat{\sigma}^2),$$

where $\hat{\sigma}^2$ is the maximum likelihood estimated variance of σ^2 and can be estimated as follows:

$$\hat{\mu} = \frac{1}{n+m}\left\{\sum_{i=1}^{n}x_i + \sum_{i=1}^{m}y_i\right\}$$

$$\hat{\sigma}^2 = \frac{1}{n+m}\left\{\sum_{i=1}^{n}(x_i-\hat{\mu})^2 + \sum_{i=1}^{m}(y_i-\hat{\mu})^2\right\}$$

where $\hat{\mu}$ is the maximum likelihood estimated mean of μ. Therefore, the description length $l_{MDL}(M1)$ of the model $M1$ based on MDL principle is

$$l_{MDL}(M1) = -MLL_1 + \log(n+m).$$

On the contrary, if $\boldsymbol{x}^n$ and $\boldsymbol{y}^m$ follow the model $M2$, the probabilistic density function of these sampled data can be written by $\Pi_{i=1}^n f(x_i:\mu_1,\sigma'^2) + \Pi_{i=1}^m f(y_i:\mu_2,\sigma'^2)$. The likelihood function LL_2 of the model $M2$ can be written by

$$LL_2 = \sum_{i=1}^{n}\log f(x_i:\mu_1,\sigma'^2) + \sum_{i=1}^{m}\log f(y_i:\mu_2,\sigma'^2).$$

Therefore, the maximum likelihood MLL_2 can be derived as follows:

$$MLL_2 = -\frac{n+m}{2}(1+\log 2\pi\hat{\sigma'}^2),$$

where $\hat{\sigma_1}^2$ and $\hat{\sigma_2}^2$ is the maximum likelihood estimated variance of σ_1^2 and σ_2^2, respectively. These statistics can be estimated as follows:

$$\hat{\mu}_1 = \frac{1}{n}\sum_{i=1}^{n}x_i, \quad \hat{\mu}_2 = \frac{1}{m}\sum_{i=1}^{m}y_i,$$

$$\hat{\sigma'}^2 = \frac{1}{n+m}\left\{\sum_{i=1}^{n}(x_i-\hat{\mu}_1)^2 + \sum_{i=1}^{m}(y_i-\hat{\mu}_2)^2\right\}.$$

where $\hat{\mu}_1$ and $\hat{\mu}_2$ is the maximum likelihood estimated mean of μ_1 and μ_2, respectively. Therefore, the description length $l_{MDL}(M2)$ of the model $M2$ based on MDL principle is

$$l_{MDL}(M2) = -MLL_2 + \frac{3}{2}\log(n+m).$$

We can recognize the difference between the distributions based on the following condition: If $l_{MDL}(M1) > l_{MDL}(M2)$, $\boldsymbol{x}$ and $\boldsymbol{y}$ arise from the different normal distributions. Otherwise, $\boldsymbol{x}$ and $\boldsymbol{y}$ arise from the same normal distribution.

3 Experimental Results

The experiment consists of two phases: first, learn the optimal policy through the computer simulation, then apply the learned policy to a real situation. To show that our algorithm has a generalization capability, we apply it to acquire two different behaviors: one is a shooting behavior and the other is a defending behavior for soccer robots. In this work, we assume that our robot does not know the location and the size of the goal, the size and the weight of the ball, any camera parameters such as focal length and tilt angle, or kinematics/dynamics of itself.

3.1 Simulation

We consider an environment shown in Figure 3 (a) where the task for a mobile robot is to shoot a ball into a goal or to defend a shot ball. The environment consists of a ball and a goal, and the mobile robot has a single CCD camera. The robot does not know the location and the size of the goal, the size and the weight of the ball, any camera parameters such as focal length and tilt angle, or kinematics/dynamics of itself.

We performed the computer simulation with the following specifications. The field is 1.52m × 2.74m. The

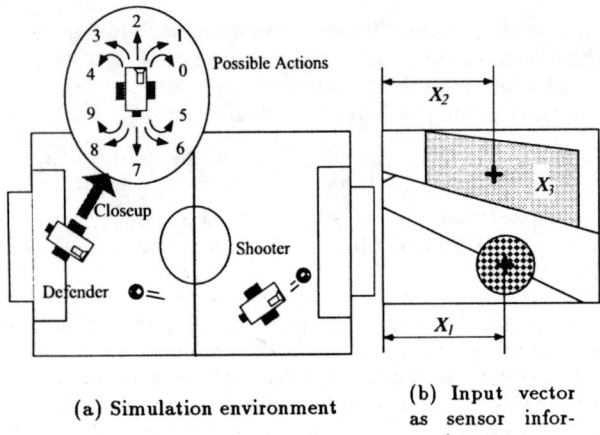

(a) Simulation environment

(b) Input vector as sensor information

Figure 3: Computer Simulation

goal posts are located at the center of the left and right line of the rectangle (see Figure 3 (a)) and its height and width are 0.20m and 0.5m, respectively. The robot is 0.10m wide and 0.18m long and kicks a ball of diameter 0.05m. The maximum translation velocity is 5cm/s. The camera is horizontally mounted on the robot (no tilt) and is in off-centered position. Its visual angle is 36 degrees. The velocities of the ball before and after being kicked by the robot is calculated by assuming that the mass of the ball is negligible compared to that of the robot. The speed of the ball is temporally decreased by a factor 0.8 in order to reflect the so-called "viscous friction." The values of these parameters are determined so that they can roughly simulate the real world.

The robot is driven by two independent motors and steered by front and rear wheels which is driven by one motor. Since we can send the motor control commands such as "move forward or backward in the given direction," all together, we have 10 actions in the action primitive set A as shown in **Fig.** 3 (a). The robot continues to take one action primitive at a time until the current state changes. This sequence of the action primitives is called an action. Actually, a stop motion does not causes any changes in the environment, we do not take into account this action primitive.

The size of the image taken by the camera is 256 × 240 pixels. An input vector x to our algorithm consists of:

- x_1: the horizontal position of the ball in the image, that ranges from 0 to 256 pixels,
- x_2: the horizontal position of the goal in the image ranging from 0 to 256,
- x_3: the area of the goal region in the image, that ranges from 0 to 256×240 pixels^2

(see Figure 3 (b)). After the range of these values is normalized in such a way that the range may become the semi open interval $[0, 1)$, they are used as inputs of our method.

A discounting factor γ is used to control to what degree rewards in the distant future affect the total value of a policy. In our case, we set the value a slightly less than 1 ($\gamma = 0.9$). In this work, we set the learning rate

$\alpha = 0.25$. In case the shooting behavior tried to be acquired by our method, as a reward value, 1 is given when the robot succeeded in shooting a ball into a goal, 0.3 is given when the robot just kicked a ball, -0.01 is given when the robot went out of field, 0 is given otherwise. In the same way, in case the defending behavior tried to be acquired by our method, as a reward value, -0.7 is given when the robot failed in preventing a ball from entering a goal, 1.0 is given when the robot just kicked a ball, -0.01 is given when the robot went out of field, 0 is given otherwise.

In the learning process, Q-learning continues until the sum of estimated Q values seems to be almost convergent. When our algorithm tried to acquire the shooting behavior, our algorithms ended after it iterated the process (Q-learning + statistical test) 8 times. In this case, about $160K$ trials were required to converge our algorithm and the total number of the states is 246. In case our algorithm tried to acquire the defending behavior, our algorithms ended after it iterated the process (Q-learning + statistical test) 5 times. In this case, about $100K$ trials were required to converge our algorithm and the total number of the states is 141.

Fig. 4 shows the success ratio versus the step of trials in the two learning processes that one is for acquiring a shooting behavior the other is for a defending behavior. We define the success rate as (# of successes)/(# of trials) × 100(%). As you can see, the bigger the number of iteration is, the higher the success ratio at the final step in each iteration. This means that our algorithm gradually made better segmentation of state space for accomplishing the given task.

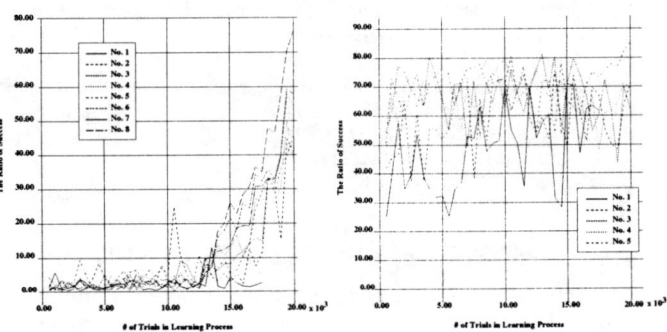

(a) In case of Shooting Behavior

(b) In case of Defending Behavior

Figure 4: The success ratio versus the step of trial

Fig. 5 shows the partitioned state spaces obtained by our method. **Fig.** 5 (a) and (b) shows the state space for the shooting behavior and for defending behavior, respectively. In each figure, $Dim.1$, $Dim.2$ and $Dim.3$ shows the position of ball, the position of goal and the area of goal region, respectively. Furthermore, in each figure, $Action2$ and $Action7$ corresponds to "moving forward" and "moving backward," respectively. One cube in the partitioned state space corresponds to one state. For example, the left in **Fig.** 5 (a) shows a group of the cube where the action 2 is assigned as an optimal action. As shown in **Fig.** 5 (a),

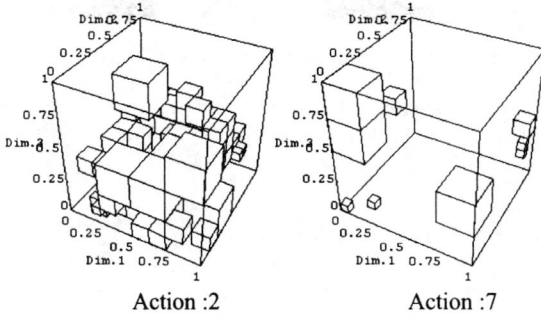

(a) The State Space for Shooting Behavior

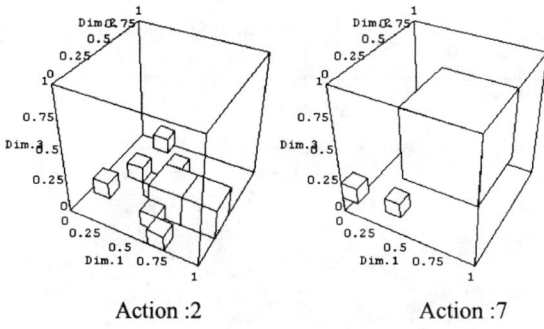

(b) The State Space for Defending Behavior

Figure 5: The Partitioned State Space

many cubes where forward actions are assigned concentrate around the center of the entire state space. This means that the robot will take an forward action if the ball and goal are observed around the center of field of its view. This shows very natural behavior for shooting a ball into a goal.

In the right of **Fig.** 5 (b), there is one large cube. This means that the the robot will take an backward action if large goal are observed around the center of field of its view. This strategy is plausible behavior for preventing a ball from entering a goal because the robot will have to go back in front of own goal after it moved out there in order to kick a ball.

3.2 Real Robot Experiments

We have developed our real robot system to take part in RoboCup-97 competition where several robotic teams are competing on a field. So, the system includes two robots which have the same structure: one for a shooter, the other for a defender. Off-board computer SGI ONYX (R4400/250MHz) perceives the environment through on-board cameras, performs the decision making based on the learned policy and sends motor commands to each robot. A CCD camera is set at bottom of each robot in off-centered position. Each robot is controlled by SGI ONYX through radio RS232C. The maximum vehicle speed is about 5cm/s. The images taken by the CCD camera on each robot are transmitted to a video signal receiver. In order to process two images (one is sent from the shooter robot, the other from the defender robot) simultaneously, two video signals are combined into one video signal by a video combiner on PC. Then, the video signal is sent to SGI ONYX for image processing. The color-based visual tracking routine is implemented for tracking and finding a ball and a goal in the image. In our current system, it takes 66 ms to do this image processing for one frame.

Fig. 9 shows how a real robot shoots a ball into a goal based on the state space obtained by our method. 6 images are shown in raster order from the top left to the bottom right in every 2.0 seconds, in which the robot tried to shoot a ball, but failed, then moved backward so as to find a position to shoot a ball, finally succeeded in shooting. Note that the backward motion for retry is just the result of learning and not hand-coded.

Fig. 10 shows how a real robot prevents a ball shot by the opponent's robot from entering a goal based on the state space obtained by our method. 6 images are shown in raster order from the top left to the bottom right in every 2.0 seconds. This sequence was appeared in the RoboCup-97 competition. In the top left image, the white circle shows the position of the defender robot and the white arrow shows the position of the ball. In the time step $1 \sim 2$, first, the opponent's robot tried to shoot a ball, then the defender robot prevent a ball from entering a goal. In the time step $3 - 6$, another opponent's robot got a ball cleared by the defender robot and tried to shoot a ball again, then the defender robot defend a goal again. As shown in these figures, the defender robot always moves in front of own goal to find out a shot ball as soon as possible. Note that this behavior is just the result of our learning algorithm and not hand-coded.

4 Concluding Remarks

We have proposed a method for partitioning the state space based on experiences, and shown the validity of the method with computer simulations and real robot experiments. We can regard the problem of quantizing the state space as "segmentation" problem. From a viewpoint of robotics, segmentation of sensory data from the environment should depend on the purpose (task), capabilities (sensing, acting, and processing) of the robot, and the complexity of its environment, and its evaluation. In this sense, the state space obtained by our method (**Fig.**5 indicates a projection of such a space) corresponds to the subjective representation of the world for the robot to accomplish a given task.

5 Acknowledgments

The main idea of this paper is thought of while I stayed at AI Lab of Comp. Sci. Dept. of Brown University. I would like to thank L. P. Kaelbling for their helpful comments during my stay. I also would like to thank M. Imai for providing research fund and S. Morita (Japan SGI Cray Corp) for lending SGI ONYX to me.

References

[1] J. H. Connel and S. Mahadevan, editors. *Robot Learning*. Kluwer Academic Publishers, 1993.

[2] D. Chapman and L. P. Kaelbling. "Input generalization in delayed reinforcement learning: An alogorithm and performance comparisons". In *Proc. of IJCAI-91*, pages 726–731, 1991.

[3] A. W. Moore and C. G. Atkeson. "The parti-game algorithm for variable resolution reinforcement learning in multidimensional state-spaces". *Machine Learning*, 21:199–233, 1995.

[4] Long-Ji Lin. "Self-improving reactive agents based on reinforcement learning, planning and teaching". *Machine Learning*, 8:293–321, 1992.

[5] R. Sato H. Ishiguro and T. Ishida. "Robot oriented state space construction". In *Proc. of IROS'96*, volume 3, 1996.

[6] K. Hori A. Ueno and S. Nakasuka. "Simultaneous learning of situation classification based on rewards and behavior selection based on the situation". In *Proc. of IROS'96*, volume 3, 1996.

[7] S. Noda M. Asada and K. Hosoda. "Action-based sensor space categorization for robot learning". In *Proc. of IROS'96*, volume 3, 1996.

[8] Y. Takahashi, M. Asada, and K. Hosoda. "Reasonable performance in less learning time by real robot based on incremental state space segmentation". In *Proc. of IROS'96*, volume 3, pages 1518–1524, 1996.

[9] J. Boyan and A. Moore. "Generalization in reinforcement learning: Safely approximating the value function". In *Proceedings of Neural Information Processings Systems 7*. Morgan Kaufmann, January 1995.

[10] C. J. C. H. Watkins. *"Learning from delayed rewards"*. PhD thesis, King's College, University of Cambridge, May 1989.

[11] G. Tesauro. "Practical issues in temporal difference learning". *Machine Learning*, 8:257–277, 1992.

[12] J. Rissanen. *Stochastic Complexity in Statistical Inquiry*. World Scientific, 1989.

[13] R. Bellman. *Dynamic Programming*. Princeton University Press, Princeton, NJ, 1957.

[14] L. P. Kaelbling. "Learning to achieve goals". In *Proc. of IJCAI-93*, pages 1094–1098, 1993.

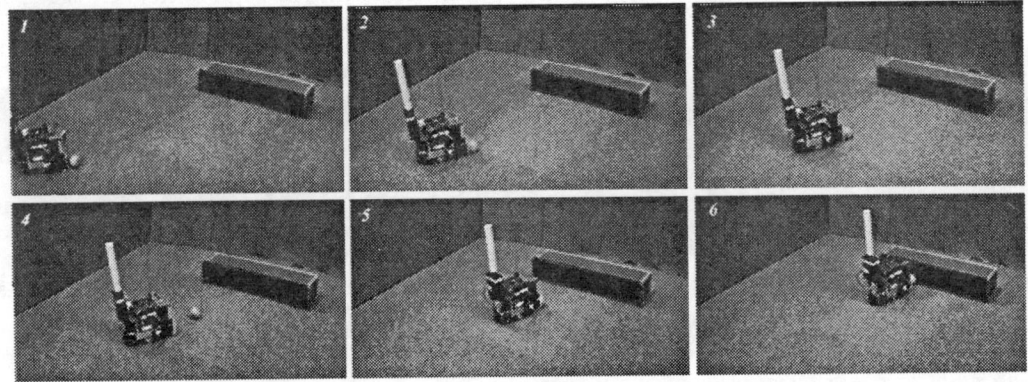

Figure 9: Shooting Behavior on Real Robot

Figure 10: Defending Behavior on Real Robot

Motion control for micro-robots playing soccer games

Sukhan Lee and Javier Bautista

Dept. of Computer Science, University of Southern California. Los Angeles, CA, 90089-0781

Abstract

The problem of two teams of micro-robots playing a soccer game constitutes an excellent testbed where various aspects of machine intelligence such as visual perception, path-planning, behavior coordination as well as learning need to be integrated. This paper describes the set of algorithms that govern the behavior of a team of three microrobots playing a soccer game under MIROSOT regulations [1]. These algorithms are structured in three levels: strategy, path-planning and motor control. At the strategy level the motions of individual team players are coordinated against those of the opponent. This is achieved based on the novel mathematical concept of 'Controllable Zone'. The arrangement of the players is optimized in such a way as to maximize the probability of possessing and shooting the ball, This concept leads to a formal mathematical framework to represent behavior-coordinated strategies for shooting, capturing and passing the ball. At the path-planning level a pseudo-optimal path is generated in real time to minimize the time required by the player to perform the desired behavior and the probability to bump into an obstacle. Finally, the motor control level computes the necessary motor commands to keep the robot on the right track. Experimental results of the system are shown.

I INTRODUCTION

The Micro-Robot Soccer Tournament (MIROSOT) establishes the rules for a soccer match between two teams of three microrobots of limited size ($7.5cm^3$). A camera placed on top of the field feeds the visual system with a continuous sequence of overhead views of the field. These are at each time processed in an off-board computer to extract the position and orientation of all the robots in the field, as well as the position of the ball. These parameters are in turn passed to each robot which will calculate their future motion in their on-board processor. Each robot will autonomously decide its next action (like running towards the ball to kick it, obstructing an opponent or saving the goal) watching some rules. For example, no more than two robots can be inside a small area around their goal and a robot is not allowed to push an opponent. Finally, each robot will have to plan in real time the path to follow, and compute the corresponding motor commands, in order to perform the desired task

Planning of coordinated motions for robots playing a soccer game opens a new area in motion planning, namely coordinated motion planning in team competition. For coordinated motion planning in team competition, planning motions of individual players in such a way as to gain advantages in offense as well as defense as a team against the opponent is of a major concern. However, this problem is quite challenging due to the fact that the game situation may vary constantly and that the complexity of search space involved is rather excessive. Therefore, the goal is how to formulate computational paradigms and strategies that allow a real-time search for optimal coordinated motions of individual players under excessive search complexity.

In this paper, we propose the concept of the ``Controllable Zone'' of a player or a team at time t, representing a set of field positions that the player or the team can occupy first, should it choose to do so at time t against the same attempt of the opponents. A systematic way of computing various forms of controllable zones is presented. Based on the controllable zones, a team can identify whether it is in offensive or defensive mode, and which strategy (shooting, capturing, dribbling, passing, or searching) is appropriate for the team. The coordinated motions of individual players are then determined based on the chosen strategy.

II STRATEGY

This level determines how the robot should behave. Possible behaviors include shooting. passing. capturing, dribbling, blocking or saving. The goal here is that each robot determines its motion in a coordinated fashion, so as to achieve a global strategy. The specific behavior is selected based on the field positions of ball and robots (input) provided by the visual system. The chosen behavior will be instructed in the form of a desired position and orientation for the robot in a future time (output). It will be the responsibility of the lower levels to achieve this configuration as fast as possible.

Because of the great amounts of uncertainty involved, we avoid the use of complicated strategies designing sophisticated maneuvers that require a high level of accuracy in order to perform them effectively. Instead we designed robust algorithms that still work under uncertainty. This algorithms also need to be fast in order to keep a high sampling rate of the visual input, that will allow to promptly detect and compensate deviations from the computed plan. Finally, It is also important that estimations made at this level allow for some degree of error. For example, when the robot is instructed to kick the ball, it is advisable to give the robot some more time than the strictly estimated it requires to reach the point where it will hit the ball.

In order for the robots to act as a team, their behaviors must be chosen in a coordinated fashion. It has to be avoided having two robots performing the same task (blocking the same opponent or running towards the ball at the same time). To achieve this a role assignment scheme is used. Based on the current game position a global strategy for the whole team is selected. Each global strategy will assign a specific role to each of the three robots depending on their position. The role assigned to a robot determines its behavior.

The main criterion for selecting a global strategy is what team posses (currently or in the next future) the ball. The team that will get the ball first is declared the offender while the other will be the defender. When designing offensive and defensive strategies it is important to note that offensive playing requires far more accuracy than defensive. It is more difficult to kick the ball with the appropriate orientation in order to hit the goal, than zone. Following this approach the arrangement of the players in the field is optimized by maximizing the team's controllable zone. This way, we maximize the probability of getting the ball.

More formally, the controllable zone of a player O_1 against an opponent X_1 at time t denoted as $C(O_1|X_1,t)$, represents a set of field positions that the player, O_1, can occupy first against the opponent, X_1. For convenience, we simplify the notation $C(O_1|X_1,t)$ as $C(O_1|X_1)$, assuming that t is understood from the context.

Fig. 1 illustrates $C(O_1|X_1)$ and $C(X_1|O_1)$ with the assumption that the time for O_1 and X_1 to reach a field position P is strictly proportional to the distances to P from their current positions. However, this assumption may not be realistic: the time for O_1 or X_1 to reach P should be a function of the initial and maximum velocity, the acceleration, as well as the initial orientation of O_1 or X_1. Should these factors be considered, a different shape and location for the border line would have resulted.

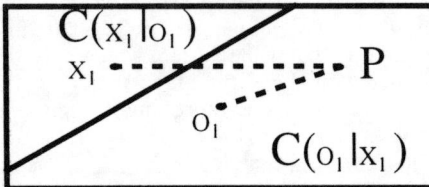

Fig 1: $C(O_1|X_1)$ and $C(X_1|O_1)$ with the assumption that the access time to P is proportional to distance.

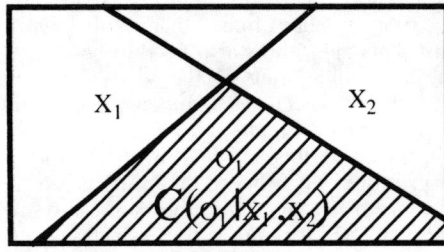

Fig. 2: $C(O_1|X_1,X_2)$

By the same token, we can define $C(O_1|X_1,X_2)$ as a set of field positions that O_1 can occupy first against X_1 and X_2, as illustrated in Fig. 2. Then, the following theorem holds:

Theorem1 $\quad C(O_1|X_1,X_2) = C(O_1|X_1) \cap C(O_1|X_2)$

Now, let us define a new notation $C(O_1|X_1,X_2,O_2)$ to represent a set of field positions that O_1 can occupy against X_1 and X_2 with the assistance of O_2. That is, O_2 can block the motion of X_1 or X_2 in such a way as to maximize the controllable zone of the team X_1 and X_2. Fig.3 illustrates $C(O_1|X_1,X_2,O_2)$.

In Fig. 3, when O_2 blocks X_1, O_1 can only be blocked by X_2 such that O_1 can extend its controllable zone to the area where the controllable zone of O_2 against X_1 intersects

or $\quad C(O_1|X_1,X_2) \cup [C(O_1|X_2) \cap C(O_2|X_1)]$

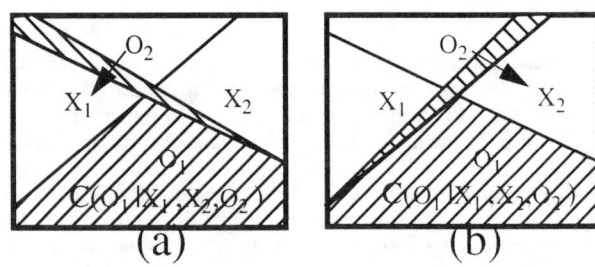

Fig. 3: (a) $C(O_1|X_1,X_2,O_2)$ when O_2 blocks X_1 (b) $C(O_1|X_1,X_2,O_2)$ when O_2 blocks X_2}

Since it is also important to describe the behavior of the ball in connection with the players, the following zones are defined:
- $C(B(O_1)|X_1,X_2)$: the zone where the Ball B possessed and kicked by O_1 can reach against X_1 and X_2 (which try to block the Ball)
- $C(O_2(B)|X_1,X_2)$: the zone where O_2 can capture the ball B against X_2 and X_2 (which try to capture the ball).

Fig. 4 illustrates $C(O_1(B)|X_1,X_2)$. In Fig. 4(a), $C(O_1(B)|X_1,X_2) = \{P\}$ since the ball is assumed dead on P and O_1 can capture it against X_1 and X_2 at P. Should the ball move in $C(O_1(B)|X_1,X_2)$ may become a set of points or nil. In Fig. 4(b), the length of L depends on the relative speed between the ball and O_1. Note that we can take the uncertainty involved in the ball motion into consideration for $C(O_1(B)|X_1,X_2)$:

Theorem3

$$C(O_1(B)|X_1,X_2) = C(O_1(B)|X_1) \cap C(O_1(B)|X_2)$$

Fig. 5 illustrates that $C(B(O_1)|X_1,X_2) = \{L\}$ where O_1 kicks the ball such that the ball moves along the trajectory, $\{L\}$, until it is interfered with either X_1 or X_2. Depending on the locations and speeds of X_1, X_2 and the ball, the ball may or may not penetrate X_1 and X_2 to reach the goal region. Again, the following theorem holds:

Theorem4

$$C(B(O_1)|X_1,X_2) = C(B(O_1)|X_1) \cap C(B(O_1)|X_2)$$

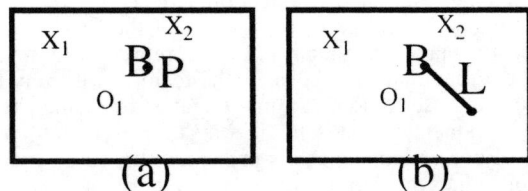

Fig. 4: (a) $C(O_1(B)|X_1,X_2) = \{P\}$ (b) $C(O_1(B)|X_1,X_2) = L$

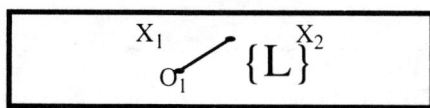

Fig. 5: $C(B(O_1)|X_1,X_2)) = C(B(O_1)|X_1)C(B(O_1)|X_2)$

Computation of Controllable Zones

First, we discretize the soccer field and represent it by nxm grid points. The number of grid points may be determined based on the trade-off between the precision and the required computational resources (such as memory and processing time).

For each grid point, (i,j), and at each sampling time for O_1, O_2, X_1, X_2, and B to reach (i,j) will be computed. As stated earlier, this time depends on the position, velocity, acceleration, and orientation of each object. For instance, the time for O_1 to reach the position P(i,j) can be calculated by generating a linear trajectory with circular blend based on the initial velocity $V_c(O_1)$, the maximum velocity, $V_{max}(O_1)$, and the average acceleration, $a(O_1)$. In Section III we show how this can be done based on our path planner. In general, the time can be computed by the formula for the trajectory generation for a 2 degrees of freedom manipulator in the Cartesian space [2]. With the time computed for O_1, O_2, X_1, X_2, B to reach (i,j), we can calculate various controllable regions defined before:

- $C(O_1|X_1)$ can be obtained as a collection of P(i,j) at which the time for O_1 to reach (i,j) is shorter than the time for X_1.
- $C(O_1|X_1,X_2)$ is a collection of P(i,j) at which O_1 precedes X_1 and X_2 in time. Or it can be obtained by the intersection of $C(O_1|X_1)$ and $C(O_1|X_2)$.
- $C(O_1|X_1,X_2,O_2)$ can be obtained by $C(O_1|X_1,X_2)$, $C(O_1|X_1)$, $C(O_1|X_2)$, $C(O_2|X_1)$, and $C(O_2|X_2)$ based on Theorem. 2.
- $C(O_1(B)|X_1)$ can be obtained as a collection of P(i,j) at which O_1 precedes B and B precedes X_1 in time.
- $C(O_1(B)|X_1,X_2)$ can be obtained as a collection of P(i,j) at which O_1 precedes B and B precedes X_1 and X_2 in time. Or, it can be obtained from $C(O_1(B)|X_1)$ and $C(O_1(B)|X_2)$ based on Theorem. 3.
- $C(B(O_1)|X_1)$ can be obtained as a collection of P(i,j) at which B precedes X_1.
- $C(B(O_1)|X_1,X_2)$ can be obtained as a collection of P(i,j) at which B precedes X_1 and X_2. Or, it can be obtained from $C(B(O_1)|X_1)$ and $C(B(O_1)|X_2)$ based on Theorem. 4.

Offensive strategy algorithm

The proposed coordinated motion strategy starts with classifying the game situation at time t as an offensive mode or a defensive mode. Let us assume that the two opposing teams, O and X, consist of $\{O_1,O_2\}$ and $\{X_1,X_2\}$ as their respective offensive and defensive excluding their goalkeepers, O_g and X_g.

The game situation at time t is declared as in an offensive mode for the team O, if O_1 or O_2 possesses the ball or $C(O_1(B)|X_1,X_2) \neq 0$, or $C(O_2(B)|X_1,X_2) \neq 0$ at time t. Otherwise, it is declared as a defensive mode. In an offensive mode, the team O can generate coordinated motions of its individual players by identifying the game pattern and applying the corresponding game strategy, as described in the following rules:

1. Shooting: If $C(B(O_1)|X_1,X_2)$ is a subset of goal region, then shoot the ball; while O_2 moves in such a way as to maximize $C(O_1|X_1,X_2) \cup C(O_2|X_1,X_2)$.
2. Capturing: If O_1 does not possess B, but $C(O_1(B)|X_1,X_2) \neq 0$, then O_1 moves to capture B, while O_2 moves in such a way as to maximum $C(O_1|X_1,X_2,O_2)$.
3. Dribbling: If O_1 possesses the ball and $C(O_1|X_1,X_2,O_2)$ is a subset of a penetration region. Then O_1 dribbles B into the penetration region; while O_2 blocks either X_1 or X_2 depending on whether $C(O_1|X_1) \cap C(O_2|X_2)$ or $C(O_1|X_2) \cap C(O_2|X_1)$ makes better in penetration. A penetration region is a collection of P(i,j) which lies on or behind the last defender (excluding the goal keeper) and is connected from any controllable region in front of the last defender.
4. Passing: If O_1 posses the ball and $C(O_1|X_1,X_2,O_2)$ is not a subset of penetration region, but $C(O_2|X_1,X_2,O_1)$ is a subset of penetration region and $C(B(O_1)|X_1,X_2) \supseteq C(O_2(B)|X_1,X_2)$, then O_1 passes the ball to O_2.
5. Searching: If the situation is none of the above classes, then move O_1 and O_2 in such a way as to maximize $C(O_1|X_1,X_2) \cup C(O_2|X_1,X_2)$.

For the determination of better choice between $C(O_1|X_1) \cap C(O_2|X_2)$ and $C(O_1|X_2) \cap C(O_2|X_1)$ as well as in maximizing $C(O_1|X_1,X_2) \cup C(O_2|X_1,X_2)$, we define a quality. To do so, we assign weights to individual grid points in such a way that the points near the goal (where the scoring chance is greater) gets higher weights. The quality is then the sum of the weights of the points included in the controllable region.

The quality indices associated with various controllable regions together with the above strategic rules allow to determine the coordinated motion of individual players.

2.2 Defensive Strategies

As opposite to offensive strategies, defensive strategies require less accuracy. To effectively block an opponent or save the ball, we just need to place the robot in the target point as soon as possible. Therefore, more deterministic approaches can be used. When a defensive position is declared, each robot in the team is assigned a specific role based on its current position in the field. We identify three roles: goalkeeper, kicker's blocker and striker's blocker.

Goalkeeper: the goalkeeper is responsible for saving the goal by intercepting the ball each time it approaches the goal. For this purpose, the goalkeeper will be placed lieing on the straight line connecting the current ball's position and the center of the goal. The exact point within this line depends (linearly) upon how far the ball is from the goal. When the ball is far away, the goalkeeper will be allowed to take some risks and move forward, thus hiding a greater portion of the goal while augmenting the field area controlled by the team (Fig.6a). However, when the ball is close to the goal, the goalkeeper will be more precautious and remain at the goal (Fig. 6b). Finally, if the ball dangerously approaches the goal, so that the goalkeeper is closer than anybody else to the ball, the goalkeeper will be instructed to run towards it and kick it away. When this happens, the robot closest to the goal among the remaining team-mates, will run to cover the goal that has been left empty by the goalkeeper (Fig. 6c). As experimental results show, this behavior resembles very much the one displayed by human goalkeepers playing a real soccer game.

The goalkeeper role, as any other, is dynamically assigned. Each time the robot that is best placed to act as the goalkeeper is determined. The robot closest to the goal is picked up as the goalkeeper.

kicker's blocker; the robot performing this role will be responsible for hindering the movements of the opponent possessing the ball, by blocking the direction of progression, shooting or passing. Eventually, this robot will capture the ball and begin the offensive strategy. These objectives are achieved by predicting where the opponent kicker is going to move next. For this purpose the point where the kicker will hit the ball is estimated. The defender will run towards this point (see Fig. 7).

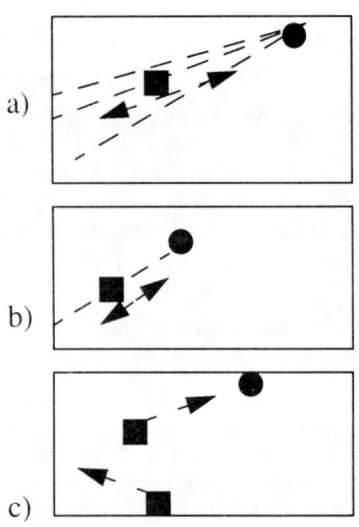

Fig. 6: Goalkeeper role.

This role is assigned to the robot closest to the estimated point where the opponent will kick the ball. If this robot has been already assigned to the goalkeeper role, the next closest is taken.

stricker's blocker: This robot will be responsible for blocking the opponent's second striker. The opponent's second striker is the closest opponent robot to the team's goal, without considering the opponent robot possessing the ball. Its defender will try to impede it to move forward. For this purpose, the stricker's blocker will run to stay in front of the stricker (see Fig. 8)

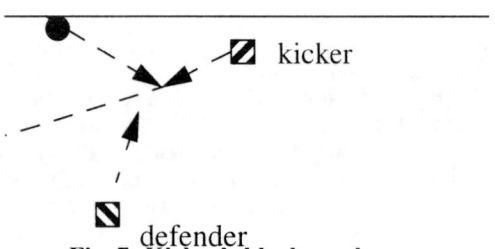

Fig. 7: Kicker's blocker role.

The robot assigned to this role is the remaining robot with no assigned role yet.

III PATH PLANNING

This module is responsible for computing the path that will bring a micro-robotic soccer player equipped with one independently controlled wheel at each lateral side, to a desired target as fast as possible. At this target the robot will kick the ball, capture it or block an opponent. Depending on the specific task to be performed, the orientation of the robot at the final point is relevant or not. For example, a striker will need to face the ball with the exact orientation in order to score a goal. On the other side, a defender will adequately block an intruder regardless of its orientation. To design a path, the path planner needs to be informed whether the orientation at the final point is relevant or not, since this places two very different requirements on the path planner.

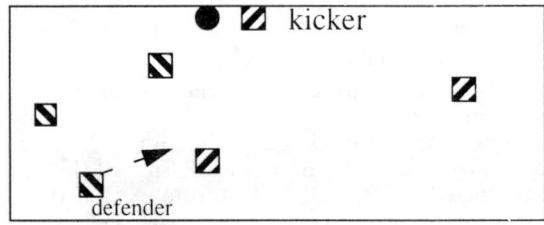

Fig. 8: stricker's blocker role.

In order to optimally meet both of these individual requirements at the same time, two different algorithms have been used. A simple, fast algorithm will compute a short path when no final orientation is required. On the other hand, a more complex, somewhat slower algorithm will optimize the design of paths where the final point has to be reached with a given orientation.

3.1 Shortest Path: Related Work

The study of shortest paths in the absence of obstacles for a system similar to our micro-robot soccer players was first performed by Dubins [3]. The vehicle he studied had a fixed linear velocity and a low-bounded turning radius. Dubins proved that the shortest paths are curves of class C1, composed of arcs of circle with minimum radius and straight line segments. Reeds and Shepp [4] extended the work of Dubins to the case of car-like vehicles. Here the linear velocity can be positive or negative, thus allowing for maneuvers, or cusps, along the path. The shortest paths are then piece-wise smooth with bounded curvature where it is defined. Between cusps the path must be of the form given by Dubins. Essentially their method provides a family of paths containing at most two cusps, that includes the shortest path. These family is characterized by a sequence of at most five pieces which are straight line segments or arcs of circle with minimum curvature. More proofs of Reeds and Shepp's result have been obtained in the framework of the optimal control theory [5] and [6]. Finally, it is possible to complete Reeds and Shepp's characterization of the shortest path by providing the synthesis of all the shortest paths [7].

However, as it might be seen, all these studies are based on the construction of a family of paths containing the shortest path. The actual shortest path is extracted by searching in this family. Such techniques are too time-consuming to be implemented in our micro robots, where fast, suboptimal algorithm are preferred over slow, optimal algorithms. On the other hand, our micro robots have slightly different characteristics from the vehicles studied in the literature depicted above. First, for our players it is irrelevant whether the goal is reached with the front or the back side ahead of the robot. The robot can kick the ball equally well regardless of the side it bounces against. This important feature eliminates the need for cusps along the path. Second, the low-bound constrain on the turning curvature can be almost removed for our micro robots. By commanding opposite turning directions to the wheels we achieve a near-zero radius of curvature. Finally, in our case, the linear velocity is not fixed, but depends on the turning radius:

$(r+a)/v_l = r/v$
$(r-a)/v_r = r/v$

where v_l and v_r are the commanded speeds for the left and right wheel respectively, v is the linear velocity, r the turning radius and 'a' half the robot's width.

Nevertheless, the work of Dubins and his followers provides very valuable insights about how to plan the paths for our robots: paths will be sequences of at most three pieces. The first part will be a turning segment where the robot will track a circle of minimum radius of curvature until it faces the target. The second segment will be a translation phase, where the robot approaches the target in a straight motion. Finally, when a desired orientation must be met at the end, a third piece will guide the robot along a low-radius (ball's radius plus half robot's width) circle to the desired goal. Fig 9 illustrates these ideas.

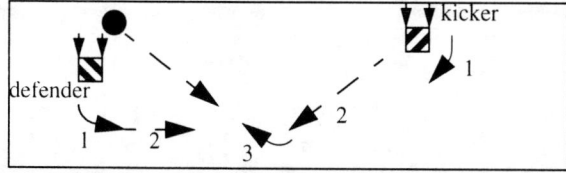

Fig. 9: Piece-wise decomposition of the planned paths.

Paths of this kind are optimal in the spatial domain. However, since the linear velocity decreases with the turning radius, they are not optimal in the temporal domain. When the target is close to the origin, it might be faster turning with a greater radius. However, by traversing in straight lines we ensure that the robot will not bump into the field boundaries. Furthermore, in general, we minimize the probability of collisions. Finally, we respect a minimum turning radius (ball's radius plus half the robot's width) in the third part of our paths, in order to avoid to touch the ball before the appropriate orientation is attained. In addition, when hitting the ball, a minimum linear velocity will be required in order to provide a strong kick.

3.2 Simple, Orientation-Free Path Planner

This simple path planner provides a fast way to move towards a target when no final orientation must be accomplished. This is rather frequent, since most tasks can be performed without considering the orientation.

These paths are made up of two pieces: first the robot rotates until it faces the target. Then it starts running in a straight motion. Since there is a great uncertainty involved, the robot will not wait until the angle to the goal is exactly zero before it starts running; this would keep the robot rotating forever. To compensate the deviation along the way, the robot will follow the arc of the circle connecting its current position and the target point, tangent to the robot direction at the robot's current position. The radius of this circle is calculated as follows:

$r = {}^r y/(2\sin(Q)\cos(Q))$
$Q = \arctan({}^r x/{}^r y)$

where ${}^r x$ and ${}^r y$ are the target coordinates in robot-based coordinate system.

The length of the path tracked by the robot's wheels can be estimated as follows:

$$l = \arctan\left(\frac{{}^r x}{{}^r y}\right) a + \sqrt{{}^r x^2 + {}^r y^2}$$

where ${}^r x$ and ${}^r y$ give the target's location in the robot-based coordinate system and 'a' is half the robot's width

3.3 Fixed-Orientation Path Planner

This algorithm is able to compute paths where the robot will reach the destination with a desired orientation. Such paths are required when the robot is instructed to appropriately shoot the ball in order to hit the goal, or when the goalkeeper has to save the goal from a ball that is already behind him.

As Fig. 10 shows, the paths are made up of three pieces. In addition to the two initial phases of the simple path planner, a final stage brings the robot along a circle to reach the target with the right orientation. However, first it is crucial to find out where the robot will enter this circle. In fact, the entry point to this circle will constitute the actual target the robot will run towards during the first two stages. Therefore the algorithm will begin drawing both possible final circles taking into account the target orientation. Depending on whether the robot is to the left or the right of the target, one or the other is chosen. Then the point where the robot will enter this circle is found by solving the following system of equations:

$(x_{cc}-x_r)\cos Q + (y_{cc}-y_r)\sin Q = r$

$(x_{cc}-x_r)(-\sin Q) + (y_{cc}-y_r)\cos Q = {}^r y$

$({}^r x^2 + {}^r y^2)^{0.5} = r$

where (x_{cc}, y_{cc}) and (x_r, y_r) are the absolute coordinates of the center of the circle and the robot position respectively. $({}^r x, {}^r y)$ are the coordinates of the entry point relative to the robot's coordinate system, Q is the angle the robot must rotate in order to face the entry point, and r is the radius of the circle.

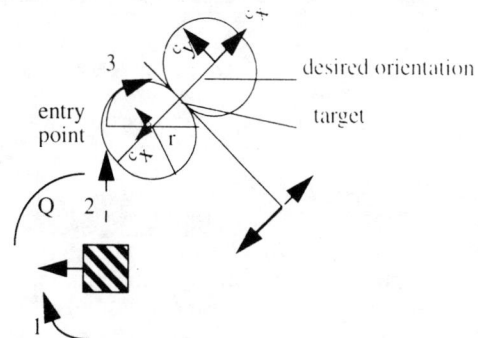

Fig. 10 Computation of fixed-orientation paths

At this point, the simple path planner is invoked to bring the robot to the entry point. This will place the robot on the circle with velocity tangent to it. Finally the circle is tracked and the goal is reached.

We estimate the length of the path traversed by the robot's outer wheel, as the length of the path to reach the entry point plus the length of the tracked final circle's arc:

where ($^r x, ^r y$) and ($^c x, ^c y$) are the coordinates of the entry point relative, respectively, to the robot and the center of the circle, and yneg is a boolean term that indicates that half circle should be added when $^c y$ is negative. The output of both path planners is a pair (v,r) giving the desired linear velocity and turning radius. This is translated into the corresponding motor commands by the following equations:

$v_l = (r+a)v/r$
$v_r = (r-a)v/r$

where v_r and v_l represent the commanded speed for the right and left wheel respectively and 'a' is again half the robot's width.

IV SIMULATION

We implemented the system in a simulation where two teams compete against each other. To make matters more realistic we added some noise to the perceptual information. We then discretized the field by 26x18 grids of size 5cm x 5cm. We calculate the time information using this formula.

$$t = \frac{l}{V_c(O)} + \frac{V_{max}(O)}{a(O)}$$

where l is the path length to the target position, given in Section III, $V_c(O)$ and $V_{max}(O)$ are, respectively, the current and maximum velocity of O, and a(O) is the average acceleration of O.

We also put higher weights around the enemy goal. We then computed controllable zones C(O|X) functions as described above. After deciding which is the offensive team and which is the defensive team, we applied for each robot the appropriate strategy. With the target position known for each robot, we then find the path to follow, and finally compute the corresponding motor commands.

Fig. 11 to 13 show different snapshots of the simulation. We see how the defenders are looking for their assigned striker (Fig. 11). In Fig. 12 the robot possessing the ball has penetrated the defender's area, while his teammate prevents a defender from hindering the penetration. Finally, in Fig. 13, the goalkeeper is instructed to kick the ball away, while the nearest defender replaces him.

V CONCLUSIONS

To achieve intelligent coordinated motion in team competition a global assessment of the game position is needed. Then, as humans do, each player determines, based on his current position, individual skills, etc., which of the required actions he can perform better than any other team-mate. The novel method based on controllable zones we presented provides powerful, rigorous yet efficient means of assessing and representing the game, allowing each robot to decide what action he is supposed to perform in order to achieve coordinated motions and global strategies. Simulation results demonstrated that complementary behaviors were accomplished. We also showed optimized paths that allow easy estimation of the time required to reach a field position; a factor that simulation results showed to be crucial.

Fig. 11: Simulation results.

Fig. 12: Simulation results

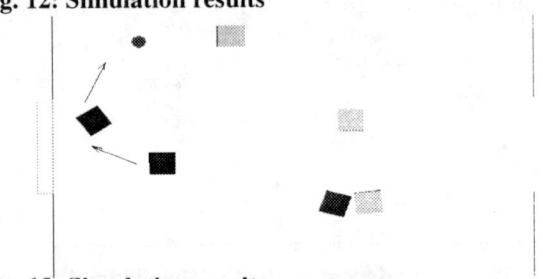

Fig. 13: Simulation results

IV REFERENCES

[1] "http://vivaldi.kaist.ac.kr/MIROSOT/".
[2] John Craig, "Introduction to Robotics: Mechanics and Control," Addison-Wesley, 1986
[3] L E Dubins; "On curves of minimal length with a constraint on average curvature and with prescribed inital and terminal position and tangents". Am. J. Math. vol. 79, pp. 497-516, 1957.
[4] J A Reeds, R A Shepp; "Optimal Paths for a robot that goes both forward and backwards". Pacific J. Mathematics, vol. 145, no. 2, pp. 367-393, 1990.
[5] J D Boissonnat, A Cerezo, J Leblong; "Shortest paths of bounded curvature in the plane". in IEEE Conf. RObotic Automat. Nice, France. pp. 2315-2320, 1992.
[6] H J Sussmann, W Tang; "Shortest paths for the Reeds-Shepp car: A worked out example of the use of geometric techniques in nonlineal optimal control". Report SYCON-91-10, Rutgers University, 1991.
[7] P Soueres, J P Lamond; "Shortest path synthesis for a car-like robot". European Control Conf. Groningen. pp 570-577, June 1993

Sony Legged Robot for RoboCup Challenge

Hiroaki Kitano[‡], Masahiro Fujita[†], Stephane Zrehen[†], and Koji Kageyama[†]

[†]D21 Laboratory, Sony Corporation
6-7-35, Kitashinagawa, Shinagawa-ku, Tokyo, 141 JAPAN
[‡]Sony Computer Science Laboratory Inc.
Takanawa Muse Building, 3-14-13 Higashi-Gotanda, Shinagawa-ku, Tokyo, 141 JAPAN

Abstract

One of the ultimate dream in robotics is to create life-like robotics systems, such as humanoid robot and animal-like legged robot. We choose to build pet-type legged robot because we believe that dog-like and cat-like legged robot has major potential for future entertainment robotics markets for personal robots. However, numbers of challenges exists before any of such robot to be fielded in the real world. Robots have to be reasonably intelligent, maintains certain level of agility, and be able to engaged in some collaborative behaviors. RoboCup is an ideal challenge to foster robotics technologies for small personal and mobile robotics system. This paper, we present Sony's legged robots system which enter RoboCup-98 Paris as a special exhibition games.

1 Introduction

Robot systems with life-like appearance and behaviors are one of the ultimate dream in robotics and AI [Inaba, 1993; Maes, 1995]. Honda's Humanoid Robot announced in early 1997 clearly demonstrated that it is technically feasible and we have all recognized that social and psychological impacts of such technologies are far reaching. A robot system described in this paper presents our effort to build reconfigurable robot system that has high degree of design flexiblilty. While many robotics system uses wheel-based driving mechanisms, we choose to develop legged robot which can be converted into wheel-based robots easily by using reconfigurable physical components. Among various types of robot configurations attainable using our architecture, our immediate focus is a legged robot configuration with four legs and a head each of which has three degree of freedom. Given that our goal is to establish Robot Entertainment industry, this configuration is attractive because it resembles dogs or cats, so that people may view it as robot pets. It also entails numbers of technical issues in controlling its motion, while avoiding difficulties of bi-pedal robot systems.

We are interested in participating in RoboCup, Robot World Cup Initiative [Kitano, et al., 1997], using our legged robots, as a part of its exhibition program. RoboCup is an ideal forum for promoting intelligent robotics technologies because it offer challenging and exciting task of playing soccer/football games by multiple robots. In order for a team of robots to play soccer game, robots should be able to identify environment in real-time under less controlled world than the laboratory set up, move quickly, and has to have certain level of intelligence. For our legged robots, RoboCup would be even more challenging than wheel-based robots because participation to RoboCup requires sophisticated control system for 15 degree-of-freedom for multiple robots. In addition, body posture and head position moves constantly due to legged locomotion. This imposes serious problem for a vision system which acts as a central sensor system to identify ball position, opponent position, and position of goals.

All these challenges actually represents basic requirements for developing robot for personal use, especially for entertainment robots. For robot entertainment, each robot has to be fully autonomous, and it should be exhibit reasonable intelligence and maneuvering capability. For example, a robot dog should be able to identify the owner and chase a ball, and play with it. If unwanted person or objects show up in front of it, a robot dog should react quickly to escape from it. Often, few robot dogs may chase an emeny as a team. Clearly, these basic elements can be found in RoboCup games. Of course, legged robot to play soccer game constitutes only a part of robot entertainment market. However, it can be a clear and present new market and it is a good starting point.

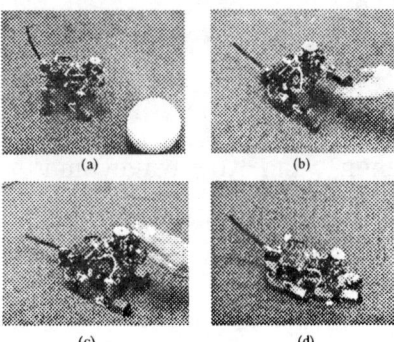

Figure 1: MUTANT: Fully autonomous pet-type robot

Figure 2: Remote-operated soccer game robot

The legged robot described in this paper is based on OPENR standard [1], which is designed as a standard architecture and interfaces for Robot Entertainment [Fujita and Kageyama, 1997a]. OPENR aims at defining a standard whereby various physical and software components which meet the OPENR standard can be assembled with no further modifications, so that various robots can be created easily. In the next section, OPENR will be described briefly, followed by description of actual implementation of legged robots for RoboCup-98.

2 OPENR Standard

2.1 Overview

OPENR is a standard for robot entertainment which enables reconfiguration of robot structure, as well as easy replacement of software and hardware components.

One of the purposes of proposing OPENR is that we wish to further promote research in intelligent robotics by providing off-the-shelf components and basic robot systems. These robots should be highly reliable and flexible, so that researchers can concentrate on the aspect of intelligence, rather than spending a substantial proportion of their time on hardware troubleshooting.

For the feasibility study of OPENR, we developed an autonomous quadruped pet-type robot named MUTANT[Fujita and Kageyama, 1997a; Fujita, et al., 1998a] (Fig.1). We also implemented and tested a remote-operated robot system for soccer game (Fig.2).

---

[1] OPENR is a trade mark of Sony Corporation.

Using these robots, software and application feasibility studies were carried out.

The legged robot for RoboCup described in is paper is a direct descendent of MUTANT and remote control soccer robots. The legged robot for RoboCup is a fully autonomous soccer robot.

2.2 Major Features

One of the salient feature of the OPENR standard is that is attains several critical dimensions of scalability. These are (1) size scalability, (2) category scalability, (3) time scalability, and (4) user expertise scalability.

Size Scalability (Extensibility): OPENR is extensible for various system configuration. For examples, in a minimum configuration, a robot may be composed of only few components, and perform a set of behaviors as a complete agent. This robot can be scaled up by adding additional physical components. It is possible to scale up such a system by having such robots as sub-systems of large robot systems.

Category Scalability (Flexibility): Category scalability ensures that various kinds of robots can be designed based on the OPENR standard. For example, two very different styles of robots, such as a wheel-based robot and a quadruped-legged robot, should be able to described by the OPENR standard. These robots may have various sensors, such as cameras, infra-red sensors, and touch sensors and motor controllers.

Time Scalability (Upgradability): OPENR can evolve together with the progress of hardware and software technologies. Thus, it must maintain a modular

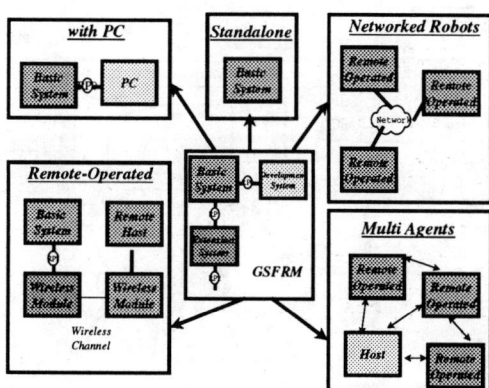
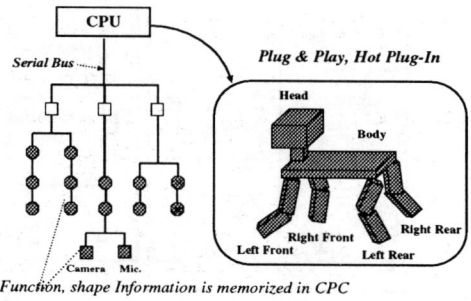

Figure 3: Generic System Functional Reference Model and Examples of Derived System Architectures

Figure 4: Configurable Physical Component

organization so that each component can be replaced with up-to-date modules.

User Expertise Scalability (Friendly Development Environment): OPENR provides a development environments, both for professional developers and for end-users who do not have technical knowledge. End users may develop or compose their own programs using the development environment. Thus, it is scalable in terms of the level of expertise that designers of the robot have.

2.3 OPENR Strategy

Our strategy to meet these requirements consists of the following:

Generic Reference Model: To meet the requirements of Extensibility and Friendly Development Environment, we define a generic system functional reference model (GSFRM) composed of Basic System, Extension System and Development System. By defining GSFRM, we are able to construct various kinds of robot systems with extensibility and development environments, as shown in Fig.3.

Configurable Physical Component: To meet the requirements of Flexibility and Extensibility, we devise a new idea of Configurable Physical Component (CPC). The physical connection between the robot components is done by a serial bus. In addition every CPC has non-volatile memory with (1) functional properties, such as an actuator and a two dimensional image sensor, and (2) physical properties, which help solve the dynamics of the robot consisting of these CPCs. Fig. 4 illustrate this concept. With this scheme, the robot will be physically reconfigurable [Fujita, et al., 1998b].

Object-Oriented Programming: To meet the requirements of Up-gradability and Flexibility, we employ an object-oriented OS, **Aperios** [Yokote, 1992], which supports the Object-Oriented Programming (OOP) paradigm from the system level with several types of message passing among objects. In addition, Aperios is capable of customizing APIs by system designers.

Layering: To meet the requirements of Up-gradability and Friendly Development Environment, we utilize the layering technique which is often used for multi-vendor open architecture. OPENR divides each functional element into three layers, Hardware Abstraction Layer (HAL), System Service Layer (SSL), and Application Layer (APL), as shown in Fig.5

In order to achieve the software component concept, an Agent Architecture for entertainment application was studied. The details is described in [Fujita and Kageyama, 1997a].

3 Legged Robot Challenge

A team of legged robot soccer players is a major challenge. It must address various issues involving control of robot with multiple degree of freedoms and stabilization and compensation of images obtained

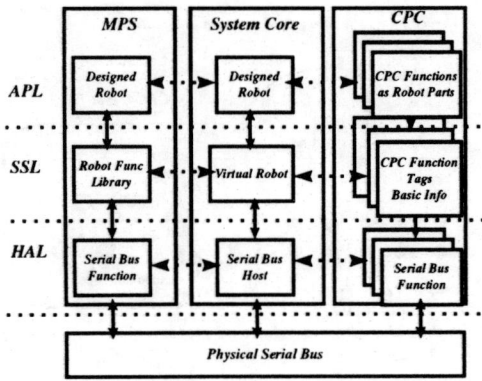

Figure 5: Layering for Basic System

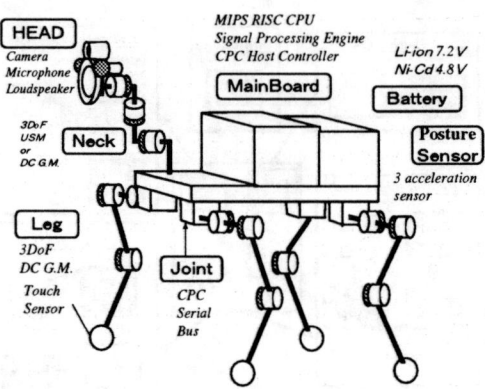

Figure 6: Mechanical Configuration

from continuously dislocating vision system due to legged locomotion, and other issues, in addition to all difficult problems offered in RoboCup Challenge for wheel-based robots.

3.1 Legged Robot Platform for RoboCup

3.1.1 Quadruped Legged Robot

For RoboCup-98 Paris, we will deploy legged robot with four legs and one head, each of which has three degree of freedom, and rich sensory channels, for example, a head has a color CCD camera, stereo microphone, touch sensors, and a loud speaker.

Most of the intelligent autonomous robots are implemented in wheel-based mechanical configuration. A wheel-based robot has advantage in their simplicity of motion control, so that researchers can concentrate on vision, planning, and other high-level issues. However, since our goal is robot entertainment, different emphasis shall be made. We believe that the capability of representation and communication using gesture and motion is very important in entertainment applications. Therefore, we choose a mechanical configuration of our robot as a quadruped-legged type, as shown in Fig.6.

The merits of the quadruped-legged configuration are, (1) walking control of a quadruped-legged is easier than that of a biped robot, and (2) when in a posture of sitting, two hands are free to move, therefore, they can be used to present emotions or to communicate with a human by the motions of the hands. Since each leg or hand has to be used for various purposes besides walking, we assign three degree of freedom (DoF) for each leg/hand. In addition, we add a tail and three

DoF for neck/head so that the robot has enough representation and communication capabilities using motions.

During the RoboCup games, legs are not necessary used for expressing emotions. However, they can be used for sophisticated control of balls, such as passing ball to the side or back, or engaged in deceptive motions.

Disadvantages of using legged robot is their moving speed is not as fast as wheel-based robots. In future, speed issue may be resolved when galloping was made possible. For now, legged robot will be played within dedicated league. Although serious hardware limitation exists, teams with efficient leg motion coordination will have major advantages in the game.

3.1.2 Standalone System

In general, it is difficult for a standalone robot system to perform these tasks in real time in a real world environment because of its limited computational power. The remote operated robot system depicted in Fig.3 can solve the computational power problem; however, in general, much computational power is necessary for image processing tasks. This implies that it is necessary for each robot to be equipped with a video transmitter. This is sometimes difficult for regulation reason.

Another solution to this problem is to set up a camera overlooking the entire field, and to distribute the image signal to all host computers. In this set up, each robot can use the huge computer power of the host computer, or a special engine such as image processing hardware. However, the image information taken

Figure 7: Micro Camera Module

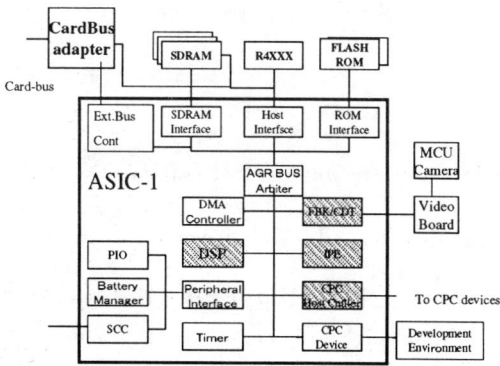

Figure 8: Electrical Block Diagram including the ASIC

by the overlooking camera is not the image taken from each robot's viewpoint.

We consider that technologies to process the image from each robot viewpoint without any global information will become very important in Robot Entertainment in future. Therefore, we decide to build RoboCup System with standaolone robots under local communication constraint.

There are two hardware issues that need to be resolved to enable full on-board vision system for small size robots: (1) camera size, and (2) processor power. We solved these problem by actually manufacturing a dedicated camera and a processor chip.

Micro-Camera-Unit: In order to make a robot small in size and weight and to reduce cost, we developed a Micro-Camera-Unit (MCU) using multi-chip-module technology ([Ueda, 1996]), as shown in Fig.7. This MCU includes a lens in the same package to achieve a single thin camera module. The size of the MCU is $23 \times 16 \times 4mm$, with pixels 362×492.

OPENR Chip: To make the robot small in size and in power consumption, we employ MIPS architecture's R4000-series CPU with more than 100 MIPS performance. We have also developed the dedicated ASICs including the peripheral controllers of the CPU, as described before. Fig.8 shows the entire electrical block diagram, where we employ the Unified Memory Architecture with synchronous DRAMs (SDRAM) as a main memory. The features of the ASIC are as follows:

CPC bus host controller: The CPC bus host controller controls all CPC devices which are connected in three structure fashion (Fig.??).

DMAC: The direct memory access controller (DMAC) copies the data with sufficient flexibility by using the list data structure which has the information of the source address, the destination address and the size of data.

FBK: The Camera data is transferred to the SDRAM through the multi-resolution filter bank (FBK). This FBK consists of three layer resolution filters. The resolutions of the filters are 360×240, 180×120 and 90×60.

CDT: The filtered image data through the FBK can be applied by the color detection engine (CDT) so that eight color can be detected in real-time.

IPE: Because the inner-product calculation is often used in signal processing, the inner product engine (IPE) is integrated in the ASIC. This IPE can be used particularly for a comparison between a template image segment and an input image.

DSP: For sound processing, the 16bit-integer DSP with about 50MHz clock is integrated in the ASIC so that the FFT or filter bank processing for sound data can be done in real-time.

Although dedicated camera module and OPENR Chip enables on-board vision processing, it still requires major efforts to recognize natural images of balls and field. However, color marking regulation of

RoboCup enable us to identify goals, a ball, and other robots using Color Detection Engine.

4 Vision Subsystem

While there are numbers of software issues exist, this paper briefly describe vision subsystem to detect and track objects, and to localize robot's own position.

4.1 Object Tracking and Distance Estimation

There are few objects which need to be identified in RoboCup games – goals, a ball, and other robots. Basically, each object can be identified with color. A ball is painted in a single color, and each goal is painted in a distinguishable color. Robots are basically painted in black or a dark color with foe-friend identification markers. Obviously, the most important object is a ball. This can be done in rather simple manner, that to find a define color for the ball, e.g. red. To keep the ball in the center of the image, the robot head direction is feedback controlled, and the walking direction is also controlled using the horizontal head direction. The advantage of having a neck is that the robot can continue to keep the ball within visual field even when the robot do not walk toward the ball. These head position control is carried out by neck-subsystem using behavior-based approach [Brooks, 1986]. Our distributed layering of agent architecutre, similar to [Firby, 1994] enables such an implementation.

Both the horizontal and vertical neck angle can be utilized to estimate the distance between the ball and the robot body (Fig.9). This information is also used as a queue to execute a kick motion.

Assume that the robot knows H, the height of its camera view point from the ground, following procedure identifies distance of the robot from the ball.

1. Find the ball.

2. Measure the neck's vertical angle facing the ball (α), and horizontal angle (θ).

3. $H tan^{-1}(\alpha)$ is the distance from the ball along the axis of body's moving direction (d1), and $H tan^{-1}(\theta)$ is the displacement of the ball from the axis of body direction (d2).

4.2 Self-localization through vision

Visual landmarks on the walls of the field can be used for self-localization purposes. Indeed, any x-y po-

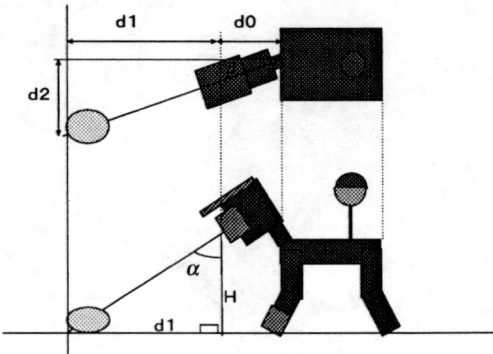

Figure 9: Ball Positioning

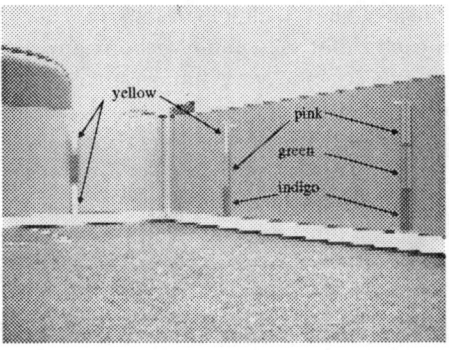

Figure 10: an image of two color landmarks pasted on the field walls (Original in color)

sition can be computed by triangulation of three landmarks azimuths, assuming that the absolute position of landmarks is known. In order to limit the computational demands on the CPU for visual processing and to make exhaustive use of the color processing hardware provided with the camera module, we will paint visual landmarks regularly along the walls of the field. These landmarks are represented on figure 10. They are about 10cm high and 3cm wide. They are composed of three color strips, and are limited on their top and their bottom by bright yellow bands. The information carrying colors are chosen among four: blue, pink, yellow and indigo. The bottom color indicates the wall on which the landmark is located.

The special color processing hardware allows the detection of 8 template colors and provides output in the form of 8 bit-planes, for every field, that is, every

Figure 11: The result of color detection of the two landmarks (Original in color)

50ms. Color detection proceeds as follows: The input image is represented in the YUV space, using 256 values for U and V and 32 values for Y. A lookup table is established for every template color in the form of a cube in the YUV space, the dimensions of which representing a matching threshold on color comparison. In our preliminary studies, using landmarks printed on a simple color printer and then pasted on the walls with yellow adhesive tapes, it appears that it is best to use two templates for each color, one for the lighter regions and one for the darker ones. This is notably due to the presence of shadows whose color is hard to model. An example of color detection is presented on Figure 11. The image is noisy, due to the existence of reflections on the plastic walls and to the absence of any noise-removing routine. However, this is not a major problem since in the localization subsystem we are only trying to identify and localize the landmarks in the visual field, and not to exhaustively segment the image. Landmark detection and identification is performed by detecting three small color squares in a vertical arrangement. On a sequence of 10 seconds where a -90 to +90 degrees motion was performed, all the landmarks but one were correctly identified. In such cases, the information extracted from the sequence can help overcome the wrong detection.

Since the micro-camera module has a limited visual angle, a head rotation movement might be necessary to identify and localize distant landmarks, for good resolution in the triangulation computation. This is time consuming and may not always be necessary. Perspective information can be used to obtain qualitative information about the robot's position in the field. For instance, on Figure 10, the relative size of the two landmarks on the right tells that the right wall of the field is on the right side of the robot. On the other hand the similar size of the two landmarks on the left tells that the robot is facing a corner since these two landmarks are known to be on different walls. In addition, it is possible to establish a lookup table of the visual size of a landmark as a function of its distance to the robot.

The use of this system will thus make self-localization possible and easy without requiring any global positioning system, which would need to be outside the robot.

5 Conclusion

Building a reliable and powerful robot platform meeting commercial-level product reliability is a challenging task. A small-size legged robot is even more challenging. We managed to produce such a robot platform by actually fabricating dedicated components, such as a micro camera unit, a special chip, and mechanical/electrical joints. These specification of robot were now being compiled as OPENR standard. OPENR will be able to supply broad range of useful mechanical and electrical hardware and software components.

Nevertheless, software to control such a robot and to make them behave to carry out meaningful tasks is major research topic. In order to facilitate research in the control of legged robot with multiple degree of freedom and rich sensor systems, we choose to developed a physical robot platform for RoboCup based on OPENR, and to participate in RoboCup.

6 Acknowledgment

The implementations of the proposed RoboCup System and robot platform are being performed by the members of Group 1, D21 laboratory. The MCM camera is developed by Sony Semiconductor Company, and Aperios is developed by Architecture Laboratory, Sony Corporation.

Authors would like to thank to Dr. Tani at Sony Computer Science Laboratory for his advice to the hand-eye coordination problem.

References

[Brooks, 1986] Brooks, R. A., 1986. A Robust Layered Control System for a Mobile Robot. *IEEE Jour-*

nal of Robotics and Automation, RA-2(1), March, pp.14–23.

[Firby, 1994] Firby, R. J., 1994, Task Networks for Controlling Continuous Processes, In *Proceedings on the Second International Conference on AI Planing Systems*, June.

[Fujita and Kageyama, 1997a] Fujita, M. and Kageyama, K., 1997, An Open Architecture for Robot Entertainment, In *Proceedings of the First International Conference on Autonomous Agents*, Marina del Ray, pp.234–239.

[Fujita, et al., 1998a] Fujita, M., Kitano, H. and Kageyama, K., 1998, "Development of an Autonomous Quadruped Robot for Robot Entertainment," *Autonomous Robots*, 1998.

[Fujita, et al., 1998b] Fujita, M., Kitano, H. and Kageyama, K., 1998, "Reconfigurable Physical Agent", *Proceedings of the Second International Conference on Autonomous Agents*, St. Paul, 1998

[Inaba, 1993] Inaba, M., 1993, "Remote-Brained Robotics: Interfacing AI with Real World Behaviors, In *Proceedings of the 6th International Symposium on Robotics Research (ISRR6)*, pp.335–344.

[Kitano, et al., 1997] Kitano, H., et al, 1997, RoboCup: A Challenge Problem for AI, *AI Magazine*, Spring, pp.73–85.

[Maes, 1995] Maes, P., 1995, Artificial Life meets Entertainment: Lifelike Autonomus Agents, *Communication of the ACM: Special Issue on New Horizons of Commercial and Industrial AI*, pp.108–114.

[Ueda, 1996] Ueda, K. and Takagi, Y., 1996, Development of Micro Camera Module, In *Proceedings of the 6th Sony Research Forum*, pp.114–119.

[Yokote, 1992] Yokote, Y., 1992, The Apertos Reflective Operating System: The Concept and Its Implementation, In *Proceeding of the 1992 International Conference of Object-Oriented Programing, System, Languages, and Applications*.

Building Integrated Mobile Robots for Soccer Competition

Wei-Min Shen, Jafar Adibi, Rogelio Adobbati, Bonghan Cho,
Ali Erdem, Hadi Moradi, Behnam Salemi, Sheila Tejada
Computer Science Department / Information Sciences Institute
University of Southern California
4676 Admiralty Way, Marina del Rey, CA 90292-6695
http://www.isi.edu/~shen

Abstract

Robot soccer competition provides an excellent opportunity for robotics research. In particular, robot players in a soccer game must perform real-time visual recognition, navigate in a dynamic field, track moving objects, collaborate with teammates, and strike the ball in the correct direction. All these tasks demand robots that are autonomous (sensing, thinking, and acting as independent creatures), efficient (functioning under time and resource constraints), cooperative (collaborating with each other to accomplish tasks that are beyond individual's capabilities), and intelligent (reasoning and planing actions and perhaps learning from experience). Furthermore, all these capabilities must be integrated into a single and complete system. To build such integrated robots, we should use different approaches from those employed in separate research disciplines. This paper describes our experience (problems and solutions) in this aspect for building soccer robots. Our robots share the same general architecture and basic hardware, but they have integrated abilities to play different roles and utilize different strategies in their behavior. Our philosophy in building these robots is to use the least possible sophistication to make them as robust as possible. In RoboCup97, our Dreamteam robots performed well (scored 8 of 9 goals of all teams in the league) and won the world championship in the middle-sized robot league.

1. Introduction

The RoboCup task is for a team of fast-moving robots to cooperatively play soccer in a dynamic environment [5,7]. Since individual skills and teamwork are fundamental factors in the performance of a soccer team, RoboCup is an excellent test-bed for integrated robots. Each soccer robot (or agent) must have the basic soccer skills— dribbling, shooting, passing, and recovering the ball from an opponent, and must use these skills to make complex plays according to the team strategy and the current situation on the field. For example, depending on the role it is playing, an agent must evaluate its position with respect to its teammates and opponents, and then decide whether to wait for a pass, run for the ball, cover an opponent's attack, or go to help a teammate.

In the "middle-sized" RoboCup league, robots are playing in a 8.22m x 4.57m green-floor area surrounded by walls of 50cm high. The ball is an official size-4 soccer ball and the size of goal is 150x50cm. (In the "small-sized" RoboCup league, the field is similar to a Ping-Pong table and the robots are playing a golf ball. There is no "large-sized" RoboCup.) The objects in the field are color coded, the ball is red, one goal is blue, the other is yellow, the lines are white, and players may have different colors. Each team can have up to five robot players with size less than 50cm in diameter. There was no height limit in 1997, so some robots were up to 100cm high. Since this was the first time for such a competition, teams were allowed to use global cameras, remote computing processors, and other remote computing devices. (We did not use any off-board resource, as you can see below, because we believe in total autonomous and integrated robots.)

Figure 1. Integrated Soccer Robots

To build agents with soccer-playing capabilities, there are a number of tasks that must be addressed. First, we must design an architecture to balance the system's performance, flexibility and resource consumption (such as power and computing cycles). This architecture, integrating hardware and software, must work in real-time. Second, we must have a fast and reliable vision system to detect various static and dynamic objects in the field, and such a system must be easy to adjust to different lighting conditions and color schema (since no two soccer fields are the same, and even in the same field, conditions may vary with time). Third, we must have an effective and accurate motor system and must deal with uncertainties (discrepancy between the motor control signals and the actual movements) in the system. Finally, we must develop a set of software strategy for robots to play different roles. This can add considerable amount of flexibility to our robots.

We realize that we are not the only nor the first to consider these problems. For example, long before the publication of [5,7], layered-controlled robots [3] and behavior-based robots [1,2] already began to address the problem of integrated robots. In a 1991 AI Spring symposium, the entire discussion [6] was centered around integrated cognitive architectures. We will have more detailed discussion on related work later.

Since building integrated robots for soccer competition requires integration of several distinct research fields, such as robotics, AI, vision, etc., we have to address some of the problems that have not been attacked before. For example, different from the small-sized league and most other teams in the middle-sized league, our robots perceive and process all visual images on-board. This will give much higher noise-ratio if one is not careful about how the pictures are taken. Furthermore, since the environment is highly dynamic, uncertainties associated with the motor system will vary with different actions and with the changes of power supply. This posts additional challenges on real-time reasoning about action than systems that are not integrated as complete and independent physical entities.

Our approach to built the robots is to use the least possible sophistication to make them as robust as possible. It is like teaching a kid to slowly improve his/her ability. Instead of using sophisticated equipment, programming very complicated algorithms, we use simple but fairly robust hardware and software (e.g., a vision system without any edge detection). This proved to be a good approach and showed its strength during the competition.

In the following sections of this paper, we will address the above tasks and problems in detail. The discussion will be organized as descriptions of component in our systems, with highlights on key issues and challenges. The related work will be discussed at the end.

2. The System Architecture

Our design philosophy for the system architecture is that we view each robot as a complete and active physical entity, who can intelligently maneuver and perform in realistic and challenging surroundings. In order to survive the rapidly changing environment in a soccer game, each robot must be physically strong, computationally fast, and behaviorally accurate. Considerable importance is given to an individual robot's ability to perform on its own without any off-board resources such as global, birds-eye view cameras or remote computing processors. Each robot's behavior must base on its own sensor data, decision-making software, and eventually communication with teammates.

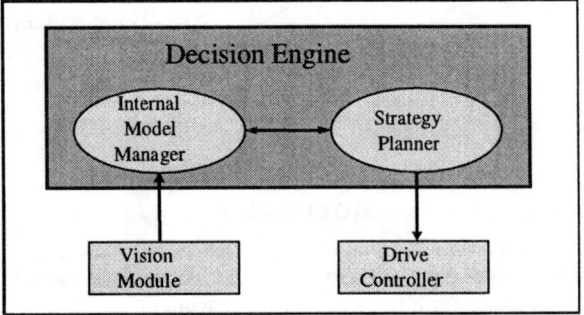

Figure 2. The System Architecture

The hardware configuration of our robot is as follows (see examples in Figure 1). The basis of each robot is a 30x50cm, 4-wheel, 2x4 drive, DC model car. The wheels on each side can be controlled independently to make the car spin fast and maneuver easily. The two motors are controlled by the on-board computer through two serial ports. The hardware interface between the serial ports and the motor control circuits on the vehicle are designed and built by ourselves. The robot can be controlled to move forward and backward, and turn left and right. The "eye" of the robot is a commercial digital color camera called QuickCam made by Connectix Corp.. The images from this camera are sent into the on-board computer through a parallel port. The on-board computer is an all-in-one 133MHz 586 CPU board extensible to connect various I/O devices. There are two batteries on board, one for the motor and the other for the computer and camera.

The software architecture of our robot is illustrated in Figure 2. The three main software components of a robot agent are the vision module, the decision engine, and the drive controller. The task of the vision module is to drive the camera to take pictures, and to extract information from the current picture. Such information contains an object's type, direction, and distance. This information is then processed by the decision engine, which is composed of two processing units - the internal model manager and the strategy planner. The model manager takes the vision module's output and maintains an internal representation of the key objects in the soccer field. The strategy planner combines the internal model with its own strategy knowledge, and decide the robot's next action. Once the action has been decided, a command is sent to the drive controller which is in charge of properly executing. Notice that in this architecture, the functionality is designed in a modular way, so that we can easily add new software or hardware to extend its working capabilities.

We use Linux as the on-board operating system and built a special kernel with 4MB file system, all compressed on a single 1.4MB floppy disk for easy down-loading. The entire software system (for vision, decision, and motor drive) consists of about 6,500 lines of C and C++ code.

One challenge we faced during the design of architecture was to draw a proper line between hardware and software. For example, to control the motors, we had to choice between using one serial port (a commercial laptop) or two serial ports (a complete all-in-one CPU board), we chose the later because we decide to solve the interface issue completely in hardware. (The former requires a complex software protocol and hardware interface). In retrospect, it seems that our decision on this issue were mainly driven by two factors: feasibility and robustness.

3. The Vision Module

Just as eyesight is essential to a human player, a soccer robot depends almost entirely on its visual input to perform its tasks, such as determining the direction and distance of objects in the visual field. These objects include the ball, the goals, other players, and the lines in the field (sidelines, end of field, and penalty area). All this information is extracted from an image of 658x496 RGB pixels, received from the on-board camera via a set of basic routines from a free package called CQCAM, provided by Patrick Reynolds from the University of Virginia.

Since the on-board computing resources for an integrated robot are very limited, it is a challenge to design and implement a vision system that is fast and reliable. In order to make the recognition procedure fast, we have developed a sample-based method that can quickly focus attention on certain objects. Depending on the object that needs to be identified, this method will automatically select certain number of rows or columns in an area of the frame where the object is most likely to be located. For example, to search for a ball in a frame, this method will selectively search only a few horizontal rows in the lower part of the frame. If some of these rows contain segments that are red, then the program will report the existence of the ball (recall that the ball is painted red). Notice that domain knowledge about soccer is useful here to determine where and how the sample pixels should be searched. For example, since the ball is often on the floor, only the lower part of the image needs to be searched when we are looking for the ball. Similarly, when the robot is looking for a goal, it will selectively search columns across the image and the search should from the floor up. Using this method, the speed to reliably detect and identify objects, including take the pictures, is greatly improved; we have reached frame rates of up to 6 images per second.

To increase the reliability of object recognition, the above method is combined with two additional processes. One is the conversion of RGB to HSV, and the other is "neighborhood checking" to determine the color of pixels. The reason we convert RGB to HSV is that HSV is much more stable than RGB when light conditions are slightly changed. Neighborhood checking is an effective way to deal with noisy pixels when determining colors. The basic idea is that pixels are not examined individually for their colors, but rather grouped together into segment windows and using a majority-vote scheme to determine the color of a window. For example, if the window size for red is 5 and the voting threshold is 3/5, then a line segment of "rrgrr" (where r is red and g is not red) will still be judged as red.

Object's direction and distance are calculated based on their relative position and size in the image. This is possible because the size of ball, goal, wall, and others are known to the robot at the outset. For example, if one image contains a blue rectangle of size 40x10 pixels (for width and height) centered at x=100 and y=90 in the image, then we can conclude that the blue goal is currently at 10 degree left and 70 inches away.

To make this vision approach more easily adjustable when environment is changed, we have kept the parameters for all objects in a table, in a separate file. This table contains the values of camera parameters such as brightness and contrast, as well as window size, voting threshold, average HSV values, and search fashion (direction, steps, and area). When the environment is changed, only this file needs to be changed and the vision program will function properly. We have tried various machine learning methods to allow program to automatically determine the values for each object in a new environment. However, they are still best determined by manually analyzing the color of each object in a new environment.

Given the speed of current processing rate of object recognition, it is now possible to track the moving direction of the ball and other players. To do so, a robot will take two consecutive pictures, and compare the locations of the ball in these two pictures. If the direction of the ball moves to left (right), then the robot concludes the real ball is moving towards left (right). In fact, this is how our goal-keeper predicts the movement of an incoming ball.

Vision modules such as the one described here also face problems that are unique for integrated robots. For example, images will have much higher noise-ratio if the robot is not careful about when and how the pictures are taken. It took us quite a long time to realize this problem. At first, we were very puzzled by the fact that although the vision system is tested well statically, our robot would sometimes behave very strangely as if it is blind. After many trials and errors, we noticed that pictures that are taken while the robot is still moving have very low quality. Such pictures are not useful at all in decision-making. Since then, special care has been given to the entire software system; furthermore, the robot takes pictures only when it is not moving.

4. Drive Controller

As specified in the system architecture, the drive controller takes commands from the decision engine, and sends the control signals to the two motors in parallel via two serial ports and a special-purpose hardware interface board. The interface provides a bridge between the two systems (the computer and the robot body) that have different power supplies.

Since the two motors (one for each side of the robot) can be controlled separately, the robot can respond to a large set of flexible commands. The basic ones include turning left and right, moving forward and backward. Others include making a big circle in the forward-left, forward-right, back-left and back-right direction. This is done by giving different amounts of drive force to the different sides. In the competition, however, we only used the basic actions for reliability reasons.

One challenge for building this simple drive controller is how to make the measured movements, such as moving forward 10 inches or turning left 35 degree. We solve this problem first by building a software mapping from the measurements of movement to the time duration of the motor running. For example, a command turning left for 30 degree would be translated by this mapping to forwarding the right-motor and backwarding the left-motor for 300ms. This solution works well when all components in the system, especially the batteries, are in perfect condition and floor material is good for wheel movement. But the accuracy of this open-loop control "deteriorates" when the power decreases or as the environment changes. Once this happens, the whole robot will behave strangely because the motor movements are no longer agreeing with the control signals.

To solve this problem, we have made all motor controls closed-loop in the entire system. Instead of saying "turning 75 degree," we also specify the termination criteria for such a turn command. For example, if the purpose of this turning is to find a goal, then the program will repeat issue smaller turnings until the goal is found. With these closed-loop control commands, the reliability of motor control has increased considerably and become more robust with respect to power fluctuation.

This closed-loop motor control also results in one of our secret weapons for well-behaved dribbling actions. Different from other team's "kamikaze" action which often lose the ball quickly in dribbling, our robot uses closed-loop control and continuously adjusts its moving direction according to the current direction of the ball. This approach worked very well in the competition, and our robots were praised for such behaviors by the audience.

5. The Decision Engine

Based on the existing theories of autonomous agents (see for example [9]), we believe that integrated robots are best to be model-driven. This principle has guided our design and implementation of the brain of our robots, namely the Decision Engine. Compared to other model-less and pure-reactive approaches, our approach could in principle demonstrate more intelligent behaviors without sacrificing the ability to quickly react to different situations.

As one can see in Figure 2, the Decision Engine receives input from the vision module and sends move commands to the drive controller. The decision engine bases its decisions on a combination of the received sensor input, the agent's internal model of its environment, and knowledge about the agent's strategies and goals. The agent's internal model and strategies are influenced by the role the agent plays on the soccer field. There are three types of agent roles or playing positions: goal keeper, defender, and forward. The team strategy is distributed into the role strategies of each individual agent. Depending on the role type, an agent can be more concerned about a particular area or object on the soccer field, e.g. a goal keeper is more concerned about its own goal, while the forward is interested in the opponent's goal. These differences are encoded into the two modules that deal with the internal model and the agent's strategies.

The decision engine consists of two sub-modules: the internal model manager and the strategy planner. These sub-modules communicate with each other to formulate the best decision for the agent's next action. The model manager converts the vision module's output into a "map" of the agent's current environment, as well as generating a set of object movement predictions. It calculates the salient features in the field and then communicates them to the strategy planner. To calculate the best action, the strategy planner uses both the information from the model manager and the strategy knowledge that it has about the agent's role on the field. It then sends this information to the drive controller and back to the model manager, so that the internal model can be properly updated.

5.1 Model Manager

For robots to know about their environment and themselves, the model manager uses the information detected by the vision module to construct or update an internal model. This model contains a map of the soccer field and location vectors for nearby objects.

A location vector consists of four basic elements; distance and direction to the object and the change in distance and direction for the object. The changes in distance and direction are used to predict a dynamic object's movement; these are irrelevant for objects that are static. Depending on the role a robot is playing, the model manager actively calls the vision module to get the information that is important to the robot and updates the internal model. For example, if the robot is playing goal keeper, then it needs to know constantly about the ball, the goal, and its current location relative to the goal.

An internal model is necessary for several reasons. First, since a robot can see only the objects within its current visual frame, a model is needed to keep information that is perceived previously. For example, a forward robot may not able to see the goal all the time. But when it sees the ball, it must decide quickly in which direction to kick. The information in the model can facilitate such decision readily. Second, the internal model adds robustness for a robot. If the camera fails for a few cycles (e.g. due to a hit or being blocked, etc.), the robot can still operate using its internal model of the environment. Third, the model is necessary for predicting the environment. For example, a robot needs to predict the movement of the ball in order to intercept it. This prediction can be computed by comparing the ball's current direction with its previous one. Fourth, the internal model can be used to provide feedback to the strategy planner to enhance and correct its actions. For example, in order to perform a turn-to-find-the-ball using the closed-loop control discussed above, the internal model provides the determination criteria to be checked with the current visual information.

5.2 Strategy Planner

In order to play a successfully soccer game, each robot must react appropriately to different situations in the field. This is accomplished by the strategy planner that resides as a part of the decision engine on each robot. Internally, a situation is represented as a vector of visual clues such as the relative direction and distance to the ball, goals, and other players. A strategy is then a set of mappings from situations to actions. For example, if a forward player is facing the opponent's goal and sees the ball, then there is a mapping to tell it to perform the kick action.

For our robots, there are five basic actions: forward, backward, stop, turn-left and turn-right. These actions can be composed to form macro actions such as kick, line-up, intercept, homing, and detour. For example, a detour action is basically a sequence of actions to turn away from the ball, move forward to pass the ball, turn back to find the ball again, and then forward to push the ball. These compound actions represent a form of simple planning. This simple reasoning and planning of actions is very effective to create an illusion that the robots are "intelligent." Indeed, during the competition, the audience cheered when they saw one of our robot make such a detour in order to protect our goal.

5.3 Role Specifications

There five roles that a robot can play for its team: left-forward, right-forward, left-defender, right- defender, and goal keeper. Each role is actually implemented as a set of mappings from situations to actions, as described above. Shown in Figure 8, each role has its own territory and home position. For example, the left-forward has the territory of the left-forward quarter of the field, and its home position is near the center line and roughly 1.5 meter from the left board line. Similarly, the left-defender is in charge of the left-back quarter of the field and its home position is at the left front of the base goal. The mappings for each role, that is goal-keeper, forward and defender, are defined briefly as follows.

For the goal-keeper, the two most important objects in the field are the ball and its own goal. Its home position is in front of the goal, and its strategy is to keep itself in line of the ball and the goal. Since most of its actions are parallel to the base line, the goal keeper's camera is mounted on the side (in the rest of the robots, the camera is mounted on the front), so that it can move sideways while keeping an eye on the ball. As we mentioned before, the goal-keeper also predicts the movement of an incoming ball in order to fulfill its strategy in time. There are four compound actions for the goal-keeper. Two actions, move to the left or right side, are used to prevent the ball from entering the goal. The third action is to search for the ball, and the fourth one is to position itself in the best location. This last action is the most difficult one to implement (there are still much room to improve our current implementation) because the goal-keeper must simultaneously track three types of information, the ball, the horizontal and vertical offset with respect to the goal. In fact, we have tried to give the goal-keeper more sensors. But due to the limit time we had before shipping everything to Japan, we could not make the additional sensors reliable enough for the competition.

The strategy for the forward role is relatively simpler. Its task is to push the ball towards the opponent's goal whenever possible. A forward must look for the ball, decide which direction to kick when the ball is found, and perform the kick or detour action appropriately. This strategy proved to be fast and effective in the competition. (In an early competition, due to a bug in our first implementation, one of our forward robot mistakenly scored on our own goal twice.)

The defender's strategy is very similar to that of the forward, except that the distance to the opponent goal is substantially larger compared to the position of the forward. Similar to the goal keeper, it also tries to position itself between the ball and its own goal. The most difficult action for a defender is to reliably come back to its position after it chases the ball away.

6. Collaboration and Learning

As we can see from the role specifications, there is no explicit collaboration built into the role strategies. This is partially due to the limited preparation time we had before competition, and our general belief that if every robot plays its role perfectly collaboration will emerge naturally. Indeed, during the competition, we saw two of our forwards helped each other to score a goal: one robot rescued the ball from two opponents, and the other robot saw the ball right in front of the goal, and pushed it in. In the future, we will improve our role specification to include passing and assisting ball dribbling.

Learning is another important issue that we have not addressed yet, although our model-based approach provides the baisc elements for its implementation. One particular area that especially needs learning is the vision calibration. In the long run, it would also be nice to have the robot learn from its own experience (such as the mistakes of scoring at one's own goal).

7. Related Work

The approach we used in RoboCup97 is descended from an earlier, integrated system called LIVE [10] for autonomous learning from the environment [9]. It also shares ideas with integrated cognitive architectures in [6], layered-controlled robots [3], behavior-based robots [1,2], as well as recent progress in Agent research [4]. The unique feature of our robots, however, is the use of internal model and closed-loop control in action planning and execution. Our earlier work along this line includes a silver medal winner robot called YODA [8] in the 1996 AAAI Robot competition for indoor navigation and problem solving.

8. Future Work and Conclusions

In building integrated robots that are autonomous, efficient, collaborative, and intelligent, we have demonstrated a simple but effective approach. In the future, we will continue following our design strategy but improving our robots to make them truly integrated. We plan to add communication and passing capacities to increase their ability to collaborate, provide better sensors to increase awareness, and allow them to learn from their own experience.

References

[1] Arbib,M. 1981. Perceptual Structures and Distributed Motor Control. Handbook of Physiology- The Nervous System, II, ed. V. B. Brooks, 1449-1465. American Physiological Society.

[2] Arkin, R.C. 1987. Motor Schema-Based Mobile Robot Navigation. International Journal of Robotics Research, 92-112

[3] Brooks, R. A. 1986. A Robust Layered Control System for a Mobile Robot. IEEE Journal of Robotics and Automation 2(1).

[4] Garcia-Alegre M. C., Recio F. Basic Agents for Visual/Motor Coordination of a Mobile Robot, Proceeding of the first International Conference on Autonomous Agents, Marina del Rey, CA, 1997, 429:434.

[5] Kitano H., Asada M. , Kuniyoshi Y., Noda I., Osawa E. RoboCup: The Robot World Cup Initiative, Proceeding of the first International Conference on Autonomous Agents, Marina del Rey, CA, 1997, 340-347.

[6] Laird, J.E. (ed) Special Issue on Integrated Cognitive Architectures. ACM SIGART Bulletin 2(4).

[7] Mackworth, A and M. Sahota. Can situated robots play soccer? In Proceedings of Artificial Intelligence, 94, Manff AB.

[8] Shen, W.H., J. Adibi, B. Cho, G. Kaminka, J. Kim, B. Salemi, and S. Tejada. YODA—The Young Observant Discovery Agent. AI Magzine, Spring 1997. 37-45.

[9] Shen, W.M. 1994. Autonomous Learning From Environment. W. H. Freeman, Computer Science Press. New York.

[10] Shen, W. M. 1991. LIVE: An Architecture for Autonomous Learning from the Environment. ACM SIGART Bulletin 2(4).

Time-scaling Control of an Underactuated Manipulator

Hirohiko Arai, Kazuo Tanie
Robotics Department
Mechanical Engineering Laboratory
1-2 Namiki, Tsukuba 305-8564 Japan
harai@mel.go.jp, tanie@mel.go.jp

Naoji Shiroma
Inst. of Eng. Mechanics
University of Tsukuba
1-1 Tennodai, Tsukuba 305-0006 Japan
naoji@melcy.mel.go.jp

Abstract

Position control of an underactuated manipulator that has one passive joint is investigated. The dynamic constraint caused by the passive joint is second-order nonholonomic. Time-scaling of the active joint trajectory and bi-directional motion planning from the initial and the desired configurations provide an exact solution of the positioning trajectory. The active and passive joints can be positioned to the desired angles simultaneously by swinging the active joints only twice. Feedback control constrains the manipulator along the planned path in the configuration space. Simulation and experimental results show the validity of the proposed methods.

1 Introduction

Given a class of robotic manipulation tasks, it might be possible to use a simpler robot mechanism (e.g. fewer joints, actuators or sensors) than ordinary ones to perform the task by considering and utilizing the task dynamics. Control of underactuated mechanisms, which have fewer actuators than the number of the generalized coordinates associated with the task, has received increasing attention from the viewpoint of nonholonomic systems in recent years. For example, if we can dexterously control an underactuated manipulator that has passive joints equipped with no actuators, the weight, cost and energy consumption can be reduced, and failure recovery of even fully-actuated manipulators can be facilitated.

An underactuated manipulator is under a dynamic constraint caused by the zero torque at the passive joints. The constraint is generally a nonholonomic constraint as nonintegrable differential equations, unless the joints are placed in some special way [1]. Exploiting this constraint, the positioning of the manipulator to the desired configuration can be achieved by the motions of the active joints, even if the passive joints are completely free joints with no brakes, etc. [2]-[7].

The nonholonomic constraint of the underactuated manipulator has different characteristics from those of wheeled vehicles and space robots, which have been treated as typical nonholonomic systems. The constraints of these examples are caused by the rolling contact or the conservation of angular momentum, and are represented as a first-order nonintegrable differential equation, $H(q)\dot{q} = 0$, where q is the generalized coordinate and $\dot{q}$ is the generalized velocity. The state equation is written as a drift-free symmetrical affine system, $\dot{q} = G(q)u$, with the velocity input u. On the other hand, the dynamic constraint of the underactuated manipulator is described as a second-order nonintegrable differential equation, $M_p(q)\ddot{q} + b_p(q,\dot{q}) = 0$, including the generalized acceleration $\ddot{q}$, and is called a second-order nonholonomic constraint. The state equation, $\frac{d}{dt}[q^T, \dot{q}^T]^T = f(q,\dot{q}) + G(q)u$, with the acceleration or torque input u, has a drift term $f(q,\dot{q})$.

There have been many studies on the conversion of symmetrical affine systems to standard forms such as the chained system or Caplygin system, and also on motion planning and feedback control based upon those forms. However, those methods cannot be directly applied to the affine system with a drift term. Thus there are no unified control methods for underactuated manipulators yet, and most of the methods so far proposed rely on the specific dynamics of the individual mechanisms.

Here, we propose a motion planning and feedback control method for an underactuated manipulator which has one passive joint with no gravity applied. Ref. [2]-[7] already proposed control methods for manipulators belonging to this class. However, the methods in [2]-[4] resort to repetitive motion of the active joint and usually take a long time for the positioning. The methods in [5]-[7] can be applied only when the passive joint is the final axis and can move freely in the horizontal plane. In this paper, the trajectory for the positioning is planned by time-scaling of the active joint trajectory and bi-directional planning from the initial and final configurations. This method can provide an exact solution (not an approximated one) for the position control. The manipulator can reach the desired configuration by swinging the active joints only twice.

The rest of this paper is organized as follows. In Section 2, the manipulator is modeled to show the dynamic constraint caused by the passive joint. In Section 3, the behavior of the manipulator is considered when the time-axis of the active joint trajectory stretches or shrinks. Then the trajectory for the positioning is designed by the bi-directional planning from the initial and the desired configurations. The feedback control for tracking of the desired path is proposed in Section 4. The simulations and experiments in Section 5 demonstrate that the manipulator can be positioned to the desired configuration by the proposed method.

2 Model of Manipulator

We consider an n-axis serial underactuated manipulator which has $n-1$ active joints and one passive joint. The passive joint is a revolute joint without angle limit, and neither gravity nor friction torque acts on it. A horizontal underactuated manipulator with a passive revolute joint around a vertical axis is a typical example. We define $\theta_a = [\theta_1, ..., \theta_{n-1}]^T \in \Re^{n-1}$ as the active joint angle and $\theta_p = \theta_n \in \Re$ as the passive joint angle. The equation of motion of the manipulator is:

$$M_{aa}(\theta)\ddot{\theta}_a + m_{pa}^T(\theta)\ddot{\theta}_p + b_a(\theta,\dot{\theta}) = \tau_a \quad (1)$$
$$m_{pa}(\theta)\ddot{\theta}_a + m_{pp}(\theta)\ddot{\theta}_p + b_p(\theta,\dot{\theta}) = 0. \quad (2)$$

The passive joint is not necessarily at the n-th axis. The elements of the vectors and matrices can be rearranged to get the above representation. Eq.(2) means the dynamic constraint caused by the zero torque at the passive joint. $b(\theta,\dot{\theta}) = [b_a^T, b_p]^T$ is a Coriolis and centrifugal term, and generally has a form of

$$b(\theta,\dot{\theta}) = \sum_{j=1}^{n}\sum_{k=1}^{j} c_{jk}(\theta)\dot{\theta}_j\dot{\theta}_k. \quad (3)$$

This manipulator is assumed not to satisfy the condition for integrability in Ref. [1]. That is, the inertia matrix explicitly includes the passive joint angle θ_p. Then the constraint (2) does not have the first integral, which is a function of the joint angle θ and the angular velocity $\dot{\theta}$, and is a second-order nonholonomic constraint that includes the angular acceleration $\ddot{\theta}$.

Since $m_{pp}(\theta) \neq 0$ from the property of the inertia matrix, the angular acceleration $\ddot{\theta}_p$ of the passive joint is,

$$\ddot{\theta}_p = -m_{pp}(\theta)^{-1}\{m_{pa}(\theta)\ddot{\theta}_a + b_p(\theta,\dot{\theta})\} \quad (4)$$

from Eq.(2).

3 Motion Planning

We develop a motion planning method for the position control in this section. Namely, we find the trajectory and input to transfer the manipulator from the initial configuration to the desired configuration. The manipulator should have zero velocity at the initial and final state. Though it is easy to stop the active joints, the passive joint usually continues moving due to the drift. Therefore, we must choose the trajectory along which the active and passive joints simultaneously stop at the desired angles.

3.1 Time-scaling of Trajectory

First, we consider the motion of the passive joint when the trajectory profile of the active joints is scaled along the time-axis. It is shown that the angle of the passive joint at the final point of the trajectory does not vary however the time-axis is scaled, and that the angular velocities of the passive joint at both ends of the trajectory are proportional to the scaling factor.

The active joints are given the following trajectory,

$$\theta_a = f(\kappa t) = [f_1(\kappa t), ..., f_{n-1}(\kappa t)]^T, \quad (5)$$

which is a function of time t ($0 \leq t \leq 1/\kappa$). $f_1(\cdot), ..., f_{n-1}(\cdot)$ are twice differentiable scalar functions. The constant $\kappa > 0$ is the time-scaling factor. Suppose $s = \kappa t$ is a new time-variable, then

$$\dot{\theta}_i = \kappa \frac{d\theta_i}{ds}, \quad \ddot{\theta}_i = \kappa^2 \frac{d^2\theta_i}{ds^2} \quad (i=1,...,n). \quad (6)$$

Substituting Eq.(6) to Eq.(4), and considering Eq.(3), we obtain

$$\frac{d^2\theta_p}{ds^2} = -m_{pp}(\theta)^{-1}\{m_{pa}(\theta)\frac{d^2\theta_a}{ds^2} + b_p(\theta,\frac{d\theta}{ds})\}. \quad (7)$$

Eq.(7) is a differential equation with s being the independent variable and includes neither t nor κ explicitly. The input, $d^2\theta_a/ds^2$, is the second-order derivative of $f(s)$ with regard to s, and depends on s only.

Hence the solution of this differential equation can be represented as a function of s, $\theta_p(s)$. In other words, the manipulator moves along the same path [1] irrespective of κ. When the passive joint starts from the same initial state $\theta_p|_{s=0}, d\theta_p/ds|_{s=0}$ and the active joints are given the same trajectory $\theta_a = f(s)$ ($0 \leq s \leq 1$), the passive joint reaches the same final state $\theta_p|_{s=1}, d\theta_p/ds|_{s=1}$. Thus θ_p and $\dot{\theta}_p/\kappa$ at $t = 1/\kappa$ are independent of the scaling factor κ and have constant values.

Furthermore, we add the following boundary condition,

$$f(0) = \theta_{a0}, f(1) = \theta_{a1}, \frac{df}{ds}(0) = \frac{df}{ds}(1) = 0$$

on the active joint trajectory $f(s)$. The active joints start from θ_{a0} with zero velocity, and stop again at θ_{a1} for any time-scaling factor κ. We call such a motion as a "swing" of the active joints. Suppose the passive joint takes the initial state ($s = 0$) as zero velocity at the angle θ_{p0}. The active joints stop at $s = 1$, but the passive joint does not stop due to the drift term. It usually has non-zero velocity $\dot{\theta}_{p1}$ at the angle θ_{p1}. Those values can be calculated by giving the active joint acceleration $\ddot{\theta}_a$ to Eq.(4) and integrating the equation numerically from the initial state. The passive joint has the same angle θ_{p1} at the end of the trajectory for arbitrary scaling factor κ, while the angular acceleration $\dot{\theta}_{p1}$ is proportional to κ.

Conversely, we can consider such angle θ_{p0} and angular velocity $\dot{\theta}_{p0}$ at $s = 0$ that the passive joint has zero velocity at the end of the swing ($s = 1$). θ_{p0} and $\dot{\theta}_{p0}$ can be calculated backwards by integrating Eq.(4) in the reverse direction from $t = 1/\kappa$ to $t = 0$. The passive joint angle θ_{p0} at the initial state does not change regardless of the scaling factor κ, and the angular velocity $\dot{\theta}_{p0}$ is proportional to κ.

3.2 Bi-directional Motion Planning

The trajectory for the positioning between two configurations is planned utilizing the property of the time-scaling described above. It is assumed that the active joints stop at a certain intermediate position besides the initial and the desired positions. The positioning trajectory is composed of three trajectory primitives. The first one is the trajectory from the initial state which can be realized by the swing of the active joints from the initial to the intermediate position. The next one is the free rotation of the passive joint while the active joints stay at the intermediate position. The last one is the trajectory that can reach the desired state by the swing from the intermediate to the desired position. Those trajectories are parameterized by the time-scaling factor for each swing and the intermediate position of the active joints. The parameters of the swings are chosen so that the trajectory from the initial state and the trajectory to the desired state can be smoothly connected by the trajectory of free rotation. The idea of bi-directional trajectory planning from both the initial and desired states was suggested by Ref. [8], which proposed a bi-directional approach to the motion planning of a space robot.

The initial and the desired configurations of the manipulator are denoted by $[\theta_{a0}^T, \theta_{p0}]^T$ and $[\theta_{ad}^T, \theta_{pd}]^T$, respectively. The intermediate position of the active joints is represented as θ_{am}. First, the active joints move from the initial position θ_{a0} to the intermediate position θ_{am} according to the trajectory

$$\theta_a = f_A(\kappa_A t) \quad (0 \leq t \leq 1/\kappa_A)$$

which satisfies the following boundary conditions,

$$f_A(0) = \theta_{a0}, f_A(1) = \theta_{am}, \frac{df_A}{ds}(0) = \frac{df_A}{ds}(1) = 0$$

(Swing A). Then the active joints stay there for a while. Finally, the active joints move to the desired position θ_{ad} according to the trajectory

$$\theta_a = f_B(\kappa_B t) \quad (0 \leq t \leq 1/\kappa_B)$$

satisfying the boundary conditions,

$$f_B(0) = \theta_{am}, f_B(1) = \theta_{ad}, \frac{df_B}{ds}(0) = \frac{df_B}{ds}(1) = 0$$

(Swing B). **Fig.1** and **Fig.2** show the trajectories of the active and passive joints in the state space for a two-axis manipulator.

When the time-scaling factor κ_A for Swing A equals 1, the angle and angular velocity of the passive joint at the instant the active joints reach θ_{am} are denoted by $\theta_{pA}^{\kappa_A=1}$ and $\dot{\theta}_{pA}^{\kappa_A=1}$, respectively. These values can be easily obtained by numerically integrating Eq.(7) over $0 \leq s \leq 1$. In the same way, the time-scaling factor κ_B for Swing B is assumed to be 1. Then the initial passive joint angle $\theta_{pB}^{\kappa_B=1}$ and the angular velocity $\dot{\theta}_{pB}^{\kappa_B=1}$ at which the manipulator stops at the desired configuration θ_{ad}, θ_{pd} after Swing B can be calculated back by the integration of Eq.(7), where the time-axis

---
[1] A *path* means a geometrical curve in the configuration space that is not associated with time. We make a distinction from a *trajectory*, which represents the motion with the progress of time.

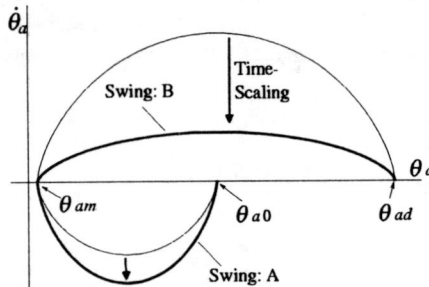

Fig. 1: Trajectory of active joint

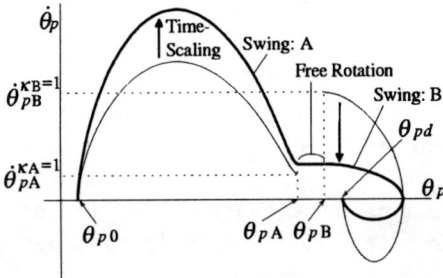

Fig. 2: Trajectory of passive joint

is reversed in this case. See the thin lines of **Fig.1** and **Fig.2**.

When the time-scaling factor κ_A for Swing A has an arbitrary positive value, the passive joint angle θ_{pA} at the end of the swing does not depend on κ_A and is constant. The angular velocity $\dot{\theta}_{pA}$ is proportional to κ_A.

$$\theta_{pA} = \theta_{pA}^{\kappa_A=1}, \quad \dot{\theta}_{pA} = \kappa_A \dot{\theta}_{pA}^{\kappa_A=1} \qquad (8)$$

The passive joint angle θ_{pB} at the start of Swing B does not change with κ_B, and the angular velocity $\dot{\theta}_{pB}$ is proportional to κ_B, too.

$$\theta_{pB} = \theta_{pB}^{\kappa_B=1}, \quad \dot{\theta}_{pB} = \kappa_B \dot{\theta}_{pB}^{\kappa_B=1} \qquad (9)$$

While the active joints stand still at the intermediate position, the manipulator is equivalent to a rigid link pivoted around the passive joint at a fixed location in the inertial coordinate frame. Therefore the passive joint rotates with constant angular velocity. The passive joint passes through θ_{pB} if $\dot{\theta}_{pA} \neq 0$, since the passive joint is a limitless revolute axis. If $\dot{\theta}_{pA} = \dot{\theta}_{pB}$ and Swing B starts at the moment the passive joint reaches θ_{pB}, Swing A and B can be smoothly connected. The angular velocity of the passive joint at the end of Swing A and the start of Swing B can be equalized by the time-scaling, if we can find such intermediate position of the active joints, θ_{am}, that $\dot{\theta}_{pA}^{\kappa_A=1}$

and $\dot{\theta}_{pB}^{\kappa_B=1}$ have the same turning directions. From Eq.(8) and (9), the time-scaling factors should satisfy,

$$\kappa_A/\kappa_B = \dot{\theta}_{pB}^{\kappa_B=1}/\dot{\theta}_{pA}^{\kappa_A=1}. \qquad (10)$$

The time-axis of the active joint trajectory in each swing is scaled using these κ_A and κ_B, like the thick line in **Fig.1**. Then the state of the passive joint traces the thick line in **Fig.2** and stops at the desired angle. The pause at the intermediate position θ_{am} is the smallest positive value of $(\theta_{pB} - \theta_{pA} + 2n\pi)/\dot{\theta}_{pA}$, where n is an integer.

The necessary condition to plan the desired trajectory with the above method is that there exists such intermediate position, θ_{am}, that $\dot{\theta}_{pA}^{\kappa_A=1}$ and $\dot{\theta}_{pB}^{\kappa_B=1}$ have the same signs. It is neither obvious nor easy to show analytically for what combination of the initial and the desired configurations there exists such θ_{am}. Instead, we numerically show later that such intermediate positions exist in a wide area for a two-axis planar manipulator.

Provided the above condition is satisfied, the intermediate position θ_{am}, the angular velocity of the passive joint during the free rotation, and the trajectory profile of the active joints $f_A(\cdot)$, $f_B(\cdot)$ remain unconstrained as the freedoms in the trajectory design, and can be exploited for optimizing the trajectory (e.g. minimum-time).

4 Feedback Control

Since no potential force such as gravity acts on the passive joint, this manipulator cannot be asymptotically stabilized to an equilibrium point with any smooth state feedback [1]. However, the open-loop trajectory for the positioning from the initial configuration to the desired configuration is rigorously calculated by the motion planning in the previous section. The manipulator is expected to reach the neighborhood of the desired configuration if some feedback control can be assembled in accordance with this nominal trajectory.

The manipulator moves along the geometrically identical path in the configuration space when the time-axis of the active joint trajectory is scaled uniformly. If the posture of the manipulator is constrained along the desired path, the time-axis might be automatically scaled and the velocity profile along the path could be what the desired trajectory is scaled to be. Since the desired trajectory is planned so that the velocity of each joint becomes zero at both ends of the trajectory, the manipulator is expected to stop at the desired configuration however the time-axis is scaled.

In this section, we first show that the swing of $n-1$ active joints can be represented by one coordinate. Next, the feedback control is developed to make the manipulator follow the desired path using the acceleration of this coordinate as the input. We proposed the path tracking control of an underactuated manipulator in Ref.[9]. However, in that method the path coordinate frame had to be defined based on the desired path. The feedback control law in this paper does not use the path coordinate frame and requires the planned trajectory data only.

The motions of the $n-1$ active joints are constrained by each other according to the function $\boldsymbol{f}(\cdot)$ as in Eq.(5). These motions can be combined to one degree-of-freedom motion. Suppose the coordinate which represents this motion is $x_a \in \Re$, then the active joint position $\boldsymbol{\theta}_a$ is represented as,

$$\boldsymbol{\theta}_a = \boldsymbol{g}(x_a) = [g_1(x_a), ..., g_{n-1}(x_a)]^T. \quad (11)$$

The active joint velocity $\dot{\boldsymbol{\theta}}_a$, and acceleration $\ddot{\boldsymbol{\theta}}_a$ are

$$\dot{\boldsymbol{\theta}}_a = \frac{d\boldsymbol{g}}{dx_a}\dot{x}_a, \quad \ddot{\boldsymbol{\theta}}_a = \frac{d^2\boldsymbol{g}}{dx_a^2}\dot{x}_a^2 + \frac{d\boldsymbol{g}}{dx_a}\ddot{x}_a. \quad (12)$$

Substituting them to the constraint (2) results in

$$m_a(x_a, \theta_p)\ddot{x}_a + m_p(x_a, \theta_p)\ddot{\theta}_p + b(x_a, \theta_p, \dot{x}_a, \dot{\theta}_p) = 0 \quad (13)$$

where,

$$m_a = \boldsymbol{m}_{pa}\frac{d\boldsymbol{g}}{dx_a}, \quad m_p = m_{pp}, \quad b = b_p + \boldsymbol{m}_{pa}\frac{d^2\boldsymbol{g}}{dx_a^2}\dot{x}_a^2. \quad (14)$$

Eq.(13) represents the dynamic constraint between the coordinate x_a describing the active joint motion, and the passive joint angle θ_p. It has the same form as the constraint of a manipulator with one active and one passive joint.

We assume the desired geometrical path in the configuration space is represented as

$$\theta_p = \theta_{pd}(x_a). \quad (15)$$

In other words, the passive joint angle θ_p is uniquely determined for the position of the active joints, x_a. Eq.(15) defines the geometrical relationship between θ_p and x_a, and does not depend on time. This relationship can be obtained by eliminating s from the desired trajectory

$$\theta_p = \theta_{pd}(s), \quad x_a = x_{ad}(s)$$

planned for each swing. x_a should be monotonously increasing or decreasing, and x_a and s should have a one-to-one correspondence, in order to represent the path as Eq.(15). x_a is not allowed to stop or go back along the trajectory. Such swings of the active joints must be given in the motion planning. The angle error e_p of the passive joint from the desired path is defined as

$$e_p = \theta_p - \theta_{pd}(x_a). \quad (16)$$

Differentiating the above equation with respect to time,

$$\dot{e}_p = \dot{\theta}_p - \frac{d\theta_{pd}}{dx_a}\dot{x}_a \quad (17)$$

$$\ddot{e}_p = \ddot{\theta}_p - \frac{d\theta_{pd}}{dx_a}\ddot{x}_a - \frac{d^2\theta_{pd}}{dx_a^2}\dot{x}_a^2 \quad (18)$$

The angular acceleration of the passive joint, $\ddot{\theta}_p$, is,

$$\ddot{\theta}_p = -m_p^{-1}(m_a\ddot{x}_a + b)$$

from the dynamic constraint (13). Substituting it to Eq.(18),

$$\ddot{e}_p = -(m_p^{-1}m_a + \frac{d\theta_{pd}}{dx_a})\ddot{x}_a - m_p^{-1}b - \frac{d^2\theta_{pd}}{dx_a^2}\dot{x}_a^2 \quad (19)$$

When the acceleration of the error is $\ddot{e}_p$, the acceleration of the active joints is,

$$\ddot{x}_a = -\frac{\ddot{e}_p}{m_p^{-1}m_a + \frac{d\theta_{pd}}{dx_a}} - \frac{m_p^{-1}b + \frac{d^2\theta_{pd}}{dx_a^2}\dot{x}_a^2}{m_p^{-1}m_a + \frac{d\theta_{pd}}{dx_a}} \quad (20)$$

if $m_p^{-1}m_a + \frac{d\theta_{pd}}{dx_a} \neq 0$. $\ddot{e}_p$ is determined by the following PD feedback,

$$\ddot{e}_p = -k_V\dot{e}_p - k_P e_p \quad (21)$$

where k_P and k_V are position and velocity feedback gains, respectively. The acceleration $\ddot{x}_a$ of the active joints is calculated by substituting Eq.(21) to Eq.(20). Then the error converges to zero as $\ddot{e}_p + k_V\dot{e}_p + k_P e_p = 0$. The manipulator is thus constrained to the desired path.

The second term of the right side in Eq.(20) is the feedforward term in the sense that it gives the motion along the desired path if $e_p = 0$. However, it is actually a state feedback which does not depend on time, since it consists of the functions of x_a, θ_p, $\dot{x}_a$ and $\dot{\theta}_p$ only. The acceleration of the active joints is not given as a function of time.

The calculations of Eq.(16), (17), (20) and (21) require $\theta_{pd}(x_a)$, $\frac{d\theta_{pd}}{dx_a}(x_a)$ and $\frac{d^2\theta_{pd}}{dx_a^2}(x_a)$ as functions of

x_a in real-time. $\theta_{pd}(s)$, $\frac{d\theta_{pd}}{ds}(s)$ and $\frac{d^2\theta_{pd}}{ds^2}(s)$ can be obtained in advance as functions of s from Eq.(7) and its integration, while the desired trajectory is planned. $x_a(s)$, $\frac{dx_a}{ds}(s)$ and $\frac{d^2x_a}{ds^2}(s)$ are calculated as the active joint motion. If these values are stored in memory arrays, $\frac{d\theta_{pd}}{dx_a}(s)$ and $\frac{d^2\theta_{pd}}{dx_a^2}(s)$ can be calculated as follows:

$$\frac{d\theta_{pd}}{dx_a} = \frac{\frac{d\theta_{pd}}{ds}}{\frac{dx_a}{ds}}, \quad \frac{d^2\theta_{pd}}{dx_a^2} = \frac{\frac{d^2\theta_{pd}}{ds^2} - \frac{d\theta_{pd}}{dx_a}\frac{d^2x_a}{ds^2}}{\frac{dx_a}{ds}^2}.$$

As x_a is strictly increasing or decreasing according to s, s can be uniquely determined for x_a. θ_{pd}, $\frac{d\theta_{pd}}{dx_a}$ and $\frac{d^2\theta_{pd}}{dx_a^2}$ are obtained by table-lookup using this s.

The velocity and acceleration of the manipulator along the desired path are determined by the shape of the path and the initial velocity. Substituting $\ddot{e}_p = 0$ to Eq.(20), the acceleration $\ddot{x}_a$ of the active joints is,

$$\ddot{x}_a = -\frac{m_p^{-1}b + \frac{d^2\theta_{pd}}{dx_a^2}\dot{x}_a^2}{m_p^{-1}m_a + \frac{d\theta_{pd}}{dx_a}}. \tag{22}$$

$\frac{d\theta_{pd}}{dx_a}$ and $\frac{d^2\theta_{pd}}{dx_a^2}$ are functions of x_a. The angle θ_p of the passive joint is also a function of x_a, and then m_a and m_p are functions of x_a, too. Considering $\dot{\theta}_p = \frac{d\theta_{pd}}{dx_a}\dot{x}_a$ and Eq.(3)(12)(14), b is a product of $\dot{x}_a^2$ and a function of x_a. Hence Eq.(22) can be rewritten as,

$$\ddot{x}_a = a(x_a)\dot{x}_a^2. \tag{23}$$

$a(x_a)$ is a function of x_a determined by the shape of the path. If the time-axis is scaled by the factor $\kappa > 0$ as $s = \kappa t$,

$$\dot{x}_a = \kappa\frac{dx_a}{ds}, \quad \ddot{x}_a = \kappa^2\frac{d^2x_a}{ds^2}$$

then

$$\frac{d^2x_a}{ds^2} = a(x_a)\left(\frac{dx_a}{ds}\right)^2 \tag{24}$$

Eq.(24) explicitly includes neither t nor κ, and has the equivalent form of Eq.(23). Therefore, if the time-axis of the trajectory that satisfies Eq.(23) is scaled by a constant factor, it is also a solution of Eq.(23). It is the only solution for that scaling factor because of the uniqueness of the solution. Since the planned trajectory is one of the solutions of Eq.(23), all the trajectories when the manipulator is constrained to the desired path coincide with the trajectories to which the desired trajectory is time-scaled.

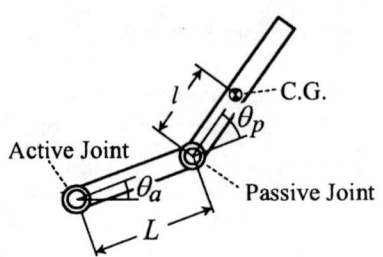

Fig. 3: Two-axis planar underactuated manipulator

5 Simulations and Experiments

We tested the proposed motion planning and feedback control method with the simulations and experiments using a two-axis planar manipulator (**Fig.3**) in which the first axis is active and the second axis is passive. m is the mass of the second link, I is the moment of inertia of the second link around the center of mass, l is the distance between the passive joint and the center of mass of the second link, and L is the length of the first link. The dynamic constraint due to the passive joint is,

$$(L\cos\theta_p + \lambda)\ddot{\theta}_a + \lambda\ddot{\theta}_p + L\sin\theta_p\dot{\theta}_a^2 = 0 \tag{25}$$

where $\lambda \equiv l + I/(ml)$. λ is the distance between the passive joint and the center of percussion of the second link.

First, the trajectories for the positioning are planned by the method of Section 3. The active joint trajectory of each swing, which corresponds to $f(\kappa t)$ in Eq.(5), is given by

$$\begin{cases} \theta_a(t) &= \theta_{a1} + (\theta_{a2} - \theta_{a1})(\kappa t - \frac{1}{2\pi}\sin 2\pi\kappa t) \\ \dot{\theta}_a(t) &= \kappa(\theta_{a2} - \theta_{a1})(1 - \cos 2\pi\kappa t) \\ \ddot{\theta}_a(t) &= 2\pi\kappa^2(\theta_{a2} - \theta_{a1})\sin 2\pi\kappa t \end{cases}$$

In Swing A, θ_{a1} is the initial angle θ_{a0}, and θ_{a2} is the intermediate angle θ_{am}. In Swing B, θ_{a1} is the intermediate angle θ_{am}, and θ_{a2} is the desired angle θ_{ad}. The maximum acceleration of the active joints is limited. The time-scaling factor, κ_A and κ_B, for each swing is determined so that the peak angular acceleration, $2\pi\kappa^2(\theta_{a2} - \theta_{a1})$, for one swing equals the maximum and the acceleration for the other swing is within the limit.

The positioning trajectories are planned from the initial configuration $\theta_a = 0$[rad], $\theta_p = 0$[rad] to various desired configurations, and the intervals for the positionings are examined. The desired configurations are in the area of $0 \leq \theta_a \leq \pi$, $-\pi \leq \theta_p \leq \pi$ and the

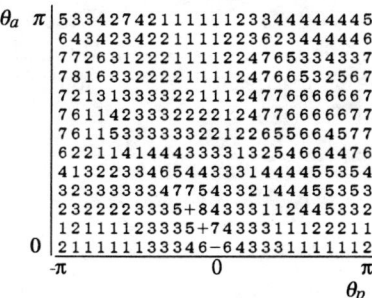

Fig. 4: Positioning intervals

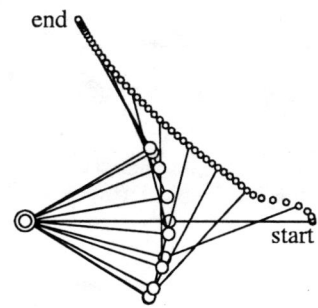

Fig. 6: Planned motion of manipulator

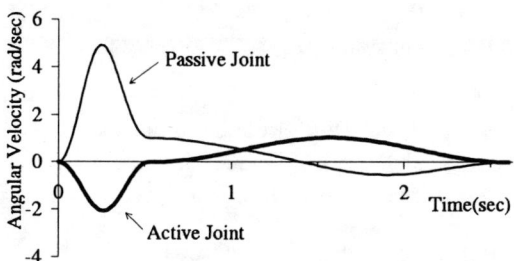

Fig. 5: Planned trajectory

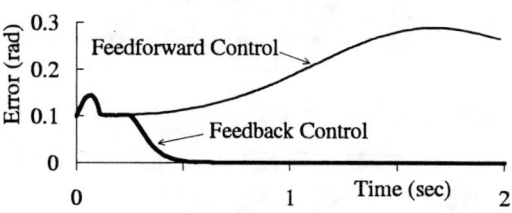

Fig. 7: Tracking error (simulation)

step angle is $\pi/12$. The intermediate angle of the active joint is searched within the initial angle $\pm\pi$ with $\pi/180$ step to find the angle where the positioning interval becomes minimum. The parameters of the manipulator are $L = 0.3$[m] and $\lambda = 0.2$[m]. The angular acceleration of the active joint is limited within $|\ddot{\theta}_a| \leq 4\pi$[rad/sec^{-2}]. **Fig.4** shows the positioning intervals ([sec]) to the desired configurations. To the configuration marked as +, the positioning takes more than 10 seconds. The positioning trajectory cannot be found for the configuration marked as −. Positioning to most of the desired configurations can be achieved within 10 seconds.

Fig.5 and **Fig.6** show an example of the planned trajectories. The initial configuration is $\theta_a = 0$[rad], $\theta_p = 0$[rad] and the desired configuration is $\theta_a = 0.524$[rad], $\theta_p = 1.571$[rad]. The intermediate angle of the active joint is $\theta_a = -0.541$[rad]. The Swing A, the free rotation period, and Swing B take 0.520, 0.018 and 2.077 [sec], respectively. The total positioning interval is 2.615 seconds.

Next, we verified the path tracking by the feedback control in Section 4. As the manipulator has one active joint, x_a coincides with θ_a. The desired path is the Swing B part of the positioning trajectory in **Fig.6**. An angle error of 0.1 [rad] is given to the passive joint at the start of Swing B. **Fig.7** shows the plot of the tracking error e_p. In the case of the feedforward control (thin line), the error increases and the manipulator cannot stop at the end of the trajectory. On the other hand, the feedback control suppresses the error and the manipulator follows the path (thick line). The manipulator stops at $\theta_a = 0.524$[rad], $\theta_p = 1.571$[rad] and positioning is achieved in spite of the initial error.

Then, the positioning of an experimental manipulator (**Fig.8**) is actually performed. The manipulator has two active joints and one passive joint. The base joint is fixed and the experiment is conducted using the remaining two joints. The feedback for the path tracking is applied during Swing B only. **Fig.9** shows the experimental result of the positioning from the initial configuration $\theta_a = 0.000$[rad], $\theta_p = 0.000$[rad] to the desired configuration $\theta_a = 0.524$[rad], $\theta_p = 1.571$[rad]. The manipulator stops at $\theta_a = 0.523$[rad], $\theta_p = 1.594$[rad] in 2.55 seconds.

6 Conclusions

We proposed a motion planning and feedback control method for the position control of an underactuated manipulator with one passive joint on which no gravity acts. The trajectory for positioning is planned

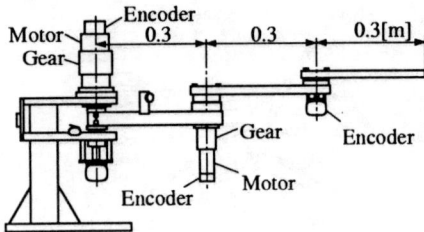

Fig. 8: Experimental setup

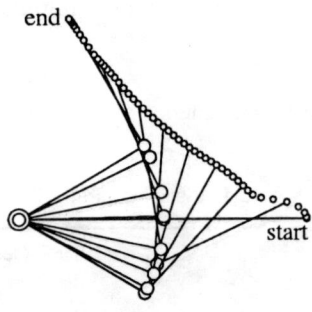

Fig. 9: Motion of manipulator (experiment)

by time-scaling of the trajectories that are numerically calculated from the initial and the desired trajectory, and connecting them by the free rotation trajectory. This method gives the exact solution for the positioning, and the active and passive joints reach the desired angles simultaneously by only two swings of the active joints. Feedback control that makes the manipulator follow the desired path is also proposed. We experimentally demonstrated that position control to the desired configuration can be achieved by the proposed methods.

References

[1] G. Oriolo and Y. Nakamura, "Free-Joint Manipulators: Motion Control under Second-Order Nonholonomic Constraints," Proc. IEEE/RSJ Int. Workshop on Intelligent Robots and Systems (IROS'91), pp. 1248–1253, 1991.

[2] Y. Nakamura, T. Suzuki and M. Koinuma, "Nonlinear Behavior and Control of A Nonholonomic Free-Joint Manipulator," IEEE Trans. Robotics and Automation, vol. 13, no. 6, pp. 853–862, 1997.

[3] T. Suzuki and Y. Nakamura, "Nonlinear Control of A Nonholonomic Free Joint Manipulator with the Averaging Method," Proc. 35th IEEE Int. Conf. on Decision and Control, pp. 1694–1699, 1996.

[4] A. De Luca, R. Mattone and G. Oriolo, "Control of Underactuated Mechanical Systems: Application to the Planar 2R Robot," Proc. 35th IEEE Int. Conf. on Decision and Control, pp. 1455–1460, 1996.

[5] H. Arai, "Controllability of a 3-DOF Manipulator with a Passive Joint under a Nonholonomic Constraint," Proc. 1996 IEEE Int. Conf. on Robotics and Automation, pp. 3707-3713, 1996.

[6] H. Arai, K. Tanie and N. Shiroma, "Feedback Control of a 3-DOF Planar Underactuated Manipulator," Proc. 1997 IEEE Int. Conf. Robotics and Automation, pp. 703–709, 1997.

[7] J. Imura, K. Kobayashi and T. Yoshikawa, "Nonholonomic Control of 3 Link Planar Manipulator with a Free Joint," Proc. 35th IEEE Int. Conf. on Decision and Control, pp.1435-1436, 1996.

[8] Y. Nakamura and R. Mukherjee, "Nonholonomic Path Planning of Space Robots via a Bidirectional Approach," IEEE Trans. Robotics and Automation, vol. 7, no. 4, pp. 500–514, 1991.

[9] H. Arai, K. Tanie and S. Tachi, "Dynamic Control of a Manipulator with Passive Joints in Operational Space," IEEE Trans. Robotics and Automation, Vol. 9, No. 1, pp. 85-93, 1993.

Adjustable Manipulability of Closed-Chain Mechanisms through Joint Freezing and Joint Unactuation

Sungbok Kim

Dept. of Control & Instrumentation Engineering
Hankuk University of Foreign Studies
Kyoungki-Do 449-791, Korea

Abstract

This paper presents the adjustment of the manipulability of a closed-chain mechanism using joint freezing/releasing and joint unactuation/actuation, leading to the improved task adaptability. A closed-chain mechanism under consideration consists of multiple redundant subchains and individual joints can be put into frozen, actuated or unactuated mode, as needed. First, joint freezing and joint unactuation allowed for a closed-chain mechanism are examined, which limits the possible combinations of joint modes. Second, the kinematics of a closed-chain mechanism is formulated, taking into account joint freezing and joint unactuation. Third, the manipulability ellipsoids of a closed-chain mechanism with/without joint freezing and/or joint unactuation are defined and compared. Finally, the wide range of the manipulability of a closed-chain mechanism due to the changes of joint modes is exemplified.

1 Introduction

Closed-chain mechanisms are of the parallel structure of multiple subchains consisting of passive as well as active joints, while open-chain mechanisms are of the serial structure of active joints only. To expand the capability of closed-chain mechanisms, two types of redundancies can be considered: the redundancy in serial-chain and the redundancy in joint actuation [1]. The former is introduced by increasing the number of joints per subchain, and the latter introduced by increasing the total number of active joints of a closed-chain mechanism.

There have been several works on the redundancy in joint actuation of a closed-chain mechanism: for example, the optimal load distribution among active joints was discussed in [2, 3], the manipulability/force applicability was defined in [1], and the stiffness/compliance was analyzed in [4, 5]. However, a little attention has been paid to the redundancy in serial chain of a closed-chain mechanism, and few works can be found on the coexistence of the two types of redundancies in a closed-chain mechanism.

Suppose that individual subchains of a closed-chain mechanism are redundant and all the joints are released and actuated. In such a mechanism, some joints may be enforced to be locked using joint freezing, and some joints may be left to be passive using joint unactuation. Accordingly, individual joints of a closed-chain mechanism operate in frozen, actuated, or unactuated mode. Certain restrictions should be placed on joint freezing and joint unactuation to maintain the mobility and controllability of the mechanism, however, there can be a large number of combinations of joint modes, in general.

Fig. 1 shows a 3 d.o.f. planar closed-chain mechanism consisting of two subchains, each having three joints. Without joint freezing, Fig. 1a) and 1b) show the mechanisms without and with actuation redundancy (due to joint unactuation), respectively. With joint freezing, Fig. 1c) and 1d) show the mechanisms without and with actuation redundancy, respectively. Note that there are many possible joint mode combinations resulting from different joint freezing and joint unactuation, besides the examples given in Fig. 1.

The manipulability ellipsoid has been most widely accepted for the representation of the kinematic performance of the mechanisms of serial, parallel, and serial-parallel structures [1, 6]. The manipulability ellipsoid of a closed-chain mechanism is defined by the range of task velocities mapped from the unit sphere of active joint velocities. Note that the reciprocal of the manipulability ellipsoid represents the force applicability as a result of the duality of velocity and force [1]. An appropriate metric can be incorporated in defining the manipulability ellipsoid to avoid the physical inconsistency incurred due to mixed types of joints [7].

The operational characteristics of a closed-chain mechanism needs to be changed depending on the requirements of a given task. For instance, higher manipulability (lower force applicability) may be needed in some directions, and higher force applicability (lower manipulability) may be needed in other directions. The manipulability of a closed-chain mechanism can be adjusted through joint freezing/releasing and joint unactuation/actuation, which leads to the improvement in task adaptability. Notice that joint freezing and joint unactuation are simple to realize: the former using joint brake and the latter using circuit breaker.

It is a major concern of this paper to demonstrate the effectiveness of joint freezing and joint unactuation for improving the task adaptability of a closed-chain mechanism. In this paper, it is assumed that a closed-chain mechanism consists of two redundant subchains, referred here to as subchain 1 and subchain 2, and all the joints of each subchain are independent and of the same type. However, the results of this paper can be extended to a closed-chain mechanism having more than two subchains of the mixed types of joints.

2 Joint Mode Combination

Joint freezing and joint unactuation in a closed-chain mechanism may be attempted for individual joints, however, the allowable combinations of joint modes are limited by the mobility and controllability of the mechanism.

Let m is the dimension of a task assigned to a closed-chain mechanism. For subchain i, $i=1,2$, let n_i ($\geq m$) be the number of all joints, and n_{ir} and n_{if} be the numbers of the joints to be released and to be frozen, such that

$$n_i = n_{ir} + n_{if}, \quad i=1,2 \qquad (1)$$

To maintain the mobility of each subchain,

$$n_{ir} \geq m, \quad i=1,2 \qquad (2)$$

and, to maintain the mobility of the mechanism,

$$n_{1r} + n_{2r} - K \geq m \qquad (3)$$

where K ($\geq m$) is equal to 3 for planar cases and 6 for spatial cases.

Using (1), from (2) and (3), we have

$$n_{if} \leq n_i - m, \quad i=1,2 \qquad (4)$$
$$n_{1f} + n_{2f} \leq (n_1 + n_2) - (m + K) \qquad (5)$$

which restrict joint freezing in a closed-chain mechanism. The number of the frozen joints of individual subchains is limited by (4), and the total number of the frozen joints of a closed-chain mechanism is limited by (5). Note that (4) implies (5), if $K = m$.

For subchain i, $i=1,2$, let n_{ip} ($\leq n_{ir}$) be the number of the joints to be unactuated. To maintain the controllability of a closed-chain mechanism, we have

$$n_{1p} + n_{2p} \leq K \qquad (6)$$

which restricts joint unactuation in the mechanism. The limitation on the number of the unactuated joints is placed on a closed-chain mechanism as a whole. Note that $n_{ip} \leq K$, $i=1,2$, if (6) holds. There may be multiple choices in selecting the joints to be frozen and to be unactuated, as far as (4), (5) and (6) are satisfied simultaneously. Under these restrictions, a variety of combinations of joint modes are possible, which differ in the numbers and the locations of the frozen and the unactuated joints.

Consider a planar closed-chain mechanism shown in Fig. 1, for which $n_1 = n_2 = 3$, $m = 2$, and $K = 3$. From (4) and (5), we have $n_{1f} \leq 1$, $n_{2f} \leq 1$, and $n_{1f} + n_{2f} \leq 1$, which tell that only a single joint out of subchain 1 or 2 can be frozen. On the other hand, from (6), we have $n_{1p} + n_{2p} \leq 3$, which tells that at most three joints of the mechanism can be unactuated. Without joint freezing, the number of possible joint unactuation amounts to

$$_6C_3 + {_6C_2} + {_6C_1} + {_6C_0} = 42 \qquad (7)$$

With joint freezing, the number of possible joint unactuation amounts to

$$6 \times (\,_5C_3 + {_5C_2} + {_5C_1} + {_5C_0}\,) = 156 \qquad (8)$$

There are 198 allowable combinations of joint modes even for a simple form of closed-chain mechanisms. Such a large number of the joint mode combinations anticipates the improvement of the task adaptability of a closed-chain mechanism.

3 Kinematic Formulation

The kinematics of a closed-chain mechanism can be formulated by imposing the kinematic constraints of the mechanism on the kinematics of individual subchains. Joint freezing is incorporated into the forward kinematics of individual subchains, and the joint unactuation incorporated into the inverse kinematics of a closed-chain mechanism.

For subchain i, $i=1,2$, let $\boldsymbol{\theta}_i$ be all joints, let $\boldsymbol{\theta}_{if}$ and $\boldsymbol{\theta}_{ir}$ be the joints to be frozen and to be released among $\boldsymbol{\theta}_i$, and let $\boldsymbol{\theta}_{ia}$ and $\boldsymbol{\theta}_{ip}$ be the joints to be actuated and to be unactuated among $\boldsymbol{\theta}_{ir}$. The Cartesian velocity at a task point of subchain i, $\dot{\mathbf{x}}_i$, $i=1,2$, is given by

$$\begin{aligned}
\dot{\mathbf{x}}_i &= \mathbf{J}_i \dot{\boldsymbol{\theta}}_i \\
&= \mathbf{J}_{if} \dot{\boldsymbol{\theta}}_{if} + \mathbf{J}_{ir} \dot{\boldsymbol{\theta}}_{ir} \\
&= \mathbf{J}_{if} \dot{\boldsymbol{\theta}}_{if} + \mathbf{J}_{ia} \dot{\boldsymbol{\theta}}_{ia} + \mathbf{J}_{ip} \dot{\boldsymbol{\theta}}_{ip}, \quad i=1,2
\end{aligned} \quad (9)$$

where for $i=1,2$, $\mathbf{J}_i$ represents the Jacobian of subchain i, $\mathbf{J}_{if}$ and $\mathbf{J}_{ir}$ represent the submatrices of $\mathbf{J}_i$, corresponding to $\dot{\boldsymbol{\theta}}_{if}$ and $\dot{\boldsymbol{\theta}}_{ir}$, and $\mathbf{J}_{ia}$ and $\mathbf{J}_{ip}$ represent the submatrices of $\mathbf{J}_{ir}$, corresponding to $\dot{\boldsymbol{\theta}}_{ia}$ and $\dot{\boldsymbol{\theta}}_{ip}$.

Considering that the velocities of the frozen joints become zero, (9) can be written as follows: Without joint freezing,

$$\dot{\mathbf{x}}_i = \begin{bmatrix} \mathbf{J}_{if} & \mathbf{J}_{ia} & \mathbf{J}_{ip} \end{bmatrix} \begin{bmatrix} \dot{\boldsymbol{\theta}}_{if} \\ \dot{\boldsymbol{\theta}}_{ia} \\ \dot{\boldsymbol{\theta}}_{ip} \end{bmatrix}, \quad i=1,2 \quad (10)$$

With joint freezing,

$$\dot{\mathbf{x}}_i = \begin{bmatrix} \mathbf{J}_{ia} & \mathbf{J}_{ip} \end{bmatrix} \begin{bmatrix} \dot{\boldsymbol{\theta}}_{ia} \\ \dot{\boldsymbol{\theta}}_{ip} \end{bmatrix}, \quad i=1,2 \quad (11)$$

(10) and (11) represent the forward kinematics of individual subchains without and with joint freezing, respectively. Note that the reduced Jacobian due to joint freezing is constructed from the original Jacobian by deleting the columns corresponding to the frozen joints.

Subject to the minimization of the norm of the released joint velocities, the inverse kinematics of individual subchains can be obtained as follows: Without joint freezing, from (10),

$$\begin{bmatrix} \dot{\boldsymbol{\theta}}_{if} \\ \dot{\boldsymbol{\theta}}_{ia} \\ \dot{\boldsymbol{\theta}}_{ip} \end{bmatrix} = \begin{bmatrix} \mathbf{Q}_{irf} \\ \mathbf{Q}_{ira} \\ \mathbf{Q}_{irp} \end{bmatrix} \dot{\mathbf{x}}_i, \quad i=1,2 \quad (12)$$

where

$$\mathbf{Q}_{ir\beta} = \mathbf{J}_{i\beta}^t (\mathbf{J}_{if} \mathbf{J}_{if}^t + \mathbf{J}_{ia} \mathbf{J}_{ia}^t + \mathbf{J}_{ip} \mathbf{J}_{ip}^t)^{-1},$$
$$i=1,2, \quad \beta=f,a,p \quad (13)$$

With joint freezing, from (11),

$$\begin{bmatrix} \dot{\boldsymbol{\theta}}_{ia} \\ \dot{\boldsymbol{\theta}}_{ip} \end{bmatrix} = \begin{bmatrix} \mathbf{Q}_{ifa} \\ \mathbf{Q}_{ifp} \end{bmatrix} \dot{\mathbf{x}}_i, \quad i=1,2 \quad (14)$$

where

$$\mathbf{Q}_{if\beta} = \mathbf{J}_{i\beta}^t (\mathbf{J}_{ia} \mathbf{J}_{ia}^t + \mathbf{J}_{ip} \mathbf{J}_{ip}^t)^{-1},$$
$$i=1,2, \quad \beta=a,p \quad (15)$$

Note that (12) and (14) represent the minimum norm solutions to (10) and (11), respectively.

The Cartesian velocity at a task point of a closed-chain mechanism, $\dot{\mathbf{x}}_o$, is determined under the kinematic constraints between two subchains:

$$\dot{\mathbf{x}}_o = \dot{\mathbf{x}}_1 = \dot{\mathbf{x}}_2 \quad (16)$$

The inverse kinematics of a closed-chain mechanism represents the relationship of the actuated joint velocities and the Cartesian velocity of the mechanism, and can be obtained as follows:

i) Without joint freezing and joint unactuation ($\overline{\mathrm{F}}\ \overline{\mathrm{U}}$), from (12) and (16),

$$\begin{bmatrix} \dot{\boldsymbol{\theta}}_{if} \\ \dot{\boldsymbol{\theta}}_{ia} \\ \dot{\boldsymbol{\theta}}_{ip} \end{bmatrix} = \begin{bmatrix} \mathbf{Q}_{irf} \\ \mathbf{Q}_{ira} \\ \mathbf{Q}_{irp} \end{bmatrix} \dot{\mathbf{x}}_o, \quad i=1,2 \quad (17)$$

ii) Without joint freezing but with joint unactuation ($\overline{\mathrm{F}}\ \mathrm{U}$), from (12) and (16),

$$\begin{bmatrix} \dot{\boldsymbol{\theta}}_{if} \\ \dot{\boldsymbol{\theta}}_{ia} \end{bmatrix} = \begin{bmatrix} \mathbf{Q}_{irf} \\ \mathbf{Q}_{ira} \end{bmatrix} \dot{\mathbf{x}}_o, \quad i=1,2 \quad (18)$$

iii) With joint freezing but without joint unactuation ($\mathrm{F}\ \overline{\mathrm{U}}$), from (14) and (16),

$$\begin{bmatrix} \dot{\boldsymbol{\theta}}_{ia} \\ \dot{\boldsymbol{\theta}}_{ip} \end{bmatrix} = \begin{bmatrix} \mathbf{Q}_{ifa} \\ \mathbf{Q}_{ifp} \end{bmatrix} \dot{\mathbf{x}}_o, \quad i=1,2 \quad (19)$$

iv) With joint freezing and joint unactuation ($\mathrm{F}\ \mathrm{U}$), from (14) and (16),

$$\dot{\boldsymbol{\theta}}_{ia} = \mathbf{Q}_{ifa} \dot{\mathbf{x}}_o, \quad i=1,2 \quad (20)$$

Note that the reduced inverse Jacobian due to joint unactuation is constructed from the original inverse Jacobian by deleting the rows corresponding to the unactuated joints.

From (17), (18), (19) and (20), there can be 16 different expressions of the inverse kinematics of a closed-chain mechanism, depending on how joint freezing and joint unactuation are combined for two subchains. A general form of the inverse kinematics of a closed-chain mechanism can be given as

$$\begin{bmatrix} \dot{\boldsymbol{\theta}}_{1a} \\ \dot{\boldsymbol{\theta}}_{2a} \end{bmatrix} = \begin{bmatrix} \mathbf{Q}_{1a} \\ \mathbf{Q}_{2a} \end{bmatrix} \dot{\mathbf{x}}_o \quad (21)$$

Table 1 summarizes the expressions of $\dot{\boldsymbol{\theta}}_{ia}$ and $\mathbf{Q}_{ia}$, $i=1,2$, depending on joint freezing and joint unactuation of subchain i. For instance, in the case of subchain 1 with $\overline{\mathrm{F}}\ \mathrm{U}$ and subchain 2 with $\mathrm{F}\ \overline{\mathrm{U}}$,

$$\begin{bmatrix} \dot{\boldsymbol{\theta}}_{1f} \\ \dot{\boldsymbol{\theta}}_{1a} \\ \dot{\boldsymbol{\theta}}_{2a} \\ \dot{\boldsymbol{\theta}}_{2p} \end{bmatrix} = \begin{bmatrix} \mathbf{Q}_{1rf} \\ \mathbf{Q}_{1ra} \\ \mathbf{Q}_{2fa} \\ \mathbf{Q}_{2fp} \end{bmatrix} \dot{\mathbf{x}}_o \quad (22)$$

4 Manipulability Ellipsoid

The manipulability ellipsoid of a closed-chain mechanism is defined as the range of the Cartesian velocities mapped from the unit sphere of the actuated joint velocities through the inverse kinematics of the mechanism.

Based on (21), the manipulability ellipsoid, $R_{\dot{x}_o}$, of a closed-chain mechanism is obtained from $\dot{\Theta}_{1a}^t \dot{\Theta}_{1a} + \dot{\Theta}_{2a}^t \dot{\Theta}_{2a} \leq 1$, as

$$R_{\dot{x}_o}: \dot{x}_o^t [Q_{1a}^t Q_{1a} + Q_{2a}^t Q_{2a}] \dot{x}_o \leq 1 \quad (23)$$

Using Table 1, (23) can be tailored for 16 different cases of joint freezing and joint unactuation. For instance, in the case of subchain 1 with $\overline{F}\,U$ and subchain 2 with $F\,\overline{U}$,

$$Q_{1a}^t Q_{1a} = Q_{1rf}^t Q_{1rf} + Q_{1ra}^t Q_{1ra} \quad (24)$$
$$Q_{2a}^t Q_{2a} = Q_{2fa}^t Q_{2fa} + Q_{2fp}^t Q_{2fp} \quad (25)$$

The geometry of the manipulability ellipsoid, given by (23), can be found in the following steps:
1) Compute the Jacobian J_i, $i=1,2$, of subchain i.
2) Construct J_{ir}, $i=1,2$, from J_i by deleting the columns corresponding to the frozen joints.
3) Compute Q_i, $i=1,2$, by inverting J_{ir}.
4) Construct Q_{ia}, $i=1,2$, from Q_i by deleting the rows corresponding to the unactuated joints.
5) Find the SVD (Singular Value Decomposition) of $[\; Q_{1a}^t \quad Q_{2a}^t \;]^t$.

As seen from (23), the contribution of individual subchains to the manipulability ellipsoid of a closed-chain is independent of each other. To simplify the analysis of the effects of joint freezing and joint unactuation, the joint mode combinations for two subchains are assumed to be the same. Let the manipulability ellipsoids of a closed-chain mechanism with $\overline{F}\,\overline{U}$, $\overline{F}\,U$, $F\,\overline{U}$, and $F\,U$ be denoted by $R_{\dot{x}_o}(\overline{F}\,\overline{U})$, $R_{\dot{x}_o}(\overline{F}\,U)$, $R_{\dot{x}_o}(F\,\overline{U})$, and $R_{\dot{x}_o}(F\,U)$, respectively.

i) $R_{\dot{x}_o}(\overline{F}\,\overline{U})$ and $R_{\dot{x}_o}(\overline{F}\,U)$: Using Table 1, with $\overline{F}\,\overline{U}$,

$$Q_{ia}^t Q_{ia} = Q_{irf}^t Q_{irf} + Q_{ira}^t Q_{ira} + Q_{irp}^t Q_{irp} \quad (26)$$

and, with $\overline{F}\,U$,

$$Q_{ia}^t Q_{ia} = Q_{irf}^t Q_{irf} + Q_{ira}^t Q_{ira} \quad (27)$$

Thus, we have

$$R_{\dot{x}_o}(\overline{F}\,\overline{U}) \subset R_{\dot{x}_o}(\overline{F}\,U) \quad (28)$$

(28) tells that the manipulability ellipsoid is expanded due to joint unactuation. This is because the unactuated joints are of infinite dexterity, while the actuated joints have finite dexterity within their velocity limits.

ii) $R_{\dot{x}_o}(\overline{F}\,U)$ and $R_{\dot{x}_o}(F\,U)$: Using Table 1, with $\overline{F}\,U$, from (13),

$$Q_{ia}^t Q_{ia} = Q_{irf}^t Q_{irf} + Q_{ira}^t Q_{ira} + Q_{irp}^t Q_{irp}$$
$$= (J_{if} J_{if}^t + J_{ia} J_{ia}^t + J_{ip} J_{ip}^t)^{-1} \quad (29)$$

and, with $F\,U$, from (15),

$$Q_{ia}^t Q_{ia} = Q_{ifa}^t Q_{ifa} + Q_{ifp}^t Q_{ifp}$$
$$= (J_{ia} J_{ia}^t + J_{ip} J_{ip}^t)^{-1} \quad (30)$$

Thus, we have

$$R_{\dot{x}_o}(\overline{F}\,U) \supset R_{\dot{x}_o}(F\,U) \quad (31)$$

(31) tells that the manipulability ellipsoid is shrunk due to joint freezing, which complies with our intuition.

iii) $R_{\dot{x}_o}(\overline{F}\,U)$ and $R_{\dot{x}_o}(F\,U)$: From (28) and (31),

$$R_{\dot{x}_o}(\overline{F}\,U) \supset R_{\dot{x}_o}(\overline{F}\,U) \supset R_{\dot{x}_o}(F\,U) \quad (32)$$

Thus, we have

$$R_{\dot{x}_o}(\overline{F}\,U) \supset R_{\dot{x}_o}(F\,U) \quad (33)$$

(33) tells that the manipulability ellipsoid is shrunk due to joint freezing and joint unactuation.

iv) $R_{\dot{x}_o}(\overline{F}\,U)$ and $R_{\dot{x}_o}(F\,U)$: Using Table 1, with $\overline{F}\,U$, from (13),

$$Q_{ia}^t Q_{ia} = Q_{irf}^t Q_{irf} + Q_{ira}^t Q_{ira} + Q_{irp}^t Q_{irp}$$
$$= (J_{if} J_{if}^t + J_{ia} J_{ia}^t + J_{ip} J_{ip}^t)^{-1}$$
$$(J_{if} J_{if}^t + J_{ia} J_{ia}^t + J_{ip} J_{ip}^t) \quad (34)$$
$$(J_{if} J_{if}^t + J_{ia} J_{ia}^t + J_{ip} J_{ip}^t)^{-1}$$

and, with $F\,U$, from (15),

$$Q_{ia}^t Q_{ia} = Q_{ifa}^t Q_{ifa}$$
$$= (J_{ia} J_{ia}^t + J_{ip} J_{ip}^t)^{-1} J_{ia} J_{ia}^t \quad (35)$$
$$(J_{ia} J_{ia}^t + J_{ip} J_{ip}^t)^{-1}$$

The inclusion of $R_{\dot{x}_o}(\overline{F}\,U)$ and $R_{\dot{x}_o}(F\,U)$ is not fixed and uniform, but varies depending on how joint freezing and joint unactuation are combined. Notice that the kinematic formulation in Section 3 plays a key role in the above analysis.

5 Task Adaptability

With joint freezing and joint unactuation, individual joints of a closed-chain mechanism are allowed to operate in frozen, actuated, or unactuated mode, under the restrictions given by (4), (5), and (6). The changes of the joint modes affect the manipulability of the mechanism, which leads to the improvement in task adaptability. The following discussions are made based on the simulation results for a planar closed-chain mechanism shown in Fig. 1.

For simplicity, the variations of the manipulability

ellipsoid according to the changes of joint modes are expressed in terms of the volume of the ellipsoid, so-called manipulability measure. Two subchains are identical with link length of [1.0, 0.75, 0.5] [m] and their bases are 2.58 [m] apart. Without joint freezing, Fig. 2a) shows the variation of the manipulability measure with respect to 42 different joint unactuation. With joint freezing, Fig. 2b) and 2c) show the variations of the manipulability measure with respect to 26 different joint unactuation with the 1st and the 3rd (from the base) joints of subchain 1 frozen, respectively. In Fig. 2, the joint mode combinations including 3, 2, 1, and 0 unactuated joint(s) are marked as 'o', 'x', '+' and '*', respectively.

From Fig. 2, the following can be observed:
1) The range of the manipulability measure is significantly extended as a result of joint freezing/releasing and joint unactuation/actuation. This demonstrates the effectiveness of joint freezing and joint unactuation as a means of improving the task adaptability of the mechanism.
2) With joint freezing, the manipulability measure decreases but such a decrease depends on the location of the frozen joint.
3) For fixed joint freezing, the manipulability measure tends to decrease as joint actuation spreads over, but the rate of decrease becomes slower as the number of the actuated joints increases. The manipulability measure varies depending on the location of the actuated joints, and the variation becomes smaller as the number of the actuated joints increases.
4) With joint freezing and joint unactuation combined, the manipulability measure may increase or decrease depending on the numbers and locations of the frozen and the unactuated joints. Note that joint freezing and joint unactuation are opposite to each other in their effects on the manipulability measure.

The changes of the joint modes through joint freezing/releasing and joint unactuation/actuation provide a wide range of the manipulability, resulting in the improved task adaptability of a closed-chain mechanism.

6 Conclusion

This paper demonstrated the effectiveness of joint freezing/releasing and joint unactuation/actuation to adjust the operational characteristics of a closed-chain mechanism consisting of multiple redundant subchains. Specifically, the manipulability of a closed-chain mechanism is shown to vary widely through the joint mode conversion, which enables the mechanism to be adapted to given task requirements. The main contributions of this paper include 1) the kinematic analysis of a closed-chain mechanism with the redundancy in serial-chain as well as the redundancy in joint actuation, and 2) the provision of a simple but effective means of improving the task adaptability of a closed-chain mechanism. It is hoped that the results of this paper open the door to the development of advanced closed-chain mechanisms with high versatility and high performance.

References

[1] S. Lee and S. Kim, "Kinematic Feature Analysis of Parallel Manipulator Systems," Proc. IEEE Int. Conf. Robotics and Automation, pp. 77-82, 1994.
[2] Y. Nakamura and M. Ghodoussi, "Dynamics Computation of Closed-Link Robot Mechanisms with Nonredundant and Redundant Actuators," IEEE Trans. Robotics and Automation, Vol. 5, pp. 294-302, 1989.
[3] V. Kumar and J. F. Gardner, "Kinematics of Redundantly Actuated Closed Chains," IEEE Trans. Robotics and Automation, Vol. 6, pp. 269-274, 1990.
[4] B. J. Yi and R. A. Freeman, "Synthesis of Actively Adjustable Springs by Antagonistic Redundant Actuation," Trans. ASME Jour. Dynamic Systems, Measurements, and Control, Vol. 114, pp. 454-461, 1992.
[5] S. Kim, "Operational Quality Analysis of Parallel Manipulators with Actuation Redundancy," Proc. IEEE Int. Conf. Robotics and Automation, pp. 2651-2656, 1997.
[6] T. Yoshikawa, "Analysis and Control of Robot Manipulator with Redundancy," Proc. 1st Int. Symp. Robotics Research, MIT Press, pp. 735-748, 1984.
[7] K. L. Doty, C. Melchiorri, E. M. Schwartz, and C. Bonivento, "Robot Manipulability," IEEE Trans. Robotics and Automation, Vol. 11, pp. 462-468, 1995.

Table 1. The expressions of $\dot{\Theta}_{ia}$ and $\mathbf{Q}_{ia}$, $i = 1, 2$, depending on joint freezing and joint unactuation of subchain i.

	$\dot{\Theta}_{ia}$	$\mathbf{Q}_{ia}$
$\overline{F}\ \overline{U}$	$\begin{bmatrix} \dot{\theta}_{if} \\ \dot{\theta}_{ia} \\ \dot{\theta}_{ip} \end{bmatrix}$	$\begin{bmatrix} \mathbf{Q}_{irf} \\ \mathbf{Q}_{ira} \\ \mathbf{Q}_{irp} \end{bmatrix}$
$\overline{F}\ U$	$\begin{bmatrix} \dot{\theta}_{if} \\ \dot{\theta}_{ia} \end{bmatrix}$	$\begin{bmatrix} \mathbf{Q}_{irf} \\ \mathbf{Q}_{ira} \end{bmatrix}$
$F\ \overline{U}$	$\begin{bmatrix} \dot{\theta}_{ia} \\ \dot{\theta}_{ip} \end{bmatrix}$	$\begin{bmatrix} \mathbf{Q}_{ifa} \\ \mathbf{Q}_{ifp} \end{bmatrix}$
$F\ U$	$\dot{\theta}_{ia}$	$\mathbf{Q}_{ifa}$

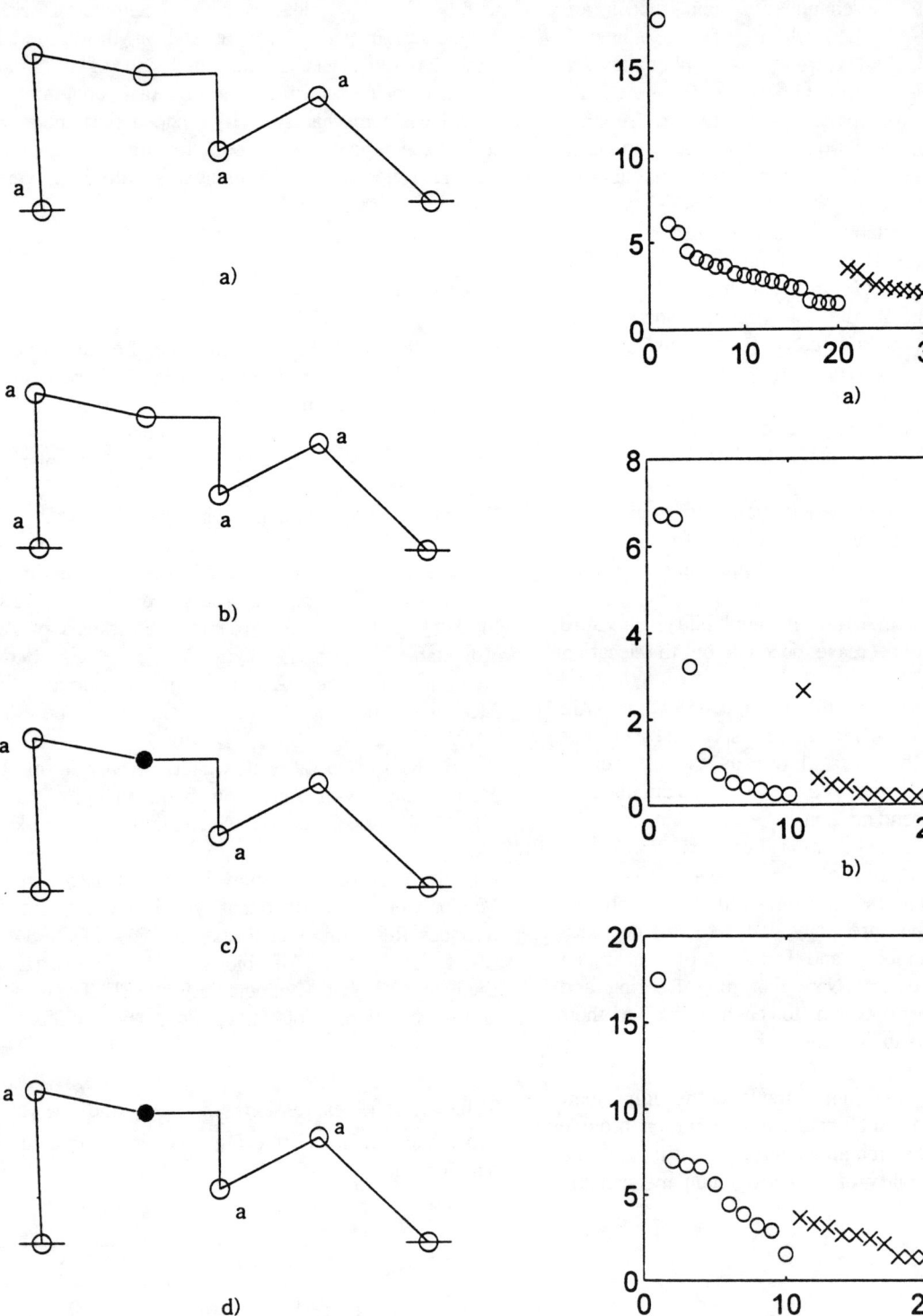

Fig. 1 Examples of a 3 d.o.f. planar closed-chain mechanism: a) without joint freezing and actuation redundancy, b) without joint freezing but with actuation redundancy, c) with joint freezing but without actuation redundancy, and d) with joint freezing and actuation redundancy (the frozen and the actuated joints are indicated by a black circle and 'a', respectively).

Fig. 2 The variations of the manipulability measure with respect to different joint mode combinations: a) with no joint frozen, b) with the 1st joint of subchain 1 frozen, and c) with the 3rd joint of subchain 1 frozen (the joint actuations with 3, 2, 1, and 0 unactuated joint(s) are marked as 'o', 'x', '+' and '*', respectively).

Scaling Laws for Nonlinear Controllers of Dynamically Equivalent Rigid-Link Manipulators

Milind Ghanekar
AlliedSignal Aerospace Canada Ltd.
Etobicoke, Ontario, CANADA

David W.L. Wang
Electrical & Computer Engineering
University of Waterloo
Waterloo, Ontario, CANADA

Glenn R. Heppler
Systems Design Engineering
University of Waterloo
Waterloo, Ontario, CANADA

Abstract

Dimensional analysis is used to determine the nondimensional groups which define the dynamic equivalence conditions for rigid-link manipulators. Scaling laws for general actuators and nonlinear controllers are also presented. As an example, simulation results are presented which illustrate how a large two-link elbow manipulator operating on the Moon can be designed by constructing a dynamically equivalent prototype in Earth gravity, and also, how a sliding mode controller designed for the prototype Earth robot can be scaled for implementation on the Moon-based manipulator.

1 Introduction

As the twenty-first century approaches the use of large robotic manipulators is increasing. Large manipulators are being used to clean hazardous waste from underground storage tanks. In space, the Canadarm is already being used on the space shuttle, and soon more robots that are designed and tested *on Earth*, will be sent to the Moon and to Mars which have significantly different gravitational environments than the Earth.

Many of these large manipulators will not fit inside a conventional research lab. In order to design and test them efficiently, *smaller scale-model* prototype robots will have to be constructed. By constructing prototypes instead of solely relying on computer simulations, the engineer can obtain physical insight into the system dynamics, implement appropriate hardware interfaces, and capture unmodelled dynamics.

Building scale-model prototypes is standard practice in many engineering disciplines, such as shipbuilding and aeronautics. The prototype is a dynamic scale-model of the actual system being modelled, and therefore all dynamic aspects of the system are scaled, including gravitational effects. Such a prototype is termed a *dynamically equivalent* prototype, because the dynamic behaviour of the prototype can be *directly scaled* to quantitatively predict the behaviour of the actual manipulator.

Dimensional analysis provides one method of determining the conditions which define dynamic equivalence. With this technique, the *nondimensional groups* for robotic manipulators, controllers, and actuators can be found, and then used to provide scaling information about the robot manipulator system.

Relevant research applying dimensional analysis and scaling theory to robotic maniplulators has been done previously. Hollerbach identified a time scaling property for single manipulator dynamics which allows planned manipulator trajectories to be analyzed and modified if unrealizable[2]. Youcef-Toumi et al. used dimensionless groups to characterize impact phenomena in force feedback control applications for robot manipulators and drive systems[3]. The dimensionless groups characterizing single flexible link manipulators and linear controllers were investigated by Ghanekar et al. in [1]. The topic of interest in this paper is to determine the dimensionless groups describing general manipulator dynamics. In addition, the scaling laws for general nonlinear controllers are also determined. Although not the focus of this paper, results for general flexible link manipulators have also been determined by Ghanekar[4], and are summarized at the end of the paper.

2 Dimensional Analysis

All physical quantities can be characterized using seven fundamental dimensions[5]: mass$[M]$, length$[L]$, time$[T]$, electric current$[A]$, temperature$[\Theta]$, amount of substance$[mol]$, and luminous intensity$[cd]$. For example, the dimensions of a variable of force are expressed by $F \equiv [M][L][T]^{-2}$, where the notation ($\equiv$) is used to indicate that the dimensions of a quantity are being given.

By examining the *fundamental dimensions* of the

variables describing a physical phenomenon, dimensional analysis can be used to find the nondimensional groups which characterize the system behaviour. A nondimensional group is a set of variables which combine to produce a quantity which is dimensionless, in that it has exponent zero for each fundamental dimension. The Buckingham Pi method, perhaps the best known dimensional analysis tool, provides a systematic method of determining the minimum number of nondimensional Pi groups which characterize the dynamics of a system.

Theorem 1 (Buckingham Pi Theorem[6])
Let $\phi(v_1, v_2, \ldots, v_n) = 0$ be a set of equations in n parameters, v_j, and m fundamental dimensions. Then ϕ can be equivalently written as a set of equations, $\mathbf{F}$, in $n - m$ parameters which are dimensionless in the fundamental dimensions; i.e. as $\mathbf{F}(\Pi_1, \Pi_2, \ldots, \Pi_{n-m}) = 0$, where the Π_i are the nondimensional groups which characterize the system behaviour.

A proof of the Buckingham Pi theorem can be found in Bridgman[6]. These nondimensional groups provide scaling information about the system. If the parameters of two physically *different* systems (e.g. different size, different material) combine to give the *same* values for these dimensionless quantities, then this means that the parameters of one system can be scaled, via the nondimensional groups, to produce the parameters of the second system. One can do this by identifying the essential physical parameters of the system, assuming that the variables appear in the nondimensional groups with arbitrary powers, and solving for these powers so that the overall result is dimensionless. In this paper, the Buckingham Pi method is used to determine the nondimensional groups characterizing the manipulator, the actuators, and the controllers.

3 General Rigid-Link Manipulators

In this paper, *general rigid-link manipulator* refers to the class of manipulators with n rigid-links, p actuators, and having any link topology in unconstrained motion.

3.1 Nondimensional Pi Groups

In order to perform dimensional analysis on the system, the variables characterizing the system dynamics must be identified. The dynamic behaviour is a function of the link parameters, the joint parameters, gravity, and time. The dynamics of the ith link are characterized by its mass m_i, and the mass moments of inertia about each of its three principle axes: J_i^{xx}, J_i^{yy}, and J_i^{zz}. The jth joint is assumed to have mass M_j, and rotational inertia I_j^h about the joint axis (rotary actuators only). The acceleration due to gravity is denoted by g, and time will be measured relative to a *time scaling frequency*, Ω. The fundamental units associated with these variables are:

i^{th} link : $m_i \equiv [M]$, $J_i^{xx} \equiv J_i^{yy} \equiv J_i^{zz} \equiv [M][L]^2$,
j^{th} joint : $M_j \equiv [M]$, $I_j^h \equiv [M][L]^2$,
gravity : $g \equiv [L][T]^{-2}$, time scale freq. : $\Omega \equiv [T]^{-1}$.

Define J_1 as any one of $\{J_1^{xx}, J_1^{yy}, \text{or } J_1^{zz}\}$, and consider the axis about which the inertia J_1 is computed. The *radius of gyration* of link 1 about this axis is denoted by k_1, and is defined[7] as $k_1 \triangleq \sqrt{\frac{J_1}{m_1}} \equiv [L]$. Applying the Buckingham Pi method with m_1, J_1, and Ω as base variables, gives the Pi groups:

$$\Pi_{m_i} = \frac{m_i}{m_1},\ \Pi_{J_i^{xx}} = \frac{J_i^{xx}}{J_1},\ \Pi_{J_i^{yy}} = \frac{J_i^{yy}}{J_1},\ \Pi_{J_i^{zz}} = \frac{J_i^{zz}}{J_1}$$
$$\Pi_{M_j} = \frac{M_j}{m_1},\ \Pi_{I_j^h} = \frac{I_j^h}{J_1},\ \Pi_G = \frac{g}{k_1 \Omega^2}, \quad (1)$$

where $i \in [1, n]$ and $j \in [1, p]$.

Notice that each link is characterized by four Pi groups. However, there are only *two* Pi groups for the first link. Hence, the dimensional dynamic equations, ϕ, have been reduced to the nondimensional equations, $\mathbf{F}$, as per,

$$\phi \underbrace{(m_i, J_i^{xx}, J_i^{yy}, J_i^{zz}, M_j, I_j^h, g, \Omega)}_{4n+2p+2 \text{ variables}} =$$
$$\mathbf{F} \underbrace{(\Pi_{m_i}, \Pi_{J_i^{xx}}, \Pi_{J_i^{yy}}, \Pi_{J_i^{zz}}, \Pi_{M_j}, \Pi_{I_j^h}, \Pi_G)}_{4n+2p-1 \text{ variables}}.$$

The Pi groups in (1) have been verified by directly nondimensionalizing the manipulator dynamic equations. The Pi groups are present in the nondimensional equations of motion[1, 4].

3.2 Nondimensionalization Theorem

For manipulator systems a theorem can be stated which states the form of a nondimensional Pi group for *any* parameter of the manipulator system. Before stating the theorem, the notion of *fundamentally distinct* parameters must be introduced.

A set of parameters is called *fundamentally distinct* if each parameter in the set represents only one fundamental dimension. For rigid manipulator systems, the set $\{m_1, k_1, \Omega\}$ forms a fundamentally distinct set, since each parameter represents only one fundamental dimension.

Theorem 2 (Pi Group for System Parameter)

Let the m fundamental dimensions be $[u_1], \ldots, [u_m]$, and let ϕ be any parameter of the overall system. If the dimensions of ϕ are

$$\phi \equiv [u_1]^{p_1}[u_2]^{p_2} \cdots [u_m]^{p_m}, \quad p_i \in \mathbf{R}, \ i \in [1, m],$$

and the system has m fundamentally distinct parameters $\{v_1, \ldots, v_m\}$ (with $v_i \equiv [u_i]$), then a nondimensional group for ϕ is

$$\Pi_\phi = \frac{\phi}{v_1^{p_1} \cdots v_m^{p_m}}.$$

Proof

The Buckingham Pi method is used to prove the theorem. In order to apply dimensional analysis to a system with m fundamental dimensions, it is required that m parameters which are independent in the fundamental dimension be selected as base parameters[6]. One such choice of base parameters would be a set of m fundamentally distinct parameters: $\{v_1, \ldots, v_m\}$. For example, in (1) the base parameters were chosen as m_1, J_1, and Ω. With the base parameters v_j, the Pi group for the parameter ϕ will be of the form

$$\Pi_\phi = \phi v_1^{e_1} \cdots v_m^{e_m}, \quad e_i \in \mathbf{R}, \ i \in [1, m],$$

where the e_i are the exponents that have to be solved in order to make the Pi group nondimensional. However, because the base parameters are fundamentally distinct by assumption, then only one fundamental dimension $[u_i]$ is associated with each parameter v_j. In terms of fundamental dimensions, the Pi group is

$$\begin{aligned}\Pi_\phi &\equiv ([u_1]^{p_1} \cdots [u_m]^{p_m})([u_1])^{e_1} \cdots ([u_m])^{e_m}\\ &\equiv [u_1]^0 \cdots [u_m]^0,\end{aligned}$$

since the Pi group is nondimensional. By inspection, it is obvious, that the values of the exponents e_i must satisfy

$$e_i = -p_i, \quad i \in [1, m].$$

Hence, as requried, the nondimensional Pi group for the parameter ϕ is

$$\Pi_\phi = \frac{\phi}{v_1^{p_1} \cdots v_m^{p_m}}.$$

The application of dimensional analysis to manipulators was examined extensively by Ghanekar et al. in [1]. With this theorem, the nondimensional Pi groups for *any* aspect of the manipulator system can now be found. In this paper, this theorem will be applied to general actuators and general nonlinear controllers.

4 Actuator Dynamics

In the manipulator model, only the generalized actuation forces are modelled. However, in scaling robotic manipulators, the dynamics behind the generation of the actuation forces must also be scaled. In this section, Theorem 2 is used to determine the nondimensional Pi groups for arbitrary joint actuators.

To scale the actuator dynamics, the physical parameters which characterize the actuator must be scaled. For example, the physical parameters characterizing the dynamics of a DC motor include the motor inertia I^h, and motor torque constant K_i. In general, let the set Φ contain the physical parameters of the actuator. Therefore, if the actuator is characterized by m parameters, ϕ_i, then

$$\Phi \triangleq \{\phi_1, \phi_2, \ldots, \phi_m\}.$$

Although the manipulator is a mechanical system defined solely by the dimensions of mass, length, and time, the actuators themselves may be defined by electrical and thermal dependent quantities. Two examples are photodiode sensors and SMA actuators. Therefore, the fundamental dimensions of an arbitrary actuator parameter ϕ_i can be expressed as

$$\phi_i \equiv [M]^{a_i}[L]^{b_i}[T]^{c_i}[A]^{d_i}[\Theta]^{e_i}, \quad a_i, b_i, c_i, d_i, e_i \in \mathbf{R}.$$

To determine the Pi groups for the actuator parameters using Theorem 2, a set of five fundamentally distinct parameters is required. Now, it is imperative that for a particular manipulator *system* (manipulator, actuator, controller), the *same* base parameters are always used for the Pi groups. This is required in order to maintain the dimensional homogeneity of the entire system. Hence, select the three parameters m_1, k_1, Ω, and introduce reference values for current A_0 and temperature θ_0. Therefore, applying Theorem 2, the nondimensional Pi group for the actuator parameters can be described by,

$$\Pi_{\phi_i} \triangleq \frac{\phi_i}{m_1^{a_i} k_1^{b_i} \Omega^{-c_i} A_0^{d_i} \theta_0^{e_i}}.$$

The form of this Pi group can be verified by deriving the nondimensional equations of motion for a general actuator[4].

5 Controller Scaling Laws

Multi-link robots are often controlled using *independent joint control*[8], where each joint actuator has its own independent controller. In this section, Theorem 2 will be used to determine the scaling laws for a general controller.

Every controller is a function of certain parameters whose values define the effect of the control law. Furthermore, these parameters can be *tuned* to adjust the controller performance. For example, the characteristic parameters for a PID controller are the gains K_p, K_i, and K_d.

Let Φ define the set of controller parameters for a general controller. Suppose that the controller is defined by m parameters, then,

$$\Phi \triangleq \{\phi_1, \phi_2, \ldots, \phi_m\},$$

where the fundamental dimensions of ϕ_i are

$$\phi_i \equiv [M]^{a_i}[L]^{b_i}[T]^{c_i}[A]^{d_i}[\Theta]^{e_i}, \quad a_i, b_i, c_i, d_i, e_i \in \mathbf{R}.$$

As was stressed in the previous section, in order to maintain dimensional consistency of the system, it is important that the same base parameters are used as the fundamentally distinct set. Therefore, applying Theorem 2, the form of the scaling law for a general controller parameter ϕ_i is

$$\Pi_{\phi_i} \triangleq \frac{\phi_i}{m_1^{a_i} k_1^{b_i} \Omega^{-c_i} A_0^{d_i} \theta_0^{e_i}}. \tag{2}$$

5.1 Example: Sliding Mode Control

It is not clear that the above methodology would apply to discontinuous nonlinear controllers. To illustrate the versatility of the scaling theory to nonlinear and switching controllers, this example will determine the scaling conditions for a sliding mode control law.

In the sliding control technique, the tracking error is represented using a *sliding surface*. Then, two control inputs are designed: the *switching control* input which drives all trajectories towards the sliding surface, and the *equivalent control* input, which maintains all trajectories on the sliding surface. Both of the control outputs are highly nonlinear, and the transition from one input to the other is often discontinuous. If the scaling theory works for this complicated controller, then it will work for simpler control strategies. The sliding mode control theory will not be covered in detail in this paper. For more information see the tutorial by DeCarlo et al.[9].

5.1.1 Control Design

Let the manipulator dynamics be expressed as a system of first order differential equations

$$\dot{x} = f(x) + Bu,$$

where x is the $2n \times 1$ state vector, and u is the $n \times 1$ input vector. States x_1 to x_n represent joint positions, whereas states x_{n+1} to x_{2n} represent the corresponding joint velocities.

Typically, a sliding surface is selected to be a weighted sum of joint position error and velocity error. Define the tracking error by the vector $\tilde{x} = x - x_{ref}$. Then, in terms of the parameters λ_i, the sliding surfaces σ_i, are defined as follows:

$$\sigma_i(\tilde{x}, t) = \lambda_i \tilde{x}_i + \tilde{x}_{n+i}, \quad i \in [1, n].$$

The parameters λ_i determine the rate at which the trajectories approach the desired trajectory on the sliding surface. Define

$$S \triangleq [\text{diag}\{\lambda_1, \ldots, \lambda_n\}, I_{n \times n}],$$

then the equivalent control input is[10]:

$$\hat{u} \triangleq (SB)^{-1}(-Sf - S\dot{x}_{ref}).$$

The switching control input $u^{\pm}$, drives the trajectories toward the sliding surface. In terms of the parameters k_i, this control input is defined as[10]:

$$\begin{aligned} u^{\pm} &\triangleq -(SB)^{-1}[k_1 \cdots k_n]^T \text{sgn}(\sigma) \\ &= -(SB)^{-1} k \text{sgn}(\sigma). \end{aligned}$$

The sliding control input is given by $u = \hat{u} + u^{\pm}$:

$$u(t) \triangleq -(SB)^{-1}(Sf(x) + S\dot{x}_{ref} + k\text{sgn}(\sigma)). \tag{3}$$

A common artifact of switching controllers is a phenomenon known as chattering. The chattering effect can be smoothed out by replacing the sgn operator with a saturation operator of width Δ_i[10].

5.1.2 Scaling Conditions

The controller parameters which can be tuned to adjust the controller performance are λ_i, k_i, and Δ_i. Therefore

$$\Phi \triangleq \{\lambda_1, \ldots, \lambda_n, k_1, \ldots, k_n, \Delta_1, \ldots, \Delta_n\},$$

where,

$$\lambda_i \equiv [T]^{-1}, \quad k_i \equiv [T]^{-2}, \quad \Delta_i \equiv [T]^{-1}.$$

Using the controller scaling laws (2), the Pi groups for the controller parameters are,

$$\Pi_{\lambda_i} = \frac{\lambda_i}{\Omega}, \quad \Pi_{k_i} = \frac{k_i}{\Omega^2}, \quad \Pi_{\Delta_i} = \frac{\Delta_i}{\Omega}. \tag{4}$$

6 Numerical Example

To illustrate the use of the dynamic equivalence conditions, a two-link rigid elbow manipulator with rotary joints will be scaled from a lunar gravity environment to an Earth-based environment. The Moon-based and Earth-based manipulator will be denoted "system 1" and "system 2" respectively. To illustrate the controller scaling laws, a sliding mode controller designed on the prototype Earth-based manipulator will be scaled to apply to the Moon-based robot.

6.1 Moon-Based Manipulator: System 1

For system 1, each link is a solid cylinder of length $l_i = 5$ m, radius $r_i = 0.04$ m and mass $m_i = 60$ kg. From [7], the mass moments of inertia of each link about principle axes at the centre of mass can be calculated to be $J_i^{xx} = 0.044$ kgm^2 and $J_i^{yy} = J_i^{zz} = 125.02$ kgm^2, where the principle x-axis is lined up with the link. With $J_1 = J_1^{xx}$, compute $k_1 = 0.0272$ m. The two rotary joints have the following parameters: $M_1 = 12$ kg, $I_1^h = 3$ kgm^2, and $M_2 = 6$ kg, $I_2^h = 3$ kgm^2. For the Moon, $g = 1.635$ ms^{-2}, and the time scale frequency is set to $\Omega = 1.0$ s^{-1}. Using (1), the Pi groups for system 1 are,

$$\Pi_{m_2} = 1.0, \quad \Pi_{J_1^{yy}} = 2813.0, \quad \Pi_{J_1^{zz}} = 2813.0$$
$$\Pi_{J_2^{xx}} = 1.0, \quad \Pi_{J_2^{yy}} = 2813.0, \quad \Pi_{J_2^{zz}} = 2813.0$$
$$\Pi_{I_1^h} = 67.5, \quad \Pi_{I_2^h} = 67.5$$
$$\Pi_{M_1} = 0.2, \quad \Pi_{M_2} =, 0.1, \quad \Pi_G = 60.074. \quad (5)$$

6.2 Earth-Based Manipulator: System 2

In designing the Earth-based manipulator, the values for three dimensionally independent parameters can be chosen, and then the remaining parameters are solved to satisfy the dynamic equivalence conditions(5). In this example, because the system is operating on Earth, $g = 9.81$ ms^{-2}. Also, suppose that the desired link 1 mass of the robot is $m_1 = 1.27$ kg, and the desired inertia for link 1 is $J_1 = J_1^{xx} = 7.2 \times 10^{-5}$ kgm^2. For dynamic equivalence, the remaining system parameters can be calculated by rearranging the Pi groups in (1) and using the three base parameters along with the Pi values in (5) to obtain,

$$\Omega = 4.66 \text{ s}^{-1}, \quad m_2 = 1.27 \text{ kg}$$
$$J_1^{yy} = 0.20 \text{ kgm}^2, \quad J_1^{zz} = 0.20 \text{ kgm}^2$$
$$J_2^{xx} = 7.2 \times 10^{-5} \text{ kgm}^2, \quad J_2^{yy} = J_2^{zz} = 0.20 \text{ kgm}^2$$
$$M_1 = 0.26 \text{ kg}, \quad M_2 = 0.13 \text{ kg}$$
$$I_1^h = I_2^h = 0.0049 \text{ kgm}^2.$$

From the link inertia values $(J_i^{xx}, J_i^{yy}, J_i^{zz})$, it can be determined that the cylindrical links of the Earth-based robot have length $l_1 = 1.38$ m and radius $r_i = 0.01$ m. Hence the physical dimensions of the links have been scaled down by approximately a factor of 5.

Notice that for this Earth-based system, the value of the time scaling parameter is $\Omega = 4.66$ s^{-1}. The implication of this is that the motion of the Earth-based system is approximately 4.7 times *faster* than the dynamically equivalent Moon-based counterpart.

6.3 Scaling the Sliding Mode Controller

A sliding mode controller was designed to move the prototype manipulator from the initial position of $(q_1, q_2) = (0,0)$ to the reference position: $(q_1, q_2) = (\frac{\pi}{2}, -\frac{\pi}{2})$, and the following parameters were found to give good performance:

$$\lambda_i = 10 \text{ s}^{-1}, \quad k_i = 10 \text{ s}^{-2}, \quad \Delta_i = 0.17 \text{ s}^{-1}, \quad i \in [1,2].$$

From (4), the values for the controller Pi groups can be calculated to be:

$$\Pi_{\lambda_i} = 1.76, \quad \Pi_{k_i} = 0.31, \quad \Pi_{\Delta_i} = 0.03, \quad i \in [1,2].$$

For the Moon-based system, the sliding mode controller parameters which will produce the same Pi values are:

$$\lambda_i = 1.76 \text{ s}^{-1}, \quad k_i = 0.31 \text{ s}^{-2}, \quad \Delta_i = 0.03 \text{ s}^{-1}, \quad i \in [1,2]$$

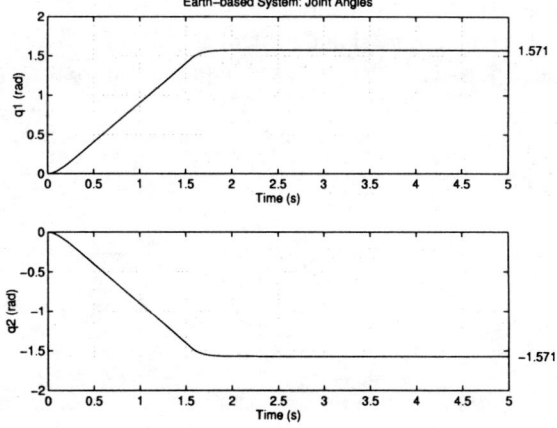

Figure 1: Earth System: Joint Angles $q_{ref} = (\frac{\pi}{2}, -\frac{\pi}{2})$

6.4 Observations

In Figures 1 to 4, the responses for both the Earth-based and the Moon-based manipulators with $q_{ref} = (\frac{\pi}{2}, -\frac{\pi}{2})$ are given. Figures 1 and 2 show the joint angle histories, and Figures 3 and 4 show the torque histories for each system.

For each pair of plots (one from the Earth-based system and one from the Moon-based system), it is clear that although the axes sizes are different, the response of the Moon-based manipulator is an *exact scaled version* of the Earth-based manipulator response. In other words, the response of the Moon-based manipulator could have been *predicted* directly from the Earth-based prototype.

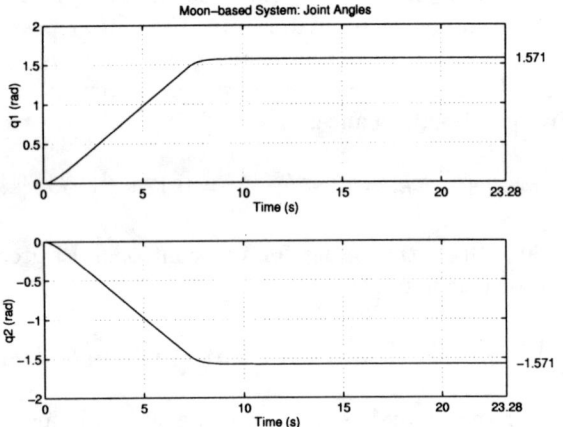

Figure 2: Moon System: Joint Angles $q_{ref} = (\frac{\pi}{2}, -\frac{\pi}{2})$

This can be easily explained by considering the scaling factors involved. Let the subscripts e and m denote the Earth-based and Moon-based systems respectively. Let the nondimensional time variable be τ, and the nondimensional control effort be $\hat{H}$. Since the Earth-based and Moon-based manipulator systems are dynamically equivalent, then all the nondimensional values for both systems must be equal. Therefore,

time: $\tau = (\Omega)_m t_m = (\Omega)_e t_e \Rightarrow t_m = 4.66 t_e$

torque: $\hat{H} = \dfrac{H_m}{(m_1 k_1^2 \Omega^2)_m} = \dfrac{H_e}{(m_1 k_1^2 \Omega^2)_e}$
$\Rightarrow H_m = 28.46 H_e,$

where the actual scaling values are obtained from the system parameters for this example. Therefore, stretching the time axis of the Earth-based system responses by a factor of 4.66 will give the time axis for the Moon-based system responses. Similarly, the torques are scaled by a factor of 28.46.

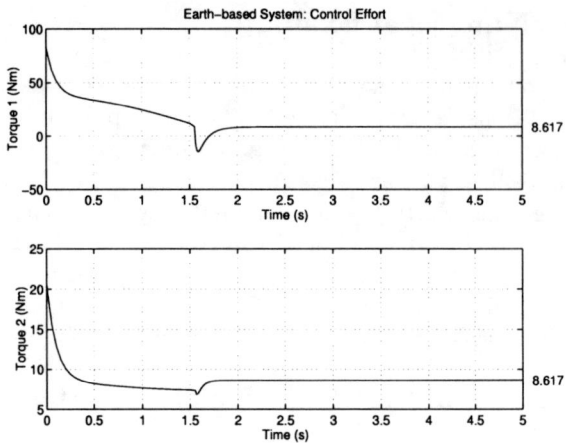

Figure 3: Earth System: Ctrl. Effort $q_{ref} = (\frac{\pi}{2}, -\frac{\pi}{2})$

7 Extensions

7.1 Flexible Manipulators

The scaling theory has also been applied successfully to flexible-link manipulators[4]. The major difference between the flexible-link Pi groups and the rigid-link Pi groups (1), is the presence of the material properties of the link in the Pi groups. The Pi groups listed below contain scaling conditions for the elastic modulus E, and the linear mass density of the material ρ.

$$\Pi_{\rho_i} = \frac{\rho_i}{\rho_1}, \quad \Pi_{l_i} = \frac{l_i}{l_1}, \quad \Pi_{I_{x_i}} = \frac{I_{x_i}}{l_1^4}, \quad \Pi_{I_{y_i}} = \frac{I_{y_i}}{l_1^4},$$

$$\Pi_{E_i} = \frac{E_i}{m_1 l_1^{-1} \Omega^2}, \quad \Pi_{M_j} = \frac{M_j}{m_1}, \quad \Pi_{I_j^h} = \frac{I_j^h}{m_1 l_1^2},$$

$$\Pi_G = \frac{g}{l_1 \Omega^2}.$$

7.2 Friction

Scaling laws for friction effects can also be found using Theorem 2[4]. As an example, consider a viscous friction model, in which the friction force F is proportional to velocity $\dot{x}$ through the friction parameter c_v: $F = c_v \dot{x}$. The fundamental dimensions of the friction parameter are $c_v = [M][T]^{-1}$. For a rigid-link manipulator with the fundamentally distinct parameters m_1, k_1, and Ω, the Pi group for c_v is given by $\Pi_{c_v} = \frac{c_v}{m_1 \Omega}$. Using the scaling condition, the required friction constant for dynamic equivalence can be determined.

Practically, it is impossible to achieve a preset friction value, and hence for the friction constant, the

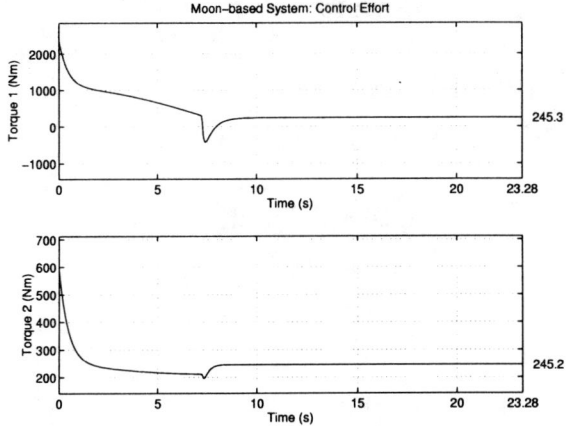

Figure 4: Moon System: Ctrl. Effort $q_{ref} = (\frac{\pi}{2}, -\frac{\pi}{2})$

Pi value required for dynamic equivalence will not be achieved. This implies that the actual manipulator system may not exhibit the same dynamics as the "desired" dynamically equivalent system. Because of this, friction effects are the ideal candidates for techniques which can compensate for errors in the Pi values, thereby making the actual system exhibit the desired dynamic behaviour. Such a compensation technique using a double inverse dynamics loop was proposed and verified in simulation in [4].

8 Conclusions

The dynamic equivalence conditions for general rigid-link manipulators were derived and the scaling laws for general actuators and general nonlinear controllers were also presented. To illustrate the application of the scaling theory a two-link rigid manipulator under sliding mode control was considered. For a manipulator operating in lunar gravity, a dynamically equivalent manipulator was designed for operation on Earth. In addition a sliding mode controller designed for the Earth robot was scaled to operate on the Moon manipulator. This example was verified via simulation.

It can be argued that this methodology seems to require that the nondimensional groups be exactly the same for the two systems. In [4], a sensitivity analysis was conducted to examine the robustness of this technique to parameter uncertainties, and a feedback linearization technique was proposed to compensate for small parameter uncertainties. This work is beyond the scope of this paper and can be found in [4].

References

[1] M. Ghanekar, D.W.L. Wang, and G.R. Heppler. Scaling Laws for Linear Controllers of Flexible Link Manipulators Characterized by Nondimensional Groups. *IEEE Transactions on Robotics and Automation*, **13**(1):117–127, February 1997.

[2] J.M. Hollerbach. Dynamic Scaling of Manipulator Trajectories. *Journal of Dynamic Systems, Measurement, and Control*, **106**(1):102–106, March 1984.

[3] K. Youcef-Toumi and D.A. Gutz. Impact and Force Control: Modelling and Experiments. *Journal of Dynamic Systems, Measurement, and Control*, **116**(1):89–98, March 1994.

[4] Milind Ghanekar. *Dynamic Equivalence Conditions and Controller Scaling Laws for Robotic Manipulators*. PhD thesis, University of Waterloo, 1997.

[5] Canadian Standards Association. *The International System of Units (SI)*. Canadian Standards Association, Rexdale, Ontario, Canada, 1973.

[6] P.W. Bridgman. *Dimensional Analysis*. Yale University Press, New Haven, 1931.

[7] Donald T. Greenwood. *Principles of Dynamics*. Prentice-Hall, Inc., Englewood Cliffs, 1988.

[8] M.W. Spong and M. Vidyasagar. *Robot Dynamics and Control*. John Wiley & Sons, New York, 1989.

[9] R.H. DeCarlo, S.H. Zak, and G.P. Matthews. Variable Structure Control of Nonlinear Multivariable Systems. *Proceedings of the IEEE*, **76**(3):212–232, March 1988.

[10] J. Slotine and W Li. *Applied Nonlinear Control*. Prentice-Hall, Inc., New Jersey, 1991.

Robust Global Stabilization of the Underactuated 2-DOF Manipulator R2D1

Jörg Mareczek   Martin Buss   Günther Schmidt

Institute of Automatic Control Engineering, Technische Universität München
D-80290 München, Germany, E-mail: {M.Buss,G.K.Schmidt}@ieee.org

Abstract

In this paper a switching control strategy for robust stabilization of the 2nd-order non-holonomic 2-DOF SCARA robot R2D1 is presented. The 1st joint is actuated by a direct drive motor, whereas the 2nd joint is equipped with a brake. The unactuated 2nd joint is controlled by non-collocated linearization and a PD-controller. A stability region is derived and robustness is achieved by exploiting the contractive character of the perturbed stability region. The proposed switching control strategy assures global and robust position control of the 2nd joint. Experimental results confirm the efficiency of the proposed approach.

1 Introduction

Underactuated robots have been a recent topic of interest [1–7]. Applications include the folding of robot arms in space when some of the actuators fail. In future cost reductions could be a motivation to build robots with fewer actuators than joints and replacing actuators with holding brakes.

It is known that underactuated robots are with second-order nonholonomic constraints [3]. Most control algorithms are based on *strong inertial coupling* [4] to control as many joints (unactuated or actuated) as the number of available actuators. Other methods use a holding brake [1] or periodic inputs to control both joints [5].

Especially underactuated systems are highly sensible to inexact parameter estimates due to nonholonomic constraints and the linearizing control law. Classical methods for nonlinear systems [8] use Lyapunov control, achieving global asymptotic stability if the uncertainties satisfy the matching condition (enter the system through the same channel as the control input). For the underactuated robot R2D1 under consideration in this paper the matching condition is not satisfied. The approach presented in [9] takes uncertainties into account when calculating a global stabilizing control Lyapunov function recursively. However, this method is only applicable to a system class not including underactuated systems.

In this paper we present a global and robust switching control law for the underactuated SCARA-type robot R2D1 (2 rotational-degrees-of-freedom, 1 drive). The robot R2D1 has a DD-motor in the first and a holding brake in the second joint, see Figure 1. The rotational plane can be inclined versus the gravitational field at an angle $-\pi/6 \leq \alpha \leq 0$.

The organization of the paper is as follows: In Section 2 we present the dynamic equations of R2D1 and estimated physical parameters. Following Spong [4], the method of *non-collocated linearization* is derived in Section 3 also showing the characteristics of the zero dynamics. A PD-controller for the linearized 2nd joint (NCL-PD-control) is proposed and the uncertain parameter case is examined. Additionally, a stability analysis for the NCL-PD-controlled system using a Lyapunov-function for the 2nd joint is presented. Section 4 proposes a switching control law to achieve a quasi-invariant stability area and therewith robust stability in the case of contractive stability regions. Experimental results in Section 5 validate the robust stabilizing effect of the switching control algorithm.

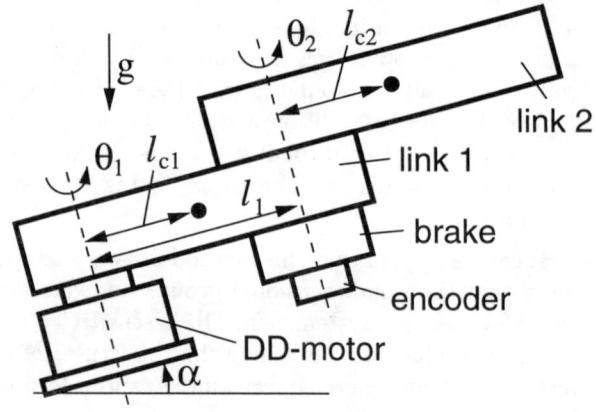

Figure 1: Scheme of R2D1.

2 Kinematics and Dynamic Model

Figure 1 shows the 2-DOF-SCARA-type robot R2D1. The rotational plane can be inclined against gravitation by the angle α. The dynamic equations of R2D1 are

$$M(\theta)\ddot{\theta} + n(\theta,\dot{\theta}) + g(\theta) + f(\dot{\theta}) = \tau , \quad (1)$$

where $\theta = [\theta_1, \theta_2]$ are the joint rotational angles, $\tau = [\tau_1, \tau_b]$ is the motor/brake torque, $n = [n_1, n_2]$ computes the coriolis and centrifugal torque, $g = [g_1, g_2]$ is the gravitational torque, $M = [m_{i,j}]$, $i,j \in \{1,2\}$ is the inertia matrix and $f = [f_1, f_2]$ is the frictional torque. With the brake at joint 2 applied a brake torque τ_b acts; otherwise we have $\tau_b \equiv 0$. Using the abbreviations $s_\alpha = \sin(\alpha)$, $c_2 = \cos(\theta_2)$, $s_2 = \sin(\theta_2)$, $s_{12} = \sin(\theta_1 + \theta_2)$, the dynamic parameters of (1) are given as

$$\begin{aligned} m_{11} &= A + 2Bc_2 & n_1 &= -B(\dot{\theta}_2^2 + 2\dot{\theta}_1\dot{\theta}_2)s_2 \\ m_{12} &= m_{22} + Bc_2 & n_2 &= Bs_2\dot{\theta}_1^2 \\ m_{21} &= m_{12} & g_1 &= -Cs_\alpha s_1 + g_2 \\ m_{22} &= m_2 l_{c2}^2 + I_2 & g_2 &= -Ds_\alpha s_{12} , \end{aligned}$$

with

$$\begin{aligned} A &= m_1 l_{c1}^2 + m_2(l_1^2 + l_{c2}^2) + I_1 + I_2 & B &= m_2 l_1 l_{c2} \\ C &= g_0(m_1 l_{c1} + m_2 l_1) & D &= g_0 m_2 l_{c2} , \end{aligned}$$

where the physical parameters are estimated in standard units to $l_1 = 0.300$, $l_{c1} = 0.206$, $l_{c2} = 0.092$, $I_1 = 0.430$, $I_2 = 0.127$, $m_1 = 10.2$, $m_2 = 5.75$. This yields $A = 1.55$, $B = 0.16$, $C = 37.5$, $D = 5.20$, $m_{22} = 0.18$. A viscous and static friction model for both joints established from experiments is $f_i = a_i \text{sign}(\dot{\theta}_i) + b_i \dot{\theta}_i$, $i \in \{1,2\}$, where $\text{sign}(0) = 0$ and $a_1 = 2.3$, $a_2 = 0.32$, $b_1 = 0.15$, $b_2 = 0.01$.

3 NCL-PD-control

It is known that we can control either one of the joint angles θ_1 or θ_2 of R2D1 with the actuator in joint 1 exploiting strong inertial coupling [1,4,6]. In this paper we examine the case of controlling the unactuated joint θ_2 applying non-collocated linearization [4,6]. A PD-controller is applied to the resulting 2nd order integrator.

3.1 Non-Collocated Linearization

Solving the second row of (1) for $\ddot{\theta}_1$ and substituting the result into the first row of (1) yields

$$\begin{aligned} m_{12}^* \ddot{\theta}_2 + n_1^* + g_1^* + f_1^* &= \tau_1 & (2) \\ m_{12} \ddot{\theta}_1 + m_{22} \ddot{\theta}_2 + n_2 + g_2 + f_2 &= 0 , & (3) \end{aligned}$$

with the following abbreviations

$$\begin{aligned} m_{12}^* &= m_{12} - \tfrac{m_{11} m_{22}}{m_{12}} & g_1^* &= g_1 - \tfrac{m_{11}}{m_{12}} g_2 \\ n_1^* &= n_1 - \tfrac{m_{11}}{m_{12}} n_2 & f_1^* &= f_1 - \tfrac{m_{11}}{m_{12}} f_2 . \end{aligned} \quad (4)$$

With v_2 as the new control input, the computed torque method with the motor moment at joint 1 being

$$\tau_1 = v_2 \tilde{m}_{12}^* + \tilde{n}_1^* + \tilde{g}_1^* + \tilde{f}_1^* , \quad (5)$$

is applied, where $\tilde{\bullet}$ denote estimated and therefore uncertain model parameters. Substitution of (5) into (2), solving for $\ddot{\theta}_2$ and then inserting the resulting expression in (3) yields

$$\begin{aligned} \ddot{\theta}_1 &= -\tfrac{1}{m_{12}}\left((v_2 \tilde{m}_{12}^* + \Delta)\tfrac{m_{22}}{m_{12}^*} + n_2 + g_2 + f_2\right) \\ \ddot{\theta}_2 &= \tfrac{1}{m_{12}^*}(v_2 \tilde{m}_{12}^* + \Delta) , \end{aligned} \quad (6)$$

where $\Delta = \tilde{n}_1^* - n_1^* + \tilde{g}_1^* - g_1^* + \tilde{f}_1^* - f_1^*$.

Remark 1 *Under the assumption of incorrect model parameters it is not possible to decouple (1) by means of non-collocated linearization. This is obvious as the right hand side of (6) depends on θ and $\dot{\theta}$.*

In case of exactly known physical parameters, complete *non-collocated linearization* [4] and decoupling is obtained as

$$\begin{aligned} \ddot{\theta}_1 &= -\tfrac{m_{22}}{m_{12}} v_2 - \tfrac{n_2 + g_2 + f_2}{m_{12}} \\ \ddot{\theta}_2 &= v_2 . \end{aligned} \quad (7)$$

We may use the control law (5) only if (1) is *strongly inertially coupled*. This requires $m_{12}(\theta_2) \neq 0$ for all θ_2. For the experimental system R2D1 the strong inertial coupling property is valid globally, i.e.

$$0.02 \leq m_{12} \leq 0.34 . \quad (8)$$

3.2 PD-Control

For the outer loop term v_2 a PD-control algorithm is implemented. With the desired angle of the 2nd joint θ_2^d the control error is defined as $e_2 = \theta_2^d - \theta_2$. The PD-control law is given by

$$v_2 = k_d \dot{e}_2 + k_p e_2 . \quad (9)$$

Substitution of (9) into the second row of (7) yields the error differential equation for the non-perturbed system as

$$\ddot{e}_2 + k_d \dot{e}_2 + k_p e_2 = 0 , \quad (10)$$

where the choice of the parameters k_d and k_p determines the dynamic behavior of the error e_2.

In case of faulty parameter estimates, the error dynamics are

$$\ddot{e}_2 + \tilde{k}_d \dot{e}_2 + \tilde{k}_p e_2 = z, \tag{11}$$

with the configuration and joint angle dependent control parameters $\tilde{k}_d = k_d \frac{\tilde{m}_{12}^*}{m_{12}^*}$, $\tilde{k}_p = k_p \frac{\tilde{m}_{12}^*}{m_{12}^*}$ and the forcing term $z = -\Delta/m_{12}^*$.

In the non-perturbed case, the control error e_2 vanishes completely and the system dynamics are governed by the *zero dynamics*. Because of the importance for stability, a brief discussion of the zero dynamics is given next.

3.3 Zero Dynamics

If the control law (9) stabilizes the system such that the control error e_2 vanishes, the system behavior is governed by the zero dynamics, obtained from (7) with $v_2 = 0$ as

$$\ddot{\theta}_1 = -\frac{n_2 + g_2}{m_{12}}. \tag{12}$$

Figure 2 shows some characteristic stable and unstable zero dynamics for $\alpha = -\pi/6$. Note that with $v_2 = 0$ the joint velocity of the 2nd joint also becomes zero and therefore $f_2 = 0$.

In case of a parameter disturbed system, NCL results in (6). It is not possible anymore to linearize the system by the computed torque method, as the system remains coupled (cf. Remark 1) and therefore $\ddot{\theta}_2$ and the control input v_2 are not linearly related. As a consequence, a PD-controller is not sufficient for zeroing the error e_2.

Remark 2 *In case of faulty parameter estimates, the NCL-PD-controlled system does not display zero dynamics.*

Besides the unstable zero dynamics, there are additional effects, causing instability.

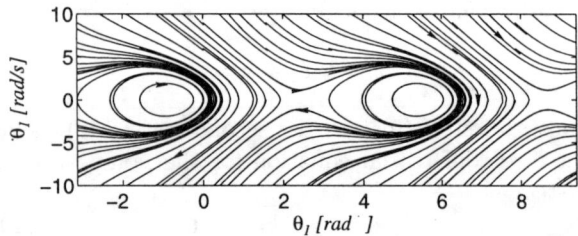

Figure 2: Zero dynamics for $\alpha = -\pi/6$, $\theta_2^d = \pi/4$.

3.4 Stability of the NCL-PD-controlled system

In case of exact parameter estimates, the 2nd joint is stable for $k_p > 0$, $k_d > 0$. However, this does not guarantee stability of the total system. Due to the nonholonomic constraint of the 1st joint, the system may approach unstable zero dynamics. Another reason for instability is caused by the control output limitation as the system looses output controllability.

Looking at the perturbed system, there is one more reason for instability. As seen in (11), the error dynamics of the 2nd joint depend on the state variable dependent control parameters $\tilde{k}_p$ and $\tilde{k}_d$ and the forcing term z. Therefore the NCL-PD-control can become unstable. This motivates a stability analysis by means of a Lyapunov-function for the 2nd joint.

3.5 Lyapunov-function for the 2nd joint

For simplification we introduce

$$\gamma(\theta_2) := \frac{\tilde{m}_{12}^*}{m_{12}^*} = \frac{|\tilde{M}| m_{12}}{|M| \tilde{m}_{12}}. \tag{13}$$

Inserting (9) into (6) and using (13) we obtain

$$\ddot{\theta}_2 = \gamma(\theta_2)\left(k_p\left(\theta_2^d - \theta_2\right) - k_d \dot{\theta}_2\right) + \frac{\Delta}{m_{12}^*}. \tag{14}$$

With the constant λ_4 and a positive definite function $\lambda_3 = \lambda_3(\theta_2)$ a Lyapunov-function for the 2nd joint is defined by

$$V_2 = \frac{1}{2}\left(\lambda_3(\theta_2)(\theta_2^d - \theta_2)^2 + \lambda_4 \dot{\theta}_2^2\right), \tag{15}$$

with the time derivative along the trajectories being

$$\dot{V}_2 = -\lambda_3(\theta_2)(\theta_2^d - \theta_2)\dot{\theta}_2 + \frac{1}{2}\left(\theta_2^d - \theta_2\right)^2 \dot{\lambda}_3(\theta_2) + \lambda_4 \dot{\theta}_2 \ddot{\theta}_2. \tag{16}$$

Substituting (14) into (16) and using the fact that $\dot{\lambda}_3(\theta_2) = \lambda_3' \dot{\theta}_2$ one obtains

$$\dot{V}_2 = -\lambda_3(\theta_2)(\theta_2^d - \theta_2)\dot{\theta}_2 + \frac{(\theta_2^d - \theta_2)^2}{2}\lambda_3'(\theta_2)\dot{\theta}_2 + \lambda_4 \dot{\theta}_2\left(\gamma(\theta_2)\left(k_p\left(\theta_2^d - \theta_2\right) - k_d \dot{\theta}_2\right) + \frac{\Delta}{m_{12}^*}\right), \tag{17}$$

which is simplified by the following lemma.

Lemma 1 *If $\gamma(\theta_2)$ is continuous, then there exists a positive function $\lambda_3(\theta_2)$ such that*

$$-\lambda_3(\theta_2)\left(\theta_2^d - \theta_2\right)\dot{\theta}_2 + \frac{(\theta_2^d - \theta_2)^2}{2}\lambda_3'(\theta_2)\dot{\theta}_2 + \lambda_4 \dot{\theta}_2 \gamma(\theta_2) k_p \left(\theta_2^d - \theta_2\right) = 0. \tag{18}$$

Proof: The proof is based on the existence of a solution of the ordinary first order differential equation (18) by means of *variation of constants*. Within the manifold of solutions, the constant can be chosen such that $\lambda_3(\theta_2)$ is positive definite. Further, a necessary and sufficient condition for the existence of a solution of (18) is the continuity of $\gamma(\theta_2)$ which follows directly from (13) as $\tilde{m}_{12}^*$ and m_{12}^* are continuous and the denominator $|M|\tilde{m}_{12} \neq 0$, $\forall \theta_2$, as strong inertial coupling is assured globally for R2D1. ∎

Inserting (18) into (17) yields

$$\dot{V}_2 = \frac{\lambda_4}{m_{12}^*}\left(-\tilde{m}_{12}^* \dot{\theta}_2^2 k_d + \dot{\theta}_2 \Delta\right) \quad (19)$$

and in case of no perturbation $\dot{V}_2 = -\dot{\theta}_2^2 k_d$.

Lemma 2 V_2 *of (15) is a Lyapunov-function for (6) within the state space area*

$$\mathcal{G} = \left\{\boldsymbol{\theta}, \dot{\boldsymbol{\theta}} \mid \dot{V}_2(\boldsymbol{\theta}, \dot{\boldsymbol{\theta}}) \leq 0\right\}.$$

Proof: As $\lambda_4 > 0$ and with $\lambda_3(\theta_2)$ being positive definite, V_2 is also positive definite. Choosing $\boldsymbol{\theta}, \dot{\boldsymbol{\theta}}$ such that $\dot{V}_2$ is negative definite, the result follows. ∎

4 Robust stabilization of R2D1

Applying NCL-PD-control, the 2nd joint is linearized and stabilized by a PD-controller in case of no perturbation. Next, for inexact parameters, a stability region for R2D1 is derived.

4.1 Stability region

The stability region $\mathcal{S} \subseteq \mathcal{G}$ is the region of the state space, where stability of R2D1 can be assured in case of perturbations.

As $\mathcal{S}$ depends on the parameter estimates, we introduce the relative error r_{x_i} of each uncertain physical parameter x_i as $x_i := (1 + r_{x_i})\tilde{x}_i$. The error vector

$$\boldsymbol{r} = [r_{l_1}\ r_{l_{c1}}\ r_{l_{c2}}\ r_{I_1}\ r_{I_2}\ r_{m_1}\ r_{m_2}\ r_{a_1}\ r_{b_1}\ r_{a_2}\ r_{b_2}]^T$$

combines all possible relative errors inside the hypercuboid $|r_{l_1}^{max}| \leq 0.05$, $|r_{l_{c1}}^{max}| \leq 0.1$, $|r_{l_{c2}}^{max}| \leq 0.1$, $|r_{I_1}^{max}| \leq 0.1$, $|r_{I_2}^{max}| \leq 0.2$, $|r_{m_1}^{max}| \leq 0.08$, $|r_{m_2}^{max}| \leq 0.1$, $|r_{a_1}^{max}| \leq 0.4$, $|r_{b_1}^{max}| \leq 0.4$, $|r_{a_2}^{max}| \leq 0.4$ and $|r_{b_2}^{max}| \leq 0.4$.

The error space, arising from the above error-constraints is called $\mathcal{R} \subset \mathbb{R}^{11}$. For each $\boldsymbol{r} \in \mathcal{R}$ there exists a corresponding smallest dynamical system $\Sigma(\boldsymbol{r})$. A system Σ is robustly stable, if it is stable for all $\boldsymbol{r} \in \mathcal{R}$.

To determine the area $\mathcal{G}$ we need to compute the intersection of all $\mathcal{G}(\boldsymbol{r})$. Dividing each element of $\mathcal{R} \in \mathbb{R}^{11}$ into n intervals, one has to calculate n^{11} different points in $\mathcal{G}$. This method is not sufficient for robustness and the computational effort is high. Therefore a simpler method to determine a conservative (smallest) stability area is required. Under the assumption that there exists one error vector $\boldsymbol{r}^*$, rendering the corresponding area $\mathcal{G}^*$ stable for all other possible error vectors $\boldsymbol{r} \in \mathcal{R}$, we can formulate the following optimization problem

$$\mathcal{G}^* = \min_{\boldsymbol{r} \in \mathcal{R}} \mathcal{G}(\boldsymbol{r}). \quad (20)$$

Calculations have shown, that for R2D1 the areas $\mathcal{G}(\boldsymbol{r})$ have a *contractive character*, see Figure 3.

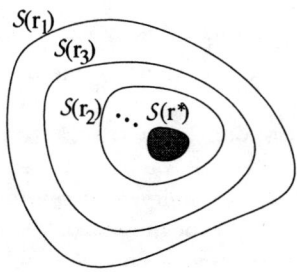

Figure 3: Contractive stability areas.

To compute the conservative (smallest) stability area $\mathcal{G}^*$ we need to determine the corresponding worst-case relative error vector $\boldsymbol{r}^*$. This yields the optimization problem

$$\boldsymbol{r}^* = \max_{\boldsymbol{r}} J(\boldsymbol{r}, \boldsymbol{\theta}, \dot{\boldsymbol{\theta}}, k_d),$$

with $J = \text{sign}(\dot{V}_2(\boldsymbol{r}, k_d, \boldsymbol{\theta}, \dot{\boldsymbol{\theta}}))$ for a predetermined control parameter k_d. For a numerical solution, the state space is discretizised and a genetic algorithm is applied, yielding the error vector

$$\boldsymbol{r}^* = [-0.0349, -0.0661, -0.0982, -0.0584,$$
$$0.1614, 0.0009, -0.0938, -0.0007, 0.3080,$$
$$-0.0861, -0.0550]^T,$$

and the smallest stability area $\mathcal{G}^*$ and therewith $\dot{V}_2^*$ from (19).

Experiments have shown, that there exists another stable area outside $\mathcal{G}^*$. Due to the perturbed error dynamics (11), the system performs stable limit cycles within the area $|\dot{\theta}_2| \leq \dot{\theta}_2^{max} \leq 0.2\text{rad/s}$. Therefore the stability area contains this area plus the area $\mathcal{G}^*$ and the area of assured output controllability, hence

$$\mathcal{S} = \left\{\boldsymbol{\theta}, \dot{\boldsymbol{\theta}} \mid (\dot{V}_2^*(\boldsymbol{\theta}, \dot{\boldsymbol{\theta}}) < 0 \vee |\dot{\theta}_2| < \dot{\theta}_2^{max}) \wedge |\dot{\theta}_1| < \dot{\theta}_1^{max}\right\}.$$

4.2 Switching control strategy

For the Lyapunov-function V_2 in (15) it is not possible to find a constant of motion, covered completely in the area $\mathcal{G}$, where $\dot{V}_2$ is negative definite. Therefore the system state may leave $\mathcal{G}$ and become unstable. If this occurs the control is switched from NCL-PD-mode to a damping control mode forcing the state to reenter the stable area $\mathcal{G}$. Hence, by switching the control the stability area $\mathcal{G}$ becomes *quasi-invariant*.

Definition 1 *If system trajectories leaving a stability area $\mathcal{G}$ cause a controller to be switched such that the trajectories reenter $\mathcal{G}$ again, this switching controller makes the stability area $\mathcal{G}$ quasi-invariant.*

The control law in damping mode is a sliding-mode position control of the 1st joint: $\tau_1 = \tau_{max} \,\text{sign}(e_1 + \dot{e}_1)$, $e_1 = \theta_1^d - \theta_1, \dot{e}_1 = \dot{\theta}_1^d - \dot{\theta}_1$, where $\text{sign}(0) = 0$ and $\theta_1^d, \dot{\theta}_1^d$ is chosen as a stable point of the zero dynamics. Additionally, the brake in the 2nd joint is applied.

To achieve a stabilizing switching strategy a pseudo Lyapunov-function (PLF) is constructed for each control mode [10]. For stability a decreasing sequence of the PLF value at the beginning of the time interval in each mode is required. The PLF for the NCL-PD mode is V_2 of (15) and a PLF for the damping mode is

$$V^{da} = \frac{1}{2}\left(e_1^2 + \lambda_1^{da}\dot{e}_1^2 + \theta_2^2 + \lambda_2^{da}\dot{\theta}_2^2\right),$$

where λ_1^{da} and λ_2^{da} are positive constants. The requirement of the values of V_2 at the beginning of each time interval forming a decreasing sequence [10] is fulfilled, however, for V^{da} only the values at the end of each time interval form a decreasing sequence. Nevertheless, stability can be proofed as the system is positioned to a stable point of the zero dynamics at the end of each damping mode time interval. From this and the fact that the control error of the 2nd joint asymptotically approaches zero it is clear that the system is driven to stable zero dynamics after a finite number of controller switches.

Figure 4 illustrates the stability requirements of the values of the PLFs. The small circles mark the values of the PLF for the NCL-PD mode and the small squares mark those of the damping mode. The values of V_2 at the beginning of subsequent NCL-PD-control intervals and the values of V^{da} at the end of damping mode intervals, each form a decreasing sequence.

The switching strategy to achieve this is:

IF $[\theta, \dot{\theta} \notin \mathcal{S}]$ THEN switch from NCL-PD to damping.

IF $[(\theta, \dot{\theta} \in \mathcal{S}) \wedge (e_1 = \dot{e}_1 = 0) \wedge (\Delta V_2^* < 0)]$ THEN switch from damping to NCL-PD,

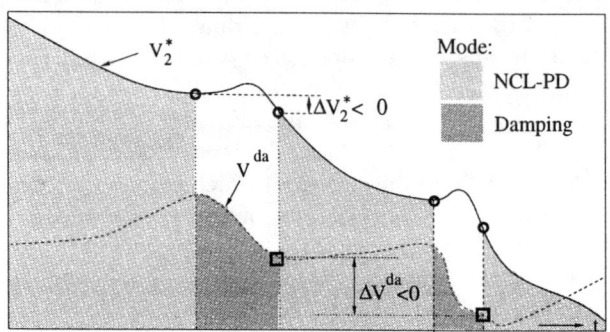

Figure 4: $V_2^*(t)$ and $V^{da}(t)$ when switching between the control modes.

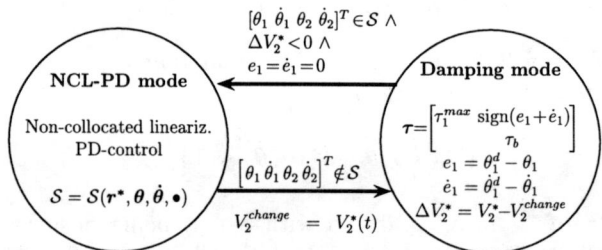

Figure 5: Hybrid automaton of the switching control strategy.

see Figure 4 for a definition of ΔV_2^*. This switching strategy is shown as a hybrid automaton in Figure 5.

5 Experimental Validation

The experiments were performed with the parameters $\alpha = -30°$, $k_p = 221, k_d = 21$, the joint velocity limitations were $\dot{\theta}_1^{max} = 10 rad/s$, $\dot{\theta}_2^{min} = 0.12 rad/s$. The desired value θ_2^d is a square-wave function with a period of $10s$, defined as $\theta_2^d(t) = \sqcap(t) \cdot \pi/2$. It should be stressed, that in this experiment every 5s the system starts from a different initial value. Therefore the trajectories for each of the time intervals (of 5s) are different from each other.

In addition to the changing value of θ_2^d, the system was sometimes disturbed by an external force. $[\theta_1^d, \dot{\theta}_1^d]$ is chosen as the equilibrium point of the zero dynamics, existing for $\alpha \neq 0$. Furthermore, a minimal closing time of 30ms for the brake has to be assured.

Figure 6 (a) shows within the first 10 seconds the case of unstable zero dynamics. In the second period, the system is disturbed twice manually such that the system leaves the stability region and control needs to switch. During the last period, the system approaches stable zero dyanamics on its own without the use of the damping mode.

Figure 6 (b) shows the mode of the system during

the experiment. *ON* stands for active damping mode, *OFF* for active NCL-PD mode. The solid line shows the case, where the system leaves the negative definite area of $\dot{V}_2$ with the 2nd joint velocity exceeding $\dot{\theta}_2^{max}$. The dashed line shows, when the system hits the control output limitations and therefore is unstable.

The same experiment without the switching strategy gets unstable from the very beginning. An MPEG-video of R2D1 can be found under [11].

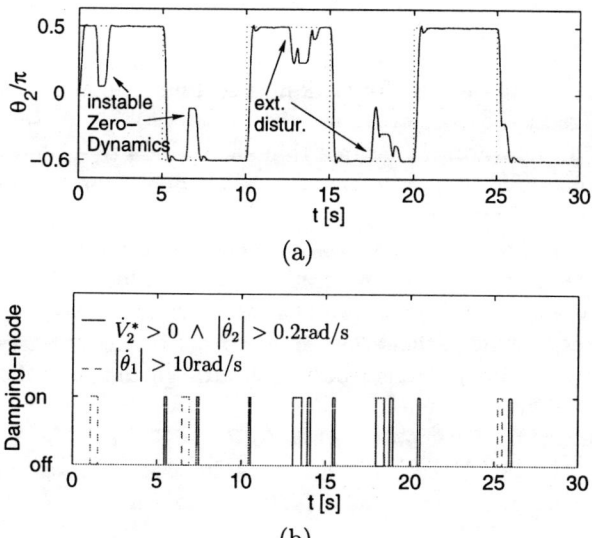

Figure 6: (a) $\theta_2(t)$, following a square-wave signal (b) Control-modes

6 Conclusions

A switching control strategy for the SCARA-type underactuated robot R2D1 using NCL-PD-control and a damping control mode has been presented. During damping control the holding brake in the 2nd joint is applied. Robust stability is achieved by an analysis of the conservative stability region and a controller switching strategy rendering this stability area quasi-invariant. The argument for robust stability assumes that the stability areas are contractive, which is the case for R2D1. Experiments have been presented, validating the robust stability of the switching control strategy to global position control of the unactuated 2nd joint angle of R2D1.

References

[1] H. Arai and S. Tachi, "Position Control of a Manipulator with Passive Joints Using Dynamic Coupling," *IEEE Transactions on Robotics and Automation*, vol. 7, pp. 528–534, August 1991.

[2] P. Chodavarapu and M. Spong, "On Noncollocated Control of a Single Flexible Link," in *Proceedings of the IEEE International Conference on Robotics and Automation*, (Minneapolis, Minnesota), pp. 1101–1106, 1996.

[3] G. Oriolo and Y. Nakamura, "Control of Mechanical Systems with Second-Order Nonholonomic Constraints: Underactuated Manipulators," in *Proceedings of the IEEE Conference on Decision and Control*, (Brighton, England), pp. 2398–2403, 1991.

[4] M. W. Spong, "Partial Feedback Linearization of Underactuated Mechanical Systems," in *Proceedings of the IEEE/RSJ/GI International Conference on Intelligent Robots and Systems IROS*, (München), pp. 314–321, 1994.

[5] T. Suzuki, M. Koinuma, and Y. Nakamura, "Chaos and Nonlinear Control of a Nonholonomic Free-Joint Manipulator," in *Proceedings of the IEEE International Conference on Robotics and Automation*, (Minneapolis, Minnesota), pp. 2668–2675, 1996.

[6] J. Mareczek, M. Buss, and G. Schmidt, "Comparison of Control Algorithms for a Nonholonomic Underactuated 2-DOF Robot," in *Proceedings of the IEEE/ASME International Conference on Advanced Intelligent Mechatronics AIM'97*, (Tokyo, Japan, Paper No. 96), 1997.

[7] M. Bergerman and Y. Xu, "Robust joint and Cartesian control of underactuated manipulators," *Transactions of the ASME: Journal of Dynamic Systems, Measurement and Control*, vol. 118, pp. 557–565, Sep. 1996.

[8] F. Najson and E. Kreindler, "On the Lyapunov Approach to Robust Stabilization of Uncertain Nonlinear Systems," in *International Journal of Robust and Nonlinear Control*, (Haifa, Israel), pp. 41–63, 1996.

[9] R. Freeman and P. Kokotović, *Robust Nonliner Control Design*. Berlin: Birkhäuser-Verlag, 1 ed., 1996.

[10] M. S. Branicky, "Stability of switched and hybrid systems," in *Proc. 33rd IEEE Conf. Decision Control*, (Lake Buena Vista), pp. 3498–3503, 1994.

[11] http://www.lsr.e-technik.tu-muenchen.de/movies/dd32.mpg. 1997.

Predictive Vision Based Control of High Speed Industrial Robot Paths

Friedrich Lange   Patrick Wunsch   Gerhard Hirzinger

DLR Institute of Robotics and System Dynamics
P. O. Box 1116, D-82230 Wessling, Germany
e-mail: Friedrich.Lange@dlr.de

Abstract

A predictive architecture is presented to react on sensor data in the case of high speed motion and low bandwidth sensor data. This concept is used for the vision based control of an industrial robot (KUKA) to track a contour at a speed of 1.6 m/s. The vision task can be performed very fast since only 2 rows of the image are analyzed. In this way an accuracy of 0.3 mm is reached in spite of uncertainties in the robot's kinematic parameters. Vision and control work asynchronously so that even delay times are tolerable during sensing as long as the time-instant of the exposure is known.

1 Introduction

It is well known that sensor control of robots is limited by the bandwidth of both signal processing and robot dynamics. Therefore, regardless of the sensor system, tracking of unknown contours requires low speed. This may be one reason why in industrial applications online sensor control predominantly has been avoided.

However a camera can provide more information than the current deviation from a nominal position at the sample instant. It can also provide some look-ahead along the desired trajectory. This property can be exploited to overcome the bandwidth limitation of traditional vision based control to execute high speed sensor controlled movements. This paper presents such an approach.

In contrast to other visual servoing problems [HHC96] we do not track a target but we follow an edge with predetermined speed. This is motivated by industrial applications where the desired path is only inaccurately known because of tolerances of the workpiece dimensions or an uncertain positioning of the workpiece. A typical example is a robot which has to distribute the glue for sticking a sealing strip along an edge at the door of a car.

In addition we assume that the robot's tool center point (TCP) is only roughly known due to uncertain kinematic parameters.

In this paper, the task is to control a 6-axis robot with an endeffector integrated camera to follow a planar edge which is known coarsely (see figure 1). The prespecified path is only used to define the velocity profile. Information about the path geometry is derived from the sensor values. So the camera is used to detect displacements normal to the edge direction.

In general we see 3 steps to attain accurate tracking during high speed sensor controlled movements:

1. Localization of the desired position with respect to the actual position and the coordinate system (joint angles) of the robot.

2. Prediction of future steps of the desired path.

3. Control of the desired path in spite of disturbances due to the robot dynamics.

Step 1 means the evaluation of the sensor device to obtain the transformation between the actual and

Figure 1: Experimental setup with endeffector mounted camera

the desired position. For vision based sensors features are extracted from an image. Then the corresponding positions are calculated using the known robot and sensor characteristics. Unknown kinematics require learning of the hand-eye-transformation (see e.g. [vdSG97, JFN97]). In industrial applications this is not a problem since the kinematic parameters are known and the deviations between the tool center point and the sensed desired position will be small. So step 1 is a standard problem which will not be outlined here (see e.g. [HHC96]).

Step 2 requires the localization of future desired positions as in step 1 for the current timestep. So the robot motion along the contour has to be mapped into the image. Then the sensed desired motion has to be transformed back to the robot coordinate system. In this case the transformation is not trivial since small uncertainties in the kinematic parameters of the robot or the camera produce orientational errors which yield significant positional differences. The problem arises from the fact that the position has to be determined with an accuracy of about .001 of the distance to the current TCP. So this step will be discussed in section 3.

Step 3 means the compensation of dynamical path errors. This problem has been solved for position controlled robots in [LH94]. The learned feedforward controller requires the desired joint values for the current and n_d future controller timesteps. This corresponds to a prediction of the desired path over at least a period equivalent to the time constant of the robot, the motor-drives, and the feedback control loops. These future desired positions are obtained by step 2.

Known approaches for high speed tracking of a target [Cor95, CG96] use feedforward control as well, exploiting only estimations about the target speed in contrast to the approach taken here, where feedforward control will be based on learned dynamic characteristics of the robot.

2 Architecture for accurate control of robots with positional interface

Figure 2 shows the proposed architecture. Thin lines mean positions, e.g. the commanded and the actual position of the robot at timestep k. In contrast thick lines mean trajectories, e.g. the desired path from timestep k to timestep $k + n_d$.

The vision system can not only measure the current control error, i.e. the difference between the actual position and the edge. In addition it can detect a spatially extended part of the desired path. So at

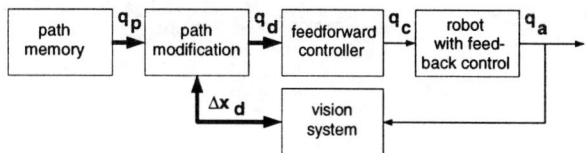

Figure 2: Indirect control architecture

timestep k the following n_d timesteps of the originally programmed path can be modified as well. This allows the feedforward controller with the learned parameters r_i to output a commanded position

$$q_c(k) = q_d(k) + r_1 \cdot (q_d(k+1) - q_d(k)) \quad (1)$$
$$+ \cdots + r_{n_d} \cdot (q_d(k+n_d) - q_d(k))$$

which in combination with the robot feedback control loops yields accurate following of the desired path q_d. The learning procedure is shown in [LH94]. It can be performed in advance without any sensor feedback.

Please note that all positions q are expressed in the coordinate system of the robot, the joint space. Cartesian measurements x have to be transformed therefore. Thus at timestep k the path modification module calculates the modified desired path as

$$q_d(k+i) = inv\_kin[kin[q_p(k+i)] + \Delta x_d(k+i,k)] \quad (2)$$

or

$$q_d(k+i) = inv\_kin[x_p(k+i) + \Delta x_d(k+i,k)] \quad (3)$$

from the original desired path q_p or x_p and the cartesian difference $\Delta x_d(k+i,k)$ between the programmed and the sensed desired path at timestep $k+i$, measured at timestep k. $kin[.]$ and $inv\_kin[.]$ denote the forward and inverse kinematic transformation, respectively.

The path modifications Δx_d are calculated from the actual position x_a (which is transformed from q_a), the sensor values Δx_s, and the nominal edge positions x_n (see figure 3). This approach allows the control of (curved) paths which do not coincide with an edge.

The positions x are expressed in the cartesian sensor coordinate system which is chosen such that the robot moves along for instance the y-direction. So the y-direction is not sensor controlled. However deviations due to the robot dynamics are compensated in all components.

Hence,

$$\Delta x_d(k+i,k) = \begin{pmatrix} \Delta x_{dx}(k+i,k) \\ 0 \\ \Delta x_{dz}(k+i,k) \\ \vdots \end{pmatrix} \quad (4)$$

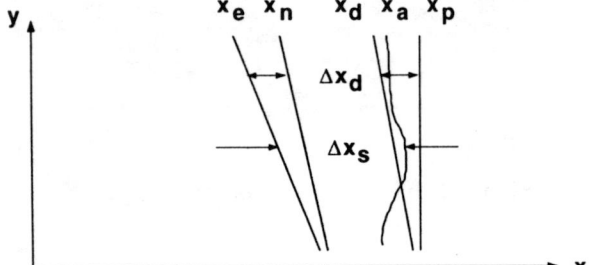

Figure 3: Notation of different paths (x_p = original desired path, x_d = sensed desired path, x_a = actual path, x_e = actual edge, x_n = nominal edge, Δx_s = sensor value, Δx_d = modification of the desired path due to a difference between the nominal and the actual edge)

In this notation indices x, y or z as second indices represent the corresponding components of the vector.

At timestep k the sensor values $\Delta x_s(k+i,k)$ are taken which look at the positions $x_p(k+i)$ by measurements from the current position $x_a(k)$. Then $\Delta x_d(k+i,k)$ can be revised by

$$\Delta x_d(k+i,k) = x_a(k) + \Delta x_s(k+i,k) - x_n(k+i) \quad (5)$$

The generation of sensor values $\Delta x_s(.)$ from the camera data will be explained in section 3. The nominal edge position $x_n(.)$ is given. For an edge parallel to the y-axis x_n is constant.

In contrast to direct feedback of the sensor values (see [HHC96]) the separation of sensor evaluation and control is insensitive to delays during the signal processing or the execution of the positional commands. This superiority with respect to stability has already been discussed in [LH96], there concerning indirect force control. It allows high bandwidth of the desired motion in spite of low bandwidth sensing which is typical for vision based systems. The only requirement is that x_a and Δx_s are measured at the same time. This will be studied in subsection 3.3.

For very low vision bandwidth the sensed deviations may become so big that instantaneous return to the desired path exceeds the allowed accelerations of the robot. In that case it is not sufficient to modify some sampled values of the desired path. Instead, a new trajectory has to be generated which leads the robot from the actual state (position and velocity) to the desired path. Such a path planning module is required for large time delays as in [Hir93] or immediately after the sensor control loop has been started. The experiments of section 4 were chosen such that explicit path planning was not necessary.

3 Vision based prediction of future control errors

3.1 General setup

This section explains the determination of the n_d sensor values $\Delta x_s(k+i,k)$ for the timesteps $k+i$ from the camera data of timestep k. It is assumed that the x-component of the sensor coordinate system is the only component to be controlled by the sensor. In this simple case, a single camera that tracks one single edge provides sufficient information. Additional sensor-controlled DOFs require an extended sensor system, e.g. a stereo camera system or a second edge which can be tracked. The x-component corresponds to the horizontal image coordinate u whereas the y-component, i.e. the direction of the fast motion, affects the vertical image coordinate v.

The camera images are filtered by a 3×3 matrix to extract vertical edges (figure 4), which allows robust localization of the edge pixels.

Then two image rows v_0 and v_1 (orthogonal to the edge) are selected in which the positions u_0 and u_1 of the edge have to be determined. These rows correspond to the TCP (more specific, the center of the sensor coordinate system) at the current timestep $x_{ay}(k)$ (see "o" in figure 4) and to a future programmed position $x_{py}(k+n_d)$ (see "+" in figure 4) if the visual angle is big enough. This means that the choice of the rows is dependent on the programmed path. The rows are computed as

$$v_0(k) = f_y^{-1}(x_{ay}(k) - x_{ay}(k)) = f_y^{-1}(0), \quad (6)$$

and

$$v_1(k) = f_y^{-1}(x_{py}(k+n_d) - x_{ay}(k)), \quad (7)$$

with extra consideration for the periphery of the image. So the "+" in the left part of figure 4 corresponds to the maximal value whereas the "+" on the

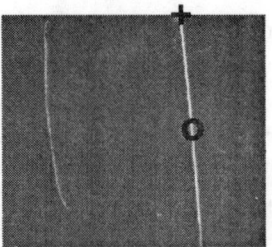 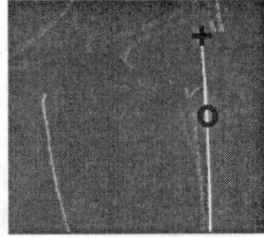

Figure 4: Image filtered for extraction of vertical edges (left shot: in the middle of the path, right shot: near the end of the path, o = edge at current TCP, + = edge at future desired TCP)

right hand side marks the end of the path which at the same time is the end of the vertical edge.

The image transformations $f_x(.)$ and $f_y(.)$ describe the mapping from pixels to positions which is defined by a calibration motion which is executed at the beginning of the path. This mapping is time-invariant if the distance between the camera and the edge is approximately constant in time. It is linear if the visual angle is small enough i.e. if the distance between the camera and the different edge positions is almost the same. For cameras with a large visual angle lens distortions may become significant if the edge is distant from the center of the rows. Then camera calibration methods as in [WCH92] have to be used to get at least $f_x(.)$. This is required when tracking different edges for sensor control of multiple degrees of freedom.

The limitation of the interpretation of the image with only two rows allows to compute the sensor values within 20 ms for each image (50 images per second) even with inexpensive standard vision hardware, as computation intensive edge filtering can be limited to the two selected rows.

The localization of the edge at the current and the future position are used to calculate a linear approximation of the edge between these two positions. This finally yields the n_d sensor values Δx_{sx} which are required in equation (5). Thus

$$\Delta x_{sx}(k+i, k) = f_x(u_0(k))$$
$$+(x_{py}(k+i) - x_{ay}(k)) \cdot \frac{f_x(u_1(k)) - f_x(u_0(k))}{f_y(v_1(k)) - f_y(v_0(k))}$$
$$= a(k) + (x_{py}(k+i) - x_{ay}(k)) \cdot b(k) \quad (8)$$

The number n_d of calculated edge positions is between 10 and 20 for usual industrial robots. But the edge is computed by a linear approximation from two points in the image. So the number of points to be localized in the image might have to be increased for edges with high curvatures as long as this does not change the image sampling rate.

Originally, the focal length of the camera system should allow a prediction length of n_d timesteps even at full speed. However this yields a very coarse resolution of the edge position. So a compromise has to be found. In this sense it is advantageous to tilt the camera towards the direction of motion for a fine resolution nearby and a large visual range.

3.2 Calibration

Equation (8) is sufficient if the kinematic parameters of the robot and the sensor are exactly known and the distance of the edge is time-invariant. Strictly speaking, the kinematic parameters of the TCP should not point to the actual robot's TCP but to the point where the optical axis of the camera systems hits the plane in which the edge lies. Otherwise a tilted end-effector would give an error in Δx_d.

If the kinematic parameters are not exactly known, calibration is required since at least orientational errors ϕ within the plane of the edge lead to big errors of the sensor values of v_1. In contrast to the detection of $f_x(.)$ and $f_y(.)$ (see sect. 3.1) the calibration of the orientation has to be repeated online because the effect of uncertain kinematic parameters or badly referenced joints is variable in space. This means that no extra motion is tolerable.

This calibration is done by comparing two positions from the real motion, measured by the joint values q_a, and from the observed motion which is tracked in the image.

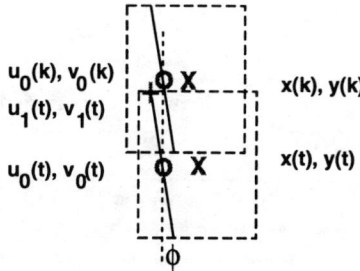

Figure 5: Calculation of the sensor orientation $\phi(k)$ from two images at time instants k and t (x = TCP, o = edge at TCP, + = edge at future position)

The first task for the calibration is to find the timestep t in which the part of the edge at the future position $x_{ay}(t) + f_y(v_1(t))$ corresponds to the current position $x_{ay}(k) + f_y(v_0(k))$ at timestep k (see figure 5).

The orientational error of the image is calculated as the difference between the slope b of the edge as it has been sensed at time-instant t (lower left solid line) and the actual slope which is represented by the difference of the current edge positions in the two images (dotted line). This approach assumes that the measurements u_0, v_0 of the current edge positions are not affected by small orientational errors.

The rotation $\phi(k)$ between the real image and the nominal sensor coordinate system is then

$$sin(\phi(k)) = \frac{x_{ax}(k) + f_x(u_0(k)) - x_{ax}(t) - f_x(u_0(t))}{x_{ay}(k) + f_y(v_0(k)) - x_{ay}(t) - f_y(v_0(t))}$$
$$- \frac{f_x(u_1(t)) - f_x(u_0(t))}{f_y(v_1(t)) - f_y(v_0(t))} \quad (9)$$

Beyond that, if the kinematic transformation of q_a

calculates changes in the orientation of the camera, $x_{a\phi}(k) - x_{a\phi}(t)$ has to be added to $\phi(k)$.

So equation (8) is changed to

$$\Delta x_{sx}(k+i, k) \qquad (10)$$
$$= a(k) + (x_{py}(k+i) - x_{ay}(k)) \cdot (b(k) + sin(\phi(k)))$$

This means that Δx_{sx} is expressed in the nominal sensor coordinate system and not in the real sensor system.

3.3 Fusion of asynchronuous sensor data

Equation (5) requires measurements x_a and Δx_s to be simultaneous. In practice however, the robot and the vision system have their own time bases and different sampling rates. In addition, delays may occur during sending of vision data to the robot controller since in our experiments we use a non-dedicated ethernet-based network to communicate measurements to the robot.

Therefore the communication is organized in a bidirectional fashion. First, the robot controller sends a request which by the way includes the arguments 0 and $x_{py}(k + n_d) - x_{ay}(k)$ of equations (6) and (7). When this request reaches the vision system, the next image will be taken for computation. So the image is at most one frame of 20 ms plus 20 ms for the exposure older than the request. After the computation of the sensor values the data are sent to the robot controller which then examines if the data are valid. Two tests are proposed when the data are received by the robot controller in timestep l:

- The sensor values are not more accurate than the difference between the positions $x_{ax}(l)$ and $x_{ax}(k - 40ms)$. So for big positional changes the sensor values have to be discarded and a new request has to be sent. The choice of a suitable limit for this turns out to be difficult since high bandwidth sensor values may cause big changes in the x-component of the robot position. Strictly speaking not the component x has to be constant but the component normal to the edge.

- The time difference $l - k$ between request and reception of data has to be limited to a maximum of about 40 ms to prevent accidentally small differences between $x_{ax}(l)$ and $x_{ax}(k - 40ms)$ (e.g. for paths with high curvature).

If a set of sensor data is valid, equation (10) is evaluated with $k-2$ instead of k and $l+i$ instead of $k+i$, since the image which has been taken about 20 ms $\approx 2 \cdot 12$ ms before timestep k is used to predict the edge for i steps after the current timestep l. If no valid sensor data is available, the edge positions are predicted as well from equation (10) with k according to the the last valid sensor data.

So a delayed or missing dataset is no problem for the stability of the servo loop since the control loop is not concerned. In the same way large delay times can be considered and, apart from the fact that the information arrives later, do not affect the performance.

4 Experiments

For the experiments a straight line of 80 cm is programmed. It decribes coarsely the edge (see figure 1). The path is executed by the robot with a maximal speed of 1.6 m/s. It is sensed by an endeffector mounted camera in a distance of about 30 cm from the edge.

Figure 6 shows the result using a straight edge. The programmed line (solid curve with an RMS deviation of 10 mm) turns out to be far away from the edge at the end of the path. Sensor control according to equations (1), (2) and (5), but without the prediction of equation (8) (dashed curve with an RMS deviation of 3.4 mm) can reduce the path error substantially but not totally. The errors come from fact that the actual edge position is sensed too late. Normal visual servoing with a specially tuned PD-type controller according to [HHC96] (dash-dotted curve with an RMS-value of 1.9 mm) reduces the path error somewhat better. Here, the accuracy is futher limited by the dynamical delays of the robot which affect the different joints differently. Both types of errors can be compensated by the predictive control approach with online calibration of the orientation (dotted curve with an RMS deviation of 0.3 mm). In contrast to the slope of the edge b

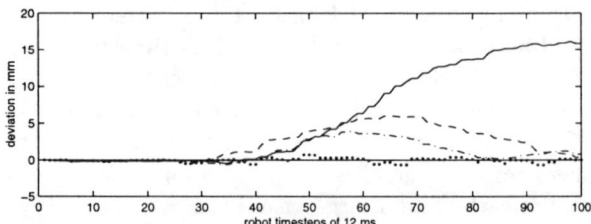

Figure 6: Control error at a straight edge (solid = without sensor control, dashed = with indirect sensor control but without prediction, dotted = with predictive sensor control, dash-dotted = normal visual servoing)

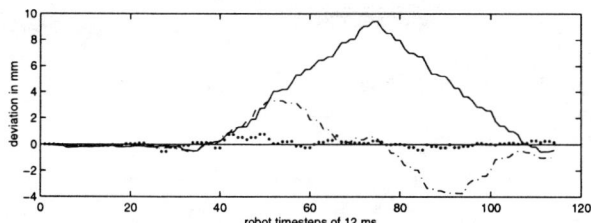

Figure 7: Control error at an edge with an unexpected corner (solid = without sensor control, dotted = with predictive sensor control, dash-dotted = normal visual servoing)

the orientation of the camera ϕ has been maintained after the previous experiement since calibration is not possible at the very beginning of the path.

The method is limited by joint elasticity which at least in some timesteps inhibits the correct measurement of x_a. So the method can be improved if the deflection of the joints can be estimated thus yielding the actual position and orientation of the camera.

The accuracy further depends on the ability of the feeedforward controller to follow a given path without dynamical deviations. By this, the reachable accuracy is restricted to about 0.3 mm as well.

On the other side, the resolution of the images seems adequate though 1 pixel corresponds to 0.45 mm. This means that the control error reached is less than 1 pixel.

The experiment is repeated in figure 7 with reduced speed. This time there is a non-modelled corner in the edge. Again, the proposed method (RMS = 0.25 mm) is superior to the standard visual servoing method (RMS = 1.7 mm).

5 Conclusion

Vision based control of an industrial robot at full speed highlights the performance of the proposed predictive control approach. In each image two points are localized to compute the next controller timesteps of the desired path and to calibrate the actual sensor coordinate system. This is used for a learned feedforward controller to follow the sensed path in spite of deviations due to the robot dynamics. Prediction and control is realized in a position based fashion. At the same time the vision system asynchronuously provides the controller with new sensory data.

Future experiments will focus on implementation aspects as filtering or the problem dependent selection of the camera's focal length or tilt, respectively. In addition, representations of the edge as higher order polynomials are tested with curved edges and more DOFs.

References

[CG96] P. I. Corke and M. C. Good. Dynamic effects in visual closed-loop systems. *IEEE Trans. on Robotics and Automation*, 12(5):671–683, Oct. 1996.

[Cor95] P. I. Corke. Dynamic issues in robot visual-servo systems. In *Int. Symp. on Robotics Research ISRR'95*, pages 488–498, Herrsching, Germany, 1995. Springer.

[HHC96] S. Hutchinson, G. D. Hager, and P. I. Corke. A tutorial on visual servo control. *IEEE Trans. on Robotics and Automation*, 12(5):651–670, Oct. 1996.

[Hir93] G. Hirzinger. ROTEX - the first robot in space. In *6th International Conference on Advanced Robotics ICAR '93*, Tokyo, Japan, Nov. 1993.

[JFN97] M. Jägersand, O. Fuentes, and R. Nelson. Experimental evaluation of uncalibrated visual servoing for precision manipulation. In *Proc. IEEE Int. Conference on Robotics and Automation*, Albuquerque, New Mexico, April 1997.

[LH94] F. Lange and G. Hirzinger. Learning to improve the path accuracy of position controlled robots. In *IEEE/RSJ/GI Int. Conference on Intelligent Robots and Systems*, München, Germany, Sept. 1994.

[LH96] F. Lange and G. Hirzinger. Learning force control with position controlled robots. In *Proc. IEEE Int. Conference on Robotics and Automation*, pages 2282–2288, Minneapolis, Minnesota, April 1996.

[vdSG97] P. van der Smagt and F. Groen. *Visual feedback in motion*, pages 37–73. Academic Press, Boston, Massachusetts, 1997.

[WCH92] J. Weng, P. Cohen, and M. Herniou. Camera calibration with distortion models and accuracy analysis. *IEEE Trans. on Pattern Analysis and Machine Intelligence*, 14(10):965–980, Oct. 1992.

Tracking a Moving Target with Model Independent Visual Servoing: a Predictive Estimation Approach

Jenelle Armstrong Piepmeier
George W. Woodruff
School of Mechanical Engineering
Georgia Institute of Technology

Gary V. McMurray
Georgia Tech
Research Institute

Harvey Lipkin
George W. Woodruff
School of Mechanical Engineering
GeorgiaInstitute of Technology

Abstract

Target tracking by model independent visual servo control is achieved by augmenting quasi-Newton trust region control with target prediction. Model independent visual servo control is defined as using visual feedback to control the robot without precise kinematic and camera models. While a majority of the research assumes a known robot and camera model, there is a paucity of literature addressing model independent control. In addition, that research has focused primarily on static targets. The work presented here demonstrates the use of predictive filters to improve performance of the control algorithm for linear and circular target motions. The results show a performance of the same order of magnitude as compared to some model based visual servo control research. Certain limitations to the algorithm are also discussed.

1 Introduction

The objective of this research is to achieve model independent, vision guided robotic control. Model independent means that the control does not require a precisely calibrated kinematic or camera models. The control method should be independent of robot type and location as well as camera type and location. The robot should be able to reach a static target and to track a moving target. Previous model independent research [8][10] has focused primarily on static target scenarios. Specifically, this paper addresses model independent control of a robot tracking a moving target.

A thorough overview of visual servoing architectures can be found in [9]. Most visual servoing research falls in two categories: position based and image based. In position based control, vision information is interpreted and transformed into information with respect to a base or world coordinate system. Image based control uses feature vectors from the camera image plane as input into the controller.

By definition, position based visual servoing requires a precise knowledge of the kinematic robot model, the exact location of the target in world coordinates, and a precise camera calibration model. Errors in these models can result in large positioning errors[17]. The *a priori* knowledge required makes position based visual servoing unsuitable for a model independent approach.

Image based visual servo control acts on an error signal which is defined by image features, such that when the servoing goal is reached, the error is zero. The error may be calculated by comparison with a pre-recorded image, i.e. a "teach by showing" approach, or by comparing the target position with the end-effector position. Most feature-based control uses either an image Jacobian or a composite Jacobian, the product of the image and robot Jacobians. A composite Jacobian relates differential changes in joint angles to differential changes in image features. Jacobians are determined by various methods: evaluated analytically[6], determined by estimation[8][10][13], updated with predictive methods[16], or solved for experimentally[2][7].

Much of the recent image based work implements *a priori* knowledge of the system structure, e.g. the camera model, kinematic model, and system parameters. The need for off-line parameter identification limits the robustness of the algorithms to disturbances in the system configuration which are common in industrial settings.

On-line Jacobian estimation presents a method of arriving at a Jacobian without using precise models of the camera and robotic systems. Hosoda[8] and Jägersand[10] have demonstrated the use of a rank one Broyden update for Jacobian estimation. The estimated Jacobian is used as part of an image based control law to servo the end-effector to the target. It should be emphasized previous model independent visual servoing research has focused on static targets.

This paper investigates the use of Jacobian estimation for model independent visual servo control to track a moving target.

2 Control Algorithms

The controller is based on a nonlinear least squares algorithm employing Jacobian estimation. A stereo vision system whose field of view includes both the end-effector and the target is assumed. The objective is to minimize the feature error between the end-effector and the desired target as seen in the image plane. Finding the joint variables that minimizes the feature error is a multivariate optimization problem that can be formulated as a nonlinear least squares problem. Jacobian estimation using a Broyden update law is a strategy employed in the solution of nonlinear least squares.

2.1 Nonlinear Least Squares

For a static target y^*, and a end-effector $y(\theta)$, as seen in the image plane, the residual error between the two points can be expressed as $f(\theta) = y(\theta) - y^*$. The objective function to be minimized, F, is a function of squared error.

$$F = \frac{1}{2} f^T f$$

The robot's forward kinematics and the camera's imaging geometry render $f(\theta)$ a highly nonlinear function. Thus the problem of finding the θ^* that minimizes F is a nonlinear least-squares problem.

Let $x = \begin{bmatrix} \theta_1 & \theta_2 & \cdots & \theta_n \end{bmatrix}^T$ represent the n joint angles, and let x_k represent x at time k. In the region of a point x_k, the value of F at a point $x_k + p$ can be modeled using a Taylor series expansion,

$$M_k(x_k + p) = F(x_k) + \nabla F p + \frac{1}{2} p^T \nabla^2 F p \quad (1)$$

where M_k is an affine model of F around the point x_k. A necessary condition for the minimizer of M_k is $\nabla M_k = 0$. If the point $x_k + p$ is a minimizer, then

$$\nabla M_k(x_k + p) \approx \nabla F + \nabla^2 F p = 0 \quad (2)$$

Manipulating (2) results in the incremental change in joint angles required to move to point $x_{k+1} = x_k + p$.

$$p = -(\nabla^2 F)^{-1} \nabla F \quad (3)$$

For the static case, (4) and (5) are the derivatives needed to compute the desired joint angles for the next iteration, x_{k+1}, which is given in (7).

$$\nabla F = J^T f \quad (4)$$
$$\text{where } J = \nabla(y(\theta))$$
$$\nabla^2 F = J^T J + \sum_i \nabla^2(f_i) f \quad (5)$$
$$\text{let } S = \sum_i \nabla^2(f_i) f \quad (6)$$
$$x_{k+1} = x_k - (J^T J + S)^{-1} J^T f \quad (7)$$

In the case of a moving target $y^*(t)$, the function to minimize is a time-varying surface $F(\theta, t)$, and (4) and (5) do not include all of the appropriate terms. Equation (7) may still be valid for this case; however, a lag between $\theta^*(t)$ and $\theta(t)$ is possible consequence when the target is moving.

Equation 6 is a difficult term to calculate or estimate. There exists two approaches in nonlinear optimization literature for dealing with this term: assume it is zero, or approximate it. Nonlinear least squares algorithms are known as either small residual algorithms or large residual algorithms, respectively, since the size of the residual term f influences the size of S. Small residual methods are typically modifications of the Gauss-Newton approach, as seen in (8), which omits the S altogether.

$$x_{k+1} = x_k - (J^T J)^{-1} J^T f \quad (8)$$

In previous literature, [8] and [11], small residual algorithms have been implemented for model independent control.

2.2 Jacobian Estimation

The composite Jacobian J relates the image feature vector y to the joint variable vector θ, by $\dot{y} = J\dot{\theta}$. The Jacobian is defined by the camera and kinematic models. As part of model independent visual servoing, an exact analytical model of the Jacobian is unavailable. Broyden's rank one update method [5] allows on-line estimation of the Jacobian based on changes in the image plane, eliminating the need for a precise analytical model.

When an estimated Jacobian is used as part of a nonlinear least squares minimization technique, it is called a quasi-Newton method. Let $\hat{J}_k$ denote the current approximation to the Jacobian, $\hat{J}^+$ denote the pseudoinverse of $\hat{J}$, and x_k the current state of the joint variables. The desired joint position x_{k+1} can be approximated by taking a quasi-Newton step.

$$\theta_{k+1} = \theta_k - \hat{J}_k^+ f(\theta) \quad (9)$$

Broyden's method is used to compute the update $\hat{J}_{k+1}$

$$\hat{J}_{k+1} = \hat{J}_k + \frac{\left(y - \hat{J}_k s\right) s^T}{s^T s} \quad (10)$$
$$\text{where} \quad y = f(\theta_{k+1}) - f(\theta_k)$$
$$s = \theta_{k+1} - \theta_k$$

If the initial Jacobian estimate $\hat{J}_0$ and the initial joint angles θ_0 are sufficiently close to the actual J and the desired point θ^*, then (9) will converge q-superlinearly when used in conjunction with (10) [15]. However, convergence is not guaranteed for larger initial residuals.

2.3 Trust Region

When the initial Jacobian estimate differs significantly from the actual system model, additional techniques, such as the trust region method, are required to assure convergence from some starting points[5]. The trust region method given by (11) limits the incremental change in joint variables.

$$\theta_{k+1} = \theta_k - \alpha_k \hat{J}_k^+ f(\theta) \quad (11)$$

There are various methods for calculating the region of trust, α_k, usually based on some confidence in the current model of the system[15]. The idea is to limit the step taken so that it is within a region that the model, $\hat{J}_k$, is valid.

2.4 Powell's Hybrid Method

The control law used here is Powell's hybrid method, a small residual method. It incorporates Broyden's rank-one method and the trust region method for simultaneous nonlinear least squares[15]. Similar to the well-known Levenberg-Marquardt, the algorithm is a combined strategy implementing either steepest descent or quasi-Newton steps, p_k.

3 Simulation Parameters

Visual servo control implementing Powell's method has been investigated in simulation using MATLAB[14]. The simulations are a dynamic look-and-move, image based model independent visual servoing system. The robot is a simple three degree-of-freedom, shoulder-elbow, two link manipulator as described in [8]. The stereo camera system is fixed in the world coordinate system approximately $2m$ from the robot base. The digital sampling time is $50ms$. Band-limited process noise is added to the target motion, and uniformly distributed $\pm \frac{1}{2}$ pixel quantization measurement noise is assumed. The initial Jacobian is calculated using kinematic parameters perturbed slightly from their actual values. Two features are tracked, the endpoint of the end-effector and the target, in each of the two cameras image planes

4 Tracking Performance

The vision-based controller easily converges on a static target. Indeed, Jägersand has demonstrated improved repeatability using visual feedback instead of joint feedback [11]. For robot trajectory planning, Jägersand implements a series of way-points or static targets. Since the controller is designed to converge on a static target, tracking a moving target presents additional challenges for the controller. With each vision update, the algorithm is provided with a new target.

A straight line target trajectory has been chosen to evaluate controller tracking abilities. The trajectory is in clear view of the cameras and the manipulator does not need to approach singularity points to track the target. The initial end-effector and target positions are collocated. Along this trajectory, 1 pixel corresponds to a Cartesian space approximately $3.3mm$ by $1.5mm$. Figure 1 gives Cartesian tracking error means and 95% confidence intervals for a target moving along a Z-directed trajectory at $0.1m/s$ and $0.5m/s$. Negative averages indicate a tendency to lag and positive values indicate a tendency to lead. At the lower velocity, Z-direction lag on the order of millimeters is evident. Tracking precision, as measured by the size of the confidence interval, is also degraded in the X and Y dimensions.

Clearly, faster target speeds result in the end-effector lagging the target. There are two possible sources of this delay. As the controller attempts to converge on the moving target each iteration, it may be consistently behind. A more likely cause is system latencies, including vision delays.

5 Target Prediction Schemes

While it is suspected that some tracking lag may occur due to controller design, latency is expected in the presence of vision delays. Estimating the actual position of the target in the image plane is proposed to investigate the source of tracking delays. The prediction of moving targets has been presented by several authors [1][3] using prediction to augment model based visual servoing algorithms. Several different target state estimators are considered here: a first order difference scheme, an $\alpha - \beta$ filter, and a Kalman filter as suggested in [4].

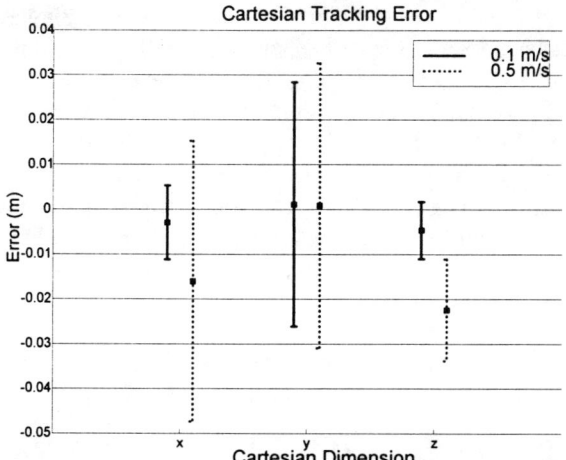

Figure 1: Cartesian tracking error is shown with 95% confidence intervals. Tracking performance degrades for faster targets.

5.1 First Order

The simplest prediction scheme is based on first order difference equations. The predicted target position is based on the current observed position and velocity of the target. This method will not compensate for process or measurement noise.

$$\widetilde{x}_{\text{target}}(k+1) = x_{\text{target}}(k) + v_{\text{target}}(k)T \quad (12)$$

5.2 $\alpha - \beta$ Tracking Filter Prediction

The $\alpha-\beta$ filter was developed for radar/target tracking systems. This filter tracks a one dimensional state assuming a constant velocity and zero-mean Gaussian acceleration errors. The estimate $\widetilde{x}$ of the target state given measurement y is as follows. The optimal values for α and β are chosen according to the tracking index as described in [12].

$$\widetilde{x}_p(k+1) = x_{\text{target}}(k) + v_{\text{target}}(k)T \quad (13)$$
$$\widetilde{x}_{\text{target}}(k+1) = x_p(k) + \quad (14)$$
$$\alpha(y(k+1) - \widetilde{x}_p(k+1)) \quad (15)$$
$$\widetilde{v}_{\text{target}}(k+1) = \widetilde{v}_{\text{target}}(k)$$
$$+ \frac{\beta}{T}(y(k+1) - \widetilde{x}_p(k+1)) \quad (16)$$

5.3 Kalman Filter

A more sophisticated prediction strategy employs a Kalman filter. This filter is an optimal estimator of state x, given measurement y. It assumes zero mean Gaussian process and measurement noise sequences.

The state vector for a single feature element from one camera consists of the position and velocities of the target. The target prediction problem can be described by the following discrete time state and output equations.

$$x(k+1) = Ax(k) + Bw(k) \quad (17)$$
$$y(k) = Cx(k) + \nu(k) \quad (18)$$

The vectors $w(k)$ and $\nu(k)$ are the process noise sequence with covariance Q and the measurement noise sequence with covariance R respectively. The state vector and system model are extended for the stereo camera system. For this system, the Kalman predictor is given by the following:

$$K_e(k) = \frac{AP(k)C^T}{R + CP(k)C^T}$$
$$\widetilde{x}(k+1) = A\widetilde{x}(k) + K_e(k)[y(k) - C\widetilde{x}(k)]$$
$$P(k+1) = BQB^T + [A - K_e(k)C]PA$$

where K_e is the Kalman gain, $\widetilde{x}(k+1)$ is the state prediction, and $P(k+1)$ is the error covariance update. This model assumes the target is moving with zero acceleration. The measurement error covariance matrix R is calculated analytically from the uniformly distributed quantization noise, and the process error covariance matrix Q is calculated using Monte Carlo trials. For straight line trajectories, Q is the covariance of the process noise. For maneuvering targets, Q represents the covariance of the accelerations.

6 Simulation Results and Discussion

6.1 Constant Velocity Target Trajectories

Simulations were run with a target moving at a velocity of $0.5 \frac{m}{s}$ from the point $[0.52, 0.0, 0.1]$ to the point $[0.52, 0.0, 0.6]$. By simply moving in the z dimension, the tracking lag is emphasized. Four simulations were run: no prediction, first-order prediction, $\alpha - \beta$ prediction, and Kalman filter prediction. Figure 2 gives Cartesian tracking error means in the 'Z' dimension with 95% confidence intervals. Without any type of prediction, the end-effector lags the target by more than two centimeters. While the first-order predictor introduces an effective lead and increased error variation, the addition of $\alpha - \beta$ filter or a Kalman filter results in near zero error means. Thus Powell's method with either a $\alpha-\beta$ or Kalman target prediction scheme has significantly reduced tracking error accuracy over Powell's method without a predictor. However, the similar error variances indicate a need to improve controller precision. While Cartesian precision can be improved by increasing image resolution, the current controller is unable to produce single pixel precision while tracking. In other words, tracking precision on the order of the vision system's quantization noise has not been achieved. Despite these limitations, zero mean tracking of a constant linear velocity target has been demonstrated.

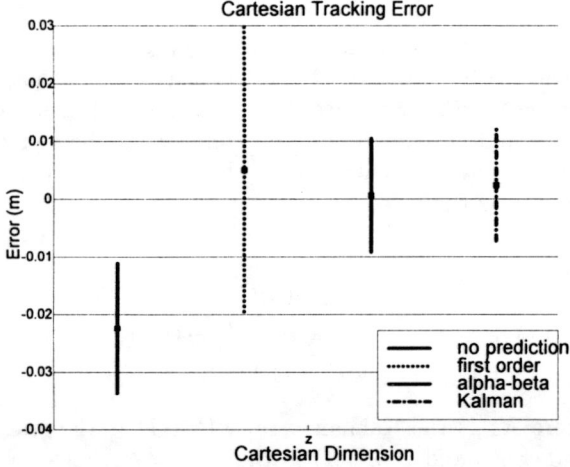

Figure 2: 'Z' dimension Cartesian tracking error in meters showing mean and 95% confidence intervals for four different prediction scenarios. Target is moving at 0.5 m/s.

6.2 Circular Target Trajectory

A circular target trajectory presents additional challenges to Powell's method since the target is moving in three dimensions with variable accelerations. However, as Figure 3 shows, the controller is capable of tracking a circle $60cm$ in diameter over a period of $10s$. Again, without prediction, the end-effector lags the target; Figure 4A shows the lag in the Z dimension. The lag is sinusoidal in nature because of the circular target motion. Predicting this type of target motion is more difficult than a straight line constant velocity motion. By modeling the accelerations of the motion as the noise signal w in Equation 17, a useful prediction can be made. With the addition of the Kalman filter, prediction error is ±1 pixel. In the image plane, the tracking error norm between the end-effector and the target is less than 10 pixels. Similar prediction and tracking errors are given in [3] for model-based visual servoing. Figure 4B shows the Cartesian tracking error in the 'Z' dimension. By comparing this with Figure 4A, we can see that the sinusoidal lag is removed. Similar results are evident in the 'Y' and 'Z' dimensions.

These results further confirm that a model independent visual servo control scheme is capable of tracking a moving target undergoing acceleration.

The tracking success demonstrated here validates the abilities of the quasi-Newton algorithm. It should be noted however, that the S term omitted in (8) is not always negligible. As the robot tracks a moving target, S can become much larger than $J^T J$, and the validity of the small-residual approach is diminished.

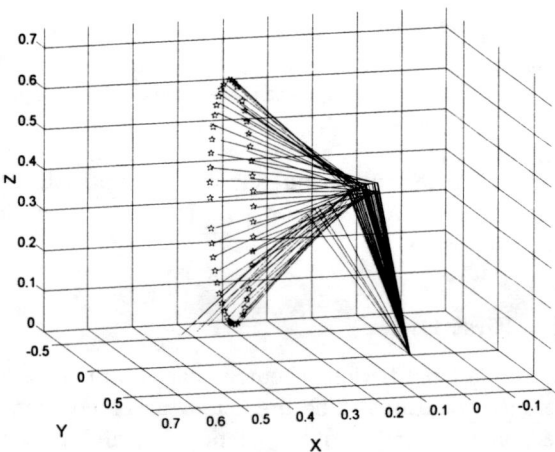

Figure 3: Circular target trajectory represented by stars. The target moves around a circle 30 cm in diameter over a period of 10 seconds starting at the lowest point of the circle in the 'Z' dimension. The end-effector starts approximately 12 cm away from the target.

7 Conclusion

This paper has augmented a model independent visual servoing control scheme with predictive elements. While Powell's method using Jacobian estimation is designed to converge on a static target, this paper demonstrates the tracking capabilities of the algorithm for both linear and circular trajectories. With the addition of prediction, the controller achieves zero-mean tracking. Current tracking abilities are coarse, but they are of the same order of magnitude as some model based algorithms. Further research will focus on improving tracking precision and incorporating appropriate algorithms for large residual scenarios.

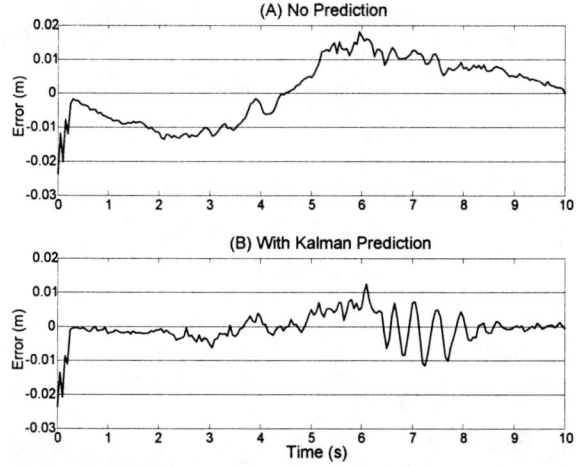

Figure 4: Cartesian error in the 'Z' dimension for the circular trajectory. (A) Without predition, a sinusoidal lag is evident. (B) The addition of a Kalman filter to the simulation eliminates lag.

References

[1] P.K. Allen, A. Timcenko, B. Yoshimi, and P. Michelman. Automated tracking and grasping of a moving object with a robotic hand-eye system. *IEEE Transactions on Robotics and Automation*, 2(2):152–165, April 1993.

[2] D. Bennett, D. Geiger, and J.M. Hollerbach. Autonomous robot calibration for hand-eye coordination. *The International Journal of Robotics Research*, 10(5):550–559, October 1991.

[3] F. Chaumette and A. Santos. Tracking a moving object by visual servoing. In *IFAC 12th Triennial World Congress*, pages 643–648, Sydney, Australia, July 1993.

[4] P.I. Corke. *Visual Control of Robots: high-performance visual servoing*. Research Studies Press Ltd., 1996.

[5] J.E. Dennis and R.B. Schnabel. *Numerical Methods for Unconstrained Optimization and Nonlinear Equations*. Prentice-Hall, Englewood Cliffs, New Jersey, 1983.

[6] B. Espiau, F. Chaumette, and P. Rives. A new approach to visual servoing in robotics. *IEEE Transactions on Robotics and Automation*, 8(3):313–326, June 1992.

[7] N. Hollinghurst and R. Cipolla. Uncalibrated stereo hand-eye coordination. Technical Report CUED/F-INFENG/TR126, Department of Engineering, University of Cambridge.

[8] K. Hosoda and M. Asada. Versatile visual servoing without knowledge of true jacobian. In *IEEE/RSJ/GI International Conference on Intelligent Robots and Systems*, pages 186–193, 1994.

[9] S. Hutchison, G.D.Hager, and P.I. Corke. A tutorial on visual servo control. *IEEE Transactions on Robotics and Automation*, 12(5):651–670, October 1996.

[10] M. Jagersand. Visual servoing using trust region methods and estimation of the full coupled visual-motor jacobian. In *IASTED Applications of Robotics and Control*, 1996.

[11] M. Jagersand, O. Fuentes, and R. Nelson. Experimental evaluation of uncalibrated visula servoing for precision manipulation. In *Proceedings of International Conference on Robotics and Automation*, 1997.

[12] P.R. Kalata. The tracking index: A generalized parameter for $\alpha - \beta$ and $\alpha - \beta - \gamma$ target trackers. *IEEE Transactions on Aerospace and Electronic Systems*, pages 174–182, 1984.

[13] N. Papanikolopoulos and P. Khosla. Robotic visual servoing around a static target: An example of controlled active vision. In *Proceedings of the 1992 American Control Conference*, pages 1489–1494, 1992.

[14] J.A. Piepmeier, H. Lipkin, and G.V. McMurray. A predictive estimation approach to model independent visual servoing. In *To Appear in Proceedings of Robotics98*, Alburquerque, New Mexico, 1998.

[15] L.E. Scales. *Introduction to Non-Linear Optimization*. Springer-Verlag, New York, 1985.

[16] R. Sharma, J-Y. Herve, and Peter Cucka. Dynamic robot manipulation using visual tracking. In *Proceedings of the 1992 EEE International Conference OnRobotics and Automation*, pages 1844–1849, Nice, France, 1992.

[17] S.W. Wijesoma, D.F.H. Wolfe, and R.J. Richards. Eye-to-hand coordination for vision-guided robot control applications. *International Journal of Robotics Research*, 12(1):65–78, February 1993.

Toward Global Visual Servos and Estimators for Rigid Bodies*

Noah J. Cowan and Daniel E. Koditschek

Electrical Engineering and Computer Science, The University of Michigan
Ann Arbor, MI 48105; E-mail: {ncowan, kod}@eecs.umich.edu

Abstract

We describe work-in-progress toward a nonlinear image-based rigid body *dynamic triangulator* which we believe tracks a moving target from "essentially all" initial conditions (all initial conditions except a set of measure zero.) The dynamic triangulator depends on the goal state only through its image plane position and velocity and requires a *navigation function*, imposed directly upon image features, to serve as a regressor for a gradient-like state update law.

1 Introduction

The control and vision literature loosely define *visual servoing* and *visual estimation* as computer-vision-based closed-loop servo control and state estimation, respectively. Sanderson and Weiss [14] propose two classifications for visual servos, *position-based*, in which the objective is to minimize a positioning error defined in the robot's Cartesian task space, and *image-based* in which the controller directly minimizes the perceived error. The same taxonomy applies to visual estimators, i.e. a position based estimator minimizes the task space tracking error and an image based estimator dynamically updates the estimate to drive the internal model to visually align with the observation.

Generically, all vision-based estimators and servos are *triangulators* in the sense that they (either explicitly or implicitly) "compute" the task space coordinates of the objects observed by cameras. Position-based systems are *algebraic triangulators* since they explicitly compute task-space information from image features and parametric knowledge of the world. Image based systems are *dynamic triangulators* since they do not require the explicit inversion of perceptual models to recover task space coordinates of the goal. Consequently, they often require less computation and are thought to be more robust with respect to calibration uncertainty [7, 8, 16, 11, 5].

When the object being tracked has rotational degrees of freedom, dynamic vision is greatly complicated. Many researchers in object tracking literature have addressed this problem by using local linearizations, e.g. Extended Kalman Filters [16, 15], which provide good results for incremental tracking but do not address the issue of large initial error.

1.1 Motivation

The long term aim of our research seeks to develop a system that couples visual estimation of a dynamical rigid body with visual servoing of a robot manipulator in order to achieve a dynamical task, such as catching

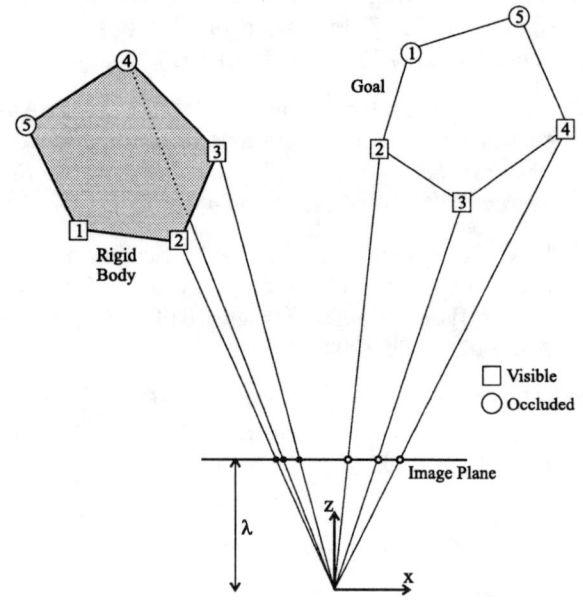

Figure 1: The objective is to drive the rigid body so that each corner aligns with the respective corner on the goal. The algorithm proposed requires three common visible corners, a condition not always satisfied. The "simple-minded" workaround is to "hallucinate" the occluded feature for the controller. We prefer to work only with the visible features available at each position, as depicted in Figure 3.

---

*This work was supported in part by the NSF under grant IRI-9510673

an otherwise unsensed falling body, or snatching an object from a conveyor. Our approach to such problems presupposes well designed robust "early vision" algorithms [9, 4] that track features such as corners and edges of the objects being observed. This affords the use of a growing body of signal processing algorithms designed to identify such features of an image and models the camera as a *virtual sensor* providing *image plane coordinates* for the objects being observed. Hutchinson *et. al.* provide a tutorial introduction to this approach [7].

As a rigid body moves in space, actuated or not, its corners and edges typically cycle into and out of the view of the cameras. Consequently it will often be necessary to switch the focus of attention during motion, introducing a hybrid aspect to the problem. To achieve stable systems, therefore, it is desirable to develop dynamic triangulators with very large domains of attraction in order to simplify the very challenging switching problem that inevitably results from the image-based approach.

1.2 Relation to Existing Literature

Much of the recent literature [15, 6, 16] uses local linearizations to solve tracking and servoing problems for both points and 3D objects. Some recent papers from our laboratory [13, 11] present algorithms, stability analysis and a working implementation [11] of systems with provably large domains of attraction for point positioning and estimation without local linearizations. This paper proposes extensions of that work to rigid bodies.

1.3 Organization and Contributions

The next section introduces our virtual sensor. Section 3 describes an approach to dynamic triangulation that imposes a cost function directly upon image features, and uses that cost as a regressor in a gradient-based state update. If one can show that the cost is in fact a *navigation function*,[1] then convergence to a static goal is guaranteed [12]. Similarly, using an image plane "tachometer" (Section 3.1) we might achieve asymptotic tracking of a moving target as well, subject to the extension of nonlinear time-varying stability theory to time-varying navigation functions. After presenting our triangulator we discuss the analytical properties for a specific objective function for a planar monocular camera in Section 3.2. In the planar case ($n = 2$) our nearly complete characterization of the critical points suggests that the cost function is indeed a navigation function. We also present a statistical summary of our simulation results suggesting that the servoing system is reasonably efficient. Finally we speculate on the implications of this paper and discuss future directions of our work in Section 4.

2 Virtual Sensor

As stated, we wish to pose an objective function in "camera-space", $\mathcal{C}$, and therefore we must construct a virtual sensor, $c : \text{SE}(n) \to \mathcal{C}$, from the camera data using knowledge of the rigid body. First we introduce some notation, and then we present the system output model.

2.1 Rigid Transformations

The group of rigid transformations, SE(n), may be embedded in GL($n+1$) (nonsingular matrices) by writing the transformation in homogeneous representation

$$\begin{aligned} \text{SE(n)} &= \left\{ H \in \text{GL}(n+1) \,\middle|\, R^T R = I, |R| = 1 \right\} \\ H &= \begin{bmatrix} R & r \\ 0^T & 1 \end{bmatrix} \end{aligned} \quad (1)$$

where $R \in \text{SO}(n)$ is an $n \times n$ rotation matrix and $r \in \mathbb{R}^n$ is a translation vector. Let $h : \mathbb{R}^n \to \mathbb{R}^n$ be a rigid transformation. Then, using the notation above, if $p \in \mathbb{R}^n$ then the point $b = h(p)$ is given by

$$\begin{bmatrix} b \\ 1 \end{bmatrix} = H \begin{bmatrix} p \\ 1 \end{bmatrix}$$

where H is the homogeneous representation of h.

2.2 Camera Model

The map, $\pi : \mathbb{R}^n \to \mathbb{R}^{(n-1)}$, maps a point in space to a point on the image plane. It is assumed that π takes an argument in a local camera coordinate frame. Figure 1 depicts the planar case ($n = 2$), corresponding to a "one dimensional" camera and a planar world. The specific form of π depends on the parametric camera model chosen. The pinhole camera model, reviewed in Appendix A, has lent theoretical and practical utility to previous work by our laboratory [13, 11] and we have chosen to exploit its simple structure when analyzing the specific cost function presented in Section 3.2.

Let $^i h$ denote the rigid transformation from world coordinates into the ith camera coordinate frame. The total camera map is then given by

$$g(b) = \begin{bmatrix} \pi \circ {}^1 h(b) \\ \vdots \\ \pi \circ {}^k h(b) \end{bmatrix} \quad (2)$$

where k is the total number of cameras.

---
[1] A navigation function has a unique global minimum, and all other critical points are nondegenerate saddles and maximums.

2.3 System Output

Consider a rigid body and let $H \in \mathrm{SE}(n)$ denote the homogeneous representation of h, the change from body to world coordinates, i.e. if p is a point in body coordinates, then $b = h(p)$ is the same point in world coordinates. Let $P = [p_1, \ldots, p_m] \in \mathbb{R}^{n \times m}$ denote m distinguishable points ("corners") fixed on the rigid body, expressed with respect to the body reference frame. Let B denote the same set of m feature points as expressed in world coordinates, i.e.

$$\begin{bmatrix} B \\ \mathbf{1}^T \end{bmatrix} = H \begin{bmatrix} P \\ \mathbf{1}^T \end{bmatrix}$$

where $\mathbf{1} = [1, \ldots, 1]^T$. The constant matrix P is assumed known *a priori*, i.e. we know the block geometry exactly.

Let $c : \mathrm{SE}(n) \to \mathcal{C}$ denote the camera image of the m points in the rigid body, i.e.

$$c(H) := \begin{bmatrix} g(b_1) \\ \vdots \\ g(b_m) \end{bmatrix} =: \begin{bmatrix} c_1 \\ \vdots \\ c_m \end{bmatrix}, \quad (3)$$

where g is given in (2). The camera space $\mathcal{C}$ is $\mathbb{R}^{mk(n-1)}$, where k is the number of cameras.

3 Dynamic Triangulation

Suppose there is a target, whose position and orientation is given by $H^* \in \mathrm{SE}(n)$, which cannot be directly measured; instead we measure $c^* = c(H^*)$. In effect, our objective is to solve

$$c^* - c(H) = 0, \quad (4)$$

for H where the parameters of c, such as the focal length and P, are assumed known.

We distinguish two types of triangulation: dynamic and algebraic. The aim of both is to solve Equation (4). Algebraic triangulation provides a "pseudo-inverse" $c^\dagger : \mathcal{C} \to \mathrm{SE}(n)$, whereas dynamic triangulation uses an iterative method such as gradient or Newton descent to dynamically solve for the minimum of an objective function on the perceived output $c(H)$ and the perceived goal c^*. For example $\|c^* - c(H)\|$ is a candidate objective function with a global minimum at $H = H^*$. Of course, in our research agenda the recourse to dynamical triangulation is motivated in part by a real-time servo implementation wherein the descent step is executed in the physical world by the direct manipulation of the observed object. Alternatively, we might wish to obtain an asymptotic estimate of the position, orientation and velocity of an object which may be moving according to some dynamic equation.

Whether for estimation or for servoing, we posit a purely kinematic model of the form

$$\begin{aligned} \dot{H} &= u \\ y &= c(H) \end{aligned} \quad (5)$$

where $u \in T\mathrm{SE}(n)$ is the input variable.[2] In terms of local coordinates ((16) in Appendix B), we have

$$\begin{aligned} \dot{q} &= u \\ y &= c(q) \end{aligned} \quad (6)$$

where we associate with u its local coordinate representation, and by $c(q)$ it is understood that we mean $c \circ \phi_{H_0}^{-1}(q)$, although an abuse of notation. If we choose $H_0 = I$ in (16), then our local coordinates are given by $q = [\theta, r^T]^T$, the rotation and translation of the body, and $u = [\omega, v^T]^T$, the angular and translational velocity.

3.1 Generic Image-Based Tracking

We wish to triangulate a part moving on a conveyor, or a falling body, with an input dependent on the goal only through its image plane positions and velocities, and yet still guarantee convergence. To achieve this we pose the cost $\varphi : \mathrm{SE}(n) \times \mathrm{SE}(n) \to \mathbb{R}$ on image plane measurements, that is, φ admits of a factorization

$$\varphi(H, H^*) = \overline{\varphi}(c(H), c(H^*)). \quad (7)$$

Furthermore, we suppose the possibility of taking numerical derivatives of our image plane motions, and assume that we have perfect measurement of the image plane coordinates, $c^* = c(H^*)$ of the body we are tracking, and the image plane velocities, $\dot{c}^*$ (motivating the term image plane "tachometer").

Our input is given in local coordinates (16) by

$$u = -M^{-1}(q)\, D_q\varphi(q, q^*)^T - u_2 D_{c^*}\overline{\varphi}(c, c^*)\, \dot{c}^* \quad (8)$$

where

$$u_2 = D_q\varphi(q, q^*)^T (D_q\varphi(q, q^*) D_q\varphi(q, q^*)^T)^{-1}$$

and $D_x f$ denotes the Jacobian of f with respect to x. M is an arbitrary Riemannian metric.

The reader can check that u depends on $(H^*, \dot{H}^*)$ only through $(c^*, \dot{c}^*)$. Furthermore, since

$$\dot{\varphi} = D_q\varphi(q, q^*)\, u + D_{c^*}\overline{\varphi}(c, c^*)\, \dot{c}^*,$$

substituting for u from (8)

$$\dot{\varphi} = -\|\mathrm{grad}_q \varphi(q, q^*)\|_M^2 \leq 0.$$

---
[2] $T\mathrm{SE}(n)$ denotes the tangent space of $\mathrm{SE}(n)$.

If $\varphi(\cdot, H^*)$ is a navigation function and $\dot{H}^* = 0$ then we achieve asymptotic tracking for "essentially all" initial conditions [12]. When the goal is moving then $\varphi(\cdot, H^*(t))$ is time varying, and the convergence result is slightly more elusive.[3]

3.2 Navigation Function Candidate

We now investigate the critical points of a novel cost function, based on "angular error", using a planar monocular pinhole camera (see Appendix A).

Note that given the projection of two points on the image plane one can deduce the angle subtended between them (measured in the camera frame) purely from image plane coordinates, and let

$$\gamma_i = \frac{b_i^T b_i^*}{\|b_i\| \|b_i^*\|} = \frac{c_i^T c_i^* + \lambda^2}{\sqrt{(\|c_i\|^2 + \lambda^2)(\|c_i^*\|^2 + \lambda^2)}} \quad (9)$$

denote the cosine of the angle between the ith corner and its goal, where $c = c(H)$, $c^* = c(H^*)$ and λ is the focal length. This serves as the primary building block for our cost function

$$\varphi(H, H^*) = \sum_{i=1}^{m} 1 - \gamma_i. \quad (10)$$

Note that φ has been factored according to (7). We now wish to verify that φ is a navigation function by investigating the critical points.

3.2.1 Gradient

In reporting the progress of our analysis, we restrict attention to the case in which the rigid body is a triangle ($m = 3$) on the plane ($n = 2$).

To investigate the gradient at a particular H_0, we simply compute the gradient in local coordinates (16) and evaluate at $q = 0$

$$\mathrm{grad}_q \varphi(q, q^*)\big|_{q=0} = M^{-1}(0) \left(D_q \varphi(q, q^*)\right)^T \big|_{q=0} \quad (11)$$

where M is a Riemannian metric. For convenience, let $M(0) = I$, as the choice of metric does not change the limiting behavior. Simplifying, yields

$$\left(D_q \varphi(q, q^*)\right)^T_{q=0} = A s$$

where

$$A = \begin{bmatrix} \frac{1}{\|b^1\|^2} & \frac{1}{\|b^2\|^2} & \frac{1}{\|b^3\|^2} \end{bmatrix}, \quad J = \begin{bmatrix} 0 & -1 \\ 1 & 0 \end{bmatrix},$$

---

[3] We are presently developing a time-varying extension to navigation functions (similar to the time-varying extension to Lyapunov theory presented, for example, in Khalil [10]) that will guarantee convergence under (presumably reasonable) restrictions on H^*.

$$s_i = \frac{\lambda(c_i^* - c_i)}{\sqrt{(\|c_i\|^2 + \lambda^2)(\|c_i^*\|^2 + \lambda^2)}}, \quad i = 1, 2, 3.$$

Assuming that (4) has a unique solution in front of the camera,[4] we believe but have not yet shown formally that $\varphi(\cdot, H^*)$ has exactly two critical points. One of the critical points is the goal, which is a minimum by design. Another is the "ghost goal" behind the camera and it is a maximum, as shown in the next section. The details of this conjecture (the final proof of which is in progress) may be found in our technical report [2].

3.2.2 Hessian and Stability

The Hessian is calculated by taking the Jacobian matrix of the vector field which we may evaluate at $q = 0$ to study the stability properties at H_0. The calculations are further simplified if we evaluate the Hessian at a critical point, which implies that $s = 0$:

$$\mathbf{H} = D_q (D_q \varphi(q, q^*))^T \big|_{q=0,\, s=0}. \quad (12)$$

It is straightforward to show that

$$\mathbf{H} = A \Gamma A^T \quad (13)$$

where $\Gamma = \mathrm{diag}\{\gamma_1, \gamma_2, \gamma_3\}$.

At the goal $\Gamma = I$ and at the "ghost goal" $\Gamma = -I$. Hence the goal is a local minimum and the remaining critical point is a maximum. Subject to the verification that there are no additional critical points, we have shown $\varphi(\cdot, H^*)$ satisfies the requirements of a navigation function. In addition to the analytical evidence, numerical simulations suggest that this is true.

3.2.3 Numerical Results

Table 1 summarizes the results of 125 simulations of (8) assuming a static goal and no occlusions, wherein 25 initial conditions were spaced evenly around each of five different initial distance balls. The maximum initial error[5] corresponds to about $100°$ of angular error or about three times the body radius of translational error. The camera was assumed to have unit focal length, and the rigid body is an equilateral triangle inscribed in a circle of radius 1. The goal is centered at $(0,5)$ and was chosen as the local coordinates for the gradient calculation, i.e. $H_0 = H^*$. Of course practical settings will vary greatly in detail and the table is merely qualitative.

---

[4] Such a solution exists provided the circle which passes through the three features of the goal does not pass through the camera's focus.

[5] We have arbitrarily scaled orientation angles by body radius to fix a unique metric for $SE(2)$.

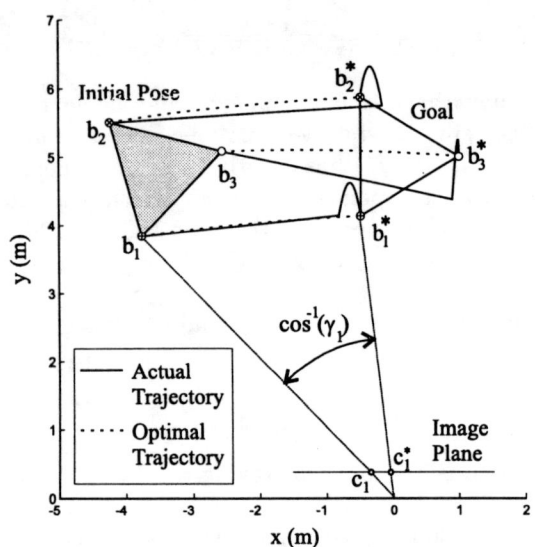

Figure 2: Typical simulated trajectory.

Distance to Goal ($\|q(0)\|$)	Avg Distance Traveled	Average Inefficiency	Standard Dev
0.25	0.37	48%	25%
0.50	0.73	46%	25%
1.00	1.44	44%	25%
1.25	1.80	41%	25%
1.50	1.39	39%	25%

Table 1: Summary of 125 simulations, wherein 25 initial conditions were spaced evenly around each of 5 different initial distances to the goal. "Average Inefficiency" is the extra distance traveled by the vision-based algorithm compared to the "straight-line" trajectory.

The algorithm seems to be reasonably "efficient" in the sense that the distance traveled is generally about one and a half times the shortest possible distance. Of course gradient algorithms do not guarantee optimality. A typical trajectory is shown in Figure 2.

4 Conclusions

4.1 Summary

The two central contributions of this paper to the field of visual servoing estimation are the application of gradient techniques to the dynamic triangulation problem for rigid bodies, including the introduction of a novel angle-based candidate navigation function, and the use of an image-based velocity estimate – image plane "tachometer" – to track nonstationary objects.

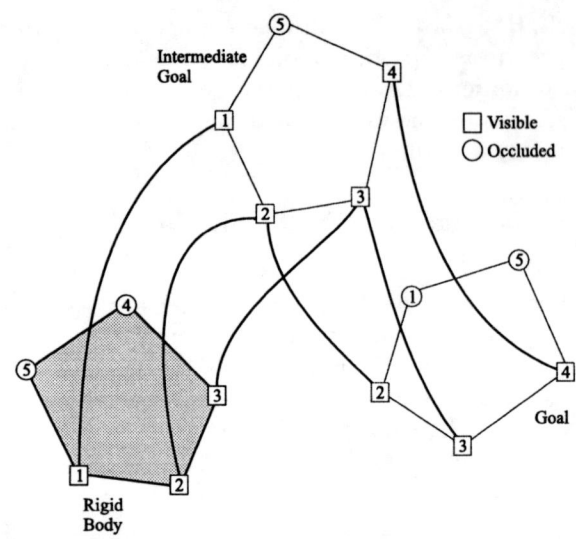

Figure 3: The camera (not shown), rigid body and goal are in the same positions as in Figure 1, and, as before, the objective is to drive the rigid body corner-wise to the goal. Here, there is a hierarchy of two controllers for backchaining. The first controller "prepares" the second.

4.2 Work-in-Progress

Our near-term research agenda addresses dynamic estimation and control in which the accelerations of the bodies being triangulated change in some known way, e.g. according to Newton's laws or a control law, so we can advance from the quasi-static approach we have used thus far.

We have also begun to address the problem of "switching." As the rigid body is servoed, it will generally be necessary to switch the focus of attention (due to occlusions), as depicted in Figure 3. Consequently a systematic switching method must be used to guarantee convergence to the goal. Back-chaining [1], wherein one constructs a set of intermediate goals which satisfy an appropriate "prepares" relationship, is one such approach. A "cartoon" version of this is shown in Figure 3. The initial condition and goal are the same as in Figure 1, however there is an intermediate goal, H' constructed *a priori* that shows corners $1,2,3,4$. One first navigates according to $\varphi(\cdot, H')$, to the *intermediate* goal until corner 4 is visible, at which time the controller switches to the navigation function $\varphi(\cdot, H^*)$. In this sense, the first navigation function "prepares" the second. Of course one seeks to plan out a complete set of these intermediate "preparatory" subgoals in such a fashion that the entire configuration space, $SE(2)$, is "covered".

Finally, and most importantly, we have begun constructing research apparatus for validating the empirical utility of the analytical results and conjectures set forth in this paper.

A Pinhole Camera

The spatial pinhole camera transformation $\pi_\lambda : \mathbb{R}^3 \to \mathbb{R}^2$ is given by

$$\pi_\lambda(b) := \frac{\lambda}{b_z} \begin{bmatrix} b_x \\ b_y \end{bmatrix}, \qquad (14)$$

where λ is the focal length (assumed known) of the camera and $b = [b_x, b_y, b_z]^T$ is a point specified in camera coordinates. Note that the z-axis is perpendicular to the image plane.

For the planar case, SE(2), there are only two frame axes which we denote $\{x, z\}$. The z-axis is chosen orthogonal to the image line, as before.[6] See Figure 1. The planar pinhole camera map $\pi_\lambda : \mathbb{R}^2 \to \mathbb{R}$ is simply

$$\pi_\lambda(b) = \frac{\lambda}{b_z}[b_x]. \qquad (15)$$

Note that the same symbol, π_λ is used for both the planar and spatial camera. It will be clear from context which is being used.

B Local Coordinates on SE(n)

The local coordinate charts used in this paper were chosen to simplify the analysis, and are defined below.

The skew symmetric operator $J : \mathbb{R}^m \to \text{Skew}(m)$ is given by

$$J(v) := \begin{bmatrix} 0 & -v_3 & v_2 \\ v_3 & 0 & -v_1 \\ -v_2 & v_1 & 0 \end{bmatrix}.$$

for $m = 3$ and

$$J(v) := \begin{bmatrix} 0 & -v \\ v & 0 \end{bmatrix}$$

for $m = 1$. Let $q_1 \in \mathbb{R}^{n(n-1)/2}$ and $q_2 \in \mathbb{R}^n$. Define the map $\phi_{H_0}^{-1} : \mathbb{R}^{n(n+1)/2} \to \text{SE}(n)$ by

$$\phi_{H_0}^{-1}(q) := \begin{bmatrix} \exp(J(q_1)) & q_2 \\ 0^T & 1 \end{bmatrix} H_0. \qquad (16)$$

In a neighborhood of H_0, $\phi_{H_0}^{-1}$ is invertible. In particular, $\phi_{H_0} \circ \phi_{H_0}^{-1}$ is the identity on $\mathbb{R}^{n(n+1)/2}$ in a neighborhood of 0, and $\phi_{H_0}^{-1} \circ \phi_{H_0}$ is the identity on SE(n) in a neighborhood of H_0 [3].

---

[6]Since the "world" is chosen to be a plane, the camera image is one dimensional.

References

[1] R. R. Burridge, A. A. Rizzi, and D. E. Koditschek. Sequential composition of dynamically dexterous robot behaviors. *Int. J. Rob. Res.*, (to appear).

[2] N. J. Cowan and D. E. Koditschek. Visual servos and estimators for rigid bodies. Technical Report CGR 98-07, University of Michigan, 1998.

[3] M. Curtis. *Matrix Groups*. Springer Verlag, New York, 1970.

[4] G. D. Hagar. Xvision visual tracking software, 1996.

[5] G. D. Hager. Calibration-free visual control using projective invariance. In *Proceedings of 5th ICCV*, 1995.

[6] K. Hashimoto, T. Elbine, and H. Kimura. Visual servoing with hand-eye manipulator- optimal control approach. *IEEE Transactions on Robotics and Automation*, pages 651–670, October 1996.

[7] S. Hutchinson, G. D. Hager, and P. I. Corke. A tutorial on visual servo control. *IEEE Transactions on Robotics and Automation*, pages 651–670, October 1996.

[8] M. Jägersand, O. Fuentes, and R. Nelson. Experimental evaluation of uncalibrated visual servoing for precision manipulation. In *International Conference on Robotics and Automation*, Albuquerque, NM, April 1997. IEEE.

[9] R. Jain, R. Kasturi, and B. Schunck. *Machine Vision*. McGraw-Hill, Inc., 1995.

[10] H. K. Khalil. *Nonlinear Systems*. Prentice Hall, 1996.

[11] D. Kim, A. A. Rizzi, G. D. Hager, and D. E. Koditschek. A "robust" convergent visual servoing system. In *International Conf. on Intelligent Robots and Systems*, Pittsburgh, PA, 1995. IEEE/RSJ.

[12] E. Rimon and D. E. Koditschek. Exact robot navigation using artificial potential fields. *IEEE Transactions on Robotics and Automation*, 8(5):501–518, Oct 1992.

[13] A. A. Rizzi and D. E. Koditsckek. An active visual estimator for dexterous manipulation. *IEEE Transactions on Robotics and Automation*, pages 697–713, October 1996.

[14] A. C. Sanderson and L. E. Weiss. Image-based visual servo control using relational graph error signals. In *Proceedings of the IEEE*, pages 1074–1077. IEEE, 1980.

[15] S. Soatto, R. Frezza, and P. Perona. Motion estimation via dynamic vision. *IEEE Transactions on Automatic Control*, 41(3):393–413, March 1996.

[16] P. Wunsch and G. Hirzinger. Real-time visual tracking of 3-d objects with dynamic handling of occlusions. In *International Conference on Robotics and Automation*, pages 2868–2873, Albuquerque, NM, 1997. IEEE.

Toward 3D Uncalibrated Monocular Visual Servo*

Bradley E. Bishop

Weapons and Systems Engineering
United States Naval Academy
Annapolis, MD 21402

Mark W. Spong

Coordinated Science Lab
University of Illinois
Urbana, IL 61801

Abstract

This work is an initial step toward combining control theory with computer vision for the case of uncalibrated monocular (or single-camera) three-dimensional manipulation. We investigate an achievable control goal using a novel image measure and derive a sampled-data control system that results in quick convergence of the end-effector trajectory to a depth-invariant velocity subspace in the camera coordinate frame. We also discuss how this uncalibrated behavior can be used to perform on-line calibration.

1 Introduction

Full three-dimensional measurement using only camera systems requires either multiple cameras or multiple views, as the inherent 3D-to-2D projection of the scene onto the image plane creates a depth ambiguity for a single image. The study of achievable three-dimensional tasks under uncalibrated monocular visual servo is interesting from a variety of standpoints. It is a study in fault tolerance for stereo vision systems, as well as a fundamental investigation into the nature of the visual servo task in three dimensions. Visual servo systems designed for use in full 3D settings have typically relied on the use of two cameras for full depth estimation of feature points in the scene. It is easy to imagine situations where a single camera of a stereo rig might fail, leaving only one camera for sensing.

In this work, we will consider achievable goals under the framework of general uncalibrated three-dimensional monocular visual servo in the fixed-camera configuration (as opposed to eye-in-hand).

---

*This research was partially supported by the National Science Foundation under grants IRI-9216428, and MSS-9212376, and by the Energy Power Research Institute (EPRI) under contract RP 8030-14.

The underlying camera model that we will utilize is the pinhole lens. Under this model, a scene point whose coordinates in the camera frame are $x_c = (x_{c_1}, x_{c_2}, x_{c_3})^T$ projects onto the image plane as follows

$$u_i = f x_{c_1}/x_{c_3} \qquad (1)$$
$$v_i = f x_{c_2}/x_{c_3} \qquad (2)$$

where f is the focal length of the camera lens and the subscript i indicates the ideal image coordinates, with no distortion. Note that, for real CCD cameras, an additional factor α must be included to convert image plane units to pixels. In many systems, the value of α is distinct for each of the imaging axes. We call a parameter of the camera system *intrinsic* if it does not depend on the camera's position and orientation in space. For our model, the intrinsic parameters are f, α_u and α_v. We will not be concerned with other intrinsic parameters, such as the offset of the center of the CCD array [1].

In visual servo systems, the translation and rotation matrices that transform the robot coordinate frame into the camera coordinate frame contain the *extrinsic* parameters. They are defined by the equation:

$$x_c = R x_r + O_r \qquad (3)$$

where x_r is a sensing point in the robot's coordinate frame, O_r is the three-dimensional translation vector relating the origin of the camera coordinate frame to that of the robot, and R is a rotation matrix given by [1]:

$$\begin{bmatrix} c(\phi)c(\theta) & s(\phi)c(\theta) & -s(\theta) \\ -s(\phi)c(\psi) & c(\phi)c(\psi) & c(\theta)s(\psi) \\ +c(\phi)s(\theta)s(\psi) & +s(\phi)s(\theta)s(\psi) & \\ s(\phi)s(\psi) & -c(\phi)s(\psi) & c(\theta)c(\psi) \\ +c(\phi)s(\theta)c(\psi) & +s(\phi)s(\theta)c(\psi) & \end{bmatrix} \qquad (4)$$

where the angles (ψ, θ, ϕ) are yaw, pitch and tilt and $c(\cdot) = \cos(\cdot)$, $s(\cdot) = \sin(\cdot)$.

Calibration of monocular systems can be carried out using numerical techniques that are well understood and characterized [2]. In this paper we are considering methodologies by which we can quantify and control the behavior of an uncalibrated monocular visual servo system. We will consider a sampled–data controller that will guarantee (given appropriate selection of various gain matrices) convergence of the end–effector of the robot system to a depth–invariant velocity subspace in the camera coordinate frame. This convergence can be thought of as an implicit form of calibration. In fact, the extension of this work to a calibration framework is straightforward and will be discussed in Section 4.

While uncalibrated visual servo control is an area of intense research (see, e.g., [3]), the use of control theoretic techniques in on–line calibration is also an interesting topic [4]. Additional camera calibration background (outside the framework of visual servo control) can be found in [1], [5], and [6]. A discussion of the problems that arise when using standard calibration techniques on the 2D–3D calibration problem generated by a planar robotic system can be found in [7]. Background on general monocular vision is found in [8].

In Section 2 we present a novel image measure for use in uncalibrated monocular visual servo systems. In Section 3 we present a sampled–data controller that achieves an example control objective. Section 4 discusses the extension of this work to an on–line calibration system. Conclusions and further research directions are given in Section 5.

2 Linearity Measure

Inherent to the study of uncalibrated visual servo is the ability to extract information that has meaning in three dimensions from the sequence of two dimensional images. In this section, we will consider a measure that relates the general nonlinear projection equations for a pinhole lens to the fixed–depth case, which is linear in the projection parameters. This measure will allow us to determine when the end–effector of the robot is moving on a line that is depth–invariant with respect to the camera coordinate frame.

As a first step toward autonomous control of general robot–camera systems, we have considered the case of general fixed camera placement (with no lens distortions). We will assume that we know only the offset of the center of the CCD array with respect to the optical axis of the lens system and the relative scale factor K_α between α_u and α_v. We will write the camera equations using α_i to indicate a re–scaled image vector. We do not assume knowledge of α_i, where $\alpha_u = \alpha_i$ and $\alpha_v = K_\alpha \alpha_i$.

Consider a plane in the robot's workspace. For a planar manipulator, this obviously contains the entire workspace, but for an arbitrary 3D manipulator the plane chosen may contain only a subset of the workspace. We will call this plane the *robot plane*.

There is a vector $v_z \in \mathcal{R}^3$ such that any end–effector motions along this direction result in no net change of the depth of the end–effector in the camera coordinate frame. Specifically, we can see that if we translate the image plane along its optical axis toward the workspace until the origin of the image plane coincides with the intersection of the optical axis and the robot plane, the intersection of the two planes results in either a plane (on a degenerate calibration set) or a line.

The line generated by the intersection of the robot plane and the translated image plane is, by virtue of construction, at a fixed depth from the image plane, as is each line parallel to it. Thus, any motion on the robot plane that is parallel to the constructed direction will result in no motion along x_{c_3}.

The principle of the proposed approach to uncalibrated visual control is to measure the ratio of the image plane position of a sensing point to the workspace position of the same point and compare that to the ratio of the image plane velocity to the workspace velocity, adjusting the desired end–effector motion appropriately to guarantee convergence to a depth–invariant velocity subspace. We will assume, for simplicity, that the image plane measurements are continuous, and that an instantaneous image plane velocity can be approximated at each sample instant.

We will henceforth consider x_r and p_r to be the workspace and image plane locations of the sensing point, typically the end–effector of the manipulator. We can express the norm of p_r by combining (1) – (4) as:

$$\|p_r\|^2 = \left(\frac{f\alpha_i}{x_{c_3}}\right)^2 x_r^T R^T \begin{bmatrix} 1 & 0 & 0 \\ 0 & 1 & 0 \\ 0 & 0 & 0 \end{bmatrix} R x_r +$$
$$\left(\frac{f\alpha_i}{x_{c_3}}\right)^2 O_r^T \begin{bmatrix} 1 & 0 & 0 \\ 0 & 1 & 0 \\ 0 & 0 & 0 \end{bmatrix} O_r +$$
$$\left(\frac{f\alpha_i}{x_{c_3}}\right)^2 2 x_r^T R^T \begin{bmatrix} 1 & 0 & 0 \\ 0 & 1 & 0 \\ 0 & 0 & 0 \end{bmatrix} O_r$$

The important observation to make concerning this equation is that the term O_r is not known *a priori*. This results in a difficulty when attempting to write the equation for $\|p_r\|$ as $A\|x_r\|$ for some unknown A.

In order to alleviate this problem, we will consider a new measurement vector defined at time t by:

$$p_n(t) = \frac{f\alpha_i}{x_{c_3}(t)} \left[\begin{array}{c} (Rx_r(t))_1 + O_{r_1} \\ (Rx_r(t))_2 + O_{r_2} \end{array} \right] - \frac{f\alpha_i}{x_{c_3}(T)} \left[\begin{array}{c} (Rx_r(T))_1 + O_{r_1} \\ (Rx_r(T))_2 + O_{r_2} \end{array} \right]$$

where will will select $T < t$, possibly updating at discrete points. Henceforth, we will suppress the dependence on t. We notice that when $x_{c_3} = x_{c_3}(T) = Z$, the equation becomes:

$$p_n = \frac{f\alpha_i}{Z} \left[\begin{array}{c} (Rx_r)_1 - (Rx_r(T))_1 \\ (Rx_r)_2 - (Rx_r(T))_2 \end{array} \right]$$
$$= \frac{f\alpha_i}{Z} \left[\begin{array}{ccc} 1 & 0 & 0 \\ 0 & 1 & 0 \end{array} \right] R(x_r - x_r(T))$$

Because we are moving in a plane, we know that we can set our robot coordinate frame such that $(x_r - x_r(T))_3$ is exactly zero. Further, when $x_{c_3} = x_{c_3}(T) = Z$, the term $(R(x_r - x_r(T)))_3$ is identically zero. In this circumstance, we can write the equations with a "phantom" variable w (that we cannot measure directly) in the following form:

$$\left[\begin{array}{c} p_n \\ w \end{array} \right] = \frac{f\alpha_i R}{Z}(x_r - x_r(T)) \quad (5)$$

where w is always zero by construction. We then write the norm $\|(p_n, w)\|^2 = \|p_n\|^2$ as:

$$\left(\frac{f\alpha_i}{Z}\right)^2 (x_r - x_r(T))^T R^T R(x_r - x_r(T)) \quad (6)$$
$$= \left(\frac{f\alpha_i}{Z}\right)^2 \|(x_r - x_r(T))\| \quad (7)$$

Using this vector p_n as an output with input $x_n = x_r - x_r(T)$, we can accurately compare the magnitude of the ratios $\frac{\|p_n\|}{\|x_n\|}$ and $\frac{\|\dot{p}_n\|}{\|\dot{x}_n\|}$ to determine if the depth is changing. If x_{c_3} is not constant, terms in $\dot{x}_{c_3}$ appear in the equations for $\dot{p}_n$.

What we have shown is that if the depth is constant the ratios of $\frac{\|p_n\|}{\|x_n\|}$ and $\frac{\|\dot{p}_n\|}{\|\dot{x}_n\|}$ will be identical. What we have not shown is that the equality of these ratios is a sufficient condition to guarantee that the depth is fixed. To accomplish this, we fall back on a geometric argument. For simplicity, let us consider those times when the velocity of the sensing point is constant. It can then be shown that for the two ratios to remain equal over time intervals of nonzero length, the velocity $\dot{p}_n$ must remain constant. Let us consider the equations for the velocity and acceleration of p_n:

$$\dot{p}_n = \left[\begin{array}{ccc} 1 & 0 & 0 \\ 0 & 1 & 0 \end{array} \right] \left(\frac{f\alpha_i}{x_{c_3}} R\dot{x}_r - \frac{f\alpha_i \dot{x}_{c_3}}{x_{c_3}^2}(Rx_r + O_r) \right) \quad (8)$$

$$\ddot{p}_n = \left[\begin{array}{ccc} 1 & 0 & 0 \\ 0 & 1 & 0 \end{array} \right] \frac{f\alpha_i}{x_{c_3}} A_X \quad (9)$$

where

$$A_X = \left(R(\ddot{x}_r - \frac{2\dot{x}_{c_3}}{x_{c_3}}\dot{x}_r) - (\frac{\ddot{x}_{c_3}}{x_{c_3}} - \frac{2\dot{x}_{c_3}^2}{x_{c_3}^2})(Rx_r + O_r) \right) \quad (10)$$

Due to the linear nature of the transformation from x_r to x_{c_3}, the constant velocity $\dot{x}_r$ forces $\ddot{x}_{c_3} = 0$. Looking at the equation for $\ddot{p}_n$, we see that this results in net acceleration of zero only when $\dot{x}_{c_3} = 0$ or when the first two elements of $(R\dot{x}_r - \frac{\dot{x}_{c_3}}{x_{c_3}}(Rx_r + O_r))$ are zero.

The latter case is easier to analyze when we realize that a constant $\dot{x}_r$ implies a constant $\dot{x}_{c_3}$, which results in an equation of the form:

$$L = \left[\begin{array}{ccc} 1 & 0 & 0 \\ 0 & 1 & 0 \end{array} \right] \frac{1}{x_{c_3}}(Rx_r + O_r) \quad (11)$$

where L is a constant vector.

The right hand side of (11) is exactly the projection equation scaled by $\frac{1}{f\alpha_i}$. Thus, (11) is satisfied only when the image point remains stationary. This can only happen if $\dot{x}_r = 0$ or the sensing point is moving along a ray through the center of the lens. As this results in $\dot{p}_n = 0$, no fundamental difficulties will arise from this case. We must be sure, however, to avoid indicating that the system is moving in the depth-invariant subspace when $\dot{p}_n = p_n = 0$.

We have shown the following proposition:

Proposition 2.1 *Given a constant, nonzero end-effector velocity $\dot{x}_n$ that results in a nonzero $\dot{p}_n$, the equality given by:*

$$\|p_n\|\|\dot{x}_n\| = \|\dot{p}_n\|\|x_n\| \quad (12)$$

will hold over intervals of nonzero length if and only if x_{c_3} is constant.

3 Sampled–Data Control

Of paramount interest in the uncalibrated visual servo problem is the determination of a class of achievable objectives and control primitives that can be used to generate complex behaviors for uncalibrated systems. As such, we now present a sampled–data control system that demonstrates one achievable control task utilizing the measure defined in Section 2.

Consider a sampled–data system in which the continuous plant dynamics are discretized for purposes of high–level control objective selection. The plant (a robot system) is controlled in continuous time using a computed torque scheme [9] with end–effector control given by:

$$R_\alpha = \begin{bmatrix} \cos\alpha & -\sin\alpha \\ \sin\alpha & \cos\alpha \end{bmatrix} \quad (13)$$

$$\dot{x}_r^d = R_\alpha \begin{bmatrix} 1 \\ 0 \end{bmatrix} \quad (14)$$

$$\ddot{x}_r = K_v(\dot{x}_r^d - \dot{x}_r) \quad (15)$$

where we note that $\dot{x}_n = \dot{x}_r$.

The input to this controller is an angle α that corresponds to the velocity subspace of the desired motion. Using k as the discrete index for sample period $(k-1)T < t \leq kT$, the algorithm for the proposed controller (based on Proposition 2.1) is given by:

Step 1. Converge to the initial desired $\alpha(0)$ to a predetermined degree, setting $l = k$ at that point.

Step 2. Choose a new desired end–effector velocity subspace by selecting $\alpha(k)$ based on the following system (with $k > l$):

$$x_n(k) = x_r(k) - x_r(l) \quad (16)$$

$$p_n(k) = p_r(k) - p_r(l) \quad (17)$$

$$\kappa(k) = \sqrt{\frac{p_n(k)^T p_n(k)}{x_n(k)^T x_n(k)}} \quad (18)$$

$$\gamma_1(k) = \sqrt{\dot{p}_n(k)^T \dot{p}_n(k)} \quad (19)$$

$$\gamma_2(k) = \sqrt{\dot{x}_n(k)^T \dot{x}_n(k)} \quad (20)$$

$$\beta(k) = \kappa(k)\gamma_2 - \gamma_1 \quad (21)$$

$$\alpha(k) = \alpha(k-1) + K\beta(k) \quad (22)$$

where it is assumed that $\alpha(k) = \alpha(k-1)$ during end–effector motion convergence.

Step 3. Use the continuous–time end–effector controller to converge to a desired degree to the velocity subspace defined by α. This guarantees that the end–effector motion has approximately constant velocity, validating our measure except for (possibly) an isolated point on each trajectory. During this convergence, all of the discrete–time parameters remain fixed.

Step 4. When the end–effector motion has sufficiently converged set $l = k$ and loop to Step 2.

The advantages of this approach are that it takes into account the unknown O_r and guarantees constant velocity over measuring intervals (as required in our discussion of the sufficiency of norm–ratio equality in Section 2). The disadvantages of this method lie in the analysis. We cannot say, a priori, how long each convergence step (Step 3) will take. Thus, the actual update law for α will have a varying sample period that is lower bounded by the vision sample period T. Even so, as the update law for α is based on subsequent vision samples, the design of the discrete–time system that is utilized for this update is unaffected by the actual time required for converge of the end–effector motion. By abstracting the discrete–time dynamical system for α to the point of ignoring the continuous–time dynamics, we can see that so long as the α does not overshoot the depth–invariant velocity subspace, the system will converge correctly.

We do not wish to allow overshoot due to the fact that we cannot say a priori that the sign of β changes as α overshoots the appropriate value. The nonlinear measure β is a complex, unknown function with possible local minima but only one solution $\beta = 0$ over time intervals of nonzero length. As such, converging to the situation where $\beta = 0$ requires care in selection of the appropriate gain K. Algorithmic solution methods can be utilized to guarantee convergence even if the α overshoots the $\beta = 0$ case.

We have simulated the system with $O_r = (0.5, 0.5, 3.0)'$, $(\theta, \psi, \phi) = (\pi/8, -\pi/12, \pi/4)$ and $f\alpha_i = 0.08$ image plane units. The initial state of the system was $x_r(0) = (0.5, 0.5, 0)^T$. The vision sample rate was taken to be 60Hz (standard for field–shuttered CCD cameras) and the gain K was set at 4000. In this case, seen in Figure 1, the measure of convergence β actually increases before converging to a sufficiently low value. We could have chosen to decrease α by selecting a negative K, but this information is not available a priori. We further note that the curve generated by the measure β over time has a local minimum that prevents us from using a simple heuristic based on increase or decrease of the measure to determine the desired sign of K. Additionally, as $\beta \geq 0$ over the entire workspace for this configuration, overshoot of the desired α would result in no convergence.

The tolerance level for convergence of β (determining when we stop updating α to avoid possible overshoot) was taken to be $(1 \times 10^{-6})\|\dot{x}_n\|$. To select this tolerance, we must consider the magnitude of K

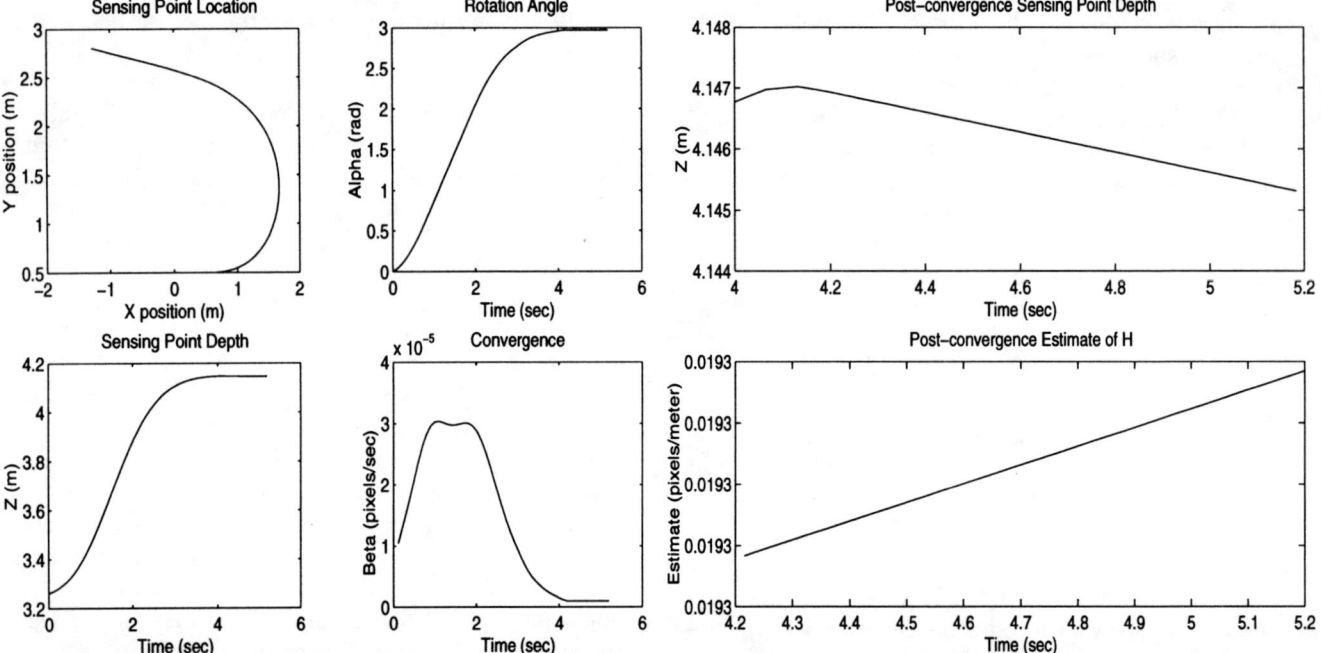

Figure 1: Sampled–data Depth–invariant controller

Figure 2: Post–convergence behavior of the sampled-data depth–invariant controller

compared with the desired convergence time and the function β for all possible configurations. Appropriate selection of K to optimize convergence time and accuracy is an important open issue.

In Figure 2 we see a closeup view of the depth of the sensing point from the camera over an interval of 1.2 seconds after sufficient convergence of α. The total end–effector excursion in the workspace during this interval was 1.2 meters, and the depth changed a total of 0.002 meters. We can determine $H = \frac{f\alpha_i}{Z}$ along the line corresponding to the depth–invariant path of the end–effector by simply taking the ratio of $\|p_n\|$ to $\|x_n\|$. These results are also shown in Figure 2, and agree with the actual value within 1×10^{-6} image plane units/meter. This is acceptable performance.

Having seen that we are able to guarantee convergence of end–effector motions to a depth–invariant velocity subspace in the camera coordinate frame, we now discuss an application of this controller for on–line calibration of general monocular visual servo systems.

4 Application: Calibration

Full calibration of a monocular visual servo system requires information beyond that available from a single depth–invariant line. After the initial system convergence, we have an estimate of the constant $\frac{f\alpha_i}{x_{c3}}$. The orientation of the depth–invariant line in the workspace and the image plane gives us two calibration angles. The final degree of uncertainty is the possible rotation of the workspace plane about the isolated invariant line. This parameter can be determined by moving to another line parallel to the first (in the image plane) and comparing the resulting (scaled) depth information.

The methodology for using the second line to determine the full calibration requires some analysis of the geometry of the problem. Given two line segments in the image plane and the extension of these segments to infinite lines, there are points on the two lines such that the ray connecting these two points is perpendicular to both lines and, when extended, includes the origin of the image plane. At these points we are guaranteed that the connecting ray in the image plane corresponds to a ray in the robot plane that is also perpendicular to the two extended line segments. Thus, we can measure the image plane distance between the two lines and compare this to the real distance between the lines in the robot space.

Making this geometric argument allows us to simplify the projection in order to solve for the full calibration. As we are considering projections on a plane through the optical axis, the projection becomes two–

dimensional, and is illustrated in Figure 3. In this figure, we have performed a coordinate transformation on the image plane to simplify presentation. We have estimates of $k_1 = \frac{\alpha_i f}{Z_1}$ and $k_2 = \frac{\alpha_i f}{Z_2}$, as well as measurements of u_1, u_2 and the real–world distance $d_r = ((Z_1 - Z_2)^2 + (x_1 - x_2)^2)^{1/2}$. From this information, we can determine the depths Z_1 and Z_2, thereby completing the calibration. Note that we do not need to actually make an extension of the lines to the radial point described above, as it is the ratio of the distances that is important, and the distance between the two line segments in the image plane is constant. Note that the vector O_r is easily determined after the workspace plane is calibrated with respect to the camera coordinate frame.

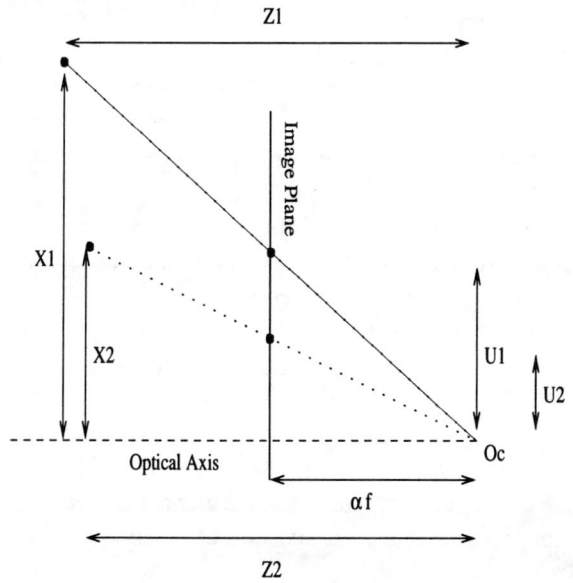

Figure 3: Transformed projection for calibration using parallel lines

5 Conclusions

In this paper we have demonstrated an achievable objective for uncalibrated monocular visual servo using sampled–data control, and have extended this result to calibration of monocular visual servo systems. The eventual goal of this approach is to design a complete visual control scheme, achieving regulation or trajectory tracking and (possibly) calibration in a single step (similar work for the 2D case is discussed in [4]).

A major obstacle to implementation of this technique is generation of an accurate instantaneous image plane velocity vector $\dot{p}_r$. Various methods exist that would allow such estimation from a single image. Generally, these techniques rely on the use of intensity–based image analysis and require either carefully structured lighting, a great deal of processing time, or both. Methods for generating a rapid and accurate estimate of $\dot{p}_r$ for general lighting, background and robot geometry are currently under investigation, as are techniques to make the process robust with regards to lens distortions and aberrations.

References

[1] R. Y. Tsai. A versatile camera calibration technique for high-accuracy 3D machine vision metrology using off-the-shelf tv cameras and lenses. *IEEE Journal of Robotics and Automation*, 3(4):323–344, August 1987.

[2] Olivier Faugeras. *Three–Dimensional Computer Vision: A Geometric Viewpoint*. The MIT Press, Cambridge, Massachusetts, 1993.

[3] B. Yoshimi and P. K. Allen. Active, uncalibrated visual servoing. In *Proc. IEEE Int'l Conference on Robotics and Automation*, pages 156 – 161, San Diego, CA, 1994.

[4] B. E. Bishop and M. W. Spong. Adaptive calibration and control of 2d monocular visual servo systems. In 9th *IFAC Symposium on Robot Control*, pages 525–530, Nantes, France, 1997.

[5] G. V. Puskorius and L. A. Feldkamp. Camera caliabration methodology based on a linear perspective transformation error model. In *Proc. IEEE Int'l Conference on Robotics and Automation*, pages 1858–1860, 1988.

[6] G. Q. Wei and S. D. Ma. Implicit and explicit camera calibration: Theory and experiments. *PAMI*, 16(5):469–480, 1994.

[7] R. J. Holt and A. N. Netravali. Camera calibration problem: Some new results. *CVGIP: Image Understanding*, 54(3):368–383, 1991.

[8] E. D. Dickmanns and V. Graefe. Dynamic monocular machine vision. *Machine Vision and Applications*, 1(1):223–240, 1988.

[9] M. W. Spong and M. Vidyasagar. *Robot Dynamics and Control*. John Wiley & Sons, New York, 1989.

Control of Autonomous Motion of Two-Wheel Bicycle with Gyroscopic Stabilisation

A.V.Beznos, A.M.Formal'sky, E.V.Gurfinkel, D.N.Jicharev, A.V.Lensky, K.V.Savitsky, L.S.Tchesalin

Institute of Mechanics, Moscow Lomonosov State University,
1, Mitchurinskii prospect, Moscow 119899, Russia.
E-mail: formal@inmech.msu.su, gurf@inmech.msu.su

Abstract

A bicycle with a gyroscopic stabilisation capable of autonomous motion along a straight line as well as along a curve is described. The stabilisation unit consists of two coupled gyroscopes spinning in opposite directions. It makes use of a gyroscopic torque due to the precession of gyroscopes. This torque counteracts the destabilising torque due to gravity forces. The control law of the actuator drive making the gyroscopes to precess is described.

1 Introduction

A bicycle equipped with a system of gyroscopic stabilisation of its upright position has been designed at the Institute of Mechanics of the Moscow Lomonosov State University (Fig. 1). The scheme of stabilisation is similar to that of Sherle-Shilovsky proposed in 1909 for stabilisation of a monorail car [1, 2]. The stabiliser contains two gyroscopes with rotors spinned in opposite directions. A torque delivered by a DC motor may be applied about the axes of the gyroscope frames. This control torque is the function of the following variables: the declination angle of the bicycle from the vertical, the orientation angle of the gyroscopes and its derivative.

Brown and Yangsheng [3] previously described a single-wheel vehicle with an internal gyroscope.

A control is presented here that provides the stability of the vertical position of the bicycle during its rectilinear motion and of an inclined position when moving over a curve. In comparison to the control in [1, 2], the feedback also incorporates the steering angle and the velocity of the bicycle. Because of this, the bicycle is able to perform turns with radii approaching its length.

In a conventional bicycle the front wheel not only controls the direction, but also contributes to the stability of the bicycle. To prevent a fall, the fork must be turned to the side of inclination. When riding without hands, the handle bar turns to the correct side by itself due to: 1) the gyroscopic effect because of the rotation of the front wheel, 2) the steering axis being inclined with respect to vertical and 3) the centre of the front wheel being placed slightly forward of the steering axis [1].

To the contrary of a conventional bike, this one is stabilised by a special gyroscopic stabiliser and does not make use of the factors mentioned above. Furthermore, to eliminate their influence, the steering tube is made vertical and the centre of the front wheel lies on its continuation (Fig. 1, 2). Another major difference is that the bicycle has the front wheel drive.

Three control loops control the motion of the bicycle along a curved path and stabilise it. Each loop has its own task.

The first loop controls the steering angle, that is, the direction of motion. An incremental encoder and a rotary drive are installed on the steering tube to measure and control the orientation of the fork.

The second servo loop controls the angular velocity of the front wheel, that is, the speed of the bicycle. Another encoder is installed on the axis of the wheel to measure its angular position and velocity. The rear wheel is passive.

The third loop controls the precession of the gyroscopes, preventing the bicycle from falling over. Two more encoders are related to this particular actuator: one, to measure the orientation of gyroscope frames, and another, to measure the declination angle of the bicycle from the vertical. To accomplish this, a free gyroscope with three DOF's is used, and its orientation relative to the frame is measured. The control law in this loop takes into account signals from all four mentioned encoders.

2 Gyroscopic Stabiliser

The stabiliser consists of two identical gyroscopes located between the wheels (Fig. 2). Each rotor is sealed within a frame, that can pivot relative to the chassis about the axis perpendicular to the sagittal plane of the bicycle (the centres of the wheels and the steering axis lie in this plane). In a normal situation, when this plane is vertical, this axis is horizontal and the rotor axis is vertical; the rotation axes of the rotor frames, parallel to each other, lie within the horizontal plane.

The rotations of gyroscope frames are not independent: they are interlinked by a gear train such that when one frame turns by some angle β, the other turns to the opposite direction by the same angle (Fig. 2).

The rotors spin with the same speed in opposite directions; hence their vectors of kinetic momentum H point in opposite directions. If a precession about the frame axes takes place, a gyroscopic torque is developed that counteracts the disbalancing torque due to the gravity.

A torque Q is applied to the axis of frame rotation by a DC motor through a reductor. The control voltage is constructed as a function of the angle ψ of the declination from the vertical (angle of bank), precession angle β, its angular velocity $\dot{\beta}$, steering angle δ, and the linear speed of the center of the front wheel V. The angle of bank ψ is measured by the free 3-DOF gyroscope - gyrovertical which is installed on the bicycle chassis.

3 Mathematical Model

In the sketch of the bicycle in the Fig. 3, the angle of bank ψ and wheel-to-ground contact points K_1 and K_2 are shown. The plane of the front wheel intersects the support surface over some straight line; for small angles of bank ψ the steering angle δ is approximately equal to the angle between this line and the segment $K_1 K_2$ (Fig. 4). The following kinematic relationship may be derived from this figure:

$$l\dot{\alpha} = V \sin \delta \qquad (1)$$

where $l = K_1 K_2$, α is the angle between the segment $K_1 K_2$ and axis OX (Fig. 4). Within the linear approximation the distance l does not depend on angles ψ and δ.

The oscillation of the bicycle on the angle of bank and the precession of the gyroscopes are described by a rather complicated set of equations. We show here only linear equations. They were obtained by linearisation of complete equations around state $\psi = \beta = \dot{\psi} = \dot{\beta} = 0$. This way we obtain the following:

$$B\ddot{\beta} - 2H\dot{\psi} = Q \qquad (2)$$

$$D\ddot{\psi} + 2H\dot{\beta} - Eg\psi = E(\ddot{y}\cos\alpha - \ddot{x}\sin\alpha) \qquad (3)$$

Here B and D are the moments of inertia of the rotors of the gyroscopes together with their frames and of the whole bicycle about the appropriate axes, $H \approx 10\ kg \cdot m^2 /s$ is the kinetic momentum of the rotor (it is considered constant), Q is the torque, developed by the drive, $E = mb$, where $m \approx 20\ kg$ is the total mass of the bicycle and $b \approx 0.2\ m$ is the distance between the common centre of mass and the segment $K_1 K_2$, x, y are the co-ordinates of middle point of this segment, where the centre of mass is assumed to map. The distance between the wheel axes or, otherwise said, the length l of the segment $K_1 K_2$ is approximately $0.75\ m$, the radius of the wheels is $0.15\ m$. The term $Eg\psi$ describes the destabilising torque component due to the gravity, the term $2H\dot{\beta}$ is a gyroscopic torque impeding the fall of the bicycle, and the right-side term in (3) is the moment of inertia forces emerging during the manoeuvres of the bicycle.

In Eq. (2) the torque due to friction in the axis of precession is neglected. We will also neglect the back emf in the DC motor and the inductance of the motor winding.

This way in composing the mathematical model (1), (2), (3), the motion of the bicycle is decomposed into two parts. Eq. (1) describes the motion of the axis K_1K_2 (this axis contains the points of wheel contacts with the support surface), which is the motion along the trajectory. Given this motion, Eqs. (2) and (3) describe the motion (inclination) of the bicycle relative to this axis.

Using relation (1), we rewrite Eq. (3) as follows:

$$D\ddot{\psi} + 2H\dot{\beta} - Eg\psi = \frac{1}{2}E\left[\frac{V^2}{l}\sin 2\delta + \frac{d}{dt}(V\sin\delta)\right] \quad (4)$$

The right-hand side of this equation depends only on the trajectory and velocity of the front wheel motion.

Let us represent the desired trajectory of the bicycle and the velocity of the front wheel V in the following way:

$$\sigma = \sigma(s), \quad \frac{ds}{dt} = V \quad (5)$$

where s is the natural parameter of the trajectory (current path length), σ is the angle between a tangent to the trajectory and some fixed direction, and the velocity V is function of time t, path s or of some other variable. If the front wheel moves exactly along the prescribed trajectory, the angle δ changes according to the following relation:

$$\delta = \sigma - \alpha \quad (6)$$

Solving (1), (5), (6), we get the functions $\delta(t)$ and $V(t)$. Substituting them into (2) and (4), it is possible for a known Q to solve them for bank angle oscillations and for the motion of the gyroscopes.

4 Stationary Regimes and Control

The equations (2) and (4) describe a system with two degrees of freedom, but having only one control parameter Q. Let $\delta = 0$, i.e., the bicycle moves straight ahead. The Eq. (4) becomes homogeneous.

The verification of the Kalman controllability criterion [4] gives that the system is quite controllable if and only if $H \neq 0$ and $b \neq 0$. This is, of course, true in our case, and the equilibrium position

$$\psi = \beta = \dot{\psi} = \dot{\beta} = 0 \quad (7)$$

corresponding to the desired upright position of the bicycle and of the axes of gyroscope rotors, can be stabilised by a linear feedback including all four state variables ψ, β, $\dot{\psi}$, $\dot{\beta}$. In [1, 2], it is shown, that the state (7) can also be stabilised by a feedback including only three variables:

$$Q = k_\beta \beta - k\dot{\beta} + k_\psi \psi \quad (8)$$

In other words, it is shown that for a sufficiently large kinetic momentum H the coefficients k_β, k, k_ψ can be selected such that the Hurwitz criterion of the asymptotic stability of solution (7) is satisfied. The coefficient k_β proves to be positive in accordance with the Kelvin's theorem, stating that an unstable system can be stabilised by gyroscopic forces only if the number of unstable degrees of freedom is even [2, 5].

If $k_\beta > 0$, the torque $k_\beta \beta$ is directed so as to turn the gyroscopes over, making them statically unstable similar to the Lagrange's gyroscope with the upper location of the mass center.

That is, the position and derivative term gains in the control (8) have opposite signs.

Assume now that the bicycle is turning with a constant velocity and a constant steering angle, that is,

$$\dot{V} = 0, \quad \dot{\delta} = 0 \quad (9)$$

Under these conditions the system (2), (4), (8) has a stationary solution

$$\psi = \psi_s = \frac{V^2}{2gl}\sin 2\delta, \quad Q = 0, \quad \beta = \beta_s = -\frac{k_\psi \psi_s}{k_\beta} \quad (10)$$

As it follows from (10), the stationary precession angle is different from zero. But it is known that the gyroscopes are most effective around the zero precession angle. This statement cannot be proven within the scope of linear model (2), (4) as it follows from the consideration of nonlinear model. This is because instead of the terms $2H\dot\psi$ and $2H\dot\beta$, describing the components of the stabilising gyroscopic torque in the linear equations (2) and (4), the nonlinear equations contain terms $2H\dot\psi\cos\beta$ and $2H\dot\beta\cos\beta$. They attain their maxima at $\beta=0$ and go to zero as $|\beta|\to\pi/2$. It would be possible to reduce the value of β_s by increasing the coefficient k_β, but its upper limit is imposed by the conditions of stability. For this reason we replace the control (8) by the following:

$$Q = k_\beta \beta - k\dot\beta + k_\psi(\psi - \psi_s) \qquad (11)$$

In this case the stationary value of β becomes zero: $\beta = \beta_s = 0$.

Assume that the angle of bank is measured by the gyrovertical with some error $\Delta\psi$. In this case, the sum $\psi+\Delta\psi$ instead of ψ is used in the feedback (11). Let $\Delta\psi = const$. This error may be caused by slightly inaccurate measurement of initial position of the gyrovertical and/or of the bicycle. This constant error would result in a stationary value

$$\beta = \beta_s = -\frac{k_\psi \Delta\psi}{k_\beta} \qquad (12)$$

To get rid of the offset (12) in the precession angle β, consider instead of (11) the following control, containing the integral of the precession angle:

$$Q = k_\beta \beta - k\dot\beta + k_\psi(\psi + \Delta\psi - \psi_s) + k_\sigma \sigma, \quad \dot\sigma = \beta \qquad (13)$$

In a stationary case for the control (13) we will have

$$\psi = \psi_s = \frac{V^2}{2gl}\sin 2\delta, \quad \beta = \beta_s = 0,$$

$$Q = 0, \quad \sigma = \sigma_s = -\frac{k_\psi \Delta\psi}{k_\beta} \qquad (14)$$

Therefore in the stationary solution (14) the precession angle β equals zero despite an unknown error $\Delta\psi$ in the measured inclination angle of the bicycle.

Using Hurwitz criterion it is possible to show that the following inequalities are the necessary conditions of the asymptotic stability of the solution (14):

$$k_\beta > 0, \quad k > 0, \quad k_\phi > 0, \quad k_\sigma > 0$$

That is, to the contrary of the common *PID*-controller the sign of the coefficient at the derivative $\dot\beta$ in the control (13) is opposite to that of the position and integral terms.

Finally we come to the following control:

$$Q = k_\beta \beta - k\dot\beta + k_\psi(\psi + \Delta\psi - \frac{V^2}{2gl}\sin 2\delta) +$$

$$+ k_\sigma \int_0^t \beta(\tau)d\tau \qquad (15)$$

It is assumed here that the bicycle is moving along a circle with a constant speed. This type of control was used in simulations as well as in real experiments.

5 Implementation of the Control Algorithm

During the motion the bicycle is fully autonomous. The stabilising gyroscopes and gyro of the inclinometer (gyrovertical) are spinned up before the start using an external power supply. All control functions are accomplished by an on-board control system.

The control system consists of two programmable controllers built around Intel 80C196KC chips with a clock frequency of 20 MHz.

The controllers communicate via CAN bus, the software is downloaded from an external PC over the same bus.

The control programmes implement the servosystems described in the Introduction. One of them controls the motion of the bicycle along the track (this is a velocity servosystem on the speed V of the front wheel) as well as the steering speed servosystem depending on the shift from the track. Both servosystems are implemented within one loop with a period of *5 ms*.

The other controller drives the gyroscopes of the stabiliser. This is in fact a servosystem including multiple variables according to Eq. (15). The values β, $\dot{\beta}$, ψ are taken from the incremental encoders and V and δ are received from the first controller. The calculations are performed with a period of *1 ms*.

To reduce the maximal precession angle of the stabiliser gyroscopes when moving over an arc, the linear speed *V* is then somewhat reduced.

6 Experiments

The experiments were carried out on the track of the mobile robot contest of the International Festival of Sciences and Technologies (France, May 1997). The track is presented in Fig. 5. Each square is 2 by 2 m in size. To follow the designated track, the control system must be informed of the current bias of the front wheel from the track. A track sensor has been installed on the front wheel to produce this signal. It was used to program the steering angle of the front wheel. For the time being the bicycle can move along the track at a speed of about *1 m/s*.

Control (15) ensures asymptotic stability of the stationary motion along the straight line or circle. However, during the motion on the experimental track the bicycle does not quite reach a stationary mode with a constant steering angle δ. The control system works in a transient mode, but nevertheless, the control (15) ensures the stable motion of the bicycle, and the maximal precession angle is small.

To further improve the quality of stabilisation, we plan to make the control (15) more complex, taking into account not only the first "centrifugal" term from the right-hand side of the equation (4), but also the second one describing the torque of the inertia force because of the changing velocity V and/or steering angle δ.

References

[1] R.Grammel, "Der Kreisel, seine Theorie und seine Anwendungen", 2 Bde., Berlin-Goettingen-Heidelberg, Springer, 1950.

[2] K.Magnus, "Kreisel, Theorie und Anwendungen", Springer-Verlag, Berlin-Heidelberg-NewYork, 1971.

[3] H.Benjamin Brown, Jr. and Yangsheng Xu, "A Single-Wheel, Gyroscopically Stabilized Robot", Proc. of the 1996 IEEE Int. Conf. on Rob. and Autom., Minneapolis. Minnesota-April 1996.

[4] N.N.Krasovskii, "Theory of Motion Control", Moscow, "Nauka", 1968. (In Russian).

[5] N.G.Tchetaev, "Motion Stability", Moscow, "Nauka", 1965 (In Russian).

Figure 1. Photo of the bicycle.

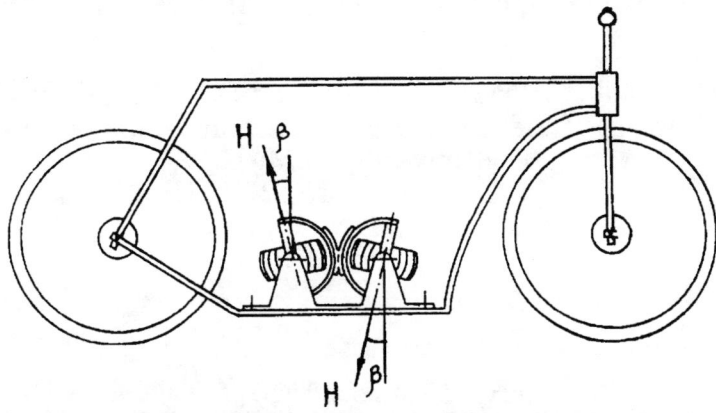

Figure 2. The scheme of the bicycle with gyroscopic stabiliser (side view).

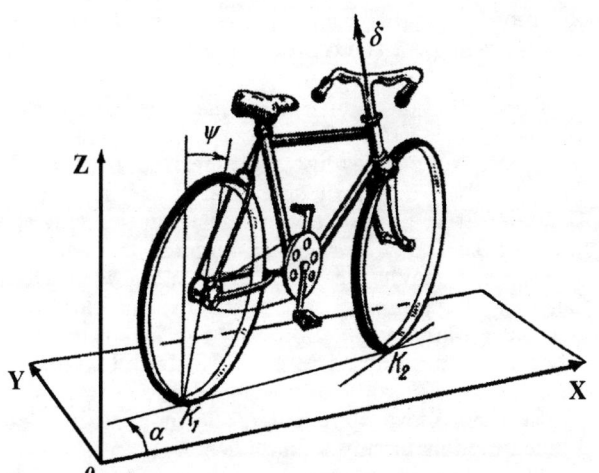

Figure 3. Sketch of the bicycle.

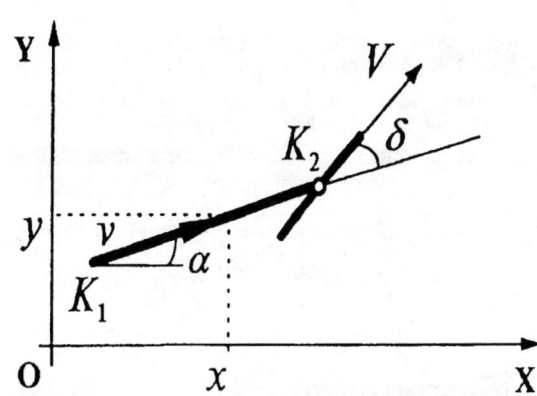

Figure 4. Kinematic scheme of the bicycle motion over the support surface.

Figure 5. Experimental track.

Toward the Control of a Multi-Jointed, Monoped Runner

Uluç Saranli, William J. Schwind*and Daniel E. Koditschek[†]
Department of Electrical Engineering and Computer Science
The University of Michigan
Ann Arbor, MI 48109-2110, USA

Abstract

In this paper, we propose a new family of controllers for multi-jointed planar monoped runners, based on approximate but accurate models of the stance phase dynamics of a two degree of freedom "SLIP" leg. Unlike previous approaches, the new scheme gives control over all parameters of the system including the hopping height, forward speed and duty cycle. The control laws are "deadbeat" in nature, derived by computing the inverse of an approximate return map and corrected by integral compensation. We use the expressions obtained in this way to control the original SLIP leg as well as radically different, more realistic four degree of freedom legs. In each case, the performance of the deadbeat scheme in controlling forward running velocity is compared to a modified Raibert control strategy, whose experimental stability properties have been analyzed carefully in the low degree of freedom setting.

1 Introduction

Biomechanists have gained great leverage in understanding basic principles of locomotion in creatures as diverse as humans and cockroaches by considering the "simple" SLIP model shown in Figure 1 as a metaphor for running and hopping [1, 3, 4, 5]. While simple to the biomechanist, even this model presents difficulties to the engineer wishing to pursue formal analysis and control since it is a hybrid system with nonlinear stance dynamics which are not closed-form integrable. Even so, previous work by two of the authors [15, 16] provides approximate functional relationships for the SLIP dynamics, enabling a consideration of control via established techniques.

The question remains, however, whether such consideration is warranted. Is the SLIP model any more than a metaphor for running and hopping? Is it actually a control target aimed for by humans and animals in spite of their greater degrees of freedom? If so, will the careful consideration of such a simple model allow the engineer to create robots with dexterity reminiscent of humans and insects, or is this a "zoomorphic fallacy" tantamount to building a flying machine with flapping wings?

In answer to the former two questions, growing biological evidence, including recent work in our lab with human running data, suggests that the SLIP dynamics are more than just a metaphor. They are the literal control target for the center of mass of the subjects we have studied to date [14].

The latter question was in one sense answered by the landmark work of Raibert and his students [12] who used robots readily characterized by the SLIP model. The power of such simple leg models was demonstrated by the extensibility of the single leg ideas to two and four legged runners as well as the variety of behaviors generated: running with a number of gaits, jumping over obstacles, and performing acrobatic maneuvers. However, the legs used in this work were constructed to be SLIP-like. The question remains: Is it possible to use the simple SLIP model to characterize more complicated and biologically plausible leg models having ankle, knee and hip joints?

The biological evidence seems to provide a proof by existence. Additionally, intuition regarding the Lagrangian dynamics suggests that a "heavily-laden" higher degree of freedom leg will behave "almost identically" to a 2 DOF SLIP leg [16].

Given this evidence, this paper reports on our preliminary efforts to investigate the extensibility of SLIP based controllers to more complicated leg models.

1.1 Scope of the Paper: Coupled Controller for a "Special" SLIP Runner

The first work in the control of SLIP runners was the successful implementation by Raibert and his students [12] of simple, roughly decoupled controllers to independently control the hopping height and forward velocity of their robots. This stunning success motivated a series of papers [10, 17, 11, 15] characterizing the stability of these decoupled controllers.

In this paper we present a new coupled approximate deadbeat controller for a SLIP runner having a "special" spring potential model which makes a simplified version of the stance dynamics closed-form integrable.

---

*Supported in part by National Science Foundation Grant IRI-9612357

[†]Supported in part by National Science Foundation Grant IRI-9510673

We then explore the applicability of the decoupled controller (that we will term Raibert-like) and the new coupled controller in more biologically plausible legs.

1.2 Contributions of the Paper: The Power of the SLIP Model

In this paper we use simulation to suggest the possibility that control laws designed for SLIP leg, can be extended more biologically plausible leg models. As far as we know, this represents the first attempt to apply any 2 DOF derived return map controller to more complex single legs. We contrast a "deadbeat" and a Raibert-like controller in so doing.

It is not surprising to find that the approximate deadbeat controller outperforms the decoupled controller in the 2 DOF leg for which they were both developed.[1] It is surprising to find that the decoupled controller continues to function well in the 4 DOF leg. However, it seems to us truly noteworthy that the aggressive 2 DOF coupled controller can be adopted in the same way to the 4 DOF leg as well, even to the point of outperforming the decoupled algorithm. This significantly bolsters our suspicion that the "collapse of dimension" observed in biological control hierarchies might be explained in terms of isometries of the kind we have explored in [16].

Good performance can be achieved in the decoupled scheme when the gain parameters are tuned, whereas in contrast, the deadbeat controller is tuned automatically in its defining formula. Moreover, it allows for explicit control over duty factor[2].

Introducing the ability to explicitly command duty factor in addition to forward speed and hopping height may be useful when considering higher level control problems in dynamic locomotion such as foot placement on irregular terrain. Hodgins [7] studied the use of three different techniques for foot placement on irregular terrain: controlling forward speed, flight duration and stance duration. While we have not explored the implications of this work on foot placement in irregular terrain, the coupled controller's ability to explicitly control forward speed, hopping height and duty factor will prove advantageous in such contexts.

2 The "Special" SLIP Runner

2.1 Model and Assumptions

The SLIP model considered in this paper is shown in Figure 1. The leg is assumed massless and the body

	Lift-Off point
q_{rl}	Leg Length at Lift-Off
$q_{\theta l}$	Leg Angle at Lift-Off
$\dot{q}_{rl}$	Radial Velocity at Lift-Off
$\dot{q}_{\theta l}$	Angular Velocity at Lift-Off
	Apex Point
$\bar{b}_y$	Apex Hopping Height
$\bar{b}_x$	Apex Forward Velocity
ϕ	$\frac{TimeFlight}{TimeStance}$
	Touch-Down Point
q_{rt}	Leg Length at Touch-Down
$q_{\theta t}$	Leg Angle at Touch-Down
$\dot{q}_{rt}$	Radial Velocity at Touch-Down
$\dot{q}_{\theta t}$	Angular Velocity at Touch-Down

Table 1: Notation for the SLIP Leg Model

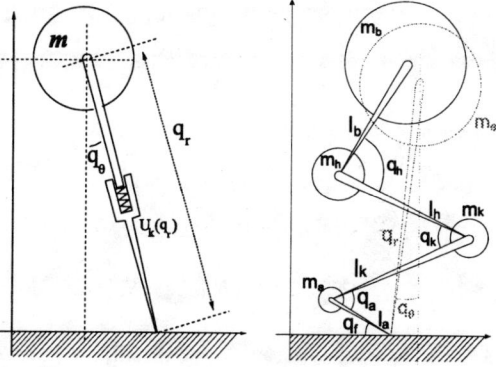

Figure 1: The spring loaded inverted pendulum(SLIP) leg model(left) and the "ankle-knee-hip"(AKH) leg model(right).

a point mass at the hip joint. During stance the leg is free to rotate around its toe and the mass is acted upon by a radial spring with potential $U(q_r)$. In flight, the mass is considered as a projectile acted upon by gravity. We assume there are no losses in either the stance or flight phases.

Despite its structural simplicity, the stance dynamics of this system are not integrable. Therefore, we begin our formal consideration by eliminating gravity from the stance dynamics yielding a simple central force problem wherein energy and angular momentum are both constants of motion and can be used to integrate the stance dynamics. The structure of the integrals suggest certain forms for the spring law which are physically realistic and also admit closed form integration [15, 16]. In particular, as in [15], we have chosen to work with the compressed air spring $U_A(q_r) := k/2(1/q_r^2 - 1/q_{r0}^2)$.

Before formulating the return map, we discuss the control inputs available for the SLIP runner. The first control input is the leg angle at touchdown, $q_{\theta t}$. We assume that during flight we are able to swing the leg to any desired angle relative to the ground. The other control inputs come from the ability to tune the spring

---

[1] The tradeoffs between deadbeat and less model dependent controllers are well understood. The relative benefits in performance promised by the former can evaporate in the presence of noise and model mismatch that might not significantly undermine the latter.

[2] In fact, the introduction of the duty factor (the ratio of time a leg is on the ground over a complete cycle of leg movements) as a control objective is also a novelty of this work. While commonly considered in the biomechanics literature for either it's power in classifying gait [2] or in its effect on metabolic efficiency, it has been all but ignored in the robotics literature.

potential. In this work, we choose to tune the spring potential via choice of the stance compression and decompression spring constants, k_1 and k_2, respectively.

2.2 The Control Objective

In formulating the control problem it is natural to work in the set of apex states (see Table 2 for state definitions),

$$\mathcal{X}_a = \{X_a \mid X_a = [\,\bar{b}_x,\, \bar{b}_y,\, \dot{\bar{b}}_x,\, \phi\,]^T\}$$

since its elements are easily observable and represent directly natural control specifications such as " jump this high" or "run this fast".

Given this perspective, an obvious next step is to introduce the apex return map, $f_a : \mathcal{X}_a \times \mathcal{U}_k \mapsto \mathcal{X}_a$ where

$$\mathcal{U} = \{u \mid u = [\,q_{\theta t},\, k_1,\, k_2\,]^T\}$$

is the set of control inputs. We are now in position to consider the coupled control problem.

That is, suppose we want to achieve the desired apex state (control objective),

$$X_a^* = [\,\bar{b}_x^*,\, \bar{b}_y^*,\, \dot{\bar{b}}_x^*,\, \phi^*\,]^T \quad (1)$$

One possible solution is the deadbeat control, that is, the control input $u^* = [\,q_{\theta t}^*,\, k_1^*,\, k_2^*\,]^T$ such that $X_a^* = f_a(X_a, u^*)$, effectively taking the current apex state X_a to the desired state X_a^* in one cycle.

The most direct way to find the deadbeat control u^* would be to invert the map f_a. However, the control inputs appear in the apex return map in a complicated manner making a direct computation of the inverse map difficult. In consequence, we introduce a new coordinate system, which affords an almost completely closed form inverse to an approximate return map.

2.3 The Liftoff Return Map

Consider the new state and control sets,

$$\mathcal{Z}_l = \{Z_l \mid Z_l = [\,q_{\theta l},\, E_l,\, \psi_l,\, \phi\,]^T\}$$

$$\overline{\mathcal{U}} = \{\overline{u} \mid \overline{u} = [\,q_{\theta t},\, a_1,\, \alpha\,]^T\}$$

where E_l is the energy at liftoff, ψ_l is the ratio of forward velocity to vertical velocity at liftoff and

$$a_i^2 := \frac{q_{rt}^2 \dot{q}_{\theta t}^2 + k_i/(m q_{rt}^2)}{\dot{q}_{rt}^2}; i = 1, 2 \quad (2)$$

$$\alpha^2 := \frac{a_2^2}{a_1^2} = \frac{q_{rt}^2 \dot{q}_{\theta t}^2 + k_1/(m q_{rt}^2)}{q_{rt}^2 \dot{q}_{\theta t}^2 + k_2/(m q_{rt}^2)} \quad (3)$$

Assuming $q_{rl} = q_{rt} = q_{r0}$, the liftoff return map $f_l : \mathcal{Z}_l \times \overline{\mathcal{U}} \mapsto \mathcal{Z}_l$ can be written as[3]

$$f_l(Z_l, \overline{u}) := \begin{bmatrix} \vartheta_l \\ E_l + \Delta E_U + \Delta E_g \\ \mathbf{t}_{(1,-\vartheta_l)} \circ \mathbf{t}_{(\frac{-1}{\alpha}, q_{\theta t})}(\psi_t) \\ \frac{t_f}{t_s} \end{bmatrix} \quad (4)$$

where

$$\vartheta_l = q_{\theta t} - \mathbf{t}_{(1, q_{\theta t})}(\psi_t)\left(\frac{\alpha+1}{\alpha}\right)\frac{1}{a_1}\mathrm{acot}(a_1) \quad (5)$$

$$\Delta E_U = U_{k_2}(q_r) - U_{k_1}(q_r) = \frac{1}{2}m\dot{q}_{rt}^2(\alpha^2 - 1) \quad (6)$$

$$\Delta E_g = mg(q_{rl}\cos\vartheta_l - q_{rt}\cos q_{\theta t}) \quad (7)$$

$$t_f = \frac{1}{g}\sqrt{\frac{2}{m}\left(\frac{1}{\psi_l^2+1}\right)}\left(\sqrt{E_l - mgq_{rl}\cos q_{\theta l}} - \sqrt{E_l + \psi_l^2 mgq_{rl}\cos q_{\theta l} - (1+\psi_l^2)mgq_{rt}\cos q_{\theta t}}\right) \quad (8)$$

$$t_s = \sqrt{\frac{q_{rt}^2}{\dot{q}_{rt}^2}}\left(\frac{1}{1+a_1^2}\right)\left(\frac{\alpha+1}{\alpha}\right) \quad (9)$$

and we define the following two parameter family of functions,

$$\mathbf{t}_{(\sigma_1, \sigma_2)}(\chi) := \tan(\sigma_1 \mathrm{atan}(\chi) + \sigma_2). \quad (10)$$

Notice that apart from certain values of the parameters (e.g. $\sigma_1 = 1$ and $\sigma_2 = 0$) this family cannot be expressed in terms of a single elementary function. Finally note that both ψ_t and $\dot{q}_{rt}$, which appear in (4) can be expressed in terms of Z_l and $q_{\theta t}$.

3 The SLIP Deadbeat Controller

We want the ability to control the SLIP hopper to achieve a goal state,[4]

$$Z_l^* = [\,q_{\theta l},\, E_l^*,\, \psi_l^*,\, \phi^*\,]^T \quad (11)$$

We are looking for the the deadbeat control, $\overline{u}^*$, such that

$$Z_l^* = f_l(Z_l, \overline{u}^*) \quad (12)$$

---

[3] Please refer to [13] for more details on the derivation of the liftoff map.

[4] As in Section 2.2 we can only choose three independent control objectives, here we select E_l, ψ_l and ϕ

3.1 Inverting the Return Map to Find Deadbeat Control

The simple form of the liftoff return map makes it possible, under a reasonable assumption, to reduce the inversion of f_l to the solution of a single equation in a single variable. The assumption that makes this possible is

$$\Delta E_g \equiv 0 \qquad (13)$$

This assumption is reasonable in practice since ΔE_g appears in (4) only as a result of the unnatural energy discontinuities at touchdown and liftoff due to our no-gravity stance model, and does not appear in the stance dynamics with gravity.

Given this assumption, solution of the E_l and ϕ equations of (4) yields

$$\alpha^2(Z_l, Z_l^*, q_{\theta t}) = \frac{2}{m} \frac{E_l^* - E_l}{\dot{q}_{rt}^2(Z_l, q_{\theta t})} + 1 \qquad (14)$$

$$a_1^2(Z_l, Z_l^*, q_{\theta t}) = \qquad (15)$$
$$\sqrt{\frac{q_{rt}^2}{\dot{q}_{rt}^2(Z_l, q_{\theta t})} \left(\frac{\alpha(Z_l, Z_l^*, q_{\theta t}) + 1)}{\alpha(Z_l, Z_l^*, q_{\theta t})} \right) \frac{\phi^*}{t_f(Z_l, q_{\theta t})} - 1}$$

We then substitute both (14) and (15) into the ψ_l equation of (4) to arrive at a single equation in a single unknown variable, $q_{\theta t}$. Namely the equation

$$\psi_l^* = \mathbf{t}_{(1, -\vartheta_l(Z_l, Z_l^*, q_{\theta t}))} \circ \qquad (16)$$
$$\mathbf{t}_{(\frac{-1}{\alpha(Z_l, Z_l^*, q_{\theta t})}, q_{\theta t})}(\psi_t(Z_l, q_{\theta t}))$$

The function of $q_{\theta t}$ on the right hand side of the equation behaves nicely (e.g. it is monotone for most choices of Z_l, Z_l^*) and can be easily solved using numerical methods.

After solving for $q_{\theta t}$ from (16), we substitute the result into (14) and (15) to obtain α and a_1. From here, it is trivial to go back to k_1 and k_2, completing the inversion.

Finally, we can express the desired liftoff state, Z_l^* in terms of X_a^* and the control inputs [13]. Substituting the appropriate relationships, (16) becomes

$$\mathbf{t}_{(1, \vartheta_l(X_a, X_a^*, q_{\theta t}))}(\psi_l^*(X_a, X_a^*, q_{\theta t})) =$$
$$\mathbf{t}_{(\frac{-1}{\alpha(X_a, X_a^*, q_{\theta t})}, q_{\theta t})}(\psi_t(X_a, q_{\theta t})) \qquad (17)$$

Equation (17) is used in the remainder of the paper to solve for $q_{\theta t}$ numerically (since no closed form expression involving elementary functions is available). This is in turn used to find k_1 and k_2 using the closed form expressions (2), (3), (15) and (14).

3.2 The Deadbeat and Modified Raibert Controllers

The procedure outlined in Section 3.1 gives an open loop approximate deadbeat controller for the ideal case where the plant exactly matches (save the omission of the ΔE_g term) the SLIP model with the compressed air spring introduced in Section 2.1.

Previous work by two of the authors [16] investigated the impact of the omission of gravity during stance on the accuracy of the approximations and suggested possible corrections to the model. To minimize the effect of the prediction errors to controller performance, we augment the inverse apex map with a gravity correction policy, increasing the stance spring constants as a function of the gravitational potential at bottom [13]. The resulting control law is the approximate deadbeat controller we have been discussing.

For the purposes of comparison, we propose a decoupled alternative to this strategy based on Raibert's original control ideas. First, the forward velocity control is achieved by approximating a neutral leg placement and adjusting it with a proportional error term, yielding

$$q_{\theta t} = \operatorname{asin}\left(\frac{\dot{x} t_s}{2 q_{rt}} + k_{\dot{x}}(\overline{\dot{b}_x}^* - \overline{\dot{b}_x}) \right) \qquad (18)$$

where $k_{\dot{x}}$ and the choice of $\dot{x}$ are controller parameters. Next, we implement a Raibert-like hopping height controller by supplying the appropriate energy at bottom, via a change in spring constant $\Delta E_U = U_{k_2}(r_b) - U_{k_1}(r_b)$, in order to provide the energy difference between two successive apex points. In the absence of an estimate for r_b, we use measurements from previous strides. Similar to $\dot{x}$ and $k_{\dot{x}}$, this is an estimation parameter which requires careful tuning for best performance.

Since both controllers, by their nature, will have tracking errors, we use integral feedback compensation, yielding a discrete closed loop system of the form

$$X_a[k+1] = f_a(X_a[k], u_c(X_a[k], X_a^*[k+1] + e[k]))$$
$$e[k+1] = e[k] + \frac{1}{c_i}(X_a^*[k] - X_a[k])$$

where $e[k]$ is the integral of the apex state error, $X_a^*[k]$ is the "reference" trace and $u_c(X_a, X_a^*)$ is a particular gait-level controller, in this paper, either the deadbeat or the modified Raibert controller.

3.3 Performance of the Deadbeat Controller

Even with integral compensation deadbeat control is an aggressive approach, imposing strong model dependence on the control law. In the absence of analytical results for the stability of the proposed controller in the presence of model mismatches, we explore in simulation the performance of the deadbeat controller and compare it to the benchmark of a modified Raibert control strategy. In particular, in Section 3.3.2, we begin by studying a simple SLIP, removing the assumption that gravity can be ignored during stance. We continue in Section 4.2 by considering two different four DOF legs having ankle, knee and hip joints and mass distributed throughout the leg.

Due to lack of space, this comparative study primarily focuses on the forward velocity behavior resulting from the control strategy. However, similar results are seen when considering the hopping height and duty factor behaviors [13].

3.3.1 Simulation Strategy

In this simulation study, we consider two families of waveforms we wish the apex velocity trace to track: one of step references and another of sinusoid references.

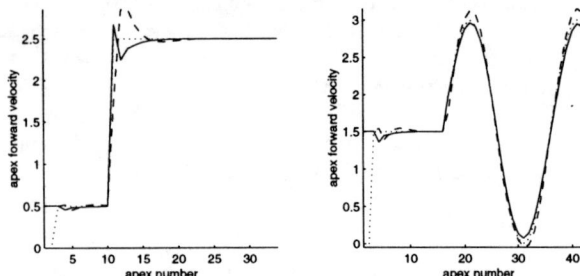

Figure 2: Sample runs of the deadbeat controller(solid lines) and modified Raibert(dashed lines) controllers applied to the 2 DOF SLIP leg for step and sinusoid references over 35 strides. Dotted lines represent the reference trace, while solid and dashed lines represent the actual performance of the SLIP runner.

Examples of both are shown in Figure 2. In each case, the hopper stabilizes around an initial running speed and the desired reference waveform is introduced at the end of 15 gait cycles.

When representing these references, we parameterize a step by its initial value and step amplitude and a sinusoid by its period and amplitude. Simulations are run over a range of these two dimensional parameter spaces. For a particular reference command, we summarize the control performance by the mean square error (MSE),

$$\text{MSE} = \frac{1}{N} \sum_{k=15}^{N} \| \dot{\bar{b}}_x^*[k] - \dot{\bar{b}}_x[k] \|^2$$

where N is the number of strides taken.

In presenting responses to these step and sinusoid reference command spaces, we collapse the initial velocity and sinusoid amplitude dimensions by averaging. In each case, 10 data points in the collapsed dimensions are chosen such that the forward velocity command always remains in the range $[0, 3]m/s$.

3.3.2 Simulation Results

Figure 3, summarizes the simulation data for step and sinusoid reference commands in forward velocity where we fix $\bar{b}_y^* = 1.2m$ and $\phi^* = 3$. The plots show the mean and variance of MSE for both controllers as a function of step amplitude(left) and sinusoid frequency(right). The results show that for this plant, the deadbeat controller provides better tracking than a modified Raibert controller. This observation about the control performance in not particular to the 2 DOF SLIP model, for we will see similar results for a 4 DOF AKH leg model in Section 4.2.

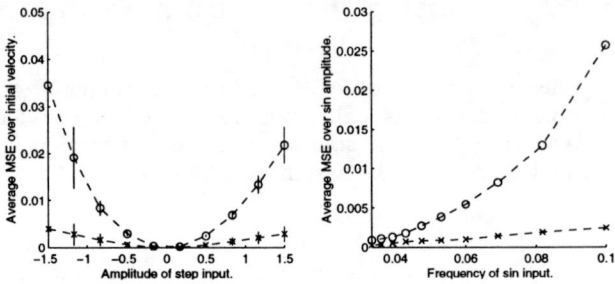

Figure 3: Step(left) and Sinusoid(right) References: The mean and variance of MSE as a function of the step amplitude(left) and sinusoid frequency(right), for the deadbeat(x) and modified Raibert(o) controllers. For this plant, $m = 50.48kg$, $\bar{b}_y^* = 1.2m$, $\phi^* = 3$.

Simulations with sinusoid reference commands reveal another property of the deadbeat controller. Due to its long settling time, the tracking error of the decoupled controller increases significantly for high frequency reference commands. The deadbeat controller, however, has shorter settling times — it ideally reaches the desired trajectory in one cycle — and consequently displays better tracking over a wide range of frequencies.

4 A More Realistic Leg Model

In this section, the application of the SLIP deadbeat controller to a much more complex dynamical leg structure, the four degree of freedom ankle/knee/hip model (Figure 1) is investigated. We consider two considerably different configurations of the four degree of freedom model: one with human-like and one with kangaroo-like kinematics and mass distribution. We present simulation evidence for the efficacy of the same approach as was used in Section 3.3.2 for the 2 DOF SLIP.

4.1 The 4 DOF AKH Leg Model

To simplify our thinking about this problem and make the application of the SLIP deadbeat controller as straightforward as possible, we consider a virtual SLIP leg connecting the toe of the 4 DOF leg to its center of mass (COM). The control objectives will remain the same as for the 2 DOF leg: the achievement of desired apex height, forward velocity and duty factor. The control implementation, however, will be considerably different, since the control inputs specified by the deadbeat controller, $u = [\, q_{\theta t},\ k_1,\ k_2\,]^T$ are not directly transferable to the control inputs of the 4 DOF leg. Furthermore there is not a one to one correspondence between the 4 DOF leg angles and $q_{\theta t}$ nor between the joint torques and the virtual leg force.

Consequently, we must develop rules for choosing posture (the leg configuration) at touchdown to achieve the desired $q_{\theta t}$ and the joint torques during stance to achieve the desired virtual leg stiffnesses, k_1 and k_2. The manner in which we use biological evidence to guide the mathematical considerations used in forming these rules is presented in the next section.

4.2 Control of the AKH Leg

In controlling the four-jointed leg, we identify two levels, a joint level torque control, and an apex level virtual leg control.

Our controller attempts to force the COM trajectory of the 4 DOF leg to mimic a SLIP leg by proper choice of touchdown joint configuration and stance torques [5]. Our objective is to develop by closed loop joint controllers a "target leg" dynamics, yielding virtual leg dynamics as close as possible to SLIP dynamics. We accomplish this by constraining the work done by the joint torques to equal the work that would be done by a virtual spring between the toe and the center of mass, yielding

$$\overline{F}^T \dot{\overline{b}} = \tau^T \dot{q} \qquad (19)$$

where $\overline{F}^T$ and $\dot{\overline{b}}$ are the virtual spring force and the center of mass velocities respectively. Note that this is substantially different from forcing the center of mass to follow a prespecified target trajectory. The actual stance trajectory is still governed by AKH dynamics.

We then combine the torque constraint of (19) with a set of symmetry constraints of the form

$$\begin{bmatrix} 1 & -1 & 1 & -1 \\ 0 & \beta & -1 & 0 \end{bmatrix} q = \begin{bmatrix} -\gamma \\ 0 \end{bmatrix} \qquad (20)$$

where β and γ are symmetry parameters, fixed for any particular locomotor. Intuitively, Equation 20 constrains the body link angle with respect to the ground to be γ, and the knee angle to be proportional to the ankle angle. In our simulations, the human-like leg has $\gamma = \pi/2$ and $\beta = 1$ and the kangaroo-like leg, has $\gamma = \pi/4$ and $\beta = 1$.

The leg configuration at touchdown is now completely specified, bridging the gap between the 4 dof leg model and the SLIP controller. As a consequence, we are able to use the controller principles explained in the preceding sections without any modifications. From the point of view of the apex controller, the combination of the torque control compensated leg dynamics are very close to SLIP dynamics.

We investigate the validity of this approach in simulation on two different 4 DOF legs, one human-like and one kangaroo-like whose structural parameters are given in Table 2.

As in Section 3.3.2 we issue step and sinusoid reference forward velocity commands and measure the tracking performance with the results shown in Figures 4 and 5, respectively. They support the validity of two major assumptions in the paper. First, they confirm that the SLIP model for running is applicable to significantly different kinematics and dynamics.

[5]Please refer to [13] for a detailed discussion.

	$[m_a, m_k, m_h, m_b]$	$[l_a, l_k, l_h, l_b]$
human	[26.4, 19.3, 3.5, 1.28]	[0.15, 0.35, 0.40, 0.35]
kangaroo	[30, 30, 5, 4]	[0.5, 0.7, 0.6, 0.5]

Table 2: Structural simulation parameters for human-like and kangaroo-like four degree of freedom legs [9].

Second, they suggest that, the connection between the SLIP model and the four-jointed complex model we consider does not rely on the particular "target pose".

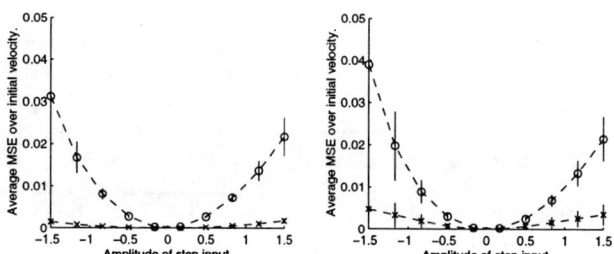

Figure 4: Step Reference: The mean and variance of MSE for human-like (left) and kangaroo-like (right) legs as a function of the step amplitude with the deadbeat(x) and modified Raibert(o) controllers. For this plant, $\overline{b}_y^* = 1.2m$, $\phi^* = 3$.

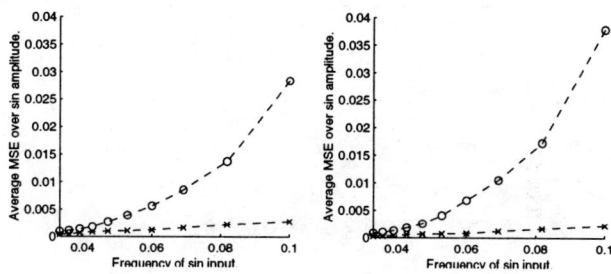

Figure 5: Sinusoid Reference: The mean of MSE for human-like (left) and kangaroo-like (right) legs as a function of the sinusoid frequency with the deadbeat(x) and modified Raibert(o) controllers. For this plant, $\overline{b}_y^* = 1.2m$, $\phi^* = 3$.

5 Conclusion

The present work serves as a tribute to the foresight of both those in the biomechanics community and those in the engineering community, such as Raibert, who have insisted that the SLIP model is the right place to begin thinking about dynamic locomotion. For not only is this model useful in describing the COM behavior of a multi-joint monoped runner as the biomechanists have claimed, but also for prescribing the control needed to achieve some desired behavior as Raibert originally intuited. In particular, in this paper the control prescription arises from the extension of the 2 DOF SLIP deadbeat control to the higher degree of freedom AKH leg.

As far as we are aware, this is the first time that the SLIP model has been shown to be applicable to more zoomorphically realistic legs. Therefore, we believe this work will be of interest to both the engineering and biomechanics communities.

5.1 Relevance to Engineering

We witness in nature that advantage is conveyed to walkers and runners with higher degree of freedom legs. As such, while Raibert's robots demonstrated remarkable abilities, it seems certain in the long term that walking and running robots must be designed with higher degree of freedom legs. But not much work has been undertaken in building multi-degree of freedom runners, presumably because of the difficulty in "getting it right". Instead, research has progressed more rapidly in the direction of high degree of freedom dynamic animations, such as the exciting work by Hodgins and her students [8]. In either case, it would be useful to design easily tunable controllers in terms of high level behaviors, such as desired speed and hopping height.

We feel that the work presented in this paper is the first step in the direction of easily implementable, provably correct task based controllers for the high degree of freedom, zoomorphically realistic problem. We are encouraged by our current successes and hope to pursue the implementation of these deadbeat inspired controllers into dynamic simulations and experimental platforms with increasing degrees of freedom.

5.2 Relevance to Biomechanics

Given the almost universal ability to characterize an animal's COM behavior by the simple SLIP model, biomechanists are beginning to question how the many degrees of freedom are coordinated to mimic the 2 DOF SLIP [6]. In other words, they would like to identify the joint level controllers that in combination give the SLIP-like behavior of the COM. Given the difficulties of such a task and the absence of any other control strategies, we feel that the multi-joint deadbeat control strategy presented in this paper may serve as a good initial guide for addressing this problem.

Acknowledgements

We thank Prof. Claire Farley for a number of informative tutorial discussions on the biomechanics of human running. We also thank Prof. Jessica Hodgins and Dr. Nancy Pollard for their help with the 4 DOF simulations, in particular for providing the mass and kinematic data needed to make the human simulations more realistic.

References

[1] R. M. Alexander. Three uses for springs in legged locomotion. *International Journal of Robotics Research*, 9(2):53–61, 1990.

[2] R. M. Alexander and A. S. Jayes. Vertical movement in walking and running. *Journal of Zoology, London*, 185:27–40, 1978.

[3] R. Blickhan. The spring-mass model for running and hopping. *Journal of Biomechanics*, 22:1217–1227, 1989.

[4] R. Blickhan and R. J. Full. Similarity in multilegged locomotion: Bouncing like a monopode. *Journal of Comparative Physiology*, 173:509–517, 1993.

[5] C. T. Farley, J. Glasheen, and T. A. McMahon. Running springs: Speed and animal size. *Journal of Experimental Biology*, 185:71–86, 1993.

[6] C. T. Farley, H. P. Houdijk, C. van Strien, and M. Louie. Mechanisms for leg stiffness adjustment during bouncing gaits. 1997. In Review.

[7] J. K. Hodgins. *Legged Robots on Rough Terrain: Experiments in Adjusting Step Length*. PhD thesis, Carnegie Mellon University, November 1989. CMU-CS-89-151.

[8] J. K. Hodgins. Three-dimensional human running. In *ICRA*, Minneapolis, MN, May 1996.

[9] J. K. Hodgins and N. S. Pollard. Typical human and kangaroo leg characteristics. Personal Communication.

[10] D. E. Koditschek and M. Bühler. Analysis of a simplified hopping robot. *International Journal of Robotics Research*, 10(6):587–605, December 1991.

[11] R. T. M'Closkey and J. W. Burdick. Periodic motions of a hopping robot with vertical and forward motion. *International Journal of Robotics Research*, 12(3):197–218, 1993.

[12] M. H. Raibert. *Legged Robots That Balance*. MIT Press, Cambridge, MA, 1986.

[13] U. Saranli and D. E. Koditschek. Analysis and control of slip and multi-jointed monoped planar hoppers. Technical report, EECS, UM, Ann Arbor, MI, 1998. In Preparation.

[14] W. J. Schwind, C. T. Farley, and D. E. Koditschek. Identification of springs in human running, 1997. Paper in Progress.

[15] W. J. Schwind and D. E. Koditschek. Control of forward velocity for a simplified planar hopping robot. In *Proceedings of the IEEE International Conference On Robotics and Automation*, Nagoya, Japan, May 1995.

[16] W. J. Schwind and D. E. Koditschek. Characterization of monoped equilibrium gaits. In *Proceedings of the IEEE International Conference On Robotics and Automation*, Albuquerque, NM, April 1997.

[17] A. F. Vakakis, J. W. Burdick, and T. K. Caughy. An 'interesting' strange attractor in the dynamics of a hopping robot. *International Journal of Robotics Research*, 10(6):606–618, December 1991.

Dynamic Model of A Gyroscopic Wheel

Gora C. Nandy[1,2] and Yangsheng Xu[1,3]

[1]Department of Mechanical and Automation Engineering, The Chinese University of Hong Kong, Hong Kong
[2]Mechanical Engineering Deparment, REC, Durgapur-713 209, India
[3]The Robotics Institute, Carnegie Mellon University, Pittsburgh, PA 15213, USA

Abstract

In this paper, we develop a dynamic model of a gyroscopic wheel, an important component of Gyrover, a single-wheel robot developed at Carnegie Mellon University. The Gyrover robot consists of a single wheel, and is actuated through a spinning flywheel attached through a two-link manipulator at the wheel bearing. The flywheel can be tilted to achieve steering, and can be driven forwards and backwards to accelerate the robot. As a first step in modeling this highly coupled, dynamically stable system, this paper focuses on developing a 3D model of the wheel part of the Gyrover. In this paper, we first describe the Gyrover robot. We then develop the dynamic model of the wheel through the Lagrangian constrained generalized formulation. Finally, we implement the resulting equations of motion and present simulation results for the unactuated Gyrover in the different gravitational environments of earth, the moon, and Mars.

1. Introduction

Gyrover is a novel, single wheel gyroscopically stabilized robot, originally developed at Carnegie Mellon University [1]. Two prototypes have already been developed, with a third currently under construction; Figures 1 and 2 show a schematic and photograph of the first prototype. Essentially, Gyrover is a sharp-edged wheel, with an actuation mechanism fitted inside the wheel. The actuation mechanism consists of three separate actuators: (1) a *spin motor*, which spins a suspended flywheel at a high rate, im-

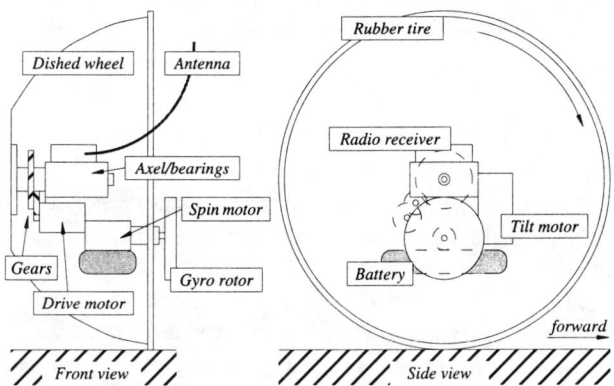

Fig. 1: A diagram of the first prototype of the Gyrover.

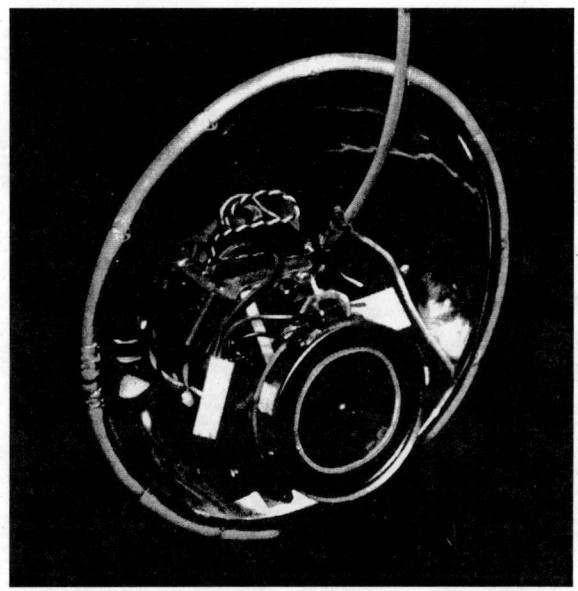

Fig. 2: Photograph of the first prototype of the Gyrover.

parting dynamic stability to the robot; (2) a *tilt motor*, which controls the angle of the spinning flywheel, and ultimately controls the steering of Gyrover; and (3) a *drive motor*, which causes forward and/or backward acceleration, by shifting Gyrover's internal pendulum mass.

The behavior of Gyrover is based on the principle of gyroscopic precession as exhibited in the stability of a rolling wheel. Because of its angular momentum, a spinning wheel tends to precess at right angles to an applied torque. Therefore, when a rolling wheel leans to one side, rather than just fall over, the gravitationally induced torque causes the wheel to precess so that it turns in the direction that it is leaning.

As a concept, Gyrover has a number of potential advantages over multi-wheeled vehicles:

1. The entire system can be enclosed within the wheel to provide mechanical and environmental protection for the equipment and actuation mechanism.

2. Gyrover is resistant to getting stuck on obstacles because it has no body to hang up, no exposed appendages (in principle), and the entire exposed surface is driven.

3. In manual control experiments, we have shown that

the tiltable flywheel can be used to right the vehicle from its statically stable rest position (on its side).
4. Gyrover can turn in place by simply leaning and precessing in the desired direction, with no special steering mechanism, thus enhancing maneuverability.
5. Single-point contact with the ground eliminates the need to accommodate uneven surfaces and potentially simplifies control.
6. Full drive traction is available because all the weight is on the single drive wheel.

Potential applications for Gyrover are numerous. We have shown that Gyrover can travel on both land and water; thus, it may find amphibious use on beaches or swampy areas, for general transportation, exploration, rescue or recreation. Similarly, with appropriate tread, it should travel well over soft snow with good traction and minimal rolling resistance. As a surveillance robot, Gyrover could use its slim profile to pass through doorways and narrow passages, and also use its ability to turn in place to maneuver in tight quarters. Another potential application is as a high-speed lunar vehicle, where the absence of aerodynamic disturbances and low gravity would permit efficient, high-speed mobility.

Thus far, the Gyrover robots have been controlled only manually, using two joysticks to control the drive and spin motors through a radio link. This is the case in part because no complete dynamic model of Gyrover has as of yet been developed. Developing such a dynamic model, while apparently tractable, involves the inherent complexities of modeling two coupled, rotating masses whose axes of rotation are misaligned in 3-dimensional space.

This paper takes the important first step in developing a complete 3-dimensional dynamic model of Gyrover by first deriving the equations of motion of an unactuated Gyrover, or in other words, a single gyroscopic wheel. A companion paper [2] subsequently incorporates the actuation mechanism with the dynamic model developed herein.

Thus, in this paper, we first provide a description of the Gyrover robot. Next, assuming rolling without slip, and point contact on a flat surface, we derive the nonholonomic constraints for a single gyroscopic wheel without actuation. We then derive the equations of motions for the wheel using the constrained generalized Lagrangian principle. Finally, we implement the dynamic equations in a real-time graphic simulator and report results for different gravitational environments, including earth, the moon and Mars.

2. Constraints of motions

2.1 Nonholonomic constraints

Rolling without slipping is a typical example of a nonholonomic system, since in most cases, some of the constrained equations for the system are nonintegrable. Thus, Gyrover is a nonholonomic system. A number of techniques for analyzing nonholonomic systems have been developed [3-7]; the *Lagrangian constrained generalized principle* is one of the better know of these methods, and we use it below for deriving the dynamic model of a gyroscopic wheel.

Let us represent a system with m generalized coordinates as,

$$q_j = (q_1, q_2, ..., q_m), j \in \{1, 2, ..., m\} \quad (1)$$

acted on by a set of m generalized forces given by,

$$Q_j = (Q_1, Q_2, ..., Q_m), j \in \{1, 2, ..., m\} \quad (2)$$

In nonholonomic systems the number of generalized coordinates exceeds the number of degrees of freedom. If there are n degrees of freedom and m generalized coordinates, there will be $(m-n)$ constraint conditions that must explicitly be satisfied by the system. In functional form, the constrained equations can be written as,

$$f_s = f_s(q) = f_s(q_1, q_2, ..., q_m, t) = 0. \quad (3)$$

We can write the corresponding time derivatives as,

$$\dot{f}_s = \sum_{j=1}^{m} \frac{\partial}{\partial q_j} f_s(q_1, q_2, ..., q_m, t) \dot{q}_j + \frac{\partial}{\partial t} f_s(q) = 0, \quad (4)$$

or, in more general form,

$$\sum_{j=1}^{m} A_{sj} \dot{q}_j + a_s = 0, s \in \{1, 2, ..., m-n\}. \quad (5)$$

For a nonholonomic system, the set of Lagrangian equations are then given by,

$$\frac{d}{dt}\left(\frac{\partial L}{\partial \dot{q}_j}\right) - \frac{\partial L}{\partial q_j} = \sum_{s=1}^{m-n} \lambda_s A_{sj}, j \in \{1, 2, ..., m\} \quad (6)$$

where $L = T - P$ is the Lagrangian function, T is the total kinetic energy of the system, P is the total potential energy of the system, and each λ_s is a Lagrangian multiplier which accounts for the system constraints.

To represent the Gyrover wheel, we require six coordinates, three for position (X, Y, Z) and three for orientation (α, β, γ). The Euler angles (α, β, γ) represent the precession, lean and spin angles of the wheel, respectively, and are illustrated in Figure 3.

2.2 Coordinate transformation

For the discussion below, let the inertial frame $\{X, Y, Z\}$ be attached to the ground x-y plane, which represents a perfectly flat surface upon which the Gyrover wheel rolls (see Figure 4). Let the body coordinate frame $\{x_B, y_B, z_B\}$ be attached to the mass center of the wheel, where z_B represents axis of rotation for the wheel. The composite rotation matrix which transforms the wheel from *state* 1 to *state* 2 in Figure 4 is given by R_c,

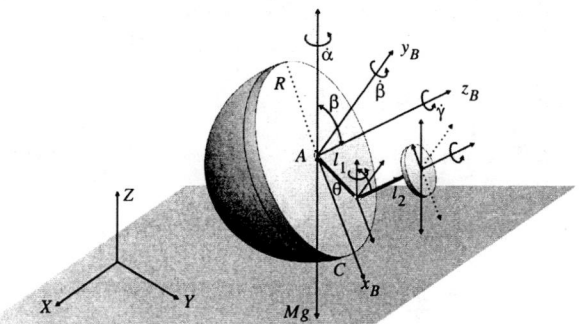

Fig. 3: Definition of system variables for Gyrover.

$$R_c = R_{(X, -\pi/2)}R_{(z_B, \pi)}R_{(x_B, -\alpha)}R_{y_B, -(\pi/2-\beta)} \quad (7)$$

$$R_c = \begin{bmatrix} -s\alpha c\beta & -c\alpha & -s\alpha s\beta \\ c\alpha c\beta & -s\alpha & c\alpha s\beta \\ -s\beta & 0 & c\beta \end{bmatrix} \quad (8)$$

where $c\alpha = \cos\alpha$, $s\alpha = \sin\alpha$, $c\beta = \cos\beta$ and $s\beta = \sin\beta$.

Now, denote $\{i, j, k\}$ and $\{l, m, n\}$ as the unit vectors along the x, y and z axes of the inertial and body frames, respectively. Then, the relationship $\{i, j, k\}$ between $\{l, m, n\}$ is given by,

$$l = -(s\alpha c\beta)i + (c\alpha c\beta)j - s\beta k \quad (9)$$

$$m = -(c\alpha)i - (s\alpha)j \quad (10)$$

$$n = -(s\alpha s\beta)i + c\alpha s\beta j + c\beta k. \quad (11)$$

2.3 Velocity constraints

Below, we derive a general expression for v_A, the velocity of the wheel's center. First, we note that,

$$\omega_B = \omega_x l + \omega_y m + \omega_z n = -\dot\alpha s\beta l + \dot\beta m + (\dot\gamma + \dot\alpha c\beta)n \quad (12)$$

where ω_B is the angular velocity of the wheel. Now, we can express v_A as,

$$v_A = \omega_B \times r_{A \leftarrow C} + v_C \quad (13)$$

where v_C is the velocity of the contact point and $r_{A \leftarrow C} = \{-Rl\}$ represents the vector from C to A in Figure 3. If we assume perfect rolling without slip, $v_C = 0$ and (13) reduces to,

$$v_A = \omega_B \times r_{A \leftarrow C} \quad (14)$$

$$v_A = \{-\dot\alpha s\beta l + \dot\beta m + (\dot\gamma + \dot\alpha c\beta)n\} \times \{-Rl\} \quad (15)$$

$$v_A = -R(\dot\gamma + \dot\alpha c\beta)m + R\dot\beta n \quad (16)$$

Transforming (16) to the inertial frame, we get the following expression for v_A:

$$v_A = \dot X i + \dot Y j + \dot Z k, \quad (17)$$

where,

$$\dot X = R(\dot\gamma c\alpha + \dot\alpha c\alpha c\beta - \dot\beta s\alpha s\beta) \quad (18)$$

$$\dot Y = R(\dot\gamma s\alpha + \dot\alpha c\beta s\alpha + \dot\beta c\alpha s\beta) \quad (19)$$

$$\dot Z = R\dot\beta c\beta \quad (20)$$

Equations (18) through (20) represent the three velocity constraint equations. The first two equations are nonintegrable and therefore nonholonomic constraint equations of the type given in (5). The last equation, however, is integrable and leads to the simple geometric or holonomic constraint,

$$Z = Rs\beta, \quad (21)$$

assuming initial conditions $Z_0 = 0$ and $\beta_0 = 0$. Therefore, we can represent Gyrover through five, rather than six independent coordinates (e.g. $\{X, Y, \alpha, \beta, \gamma\}$ or $\{X, Y, Z, \alpha, \gamma\}$).

3. Dynamic model

3.1 Equations of motion

We now derive the equations of motion by calculating the Lagrangian $L = T - P$ of the system, where T is the total kinetic energy of the system given by,

$$T = \frac{1}{2}M(\dot X^2 + \dot Y^2 + \dot Z^2) + \frac{1}{2}(I_{xx}\omega_x^2 + I_{yy}\omega_y^2 + I_{zz}\omega_z^2) \quad (22)$$

and P represents the potential energy of the system given by,

$$P = MgR\sin\beta. \quad (23)$$

Assuming the Gyrover wheel to be hemispherical,

$$I_{xx} = I_{yy} = (5/8)I_{zz}, \quad I_{zz} = (2/5)MR^2, \quad (24)$$

where M is the mass of the wheel and R is its radius. The Lagrangian function in terms of the constrained generalized coordinates then becomes,

$$L = \frac{1}{2}M\left[\dot X^2 + \dot Y^2 + \dot Z^2 + \frac{1}{4}(R\dot\alpha s\beta)^2 + \frac{1}{4}(R\dot\beta)^2 + \frac{2}{5}R^2(\dot\alpha c\beta + \dot\gamma)^2\right] - MgRs\beta \quad (25)$$

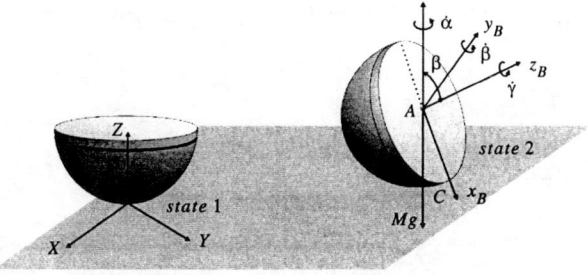

Fig. 4: Coordinate transformation from the inertial to the body coordinate frames.

From equations (5) and (6), we determine the Lagrangian equations of motion by evaluating the following expressions:

$$\frac{\partial L}{\partial \dot{X}} = M\dot{X}, \quad \frac{d}{dt}\left(\frac{\partial L}{\partial \dot{X}}\right) = M\ddot{X}, \quad \frac{\partial L}{\partial X} = 0 \quad (26)$$

$$\frac{\partial L}{\partial \dot{Y}} = M\dot{Y}, \quad \frac{d}{dt}\left(\frac{\partial L}{\partial \dot{Y}}\right) = M\ddot{Y}, \quad \frac{\partial L}{\partial Y} = 0 \quad (27)$$

$$\frac{\partial L}{\partial \dot{\alpha}} = MR^2\left[\frac{1}{4}\dot{\alpha}s\beta^2 + \frac{2}{5}(\dot{\gamma}+\dot{\alpha}c\beta)c\beta\right],$$

$$\frac{d}{dt}\left(\frac{\partial L}{\partial \dot{\alpha}}\right) = MR^2\left[\ddot{\alpha}\left(\frac{2}{5}c\beta^2+\frac{1}{4}s\beta^2\right) + \frac{2}{5}\ddot{\gamma}c\beta - \frac{2}{5}\dot{\beta}\dot{\gamma}s\beta - \frac{2}{5}\dot{\alpha}\dot{\beta}c\beta s\beta\right],$$

$$\frac{\partial L}{\partial \alpha} = 0 \quad (28)$$

$$\frac{\partial L}{\partial \dot{\beta}} = \frac{1}{4}MR^2\dot{\beta}, \quad \frac{d}{dt}\left(\frac{\partial L}{\partial \dot{\beta}}\right) = \frac{1}{4}MR^2\ddot{\beta},$$

$$\frac{\partial L}{\partial \beta} = -MR^2\dot{\alpha}\left[\frac{3}{20}\dot{\alpha}c\beta + \frac{2}{5}\dot{\gamma}\right]s\beta - MgRc\beta \quad (29)$$

$$\frac{\partial L}{\partial \dot{\gamma}} = \frac{2}{5}MR^2(\dot{\alpha}c\beta + \dot{\gamma}),$$

$$\frac{d}{dt}\left(\frac{\partial L}{\partial \dot{\gamma}}\right) = \frac{2}{5}MR^2(\ddot{\gamma}-\dot{\alpha}\dot{\beta}s\beta+\ddot{\alpha}c\beta), \quad \frac{\partial L}{\partial \gamma} = 0 \quad (30)$$

$$M\ddot{x} = \lambda_1 A_{11} + \lambda_2 A_{21} \quad (31)$$

$$M\ddot{y} = \lambda_1 A_{12} + \lambda_2 A_{22} \quad (32)$$

$$MR^2\left[\ddot{\alpha}\left(\frac{2}{5}c\beta^2+\frac{1}{4}s\beta^2\right) + \frac{2}{5}\ddot{\gamma}c\beta - \frac{2}{5}\dot{\beta}\dot{\gamma}s\beta - \frac{2}{5}\dot{\alpha}\dot{\beta}c\beta s\beta\right]$$
$$= \lambda_1 A_{13} + \lambda_2 A_{23} \quad (33)$$

$$\frac{1}{4}MR^2\ddot{\beta} + MR^2\dot{\alpha}\left[\frac{3}{20}\dot{\alpha}c\beta + \frac{2}{5}\dot{\gamma}\right]s\beta + MgRc\beta$$
$$= \lambda_1 A_{14} + \lambda_2 A_{24} \quad (34)$$

$$\frac{2}{5}MR^2(\ddot{\gamma}-\dot{\alpha}\dot{\beta}\sin\beta+\ddot{\alpha}\cos\beta) = \lambda_1 A_{15} + \lambda_2 A_{25} \quad (35)$$

$$A_{11}\dot{X} + A_{12}\dot{Y} + A_{13}\dot{\alpha} + A_{14}\dot{\beta} + A_{15}\dot{\gamma} + a_1 = 0 \quad (36)$$

$$A_{21}\dot{X} + A_{22}\dot{Y} + A_{23}\dot{\alpha} + A_{24}\dot{\beta} + A_{25}\dot{\gamma} + a_2 = 0 \quad (37)$$

Comparing the constraint equations (18) and (19) with (36) and (37) we have that,

$$A_{11} = 1, \quad A_{12} = 0, \quad A_{13} = -Rc\alpha c\beta,$$
$$A_{14} = Rs\alpha s\beta, \quad A_{15} = -Rc\alpha, \quad a_1 = 0 \quad (38)$$
$$A_{21} = 0, \quad A_{22} = 1, \quad A_{23} = -Rc\alpha s\beta,$$
$$A_{24} = -Rc\alpha s\beta, \quad A_{25} = -Rs\alpha, \quad a_2 = 0 \quad (39)$$

Thus from (6), the Lagrangian's equations of motion are,

$$M\ddot{X} = \lambda_1 \quad (40)$$

$$M\ddot{Y} = \lambda_2 \quad (41)$$

$$MR^2\left[\ddot{\alpha}\left(\frac{2}{5}c\beta^2-\frac{1}{4}s\beta^2\right) + \frac{2}{5}\ddot{\gamma}c\beta - \frac{2}{5}\dot{\beta}\dot{\gamma}s\beta - \frac{2}{5}\dot{\alpha}\dot{\beta}c\beta s\beta\right]$$
$$= -\lambda_1 Rc\alpha c\beta - \lambda_2 Rs\alpha c\beta \quad (42)$$

$$\frac{1}{4}MR^2\ddot{\beta} + MR^2\dot{\alpha}\left[\frac{3}{20}\dot{\alpha}c\beta + \frac{2}{5}\dot{\gamma}\right]s\beta + MgRc\beta$$
$$= -\lambda_1 Rs\alpha s\beta + \lambda_2 Rc\alpha s\beta \quad (43)$$

$$\frac{2}{5}MR^2(\ddot{\gamma}-\dot{\alpha}\dot{\beta}\sin\beta+\ddot{\alpha}\cos\beta) = -\lambda_1 Rc\alpha - \lambda_2 Rs\alpha \quad (44)$$

The nonlinear differential equations (40) through (44), along with the constraint equations (18) and (19) completely describe the motion for the Gyrover wheel. Below, we analyze as well as numerically simulate these equations of motion.

3.2 Precession rate

Here we derive the effective steady precession rate for Gyrover as a function of the radius of curvature ρ, the radius of the wheel R, the lean angle β, and the gravitational acceleration g. For different gravitational environments, such as on earth, the moon, or Mars, this information can be important for the automatic control of Gyrover.

To achieve steady precession (i.e. $\ddot{\alpha} = 0$), the center of the wheel must follow a circular path with radius of curvature ρ and constant lean angle β, such that $\dot{\beta} = 0$. Assume that the motion is centered about the z-axis of the inertial frame, such that,

$$X = -\rho s\alpha, \quad Y = \rho c\alpha. \quad (45)$$

Letting $\ddot{\alpha} = 0$ and $\dot{\beta} = 0$, the conditions for steady precision, the constraint equations (18) and (19) reduce to,

$$\dot{X} = R(\dot{\gamma}+\dot{\alpha}c\beta)c\alpha, \quad \dot{Y} = R(\dot{\gamma}+\dot{\alpha}c\beta)s\alpha \quad (46)$$

Differentiating equations (45) with respect to time, we get,

$$\dot{X} = -\rho\dot{\alpha}c\alpha, \quad \dot{Y} = -\rho\dot{\alpha}s\alpha. \quad (47)$$

Setting equations (46) and (47) equal to one another,

$$\left(-\frac{\rho}{R}\right)\dot{\alpha} = \dot{\gamma}+\dot{\alpha}c\beta \quad (48)$$

$$-\left(\frac{\rho}{R}+\cos\beta\right)\dot{\alpha} = \dot{\gamma} \quad (49)$$

Thus, equations (45) will be satisfied for the Lagrangian equations of motion under the condition given by equation (49). Next, differentiating equations (46) with respect to time and combining with equations (40) and (41), we get that,

$$\lambda_1 = M\rho(-\ddot{\alpha}c\alpha + \dot{\alpha}^2 s\alpha) \tag{50}$$

$$\lambda_2 = M\rho(-\ddot{\alpha}s\alpha - \dot{\alpha}^2 c\alpha). \tag{51}$$

Finally, combining equations (42), (50) and (51) and solving for $\dot{\alpha}$, we derive the desired relationship to be,

$$\dot{\alpha}^2 = \frac{20g\cot\beta}{5R\cos\beta + 28\rho} \tag{52}$$

The corresponding velocity of the wheel's center is given by $v^2 = (\rho\dot{\alpha})^2$,

$$v^2 = \frac{20g\cot\beta\rho^2}{5R\cos\beta + 28\rho} \tag{53}$$

Figures 5 and 6, plot the velocity v as a function of the radius of curvature ρ and the lean angle β, respectively, for gravitational accelerations corresponding to those of earth, Mars, and the moon. Note that for Mars and moon, whose gravitational accelerations are approximately one third and one sixth that of earth, respectively, the Gyrover wheel can be driven at significantly higher velocities through tighter turns. Also, the wheel will be relatively more stable in a lunar or Martian environment, as demonstrated by the smaller lean angles as shown in Figure 6.

4. Simulation experiments

We have solved the equations of motion numerically and simulated them for $M = 2\text{kg}$ and $R = 17\text{cm}$, and initial conditions $x_0 = 0$, $y_0 = 0$, $\alpha_0 = 0$, $\beta_0 = 80\text{deg}$, $\gamma_0 = 0$, $\dot{x}_0 = 0$, $\dot{y}_0 = 0$, $\dot{\alpha}_0 = 0$, $\dot{\beta}_0 = 0$, $\dot{\gamma}_0 = 6\text{ rad/s}$. We compare the resulting motion for gravitational accelerations g corresponding to earth, Mars and the moon.

Figures 7, 8, and 9 plot the state trajectories for a period of 20 seconds on earth, Mars, and the moon, respectively, in order of decreasing gravitational acceleration g. First, note the trajectory of the wheel on earth (Figure 7). The wheel wobbles significantly back and forth, as shown by the large variations in the lean angle β, and the irregularly shaped x-y trajectory. Now compare the trajectory on earth to that on Mars (Figure 8) and the moon (Figure 9). On Mars, the wheel precesses in a nearly perfect circular path of larger radius than on earth, and will take significantly longer than 20 seconds to eventually fall down. The x-y trajectory on the moon traces an even larger circle (twice the diameter of Mars' circle) and the wheel will take even longer to fall down in that gravitational field. Thus, we expect that a single-wheel robot such as Gyrover will have significantly greater dynamic stability as a planetary rover on either Mars or the moon rather than earth.

5. Conclusion

In this paper, we have developed a dynamic model of a gyroscopic wheel, utilizing the constrained generalized Lagrangian principle for nonholonomic systems. We have implemented the equations of motion in a real-time graphic simulator, and have simulated the dynamic behavior of the wheel for different initial conditions and different gravitational environments, such as those seen on the moon, and Mars. From these simulations, we have seen that the unactuated gyroscopic wheel has greater stability in lower gravitational fields, but ultimately needs additional actuation to maintain indefinite stability. The work in this paper is an important first step in the theoretical analysis and control of the single wheel Gyrover robot, developed at Carnegie Mellon University.

Acknowledgments

We thank Michael C. Nechyba for his help in simulating the equations of motion. We also thank Joseph Chan and W. Fung for their help in generating Figures 3 and 4.

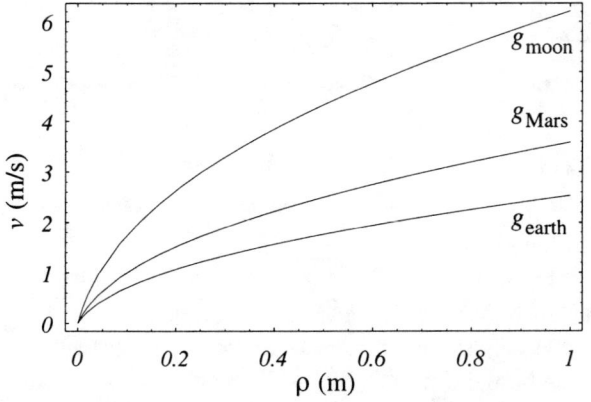

Fig. 5: Variations in the velocity as a function of the radius of curvature and different gravitational forces.

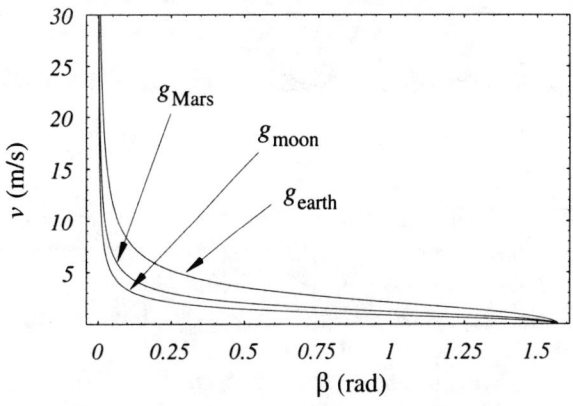

Fig. 6: Variations in the velocity for different lean angles and different gravitational forces (earth, Mars, and moon).

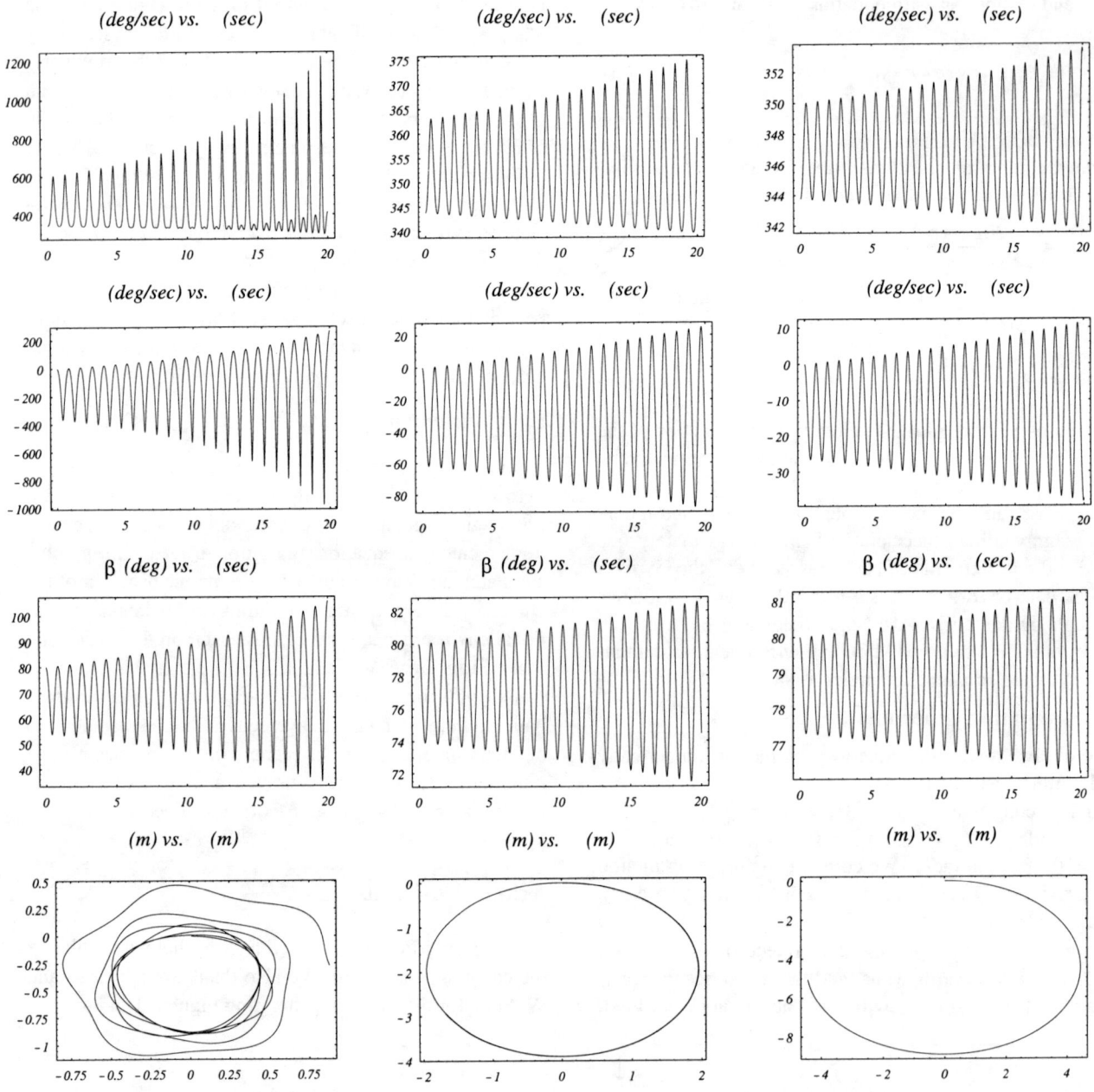

Fig. 7: State trajectories for earth's gravity.

Fig. 8: State trajectories for Mars' gravity.

Fig. 9: State trajectories for the moon's gravity.

References

[1] H. B. Brown and Y. Xu, "A single wheel gyroscopically stabilized robot," *Proc. IEEE Int. Conf. on Robotics and Automation*, vol. 4, pp. 3658-63, 1996.

[2] Y. Xu, G. C. Nandy, M. C. Nechyba and H. B. Brown, "Analysis of Actuation and Dynamic Balancing For a Single Wheel Robot," submitted to *Int. Conf. on Robotics and Automation*, 1998.

[3] G. Bianchi and W. Schiehlen, *Dynamics of Multibody Systems*, Springer Verlag, Berlin, 1986.

[4] T.R. Kane and D. A. Levinson, *Dynamics, Theory and Applications*. McGraw-Hill, New York, 1985.

[5] R. L. Huston, *Multibody Dynamics*, Butterworth-Heinemann, Boston, 1990.

[6] J. H. Ginsberg, *Advanced Engineering Dynamics*, Cambridge University Press, Cambridge, 1995.

[7] H. Goldstein, *Classical Mechanics*, 2nd ed., Addison-Wesley, Reading, Massachusetts, 1980.

OMNI-DIRECTIONAL SELF-PROPULSIVE TROWELLING ROBOT

DONG HUN SHIN[1], HO JOONG KIM[2], HO GIL LEE[3], HONG SEOK KIM[4]

[1,2] Dept. of Mechanical Engineering, the University of Seoul,
Jeonnong-Dong 90, Dongdaemun-Gu, Seoul, 130-743, Korea

[3,4] Electronics & Information Technology Research Team, KITECH,
San 17-1, Hongcheon-Ri, Ipjang-Myon, Cheon-An, 330-820, Korea

ABSTRACT

A omni-directional self-propulsive trowelling robot is proposed. The proposed robot with two rotary trowels does not require any mechanism such as wheels to obtain driving forces. When the robot flattens a concrete floor with its two rotating trowels, the unbalanced friction forces occur between the trowels and the concrete floor. These friction forces are used to move the robot. Thus, the robot can move in any direction by controlling the two rotary trowels properly.

In this paper, firstly the driving force for each trowel is computed by deriving the friction force between the trowel and the concrete floor. Secondly, the relationship between the driving force for the robot and the control variable of the robot is derived. Finally, the basic motions of the robot are realized by using the obtained relationship.

This paper figures out how the concrete floor finishing robot with two trowels moves and will contribute to realizing it.

KEYWORDS

Rotary trowel, Unbalanced friction force,
Omni-directional motion, Self-propulsive force.

1. INTRODUCTION

The concrete floor finishing robot, which is demanded mostly among the construction robots, has been developed in Japan since 1980's. It consists of the trowel and the traveling unit. The trowel has three or four rotary plastering blades and the traveling unit has the wheels or the endless tracks. Thus, the traveling control of the robot is similar to that of the mobile robot and is not so difficult. However, the semi-automatic machine developed in the United States in 1990's shows much more efficiency for concrete floor finishing, as in Fig.1.[3] It consists only of the two trowels and can move in any direction and rotates using the unbalanced frictional force between the trowel and concrete floor, while it is flattening the concrete floor.

FIGURE 1. SEMI-AUTOMATIC TROWEL MACHINE

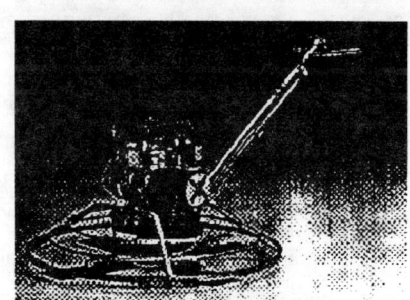

FIGURE 2. CONVENTIONAL TROWEL MACHINE

Its traveling method is similar to that of the conventional trowelling machine, as in Fig.2. When the operator wants to travel the machine, he does not need to push or pull the machine. Instead, he turns up, down, left, or right the operational handle and then, the trowel is tilted in that direction. Then, the frictional force of the tilted half of the trowel is bigger than that of the other half, thus the unbalanced force generated by the difference of the frictional forces makes the machine travel. For example, in Fig.3, if we tilt the trowel by an angle θ about the axis Y, then the trowel will move by the force $f_1 - f_2$ in the positive direction of the axis Y.

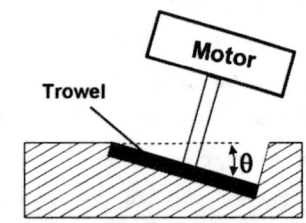

(a) FRONT VIEW OF THE TROWEL

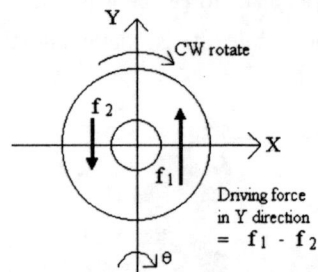

(b) BOTTOM OF THE TROWEL SEEN FROM UPSIDE

FIGURE 3. TROWEL TILTED BY θ RADIAN

Similarly, the two trowels of the recent semiautomatic machine rotate in opposite directions each other and generates the traveling frictional force by tilting themselves. However, since the operator riding on the machine experiences the vibration of engine and trowel, the fully automatic machine (robot) is very demanded now. However, it is difficult to control the trowel so that the trowel finishes the concrete floor and makes the robot to travel as wanted, simultaneously. Thus, the control problem of this robot is rather new and thus, much has not been reported, yet.

The polisher uses the rotating brush and the operational handle, and thus, is similar to the conventional trowelling machine. Furiya and Kiyohiro modeled the brush as linear springs and computed the unbalanced friction force of the polisher in 1995.[4] Shin, etc. showed the dynamics modeling and the motion control of self-propulsive polishing robot, which has the two rotating brushes and travels using the unbalanced frictional forces of the brushes.[5]

In this paper, the following are studied. Firstly, the driving force of one trowel tilted by θ radian is computed by modeling the friction force between the trowel and a concrete floor. Secondly, the relationship between the driving force acting on the trowelling robot with two rotary trowels and the control variables of the robot is derived. Finally, the basic motions, such as the forward, backward, leftward, rightward motions and the pure rotation, are realized by using the obtained relationship.

II. FRICTION FORCE MODELLING OF A TROWEL

In order to simplify the model of the friction force of a trowel, the following are assumed.

Firstly, the rotary trowel is either made of three or four plastering blades, or made of a circular steel plate of which the center part is hollow. It rotates very fast to finish plastering the concrete floor. Therefore, we assume the rotary trowel to be a center-hollowed circular steel plate.

Secondly, since the rotary trowel is made from steel and is strong enough, the amount of deformation of the trowel such as bending are negligibly small. Thus, as we tilt the trowel asymmetrically, the trowel dose not deform, but the concrete floor does.

Thirdly, the concrete deforms elastically by the following manner ;

$$N = C\delta^2 \qquad (1)$$

Where, C is a constant, and N is the compression stress, and δ is the amount of deformation. The compression strength of the concrete proper to trowelling is known to be from 0.08 to 0.4 MPa, and the concrete is not completely dried up, yet. If someone steps on this concrete floor, footmarks will be printed off distinctly on the floor. However, there has been found very few references about the wet concrete, yet. Since it is wet, the property of the concrete floor is presumed to be similar to that of a soil. Therefore, we use property of the stress-strain behavior of soils for the model of concrete floor which is trowelled. According to the result of compression tests on several soils in [6], for small stresses up to about 15MPa (2000psi), the stress-strain curves are the type of quadratic functions. Thus, we use the model for the relationship between the compression stress N and deformation δ of the concrete floor for trowelling as (1).

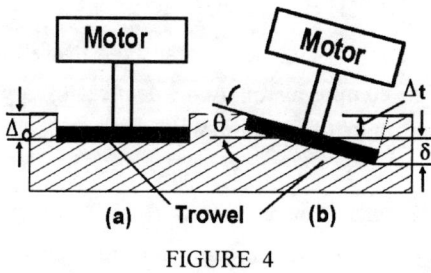

FIGURE 4

Now, let us compute the driving force generated by the friction between the trowel and the concrete floor, when we tilt the trowel by θ radian.

Fig.4(a) shows the rotating trowel which is not tilted. The floor deforms uniformly by the amount of Δ_0 due to the weight of the trowel. When we tilt the trowel by θ radian as in Fig.4(b), Δ_t is the deformation of the concrete floor at the center of the trowel. Note that the value of Δ_t is different from Δ_0, because the force-deformation relationship of the concrete floor is not linear, as shown in (1).

The neutral deformation Δ_0 of the floor can be computed as the following from (1):

$$\Delta_0 = \sqrt{\frac{Mg}{AC}}, \quad A = \pi(R_2^2 - R_1^2) \quad (2)$$

Where, A denotes the area of the floor surface which the trowel contacts, and M is the mass of the trowel, and g is the acceleration of gravity.

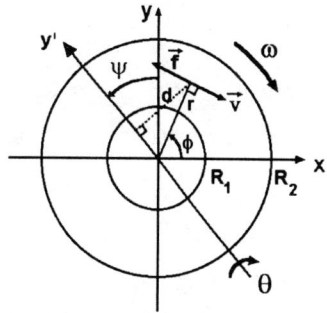

FIGURE 5

Fig.5 shows the top view of the floor contact surface of the trowel. The trowel is tilted by θ radian about the axis y'. and the deformation of the concrete floor at a point (r, Φ) is denoted as δ, and the normal force N at the point is computed as

$$N = C(\Delta_t + \delta)^2 \quad (3)$$

The length d, from the tilting axis y' to the point (r, Φ), is computed as

$$d = r\cos(\Phi - \psi) \quad (4)$$

And, since the tilting angle θ is small, the concrete deformation amount δ is expressed as

$$\delta = d\theta = r\theta\cos(\Phi - \psi) \quad (5)$$

Then, (3) can be written as

$$N = C(\Delta_t + r\theta\cos(\Phi - \psi))^2 \quad (6)$$

And the total normal force N_{total} exerted on the trowel is computed by integrating the normal force N over the area A:

$$N_{total} = C \int_{R_1}^{R_2} \int_0^{2\pi} N \, r d\Phi dr$$
$$= \pi(R_2^2 - R_1^2)C\Delta_t^2 + \frac{\pi C}{4}(R_2^4 - R_1^4)\theta^2 \quad (7)$$

Unless there is the external force except the weight of the trowel, N_{total} is constant and is equal to the weight of the trowel:

$$\pi(R_2^2 - R_1^2)C\Delta_t^2 + \frac{\pi C}{4}(R_2^4 - R_1^4)\theta^2 = \pi(R_2^2 - R_1^2)C\Delta_0^2 \quad (8)$$

Then, Δ_t is computed as the following function of θ :

$$\Delta_t = \sqrt{\Delta_0^2 - \frac{(R_2^2 + R_1^2)}{4}\theta^2} \quad (9)$$

Since the trowel rotates clockwise at a speed of ω, the friction force f at a point $(r, \varPhi)$ is computed as

$$\begin{aligned}f &= \mu N(-\sin\varPhi\ \boldsymbol{i} + \cos\varPhi\ \boldsymbol{j}) \\ &= \mu C(\Delta_t + r\theta\cos(\varPhi - \psi))^2(-\sin\varPhi\ \boldsymbol{i} + \cos\varPhi\ \boldsymbol{j})\end{aligned} \quad (10)$$

where, μ is the coefficient of friction between the trowel and the floor.

The x, y components of the entire friction force $\boldsymbol{F}$ on the trowel are computed as :

$$\begin{aligned}F_x &= \mu C \int_{R_1}^{R_2}\int_0^{2\pi} [\Delta_t + r\theta\cos(\varPhi - \psi)]^2(-\sin\varPhi)\ rd\varPhi dr \\ &= -\frac{2}{3}\pi\mu C\Delta_t(R_2^3 - R_1^3)\theta\sin\psi\end{aligned} \quad (11)$$

$$\begin{aligned}F_y &= \mu C \int_{R_1}^{R_2}\int_0^{2\pi} [\Delta_t + r\theta\cos(\varPhi - \psi)]^2 \cos\varPhi\ rd\varPhi dr \\ &= \frac{2}{3}\pi\mu C\Delta_t(R_2^3 - R_1^3)\theta\cos\psi\end{aligned} \quad (12)$$

Substituting Δ_t of (9) into (13) and (14), we can write the driving force $\boldsymbol{F}$ as :

$$\boldsymbol{F} = K_1\theta\sqrt{\Delta_0^2 - K_2\theta^2}\ (-\sin\psi\ \boldsymbol{i} + \cos\psi\ \boldsymbol{j}) \quad (13)$$

Where, $K_1 = \frac{2}{3}\pi\mu C(R_2^3 - R_1^3)$, $K_2 = \frac{R_2^2 + R_1^2}{4}$

From (13), note that when we tilt the clockwise rotary trowel by θ, the friction force on the trowel is approximately proportional to θ^2 in the direction of tilting axis y'.

Next, from (9), (10) and Fig.5, the moment T about the center of the rotary trowel by the friction forces is computed as :

$$\begin{aligned}T &= \int_{R_1}^{R_2}\int_0^{2\pi} (f \cdot r)\ rd\varPhi dr \\ &= \mu C\left[\frac{2\pi}{3}\Delta_t^2(R_2^3 - R_1^3) + \frac{\pi}{5}(R_2^5 - R_1^5)\theta^2\right] \\ &= K_1\Delta_0^2 + K_3\theta^2\end{aligned} \quad (14)$$

Where, $K_3 = \mu C\left[\frac{\pi}{5}(R_2^5 - R_1^5) - \frac{\pi}{6}(R_2^2 + R_1^2)(R_2^3 - R_1^3)\right]$

From (14), note that the moment T is proportional to θ^2. Fig.6 shows the force and the moment when the trowel is tilted about the axis y'.

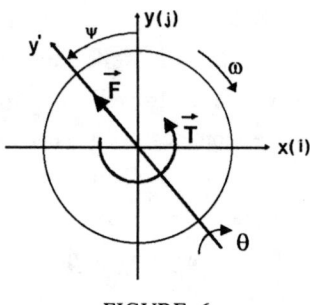

FIGURE 6

III. DYNAMICS MODELING OF THE TROWELLING ROBOT

Fig.7 shows the self-propulsive trowelling robot which is composed of two identical rotary trowels for omni-directional motion. Note that the two trowels rotates at the same speed in opposite directions each other. Thus, if the two trowels are tilted by the same amount, the moments generated by the two trowels are canceled and do not affect the motion of the robot.

The coordinate system of the robot is defined as in Fig.7(a), and its origin G is attached to the mass center of the robot. The coordinate system $x_1y_1z_1$ and $x_2y_2z_2$ are attached to the trowel 1 and trowel 2 respectively. And the fixed coordinate system $X_0 Y_0$ is defined as Fig.7(b), which shows the robot rotated by an angle Θ about the center of the robot.

In Fig.7(a), the trowel 1 is tilted by the angle θ_1 about the axis y_1 which rotates by the angle ψ_1 about the axis z_1, and similarly, the trowel 2 is tilted by the angle θ_2 about the axis y_2 which rotates by the angle ψ_2 about the axis z_2. Then, the resulting force $\boldsymbol{F}_1$ is generated in the negative direction of the axis y_1, and the resulting force $\boldsymbol{F}_2$

is generated in the positive direction of the axis y_2. And each trowel is torqued by the moments T_1 and T_2 respectively.

The entire driving force acting on the robot can be obtained by summing these F_1, F_2, T_1, and T_2. As explained in section II, the magnitude of the F_1, F_2 can be expressed as the function of θ_1, θ_2 and the direction of F_1, F_2 depends on ψ_1, ψ_2. Thus, we can change the direction of the driving force acting on each trowel by controlling ψ_1, ψ_2, and adjust the magnitude of the driving force by controlling θ_1, θ_2.

ψ_1, ψ_2. Thus, we can change the direction of the driving force acting on each trowel by controlling ψ_1, ψ_2, and adjust the magnitude of the driving force by controlling θ_1, θ_2.

Now, let us derive the relationship between the control variables of the robot, θ_1, θ_2, ψ_1, ψ_2 and the force and the moment which drive the robot.

The forces F_1 and F_2 can be expressed as the following by using (13):

$$F_1 = K_1 \theta_1 \sqrt{\Delta_0^2 - K_2 \theta_1^2} (\sin \psi_1 \, i - \cos \psi_1 \, j)$$
$$= K_1 \Gamma(\theta_1)(\sin \psi_1 \, i - \cos \psi_1 \, j)$$
$$F_2 = K_1 \theta_2 \sqrt{\Delta_0^2 - K_2 \theta_2^2} (-\sin \psi_2 \, i + \cos \psi_2 \, j) \quad (15)$$
$$= K_1 \Gamma(\theta_2)(-\sin \psi_2 \, i + \cos \psi_2 \, j)$$

$$\Delta_0 = \sqrt{\frac{M_{total} g}{2AC}}, \quad A = \pi(R_2^2 - R_1^2)$$

where, M_{total} is the total mass of the robot and Δ_0 is the neutral deflection of the concrete floor, on the assumption that each trowel is loaded by the same weight, $(M_{total} g)/2$. Since the total force F_{robot} which drive the robot can be computed by summing F_1 and F_2, its X, Y components are:

$$F_{robot\,X} = K_1(\Gamma(\theta_1)\sin\psi_1 - \Gamma(\theta_2)\sin\psi_2)$$
$$F_{robot\,Y} = K_1(-\Gamma(\theta_1)\cos\psi_1 + \Gamma(\theta_2)\cos\psi_2) \quad (16)$$

The moment T_{robot} about the center of the robot, G, can be expressed as the following:

$$T_{robot} = T_1 + T_2 + (sF_{1Y} + sF_{2Y}) \quad (17)$$

where s denotes the length from G to the center of each trowel in Fig.7(a). From (14), the moments of each trowel T_1, T_2 can be expressed as

$$T_1 = -(K_1 \Delta_0^2 + K_3 \theta_1^2), \quad T_2 = (K_1 \Delta_0^2 + K_3 \theta_2^2) \quad (18)$$

Now, from (15),(17), and (18), we can derive the moment T_{robot} about the mass center of the robot, G, as the following (19).

$$T_{robot} = [K_3(-\theta_1^2 + \theta_2^2) + sK_1(\Gamma(\theta_1)\cos\psi_1 + \Gamma(\theta_2)\cos\psi_2)] \quad (19)$$

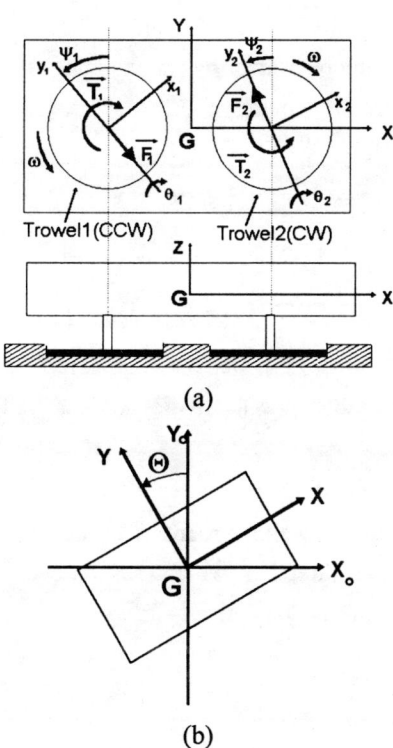

FIGURE 7. SELF-PROPULSIVE TROWELLING ROBOT

The driving force and the moment acting on the robot is shown in Fig.8. And the equations of motion for the robot are expressed as

$$M_{total}\ddot{X} = K_1(\Gamma(\theta_1)\sin\phi_1 - \Gamma(\theta_2)\sin\phi_2) \quad (20\text{-}1)$$

$$M_{total}\ddot{Y} = K_1(-\Gamma(\theta_1)\cos\phi_1 + \Gamma(\theta_2)\cos\phi_2) \quad (20\text{-}2)$$

$$I_G\ddot{\Theta} = [K_3(-\theta_1^2 + \theta_2^2) + sK_1(\Gamma(\theta_1)\cos\phi_1 + \Gamma(\theta_2)\cos\phi_2)] \quad (20\text{-}3)$$

Where I_G denotes the mass moment of inertia of the robot, and

$$K_1 = \frac{2}{3}\pi\mu C(R_2^3 - R_1^3), \quad K_2 = \frac{R_2^2 + R_1^2}{4}$$

$$K_3 = \mu C[\frac{\pi}{5}(R_2^5 - R_1^5) - \frac{\pi}{6}(R_2^2 + R_1^2)(R_2^3 - R_1^3)]$$

$$\Gamma(\theta) = \theta\sqrt{\Delta_0^2 - K_2\theta_2^2} \quad (21)$$

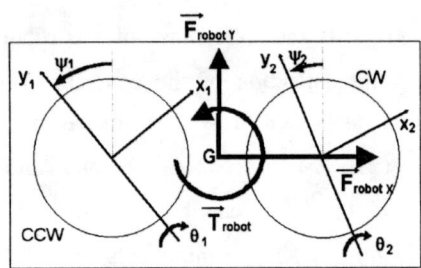

FIGURE 8. DRIVING FORCE AND MOMENT

Note that the driving force and the moment are the function of variables ϕ_1, ϕ_2, θ_1, and θ_2, and that we can drive the robot in any direction by controlling the variables ϕ_1, ϕ_2, θ_1, and θ_2. Therefore, the proposed omni-directional self-propulsive trowelling robot with 2 trowels does not need any driving mechanism such as wheels.

IV. BASIC MOTIONS OF THE TROWELLING ROBOT

In this section, we compute the values of the control variables for the basic motions, such as the pure translation and the pure rotation, by using the equations of motion for the robot, (20). In addition,
we compute the values of the control variables, ϕ_1, ϕ_2, θ_1 and θ_2 for the given arbitrary accelerations $\ddot{X}$, $\ddot{Y}$ and $\ddot{\Theta}$.

The motion of the robot is the 3 DOF motion including translation in X, Y directions and rotation. And, we have three equations in (20). But we have four unknown variables. Thus, there exist a lot of solutions. Therefore, we need one more constraint in order to obtain a unique solution. Our solution suggests the following condition.

$$-\theta_1 = \theta_2 = \theta \quad (22)$$

It means that we tilt the trowels by the same angle in opposite directions each other. The condition (22) makes the control of the robot more simpler by tilting the same magnitude, and makes the motion of the robot more stable by maintaining the symmetry of the tilts of two trowels.

Substituting (22) into (20), we can write

$$M_{total}\ddot{X} = -K_1\Gamma(\theta)[\sin\phi_1 + \sin\phi_2]$$
$$M_{total}\ddot{Y} = K_1\Gamma(\theta)[\cos\phi_1 + \cos\phi_2] \quad (23)$$
$$I_G\ddot{\Theta} = -sK_1\Gamma(\theta)[\cos\phi_1 - \cos\phi_2]$$

Now, using (23), let us find the values of the control variables for the basic motions of the robot.

1. Pure Translation

When the trowelling robot is accelerated straight in the oblique direction which has turned through η radian from Y axis about Z axis as in Fig.9, the required accelerations are :

$$\ddot{\Theta} = 0, \quad \frac{\ddot{X}}{\ddot{Y}} = -\tan\eta$$

Solving (23) under the conditions, we can obtain one of the various solutions as the following.

$$\phi_1 = \phi_2 = \eta, \quad \theta_1 = -\theta, \quad \theta_2 = \theta > 0$$

And, we compute the forces acting on the trowels as the following.

$$F_1 = F_2 = K_1\Gamma(\theta)[-\sin\eta\,\boldsymbol{i} + \cos\eta\,\boldsymbol{j}]$$

Fig.9 shows the case. The solution means that if

we tilt trowel 1 and trowel 2 by the angles $-\theta$, θ respectively about axes y_1, y_2 of the same η direction, each trowel is driven by the same forces F_1, F_2 respectively in the same η direction. This case includes the forward and sideward motions. If the robot translates to the forward, the solution is $\eta=0$, and if the robot translates to the left, the solution is $\eta=\frac{\pi}{2}$.

2. Pure Rotation

The required accelerations for this case are $\ddot{\Theta}\neq 0$, $\ddot{X}=\ddot{Y}=0$. Then, the following is one of the solutions of (23), and T_{robot} denotes the corresponding moment acting on the robot.

CCW Rotation :
$$\phi_1=\pi, \phi_2=0, \theta_1=-\theta, \theta_2=\theta > 0$$
$$T_{robot}=2sK_1\Gamma(\theta)$$

CW Rotation :
$$\phi_1=0, \phi_2=\pi, \theta_1=-\theta, \theta_2=\theta > 0$$
$$T_{robot}=-2sK_1\Gamma(\theta)$$

Fig.10 shows the counterclockwise(CCW) rotation. This means that the forces F_1, F_2 of the same magnitude act on trowel 1 and trowel 2 respectively in opposite direction each other, thus the robot rotates about the mass center of the robot.

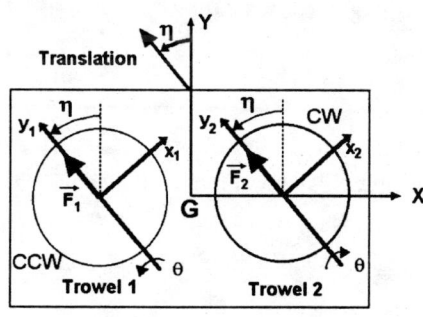

FIGURE 9. PURE TRANSLATION

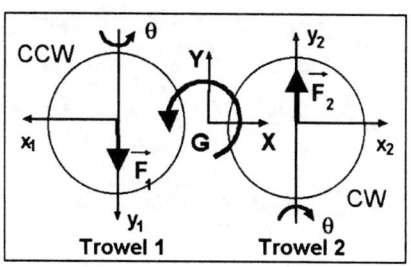

FIGURE 10. PURE ROTATION

3. Backward Solution

Now, let us compute the values of the control variables ϕ_1, ϕ_2, θ_1 and θ_2, when we are given arbitrary set of accelerations $\ddot{X}$, $\ddot{Y}$ and $\ddot{\Theta}$.

We can transform (23) to the following (24) using the rule of trigonometry.

$$\ddot{X} = -\frac{2K_1}{M_{total}}\Gamma(\theta)\sin(\frac{\phi_1+\phi_2}{2})\cos(\frac{\phi_1-\phi_2}{2})$$
$$\cdots (24\text{-}1)$$
$$\ddot{Y} = \frac{2K_1}{M_{total}}\Gamma(\theta)\cos(\frac{\phi_1+\phi_2}{2})\cos(\frac{\phi_1-\phi_2}{2})$$
$$\cdots (24\text{-}2)$$
$$\ddot{\Theta} = \frac{2sK_1}{I_G}\Gamma(\theta)\sin(\frac{\phi_1+\phi_2}{2})\sin(\frac{\phi_1-\phi_2}{2})$$
$$\cdots (24\text{-}3)$$

Dividing (24-1) and (24-3) by (24-2) and (24-1) respectively, we can eliminate θ in (24) as follows.

$$\frac{\ddot{X}}{\ddot{Y}} = -\tan(\frac{\phi_1+\phi_2}{2})$$
$$\frac{I_G\ddot{\Theta}}{sM_{total}\ddot{X}} = -\tan(\frac{\phi_1-\phi_2}{2})$$
(25)

Then, from (25), ϕ_1 and ϕ_2 are obtained as

$$\frac{\phi_1+\phi_2}{2} = \arctan(-\frac{\ddot{X}}{\ddot{Y}})$$
$$\frac{\phi_1-\phi_2}{2} = \arctan(-\frac{I_G\ddot{\Theta}}{sM_{total}\ddot{X}})$$
(26)

$$\phi_1 = \arctan(-\frac{\ddot{X}}{\ddot{Y}}) + \arctan(-\frac{I_G \ddot{\Theta}}{sM_{total}\ddot{X}})$$
$$\phi_2 = \arctan(-\frac{\ddot{X}}{\ddot{Y}}) - \arctan(-\frac{I_G \ddot{\Theta}}{sM_{total}\ddot{X}}) \quad (27)$$

Then, after computing the value $\Gamma(\theta)$ by substituting ϕ_1 and ϕ_2 in (27) into one of the three equations (23), we can find the value of θ by solving the equation (21). For example, if we substitute the values of ϕ_1 and ϕ_2 in (27) into the second equation of (23), θ can be computed as follows. Here, $\Gamma^{-1}(x)$ denotes the inverse function of $\Gamma(\theta)$ in (21).

$$-\theta_1 = \theta_2 = \theta = \Gamma^{-1}(\frac{M_{total}\ddot{Y}}{K_1(\cos\phi_1 + \cos\phi_2)}) \quad (28)$$

V. CONCLUSION

In this paper, we proposed the omni-directional self-propulsive trowelling robot with two rotary trowels, which can move in any direction and rotate by controlling the friction between the rotating trowels and the concrete floor. Thus, the proposed robot doesn't require any mechanism such as wheels which generate driving forces.

For the control of the proposed robot, the following were studied. Firstly, when a trowel rotates at a constant angular speed and it is tilted by an angle θ, we derived the driving force acting on the trowel, which is generated by the friction between the trowel and the concrete floor. The magnitude of the driving force is approximately proportional to the square of the tilting angle θ, and the direction of the force is in the direction of the tilting axis. Also, we derived the moment about the center of rotation of the trowel, and we found that the magnitude of the moment is proportional to the square of the tilting angle θ of the trowel.

Secondly, we found that the driving force of the proposed robot composed of the two rotary trowels can be controlled by adjusting the variables ϕ_1, ϕ_2, θ_1 and θ_2, and that the variables θ_1 and θ_2 are related to the magnitude of the driving force and the variables ϕ_1 and ϕ_2 are related to the direction of the driving force. And we have derived the relationship between the driving force acting on the robot and the control variables.

Finally, by using the obtained relationship, we exampled the values of the control variables for the basic omni-directional motions of the robot, such as pure translation and pure rotation.

This paper figures out how the concrete floor finishing robot with two trowels moves and will contribute to realizing it.

The real trowelling-robot system is now being built and we will be able to make actual implementation, and to show the effectiveness of the proposed robot in near future.

REFERENCES

[1] Ho Gil Lee, "The Trend of Constructional Automation", ICASE Magazine, Vol.1, No.2, pp43-48, 1995.

[2] Marvin E. Whiteman Jr., "Twin Trowel Cement Finishing Machine", U.S. Patent No. 4312603, 1982.

[3] Dewayne Allen, "Riding Trowel For Concrete Finishing", U.S. Patent No.5238323, 1993.

[4] Hiroshi Furiya and Noriaki Kiyohiro, "Floor Polishing Robot Driven by Self Propulsive Force", JRSJ, Vol. 13, No 16, pp120-125, Sept. 1995.

[5] Dong Hun Shin and Ho Joong Kim, "Self-propulsive Floor Polishing Robot", Proceedings of IASTED Int. Conf. on Applied Modeling and Simulation., pp284-290, July, 1997.

[6] T. William Lambe, Robert V. Whitman, "Soil Mechanics", John Wiley & Sons, Inc., pp122-125, 1969

Frequency Modulation In Anthropomorphic Robots With Kinematic And Force Redundancies

Byung-Ju Yi[1], Sang-Rok Oh[2], Il Hong Suh[3], Whee Kuk Kim[4]

[1]Dept. of Control & Instrumentation Eng., Hanyang Univ. Korea
[2]Div. of Electronics and Information Tech. KIST, Korea
[3]Dept. of Electronics Eng., Hanyang Univ. Korea
[4]Dept. of Control & Instrumentation Eng., Hanyang Univ. Korea

Abstract

Typical biomechanical systems such as human body and mammals possess abundant muscles which are more than required for motion generation of such systems. The purpose of this work is to verify a biological phenomenon, the so called frequency modulation, in a mathematical manner. The frequency modulation represents a simultaneous control of force and kinematic redundancies. The phenomenon of frequency modulation is explained through a human-like anthropomorphic robot. A load distribution method for frequency modulation via redundant actuation is also introduced. To show the effectiveness of the proposed algorithms, several simulation results are illustrated.

1. Introduction

Mobility (M) of a system is defined as the number of independent variables which must be specified in order to locate its elements relative to another. When M is greater than the degree-of-freedom of the operational (or task) space, the system is called *a kinematically redundant system*[1-10]. On the other hand, when the number of actuators is greater than M (this situation usually happens in closed-chain linkage systems), the system is called *redundantly actuated* For instance, five-bar mechanisms require at least two actuators (that is, $M=2$) to control the motion at the end-point. When this mechanism is driven by more than two actuators for the same motion generation, the system becomes redundantly actuated. General biomechanical systems including the human body as well as the bodies of mammals and insects are redundantly actuated. For example, mobility of the human upper-extremity (arm) shown in Figure 1 has 6 human actuators (i.e., muscles)[4], while mobility of this system is two. Accordingly, it has 4 redundant actuators. Now, a question may arise for the necessity of those hyper-redundant actuators. Hogan [12] explained the existence of the hyper-redundant muscles of the human body in terms of the spring-like impedance property. Since the neural feedback control of dynamic behaviour is severely curtailed by the inevitable time delays associated with the neural transmission, neural feedback is only effective below a certain frequency. Thus, he mentioned that synergistic activation of antagonistic muscles might be a single unified approach for controlling dynamic interaction at all frequencies. Conclusively, the spring-like impedance property comes from the antagonistic activation of hyper-redundant muscles. He also pointed out that the inertia property could be another important impedance property. Since the inertia property at the end-point of the human arm is a function of the joint angles of the system, the human arm with kinematically redundant structure is able to modulate the inertia property by its self-motion without changing the end-position of the arm. The damping impedance property is also a function of the joint angles, and thus modulation of the damping impedance is similar to that of the inertia property. Inspired by those ideas stemming from the biomechanics research fields, Yi and Freeman [1] proposed a general methodology for actively adjustable springs, and conditions for actively adjustable springs are derived for general redundantly actuated closed-chain mechanisms. Also, they successfully applied those conditions to the analysis of redundantly actuated linkage systems [2, 3, 6] and some of the biomechanical models in terms of the spring-like impedance property[5]. However, previous research was restricted to analysis of kinematically nonredundant structures such as five-bar mechanisms.

In this work, we will consider a kinematically redundant anthropomorphic robotic mechanism as shown in Fig. 2, which only controls the x- and y-directional motions at the task space. The system is driven by six linear actuators around the first- and the second-joints and one rotary actuator at the third joint. In this case, the effective inertia property at the task point can be controlled by self-motion at the joint space.

The organization of this paper is as follows : Initially, we introduce the kinematic and dynamic modeling methodology for the proposed anthropomorphic robotic mechanism, in section 2. In section 3, the frequency modulation algorithm and the associated load distribution algorithm will be proposed. In section 4, the effectiveness of the of frequency modulation is proven through simulation. Finally, we draw conclusions.

2. Kinematic Modeling

The modeling methodology integrates the Generalized Principle of D'Alembert with the method of kinematic influence coefficients(*KIC*) resulting in closed form vector expressions[11].

2.1 Open-chain kinematics

Assume that a closed-chain mechanism consists of R chains. Adopting the standard Jacobian representation for the velocity of a vector of N dependent (output) parameters u in terms of a set of M independent input coordinates $_r\dot{\theta}$ of rth open-chain, one has

$$\dot{u} = [\,_r G^u_\theta\,]\,_r\dot{\theta}_a. \tag{1}$$

Here,

$$[\,_r G^u_\theta\,] = \left[\,\frac{\partial u}{\partial_r\theta_1},\frac{\partial u}{\partial_r\theta_2},\cdots,\frac{\partial u}{\partial_r\theta_N}\,\right] \tag{2}$$

is the Jacobian relating the coordinates u and $_r\dot{\theta}$ and is of dimension of $N\times M$. Generally, the acceleration vector $\ddot{u}$ of a set of N dependent parameters u is represented in terms of the M independent coordinates ϕ as

$$\ddot{u} = [\,_r G^u_\theta\,]\,_r\ddot{\theta} + \,_r\dot{\theta}^T[\,_r H^u_{\theta\theta}\,]\,_r\dot{\theta}, \tag{3}$$

where the second-order *KIC* array $[\,_r H^u_{\theta\theta}\,]$ is of dimension of $N\times M\times M$, and the ith plane of $[\,_r H^u_{\theta\theta}\,]$, $[\,_r H^u_{\theta\theta}\,]_{i;:}$ with dimension of $M\times M$ is defined as

$$[\,_r H^u_{\theta\theta}\,]_{i;:} = \left[\frac{\partial^2 u_i}{\partial_r\theta\partial_r\theta}\right]. \tag{4}$$

2.2 Internal kinematics for six closed-chains

Consider a human-like anthropomorphic robotic manipulator shown in Fig. 2. This system has two typical features. As the first feature, it has six closed-kinematic chain as shown in Fig. 3. As the second feature, this system has one kinematic redundancy. Since the mobility of this mechanism is given as three, three actuators are minimally required to control the mechanism. However, we assume that all linear actuators are activated.

For the six closed-chains, internal kinematic relationships between the dependent joints and the independent joints is required to deal with our further analysis of the given robotic mechanism. Specifically, consider the second five-bar of Fig. 3. Note that the two open-chains of the 5-bar mechanism have a common kinematic relation at the end-location (x, y) of the linear actuator d_3. Then, the kinematic constraint equations are written by

$$x = l_1 c_1 + b_1 c_{12} = a_1 + d_2 c_5, \tag{5}$$

$$y = l_1 s_1 + b_1 s_{12} = d_2 s_5. \tag{6}$$

Differentiating Eqs. (5) and (6) with respect to time, the equivalent velocity and acceleration relations are, respectively, given by

$$[\,_1 G^u_\theta\,]\,_1\dot{\theta} = [\,_2 G^u_\theta\,]\,_2\dot{\theta}, \tag{7}$$

and

$$[\,_1 G^u_\theta\,]\,_1\ddot{\theta} + \,_1\dot{\theta}^T[\,_1 H^u_{\theta\theta}\,]\,_1\dot{\theta}$$
$$= [\,_2 G^u_\theta\,]\,_2\ddot{\theta} + \,_2\dot{\theta}^T[\,_2 H^u_{\theta\theta}\,]\,_2\dot{\theta}, \tag{8}$$

where $[\,_1 G^u_\theta\,]$ and $[\,_2 G^u_\theta\,]$, respectively, imply the Jacobians of the first and second open-chain, and $[\,_1 H^u_{\theta\theta}\,]$ and $[\,_2 H^u_{\theta\theta}\,]$ denote the Hessian arrays of the first and second open-chain of the system.

Selecting the joints θ_1 and θ_2 as the independent joints (θ_a) and the joints θ_5 and d_3 as the dependent joints (θ_b), Eq. (8) can be rewritten as

$$[A]\,\dot{\theta}_b = [B]\,\dot{\theta}_a, \tag{9}$$

where

$$[A] = [\,_2 G^u_\theta\,], \tag{10}$$

$$[B] = [\,_1 G^u_\theta\,], \tag{11}$$

$$\dot{\theta}_a = \begin{pmatrix}\dot{\theta}_1 \\ \dot{\theta}_2\end{pmatrix},\quad \dot{\theta}_b = \begin{pmatrix}\dot{\theta}_5 \\ \dot{d}_3\end{pmatrix}. \tag{12}$$

Now, premultiplying the inverse of the matrix [A] to both sides of Eq. (9) yields

$$\dot{\theta}_b = [G^b_a]\,\dot{\theta}_a, \tag{13}$$

where $[G^b_a]_{2\times 2}$ denotes the first-order *KIC* matrix relating $\dot{\theta}_b$ to $\dot{\theta}_a$.

Now, the velocity of the linear actuator is extracted from Eq. (14) as follows

$$\dot{d}_2 = [G^b_a]_{2:}\,\dot{\theta}_a, \tag{14}$$

where $[G^b_a]_{2:}$ denotes the *2nd* row of $[G^b_a]$. The same procedure is applied to the rest of the closed-chains to obtain the relationship similar to Eq. (14). Finally, the relationship between the independent joints and the six linear joints is constructed as

$$\dot{d} = [G^d_a]\,\dot{\theta}_a, \tag{15}$$

where

$$\dot{d} = \begin{pmatrix}\dot{d}_1 & \dot{d}_2 & \dot{d}_3 & \dot{d}_4 & \dot{d}_5 & \dot{d}_6\end{pmatrix}^T \tag{16}$$

and

$$[G^d_a]^T = \begin{bmatrix} g_1^{d1} & g_1^{d2} & g_1^{d3} & g_1^{d4} & 0 & 0 \\ g_2^{d1} & g_2^{d2} & 0 & 0 & g_2^{d5} & g_2^{d6} \end{bmatrix}^T. \tag{17}$$

In Eq. (17), g_i^{dj} denotes $\dfrac{\partial d_j}{\partial \theta_i}$.

The second-order, three-dimensional *KIC* array $[H^d_{aa}]_{6\times 2\times 2}$ relating $\ddot{d}$ to $\dot{\theta}_a$ can be easily obtained in a similar manner as follows [1] :

$$\ddot{d} = [G^d_a]\,\ddot{\theta}_a + \dot{\theta}_a^T[H^d_{aa}]\,\dot{\theta}_a. \tag{18}$$

Also, $[H^d_{aa}]$ is defined as

$$[H^d_{aa}]_{1;:} = \begin{bmatrix} h^{d1}_{11} & h^{d1}_{12} \\ h^{d1}_{12} & h^{d1}_{22} \end{bmatrix}, \quad [H^d_{aa}]_{2;:} = \begin{bmatrix} h^{d2}_{11} & h^{d2}_{12} \\ h^{d2}_{12} & h^{d2}_{22} \end{bmatrix},$$

$$[H^d_{aa}]_{3;:} = \begin{bmatrix} h^{d3}_{11} & 0 \\ 0 & 0 \end{bmatrix}, \quad [H^d_{aa}]_{4;:} = \begin{bmatrix} h^{d4}_{11} & 0 \\ 0 & 0 \end{bmatrix},$$

$$[H^d_{aa}]_{5;:} = \begin{bmatrix} 0 & 0 \\ 0 & h^{d5}_{22} \end{bmatrix}, \quad [H^d_{aa}]_{6;:} = \begin{bmatrix} 0 & 0 \\ 0 & h^{d6}_{22} \end{bmatrix}, \quad (19)$$

where

$$h^{dk}_{ij} = \frac{\partial^2 d_k}{\partial \theta_i \partial \theta_j}. \quad (20)$$

According to the duality between the velocity vector and force vector, the force relation between the independent joints and the linear joints is described by

$$T_a = [G^d_a]^T F_d, \quad (21)$$

where

$$T_a = (T_1 T_2)^T, \quad F_d = (F_1 F_2 F_3 F_4 F_5 F_6)^T. \quad (22)$$

Then, when activating all linear actuators, the effective load referenced to the independent joints is expressed by

$$T_a^* = T_a + [G^d_a]^T F_d = [G^d_a]^T F_d, \quad (23)$$

where T_a is equal to zero vector since the independent joints are passive joints.

2.3 Forward Kinematics

For the system given in Fig. 2, three revolute joints θ_1, θ_2 and θ_3 have been decided as the independent joints. If only the output positions are vector is controlled, they are given in terms of the independent input vector as follows

$$x = l_1 c_1 + l_2 c_{1+2} + l_3 c_{1+2+3}, \quad (24)$$

$$y = l_1 s_1 + l_2 s_{1+2} + l_3 s_{1+2+3}. \quad (25)$$

Now, the first-order forward kinematics is obtained as

$$\dot{u} = [G^u_a] \dot{\theta}_a, \quad (26)$$

where the dimension of $[G^u_a]$ is 2×3. One kinematic redundancy existing in the joint space can be employed to change the kinematic and dynamic characteristics. In the following discussion, we employ the kinematic redundancy to adjust the inertia property at the task space by changing the manipulator configuration. The second-order forward kinematic array $[H^u_{aa}]$ is also straightforward [11].

2.4 Dynamic modeling for anthropomorphic manipulator

We assume that the last revolute joint is locked once a desired manipulator configuration is achieved. In that case, the dimension of $[G^u_a]$ becomes 2×2 by extracting the third column of itself. Then, the inverse of $[G^u_a]$ (i.e., $[G^a_u]$) from Eq. (26) is uniquely decided.

Using the principle of virtual work, the generalized inertial loads of an M-link open-chain as referenced to the M relative joint parameters are given as [11]

$$_rT_\theta = [_rI^*_{\theta\theta}]_r\ddot{\theta} + _r\dot{\theta}^T[_rP^*_{\theta\theta\theta}]_r\dot{\theta}, \quad (r=1\text{-}6) \quad (27)$$

where $[_rI^*_{\theta\theta}]$ and $[_rP^*_{\theta\theta\theta}]$ denote the effective inertia matrix and the inertia power array, respectively.

Now, employing the principle of virtual work, the open-chain dynamics can be directly incorporated into closed-chain dynamics according to

$$T_\theta^T \delta\theta = T_a^T \delta\theta_a. \quad (28)$$

The total system dynamics is obtained using Eqs. (27) and (28) as follows:

$$T_a^* = [G^\theta_a]^T T_\theta$$
$$= [I^*_{aa}] \ddot{\theta}_a + \dot{\theta}_a^T [P^*_{aaa}] \dot{\theta}_a, \quad (29)$$

where the inertial matrix $[I^*_{aa}]$ and inertia power array $[P^*_{aaa}]$ defined in the independent joint set are given by

$$[I^*_{aa}] = \sum_{r=1}^{2} [_rG^\theta_a]^T [_rI^*_{\theta\theta}][_rG^\theta_a], \quad (30)$$

$$[P^*_{aaa}] = \sum_{r=1}^{2} \{([_rG^\theta_a]^T[_rI^*_{\theta\theta}]) o [_rH^\theta_{aa}] \quad (31)$$
$$+ [_rG^\theta_a]^T([_rG^\theta_a]^T o[_rP^*_{\theta\theta\theta}])[_rG^\theta_a]\},$$

and $[_rG^\theta_a]$ and $[_rH^\theta_{aa}]$, respectively, denote the first-order and the second-order kinematic influence coefficient matrices relating the joints of rth serial chain to the independent joints of the system.

Now, the dynamic formulation with respect to the output(task or operational) coordinates is obtained by employing the coordinate transformation technique between the minimum coordinates and the task coordinates:

$$T_u = [I^*_{uu}] \ddot{u} + \dot{u}^T [P^*_{uuu}] \dot{u}, \quad (32)$$

where

$$[I^*_{uu}] = [G^a_u]^T [I_{aa}][G^a_u], \quad (33)$$

$$[P^*_{uuu}] = [G^a_u]^T([G^a_u]^T o[P_{aaa}])[G^a_u]$$
$$+ ([G^a_u]^T[I_{aa}]) o[H^a_{uu}] \quad (34)$$

and T_u denotes the load vector at the output position, and $[I^*_{uu}]$, $[P^*_{uuu}]$ represent the inertial matrix, inertia power array defined in the output position, respectively.

3. Feedforward Frequency Modulation

3.1 Frequency modeling

In a state of static equilibrium, Eq. (23) can be described by

$$T_a^* = [G^d_a]^T F_d = 0. \quad (35)$$

Given a disturbance to the system under force equilibrium, a spring-like behaviour occurs to the system. Assuming that the magnitude of F_d remains constant, the effective stiffness matrix $[K_{aa}]$ with respect to the independent coordinates is obtained by differentiating Eq. (35) with respect to the independent coordinate set θ_a [1]

$$[K_{aa}] = (-F_d)^T o[H^d_{aa}]^T, \quad (36)$$

where $[H^d_{aa}]$ is given in Eq. (19). Noting that the stiffness relationship between the output coordinates and the independent coordinates is given by

$$[K_{uu}] = [G_u^a]^T[K_{aa}][G_u^a], \quad (37)$$

substituting Eq. (36) into Eq. (37) yields the following stiffness matrix expressed with respect to the output space

$$[K_{uu}] = (-\boldsymbol{F_d})^T o[H_{uu}^d], \quad (38)$$

where

$$[H_{uu}^d] = [G_u^a]^T[H_{aa}^d][G_u^a]. \quad (39)$$

Given a small displacement to the system in a state of static equilibrium ($\dot{\boldsymbol{u}} = 0$), the dynamic equation of the system is given, from Eq. (40), as

$$[I_{uu}^*]\delta\ddot{\boldsymbol{u}} = \boldsymbol{T_u}, \quad (40)$$

where

$$\boldsymbol{T_u} = \Delta([G_u^d]^T\boldsymbol{F_d})$$
$$= (\boldsymbol{F_d}^T o[H_{uu}^d]^T)\delta\boldsymbol{u}. \quad (41)$$

The above equation is rearranged as

$$[I_{uu}^*]\delta\ddot{\boldsymbol{u}} + [K_{uu}]\delta\boldsymbol{u} = 0, \quad (42)$$

where $[K_{uu}]$ is the stiffness matrix given in Eq. (38). Premultiplying $[I_{uu}^*]^{-1}$ to both sides of Eq. (43) yields

$$\delta\ddot{\boldsymbol{u}} + [I_{uu}^*]^{-1}[K_{uu}]\delta\boldsymbol{u} = 0, \quad (43)$$

where the frequency matrix $[w_{uu}]$ is defined according to

$$[w_{uu}][w_{uu}]^T = [I_{uu}^*]^{-1}[K_{uu}] \quad (44)$$
$$= (-\boldsymbol{F_d})^T o([I_{uu}^*][H_{uu}^d]).$$

Since the frequency matrix is diagonal, Eq. (44) can be written in a matrix form

$$\boldsymbol{w_u} = [W_u^d]\boldsymbol{F_d}, \quad (45)$$

where $\boldsymbol{w_u}$ and $[W_u^d]$ are defined as

$$\boldsymbol{w_u} = \begin{bmatrix} w_{xx}^2 \\ w_{xy}^2 \\ w_{yy}^2 \end{bmatrix} = \begin{bmatrix} [w_{uu}]_{1;1}^2 \\ [w_{uu}]_{1;2}[w_{uu}]_{1;2}^T \\ [w_{uu}]_{2;2}^2 \end{bmatrix} \quad (46)$$

and

$$[W_u^d] = \begin{bmatrix} \{[I_{uu}^*]^{-1}([G_u^a]^T[H_{aa}^d][G_u^a])\}_{;1;1} \\ \{[I_{uu}^*]^{-1}([G_u^a]^T[H_{aa}^d][G_u^a])\}_{;1;2} \\ \{[I_{uu}^*]^{-1}([G_u^a]^T[H_{aa}^d][G_u^a])\}_{;2;2} \end{bmatrix}. \quad (47)$$

3.2 Load Distribution Algorithm

Yi and Freeman[1] derived necessary conditions for stiffness modulation by antagonistic preloading in redundantly actuated systems. According to those conditions, a planar closed-chain system having one closed-loop has two nonlinear constraint equations, which allows modulation of the same number of stiffness elements at least. Since the anthropomorphic manipulator has six independent closed-chains, twelve nonlinear holonomic equations satisfies full modulation of three stiffness elements (that is, $k_{xx}, k_{xy},$ and k_{yy}) in the task space.

Note that the frequency matrix is nothing but a stiffness matrix weighted by the inertia matrix as seen from Eq. (44). Therefore, three components (that is, ω_{xx}, ω_{xy}, and ω_{yy}) of $\boldsymbol{w_u}$ vector can be independently controlled. In regard of the number of actuators, at least, five actuators are necessary to control the motion in the x- and y-directions and the three frequency components. In simulation, the magnitudes of ω_{xx}, ω_{xy}, and ω_{yy} are to be controlled simultaneously under static equilibrium.

Now, in order to modulate the desired $\boldsymbol{w_u}$ in static equilibrium, a load distribution method is introduced in this section. Initially, combine Eqs. (35) and (45) in a matrix form, given by

$$\begin{bmatrix} [G_a^d]^T \\ -[W_u^d] \end{bmatrix} \boldsymbol{F_d} = \begin{bmatrix} 0 \\ \boldsymbol{w_u} \end{bmatrix}. \quad (48)$$

Then, the general solution of Eq. (48) is described by

$$\boldsymbol{F_d} = [G_{com}]^+ \boldsymbol{a} + ([I] - [G_{com}]^+[G_{com}])\boldsymbol{\varepsilon}, \quad (49)$$

where $[G_{com}]^+$ denotes a pseudo-inverse solution of $[G_{com}]$, and $[G_{com}]$ and $\boldsymbol{a}$ are given as

$$[G_{com}] = \begin{bmatrix} [G_a^d]^T \\ -[W_u^d] \end{bmatrix} \quad (50)$$

and

$$\boldsymbol{a} = \begin{bmatrix} 0 \\ \boldsymbol{w_u} \end{bmatrix}. \quad (51)$$

Also, the second-term of Eq. (49) represents a homogeneous solution which creates an internal loading. This internal loading can be utilized for additional subtasks.

4. Simulation

The oscillation of the anthropomorphic manipulator with respect to a desired position can be achieved by activating the system actuators such that the system has the desired frequency characteristics at the equilibrium position, with the initial position of the end-point being deviated from the equilibrium position by the amount of the desired amplitude of oscillation. One kinematic redundancy can be utilized to change the configuration(i.e., self-motion) at the same task position, which is required to avoid obstacles or change the inertia property at the task position which enables the modulation of the motion frequency of the system in a more active manner since the inertia property can be possibly modulated by the self-motion using the kinematic redundancy of the kinematically redundant structure. Here, we assume that the last revolute joint is locked once a desired manipulator configuration is achieved.

The kinematic and dynamic parameters for the anthropomorphic manipulator are given by

$l_1 = 0.2\,m, \quad l_2 = 0.1\,m, \quad l_3 = 0.05\,m$

$a_1 = a_2 = 0.1\,m, a_3 = a_5 = 0.2\,m$

$a_4 = a_6 = 0.15\,m$

$b_1 = b_2 = b_4 = b_5 = 0.01 m$, $b_3 = 0.1 m$,
$b_6 = 0.2 m$
$m_1 = 0.5 kg$ $m_2 = 0.3 kg$ $m_3 = 0.1 kg$
$I_{z1} = 0.0017 kg \cdot m^2$ $I_{z2} = 0.00056 kg \cdot m^2$
$I_{z3} = 0.000021 kg \cdot m^2$, (52)

where the origin and insertion points for each linear actuator are determined based on the observation of the structure of the human upper extremity.

Given an initial displacement to the x-direction by $1 cm$, Fig. 4 shows the vibration response about a equilibrium position $(x, y) = (0.1\ 0.15) m$ with a hand orientation $\Phi = -90°$. In this simulation, w_{xy} is modulated as zero, and ω_{xx} and ω_{yy} are modulated identically as 4 rad/sec steadily. On the other hand, Fig. 5 shows the vibration response about the same equilibrium position $(x, y) = (0.1\ 0.15) m$ and the same frequency modulation with a different hand orientation $\Phi = -45°$. A slight oscillation in the y direction is observed. It tells us that the amount of dynamic coupling between the x- and y-directions depends upon the manipulator configuration. Therefore, in order to use the frequency modulator in a wide range of workspace, an optimization procedure to maximize the frequency modulation throughout the workspace should be performed.

Now, Fig. 6 illustrates the frequency modulation with doubled magnitude of the motion frequency. Just as expected, the frequency content of Fig. 6 is two times of that of Fig. 4 and 5.

It is remarked that frequency modulation will be useful in several complex assembly applications. For example, a certain assembly work such as peg-in-hole problem can be easily performed by inducing vibration to the grasped object. Also, the concept of frequency modulation can be employed for virtual trajectory planning. The spring-like property coupled with the inertia of the human body defines the property of frequency content. Thus, a certain frequency content of the system behaviour is directly converted to the spring-like property. It has been mentioned [12] that modulation of the spring-like property could be employed to produce movement of the system. The production of the movement can be accomplished by a progressive movement of the equilibrium position, which is called a virtual trajectory. The principle significance of the virtual trajectory is that it implies a drastic reduction in the computational effort required to obtain the inverse dynamics for movement generation of robot manipulators. Besides, the applications of motion frequency modulation will be diverse, which will be a future research topic.

5. Conclusions

Typical bio-systems are known to have abundant actuators. In this work, we investigate an anthropomorphic robotic manipulator which has the human-like musculoskeletal structure driven by abundant redundant actuators. The phenomenon of actively adjustable frequency modulation is explained through the example model. The concept of the frequency modulation has been presumed by biomechanics area., but has not been mathematically formulated yet. Possible applications of frequency modulator can be found in robotic fields, which needs to be further investigated. We believe that exploration of existing biological systems provides us with many beneficial ideas in the design and control of advanced robotic systems.

Acknowledgement

This work has been supported by Korean National Science Foundation.

References

[1] Yi, B-J. and Freeman, R.A., Geometric analysis of antagonistic stiffness in redundantly actuated parallel mechanisms. *Special Issues on Parallel Closed-Chain Mechanism, Journal of Robotic systems* **Vol. 10**, 581-603 (1993).

[2] Yi, B-J., Suh, I.H., and Oh, S-R., Analysis of a five-bar finger mechanism having redundant actuators with applications to stiffness and frequency modulation. *IEEE Proceeding on Robotics and Automation Conference*, (1997).

[3] Yi, B-J., Oh, S-R., and Suh, I.H., Synthesis of actively adjustable frequency modulators : The case for a five-bar finger mechanism. *IEEE/RSJ Proceeding on IROS*, (1997).

[4] Spence, P.A., *Basic human anatomy.* The Benjamin/Cummings Publishing Co. Inc. (1986).

[5] Yi, B-J. and Freeman, R.A., Feedforward spring-like impedance modulation in human arm models, *IEEE Proceeding on Robotics and Automation Conference*, pp. 3121-3128, (1995).

[6] Yi, B-J. and Freeman, R.A,. Synthesis of actively adjustable springs by antagonistic redundant actuation. *Trans. of the ASME, Journal of Dynamic Systems, Measurement, and Controls*, **Vol. 114**, pp. 454-461, (1992).

[7] Nakamura, Y. and Ghodoussi, M., Dynamic computation of closed-link robot mechanisms with nonredundant and redundant actuators. *IEEE Journal of Robotics and Automation* **Vol. 5**, 294-302 (1989).

[8] Nahon, M.A. and Angeles, J., Force optimization in redundantly-actuated closed kinematic chains. *IEEE Proceeding on Robotics and Automation Conference*, pp. 951-956 (1989).

[9] Kumar, V.J. and Gardner, J., Kinematics of redundantly actuated closed chain. *IEEE Journal of Robotics and Automation* **Vol. 6**, 269-273 (1990).

[10] Kurz, R. and Hayward, W., Multiple-goal kinematic optimization of a parallel spherical mechanism with actuator redundancy. *IEEE Journal of Robotics and Automation* **Vol. 8**, 644-651 (1992).

[11] Freeman, R.A. and Tesar, D., Dynamic modeling of serial and parallel mechanisms/robotic systems, Part I-Methodology, Part II-Applications, Proc. of 20th ASME Mechanisms Conference, Orlando, FL, (1988).

[12] Hogan, N., Impedance Control : An approach to manipulation : Part I - Theory, Part II : Implementation, Part III - Applications, J. of Dynamic Systems, Measurement, and Control, **Vol. 107**, pp. 1-24, (1985).

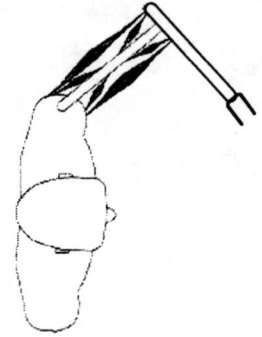

Figure 1 : Human Upper Extremity

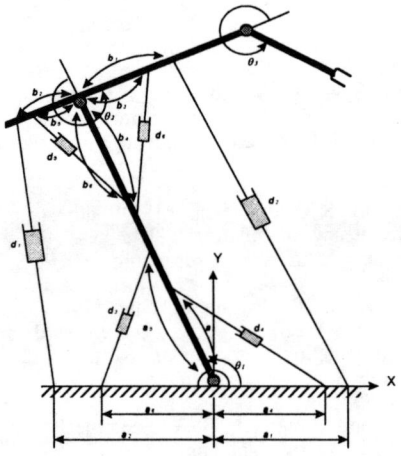

Figure 2 : Anthropomorphic Robot

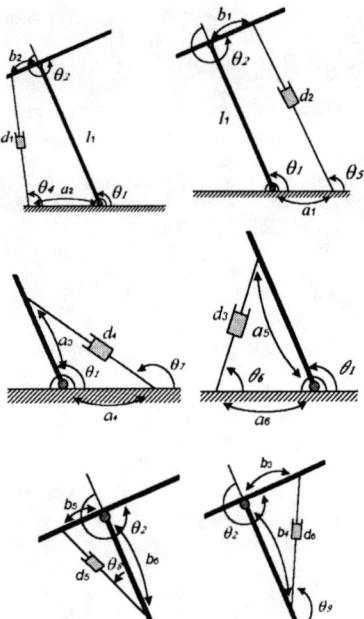

Figure 3 : Six Closed-chains

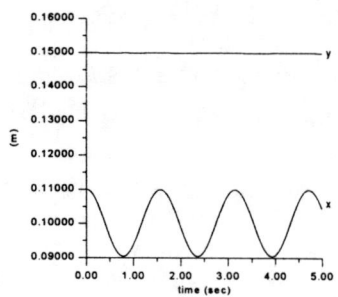

Figure 4 : Frequency Modulation
($\Phi = -90°$, $\omega_{xx} = \omega_{yy} = 4\,rad/s$, $\omega_{xy} = 0\,rad/s$)

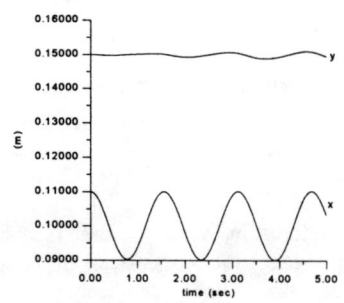

Figure 5 : Frequency Modulation
($\Phi = -45°$, $\omega_{xx} = \omega_{yy} = 4\,rad/s$, $\omega_{xy} = 0\,rad/s$)

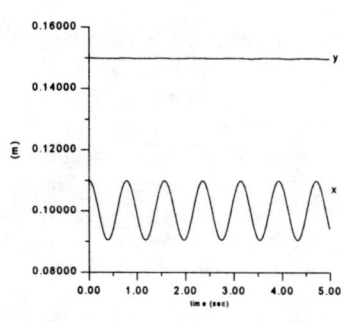

(a) $\Phi = -90°$

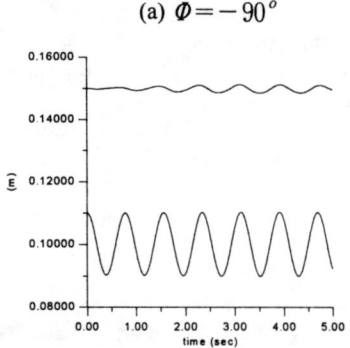

(b) $\Phi = -45°$

Figure 6 : Frequency Modulation
($\omega_{xx} = \omega_{yy} = 8\,rad/s$, $\omega_{xy} = 0\,rad/s$)

Proceedings of the 1998 IEEE
International Conference on Robotics & Automation
Leuven, Belgium • May 1998

FAULT-TOLERANT CONTROL AND OPTIMAL OPERATION
OF
REDUNDANT ROBOTIC MANIPULATORS

A. J. Koivo   M. Ramos

School Electrical and Computer Engineering
Purdue University
West Lafayette, Indiana 47907-1285

Abstract

The transformations between the (Cartesian) p-space, the (joint) q-space of a manipulator and a (reduced dimensional) µ-space are first presented. On the basis of the reduced order model in the µ-space, a low-dimensional torque control is determined. For implementation, it is transformed into the q-space while constraints are taken into account. It is then demonstrated that the foregoing transformation can be determined so that a well-defined performance criterion is optimized. The proposed approaches are illustrated by simulations.

I Introduction

A redundant manipulator working in a constraint environment offers several advantages in its operation due to its extra DOF relative to the DOF of the task. Indeed, the additional links/joints of such a manipulator make it possible, for example, to reach points in the environment which may be inaccessible for a regular manipulator, to avoid obstacles [5,6], to make the joint variables and the joint torques stay within admissible limits [7,8], to avoid overloading joint actuators, and to continue a normal operation even though a joint actuator has failed partially or completely.

The resolution of redundancy for single serial-link manipulators has been studied extensively in the framework of the kinematic equations, relating the position, velocity and acceleration of the end-effector expressed in the (Cartesian) base coordinate system to the joint variables [2,3,4]. Analytical problems arise in solving these equations because the number of the unknowns exceeds the number of the equations. To make the number of the equations equal to that of the unknowns, additional equations may be obtained, for example, from necessary conditions of an optimization problem [1].

The resolution of the redundancy using the Jacobian matrix of a manipulator in the velocity equations has attracted considerable attention of researchers. A possible approach is to use additional constraint equations expressed in the form that they can be combined with basic velocity relations to obtain a extended square Jacobian matrix. [8]. A more common approach involves the use of the pseudo-inverse of the Jacobian matrix, and a null-space vector selected as the gradient of some well-defined performance criterion. This approach offers several attractive designs, such as the optimization of a (secondary) objective function, the utilization of self-motion and a task-priority based control [4,9]. However, the dynamics of the system are not usually taken into account in resolving the redundancy on the position or velocity level.

The resolution of redundancy on the acceleration level incorporates the dynamical equations in the solution. This approach usually includes the vector in the null-space of the Jacobian matrix. The null-space vector can be chosen, for example, so as to force the torque values of the joint actuators to stay within specified limits [6]. While the redundancy is resolved on the kinematic level, the basic control of the redundant manipulator is performed by the traditional servoing methods [7].

The basic issue of controlling redundant manipulators has been addressed in [10,11,12] in which the full-dimensional redundant manipulator model is used. An alternative approach is to reduce the dimensionality of the model to that of the task space using constraint equations, and then to design a controller for the reduced order model [1]. The latter method is applied here.

The organization of the paper is as follows: the basic equations of the dynamical models and the transformation equations for the positions, the velocities and the generalized torques between the joint and reduced dimensional spaces are first presented. Then a controller for the normal and (partly or completely) faulty operations are described. Examples are presented to illustrate the design and the performance controllers. Finally, a procedure for the optimal operation of redundant manipulators is proposed.

II Relationships between Spaces

The dynamical model of a serial-link robotic manipulator is assumed to be in the familiar form

$$D(q)\ddot{q} + C(q,\dot{q})\dot{q} + G(q) = \tau^q \qquad (1)$$

where the joint variable q is an n-dimensional vector, $D(q)$ is the symmetric pseudo-inertia matrix, $C(q,\dot{q})\dot{q}$ signifies the Coriolis and centripetal terms, $G(q)$ the gravitational effects, and the n-dimensional vector τ^q represents the actuator generalized torques (i.e., torques and forces). Equation (1) describes the motion of the robotic manipulator.

The task to be performed by the robotic manipulator is specified in an m - dimensional space. If $n > m$, the robotic manipulator is called redundant with the degree of redundancy being (n - m).

An m-dimensional pseudo-position vector $\mu(t)$ in the task space is related to the joint position vector q as

$$q(t) = \Pi\mu(t) \quad (2)$$

where the n x m matrix Π of full rank m is chosen by the designer. It is called the operation matrix. It is assumed to have constant rational elements. It also describes the corresponding velocity relationship.

The Cartesian velocity $\dot{p}$ is related to the joint velocity $\dot{q}$ by the m x n Jacobian matrix J(q):

$$\dot{p} = J(q)\dot{q} \quad (3)$$

Assuming that the power (energy) is invariant under the mappings described by equations (2) and (3), the generalized torques are related as

$$\tau^q = J^T \tau^p \quad (4)$$
$$\tau^\mu = \Pi^T \tau^q \quad (5)$$

where the superscript T signifies the transposition, τ^p represents the m-dimensional external generalized torque, τ^q is the n-dimensional generalized joint torque produced by the actuators and τ^μ is the m - dimensional generalized torque in the μ - space. By solving for τ^q, equation (5) can be expressed as

$$\tau^q = (\Pi^T)^\dagger \tau^\mu \quad (6)$$

where $\Pi^\dagger$ signifies the pseudo-inverse of Π.

Equations (2) and (5) can be then applied to equation (1) to obtain the a reduced-order model for the motion in the m-dimensional μ- space [1]:

$$D_{eq}\ddot{\mu} + C_{eq}\dot{\mu} + G_{eq} = \tau^\mu \quad (7)$$

where $D_{eq} = \Pi^T D \Pi$, $C_{eq} = \Pi^T C \Pi$, and $G_{eq} = \Pi^T G$. Equation (7) represents m second-order differential equations for the manipulator motion in the μ-space.

III Controller Design in μ-Space

The μ-space torque τ^μ for the manipulator is designed so that it consists of a primary controller determined by the inverse dynamics and a secondary controller chosen as a PI -controller. Thus,

$$\tau^\mu - \hat{D}_{eq}\ddot{\mu} - \hat{C}_{eq}\dot{\mu} - \hat{G}_{eq} = \hat{D}_{eq}(-\ddot{e}^\mu - K_v \dot{e}^\mu - K_p e^\mu) \quad (8)$$

where $\hat{D}_{eq} = \Pi^T \hat{D} \Pi$, $\hat{C}_{eq} = \Pi^T \hat{C} \Pi$, $\hat{G}_{eq} = \Pi^T \hat{G}$, the tracking position error $e^\mu(t) = \mu(t) - \mu^d(t)$, and the hat refers to the matrix or vector when estimated numerical values are used for the parameters. It follows that

$$\ddot{e}^\mu + K_v \dot{e}^\mu + K_p e^\mu = \xi(t) \quad (9)$$

where $\xi(t)$ is assumed to be a bounded disturbance depending on the parameter estimates and $\hat{D}_{eq}^{-1}$. The tracking error $e^\mu(t)$ approaches zero asymptotically, if K_v and K_p are chosen so that the roots of the characteristic equation have negative real parts, for example, by the eigenvalue assignment method.

IV Dynamic Fault-tolerant Torque Design

When a joint actuator in a redundant manipulator partially fails, the joint actuator can be expected to produce only a fraction of the torque that is originally determined. The joint would still be moving due to the small input and coupling effects. Such a joint is here called ϵ - passive. Alternatively, if the joint actuator fails completely, the motion of the joint may be prevented by locking the joint with a break.

When the degree of redundancy (n - m) in a manipulator is one and a complete failure of one joint actuator occurs, then equation (5) will have n - 1 = m unknowns. Consequently, it can be solved explicitly for τ^q, since Π^T has rank m. A joint actuator in a partial failure is assumed to be able to generate a small value of torque. Since τ^μ stays the same, the components of τ^q can still be determined explicitly. In this regard, the *control* τ^μ *is dynamically fault-tolerant.*

If a joint in a complete failure is locked to a certain position, the value of the locked joint variable can be chosen in an optimal manner. The optimal solution can be calculated by a search algorithm.

If the degree of redundancy is greater than one, i.e., n - m >1, then the failure of one joint does not change the state of equation (5); that is, it still contains more unknowns than the number of the equations. The pseudo-inverse matrix can be used, or an additional secondary objective can be optimized. The *control* τ^μ *is dynamically fault-tolerant* also in this case.

It should be noted that τ^q may not be able to accomplish the control objective because the accessability (controllability) condition is not met.

The controller design proposed will be illustrated next.

Example 1: A Fault-tolerant Controller

The motion of a three-joint manipulator (Figure 1a) with an actuator at each joint takes place in a horizontal plane. It is governed by the dynamical model presented in Appendix A. The center point p(t) of the end-effector is to track the straight line that connects points (1,0,0) and (2,1,0) in the xyz coordinate system. The velocity profile $\dot{p}^d(t)$ vs time for the motion in the base coordinate system is displayed in Figure 1b. The corresponding velocity profile $\dot{\mu}^d(t)$ in the μ–space is calculated on-line for the specified parameter values of the elements of Π by equation $\dot{\mu}^d = (J\Pi)^{-1}\dot{p}^d$, and the acceleration $\ddot{\mu}^d(t)$ is obtained after differentiating the foregoing equation. Equation (8) specifies $\tau^\mu = \tau^\mu_{ps}$. The duration of the motion is 5.0 s, i.e., $t \in [0, t_f]$, and $t_f = 5s$. It is assumed that joint 2 actuator after time t_1 can generate only a small torque ϵ, i.e., $\tau^q_2(t) = \epsilon$ for $t \in [t_1, t_f]$.

Since the redundancy of the manipulator in this task is one, the μ-space is two-dimensional. The relationship between $\tau^\mu = [\tau^\mu_1 \ \tau^\mu_2]^T$ and

$\tau^q = [\tau_1^q \ \tau_2^q \ \tau_3^q]^T$ is determined by equation (5), where Π has been chosen by the designer as

$$\Pi^T = \begin{bmatrix} 1 & 1 & 0 \\ 0 & 1 & 1 \end{bmatrix} \quad (10)$$

The controller is determined by equation (8). The experimental tuning of the secondary controller for an acceptable transient response resulted in $K_v = 5.0$ and $K_p = 10.0$. For the implementation of the controller, τ^q is to be determined on the basis of equation (5) when τ^μ is known.

(i) Passive Joint

The manipulator motion starts at $t = 0$. The dynamically fault-tolerant torque τ^μ is determined by equation (8). The joint torque τ^q is governed by equation (5):

$$\tau_1^\mu = \tau_1^q + \tau_2^q \qquad \tau_2^\mu = \tau_2^q + \tau_3^q \quad (11)$$

Torque τ^q needed for the implementation is calculated using the Moore - Penrose pseudo-inverse matrix that minimizes $\|\tau^q\|^2$ subject to the constraint (11).

The solution τ^q for $t \in [0, t_1)$ is

$$\tau_1^q = \frac{2}{3}\tau_1^\mu - \frac{1}{3}\tau_2^\mu$$
$$\tau_2^q = \frac{1}{3}\tau_1^\mu + \frac{1}{3}\tau_2^\mu \quad (12)$$
$$\tau_3^q = -\frac{1}{3}\tau_1^\mu + \frac{2}{3}\tau_2^\mu$$

At time t_1, joint 2 partially fails, and it starts producing only a small torque ε.

When joint 2 becomes a ε - passive joint, the solution to equation (11) is now unique for $t \geq t_1$ when $\tau_2^q = \varepsilon$. Thus, the torques to be applied to the joint shafts are specified for $0 \leq t < t_1$ by equation (12) and for $t_1 \leq t \leq 5.0s$ by the solution to equation (11):

$$\tau_1^q(t) = \tau_1^\mu(t) - \varepsilon \qquad \tau_2^q(t) = \varepsilon \qquad \tau_3^q(t) = \tau_2^\mu(t) - \varepsilon \quad (13)$$

It is emphasized that the position and velocity vectors, $q(t)$ and $\dot{q}(t)$, are assumed to be accessible (controllable) by $\tau_1^q(t), \tau_2^q(t) = \varepsilon$ and $\tau_3^q(t)$ in order to achieve the tracking of the desired trajectory.

The motion of the redundant manipulator was simulated under ideal conditions. Actuator 2 partially fails at $t_1 = 1.5s$ and $\varepsilon = 0.01$. The torque τ^q is specified for $0 \leq t < 1.5s$ by equation (12) and for $1.5s \leq t \leq 5.0s$ by equation (13). The simulations gave the graphs shown in Figure 2a for τ_2^q and τ_3^q vs time and in Figure 2b for $\dot{p}_y$ vs time. In spite of the partial failure, the tracking is still excellent; the mean integrated squared position error is 6.0×10^{-5}. The total energy consumed is 2.745(J).

(ii) Locked joint after failure

It is now assumed that the actuator of joint 2 has failed (completely) before the motion starts, i.e., $\tau_2^q = 0$ for all $t \geq 0$. Joint 2 will be locked so that q_2 is constant for the duration of the motion. The problem to be studied is to determine the value q_2^* that minimizes the energy

$$EN = \int_0^{t_f} \dot{q}^T(\sigma)\tau^q(\sigma)d\sigma = \int_0^{t_f} \dot{\mu}^T(\sigma)\tau^\mu(\sigma)d\sigma \quad (14)$$

The nature of the solution to the minimization problem is illustrated by displaying EN vs q_2 as shown in Figure 3. The value of q_2 that gives a local minimum can be found using a one-dimensional search algorithm. In the range of (-1.5 rad, 1.5 rad), two local minima exist: at $q_2^* = -1.405$ (rad) the minimum energy is 16.543 (J), and at $q_2^* = 0.579$ (rad) the minimum energy is 18.901(J). The mean integrated squared position error for the former case is 2.40×10^{-5}.

V Optimal Operation Matrix

The transformation matrix Π^T in equation (5) determines the torque that each joint should contribute to the total torque needed for the motion. Thus, the elements of Π specify the responsibility of each joint. Instead of having the designer specify the Π - matrix heuristically, the elements of Π can be determined so that a well-defined performance criterion will be minimized. The problem to be studied is to determine Π^* that minimizes the energy EN in equation (14).

Two examples will next be presented by restricting the elements of Π to be 0's and 1's and then to positive real numbers.

Example 2: Binary-valued Optimal Operation Matrix

The redundant robotic manipulator of Example 1 is operated in a vertical plane. The center point of the end-effector is to move along a straight line connecting points (0,1,0) and (0,2,1) in the xyz - coordinate system. The desired trajectories in the μ–space and in the q–space are generated as described in Example 1. The controller τ^μ is constructed in the same way as in Example 1. The duration of the motion is 5.0s.

The problem is to determine the elements of Π so that EN in equation (14) is minimized. The admissible elements of Π are 0's and 1's.

By an exhaustive search, the solution is found: the minimum energy EN* = 14.812(J) and the corresponding minimizing binary-valued matrix Π^* is represented by two matrices

$$\Pi^* = \begin{bmatrix} 0 & 1 & 1 \\ 1 & 1 & 0 \end{bmatrix}^T \quad (15) \qquad \Pi^* = \begin{bmatrix} 1 & 1 & 0 \\ 0 & 1 & 1 \end{bmatrix}^T \quad (16)$$

The elements of Π^* specify the contributions of the joints to the realization of torque τ^μ. If equation (16) is accepted, then equation (5) gives

$$\tau_1^\mu = \tau_1^q + \tau_2^q \qquad \tau_2^\mu = \tau_2^q + \tau_3^q \quad (17)$$

The joint torque τ^q is solved from equation (17) in which τ^μ is known.

The results of two simulated cases are next presented: (i) ideal system and (ii) bounded torque input.

(i) *Ideal case*

Torque τ^q is calculated using equation (6). The result is given in equation (12) for $t \in [0, t_f]$. Torques $\tau_i^q, i = 1, 2, 3$ are then applied to the joint shafts.

Typical graphs for unbounded $\tau_i^q(t)$ $i=1,2$ obtained by simulating the motion of the manipulator are displayed in Figures 4a-4b. The minimum value of the energy is $EN^* = 14.870$ (J). The mean integrated squared position error is 15.20×10^{-5}.

(ii) *Bounded torque input*

The hard constraint here is $|\tau_1^q| \leq 20.5$ Nm. The computed torque τ^μ in the μ-space is assumed to be uneffected by the constraint. The designer could determine τ^q by minimizing the squared norm of τ^q subject to the limit constraint and equation (17).

An alternative approach, however, is used here. Because of nonlinear dynamics, the components of τ^q in equation (17) are determined so as to minimize

$$E(\tau^q) = \|\Delta p(t)\|^2 + w\|\Delta \dot{p}(t)\|^2 \quad (18)$$

where $\Delta p(t) = p(t) - p^d(t)$, and w is a positive weighting factor used to weigh the importance of the velocity errors relative to the position errors. The desired trajectory $[p^d(t), \dot{p}^d(t)]$ is specified by the planning.

The numerical solution for τ^q is calculated using MATLAB. The resulting torques τ_i^q, $i = 1, 2, 3$ are then applied to the joint shafts.

The simulations of the system model with w = 0.1 gave the torque components shown in Figures 4a - 4b, where the dashed curves represent the unconstrained torques and the dashed curves the corresponding constrained torques. The mean squared position error is 24.63×10^{-5}.

It should be noted that the solution to equation (5) may not exist depending on the constraint limits.

Example 3: Real-Valued Optimal Operation Matrix

The three-joint manipulator of Example 1 is operated on a vertical plane (x = 0). The specifications for the motion are the same as in Example 2. The controller is the same as described in Example 2. The additional objective here is to minimize energy EN in equation (14) with respect to the elements of Π given as:

$$\Pi = \begin{bmatrix} \alpha_1/(\alpha_1+\alpha_2) & 0 \\ \alpha_2/(\alpha_1+\alpha_2) & \alpha_2/(\alpha_1+\alpha_2) \\ 0 & \alpha_1/(\alpha_1+\alpha_2) \end{bmatrix} \quad (19)$$

where α_1 and α_2 are real positive constant design parameters to be determined.

In order to illustrate the nature of the solution, constant $EN(\alpha_1, \alpha_2)$-curves are displayed on the (α_1, α_2)-plane in Figure 5. The optimal values α_1^* and α_2^* can be found by applying any standard search algorithm, such as a steep-descent technique, or the conjugate gradient method. It leads to the minimizing values: $\alpha_1^* = 0.5$ and $\alpha_2^* = 0.6$. Thus, matrix Π is now specified. The minimum value of the energy is $EN(\alpha_1^*, \alpha_2^*) = 14.825$(J). The integrated mean squared position error in the base coordinate system is 2.831×10^{-5}.

Torque τ^μ causing the motion in the μ-space is related to τ^q by equation (5):

$$\tau_1^\mu = 0.6\tau_1^q + 0.4\tau_2^q \qquad \tau_2^\mu = 0.5\tau_2^q + 0.5\tau_3^q \quad (20)$$

It should be noticed that according to equation (20) the torque of joint 2 is contributing to torque τ_1^μ at the activity level of 0.4 and to torque τ_2^μ at the activity level 0.5. The activity levels of joints 1 and 3 can be expressed similarly.

Equation (20) can be solved for τ_i^q, $i = 1, 2, 3$ using the Moore-Penrose pseudo-inverse to obtain

$$\tau_1^q = \tau_1^\mu \qquad \tau_2^q = \tau_2^\mu \qquad \tau_3^q = \tau_2^\mu \quad (21)$$

Torques τ_i^q $i = 1, 2, 3$ are then applied to the joint shafts.

VI Conclusions

Using a reduced-order model for a redundant manipulation dynamics, a torque independent of possible actuator failures is determined. It is transformed into the joint space while the constraint is taken into account. The approach can be used in the case of a joint failure as well as in the case of bounded input.

The transformation matrix relating the low dimensional μ-space to the q-space can be chosen so as to minimize a well-defined performance criterion, e. g., the energy. The resulting matrix specifies the contributions of the joints to the total torque needed to produce the motion of the manipulator.

The proposed approaches are illustrated by simulation examples.

Appendix A: Model of 3-Joint Manipulator

The mathematical model used in simulations is

$$D(q)\ddot{q} + C(q,\dot{q})\dot{q} + G(q) = \tau^q \quad (A.1)$$

Vectors q and τ^q are defined as $q = [q_1 \ q_2 \ q_3]^T$, $\tau^q = [\tau_1^q \ \tau_2^q \ \tau_3^q]^T$ and

$$D(q) = \begin{bmatrix} D_{11} & D_{12} & D_{13} \\ D_{21} & D_{22} & D_{23} \\ D_{31} & D_{32} & D_{33} \end{bmatrix} \qquad C(q,\dot{q})\dot{q} = \begin{bmatrix} C_{11} \\ C_{21} \\ C_{31} \end{bmatrix}$$

$$G(q) = [G_{11} \ G_{21} \ G_{31}]^T$$

The numerical values of the parameters (in consistent units) are:
$g = -9.81$, $m_1 = 1$, $m_2 = 1$, $m_3 = 1$, $\ell_1 = 1$, $\ell_2 = 1$, $\ell_3 = 1$.

Also, $I_i = m_i(\ell_i^2)/12$, $i = 1, 2, 3$

$a_1 = (m_1/8 + m_2/2 + m_3/2)\ell_1^2 + I_1/2;$

$a_2 = (m_2/8 + m_3/2)\ell_2^2 + I_2/2;$ $a_3 = (m_3/8)\ell_3^2 + I_3/2$

$a_4 = a_1 + a_2 + a_3;$ $a_5 = (m_2/2 + m_3)\ell_1\ell_2;$

$a_6 = (m_3/2)\ell_1\ell_3;$ $a_7 = (m_3/2)\ell_2\ell_3$

$b_2 = (m_2/2 + m_3)g\ell_2$

$b_1 = (m_1/2 + m_2 + m_3)g\ell_1,$ $b_3 = (m_3/2)g\ell_3$

$D_{11} = 2a_4 + 2a_5c_2 + 2a_6c_{23} + 2a_7c_3;$ $D_{21} = D_{12}$

$D_{12} = 2a_2 + 2a_3 + a_5c_2 + a_6c_{23} + 2a_7c_3;$ $D_{31} = D_{13}$

$D_{13} = 2a_3 + a_7c_3 + a_6c_{23},$ $D_{23} = 2a_3 + a_7c_3$

$D_{22} = 2a_2 + 2a_3 + 2a_7c_3;$ $D_{32} = D_{23};$ $D_{33} = 2a_3$

$C_{11} = (-a_5s_3 - a_6s_{23})\dot{q}_2^2 + (-a_7s_3 - a_6s_{23})\dot{q}_3^2 + (-2a_5s_3$
$-2a_6s_{23})\dot{q}_1\dot{q}_2 + (-2a_6s_{23} - 2a_7s_3)\dot{q}_1\dot{q}_3$
$+(-2a_6s_{23} - 2a_7s_3)\dot{q}_2\dot{q}_3$

$C_{21} = (a_5s_2 + a_6s_{23})\dot{q}_1^2 + (-a_7s_3)\dot{q}_3^2 + (-2a_7s_3)\dot{q}_1\dot{q}_3$
$+(-2a_7s_3)\dot{q}_2\dot{q}_3$

$C_{31} = (a_6s_{23} + a_7s_3)\dot{q}_1^2 + (a_7s_3)\dot{q}_2^2 + (2a_7s_3)\dot{q}_1\dot{q}_2$

$G_{11} = -b_1c_1 - b_2c_{12} - b_3c_{123}$

$G_{21} = -b_2c_{12} - b_3c_{123};$ $G_{31} = -b_3c_{123}$

References

[1] A. J. Koivo and S. Arnautovic, "Control of Redundant Manipulators with Constraints Using a Reduced Order Model", *Automatica,* Vol. 30, No. 4, 1994, pp. 665-677.

[2] D. E. Whitney, "The Mathematics of Coordinated Control of Prosthetic Arms and Manipulators", *Trans. ASME J. Dynamical Systems, Measurements and Control,* Vol. 94, pp. 303-309, Dec. 1972.

[3] A. Liegeois, "Automatic Supervisory Control of the Configuration and Behavior of Multibody Mechanisms", *IEEE Trans. Systems, Man, Cybernetics,* Vol. SMC-7, no. 12, pp. 868-871, Dec. 1977.

[4] C. A. Klien and C. H. Huang, "Review of Pseudoinverse Control for Use with Kinematically Redundant Manipulators", *IEEE Trans. Systems, Man and Cybernetics,* Vol. SMC-13, No. 3, pp. 245-250, Mar./Apr. 1983.

[5] M. Kircanski and M. Vukobratovic, "Trajectory Planning for Redundant Manipulators in the Presence of Obstacles", in *Proc. 5th CISM-IFToMM Ro. Man. Syst.,* Udine, Italy, June 1984.

[6] A. A. Maciejewski and C. A. Klein, "Obstacle Avoidance for Kinematically Redundant Manipulators in Dynamically Varying Environments", *Int. J. of Robotics Research,* Vol. 4, No. 3, pp. 109-117, Fall 1985.

[7] J. Hollerbach and K. C. Suh, "Redundancy Resolution of Manipulators through Torque Optimization", *IEEE J. of Robotics and Automation,* Vol. RA-3, No. 4, August 1987, pp. 308-316.

[8] Y. Ting, S. Tosunoglu and R. Freeman, "Torque Redistribution Method for Fault Recovery in Redundant Serial Manipulators", *Proc. of IEEE Int. Conference on Robotics and Automation,* May 1994, pp. 1396-1401.

[9] J. Baillieul, "A Constraint Oriented Approach to Inverse Problems for Kinematically Redundant Manipulators", *Proc. IEEE Int. Conference on Robotics and Automation,* Raleigh, North Carolina.

[10] Y. Nakamura, H. Hanafusa and T. Yoshikawa, "Task-Priority Based Redundancy Control of Robot Manipulators", *Int. J. of Robotics Research,* Vol. 6, No. 2, pp 3 - 15, Summer 1987.

[11] P. Hsu, J. Hauser and S. Sastry, "Dynamic Control of Redundant Manipulators", *Proc. of IEEE Int. Conference on Robotics and Automation,* Philadelphia, PA, April 1988, pp. 183-187.

[12] R. Kankaanranta and H. N. Koivo, "Dynamics and Simulation of Compliant Motion of a Manipulator", *IEEE J. of Robotics and Automation,* Vol. 4, No. 2, April 1988, pp. 163-173.

[13] O. Khatib, "The Operational Space Formulation in the Analysis, Design and Control of Robot Manipulators", *Proc. 3rd Int. Symposium: Robotics Research,* pp. 263-270.

[14] H. Arai and S. Tachi, "Position Control of a Manipulator with Passive Joints Using Dynamic Coupling", *IEEE Trans. on Robotics and Automation,* Vol. 7, No. 4, August 1991, pp. 528-534.

[15] F. L. Lewis, C. T. Abdallah, D. M. Dawson, *Control of Robot Manipulators,* book, Macmillan Publishing Company, New York, NY, 1993.

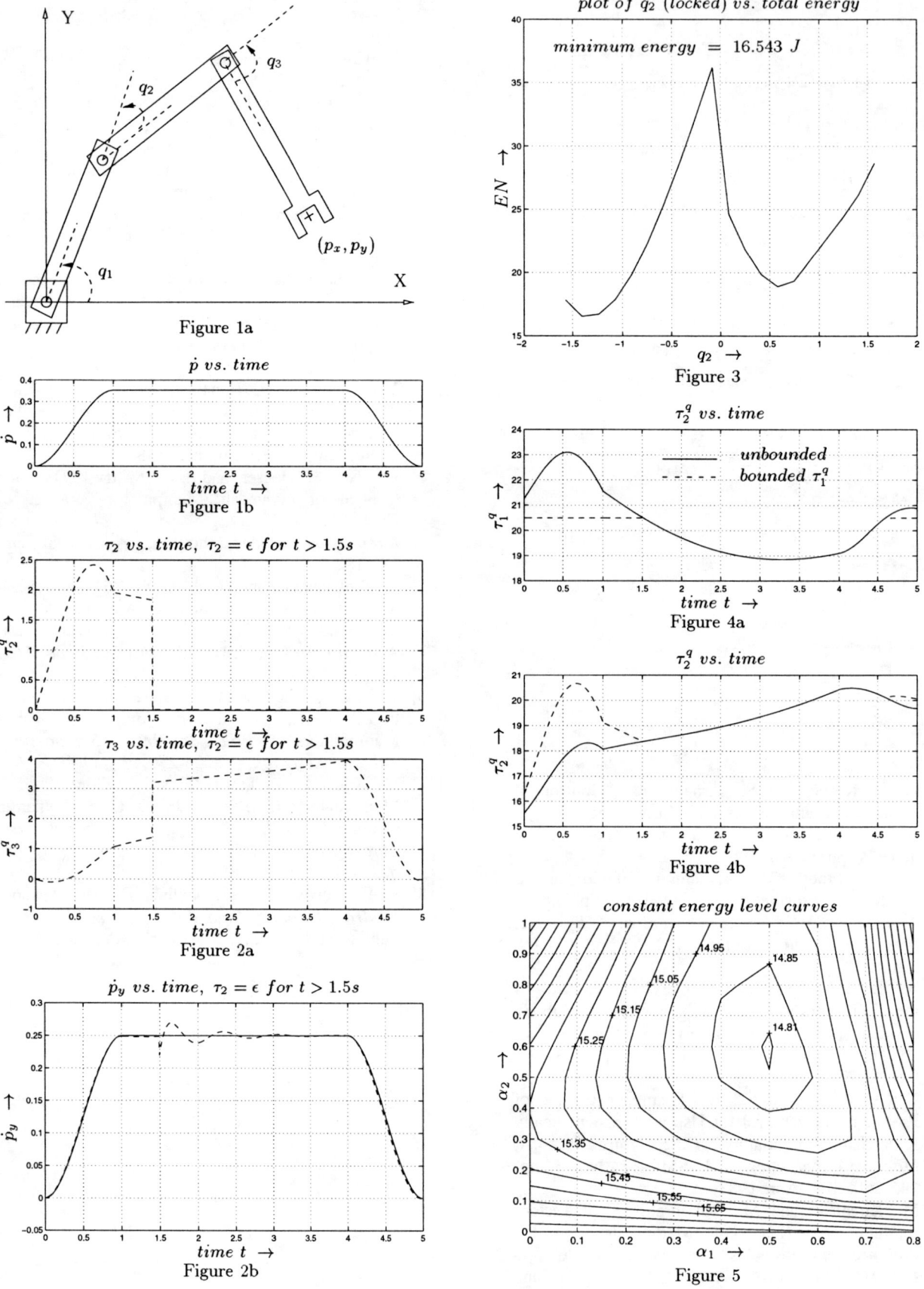

Analytic Nonlinear $\mathcal{H}_\infty$ Optimal Control for Robotic Manipulators

Jonghoon Park, Wankyun Chung, Youngil Youm,
Robotics Lab., Department of Mechanical Engineering
Pohang University of Science & Technology (POSTECH)
Pohang, 790-784, Republic of Korea

Abstract

Recent successes in nonlinear $\mathcal{H}_\infty$ control design [5, 10] has been applied to control of robot manipulator systems [1, 2, 3, 7]. It is known that the existence of $\mathcal{H}_\infty$ optimal control reduces to solvability of the Hamilton-Jacobi-Isaccs(HJI) partial differential equation, which is very difficult to solve. In this article, a robust control in the sense of L_2-gain attenuation from an external disturbance and one due to model uncertainties is designed for a class of Euler-Lagrange system based on a class of analytic solution to the associated HJI equation.

1 Introduction

Control of a nonlinear system should be evaluated in terms of the three requirements: i) stabilization, ii) performance and iii) robustness. Consider a nonlinear system, with local coordinates $x = (x_1, \cdots, x_n)^T$ for the state space manifold,

$$\dot{x} = f(x,t) + g(x,t)u + p(x,t)w, \tag{1}$$

where $u \in \Re^m$ is the control, and $w \in \Re^w$ is the disturbance. Performance can be defined using the cost variable $z \in \Re^z$,

$$z = h(x,t) + k(x,t)u, \quad h^T k = 0, k^T k > 0. \tag{2}$$

The nonlinear state-feedback $\mathcal{H}_\infty$ control aims at finding a *stabilizing*[1] nonlinear state feedback

$$u = \alpha(x,t), \quad \alpha(0,t) = 0 \tag{3}$$

such that the closed-loop system has L_2-gain $\leq \gamma$ from w to z [5, 10], i.e.

$$\int_0^t z^T z \, dt \leq \gamma^2 \int_0^t w^T w \, dt \tag{4}$$

for all $t \geq 0$, for $x(0) = 0$, and for all w such that the integral in the right hand side of the equation is well-defined.

[1] i.e. internally stabilizing which means that the closed-loop system with $w = 0$ is uniformly asymptotically stable

We provide one version of theorem [5] concerning the solution of $\mathcal{H}_\infty$ disturbance attenuation problem described above. The following definition simplifies the discussion.

Definition 1. *Given a C^1 function $V(x,t)$ and a positive number γ, the following expression denoted by $HJI_\gamma(x,t;V)$ is called the Hamilton-Jacobi-Isaccs(HJI)*

$$V_t + V_x f - \frac{1}{2} V_x \left\{ g[k^T k]^{-1} g^T - \frac{1}{\gamma^2} pp^T \right\} V_x^T + \frac{1}{2} h^T h \tag{5}$$

for the system (1) and (2), where $V_t = \dfrac{\partial V}{\partial t}$ and $V_x = \dfrac{\partial V}{\partial x^T}$.

Lemma 1. *If there exists a C^1 positive definite function $V(x,t)$ with $V(0,t) = 0$ and $V_x(0,t) = 0$ that satisfies*

$$HJI_\gamma(x,t;V) \leq 0 \tag{6}$$

then the control

$$u = -[k^T k]^{-1} g^T V_x^T \tag{7}$$

satisfies the L_2-gain attenuation as well as the closed-loop stability.

The above remarkable theorem shows that an $\mathcal{H}_\infty$ control can be designed by solving the associated HJI equation. Then, the solution, if any, will guarantee the stability as well as the disturbance attenuation in L_2 sense.

Unfortunately, however, the HJI equation is known to be very difficult to solve even for a specific nonlinear system, say robot manipulator system. This is the motivation of this article. In this work, we will derive a class of analytic solutions to the HJI equation associated to a class of systems described by Euler-Lagrange equations of motion. Note that the systems covered by the Euler-Lagrange equations are very broad and most of mechanical systems are included in this class. This paper is constructed with special focus on robot manipulator system, and much part of mathematical proofs are omitted due to space limitation.

2 Closed-Loop System for the Euler-Lagrange System

Many of physical systems can be modeled by Lagrange equations of motion. Denoting the generalized coordinates of system by $q = (q_1, q_2, \cdots, q_n)^T$, the equation yields

$$\tau \triangleq M(q)\ddot{q} + C(q,\dot{q})\dot{q} + G(q) + d, \qquad (8)$$

where $\tau \in \Re^n$ is the input, d is the external disturbance, and M, C, and G denotes the dynamic parameters of the system.

The EL (Euler-Lagrange) system satisfies the following properties.

(EL1) The EL system can be modeled as a rn-th order ordinary differential equation for some integer $r > 0$ by setting the state $x = (x_1^T, \cdots, x_r^T)^T$ with $x_i = \underbrace{\int \cdots \int}_{(r-1-i)\text{times}} q$. Then the state space equation can be represented as

$$\dot{x} = A(x,t)x + B(x,t)u + D(x,t)w, \qquad (9)$$

where

$$A(x,t) = \begin{bmatrix} 0 & I & \cdots & 0 \\ 0 & 0 & \cdots & 0 \\ \vdots & \vdots & \ddots & \vdots \\ 0 & 0 & \cdots & I \\ a_1(x,t) & a_2(x,t) & \cdots & a_r(x,t) \end{bmatrix}$$

$$B(x,t) = \begin{bmatrix} 0 \\ 0 \\ \vdots \\ b(x,t) \end{bmatrix}, \quad D(x,t) = \begin{bmatrix} 0 \\ 0 \\ \vdots \\ c(x,t) \end{bmatrix}.$$

(EL2) The Corioli's and centrifugal matrix $C(q,\dot{q})$ satisfies

$$\dot{M}(q,\dot{q}) - \left(C(q,\dot{q}) + C^T(q,\dot{q})\right) = 0.$$

Since the tracking of the coordinate q to some desired trajectory $q_d(t)$ is of primary concern, let us define

$$e = q_d - q. \qquad (10)$$

Using this transformed coordinates, consider the following closed-loop system, called the reference error output passive system,

$$M\{\ddot{e} + k_v\dot{e} + k_p e\} + C\left\{\dot{e} + k_v e + k_p \int e\right\} = u + w \qquad (11)$$

which is related to the EL system (8) by suitable choice of control input τ and of the disturbance w. As a matter of fact, the system has the property of (EL1) and (EL2). Expressing the system in the state space with $x = (\int e^T, e^T, \dot{e}^T)^T \in \Re^{3n}$ yields the system (9) with

$$A = \begin{bmatrix} 0 & I & 0 \\ 0 & 0 & I \\ -k_p M^{-1} C & -k_v M^{-1} C - k_p I & -M^{-1} C - k_v I \end{bmatrix}$$

$$B = D = \begin{bmatrix} 0 \\ 0 \\ M^{-1} \end{bmatrix}. \qquad (12)$$

Since any EL system can be rendered to the reference error output passive system with suitable choice of a control input, which will be shown in Section 4, an $\mathcal{H}_\infty$ disturbance attenuating solution for this system will be derived in the next section.

3 $\mathcal{H}_\infty$ Solution for EL System

The main result on the $\mathcal{H}_\infty$ control for the system (11) derived from the Euler-Lagrange system is summarized by the following theorem, whose proof will be given in step-by-step manner in the sequel development.

Theorem 1. *The EL reference error output passive system (11) allows the following nonlinear $\mathcal{H}_\infty$ disturbance attenuating control with the closed-loop stability*

$$u^* = -\alpha K(x,t)\left(\dot{e} + k_v e + k_p \int e\right) \qquad (13)$$

if for a given $\gamma > 0$ and a set of (k_p, k_v), the following three requirements hold: K should satisfy

$$(R1) \quad K_\gamma = K - \frac{1}{\gamma^2}I > 0,$$

and there exist $n \times n$ constant symmetric matrices $\overline{Q}_{12}$, $\overline{Q}_{13}$, $\overline{Q}_{23}$ and scalar $\alpha > 0$ such that

$$(R2) \quad \alpha^2 \begin{bmatrix} k_p k_v \overline{K}_\gamma & k_p \overline{K}_\gamma \\ k_p \overline{K}_\gamma & k_v k_p \overline{K}_\gamma \end{bmatrix} - \begin{bmatrix} \overline{Q}_{12} & \overline{Q}_{13} \\ \overline{Q}_{13} & \overline{Q}_{23} \end{bmatrix} > 0$$

$$(R3) \quad \begin{bmatrix} \alpha^2 k_p^2 \overline{K}_\gamma & \overline{Q}_{12} & \overline{Q}_{13} \\ \overline{Q}_{12} & \alpha^2(k_v^2 - 2k_p)\overline{K}_\gamma + 2\overline{Q}_{13} & \overline{Q}_{23} \\ \overline{Q}_{13} & \overline{Q}_{23} & \alpha^2 \overline{K}_\gamma \end{bmatrix} > 0$$

where $\overline{K}_\gamma = \overline{K} - \frac{1}{\gamma^2}I$ with $\overline{K}$ being a constant portion of K such that $\tilde{K} = K - \overline{K} \geq 0$.

The analytic derivation of the above $\mathcal{H}_\infty$ optimal control consists of two steps:

1. Conversion of the HJI partial differential equation (6) with (5) into a nonlinear ordinary differential Riccati equation (NDRE)

$$\dot{P} + A^T P + PA - PBR_\gamma^{-1}B^T P + Q = 0 \quad (14)$$

under the assumed solution

$$V(x,t) = \frac{1}{2}x^T P(x,t)x \quad (15)$$

using (EL1)

2. Analytic solution of the NDRE (14) using (EL2)

3.1 Derivation of NDRE for EL system (1st step)

The assumed solution (15) is a quadratic form with nonlinear time-9 matrix $P(x,t) = P(x_1, x_2, x_3, t)$ which will be defined later to be positive definite. The following lemma provides one case where the HJI equation turns into the NDRE for the system satisfying (EL1).

Lemma 2. *Let the system (1) satisfies (EL1) and $h^T(x,t)h(x,t) = \frac{1}{2}x^T Q(x,t)x$ in (2). If P is not an explicit function of x_r, that is $\frac{\partial P}{\partial x_r^T} = 0$, then the HJI equation (5) with (6) reduces to the following NDRE*

$$0 = \dot{P} + A^T P + PA - P\left(BKB^T - \frac{1}{\gamma^2}DD^T\right)P + Q, \quad (16)$$

and the $\mathcal{H}_\infty$ optimal control (7) is given by

$$u = -KB^T Px. \quad (17)$$

The proof is shown in Appendix A.

Note that for the EL reference error output system (11), (16) is further reduced to (14) by letting

$$R_\gamma = \left[K - \frac{1}{\gamma^2}I\right]^{-1} = K_\gamma^{-1}$$

since $B = D$ in (12).

3.2 Analytic solution to NDRE for EL system (2nd step)

In this section, we are to find an analytic solution to the NDRE(14) exploiting the property (EL2) of the EL system. By partitioning the $3n \times 3n$ P matrix as

$$P = \begin{bmatrix} P_{11} & P_{12} & P_{13} \\ P_{12} & P_{22} & P_{23} \\ P_{13} & P_{23} & P_{33} \end{bmatrix}$$

and Q similarly, we have a set of six simultaneous matrix differential equations, each of dimension $n \times n$. It seems that it is almost impossible to solve them analytically. However, the following lemma proposes one class of analytic solutions.

Lemma 3. *The NDRE for the system (12) has a solution $P(x,t)$ of the form*

$$P(x,t) = \begin{bmatrix} \alpha k_p^2 M + \alpha^2 k_p k_v \overline{K}_\gamma - \overline{Q}_{12} & \alpha k_p k_v M + \alpha^2 k_p \overline{K}_\gamma - \overline{Q}_{13} & \alpha k_p M \\ \alpha k_p k_v M + \alpha^2 k_p \overline{K}_\gamma - \overline{Q}_{13} & \alpha k_v^2 M + \alpha^2 k_v \overline{K}_\gamma - \overline{Q}_{23} & \alpha k_v M \\ \alpha k_p M & \alpha k_v M & \alpha M \end{bmatrix}, \quad (18)$$

if the state weighting matrix $Q(x,t)$ in Lemma 2 is given by the form

$$Q(x,t) = \begin{bmatrix} \alpha^2 k_p^2 \overline{K}_\gamma & \overline{Q}_{12} & \overline{Q}_{13} \\ \overline{Q}_{12} & \alpha^2(k_v^2 - 2k_p)\overline{K}_\gamma + 2\overline{Q}_{13} & \overline{Q}_{23} \\ \overline{Q}_{13} & \overline{Q}_{23} & \alpha^2 \overline{K}_\gamma \end{bmatrix}$$

$$+ \begin{bmatrix} \alpha^2 k_p^2 \tilde{K}_\gamma & \alpha^2 k_p k_v \tilde{K}_\gamma & \alpha^2 k_p \tilde{K}_\gamma \\ \alpha^2 k_p k_v \tilde{K}_\gamma & \alpha^2 k_v^2 \tilde{K}_\gamma & \alpha^2 k_v \tilde{K}_\gamma \\ \alpha^2 k_p \tilde{K}_\gamma & \alpha^2 k_v \tilde{K}_\gamma & \alpha^2 \tilde{K}_\gamma \end{bmatrix} \quad (19)$$

and the control weighting $R(x,t) = K^{-1}(x,t)$, where $K_\gamma(x,t) = \overline{K}_\gamma + \tilde{K}_\gamma(x,t)$ with $\overline{K}_\gamma = K - \frac{1}{\gamma^2}I$. Also, the control (17) reduces to the reference error feedback (REF) defined by

$$u = -\alpha K(x,t)\left(\dot{e} + k_v e + k_p \int e\right). \quad (20)$$

The proof is omitted, but essentially not difficult. It should be noted that the control (20) does not depend on any dynamic parameters of the system (8), although the control (17) does. Due to this property, the REF $\mathcal{H}_\infty$ control can be applied generally without any knowledge of the system dynamics, only if it is the EL system.

Now a series of simple lemmas to ensure the positive definiteness of P, Q, and R will be provided.

Lemma 4. *The matrix $K_\gamma(x,t)$ satisfies (R1) if*

$$\lambda_{min}(\overline{K}) > \frac{1}{\gamma^2} \quad \text{and} \quad \widetilde{K}(x,t) \geq 0. \quad (21)$$

where $\lambda(A)$ denotes the eigenvalue of A.

Note that if (21) holds, then $R(x,t) = K^{-1}(x,t) > 0$.

<u>REMARK 1</u> If $R(x,t)$ is constant, then $\overline{K} = R^{-1}$. Since $\lambda_{min}(K) = \lambda_{max}^{-1}(R)$, the condition is expressed as $\lambda_{max}(R) < \gamma^2$.

Lemma 5. *For the $3n \times 3n$ $\boldsymbol{P}$ matrix given in (18) to be positive definite, it is necessary and sufficient that the $2n \times 2n$ constant portion matrix given in (R2) is positive definite.*

Lemma 6. *Similarly, the matrix $\boldsymbol{Q}(\boldsymbol{x},t)$ is positive-definite if the constant portion $\overline{\boldsymbol{Q}}$ given in (R3) is positive-definite for $\widetilde{\boldsymbol{K}} \geq 0$.*

REMARK 2 The proposed $\mathcal{H}_\infty$ control has $n \times n$ matrices (nonlinear and time-varying) $\boldsymbol{K}$, $\overline{\boldsymbol{Q}}_{12}$, $\overline{\boldsymbol{Q}}_{13}$, and $\overline{\boldsymbol{Q}}_{23}$ as free parameters which should be tuned, if γ and (k_p, k_v) are fixed. Then, for a prespecified γ, the chosen $\boldsymbol{R}$ should satisfy (21), or (R1). Next, the free parameter α can be determined so that (R2) and (R3) can hold. Otherwise, the parameters should be re-selected, or α should be larger.

REMARK 3 The tuning procedure in the previous Remark can be simplified by setting $\boldsymbol{R} > 0$ as constant with $\lambda_{max}(\boldsymbol{R}) < \gamma^2$. Then $\overline{\boldsymbol{K}} = \boldsymbol{R}^{-1}, \widetilde{\boldsymbol{K}} = 0$, and (R1) holds. Next choose $\overline{\boldsymbol{Q}}_{ij} = \boldsymbol{0}$. Then (R2) reduces to

$$k_v k_p \boldsymbol{K} - (k_p \boldsymbol{K})\left(\frac{1}{k_p k_v}\boldsymbol{R}\right)(k_p \boldsymbol{K}) = (k_v k_p - \frac{k_p}{k_v})\boldsymbol{K} > 0,$$

and (R3) is simplified to $k_v^2 > 2k_p$. Hence, the gain inequality $k_v^2 > 2k_p > 1$ is sufficient for $\mathcal{H}_\infty$ control. Then the performance of L_2-gain attenuation can be tuned easily only by changing $\boldsymbol{R}$.

3.3 Closed-loop stability

Choosing the solution $V(\boldsymbol{x},t) > 0$ as the Lyapunov function, it is easy to prove that the closed-loop system is in fact *globally exponentially stable* when $\boldsymbol{w} = \boldsymbol{0}$.

4 Robust Reference Motion Compensation Control

To derive the reference error output passive closed-loop system (11) for a general EL system (8), the following control, termed the *reference motion compensation control* (RMC) is proposed[1]

$$\boldsymbol{\tau} = \widehat{\boldsymbol{M}}(\ddot{\boldsymbol{q}}_d + k_v \dot{\boldsymbol{e}} + k_p \boldsymbol{e}) + \widehat{\boldsymbol{C}}\left(\dot{\boldsymbol{q}}_d + k_v \boldsymbol{e} + k_p \int \boldsymbol{e}\right) + \widehat{\boldsymbol{G}} - \boldsymbol{u}, \qquad (22)$$

---
[1] A similar control was proposed in many literatures with the name of passive control [11] and of modified computed torque control [4], etc. Specifically, Dawson et al. [4] applied this control to solve the Hamilton-Jacobi-Bellman equation arising in the linear quadratic optimal control.

where $(\widehat{\boldsymbol{M}}, \widehat{\boldsymbol{C}}, \widehat{\boldsymbol{G}})$ is the estimated or nominal dynamic parameters of the exact $(\boldsymbol{M}, \boldsymbol{C}, \boldsymbol{G})$. By closing the control around the system (8), the reference error output system (11) is obtained by formulating the disturbance $\boldsymbol{w}$ to include the internal disturbances due to model uncertainties

$$\boldsymbol{w} = \widetilde{\boldsymbol{M}}(\ddot{\boldsymbol{q}}_d + k_v \dot{\boldsymbol{e}} + k_p \boldsymbol{e}) + \widetilde{\boldsymbol{C}}\left(\dot{\boldsymbol{q}}_d + k_v \boldsymbol{e} + k_p \int \boldsymbol{e}\right) + \widetilde{\boldsymbol{G}} + \boldsymbol{d}, \qquad (23)$$

where $\widetilde{(\cdot)} = (\cdot) - \widehat{(\cdot)}$.

Now by applying the REF (20) to the RMC (22) the robustness to the model uncertainty in the sense of L_2-gain attenuation is 1. The property of the independence of the REF of the actual dynamic parameters of the system (8) is the crucial part.

REMARK 4 [Decentralized $\mathcal{H}_\infty$ control] If the nominal dynamics is given of the form $(\widehat{\boldsymbol{M}}, \widehat{\boldsymbol{C}}, \widehat{\boldsymbol{G}}) = (\text{diag}\{\hat{m}_i\}, \boldsymbol{0}, \boldsymbol{0})$, and the reference error feedback gain matrix $\boldsymbol{K}(\boldsymbol{x},t)$ is set to $\text{diag}\{k_i\}$, then a decentralized control is obtained

$$\tau_i = \hat{m}_i \ddot{p}_{d,i} + (k_v \hat{m}_i + \alpha k_i)\dot{e}_i + (k_p \hat{m}_i + \alpha k_v k_i)e_i + \alpha k_p k_i \int e_i, \qquad (24)$$

which is a kind of PID control. Note that the control still has the property of $\mathcal{H}_\infty$ disturbance attenuation.

4.1 Task space RMC control with REF

In real applications, the trajectory to be tracked is defined using a coordinate $\boldsymbol{p} \in \Re^m$ different from the generalized coordinate $\boldsymbol{q} \in \Re^n$. The former constitutes the task space, and the latter the joint space in robotic control area. They are transformed using the kinematics

$$\boldsymbol{p} = \boldsymbol{k}(\boldsymbol{q}), \quad \dot{\boldsymbol{p}} = \boldsymbol{J}(\boldsymbol{q})\dot{\boldsymbol{q}} \qquad (25)$$

where the $m \times n$ matrix $\boldsymbol{J} = \dfrac{\partial \boldsymbol{k}}{\partial \boldsymbol{q}^T}$ is called the Jacobian matrix. We can design a direct task space $\mathcal{H}_\infty$ control making use of the well-known fact that the joint dynamics is transformed to the unique task dynamics which is also an EL system summarized in the following lemma [6].

Lemma 7. *If $m = n$, the joint dynamics of the EL system, given in (8), is transformed to the EL system in the task space, defined by (25). The dynamics is written as*

$$\boldsymbol{f} = \boldsymbol{\Lambda}(\boldsymbol{p})\ddot{\boldsymbol{p}} + \boldsymbol{\Gamma}(\boldsymbol{p},\dot{\boldsymbol{p}})\dot{\boldsymbol{p}} + \boldsymbol{\Upsilon}(\boldsymbol{p}) + \boldsymbol{\delta} \qquad (26)$$

Table 1: Dynamic parameters of a 2 DOF manipulator

link	$l(m)$	$m(kg)$	$r(m)$	$i(kgm^2)$
1	0.4	10.0	0.2	0.15
2	0.4	10.0	0.2	0.15

where
$$\Lambda = J^{-T}MJ^{-1}, \quad \Gamma = J^{-T}(C - MJ^{-1}\dot{J})J^{-1}$$
$$\Upsilon = J^{-T}G, \quad \delta = J^{-T}d.$$

where $f \in \Re^m$ is the equivalent input defined by $\tau = J^T f$.

Hence, defining $e_p = p_d - p$, the following RMC control
$$f = \hat{\Lambda}(\ddot{p}_d + k_v \dot{e}_p + k_p e_p) + \hat{\Gamma}\left(\dot{p}_d + k_v e_p + k_p \int e_p\right) + \hat{\Upsilon} - u_p \quad (27)$$

with the REF in the task space
$$u_p = -\alpha_p R_p^{-1}\left(\dot{e}_p + k_v e_p + k_p \int e_p\right) \quad (28)$$

achieves the direct L_2-gain attenuation property from the disturbance
$$w_p = \tilde{\Lambda}(\ddot{p}_d + k_v \dot{e}_p + k_p e_p) + \tilde{\Gamma}\left(\dot{p}_d + k_v e_p + k_p \int e_p\right) + \tilde{\Upsilon} + \delta. \quad (29)$$

REMARK 5 [Joint space implementation of task space RMC] When $(\hat{M}, \hat{C}, \hat{G})$ are available, the above task space RMC can be implemented directly in the joint space by
$$\tau = \hat{M}(q)\left\{\ddot{p}_{ref} - \dot{J}(q,\dot{q})J^{-1}(q)\dot{p}_{ref}\right\}$$
$$+ \hat{C}(q,\dot{q})J^{-1}(q)\dot{p}_{ref} + \hat{G}(q) - J^T(q)u_p, \quad (30)$$

which is equivalent to (27).

5 Numerical Simulation

Consider a two degrees of freedom planar manipulator, which falls into the Euler-Lagrange system. The exact set of dynamic parameters is summarized in Table 1.

The manipulator is commanded to trace a straight line connecting $(0.7, 0.0)^T (m)$ and $(0.0, 0.7)^T (m)$ in $1(sec)$, and trace it backwards, each in the bang-bang type interpolation. The initial configuration is $q = (-29.0, 57.9)^T (°)$.

The objective of the simulation is two-fold: one is to show the enhancement of robustness of the control as the L_2-gain γ decreases, and the other is to exhibit the easiness of controller tuning, since both controllers are of the decentralized type in the joint and the task space, and the tuning can be done only by the L_2-gain γ. The control gain used in reference motion was set as $k_p = 100$ and $k_v = 20$ at frequency of $1000(Hz)$. [2]

The first simulation employs the joint space implementation of the task space RMC with REF in decentralized form, that is (30) with $(\hat{M}, \hat{C}, \hat{G}) = (I, 0, 0)$. The position, velocity error, and torque are shown in Fig. 1, with γ decreased by half. It should be noted that the position and velocity error can be made arbitrary small with effectively same amount of control effort, as shown in (c1), (c2), and (c3) which shows that the control trajectory is almost similar whereas the error suppression is more and more effective. The comparison with the control performance by the conventional CTM (computed torque method) with exact parameters confirmed that there was no degradation of the performance.

The second control is based on the direct task space RMC with REF in 0 form, that is (27) with $(\hat{\Lambda}, \hat{\Gamma}, \hat{\Upsilon}) = (I, 0, 0)$. The results are shown in Fig. 2 with two values of γ, i.e. 0.025, and 0.0125. In fact, this type of control is very efficient in computation, since no matrix inverse is required. Although the control structure is very simple, it shows similarly good performances. For both implementations, the control efforts are almost the same.

Through these simulations, one can confirm superior robustness performance and the efficiency and easiness of the controller setup and tuning. The nonlinear $\mathcal{H}_\infty$ controller proposed in this article can be applied to kinematically redundant manipulators [9] by the kinematically decoupled joint space decomposition [8].

6 Conclusion

We designed a robust tracking controller for the EL system, especially for robot manipulators, by deriving the analytic solution to the nonlinear $\mathcal{H}_\infty$ disturbance attenuation control problem. The control based on the analytic solution consists of the RMC (reference motion compensation) control with the REF (reference error feedback). The robustness to model uncertainties is achieved in the sense of L_2-gain attenuation from the disturbance which includes the model uncertainties. Compared to those previously proposed nonlinear $\mathcal{H}_\infty$ controls [1, 2, 3, 7], the current controller seems to be more general and analytic, and less limited in validity

---

[2] The frequency of $1000(Hz)$ seems indeed possible because the controllers simulated are of the decentralized form.

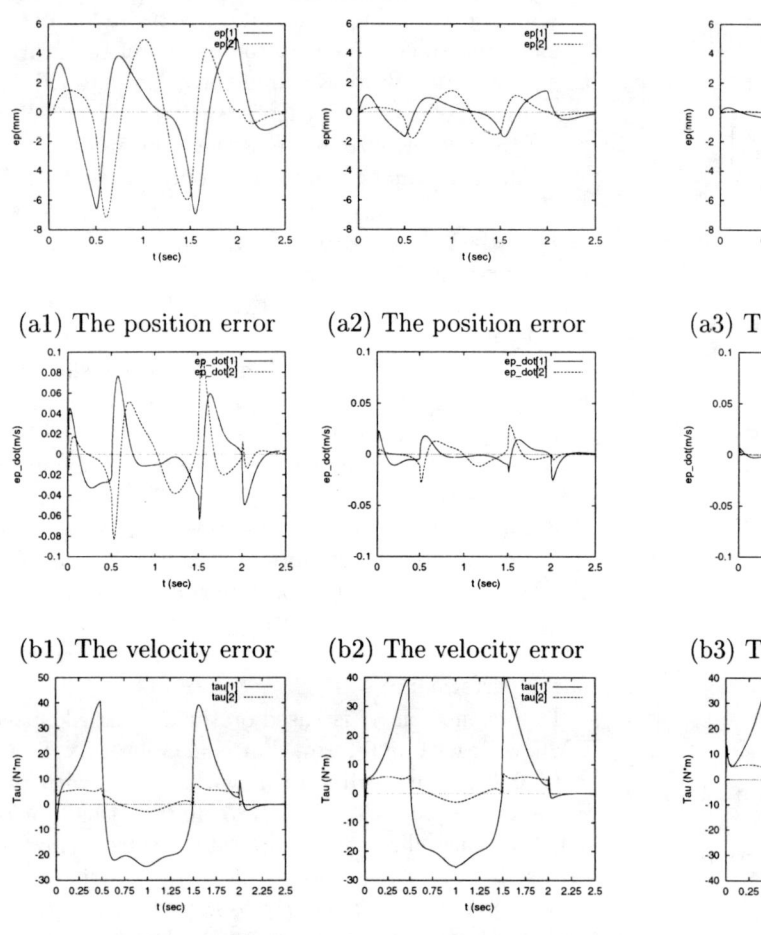

(a1) The position error (a2) The position error (a3) The position error

(b1) The velocity error (b2) The velocity error (b3) The velocity error

(c1) The control torque (c2) The control torque (c3) The control torque

Fig. 1: The joint space implementation of the task space RMC with REF $\mathcal{H}_\infty$ control ($\{$(a1),(b1),(c1)$\}$:$\gamma = 0.05\}$, $\{$(a2),(b2),(c2)$\}$:$\gamma = 0.025\}$)

Fig. 1: (Cont'd) The joint space implementation of the task space RMC with REF $\mathcal{H}_\infty$ control ($\{$(a3),(b3),(c3)$\}$:$\gamma = 0.0125\}$)

and application, thanks to easiness of controller tuning. (Consider the decentralized controller (24) with the tuning method discussed in Remark 3.2.) The control was applied to a two degrees-of-freedom manipulator and the result shows excellent performances.

References

[1] A. Astolfi, and L. Lanari, "Disturbance attenuation and set-point regulation of rigid robots via $\mathcal{H}_\infty$ control", *Proc. 33rd IEEE Int. Conf. on Decision and Control*

[2] S. Battilotti, and L, Lanari, "Tracking with disturbance attenuation for rigid robots", *Proc. 1996 IEEE Int. Conf. on Robotics and Automation*, pp. 1578-1583, 1996

[3] B.-S. Chen, T.-S. Lee, and J.-H. Feng, "A nonlinear $\mathcal{H}_\infty$ control design in robotics systems under parametric perturbation and external disturbance", *Int. J. Control*, vol. 59, no. 12, pp. 439-461, 1994

[4] D. Dawson, M. Grabbe, and F. L. Lewis, "Optimal control of a computed-torque controller for a robot manipulator", *Int. J. of Robotics and Automation*, vol. 6, no. 3, pp. 161-165, 1991

[5] A. Isidori, "Feedback control of nonlinear systems", *Int. J. Robust and Nonlinear Control* vol. 2, pp. 291-311, 1992

[6] O. Khatib, "A unified approach for motion and force control of robot manipulators: The operational space formulation", *IEEE Tr. Robotics and Automation*, vol. 3, no. 1, pp. 43-53, 1987

[7] T. Nakayama, and S Arimoto. "$\mathcal{H}_\infty$ control for robotic systems using the passivity concept",

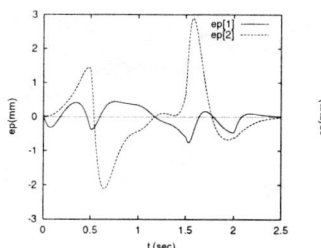
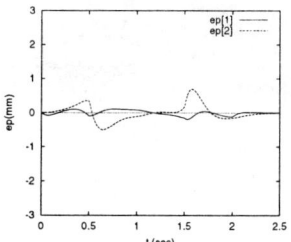

(a1) The position error   (a2) The position error

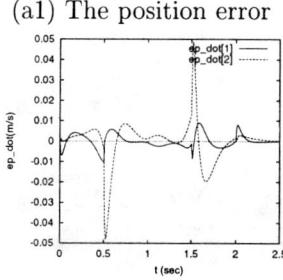
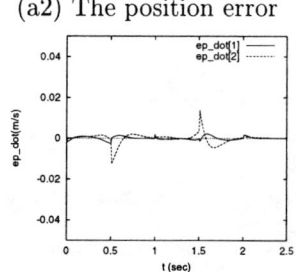

(b1) The velocity error   (b2) The velocity error

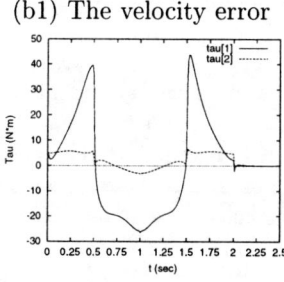
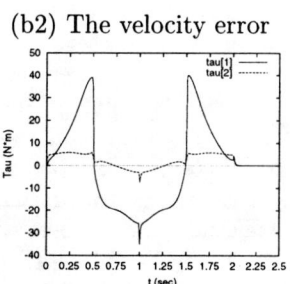

(c1) The control torque   (c2) The control torque

Fig. 2: The task space RMC with REF $\mathcal{H}_\infty$ control ({(a1),(b1),(c1):$\gamma = 0.025$}, {(a2),(b2),(c2):$\gamma = 0.0125$})

Proc. 1996 IEEE Int. Conf. on Robotics and Automation pp. 1584-1589, 1996

[8] J. Park, W. Chung, and Y. Youm, "Weighted decomposition of kinematics and dynamics of kinematically redundant manipulators", *Proc. 1996 IEEE Int. Conf. Robotics and Automation*, pp. 480–486, 1996

[9] J. Park, W. Chung, and Y. Youm, "$\mathcal{H}_\infty$ robust motion control of kinematically redundant manipulators", submitted to *Proc. 1998 IEEE/RSJ Int. Conf. Intelligent Robots and Systems*, 1998

[10] A. J. van der Schaft, "L_2-gain analysis of nonlinear systems and nonlinear state feedback $\mathcal{H}_\infty$ control", *IEEE Tr. Automatic Control*, vol. 37, no. 6, pp. 770-784, 1992

[11] J. J. E. Slotine, and W. Li, "On the adaptive control of robot manipulators", *Int. J. Robotics Research*, vol. 6, no. 3, pp. 49-59, 1987

A Proof of Lemma 2

With the candidate solution

$$V(\boldsymbol{x},t) = \frac{1}{2}\boldsymbol{x}^T \boldsymbol{P}(\boldsymbol{x}_1, \cdots, \boldsymbol{x}_{r-1}, t)\boldsymbol{x}$$

for the system (9) one gets $\left[\frac{\partial \boldsymbol{P}}{\partial \boldsymbol{x}^T}\boldsymbol{x}\right]\boldsymbol{B} = \left[\frac{\partial \boldsymbol{P}}{\partial \boldsymbol{x}^T}\boldsymbol{x}\right]\boldsymbol{D} = 0$, since

$$\left[\frac{\partial \boldsymbol{P}}{\partial \boldsymbol{x}^T}\boldsymbol{x}\right]\boldsymbol{B} = \left[\frac{\partial \boldsymbol{P}}{\partial \boldsymbol{x}_1^T}\boldsymbol{x} \cdots \frac{\partial \boldsymbol{P}}{\partial \boldsymbol{x}_{r-1}^T}\boldsymbol{x}\,0\right]\begin{bmatrix} 0 \\ \vdots \\ 0 \\ \boldsymbol{b}(\boldsymbol{x},t) \end{bmatrix}.$$

Also, it is easy to show

$$\left[\frac{\partial \boldsymbol{P}}{\partial \boldsymbol{x}^T}\boldsymbol{x}\right]\boldsymbol{A}\boldsymbol{x} = \left[\frac{\partial \boldsymbol{P}}{\partial \boldsymbol{x}^T}\boldsymbol{x}\right]\begin{bmatrix} \boldsymbol{x}_2 \\ \vdots \\ \boldsymbol{x}_r \\ \boldsymbol{a}(\boldsymbol{x},t) \end{bmatrix} = \sum_{k=1}^{r-1}\left[\frac{\partial \boldsymbol{P}}{\partial \boldsymbol{x}_k^T}\boldsymbol{x}\right]\frac{\partial \boldsymbol{x}_k}{\partial t}.$$

Since

$$\left[\frac{\partial \boldsymbol{P}}{\partial \boldsymbol{x}_k^T}\boldsymbol{x}\right]\frac{\partial \boldsymbol{x}_k}{\partial t} = \left[\frac{\partial \boldsymbol{P}}{\partial \boldsymbol{x}_{k,1}}\boldsymbol{x} \cdots \frac{\partial \boldsymbol{P}}{\partial \boldsymbol{x}_{k,n}}\boldsymbol{x}\right]\dot{\boldsymbol{x}}_k$$

$$= \sum_{i=1}^{rn}\frac{\partial \boldsymbol{P}}{\partial x_{k,j}}\dot{x}_{k,j}\boldsymbol{x},$$

we can see that

$$\left[\frac{\partial \boldsymbol{P}}{\partial \boldsymbol{x}^T}\boldsymbol{x}\right]\boldsymbol{A}\boldsymbol{x} = \left[\sum_{i=1}^{(r-1)n}\frac{\partial \boldsymbol{P}}{\partial x_i}\dot{x}_i\right]\boldsymbol{x} = \left[\sum_{i=1}^{rn}\frac{\partial \boldsymbol{P}}{\partial x_i}\dot{x}_i\right]\boldsymbol{x}.$$

Then,

$$V_t + \boldsymbol{V}_x\boldsymbol{A}\boldsymbol{x} = \frac{1}{2}\boldsymbol{x}^T\left\{\frac{\partial \boldsymbol{P}}{\partial t} + \sum_{i=1}^{rn}\frac{\partial \boldsymbol{P}}{\partial x_i}\dot{x}_i + \boldsymbol{P}\boldsymbol{A} + \boldsymbol{A}^T\boldsymbol{P}\right\}\boldsymbol{x}$$

$$\boldsymbol{V}_x\left\{\boldsymbol{B}\boldsymbol{K}\boldsymbol{B}^T - \frac{1}{\gamma^2}\boldsymbol{D}\boldsymbol{D}^T\right\}\boldsymbol{V}_x^T$$
$$= \boldsymbol{x}^T\boldsymbol{P}\left\{\boldsymbol{B}\boldsymbol{K}\boldsymbol{B}^T - \frac{1}{\gamma^2}\boldsymbol{D}\boldsymbol{D}^T\right\}\boldsymbol{P}\boldsymbol{x}$$

Hence, (5) reduces to (16). Also, (17) follows by noting that

$$\boldsymbol{V}_x\boldsymbol{B} = \boldsymbol{x}^T\boldsymbol{P}\boldsymbol{B}.$$

□

High Speed Tracking Control of Stewart Platform Manipulator via Enhanced Sliding Mode Control

Nag-In Kim* and Chong-Won Lee**

* Daewoo Heavy Industries Ltd., 6, Manseok-dong, Tong-gu, Incheon, South Korea, 401-010
Tel:+82-42-869-3056;Fax:+82-42-869-8220; E-mail:s_nikim@cais.kaist.ac.kr
** Center for Noise and Vibration Control, Department of Mechanical Engineering
KAIST, Science Town, Taejon, South Korea, 305-701
Tel:+82-42-869-3016;Fax:+82-42-869-8220; E-mail:cwlee@hanbit.kaist.ac.kr

Abstract: High speed tracking control of a 6-6 Stewart platform manipulator is performed by employing an enhanced sliding mode control with reduced manipulator dynamics when the manipulator is assumed to be operated on a low frequency planar motion unit. The high-performance tracking control strategy normally requires the complex full dynamics of 6-6 Stewart platform manipulator. The dynamics becomes even more complicated in the presence of the base motion of the manipulator, requiring additional sensors for measurement of the base motion. It is shown that enhanced sliding mode control implemented with perturbation compensation and reaching phase alleviation functions can effectively remove the use of the complex full dynamics and the additional sensors in the control system for high performance tracking control of the Stewart platform manipulator under the effects of virtual base motion.

1. INRODUCTION

In recent years, there has been considerable interest in the area of parallel manipulator, which provides better accuracy, rigidity, load-to-weight ratio, and load distribution than serial manipulator. Such advantages of fully parallel manipulators[1] originate from the fact that the actuators act in parallel sharing the common payload. Thus, parallel manipulators have been often used for high speed, high precision and large payload environments. Practical usage of the Stewart platform manipulator(SPM), the most famous 6 degree-of-freedom parallel manipulator, has generally been in the area of low speed and large payload conditions such as motion base of the classical automobile[2] or flight simulator, and motion bed of a machine tool[3]. In such applications, simple independent joint-axis PID or/and adaptive control[4], without knowledge of the complex dynamics in the design of the controller, have been widely adopted because the driving force required for generation of the desired motion is mostly consumed to compensate for the gravity force associated with the payload and external disturbance forces. However, high-performance tracking control of the high speed motion base used in enhanced automobile simulators and multi-axes vibration isolators generally necessitates the knowledge of the complex dynamics and the model-based controller.

The real time calculation of forward kinematics[5] and dynamics of SPM may be a difficult task owing to the time consuming nature. Lee and Kim[6] proposed the dual-processors based computing architecture to resolve the problem and introduced the joint-axis sliding mode control(SMC). On the other hand, SPM may be operated on a moving vehicle[7] or a low speed motion unit while the dynamics of the SPM is derived with the assumption that the lower base of the SPM is fixed to the inertia coordinates. Although such manipulators may normally experience low frequency base motion, large unpredictable dynamic force may exert to the system and the induced large tracking errors can not be avoided by the conventional SMC.

In this work, large tracking errors are effectively reduced by incorporating the reaching phase alleviation[8] and perturbation compensation[9] functions into the conventional SMC based on fixed base motion dynamics of the SPM. The reaching phase alleviator, which changes the conventional sliding surface to an augmented one during initial transient period, keeps the sliding function inside the boundary layer, or on the sliding surface, from start time, and thus it may alleviate the reaching phase of sliding mode control. The perturbation estimator effectively estimates the low frequency perturbation when the sliding function stays inside the boundary layer and then continuously compensates them in the control law. Using this enhanced SMC, the tracking control based on the fixed base motion dynamics is performed with the SPM and its high tracking control performance is proven experimentally.

2. CONTROL SYSTEM DESIGN

SPM configurations

The linear actuator system consisting of an AC servo motor and a linear ball screw system is employed in the laboratory SPM. The AC servo motor(Max. power: 200W) equipped with a 3000 pulse encoder generates torque and rotational motion, and the linear ball screw system(lead: 25mm) converts them into linear force(Max. rated force:240N; Peak force:720N) and linear motion. The motion range and dimensions of the SPM are given in Table 1 and Fig.1, respectively.

Table 1 Maximum one degree-of-freedom displacements of SPM

	Displacement, m		Displacement, °
x	+0.13 ◁▷ −0.12	α	+9.8 ◁▷ −9.8
y	+0.11 ◁▷ −0.11	β	+11.2 ◁▷ −9.2
z	+0.82 ◁▷ +0.72	γ	+21.5 ◁▷ −21.5

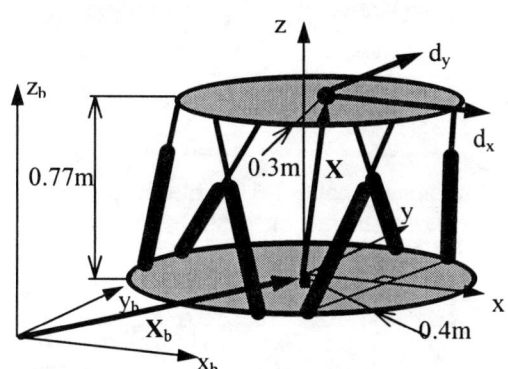

Figure 1 Stewart platform manipulator and the coordinates

Kinematics and Dynamics of SPM system

For the linear actuator system modeling, we consider only inertia and proportional friction terms, leaving other complex nonlinear behavior parts to the unknown perturbation. Lee and Kim[6] proposed the control system for the SPM that includes the iterative Newton-Raphson(NR) method[5] for the forward kinematics and the algebraically operated SPM dynamics including actuator dynamics, which basic form was proposed by Zhang and Song[10] expressed as

$$u_j(t) = \sum_{k=1}^{6}\{m_{jk}(\mathbf{X}(t))\ddot{l}_k(t)\} + V_j(\mathbf{X}(t),\dot{\mathbf{X}}(t),\dot{l}_j(t))$$
$$+ G_j(\mathbf{X}(t)), \quad j=1,2,\cdots,6 \quad (1)$$

which can be used for design of the joint-axis control system. Here u_j denotes the actuated force; $\mathbf{X} = [x\ y\ z\ \alpha\ \beta\ \gamma]^T$ and $\dot{\mathbf{X}}$ are the coordinate and velocity vectors of the upper centroid, respectively; α, β and γ are the rotational angles about x, y and z axis, respectively; $\dot{l}_j$ and $\ddot{l}_j$ are the velocity and acceleration of actuator length, respectively; $[m_{jk}] \in R^{6\times 6}$ represents the inertia mass matrix of the SPM system, which is symmetric and nonsingular; $[V_j] \in R^{6\times 1}$ corresponds to frictional force of the actuator system, the centrifugal and Coriolis force vectors of the SPM system; $[G_j] \in R^{6\times 1}$ is the gravity force.

Control hardware and software

A model based control strategy generally needs calculation of system dynamics at every fixed sampling time interval. However, the SPM dynamics including numerical forward kinematics is generally difficult to calculate within a given sampling time interval. Thus, to reduce this difficulty, control architecture for the SPM control has been proposed as shown in Fig. 2[6]. The forward kinematic and dynamic equations are calculated on a PC(Pentium 133MHz) asynchronously to a DSP that is a general digital servo controller associated with TMS320C40, ADCs, DACs and counters. The PC provides the dynamic properties of the SPM, with a little uncertainty induced by the asynchronous calculation, to the digital servo controller without processing interference. Thus it can achieve a high speed feedback loop and design the model based control system for the SPM

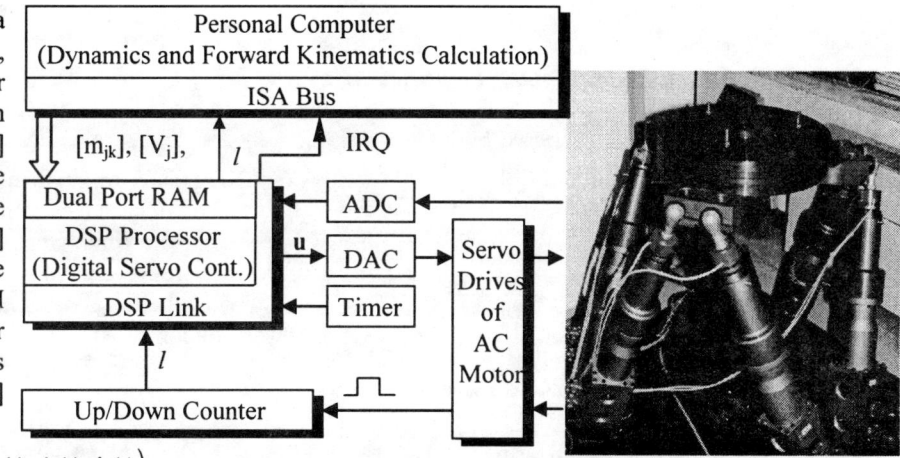

Figure 2 Control block diagram of SPM

Design of the conventional SMC

The dynamic system (1) can be re-written, in the presence of perturbation, as

$$u_j(t) = \sum_{k=1}^{6}\{m_{jk}(\mathbf{X}(t))\ddot{l}_k(t)\} + V_j(\mathbf{X}(t),\dot{\mathbf{X}}(t),\dot{l}_j(t)) + G_j(\mathbf{X}(t))$$
$$\sum_{k=1}^{6}\{\Delta m_{jk}\ddot{l}_k\} + \Delta V_j(\mathbf{X}(t),\dot{\mathbf{X}}(t),\dot{l}_j) + \Delta G_j(\mathbf{X}(t)) + d_j(t)$$
$$j=1,2,\cdots,6 \quad (3)$$

Here Δm_{jk}, ΔV_j and ΔG_j are the uncertainties of m_{jk}, V_j and G_j, respectively, and d_j denotes the external disturbance. The sliding function s is defined by[8,11]

$$s_j = \dot{e}_j + \lambda_j e_j - g_j(t) \quad j=1,2,\cdots,6 \quad (4)$$

where $e_j = l_j - l_{d,j}$, the positive constant λ_j is the desired control bandwidth, l_j and $l_{d,j}$ are the measured and desired actuator lengths, and $g_j(t)$ is the reaching phase alleviation term defined as[8]

$$g_j(t) = \{\dot{e}_j(0) + \lambda_j e_j(0)\} exp(-\varphi_j t) \quad (5)$$

where the positive definite φ_j is a time constant. Let the time derivative of the Lyapunov function candidate be given by $\frac{1}{2}\frac{d(s_j^2)}{dt} < 0$ to satisfy the boundary layer attraction condition, which gives the conventional SMC law defined as

$$u_j = \sum_{k=1}^{6}\{m_{jk}(\ddot{l}_{d,k} - \lambda_k \dot{e}_k + \dot{g}_k)\} - k_j sat(s_j, s_{o,j}) + V_j + G_j,$$
$$j = 1,2,\cdots,6 \quad (6)$$

where

$$k_j > max\left|\sum_{k=1}^{6}(\Delta m_{jk}\ddot{l}_k) + \Delta V_j + \Delta G_j + d_j\right| \quad (7)$$

Here the positive constant s_o is the thickness of boundary layer[11].

Sliding mode control with perturbation compensation

Now, we introduce the SMC with perturbation compensation which was first proposed by Kim *et al.*[9]. The actual perturbation, $P_{E,j}$, may be divided into known, or estimated, and residual perturbations, Φ_j and $\Delta P_{E,j}$, *i.e.*

$$P_{E,j} = \Phi_j + \Delta P_{E,j} \quad (8)$$

where

$$P_{E,j} = \sum_{k=1}^{6}\left\langle [b_{jk}]\left\{\sum_{i=1}^{6}(\Delta m_{ki}\ddot{l}_i) + \Delta V_k + \Delta G_k + d_k\right\}\right\rangle$$
$$\text{with } [b_{jk}] = [m_{jk}]^{-1}$$

Then equation (6) becomes

$$u_j = \sum_{k=1}^{6} m_{jk}\{\ddot{l}_{d,k} - \lambda_k \dot{e}_k + \dot{g}_k - \Phi_j\} - k_{r,j} sat(s_j, s_{o,j}) + V_j + G_j$$
$$(9)$$

with the boundary layer attraction condition defined as

$$k_{r,j} > max|\Delta P_{E,j}| \quad (10)$$

The structure of the controlled system is schematically shown in Fig. 3.

Dynamic system (3) may be divided into two parts: modeled and perturbation dynamic parts given by

$$u_{n,j} = \sum_{k=1}^{6}(m_{jk}\ddot{l}_{n,k}) + V_j + G_j \quad (11.a)$$

$$u_{p,j} = \sum_{k=1}^{6}\{m_{jk}(\ddot{l}_{p,k} + P_{E,k})\} \quad (11.b)$$

Here $l_j = l_{n,j} + l_{p,j}$, $l_{p,j}$ and $l_{n,j}$ are the coordinate vectors associated with the modeled and perturbation dynamics, respectively, $u_{n,j}$ and $u_{p,j}$ are the corresponding control forces. The perturbation dynamics does not allow the state of dynamic system to stay on the sliding surface, *i.e.* $s \neq 0$, whereas the modeled dynamics can be completely compensated by the Filippov's equivalent dynamics[11] inside boundary layers, *i.e*, $\dot{s} = 0$. Thus the s of the original system can be expressed by the state of perturbed dynamics given, inside boundary layers, as

$$s_j = \dot{l}_{p,j} + \lambda_j l_{p,j} - g_j, \quad j = 1,2,\cdots,6. \quad (12)$$

If a well-designed controller for the perturbation dynamic system can generate the control force $u_{p,j}$ that

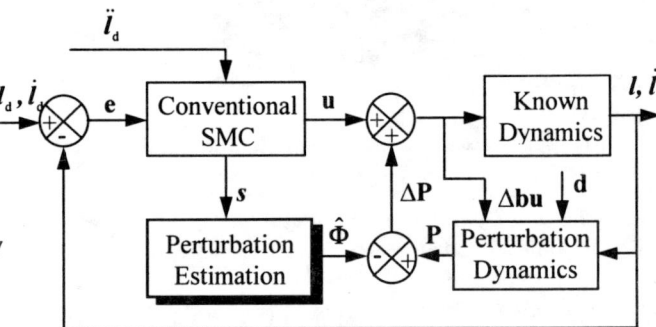

Figure 3. Structure of the proposed SMC system

drives the control states to its origin allowing only high-frequency errors, then most of low frequency perturbation acted upon the system is likely to be compensated by the control action. Thus the control force essentially reflects the relatively low frequency components of actual perturbations. It will prove convenient to introduce the relation

$$\ddot{l}_{p,j}(t) = -P_{E,j}(t) - \Phi_j(t), \quad j = 1,2,\cdots,6 \quad (13)$$

where $\Phi_j = -\sum_{k=1}^{6}\{[b_{jk}]u_{p,k}\}$. Since the terms Φ_j also contain undesirable high frequency components owing to control action, an observer is proposed as

$$\dot{\hat{\Phi}}_j(t) + \omega_{n,j}\hat{\Phi}_j(t) = \omega_{n,j}\Phi_j(t), \quad j = 1,2,\cdots,6 \quad (14)$$

Here $\hat{\Phi}_j$ is the filtered output of Φ_j and $\omega_{n,j}(rad/sec)$ is the filter frequency which indicates the frequency bandwidth for perturbation estimation.

The controlled dynamics of the form, for Φ_j,

$$\ddot{l}_{p,j}(t) + a_{j1}\dot{l}_{p,j}(t) + a_{j2}l_{p,j}(t) + g_{m,j}(t) = 0, \quad j = 1,2,\cdots,6 \quad (15)$$

The coefficients, a_{j1} and a_{j2}, make the perturbation dynamic response asymptotically decay to its origin on the sliding surface, and $g_{m,j}$ is a time variable function. Substituting equation (13) into equation (15), we obtain

$$\Phi_j(t) = -P_{E,j}(t) + a_{j1}\dot{l}_{p,j}(t) + a_{j2}l_{p,j}(t) + g_{m,j}(t),$$
$$j = 1,2,\cdots,6 \quad (16)$$

Note here that we need the information of $P_{E,j}(t)$ to obtain $\Phi_j(t)$. TDC for slowly varying perturbations and system dynamics[12] essentially assumes that, for a sufficiently small time delay Δt,

$$P_{E,j}(t) \cong P_{E,j}(t - \Delta t) = -\ddot{l}_{p,j}(t - \Delta t) - \Phi_j(t - \Delta t)$$
$$j = 1, 2, \cdots, 6 \quad (17)$$

The manipulator dynamics of interest possesses wide band frequency characteristics, so that the approximation (17) may not be valid. One of the simple, yet effective, remedies is to remove the high frequency components in $P_{E,j}$ by low-pass filtering. In other words, equation (17) may be replaced by

$$\hat{P}_{E,j}(t) \cong \hat{P}_{E,j}(t - \Delta t) = -\hat{\ddot{l}}_{p,j}(t - \Delta t) - \hat{\Phi}_j(t - \Delta t) \quad (18)$$

where '^' indicates the low-pass filtered value. Substituting equations (18) and (16) into equation (14), we obtain[9]

$$\hat{\Phi}_j(t) \cong k_{p,j}\left[s_j(t) + h_j \int s_j(t) dt\right] \quad (19)$$

for sufficiently small Δt and slow $\hat{\Phi}_j$. Here $k_{p,j} = \omega_{n,j}$, $a_{j1} = -(\lambda_j + h_j)$, $a_{j2} = -\lambda_j h_j$ and $g_{m,j} = -\dot{g}_j - h_j g_j$.

3. PROBLEM FORMULATION

In this study, SPM is to be used primarily as the motion generator of an aircraft or driving simulator. The driving simulator is known to have the smaller amplitude and higher frequency characteristics in the simulated motion than the aircraft simulator. Figure 1 shows the schematic layout and coordinates of a driving simulator running on the two degree-of-freedom planar motion unit. Thus, the two motion units share the desired motion, which is required by the automobile simulator, according to the system characteristics such that the SPM and the planar motion unit simulate the high and low frequency motions, respectively. Although the SPM simulates the high frequency motion, the inertia parts are exposed under the low frequency motion by the planar motion unit so that it leads the SPM dynamics to be a more complex one than equation (1).

Let us define the dynamics of the platform given as

$$\mathbf{u}_N(t) = \mathbf{u}(t) + \mathbf{M}_X(\mathbf{X})\ddot{\mathbf{X}}_b(t) + \Delta \mathbf{V}_X(\mathbf{X}, \dot{\mathbf{X}}, \dot{\mathbf{X}}_b) \quad (20)$$

Here $\mathbf{u}_N(t)$ represents the actuated force vector considering base motion in deriving the SPM dynamic equation, $\dot{\mathbf{X}}_b = [\dot{x}_b \ \dot{y}_b \ 0 \ 0 \ 0 \ 0]^T$ and $\ddot{\mathbf{X}}_b = [\ddot{x}_b \ \ddot{y}_b \ 0 \ 0 \ 0 \ 0]^T$ represent the base velocity and acceleration vectors of the SPM, respectively, and $\Delta \mathbf{V}_X(\mathbf{X}, \dot{\mathbf{X}}, \dot{\mathbf{X}}_b) = \mathbf{V}_{NX}(\mathbf{X}, \dot{\mathbf{X}}, \dot{\mathbf{X}}_b) - \mathbf{V}(\mathbf{X}, \dot{\mathbf{X}})$, $\mathbf{V}_{NX}(\mathbf{X}, \dot{\mathbf{X}}, \dot{\mathbf{X}}_b)$ is the Coriolis and centrifugal force vector in the presence of base motion, which may be a very small in value relative to other terms since the low dexterity of SPM. The term $\mathbf{M}_X(\mathbf{X})\ddot{\mathbf{X}}_b$, whose frequency characteristics closely resembles $\ddot{\mathbf{X}}_b$ that is sometimes difficult or/and expensive to measure, is the dominant dynamic force in dynamic equation (20) except $\mathbf{u}$. Thus we treat $\mathbf{M}_X(\mathbf{X})\ddot{\mathbf{X}}_b$ as an unmodelled dynamics in this study to simplify the SPM modeling, without using feedback sensors to measure $\ddot{\mathbf{X}}_b$.

4. EXPERIMENT

To check the controllability of the proposed controller, control experiments were performed under the given command trajectory:

$$\begin{cases} x(t) = 0.03\{1 - exp(-\pi t)\} cos(1.88\pi t), \text{ m} \\ y(t) = 0.04\{1 - exp(-\pi t)\} sin(1.88\pi t), \text{ m} \\ z(t) = \dfrac{0.02}{1 + 0.9t} sin\left\{2\pi t\left(\dfrac{0.1 + 5.9t}{10.5}\right) + \dfrac{\pi}{24}\right\}, \text{ m} \\ \alpha(t) = 0, \text{ deg} \\ \beta(t) = 4\{1 - exp(-\pi t)\} sin(0.86\pi t), \text{ deg} \\ \gamma(t) = 5\{1 - exp(-\pi t)\} sin(0.74\pi t), \text{ deg} \end{cases} \quad (21)$$
$$; 0 \leq t \leq 10$$
$$x(t) = y(t) = z(t) = \alpha(t) = \beta(t) = \gamma(t) = \gamma(t) = 0$$
$$; \text{otherwise}$$

The control parameter values throughout the tests are: $\lambda_j = 14\pi \text{ rad/sec}$, $s_{o,j} = 0.06 \text{ m/sec}$, $k_j = k_{r,j} = 160 \text{ N}$, $\omega_{n,j} = 3\pi \text{ rad/sec}$, and $h_j = \lambda_j = \varphi_j$, thus $g_{m,j} = 0$, for j=1,2, ..., 6. The payload of SPM is 64 kg. To simulate the effects of the base motion, the centroid of upper platform of the manipulator is assumed to be subjected to the disturbance forces, d_x and d_y, along the x and y directions, respectively, given by

$$\begin{Bmatrix} d_x \\ d_y \end{Bmatrix} = \begin{bmatrix} 180 sin(0.86\pi t) + 120 sin(1.26\pi t) \\ 150 sin(1.06\pi t) + 80 sin(1.74\pi t) \end{bmatrix}, \text{ N} \quad (22.a)$$

In fact, such disturbance forces can be experienced by the manipulator, for example, when the base motion of SPM is exposed to the disturbance motions, x_b and y_b, given by

$$\begin{Bmatrix} x_b \\ y_b \end{Bmatrix} = \begin{bmatrix} -0.385 sin(0.86\pi t) - 0.12 sin(1.26\pi t) \\ -211 sin(1.06\pi t) - 0.048 sin(1.74\pi t) \end{bmatrix}, \text{ m} \quad (22.b)$$

Because most of time consuming routines are executed asynchronously by a PC, we can carry out the control task within the sampling time interval of 1 m*sec*. The asynchronous task executed by the PC required about 2-3 m*sec* to finish its all routines. Figures 4-7 show the tracking errors in the operating coordinates, the typical sliding functions associated with the tracking errors in the actuator direction coordinates, the typical actuator control forces and the estimated virtual perturbations. Although both the conventional and enhanced SMCs did not induce chattering in the control input, the former failed in effectively rejecting the low frequency perturbations. Note that the enhanced SMC resulted in relatively smaller tracking errors and s values than the conventional SMC owing to its capability of rejecting the low frequency perturbation. Figure 6 shows that the time-varying sliding surface enables the estimated perturbation to exponentially track the actual low frequency perturbation in the initial start time period, and the estimated perturbations successfully track the large external disturbances that are the major source of tracking errors during the steady state. It implies that the control system was eventually subjected only to the small residual perturbation. The s value associated with the enhanced SMC remained inside the pre-defined boundary layer throughout the control test whereas the s value for the conventional SMC often violated the boundary layer attraction condition owing to the virtually low frequency and large external disturbances. It indicates that the low frequency external disturbances do not affect the tracking performance of the enhanced SMC, and the upper bound of the perturbation k_j in the conventional SMC should be increased until the boundary layer attraction condition is satisfied, that, in turn, may cause poor tracking performance and undesirable control oscillations.

5. SUMMARY AND CONCLUSIONS

The enhanced SMC and proposed control system was shown to be an effective approach for model based tracking control of the SPM, which is exposed to a low frequency base motion. Employing the enhanced SMC, we treated the dynamic part induced by low frequency base motion as an unmodelled dynamics of the SPM, which was then effectively compensated in the controller by the augmented sliding surface and perturbation estimation. It lead us to use the simple SPM dynamics in the control law, so that the control system became simple when the SPM runs on a low frequency planar motion unit. Experimental results confirmed that the control system with implemented the enhanced SMC allowed us to design a simple high-performance tracking control system for the SPM system under the high payload and large disturbance conditions.

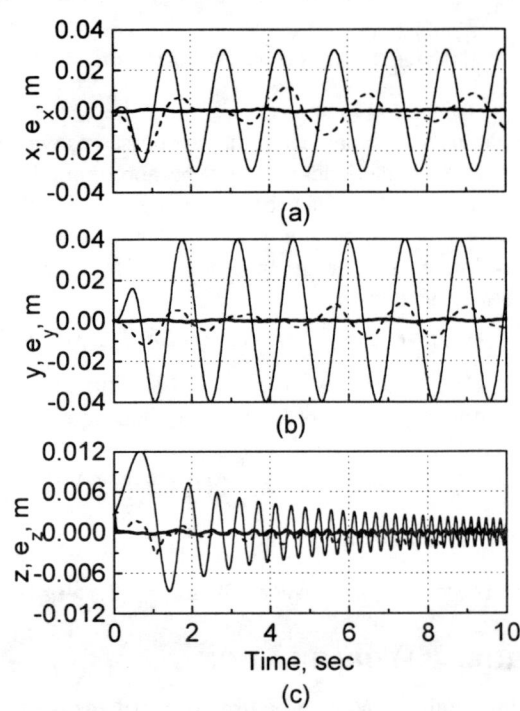

Figure 4 Typical tracking Errors: —— Command, – – Conventional SMC ($\hat{\Phi} = 0$), —— Enhanced SMC

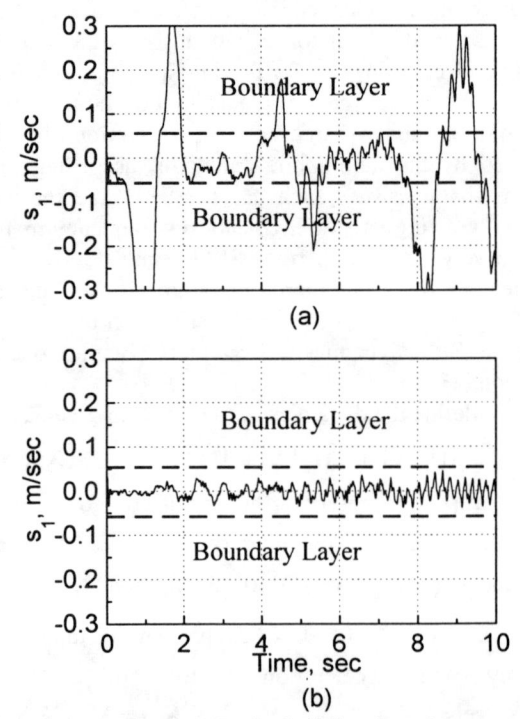

Figure 5 Typical sliding functions:
(a) Conventional SMC $\left(\hat{\Phi} = 0\right)$, (b) Enhanced SMC

REFERENCES

[1] D. Stewart, A Platform with six degree of freedom, Proceedings of the Institution of Mechanical Engineering, Vol. 180, Part 1, No. 5, 1965, pp. 371-386.

[2] S. Freeman et al., The Iowa driving simulator: An implementation and application overview, 1995, SAE 950174, pp. 1-10.

[3] G. Lebret, K. Liu and F. L. Lewis, Dynamic analysis and control of a Stewart platform manipulator, Journal of Robotic System, 1993, Vol. 10, No. 5, pp. 629-656.

[4] C. C. Nguyen, S. S. Antrazi, X.-L. Xhou and C. E. campbell, Jr., Adaptive control of a Stewart platform-based manipulator, Journal of Robotic System, 1993, Vol. 10, No. 5, pp. 657-668.

[5] R. V. Parrish, J. E. Dieudenne and D. J. Martin, Jr., Motion software for a synergistic six-degree-of-freedom motion base, NASA TN D-7350, 1973.

[6] C. W. Lee and N. I. Kim, Model based sliding mode control of Stewart platform manipulator, Korea Automatic Control Conference 97.

[7] J. Bormann and H. Ulbrich, Isolation of vibrations to avoid dynamic interactions between a telescope and its foundation by active control, Third International Conference on Motion and Vibration Control, 1996, pp. 88-93.

[8] K. B. Park and J. J. Lee, Variable structure controller for robot manipulators using time-varying sliding surface, IEEE Conference on Robotics and Automation, 1993, pp. 89-93.

[9] N. I. Kim, C. W. Lee and P. H. Chang, Tracking control of the parallel manipulator by sliding mode control with perturbation estimation, Submitted to Control Engineering Practice.

[10] C.-D. Zhang and S.-N. Song, An efficient method for inverse dynamics of manipulator based on virtual work principle, Journal of Robotic System, 1993, Vol. 10, No. 5, pp. 605-628.

[11] J.-J. E. Slotine, Sliding controller design for non-linear systems, International Journal of Control, Vol. 40, No. 2, 1984, pp. 421-434.

[12] K. Youcef-Toumi and O. Ito, A time delay controller for systems with unknown dynamics, ASME Journal of Dynamic Systems, Measurement, and Control, Vol. 112, 1990, pp. 133-142.

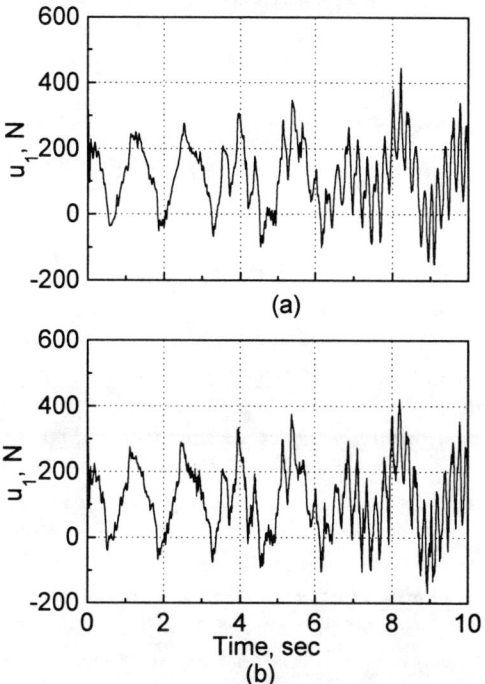

Figure 6 Typical control forces:
(a) Conventional SMC $(\hat{\Phi} = 0)$, (b) Enhanced SMC

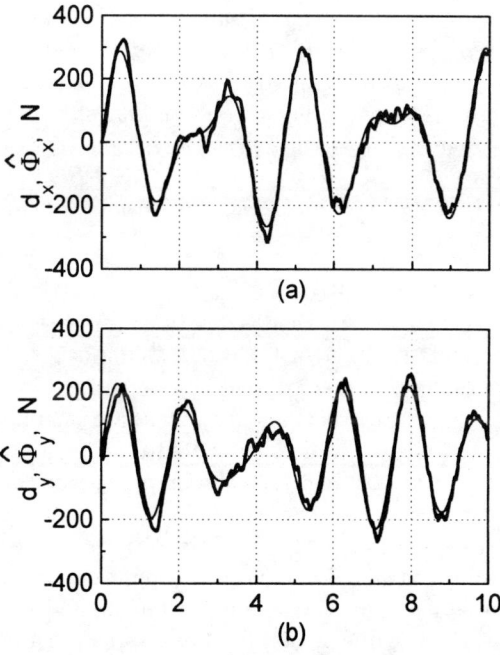

Figure 7 Applied and estimated forces: (a)x direction, (b)y direction forces
—— Applied disturbance (d_x, d_y), ▬ Estimated disturbance $(\hat{\Phi})$

Caging Planar Objects with a Three-Finger One-Parameter Gripper

Colin Davidson and Andrew Blake

University of Oxford, Oxford OX1 3PJ, UK.

davidson@robots.ox.ac.uk, ab@robots.ox.ac.uk

Abstract

This paper extends the caging theory of Rimon and Blake [11] for one-parameter two-finger grippers to one-parameter three-finger grippers. The caging theory describes how the fingers of a robot gripper can be placed around an object so that it cannot escape. As the fingers close from such a configuration, the freedom of the object to move is gradually restricted, until, in the absence of friction, it is completely immobilized by the fingers at an immobilizing grasp. The extension of the caging theory to three-fingered grippers is important because convex objects cannot be caged by two-finger grippers. The computation of the set of caging formations requires the identification of both the two- and three-finger frictionless grasps. These grasps correspond to critical points *of the opening parameter in the gripper's configuration space. There are two main problems here. Firstly, a method is needed to compute the critical points of the opening parameter in the three-finger contact space. This is complicated by the fact that this space may have more that one connected component. The second problem is how to associate the immobilizing grasps with punctures. In this paper we solve both these problems, presenting efficient algorithmic solutions.*

1 Introduction

In this paper a method is described for grasping planar objects in an error-tolerant fashion using a three-finger one-parameter gripper mounted on a robot arm. An image of the object and its immediate surroundings is provided by a camera, mounted directly above the object. It is assumed that the mapping from camera coordinates to real-world coordinates is known, and that the apparent contour of the object can be automatically extracted as a closed B-spline curve $\mathbf{r}(s)$, for example using an active contour technique [1, 5] (see figure 1). The fingers of the robot gripper are modelled as identical cylinders with their axis perpendicular to the image plane. Thus the fingers appear as discs in the plane. Without loss of generality we can assume point fingers as the original object contour can be replaced in the usual manner, via a Minkowski sum with the disc. The geometry of the object has been previously used to find optimal point grasps [9, 4, 2, 10]. A novel method of computing optimal three-finger grasps with a one-parameter gripper is presented here. We also present a method of computing an approach of the gripper which is guaranteed to reach the chosen optimal grasp. This succeeds even if the object is displaced from its observed configuration either by mechanical vibration or by measurement error. The technique is somewhat similar to those employed in *part feeder design*, where the objective is to design fixed obstacles to automatically orient parts on a conveyor belt into a stable known configuration [3]. Kinematic systems in which energy dissipates over time tend to reach configurations where the potential energy is at a local minimum, which correspond to stable poses of the object [7]. By choosing the right approach, the set of configurations which inevitably lead to a particular stable pose can be maximised, producing a highly error-tolerant system.

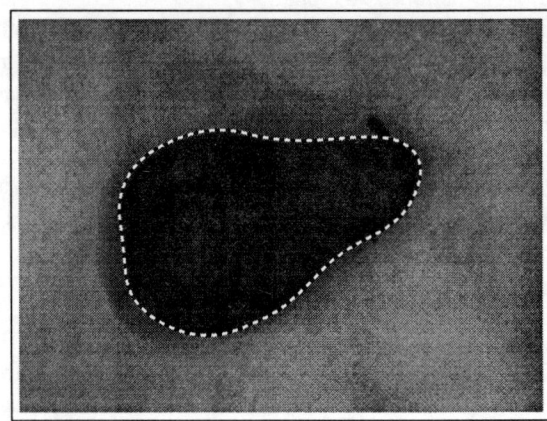

Figure 1: **A quadratic B-spline is automatically fitted to the apparent contour of a pear.**

The *caging problem* is to fix the fingers of a robot gripper around a stationary object in such a way that though the object may have some freedom to move, it cannot be removed completely. This is a *caging configuration*, and the problem of finding the *maxi-*

mal caging set of caging configurations was solved in the case of two-finger one-parameter grippers by Rimon and Blake [11]. This caging set has the additional property that closing the fingers from any configuration in the set is guaranteed to reach a particular immobilizing grasp. Here, we extend caging theory to the case of one-parameter three-finger grippers.

In the following, gripper configurations are thought of as points in configuration space, or *C-space*, denoted $\mathcal{C}$, which is four-dimensional. The four dimensions are the (x, y) translation of the arm in the plane, the rotation θ relative to the plane, and the opening parameter of the gripper σ (which is strictly positive). The function $\sigma(\mathbf{x})$ projects configurations $\mathbf{x} = (x, y, \theta, \sigma)$ of the robot gripper onto the opening parameter σ. The subset of $\mathcal{C}$ for which none of the fingers intersects the stationary object is *freespace*, denoted $\mathcal{F}$.

1.1 The caging set

A caging set $\mathcal{K} \subset \mathcal{F}$ is defined such that, given $\mathbf{x} \in \mathcal{K}$, any configuration connected to $\mathbf{x}$ by a path in freespace on which σ is non-increasing is also in $\mathcal{K}$. Thus, closing the fingers on the object from a caging configuration in $\mathcal{K}$ leads only to other caging configurations in $\mathcal{K}$. The set $\mathcal{K}$ is defined as follows. Consider the set of configurations

$$\mathcal{F}_c = \{\mathbf{x} \in \mathcal{F} \, : \, \sigma(\mathbf{x}) \leq c\}.$$

Let $\mathcal{K}_c$ be the connected component of $\mathcal{F}_c$ containing the immobilizing grasp $\mathcal{I}$, and let $\sigma(\mathcal{I}) = \sigma_0$. Then $\mathcal{K}_{\sigma_0}$ contains only the single point $\mathcal{I}$. As c increases from σ_0, $\mathcal{K}_c$ grows as a region and is the set of configurations reachable from $\mathcal{I}$ without the opening parameter exceeding c. At a certain critical value of $c = \sigma_1$, this region joins another component of $\mathcal{F}_c$ and the critical point where this happens is termed a *puncture point*. Thus any set $\mathcal{K}_c$ with $c < \sigma_1$ is a caging set and the maximal caging set $\mathcal{K}$ is

$$\mathcal{K} = \bigcup_{c < \sigma_1} \mathcal{K}_c.$$

Caging theory uses the mathematical tool *Stratified Morse Theory* [6] to prove that puncture points are frictionless equilibrium grasps which are critical points of σ with Morse index 1. (A critical point of a Morse function f on a manifold $\mathcal{M}$ is a point where $\nabla f = \mathbf{0}$ [6]. The Morse index of a critical point is the number of negative eigenvalues of the Hessian.) The puncture point of the caging set $\mathcal{K}$ is the puncture point with least σ value that also passes the *topological check*, a test which associates puncture points with immobilizing grasps. This gives rise to the algorithm in [11] for a two fingered gripper, where the set of all critical points is calculated and tested by the topological check in order of increasing σ-value. In order to use this algorithm, all the puncture points must be accounted for. In the case of the three-finger gripper, the punctures may include three-finger contact points, an example of which is illustrated in figure 2. Later, the alternate possibility of a two-finger puncture for a three-finger grasp is demonstrated, and it is shown that the three-finger-contact space can be complicated, even for a simple object.

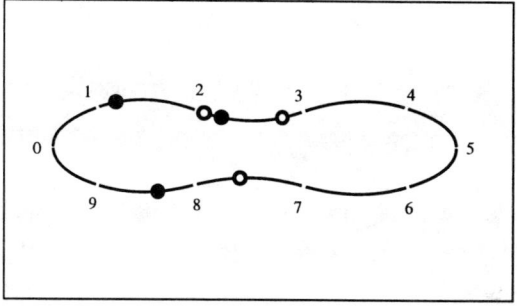

Figure 2: *A three-finger puncture point (filled circles) associated with a three-finger immobilizing grasp (unfilled circles), for a gripper whose fingers form the vertices of an equilateral triangle.*

2 The three-finger gripper

For three-finger grippers, the following problems must be solved. First, a method for efficiently searching $\mathcal{F}$ for all the critical points of σ is required. Second, these critical points must be classified as puncture points, immobilizing grasps and *false critical points* (*i.e.* critical points which are not grasps). Third, the topological check needs to be extended to deal with both two-finger and three-finger contacts.

One restriction is made on the choice of one-parameter three-finger gripper, namely that the Euclidean distance between any two of the fingers in the plane $|\mathbf{f}_i - \mathbf{f}_{i+1}|$ (where index arithmetic is modulo 3) is a strictly increasing function of σ. This restriction is acceptable as all standard three-finger grippers satisfy this property. It is important because, if the distance between fingers i and $i+1$ varies monotonically with σ, then knowing the finger positions $\mathbf{f}_i$ and $\mathbf{f}_{i+1}$ uniquely determines the position of the third finger, so we can write $\mathbf{f}_{i+2} = \mathbf{f}_{i+2}(\mathbf{f}_i, \mathbf{f}_{i+1})$. This defines three functions $\mathbf{f}_1(\mathbf{f}_2, \mathbf{f}_3)$, $\mathbf{f}_2(\mathbf{f}_3, \mathbf{f}_1)$, and $\mathbf{f}_3(\mathbf{f}_1, \mathbf{f}_2)$ which fully describe the geometry of the gripper.

This allows the definition of the *contact C-space* $\mathcal{X}_{ij} := \mathcal{S} \times \mathcal{S}$ for any two fingers i and $j = i+1$ (modulo 3), where $\mathcal{S}$ is the periodic interval $[0, L]$, L is the number of B-spline spans, and $(s_i, s_j) \in \mathcal{X}_{ij}$ corresponds to a two-finger grasp at $\mathbf{f}_i = \mathbf{r}(s_i)$ and $\mathbf{f}_j = \mathbf{r}(s_j)$. Since the location of the third finger $k = i + 2$ is uniquely defined by $\mathbf{f}_k(\mathbf{r}(s_i), \mathbf{r}(s_j))$, every two-finger contact involving fingers i and j is represented by a point in $\mathcal{X}_{ij}$. This implies that every three-finger contact is represented by a point in each of the $\mathcal{X}_{ij}$.

3 Searching for critical points

The freespace $\mathcal{F}$ is a stratified set, consisting of a collection of manifolds called *strata*. In this section, the strata are determined, and then searched for critical points.

3.1 The strata of $\mathcal{F}$

The freespace $\mathcal{F}$ is formed by the removal of three *finger C-obstacles* from $\mathcal{C}$. These are the configurations where each finger intersects with the object. It follows that points on the boundary of each finger C-obstacle are configurations where that finger touches the object. $\mathcal{F}$ can therefore be partitioned into the following eight strata:

- One four-dimensional manifold, the interior of freespace, $\mathcal{F}_0 = I(\mathcal{F})$. This is the no-finger contact stratum and contains no critical points of $\sigma(\mathbf{x})$.

- Three three-dimensional manifolds, $\mathcal{F}_1$, $\mathcal{F}_2$, and $\mathcal{F}_3$. These are the one-finger contact strata, containing configurations where exactly one finger touches the object. These strata also do not contain any critical points of $\sigma(\mathbf{x})$.

- Three two-dimensional manifolds, $\mathcal{F}_{12}$, $\mathcal{F}_{23}$, and $\mathcal{F}_{31}$. These are the two-finger contact strata, containing configurations of the gripper where exactly two fingers touch the object. These strata do contain critical points of $\sigma(\mathbf{x})$.

- One one-dimensional manifold, $\mathcal{F}_{123}$. This is the three-finger contact stratum. This stratum also contains critical points of $\sigma(\mathbf{x})$.

These sets form a disjoint partition of $\mathcal{F}$, and satisfy $\overline{\mathcal{F}_{ij}} = \overline{\mathcal{F}_i} \cap \overline{\mathcal{F}_j}$ and $\overline{\mathcal{F}_{123}} = \overline{\mathcal{F}_1} \cap \overline{\mathcal{F}_2} \cap \overline{\mathcal{F}_3}$, where $\overline{S}$ denotes the closure of S.

Let $\phi_{ij} : \mathcal{F}_{ij} \to \mathcal{X}_{ij}$ be the smooth 1-1 function, mapping configurations of the gripper in which fingers i and j are touching the object and k is outside the object, to the corresponding contact configurations. Consider $\phi_{ij}(\mathcal{F}_{123})$; this is the set of three-finger contacts expressed as a subset of $\mathcal{X}_{ij}$. Since ϕ_{ij} is 1-1 we can write

$$\mathcal{F}_{123} = \phi_{ij}^{-1}(\{(s_i, s_j) \in \mathcal{X}_{ij} : D(\mathbf{f}_k(\mathbf{r}(s_i), \mathbf{r}(s_j))) = 0\})$$

where D is the minimum Euclidean distance of a point in the plane from the object. Since all these functions are continuous, it follows that $\mathcal{F}_{123}$ is the level set of a continuous real-valued function on a two-dimensional space. Therefore it consists of closed loops and arcs which terminate at $\sigma = 0$.

3.2 The two-finger contact strata

Since ϕ_{ij} is smooth, the critical points of $\sigma(\mathbf{x})$ on $\mathcal{F}_{ij}$ can be sought in $\mathcal{X}_{ij}$ as the critical points of $\sigma(s_i, s_j)$ just as in the case of the two-finger gripper, by tracking along the antisymmetry set searching for intersections of the symmetry set [2]. For the two-finger gripper, $\phi_{ij}(\mathcal{F}_{ij})$ fills the entire contact C-space $\mathcal{X}_{ij}$ except for the axis $s_i = s_j$, where $\sigma = 0$. In the case of the three-finger gripper, it is not necessarily true that all three fingers meet at $\sigma = 0$. Hence $\mathcal{X}_{ij}$ may contain regions which do not correspond to any gripper configuration.

In addition, $\mathcal{X}_{ij}$ contains points where the third finger lies inside the object. Such configurations are not members of $\mathcal{F}$, and form *forbidden regions* in $\mathcal{X}_{ij}$ which correspond to the third finger C-obstacle. Critical points in forbidden regions are ignored for this reason. The edges of forbidden regions are made up of points on the boundary of the third finger C-obstacle, which implies that they correspond to points in $\mathcal{F}_{123}$, the three finger contact stratum.

3.3 The three-finger contact stratum

$\mathcal{F}_{123}$ is a one-dimensional set. When projected into $\mathcal{X}_{ij}$, it consists of a number of closed loops (which surround forbidden regions) and a number of curved arcs which, together with the curve $\sigma = 0$, bound forbidden regions (see figure 4). The critical points of $\sigma(\mathbf{x})$ could be easily found by tracking along these curves. Unfortunately, there is no obvious way of calculating the number of connected components in this stratum, and so we cannot easily obtain sample points on every component from which to begin tracking.

Therefore, the following alternative strategy is adopted. Consider the space of all three-finger contacts, $\mathcal{S} \times \mathcal{S} \times \mathcal{S}$. Fixing one of the fingers at a particular value, say $s_1 = t$, is equivalent to intersecting this space with a plane, containing points of the form (t, s_2, s_3). The position of fingers 1 and 2 uniquely determines the position of finger 3, $\mathbf{f}_3(\mathbf{r}(s_1), \mathbf{r}(s_2))$. Therefore, any three-finger contact in the plane (t, s_2, s_3) is also a solution of the equation

$$D(\mathbf{f}_3(\mathbf{r}(t), \mathbf{r}(s_2))) = 0 \qquad (1)$$

where as before, $D(\mathbf{x})$ is the minimum distance of a point from the object. By considering single spans of the B-spline, this equation reduces to a polynomial in

s_2. The degree of this polynomial is kept to a minimum by using quadratic B-splines, and by using a gripper of simple geometry. In any case, for higher order polynomials, all the roots can be found by a technique such as homotopy continuation [8]. Values of s_1 are sampled at N points per span, on L spans of the B-spline in $O(NL^3)$ time. This is then repeated, sampling in the s_2 and then s_3 directions. In this way, the space of three-finger contacts $\mathcal{S} \times \mathcal{S} \times \mathcal{S}$ is effectively divided into cubes of side $1/N$, and all the intersections of the surface of these cubes with $\mathcal{F}_{123}$ are calculated. As N becomes large, there will be at most 2 such intersections per cube, which can be linked to produce connected chains of points (see figure 3).

form the vertices of an equilateral triangle. In this case, the functions $\mathbf{f}_{i+2}(\mathbf{f}_i, \mathbf{f}_{i+1})$ for $i = 1, 2, 3$ are linear and so, for a quadratic B-spline, the function $D(\mathbf{f}_{i+2}(\mathbf{r}(t), \mathbf{r}(s_{i+1})))$ is only quartic in s_{i+1}. As a result, the roots can be calculated quickly.

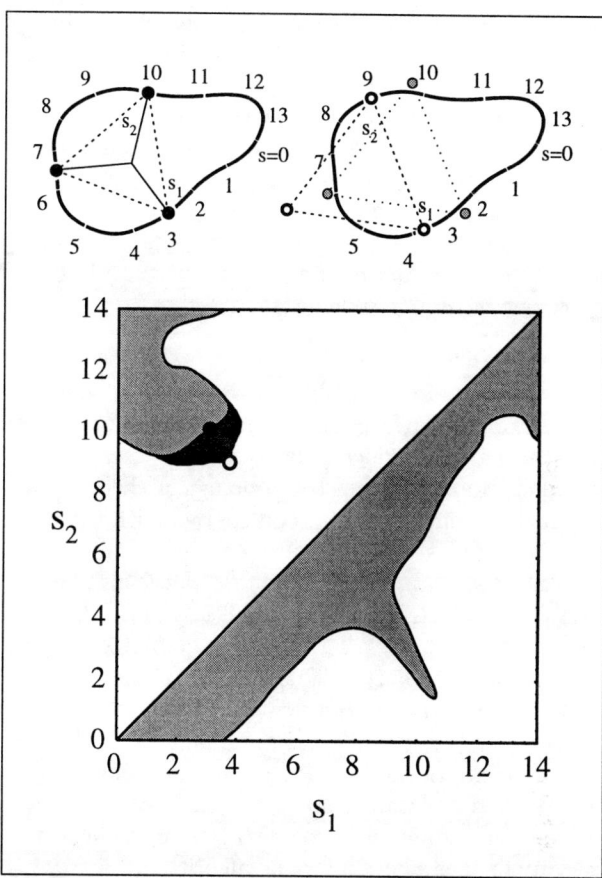

Figure 4: **Grasping a pear.** *(Top left) A three-finger immobilizing grasp. (Top right) The puncture grasp of the cage, and (grey circles) a configuration inside the cage. (Bottom) The contact C-space $\mathcal{X}_{12}$ (black curve). The cage is shaded dark. Light grey regions are forbidden regions, and the black curve boundary is the three finger contact space. The immobilizing grasp and the puncture point are shown.*

---

compute 3-finger contact stratum

1 **For** $f = 1, 2, 3$; *each finger*
  For $i = 0, ..., L-1$; *for s_f in each span*
  For $t = 0, ..., N-1$; *N samples of s_f per span*
  For $j = 0, ..., L-1$; *for s_{f+1} in each span*
  For $k = 0, ..., L-1$; *for s_{f+2} in each span*
  Set $s_f = i + t/N$
  For each root s of
  $$D_k(\mathbf{f}_{f+2}(\mathbf{r}(s_f), \mathbf{r}_j(s))) = 0$$
  If $(s \in \mathbb{R}$ and $s \in (0, 1))$ **Then**
  Set $s_{f+1} = j + s$
  Solve $\mathbf{r}_k(r) = \mathbf{f}_{f+2}(\mathbf{r}(s_f), \mathbf{r}(s_{f+1}))$ for r
  If $(r \in \mathbb{R}$ and $r \in (0, 1))$ **Then**
  Set $s_{f+2} = k + r$
  Append (s_1, s_2, s_3) to $PLIST$

2 **For** $i = 1, ..., |PLIST|$
  Calculate the centre-points (c_{11}, c_{12}, c_{13}), (c_{21}, c_{22}, c_{23}) of the two cubes of side $1/N$ to which $PLIST[i]$ belongs
  Append the vectors $(c_{11}, c_{12}, c_{13}, i)$ and $(c_{21}, c_{22}, c_{23}, i)$ to QLIST

3 **Sort** QLIST lexicographically

4 **For** $i = 1, ..., |QLIST|$
  If $QLIST[i]_j = QLIST[i+1]_j$ for $j = 1, 2, 3$ **Then**
  Mark $QLIST[i]_4$ connected to $QLIST[i+1]_4$

Figure 3: **Algorithm for computing the three-finger contact stratum.**

A demonstration of the output of the algorithm is shown in figure 4, for a gripper whose fingers

Having found all of the closed loops and arcs which make up $\mathcal{F}_{123}$, the critical points on this stratum can be determined as the local minima and maxima of σ along the curves of $\mathcal{F}_{123}$.

4 Classification of the critical points

The classification of two-finger-contact critical points is exactly as in the two-finger gripper case [11]. Local minima (index 0) give immobilizing grasps; saddles (index 1) give puncture points; local maxima (index 2) are ignored. False critical points must be pruned. The three-finger critical points are straightforward; local minima (index 0) give immobilizing grasps, and local maxima (index 1) give puncture points. All that remains is to prune false critical points, using the following standard test for a three-finger equilibrium-grasp.

At a frictionless grasp, contact forces of the three fingers are in the direction of the inward-pointing unit normals, $\mathbf{n}_i$. For equilibrium, the forces must sum to zero. Therefore, there must exist scalars $\lambda_1, \lambda_2 > 0$ such that $\lambda_1 \mathbf{n}_1 + \lambda_2 \mathbf{n}_2 + \mathbf{n}_3 = 0$. In addition, these forces must produce no net moment. This gives the equation $\lambda_1 (\mathbf{r}_1 - \mathbf{r}_3) \cdot \mathbf{t}_1 + \lambda_2 (\mathbf{r}_2 - \mathbf{r}_3) \cdot \mathbf{t}_2 = 0$ where $\mathbf{t}_i$ is the tangent to the curve. A false critical point will fail to satisfy one of these equations because it will require one of the λ_i to be negative.

5 The topological check

The purpose of the topological check is to associate puncture points with immobilizing grasps by demonstrating a path on which σ is monotonic decreasing. The key to implementing the topological check is to keep track of which fingers are in contact with the object. The path is then just gradient descent on σ *in the relevant stratum* until either (a) a local minimum of $\sigma(\mathbf{x})$ on that stratum is reached, or (b) a free finger comes into contact with the object (detected by a change in sign of the distance of the finger from the B-spline). The track stops if an immobilization is reached. Otherwise, the local minimum is false and the track continues in case (a) by allowing one finger to break contact with the object. In case (b) the trace continues in a different stratum by maintaining the new contact. See figure 5 for the algorithm.

6 The parallel-jaw gripper

A particular gripper for which the three-finger contact space can be easily computed is the three-finger parallel-jaw gripper. It consists of two fingers ($\mathbf{f}_1$ and $\mathbf{f}_2$) and an opposing thumb ($\mathbf{f}_3$). The thumb moves along the perpendicular bisector of the line segment joining the two fingers, and the two fingers remain a fixed distance apart: $|\mathbf{f}_1 - \mathbf{f}_2| = c$. Note that this technically breaks the assumption that the distance between any two fingers varies strictly monotonically with σ. In this special case, the stratum $\mathcal{F}_{12}$ cannot be searched but contains no critical points anyway. The strata $\mathcal{F}_{23}$ and $\mathcal{F}_{31}$ can be searched in the usual way as these fingers do satisfy the monotonicity assumption. Therefore, the problem reduces to calculating the three-finger contact stratum $\mathcal{F}_{123}$.

For a given contact point of either finger 1 or 2, three-finger contacts can be constructed as follows. Consider $\mathbf{f}_1 = \mathbf{r}(s_1)$ for some fixed s_1. Calculate all intersections of the circle centred on $\mathbf{f}_1$ having radius c with the object outline $\mathbf{r}(s_2)$. These are the possible positions of the other finger, $\mathbf{f}_2$. Now construct the perpendicular bisector of the line segment joining $\mathbf{f}_1$ and $\mathbf{f}_2$ and calculate its intersections with the object outline to locate the thumb, $\mathbf{f}_3$. This set of points

---

find puncture associated with $\mathcal{I}$

1 **Calculate** the set $\mathcal{A} = \{A_1, ..., A_a\}$ of possible puncture points (critical points of σ in freespace with index 1 which are also grasps and satisfy $\sigma > \sigma(\mathcal{I})$)

2 **Sort** the A_i by increasing σ-value

3 **For** $i = 1, ..., a$

  Calculate the negative eigenvalue λ of the Hessian at $\mathcal{A}_i$ and the corresponding eigenvector $\mathbf{e}$

  For $j = 0, 1$

   If (A_i is a two-finger puncture) **Then**

    Set $t_0 = A_i + (-1)^j \sqrt{\frac{2\varepsilon}{\lambda}} \mathbf{e}$ (where ε is a small constant)

   If (A_i is a three-finger puncture) **Then**

    Set t_0 to be the previous ($j = 0$) or next ($j = 1$) three-finger contact in the chain

   While (t_0 is not an immobilizing grasp and $\sigma(t_0) > \sigma(\mathcal{I})$)

    For $n = 0, ...$ **Until** (t_n is a critical point or a free finger comes into contact with the object)

     Set $t_{n+1} = t_n - (\Delta \sigma) \nabla \sigma(t_n)$ (where $\nabla \sigma$ is the gradient of the function σ restricted to the current stratum and $\Delta \sigma$ is chosen so that σ is monotonic decreasing on the line segment (t_n, t_{n+1}))

    If (free finger f has touched) **Then**

     Set t_0 to the point nearest to t_n on the stratum in which finger f is in contact

    If (t is a false critical point) **Then**

     Set t_0 to the point nearest to t_n on a stratum in which continued descent is possible

    If ($t_n = \mathcal{I}$) **Then Return** $\mathcal{A}_i$

Figure 5: **Algorithm for finding the puncture.**

PLIST can be turned into a chain in a similar way to the algorithm in figure 3, calculating the centre-points of the squares of side $1/N$ in (s_1, s_2) space to which the points belong and inserting the actual s_3 value. Nearest neighbours in the s_3 direction in each square are connected to form chains. See figure 6 for an example of this gripper.

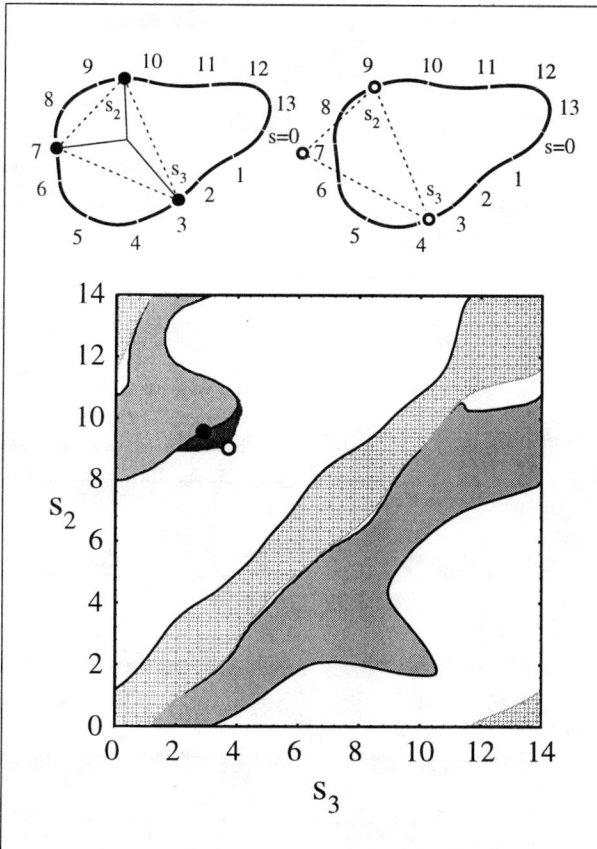

Figure 6: **A three-finger parallel-jaw gripper grasping a pear.** *(Top Left) The 3-finger immobilizing grasp. (Top Right) The puncture point, a two-finger saddle. (Bottom) The two-finger contact C-space $\mathcal{X}_{23}$ showing the forbidden regions in light grey and $\mathcal{F}_{123}$ as black curves. The hatched regions do not correspond to gripper configurations. The white region is freespace and the dark region is the caging set, with the immobilizing grasp and the puncture on its boundary.*

7 Conclusion

We have shown how to construct a cage with a one-parameter three-finger gripper. Closing the fingers on the object from any configuration in the caging set gradually restricts the possible movements of the object until it is completely immobilized. The object is thus held in a known frictionless immobilizing grasp. This paper has also proposed algorithms to efficiently compute the caging set, with run times typically under 10 seconds.

We made the assumption that the distance between any two of the fingers varies strictly monotonically with σ. We also assumed that the geometry of the gripper permits equation (1) to be written as a polynomial. However, we also demonstrated the solution in the special case where two of the fingers are a constant distance apart, breaking these assumptions. This suggests that the assumption can be relaxed to include a larger class of grippers. Another extension may be to deal with multiple degree of freedom grippers. Industrially, one-parameter grippers are preferred for their simplicity. However, it is possible that the problem of grasping curved objects with more complicated grippers may be greatly simplified by caging theory.

Finally, although it is easy to construct configurations in the caging set, it is not clear how to find a configuration in the caging set that maximises the clearance of the three fingers from the object. Such a configuration would be the ideal target to aim for in an automatic gripping system.

8 Acknowledgements

The authors gratefully acknowledge Dr Elon Rimon for introducing this subject and for many helpful conversations. Financial support for this research was provided by an EPSRC grant to CD.

References

[1] A. Blake and M. Isard. *Active contours.* Springer, 1998.

[2] A. Blake and M. Taylor. Planning planar grasps of smooth contours. In *Proc. IEEE Int. Conf. Robotics and Automation*, volume 2, pages 834–839, 1993.

[3] M. Brokowski, M. Peshkin, and K. Goldberg. Curved fences for part alignment. In *Proc. IEEE Int. Conf. Robotics and Automation*, volume 3, pages 467–473, 1993.

[4] I.M. Chen and J.W. Burdick. Finding antipodal point grasps on irregularly shaped objects. In *Proc. IEEE Int. Conf. Robotics and Automation*, pages 2278–2283, 1992.

[5] C. Davidson and A. Blake. Error-tolerant visual planning of planar grasp. In *Proc. 6th Int. Conf. on Computer Vision*, pages 911–916, 1998.

[6] M. Goresky and R. MacPherson. *Stratified Morse Theory.* Springer-Verlag, 1988.

[7] D.J. Kriegman. Let them fall where they may: Capture regions of curved objects and polyhedra. Int. J. Robotics Research, June 1997.

[8] A. Morgan. Computing all solutions to polynomial systems. *Applied Mathematics and Computation*, 24:115–138, 1987.

[9] V.D. Nguyen. Constructing force-closure grasps. *Int. J. Robotics Research*, 7(3):3–16, 1988.

[10] J. Ponce, J.W. Burdick, and E. Rimon. Computing the immobilizing three-finger grasps of planar objects. In *Proc. of the 1995 Workshop on Computational Kinematics*, pages 281–300, 1995.

[11] E. Rimon and A. Blake. Caging 2D bodies by one-parameter, two-fingered gripping systems. In *Proc. IEEE Int. Conf. Robotics and Automation*, pages 1458–1464. IEEE Press, 1996.

Geometric Formulation of Orientation Tolerances

J.B. Gou, Y.X. Chu, H. Wu and Z.X. Li*
Department of Electrical and Electronic Engineering
Hong Kong University of Science and Technology
Clearwater Bay, Kowloon, Hong Kong
(e-mail: eejbgou@ee.ust.hk, fax: 23581485)

Abstract

The rapid proliferation of coordinate measuring machines (CMMs) triggered the need for precise and rigorous formulations of each tolerance concept. In this paper, we employ the concept of configuration space of symmetric features to define each type of datum features and orientation tolerances. The establishment of datum features and orientation tolerances are formulated in an linear programming approach. Using properties of Lie groups and Lie algebra, a simple, unified and coordinate free algorithm for datum establishment and orientation tolerance is developed. The results of the algorithm comply to the definitions stipulated in the standard ANSI Y14.5M.

1 Introduction

Orientation tolerance controls orientation relationship between features. In order to evaluate deviation of an oriented feature, the datum reference frame should be first established, i.e., the location and orientation of features in the datum reference frame be determined. Two types of algorithms are ordinarily used for such purposes. The first and the most common approach is to apply a least-squares method. The objective of this method is to minimize *the sum of squares of deviation of measurement points from nominal features*. Several geometric algorithms applicable for non-symmetric features were analyzed in [4]. However, when the algorithms apply to a symmetric feature, they become degenerate. It should not be a surprise since we can not recover the symmetry of a symmetric feature. If we identify the general rigid motion with the Special Euclidean group $SE(3)$, then the configuration space of a symmetric feature is a *homogeneous space* of $SE(3)$.

However, formulation with the least-squares approach is inaccurate for tolerance evaluation purpose. The resulting tolerance zone is not in conformance to the standard ANSI/Y14.5M ([1]). It is desirable to use min-max method instead. Some algebraic algorithms to evaluate special types of form tolerances such as circularity and cylindricity can be found in ([10], [15] and [3]). P. Ikonomov al et. [9] proposed a virtual gauge algorithm to evaluate the geometric relationship between feature and datum features, the small displacement screw method was employed to solve the min-max problem.

In this paper, we first formulate datum features and oriented features with mini-max approach using the theory of configuration space of symmetric features. The configuration space of a symmetric feature is identified with the coset space $SE(3)/G_0$ of the Euclidean space $SE(3)$, where G_0 is the symmetry group of the feature. Then, using properties of Euclidean groups, especially its exponential coordinate, we convert the mini-max problem into sequential linear programs.

The paper is organized as follows: In Section 2, we formulate datum establishment and orientation tolerances as minimization or constrained minimization problems on $SE(3)/G_0$. We transform the non-differentiable minimization problem into a differentiable minimization problem on an extended configuration space. In Section 3, we use the geometric properties of $SE(3)/G_0$ to derive a a sequence of linear programming problems whose solutions can be used to approximate the actual tolerances. We present the details of the Symmetric Minimum Zone algorithm. In section 4, we conclude the paper by highlighting some future problems in this area.

2 Formulation of Orientation Tolerances

We will present in this section the formulation of datum reference frame using the theory of configuration space of symmetric features. The geometric properties of the configuration space and related notations can be find in our another paper in this proceedings ([7]).

According to definitions in the standard ANSI Y14.5M, the datum feature is determined by the envelopment principle, i.e., the tolerance zone of the datum feature reaches minimum. For the primary datum, the definition is equivalent to that of the corresponding form tolerance. For the secondary datum, the datum feature is established by the formulation of orientation tolerance related to only the primary datum.

Orientation tolerances include *angularity, parallelism* and *perpendicularity*. These tolerances control the orientation of features to one another, and are defined using three types of tolerance zones: (i) planar tolerance zone of oriented planar features; (ii) cylindrical tolerance zone of oriented axis of cylindrical features; and (iii) linear tolerance zone of line element of oriented cross-sections.

Orientation tolerances can be formulated as *constrained* minimization problems with orientation constraints imposed on the configuration space $SE(3)/G_0$, where G_0 is the symmetry subgroup of the toleranced feature.

To define the orientation constraints, we let the direction vector of a plane to be its unit normal vector and that of a cylinder to be its axis direction vector. We assume that the datum features (either a planar or a cylindrical feature) have been established using, say the form tolerance verification procedure formulated in the previous subsection and solved in the later section. Thus, the direction vectors of the primary datum and the secondary datum (if necessary) are available, and are denoted $u_1, u_2 \in \mathbb{R}^3$.

Definition 2.1. (Orientation constraints)
Let v be the direction vector of a toleranced feature. Then, the orientation constraint for a planar feature is given by

$$|\langle v, u_1 \rangle| = \begin{cases} |\cos(\theta)| & \text{for a primary datum axis} \\ |\sin(\theta)| & \text{for a primary datum plane} \end{cases}$$
$$=: c_1,$$

where θ is the basic angle between the primary datum and the feature, and, if a secondary datum is specified,

$$|\langle v', u_2 \rangle| = \begin{cases} |\cos(\alpha)| & \text{for a secondary datum axis} \\ |\sin(\alpha)| & \text{for a secondary datum plane} \end{cases}$$
$$=: c_2,$$

where v' is the projection of v to a plane orthogonal to u_1,

$$v' = \frac{v - \langle v, u_1 \rangle u_1}{\|v - \langle v, u_1 \rangle u_1\|} \qquad (1)$$

and α the basic angle between v' and the secondary datum.

Similarly, if the toleranced feature is a cylindrical feature then the orientation constraints are given by

$$|\langle v, u_1 \rangle| = \begin{cases} |\sin(\theta)| & \text{for a primary datum axis} \\ |\cos(\theta)| & \text{for a primary datum plane} \end{cases}$$
$$=: c_1,$$

and

$$|\langle v', u_2 \rangle| = \begin{cases} |\sin(\alpha)| & \text{for a secondary datum axis} \\ |\cos(\alpha)| & \text{for a secondary datum plane} \end{cases}$$
$$=: c_2,$$

where v' is the same as in (1).

Let $SE(3)/G_0$ be the configuration space of a toleranced feature F_0, and $v_0 \in \mathbb{R}^3$ the direction vector of F_0 at a nominal configuration. At a location $g \in SE(3)/G_0$, the direction of F_0 is given by $v = gv_0$, and the orientation constraints of Definition 2.1 define a submanifold of $SE(3)/G_0$,

$$Q = \{g \in SE(3)/G_0 | \langle gv_0, u_1 \rangle = c_1, \text{ and } \langle gv_0', u_2 \rangle = c_2\} \qquad (2)$$

$Q \subset SE(3)/G_0$ gives the set of Euclidean transformations of the toleranced feature satisfying the orientation constraints. Since one orientation constraint deduces one degree of freedom of $SE(3)/G_0$, the dimension of Q is determined by

$$\dim Q = \dim SE(3)/G_0 - m$$

where m is the number of orientation constraints.

Definition 2.2. (Orientation tolerance zone)
The orientation tolerance zone of size t associated with $g \in Q$ is defined by all points $y \in \mathbb{R}^3$ such that

$$|d(y, g)| \leq t$$

where

$$d(y, g) = \langle g^{-1}y - x, n \rangle$$

$x \in \mathbb{R}^3$ a corresponding point to y on F_0 and $n \in \mathbb{R}^3$ the unit outward normal vector of F_0 at x.

Let $Y = \{y_i \in \mathbb{R}^3, i = 1, \cdots n\}$ be a set of points sampled from the feature surface. The orientation tolerance value t associated with Y as defined by the above *orientation tolerance zone* is given by

$$t = \min_{g \in Q}(\max_{y_i \in Y} d(y_i, g) - \min_{y_i \in Y} d(y_i, g)). \qquad (3)$$

We give in the following the precise formulations of the planar and the cylindrical tolerance zones and leave that of the linear tolerance zone for line elements of oriented cross-sections to the reader.

Problem 1. (Planar orientation zone)
An orientation tolerance specifies that the toleranced plane must lie in a zone bounded by two parallel planes separated by a specified tolerance and basically oriented to the primary datum and, if specified, to the secondary datum as well.

Planar orientation tolerance t is determined by

$$t = 2 \min_{g \in Q} \max_{y_i \in Y} d(y_i, g)$$

This can be transformed into an ordinary minimization problem with objective function

$$f: Q \times \mathbb{R} \longrightarrow : (g, s) \longmapsto 2s$$

and the constraints

$$\begin{aligned} 0 &\leq s \\ -s &\leq d(y_i, g) \leq s, \quad i = 1, \cdots n \end{aligned}$$

where

$$d(y_i, g) = \langle g^{-1} y_i - x_i, v_0 \rangle,$$

$v_0 \in \mathbb{R}^3$ the unit normal vector of the plane at its nominal configuration, and $x_i \in \mathbb{R}^3$ the corresponding point of y_i.

Problem 2. (Cylindrical orientation zone)
An orientation tolerance specifies that the toleranced axis must lie in a zone bounded by a cylinder with a diameter equal to the specified tolerance and basically oriented to the primary datum and, if specified, to the secondary datum as well.

The value of cylindrical orientation tolerance t is determined by

$$t = 2 \min_{g \in Q} \max_{y_i \in Y} d(y_i, g)$$

and the objective function of the transformed minimization problem is given by

$$f: Q \times \mathbb{R} \longrightarrow : (g, s) \longmapsto 2s$$

and the constraints are

$$\begin{aligned} 0 &\leq s \\ d(y_i, g) &\leq s, \quad i = 1, \cdots n \end{aligned}$$

where

$$d(y_i, g) = \langle g^{-1} y_i - x_i, v_i \rangle,$$

$x_i \in \mathbb{R}^3$ a point on the axis nearest to $g^{-1} y_i$ and $v_i \in \mathbb{R}^3$ a unit vector in the direction of $g^{-1} y_i - x_i$.

The tolerance values for orientation tolerances can be computed using either the Lagrangian multiplier's technique and the Symmetric Minimium Zone (SMZ) algorithm of the next section or the SMZ algorithm alone with a proper parameterization of the constrained submanifold Q in (2). In the following example, we discuss a simple parameterization of Q using the exponential coordinates of $SE(3)/G_0$.

Example 2.1. (The constrained submanifold Q)
Consider first the case of a feature oriented to a primary datum. The direction of the transformed feature is given by

$$v = g v_0.$$

Let $\mathcal{M}_0$ be spanned by

$$\mathcal{M}_0 = \mathrm{span}\{\widehat{\eta}_1, \cdots \widehat{\eta}_{r_1}, \widehat{\eta}_{r_1+1}, \cdots \widehat{\eta}_r\}$$

where $(\widehat{\eta}_1, \cdots \widehat{\eta}_{r_1})$ represent infinitesimal translations, and $(\widehat{\eta}_{r_1+1}, \cdots \widehat{\eta}_r)$ infinitesimal rotations. For each $g \in SE(3)/G_0$, write

$$g = e^{\widehat{m}} g_0$$

where $\widehat{m} \in \mathcal{M}_0$ and $g_0 \in G_0$. Let

$$\widehat{m} = m_1 \widehat{\eta}_1 + \cdots m_r \widehat{\eta}_r$$

for some $(m_1, \cdots, m_r) \in \mathbb{R}^r$. Since $g_0 v_0 = v_0$, we have

$$v = g v_0 = e^{\widehat{m}} v_0.$$

The direction v should satisfy the orientation constraint

$$\langle v, u_1 \rangle = c_1$$

or,

$$\langle e^{\widehat{m}} v_0, u_1 \rangle = c_1. \qquad (4)$$

Note that combination of infinitesimal translations has no effect on feature orientation we can neglect the translational component of $\widehat{m}$ in Equation (4). On the other hand, if we choose a set of $(m_{r_1+1}, \cdots m_r)$ such that

$$m_{r_1+1} \widehat{\eta}_{r_1+1} + \cdots m_r \widehat{\eta}_r =: m' \widehat{u}_1$$

which indicates the combination of effective rotations being equivalent to rotations about the axis u_1. Since a vector preserves the direction under rotations about the parallel axis, i.e.,

$$e^{\widehat{u}_1 t} u_1 = u_1, \quad u_1 \in \mathbb{R}^3.$$

We have

$$\langle e^{m_1 \widehat{\eta}_1 + \cdots m_{r_1+1} \widehat{\eta}_{r_1+1} + m' \widehat{u}_1} v_0, u_1 \rangle = \langle v_0, u_1 \rangle = c_1,$$

that is, the orientation constraint is always satisfied under this parametrization. The vector $(m_1, \cdots m_{r_1}, m')$ is then an exponential coordinate of the constrained submanifold Q relative to the basis $\{\widehat{\eta}_1, \cdots \widehat{\eta}_{r_1}, \widehat{u}_1\}$, i.e.,

$$Q = \{g = e^{m_1\widehat{\eta}_1 + \cdots m_{r_1}\widehat{\eta}_{r_1} + m'\widehat{u}_1} \in SE(3) \\ |m_1, \cdots m_{r_1}, m' \in \mathbb{R}\}.$$

3 Geometric Algorithms for Tolerance Verification

In this section, we use the exponential coordinates of $SE(3)/G_0$ to develop a simple geometric algorithm for minimization of the function

$$f(g, s_1, s_2) = s_1 - s_2 \qquad (5)$$

subject to the nonlinear constraints

$$s_2 \leq d(y_i, g) \leq s_1, \quad i = 1, \cdots n \qquad (6)$$

where

$$d(y_i, g) = \langle g^{-1}y_i - x_i, n_i \rangle. \qquad (7)$$

In general, the solution of a constrained minimization problem of the form

$$\min_{q \in \mathbb{R}^n} \{\psi(q) | \phi(q) \leq 0\} \qquad (8)$$

where $\psi : \mathbb{R}^n \to \mathbb{R}$ is a C^1-function and $\phi : \mathbb{R}^n \to \mathbb{R}^m$ a set of nonlinear constraints, can be obtained by solving a sequence of linear programming (LP) problems with properly chosen initial conditions. To derive the corresponding LP problem, let $q^k \in \mathbb{R}^n$ be an initial condition satisfying the constraints and consider

$$q^{k+1} = q^k + \tilde{q} \qquad (9)$$

where $\tilde{q} \in \mathbb{R}^n$ is a perturbation term. Computing the Taylor series expansion of $\psi(\cdot)$ and $\phi(\cdot)$ at q^k and retaining the first-order terms yield

$$\psi(q^{k+1}) \approx \psi(q^k) + \langle d\psi(q^k), \tilde{q} \rangle$$

and

$$\phi(q^{k+1}) \approx \phi(q^k) + D\phi(q^k) \cdot \tilde{q}$$

where $d\psi(q^k) \in \mathbb{R}^{1 \times n}$ and $D\phi(q^k) \in \mathbb{R}^{m \times n}$ are, respectively, the differential of ψ and the Jacobian of ϕ at q^k. The solution of the LP problem

$$\min_{\tilde{q} \in \mathbb{R}^n} \{\langle d\psi(q^k), \tilde{q} \rangle | D\phi(q^k) \cdot \tilde{q} \leq 0\} \qquad (10)$$

in (9) ensures that the constraints be satisfied while the the function be minimized.

To apply the above method to tolerance verification, we need to take into account of the fact that the underlying configuration space is not Euclidean but the homogeneous space $SE(3)/G_0$. In view of Section 2, we consider perturbations of the form

$$(g^{k+1}, s_1^{k+1}, s_2^{k+1}) = (g^k e^{\widehat{m}}, s_1^k + \tilde{s}_1, s_2^k + \tilde{s}_2) \quad (11)$$

where $\widehat{m} \in \mathcal{M}_0$ and $\mathcal{M}_0$ is a complementary subspace to the Lie algebra $\mathcal{G}_0$ of G_0.

Choose a basis $(\widehat{\eta}_1, \cdots \widehat{\eta}_k)$ of $\mathcal{M}_0$ and write

$$\widehat{m} = \sum_{i=1}^{k} \widehat{\eta}_i m_i$$

for some $m = (m_1, \cdots m_k) \in \mathbb{R}^m$. Then, for small values of $m \in \mathbb{R}^k$ we have

$$e^{\widehat{m}} \approx I + \widehat{m} \qquad (12)$$

Substituting (12) into (7) and linearizing the constraints in (6) yield

$$\tilde{s}_2 \leq \langle -\widehat{m}(g^k)^{-1}y_i, n_i \rangle \leq \tilde{s}_1, \quad i = 1, \cdots n. \quad (13)$$

The linearized objective function is

$$\tilde{f}(m, \tilde{s}_1, \tilde{s}_2) = \tilde{s}_1 - \tilde{s}_2. \qquad (14)$$

Thus, the corresponding LP problem becomes

$$\min_{(m, \tilde{s}_1, \tilde{s}_2) \in \mathbb{R}^{k+2}} \{\tilde{f}(m, \tilde{s}_1, \tilde{s}_2) | \tilde{s}_2 \leq \langle -\widehat{m}g^{-k}y_i, n_i \rangle \leq s_2, \\ i = 1 \cdots n\} \qquad (15)$$

It is also useful to include following limit constraints in the LP problem.

$$m_i^- \leq m_i \leq m_i^+$$

where $m_i^-, m_i^+ \in \mathbb{R}$ are appropriately chosen limit values.

The proceeding discussions are summarized into the following algorithm for tolerance verification:

Algorithm 1. (Symmetric minimum zone (SMZ) algorithm)

Input: Measurement data set $Y = \{y_i \in \mathbb{R}^3, i = 1, \cdots n\}$;

Output: (a) Optimal location $g^* \in SE(3)/G_0$ of the nominal feature;

(b) Form tolerance value t^*.

Step 0: (a) Set $k = 0$;

(b) Initialize g_0;
Set $s_1^0 = \max_i d(y_i, g^0)$ and $s_2^0 = \min_i d(y_i, g^0)$;

(c) Solve for $x_i^0, i = 1, \cdots n$;

(d) Compute $f^0 = s_1^0 - s_2^0$;

Step 1: (a) Solve the LP problem (15) to obtain the optimal $(m, \tilde{s}_1, \tilde{s}_2)$;

(b) Update $(g^{k+1}, s_1^{k+1}, s_2^{k+1})$ according to (11);

(c) Solve for $x_i^{k+1}, i = 1, \cdots n$;

(d) Compute f^{k+1};

(e) If $(1 - f^{k+1}/f^k) > \epsilon$, then set $k = k+1$ and return to Step 1(a); Else exit and report results.

Remark 1. Proper choice of g^0 is important for convergence of the algorithm. We suggest that the centroid of the measurement data be used for the translational component of g^0, and R^0 be the identity matrix. Otherwise, one should use solutions computed by a least square algorithm such as the symmetric localization algorithm in ([6]) as the initial condition.

Example 3.1. (Planar orientation tolerance zone)
A planar orientation tolerance is specified as in Figure (1). Let $Y = \{y_i \in \mathbb{R}^3, i = 1, \cdots n\}$ be a set of

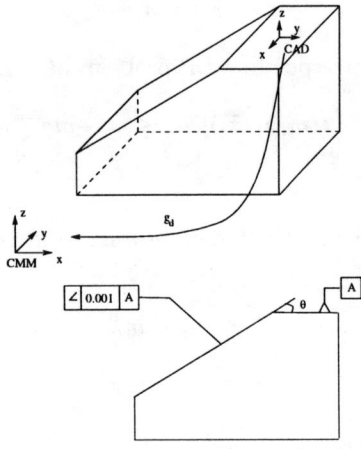

Figure 1: Planar orientation tolerance zone

measurement points from the datum planar feature A and $Z = \{z_j \in \mathbb{R}^3, j = 1, \cdots m\}$ a set of points measured from the oriented planar feature. Assume that the nominal configuration of Feature A is xy-plane. The basic angle between the oriented feature and the datum feature is θ. The orientation tolerance zone is computed in two steps.

1. **Datum establishment:** The planar datum A is established by the procedure proposed in ([7]). The right solution of the Euclidean transformation is denoted $g_d \in SE(3)$.

2. **Orientation tolerance zone:** Let a nominal configuration of the oriented feature be represented by

$$g = \left\{ \begin{bmatrix} v_1 & v_2 & v_3 & 0 \\ 0 & 0 & 0 & 1 \end{bmatrix} | v_1, v_2, v_3 \in \mathbb{R}^3 \right\}$$

where $v_1 = [\cos(\theta) \ 0 \ -\sin(\theta)]^T$, $v_2 = [0 \ 1 \ 0]^T$ and $v_3 = [\sin(\theta) \ 0 \ \cos(\theta)]^T$. A choice of a basis for the constrained submanifold Q is given by

$$Q = \text{span}\{\hat{\eta}_1, \hat{\eta}_2\}$$

where $\eta_1 = [v_3 \ 0]^T, \eta_2 = [0 \ e_3]^T \in \mathbb{R}^6$. Note that the point set Z should be first transformed to the coordinate system established by the datum feature A, i.e.,

$$z_j' = g_d^{-1} z_j, \ j = 1, \cdots m.$$

The linearized constraints is then given by

$$\begin{aligned} 0 &\leq \tilde{s} \\ -m_i^+ &\leq m_i \leq m_i^+ \\ -\tilde{s} &\leq \langle -\widehat{m} g^{-k} z_j', v_3 \rangle \leq \tilde{s}, \ j = 1, \cdots m. \end{aligned}$$

4 Conclusion

In this paper, we presented a geometric theory which unified the formulation and evaluation of datum establishment and orientation tolerances. Using orientation constraints, we showed that every case of orientation tolerances could be formulated as a constrained minimization problem on $SE(3)/G_0$, and a simple approach to parameterize the constrained submanifold was given.

To solve the derived mini-max problem we converted the non-differentiable minimization problem into a differentiable minimization problem with an extended configuration space. By exploring the special geometries of $SE(3)/G_0$, we developed the Symmetric Minimium Zone (SMZ) algorithm to unify the computation of orientation tolerances. The SMZ algorithm computes the minimum zone solutions by solving a sequence of linear programming problems, each of which could be easily derived using the geometric formulation. The SMZ algorithm is seen to possess several important features: (a) it is simple to implement; (b) it computes solutions which are accurate and consistent with the ANSI Y14.5M standard.

Future problems in this area which we are currently studying include:

1. Generalize the present framework to datum establishment and formulation and verification of location tolerances, thus complete a geometric theory for the important subject of geometric tolerancing;

2. Develop and implement on a flexible machining platform a computer-aided inspection system based on the tolerance verification algorithms proposed in this paper and the above study.

3. Generalize the geometric theory to the hybrid localization/envelopment problems and develop a computer-aided setup system to reduce the cost of workpiece setup, refixturing and dimensional inspections.

References

[1] American National Standard Institute. *Dimensioning and Tolerancing, ANSI Standard Y14.5M*. The American Society of Mechanical Engineers, 1982.

[2] American National Standard Institute. *Mathematical Definition of Dimensioning and Tolerancing Principles, ANSI Standard Y14.5.1M*. The American Society of Mechanical Engineers, 1994.

[3] W. Choi, T.R. Kurfess. Dimenstional Measurement Data Analysis Part II, Minimum Zone Evaluation. *MED-Vol.4, Manufacturing Science and Engineering ASME*, pages 457-462, 1996, New York.

[4] Y.X. Chu, J.B. Gou and Z.X. Li. Performance analysis of localization algorithms. In *IEEE Intl. Conf. on Robotics and Automation*, pages 1247-1252, 1997.

[5] A.B. Forbes. Least-squares best-fit geometric elements. Technical report, National Physical Laboratory, UK, 1989.

[6] J.B. Gou, Y.X. Chu and Z.X. Li. On the Symmetric Localization Problem. *IEEE Trans. on Robotics and Automation* (to appear),

[7] J.B. Gou, Y.X. Chu and Z.X. Li. A Geometric Approach of Form Tolerance Formulation and Evaluation. In *IEEE Intl. Conf. on Robotics and Automation*, 1998.

[8] J. Hong and X. Tan. Method and apparatus for determining position and orientation of mechanical objects. *U.S. Patent No. 5208762*, 1990.

[9] P. Ikonomov, at el, Inspection Method for Geometrical Tolerance using Virtual Gauges. *IEEE Int. Conf. on Robotics and Automation*, pages 550-555, 1995.

[10] J.Y. Lai, I.H. Chen. Minimum Zone Evaluation of Circles and Cylinders. *Int. J. Mach. Tools Manufact.*, Vol.36(4):435-451, 1996.

[11] X.M Li, M. Yeung, and Z.X. Li. An algebraic algorithm for workpiece localization. In *IEEE Intl. Conf. on Robotics and Automation*, pages 152–158, 1996.

[12] C.H. Menq, H. Yau, and G. Lai. Automated precision measurement of surface profile in CAD-directed inspection. *IEEE Transactions on Robotics and Automation*, 8(2):268–278, 1992.

[13] R. Murray, Z.X. Li, and S. Sastry. *A Mathematical Introduction to Robotic Manipulation*. CRC Press, 1994.

[14] A. Requiccha. Toward a theory of geometric tolerancing. *International Journal of Robotics Research*, pages 45-60, 1983.

[15] U. Roy, Y. Xu. Form and Orientation tolerance analysis for cylindrical surface in computer-aided inspection. *Computer in Industry*, Vol.26:127-134, 1995

[16] Z.C. Yan and C.H. Menq. Evaluation of geometric tolerances using discrete measurement data. *Journal of Design and Manufacturing*, 4:215–228, 1994.

[17] H.T. Yau and C. H. Menq. A unified least-squares approach to the evaluation of geometric errors using discrete measurement data. *Intl. J. Mach. Tools Manufact.*, 36:1269–1290, 1996.

Computing n-Finger Force-Closure Grasps on Polygonal Objects*

Yun-Hui Liu

Dept. of Mechanical and Automation Engineering
The Chinese University of Hong Kong
Shatin, N. T., Hong Kong
Email: yhliu@mae.cuhk.edu.hk

Abstract

This paper presents an efficient algorithm for computing all n-finger force-closure grasps on a polygonal object. This algorithm is based on a new qualitative test algorithm for force-closure grasps which recursively transforms the problem in the three dimensional wrench space to a problem in an one dimensional space. We demonstrate that non-force-closure grasps are two convex polytopes in the space of n parameters that represent grasp points on sides of the polygon. Therefore, the force-closure grasp region is calculated by subtracting the convex polytopes from the parameter space. The qualitative test algorithm takes $O(n^3)$ time and the grasp computation algorithm takes $O(n^3 \log n)$ time for $n \leq 3$ and $O(n^{3n/2})$ time for $n > 3$, where n is the number of the fingers. The efficiency of the algorithms is confirmed with simulations.

1 Introduction

The qualitative test of force-closure grasps of multi-fingered hands have been studied extensively since the middle 80's. Nguyen[7] proposed a simple test algorithm for 2-finger planar grasps, and Ponce and Faverjon[8] developed several sufficient conditions for 3-finger planar force-closure grasps. The aforementioned algorithms, however, are not applicable to 2D grasps with more fingers. The qualitative test algorithms for 2D grasps developed by Ferrari[4] and by Chen and Burdick[1] are suitable for any number of contacts.

A qualitative test algorithm only judges whether a given grasp is force-closure. It is more important to compute force-closure grasps from geometric model of an object. Nguyen[7] extended his test algorithm to computing all 2-finger force-closure grasps. Ponce and Faverjon[8] also presented an algorithm for computing all grasps satisfying their sufficient conditions. Similar to their qualitative algorithms, the algorithms cannot be applied to grasps with over three fingers. It should be noted that the algorithm in [8] computes only a sub-region not the entire region of the force-closure grasp.

In this paper, we propose an algorithm for computing all n-fingers force-closure grasps on a polygonal object. This algorithm works on the basis of a new qualitative test algorithm for n-finger force-closure grasps developed based on a recursive reduction technique. This technique recursively slices the convex hull of the primitive contact wrenches in the 3D wrench space by planes passing the origin point and transforms the problem in the 3D wrench space to one in an one dimensional space, which can be readily solved. To represent force-closure grasp region, for each finger i we introduce a parameter u_i representing its grasping point on the side of the object. The n parameters u_i defines an n-dimensional parameter space. We will demonstrate that in the parameter space the non-force-closure grasp region consists of two convex polytopes. The force-closure grasp region is obtained by subtracting the two convex polytopes from the parameter space. The incremental algorithm in [5] is employed to construct the convex polytopes. The algorithm for computing the entire force-closure grasp region takes $O(n^3 \log n)$ time for $n \leq 3$ and $O(n^{3n/2})$ time for $n > 3$, where n is the number of the fingers. The computational cost of the qualitative test algorithm is $O(n^3)$. The proposed algorithms have been implemented and their performance has been confirmed with simulations.

2 Force-Closure Grasp

Suppose that n hard fingers are grasping a rigid polygonal object in a 2D workspace (Figure 1a). Assume that the Coulomb friction with coefficient μ exists at the contact points. To hold the object and balance any external forces and moments, each finger must apply to the object a force f_i called *grasp force*. To assure non-slipping at the contact

---

*This work is supported in part by the Hong Kong Research Grant Council under grant CUHK4151/97E and in part by the CUHK strategic research grant.

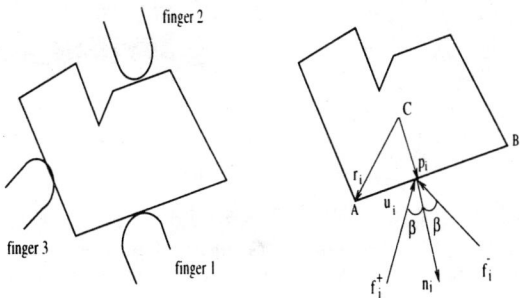

Figure 1: A planar 3-finger grasp on a polygon (a) and the friction sector (b).

point, the grasp force f_i must lie inside the friction sector defined by f_i^+ and f_i^- (Figure 1b). Here $\beta = \arctan \mu$. If it lies inside the friction sector, the grasp force f_i can be represented as follows:

$$f_i = \alpha_i^+ f_i^+ + \alpha_i^- f_i^- \qquad (1)$$

where coefficients α_i^+ and α_i^- are nonnegative constants. The force and moment, corresponding to the grasp force f_i, applied at the center of mass of the object is given by

$$\underline{w}_i = \begin{pmatrix} f_i \\ p_i \times f_i \end{pmatrix} \qquad (2)$$

where p_i denotes the position vector of the i-th grasping point w.r.t. the object coordinate frame. The force and moment $\underline{w}_i$ is called *wrench*. Substituting eq. (1) into eq. (2) derives

$$\underline{w}_i = \alpha_i^+ \underbrace{\begin{pmatrix} f_i^+ \\ p_i \times f_i^+ \end{pmatrix}}_{w_i^+} + \alpha_i^- \underbrace{\begin{pmatrix} f_i^- \\ p_i \times f_i^- \end{pmatrix}}_{w_i^-} \qquad (3)$$

w_i^+ and w_i^- are called *primitive contact wrenches*. The net wrench applied at the object by the fingers is

$$\underline{w}_{net} = \sum_{i=1}^{n} \alpha_i^+ w_i^+ + \alpha_i^- w_i^- = W\alpha \qquad (4)$$

where $W = (w_1^+, w_1^-, w_2^+, w_2^-, ..., w_n^+, w_n^-)$ and $\alpha = (\alpha_1^+, \alpha_1^-, \alpha_2^+, \alpha_2^-, ..., \alpha_n^+, \alpha_n^-)^T$. The $3 \times 2n$ matrix W is called *wrench matrix* and its column vectors are the primitive contact wrenches. α is a $2n$ by 1 vector with nonnegative elements. For convenience, in the following we use w_i to denote the i-th column vector of matrix W, and use α_i to represent the i-th component of the vector α. Let $N = 2n$. N is the total number of the primitive contact wrenches.

Definition 1 *Suppose that n hard frictional fingers are grasping an object. For any external wrench $\underline{w}_{ext}$ applied at the object, if it is always possible to find an α with $\alpha_i \geq 0$ such that $\underline{w}_{net} + \underline{w}_{ext} = 0$, the grasp is said to be force-closure.*

It is well-known that a force-closure grasp is equivalent to that the origin point of the wrench space R^3 lies exactly inside the convex hull of the primitive contact wrenches. As constructing a convex hull in R^3 is computationally expensive, research efforts have been directed to development of algorithms that do not compute the convex hull.

3 Recursive Qualitative Test

This section presents a new qualitative test algorithm, which forms the basis of the algorithm computing all force-closure grasps.

3.1 Recursive Reduction Technique

In order to perform the qualitative test for n-finger force-closure grasp efficiently, a recursive method is developed based on the following theorem:

Theorem 1 *A convex hull $H(N)$ of N points in $R^d(x_1, x_2, ..., x_d)$ contains the origin point if and only if there is such an i that,*

(a) in the given N points there are points with strictly positive x_i-coordinate and points with strictly negative x_i-coordinate as well;

(b) the intersection of the convex hull $H(N)$ with the hyperplane $x_i = 0$ contains the origin of the R^{d-1} defined by $x_i = 0$.

Proof: (a) Necessary condition: First, when the convex hull contains the origin, the N point cannot lie on one side of any hyperplane passing the origin. Therefore, for any i, there are points with strictly positive x_i-coordinate and points with strictly negative x_i-coordinate. Second, since the origin is contained by the convex hull, the origin is also contained by the intersection of $H(N)$ with any plane passing the origin point.

(b) Sufficient condition: To prove the sufficient condition, we demonstrate that any vector $v = (v_1, v_2, ..., v_d)$ in R^d can be represented by a nonnegative linear combination of the N points when the two conditions are satisfied. With the first condition, there exists a point $q = (q_1, q_2, ..., q_d)$ whose x_i-coordinate has the same sign as v_i. The vector v can be clearly represented as follows:

$$v = \frac{|v_i|}{|q_i|} \begin{pmatrix} q_1 \\ ... \\ q_i \\ ... \\ q_d \end{pmatrix} - \frac{|v_i|}{|q_i|} \begin{pmatrix} q_1 \\ ... \\ 0 \\ ... \\ q_d \end{pmatrix} + \begin{pmatrix} v_1 \\ ... \\ 0 \\ ... \\ v_d \end{pmatrix}$$

where $|\cdot|$ means the absolute value. It must be noted that the sum of the last two vectors is on the hyperplane $x_i = 0$, which spans a R^{d-1} space.

Denote the intersection of $H(N)$ with $x_i = 0$ by $H(V)$. With the second condition, any point on the hyperplane $x_i = 0$ can be represented a nonnegative linear combination of vertices of $H(V)$. Note that all vertices of $H(V)$ belong to the convex hull $H(N)$, and thus they are convex combinations of the N original points. Therefore, any point on the hyperplane $x_i = 0$ can be represented by a nonnegative linear combination of the N points and so can the point v. This implies that the convex hull $H(N)$ contains the origin point of R^d. □

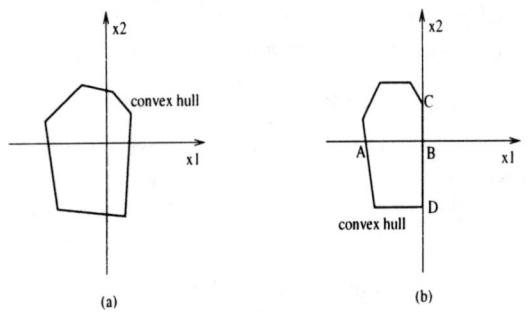

Figure 2: Cases when the conditions are and are not satisfied.

In Figure 2a, the two conditions are satisfied for all i. In Figure 2b, the first condition is satisfied but the second is not for $i = 2$; the second condition is satisfied but the first is not for $i = 1$. If for an i the two conditions are satisfied, the conditions hold for all other i. Similarly, if for an i the two conditions are not simultaneously satisfied, for any other i the conditions cannot be simultaneously satisfied.

From Theorem 1, it is possible to recursively transform the problem of qualitative test in R^d to a problem in R^1. In detail, we first slice the convex hull $H(N)$ by the hyperplane $x_d = 0$. Denote the slice of the convex hull by $H(V)$, which is a convex region in R^{d-1}. Then, the problem is to check whether $H(V)$ contains the origin of R^{d-1}. We further slice $H(V)$ by the hyperplane $x_{d-1} = 0$ so that the qualitative testing is transformed to a problem in R^{d-2}. If the slicing is recursively performed, the qualitative testing is eventually transformed to a problem in R^1. In R^1, the convex hull of a set of points is the closed interval from the minimum point to the maximum point, and whether it contains the origin point can be readily identified. This procedure is called *recursive reduction technique*.

3.2 The Algorithm

The recursive reduction technique needs to calculate the slice of a convex hull by a hyerplane. Denote intersection points of the hyperplane with edges of the convex hull by set V. The slice is the convex hull $H(V)$ of points in set V. The following fact is well-known in Computational Geometry:

Proposition 1 *For any N points in R^d, the vertices of their convex hull belong to the N points and the edges belong to segments connecting them.*

From Proposition 1, we clearly obtain

Proposition 2 *For N points in R^d, denote intersection points of the segments connecting them with the hyperplane $x_i = 0$ by set E. The convex hull of points in set E is the intersection of the convex hull $H(N)$ with the hyperplane.*

From Proposition 2, instead of explicitly calculating intersection of the convex hull with a hyperplane, we calculate intersections of a hyperplane with segments connecting the points. Following are details of the qualitative test algorithm:

Algorithm 1

Step 1 Calculate all the primitive contact wrenches w_i^+ and w_i^- and denote them by set W. Let $d = 3$ and $E = W$.

Step 2 According to signs of their d-th coordinates, divide the points in set E into E_+, E_0, and E_-, where set E_+ and E_- include points with positive and negative d-th coordinates, respectively. E_0 denotes the set of points with zero d-th coordinates. If either set E_+ or E_- is null, the grasp is not force-closure and the algorithm ends.

Step 3 For any pair of points combined from sets E_+ and E_-, calculate the intersection point of the segment connecting them with the hyperplane $x_d = 0$. Update the set E by the intersection points and set E_0.

Step 4 $d = d - 1$. If $d \neq 1$, go back to step 2; otherwise proceed to step 5.

Step 5 If in the set V there exist not only points with positive d-th coordinates but also points with negative d-th coordinates, the grasp is a force-closure grasp; otherwise it is not. The algorithm ends.

This algorithm takes $O(N^4)$ time because reducing one dimension of the space produces at most $m^2/4$ points if there are m points in the original space. Due to the fact that existing robotic hands do not have more than 5 fingers, existing computers are powerful enough to fulfill the computation in real-time. As to be described in Section 4, however, the combination number undermines computational efficiency of computing all force-closure grasps. Fortunately, the combination number can be reduced to $O(N^3)$ with a minor modification.

3.3 Improving the Efficiency

Algorithm 1 first calculates the intersections of the plane $x_3 = 0$ with segments connecting the primitive contact wrenches (we call them points here). Suppose that the segment between points A and B intersects the plane $x_3 = 0$ in point a. The points A and B are *parents* of point a. Point a is said to be *originated* from points A and B.

Proposition 3 *Suppose that two points a and c in R^2 are originated from four different points in R^3. When we transform the problem from R^2 to R^1, it is not necessary to consider the intersection of $x_2 = 0$ with the segment connecting points a and c.*

Proof: Suppose that point a is originated from points A and B, and point c is originated from points C and D. Denote the convex hull of the primitive contact wrenches in R^3 by $H(N)$. Let $H(V)$ represent the intersection of $H(N)$ with $x_3 = 0$. The following three cases may occur:

Case a: At least one of the four points is not the vertex of $H(N)$. Without loss of generality, assume that point D is an interior point of $H(N)$. Then, the intersection point c is an interior point of $H(V)$, and thus the segment ac is not an edge of the convex hull $H(V)$. Therefore, it is not necessary to consider segment ac.

Case b: All the four points are on the facets of $H(N)$ but not on the same facet. In this case, points a and c belong to different facets or are interior points of $H(N)$, and thus the segment ac is not on the boundary of $H(N)$. Therefore, it is not necessary to process segment ac.

Case c: All the four points are on the same facet of the convex hull $H(N)$. In this case, it must be noted that the hyperplane $x_3 = 0$ also intersects with segments AC and DB in points b and d, respectively(Figure 3). Obviously, the sum $ad + bc$ of two segments ad and bc covers the segment ac entirely. Note that points a and d are originated from points A, B and D. Points b and c are originated from points A, C, and D. Once we check intersections of $x_2 = 0$ with segments ad and bc, it is not necessary to consider segment ac.

Therefore, if two points in R^2 are originated from four different points in R^3, it is not necessary to consider the segment connecting them when slicing $H(V)$ by $x_2 = 0$. $\square$

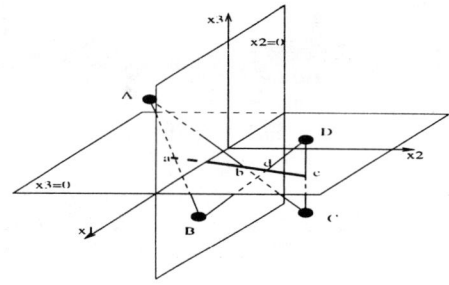

Figure 3: The case when the four points are on the same facet.

From Proposition 3, step 3 of Algorithm 1 is modified as follows:

Step 3 If d=3, combine all points from sets E_+ and E_-; otherwise, from sets E_+ and E_- combine pair of points that are not originated from four different primitive contact wrenches. For each pair of combined points, calculate the intersection of the segment connecting them with the hyperplane $x_d = 0$. Update the set E by all the intersection points and points in set E_0.

The combination number is clearly reduced to $O(N^3)$. It should be noted that the combination number is also $O(n^2)$ since $N = 2n$.

4 Computing All Force-Closure Grasps

This section addresses a more challenging problem of computing all n-finger force-closure grasps on a polygonal object. Suppose that the object's sides to be grasped by the fingers have been specified. The cases are also considered when multiple fingers are located on same side.

Suppose that finger i is in a contact with a side AB (Figure 1). Denote the position vector of the endpoint A by r_i and the direction vector of the side by vector s_i. To represent the grasp point p_i on the side, a scalar parameter u_i is introduced so that

$$p_i = r_i + s_i u_i. \quad (5)$$

The parameter u_i is constrained by

$$0 \leq u_i \leq l_i \quad (6)$$

where l_i denotes the length of the side AB. The n parameters u_i define an n-dimensional space. The two primitive contact wrenches at the contact point are represented by

$$w_i^\pm = \begin{pmatrix} f_i^\pm \\ (r_i + s_i u_i) \times f_i^\pm \end{pmatrix} \quad (7)$$

It should be noted that the moment depends linearly on the parameter u_i and the forces are constants.

The proposed qualitative test algorithm recursively reduces the dimension of the problem by successively computing intersections between segments connecting the primitive contact wrenches and the plane $x = 0$ then the plane $y = 0$. For any two points $w_i(w_{i_x}, w_{i_y}, w_{i_z})$ and $w_j(w_{j_x}, w_{j_y}, w_{j_z})$ in R^3, if the segment connecting them intersects the plane $x = 0$, the intersection point e_k is:

$$e_k = \begin{pmatrix} 0 \\ w_{i_y} - \frac{w_{i_x}}{w_{j_x} - w_{i_x}}(w_{j_y} - w_{i_y}) \\ w_{i_z} - \frac{w_{i_x}}{w_{j_x} - w_{i_x}}(w_{j_z} - w_{i_z}) \end{pmatrix} \quad (8)$$

Since the x and y coordinates of w_i and w_j are constants, the y-coordinate of the intersection point e_k is a constant while the z coordinate is a linear function of parameters u_i and u_j. For any such two intersection points e_k and e_s, the hyperplane $y = 0$ intersects the segment connecting them in the point q_j if the intersection occurs:

$$q_j = \begin{pmatrix} 0 \\ 0 \\ e_{k_z} - \frac{e_{k_y}}{e_{s_y} - e_{k_y}}(e_{s_z} - e_{k_z}) \end{pmatrix} \quad (9)$$

The z-coordinate of the intersection point q_j is obviously a linear function of the parameters u_i. Denote all the intersection points by set Q. For convenience, we re-denote the z-coordinate of point q_j in set Q by z_j, which has the following form:

$$z_j = \sum_{i=1}^{N} a_{j_i} u_i + b_j \quad (10)$$

where a_{j_i} and b_j are constants. According to Theorem 1, a non-force-closure grasp implies that

$$\sum_{i=1}^{n} a_{j_i} u_i + b_j \geq 0 \quad \forall q_j \in Q \quad (11)$$

or

$$\sum_{i=1}^{n} a_{j_i} u_i + b_j \leq 0 \quad \forall q_j \in Q \quad (12)$$

Eqs. (11) and (12) define two convex polytopes in the parameter space. The convex polytopes might not be closed regions. The closeness can be guaranteed by adding the constraints imposed by eq. (6). Denote the convex polytope defined by eqs. (6) and (11) by $CP(+)$. $CP(-)$ is employed to represent the convex polytope defined by eqs. (6) and (12).

Theorem 2 *Suppose that the sides to be grasped by the n fingers have been specified. Denote the region defined by the n inequalities in (6) in the parameter space $(u_1, u_2, ..., u_n)$ by U.*

(1) The non-force-closure region consists of two convex polytopes $CP(+)$ and $CP(-)$.

(2) Any point inside the region $U - CP(+) - CP(-)$ is a force-closure grasp.

Many algorithms are available in Computational Geometry for computing the convex polytope defined by a set of inequalities. In this paper, we employ the incremental construction algorithm developed in [5]. Following are details of the algorithm computing the force-closure grasp region:

Algorithm 2

Step 1 Compute all the primitive contact wrenches $w_i^{\pm}$ in the representation of the parametric variables u_i and denote them by set E.

Step 2 According to signs of their x coordinates, divide the primitive contact wrenches into groups E_+, E_0 and E_-, which contain points with positive, zero and negative x coordinates, respectively. If either E_+ or E_- is null, the force-closure grasp region is null and then the algorithm ends.

Step 3 Calculate intersections e_j of the plane $x = 0$ with segments connecting points in set E_+ to those in set E_-. Update the set E by all the intersection points and points in set E_0.

Step 4 According to signs of their y coordinates, divide points in set E into groups E_+, E_0 and E_-, which contain points with positive, zero and negative y coordinates, respectively. If either E_+ or E_- is null, the force-closure region is null and then the algorithm ends.

Step 5 For all pairs of points in sets E_+ and E_- that are not originated from four different primitive contact wrenches, calculate intersections of the plane $y = 0$ with segments connecting them. Denote all the intersection points by set Q. A point q_j of Q is characterized by a hyperplane $a_j^T u + b_j$ in the R^n parameter space.

Step 6 Using the incremental algorithm, calculate the convex polytope $CP(+)$ defined by eqs. (11) and (6) and then the convex polytope $CP(-)$ defined by eqs. (12) and (6). The force-closure grasp region is equal to $U - CP(+) - CP(-)$. The algorithm ends.

The computation efficiency of Algorithm 2 is mainly determined by the cost constructing the two convex polytopes $CP(+)$ and $CP(-)$. The computation cost of the incremental algorithm depends on the number of the inequality constraints. The recursive reduction technique produces $O(n^3)$ points in the R^1 space, which subsequently means $O(n^3)$ inequality constraints in the parameter space. The

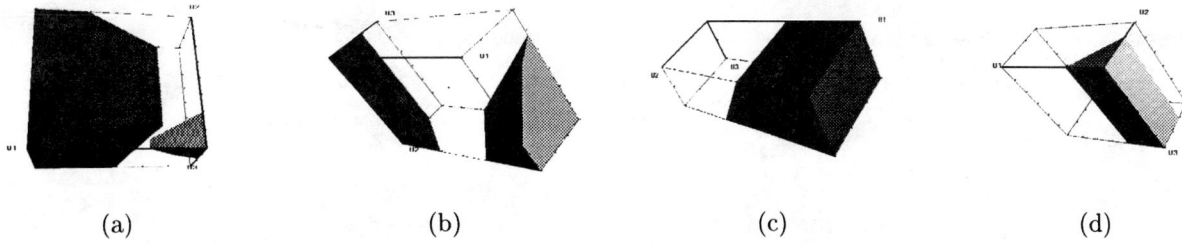

(a) (b) (c) (d)

Figure 4: The computed force-closure regions.

incremental algorithm constructs the common space of K inequality constraints (half-spaces) in R^d in $O(K^{d/2})$ time for $d > 3$ and in $O(K \log K)$ time for $d \leq 3$. Therefore, for 2 or 3-finger grasps, the n-finger force-closure grasp region can be computed in $O(n^3 \log n)$ time. For grasps with more fingers, the computation time is $O(n^{3n/2})$.

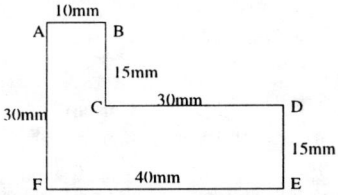

Figure 5: The polygon used.

5 Implementation

We have implemented the algorithms and examined their performance by an example. In the example, three fingers are to grasp this polygon shown in Figure 5. Four grasp configurations are considered. Figure 4a plots the computed force-closure grasp region when the three fingers are placed on sides AF, BC, and FE. The dark regions represent regions $CP(+)$ and $CP(-)$ and the white region is the force-closure grasp region. By the qualitative test algorithm, we examined that the grasp is not force-closure when the fingers are located at the center points of the sides. Figure 4b shows the results when the fingers are on sides AF, BC and CD. Figure 4c is the region when the fingers are on sides AB, BC and FE. Figure 4 depicts the case when the fingers are on sides AF, CD and DE. The friction coefficient is 0.5.

6 Conclusions

We proposed a qualitative test algorithm for n-finger planar force-closure grasps based on a recursive reduction technique developed in this paper. Furthermore, we presented an algorithm for computing all n-finger force-closure grasps on polygonal objects. It has been demonstrated that the non-force-closure grasp region consists of two convex polytopes in the parameter space. The algorithms are implemented and their performance has been examined with simulation study. Finally, it should be pointed out that the qualitative test algorithm is applicable to 3D grasps. The extension of the grasp computation algorithm to 3D grasps is being studied.

Acknowledgment: The author would thank W. K. Yu and Joseph Chan for their helps in the programming.

References

[1] I-M. Chen and J. W. Burdick, "A qualitative test for N-Finger force-closure grasps on planar objects with applications to manipulation and finger Gaits," ICRA, pp. 814-820, 1993.

[2] M. R. Cutkosky, "On grasp choice, grasp model, and the design of hands for manufacturing task," IEEE Trans. Rob. and Automat., vol 5, no. 3, pp. 269-279, 1989.

[3] B. Mishra, et al., "On the existence and synthesis of multifinger positive grips," Algorithmica, vol. 2, no. 4, pp. 541-558, 1987.

[4] C. Ferrari and J. F. Canny, "Planning optimal grasps," ICRA, pp. 2290-2295, 1992.

[5] K. Mulmuley, "Computational Geometry," Prentice Hall, 1994.

[6] R. M. Murray, et al., "A mathematical introduction to robotic manipulation," CRC Press, 1994.

[7] V. D. Nguyen, "Constructing force-closure grasps," Int. J. Rob. Res., vol. 7, no. 3, pp. 3-16, 1988.

[8] J. Ponce and B. Faverjon, "On Computing Three Finger Force-Closure Grasp of Polygonal Objects," IEEE Trans. Rob. and Automat., vol. 11, no. 6, pp. 868-881, 1995.

[9] J. K. Salisbury and B. Rotch, "Kinematic and force analysis of articulated hands," ASME J. Mech., Transmissions, Automat., Design, vol. 105, pp. 33-41, 1982.

ON GRASPING AND MANIPULATING POLYGONAL OBJECTS WITH DISC-SHAPED ROBOTS IN THE PLANE

Attawith Sudsang and Jean Ponce

Department of Computer Science and Beckman Institute
University of Illinois, Urbana, IL 61801, USA

Abstract: This paper addresses the problem of grasping and manipulating a polygonal object with three disc-shaped robots capable of translating in arbitrary directions in the plane. The main novelty of the proposed approach is that it does not assume that contact is maintained during the execution of the grasping/manipulation task. Nor does it rely on detailed (and a priori unverifiable) models of friction or contact dynamics. Instead, the range of possible object motions for a given position of the robots is characterized in configuration space. This allows the construction of manipulation plans guaranteed to succeed under the weaker assumption that jamming does not occur during the task execution.

1 Introduction

This paper addresses the problem of manipulating a planar polygonal object with three disc-shaped robots capable of translating in arbitrary directions in the plane. In practice, the discs may be the fingertips of a robot hand or independently-moving mobile platforms.

We propose an algorithm for grasping the object and bringing it to a desired position and orientation through sequences of individual straight-line robot motions. This algorithm guarantees that the object will never escape from the robots' grasp, even when contact is broken during the initial grasping phase or the subsequent manipulation stage. It does not require synchronizing the motion of the discs, and only assumes that each one of them can be moved in turn along a given straight line trajectory.

The proposed approach is based on a detailed analysis of the geometry of the joint object/robot configuration space. Instead of trying to predict the exact motion of the object, we characterize the range of possible motions associated with each position of the robots and identify the "minimal" robot configurations for which the object is totally immobilized as well as the "maximal" ones for which there is a non-empty open set of object motions within the grasp, but no escape path to infinity.

2 Background and Approach

When a hand holds an object at rest, the forces and moments exerted by the fingers should balance each other so as not to disturb the position of this object. We say that such a grasp achieves *equilibrium*. For the hand to hold the object securely, it should also be capable of preventing any motion due to external forces and torques. This is captured by the dual notions of *form and force closure* from screw theory [2, 6], that constitute the traditional theoretical basis for grasp planning (see, for example, [8, 10, 11]). Recently, Rimon and Burdick have introduced the notion of *second-order immobility* [14] and shown that certain equilibrium grasps of a part which do not achieve form closure effectively prevent any *finite* motion of this part through curvature effects in configuration space. Algorithms for computing immobilizing grasps of three-dimensional objects can be found in [15].

We introduced in [15] the notion of *inescapable configuration space* (ICS) region for a grasp. This notion generalizes the concept of immobility: an object is immobilized when it rests at an isolated point of its free configuration space. By moving the fingers in an appropriate way, this isolated point transforms into a compact region of free space (the ICS) that cannot be escaped by the object. ICS regions were first introduced in the context of in-hand manipulation with a multi-fingered reconfigurable gripper [15] (see [13] for a related notion in the two-finger case, and [1, 3, 4, 7, 9, 12] for other approaches to pushing and manipulation). Here, ICS regions allow us to move an object by pushing it with three disc-shaped robots moving along straight lines: starting from some immobilizing configuration, we move the robots one at a time in some direction, then choose another direction etc.. to achieve the desired translation and/or rotation. The object remains at all times in the ICS region associated with the discs, and the planned manipulation is guaranteed to succeed as long as the friction forces associated with contacts between the robots, the object and its supporting plane are not large enough to cause jamming. In particular, our approach does not require that finger/object contact be maintained during grasping or manipulation, nor does it rely on any particular model of friction or contact dynamics.

Let us show an example to illustrate this idea (Fig. 1). The polygon shown in Fig. 1(a) is immobilized by the three discs since the three inward normals at the contacts intersect [14]. In Fig. 1(c), we translate one of the discs to a new position along the vector v. During this motion, there is no path that will allow the polygon to escape the grasp of the three discs: the polygon can move, but is constrained to remain within the corresponding ICS region of free space. Figure 1(d)-(e) shows unsuccessful attempts to take the polygon out of the grasp.

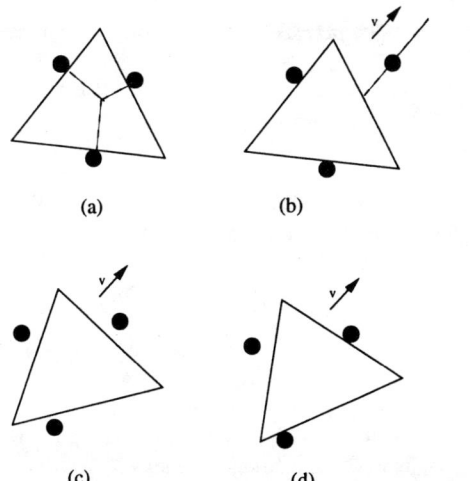

Figure 1: A polygon constrained to remain in the ICS region associated with the three robots.

3 Computing Maximum ICS Regions

We introduce formally in this section the concept of an inescapable configuration space region. The analysis proceeds along the lines of [15] by identifying the constraints imposed by the robots in the configuration space of the polygon. The general approach is the same as in [15] but the setting and the corresponding constraints are of course different.

3.1 Contact

We reduce the problem of achieving contact between a disc and a line to the problem of achieving point contact with a line. This is done without loss of generality by growing the object to be grasped by the disc radius and shrinking each disc into its center.

We attach a coordinate system (u, v) to the polygon, and write in this coordinate system the equations of the line supporting the edge e_i ($i = 1, 2, 3$) as $u \cos \alpha_i + v \sin \alpha_i - d_i = 0$, where α_i is the angle between the u axis and the *internal* normal n_i to the edge, and d_i is the distance between the origin of the (u, v) coordinate system and the edge.

Without loss of generality, we also define a world coordinate system (q, r) such that the r axis is parallel to the motion direction v and goes through the center of the first (moving) disc. We denote by $q_i = (q_i, r_i)^T$ the position of the center of disc number i in this coordinate system. In particular, $q_1 = 0$ and $r_1 = \delta$.

We can write the condition for contact between disc number i and the corresponding line as

$$q_i = \mathcal{R} p_i + t, \quad (1)$$

where $p_i = (u_i, v_i)^T$ and $q_i = (q_i, r_i)^T$ denote the positions of the contact point in the two coordinate systems, $\mathcal{R}$ is a rotation matrix of angle θ and $t = (x, y)^T$ is the translation between the two coordinate frames. Let $c_i = \cos(\theta + \alpha_i)$

and $s_i = \sin(\theta + \alpha_i)$, the above equation can be rewritten as

$$(x - q_i)c_i + (y - r_i)s_i + d_i = 0, \quad (2)$$

When the three contacts are achieved simultaneously, we have

$$\begin{pmatrix} c_1 & s_1 & \delta s_1 - d_1 \\ c_2 & s_2 & q_2 c_2 + r_2 s_2 - d_2 \\ c_3 & s_3 & q_3 c_3 + r_3 s_3 - d_3 \end{pmatrix} \begin{pmatrix} x \\ y \\ -1 \end{pmatrix} = 0.$$

For this equation to be satisfied, the determinant of the 3×3 matrix must be zero, which yields (after some simple algebraic manipulation):

$$\delta \sin(\theta + \alpha_1) + A_2 \cos(\theta + \beta_2) + A_3 \cos(\theta + \beta_3) - B = 0, \quad (3)$$

where β_2, β_3 and A_2, A_3, B are appropriate constants.

This condition defines a curve in θ, δ space, called the *contact curve*. This curve is defined on the $[0, 2\pi]$ interval, but an actual contact between the first disc and the corresponding edge can only occur when the angle between v and the internal normal to the edge is obtuse, i.e., when $\theta + \alpha_i \in [\pi, 2\pi]$. It follows from the form of its equation that the contact curve is in fact bounded by two vertical asymptotes on that interval.

3.2 Equilibrium

At equilibrium, the various forces and moments exerted at the contacts balance each other. This can be written in the object's coordinate system as

$$\sum_{i=1}^{3} \lambda_i \begin{pmatrix} n_i \\ p_i \times n_i \end{pmatrix} = 0, \quad \text{where} \quad \begin{cases} \lambda_1, \lambda_2, \lambda_3 > 0, \\ \lambda_1 + \lambda_2 + \lambda_3 = 1. \end{cases}$$

Using the change of coordinates (1) and taking advantage of the fact that $\sum_{i=1}^{3} \lambda_i n_i = 0$ allows us to rewrite this equation as

$$\sum_{i=1}^{3} \lambda_i \begin{pmatrix} n_i \\ (\mathcal{R}^{-1} q_i) \times n_i \end{pmatrix} = 0,$$

which can be interpreted as a 3×3 homogeneous equation in the coefficients $\lambda_1, \lambda_2, \lambda_3$. A necessary and sufficient condition for this equation to have a non-trivial solution is that its determinant be zero, i.e.,

$$\begin{vmatrix} n_1 & n_2 & n_3 \\ (\mathcal{R}^{-1} q_1) \times n_1 & (\mathcal{R}^{-1} q_2) \times n_2 & (\mathcal{R}^{-1} q_3) \times n_3 \end{vmatrix} = 0.$$

Expanding the determinant yields, after some additional algebraic manipulation, the condition

$$\delta \cos(\theta + \alpha_1) - A_2 \sin(\theta + \beta_2) - A_3 \sin(\theta + \beta_3) = 0,$$

and eliminating δ between this equation and the contact constraint (3) yields an equation in θ only:

$$\cos(\theta + \alpha_1) = \frac{A_2}{B} \cos(\beta_2 - \alpha_1) + \frac{A_3}{B} \cos(\beta_3 - \alpha_1). \quad (4)$$

There are (at most) two solutions for this equation in the $[0, 2\pi]$ interval. When they exist, exactly one of them is in the interval of physically achievable contacts. It is also easy to show that the corresponding solution is a minimum of the contact curve. As in [15], this minimum corresponds to an immobilizing configuration [14].[1]

Figure 2 shows an actual example in the object's and disc's coordinate frames. The triangle has to rotate 60 degrees counterclockwise to be immobilized by the matching discs (Fig. 2(c)). This is verified on the contact curve shown in Fig. 2(e) where the minimum occurs at 60 degrees in the physically realizable interval. The maximum of the curve corresponds to the configuration shown in Fig. 2(d), and it cannot be achieved in reality: the first disc would have to lie inside the triangle.

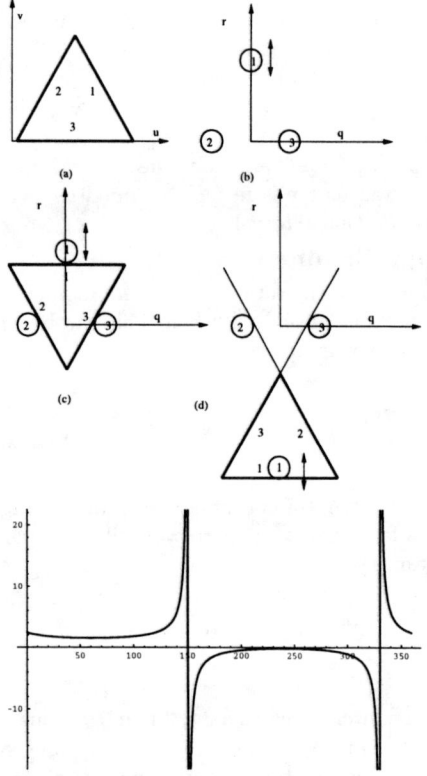

Figure 2: A grasp and the corresponding contact curve.

3.3 Free Configuration Space Regions

Let us consider an immobilizing configuration of the robots, and denote by $x_0, y_0, \theta_0, \delta_0$ the corresponding values of x, y, θ, δ. Let us also assume that the positions of robots 2 and 3 are held constant while the δ coordinate of the first robot may change.

---

[1]The object will be immobilized even if there is no friction: although this appears to contradict classical screw theory, which states that three fingers are not sufficient to immobilize a two-dimensional object in that case [6], recall that screw theory is concerned with *infinitesimal* motions: there exists an escape velocity but no finite escape motion. See [14] for details.

We denote by S_i the set of object configurations (x, y, θ) for which contact between disc number i and the corresponding object edge is achieved. From (2), this is a ruled surface in (x, y, θ) space, whose intersection with a plane $\theta =$ constant is a line $L_i(\theta)$ at distance $-d_i$ from the fixed point (q_i, r_i) of the x, y plane, and the angle between the x axis and the normal to this line is $\theta + \alpha_i$. Changing θ corresponds to rotating each line about the point (q_i, r_i). Changing δ amounts to translating the line $L_1(\theta)$.

Together, the three ruled surfaces S_1, S_2 and S_3 bound a volume V of free configuration space. Given the setup of the robots, it is obvious that if a configuration lies in free space for some value δ_1 of δ, it also lies in free space for any other value $\delta_2 \geq \delta_1$. In other words, $V(\delta_1) \subset V(\delta_2)$ when $\delta_2 \geq \delta_1$, and it follows that the immobilizing configuration (x_0, y_0, θ_0) is always in free space for $\delta \geq \delta_0$.

In addition, the intersection of V with a plane $\theta =$ constant is a triangle $T(\theta)$ that may contain an open subset, be reduced to a point, or be empty. In the second case, the three contacts are simultaneously achieved, and (3) is satisfied.

It is easy to show that a necessary and sufficient condition for the triangle $T(\theta)$ to contain at least one point is that the point (θ, δ) be *above* the contact curve. This allows us to characterize qualitatively the range of orientations θ for which $T(\theta)$ is not empty: for a given δ, the condition (3) is an equation in θ that may have zero, one, or two real solutions, with a double root at the minimum $\delta = \delta_0$ of the curve. In this case, the range of orientations reduces to a single point. For any value $\delta_1 > \delta_0$, there are two distinct roots θ', θ'', and the range of orientations is the arc bounded by these roots and containing θ_0.

In particular, since the volume V is a stack of contiguous triangles $T(\theta)$, it is clear at this point that, for $\delta \geq \delta_0$, V is a non-empty, connected, compact region of $\mathbb{R}^2 \times S^1$. The analysis confirms that the minimum point (θ_0, δ_0) of the contact curve corresponds to an isolated point of configuration space or equivalently to an immobilizing configuration: indeed, for $\delta = \delta_0$, the triangle $T(\theta_0)$ is reduced to a point, and $T(\theta)$ is empty for any $\theta \neq \theta_0$.

3.4 ICS Regions

The discussion so far has characterized the contacts between the discs and the lines supporting the corresponding edges, ignoring the fact that each edge is a compact line segment. For a given value of δ, let us construct a parameterization of the set $E_i(\theta)$ of configurations (x, y) for which disc number i touches the edge e_i. Obviously, $E_i(\theta)$ is itself a line segment supported by the line $L_i(\theta)$.

We first parameterize the corresponding edge e_i by

$$\begin{pmatrix} u_i \\ v_i \end{pmatrix} = d_i \begin{pmatrix} \cos \alpha_i \\ \sin \alpha_i \end{pmatrix} + \eta_i \begin{pmatrix} -\sin \alpha_i \\ \cos \alpha_i \end{pmatrix},$$

with η_i in some interval $[\eta_{i1}, \eta_{i2}]$. The segment $E_i(\theta)$ can now be parameterized by

$$\begin{pmatrix} x - q_i \\ y - r_i \end{pmatrix} = -d_i \begin{pmatrix} c_i \\ s_i \end{pmatrix} - \eta_i \begin{pmatrix} -s_i \\ c_i \end{pmatrix}. \quad (5)$$

The constraints $\eta_{i1} \leq \eta \leq \eta_{i2}$ ($i = 1, 2, 3$) define the regions of configuration space where actual contact will occur. When $E_i(\theta)$ and $E_j(\theta)$ intersect for all $i \neq j$, the three segments completely enclose the triangle $T(\theta)$, and we will say that the corresponding configuration satisfies the *enclosure condition* since there is no escape path for the object in the x, y plane with the corresponding orientation θ. More generally, when all triples of segments in the range of orientations associated with a given δ satisfy the enclosure condition, V itself is an *inescapable configuration space* (ICS) region: in other words, the object is free to move within the region V, but remains imprisoned by the grasp and cannot escape to infinity.

3.5 Maximum ICS Regions

We now address the problem of characterizing the maximum value δ^* for which $V(\delta)$ forms an ICS region for any δ in the $[\delta_0, \delta^*]$ interval. We know that at $\delta = \delta_0$ the three segments intersect at the immobilizing configuration, forming an ICS region reduced to a single point. Thus the enclosure condition holds at $\delta = \delta_0$. On the other hand, as $\delta \to +\infty$, the whole configuration space becomes free of obstacles, thus there must exist a critical point for some minimal value of δ greater than δ_0. This guarantees that δ^* has a finite value.

A critical configuration occurs when an endpoint of the segment $E_i(\theta)$ lies on the line $L_j(\theta)$, $j \neq i$. We intersect the lines $L_i(\theta)$ and $L_j(\theta)$ by substituting the parameterization (5) in the contact equation (2). Writing $\eta_i = \eta_{ik}$ ($k = 1, 2$) yields

$$\eta_{ik} = -\frac{d_j - d_i \cos(\alpha_i - \alpha_j) + (q_i - q_j)c_j + (r_i - r_j)s_j}{\sin(\alpha_i - \alpha_j)}. \quad (6)$$

It follows that critical points lie on one of the six *critical curves* of (θ, δ) space defined by (6) for $i, j \in \{1, 2, 3\}$ ($i \neq j$) and $k = 1, 2$. Note that when $i, j \in \{2, 3\}$, (6) is a function of θ only, and the corresponding critical curves are vertical.

We seek the minimum value of $\delta^* > \delta_0$ for which the range of possible object orientations defined by the contact curve includes one of the critical configurations. Let us suppose first that a critical value lies in the interior of the orientation range associated with some $\delta_1 \geq \delta_0$, and denote by $\delta_{\min}$ the minimum value of δ on the critical curve. By definition, we have $\delta_1 \geq \delta_{\min}$. Suppose that $\delta_1 > \delta_{\min}$. Then by continuity, there exists some δ_2 such that $\delta_{\min} < \delta_2 < \delta_1$ and the corresponding range of orientations also contains a critical orientation. The argument holds for any value $\delta > \delta_{\min}$. In other words, either the range of orientations of $\delta_{\min}$ contains a critical orientation, in which case $\delta^* = \delta_{\min}$, or it does not, in which case the critical value associated with δ^* must be one of its range's endpoints. This is checked by intersecting the contact curve and the critical curve. Note that this process must be repeated six times (once per each segment/vertex pair) to select the minimum value of δ^*.

Figure 3 shows an example, where the contact and critical curves have been constructed for some sample object (the contact curve is drawn with a thicker brush). In this case, the minimum of the critical curve occurs just below the contact curve, and the critical configuration is the intersection of the two curves, lying at the right endpoint of the corresponding range of orientations.

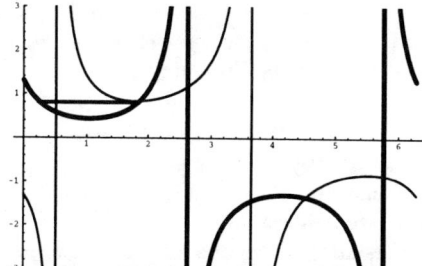

Figure 3: The contact and critical curves for a sample object. The critical range of orientations is shown as a horizontal line. See text for details.

Computing the minimum of the critical curve amounts to solving a trigonometric equation. It is easily shown that intersecting the critical curve and the intersection curve amounts to solving a quadratic equation in $\tan(\theta/2)$ when $i = 2, 3$ and $j = 1$, and a quartic equation in the same variable when $i = 1$, $j = 2, 3$. The intersection can be computed in closed form in both cases.

4 Planning Grasping and Manipulation Sequences

It is easily shown that the set of equilibrium (hence immobilizing) grasps of a polygon can be identified through linear programming, and various grasp optimality criteria (e.g., [8]) can be defined to choose a particular immobilizing grasp among this set. In this section, we will assume that an immobilizing grasp has been selected and that the initial object position and orientation are known, and we will show how to actually execute the grasp and then manipulate the polygon, moving the three robots one at a time while guaranteeing that the object will not escape.

In the rest of this section, a joint configuration of the polygon and the robots will be denoted by $q = (q_1, q_2, q_3, x, y, \theta)$, where q_i is as before the position of disc number i and (x, y, θ) denotes the polygon configuration. Given an immobilizing configuration q, MaxICS(q, i, v) will denote the maximum distance that robot number i can travel in the direction v while guaranteeing that the object cannot escape.

4.1 Capturing and Grasping a Polygon

Given some input grasp configuration q, we choose one of the discs (say the first one) and some direction v, say the external normal to the corresponding edge, and compute $\delta = \text{MaxICS}(q, 1, v)$ as described in the previous section. To capture the object, we first move the robots one by one from their home position to $q_1 + \frac{1}{2}\delta v$, $q_2 - \frac{1}{2}\delta v$, and $q_3 - \frac{1}{2}\delta v$. The polygon is now guaranteed to be in the maximum ICS region associated with the robots. We then translate the first robot by $-\delta v$.

Although the object may (and indeed will) move when contact occurs, it will end up in the planned immobilized configuration. Note that this approach is robust to uncertainty in the position of the object, but that it requires precise relative motions of the robots.

In the next two sections, we show how to achieve arbitrary translations and rotations of the object once it has been grasped. The overall motion will be decomposed into atomic translations of the three fingers along appropriate directions. The object will remained imprisoned in the grasp of the three robots during each motion.

4.2 Translating a Polygon

Let us assume that the object is currently immobilized by the discs in configuration q, and let us show how to apply the translation dv to the polygon. The immobilizing configuration is shown in Fig. 4(a). To translate the polygon, we will apply a translation δv to discs 2, 3, and 1 in succession. The problem is to compute the maximum value of δ guaranteeing that the polygon cannot escape at any time. If $d < \delta$, we will reset δ to d before applying the translation. If $d > \delta$, we will simply apply the same translation steps as many times as necessary.

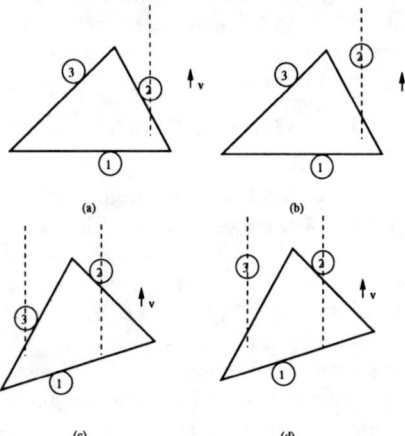

Figure 4: Translating a polygon: failed bisection step. The net translation $\delta' v$ of disc 3 in (d) is smaller than the translation δv of disc 2 in (b).

The first thing to note is that the object should not escape the grasp when we move disc 2, so δ must be smaller than or equal to $\delta_2 = \text{MaxICS}(q, 2, v)$. Likewise, once we have moved discs 2 and 3 by δv, the polygon should still be unable to escape, which implies that δ must be smaller than $\delta_1 = \text{MaxICS}(q, 1, -v)$.

Using the value $\delta = \min(\delta_1, \delta_2)$ is not sufficient because the polygon may move when the contact with discs 2 and 3 is broken. We use bisection to compute the maximum value of δ in the $[0, \min(\delta_1, \delta_2)]$ range such that both discs 2 and 3 can undergo a δv translation while maintaining inescapability (Fig. 4): For a given value of δ, we suppose first that disc 2 has already moved to its new position (Fig. 4(b)), then find the translation γ of disc 3 along v that will yield a new immobilizing configuration, say

q' (Fig. 4(c)). Consider now the net maximum translation $\delta' = \text{MaxICS}(q', 3, v) - \gamma$ that disc 3 may undergo while guaranteeing that the polygon cannot escape. If $\delta' \geq \delta$, we know that disc 2, then disc 3 can safely undergo the translation δv: the bisection step is successful and we increase the value of δ. If $\delta' \leq \delta$, then the bisection step has failed, and we try again with a lower value of δ. Figures 4 and 5 show respectively a failed bisection step and a successful one.

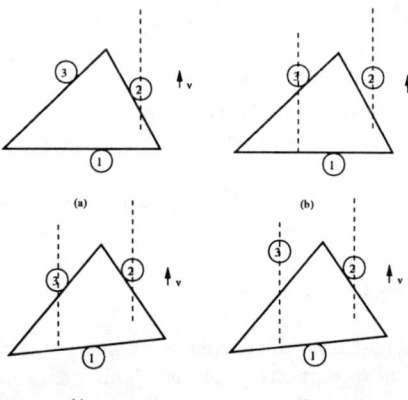

Figure 5: Translating a polygon: successful bisection step. The net translation $\delta' v$ of disc 3 in (d) is greater than the translation δv of disc 2 in (b).

A pseudocode version of the algorithm is given below. In it, q denotes the initial configuration, the function $\text{ImTransl}_3(q_1, q_2, q_3, v)$ returns the distance that q_3 must travel along v to reach an immobilizing configuration, and the function $\text{ImConfig}(q_1, q_2, q_3)$ returns the object configuration associated with the contacts q_i. The output of these two functions is easily computed using the contact and equilibrium conditions (2), (3) and (4). The variables ν and δ below are respectively the low and high values used to bound the bisection.

```
1:    q = (q_1, q_2, q_3, x, y, θ);
2:    δ_1 = MaxICS(q, 1, -v);
3:    δ_2 = MaxICS(q, 2, v);
4:    ν = 0.0;
5:    if δ_2 > δ_1 then δ = δ_1
6:        else δ = δ_2;
7:    do {
8:        q'_2 = q_2 + δv;
9:        γ = ImTransl_3(q_1, q'_2, q_3, v);
10:       q'_3 = q_3 + γv;
11:       (x', y', θ') = ImConfig(q_1, q'_2, q'_3);
12:       q' = (q_1, q'_2, q'_3, x', y', θ');
13:       δ_3 = MaxICS(q', 3, v);
14:       if δ_3 - γ ≥ δ
15:           then ν = (δ + ν)/2
16:           else δ = (δ + ν)/2
17:   } until (δ - ν < ε);
18:   return δ.
```

So far, we have shown how to translate the polygon in one direction. We can translate the polygon in arbitrary directions using plans computed for three directions only by switching the roles of the discs and alternating between the three directions. If the chosen directions positively span the plane, it is easy to see that we can arrange a sequence of translations to bring the polygon to any position. One simple choice for these directions is the inward normals at the contacts.

4.3 Rotating a Polygon

Some of the steps involved in translating a polygon also prove useful in rotating it (Fig. 4): Starting in some configuration q, we rotate the polygon counterclockwise by first translating disc 2 in direction v for some distance δ (Fig. 4(b)), before translating disc 3 in the direction $-v$ until the polygon is immobilized in a new configuration q' (Fig. 4(c)). To guarantee that the polygon will not escape, we must have $\delta \leq \text{MaxICS}(q, 2, v)$ and $\delta \leq \text{MaxICS}(q', 3, v)$. Note that in this case the translation of disc 2 is actually performed, but disc 1 does not move, and thus does not constrain the value of δ. The maximum value of δ is found as before through bisection. Clearly, for a given vector v, this value yields the maximum possible rotation. Achieving a smaller rotation is done by using a smaller δ (the corresponding positions of discs 2 and 3 are once again easily calculated in this case using the equilibrium and contact conditions). Achieving a larger one involves repeating the same elementary rotation several times.

Note that the rotation steps will change the positions of the discs along the edges of the polygon (Fig. 6(a)). We would like these positions to remain unchanged so that we can apply the same rotating steps repeatedly and achieve a pure rotation (Fig. 6(b)).

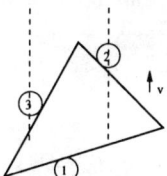 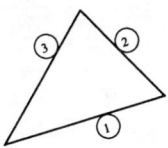

Figure 6: Effect of a rotation: (a) the position of the discs along the edges has changed; (b) corrected disc positions.

To correct the disc positions, we move them one by one toward their initial configuration along the edges. The object may move during this correction stage, but it will eventually come back to its original position and orientation, which are uniquely determined by the disc positions in their initial configurations. The difficulty is to guarantee that the polygon will remain imprisoned by the discs during the correction steps. This corresponds to computing the maximum ICS region in the direction of the corresponding edges. The ICS computation degenerates in this case to the following construction (Fig. 7): In Fig. 7(a), the polygon is immobilized, and disc 2 touches the edge E in A. Let L denote the line supporting E in this configuration (L is fixed but E will move when the position of disc 2 changes), and let w denote the direction in which we want to move disc 2 along L. We want to find the furthermost point on L to which we can move disc 2 from A in the direction w.

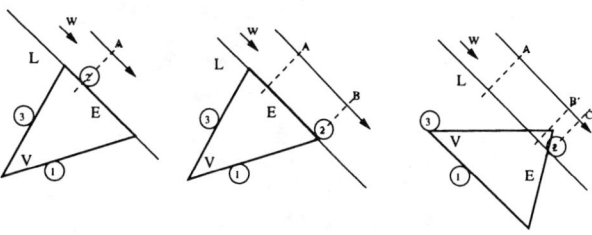

Figure 7: Moving a disc along a line.

If this point is further than the edge endpoint B, the polygon will obviously be able to escape (Fig. 7(b)). In addition, moving disc 2 in the direction w will allow the polygon to rotate clockwise. Figure 7(c) shows the maximum clockwise rotation preventing the polygon to escape (vertex V touches disc 3). Let C denote the intersection of L and E in that case. Clearly, the point we seek must not be further than C, since this would also allow the polygon to escape. In general, it can be shown that we can safely move disc 2 along L anywhere from A to the closest of the two points B and C.

With this method, we move a disc toward its initial position along the corresponding edge (if the position is not in the allowable range, we move to the point closest to the position), update the current position of the disc, apply the method to another disc, and continue this procedure until all three discs reach the desired positions.

Note that the rotation steps also affect the overall position of the polygon. To perform a pure rotation, a final translation stage has to be performed, using the technique presented in the previous section.

5 Implementation and Results

We have implemented the algorithm for planning manipulation sequences described in the previous section. Figure 8 shows intermediate immobilizing configurations in a manipulation sequence that brings an equilateral triangle with edges of unit length from some initial configuration (lower right) to a goal configuration (upper left). The triangle is first rotated to the desired orientation, and then translated to the desired position. The entire sequence has 28 steps. By choosing a grasp with contacts at the center of the edges, and the inward normals as the translation directions, we obtain a maximum translation distance of 0.092, and a maximum rotation angle of 14.98 degrees. The program takes less than 1 second to compute the sequence on a 200MHz PC.

Figure 9 shows snapshots of some of the translation and rotation steps used in the experiment of Fig. 8. Figure 9(a) shows the input grasp. Figure 9(b)-(d) shows the

Figure 8: A manipulation sequence.

elementary stages of a translation step, and Fig. 9(e)-(i) shows the elementary stages of a rotation step, including the disc motions along the edges.

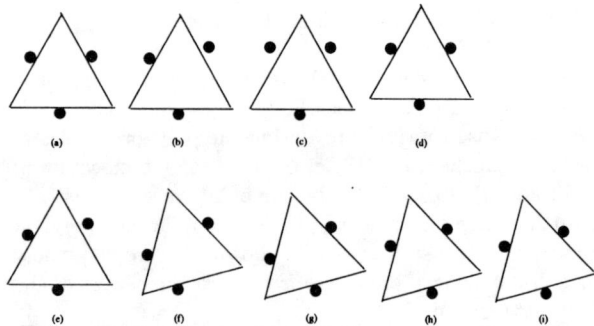

Figure 9: Snapshots of translation and rotation steps.

Figure 10 shows another example where a simple potential-field method [5] has been used to plan a motion avoiding polygonal obstacles. At each step, virtual attractive and repulsive forces acting on the object are computed to determine the next free-space configuration of the object. To move the object from the current configuration to the next one, a local plan composed of elementary translation and rotation steps is computed. If the local plan is collision free, it is used, otherwise a new candidate configuration is generated and the corresponding local plan is computed. The process is iterated until a collision-free global plan is found. The run time for the example shown in the figure is 11.0 seconds on a 200 MHz PC.

Acknowledgments. This research was supported in part by the National Science Foundation under grant IRI-9634393, a UIUC Research Board grant, a UIUC Critical Research Initiative planning grant, and an equipment grant from the Beckman Institute for Advanced Science and Technology.

References

[1] S. Akella and M.T. Mason. Parts orienting by push-aligning. In *Proc. ICRA*, pp. 414–420, Nagoya, Japan, 1995.

[2] R.S. Ball. *A treatise on the theory of screws*. Cambridge University Press, 1900.

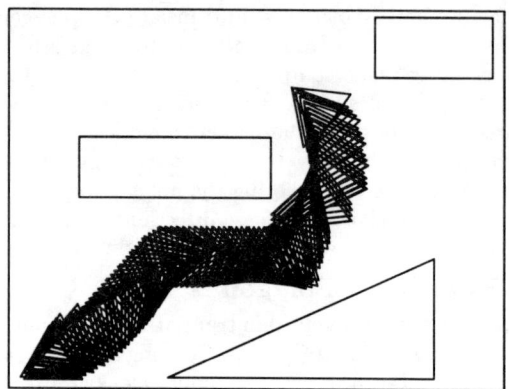

Figure 10: Obstacle avoidance example.

[3] M.A. Erdmann and M.T. Mason. An exploration of sensorless manipulation. *IEEE J. Robot. Autom.*, 4:369–379, 1988.

[4] K.Y. Goldberg. Orienting polygonal parts without sensors. *Algorithmica*, 10(2):201–225, 1993.

[5] O. Khatib. Real-time obstacle avoidance for manipulators and mobile robots. *IJRR*, 5(1):90–98, 1986.

[6] K. Lakshminarayana. Mechanics of form closure. Tech. Rep. 78-DET-32, ASME, 1978.

[7] K.M. Lynch and M.T. Mason. Stable pushing: mechanics, controllability, and planning. In K.Y. Goldberg, D. Halperin, J.-C. Latombe, and R. Wilson, eds., *Algorithmic Foundations of Robotics*, pp. 239–262. A.K. Peters, 1995.

[8] X. Markenscoff and C.H. Papadimitriou. Optimum grip of a polygon. *IJRR*, 8(2):17–29, 1989.

[9] M.T. Mason. Mechanics and planning of manipulator pushing operations. *IJRR*, 5(3):53–71, 1986.

[10] B. Mishra, J.T. Schwartz, and M. Sharir. On the existence and synthesis of multifinger positive grips. *Algorithmica*, 2(4):541–558, 1987.

[11] V-D. Nguyen. Constructing force-closure grasps. *IJRR*, 7(3):3–16, 1988.

[12] M.A. Peshkin and A.C. Sanderson. Planning robotic manipulation strategies for workpieces that slide. *IEEE J. Robot. Autom.*, 4(5), 1988.

[13] E. Rimon and A. Blake. Caging 2D bodies by one-parameter two-fingered gripping systems. In *Proc. ICRA*, pp. 1458–1464, Minneapolis, MN, 1996.

[14] E. Rimon and J.W. Burdick. Towards planning with force constraints: On the mobility of bodies in contact. In *Proc. ICRA*, pp. 994–1000, Atlanta, GA, 1993.

[15] A. Sudsang, J. Ponce, and N. Srinivasa. Algorithms for constructing immobilizing fixtures and grasps of three-dimensional objects. In J.-P. Laumont and M. Overmars, editors, *Algorithmic Foundations of Robotics II*, pp. 363–380. AK Peters, Ltd., 1997.

Schedule Execution using Perturbation Analysis

Luc Bongaerts, Hendrik Van Brussel, Paul Valckenaers

Katholieke Universiteit Leuven - Mechanical Engineering Department
Celestijnenlaan 300B - B-3001 Leuven, Belgium - Tel: 32-16-32 24 80 ; fax: 32-16-32 29 87
e-mail: Luc.Bongaerts@mech.kuleuven.ac.be - http://www.mech.kuleuven.ac.be/pma/pma.html

Abstract

In a holonic shop floor control (SFC) architecture, a reactive scheduler co-operates with an autonomous on-line SFC system to combine high performance with reactivity against disturbances. To take its decisions, the on-line SFC system uses the existing schedule as advice. This "schedule execution" problem (SE) is not trivial, because the decisions taken by the on-line SFC system are local decisions in a combinatorial optimisation problem. By representing the schedule as a graph, this paper presents an algorithm for SE, based on perturbation analysis. It tackles the non-linearities in the production system by a combination of linearisation and feedback control.

1. Introduction

In industry, high performance schedulers currently are hardly used, due to the lack of schedule robustness [14]. While a near-optimal schedule is executed on the shop floor, several disturbances occur: rush orders, machine breakdowns, etc. It turns out that the performance of a schedule is very sensible to these disturbances and it is difficult to execute that schedule [14]. The problem is partly due to the fact that the input parameters for the scheduling algorithm are changing while the schedule is being calculated. Hence, the resulting schedule may not be feasible nor optimal anymore.

To cope with disturbances, researchers have proposed two main approaches [14]: reactive scheduling and heterarchical control. The reactive scheduling approach [12] uses schedule repair techniques [6, 11], feed-back [10] and feed-forward [13] to address the dynamic and stochastic aspects of the problem. The idea of schedule repair is very valid for adapting a schedule to a new situation, but neglects the opportunities to immediately react with an intelligent dispatcher and have a second line reaction in the reactive scheduler. In heterarchical control [7, 8, 9], reactivity to disturbances emerges from the co-operation of the individual agents. It does not address any optimisation either and confines itself to fast heuristics. Since, usually, advice from specific scheduling agents is not used, the opportunities to immediately react with an intelligent dispatcher and have a second line reaction in the reactive scheduler, allowing for optimisation, are totally neglected. Further references can be found in [4].

In previous work on holonic manufacturing [3], we have proposed the concept of concurrent scheduling and schedule execution (SE) to combine schedule optimisation and robustness against disturbances. While a reactive scheduler periodically calculates an (updated) schedule, an on-line shop floor control (SFC) system (like an intelligent dispatcher) executes this schedule, autonomously reacting to disturbances. This task of the on-line SFC system, defined as *autonomous schedule execution*, consists of real time monitoring and control of the manufacturing resources, reacting to disturbances if necessary, but following the schedule if possible, considering it as advice [2]. (In this way, it is similar to a robot controller following a trajectory generated by a trajectory planner.) Both scheduler and on-line SFC system have their own intelligence, and work concurrently and asynchronously, but co-operate on a regular basis to maintain performance at a high level. Thereby, the scheduler exploits the available time for the optimisation of the schedule, providing the system with a high performance. Meanwhile the on-line SFC system reacts immediately to disturbances, providing the system with a quick response to disturbances.

This paper describes an algorithm for SE that implements the concepts behind holonic manufacturing. While reactive scheduling algorithms have already been studied in detail, SE is a new concept and far from trivial, because the decisions taken by the on-line SFC system are local decisions, which affect global performance in a non-linear way. The algorithm is based on perturbation analysis (PA). PA explicitly models the effect of local decisions to the global performance. Since on-line control consists mainly of taking local decisions, PA is particularly suited for SE.

The paper is structured as follows. The second section shows why SE is not trivial. The third section presents a different way to represent a schedule, such that it is more suited for SE with perturbation analysis. The next two sections then present the resulting algorithms in the scheduler and the on-line SFC system. The conclusion states that this algorithm explicitly considers the final goal of the research: to combine a good schedule with reactivity to disturbances.

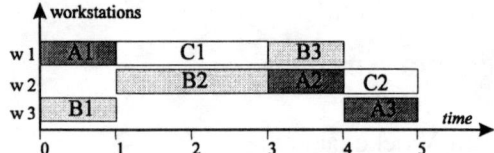

Figure 1: An example of problems that occur when executing a schedule. If operation A1 takes 1 time unit more to complete than expected, it is best to skip operation C1, but only if order B is more important than order C.

2. Concepts

2.1 Schedule execution problem

It is impossible to find a good SE algorithm that deals with disturbances autonomously and does not have to perform rescheduling itself, if the Gantt chart is the only available information of the schedule.

This is shown in an example for a job-shop scheduling problem. While the on-line SFC system is executing the schedule given in figure 1, operation A1 is delayed. The on-line SFC system should decide whether to keep the scheduled sequence of operations on workstation 1 (namely C1 - B3, i.e. follow the schedule) or to swap the operations (i.e. react autonomously). The decision whether to swap or not will influence the global performance (like the weighted mean tardiness) via precedence constraints within orders and sequencing on workstations and other resources, and via the finish times of the all orders. Therefore, if order C is more urgent than B, operations C1 and B3 probably cannot be swapped. However, if operation A2 takes a lot of time, the swap would again be possible, because C2 would not be delayed by the swap. However, for this kind of reasoning, the information contained in the Gantt-chart was not sufficient, and it was necessary to consider how local decisions affect global performance.

2.2 Solution

Since a Gantt chart does not contain sufficient information to execute a schedule autonomously, additional advice has to be given: the on-line SFC system needs to know how local decisions affect the global performance.

A mathematical technique designed specifically for this problem is Perturbation Analysis (PA — [5] describes the application of this technique to discrete event systems). In PA, the global performance is expressed in function of local parameters. Partial derivatives of the global performance to the local parameters are calculated. Then, in standard PA, the global performance is linearised in function of the local parameters, using these partial derivatives. As such, local decisions explicitly conform to the global goal.

These partial derivatives need to be defined and calculated off-line (by the reactive scheduler) and in such a way that the on-line SFC system can instantly derive the correct local decision from this advice and the on-line manufacturing data. If a schedule is represented as a Gantt chart, it is not even obvious to clearly define these partial derivatives. Therefore, the next section gives another representation for a schedule, based on graph theory. This allows to define the local decision parameters ε_e, and formulate the global performance Λ in function of the ε_e. This way, section 4 can define the partial derivatives $\delta\Lambda/\delta\varepsilon_e$ exactly and describe an efficient calculation schema for the scheduler. During the on-line control (see section 5), a number of local decision alternatives ('choices') are defined and evaluated according to the local decision parameters and the partial derivatives, thereby making local decisions with a global view, but based on real time data. For each alternative, the effect on the values ε_e is calculated in a simple but non-generic way and as such, $\Delta\Lambda$ is calculated. The alternative with the best $\Delta\Lambda$ is selected.

3. Schedule Represented as a Graph

The global performance Λ is function of the start times of all operations: the done ones, and the ones that still have to be executed. (The estimated value of) Λ therefore depends on done operations, on local decisions (for the operations that are about to be executed) and on the schedule (for those operations that still have to be executed later on). This section shows how to express Λ in an unambiguous way.

3.1 Precondition on the schedule

Usually, in short-range scheduling in an environment full of disturbances, WIP minimisation is of less importance. Hence, the proposed approach is limited to the execution of **active schedules**. An active schedule is defined as a schedule in which no operation can be executed on an earlier time instant without delaying another operation. If this constraint is not satisfied, the schedule can still be executed, but delays for operations as imposed by the schedule cannot be followed in the suggested approach.

3.2 Operation start times

The resource allocation problem addressed here, is job shop scheduling extended with alternative operations and alternative routings (represented by precedence graphs). The schedule consists of an allocation of a workstation and a starting time for each operation. Since the schedule is active, the starting time of an operation can be represented as the earliest time the operation can execute, re-

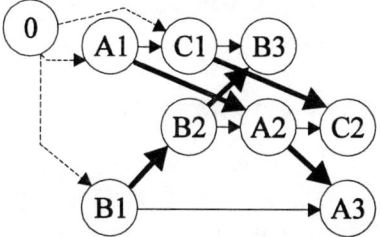

Figure 2: Precedence constraints representing a schedule. The thick arrows represent the technological precedence constraints, the thin ones the sequencing of operations on machines, and the broken line ones the release date of an order.

specting the (scheduled) sequence of the operations of the workstation, respecting the (scheduled) sequence of operations for auxiliary equipment (like a pallet), respecting its (technological) precedence constraints with other operations of the same order and respecting eventual transport and set-up times. Thus, the start time of an operation can be expressed mathematically in function of earlier operations:

$$t_{ij} = \max \begin{pmatrix} \max_{p=1..P}(t_{i,j_p} + d_{i,j_p} + T_{i,j_p,j}), \\ t_{i_a,j_a} + d_{i_a,j_a} + T_{i_a,j_a,i,j}, \\ t_{i_k,j_k} + d_{i_k,j_k} + S_{i_k,j_k,i,j}, \\ R_i \end{pmatrix}, \quad (1)$$

with the following definitions for the symbols used:
t_{ij} : starting time of operation j of order i (oper. i,j),
d_{ij} : duration of operation i,j ,
j_p : predecessor operations of oper. j, p = 1 ... P ,
i_k,j_k : previous operation scheduled on workstation k,
i_a,j_a : previous operation scheduled on auxiliary resource a,
R_i : release date of order i,
$T_{i,j_p,j}$: transport time between oper. i,j_p and oper. i,j,
$T_{i_a,j_a,i,j}$: transport time between oper. i_a,j_a and oper. i,j,
$S_{i_k,j_k,i,j}$: set-up time between oper. i_k,j_k and oper. i,j.

This implies that the start time of all operations can be calculated, starting from R_i, the workstation allocations and the sequencing of all operations on each resource (workstation and auxiliary resources). By definition, exactly following the schedule means keeping the workstation allocation and operation sequencing on each resource.

3.3 Graph

Using formula (1), the schedule can be modelled as a directed graph G [1], where every node represents an operation and where every edge represents a precedence constraint. Some of the precedence constraints represent the technological precedence constraints, while other ones represent the scheduling decisions. The first operation of each order also has a release date. This is represented as a precedence constraint between a dummy operation with duration 0 on the time instant of the release date and the first operation of each order. For instance, figure 2, shows a graph of the schedule shown in figure 1. The graph shows the constraints given by the original (technological) precedence constraints and the scheduling decisions. In this example, auxiliary resources are not taken into account, and the precedence constraints represent a fixed sequence. All orders have to start at time 0 or later.

Figure 3 illustrates some additional definitions. On each instant of time t_{cut}, the schedule can be partitioned in a set of operations that start before t_{cut} and a set of operations that start at time t_{cut} or later. The graph G_h, that is a subgraph of G and consists only of operations starting before t_{cut} is called the "schedule head". The graph G_b, that is a subgraph of G and consists of the other operations is called the "schedule body". The set E of edges (the arrows) of the graph that connect a node (an operation) of the schedule head with a node of the schedule body, is called the "cut" E. The union of the two subgraphs G_h and G_b and the cut E is again the original graph G. If the schedule is feasible, then all edges between the schedule head and the schedule body are oriented in the direction from schedule head to schedule body. In graph theory, one can prove that a cut is sufficient to define the schedule head and the schedule body, if the graph G is connected.

3.4 Definition of local decision parameters ε_e

Given all starting times of the operations of the schedule head, it should be possible to calculate the starting times of all operations of the schedule body. Therefore, for every edge $e = (n_1, n_2)$ in the cut E, a new variable is defined: the earliest start time ε_e of an operation n_2 is the earliest time operation n_2 could start if it would respect the precedence constraint represented by the edge e (an arrow of the graph). Therefore,

$$\varepsilon_e = t_{i_1,j_1} + d_{i_1,j_1} + A_{i_1,j_1,i_2,j_2}, \quad (2)$$

where t_{i_1,j_1} is the start time of operation (i_1,j_1), d_{i_1,j_1} is the duration of operation (i_1,j_1) and A_{i_1,j_1,i_2,j_2} is the auxiliary operation time (i.e. set-up time or transport time) between operation (i_1,j_1) and operation (i_2,j_2).

3.5 Performance in function of ε_e

If the values of the earliest start times ε_e belonging to every edge of the cut are known, it is possible to calculate the start time of all operations of the schedule body. After extending subgraph G_b with the dummy nodes for each ε_e and an arrow between each dummy node and its respective operation, this extended subgraph contains all information necessary to derive all start times as they were in the original schedule (figure 4).

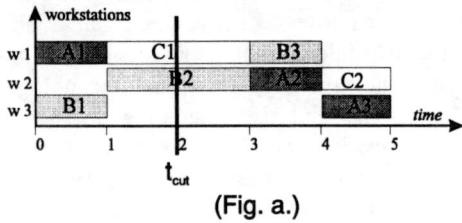

(Fig. a.)

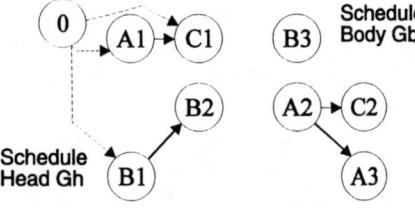

(Fig. b.)

Figure 3: (a) The time t_{cut} is chosen equal to 2 time units. (b) The first graph G_h is the schedule head. The second graph G_b is the schedule body. The cut E = {(C1,B3), (B2,A2), (B1,A3), (A1,A2), (C1,C2), (B2,B3)}, but it is not drawn.

The objective function Λ (the schedule performance) is a function of the start date of all operations: for instance, for the weighted mean tardiness objective function, Λ is a linear function of the tardiness of each order, and the tardiness is a piece-wise linear function of the start time of the last operation of each order. Thus, for the example already shown, the global performance Λ of the schedule can be written as:

$$\Lambda = w_A \cdot \max(0, t_{A3} + d_{A3} - D_A) + \\ w_B \cdot \max(0, t_{B3} + d_{B3} - D_B) + \qquad (3), \\ w_C \cdot \max(0, t_{C2} + d_{C2} - D_C)$$

where w_i is the weight of order i, representing its relative importance; and D_i is the due date of order i.

If the cut that partitions the schedule in a schedule head and a schedule body, is defined, it is possible to express the performance Λ of the schedule in function of the earliest start times ε_e of the cut E and the start times $t_{ij,head}$ of the operations of the schedule head:

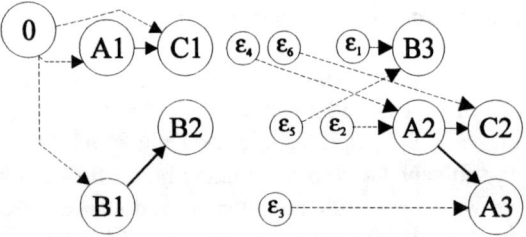

Figure 4: The subgraph G_b is extended with dummy nodes such that the ε_e are the only parameters to be known to derive the start time of all operations of G_b.

$$\Lambda = \lambda_{t_{cut}}(\{t_{ij,head}\}, \{\varepsilon_e\}).$$

For instance, for the example given above, using formula (1) and (3),

$$\Lambda = \begin{aligned} & w_A \cdot \max(0, \max(\max(\varepsilon_2, \varepsilon_4) + d_{A2}, \varepsilon_3) + d_{A3} - D_A) + \\ & w_B \cdot \max(0, \max(\varepsilon_1, \varepsilon_6) + d_{B3} - D_B) + \\ & w_C \cdot \max(0, \max(\varepsilon_5, \max(\varepsilon_2, \varepsilon_4) + d_{A2}) + d_{C2} - D_C) \end{aligned} \quad (4)$$

It shows that it is possible to express the performance of an entire schedule in function of a reasonably small amount of parameters that represent local aspects of the schedule. It also shows that the parameters $t_{ij,head}$ only appear in the equation if they directly affect the performance, because all starting times of operations of the schedule body are expressed in function of the earliest start time variables only.

4. Calculations of Reactive Scheduler

The reactive scheduler has to calculate the partial derivatives of the global performance to the local decision parameters ε_e. The rest of the calculations are performed on-line by the on-line SFC system.

If the performance Λ is a continuous function of the start times, with a continuous derivative except in a limited number of points, then the functions $\lambda_{t_{cut}}(\varepsilon_{e_1}, \varepsilon_{e_2}, ..., \varepsilon_{e_N}, t_{(ij,head)_1}, t_{(ij,head)_2}, ..., t_{(ij,head)_{N_{head}}})$ also have a continuous derivative except in a limited number of points (a lower dimensional hyperplane in the hyperspace of ε_e). As an example, figure 5 shows $\lambda_3(\varepsilon_2)$, the function that expresses how the global performance Λ changes due to changes of the resource driven earliest start time for operation A2. In other words, $\lambda_3(\varepsilon_2)$ describes the consequences of a delayed (or earlier) availability of workstation w2 for operation A2. It can be seen that, for this example with weighted mean tardiness as objective function, the function $\lambda_3(\varepsilon_2)$ is a piece-wise linear function of ε_2.

Consequently, where the function has a derivative, the effect can be expressed by a partial derivative of Λ to ε_e:

$$\frac{\partial \lambda_{t_{cut}} \begin{pmatrix} \varepsilon_{e_1}, \varepsilon_{e_2}, ..., \varepsilon_{e_N}, \\ t_{(ij,head)_1}, t_{(ij,head)_2}, ..., t_{(ij,head)_{N_{head}}} \end{pmatrix}}{\partial \varepsilon_e}, \quad (5)$$

where N is the number of edges of the cut E and N_{head} is the number of operations in the schedule head. Where the function is non-differentiable, it is possible to calculate the left hand and right hand partial derivative.

The reactive scheduler calculates off-line the partial derivative of the global performance measure to the earliest start time defined by each arrow in the schedule graph. After the schedule has been generated, it calculates the function $\lambda_{t_{cut}}$ for each $t_{cut}(i,j)$ corresponding with the scheduled start time of an operation (i,j). For each edge e of the cut E, the value of (5) is calculated. Along with the schedule in Gantt chart format, the graph representation and the values of the partial derivatives (5) are sent to the on-line SFC system. The partial derivatives are calculated efficiently using linear programming techniques [5].

The following example demonstrates how to calculate the partial derivative for time $t_{cut} = 3$ for all edges e of the cut. Using (2) and (4), the nominal values of ε_e and the partial derivatives $\delta\Lambda/\delta\varepsilon_e$ can be calculated (see figures 3a and 3b and table 1). For instance, using formula (4), $\delta\Lambda/\delta\varepsilon_4$ is calculated:

$$\frac{\partial \Lambda}{\partial \varepsilon_4} = w_A \cdot \frac{\partial \max\begin{pmatrix} 0, \varepsilon_2 + d_{A2} + d_{A3} - D_A, \\ \varepsilon_4 + d_{A2} + d_{A3} - D_A, \varepsilon_3 + d_{A3} - D_A \end{pmatrix}}{\partial \varepsilon_4} + w_C \cdot \frac{\partial \max\begin{pmatrix} 0, \varepsilon_5 + d_{C2} - D_C, \\ \varepsilon_2 + d_{A2} + d_{C2} - D_C, \varepsilon_4 + d_{A2} + d_{C2} - D_C \end{pmatrix}}{\partial \varepsilon_4} \quad (6).$$

Given the other values of ε_e (for e not equal to 4), and given the durations and due dates, the partial derivative becomes:

$$\frac{\partial \Lambda}{\partial \varepsilon_4} = w_A \cdot \frac{\partial \max(1, \varepsilon_4 - 2)}{\partial \varepsilon_4} + w_C \cdot \frac{\partial \max(0, \varepsilon_4 - 4)}{\partial \varepsilon_4}.$$

5. On-line Control

5.1 Algorithm

During on-line control of the manufacturing system, local decisions should be made, like allocating the resource to an operation, or sequencing the operations on a resource. These decisions are taken in three phases: 1) alternative local decisions are proposed; 2) they are evaluated and 3) the best one is chosen.

To find alternative decisions, three kind of scheduling decisions can be modified: the allocation of operations to workstations; the sequencing of operations on a workstation (a 'swap'); and the sequencing of operations on secondary resources. In the simple version of the algorithm, (as it is presented here) only re-allocation to another workstation and the swap of the first two operations scheduled on a resource are considered.

In the second phase, for each alternative decision, an estimate of the resulting global performance is calculated. Therefore, the on-line SFC system maintains a set N_D of done operations, a set N_{Bu} of busy operations, a set N_P of operations it is about to take local resource allocation decisions about (the 'pending' operations), and finally the set N_B of operations contained in the schedule body. (In other words, the first three sets N_D, N_{Bu}, and N_P together form the schedule head.) For the set of pending operations N_P,

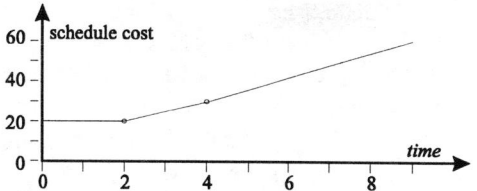

Figure 5: The function $\lambda_3(\varepsilon_2)$ expresses how the global performance Λ changes due to changes of the resource driven earliest start time ε_2 for operation A2. In other words, $\lambda_3(\varepsilon_2)$ describes the consequences of a delayed (or earlier) availability of workstation w2 for operation A2.

the start and finish times are determined by the local decisions. The set N_P itself is constructed by taking the next few operations of the schedule, until a given time t_{cut}. For each alternative, the on-line SFC system calculates the effect on all ε_e of the cut at time t_{cut}, such that it can also estimate the effect on Λ (linear estimation, using the partial derivatives).

In the third phase, the alternative with the lowest cost increase is selected as the best decision.

In the next subsection, an example shows how to react to a delayed operation. Examples on operations finished earlier, machine breakdowns, tool breakdowns, and the use of alternative operations are shown in [4].

5.2 Example: delayed operation

The example is the same one as explained in figure 1. Suppose operation A1 is delayed with half a time unit and could only be finished at time 1.5. In this very simple case, the alternatives to be considered are swapping operations C1 and B3 or keeping the sequence. Alternative workstations or secondary resources are not considered.

To evaluate the first alternative, t_{cut} is chosen to be 3. Then $N_D=\{A1, B1\}$; $N_{Bu} = \{B2\}$, $N_P=\{C1\}$, and $N_B =\{A2, A3, B3, C2\}$. The performance depends on the schedule

	$(\varepsilon_e)_{nom}$	$\delta\Lambda/\delta\varepsilon_e$	$(\delta\Lambda/\delta\varepsilon_e)_{nom}$
ε_1	3	0 if $\varepsilon_1 < 3$; w_B if $\varepsilon_1 > 3$	$\delta\Lambda/\delta\varepsilon_1(3^-) = 0$ $\delta\Lambda/\delta\varepsilon_1(3^+) = 10$
ε_2	3	0 if $\varepsilon_2 < 2$; w_A if $2 < \varepsilon_2 < 4$; w_A+w_C if $\varepsilon_2 > 4$	$\delta\Lambda/\delta\varepsilon_2(3) = 5$
ε_3	1	0 if $\varepsilon_3 < 4$; w_A if $\varepsilon_3 > 4$	$\delta\Lambda/\delta\varepsilon_3(1) = 0$
ε_4	1	0 if $\varepsilon_4 < 3$; w_A if $3 < \varepsilon_4 < 4$; w_A+w_C if $\varepsilon_4 > 4$	$\delta\Lambda/\delta\varepsilon_4(1) = 0$
ε_5	3	0 if $\varepsilon_5 < 3$; w_B if $\varepsilon_5 > 3$	$\delta\Lambda/\delta\varepsilon_5(3^-) = 0$ $\delta\Lambda/\delta\varepsilon_5(3^+) = 10$
ε_6	3	0 if $\varepsilon_6 < 5$; w_C if $\varepsilon_6 > 5$	$\delta\Lambda/\delta\varepsilon_6(3) = 0$

Table 1: The nominal values of ε_e and $\delta\Lambda/\delta\varepsilon_e$, for $t_{cut} = 3$.

body, and thus indirectly on the values of ε_e, as defined and calculated in the example above. If C1 is executed immediately after A1, then C1 has finished at time 3.5, and $\varepsilon_1 = 3.5$; $\varepsilon_4 = 1.5$; $\varepsilon_6 = 3.5$. The linear approximation of the new Λ (see table 1 for numerical values)

$= \Lambda_0 + \delta\Lambda/\delta\varepsilon_1.\Delta\varepsilon_1 + \delta\Lambda/\delta\varepsilon_4.\Delta\varepsilon_4 + \delta\Lambda/\delta\varepsilon_6.\Delta\varepsilon_6 = \Lambda_0 + 5$,

where Λ_0 is the nominal value of Λ. In this simple case, the linear approximation equals the correct value of Λ.

To evaluate the second strategy, t_{cut} is chosen to be 4, such that operation B3 can be included in the rescheduling process. In a similar way as in section 4, the nominal values for ε_e and the partial derivatives can be calculated. For this alternative, $N_D=\{A1, B1\}$; $N_{Bu} = \{B2\}$, $N_P=\{C1, A2, B3\}$, and $N_B =\{A3, C2\}$. Since this alternative wants to swap operations C1 and B3, $t_{B3} = 3$, $t_{C1} = 4$ and t_{A2} remains = 3. Defining ε_1, ε_2 and ε_3 again for the workstation schedule and ε_4, ε_5 and ε_6 again for the precedence constraints, calculations show that $\varepsilon_2 = 4$; $\varepsilon_3 = 1$; $\varepsilon_4 = 4$; and $\varepsilon_6 = 6$. In other words, only ε_6 has changed from 3 till 6. The linear approximation of the new $\Lambda = \Lambda_0 + \delta\Lambda/\delta\varepsilon_6. \Delta\varepsilon_6 = \Lambda_0$, which is close to the correct value $\Lambda_0 + 1$.

Therefore, the on-line SFC system concludes that it should select the second alternative and swap C1 and B3.

5.3 Evaluation and extensions

Examples have shown that this yields a good overall behaviour for the SE problem [4]. However, they have also shown that the results can still be improved in several ways: considering finite differences for the PA; considering more operations in the set N_P; and triggering additional events [4].

This algorithm computationally outperforms a similar approach based on simulations for each alternative instead of using partial derivatives, by an order of magnitude $O(N_{oper})$, where N_{oper} is the total number of operations. Moreover, this approach effectively controls nervousness to stay below reasonable bounds, by keeping a good homeostasis of the schedule when few disturbances occur.

6. Conclusion

Summarising, this paper presents an algorithm for executing a schedule, based on a graph representation of a schedule and the use of partial derivatives for estimating the effect of local decisions to the global goal. As such, it provides a generic concept that combines fast reaction to disturbances with optimisation by a combination of feedback control (reactive scheduling) and linearised feedforward. Reaction to disturbances not only is quick, it also explicitly considers the optimisation of the global goal. The algorithms are currently being implemented in the control software for an FAS [2], based on the concepts of holonic manufacturing. Future work include [1] the extension of the existing communication protocol to accommodate advice; [2] the exhaustive testing of the algorithm; and [3] the tuning of the right parameters to obtain an agile and high performance manufacturing system.

References

[1] Bollobas, B., *Graph Theory, An Introductory Course*, Springer-Verlag, New York, 1979

[2] Bongaerts, L., P. Valckenaers, H. Van Brussel, P. Peeters, "Schedule Execution in Holonic Manufacturing Systems," *Proc. of the 29th CIRP Int. Sem. on Manufacturing Systems*, May 11-13, 1997, Osaka, Japan, pp. 209-214.

[3] Bongaerts, L., H. Van Brussel, P. Valckenaers, P. Peeters, "Reactive Scheduling in Holonic Manufacturing Systems: Architecture, Dynamic Model and Co-operation Strategy," *Proc. of ASI 97*, Budapest, 14-17 July 1997.

[4] Bongaerts, L., H. Van Brussel, P. Valckenaers, "Schedule Execution using Perturbation Analysis," *Int. Symp. on non-linear dynamics in production processes and systems*, Hannover, Germany, September 17-18, 1997.

[5] Cassandras, C.G., *Discrete Event Systems, Modeling and Performance Analysis*, Irwin, Boston, 1993.

[6] Czerwinsky and P. Luh, "Scheduling Products with Bill of Material Using an Improved Lagrangian Relaxation Technique," *Trans. on Rob. and Autom.*, V.10 N.2 p.99.

[7] Dilts, D.M., N.P. Boyd, H.H.Whorms, "The Evolution of Control Architectures for Automated Manufacturing Systems," *J. of Manufacturing Systems*, Vol. 10, N. 1, 1991.

[8] Duffie, N.A., V.V. Prabhu, "Real Time Distributed Scheduling of Heterarchical Manufacturing," *J. of Manufacturing Systems* V.13 N.2 p.94, 1994.

[9] Márkus, A., T. Kis, J. Váncza, and L. Monostori, "A Market Approach to Holonic Manufacturing," *CIRP Annals*, Vol. 45/1, pp.433-436.

[10] Schmidt, G., "Case-Based Reasoning for Production Scheduling," *Pre-prints, 9th Intern. Working Sem. on Production Economics*, Feb. 19-23, 1996, Innsbruck, Austria.

[11] Smith, S. F., "Reactive Scheduling Systems", appeared in *Intelligent scheduling systems*, Brown, Donald E. (ed), Kluwer Academic publ. Boston (Mass.), 1995.

[12] Szelke, E., R.M. Kerr, "Knowledge Based Reactive Scheduling - State of the Art," *Int. J. of Production Planning and Control*, Vol. 5, March-April, 1994, pp. 124-145.

[13] Szelke, E., L. Monostori, "Reactive and proactive scheduling with learning in reactive operation management," *Proc. of IFIP Int. Working Conf. on Managing Concurrent Manufacturing to Improve Industrial Performance*, Sept. 11-15, 1995, Seattle, Washington, USA, pp. 456-483.

[14] Van Dyke Parunak, H., "Characterising the Manufacturing Scheduling Problem," *J. of Manufacturing Systems*, Vol 10, No 3, pp. 241-258, 1991.

A dynamic control problem for a two part-type pull manufacturing system

Francesco Martinelli and Paolo Valigi
Dipartimento di Informatica, Sistemi e Produzione
Università di Roma "Tor Vergata"
Via della Ricerca Scientifica, 00133 Roma – Italy
e-mail: {martinelli|valigi}@disp.uniroma2.it

Abstract In this paper it is studied the problem of dynamic part scheduling and Kanban allocation on a pull manufacturing system, with finite inventory space, and stochastic demand. It is assumed a finite number of requests can be backlogged, and requests finding the system full are dropped. The system is a single machine, two part-type one, and the control objective is that of minimizing an infinite horizon discounted cost in the inventory and backlog level, and demand losses. No set-up times and costs are considered.

The original scheduling and Kanban allocation problem is reformulated as a "pure" scheduling problem for constant kanban systems. A continuous-time Markov decision problem is formulated, and solved by means of a uniformization procedure, and discrete-time dynamic programming. Some initial analytical results are given, while the solution structure is derived numerically.

1 Introduction

The problem of dynamic scheduling for manufacturing systems has been studied for decades, with a variety of approaches (see [1]-[4], just to cite a few). The dynamic scheduling problem also arises in several other fields, such as computer networks, telecommunication systems, traffic networks.

For a broad class of infinite capacity queueing networks, under fairly general assumptions on the arrival processes, a well known optimal solution of the scheduling problem is the μc-rule. The optimality of such a rule has been initially proved via dynamic programming arguments, hence requiring a Markovian structure for the problem, and then via an elegant "interchange" argument, thus relaxing considerably the assumptions on the arrival processes (see, among many others, [5, 6, 7]).

The dynamic scheduling problem for queueing networks, for which the μc rule is not optimal, has been studied by several authors, and it is well known to be a difficult one.

In this paper, it is studied the problem of optimal dynamic scheduling and Kanban allocation on a single machine, two part-type pull manufacturing system, with finite inventory space, stochastic service times and stochastic demand. It is assumed only a finite number of requests can be backlogged, and requests finding the system full are dropped. The control objective is that of minimizing an infinite horizon discounted cost in the inventory/backlog level and demand loss. No set-up times and costs are considered. For such a system, the μc rule is not optimal.

The Kanban allocation problem consists of dynamically adjusting the number of Kanban assigned to each part type during system operation. The optimal scheduling policy is sought in the class of state feedback, non-preemptive and possibly idling policies.

The combined problem of dynamic Kanban allocation and scheduling, at the best of author's knowledge, has not been studied before. An analysis of some classical scheduling policies in a Kanban system (constant allocation) has been studied in [8], and the effect of sequencing rules has been considered in [9].

As for the dynamic scheduling problem, in [3] systems with an infinite demand capacity, non null set-up times and costs, and constant production times, are considered. The overall system is in the framework of semi-Markov decision models; approximate solutions are proposed, and error estimates are given. In [1] and [2], for infinite capacity deterministic systems, control policies are proposed aimed at achieving system stability, in terms of limited buffer levels. In [10] the single server multi-class problem for push systems is considered, and an approximate solution, the "overflow scheduling policy", is proposed, aimed at optimizing the weighted throughput. The problem is studied in the framework of Markov decision problems using the uniformization technique. In [11], for infinite capacity multiqueue single server, the problem of minimizing the number of customers in each queue, is considered. The proposed policy is the "most customers first": at each control epoch, the queue with the largest number of pending requests is serviced. In [12], a station comprising four infinite capacity single-server is considered, and two customer classes. The problem is solved by making use of a Brownian motion approximation. In [13], for the class of finite capacity queueing networks, the duality between routing and scheduling systems (with respect to queue lenght and admitted jobs vs residual capacity and serviced customers), is studied. Then, a routing policy optimal with respect to losses clearly identifies a scheduling policy which is optimal with respect to throughput. Related results are presented in [14], where optimality of routing/scheduling policies with respect to losses/throughput is studied, for multiclass finite capacity parallel server systems.

The first contribution of this paper is that of showing that the problem of jointly scheduling and allocating Kanbans can be reformulated as an equivalent "pure" dynamic scheduling problem, for a system with fixed Kanban allocation, in the sense that the optimal solutions of the two problems have the same cost.

The second contribution of the paper consists in providing the structure of the optimal solution for such a "pure" scheduling problem, modeled by means of a continuous-time Markov decision problem, and solved by means of a uniformization procedure, and a discrete-time dynamic programming approach. A numerical solution to the problem is presented together with a few initial analytical results.

2 Problem formulation and preliminary results

The pull manufacturing system $\mathcal{S}$ considered in this paper, and depicted in Figure 1, comprises a reliable machine $\mathcal{R}$, and two "matching nodes" $\mathcal{M}_a$ and $\mathcal{M}_b$. Machine $\mathcal{R}$ can provide service to two different part-types a and b, with mutually independent exponentially distributed service times, with rates λ_a and λ_b, respectively. Requests of finished parts arrive to the system according to mutually independent Poisson processes, with parameter μ_a and μ_b, independent of the service processes. A new request for part type a or b is serviced immediately if finished part inventory of the proper part is available in the *finished part queue*, otherwise the request is queued in the corresponding *request queue* or rejected if there is no room in this queue, whose capacity is finite and equal to L_a for part type a and L_b for part type b. Notice that, for each matching node, only one queue at a time can be non-empty.

Machine $\mathcal{R}$ is controlled by means of a Kanban mechanism: each finished part is attached a card (Kanban), which is detached by the matching node upon request fulfilling, and placed in the card board of the corresponding type. A finite number n_a and n_b of cards is available for part-types a and b, respectively. Machine $\mathcal{R}$ can start service on a new part type only if at least a card is available in the corresponding board. Hence, the maximum number of finished parts of each type allowed in the system is given by the corresponding number of Kanbans. It is assumed that an infinite source of raw material is available and an infinite capacity sink is placed downstream of each matching node. The Kanban mechanism makes the considered system a "pull" system.

Let $x_a > 0$ represent the number of type a parts in the finished part queue, $-x_a > 0$ the number of type a requests in the request queue (backlog), and x_b be similarly defined. This can be seen also as a single queue whose content is denoted by x_a [x_b], where both positive and negative arrivals are possible. Let $s = 0$ indicate that $\mathcal{R}$ is idle, $s = a$ [$s = b$] indicate that $\mathcal{R}$ is servicing part-type a [b]. Then, the system state is given by $x = (x_a, x_b, s)^T$; the state at time t will be denoted by $x(t)$.

The overall cost per time unit $c_c(x)$ incurred while the state is $x = (x_a, x_b, s)^T$, is given by the combination of inventory/backlog cost $g(x_a, x_b)$, production cost $h(s)$ and demand loss cost $d(x_a, x_b)$:

$$c_c(x) = g(x_a, x_b) + h(s) + d(x_a, x_b), \quad (1)$$

The form of these functions will be better specified in the following.

Control actions can be exerted on the system at some *control epochs*, namely upon service completion at machine $\mathcal{R}$ and upon arrival of a new request, provided that machine $\mathcal{R}$ is idle. In this paper, two different types of control actions are considered: *dynamic scheduling* actions and *dynamic Kanban allocation* actions. At each control epoch, both types of control actions can be exerted. Notice that the order of implementation of the actions may have an influence on system behavior. In the following the kanban allocation will be always supposed to precede the scheduling action.

A *dynamic Kanban allocation policy* is a state feedback policy which, at each control epoch, determines the number of kanbans to assign to each part type, and satisfies the following constraint:

$$n_a + n_b = n, \quad (2)$$

where n is a given system parameter.

A *dynamic scheduling policy* is a state feedback, non-preemptive, possibly idling policy which, at each control epoch, selects the new state for machine $\mathcal{R}$ (i.e., whether the machine will remain idle, or which part-type will provide service to) and complies with the following Kanban constraint on inventory levels:

$$x_a \leq n_a, \quad x_b \leq n_b. \quad (3)$$

In the following the class of policies resulting from the joint use of a dynamic Kanban allocation policy and a dynamic scheduling policy will be denoted by Π.

Now, the *Dynamic Kanban Allocation and Scheduling* (DKAS) problem considered in this paper can be formally stated as follows, where $\beta \in \mathbb{R}, \beta > 0$, is a discount factor, x_0 is a given system initial state, and $x_\eta(\cdot)$ is the state evolution under a generic policy η.

DKAS Problem *For system $\mathcal{S}$, find the control policy $\pi^* \in \Pi$ minimizing, for all initial states x_0, the total expected β-discounted cost criterion:*

$$J_\eta(x_0) = E\left[\int_0^\infty e^{-\beta t} c_c(x_\eta(t)) dt \,|\, x(0) = x_0\right], \quad (4)$$

over all $\eta \in \Pi$.

Consider now the class of state-feedback, non-preemptive, possibly idling dynamic scheduling policies complying with the constraint

$$x_a + x_b \leq n, \qquad (5)$$

instead of the stronger constraint (3), and denote this class by Σ.

Minimizing the cost function (4) over all $\eta \in \Sigma$ corresponds to solving the following *Dynamic Scheduling* (DS) problem:

DS Problem *For system $\mathcal{S}$, find the optimal scheduling policy $\sigma^* \in \Sigma$ minimizing, for all initial states x_0, the cost criterion (4) over all $\eta \in \Sigma$.*

The solution of the two problems is different, namely a whole control policy $\pi \in \Pi$ in case of the DKAS problem and a "pure" scheduling policy $\sigma \in \Sigma$ in case of the DS problem are provided. Nevertheless, according to the following lemma, the corresponding state evolution and total cost are exactly the same.

Lemma 1 *If Kanban allocation actions can be performed in null time, then for any policy $\pi \in \Pi$ $[\sigma \in \Sigma]$ there exists a policy $\sigma' \in \Sigma$ $[\pi' \in \Pi]$ such that:*

$$\begin{aligned} x_\pi(t) &= x_{\sigma'}(t) \; [x_\sigma(t) = x_{\pi'}(t)] \quad \forall \, t \geq 0, \quad (6a) \\ J_\pi(x_0) &= J_{\sigma'}(x_0) \; [J_\sigma(x_0) = J_{\pi'}(x_0)] \quad \forall x_0 \quad (6b) \end{aligned}$$

where x_0 is any admissible initial state.

Proof. Since constraint (5) is weaker than constraint (3), it is enough to prove that, for any $\sigma \in \Sigma$ there exists a $\pi' \in \Pi$ such that (6) holds. This is a direct consequence of instantaneous kanban allocation hypothesis. □

In view of Lemma 1, in the following only the solution for Problem DS will be provided. It is remarked that constraint (5) corresponds to a system with a constant number of Kanban, shared by all the part-types.

Notice that, if Kanban allocation actions are not performed in null time, and/or if they have a pace slower than the scheduling actions, then Lemma 1 does not hold, and one has to cope with the more complex DKAS problem. In this case, since parameters n_a and n_b may change over time, it may turn out that several scheduling problems have to be solved. Indeed, it has been proved in [15] that, for push systems with finite capacities, scheduling policies which are extremal when all the capacities are equal, may be not extremal with unequal capacities.

3 Optimal scheduling policy

System $\mathcal{S}$ is a finite capacity system, hence its state space $\mathcal{X}$ is finite, and given by $\mathcal{X} = \{(x_a, x_b, s)^T, (x_a, x_b) \in \mathbb{Z}^2, -L_a \leq x_a \leq n, -L_b \leq x_b \leq n,$ $x_a + x_b \leq n, s \in \{0, a, b\}\}$. The DS problem will be studied under the following cost structure:

$$g(x_a, x_b) = g_{a+} x_a^+ + g_{b+} x_b^+ + g_{a-} x_a^- + g_{b-} x_b^- \qquad (7)$$

$$h(s) = \begin{cases} 0 & \text{if } s = 0 \\ h & \text{otherwise} \end{cases} \qquad (8)$$

$$d(x_a, x_b) = d_a I(x_a = -L_a) + d_b I(x_b = -L_b) \qquad (9)$$

where $x_\ell^+ = \max[x_\ell, 0]$, $x_\ell^- = \max[-x_\ell, 0]$, $\ell = a, b$; g_{a+}, g_{b+}, g_{a-}, g_{b-}, h, d_a and d_b are positive constant cost coefficients and $I(\cdot)$ is 1 if the argument is true and 0 otherwise.

The considered control system can be modeled by means of a Markov controlled process, and the DS problem can be recasted in the framework of Markov decision problems and solved by using dynamic programming techniques. Transforming the original continuous time problem into a discrete time one by means of a well known uniformization procedure (see [16, 17]), the solution of the DS problem, which exists because the cost is positive and bounded and the action space is finite, is given by the following optimality equation, where the dependence on policy η has been omitted for simplicity:

$$J(x) = \min_{u \in \mathcal{U}_x} \left[c(x) + \alpha \sum_{x' \in \mathcal{X}} p(x'; x, u) J(x') \right] \qquad (10)$$

where the one-step-cost $c(x)$, the transition probability $p(\cdot; \cdot, \cdot)$ and the discount factor α, derived through the uniformization procedure, are defined in the following, together with the set $\mathcal{U}_x$ of *admissible control actions in state x*. In particular, a control $u = a[b]$ means that a service on a new part of type $a[b]$ must be started, while $u = 0$ leaves the machine idle for the next control epoch.

Let $\gamma := \mu_a + \mu_b + \max\{\lambda_a, \lambda_b\}$, then the cost and discount factor for the discrete-time problem are given by $c(x) := c_c(x)/(\beta + \gamma)$, $\alpha := \gamma/(\beta + \gamma)$.

For a given inventory/backlog state (x_a, x_b), the control function u considered in this paper has four different components, depending on machine state: $u_a(x_a, x_b)$, $u_b(x_a, x_b)$, $u_{0a}(x_a, x_b)$ and $u_{0b}(x_a, x_b)$. The components u_a and u_b are used if the control decision is triggered by service completion at machine $\mathcal{R}$. Viceversa, the components u_{0a} and u_{0b} are used if the control decision is triggered by a new request arrival, occurring when machine $\mathcal{R}$ is idle, and therefore they also depend on the type of the newly arrived request.

The set $\mathcal{U}_x$ is defined as $\mathcal{U}_x := \{0, a, b\}$, except when the backlog/inventory level hits (or is close to hit) the boundary, making some control actions not possible. As an example, consider the case $x = (x_a, x_b, a)$, with $x_a = n - 1$ or $x_a + x_b = n - 1$. Then $\mathcal{U}_x = \{0\}$, i.e., the only admissible control action is $u_a(x_a, x_b) = 0$. A similar reasoning can be followed to derive the set $\mathcal{U}_x$ in all the other cases.

The probability $p(x'; x, u)$ of a transition from state x to state x', under the value u of the control action, is given in the following, for some typical cases.

(i) If the state is $x = (x_a, x_b, 0)$, with $-L_a < x_a < n$, $-L_b < x_b < n$, $x_a + x_b < n$, then only a new request arrival is possible, namely an arrival of a part of type a, with probability $p_a := \mu_a/\gamma$, and an arrival of a part of type b, with probability $p_b := \mu_b/\gamma$. This implies that the possible new states reached by the system are $x' = (x_a - 1, x_b, s')$ for a type a arrival and $x'' = (x_a, x_b - 1, s'')$ for a type b arrival, where s' and s'' denote the new state of machine $\mathcal{R}$. Because of the uniformization procedure, it is also possible for the system to remain in state x, with probability $1 - p_a - p_b$. Then, $prob(x'; x, u_{0a}) = p_a$ if $u_{0a} = s'$, and $prob(x'; x, u_{0a}) = 0$ otherwise. Similarly, $prob(x''; x, u_{0b}) = p_b$ if $u_{0b} = s''$, and $prob(x''; x, u_{0b}) = 0$ otherwise. Finally, $prob(x; x, \cdot) = 1 - p_a - p_b$, and the transition does not depend on the control function.

(ii) Let the state be $x = (x_a, x_b, s)$, with $-L_a < x_a < n - 1$, $-L_b < x_b < n - 1$, $x_a + x_b < n - 1$ and $s = a$ (the case $s = b$ is similar). In this case, both a service completion event may occur, with probablity $q_a := \lambda_a/\gamma$, and an arrival event, with probability p_a (or p_b). The possible new states that may be reached by the system are: $x' = (x_a + 1, x_b, s')$ if a service completion occurs, $x'' = (x_a - 1, x_b, s)$ for a type a arrival, $x''' = (x_a, x_b - 1, s)$ for a type b arrival. In addition, the system may remain in state $x = (x_a, x_b, s)$ with probability $1 - q_a - p_a - p_b$. Then, $prob(x'; x, u_a) = q_a$ if $u_a = s'$, and $prob(x'; x, u_a) = 0$ otherwise. In addition, independently of the control, $prob(x''; x, \cdot) = p_a$, $prob(x'''; x, \cdot) = p_b$ and $prob(x; x, \cdot) = 1 - p_a - p_b - q_a$.

Special care is required for transitions involving states in which the backlog/inventory level hits (or is close to hit) the boundary, i.e., states for which one or more of the following conditions hold: $x_a = -L_a$, $x_b = -L_b$, $x_a = n - 1$, $x_b = n - 1$, $x_a = n$, $x_b = n$, $x_a + x_b = n - 1$, $x_a + x_b = n$. In these situations, some control actions may be not possible, and/or some requests may be rejected. For example, let the state be $x = (-L_a, x_b, 0)$, with $x_b > -L$. The transition to the state $x' = (x_a - 1, x_b, s')$ cannot occur, for any s', while the other transition probabilities are similar to those in (i), with the self-loop transition having probability $1 - p_b$ instead of $1 - p_a - p_b$.

In this paper, the interest is in the dynamic behavior of the considered system, and on basic properties of the optimal scheduling policy. Therefore, in order to avoid a natural or intrinsic priority between the two part-types, a symmetrical system is considered, i.e., a system characterized by the following set of parameters: $\mu := \mu_a = \mu_b$, $\lambda := \lambda_a = \lambda_b$, $g_+ := g_{a+} = g_{b+}$, $g_- := g_{a-} = g_{b-}$, $d := d_a = d_b$, $q := q_a = q_b$, $p := p_a = p_b$, and, finally, $L := L_a = L_b$.

For the problem to be sensible, it is necessary that:

$$\lambda \geq 2\mu. \quad (11)$$

If this condition does not hold, machine $\mathcal{R}$ does not have enough capacity to meet demand and many requests will be dropped at steady state. Condition (11) will be assumed to hold throughout the paper, although not explicitly mentioned.

Some initial analytical results are reported in the following, to clarify the structure of the optimal solution of the DS problem, which has been derived numerically. Given the optimality equation (10), the optimal scheduling policy $\sigma^* \in \Sigma$ solving the DS problem has been found by means of the successive approximation procedure:

$$J^{(n+1)}(x) = \min_{u \in \mathcal{U}_x} [c(x) + \alpha \sum_{x' \in \mathcal{X}} p(x'; x, u) J^{(n)}(x')] \quad (12a)$$

$$\forall n \geq 0, \forall x \in \mathcal{X}, \quad (12b)$$

$$J^{(0)}(x) = 0, \quad \forall x \in \mathcal{X}, \quad (12c)$$

which, for the case considered here, converges to the optimal value $J(x)$ as $n \to \infty$.

The structure of the optimal solution is strongly influenced by the relative magnitude of the cost parameters. In particular, two major issues arise: a) *production vs no-production* – if production cost h is large with respect to the inventory/backlog cost, the optimal solution turns out to be that of avoiding production; and b) *part-type a vs part-type b* – given that h is small enough to make production the optimal choice, the relative magnitude of the other parameters determines which part type has to be produced in each state.

These two issues will be now better clarified.

In order to better explain the first issue (production vs no-production), consider a single part-type system $\mathcal{S}_1$, similar to system $\mathcal{S}$, i.e., a system comprising only machine $\mathcal{R}$ and a matching node $\mathcal{M}$. Assume λ_1 is the service rate of machine $\mathcal{R}$, μ_1 the rate of the Poisson arrival process (of requests), g_{1+} and g_{1-} the inventory and backlog costs, respectively, h_1 the production cost, and d_1 the demand loss cost. Finally, let $\gamma_1 = \lambda_1 + \mu_1$, $q_1 = \lambda_1/\gamma_1$, $p_1 = \mu_1/\gamma_1$, let $x \in \{-L_1, \ldots, n_1\}$, $L_1, n_1 \in \mathbb{Z}^+$, be the inventory/backlog state, let $s \in \{0, 1\}$ the machine state (with $s = 0$ denoting machine idle, and $s = 1$ denoting machine in service), and let $J^1(x, s)$ be the total expected β_1-discounted cost, defined similarly to (4). For this system, a scheduling policy is a policy that, at each control epoch, decides whether to operate machine (i.e., to start a service) or to let $\mathcal{R}$ stay idle, and the DS problem reduces to find a scheduling policy minimizing $J^1(x, s)$, for all (x, s).

Conditions under which the optimal policy for system $\mathcal{S}_1$ is to operate machine $\mathcal{R}$ whenever there is a backlog, are presented in the following two lemmas, whose proofs, omitted for space limitations, can be found in [17]. In particular, Lemma 2 gives a sufficient condition, while

Lemma 3 provides a necessary condition. Based on numerical computations, the condition of Lemma 3 turns out to be necessary also for the two part-type system.

Lemma 2 *Assume that $\beta_1 g_{1-} > \lambda_1 g_{1+}$. Then, the total expected β_1-discounted cost $J^1(x,s)$ for system S_1 satisfies the condition*

$$J^1(x,1) < J^1(x,0), \quad \forall x \in \{-L_1, \ldots, -1\}, \quad (13)$$

if

$$h_1 < B_{suf} = \frac{\lambda_1 g_{1-}}{\beta_1 + \lambda_1} - \frac{\lambda_1^2 g_{1+}}{\beta_1(\beta_1 + \lambda_1)}. \quad (14)$$

The proof of Lemma 2 can be obtained using the uniqueness of the optimal cost $J^1(x,s)$ and the convergence to this cost of the recursive form of Hamilton Bellman Jacobi equation (12) [17]. Condition (13) implies that, whenever backlog level is non zero, it is convenient to produce.

Lemma 3 *Assume that :*

$$J^1(x+1,1) - J^1(x,1) < 0 \ \forall \ x \in \{-L_1, \ldots, -1\}, \quad (15)$$

then, condition (13) holds only if

$$h_1 < B_{nec} = \frac{\lambda_1 g_{1-}}{\beta_1}. \quad (16)$$

The proof of Lemma 3 can be obtained using arguments similar to the ones used for Lemma 2. Assumption (15) is motivated by numerical results, and by the following consideration: the cost function $J^1(x,s)$ is decreasing as x approaches the origin. Observe that, as expected, $B_{suf} < B_{nec}$.

As for the original two part-type system S considered in this paper, preliminary results, as well as numerical computations, show that there is a sudden change in the structure of the control functions around the value $h = \lambda g_-/\beta$ of the production cost, i.e., (16) rewritten for system S.

As for issue b), i.e., for the policy for part selection, provided that production is the best choice, numerical computations indicate that the optimal solution has two possible basic structures (see fig. 2 and 3), depending on system parameters. In particular, the dependence is very strong on demand loss cost parameter d.

The two basic structures are quite different: in one case, the optimal solution is of type *Largest Queue First* (LQF), while the other structure follows a kind of inverse rule, i.e., *Shortest Queue First* (SQF). Analytical computations on a simpler system, obtained from system S by setting $n = 0$, show that there is an interval $[d_m, d_M]$ on demand loss cost, such that the optimal solution is of type SQF for cost values lower than d_m, and of type LQF for values greater than d_M. For cost values in the above interval, the optimal solution is a mixed SQF/LQF policy [17]. The optimality of the inverse rule when the demand loss cost is smaller than d_m, arises from the fact that state motion along the border $x_a = -L$ or $x_b = -L$ (i.e., when one of the two arrival processes is not effective) is faster than along any trajectory inside the backlog region.

The optimal scheduling policy solving the DS problem for a set of parameters for which the solution structure is of the SQF type, is shown in Figure 2. In particular, the figure reports the function $u_a(x_a, x_b)$, for $L = 10$, $n = 4$, and with the other system parameters given by: $\lambda = 3$, $\mu = 1$, $\beta = 0.1$, $g_+ = 10$, $g_- = 100$, $h = 1$, and $d = 500$.

The major characteristics of the solution are: i) it is a truly feedback policy, not an index policy; and ii) the policy is based on a kind of "inverse rule": the part type with the smaller backlog is produced first, whenever both parts have a backlog.

It is recalled that the system state, at the epoch of occurrence of an event, does not take into account the event. This has to be considered in examining the structure of the control functions. For example, the raw $x_a = 3$ in the u_a control function reported in Figure 2, corresponds to the situations in which three type a parts are already queued, and an additional part a is completed, hence, since $n = 4$, no production is possible. The same occurs for the diagonal zeros on the $x_a > 0$, $x_b > 0$ quadrant.

The structure of the optimal solution, for the same set of parameters, but demand loss cost increased to $d = 1000$, is shown in Figure 3. The solution is now of type LQF.

As for the dependence of the optimal solution on demand loss cost, numerical computations indicate that, for a significantly large set of parameter values, the form SQF of the solution remains optimal also for a "quadratic" form of the inventory/backolg cost, instead of the "linear" form (7) [17].

4 Conclusions

The problem of dynamic part scheduling and Kanban allocation on a pull, manufacturing system, with finite inventory space, and stochastic demand has been studied and solved numerically. The system is single machine, two part-type, and the control objective is that of minimizing an infinite horizon discounted cost in the inventory and backlog level.

As part of the proposed solution, the original scheduling and Kanban allocation problem has been reduced to a "pure" scheduling one, for a system with constant number of Kanban assigned to each part-type. Then, such a problem has been modeled by means of a continuous-time Markov decision problem, and solved by means of a uniformization procedure, and a discrete-time dynamic programming approach.

Some initial analytical results have been reported in order to better understand the structure of the optimal solution, numerically derived.

References

[1] J. R. Perkins and P. R. Kumar, "Stable, distributed, real-time scheduling of flexible manufacturing/assembly/disassemble systems," *IEEE Trans. on Automatic Control*, vol. 34, no. 2, pp. 139–148, 1989.

[2] C. Chase and P. J. Ramadge, "On real-time scheduling policies for flexible manufacturing systems," *IEEE Trans. on Automatic Control*, vol. 37, pp. 491–496, April 1992.

[3] J. Qiu and R. Loulou, "Multiproduct production/inventory control under random demands," *IEEE Trans. on Automatic Control*, vol. 40, pp. 350–355, February 1995.

[4] S. Stidham Jr. and R. Weber, "A survey of Markov decision models for control of networks of queues," *Queueing Systems: Theory and Applications*, vol. 13, pp. 291–314, May 1993.

[5] P. P. Varaiya, J. C. Walrand, and C. Buyukkoc, "Extensions of the multiarmed bandit problem: The discounted case," *IEEE Trans. on Automatic Control*, vol. 30, pp. 426–439, May 1985.

[6] J. Baras, D. Ma, and A. Makowski, "K competing queues with geometric service requirements and linear costs: The μc rule is always optimal," *Systems & Control Letters*, vol. 6, pp. 173–180, 1985.

[7] P. Nain, "Interchange arguments for classical scheduling problems in queues," *Systems & Control Letters*, vol. 12, pp. 177–184, 1989.

[8] C. Duri, Y. Frein, and M. Di Mascolo, "Performance evaluation of Kanban multiple-product production systems," in *Proc. IEEE Conference*, pp. 557–566, 1995.

[9] B. Berkley, "Effect of buffer capacity and sequencing rules on a single-card Kanban system performance," *Int. J. Prod. Res.*, vol. 31, no. 12, pp. 2875–2893, 1993.

[10] Z. Rosberg and P. Kermani, "Customer scheduling under queueing constraints," *IEEE Trans. on Automatic Control*, vol. 37, pp. 252–257, February 1992.

[11] Z. Liu and P. Nain, "Optimal scheduling in some multiqueue single-server systems," *IEEE Trans. on Automatic Control*, vol. 37, pp. 247–252, February 1992.

[12] C. Laws and G. Louth, "Dynamic scheduling of a four-station queueing network," *Probability in the Engineering and Informational Science*, vol. 4, pp. 131–156, 1990.

[13] D. Sparaggis, C. G. Cassandras, and D. Towsley, "On the duality between routing and scheduling systems with finite buffer space," in *31st Conference on Decision and Control*, (Tucson, Arizona), pp. 2364–2365, 1992.

[14] D. Sparaggis, D. Towsley, and C. G. Cassandras, "Optimal control of multiclass parallel service systems," *Discrete Event Systems: Theory and Applications*, vol. 6, pp. 139–158, 1996.

[15] D. Sparaggis, D. Towsley, and C. G. Cassandras, "Extremal properties of the shortest/longest non-full queue policies in finite-capacity systems with state-dependent service rates," *J. Appl. Probability*, vol. 30, pp. 223–236, 1993.

[16] C. Cassandras, *Discrete Event Systems: Modeling and Performance Analysis*. Boston, MA: Irwin & Aksen, 1993.

[17] F. Martinelli and P. Valigi, "Dynamic scheduling policies for push and pull finite capacity queueing networks," tech. rep., Dipartimento di Informatica, Sistemi e Produzione, Università di Roma "Tor Vergata", Roma, 1997.

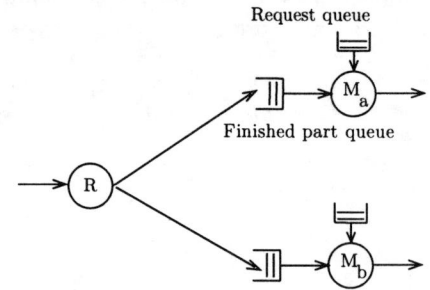

Figure 1: The manufacturing system

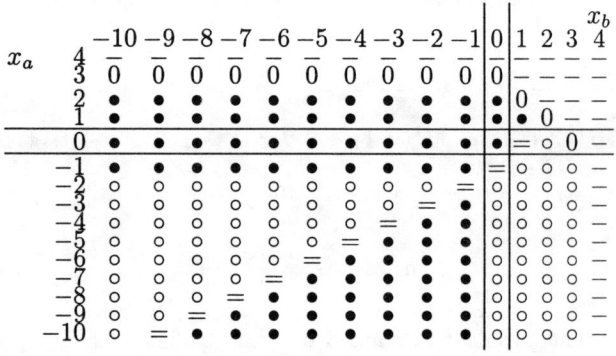

Figure 2: Structure of the SQF type optimal scheduling policy: control function component $u_a(x_a, x_b)$ [$\circ \to u_a = a$; $\bullet \to u_a = b$; $= \to u_a = a$ or $u_a = b$; $0 \to u_a = 0$; $- \to u_a$ not defined]

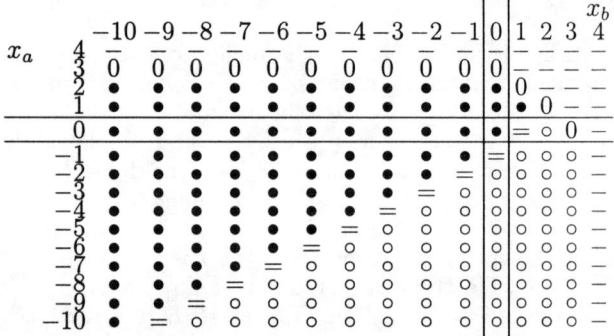

Figure 3: Structure of the LQF type optimal scheduling policy: control function component $u_a(x_a, x_b)$

Multiple Control Policies for Two-Station Production Networks with Two Types of Parts Using Fuzzy Logic

Runtong Zhang and Yannis A. Phillis

Department of Production Engineering and Management
Technical University of Crete, Chania 73100, Greece
Email: zhang@dssl.tuc.gr and phillis@dpem.tuc.gr

Abstract: This paper considers the problem of optimal control to tandem two-station production networks with two types of parts. Part routing, service rate selection and flow control, are simultaneously considered. We use fuzzy control to solve the problem. Simulation shows that the approach is efficient and promising.

1. Introduction

It is generally very difficult to dynamically and simultaneously determine in an optimal fashion multiple control policies to production networks. For example, it is quite possible in practice to have to select the service rate and at the same time to schedule a part among different machines with the choice of rejecting it.

In this paper, we consider two models concerning production control. Both are tandem two-station production networks with two types of parts, in which multiple control policies are studied simultaneously, i.e., part routing, service rate selection and flow control. Problems of part routing and service rate selection as well as routing and admission have been studied in recent years, but the determination of an explicit solution is still an open problem (see Chen et al [1], Harrison and Wein [3], Wein [7] and [8] and Shioyama [4]). Routing and service rate selection in the presence of the service costs is new in the literature. In this paper, we tackle the problems using fuzzy logic techniques [2]. The fuzzy control approach is explicitly determines the optimal control policies to all three cases.

Optimal control of production systems has been extensively investigated in the literature. A comprehensive discussion on optimal control of queueing networks can be found in the survey papers of Stidham and Weber [5] and Teghem [6]. Recently, Zhang and Phillis [9]-[12] proposed a new method using fuzzy logic to solve production or more general queueing control problems and show via simulation that the new method generalizes existing solutions and also efficiently solves cases intractable with classical methods. This approach signals a departure from classical techniques.

Based on the current state of the system, an inference engine equipped with fuzzy rule bases determines on-line decisions to adjust the system behavior in order to guarantee that the system is optimal in some specific sense. This is the *fuzzy logic controller* (see Figure 1).

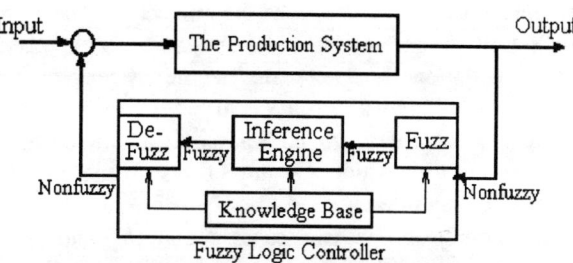

Figure 1: The fuzzy production control system

2. Part Routing and Service Rate Selection (Case 1)
2.1 Problem Description

The tandem two-station network under consideration is illustrated in Figure 2.

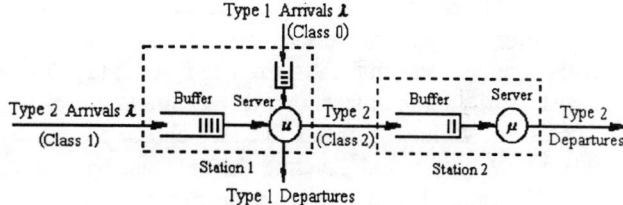

Figure 2: A production network

Station 1 is fed by two independent Poisson streams i, $i=1,2$, of parts with rate λ. Parts of stream i are referred to as type i parts. Type 1 parts visit station 1 only and type 2 parts visit both stations 1 and 2 in series. To simplify the notation we categorize the two part types into three classes: class 0 refers to type 1 parts whereas class 1 and class 2 refer to type 2 in station 1 and station 2 respectively. There is one exponential machine in each station, where the service rate in station 1 is a control variable u taking any value in $[0,a]$ but in station 2 the rate μ is constant. The buffer in either station has unlimited capacity and the order of service is irrelevant. The instantaneous cost is $h_1(s_0+s_1)+h_2 s_2$, where $h_1>0$ and $h_2>0$ are the holding costs per part per unit time in stations 1 and 2 respectively, and s_j, $j=0,1,2$, is the number of class j parts. For stability it is assumed that

$\lambda<\mu$ and $2\lambda<a$. The problem is to dynamically determine the service rate in station 1 and schedule the machine in station 1 between the two types of parts in order to minimize the average cost of the system.

Chen et al [1], Harrison and Wein [3] and Wein [7] and [8], have examined this problem and provided characterizations of solutions. However, the explicit determination of the optimal polices is "extremely tedious, if possible at all" and hence it still remains an open problem.

2.2 Architecture of the Fuzzy Logic Controller

The state of the system is described by (s_1, s_2, s_3, u), where $s_j=0,1,2,...,j=1,2,3$, is the number of class j parts including the one in service (if any) and u is the service rate in station 1. The state of the system changes whenever an arrival or a departure in either station occurs. Since there are no service costs involved, the service rate u does not bear any effects on the course of fuzzy decision inference.

We now examine two cases for $h_1 \geq h_2$ and $h_1 < h_2$, because, as we shall see, each produces different results.

2.2.1 Case 1-1: $h_1 \geq h_2$

To build the fuzzy logic controller, we choose as fuzzy inputs the numbers of class j parts $s_j \in [0,+\infty)$, $j=0,1,2$. The decision $d=1,0$, which indicates that the machine in station 1 should serve a class 1 or 0 part is the fuzzy output. We develop the fuzzy rule base in Table 1. According to the above discussions, when there are no parts of either class in the system, i.e., $s_1=0$ and $s_2=0$, the decisions d in the rule base are meaningless and the machine in station 1 remains idle. We choose four fuzzy sets for each input and therefore the rule base has a total of $4^3=64$ rules. The membership functions for s_0 and s_1 are both shown in Figure 3(a), and for s_2 and d are shown in Figures 3(b) and 3(c), respectively.

Table 1: Fuzzy rule base

s_0	s_1	s_2	d	s_0	s_1	s_2	d
ZO	ZO	ZO	1	...	...	...	...
PS	ZO	ZO	0	...	...	...	...
...	...	...	...	PM	PB	PB	0
...	...	...	...	PB	PB	PB	0

Because of the condition that the machine in station 2 should be starved as little as possible, priority of service in station 1 is given to class 0 parts. It is desirable that, as long as the machine in station 2 is busy, the machine in station 1 always serve a class 0 part if asked. However, this policy has the risk of starving the machine in station 2 even when there are type 2 parts in the system. On the other hand, if the machine in station 1 serves class 1 parts too often, the philosophy of the $c\mu$ rule will be violated. The fuzzy rule base is developed by seeking a balance between these two situations. (a) When s_2 is relatively small, to avoid both risks, the machine in station 1 receives a class 1 part, otherwise it does not. (b) If s_1 is relatively small, the machine in station 1 receives a class 0 part otherwise it does not. (c) If s_0 is relatively small, the machine in station 1 receives a class 1 part. Based on (a)-(c), the rule base is ready.

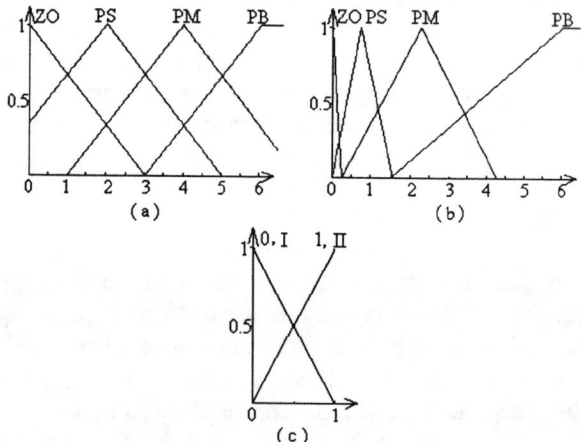

Figure 3: Membership functions

We adjust s_2 from s_1 to 1 as the difference h_2-h_1 ranges accordingly from 0 to $+\infty$. If $h_1 \approx h_2$, the membership functions for s_1 and s_2 should be similar but if $h_1 << h_2$, PB for s_2 should reach 1 from above in the physical domain. Besides h_1 and h_2, the relationship between the service rates in both stations a and μ also has a similar effect. Also, if $a \approx \mu$, the membership functions for s_1 and s_2 should be similar but if $a << \mu$, PB for s_2 should reach 1 from above in the physical domain, whereas if $a >> \mu$, PS for s_2 should reach $+\infty$ from below in the physical domain. Since the membership functions of s_0, s_1 and s_2 are interrelated, knowledge of the membership functions for s_2 implies knowledge of the others. The membership functions for s_2 are devised based on the previous discussions. Finally, the above discussion implies that PS for s_2 with membership grade 1.0 is fixed at 1.

2.2.2 Case 1-2: $h_1 < h_2$

Because $h_1 < h_2$ and the machine in station 2 operates always at a constant service rate, it is reasonable to keep the number of parts in station 2 as small as possible under the condition that the machine in station 2 should never be starved as long as station 1 is not empty. It would be desirable to always have only one part (in service) in station 2, unless there are no parts in station 1. Such a policy allows the parts to stay in the lower cost place (station 1) for the longest possible time and in the higher cost place (station 2) for the shortest possible time, if they have to wait somewhere. However, the memoryless property of this process, which means that neither the interarrival nor the service times can be conditioned on the present observable state, makes it impossible to implement this policy. The fuzzy rule base as well as the corresponding membership functions for this special case are developed by seeking a balance between these two situations. When s_1 is relatively small, we turn off the machine in station 1 earlier. If s_1 is relatively large, we turn off the machine in station 1 a little later. Since $h_1 < h_2$, we should avoid situations where $s_1 < s_2$. We choose the numbers $s_i \in [0, +\infty)$ of class i parts, $i=0,1,2$, as fuzzy inputs and the decision $d=1,0$, which indicates that the machine in station 1 should serve a class 1 or 0 part as fuzzy output. The membership functions for s_1, s_2 and d are shown in Figures 3(a), 3(b) and 3(c), respectively. Based on the previous discussions, we first write a rule base shown in Table 2 for the special case where parts of type 1 are absent. The fuzzy output here has two sets, PB and ZO, which represent the service rate a and 0 respectively in station 1. Hence, PB and ZO indicate whether or not a class 1 part is accepted.

Table 2: Fuzzy rule base

s_1 \ s_2	ZO	PS	PM	PB
ZO	PB	ZO	ZO	ZO
PS	PB	PB	ZO	ZO
PM	PB	PB	PB	ZO
PB	PB	PB	PB	PB

Based on Table 2, we develop the fuzzy rule base in Table 3, which consists of 64 rules. When no class 0 or 1 parts are present in the system, the decisions d in the rule base are meaningless and the machine in station 1 simply idles. It is easy to see that, when s_0 is ZO, the rules in Table 3 correspond to Table 2. The membership functions for s_0 are shown in Figure 3(a).

Table 3: Fuzzy rule base

s_0	s_1	s_2	d	s_0	s_1	s_2	d
ZO	ZO	ZO	1	...	...	...	...
PS	ZO	ZO	0	...	...	...	...
...	...	...	...	PM	PB	PB	0
...	...	...	...	PB	PB	PB	0

The membership functions are determined following the ideas of Case 1-1.

2.3 A Numerical Example

Example 1: We examine a two-stage tandem production network with two types of parts and no service costs. The service rate in station 2 is $\mu=0.1$, and service rate in station 1 is either $u_0=0$ or $u_1=a=0.1$. The holding cost per part per unit time in each station $h_1=2.8$ and $h_2=1.4$. This is Case 1-1.

We determine the optimal policy from the architecture of the fuzzy logic controller (refer to Figure 1). The algorithm is outlined as follows.

(a) We determine the fuzzy membership functions for s_2 according to the given parameters μ, a, h_1 and h_2.

(b) We start the algorithm from an initial state $s_0=s_1=s_2=0$.

(c) Using the current s_0, s_1 and s_2 as crisp inputs, we determine the decision d via fuzzification, fuzzy inference and de-fuzzification.

(d) Plot the decision d in the three dimensional space of s_0, s_1 and s_2,

(e) If the decision d is different from the one for $s_2=0$, go to (f), otherwise let $s_2=s_2+1$ and go to (c).

(f) If the decision d is different from the one for $s_1=0$, go to (g), otherwise let $s_1=s_1+1$ and go to (c).

(g) If the decision d is different from the one for $s_0=0$, the calculations stop, otherwise let $s_0=s_0+1$ and go to (c).

Each rule in the rule base expresses an optimal decision in a certain situation. Therefore the decisions that result from all possible individual situations are optimal in the intuitive sense of fuzzy control. Following this algorithm, we obtain the optimal control policy shown in Figure 4.

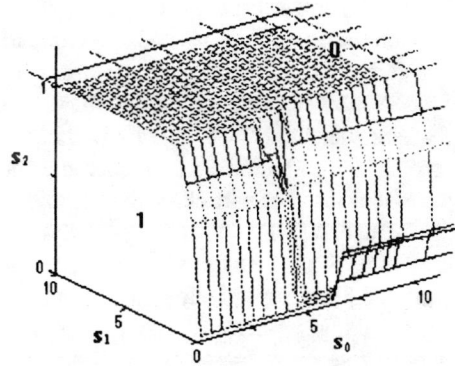

Figure 4: Optimal switching-curve policy

The optimal policy is of a switching-surface structure in the three dimensional state space of s_1, s_2 and s_0, which is essentially the same as the switching-curve structure proved in [1]. The switching-surface divides the state space into two areas, 0 and 1, in which parts of class 0 and 1 should be served by the machine in station 1. When there are no parts present in the system, the machine in station 1 will simply idle. Our work not only verifies the characterizations of the optimal policy, but also explicitly determines the optimal solutions.

3. An Extension of Case 1 (Case 2)
3.1 Problem Description

Now we extend the previous case by introducing service costs whenever the machine in station 1 is used. To take into account the service costs, we let the service rate u of the machine in station 1 assume any value from a finite *denumerable* set u_k, $k=1,2,....,m$, with $0 \leq u_1 < u_2 < .... < u_m < +\infty$. When $u=u_k$, the service cost per unit time is r_k, where $0 \leq r_1 < r_2 < .... < r_m < +\infty$. Again we assume that $\lambda < \mu$ and $2\lambda < u_m$. The system objective is to dynamically determine the service rate in station 1 and schedule the machine in station 1 among the two types of parts in order to minimize the average holding and service costs of the system.

This model is new in the literature.

3.2 Architecture of the Fuzzy Logic Controller

The state of the system is described by four variables (s_1,s_2,s_3,k), where s_1, s_2 and s_3 have the same meaning as in Case 1, and k is the service class. As previously, the decision epochs are the times when a part arrives at the empty station 1 and/or leaves from station i, $i=1,2$. A significant difference from the previous work is that, because service costs have been introduced, the service class k does play an important role in the course of the fuzzy decision inference.

Again, we consider two subcases $h_1 \geq h_2$ and $h_1 < h_2$.

3.2.1 Case 2-1: $h_1 \geq h_2$

It is obvious that this subcase is related to Case 1-1 due to the common assumption $h_1 \geq h_2$. All previous ideas on how to schedule the machine in station 1 are valid in the present subcase. Hence, we have a rule base as shown in Table 1, where the numbers of class i parts $s_i \in [0,+\infty)$, $i=0,1,2$, are chosen to be the fuzzy inputs and the fuzzy output is the decision d of part class, 0 or 1, to be served. The fuzzy membership functions for s_0 and s_1 are both shown in Figure 3(a), and for s_2 and d are shown in Figures 3(b) and 3(c), respectively.

Formally speaking, we select the fuzzy inputs to be the current service rate class k, $k=1,2,....,m$, and the current holding cost difference between the two stations per unit time which is

$$hs = h_1(s_0+s_1) - \mathbf{1}(d)h_2s_2 \quad (1)$$

where $\mathbf{1}(\bullet)$ is an indicator function defined $\mathbf{1}(0)=0$ and $\mathbf{1}(1)=1$, and $d=0,1$ is the decision fired by the rule base in Table 1 concerning the class of part, 0 or 1, to be served.

The fuzzy output is the variation (dk) of the service rate class. The machine amends its service rate class by simply adding the defuzzified crisp output $dk^*=-(m-1),....,-2,-1,0,1,2,....m-1$, to the current class k. The fuzzy rule base is shown in Table 4. We set the fuzzy input k at ZO when the service is of class 1, because this is the basic class even when there are no parts present in the system.

Table 4: Fuzzy rule base

k \ hs	ZO	PS	PM	PB
ZO	ZO	PS	PM	PB
PS	NS	ZO	PS	PM
PM	NM	NS	ZO	PS
PB	NB	NM	NS	ZO

The fuzzy membership functions for k, hs and dk are shown in Figures 5(a), 3(b) and 5(b).

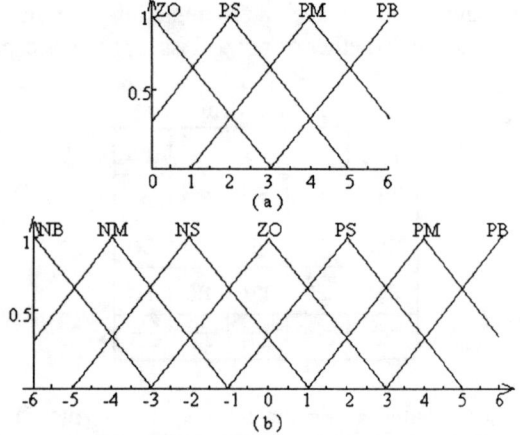

Figure 5: Membership functions

The accumulated holding cost of parts increases in proportion to the number of parts in the system. This is the basis for determining the shape of the fuzzy membership functions for hs. PB for hs should be equivalent to the difference of the service costs per unit time between those of the lowest and the highest service rate. Hence, the fuzzy set PB for hs with membership grade 1.0 is fixed at r_m-r_1. Using a scaling factor we change each physical domain into its normalized counterpart, e.g. r_m-r_1 corresponds to 6 in Figure 3(b).

3.2.2 Case 2-2: $h_1 \geq h_2$

As previously, the decisions for machine scheduling and service rate control are fired independently by their own input parameters and rule bases. Specifically, the fuzzy outputs for this case d and dk correspond to those in Cases 1-2 and 2-1, respectively. It is immediate that all the fuzzy variables as well as corresponding membership functions are the same as in Case 2-1, i.e., the five fuzzy inputs and two fuzzy outputs. The fuzzy rule bases are shown in Tables 2 and 4.

4. Simultaneous Routing and Admission of Parts (Case 3)

4.1 Problem Description

The production network under consideration which consists of two stations 1 and 2 in tandem is illustrated in Figure 6. Station 1 consists of one machine called machine 0 and station 2 consists of two machines called 1 and 2. The system processes two types of parts, $i=1,2$, which are first served in station 1 and then proceed to the corresponding machine in station 2. The service times of the machines are exponential with parameter μ_j, $j=0,1,2$, where $\mu_0 > \mu_1 + \mu_2$. There is an infinite supply of parts of type i, $i=1,2$, in front of machine 0 and each machine in station 2 has its own infinite queue called queue i. A reward r_i, $i=1,2$, is earned whenever machine 0 completes a service on a type i part, and a holding cost h_i per part of type i in queue per unit time is incurred. When machine 0 completes a service, the type of part to be next served is determined based on the number of parts in the buffers of station 2. The problem is to dynamically select the type of part to be next served in order to maximize the average benefit of the system.

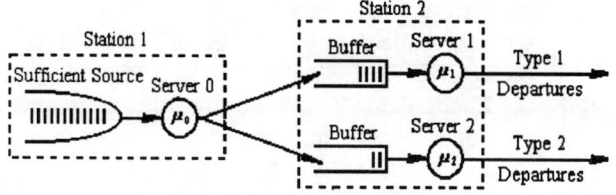

Figure 6: A production network

The control policy at each decision epoch will be one of the following three choices, type 1, type 2, neither type 1 nor type 2. Besides the scheduling of type 1 and type 2 of parts, machine 0 may deny any entrance. This is a problem of simultaneous part scheduling and admission. Shioyama [4] has examined this problem with regard to the monotonicity of the optimal policy. However the explicit determination of the optimal policy remains an open problem.

4.2. Architecture of the Fuzzy Logic Controller

The state of the system is described by (s_1, s_2), where $s_i = 0,1,2,...$, is the number of type i parts in queue i. The decision epochs correspond to the time instants when machine 0 has just completed a service.

To construct the fuzzy controller, we choose the number of parts in queue i, s_i, as the fuzzy inputs and the decision d about which type of part to be served next as the output. We develop a rule base in Table 5. The fuzzy set PVB for the fuzzy inputs s_1 and s_2 in the rule base indicates "positive very big", and 0, 1 and 2 for d indicate machine 0 selects nothing, type 1 and type 2, respectively. The sign "*" represents 1 or 2, but not 0.

Table 5: Rule base

s_1 \ s_2	ZO	PS	PM	PB	PVB
ZO	*	1	1	1	1
PS	2	*	1	0	0
PM	2	2	0	0	0
PB	2	0	0	0	0
PVB	2	0	0	0	0

The membership functions for the fuzzy inputs s_i are shown in Figure 7, and for the output d in Figure 3(c).

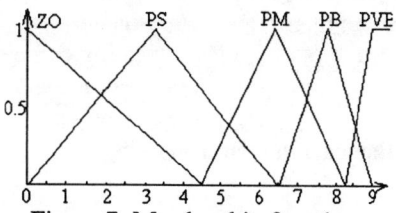

Figure 7: Membership functions

It is always beneficial to admit a part of type i, $i=1,2$, in machine 0 when queue i is empty. Therefore we have the rules for the state $s_i=0$, *if there are no parts in queue i, then a type i part is admitted in machine 0*. Next we answer the question how big is "big" for the queue length in the rule base.

We now determine that scaling factors for s_1 and s_2. In the absence of type 2 parts in the system, that is $s_2=0$, machine 0 rejects a type 1 part whenever the length of queue 1 is greater or equal to

$$n_1 = \frac{r_1 \mu_1}{h_1} + \frac{\mu_1}{\mu_0}. \quad (2)$$

This proposition says that a part is denied entry because its reward does not compensate its expected holding cost, which is $r_1 \leq (n_1 - \mu_1/\mu_0)(h_1/\mu_1)$. The fuzzy set PVB for s_1 with membership grade 1.0 in the rule base is fixed at the value of n_1, which is not necessarily an integer. In a similar fashion, we obtain the crisp value

$$n_2 = \frac{r_2 \mu_2}{h_2} + \frac{\mu_2}{\mu_0} \quad (3)$$

for the fuzzy set PVB for s_2 with membership grade 1.0.

Observing the rule base in Table 5, the output d has three fuzzy sets 0, 1 and 2. However, these three fuzzy sets have no numerical relationships ranking from small to big etc. and hence can not be expressed in one figure, but they are related as follows: (i) if machine 0 selects nothing, the decision d is 0, otherwise it is 1 or 2, and (ii) on the basis of (i) if machine 0 selects type 1, the decision d is 1 otherwise it is 2. In other words, the fuzzy inference is completed in two stages. Therefore we device the fuzzy membership functions for d shown in Figure 3(c). In stage (i), the fuzzy sets I and II represent the fuzzy decisions 0 and 1 or 2, respectively, whereas in stage (ii) they represent 1 and 2.

4.3 A Numerical Example

Example 3: We examine a tandem two-station production network with two types of parts. The parameters are as follows: service rates $\mu_0=4$, $\mu_1=1$ and $\mu_2=2$, reward $r_1=r_2=3$, and holding cost per part per unit time $h_1=h_2=1$.

The optimal control policy is shown in Figure 8, which is of a switching-curve structure in the two dimensional state space of s_1 and s_2. The switching-curve divides the state space into three areas. In areas 1 and 2 machine 0 selects a part of type 1 and type 2 respectively, whereas in area 0 the machine selects nothing.

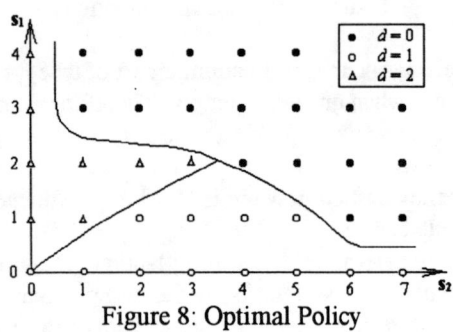

Figure 8: Optimal Policy

5. Conclusions

In this paper we apply fuzzy logic to the optimal control problem of tandem two-station production networks with two types of parts. We examine two problems, Cases 1 and 3, whose optimal control structure has been studied analytically. We verify this structure but we also specify explicitly the optimal control. Finally, an entirely new problem, Case 2, is introduced and again, the optimal strategy is given using fuzzy logic.

Our methodology seems promising in solving problems where analysis is rather hopeless such as non-Markovian systems or complex networks.

References

[1] H. Chen, P. Yang and D. D. Yao, "Control and scheduling in a two-station queueing network: Optimal policies and heuristics", *Queueing Syst.*, vol. 18, pp. 301-332, 1994.

[2] D. Driankov, H. Hellendoorn and M. Reinfrank, *An Introduction to Fuzzy Control*. New York: Springer-Verlag, 1993.

[3] J. M. Harrison and L. M. Wein, "Scheduling networks of queues: Heavy traffic analysis of a simple open network", *Queueing Syst.*, vol. 5, pp. 265-280, 1989.

[4] T. Shioyama, "Optimal control of a queueing network system with two types of customers," *Euro. J. Operat. Res.*, vol. 52, pp. 367-372, 1991.

[5] S. Stidham Jr and R. Weber, "A survey of Markov decision models for control of networks of queues", *Queueing Syst.*, vol. 13, pp. 291-314, 1993.

[6] J. Teghem Jr., "Control of the service process in a queueing system", *Euro. J. Operat. Res.*, vol. 23, pp. 141-158, 1986.

[7] L. M. Wein, "Optimal control of a two-station Brownian network," *Math. Operat. Res.*, vol. 15, pp. 215-242, 1990.

[8] L. M. Wein, "Scheduling networks of queues: Heavy traffic analysis of a two-station network with controllable inputs," *Operat. Res.*, vol. 38, pp. 1065-1078, 1990.

[9] R. Zhang and Y. A. Phillis, "Fuzzy service control of M/M/1 queues", *Proc. Second World Automation Congress*, Montpellier, France, vol. 5, pp. 731-736, 1996.

[10] R. Zhang and Y. A. Phillis, "Fuzzy service rate control of queueing systems", *Proc. 1996 IEEE International Conference on Systems, Man and Cybernetics on Information, Intelligence and Systems*, Beijing, China, vol. 4, pp. 3045-3050, 1996.

[11] R. Zhang and Y. A. Phillis, "Fuzzy routing of queueing systems with heterogeneous severs", *Proc. 1997 IEEE International Conference on Robotics and Automation*, vol. 3, pp. 2340-2345 Albuquerque, NM, USA, 1997.

[12] R. Zhang and Y. A. Phillis, "A fuzzy approach to flow control problems" To appear in *J. Intelligent and Fuzzy Systems*.

Analysis of Robot Motion Performance and Implications to Economy Principles

S. Shoval, J. Rubinovitz, S. Nof
Faculty of Industrial Engineering & Management
Technion - Israel Institute of Technology, Haifa, Israel 32000.

Abstract

This paper utilizes the idea of "robot ergonomics" in which robotic tasks are evaluated according to ergonomics criteria. Robot performance is continuously recorded in terms of total motion for each robotic link, the load applied on each joint, and the accuracy of each motion. Evaluating robot motion can assist in the design of a robotic cell, where robot position and layout of peripheral equipment are determined. In previous work, the main measure has been AJU (Arm Joint Utilization). Two new measures are developed here: AJA (Arm Joint Accuracy) and AJL (Arm Joint Load). A test case of an articulated industrial robot, performing a kitting task is presented. Robot performance is evaluated for various bin positions, and based on these results, the bin position relative to the robot is determined based on all three measures, AJA, AJL and AJU.

1. Introduction

Robot performance depends on the particular tasks required by the application. Commonly, performance is defined by several parameters such as speed, accuracy and payload. Robot performance is not uniform across the entire working volume, therefore, some areas are more suitable for particular tasks than others. Furthermore, in most robot types there is a trade-off between performance parameters (speed vs. accuracy, accuracy vs. payload etc.) [Rubinovitz and Wysk, 1989].

In the design of a robotic cell, the layout of peripheral equipment (feeders, clamps, tools etc.), as well as the position of the robot relative to other cells in the process, are determined. Nof [1985] describes various techniques that can be used for robotic cell design. These techniques are described within the context of robot ergonomics, and have lead to general systematic motion economy principles (MEP).

Nnaji and Asano [1989] describe evaluation techniques of trajectories for various robot types. They conclude that trajectories are robot-independent, and that different trajectories are optimal for different working ranges. Nof and Robinson [1986] developed and analyzed principles for Arm Joint Utilization (AJU) and cycle times for specific tasks of bin picking and part kitting performed by two different robots. Edan and Nof [1996] used graphic robot simulation to derive MEP for various robot classes in different size group. In particular, they concentrated on the effect of bin orientation and location on robot performance, and compared different robot classes for specific cell set up in a kitting process. In their work, Edan and Nof have described the use of Arm Joint Utilization (AJU) measure, which is continuously calculated based on the total movement of all joints during the task performance. The AJU measure is calculated for the first three links of the robot, and is derived according to eq. 1

$$AJU = \sum_{j=1}^{N}(0.5 d\theta_1 + 0.33 d\theta_2 + 0.17 d\theta_3) \quad (1)$$

where N - the number of motions performed by the robot
$d\theta_1$, $d\theta_2$, $d\theta_3$ - the incremental motions of links 1, 2 and 3 from position j-1 to j.

The AJU measure provides a weighted mean such that the first joint is given 3 times more weight than the third joint, and the second is given twice the weight of the third joint. According to the AJU measure definition, a lower value is more desirable as it indicates that less motion is required by the driving system, wear is reduced, less maintenance is required, and operations are more accurate and consistent. In previous work, AJU has been calculated for six degrees of freedom using a similar weighting method for the individual joints. In this paper, only the three more significant joints are considered.

The AJU principle is based on the kinematics of the robot motion. In the work described in this paper, we suggest additional measures which consider kinematics as well as dynamic parameters that affect robot performance. The first measure - the Arm Joint Accuracy (AJA) measure, determines the accuracy of the robotic arm during the motion to a specific location. The Arm Joint Load (AJL) measures the load (in terms of torques) that the driving systems is subjected to. These new measures, combined with the AJU measure provide a broad measure for estimation of robot motion performance, and can be an effective tool in the design and set up of robotic cells.

2. Definitions and Assumptions

The Arm Joint Accuracy (AJA) Measure
As previously mentioned, the AJA measure measures the accuracy of the robotic arm during motion. In Point-to-Point tasks the accuracy is important only at the end of the motion (i.e. simple assembly tasks, spot welding, drilling etc.). However, many processes require continuos motion control (i.e. arc welding, painting, gluing etc.), therefore, the accuracy during motion is an important measure in the overall performance. The calculation of the AJA measure is given by (2):

$$AJA = \sum_{j=1}^{N} \int_{t_{j-1}}^{t_j} (K_1\varepsilon_1 + K_2\varepsilon_2 + K_3\varepsilon_3)dt \quad (2)$$

where N is the number of motions performed by the arm.
K_1, K_2, K_3 - kinematics normalized coefficients for joints 1,2 and 3 respectively.
$t_{j-1}-t_j$ - the time through which the arm travels from point j-1 to point j.
$\varepsilon_1, \varepsilon_2, \varepsilon_3$ - the difference between the reference and actual position of joint 1, 2 and 3 respectively given in (2a).

$$\varepsilon_i = \theta_{ir} - \theta_{im} \quad i=1,2,3 \quad (2a)$$

θ_{ir} - the reference value for joint i, as calculated by the robot's controller.
θ_{im} - the measured value for joint i, as measured by the encoder.

An alternative measure for Arm Joint Accuracy measure, the AJA_{abs} considers the absolute values of $d\theta_1$, $d\theta_2$ and $d\theta_3$. This measure expresses a more precise data about motion oscillations around the reference value which will not be indicated by the AJA measure. The AJA_{abs} is given by Eq. (2b):

$$AJA_{abs} = \sum_{j=1}^{N} \int_{t_{j-1}}^{t_j} (K_1|\varepsilon|_1 + K_2|\varepsilon|_2 + K_3|\varepsilon|_3)dt \quad (2b)$$

where $|\varepsilon_1|$, $|\varepsilon_2|$ nd $|\varepsilon_3|$ - the difference between the reference and actual position of joint 1 2 and 3 respectively.

As with the AJU measure, only the first three joints are considered in our research, and they are weighted by the K_1, K_2, K_3 coefficients. While the AJU measure assumes that the "contribution" of each link to the total arm utility is given constant and normalized coefficients (0.5, 0.33 and 0.17 respectively), this assumption is not correct for AJA calculation. A careful measurement of a robotic arm accuracy requires to include the full direct kinematics calculation of the particular arm. For example, in Cartesian robotic arm, each link has an equal "contribution" to the total accuracy, while in SCARA type robot the first link has a larger affect on the accuracy in the X-Y plane compared with the second joint, and the third joint affects the accuracy along the Z axes only. Equation 2 expresses the kinematics of the arm by the K_1, K_2, K_3 coefficients, which are proportional not only to the arm type, but also depend on the links' geometry. The full kinematics calculation requires that these coefficients are continuously adjusted according to the robot and current arm configuration. To simplify the calculation of the AJA measure, the coefficients are determined in advance for each robot type and are assumed constant throughout the overall working envelope. This assumption, although reducing the accuracy of the AJA measure, provides a simple tool for expressing the robotic arm accuracy.

The Arm Joint Load (AJL) Measure
The AJL measure measures the load to which the arm is subjected during motion. In other terms, this measure represents the "effort" that the arm's driving system is applying to perform the required motion. The AJL measure is calculated similar to the AJU and AJA measures, and is given by (3):

$$AJL = \sum_{j=1}^{N} \int_{t_{j-1}}^{t_j} (T_1 + T_2 + T_3)dt \quad (3)$$

where N is the number motions performed by the arm
$t_{j-1}-t_j$ - the time through which the arm travels from point j-1 to point j
T_1, T_2, T_3 - the moments applied on joint 1, 2 3 respectively.

It should be noted that no coefficients are included in the AJL measure as the moment produced by each link is independent of other links.

Similar to the AJA_{abs}, the AJL_{abs} is an additional measure that calculates the absolute values of torques applied by the driving motors. This measure is given by Eq. (3a) as followes:

$$AJL_{abs} = \sum_{j=1}^{N} \int_{t_{j-1}}^{t_j} (|T_1| + |T_2| + |T_3|) dt \qquad (3a)$$

Using the above definition, together with the AJU measure, provides a means for qualitative, as well as quantitative measures for estimating robotic arm performance. In our work we analyze the effect of process parameters on the performance of an articulated robot. In particular, we analyze the effect of bin location and orientation under different loads.

3. The Experimental Environment and Evaluation Methods

Kitting task is common to many robot applications. In this task the cycle time is determined directly by the robot motion, and it does not depend on extraneous factors. Bin position and orientation can be easily modified, with no major changes to the application software and hardware, therefore a reliable comparison is available. The robotic arm selected for our experiment is Eshed/Yaskawa MK-3 articulated robot. This is a five Degrees of Freedom (DOF) robot, with an AC servo drive mechanisms. The controller is an ACL AC-type controller with an asynchronous computer link to a PC. The application is similar to the one suggested by Edan and Nof [1996], where a 4X4 bin of 16 cells is placed in different positions and orientations inside the robot working volume. Since the robot size is relatively small, each cell in the bin is set to 150X150X200 mm.

The ACL programming language provide tools for monitoring the arm's motion in terms of speed, accuracy and load. Reference values for each joint and for the tool center point (TCP) are accessible to the application program. Also, actual values of the joints' encoders can be recorded in real-time during motion at a rate of 100Hz. The difference between the reference and actual encoder values determine the AJA measure according the Eq. 2. The load on each driving system is available to the application program by measurement of the current consumed by each motor driver.

As the ACL controller has a multi-task processor, one process records the required data during the arm's motion, while the main process controls the motion according to the application. Data is recorded and transmitted through an asynchronous link to a PC. This data is then processed by the PC using Microsoft Excel. Figure 1 illustrates the experimental environment used in our experiments.

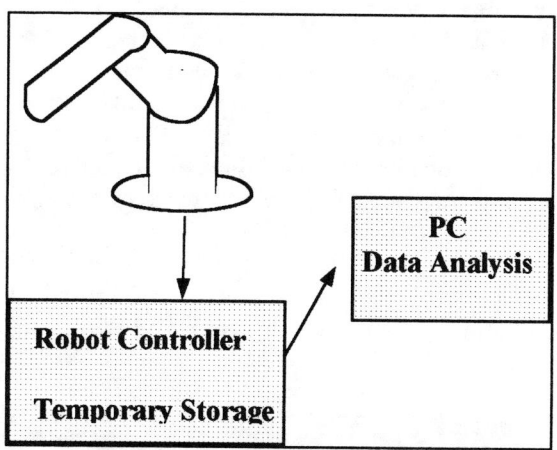

Figure 1: The exprimental environment

The application

As described, the task that the arm performs is kitting, where the robot picks 16 "components" from a 4X4 bin. The location and orientation of the first cell (bottom left cell in the matrix) in the bin is indicated at the beginning of each test by the operator. The controller than calculates the position of all other cells, and the robot kits through all bin cells in the following order (fig. 2): $1 \Rightarrow 2 \Rightarrow 3 \Rightarrow 4 \Rightarrow 5 \Rightarrow 6 \Rightarrow 7 \Rightarrow 8 \Rightarrow 9 \Rightarrow 10 \Rightarrow 11 \Rightarrow 12 \Rightarrow 13 \Rightarrow 14 \Rightarrow 15 \Rightarrow 16$.

4	8	12	16
3	7	11	15
2	6	10	14
1	5	9	13

Figure 2

Data Type

The data recorded during the arm's motion consists of three values:

- **Link angular displacement** - this value is calculated at the beginning and the end of each motion. A single motion is defined as the movement of the arm from one cell to the next one. Each experiment includes 17 motions (15 motion between the cells plus one motion from start position to the first cell and one motion from the last cell to the start position). In total 51 values are recorded in each experiment (17 motions X 3 joints) to be used by the AJU measure.

- **Positional error** - recorded continuously during motion. This value is the difference between the reference angular value of each link, calculated by the controller and the actual joint position measured from the joint's encoder. Data is recorded at a rate of 100Hz per each link, and therefore the amount of data data is proportional to the time required to complete the motion.
- **Link load** - the load on each link is proportional to the current consumed by the motor's driver. As with the positional error, the link load is recorded at a rate of 100Hz for each link.

Bin Position

To examine the performance of the robot under various conditions, 27 bin positions were selected. These positions include three locations at the lower level of the working space, on the robot base level, three in mid-height level, and three positions at the robot's highest level of the working space (Fig. 3b). At each level one position is in close proximity to the robot base, one is far from the robot base (extending the arm to the limits of the working space) and one position is in mid-range (Fig 3a). At each position the bin is arranged in three different orientations creating 0°, 45° and 90° angle with the horizontal plane (Fig 3c).

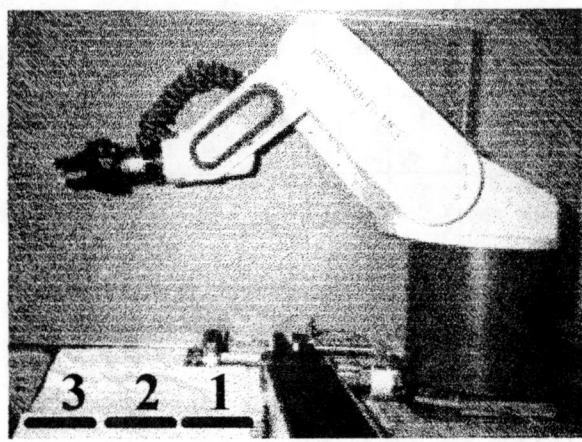

Figure 3a: Bin positions in lower level

Figure 3(b): Bin position in three levels

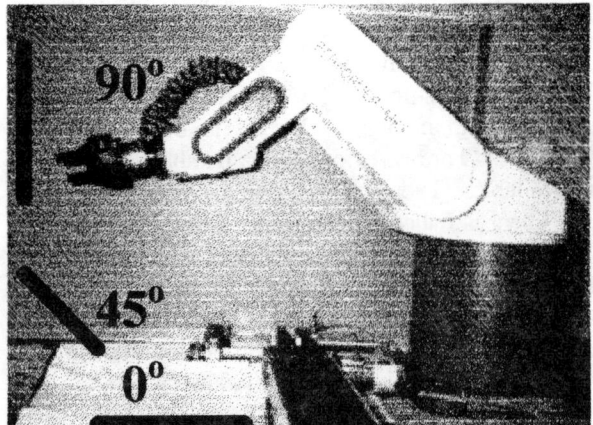

Figure 3c: Bin position in three orientations

4. Experimental results

At the end of each experiment the three measures (AJU, AJA and AJL) are calculated both for regular and absolute values. Figure 4(a) shows the torque values recorded during a typical experiment. As shown, the torques for all three joints increase from a zero value at the beginning of the motion to a maximum value at the middle of the motion, and than decreases towards the end of the motion. However, the torques applied by some joints do not decrease to zero as a moment is required even during steady state to overcome gravitational and other external forces. Similarly, the values for joint's accuracy are shown in figure 4(b). In this case, however, the values increase sharply at the beginning and the end of the motion. This is expected as larger errors occur during acceleration and declaration stages.

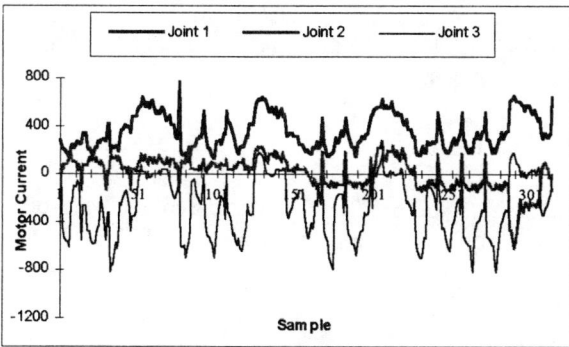

Figure 4(a): Joints torques during motion

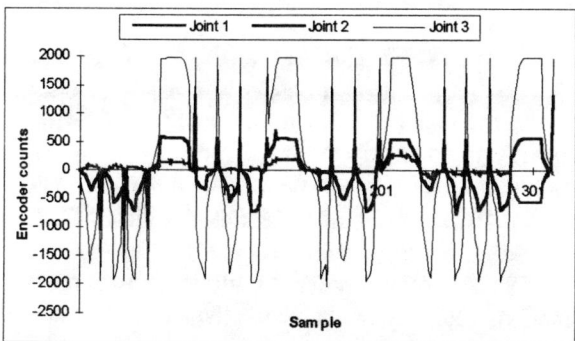

Figure 4(b): Joint error during motion

A summary of the experiments' results is shown in Table 1. Each experiment is coded as followed: The first digit identifies the bin height level according to figure 3(a) where 1 represents the base level, 2- mid height and 3- highest level. The second digit represents the bin proximity to the robot's base according to fig 3(b) where 1 - closest, 2- mid range, and 3 - the largest distance. The last digit identifies the bin orientation (Fig 3c) with 0 represents 0°, 45 - 45°, and 90 - 90°, all relative to the horizontal plane. It should be noted that in all experiments robot motion was well within the working envelope, and not within close proximity to singular positions.

For each experiment the three measures, AJU, AJA and AJL are shown. The dimensions for the AJA and AJU measures are in encoder counts. The AJL measures are shown in driving motor current values. Although these values can be translated to real dimensions (angular values for AJU and AJA, and torque values for AJL), only the relations between the results are of importance for our research. The values for AJA and AJL, shown in Table 1, are normalized by a time factor to eliminate the effect of calculation and data transmission delays.

Table 1 Experimental Results

No.	Code	AJL Abs.	AJL Reg.	AJA Abs.	AJA Reg.	AJU
1	110	815	184	2272	149	205211
2	210	947	210	2413	136	156045
3	310	981	7	2461	-252	136611
4	120	775	23	2122	-102	185754
5	220	949	150	2591	-303	157184
6	320	935	-61	2392	-267	141726
7	130	797	17	2166	-164	167411
8	230	840	37	2376	-127	150585
9	330	907	-91	2266	-99	146417
10	1145	873	115	2160	-243	77034
11	2145	865	199	2116	133	96996
12	3145	1043	118	2318	-134	65735
13	1245	868	91	2354	-296	129120
14	2245	957	139	2420	-123	115694
15	3245	988	7	2400	-211	112211
16	1345	863	49	2445	-120	161508
17	2345	954	86	2589	-154	175550
18	3345	**	**	**	**	199040
19	1190	933	92	2117	203	130516
20	2190	1084	225	1805	167	124570
21	3190	1156	207	2086	-8	114730
22	1290	987	73	2236	34	76285
23	2290	1045	128	1970	228	69391
24	3290	1106	70	1999	-128	64152
25	1390	913	59	2250	-74	84570
26	2390	1029	108	2188	225	79372
27	3390	1009	13	2103	-256	69427

** no data recorded

5. Discussion

Analysis of the results shows the following observations:

- Minimum AJL_{abs} occurs at experiments No. 4,7,and 1 respectively. All these experiments include bin position No.1 - closest to the robot base, bin orientation of 0°, and varying height levels.
- Minimum AJU occur at experiments 24, 23, 22 and 27. These results can be directly derived from the inverse kinematics solutions, and are similar to the results obtained by Edan and Nof [1996], where AJU for articulated robot decreases as distance from the robot base increases.
- Minimum AJA_{abs} occur at experiments No. 20, 23, 24 and 27. In all these experiments the bin orientation is at 90°, and bin position is No. 2 and 3 (largest distance from robot base or mid range for experiment). As with the AJL measure, bin's height has minimal effect on and AJA measure.

The AJL_{abs} results are expected as the AJL_{abs} measure represents the "effort" exerted by the arm to perform the task in terms of motor torques. As torque is directly proportional to the arm length, bin positions which are closer to the robot base require less moment compared with further positions, resulting in lower AJL_{abs} values for positions close to the robot base.

Although the AJA_{abs} results are not identical to the minimum AJU results, it is clear that low AJA_{abs} occurs when the AJU measure is also low. This can be related to the fact that shorter joint motions (low AJU) create smaller errors (low AJA_{abs}) as errors during the transient response period, as well as steady state errors, are proportional to the differential input to the control system of each robot joint. This differential input is evidently smaller for smaller joint motions, resulting in low AJA_{abs} for low AJU.

Linear regression of the results between the AJA_{abs} and AJU shows high correlation factor ($R^2=0.9$). However, similar analysis between the AJA_{abs} and AJL_{abs} results in a much lower correlation factor, indicating low interdependence between the moment and the positional error.

One of the most distinct observation from all the experiments is that the height level of the bin has little effect on all measures compared with the horizontal bin position and orientation. This may be due to the fact that the MK-3 robot is a small articulated type robot, and all bin positions examined in this research are within the working volume. As a result, the height range is smaller compared with the horizontal range.

6. Conclusions

Bin position relative to the robot base affects robot performance in terms of motion, accuracy and load. For the robot examined (small articulated type), load on the driving motors (as expressed by the AJL_{abs} measure) increases as bin distance from the robot base increases. Accuracy, on the other hand, improves as bin distance increases (lower AJA_{abs}). However, it should be noted that these results were obtained for relatively low payload compared with the maximum allowed payload.

The results of this research provide tools for the design of robotic cells and processes in terms of bin position relative to the robot base. Following the guidelines from the measures developed can result in lower load on the driving motors, therefore increasing their life span, as well as protecting gears, pulleys, and other driving mechanisms. Also, these tools assist in determining bin positions for tasks requiring continuous accurate motion. Future work will cover other aspects of performance such as cycle time, effects of varying payloads, and effects of motion speed direction and distance.

7. References

1. Edan Y., Nof S. Y., "Graphic-based analysis of robot motion economy principles", " Robotics & Computer Integrated Manufacturing, Vol. 12, N0. 2, pp. 185-193, 1996.
2. Edan Y., Nof S. Y., "Motion Economy Analysis for Robotic Kitting Tasks", International Journal of Production Research", Vol. 33No. 5, 1995, pp. 1213-1227.
3. Nnaji B. O., Asano D. K.., "Evaluation of trajectories for different classes of robots" Robotics & Computer Integrated Manufacturing 6(1):25-35,1989.
4. Nof S. Y., "Robot ergonomics: optimizing robot work", In Handbook of Industrial Robotics, Nof S. Y. (Ed.). New York, John Wiley, 1985.
5. Nof S. Y., Robinson P., "Analysis of two robot motion economy principles", Israel J. Tehnol. 23::125-128, 1986.
6. Rubinovitz J., Wysk R. A., "Methodology for robot capability and performance characterization - the basis for robot task planning", Proceedings of the International Conference on CAD/CAM and AMT, 1989, Vol. F-2-3, pp. 1-9.